Foundations of Image Science

WILEY SERIES IN PURE AND APPLIED OPTICS

Founded by Stanley S. Ballard, University of Florida

EDITOR: Bahaa E. A. Saleh

BARRETT AND MYERS · *Foundations of Image Science*
BEISER · *Holographic Scanning*
BERGER-SCHUNN · *Practical Color Measurement*
BOYD · *Radiometry and The Detection of Optical Radiation*
BUCK · *Fundamentals of Optical Fibers*
CATHEY · *Optical Information Processing and Holography*
CHUANG · *Physics of Optoelectronic Devices*
DELONE AND KRAINOV · *Fundamentals of Nonlinear Optics of Atomic Gases*
DERENIAK AND BOREMAN · *Infrared Detectors and Systems*
DERENIAK AND CROWE · *Optical Radiation Detectors*
DE VANY · *Master Optical Techniques*
GASKILL · *Linear Systems, Fourier Transform, and Optics*
GOODMAN · *Statistical Optics*
HOBBS · *Building Electro-Optical Systems: Making It All Work*
HUDSON · *Infrared System Engineering*
IIZUKA · *Elements of Photonics, Volume I: In Free Space and Special Media*
IIZUKA · *Elements of Photonics, Volume II: For Fiber and Integrated Optics*
JUDD AND WYSZECKI · *Color in Business, Science, and Industry,* Third Edition
KAFRI AND GLATT · *The Physics of Moire Metrology*
KAROW · *Fabrication Methods for Precision Optics*
KLEIN AND FURTAK · *Optics,* Second Edition
MALACARA · *Optical Shop Testing,* Second Edition
MILONNI AND EBERLY · *Lasers*
NASSAU · *The Physics and Chemistry of Color: The Fifteen Causes of Color,* Second Edition
NIETO-VESPERINAS · *Scattering and Diffraction in Physical Optics*
OSCHE · *Optical Detection Theory for Laser Applications*
O'SHEA · *Elements of Modern Optical Design*
OZAKTAS · *The Fractional Fourier Transform*
SALEH AND TEICH · *Fundamentals of Photonics,* Second Edition
SCHUBERT AND WILHELMI · *Nonlinear Optics and Quantum Electronics*
SHEN · *The Principles of Nonlinear Optics*
UDD · *Fiber Optic Sensors: An Introduction for Engineers and Scientists*
UDD · *Fiber Optic Smart Structures*
VANDERLUGT · *Optical Signal Processing*
VEST · *Holographic Interferometry*
VINCENT · *Fundamentals of Infrared Detector Operation and Testing*
WILLIAMS AND BECKLUND · *Introduction to the Optical Transfer Function*
WYSZECKI AND STILES · *Color Science: Concepts and Methods, Quantitative Data and Formulae,* Second Edition
XU AND STROUD · *Acousto-Optic Devices*
YAMAMOTO · *Coherence, Amplification, and Quantum Effects in Semiconductor Lasers*
YARIV AND YEH · *Optical Waves in Crystals*
YEH · *Optical Waves in Layered Media*
YEH · *Introduction to Photorefractive Nonlinear Optics*
YEH AND GU · *Optics of Liquid Crystal Displays*

To

Cathy, David, Mindy, Ryan, Justin

Matt, Ben, Andrew

Preface

Images are ubiquitous in the modern world. We depend on images for news, communication and entertainment as well as for progress in medicine, science and technology. For better or worse, television images are virtually a *sine qua non* of modern life. If we become ill, medical images are of primary importance in our care. Satellite images provide us with weather and crop information, and they provide our military commanders with timely and accurate information on troop movements. Biomedical research and materials science could not proceed without microscopic images of many kinds. The petroleum reserves so essential to our economy are usually found through seismic imaging, and enemy submarines are located with sonic imaging. These examples, and many others that readily come to mind, are ample proof of the importance of imaging systems.

While many of the systems listed above involve the latest in high technology, it is not so obvious that there is an underlying intellectual foundation that ties the technologies together and enables systematic design and optimization of diverse imaging systems. A substantial literature exists for many of the subdisciplines of image science, including quantum optics, ray optics, wave propagation, image processing and image understanding, but these topics are typically treated in separate texts without significant overlap. Moreover, the practitioner's goal is to make better images, in some sense, but little attention is paid to the precise meaning of the word "better." In such circumstances, can imaging be called a science?

There are three elements that must be present for a discipline to be called a science. First, the field should have a common language, an agreed-upon set of definitions. Second, the field should have an accepted set of experimental procedures. And finally, the field should have a theory with predictive value. It is the central theme of this book that there is indeed a science of imaging, with a well-defined theoretical and experimental basis. In particular, we believe that image quality can be defined objectively, measured experimentally and predicted and optimized theoretically.

Our goal in writing this book, therefore, is to present a coherent treatment of the mathematical and physical foundations of image science and to bring image evaluation to the forefront of the imaging community's consciousness.

ORGANIZATION OF THE BOOK

There are a number of major themes that weave their way throughout this book, as well as philosophical stances we have taken, so we recommend that the reader begin with the prologue to get an introduction to these themes and our viewpoint. Once this big picture is absorbed, the reader should be ready to choose where to jump into the main text for more detailed reading.

Mathematical Foundations

The first six chapters of this book represent our estimation of the essential mathematical underpinnings of image science. In our view, anyone wishing to do advanced research in this field should be conversant with all of the main topics presented there.

The first four chapters are devoted to the important tools of linear algebra, generalized functions, Fourier analysis and other linear transformations. Chapter 5 treats a class of mathematical descriptions called mixed representations, that is, descriptions that mix seemingly incompatible variables such as spatial position and spatial frequency. Chapter 6 presents the basic concepts of group theory, the mathematics of symmetry, which will be applied to the description of imaging systems in later chapters.

It was our objective in writing these introductory chapters to present the mathematical foundations of image science at a level that will be accessible to graduate students and well-motivated undergraduates. At the same time, we have attempted to include sufficient advanced material so that the material will be beneficial to established workers in the field. This dual goal requires examining many concepts at different levels of sophistication. We have attempted to do this by providing both elementary explanations of the key points and more detailed mathematical treatments. The reader will find that the level of mathematical rigor is not uniform throughout these chapters or even within a particular chapter. We hope that this approach allows each reader to extract from the book insights appropriate to his or her individual interests and mathematical preparation.

Image Formation: Models and Mechanisms

A quick perusal of the of Contents will reveal that a significant portion of the book is devoted to the subject of image formation. We have strived to present a comprehensive and unified treatment of the mathematical and statistical principles of imaging. We hope this serves the image-science community by giving a common language and framework to our many disciplines. Additionally, a thorough understanding of the image-formation process is a prerequisite for the image-evaluation methodology we advocate.

The deterministic analysis of imaging systems begins in Chap. 7, where we present a wide variety of mathematical descriptions of objects and images and mappings from object to image. We argue in Chap. 7 (and briefly also in the Prologue) that digital imaging systems are best described as a mapping from a function to a discrete set of numbers, so much of the emphasis in that chapter will be on such mappings. More conventional mappings such as convolutions are, however, also treated in a unified way. An important tool in Chap. 7 is singular-value decomposition, which is introduced mathematically in Chap. 1.

The deterministic mappings are not a complete description of image formation. Repeated images of a single object will not be identical because of electronic noise in detectors and amplifiers, as well as photon noise, which arises from the discrete nature of photoelectric interactions. In addition, the object itself can often be usefully regarded as random. Object statistics are important in pattern recognition, image reconstruction and evaluation of image quality. Chapter 8 provides a general mathematical framework for the description of random vectors and processes. Particular emphasis is given to Gaussian random vectors and processes, which often arise as the result of the central-limit theorem.

The next two chapters go more deeply into specific mechanisms of image formation. Chapter 9 develops the theory of wave propagation from first principles and treats diffraction and imaging with waves within this framework. Though the objective of the discussion is to develop deterministic models of wave-optical imaging systems, we cannot avoid discussing random processes when we consider the coherence properties of wave fields, so an understanding of the basics of random processes, as presented in Chap. 8, is needed for a full understanding of Chap. 9. The reader with previous exposure to such topics as autocorrelation functions and complex Gaussian random fields can, however, skip Chap. 8 and move directly to Chap. 9.

Chapter 10 is ostensibly devoted to radiometry and radiative transport, but actually it covers a wide variety of topics ranging from quantum electrodynamics to tomographic imaging. A key mathematical tool developed in that chapter is the Boltzmann equation, a general integro-differential equation that is capable of describing virtually all imaging systems in which interference and diffraction play no significant role. The Boltzmann equation describes a distribution function that can loosely be interpreted as a density of photons in phase space, so it is necessary to discuss in that chapter just what we mean by the ubiquitous word *photon*.

In Chap. 10 we discuss only the mean photon density or the mean rate of photoelectric interactions, but in Chap. 11 we begin to discuss fluctuations about these means. In particular, we present there an extensive discussion of the Poisson probability law and its application to simple counting detectors and imaging arrays. Included is a discussion of photon counting from a quantum-mechanical perspective. Many of the basic principles of random vectors and processes enunciated in Chap. 8 are used in Chap. 11.

Chapter 12 goes into more detail on noise mechanisms in various detectors of electromagnetic radiation. The implications of Poisson statistics are discussed in practical terms, and a number of noise mechanisms that are not well described by the Poisson distribution are introduced. A long section is devoted to x-ray and gamma-ray detectors, not only because of their practical importance in medical imaging, but also because they illustrate some important aspects of the theory developed in Chap. 11.

Inferences from Images

With the background developed in Chaps. 1 – 12, we can discuss ways of drawing inferences from image data. The central mathematical tool we need for this purpose is statistical decision theory, introduced in Chap. 13. This theory allows a systematic approach to estimation of numerical parameters from image data as well as classifying the object that produced a given image, and it will form the cornerstone of our treatment of image quality. In accordance with this theory, we shall define image quality in terms of how well an observer can extract some desired information from an image.

Chapter 14 is a nuts-and-bolts guide to objective assessment of image quality for both hardware and software. Particular attention is paid to evaluating the performance of human observers, for whom most images are intended.

Chapter 15 provides a general treatment of inverse problems or image reconstruction, defined as inferring properties of an object from data that do not initially appear to be a faithful image of that object. Considerable attention is given to what information one can hope to extract and what aspects of an object are intrinsically inaccessible from data obtained with a specific imaging system. A wide variety of image-reconstruction algorithms will be introduced, and special attention will be given to the statistical properties of the resulting images.

Applications

Chapters 16 – 19 are intended as detailed case studies of specific imaging systems, with the goal of providing examples of how various mathematical tools developed earlier in the book can be applied. Two of these chapters (16 and 18) cover direct imaging systems in which the image is formed without the need for a processing or reconstruction algorithm, and two of them (17 and 19) cover indirect systems in which the initial data set is not a recognizable image. By a different dichotomy, two of the chapters (16 and 17) relate to imaging with x rays and gamma rays, and two of them (18 and 19) relate to imaging with light. The key physical and pedagogical difference is that x rays and gamma rays have such short wavelengths that interference and diffraction can be neglected, and the Boltzmann transport equation of Chap. 10 is applicable. With light, the diffraction theory developed in Chap. 9 takes a central role.

Appendices

Three appendices are provided: one on matrix algebra, a second on complex variables and a third on the fundamentals of probability theory. The material contained there is expected to have been seen by most readers during their undergraduate training. In writing the appendices, we tried to provide a self-contained treatment of the prerequisite material necessary for the understanding of the material in the main text.

SUGGESTIONS FOR COURSE OUTLINES

Drafts of this book have been used as text material for three different courses that have been taught at the Optical Sciences Center of the University of Arizona. Each course has been taught several times, and there has been some experimentation with the course outlines as the book evolved.

The first six chapters of the book were developed for a one-semester course called Mathematical Methods for Optics. This course was originally intended for first-year graduate students but has proved to be more popular with advanced students. It is basically an introductory course in applied mathematics with emphasis on topics that are useful in image science. Expected preparation includes calculus and differential equations and an elementary understanding of matrix algebra and complex analysis. Appendices A and B were originally used as introductory units in the course but are now considered to define the prerequisites for the course. The current syllabus covers Chaps. 1 – 6 of the book. Earlier, however, a more optics-oriented course was offered based on Chaps. 1 – 3, 9 and 10. For this course it was necessary to assume that the students had some elementary understanding of random processes.

For advanced graduate students, especially ones who will pursue dissertation research in image science, there is a two-course sequence: Principles of Image Science, taught in the Fall semester, and Noise in Imaging Systems, taught in the Spring. The Principles course begins with Chap. 1; this is a review for those who have previously taken the Mathematical Methods course, but there have been no complaints about redundancy. Chapters 7, 9, 10 and 15 are then covered sequentially. Occasionally it is necessary to review material from Chaps. 2 – 5, but basically these chapters are assumed as prerequisites. Appendices A and B are available for reference.

Noise in Imaging Systems covers Chaps. 8 and 11 – 14. Appendix C defines the prerequisite knowledge of probability and statistics, but a general acquaintance with Chaps. 1 – 3 is also presumed. Neither Mathematical Methods nor Principles of Image Science is a formal prerequisite for the Noise course.

Alternatively, a one-year advanced sequence could be taught by covering Chap. 1 and then 7 – 15 in sequence. Prerequisite material in this case would be defined by Chaps. 2 and 3 and the three appendices. Necessary topics in Chaps. 4 – 6 could be sketched briefly in class and then assigned for reading.

The applications chapters, 16 – 19, have not been used in teaching, although they have been reviewed by graduate students working in image science. They could form the basis for an advanced seminar course.

Acknowledgments

The seeds of this project can be found in the interactions of the authors with Robert F. Wagner, who more than anyone else founded the field of objective assessment of image quality, especially in regard to radiological imaging. Without his insights and guidance, neither that field nor this book would have been born.

Many people have read parts of this book and provided invaluable feedback, but two in particular must be singled out. Matthew Myers may be the only person other than the authors who has read every word of the book (to date, we hope!), and Eric Clarkson has read large portions. Both have provided continuing guidance on mathematics, physics and pedagogy; our debt to them is enormous.

A bevy of students at the University of Arizona also struggled through many parts of the book, sometimes in early draft form, and their diligence and insightful feedback have been invaluable. It is almost unfair to make a list of those who have helped in this respect, since we will surely leave off many who should not be overlooked, but we thank in particular Rob Parada, Jim George, Elena Goldstein, Angel Pineda, Andre Lehovich, Jack Hoppin, Kit-Iu Cheong, Dana Clarke, Liying Chen and Bill Hunter. Former students, too, have been very helpful, especially Brandon Gallas, Craig Abbey, John Aarsvold and Jannick Rolland. Colleagues who have provided invaluable review and guidance include Rolf Clackdoyle, Jeffrey Fessler, Charles Metz, Keith Wear, Robert Gagne, Xiaochuan Pan, Mike Insana, Roger Zemp, Steve Moore, Adriaan Walther, Jim Holden, Todd Peterson, Elizabeth Krupinski, Jack Denny, Donald Wilson and Matthew Kupinski.

Staff at the Radiology Department of the University of Arizona, especially Debbie Spargur, Lisa Gelia and Jane Lockwood, have been a continuing source of cheerful and highly competent assistance in myriad details associated with this project. We also thank Brian W. Miller and Meredith Whitaker for their assistance with figures and Bo Huang for his assistance in converting parts of the book from another word processor to LaTeX. The authors have benefited significantly from the help of staff at the Center for Devices and Radiological Health of the FDA as well, especially Phil Quinn and Jonathan Boswell.

Special thanks are owed to Stefanie Obara, also known as LEXI, for her diligence and care in polishing up our LaTeX and producing the final camera-ready text. LEXI describes herself as "anal and proud of it," and we are proud of her production. She participated very capably in formatting, indexing and preparing the bibliography as well as meticulous editing.

Finally, we thank our loving families, who supported and encouraged us during the many years it took to bring this project to fruition.

HARRISON H. BARRETT

KYLE J. MYERS

Tucson, Arizona
July 1, 2003

Prologue

We shall attempt here to provide the reader with an overview of topics covered in this book as well as some of the interrelationships among them. We begin by surveying and categorizing the myriad imaging systems that might be discussed and then suggest a unifying mathematical perspective based on linear algebra and stochastic theory. Next we introduce a key theme of this book, objective or task-based assessment of image quality. Since this approach is essentially statistical, we are led to ruminate on Bayesian and frequentist interpretations of probability. In discussing image quality, probability and statistics, our personal views, developed as we have worked together on imaging issues for two decades, will be much in evidence. The viewpoints presented here are, we hope, more firmly given mathematical form and physical substance in the chapters to follow.

KINDS OF IMAGING SYSTEMS

There are many kinds of objects to be imaged and many mechanisms of image formation. Consequently, there are many ways in which imaging systems can be classified. One such taxonomy, represented by Table I, classifies systems by the kind of radiation or field used to form an image. The most familiar kind of radiation is electromagnetic, including visible light, infrared and ultraviolet radiation. Also under this category we find long-wavelength radiation such as microwaves and radio waves and short-wavelength radiation in the extreme ultraviolet and soft x-ray portions of the spectrum. Of course, the electromagnetic spectrum extends further in both directions, but very long wavelengths, below radio frequencies, do not find much use in imaging, while electromagnetic waves of very short wavelength, such as hard x rays and gamma rays, behave for imaging purposes as particles. Other particles used for imaging include neutrons, protons and heavy ions.

Other kinds of waves are also used in various imaging systems. Mechanical waves are used in seismology, medical ultrasound and even focusing of ocean waves. The DeBroglie principle tells us that matter has both wave-like and particle-like characteristics. The wave character of things we usually call matter is exploited for imaging in scanning tunneling microscopes and in recent work on diffraction of atoms.

Not only radiation, in the usual sense, but also static or quasistatic fields may be the medium of imaging. Magnetic fields are of interest in geophysics and in biomagnetic imaging, while electric fields are imaged in some new medical imaging modalities.

Table I. CLASSIFICATION BY KIND OF RADIATION OR FIELD

Electromagnetic waves	Other waves	Particles	Quasistatic fields
Radio waves	Seismic waves	Neutrons	Geomagnetic fields
Microwaves	Water waves	Protons	Biomagnetic fields
Infrared	Ultrasound	Heavy ions	Bioelectric fields
Visible light	DeBroglie waves	Hard x rays	Electrical impedance
Ultraviolet		Gamma rays	
Soft x rays			

A second useful taxonomy of imaging systems groups them according to the property of the object that is displayed in the image (see Table II). In other words, what does the final image represent?

Table II. CLASSIFICATION BY PROPERTY BEING IMAGED

Optical reflectance	Microwave reflectance	Acoustic reflectance
Photography	Radar	Medical ultrasound
Remote sensing		Sonar
LIDAR		
Source strength	**Concentration**	**Wave amplitude**
Astronomical imaging	Nuclear medicine	Interferometry
Fluorescence microscopy	MRI (spin density)	Seismology
Attenuation	**Index of refraction**	**Scattering properties**
Film densitometry	Phase-contrast microscope	Medical ultrasound
Transmission x ray		Weather radar
Field strength	**Electric, magnetic properties**	**Surface height**
Biomagnetic imaging	Impedance tomography	Optical profilometry
Geomagnetic imaging	MRI (magnetization)	Laser ranging
	MRI (spin relaxation)	Moiré topography

In an ordinary photographic camera, the snapshot usually represents the light reflected from a scene. More precisely, the image reaching the film is related to the product of the optical reflectance of the object and its illumination. Other imaging techniques that essentially map object reflectance include radar imaging and medical ultrasound.

In some instances, however, a photograph measures not reflectance but the source strength of a self-luminous source; a snapshot of a campfire, an astronomical image and a fluorescence micrograph are all examples of emission images. The source strength, in turn, is often related to some concentration or density of physical interest. For example, in nuclear medicine one is interested ultimately in the concentration of some pharmaceutical; if the pharmaceutical is radioactive, its concentration is directly related to the strength of a gamma-ray-emitting source.

Other optical properties can also be exploited for imaging. The index of refraction is used in phase-contrast microscopy, while attenuation or transmissivity of radiation is used in film densitometry and ordinary x-ray imaging. The complex amplitude of a wave is measured in many kinds of interferometry and some forms of seismology, and scattering properties are used in medical ultrasound and weather radar. Electrical and magnetic properties such as impedance and magnetization are of increasing interest, especially in medicine.

We might also classify systems by the imaging mechanism. In other words, how is the image or data set formed? Included in this list (see Table III) are simple refraction and reflection, along with the important optical effects of interference and diffraction. Some imaging systems, however, make use of less obvious physical mechanisms, including scattering and shadow casting. Perhaps the least obvious mechanism is what we shall designate as modulation imaging. In this technique, the imaging system actively modulates the properties of the object being imaged in a space-dependent manner. Examples include the important medical modality of magnetic resonance imaging (MRI) and the lesser-known method of photothermal imaging, which originated with Alexander Graham Bell.

Table III. CLASSIFICATION BY IMAGING MECHANISM

Refraction	Reflection	Diffraction
Eyes	Reflecting telescope	Holographic elements
Microscopes	Wolters x-ray telescope	Kinoforms
Cameras		Binary optics
Refracting telescopes		Fresnel zone plates
Fresnel lenses		

Interference	Scattering	Modulation
Holography	Compton telescope	MRI
Synthetic-aperture radar		Photothermal imaging
Stellar interferometers		
Hanbury Brown/Twiss		

Shadow casting
X-ray computed tomography
Pinhole imaging of x rays
Collimator for gamma rays
Coded apertures

The next dichotomy to consider is direct vs. indirect imaging. By direct imaging we mean any method where the initial data set is a recognizable image. In indirect imaging, on the other hand, a data-processing or reconstruction step is required to obtain the image. Examples of direct and indirect imaging systems are provided in Table IV.

Direct imaging techniques may be divided into serial-acquisition systems or scanners, in which one small region of the object is interrogated at a time, and parallel-acquisition systems where detector arrays or continuous detectors are used to capture many picture elements or pixels in the object simultaneously. Hybrid serial/parallel systems are also possible.

Table IV. DIRECT VS. INDIRECT IMAGING

Direct – serial acquisition	Direct – parallel acquisition
Scanning microdensitometer	Human eye
Medical gamma-ray scanner	Photographic camera
Confocal scanning microscope	Electronic camera
Scanning-tip microscopes	Optical microscope with CCD
Image dissector	Scintillation camera

Indirect
X-ray CT
SPECT and PET
MRI
Holography
Synthetic-aperture radar

Perhaps the most common type of indirect imaging is tomography in all its varied forms, including the now-familiar x-ray computed tomography (CT), emission tomography such as single-photon emission computed tomography (SPECT) and positron emission tomography (PET), as well as MRI and certain forms of ultrasonic and optical imaging. In all of these methods, the data consist of a set of line integrals or plane integrals of the object, and a reconstruction step is necessary to obtain the final image.

The indirect method of coded-aperture imaging is a shadow-casting method used in x-ray astronomy and nuclear medicine. The shadows represent integrals of the object but here the path of integration depends on the shape of the aperture. As above, tomographic information can be retrieved from the data set following a reconstruction step.

An earlier example of indirect imaging is holography, in which the initial data include information, in coded form, about the amplitude and phase of a diffraction pattern. As is implicit in the name ($holo$ = entire), the holographic data are complete in some sense, but are virtually useless to the human observer; again a reconstruction step is required. The holographic principle finds use in nonoptical techniques such as acoustic holography, microwave holography and synthetic-aperture radar or SAR, all indirect imaging methods.

Another principle that leads to specific indirect imaging systems is embodied in the van Cittert – Zernike theorem, relating the intensity distribution of an incoherent source to the coherence properties of the field it produces. Systems that exploit this theorem include the Michelson stellar interferometer and the Hanbury Brown – Twiss interferometer.

The final dichotomy we shall consider is passive vs. active imaging (see Table V). In passive imaging, measurements are made without interacting with a source. Familiar examples include ordinary photography of self-luminous sources or of a reflecting source with natural illumination as well as astronomical imaging and medical thermography. By contrast, an active imaging system supplies the radiation being imaged. Systems in this category include flash photography, transmission imaging (x rays, microscopy, etc.), radar, active SONAR and medical ultrasound.

Table V. PASSIVE VS. ACTIVE IMAGING

Passive systems	Active systems
Fluorescent microscopy	Conventional transmission microscopy
Nuclear medicine	Diagnostic radiology
Lunar imaging with a telescope	Radar ranging of the moon
IR thermography	Photoacoustic imaging
Seismology	Geophysical imaging with explosives
Natural-light photography	Flash photography
Biomagnetic imaging	Magnetic resonance imaging

OBJECTS AND IMAGES AS VECTORS

As we have just seen, many different physical entities can serve as objects to be imaged. In most cases, these objects are functions of one or more continuous variables. In astronomy, for example, position in the sky can be specified by two angles, so the astronomical object being imaged is a scalar-valued function of two variables, or a two-dimensional (2D) function for short. In nuclear medicine, on the other hand, the object of interest is the three-dimensional (3D) distribution of some radiopharmaceutical, so mathematically it is described as a 3D function. Moreover, if the distribution varies with time — not an uncommon situation in nuclear medicine — then a 4D function is required (three spatial dimensions plus time).

Even higher dimensionalities may be needed in some situations. For example, multispectral imagers may be used on objects where wavelength is an important variable. An accurate object description might then require five dimensions (three spatial dimensions plus time and wavelength).

Sometimes the function is vector-valued. In magnetic resonance imaging, for example, the object is characterized by the proton density and two relaxation times, so a complete object description consists of a 3D vector function of space and time.

Images, too, are often functions. A good example occurs in an ordinary camera, where the image is the irradiance pattern on a piece of film. Even if this pattern is time varying, usually all we are interested in is its time integral over the exposure time, so the most natural description of the image is as a continuous 2D spatial function. Similar mathematics applies to the developed film. The image might then be taken as the optical density or transmittance of the film, but again it is a 2D function. A color image is a vector-valued function; the image is represented by the density of the three color emulsions on the film.

Sometimes images are not functions but discrete arrays of numbers. In the camera example, suppose the detector is not film but an electronic detector such as a charge-coupled device (CCD). A CCD is a set of, say, M discrete detector elements, each of which performs a spatial and temporal integration of the image irradiance. The spatial integral extends over the area of one detector element, while the temporal integration extends over one frame period, typically 1/30 sec. As a result of these two integrations, the image output from this detector is simply M numbers per frame. In this example, the object is continuous, but the image is discrete. In fact, any digital data set consists of a finite set of numbers, so a discrete representation is virtually demanded.

Another example that requires a discrete representation of the image is indirect imaging such as computed tomography (CT). This method involves reconstruction of an image of one or more slices of an object from a set of x-ray projection data. Even if the original projection data are recorded by an analog device such as film, a digital computer is usually used to reconstruct the final image. Again, the use of the computer necessitates a discrete representation of the image.

In this book and throughout the imaging literature, mathematical models or representations are used for objects and images, and we need to pay particular attention to the ramifications of our choice of model. Real objects are functions, but the models we use for them are often discrete. Familiar examples are the digital simulation of an imaging system and the digital reconstruction of a tomographic image; in both cases it is necessary to represent the actual continuous object as a discrete set of numbers. A common way to construct a discrete representation of a continuous object is to divide the space into N small, contiguous regions called *pixels* (picture elements) or *voxels* (volume elements). The integral of the continuous function over a single pixel or voxel is then one of N discrete numbers representing the continuous object. As discussed in detail in Chap. 7, many other discrete object representations are also possible, but the pixel representation is widespread.

There are rare circumstances where the object to be imaged is more naturally described by a discrete set of numbers rather than by a continuous function. For example, in some kinds of optical computing systems, data are input by modulating a set of point emitters such as light-emitting diodes. If we regard the optical computer itself as a generalized imaging system, then this array of luminous points is the object being imaged. If there are N emitters in the array, we can consider the object to be defined by a set of N numbers. Even in this case, however, we are free to adopt a continuous viewpoint, treating the point emitters as Dirac delta functions. After all, the object is not a set of discrete numbers but the radiance distribution at the diode array face. In our view, then, any finite, discrete object representation is, at best, an approximation to the real world.

To summarize to this point, both objects and images can be represented as either continuous functions in some number of dimensions or as sets of discrete numbers. Discrete representations for *objects* are not an accurate reflection of the real world and should be used with caution, while discrete representations of *images* may be almost mandatory if a computer is an integral part of the imaging system.

These diverse mathematical descriptions of objects and images can be unified by regarding all of them as *vectors* in some *vector space*. A discrete object model consisting of N pixels can be treated as a vector in an N-dimensional Euclidean space, while a continuous object is a vector in an infinite-dimensional Hilbert space. We shall refer to the space in which the object vector resides as *object space*, denoted $\mathbb{U}$, and we shall consistently use the designation $\mathbf{f}$ to denote the object vector. Similarly, the space in which the data vector is defined will be called *data space* and denoted $\mathbb{V}$. This space will also be referred to as *image space* when direct imaging is being discussed.

IMAGING AS A MAPPING OPERATION

A unifying theme of this book is the treatment of the image formation process as a mapping between object space $\mathbb{U}$ and image space $\mathbb{V}$. If we ignore the statistical nature of the imaging process, this mapping is unique in the sense that a particular object $\mathbf{f}$ maps to a single image $\mathbf{g}$, though it may well be true that many different $\mathbf{f}$ can produce the same $\mathbf{g}$. We shall refer to the mapping operator as $\boldsymbol{\mathcal{H}}$, so that $\mathbf{g} = \boldsymbol{\mathcal{H}}\mathbf{f}$.

The mapping operator $\boldsymbol{\mathcal{H}}$ can be either linear or nonlinear. For many reasons, linear systems are easier to analyze than nonlinear ones, and it is indeed fortunate that we can often get away with the assumption of linearity. One common exception to this statement is that many detectors are nonlinear, or at best only approximately linear over a restricted range of inputs.

Chapter 1 provides the mathematical foundation necessary to describe objects and images as vectors and imaging systems as linear operators. A particular kind of operator will emerge as crucial to the subsequent discussions: *Hermitian operators*, better known in quantum mechanics than in imaging. Study of the eigenvectors and eigenvalues of Hermitian operators will lead to the powerful mathematical technique known as *singular-value decomposition* (SVD). SVD provides a set of basis vectors such that the mapping effect of an arbitrary (not necessarily Hermitian) linear operator reduces to simple multiplication.

If the object and the image are both continuous functions, $\boldsymbol{\mathcal{H}}$ is referred to as a continuous-to-continuous, or CC, operator. If, in addition, $\boldsymbol{\mathcal{H}}$ is linear, the relation between object and image is an integral. Similarly, if both object and image are discrete vectors, $\mathcal{H}$ is referred to as a discrete-to-discrete, or DD, operator. If this DD operator is linear, the relation between object and image is a matrix-vector multiplication.

While both linear models, CC and DD, are familiar and mathematically tractable, neither is really a good description of real imaging systems. As noted above, real objects are continuous functions while digital images are discrete vectors. The correct description of a digital imaging system is thus a continuous-to-discrete, or CD, operator. While such operators may be unfamiliar, they can nevertheless be analyzed by methods similar to those used for CC and DD operators provided the assumption of linearity is valid.

Choice of basis When we describe objects and images as functions, which are vectors in a Hilbert space, we have many options for the basis vectors in this space. One very important basis set consists of plane waves or complex exponentials, and the resulting theory is known broadly as Fourier analysis. When a discrete sum of complex exponentials is used to represent a function, we call the representation a Fourier series, while a continuous (integral) superposition is called a Fourier transform.

A Fourier basis is a natural way to describe many imaging systems. If a spatial shift of the object produces only a similar shift of the image and no other changes, the system is said to be *shift invariant* or to have *translational symmetry*. The mapping properties of these systems are described by an integral operator known as *convolution*, but when object and image are described in the Fourier basis, this mapping reduces to a simple multiplication. In fact, Fourier analysis is equivalent to SVD for linear, shift-invariant systems.

When we use pixels or some other approximate representation of an object, we shall refrain from calling the expansion functions a basis since they do not form a basis for object space $\mathbb{U}$. Of course, any set of functions is trivially a basis for *some* space (the space of all linear combinations of functions in the set), but it is too easy to lose sight of the distinction between true basis functions for objects and the approximate models we construct.

DETECTORS AND MEASUREMENT NOISE

Every imaging system must include a detector, either an electronic or a biological one. Most detectors exhibit some degree of nonlinearity in their response to incident radiation. Some detectors, such as photographic film, are intrinsically very nonlinear, while others, such as silicon photodiodes, are quite linear over several orders of magnitude if operated properly. All detectors, however, eventually saturate at high radiation levels or display other nonlinearities.

Nonlinearities may be either *global* or *local*. With respect to imaging detectors, a global nonlinearity is one in which the response at one point in the image depends nonlinearly on the incident radiation at another (perhaps distant) point. An example would be the phenomenon known as blooming in various kinds of TV camera tubes. In a blooming detector, a bright spot of light produces a saturated image, the diameter of which increases as the intensity of the spot increases.

A simpler kind of nonlinearity is one in which the output of the detector for one image point or detector element depends nonlinearly on the radiation level incident on that element but is independent of the level on other detector elements. This kind of nonlinearity is referred to as a local or point nonlinearity. To a reasonable approximation, film nonlinearities are local. A local nonlinearity may be either *invertible* or *noninvertible*, depending on whether the nonlinear input-output characteristic is monotonic. If the characteristic is monotonic and known, then it can be corrected in a computer with a simple algorithm.

Randomness due to noise is an essential limitation in any measurement system. We can include noise in our basic imaging equation by writing $\mathbf{g} = \mathcal{H}\mathbf{f} + \mathbf{n}$, where $\mathbf{n}$ is a random perturbation added to the noise-free image $\mathcal{H}\mathbf{f}$. Perhaps surprisingly, this additive form is completely general; we can think of $\mathcal{H}\mathbf{f}$ as the mean value of the random data vector $\mathbf{g}$, and $\mathbf{n}$ is then just *defined* by $\mathbf{n} = \mathbf{g} - \mathcal{H}\mathbf{f}$. With this definition, $\mathbf{n}$ has zero mean, but its other statistical properties can depend in a complicated way on the object and the imaging system.

In imaging, the two main noise sources are photon noise, which arises from the discrete nature of photoelectric interactions, and electronic noise in detectors or amplifiers. Photon noise usually obeys the Poisson probability law and electronic noise is almost always Gaussian.

IMAGE RECONSTRUCTION AND PROCESSING

The mapping from object to image is called the *forward problem*: given an object and knowledge of the imaging system, find the image. We are often interested also in the *inverse problem*: given an image (or some other data set), learn as much as we can about the object that produced it. Note that we do not say: given the image, find the object. Except in rare and usually highly artificial circumstances, it will not be possible to determine the object exactly.

An inverse problem is fundamental to indirect imaging systems, where an image-reconstruction step is needed in order to produce the final useful image. Even in direct imaging, a post-detection processing step may be used for image enhancement. For example, it may be desirable to smooth the image before display or to manipulate its contrast. We shall refer to all such manipulations, whether for purpose of image reconstruction or enhancement, as *post-processing*.

When we are explicitly discussing post-processing, it will often be necessary to distinguish the detector output from the final, processed image. We shall reserve the notation $\mathbf{g}$ for the detector output. The vector $\mathbf{g}$ may be the final image in a direct imaging system, but in indirect imaging it will refer to a data set to be processed further by an operator which we can call $\mathcal{O}$. In the latter case, we shall use the notation $\hat{\boldsymbol{\theta}} = \mathcal{O}\mathbf{g}$ to denote the final image, a vector of coefficients in a finite-dimensional (hence approximate) object representation. Since we are virtually never able to find these coefficients exactly, we use the caret to denote an estimate.

With post-processing, we have three vectors to deal with: $\mathbf{f}$, $\mathbf{g}$ and $\hat{\boldsymbol{\theta}}$. These vectors are defined in, respectively, object space, data space and image or reconstruction space.

OBJECTIVE ASSESSMENT OF IMAGE QUALITY

In scientific or medical applications, the goal of imaging is to obtain information about the object. Aesthetic considerations such as bright colors, high contrast or pleasing form play no obvious role in conveying this information, and subjective impressions of the quality of an image without consideration of its ability to convey information can be very misleading. We adopt the view, therefore, that any meaningful definition of image quality must answer these key questions:

1. What information is desired from the image?
2. How will that information be extracted?
3. What objects will be imaged?
4. What measure of performance will be used?

We refer to the desired information as the *task* and the means by which it is extracted as the *observer*. For a given task and observer and a given set of objects, it is possible to define quantitative figures of merit for image quality. We call this approach *objective* or *task-based* assessment of image quality.

Tasks Two kinds of information might be desired from an image: labels and numbers. There are many circumstances in which we want merely to label or identify the object. For example, a Landsat image is used to identify crops and a screening mammogram is used to identify a patient's breast as normal or abnormal.

More and more commonly in modern imaging, however, the goal is to extract quantitative information from images. A cardiologist might want to know the volume of blood ejected from a patient's heart on each beat, for example, or a military photointerpreter might want to know the number of planes on an airfield or the area of an encampment.

Different literatures apply different names to these two kinds of task. In medicine, labeling is *diagnosis* and extraction of numbers is *quantitation*. In statistics, the former is *hypothesis testing* and the latter is *parameter estimation*. In radar, detection of a target is equivalent to hypothesis testing (signal present vs. signal absent), but determination of the range to the target is parameter estimation.

We shall use the term *estimation* to refer to any task that results in one or more discrete numbers, while a task that assigns the object to one of two or more categories will be called *classification*. Various parts of the imaging literature discuss pattern recognition, character recognition, automated diagnosis, signal detection and image segmentation, all of which fall under classification, while other parts of the literature discuss metrology, image reconstruction and quantitation of gray levels in regions of interest, all of which are estimation tasks.

This dichotomy is not absolute, however, since often the output of an estimation procedure is used immediately in a classification, for example, when features are extracted from an image and passed to a pattern-recognition algorithm. Also, both aspects may be desired from a single image; radar, for example, is an acronym for Radio Detection and Ranging, implying that we want both to detect a signal (classification) and to determine its range (estimation). Often the very same image-analysis problem may be logically cast as either classification or estimation. In functional magnetic resonance imaging (fMRI), for example, the task can be formulated as detecting a change in signal as a result of some stimulus or estimating the strength of the change.

The conceptual division of tasks into classification and estimation will serve us well when we discuss image quality in Chaps. 13 and 14, but we shall also explore there the interesting cases of hybrid estimation/classification tasks and the implications of performing one kind of task before the other.

Observers We call the means by which the task gets done, or the strategy, the *observer*. In spite of the anthropomorphic connotation of this term, the observer need not be a human being. Often computer algorithms or mathematical formulas serve as the observer.

In many cases the observer for a classification task *is* a human, a physician viewing a radiograph or a photointerpreter studying an aerial photo, for example. In these cases the label results from the judgment and experience of the human, and the output is an oral or written classification.

It is, however, becoming increasingly common in radiology and other fields of imaging to perform classification tasks by a computer. If not fully computerized diagnosis, at least computer-aided diagnosis is poised to make an important contribution to clinical radiology, and computer screening of specimen slides is a great help to overworked pathologists. Satellite images generate such an enormous volume of data that purely human classification is not practical. In these cases, the computer that either supplements or supplants the human observer can be called a *machine observer*.

A special observer known as the *ideal observer* will receive considerable attention in this book. The ideal observer is defined as the observer that utilizes all statistical information available regarding the task to maximize task performance (see below for measures of performance). Thus the performance of the ideal observer is a standard against which all other observers can be compared.

Estimation tasks using images as inputs are most often performed by computer algorithms. These algorithms can run the gamut from very ad hoc procedures to ones based on stated optimality criteria that have to do with the bias and/or variance of the resulting estimates.

Objects The observer's strategy will depend on the source of the signal, that is, the parameters of the objects that distinguish one class from another if the task is classification, or the parameter to be estimated if the task is estimation. Table II lists possible sources of signal for a variety of imaging mechanisms.

Given a particular object, there will be a fixed signal and fixed background, or nonsignal component, at the input to the imaging system. Real imaging systems collect data from multiple objects, though, and there will therefore be a distribution of the signal-carrying component and background component across the full complement of objects in each class. The observer's strategy will depend on these distributions of signal and background in the object space.

The observer must also make use of all available knowledge regarding the image-formation process, including the deterministic mapping from object space to image space described above, and knowledge of any additional sources of variability from measurement noise, to generate as complete a description of the data as possible. The more accurately the observer's knowledge of the properties of the data, the better the observer will be able to design a strategy for performing the task.

Measures of task performance For a classification task such as medical diagnosis, the performance is defined in terms of the average error rate, in some sense, but we must recognize that there are different kinds of errors and different costs attached to them. In a binary classification task such as signal detection, where there are just two alternatives (signal present and signal absent), there are two kinds of error. The observer can assert that the signal is present when it is not, which is a false alarm or false-positive decision; or the assertion can be that the signal is not present when in fact it is, which would be a missed detection or a false-negative decision. The trade-off between these two kinds of error is depicted in the receiver operating characteristic (ROC) curve, a useful concept developed in the radar literature but now widely used in medicine. Various quantitative measures of performance on a binary detection task can be derived from the ROC curve (see Chap. 13).

Another approach to defining a figure of merit for classification tasks is to define costs associated with incorrect decisions and benefits (or negative costs) associated with correct decisions. Then the value of an imaging system can be defined in terms of the average cost associated with the classifications obtained by some observer on the images.

For an estimation task where the output is only a single number, we can define the performance in terms of an average error, but we must again recognize that there are two types of error: random and systematic. In the literature on statistical parameter estimation, the random error is defined by the variance of the estimate, while systematic error is called bias. Usually in that literature, bias is computed on the basis of a particular model of the data-generation process, but an engineer will recognize that systematic error can also arise because the model is wrong or the system is miscalibrated. Both kinds of bias will be discussed in this book.

Often more than one quantitative parameter is desired in an estimation task. An extreme example is image reconstruction where an attempt is made to assign one parameter to each pixel in the image. In this case some very subtle issues arise in attempting even to define bias, much less to minimize it. These issues get considerable attention in Chap. 15.

PROBABILITY AND STATISTICS

The figures of merit for image quality just mentioned are inherently statistical. They involve the random noise in the images and the random collection of objects that might have produced the images. In our view, any reasonable definition of image quality must take into account both of these kinds of randomness, so probability and statistics play a central role in this book.

The conventional distinction is that probability theory deals with the description of random data while statistics refers to drawing inferences from the data. From a more theoretical viewpoint, probability theory is essentially *deductive logic*, where a set of axioms is presented and conclusions are drawn, but these conclusions are, in fact, already inherent in the axioms. Statistics, with this dichotomy, is *inductive logic*, where the conclusions are drawn from the data, not just the axioms. Thus, as we shall see, we use probability theory to describe the randomness in objects and images. Statistical decision theory is the tool we use in the objective assessment of images formed by an imaging system for some predefined task.

Bayesians, frequentists and pragmatists We expect that our readers will have some previous acquaintance with basic definitions of probability and the associated operational calculus. There are, however, many subtleties and points of contention in the philosophy of probability and statistical inference. The divisions between different schools, classified broadly as frequentists and Bayesians, are profound and often bitter. We do not propose to enter seriously into this fray, but we cannot avoid adopting a point of view and an operational approach in this book. In this section we give a short summary of what this point of view is and why we have adopted it for various problems in image science.

Of the various definitions of probability given in App. C, perhaps the most intuitively appealing is the one that defines the probability of an event as its relative frequency of occurrence in a large number of trials under identical conditions. Because of the historical origins of probability theory, the concept of probability as relative frequency is usually illustrated in terms of games of chance; *e.g.*, the probability of a coin showing heads is the relative number of times it does so in a very large number of trials. In an optical context, it is easy to conceive of an experiment where the number of photoelectric events in some detector, say a photomultiplier, is recorded for many successive one-second intervals. The limit of the histogram of these counts as the number of trials gets very large can be regarded as the probability law for the counts.

To a Bayesian, probability is a measure of degree of belief, an element of inductive logic and not a physical property or an observable quantity. To illustrate this concept and its relation to frequency, we let some well-known Bayesians speak for themselves:

> We ... regard probability as a mathematical expression of our degree of belief in a certain proposition. In this context the concept of verification of probabilities by repeated experimental trials is regarded as *merely* [emphasis added] a means of calibrating a subjective attitude. (Box and Tiao, 1992)

> ... long-run frequency remains a useful notion for some applications, as long as we remain prepared to recognize that the notion must be abandoned if it contradicts our degree of belief .... (Press, 1989)

> The essence of the present theory is that no probability, direct, prior, or posterior, is simply a frequency. (Jeffreys, 1939)

The interpretation we shall intend in this book when we use the word *probability* will depend somewhat on context. We are pragmatists — many times a relative-frequency interpretation will serve our needs well, but we shall not hesitate to use Bayesian methods and a subjective interpretation of probability when it is useful to do so. In particular, we can often present certain conditional probabilities of the data that we can be "reasonably certain" (a measure of belief, to be sure) would be verified by repeated experiments. Other times, as we shall argue below, there is no conceivable way to experimentally verify a postulated probability law, and in those cases we must be content to regard the probabilities in a Bayesian manner.

Our way of resolving this ambivalence will be presented below after we have seen in more detail how some probabilities of interest in imaging can be interpreted as frequencies and others cannot.

Conditional probability of the image data A statistical entity that plays a major role in the description of the randomness in an image is the conditional probability (or probability density) of an image data set obtained from a particular object. Since we usually refer to a data set as **g** and an object as **f**, this conditional probability is denoted $\mathrm{pr}(\mathbf{g}|\mathbf{f})$. Here each component of the image vector **g** refers to a particular measurement; for example, the component g_m could be the measured gray level in the mth pixel on an image sensor. The object **f**, on the other hand, is best conceptualized as a vector (or function) in an infinite Hilbert space.

The conditional probability $\mathrm{pr}(\mathbf{g}|\mathbf{f})$ is the basic description of the randomness in the data, and it is paramount to compute it if we want to give a full mathematical description of the imaging system. Moreover, all inferences we wish to draw about the object **f**, whether obtained by frequentist or Bayesian methods, require knowledge of $\mathrm{pr}(\mathbf{g}|\mathbf{f})$.

Fortunately, $\mathrm{pr}(\mathbf{g}|\mathbf{f})$ is usually quite simple mathematically. For example, with so-called photon-counting imaging systems, it follows from a few simple axioms (enumerated in Chap. 11) that each component of **g** is a Poisson random variable and that different components are statistically independent. As a second example, if there are many independent sources of noise in a problem, the central-limit theorem (derived in Chap. 8) allows us to assert that each component of **g** is a normal or Gaussian random variable, and often the statistical independence of the different components (conditional on a fixed **f**) can be justified as well. In these circumstances, we see no difficulty in regarding $\mathrm{pr}(\mathbf{g}|\mathbf{f})$ as a relative frequency of occurrence. We could easily repeat the image acquisition many times with the same object and accumulate a histogram of data values g_m for each sensor element m. If we have chosen the correct model for the imaging system and all experimental conditions (including the object) are held constant, each of these histograms should approach the corresponding marginal probability $\mathrm{pr}(g_m|\mathbf{f})$ computed on the basis of that model. If we can also experimentally verify the independence of the different values of g_m, we will have shown that the histogram, in the limit of a large number of data sets, is well described by the probability law $\mathrm{pr}(\mathbf{g}|\mathbf{f})$, which in this case is just the product of the individual $\mathrm{pr}(g_m|\mathbf{f})$. Of course, this limiting frequency histogram will depend on the nature of the object and properties of the imaging system, but the mathematical form should agree with our calculation.

The conditional probability $\mathrm{pr}(\mathbf{g}|\mathbf{f})$ is naturally interpreted as a relative frequency since it describes the random variation of the image for repeated observations on the same object. This conditional probability is as much a part of our mathematical description of an imaging system as the system operator $\boldsymbol{\mathcal{H}}$. The distinction is that $\boldsymbol{\mathcal{H}}$ specifies the average or deterministic characteristics of the system, while $\mathrm{pr}(\mathbf{g}|\mathbf{f})$ specifies the randomness in the image, in a relative-frequency sense. Both are needed for a full description.

Priors When we go beyond the statistical description of the data to the task of drawing inferences from the data, we quickly find that the relative-frequency interpretation of probability is quite limiting. The frequentist approach is to disavow all prior knowledge of the object and form inferences using only the data **g**. A little experience with this approach, however, soon convinces one that prior knowledge is essential. If we have only the data, a parameter estimate would logically have to be chosen to be maximally consistent with the data. But the data are often very noisy, so the resulting estimate must be noisy as well. Presented with an image reconstructed by any method that enforces strict agreement with the data, and which incorporates no prior object information, almost everyone would reject it as virtually useless because of the enormous noise amplification that results. This subjective judgment is really an assertion that we have not incorporated prior knowledge of the objects being imaged.

Bayesian methodology allows us to use prior knowledge about the object being imaged in a logically consistent manner in forming a reconstruction or drawing other inferences. In the image-reconstruction problem, incorporation of prior information results in a smoother image that most observers would assess as more in keeping with prior beliefs about the object.

The difficulty is that the Bayesian method requires a specific functional form for the prior object probability. For hypothesis testing, a conditional probability $\mathrm{pr}(\mathbf{f}|H_j)$ is required, where H_j is the jth hypothesis. For estimation of a parameter $\theta(\mathbf{f})$ related to the object, we need $\mathrm{pr}[\theta(\mathbf{f})]$, which might be calculated from $\mathrm{pr}(\mathbf{f})$. But how do we get such prior object information and what does it mean?

The Bayesian would argue that it is illogical to even try to think of $\mathrm{pr}(\mathbf{f})$ in frequency terms. His contention would be that there is but a single object of interest at any one time, so $\mathrm{pr}(\mathbf{f})$ cannot represent a frequency. Rather, it is a logical element representing our state of knowledge about the (unvarying) object. But in fact this knowledge, like all human knowledge, must have been acquired by experience and observation. We can have no conception of what a liver image should look like without seeing at least a few liver images. Different livers, belonging to different patients, *do* vary, so there should be no objection to using this known variation as an element of an inference. Often the crucial determinant in a diagnosis of liver disease will be whether the observed shape represents a normal anatomical variant or a pathological condition, and this question can be addressed only with frequency information about normal livers.

Because of the sheer dimensionality of the problem, however, the full prior $\mathrm{pr}(\mathbf{f})$ cannot be defined *ab initio* in relative-frequency terms. Even for a relatively coarse discrete model, the number of possible vectors $\mathbf{f}$ is astronomical. For example, if we sample $\mathbf{f}$ on a 64×64 grid and allot 256 gray levels (8 bits) to each sample, there are 256^{4096} or some $10^{10,000}$ possible vectors $\mathbf{f}$. No amount of experimentation would give an estimate of the relative frequency of occurrence of these vectors. Moreover, in contrast to the situation with $\mathrm{pr}(\mathbf{g}|\mathbf{f})$, there is no simple mathematical model for $\mathrm{pr}(\mathbf{f})$.

Thus, our view coincides with the Bayesian one at this point: the role of the prior $\mathrm{pr}(\mathbf{f})$ in imaging problems is to incorporate what knowledge we do have about $\mathbf{f}$, not to devise a density from which a sample would resemble the object. A prior on $\mathbf{f}$ must be interpreted as a statement about our knowledge of $\mathbf{f}$, not a relative frequency. Our strongest argument for this conclusion is, however, one of complexity and dimensionality; it does not stem from a belief that it is somehow inappropriate to consider other objects when making inferences about a particular one.

On the other hand, many priors *can* be interpreted as frequencies. Suppose we want to estimate a single scalar parameter such as a cardiac ejection fraction. It is not at all unreasonable to collect data on the frequency of occurrence of different values of this parameter in some population of interest, and it is quite appropriate to regard these data as a prior probability when estimating the parameter for a particular patient (Mr. Jones, say). If we are "confident" (again, a statement of belief) that Mr. Jones has many characteristics in common with the population used to construct the prior, then we can improve the estimate of ejection fraction for Mr. Jones by considering the population data.

Similarly, in hypothesis testing there may well be useful information available about the frequency with which each hypothesis is true. In medical terms, this information is called disease prevalence, and it should certainly be incorporated into any diagnosis. Of course, there may also be information about a particular patient, say history or clinical indications, which would lead the Bayesian diagnostician to alter the prior probability of disease for that particular patient. In this case, subjective and objective (frequency) information can be combined in a single inference.

Our overall attitude to priors is that they should be justified as relative frequencies when they can, but that Bayesian methods should not be avoided when they cannot. The best priors are the most realistic ones, in the sense that they best predict measurable sampling distributions, but less detailed prior information requires a more broadly construed prior probability law.

Image quality revisited Let us now return to the subject of objective, task-based assessment of image quality. Our basic premise is that image quality must be defined in terms of the *average* performance of some observer on some task of practical interest, but there is considerable room for discussion on the choice of task and observer as well as on the meaning of the word "average."

The fact that measures of image quality depend on task and observer is simply a fact of life. We cannot expect one system to be better than another for all possible applications. One system might be superior in terms of quantitative accuracy and another in presenting subtle signals to a human for optimal detection. A diligent reader of this book will learn how to compute figures of merit under a wide variety of conditions for many different tasks and observers.

The meaning of the word average is more subtle. An average is always computed with respect to some probability law, and in imaging every probability law rests on statistical knowledge of the object. As we saw above, that knowledge can be partially frequentist but will always have a significant subjective or Bayesian component.

Even so, the philosophy we have espoused for many years, and adopted for this book, is applicable. Image quality must be assessed on the basis of average performance of some task of interest by some observer or decision maker. Whether we are considering a classification or an estimation task, the performance measures are long-run averages, taking into account both conditional randomness of the data, as specified by $\mathrm{pr}(\mathbf{g}|\mathbf{f})$, and randomness, in the sampling sense, of the object. These averages are unabashedly frequentist concepts, but they seem to us to be unavoidable. An image might be reconstructed or a diagnosis made by a Bayesian method, but ultimately we must keep score. How well does the method perform in the long run? We want to know not only our doctor's personal certainty about our diagnosis, but also his batting average on similar cases. And if the batting averages for some diagnostic task are higher with imaging system A than with system B, we feel justified in claiming that system A is superior to system B for this task.

This is the point at which our approach differs from the strict Bayesian approach. We allow the use of prior information, and even subjective priors, in situations where sampling priors are unavailable or unreliable, but we compute the final performance of the procedure by averaging in a frequentist sense. This pragmatic approach, which will probably earn us the scorn of both camps, strikes us as a sensible middle ground. To a Bayesian who challenges this approach, we pose a question: How else would you compare two priors, two reconstruction algorithms or two imaging systems? The frequentist, on the other hand, can hardly object to our allowance of subjective priors; they are, after all, justified in a frequentist sense if they give improved long-run performance.

Accounting for the object randomness in performance assessment could be done analytically if we had the full probability density function $\mathrm{pr}(\mathbf{f})$, but we have already seen that this is difficult, if not impossible, even for the Bayesian. In practice, we must estimate the performance rather than compute it analytically. This estimate can be arrived at by using actual or simulated images, *i.e.*, samples drawn from $\mathrm{pr}(\mathbf{f})$. The theoretical expressions for various diagnostic performance measures will be expressed as integrals over an object space of huge dimensionality, but in practice the integrals can be accurately estimated with an amazingly small number of samples. Procedures for carrying out this performance estimate for hypothesis-testing tasks will be detailed in Chap. 14.

Similarly, for an estimation task, we can compute the conditional estimation performance, for example the bias and variance of the estimate for some assumed object, then average the result over possible objects by sampling. In this procedure, any prior whatsoever can be used to find the parameter estimate in the first place, but the performance of that estimate is itself estimated with respect to a realistic sampling prior.

To return to a point raised in the Preface, we believe (however subjectively) that this approach is essential if imaging is to qualify as a science. One of the required elements of a science is that its theories have predictive value. For decades, centuries even, imaging has had its predictive theories for the propagation of light and the formation of images, but only much more recently has it been possible to predict the usefulness of those images. There is now a science of image evaluation that completes the picture by providing a formal, predictive theory of imaging and image quality. We hope after reading this book that you will be able to apply this theory toward the objective evaluation of the imaging systems that are of interest to you.

Contents

1

Vectors and Operators

As discussed in the Prologue, an imaging system is a transformation or mapping from an object to an image. It facilitates the analysis of imaging systems to regard the object as a vector in some space and the image as a vector in some other space, so that the system is an operator mapping from one space to another. To do this, we need a concept of a vector that is more general than the familiar one of conventional 3D vector analysis. For digital images, the extension is easy: we simply define a vector with the number of components equal to the number of pixels in the image and presume that the usual rules of conventional vector analysis still apply. It is often necessary, however, to regard the object, image or both as a function of one or more continuous[1] variables. To preserve the notion of an imaging system as an operator mapping between an object space and an image space in these cases, we must have a more general notion of a vector and the space in which it is defined.

Several different branches of mathematics, including linear algebra, functional analysis and integral transforms, relate to description of these general spaces and mappings between them. Linear algebra, at least as taught at an elementary level, deals with matrices, which can be viewed as mappings from one finite-dimensional space to another: an $M \times N$ matrix maps or transforms an N-dimensional vector to an M-dimensional one (Smith, 1984). A function is another kind of mapping. A scalar-valued function of a real variable x maps one point on the real line to another, thus transforming a scalar to another scalar. A *functional*, on the other hand, maps a function to a scalar; a simple example is a definite integral of the function. An integral transform is a functional that depends on a continuous parameter, so that

[1]The word *continuous* has many meanings in mathematics and engineering. The reader should be careful to distinguish a *continuous variable* from a *continuous function.* A scalar that can assume any value over some segment of the real line is called a continuous variable to distinguish it from a discrete index. The elementary meaning of *continuous* for functions is presumed to be known to reader, and the extension of this concept to other mappings is discussed in Sec. 1.3.2.

the result can be thought of as a function itself. In other words, integral transforms, such as the Fourier transform, map a function to a function.

These different kinds of mappings are often treated separately in elementary courses, but there is an essential unity among them. In all cases we are dealing with vectors in some generalized sense and with operators that map one vector to another. Our approach in this chapter is to build a generalized theory on the foundation of ordinary 3D vector analysis with which the reader is presumably familiar.

At a few places in this chapter, we anticipate some simple properties of Dirac delta functions and Fourier transforms, discussed in more detail in Chaps. 2 and 3, respectively. The reader who is unfamiliar with these topics may wish to read at least Secs. 2.2.1, 2.2.4, 3.2.1 and 3.3.1 as background for this chapter.

1.1 LINEAR VECTOR SPACES

1.1.1 Vector addition and scalar multiplication

The simplest definition of vectors is that they are objects that can be added to each other and multiplied by numbers (Gel'fand, 1961). The familiar vectors from 3D vector analysis obviously satisfy this definition. We know that the multiplication of a vector $\mathbf{f}$, with Cartesian components (f_x, f_y, f_z), by the scalar λ yields a new vector $\lambda\mathbf{f}$, the components of which are $(\lambda f_x, \lambda f_y, \lambda f_z)$. Similarly, addition of two vectors $\mathbf{f}_1$ and $\mathbf{f}_2$ is defined as the operation that yields the new vector $\mathbf{f}_1 + \mathbf{f}_2$, with components given by the sum of the corresponding components of $\mathbf{f}_1$ and $\mathbf{f}_2$.

These definitions extend easily to many other kinds of vectors. Any ordered set of N numbers, called an N-tuple, can be thought of as an N-dimensional vector, with rules for scalar multiplication and addition analogous to the ones above. The individual numbers in the N-tuple are called the components of the vector, and the collection of all such vectors is called a *linear vector space*. If each component is a real number in the range $(-\infty, \infty)$, the space is designated $\mathbb{R}^N$. Hence the usual 3D space of vector analysis is $\mathbb{R}^3$, and each vector in that space is a 3-tuple or triplet of real numbers. Similarly, if each vector is an N-tuple of complex numbers, and each component can assume any value in the complex plane, the space is denoted $\mathbb{C}^N$.

As a less familiar example, consider a quadratic function $f_1(x)$ defined by

$$f_1(x) = a_1x^2 + b_1x + c_1\,. \tag{1.1}$$

Multiplication of this function by the scalar λ gives a new quadratic $\lambda f_1(x) = \lambda a_1x^2 + \lambda b_1x + \lambda c_1$. Addition of $f_1(x)$ to a similar quadratic $f_2(x) = a_2x^2 + b_2x + c_2$ yields a third quadratic $f_3(x)$ with coefficients $a_3 = a_1 + a_2$, etc. The collection of all quadratics of the form of (1.1) thus constitutes a vector space, and the functions themselves are the vectors.

An easy extension of this argument shows that any polynomial qualifies as a vector, and the collection of all polynomials is another vector space. The space of polynomials of degree N or less is formally $\mathbb{R}^N$ or $\mathbb{C}^N$, depending on whether the coefficients are real or complex. Other vector spaces where the vectors are functions will be defined below. Throughout this book, we shall designate vectors by boldface type, even when they are also functions, so $f(x)$ will also be written as $\mathbf{f}$.

A formal definition of a linear vector space is that it is a set of elements, called vectors, in which the operations of addition and multiplication by a scalar are defined and satisfy the following conditions (Gellert *et al.*, 1977; Stakgold, 1979):

(a) Commutative property of scalar multiplication: $\mu(\lambda\mathbf{f}) = \lambda(\mu\mathbf{f})$, where λ and μ are arbitrary real or complex numbers.

(b) Distributive property of scalar multiplication: $(\lambda+\mu)\mathbf{f} = \lambda\mathbf{f}+\mu\mathbf{f}$.

(c) Existence of identity operator for scalar multiplication: $1\mathbf{f} = \mathbf{f}$.

(d) Commutative property of addition: $\mathbf{f}_1+\mathbf{f}_2 = \mathbf{f}_2+\mathbf{f}_1$.

(e) Associative property of addition: $(\mathbf{f}_1+\mathbf{f}_2)+\mathbf{f}_3 = \mathbf{f}_1+(\mathbf{f}_2+\mathbf{f}_3)$.

(f) Distributive property of scalar multiplication with respect to addition: $\lambda(\mathbf{f}_1+\mathbf{f}_2) = \lambda\mathbf{f}_1+\lambda\mathbf{f}_2$.

(g) Existence of the zero vector, denoted $\mathbf{0}$ and defined such that $0\mathbf{f} = \mathbf{0}$ and $\mathbf{0}+\mathbf{f} = \mathbf{f}$ for all $\mathbf{f}$, where 0 (without boldface) is the ordinary scalar zero.

(h) Continuity of addition: If $\lim_{n\to\infty}\mathbf{f}_n = \mathbf{f}$ and $\lim_{n\to\infty}\mathbf{g}_n = \mathbf{g}$, then $\lim_{n\to\infty}(\mathbf{f}_n+\mathbf{g}_n) = \mathbf{f}+\mathbf{g}$.

(i) Continuity of multiplication: If $\lim_{n\to\infty}\mathbf{f}_n = \mathbf{f}$ and $\lim_{n\to\infty}\lambda_n = \lambda$, then $\lim_{n\to\infty}\lambda_n\mathbf{f}_n = \lambda\mathbf{f}$.

1.1.2 Metric spaces and norms

A familiar concept from 3D vector analysis is that of a distance or metric. The 3D vector $\mathbf{f}_1$ defines a point with Cartesian coordinates[2] (f_{1x}, f_{1y}, f_{1z}) in the 3D space, while $\mathbf{f}_2$ defines the point (f_{2x}, f_{2y}, f_{2z}). One common way of defining the distance $d(\mathbf{f}_1,\mathbf{f}_2)$ between $\mathbf{f}_1$ and $\mathbf{f}_2$ is

$$d(\mathbf{f}_1,\mathbf{f}_2) = \left[(f_{1x}-f_{2x})^2+(f_{1y}-f_{2y})^2+(f_{1z}-f_{2z})^2\right]^{\frac{1}{2}}. \qquad (1.2)$$

It is easy to show that this definition satisfies the following conditions:

(a) $d(\mathbf{f}_1,\mathbf{f}_2) = d(\mathbf{f}_2,\mathbf{f}_1)$;

(b) $d(\mathbf{f}_1,\mathbf{f}_2) \geq 0$;

(c) $d(\mathbf{f}_1,\mathbf{f}_2) = 0$ if and only if $\mathbf{f}_1 = \mathbf{f}_2$;

(d) $d(\mathbf{f}_1,\mathbf{f}_3) \leq d(\mathbf{f}_1,\mathbf{f}_2)+d(\mathbf{f}_2,\mathbf{f}_3)$ (triangle inequality).

In more general settings, we accept as a possible distance or metric any quantity that satisfies these four relations. A linear vector space for which the distance

[2]Do not confuse the notations $\mathbf{f}_n$ and f_n. The boldface $\mathbf{f}_n$ denotes the nth vector in a set, while f_n denotes the nth component of the vector $\mathbf{f}$. Thus, by this notation, the jth component of the vector $\mathbf{f}_n$ is denoted f_{nj}.

between any two vectors is defined, in a way consistent with these four conditions, is called a *metric space*. Different metric spaces are distinguished by the definition of distance employed.

Closely related to distance is the concept of a *norm*. In ordinary vector analysis, the norm or length of a vector is the distance between the vector and the origin (or the zero vector). In general, we denote the norm of $\mathbf{f}$ as $||\mathbf{f}||$ and require that it satisfy

(*a*) $||\mathbf{f}|| \geq 0$, with equality if and only if $\mathbf{f} = \mathbf{0}$;

(*b*) $||\lambda \mathbf{f}|| = |\lambda| \cdot ||\mathbf{f}||$, λ an arbitrary real or complex number;

(*c*) $||\mathbf{f}_1 + \mathbf{f}_2|| \leq ||\mathbf{f}_1|| + ||\mathbf{f}_2||$ (triangle inequality).

Again, it is easy to show that these requirements are met by the conventional definition of length in vector analysis, but many other definitions are also consistent with them.

If we have any definition of norm, it leads immediately to a corresponding definition of distance. The distance $d(\mathbf{f}_1, \mathbf{f}_2)$ between $\mathbf{f}_1$ and $\mathbf{f}_2$ can be taken as the norm of $\mathbf{f}_1 - \mathbf{f}_2$.

In an N-dimensional vector space $\mathbb{R}^N$ or $\mathbb{C}^N$, where a vector $\mathbf{f}$ is an ordered set of N numbers $\{f_n\}$, some possible definitions of the norm are:

$$||\mathbf{f}||_2 = \left[\sum_{n=1}^{N} |f_n|^2\right]^{\frac{1}{2}} \qquad (\mathbb{L}_2 \text{ norm}), \tag{1.3}$$

$$||\mathbf{f}||_1 = \sum_{n=1}^{N} |f_n| \qquad (\mathbb{L}_1 \text{ norm}), \tag{1.4}$$

$$||\mathbf{f}||_p = \left[\sum_{n=1}^{N} |f_n|^p\right]^{\frac{1}{p}} \qquad (\mathbb{L}_p \text{ norm}), \tag{1.5}$$

$$||\mathbf{f}||_\infty = \lim_{p\to\infty} ||\mathbf{f}||_p = \max_{n=1,\ldots,N} |f_n| \qquad (\mathbb{L}_\infty \text{ norm}). \tag{1.6}$$

In the $\mathbb{L}_p$ norm, p can be any positive, real number, not necessarily an integer. The $\mathbb{L}_\infty$ norm is also called the *sup norm* (for *supremum*).

If the $\mathbb{L}_2$ norm is adopted in an N-dimensional space, we refer to the space as *Euclidean* and denote it as $\mathbb{E}^N$, regardless of whether the components are real or complex.

1.1.3 Sequences of vectors and complete metric spaces

Another important concept in a vector space is the notion of the limit of a sequence of vectors, written as

$$\lim_{j\to\infty} \mathbf{f}_j = \mathbf{a}. \tag{1.7}$$

Several interpretations of this limit can be given. As a simple example, consider a sequence of ordinary 3D vectors $\mathbf{f}_j$ with components (f_{jx}, f_{jy}, f_{jz}). We can say that $\mathbf{f}_j \to \mathbf{a}$ as $j \to \infty$, provided each of the three components $f_{jn} (n = x, y, z)$ approaches the corresponding component a_n of $\mathbf{a}$. Similarly, a sequence of vectors in either $\mathbb{R}^N$ or $\mathbb{C}^N$ approaches a limiting vector $\mathbf{a}$ if each component of the vectors in the sequence limits to the corresponding component of $\mathbf{a}$, *i.e.*, if $f_{jn} \to a_n$ as $j \to \infty$ for $n = 1, ..., N$.

There is another way to interpret (1.7) if we have a distance measure in our space. A sequence of vectors $\mathbf{f}_j$ in a metric space converges to the fixed vector $\mathbf{a}$ if the distance from $\mathbf{f}_j$ to $\mathbf{a}$, *i.e.*, $||\mathbf{f}_j - \mathbf{a}||$, tends to zero. The precise interpretation of the limit then depends on the metric chosen.

If the distance from one vector in the sequence, $\mathbf{f}_j$, to another, $\mathbf{f}_k$, approaches zero as j and k both go to infinity, the sequence is called a *Cauchy sequence.* Every convergent sequence is a Cauchy sequence, but the converse is not true in general because the limit vector may not be in the space (Stakgold, 1967).

If the limit vector $\mathbf{a}$ is an element of the same space in which the $\mathbf{f}_j$ are defined, for every Cauchy sequence, then the space is said to be *complete.* Several examples of complete and incomplete metric spaces are given by Kreysig (1978). For example, the real line and the complex plane are complete, but they become incomplete if even a single isolated point is omitted, or if only rational numbers are considered. The N-dimensional Euclidean space is complete for any finite N. The space of all polynomials of degree M or less is complete when the $\mathbb{L}_2$ norm is used, but not with $\mathbb{L}_\infty$. The space of all continuous, absolutely integrable functions on $[0, 1]$ is not complete, but the space of all absolutely integrable functions on that interval is complete.

A complete normed linear space is called a *Banach space.* The appellation honors Stefan Banach (1892–1945), a founder of functional analysis and member of the illustrious group of mathematicians who met at the Scottish Cafe in Lvov (now Lviv, Ukraine) in the 1930s (Kaluza, 1996).

1.1.4 Scalar products and Hilbert space

The space of most use to us in this book is Hilbert space, after David Hilbert (1862–1943). Hilbert launched the twentieth century (for mathematicians, at least) with a talk entitled "Mathematical Problems," delivered to the Second International Congress of Mathematicians in Paris in 1900. In it, he listed 23 unsolved problems that would occupy the best mathematical minds far into the future. Problem no. 6, in which Hilbert himself took the lead, called for a rigorous mathematical treatment of the axioms of physics. In the first decade of the century, Hilbert and his student Erhard Schmidt developed a theory of integral equations and functions of infinitely many variables. Hilbert showed that many obscure analytic relations in this field become almost intuitively obvious when stated in geometric terms (Reed, 1996). This work, which Hilbert called "spectral theory," can be regarded as the birth of modern linear algebra, while Hilbert space is the mathematical underpinning of quantum mechanics and hence all of modern physics (Boyer and Merzbach, 1989).

A Hilbert space, an important special case of a normed linear space, is a Banach space in which a scalar product is defined. Again we introduce the subject by appeal to ordinary vector analysis, where the scalar or dot product of two 3D vectors $\mathbf{a}$ and $\mathbf{b}$ is defined as

$$(\mathbf{a}, \mathbf{b}) = a_x b_x + a_y b_y + a_z b_z = ||\mathbf{a}|| \cdot ||\mathbf{b}|| \cos[\theta(\mathbf{a}, \mathbf{b})] \,, \tag{1.8}$$

where $\theta(\mathbf{a}, \mathbf{b})$ is the angle between $\mathbf{a}$ and $\mathbf{b}$. Thus, for ordinary 3D vectors, both norm and scalar product are defined. The norm expresses the length of a vector, while the scalar product of two vectors depends not only on their lengths, but also on the angle between them. Since we have already noted that the 3D Euclidean space is complete, the existence of a scalar product as defined above makes it a Hilbert space.

Scalar products In a general Hilbert space, the scalar or inner product $(\mathbf{f}_1, \mathbf{f}_2)$ is a complex-valued functional of the vectors $\mathbf{f}_1$ and $\mathbf{f}_2$ with the following properties:

(a) $(\mathbf{f}_1, \mathbf{f}_2) = (\mathbf{f}_2, \mathbf{f}_1)^*$;

(b) $(\mathbf{f}_1, \lambda\mathbf{f}_2) = \lambda(\mathbf{f}_1, \mathbf{f}_2)$;

(c) $(\lambda\mathbf{f}_1, \mathbf{f}_2) = \lambda^*(\mathbf{f}_1, \mathbf{f}_2)$;

(d) $(\mathbf{f}_1 + \mathbf{f}_2, \mathbf{f}_3) = (\mathbf{f}_1, \mathbf{f}_3) + (\mathbf{f}_2, \mathbf{f}_3)$;

(e) $(\mathbf{f}_1, \mathbf{f}_1) \geq 0$, with equality if and only if $\mathbf{f}_1$ is the zero vector.

Property (c) follows from (a) and (b) but is listed separately since it is an important operational rule. Slightly different versions of (b) and (c) are often used; many authors, especially in the mathematics literature, require that $(\mathbf{f}_1, \lambda\mathbf{f}_2) = \lambda^*(\mathbf{f}_1, \mathbf{f}_2)$. The convention given above, however, is common in the physics literature and is the one used throughout this book.

A Hilbert space is always normed, with the norm $||\mathbf{f}||$ defined as $\sqrt{(\mathbf{f}, \mathbf{f})}$, so all Hilbert spaces are Banach spaces. The converse, however, does not hold; not all norms are compatible with the required properties of scalar products. For example, the $\mathbb{L}_p$ norm with $p \neq 2$, defined in (1.5), cannot be generated from a scalar product (Stakgold, 1979). If we choose to work with $\mathbb{L}_2$, we can therefore take advantage of many useful properties of Hilbert spaces, but the choice of $\mathbb{L}_p$ with $p \neq 2$ confines us to the more general Banach space. Since many of the results in the remainder of this chapter are derived for Hilbert spaces, this is a powerful incentive to choose the $\mathbb{L}_2$ norm.

Euclidean and $\mathbb{L}_2$ spaces An important example of a Hilbert space is the N-dimensional Euclidean space $\mathbb{E}^N$, where each vector $\mathbf{f}$ is an ordered set of complex numbers $\{f_n,\ n = 1, ..., N\}$, and the scalar product is defined as

$$(\mathbf{f}_1, \mathbf{f}_2) = \sum_{n=1}^{N} f_{1n}^* \, f_{2n} \, . \tag{1.9}$$

It is straightforward to show that this definition satisfies properties $(a)-(e)$ above, and the $\mathbb{L}_2$ norm of (1.3) follows at once by setting $\mathbf{f}_1 = \mathbf{f}_2$ and using $||\mathbf{f}|| = \sqrt{(\mathbf{f}, \mathbf{f})}$.

Another important example of a Hilbert space, usually denoted $\mathbb{L}_2(\alpha, \beta)$, or just $\mathbb{L}_2$ for short, is the space of complex-valued functions $f(x)$ with a scalar product defined by

$$(\mathbf{f}_1, \mathbf{f}_2) = \int_{\alpha}^{\beta} dx \; f_1^*(x) \, f_2(x) \, . \tag{1.10}$$

In this space, the norm is defined by

$$||\mathbf{f}|| = \sqrt{(\mathbf{f}, \mathbf{f})} = \left[\int_{\alpha}^{\beta} dx \; |f(x)|^2 \right]^{\frac{1}{2}} . \tag{1.11}$$

For this norm to exist, the integral of the squared modulus of the function over the range (α, β) must exist, so $\mathbb{L}_2(\alpha, \beta)$ is frequently referred to as the space of square-integrable functions. The integral may have to be interpreted in the Lebesgue sense in some cases, but we do not need to go into that detail here. For a concise introduction to Lebesgue integration, see Champeney (1987) or Friedman (1991).

Comparison of (1.9) and (1.10) shows the analogy between the discrete N-dimensional Euclidean space and the space $\mathbb{L}_2(\alpha, \beta)$. In the latter, scalar products and norms have the same structure as in the former, but the discrete sums are replaced with continuous integrals. In a sense, each x specifies a component of $\mathbf{f}$. Thus we may loosely view $\mathbb{L}_2(\alpha, \beta)$ as a Euclidean space with infinitely many dimensions.

Often we shall have occasion to consider functions of the form $f(\mathbf{r})$, where $\mathbf{r}$ is itself a vector. If $\mathbf{r}$ is defined in two spatial dimensions, it is a vector in $\mathbb{R}^2$. The function $f(\mathbf{r})$, on the other hand, if it is square-integrable, is a vector in a Hilbert space in which norms and scalar products are defined by 2D integrals. If the range of integration in each dimension is $(-\infty, \infty)$, we denote the Hilbert space of $f(\mathbf{r})$ as $\mathbb{L}_2[(-\infty, \infty) \times (-\infty, \infty)]$ or $\mathbb{L}_2(\mathbb{R}^2)$. In the same spirit, $f(\mathbf{r})$ may be a square-integrable function defined on the unit disc, *i.e.*, the 2D region interior to the unit circle, which we shall denote as $\mathbb{D}$. In this case, the function $f(\mathbf{r})$ is said to be a vector in $\mathbb{L}_2(\mathbb{D})$. Where no confusion is likely to result, we shall just use $\mathbb{L}_2$ to denote the Hilbert space of square-integrable functions $f(\mathbf{r})$, regardless of the space in which $\mathbf{r}$ is defined.

Weighted scalar products Sometimes it is useful to define scalar products with a weighting function in the integrand. For example, we might define

$$(\mathbf{f}_1, \mathbf{f}_2)_w = \int_\alpha^\beta dx\; w(x)\, f_1^*(x)\, f_2(x)\,, \tag{1.12}$$

where we must require $w(x)$ to be real and ≥ 0 to avoid negative or imaginary norms. In this space, denoted $\mathbb{L}_2(\alpha, \beta; w(x))$, the norm is defined by

$$||\mathbf{f}||_w = \left[\int_\alpha^\beta dx\; w(x)\, |f(x)|^2 \right]^{\frac{1}{2}} . \tag{1.13}$$

The advantage of this space is that we can deal with functions that are not, by themselves, square-integrable. For example, $\cos x$ is not in $\mathbb{L}_2(-\infty, \infty)$ but it is in $\mathbb{L}_2(-\infty, \infty; \exp(-x^2))$.

Sobolev space *Sobolev space* is the general term for a function space (not necessarily a Hilbert space) where the norm involves derivatives. The *Sobolev-Hilbert space of order* N, denoted $\mathbb{W}_N$, consists of functions $f(x)$ such that $f(x)$ and all of its derivatives up to order $N-1$ are absolutely continuous, while the Nth derivative lies in $\mathbb{L}_2$ (Wahba, 1990).

Schwarz inequality It follows from the definitions of scalar products and norms that the Schwarz inequality holds in every Hilbert space. This important inequality states that

$$|(\mathbf{f}_1, \mathbf{f}_2)|^2 \leq ||\mathbf{f}_1||^2 \cdot ||\mathbf{f}_2||^2\,, \tag{1.14}$$

where the equality holds if and only if $\mathbf{f}_2 = \gamma\mathbf{f}_1$, with γ a constant. Once again, this result is familiar from ordinary 3D vector analysis, where it merely states that the cosine of the angle between two vectors is less than or equal to unity. For real functions, it is reasonable to interpret $(\mathbf{f}_1, \mathbf{f}_2)/(||\mathbf{f}_1|| \cdot ||\mathbf{f}_2||)$ as the cosine of the angle between $\mathbf{f}_1$ and $\mathbf{f}_2$ in spaces of any dimensionality.

1.1.5 Basis vectors

In conventional vector analysis, we are familiar with the concept of basis vectors or unit vectors along the coordinate axes. Essentially the same concept is applicable to the more general vectors we are discussing here.

If $\mathbf{f}$ is a vector in an N-dimensional space, it is possible to find a set of N vectors $\{\mathbf{u}_n, n = 1, ..., N\}$ such that $\mathbf{f}$ can be represented as

$$\mathbf{f} = \sum_{n=1}^{N} \alpha_n \mathbf{u}_n \, , \tag{1.15}$$

where the $\{\alpha_n\}$ are a set of scalar coefficients particular to the vector $\mathbf{f}$ being represented. If all vectors in the space can be represented in terms of the set $\{\mathbf{u}_n\}$ in this way, the set is said to be *complete* or to form a basis, and the space is said to be spanned by $\{\mathbf{u}_n\}$. (Do not confuse this meaning of the word complete with its meaning in complete metric spaces; unfortunately, both usages are common.) Conversely, if it is possible to find a set of N linearly independent vectors such that any $\mathbf{f}$ in the space can be represented as in (1.15), then we say that the space is N-dimensional.

The choice of a basis is not unique; infinitely many different sets $\{\mathbf{u}_n\}$ can be found to represent $\mathbf{f}$ in the form of (1.15). When a particular set has been chosen, however, the set of components $\{\alpha_n\}$ is a representation of the vector in the given basis. For finite-dimensional spaces, it is conventional to write these numbers in a column and refer to them as a column vector. It must be remembered, however, that this set of numbers will be different for different choices of the basis vectors, while the vector $\mathbf{f}$ has a meaning and algebraic properties quite independent of the basis.

The minimum requirement to impose on a set of basis vectors is that they be linearly independent, so that it is not possible to write one of them as a linear combination of the others. It is very convenient to require further that they be *orthonormal*, which means that they are *orthogonal* and *normalized*. Two vectors $\mathbf{u}_m$ and $\mathbf{u}_n$ $(m \neq n)$ are said to be orthogonal if their scalar product vanishes. If every vector in the set is orthogonal to every other, the set is linearly independent. A vector $\mathbf{u}_n$ is normalized if its norm $||\mathbf{u}_n|| = 1$. Since $||\mathbf{u}_n||^2 = (\mathbf{u}_n, \mathbf{u}_n)$, these two requirements can be combined into a single orthonormality condition, written as

$$(\mathbf{u}_m, \mathbf{u}_n) = \delta_{mn} \, . \tag{1.16}$$

Here the symbol δ_{mn}, with value 1 when $m = n$ and 0 if $m \neq n$, is known as the *Kronecker delta*, after Leopold Kronecker (1823–1891), a number theorist who amassed a fortune in Bismarck's Germany. It was Kronecker who said, "God made the integers; all the rest is the work of man" (Bell, 1937).

If the basis vectors satisfy the orthonormality condition, we can easily determine the expansion coefficients $\{\alpha_n\}$. Taking the scalar product of both sides of

(1.15) with $\mathbf{u}_m$, we find

$$(\mathbf{u}_m, \mathbf{f}) = \sum_{n=1}^{N} \alpha_n (\mathbf{u}_m, \mathbf{u}_n) = \sum_{n=1}^{N} \alpha_n \, \delta_{mn} = \alpha_m \,. \tag{1.17}$$

So the expansion coefficient α_m is just the scalar product of the basis vector $\mathbf{u}_m$ with the vector $\mathbf{f}$ being expanded, provided, of course, that a scalar product is defined and the orthonormality condition holds.

For functions (vectors) in an infinite-dimensional Hilbert space, the situation is more delicate. It is not possible to represent an arbitrary vector in such a space by a finite sum such as (1.15), and it is not obvious that it is possible to salvage the formalism by considering an infinite sum.

Fortunately, for a wide class of Hilbert spaces with infinite dimensionality, it *is* possible to expand an arbitrary vector in terms of a countably infinite set of basis vectors, so that the expansion of (1.15) is valid with the simple modification of setting the upper summation limit to ∞. Hilbert spaces for which such a countable basis exists are called *separable.* Almost all Hilbert spaces used in analysis are separable, including $\mathbb{L}_2(\alpha, \beta)$ for α and β finite or infinite (Stakgold, 1979). For basis functions on these spaces, orthonormality is still specified by (1.16), and the expansion coefficients are still given by (1.17), though the indices run from 1 to ∞.

1.1.6 Continuous bases

While it is always possible to use a denumerable basis set for any of the Hilbert spaces considered here, it may nevertheless be convenient or desirable to use a nondenumerably infinite set, especially for $\mathbb{L}_2(-\infty, \infty)$. In that case, the basis function is designated $u_\nu(x)$, where ν is a continuous index, and we refer to the set $\{u_\nu(x)\}$ for all ν in some range as a *continuous basis.*

To demonstrate the utility of continuous bases, we anticipate a few results from Chap. 3 on Fourier analysis. A function $f(x)$ defined on $\mathbb{L}_2(-\frac{1}{2}a, \frac{1}{2}a)$ can be expanded in a Fourier series of the form

$$f(x) = \frac{1}{\sqrt{a}} \sum_{n=-\infty}^{\infty} F_n \exp\left(\frac{2\pi i n x}{a}\right), \tag{1.18}$$

where the expansion coefficient F_n is given by

$$F_n = \frac{1}{\sqrt{a}} \int_{-a/2}^{a/2} dx \; f(x) \exp\left(-\frac{2\pi i n x}{a}\right). \tag{1.19}$$

These equations have the same structure as (1.15) and (1.17) if we take $u_n(x) = [1/\sqrt{a}\,] \exp(2\pi i n x/a)$, so these functions comprise a countable orthonormal basis on $\mathbb{L}_2(-\frac{1}{2}a, \frac{1}{2}a)$. But what happens if we let $a \to \infty$? We cannot continue to define $u_n(x)$ with the factor of $1/\sqrt{a}$ since then $u_n(x)$ would tend to zero for all x. From the theory of Fourier *transforms*, however, we know that we can use as a basis the functions $u_\nu(x) = \exp(2\pi i \nu x)$, where the label ν ranges continuously from $-\infty$ to ∞. The expansion of $f(x)$ is then given by

$$f(x) = \int_{-\infty}^{\infty} d\nu \; F(\nu) \, u_\nu(x) \,. \tag{1.20}$$

The function $F(\nu)$ is known as the Fourier transform of $f(x)$, but in this discussion it plays the role of a set of expansion coefficients, analogous to F_n except that $F(\nu)$ depends on the continuous variable ν. To find the expansion coefficients, we follow a prescription similar to (1.17) and take the scalar product of both sides of (1.20) with $u_{\nu'}(x)$. The result is

$$F(\nu) = \int_{-\infty}^{\infty} dx\; u_\nu^*(x)\, f(x)\,, \tag{1.21}$$

provided

$$(\mathbf{u}_\nu, \mathbf{u}_{\nu'}) = \int_{-\infty}^{\infty} dx\; u_\nu^*(x)\, u_{\nu'}(x) = \delta(\nu - \nu')\,, \tag{1.22}$$

where $\delta(\nu - \nu')$ is the Dirac delta function defined in Chap. 2. Equation (1.22) is in fact valid for $u_\nu(x) = \exp(2\pi i \nu x)$, but it also holds for many other sets of functions. It can be regarded as the generalization of the orthonormality relation, (1.16), when the functions are indexed by a continuous variable ν rather than a discrete index n.

Though developed in the context of Fourier analysis, (1.20) – (1.22) have much wider applicability. If all functions $f(x)$ in our space can be expanded in terms of some set of functions $u_\nu(x)$ satisfying (1.22), we say that the functions form a complete, continuous, orthonormal basis for the space. The expansion coefficients are found by (1.21), which looks very much like the scalar product $(\mathbf{u}_\nu, \mathbf{f})$ except that $\mathbf{u}_\nu$ is not a vector in the same space as $\mathbf{f}$. In the Fourier example, $\exp(2\pi i \nu x)$ is not square-integrable and hence not contained in $\mathbb{L}_2(-\infty, \infty)$.

1.2 TYPES OF OPERATORS

1.2.1 Functions and functionals

A function $f(x)$ is a transformation or mapping from one linear vector space to another. For simplicity, we assume throughout this section that $f(x)$ is a real, scalar-valued function, defined on $(-\infty, \infty)$. Therefore it associates each value x on the real line with another value on the real line. It maps $\mathbb{R}^1$ to $\mathbb{R}^1$ since the particular value $f(x_0)$ for some specific value x_0 of the variable x is a single real number. The complete function $f(x)$ is a vector in $\mathbb{L}_2$ but the particular value $f(x_0)$ is a vector or point in $\mathbb{R}^1$. A scalar-valued function can thus be thought of as an algorithm in which one number, x_0, is the input and another number, $f(x_0)$, is the output. The space in which the input is defined is referred to as the *domain* of the function, and the space of possible outputs is its *range*. For example, if $f(x) = x^2$, the domain is $\mathbb{R}^1(-\infty, \infty)$ and the range is $\mathbb{R}^1[0, \infty)$.

By contrast, a functional is a mapping from a function space to a scalar (or more generally, from a vector space to a scalar). Like a function, a functional can be regarded as an algorithm; its input is the function $f(x)$ and the output is the functional, a scalar denoted $\Phi[f(x)]$. The complete function $f(x)$ is required for the computation, even though the output is a single number. If $f(x)$ is square-integrable and $\Phi[f(x)]$ is real, the mapping is from $\mathbb{L}_2$ to $\mathbb{R}^1$. As with functions, we can refer to the input space (here $\mathbb{L}_2$) as the domain of the functional. The range of the functional is the output space $\mathbb{R}^1$ (or some subset of it if not all real numbers can be generated by the functional).

The functional $\Phi[f(x)]$ is said to be linear if it satisfies

$$\Phi[f_1(x) + cf_2(x)] = \Phi[f_1(x)] + c\Phi[f_2(x)], \tag{1.23}$$

where c is an arbitrary scalar.

According to the *Riesz Representation Theorem* (Stakgold, 1979), the most general form of a linear functional on a Hilbert space is a scalar product of the form[3]

$$\Phi[f(x)] = (\mathbf{a}, \mathbf{f}) = \int_{-\infty}^{\infty} dx\ a^*(x)\, f(x). \tag{1.24}$$

In this case the domain of the functional is $\mathbb{L}_2$ but the range can be $\mathbb{C}^1$ or a subset of it if $a(x)$ is complex.

A simple example of a nonlinear functional is the norm, as defined, for example, by (1.11). Any definite integral in which $f(x)$ appears nonlinearly in the integrand is a nonlinear functional.

1.2.2 Integral transforms

Just as a function maps a scalar to a scalar and a functional maps a function to a scalar, an integral transform maps a function to a function. Once again we regard the mapping as an algorithm, but now both input and output are functions. Examples important in image science include the convolution and the Fourier transform.

The general form of a linear integral transform connecting 1D functions is

$$g(x') = \int_{\alpha}^{\beta} dx\ h(x', x)\, f(x), \tag{1.25}$$

where $h(x', x)$ is called the kernel of the transform. If both $f(x)$ and $g(x')$ are square-integrable over (α, β), this integral transform maps $\mathbb{L}_2(\alpha, \beta)$ to the same space, $\mathbb{L}_2(\alpha, \beta)$. Note that the integral transform can be considered as a nondenumerably infinite set of functionals, one for each value of x'.

An example of a nonlinear integral transform from $\mathbb{L}_2(\alpha, \beta)$ to itself is

$$g(x') = \int_{\alpha}^{\beta} dx\ h(x', x)\, |f(x)|^2. \tag{1.26}$$

Integral transforms may also involve functions of several variables. For example, the 2D Fourier transform has the structure

$$F(\xi, \eta) = \int_{-\infty}^{\infty} dx \int_{-\infty}^{\infty} dy\ f(x, y) \exp[-2\pi i(\xi x + \eta y)]. \tag{1.27}$$

It is thus a linear integral transform from $\mathbb{L}_2(\mathbb{R}^2)$ to $\mathbb{L}_2(\mathbb{R}^2)$ if both $f(x, y)$ and $F(\xi, \eta)$ are square-integrable (a condition we frequently violate!).

As yet another example of a linear integral transform, consider the integral

$$g(x) = \int_{-\infty}^{\infty} dy\ f(x, y). \tag{1.28}$$

[3]Strictly speaking, this equation gives the general form of a *bounded, continuous* linear functional, where these terms are defined below.

This transform maps from $\mathbb{L}_2(\mathbb{R}^2)$ to $\mathbb{L}_2(\mathbb{R}^1)$ if $g(x)$ and $f(x, y)$ are square-integrable in their respective spaces.

We shall often use the term *continuous-to-continuous mapping*, (or *CC mapping*) to describe integral transforms. The reason for this term is that an integral transform maps a function of a continuous variable (as opposed to a discrete index) to another function of a continuous variable. There is no implication that either function is itself continuous or even that the mapping itself is continuous in the technical sense discussed in Sec. 1.3.2.

1.2.3 Matrix operators

An $M \times N$ real matrix $\mathbf{H}$ has M rows and N columns. This matrix acting on a real N-dimensional vector $\mathbf{f}$ yields a real M-dimensional vector $\mathbf{g}$. It therefore maps $\mathbb{R}^N$ to $\mathbb{R}^M$. If we think of both vectors as elements of a Hilbert space with $\mathbb{L}_2$ norm, the mapping is from $\mathbb{E}^N$ to $\mathbb{E}^M$. The mapping rule is the usual rule of matrix multiplication,

$$g_m = \sum_{n=1}^{N} H_{mn} f_n \,, \tag{1.29}$$

or in matrix form, $\mathbf{g} = \mathbf{Hf}$. This mapping is linear since $\mathbf{H}(\mathbf{f}_1 + \alpha \mathbf{f}_2) = \mathbf{Hf}_1 + \alpha \mathbf{Hf}_2$.

In many applications of this equation, $\mathbf{g}$ represents a vector of physical measurements (such as a digital image) and $\mathbf{f}$ is a vector of unknown parameters characteristic of some physical object, and the objective is to find or estimate $\mathbf{f}$. Success in this endeavor depends critically on the number of independent measurements. This number is not necessarily M since it may be possible to express some of the measurements as linear combinations of the others. The number of linearly independent measurements is the same as the number of linearly independent rows of $\mathbf{H}$. This number is called the *rank* of $\mathbf{H}$ and denoted $R(\mathbf{H})$, or simply R if only one matrix is under discussion. It is not an obvious point, but it can be shown that the number of linearly independent columns is also R. An important rule is that the rank is less than or equal to the smaller of M and N, *i.e.*, $R \leq \min(M, N)$.

1.2.4 Continuous-to-discrete mappings

In Sec. 1.2.1 we saw that functionals map a function, for example in $\mathbb{L}_2(\alpha, \beta)$, to a scalar in $\mathbb{R}^1$ or $\mathbb{C}^1$. Many physical measurement systems yield not one but a finite set of scalars. If the object being measured by such systems is a function in $\mathbb{L}_2(\alpha, \beta)$ and M real measurements are obtained, the system is a mapping from $\mathbb{L}_2(\alpha, \beta)$ to $\mathbb{R}^M$. If this mapping is linear, its most general form is

$$g_m = \int_{\alpha}^{\beta} dx \; f(x) \, h_m(x) \,, \tag{1.30}$$

where g_m is the mth component of the measurement vector $\mathbf{g}$. The kernel $h_m(x)$ can be called a *sensitivity function* since it describes the sensitivity of the mth measurement to the value of the function $f(x)$ at point x. In an imaging context, $h_m(x)$ is also called the *point response function*.

We refer to the mapping of (1.30) as a *continuous-to-discrete* mapping since it maps a function of a continuous variable x to a discrete set of numbers. The function $f(x)$ itself need not be continuous, however, so long as it is square-integrable.

As noted in the Prologue, continuous-to-discrete mappings are the appropriate descriptions of digital imaging systems viewing real-world objects.

1.2.5 Differential operators

For completeness we mention that differential operators also fit into the general framework of this section. The operator d/dx, for example, is a linear operator whose domain is the set of all differentiable functions. The range is not the same as the domain since the derivative of a differentiable function may not be itself differentiable.

In physical applications, it is often desirable to consider functions that are square-integrable but not differentiable. For example, we might want to consider a function with an abrupt discontinuity, perhaps representing the edge of an object to be imaged. Such a function is in $\mathbb{L}_2(-\infty, \infty)$ since the discontinuity does not prohibit it from being square-integrable. The derivative across the discontinuity does not exist in an elementary sense but can be defined in terms of a generalized function, the delta function (see Chap. 2). By allowing generalized functions, we can thus consider discontinuous, square-integrable functions to be in the domain of d/dx. Note, however, that the derivative (a delta function) is not square-integrable, so again the range is not the same as the domain.

The use of generalized functions may also allow us to treat differential operators as integral operators. For example, the differential operator d/dx has the same effect on functions in $\mathbb{L}_2(-\infty, \infty)$ as the integral operator with kernel $h(x', x) = \delta'(x' - x)$, where $\delta'(x)$ denotes the derivative of a delta function as defined in Chap. 2.

Partial differential operators are of considerable importance in physics and image science. Virtually all of modern physics is based on linear, second-order partial differential operators. The Schrödinger equation, the Poisson equation, the time-dependent wave equation and the time-independent wave equation all fit into this category. An example of prime importance in optics and imaging is the Helmholtz operator, $\nabla^2 + k^2$, where ∇^2 is the 3D Laplacian. This operator appears in the time-independent wave equation, which describes the propagation of monochromatic light, so it is fundamental in the mathematical description of many kinds of imaging systems (see Chap. 9)

1.3 HILBERT-SPACE OPERATORS

1.3.1 Range and domain

Consider a linear operator $\mathcal{H}$ that maps a vector $\mathbf{f}$ in the Hilbert space $\mathbb{U}$ to a vector $\mathbf{g}$ in the Hilbert space $\mathbb{V}$ (see Fig. 1.1). In operator form we write

$$\mathbf{g} = \mathcal{H}\mathbf{f}. \tag{1.31}$$

The space $\mathbb{U}$ is the set of vectors on which $\mathcal{H}$ can act, referred to as the domain of $\mathcal{H}$. The vector $\mathbf{g}$ will be referred to as the *image* of $\mathbf{f}$, a term that is used in both mathematics and image science. Though $\mathbf{g}$ is in $\mathbb{V}$, not all vectors in $\mathbb{V}$ are necessarily images through $\mathcal{H}$. Thus the range of $\mathcal{H}$ might be a subspace of $\mathbb{V}$.

Scalar products in $\mathbb{U}$ and $\mathbb{V}$ are denoted $(\,,\,)_{\mathbb{U}}$ and $(\,,\,)_{\mathbb{V}}$, respectively. The norms of $\mathbf{f}$ and $\mathbf{g}$ are, respectively, $||\mathbf{f}|| = \sqrt{(\mathbf{f},\mathbf{f})_{\mathbb{U}}}$ and $||\mathbf{g}|| = \sqrt{(\mathbf{g},\mathbf{g})_{\mathbb{V}}}$.

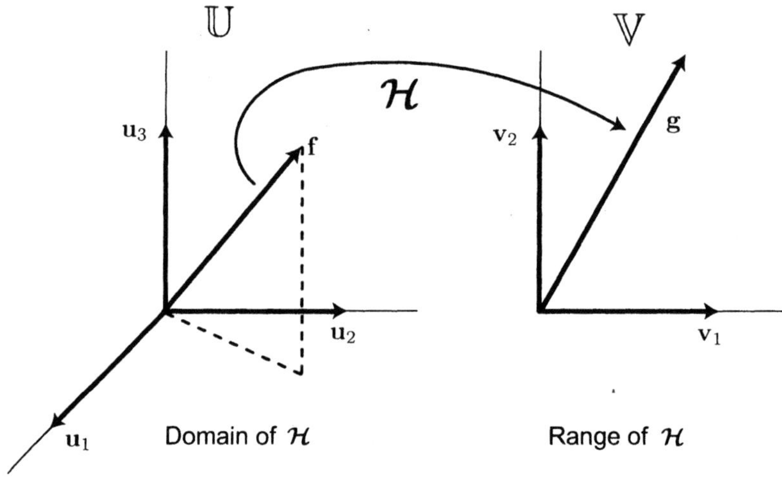

Fig. 1.1 Illustration of the mapping by an operator $\mathcal{H}$ from a Hilbert space $\mathbb{U}$ to a Hilbert space $\mathbb{V}$. For this illustration, $\mathbb{U}$ is $\mathbb{E}^3$ and $\mathbb{V}$ is $\mathbb{E}^2$. The basis vectors $\{\mathbf{u}_n\}$ and $\{\mathbf{v}_n\}$ in $\mathbb{U}$ and $\mathbb{V}$, respectively, are also shown.

1.3.2 Linearity, boundedness and continuity

The operator $\mathcal{H}$ is linear if

$$\mathcal{H}[\mathbf{f}_1 + c\mathbf{f}_2] = \mathcal{H}\mathbf{f}_1 + c\mathcal{H}\mathbf{f}_2 \tag{1.32}$$

for all scalars c and all vectors in the domain of $\mathcal{H}$.

A linear operator $\mathcal{H}$ is bounded if there exists a positive number Q such that $||\mathcal{H}\mathbf{f}|| < Q||\mathbf{f}||$ for all $\mathbf{f}$ in $\mathbb{U}$. In other words, a bounded operator cannot produce an image of infinite norm when operating on a vector of finite norm. The smallest constant Q for which this inequality holds for all $\mathbf{f}$ in $\mathbb{U}$, called the norm of $\mathcal{H}$, is sometimes denoted $||\mathcal{H}||$.

A simple example of an unbounded operator on $\mathbb{L}_2(-\infty, \infty)$ is the differential operator d/dx. A discontinuous function $f(x)$ might have finite norm, but its derivative has infinite norm.

The operator $\mathcal{H}$ is continuous if infinitesimal perturbations in $\mathbf{f}$ lead to infinitesimal perturbations in $\mathbf{g}$. More formally, for each $\epsilon > 0$ there exists a $\delta > 0$ such that $||\mathcal{H}\mathbf{f}_2 - \mathcal{H}\mathbf{f}_1|| < \epsilon$ whenever $||\mathbf{f}_2 - \mathbf{f}_1|| < \delta$. Equivalently, if a sequence $\{\mathbf{f}_n\}$ converges to $\mathbf{f}$ as $n \to \infty$, then $\{\mathcal{H}\mathbf{f}_n\}$ converges to $\mathcal{H}\mathbf{f}$ if and only if $\mathcal{H}$ is continuous. For linear operators, boundedness implies continuity and vice versa. If the range and domain are both finite-dimensional Euclidean spaces, all linear operators are bounded.

1.3.3 Compactness

Another concept related to continuity is *compactness* of an operator. The importance of compactness will become evident in Sec. 1.4 when we discuss eigenvalue spectra, but in this section we discuss the meaning of compactness and conditions under which an operator is compact.

To define a compact operator, we return to the notion of an infinite sequence of input vectors $\{\mathbf{f}_j\}$. We presume the sequence is bounded, so that $||\mathbf{f}_j|| < C$ for all j and some finite number C. Note that this sequence need not have a limit. It may, however, contain a subsequence that does have a limit. To illustrate, consider a sequence of vectors in the 2D plane ($\mathbb{R}^2$) where $\mathbf{f}_{j+1}$ is obtained from $\mathbf{f}_j$ by a rotation $\Delta\theta$. For simplicity, assume that $||\mathbf{f}_j|| = 1$ for all j. As $j \to \infty$, the vector continues to rotate around the unit circle *ad infinitum*, and there is no limit to the sequence. If $2\pi/\Delta\theta$ is an integer k, then every k^{th} vector forms a convergent subsequence; since $\mathbf{f}_{j+k} = \mathbf{f}_j$, this subsequence has a limit. If $2\pi/\Delta\theta$ is not an integer, one can make any desired vector on the unit circle the limit point simply by choosing from the sequence vectors that lie successively closer to the chosen limit vector. In both cases, therefore, a convergent subsequence exists.

In a finite-dimensional space, such as the 2D space in the example above, every bounded infinite sequence contains a convergent subsequence, but in infinite-dimensional spaces that is not necessarily so, as a simple counterexample will show. Consider a separable Hilbert space spanned by the countably infinite orthonormal basis $\{\mathbf{u}_n\}$. This basis set can be thought of as a sequence, which is bounded since $||\mathbf{u}_n|| = 1$ for all n. Since the $\mathbf{u}_n$ are distinct (in fact, orthogonal), no subset of them will form a convergent subsequence.

We are now in a position to define compactness: A compact operator is one that maps a bounded sequence into one having a convergent subsequence. More precisely, whenever $\{\mathbf{f}_j\}$ is a sequence in $\mathbb{U}$ with $||\mathbf{f}_j|| < C$, then $\{\boldsymbol{\mathcal{H}}\mathbf{f}_j\}$ contains a convergent subsequence if $\boldsymbol{\mathcal{H}}$ is compact. Compact operators are also called *completely continuous*. It can be shown that all bounded operators with a finite-dimensional range are compact (Stakgold, 1967). The interesting situation thus arises with operators such as integral transforms that have a range of infinite dimensionality.

The Hilbert-Schmidt theorem (Stakgold, 1979) provides an important way of determining if an integral transform is compact. A linear integral operator of the form of (1.25) is compact if its kernel satisfies the condition

$$\int_\alpha^\beta dx \int_\alpha^\beta dx' \, |h(x',x)|^2 < \infty \,. \tag{1.33}$$

Integral operators for which this condition is satisfied are called *Hilbert-Schmidt operators*. Though all Hilbert-Schmidt operators are compact, not all linear, compact integral operators are Hilbert-Schmidt.

If $\beta \quad \alpha$ is finite, the Hilbert-Schmidt condition is satisfied whenever $h(x',x)$ is bounded. If, on the other hand, (α,β) is $(-\infty,\infty)$, it is not sufficient that the kernel be bounded. Consider, for example, the convolution operator, where the kernel $h(x',x)$ is a function only of the difference $x'-x$ and (α,β) is $(-\infty,\infty)$. We can rewrite $h(x',x)$ as $h(x'-x)$ and, through a change of variables, perform the inner integral in (1.33) if $h(x)$ is square-integrable. The result is some finite positive number, but the outer integral still remains. Since the integral of a constant over an infinite range is infinite, the Hilbert-Schmidt condition is not satisfied.

A class of integral operators for which we can establish compactness, even when (α,β) is $(-\infty,\infty)$, consists of operators with kernels that can be written as a finite sum of products of the form

$$h(x', x) = \sum_{j=1}^{J} p_j(x')\, q_j(x)\,, \tag{1.34}$$

where J is finite and $p_j(x')$ and $q_j(x)$ are square-integrable for all j (Stakgold, 1979). Such kernels are referred to as *degenerate*. With this form of the kernel, the double integral in (1.33) becomes a sum of terms, each of which factors into a product of two convergent single integrals, and the Hilbert-Schmidt condition is satisfied. Thus all linear operators with kernels of the form of (1.34) are compact. For example, the operator with kernel $h(x', x) = \exp(-x^2 - x'^2)$ is compact, but one with kernel $h(x', x) = \exp[-(x - x')^2]$ is not.

1.3.4 Inverse operators

The operator $\mathcal{H}$ is said to be *one-to-one* (or *injective*, in the older literature) if each vector $\mathbf{g}$ in the range of $\mathcal{H}$ is the image of exactly one vector $\mathbf{f}$ in $\mathbb{U}$. Conversely, if $\mathcal{H}$ is not one-to-one, two or more vectors in $\mathbb{U}$ may produce the same image $\mathbf{g}$. Let $\mathbf{f}_1$ and $\mathbf{f}_2$ both have the same image, so that $\mathcal{H}\mathbf{f}_1 = \mathbf{g}$ and $\mathcal{H}\mathbf{f}_2 = \mathbf{g}$. It follows that $\mathcal{H}\{\mathbf{f}_2 - \mathbf{f}_1\} = \mathbf{0}$, where $\mathbf{0}$ is the zero vector. We say that $\mathbf{f}_2 - \mathbf{f}_1$ is a *null vector* of $\mathcal{H}$ or a vector in the *null space* of $\mathcal{H}$. Only one-to-one operators have no (nontrivial) null space.[4]

If $\mathcal{H}$ is one-to-one, we can define an inverse operator $\mathcal{H}_L^{-1}$ such that

$$\mathcal{H}_L^{-1}\mathcal{H}\mathbf{f} = \mathbf{f} \tag{1.35}$$

for all $\mathbf{f}$ in $\mathbb{U}$. That it is possible to define such an operator follows because $\mathcal{H}\mathbf{f}$ is, by definition, in the range of $\mathcal{H}$, and since $\mathcal{H}$ is one-to-one, only a single $\mathbf{f}$ can have this image. The subscript L on $\mathcal{H}_L^{-1}$ indicates that it appears to the *left* of $\mathcal{H}$ in (1.35), and $\mathcal{H}_L^{-1}$ is referred to as the *left inverse* of $\mathcal{H}$. One-to-one operators always possess a left inverse.

Equation (1.35) can also be written in pure operator form without a specific operand as

$$\mathcal{H}_L^{-1}\mathcal{H} = \mathcal{I}_{\mathbb{U}}\,, \tag{1.36}$$

where $\mathcal{I}_{\mathbb{U}}$ is the unit operator in $\mathbb{U}$, mapping any vector $\mathbf{f}$ into itself. Equation (1.35) follows from (1.36) by operating with both sides of the latter equation on an arbitrary $\mathbf{f}$ in $\mathbb{U}$.

If all vectors in $\mathbb{V}$ are in the range of $\mathcal{H}$, the mapping of $\mathcal{H}$ is said to be *onto* $\mathbb{V}$, or simply *onto*. (The term *surjective* is also used.) We can, of course, simply define $\mathbb{V}$ as the range of $\mathcal{H}$, in which case $\mathcal{H}$ is automatically onto, but that is not always convenient. We might, for example, want to use some general Hilbert space such as $\mathbb{L}_2$ for the space $\mathbb{V}$, but there is no guarantee that $\mathcal{H}$ can produce every vector in this space as an image.

If $\mathcal{H}$ is onto (but not necessarily one-to-one), any $\mathbf{g}$ in $\mathbb{V}$ is the image of one or more vectors in $\mathbb{U}$. It is therefore possible to find an inverse operator that maps

[4] Do not confuse the terms *zero vector* and *null vector*. A zero vector has zero norm, while a null vector for some operator has finite norm itself, but its image through that operator has zero norm. The zero vector is, of course, a null vector of any linear operator, but it is a trivial one.

an arbitrary $\mathbf{g}$ back to one of the vectors in $\mathbb{U}$ that could have produced it as an image. If we operate on this $\mathbf{f}$ with $\mathcal{H}$, we must get back to our original $\mathbf{g}$. We can thus define a *right inverse* $\mathcal{H}_R^{-1}$ such that

$$\mathcal{H}\mathcal{H}_R^{-1}\mathbf{g} = \mathbf{g}, \tag{1.37}$$

or, in operator form,

$$\mathcal{H}\mathcal{H}_R^{-1} = \mathcal{I}_{\mathbb{V}}, \tag{1.38}$$

where $\mathcal{I}_{\mathbb{V}}$ is the unit operator in $\mathbb{V}$.

In summary, a one-to-one operator always possesses a left inverse satisfying (1.36), while an operator that is onto always possesses a right inverse satisfying (1.38). We shall use the term *inverse*, without further qualification, only when (1.36) and (1.38) are both satisfied. If an inverse in this stricter sense exists, the operator is both one-to-one and onto, or *invertible* for short. The more mathematical literature uses the term *bijective* (injective plus surjective) for invertible. An invertible operator has no null space (except the trivial one containing only the zero vector). An operator that is not invertible is said to be *singular*. Thus, nonsingular, invertible and bijective are synonymous.

In an imaging context, real-world imaging operators (as opposed to computer simulations) are *always* singular.

1.3.5 Adjoint operators

Another operator that maps from $\mathbb{V}$ to $\mathbb{U}$ is the adjoint of $\mathcal{H}$, denoted $\mathcal{H}^\dagger$. The adjoint of a bounded operator is defined in terms of scalar products. Consider a particular vector in $\mathbb{U}$ and call it $\mathbf{f}_1$. The image of $\mathbf{f}_1$ is $\mathbf{g}_1 = \mathcal{H}\mathbf{f}_1$. The scalar product between $\mathbf{g}_1$ and some other vector $\mathbf{g}_2$ in $\mathbb{V}$, denoted $(\mathbf{g}_2, \mathcal{H}\mathbf{f}_1)_{\mathbb{V}}$, is easily expressed as a sum or integral, as appropriate, over the space $\mathbb{V}$. It is often useful, however, to express the *same* scalar product as a sum or integral in the original space $\mathbb{U}$. We can do so by defining $\mathcal{H}^\dagger$ such that

$$(\mathbf{g}_2, \mathcal{H}\mathbf{f}_1)_{\mathbb{V}} = (\mathcal{H}^\dagger\mathbf{g}_2, \mathbf{f}_1)_{\mathbb{U}} \tag{1.39}$$

for all $\mathbf{g}_2$ and $\mathbf{f}_1$. We are thus free, by definition, to shift an operator $\mathcal{H}$ from the right-hand side of a scalar product to the left so long as we replace the operator by its adjoint. This definition can be used to find explicit forms for adjoints, as demonstrated by several examples below.

The following properties of adjoints follow easily from the definition and properties of scalar products (Messiah, 1961):

(*a*) $(c\mathcal{H})^\dagger = c^*\mathcal{H}^\dagger$;

(*b*) $(\mathcal{H}_1 + \mathcal{H}_2)^\dagger = \mathcal{H}_1^\dagger + \mathcal{H}_2^\dagger$;

(*c*) $(\mathcal{H}_1\mathcal{H}_2)^\dagger = \mathcal{H}_2^\dagger\mathcal{H}_1^\dagger$;

(*d*) $(\mathcal{H}^\dagger)^\dagger = \mathcal{H}$,

where c is a scalar and $\mathcal{H}_1$ and $\mathcal{H}_2$ are two different operators mapping $\mathbb{U}$ to $\mathbb{V}$.

If $\mathcal{H} = \mathcal{H}^\dagger$, which is possible only if $\mathbb{U} = \mathbb{V}$, the operator is said to be self-adjoint or *Hermitian*. If $\mathcal{H}^\dagger = \mathcal{H}^{-1}$, again possible only if $\mathbb{U} = \mathbb{V}$, the operator is said to be *unitary*.

Adjoint of a matrix operator To understand how to compute adjoints, we begin with the simple case of a matrix operator. If $\mathcal{H}$ maps $\mathbb{E}^N$ to $\mathbb{E}^M$, it is represented by an $M \times N$ matrix $\mathbf{H}$. Hence $\mathcal{H}^\dagger$ maps $\mathbb{E}^M$ to $\mathbb{E}^N$ and is represented by an $N \times M$ matrix $\mathbf{H}^\dagger$. To determine the explicit form of the matrix elements of $\mathbf{H}^\dagger$, we use the definition of scalar products in Euclidean space, (1.9), in the definition of the adjoint, (1.39). From the left-hand side of (1.39), we obtain

$$(\mathbf{g}_2, \mathcal{H}\mathbf{f}_1)_{\mathbb{V}} = \sum_{m=1}^{M} g_{2m}^* \sum_{n=1}^{N} H_{mn} f_{1n} = \sum_{n=1}^{N} \sum_{m=1}^{M} g_{2m}^* H_{mn} f_{1n} \,. \tag{1.40}$$

Similarly, the right-hand side of (1.39) yields

$$(\mathcal{H}^\dagger \mathbf{g}_2, \mathbf{f}_1)_{\mathbb{U}} = \sum_{n=1}^{N} \left[\sum_{m=1}^{M} \left[\mathbf{H}^\dagger\right]_{nm} g_{2m} \right]^* f_{1n} = \sum_{n=1}^{N} \sum_{m=1}^{M} \left[\mathbf{H}^\dagger\right]_{nm}^* g_{2m}^* f_{1n} \,. \tag{1.41}$$

Comparison of the final forms of (1.40) and (1.41) shows that $[\mathbf{H}^\dagger]_{nm} = H_{mn}^*$, so the adjoint of $\mathbf{H}$ is obtained by transposing it (interchanging rows and columns) and taking the complex conjugate of each element. For real matrices, adjoint and transpose are synonymous, and we can write $\mathbf{H}^\dagger = \mathbf{H}^t$, where the superscript t denotes transpose.

Adjoint of an integral operator Similar results hold for integral operators. Let $\mathcal{H}$ now represent a linear integral operator as in (1.25). Then a scalar product $(\mathbf{g}_2, \mathcal{H}\mathbf{f}_1)$ can be written as

$$(\mathbf{g}_2, \mathcal{H}\mathbf{f}_1) = \int_\alpha^\beta dx \; f_1(x) \int_\alpha^\beta dx' \; g_2^*(x') \, h(x', x) \,. \tag{1.42}$$

By the definition of the adjoint, this expression must also equal

$$(\mathcal{H}^\dagger \mathbf{g}_2, \mathbf{f}_1) = \int_\alpha^\beta dx \; \left[\int_\alpha^\beta dx' \, h^{(\dagger)}(x, x') \, g_2(x') \right]^* f_1(x) \,, \tag{1.43}$$

where $h^{(\dagger)}(x, x')$ is the kernel of the adjoint operator. Comparison of (1.42) and (1.43) shows that

$$h^{(\dagger)}(x, x') = h^*(x', x) \,. \tag{1.44}$$

Thus the kernel for $\mathcal{H}^\dagger$ is obtained from the kernel for $\mathcal{H}$ by interchanging x and x' and taking the complex conjugate. This result is the continuous generalization of the adjoint of a matrix.

In using (1.44), it is important to pay attention to which of the two arguments of the kernel is the variable of integration. By (1.25), we know that $g(x') = [\mathcal{H}\mathbf{f}](x') = \int dx \; h(x', x) \, f(x)$, so the integral is over the second argument of $h(x', x)$. Similarly, $[\mathcal{H}^\dagger \mathbf{g}](x)$ would be written as $\int dx' \, h^{(\dagger)}(x, x') \, g(x')$; again, the integral is over the second argument of the kernel, which here is $h^{(\dagger)}(x, x')$. With (1.44), however, we can also write $[\mathcal{H}^\dagger \mathbf{g}](x) = \int dx' \, h^*(x', x) \, g(x')$. The integral is now over the *first* argument of $h^*(x', x)$, which is just the conjugate of the kernel needed to compute $[\mathcal{H}\mathbf{f}](x')$; no explicit interchange is needed in moving from $\mathcal{H}$ to $\mathcal{H}^\dagger$ if we are careful to associate x with space $\mathbb{U}$ and x' with $\mathbb{V}$ consistently. The situation is exactly the same as in the DD case, where we can write $[\mathbf{H}^\dagger\mathbf{g}]_n$ as either $\sum_m [\mathbf{H}^\dagger]_{nm} g_m$ or $\sum_m [\mathbf{H}]^*_{mn} g_m$.

The integral operator $\mathcal{H}$ is Hermitian or self-adjoint if $h(x, x') = h^*(x', x)$. Integral operators with real kernels are Hermitian if the kernel is symmetric in interchange of the arguments.

Adjoint of a continuous-to-discrete mapping If the operator $\mathcal{H}$ is the continuous-to-discrete mapping of (1.30), then its adjoint is a discrete-to-continuous mapping. This mapping yields a function of x denoted $[\mathcal{H}^\dagger\mathbf{g}](x)$, where $\mathbf{g}$ is the vector with m^{th} component given by (1.30). An argument similar to the ones given for matrix and integral operators shows that

$$\left[\mathcal{H}^\dagger\mathbf{g}\right](x) = \sum_{m=1}^{M} g_m h_m^*(x)\,. \tag{1.45}$$

Thus the adjoint operator in this case consists of a superposition of the complex conjugates of the sensitivity functions $h_m(x)$ with weights g_m.

1.3.6 Projection operators

An important class of operators is known in the linear-algebra literature as *projection operators*, though the reader is cautioned that this same term is used in other arenas (even elsewhere in this book!) with different meanings.

Projection operators have a property known as *idempotency*. An operator $\mathcal{P}$ is said to be idempotent if

$$\mathcal{P}^2 = \mathcal{P}. \tag{1.46}$$

In other words, acting a second time with the same operator produces no further effect.

An idempotent operator that is also Hermitian is called a projection operator or *projector*.[5] To see the reason for this designation, we consider a simple example in $\mathbb{E}^N$. Suppose that the vector $\mathbf{f}$ is expanded in the orthonormal basis $\{\mathbf{u}_n, n = 1, ..., N\}$ as in (1.15). The projection operator $\mathcal{P}_n$ is defined so that

$$\mathcal{P}_n\mathbf{f} = \alpha_n \mathbf{u}_n\,, \tag{1.47}$$

for all $\mathbf{f}$, where α_n is the component of $\mathbf{f}$ along direction $\mathbf{u}_n$. Thus the operator $\mathcal{P}$ singles out one component of $\mathbf{f}$ and produces a vector in the direction of the corresponding basis vector; all other components of $\mathbf{f}$ are set to zero by $\mathcal{P}_n$. Once the components are set to zero, a second application of the operator has no further effect, so its idempotency is obvious. The resulting vector $\mathcal{P}_n\mathbf{f}$ is said to be the projection of $\mathbf{f}$ along $\mathbf{u}_n$ or onto the 1D subspace spanned by $\mathbf{u}_n$.

As a second example, consider the operator $\mathcal{P}_{nm}$ defined so that

$$\mathcal{P}_{nm}\mathbf{f} = \alpha_n \mathbf{u}_n + \alpha_m \mathbf{u}_m\,, \tag{1.48}$$

[5]Some books use the term projector for any idempotent operator. The more specific term *orthogonal projector* is then used for an idempotent Hermitian operator. Our usage implies that a projector is both Hermitian and idempotent.

where $\mathbf{f}$ is expressed by (1.15). In this case, the resulting vector lies in the 2D subspace spanned by $\mathbf{u}_n$ and $\mathbf{u}_m$, and the operator $\mathcal{P}_{nm}$ is said to project $\mathbf{f}$ onto the nm-plane (see Fig. 1.2).

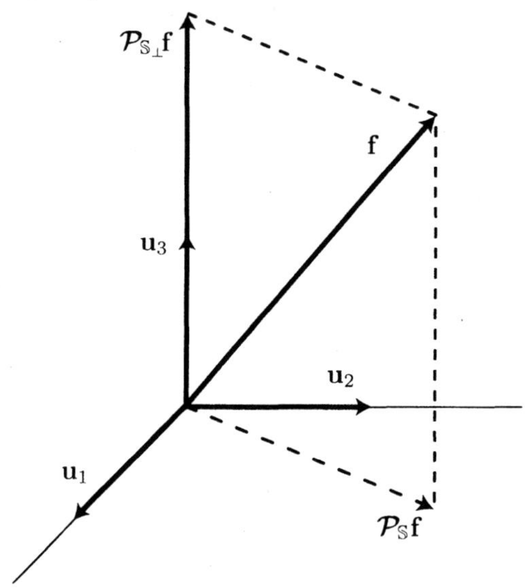

Fig. 1.2 Illustration of the meaning of the projection operators $\mathcal{P}_{\mathbb{S}}$ and $\mathcal{P}_{\mathbb{S}_\perp}$. The Hilbert space $\mathbb{U}$ is $\mathbb{E}^3$ as in Fig. 1.1, subspace $\mathbb{S}$ is the 1-2 plane, and subspace $\mathbb{S}_\perp$ is the 3-axis.

More generally, if $\mathbf{f}$ is defined in a Hilbert space $\mathbb{U}$, we can consider two subspaces $\mathbb{S}$ and $\mathbb{S}_\perp$ that together comprise the entire space $\mathbb{U}$. The subspaces are said to be orthogonal if any vector in $\mathbb{S}$ is orthogonal to any vector in $\mathbb{S}_\perp$. The subspace $\mathbb{S}_\perp$ is said to be the *orthogonal complement* of $\mathbb{S}$ if any vector $\mathbf{f}$ in $\mathbb{U}$ can be written uniquely as a sum of two vectors, one in $\mathbb{S}$ and one in $\mathbb{S}_\perp$. The required decomposition is

$$\mathbf{f} = \mathcal{P}_{\mathbb{S}}\,\mathbf{f} + \mathcal{P}_{\mathbb{S}_\perp}\mathbf{f}, \tag{1.49}$$

where $\mathcal{P}_{\mathbb{S}}$ is the projector onto $\mathbb{S}$ and $\mathcal{P}_{\mathbb{S}_\perp}$ is the projector onto $\mathbb{S}_\perp$. Since $\mathbf{f} = \mathcal{I}\mathbf{f}$, where $\mathcal{I}$ is the identity operator, it follows that

$$\mathcal{P}_{\mathbb{S}_\perp} = \mathcal{I} - \mathcal{P}_{\mathbb{S}}\,, \tag{1.50}$$

provided $\mathbb{S}_\perp$ is the orthogonal complement of $\mathbb{S}$.

Explicit forms for these projectors can be given in terms of basis vectors for the two subspaces. Assume that $\mathbb{S}$ is spanned by the orthonormal basis set $\{\phi_k, k = 1, ..., K\}$ while $\mathbb{S}_\perp$ is spanned by the orthonormal set $\{\psi_m, m = 1, ..., M\}$. If the original space is $\mathbb{E}^N$, we must have $K + M = N$, but in general K or M or both can be infinite. One possible choice of bases would be for $\{\phi_k\}$ and $\{\psi_m\}$ to be disjoint subsets of $\{\mathbf{u}_n\}$, but any other bases would suffice as well.

In terms of the subspace basis vectors, the projections of $\mathbf{f}$ are

$$\mathcal{P}_{\mathbb{S}}\mathbf{f} = \sum_{k=1}^{K} (\phi_k, \mathbf{f})\,\phi_k\,, \tag{1.51}$$

$$\mathcal{P}_{\mathbb{S}_\perp}\mathbf{f} = \sum_{m=1}^{M} (\psi_m, \mathbf{f})\,\psi_m\,. \tag{1.52}$$

1.3.7 Outer products

A convenient way of writing the projection operators of the last section, without reference to a particular vector **f**, is in terms of *outer products* (also known as *tensor products*). Like the inner or scalar product, the outer product involves two vectors. Unlike the scalar product, however, an outer product is not a scalar but instead an operator. This concept will be explained first in terms of vectors in finite-dimensional Euclidean spaces.

Consider an M-dimensional vector $\mathbf{b}$ and an N-dimensional vector $\mathbf{a}$. The outer product of these two vectors is the $M \times N$ matrix $\mathbf{ba}^\dagger$ with components given by

$$\left[\mathbf{ba}^\dagger\right]_{mn} = b_m a_n^* \,. \tag{1.53}$$

One way to view this definition is to regard $\mathbf{b}$ as an $M \times 1$ matrix (or column vector) and $\mathbf{a}$ as an $N \times 1$ matrix. Then $\mathbf{a}^\dagger$ is the $1 \times N$ matrix (row vector) obtained from $\mathbf{a}$ by the rule derived above for forming the adjoint of a matrix operator: interchange rows and columns and take the complex conjugate. With this view, (1.53) follows from the usual rule for matrix-matrix multiplication (see Fig. 1.3).

$$\mathbf{b} = \begin{pmatrix} b_1 \\ b_2 \\ b_3 \\ b_4 \end{pmatrix} \qquad \mathbf{a} = \begin{pmatrix} a_1 \\ a_2 \\ a_3 \end{pmatrix}$$

$$4 \times 1 \qquad\qquad 3 \times 1$$

$$\mathbf{ba}^\dagger = \begin{bmatrix} b_1 \\ b_2 \\ b_3 \\ b_4 \end{bmatrix} \begin{bmatrix} a_1^* & a_2^* & a_3^* \end{bmatrix} = \begin{bmatrix} b_1 a_1^* & b_1 a_2^* & b_1 a_3^* \\ b_2 a_1^* & b_2 a_2^* & b_2 a_3^* \\ b_3 a_1^* & b_3 a_2^* & b_3 a_3^* \\ b_4 a_1^* & b_4 a_2^* & b_4 a_3^* \end{bmatrix}$$

$$4 \times 3$$

Fig. 1.3 Illustration of the outer product of two vectors $\mathbf{b}$ and $\mathbf{a}$, where $\mathbf{b}$ has 4 elements and is represented as a 4×1 column vector, while $\mathbf{a}$ has 3 elements and is represented as a 3×1 column vector. The outer product $\mathbf{ba}^\dagger$ is a 4×3 matrix or an operator that can map from $\mathbb{E}^3$ to $\mathbb{E}^4$.

Another view of $\mathbf{ba}^\dagger$ is to regard it as an operator mapping an $N \times 1$ vector to an $M \times 1$ vector. Consider the action of this operator on some $N \times 1$ vector $\mathbf{c}$. The result is an $M \times 1$ vector, the mth element of which is

$$\left[\mathbf{ba}^\dagger\mathbf{c}\right]_m = \sum_{n=1}^{N} \left[\mathbf{ba}^\dagger\right]_{mn} c_n = \sum_{n=1}^{N} b_m a_n^* c_n = b_m \sum_{n=1}^{N} a_n^* c_n = (\mathbf{a}, \mathbf{c})\, b_m \,, \tag{1.54}$$

or, in vector form,

$$\mathbf{ba}^\dagger\mathbf{c} = (\mathbf{a}, \mathbf{c})\mathbf{b} \,. \tag{1.55}$$

Thus the result of this operator acting on $\mathbf{c}$ is the vector $\mathbf{b}$ times the scalar $(\mathbf{a}, \mathbf{c})$. The notation makes this result evident[6] when we realize that $\mathbf{a}^\dagger\mathbf{c}$ is just another

[6]Readers familiar with quantum mechanics will recognize that $\mathbf{ba}^\dagger$ in Dirac notation is $|\mathbf{b}\rangle\langle\mathbf{a}|$ and $\mathbf{a}^\dagger\mathbf{c}$ is $\langle\mathbf{a}|\mathbf{c}\rangle$.

way of writing the scalar product,

$$\mathbf{a}^\dagger\mathbf{c} = (\mathbf{a}, \mathbf{c}) = \sum_{n=1}^{N} a_n^* c_n \,. \tag{1.56}$$

Thus (1.55) can be viewed as $(\mathbf{b}\mathbf{a}^\dagger)\mathbf{c} = \mathbf{b}\mathbf{a}^\dagger\mathbf{c} = \mathbf{b}(\mathbf{a}^\dagger\mathbf{c}) = (\mathbf{a}^\dagger\mathbf{c})\mathbf{b}$, where the last step is valid since $(\mathbf{a}^\dagger\mathbf{c})$ is just a scalar, and there is no problem with rewriting it on the other side of the vector $\mathbf{b}$.

We can readily generalize the concept of outer product to include continuous functions. We simply regard (1.55) as *defining* the operator $\mathbf{b}\mathbf{a}^\dagger$, with whatever form of the scalar product is appropriate. Thus either $\mathbf{a}$ or $\mathbf{b}$ can be either discrete or continuous, though of course $\mathbf{a}$ and $\mathbf{c}$ must be in the same space so that the scalar product can be defined.

In this new notation, one specific form of projection operator is

$$\mathcal{P} = \mathbf{p}\mathbf{p}^\dagger \,, \tag{1.57}$$

where $\mathbf{p}$ is some vector with unit norm. The actions of this projector and its square on an arbitrary $\mathbf{f}$ are

$$\mathcal{P}\mathbf{f} = \mathbf{p}\mathbf{p}^\dagger\mathbf{f} = (\mathbf{p}, \mathbf{f})\,\mathbf{p} \tag{1.58}$$

$$\mathcal{P}^2\mathbf{f} = \mathbf{p}\mathbf{p}^\dagger\mathcal{P}\mathbf{f} = (\mathbf{p}, \mathbf{f})(\mathbf{p}, \mathbf{p})\mathbf{p} = (\mathbf{p}, \mathbf{f})\mathbf{p} = \mathcal{P}\mathbf{f}, \tag{1.59}$$

where the last line follows since $(\mathbf{p}, \mathbf{p}) = ||\mathbf{p}||^2 = 1$.

If $\mathbf{p} = \mathbf{u}_n$, then $\mathbf{p}\mathbf{p}^\dagger$ is the same as the simple projector defined in (1.47). The general form of a projector is

$$\mathcal{P} = \sum_n \mathbf{p}_n\mathbf{p}_n^\dagger \,, \tag{1.60}$$

where $\{\mathbf{p}_n\}$ is any set of mutually orthogonal vectors with unit norm. The idempotency follows since

$$\mathcal{P}^2\mathbf{f} = \sum_n \mathbf{p}_n\mathbf{p}_n^\dagger\mathcal{P}\mathbf{f} = \sum_n\sum_m (\mathbf{p}_m, \mathbf{f})(\mathbf{p}_n, \mathbf{p}_m)\mathbf{p}_n = \sum_n (\mathbf{p}_n, \mathbf{f})\mathbf{p}_n = \mathcal{P}\mathbf{f}, \tag{1.61}$$

and we have used the orthonormality of the $\mathbf{p}_n$.

We can also use this notation to express the completeness of a set of orthonormal basis vectors $\{\mathbf{u}_n, n = 1, ..., N\}$. We have defined this set to be complete for representation of the space $\mathbb{U}$ if any vector $\mathbf{f}$ in the space can be expanded in terms of the $\mathbf{u}_n$. In terms of outer products, this requires that

$$\sum_{n=1}^{N} \mathbf{u}_n\mathbf{u}_n^\dagger = \mathcal{I}_{\mathbb{U}} \,, \tag{1.62}$$

where $\mathcal{I}_{\mathbb{U}}$ is the identity operator in $\mathbb{U}$. This completeness condition, illustrated in Fig. 1.4, is referred to as *closure* of the set $\{\mathbf{u}_k\}$ or the decomposition of the unit operator. If it is satisfied, the component expansion is

$$\mathbf{f} = \sum_{n=1}^{N} \mathbf{u}_n\mathbf{u}_n^\dagger\mathbf{f} = \sum_{n=1}^{N} (\mathbf{u}_n^\dagger\mathbf{f})\,\mathbf{u}_n \,, \tag{1.63}$$

so that the expansion coefficients are given by $\alpha_n = \mathbf{u}_n^\dagger\mathbf{f}$, which is just another notation for the scalar product $(\mathbf{u}_n, \mathbf{f})$.

A similar closure relation can also be given for a continuous basis. If the set of functions $\{u_\nu(x)\}$, where ν is a continuous index, forms a basis for some Hilbert space $\mathbb{U}$, then any function in $\mathbb{U}$ can be expanded as in (1.20). For this to be possible, we must have

$$\int_{-\infty}^{\infty} d\nu\ u_\nu^*(x')\, u_\nu(x) = \delta(x - x')\,, \tag{1.64}$$

which is the continuous closure relation analogous to (1.62). One way to interpret this equation is that it represents the expansion of the function $\delta(x - x')$ as a continuous superposition of the basis functions $u_\nu(x)$. Since x' is arbitrary, a delta function at any location can be synthesized in this way. Since, as we shall see in Chap. 2, any function in $\mathbb{L}_2$ can be expanded in terms of delta functions, (1.64) guarantees that any function in $\mathbb{L}_2$ can be expanded in terms of the full set of $u_\nu(x)$.

$$\mathbf{u}_1 = \begin{bmatrix} 1 \\ 0 \\ 0 \end{bmatrix} \qquad \mathbf{u}_2 = \begin{bmatrix} 0 \\ 1 \\ 0 \end{bmatrix} \qquad \mathbf{u}_3 = \begin{bmatrix} 0 \\ 0 \\ 1 \end{bmatrix}$$

$$\mathbf{u}_1\mathbf{u}_1^\dagger = \begin{bmatrix} 1 & 0 & 0 \\ 0 & 0 & 0 \\ 0 & 0 & 0 \end{bmatrix} \quad \mathbf{u}_2\mathbf{u}_2^\dagger = \begin{bmatrix} 0 & 0 & 0 \\ 0 & 1 & 0 \\ 0 & 0 & 0 \end{bmatrix} \quad \mathbf{u}_3\mathbf{u}_3^\dagger = \begin{bmatrix} 0 & 0 & 0 \\ 0 & 0 & 0 \\ 0 & 0 & 1 \end{bmatrix}$$

$$\sum_{j=1}^{3} \mathbf{u}_j\mathbf{u}_j^\dagger = \begin{bmatrix} 1 & 0 & 0 \\ 0 & 1 & 0 \\ 0 & 0 & 1 \end{bmatrix} = \mathbf{I}$$

Fig. 1.4 Illustration of the completeness of the three basis vectors in $\mathbb{E}^3$. The sum of the 3 outer products is the identity operator.

1.4 EIGENANALYSIS

1.4.1 Eigenvectors and eigenvalue spectra

If the range and domain of a linear operator $\boldsymbol{\mathcal{A}}$ are the same space, it may be possible to find vectors $\boldsymbol{\psi}$ such that $\boldsymbol{\mathcal{A}}\boldsymbol{\psi}$ is the same as $\boldsymbol{\psi}$ except for a multiplicative constant. In other words, the length of the vector is changed but its angle is not. Since most vectors are changed into another vector in a different direction by action of the operator, these vectors are special ones characteristic of the operator. They are called *eigenvectors*, from the German for characteristic vectors, and the constant λ is an *eigenvalue*. A central theme running throughout this book is that eigenvectors and eigenvalues are essential tools for analyzing imaging systems and solving inverse problems.

The eigenvectors and eigenvalues satisfy an *eigenvalue equation* of the form

$$\boldsymbol{\mathcal{A}}\boldsymbol{\psi} = \lambda\boldsymbol{\psi}\,, \tag{1.65}$$

or

$$[\boldsymbol{\mathcal{A}} - \lambda\boldsymbol{\mathcal{I}}]\boldsymbol{\psi} = \mathbf{0}\,, \tag{1.66}$$

where $\mathbf{0}$ is the zero vector and $\mathcal{I}$ is the unit operator.[7]

If the operator $\mathcal{A} - \lambda\mathcal{I}$ is invertible, (1.66) has only trivial solutions, $[\mathcal{A} - \lambda\mathcal{I}]^{-1}\mathbf{0} = \mathbf{0}$. Nontrivial solutions exist only if $\mathcal{A} - \lambda\mathcal{I}$ is singular. This conclusion leads to another interpretation of the eigenvalues: $\lambda\mathcal{I}$ is a particular shift of the operator $\mathcal{A}$ that makes it singular. If $\mathcal{A}$ itself is singular, $\lambda = 0$ is an eigenvalue, and conversely.

Depending on the operator, the eigenvalues may form a discrete set, or there might be some continuous range of possible values. In the former case we can use a discrete index n to label different eigenvalues and their associated eigenvectors. We say that $\mathcal{A}$ has a *discrete spectrum* of eigenvalues, and (1.65) becomes

$$\mathcal{A}\psi_n = \lambda_n \psi_n \,. \tag{1.67}$$

If the eigenvalues fall in some continuous range, we need a continuous label ν to distinguish them, so the eigenvalue equation is written

$$\mathcal{A}\psi_\nu = \lambda_\nu \psi_\nu \,. \tag{1.68}$$

Such operators are said to have a *continuous spectrum*. Some operators have spectra with both continuous and discrete components.

Compact operators have only discrete spectra. In a finite-dimensional space, all linear operators are compact, so continuous spectra occur only for operators in infinite-dimensional spaces such as $\mathbb{L}_2$. This class of operators includes both integral and differential operators, but we shall be concerned mainly with the former in this book.

In Sec. 1.3.3 we noted that a linear integral operator in $\mathbb{L}_2(\alpha, \beta)$ is compact if the Hilbert-Schmidt condition, (1.33), is satisfied. If α and β are finite, this condition requires only that the kernel be bounded. If, however, $(\alpha, \beta) = (-\infty, \infty)$, the only general rule we have is that the Hilbert-Schmidt condition is valid if the kernel is degenerate in the sense expressed by (1.34). Since even so common an operator as the convolution violates the Hilbert-Schmidt condition, we shall frequently have to account for continuous spectra.

1.4.2 Similarity transformations

We can operate on the eigenvalue equation (1.67) with some invertible operator $\mathcal{W}$ and obtain a new eigenvalue equation:

$$\mathcal{W}\mathcal{A}\psi_n = \lambda_n \mathcal{W}\psi_n \,. \tag{1.69}$$

Since $\mathcal{W}$ is invertible, we can insert the unit operator $\mathcal{W}^{-1}\mathcal{W}$, obtaining

$$\mathcal{W}\mathcal{A}\mathcal{W}^{-1}\mathcal{W}\psi_n = \lambda_n \mathcal{W}\psi_n \,. \tag{1.70}$$

This equation shows that the new operator $\mathcal{W}\mathcal{A}\mathcal{W}^{-1}$ has the same eigenvalue spectrum as the original operator $\mathcal{A}$, but with the new eigenvectors $\mathcal{W}\psi_n$. A

[7] A slightly different definition of eigenvalue is often used in treatises on integral equations. There it is common to write the eigenvalue equation as $\lambda\mathcal{A}\psi = \psi$. This convention yields eigenvalues that are the reciprocals of the ones used here.

transformation of operators and eigenvectors in this way by means of an invertible operator is called a *similarity transformation*, and $\mathcal{A}$ and $\mathcal{W}\mathcal{A}\mathcal{W}^{-1}$ are said to be *similar operators*. Equation (1.70) shows that a discrete eigenvalue spectrum is invariant under a similarity transformation, and the same result for a continuous spectrum follows from an analogous derivation. In summary, similar operators have the same eigenvalue spectrum.

A particular form of similarity transformation is the *unitary transformation*. If $\mathcal{W}$ is a unitary operator, its inverse equals its adjoint. From the defining property of the adjoint, (1.39), it is easy to show that the unitary transformation preserves the norm of a vector, *i.e.*, $||\mathcal{W}\psi|| = ||\psi||$. A unitary transformation is thus a generalized rotation of a vector, changing its direction but not its length.

Similarity transformations, basis vectors and coordinate systems It is evident from (1.70) that the transformed operator $\mathcal{W}\mathcal{A}\mathcal{W}^{-1}$ has the same effect when acting on $\mathcal{W}\psi_n$ that $\mathcal{A}$ does when acting on ψ_n. Since this is true for any vector in the domain of $\mathcal{A}$, it means we can always replace any operator equation of the form $\mathcal{A}\mathbf{f} = \mathbf{g}$ with an equivalent relation $\mathcal{A}'\mathbf{f}' = \mathbf{g}'$, where $\mathcal{A}' = \mathcal{W}\mathcal{A}\mathcal{W}^{-1}$, $\mathbf{f}' = \mathcal{W}\mathbf{f}$, and $\mathbf{g}' = \mathcal{W}\mathbf{g}$ (provided $\mathbf{f}$ and $\mathbf{g}$ are both in the domain of $\mathcal{W}$). The two equations $\mathcal{A}'\mathbf{f}' = \mathbf{g}'$ and $\mathcal{A}\mathbf{f} = \mathbf{g}$ represent the same mathematical relation expressed in terms of different basis vectors; the new basis vectors are related to the old ones by $\mathbf{u}'_n = \mathcal{W}\mathbf{u}_n$. A similarity transformation can be regarded as a change of the coordinate axes in terms of which a vector or operator is expressed. If $\mathcal{A}$ is a matrix, then the specific matrix elements are changed, $A_{mn} \neq A'_{mn}$, but the effect of the matrix as an operator is unchanged if the relevant vectors are also changed appropriately. Thus a matrix is dependent on choice of basis, while the underlying operator is not. Similarly, for integral operators like (1.25), a change of variables will change the functional forms of the functions and the kernel, but not the meaning of the underlying operator equation.

1.4.3 Eigenanalysis in finite-dimensional spaces

In this section we restrict attention to operators for which the range and domain are both $\mathbb{E}^N$. Then, with respect to some basis, the operator $\mathcal{A}$ is the $N \times N$ matrix $\mathbf{A}$, and the eigenvector ψ_n is an $N \times 1$ column vector. In that case, $\mathbf{A}$ is necessarily bounded, continuous and compact, and we can make some general statements about the eigenvalues and eigenvectors.

The eigenvectors are the solution of the set of N simultaneous, homogeneous equations, one for each component of the following vector equation:

$$[\mathbf{A} - \lambda\mathbf{I}]\,\psi = \mathbf{0}\,, \tag{1.71}$$

where $\mathbf{I}$ is the $N \times N$ unit matrix. This set of equations has no nontrivial solutions unless $\mathbf{A}-\lambda\mathbf{I}$ is singular, or equivalently, its determinant vanishes (see App. A). This determinant, called the characteristic determinant for $\mathbf{A}$, is an Nth-order polynomial of the form (Eves, 1966)

$$\chi(\lambda) = \det[\mathbf{A} - \lambda\mathbf{I}] = (-1)^N\left[\lambda^N + a_1\lambda^{N-1} + a_2\lambda^{N-2} + \ldots + (-1)^N a_N\right]\,, \tag{1.72}$$

where $\det(\cdot)$ denotes determinant. By the Fundamental Theorem of Algebra (Gellert *et al.*, 1977), this polynomial has N roots, at least one of which is nonzero if a_N is nonzero. It is only when λ is equal to one of these roots that (1.71) has a nontrivial solution. The roots of $\chi(\lambda)$ are thus the eigenvalues of $\mathbf{A}$ for which a solution of the eigenvalue equation exists. The eigenvalues are not necessarily distinct, and if a particular eigenvalue occurs k times, it is said to be an eigenvalue of *multiplicity* k. In the physics literature, the term *degeneracy* is used for multiplicity.

One consequence of multiplicity of eigenvalues is that the eigenvectors are not unique. Any linear combination of two eigenvectors with the same eigenvalue is still an eigenvector associated with that same eigenvalue. Suppose $\mathbf{A}\psi_1 = \lambda\psi_1$ and $\mathbf{A}\psi_2 = \lambda\psi_2$ (same λ). It follows that $\mathbf{A}(\alpha\psi_1 + \beta\psi_2) = \lambda(\alpha\psi_1 + \beta\psi_2)$, with α and β arbitrary scalars, so $(\alpha\psi_1 + \beta\psi_2)$ is also an eigenvector with eigenvalue λ.

In principle (and in practice for very small matrices), one can find the eigenvalues by solving the characteristic equation $\chi(\lambda) = 0$. These eigenvalues can then be plugged back into (1.71) to find the corresponding eigenvectors. As a practical matter, however, the characteristic polynomial is of no use in finding eigenvalues of matrices with $N > 4$, since there are no formulas for factoring polynomials of order 5 or higher. Actual calculation of the $\{\lambda_n\}$ and the corresponding $\{\psi_n\}$ is an enterprise best left to the computer. Many computer packages exist for this purpose. An invaluable resource for anyone wanting to write new code or understand the packages is Numerical Recipes (Press *et al.*, 1992).

Both the sum and the product of the eigenvalues can be related back to properties of the matrix (Strang, 1980). Let $\lambda_n (n = 1, ..., N)$ denote the (not necessarily distinct) eigenvalues of $\mathbf{A}$. The sum of these eigenvalues is the *trace* (sum of the diagonal elements) of $\mathbf{A}$:

$$\sum_{n=1}^{N} \lambda_n = \sum_{n=1}^{N} A_{nn} \equiv \mathrm{tr}(\mathbf{A}), \tag{1.73}$$

where $\mathrm{tr}(\cdot)$ stands for trace. The product of the eigenvalues is the determinant of the matrix:

$$\prod_{n=1}^{N} \lambda_n = \det(\mathbf{A}). \tag{1.74}$$

This result shows that $\mathbf{A}$ is singular if any eigenvalue is zero, since then $\det(\mathbf{A}) = 0$. Since the eigenvalues are unchanged by a similarity transformation, the trace and determinant are also unchanged.

As we noted in Sec. 1.2.3, the rank of a general (not necessarily square) matrix is defined as the number of linearly independent rows or columns. For a square matrix, it can be shown that the rank is also the number of nonzero eigenvalues (counting multiplicity). Thus the multiplicity of the zero eigenvalue is $N - R$, where R is the rank and N is the dimension. If the matrix is nonsingular, no eigenvalues are zero, so $R = N$. A nonsingular matrix is sometimes referred to as a *full-rank* matrix.

There is one form of matrix for which it is possible to determine the eigenvalues without any computation at all. If the matrix is diagonal, then the diagonal elements are just the eigenvalues. For matrices in this form, (1.73) and (1.74) are easily verified. In fact, a somewhat broader statement can also be made. If the matrix is in upper (or lower) triangular form, where all elements below (above) the diagonal are zero, then the diagonal elements are the eigenvalues (Strang, 1980).

In some cases it is possible to find a similarity transformation that will reduce a matrix to diagonal form. This process is known as *diagonalization.* We noted above that a similarity transformation preserves the eigenvalue spectrum, so diagonalization is one way of solving the eigenvalue problem. Unfortunately, not all square matrices are diagonalizable (Strang, 1980; Smith, 1984).

1.4.4 Eigenanalysis of Hermitian operators

In this section we consider Hermitian operators in some Hilbert space $\mathbb{U}$. This space may have either finite or infinite dimensionality, and in the latter case the spectrum may be either discrete or continuous. The goal of the section is to see what additional properties of the eigenvalues and eigenvectors follow from the Hermiticity. An excellent reference for this section is Messiah (1961).

One key point is that the eigenvalues of a Hermitian operator are real. To show this, let $\mathcal{A}$ be a Hermitian operator satisfying the eigenvalue equation (1.65). Taking the scalar product of this equation with ψ and making use of properties of the scalar product (see Sec. 1.1.4), we find

$$(\psi, \mathcal{A}\psi) = (\psi, \lambda\psi) = \lambda(\psi, \psi) = \lambda||\psi||^2 . \tag{1.75}$$

With the definition of the adjoint, we also have

$$(\psi, \mathcal{A}\psi) = (\mathcal{A}^\dagger\psi, \psi) = (\lambda\psi, \psi) = \lambda^*(\psi, \psi) = \lambda^*||\psi||^2 . \tag{1.76}$$

Since $||\psi||$ is nonzero, comparison of (1.75) and (1.76) shows that $\lambda^* = \lambda$ if $\mathcal{A}$ is Hermitian.

An interesting special case of this result holds for operators of the form $\mathcal{B}^\dagger\mathcal{B}$. Any operator of this form is Hermitian since $(\mathbf{u}, \mathcal{B}^\dagger\mathcal{B}\mathbf{v}) = (\mathcal{B}\mathbf{u}, \mathcal{B}\mathbf{v}) = (\mathcal{B}^\dagger\mathcal{B}\mathbf{u}, \mathbf{v})$ for arbitrary $\mathbf{u}$ and $\mathbf{v}$. As we shall see in the next section, this result is very useful since it allows us to construct a Hermitian operator $\mathcal{B}^\dagger\mathcal{B}$ from any arbitrary operator $\mathcal{B}$. Moreover, for operators of this form, the eigenvalues are real and nonnegative. To show this, let $\mathcal{B}^\dagger\mathcal{B}\phi = \mu\phi$ and again take the scalar product with ϕ. The calculation proceeds as in (1.75) and (1.76):

$$(\phi, \mathcal{B}^\dagger\mathcal{B}\phi) = \mu||\phi||^2 = (\mathcal{B}\phi, \mathcal{B}\phi) = ||\mathcal{B}\phi||^2 , \tag{1.77}$$

from which we find that $\mu = ||\mathcal{B}\phi||^2/||\phi||^2$. The denominator in this expression is always a positive, real number, while the numerator is either zero or positive and real. Thus μ cannot be negative and is zero only if $\mathcal{B}\phi = \mathbf{0}$. We say that an operator of the form $\mathcal{B}^\dagger\mathcal{B}$ is *nonnegative-definite* or *positive-semidefinite.* If it has no null space, all eigenvalues are strictly greater than zero, and the operator is *positive-definite.*

Next we show that eigenvectors corresponding to different eigenvalues are orthogonal. Suppose $\mathcal{A}\psi_1 = \lambda_1\psi_1$ and $\mathcal{A}\psi_2 = \lambda_2\psi_2$, where $\mathcal{A}$ is Hermitian and $\lambda_1 \neq \lambda_2$. Then

$$(\psi_1, \mathcal{A}\psi_2) = \lambda_2(\psi_1, \psi_2) = (\mathcal{A}\psi_1, \psi_2) = \lambda_1(\psi_1, \psi_2) , \tag{1.78}$$

where we have used the facts that $\mathcal{A}$ is Hermitian and the eigenvalues are real. Subtracting the second and fourth forms of (1.78) yields

$$(\lambda_2 - \lambda_1)(\psi_1, \psi_2) = 0 . \tag{1.79}$$

Hence,

$$(\psi_1, \psi_2) = 0 \qquad \text{if } \lambda_2 \neq \lambda_1 \,. \tag{1.80}$$

Since we can always normalize the eigenvectors to unit norm without changing the eigenvalues, (1.80) says that ψ_1 and ψ_2 are orthonormal, just the property we desire in a set of basis vectors.

If two or more eigenvectors share the same eigenvalue, we cannot be guaranteed that they are orthonormal, but we can always construct orthonormal linear combinations of them. As noted in Sec. 1.4.3, any linear combination of eigenvectors with the same eigenvalue is still an eigenvector associated with that same eigenvalue, and the coefficients in the linear combinations can be chosen to ensure orthonormality. This procedure, known as *Gram-Schmidt orthogonalization*, is detailed in App. A.

Consider a Hermitian operator in $\mathbb{E}^N$. It has exactly N eigenvectors (some of which may be null vectors), and with Gram-Schmidt orthogonalization, they form an orthonormal set. Since *any* set of N orthonormal vectors constitutes a complete basis for the N-dimensional space, we can be assured that the full set of eigenvectors forms a basis.

In infinite-dimensional spaces, as usual, we cannot make such broad statements. If we restrict attention to separable Hilbert spaces and compact, Hermitian operators, it can be shown that the denumerably infinite set of eigenvectors does indeed form a basis (Stakgold, 1979). As far as the authors know, no general statements are possible for noncompact operators or nonseparable Hilbert spaces. Fortunately, nonseparable spaces are seldom encountered, but noncompact operators are common.

A noncompact Hermitian operator may have a continuous spectrum; if it does, the eigenvectors will not be in the Hilbert space (Messiah, 1961). Nevertheless, the eigenvectors may form a continuous basis for the space. An important example is the Fourier basis discussed previously. It will be shown in Chap. 7 that the basis functions $\exp(2\pi i\nu x)$ are eigenfunctions of convolution operators, which we know to be noncompact. The main result of Fourier theory is that this continuous set of eigenfunctions can be used to expand any function in the separable Hilbert space $\mathbb{L}_2(-\infty, \infty)$.

Except for the Fourier example and one or two others, it is difficult to make precise mathematical statements about completeness of the eigenvectors of a noncompact Hermitian operator. This difficulty is apparent in the literature on quantum mechanics, where it is essential to deal with noncompact Hermitian operators. A central tenet of quantum mechanics is that all physically measurable quantities are represented by Hermitian operators in a Hilbert space. Many of these operators, including the important position and momentum operators, are noncompact. Messiah (1961) comments that it is a difficult mathematical problem to prove that the eigenfunctions of any particular operator form a basis for the space, but he says it has been done for the position and momentum operators.[8] He goes on to say, "In fact, the completeness property is so closely related to the physical interpretation (of quantum mechanics) that the whole theory would have to be profoundly revised

[8]The eigenfunctions of the momentum operator are the Fourier basis functions, and the eigenfunctions of the position operator are the delta functions, both of which can be used as a basis for $\mathbb{L}_2(-\infty, \infty)$.

if it did not hold true." Messiah and other books (*e.g.*, Cohen-Tannoudji *et al.*, 1977) duck this problem by defining an *observable* as a Hermitian operator whose eigenvectors form a basis in the Hilbert space and simply postulating that all physically measurable quantities are observables in this sense. Mathematical research into this important question is actively continuing (Dubin and Hennings, 1990).

1.4.5 Diagonalization of a Hermitian operator

From Sec. 1.4.2, we know that the eigenvalue spectrum of an operator is unchanged by a similarity transformation, a special case of which is a unitary transformation. We noted also in Sec. 1.4.3 that one way of finding the eigenvalues is to diagonalize the operator by means of a similarity transformation. In this section we pursue this approach for Hermitian operators, beginning with Hermitian matrices.

Hermitian matrices An important property of any Hermitian matrix is that it can be diagonalized by a suitable unitary transformation (Strang, 1980). This property is a direct consequence of the fact that the eigenvectors $\{\boldsymbol{\psi}_n\}$ of an $N \times N$ Hermitian matrix $\mathbf{A}$ can be chosen (using the Gram-Schmidt procedure if needed) to form an orthonormal basis in $\mathbb{E}^N$. From these eigenvectors we can construct an $N \times N$ matrix $\boldsymbol{\Psi}$, the mn^{th} element of which is defined to be the mth component of $\boldsymbol{\psi}_n$, or

$$\Psi_{mn} = \psi_{nm} \,. \tag{1.81}$$

The reversal of indices may look peculiar, but it has an easy interpretation. Since Ψ_{mn} is the element in the mth row and n^{th} column of $\boldsymbol{\Psi}$, the entire n^{th} column is just the column vector $\boldsymbol{\psi}_n$. In other words, we form the matrix $\boldsymbol{\Psi}$ by arraying the column vectors $\boldsymbol{\psi}_n$ side by side.

It follows from the completeness and orthonormality of the eigenvectors that $\boldsymbol{\Psi}$ is unitary. In particular, $\boldsymbol{\Psi}^\dagger \boldsymbol{\Psi} = \mathbf{I}$ is a statement of the orthonormality condition, (1.16), since

$$\begin{aligned} \left[\boldsymbol{\Psi}^\dagger \boldsymbol{\Psi}\right]_{nn'} &= \sum_{k=1}^{N} \left[\Psi^\dagger\right]_{nk} \Psi_{kn'} = \sum_{k=1}^{N} \Psi_{kn}^* \Psi_{kn'} \\ &= \sum_{k=1}^{N} \psi_{nk}^* \psi_{n'k} = (\boldsymbol{\psi}_n, \boldsymbol{\psi}_{n'}) = \delta_{nn'} \,. \end{aligned} \tag{1.82}$$

A similar analysis shows that $\boldsymbol{\Psi}\boldsymbol{\Psi}^\dagger = \mathbf{I}$ is a statement of the completeness or closure condition, (1.62). Hence $\boldsymbol{\Psi}^\dagger = \boldsymbol{\Psi}^{-1}$.

Since $\boldsymbol{\psi}_n$ is an eigenvector of $\mathbf{A}$, $\boldsymbol{\Psi}$ is precisely the unitary matrix that diagonalizes $\mathbf{A}$:

$$\left[\boldsymbol{\Psi}^\dagger \mathbf{A} \boldsymbol{\Psi}\right]_{nn'} = \sum_{j=1}^{N} \sum_{k=1}^{N} \psi_{nj}^* A_{jk} \psi_{n'k} = \lambda_{n'} \sum_{j=1}^{N} \psi_{nj}^* \psi_{n'j} = \lambda_n \, \delta_{nn'} \,. \tag{1.83}$$

This result can be summarized succinctly as

$$\boldsymbol{\Psi}^\dagger \mathbf{A} \boldsymbol{\Psi} = \boldsymbol{\Lambda} \,, \tag{1.84}$$

where $\boldsymbol{\Lambda}$ is a diagonal matrix with the nth diagonal element equal to λ_n.

Spectral decomposition Equation (1.84) leads to a useful representation of the Hermitian operator $\mathbf{A}$. Since $\boldsymbol{\Psi}^\dagger = \boldsymbol{\Psi}^{-1}$, we have

$$\mathbf{A} = \boldsymbol{\Psi}\boldsymbol{\Lambda}\boldsymbol{\Psi}^\dagger . \tag{1.85}$$

This representation can also be expressed in terms of outer products [see (1.55)] as

$$\mathbf{A} = \sum_{n=1}^{N} \lambda_n \boldsymbol{\psi}_n \boldsymbol{\psi}_n^\dagger . \tag{1.86}$$

This equation, called the *spectral decomposition* of $\mathbf{A}$, shows that the Hermitian matrix $\mathbf{A}$ can be expressed as a weighted sum of projection operators, with the weighting coefficients just being the eigenvalues (Lorch, 1962). Moreover, any operator that can be expressed in the form of the right-hand side of (1.86) with real λ_n is necessarily Hermitian since each individual term $\boldsymbol{\psi}_n \boldsymbol{\psi}_n^\dagger$ is Hermitian (see Sec. 1.3.5).

There is a similar spectral decomposition for the inverse of $\mathbf{A}$, if it exists:

$$\mathbf{A}^{-1} = \sum_{n=1}^{N} \frac{1}{\lambda_n} \boldsymbol{\psi}_n \boldsymbol{\psi}_n^\dagger . \tag{1.87}$$

The validity of this representation can be demonstrated by forming the product $\mathbf{A}^{-1}\mathbf{A}$ or $\mathbf{A}\mathbf{A}^{-1}$ using the right-hand sides of (1.86) and (1.87), yielding

$$\mathbf{A}^{-1}\mathbf{A} = \mathbf{A}\mathbf{A}^{-1} = \sum_{n=1}^{N}\sum_{m=1}^{N} \frac{\lambda_n}{\lambda_m} \boldsymbol{\psi}_n \boldsymbol{\psi}_n^\dagger \boldsymbol{\psi}_m \boldsymbol{\psi}_m^\dagger = \sum_{n=1}^{N} \boldsymbol{\psi}_n \boldsymbol{\psi}_n^\dagger = \mathbf{I} , \tag{1.88}$$

where we have used the representation of the unit operator $\mathbf{I}$ from (1.62) and the orthonormality of the eigenvectors, expressed in the present notation as

$$\boldsymbol{\psi}_n^\dagger \boldsymbol{\psi}_m = \delta_{mn} . \tag{1.89}$$

Other Hermitian operators Though derived for a Hermitian operator in $\mathbb{E}^N$, the results given above apply, with some modification, to any Hermitian operator $\boldsymbol{\mathcal{A}}$. If the operator is compact, it has a discrete spectrum, and (1.86) applies with the simple modification of setting N to ∞. In that case $\boldsymbol{\Psi}$ and $\boldsymbol{\Lambda}$ are infinite square matrices.

Even if $\boldsymbol{\mathcal{A}}$ is a noncompact integral operator, a representation analogous to (1.86) can be obtained. Let $a(x,\ x')$ be the kernel of $\boldsymbol{\mathcal{A}}$ and assume that the eigenfunctions form a complete continuous basis. The counterpart of (1.86) is

$$a(x, x') = \int_{-\infty}^{\infty} d\nu \; \lambda_\nu \, \psi_\nu(x) \, \psi_\nu^*(x') . \tag{1.90}$$

To demonstrate that this representation is correct, note that it yields

$$[\boldsymbol{\mathcal{A}}\psi_\nu](x) = \int_{-\infty}^{\infty} d\nu' \int_{-\infty}^{\infty} dx' \, \lambda_{\nu'} \, \psi_{\nu'}(x) \, \psi_{\nu'}^*(x') \, \psi_\nu(x') = \lambda_\nu \, \psi_\nu(x) , \tag{1.91}$$

where the last step follows from the continuous orthonormality [*cf.* (1.22)]. Equation (1.91) shows that representation (1.90) gives the right answer when $\boldsymbol{\mathcal{A}}$ operates on one of its eigenfunctions. Since the eigenfunctions are assumed to be complete, any function in $\mathbb{U}$ can be written as a (continuous) linear superposition of the eigenfunctions, and the representation is correct in general.

In the discrete case, we constructed a unitary operator by arraying the column eigenvectors side by side; we can do much the same in the continuous case as well. We simply define the operator $\boldsymbol{\Psi}$ as an integral operator mapping a function of ν to a function of x by means of the kernel $\psi_\nu(x)$, so that

$$[\boldsymbol{\Psi}\mathbf{F}](x) = \int_{-\infty}^{\infty} d\nu \; \psi_\nu(x)\, F(\nu)\,. \tag{1.92}$$

(The analogy to Fourier transformation should be apparent.) The adjoint of $\boldsymbol{\Psi}$ is given by

$$\left[\boldsymbol{\Psi}^\dagger\mathbf{f}\right](\nu) = \int_{-\infty}^{\infty} dx \; \psi_\nu^*(x)\, f(x) \equiv F(\nu)\,, \tag{1.93}$$

and it follows from the orthonormality of the $\{\psi_\nu\}$ that $\boldsymbol{\Psi}$ is unitary.

From the discussion above in the discrete case, we would expect $\boldsymbol{\Psi}$ to diagonalize $\boldsymbol{\mathcal{A}}$ in some sense. To make this statement more precise, note that $\boldsymbol{\Psi}^\dagger\boldsymbol{\mathcal{A}}\boldsymbol{\Psi}$ is an integral operator, the kernel of which we can denote (rather cumbersomely) as $[\boldsymbol{\Psi}^\dagger\boldsymbol{\mathcal{A}}\boldsymbol{\Psi}](\nu,\nu')$. From the orthonormality and completeness of the eigenfunctions, we can show that

$$\left[\boldsymbol{\Psi}^\dagger\boldsymbol{\mathcal{A}}\boldsymbol{\Psi}\right](\nu,\nu') = \lambda_\nu\, \delta(\nu-\nu')\,. \tag{1.94}$$

By comparing (1.94) and (1.83), we see that in the continuous case the effect of the unitary transformation is again to reduce $\boldsymbol{\mathcal{A}}$ to diagonal form, which here means that the kernel is proportional to the delta function $\delta(\nu-\nu')$. One can think of a matrix with continuous indices where the matrix elements are zero if the indices are not equal and infinite if they are.

1.4.6 Simultaneous diagonalization of Hermitian matrices

Many problems in imaging involve two or more Hermitian operators. For example, the deterministic properties of an imaging system are described by the Hermitian operators $\boldsymbol{\mathcal{H}}^\dagger\boldsymbol{\mathcal{H}}$ and $\boldsymbol{\mathcal{H}}\boldsymbol{\mathcal{H}}^\dagger$ (for more on this point, see Sec. 1.5.1), and the stochastic properties are described by the covariance matrix, a Hermitian matrix to be defined in Chap. 8. In addition, various Hermitian symmetry and projection operators arise, so it is of considerable value to be able to find a representation in which two or more Hermitian operators are simultaneously diagonalized.

We address that problem in this section specifically for Hermitian matrices, but essentially the same results hold for other Hermitian operators. As we shall see, the nature of the transformation to the simultaneously diagonal representation depends critically on whether the matrices commute.

Commuting matrices A well-known theorem from quantum mechanics is that two Hermitian operators can be simultaneously diagonalized by a unitary transformation if and only if they commute. We shall now learn how to find that transformation for the special case of Hermitian matrices.

Let $\mathbf{A}$ and $\mathbf{B}$ be commuting Hermitian matrices. We seek a unitary matrix $\boldsymbol{\Psi}$ such that $\boldsymbol{\Psi}^\dagger\mathbf{A}\boldsymbol{\Psi}$ and $\boldsymbol{\Psi}^\dagger\mathbf{B}\boldsymbol{\Psi}$ are both diagonal. We begin with the eigenvalue equation for $\mathbf{A}$:

$$\mathbf{A}\psi_n = \lambda_{An}\psi_n \, . \tag{1.95}$$

For simplicity we assume that the eigenvalues are not degenerate; for a detailed discussion of the degenerate case, see Secs. 6.7.3–6.7.6. Now apply $\mathbf{B}$ to both sides of (1.95) and use the commutativity:

$$\mathbf{B}\mathbf{A}\psi_n = \lambda_{An}\mathbf{B}\psi_n = \mathbf{A}\mathbf{B}\psi_n \, . \tag{1.96}$$

Thus $\mathbf{B}\psi_n$ is also an eigenvector of $\mathbf{A}$ with eigenvalue λ_{An}. We have assumed, however, that the eigenvalues are nondegenerate, so there cannot be two linearly independent eigenvectors with the same eigenvalue. Thus $\mathbf{B}\psi_n$ must be just a constant times ψ_n; calling that constant λ_{Bn}, we see that

$$\mathbf{B}\psi_n = \lambda_{Bn}\psi_n \, . \tag{1.97}$$

Thus $\mathbf{A}$ and $\mathbf{B}$ share the same eigenvectors, but not necessarily the same eigenvalues. Since any Hermitian matrix is diagonalized by the unitary matrix with its normalized eigenvectors as columns, we can write

$$\boldsymbol{\Psi}^\dagger \mathbf{A} \boldsymbol{\Psi} = \boldsymbol{\Lambda}_A \, , \qquad \boldsymbol{\Psi}^\dagger \mathbf{B} \boldsymbol{\Psi} = \boldsymbol{\Lambda}_B \, , \tag{1.98}$$

where $\boldsymbol{\Lambda}_A$ and $\boldsymbol{\Lambda}_B$ are diagonal, with the former having λ_{An} as its diagonal elements and the latter having λ_{Bn}.

Prewhitening If a Hermitian matrix $\mathbf{A}$ is positive-definite (see Secs. 1.4.4 and A.8), not only can we transform it to diagonal form, we can also transform it to the unit matrix. The tool we need for this purpose is the square-root matrix defined in Sec. A.8.3. Specifically, we need $\boldsymbol{\Lambda}_A^{\frac{1}{2}}$, defined as the diagonal matrix with $\lambda_{An}^{\frac{1}{2}}$ as the diagonal elements. Since $\mathbf{A}$ is positive-definite, its eigenvalues are positive, and we do not have to worry about taking square roots of negative numbers. Moreover, since positive-definite matrices are nonsingular, the eigenvalues are not zero, and we can define $\boldsymbol{\Lambda}_A^{-\frac{1}{2}}$ as the diagonal matrix with $\lambda_{An}^{-\frac{1}{2}}$ as the diagonal elements.

Applying $\boldsymbol{\Lambda}_A^{-\frac{1}{2}}$ to the diagonal representation of $\mathbf{A}$, we obtain

$$\boldsymbol{\Lambda}_A^{-\frac{1}{2}} \boldsymbol{\Psi}^\dagger \mathbf{A} \boldsymbol{\Psi} \boldsymbol{\Lambda}_A^{-\frac{1}{2}} = \boldsymbol{\Lambda}_A^{-\frac{1}{2}} \boldsymbol{\Lambda}_A \boldsymbol{\Lambda}_A^{-\frac{1}{2}} = \mathbf{I} \, . \tag{1.99}$$

Because of a connection with white noise, to be clarified in Chap. 8, this transformation is known as *prewhitening*. We shall have several opportunities to apply this transformation later in the book, and immediately below we apply it to simultaneous diagonalization of noncommuting Hermitian matrices.

Noncommuting matrices Consider two Hermitian matrices $\mathbf{A}$ and $\mathbf{B}$ that do not necessarily commute, and assume that $\mathbf{A}$ is positive-definite and hence nonsingular. We seek a matrix $\mathbf{W}$ such that

$$\mathbf{W}^\dagger \mathbf{A} \mathbf{W} = \mathbf{I} \qquad \text{and} \qquad \mathbf{W}^\dagger \mathbf{B} \mathbf{W} = \mathbf{D} \, , \tag{1.100}$$

where $\mathbf{D}$ is diagonal. We do not require that $\mathbf{W}^\dagger = \mathbf{W}^{-1}$, so (1.100) is not a similarity transformation. Many books use the term *diagonalization* only for similarity transformations, but our usage is more general.

Following Fukunaga (1990), we first diagonalize $\mathbf{A}$ by a unitary transformation, then apply a nonunitary prewhitening transformation to convert it to the unit matrix, and finally apply another unitary transformation to diagonalize $\mathbf{B}$. This second unitary transformation leaves the unit matrix unchanged, so both matrices have the desired form after this sequence of three transformations.

After the first step, we have

$$\mathbf{A}' \equiv \mathbf{\Psi}^\dagger \mathbf{A} \mathbf{\Psi} = \mathbf{\Lambda}_A , \qquad \mathbf{B}' \equiv \mathbf{\Psi}^\dagger \mathbf{B} \mathbf{\Psi} , \tag{1.101}$$

where $\mathbf{\Lambda}_A$ is diagonal but $\mathbf{B}'$ is not.

After we apply the prewhitening transform, we obtain

$$\mathbf{A}'' \equiv \mathbf{\Lambda}_A^{-\frac{1}{2}} \mathbf{A}' \mathbf{\Lambda}_A^{-\frac{1}{2}} = \mathbf{\Lambda}_A^{-\frac{1}{2}} \mathbf{\Psi}^\dagger \mathbf{A} \mathbf{\Psi} \mathbf{\Lambda}_A^{-\frac{1}{2}} = \mathbf{I} , \tag{1.102}$$

$$\mathbf{B}'' \equiv \mathbf{\Lambda}_A^{-\frac{1}{2}} \mathbf{B}' \mathbf{\Lambda}_A^{-\frac{1}{2}} = \mathbf{\Lambda}_A^{-\frac{1}{2}} \mathbf{\Psi}^\dagger \mathbf{B} \mathbf{\Psi} \mathbf{\Lambda}_A^{-\frac{1}{2}} . \tag{1.103}$$

The matrix $\mathbf{B}''$ is still Hermitian, so it can be diagonalized by a unitary transformation. As in Sec. 1.4.5, this transformation matrix can be found by solving an eigenvalue problem:

$$\mathbf{B}'' \mathbf{\Phi} = \mathbf{\Phi} \mathbf{D} , \tag{1.104}$$

where $\mathbf{D}$ is the diagonal matrix of the eigenvalues of $\mathbf{B}''$, and the eigenvectors form the columns of $\mathbf{\Phi}$.

Now apply the final transformation to both matrices:

$$\mathbf{A}''' \equiv \mathbf{\Phi}^\dagger \mathbf{A}'' \mathbf{\Phi} = \mathbf{\Phi}^\dagger \mathbf{\Phi} = \mathbf{I} , \tag{1.105}$$

$$\mathbf{B}''' \equiv \mathbf{\Phi}^\dagger \mathbf{B}'' \mathbf{\Phi} = \mathbf{\Phi}^\dagger \mathbf{\Lambda}_A^{-\frac{1}{2}} \mathbf{\Psi}^\dagger \mathbf{B} \mathbf{\Psi} \mathbf{\Lambda}_A^{-\frac{1}{2}} \mathbf{\Phi} = \mathbf{D} . \tag{1.106}$$

The desired overall transformation matrix $\mathbf{W}$ is thus

$$\mathbf{W} = \mathbf{\Psi} \mathbf{\Lambda}_A^{-\frac{1}{2}} \mathbf{\Phi} . \tag{1.107}$$

Because of the presence of $\mathbf{\Lambda}_A^{-\frac{1}{2}}$, the transformation matrix $\mathbf{W}$ cannot be unitary, even if $\mathbf{A}$ and $\mathbf{B}$ do commute; two commuting Hermitian matrices can be simultaneously diagonalized by a unitary matrix, but converting one of them to the unit matrix requires a nonunitary transformation (except for the trivial case $\mathbf{A} = \mathbf{I}$), and it also requires that at least one of the matrices be positive-definite.

Generalized eigenvalue problems Since $\mathbf{W}$ as given by (1.107) satisfies (1.100), we can write

$$\mathbf{W}^\dagger \mathbf{B} \mathbf{W} = \mathbf{D} = \mathbf{I} \mathbf{D} = \mathbf{W}^\dagger \mathbf{A} \mathbf{W} \mathbf{D} . \tag{1.108}$$

Since $\mathbf{W}^\dagger$ is nonsingular, we can multiply through by its inverse in (1.108) and obtain

$$\mathbf{B} \mathbf{W} = \mathbf{A} \mathbf{W} \mathbf{D} . \tag{1.109}$$

This is the *generalized eigenvalue equation* for $\mathbf{A}$ and $\mathbf{B}$.

Since $\mathbf{A}$ has been assumed to be nonsingular, we can apply its inverse to both sides of (1.109) and get

$$\mathbf{A}^{-1} \mathbf{B} \mathbf{W} = \mathbf{W} \mathbf{D} . \tag{1.110}$$

Thus the columns of $\mathbf{W}$ are eigenvectors of $\mathbf{A}^{-1}\mathbf{B}$. Sometimes it will be easier to find $\mathbf{W}$ by solving a single eigenvalue problem, (1.109) or (1.110), rather than the two separate eigenvalue problems implied by (1.101) and (1.104).

1.5 SINGULAR-VALUE DECOMPOSITION

The spectral decomposition of (1.86) is a powerful tool for working with Hermitian operators, but it is not directly applicable to other kinds of linear operators. This is a significant drawback since only rarely can imaging systems be described by Hermitian operators. Fortunately, there is a closely related decomposition, known as *singular-value decomposition* or *SVD*, that is much more widely applicable.

1.5.1 Definition and properties

Consider a general linear operator $\mathcal{H}$ that maps a vector $\mathbf{f}$ in the Hilbert space $\mathbb{U}$ to a vector $\mathbf{g}$ in the Hilbert space $\mathbb{V}$. From this operator we can form two new operators, $\mathcal{H}^\dagger\mathcal{H}$ and $\mathcal{H}\mathcal{H}^\dagger$. As discussed in Sec. 1.4.4, both of these operators are Hermitian and nonnegative-definite, regardless of the nature of $\mathcal{H}$ itself. In addition, we assume that both $\mathcal{H}^\dagger\mathcal{H}$ and $\mathcal{H}\mathcal{H}^\dagger$ are compact and hence have discrete spectra. The eigenvalue equation for $\mathcal{H}^\dagger\mathcal{H}$ is then

$$\mathcal{H}^\dagger\mathcal{H}\mathbf{u}_n = \mu_n\mathbf{u}_n\,, \tag{1.111}$$

where we know that μ_n is nonnegative and real.

Since $\mathcal{H}^\dagger\mathcal{H}$ is Hermitian, and we assume it is compact, we know from Sec. 1.4.4 that the set $\{\mathbf{u}_n\}$ can be taken as a complete, orthonormal basis in $\mathbb{U}$ (invoking the Gram-Schmidt procedure if needed). The orthonormality is expressed as

$$\mathbf{u}_n^\dagger\mathbf{u}_m = \delta_{nm}\,, \tag{1.112}$$

where, in the notation of Sec. 1.3.7, $\mathbf{u}_n^\dagger\mathbf{u}_m$ denotes the scalar or inner product of $\mathbf{u}_n$ and $\mathbf{u}_m$. Completeness means that

$$\sum_{n=1}^{N}\mathbf{u}_n\mathbf{u}_n^\dagger = \mathcal{I}_{\mathbb{U}}\,, \tag{1.113}$$

where $\mathbf{u}_n\mathbf{u}_n^\dagger$ denotes an outer product (see Sec. 1.3.7) and $\mathcal{I}_{\mathbb{U}}$ denotes the identity operator in $\mathbb{U}$.

It is conventional to order the eigenvalues by decreasing value, so that

$$\mu_1 \geq \mu_2 \geq .... \geq \mu_R > 0\,, \tag{1.114}$$

where R is the number of nonzero eigenvalues of $\mathcal{H}^\dagger\mathcal{H}$ (counting multiplicity, as discussed in Sec. 1.4.3). Equation (1.114) takes advantage of the fact that all of the eigenvalues are nonnegative, which is true for the nonnegative-definite operator $\mathcal{H}^\dagger\mathcal{H}$ but not for an arbitrary Hermitian operator.

If $\mathbb{U} = \mathbb{E}^N$, then $\mathcal{H}^\dagger\mathcal{H}$ is an $N \times N$ matrix and, as noted in Sec. 1.4.3, R is its rank, which must be less than or equal to N. Though R was defined for a general matrix as the number of linearly independent rows or columns, for a Hermitian operator it is also equal to the number of nonzero eigenvalues.

If we have solved the eigenvalue problem for $\mathcal{H}^\dagger\mathcal{H}$, we can get some solutions of the corresponding problem for $\mathcal{H}\mathcal{H}^\dagger$ for free. Operating on both sides of (1.111) with $\mathcal{H}$ yields

$$\mathcal{H}\mathcal{H}^\dagger\mathcal{H}\mathbf{u}_n = \mu_n\mathcal{H}\mathbf{u}_n\,, \tag{1.115}$$

which shows that $\mathcal{H}\mathbf{u}_n$ is an eigenvector of $\mathcal{H}\mathcal{H}^\dagger$ with eigenvalue μ_n. It is straightforward to show that the vectors $\{\mathcal{H}\mathbf{u}_n\}$ are orthogonal if the $\{\mathbf{u}_n\}$ are, but it does not follow that they are correctly normalized. In fact, $||\mathcal{H}\mathbf{u}_n||^2 = \mu_n$ rather than unity. We can, however, multiply an eigenvector by any constant and it will remain an eigenvector with the same eigenvalue. Thus, if $\mathbf{u}_n$ is not a null vector, we can define

$$\mathbf{v}_n = \frac{1}{\sqrt{\mu_n}}\mathcal{H}\mathbf{u}_n\,, \qquad (\mu_n \neq 0)\,, \tag{1.116}$$

where $\sqrt{\mu_n}$ is positive and real since μ_n is. With this definition, we have

$$\mathcal{H}\mathcal{H}^\dagger\mathbf{v}_n = \mu_n\mathbf{v}_n\,. \tag{1.117}$$

Thus $\mathbf{v}_n$ is an eigenvector of $\mathcal{H}\mathcal{H}^\dagger$ with eigenvalue μ_n, so $\mathcal{H}^\dagger\mathcal{H}$ and $\mathcal{H}\mathcal{H}^\dagger$ have the same eigenvalue spectra.

The $\{\mathbf{v}_n\}$ as defined by (1.116) are an orthonormal set in $\mathbb{V}$, but they are not necessarily complete. If some of the $\{\mu_n\}$ are zero, the corresponding $\{\mathbf{v}_n\}$ cannot be constructed by (1.116) but must be obtained by solving (1.117) directly.

If all of the eigenvectors of $\mathcal{H}\mathcal{H}^\dagger$ have been found, including those corresponding to zero eigenvalue, they form a complete orthonormal basis in $\mathbb{V}$ and satisfy

$$\mathbf{v}_n^\dagger\mathbf{v}_m = \delta_{nm}\,, \tag{1.118}$$

$$\sum_{n=1}^{N}\mathbf{v}_n\mathbf{v}_n^\dagger = \mathcal{I}_{\mathbb{V}}\,. \tag{1.119}$$

The set $\{\mathbf{u}_n, \mathbf{v}_n, \mu_n\}$ is called the *singular system* of $\mathcal{H}$. Knowledge of this singular system allows us to construct a representation of $\mathcal{H}$ analogous to the spectral decomposition. We shall show that

$$\mathcal{H} = \sum_{n=1}^{R}\sqrt{\mu_n}\,\mathbf{v}_n\mathbf{u}_n^\dagger\,, \tag{1.120}$$

where each term $\mathbf{v}_n\mathbf{u}_n^\dagger$ is to be interpreted as an outer-product operator as discussed in Sec. 1.3.7 [*cf.* (1.54)]. The sum in this equation runs over all n for which $\mu_n \neq 0$, which means $n \leq R$ with the ordering of (1.114). We could, of course, also run the sum in (1.120) from 1 to N, but the factor $\sqrt{\mu_n}$ would set to zero all terms for $n > R$ anyway.

To prove the validity of (1.120), all we have to do is show that both sides give the same result when acting on an arbitrary vector $\mathbf{f}$ in the domain of $\mathcal{H}$. Since the $\{\mathbf{u}_n\}$ form a basis in $\mathbb{U}$, we can write the arbitrary $\mathbf{f}$ as

$$\mathbf{f} = \sum_{n=1}^{N}\alpha_n\mathbf{u}_n\,, \tag{1.121}$$

where the coefficient α_n is given by the scalar product,

$$\alpha_n = \mathbf{u}_n^\dagger\mathbf{f}. \tag{1.122}$$

The sum in (1.121) must include all vectors in the basis, not just ones with nonzero eigenvalue; it therefore runs up to N rather than R.

Operating on this representation of $\mathbf{f}$ with $\mathcal{H}$ gives

$$\mathcal{H}\mathbf{f} = \sum_{n=1}^{N} \alpha_n \mathcal{H}\mathbf{u}_n = \sum_{n=1}^{R} \alpha_n \sqrt{\mu_n}\, \mathbf{v}_n \,, \tag{1.123}$$

where we have used the linearity of $\mathcal{H}$ and (1.116). Applying the right-hand side of (1.120) to the same representation of $\mathbf{f}$ yields

$$\mathcal{H}\mathbf{f} = \sum_{n=1}^{R} \sqrt{\mu_n}\, \mathbf{v}_n \mathbf{u}_n^\dagger \sum_{m=1}^{N} \alpha_m \mathbf{u}_m = \sum_{n=1}^{R} \sum_{m=1}^{N} \alpha_m \sqrt{\mu_n}\, \mathbf{v}_n \mathbf{u}_n^\dagger \mathbf{u}_m$$

$$= \sum_{n=1}^{R} \sum_{m=1}^{N} \alpha_m \sqrt{\mu_n}\, \mathbf{v}_n \,\delta_{nm} = \sum_{n=1}^{R} \alpha_n \sqrt{\mu_n}\, \mathbf{v}_n \,, \tag{1.124}$$

where we have used the orthonormality of the $\{\mathbf{u}_n\}$ to set $\mathbf{u}_n^\dagger \mathbf{u}_m = \delta_{nm}$. The agreement of the final forms of (1.123) and (1.124) establishes the validity of (1.120).

Equation (1.120) is the singular-value decomposition of $\mathcal{H}$. Like the spectral decomposition of a Hermitian operator, it expresses the operator as a sum of outer-product operators. This sum has R nonzero terms, where R is the rank of both $\mathcal{H}^\dagger\mathcal{H}$ and $\mathcal{H}\mathcal{H}^\dagger$. We shall refer to R as the rank of $\mathcal{H}$ as well, where the rank of a general linear operator is the number of nonzero *singular* values. This definition is consistent with our two earlier definitions of rank: the number of linearly independent rows or columns of a general matrix or the number of nonzero eigenvalues of a Hermitian operator.

In a similar fashion, $\mathcal{H}^\dagger$ can be represented as

$$\mathcal{H}^\dagger = \sum_{k=1}^{R} \sqrt{\mu_k}\, \mathbf{u}_k \mathbf{v}_k^\dagger \,. \tag{1.125}$$

Again we have a sum of R outer products. If $\mathcal{H}$ is an $M \times N$ matrix $\mathbf{H}$, then $\mathbf{H}^\dagger$ is an $N \times M$ matrix, $\mathbf{u}_k$ is an $N \times 1$ vector, $\mathbf{v}_k$ is an $M \times 1$ vector, and each outer product in the sum is an $N \times M$ matrix, as it must be if the sum is to represent $\mathbf{H}^\dagger$.

1.5.2 Subspaces

It is often true that $R < N$, where R is the rank of $\mathcal{H}$ and N is the (possibly infinite) dimension of $\mathbb{U}$. Under this condition, $\mathbb{U}$ can be divided into two orthogonal subspaces[9] (see Fig. 1.5).

[9] Recall from Sec. 1.3.6 that two subspaces are orthogonal if all vectors in one are orthogonal to all vectors in the other.

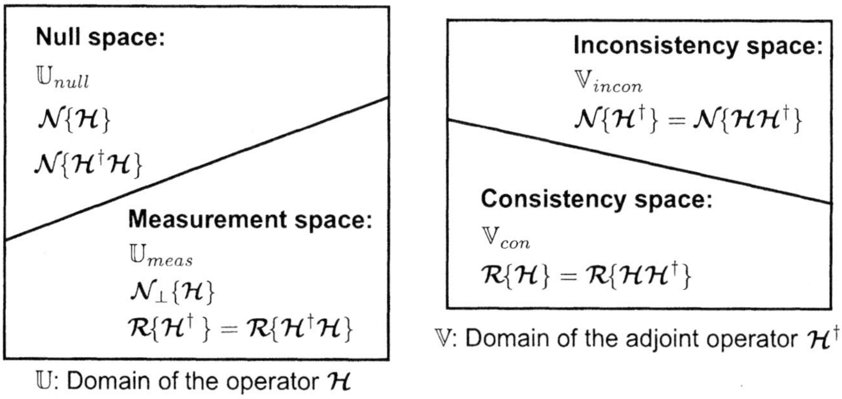

Fig. 1.5 Division of the domain of $\mathcal{H}$ and the domain of $\mathcal{H}^\dagger$ into orthogonal subspaces. Several alternative designations are given for each subspace.

The null space of $\mathcal{H}$ is an $(N-R)$-dimensional Euclidean space formally denoted $\mathcal{N}\{\mathcal{H}\}$. We shall also use the notation $\mathbb{U}_{null}$ for $\mathcal{N}\{\mathcal{H}\}$, emphasizing that it is a subspace of $\mathbb{U}$. A vector $\mathbf{f}_{null}$ in this space satisfies

$$\mathcal{H}\mathbf{f}_{null} = 0\,. \tag{1.126}$$

If $R = N$, $\mathcal{N}\{\mathcal{H}\}$ is trivial, in the sense that it contains only the zero vector. In that case $\mathcal{H}$ is *one-to-one*, which means that, in the absence of noise, every $\mathbf{g}$ is produced by exactly one $\mathbf{f}$.

The orthogonal complement of $\mathcal{N}\{\mathcal{H}\}$ is an R-dimensional Euclidean space formally denoted $\mathcal{N}_\perp\{\mathcal{H}\}$. We shall also refer to this subspace as *measurement space* since vectors in this space produce nonzero measurements if $\mathcal{H}$ describes a measurement system. To emphasize this point, we denote the space by $\mathbb{U}_{meas}$. An arbitrary vector $\mathbf{f}$ in $\mathbb{U}$ can be decomposed into a sum of two orthogonal vectors, one in $\mathbb{U}_{meas}$ and one in $\mathbb{U}_{null}$:

$$\mathbf{f} = \mathbf{f}_{meas} + \mathbf{f}_{null}\,. \tag{1.127}$$

If $R < M$, the data space $\mathbb{V}$ can also be divided into two nontrivial orthogonal subspaces. All possible data vectors $\mathbf{g}$ lie in $\mathbb{V}$, but not all vectors in $\mathbb{V}$ can be written as $\mathcal{H}\mathbf{f}$ for some $\mathbf{f}$ in $\mathbb{U}$. In other words, if $R < M$, $\mathcal{H}$ is not *onto* and the range of $\mathcal{H}$ is not the whole space $\mathbb{V}$ but some subspace of it.

Though the range of $\mathcal{H}$ is denoted formally as $\mathcal{R}\{\mathcal{H}\}$, we shall refer to it as *consistency space* and denote it $\mathbb{V}_{con}$. Any vector $\mathbf{g}$ in $\mathbb{V}_{con}$ is consistent in the sense that there is some vector $\mathbf{f}$ in $\mathbb{U}$ such that $\mathbf{g} = \mathcal{H}\mathbf{f}$. In an imaging context, $\mathcal{H}\mathbf{f}$ is a noise-free data vector, but a real, measured data vector, $\mathbf{g} = \mathcal{H}\mathbf{f} + \epsilon$, contains noise which creates inconsistency in the data and requires the use of the entire space $\mathbb{V}$ to describe an arbitrary data vector.

The orthogonal complement of $\mathbb{V}_{con}$ will be called *inconsistency space* and denoted $\mathbb{V}_{incon}$. An arbitrary vector $\mathbf{g}$ in $\mathbb{V}$ can be decomposed into a sum of two orthogonal vectors, one in $\mathbb{V}_{con}$ and one in $\mathbb{V}_{incon}$:

$$\mathbf{g} = \mathbf{g}_{con} + \mathbf{g}_{incon}\,. \tag{1.128}$$

If $R = M$, $\mathbb{V}_{incon}$ is trivial and any $\mathbf{g}$ is automatically consistent.

The division of $\mathbb{U}$ and $\mathbb{V}$ into orthogonal subspaces can also be viewed another way. The entire space $\mathbb{V}$ is the domain of the adjoint operator $\mathcal{H}^\dagger$, but that operator can have a null space if $R < M$. The null space of $\mathcal{H}^\dagger$, denoted $\mathcal{N}\{\mathcal{H}^\dagger\}$, is a subspace of $\mathbb{V}$. Moreover, if $R < N$, the range of $\mathcal{H}^\dagger$ is a subspace of $\mathbb{U}$.

It can be shown (Campbell and Meyer, 1979) that the null space of $\mathcal{H}^\dagger$ is just $\mathbb{V}_{incon}$, while its orthogonal complement is $\mathbb{V}_{con}$. Also, the range of $\mathcal{H}^\dagger$ is $\mathbb{U}_{meas}$ and its orthogonal complement is $\mathbb{U}_{null}$. Furthermore, the range of $\mathcal{H}$ is the same as the range of $\mathcal{H}\mathcal{H}^\dagger$ and the range of $\mathcal{H}^\dagger$ is the same as the range of $\mathcal{H}^\dagger\mathcal{H}$. Also, the null space of $\mathcal{H}$ is the same as the null space of $\mathcal{H}^\dagger\mathcal{H}$ and the null space of $\mathcal{H}^\dagger$ is the same as the null space of $\mathcal{H}\mathcal{H}^\dagger$. In summary,

$$\mathcal{R}\{\mathcal{H}\} = \mathcal{R}\{\mathcal{H}\mathcal{H}^\dagger\} = \mathbb{V}_{con} \,; \tag{1.129a}$$

$$\mathcal{R}\{\mathcal{H}^\dagger\} = \mathcal{R}\{\mathcal{H}^\dagger\mathcal{H}\} = \mathbb{U}_{meas} \,; \tag{1.129b}$$

$$\mathcal{N}\{\mathcal{H}\} = \mathcal{N}\{\mathcal{H}^\dagger\mathcal{H}\} = \mathbb{U}_{null} \,; \tag{1.129c}$$

$$\mathcal{N}\{\mathcal{H}^\dagger\} = \mathcal{N}\{\mathcal{H}\mathcal{H}^\dagger\} = \mathbb{V}_{incon} \,. \tag{1.129d}$$

These various designations of spaces and subspaces are illustrated in Fig. 1.5.

1.5.3 SVD representations of vectors and operators

The great advantage of the SVD approach is that it provides a consistent set of representations for all of the vectors and operators that arise in connection with a particular linear system. We have already encountered some of these representations, and the full set is summarized in Tables 1.1 and 1.2 for reference.

Table 1.1 lists the representations of vectors in either $\mathbb{U}$ or $\mathbb{V}$, including $\mathbf{f}$ and $\mathbf{g}$ as well as their components $\mathbf{f}_{meas}$, $\mathbf{f}_{null}$, $\mathbf{g}_{con}$ and $\mathbf{g}_{incon}$. Other vectors that will be important in later discussion include the vector $\boldsymbol{\epsilon}$ in $\mathbb{V}$ describing noise in the data and an estimated or reconstructed version of the object, denoted $\hat{\mathbf{f}}$. For completeness these representations are also given in Table 1.1.

Table 1.2 lists the singular-value decompositions for $\mathcal{H}$ and $\mathcal{H}^\dagger$ as well as the spectral decompositions for $\mathcal{H}^\dagger\mathcal{H}$ and $\mathcal{H}\mathcal{H}^\dagger$. This table also gives forms for $[\mathcal{H}^\dagger\mathcal{H}]^{-1}$, which exists if $R = N$, and $[\mathcal{H}\mathcal{H}^\dagger]^{-1}$, which exists if $R = M$. The correctness of these expressions can easily be verified from the orthonormality and completeness relations. Also listed in Table 1.2 are some additional operators to be discussed in Sec. 1.6.

1.6 MOORE-PENROSE PSEUDOINVERSE

A powerful tool for dealing with linear systems is the Moore-Penrose pseudoinverse, originally proposed by Moore (1920) and then apparently forgotten. Twenty years later the same concept was independently rediscovered by Sir Roger Penrose, British physicist and cosmologist and frequent collaborator of Stephen Hawking. Remarkably, Penrose published his classic paper on generalized inverses when he was in his early 20s and working toward his doctorate at Cambridge.

The literature on pseudoinverses is vast; Nashed (1976) gives a total of 1,775 references. Excellent accounts are given by Albert (1972), Ben-Israel and Greville (1974) and Campbell and Meyer (1979).

1.6.1 Penrose equations

A linear operator $\mathcal{H}^{\#}$ is called a *generalized inverse* or *pseudoinverse* of $\mathcal{H}$ if it satisfies $\mathcal{H}\mathcal{H}^{\#}\mathcal{H} = \mathcal{H}$. This equation, which is satisfied by $\mathcal{H}^{-1}$ if it exists, is now commonly referred to as the first *Penrose equation*. The full set of Penrose equations is:

$$\text{Penrose Eq. 1: } \mathcal{H}\mathcal{H}^{\#}\mathcal{H} = \mathcal{H}\,. \tag{1.130a}$$

$$\text{Penrose Eq. 2: } \mathcal{H}^{\#}\mathcal{H}\mathcal{H}^{\#} = \mathcal{H}^{\#}\,. \tag{1.130b}$$

$$\text{Penrose Eq. 3: } (\mathcal{H}\mathcal{H}^{\#})^{\dagger} = \mathcal{H}\mathcal{H}^{\#}\,. \tag{1.130c}$$

$$\text{Penrose Eq. 4: } (\mathcal{H}^{\#}\mathcal{H})^{\dagger} = \mathcal{H}^{\#}\mathcal{H}\,. \tag{1.130d}$$

If $\mathcal{H}$ has a true inverse, it satisfies all four of the Penrose equations. A matrix that satisfies Penrose Eq. 1 but not the other three is called a 1-inverse of $\mathcal{H}$, one that satisfies Eqs. 1 and 2 is called a (1,2)-inverse, etc. For matrices, a 1-inverse always exists and can be found by Gaussian elimination.

The Moore-Penrose pseudoinverse, denoted $\mathcal{H}^{+}$, is the generalized inverse that satisfies all four Penrose equations. Thus the Moore-Penrose pseudoinverse is a (1,2,3,4)-inverse and is equal to the true inverse if one exists. When we use the term pseudoinverse without proper names or other qualifiers, the Moore-Penrose pseudoinverse will be understood.

It can be shown that the Moore-Penrose pseudoinverse of a matrix always exists and is always unique (Albert, 1972). If $\mathcal{H}$ is an $M \times N$ matrix $\mathbf{H}$, then $\mathbf{H}^{+}$ is an $N \times M$ matrix, but the concept of pseudoinverse is not restricted to matrices. Under broad conditions, it is applicable to any bounded, linear operator $\mathcal{H}$ mapping one separable Hilbert space to another.[10]

1.6.2 Pseudoinverses and SVD

We shall show that the Moore-Penrose pseudoinverse can be represented as

$$\mathcal{H}^{+} = \sum_{k=1}^{R} \frac{1}{\sqrt{\mu_k}} \mathbf{u}_k \mathbf{v}_k^{\dagger}\,. \tag{1.131}$$

Since $\mu_k \neq 0$ if $k \leq R$, there is no worry about dividing by zero.

To show that (1.131) is a valid representation of $\mathcal{H}^{+}$, we need to prove that the right-hand side satisfies the four Penrose equations. To prove the first one, we use (1.120) and (1.131) to write

$$\begin{aligned}\mathcal{H}\mathcal{H}^{+}\mathcal{H} &= \sum_{k=1}^{R} \sqrt{\mu_k}\, \mathbf{v}_k \mathbf{u}_k^{\dagger} \sum_{m=1}^{R} \frac{1}{\sqrt{\mu_m}} \mathbf{u}_m \mathbf{v}_m^{\dagger} \sum_{n=1}^{R} \sqrt{\mu_n}\, \mathbf{v}_n \mathbf{u}_n^{\dagger} \\ &= \sum_{k=1}^{R} \sum_{m=1}^{R} \sum_{n=1}^{R} \frac{\sqrt{\mu_k}\sqrt{\mu_n}}{\sqrt{\mu_m}} \mathbf{v}_k \mathbf{u}_k^{\dagger} \mathbf{u}_m \mathbf{v}_m^{\dagger} \mathbf{v}_n \mathbf{u}_n^{\dagger}\,. \end{aligned} \tag{1.132}$$

[10] Technically, the condition is that the range of the operator be closed (Ogawa, 1988). This condition is satisfied for matrices and other operators with finite-dimensional range, including especially continuous-to-discrete operators. See also Caradus (1978) and Groetsch (1977).

Note that no parentheses are needed in the last line; the following forms are all equivalent:

$$(\mathbf{v}_k\mathbf{u}_k^\dagger)(\mathbf{u}_m\mathbf{v}_m^\dagger)(\mathbf{v}_n\mathbf{u}_n^\dagger) = \mathbf{v}_k\mathbf{u}_k^\dagger\mathbf{u}_m\mathbf{v}_m^\dagger\mathbf{v}_n\mathbf{u}_n^\dagger = \mathbf{v}_k(\mathbf{u}_k^\dagger\mathbf{u}_m)(\mathbf{v}_m^\dagger\mathbf{v}_n)\mathbf{u}_n^\dagger\,, \tag{1.133}$$

as one can demonstrate by patiently writing out each expression in component form. The advantage of the last expression in (1.133), however, is that we can recognize the inner products $(\mathbf{u}_k^\dagger\mathbf{u}_m)$ and $(\mathbf{v}_m^\dagger\mathbf{v}_n)$, allowing us to use the orthogonality relations (1.112) and (1.118). The resulting Kronecker deltas allow us to perform two of the three sums in (1.132), and we obtain

$$\mathcal{H}\mathcal{H}^+\mathcal{H} = \sum_{k=1}^{R}\sum_{m=1}^{R}\sum_{n=1}^{R}\frac{\sqrt{\mu_k}\sqrt{\mu_n}}{\sqrt{\mu_m}}\,\delta_{km}\,\delta_{mn}\mathbf{v}_k\mathbf{u}_n^\dagger = \sum_{k=1}^{R}\sqrt{\mu_k}\,\mathbf{v}_k\mathbf{u}_k^\dagger\,, \tag{1.134}$$

which, by (1.120), is just $\mathcal{H}$, verifying the first Penrose equation. A similar procedure shows that the remaining Penrose equations are also valid when we represent $\mathcal{H}^+$ by means of (1.131).

1.6.3 Properties of the pseudoinverse

The SVD representations of Sec. 1.5.3 enable us to derive many additional properties of the Moore-Penrose pseudoinverse. We begin by deriving two additional representations for it.

Limiting representations An important limiting representation of the pseudoinverse, which will prove useful in Sec. 1.7.6, is

$$\mathcal{H}^+ = \lim_{\eta\to 0^+}\left[\mathcal{H}^\dagger\mathcal{H} + \eta\mathcal{I}_{\mathrm{U}}\right]^{-1}\mathcal{H}^\dagger\,. \tag{1.135}$$

To prove that the right-hand side of (1.135) is indeed $\mathcal{H}^+$, we first represent $\mathcal{H}^\dagger\mathcal{H}$ and $\mathcal{I}_{\mathrm{U}}$ in terms of the SVD basis vectors (see Table 1.2) as

$$\left[\mathcal{H}^\dagger\mathcal{H} + \eta\mathcal{I}_{\mathrm{U}}\right] = \sum_{n=1}^{R}\mu_n\mathbf{u}_n\mathbf{u}_n^\dagger + \eta\sum_{n=1}^{N}\mathbf{u}_n\mathbf{u}_n^\dagger = \sum_{n=1}^{N}(\mu_n+\eta)\mathbf{u}_n\mathbf{u}_n^\dagger\,, \tag{1.136}$$

where the first sum could be extended to N since $\mu_n = 0$ for $n > R$. The inverse of the operator in (1.136) is given by

$$\left[\mathcal{H}^\dagger\mathcal{H} + \eta\mathcal{I}_{\mathrm{U}}\right]^{-1} = \sum_{n=1}^{N}(\mu_n+\eta)^{-1}\mathbf{u}_n\mathbf{u}_n^\dagger\,, \tag{1.137}$$

where $\mu_n + \eta$ cannot vanish since $\mu_n \geq 0$ and $\eta > 0$. The correctness of (1.137) can be checked by multiplying it by (1.136) and using the orthogonality relation (1.112). The result will be the SVD representation for the identity operator as given in Table 1.2.

Using (1.137), (1.125) and (1.112) again, we find

$$\left[\mathcal{H}^\dagger\mathcal{H} + \eta\mathcal{I}_{\mathrm{U}}\right]^{-1}\mathcal{H}^\dagger = \sum_{n=1}^{N}\frac{\sqrt{\mu_n}}{\mu_n+\eta}\,\mathbf{u}_n\mathbf{v}_n^\dagger\,. \tag{1.138}$$

We can pass to the limit $\eta \to 0$ by noting that

$$\lim_{\eta \to 0^+} \frac{\sqrt{\mu_n}}{\mu_n + \eta} = \begin{cases} 1/\sqrt{\mu_n} & \text{if} \quad \mu_n \neq 0 \\ 0 & \text{if} \quad \mu_n = 0 \end{cases} . \tag{1.139}$$

Hence,

$$\lim_{\eta \to 0^+} \left[\mathcal{H}^\dagger \mathcal{H} + \eta \mathbf{I}_{\mathbb{U}}\right]^{-1} \mathcal{H}^\dagger = \sum_{n=1}^{R} \frac{1}{\sqrt{\mu_n}} \mathbf{u}_n \mathbf{v}_n^\dagger , \tag{1.140}$$

which, by (1.131), is just $\mathcal{H}^+$.

By a similar procedure, we can also show that

$$\mathcal{H}^+ = \lim_{\eta \to 0^+} \mathcal{H}^\dagger \left[\mathcal{H}\mathcal{H}^\dagger + \eta \mathcal{I}_{\mathbb{V}}\right]^{-1} . \tag{1.141}$$

Special cases If $\mathbf{H}$ is an $N \times N$ Hermitian matrix of rank R, it can be written in terms of its spectral decomposition (see Sec. 1.4.5) as

$$\mathbf{H} = \sum_{j=1}^{R} \lambda_j \mathbf{u}_j \mathbf{u}_j^\dagger . \tag{1.142}$$

This expansion is a special case of SVD with $\mu_j = \lambda_j^2$ and $\mathbf{v}_j = \mathbf{u}_j$. Of course, λ_j must be real since $\mathbf{H}$ is Hermitian. Moreover, all of the λ_j are ≥ 0 if $\mathbf{H}$ is nonnegative-definite. In this case, the pseudoinverse of $\mathbf{H}$ is

$$\mathbf{H}^+ = \sum_{j=1}^{R} \frac{1}{\lambda_j} \mathbf{u}_j \mathbf{u}_j^\dagger . \tag{1.143}$$

As a further specialization, suppose that $\mathbf{H}$ is Hermitian and diagonal, so we can write

$$\mathbf{H} = \text{diag}(\lambda_1, \lambda_2, ..., \lambda_N) . \tag{1.144}$$

The notation indicates that $\mathbf{H}$ is diagonal with elements $\lambda_1, \lambda_2, ..., \lambda_N$ along the diagonal, and we assume that all of the λ_j are real and nonnegative. In this notation, the pseudoinverse is

$$\mathbf{H}^+ = \text{diag}(\lambda_1^+, \lambda_2^+, ..., \lambda_N^+) , \tag{1.145}$$

where

$$\lambda_j^+ = \begin{cases} 1/\lambda_j & \text{if} \quad \lambda_j \neq 0 \\ 0 & \text{if} \quad \lambda_j = 0 \end{cases} . \tag{1.146}$$

Other useful identities The pseudoinverse obeys the following identities (Albert, 1972):

$$\left[\mathcal{H}^+\right]^+ = \mathcal{H} ; \tag{1.147}$$

$$\mathcal{H}^+ = \left(\mathcal{H}^\dagger \mathcal{H}\right)^+ \mathcal{H}^\dagger ; \tag{1.148}$$

$$\mathcal{H}^+ = \mathcal{H}^\dagger \left[\mathcal{H}\mathcal{H}^\dagger\right]^+ ; \tag{1.149}$$

$$\left[\mathcal{H}^\dagger\right]^+ = \left[\mathcal{H}^+\right]^\dagger ; \tag{1.150}$$

$$\left[\mathcal{H}^\dagger\right]^+ = \left[\mathcal{H}\mathcal{H}^\dagger\right]^+ \mathcal{H} ; \tag{1.151}$$

$$[\mathcal{H}^+\mathcal{H}]^+ = \mathcal{H}^+\mathcal{H}\,; \tag{1.152}$$

$$\mathcal{H}^+\mathcal{H}\mathcal{H}^\dagger = \mathcal{H}^\dagger\,; \tag{1.153}$$

$$\mathcal{H}^\dagger\mathcal{H}\mathcal{H}^+ = \mathcal{H}^\dagger\,; \tag{1.154}$$

$$\mathcal{H}^+\mathcal{H} = \left(\mathcal{H}^\dagger\mathcal{H}\right)^+\left(\mathcal{H}^\dagger\mathcal{H}\right) = \left(\mathcal{H}^\dagger\mathcal{H}\right)\left(\mathcal{H}^\dagger\mathcal{H}\right)^+\,; \tag{1.155}$$

$$\mathcal{H}\mathcal{H}^+ = \left(\mathcal{H}\mathcal{H}^\dagger\right)^+\mathcal{H}\mathcal{H}^\dagger = \mathcal{H}\mathcal{H}^\dagger\left(\mathcal{H}\mathcal{H}^\dagger\right)^+\,; \tag{1.156}$$

$$[\mathcal{H}^\dagger\mathcal{H}]^+ = \mathcal{H}^+[\mathcal{H}^\dagger]^+\,; \tag{1.157}$$

$$[\mathcal{H}^\dagger\mathcal{H}]^+ = \mathcal{H}^+[\mathcal{H}\mathcal{H}^\dagger]^+\mathcal{H} = \mathcal{H}^\dagger[\mathcal{H}\mathcal{H}^\dagger]^+[\mathcal{H}^\dagger]^+\,; \tag{1.158}$$

$$[\mathcal{H}\mathcal{H}^\dagger]^+ = [\mathcal{H}^\dagger]^+\mathcal{H}^+\,. \tag{1.159}$$

All of these identities can be proved either directly from the Penrose equations or from the SVD representations of $\mathcal{H}$ and $\mathcal{H}^+$.

Pseudoinverse of a product It is not true in general that $(\mathbf{AB})^+ = \mathbf{B}^+\mathbf{A}^+$, but there is one special case in which this useful relation is valid (Harville, 1997). It holds for any $M \times N$ matrix $\mathbf{A}$ of full column rank (*i.e.*, rank N) and any $N \times K$ matrix $\mathbf{B}$ of full row rank (rank N again). More generally, all that can be said is that $(\mathbf{AB})^+ = (\mathbf{A}^+\mathbf{AB})^+(\mathbf{ABB}^+)^+$ (Campbell and Meyer, 1979, theorem 1.4.1).

1.6.4 Pseudoinverses and projection operators

In this section we relate the pseudoinverse to projection operators for the subspaces discussed in Sec. 1.5.2. Consider first the division of $\mathbb{U}$ into $\mathbb{U}_{meas}$ and $\mathbb{U}_{null}$. An arbitrary vector $\mathbf{f}$ in $\mathbb{U}$ can be expanded in terms of the basis $\{\mathbf{u}_k\}$ as

$$\mathbf{f} = \sum_{k=1}^{N} \alpha_k \mathbf{u}_k\,, \qquad \alpha_k = \mathbf{u}_k^\dagger \mathbf{f}. \tag{1.160}$$

The vector $\mathcal{H}\mathbf{f}$ can be computed from (1.120) and (1.160) as follows:

$$\mathcal{H}\mathbf{f} = \sum_{k=1}^{R} \sqrt{\mu_k}\,\mathbf{v}_k\mathbf{u}_k^\dagger \sum_{n=1}^{N} \alpha_n \mathbf{u}_n = \sum_{k=1}^{R}\sum_{n=1}^{N} \sqrt{\mu_k}\,\alpha_n \mathbf{v}_k \mathbf{u}_k^\dagger \mathbf{u}_n = \sum_{n=1}^{R} \sqrt{\mu_n}\,\alpha_n \mathbf{v}_n\,, \tag{1.161}$$

where the last step made use of the orthogonality relation (1.112). The key point to notice here is that (in this particular basis) components f_n for which $n > R$ make no contribution to $\mathcal{H}\mathbf{f}$ since they correspond to zero singular values μ_n. Thus these components define $\mathbf{f}_{null}$, and we can write

$$\mathbf{f}_{null} = \sum_{n=R+1}^{N} \alpha_n \mathbf{u}_n\,. \tag{1.162}$$

The measurement component is given by

$$\mathbf{f}_{meas} = \mathbf{f} - \mathbf{f}_{null} = \sum_{m=1}^{R} \alpha_m \mathbf{u}_m\,. \tag{1.163}$$

Since $\mathbf{u}_n$ and $\mathbf{u}_m$ are orthogonal if $n > R$ and $m \leq R$, it follows that $\mathbf{f}_{meas}$ and $\mathbf{f}_{null}$ are orthogonal.

Now we bring the pseudoinverse into the discussion. Consider the operator $\mathcal{H}^+\mathcal{H}$, given from (1.120) and (1.131) as

$$\mathcal{H}^+\mathcal{H} = \sum_{k=1}^{R} \frac{1}{\sqrt{\mu_k}} \mathbf{u}_k \mathbf{v}_k^\dagger \sum_{m=1}^{R} \sqrt{\mu_m}\, \mathbf{v}_m \mathbf{u}_m^\dagger = \sum_{k=1}^{R} \mathbf{u}_k \mathbf{u}_k^\dagger . \tag{1.164}$$

The last sum is in the form of a general projection operator as in (1.60), and in fact it is just the projector onto the measurement space:

$$\mathcal{H}^+\mathcal{H}\mathbf{f} = \sum_{k=1}^{R} \mathbf{u}_k \mathbf{u}_k^\dagger \mathbf{f} = \sum_{k=1}^{R} \alpha_k \mathbf{u}_k = \mathbf{f}_{meas} , \tag{1.165}$$

where we have used (1.160) and (1.163). We can thus define a projection operator by

$$\mathcal{P}_{meas} = \mathcal{H}^+\mathcal{H} , \tag{1.166}$$

and it follows that

$$\mathbf{f}_{meas} = \mathcal{P}_{meas}\mathbf{f} . \tag{1.167}$$

The null component of $\mathbf{f}$ is given by

$$\mathbf{f}_{null} = \mathbf{f} - \mathbf{f}_{meas} = (\mathcal{I}_{\mathbb{U}} - \mathcal{P}_{meas})\mathbf{f} = \mathcal{P}_{null}\mathbf{f} . \tag{1.168}$$

The projector $\mathcal{P}_{null}$ is thus given by

$$\mathcal{P}_{null} = \mathcal{I}_{\mathbb{U}} - \mathcal{P}_{meas} = \mathcal{I}_{\mathbb{U}} - \mathcal{H}^+\mathcal{H} . \tag{1.169}$$

As a check on these formulas, note that

$$\mathcal{P}_{meas} + \mathcal{P}_{null} = \sum_{k=1}^{N} \mathbf{u}_k \mathbf{u}_k^\dagger = \mathcal{I}_{\mathbb{U}} , \tag{1.170}$$

where the sum is equal to the identity operator in $\mathbb{U}$ by (1.113).

A similar decomposition holds for any vector $\mathbf{g}$ in $\mathbb{V}$. We can write

$$\mathbf{g} = \sum_{k=1}^{M} \beta_k \mathbf{v}_k = \mathbf{g}_{con} + \mathbf{g}_{incon} ; \tag{1.171}$$

$$\beta_k = \mathbf{v}_k^\dagger \mathbf{g} ; \tag{1.172}$$

$$\mathbf{g}_{con} = \mathcal{P}_{con}\mathbf{g} = \sum_{k=1}^{R} \beta_k \mathbf{v}_k ; \tag{1.173}$$

$$\mathbf{g}_{incon} = \mathcal{P}_{incon}\mathbf{g} = \sum_{k=R+1}^{M} \beta_k \mathbf{v}_k . \tag{1.174}$$

By a procedure analogous to the one used for $\mathcal{P}_{meas}$ and $\mathcal{P}_{null}$, we can show that

$$\mathcal{P}_{con} = \mathcal{H}\mathcal{H}^+ ; \tag{1.175}$$

$$\mathcal{P}_{incon} = \mathcal{I}_{\mathbb{V}} - \mathcal{H}\mathcal{H}^+ . \tag{1.176}$$

Though we have derived all of these projection operators from the SVD, they also follow directly from the Penrose equations and the definitions of the subspaces. For example, to show that (1.169) is correct, we operate on $\mathcal{P}_{null}\mathbf{f}$ with $\mathcal{H}$, yielding

$$\mathcal{H}\mathcal{P}_{null}\mathbf{f} = [\mathcal{H} - \mathcal{H}\mathcal{H}^+\mathcal{H}]\mathbf{f} . \tag{1.177}$$

The right-hand side is identically zero, as expected, because of the first Penrose equation, (1.130a). This proves that $[\mathcal{I}_{\mathbb{U}} - \mathcal{H}^+\mathcal{H}]\mathbf{f}$ is indeed a null vector of $\mathcal{H}$ and hence that $\mathcal{I}_{\mathbb{U}} - \mathcal{H}^+\mathcal{H}$ is a valid form for $\mathcal{P}_{null}$. We used only the first Penrose equation in this argument, but any projector must be Hermitian so the fourth equation must also be satisfied. Thus

$$\mathcal{P}_{null} = \mathcal{I}_{\mathbb{U}} - \mathcal{H}^{\#}\mathcal{H} \tag{1.178}$$

if $\mathcal{H}^{\#}$ is a (1,4)-inverse.

Similar arguments can be used to derive the other three projectors from the Penrose equations. It is found that (1.166) and (1.169) hold if $\mathcal{H}^+$ is replaced by a (1,4)-inverse, while (1.175) and (1.176) hold if $\mathcal{H}^+$ is replaced by a (1,3)-inverse. Since the Moore-Penrose operator is a (1,2,3,4)-inverse, however, we are always safe in using $\mathcal{H}^+$ itself.

It is also possible to go in the other direction. We can *define* $\mathcal{H}^+$ by the projector equations (1.166) and (1.175) and use this definition to derive all four Penrose equations[11] and other properties of the pseudoinverse. This was the approach taken by Moore in 1935.

1.7 PSEUDOINVERSES AND LINEAR EQUATIONS

An important use of the pseudoinverse operator is to solve systems of linear equations. In imaging, such systems arise when we represent an object by a discrete vector and then try to relate the elements of this vector to a discrete image data set.

The goal of this section is to discuss two particular kinds of solutions to systems of linear equations, so-called *exact* solutions and approximate *least-squares* solutions (Lawson and Hanson, 1995), and to show how each is related to pseudoinverses. Many other kinds of solutions of linear equations will be discussed in Chap. 15, where we deal with inverse problems in general. That chapter will also discuss in detail the problems that can arise with discrete object representations, but for now we shall accept them uncritically.

1.7.1 Nature of solutions of linear equations

Noisy data So far, we have considered only ideal linear mappings from a vector $\mathbf{f}$ to a vector $\mathbf{g}$. In real life, however, there is always some randomness or noise in any set of measurements, so $\mathbf{g}$ is a random vector. There can also be systematic

[11] It might be surprising that we can get all four equations this way, since we had to use only Eqs. 1, 3 and 4 to go from the Penrose equations to the projectors, but it can be shown that a (1,3,4)-inverse is also a 2-inverse.

errors in $\mathbf{g}$. To account for these errors, we must modify the basic matrix equation $\mathbf{g} = \mathbf{Hf}$. If we assume that $\mathbf{f}$ is a finite vector with components $\{f_n, n = 1, ..., N\}$, the mth component of $\mathbf{g}$ is given by

$$g_m = \sum_{n=1}^{N} H_{mn} f_n + \epsilon_m \,, \qquad m = 1, ..., M \,, \tag{1.179}$$

where the random numbers $\{\epsilon_m\}$ represent noise or other errors in the measurements. In matrix-vector form, we now have

$$\mathbf{g} = \mathbf{Hf} + \boldsymbol{\epsilon} \,, \tag{1.180}$$

where $\boldsymbol{\epsilon}$ is an $M \times 1$ column vector with elements $\{\epsilon_m\}$. We shall refer to $\boldsymbol{\epsilon}$ as a noise vector, though it can have both random and systematic components. The specific properties of $\boldsymbol{\epsilon}$ are not needed here but will be discussed in detail in later chapters.

Exact and approximate solutions One sense in which one might attempt to solve (1.180) would be to try to find the unknown vectors $\mathbf{f}$ and $\boldsymbol{\epsilon}$, but there are two difficulties with this approach. One is that determination of both vectors requires solving the M equations for $M + N$ unknowns, a clear impossibility. The other difficulty is that we aren't really interested in the components of $\boldsymbol{\epsilon}$ since they are not characteristic of the object.

A more sensible approach is to attempt to find solutions to

$$\mathbf{g} = \mathbf{H}\hat{\mathbf{f}} \,, \tag{1.181}$$

where the caret indicates that $\hat{\mathbf{f}}$ is some approximation to $\mathbf{f}$. In the language of statistical decision theory, to be introduced Chap. 13, $\hat{\mathbf{f}}$ is called an *estimate* of $\mathbf{f}$. Since $\mathbf{H}$ is an $M \times N$ matrix, $\hat{\mathbf{f}}$ must be an $N \times 1$ vector like $\mathbf{f}$ itself. Thus, solution of (1.181) requires finding the N unknown values $\{\hat{f}_n\}$ from M equations (one for each component of $\mathbf{g}$). An exact solution of (1.181) amounts to finding a vector $\hat{\mathbf{f}}$ that would exactly reproduce $\mathbf{g}$ if it were mapped through $\mathbf{H}$ without noise, even though $\mathbf{g}$ itself actually contains noise.

1.7.2 Existence and uniqueness of exact solutions

We now state several formal conditions for the existence and uniqueness of solutions of (1.181) and then discuss them from various perspectives. All of the conditions listed under each theorem below are mathematically equivalent. A good general reference to this section is Strang (1980).

Existence theorems Equation (1.181) has *at least* one exact solution for *a particular* $\mathbf{g}$ if and only if

(*a*) $\mathbf{g}$ can be expressed as a linear combination of the columns of $\mathbf{H}$;

(*b*) The data are consistent;

(*c*) $\mathbf{g}$ lies in $\mathbb{V}_{con}$;

(*d*) $\mathbf{HH}^{+}\mathbf{g} = \mathbf{g}$.

Equation (1.181) has *at least* one exact solution for *every* $\mathbf{g}$ if and only if

(a) The columns of $\mathbf{H}$ span $\mathbb{V} = \mathbb{E}^M$;

(b) $R = M$;

(c) A right inverse of $\mathbf{H}$ exists;

(d) $\mathbf{H}$ is *onto*;

(e) $\mathbb{V}_{incon}$ is trivial;

(f) $\mathbf{HH}^\dagger$ is nonsingular.

These conditions can be satisfied only if $M \leq N$.

Uniqueness theorems Equation (1.181) has *at most* one solution for every $\mathbf{g}$ if and only if

(a) The columns of $\mathbf{H}$ are linearly independent;

(b) $R = N$;

(c) A left inverse of $\mathbf{H}$ exists;

(d) $\mathbf{H}$ is *one-to-one*;

(e) $\mathbb{U}_{null}$ is trivial;

(f) $\mathbf{H}^\dagger\mathbf{H}$ is nonsingular.

These conditions can be satisfied only if $M \geq N$. Putting the existence and uniqueness conditions together, we see that a unique, exact solution exists only if $\mathbf{H}$ is square and full rank, $M = N = R$.

Discussion A system of linear equations is said to be *underdetermined* if the number of measurements is less than the number of unknowns, or $M < N$ in our notation. There are not enough measurements to determine all of the unknown values f_n in this case. Conversely, the system is said to be *overdetermined* if $M > N$. The implication of the word overdetermined is that the data are inconsistent so that no exact solution exists.

In fact, it is possible to have an exact solution to (1.181) even if $M > N$. All that is required is that the particular $\mathbf{g}$ lie entirely in $\mathbb{V}_{con}$ as expressed by the first set of existence conditions above. Suppose, for example, that we somehow generate noise-free measurements (as in computer simulations). Then the data vector is given precisely by $\mathbf{g} = \mathbf{Hf}$, so (1.181) has an exact solution, namely $\hat{\mathbf{f}} = \mathbf{f}$. Thus the first set of existence conditions can be satisfied if there is no noise in the data.

Of course, the assumption of noise-free measurements is highly artificial. Real, noisy measurements are unlikely to be confined to any subspace of $\mathbb{V}$. If there is a nontrivial inconsistency space, chances are that noisy data will have a component in it. Thus the only practical way to avoid inconsistent data is to have no inconsistency space, which is a paraphrase of the second set of existence conditions above. There is no inconsistency space if $R = M$. Since $R \leq \min(M, N)$, the condition $M > N$ implies $M > R$. An overdetermined system of equations thus necessarily has an inconsistency space. Barring noise-free data or other bizarre circumstances, there will be no exact solution to (1.181) in that case.

Next we discuss the uniqueness theorems. If a solution to (1.181) exists but $\mathbf{H}$ has a null space, as it will if $R < N$, the solution is not unique. All we have to do to construct another solution is to add any null vector to the original solution. If $\hat{\mathbf{f}}$ satisfies (1.181), then so does $\hat{\mathbf{f}} + \mathbf{f}_{null}$, where $\mathbf{f}_{null}$ is any vector such that $\mathbf{H}\mathbf{f}_{null} = \mathbf{0}$. All of the uniqueness conditions listed above thus amount to saying that there is no nontrivial null space.

1.7.3 Explicit solutions for consistent data

Right inverse If $R = M$, at least one exact solution to (1.181) exists, which is the same thing as saying that the right inverse of $\mathbf{H}$ exists. In the present problem, the right inverse $\mathbf{H}_R^{-1}$ is an $N \times M$ matrix that satisfies (1.38), or

$$\mathbf{H}\mathbf{H}_R^{-1} = \mathbf{I}_M \,, \tag{1.182}$$

where $\mathbf{I}_M$ is the $M \times M$ unit matrix. In terms of $\mathbf{H}_R^{-1}$, an exact solution to (1.181) is given by

$$\hat{\mathbf{f}} = \mathbf{H}_R^{-1}\mathbf{g} \,. \tag{1.183}$$

With (1.182), it is easy to see that this $\hat{\mathbf{f}}$ satisfies (1.181).

To construct the right inverse explicitly, we note that the rank of the $M \times M$ matrix $\mathbf{H}\mathbf{H}^\dagger$ is also R, a point which follows from (1.129d). Thus, if $R = M$, $\mathbf{H}\mathbf{H}^\dagger$ is invertible, and the right inverse of $\mathbf{H}$ is given by

$$\mathbf{H}_R^{-1} = \mathbf{H}^\dagger \left(\mathbf{H}\mathbf{H}^\dagger\right)^{-1}, \qquad (R = M) \,. \tag{1.184}$$

Direct substitution shows that (1.184) satisfies (1.182).

Consistency and pseudoinverses An exact solution to (1.181) exists if the data are consistent, or $\mathbf{g}$ lies entirely in $\mathbb{V}_{con}$. That will be the case either if we have somehow generated noise-free data or if there is no nontrivial inconsistency space. A mathematical statement that covers both contingencies is that $\mathbf{g}$ is consistent if

$$\mathbf{H}\mathbf{H}^+\mathbf{g} = \mathbf{g} \,, \tag{1.185}$$

where we recall from (1.175) that $\mathbf{H}\mathbf{H}^+ = \mathbf{P}_{con}$ (and we use $\mathbf{P}$ instead of $\mathcal{P}$ since we are dealing explicitly with matrices).

If the data are consistent, one solution to (1.181) is

$$\hat{\mathbf{f}} = \mathbf{H}^+\mathbf{g} \,. \tag{1.186}$$

To show that this is a solution, operate on it with $\mathbf{H}$. The result is

$$\mathbf{H}\hat{\mathbf{f}} = \mathbf{H}\mathbf{H}^+\mathbf{g} = \mathbf{g} \,, \tag{1.187}$$

where the last step follows from the consistency condition (1.185).

If the data are consistent because there is no nontrivial inconsistency space, then $R = M$ and the right inverse exists. In this case, right inverse and Moore-Penrose pseudoinverse are identical, but (1.186) is more general than (1.183). The pseudoinverse provides a solution to (1.181) even when there *is* an inconsistency space but the particular $\mathbf{g}$ happens to have no inconsistent component.

General exact solution While (1.186) provides a solution to (1.181), it is not the most general solution if $\mathbf{H}$ has a nontrivial null space. The general solution to (1.181) is

$$\hat{\mathbf{f}} = \mathbf{H}^{+}\mathbf{g} + [\mathbf{I}_N - \mathbf{H}^{+}\mathbf{H}]\mathbf{y}, \tag{1.188}$$

where $\mathbf{y}$ is an arbitrary vector in $\mathbb{U}$. Since $\mathbf{I}_N - \mathbf{H}^{+}\mathbf{H}$ is just $\mathbf{P}_{null}$, the second term in (1.188) is a null vector of $\mathbf{H}$. Thus the solution to (1.181) is unique if and only if $\mathbf{P}_{null} = \mathbf{0}$ or $\mathbf{H}^{+}\mathbf{H} = \mathbf{I}_N$.

To demonstrate that (1.188) is a solution to (1.181) for consistent data, we again operate on it with $\mathbf{H}$, yielding

$$\mathbf{H}\hat{\mathbf{f}} = \mathbf{H}\mathbf{H}^{+}\mathbf{g} + [\mathbf{H} - \mathbf{H}\mathbf{H}^{+}\mathbf{H}]\mathbf{y}. \tag{1.189}$$

The first term is $\mathbf{g}$ by the consistency condition (1.185), while the second term is zero by the first Penrose equation, so (1.188) is indeed a solution to (1.181) if the data are consistent. Furthermore, (1.188) must be the most general solution to (1.181); we cannot add any vector in $\mathbb{U}_{meas}$ to this $\hat{\mathbf{f}}$ and still have it satisfy (1.181), and $[\mathbf{I}_N - \mathbf{H}^{+}\mathbf{H}]\mathbf{y}$ is the general form for a vector in $\mathbb{U}_{null}$.

A way of selecting among the various solutions specified by (1.188) will be given in Sec. 1.7.5.

Unique, exact solutions Since $\mathbf{H}$ has a null space unless $R = N$, and we have already said that we must have $R = M$ for *any* solution to exist in general, a unique, exact solution to (1.181) exists for all $\mathbf{g}$ only if $R = M = N$. In this case $\mathbf{H}$ is a square, nonsingular matrix, and the unique solution is given by

$$\hat{\mathbf{f}} = \mathbf{H}^{-1}\mathbf{g}, \qquad (R = M = N). \tag{1.190}$$

Only rarely will the conditions necessary for the applicability of (1.190) be satisfied in real imaging problems.

1.7.4 Least-squares solutions

The method of least squares is a powerful and widely applicable method for drawing inferences from incomplete or noisy data (Lawson and Hanson, 1995). As with so much of the mathematics in this book, this method originated with Gauss. In 1801 an astronomer named G. Piazzi briefly observed and then lost the asteroid we now know as Ceres. Thinking it was a new planet, Piazzi and other astronomers tried in vain to locate this elusive heavenly body. Gauss assumed that it travelled in an elliptical orbit, and he found the parameters of the ellipse by least-squares fitting to Piazzi's data. He astounded the astronomy community, not only by telling them where to find the asteroid, but also by predicting its future path. He then waited another eight years before he revealed how he had done it (Campbell and Meyer, 1979).

In this section we discuss least-squares solutions to (1.181) for the case $R < M$ where no exact solution exists. We shall refer to the difference $\mathbf{g} - \mathbf{H}\hat{\mathbf{f}}$ as the *residual vector* or simply the residual.[12] It represents the amount by which a

[12]Do not confuse the residual with the measurement error ϵ. The latter is the difference between the data $\mathbf{g}$ and the image of the unknown vector $\mathbf{f}$ while the former is the difference between the data and the image of the estimate.

particular estimate $\hat{\mathbf{f}}$ fails to reproduce the data when imaged through $\mathbf{H}$. A least-squares solution to (1.181) is one for which the $\mathbb{L}_2$ norm of the residual vector is the smallest. Formally, we write

$$\hat{\mathbf{f}}_{LS} = \underset{\hat{\mathbf{f}}}{\text{argmin}} \ ||\mathbf{g} - \mathbf{H}\hat{\mathbf{f}}||^2 . \tag{1.191}$$

This notation means that $\hat{\mathbf{f}}_{LS}$ (where LS denotes least-squares) is the $\hat{\mathbf{f}}$ argument for which the norm is minimum. Since we are using the $\mathbb{L}_2$ norm, this means that the sum of the squares of the components of the residual is minimum. A statistical justification for the least-squares solution will be given in Chap. 13, but for now we merely investigate its mathematical properties.

Normal equation To find an explicit equation for $\hat{\mathbf{f}}_{LS}$, we write the squared norm in (1.191) as the scalar product of the residual with itself. Various equivalent forms are:

$$\begin{aligned} ||\mathbf{g} - \mathbf{H}\hat{\mathbf{f}}||^2 &= (\mathbf{g} - \mathbf{H}\hat{\mathbf{f}}, \mathbf{g} - \mathbf{H}\hat{\mathbf{f}}) \\ &= (\mathbf{g}, \mathbf{g}) - (\mathbf{H}\hat{\mathbf{f}}, \mathbf{g}) - (\mathbf{g}, \mathbf{H}\hat{\mathbf{f}}) + (\mathbf{H}\hat{\mathbf{f}}, \mathbf{H}\hat{\mathbf{f}}) \\ &= ||\mathbf{g}||^2 - 2\,\mathrm{Re}(\mathbf{g}, \mathbf{H}\hat{\mathbf{f}}) + ||\mathbf{H}\hat{\mathbf{f}}||^2 \\ &= ||\mathbf{g}||^2 - 2\,\mathrm{Re}(\mathbf{H}^\dagger\mathbf{g}, \hat{\mathbf{f}}) + (\mathbf{H}^\dagger\mathbf{H}\hat{\mathbf{f}}, \hat{\mathbf{f}}) , \end{aligned} \tag{1.192}$$

where we have made use of several properties of scalar products and adjoints (see Secs. 1.1.4 and 1.3.5), and Re denotes real part. If $\mathbf{H}$ and $\hat{\mathbf{f}}$ are real, the Re is irrelevant.

In some problems there may be constraints on the possible values of $\hat{f}_n$; for example, the values may have to be nonnegative if $\mathbf{f}$ represents an irradiance or other intrinsically nonnegative quantity. Such constraints are discussed in detail in Chap. 15, but for now we assume that each $\hat{f}_n$ can take on any value in $(-\infty, \infty)$.

With that assumption, the least-squares solution must occur at a point in $\mathbb{U}$ where all derivatives of the residual norm vanish. If a component $\hat{f}_n$ can be complex, the residual norm is a function of $2N$ variables (the real and imaginary parts of $\hat{f}_n$ for $n = 1, ..., N$), so $2N$ derivatives must vanish. It is convenient, however, to take $\hat{f}_n$ and its complex conjugate $\hat{f}_n^*$ as the independent variables instead of the real and imaginary parts. These quantities can be regarded as components of two *independent* vectors $\hat{\mathbf{f}}$ and $\hat{\mathbf{f}}^\dagger$. Rules for differentiating with respect to these vectors are discussed in App. A, Sec. A.9.5. Applying these rules to (1.192), we find[13]

$$\frac{\partial}{\partial \hat{\mathbf{f}}^\dagger} ||\mathbf{g} - \mathbf{H}\hat{\mathbf{f}}||^2 = \left[-\mathbf{H}^\dagger\mathbf{g} + \mathbf{H}^\dagger\mathbf{H}\hat{\mathbf{f}}\right]^\dagger . \tag{1.193}$$

For a least-squares solution, this vector derivative must vanish, which requires that

$$\mathbf{H}^\dagger\mathbf{H}\hat{\mathbf{f}} = \mathbf{H}^\dagger\mathbf{g} . \tag{1.194}$$

[13]If all quantities are real, we can forgo the subtleties of differentiating with respect to a complex vector and use the conventions of Sec. A.9.2 rather than those of Sec. A.9.5. The result is that adjoint is replaced by transpose and an additional factor of 2 appears in (1.193). The factor of 2 can be cancelled in (1.194) when the derivative is set to zero.

This important result is called the *normal equation*. Solving the normal equation is equivalent to finding a least-squares solution to the original equation, (1.181). A least-squares solution to (1.181) must satisfy (1.194). The same conclusion is reached by differentiating (1.192) with respect to $\hat{\mathbf{f}}$ rather than $\hat{\mathbf{f}}^\dagger$.

Existence and uniqueness We can now pose two important questions: Does a least-squares solution to (1.181) exist, and is it unique?

The answer to the first question is straightforward. If no $\hat{\mathbf{f}}$ drives the norm of the residual to zero, then that norm must have some minimum value, and an estimate $\hat{\mathbf{f}}$ for which it attains the minimum value is a least-squares solution. In other words, there is *always* at least one exact solution to (1.194). One way to think about this result is that the modified data $\mathbf{H}^\dagger\mathbf{g}$ are always consistent. The operator $\mathbf{H}^\dagger$ wipes out the inconsistent part of $\mathbf{g}$ since the inconsistency space is precisely the null space of $\mathbf{H}^\dagger$.

Next consider the second question. The least-squares solution is certainly not unique if $\mathbf{H}$ has a null space $(R < N)$. If it does, we can generate an infinite set of least-squares solutions simply by adding arbitrary null vectors $\mathbf{f}_{null}$. Since $\mathbf{H}\mathbf{f}_{null} = \mathbf{0}$, the norm of the residual in (1.191) is unaffected by the null vectors. If the residual achieves its minimum norm for some $\hat{\mathbf{f}}$, it must have exactly the same norm for $\hat{\mathbf{f}} + \mathbf{f}_{null}$.

Another way to make the same point is to note that $\mathbf{H}^\dagger\mathbf{H}$ is singular if $R < N$. The null space of $\mathbf{H}^\dagger\mathbf{H}$ is the same as the null space of $\mathbf{H}$, so if (1.194) holds for some $\hat{\mathbf{f}}$, it holds also for $\hat{\mathbf{f}} + \mathbf{f}_{null}$. Thus the least-squares solution is unique if and only if $\mathbf{H}^\dagger\mathbf{H}$ is nonsingular or $R = N$, in which case the solution is given by

$$\hat{\mathbf{f}}_{LS} = \left(\mathbf{H}^\dagger\mathbf{H}\right)^{-1}\mathbf{H}^\dagger\mathbf{g}, \qquad (R = N). \tag{1.195}$$

General least-squares solutions We have seen that any $\hat{\mathbf{f}}$ that minimizes $||\mathbf{g} - \mathbf{H}\hat{\mathbf{f}}||$ also satisfies the normal equation (1.194). As we shall now show, the general solution to (1.194) has exactly the same form as (1.188):

$$\hat{\mathbf{f}}_{LS} = \mathbf{H}^+\mathbf{g} + \left[\mathbf{I}_N - \mathbf{H}^+\mathbf{H}\right]\mathbf{y}, \tag{1.196}$$

where again $\mathbf{y}$ is an arbitrary vector in $\mathbb{U}$. Operating on $\hat{\mathbf{f}}_{LS}$ with $\mathbf{H}^\dagger\mathbf{H}$ yields

$$\mathbf{H}^\dagger\mathbf{H}\hat{\mathbf{f}}_{LS} = \mathbf{H}^\dagger\mathbf{H}\mathbf{H}^+\mathbf{g} + \left[\mathbf{H}^\dagger\mathbf{H} - \mathbf{H}^\dagger\mathbf{H}\mathbf{H}^+\mathbf{H}\right]\mathbf{y}. \tag{1.197}$$

From (1.154), $\mathbf{H}^\dagger\mathbf{H}\mathbf{H}^+ = \mathbf{H}^\dagger$, and from the first Penrose equation $\mathbf{H}^\dagger\mathbf{H}\mathbf{H}^+\mathbf{H} = \mathbf{H}^\dagger\mathbf{H}$, so the right-hand side of (1.197) reduces to $\mathbf{H}^\dagger\mathbf{g}$, proving that (1.196) indeed satisfies (1.194). Moreover, it must be the most general solution. We cannot add any vector in $\mathbb{U}_{meas}$ and still have (1.194) satisfied, and we have already added the most general vector in $\mathbb{U}_{null}$.

1.7.5 Minimum-norm solutions

There is an interesting analogy between (1.188) or (1.196) and the general solution to an inhomogeneous differential equation. The first term, $\mathbf{H}^+\mathbf{g}$, is a particular solution to (1.181) or (1.194), while the second term is the general solution to the appropriate homogeneous equation, $\mathbf{H}\hat{\mathbf{f}} = \mathbf{0}$ or $\mathbf{H}^\dagger\mathbf{H}\hat{\mathbf{f}} = \mathbf{0}$. What we are here calling a null vector would be called a solution to the homogeneous equation in the theory of differential equations. That solution would usually be adjusted to satisfy boundary conditions, and we must look for an analogous subsidiary condition in our problem.

Since the various solutions in (1.188) or (1.196) differ by null vectors of $\mathbf{H}^\dagger\mathbf{H}$, one approach is to choose the solution that contains no such null vector, *i.e.*, one that lies entirely in measurement space. Since the measurement and null spaces are orthogonal, the norm of the solution is given by

$$||\hat{\mathbf{f}}||^2 = ||\hat{\mathbf{f}}_{meas}||^2 + ||\hat{\mathbf{f}}_{null}||^2 \,. \tag{1.198}$$

The first term on the right has the same value for all solutions. Since the second term is nonnegative, $||\hat{\mathbf{f}}||$ is a minimum if $||\hat{\mathbf{f}}_{null}|| = 0$. An exact solution with no null component is called a *minimum-norm exact* solution and denoted $\hat{\mathbf{f}}_{MN}$. Similarly, a least-squares solution with no null component is called a *minimum-norm least-squares* (MNLS) solution and denoted $\hat{\mathbf{f}}_{MNLS}$. The norm referred to here should not be confused with the one given in (1.191), where a least-squares solution is defined as the one with minimum norm of the residual. The norm implied in the designation MN is the norm of $\hat{\mathbf{f}}$ itself, not the norm of $\mathbf{g} - \mathbf{H}\hat{\mathbf{f}}$.

To apply the minimum-norm condition to (1.181), we note that $\mathbf{H}^+\mathbf{g}$ lies in $\mathbb{U}_{meas}$ and $[\mathbf{I}_N - \mathbf{H}^+\mathbf{H}]\mathbf{y}$ lies in $\mathbb{U}_{null}$. The choice of $\mathbf{y}$ that minimizes the norm of $\hat{\mathbf{f}}$ is the one that makes the null component vanish, namely, $\mathbf{y} = \mathbf{0}$. Thus the minimum-norm solution to (1.181) for consistent data is

$$\hat{\mathbf{f}}_{MN} = \mathbf{H}^+\mathbf{g} \,. \tag{1.199}$$

Since the general least-squares solution has the same structure as the general exact solution, the MNLS solution is also the pseudoinverse solution:

$$\hat{\mathbf{f}}_{MNLS} = \mathbf{H}^+\mathbf{g} \,. \tag{1.200}$$

Since (1.199) and (1.200) have the same form, we can simply forget about the distinction between consistent and inconsistent data. In either case we can seek the unique, minimum-norm solution $\mathbf{H}^+\mathbf{g}$.

Explicit solutions in the SVD domain Since the eigenvectors of $\mathbf{H}\mathbf{H}^\dagger$ form a basis in $\mathbb{V}$, we can represent an arbitrary data vector $\mathbf{g}$ as (see Table 1.1)

$$\mathbf{g} = \sum_{k=1}^{M} \beta_k \mathbf{v}_k \,, \tag{1.201}$$

where

$$\beta_k = \mathbf{v}_k^\dagger \mathbf{g} \,. \tag{1.202}$$

Similarly, an arbitrary estimate of the object can be represented as

$$\hat{\mathbf{f}} = \sum_{k=1}^{N} \hat{\alpha}_k \mathbf{u}_k \,, \tag{1.203}$$

where

$$\hat{\alpha}_k = \mathbf{u}_k^\dagger \hat{\mathbf{f}} \,. \tag{1.204}$$

With the SVD representation for $\mathbf{H}^+$ and the orthonormality relations, the expansion coefficients for $\hat{\mathbf{f}}_{MN}$ or $\hat{\mathbf{f}}_{MNLS}$ are given by

$$\hat{\alpha}_k = \begin{cases} \beta_k/\sqrt{\mu_k} & \text{if} \quad \mu_k \neq 0 \\ 0 & \text{if} \quad \mu_k = 0 \end{cases} . \tag{1.205}$$

The coefficients in the estimate are thus obtained by an operation called *inverse filtering*, where each β_k is divided by $\sqrt{\mu_k}$, no matter how small μ_k is. The process is truncated at $k = R$, so we never divide by zero, but we can divide by a very small number in some cases.

With these coefficients, the explicit form of the minimum-norm estimates is

$$\hat{\mathbf{f}}_{MN} = \hat{\mathbf{f}}_{MNLS} = \sum_{k=1}^{R} \frac{\beta_k}{\sqrt{\mu_k}} \mathbf{u}_k . \tag{1.206}$$

An algorithm to obtain these estimates is:

(*a*) Perform an SVD on $\mathbf{H}$, obtaining $\{\mathbf{u}_k, \mathbf{v}_k, \mu_k\}$;

(*b*) Use the vectors $\{\mathbf{v}_k\}$ to compute $\{\beta_k\}$;

(*c*) Compute $\{\hat{\alpha}_k, k = 1, ..., R\}$ via (1.205);

(*d*) Form the weighted sum in (1.203) to compute $\hat{\mathbf{f}}$.

Unfortunately this algorithm is not practical for large matrices. If there are more than a few thousand elements (pixels, say) in the vectors $\mathbf{f}$ and $\mathbf{g}$, the SVD cannot be performed on current computers. In such cases we can use iterative algorithms, to be introduced in Sec. 1.7.6.

Noise amplification Another serious problem with the SVD algorithm is that (1.205) requires dividing by all nonzero singular values, no matter how small they may be. This step has the effect of amplifying the errors in data coefficients β_k for which μ_k is small. A full treatment of this problem must wait until we have established statistical models for the noise, but a simple analysis will illustrate the problem.

If we represent the actual vector $\mathbf{f}$ (not its estimate) by

$$\mathbf{f} = \sum_{k=1}^{N} \alpha_k \mathbf{u}_k , \qquad \alpha_k = \mathbf{u}_k^{\dagger} \mathbf{f}, \tag{1.207}$$

and the noise by

$$\boldsymbol{\epsilon} = \sum_{k=1}^{M} \gamma_k \mathbf{v}_k , \qquad \gamma_k = \mathbf{v}_k^{\dagger} \boldsymbol{\epsilon} , \tag{1.208}$$

then the expansion coefficients for $\mathbf{g}$ are given by [*cf.* (1.161)]

$$\beta_k = \sqrt{\mu_k}\, \alpha_k + \gamma_k . \tag{1.209}$$

Substituting (1.209) into (1.206), we find

$$\hat{\mathbf{f}}_{MN} = \hat{\mathbf{f}}_{MNLS} = \sum_{k=1}^{R} \left[\alpha_k + \frac{\gamma_k}{\sqrt{\mu_k}} \right] \mathbf{u}_k = \mathbf{f}_{meas} + \sum_{k=1}^{R} \frac{\gamma_k}{\sqrt{\mu_k}} \mathbf{u}_k , \tag{1.210}$$

where we have used (1.165) in the last step.

Were it not for the noise, the pseudoinverse solution would thus exactly reproduce $\mathbf{f}_{meas}$. Noise in the data causes error, of course, and (1.210) shows that the error can become very large if μ_k is near zero. This large noise amplification for small singular values is the major drawback to pseudoinverse solutions. Methods of dealing with this problem will be discussed in Chap. 15.

1.7.6 Iterative calculation of pseudoinverse solutions

In this section we discuss an important iterative technique for finding the pseudoinverse solution of a system of linear equations. This method will work when the matrix is too large for direct SVD computation, but it does not actually yield the pseudoinverse matrix; instead, for any given $\mathbf{g}$, it computes $\mathbf{H}^+\mathbf{g}$ iteratively. If we actually knew $\mathbf{H}^+$, computation of $\mathbf{H}^+\mathbf{g}$ would be just a matrix-vector multiply, but the iterative methods have to be run anew for each $\mathbf{g}$.

The starting point for developing an iterative algorithm is the limiting representation (1.135). If $\boldsymbol{\mathcal{H}}$ is the $M \times N$ matrix $\mathbf{H}$, that representation can be rewritten as

$$\mathbf{H}^+ = \lim_{\eta \to 0^+} \left[\mathbf{H}^\dagger\mathbf{H} + \eta\mathbf{I}\right]^{-1}\mathbf{H}^\dagger = \lim_{\eta \to 0^+} \left[\mathbf{I} - \boldsymbol{\Omega}\right]^{-1}\mathbf{H}^\dagger , \tag{1.211}$$

where $\mathbf{I}$ is the $N \times N$ identity matrix and $\boldsymbol{\Omega}$ is an $N \times N$ Hermitian matrix given by

$$\boldsymbol{\Omega} = (1-\eta)\mathbf{I} - \mathbf{H}^\dagger\mathbf{H} . \tag{1.212}$$

From App. A, we know that we can express $[\mathbf{I} - \boldsymbol{\Omega}]^{-1}$ by a *Neumann series.* Provided the series converges uniformly, we can write [see (A.59)]

$$[\mathbf{I} - \boldsymbol{\Omega}]^{-1} = \sum_{j=0}^{\infty} \boldsymbol{\Omega}^j . \tag{1.213}$$

Convergence To determine the conditions under which the Neumann series converges, we use the representations for $\mathbf{I}$ and $\mathbf{H}^\dagger\mathbf{H}$ from Table 1.2 and write $\boldsymbol{\Omega}$ in SVD form as

$$\boldsymbol{\Omega} = (1-\eta)\mathbf{I} - \mathbf{H}^\dagger\mathbf{H} = \sum_{n=1}^{N} (1 - \eta - \mu_n)\mathbf{u}_n\mathbf{u}_n^\dagger . \tag{1.214}$$

The SVD representation for the jth power of a matrix is obtained simply by raising each coefficient to the jth power:

$$\boldsymbol{\Omega}^j = \sum_{n=1}^{N} (1 - \eta - \mu_n)^j\mathbf{u}_n\mathbf{u}_n^\dagger . \tag{1.215}$$

The proof that (1.215) is correct makes use of the orthonormality of the $\{\mathbf{u}_n\}$ j times.

Now consider the effect of this matrix operating on an arbitrary vector $\mathbf{f}$ in $\mathbb{U}$.

The squared norm of $\mathbf{\Omega}^j \mathbf{f}$ is

$$||\mathbf{\Omega}^j \mathbf{f}||^2 = \Big|\Big|\sum_{n=1}^{N} (1-\eta-\mu_n)^j \mathbf{u}_n \mathbf{u}_n^\dagger \sum_{m=1}^{N} \alpha_m \mathbf{u}_m\Big|\Big|^2$$

$$= \Big|\Big|\sum_{n=1}^{N} (1-\eta-\mu_n)^j \alpha_n \mathbf{u}_n\Big|\Big|^2 = \sum_{n=1}^{N} |1-\eta-\mu_n|^{2j}\, |\alpha_n|^2 . \tag{1.216}$$

By the triangle inequality (see Sec. 1.1.2),

$$\Big|\Big|\sum_{j=0}^{J} \mathbf{\Omega}^j \mathbf{f}\Big|\Big| \le \sum_{j=0}^{J} ||\mathbf{\Omega}^j \mathbf{f}|| = \sum_{j=0}^{J} \left[\sum_{n=1}^{N} |1-\eta-\mu_n|^{2j} |\alpha_n|^2\right]^{\frac{1}{2}} . \tag{1.217}$$

But

$$|1-\eta-\mu_n|^{2j} \le \max_n |1-\eta-\mu_n|^{2j} , \tag{1.218}$$

so

$$\Big|\Big|\sum_{j=0}^{J} \mathbf{\Omega}^j \mathbf{f}\Big|\Big| \le \sum_{j=0}^{J} \max_n |1-\eta-\mu_n|^j \left[\sum_{n=1}^{N} |\alpha_n|^2\right]^{\frac{1}{2}}$$

$$= ||\mathbf{f}|| \max_n \sum_{j=0}^{J} |1-\eta-\mu_n|^j . \tag{1.219}$$

The remaining sum over j, an ordinary geometric series, converges absolutely if

$$|1-\eta-\mu_n| < 1 . \tag{1.220}$$

Small singular values cause no problem since $\mu_n \ge 0$ and $\eta > 0$. The geometric series converges for all n if the *maximum* singular value satisfies

$$\mu_{max} < 2-\eta . \tag{1.221}$$

If (1.221) is satisfied, we can pass to the limit $J \to \infty$, and (1.219) becomes

$$\Big|\Big|\sum_{j=0}^{\infty} \mathbf{\Omega}^j \mathbf{f}\Big|\Big| \le ||\mathbf{f}|| \max_n \sum_{j=0}^{\infty} |1-\eta-\mu_n|^j = \max_n \frac{||\mathbf{f}||}{1-|1-\eta-\mu_n|} . \tag{1.222}$$

In the limit $\eta \to 0$, the right-hand side is finite provided $\mu_{max} < 2$. Thus, under this condition, the pseudoinverse can be represented as

$$\mathbf{H}^+ = \lim_{\eta \to 0^+} \sum_{j=0}^{\infty} \mathbf{\Omega}^j \mathbf{H}^\dagger , \tag{1.223}$$

where $\mathbf{\Omega}$ is given by (1.212). This form for $\mathbf{H}^+$ will be manipulated into an iterative algorithm below.

A trick Suppose, however, that the condition (1.221) is not satisfied. With a simple trick, we can still use the Neumann series. Define a new matrix $\mathbf{H}'$ by

$$\mathbf{H}' = C\mathbf{H}\,, \tag{1.224}$$

where C is a real constant. All of the eigenvalues μ_n are scaled by $|C|^2$, so if C is chosen such that $|C|^2\mu_{max} < 2$, the Neumann series for $\mathbf{H}'$ converges, and (1.223) becomes

$$[\mathbf{H}']^+ = \lim_{\eta\to 0^+} \sum_{j=0}^{\infty} [\boldsymbol{\Omega}']^j \, [\mathbf{H}']^\dagger \,, \tag{1.225}$$

where $\boldsymbol{\Omega}'$ is defined like $\boldsymbol{\Omega}$ but with $\mathbf{H}'$ in place of $\mathbf{H}$. The pseudoinverse of the original $\mathbf{H}$ is then found from

$$[\mathbf{H}']^+ = \frac{1}{C}\mathbf{H}^+ . \tag{1.226}$$

This trick can be used even if μ_{max} is not known. If the series diverges, simply pick a $C < 1$, scale the matrix by (1.224), and try again. If it still diverges, use a smaller C. When it converges, $|C|^2\mu_{max}$ will be less than 2.

Landweber algorithm In order to convert the Neumann series into an iterative algorithm, we define partial sums of the series as

$$\hat{\mathbf{f}}^{(k)} = \lim_{\eta\to 0^+} \sum_{j=0}^{k} \boldsymbol{\Omega}^j \mathbf{H}^\dagger \mathbf{g} . \tag{1.227}$$

If the series converges, then

$$\hat{\mathbf{f}}^{(\infty)} = \mathbf{H}^+\mathbf{g} = \hat{\mathbf{f}}_{MNLS} . \tag{1.228}$$

Now consider $\hat{\mathbf{f}}^{(k+1)}$, given by

$$\hat{\mathbf{f}}^{(k+1)} = \lim_{\eta\to 0} \sum_{j=0}^{k+1} \boldsymbol{\Omega}^j \mathbf{H}^\dagger \mathbf{g} = \lim_{\eta\to 0} \left\{ \mathbf{H}^\dagger \mathbf{g} + \sum_{j=1}^{k+1} \boldsymbol{\Omega}^j \mathbf{H}^\dagger \mathbf{g} \right\} . \tag{1.229}$$

If we let $m = j - 1$, this equation becomes

$$\hat{\mathbf{f}}^{(k+1)} = \lim_{\eta\to 0} \left\{ \mathbf{H}^\dagger \mathbf{g} + \sum_{m=0}^{k} \boldsymbol{\Omega}^{m+1} \mathbf{H}^\dagger \mathbf{g} \right\} = \lim_{\eta\to 0} \left\{ \mathbf{H}^\dagger \mathbf{g} + \boldsymbol{\Omega}\hat{\mathbf{f}}^{(k)} \right\} . \tag{1.230}$$

We can now insert the definition of $\boldsymbol{\Omega}$ from (1.212) and pass to the limit $\eta \to 0$; the result is

$$\hat{\mathbf{f}}^{(k+1)} = \hat{\mathbf{f}}^{(k)} + \mathbf{H}^\dagger \left[\mathbf{g} - \mathbf{H}\hat{\mathbf{f}}^{(k)} \right] . \tag{1.231}$$

Thus the $(k+1)^{th}$ partial sum (or estimate of $\mathbf{f}$) can be calculated from the kth by adding a correction term $\mathbf{H}^\dagger \left[\mathbf{g} - \mathbf{H}\hat{\mathbf{f}}^{(k)}\right]$. The algorithm converges under the same conditions as required for the convergence of the Neumann series.

Moreover, if the algorithm converges at all, it converges to $\hat{\mathbf{f}}_{MNLS}$, provided the initial estimate contains no null functions. One way to see this is to note that the correction term is zero if $\hat{\mathbf{f}}^{(k)}$ satisfies the normal equation, (1.194), so the convergent estimate is at least an LS solution. The MN part comes about since each correction step makes use of $\mathbf{H}^\dagger$, an operator that erases null functions. If the initial estimate is chosen by (1.227) with $k = 0$, it is given by $\hat{\mathbf{f}}^{(0)} = \mathbf{H}^\dagger\mathbf{g}$; since this vector contains no null functions, neither will any subsequent estimates, and the convergence point will be $\hat{\mathbf{f}}_{MNLS}$.

The algorithm described by (1.231) has been frequently rediscovered in various communities. In image processing it is known as the van Cittert or Landweber algorithm (Landweber, 1951), and in tomography it is called SIRT (simultaneous iterative reconstruction technique). As we shall see below, it is also closely related to the Jacobi method from numerical analysis.

Other iterative methods One approach (Golub and van Loan, 1989) to devising a whole family of iterative pseudoinversion algorithms is to define a *splitting* of the matrix $\mathbf{H}^\dagger\mathbf{H}$ by

$$\mathbf{H}^\dagger\mathbf{H} = \mathbf{A} - \mathbf{B}\,, \tag{1.232}$$

where $\mathbf{A}$ can be any $N \times N$ matrix so long as $\mathbf{A}^{-1}$ exists. Some simple matrix algebra is then used to rewrite the normal equation (1.194):

$$(\mathbf{A} - \mathbf{B})\hat{\mathbf{f}} = \mathbf{H}^\dagger\mathbf{g}\,; \tag{1.233}$$

$$\mathbf{A}\hat{\mathbf{f}} = \mathbf{H}^\dagger\mathbf{g} + \mathbf{B}\hat{\mathbf{f}} = \mathbf{H}^\dagger\mathbf{g} + \mathbf{A}\hat{\mathbf{f}} - \mathbf{H}^\dagger\mathbf{H}\hat{\mathbf{f}}\,; \tag{1.234}$$

$$\hat{\mathbf{f}} = \hat{\mathbf{f}} + \mathbf{A}^{-1}\left[\mathbf{H}^\dagger\mathbf{g} - \mathbf{H}^\dagger\mathbf{H}\hat{\mathbf{f}}\right]\,. \tag{1.235}$$

So far, this equation is algebraically equivalent to (1.194); to convert it into an iterative algorithm, we assume that $\hat{\mathbf{f}}$ on the right-hand side refers to the current estimate $\hat{\mathbf{f}}^{(k)}$, and that the entire right-hand side is then the rule for forming the next estimate $\hat{\mathbf{f}}^{(k+1)}$. This procedure, known as a *fixed-point iteration*, is discussed in more detail in Sec. 15.4. Here it yields

$$\hat{\mathbf{f}}^{(k+1)} = \hat{\mathbf{f}}^{(k)} + \mathbf{A}^{-1}\left[\mathbf{H}^\dagger\mathbf{g} - \mathbf{H}^\dagger\mathbf{H}\hat{\mathbf{f}}^{(k)}\right]\,. \tag{1.236}$$

If the algorithm converges, then $\left[\mathbf{H}^\dagger\mathbf{g} - \mathbf{H}^\dagger\mathbf{H}\hat{\mathbf{f}}^{(k)}\right]$ is zero and (1.194) is satisfied. We can thus write

$$\hat{\mathbf{f}}^{(\infty)} = \hat{\mathbf{f}}_{LS}\,, \tag{1.237}$$

regardless of the choice of $\mathbf{A}$, so long as the algorithm converges. It does not follow, however, that $\hat{\mathbf{f}}^{(\infty)} = \hat{\mathbf{f}}_{MNLS} = \mathbf{H}^+\mathbf{g}$ since the matrix $\mathbf{A}^{-1}$ can introduce null functions. It can be shown that (1.236) does indeed converge so long as the maximum eigenvalue of $\mathbf{A}^{-1}\mathbf{B}$ is less than 1 (Golub and van Loan, 1989, p. 508).

Different iterative algorithms are generated by different choices of $\mathbf{A}$. If $\mathbf{A} = \mathbf{I}$ and $\mathbf{B} = \mathbf{I} - \mathbf{H}^\dagger\mathbf{H}$, then (1.236) reverts to the Landweber algorithm. A classical method called the *Jacobi iteration* is obtained by letting $\mathbf{A}$ be a diagonal matrix with the same diagonal elements as $\mathbf{H}^\dagger\mathbf{H}$, or

$$A_{mn} = \left[\mathbf{H}^\dagger\mathbf{H}\right]_{mn}\delta_{mn}\,. \tag{1.238}$$

The resulting iteration rule, in component form, is

$$\hat{f}_n^{(k+1)} = \hat{f}_n^{(k)} + \frac{1}{A_{nn}}\left[\mathbf{H}^\dagger\mathbf{g} - \mathbf{H}^\dagger\mathbf{H}\hat{\mathbf{f}}^{(k)}\right]_n$$

$$= \frac{1}{A_{nn}}\left[(\mathbf{H}^\dagger\mathbf{g})_n - \sum_{m=1}^{n-1}(\mathbf{H}^\dagger\mathbf{H})_{nm}\hat{f}_m^{(k)} - \sum_{m=n+1}^{N}(\mathbf{H}^\dagger\mathbf{H})_{nm}\hat{f}_m^{(k)}\right]\,. \tag{1.239}$$

It is only the normalizing factor $1/A_{nn}$ that distinguishes the Jacobi and Landweber algorithms.

Another classical method, the *Gauss-Seidel iteration*, corresponds to taking $\mathbf{A}$ as the lower-triangular part of $\mathbf{H}^\dagger\mathbf{H}$, *i.e.*,

$$A_{mn} = \begin{cases} [\mathbf{H}^\dagger\mathbf{H}]_{mn} & \text{if} \quad n \leq m \\ 0 & \text{if} \quad n > m \end{cases} . \tag{1.240}$$

The Gauss-Seidel iteration rule is

$$\hat{f}_n^{(k+1)} = \frac{1}{A_{nn}} \left[(\mathbf{H}^\dagger\mathbf{g})_n - \sum_{m=1}^{n-1} (\mathbf{H}^\dagger\mathbf{H})_{nm} \hat{f}_m^{(k+1)} - \sum_{m=n+1}^{N} (\mathbf{H}^\dagger\mathbf{H})_{nm} \hat{f}_m^{(k)} \right] . \tag{1.241}$$

The only difference between Jacobi and Gauss-Seidel is that the latter makes use of an updated component $\hat{f}_m^{(k+1)}$ as soon as it is available, while the former waits until all components have been computed and then updates the entire vector $\hat{\mathbf{f}}^{(k)}$ to $\hat{\mathbf{f}}^{(k+1)}$ simultaneously [compare the first sums in (1.241) and (1.239)].

There are many other iterative algorithms for solving (1.181) or (1.194). One famous family is the Kaczmarz method/algebraic reconstruction technique (ART) and its variations. The Kaczmarz method was proved to converge for nonsingular matrices by Kaczmarz (1937). It was rediscovered in a tomography context by Gordon et al. (1970), who gave it the name ART, and it has been widely used in that field (and widely called by that name) ever since.

For very large and sparse matrices, an important class of algorithms (of which ART is an example) is called row-action methods. The distinguishing features of these methods are that they do not make any modifications to the original matrix (unlike Gaussian elimination, for example), that they require access to only a single row of the matrix at a time, and that, when a new iterate is computed, only the immediately preceding iterate is required. For a comprehensive survey of row-action algorithms, see Censor (1981).

1.8 REPRODUCING-KERNEL HILBERT SPACE

We close this chapter with a discussion of a kind of vector space called a *reproducing-kernel Hilbert space*, which is particularly useful for representing functions that are smooth in some sense. Since smooth functions are very important in imaging applications, reproducing-kernel Hilbert spaces will arise in many different contexts in this book.

A formal definition of a reproducing-kernel Hilbert space is that it is a function space in which each function can be evaluated at a point by use of a bounded linear functional (Weinert, 1983; Wahba, 1990; Daubechies, 1992; Zayed, 1993). Our most familiar function space, $\mathbb{L}_2(\mathbf{R})$, does not have this property since it contains discontinuous or even mildly singular functions. In fact, functions in $\mathbb{L}_2(\mathbf{R})$ are not necessarily even defined pointwise. There are, however, many ways to select a subspace of $\mathbb{L}_2(\mathbf{R})$ that is a reproducing-kernel Hilbert space.

By the Riesz representation theorem (1.24), a bounded linear functional can be written as a scalar product. Thus, in a reproducing-kernel Hilbert space there exists a vector $\mathbf{h}(x')$ such that, for all $\mathbf{f}$ in the space,

$$(\mathbf{h}(x'), \mathbf{f})_{rk} = f(x') , \tag{1.242}$$

where $(\cdot,\cdot)_{rk}$ denotes a scalar product in the space. Here $\mathbf{h}(x')$ and $\mathbf{f}$ are both vectors in the space, though $\mathbf{h}(x')$ is characterized by the scalar parameter x'. Any vector in a function space can be identified with a function; in (1.242), $\mathbf{f}$ is identified with $f(x)$ and $\mathbf{h}(x')$ is identified with the function of x denoted $h(x, x')$, with x' regarded as a fixed parameter. To fully specify the reproducing-kernel Hilbert space, we must define the scalar product and find the kernel $h(x, x')$ such that (1.242) is satisfied.

1.8.1 Positive-definite Hermitian operators

We can start with a compact, positive-definite Hermitian operator $\mathcal{A}$ on $\mathbb{L}_2(\mathbf{R})$ and use it to identify the subspace of $\mathbb{L}_2(\mathbf{R})$ that constitutes a reproducing-kernel Hilbert space related to the operator. Since $\mathcal{A}$ is compact, it satisfies an eigenvalue equation like (1.67) with a discrete spectrum $\{\lambda_n, n = 1, ..., \infty)$, and the eigenvectors $\{\boldsymbol{\psi}_n\}$ form an orthonormal basis in $\mathbb{L}_2(\mathbf{R})$. Thus any vector $\mathbf{f}$ in $\mathbb{L}_2(\mathbf{R})$ can be expressed as [*cf.* (1.63)]

$$\mathbf{f} = \sum_{n=1}^{\infty} \alpha_n \boldsymbol{\psi}_n \,, \qquad \alpha_n = \boldsymbol{\psi}_n^\dagger \mathbf{f}. \tag{1.243}$$

An equivalent representation of $\mathbf{f}$ as a function rather than as an abstract vector is

$$f(x) = \sum_{n=1}^{\infty} \alpha_n \, \psi_n(x) \,, \qquad \alpha_n = \int_{-\infty}^{\infty} dx \; \psi_n^*(x) \, f(x) \,. \tag{1.244}$$

The $\mathbb{L}_2$ norm of $\mathbf{f}$ is

$$||\mathbf{f}||_2 = \left[\sum_{n=1}^{\infty} |\alpha_n|^2\right]^{\frac{1}{2}} = \left[\sum_{n=1}^{\infty} \left| \int_{-\infty}^{\infty} dx \; \psi_n^*(x) \, f(x) \right|^2\right]^{\frac{1}{2}} = \left[\int_{-\infty}^{\infty} dx \; |f(x)|^2\right]^{\frac{1}{2}} , \tag{1.245}$$

where the last step follows from (1.62) with the recognition that the unit operator in $\mathbb{L}_2$ has the kernel $\delta(x - x')$.

To construct a reproducing-kernel Hilbert space associated with $\mathcal{A}$, we define the norm in the space as

$$||\mathbf{f}||_{rk}^2 = \sum_{n=1}^{\infty} \frac{1}{\lambda_n} \, |\alpha_n|^2 \,. \tag{1.246}$$

Since we have assumed that $\mathcal{A}$ is positive-definite, none of the eigenvalues is zero for finite n, but they may approach zero rapidly as $n \to \infty$. It is possible for the sum in (1.246) to diverge if $|\alpha_n|^2$ falls off more slowly with increasing n than λ_n does. A vector $\mathbf{f}$ (or equivalently, a function $f(x)$) is said to be in this reproducing-kernel Hilbert space if $||\mathbf{f}||_{rk}$ is finite.

For a normed space to be a Hilbert space, the norm must be derivable from a scalar product. A suitable definition of scalar product is

$$(\mathbf{f}_1, \mathbf{f}_2)_{rk} = \left(\mathbf{f}_1, \mathcal{A}^{-1}\mathbf{f}_2\right)_2 = \left(\mathcal{A}^{-1}\mathbf{f}_1, \mathbf{f}_2\right)_2 \,, \tag{1.247}$$

where $(\cdot,\cdot)_2$ denotes the $\mathbb{L}_2$ scalar product. The equality of the two $\mathbb{L}_2$ scalar products in (1.247) holds because $\mathcal{A}^{-1}$ is a Hermitian operator in $\mathbb{L}_2$.

With (1.87), it is straightforward to show that the norm (1.246) is compatible with the scalar product (1.247) in the sense that

$$||\mathbf{f}||_{rk}^2 = (\mathbf{f}, \mathbf{f})_{rk} \,. \tag{1.248}$$

The kernel for this reproducing-kernel Hilbert space is simply the kernel of $\mathcal{A}$ when it is expressed as an integral operator. Combining (1.25) and (1.86), we find

$$[\mathcal{A}\mathbf{f}](x) = \int_{-\infty}^{\infty} dx' \, h(x, x') \, f(x') = \sum_{n=1}^{\infty} \lambda_n (\boldsymbol{\psi}_n, \mathbf{f}) \, \psi_n(x)$$

$$= \int_{-\infty}^{\infty} dx' \left[\sum_{n=1}^{\infty} \lambda_n \, \psi_n(x) \, \psi_n^*(x') \right] f(x') \,. \tag{1.249}$$

Hence,

$$h(x, x') = \sum_{n=1}^{\infty} \lambda_n \, \psi_n(x) \, \psi_n^*(x') \,, \tag{1.250}$$

which is the functional counterpart of the spectral decomposition (1.86). The corresponding vector in the reproducing-kernel Hilbert space is given by

$$\mathbf{h}(x') = \sum_{n=1}^{\infty} [\lambda_n \, \psi_n^*(x')] \, \boldsymbol{\psi}_n \,, \tag{1.251}$$

where the quantity in square brackets is the expansion coefficient.

Next consider the scalar product, defined as in (1.247), between $\mathbf{h}(x')$ and an arbitrary $\mathbf{f}$ in the reproducing-kernel space. We have

$$(\mathbf{h}(x'), \mathbf{f})_{rk} = \left(\mathcal{A}^{-1} \, \mathbf{h}(x'), \mathbf{f} \right)_2 \,. \tag{1.252}$$

But, from (1.87), (1.89) and (1.251), we find

$$\mathcal{A}^{-1} \, \mathbf{h}(x') = \sum_{n=1}^{\infty} \psi_n^*(x') \, \boldsymbol{\psi}_n \,. \tag{1.253}$$

After some algebra of the sort that by now should be standard, we find

$$(\mathbf{h}(x'), \mathbf{f})_{rk} = f(x') \,, \tag{1.254}$$

so (1.242) is satisfied.

1.8.2 Nonnegative-definite Hermitian operators

So far, we have considered reproducing-kernel Hilbert spaces based on positive-definite operators, but (1.250) and (1.251) suggest a more general approach. Consider an operator $\mathcal{B}$ defined by

$$\mathcal{B} = \sum_{n=1}^{N} \mu_n \mathbf{u}_n \mathbf{u}_n^\dagger \,, \tag{1.255}$$

where $\{\mathbf{u}_n, n = 1, ..., N\}$ is any orthonormal set in $\mathbb{L}_2(\mathbb{R})$ and $\{\mu_n\}$ is any set of positive real numbers. The operator $\mathcal{B}$ constructed in this way is nonnegative-definite and Hermitian. Since N is not infinite, however, $\mathcal{B}$ is singular in $\mathbb{L}_2(\mathbb{R})$. For the same reason, $\{\mathbf{u}_n\}$ is not a basis in $\mathbb{L}_2(\mathbb{R})$.

The reproducing-kernel Hilbert space associated with this operator is the ND subspace of $\mathbb{L}_2(\mathbb{R})$ spanned by $\{\mathbf{u}_n, n = 1, ..., N\}$. Any vector in this space can be represented as

$$\mathbf{f} = \sum_{n=1}^{N} \beta_n \mathbf{u}_n , \qquad \beta_n = \mathbf{u}_n^\dagger \mathbf{f}. \tag{1.256}$$

The norm is now defined by

$$||\mathbf{f}||_{rk}^2 = \sum_{n=1}^{N} \frac{1}{\mu_n} |\beta_n|^2 . \tag{1.257}$$

None of the factors $|\beta_n|^2$ can be infinite since $\mathbf{f}$ is also in $\mathbb{L}_2(\mathbb{R})$, none of the μ_n is zero by definition, and there is a finite number of terms, so the norm in (1.257) is finite for any $\mathbf{f}$ in the space.

The corresponding scalar product is given by

$$(\mathbf{f}_1, \mathbf{f}_2)_{rk} = \sum_{n=1}^{N} \frac{1}{\mu_n} \beta_{1n}^* \beta_{2n} , \tag{1.258}$$

where β_{1n} and β_{2n} are expansion coefficients for $\mathbf{f}_1$ and $\mathbf{f}_2$, respectively. The rk scalar product can be related to an $\mathbb{L}_2$ scalar product by (Helstrom, 1995)

$$(\mathbf{f}_1, \mathbf{f}_2)_{rk} = (\mathbf{f}_1, \tilde{\mathbf{f}}_2)_2 , \tag{1.259}$$

where $\tilde{\mathbf{f}}_2$ is any vector in $\mathbb{L}_2(\mathbb{R})$ related to $\mathbf{f}_2$ by

$$\mathbf{f}_2 = \mathcal{B}\tilde{\mathbf{f}}_2 . \tag{1.260}$$

Since $\mathcal{B}$ is singular in $\mathbb{L}_2(\mathbb{R})$, there will be an infinity of vectors in that space that can serve as $\tilde{\mathbf{f}}_2$; it does not matter which we choose since the scalar product of $\mathcal{B}\tilde{\mathbf{f}}_2$ with an $\mathbf{f}_1$ in the span of $\{\mathbf{u}_n\}$ will not be affected by null vectors of $\mathcal{B}$ in $\tilde{\mathbf{f}}_2$.

The reproducing kernel of the space associated with $\mathcal{B}$ is given by

$$h(x, x') = \sum_{n=1}^{N} \mu_n \, u_n(x) \, u_n^*(x') . \tag{1.261}$$

The relations in (1.258) – (1.261) agree with those given earlier if $N \to \infty$ and $\mathcal{B}$ is nonsingular.

An important special case of (1.258) – (1.261) is where N is finite and all $\mu_n = 1$. In this case, $\mathcal{B}$ is the projector on the ND subspace spanned by $\{\mathbf{u}_n, n = 1, ..., N\}$ and the reproducing kernel is the kernel of the projection operator.

Table 1.1 SVD Representations of Vectors

$$\mathbf{f} = \sum_{n=1}^{N} \alpha_n \mathbf{u}_n \,, \qquad \alpha_n = \mathbf{u}_n^\dagger \mathbf{f}$$

$$\hat{\mathbf{f}} = \sum_{n=1}^{N} \hat{\alpha}_n \mathbf{u}_n \,, \qquad \hat{\alpha}_n = \mathbf{u}_n^\dagger \hat{\mathbf{f}}$$

$$\mathbf{g} = \sum_{k=1}^{M} \beta_k \mathbf{v}_k \,, \qquad \beta_k = \mathbf{v}_k^\dagger \mathbf{g}$$

$$\boldsymbol{\epsilon} = \sum_{k=1}^{M} \gamma_k \mathbf{v}_k \,, \qquad \gamma_k = \mathbf{v}_k^\dagger \boldsymbol{\epsilon}$$

$$\mathbf{v}_n = \frac{1}{\sqrt{\mu_n}} \boldsymbol{\mathcal{H}} \mathbf{u}_n \,, \qquad (\mu_n \neq 0)$$

$$\mathbf{f}_{meas} = \sum_{n=1}^{R} \alpha_n \mathbf{u}_n \,, \quad \mathbf{f}_{null} = \sum_{n=R+1}^{N} \alpha_n \mathbf{u}_n$$

(where $\mathbf{f} \equiv \mathbf{f}_{meas} + \mathbf{f}_{null}$ and $\boldsymbol{\mathcal{H}} \mathbf{f}_{null} \equiv \mathbf{0}$)

$$\mathbf{g}_{con} = \sum_{k=1}^{R} \beta_k \mathbf{v}_k \,, \quad \mathbf{g}_{incon} = \sum_{k=R+1}^{M} \beta_k \mathbf{v}_k$$

(where $\mathbf{g} \equiv \mathbf{g}_{con} + \mathbf{g}_{incon}$ and $\boldsymbol{\mathcal{H}}^\dagger \mathbf{g}_{incon} \equiv \mathbf{0}$)

Table 1.2 SVD Representations of Operators

$\mathcal{H}^\dagger\mathcal{H} = \sum_{k=1}^{R} \mu_k \mathbf{u}_k \mathbf{u}_k^\dagger$	$\mathcal{H}\mathcal{H}^\dagger = \sum_{k=1}^{R} \mu_k \mathbf{v}_k \mathbf{v}_k^\dagger$
$\mathcal{H} = \sum_{k=1}^{R} \sqrt{\mu_k}\, \mathbf{v}_k \mathbf{u}_k^\dagger$	$\mathcal{H}^\dagger = \sum_{k=1}^{R} \sqrt{\mu_k}\, \mathbf{u}_k \mathbf{v}_k^\dagger$
$\left[\mathcal{H}^\dagger\mathcal{H}\right]^{-1} = \sum_{k=1}^{N} \frac{1}{\mu_k} \mathbf{u}_k \mathbf{u}_k^\dagger$	(provided $N = R$)
$\left[\mathcal{H}\mathcal{H}^\dagger\right]^{-1} = \sum_{k=1}^{M} \frac{1}{\mu_k} \mathbf{v}_k \mathbf{v}_k^\dagger$	(provided $M = R$)
$\mathcal{H}^+ = \sum_{k=1}^{R} \frac{1}{\sqrt{\mu_k}} \mathbf{u}_k \mathbf{v}_k^\dagger$	(Moore-Penrose pseudoinverse)
$\mathcal{P}_{meas} = \mathcal{H}^+\mathcal{H} = \sum_{k=1}^{R} \mathbf{u}_k \mathbf{u}_k^\dagger$	(Projector of $\mathbf{f}$ onto measurement space)
$\mathcal{I}_{\mathbb{U}} = \sum_{k=1}^{N} \mathbf{u}_k \mathbf{u}_k^\dagger$	(Identity operator in $\mathbb{U}$)
$\mathcal{P}_{null} = \mathcal{I}_{\mathbb{U}} - \mathcal{H}^+\mathcal{H} = \sum_{k=R+1}^{N} \mathbf{u}_k \mathbf{u}_k^\dagger$	(Projector of $\mathbf{f}$ onto null space)
$\mathcal{P}_{con} = \mathcal{H}\mathcal{H}^+ = \sum_{k=1}^{R} \mathbf{v}_k \mathbf{v}_k^\dagger$	(Projector of $\mathbf{g}$ onto consistency space)
$\mathcal{I}_{\mathbb{V}} = \sum_{k=1}^{M} \mathbf{v}_k \mathbf{v}_k^\dagger$	(Identity operator in $\mathbb{V}$)
$\mathcal{P}_{incon} = \mathcal{I}_{\mathbb{V}} - \mathcal{H}\mathcal{H}^+ = \sum_{k=R+1}^{M} \mathbf{v}_k \mathbf{v}_k^\dagger$	(Projector of $\mathbf{g}$ onto inconsistency space)

2

Dirac Delta and Other Generalized Functions

Optical imaging systems are often experimentally characterized by imaging a small, bright light source. If the dimensions of this source are much smaller than the spatial resolution capability of the imaging system, the object is essentially a mathematical point. Such an object is called a *point source*, and its image is the *point spread function* or PSF of the system. A more general source can be decomposed into a collection of points; if the system is linear, the response to the general object is obtained by adding up the responses to the individual points.

The mathematical construct that corresponds to a physical point source is the *delta function*, introduced by P. A. M. Dirac in quantum mechanics. There are three basic ways of defining a delta function. The first is an intuitive approach, where we simply postulate a function that is zero except at a single point, but infinite at that point in such a way that the integral of the function is constrained to be unity. A second, somewhat better approach is to consider a sequence of well-behaved functions, all of which integrate to unity. The width of these functions tends to zero and the amplitude tends to infinity in order to hold the integral constant. The delta function is defined as the limit of this sequence.

Both of these approaches leave many mathematical questions unanswered. They do not make it clear just when the definition is valid or how the delta function can legitimately be manipulated in practice. The third approach, which puts the delta function on a firmer mathematical footing, requires the *theory of distributions*, originally developed by Laurent Schwartz (1950). We give a brief summary of this theory in the next section, but then make use of all three approaches to delta functions in what follows. The hope is that this strategy will allow delta functions to be understood at several different levels of rigor. The reader who is content with a less rigorous development can jump to Sec. 2.2 without loss of continuity.

2.1 THEORY OF DISTRIBUTIONS

2.1.1 Basic concepts

In brief, a distribution is a linear, continuous functional that maps a function $t(x)$ to a real or complex number. For simplicity, we assume initially that $t(x)$ is a real, scalar-valued function of a single real variable x. We have seen in Chap. 1 that the general form of a bounded, continuous, linear functional on a Hilbert space is given by the Riesz representation theorem, (1.24), as

$$\Phi\{t(x)\} = \int_{-\infty}^{\infty} dx\; g(x)\, t(x)\,, \tag{2.1}$$

where $g(x)$, known as the kernel of the functional, must lie in the Hilbert space (Gohberg and Goldberg, 1981, p. 61). If the Hilbert space is $\mathbb{L}_2$, $g(x)$ must be square-integrable.

If we give up on the requirement that the kernel lie in a Hilbert space, we can define a wider class of functionals known as distributions. When we do so, we shall always write the distribution in the Riesz form, like (2.1), but then $g(x)$ need not be a square-integrable function or even a function at all in the conventional sense. All we have to do is specify the mapping rule of the functional, and that, in turn, will give meaning to the kernel function, which we then term a *generalized function.*

As a simple example, consider the function $g(x) = 1/x$. We might be tempted to define a functional using this kernel as

$$\Phi_{1/x}\{t(x)\} \stackrel{?}{=} \int_{-\infty}^{\infty} dx\; \frac{t(x)}{x}\,, \tag{2.2}$$

but this integral is not well defined (unless $t(0) = 0$) because of the singularity at $x = 0$. Converting the integral to a contour integral in the complex plane, we have three options in dealing with this pole. We can indent the contour above the pole, indent it below the pole, or take the Cauchy principal value (see App. B) at the singularity. Since these three options can lead to three different numerical values for the integral, we must make an arbitrary choice. In some practical applications, the physics of the situation dictates the use of the Cauchy principal value, and in that case it is useful to define a functional by

$$\Phi_{1/x}\{t(x)\} = \lim_{\epsilon \to 0} \left\{ \int_{-\infty}^{-\epsilon} dx\; \frac{t(x)}{x} + \int_{\epsilon}^{\infty} dx\; \frac{t(x)}{x} \right\}. \tag{2.3}$$

Now we have a clearly specified functional, provided that $t(x)$ is sufficiently well behaved that the limit exists. This functional gives meaning to the generalized function $\mathcal{P}\{1/x\}$, where $\mathcal{P}$ denotes the Cauchy principal value, and we can write

$$\Phi_{1/x}\{t(x)\} \equiv \int_{-\infty}^{\infty} dx\; t(x)\, \mathcal{P}\left\{\frac{1}{x}\right\}, \tag{2.4}$$

where the right-hand side must be interpreted according to (2.3). We have thus succeeded in defining a distribution and, simultaneously, a generalized function. It is important to note, however, that the generalized function is defined only by its action within an integral, and then only when certain restrictions are placed on the function $t(x)$ that appears in the integrand.

One caution regarding terminology: The distinction between a distribution (a functional) and its kernel (a generalized function) is often blurred. For example, it is often said that a delta function is a distribution. What this means is that the delta function is a generalized function that can be used as the kernel in a functional.

2.1.2 Well-behaved functions

As the example above shows, definition of a distribution as a functional places some restrictions on the functions in the domain of the functional. For the distribution associated with $\mathcal{P}\{1/x\}$, the requirement on $t(x)$ is that the limit in (2.3) exist.

Different kinds of distributions require different properties for the functions $t(x)$. The strongest requirements are that the nonzero values of $t(x)$ be confined to a finite interval $a < x < b$ and that the function and all of its derivatives be bounded and continuous within this region. If these conditions are satisfied, we say that $t(x)$ is infinitely differentiable and has compact support. Such functions are called *test functions*. The space of all test functions with support (a, b) will be denoted by $\mathbb{T}(a, b)$. Because $t(x)$ is bounded and has compact support, it necessarily lies in $\mathbb{L}_2(a, b)$, but not all functions in $\mathbb{L}_2(a, b)$ are test functions, so $\mathbb{T}(a, b)$ is a subspace of $\mathbb{L}_2(a, b)$.

It is actually fairly difficult to construct test functions since infinitely differentiable functions such as polynomials or Gaussians tend not to have compact support. One way to form test functions meeting the necessary conditions is to use as a building block the function $h(x)$ defined by (Richards and Youn, 1990)

$$h(x) = \begin{cases} \exp(-1/x) & \text{if} \quad x > 0 \\ 0 & \text{if} \quad x \leq 0 \end{cases} , \tag{2.5}$$

which is plotted in Fig. 2.1a. It is not difficult to show that all derivatives of this function exist everywhere, even at the transition point $x = 0$.

Amusingly, in the theory of functions of a complex variable, $\exp(-1/z)$ is the archetype of a function that has an essential singularity (see App. B). Within an arbitrarily small radius of the origin in the complex plane, $\exp(-1/z)$ takes on all finite values an infinite number of times. The complex function $\exp(-1/z)$ is as ill-behaved as a function can be, while the real function $\exp(-1/x)$ is the paragon of good behavior.

By use of $h(x)$ we can construct a test function on (a, b) as

$$t(x) = h\left(\frac{x-a}{b-a}\right) h\left(\frac{b-x}{b-a}\right) . \tag{2.6}$$

As required, this function vanishes unless x is in the interval (a, b), and it is infinitely differentiable within this interval (see Fig. 2.1b). More general test functions can be constructed by superimposing shifted and scaled versions of the function defined by (2.5) (see Richards and Youn, 1990, Chap. 1).

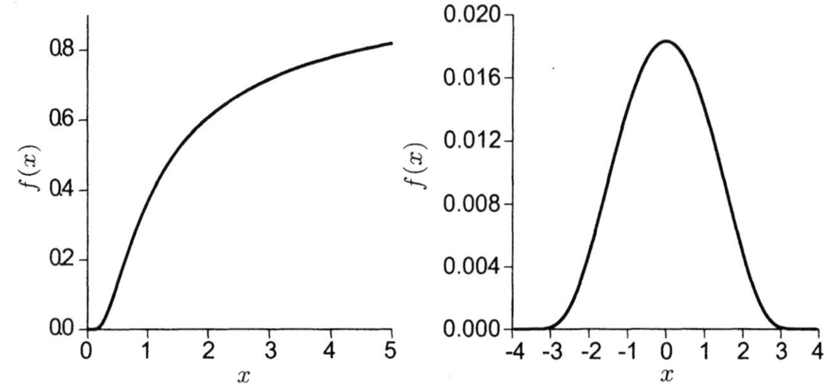

(a) (b)

Fig. 2.1 (a) Plot of the building-block function $h(x)$ from (2.5); (b) plot of the test function $t(x)$ from (2.6), with $(a, b) = (-4, 4)$.

The requirement of compact support is often onerous, and it is useful to define distributions based on infinitely differentiable functions without compact support. Following Lighthill (1958), we define a *good function* as one that is everywhere differentiable any number of times and such that it and all of its derivatives vanish faster than $|x|^{-N}$ for all N as $x \to \pm\infty$. In other words, $t(x)$ is a good function if all derivatives exist and

$$\lim_{x \to \pm\infty} \left\{ |x|^N t(x) \right\} = 0 \qquad \text{for all } N\,. \tag{2.7}$$

An example of a good function is the Gaussian $\exp(-\pi x^2)$. Good functions are also called *open-support test functions* (Richards and Youn, 1990) or Schwartz functions (Strichartz, 1994). The space of all good functions is often called *Schwartz space*.

A *fairly good function* (Lighthill, 1958) is infinitely differentiable but may be unbounded as $x \to \pm\infty$. The requirement is that it must not blow up faster than some power of x. Specifically, $t(x)$ is a fairly good function if

$$\lim_{x \to \pm\infty} \left\{ |x|^{-N} t(x) \right\} = 0 \qquad \text{for some } N\,. \tag{2.8}$$

Any polynomial is a fairly good function, since it is infinitely differentiable and (2.8) is satisfied for N greater than the degree of the polynomial, but $\exp(x)$ is not a fairly good function. Fairly good functions are also called *test functions of slow growth*.

2.1.3 Approximation of other functions

Many functions of practical interest do not fit into any of the categories of well-behaved functions defined above. In optics, for example, the transmission of light through a rectangular slit is described by the *rect function*, defined by

$$\operatorname{rect}\left(\frac{x}{L}\right) \equiv \begin{cases} 1 & \text{if} \quad |x| < L/2 \\ 0 & \text{if} \quad |x| > L/2 \end{cases}\,. \tag{2.9}$$

This function has compact support but is not continuous at $x = \pm L/2$, so it is not a test function. As another example, the *Heaviside unit step function*, defined by

$$\text{step}(x) \equiv \begin{cases} 1 & \text{if} \quad x > 0 \\ 0 & \text{if} \quad x < 0 \end{cases} , \tag{2.10}$$

satisfies (2.8) but is not differentiable at $x = 0$, so it is not a fairly good function.

Nevertheless, these and other functions of practical importance can be approximated arbitrarily closely by test functions or good functions, depending on their behavior at infinity, so the restrictions of differentiability are not so demanding as might appear (Zemanian, 1965, p. 3, 1987).

To illustrate this point, we note that the building-block function $h(kx)$ is a fairly good function for all k. (In fact, it is the best of the fairly good functions since it remains bounded at infinity.) In terms of this function,

$$\text{step}(x) = \lim_{k\to\infty} h(kx)\,, \tag{2.11}$$

as illustrated in Fig. 2.2. The limit implies uniform convergence, which means that the maximum value of the difference between $h(kx)$ and $\text{step}(x)$ tends to zero:

$$\lim_{k\to\infty} \max_{x} \left| \text{step}(x) - h(kx) \right| = 0 \qquad \text{for all } x\,, \tag{2.12}$$

except possibly at the isolated point $x = 0$, where the step function has been deliberately left undefined. The technical term is that it *converges uniformly almost everywhere*. The proof of this result is elementary, but see Sec. 2.3.2 for some cautions about using it.

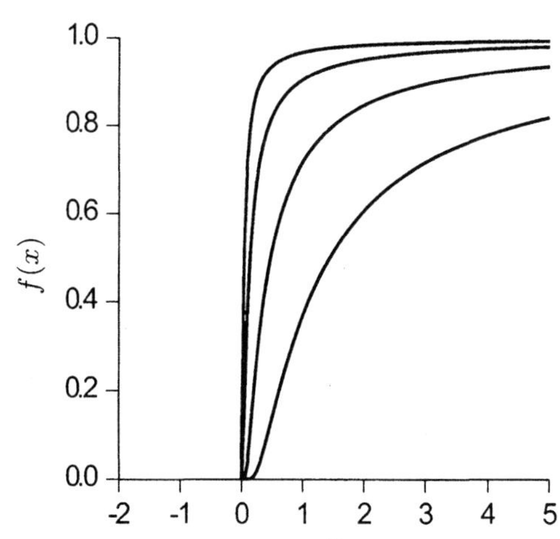

Fig. 2.2 Plot of the limiting representation of a step function from (2.11); The four curves are for $k = 0$, 3, 10 and 30.

A test function approximating $\text{rect}(x/L)$ can also be obtained from $h(x)$:

$$\text{rect}\left(\frac{x}{L}\right) = \lim_{k\to\infty} h\left[k\left(x+\tfrac{L}{2}\right)\right] h\left[-k\left(x+\tfrac{L}{2}\right)\right] . \tag{2.13}$$

This function vanishes identically unless $-L/2 < x < L/2$, and it is infinitely differentiable inside that interval, so it qualifies as a test function. Again, the representation converges uniformly to $\mathrm{rect}(x/L)$ except at the isolated points $x = \pm L/2$.

A general procedure for approximating integrable functions by good functions will be given in Sec. 2.2.5.

2.1.4 Formal definition of distributions

In order to define distributions, we must first define the notion of convergence of test functions, which is stronger than the uniform convergence used above. Our development follows Richards and Youn (1990).

A sequence of test functions $\{t_n(x)\}$ is said to converge to a specific test function $t(x)$ as $n \to \infty$ if:

(a) The functions $t(x)$ and all of the $t_n(x)$ have a common region of support (a, b);

(b) For each k, the sequence of kth derivatives $\left\{t_n^{(k)}(x)\right\}$ converges to $t^{(k)}(x)$.

A distribution Φ is defined as a linear, continuous mapping from the space $\mathbb{T}$ of test functions to the real or complex numbers. Linearity and continuity are defined in Sec. 1.3.2 of Chap. 1; their interpretation in the present notation is:

(a) (Linearity) $\Phi\{\alpha s(x)+\beta t(x)\} = \alpha\Phi\{s(x)\}+\beta\Phi\{t(x)\}$ for all test functions $s(x)$ and $t(x)$ and all constants α and β;

(b) (Continuity) If $t_n(x) \to t(x)$ in the sense defined above, then $\Phi\{t_n(x)\} \to \Phi\{t(x)\}$.

The continuity condition will allow us to interchange distribution operations and limits at will; the interchange is valid *by definition.*

A specific, if not very interesting, distribution is the one defined by an ordinary function. If $f(x)$ is a piecewise-continuous function on the real line, it defines a distribution Φ_f given by

$$\Phi_f\{t(x)\} = \int_{-\infty}^{\infty} dx\ f(x)\, t(x)\,. \tag{2.14}$$

It is easy to demonstrate that this distribution satisfies the requirements of linearity and continuity.

2.1.5 Properties of distributions

In this section we give several important properties of distributions. The proofs can be found in Messiah (1961) or Richards and Youn (1990).

The linear combination of two distributions is a distribution. If Φ_1 and Φ_2 are distributions, $\Phi = \alpha\Phi_1 + \beta\Phi_2$ is a distribution defined such that

$$\Phi\{t(x)\} = \alpha\Phi_1\{t(x)\} + \beta\Phi_2\{t(x)\}\,. \tag{2.15}$$

It follows from this definition that the product of a generalized function and a constant behaves just as one would expect from (2.1). When the product is used in an integral, we can simply take the constant out of the integral.

As an extension of (2.15), if the infinite series $\sum_i \Phi_i\{t(x)\}$ is summable, its sum defines a distribution.

Integrals of distributions can also be defined. If Φ_λ is a distribution that depends on a parameter λ that can vary continuously in some domain Λ, and if the integral

$$I\{t(x)\} = \int_\Lambda \Phi_\lambda\{t(x)\}\, d\lambda \tag{2.16}$$

converges for all test functions $t(x)$, then it defines a distribution (Messiah, 1961) I given by

$$I = \int_\Lambda \Phi_\lambda\, d\lambda\,. \tag{2.17}$$

Derivatives of distributions (or, more correctly, derivative distributions) can be defined by analogy with the familiar operation of integration by parts. If Φ_g is a distribution corresponding to the kernel $g(x)$, which may be an ordinary function or a generalized one, then a distribution $\Phi_{g'}$, where prime denotes derivative, can be defined by

$$\Phi_{g'}\{t(x)\} \equiv -\Phi_g\left\{\frac{dt(x)}{dx}\right\} = -\int_{-\infty}^{\infty} dx\; g(x)\,\frac{dt(x)}{dx} = \int_{-\infty}^{\infty} dx\; g'(x)\,t(x)\,. \tag{2.18}$$

If $g(x)$ is an ordinary differentiable function, then (2.18) is simply a statement of integration by parts. If $g(x)$ is a generalized function, the last form in this equation can be regarded as a definition of the new generalized function $g'(x)$. It is straightforward to verify that the definition in (2.18) satisfies the requirements of linearity and continuity, so we have indeed defined a legitimate distribution.

Next we inquire about the product of two generalized functions. Suppose the kernel $g(x) = f(x)\,h(x)$, where $f(x)$ or $h(x)$ or both may be generalized functions. The distribution Φ_g corresponding to $g(x)$ does not necessarily exist, and there are only a few conditions where we can be sure that it does (Messiah, 1961). If $f(x)$ has derivatives of all orders, then Φ_g exists for all $h(x)$. This statement can easily be proven by noting that $f(x)\,t(x)$ is a test function in that case. Another case where the product distribution is guaranteed to exist is when both $f(x)$ and $h(x)$ are square-integrable functions, since in that case the product kernel is also square-integrable and the Riesz representation theorem applies. With these exceptions, however, we cannot be sure that products of generalized functions (or, loosely, products of distributions) make any sense. For example, we shall see that the square of a delta function is not defined.

2.1.6 Tempered distributions

Tempered distributions are defined exactly as other distributions except that the conditions on the test functions are relaxed. In particular, we require infinite differentiability but not compact support. Thus *tempered distributions are continuous, linear functionals acting on good functions* as defined above. Moreover, it will be useful to allow the number returned by the functional to be complex, so a tempered distribution is a mapping from the space of good functions to $\mathbb{C}^1$.

Tempered distributions are a subset of all distributions. Since a tempered distribution is well defined for differentiable functions of open support, it is also well defined for test functions with compact support. Thus a tempered distribution is a distribution, but not all distributions are tempered distributions. All of the distributions discussed in this chapter are, however, tempered distributions.

The main motivation for introducing tempered distributions is in connection with Fourier theory, as explored further in the next chapter. A wide variety of other distributions, based on others kinds of test functions, can also be defined (Zemanian, 1965, 1987), but we shall not need them.

2.2 ONE-DIMENSIONAL DELTA FUNCTION

2.2.1 Intuitive definition and elementary properties

As noted in the introduction, the simplest — and least rigorous — approach to defining the delta function is simply to require it to vanish everywhere except at a single point, say $x = x_0$:

$$\delta(x - x_0) = 0 \qquad \text{if } x \neq x_0 \,. \tag{2.19}$$

If this function were to be finite at $x = x_0$, its integral would vanish, no matter how the integral was defined. Thus, for $\delta(x - x_0)$ to have nontrivial properties, it must be infinite at x_0, or at least in an infinitesimal neighborhood around x_0. We require that this infinity be such that the integral of the function is unity, *i.e.*,

$$\int_{-\infty}^{\infty} dx\ \delta(x - x_0) = 1 \,. \tag{2.20}$$

Because of (2.19), we also have

$$\int_{x_0-\epsilon}^{x_0+\epsilon} dx\ \delta(x - x_0) = 1 \,, \tag{2.21}$$

where ϵ is any finite positive number.

A number of elementary properties follow from these equations (Gaskill, 1978; Barrett and Swindell, 1981, 1996). The most important is the so-called *sifting property* which arises when the delta function $\delta(x - x_0)$ is multiplied by another function $f(x)$ in the integrand of an integral. Since the delta function is zero for $x \neq x_0$, the behavior of $f(x)$ for $x \neq x_0$ is irrelevant. We can thus write

$$\int_{-\infty}^{\infty} dx\ f(x)\,\delta(x - x_0) = \int_{x_0-\epsilon}^{x_0+\epsilon} dx\ f(x)\,\delta(x - x_0) \,, \tag{2.22}$$

where ϵ is arbitrarily small. Then, if $f(x)$ is continuous at x_0, it does not vary over the range of integration in the limit as ϵ approaches zero, and we can replace it by $f(x_0)$, obtaining

$$\begin{aligned} \int_{-\infty}^{\infty} dx\ f(x)\,\delta(x - x_0) &= \lim_{\epsilon \to 0} \int_{x_0-\epsilon}^{x_0+\epsilon} dx\ f(x)\,\delta(x - x_0) \\ &= f(x_0) \lim_{\epsilon \to 0} \int_{x_0-\epsilon}^{x_0+\epsilon} dx\ \delta(x - x_0) = f(x_0) \,. \end{aligned} \tag{2.23}$$

This result is called the *sifting property* because integration of $f(x)\,\delta(x - x_0)$ sifts out the value of $f(x)$ at $x = x_0$. A more general statement of the property is that

$$\int_a^b dx\ f(x)\,\delta(x - x_0) = \begin{cases} f(x_0) & \text{if} \quad a < x_0 < b \\ 0 & \text{if} \quad x_0 < a \ \text{ or } \ x_0 > b \end{cases} \,. \tag{2.24}$$

Another way to state the same property is to write

$$f(x)\,\delta(x - x_0) = f(x_0)\,\delta(x - x_0)\,, \tag{2.25}$$

which is true in the sense that both sides of the equation yield the same result when integrated over an arbitrary interval (a, b). A special case of (2.25) is

$$x\,\delta(x) = 0\,. \tag{2.26}$$

Additional properties may be derived by simple changes of variable in (2.24). Letting $x' = -x$ and setting x_0 to zero shows that

$$\delta(x) = \delta(-x)\,. \tag{2.27}$$

Thus the delta function is an inherently even function. Similarly, the transformation $x' = ax$, where a is a real constant, leads to

$$\delta(ax - x_0) = |a|^{-1}\,\delta(x - x_0/a) \tag{2.28}$$

and the special case

$$\delta(ax) = |a|^{-1}\,\delta(x)\,. \tag{2.29}$$

In these equations, the absolute value arises since the direction of integration is reversed if a is negative.

Now suppose the argument of a delta function is itself a function, as in $\delta[g(x)]$. This delta function is zero except where its argument vanishes. Suppose that

$$g(x) = 0 \text{ at } x = x_n\,, \qquad n = 1, ..., N\,. \tag{2.30}$$

Then $\delta[g(x)] = 0$ unless $x = x_n$ and we can write

$$\int_{-\infty}^{\infty} dx\; f(x)\,\delta[g(x)] = \sum_{n=1}^{N} \int_{x_n-\epsilon}^{x_n+\epsilon} dx\; f(x)\,\delta[g(x)]\;. \tag{2.31}$$

If $g(x)$ is differentiable at the points x_n, we can expand it in a Taylor series about that point. By definition, $g(x_n) = 0$, and we shall assume that the first derivative $g'(x_n) \neq 0$. Since ϵ is arbitrarily small, all higher terms in the Taylor series can be neglected without approximation. We thus have

$$\int_{-\infty}^{\infty} dx\; f(x)\,\delta[g(x)] = \sum_{n=1}^{N} \int_{x_n-\epsilon}^{x_n+\epsilon} dx\; f(x)\,\delta[g'(x_n)\,(x - x_n)]$$

$$= \sum_{n=1}^{N} \frac{1}{|g'(x_n)|} \int_{x_n-\epsilon}^{x_n+\epsilon} dx\; f(x)\,\delta(x - x_n) = \sum_{n=1}^{N} \frac{f(x_n)}{|g'(x_n)|}\,, \tag{2.32}$$

where we have used (2.29), recognizing that $g'(x_n)$ is a constant. This result can be summarized succinctly as

$$\delta[g(x)] = \sum_{n=1}^{N} \frac{\delta(x - x_n)}{|g'(x_n)|}\,. \tag{2.33}$$

Again, the sense of this equation, as with all equations involving delta functions, is that it is true when both sides are multiplied by a suitable function and integrated over some interval.

2.2.2 Limiting representations

The delta function can be put on a somewhat more rigorous footing if it is defined as the limit of an ordinary function. We define a *Dirac sequence*[1] as a set of functions $\psi_k(x)$ with the following properties:

(a) $\psi_k(x) \geq 0$ for all k;

(b) $\psi_k(x) = \psi_k(-x)$ for all k;

(c) $\int_{-\infty}^{\infty} dx\ \psi_k(x) = 1$ for all k;

(d) The width of $\psi_k(x)$ decreases uniformly with increasing k.

This set of requirements follows Lang (1993, pp. 227–228), except that he does not include property (b), so his results are more general. Other authors omit property (a), and we shall also relax it eventually.

The width in property (d) is defined in terms of the integral of ψ_k over a finite interval. By property (c), if the function is integrated over $(-\infty, \infty)$, the result is unity. If it is integrated over an interval $(-\beta, \beta)$, where $0 < \beta < \infty$, the result is $1-\epsilon$, where $0 \leq \epsilon \leq 1$. The value ϵ is the fraction of the integral contributed by the part of ψ_k that lies outside $(-\beta, \beta)$, and 2β is the width of ψ_k at level ϵ as defined by

$$\int_{-\beta}^{\beta} dx\ \psi_k(x) = 1 - \epsilon\,. \tag{2.34}$$

The width 2β can obviously depend on both ϵ and the index k. The formal definition of width in property (d) is basically that, whatever ϵ is chosen as a reference, the resulting width 2β decreases as the index k is increased.

The key result from formal analysis that we need is that the Dirac sequence can be used to approximate a bounded, continuous function $f(x)$ arbitrarily closely. That is,

$$\lim_{k\to\infty} \int_{-\infty}^{\infty} dx\ f(x)\,\psi_k(x - x_0) = f(x_0)\,, \tag{2.35}$$

where the limit is to be understood in the sense of uniform convergence[2] (Lang, 1993, p. 228). The particular form of integral in (2.35) is called a *convolution*, and we say that $f(x)$ *is convolved with* $\psi_k(x)$ in this integral. Equation (2.35) says that this convolution converges to $f(x_0)$ as $k \to \infty$ provided $\psi_k(x)$ satisfies the conditions for a Dirac sequence.

Another way to write the result of (2.35) is

$$\delta(x - x_0) = \lim_{k\to\infty} \psi_k(x - x_0)\,, \tag{2.36}$$

by which we mean that use of the right-hand side will reproduce the sifting property, (2.24), for all $f(x)$ that are bounded and continuous at $x = x_0$.

[1] The term *Dirac family* is also used, especially when k is not restricted to be an integer, but we shall not make this distinction.

[2] Actually the function $f(x)$ need not be continuous everywhere for (2.35) to hold. It is sufficient if it is measurable and continuous in the neighborhood of x_0. Specifically, if $f(x)$ is in $\mathbb{L}_p(\mathbb{R}^1)$, with $1 \leq p < \infty$, and is continuous over some range $x_0 - \epsilon < x < x_0 + \epsilon$, then the integral in (2.35) converges in the $\mathbb{L}_p$ sense to $f(x_0)$ (Lang, 1993, pp. 234–235).

There are many possible choices for the functions $\{\psi_k(x)\}$, and each choice yields a different limiting representation for the delta function. One choice is a sequence of Gaussians, $\psi_k(x) = k\exp(-\pi k^2 x^2)$, so that

$$\delta(x) = \lim_{k\to\infty} k\exp(-\pi k^2 x^2) = \lim_{\epsilon\to 0^+} \frac{1}{\epsilon}\exp\left[-\pi(x/\epsilon)^2\right], \tag{2.37}$$

where $\epsilon = 1/k$ [not to be confused with the ϵ in (2.34)]. This particular Dirac sequence is illustrated in Fig. 2.3.

Other choices for the $\{\psi_k\}$ yield

$$\delta(x) = \lim_{\epsilon\to 0^+} \frac{1}{\epsilon}\,\text{rect}\left(\frac{x}{\epsilon}\right), \tag{2.38}$$

$$\delta(x) = \lim_{\epsilon\to 0^+} \frac{\epsilon/\pi}{x^2+\epsilon^2}, \tag{2.39}$$

$$\delta(x) = \lim_{\epsilon\to 0^+} \frac{1}{2\epsilon}\exp\left(-\frac{|x|}{\epsilon}\right), \tag{2.40}$$

$$\delta(x) = \lim_{\epsilon\to 0^+} \frac{1}{2\epsilon}\,\text{sech}^2\left(\frac{|x|}{\epsilon}\right). \tag{2.41}$$

With all of these forms, it is straightforward to show that $\delta(x) = 0$ if $x \neq 0$ and that

$$\int_{-\infty}^{\infty} dx\;\delta(x) = 1\,, \tag{2.42}$$

as required. Furthermore, for all of these examples, $\psi_k(0)$ tends to infinity, though limiting representations of the delta function can be constructed for which this is not the case (see, for example, Kanwal, 1983, pp. 10–12).

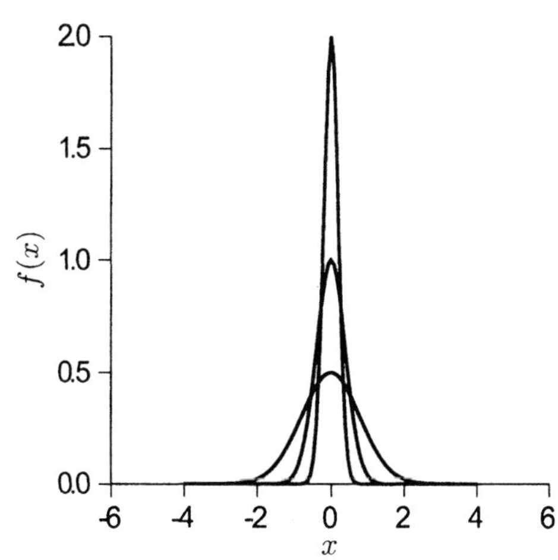

Fig. 2.3 Plot of the limiting representation of a delta function as a sequence of Gaussians from (2.37). The three curves shown are for $k = 0.5$, 1.0 and 2.0.

A somewhat more subtle representation, which will turn out to be very useful, is based on the sinc function, defined as

$$\text{sinc}(u) \equiv \frac{\sin(\pi u)}{\pi u}, \tag{2.43}$$

and plotted in Fig. 2.4. The sinc function is frequently referred to as the *Dirichlet kernel*, after the Prussian mathematician Peter Gustave Lejeune Dirichlet (1805–1859), disciple of Gauss and mentor of Riemann (Bell, 1937).

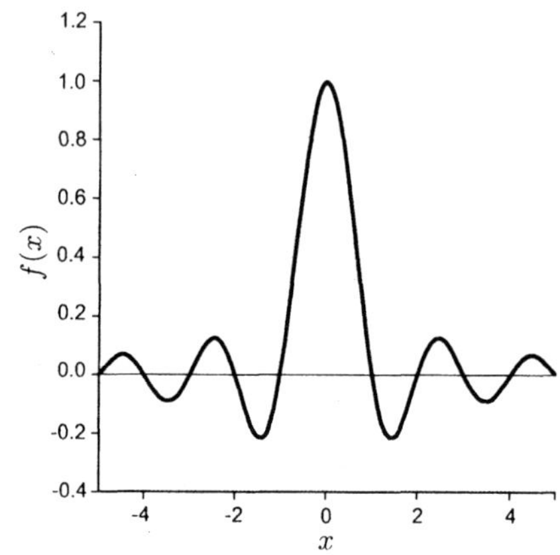

Fig. 2.4 Plot of the sinc function defined in (2.43).

We could consider a sequence of functions $\psi_k(x) = k\,\mathrm{sinc}(kx)$, but since the sinc function can go negative, this choice for ψ_k does not define a Dirac sequence. Nevertheless, $\psi_k(0) \to \infty$ as $k \to \infty$ and the integral of $\psi_k(x)$ over $(-\infty, \infty)$ is unity for all k, leading us to expect that the limit behaves as a delta function and that it is possible to write

$$\delta(x) = \lim_{k\to\infty} k\,\mathrm{sinc}(kx) = \lim_{k\to\infty} \frac{\sin(\pi k x)}{\pi x}\,. \tag{2.44}$$

This statement is true, in the sense that the right-hand side yields the sifting property, (2.24), if suitable restrictions are placed on the function $f(x)$ that appears in that equation. For example, it is sufficient to require that $f(x)$ be differentiable, with $f'(x)$ bounded and continuous, and $f(\pm\infty) = 0$ (Stakgold, 1967, p. 27; Kanwal, 1983, pp. 6–7). Moreover, it works for all functions in the space $\mathbb{L}_p(-\infty, \infty)$, with $1 < p < \infty$, provided the limit is interpreted as limit in the mean (see Champeney, 1987, p. 35, and Sec. 3.2.2 in Chap. 3 of this book).

Another useful way to write (2.44) is to recognize that the sinc function is the integral of an exponential, so that we also have

$$\delta(x) = \lim_{k\to\infty} \int_{-k/2}^{k/2} d\nu \, \exp(2\pi i \nu x)\,. \tag{2.45}$$

We shall often write this equation in the simplified form,

$$\delta(x) = \int_{-\infty}^{\infty} d\nu \, \exp(2\pi i \nu x)\,, \tag{2.46}$$

where the infinite integral is to be understood in the sense of the limit in (2.45).

A useful form akin to (2.44) is the *comb function*, defined by

$$\mathrm{comb}(x) = \lim_{N\to\infty} \frac{\sin(\pi N x)}{\sin(\pi x)}\,, \tag{2.47}$$

where N is an integer that tends to ∞. The reason for the name of this function can be discerned from Fig. 2.5. For finite N, the function has sharp peaks like teeth on a comb for all integer values of x; as N tends to infinity, the heights of the peaks tend to infinity while the widths tend to zero. That these peaks tend to delta functions can be seen by examining the one at $x = 0$. For a small neighborhood of $x = 0$, the denominator of (2.47) is approximately πx, which is also the denominator in (2.44). Thus, for $|x| < 1/2$, $\text{comb}(x)$ is indistinguishable from $\delta(x)$. Furthermore, for integer N, $\sin(\pi N x)/\sin(\pi x)$ is periodic with period 1, so there must be other delta peaks at all integer values of x. We can therefore write

$$\text{comb}(x) = \sum_{n=-\infty}^{\infty} \delta(x - n) . \tag{2.48}$$

Another way to express the comb function is to recognize that $\sin(\pi N x)/\sin(\pi x)$ is the sum of a geometric series:

$$\sum_{n=-M}^{M} \exp(2\pi i n x) = \frac{\sin(\pi N x)}{\sin(\pi x)} , \tag{2.49}$$

where $N = 2M + 1$. Thus we can write

$$\text{comb}(x) = \sum_{n=-\infty}^{\infty} \exp(2\pi i n x) . \tag{2.50}$$

The difference between $\text{comb}(x)$ from (2.50) and $\delta(x)$ from (2.46) is only in the appearance of the sum rather than the integral; the replacement of integral by sum results in an infinite comb of delta functions.

With any of the limiting forms discussed in this section, the various properties listed in Sec. 2.2.1 can be derived by ordinary rules for manipulating integrals.

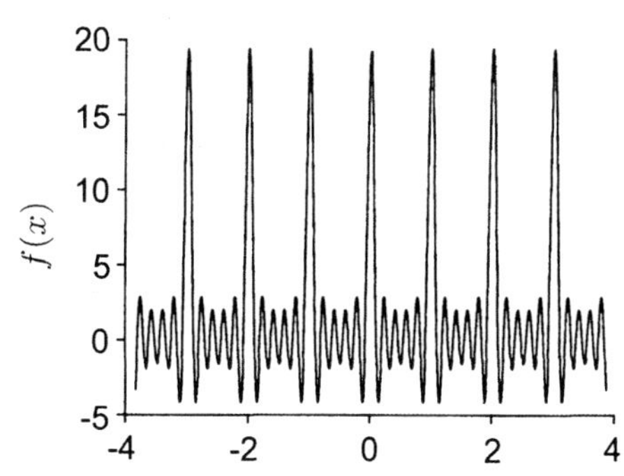

Fig. 2.5 Plot of a representation of the comb function from (2.49).

2.2.3 Distributional approach

In Sec. 2.2.1 we simply postulated, in a very nonrigorous way, the existence of a function that vanished except for a single point and became infinite at that point in such a way as to maintain an integral equal to one. The limiting representations of Sec. 2.2.2 remove some, but not all, of the mathematical uncertainties associated with this definition. One question left unanswered is when it is valid to interchange the order of an integral and a limit; the limit of an integral is not necessarily the integral of the limit. Also, we have not clearly specified the conditions that have to be placed on $f(x)$ in order for our expressions to be valid.

These uncertainties are removed by using the theory of distributions to define a delta function. In this view, $\delta(x)$ is a generalized function associated with the distribution Φ_δ in such a way that

$$\Phi_\delta\{t(x)\} = \int_{-\infty}^{\infty} dx\ \delta(x)\, t(x) \equiv t(0)\,, \tag{2.51}$$

where $t(x)$ is a test function. The same equation can be used to define a tempered distribution if $t(x)$ is a good function as defined in Sec. 2.1.2.

To get a more general sifting property analogous to (2.24), we simply define another distribution, rather clumsily designated as $\Phi_{\delta_{x_0}}$, specified by

$$\Phi_{\delta_{x_0}}\{t(x)\} = \int_{-\infty}^{\infty} dx\ \delta(x-x_0)\, t(x) \equiv t(x_0)\,. \tag{2.52}$$

There can be no question of the validity of this equation since it is just a definition, not something derived from (2.51). We define the result of the distribution to be $t(x_0)$, and the integral expression involving $\delta(x-x_0)$ is just a convenient mnemonic for this definition.

We can proceed in this way, avoiding all questions of validity by just defining whatever distributions we need. We do not do so willy-nilly, however, but try to choose definitions consistent with ordinary functionals in the Riesz form, (2.1). Moreover, the resulting distribution must satisfy the requirements of linearity and continuity discussed in Sec. 2.1.4. We illustrate this approach in the next section where we define the derivative of a delta function.

2.2.4 Derivatives of delta functions

If $\delta(x)$ is defined by the limit of a Dirac sequence as in (2.36) and the functions $\psi_k(x)$ are all differentiable, it is natural to define the derivative of a delta function as

$$\delta'(x-x_0) = \lim_{k\to\infty} \psi_k'(x-x_0)\,, \tag{2.53}$$

where the prime denotes derivative. With this definition and the assumption that $f(x)$ is differentiable, we have

$$\int_{-\infty}^{\infty} dx\ f(x)\,\delta'(x-x_0) = \lim_{k\to\infty}\int_{-\infty}^{\infty} dx\ f(x)\,\psi_k'(x-x_0)$$

$$= -\lim_{k\to\infty}\int_{-\infty}^{\infty} dx\ f'(x)\,\psi_k(x-x_0) = -\int_{-\infty}^{\infty} dx\ f'(x)\,\delta(x-x_0) = -f'(x_0)\,, \tag{2.54}$$

where we have performed an integration by parts and made use of the fact that all of the ψ_k vanish at infinity.

Repetition of this procedure and consideration of finite limits of integration lead to the conclusion that

$$\int_a^b dx\ f(x)\,\delta^{(n)}(x-x_0) = \begin{cases} (-1)^n f^{(n)}(x_0) & \text{if}\quad a < x_0 < b \\ 0 & \text{if}\quad x_0 < a \ \text{ or }\ x_0 > b \end{cases}\,, \tag{2.55}$$

where $f^{(n)}(x)$ is the nth derivative of $f(x)$.

The development of (2.54) and (2.55) raises a number of mathematical questions about differentiability, support and the validity of interchanging limits and integrals. All of these problems are avoided by use of distributions. In Sec. 2.1.5, we discussed derivatives of distributions, and it is easy to apply this idea here. By analogy to (2.18), we define a distribution

$$\Phi_{\delta'}\{t(x)\} = \int_{-\infty}^{\infty} dx\ \delta'(x)\, t(x) \equiv -\Phi_{\delta}\{t'(x)\} = -t'(0)\,, \tag{2.56}$$

where the generalized function $\delta'(x)$ is defined by this equation. Similarly, higher derivatives of the delta function are associated with other new distributions, and we have in general

$$\int_{-\infty}^{\infty} dx\ t(x)\, \delta^{(n)}(x - x_0) = (-1)^n t^{(n)}(x_0)\,, \tag{2.57}$$

where we have abandoned the Φ notation since it was getting impossibly clumsy.

From either the Dirac-sequence approach or the distributional one, the following operational properties of derivatives of the delta function can be derived:

$$\delta^{(n)}(x - x_0) = 0 \qquad \text{if } x \neq x_0\,, \tag{2.58}$$

$$\int_{-\infty}^{\infty} dx\ \delta^{(n)}(x - x_0) = 0 \qquad \text{if } n > 0\,, \tag{2.59}$$

$$\delta^{(n)}(x) = (-1)^n\, \delta^{(n)}(-x)\,, \tag{2.60}$$

$$x\, \delta'(x) = -\delta(x)\,. \tag{2.61}$$

All of these equations are true in the sense that, when multiplied by a test function and integrated, both sides yield the same result.

2.2.5 A synthesis

As we have seen, the distributional approach to delta functions avoids many mathematical difficulties, basically by defining them away. Anytime we need a previously undefined result, such as the derivative of a delta function, we simply define a new distribution. The rigor of equations such as (2.56) and (2.57) is beyond reproach since they are merely definitions; virtually any definition of distribution is allowed, so long as it satisfies the basic requirements of continuity and linearity. The price we have paid is that the definitions are restricted to test functions (or at least good functions), and we might want to use the results more widely. From the viewpoint of Dirac sequences, on the other hand, (2.56) and (2.57) can be derived directly since there is no difficulty in differentiating a delta function so long as the $\{\psi_k\}$ are themselves differentiable. The difficulty in that case is that it is not always clear what restrictions have to be placed on the function $f(x)$ that appears along with the generalized function $g(x)$ in an integral of the form $\int f(x)\, g(x)\, dx$.

To illustrate this dichotomy, consider an integral where the function $f(x)$ is bounded but discontinuous at $x = x_0$. Then, from the Dirac-sequence view, there is no problem in showing that

$$\int_{-\infty}^{\infty} dx\ f(x)\, \delta(x - x_0) = \tfrac{1}{2} \lim_{x \to x_0^+} f(x) + \tfrac{1}{2} \lim_{x \to x_0^-} f(x) = \tfrac{1}{2}[f(x_0^+) + f(x_0^-)]\,. \tag{2.62}$$

This very useful result would be meaningless in a strict distributional treatment since the distribution is defined only with respect to continuous test functions. What we need is an approach that will allow more flexibility in the choice of $f(x)$ yet retain the rigor and clarity of distributions.

The way to achieve this synthesis is to use test functions to approximate more general functions. We have seen two examples of how this can be done in Sec. 2.1.2, but we shall now make use of properties of Dirac sequences to develop a general procedure.

The functions $\psi_k(x)$ used in a Dirac sequence need not be test functions, but they can be. A Dirac sequence of test functions can be constructed from our building-block function, $h(x)$, as defined in (2.5). The functions

$$s_k(x) \equiv kh\left(kx + \tfrac{1}{2}\right) h\left(-kx + \tfrac{1}{2}\right) \tag{2.63}$$

are test functions with support $(-\epsilon, \epsilon)$, where $\epsilon = 1/2k$, and with a proper normalizing constant to satisfy condition (c) of Sec. 2.2.2, they also satisfy all of the requirements for a Dirac sequence. Using these functions or any similar test functions that qualify as a Dirac sequence, we define the kth approximant of a function $f(x)$ as

$$\hat{f}_k(x) \equiv \int_{-\infty}^{\infty} dx'\, f(x')\, s_k(x - x')\,. \tag{2.64}$$

If $f(x)$ has support $(-a, a)$, then $\hat{f}_k(x)$ has support $(-a-\epsilon, a+\epsilon)$, which approaches $(-a, a)$ as $k \to \infty$.

To complete the demonstration that $\hat{f}_k(x)$ is a test function, we need to show that all of its derivatives exist. For this purpose, we must assume that it is legitimate to differentiate under the integral sign, which it is if the resulting integral exists (perhaps in the Lebesgue sense[3]). For example, it is legitimate if $f(x)$ has a finite number of finite discontinuities. Under these conditions, we have

$$\frac{d^n \hat{f}_k(x)}{dx^n} = \int_{-\infty}^{\infty} dx'\, f(x')\, \frac{d^n\, s_k(x - x')}{dx^n}\,. \tag{2.65}$$

Since s_k is a test function for all k, all of its derivatives are guaranteed to exist everywhere. Thus $\hat{f}_k(x)$ is a test function of support $(-a - \epsilon, a + \epsilon)$ for all k and, since $\epsilon = 1/2k$, $\hat{f}_\infty(x)$ is a test function of support $(-a, a)$. Moreover, we know from the discussion in Sec. 2.2.2 that $\hat{f}_k(x)$ converges uniformly to $f(x)$ at every point where $f(x)$ is continuous. If $f(x)$ is bounded but discontinuous at $x = x_0$, then $\hat{f}_k(x_0)$ can be shown to converge to the average as in (2.62) since $s_k(x) = s_k(-x)$.

Thus integrable functions of compact support can be uniformly approximated by test functions $\hat{f}_k(x)$. If $f(x)$ does not have compact support, but interchange of differentiation and integration is still legitimate, $\hat{f}_k(x)$ is a good function instead of a test function.

To summarize, all of the functionals considered here can be written in the form $\int f(x)\, g(x)\, dx$. In the distributional approach, $g(x)$ is a generalized function and $f(x)$ is restricted to be a test function (or a good function for a tempered distribution). The generalized function itself has no restrictions at all placed on it

[3]For a concise introduction to Lebesgue integration, see Champeney (1987) or Friedman (1991).

since it is defined by the functional. In the Dirac-sequence approach, $g(x)$ is the limit of a sequence of functions obeying conditions $(a)-(d)$ of Sec. 2.2.2, while $f(x)$ can be any function for which the product $f(x)\,g(x)$ is integrable. The synthesis suggested in this section combines these two approaches by allowing $g(x)$ to be a generalized function, defined by its distribution, but representing $f(x)$ as the limit of a Dirac sequence *of test functions*. Since a wide variety of functions can be so represented, the distributional results derived for test functions can be more widely applied. The reader is cautioned, however, not to use limiting representations for *both* $f(x)$ and $g(x)$; see Sec. 2.3.2 for an example where the order of the limits cannot be interchanged.

2.2.6 Delta functions as basis vectors

The sifting property of delta functions has a useful interpretation in terms of vector spaces. If $f(x)$ is a square-integrable function and hence a vector in $\mathbb{L}_2(-\infty,\infty)$ (see Chap. 1), then it can be expanded in terms of various sets of basis vectors for that space. As discussed in Sec. 1.1.6, the basis vectors need not lie in the space for which they form the basis. The example given in that section was the Fourier basis, in which the basis vectors are the set $\{\exp(2\pi i\nu x)\}$, indexed by the continuous variable ν. The delta functions $\{\delta(x-a)\}$ can also be considered as a continuous basis, this time indexed by the continuous variable a. Since the delta functions are not square-integrable, they are not in $\mathbb{L}_2(-\infty,\infty)$, but they can be used to expand a continuous function in that space. The required expansion is just the sifting property,

$$f(x) = \int_{-\infty}^{\infty} da\ f(a)\,\delta(x-a)\,, \tag{2.66}$$

from which we see that the expansion coefficients are just the values of the function.

This viewpoint runs into some difficulties when $f(x)$ is not continuous since then $f(a)$ is not uniquely defined. We can avoid these difficulties by confining $f(x)$ to the space of test functions, or we can appeal to the argument of the last section that certain discontinuous functions can be approximated arbitrarily closely by test functions.

2.3 OTHER GENERALIZED FUNCTIONS IN 1D

2.3.1 Generalized functions as limits

As we have seen, the generalized function $\delta(x)$ can be defined in two mathematically defensible ways: either in terms of a distribution or as the limit of an ordinary function. Either of these approaches can be extended to define a wide variety of other generalized functions. The distributional approach merely defines whatever generalized functions we need; suitable definitions will be given in the sections that follow. The limiting-sequence approach, however, requires that we be able to place suitable restrictions on the functions in the sequence. We have done this in Sec. 2.2.2 for the delta function, but we now need to examine how to extend the method to other generalized functions.

The key mathematical result justifying the limiting-sequence approach is that any generalized function $g(x)$, no matter how wild, can be written as the limit of a sequence of good functions (Lighthill, 1958; Richards and Youn, 1990):

$$g(x) = \lim_{k\to\infty} w_k(x)\,, \tag{2.67}$$

where $w_k(x)$ is a good function for all k. The distribution associated with $g(x)$ is, of course,

$$\Phi_g\{t(x)\} = \int_{-\infty}^{\infty} dx\; g(x)\, t(x) = \lim_{k\to\infty} \int_{-\infty}^{\infty} dx\; w_k(x)\, t(x)\,, \tag{2.68}$$

where $t(x)$ is a test function.

In the sections below, we shall show how to define new generalized functions distributionally and to devise suitable limiting representations of them.

2.3.2 Generalized functions related to the delta function

Step function Consider the function step(x), which we defined as an ordinary discontinuous function in (2.10). A distributional definition of the step function is

$$\Phi_{\text{step}}\{t(x)\} \equiv \int_0^{\infty} dx\; t(x) = \int_{-\infty}^{\infty} dx\; t(x)\,\text{step}(x)\,, \tag{2.69}$$

where the first integral defines the functional and the second defines the generalized function step(x). We have already seen in (2.11) one way in which this generalized function can be written as the limit of a sequence of fairly good functions. Other representations can be derived from any of the Dirac sequences used to define $\delta(x)$. If we have a Dirac sequence of functions $\{\psi_k(x)\}$ that converges to $\delta(x)$, we can define step(x) as

$$\text{step}(x) = \lim_{k\to\infty} \int_{-\infty}^{x} dx'\; \psi_k(x')\,. \tag{2.70}$$

Each integral in this limit is a fairly good function if $\psi_k(x)$ is a test function.[4]

The shifted step function step$(x - x_0)$ is defined distributionally (without the cumbersome Φ notation) as

$$\int_{-\infty}^{\infty} dx\; t(x)\,\text{step}(x - x_0) = \int_{x_0}^{\infty} dx\; t(x)\,. \tag{2.71}$$

It is easy to modify the limiting representation of step(x) into one for step$(x - x_0)$.

Derivative of a step function We can define a derivative of step(x), even though it is not differentiable in the conventional sense. From the discussion in Sec. 2.1.5, and especially (2.18), we have

$$\int_{-\infty}^{\infty} dx\; t(x)\frac{d}{dx}\,\text{step}(x) = -\int_{-\infty}^{\infty} dx\; t'(x)\,\text{step}(x) = -\int_0^{\infty} dx\; t'(x) = t(0)\,, \tag{2.72}$$

[4]The result in (2.70) is a bit weaker than the theorem of (2.67), which says that any generalized function can be expressed as the limit of a sequence of good functions. The integrals in (2.70) are not good functions since they do not approach zero as $x \to \infty$, but they are better than most fairly good functions since they do not grow at infinity either. If we wanted to write step(x) as the limit of a sequence of good functions, we could do so by multiplying (2.70) by a factor such as $\exp(-x^2/k^2)$. This factor becomes unity for all finite x in the limit as $k \to \infty$, but meets the requirements for a good function at all k.

since $t(\infty) = 0$ for any test function $t(x)$. Comparison of the first and last forms of (2.72) shows that

$$\frac{d}{dx}\,\text{step}(x) = \delta(x)\,, \tag{2.73}$$

since both sides of this equation yield the same result when multiplied by a test function and integrated.

We can get the same result from a limiting sequence. If step(x) is represented by (2.70), its derivative is

$$\frac{d}{dx}\,\text{step}(x) = \lim_{k\to\infty} \frac{d}{dx} \int_{-\infty}^{x} dx'\, \psi_k(x') = \lim_{k\to\infty} \psi_k(x) = \delta(x)\,. \tag{2.74}$$

Thus we see from this viewpoint also that the derivative of a step function is the delta function.

This is a good point at which to illustrate a potential pitfall when manipulating generalized functions. Since we have two different limiting representations for step(x), (2.11) and (2.70), it might be expected that the derivative of either one could be used to represent $\delta(x)$ in an integral of the form $\int \delta(x)\, f(x)\, dx$. That expectation is correct if $f(x)$ is continuous at $x = 0$, as it must be if it is a test function, but the two representations give different results if $f(x)$ has a discontinuity at $x = 0$. Use of the derivative of (2.70) to represent $\delta(x)$ yields $[f(0^+) + f(0^-)]/2$ as in (2.62), but use of the derivative of $h(kx)$ to represent $\delta(x)$, as in (2.11), can give only $f(0^+)$ since $h(kx)$ and all of its derivatives are zero for $x \le 0$.

The reason for the discrepancy between these two results is that *both* $f(x)$ and $g(x)$ are being represented as limits in $\int f(x)\, g(x)\, dx$, and it matters in which order the limits are taken. No problem would arise if either $f(x)$ or $g(x)$ were continuous. The limiting representations are useful, but two of them should not be used in the same integral.

Signum function Closely related to step(x) is the function sgn(x), pronounced signum of x. Treated as a conventional (though discontinuous) function, sgn(x) is defined as

$$\text{sgn}(x) \equiv \begin{cases} 1 & \text{if} \quad x > 0 \\ -1 & \text{if} \quad x < 0 \end{cases}\,. \tag{2.75}$$

Distributionally, it is defined by

$$\Phi_{\text{sgn}}\{t(x)\} = \int_{-\infty}^{\infty} t(x)\, \text{sgn}(x)\, dx \equiv \int_{0}^{\infty} t(x)\, dx - \int_{-\infty}^{0} t(x)\, dx\,. \tag{2.76}$$

Thus

$$\text{sgn}(x) = 2\,\text{step}(x) - 1\,, \tag{2.77}$$

which, when combined with the distributional definition of step(x) and the rules for manipulating distributions given in Sec. 2.1.5, will reproduce (2.76). Any valid limiting representation for the step function gives one for the signum function via (2.77). Similarly, we can show either distributionally or by a limiting representation that

$$\frac{d}{dx}\,\text{sgn}(x) = 2\,\delta(x)\,. \tag{2.78}$$

Products of delta functions Consider an integral of the form

$$I(x_0, x_1) = \int_{-\infty}^{\infty} dx\; t(x)\, \delta(x - x_0)\, \delta(x - x_1)\,, \tag{2.79}$$

where $t(x)$ is a test function. As noted above, we can represent $\delta(x - x_0)$ as the limit of a Dirac sequence of test functions, and the product of two test functions is a test function. Thus there is no problem in passing to the limit of the sequence and writing

$$I(x_0, x_1) = t(x_1)\, \delta(x_1 - x_0)\,, \tag{2.80}$$

which is identically zero if $x_0 \neq x_1$. If $x_0 = x_1$, $I(x_0, x_1)$ is infinite, so the square of a delta function is not defined directly, but $I(x_0, x_1)$ itself may be treated as a generalized function. It is easy to establish that it corresponds to the functional

$$\Phi_I\{s(x_0)\} = \int_{-\infty}^{\infty} dx_0\; I(x_0, x_1)\, s(x_0) = s(x_1)\, t(x_1)\,, \tag{2.81}$$

where $s(x)$ is a test function. In other words, the product of two delta functions, $\delta(x-x_0)\,\delta(x-x_1)$, behaves as a double generalized function; it has to be integrated twice before arriving at a simple number. In the example above, the two integrals were over x and x_0, but in fact any two of the three variables can be chosen. The following operational rule holds:

$$t(x)\, \delta(x-x_0)\, \delta(x-x_1) = t(x_0)\, \delta(x-x_0)\, \delta(x-x_1) = t(x_1)\, \delta(x-x_0)\, \delta(x-x_1)\,. \tag{2.82}$$

Although this rule does not work if $x_0 = x_1$, there is one way we can define something akin to the square of a delta function. The integral in (2.45) is a useful limiting form for the delta function that often arises in spectral analysis. If we denote this integral by $d_k(x)$, its square can be written as

$$[d_k(x)]^2 = \int_{-k/2}^{k/2} d\nu \int_{-k/2}^{k/2} d\nu'\; \exp[2\pi i(\nu - \nu')x]\,. \tag{2.83}$$

For k large, $[d_k(x)]^2$ is sharply peaked like a delta representation; its width varies as $1/k$, but its peak amplitude grows as k^2, so it does not integrate to unity as desired in a Dirac sequence. If integrated against a slowly varying function $f(x)$, a useful approximation might be

$$\int_{-\infty}^{\infty} dx\; [d_k(x)]^2\, f(x) \simeq k\, f(0)\,, \tag{2.84}$$

provided $f(x) \simeq f(0)$ for $|x| < 1/k$. Since $d_k(x) \to \delta(x)$ as $k \to \infty$, we can express this result loosely as (Bjorken and Drell, 1964, p. 101)

$$[\delta(x)]^2 \simeq k\, \delta(x)\,, \qquad k \to \infty\,. \tag{2.85}$$

A better way to express the same mathematics, however, is

$$\lim_{k\to\infty} \frac{1}{k}[d_k(x)]^2 = \lim_{k\to\infty} \frac{1}{k} \int_{-k/2}^{k/2} d\nu \int_{-k/2}^{k/2} d\nu'\; \exp[2\pi i(\nu - \nu')x] = \delta(x)\,. \tag{2.86}$$

This equation can also be written as

$$\lim_{k\to\infty} k\, \mathrm{sinc}^2(kx) = \delta(x)\,, \tag{2.87}$$

providing yet another useful limiting representation for the delta function.

2.3.3 Other point singularities

$1/x$ As an illustration of the concept of distributions, we introduced the generalized function $\mathcal{P}\{1/x\}$ in (2.4). The use of this function in complex integrals was also discussed in App. B. Henceforth, we drop the designation $\mathcal{P}$ and write $1/x$, with the understanding that the function is defined by a principal-value integral.

An alternative definition of $1/x$ (Lighthill, 1958) is that it is the odd generalized function $g(x)$ that satisfies $x\,g(x) = 1$. By this definition, $g(x)$ coincides with the ordinary function $1/x$ for $x \neq 0$, while the requirement that $g(x)$ be odd guarantees that the value at $x = 0$ is zero.

As with any other generalized function, $1/x$ can be represented as the limit of good or fairly good functions. One such representation can be derived by using (2.39) to represent a delta function and convolving that representation with $1/x$:

$$\frac{1}{x} = \lim_{\epsilon \to 0^+} \frac{\epsilon}{\pi} \int_{-\infty}^{\infty} dx' \left(\frac{1}{x - x'}\right)\left(\frac{1}{x'^2 + \epsilon^2}\right) = \lim_{\epsilon \to 0} \frac{x}{x^2 + \epsilon^2}, \tag{2.88}$$

where the integral has been performed by contour methods, taking advantage of the distributional definition of $1/x$, (2.4), for guidance in handling the pole on the contour. The representation of (2.88) is illustrated in Fig. 2.6a.

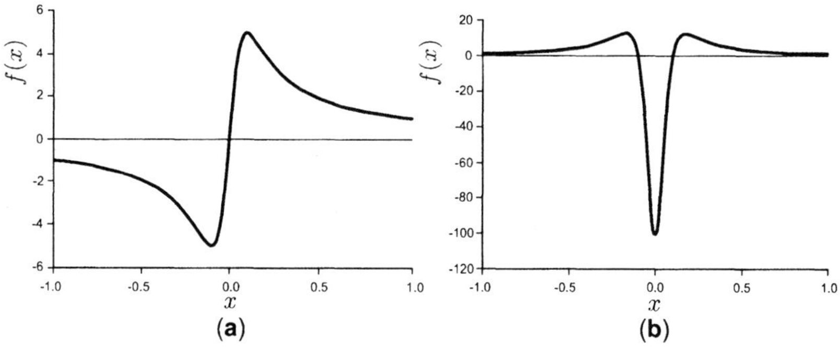

Fig. 2.6 (a) Plot of a representation of the generalized function $1/x$ from (2.88); (b) plot of a similar representation of the generalized function $1/x^2$ from (2.91).

The final form of (2.88) shows again that the generalized function $1/x$ agrees exactly with the ordinary function $1/x$ except at $x = 0$, where the ordinary function is not defined. The limit of the ordinary function $1/x$ as $x \to 0$ is $\pm\infty$, depending on the direction of approach. The generalized $1/x$, on the other hand, is zero at $x = 0$, corresponding to the deletion of the neighborhood of $x = 0$ from any integral in which it appears [*cf.* (2.3)].

$1/x^m$ From the generalized function $1/x$, we can use the rule for forming derivatives of generalized functions, (2.18), to define new generalized functions $1/x^m$ by (Lighthill, 1958)

$$\frac{1}{x^m} \equiv \frac{(-1)^{m-1}}{(m-1)!} \frac{d^{m-1}}{dx^{m-1}} \frac{1}{x}, \tag{2.89}$$

where m is an integer greater than one. Limiting representations can be obtained by differentiating (2.88) the appropriate number of times.

The generalized function $1/x^2$ will prove to be important in analysis of tomographic imaging systems. By (2.89) it should be interpreted as

$$\int_{-\infty}^{\infty} dx\, \frac{t(x)}{x^2} = -\mathcal{P}\int_{-\infty}^{\infty} dx\, \frac{t'(x)}{x}. \tag{2.90}$$

Again, the generalized function $1/x^2$ coincides with the ordinary function if $x \neq 0$, but, unlike $1/x$, it is not correct to say that the generalized $1/x^2$ is zero at $x = 0$. To see what does happen at the origin, we take the first derivative of (2.88), obtaining

$$\frac{1}{x^2} = \lim_{\epsilon\to 0}\left[\frac{2x^2}{(x^2+\epsilon^2)^2} - \frac{1}{x^2+\epsilon^2}\right], \tag{2.91}$$

which is depicted in Fig. 2.6b. For $x \neq 0$, the first term limits to $2/x^2$ and the second to $-1/x^2$, so the sum has the proper behavior. For $x \equiv 0$, however, the first term is zero and the second term limits to $-\infty$. Moreover, this singularity has an infinite integral over any vanishingly small region, *i.e.*,

$$\lim_{b\to 0}\int_{-b}^{b} \frac{dx}{x^2} = -\infty. \tag{2.92}$$

We can say, very loosely, that $1/x^2$ behaves like a delta function *of weight negative infinity* at the origin (Richards and Youn, 1990, pp. 76–77). One way to understand this wild singularity is to say that it arises from differentiating across the infinite discontinuity of $1/x$ at $x = 0$ [*cf.* (2.79)]. Nevertheless, when integrated from $-\infty$ to ∞, the infinities from the two terms in (2.91) cancel, giving

$$\int_{-\infty}^{\infty} \frac{dx}{x^2} = 0. \tag{2.93}$$

This result could be anticipated from (2.90) by letting $t(x) \to 1$ (or $t'(x) \to 0$), but it is definitely not what one would expect by taking the integrand at face value. A function denoted $1/x^2$ might be expected to be positive everywhere (which, in fact, is true except for the isolated point $x = 0$), so a zero integral is a surprise.

These subtleties illustrate some of the potential difficulties in interpreting generalized functions and the need for anchoring the theory in clear definitions. In the present example, the basic definition from which all interpretations must be derived is (2.90).

Noninteger powers Noninteger powers of x present some new subtleties. In what follows, we shall consider functions of the form $|x|^\alpha$ to avoid complications with noninteger powers of negative numbers. As the notation implies, $|x|^\alpha$ is an even generalized function. For $\alpha > 0$, $|x|^\alpha$ is a continuous function, bounded for all finite x, so there is no difficulty in interpreting integrals of the form $\int |x|^\alpha\, t(x)\, dx$, where $t(x)$ is a test function. For $-1 < \alpha < 0$, there is also no problem since $|x|^\alpha$ is singular but integrable at $x = 0$. To extend the definition of $|x|^\alpha$ to noninteger $\alpha < -1$, we make use of the derivatives of $|x|^\alpha$. In particular, if $x \neq 0$,

$$\frac{d}{dx}|x|^\alpha = \alpha|x|^{\alpha-1}\,\mathrm{sgn}(x), \tag{2.94}$$

suggesting that we define the generalized function $|x|^{\alpha-1}$ by

$$|x|^{\alpha-1} = \frac{1}{\alpha}\frac{d}{dx}\left\{|x|^{\alpha}\operatorname{sgn}(x)\right\}, \qquad -1 < \alpha < 0 . \tag{2.95}$$

The distribution corresponding to $|x|^{\alpha-1}$ is, by (2.18),

$$\int_{-\infty}^{\infty} dx\ t(x)\,|x|^{\alpha-1} = -\frac{1}{\alpha}\int_{-\infty}^{\infty} dx\ \frac{dt(x)}{dx}|x|^{\alpha}\operatorname{sgn}(x) \qquad \text{for } \alpha > -1 . \tag{2.96}$$

Since $t(x)$ need only be a good function, and not strictly a test function, for this integral to converge, (2.96) defines a tempered distribution.

This process can be extended to include all noninteger negative α by defining (Lighthill, 1958)

$$|x|^{\alpha} = \frac{1}{(\alpha+1)(\alpha+2)\cdots(\alpha+n)}\frac{d^n}{dx^n}\left\{|x|^{\alpha+n}\operatorname{sgn}(x)\right\}, \tag{2.97}$$

where n is an integer such that $\alpha + n > -1$.

To illustrate this result, consider the generalized function $|x|^{-3/2}$, defined according to (2.97) by

$$|x|^{-3/2} = -2\frac{d}{dx}\left\{\frac{\operatorname{sgn}(x)}{\sqrt{|x|}}\right\}. \tag{2.98}$$

Distributionally, this equation means

$$\int_{-\infty}^{\infty} dx\ t(x)\,|x|^{-3/2} = 2\int_{-\infty}^{\infty} dx\ \frac{t'(x)\operatorname{sgn}(x)}{\sqrt{|x|}} . \tag{2.99}$$

There is no problem with this definition at $x = 0$ since $t'(x)$ is necessarily finite and continuous there and $1/\sqrt{x}$ is an integrable singularity.

Though (2.98) and (2.99) are the formal definition of $|x|^{-3/2}$, further insight can be obtained by use of limiting representations. We choose

$$\frac{1}{\sqrt{|x|}} = \lim_{\epsilon\to 0}\frac{1}{(x^2+\epsilon^2)^{1/4}} \tag{2.100}$$

and

$$\operatorname{sgn}(x) = \lim_{\epsilon\to 0}\frac{x}{(x^2+\epsilon^2)^{1/2}}, \tag{2.101}$$

from which we find

$$\frac{1}{|x|^{3/2}} = -2\lim_{\epsilon\to 0}\frac{d}{dx}\frac{x}{(x^2+\epsilon^2)^{3/4}} = \lim_{\epsilon\to 0}\left\{\frac{3x^2}{(x^2+\epsilon^2)^{7/4}} - \frac{2}{(x^2+\epsilon^2)^{3/4}}\right\}. \tag{2.102}$$

As with the function $1/x^2$, this form shows a strong negative singularity at the origin [*cf.* (2.91)].

$x^{-m}\operatorname{sgn}(x)$ *and* $\ln|x|$ Though singular at the origin, $\ln|x|$ is an ordinary, integrable function, and functionals of the form $\int t(x)\ln|x|\,dx$ are easily evaluated. Away from the origin, the following derivative relations hold:

$$\frac{d}{dx}\ln|x| = \frac{1}{x}, \qquad (x \neq 0), \tag{2.103}$$

and

$$\frac{d}{dx}\{(\ln|x|)\,\mathrm{sgn}(x)\} = \frac{1}{|x|} = \frac{\mathrm{sgn}(x)}{x}\,, \qquad (x \neq 0)\,. \tag{2.104}$$

The first of these equations is valid without the exclusion of the origin if $1/x$ is understood as the generalized function (Richards and Youn, 1990, p. 72), but the second is more problematical. The generalized functions $x^{-1}\,\mathrm{sgn}(x)$ and $x^{-m}\,\mathrm{sgn}(x)$ turn out not to be uniquely definable.

We could simply define a generalized function $g(x)$ by $d\{(\ln|x|)\,\mathrm{sgn}(x)\}/dx$ and call it $1/|x|$, but it would obey some very strange manipulation rules (Lighthill, 1958, p. 38). For example, we would hope that $g(ax)$ would equal $g(x)/|a|$, but instead we find

$$g(ax) = \frac{d\{(\ln|x| + \ln|a|)\,\mathrm{sgn}(a)\,\mathrm{sgn}(x)\}}{a\,dx} = \frac{1}{|a|}\frac{d}{dx}\{(\ln|x|)\,\mathrm{sgn}(x) + (\ln|a|)\,\mathrm{sgn}(x)\}$$

$$= \frac{1}{|a|}\{g(x) + 2(\ln|a|)\,\delta(x)\}\,. \tag{2.105}$$

Since a is arbitrary, we must conclude that $1/|x|$ is not uniquely defined. The best we can say is that it is *any* generalized function that satisfies

$$\frac{1}{|x|} = \frac{d}{dx}\{(\ln|x|)\,\mathrm{sgn}(x)\} + 2C\,\delta(x) = \frac{d}{dx}\{(\ln|x| + C)\,\mathrm{sgn}(x)\}\,, \tag{2.106}$$

where C is an arbitrary constant (Lighthill, 1958; Champeney, 1987). Similarly,

$$x^{-m}\,\mathrm{sgn}(x) = \frac{(-1)^{m-1}}{(m-1)!}\frac{d^m}{dx^m}\{(\ln|x| + C)\,\mathrm{sgn}(x)\}$$

$$= \frac{(-1)^{m-1}}{(m-1)!}\frac{d^m}{dx^m}\{(\ln|x|)\,\mathrm{sgn}(x)\} + C'\,\delta^{m-1}(x)\,, \tag{2.107}$$

where m is a positive integer and C and C' are arbitrary.

2.4 MULTIDIMENSIONAL DELTA FUNCTIONS

2.4.1 Multidimensional distributions

The theory of distributions is easily extended to functions of two or more variables. We shall refer to a scalar-valued function of n scalar variables as an n-dimensional (or nD) function for short.

A function $t(x, y)$ is a 2D test function if all partial derivatives exist and it has compact support in the x-y plane. For example, a suitable support region would be the square of side $2L$ centered on the origin, in which case $t(x, y) = 0$ unless $L < x < L$ and $-L < y < L$. More generally, we denote the support region by $\mathbf{S}$ and say that $t(x, y) = 0$ unless (x, y) lies in $\mathbf{S}$. The test functions themselves lie in a subset of $\mathbb{L}_2(\mathbb{R}^2)$ denoted $\mathbb{T}(\mathbf{S})$ for test functions of support $\mathbf{S}$.

Good functions and fairly good functions in 2D are defined by analogy with (2.7) and (2.8), respectively. A good function in 2D is one for which all partial derivatives exist and

$$\lim_{x\to\pm\infty}\{|x|^N t(x, y)\} = 0 \qquad \text{for all } N \text{ and all } y;$$

$$\lim_{y\to\pm\infty}\{|y|^N t(x,y)\}=0 \qquad \text{for all } N \text{ and all } x\,. \tag{2.108}$$

A fairly good function in 2D is one for which all partial derivatives exist and

$$\lim_{x\to\pm\infty}\left\{|x|^{-N} t(x,y)\right\}=0 \qquad \text{for some } N \text{ and all } y;$$

$$\lim_{y\to\pm\infty}\left\{|y|^{-N} t(x,y)\right\}=0 \qquad \text{for some } N \text{ and all } x\,. \tag{2.109}$$

Similar definitions apply in three or more dimensions.

A 2D distribution is a linear, continuous functional that maps a test function $t(x, y)$ to a real or complex number. A 2D tempered distribution is defined the same way except that $t(x, y)$ need only be a good function. In both cases, the functional can be written in terms of a generalized function $g(x, y)$ as

$$\Phi_g\{t(x,y)\}=\int_{-\infty}^{\infty}dx\int_{-\infty}^{\infty}dy\; t(x,y)\,g(x,y)\,. \tag{2.110}$$

Again, the extension to higher dimensions is obvious.

Vector notation In 2D problems we can define a vector $\mathbf{r}$ with Cartesian components (x, y), so that (2.110) can be written more compactly as

$$\Phi_g\{t(\mathbf{r})\}=\int_{\infty} d^2r\; t(\mathbf{r})\,g(\mathbf{r})\,, \tag{2.111}$$

where $d^2r = dx\,dy$ and the subscript ∞ on the integral sign denotes an integral over the infinite x-y plane.

To extend this notation to n-dimensional (nD) functionals, all we have to do is to replace d^2r by d^nr in (2.111). For example, in 3D, $\mathbf{r}$ is a vector with Cartesian components (x, y, z), $d^3r = dx\,dy\,dz$ and the integral runs over the infinite 3D volume. No notational distinction will be made between 2D and 3D vectors unless both appear in the same problem; $\mathbf{r}$ will denote the general position vector in any number of dimensions, and the dimensionality will usually be clear by context (*e.g.*, from d^nr).

We also adopt the convention that r, without the boldface type, is the magnitude of the vector $\mathbf{r}$ in any number of dimensions. Thus, $r = \sqrt{x^2+y^2}$ in 2D or $\sqrt{x^2+y^2+z^2}$ in 3D.

2.4.2 Multidimensional delta functions

The 2D delta function $\delta(\mathbf{r})$ is the generalized function associated with the distribution Φ_δ, defined by

$$\Phi_\delta\{t(\mathbf{r})\}=\int_{\infty} d^2r\; t(\mathbf{r})\,\delta(\mathbf{r})\equiv t(\mathbf{0})\,, \tag{2.112}$$

where $t(\mathbf{0})$ is $t(\mathbf{r})$ evaluated at $\mathbf{r} = \mathbf{0}$, (*i.e.*, $x = 0$, $y = 0$). It follows from this definition that

$$\delta(\mathbf{r})=\delta(x)\,\delta(y)\,, \qquad (\mathbf{r} \text{ a 2D vector })\,, \tag{2.113}$$

since

$$\int_{-\infty}^{\infty}dx\int_{-\infty}^{\infty}dy\; t(x,y)\,\delta(x)\,\delta(y)=\int_{-\infty}^{\infty}dy\; t(0,y)\,\delta(y)=t(0,0)\,. \tag{2.114}$$

In performing these integrals, we have used the fact that $t(x, y)$ for fixed y is a test function of x, and vice versa.

The extension to nD is straightforward, and the following properties of nD delta functions are easily established:

$$\delta(\mathbf{r} - \mathbf{r}_0) = 0 \qquad \text{if } \mathbf{r} \neq \mathbf{r}_0\,; \tag{2.115}$$

$$\int_\infty d^n r\ \delta(\mathbf{r} - \mathbf{r}_0) = 1\,; \tag{2.116}$$

$$\delta(\mathbf{r}) = \delta(-\mathbf{r})\,; \tag{2.117}$$

$$\int_\infty d^n r\ t(\mathbf{r})\,\delta(\mathbf{r} - \mathbf{r}_0) = t(\mathbf{r}_0)\,; \tag{2.118}$$

$$\int_S d^n r\ t(\mathbf{r})\,\delta(\mathbf{r} - \mathbf{r}_0) = t(\mathbf{r}_0) \qquad \text{if } \mathbf{r}_0 \text{ is in region } \mathbf{S}\,; \tag{2.119}$$

$$t(\mathbf{r})\,\delta(\mathbf{r} - \mathbf{r}_0) = t(\mathbf{r}_0)\,\delta(\mathbf{r} - \mathbf{r}_0)\,; \tag{2.120}$$

$$\delta(a\mathbf{r} - \mathbf{r}_0) = |a|^{-n}\,\delta(\mathbf{r} - \mathbf{r}_0/a)\,. \tag{2.121}$$

Multidimensional limiting representations Any of the Dirac sequences introduced in Sec. 2.2.2 can be used to construct a limiting representation for a multidimensional delta function in Cartesian coordinates. In 2D, we have

$$\delta(\mathbf{r}) = \delta(x)\,\delta(y) = \lim_{k\to\infty} \psi_k(x)\,\psi_k(y)\,. \tag{2.122}$$

For example, the 2D generalization of (2.40) is

$$\begin{aligned} \delta(\mathbf{r}) &= \lim_{k\to\infty} k^2 \exp\left[-\pi k^2(x^2+y^2)\right] = \lim_{\epsilon\to 0} \frac{1}{\epsilon^2} \exp\left[-\pi(x^2+y^2)/\epsilon^2\right] \\ &= \lim_{\epsilon\to 0} \frac{1}{\epsilon^2} \exp\left[-\frac{\pi r^2}{\epsilon^2}\right], \end{aligned} \tag{2.123}$$

where $\epsilon = 1/k$. The last form follows since $x^2 + y^2 = r^2$ in 2D.

The extension of (2.122) to nD is

$$\delta(\mathbf{r}) = \lim_{k\to\infty} \prod_{j=1}^{n} \psi_k(x_j)\,, \tag{2.124}$$

where x_j, $j = 1, ..., n$, are the Cartesian components of $\mathbf{r}$. In particular, (2.123) generalizes to

$$\delta(\mathbf{r}) = \lim_{\epsilon\to 0^+} \frac{1}{\epsilon^n} \exp\left[-\frac{\pi r^2}{\epsilon^2}\right]. \tag{2.125}$$

2.4.3 Delta functions in polar coordinates

The reader is cautioned that the product form of (2.124) is valid in Cartesian coordinates only. In 2D polar coordinates (r, θ), the following is true:

$$\delta(\mathbf{r}) = \frac{\delta(r)}{\pi r}, \tag{2.126}$$

where $r = |\mathbf{r}| = \sqrt{x^2 + y^2}$. As with all equations involving delta functions, (2.126) must hold if both sides are multiplied by a test function and integrated. We convert the test function $t(x, y)$ to polar coordinates by defining $t_p(r, \theta) = t(x, y)$ and using the usual transformation formulas from (x, y) to (r, θ). We have two options for computing the integral in question. One is to write

$$\int_\infty d^2r\ t(\mathbf{r})\,\delta(\mathbf{r}) = \int_0^{2\pi} d\theta \int_0^\infty r\,dr\ t_p(r,\theta)\frac{\delta(r)}{\pi r} = \frac{1}{\pi}\int_0^{2\pi} d\theta \int_0^\infty dr\ t_p(r,\theta)\,\delta(r)\,. \tag{2.127}$$

At this point we can treat r as a dummy 1D variable of integration and use (2.62) to obtain

$$\int_0^\infty dr\ t_p(r,\theta)\,\delta(r) = \int_{-\infty}^\infty dr\ t_p(r,\theta)\,\mathrm{step}(r)\,\delta(r) = \tfrac{1}{2}t_p(0,\theta)\,. \tag{2.128}$$

Since $t_p(r, \theta)$ is a test function, it must be independent of θ at $r = 0$, so we have

$$\int_\infty d^2r\ t(\mathbf{r})\,\delta(\mathbf{r}) = \frac{t_p(0,0)}{2\pi}\int_0^{2\pi} d\theta = t(\mathbf{0})\,, \tag{2.129}$$

as required.

The other option for performing this integral without appealing to (2.62) is to allow r to assume negative values, which we do by redefining r as $|\mathbf{r}|$ if $\mathbf{r}$ is in quadrants 1 or 2 and $-|\mathbf{r}|$ if it is in quadrants 3 or 4. Then $d^2r = |r|dr\,d\theta$, and the infinite plane is traversed by letting θ run from 0 to π while r runs from $-\infty$ to ∞. This yields

$$\int_\infty d^2r\ t(\mathbf{r})\,\delta(\mathbf{r}) = \int_0^\pi d\theta \int_{-\infty}^\infty |r|dr\ t_p(r,\theta)\,\frac{\delta(r)}{\pi}|r| = \frac{t_p(0,0)}{\pi}\int_0^\pi d\theta = t(\mathbf{0})\,. \tag{2.130}$$

Both of these approaches establish the validity of (2.126) in 2D. The 3D counterpart is

$$\delta(\mathbf{r}) = \frac{\delta(r)}{2\pi r^2}, \tag{2.131}$$

where now $r = |\mathbf{r}| = \sqrt{x^2 + y^2 + z^2}$.

Dimensions and dimensions Equations (2.126) and (2.131) afford an opportunity to comment on some semantic and notational points. In the former equation, $\mathbf{r}$ is a 2D vector while r, its magnitude, is a scalar. We refer to $\delta(\mathbf{r})$ as a 2D delta function, meaning that it can be used to perform a 2D integral. The right-hand side of the equation involves $\delta(r)$, a delta function with a scalar argument, which we refer to as a 1D delta function since it can be used to perform only a single integral, namely the one over the scalar r (or $|\mathbf{r}|$). That disparity in dimensions caused no problem in the calculation above since the integral over θ was trivial once the one over r was performed.

We adopt the terminology that an nD delta function is one that can be used to perform n integrals. Some authors use a notation such as $\delta_n(\mathbf{r})$ to state explicitly the dimensionality of the delta function, but we shall not do so. Instead, we use the convention that the dimensionality of the delta function is the same as the dimensionality of the vector (or scalar) in its argument. By this rule, $\delta(\mathbf{r})$ is an nD delta function if $\mathbf{r}$ is an nD vector, but $\delta(r)$ is always a 1D delta function.

Having said this, we now introduce a completely different usage of the word dimension. In physical problems, *dimensional analysis* is a very useful tool for ensuring that equations are consistent and independent of the particular system of units chosen. One way to perform a dimensional analysis is to note that all physical quantities can be expressed in terms of mass (M), length (L), time (T) and charge (Q). Different *units* can be used, but the *dimensions* of a physical quantity are always the same. Speed, for example, can be measured in cm/sec or furlongs/fortnight, but it always has dimensions of L/T. An equation like $v = dx/dt$ (where v = speed, x = position and t = time) is dimensionally consistent since dx has dimensions of length and dt dimensions of time, so speed has dimensions of L/T as required. We shall often use square brackets to denote dimension. Here $[dx] = L$, $[dt] = T$ and $[v] = L/T$. Note that the fact that dx and dt are infinitesimal is of no import for purposes of dimensional analysis. Neither is the distinction between a vector and a scalar; speed and velocity have the same dimensions.

These same concepts can be applied to equations involving delta functions. The equation $\int dx\, t(x)\,\delta(x) = t(0)$ must be dimensionally consistent, in the sense of dimensional analysis. To be concrete, assume that x is a spatial position and hence has dimensions of length. The test function $t(x)$ can represent any physical quantity, so we denote its (unknown) dimensions as $[t(x)]$. The dimensions of the right-hand side of the equation, namely, $[t(x)]$, must match those of the left-hand side. On the left, however, dimensional analysis yields $[dx][\delta(x)][t(x)]$, so consistency demands that

$$[\delta(x)] = \frac{1}{[dx]} = \frac{1}{[x]}\,, \tag{2.132}$$

or $1/L$ if x is a length. In (confusing) words, the one-dimensional delta function has dimensions equal to the reciprocal of the dimensions of its argument.

If delta functions have dimensions, their limiting representation must as well. To apply dimensional analysis to (2.40), for example, we must first determine the dimensions of ϵ. Recall that an exponential can be expressed as a power series, and one cannot add quantities with different dimensions. Thus each term in the power series must be dimensionless, requiring that ϵ have the same dimensions as r, say L. The exponential function itself, being a sum of powers of the exponent, is also dimensionless, and the delta function represented by (2.40) has dimensions of $1/[\epsilon]$ or $1/L$ as required.

Similarly, an nD delta function has dimensions of

$$[\delta(\mathbf{r})] = \frac{1}{[d^n r]} = \frac{1}{[\mathbf{r}]^n}\,. \tag{2.133}$$

Thus, if $\mathbf{r}$ is a 3D position vector, then $\delta(\mathbf{r})$ must be assigned the dimensions of $1/L^3$ for purposes of dimensional analysis. The 1D delta function $\delta(r)$, on the other hand, has dimensions of just $1/L$. Equations (2.126) and (2.131) are dimensionally consistent since the denominators serve to balance the dimensions of the two sides. (Note that an angle, being the ratio of an arc length to a radius, is dimensionless.)

To add further to the confusion, we remind the reader that yet another usage of dimension was introduced in the last chapter. The test functions $t(\mathbf{r})$, which map from $\mathbb{R}^n$ to $\mathbb{T}(\mathbf{S})$, are themselves vectors in a Hilbert space of infinite dimensionality. Here are all three usages in a single sentence: An n-dimensional delta function, which has dimensions $1/L^n$, is a basis vector for an infinite-dimensional space. Caveat lector.

2.4.4 Line masses and plane masses

The 1D delta function $\delta(r)$, where r is the magnitude of an nD vector, is defined for all points in an nD space; it just happens that it is nonzero only at a single point in that space, namely the origin. One can devise other 1D delta functions that are also defined at all points in an nD space but are nonzero over lines or planes. Consider, for example, a delta function of the form $\delta(\mathbf{r} \cdot \hat{\mathbf{n}})$ where $\mathbf{r}$ is a 2D vector, $\hat{\mathbf{n}}$ is a 2D unit vector and the dot indicates the usual 2D scalar product. This delta function is nonzero everywhere that its argument vanishes, or for the set of points $\mathbf{r}$ such that $\mathbf{r} \cdot \hat{\mathbf{n}} = 0$. Since this equation requires that $\mathbf{r}$ be perpendicular to $\hat{\mathbf{n}}$, the delta function is nonzero along a line through the origin and normal to $\hat{\mathbf{n}}$. Similarly, the delta function $\delta(\mathbf{r} \cdot \hat{\mathbf{n}} - p)$ is nonzero along a line normal to $\hat{\mathbf{n}}$ and a distance p from the origin as shown in Fig. 2.7. To use either of these delta functions in an integral, it is convenient to use a rotated coordinate system where $\mathbf{r}$ has Cartesian coordinates (x', y'), with the x'-axis parallel to $\hat{\mathbf{n}}$. In this system, a test function $t(\mathbf{r})$ is given by $t_{rot}(x', y')$ and $\mathbf{r} \cdot \hat{\mathbf{n}} = x'$. We therefore have

$$\int_{\infty} d^2r \; t(\mathbf{r})\, \delta(\mathbf{r} \cdot \hat{\mathbf{n}} - p) = \int_{-\infty}^{\infty} dx' \int_{-\infty}^{\infty} dy' \; t_{rot}(x'y')\, \delta(x' - p) = \int_{-\infty}^{\infty} dy' \; t_{rot}(p, y') \,. \tag{2.134}$$

Since $\delta(\mathbf{r} \cdot \hat{\mathbf{n}} - p)$ is 1D, we can go no further. The result of using this 1D delta function in a 2D integral is a specific number, namely the line integral of the test function along the line $\mathbf{r} \cdot \hat{\mathbf{n}} = p$. Thus $\delta(\mathbf{r} \cdot \hat{\mathbf{n}} - p)$ qualifies as a generalized function

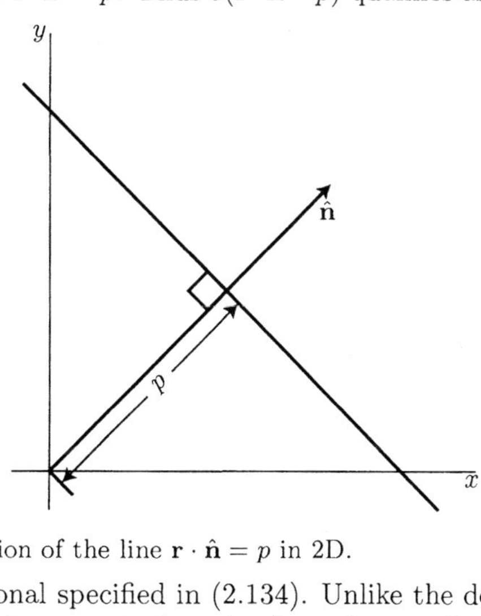

Fig. 2.7 Illustration of the line $\mathbf{r} \cdot \hat{\mathbf{n}} = p$ in 2D.

and defines the functional specified in (2.134). Unlike the delta function $\delta(\mathbf{r} - \mathbf{r}_0)$, however, it does not sift out a single point from the test function.

In 3D, the delta function $\delta(\mathbf{r}\cdot\hat{\mathbf{n}}-p)$ has a different interpretation. Again it is nonzero for those points $\mathbf{r}$ for which the argument vanishes, or $\mathbf{r}\cdot\hat{\mathbf{n}}=p$. In 3D, however, this equation specifies a plane, so the result of using the generalized function $\delta(\mathbf{r}\cdot\hat{\mathbf{n}}-p)$ in a 3D integral is the integral of the test function over the plane normal to $\hat{\mathbf{n}}$ and a distance p from the origin. We refer to $\delta(\mathbf{r}\cdot\hat{\mathbf{n}}-p)$ as a line mass or line delta function in 2D and as a plane mass or plane delta function in 3D. Both of these constructs will prove to be very important when we discuss tomography in Chap. 16.

If we wish to construct a line mass in 3D, one way to do so is to define a generalized function $g_{\hat{\mathbf{n}}}(\mathbf{r})$ by

$$g_{\hat{\mathbf{n}}}(\mathbf{r})=\int_{-\infty}^{\infty}d\ell\;\delta(\mathbf{r}-\hat{\mathbf{n}}\ell)\,. \tag{2.135}$$

Since the integrand is a 3D delta function and a single integral is performed, $g_{\hat{\mathbf{n}}}(\mathbf{r})$ is a 2D delta function or, equivalently, a line delta function in 3D. It is nonzero for all points along a line through the origin and parallel to $\hat{\mathbf{n}}$.

The line delta functions discussed above are all nonzero along *straight* lines. An example of a line delta function in 2D where the line is not straight is the ring delta function,

$$\delta(r-R)=\delta(|\mathbf{r}|-R)\,,\qquad \mathbf{r}\text{ a 2D vector}\,. \tag{2.136}$$

Not to be confused with $\delta(\mathbf{r}-\mathbf{R})$, the ring delta function is nonzero along a circle of radius R centered on the origin in the 2D plane. The same expression $\delta(r-R)$ but with $\mathbf{r}$ a 3D vector would be nonzero along a shell of radius R and infinitesimal thickness in the 3D space.

2.4.5 Multidimensional derivatives of delta functions

Various partial derivatives of multidimensional delta functions can be defined by analogy to the 1D derivatives discussed in Sec. 2.2.4. If $\mathbf{r}$ is an nD vector with Cartesian coordinates $\{x_j : j=1,...,n\}$, then the first partial derivatives of $\delta(\mathbf{r}-\mathbf{r}_0)$ are defined by

$$\int_\infty d^n r\; t(\mathbf{r})\frac{\partial}{\partial x_j}\,\delta(\mathbf{r}-\mathbf{r}_0)=-\left.\frac{\partial t(\mathbf{r})}{\partial x_j}\right|_{\mathbf{r}=\mathbf{r}_0}. \tag{2.137}$$

Higher partial derivatives are defined similarly. For example, gradients and Laplacians of a delta function are defined by

$$\int_\infty d^n r\; t(\mathbf{r})\,\nabla\,\delta(\mathbf{r}-\mathbf{r}_0)=-\nabla t(\mathbf{r}_0)\,; \tag{2.138a}$$

$$\int_\infty d^n r\; t(\mathbf{r})\,\nabla^2\,\delta(\mathbf{r}-\mathbf{r}_0)=\nabla^2 t(\mathbf{r}_0)\,. \tag{2.138b}$$

A general differential operator acting on a delta function is defined in terms of its adjoint acting on the test function (see Sec. 1.3.5). The adjoint of ∇ is $-\nabla$ (which is a statement of integration by parts), while ∇^2 is self-adjoint or Hermitian.

2.4.6 Other point singularities

In discussing various imaging applications in later chapters, we shall encounter some multidimensional point singularities similar to the 1D ones defined in Sec. 2.3.3. A very common one will be the function $1/|\mathbf{r} - \mathbf{r}_0|$, where $\mathbf{r}$ and $\mathbf{r}_0$ are vectors in 2D or 3D. Unlike the 1D function $1/|x - x_0|$, $1/|\mathbf{r} - \mathbf{r}_0|$ is an integrable singularity, and no special definition of a generalized function is needed. To prove this contention in 2D, we make the change of variables, $\mathbf{r}' = \mathbf{r} - \mathbf{r}_0$, and write

$$\int_\infty d^2r \, \frac{t(\mathbf{r})}{|\mathbf{r} - \mathbf{r}_0|} = \int_\infty d^2r' \, \frac{t(\mathbf{r}' + \mathbf{r}_0)}{|\mathbf{r}'|} = \int_0^{2\pi} d\theta' \int_0^\infty r' \, dr' \, \frac{t(\mathbf{r}' + \mathbf{r}_0)}{r'}$$

$$= \int_0^{2\pi} d\theta' \int_0^\infty dr' \, t(\mathbf{r}' + \mathbf{r}_0) \,, \tag{2.139}$$

which must converge if $t(\mathbf{r})$ is a test function. Similarly, $1/|\mathbf{r} - \mathbf{r}_0|^2$ is an integrable singularity in 3D.

The Laplacian of $1/|\mathbf{r} - \mathbf{r}_0|$ in 3D is an important generalized function with applications in coherent imaging and tomography. A change of variables as in (2.139) allows us to set $\mathbf{r}_0$ to zero without loss of generality, and the Laplacian can be calculated in spherical coordinates as follows:

$$\nabla^2 \frac{1}{r} = \frac{1}{r^2} \frac{\partial}{\partial r} r^2 \frac{\partial}{\partial r} \frac{1}{r} = 0 \qquad \text{if } r \neq 0 \,. \tag{2.140}$$

To see the behavior at $r = 0$, we make use of the divergence theorem (Gauss's theorem), which states that

$$\int_S \mathbf{D} \cdot \hat{\mathbf{n}} \, da = \int_V \nabla \cdot \mathbf{D} \, d^3r \,, \tag{2.141}$$

where S is a closed surface surrounding volume V, da is the area element on S, $\hat{\mathbf{n}}$ is an outwardly directed unit normal to S, and $\mathbf{D}$ is an arbitrary vector field. In the present problem, $\mathbf{D}$ can be taken as the gradient of $1/r$, which is a vector directed radially away from the origin, and S can be taken as a sphere of radius R centered on the origin so that $\hat{\mathbf{n}}$ is in the same direction as $\nabla(1/r)$. Since $\nabla^2 f(\mathbf{r}) = \nabla \cdot \nabla f(\mathbf{r})$, we have

$$\int_V \nabla \cdot \nabla \frac{1}{r} \, d^3r = \int_S \hat{\mathbf{n}} \cdot \nabla \frac{1}{r} \, da = \int_S \left[\frac{\partial}{\partial r} \frac{1}{r} \right]_{r=R} da = -\frac{1}{R^2} \int_S da = -\frac{1}{R^2} 4\pi R^2 = -4\pi \,, \tag{2.142}$$

independent of R. Comparison of the first and last forms establishes that

$$\nabla^2 \frac{1}{r} = -4\pi \, \delta(\mathbf{r}) \,, \qquad \mathbf{r} \text{ a 3D vector} \,, \qquad r = |\mathbf{r}| \,, \tag{2.143}$$

or, more generally,

$$\nabla^2 \frac{1}{|\mathbf{r} - \mathbf{r}_0|} = -4\pi \, \delta(\mathbf{r} - \mathbf{r}_0) \,. \tag{2.144}$$

The same result can also be obtained more rigorously either by using a limiting representation of $1/r$ or by using distribution theory.

An analogous result in 2D is

$$\nabla^2 \ln |\mathbf{r} - \mathbf{r}_0| = 2\pi \, \delta(\mathbf{r} - \mathbf{r}_0) \,, \qquad \mathbf{r} \text{ a 2D vector} \,. \tag{2.145}$$

Both (2.144) and (2.145) have important applications in the theory of wave propagation and diffraction, and they show up in surprising ways in tomographic imaging as well.

2.4.7 Angular delta functions

So far in this chapter, the variable in the argument of a delta function has been a spatial variable in some number of dimensions, but it is also frequently useful to consider angular variables.

In 2D problems in polar coordinates, there is no difficulty in defining an angular delta function. We can simply define $\delta(\theta-\theta_0)$ by

$$\int_0^{2\pi} d\theta\; \delta(\theta-\theta_0)t(\theta) = t(\theta_0)\,, \tag{2.156}$$

where $t(\theta)$ is a test function.

In 3D spherical coordinates, we need two angles; the usual choices are the colatitude θ, measured from the z axis, and the longitude or azimuth ϕ, defined as the rotation about the z axis as measured from the x axis. A unit vector $\hat{\mathbf{n}}$ in the direction determined by these angles is given in Cartesian coordinates by

$$\hat{\mathbf{n}} = (\sin\theta\cos\phi,\ \sin\theta\sin\phi,\ \cos\theta)\,. \tag{2.157}$$

With these variables, we can define

$$\int_{4\pi} d\Omega_n\; \delta(\hat{\mathbf{n}}-\hat{\mathbf{n}}_0)\,t(\hat{\mathbf{n}}) = t(\hat{\mathbf{n}}_0)\,, \tag{2.158}$$

where $d\Omega_n = \sin\theta\, d\theta\, d\phi$ is the element of solid angle associated with $\hat{\mathbf{n}}$, and $t(\hat{\mathbf{n}})$ is a test function in the angular variables.

3

Fourier Analysis

Jean Baptiste Joseph Fourier was born in a working-class French family in 1768. Orphaned at age eight, he entered the military school at Auxerre, where his aptitude for mathematics was evident. He was denied a commission in the artillery because of his low birth and decided instead to become a monk (Whitrow, 1965). He entered the Benedictine novitiate, but his plans were interrupted in 1789; the French Revolution changed both Fourier's life and the future of mathematics. Fourier was an early supporter of the revolution but an outspoken opponent of the Terror. He caught the eye of Napoleon and accompanied him to Egypt, where he became governor of lower Egypt. On his return to France in 1801 he became prefect of Isère, and during that time he conducted the investigations into heat flow that were to lead to the methodology that we now know as Fourier analysis.

Lord Kelvin called Fourier's Théorie Analytique de la Chaleur "a great mathematical poem," and Kelvin and P. G. Tait called Fourier's theorem "one of the most beautiful results of modern analysis ... an indispensable instrument in the treatment of nearly every recondite question in modern physics" (quoted in Bell, 1937).

Today, Fourier analysis is the cornerstone of many branches of science, engineering and mathematics. The modern field of image science can be traced to the turn-of-the century investigations of Ernst Abbé on the Fourier-transforming properties of lenses. Later work by P. M. Duffieux, Otto Schade, H. H. Hopkins, Edward O'Neill and others laid a firm foundation for the concepts of modulation transfer function and optical transfer function, which amount to analyzing an imaging system in terms of its Fourier decomposition. The full power of this approach to image science was evident by 1961 when Emmett Leith and Juris Upatnieks used Fourier methods to analyze holographic imaging systems.

The term Fourier analysis actually covers a diverse range of mathematical methods. We need to distinguish *Fourier series*, *Fourier transforms* and *discrete Fourier transforms*. The Fourier series is an expansion of a continuous function into a weighted sum of sines and cosines. We can view the Fourier-series expansion as a continuous-to-discrete mapping; a function of a continuous variable is transformed into a discrete set of numbers. The Fourier transform, on the other hand, maps one function to another, so it is a continuous-to-continuous mapping or integral transform. The discrete Fourier transform (DFT), an often-used approximation to the Fourier integral, maps a discrete vector to another discrete vector, so it is a discrete-to-discrete mapping or matrix operator. Finally, the discrete-space Fourier transform, the spatial analog of the discrete-time Fourier transform often used in digital signal processing, maps an infinite sequence of data samples to a function, so it is a DC mapping. All four of these transforms are linear, so we can build on what we learned in Chap. 1 about linear operators.

Since Fourier analysis is all about sines and cosines, we begin in Sec. 3.1 with a short overview of some mathematical properties of these familiar trigonometric functions. Then in Sec. 3.2 we introduce the Fourier series and discuss its properties in some depth. In Secs. 3.3 and 3.4, the Fourier transform in one or several dimensions is developed as a generalization of the Fourier series, though it is also possible to go in the other direction and regard the Fourier series as a special case of the Fourier transform.

In Sec. 3.5 we discuss sampling, which can be viewed as a transformation from a function to a discrete set of numbers. Conditions under which this transformation is invertible and some specific inversion formulas are derived. In Sec. 3.6 we apply sampling theory and derive the discrete Fourier transform, which can be viewed either as an interesting matrix transformation in its own right or as an approximation to the continuous Fourier transform.

The available books on Fourier analysis tend to cluster into two classes. Engineering-oriented texts such as Gaskill (1978) and Bracewell (1965) are excellent compendia of practical properties of the transforms but spend little time discussing the underlying mathematical issues such as convergence and the applicability of the transforms to strange functions. At the opposite extreme, mathematical texts such as Champeney (1987) and Körner (1988) discuss these points at length but provide little guidance to the practitioner. To make matters worse, the mathematical literature itself is divided into two camps, depending on whether or not generalized functions are admitted.

In this chapter we attempt to take a middle ground. It is our hope that the chapter will indeed serve as a useful reference on practical properties of the transforms, but we also discuss conditions that must be imposed on the functions for the various results to hold in specified senses. Most importantly, we attempt to integrate generalized functions into the overall framework and to relate the transform theory to the Hilbert-space viewpoint that pervades this book.

As with the previous chapters, readers can skip over some of the more mathematical sections without loss of continuity. For example, any section with the word 'convergence' in its title can be skipped on a first reading. A few key results from these sections are used later, but they are called to the attention of the reader when needed.

3.1 SINES, COSINES AND COMPLEX EXPONENTIALS

3.1.1 Orthogonality on a finite interval

The basic building blocks of Fourier analysis are trigonometric functions of the form $\cos(2\pi nx/L)$ or $\sin(2\pi nx/L)$, where n is an integer and x is a real variable. The period of each of these functions is L/n, so the interval $-\frac{1}{2}L \le x \le \frac{1}{2}L$ contains exactly n cycles of both functions. The cosine is an even function of x on this interval and the sine is an odd function.

The reciprocal of the period, n/L, is the frequency associated with the periodic function $\cos(2\pi nx/L)$ or $\sin(2\pi nx/L)$. If x is a spatial variable, as it usually is in imaging applications, we shall refer to n/L as a spatial frequency and denote it ξ_n. Thus the functions $\{\cos(2\pi nx/L)\}$ and $\{\sin(2\pi nx/L)\}$ are sets of periodic functions with different spatial frequencies ξ_n indexed by the discrete parameter n. The numerical value of ξ_n is the number of cycles of $\cos(2\pi nx/L)$ or $\sin(2\pi nx/L)$ per unit length. For purposes of dimensional analysis, spatial frequency has dimensions $(\text{length})^{-1}$.

The following integrals involving these functions are well known from elementary calculus:

$$\frac{2}{L}\int_{-\frac{1}{2}L}^{\frac{1}{2}L} dx \cos\left(\frac{2\pi nx}{L}\right)\cos\left(\frac{2\pi mx}{L}\right) = \delta_{mn}\,, \qquad m \neq 0\,, \qquad n \neq 0\,, \tag{3.1}$$

$$\frac{2}{L}\int_{-\frac{1}{2}L}^{\frac{1}{2}L} dx \sin\left(\frac{2\pi nx}{L}\right)\sin\left(\frac{2\pi mx}{L}\right) = \delta_{mn}\,, \qquad m \neq 0\,, \qquad n \neq 0\,, \tag{3.2}$$

$$\int_{-\frac{1}{2}L}^{\frac{1}{2}L} dx \cos\left(\frac{2\pi nx}{L}\right)\sin\left(\frac{2\pi mx}{L}\right) = 0\,, \tag{3.3}$$

where m and n are integers and δ_{mn} is the Kronecker delta symbol, with value 1 if $n = m$ and 0 if $n \neq m$. The factor of 2 is required in (3.1) and (3.2) since, for $n = m \neq 0$, the integrals are L times the spatial averages of $\sin^2$ or $\cos^2$ over an integer number of periods. The average of $\sin^2$ or $\cos^2$ must be $\frac{1}{2}$ since the corresponding average of $\sin^2 + \cos^2$ is 1 and the average of $\sin^2$ must equal the average of $\cos^2$ over an integer number of periods.

The case $n = m = 0$ must be treated separately. The left-hand side of (3.1) is 2 in this case since both cosines are one, while the left-hand side of (3.2) is 0 since the sines are zero.

These sine and cosine functions are square-integrable over the interval $[-\frac{1}{2}L, \frac{1}{2}L]$, so they are vectors in the Hilbert space $\mathbb{L}_2(-\frac{1}{2}L, \frac{1}{2}L)$. From the discussion in Chap. 1, the integrals in the three equations above will be recognized as scalar products in that space. Equations (3.1) and (3.2) say that $\{\sqrt{2/L}\cos(2\pi nx/L)\}$ and $\{\sqrt{2/L}\sin(2\pi nx/L)\}$ are two orthonormal sets in $\mathbb{L}_2(-\frac{1}{2}L, \frac{1}{2}L)$, while (3.3) says that any member of the first set is orthogonal to any member of the second.

We know from Chap. 1 that *complete* orthonormal sets are useful since they form a basis in terms of which any vector in the space can be expanded. The central question in Fourier analysis is whether the sets $\{\sqrt{2/L}\cos(2\pi nx/L)\}$ and $\{\sqrt{2/L}\sin(2\pi nx/L)\}$ are complete in that sense. This question is addressed in Sec. 3.2.

3.1.2 Complex exponentials

A more compact way to express the results of the last section uses complex exponentials. By DeMoivre's theorem (see App. B)

$$e^{i\theta} = \cos\theta + i\sin\theta\,. \tag{3.4}$$

Equation (3.4), like any equation involving complex variables, is really two equations, one for the real part and one for the imaginary part. These two simultaneous equations can be inverted to express the sine and cosine in terms of complex exponentials, with the result

$$\cos\theta = \frac{e^{i\theta} + e^{-i\theta}}{2}\,, \tag{3.5}$$

$$\sin\theta = \frac{e^{i\theta} - e^{-i\theta}}{2i}\,. \tag{3.6}$$

In terms of complex exponentials, the three orthogonality relations, (3.1) – (3.3), can be collapsed into a single equation:

$$\frac{1}{L}\int_{-\frac{1}{2}L}^{\frac{1}{2}L} dx \exp\left(-\frac{2\pi inx}{L}\right)\exp\left(\frac{2\pi imx}{L}\right) = \delta_{mn}\,, \tag{3.7}$$

which is readily proved by separating real and imaginary parts and using (3.1) – (3.3). The ugly $\sqrt{2}$ has disappeared since the complex exponentials have modulus unity; if $n = m$, the integral is L. Also, it is no longer necessary to treat $n = 0$ or $m = 0$ separately.

In (3.7) we introduced a complex conjugate in the first exponential in keeping with the definition of a scalar product in $\mathbb{L}_2$. If we denote $(1/\sqrt{L})\exp(2\pi inx/L)$ as $u_n(x)$, (3.7) says that the $\{u_n(x)\}$ constitute an orthonormal set in $\mathbb{L}_2(-\frac{1}{2}L, \frac{1}{2}L)$. Again, the key question is whether they are a complete set.

3.1.3 Orthogonality on the infinite interval

In many applications of Fourier analysis, it is necessary to consider the infinite interval $-\infty < x < \infty$. If we simply let $L \to \infty$ in the treatment above, we run into difficulty because of the factor $1/\sqrt{L}$ in the definition of the orthonormal functions. Orthonormal functions with zero amplitude everywhere are of little use. A better approach is to designate the functions by a continuous index rather than a discrete one, so that n/L and m/L become the real continuous variables ξ and ξ', respectively. Then the counterpart of (3.7) is

$$\int_{-\infty}^{\infty} dx \exp(-2\pi i\xi x)\exp(2\pi i\xi' x) = \delta(\xi - \xi')\,, \tag{3.8}$$

which is (2.45) in Chap. 2 with a notational change. In the limit $L \to \infty$, L times the Kronecker delta has become the Dirac delta.

Equation (3.8) says that the functions $\exp(2\pi i\xi x)$ and $\exp(2\pi i\xi' x)$ are orthogonal in the sense that their scalar product, defined appropriately for $\mathbb{L}_2(-\infty,\infty)$, vanishes if $\xi \neq \xi'$. If $\xi = \xi'$, however, the scalar product is infinite, so $\exp(2\pi i\xi x)$ is not itself a vector in $\mathbb{L}_2(-\infty,\infty)$. Nevertheless, we can expect from the discussions in Chap. 1 (see Sec. 1.1.6) that these complex exponentials will prove useful as a basis for $\mathbb{L}_2(-\infty, \infty)$. Anticipating that result, we denote the set $\{\exp(2\pi i \xi x)\}$ as $\{u_\xi(x)\}$.

The corresponding orthogonality relations for sines and cosines are

$$2\int_{-\infty}^{\infty} dx \cos(2\pi\xi x)\cos(2\pi\xi' x) = \delta(\xi+\xi') + \delta(\xi-\xi'), \tag{3.9}$$

$$2\int_{-\infty}^{\infty} dx \sin(2\pi\xi x)\sin(2\pi\xi' x) = \delta(\xi-\xi') - \delta(\xi+\xi'). \tag{3.10}$$

Thus $\cos(2\pi\xi x)$ and $\cos(2\pi\xi' x)$ are orthogonal unless $\xi = \xi'$ or $\xi = -\xi'$, and similarly for the sines.

3.1.4 Discrete orthogonality

In the last section we converted the discrete index n in the function $\exp(2\pi inx/L)$ into a continuous variable. We can also do the reverse and convert the continuous variable x into a discrete index k. In this case it is more convenient to work with the range $0 \leq x < L$ rather than the centered range used above. We get a discrete variable simply by dividing the range $[0, L)$ into K equal steps of size $\Delta x = L/K$ and defining

$$x_k = k\Delta x = \frac{kL}{K}, \qquad k = 0, ..., K-1. \tag{3.11}$$

The function $\exp(2\pi inx/L)$ evaluated at $x = x_k$ is $\exp(2\pi ink/K)$, which, for fixed n, is a set of K complex numbers or a $K \times 1$ vector. Of course, the index n specifies *which* $K \times 1$ vector we are considering. We denote by $\mathbf{u}_n$ the vector for which the kth component is given by $(1/\sqrt{K})\exp(2\pi ink/K)$. With the scalar product appropriate to the K-dimensional Euclidean space $\mathbb{E}^K$, we have the following orthogonality relation:

$$(\mathbf{u}_m, \mathbf{u}_n) = \frac{1}{K}\sum_{k=0}^{K-1} e^{2\pi i(n-m)k/K} = \delta_{mn}\,. \tag{3.12}$$

This result can be derived algebraically by recognizing the sum as a geometric series, with terms of the form t^k, where $t = \exp[2\pi i(n-m)/K]$. The usual expression for the sum of a geometric series then gives (3.12). A more intuitive view of the same result is given in the next section.

3.1.5 View from the complex plane

For fixed x, the function $u_n(x) = \exp(2\pi inx/L)$ can be represented as a point on the unit circle in the complex plane or as a vector from the origin to that point. The axes in this complex representation are the real and imaginary parts of $u_n(x)$. The angle between the vector and the real axis is the phase, $2\pi nx/L$, which increases linearly as x increases. For this reason, the complex exponential is often referred to as a *linear phase factor*. The term *phasor* is also frequently encountered, especially in the electrical engineering literature.

A useful mental image is that there is a vector in the complex plane associated with each value of x, with the vector rotating linearly as x increases (see Fig. 3.1). The spatial frequency n/L specifies the rate of rotation.

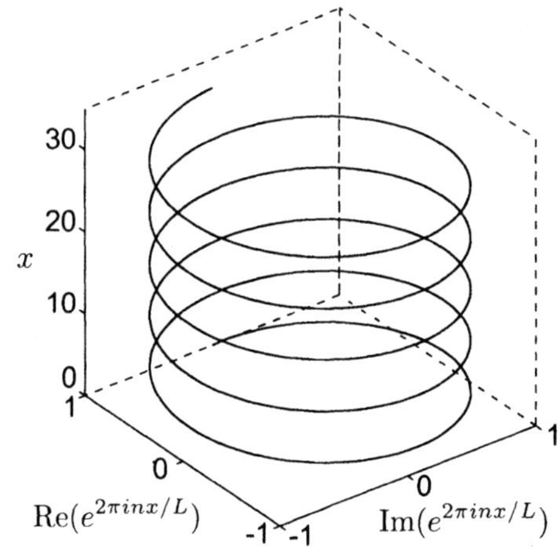

Fig. 3.1 Illustration of a complex exponential or linear phase factor $\exp(2\pi i \xi_0 x)$.

This viewpoint can be used to give geometric interpretations of the various orthogonality relations derived above. Consider first (3.7). The integrand is the product of two complex exponentials, which is itself a complex exponential, $\exp[2\pi i(m-n)x/L]$. The resulting vector in the complex plane makes an angle $2\pi(m-n)x/L$ with the real axis. Again, this angle increases linearly as x varies, with the rate of rotation now specified by the *difference* in spatial frequencies, $(m-n)/L$. As x varies over the range of integration from $-\frac{1}{2}L$ to $\frac{1}{2}L$, the vector rotates through a total angle of $2\pi(m-n)$, which is an integer number of turns since $m-n$ is an integer. Each x in the range of integration is paired with another x such that the vector is oppositely directed. The integral, being just the sum of all these vectors, is zero unless $m=n$.

Similarly, in (3.12), the complex number $\exp[2\pi i(m-n)k/K]$ is a vector in the complex plane, making an angle $2\pi(m-n)k/K$ with the real axis. If $m=n$, the angles are zero for each k, and the sum is K, but if $m \neq n$, the vectors are uniformly spaced in angle and the vector sum is zero.

3.2 FOURIER SERIES

3.2.1 Basic concepts

Consider a one-dimensional (1D) function $f(x)$ defined on the interval $-\frac{1}{2}L \leq x < \frac{1}{2}L$. In the next section we shall be more precise about the requirements on $f(x)$, but for now we simply assume that it can be expanded in terms of sines and cosines in the form

$$f(x) = F_0 + \sum_{n=1}^{\infty} F_n^{(c)} \cos\left(\frac{2\pi nx}{L}\right) + \sum_{n=1}^{\infty} F_n^{(s)} \sin\left(\frac{2\pi nx}{L}\right). \tag{3.13}$$

Discussion of the validity of this form and the sense in which the series converges to $f(x)$ is postponed to the next section.

How do we find the coefficients of the expansion in (3.13)? The first term, F_0, is found by integrating both sides of (3.13) from $-\frac{1}{2}L$ to $\frac{1}{2}L$, with the result

$$F_0 = \frac{1}{L}\int_{-\frac{1}{2}L}^{\frac{1}{2}L} f(x)\,dx\,, \tag{3.14}$$

since all other terms in (3.13) yield integrals of a sine or cosine over an integer number of periods and hence integrate to zero. Thus F_0 is simply the average of $f(x)$ over the interval. The coefficient F_0 is often referred to colloquially as the *DC term*; if x is a temporal variable and $f(x)$ is an electrical current, F_0 is the average or direct-current component, and the terminology has carried over to spatial variables as well.

To find the other coefficients, we multiply (3.13) by either $\cos(2\pi mx/L)$ or $\sin(2\pi mx/L)$, integrate over the same interval as before, and make use of the orthogonality relations, (3.1) – (3.3). The results are

$$F_n^{(c)} = \frac{2}{L}\int_{-\frac{1}{2}L}^{\frac{1}{2}L} dx\, f(x)\cos\left(\frac{2\pi nx}{L}\right)\,, \tag{3.15}$$

$$F_n^{(s)} = \frac{2}{L}\int_{-\frac{1}{2}L}^{\frac{1}{2}L} dx\, f(x)\sin\left(\frac{2\pi nx}{L}\right)\,. \tag{3.16}$$

It follows from these equations that $F_n^{(c)}$ and $F_n^{(s)}$ are real numbers if $f(x)$ is a real-valued function (or simply a real function for short).

If $f(x)$ is an even function, so that $f(-x) = f(x)$, then $F_n^{(s)} = 0$ and only the cosine series is needed. Similarly, if $f(x)$ is odd, so that $f(-x) = -f(x)$, then $F_n^{(c)} = 0$ and only the sine series is needed.

Fourier series in terms of complex exponentials A more compact way to state these results is by use of complex exponentials, so that the Fourier series takes the form

$$f(x) = \sum_{n=-\infty}^{\infty} F_n\, e^{2\pi inx/L}\,. \tag{3.17}$$

To find the coefficients $\{F_n\}$, we multiply both sides of (3.17) by $\exp(-2\pi imx/L)$ and integrate from $-\frac{1}{2}L$ to $\frac{1}{2}L$. With the orthogonality relation of (3.7), we find

$$\int_{-\frac{1}{2}L}^{\frac{1}{2}L} dx\, e^{-2\pi imx/L} f(x) = \sum_{n=-\infty}^{\infty} F_n \int_{-\frac{1}{2}L}^{\frac{1}{2}L} dx\, e^{-2\pi imx/L}\, e^{2\pi inx/L}$$

$$= \sum_{n=-\infty}^{\infty} F_n L\,\delta_{mn} = LF_m\,. \tag{3.18}$$

Solving for F_m and changing the dummy index m to n yields

$$F_n = \frac{1}{L}\int_{-\frac{1}{2}L}^{\frac{1}{2}L} dx\, f(x)\, e^{-2\pi inx/L}\,. \tag{3.19}$$

Thus, if we know *a priori* that $f(x)$ can be expanded in the form of (3.17), the coefficients in the expansion are given by (3.19). The question of when $f(x)$ can be so represented will be treated in Sec. 3.2.2.

Periodicity We assumed above that $f(x)$ was defined on the interval $[-\frac{1}{2}L, \frac{1}{2}L]$, and of course the coefficients were fully determined by the values of $f(x)$ in this interval. It is, however, interesting to inquire about the behavior of the series outside that interval. Since the function $\exp(2\pi inx/L)$ is periodic with period L/n, where n is an integer, it is also periodic with period L. Since this is true for every term in the Fourier series, it is true for the sum as well (if the sum converges). That is,

$$f(x + mL) = f(x), \qquad m = 0, \pm 1, \pm 2, \dots, \tag{3.20}$$

provided $f(x)$ is expressed by (3.17). We thus have two different ways of looking at the Fourier series. It is a representation of an arbitrary function on the interval $[-\frac{1}{2}L, \frac{1}{2}L]$, and it is a representation of a periodic function[1] with period L on the interval $(-\infty, \infty)$.

This view gives us alternative ways of expressing the Fourier coefficients. Since both factors in the integrand of (3.19) are periodic with period L, we can perform the integration over any period we choose. For example, it is straightforward to show that (3.19) is equivalent to

$$F_n = \frac{1}{L} \int_0^L dx\, f(x)\, e^{-2\pi inx/L} \tag{3.21}$$

if $f(x)$ has the periodic symmetry stated by (3.20).

Relation to the Laurent series For x real, the complex exponential $\exp(2\pi inx/L)$ lies on the unit circle in the complex plane. We can use this observation to relate the Fourier series to the Laurent series for functions of a complex variable (see App. B).

We define $w(x) \equiv \exp(2\pi ix/L)$, but we shall immediately drop the argument and write simply w for $w(x)$. Then the basic Fourier kernel is related to w by

$$\exp(2\pi inx/L) = w^n\,. \tag{3.22}$$

To get to the Laurent series, $f(x)$, must also be expressed in terms of w. To this end we define

$$f(x) = f_w(w) = f_w\left(e^{2\pi ix/L}\right). \tag{3.23}$$

The *Fourier* series for $f(x)$ is then the *Laurent* series for $f_w(w)$ around the point $w = 0$, *i.e.*,

$$f(x) = f_w(w) = \sum_{n=-\infty}^{\infty} F_n w^n\,. \tag{3.24}$$

The Laurent coefficients are identical to the Fourier coefficients. A change of variables from x to w in (3.19) allows us to write

$$F_n = \frac{1}{L} \int_{-\frac{1}{2}L}^{\frac{1}{2}L} dx\, f(x)\, e^{-2\pi inx/L} = \frac{1}{2\pi i} \oint_C dw\, \frac{f_w(w)}{w^{n+1}}\,, \tag{3.25}$$

where the contour integral is around the unit circle, denoted C. Equation (3.25) is identical (except for notation) to the relation given in App. B for Laurent coefficients.

[1] If the original definition of $f(x)$ on $[-\frac{1}{2}L, \frac{1}{2}L]$ does not have $f(-\frac{1}{2}L) = f(\frac{1}{2}L)$, then the corresponding periodic function on $(-\infty, \infty)$ will have periodically placed discontinuities.

A potential advantage of the Laurent series is that it would allow us to analytically extend $f(x)$ from a function of a real variable to a function of a complex variable, and the theory of Laurent series would enable us to make statements about convergence. Though we shall not have use for this extension, a related extension will prove useful when we discuss Fourier transforms in Sec. 3.3. For the remainder of the chapter, x will continue to denote a real variable.

3.2.2 Convergence of the Fourier series

In Sec. 3.2.1 we assumed that $f(x)$ could be represented by a Fourier series and used the orthogonality relations to derive expressions for the coefficients. In this section we pose the opposite question: Given the series and the coefficients, to what function does it converge?

To make this question more precise, we define

$$S_N(x) = \sum_{n=-N}^{N} F_n \, e^{2\pi i n x/L} , \tag{3.26}$$

where the $\{F_n\}$ are now *defined* by (3.19). This sum is called the Nth partial sum of the Fourier series of $f(x)$. For any finite N, $S_N(x)$ is a continuous, periodic function of x with period L (Champeney, 1987). We discuss below several ways in which S_N might converge to $f(x)$.

Pointwise convergence If $f(x)$ is smooth in some sense, there is no difficulty in showing that $S_N(x) \to f(x)$ for all x as $N \to \infty$. One way we could demonstrate this point is by use of the Laurent series introduced above. As we have used it, the Laurent series is convergent if $f_w(w)$ is analytic on the unit circle. As discussed in App. B, all derivatives of analytic functions exist. From this we can deduce that $S_N(x)$ converges to $f(x)$ at all x if all derivatives of $f_w(w)$ exist for w on the unit circle, which implies that all derivatives of $f(x)$ exist for real x. This result is of extremely limited usefulness.

A somewhat less restricted convergence statement is a classical theorem due to Dirichlet (Körner, 1988; Stakgold, 1979). If $f(x)$ is continuous and has a bounded continuous derivative on $(-\frac{1}{2}L, \frac{1}{2}L)$, except possibly at a finite number of points, then $S_N(x)$ converges to $f(x)$ for all points where $f(x)$ is continuous in that interval. This is an improvement over the previous paragraph, but it still does not encompass all of the functions for which we might want to construct a Fourier series. For more interesting functions, $S_N(x)$ may converge but not to $f(x)$, or it may not converge at all.

Consider two functions $f_1(x)$ and $f_2(x)$ that differ only at an isolated point x_0. If the values of both functions at x_0 are finite, these functions have the same set of Fourier coefficients since the isolated point does not affect the integrals. Hence the two functions have the same $S_N(x)$ for all N, so $S_N(x)$, while it may converge, cannot converge to both functions at every point. A simple way to deal with this problem is to say that $f_1(x)$ and $f_2(x)$ are really the same function *almost everywhere*. This disclaimer, often abbreviated *a.e.*, allows us to disregard the isolated point. The isolated point (or any countable set of points) is, in mathematical jargon, a set of measure zero, which simply means it doesn't affect any integrals so we don't worry about it.

Having dismissed problems arising from isolated points, we turn next to discontinuities in $f(x)$. Consider a point x_0 where $f(x)$ is bounded but discontinuous and assume that the limit of $f(x)$ exists as $x \to x_0$ from both directions. Denote the limit from above by $f(x_0^+)$ and from below by $f(x_0^-)$. A classical result (Stakgold, 1979) is that $S_N(x_0) \to \frac{1}{2}[f(x_0^+) + f(x_0^-)]$ as $N \to \infty$. A similar conclusion follows if $f(x)$ has a finite number of finite discontinuities. In other words, if we simply redefine $f(x)$ such that its value at every discontinuity is the average of its values on the two sides, we can say that $S_n(x)$ converges to $f(x)$ everywhere; without redefinition, we still need the *almost everywhere* disclaimer to exclude the discontinuity.

Many other theorems relate to the pointwise convergence of Fourier series (Champeney, 1987; Titchmarsh, 1948). The following theorem, quoted from Champeney, provides some useful guidelines:

> Suppose that $f(x)$ is periodic of period L, defined and bounded on $[-\frac{1}{2}L, \frac{1}{2}L]$, and that at least one of the following four conditions is satisfied: (*i*) $f(x)$ is piecewise monotonic on $[-\frac{1}{2}L, \frac{1}{2}L]$; (*ii*) $f(x)$ has a finite number of maxima and minima on $[-\frac{1}{2}L, \frac{1}{2}L]$ and a finite number of discontinuities on $[-\frac{1}{2}L, \frac{1}{2}L]$; (*iii*) $f(x)$ is of bounded variation on $[-\frac{1}{2}L, \frac{1}{2}L]$; (*iv*) $f(x)$ is piecewise smooth on $[-\frac{1}{2}L, \frac{1}{2}L]$. Then the Fourier series coefficients can be defined by (3.19), using proper Riemann integrals, and the Fourier series given by (3.26) converges to $f(x)$ at each point of continuity of $f(x)$ and to the value $\frac{1}{2}[f(x_0^+) + f(x_0^-)]$ at all x.

While the conditions listed above allow considerable latitude in the function $f(x)$, clever mathematicians can always find pathological cases. Certain peculiar functions with a fractal character may be everywhere continuous but nowhere differentiable (Walker, 1991; Körner, 1988). These functions are not covered by the theorem above; for example, they have an infinite number of maxima and minima in a finite interval. In such cases, it is possible that the Fourier series for a periodic and everywhere continuous function can diverge for some x and even for a nondenumerable and dense set of x values. About all that can be proved unequivocally is that the Fourier series of a continuous function cannot diverge everywhere (Champeney, p. 157).

Uniform convergence After pointwise convergence, the next kind to consider is uniform convergence (Champeney, p. 26). Even though $S_N(x)$ may converge pointwise almost everywhere to $f(x)$, the rate of convergence may be different for different x, so that even at large N, $f(x) - S_N(x)$ may not be small for some x. A formal definition of uniform convergence is that the least upper bound of $|f(x) - S_N(x)|$ over any interval of x tends to 0 as N goes to ∞.

One simple way to ensure uniform convergence of the Fourier series is to require that the function and its derivative be continuous (Stakgold, 1979). If the function is discontinuous, however, there is a well known failure of the condition of uniform convergence. As illustrated in Fig. 3.2, there is an oscillation and a significant overshoot near the point of discontinuity. As N gets larger, the spatial extent of the oscillation decreases, but the peak amplitude of the overshoot does not tend to zero. This finite limiting overshoot is known as the *Gibbs phenomenon* (after the American physicist J. Willard Gibbs). The Gibbs phenomenon does not contradict the fact that the Fourier series converges pointwise, but it does show that the convergence is not uniform in an interval surrounding the discontinuity (Stakgold, 1979, p. 135). A good discussion of the Gibbs phenomenon is given in Carslaw (1930).

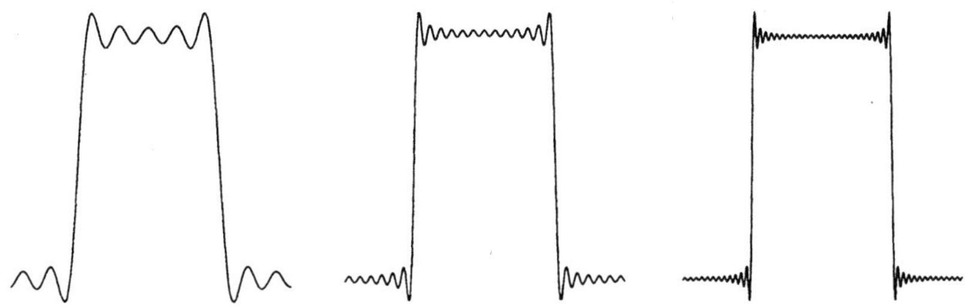

Fig. 3.2 Illustration of the Gibbs phenomenon. The function $f(x)$ is a rect function, and the partial Fourier sum $S_N(x)$ from (3.26) is shown for three values of N. Though the oscillations become more rapid as N increases, the amount of overshoot error does not change.

Convergence in the $\mathbb{L}_2$ sense Another important kind of convergence is *convergence in the $\mathbb{L}_2$ sense*, which is often referred to in the literature as *convergence in the mean.* If we approximate $f(x)$ by $S_N(x)$, the mean-square error in the approximation is just $1/L$ times the square of the $\mathbb{L}_2$ norm of $f(x) - S_N(x)$. Convergence in the mean states that this mean-square error tends to zero as N tends to ∞. To be precise, if we assume that $f(x)$ is in $\mathbb{L}_2(-\frac{1}{2}L, \frac{1}{2}L)$, then

$$\lim_{N\to\infty} \int_{-\frac{1}{2}L}^{\frac{1}{2}L} dx \, |f(x) - S_N(x)|^2 = 0 \,. \tag{3.27}$$

A formal proof of this theorem is given in Stakgold (1979), and a less formal one based on properties of delta functions is given below.

This kind of convergence says that if we use the Fourier series in place of the original function $f(x)$, the error we make is a function with zero length, in the sense of its $\mathbb{L}_2$ norm. In particular, if we form a scalar product with any function $h(x)$ in $\mathbb{L}_2(-\frac{1}{2}L, \frac{1}{2}L)$, it does not matter whether we use $f(x)$ or its Fourier-series representation, *i.e.*,

$$\int_{-\frac{1}{2}L}^{\frac{1}{2}L} dx \, h^*(x) \, f(x) = \lim_{N\to\infty} \int_{-\frac{1}{2}L}^{\frac{1}{2}L} dx \, h^*(x) \, S_N(x) \,. \tag{3.28}$$

Equation (3.27) says that $f(x) - S_N(x)$ tends to the zero vector in $\mathbb{L}_2$, and (3.28) is the extension that any scalar product with the zero vector is zero.

A proof of (3.27) and (3.28) based on properties of delta functions is instructive. If we use (3.26) for $S_N(x)$ and (3.19) for the coefficients, the right-hand side of (3.28) becomes

$$\lim_{N\to\infty} \frac{1}{L} \int_{-\frac{1}{2}L}^{\frac{1}{2}L} dx \int_{-\frac{1}{2}L}^{\frac{1}{2}L} dx' h^*(x) \, f(x') \sum_{n=-N}^{N} e^{2\pi i n(x-x')/L} \,. \tag{3.29}$$

The sum is a geometric series with terms of the form $s^n = \exp(i\alpha n)$, where $\alpha = 2\pi(x - x')/L$. From (2.50) we know that

$$\lim_{N\to\infty} \sum_{n=-N}^{N} e^{2\pi i n(x-x')/L} = \operatorname{comb}\left(\frac{x - x'}{L}\right) . \tag{3.30}$$

As discussed in Chap. 2, the comb function is an infinite series of delta functions, but in the present problem only one of them lies in the range of integration, $(-\frac{1}{2}L, \frac{1}{2}L)$, so $\text{comb}(x)$ is equivalent to $\delta(x)$. Moreover, this delta function has the sifting property (2.24) not only for test functions but also for all functions in $\mathbb{L}_2(-\frac{1}{2}L, \frac{1}{2}L)$ if convergence in the mean is understood (Champeney, 1987, p. 35). Thus (3.29) becomes

$$\frac{1}{L}\int_{-\frac{1}{2}L}^{\frac{1}{2}L} dx \int_{-\frac{1}{2}L}^{\frac{1}{2}L} dx' h^*(x)\, f(x')\, \delta\left(\frac{x-x'}{L}\right) = \int_{-\frac{1}{2}L}^{\frac{1}{2}L} dx\, h^*(x)\, f(x)\,, \tag{3.31}$$

again establishing (3.28) and the conclusion that $S_N(x) - f(x)$ converges to the zero vector in $\mathbb{L}_2(-\frac{1}{2}L, \frac{1}{2}L)$.

Having established that $S_N(x)$ converges to $f(x)$ and recalling our previous discussion of orthogonality, we are now justified in claiming that the complex exponentials $\{(1/\sqrt{L})\exp(2\pi inx/L)\}$ form a complete, orthonormal basis in $\mathbb{L}_2(-\frac{1}{2}L, \frac{1}{2}L)$. This basis set is infinite but denumerable. Note that neither $\{\cos(2\pi nx/L)\}$ nor $\{\sin(2\pi nx/L)\}$ by itself is a complete basis (unless we restrict attention to even or odd functions). Both sines and cosines, or equivalently complex exponentials, are needed to expand an arbitrary function in $\mathbb{L}_2(-\frac{1}{2}L, \frac{1}{2}L)$.

Fourier series divergent at a point So far we have been discussing situations where $S_N(x)$ converges to *something* at all x, though not necessarily to $f(x)$. It is also possible that $S_N(x)$ does not converge at all for one or more values of x.

One way to deal with such situations is to treat the resulting series not as an ordinary function but as a generalized function. As discussed in Chap. 2, a generalized function is defined by multiplying it by a well behaved test function $t(x)$ and integrating. Thus, if we can assign a meaning to the integral

$$\int_{-\frac{1}{2}L}^{\frac{1}{2}L} dx\, t(x)\, f(x) = \lim_{N\to\infty} \int_{-\frac{1}{2}L}^{\frac{1}{2}L} dx\, t(x) \sum_{n=-N}^{N} F_n\, e^{2\pi inx/L}\,, \tag{3.32}$$

then we shall have defined the generalized function $f(x)$ associated with the Fourier coefficients $\{F_n\}$, even if the series by itself does not converge.

As a simple example, consider the case where $F_n = 1$ for all n. The Fourier series diverges at $x = 0$, but the right-hand side of (3.32) can be written as

$$\lim_{N\to\infty} \sum_{n=-N}^{N} \int_{-\frac{1}{2}L}^{\frac{1}{2}L} dx\, t(x)\, e^{2\pi inx/L} = \lim_{N\to\infty} L \sum_{n=-N}^{N} T_{-n} = \lim_{N\to\infty} L \sum_{n=-N}^{N} T_n = Lt(0)\,, \tag{3.33}$$

where T_n is the well-defined Fourier coefficient of the continuous test function $t(x)$. Since exactly this same result would been obtained by integrating $Lt(x)\,\delta(x)$, we can say

$$\sum_{n=-\infty}^{\infty} e^{2\pi inx/L} = \delta(x/L) = L\,\delta(x)\,, \qquad -\tfrac{1}{2}L < x < \tfrac{1}{2}L\,. \tag{3.34}$$

Furthermore, since any Fourier series is periodic, we can remove the restriction on x and write

$$\sum_{n=-\infty}^{\infty} e^{2\pi inx/L} = \text{comb}(x/L)\,. \tag{3.35}$$

Equation (3.30) provides another example of a divergent Fourier series. It can be regarded as a Fourier series with coefficients given by $F_n = \exp(-2\pi inx'/L)$, where x' is a constant. The resulting Fourier series diverges if $x = x' + k/L$, where k is any integer, but it can be treated as the generalized function $\mathrm{comb}[(x - x')/L]$. Some other examples will be introduced below when we discuss derivatives of a Fourier series.

Mollifying a divergent series A common trick in mathematical physics is to deal with a divergent sum or integral by introducing a *convergence factor* in the summand or integrand. This factor is chosen so that, in some limit, it becomes unity, but the new sum or integral nevertheless converges. Fortunately, there is a good mathematical justification for this trick, as we shall see here for the case of a divergent Fourier series.

Consider a function $f(x)$ with Fourier coefficients given by the usual integral, (3.19). For this integral to exist, it is sufficient for $f(x)$ to be absolutely integrable, *i.e.*, in $\mathbb{L}_1(-\frac{1}{2}L, \frac{1}{2}L)$. If that is the case, then F_n is bounded since

$$|F_n| = \frac{1}{L}\left|\int_{-\frac{1}{2}L}^{\frac{1}{2}L} dx\, f(x)\, e^{-2\pi i\xi x}\right| \le \frac{1}{L}\int_{-\frac{1}{2}L}^{\frac{1}{2}L} dx\, \left|f(x)\, e^{-2\pi i\xi x}\right| = \frac{1}{L}\int_{-\frac{1}{2}L}^{\frac{1}{2}L} dx\, |f(x)|\,. \tag{3.36}$$

This condition is not, however, sufficient to guarantee the convergence of the Fourier series, so it is not clear to what extent a function in $\mathbb{L}_1(-\frac{1}{2}L, \frac{1}{2}L)$ can be reconstructed from its Fourier coefficients. (A function in $\mathbb{L}_1(-\frac{1}{2}L, \frac{1}{2}L)$ need not be in $\mathbb{L}_2(-\frac{1}{2}L, \frac{1}{2}L)$, as shown by the example of a delta function.)

The key question is the following: Given a set of Fourier coefficients of a function in $\mathbb{L}_1(-\frac{1}{2}L, \frac{1}{2}L)$, can we recover the function even when the series diverges? Surprisingly, the answer to this question is yes. Consider two new partial sums defined by

$$S_N^{(C)}(x) = \sum_{n=-N}^{N} \left(1 - \frac{|n|}{N}\right) F_n\, e^{2\pi inx/L}\,, \tag{3.37}$$

$$S_\lambda^{(A)}(x) = \sum_{n=-\infty}^{\infty} e^{-|n|/\lambda}\, F_n\, e^{2\pi inx/L}\,. \tag{3.38}$$

The first of these sums was devised by Cesaro and the second by Abel, hence the superscripts. The additional factor in the summand in each case is the convergence factor. If $N \to \infty$ or $\lambda \to \infty$, these factors approach unity for any fixed, finite n, so they would not appear to influence the sums in the limit; in fact, they don't if the original Fourier series converges. The nice result, however, is that these partial sums can converge to $f(x)$ in some sense even when the Fourier series diverges, *i.e.*,

$$\lim_{N\to\infty} S_N^{(C)}(x) = f(x)\,, \tag{3.39}$$

$$\lim_{\lambda\to\infty} S_\lambda^{(A)}(x) = f(x)\,. \tag{3.40}$$

Convergence proofs can be found in Körner (1988) and Champeny (1987). For example, it can be shown that the sums in (3.37) and (3.38) converge in all of the following senses (Champeney, p. 159):

(a) To $f(x)$ almost everywhere;

(b) To $f(x)$ at each point of continuity;

(c) Uniformly to $f(x)$ on an interval $[a, b]$ whenever $f(x)$ is continuous on $[a-\epsilon, b+\epsilon]$ for some $\epsilon > 0$;

(d) To $\frac{1}{2}[f(x^+)+f(x^-)]$ at each point where $f(x^+)$ and $f(x^-)$ exist, where $f(x^{\pm}) = \lim_{\epsilon\to 0} f(x \pm \epsilon)$;

(e) As a limit in the mean in the $\mathbb{L}_1$ sense over $(-\frac{1}{2}L, \frac{1}{2}L)$;

(f) To the Lebesgue value of $f(x)$ at each Lebesgue point.

A Lebesgue value is simply the local average of $f(x)$ over some small region, where the average is defined by an integral in the Lebesgue sense (Champeney, p. 31). A Lebesgue point is one where such an integral exists.

The uniform convergence in point (c) is interesting. It says that the Gibbs phenomenon is eliminated by inclusion of the convergence factors.

The use of a convergence factor can be illustrated by returning to our example of a delta function, where $f(x) = \delta(x)$ and $F_n = 1/L$ for all n. In this case the limit of the Cesaro partial sum can be written

$$\lim_{N\to\infty} S_N^{(C)}(x) = \lim_{N\to\infty} \frac{1}{L} \sum_{n=-N}^{N} \left(1 - \frac{|n|}{N}\right) e^{2\pi i n x/L} . \tag{3.41}$$

Using results to be derived later [the Poisson summation formula (3.197) and the expression for the Fourier transform of a triangle function, (3.141)], we find

$$\begin{aligned} \lim_{N\to\infty} S_N^{(C)}(x) &= \lim_{N\to\infty} \frac{N}{L} \sum_{n=-N}^{N} \operatorname{sinc}^2\left[\frac{N}{L}(x - nL)\right] \\ &= \sum_{n=-\infty}^{\infty} \delta(x - nL) = \frac{1}{L} \operatorname{comb}\left(\frac{x}{L}\right) , \end{aligned} \tag{3.42}$$

where we have used (2.48) and (2.87). This is the correct answer since the function $f(x) = \delta(x)$ in $(-\frac{1}{2}L, \frac{1}{2}L)$ corresponds to $(1/L)\operatorname{comb}(x/L)$ when periodically extended.

The relation between this subsection and the previous one should not be overlooked. Above we didn't worry about the divergence of the Fourier series at a point but simply used the divergent series to define a generalized function. Here we fixed up the divergence with a convergence factor, arriving at a limiting representation for the same generalized function. Different choices of convergence factors lead to different limiting representations. For example, the reader may wish to show that the Abel convergence factor leads to (2.39).

3.2.3 Properties of the Fourier coefficients

Having found expressions for the Fourier coefficients and discussed the convergence properties of the series, we now examine some important practical properties of the coefficients.

Linearity The mapping from a function $f(x)$ to its Fourier coefficients is a linear operator, as defined in Chap. 1. If we define

$$h(x) = \alpha\, f(x) + \beta\, g(x)\,, \tag{3.43}$$

where α and β are constants, then

$$H_n = \alpha F_n + \beta G_n\,, \tag{3.44}$$

where F_n, G_n, and H_n are Fourier coefficients of $f(x)$, $g(x)$ and $h(x)$, respectively.

Real functions Up until now, we have not imposed the requirement that $f(x)$ be real-valued; all of our results hold even if it is complex-valued. In practical applications, $f(x)$ will often be real, so of course its Fourier series must be real as well. By a change of variables, $x' = -x$, in (3.19), it can be seen that

$$F_n^* = F_{-n} \qquad \text{if } f(x) \text{ is real}\,. \tag{3.45}$$

A consequence of this relation is that F_0 must be real if $f(x)$ is.

Equation (3.45) is sometimes referred to as the Hermiticity of the Fourier coefficients, and the series itself is said to be Hermitian. This terminology derives from properties of Hermitian matrices, where interchange of rows and columns followed by complex conjugation leaves the matrix element unchanged. Here we see that replacement of n with $-n$ followed by complex conjugation leaves F_n unchanged if it is a Fourier coefficient of a real function, so many writers use the term Hermitian to describe this symmetry. An unfortunate aspect of this terminology is that it is likely to foster the impression that we are dealing with Hermitian operators, which we are not.

Even and odd functions Additional restrictions on the Fourier coefficients arise from symmetry of $f(x)$. In Sec. 3.2.1, we noted that an even function could be described completely by a Fourier cosine series, while an odd function required only the sine series. In terms of the coefficients in the complex exponential series, this implies

$$F_n = F_{-n} \qquad \text{if } f(x) = f(-x)\,, \tag{3.46a}$$

$$F_n = -F_{-n} \qquad \text{if } f(x) = -f(-x)\,. \tag{3.46b}$$

Thus even (odd) functions have Fourier coefficients that are even (odd) when we replace n with $-n$.

Combining (3.45) and (3.46), we can make stronger statements: If $f(x)$ is real and even, its Fourier coefficients are also real and even, while if $f(x)$ is odd, its coefficients are pure imaginary and odd.

Asymptotic behavior It is of considerable importance to know how F_n behaves as n tends to $\pm\infty$. The answer depends on the properties of $f(x)$, but the following theorem, proved by Stakgold (1979) is often useful: If $f(x)$ is in $\mathbb{L}_2(-\frac{1}{2}L, \frac{1}{2}L)$ and the $\{F_n\}$ are defined by (3.19), then

$$\lim_{n\to\pm\infty} F_n = 0 \tag{3.47}$$

and the decay of F_n is sufficiently rapid so that

$$\sum_{n=-\infty}^{\infty} |F_n|^2 < \infty \,. \tag{3.48}$$

Moreover, the rate of decay is dependent on the smoothness of $f(x)$. If we know that $f(x)$ is continuous and has continuous derivatives up to at least order q, then (Stakgold, 1979)

$$\lim_{n\to\pm\infty} n^q F_n = 0 \,. \tag{3.49}$$

We can understand these results intuitively by envisioning the real and imaginary parts of the integrand in (3.19). When a sine or cosine is multiplied by a smooth function, adjacent positive and negative half-cycles tend to cancel out. This cancellation becomes more complete as the frequency of the sine or cosine increases or as the function becomes smoother.

Parseval's relations Equation (3.48) says that the sum of squared moduli of the Fourier coefficients converges if $f(x)$ is in $\mathbb{L}_2(-\frac{1}{2}L, \frac{1}{2}L)$. *What* it converges to is interesting. An important result called Parseval's identity states that

$$\sum_{n=-\infty}^{\infty} |F_n|^2 = \frac{1}{L}\int_{-\frac{1}{2}L}^{\frac{1}{2}L} dx\, |f(x)|^2 \,. \tag{3.50}$$

This equation is a special case of a more general formula, known as the power theorem or generalized Parseval's theorem, which relates to products of two different functions, $f(x)$ and $g(x)$:

$$\sum_{n=-\infty}^{\infty} F_n^* G_n = \frac{1}{L}\int_{-\frac{1}{2}L}^{\frac{1}{2}L} dx\, f^*(x)\, g(x) \,. \tag{3.51}$$

There are several ways to derive these relations. A one-liner adapted from Marks (1991) starts by expressing $f^*(x)$ in (3.51) in terms of its Fourier series. From the complex conjugate of (3.17), we have

$$\frac{1}{L}\int_{-\frac{1}{2}L}^{\frac{1}{2}L} dx\, f^*(x)\, g(x) = \frac{1}{L}\int_{-\frac{1}{2}L}^{\frac{1}{2}L} dx \sum_{n=-\infty}^{\infty} F_n^*\, e^{-2\pi i n x/L} g(x) \,. \tag{3.52}$$

Interchanging sum and integral and recognizing the integral as G_n completes the proof of (3.51), while setting $f(x) = g(x)$ establishes (3.50).

An alternative derivation of the generalized Parseval relation parallels the derivation given for (3.28). We begin with the left-hand side of (3.51) and insert the integral expressions for both Fourier coefficients. Regarding the infinite sum as the limit of partial sums, we obtain

$$\begin{aligned}\sum_{n=-\infty}^{\infty} F_n^* G_n &= \lim_{N\to\infty} \frac{1}{L^2} \sum_{n=-N}^{N} \int_{-\frac{1}{2}L}^{\frac{1}{2}L} dx\, f^*(x)\, e^{2\pi i n x/L} \int_{-\frac{1}{2}L}^{\frac{1}{2}L} dx' g(x')\, e^{-2\pi i n x'/L} \\ &= \lim_{N\to\infty} \frac{1}{L^2} \int_{-\frac{1}{2}L}^{\frac{1}{2}L} dx \int_{-\frac{1}{2}L}^{\frac{1}{2}L} dx'\, f^*(x)\, g(x') \sum_{n=-N}^{N} e^{2\pi i n (x-x')/L} \,.\end{aligned} \tag{3.53}$$

By arguments identical to those used below (3.29), the sum becomes a delta function, and (3.51) follows.

Fourier series and unitary transformations Parseval's relations have an important interpretation in terms of scalar products. The right-hand side of (3.51) is $1/L$ times the $\mathbb{L}_2$ scalar product of $g(x)$ and $f(x)$, while the left-hand side can be interpreted as a scalar product in an infinite-dimensional Euclidean space $\mathbb{E}^\infty$. If we regard the process of constructing a Fourier series as a linear transformation from $\mathbb{L}_2$ to $\mathbb{E}^\infty$, the generalized Parseval relation says that the transformation preserves the scalar product except for the factor of $1/L$. Similarly, (3.50) says that the norm is preserved within $1/L$.

In Chap. 1 we saw that *unitary* transformations preserve norms and scalar products and that transformations based on orthonormal functions are unitary (see Sec. 1.4.5). The only reason the transformation from $f(x)$ to the set of Fourier coefficients $\{F_n\}$ is *not* unitary is that we did not define the coefficients with properly normalized orthonormal functions. As we have seen, the orthonormal set is $\{(1/\sqrt{L})\exp(2\pi inx/L)\}$, while the Fourier coefficients are defined with $1/L$ rather than $1/\sqrt{L}$. Except for the gratuitous factor of $\sqrt{L}$, the transformation from $f(x)$ to $\{F_n\}$ is unitary, and the Parseval relations follow.

Differentiation and integration of a Fourier series Suppose $f(x)$ is continuous and has a continuous derivative $f'(x)$ for some interval of x. Then both $f(x)$ and $f'(x)$ can be represented by Fourier series in this interval. Direct differentiation of the series for $f(x)$ yields

$$f'(x) = \sum_{n=-\infty}^{\infty} \frac{2\pi inF_n}{L}\, e^{2\pi inx/L}\,. \tag{3.54}$$

Hence the Fourier coefficients for $f'(x)$ are $\{2\pi inF_n/L\}$, where $\{F_n\}$ are the coefficients for $f(x)$.

Several comments about this result are in order. First, in spite of the factor of i, $f'(x)$ is real if $f(x)$ is real since replacing n with $-n$ and taking the complex conjugate leaves the product in unchanged [see (3.32)]. Second, if $f'(x)$ exists, the series in (3.54) will converge. Finally, if the series in (3.54) does not converge, we can nevertheless interpret it as a series representation of a generalized function.

For example, consider the function $f(x) = \text{rect}(2x/L)$ defined in (2.9). As discussed in Chap. 2, this function is discontinuous at $x = \pm L/4$, and its derivative, interpreted as a generalized function, has a delta function of weight $+1$ at $x = -L/4$ and another of weight -1 at $x = +L/4$. To show the same result in terms of Fourier series, we first expand $f(x)$ and then use (3.54). It follows that $F_0 = \frac{1}{2}$ and $F_n = (1/\pi n)\sin(\pi n/2)$ for $n \neq 0$. Hence the coefficients for $f'(x)$ are $(2i/L)\sin(\pi n/2)$ for all n. The series for $f'(x)$ does not converge at $x = \pm L/4$, but we can write

$$f'(x) = \sum_{n=-\infty}^{\infty} \frac{2i}{L}\sin(\pi n/2)\, e^{2\pi inx/L} = \delta\big(x + \tfrac{L}{4}\big) - \delta\big(x - \tfrac{L}{4}\big)\,, \qquad -\tfrac{1}{2}L \le x < \tfrac{1}{2}L\,. \tag{3.55}$$

The validity of this form can be checked by multiplying by a test function and integrating. Starting with the right-hand side, we obtain

$$\int_{-\frac{1}{2}L}^{\frac{1}{2}L} dx\, t(x)\left[\delta\left(x + \tfrac{L}{4}\right) - \delta\left(x - \tfrac{L}{4}\right)\right] = t\left(-\tfrac{L}{4}\right) - t\left(\tfrac{L}{4}\right)\,. \tag{3.56}$$

This result can be written in terms of the Fourier series for $t(x)$ evaluated at $x = \pm L/4$, with the result

$$t\left(-\tfrac{L}{4}\right) - t\left(\tfrac{L}{4}\right) = \sum_{n=-\infty}^{\infty} T_n \left[e^{-i\pi n/2} - e^{i\pi n/2}\right] . \tag{3.57}$$

The same result is obtained by multiplying the left-hand side of (3.55) by $t(x)$ and integrating, so (3.55) is a valid representation of the divergent series for the derivative of the rect function.

Integration of a Fourier series is much simpler than differentiation. A term-by-term indefinite integral introduces a factor of $1/(2\pi i n)$, which can only help the convergence as $n \to \infty$. It can be proved that any Fourier series, whether convergent or not, can be integrated term by term between any limits. The integrated series converges to the integral of the periodic function corresponding to the original series (Sokolnikoff and Redheffer, 1958).

3.3 1D FOURIER TRANSFORM

3.3.1 Basic concepts

The Fourier series is an expansion of a function on a finite interval, or of a periodic function on the infinite interval. For many applications, it is necessary to have an expansion on the infinite interval without the requirement of periodicity. Development of such an expansion is the goal of this section.

Formal limit of a Fourier series The simplest route to this goal is the one travelled by Fourier himself, simply passing formally to the limit $L \to \infty$. In this limit, the spatial frequency n/L goes over to a continuous variable ξ. A change of one in n is equivalent to a change of ξ by an amount $\Delta\xi = 1/L$. The Fourier series thus limits to

$$\lim_{L\to\infty} \sum_{n=-\infty}^{\infty} F_n \, e^{2\pi i n x/L} = \lim_{L\to\infty} \frac{1}{\Delta\xi} \int_{-\infty}^{\infty} d\xi \, F_n \, e^{2\pi i \xi x} . \tag{3.58}$$

Now we simply define the limit of $F_n/\Delta\xi$ as $F(\xi)$, obtaining

$$F(\xi) \equiv \lim_{L\to\infty} \frac{F_n}{\Delta\xi} = \lim_{L\to\infty} \int_{-\frac{1}{2}L}^{\frac{1}{2}L} dx \, f(x) \, e^{-2\pi i \xi x} = \int_{-\infty}^{\infty} dx \, f(x) \, e^{-2\pi i \xi x} . \tag{3.59}$$

The function $F(\xi)$ is called the *Fourier transform* of $f(x)$. Using an operator notation similar to that introduced in Chap. 1, we denote this integral transform by the operator $\mathcal{F}$, with a subscript to denote the dimensionality of the integral. Hence, we write

$$F(\xi) = \mathcal{F}_1 \{f(x)\} = \int_{-\infty}^{\infty} dx \, f(x) \, e^{-2\pi i \xi x} . \tag{3.60}$$

The subscript will often be omitted when it is obvious from the context.

The limit in (3.58) suggests that the inverse transform is

$$f(x) = \mathcal{F}_1^{-1} \{F(\xi)\} = \int_{-\infty}^{\infty} d\xi \, F(\xi) \, e^{2\pi i \xi x} . \tag{3.61}$$

Showing that this operator $\mathcal{F}_1^{-1}$ is indeed the inverse to $\mathcal{F}_1$, and defining the sense in which the limits hold, requires a detailed discussion of convergence issues, which we give in Sec. 3.3.2.

Fourier integral theorem Before launching into a formal discussion of convergence, we present an alternative statement of the inverse Fourier transform. If the integral transform in (3.61) is indeed the inverse of the one in (3.60), we can write

$$f(x) = \int_{-\infty}^{\infty} d\xi \, e^{2\pi i \xi x} \left[\int_{-\infty}^{\infty} dx' f(x') \, e^{-2\pi i \xi x'} \right] . \tag{3.62}$$

This equation is commonly known as the *Fourier integral theorem*, though it is not yet a proper theorem since we have not stated the conditions that $f(x)$ must satisfy or the sense in which equality holds.

Taking a hint from the discussion above on Fourier series, we can also express the Fourier integral theorem using the limit of a finite integral rather than an infinite integral. We define

$$f_K(x) = \int_{-\frac{1}{2}K}^{\frac{1}{2}K} d\xi \, e^{2\pi i \xi x} \left[\int_{-\infty}^{\infty} dx' f(x') \, e^{-2\pi i \xi x'} \right] . \tag{3.63}$$

The advantage of this form is that, with a finite integral, we might be able to interchange order of integration and write

$$\begin{aligned} f(x) = \lim_{K\to\infty} f_K(x) &= \lim_{K\to\infty} \int_{-\infty}^{\infty} dx' f(x') \int_{-\frac{1}{2}K}^{\frac{1}{2}K} d\xi \, e^{2\pi i \xi (x-x')} \\ &= \lim_{K\to\infty} \int_{-\infty}^{\infty} dx' f(x') K \operatorname{sinc}\left[K(x-x')\right] , \end{aligned} \tag{3.64}$$

which is plausible since $K \operatorname{sinc}[K(x-x')]$ is one of our limiting representations of $\delta(x-x')$. Proving the theorem in this form requires justifying the interchange of order of integration, stating the conditions under which $K \operatorname{sinc}[K(x-x')] \to \delta(x-x')$, and specifying the sense in which the right-hand side converges to $f(x)$. We undertake this enterprise in the next section.

3.3.2 Convergence issues

There is a vast literature on determination of conditions on $f(x)$ sufficient for one or more of the forms of the Fourier integral theorem to be valid. Papers and books on this subject can be divided into two categories, which we might call classical and distributional. The most influential book on the classical approach is, no doubt, Titchmarsh (1948), while the distributional approach was most forcefully advocated by Lighthill (1958).

The classical approach requires that we place restrictions on the function $f(x)$, just as we did with Fourier series. One possible restriction is that $f(x)$ be in one of the spaces $\mathbb{L}_p(-\infty,\infty)$, with $1 \le p < \infty$. As we shall see, one immediate difficulty with this approach is that $f(x)$ and $F(\xi)$ need not be in the same space. Another problem is that some rather ordinary functions are not in a convenient $\mathbb{L}_p$ space. For example, $\operatorname{sinc}(ax)$ is in $\mathbb{L}_p(-\infty,\infty)$ for $p > 1$, but is not in $\mathbb{L}_1(-\infty,\infty)$, while a delta function, which we would certainly like to include in the theory, is absolutely integrable and hence can be regarded as a function in $\mathbb{L}_1(-\infty,\infty)$, but it is not in $\mathbb{L}_p(-\infty,\infty)$ for $p > 1$. The function $|x|^{-\frac{1}{2}}$ is not in $\mathbb{L}_p(-\infty,\infty)$ for *any* p. All of these problems will be circumvented by the use of distribution theory and generalized functions in Sec. 3.3.4.

Before proceeding, we need to make a distinction between the Hilbert space $\mathbb{L}_2$ in which $f(x)$ is a vector and the domain of the function itself. The function $f(x)$ maps the real line (the x axis) to the complex plane. The Fourier transform $F(\xi)$ also maps the real line — this time the ξ axis — to the complex plane. Colloquially, $f(x)$ is often said to be a function in *real space* or *coordinate space*, while $F(\xi)$ is a function in *reciprocal space* or *frequency space*. Alternatively, the terms *space domain* and *Fourier domain* are often encountered for the x and ξ axes, respectively. All of these designations, however, refer merely to the real line $\mathbb{R}$. The space $\mathbb{L}_p$ is not the same as either real space or frequency space.

Absolutely integrable functions Much of the classical literature on Fourier's theorem centers on functions in $\mathbb{L}_1(-\infty,\infty)$, the space of functions whose absolute value is integrable on the real line. The reason for this is simple; if $f(x)$ is in $\mathbb{L}_1(-\infty,\infty)$, then $F(\xi)$ is finite for all ξ:

$$|F(\xi)| = \left| \int_{-\infty}^{\infty} dx\, f(x)\, e^{-2\pi i \xi x} \right| \leq \int_{-\infty}^{\infty} dx\, |f(x)\, e^{-2\pi i \xi x}| = \int_{-\infty}^{\infty} dx\, |f(x)| < \infty\,. \tag{3.65}$$

Moreover, it can be shown that $F(\xi)$ is continuous everywhere and vanishes at infinity if $f(x)$ is in $\mathbb{L}_1(-\infty,\infty)$, *i.e.*,

$$\lim_{\xi \to \pm\infty} F(\xi) = 0\,. \tag{3.66}$$

This theorem, known as the Riemann-Lebesgue lemma, is proved in many books, such as Lang (1993).

Note, however, that having $f(x)$ an $\mathbb{L}_1$ function does not necessarily mean that $F(\xi)$ is also an $\mathbb{L}_1$ function. Since the maximum absolute value of a continuous function is its $\mathbb{L}_\infty$ norm, all we can say about $F(\xi)$ so far is that it is in $\mathbb{L}_\infty(-\infty,\infty)$. In spite of the similarity of (3.60) and (3.61), we cannot be guaranteed that the integral in the inverse transform exists just because the forward one does.

To proceed, we need to rule out certain pathological functions that, while absolutely integrable and perhaps even differentiable everywhere, neverthelesss vary arbitrarily rapidly. Such functions are said to be *of unbounded variation*. Roughly speaking, a function of bounded variation is one where the graph of the function has a finite length for any finite range of the variable (Strichartz, 1994, p. 39). Functions of unbounded variation include any function that goes to infinity as well as functions like $\cos(1/x)$ that oscillate infinitely rapidly. Differentiability and bounded variation are independent conditions, neither implying the other (Champeney, 1987).

For functions of bounded variation that are also in $\mathbb{L}_1(-\infty,\infty)$, there is a useful form of the Fourier integral theorem. Such functions cannot go to infinity and can have at most a finite number of discontinuities. These conditions are known as the *Dirichlet conditions*; if they hold, it can be shown that (Lang, 1993, pp. 289–290)

$$\lim_{K \to \infty} f_K(x) = \tfrac{1}{2}[f(x^+) + f(x^-)] = f(x)\,, \tag{3.67}$$

where $f_K(x)$ is defined in (3.63) and $f(x^{\pm})$ is defined below (3.40). The last form in (3.67) holds either if $f(x)$ is continuous at x or if we simply define $f(x)$ as the average of its values on either side of a finite discontinuity.

The reason that (3.67) works for $\mathbb{L}_1$ functions of bounded variation is that the interchange of order of integration in (3.64) is legal under these conditions. Basically, a double integral can be written in either order if both single integrals are absolutely convergent, which means that the integral of the absolute value of the integrand remains finite. An $\mathbb{L}_1$ function, by definition, is absolutely integrable, and any integral over a finite range is finite unless the integrand goes to infinity somewhere. Thus the twin conditions of an $\mathbb{L}_1$ function and bounded variation ensure that the interchange in (3.64) is allowed.

Comparison of (3.64) and (3.67) shows that it is legitimate to say that $K\,\text{sinc}(Kx)$ limits to $\delta(x)$ when used in an integral with a function satisfying the Dirichlet conditions. Moreover, should that function happen to be discontinuous at $x = 0$, the interpretation of the delta function is that it sifts out the average of the two values on either side of the discontinuity, as in (2.62).

Square-integrable functions As discussed in Chap. 1, a Hilbert space has advantages over the more general Banach space, and $\mathbb{L}_2$ is a Hilbert space but $\mathbb{L}_p$ for $p \neq 2$ is not. Thus we turn next to Fourier transforms of functions in $\mathbb{L}_2(-\infty, \infty)$, the space of square-integrable functions on the real line.

The theory of Fourier transforms of functions in $\mathbb{L}_2(-\infty, \infty)$ was developed by Plancherel in the early part of the twentieth century. Plancherel's main result is stated in terms of two limiting functions, $f_K(x)$ defined in (3.63) and $F_L(\xi)$, defined by

$$F_L(\xi) = \int_{-\frac{1}{2}L}^{\frac{1}{2}L} dx\, f(x)\, e^{-2\pi i \xi x} . \tag{3.68}$$

Plancherel's theorem (Titchmarsh, 1948, p. 69) states that if $f(x)$ is in $\mathbb{L}_2(-\infty, \infty)$, then $F_L(\xi)$ converges in the mean to a function $F(\xi)$ that is also in $\mathbb{L}_2(-\infty, \infty)$, and conversely if $F(\xi)$ is in $\mathbb{L}_2(-\infty, \infty)$, $f_K(x)$ converges in the mean to a function $f(x)$ in $\mathbb{L}_2(-\infty, \infty)$. The limits $f(x)$ and $F(\xi)$ are related by the usual Fourier transform relations. Loosely speaking, $\mathbb{L}_2$ functions transform to $\mathbb{L}_2$ functions. There is now a satisfying symmetry between the transform and its inverse. The only catch is that some of the functions we might wish to transform are not in $\mathbb{L}_2(-\infty, \infty)$.

It is important to note that Plancherel's theory is couched in terms of convergence in the mean. The mean-square error we make in computing the inverse Fourier transform via (3.61) is a function with zero $\mathbb{L}_2$ norm. More precisely,

$$\lim_{K\to\infty} \int_{-\infty}^{\infty} dx\, |f_K(x) - f(x)|^2 = 0 , \tag{3.69}$$

and similarly,

$$\lim_{L\to\infty} \int_{-\infty}^{\infty} d\xi\, |F_L(\xi) - F(\xi)|^2 = 0 . \tag{3.70}$$

One way to look at these results is that the functions $\{\exp(2\pi i \xi x)\}$ form a continuous basis for $\mathbb{L}_2(-\infty, \infty)$. To expand an $\mathbb{L}_2$ function $f(x)$ in this basis, we write

$$f(x) = \int_{-\infty}^{\infty} d\xi\, F(\xi)\, u_\xi(x) , \tag{3.71}$$

where $u_\xi(x) = \exp(2\pi i \xi x)$, so the expansion is precisely the *inverse* Fourier transform of $F(\xi)$. Similarly, an expansion of the $\mathbb{L}_2$ function $F(\xi)$ is given by

$$F(\xi) = \int_{-\infty}^{\infty} dx\, f(x)\, u_x(\xi)\,, \tag{3.72}$$

where $u_x(\xi) = \exp(-2\pi i \xi x)$. The expansion is now just the Fourier transform of $f(x)$. Except for the minus sign, the same basis functions are used in both expansions. In (3.71) the basis function $\exp(2\pi i \xi x)$ can be viewed as a function of x indexed by the continuous index ξ, while in (3.72) the basis function $\exp(-2\pi i \xi x)$ can be viewed as a function of ξ indexed by x. Because of the continuous index, the basis functions are not themselves in the space $\mathbb{L}_2$ (see Chap. 1, Sec. 1.4.4).

Convergence factors While the picture of a Fourier transform as a mapping from $\mathbb{L}_2(-\infty, \infty)$ to $\mathbb{L}_2(-\infty, \infty)$ is appealing, there are many functions of practical interest that are not in $\mathbb{L}_2(-\infty, \infty)$. We can deal with such cases with a trick similar to the one used with Fourier series in (3.37) and (3.38); we can multiply the integrand by a convergence factor that tends to unity in some limit.

There are many theorems that justify this procedure, but we shall be content with stating just one of them here (Champeney, 1987, p. 63). Suppose two functions $f(x)$ and $F(\xi)$ are related almost everywhere as follows:

$$F(\xi) = \lim_{\lambda\to\infty} \int_{-\infty}^{\infty} dx\, f(x)\, e^{-|x|/\lambda}\, e^{-2\pi i \xi x}\,, \tag{3.73}$$

$$f(x) = \lim_{\lambda\to\infty} \int_{-\infty}^{\infty} d\xi\, F(\xi)\, e^{-|\xi|/\lambda}\, e^{2\pi i \xi x}\,. \tag{3.74}$$

If at least one of the functions $f(x)$ or $F(\xi)$ is in $\mathbb{L}_p(-\infty, \infty)$ for some p in the range $1 \le p \le 2$, then the other function will be in $\mathbb{L}_q(-\infty, \infty)$, where $p^{-1} + q^{-1} = 1$, and both integrals will be well behaved. The special case $p = q = 2$ gets us back to the statement that the Fourier transform of an $\mathbb{L}_2$ function is an $\mathbb{L}_2$ function, while $p = 1$, $q = \infty$ corresponds to the asymmetric situation where the transform of an $\mathbb{L}_1$ function is in $\mathbb{L}_\infty$, or conversely.

An important example of the use of this theorem is the delta function, $f(x) = \delta(x)$. Any limiting representation of the delta function is in $\mathbb{L}_1$, and it follows from (3.73) and the definition of a delta function that $F(\xi) = 1$, which is a function in $\mathbb{L}_\infty$. We can also show that (3.74) is correct for this example. With $F(\xi) = 1$, (3.74) becomes

$$f(x) = \lim_{\lambda\to\infty} \int_{-\infty}^{\infty} d\xi\, e^{-|\xi|/\lambda}\, e^{2\pi i \xi x} = \lim_{\lambda\to\infty} \frac{2\lambda}{4\pi^2 x^2 \lambda^2 + 1}\,, \tag{3.75}$$

which is one of the limiting representations for a delta function, (2.39) with $\lambda = 1/(2\pi\epsilon)$.

Many different convergence factors can be used in similar theorems. When applied to $F(\xi) = 1$, they lead to different limiting representations of $\delta(x)$. In fact, (3.64) is analogous to (3.74) but with convergence factor $\mathrm{rect}(\xi/K)$.

3.3.3 Unitarity of the Fourier operator

As discussed in the last section, if $f(x)$ is in $\mathbb{L}_2(-\infty,\infty)$, so is $F(\xi)$ except possibly for a function of zero $\mathbb{L}_2$ norm; thus the Fourier transform is an integral operator that maps $\mathbb{L}_2$ to itself. In Chap. 1 we discussed operators that map a Hilbert space $\mathbb{U}$ to another Hilbert space $\mathbb{V}$, but here $\mathbb{U} = \mathbb{V} = \mathbb{L}_2$. Even though functions in the range and domain of the Fourier operator are denoted by different variables, the two spaces are really the same.

The Fourier mapping can be given in abstract operator form by

$$\mathbf{F} = \boldsymbol{\mathcal{F}}\mathbf{f}, \tag{3.76}$$

where $\mathbf{f}$ and $\mathbf{F}$ are the $\mathbb{L}_2$ vectors corresponding to the functions $f(x)$ and $F(\xi)$, respectively. The operator $\boldsymbol{\mathcal{F}}$ transforms $\mathbf{f}$ to $\mathbf{F}$.

The mapping properties of an integral transform are specified by its kernel function. The kernel of the Fourier transform is $\exp(-2\pi i\xi x)$ while the kernel of the inverse transform is $\exp(2\pi i\xi x)$. In Chap. 1 we saw that the kernel for the adjoint of an integral operator is obtained from the kernel for the operator itself by interchanging the variables and taking the complex conjugate. Here the variables are x and ξ, so the kernel for the adjoint is precisely the same as the kernel for the inverse. A unitary operator is one for which the adjoint equals the inverse, so the Fourier transform is a unitary operator. The analogous conclusion was found in Sec. 3.2.3 for a Fourier series.

Recall from Chap. 1 that an operator can be moved from one factor in a scalar product to the other if it is replaced by its adjoint (see Sec. 1.3.5). In the present context, that means

$$(\mathbf{F}_2, \boldsymbol{\mathcal{F}}\mathbf{f}_1) = (\boldsymbol{\mathcal{F}}^{\dagger}\mathbf{F}_2, \mathbf{f}_1)\,, \tag{3.77}$$

where the first scalar product is an integral over ξ and the second is over x. Since $\boldsymbol{\mathcal{F}}$ is unitary, we have

$$(\mathbf{F}_2, \mathbf{F}_1) = (\boldsymbol{\mathcal{F}}^{-1}\mathbf{F}_2, \mathbf{f}_1) = (\mathbf{f}_2, \mathbf{f}_1)\,. \tag{3.78}$$

In other words, scalar products are preserved under unitary transformations.

Parseval's theorems Writing out (3.78) in terms of explicit integrals, we find

$$\int_{-\infty}^{\infty} d\xi\, [F_2(\xi)]^*\, F_1(\xi) = \int_{-\infty}^{\infty} dx\, [f_2(x)]^*\, f_1(x)\,. \tag{3.79}$$

The special case where $f_1(x) = f_2(x)$ gives

$$\int_{-\infty}^{\infty} d\xi\, |F(\xi)|^2 = \int_{-\infty}^{\infty} dx\, |f(x)|^2\,. \tag{3.80}$$

The left-hand side of this equation is the $\mathbb{L}_2$ norm of $F(\xi)$ and the right-hand side is the $\mathbb{L}_2$ norm of $f(x)$.

Equation (3.80) is commonly called Parseval's theorem for Fourier transforms, while (3.79) is called the generalized Parseval theorem. Other terms such as Rayleigh's theorem or Plancherel's theorem are occasionally encountered, especially in the older literature. In essence, Parseval's theorems are just a statement of the unitarity of the Fourier operator and the fact that unitary transformations preserve norms and scalar products.

3.3.4 Fourier transforms of generalized functions

All of the Fourier theorems presented so far place restrictions on $f(x)$ and $F(\xi)$, often with different restrictions on each. It would be highly desirable to be able to choose a single class of functions broad enough to cover all functions of practical interest and their transforms. Fortunately, that goal can be reached by means of the theory of distributions and generalized functions.

Excellent pedagogical accounts of the distributional approach to Fourier transforms are given by Lighthill (1958), Strichartz (1994), and Richards and Youn (1990). For a much broader and more detailed treatment, see Zemanian (1965) and Zemanian (1987).

Tempered distributions The specific generalized functions we shall use are those corresponding to tempered distributions. Recall from Chap. 2 that tempered distributions are continuous, linear functionals acting on good functions, and that a good function is an open-support test function. More specifically, a good function is everywhere differentiable any number of times, and it and all of its derivatives vanish at infinity faster than $|x|^{-N}$ for all N. The space of good functions on the real line is sometimes called the Schwartz space and denoted $\mathbb{S}(-\infty, \infty)$ or simply $\mathbb{S}$. Note that good functions are necessarily square-integrable, so $\mathbb{S}$ is a subspace of $\mathbb{L}_2$.

The general form for a tempered distribution is

$$\Phi_g\{t(x)\} = \int_{-\infty}^{\infty} dx\, g(x)t(x)\,, \tag{3.81}$$

where $t(x)$ is a good function and $g(x)$ is a generalized function, defined by specifying the number returned by the functional for each $t(x)$. For example, if $g(x) = \delta(x)$, $\Phi_g\{t(x)\} = t(0)$.

As its name implies, a good function is indeed very good as far as Fourier transforms are concerned. It can be shown that, if $t(x)$ is in $\mathbb{S}$, then its Fourier transform $T(\xi)$ is also in $\mathbb{S}$; Fourier transforms of good functions are good functions (Strichartz, 1994, pp. 30–38). The same cannot be said of test functions in general; the transform of a function of compact support cannot have compact support, so it is important that we deal here with good functions and tempered distributions.

It might appear that this result is of little use since the conditions on good functions are so stringent. It would be much more interesting if we could make a similar statement about generalized functions. To do so, we must first define the Fourier transform of a generalized function.

Definition of Fourier transform The key to the definition of the Fourier transform of a generalized function is the unitary nature of the Fourier-transform operator. We note that the functional in (3.81) resembles a scalar product in $\mathbb{L}_2$, even though $g(x)$ is almost never in $\mathbb{L}_2$. Of course, the scalar product usually includes an asterisk indicating complex conjugate on the first factor in the integrand, but that is not needed since the generalized functions we consider are usually real. If $t(x)$ or $\mathcal{F}^{-1}\{T(\xi)\}$ is in $\mathbb{S}$, the functional $\Phi_g\{\mathcal{F}^{-1}T(\xi)\}$ is defined by

$$\Phi_g\{\mathcal{F}^{-1}[T(\xi)]\} = \int_{-\infty}^{\infty} dx\, g(x)\left[\mathcal{F}^{-1}\{T(\xi)\}\right]\,, \tag{3.82}$$

which we can think of as the scalar product of $g(x)$ and the good function $\mathcal{F}^{-1}\{T(\xi)\}$. Since $\mathcal{F}$ is a unitary operator in $\mathbb{L}_2$ and $\mathbb{S}$ is a subspace of $\mathbb{L}_2$, we can write

$$\Phi_g\{\mathcal{F}^{-1}[T(\xi)]\} = \int_{-\infty}^{\infty} dx\, g(x)\, [\mathcal{F}^{\dagger}\{T(\xi)\}] \,. \tag{3.83}$$

By analogy to scalar products in $\mathbb{L}_2$, we now *define*

$$\Phi_g\{\mathcal{F}^{-1}[T(\xi)]\} \equiv \int_{-\infty}^{\infty} d\xi\, [\mathcal{F}\{g(x)\}]^* \, T(\xi) = \int_{-\infty}^{\infty} d\xi\, [G(\xi)]^* \, T(\xi) \equiv \Phi_{\mathbf{G}}\{T(\xi)\}\,, \tag{3.84}$$

where $\mathcal{F}\{g(x)\} \equiv G(\xi)$, and the last distribution exists since $T(\xi)$ is a good function. In essence, the basic definition [for real $g(x)$] is

$$\int_{-\infty}^{\infty} dx\, g(x)\, t(x) = \int_{-\infty}^{\infty} d\xi\, [G(\xi)]^* \, T(\xi) \,. \tag{3.85}$$

Note that the complex conjugate on $G(\xi)$ is needed; even though $g(x)$ is real, its Fourier transform need not be.

We emphasize that (3.85) is a definition, but it is a reasonable one. As with other distributions defined in Chap. 2, this one holds for ordinary functions as well as generalized functions. If $g(x)$ were in $\mathbb{L}_2$, the transition from (3.83) to (3.85) would follow from the definition of the adjoint (see Sec. 1.3.5 in Chap. 1). The definition in (3.85) is tantamount to asserting that the generalized Parseval theorem holds even for generalized functions.

Thus, so long as the action of the functional on any good function is defined, (3.85) provides a definition of the Fourier transform of the associated generalized function. Moreover, the Fourier transform of a generalized function defined this way is itself a generalized function, since $T(\xi)$ in (3.85) is a good function. So long as we deal with tempered distributions, *the Fourier transform of a generalized function always exists and is always a generalized function.*

This definition takes care of any doubts about the existence of the Fourier transform of a generalized function, but we still need an operational way to calculate it. To that end, we use the fact that a generalized function can always be written as the limit of a regular sequence of good functions (Lighthill, 1958). With the representation,

$$g(x) = \lim_{n\to\infty} g_n(x)\,, \tag{3.86}$$

we have

$$\int_{-\infty}^{\infty} dx\, g(x)\, t(x) = \lim_{n\to\infty} \int_{-\infty}^{\infty} dx\, g_n(x)\, t(x) = \lim_{n\to\infty} \int_{-\infty}^{\infty} d\xi\, [G_n(\xi)]^* \, T(\xi)\,, \tag{3.87}$$

where the interchange of limit and integral is allowed since the integrand is so well behaved (Lighthill, 1958, p. 18). Comparing (3.87) to (3.85) establishes the important result that

$$\mathcal{F}\{g(x)\} = \lim_{n\to\infty} \mathcal{F}\{g_n(x)\} = \lim_{n\to\infty} G_n(\xi)\,. \tag{3.88}$$

Example: Transform of the derivative of a delta function We have already seen that the Fourier transform of a delta function is unity (*i.e.*, the function $F(\xi) = 1$), so we turn to a slightly more complicated example to illustrate the transform of a generalized function. Suppose $g(x) = \delta'(x)$, the derivative of a delta function as defined by [see (2.56)]

$$\Phi_{\delta'}\{t(x)\} = \int_{-\infty}^{\infty} dx\, \delta'(x)\, t(x) \equiv -t'(0)\,. \tag{3.89}$$

From (3.82) and (3.84), we have

$$\begin{aligned}\int_{-\infty}^{\infty} d\xi \left[\mathcal{F}\{\delta'(x)\}\right]^* T(\xi) &= \int_{-\infty}^{\infty} dx\, \delta'(x) \left[\mathcal{F}^{-1}\{T(\xi)\}\right] \\ &= \int_{-\infty}^{\infty} dx\, \delta'(x)\, t(x) = -t'(0)\,. \end{aligned} \tag{3.90}$$

But note that

$$-t'(0) = -\left[\frac{\partial}{\partial x}\int_{-\infty}^{\infty} d\xi\, T(\xi)\, e^{2\pi i\xi x}\right]_{x=0} = \int_{-\infty}^{\infty} d\xi\, (-2\pi i\xi)\, T(\xi)\,, \tag{3.91}$$

where differentiation under the integral sign is legal since the good function $T(\xi)$ is so well behaved. Comparison of (3.90) and (3.91) allows us to write

$$\mathcal{F}\{\delta'(x)\} = 2\pi i\xi\,. \tag{3.92}$$

Since this function increases without bound as $\xi \to \infty$, it certainly isn't in any of the $\mathbb{L}_p$ spaces, but as a generalized function corresponding to a tempered distribution, it is well defined.

The result in (3.92) could also be obtained from the limiting definition (3.88), along with any of the limiting representations of delta functions in Chap. 2.

3.3.5 Properties of the 1D Fourier transform

In this section we list several important properties of the Fourier transform. Many of these properties are implicit in the discussion above or follow from simple algebraic manipulation of the definitions, so we shall merely indicate the method of derivation without many details. Some comments will be made, however, on the extension of classical theorems to generalized functions.

Linearity The Fourier transform is a linear operator, so

$$\mathcal{F}\{\alpha\, f(x) + \beta\, g(x)\} = \alpha\, F(\xi) + \beta\, G(\xi)\,, \tag{3.93}$$

where $F(\xi)$ and $G(\xi)$ are the Fourier transforms of $f(x)$ and $g(x)$, respectively.

Symmetry properties Fourier transforms have symmetry properties similar to those discussed in Sec. 3.2.3 for Fourier coefficients. It is straightforward to show from the definition of the transform that

$$F(-\xi) = [F(\xi)]^*\,, \qquad \text{if } f(x) \text{ is real}\,. \tag{3.94}$$

As with Fourier coefficients, this relation is often referred to as Hermiticity, but note that the Fourier operator is unitary, not Hermitian.

Additional restrictions on the Fourier transform arise if $f(x)$ is an even or odd function. In particular,

$$F(\xi) = F(-\xi)\,, \qquad \text{if } f(x) = f(-x)\,, \tag{3.95a}$$

$$F(\xi) = -F(-\xi)\,, \qquad \text{if } f(x) = -f(-x)\,. \tag{3.95b}$$

Thus even (odd) functions have Fourier transforms that are even (odd). If $f(x)$ is real and even, its Fourier transform is also real and even, while if $f(x)$ is odd, its transform is pure imaginary and odd.

These results apply also to generalized functions. For example, $\delta(x)$ is real and even, so its Fourier transform, unity, is real and even. The derivative of a delta function is odd, and we have seen in (3.92) that its transform is pure imaginary and odd.

Derivatives To discover an expression for the Fourier transform of the first derivative of a function $f(x)$, we first represent the function in terms of its transform and then apply the derivative operator:

$$\frac{df}{dx} = \frac{d}{dx}\int_{-\infty}^{\infty} d\xi\, F(\xi)\, e^{2\pi i \xi x}\,. \tag{3.96}$$

If the derivative is bounded, it is legitimate to differentiate under the integral sign, yielding

$$\frac{df}{dx} = \int_{-\infty}^{\infty} d\xi\, F(\xi)\, 2\pi i \xi\, e^{2\pi i \xi x}\,. \tag{3.97}$$

Thus the Fourier transform of df/dx is $2\pi i \xi\, F(\xi)$. In operator form, we can write

$$\mathcal{F}\left\{\frac{d}{dx} f(x)\right\} = 2\pi i \xi\, F(\xi)\,. \tag{3.98}$$

In words, taking the derivative of a function is equivalent to multiplying its Fourier transform by $2\pi i \xi$.

We can repeat the process with impunity so long as the derivatives remain bounded, yielding

$$\mathcal{F}\left\{\left(\frac{d}{dx}\right)^q f(x)\right\} = (2\pi i \xi)^q F(\xi)\,. \tag{3.99}$$

In fact, the restriction to bounded derivatives is quite unnecessary. We have already seen in (3.92) that $\mathcal{F}\{\delta'(x)\} = 2\pi i \xi\, \mathcal{F}\{\delta(x)\} = 2\pi i \xi$, which is just (3.98) with $f(x) = \delta(x)$. Equations (3.98) and (3.99) can be derived from the definitions of derivatives and Fourier transforms of generalized functions, (2.18) and (3.85).

We can often discover useful dual relations by interchanging x and ξ and $\mathcal{F}$ and $\mathcal{F}^{-1}$. For example, the dual to (3.99) is

$$\mathcal{F}^{-1}\left\{\left(\frac{d}{d\xi}\right)^q F(\xi)\right\} = (-2\pi i x)^q f(x)\,. \tag{3.100}$$

This relation can be derived for ordinary functions with bounded derivatives by differentiating under the integral sign; for generalized functions it follows from the definitions of derivatives and Fourier transforms.

Moments The qth moment of a function $f(x)$ on $(-\infty, \infty)$ is defined by

$$m_q \equiv \int_{-\infty}^{\infty} dx\, x^q f(x)\,. \tag{3.101}$$

This integral can be thought of as the Fourier transform of $x^q f(x)$ at zero frequency, *i.e.*,

$$m_q = \left[\int_{-\infty}^{\infty} dx\, x^q f(x)\, e^{-2\pi i \xi x}\right]_{\xi=0}\,. \tag{3.102}$$

From (3.100), we now have

$$m_q = (-2\pi i)^{-q} F^{(q)}(0)\,, \tag{3.103}$$

where $F^{(q)}(\xi)$ is the qth derivative of $F(\xi)$. Thus, except for a constant, the qth moment of a function is the qth derivative of its Fourier transform evaluated at the origin.

The special case $q = 0$ gives

$$m_0 = \int_{-\infty}^{\infty} dx\, f(x) = F(0)\,. \tag{3.104}$$

This result is known as the *central-ordinate theorem* since a graph of $F(\xi)$ crosses the ordinate $\xi = 0$ at the value m_0.

The dual theorem to (3.103) relates to moments of $F(\xi)$:

$$M_q \equiv \int_{-\infty}^{\infty} d\xi\, \xi^q F(\xi) = (2\pi i)^{-q} f^{(q)}(0)\,. \tag{3.105}$$

The dual central-ordinate theorem is

$$M_0 = \int_{-\infty}^{\infty} d\xi\, F(\xi) = f(0)\,. \tag{3.106}$$

Asymptotic behavior The asymptotic behavior of $F(\xi)$ as $\xi \to \infty$ is analogous to the behavior of a Fourier coefficient as $n \to \infty$. The classical Riemann-Lebesgue lemma, (3.66), states that if $f(x)$ is an ordinary function of bounded variation in $\mathbb{L}_1(-\infty, \infty)$ and $F(\xi)$ is its Fourier transform, then $F(\xi) \to 0$ as $|\xi| \to \infty$. Further, if $f(x)$ is continuous and has bounded derivatives up to at least order q, then (Stakgold, 1979)

$$\lim_{\xi \to \pm\infty} \xi^q F(\xi) = 0\,. \tag{3.107}$$

This result follows from the Riemann-Lebesgue lemma when we recognize that $\xi^q F(\xi)$ is just a constant times the Fourier transform of the qth derivative of $f(x)$ [see (3.99)]. If that derivative is bounded and absolutely integrable, its Fourier transform must tend to zero as $\xi \to \infty$.

This time the condition of boundedness is important, however. We know that $\delta(x)$, though absolutely integrable, has a Fourier transform that does not vanish at infinity. Equation (3.107) is one of the few results in Fourier theory where it is really necessary to make a distinction between ordinary and generalized functions.

As an important example of this theorem, suppose $f(x)$ is a good function. Since all derivatives of a good function exist, (3.107) must hold for *all* q, which is equivalent to saying that $F(\xi)$ is also a good function.

Shifts and linear phase factors It follows readily from the definition of the Fourier transform that, for ordinary functions,

$$\mathcal{F}\{f(x - x_0)\} = e^{-2\pi i \xi x_0} F(\xi)\,. \tag{3.108}$$

This equation holds also for generalized functions if we define shifted generalized functions in a sensible way [for example, by analogy to Eq. (2.52)].

Equation (3.108) shows that shifting a function by x_0 is equivalent to multiplying its Fourier transform by the complex exponential (or linear phase factor) $\exp(-2\pi i \xi x_0)$. This result can be understood by recalling the viewpoint presented in Sec. 3.1.5 and Fig. 3.1. When we express a function $f(x)$ in terms of its Fourier transform, we are resolving it into a sum of vectors, $F(\xi)\exp(2\pi i \xi x)$, in the complex plane. Replacing x by $x - x_0$ changes the angle of each vector by an amount $-2\pi\xi x_0$ linearly dependent on ξ. Since the angle of the vector is the phase of the complex number, this is equivalent to multiplying $F(\xi)$ by $\exp(-2\pi i \xi x_0)$.

The dual theorem is

$$\mathcal{F}\left\{e^{+2\pi i \xi_0 x} f(x)\right\} = F(\xi - \xi_0)\,, \tag{3.109}$$

which can also be derived from basic definitions. Thus multiplying a function by a linear phase factor (this time linear in x) is equivalent to shifting its Fourier transform.

Scaled functions If a is a real number, $f(x/a)$ is a scaled version of the function $f(x)$. For $|a| > 1$, a graph of $f(x/a)$ is wider than a graph of $f(x)$, while for $|a| < 1$ it is narrower. Put another way, $f(x)$ takes on the same value at $x = x_0$ as $f(x/a)$ does at $x = ax_0$. Negative values of a correspond to reversal of the sense of the function along the x axis as well as scaling.

For ordinary functions, the Fourier transform of $f(x/a)$ is readily computed by a change of variables:

$$\mathcal{F}\{f(x/a)\} = \int_{-\infty}^{\infty} dx\, f(x/a)\, e^{-2\pi i \xi x} = |a| \int_{-\infty}^{\infty} dx'\, f(x')\, e^{-2\pi i \xi a x'}\,, \tag{3.110}$$

where $x' \equiv x/a$. The absolute value of a is needed since, for x negative, the direction of integration is reversed; if the integral is to always run from $-\infty$ to ∞, we must write $dx = |a| dx'$. The right-hand side of this equation is recognized as a scaled version of $F(\xi)$,

$$\mathcal{F}\{f(x/a)\} = |a| F(a\xi)\,. \tag{3.111}$$

Note that the scaling factor a appears in the numerator of the argument of $F(\xi)$, so a function that is scaled so that it becomes wider has its transform scaled so that it is narrower.

One way to understand the constant $|a|$ in (3.111) is via the central-ordinate theorem. The functions $|a|^{-1} f(x/a)$ and $f(x)$ have the same integral since the former has been increased in width by a and decreased in height by $1/a$ relative to the latter. The integral of both functions is given by $F(0)$ since $F(a \cdot 0) = F(0)$.

A derivation of (3.111) for generalized functions necessitates use of a scaled generalized function, which is defined such that

$$\int_{-\infty}^{\infty} dx\, g(x/a)\, t(x) = |a| \int_{-\infty}^{\infty} dx'\, g(x')\, t(ax')\,, \tag{3.112}$$

where $g(x)$ is a generalized function, $g(x/a)$ is its scaled counterpart and $t(x)$ is a good function. This definition is not directly applicable to the integral in (3.110) since the complex exponential is not a good function, but we can construct a good function such as $\exp[-2\pi i\xi x - (x/L)^2]$ and pass to the limit $L \to \infty$, yielding (3.111). Alternatively, we can get (3.111) from (3.112) by appealing to (3.87).

Powers of the Fourier operator Application of the inverse Fourier transform to $F(\xi)$ yields $f(x)$, but what happens if we take a Fourier *transform*, not the inverse transform, of $F(\xi)$? It is straightforward to show from basic definitions that

$$[\mathcal{F}F](x) = [\mathcal{F}\mathcal{F}f](x) = f(-x) . \tag{3.113}$$

Thus the square of the Fourier operator is the coordinate-inversion operator which maps $f(x)$ to $f(-x)$. If $f(x) = f(-x)$, there is no distinction between $\mathcal{F}$ and $\mathcal{F}^{-1}$. Since coordinate inversion applied twice returns us to the original function, the fourth power of the Fourier operator is the identity operator.

3.3.6 Convolution and correlation

We briefly encountered the mathematical operation of convolution in Chap. 2 (Sec. 2.2.2), and we shall have many occasions to use it later in this book. As we shall see in Chap. 7, convolution is a common and fruitful model for imaging systems. Systems for which the input–output relation is a convolution are called *linear, shift-invariant systems.* In Chap. 7, we shall examine in detail the conditions under which the convolution model is applicable and develop alternative mathematical descriptions when it is not. Here we confine our attention to some basic mathematical properties of the convolution operation, especially its Fourier transform. Also included here is a brief discussion of the closely related operation of correlation.

Definition of convolution The convolution of two functions $f(x)$ and $h(x)$, denoted $[f * h](x)$ or simply $f * h$, is defined by

$$[f * h]\,(x) = \int_{-\infty}^{\infty} dx' f(x')\, h(x - x') = \int_{-\infty}^{\infty} dx'' f(x - x'')\, h(x'') . \tag{3.114}$$

The second form follows from the first by the change of variable $x' = x - x''$, so $[f * h](x) = [h * f](x)$.

It can be shown that $[f * h](x)$ is bounded for all x (*i.e.*, is in $\mathbb{L}_\infty$) if $f(x)$ and $h(x)$ are both in $\mathbb{L}_2(-\infty, \infty)$. Moreover, if one function, say $f(x)$, is in $\mathbb{L}_1(-\infty, \infty)$ and the other is in $\mathbb{L}_p(-\infty, \infty)$, then $[f * h](x)$ is also in $\mathbb{L}_p(-\infty, \infty)$. In this sense, convolution with an $\mathbb{L}_1$ function or between two $\mathbb{L}_2$ functions is always possible (Richards and Youn, 1990, p. 128).

There are several ways to look at the operation of convolution. If we regard $h(x)$ as a fixed kernel, then convolution is a linear integral transform mapping $f(x)$ to a new function $[f * h](x)$. The problem with this view is that it overlooks the essential symmetry between $f(x)$ and $h(x)$. We could equally well fix $f(x)$ and regard the convolution as a transform mapping $h(x)$ to $[f*h](x)$. The convolution is really a *bilinear* transform with two inputs, $f(x)$ and $h(x)$, and one output, $[f*h](x)$. Still another way to look at convolution is as a linear functional mapping either $f(x)$ or $h(x)$ to a number, $[f * h](x)$, which happens to depend on the continuous variable x.

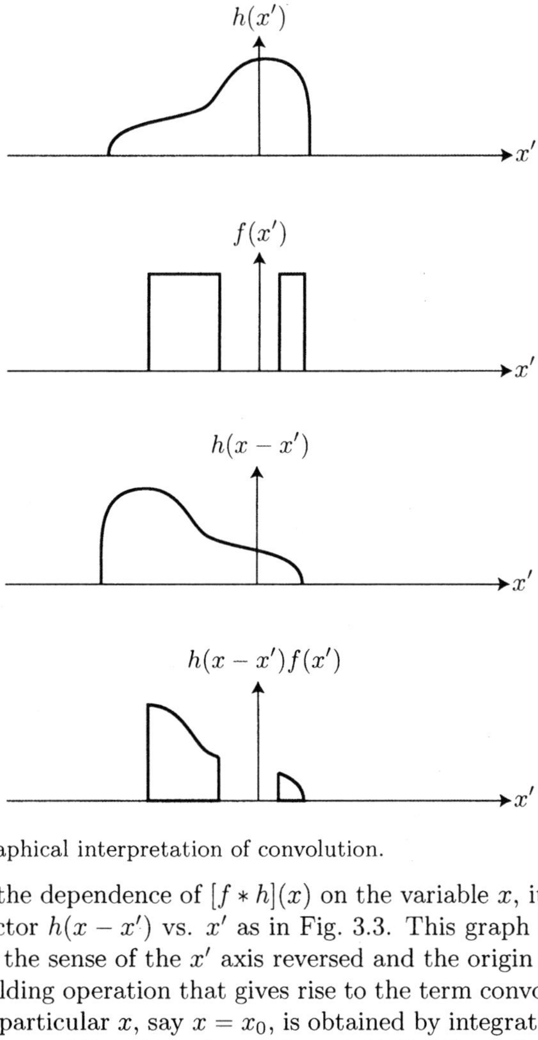

Fig. 3.3 Graphical interpretation of convolution.

To envision the dependence of $[f * h](x)$ on the variable x, it is useful to draw a graph of the factor $h(x - x')$ vs. x' as in Fig. 3.3. This graph looks like a graph of $h(x')$ but with the sense of the x' axis reversed and the origin shifted by x. It is this reversal or folding operation that gives rise to the term convolution. The value of $[f * h](x)$ for a particular x, say $x = x_0$, is obtained by integrating the product of the shifted and reversed function $h(x_0 - x')$ and $f(x')$. The full function $[f * h](x)$ is obtained by repeating the process for all values of the shift variable x. For a lucid introduction to convolution with copious examples, see Gaskill (1978).

Notation A common alternative notation for the convolution is $f(x) * h(x)$ instead of $[f * h](x)$. Since x is the shift variable, and hence the *only* variable of which $f * h$ is a function, the notation $f(x) * h(x)$ would appear to be redundant. Nevertheless, expressions of the form $f(x_1) * h(x_2)$ can often be useful. For example, suppose we wanted to convolve the shifted function $f(x - x_0)$ with $h(x)$. We could define $f_{x_0}(x) = f(x - x_0)$ and write the convolution as $[f_{x_0} * h](x)$ or $f_{x_0}(x) * h(x)$, but the more straightforward notation is $f(x - x_0) * h(x)$. In practice, this notation will not lead to difficulty since convolution commutes with translation [see (3.121) below].

Scale factors are a bit more problematical. If we wrote an expression like $f(x/a) * h(x/b)$, we would not know how to convert it to an integral. Is the shift variable x, x/a or x/b? Even $f(x/a) * h(x/a)$ is inherently ambiguous and best avoided [see, *e.g.*, (3.143) below]. We shall, however, stipulate that expressions like $f(x) * h(x/b)$ are allowed and that the shift variable will be the unscaled one,

x rather than x/b. With these caveats, we shall use $f(x) * h(x)$ and $[f * h](x)$ interchangeably.

Correlation Correlation (sometimes called cross-correlation) is a bilinear transform like convolution but without the reversal. It is defined by

$$[f \star h](x) = \int_{-\infty}^{\infty} dx' f(x + x')\, h(x') = \int_{-\infty}^{\infty} dx'' f(x'')\, h(x'' - x)\,. \tag{3.115}$$

The notation $f(x) \star h(x)$ is also common. Note that, unlike convolution, the order of the functions matters in correlation, and $f \star h$ does not necessarily equal $h \star f$.

The relation between convolution and correlation is often written

$$f(x) \star h(x) = f(x) * h(-x)\,, \tag{3.116}$$

but this notation must be used with caution. The right-hand side really means $[f * h_r](x)$, where $h_r(x) \equiv h(-x)$. The shift variable in the convolution is x, not $-x$.

Basic properties of convolutions As already stated, convolution is commutative and linear in both inputs, *i.e.*,

$$f * g = g * f\,, \tag{3.117}$$

$$f * [\alpha h_1 + \beta h_2] = \alpha f * h_1 + \beta f * h_2\,, \tag{3.118}$$

$$[\alpha f_1 + \beta f_2] * h = \alpha f_1 * h + \beta f_2 * h\,, \tag{3.119}$$

where α and β are arbitrary complex numbers. Convolution is also associative,

$$f * [g * h] = [f * g] * h\,, \tag{3.120}$$

which can be shown by writing out both sides as a double integral.

It is straightforward to show that convolution commutes with translation and differentiation (Richards and Youn, 1990, p. 23):

$$f(x - x_0) * h(x) = f(x) * h(x - x_0) = [f * h]\,(x - x_0)\,, \tag{3.121}$$

$$\frac{d}{dx}\,[f(x) * g(x)] = f'(x) * g(x) = f(x) * g'(x)\,. \tag{3.122}$$

An interesting observation that follows from this last equation is that $f * g$ is differentiable if *either* f or g is differentiable. Convolution is a smoothing operation in this sense (Richards and Youn, 1990, pp. 23–24).

Convolutions involving generalized functions Convolutions are bilinear transforms with two input functions; in terms of distribution theory, either or both of these functions could be a test function, a good function or a generalized function. We shall examine each of these possibilities in turn.

If both f and g are test functions (infinitely differentiable functions of compact support), so is $f * g$. This statement follows from the comments above on differentiability and support; the differentiability of f or g ensures the differentiability of $f * g$, while the support of $f * g$ is the sum of the supports of f and g, so $f * g$ must be a function of compact support if f and g are. Thus the convolution of two test functions is a test function. Likewise, the convolution of two good functions or a test function and a good function is a good function.

The convolution of a test function or a good function with an appropriate generalized function presents no special difficulties (see Richards and Youn, 1990, p. 25, or Strichartz, 1994, p. 52). If $t(x)$ is a test function and $g(x)$ is a generalized function as defined by (3.81), then

$$t(x) * g(x) = \int_{-\infty}^{\infty} dx' g(x')\, t(x - x') = \Phi_g\{t(x - x')\}\,. \tag{3.123}$$

Since $t(x - x')$ is a test function if $t(x)$ is, the distribution on the right is well defined. The analogous result holds for good functions and tempered distributions.

As a simple but important example, let $g(x) = \delta(x)$ and $t(x)$ be a good function. Then

$$t(x) * \delta(x) = \int_{-\infty}^{\infty} dx' \delta(x')\, t(x - x') = t(x)\,. \tag{3.124}$$

Thus convolution with $\delta(x)$ is the identity operator. Similarly,

$$f(x) * \delta(x - x_0) = f(x - x_0)\,, \tag{3.125}$$

so convolution of a good function with $\delta(x - x_0)$ produces a shifted version of the function. Equations (3.124) and (3.125) provide an illustration of the statement above that convolution commutes with translation.

Another important example of convolution with a generalized function involves the derivative of a delta function. Either from the definition of that generalized function or from (3.122) and (3.125), we see that

$$f(x) * \delta'(x - x_0) = \int_{-\infty}^{\infty} dx' f(x')\, \delta'(x - x_0 - x') = f'(x - x_0)\,. \tag{3.126}$$

Note that there is no overall minus sign on the right-hand side, even though there is one in the basic sifting property of $\delta'(x)$; the minus sign disappears when we make a change of variables to put the integral in (3.126) into the form of (3.89).

Convolution of two generalized functions is a more delicate topic. To discuss it, we must introduce the idea of support of a generalized function (Richards and Youn, 1990, p. 25). The intuitive notion is that some generalized functions like the delta function are zero outside a compact set, while others like the step function are not. To get a more formal definition, consider a test function with support entirely *outside* the interval $[a, b]$. If the action of a distribution Φ_g on this test function is identically zero, we say that the generalized function $g(x)$ has support *inside* $[a, b]$. For example, for all test functions that vanish in $[-\epsilon, \epsilon]$, where ϵ is a small positive number, the distribution associated with $\delta(x)$ yields zero, so we can say that $\delta(x)$ has support $[-\epsilon, \epsilon]$. If the support defined this way is finite, we say that the generalized function has compact support.

An important property of generalized functions of compact support is that they can always be represented as the limit of a sequence of test functions, which allows us to define convolutions involving such functions. Richards and Youn (pp. 29 – 32) use this theorem as the starting point to show that all of the basic properties of convolutions discussed above, (3.117) – (3.122), hold even if both functions are generalized functions of compact support. For example,

$$\delta(x - x_1) * \delta(x - x_2) = \delta(x - x_1 - x_2) \tag{3.127}$$

and

$$\delta^{(m)}(x) * \delta^{(n)}(x) = \delta^{(m+n)}(x)\,, \tag{3.128}$$

where $\delta^{(m)}(x)$ is the mth derivative of a delta function.

Richards and Youn state (p. 29) that no one has yet given a definition of the convolution of two generalized functions that do not have compact support, and they speculate that there probably is none.

Fourier transforms of convolutions and correlations The Fourier transform of a convolution $[f * h](x)$ will turn out to play a pivotal role in the analysis of linear, shift-invariant imaging systems. We shall compute the transform indirectly by writing the convolution in a form that looks like the inverse transform of something.

Representing each of the functions $f(x)$ and $h(x)$ by its Fourier transform, we obtain

$$[f * h]\,(x) = \int_{-\infty}^{\infty} dx' f(x')\, h(x - x')$$

$$= \int_{-\infty}^{\infty} dx' \int_{-\infty}^{\infty} d\xi \int_{-\infty}^{\infty} d\xi' F(\xi)\, e^{2\pi i \xi x'}\, H(\xi')\, e^{2\pi i \xi'(x-x')}\,. \tag{3.129}$$

We shall assume that $f(x)$ and $h(x)$ are sufficiently well behaved that we can interchange the order of integration. This assumption holds for test functions, good functions and $\mathbb{L}_2$ functions, for example, but it also holds for generalized functions of compact support since they are the limit of a sequence of test functions. With this assumption, we can write

$$[f * h]\,(x) = \int_{-\infty}^{\infty} d\xi \int_{-\infty}^{\infty} d\xi' F(\xi)\, H(\xi')\, e^{2\pi i \xi' x} \int_{-\infty}^{\infty} dx' e^{2\pi i(\xi - \xi')x'}\,. \tag{3.130}$$

The integral over x' yields $\delta(\xi - \xi')$, so

$$[f * h]\,(x) = \int_{-\infty}^{\infty} d\xi \int_{-\infty}^{\infty} d\xi' F(\xi)\, H(\xi')\, e^{2\pi i \xi' x} \delta(\xi - \xi')$$

$$= \int_{-\infty}^{\infty} d\xi F(\xi)\, H(\xi)\, e^{2\pi i \xi x} = \mathcal{F}^{-1}\left\{F(\xi)\, H(\xi)\right\}\,. \tag{3.131}$$

Thus

$$\mathcal{F}\{[f * h]\,(x)\} = F(\xi)\, H(\xi)\,. \tag{3.132}$$

This result is often referred to as the *convolution theorem*. In words, the Fourier transform of the convolution of two functions is the product of the transforms of the two functions. The functions in question can be any functions for which both sides of the equation are defined, so generalized functions of compact support are allowed. One of the two functions can even be a generalized function of infinite support, but problems arise if both are.

An interesting point follows by combining the convolution theorem with the central-ordinate theorem, (3.104). If $g(x) = [f * h](x)$, then its integral is given by

$$\int_{-\infty}^{\infty} dx\, g(x) = G(0) = F(0)\, H(0) = \int_{-\infty}^{\infty} dx\, f(x) \int_{-\infty}^{\infty} dx\, h(x)\,. \tag{3.133}$$

Thus the infinite integral of a convolution is the product of the integrals of the two functions being convolved.

From (3.116) and (3.132), we find that the Fourier transform of a correlation is

$$\mathcal{F}\{[f \star h](x)\} = F(\xi)\,H(-\xi) = F(\xi)\,H^*(\xi)\,, \tag{3.134}$$

where the last form is valid only if $h(x)$ is real [see (3.94)]. If $h(x)$ is even, there is no difference between correlation and convolution, and (3.132) and (3.134) are identical.

In incoherent optical imaging, the complex autocorrelation will turn out to be a very important function (Papoulis, 1968; Gaskill, 1978). It is defined by (3.115) with $h(x) = f^*(x)$, and its Fourier transform is given by

$$\mathcal{F}\{[f \star f^*](x)\} = |F(\xi)|^2\,. \tag{3.135}$$

3.3.7 Fourier transforms of some special functions

Useful tables of Fourier transforms are found in Campbell and Foster (1948), Magnus and Oberhettinger (1949), Erdélyi (1954), Papoulis (1962), Bracewell (1965), and Gaskill (1978). We discuss here a few transforms that we shall need frequently in this book. We shall also use the opportunity to discuss further the basic Fourier theorems given above.

Rect and sinc functions The rect and sinc functions were defined in Chap. 2 [see (2.9) and (2.43)]. An elementary integration shows that these functions form a Fourier transform pair and, in fact, we have already implicitly used this fact in (2.44) and (2.45). The explicit relations are:

$$\mathcal{F}\{\text{rect}(x/L)\} = L\,\text{sinc}(L\xi)\,, \tag{3.136}$$

$$\mathcal{F}\{\text{sinc}(x/L)\} = L\,\text{rect}(L\xi)\,. \tag{3.137}$$

This Fourier-transform pair provides a nice illustration of many of the properties of Fourier transforms discussed above. For example, note that both the transform and the inverse transform of a rect is a sinc, and vice versa. This symmetry between forward and inverse transforms holds because both the sinc and rect are real, even functions.

Note also that the rect function is discontinuous and hence its first derivative is not bounded. Its transform, the sinc function, falls off only as $1/\xi$ as $\xi \to \pm\infty$, in accordance with (3.107), which is valid for $q = 0$ but not $q = 1$.

The central-ordinate theorem of (3.104) can also be verified here. The integral of $\text{rect}(x/L)$ is trivially found to be L, and $L\ \text{sinc}(L\xi)$ evaluated at $\xi = 0$ is also L since $\sin(u)/u \to 1$ as $u \to 0$. Thus the integral of $f(x) = \text{rect}(x/L)$ is the same as $F(0)$ as required by (3.104).

One way to remember the normalizations of the rect and sinc functions is to note that $\text{rect}(x)$ and $\text{sinc}(x)$, without the scale factors of L, have unit integrals, *i.e.*,

$$\int_{-\infty}^{\infty} dx\,\text{rect}(x) = \int_{-\infty}^{\infty} dx\,\text{sinc}(x) = 1\,. \tag{3.138}$$

By a simple change of variables, the integrals of $\text{rect}(x/L)$ and $\text{sinc}(x/L)$ are each L, and the Fourier transform of each function evaluated at the origin is thus also L. Hence the normalized functions $L^{-1}\,\text{rect}(x/L)$ and $L^{-1}\,\text{sinc}(x/L)$ integrate to unity over $(-\infty, \infty)$ and their Fourier transforms are unity at the origin.

Triangle functions The triangle function is defined as

$$\operatorname{tri}\left(\frac{x}{L}\right) \equiv \begin{cases} 1 - |x|/L & \text{if } |x| \leq L \\ 0 & \text{if } |x| > L \end{cases} . \tag{3.139}$$

Like the rect and sinc functions, $\operatorname{tri}(x)$ without the $1/L$ scaling has unit integral when integrated from $-\infty$ to ∞.

To find the Fourier transform of a triangle function, we recognize that it is the convolution of a rect with itself, *i.e.*,

$$\operatorname{tri}(x) = \operatorname{rect}(x) * \operatorname{rect}(x)\,, \tag{3.140}$$

a relation that follows from the graphical interpretation of convolution. Hence, by (3.132) and (3.136),

$$\mathcal{F}\{\operatorname{tri}(x)\} = \operatorname{sinc}^2(\xi)\,. \tag{3.141}$$

From the scaling property of Fourier transforms, (3.111), we have

$$\mathcal{F}\{\operatorname{tri}(x/L)\} = L\operatorname{sinc}^2(L\xi)\,. \tag{3.142}$$

This development affords us an opportunity to comment on a potential pitfall with the shorthand notation for convolutions. It might be tempting to substitute x/L for x in (3.140), but it would not give the right answer. To discover the form for the convolution of $\operatorname{rect}(x/L)$ with itself, we define $r_L(x) \equiv \operatorname{rect}(x/L)$ and write

$$\begin{aligned} [r_L * r_L](x) &= \int_{-\infty}^{\infty} dx' r_L(x')\, r_L(x - x') \\ &= \int_{-\frac{1}{2}L}^{\frac{1}{2}L} dx' \operatorname{rect}\left[\frac{x - x'}{L}\right] . \end{aligned} \tag{3.143}$$

Note that $r_L(x - x')$ is just $\operatorname{rect}[\frac{x-x'}{L}]$, which results from the substitution of $x - x'$ for x in the definition of $r_L(x)$; it is not the same as $\operatorname{rect}[\frac{x}{L} - x']$ or $\operatorname{rect}[x - \frac{x'}{L}]$, neither of which is even dimensionally consistent.

Now the change of variable $s = x'/L$ gives

$$[r_L * r_L](x) = L \int_{-\frac{1}{2}}^{\frac{1}{2}} ds \operatorname{rect}\left(\frac{x}{L} - s\right) = L\,[r_1 * r_1](x/L)\,, \tag{3.144}$$

where r_1 is r_L for $L = 1$, or simply $\operatorname{rect}(x)$. Taking a Fourier transform of (3.144) gives

$$\mathcal{F}\{[r_L * r_L](x)\} = L^2 \operatorname{sinc}^2(L\xi)\,, \tag{3.145}$$

which is not the same as $\mathcal{F}\{\operatorname{tri}(x/L)\}$ in (3.142) because of the extra L. If we now insist on going back to the rect notation, we are forced to write (see Gaskill, 1978, p. 166)

$$\operatorname{rect}(x/L) * \operatorname{rect}(x/L) = L\operatorname{tri}(x/L)\,. \tag{3.146}$$

This expression violates an elementary rule for functional equations: If $f(x) = g(x)$ for all x, then $f(x/L) = g(x/L)$. We can either say that a convolution is not a function and this rule doesn't apply or avoid convolution expressions with scale factors by defining auxiliary functions like r_L. We prefer the latter.

As a practical matter, if one has difficulty with scale factors in convolutions or Fourier transforms, dimensional analysis can be used. For example, in (3.145) above, if we assume that x has dimensions of length, then ξ must have dimensions of (length)$^{-1}$ since the argument of a trigonometric function, and hence of a sinc function, must be dimensionless. Similarly, the argument of the rect function must be dimensionless, and hence L must have dimensions of length. The convolution and Fourier-transform operations are both integrals over x, and the dx contributes an additional factor of length. The asterisk $*$ and the Fourier operator $\mathcal{F}$ have the same dimensions as x, at least in one dimension. Thus (3.146) is dimensionally consistent since both sides have dimensions of length, while both sides of (3.145) have dimensions of (length)^2.

The central-ordinate theorem can also help resolve difficulties with scale factors. A graph of the function $\text{tri}(x/L)$ is an isosceles triangle with height unity and base $2L$, so its area is L. The central-ordinate theorem applied to (3.142) also shows that the integral of $\text{tri}(x/L)$ is L since $\text{sinc}^2(0) = 1$.

The asymptotic properties of the Fourier transform of a triangle function are worthy of note. While a rect function is discontinuous and has a transform that falls off as $1/\xi$, a triangle is continuous but has a discontinuous derivative. It can be differentiated once but not twice, and its transform falls off as $1/\xi^2$, consistent with (3.107) for $q = 1$. As noted above, a convolution is a smoothing operation which increases the differentiability of the function and thus the rate of decay of its transform.

Delta functions, sines and cosines We have discussed various Fourier transforms involving delta functions above, but the main results are collected here for ease of reference:

$$\mathcal{F}\{\delta(x)\} = 1\,, \tag{3.147}$$

$$\mathcal{F}\{\delta(x - x_0)\} = e^{-2\pi i x_0 \xi}\,, \tag{3.148}$$

$$\mathcal{F}\left\{\delta^{(k)}(x)\right\} = (2\pi i \xi)^k\,, \tag{3.149}$$

$$\mathcal{F}\left\{\delta^{(k)}(x - x_0)\right\} = (2\pi i \xi)^k\, e^{-2\pi i x_0 \xi}\,, \tag{3.150}$$

where $\delta^{(k)}(x)$ is the kth derivative of $\delta(x)$.

Since a shifted delta function transforms into a linear phase factor (or complex exponential) as in (3.148), the converse also holds:

$$\mathcal{F}\{e^{2\pi i \xi_0 x}\} = \delta(\xi - \xi_0) \tag{3.151}$$

and

$$\mathcal{F}\{(-2\pi i x)^k\, e^{2\pi i \xi_0 x}\} = \delta^{(k)}(\xi - \xi_0)\,. \tag{3.152}$$

A linear phase factor transforms to a single delta function, which can be paraphrased by saying it consists of a single spatial-frequency component.

The linear phase factors can be used to construct sines and cosines via (3.5) and (3.6), and the following transforms are obtained:

$$\mathcal{F}\{\cos(2\pi\xi_0 x)\} = \frac{1}{2}\left[\delta(\xi - \xi_0) + \delta(\xi + \xi_0)\right]\,, \tag{3.153}$$

$$\mathcal{F}\{\sin(2\pi\xi_0 x)\} = \frac{1}{2i}\left[\delta(\xi - \xi_0) - \delta(\xi + \xi_0)\right]\,. \tag{3.154}$$

Sines and cosines transform to pairs of delta functions. First appearances aside, these functions consist of two spatial frequencies each; positive and negative frequencies must be considered distinct. For the cosine, the weights of the two delta functions are real and equal, while for the sine they are pure imaginary and differ by a minus sign.

Sines and cosines are infinitely differentiable and their transforms vanish identically at $\xi = \pm\infty$. Equation (3.107) holds for all q.

Comb function The comb function is defined as an infinite sum of delta functions in (2.48), but (2.50) shows that it is also an infinite sum of linear phase factors. Hence its Fourier transform is

$$\mathcal{F}\{\text{comb}(x)\} = \sum_{n=-\infty}^{\infty} \mathcal{F}\{e^{2\pi inx}\} = \sum_{n=-\infty}^{\infty} \delta(\xi - n) = \text{comb}(\xi)\,. \tag{3.155}$$

Thus the Fourier transform of a comb is a comb. The scaled version of a comb function is

$$\text{comb}\left(\frac{x}{L}\right) = \sum_{n=-\infty}^{\infty} \delta\left(\frac{x}{L} - n\right) = L \sum_{n=-\infty}^{\infty} \delta(x - nL)\,, \tag{3.156}$$

where the last step follows from (2.28). Thus $\text{comb}(x/L)$ is a sum of delta functions of weight L at the points $x = nL$, $n = 0, \pm1, \pm2, ..., \pm\infty$. The Fourier transform of the scaled comb is

$$\mathcal{F}\{\text{comb}(x/L)\} = L\ \text{comb}(L\xi) = L \sum_{n=-\infty}^{\infty} \delta(L\xi - n) = \sum_{n=-\infty}^{\infty} \delta\left(\xi - \frac{n}{L}\right), \tag{3.157}$$

which is a sum of delta functions of unit weight at points $\xi = n/L$.

It is worth looking at these results from the viewpoint of dimensional analysis. In (3.155) x must be dimensionless since otherwise the exponent $2\pi inx$ would be meaningless. Hence ξ is also dimensionless, as are the delta functions, and dimensional analysis gets us nowhere. In (3.156) and (3.157) on the other hand, x and L can be physical lengths. In this case ξ has dimensions of $(\text{length})^{-1}$, $\delta(\xi - n/L)$ has dimensions of length and $\delta(x - nL)$ has dimensions of $(\text{length})^{-1}$. (See Sec. 2.4.3 for a discussion of dimensional analysis of delta functions.) All forms in (3.156) are thus dimensionless, while all forms in (3.157) have dimensions of length.

Step and signum functions In Sec. 2.3.2 we discussed two generalized functions related to the delta function, namely the step and signum functions. An important property of $\text{step}(x)$ is that its derivative is $\delta(x)$. Thus we know at once from (3.98) that

$$\mathcal{F}\left\{\frac{d}{dx}\,\text{step}(x)\right\} = 2\pi i\xi\,\mathcal{F}\{\,\text{step}(x)\} = \mathcal{F}\{\delta(x)\} = 1\,. \tag{3.158}$$

One might be tempted to conclude from this result that $\mathcal{F}\{\text{step}(x)\} = 1/(2\pi i\xi)$, but this is not correct. For one thing, we wouldn't know how to interpret the singularity $1/\xi$. For another, specifying the derivative of a function does not uniquely specify the function itself or its Fourier transform; we must account for the constant of integration.

To fix the constant, let us retreat to basic definitions. From (3.85), we have

$$\int_{-\infty}^{\infty} dx\, \text{step}(x)\, t(x) = \int_{-\infty}^{\infty} d\xi\, [\mathcal{F}\{\text{step}(x)\}]^*\, T(\xi)\,, \tag{3.159}$$

where $t(x)$ is any good function and $T(\xi)$ is its Fourier transform. Expressing $t(x)$ on the left-hand side in terms of $T(\xi)$ and using the definition of the step function gives

$$\int_{-\infty}^{\infty} d\xi\,[\mathcal{F}\{\text{step}(x)\}]^*\,T(\xi) = \int_0^{\infty} dx \int_{-\infty}^{\infty} d\xi\,T(\xi)\,e^{2\pi i\xi x}$$
$$= \int_{-\infty}^{\infty} d\xi\,T(\xi) \int_0^{\infty} dx\,e^{2\pi i\xi x}\,, \tag{3.160}$$

where the interchange of order of integration is legal since $T(\xi)$ is a good function. Comparison of the first and last integrals now reveals that

$$\mathcal{F}\{\text{step}(x)\} = \int_0^{\infty} dx\,e^{-2\pi i\xi x}\,, \tag{3.161}$$

which we might have written immediately if we had treated the step function as an ordinary function instead of defining it distributionally.

All that remains is to deal with the improper integral in (3.161), for which purpose we appeal to (3.88). Introducing the convergence factor $\exp(-x/L)$ in order to generate an integrable sequence, we find

$$\mathcal{F}\{\text{step}(x)\} = \lim_{L\to\infty} \int_0^{\infty} dx\,e^{-2\pi i\xi x - x/L} = \lim_{L\to\infty} \frac{L}{1+2\pi i\xi L}$$
$$= \lim_{L\to\infty} \frac{L}{1+4\pi^2\xi^2L^2} - \lim_{L\to\infty} \frac{2\pi i\xi L^2}{1+4\pi^2\xi^2L^2}\,. \tag{3.162}$$

Letting $\epsilon = 1/(2\pi L)$ and comparing to (2.39), we recognize the first limit in (3.162) as $\frac{1}{2}\,\delta(\xi)$. The second limit is a representation of $\mathcal{P}\{1/(2\pi i\xi)\}$, where $\mathcal{P}$ denotes Cauchy principal value [see (2.88)]. Thus we have, finally,

$$\mathcal{F}\{\text{step}(x)\} = \tfrac{1}{2}\,\delta(\xi) + \mathcal{P}\left\{\frac{1}{2\pi i\xi}\right\}. \tag{3.163}$$

Note that this result is consistent with (3.158) since $\xi\,\delta(\xi) = 0$. The constant of integration turned out to be $\frac{1}{2}$, which gave $\frac{1}{2}\,\delta(\xi)$ in the Fourier domain.

The Fourier transform of the signum function can now be derived. Since $\text{sgn}(x) = 2\,\text{step}(x) - 1$ by (2.77), we have

$$\mathcal{F}\{\text{sgn}(x)\} = \mathcal{P}\left\{\frac{1}{i\pi\xi}\right\}. \tag{3.164}$$

We can verify this transform by taking the inverse, *i.e.*, by performing the integral

$$\mathcal{F}^{-1}\left\{\mathcal{P}\left(\frac{1}{i\pi\xi}\right)\right\} = \mathcal{P}\int_{-\infty}^{\infty} d\xi\,\frac{e^{2\pi i\xi x}}{i\pi\xi}\,. \tag{3.165}$$

This integral is a standard exercise in complex integration (see App. B), with the effect of the principal-value designation being that only half of the residue at $\xi = 0$ contributes (Friedman, 1991). The result is

$$\mathcal{F}^{-1}\left\{\mathcal{P}\left(\frac{1}{i\pi\xi}\right)\right\} = \text{sgn}(x)\,, \tag{3.166}$$

as expected. The dual to this transform is

$$\mathcal{F}\left\{\mathcal{P}\left(\frac{1}{x}\right)\right\} = -i\pi\,\text{sgn}(\xi)\,. \tag{3.167}$$

A derivation of this result by contour integration is given in App. B.

Powers and logarithms The generalized functions x^m (m an integer), $|x|^\alpha$ (α any real number not an integer) and $\ln|x|$ are defined in Sec. 2.3.3. These functions play an important role in tomographic imaging. Their Fourier transforms are derived from the basic definitions in Lighthill (1958). The results are

$$\mathcal{F}\{|x|^\alpha\} = \left\{2\cos\left[\tfrac{1}{2}\pi(\alpha+1)\right]\right\} \alpha!(2\pi|\xi|)^{-\alpha-1}, \qquad \alpha \text{ real, noninteger}, \tag{3.168}$$

$$\mathcal{F}\{x^{-m}\} = -\pi i\, \frac{(-2\pi i\xi)^{m-1}}{(m-1)!}\, \mathrm{sgn}(\xi), \qquad m > 0, \text{ integer}, \tag{3.169}$$

$$\mathcal{F}\{x^m\} = (-2\pi i)^{-m}\, \delta^{(m)}(\xi), \qquad m \geq 0, \text{ integer}, \tag{3.170}$$

$$\mathcal{F}\{x^m \ln|x|\} = \pi i\, \frac{m!}{(2\pi\xi)^{m+1}}\, \mathrm{sgn}(\xi), \qquad m \geq 0, \text{ integer}. \tag{3.171}$$

In these equations, $m!$ denotes the ordinary factorial function for the integer m, while $\alpha!$ is defined for noninteger α in terms of the gamma function [see Abramowitz and Stegun, 1965, or (C.123) in App. C] as $\alpha! = \Gamma(\alpha+1)$. Transforms involving products of these generalized functions and signum and step functions can be found in Lighthill.

An interesting special case of (3.168) is

$$\mathcal{F}\left\{|x|^{-\frac{1}{2}}\right\} = |\xi|^{-\frac{1}{2}}. \tag{3.172}$$

Thus $|x|^{-\frac{1}{2}}$ is its own Fourier transform.

Gaussians and quadratic phase factors The Gaussian function is important in probability theory and as a model for the point spread function of imaging systems. It is also the prototype of a good function (see Sec. 2.1.2). One convenient form for the Gaussian is (Gaskill, 1978)

$$\mathrm{gaus}(x) \equiv \exp(-\pi x^2). \tag{3.173}$$

An advantage of this definition is that

$$\int_{-\infty}^{\infty} dx\, \mathrm{gaus}(x) = 1. \tag{3.174}$$

One way to derive this result, which will be important below, is to compute not the integral, denoted I, but its square:

$$I^2 = \left[\int_{-\infty}^{\infty} dx\, \mathrm{gaus}(x)\right]^2 = \int_{-\infty}^{\infty} dx\, e^{-\pi x^2} \int_{-\infty}^{\infty} dy\, e^{-\pi y^2} = \int_{-\infty}^{\infty} dx \int_{-\infty}^{\infty} dy\, e^{-\pi(x^2+y^2)}. \tag{3.175}$$

In this form the double integral can be interpreted as a 2D integral over the infinite plane. This integral can be performed in polar coordinates; letting $r^2 = x^2 + y^2$, we find

$$I^2 = 2\pi \int_0^{\infty} r\, dr\, e^{-\pi r^2} = \int_0^{\infty} du\, e^{-u} = 1, \tag{3.176}$$

confirming (3.174).

The Fourier transform of gaus(x) is given by

$$\mathcal{F}\{\mathrm{gaus}(x)\} = \int_{-\infty}^{\infty} dx\, e^{-2\pi i\xi x - \pi x^2}. \tag{3.177}$$

The integral will be evaluated by forcing it to look like (3.174). To do so, we complete the square by adding and subtracting $\pi\xi^2$ in the exponent, yielding

$$\mathcal{F}\{\text{gaus}(x)\} = e^{-\pi\xi^2}\int_{-\infty}^{\infty} dx\, e^{-\pi(x+i\xi)^2} = e^{-\pi\xi^2}\int_{C_1} dz\, e^{-\pi z^2}. \tag{3.178}$$

where z is a complex variable and the (open) contour integral is along the path C_1 parallel to the real axis and displaced upward by $i\xi$; for points on this contour, $z = x + i\xi$. Note, however, that $\exp(-\pi z^2)$ is entire (analytic for all finite z) and vanishes at infinity on both C_1 and the real axis. We can thus deform the contour to be the real axis, denoted C_2, without changing the value of the integral (see App. B), so

$$\mathcal{F}\{\text{gaus}(x)\} = e^{-\pi\xi^2}\int_{C_2} dz\, e^{-\pi z^2} = e^{-\pi\xi^2}\int_{-\infty}^{\infty} dx\, e^{-\pi x^2} = e^{-\pi\xi^2}, \tag{3.179}$$

where the last step follows from (3.174). Like the function $|x|^{-\frac{1}{2}}$ encountered above, gaus(x) and its Fourier transform have the same form.

By the scaling theorem, (3.111), we now have

$$\mathcal{F}\left\{|a|^{-1}\text{gaus}(x/a)\right\} = e^{-\pi a^2\xi^2}, \tag{3.180}$$

where a is real. It follows from this equation and the central ordinate theorem that $|a|^{-1}\text{gaus}(x/a)$ integrates to unity over $(-\infty, \infty)$.

Closely related to the Gaussian is the quadratic phase factor $\exp(i\pi\beta x^2)$, where β is real. In fact, one might be tempted to regard this as a Gaussian of the form gaus(bx), where $b = \sqrt{i\beta}$, but that view is dangerous. While gaus(bx) for b real is very well behaved (in fact, the prototype of a good function), $\exp(i\pi\beta x^2)$ for β real is not even in $\mathbb{L}_1(-\infty, \infty)$. A safer procedure is to regard the quadratic phase factor as a generalized function and compute its transform from first principles.

As in (3.88), we shall regard the generalized function $\exp(i\pi\beta x^2)$ as the limit of a sequence of good functions. A convenient form for the good functions is $\exp(i\pi\beta x^2 - \pi\alpha x^2)$, where α is real and positive. We then have

$$\mathcal{F}\left\{e^{i\pi\beta x^2}\right\} = \lim_{\alpha\to 0}\int_{-\infty}^{\infty} dx\, e^{i\pi\beta x^2 - \pi\alpha x^2 - 2\pi i\xi x}. \tag{3.181}$$

As above, the approach is to complete the square. A little algebra yields

$$\mathcal{F}\left\{e^{i\pi\beta x^2}\right\} = \lim_{\alpha\to 0} e^{\pi b^2}\int_{-\infty}^{\infty} dx\, e^{-\pi(\gamma x+b)^2}, \tag{3.182}$$

where $b = i\xi/\gamma$ and $\gamma = \sqrt{\alpha - i\beta}$. Note that γ is a complex number with phase $\phi_\gamma = \frac{1}{2}\tan^{-1}(-\beta/\alpha)$, which approaches $\pi/4$ as $\alpha \to 0$. The change of variables $z = \gamma x + b$ allows us to write

$$\mathcal{F}\{e^{i\pi\beta x^2}\} = \lim_{\alpha\to 0} e^{\pi b^2}\frac{1}{\gamma}\int_C dz\, e^{-\pi z^2}, \tag{3.183}$$

where the contour C makes an angle ϕ_γ to the real axis. The key point, however, is that there are no singularities of the integrand anywhere between C and the real axis, and the integrand vanishes rapidly at infinity along both C and the real axis. Therefore, we can again deform the contour to the real axis and write

$$\mathcal{F}\{e^{i\pi\beta x^2}\} = \lim_{\alpha\to 0} e^{\pi b^2}\frac{1}{\gamma}\int_{-\infty}^{\infty} dx\, e^{-\pi x^2} = \lim_{\alpha\to 0} e^{\pi b^2}\frac{1}{\gamma}. \tag{3.184}$$

Passing to the limit, we have, finally

$$\mathcal{F}\{e^{i\pi\beta x^2}\} = \sqrt{\frac{i}{\beta}}\, e^{-i\pi\xi^2/\beta}. \tag{3.185}$$

The exponents in this expression are the same as we would have obtained by use of (3.180) with complex a, but that *ad hoc* approach would have left us unsure how to handle the factor $|a|$.

The quadratic phase factor and its Fourier transform play a crucial role in diffraction theory (see Chap. 9) and radar signal processing (Chap. 18). Additional properties are discussed in Sec. 4.3.

3.3.8 Relation between Fourier series and Fourier transforms

Most treatments of Fourier analysis introduce the Fourier series as a way of representing a periodic function by a discrete sum of complex exponentials. The Fourier transform is then used to represent a general, nonperiodic function by a continuous superposition or integral of complex exponentials. In a classical approach it would not be possible to use the Fourier transform for a periodic function which cannot be in $\mathbb{L}_1(-\infty,\infty)$. The use of generalized functions, however, frees us of that restriction and makes it possible to look at the Fourier *transform* of a periodic function.

Consider a periodic function $f(x)$ of bounded variation and period L, so that $f(x+mL) = f(x)$, where m is any integer. Let $f_0(x)$ be identical with $f(x)$ for x in $(-\frac{1}{2}L, \frac{1}{2}L)$ and zero outside this interval, *i.e.*,

$$f_0(x) = f(x)\,\mathrm{rect}(x/L). \tag{3.186}$$

Since $f_0(x)$ has bounded support, it is either a test function or the limit of a sequence of test functions. Then we can write $f(x)$ as

$$f(x) = \sum_{n=-\infty}^{\infty} f_0(x-nL) = f_0(x) * \sum_{n=-\infty}^{\infty} \delta(x-nL), \tag{3.187}$$

where we have used (3.125) and the linearity of the convolution operation. With some trepidation, we rewrite this equation in terms of the shorthand notation for the convolution as

$$f(x) = f_0(x) * \left[\frac{1}{L}\,\mathrm{comb}\left(\frac{x}{L}\right)\right]. \tag{3.188}$$

Recall that we stipulated in Sec. 3.3.6 that, in expressions of this form with one unscaled variable and one scaled one, the shift variable in the convolution would be the unscaled one. Thus the interpretation of $f_0(x) * L^{-1}\,\mathrm{comb}(x/L)$ is that it is the same thing as $f_0(x) * c_L(x)$, where $c_L(x) = L^{-1}\,\mathrm{comb}(x/L)$.

With this notation, we can now take the Fourier transform of $f(x)$ by use of the convolution theorem, (3.132), along with (3.157) for the Fourier transform of a comb. The result is

$$F(\xi) = \mathcal{F}\{f(x)\} = \mathcal{F}\left\{f_0(x) * \left[\frac{1}{L}\operatorname{comb}\left(\frac{x}{L}\right)\right]\right\} = F_0(\xi)\operatorname{comb}(L\xi)\,, \tag{3.189}$$

where $F_0(\xi)$ is the Fourier transform of $f_0(x)$. To be more explicit,

$$F(\xi) = F_0(\xi)\sum_{n=-\infty}^{\infty}\delta(\xi L - n) = \frac{1}{L}\sum_{n=-\infty}^{\infty}F_0\left(\frac{n}{L}\right)\delta\left(\xi - \frac{n}{L}\right), \tag{3.190}$$

where the last form has made use of (2.25) and (2.28).

Equation (3.190) says that the Fourier transform of a periodic function is defined for all ξ but is zero unless $\xi = n/L$, where n is an integer. The behavior at these discrete frequencies is that of a delta function, and the weight of the delta function at $\xi = n/L$ is given by $F_0(\xi)$ sampled (evaluated) at this discrete value. The key point is that the *periodicity of the function f(x) restricts its Fourier transform to a discrete set of frequencies.*

An interesting result is obtained by simply taking the inverse transform of (3.190). We find

$$f(x) = \frac{1}{L}\sum_{n=-\infty}^{\infty}F_0\left(\frac{n}{L}\right)e^{2\pi i n x/L}\,, \tag{3.191}$$

which we recognize as a Fourier *series* with coefficients $L^{-1}F_0(n/L)$. In other words, the *Fourier coefficients of a periodic function are sampled values of the Fourier transform of one period of the function.* Of course, the Fourier-series representation inherently has only a discrete set of frequencies present, and these are the same frequencies, n/L, as selected by the comb of delta functions in the Fourier transform.

Poisson summation formula The starting point for the discussion above was the representation of a periodic function $f(x)$ as the convolution of its values in one period, $f_0(x)$, with a comb function. Now let us consider a slightly different problem. Consider a function $g(x)$ that is *not* periodic and does not necessarily have bounded support. From this function we can *create* a periodic function $g_p(x)$ of period L by forming all possible translated replicas of the original function and adding them together. Mathematically, we define

$$g_p(x) = \sum_{n=-\infty}^{\infty} g(x - nL) = g(x) * \frac{1}{L}\operatorname{comb}\left(\frac{x}{L}\right). \tag{3.192}$$

The important difference between this equation and (3.188) is that here the different terms can overlap spatially.

The new function $g_p(x)$ is periodic with period L and hence can be expanded in a Fourier series:

$$g_p(x) = \sum_{k=-\infty}^{\infty} C_k\, e^{2\pi i k x/L}\,. \tag{3.193}$$

The coefficients are given by

$$C_k = \frac{1}{L} \int_{-\frac{1}{2}L}^{\frac{1}{2}L} dx \sum_{n=-\infty}^{\infty} g(x - nL)\, e^{-2\pi ikx/L} . \tag{3.194}$$

The change of variables $x' = x - nL$ leads to

$$C_k = \frac{1}{L} \sum_{n=-\infty}^{\infty} \int_{-\frac{1}{2}L-nL}^{\frac{1}{2}L-nL} dx' g(x')\, e^{-2\pi ikx'/L}\, e^{-2\pi ink} . \tag{3.195}$$

The last exponential is identically one since nk is an integer, and the sum of integrals over contiguous intervals of width L is equivalent to one integral over $(-\infty, \infty)$. Thus we have

$$C_k = \frac{1}{L} \int_{-\infty}^{\infty} dx' g(x')\, e^{-2\pi ikx/L} = \frac{1}{L} G\left(\frac{k}{L}\right) , \tag{3.196}$$

where, as usual, $G(\xi)$ is the Fourier transform of $g(x)$. Note that, in contrast to (3.190) or (3.191), the transform that appears here is not the transform of one period of a periodic function but rather of the original, nonperiodic function from which we constructed the periodic one, g_p.

Combining (3.192), (3.193) and (3.196), we obtain the *Poisson summation formula*,

$$\sum_{n=-\infty}^{\infty} g(x - nL) = \frac{1}{L} \sum_{k=-\infty}^{\infty} G\left(\frac{k}{L}\right) \exp\left(\frac{2\pi ikx}{L}\right) . \tag{3.197}$$

The special case $x = 0$ gives

$$\sum_{n=-\infty}^{\infty} g(nL) = \frac{1}{L} \sum_{k=-\infty}^{\infty} G\left(\frac{k}{L}\right) . \tag{3.198}$$

This remarkable theorem says that, with suitable scale factors, *the sum of the samples of any function $g(x)$ equals the sum of the samples of its Fourier transform.* Note that L is now simply the sample spacing in x, not a period. The sample spacing in ξ is $1/L$.

If $g(x)$ has compact support, the Poisson summation formula takes a familiar form. Suppose $g(x)$ vanishes outside the range $-\frac{1}{2}L < x < \frac{1}{2}L$. Then the left-hand side of (3.197) is a sum of periodically repeated replicas of $g(x)$, while on the right side, the sampled Fourier transform $G(k/L)$ is identical to G_k, the kth Fourier coefficient of $g(x)$. Thus, when $g(x)$ has support $(-\frac{1}{2}L, \frac{1}{2}L)$, (3.197) is just the usual Fourier-series representation. As it stands, however, (3.197) places no restriction on $g(x)$.

A practical advantage of the Poisson summation formula is that it converts a slowly converging series into a rapidly converging one (Kanwal, 1983). For example, if $g(x)$ is a Gaussian, we find (Strichartz, 1994)

$$\sum_{k=-\infty}^{\infty} e^{-4\pi^2 tk^2} = \frac{1}{\sqrt{4\pi t}} \sum_{n=-\infty}^{\infty} e^{-n^2/4t} . \tag{3.199}$$

The left-hand side converges rapidly for large t, while the right-hand side does so for small t.

3.3.9 Analyticity of Fourier transforms

Up until now, the spatial frequency ξ has been a real variable, but it is also of interest to consider complex frequencies. For this purpose, we write

$$\xi = \xi_r + i\xi_i \,, \tag{3.200}$$

where subscripts r and i denote real and imaginary parts, respectively. The Fourier transform is still defined by (3.60), which now reads

$$F(\xi) = \int_{-\infty}^{\infty} dx\, f(x)\, e^{-2\pi i\xi_r x}\, e^{2\pi\xi_i x} \,. \tag{3.201}$$

It is not obvious that $F(\xi)$ exists for all complex ξ. For example, if $\xi_i > 0$, the factor $\exp(2\pi\xi_i x)$ in the integrand blows up exponentially as $x \to \infty$. This problem is avoided, however, if $f(x)$ is square-integrable and has compact support. In that case, as we shall now show, $F(\xi)$ is an analytic function of the complex variable ξ.

Suppose the support of $f(x)$ is $(-\frac{1}{2}L, \frac{1}{2}L)$ and denote $F(\xi)$ by

$$F(\xi) = F_r(\xi_r, \xi_i) + iF_i(\xi_r, \xi_i) \,, \tag{3.202}$$

where, if $f(x)$ is real,

$$F_r(\xi_r, \xi_i) = \int_{-\frac{1}{2}L}^{\frac{1}{2}L} dx\, f(x) \cos(2\pi\xi_r x)\, e^{2\pi\xi_i x} \,; \tag{3.203}$$

$$F_i(\xi_r, \xi_i) = -\int_{-\frac{1}{2}L}^{\frac{1}{2}L} dx\, f(x) \sin(2\pi\xi_r x)\, e^{2\pi\xi_i x} \,. \tag{3.204}$$

To demonstrate the analyticity of $F(\xi)$, we must show that the Cauchy-Riemann conditions are satisfied (see App. B). If $f(x)$ is in $\mathbb{L}_2(-\frac{1}{2}L, \frac{1}{2}L)$, there is no problem in differentiating under the integral sign, so

$$\frac{\partial F_r}{\partial \xi_r} = \int_{-\frac{1}{2}L}^{\frac{1}{2}L} dx\, f(x)\, (-2\pi x) \sin(2\pi\xi_r x)\, e^{2\pi\xi_i x} = \frac{\partial F_i}{\partial \xi_i} \,, \tag{3.205}$$

verifying the first of the Cauchy-Riemann conditions; the second one follows similarly. The conclusion is that $F(\xi)$ is *entire* (analytic for all finite ξ) if $f(x)$ is square-integrable and has compact support. This statement is frequently referred to as the *Paley-Wiener theorem* (Paley and Wiener, 1934).

The Paley-Wiener theorem can also be derived if $f(x)$ is a generalized function of compact support (Strichartz, 1994), though in this case $F(\xi)$ might grow exponentially as $\xi \to \infty$. The general statement of the theorem is that the Fourier transform of a function or generalized function of compact support is an entire function of exponential type.

The Paley-Wiener theorem should not be surprising when we recall the relation between the smoothness of a function and the asymptotic behavior of its Fourier transform. Compact support is the ultimate in decay at infinity and analyticity is the ultimate in smoothness (Strichartz, 1994).

An interesting consequence of the Paley-Wiener theorem is that the Fourier transform of a function of compact support cannot itself have compact support. That is the case since an analytic function that is zero over any finite region of the complex plane must be zero everywhere (see App. B).

3.3.10 Related transforms

Fourier cosine transform We saw in (3.95) that the Fourier transform of an even function is an even function. If the function is also real, then

$$F(\xi) = F(-\xi) = 2\int_0^\infty dx\, f(x)\cos(2\pi\xi x)\,, \qquad f(x) \text{ real and even}\,. \tag{3.206}$$

Even if $f(x)$ is not even, however, we can use this form as the definition of the *Fourier cosine transform* of $f(x)$. Denoting this transform by $F_c(\xi)$, we write

$$F_c(\xi) \equiv 2\int_0^\infty dx\, f(x)\cos(2\pi\xi x)\,. \tag{3.207}$$

Since the cosine is an even function, it follows at once that

$$F_c(\xi) = F_c(-\xi)\,. \tag{3.208}$$

Since $F_c(x)$ is computed from knowledge of $f(x)$ for $x \geq 0$, and negative x values never enter into (3.207), we cannot expect to recover $f(x)$ for $x < 0$ from $F_c(\xi)$. The inversion formula for $x \geq 0$ follows from (3.153), with the result

$$f(x) = 2\int_0^\infty d\xi\, F_c(\xi)\cos(2\pi\xi x)\,, \qquad x \geq 0\,. \tag{3.209}$$

This result is valid for $x \geq 0$ without any assumptions about the symmetry of $f(x)$. On the other hand, should we ignore the condition on x and evaluate the right-hand side of (3.209) for negative x, we would find that we had a representation of an even function. The situation is analogous to a Fourier series, which can be viewed as a representation of a periodic function for $-\infty \leq x \leq \infty$ or of an arbitrary function for $-\frac{1}{2}L \leq x \leq \frac{1}{2}L$. Equation (3.209) is a representation of an even function for $-\infty \leq x \leq \infty$ or of an arbitrary function for $x \geq 0$.

Fourier sine transform The Fourier sine transform is defined as

$$F_s(\xi) \equiv 2\int_0^\infty dx\, f(x)\sin(2\pi\xi x)\,. \tag{3.210}$$

Since the sine is an odd function,

$$F_s(\xi) = -F_s(-\xi)\,, \tag{3.211}$$

no matter the symmetry of $f(x)$. The inversion formula, which follows from (3.154), is

$$f(x) = 2\int_0^\infty d\xi\, F_s(\xi)\sin(2\pi\xi x)\,, \qquad x \geq 0\,. \tag{3.212}$$

Again, this result is valid for $x \geq 0$ without any assumptions about the symmetry of $f(x)$, but the right-hand side of the equation is an odd function of x. Equation (3.212) is a representation of an odd function for $-\infty \leq x \leq \infty$ or an arbitrary function for $x \geq 0$.

3.4 MULTIDIMENSIONAL FOURIER TRANSFORMS

3.4.1 Basis functions

One-dimensional (1D) Fourier analysis involves the complex-exponential basis function $u_\xi(x) = \exp(2\pi i \xi x)$. To perform a similar analysis for a 2D or 3D function, we must construct 2D and 3D complex exponentials. In 2D a reasonable generalization of $u_\xi(x)$ is

$$u_{\xi\eta}(x, y) \equiv \exp[2\pi i(\xi x + \eta y)] . \tag{3.213}$$

Regarded as a function of x for fixed y, this function is periodic with period $1/\xi$, and as a function of y for fixed x it is periodic with period $1/\eta$ (see Fig. 3.4). Thus ξ is the spatial frequency for the x variation and η is the spatial frequency for the y variation.

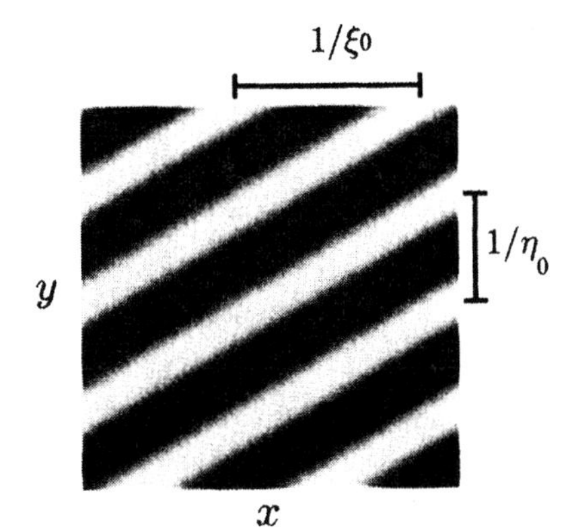

Fig. 3.4 Illustration of the 2D basis function, $\exp[2\pi i(\xi_0 x + \eta_0 y)]$.

Vector notation It is convenient to think of ξ and η as Cartesian coordinates of a 2D vector $\boldsymbol{\rho}$. Similarly, x and y are the components of a 2D spatial-position vector $\mathbf{r}$. We can then recognize the exponent in (3.213) as a simple 2D scalar product and write

$$u_{\boldsymbol{\rho}}(\mathbf{r}) = \exp(2\pi i \boldsymbol{\rho} \cdot \mathbf{r}) . \tag{3.214}$$

In the same fashion, we can define a 3D complex-exponential basis function by

$$u_{\xi\eta\zeta}(x, y, z) \equiv \exp[2\pi i(\xi x + \eta y + \zeta z)] . \tag{3.215}$$

Now ξ, η and ζ are the spatial frequencies for variation along x, y and z, respectively. We shall usually use the same notation in 3D as in 2D; that is, $\boldsymbol{\rho}$ is now a 3D spatial-frequency vector with Cartesian components (ξ, η, ζ) while $\mathbf{r}$ is a 3D spatial position vector with Cartesian components (x, y, z). With these conventions, (3.214) is valid in 3D as well as 2D. In fact, it holds in any number of dimensions.

Sometimes we shall encounter 2D and 3D vectors in the same problem and must make a notational distinction between them. In such cases, we shall reserve $\boldsymbol{\rho}$ and $\mathbf{r}$ for the 2D vectors and use $\boldsymbol{\sigma}$ and $\mathbf{t}$ for the 3D ones. For the remainder of this chapter, however, that distinction is not needed and dimensionality will be determined by context.

Orthonormality and completeness The orthonormality and completeness properties of the multidimensional basis functions follow from their 1D counterparts. The statement of orthonormality in two dimensions is [*cf.* (3.8)]

$$\int_{-\infty}^{\infty} dx \int_{-\infty}^{\infty} dy\, [u_{\xi\eta}(x,y)]^* \, u_{\xi'\eta'}(x,y)$$

$$= \int_{-\infty}^{\infty} dx \int_{-\infty}^{\infty} dy \exp[-2\pi i(\xi x + \eta y)] \exp[2\pi i(\xi' x + \eta' y)]$$

$$= \delta(\xi - \xi')\, \delta(\eta - \eta') \,. \tag{3.216}$$

A similar expression holds in 3D, but it is much cleaner to use the vector notation. In any number of dimensions we can write

$$\int_\infty d^n r\, [u_{\boldsymbol{\rho}'}(\mathbf{r})]^* \, u_{\boldsymbol{\rho}}(\mathbf{r}) = \int_\infty d^n r \exp[2\pi i(\boldsymbol{\rho} - \boldsymbol{\rho}') \cdot \mathbf{r}] = \delta(\boldsymbol{\rho} - \boldsymbol{\rho}') \,. \tag{3.217}$$

See Sec. 2.4.1 for a review of notational conventions in nD and Sec. 2.4.2 for a discussion of nD delta functions.

The closure or completeness integral in nD is

$$\int_\infty d^n\rho\, [u_{\boldsymbol{\rho}}(\mathbf{r})]^* \, u_{\boldsymbol{\rho}}(\mathbf{r}') = \int_\infty d^n\rho \exp[-2\pi i \boldsymbol{\rho} \cdot (\mathbf{r} - \mathbf{r}')] = \delta(\mathbf{r} - \mathbf{r}') \,. \tag{3.218}$$

Thus the nD complex exponentials form a complete orthonormal set in any of the senses that their 1D counterparts do. In particular, they are complete in $\mathbb{L}_2(\mathbb{R}^n)$ if convergence in the mean is understood.

3.4.2 Definitions and elementary properties

Definitions The 2D Fourier transform of a suitably well behaved function in Cartesian coordinates is defined as

$$F(\xi,\eta) = \mathcal{F}_2\{f(x,y)\} = \int_{-\infty}^{\infty} dx \int_{-\infty}^{\infty} dy\, f(x,y)\, e^{-2\pi i(\xi x + \eta y)} \,. \tag{3.219}$$

We can look at this equation as two consecutive 1D transforms, first on y then on x. In operator notation,

$$\mathcal{F}_2 = \mathcal{F}_{1(x\to\xi)} \mathcal{F}_{1(y\to\eta)} \,. \tag{3.220}$$

These operators commute so long as both integrals remain finite. To obtain the inverse of the 2D transform, we simply perform two consecutive 1D inverse transforms (in either order), giving

$$f(x,y) = \mathcal{F}_2^{-1}\{F(\xi,\eta)\} = \int_{-\infty}^{\infty} d\xi \int_{-\infty}^{\infty} d\eta\, F(\xi,\eta)\, e^{2\pi i(\xi x + \eta y)} \,. \tag{3.221}$$

This result is valid in any circumstance where both of the individual 1D operations are valid.

A similar argument holds in any number of dimensions; the nD Fourier transform in Cartesian coordinates can always be decomposed into n consecutive 1D transforms, and the inverse transform has the expected form so long as all of the constituent 1D operations are legal.

In vector form, the general nD expressions for the Fourier transform and its inverse are

$$F(\boldsymbol{\rho}) = \mathcal{F}_n\{f(\mathbf{r})\} = \int_\infty d^n r\, f(\mathbf{r}) \exp(-2\pi i \boldsymbol{\rho} \cdot \mathbf{r}), \tag{3.222}$$

$$f(\mathbf{r}) = \mathcal{F}_n^{-1}\{F(\boldsymbol{\rho})\} = \int_\infty d^n \rho\, F(\boldsymbol{\rho}) \exp(2\pi i \boldsymbol{\rho} \cdot \mathbf{r}). \tag{3.223}$$

Multidimensional tempered distributions Multidimensional test functions and good functions were defined in Sec. 2.4.1. By decomposing the nD transform of a good function into n 1D transforms as above, it can be shown that the Fourier transform of a good function is a good function in any number of dimensions (Strichartz, 1994). From this result we can define the nD Fourier transform of a generalized function analogously to (3.85). Let $g(\mathbf{r})$ be a generalized function corresponding to an nD tempered distribution, $t(\mathbf{r})$ be an nD good function and $T(\boldsymbol{\rho})$ be its nD Fourier transform. Then the defining equation for $G(\boldsymbol{\rho})$ is

$$\int_\infty d^n r\, g(\mathbf{r})\, t(\mathbf{r}) = \int_\infty d^n \rho\, [G(\boldsymbol{\rho})]^*\, T(\boldsymbol{\rho}). \tag{3.224}$$

Both sides of this equation have the look and feel of a scalar product in $\mathbb{L}_2(\mathbb{R}^n)$, but of course they are not since $g(\mathbf{r})$ is not necessarily in that space.

From this definition we can derive nD counterparts of most of the Fourier theorems developed in 1D with the assurance that they are valid for generalized functions as well as ordinary ones. In particular, (3.222) and (3.223) work for generalized functions so long as we interpret any improper integrals by expressing the generalized function as the limit of good functions.

Below we present the nD extensions of several other theorems from Sec. 3.3. Derivations are omitted, but in most cases they involve little more than the notational change, $x \to \mathbf{r}$ and $\xi \to \boldsymbol{\rho}$, in the corresponding 1D derivation. Except as noted, all theorems given are valid for generalized functions associated with nD tempered distributions (Strichartz, 1994).

Multidimensional Parseval's relations If both $f(\mathbf{r})$ and $F(\boldsymbol{\rho})$ are in $\mathbb{L}_2(\mathbb{R}^n)$, then the nD Fourier transformation is a unitary operator in that space. Since unitary transformations preserve norms and scalar products, we can write at once

$$\int_\infty d^n r\, [f_1(\mathbf{r})]^*\, f_2(\mathbf{r}) = \int_\infty d^n \rho\, [F_1(\boldsymbol{\rho})]^*\, F_2(\boldsymbol{\rho}), \tag{3.225}$$

$$\int_\infty d^n r\, |f(\mathbf{r})|^2 = \int_\infty d^n \rho\, |F(\boldsymbol{\rho})|^2, \tag{3.226}$$

provided both $f_1(\mathbf{r})$ and $f_2(\mathbf{r})$ are in $\mathbb{L}_2(\mathbb{R}^n)$. If one of these functions is an nD good function and the other is a generalized function, (3.225) also holds, virtually by definition [see (3.224)]. Note, however, that (3.226) may not work since the square of a generalized function is not necessarily defined.

Symmetry All of the 1D symmetry properties derived in Sec. 3.3.5 carry over to nD essentially unchanged. Directly from the definition we have the so-called Hermiticity property:

$$F(-\boldsymbol{\rho}) = [F(\boldsymbol{\rho})]^* \qquad \text{if } f(\mathbf{r}) \text{ is real}. \tag{3.227}$$

To clarify the notation, let us consider specifically the 2D case. If $F(\boldsymbol{\rho})$ is a shorthand for $F(\xi, \eta)$, then $F(-\boldsymbol{\rho})$ signifies $F(-\xi, -\eta)$. If we represent $F(\boldsymbol{\rho})$ as a gray-scale image, then $F(-\boldsymbol{\rho})$ is the same image with the ξ and η axes inverted.

Even (odd) functions transform to even (odd) functions in nD, *i.e.*,

$$F(\boldsymbol{\rho}) = \pm F(-\boldsymbol{\rho}) \qquad \text{if } f(\mathbf{r}) = \pm f(-\mathbf{r}). \tag{3.228}$$

Central-ordinate and central-slice theorems The central-ordinate theorems in nD follow directly from the definitions of the transform and its inverse. The Fourier kernel $\exp(2\pi i \boldsymbol{\rho} \cdot \mathbf{r}) = 1$ if either $\boldsymbol{\rho} = \mathbf{0}$ or $\mathbf{r} = \mathbf{0}$ (but note that these are vector equations, so *all* components must be zero). Thus

$$F(\mathbf{0}) = \int_\infty d^n r \, f(\mathbf{r}), \tag{3.229}$$

$$f(\mathbf{0}) = \int_\infty d^n \rho \, F(\boldsymbol{\rho}). \tag{3.230}$$

As in 1D, these relations are very useful for checking the constants in a Fourier calculation.

An important generalization of the central-ordinate theorem ensues if we use Cartesian coordinates and set some but not all of the corresponding frequency variables to zero. For example, if we write $f(\mathbf{r}) = f(x, y)$ and $F(\boldsymbol{\rho}) = F(\xi, \eta)$ in 2D, we find that

$$F(0, \eta) = \int_{-\infty}^{\infty} dx \int_{-\infty}^{\infty} dy \, f(x, y) \, e^{-2\pi i \eta y}; \tag{3.231}$$

$$F(\xi, 0) = \int_{-\infty}^{\infty} dx \int_{-\infty}^{\infty} dy \, f(x, y) \, e^{-2\pi i \xi x}. \tag{3.232}$$

These equations are special cases of the *central-slice theorem*, which plays a key role in the theory of tomographic imaging. This theorem is explored further in Chap. 4, but to get a glimmering of its content we rewrite (3.231) as

$$F(0, \eta) = \int_{-\infty}^{\infty} dy \left[\int_{-\infty}^{\infty} dx \, f(x, y) \right] e^{-2\pi i \eta y}. \tag{3.233}$$

The quantity in square brackets is a 1D function of y obtained by integrating $f(x, y)$ along lines parallel to the x axis. This 1D function is called the *line-integral projection* (or simply *projection*) of $f(x, y)$, with the projection direction being parallel to the x axis. The remaining integral in (3.233) is just a 1D Fourier transform with respect to y. Thus (3.233) says that the 1D Fourier transform of a projection of a 2D function is the same as the 2D Fourier transform of the function, but evaluated along a line through the origin ($\xi = 0$) in the 2D frequency domain.

Derivatives The Fourier transform of the derivative of a function is essentially the same thing in 1D and nD, except that in nD we must distinguish the various possible partial derivatives. For example, the rationale that led to (3.98) shows also that

$$\mathcal{F}_n\left\{\frac{\partial}{\partial x} f(\mathbf{r})\right\} = 2\pi i\xi\, F(\boldsymbol{\rho})\,. \tag{3.234}$$

Here ξ is the x-component of the spatial-frequency vector in any number of dimensions. Similar results hold for other components, higher derivatives and mixed partial derivatives.

On the other hand, there is an additional richness to the subject of derivatives in higher dimensions. Since $f(\mathbf{r})$ is a scalar function, we can compute its gradient, which is a vector in the direction of maximum rate of change of $f(\mathbf{r})$. Proceeding from the expression for a gradient in Cartesian components and using (3.234), we can show that

$$\mathcal{F}_n\{\nabla f(\mathbf{r})\} = 2\pi i\boldsymbol{\rho}\, F(\boldsymbol{\rho})\,. \tag{3.235}$$

Since $\nabla f(\mathbf{r})$ is a vector, so is its Fourier transform; (3.235) is really n equations, one for each component of the vector. The Fourier transform of a gradient always points in the same direction as $\boldsymbol{\rho}$.

Similarly, the Fourier transform of a Laplacian (divergence of the gradient) is given by

$$\mathcal{F}_n\{\nabla^2 f(\mathbf{r})\} = (2\pi i\rho)^2 F(\boldsymbol{\rho})\,, \tag{3.236}$$

where $\rho = |\boldsymbol{\rho}|$ is the magnitude of the spatial frequency vector. For example, in 3D, $\rho = \sqrt{\xi^2 + \eta^2 + \zeta^2}$. Since the Laplacian is a scalar, so is its Fourier transform.

Shifts and linear phase factors Just as in 1D, multiplying an nD function by a linear phase factor produces a translation or shift of its Fourier transform, and vice versa. It follows from the definition of the nD transform that

$$\mathcal{F}_n\{\exp(+2\pi i\boldsymbol{\rho}_0\cdot\mathbf{r}) f(\mathbf{r})\} = F(\boldsymbol{\rho} - \boldsymbol{\rho}_0)\,. \tag{3.237}$$

$$\mathcal{F}_n\{f(\mathbf{r} - \mathbf{r}_0)\} = \exp(-2\pi i\boldsymbol{\rho}\cdot\mathbf{r}_0)\, F(\boldsymbol{\rho})\,. \tag{3.238}$$

Scaled functions In 2D, $f(\mathbf{r})$ is the vector notation for the function $f(x, y)$. We shall use the notation $f(\mathbf{r}/a)$ (with a real) to mean $f(x/a, y/a)$; the extension to any number of dimensions is straightforward. By a change of variables in the definition of the nD transform, we find, analogously to (3.111),

$$\mathcal{F}_n\{f(\mathbf{r}/a)\} = |a|^n F(a\boldsymbol{\rho})\,. \tag{3.239}$$

This is one of the few Fourier theorems where the number of dimensions n enters explicitly.

A consequence of this result and the central-ordinate theorem is that the integral of $f(\mathbf{r})$ and $|a|^{-n} f(\mathbf{r}/a)$ are the same. In nD, if we wish to scale all n coordinate axes of a function by a factor a, we must scale its amplitude by $|a|^{-n}$ to maintain a constant integral.

3.4.3 Multidimensional convolution and correlation

The vector notation makes it simple to generalize the definitions of convolution and correlation to nD. An nD convolution is defined by [*cf.* (3.114)]

$$[f * h](\mathbf{r}) = f(\mathbf{r}) * h(\mathbf{r}) = \int_{\infty} d^n r' f(\mathbf{r}')\, h(\mathbf{r}-\mathbf{r}') = \int_{\infty} d^n r'' f(\mathbf{r}-\mathbf{r}'')\, h(\mathbf{r}''). \quad (3.240)$$

Explicitly in 2D, we have

$$[f * h](x, y) = \int_{-\infty}^{\infty} dx' \int_{-\infty}^{\infty} dy' f(x', y')\, h(x - x', y - y'). \quad (3.241)$$

Similarly, the nD correlation is given by [*cf.* (3.115)]

$$[f \star h](\mathbf{r}) = \int_{\infty} d^n r' f(\mathbf{r} + \mathbf{r}') h(\mathbf{r}') = \int_{\infty} d^n r'' f(\mathbf{r}'')\, h(\mathbf{r}'' - \mathbf{r}). \quad (3.242)$$

The nD Fourier transforms of the convolution and correlation are, respectively,

$$\mathcal{F}_n\{[f * h](\mathbf{r})\} = F(\boldsymbol{\rho})\, H(\boldsymbol{\rho}), \quad (3.243)$$

$$\mathcal{F}_n\{[f \star h](\mathbf{r})\} = F(\boldsymbol{\rho})\, H(-\boldsymbol{\rho}) = F(\boldsymbol{\rho})\, [H(\boldsymbol{\rho})]^*, \quad (3.244)$$

where the last step in (3.244) is valid if $h(\mathbf{r})$ is real.

An important special case of (3.244) occurs when $h(\mathbf{r}) = f^*(\mathbf{r})$. The correlation is referred to as a complex autocorrelation in that case, and we have

$$\mathcal{F}_n\{\, [f \star f^*](\mathbf{r})\} = |F(\boldsymbol{\rho})|^2, \quad (3.245)$$

which is the generalization of (3.135). This equation will prove to be essential in discussing the properties of incoherent optical systems.

3.4.4 Rotationally symmetric functions

So far we have discussed nD Fourier transforms as n-fold integrals in Cartesian coordinates, but if the function being transformed has symmetry properties, other coordinate systems could be preferable. Consider, for example, a 2D function with rotational symmetry about the origin. The natural coordinate system is polar coordinates (r, θ), with $r = |\mathbf{r}|$ being the distance from the origin and θ being the angle measured from the x axis. In these coordinates, $f(\mathbf{r})$ is independent of θ and can be written as $f(r)$.

In 2D polar coordinates, $\boldsymbol{\rho} \cdot \mathbf{r} = \rho r \cos(\theta - \theta_\rho)$, where (ρ, θ_ρ) are the polar coordinates of $\boldsymbol{\rho}$, and we can write

$$\mathcal{F}_2\{f(r)\} = \int_0^{\infty} r\, dr\, f(r) \int_0^{2\pi} d\theta\, e^{-2\pi i \rho r \cos(\theta - \theta_\rho)}. \quad (3.246)$$

Since $f(r)$ is independent of θ, the integral over θ can be done once and for all. A well known integral (and a very important one in optics) is

$$\mathrm{J}_k(z) = \frac{i^{-k}}{\pi} \int_0^{\pi} d\theta\, e^{iz\cos\theta} \cos(k\theta), \quad (3.247)$$

where $\mathrm{J}_k(z)$ is the Bessel function of the first kind of order k (Abramowitz and Stegun, 1965). This integral is often used as the definition of this Bessel function.

With a little algebra, the integral in (3.246) can be put into the form of (3.247) with $k = 0$, yielding

$$\mathcal{F}_2\{f(r)\} = 2\pi \int_0^\infty r\,dr\,\mathrm{J}_0(2\pi\rho r)\,f(r)\,. \tag{3.248}$$

This form is often referred to as the *Hankel transform of zeroth order*, but we emphasize that it is also the 2D *Fourier* transform of a function with rotational symmetry.

A similar calculation applies in 3D for spherically symmetric functions. The natural coordinate system is spherical polar coordinates (r, θ, ϕ), where θ is the co-latitude, measured from the z axis, and ϕ is the azimuth or longitude. A spherically symmetric function $f(\mathbf{r})$ is independent of θ and ϕ, so we can write $f(\mathbf{r}) = f(r)$, where $r = |\mathbf{r}|$. The exponent in the Fourier kernel is simplified, without loss of generality, by taking the z axis to be parallel to $\boldsymbol{\rho}$, so that $\boldsymbol{\rho}\cdot\mathbf{r} = \rho r\cos\theta$. We then have

$$\mathcal{F}_3\{f(r)\} = \int_0^\infty r^2 dr\, f(r) \int_0^{2\pi} d\phi \int_0^\pi d\theta \sin\theta \exp(-2\pi i \rho r\cos\theta)\,, \tag{3.249}$$

and some algebra gives

$$\mathcal{F}_3\{f(r)\} = 4\pi \int_0^\infty r^2 dr\,\mathrm{sinc}(2\rho r)\,f(r)\,. \tag{3.250}$$

The constants in (3.248) and (3.250) can be checked by the central-ordinate theorem; since $\mathrm{J}_0(0) = \mathrm{sinc}(0) = 1$, $F(\mathbf{0})$ is given by the integral of the function in both cases.

3.4.5 Some special functions and their transforms

Delta functions The definitions of the nD delta function and the nD Fourier transform show at once that

$$\mathcal{F}_n\{\delta(\mathbf{r})\} = 1\,, \tag{3.251}$$

$$\mathcal{F}_n\{\delta(\mathbf{r}-\mathbf{r}_0)\} = \exp(-2\pi i\mathbf{r}_0\cdot\boldsymbol{\rho})\,. \tag{3.252}$$

The transforms of various nD derivatives of the delta function follow from these results and (3.234) – (3.236). For example,

$$\mathcal{F}_n\{\nabla^2\delta(\mathbf{r})\} = -4\pi^2\rho^2\,. \tag{3.253}$$

Complex exponentials, sines and cosines A modest change of variables in the expression for the nD Fourier transform shows that

$$\mathcal{F}_n\{\exp(2\pi i\boldsymbol{\rho}_0\cdot\mathbf{r})\} = \delta(\boldsymbol{\rho}-\boldsymbol{\rho}_0)\,. \tag{3.254}$$

From this result and (3.5) and (3.6), we find

$$\mathcal{F}_n\{\cos(2\pi\boldsymbol{\rho}_0\cdot\mathbf{r})\} = \frac{1}{2}\left[\delta(\boldsymbol{\rho}-\boldsymbol{\rho}_0) + \delta(\boldsymbol{\rho}+\boldsymbol{\rho}_0)\right]\,, \tag{3.255}$$

$$\mathcal{F}_n\{\sin(2\pi\boldsymbol{\rho}_0\cdot\mathbf{r})\} = \frac{1}{2i}\left[\delta(\boldsymbol{\rho}-\boldsymbol{\rho}_0) - \delta(\boldsymbol{\rho}+\boldsymbol{\rho}_0)\right]\,. \tag{3.256}$$

As in 1D, sines and cosines transform to pairs of delta functions. These delta functions are located at $\boldsymbol{\rho} = \pm\boldsymbol{\rho}_0$, two points oppositely displaced from the origin in the nD frequency space.

Cylinders and besincs A convenient special function in 2D is the cylinder function, which is unity inside a disc of diameter D and zero outside:

$$\operatorname{cyl}\left(\frac{r}{D}\right) \equiv \begin{cases} 1 & \text{if } r < D/2 \\ 0 & \text{if } r > D/2 \end{cases} . \tag{3.257}$$

Note that r is $|\mathbf{r}|$, where $\mathbf{r}$ is a 2D vector, so $\operatorname{cyl}(r/D)$ is specifically a 2D function. The name cylinder evokes a surface relief plot of the function, which looks like a cylinder (see Fig. 3.5). The cylinder function is a useful representation of a circular aperture in optics.

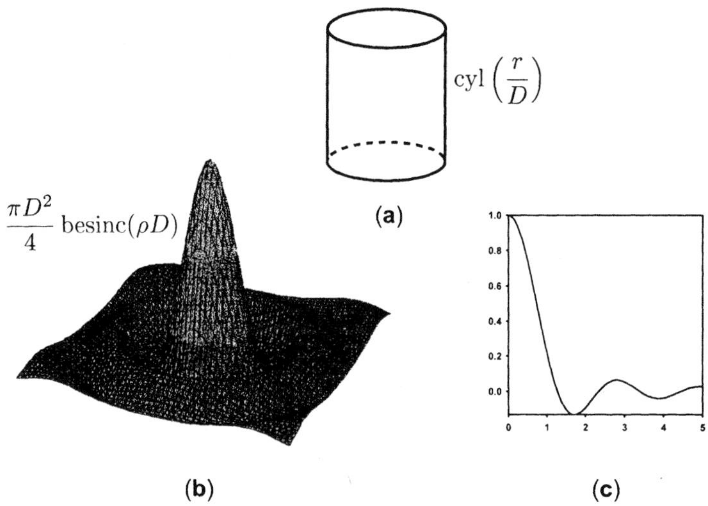

Fig. 3.5 (a) The cylinder function $\operatorname{cyl}(r)$; (b) The 2D Fourier transform of the cylinder, $\operatorname{besinc}(\rho)$ presented as a relief plot; (c) Normalized radial profile of the 2D function in (b).

Since the cylinder function is rotationally symmetric, its Fourier transform can be found from (3.248):

$$\mathcal{F}_2\left\{\operatorname{cyl}(r/D)\right\} = 2\pi \int_0^{\frac{1}{2}D} r\,dr\, \mathrm{J}_0(2\pi\rho r) = \frac{\pi D^2}{4} \cdot \frac{2\mathrm{J}_1(\pi\rho D)}{\pi\rho D}, \tag{3.258}$$

where $\mathrm{J}_1(\cdot)$ is the first-order Bessel function of the first kind (Abramowitz and Stegun, 1965). A succinct way to write this equation is

$$\mathcal{F}_2\{\operatorname{cyl}(r/D)\} = \frac{\pi D^2}{4} \operatorname{besinc}(\rho D), \tag{3.259}$$

where

$$\operatorname{besinc}(t) \equiv \frac{2\mathrm{J}_1(\pi t)}{\pi t}. \tag{3.260}$$

The besinc function, also referred to as a jinc or sombrero function, is the 2D counterpart of the sinc function. It is plotted in Fig. 3.5. Since $\operatorname{besinc}(0) = 1$, the factor of $\pi D^2/4$ in (3.259) is in accord with the 2D central-ordinate theorem; the area of the disc is the integral of the cylinder function.

Gaussians and quadratic phase factors The nD Gaussian is just a product of 1D Gaussians. For example, in 3D,

$$\text{gaus}\left(\frac{r}{a}\right) = e^{-\pi(x^2+y^2+z^2)/a^2} = e^{-\pi x^2/a^2} e^{-\pi y^2/a^2} e^{-\pi z^2/a^2} . \tag{3.261}$$

Since the nD function factors into a product of n 1D functions, it follows that

$$\mathcal{F}_n \{\text{gaus}(r/a)\} = |a|^n \, e^{-\pi a^2 \rho^2} = |a|^n \, \text{gaus}(a\rho) , \tag{3.262}$$

where we have used (3.180) and (3.239).

By the same argument, an nD quadratic phase factor is a product of n 1D ones, and we have [*cf.* (3.185)]

$$\mathcal{F}_n\{e^{i\pi\beta r^2}\} = \left(\frac{i}{\beta}\right)^{\frac{1}{2}n} e^{-i\pi\rho^2/\beta} . \tag{3.263}$$

3.4.6 Multidimensional periodicity

Comb functions Consider the 2D function comb($\mathbf{r}/a$) defined by

$$\text{comb}(\mathbf{r}/a) \equiv \text{comb}(x/a)\,\text{comb}(y/a) = \sum_{k=-\infty}^{\infty} \sum_{m=-\infty}^{\infty} \delta\left(\frac{x}{a} - k\right) \delta\left(\frac{y}{a} - m\right) . \tag{3.264}$$

Regarded as a 2D function, $\delta(x/a - k)$ is a line mass (see Sec. 2.4.4) located at $x = ka$. The product $\delta(x/a - k)\,\delta(y/a - m)$ is the intersection of two line masses or a point mass located at $(x, y) = (ka, ma)$, so comb($\mathbf{r}/a$) is a set of 2D delta functions located on a square lattice. Similarly, in nD, comb($\mathbf{r}/a$) is a product of n 1D combs or a set of delta functions on a regular lattice in $\mathbb{R}^n$.

Since the nD comb function is a product of n 1D functions, we have

$$\mathcal{F}_n\{\text{comb}(\mathbf{r}/a)\} = |a|^n \, \text{comb}(a\boldsymbol{\rho}) . \tag{3.265}$$

If the lattice has different spacings in different directions, the nD comb function can be expressed in product form and transformed one dimension at a time. For example, in 2D,

$$\mathcal{F}_2\{\text{comb}(x/a)\,\text{comb}(y/b)\} = |ab| \, \text{comb}(a\xi)\,\text{comb}(b\eta) . \tag{3.266}$$

The delta functions in frequency space are now spaced by $1/a$ along ξ and $1/b$ along η.

If the lattice is not orthogonal, more care is needed. Consider a 2D lattice of delta functions located at the points $\mathbf{r} = m_1\mathbf{a} + m_2\mathbf{b}$, where $\mathbf{a}$ and $\mathbf{b}$ are arbitrary noncollinear vectors and m_1 and m_2 are integers. An array of delta functions on these points can be written

$$\sum_{m_1=-\infty}^{\infty} \sum_{m_2=-\infty}^{\infty} \delta(\mathbf{r} - m_1\mathbf{a} - m_2\mathbf{b}) . \tag{3.267}$$

A more compact way of writing this expression (and one more easily generalized to higher dimensions) makes use of a matrix $\mathbf{P}$, defined by

$$\mathbf{P} = \begin{bmatrix} a_x & b_x \\ a_y & b_y \end{bmatrix} , \tag{3.268}$$

and a column vector $\mathbf{m} = (m_1, m_2)^t$, called a *multi-index*. With this notation,

$$\sum_{m_1=-\infty}^{\infty} \sum_{m_2=-\infty}^{\infty} \delta(\mathbf{r} - m_1\mathbf{a} - m_2\mathbf{b}) = \sum_{\mathbf{m}} \delta(\mathbf{r} - \mathbf{Pm}), \tag{3.269}$$

where the sum over $\mathbf{m}$ implies the double sum over m_1 and m_2 from $-\infty$ to ∞. This matrix-vector form can be generalized to nD, in which case $\mathbf{P}$ becomes an $n \times n$ matrix and $\mathbf{m}$ an $n \times 1$ column vector of integers. With this notation, it can be shown that (Marks, 1991)

$$\mathcal{F}_n \left\{ |\det(\mathbf{P})| \sum_{\mathbf{m}} \delta(\mathbf{r} - \mathbf{Pm}) \right\} = \sum_{\mathbf{m}} \delta(\boldsymbol{\rho} - \mathbf{Qm}), \tag{3.270}$$

where $\mathbf{Q}^t = \mathbf{P}^{-1}$, and $\det(\cdot)$ denotes determinant.

This equation shows that the Fourier transform of a general nD comb of delta functions on the nonorthogonal lattice $\mathbf{Pm}$ is a similar comb on the *reciprocal lattice* $\mathbf{Qm}$. The reciprocal lattice is an important tool in solid-state physics (see, for example, Ashcroft and Mermin, 1976).

In 2D, the points $\mathbf{Qm}$ are located at $\boldsymbol{\rho} = m_1\mathbf{A} + m_2\mathbf{B}$, where $\mathbf{A}$ and $\mathbf{B}$ are the basis vectors for the reciprocal lattice, defined such that $\mathbf{a} \cdot \mathbf{A} = \mathbf{b} \cdot \mathbf{B} = 1$ and $\mathbf{a} \cdot \mathbf{B} = \mathbf{b} \cdot \mathbf{A} = 0$. Equation (3.266) is the special case of (3.270) for diagonal matrices $\mathbf{P}$ and $\mathbf{Q}$, in which case $\mathbf{a} \perp \mathbf{b}$ and $\mathbf{A} \perp \mathbf{B}$.

Multidimensional periodic functions An nD function $f(\mathbf{r})$ satisfies the periodicity condition

$$f(\mathbf{r} + \mathbf{Pm}) = f(\mathbf{r}), \tag{3.271}$$

where $\mathbf{P}$ is some $n \times n$ matrix $\mathbf{P}$ and $\mathbf{m}$ is again an $n \times 1$ vector of integers. As a simple example, consider $n = 3$ and $\mathbf{P} = L\mathbf{I}_3$, where $\mathbf{I}_3$ is the 3×3 unit matrix. With this choice, (3.271) becomes

$$f(x + m_1L,\, y + m_2L,\, z + m_3L) = f(x, y, z). \tag{3.272}$$

Since m_1, m_2 and m_3 are integers, this equation describes a 3D function that is periodic on a cubic lattice of spacing L. It will be useful to keep this simple example in mind, but the remainder of this section will use the more general formula (3.271).

To proceed, we need to define a *unit cell*, just as in solid state physics (Ashcroft and Mermin, 1976). One definition of a unit cell is that it is the set of points that lies closer to the origin than to any other point in the lattice. This particular choice is known as the *Wigner-Seitz unit cell.* A geometrical method to construct a Wigner-Seitz unit cell is shown in Fig. 3.6.

By analogy to (3.188), the periodic function of (3.271) can be expressed as

$$f(\mathbf{r}) = f_0(\mathbf{r}) * \sum_{\mathbf{m}} \delta(\mathbf{r} - \mathbf{Pm}), \tag{3.273}$$

where

$$f_0(\mathbf{r}) = \begin{cases} f(\mathbf{r}) & \text{if } \mathbf{r} \text{ is in unit cell} \\ 0 & \text{otherwise} \end{cases}. \tag{3.274}$$

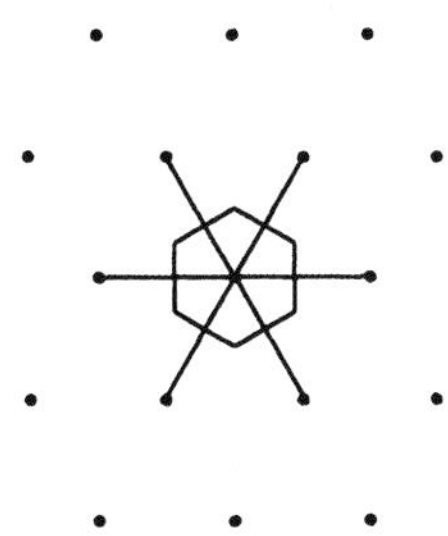

Fig. 3.6 Method of constructing the Wigner-Seitz unit cell in 2D. The perpendicular bisector is constructed for each line connecting the origin to another point in the lattice, and the unit cell is the smallest area bounded by all of these lines. Hence points in the unit cell lie closer to the origin than to any other lattice point.

We can now take the Fourier transform of $f(\mathbf{r})$ by use of (3.243) and (3.270). The result is [*cf.* (3.189)]

$$F(\boldsymbol{\rho}) = \mathcal{F}_n\{f(\mathbf{r})\} = F_0(\boldsymbol{\rho})\frac{1}{|\det(\mathbf{P})|}\sum_{\mathbf{m}}\delta(\boldsymbol{\rho} - \mathbf{Qm}), \tag{3.275}$$

where $\mathbf{Q}^t = \mathbf{P}^{-1}$ and $F_0(\boldsymbol{\rho})$ is the Fourier transform of $f_0(\mathbf{r})$. Thus, just as in the 1D case, the Fourier transform of a periodic function is the Fourier transform of its values in the unit cell times a sum of delta functions. Again the periodicity of the function restricts its Fourier transform to a discrete set of frequencies. Here the discrete frequencies are the points of the reciprocal lattice. In fact, it is precisely this property that gives rise to the term reciprocal lattice; many books refer to Fourier space as *reciprocal space.*

Still pursuing the analogy to the 1D case, we take the inverse transform of (3.275), yielding [*cf.* (3.191)]

$$f(\mathbf{r}) = \frac{1}{|\det(\mathbf{P})|}\sum_{\mathbf{m}} F_0(\mathbf{Qm})\exp[2\pi i(\mathbf{Qm})\cdot\mathbf{r}]. \tag{3.276}$$

This is the nD Fourier series for the periodic function $f(\mathbf{r})$. As in the 1D case, the Fourier coefficients of a periodic function are sampled values of the Fourier transform of one period (unit cell) of the function.

As noted in Sec. 3.2.1, there are two ways to view a Fourier series. It is a representation of a periodic function in all space or a representation of an arbitrary function in a finite region. We now investigate the latter view in the multidimensional case.

Consider a function $f(\mathbf{r})$ that vanishes identically unless $\mathbf{r}$ is in a region $\mathbf{S}_f$ of the nD space. We choose an arbitrary lattice, defined by a matrix $\mathbf{P}$, with no restrictions except that $\mathbf{S}_f$ must fit entirely within the Wigner-Seitz unit cell of the lattice. We define a *support function* $S_f(\mathbf{r})$ via

$$S_f(\mathbf{r}) = \begin{cases} 1 & \text{if } \mathbf{r} \text{ is in } \mathbf{S}_f \\ 0 & \text{otherwise} \end{cases}. \tag{3.277}$$

Then an exact representation of $f(\mathbf{r})$, valid in all space, is[2]

[2]The possibility of an exact representation like (3.278) is a consequence of sampling theory, a topic to be introduced in the next section. The requirement that the function fit within the unit cell

$$f(\mathbf{r}) = \frac{1}{|\det(\mathbf{P})|} \sum_{\mathbf{m}} F(\mathbf{Qm}) \exp[2\pi i(\mathbf{Qm}) \cdot \mathbf{r}] \, S_f(\mathbf{r}) \,. \tag{3.278}$$

Without the support function, the series in (3.278) would represent an infinite set of replicas of $f(\mathbf{r})$, but the presence of $S_f(\mathbf{r})$ sets all replicas but one to zero.

A suggestive way to rewrite (3.278) is:

$$f(\mathbf{r}) = \sum_{\mathbf{m}} F(\mathbf{Qm}) \Phi_{\mathbf{m}}(\mathbf{r}) \,, \tag{3.279}$$

where $\Phi_{\mathbf{m}}(\mathbf{r})$ is a basis function defined by

$$\Phi_{\mathbf{m}}(\mathbf{r}) \equiv \frac{1}{|\det(\mathbf{P})|} \exp[2\pi i(\mathbf{Qm}) \cdot \mathbf{r}] \, S_f(\mathbf{r}) \,. \tag{3.280}$$

These basis functions can thus be used to represent exactly an arbitrary function of compact support.

3.5 SAMPLING THEORY

So far we have considered Fourier expansions of functions of a continuous variable x, and most of our results have been couched in terms of integrals with respect to this variable. When it comes time to perform real calculations, however, we often must have recourse to a digital computer, and computers usually deal with discrete sets of numbers. It is straightforward to convert a continuous function $f(x)$ to such a discrete set — we can simply evaluate it at a discrete set of points. The resulting numbers $f_j \equiv f(x_j)$ are referred to as *samples* of $f(x)$, and the process of evaluation is called *sampling*. We shall express this operation via the *sampling operator* $\boldsymbol{\mathcal{S}}$, a continuous-to-discrete operator that derives sample values from a function. The inverse process — determination of a continuous function from its discrete samples — is more difficult and in fact impossible without placing stringent conditions on the function. If there were no restrictions, the function could vary in an arbitrary manner between the samples and we would have no way of knowing it. Some sort of smoothness constraint is required.

3.5.1 Bandlimited functions

One way of saying that a function is smooth is to say that its Fourier transform vanishes identically outside a finite interval; such a function is said to be *bandlimited*. One way to construct a bandlimited function, at least in principle, is to start with an arbitrary function and pass it through an ideal low-pass filter. As we shall see in Chap. 9, certain kinds of optical systems act as low-pass filters with a sharp cutoff, so the image with such systems is bandlimited; its Fourier transform has compact support. As we noted in Sec. 3.3.9, however, the Fourier transform of a function of compact support cannot itself have compact support, so the output of

is equivalent to saying that the reciprocal lattice satisfies the Nyquist condition for sampling in frequency space. See Sec. 3.5.4.

an ideal low-pass filter must in principle have infinite spatial extent. If we truncate the spatial function (*e.g.*, record the image on a piece of film of finite size), we induce higher spatial frequencies, so the image is no longer strictly bandlimited.

Nevertheless, the Fourier transform may drop to an insignificant level for $|\xi|$ larger than some frequency ξ_{max}, in which case we can proceed as if the function were bandlimited, recognizing that the resulting interpolation formula will be only approximate. In practice, if we go through the two-step procedure of low-pass filtering and truncation, an appropriate value for ξ_{max} will be just slightly greater than the cutoff frequency of the filter.

Paley-Wiener space A more formal way of discussing bandlimited functions makes use of the Paley-Wiener theorem (see Sec. 3.3.9) and a reproducing-kernel Hilbert space (see Sec. 1.8) called *Paley-Wiener space.*

The Paley-Wiener theorem shows that the Fourier transform of a function with compact support is an entire function of exponential type. Similarly, if the Fourier transform $F(\xi)$ has compact support, the function $f(x)$ itself, regarded as a function of a complex variable, is entire. That means that a bandlimited function is continuous, differentiable and expandable in a complex Taylor series (see App. B). We won't have need for complex x, but all of these nice properties hold for real x as well. Bandlimited functions of a real variable are said to be in Paley-Wiener space.

Ideal low-pass filtering with a cutoff at $\xi = \pm\xi_{max}$ corresponds to convolution with $B\,\mathrm{sinc}(Bx)$ in the spatial domain, where $B = 2\xi_{max}$. Equivalently, this convolution corresponds to multiplication in the frequency domain by the Fourier transform of the sinc. Since

$$\mathcal{F}_1\{B\,\mathrm{sinc}(Bx)\} = \mathrm{rect}(\xi/B)\,, \tag{3.281}$$

subsequent convolution with $B\,\mathrm{sinc}(Bx)$ has no further effect. (A rect function raised to any power is still the same rect function.) Thus bandlimited functions containing only frequencies for which $|\xi| \leq B/2$ must satisfy

$$\int_{-\infty}^{\infty} dx' B\,\mathrm{sinc}[B(x-x')]\, f(x') = f(x)\,. \tag{3.282}$$

From the discussion in Sec. 1.8, (3.282) will be recognized as the condition for Paley-Wiener space to be a reproducing-kernel Hilbert space. The operator $\boldsymbol{\mathcal{B}}$ used in Sec. 1.8 here corresponds to convolution with $B\,\mathrm{sinc}(Bx)$. All functions in Paley-Wiener space are invariant to this operator, and $\boldsymbol{\mathcal{B}}$ is simply the unit operator in the reproducing-kernel space (or, equivalently, the projection operator from $\mathbb{L}_2(\mathbb{R})$ to the space). Thus Paley-Wiener space is a reproducing-kernel Hilbert space with a sinc function as the kernel (Walter, 1994; Daubechies, 1992). As noted in Sec. 1.8, a function in a reproducing-kernel Hilbert space is smooth in a certain sense; the smoothness of bandlimited functions will be exploited below.

3.5.2 Reconstruction of a bandlimited function from uniform samples

Bandlimited functions are so smooth that their behavior between samples can be predicted exactly from the sample values, provided the samples are sufficiently close together. In other words, there exists an exact interpolation formula, usually referred to as the *Whittaker-Shannon sampling theorem*, for reconstructing a bandlimited function from its samples. Other names for this theorem include the *Whittaker-Shannon-Kotel'nikov theorem*, the *Whittaker-Shannon-Kotel'nikov-Kramer theorem*, the *sampling theorem*, and the *cardinal series* (Marks, 1991). We shall derive this important theorem in two different ways below.

Shannon's derivation The first derivation of the sampling theorem that we present is the one given originally by Shannon and discussed in Jerri (1977).

A bandlimited function can be represented by the truncated inverse Fourier transform,

$$f(x) = \int_{-\frac{1}{2}B}^{\frac{1}{2}B} d\xi \, F(\xi) \exp(2\pi i \xi x) \,. \tag{3.283}$$

The truncation of the range of integration to $(-\frac{1}{2}B, \frac{1}{2}B)$ entails no error since $F(\xi)$ vanishes identically outside this range. The full range B, including positive and negative frequencies, is called the *bandwidth* of the function. (Caution: In engineering texts, the term bandwidth often refers to the maximum frequency present in a signal, which is our $\frac{1}{2}B$.)

Another consequence of the bandlimited character is that $F(\xi)$ can be exactly represented on $(-\frac{1}{2}B, \frac{1}{2}B)$ by the Fourier series,

$$F(\xi) = \sum_{k=-\infty}^{\infty} C_k \exp(-2\pi i \xi k/B) \,. \tag{3.284}$$

This expression is essentially the same as (3.17) but with $x \to \xi$ and $L \to B$. The coefficients in (3.284) are given by [*cf.* (3.191)]

$$C_k = \frac{1}{B} \int_{-\frac{1}{2}B}^{\frac{1}{2}B} d\xi \, F(\xi) \exp(2\pi i \xi k/B) = \frac{1}{B} f\left(\frac{k}{B}\right) . \tag{3.285}$$

Plugging (3.285) into (3.284) and the result into (3.283), we find

$$f(x) = \int_{-\frac{1}{2}B}^{\frac{1}{2}B} d\xi \sum_{k=-\infty}^{\infty} \frac{1}{B} f\left(\frac{k}{B}\right) \exp(-2\pi i \xi k/B) \exp(2\pi i \xi x) \,. \tag{3.286}$$

Interchanging sum and integral and performing an integration yields

$$f(x) = \sum_{k=-\infty}^{\infty} f\left(\frac{k}{B}\right) \operatorname{sinc}(Bx - k) \,. \tag{3.287}$$

This is the desired interpolation formula, the Whittaker-Shannon sampling theorem. We can see that the right-hand side of (3.287) exactly reproduces $f(x)$ at the sample points as follows: If $x = j/B$ (j an integer), then $\operatorname{sinc}(Bx - k) = \operatorname{sinc}(j - k) = \delta_{jk}$, so the equation reads $f(j/B) = f(j/B)$. The interpolation function has the nice property that it is unity at one of the sampling points and zero at all others.

Of course, giving the right answer at the sample points is not sufficient for an interpolation formula; it must also work for all x. Linear or spline interpolants, for example, could reproduce $f(x)$ at $x = j/B$, but only the sinc function gives the exact values for $f(x)$ at all intermediate points, and then only if $f(x)$ is bandlimited to bandwidth B.

In spite of their approximate character, linear interpolants and splines do have one practical advantage over the Whittaker-Shannon theorem; determination of $f(x)$ at some point x requires knowledge of the sample values at only the adjacent sample points. Equation (3.287) shows that knowledge of $f(j/B)$ for *all* j is required to determine $f(x)$ for *any* x other than a sampling point.

Alternative derivation In this section, we present an alternative derivation of the Whittaker-Shannon sampling theorem based on comb functions. This derivation, found in many modern texts (*e.g.*, Marks, 1991 or Gaskill, 1978), makes it clearer what the sampling rate must be and what happens if it is not large enough.

Suppose the continuous-to-discrete operator $\boldsymbol{\mathcal{S}}$ acts on $f(x)$ to give a set of sample values $\{f(j\Delta x)\}$, $j = 0, \pm 1, \pm 2, ..., \pm\infty$, where for the moment Δx is arbitrary. We can use the adjoint operator $\boldsymbol{\mathcal{S}}^\dagger$ to construct an auxiliary function $f_s(x)$, formed by using the sample values as weights for an infinite set of delta functions. Thus $f_s(x)$ is defined by

$$f_s(x) = \boldsymbol{\mathcal{S}}^\dagger \boldsymbol{\mathcal{S}} f(x) = \sum_{j=-\infty}^{\infty} f(j\Delta x)\, \delta(x - j\Delta x) \,. \tag{3.288}$$

This function and others encountered in this derivation are illustrated in Fig. 3.7.

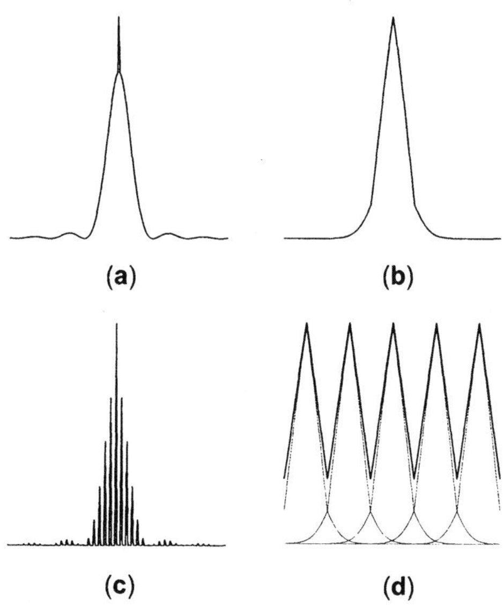

Fig. 3.7 Functions encountered in a derivation of the Whittaker-Shannon sampling theorem. (a) The original function $f(x)$, illustrated by a sinc^2 function plus a narrow Gaussian. The narrow Gaussian keeps $f(x)$ from being bandlimited. (b) The Fourier transform of $f(x)$. (c) The sampled function $f_s(x)$; the spikes represent delta functions. (d) The Fourier transform of $f_s(x)$, showing the replication induced by sampling.

Since $f(x)\,\delta(x - j\Delta x) = f(j\Delta x)\,\delta(x - j\Delta x)$ [see (2.25)], we can also write

$$f_s(x) = \sum_{j=-\infty}^{\infty} f(x)\,\delta(x - j\Delta x) = f(x) \sum_{j=-\infty}^{\infty} \delta(x - j\Delta x) = f(x) \frac{1}{\Delta x} \mathrm{comb}\left(\frac{x}{\Delta x}\right) . \tag{3.289}$$

Reconstructing $f(x)$ from the set of samples $\{f(j\Delta x)\}$ is equivalent to finding an operator that maps $f_s(x)$ to $f(x)$. To ferret out such an operator, we take the Fourier transform of $f_s(x)$, with the result

$$F_s(\xi) = \mathcal{F}_1\{f_s(x)\} = F(\xi) * \operatorname{comb}(\Delta x \xi) = \frac{1}{\Delta x} \sum_{j=-\infty}^{\infty} F\left(\xi - \frac{j}{\Delta x}\right), \tag{3.290}$$

where we have used (3.125), (3.132) and (3.157).

Thus, as illustrated in Fig. 3.7, the Fourier transform of $f_s(x)$ is a set of displaced replicas of the Fourier transform of $f(x)$. There is no overlap of the replicas if $f(x)$ is bandlimited to bandwidth B and the sample interval Δx satisfies

$$\Delta x \leq 1/B. \tag{3.291}$$

This important condition is known as the *Nyquist sampling condition* (Nyquist, 1928a). If it is satisfied, we can use a simple rect function to isolate a single replica,

$$F(\xi) = \frac{1}{B} \operatorname{rect} \frac{\xi}{B} F_s(\xi). \tag{3.292}$$

Thus, provided the Nyquist condition holds, the operator that maps $F_s(\xi)$ to $F(\xi)$ is nothing more than multiplication by a rect.

The corresponding operator in the space domain is convolution with a sinc; an inverse transform of (3.292) gives

$$\begin{aligned} f(x) = \operatorname{sinc}(Bx) * f_s(x) &= \operatorname{sinc}(Bx) * \left[\sum_{j=-\infty}^{\infty} f(j\Delta x)\, \delta(x - j\Delta x) \right] \\ &= \sum_{j=-\infty}^{\infty} f(j\Delta x) \operatorname{sinc}\left[B(x - j\Delta x)\right]. \end{aligned} \tag{3.293}$$

This result is a valid interpolation formula provided $B\Delta x \leq 1$; the special case $B\Delta x = 1$ gives us back (3.287).

How many samples are needed? If we really had a bandlimited function, it could not be strictly spatially limited, so an infinite number of samples spaced at the Nyquist interval would be needed to specify the function. In many practical situations however, the function is spatially limited, and we regard it as approximately bandlimited to a bandwidth of B. If this approximation is adequate and we sample at an interval of $\Delta x = 1/B$, then a function of spatial extent L requires $L/\Delta x = LB$ samples. The product LB, known as the *space-bandwidth product* measures the number of *degrees of freedom* of the function, or the number of independent parameters needed to specify it.

Generalizations of Whittaker-Shannon There are many generalizations and extensions of the sampling theorem. The reader who wishes to delve further is referred to Jerri (1977), Butzer (1983), Jerri (1986), Marks (1991) and Zayed (1993). Topics treated in these references include nonuniform sampling, sampling of random processes, non-bandlimited functions, implicit sampling and sampling for general integral transforms.

A very powerful generalization of the sampling theorem uses the idea of a *sampling basis* in a reproducing-kernel Hilbert space (Zayed, 1993). Consider a space of functions $f(x)$ with reproducing kernel $h(x, x')$. A basis $\{s_n(x)\}$ in this space is called a sampling basis if there exists a set of points x_n along the real line such that

$$f(x) = \sum_n f(x_n)\, s_n(x)\,. \tag{3.294}$$

For simplicity, we consider only orthonormal bases. It is proved in Zayed (1993) that an orthonormal basis is a sampling basis if and only if it is generated from the reproducing kernel by

$$s_n(x) = h(x, x_n)\,. \tag{3.295}$$

Paley-Wiener space satisfies this condition with the kernel being a sinc function, and (3.294) is just the Whittaker-Shannon sampling theorem in this case.

3.5.3 Aliasing

A crucial assumption in the derivation of the Whittaker-Shannon sampling theorem above is that there is no overlap of the displaced replicas of $F(\xi)$ in (3.290). If there is overlap, there is no way to distinguish a particular frequency ξ_0 from the frequencies $\xi_0 \pm n/\Delta x$. Such frequencies are said to be *aliased* since they produce identical sets of samples, as illustrated in Fig. 3.8.

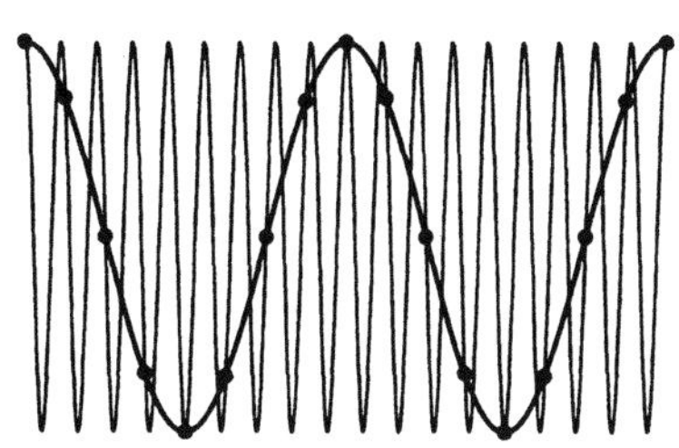

Fig. 3.8 Illustration of aliasing. The two cosines differ in frequency by the reciprocal of the sampling interval, so they agree exactly at the sample points.

One way to think about aliasing is that it results from heterodyning. In the temporal domain, heterodyne detection is the multiplication of a signal, say $\cos(2\pi\nu_1 t)$, with a local oscillator at a different frequency, say $\cos(2\pi\nu_2 t)$. By elementary trigonometry, the product has terms at the sum and difference frequencies, $\cos[2\pi(\nu_1 \pm \nu_2)t]$. Exactly the same argument holds in the spatial domain; multiplication of $\cos(2\pi\xi_1 x)$ and $\cos(2\pi\xi_2 x)$ yields terms in $\cos[2\pi(\xi_1 \pm \xi_2)x]$. In (3.289), the function $f(x)$ is sampled by multiplying it by the comb function, which contains all harmonics of the fundamental sampling frequency $1/\Delta x$. In the multiplication each frequency component ξ in $f(x)$ is heterodyned to a set of new frequencies $\xi \pm \frac{n}{\Delta x}$, as seen in (3.290).

Another way to think about aliasing is in terms of null functions of the sampling operator $\boldsymbol{\mathcal{S}}$ which maps $f(x)$ to its samples $\{f(j\Delta x)\}$. Since $\cos(2\pi\xi x)$ and $\cos[2\pi(\xi \pm \frac{n}{\Delta x})x]$ yield exactly the same samples, the difference function $\cos(2\pi\xi x) - \cos[2\pi(\xi \pm \frac{n}{\Delta x})x]$ is a null function of $\boldsymbol{\mathcal{S}}$.

The reader has, no doubt, seen examples of aliasing in everyday experience. A television image is sampled by the raster lines, and the moiré pattern seen in a TV image of empty seats in a football stadium or a performer's striped coat is due to aliasing. Here the coat or stadium is multiplied by the raster function, generating a multitude of new frequencies. An interesting point in this case is that the spatial frequencies are vectors, and heterodyning produces vector sum and difference frequencies. Both the raster pattern and the coat might correspond to very high spatial frequencies, but if the vector difference is small, it is easily seen in the TV image.

There are two ways to avoid significant aliasing in practice. One is simply to use a very fine sampling interval small (small Δx) to ensure that the Nyquist condition is satisfied. If the function being sampled is not truly bandlimited, it would be desirable to choose Δx small compared to $1/(2\xi_{max})$, where $F(\xi)$ has dropped to some very small value by $\xi = \pm\xi_{max}$.

The second way to avoid aliasing is to prefilter the function with a low-pass *anti-aliasing* filter before sampling. In the TV example above, the camera lens might constitute such a filter. If the lens cannot pass the high spatial frequencies in the performer's coat, they cannot heterodyne with the raster pattern. Some information has been lost by the low-pass filter, but the effect is usually less deleterious than aliasing.

3.5.4 Sampling in frequency space

The sampling theorem (3.293) and the Nyquist condition (3.291) are applicable when $f(x)$ is a bandlimited function, which means that its Fourier transform has finite support. Two dual relations hold if $f(x)$ itself has finite support. In that case we can reconstruct its Fourier transform $F(\xi)$ from samples taken according to a different Nyquist condition.

Suppose $f(x)$ is zero outside $(-\frac{1}{2}L, \frac{1}{2}L)$ and that $F(\xi)$ is sampled at $\xi_j = j\Delta\xi$, $j = 0, \pm 1, \pm 2, ..., \pm\infty$. Under what circumstances can $F(\xi)$ be recovered from $\{F(j\Delta\xi)\}$ and what is the explicit interpolation formula? We can answer these questions by little more than a change of notation in the derivation above. The spatial variable x becomes the frequency variable ξ, $f(x)$ becomes $F(\xi)$, the sample interval Δx becomes $\Delta\xi$, and the bandlimit B is replaced by the space limit L. The new Nyquist condition is, by analogy to (3.291),

$$\Delta\xi \leq 1/L\,, \tag{3.296}$$

and the interpolation formula analogous to (3.293) is

$$F(\xi) = \sum_{j=-\infty}^{\infty} F(j\Delta\xi)\,\mathrm{sinc}\left(\frac{\xi}{\Delta\xi} - j\right). \tag{3.297}$$

3.5.5 Multidimensional sampling

If samples are acquired in Cartesian coordinates, the Whittaker-Shannon sampling theorem extends trivially to higher dimensions. Suppose, for example, we sample a 2D function $f(x, y)$ at $x = ja$, $y = ka$, where j and k are integers, and that $F(\xi, \eta) = 0$ unless $-\frac{1}{2}B \leq \xi,\ \eta \leq \frac{1}{2}B$, with $aB \leq 1$. Under these circumstances,

(3.293) generalizes to

$$f(x,y) = \sum_{j=-\infty}^{\infty} \sum_{k=-\infty}^{\infty} f(ja, ka)\,\text{sinc}\left(\frac{x}{a} - j\right) \text{sinc}\left(\frac{y}{a} - k\right) . \tag{3.298}$$

But there are many other possible sampling theorems in two or more dimensions. The sampling need not be on a square lattice, and the support of the Fourier transform of $f(x, y)$ need not be a simple product of rects. Rather than enumerate some subset of these theorems, we shall present a single general formula from which they can all be derived. The treatment parallels the 'alternative derivation' of the 1D sampling theorem above.

Consider an nD function $f(\mathbf{r})$ with Fourier transform $F(\boldsymbol{\rho})$. Consider also a function $S(\boldsymbol{\rho})$ that takes on the value 1 within the support of $F(\boldsymbol{\rho})$ and 0 outside it, so that

$$F(\boldsymbol{\rho}) = F(\boldsymbol{\rho})\, S(\boldsymbol{\rho}) . \tag{3.299}$$

Using the notation of Sec. 3.4.6 and following the development in Marks (1991), we presume that $f(\mathbf{r})$ is sampled on the points $\mathbf{r} = \mathbf{Pm}$, where $\mathbf{P}$ is an $n \times n$ matrix and $\mathbf{m}$ is an $n \times 1$ column vector with all elements given by integers ranging from $-\infty$ to ∞. The samples of $f(\mathbf{r})$ are thus the set of numbers $\{f(\mathbf{Pm})\}$. As in (3.288), we can use the samples to define an auxiliary function $f_s(\mathbf{r})$, given by

$$f_s(\mathbf{r}) = \sum_{\mathbf{m}} f(\mathbf{Pm})\, \delta(\mathbf{r} - \mathbf{Pm}) = f(\mathbf{r}) \sum_{\mathbf{m}} \delta(\mathbf{r} - \mathbf{Pm}) . \tag{3.300}$$

An nD Fourier transform and use of (3.270) yields [*cf.* (3.290)]

$$\begin{aligned} F_s(\boldsymbol{\rho}) &= F(\boldsymbol{\rho}) * \mathcal{F}_n \left\{ \sum_{\mathbf{m}} \delta(\mathbf{r} - \mathbf{Pm}) \right\} = \frac{1}{|\det(\mathbf{P})|} F(\boldsymbol{\rho}) * \sum_{\mathbf{m}} \delta(\boldsymbol{\rho} - \mathbf{Qm}) \\ &= \frac{1}{|\det(\mathbf{P})|} \sum_{\mathbf{m}} F(\boldsymbol{\rho} - \mathbf{Qm}) , \end{aligned} \tag{3.301}$$

where $\mathbf{Q}^t = \mathbf{P}^{-1}$. As discussed in Sec. 3.4.6, the points $\mathbf{Qm}$ define the reciprocal lattice.

As in the corresponding 1D derivation, we now have a set of shifted replicas of $F(\boldsymbol{\rho})$; here the replicas are centered on the reciprocal lattice points $\mathbf{Qm}$. If these replicas do not overlap, we can multiply $F_s(\boldsymbol{\rho})$ by the support function $S(\boldsymbol{\rho})$ to isolate a single replica, giving [*cf.* (3.292)]

$$F(\boldsymbol{\rho}) = |\det(\mathbf{P})|\, S(\boldsymbol{\rho})\, F_s(\boldsymbol{\rho}) . \tag{3.302}$$

An inverse transform now gives

$$\begin{aligned} f(\mathbf{r}) &= |\det(\mathbf{P})|\, s(\mathbf{r}) * f_s(\mathbf{r}) = |\det(\mathbf{P})|\, s(\mathbf{r}) * \left[\sum_{\mathbf{m}} f(\mathbf{Pm})\, \delta(\mathbf{r} - \mathbf{Pm}) \right] \\ &= |\det(\mathbf{P})| \left[\sum_{\mathbf{m}} f(\mathbf{Pm})\, s(\mathbf{r} - \mathbf{Pm}) \right] . \end{aligned} \tag{3.303}$$

This is the desired multidimensional generalization of the Whittaker-Shannon sampling theorem. The interpolation function is, within a constant, the inverse Fourier transform of the support function $S(\boldsymbol{\rho})$. There is no simple statement of the Nyquist condition except to say that there must be no overlap of the replicas of $F(\boldsymbol{\rho})$ centered on the reciprocal-lattice points. For many special cases and applications of this formula, see Marks (1991).

3.5.6 Sampling with a finite aperture

So far we have considered only point sampling as the operator $\boldsymbol{\mathcal{S}}$. In one dimension, $\boldsymbol{\mathcal{S}}$ transforms the function $f(x)$ into a discrete set of numbers $\{f(x_j)\}$. There are, however, many other possible forms for the sampling operator. We can, for example, define $\boldsymbol{\mathcal{S}}$ as a normalized integral of the function over a small region, so that

$$f_j = \boldsymbol{\mathcal{S}} f(x) = \frac{1}{\epsilon} \int_{x_j - \frac{1}{2}\epsilon}^{x_j + \frac{1}{2}\epsilon} dx\, f(x)\,. \tag{3.304}$$

If we take $x_j = j\Delta x$ and $\Delta x = \epsilon$, then the regions are contiguous pixels. On the other hand we can also take $x_j = j\Delta x$ but pass to the limit $\epsilon \to 0$. In this case, if $f(x)$ is continuous at $x = x_j$, we have

$$\lim_{\epsilon \to 0} f_j = f(x_j) = f(j\Delta x)\,. \tag{3.305}$$

A general way of describing these transformations is in terms of continuous-to-discrete mappings as discussed in Sec. 1.2.4. By analogy to (1.30), we can write

$$f_j = \int_{-\infty}^{\infty} dx\, f(x)\, a(x_j - x)\,, \tag{3.306}$$

where $a(x)$ is called the *sampling function* or *aperture function*. The latter designation is suggestive of instruments such as scanning microdensitometers where an optical aperture is placed at a sequence of locations in an image plane.

To get back to (3.304) from this general form, we take

$$a(x) = \epsilon^{-1}\, \mathrm{rect}(x/\epsilon)\,, \tag{3.307}$$

while point sampling is recovered by taking $a(x) = \delta(x)$.

It is interesting to inquire whether there exists a counterpart of the Whittaker-Shannon theorem for the more general kind of sampling defined in (3.306). To answer this question, note that (3.306) can also be written as

$$f_j = [f * a]\,(x_j)\,, \tag{3.308}$$

where $[f * a](x)$ denotes the convolution of $f(x)$ and $a(x)$ (see Sec. 3.3.6). Thus sampling a function with a finite aperture is equivalent to first convolving it with an aperture function, then performing point sampling on the result.

Since convolution corresponds to multiplication in the frequency domain, $[f * a](x)$ is bandlimited to $(-\frac{1}{2}B, \frac{1}{2}B)$ if $f(x)$ is. Thus $[f * a](x)$ can be recovered from its samples $\{f_j\}$ if $x_j = j\Delta x$ and Δx satisfies the Nyquist condition, $B\Delta x = 1$. From (3.293), we have immediately that

$$[f * a]\,(x) = \sum_{j=-\infty}^{\infty} f_j\, \mathrm{sinc}\left(\frac{x}{\Delta x} - j\right)\,. \tag{3.309}$$

Of course, all this says is that we can recover $[f * a](x)$; we would really like to recover $f(x)$ itself. To see when that is possible, we take the Fourier transform of the convolution, yielding

$$\mathcal{F}_1\{[f * a]\,(x)\} = F(\xi)\,A(\xi)\,, \tag{3.310}$$

where, as usual, functions with capital letters are the Fourier transforms of the corresponding lower-case functions. If $A(\xi)$ does not vanish in the interval $-\frac{1}{2}B < \xi < \frac{1}{2}B$, we can find $F(\xi)$ in that interval by dividing through by $A(\xi)$. Since $F(\xi)$ is assumed to vanish outside that interval, an inverse Fourier transform then yields the desired $f(x)$.

As an example, let $a(x)$ be given by (3.307). Then, by (3.136), $A(\xi) = \text{sinc}(\epsilon\xi)$, which does not vanish in $(-\frac{1}{2}B, \frac{1}{2}B)$ provided $\epsilon < 2B^{-1}$. If the Nyquist condition is exactly satisfied, so that $B\Delta x = 1$, $A(\xi)$ will not vanish in the interval of interest if $\epsilon < 2\Delta x$. Since the usual pixel sampling corresponds to $\epsilon = \Delta x$, there is no problem in recovering $f(x)$ from its samples on contiguous pixels if the Nyquist condition is satisfied. It will be left as an exercise for the reader to develop an explicit interpolation formula for this case.

3.6 DISCRETE FOURIER TRANSFORM

3.6.1 Motivation and definitions

Though many Fourier transforms can be found analytically, there are also many circumstances where numerical methods are needed. In image processing, for example, we usually do not have an analytical description of the image, so analytical determination of its Fourier transform is out of the question. In other circumstances we might have an analytic expression for the function but not be able to express its Fourier transform in closed form.

In such cases, we are forced to represent the function to be transformed by a discrete set of numbers and to approximate the Fourier integral by a sum. One way to do so is:

$$F(\xi) \simeq \Delta x \sum_{k=-\infty}^{\infty} f(k\Delta x) \exp(-2\pi i \xi k \Delta x)\,. \tag{3.311}$$

This form is still not amenable to numerical computation for two reasons: For each ξ an infinite sum is required, and there are an infinite number of values of ξ.

To restrict the sum to a finite number of terms, we must presume that $f(x)$ vanishes, at least approximately, outside some finite interval. As in Sec. 3.1.4, we take that interval as $[0, L)$ and divide it into N equal steps of size $\Delta x = L/N$, so that the samples are at $x = x_k = k\Delta x = kL/N$, $k = 0, ..., N-1$. Since $f(x)$ now has finite support, $F(\xi)$ can be represented exactly by its samples at $\xi = \xi_n = n\Delta\xi$, provided these samples in frequency space satisfy the dual Nyquist condition (3.296), $\Delta\xi \le 1/L$. We choose the inequality to be an equality. The samples $F(n\Delta\xi)$ can be approximated by $\Delta x F_n$, where

$$F_n \equiv \sum_{k=0}^{N-1} f_k \exp(-2\pi i k n/N)\,, \tag{3.312}$$

and $f_k = f(k\Delta x)$. In the limit as $N \to \infty$ (or equivalently, $\Delta x \to 0$), $\Delta x F_n \to F(n\Delta\xi)$.

Though motivated as an approximation to the Fourier integral, (3.312) is an important linear transformation in its own right. Known as the *discrete Fourier transform* or *DFT*, it transforms the N-dimensional vector $\mathbf{f}$ with components $\{f_k\}$ to another N-dimensional vector $\mathbf{F}$ with components $\{F_n\}$. The exact inverse to this transform can be established from the orthogonality relation (3.12), yielding

$$f_k = \frac{1}{N}\sum_{n=0}^{N-1} F_n \exp(2\pi ikn/N)\,. \tag{3.313}$$

Another way to express these results is to recognize that $\exp(-2\pi i/N)$, which we can denote as W_N, is an Nth root of unity (see Sec. B.1.5), *i.e.*,

$$[W_N]^N = [\exp(-2\pi i/N)]^N = \exp(-2\pi i) = 1\,. \tag{3.314}$$

In terms of W_N, the DFT is

$$F_n = \sum_{k=0}^{N-1} f_k W_N^{kn}\,, \tag{3.315}$$

and its inverse is

$$f_k = \frac{1}{N}\sum_{n=0}^{N-1} F_n W_N^{-kn}\,. \tag{3.316}$$

We shall return to the question of how well the DFT approximates the Fourier transform in Sec. 3.6.3, but first we establish some basic properties of the DFT itself.

3.6.2 Basic properties of the DFT

The DFT has many properties analogous to those of the Fourier series and Fourier transform as discussed in Secs. 3.2 and 3.3. The one exception is that there is no need to discuss convergence since finite sums of finite values are necessarily finite.

Linearity Like its kin, the DFT is a linear transformation. If $\mathbf{f}$ and $\mathbf{g}$ are N-dimensional vectors with DFTs $\mathbf{F}$ and $\mathbf{G}$, respectively, and we define

$$\mathbf{h} = \alpha\mathbf{f} + \beta\mathbf{g}\,, \tag{3.317}$$

where α and β are constants, then

$$\mathbf{H} = \alpha\mathbf{F} + \beta\mathbf{G}\,, \tag{3.318}$$

where $\mathbf{H}$ is the DFT of $\mathbf{h}$.

Periodicity and folding Since W_N is an Nth root of unity, raising it to a power that is an integer multiple of N leaves it invariant. From this observation it follows that

$$W_N^{kn} = W_N^{k(n+N)}\,. \tag{3.319}$$

Plugging this result into (3.315) reveals that

$$F_{n+N} = F_n \,. \tag{3.320}$$

In other words, regardless of the sequence $\{f_k\}$, its DFT is periodic with period N. Though we are free to evaluate F_n for any n, there are only N independent values. It is conventional to choose $n = 0, 1, ..., N-1$, but we could equally well choose any other range of N contiguous integers.

Furthermore, if all of the $\{f_k\}$ are real, there is an additional constraint on the $\{F_n\}$. It is straightforward to show in this case that

$$F_n = F_{-n}^* \qquad \text{if } f_k \text{ is real for all } k \,. \tag{3.321}$$

This result is the DFT analog of the Hermiticity condition for Fourier series or transforms, (3.45) or (3.94). Combining (3.320) and (3.321), we also have

$$F_{\frac{1}{2}N+n} = F_{\frac{1}{2}N-n}^* \qquad \text{if } f_k \text{ is real for all } k \,, \tag{3.322}$$

where N must be even. Thus if we choose the basic range of n to run from 0 to $N-1$ and all of the $\{f_k\}$ are real, there is a folding point at $n = \frac{1}{2}N$. The real part of F_n is symmetric about this point while the imaginary part is antisymmetric. These relations are illustrated in Fig. 3.9.

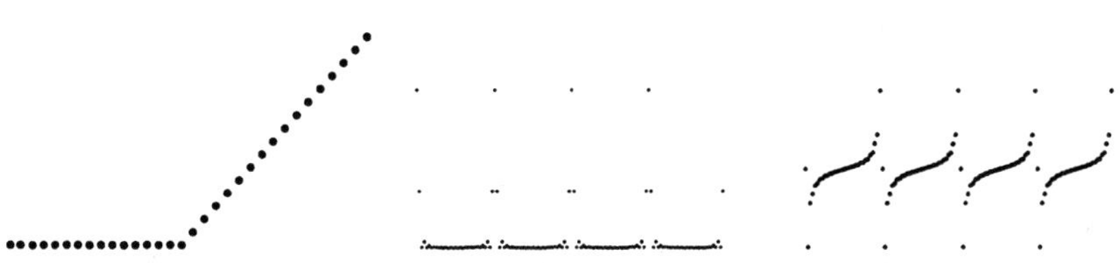

(a) (b) (c)

Fig. 3.9 Illustration of symmetry properties of the DFT of a real sequence. (a) A 32-point real sequence $\{f_n, n = 0, ..., 31\}$; (b) Real part of the DFT of this sequence plotted over five periods; (c) Imaginary part of the DFT of this sequence plotted over five periods.

The same symmetry relations apply to the inverse DFT. We can regard (3.313) as a representation of our original sequence for $n = 0, 1, ..., N-1$, but given the representation, we can plug in any n whatsoever. Doing so shows that the representation is periodic with period N. Like the Fourier series, an inverse DFT is a representation of an arbitrary function on a finite interval or of a periodic function on the infinite interval.

Shifted sequences In discussing Fourier transforms, we found that shifting the function was equivalent to multiplying its transform by a linear phase factor [see (3.108)]. A similar result holds for the DFT, but we must be careful how we define a shift.

Given a sequence $\{f_k, k = 0, ..., N-1\}$, let us define a new sequence $\{y_k, k = 0, ..., N-1\}$ by

$$y_k = f_{k-m} \,. \tag{3.323}$$

The problem with this definition is that $k-m$ may not be in the range 0 to $N-1$. When this occurs, we have two options: we can either take f_{k-m} to be zero or we can assume that the original sequence $\{f_k\}$ is defined for all k by periodically repeating the original N values. This latter view is the one that is consistent with the periodicity properties of the DFT as discussed above, and it is the choice we shall make. Thus, if $k-m$ in (3.323) lies outside $[0, N-1]$, we must add or subtract a multiple of N to get it back into this range. Technically, the index of f is $k-m$, modulo N, but to keep the notation simple we do not explicitly show the modulo convention. Shifting modulo N is also referred to as *cyclic shifting*.

The DFT of the (cyclically) shifted sequence is

$$Y_n = \exp(-2\pi i n m/N)\, F_n \,. \tag{3.324}$$

Again, as in (3.108), shifting the input to the transform operation has the effect of multiplying the output by a linear phase factor, here linear in n.

Kronecker deltas Suppose the sequence $\{f_k\}$ consists of all zeros except for a one in the mth position, *i.e.*,

$$f_k = \delta_{km} \,, \tag{3.325}$$

where δ_{km} is the Kronecker delta. The DFT of this particular sequence is

$$F_n = \sum_{k=0}^{N-1} \delta_{km} \exp(-2\pi i k n/N) = \exp(-2\pi i m n/N) \,. \tag{3.326}$$

This result is the DFT analog of (3.148); again a delta function transforms to a linear phase factor.

Since $\exp(-2\pi i m n/N)$ is periodic in n with period N, we have a simple example of the general periodicity condition (3.320). Equation (3.326) is also an illustration of the shift theorem (3.324) since $F_n = 1$ for all n if $f_k = \delta_{k0}$.

Now consider a sequence $\{f_k\}$ defined by

$$f_k = \exp(2\pi i k m/N) \,. \tag{3.327}$$

The DFT yields

$$F_n = N\, \delta_{mn} \,, \tag{3.328}$$

where we have used (3.12). This is the analog of (3.151); a linear phase factor transforms to a delta function. Here it is a Kronecker delta times a factor of N.

Discrete convolution and correlation A discrete implementation of the convolution equation $g(x) = [f * h](x)$ is [*cf.* (3.114)]

$$g_n = \frac{1}{N} \sum_{k=0}^{N-1} f_k h_{n-k} \,, \qquad n = 0, 1, 2, \ldots, N-1 \,, \tag{3.329}$$

where, as above, the shift is interpreted modulo N. The factor of $1/N$, corresponding to the dx in a convolution integral, should not be overlooked.

The DFT counterpart of the convolution theorem, (3.132), is

$$G_n = F_n H_n \,. \tag{3.330}$$

The proof follows from the definition of the DFT and (3.12). The factor of $1/N$ in the definition of discrete convolution, (3.329), eliminates the factor of N that comes from the discrete orthogonality relation, (3.12).

Discrete correlation can be defined similarly:

$$g_n = \frac{1}{N} \sum_{k=0}^{N-1} f_k h_{n+k}, \qquad n = 0, 1, 2, ..., N-1, \tag{3.331}$$

and a DFT gives the analog of (3.134),

$$G_n = F_n H_n^* . \tag{3.332}$$

Parseval's relation Parseval's relation, the analog of (3.51) or (3.79), is

$$\sum_{k=0}^{N-1} f_k h_k^* = \frac{1}{N} \sum_{n=0}^{N-1} F_n H_n^* . \tag{3.333}$$

The left-hand side of this equation is the Euclidean scalar product of the N-dimensional vectors $\mathbf{f}$ and $\mathbf{h}$, while the right-hand side is $1/N$ times the similar scalar product of $\mathbf{F}$ and $\mathbf{H}$. Had we defined the DFT with a factor of $1/\sqrt{N}$ in both the forward and inverse transforms, this extra factor of $1/N$ would not have appeared and we would be able to say that discrete Fourier transformation was unitary. The situation is completely analogous to Fourier series, where we have an asymmetric factor of $1/L$ (see Sec. 3.2.3).

3.6.3 Relation between discrete and continuous Fourier transforms

We return now to the question of how a DFT is related to a continuous Fourier transform. Our goal in this section is not only to show the mathematical relation but also to offer some practical suggestions to anyone who wishes to use the DFT to compute an approximation to a continuous transform.

The formal mathematical connection between the discrete and continuous transforms rests on two concepts: sampling and periodic replication. We saw in Sec. 3.5 that sampling a function in the space domain, as in (3.289), produces a periodic function in the frequency domain, (3.290). Conversely, periodically replicating a function as in (3.192) yields a Fourier transform that is sampled in the sense that it contains only a discrete set of frequencies. The DFT of any sequence is both discrete (sampled) and periodic, so we might suspect that it corresponds in some sense to a continuous Fourier transform from one periodic, sampled function to another.

To make this argument more precise, we begin with an arbitrary continuous function $f(x)$ and consider the sequence of steps needed to create from it a periodic, sampled function (Hayes, 1992; Hakimmashhadi, 1988). The steps involved are illustrated in Fig. 3.10.

The first step is to sample the function at points $x_n = n\Delta x$. As in (3.289) we can then define the auxiliary function $f_s(x)$ in terms of the sample values by

$$f_s(x) = f(x) \frac{1}{\Delta x} \operatorname{comb}\left(\frac{x}{\Delta x}\right) = \sum_{n=-\infty}^{\infty} f(n\Delta x)\, \delta(x - n\Delta x) . \tag{3.334}$$

From (3.290) and Fig. 3.7, we know that this operation has the effect of periodically replicating the Fourier transform $F(\xi)$. If $F(\xi)$ does not have finite support or the

Nyquist condition is not satisfied, there is aliasing or overlap of the various replicas.

Since we do not wish to assume that the original $f(x)$ itself has finite support, we now impose a finite support by multiplying $f_s(x)$ by a window function that selects $0 \leq x < L$. The appropriate window function is

$$w(x) = \text{rect}\left[\frac{x - \frac{1}{2}L + \epsilon}{L}\right], \tag{3.335}$$

where ϵ is a positive number less than Δx which ensures that the point at $x = 0$ is included in the window while the one at $x = L$ is not. We define $N = L/\Delta x$ so that exactly N points are included in the window; the index n thus ranges from 0 to $N - 1$. The resulting function after sampling and windowing is

$$f_{sw}(x) = f_s(x)\, w(x) = \sum_{n=0}^{N-1} f(n\Delta x)\, \delta\left(x - \frac{nL}{N}\right). \tag{3.336}$$

The (continuous) Fourier transform of this function is

$$F_{sw}(\xi) = F_s(\xi) * W(\xi). \tag{3.337}$$

The next (and last) step is periodic replication. As defined in (3.192), this operation is described mathematically by convolution with $L^{-1}\,\text{comb}(x/L)$, where L is the period. (Note that we take the period to be identical to the width of the window function.) The resulting sampled, windowed and periodically replicated function is given by[3]

$$f_{swp}(x) = f_{sw}(x) * \frac{1}{L}\,\text{comb}\left(\frac{x}{L}\right). \tag{3.338}$$

Substituting (3.336) into (3.338) and performing the convolution by means of (3.127), we find

$$f_{swp}(x) = \sum_{n=0}^{N-1} f(n\Delta x)\, \delta\left(x - \frac{nL}{N}\right) * \sum_{m=-\infty}^{\infty} \delta(x - mL) =$$

$$\sum_{m=-\infty}^{\infty} \sum_{n=0}^{N-1} f(n\Delta x)\, \delta\left(x - \frac{nL}{N} - mL\right). \tag{3.339}$$

Now it is straightforward to take the Fourier transform of $f_{swp}(x)$, with the result

$$F_{swp}(\xi) = \sum_{m=-\infty}^{\infty} \sum_{n=0}^{N-1} f(n\Delta x) \exp[-2\pi i \xi (nL/N + mL)]. \tag{3.340}$$

By (2.50) we can also write

$$F_{swp}(\xi) = \frac{1}{L} \sum_{m=-\infty}^{\infty} \left[\sum_{n=0}^{N-1} f(n\Delta x) \exp(-2\pi i nm/N)\right] \delta(\xi - m/L). \tag{3.341}$$

[3]The shorthand notation for convolution in (3.338) should not be misinterpreted. The shift variable is x, not x/L. See Sec. 3.3.6 and the discussion below (3.142).

Not surprisingly, $F_{swp}(\xi)$ is sampled and periodic, and the samples are given by the DFT of the sequence $\{f(n\Delta x)\}$. Thus, *performing a DFT of the samples of $f(x)$ produces the samples of the Fourier transform of a sampled, windowed and periodically replicated version of $f(x)$.*

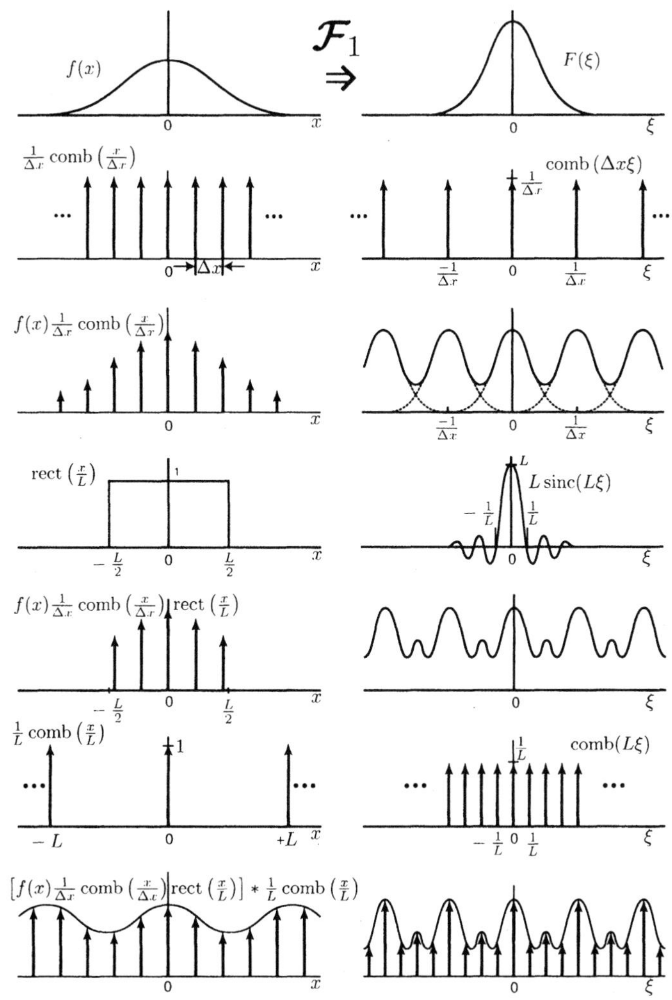

Fig. 3.10 Steps in going from the FT to the DFT.

Errors in the DFT: Their causes and cures Each of the steps leading up to (3.341) constitutes a potential source of error in computation of the continuous Fourier transform by means of the DFT. In this section we briefly discuss the nature of these errors and some ways to minimize them.

As we have seen, sampling the original function produces error if $f(x)$ is not bandlimited or the Nyquist condition is not satisfied. One solution to this problem is to use finer sampling. If we have an analytic expression for $f(x)$, the only price paid for this remedy is increased computer time. In many practical situations, however, the number of samples is limited by a finite sensor array or other technological considerations. In those cases some sort of anti-aliasing filter as discussed in Sec. 3.5.3 is desirable. In an imaging context, the bandwidth of the signal being sampled can be limited by the characteristics of the imaging lens or by finite detector size.

The windowing step produces an error if the function being sampled does not fit into the window. The nature of the error is seen from (3.337); the convolution of $F_s(\xi)$ with the transform of the window function has the effect of smoothing out sharp features in $F_s(\xi)$ as shown in Fig. 3.10. Since adjacent frequency components are blurred together in this way, the effect is sometimes referred to as *spectral leakage*. The effect can be minimized by using a larger window so that $W(\xi)$ more closely approximates a delta function. Another useful measure is to use a smooth function for $w(x)$ rather than a rect so that $W(\xi)$ falls off more rapidly with increasing ξ.

The final source of error is the sampling of $F(\xi)$ itself. The spacing of samples in the frequency domain is $1/L$, which satisfies the dual Nyquist condition (3.296) if the support of $f(x)$ is L, but nevertheless we may want finer samples in order to fully appreciate the structure of $F(\xi)$. It is often desirable to use a window function that is substantially larger than the support of $f(x)$ simply to decrease the spacing of the points in the frequency domain. In practice, this is done by appending a string of zeros to the sample values $f(n\Delta x)$, creating an artificially longer sequence as the input to the DFT. This procedure is known as *zero padding*.

For further tips on practical applications of the DFT, see Lathi (1992), Hayes (1992), or Walker (1991).

3.6.4 Discrete-Space Fourier Transform

The mathematics of the DFT are based on the assumption that the input is a finite set of N numbers. In this section we investigate the *discrete-space Fourier transform*, or DSFT, which maps an infinite number of discrete values to the spatial-frequency domain. The assumption of an infinite number of samples is often plausible in the analysis of temporal data sets such as those encountered in many communications applications. For such problems the *discrete-time Fourier transform* (Oppenheim and Schafer, 1989) is a common method of analysis. In imaging applications involving large detector arrays, where boundary effects are less important, the assumption of an infinite set of samples may also be reasonable. We present here a description of the properties of the DSFT for completeness.

Suppose we are given an infinite number of samples f_k. The discrete-space Fourier transform of this set is defined by

$$F_{DS}(\xi) \equiv \sum_{k=-\infty}^{\infty} f_k \exp(-2\pi ik\xi\Delta x)\,, \tag{3.342}$$

where the subscript DS is short for *discrete-space* and Δx is the distance between sample values. The sample values f_k can be recovered from $F_{DS}(\xi)$ by the following inverse transformation:

$$f_k \equiv \Delta x \int_{-1/(2\Delta x)}^{1/(2\Delta x)} d\xi\, F_{DS}(\xi) \exp(2\pi ik\xi\Delta x)\,. \tag{3.343}$$

As we shall see, the DSFT can be thought of as the limit of the DFT in the case of an infinite number of samples, but it has several unique properties of its own.

Relationship with DFT The DFT (3.312) transforms the set of discrete samples $\{f_k\}$ to the samples $\{F_n\}$. If we were to consider a slightly different definition, in which the summation is symmetric, we could write

$$F_n = \sum_{k=-\frac{N}{2}}^{\frac{N}{2}-1} f_k \exp(-2\pi ikn/N). \tag{3.344}$$

When we make the changes of variables, $\Delta\xi = 1/L$, $\xi = n\Delta\xi$ and $\Delta x = L/N$, we find that

$$F_n\big|_{n=\frac{\xi}{\Delta\xi}} = \sum_{k=-\frac{N}{2}}^{\frac{N}{2}-1} f_k \exp(-2\pi ik\xi\Delta x). \tag{3.345}$$

We know from our discussion of the DFT that as the number of samples N increases, the output of (3.345) becomes increasingly finely sampled. In the limit where N becomes infinite, (3.345) is equivalent to (3.342), and the output is continuous. Thus the transformation of (3.342) is a discrete-to-continuous transformation.

From Sec. 3.6.2 we also know that F_n in (3.345) is periodic with period $1/\Delta x$. Thus $F_{DS}(\xi)$ is also periodic with period $1/\Delta x$. For this reason the integral in (3.343) need only be evaluated over the central zone of the periodic function $F_{DS}(\xi)$; values of the function outside this zone are redundant.

Relationship with Fourier series By comparison with (3.17) we see that (3.342) is a Fourier-series expansion of the continuous function $F_{DS}(\xi)$. The samples of the discrete function f_k in (3.342) play the role of the series coefficients. The inverse discrete-space transform of (3.343) is directly analogous to the formula given for determination of the Fourier series expansion coefficients [*cf.* (3.19)]. Perhaps the odd thing here is that (3.342) is a Fourier series expansion of a function we think of as being in the Fourier domain!

Another way to see the relationship between the DSFT and the Fourier series is to use operator theory. We noted in Sec. 3.2.3 that the mapping from a function on a finite interval to the infinite set of Fourier coefficients is a linear operator; the DSFT, which is a mapping from an infinite set of coefficients to a function on a finite interval, is the adjoint of the Fourier-series operator. Since we also know from Sec. 3.2.3 that the Fourier-series operator is unitary (up to normalization factors), the DSFT is the inverse of the Fourier series, slightly disguised by the notation.

Relationship with sampling Suppose, as we did in Sec. 3.5.2, that the samples $\{f_k\}$ have been acquired from an underlying continuous function $f(x)$ via the sampling operator $\boldsymbol{\mathcal{S}}$. The samples can be used to form the auxiliary function $f_s(x)$, related to $f(x)$ by

$$f_s(x) = \boldsymbol{\mathcal{S}}^\dagger \boldsymbol{\mathcal{S}} f(x) = \sum_{k=-\infty}^{\infty} f(x)\,\delta(x - k\Delta x) = f(x)\frac{1}{\Delta x}\operatorname{comb}\left(\frac{x}{\Delta x}\right). \tag{3.346}$$

The Fourier transform of this function is

$$F_s(\xi) = \sum_{k=-\infty}^{\infty} \int_{-\infty}^{\infty} dx\, f(x)\,\delta(x - k\Delta x)\exp(-2\pi i\xi x) = \sum_{k=-\infty}^{\infty} f_k \exp(-2\pi ki\xi\Delta x), \tag{3.347}$$

where $f_k = f(k\Delta x)$. By comparison with (3.342) we see that $F_s(\xi)$ is equivalent to $F_{DS}(\xi)$, the DSFT of the samples of $f(x)$.

Consider again the transform pairs represented by the various rows of Fig. 3.10. We have just found that the DSFT plays a role in row 3 of the figure. The function on the left in row 3 is $f_s(x)$, the continuous auxiliary function obtained by weighting delta functions with the sample values of $f(x)$. The Fourier transform on the right of row 3 is $F_s(\xi)$, the continuous FT of $f_s(x)$. We now know this to be also the DSFT of the sample values f_k, $F_{DS}(\xi)$. The figure shows that $F_{DS}(\xi)$ is the sum of an infinite number of shifted replications of the Fourier transform $F(\xi)$. The amount of overlap of the replications is zero if the bandwidth B of $f(x)$ is such that the sampling distance $\Delta x \leq 1/B$.

In summary, the striking difference between the discrete Fourier transform (DFT) and the discrete-space Fourier transform (DSFT) is that the DFT is a transformation from a finite set of samples to another finite set of samples. In the case of the DSFT, where the number of input samples is infinite, the output is a continuous function. Another way to consider this is that the input to the DSFT is not assumed to be periodic, or equivalently, the period is infinite. The continuous-to-continuous Fourier transform can be recovered from the DSFT in the limit where the sampling distance becomes infinitesimally small.

Relationship with Poisson summation We have seen that $F_{DS}(\xi)$ is a periodic function with periodicity $1/\Delta x$. It is the function we obtain when we add an infinite number of shifted replications of the general (nonperiodic) function $F(\xi)$. That is,

$$F_{DS}(\xi) = \frac{1}{\Delta x} \sum_{n=-\infty}^{\infty} F\left(\xi - \frac{n}{\Delta x}\right), \tag{3.348}$$

which can be verified by taking the Fourier transform of (3.346). By (3.342) we can equate this periodic function with a Fourier series with coefficients f_k, which are the samples of $f(x)$, giving

$$\frac{1}{\Delta x} \sum_{n=-\infty}^{\infty} F\left(\xi - \frac{n}{\Delta x}\right) = \sum_{k=-\infty}^{\infty} f_k \exp(-2\pi i k \xi \Delta x), \tag{3.349}$$

which is equivalent to the Poisson summation formula of (3.197). For the special case $\xi = 0$ we obtain

$$\frac{1}{\Delta x} \sum_{n=-\infty}^{\infty} F\left(\frac{n}{\Delta x}\right) = \sum_{k=-\infty}^{\infty} f_k . \tag{3.350}$$

As expected, the sum of the samples of $f(x)$ is equal to the sum of the samples of its Fourier transform.

3.6.5 Fast Fourier Transform

It can be seen from (3.315) that the DFT can be regarded as a matrix-vector multiplication. We can rewrite that equation in matrix-vector form as

$$\mathbf{F} = \mathbf{D}\mathbf{f}, \tag{3.351a}$$

or, in detail,

$$F_n = \sum_{k=0}^{N-1} D_{nk} f_k \,, \tag{3.351b}$$

where $D_{nk} = W_N^{nk} = \exp(-2\pi i n k/N)$. In order to obtain the value of a particular F_n, we must multiply each f_k by the appropriate matrix element, D_{nk} and sum the results. Doing this for all N components appears to require a total of N^2 multiplications and additions, but in fact it is possible to perform a DFT with only about $N \log_2 N$ operations if N is a power of 2. Algorithms for computing a DFT in this manner are known as *fast Fourier transforms* or *FFTs*. The classic text on the FFT is Brigham (1974), and the more recent books by Ramirez (1985) and Walker (1991) are also useful. Excellent short discussions are given by Hayes (1992), Hakimmashhadi (1988), and Kraniauskas (1994).

Though often attributed to Cooley and Tukey (1965), the basic idea of the FFT is actually much older. A fascinating historical survey is given by Heidemann *et al.* (1985) who trace the algorithm back to work by Gauss, first published only posthumously in 1866. Heidemann estimates that Gauss developed the FFT in 1805, which, if accurate, puts it two years before Fourier's own work on the Fourier series. The method was also used by the German numerical analyst Karl Runge in 1905, and there were numerous other independent discoveries of the same basic algorithm. What all of these works have in common is the recognition that the N^2 elements of $\mathbf{D}$ are not independent. Rather, the elements are all powers of W_N, the Nth root of unity, so there are only N independent elements.

We now show how to take advantage of this redundancy in the matrix elements. Our treatment closely follows the lucid account given by Hayes (1992), but similar treatments can be found in all of the references given in this section. Assume that N is even,

$$N = 2M \,, \tag{3.352}$$

and divide the sum in (3.315) into even and odd parts:

$$F_n = \sum_{k=0}^{N-1} f_k (W_N)^{nk} = \sum_{m=0}^{M-1} f_{2m} (W_{2M})^{2mn} + \sum_{m=0}^{M-1} f_{2m+1} (W_{2M})^{(2m+1)n} \,. \tag{3.353}$$

Since $(W_{2M})^2 = W_M$, we can also write

$$F_n = \sum_{m=0}^{M-1} f_{2m} (W_M)^{mn} + (W_{2M})^n \sum_{m=0}^{M-1} f_{2m+1} (W_M)^{mn} \,. \tag{3.354}$$

Each of these sums will be recognized as an M-element DFT. It is useful to define separate DFTs of the even and odd terms as

$$F_n^{(e)} = \sum_{m=0}^{M-1} f_{2m} (W_M)^{mn} \tag{3.355a}$$

$$F_n^{(o)} = \sum_{m=0}^{M-1} f_{2m+1} (W_M)^{mn} \,, \tag{3.355b}$$

where the superscripts denote even and odd, respectively. Thus

$$F_n = F_n^{(e)} + (W_{2M})^n F_n^{(o)} \,. \tag{3.356}$$

Since each of the M-element DFTs returns M numbers, this formula gives only the first M values of F_n, but the remaining ones are found by adding M to each index:

$$F_{n+M} = F_{n+M}^{(e)} + (W_{2M})^{n+M} F_{n+M}^{(o)} = F_n^{(e)} - (W_{2M})^n F_n^{(o)} , \tag{3.357}$$

where the last steps follows since $(W_{2M})^M = \exp(-i\pi) = -1$.

Equations (3.356) and (3.357) show that the desired N-element DFT can be computed as two M-element ones, where $M = N/2$. This result was first derived by Danielson and Lanczos (1942). If N is a power of 2, say 2^K, we can repeat this halving process K times and eventually perform a set of single-point transforms.[4] The total number of operations is then of order NK or $N \log_2 N$ instead of N^2. Some additions are also required by (3.356) and (3.357), but if N is large, the number of additions is small compared to $N \log_2 N$. We say that the required number of operations is $\mathcal{O}(N \log_2 N)$, where $\mathcal{O}$ is to be read "on the order of."

Practical algorithms for efficient computation of the DFT are found in many places. Hayes gives programs in BASIC, and Press *et al.* (1992) give ones in FORTRAN and C. Many commercial software packages provide FFT routines integrated with excellent graphics.

3.6.6 Multidimensional DFTs

The 2D DFT The 2D DFT is defined by

$$F_{mn} = \sum_{j=0}^{N-1} \sum_{k=0}^{N-1} f_{jk} \exp[-2\pi i(mj + nk)/N] = \sum_{j=0}^{N-1} \sum_{k=0}^{N-1} f_{jk} (W_N)^{mj+nk} . \tag{3.358}$$

The indices j, k, m and n all run from 0 to $N - 1$.

An alternative form of (3.358) is obtained by regarding F_{nm} and f_{jk} as elements of the $N \times N$ matrices $\mathbf{F}$ and $\mathbf{f}$, respectively. In both matrices, the first index denotes the row and the second the column. Now we can write

$$\mathbf{F} = \mathbf{D f D} , \tag{3.359}$$

where $\mathbf{D}$ is the $N \times N$ matrix defined in (3.351). Note that this is not a similarity transformation since $\mathbf{D}$ itself, rather than $\mathbf{D}^\dagger$ or $\mathbf{D}^{-1}$, appears in both places. Thus the 2D DFT with N elements in each direction can be conceptualized as resulting from a matrix product of three $N \times N$ matrices.

To see how the 2D DFT can be computed in practice, rewrite (3.358) as:

$$F_{mn} = \sum_{j=0}^{N-1} \left\{ \sum_{k=0}^{N-1} f_{jk} \exp(-2\pi i nk/N) \right\} \exp(-2\pi i mj/N) . \tag{3.360}$$

The quantity in brackets is a 1D N-element DFT on each column, with the column replaced *in situ* by its DFT. The outer sum corresponds to similar 1D DFTs on

[4]The halving process is frequently referred to as decimation or radix-2 decimation, but the word decimation has an evident root in Latin *decem* for ten, not two. Though used more broadly, the word decimate refers to the practice of selecting by lot and killing one-tenth of a population. It does not mean to reduce by a factor of ten, and certainly not to reduce by a factor of two.

the resulting rows. Thus the 2D DFT of an $N \times N$ image can be computed in $\mathcal{O}(N^2 \log_2 N^2)$ operations.

The extension to n dimensions is straightforward: An nD DFT can be performed as a sequence of n 1D DFTs in $\mathcal{O}(N^n \log_2 N^n)$ operations.

General lattices If we regard the nD DFT as a sampled approximation to the Fourier integral, it is of interest to study the case of sampling on an arbitrary lattice as in Sec. 3.4.6. The starting point for this discussion is the nD Fourier series (3.276).

Consider a function $f_0(\mathbf{r})$ defined at all points within a unit cell of an arbitrary lattice. The lattice points are $\mathbf{r} = \mathbf{Pk}$, where $\mathbf{k}$ is a multi-index, *i.e.* an nD vector with integer components. If we extend $f_0(\mathbf{r})$ to a periodic function $f(\mathbf{r})$ as in (3.273) and sample $f(\mathbf{r})$ at the points $\mathbf{r} = \mathbf{Pk}/N$, so that N points fit along each axis in the unit cell, then (3.276) becomes

$$f(\mathbf{Pk}/N) = \frac{1}{|\det(\mathbf{P})|} \sum_{\mathbf{m}} F_0(\mathbf{Qm}) \exp(2\pi i \mathbf{Qm} \cdot \mathbf{Pk}/N) =$$

$$\frac{1}{|\det(\mathbf{P})|} \sum_{\mathbf{m}} F_0(\mathbf{Qm}) \exp(2\pi i \mathbf{m} \cdot \mathbf{k}/N) , \tag{3.361}$$

where $\mathbf{Qm}$ are the points on the reciprocal lattice and the second form of (3.361) follows from $\mathbf{Q}^t = \mathbf{P}^{-1}$.

With this motivation we define the nD inverse DFT as

$$f_{\mathbf{k}} = \frac{1}{N^n} \sum_{\mathbf{m}=0}^{N-1} F_{\mathbf{m}} \exp(2\pi i \mathbf{m} \cdot \mathbf{k}/N) , \tag{3.362}$$

where $f_{\mathbf{k}}$ corresponds to $f(\mathbf{Pk}/N)$ and $F_{\mathbf{m}}$ to $F_0(\mathbf{Qm})$. We emphasize however, that these are correspondences and not equalities; the DFT is only an approximation to the sampled Fourier series, so (3.362) is a definition, not something derived from (3.361).

Note that the specific geometry of the lattices has disappeared in (3.362); the exponent involves only the scalar product $\mathbf{m}\cdot\mathbf{k}$ of the multi-indices, not the matrices $\mathbf{P}$ and $\mathbf{Q}$. It is only when we want to interpret $f_{\mathbf{k}}$ and $F_{\mathbf{m}}$ as samples of a function and its Fourier transform that the geometry matters.

The inverse of (3.362), the DFT itself, is given by

$$F_{\mathbf{k}} = \sum_{\mathbf{m}=0}^{N-1} f_{\mathbf{m}} \exp(-2\pi i \mathbf{m} \cdot \mathbf{k}/N) . \tag{3.363}$$

Lexicographic ordering Another way to formulate a 2D DFT is to reorder the N^2 elements of $\mathbf{f}$ into an $N^2 \times 1$ column vector by use of *lexicographic ordering.* In this approach we define a new index ℓ as

$$\ell \equiv k+jN , \qquad j = 0, ..., N-1 , \qquad k = 0, ..., N-1 , \qquad \ell = 0, ..., N^2-1 . \tag{3.364}$$

This indexing amounts to starting with 0 at the upper left of the 2D matrix and counting in a raster fashion from left to right and top to bottom.

The inverse of (3.364) is

$$j = \mathrm{int}(\ell/N) , \qquad k = \ell - jN = \ell \,(\mathrm{modulo}\ N) , \tag{3.365}$$

where int(x) denotes the integer part of x.

A similar ordering of $\mathbf{F}$ is obtained by defining

$$p = n + mN\,, \qquad m = 0, ..., N-1\,, \qquad n = 0, ..., N-1\,, \qquad p = 0, ..., N^2-1\,, \tag{3.366}$$

with a similar inversion rule. With these definitions, (3.358) becomes

$$F_p^{(lex)} = \sum_{\ell=0}^{N^2-1} D_{p\ell}^{(lex)} f_\ell^{(lex)}\,, \tag{3.367}$$

or, in matrix-vector form,

$$\mathbf{F}^{(lex)} = \mathbf{D}^{(lex)} \mathbf{f}^{(lex)}\,, \tag{3.368}$$

where $\mathbf{f}^{(lex)}$ and $\mathbf{F}^{(lex)}$ are the lexicographically ordered $N^2 \times 1$ vectors corresponding to f_{jk} and F_{mn}, respectively, and $\mathbf{D}^{(lex)}$ is an $N^2 \times N^2$ matrix. The elements of $\mathbf{D}^{(lex)}$ are $\exp[-2\pi i(jm + kn)/N]$ with application of (3.365) and the analogous rule for p.

4

Series Expansions and Integral Transforms

This chapter comprises a collection of miscellaneous mathematical concepts and techniques that we shall need later in the book. Section 4.1 develops the theory of orthogonal expansions and provides us with several sets of orthonormal functions that can be used to represent objects and images. Section 4.2 surveys several classical integral transforms that are of use in describing imaging systems. The treatment in this section is brief, amounting to little more than a compendium of properties, since the transforms in question are well known and treated fully in many texts.

Section 4.3 deals with an integral transform that is not so widely known, yet which plays a fundamental role in image science. The *Fresnel transform* arises in signal processing, diffraction theory, coherent imaging and radar, so we present a relatively detailed account of its properties.

Finally, in Sec. 4.4, we introduce the *Radon transform*, an important tool in the mathematical description of tomographic imaging systems.

4.1 EXPANSIONS IN ORTHOGONAL FUNCTIONS

Fourier analysis is an expansion in complex exponentials, but there are many other sets of orthogonal functions that can be used in an exactly parallel manner. Since virtually all of the theorems that apply to conventional Fourier analysis have counterparts with these functions, the broader field of expansion in orthonormal functions is often called *generalized Fourier analysis*. We survey this field here, concentrating on families of orthonormal functions with particular usefulness in image science.

4.1.1 Basic concepts

The main thing that distinguishes one family of orthogonal functions from another is the definition of orthogonality used. The simplest definition is in terms of a scalar product on $\mathbb{L}_2(a,b)$, but we can also use the scalar product appropriate to the weighted Hilbert space $\mathbb{L}_2(a,b;w(x))$ (see Sec. 1.1.4). Norms in $\mathbb{L}_2(a,b;w(x))$ are denoted $\|\cdot\|_w$. In this space, two functions $u_n(x)$ and $u_m(x)$ are said to be orthonormal if

$$(u_n, u_m)_w = \int_a^b dx\, w(x)\, u_n^*(x)\, u_m(x) = \delta_{nm}\,, \tag{4.1}$$

where $w(x) \geq 0$ for $a < x < b$.

Suppose we are given a set of functions $\{u_n(x), n = 1, ..., \infty\}$, satisfying (4.1) and wish to use them to represent a function $f(x)$ in $\mathbb{L}_2(a,b;w(x))$. As in Sec. 3.2.1, we begin by simply assuming that $f(x)$ can be represented by a series of the form [*cf.* (3.17)]

$$f(x) = \sum_{n=1}^{\infty} \alpha_n u_n(x)\,. \tag{4.2}$$

If this expansion is valid, then the coefficients must be given by [*cf.* (3.19)]

$$\alpha_n = \int_a^b dx'\, w(x')\, u_n^*(x')\, f(x')\,. \tag{4.3}$$

This set of coefficients is optimal in the sense that

$$\left\|f(x) - \sum_{n=1}^{N} \beta_n u_n(x)\right\|_w \geq \left\|f(x) - \sum_{n=1}^{N} \alpha_n u_n(x)\right\|_w\,, \tag{4.4}$$

where $\{\alpha_n\}$ is given by (4.3) and $\{\beta_n\}$ is any other set of coefficients (Rade and Westergren, 1990).

The function set $\{u_n(x)\}$ is complete in $\mathbb{L}_2(a,b;w(x))$ if

$$\lim_{N\to\infty} \left\|f(x) - \sum_{n=1}^{N} \alpha_n u_n(x)\right\|_w = 0 \tag{4.5}$$

for all $f(x)$ in the space. If this condition is satisfied, the assumed expansion (4.2) indeed exists.

Expansions in orthogonal functions obey Parseval's relations analogous to those found in Fourier analysis (Rade and Westergren, 1990). The analog of the Parseval relation for a Fourier series, (3.50), is

$$\sum_{n=1}^{\infty} |\alpha_n|^2 = \int_a^b dx'\, w(x')\, |f(x')|^2\,, \tag{4.6}$$

where $\{u_n(x)\}$ is any complete set of orthonormal functions in $\mathbb{L}_2(a,b;w(x))$, and α_n is given by (4.3). The analog of the generalized Parseval relation (3.51) also holds.

4.1.2 Orthogonal polynomials

Some of the most important sets of orthogonal functions are real polynomials $\{\phi_n(x), n = 0, 1, 2, ...\}$, where $\phi_n(x)$ is a polynomial of degree n. For historical reasons the polynomials are usually not normalized but instead constructed to satisfy

$$\int_a^b dx\; w(x)\, \phi_n(x)\, \phi_m(x) = h_n\, \delta_{nm}\,, \tag{4.7}$$

where h_n is a normalization factor to be discussed below. The orthonormal functions required in Sec. 4.1.1 can be obtained simply by

$$u_n(x) = \frac{\phi_n(x)}{\sqrt{h_n}}\,. \tag{4.8}$$

The Gram-Schmidt procedure outlined in App. A can be used to construct a set of orthogonal polynomials satisfying (4.7). Starting by convention with $\phi_0(x) = 1$, we write $\phi_1(x) = c_{10} + c_{11}x$, choosing the coefficients so that ϕ_0 and ϕ_1 are orthogonal according to (4.7). Then we write $\phi_2(x) = c_{20} + c_{21}x + c_{22}x^2$, choosing the coefficients via the algorithm in App. A to make ϕ_2 orthogonal to both ϕ_1 and ϕ_0. Continuing in this way, we can in principle construct the infinite set; a detailed example of this procedure can be found in Arfken and Weber (1995).

The Gram-Schmidt procedure tends to be numerically unstable because it requires subtracting numbers of similar magnitude. A more practical way to compute the polynomials is by means of *recurrence relations*. It can be shown (Walter, 1994) that it is always possible to write

$$x u_n(x) = A_n u_{n+1}(x) + B_n u_n(x) + C_n u_{n-1}(x)\,, \qquad n = 1, 2, \ldots\,, \tag{4.9}$$

where the coefficients depend on the interval (a, b) and the weighting function $w(x)$.

One use of the recurrence relation is to prove the completeness of the set of polynomials. Given a function $f(x)$ in $\mathbb{L}_2(a, b; w(x))$ and assuming that the polynomials are orthonormal, we can form the partial sum $S_N(x)$ given by

$$S_N(x) = \sum_{n=0}^{N} \alpha_n u_n(x)\,, \qquad \alpha_n = \int_a^b dx'\, w(x')\, u_n(x')\, f(x')\,, \tag{4.10}$$

where $u_n(x)$ is $\phi_n(x)$ normalized as in (4.8), and we have omitted the complex conjugate on the assumption that the polynomials are real. In the limit $N \to \infty$, $S_N(x)$ converges to $f(x)$ in the sense that

$$\lim_{N\to\infty} \|S_N(x) - f(x)\|_w = 0\,, \tag{4.11}$$

so the functions $\{u_n(x)\}$ (and hence also $\{\phi_n(x)\}$) are complete in $\mathbb{L}_2(a, b; w(x))$.

We can also write $S_N(x)$ as an integral transform:

$$S_N(x) = \int_a^b dx'\, w(x')\, f(x') \sum_{n=0}^{N} u_n(x)\, u_n(x') \equiv \int_a^b dx'\, w(x')\, f(x')\, K_N(x, x')\,. \tag{4.12}$$

Walter (1994) shows that the kernel K_N is given by the *Christoffel-Darboux formula*:

$$K_N(x, x') = \frac{A_N[u_{N+1}(x)\, u_N(x') - u_N(x)\, u_{N+1}(x')]}{x - x'}\,. \tag{4.13}$$

This kernel is the analog of the Dirichlet kernel in Fourier theory [see Secs. 2.2.2 and 3.2.2].

The convergence of $S_N(x)$ can now be restated as

$$\lim_{N\to\infty} w(x')\, K_N(x, x') = \delta(x - x')\,. \tag{4.14}$$

The Dirac delta function allows the integral in (4.12) to be performed readily in the limit, showing that $S_N(x) \to f(x)$. Thus each distinct set of orthogonal polynomials has associated with it a representation of the delta function.

4.1.3 Sturm-Liouville theory

A rich source of orthonormal function sets is the literature on Hermitian operators. As discussed in detail in Chap. 1, the eigenfunctions of any Hermitian operator can be chosen to form a complete, orthonormal basis in the relevant Hilbert space. In this section we discuss orthonormal functions that arise as eigenfunctions of certain Hermitian differential operators.

Many of the classical second-order partial differential equations encountered in mathematical physics can be solved by the method of separation of variables (Morse and Feshbach, 1953, p. 719). The resulting ordinary differential equations involve differential operators called *Sturm-Liouville operators*. The general form of a Sturm-Liouville operator $\boldsymbol{\mathcal{L}}$ is (Arfken and Weber, 1995)

$$\boldsymbol{\mathcal{L}} f(x) = \frac{d}{dx}\left[p(x)\frac{df(x)}{dx}\right] + q(x)f(x)\,, \tag{4.15}$$

where $p(x)$ and $q(x)$ are real, bounded functions that are characteristic of the original partial differential equation and of the coordinates used for separation. If $p(x) \neq 0$ on an interval $[a, b]$, the operator is said to be *regular* on that interval. If $p(x) = 0$ or $q(x)$ is singular on $[a, b]$, or if the interval is infinite, the operator is said to be *singular*. The points at which $p(x) = 0$ are called the *singular points* of the operator, and the range $[a, b]$ is often taken to extend from one singular point to another.

It is a straightforward exercise in integration by parts to show that any operator in the form (4.15) is Hermitian in $\mathbb{L}_2(a, b)$ subject to certain boundary conditions. That is,

$$(f_1(x), \boldsymbol{\mathcal{L}} f_2(x)) = \int_a^b dx\ [f_1(x)]^*\, \boldsymbol{\mathcal{L}} f_2(x)$$

$$= (\boldsymbol{\mathcal{L}} f_1(x), f_2(x)) = \int_a^b dx\ [\boldsymbol{\mathcal{L}} f_1(x)]^*\, f_2(x)\,. \tag{4.16}$$

The boundary conditions are that the boundary terms produced by the integration by parts must vanish, which will happen if f_1 and f_2 satisfy

$$\left[f_1^* p\,\frac{df_2}{dx}\right]_{x=a} = \left[f_1^* p\,\frac{df_2}{dx}\right]_{x=b} \tag{4.17a}$$

and

$$\left[f_2^* p\,\frac{df_1}{dx}\right]_{x=a} = \left[f_2^* p\,\frac{df_1}{dx}\right]_{x=b}\,. \tag{4.17b}$$

These conditions are satisfied if either the functions or their derivatives vanish at the end points.

The eigenvalue problem for a Sturm-Liouville operator is usually stated as

$$\mathcal{L}\psi_n(x) + \lambda_n w(x)\, \psi_n(x) = 0\,, \tag{4.18}$$

where $w(x)$ is real, continuous and positive for $a < x < b$. Though this equation arises naturally in many physical problems (Morse and Feshbach, 1953, p. 719), it has a rather different structure from our usual eigenvalue equation $\mathcal{O}u_n(x) = \lambda_n u_n(x)$. To get the standard form, we can define a new operator $\mathcal{L}_w$ by

$$\mathcal{L}_w f(x) = -\frac{1}{w(x)}\mathcal{L}f(x)\,, \tag{4.19}$$

so that

$$\mathcal{L}_w\psi_n(x) = \lambda_n\psi_n(x)\,. \tag{4.20}$$

This new operator $\mathcal{L}_w$ is Hermitian in the space $\mathbb{L}_2(a, b; w(x))$ (see Sec. 1.1.4) if the original operator $\mathcal{L}$ is Hermitian in $\mathbb{L}_2(a, b)$ since

$$\int_a^b dx\ w(x) f_1^*(x)\,\mathcal{L}_w f_2(x) = -\int_a^b dx\ f_1^*(x)\,\mathcal{L}f_2(x)$$

$$= -\int_a^b dx\ [\mathcal{L}f_1(x)]^*\, f_2(x) = \int_a^b dx\ w(x)[\mathcal{L}_w f_1(x)]^* f_2(x)\,. \tag{4.21}$$

Thus we can work with either the original $\mathcal{L}$, considered as an operator in $\mathbb{L}_2(a, b)$, or with the modified operator $\mathcal{L}_w$ in $\mathbb{L}_2(a, b; w(x))$.

With either viewpoint, the eigenfunctions form an orthogonal set with respect to $w(x)$:

$$\int_a^b dx\ w(x)\, \psi_n^*(x)\, \psi_m(x) = h_n\, \delta_{nm}\,. \tag{4.22}$$

The eigenfunctions are often but not always polynomials. They form a complete set in the sense that any function in $\mathbb{L}_2(a, b; w(x))$ which satisfies the boundary conditions of the eigenvalue problem can be expanded as

$$f(x) = \sum_n c_n\psi_n(x)\,, \qquad c_n = \frac{1}{h_n}\int_a^b dx\ w(x)\, \psi_n^*(x)\, f(x)\,. \tag{4.23}$$

The particular choice of indexing is arbitrary, but $n = 0$ to ∞ is common.

Sturm-Liouville and Fourier Fourier analysis emerges as a special case of Sturm-Liouville theory if we take $p(x) = 1$ and $q(x) = 0$, so that $\mathcal{L} = d^2/dx^2$. For $(a, b) = (-\frac{1}{2}L, \frac{1}{2}L)$ and $w(x) = 1$, the eigenfunctions (for various boundary conditions) are

$$\psi_n(x) = \sin(2\pi nx/L)\,, \qquad \psi_n\left(\pm\tfrac{1}{2}L\right) = 0\,; \tag{4.24a}$$

$$\psi_n(x) = \cos(2\pi nx/L)\,, \qquad \psi_n'\left(\pm\tfrac{1}{2}L\right) = 0\,; \tag{4.24b}$$

$$\psi_n(x) = \exp(2\pi inx/L)\,, \qquad \psi_n\left(\tfrac{1}{2}L\right) = \psi_n\left(-\tfrac{1}{2}L\right)\,. \tag{4.24c}$$

Green's functions If zero is not an eigenvalue, a regular Sturm-Liouville operator can be inverted by use of a *Green's function* (Walter, 1994). This function, denoted $G(x, x')$, is the kernel of the integral operator that is the inverse of the differential operator $\mathcal{L}$. It satisfies

$$\mathcal{L}G(x, x') = \delta(x - x') . \tag{4.25}$$

The Green's function must also satisfy the relevant boundary conditions.

One use of a Green's function is to solve the inhomogeneous differential equation,

$$\mathcal{L}f(x) = s(x) , \tag{4.26}$$

subject to boundary conditions on $f(x)$. For example, if $f(x)$ is required to vanish at $x = \pm\frac{1}{2}L$, then the solution to (4.26) is given by

$$f(x) = \int_{-\frac{1}{2}L}^{\frac{1}{2}L} dx' \, s(x') \, G(x, x') , \tag{4.27}$$

provided $G(\pm\frac{1}{2}L, x') = 0$. That (4.27) indeed solves (4.26) can be verified by operating on both sides with $\mathcal{L}$ and differentiating under the integral sign.

An important practical application of Green's functions is diffraction theory, introduced in Chap. 9. Extension of the concept to higher dimensions and more complicated boundary conditions will be discussed there.

4.1.4 Classical orthogonal polynomials and related functions

This section is a brief summary of some useful orthogonal polynomials and orthogonal functions derived from them. Most of these functions arise from Sturm-Liouville operators. For a tutorial discussion of these functions in the context of mathematical physics, the reader can consult Morse and Feshbach (1953) or Arfken and Weber (1995). A detailed compendium of their properties can be found in Abramowitz and Stegun (1965) or Magnus and Oberhettinger (1949).

Legendre polynomials The Legendre polynomials $P_n(x)$ are perhaps the simplest and best-known of the classical orthogonal polynomials. They are not normalized but satisfy the orthogonality relation (4.7) with $(a, b) = (-1, 1)$, $w(x) = 1$ and $h_n = 2/(2n + 1)$. For n even (odd), $P_n(x)$ is a polynomial of order n with only even (odd) terms. The first few Legendre polynomials are

$$P_0(x) = 1 , \qquad P_1(x) = x , \qquad P_2(x) = \tfrac{1}{2}(3x^2 - 1) , \qquad P_3(x) = \tfrac{1}{2}(5x^3 - 3x) . \tag{4.28}$$

On the boundaries of the orthogonality interval, the Legendre polynomials satisfy

$$P_n(1) = 1 , \qquad P_n(-1) = (-1)^n . \tag{4.29}$$

Legendre polynomials are particularly useful for expanding functions of the cosine of an angle. Such functions arise frequently in physical applications when scalar products between 3D vectors occur. If we let $x = \cos\theta$, then the interval $(-1, 1)$ for x corresponds to the interval $(0, \pi)$ for θ. For any function $f(\theta)$ that depends only on $\cos\theta$, $f(2\pi - \theta) = f(\theta)$, so the full angular range $(0, 2\pi)$ is not needed, and the functions $\{P_n(\cos\theta), n = 0, ..., \infty\}$ form a complete set on $(0, \pi)$. The boundary conditions in (4.29) are also appropriate for functions of $\cos\theta$. All powers of $\cos\theta$ are 1 at $\theta = 0$ (or $\cos\theta = 1$), and the nth power of $\cos\theta$ is $(-1)^n$ at $\theta = \pi$ (or $\cos\theta = -1$).

One important function that can be expanded in Legendre polynomials is $1/R$, where $R = |\mathbf{r} - \mathbf{r}'|$ is the distance between two 3D vectors $\mathbf{r}$ and $\mathbf{r}'$. This function fits the scheme of the paragraph above since

$$R = |\mathbf{r} - \mathbf{r}'| = \sqrt{r^2 + r'^2 - 2rr'\cos\theta}\,, \tag{4.30}$$

where θ is the angle between $\mathbf{r}$ and $\mathbf{r}'$. The reciprocal of this distance, $1/R$, plays an important role in both electrostatics and optics. It is the potential of a point source, and hence the Green's function for the Poisson equation in electrostatics, and it is one factor in the Green's function for the Helmholtz equation or the time-dependent wave equation in optics (see Chap. 9).

An expansion of $1/R$ can be obtained from the *generating function* for the Legendre polynomials (Arfken and Weber, 1995), which is given by

$$g(u,x) = (1 - 2xu + u^2)^{-\frac{1}{2}} = \sum_{n=0}^{\infty} P_n(x)\, u^n\,. \tag{4.31}$$

This function is called a generating function since $P_n(x)$ can, in principle, be generated from

$$P_n(x) = \frac{1}{n!}\left[\frac{\partial^n g(u,x)}{\partial u^n}\right]_{u=0}\,. \tag{4.32}$$

To get the expansion for $1/R$, we let $x = \cos\theta$ and $u = r'/r$ in (4.31).

Associated Legendre functions Closely related to the Legendre polynomials are the *associated Legendre functions*, defined by[1]

$$P_n^m(x) = (1 - x^2)^{m/2}\frac{d^m}{dx^m}P_n(x)\,. \tag{4.33}$$

Since $P_n(x)$ is a polynomial of degree n, we must have $m \le n$. While it might be expected that only nonnegative values of m are allowed, a series representation of the Legendre polynomials permits $-n \le m \le n$ (see Arfken and Weber, 1995, p. 724). The Legendre polynomials themselves are a special case of P_n^m with $m = 0$.

The orthogonality relation for the associated Legendre functions is (Arfken and Weber, 1995, p. 727)

$$\int_{-1}^{1} dx\; P_k^m(x)\, P_n^m(x) = \int_0^{\pi} P_k^m(\cos\theta)\, P_n^m(\cos\theta)\sin\theta\; d\theta = \frac{2}{2k+1}\cdot\frac{(k+m)!}{(k-m)!}\,\delta_{kn}\,. \tag{4.34}$$

The first few associated Legendre functions are

$$P_1^1(x) = \left(1 - x^2\right)^{\frac{1}{2}} = \sin\theta\,, \qquad P_2^1(x) = 3x\left(1 - x^2\right)^{\frac{1}{2}} = 3\cos\theta\sin\theta\,,$$

$$P_2^2(x) = 3(1 - x^2) = 3\sin^2\theta\,, \quad P_3^1(x) = \tfrac{3}{2}(5x^2 - 1)(1 - x^2)^{\frac{1}{2}} = \tfrac{3}{2}(5\cos^2\theta - 1)\sin\theta\,,$$

$$P_3^2(x) = 15x\left(1 - x^2\right) = 15\cos\theta\sin^2\theta\,. \tag{4.35}$$

[1]The reader is cautioned that other normalizations for these functions can be found in the literature. Our convention follows that of Arfken and Weber (1995).

Spherical harmonics The associated Legendre functions are used to construct the *spherical harmonics*, familiar in virtually all branches of mathematical physics (Morse and Feshbach, 1953; Arfken and Weber, 1995). The spherical harmonic $Y_{\ell m}(\theta, \phi)$ is defined by

$$Y_{\ell m}(\theta, \phi) = (-1)^m \sqrt{\frac{2\ell+1}{4\pi} \frac{(\ell-m)!}{(\ell+m)!}}\, P_\ell^m(\cos\theta) \exp(im\phi) \,. \tag{4.36}$$

These functions satisfy the orthogonality relation,

$$\int_0^{2\pi} d\phi \int_0^{\pi} \sin\theta\, d\theta\, Y^*_{\ell m}(\theta, \phi)\, Y_{\ell' m'}(\theta, \phi) = \delta_{mm'}\, \delta_{\ell\ell'} \,, \tag{4.37}$$

and they form a complete set for expansions of square-integrable functions $f(\theta, \phi)$ in spherical polar coordinates.

An important property of spherical harmonics is the *addition theorem* (Arfken and Weber, 1995). If we consider two unit vectors $\hat{\mathbf{s}}$ and $\hat{\mathbf{s}}'$, with polar coordinates (θ, ϕ) and (θ', ϕ'), respectively, and denote the angle between $\hat{\mathbf{s}}$ and $\hat{\mathbf{s}}'$ as θ_0, then

$$P_\ell(\cos\theta_0) = \frac{4\pi}{2\ell+1} \sum_{m=-\ell}^{\ell} Y_{\ell m}(\theta, \phi)\, Y^*_{\ell m}(\theta', \phi') \,. \tag{4.38}$$

This result will prove beneficial in Chap. 10, where we discuss scattering processes.

Zernike polynomials and circular harmonics The circle polynomials, now known as Zernike polynomials, were introduced by Frits Zernike in a classic 1934 paper for which he was to win the Nobel Prize in physics two decades later. The Zernike polynomials $R_k^j(r)$ are orthogonal on $[0, 1]$ with respect to weight r and are therefore well suited to expressing the radial dependence of a 2D function $f(\mathbf{r})$ in polar coordinates.

When $f(\mathbf{r})$ is expressed in Cartesian coordinates, we shall denote it as $f^{(c)}(x, y)$, and in polar coordinates we shall call it $f^{(p)}(r, \theta)$. Since $f^{(p)}(r, \theta)$ is periodic in θ with period 2π, it can be expanded in an angular Fourier series of the form

$$f^{(p)}(r, \theta) = \sum_{j=-\infty}^{\infty} f_j(r) \exp(ij\theta) \,, \tag{4.39}$$

where the coefficients, which are still functions of r, are given by

$$f_j(r) = \frac{1}{2\pi} \int_0^{2\pi} d\theta\, f^{(p)}(r, \theta) \exp(-ij\theta) \,. \tag{4.40}$$

To obtain a complete expansion of $f^{(p)}(r, \theta)$ in orthogonal functions, we must find an appropriate expansion of $f_j(r)$. In order to do so, the range of the variable r must be specified. We can assume, without loss of generality, that the maximum value of r is unity; if it is any other finite value R, we merely introduce the normalized radius r/R.

Of course, any set of orthogonal polynomials on $[0, 1]$ could be used as a basis for $f_j(r)$, but such an expansion might contain terms that cannot arise physically and must therefore have zero coefficients. To discover what terms are allowed, we consider an expansion of $f^{(c)}(x, y)$ in a power series:

$$f^{(c)}(x,y) = \sum_{n=0}^{\infty} \sum_{m=0}^{\infty} A_{nm} x^n y^m . \tag{4.41}$$

Since $x = r\cos\theta$ and $y = r\sin\theta$, we can also write $f(\mathbf{r})$ in polar coordinates as

$$f^{(p)}(r,\theta) = f^{(c)}(r\cos\theta, r\sin\theta) = \sum_{n=0}^{\infty} \sum_{m=0}^{\infty} A_{nm} r^{n+m} [\cos\theta]^n [\sin\theta]^m$$

$$= \sum_{n=0}^{\infty} \sum_{m=0}^{\infty} A_{nm} r^{n+m} \left[\frac{\exp(i\theta)+\exp(-i\theta)}{2}\right]^n \left[\frac{\exp(i\theta)-\exp(-i\theta)}{2i}\right]^m . \tag{4.42}$$

The radial Fourier coefficients are thus given by

$$f_j(r) = \frac{1}{2\pi} \sum_{n=0}^{\infty} \sum_{m=0}^{\infty} A_{nm} r^{n+m} \int_0^{2\pi} d\theta \left[\frac{\exp(i\theta)+\exp(-i\theta)}{2}\right]^n$$

$$\times \left[\frac{\exp(i\theta)-\exp(-i\theta)}{2i}\right]^m \exp(-ij\theta) . \tag{4.43}$$

With the binomial theorem and a little algebra, we find that

$$\frac{1}{2\pi} \int_0^{2\pi} d\theta \left[\frac{\exp(i\theta)+\exp(-i\theta)}{2}\right]^n \left[\frac{\exp(i\theta)-\exp(-i\theta)}{2i}\right]^m \exp(-ij\theta)$$

$$= \frac{1}{2^n} \frac{1}{(2i)^m} \sum_{p=0}^{n} \sum_{q=0}^{m} \binom{n}{p} \binom{m}{q} (-1)^q \frac{1}{2\pi} \int_0^{2\pi} d\theta \exp[-2i(p+q)\theta + i(n+m)\theta - ij\theta] . \tag{4.44}$$

Because of the orthogonality of the complex exponentials [see (3.7)], the last integral is unity if

$$j = 2(p+q) - (n+m) , \qquad 0 \le p+q \le n+m , \tag{4.45}$$

and zero otherwise. This condition restricts the values of j that can be associated with any particular power of r. If $k = n+m$, we must have

$$j = -k, -k+2, ..., k . \tag{4.46}$$

Thus we must have $|j| \le k$ (where k is the power of r), and j must vary in steps of 2. Stated differently, a polynomial of degree k in r can only have terms r^j, $r^{|j|+2}, ..., r^k$ if it is to be used in expanding $f_j(r)$.

If we look for polynomials of this form and require them to be orthogonal on $[0, 1]$ with respect to weight r, we are led to the Zernike polynomials, given by (Born and Wolf, 1965)

$$R_k^j(r) = \sum_{s=0}^{\frac{1}{2}(k-|j|)} \frac{(-1)^s (k-s)!}{s! \left[\frac{k-|j|}{2} - s\right]! \left[\frac{k+|j|}{2} - s\right]!} r^{k-2s} , \tag{4.47}$$

where j and k must satisfy (4.46). The orthogonality condition for the Zernike polynomials is

$$\int_0^1 r\,dr\; R_k^j(r)\, R_{k'}^j(r) = \frac{1}{2(k+1)}\,\delta_{kk'}\,. \tag{4.48}$$

The functions $\sqrt{2(k+1)}R_k^j(r)\exp(ij\theta)$ thus form a convenient orthonormal set on the unit disk.

For a detailed treatment of Zernike polynomials, see Born and Wolf (1965), and for a practical discussion of their use in aberration theory, see Mahajan (1981).

Hermite polynomials Hermite polynomials, denoted $H_n(x)$, can be defined as

$$H_n(x) = (-1)^n \exp(x^2)\left(\frac{d}{dx}\right)^n \exp(-x^2)\,. \tag{4.49}$$

The first few Hermite polynomials are:

$$H_0(x) = 1\,, \qquad H_1(x) = 2x\,, \qquad H_2(x) = 4x^2 - 2\,, \qquad H_3(x) = 8x^3 - 12x\,. \tag{4.50}$$

The Hermite polynomials are orthogonal on $(-\infty, \infty)$ with respect to a Gaussian weighting function (Arfken and Weber, 1995, p. 768):

$$\int_{-\infty}^{\infty} dx\; H_n(x)\, H_m(x) \exp(-x^2) = 2^n n! \sqrt{\pi}\,\delta_{mn}\,. \tag{4.51}$$

The generating function for the Hermite polynomials is (Poularikas, 1996)

$$\exp(2tx - t^2) = \sum_{n=0}^{\infty} \frac{H_n(x)}{n!} t^n\,. \tag{4.52}$$

The Hermite polynomial $H_k(x)$ can be obtained from this expression by differentiating k times with respect to t and then setting $t = 0$.

Hermite-Gauss functions Instead of considering the Gaussian in (4.51) as a weight in the scalar product, we can lump it into the definition of the orthogonal functions. Thus we can consider a set of functions $\{HG_n(x)\}$ orthonormal on $(-\infty, \infty)$ with respect to *unit* weight and defined by

$$HG_n(x) = \left[2^n n! \sqrt{\pi}\right]^{-\frac{1}{2}} H_n(x) \exp\left(-\tfrac{1}{2}x^2\right)\,. \tag{4.53}$$

These functions, known as *Hermite-Gauss functions*, arise in a variety of physical applications. In quantum mechanics, they are the wavefunctions for the simple harmonic oscillator. In optics they occur in beam-like solutions of the Helmholtz equation and as eigenmodes of laser resonators.

An important property of the Hermite-Gauss functions is that they form a complete orthonormal basis for $\mathbb{L}_2(\mathbb{R}; w(x) = 1)$ (or simply $\mathbb{L}_2(\mathbb{R})$ for short). A formal proof of this statement is given by Titchmarsh (1948), but it also follows from the observation that the functions are eigenfunctions of a Hermitian operator, namely the Schrödinger operator for the simple harmonic oscillator.

Because of the completeness of the Hermite-Gauss functions in $\mathbb{L}_2(\mathbb{R})$, any function in that space can be expressed with vanishing $\mathbb{L}_2$ error as

$$f(x) = \sum_{n=0}^{\infty} \alpha_n \left[2^n n! \sqrt{\pi}\right]^{-\frac{1}{2}} H_n(x) \exp\left(-\tfrac{1}{2}x^2\right)\,, \tag{4.54}$$

and the coefficients $\{\alpha_n\}$ can be found by using (4.51).

The discrete expansion in (4.54) may be somewhat surprising. We are more used to expanding functions on $(-\infty, \infty)$ in Fourier basis functions $\{\exp(2\pi i\xi x)\}$, which can be regarded as a basis set with a *continuous* index ξ, but (4.54) shows we can also use a set with a *discrete* index n. The possibility of using a discrete (denumerably infinite) basis follows from the fact that $\mathbb{L}_2(-\infty, \infty)$ is a separable Hilbert space (see Sec. 1.1.5).

The Hermite-Gauss functions are also naturally associated with tempered distributions and Schwartz space (Walter, 1994). A polynomial times a Gaussian is, virtually by definition, a *good function* or *open-support test function* (see Sec. 2.1.2). Schwartz space is the space of good functions, and tempered distributions are linear functionals on this space.

One interesting property of Hermite-Gauss functions, occasionally exploited in imaging applications, is that they are their own Fourier transforms. Explicitly,

$$\int_{-\infty}^{\infty} dx \, \exp\left(-\tfrac{1}{2}x^2\right) H_n(x) \exp(ixy) = i^n \sqrt{2\pi} \exp\left(-\tfrac{1}{2}y^2\right) H_n(y) \,, \tag{4.55}$$

from which a change of variables shows that

$$\mathcal{F}_1 \left\{ HG_n \left(\sqrt{2\pi} x\right) \right\} = (-i)^n HG_n \left(\sqrt{2\pi}\xi\right) . \tag{4.56}$$

Thus the Hermite-Gauss functions are eigenfunctions of the Fourier operator.

Laguerre polynomials The Laguerre polynomials, orthogonal on $(0, \infty)$ with respect to the weight $\exp(-x)$, are familiar in the quantum mechanics of the hydrogen atom. They are defined by (Arfken and Weber, 1995)

$$L_n(x) = \sum_{m=0}^{n} (-1)^m \binom{n}{m} \frac{x^m}{m!} \,, \tag{4.57}$$

and the orthogonality relation is

$$\int_0^{\infty} dx \, e^{-x} L_n(x) \, L_m(x) = \delta_{nm} \,. \tag{4.58}$$

Imaging uses of Laguerre polynomials are rare, but a few authors (*e.g.*, Seger, 1993) have exploited the fact that Laguerre polynomials can be used to construct eigenfunctions of the zeroth-order Hankel transform. From Sec. 3.4.4 we know that this Hankel transform is the same as the 2D Fourier transform for rotationally symmetric functions. A tabulated integral (Gradshteyn and Ryzhik, 1980, formula 7.421.1) shows that

$$\int_0^{\infty} r \, dr \, \exp(-\pi r^2) \, L_n(2\pi r^2) \, \mathrm{J}_0(2\pi r\rho) = \frac{(-1)^n}{2\pi} \exp(-\rho^2/4\pi) \, L_n(\rho^2/2\pi) \,. \tag{4.59}$$

The similarity to (4.55) and (4.56) should be noted. Just as Hermite-Gauss functions are eigenfunctions of the 1D Fourier operator, so too are Laguerre-Gauss functions eigenfunctions of the 2D rotationally symmetric Fourier operator.

Chebyshev polynomials There are two kinds of Chebyshev polynomials, both orthogonal on $(-1, 1)$ but with different weights. Chebyshev polynomials of the first kind, denoted $T_n(x)$, satisfy the orthogonality relation

$$\int_{-1}^{1} dx \, (1-x^2)^{-\frac{1}{2}} T_n(x) \, T_m(x) = \frac{\pi}{2} \, \delta_{mn}(1+\delta_{m0}) \, . \tag{4.60}$$

The right-hand side is $\pi/2$ if $m = n \neq 0$ and π if $m = n = 0$.

Chebyshev polynomials of the second kind, denoted $U_n(x)$, satisfy

$$\int_{-1}^{1} dx \, (1-x^2)^{\frac{1}{2}} U_n(x) \, U_m(x) = \frac{\pi}{2} \, \delta_{mn} \, . \tag{4.61}$$

These functions occur most commonly when x is the cosine of some angle, say $x = \cos\theta$. Then we have

$$T_n(\cos\theta) = \cos n\theta = \cos(n \cos^{-1} x) \, , \qquad |x| \leq 1 \, , \tag{4.62}$$

$$U_n(\cos\theta) = \frac{\sin(n+1)\theta}{\sin\theta} \, . \tag{4.63}$$

As we shall see in Chap. 17, Chebyshev polynomials play a prominent role in the analysis of tomographic imaging systems; for a review, see Barrett (1984).

4.1.5 Prolate spheroidal wavefunctions

Walter (1994) refers to prolate spheroidal wavefunctions $\psi_n(x)$ as a lucky accident since they are eigenfunctions of both an integral operator and a Sturm-Liouville differential operator. The Sturm-Liouville operator arises when the wave equation is separated in prolate spheroidal coordinates, hence the name of the functions. In imaging applications, however, the integral operator is more important.

The integral operator in question corresponds to first truncating a function, so that it is spatially limited, and then passing it through an ideal low-pass filter. The prolate spheroidal wavefunction is the eigenfunction of this operator (Marks, 1991; Percival and Walden, 1993):

$$B \int_{-\frac{1}{2}L}^{\frac{1}{2}L} dx' \, \psi_n(x') \, \text{sinc}[B(x-x')] = \lambda_n \psi_n(x) \, . \tag{4.64}$$

In abstract operator form, we can rewrite (4.64) as

$$\mathcal{B}_B \mathcal{S}_L \psi_n(x) = \lambda_n \psi_n(x) \, , \tag{4.65}$$

where $\mathcal{S}_L$ is the space-limiting operator [multiplication by $\text{rect}(x/L)$] and $\mathcal{B}_B$ is the band-limiting operator (convolution with $B \, \text{sinc}(Bx)$ or equivalently, multiplication by $\text{rect}(\xi/B)$ in the frequency domain).

The eigenfunctions depend on the parameters L and B and the eigenvalues depend only on the product LB, but we shall not indicate these dependences explicitly. The first few eigenfunctions are plotted in Fig. 4.1, and the eigenvalue spectrum is shown in Fig. 4.2. The key point to note from Fig. 4.2 is the precipitous rolloff near $n = LB$ of the plot of λ_n vs. n.

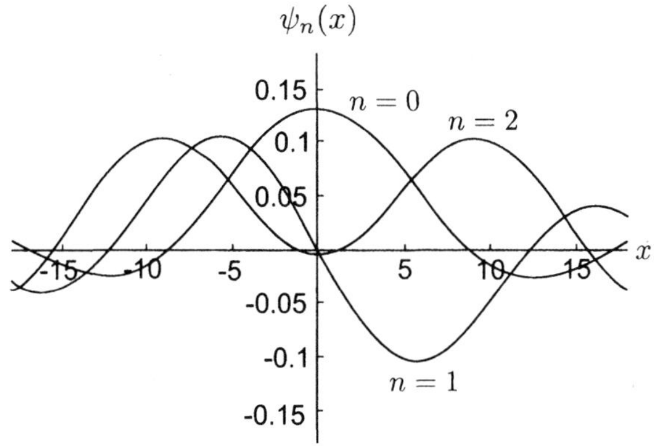

Fig. 4.1 The first few prolate spheroidal wavefunctions.

Because of the lucky accident, prolate spheroidal wavefunctions satisfy two separate orthogonality conditions. They are orthonormal on $(-\infty, \infty)$,

$$\int_{-\infty}^{\infty} dx\ \psi_n(x)\,\psi_m(x) = \delta_{nm}\,, \tag{4.66}$$

and they are orthogonal but not normalized on the finite interval $(-\frac{1}{2}L, \frac{1}{2}L)$,

$$\int_{-\frac{1}{2}L}^{\frac{1}{2}L} dx\ \psi_n(x)\,\psi_m(x) = \lambda_n\,\delta_{nm}\,. \tag{4.67}$$

Since the prolates are eigenfunctions of a Hermitian operator on $\mathbb{L}_2(-\frac{1}{2}L, \frac{1}{2}L)$, they are complete on that space. On the infinite interval, however, they are orthonormal but not complete; by (4.65) they are bandlimited so they cannot be used to represent an arbitrary function in $\mathbb{L}_2(-\infty, \infty)$. The best we can say is that they are complete in the space of bandlimited, square-integrable functions on the real line (Paley-Wiener space).

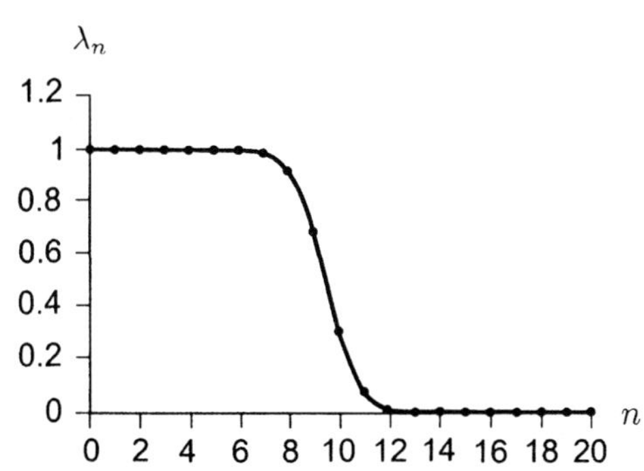

Fig. 4.2 The eigenvalue spectrum λ_n for the integral operator with eigenfunctions given by prolate spheroidal wavefunctions. Figure adapted from Marks (1991).

These properties give the prolates great utility in a variety of bandwidth-extrapolation and superresolution problems [see, *e.g.*, Mammone, (1987)].

Finite Fourier transform Another way of understanding prolate spheroidal wavefunctions is through the finite Fourier transform, performed by multiplying a function by a rect function before doing an ordinary Fourier transform. Under this operation, prolates satisfy (Slepian and Pollak, 1961; Marks, 1991):

$$\mathcal{F}\{\psi_n(x)\,\mathrm{rect}(x/L)\} = \sqrt{\frac{L\lambda_n}{B}}\,\psi_n\left(\frac{\xi L}{B}\right), \tag{4.68}$$

where both λ_n and ψ_n depend on the parameter LB. Note that this is not exactly an eigenvalue equation because of the scale factor in the argument on the right. If we use dimensionless variables and set $L/B = 1$, we can say that prolates are eigenfunctions of the finite Fourier transform.

The dual relation to (4.68) is

$$\mathcal{F}\{\psi_n(x)\} = \sqrt{\frac{L}{B\lambda_n}}\,\psi_n\left(\frac{\xi L}{B}\right)\mathrm{rect}(\xi/B). \tag{4.69}$$

From (4.68) and (4.69), it is straightforward to derive (4.64), so (4.68) can be regarded as the defining property of prolates.

Degrees of freedom One use of prolate spheroidal wavefunctions is in enumerating the degrees of freedom of a bandlimited and approximately spacelimited function. Consider a function $f_0(x)$ that is obtained by starting with an arbitrary square-integrable function $f(x)$ and applying the operator $\boldsymbol{\mathcal{B}}_B\boldsymbol{\mathcal{S}}_L$:

$$f_0(x) \equiv \boldsymbol{\mathcal{B}}_B\boldsymbol{\mathcal{S}}_L f(x) = B\int_{-\frac{1}{2}L}^{\frac{1}{2}L} dx'\, f(x')\,\mathrm{sinc}[B(x-x')]. \tag{4.70}$$

Since $\boldsymbol{\mathcal{B}}_B$ is applied last, $f_0(x)$ is exactly bandlimited, but if LB is large the function is also approximately spacelimited.

Since $f(x)$ is in $\mathbb{L}_2(-\infty,\infty)$, $\boldsymbol{\mathcal{S}}_L f(x)$ is in $\mathbb{L}_2(-\frac{1}{2}L,\frac{1}{2}L)$ and can be expressed as

$$\boldsymbol{\mathcal{S}}_L f(x) = \sum_{n=0}^{\infty} \alpha_n\psi_n(x)\,\mathrm{rect}(x/L). \tag{4.71}$$

The operator $\boldsymbol{\mathcal{S}}_L$ is idempotent, so we can write

$$f_0(x) = \boldsymbol{\mathcal{B}}_B\boldsymbol{\mathcal{S}}_L f(x) = \boldsymbol{\mathcal{B}}_B\boldsymbol{\mathcal{S}}_L\sum_{n=0}^{\infty}\alpha_n\psi_n(x) = \sum_{n=0}^{\infty}\alpha_n\boldsymbol{\mathcal{B}}_B\boldsymbol{\mathcal{S}}_L\psi_n(x) = \sum_{n=0}^{\infty}\alpha_n\lambda_n\psi_n(x). \tag{4.72}$$

From Fig. 4.2, however, we see that $\lambda_n \simeq 0$ for $n > LB$ (provided $LB > 1$), so only about LB terms are required to represent $f_0(x)$ in the expansion (4.72).[2]

We recognize LB as the space-bandwidth product, introduced in Sec. 3.5.2 as a way of estimating the number of degrees of freedom of a signal. Here the same quantity appears as the number of parameters needed to represent a function $f_0(x)$ resulting from the operator in (4.70), placing the concept of space-bandwidth product on a firm theoretical footing. For an excellent review of this topic, see Slepian (1976).

[2]It is possible to construct pathological functions for which α_n grows rapidly in the vicinity of $n = LB$, and then this argument does not hold.

4.2 CLASSICAL INTEGRAL TRANSFORMS

The Fourier transform is treated in detail in Chap. 3. Here we survey a few other integral transforms with applications in imaging. For simplicity, only 1D versions of the transforms are discussed, but all are readily extended to two or more dimensions in Cartesian coordinates.

Good general treatments of these transforms are given by Bracewell (1965), Jerri (1992) and Poularikas (1996). Extensive tables of transforms are given by Erdélyi (1954), Abramowitz and Stegun (1965), and Oberhettinger (1972, 1973, 1974).

4.2.1 Laplace transform

The Laplace transform is defined as

$$F_L(s) = \int_0^\infty dt\ f(t)\exp(-st)\,. \tag{4.73}$$

We have written the variable of integration as t and taken the range of integration as $(0,\infty)$ since the Laplace transform is most commonly used for functions of time, and we are often interested only in $t > 0$. For example, if we are dealing with a transient signal that is excited at $t = 0$, then we know that there is no signal before the excitation. Similarly, a linear shift-invariant temporal filter is described by its *impulse response* $h(t)$, which must vanish if $t < 0$. This mathematical condition simply says that there can be no output from the filter before an input is applied, or that cause must precede effect. Temporal functions that vanish for $t < 0$ are called *causal* or *one-sided.*

We shall denote the Laplace operator by $\mathcal{L}\mathbf{a}$. Thus we write

$$F_L(s) = \mathcal{L}\mathbf{a}\{f(t)\}\,. \tag{4.74}$$

If we restrict attention to causal functions, so that $f(t) = 0$ for $t < 0$, then the Laplace transform is formally the same thing as the Fourier transform with the temporal frequency being $s/2\pi i$. That is,

$$F_L(s) = \left[\int_0^\infty dt\ f(t)\exp(-2\pi i\nu t)\right]_{\nu=s/2\pi i} = F(s/2\pi i)\,, \tag{4.75}$$

where $F(\nu)$ is the 1D Fourier transform of $f(t)$. If we do not restrict the function to be causal, then the Laplace transform of $f(t)$ is the same as the Fourier transform of $f(t)\,\mathrm{step}(t)$ with $\nu = s/2\pi i$, or

$$F_L(s) = \left[\mathcal{F}_1\{f(t)\,\mathrm{step}(t)\}\right]_{\nu=s/2\pi i}\,. \tag{4.76}$$

Alternatively, for noncausal functions we can also define a *two-sided Laplace transform* $F_{L2}(s)$ and the associated operator $\mathcal{L}\mathbf{a}_2$ via

$$F_{L2}(s) = \mathcal{L}\mathbf{a}_2\{f(t)\} = \int_{-\infty}^\infty dt\ f(t)\exp(-st)\,, \tag{4.77}$$

which is identically $F(s/2\pi i)$ without any further restrictions. Unless the specification *two-sided* is given explicitly, the term Laplace transform will refer to the one-sided expression given in (4.73).

One practical difference between Fourier and Laplace transforms is that the Fourier variable ν is usually taken to be real (though we saw in Sec. 3.3.9 that complex frequencies can be useful). The Laplace variable s, on the other hand, is usually regarded as complex. The transform $F_L(s)$ is defined for any complex s for which the integral in (4.73) converges. The one-sided nature of the integral is helpful in this respect since the factor e^{-st} decays rapidly for all positive t if s has a positive real part. Even if $f(t)$ diverges exponentially, say as $\exp(\alpha t)$ with α real, then the Laplace transform will still exist whenever $\mathrm{Re}(s) > \alpha$. The point $s = \alpha$ in the complex s-plane is a singularity of $F_L(s)$ in this case.

Inverse Laplace transform By analogy to the Fourier transform, the inverse Laplace transform involves an integral over s. Since s is complex, we must specify a contour for this integral. The usual choice is the *Bromwich contour* shown in Fig. 4.3. With this contour, the inverse Laplace transform is (Titchmarsh, 1948)

$$\frac{1}{2\pi i}\int_{c-i\infty}^{c+i\infty} ds\, F_L(s)\, e^{st} = \begin{cases} f(t) & \text{if} \quad t > 0 \\ 0 & \text{if} \quad t < 0 \end{cases}, \tag{4.78}$$

where c is a real constant chosen so that the contour is to the right of all singularities of $F_L(s)$ in the complex s-plane. The integral vanishes for $t < 0$ since the integrand in (4.78) is analytic within and on the closed contour shown in Fig. 4.3. For $t > 0$, the validity of (4.78) can be shown by use of (4.73), (2.46) and (2.25).

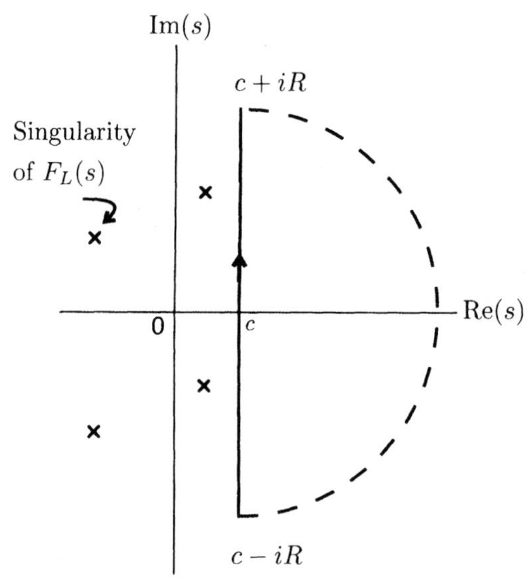

Fig. 4.3 The Bromwich contour used in the inverse Laplace transform. This contour in the complex s-plane lies to the right of all singularities of $F_L(s)$ and extends vertically from $c - iR$ to $c + iR$, with $R \to \infty$.

The inversion formula (4.78) is still valid if $F_L(s)$ is replaced by the two-sided transform of (4.75), without the causality restriction, but in that case the integral on the left agrees with $f(t)$ only for $t > 0$. Thus the inverse Laplace transform can be regarded as a representation of a one-sided function for all t or of an arbitrary function for $t > 0$. In this sense it is analogous to the Fourier series or Fourier cosine transform. We saw in Chap. 3 that a Fourier series is a representation of an arbitrary function on a finite interval or of a periodic function for all space. Similarly, a Fourier cosine transform is a representation of an arbitrary function for positive x or of an even function for all x. No matter what function was used to find the original transform, the inverse transform is a representation of a function with a certain symmetry property: one-sidedness for the Laplace transform, periodicity for the Fourier series, and evenness for the Fourier cosine transform.

Filters and Laplace convolution The Laplace transform is very important for solving equations of motion—differential equations for the time evolution of a linear system subject to boundary conditions. Electrical filters and other causal linear systems fit this description. For spatial functions, the one-sided integral of (4.73) does not usually arise naturally, and we may as well use Fourier transforms. One exception to this statement occurs with exponentially attenuated beams of radiation, for which the Laplace transform is frequently useful.

An important tool in many of these applications is the *Laplace convolution theorem*. The Laplace convolution of two functions $f(t)$ and $p(t)$ is defined by

$$[p *_L f](t) = \int_0^t dt'\, p(t')\, f(t-t')\,, \tag{4.79}$$

which is the same as the ordinary convolution if $p(t)$ and $f(t)$ are both causal.

Just as Fourier transformation converts convolution to multiplication, so too does Laplace transformation convert Laplace convolution to multiplication (Carrier *et al.*, 1966). That is,

$$\mathcal{L}\mathrm{a}\{p *_L f\} = \int_0^\infty dt\, \exp(-st) \int_0^t dt'\, p(t')\, f(t-t') = \int_0^\infty dt' \int_{t'}^\infty dt\, \exp(-st)\, p(t')\, f(t-t')$$

$$= \int_0^\infty dt'\, \exp(-st')\, p(t') \int_{t'}^\infty dt\, \exp\left[-s(t-t')\right] f(t-t') = P_L(s)\, F_L(s)\,. \tag{4.80}$$

Conversely, the inverse Laplace transform of the product $P_L(s)F_L(s)$ is the Laplace convolution (4.79).

4.2.2 Mellin transform

The Mellin transform can be derived from the two-sided Laplace transform by making a change of variables (Bracewell, 1965). If we let $x = \exp(-t)$ in (4.77), we obtain

$$F_{L2}(s) = \int_0^\infty dx\ f(-\ln x)\, x^{s-1}\,. \tag{4.81}$$

The integral on the right is the Mellin transform of the logarithmically distorted function $f(-\ln x)$. Letting $k(x) = f(-\ln x)$, we can write

$$K_M(s) = \int_0^\infty dx\ k(x)\, x^{s-1} = \mathcal{M}\{k(x)\} = \mathcal{L}\mathrm{a}_2\{f(t)\}\,, \tag{4.82}$$

where $K_M(s)$ denotes the Mellin transform of $k(x)$ and $\mathcal{M}$ is the Mellin operator.

Expressions like (4.82) were used by Riemann in the context of number theory, but the first rigorous treatment was due to Mellin (Titchmarsh, 1948).

A similar change of variables in the inverse Laplace transform shows that the inverse Mellin transform is given by (Bracewell, 1965; Carrier *et al.*, 1966)

$$k(x) = \mathcal{M}^{-1}\{K_M(s)\} = \frac{1}{2\pi i}\int_{c-i\infty}^{c+i\infty} ds\ K_M(s)\, x^{-s}\,. \tag{4.83}$$

One interpretation of the Mellin transform is in terms of moments of the function being transformed. If s is real and $k(x) = 0$ for $x < 0$, then $K_M(s)$ is the $(s-1)^{th}$ moment of $k(x)$.

Mellin and magnification The Mellin transform finds its main imaging application in dealing with magnification. If $k(x)$ represents an image, then $k(\alpha x)$ represents an image magnified by $1/\alpha$. It follows readily from (4.82) that

$$\mathcal{M}\{k(\alpha x)\} = \int_0^\infty dx\ k(\alpha x)\, x^{s-1} = \alpha^{-s} K_M(s) = \exp(-s\ln\alpha)\, K_M(s)\,. \tag{4.84}$$

Hence the magnification has the effect of multiplying the Mellin transform by $\exp(-s\ln\alpha)$, which has unit modulus if s is pure imaginary. Thus $|\mathcal{M}\{k(\alpha x)\}|$ is invariant to magnification. This property is analogous to the shifting property of Fourier transforms: If a function $f(x)$ is shifted by α, its Fourier transform $F(\xi)$ is multiplied by $\exp(2\pi i\alpha\xi)$, which has unit modulus if ξ is real, and $|F(\xi)|$ is invariant to shift. For this reason, Fourier transforms are important in the analysis of shift-invariant systems, while Mellin transforms are used for magnification-invariant or scale-invariant systems. Applications include modeling of the visual cortex of primates (Kaas, 1978; Cavanaugh, 1978), which evidently functions in a scale-invariant manner, and scale-invariant pattern recognition (Casasent and Psaltis, 1976).

Mellin convolution The Mellin convolution is defined by

$$[p *_M f]\,(x) = \int_0^\infty \frac{dx'}{x'}\ p\left(\frac{x}{x'}\right) f(x')\,. \tag{4.85}$$

This expression is similar to the usual convolution expression (3.114); one key difference, however, is that x/x' appears rather than $x - x'$. From the discussion above of Mellin and magnification, we might expect that a Mellin transform would be a useful thing to apply to (4.85). In fact, it gives

$$\begin{aligned}\mathcal{M}\{p *_M f\} &= \int_0^\infty dx\ x^{s-1}\int_0^\infty \frac{dx'}{x'}\ p\left(\frac{x}{x'}\right) f(x') = \int_0^\infty \frac{dx'}{x'}\ f(x')\int_0^\infty dx\ x^{s-1} p\left(\frac{x}{x'}\right)\\ &= \int_0^\infty dx'\ x'^{s-1} f(x')\int_0^\infty du\ u^{s-1} p(u) = F_M(s)\, P_M(s)\,,\end{aligned} \tag{4.86}$$

where $u = x/x'$. Thus Mellin transformation converts Mellin convolution to multiplication. A related result is

$$\mathcal{M}\left\{\int_0^\infty dx\ f(x)\, p(\alpha x)\right\} = \int_0^\infty d\alpha\ \alpha^{s-1}\int_0^\infty dx\ f(x)\, p(\alpha x) = F_M(1-s)\, P_M(s)\,. \tag{4.87}$$

4.2.3 z transform

The z transform is used in discrete problems where a function $f(t)$ is known only at a discrete set of sample points $\{t_n = n\Delta t,\ n = 0, 1, 2, ...\}$. Letting $f_n = f(n\Delta t)$, we can define the z transform of the sequence $f_0, f_1, ...$, by constructing a polynomial in z^{-1}, where z is a complex number, with f_n as the weight of the nth term. Explicitly,

$$F_z(z) = \mathcal{Z}\{f_n\} = \sum_{n=0}^{\infty} f_n z^{-n}\,, \tag{4.88}$$

where $\mathcal{Z}$ is the z-transform operator, which maps a sequence of numbers to a function of a complex variable.

Another way to view the z transform is that it is the Laplace transform of a function $\phi(t)$ constructed from the sample values of $f(t)$ and delta functions (Bracewell, 1965), *i.e.*,

$$\phi(t) = \sum_{n=0}^{\infty} f_n\, \delta(t - t_n)\,. \tag{4.89}$$

The Laplace transform of $\phi(t)$, denoted $\Phi_L(s)$, is given by[3]

$$\Phi_L(s) = \sum_{n=0}^{\infty} f_n \exp(-sn\Delta t) = \sum_{n=0}^{\infty} f_n[\exp(-s\Delta t)]^n\,. \tag{4.90}$$

This expression agrees with (4.88) if we let

$$z = \exp(s\Delta t)\,. \tag{4.91}$$

Since both s and z are arbitrary complex numbers, this substitution entails no loss of generality.

The main utility of the z transform is in analyzing problems involving discrete convolutions. Given two sequences $\{f_n\}$ and $\{p_n\}$, $n = 0, 1, 2, ...$, their discrete convolution is defined by analogy to (4.79) as

$$(p * f)_n = \sum_{m=0}^{n} p_m f_{n-m}\,. \tag{4.92}$$

The z transform yields

$$\mathcal{Z}\{p * f\} = \sum_{n=0}^{\infty} z^{-n} \sum_{m=0}^{n} p_m f_{n-m} = \sum_{m=0}^{\infty} \sum_{n=m}^{\infty} z^{-m} p_m z^{-(n-m)} f_{n-m} = P_z(z)\, F_z(z)\,. \tag{4.93}$$

Once again, a particular kind of convolution has been converted to a simple product by the appropriate transform.

[3]Strictly speaking, the Laplace transform used here has to be two-sided. Even though $\phi(t) = 0$ for $t < 0$, the impulse exactly at $t = 0$ has to be included.

4.2.4 Hilbert transform

The Hilbert transform of a function $f(x)$ is defined by

$$F_{Hi}(x) = [\mathcal{H}\mathbf{i}\{f\}](x) = -\frac{1}{\pi}\mathcal{P}\int_{-\infty}^{\infty} dx' \, \frac{f(x')}{x - x'}, \tag{4.94}$$

where $\mathcal{P}$ denotes the Cauchy principal value, discussed in App. B. The Hilbert transform of $f(x)$ can be regarded as the convolution of $f(x)$ with the generalized function $-\pi^{-1}\mathcal{P}\{1/x\}$, which was discussed in Secs. 2.1.1 and 3.3.7.

The inverse of the Hilbert transform can be found by use of the convolution theorem (3.132) and the expression (3.167) for $\mathcal{F}\{\mathcal{P}[1/x]\}$. These results allow us to write

$$\mathcal{F}\{F_{Hi}(x)\} = iF(\xi)\,\mathrm{sgn}(\xi), \tag{4.95}$$

where the sgn function is defined in (2.75) and $F(\xi)$ is the usual Fourier transform of $f(x)$. Thus Hilbert transformation corresponds to multiplication by $i\,\mathrm{sgn}\,\xi$ in the frequency domain, and inverse Hilbert transformation is simply multiplication by $-i\,\mathrm{sgn}\,\xi$. Use of the convolution theorem in reverse then shows that

$$f(x) = [\mathcal{H}\mathbf{i}^{-1}\{F_{Hi}\}](x) = \frac{1}{\pi}\mathcal{P}\int_{-\infty}^{\infty} dx' \, \frac{F_{Hi}(x')}{x - x'}. \tag{4.96}$$

Causality and the Kramers-Kronig relations The Hilbert transform plays an important role in the analysis of temporal filters and other causal linear systems. As noted above in Sec. 4.2.1, the temporal impulse response $h(t)$ of a linear, shift-invariant filter must vanish if $t < 0$. As we shall now show, this condition imposes a strong requirement on the Fourier transform of $h(t)$, which is the transfer function of the filter.

We can write the causality condition as

$$h(t) = 0 \qquad \text{if } t < 0, \tag{4.97}$$

but an equivalent statement is

$$h(t) = h(t)\,\mathrm{sgn}(t). \tag{4.98}$$

For $t > 0$, $\mathrm{sgn}(t) = 1$ and (4.98) reads $h(t) = h(t)$. For $t < 0$, (4.98) requires that $h(t) = -h(t)$, which can be satisfied only if $h(t) = 0$.

Taking the Fourier transform of both sides of (4.98), we obtain

$$H(\nu) * \left(-\frac{i}{\pi}\right)\mathcal{P}\left(\frac{1}{\nu}\right) = H(\nu). \tag{4.99}$$

Since $H(\nu)$ is a complex function, this equation, like any equation involving complex variables, is really two equations: one for the real part and one for the imaginary part. If we split $H(\nu)$ into real and imaginary parts as

$$H(\nu) = H_r(\nu) + iH_i(\nu), \tag{4.100}$$

then (4.99) is equivalent to

$$H_r(\nu) = \frac{1}{\pi}\int_{-\infty}^{\infty} d\nu' \, \frac{H_i(\nu')}{\nu - \nu'} = -\mathcal{H}\mathbf{i}\{H_i(\nu)\}; \tag{4.101a}$$

$$H_i(\nu) = -\frac{1}{\pi}\int_{-\infty}^{\infty} d\nu' \, \frac{H_r(\nu')}{\nu - \nu'} = \mathcal{H}\mathbf{i}\{H_r(\nu)\}\,. \tag{4.101b}$$

These equations, known as the *Kramers-Kronig relations*, show that the real and imaginary parts of the transfer function of a causal linear system are a Hilbert-transform pair. Imaging systems, however, do not usually have this nice mathematical property.

Analytic signals In practical applications we often deal with real-valued functions of time that oscillate at a single frequency ω. It is convenient to represent such functions as the real part of a complex exponential:

$$f(t) = A\cos(\omega t + \phi) = \mathrm{Re}\{A\exp(i\omega t + i\phi)\}\,. \tag{4.102}$$

But we may also be interested in amplitude-modulated waves where A is a function of time or phase-modulated waves where ϕ is a function of time. We consider, therefore, the more general real-valued function,

$$f(t) = A(t)\cos[\omega t + \phi(t)]\,. \tag{4.103}$$

We may still want to express $f(t)$ as the real part of some complex function $f_c(t)$. Trivially, we can write

$$f_c(t) = f(t) - ig(t)\,, \tag{4.104}$$

where $g(t)$ is an arbitrary real function. No matter what we choose for $g(t)$, we still have $f(t) = \mathrm{Re}\{f_c(t)\}$ since $-ig(t)$ is pure imaginary.

One way to fix $g(t)$ is to require that $f_c(t)$ be an analytic function when t is replaced by a complex variable $z = t + i\tau$. Since the real and imaginary parts of an analytic function must be related by the Cauchy-Riemann equations (see App. B), this condition removes the arbitrariness of the imaginary part $g(z)$. It will be left as an exercise for the reader to show that the Cauchy-Riemann equations are satisfied if we take $g(z)$ to be the Hilbert transform of $f(z)$. Then the *analytic signal* associated with the real function $f(t)$ is defined as

$$f_a(t) = f(t) - i\mathcal{H}\mathbf{i}\{f(t)\}\,. \tag{4.105}$$

From (4.95), the Fourier transform of the analytic signal, denoted $F_a(\nu)$, is given by

$$F_a(\nu) = F(\nu)\,[1 + \mathrm{sgn}(\nu)] = \begin{cases} 2F(\nu) & \text{if} \quad \nu > 0 \\ 0 & \text{if} \quad \nu < 0 \end{cases}, \tag{4.106}$$

and therefore $f_a(t)$ itself is given by

$$f_a(t) = 2\int_0^{\infty} d\nu \; F(\nu)\exp(2\pi i\nu t)\,. \tag{4.107}$$

An operational way to get the analytic signal is thus to take the Fourier transform of the original real signal, set negative-frequency components to zero, multiply the positive-frequency components by two, and inverse transform.

As an example, consider the single-frequency signal $f(t) = \cos(2\pi\nu_0 t)$. Since $F(\nu) = \frac{1}{2}\delta(\nu - \nu_0) + \frac{1}{2}\delta(\nu + \nu_0)$ for this signal, we find that $F_a(\nu) = \delta(\nu - \nu_0)$. Thus the analytic signal in this case is the expected $\exp(2\pi i\nu_0 t)$. The same result follows from the observation that $\mathcal{H}\mathbf{i}\{\cos(2\pi\nu_0 t)\} = -\sin(2\pi\nu_0 t)$.

4.2.5 Higher-order Hankel transforms

We encountered the zeroth-order Hankel transform in Sec. 3.4.4, where we saw that it was also the 2D Fourier transform for rotationally symmetric functions. Higher-order Hankel transforms are also important in some imaging applications. The nth-order Hankel transform of a 1D function $f(x)$ is defined by

$$F_{Hn}(\rho) = \mathcal{H}_n\{f(x)\} = 2\pi \int_0^\infty x\,dx\; \mathrm{J}_n(2\pi\rho x)\, f(x)\,, \tag{4.108}$$

where $\mathrm{J}_n(\cdot)$ in the nth-order Bessel function of the first kind. The inverse transform is

$$f(x) = \mathcal{H}_n^{-1}\{F_{Hn}(\rho)\} = 2\pi \int_0^\infty \rho\,d\rho\; \mathrm{J}_n(2\pi\rho x)\, F_{Hn}(\rho)\,. \tag{4.109}$$

Thus, for any order, the Hankel transform is its own inverse.

To see how higher-order Hankel transforms can arise in 2D imaging problems, consider the Fourier transform of $f^{(p)}(r,\theta)$ as given by (4.39). The derivation of this transform in polar coordinates parallels the derivation of (3.248). From (3.246) and (4.39), we find

$$\mathcal{F}_2\left\{f^{(p)}(r,\theta)\right\} = \sum_{n=-\infty}^{\infty} \int_0^\infty r\,dr\; f_n(r) \int_0^{2\pi} d\theta\; \exp[-2\pi i\rho r\cos(\theta-\theta_\rho)]\exp(in\theta)\,. \tag{4.110}$$

From (3.247) and a bit of algebra, the θ-integral is given by

$$\int_0^{2\pi} d\theta\; \exp[-2\pi i\rho r\cos(\theta-\theta_\rho)]\exp(in\theta) = 2\pi i^n \mathrm{J}_n(2\pi\rho r)\exp(in\theta_\rho)\,. \tag{4.111}$$

Thus

$$\mathcal{F}_2\left\{f^{(p)}(r,\theta)\right\} = \sum_{n=-\infty}^{\infty} i^n F_{Hn}(\rho)\exp(in\theta_\rho) = \sum_{n=-\infty}^{\infty} i^n \mathcal{H}_n\{f_n(r)\}\exp(in\theta_\rho)\,. \tag{4.112}$$

This expansion has the form of an angular Fourier series, but with the angular variable θ_ρ itself being a variable in Fourier space. The Fourier coefficient associated with this variable, is i^n times the nth-order Hankel transform of the original radial Fourier coefficient $f_n(r)$. In other words, if we represent both a 2D function and its 2D Fourier transform by angular Fourier series, then the two sets of coefficients are related by Hankel transforms of appropriate order.

4.3 FRESNEL INTEGRALS AND TRANSFORMS

In Sec. 3.3.7 we introduced the quadratic phase factor $\exp(i\pi\beta x^2)$ as a Gaussian function with a complex argument. Another name for a quadratic phase factor is *chirp*; the reason for this designation will be seen in the next chapter. In Chap. 9 we shall encounter the same function in Fresnel diffraction theory, where it is an approximate description of a spherical wave. Here we explore two topics related to chirps: special functions called *Fresnel integrals* and an integral transform called the *Fresnel transform*.

4.3.1 Fresnel integrals

The complex Fresnel integral $Z(x)$ is defined as the integral of a chirp:

$$Z(x) = \int_0^x dt \exp\left(i\frac{\pi}{2}t^2\right) . \tag{4.113}$$

Related functions, also called Fresnel integrals, are defined by

$$C(x) = \int_0^x dt \cos\left(\frac{\pi}{2}t^2\right) , \tag{4.114a}$$

$$S(x) = \int_0^x dt \sin\left(\frac{\pi}{2}t^2\right) . \tag{4.114b}$$

If x is real, $C(x)$ and $S(x)$ are, respectively, the real and imaginary parts of $Z(x)$. These functions are plotted vs. x in Fig. 4.4, and $C(x)$ is plotted vs. $S(x)$ in Fig. 4.5, a plot known as a *Cornu spiral*. Note that $C(0) = S(0) = 0$, and $C(\pm\infty) = S(\pm\infty) = \pm\frac{1}{2}$. Perhaps surprisingly, both $C(x)$ and $S(x)$ are odd functions.

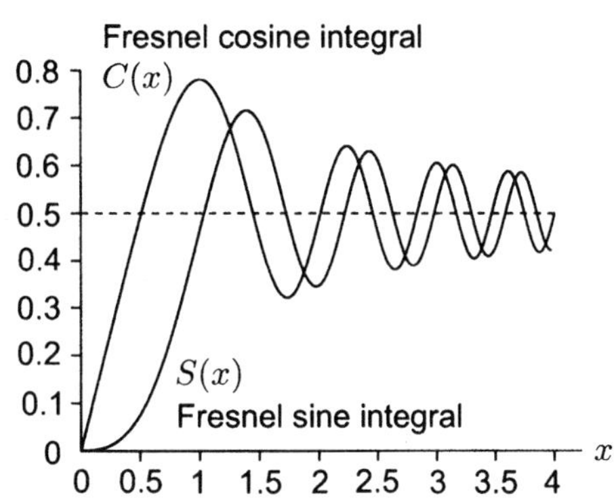

Fig. 4.4 Plots of the Fresnel integrals $C(x)$ and $S(x)$.

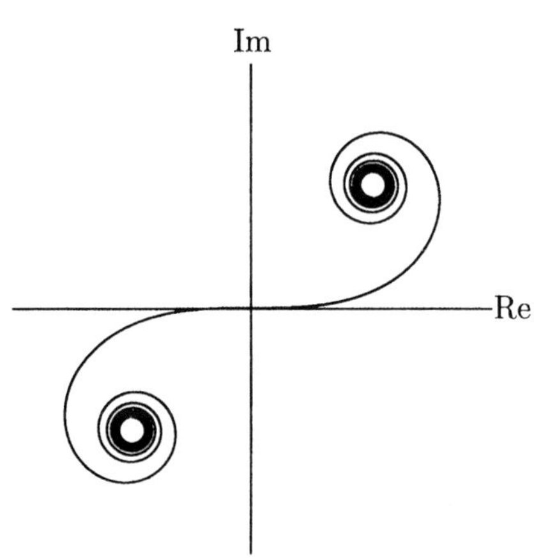

Fig. 4.5 The Cornu spiral, a plot of $C(x)$ vs. $S(x)$.

Fourier transform of a chirp segment As an illustration of the use of Fresnel integrals, we now compute the Fourier transform of a finite segment of a chirp:

$$f(x) = \begin{cases} \exp(i\pi\beta x^2) & \text{if } L_1 < x < L_2 \\ 0 & \text{otherwise} \end{cases} . \tag{4.115}$$

The Fourier transform of this function is given by

$$F(\xi) = \int_{L_1}^{L_2} dx \, \exp(i\pi\beta x^2 - 2\pi i \xi x) . \tag{4.116}$$

To evaluate this integral, we complete the square by requiring

$$\pi\beta x^2 - 2\pi\xi x = \frac{\pi}{2}[(t - t_0)^2 - t_0^2] , \tag{4.117}$$

which is achieved if

$$t = \sqrt{2\beta}\, x , \qquad t_0 = \sqrt{2/\beta}\, \xi . \tag{4.118}$$

With these substitutions,

$$F(\xi) = \frac{1}{\sqrt{2\beta}} \exp\left[-i\frac{\pi}{2}t_0^2\right] \int_{\sqrt{2\beta}L_1}^{\sqrt{2\beta}L_2} dt \, \exp\left[i\frac{\pi}{2}(t - t_0)^2\right] . \tag{4.119}$$

From the definition of the complex Fresnel integral, we find

$$F(\xi) = \frac{1}{\sqrt{2\beta}} \exp(-i\pi\xi^2/\beta) \left[Z\left(\sqrt{2\beta}L_2 - \sqrt{2/\beta}\xi\right) - Z\left(\sqrt{2\beta}L_1 - \sqrt{2/\beta}\xi\right)\right] . \tag{4.120}$$

If we let $L_1 \to -\infty$ and $L_2 \to \infty$, then, since $Z(\pm\infty) = \pm\frac{1}{2}(1 + i) = \pm\sqrt{i/2}$,

$$F(\xi) \to \sqrt{i/\beta} \exp(-i\pi\xi^2/\beta) , \qquad (L_1, L_2) \to (-\infty, \infty) , \tag{4.121}$$

which is exactly the expression (3.185) for the Fourier transform of an infinite chirp.

4.3.2 Fresnel transforms

One use of a chirp is as the kernel in a little-known integral transform, the Fresnel transform (Gori, 1994). As we shall see in Chap. 9, many diffraction problems can be formulated as Fresnel transforms.

The Fresnel transform associated with a chirp of rate β is defined as

$$F_\beta(x) = [\mathcal{F}r_\beta\{f\}](x) = \sqrt{-i\beta} \int_{-\infty}^{\infty} dx' \, f(x') \exp\left[i\pi\beta(x - x')^2\right] . \tag{4.122}$$

Different values of β lead to different functions $F_\beta(x)$, so the operator $\mathcal{F}r_\beta$ actually refers to a whole family of transforms. When we revisit Fresnel transforms in the context of diffraction, we shall see that the parameter β indexes different planes in which the diffraction pattern can be observed.

Since the integral in (4.122) is a convolution, we can find the form for the inverse Fresnel transform by taking *Fourier* transforms of both sides. With (4.121), the result is

$$\mathcal{F}_1\{F_\beta(x)\} = \exp(-i\pi\xi^2/\beta) F(\xi) . \tag{4.123}$$

Since the chirp on the right never goes to zero, we can divide through with impunity and obtain

$$F(\xi) = \mathcal{F}_1\{F_\beta(x)\}\exp(+i\pi\xi^2/\beta)\,. \tag{4.124}$$

An inverse Fourier transform yields

$$f(x) = \sqrt{i\beta}\int_{-\infty}^{\infty} dx'\, F_\beta(x')\exp\left[-i\pi\beta(x-x')^2\right] = \left[\mathcal{F}r_\beta^{-1}\{F_\beta\}\right](x)\,. \tag{4.125}$$

Thus the inverse Fresnel transform for parameter β is the same as the forward Fresnel transform with parameter $-\beta$. In operator form,

$$\mathcal{F}r_\beta^{-1} = \mathcal{F}r_{-\beta}\,. \tag{4.126}$$

When we apply this transform to diffraction, reversal of the sign of β will imply propagation of the wave in the opposite direction.

Some useful properties of the Fresnel transform are listed below. For more details, see Gori (1994) or Papoulis (1994).

Convolution Since Fresnel transformation is a convolution, and convolution is associative, it follows that

$$\mathcal{F}r_\beta\{f(x) * p(x)\} = F_\beta(x) * p(x) = f(x) * P_\beta(x)\,. \tag{4.127}$$

Derivatives Differentiation commutes with Fresnel transformation. That is,

$$\mathcal{F}r_\beta\{f'(x)\} = \frac{d}{dx}F_\beta(x)\,, \tag{4.128}$$

which can be proved by differentiating under the integral sign in (4.122), recognizing that

$$\frac{d}{dx}\exp\left[i\pi\beta(x-x')^2\right] = -\frac{d}{dx'}\exp\left[i\pi\beta(x-x')^2\right]\,, \tag{4.129}$$

and then performing an integration by parts.

Shifting and scaling Shifting and scaling of a function have a very simple effect on its Fourier transform; the situation is not quite so neat with the Fresnel transform. The scaling relation is (Gori, 1994):

$$\mathcal{F}r_\beta\{f(x/k)\} = F_{\beta k^2}(x/k)\,. \tag{4.130}$$

Thus scaling changes only the parameter β in the Fresnel transform, leaving the argument unchanged.

Thc shift rclation is

$$\mathcal{F}r_\beta\{f(x - k/2\pi\beta)\} = \exp(ik^2/4\pi\beta - ikx)\,\mathcal{F}r_\beta\{f(x)\exp(ikx)\}\,. \tag{4.131}$$

Correlation invariance The correlation operation is defined in (3.115). Since the Fourier transform of $[f\star p^*](x)$ is just $F(\xi)P^*(\xi)$ [*cf.* (3.135)], it follows from (4.123) that

$$\mathcal{F}r_\beta\{[f\star p^*](x)\} = \left[F_\beta \star P_\beta^*\right](x)\,. \tag{4.132}$$

Thus the Fresnel transform of the complex cross-correlation of two functions is the same as the complex cross-correlation of their Fresnel transforms.

Parseval's relations Since Fresnel transformation is simply multiplication by a pure phase factor in the Fourier domain [see (4.123)], it is a unitary transformation. Since unitary transformations preserve norms and scalar products, we have

$$\int_{-\infty}^{\infty} dx\, |f(x)|^2 = \int_{-\infty}^{\infty} dx\, |F_\beta(x)|^2 \,, \tag{4.133}$$

$$\int_{-\infty}^{\infty} dx\, f^*(x)\, p(x) = \int_{-\infty}^{\infty} dx\, F_\beta^*(x)\, P_\beta(x) \,. \tag{4.134}$$

Fresnel transform of a constant Suppose $f(x) = C$, a constant. Then

$$F_\beta(x) = \sqrt{-i\beta} \int_{-\infty}^{\infty} dx'\, C \exp\left[i\pi\beta(x - x')^2\right] = C \,, \tag{4.135}$$

where the last step follows from (4.121) and the central-ordinate theorem, (3.104). Thus a constant is invariant to Fresnel transformation.

Special functions Gori (1994) gives expressions for Fresnel transforms of sgn, step, rect, cos, delta, comb and Hermite-Gauss functions. The Fresnel transforms of sgn, step and rect functions are all expressible in terms of Fresnel integrals. The cosine turns out to be an eigenfunction of the Fresnel transform:

$$\mathcal{F}r_\beta\{\cos(Kx + \phi)\} = \exp(iK^2/4\pi\beta)\cos(Kx + \phi) \,. \tag{4.136}$$

This result follows immediately from (3.153) and (4.123).

The Fresnel transform of a single delta function is easy, but that of a comb is surprisingly subtle; see Gori (1994) for details.

Hermite-Gauss functions transform into different Hermite-Gauss functions with complex arguments, a property that has some use in discussing Gaussian laser beams (Siegman, 1986).

4.3.3 Chirps and Fourier transforms

Though chirps are most naturally associated with *Fresnel* transforms, they can also be used to perform *Fourier* transforms. Given a function $f(x)$, consider the following sequence of operations:

(a) Multiply $f(x)$ by a chirp, $\exp(-i\pi\beta x^2)$;

(b) Convolve with the conjugate chirp, $\exp(i\pi\beta x^2)$;

(c) Multiply again by the original chirp, $\exp(-i\pi\beta x^2)$.

This sequence results in the Fourier transform of $f(x)$, as we now demonstrate. Simply writing out the three steps gives

$$\begin{aligned} &\exp(-i\pi\beta x^2)\left\{\left[f(x)\exp(-i\pi\beta x^2)\right] * \exp(i\pi\beta x^2)\right\} \\ &= \exp(-i\pi\beta x^2)\int_{-\infty}^{\infty} dx'\, f(x')\exp(-i\pi\beta x'^2)\exp\left[i\pi\beta(x - x')^2\right] \\ &= \int_{-\infty}^{\infty} dx'\, f(x')\exp(-2\pi i\beta x x') = F(\beta x) \,. \end{aligned} \tag{4.137}$$

Thus the result of the three steps, though a function of x, is precisely the Fourier transform of $f(x)$, but evaluated at $\xi = \beta x$. As we shall see in Sec. 5.1, frequency ξ and position x are always linearly related for chirps.

The three steps listed above are often referred to as the *chirp-transform algorithm*, but this designation risks confusion with a Fresnel transform. A more precise (though somewhat awkward) term is *chirp-Fourier transform*. Even this term is ambiguous, however, since there is another algorithm that uses chirps to obtain a Fourier transform. We can also perform the following steps:

(*a*) Convolve $f(x)$ with a chirp, $\exp(i\pi\beta x^2)$;

(*b*) Multiply by the conjugate chirp, $\exp(-i\pi\beta x^2)$;

(*c*) Convolve again with the original chirp, $\exp(i\pi\beta x^2)$.

Again the result is $F(\beta x)$, though the proof is somewhat more complicated since two integrals are involved. If it is necessary to distinguish these two chirp-Fourier algorithms, the first can be called the MCM algorithm (for multiply-convolve-multiply) and the second CMC (for convolve-multiply-convolve).

A discrete version of the chirp-Fourier transform leads to the chirp-z transform, an implementation of the z-transform (see Sec. 4.2.3) by means of discrete convolutions with sampled chirps.

Fourier implementation of the Fresnel transform We have just seen that chirps can be used to perform a Fourier transform. It is also possible to go in the other direction and use Fourier transforms to perform a Fresnel transform, or convolution with a chirp.

The defining integral for the Fresnel transform, (4.122), can be rewritten as

$$F_\beta(x) = \sqrt{-i\beta}\,\exp(i\pi\beta x^2)\int_{-\infty}^{\infty} dx'\,\left\{f(x')\exp(i\pi\beta x'^2)\right\}\exp(-2i\pi\beta x x')\,. \quad (4.138)$$

The integral can now be recognized as the Fourier transform of the quantity in $\{\}$, with the role of spatial frequency ξ played by βx. The Fresnel transform of $f(x)$ can thus be calculated by the following algorithm:

(*a*) Multiply $f(x)$ by a chirp $\exp(i\pi\beta x^2)$;

(*b*) Compute the Fourier transform of the product;

(*c*) Evaluate the Fourier transform at $\xi = \beta x$;

(*d*) Multiply again by $\exp(i\pi\beta x^2)$;

(*e*) Multiply by $\sqrt{-i\beta}$.

This approach, though mathematically equivalent to (4.122), has practical implications since the Fourier transform can be performed efficiently with the FFT algorithm.

4.4 RADON TRANSFORM

In 1917, the Austrian mathematician Johann Radon published a classic paper (Radon, 1917) on the reconstruction of a multidimensional function from its integrals over lower-dimensional manifolds. In imaging applications, the concern is usually with a function $f(\mathbf{r})$ defined on a 2D plane, so the manifolds are lines. A continuous set of integrals of $f(\mathbf{r})$ over all parallel lines in a specified direction is called a 1D projection of the 2D function, and the 2D *Radon transform* is the set of 1D projections for all projection directions. This mathematical construct is often used as an idealized description of tomographic imaging systems that collect 1D data from 2D objects. Reconstruction of some representation of the object from a sequence of such measurements is called *reconstruction from projections* or *tomographic reconstruction*. A useful starting point for developing tomographic reconstruction algorithms is the *inverse Radon transform*.

In this section we shall introduce the basic mathematics of the Radon transform and its inverse, postponing until Chaps. 15 and 17 the connection between this theory and real tomographic imaging systems. The notation and approach used here follow Barrett (1984), to which the reader is referred for more details. A more comprehensive discussion at a similar level is given by Deans (1983), and advanced mathematical treatments are given by Helgason (1980) and Natterer (1986).

4.4.1 2D Radon transform and its adjoint

The 2D Radon transform is an integral transform in which the kernel is a line mass as introduced in Sec. 2.4.4. Specifically, if the equation of the line is $\mathbf{r} \cdot \hat{\mathbf{n}} = p$, where $\hat{\mathbf{n}}$ is a unit vector making an angle ϕ to the x-axis (see Fig. 2.7), then the line integral is defined by (2.134) as

$$\lambda(p, \phi) = \int_\infty d^2r \; f(\mathbf{r}) \, \delta(p - \mathbf{r} \cdot \hat{\mathbf{n}}) \,. \tag{4.139}$$

Alternative expressions for $\lambda(p, \phi)$ in rotated coordinate systems are given in (2.133). The set $\lambda(p, \phi)$ for fixed ϕ and all p is the 1D projection of $f(\mathbf{r})$ in direction ϕ, and the set of $\lambda(p, \phi)$ for all p and ϕ is the 2D Radon transform of $f(\mathbf{r})$.

Equation (4.139) can be written in operator form as

$$\lambda(p, \phi) = \mathcal{R}_2\{f(\mathbf{r})\} \,, \tag{4.140}$$

where $\mathcal{R}_2$ is the 2D Radon-transform operator as defined in (4.139).

Norms and scalar products in Radon space The Radon operator $\mathcal{R}_2$ maps the function $f(\mathbf{r})$, or $f(x, y)$, to another function $\lambda(p, \phi)$. If we restrict $f(\mathbf{r})$ to be square-integrable, the domain of $\mathcal{R}_2$ is $\mathbb{L}_2(\mathbb{R}^2)$. It is necessary, however, to impose certain smoothness or differentiability conditions on $f(\mathbf{r})$ to avoid mathematical difficulties. It suffices to assume that $f(\mathbf{r})$ is a good function as defined in Sec. 2.1.2 (see, *e.g.*, Natterer, 1986). Since any $\mathbb{L}_2$ function can be approximated arbitrarily closely with a good function, however, there is not much loss of rigor in considering the object space to be $\mathbb{L}_2(\mathbb{R}^2)$.

The range of $\mathcal{R}_2$ is usually called *Radon space*, but we need to define the characteristics of this space more carefully. All Hilbert spaces are defined in terms of scalar products. One natural way to define Radon space would be to use p and ϕ as polar coordinates of a vector $\mathbf{p}$, with area element $d^2p = p\,dp\,d\phi$, and to use integrals over area to define norms and scalar products. As we shall see below, however, it is more convenient to treat ϕ as a Cartesian coordinate for this purpose. Thus we define the norm of $\lambda(p, \phi)$ by

$$||\boldsymbol{\lambda}||^2 = \int_0^\pi d\phi \int_{-\infty}^{\infty} dp\ |\lambda(p, \phi)|^2 . \tag{4.141}$$

The function $\lambda(p, \phi)$ constitutes a vector $\boldsymbol{\lambda}$ in Radon space if this norm is finite.

Scalar products are defined similarly:

$$(\boldsymbol{\lambda}_1, \boldsymbol{\lambda}_2) = \int_0^\pi d\phi \int_{-\infty}^{\infty} dp\ \lambda_1^*(p, \phi)\, \lambda_2(p, \phi) . \tag{4.142}$$

Having defined the Hilbert spaces, we can now write (4.140) more abstractly as

$$\boldsymbol{\lambda} = \boldsymbol{\mathcal{R}}_2 \mathbf{f} . \tag{4.143}$$

Adjoint From Sec. 1.3.5, we know how to compute the adjoint of an integral operator. Since the kernel of the Radon operator is real, we can write

$$\left[\boldsymbol{\mathcal{R}}_2^\dagger \boldsymbol{\lambda}\right](\mathbf{r}) = \int_0^\pi d\phi \int_{-\infty}^{\infty} dp\ \lambda(p, \phi)\, \delta(p - \mathbf{r} \cdot \hat{\mathbf{n}}) . \tag{4.144}$$

The integral over p is easily performed, and we have

$$\left[\boldsymbol{\mathcal{R}}_2^\dagger \boldsymbol{\lambda}\right](\mathbf{r}) = \int_0^\pi d\phi\ \lambda(\mathbf{r} \cdot \hat{\mathbf{n}}, \phi) . \tag{4.145}$$

The substitution $p \to \mathbf{r} \cdot \hat{\mathbf{n}}$ converts each 1D function of p into a 2D function. As illustrated in Fig. 4.6, this function is defined for all $\mathbf{r}$, but it varies only along the direction parallel to $\hat{\mathbf{n}}$. Since the original projection direction was perpendicular to $\hat{\mathbf{n}}$, the function is smeared out in the projection direction by the substitution. For this reason, the operation $p \to \mathbf{r} \cdot \hat{\mathbf{n}}$ is called *backprojection.* The adjoint operation is then equivalent to backprojecting each 1D projection and integrating the result over all projection directions. Geometrically, the 2D Radon transform integrates over all points along a line, and its adjoint integrates over all lines passing through a point (Natterer, 1986).

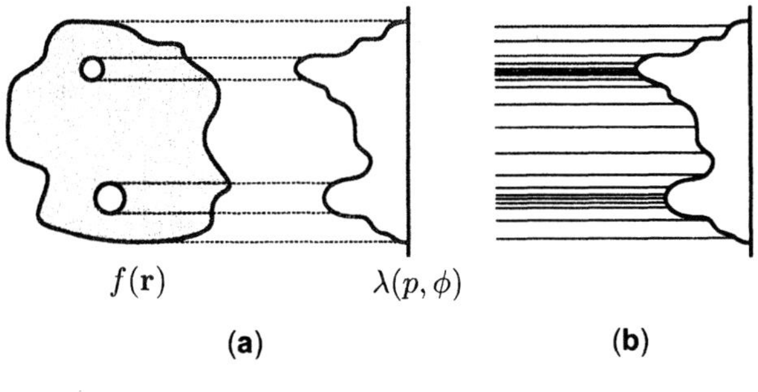

Fig. 4.6 Illustration of the operation of backprojection. (a) A 2D object and one of its 1D projections. (b) Backprojection of the 1D projection into the 2D space. (The density of lines in the backprojection suggests a gray scale).

We can now see why it was more convenient to define the scalar product with $dp\,d\phi$ rather than $p\,dp\,d\phi$; with the latter form, we would have had a factor of $\mathbf{r}\cdot\hat{\mathbf{n}}$ in the definition of backprojection.

Sinogram Since the definition of adjoint treats ϕ as a Cartesian coordinate rather than a polar one, it is useful to plot projection data in this way. A gray-level image of $[\mathcal{R}_2\mathbf{f}](p,\phi)$ with p and ϕ as the axes is known as a *sinogram* (see Fig. 4.7).

To understand the structure of a sinogram, consider first an object consisting of a single point, $f^\delta(\mathbf{r}) = \delta(\mathbf{r}-\mathbf{r}_0)$. The Radon transform of this object is

$$\left[\mathcal{R}_2\mathbf{f}^\delta\right](p,\phi) = \int_\infty d^2r\;\delta(\mathbf{r}-\mathbf{r}_0)\,\delta(p-\mathbf{r}\cdot\hat{\mathbf{n}}) = \delta(p-\mathbf{r}_0\cdot\hat{\mathbf{n}})\,. \tag{4.146}$$

Not surprisingly, the 1D projection of a 2D delta function is a 1D delta function. Explicitly, in terms of the p and ϕ coordinates,

$$\left[\mathcal{R}_2\mathbf{f}^\delta\right](p,\phi) = \delta[p - r_0\cos(\theta_0-\phi)]\,, \tag{4.147}$$

where r_0 and θ_0 are the polar coordinates of $\mathbf{r}_0$.

If we plot this delta function as a function of p for fixed ϕ, we see a spike at $p = r_0\cos(\theta_0-\phi)$. If we also vary ϕ, the spike moves along the p axis in a cosinusoidal fashion (see Fig. 4.7b). The amplitude of the cosine is the radial coordinate r_0, and the phase is the polar angle θ_0.

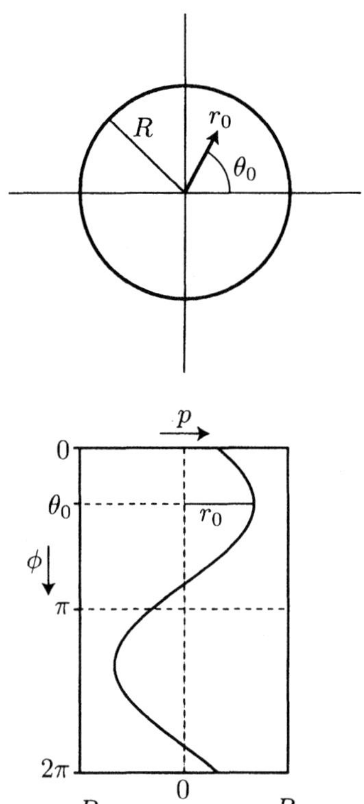

Fig. 4.7 The sinogram format, in which $\lambda(p,\phi)$ is plotted against Cartesian coordinates p and ϕ. (a) An object consisting of a single point at $\mathbf{r}=\mathbf{r}_0$. (b) Sinogram depiction of the projection data from the single-point object.

Since the Radon transform is linear, the sinogram of a more general object can be constructed by linear superposition. If we express an object as

$$f(\mathbf{r}) = \int_{\infty} d^2 r_0 \; f(\mathbf{r}_0)\, \delta(\mathbf{r} - \mathbf{r}_0) \,, \tag{4.148}$$

then its Radon transform is given by

$$[\mathcal{R}_2 \mathbf{f}]\,(p, \phi) = \int_{\infty} d^2 r_0 \; f(\mathbf{r}_0)\, \delta[p - r_0 \cos(\theta_0 - \phi)] \,. \tag{4.149}$$

The sinogram is thus the superposition of cosinusoidal patterns of the form (4.147) with the gray level of the cosine of amplitude r_0 and phase θ_0 determined by $f(\mathbf{r}_0)$.

4.4.2 Central-slice theorem

To see what information about $f(\mathbf{r})$ is contained in a single 1D projection, we can take the 1D Fourier transform (with respect to p) of $\lambda(p, \phi)$. From (4.139), we find

$$\Lambda(\nu, \phi) = [\mathcal{F}_1 \lambda(p, \phi)]\,(\nu) = \int_{-\infty}^{\infty} dp \, \exp(-2\pi i \nu p) \int_{\infty} d^2 r \; f(\mathbf{r})\, \delta(p - \mathbf{r} \cdot \hat{\mathbf{n}})$$

$$= \int_{\infty} d^2 r \; f(\mathbf{r}) \int_{-\infty}^{\infty} dp \, \exp(-2\pi i \nu p)\, \delta(p - \mathbf{r} \cdot \hat{\mathbf{n}}) = \int_{\infty} d^2 r \; f(\mathbf{r}) \exp(-2\pi i \mathbf{r} \cdot \hat{\mathbf{n}} \nu) \,. \tag{4.150}$$

The remaining integral is recognized as the 2D Fourier transform, but with the vector $\hat{\mathbf{n}}\nu$ appearing in place of the usual spatial frequency $\boldsymbol{\rho}$. Since $\hat{\mathbf{n}}$ is fixed once we have chosen the particular 1D projection $\lambda(p, \phi)$, the points in the frequency plane for which $\boldsymbol{\rho} = \hat{\mathbf{n}}\nu$ describe a line passing through the origin. For this reason, (4.150) is called the *central-slice theorem*. In words, the 1D Fourier transform of the projection $\lambda(p, \phi)$ is equal to the 2D transform of the original function $f(\mathbf{r})$ evaluated along a line through the origin of the Fourier plane. This relation is illustrated in Fig. 4.8.

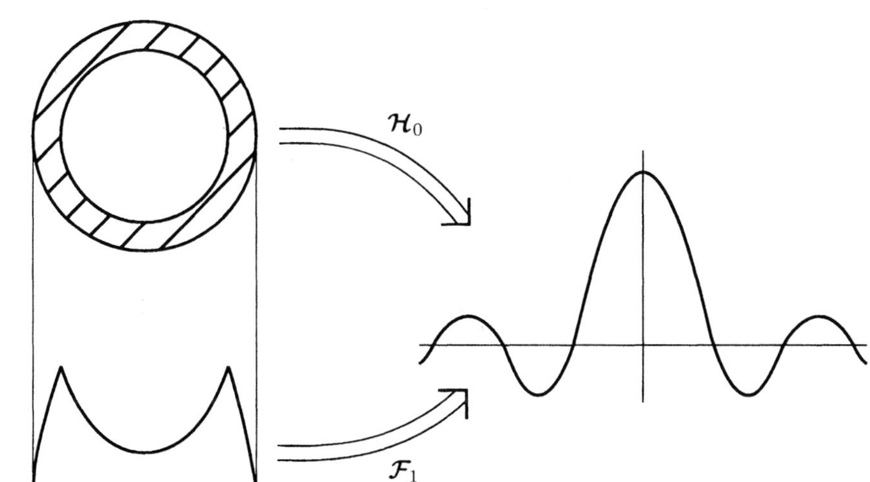

Fig. 4.8 Illustration of the central-slice theorem. A symmetric 2D function is shown so that its Fourier transform is real.

We can also express the central-slice theorem in operator form as

$$\mathcal{F}_2 = \mathcal{F}_1 \mathcal{R}_2 \,. \tag{4.151}$$

Applied to any 2D function $f(\mathbf{r})$, this relation says that forming a 1D projection and then taking the 1D Fourier transform is equivalent to taking the 2D Fourier transform. In interpreting the operators, however, we must keep in mind the need to evaluate the 2D transform at $\boldsymbol{\rho} = \hat{\mathbf{n}}\nu$.

One consequence of the central-slice theorem is that the Radon transform is invertible if 1D projection data are collected for $0 \le \phi < \pi$ since all points in the frequency plane are reached by ϕ in this range and $-\infty < \nu < \infty$.

Moreover, projections with $\pi \le \phi < 2\pi$ are redundant with those with $0 \le \phi < \pi$ since projections at ϕ and $\phi + \pi$ give information on the same line in the Fourier plane. We can reach this same conclusion by noting that $\phi \to \phi + \pi$ is equivalent to $\hat{\mathbf{n}} \to -\hat{\mathbf{n}}$. The delta function in (4.139) is unchanged if $p - \mathbf{r}\cdot\hat{\mathbf{n}} \to -p + \mathbf{r}\cdot\hat{\mathbf{n}}$, so

$$\lambda(p, \phi) = \lambda(-p, \phi + \pi) \,. \tag{4.152}$$

Thus we can always recover $f(\mathbf{r})$ if we know $\lambda(p, \phi)$ for $-\infty < p < \infty$ and ϕ in any interval of length π.

If $f(\mathbf{r})$ has finite support, we need even less data. If $f(\mathbf{r}) = 0$ for $r > \mathbb{R}_{max}$, then $\lambda(p, \phi) = 0$ for $|p| > \mathbb{R}_{max}$. We do not need to collect projections for which the line of integration does not intersect the object.

4.4.3 Filtered backprojection

One route to the inverse Radon transform starts with the operator relation (4.151). Turning that relation around, we can express the Radon transform as

$$\mathcal{R}_2 = \mathcal{F}_1^{-1} \mathcal{F}_2 \,. \tag{4.153}$$

Thus the inverse is, formally,

$$\mathcal{R}_2^{-1} = \mathcal{F}_2^{-1} \mathcal{F}_1 \,. \tag{4.154}$$

Implicit in this relation is the identification of $\boldsymbol{\rho}$ with $\hat{\mathbf{n}}\nu$, so it is necessary to express the integral over $\boldsymbol{\rho}$ in $\mathcal{F}_2^{-1}$ in terms of ν and ϕ. We could write $d^2\rho = \nu d\nu\, d\phi$ and integrate over $0 \le \phi < 2\pi$ and $0 \le \nu < \infty$, but we will usually know $\lambda(p, \phi)$ only for $0 \le \phi < \pi$. Fortunately, this suffices to give $\Lambda(\nu, \phi)$ for $-\infty < \nu < \infty$, so it is much more convenient to integrate over $0 \le \phi < \pi$ and $-\infty \le \nu < \infty$. When we do so, we must write $d^2\rho = |\nu| d\nu\, d\phi$ since the area element must be positive. Thus,

$$\begin{aligned} \left[\mathcal{F}_2^{-1}\mathcal{F}_1 \boldsymbol{\lambda}\right](\mathbf{r}) &= \int_0^\pi d\phi \int_{-\infty}^{\infty} |\nu| d\nu \, \exp(2\pi i \mathbf{r}\cdot\hat{\mathbf{n}}\nu) \int_{-\infty}^{\infty} dp \; \lambda(p,\phi) \exp(-2\pi i \nu p) \\ &= \int_0^\pi d\phi \int_{-\infty}^{\infty} d\nu \, \exp(2\pi i \mathbf{r}\cdot\hat{\mathbf{n}}\nu)\, |\nu| \Lambda(\nu,\phi) \,. \end{aligned} \tag{4.155}$$

The integral over ν can be regarded as the inverse Fourier transform of the product $|\nu|\Lambda(\nu, \phi)$. This transform would normally yield a function of p, but we see that $\mathbf{r}\cdot\hat{\mathbf{n}}$ appears where p should be, so we can write

$$\int_{-\infty}^{\infty} d\nu \, \exp(2\pi i \mathbf{r}\cdot\hat{\mathbf{n}}\nu)\, |\nu| \Lambda(\nu,\phi) = \hat{\lambda}(\mathbf{r}\cdot\hat{\mathbf{n}}, \phi) \,, \tag{4.156}$$

where $\hat{\lambda}(p,\phi)$ is a filtered version of $\lambda(p,\phi)$, with the filtering operation expressed in the 1D frequency domain as multiplication by $|\nu|$. For later convenience, we define

$$H(\nu) = |\nu| \,. \tag{4.157}$$

Putting the pieces together, we now have one form of the inverse Radon transform:

$$f(\mathbf{r}) = \left[\mathcal{R}_2^{-1}\boldsymbol{\lambda}\right](\mathbf{r}) = \int_0^{\pi} d\phi \; \hat{\lambda}(\mathbf{r}\cdot\hat{\mathbf{n}},\phi) \,. \tag{4.158}$$

Thus, one recovers $f(\mathbf{r})$ by filtering each projection (multiplying it by $H(\nu)$ in the 1D frequency domain), backprojecting the result and integrating over all projection angles. A discretized version of this inverse, known as *filtered backprojection*, is used in almost all commercial tomographic instruments at this writing.

Alternative form A shift-invariant filtering operation can be specified either as a multiplication in the frequency domain or as a convolution in the space domain. Thus the filtering operation of (4.158) can be written as a convolution of $\lambda(p,\phi)$ with the inverse Fourier transform of $H(\nu)$, which we denote by $h(p)$. The shift variable in the convolution is given by $\mathbf{r}\cdot\hat{\mathbf{n}}$, so we can write

$$\begin{aligned} f(\mathbf{r}) &= \int_0^{\pi} d\phi \left[\int_{-\infty}^{\infty} dp' \; \lambda(p',\phi)\, h(p-p')\right]_{p=\mathbf{r}\cdot\hat{\mathbf{n}}} \\ &= \int_0^{\pi} d\phi \int_{-\infty}^{\infty} dp' \; \lambda(p',\phi)\, h(\mathbf{r}\cdot\hat{\mathbf{n}}-p') \,. \end{aligned} \tag{4.159}$$

We know the functional form of $h(p)$ from Sec. 3.3.7. In particular, from (3.169) with $m=2$, we see that

$$-\frac{1}{2\pi^2}\mathcal{F}_1\left\{\frac{1}{p^2}\right\} = \nu\,\mathrm{sgn}(\nu) = |\nu| \,, \tag{4.160}$$

where $1/p^2$ is the rather bizarre generalized function[4] discussed in Sec. 2.3.3.

Combining (4.159) and (4.160) yields another form for the inverse Radon transform:

$$f(\mathbf{r}) = -\frac{1}{2\pi^2}\int_0^{\pi} d\phi \int_{-\infty}^{\infty} dp' \, \frac{\lambda(p',\phi)}{(\mathbf{r}\cdot\hat{\mathbf{n}}-p')^2} \,. \tag{4.161}$$

The structure of this result may be surprising; if $f(\mathbf{r})$ is nonnegative, as it is in most tomographic problems, then $\lambda(p,\phi)$ is also nonnegative, so the integrand appears to be nonnegative everywhere. There is an overall minus sign in front of the integral, yet the right-hand side of (4.161) is supposed to equal the nonnegative function $f(\mathbf{r})$. To resolve this paradox, recall from Sec. 2.3.3 that the generalized function $1/p^2$ has a strong negative singularity at the origin (see (2.91) and the discussion in that vicinity, esp. Fig. 2.6b). Because of this singularity, the integral is in fact negative.

[4] Strictly speaking, $\lambda(p',\phi)$ must be a good function (see Sec. 2.1.2) for this generalized function to be defined, so we are dealing with a Schwartz space rather than the more general $\mathbb{L}_2$ space. As we noted in Sec. 2.1.3, however, $\mathbb{L}_2$ functions can be approximated arbitrarily closely by good functions.

4.4.4 Unfiltered backprojection

It is instructive to see what happens if we leave out the filtering operation in the filtered-backprojection algorithm. Since backprojection plus integration over ϕ is the same as $\mathcal{R}_2^\dagger$, we wish to compute

$$\left[\mathcal{R}_2^\dagger \boldsymbol{\lambda}\right](\mathbf{r}) = \left[\mathcal{R}_2^\dagger \mathcal{R}_2 \mathbf{f}\right](\mathbf{r}). \tag{4.162}$$

If we can find a simple form for $\mathcal{R}_2^\dagger \mathcal{R}_2$, we may be able to construct $\mathcal{R}_2^{-1}$ as

$$\mathcal{R}_2^{-1} = \left[\mathcal{R}_2^\dagger \mathcal{R}_2\right]^{-1} \mathcal{R}_2^\dagger. \tag{4.163}$$

The alert reader will recognize (4.162) as the normal equation developed in Sec. 1.7.4 in the context of least-squares solutions of noisy, inconsistent equations and singular operators. Here, however, we are ignoring noise, and we argued in Sec. 4.4.2 that $\mathcal{R}_2$ is nonsingular if we know $\lambda(p, \phi)$ for all p and an angular range of π. Therefore (4.163) is just an identity which may provide another form of $\mathcal{R}_2^{-1}$.

To compute $\mathcal{R}_2^\dagger \mathcal{R}_2 \mathbf{f}$ explicitly, we combine (4.139) and (4.145), yielding

$$\left[\mathcal{R}_2^\dagger \mathcal{R}_2 \mathbf{f}\right](\mathbf{r}) = \int_0^\pi d\phi \int_\infty d^2r'\, f(\mathbf{r}')\, \delta(\mathbf{r}\cdot\hat{\mathbf{n}} - \mathbf{r}'\cdot\hat{\mathbf{n}}). \tag{4.164}$$

To simplify the integral over ϕ, we define a vector $\mathbf{R} = \mathbf{r} - \mathbf{r}'$ and express it in polar coordinates as (R, θ_R), where $R = |\mathbf{r} - \mathbf{r}'|$. The angle between $\hat{\mathbf{n}}$ and $\mathbf{R}$ is thus $\phi - \theta_R$, and we have

$$\left[\mathcal{R}_2^\dagger \mathcal{R}_2 \mathbf{f}\right](\mathbf{r}) = \int_\infty d^2r'\, f(\mathbf{r}') \int_0^\pi d\phi\, \delta[R\cos(\phi - \theta_R)]. \tag{4.165}$$

We can perform the integral over ϕ with the help of (2.33), from which we know

$$\delta(R\cos\phi) = \frac{\delta(\phi - \theta_R - \frac{\pi}{2})}{|R\sin\frac{\pi}{2}|} = \frac{1}{R}\,\delta\left(\phi - \theta_R - \frac{\pi}{2}\right). \tag{4.166}$$

Since the argument of the delta function vanishes exactly once in $(0, \pi]$, the integral over ϕ yields

$$\left[\mathcal{R}_2^\dagger \mathcal{R}_2 \mathbf{f}\right](\mathbf{r}) = \int_\infty d^2r'\, \frac{f(\mathbf{r}')}{R} = \int_\infty d^2r'\, \frac{f(\mathbf{r}')}{|\mathbf{r} - \mathbf{r}'|}. \tag{4.167}$$

The integral in (4.167) is a convolution; even though neither $\mathcal{R}_2$ nor $\mathcal{R}_2^\dagger$ describes a shift-invariant system, $\mathcal{R}_2^\dagger \mathcal{R}_2$ does. The point spread function for this operator is $1/r$, and the combined effect of Radon projection and its adjoint (unfiltered backprojection) is to convolve the object with $1/r$.

A graphical way of understanding this result is given in Fig. 4.9. Shown there is an object consisting of the single 2D delta function, $f^\delta(\mathbf{r}) = \delta(\mathbf{r} - \mathbf{r}_0)$. From (4.146) we know that the Radon transform of this object is a 1D delta function. The backprojection operation smears this 1D delta function into a line delta function: $\delta(p - \mathbf{r}_0\cdot\hat{\mathbf{n}}) \rightarrow \delta(\mathbf{r}\cdot\hat{\mathbf{n}} - \mathbf{r}_0\cdot\hat{\mathbf{n}})$. Though still 1D, in the sense that it can be used for only one integral, the line delta function is defined in the 2D space. Note that this line delta function passes through the original point $\mathbf{r}_0$ for all p and ϕ.

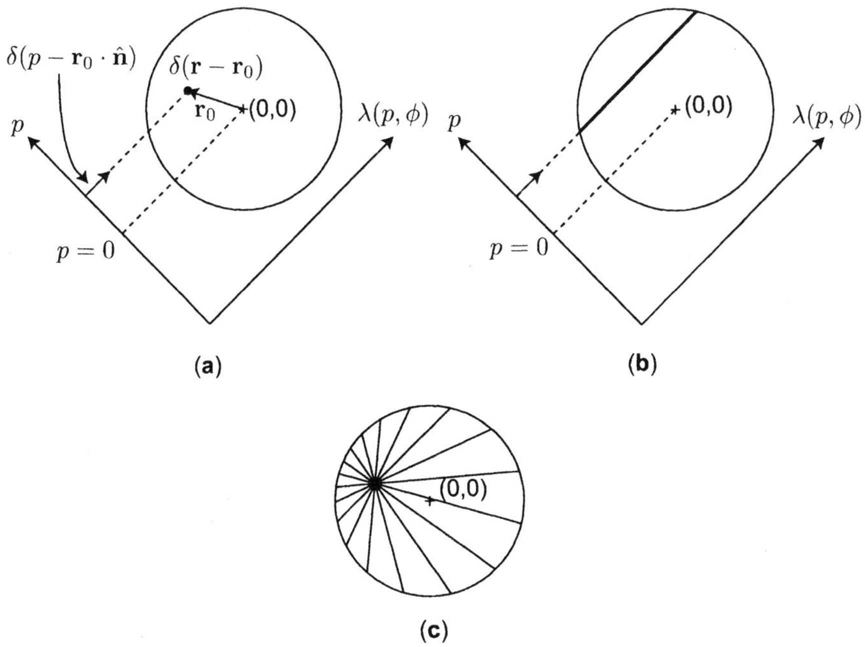

Fig. 4.9 Graphical explanation of the point spread function for $\mathcal{R}_2^\dagger \mathcal{R}_2$. (a) A single 2D delta function at $\mathbf{r} = \mathbf{r}_0$ and its projection at angle ϕ. (b) Backprojection of the 1D projection in (a). (c) Sum of many backprojections. This sum limits to a $1/r$ function as the number of projection angles approaches infinity.

We now bring in the integration over angles needed in the definition of $\mathcal{R}_2^\dagger$. If we had only a finite set of angles (as we necessarily do in any practical situation), then the integral over ϕ would be a sum, and the result would be a spoke pattern as shown in Fig. 4.9c. As the number of angles approaches infinity, the sum of line deltas becomes an integral and the spoke pattern approaches $1/|\mathbf{r} - \mathbf{r}_0|$, a cusp-like function that is singular at the original point location. We shall have more to say about finite angular sampling and the resulting PSF in Chap. 17.

If the object is more complicated than a single point, then there is a spoke pattern, or a $1/r$ pattern in the limit, associated with each object point. Because $\mathcal{R}_2^\dagger \mathcal{R}_2$ is shift invariant (if the object support is infinite), the form of this pattern is the same for each object point. The overlapping of these long-tailed point spread functions results in a considerable blurring of the object.

Deblurring Since the blurring is a simple convolution with a known blur function, it can in principle be corrected by inverse filtering in the Fourier domain. (We are, of course, neglecting practical issues like noise and discrete sampling in this section.)

To find the form of the inverse filter, we must first determine the Fourier transform of the blur function. From (3.248), we know that

$$\mathcal{F}_2 \left\{ \frac{1}{r} \right\} = 2\pi \int_0^\infty r\, dr\ \mathrm{J}_0(2\pi\rho r) \frac{1}{r} \,. \tag{4.168}$$

The change of variables $u = 2\pi\rho r$ and a well-known integral yield

$$\mathcal{F}_2\left\{\frac{1}{r}\right\} = \frac{1}{\rho}\int_0^\infty du\ \mathrm{J}_0(u) = \frac{1}{\rho}\,. \tag{4.169}$$

Thus blurring by convolution with $1/r$ is equivalent to multiplication with $1/\rho$ in the Fourier domain.

We can summarize what we have learned to this point by defining a multiplicative operator $\boldsymbol{\mathcal{M}}_{1/\rho}$ which acts on functions in the frequency domain and multiplies their value at each point by $1/\rho$. In terms of this operator, we have

$$\boldsymbol{\mathcal{R}}_2^\dagger\boldsymbol{\mathcal{R}}_2 = \boldsymbol{\mathcal{F}}_2^{-1}\boldsymbol{\mathcal{M}}_{1/\rho}\boldsymbol{\mathcal{F}}_2\,. \tag{4.170}$$

If we don't worry too much about the isolated point $\rho = 0$, the inverse of multiplication by $1/\rho$ is multiplication by ρ, so $\boldsymbol{\mathcal{M}}_{1/\rho}^{-1} = \boldsymbol{\mathcal{M}}_\rho$, and

$$\left[\boldsymbol{\mathcal{R}}_2^\dagger\boldsymbol{\mathcal{R}}_2\right]^{-1} = \boldsymbol{\mathcal{F}}_2^{-1}\boldsymbol{\mathcal{M}}_\rho\boldsymbol{\mathcal{F}}_2\,. \tag{4.171}$$

With (4.163), therefore, we have

$$\boldsymbol{\mathcal{R}}_2^{-1} = \boldsymbol{\mathcal{F}}_2^{-1}\boldsymbol{\mathcal{M}}_\rho\boldsymbol{\mathcal{F}}_2\boldsymbol{\mathcal{R}}_2^\dagger\,. \tag{4.172}$$

In words, the inverse Radon transform can be implemented by first performing an unfiltered backprojection, then transforming to the 2D Fourier domain, multiplying by the inverse filter $H(\boldsymbol{\rho}) = \rho$, and inverse-transforming back to the 2D space domain.

4.4.5 Radon transform in higher dimensions

In n dimensions, the Radon transform is an integral of an nD function over a set of $(n-1)$-dimensional hyperplanes. In particular, the 3D Radon transform is a set of integrals of a 3D function over ordinary 2D planes.

The equation of a 2D plane in a 3D space is $p = \mathbf{r}\cdot\hat{\mathbf{n}}$, which is identical in form to the equation of a line in a 2D space except that now $\mathbf{r}$ and $\hat{\mathbf{n}}$ are 3D vectors. In fact, with appropriate interpretation of the vectors, $p = \mathbf{r}\cdot\hat{\mathbf{n}}$ is the general equation of an $(n-1)$-dimensional hyperplane in an nD space, and the nD Radon transform is defined as

$$[\boldsymbol{\mathcal{R}}_n\mathbf{f}]\,(p,\hat{\mathbf{n}}) = \int_\infty d^n r\ f(\mathbf{r})\,\delta(p - \mathbf{r}\cdot\hat{\mathbf{n}})\,. \tag{4.173}$$

Note that the delta function here is 1D, so it can be used to perform only one of the n integrals implicit in $\int d^n r$. Also note that the unit vector $\hat{\mathbf{n}}$ is specified by $n-1$ variables in an nD space. For example, if $n = 3$, we can specify $\hat{\mathbf{n}}$ by two polar angles.

Norms, scalar products and adjoints By analogy to (4.142), we define the norm of the Radon transform of an nD function as

$$||\boldsymbol{\lambda}||^2 = \int_S d\Omega \int_0^\infty dp\ |\lambda(p,\hat{\mathbf{n}})|^2\,, \tag{4.174}$$

where $d\Omega$ is the element of solid angle in the nD space and the integral is over the unit sphere S. Since $\lambda(p, \hat{\mathbf{n}}) = \lambda(-p, -\hat{\mathbf{n}})$, we could equally well extend the p integral to $(-\infty, \infty)$ and integrate over half the unit sphere:

$$||\boldsymbol{\lambda}||^2 = \int_{\frac{1}{2}S} d\Omega \int_{-\infty}^{\infty} dp\ |\lambda(p, \hat{\mathbf{n}})|^2 . \tag{4.175}$$

For example, if $n = 3$, $d\Omega = \sin\theta\, d\theta\, d\phi$, where θ and ϕ are the usual spherical coordinates, and the integral is over 4π steradians if $0 \le p < \infty$ or 2π steradians if $-\infty < p < \infty$. The function $\lambda(p, \hat{\mathbf{n}})$ constitutes a vector $\boldsymbol{\lambda}$ in nD Radon space if the norm defined in (4.174) is finite.

Scalar products are defined by analogy to (4.142) as

$$(\boldsymbol{\lambda}_1, \boldsymbol{\lambda}_2) = \int_S d\Omega \int_0^{\infty} dp\ \lambda_1^*(p, \hat{\mathbf{n}})\, \lambda_2(p, \hat{\mathbf{n}}) = \int_{\frac{1}{2}S} d\Omega \int_{-\infty}^{\infty} dp\ \lambda_1^*(p, \hat{\mathbf{n}})\, \lambda_2(p, \hat{\mathbf{n}}) . \tag{4.176}$$

Note that we use $dp\, d\Omega$ as the volume element here rather than $p^{n-1} dp\, d\Omega$. With this definition of the scalar product, the adjoint of $\boldsymbol{\mathcal{R}}_n$ is given by

$$\left[\boldsymbol{\mathcal{R}}_n^{\dagger} \boldsymbol{\lambda}\right](\mathbf{r}) = \int_{\frac{1}{2}S} d\Omega \int_{-\infty}^{\infty} dp\ \lambda(p, \hat{\mathbf{n}})\, \delta(p - \mathbf{r}\cdot\hat{\mathbf{n}}) = \int_{\frac{1}{2}S} d\Omega\, \lambda(\mathbf{r}\cdot\hat{\mathbf{n}}, \hat{\mathbf{n}}) . \tag{4.177}$$

The interpretation of this equation is analogous to that of (4.145); the substitution $p \to \mathbf{r}\cdot\hat{\mathbf{n}}$ converts each 1D function of p into an nD function by assigning the value $\lambda(p, \hat{\mathbf{n}})$ uniformly to all $\mathbf{r}$ for which $\mathbf{r}\cdot\hat{\mathbf{n}} = p$. For $n = 3$, this means smearing it uniformly over the original plane of integration.

Central-slice theorem By taking the 1D transform of (4.173) as we did in Sec. 4.4.2, we can derive the nD central-slice theorem. By analogy to (4.150), we have

$$\Lambda(\nu, \hat{\mathbf{n}}) = [\boldsymbol{\mathcal{F}}_1 \lambda(p, \hat{\mathbf{n}})](\nu) = \int_{\infty} d^n r\ f(\mathbf{r}) \int_{-\infty}^{\infty} dp\ \exp(-2\pi i \nu p)\, \delta(p - \mathbf{r}\cdot\hat{\mathbf{n}})$$

$$= \int_{\infty} d^n r\ f(\mathbf{r}) \exp(-2\pi i \mathbf{r}\cdot\hat{\mathbf{n}}\nu) = F(\hat{\mathbf{n}}\nu) , \tag{4.178}$$

where $F(\boldsymbol{\rho})$ is the nD Fourier transform of $f(\mathbf{r})$. In words, the 1D Fourier transform of the projection is the same as one line through the nD of the original function $f(\mathbf{r})$. Note that the central-slice theorem in any number of dimensions always relates to a line in nD Fourier space, not a hyperplane. In operator form, we can express the central-slice theorem as

$$\boldsymbol{\mathcal{F}}_n = \boldsymbol{\mathcal{F}}_1 \boldsymbol{\mathcal{R}}_n . \tag{4.179}$$

In using this form, the implicit substitution $\boldsymbol{\rho} \to \hat{\mathbf{n}}\nu$ must always be kept in mind.

Inverse transform From the central-slice theorem, the formal inverse Radon transform is [*cf.* (4.154)]

$$\boldsymbol{\mathcal{R}}_n^{-1} = \boldsymbol{\mathcal{F}}_n^{-1} \boldsymbol{\mathcal{F}}_1 . \tag{4.180}$$

Converting the operators into explicit integrals as in Sec. 4.4.3, we find [*cf.* (4.155)]

$$\begin{aligned}\left[\mathcal{F}_n^{-1}\mathcal{F}_1\boldsymbol{\lambda}\right](\mathbf{r}) &= \int_{\frac{1}{2}S} d\Omega \int_{-\infty}^{\infty} |\nu|^{n-1} d\nu \, \exp(2\pi i \mathbf{r}\cdot\hat{\mathbf{n}}\nu) \int_{-\infty}^{\infty} dp \, \lambda(p,\hat{\mathbf{n}}) \exp(-2\pi i \nu p) \\ &= \int_{\frac{1}{2}S} d\Omega \int_{-\infty}^{\infty} d\nu \, \exp(2\pi i \mathbf{r}\cdot\hat{\mathbf{n}}\nu)\, |\nu|^{n-1} \Lambda(\nu,\hat{\mathbf{n}}) \,. \end{aligned} \tag{4.181}$$

The integral over ν can be regarded as the inverse Fourier transform of the product $|\nu|^{n-1}\Lambda(\nu,\hat{\mathbf{n}})$, so (4.156) generalizes to

$$\int_{-\infty}^{\infty} d\nu \, \exp(2\pi i \mathbf{r}\cdot\hat{\mathbf{n}}\nu)\, |\nu|^{n-1} \Lambda(p,\hat{\mathbf{n}}) = \hat{\lambda}(\mathbf{r}\cdot\hat{\mathbf{n}},\hat{\mathbf{n}}) \,, \tag{4.182}$$

where again $\hat{\lambda}(p,\hat{\mathbf{n}})$ denotes a filtered version of $\lambda(p,\hat{\mathbf{n}})$. The filter function is now

$$H(\nu) = |\nu|^{n-1} \,, \tag{4.183}$$

but note that the absolute-value signs are unnecessary if n is odd.

Odd vs. even dimensions The filter function given in (4.183) is fundamentally different for even and odd n. If n is odd, then $H(\nu)$ is just ν^{n-1}, and we know from (3.97) that multiplying a function by $(2\pi i\nu)^k$ in the frequency domain is equivalent to differentiating it k times. Thus the 1D filtering operation before backprojection is equivalent to differentiating the projection data $n-1$ times with respect to p.

The key point is that differentiation is a local operation; to compute any derivative of the projection $\lambda(p,\hat{\mathbf{n}})$ at, say, $p = p_0$, we need to know the function only in an infinitesimal neighborhood of $p = p_0$. That means that we can reconstruct an nD function $f(\mathbf{r})$ at some point $\mathbf{r}_0$ knowing only its integrals over hyperplanes passing through a neighborhood of $\mathbf{r}_0$, provided n is odd.

For n even, we must account for the absolute-value operation in $H(\nu)$, which greatly affects the behavior of its Fourier transform, $h(p)$. One way to think about this difference is in terms of the Hilbert transform, introduced in Sec. 4.2.4. We know from (4.95) that the Hilbert transform is equivalent to multiplying by $i\,\mathrm{sgn}(\nu)$ in the frequency domain, and for n even we can write

$$H(\nu) = |\nu|^{n-1} = \nu^{n-1}\,\mathrm{sgn}(\nu) \,. \tag{4.184}$$

Thus, within constant factors, the 1D filtering operation can be expressed as $n-1$ derivatives compounded with a Hilbert transform. (The differentiations and Hilbert transformation can be performed in any order.) From (4.94) we know that Hilbert transformation is essentially convolution with the principal value of $1/p$. Since $1/p$ has long tails, Hilbert transformation is decidedly nonlocal. To reconstruct $f(\mathbf{r})$ at $\mathbf{r} = \mathbf{r}_0$ for n even, we need to know its integrals over all hyperplanes, not just those passing through a neighborhood of $\mathbf{r}_0$.

Another way to think about this difference is in terms of the asymptotic behavior of $h(p)$ as $p \to \infty$. It follows from the Riemann-Lebesgue lemma [see (3.107)] that $p^k h(p) \to 0$ as $p \to \infty$ if $H(\nu)$ has bounded derivatives at least up to order k. For n odd, all derivatives of $H(\nu)$ are bounded, so $h(p) \to 0$ faster than p^{-k} for all k. In fact, we know that $h(p)$ is the $(n-1)^{th}$ derivative of a delta function in this case, so it is identically zero away from the neighborhood of the origin. For n even, however, the absolute-value signs spoil the differentiability. For $n = 2$, for example, the first derivative of $|\nu|$ is bounded but the second is not, and we know from Sec. 4.4.3 that $h(p)$ is the generalized function $1/p^2$. This function coincides with the ordinary function $1/p^2$ for $p \neq 0$, so $p\,h(p) \to 0$ as $p \to \infty$, but $p^2 h(p)$ does not.

3D case If $n = 3$, then $H(\nu) = \nu^2$, and we know that multiplication by ν^2 in a 1D frequency domain is essentially the same thing as taking a second derivative [see (3.99)], which in turn is equivalent to convolving with the second derivative of a 1D delta function. Specifically, the space-domain filter function for the 3D Radon transform is

$$h(p) = \mathcal{F}_n^{-1}\left\{\nu^2\right\} = -\frac{1}{4\pi^2}\,\delta''(p)\,. \tag{4.185}$$

Thus one form of the 3D inverse Radon transform is

$$f(\mathbf{r}) = -\frac{1}{4\pi^2}\int_{\frac{1}{2}S} d\Omega\;\lambda''(\mathbf{r}\cdot\hat{\mathbf{n}},\hat{\mathbf{n}})\,, \tag{4.186}$$

where $\lambda''(p,\hat{\mathbf{n}})$ is the second derivative of the projection with respect to p. Reconstruction in this case is just backprojection of the second derivative of the projection.

As in 2D, it is also possible to backproject first and then filter. The reader may show that [*cf.* (4.167)]

$$\left[\mathcal{R}_3^\dagger\boldsymbol{\lambda}\right](\mathbf{r}) = \int_{\frac{1}{2}S} d\Omega\;\lambda(\mathbf{r}\cdot\hat{\mathbf{n}},\hat{\mathbf{n}}) = \int_\infty d^3r'\,\frac{f(\mathbf{r}')}{|\mathbf{r}-\mathbf{r}'|^2}\,. \tag{4.187}$$

It then follows from (2.143) that

$$f(\mathbf{r}) = -\frac{1}{4\pi^2}\,\nabla^2\int_{\frac{1}{2}S} d\Omega\;\lambda(\mathbf{r}\cdot\hat{\mathbf{n}},\hat{\mathbf{n}})\,. \tag{4.188}$$

Thus the 3D filter is also a second derivative, the Laplacian.

Dipole-sheet transform The dipole-sheet transform (Barrett, 1982, 1984) is a symmetrized version of the 3D Radon transform and its inverse. The asymmetry can be seen by defining a 3D vector $\mathbf{p} \equiv p\,\hat{\mathbf{n}}$, so that $d^3p = p^2dp\,d\Omega_{\mathbf{n}}$. With this definition, we can write the inverse 3D Radon transform, (4.186), as

$$f(\mathbf{r}) = -\frac{1}{4\pi^2}\int_\infty \frac{d^3p}{p^2}\,[\mathcal{R}_3\mathbf{f}](\mathbf{p})\,\delta''(p-\mathbf{r}\cdot\hat{\mathbf{n}})\,, \tag{4.189}$$

while the forward Radon transform is

$$[\mathcal{R}_3\mathbf{f}](\mathbf{p}) = \int_\infty d^3r\;\mathbf{f}(\mathbf{r})\,\delta(p-\mathbf{r}\cdot\hat{\mathbf{n}})\,. \tag{4.190}$$

These forms differ by the second derivative and the factor of $-1/(4\pi^2p^2)$ in (4.189); to symmetrize, we can move one derivative and a factor of $i/(2\pi p)$ to the forward transform. We thus define the dipole-sheet operator $\mathcal{D}_3$ by

$$[\mathcal{D}_3\mathbf{f}](\mathbf{p}) = \int_\infty d^3r\;\mathbf{f}(\mathbf{r})\,\psi(\mathbf{p},\mathbf{r})\,, \tag{4.191}$$

where $\psi(\mathbf{p},\mathbf{r})$ is the dipole-sheet basis function, defined by

$$\psi(\mathbf{p},\mathbf{r}) = \frac{i}{2\pi p}\,\delta'(p-\mathbf{r}\cdot\hat{\mathbf{n}})\,. \tag{4.192}$$

The derivative of the delta function here has the same structure as the double layer of charge used to enforce boundary conditions in electrostatics (Jackson, 1998).

Regarded as a function in three dimensions, $\delta'(p - \mathbf{r} \cdot \hat{\mathbf{n}})$ vanishes except in the neighborhood of the plane $p = \mathbf{r} \cdot \hat{\mathbf{n}}$, and it can be regarded, loosely, as being $+\infty$ just to one side of this plane and $-\infty$ on the other side. If we think of $\delta(p - \mathbf{r} \cdot \hat{\mathbf{n}})$ as a distribution of charge confined to a plane, then $\delta'(p - \mathbf{r} \cdot \hat{\mathbf{n}})$ is a double layer of positive and negative charges, or a sheet of dipoles.

It follows from (4.189) and (4.190) that the dipole-sheet transform is unitary; the inverse is given by the adjoint. Many other interesting properties of the dipole-sheet transform are given in Barrett (1982, 1984). For example, the transform of any rotationally symmetric function is just a constant times that same function. In addition, as we shall see in Chap. 17, the dipole-sheet transform arises naturally in cone-beam tomography.

Related transforms in an arbitrary number of dimensions are discussed in Ludwig (1966).

4.4.6 Radon transform in signal processing

One application of the Radon transform is to reduce nD signal-processing operations to a sequence of 1D operations (Gmitro *et al.*, 1983; Barrett, 1984b). Though many different operations can be reduced in this way (Easton and Barrett, 1987), we shall confine our attention here to convolutions.

There are two mathematical theorems that make it possible to reduce nD convolutions to 1D. The first one, which follows directly from the central-slice theorem, states that

$$(\mathcal{R}_n \mathbf{f}_1) * (\mathcal{R}_n \mathbf{f}_2) = \mathcal{R}_n(\mathbf{f}_1 * \mathbf{f}_2)\,, \tag{4.193}$$

where the asterisk on the left denotes a 1D convolution with respect to p, while the asterisk on the right denotes an nD convolution. Thus, to convolve two nD functions, one can take the nD Radon transform of each, perform 1D convolutions for every projection angle, and then perform an nD inverse Radon transform. This approach is particularly attractive for n odd because of the local character of the inverse, as discussed above.

The second theorem (Natterer, 1986) does not directly require an inverse Radon transform; it states that

$$(\mathcal{R}_n^\dagger \mathbf{h}) * \mathbf{f} = \mathcal{R}_n^\dagger[\mathbf{h} * (\mathcal{R}_n \mathbf{f})]\,. \tag{4.194}$$

Note that $\mathbf{h}$ here denotes a set of 1D functions of p, one function for each projection direction. Therefore, the asterisk on the left denotes nD convolution but the one on the right denotes 1D convolution on the variable p. Proof of (4.194) requires writing the operators in integral form and shuffling the order of integration.

If we want to use (4.194) to compute $\mathbf{f}_1 * \mathbf{f}_2$, we must first solve the problem $\mathcal{R}_n^\dagger \mathbf{h} = \mathbf{f}_1$ to find the set of functions $\mathbf{h}$, but then no further inverse transforms are required. We merely compute a set of 1D convolutions and backproject.

Quite apart from signal processing, (4.194) can also be used to derive inverses of the Radon transform and related transforms. For example, we shall see in Chap. 17 how it is useful in analyzing attenuation problems in SPECT.

5

Mixed Representations

In this chapter we discuss a class of mathematical descriptions called *mixed representations*, since they mix apparently incompatible variables such as spatial position and spatial frequency. These descriptions can also be called *extended representations*, since they use more variables than are needed for a parsimonious specification of a function.

In Sec. 5.1 we introduce a variety of integral transforms where a function of a spatial variable is augmented, apparently redundantly, with a spatial frequency ξ. These transforms are linear, like all of the other integral transforms treated earlier in the book, but in Sec. 5.2 we consider *bilinear* transforms, where a product of the original function with itself appears in the integrand. Later, in Sec. 5.3, the spatial variable will be augmented with a scale variable.

As we shall see, these new variables, though superfluous from a strict mathematical viewpoint, offer new insights into the behavior of the functions. Moreover, when the functions are sampled in all variables, little or no increase in the number of required sample points is needed with the extended representations.

5.1 LOCAL SPECTRAL ANALYSIS

A spatial description $f(x)$ gives information about the behavior of a function at every point x, while its Fourier transform $F(\xi)$ gives information about the behavior at every spatial frequency ξ. Neither description gives any information about what frequencies are associated with what spatial location. Often it would be useful to have the spatial counterpart of a musical score, associating frequencies with positions.

5.1.1 Local Fourier transforms

One mathematical construct to accomplish this end is the *sliding-window Fourier transform* or *local Fourier transform*, defined by

$$F_b(\xi; x_0) = \int_{-\infty}^{\infty} dx \; b^*(x - x_0)\, f(x) \exp(-2\pi i \xi x) \,, \tag{5.1}$$

where $b(x)$ is a window function. The window function is usually chosen to be real-valued, but can, in general, be complex. The shape of $b(x)$ is arbitrary as far as the mathematics is concerned, but it is usually taken to be more or less concentrated around $x = 0$, so that the shifted function $b(x - x_0)$ in (5.1) is concentrated around $x = x_0$. Suppose, for example, that $b(x) = \text{rect}(x/\Delta x)$. Then the window in (5.1) extends from $x_0 - \frac{1}{2}\Delta x < x < x_0 + \frac{1}{2}\Delta x$, and only in this range does the behavior of $f(x)$ contribute to $F_b(\xi; x_0)$.

The window width Δx controls the degree of localization in x, but it also controls the resolution in ξ. To see this, suppose that $f(x)$ can be well approximated by $\exp(2\pi i \xi_0 x)$ over the window width. We would like to find that $F_b(\xi; x_0)$ is sharply peaked at $\xi = \xi_0$, approximating a delta function, but (5.1) shows that

$$F_b(\xi; x_0) = B^*(\xi_0 - \xi) \exp[2\pi i (\xi_0 - \xi) x_0] \,, \tag{5.2}$$

where $B(\xi) = \mathcal{F}\{b(x)\}$. Continuing the previous example where $b(x) = \text{rect}(x/\Delta x)$, we now have $F_b(\xi; x_0) \propto \Delta x \,\text{sinc}[(\xi - \xi_0)\Delta x]$, which has a width (measured from the peak to the first zero) of $1/\Delta x$. Thus if we make Δx smaller to improve the spatial localization, we pay the price in spectral localization.

5.1.2 Uncertainty

One of the most celebrated scientific discoveries of the twentieth century is the *Heisenberg uncertainty principle*, which says that two physical observables that are represented quantum mechanically by noncommuting operators cannot be measured simultaneously with arbitrary precision. Any reduction in the inherent uncertainty in one observable must be accompanied by an increase in the uncertainty of the other. This is a profound result, intimately related to the nature of matter and the measurement process in quantum-mechanical systems, but in a purely formal sense it is a simple consequence of basic Fourier mathematics.

The connection between Fourier analysis and quantum-mechanical uncertainty was made by the physicist C. G. Darwin, grandson of Charles Darwin (Cohen, 1995). We shall present the concept of Fourier uncertainty here in the context of the local Fourier transform and then return briefly to the quantum-mechanical analogy.

To understand the inverse relation between the width of a function like $b(x)$ and the width of its Fourier transform, we need a more general definition of width. A mathematically convenient approach is to treat $|b(x)|^2$, when properly normalized, as a probability density function (PDF) and use the associated standard deviation σ_x (see App. C) as a measure of width. The width is thus the square-root of the variance σ_x^2, where

$$\sigma_x^2 = \frac{\int_{-\infty}^{\infty} dx \; (x - \overline{x})^2 \, |b(x)|^2}{\int_{-\infty}^{\infty} dx \; |b(x)|^2} \,. \tag{5.3}$$

In this expression, $\overline{x}$ is the mean value of x with respect to the density $|b(x)|^2$, *i.e.*,

$$\overline{x} = \frac{\int_{-\infty}^{\infty} dx\; x|b(x)|^2}{\int_{-\infty}^{\infty} dx\; |b(x)|^2}\,. \tag{5.4}$$

Similarly, the width of $B(\xi)$ is taken as σ_ξ, where

$$\sigma_\xi^2 = \frac{\int_{-\infty}^{\infty} d\xi\; (\xi - \overline{\xi})^2\, |B(\xi)|^2}{\int_{-\infty}^{\infty} d\xi\; |B(\xi)|^2}\,, \tag{5.5}$$

and $\overline{\xi}$ is defined analogously to (5.4). Since a change of variables can set $\overline{x}$ and $\overline{\xi}$ to zero, we shall henceforth neglect these terms without loss of generality.

Use of the Parseval and derivative theorems, (3.80) and (3.98), respectively, lets us express σ_ξ^2 in the spatial domain as

$$\sigma_\xi^2 = \frac{\frac{1}{4\pi^2}\int_{-\infty}^{\infty} dx\; |b'(x)|^2}{\int_{-\infty}^{\infty} dx\; |b(x)|^2}\,, \tag{5.6}$$

where the prime denotes derivative. The Schwarz inequality (1.14) yields

$$\int_{-\infty}^{\infty} dx\; x^2\, |b(x)|^2 \int_{-\infty}^{\infty} dx\; |b'(x)|^2 \geq \left| \int_{-\infty}^{\infty} dx\; b^*(x)\, x b'(x) \right|^2 , \tag{5.7}$$

from which it follows that

$$\sigma_x \sigma_\xi \geq \frac{\left| \int_{-\infty}^{\infty} dx\; b^*(x)\, x b'(x) \right|}{2\pi \int_{-\infty}^{\infty} dx\; |b(x)|^2}\,. \tag{5.8}$$

A more useful form results if we use the operator relation,

$$\frac{d}{dx} x - x \frac{d}{dx} = 1\,, \tag{5.9}$$

which can be proved by operating on an arbitrary function and using the chain rule of differentiation. This relation can be used to express the function $xb'(x)$, in (5.8). With an integration by parts and a little algebra,[1] we then find that

$$\sigma_x \sigma_\xi \geq \frac{1}{4\pi}\,. \tag{5.10}$$

The inequality in (5.8) becomes an equality if $b'(x)$ is proportional to $xb(x)$, which occurs if $b(x)$ is a Gaussian function of the general form $A\exp(-\alpha x^2)$. Gaussians are therefore sometimes referred to as minimum uncertainty signals. They are the most common choice for the window function in a local Fourier transform.

It is interesting to verify the uncertainty relation for a rect function, $b(x) = \text{rect}(x)$. Then an easy integral shows that $\sigma_x = 1/(2\sqrt{3})$, but in fact $\sigma_\xi = \infty$ since the integral of $[\xi\, \text{sinc}(\xi)]^2$ diverges. The sinc has a finite width (unity, in fact) as measured from its peak to the first zero, but an infinite width by the variance measure.

[1]The derivation of (5.10) from (5.8) and (5.9) is straightforward if $b(x)$ is real, but a little tricky if it is complex. See Cohen (1995), p. 47.

Relation to quantum mechanics The connection of this Fourier math to quantum mechanics takes two distinctly different forms, depending on just which variables are involved. The most familiar form of Heisenberg uncertainty involves the position and momentum of a particle such as an electron. We discuss this position-momentum uncertainty relation first and then comment briefly on the so-called energy-time uncertainty principle.

For simplicity we consider one-dimensional motion, denoting the electron position by x and its momentum by p. One way of identifying p with a spatial frequency is to follow Prince Louis DeBroglie, who in 1923 postulated that an electron could behave either as a particle or as a wave. When it behaved as a wave, he showed, its wavelength λ would be related to the momentum by $p = h/\lambda$, where h is Planck's constant. This audacious suggestion, made several years before Schrödinger's quantum theory, was experimentally verified in 1927 by Davisson and Germer in New York and Thomson in Aberdeen.

We rewrite the DeBroglie relation as

$$p = \frac{h}{\lambda} = h\xi \,, \tag{5.11}$$

where ξ is now interpreted as the spatial frequency (reciprocal wavelength) of the electron wave. From (5.10) and (5.11), we see that

$$\sigma_x \sigma_p \geq \tfrac{1}{2}\hbar \,, \tag{5.12}$$

where $\hbar = h/2\pi$. This is the usual form of the Heisenberg position-momentum uncertainty principle.

Another way to derive this relation is to follow modern wave mechanics as pioneered by Schrödinger. This theory postulates that the state of a physical system is specified by a wavefunction and that physical observables are represented by Hermitian operators that can operate on this function. For a single particle in one dimension, the wavefunction is denoted $\psi(x)$, and the effect of the position operator, which we shall denote by $\hat{X}$, is simply multiplication by x. The momentum in this picture is represented by the Hermitian differential operator $\hat{P} \equiv -i\hbar\frac{d}{dx}$. The operator equation (5.9) can thus be reinterpreted as a commutation relation between position and momentum operators:

$$\left[\hat{P}, \hat{X}\right] = \hat{P}\hat{X} - \hat{X}\hat{P} = -i\hbar \,. \tag{5.13}$$

In quantum mechanics the expectation value of an operator $\hat{\Omega}$ for a system in state $\psi(x)$ is given by the $\mathbb{L}_2$ scalar product,

$$\langle\hat{\Omega}\rangle = (\psi, \hat{\Omega}\psi) = \int_{-\infty}^{\infty} dx\, \psi^*(x)\, \hat{\Omega}\psi(x) \,, \tag{5.14}$$

provided the wavefunction is normalized so that

$$\|\psi\|^2 = \int_{-\infty}^{\infty} dx\, |\psi(x)|^2 = 1 \,. \tag{5.15}$$

The variance of the position is now defined as $\sigma_x^2 = \langle[\hat{X} - \langle\hat{X}\rangle]^2\rangle$, and similarly for the variance of the momentum.

To derive the uncertainty relation, we take the expectation value of the operator $\hat{\Omega}^\dagger\hat{\Omega}$, where $\hat{\Omega} = \hat{X} + i\epsilon\hat{P}$ and ϵ is an arbitrary real number (Cohen-Tannoudji *et al.*, 1977). Since $\hat{X}$ and $\hat{P}$ are Hermitian, the adjoint of $\hat{X} + i\epsilon\hat{P}$ is $\hat{X} - i\epsilon\hat{P}$ (see Sec. 1.3.5), and we find

$$\langle\hat{\Omega}^\dagger\hat{\Omega}\rangle = (\psi, [\hat{X} - i\epsilon\hat{P}][\hat{X} + i\epsilon\hat{P}]\psi) = \langle\hat{X}^2\rangle + \epsilon^2\langle\hat{P}^2\rangle - i\epsilon\langle\hat{P}\hat{X} - \hat{X}\hat{P}\rangle = \langle\hat{X}^2\rangle + \epsilon^2\langle\hat{P}^2\rangle - \epsilon\hbar\,. \tag{5.16}$$

Since $\langle\hat{\Omega}^\dagger\hat{\Omega}\rangle = (\hat{\Omega}\psi, \hat{\Omega}\psi) = ||\hat{\Omega}\psi||^2$, we also have

$$||\hat{\Omega}\psi||^2 = \langle\hat{X}^2\rangle + \epsilon^2\langle\hat{P}^2\rangle - \epsilon\hbar \geq 0\,. \tag{5.17}$$

The discriminant of this quadratic form in ϵ is $\hbar^2 - 4\langle\hat{X}^2\rangle\langle\hat{P}^2\rangle$; if this discriminant is positive, then the quadratic form has two real roots, and the form itself goes negative between these two values of ϵ. To satisfy (5.17) for all ϵ, we must therefore have

$$\langle\hat{X}^2\rangle\langle\hat{P}^2\rangle \geq \tfrac{1}{4}\hbar^2\,. \tag{5.18}$$

A change of variables to remove the mean (Cohen-Tannoudji *et al.*, 1977) completes the derivation of the uncertainty relation, (5.12).

An important manifestation of Heisenberg uncertainty arises when a particle moving in three dimensions is localized in one direction (say x) by passing it through a slit. If the slit width is w, then the particle has an uncertainty in the x component of position of this amount and a corresponding minimum uncertainty in p_x of order h/w. An image scientist conversant with the DeBroglie relation, $p = h/\lambda$, would recognize the uncertainty in p_x as a natural consequence of diffraction.

Energy-time uncertainty Another relation, also frequently referred to as an uncertainty relation in quantum mechanics, involves the energy E and the time t. The energy-time uncertainty relation conveys very different physics from the position-momentum principle, however, since time is simply a classical parameter in quantum mechanics, not an operator. Time can be measured with arbitrary precision.

An electromagnetic wave can induce transitions between two energy states of an atom. If the states have an energy difference of ΔE and the wave has frequency ν, then an energy-conserving transition can be made if $\nu = \Delta E/h$. In a quantum-electrodynamical view (discussed further in Sec. 9.1), the field is quantized and the transition corresponds to absorption or emission of one quantum (photon) of energy $h\nu$. Quantization of the field is, however, not necessary to an accurate quantum-mechanical description of the transition; almost all of the salient features can be derived with a purely classical model for the field, using quantum mechanics only for the atom (see Sec. 9.1.4).

If the atom is initially in the lower-energy state and the field is applied at $t = 0$, then there is some probability of observing it in the upper state at time $t = T$ even if $h\nu \neq \Delta E$. All that is required is that $|h\nu - \Delta E|$ be less than about h/T. From a Fourier viewpoint, this result is not surprising since the windowed temporal Fourier transform of the field has a frequency spread of order $1/T$, and some frequency component in this range resonates with the atomic transition. For large T, the Fourier transform of the field over this interval is sharply peaked, and a transition can be observed only if energy is conserved, in the sense that $h\nu = \Delta E$.

5.1.3 Local frequency

Local spectrogram The local spectrogram is the squared modulus of the local Fourier transform,

$$W_b(\xi;x_0) = |F_b(\xi;x_0)|^2\,, \tag{5.19}$$

where $F_b(\xi;x_0)$ is defined by (5.1). This nonnegative real quantity has many applications in signal analysis and pattern recognition since it shows how the signal energy is distributed in both spatial position and spatial frequency.

If $W_b(\xi;x_0)$ as a function of ξ (for some choice of window function and some shift x_0) exhibits one or more well-defined peaks, then it is reasonable to say that the position of each peak is a frequency associated with position x_0. For example, suppose $f(x)$ can be approximated by $\cos(2\pi\xi_0 x)$ over the region defined by the window when it is centered at $x = x_0$. Then we can write

$$\begin{aligned} W_b(\xi;x_0) &\simeq \left|\int_{-\infty}^{\infty} dx\; b(x-x_0)\cos(2\pi\xi_0 x)\exp(-2\pi i\xi x)\right|^2 \\ &= \tfrac{1}{4}\left|B(\xi-\xi_0)\right|^2 + \tfrac{1}{4}\left|B(\xi+\xi_0)\right|^2\,. \end{aligned} \tag{5.20}$$

No cross terms appear in this expression if the window is chosen so that $B(\xi-\xi_0)$ and $B(\xi+\xi_0)$ do not overlap at any frequency.

The two terms in (5.20) show peaks when their arguments vanish, namely at $\xi = \pm\xi_0$, so frequencies ξ_0 and $-\xi_0$ are both local to x_0. Similarly, if $f(x)$ can be approximated by a square wave of fundamental frequency ξ_0 for a window centered at x_0, then $W_b(\xi;x_0)$ will exhibit peaks at all odd harmonics of $\pm\xi_0$, and all of these frequencies are local to x_0.

Pure phase functions There is one circumstance where a unique local frequency can be associated with each position x_0. That is for a *pure phase function* with slowly varying phase. A pure phase function, illustrated in Fig. 5.1, is a complex-valued function of a real variable x with the form

$$f(x) = \exp[i\phi(x)] = \cos\phi(x) + i\sin\phi(x)\,, \tag{5.21}$$

where $\phi(x)$ is real. Since $|f(x)| = 1$, both the real part and the imaginary part are fully determined by the phase function $\phi(x)$. As we shall see later in the book, functions of this form have many applications in optics and imaging. Plane waves, spherical waves and nonabsorbing optical elements such as lenses and prisms can be described by pure phase functions.

If $\phi(x)$ is differentiable to all orders at $x = x_0$, it can be represented as a Taylor series:

$$\phi(x) = \phi(x_0) + (x-x_0)\,\phi'(x_0) + \tfrac{1}{2}(x-x_0)^2\,\phi''(x_0) + \ldots\,. \tag{5.22}$$

If the phase varies sufficiently slowly for x near x_0 that this expansion can be truncated with the linear term, we can write

$$f(x) \simeq \exp[i\phi(x_0) + i(x-x_0)\,\phi'(x_0)] = const\cdot\exp[2\pi i\xi_{loc}(x_0)x]\,, \tag{5.23}$$

where

$$\xi_{loc}(x_0) = \frac{1}{2\pi}\phi'(x_0)\,. \tag{5.24}$$

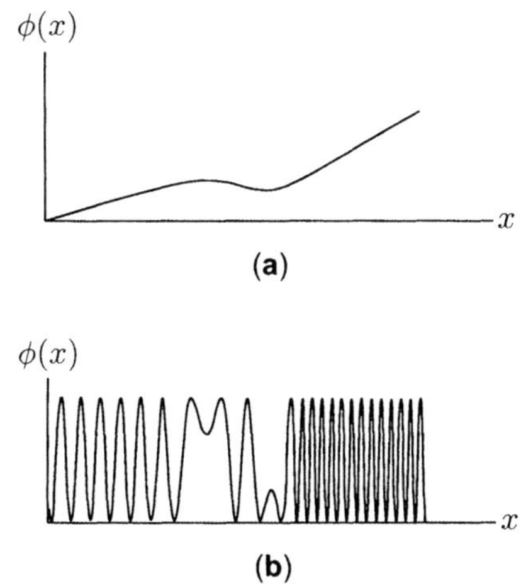

Fig. 5.1 (a) Plot of a phase $\phi(x)$; (b) Real part of the pure phase function $\exp[i\phi(x)]$.

The approximation in (5.23) thus has the effect of replacing the pure phase function $f(x)$ in the vicinity of x_0 with a simple, exactly periodic, complex exponential. The spatial frequency of this exponential is just $\xi_{loc}(x_0)$.

Local period Another way to look at spatial frequency is that it is the reciprocal of the period (or wavelength) of a periodic function. An approximate period $\Lambda(x_0)$ of $f(x)$ in the vicinity of x_0 can be defined by

$$f[x_0 + \Lambda(x_0)] = f(x_0)\,. \tag{5.25}$$

For a pure phase function, this means that

$$\phi[x_0 + \Lambda(x_0)] - \phi(x_0) = 2\pi\,. \tag{5.26}$$

If $\phi'(x_0)$ is approximately constant over a distance $\Lambda(x_0)$, a Taylor expansion shows that

$$\frac{1}{\Lambda(x_0)} \simeq \frac{1}{2\pi}\phi'(x_0) = \xi_{loc}(x_0)\,. \tag{5.27}$$

Thus $\xi_{loc}(x_0)$ can be interpreted as the reciprocal of the distance from one crest of $f(x)$ in the vicinity of x_0 to the next (see Fig. 5.2).

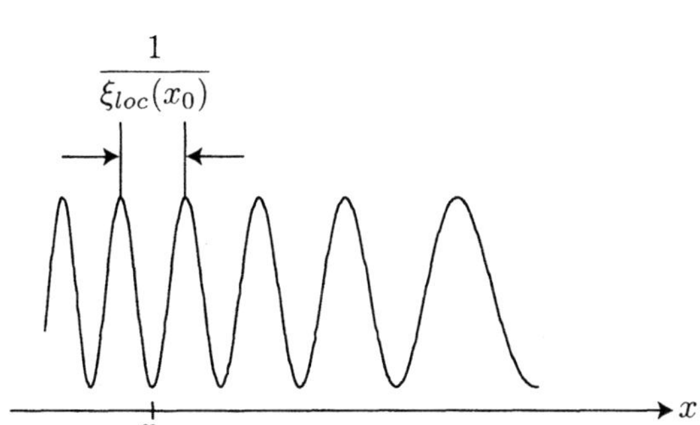

Fig. 5.2 Interpretation of local spatial frequency as reciprocal of local period.

Local frequency of a chirp In Secs. 3.3.7 and 4.3 we discussed quadratic phase factors, also known as chirp functions. Now we shall discover the reason for the latter designation.

For a 1D chirp of the form

$$f(x) = \exp(i\pi\beta x^2), \tag{5.28}$$

the local frequency is given from (5.24) by

$$\xi_{loc}(x) = \beta x. \tag{5.29}$$

Thus the local frequency increases linearly with x, reminiscent of a bird's chirp which increases with time. The parameter β is known as the *chirp rate.* Because of the linear variation of local frequency, a chirp is also referred to as *linear FM* (frequency modulation).

Pure phase functions and quantum-mechanical operators We saw in Sec. 5.1.2 that DeBroglie related momentum to spatial frequency via $p = h/\lambda = h\xi$. Later Schrödinger represented p by an operator $\hat{P} = -i\hbar\frac{d}{dx}$. For a pure phase function we also know that ξ_{loc} is given by (5.24). With $f(x) = \exp[i\phi(x)]$, we can write

$$\hat{P}f(x) = -i\hbar\frac{d}{dx}\exp[i\phi(x)] = \hbar f(x)\frac{d\phi(x)}{dx} = h\xi_{loc}(x)\,f(x) = \frac{h}{\Lambda(x)}f(x). \tag{5.30}$$

The DeBroglie relation thus holds for pure phase functions in the sense that applying the momentum operator to such functions is the same as multiplying by Planck's constant over the local wavelength.

Local frequency and analytic signals Another way to define local frequency is by means of the analytic signal, introduced in Sec. 4.2.4. For a real spatial function $f(x)$, the analytic signal $f_a(x)$ is a complex function given by the spatial counterpart of (4.107). If $f_a(x)$ is written as

$$f_a(x) = |f_a(x)|\exp[i\phi_a(x)], \tag{5.31}$$

then a local frequency can be defined as

$$\xi_a(x_0) = \frac{1}{2\pi}\frac{d\phi_a(x)}{dx}. \tag{5.32}$$

This definition applies to all functions $f(x)$, without any assumptions about the phase being slowly varying, but a number of counterintuitive features arise if the function has a large bandwidth (Cohen, 1977). For example, the local frequency may not even be contained in the Fourier transform of $f(x)$.

Extension to two or more dimensions The concept of local spatial frequency can be extended to pure phase functions of two or more real variables. In 2D Cartesian coordinates, for example, such a function would have the form

$$f(x, y) = \exp[i\phi(x, y)]. \tag{5.33}$$

By arguments similar to those used in the 1D case, we can define local spatial frequencies $\xi_{loc}(x_0, y_0)$ and $\eta_{loc}(x_0, y_0)$ for the x and y directions, respectively, by

$$\xi_{loc}(x_0, y_0) = \frac{1}{2\pi}\left[\frac{\partial}{\partial x}\phi(x, y)\right]_{x=x_0, y=y_0}, \qquad \eta_{loc}(x_0, y_0) = \frac{1}{2\pi}\left[\frac{\partial}{\partial y}\phi(x, y)\right]_{x=x_0, y=y_0}. \tag{5.34}$$

A succinct vector notation for these expressions is

$$\boldsymbol{\rho}_{loc}(\mathbf{r}_0) = \frac{1}{2\pi}\nabla\phi(\mathbf{r}_0), \tag{5.35}$$

where $\boldsymbol{\rho}_{loc}$ has Cartesian coordinates $(\xi_{loc}, j1\eta_{loc})$, $\mathbf{r}_0$ has Cartesian coordinates (x_0, y_0), and ∇ is the usual 2D gradient operator. Because of the gradient, $\boldsymbol{\rho}_{loc}(\mathbf{r}_0)$ is always normal to contours of constant phase.

The vector form of local frequency, (5.35), holds in any number of dimensions. All that is required is to interpret $\boldsymbol{\rho}_{loc}$, $\mathbf{r}_0$ and ∇ as vectors with an appropriate number of components.

As an example of a multidimensional local frequency, consider a 2D rotationally symmetric chirp defined by

$$f(\mathbf{r}) = \exp(i\pi\beta r^2), \tag{5.36}$$

where $r = |\mathbf{r}| = \sqrt{x^2 + y^2}$. For this function, illustrated in Fig. 5.3, the contours of constant phase are concentric circles, and the local frequency vector is given from (5.35) by

$$\boldsymbol{\rho}_{loc}(\mathbf{r}) = \beta\mathbf{r}. \tag{5.37}$$

This frequency vector is directed radially away from the center of the circles.

Fig. 5.3 Illustration of a 2D rotationally symmetric chirp. The function is complex, and only its real part is shown here.

5.1.4 Gabor's signal expansion

The local Fourier transform of a function $f(x)$ is a way of expressing the spatial and frequency content of that function. In this section we look at the inverse problem: recovering the function from its spatial and frequency content, as expressed by the local Fourier transform. This discussion will lead to a signal expansion developed by Dennis Gabor (1900–1979), one of the founders of modern communications theory. Gabor also brought communications theory into image science, establishing the viewpoint that pervades this book. He received the Nobel prize for physics in 1971 for the invention of holography.

Inversion of the local Fourier transform As a step toward Gabor's signal expansion, we first discuss a method of inverting the local Fourier transform, (5.1) (de Bruijn, 1973; Bastiaans, 1981). Note that (5.1) is an $\mathbb{L}_2$ scalar product of $f(x)$ with the function $b(x - x_0)\exp(2\pi i\xi x)$. We define a normalized version of this function by

$$u(x;\xi,x_0) = \frac{b(x-x_0)\exp(2\pi i\xi x)}{\left[\int_{-\infty}^{\infty} dx'\,|b(x')|^2\right]^{\frac{1}{2}}} = \frac{b(x-x_0)\exp(2\pi i\xi x)}{\|b\|}\,. \tag{5.38}$$

Consider the family of functions $\{u(x;\xi,x_0)\}$ indexed by two continuous indices ξ and x_0. Perhaps surprisingly, this family forms a complete set in $\mathbb{L}_2(\mathbb{R})$; no matter what we choose for the window function $b(x)$, any function in $\mathbb{L}_2(\mathbb{R})$ can be expressed as a linear combination of $\{u(x;\xi,x_0)\}$. The closure relation (decomposition of the unit operator) has the form [*cf.* (1.64)]

$$\int_{-\infty}^{\infty} d\xi \int_{-\infty}^{\infty} dx_0\; u^*(x';\xi,x_0)\,u(x;\xi,x_0) = \delta(x-x')\,, \tag{5.39}$$

which can be demonstrated with the help of (2.46) and (2.25). From (5.39) it follows that

$$f(x) = \frac{\int_{-\infty}^{\infty} d\xi \int_{-\infty}^{\infty} dx_0\; F_b(\xi;x_0)\,b(x-x_0)\exp(2\pi i\xi x)}{\int_{-\infty}^{\infty} dx'\,|b(x')|^2}\,, \tag{5.40}$$

which is the desired representation of $f(x)$ in terms of its local Fourier transform.

Sampling Just as the local Fourier transform is redundant, using two variables where one would suffice, so too is the set $\{u(x;\xi,x_0)\}$ redundant; it is an *overcomplete* set for $\mathbb{L}_2$, in the sense that some subset would form a basis. One way to select such a subset is to sample $u(x;\xi,x_0)$ in the variables ξ and x_0. We denote the sample interval in x_0 as δx and the interval in ξ as $\delta\xi$. A basis results if $\delta\xi = 1/\delta x$ (Bastiaans, 1981). Samples chosen in this way are said to be points on the *Gabor lattice*, illustrated in Fig. 5.4. The sampled counterpart of (5.40), known as Gabor's signal expansion, has the form

$$f(x) = \sum_{m=-\infty}^{\infty}\sum_{n=-\infty}^{\infty} a_{mn} b(x - m\,\delta x)\exp(2\pi i n x\,\delta\xi)\,, \qquad (\delta\xi = 1/\delta x)\,. \tag{5.41}$$

Gabor referred to $b(x)$ as the *elementary signal*; the basis functions in the Gabor expansion are thus the elementary signal shifted in discrete steps of δx and modulated with linear phase factors with a discrete set of frequencies $n\,\delta\xi$. To force (5.41) to look like the other expansions in Chap. 4, we define

$$b_{mn}(x) = b(x - m\,\delta x)\exp(2\pi i n x\,\delta\xi)\,, \tag{5.42}$$

so that the Gabor expansion becomes

$$f(x) = \sum_{m=-\infty}^{\infty}\sum_{n=-\infty}^{\infty} a_{mn} b_{mn}(x)\,. \tag{5.43}$$

With suitable choice of the coefficients $\{a_{mn}\}$, this expansion can represent any function in $\mathbb{L}_2(\mathbb{R})$. Unfortunately, since the expansion functions are not orthogonal in general, it is not trivial to find the coefficients.

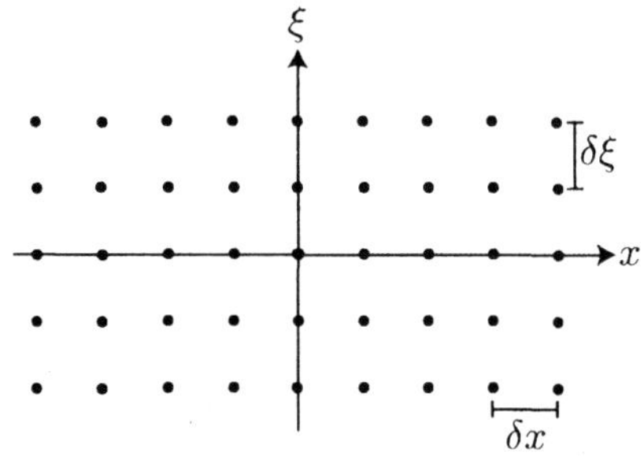

Fig. 5.4 The Gabor lattice of sample points in the x-ξ plane.

Degrees of freedom The number of coefficients $\{a_{mn}\}$ required to represent a function adequately via a Gabor expansion is another measure of the number of degrees of freedom of the function. To estimate this number, we consider, as in Sec. 5.1, a function that is spatially limited and approximately bandlimited. If we neglect the width of $b(x)$ compared to the width L of the function, then $L/\delta x$ points are required along the x axis in the Gabor lattice. Note that there is no requirement that δx satisfy the Nyquist condition, so this number is quite arbitrary. However, if the function has bandwidth B, then $B/\delta\xi = B\,\delta x$ points are required along the ξ axis. The total number of points required is simply the space-bandwidth product LB, just as with Nyquist sampling. Whatever we save by sampling more sparsely along x, we must make up by sampling more finely in ξ. The space-bandwidth product is the number of parameters required to specify the function, and all the Gabor expansion does is give us some flexibility in how we distribute the samples between space and spatial frequency.

An important consequence of this discussion is that we do not need *more* samples when we decide to represent a function of one variable with a function of two variables, in spite of the apparent redundancy of the representation.

Biorthonormality It would be easy to find the coefficients $\{a_{mn}\}$ in the Gabor expansion if we could find an auxiliary function $w(x)$ such that

$$\int_{-\infty}^{\infty} dx\; w^*_{m'n'}(x)\, b_{mn}(x) = \delta_{mm'}\,\delta_{nn'}\,, \tag{5.44}$$

where $w_{mn}(x)$ is defined by analogy to (5.42) as

$$w_{mn}(x) = w(x - m\,\delta x)\exp(2\pi i n x\,\delta\xi)\,. \tag{5.45}$$

Equation (5.44) is called a *biorthonormality relation* (Bastiaans, 1981, 1994). If it is satisfied for some $w(x)$, we can multiply both sides of (5.43) by $w^*_{m'n'}(x)$ and integrate over x, obtaining

$$a_{mn} = \int_{-\infty}^{\infty} dx\; w^*_{mn}(x)\, f(x)\,. \tag{5.46}$$

The next section describes one approach to finding a suitable $w(x)$. The concept of biorthonormality will recur in Sec. 5.3 in the context of wavelets.

Zak transform An elegant approach to finding $w(x)$ that satisfies (5.44) uses the *Zak transform*, defined for an arbitrary function $\mu(x)$ by (Bastiaans, 1994)

$$\mu_{zak}(x,\xi) = \sum_{m=-\infty}^{\infty} \mu(x + m\,\delta x) \exp(-2\pi i \xi m\,\delta x)\,. \tag{5.47}$$

There is an equivalent form of the Zak transform in the frequency domain. From the Poisson summation formula (3.197) with $g(x) = \mu(x)\exp(-2\pi i\xi x)$, we can show that

$$\mu_{zak}(x,\xi) = (\delta x)^{-1} \exp(2\pi i \xi x) \sum_{m=-\infty}^{\infty} M(\xi + m\,\delta\xi) \exp(-2\pi i m x\,\delta\xi)\,, \tag{5.48}$$

where $M(\xi)$ is the Fourier transform of $\mu(x)$.

Since (5.47) can be recognized as a Fourier series in ξ (with coefficients that depend on x), it follows at once that $\mu_{zak}(x,\xi)$ is periodic in ξ with period $\delta\xi$. Moreover, from (5.48) it follows that $\mu_{zak}(x,\xi)$ is quasiperiodic in x (periodic except for a phase factor). Thus, for n and k integers,

$$\mu_{zak}(x + k\,\delta x, \xi + n\,\delta\xi) = \exp(2\pi i \xi k\,\delta x)\,\mu_{zak}(x,\xi)\,. \tag{5.49}$$

Because of this periodicity, $\mu_{zak}(x,\xi)$ is fully determined by its behavior in a unit cell of the Gabor lattice, *i.e.*, $-\frac{1}{2}\,\delta x < x \le \frac{1}{2}\,\delta x$, $-\frac{1}{2}\,\delta\xi < \xi \le \frac{1}{2}\,\delta\xi$.

The inverse of the Zak transform follows easily from its interpretation as a Fourier series in ξ. From (5.47) and (3.19), we see that (Bastiaans, 1994)

$$\mu(x + m\,\delta x) = \frac{1}{\delta\xi} \int_{-\frac{1}{2}\delta\xi}^{\frac{1}{2}\delta\xi} d\xi\ \mu_{zak}(x,\xi) \exp(2\pi i \xi m\,\delta x)\,. \tag{5.50}$$

We can find $\mu(x')$ for any $x' = x + m\,\delta x$ from this formula by restricting x to the unit cell and letting m range over all integers.

Application of the Zak transform to the Gabor expansion We shall now apply the Zak transform to the problem of finding the coefficients in a Gabor expansion. With a change of variables and the explicit expressions (5.42) and (5.45) for $b_{mn}(x)$ and $w_{mn}(x)$, respectively, the biorthonormality relation (5.44) can be written,

$$\int_{-\infty}^{\infty} dx\ w^*[x - (m' - m)\,\delta x]\, b(x) \exp[2\pi i (n' - n) x\,\delta\xi] = \delta_{mm'}\,\delta_{nn'}\,. \tag{5.51}$$

Letting $k = m - m'$ and $l = n - n'$ and summing over k and l, we find

$$\sum_{k=-\infty}^{\infty} \sum_{l=-\infty}^{\infty} \int_{-\infty}^{\infty} dx\ w^*[x + k\,\delta x]\, b(x) \exp(-2\pi i x\,\delta\xi) = 1\,. \tag{5.52}$$

After a nontrivial amount of algebra, including the use of (2.50), we find (Bastiaans, 1994)

$$\delta x\, b_{zak}(x,\xi)\, w^*_{zak}(x,\xi) = 1\,. \tag{5.53}$$

Thus, like many of the transforms considered in the last chapter, the Zak transform converts a certain kind of convolution to a simple product; here the convolution is expressed by the integral in (5.44).

To find the auxiliary function $w(x)$ that corresponds to any given elementary signal $b(x)$, we must take the Zak transform of $b(x)$, compute its reciprocal, and perform an inverse Zak transform. The Gabor coefficients $\{a_{mn}\}$ then follow from (5.46). Even for the simple case of a Gaussian elementary signal, however, this program leads to quite complicated expressions involving theta functions.

5.2 BILINEAR TRANSFORMS

To this point, most of the integral transform we have considered have been linear, but it is also of some interest to study *bilinear forms*, where a product of two versions of the function being transformed appears in the integrand. We have already encountered one example of a bilinear form: the autocorrelation integral defined in Sec. 3.3.6 involves an integral of the product of a function with a shifted version of the same function. Two other bilinear forms, the Wigner distribution function and the Woodward ambiguity function, will be introduced here. Stochastic bilinear forms, including a stochastic version of the Wigner distribution function, will be discussed in Chaps. 8 and 10.

5.2.1 Wigner distribution function

Another candidate for a local spectrum is the Wigner distribution function (WDF), introduced by Wigner in 1932 in quantum mechanics and first applied to signal processing by Ville (1948). For an excellent review of the Wigner representation in quantum mechanics, see Tatarskii (1983), and for a detailed treatment in a signal-processing context, see Claasen and Mecklenbräuker (1980). For a snapshot of the applications of Wigner distributions and phase space in optics at the turn of the millennium, see the December, 2000 special issue of *Journal of the Optical Society of America A*.

The WDF of a 1D function $f(x)$ is defined by

$$W_f(x,\xi) = \int_{-\infty}^{\infty} dx'\, f(x+\tfrac{1}{2}x')\, f^*(x-\tfrac{1}{2}x') \exp(-2\pi i \xi x')\,. \tag{5.54}$$

Unlike the local Fourier transform, the WDF is not a linear functional of $f(x)$; it is a bilinear form since $f(x)$ appears twice with two different shifts. One way of thinking about the WDF is that it is a sliding-window Fourier transform, where the function itself serves as the window.

As the reader may demonstrate, $W_f(x,\xi)$ can also be expressed in terms of the Fourier transform $F(\xi)$ rather than $f(x)$ directly:

$$W_f(x,\xi) = \int_{-\infty}^{\infty} d\xi'\, F(\xi+\tfrac{1}{2}\xi')\, F^*(\xi-\tfrac{1}{2}\xi') \exp(2\pi i \xi' x)\,. \tag{5.55}$$

Even though (5.55) is written in terms of $F(\xi)$, it is the WDF of $f(x)$, which is not the same thing as the WDF of $F(\xi)$ itself. For more on this latter function, which we denote as $W_F(\xi, x)$, see Sec. 5.2.3.

Inversion One way to recover $f(x)$, up to a constant, from the WDF is to perform an inverse Fourier transform on the ξ variable in $W_f(x, \xi)$, yielding

$$\int_{-\infty}^{\infty} d\xi \; W_f(x, \xi) \exp(4\pi i \xi x) = f(2x) \, f^*(0) \,. \tag{5.56}$$

The modulus of the constant, $|f(0)|$, can be found since

$$\int_{-\infty}^{\infty} d\xi \; W_f(x, \xi) = |f(x)|^2 \,. \tag{5.57}$$

Thus $f(x)$ can be uniquely determined from $W_f(x, \xi)$ except for a constant phase factor; multiplying $f(x)$ by $\exp(i\alpha)$ with α constant leaves $W_f(x, \xi)$ unchanged.

Similarly,

$$\int_{-\infty}^{\infty} dx \; W_f(x, \xi) \exp(-4\pi i \xi x) = F(2\xi) \, F^*(0) \tag{5.58}$$

and

$$\int_{-\infty}^{\infty} dx \; W_f(x, \xi) = |F(\xi)|^2 \,. \tag{5.59}$$

Pure phase functions revisited Suppose that $f(x) = \exp[i\phi(x)]$. Then the Wigner distribution function is given by

$$W_f(x, \xi) = \int_{-\infty}^{\infty} dx' \; \exp[i\phi(x + \tfrac{1}{2}x')] \exp[-i\phi(x - \tfrac{1}{2}x')] \exp(-2\pi i \xi x') \,. \tag{5.60}$$

If $\phi(x \pm \frac{1}{2}x')$ is expanded in powers of x' as in (A.176), the quadratic terms cancel. If cubic and higher terms can be neglected, we find

$$W_f(x, \xi) \simeq \int_{-\infty}^{\infty} dx' \; \exp\left[-2\pi i x' \left(\xi - \frac{1}{2\pi} \frac{\partial \phi(x)}{\partial x}\right)\right] = \delta\left(\xi - \frac{1}{2\pi} \frac{\partial \phi(x)}{\partial x}\right) \,. \tag{5.61}$$

For slowly varying phase functions, therefore, the WDF is nonzero only if $\xi = \xi_{loc}(x)$. If the Taylor expansion is not strictly valid, the WDF will have some width around the line $\xi = \xi_{loc}(x)$ in the ξ-x plane. Examination of the WDF is thus a way of assessing the validity of the local-frequency approximation.

A curious nonlocality There are some decidedly counterintuitive features of the WDF as a local spectrum. For example, suppose $f(x)$ is given by

$$f(x) = \delta(x - x_1) + \delta(x - x_2) \,. \tag{5.62}$$

One might expect a local spectrum of this function to be localized spatially around the two points x_1 and x_2, but the WDF says otherwise. Some manipulations using formulas from Chap. 2 show that

$$W_f(x, \xi) = \delta(x - x_1) + \delta(x - x_2) + 2\,\delta\left(x - \frac{x_1 + x_2}{2}\right) \cos[2\pi \xi (x_1 - x_2)] \,. \tag{5.63}$$

We see that there is a concentration at the midpoint $\frac{1}{2}(x_1 + x_2)$, even though $f(x)$ is zero at that point. This nonphysical effect, resulting from the cross-term when a sum of two delta functions is multiplied by another such sum, is the price we pay for using a bilinear form rather than the square of a linear form as a spectrum. Note, however, that the cosine term is an oscillatory function of ξ; if we smooth the spectrum in the ξ direction with a filter of width greater than $1/(x_2 - x_1)$, then the cross-term will be removed.

Positivity Another objection to the use of the WDF as a spectrum is that it can take on negative values. One approach to dealing with this problem is to smooth $W_f(x,\xi)$ by convolving with a 2D Gaussian in x and ξ. It can be shown that if the width of the Gaussian is chosen properly, the smoothed WDF cannot go negative (Cohen, 1995). This smoothing operation has an important interpretation in quantum mechanics; it relates to the choice of ordering of the noncommutative operators in an expectation value (Tatarskii, 1983).

5.2.2 Ambiguity functions

Another useful extended representation of a function $f(x)$ is obtained by taking a double Fourier transform of $W_f(x,\xi)$:

$$A_f(\xi',x') = \int_{-\infty}^{\infty} dx \int_{-\infty}^{\infty} d\xi \; W_f(x,\xi)\exp[2\pi i(\xi x' - \xi' x)]\,. \tag{5.64}$$

Use of (5.54), (2.46) and some algebra shows that

$$A_f(\xi',x') = \int_{-\infty}^{\infty} dx \; f(x+\tfrac{1}{2}x')\,f^*(x-\tfrac{1}{2}x')\exp(-2\pi i\xi' x)\,. \tag{5.65}$$

This function, known as the *Woodward ambiguity function*, arises in radar signal processing, where the variable x is time and ξ is temporal frequency. In this application, the ambiguity function measures the degree of similarity between a signal and a delayed and Doppler-shifted version of the same signal. Note that if $\xi' = 0$, then the ambiguity function is essentially an autocorrelation, analogous to (3.115) but defined with a symmetric shift.

The ambiguity function is very similar in form to the WDF, both involving a Fourier transform of the product function $f(x+\frac{1}{2}x')\,f^*(x-\frac{1}{2}x')$. The difference is that the transform is over the shift variable x' in the WDF and over the center variable x in the ambiguity function.

5.2.3 Fractional Fourier transforms

An interesting line of research emerges if we examine the WDF associated with the Fourier transform $F(\xi)$ of a signal $f(x)$. This procedure should not be confused with the ambiguity function, which is the double Fourier transform of the WDF of a signal $f(x)$; here we are concerned with the WDF of a Fourier transform, not the Fourier transform of the WDF.

We define the WDF of $F(\xi)$ by analogy to (5.54) as

$$W_F(\xi,x) = \int_{-\infty}^{\infty} d\xi' \; F(\xi+\tfrac{1}{2}\xi')\,F^*(\xi-\tfrac{1}{2}\xi')\exp(-2\pi i x\xi')\,. \tag{5.66}$$

Comparing this integral to (5.55), we see that

$$W_F(\xi,x) = W_f(-x,\xi)\,. \tag{5.67}$$

Thus W_F is the same function as W_f but rotated by an angle of $\pi/2$ in the x-ξ plane. To compute a Fourier transform, therefore, one can compute the WDF of $f(x)$, rotate it by $\pi/2$, and then compute the inverse WDF. The result will be $F(\xi)$.

This observation has prompted several authors to consider rotations of the WDF by angles other than $\pi/2$ and to use such rotations to define Fourier transforms of fractional (or even complex) order. Rotation of the WDF by an angle of $p\pi/2$ corresponds to a fractional Fourier transform of order p. The case $p = 1$ corresponds to the usual Fourier transform and $p = 0$ to the identity. Since the Fourier domain allows perfect localization of a signal in frequency and the space domain allows perfect spatial localization, the fractional order p controls the tradeoff between spatial and frequency localization.

Given a function $f(x)$, we denote its fractional Fourier transform of order p by $f_p(x)$. An explicit formula for $f_p(x)$ is (Lohmann, 1993)

$$f_p(x) = \int_{-\infty}^{\infty} dx'\, f(x') \exp\left[\frac{i\pi}{\tan\phi}(x'^2 + x^2)\right] \exp\left[\frac{-2i\pi}{\sin\phi} xx'\right] \tag{5.68}$$

where ϕ is related to the fractional order p by

$$\phi = p\frac{\pi}{2}\,. \tag{5.69}$$

If $p = 1$, then $\sin\phi = 1$ and $1/\tan\phi = 0$, so the ordinary Fourier transform is obtained. If p is small, so $\sin\phi \simeq \tan\phi \simeq \phi$, then (5.68) reduces to the Fresnel transform of parameter $\beta = 1/\phi$.

Applications of the fractional Fourier transform to optics and imaging are discussed by Lohmann (1993), Ozaktis *et al.* (1994), Ozaktis and Mendlovic (1994), Alieva *et al.* (1994) and Dorsch (1995).

5.3 WAVELETS

As discussed in Sec. 5.1.4, the local Fourier transform can be computed as a sequence of $\mathbb{L}_2$ scalar products of a function $f(x)$ with expansion functions $u(x;\xi,x_0) \propto b(x - x_0)\exp(2\pi i\xi x)$. These expansion functions are parameterized by a shift x_0 and a modulation frequency ξ, and there is considerable freedom in choosing the elementary signal $b(x)$ from which the expansion functions are constructed. In this section we introduce expansion functions known as *wavelets*, where the parameters are shift and scale rather than shift and frequency. Again, as we shall see, there is considerable freedom in how these functions are constructed.

Though wavelets have many historical antecedents in mathematics and physics, their emergence as practical tools for signal analysis stems from the work of a cadre of European mathematicians in the 1980s. Notable among this group were Morlet, who coined the word wavelets (*ondelettes*), Meyer, Mallat, Grossmann and Daubechies.

Comprehensive treatments of wavelets are given by Kaiser (1994), Walter (1994), Chui (1992) and Daubechies (1992).

5.3.1 Mother wavelets and scaling functions

Just as local Fourier transforms and Gabor expansions are built on elementary signals, wavelets are built on elementary functions called *mother wavelets*. Given a mother wavelet $\psi(x)$, a wavelet $\psi_{a,b}(x)$ is defined by

$$\psi_{a,b}(x) = \frac{1}{\sqrt{|a|}} \psi\left(\frac{x-b}{a}\right). \tag{5.70}$$

Thus b is the shift of the wavelet, a specifies the scale, and the factor of $1/\sqrt{|a|}$ ensures that the $\mathbb{L}_2$ norm of $\psi_{a,b}(x)$ is independent of a. Both a and b range over the real line, with negative a corresponding to reversal of the sense of the wavelet. For even wavelets, positive a would suffice.

The Fourier transform of $\psi_{a,b}(x)$ is given by

$$\Psi_{a,b}(\xi) = \sqrt{|a|}\, \exp(-2\pi i b\xi)\, \Psi(a\xi). \tag{5.71}$$

The mother wavelet can be derived from a *scaling function* $\phi(x)$, which will usually be chosen to be smooth and compact in some sense. In addition, we shall require the scaling functions to have unit $\mathbb{L}_2$ norm, so

$$\int_{-\infty}^{\infty} dx\, |\phi(x)|^2 = 1. \tag{5.72}$$

Given a scaling function, the mother wavelet can be defined by

$$\psi(x) = \sqrt{2} \sum_{n=0}^{2N-1} c_{N-1-n}\, \phi(2x-n), \tag{5.73}$$

where the value of N and the coefficients are specific to the family of wavelets chosen. For the Haar wavelet (described below), $N = 2$ and there are only two terms in the sum.

The usefulness of scaling functions will be seen below when we discuss multiresolution analysis. A few wavelets and their scaling functions are illustrated in Fig. 5.5.

Examples: Haar and spline One specific example that has received considerable attention is the *Haar wavelet*, for which the scaling function is $\text{rect}(x - \frac{1}{2})$ and the mother wavelet is given by

$$\psi_H(x) = \text{rect}\left(2x - \tfrac{1}{2}\right) - \text{rect}\left(2x - \tfrac{3}{2}\right) = \begin{cases} +1 & \text{if } 0 < x < \frac{1}{2} \\ -1 & \text{if } \frac{1}{2} < x < 1 \\ 0 & \text{otherwise} \end{cases}. \tag{5.74}$$

For reference, the Fourier transform of this mother wavelet is

$$\Psi_H(\xi) = \frac{[1 - \exp(-i\pi\xi)]^2}{2\pi i\xi}. \tag{5.75}$$

Haar wavelets are related to a family of interpolating functions called *B-splines*.[2] The first-order B-spline is simply the scaling function for the Haar

[2]Different books give conflicting accounts of the origin of the term; B may stand for *basis* or *Bernstein*, the latter because of a connection to polynomials of the same name (Chui, 1992).

wavelet, $\mathrm{rect}(x - \frac{1}{2})$. The B-spline of order m, denoted $N_m(x)$, is the first-order function convolved with itself $m-1$ times (de Boor, 1978). Thus a B-spline of order m is a piecewise polynomial of degree $m - 1$. From any B-spline, a corresponding wavelet can be constructed by

$$\psi_m(x) = \sum_{n=0}^{3m-2} q_n N_m(2x - n) \, , \tag{5.76}$$

where an explicit expression for the coefficients $\{q_n\}$ is given in Chui (1992, Chap. 6). Wavelets constructed this way are piecewise polynomials with compact support. The higher-order spline wavelets are very similar to Gabor functions, being concentrated both spatially and in spatial frequency. The key difference is that the width and frequency are independent in a Gabor function, while they scale together for a wavelet. High-frequency wavelets are narrow, while low-frequency wavelets are broad.

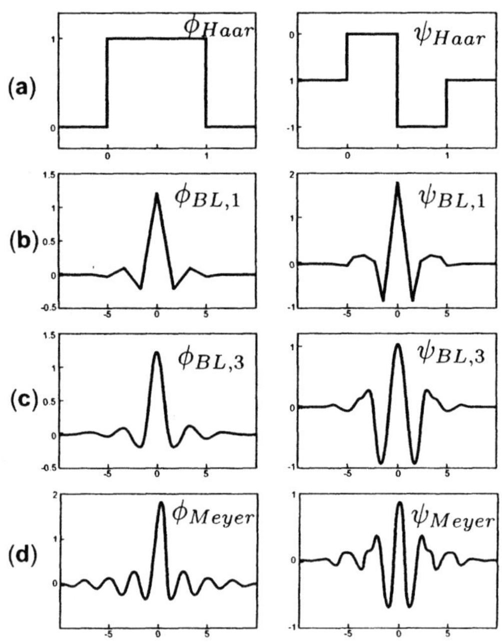

Fig. 5.5 Some wavelets and their scaling functions (adapted from Daubechies, 1992).

5.3.2 Continuous wavelet transform

For a function $f(x)$ in $\mathbb{L}_2(\mathbb{R})$, the wavelet transform associated with the mother wavelet $\psi(x)$ is given by

$$[\mathcal{W}f]\,(a, b) = (\psi_{a,b}(x), f(x)) = \frac{1}{\sqrt{|a|}} \int_{-\infty}^{\infty} dx \; \psi^* \left(\frac{x - b}{a} \right) f(x) \, , \tag{5.77}$$

where $-\infty < a < \infty$ and $-\infty < b < \infty$.

The inverse wavelet transform, to be derived below, is given by

$$f(x) = \frac{1}{C_\psi} \int_{-\infty}^{\infty} \frac{da}{a^2} \int_{-\infty}^{\infty} db \; [\mathcal{W}f](a, b) \, \psi_{a,b}(x) \, , \tag{5.78}$$

where the constant C_ψ is most easily defined in the Fourier domain:

$$C_\psi = \int_{-\infty}^{\infty} d\xi \, \frac{|\Psi(\xi)|^2}{|\xi|} \,. \tag{5.79}$$

The similarity of (5.78) to the corresponding result for the local Fourier transform, (5.40), should be noted.

For (5.78) to make sense, we must have

$$C_\psi < \infty \,. \tag{5.80}$$

This condition is known as the *admissibility condition* for the mother wavelet $\psi(x)$. To satisfy it in spite of the apparent singularity at $\xi = 0$, we must have $\Psi(0) = 0$, which implies that

$$\int_{-\infty}^{\infty} dx \; \psi(x) = 0 \,. \tag{5.81}$$

The admissibility of the Haar wavelets can be verified from (5.75); the integrand in (5.79) is linear for small ξ, so there is no divergence at the origin.

Decomposition of the unit operator The inversion formula (5.78) is equivalent to

$$\frac{1}{C_\psi} \int_{-\infty}^{\infty} \frac{da}{a^2} \int_{-\infty}^{\infty} db \; \psi^*_{a,b}(x')\, \psi_{a,b}(x) = \delta(x - x') \,. \tag{5.82}$$

This equation is the decomposition of the unit operator in $\mathbb{L}_2(\mathbb{R})$ in terms of expansion functions of the wavelet operator $\mathcal{W}$, hence a statement of the completeness of those functions.

To derive (5.82), we express $\psi_{a,b}(x)$ in terms of its inverse Fourier transform. The integral in (5.82), denoted I, is then given by

$$I = \int_{-\infty}^{\infty} \frac{da}{a^2} \int_{-\infty}^{\infty} db \; \psi^*_{a,b}(x')\, \psi_{a,b}(x)$$

$$= \int_{-\infty}^{\infty} \frac{da}{|a|} \int_{-\infty}^{\infty} db \int_{-\infty}^{\infty} d\xi \int_{-\infty}^{\infty} d\xi' \; \Psi^*(a\xi)\, \Psi(a\xi') \exp[2\pi i b(\xi - \xi')] \exp[2\pi i(\xi x - \xi' x')] \,, \tag{5.83}$$

where we have used (5.71). The integral over b yields $\delta(\xi - \xi')$ by (2.46), so

$$I = \int_{-\infty}^{\infty} \frac{da}{|a|} \int_{-\infty}^{\infty} d\xi \; |\Psi(a\xi)|^2 \exp[2\pi i \xi(x - x')] \,. \tag{5.84}$$

Letting $a' = 1/a$ and $\xi' = a\xi$, we find

$$I = \int_{-\infty}^{\infty} d\xi' \; |\Psi(\xi')|^2 \int_{-\infty}^{\infty} da' \; \exp[2\pi i a' \xi'(x - x')] = C_\psi \, \delta(x - x') \,, \tag{5.85}$$

where the last step has used (2.46), (2.29) and (5.79). From (5.85), the decomposition of the unit operator in (5.82) follows readily, and from there we get the wavelet inversion formula (5.78).

Reproducing-kernel of the continuous wavelet transform If $f(x)$ is in $\mathbb{L}_2(\mathbb{R})$, then it can be shown that $[\mathcal{W}f](a,b)$ is square-integrable with weight a^{-2} in the 2D a-b plane, so $[\mathcal{W}f](a,b)$ is in $\mathbb{L}_2(\mathbb{R}^2; a^{-2})$. Not every function in this space, however, can be realized by use of the operator $\mathcal{W}$ on a function in $\mathbb{L}_2(\mathbb{R})$, so the space of realizable wavelet transforms is a subspace of $\mathbb{L}_2(\mathbb{R}^2; a^{-2})$. The reader who has perused Sec. 1.8 should expect this subspace to be a reproducing-kernel Hilbert space (Kaiser, 1994).

To find the reproducing kernel, we simply insert the unit operator on $\mathbb{L}_2(\mathbb{R})$, denoted $\mathcal{I}$, into the wavelet transform. Symbolically, we write

$$\mathcal{W}f = \mathcal{W}\mathcal{I}f\,. \tag{5.86}$$

The unit operator has kernel $\delta(x - x')$, which of course we express by means of (5.82). With the wavelet transform from (5.77) and some shuffling of integrals, we find that

$$[\mathcal{W}f]\,(a',b') = \int_{-\infty}^{\infty} \frac{da}{a^2} \int_{-\infty}^{\infty} db\; K(a',b';a,b)\,[\mathcal{W}f]\,(a,b)\,, \tag{5.87}$$

where

$$K(a',b';a,b) = \frac{1}{C_\psi}\int_{-\infty}^{\infty} dx\; \psi_{a',b'}(x)\,\psi^*_{a,b}(x)\,. \tag{5.88}$$

Thus the space of all realizable wavelet transforms is a reproducing-kernel Hilbert space with scalar product defined by the measure $a^{-2}da\,db$ and kernel given by (5.88).

5.3.3 Discrete wavelet transform

Orthonormal wavelets The discussion above shows that the set of all wavelets $\{\psi_{a,b}(x)\}$, $-\infty < a,\ b < \infty$, is complete for $\mathbb{L}_2(\mathbb{R})$ provided the mother wavelet is admissible. Like the elementary signals in the local Fourier transform, however, this set is overcomplete in the sense that some subset will form a basis. We can sample the wavelet transform and still have an invertible mapping.

A virtually universal sampling of the wavelet transform consists of *binary dilations* ($a = 2^{-j}$) and *dyadic translations* ($b = 2^{-j}k$). The resulting functions are

$$\psi_{jk}(x) = 2^{j/2}\psi(2^j x - k)\,, \qquad j,k \text{ integers}\,. \tag{5.89}$$

With certain conditions on $\psi(x)$, discussed below, every function in $\mathbb{L}_2(\mathbb{R})$ can be written as

$$f(x) = \sum_{j=-\infty}^{\infty}\sum_{k=-\infty}^{\infty} c_{jk}\psi_{jk}(x)\,, \tag{5.90}$$

where convergence of this series is in $\mathbb{L}_2(\mathbb{R})$ (see Sec. 3.2.2). If we use the wavelet series in place of the original function $f(x)$, the $\mathbb{L}_2$ norm of the error converges to zero.

It is straightforward to find the coefficients in (5.90) if we use orthonormal wavelets, *i.e.*, a set $\{\psi_{jk}(x), j,k \text{ integers}\}$ satisfying

$$(\psi_{jk}, \psi_{j'k'}) = \int_{-\infty}^{\infty} dx\; \psi^*_{jk}(x)\,\psi_{j'k'}(x) = \delta_{jj'}\,\delta_{kk'}\,. \tag{5.91}$$

For orthonormal wavelets, the coefficients are scalar products $(\psi_{jk}, \mathbf{f})$, which are samples of the continuous wavelet transform:

$$c_{jk} = [\mathcal{W}f](2^{-j}, 2^{-j}k)\,. \tag{5.92}$$

The decomposition of the unit operator (or closure relation) for orthonormal wavelets is

$$\sum_{j=-\infty}^{\infty} \sum_{k=-\infty}^{\infty} \psi_{jk}(x)\, \psi_{jk}^{*}(x') = \delta(x - x')\,. \tag{5.93}$$

The simplest example of orthonormal wavelets is the set derived from the Haar mother wavelet, but many other complete, orthonormal sets exist as well (Daubechies, 1992; Chui, 1992).

Frames For the continuous wavelet transform, any mother wavelet satisfying the admissibility condition can be used in (5.77) and (5.78). This transform pair is symmetric in the sense that the same function $\psi_{a,b}(x)$ is used in both the forward and the inverse transforms (though of course the variables of integration are different). For the discrete wavelet transform, somewhat stronger conditions must be placed on the choice of mother wavelet, and the forward and inverse transforms may involve different functions.

The theory that determines when an expansion like (5.90) exists, and allows us to find the coefficients, is the theory of *frames.* Frames, like bases, are building blocks of Hilbert space, allowing expansion of any vector in the space (Zayed, 1993). Frames for $\mathbb{L}_2(\mathbb{R})$ are generated from a single function by translations, dilations, modulations or some combination of these operations. Unlike basis functions, the functions that form a frame are not necessarily linearly independent; a frame may be an overcomplete set.

For a set of functions to constitute a frame in $\mathbb{L}_2(\mathbb{R})$, it must satisfy the *stability condition*, which states that there exist constants A and B independent of $f(x)$, with $0 < A \leq B < \infty$, such that

$$A\|\mathbf{f}\|^2 \leq \sum_{j,k} |(\psi_{jk}, \mathbf{f})|^2 \leq B\|\mathbf{f}\|^2\,, \tag{5.94}$$

for all $f(x)$. Function sets $\{\psi_{jk}(x)\}$ satisfying this condition are called frames, and the constants A and B are the *frame bounds.* If $A = B$, the frame is said to be *tight.* If $\{\psi_{jk}(x)\}$ is a complete orthonormal set, then $A = B = 1$ by Parseval's relation, (4.6).

Whenever (5.94) is satisfied, there exists a *dual frame* of functions $\{\overline{\psi}_{jk}(x)\}$ that are biorthogonal to $\{\psi_{jk}(x)\}$, *i.e.*,

$$(\overline{\psi}_{j'k'}, \psi_{jk}) = \delta_{jj'}\, \delta_{kk'}\,. \tag{5.95}$$

The expansion coefficients in (5.90) are then given by (Chui, 1992; Zayed, 1993)

$$c_{jk} = (\overline{\psi}_{jk}, \mathbf{f})\,. \tag{5.96}$$

We have already encountered two examples of dual frames. The functions $\{w_{mn}(x)\}$ in (5.44) are a dual frame to the Gabor expansion functions, and the reciprocal lattice is a dual frame to an ordinary lattice (see Secs. 3.4.6 and 5.1.4). For orthonormal frames, the dual frame is identical to the original one.

5.3.4 Multiresolution analysis

The completeness relations (5.90) and (5.93) require summation over both translations k and dilations j in order to represent an arbitrary function in $\mathbb{L}_2(\mathbb{R})$ by a discrete wavelet series. There is, however, a subset of functions in $\mathbb{L}_2(\mathbb{R})$ that can be represented by fixing j and summing only over k. This set of functions defines a Hilbert space, which we shall denote as $\mathbb{W}_j$, with a scalar product defined by

$$(f_{1j}(x), f_{2j}(x))_j = \sum_{k=-\infty}^{\infty} c_{1jk}^* c_{2jk} \,, \tag{5.97}$$

where c_{1jk} and c_{2jk} are the wavelet coefficients for $f_{1j}(x)$ and $f_{2j}(x)$, respectively, and we are considering only orthonormal wavelets.

Since any function in $\mathbb{L}_2(\mathbb{R})$ can be represented by summing the wavelet series over both j and k, it follows that any such function can be decomposed into a sum of functions, each of which is in one of the subspaces $\mathbb{W}_j$. Moreover, for orthonormal wavelets, these functions are mutually orthogonal, so the spaces $\mathbb{W}_j$ are orthogonal subspaces of $\mathbb{L}_2(\mathbb{R})$. In formal language, this says that $\mathbb{L}_2(\mathbb{R})$ can be expressed as the *direct sum* of the subspaces $\mathbb{W}_j$. We thus write

$$\mathbb{L}_2(\mathbb{R}) = \cdots \oplus \mathbb{W}_{j-2} \oplus \mathbb{W}_{j-1} \oplus \mathbb{W}_j \oplus \mathbb{W}_{j+1} \oplus \cdots, \tag{5.98}$$

where $\oplus$ denotes direct sum. A more transparent expression of the same mathematics is

$$f(x) = \sum_{j=-\infty}^{\infty} f_j(x)\,, \qquad f(x) \text{ in } \mathbb{L}_2(\mathbb{R})\,, \qquad f_j(x) \text{ in } \mathbb{W}_j\,. \tag{5.99}$$

The component $f_j(x)$ is said to represent $f(x)$ at the scale 2^{-j}. Often in image analysis, it turns out that the important information about an object or a scene is contained in only a few scales.

Another way of decomposing $\mathbb{L}_2(\mathbb{R})$ is to lump together the subspaces $\mathbb{W}_j$, creating a cumulative space $\mathbb{V}_j$ defined by

$$\mathbb{V}_j = \cdots \oplus \mathbb{W}_{j-2} \oplus \mathbb{W}_{j-1}\,. \tag{5.100}$$

Thus $\mathbb{V}_j$ is spanned by wavelets $\{\psi_{ik}(x), i < j, -\infty < k < \infty\}$.

The function in $\mathbb{V}_j$ corresponding to $f(x)$ is

$$f_{\mathbb{V}_j}(x) = \sum_{i=-\infty}^{j-1} f_i(x) = \sum_{i=-\infty}^{j-1} \sum_{k=-\infty}^{\infty} c_{ik}\psi_{ik}(x)\,. \tag{5.101}$$

It follows from this equation that if some function $g(x)$ is in $\mathbb{V}_j$, then $g(2x)$ is in $\mathbb{V}_{j+1}$, and conversely.

The spaces $\mathbb{V}_j$, called *scaling subspaces*, are not orthogonal; $\mathbb{V}_j$ is contained entirely in $\mathbb{V}_{j+1}$. The full set of scaling subspaces $\{\mathbb{V}_j\}$, with j being an integer ranging from $-\infty$ to ∞, is said to form a *multiresolution ladder* of nested Hilbert spaces, and the sequence of functions $\{f_{\mathbb{V}_j}(x)\}$ is called the *multiresolution analysis* of

$f(x)$. As j increases, $f_{\mathbb{V}_j}(x)$ contains finer details and more closely approximates $f(x)$, *i.e.*, $f_{\mathbb{V}_j}(x) \to f(x)$ in the $\mathbb{L}_2$ sense as $j \to \infty$.

The definition (5.100) of $\mathbb{V}_j$ can also be written recursively as

$$\mathbb{V}_{j+1} = \mathbb{V}_j \oplus \mathbb{W}_j \,. \tag{5.102}$$

Since $\mathbb{W}_j$ consists of wavelets with scale index j while $\mathbb{V}_j$ consists of wavelets with scale index less than j, and we are assuming orthogonal wavelets, $\mathbb{W}_j$ is the orthogonal complement of $\mathbb{V}_j$ in $\mathbb{V}_{j+1}$. Thus any function in $\mathbb{V}_{j+1}$ can be written uniquely as the sum of a function in $\mathbb{W}_j$ and one in $\mathbb{V}_j$. This recursive characteristic leads to fast algorithms for multiresolution analysis.

Let us suppose that one of the scaling subspaces, say $j = 0$, can be generated by integer translates of a scaling function $\phi(x)$. In other words, $\mathbb{V}_0$ consists of all functions in $\mathbb{L}_2(\mathbb{R})$ that can be written as linear combinations of $\phi(x-k)$. Then, by (5.102), $\mathbb{V}_1$ consists of the functions in $\mathbb{V}_0$ supplemented by linear combinations of wavelets $\psi_{0k}(x)$. By (5.73), however,

$$\psi_{0k}(x) = \psi(x-k) = \sqrt{2} \sum_{n=0}^{2N-1} c_{N-1-n}\, \phi(2x - 2k - n) \,. \tag{5.103}$$

Thus a function $f_{\mathbb{V}_1}(x)$ in $\mathbb{V}_1$ can be expressed as a linear combination of $\{\phi(x-k)\}$ and $\{\phi(2x-k)\}$ for integer k. However, $\phi(x-k)$ itself can be expressed as a linear combination of $\{\phi(2x-k)\}$ since $\mathbb{V}_0$ is a subspace of $\mathbb{V}_1$. It follows that all functions in $\mathbb{V}_1$ are linear combinations of $\{\phi(2x-k)\}$. Continuing this process up and down the ladder, we see that every scaling subspace $\mathbb{V}_j$ is spanned by $\phi(2^j x - k)$ for integer k. We are thus led to define basis functions by analogy to (5.89) as

$$\phi_{jk}(x) = 2^{j/2} \phi(2^j x - k) \,. \tag{5.104}$$

The set $\{\phi_{jk}(x), j \text{ fixed}, k \text{ an integer}\}$ is a basis for $\mathbb{V}_j$, though not necessarily an orthonormal one.

As a trivial example, suppose we wish to approximate a function $f(x)$ with a function in $\mathbb{V}_J$ using Haar functions. The Haar scaling function is a rect function with width 1, so $\phi_{Jk}(x)$ has width 2^{-J}. Approximating $f(x)$ with a function in $\mathbb{V}_J$ thus amounts to representing it with rectangles of this width. Similarly, a Haar wavelet expansion like (5.90) truncated at $j = J-1$ allows for the finest detail to be a rectangle of width 2^{-J}.

More reproducing-kernel Hilbert spaces Each of the spaces $\mathbb{V}_j$ is a reproducing-kernel Hilbert space (see Sec. 1.8). If we assume that $\mathbb{V}_0$ is spanned by integer translates of the scaling function $\phi(x)$, then the reproducing kernel for that subspace is (Walter, 1994)

$$h_0(x, x') = \sum_{k=-\infty}^{\infty} \phi(x-k)\, \phi^*(x'-k) \,. \tag{5.105}$$

The reproducing kernel for $\mathbb{V}_j$ is obtained from $h_0(x, x')$ by (Walter, 1994)

$$h_j(x, x') = 2^j h_0(2^j x, 2^j x') \,, \tag{5.106}$$

and the reproducing kernel for the wavelet subspace $\mathbb{W}_j$ is given by

$$k_j(x, x') = 2^j \sum_{k=-\infty}^{\infty} \psi(2^j x - k)\, \psi^*(2^j x' - k) \,. \tag{5.107}$$

The derivation of these kernels parallels the derivation of (5.87) but with sums replacing integrals.

6

Group Theory

Group theory is the mathematics of symmetry, and many imaging systems have symmetry properties. An ordinary circular lens, for example, is symmetric under arbitrary rotations about its optical axis, while a tomographic system with M equally spaced projections is symmetric under rotations of $2\pi/M$. A shift-invariant system, as the name implies, is symmetric with respect to translations. In this section we develop the tools needed to describe and exploit these symmetry properties. Sections 6.1 through 6.6 survey the mathematics, and Sec. 6.7 makes the connection with physics and image science. Section 6.8 then shows how group theory leads to an expanded understanding of familiar concepts such as convolution and Fourier transformation.

An excellent place to begin reading about symmetry and groups is the classic essay by Weyl (1952). Succinct introductions to the mathematics of group theory are given by Kanatani (1990), Margenau and Murphy (1956) and Messiah (1962). More complete treatments are found in Hamermesh (1989), Armstrong (1988), Fässler and Stiefel (1992), Lomont (1959) and numerous other texts.

6.1 BASIC CONCEPTS

6.1.1 Definition of a group

Subject to certain conditions listed below, a *group* is a set of elements and a combination rule referred to as *multiplication*. This rule must allow us to combine any ordered pair of elements and obtain a unique result called the *product* of the two elements, but it need not have any relation to more familiar kinds of multiplication.

We denote the elements of a group by boldface script letters; the group itself will be denoted by boldface block letters. Thus a group $\mathbf{G}$ consists of elements $\boldsymbol{\mathcal{G}}_1$,

$\mathcal{G}_2, ..., \mathcal{G}_N$. The number of elements N is called the *order* of $\mathbf{G}$. The combination rule will be denoted by simple juxtaposition of elements, just as with ordinary algebraic multiplication. Thus $\mathcal{G}_m\mathcal{G}_n$ denotes the product of $\mathcal{G}_m$ and $\mathcal{G}_n$.

In order for the elements $\{\mathcal{G}_n\}$ to form a group, the following conditions must be satisfied:

(a) The product of any two elements in the group is an element in the group;

(b) The product of any two elements in the group is unique;

(c) Multiplication is associative: $(\mathcal{G}_i\mathcal{G}_j)\mathcal{G}_k = \mathcal{G}_i(\mathcal{G}_j\mathcal{G}_k)$;

(d) There exists a unique identity element $\mathcal{E}$ in the group such that $\mathcal{E}\mathcal{G}_n = \mathcal{G}_n\mathcal{E} = \mathcal{G}_n$ for all elements $\mathcal{G}_n$ in the group;

(e) Every element in the group has a unique inverse in the group, *i.e.*, for every element $\mathcal{G}_n$ in the group, there is another element $\mathcal{G}_n^{-1}$ in the group such that $\mathcal{G}_n\mathcal{G}_n^{-1} = \mathcal{G}_n^{-1}\mathcal{G}_n = \mathcal{E}$.

Two examples As a simple example of a group, consider the set of all integers (positive, negative and zero) and let the multiplication operation be arithmetic addition. This group satisfies the conditions listed above. The identity element is 0, the inverse of any integer n is $-n$, which is also an element of the group, and all possible products (sums) are members of the group. The order of this group is (countably) infinite.

As an example of a finite group, consider the complex numbers $\{1, i, -1, -i\}$ and let the multiplication rule be ordinary complex multiplication. Again, this set meets the requirements for a group. The identity element is the number 1, every possible product is an element of the group, and every element has an inverse in the group. For example, -1 is the inverse of 1, and $-i$ is the inverse of i. The order of this group is four, and it is frequently denoted as C_4.

6.1.2 Group multiplication tables

A convenient way to depict a finite group is by its *group multiplication table* in which an entry in row n and column m is the product $\mathcal{G}_n\mathcal{G}_m$. Since group elements are distinct, all entries in a row (or column) of a multiplication table must be distinct.

	1	i	-1	$-i$
1	1	i	-1	$-i$
i	i	-1	$-i$	1
-1	-1	$-i$	1	i
$-i$	$-i$	1	i	-1

(a)

	$\mathcal{E}$	$\mathcal{R}_1$	$\mathcal{R}_2$	$\mathcal{R}_3$
$\mathcal{E}$	$\mathcal{E}$	$\mathcal{R}_1$	$\mathcal{R}_2$	$\mathcal{R}_3$
$\mathcal{R}_1$	$\mathcal{R}_1$	$\mathcal{R}_2$	$\mathcal{R}_3$	$\mathcal{E}$
$\mathcal{R}_2$	$\mathcal{R}_2$	$\mathcal{R}_3$	$\mathcal{E}$	$\mathcal{R}_1$
$\mathcal{R}_3$	$\mathcal{R}_3$	$\mathcal{E}$	$\mathcal{R}_1$	$\mathcal{R}_2$

(b)

Fig. 6.1 Group multiplication table for the group C_4. In (a), the table is depicted for the set $\{1, i, -1, -i\}$, while in (b) it is depicted in abstract form.

The multiplication table for the group $\mathbf{C}_4$ is shown in Fig. 6.1a, where the rows and columns are labelled by the group elements $(1, i, -1, -i)$. We can also denote the elements of $\mathbf{C}_4$ abstractly as, say, $(\mathcal{E}, \mathcal{R}_1, \mathcal{R}_2, \mathcal{R}_3)$ rather than $(1, i, -1, -i)$. The group multiplication table in this notation is reproduced in Fig. 6.1b.

An advantage of the abstract notation is that very different physical entities may have the same group properties. In many applications in physics and image science, for example, group elements refer to geometrical transformations. In $\mathbf{C}_4$ we can think of $\mathcal{R}_1$ as a 90° rotation, $\mathcal{R}_2$ as a 180° rotation, and $\mathcal{R}_3$ as a 270° rotation. Then $\mathcal{E}$ is either no rotation or a 360° one. The product of two such rotations, say $\mathcal{R}_n\mathcal{R}_m$, is interpreted as the rotation $\mathcal{R}_m$ followed by $\mathcal{R}_n$. For example, $\mathcal{R}_1\mathcal{R}_3$ is a 270° rotation followed by a 90° one, which is the identity. In this way it is easy to verify that the multiplication table in Fig. 6.1b, originally derived for complex multiplications of the element $\{1, i, -1, -i\}$, holds also for physical rotations of a square. This equivalence is hardly surprising when we recall the Argand representation of a complex number, discussed in App. B, but it does emphasize that the essence of a group is the multiplication table, not the physical interpretation of the elements.

Group multiplication is not necessarily commutative. It may happen that $\mathcal{G}_n\mathcal{G}_m \neq \mathcal{G}_m\mathcal{G}_n$ for one or more pairs of elements in the group. On the other hand, if the group is commutative, then it has some special properties discussed below, and it is given a special name: a commutative group is said to be *Abelian*. The group $\mathbf{C}_4$ is Abelian.

An example of a non-Abelian group is $\mathbf{D}_3$, defined by the multiplication table of Fig. 6.2. For reasons that will appear in Sec. 6.4, the elements of this group are designated as $\{\mathcal{E}, \mathcal{R}_1, \mathcal{R}_2, \mathcal{M}_1, \mathcal{M}_2, \mathcal{M}_3\}$. The multiplication table shows that $\mathcal{M}_1\mathcal{M}_2 = \mathcal{R}_2$, for example, but $\mathcal{M}_2\mathcal{M}_1 = \mathcal{R}_1$, so the elements do not commute.

	$\mathcal{E}$	$\mathcal{R}_1$	$\mathcal{R}_2$	$\mathcal{M}_1$	$\mathcal{M}_2$	$\mathcal{M}_3$
$\mathcal{E}$	$\mathcal{E}$	$\mathcal{R}_1$	$\mathcal{R}_2$	$\mathcal{M}_1$	$\mathcal{M}_2$	$\mathcal{M}_3$
$\mathcal{R}_1$	$\mathcal{R}_1$	$\mathcal{R}_2$	$\mathcal{E}$	$\mathcal{M}_2$	$\mathcal{M}_3$	$\mathcal{M}_1$
$\mathcal{R}_2$	$\mathcal{R}_2$	$\mathcal{E}$	$\mathcal{R}_1$	$\mathcal{M}_3$	$\mathcal{M}_1$	$\mathcal{M}_2$
$\mathcal{M}_1$	$\mathcal{M}_1$	$\mathcal{M}_3$	$\mathcal{M}_2$	$\mathcal{E}$	$\mathcal{R}_2$	$\mathcal{R}_1$
$\mathcal{M}_2$	$\mathcal{M}_2$	$\mathcal{M}_1$	$\mathcal{M}_3$	$\mathcal{R}_1$	$\mathcal{E}$	$\mathcal{R}_2$
$\mathcal{M}_3$	$\mathcal{M}_3$	$\mathcal{M}_2$	$\mathcal{M}_1$	$\mathcal{R}_2$	$\mathcal{R}_1$	$\mathcal{E}$

Fig. 6.2 Group multiplication table for the group $\mathbf{D}_3$, the symmetry group of an equilateral triangle.

6.1.3 Isomorphism and homomorphism

We saw above that the set of numbers $\{1, i, -1, -i\}$ and the set of rotations of a square have the same multiplication table. We can say that the two sets are two different physical realizations of the same group $\mathbf{C}_4$, or in more formal language we can say that the two groups are *isomorphic*. Two groups $\mathbf{G}$ and $\mathbf{G}'$ of order N are isomorphic if there exists a one-to-one correspondence that preserves the multiplication table (Margenau and Murphy, 1956), *i.e.*,

(a) To each element $\mathcal{G}_j$ of $\mathbf{G}$, there corresponds one and only one element $\mathcal{G}'_j$ of $\mathbf{G}'$, and conversely;

(b) $\mathcal{G}_i\mathcal{G}_j = \mathcal{G}_k$ implies that $\mathcal{G}'_i\mathcal{G}'_j = \mathcal{G}'_k$ for all $i, j, k = 1, ..., N$, and conversely.

A related concept is *homomorphism*. A group $\mathbf{G}$ is homomorphic to $\mathbf{G}'$ if:

(a) To each element $\mathcal{G}_j$ of $\mathbf{G}$, there corresponds one and only one element $\mathcal{G}'_j$ of $\mathbf{G}'$, and to each element $\mathcal{G}'_j$ of $\mathbf{G}'$, there corresponds at least one (and perhaps more than one) element of $\mathbf{G}$.

(b) $\mathcal{G}_i\mathcal{G}_j = \mathcal{G}_k$ implies that $\mathcal{G}'_i\mathcal{G}'_j = \mathcal{G}'_k$ for all $i, j, k = 1, ..., N$, but the converse doesn't necessarily hold.

Thus isomorphism is homomorphism that is one-to-one and onto (see Sec. 1.3.4).

6.2 SUBGROUPS AND CLASSES

6.2.1 Definitions

A *subgroup* of a group $\mathbf{G}$ is a subset of the elements of $\mathbf{G}$ that itself constitutes a group under the same multiplication rule as for $\mathbf{G}$. For example, $(1, -1)$ is a subgroup of $(1, i, -1, -i)$ under complex multiplication. Since every group must contain the identity element, that element is a member of every subgroup, and the identity by itself is a subgroup of order one for any group.

Groups can be uniquely partitioned into *conjugacy classes*, or simply *classes* for short. To define a conjugacy class, we must first define conjugacy (from the Latin *conjugare*, to yoke together). Two elements $\mathcal{G}_n$ and $\mathcal{G}_m$ in a group $\mathbf{G}$ are said to be *conjugate* to each other if there exists an element $\mathcal{G}_k$ in $\mathbf{G}$ such that

$$\mathcal{G}_k^{-1}\mathcal{G}_n\mathcal{G}_k = \mathcal{G}_m\,. \tag{6.1}$$

In this equation, $\mathcal{G}_m$ is called the *transform* of $\mathcal{G}_n$ by $\mathcal{G}_k$.

Conjugate elements have the following properties (Margenau and Murphy, 1956):

(a) Every element in a group is conjugate with itself;

(b) If $\mathcal{G}_n$ is conjugate to $\mathcal{G}_m$, then $\mathcal{G}_m$ is conjugate to $\mathcal{G}_n$;

(c) If $\mathcal{G}_n$ is conjugate to both $\mathcal{G}_m$ and $\mathcal{G}_p$, then $\mathcal{G}_m$ and $\mathcal{G}_p$ are conjugate to each other.

A class is a complete set of elements that are conjugate to each other. The union of all classes is the group itself, and no element can appear in more than one class.

6.2.2 Examples

To see how to decompose a group into its classes, we examine the groups $\mathbf{C}_4$ and $\mathbf{D}_3$, with multiplication tables given by Figs. 6.1 and 6.2, respectively.

In an Abelian group like $\mathbf{C}_4$, all elements commute, and it follows at once that

$$\mathcal{G}_k^{-1}\mathcal{G}_n\mathcal{G}_k = \mathcal{G}_n\mathcal{G}_k^{-1}\mathcal{G}_k = \mathcal{G}_n\,. \tag{6.2}$$

Thus each element is transformed only into itself, so an Abelian group of order N can always be decomposed into N classes of one element each. Hence $\mathbf{C}_4$ consists of 4 classes.

Since $\mathbf{D}_3$ is not Abelian, its classes are more interesting. The identity element $\mathcal{E}$ commutes with all other elements, so it forms a class by itself. There is also a class consisting of $\mathcal{R}_1$ and $\mathcal{R}_2$ since

$$\mathcal{E}^{-1}\mathcal{R}_1\mathcal{E} = \mathcal{R}_1\,, \qquad \mathcal{R}_1^{-1}\mathcal{R}_1\mathcal{R}_1 = \mathcal{R}_1\,, \qquad \mathcal{R}_2^{-1}\mathcal{R}_1\mathcal{R}_2 = \mathcal{R}_1\,,$$

$$\mathcal{M}_1^{-1}\mathcal{R}_1\mathcal{M}_1 = \mathcal{R}_2\,, \qquad \mathcal{M}_2^{-1}\mathcal{R}_1\mathcal{M}_2 = \mathcal{R}_2\,, \qquad \mathcal{M}_3^{-1}\mathcal{R}_1\mathcal{M}_3 = \mathcal{R}_2\,. \qquad (6.3)$$

A similar argument shows that $\{\mathcal{M}_1, \mathcal{M}_2, \mathcal{M}_3\}$ forms a class. Thus $\mathbf{D}_3$ can be decomposed into three classes, one each with 1, 2 and 3 elements.

6.3 GROUP REPRESENTATIONS

6.3.1 Matrices that obey the multiplication table

Given a group multiplication table, it is always possible to find a set of square matrices (not necessarily distinct) that behave in the same way under ordinary matrix multiplication. This set of matrices is called a *representation* of the group. Explicitly, a set of nonsingular $K \times K$ matrices $\{\mathrm{M}(\mathcal{G}_n), n = 1, ..., N\}$ is a representation of the group $\mathbf{G} = \{\mathcal{G}_n, n = 1, ..., N\}$ if

$$\mathrm{M}(\mathcal{G}_n)\,\mathrm{M}(\mathcal{G}_m) = \mathrm{M}(\mathcal{G}_n\mathcal{G}_m)\,. \qquad (6.4)$$

The order K of the matrices is referred to as the *dimensionality*[1] of the representation.

If the mapping of group elements to matrices is an isomorphism, the representation is *faithful*, which is possible only if all of the matrices are distinct. If a new representation is formed by applying a similarity transformation to each matrix in a representation, the two representations are said to be *equivalent*.

A trivial KD representation of any group of order N is a set of N $K \times K$ unit matrices. This set satisfies the multiplication table for any group, but it is clearly not faithful. Since K is arbitrary, an infinite set of (trivial) group representations can be constructed in this way.

As a more interesting example, $\mathbf{C}_4$ can be represented by

$$\mathrm{M}(\mathcal{E}) = \begin{bmatrix} 1 & 0 \\ 0 & 1 \end{bmatrix}, \qquad \mathrm{M}(\mathcal{R}_1) = \begin{bmatrix} 0 & 1 \\ -1 & 0 \end{bmatrix},$$

$$\mathrm{M}(\mathcal{R}_2) = \begin{bmatrix} -1 & 0 \\ 0 & -1 \end{bmatrix}, \qquad \mathrm{M}(\mathcal{R}_3) = \begin{bmatrix} 0 & -1 \\ 1 & 0 \end{bmatrix}. \qquad (6.5)$$

It can be verified that this set of matrices follows the multiplication table of Fig. 6.1. In this example, $K = 2$, so the representation is two-dimensional.

[1] This is yet another meaning of the term *dimension*. For some others, see Sec. 2.4.6.

6.3.2 Irreducible representations

A $K \times K$ matrix $\mathbf{M}_{K\times K}$ is said to be in *lower-block-triangular form* if

$$\mathbf{M}_{K\times K} = \begin{bmatrix} \mathbf{M}_{Q\times Q} & \mathbf{0}_{Q\times P} \\ \mathbf{M}_{P\times Q} & \mathbf{M}_{P\times P} \end{bmatrix}, \tag{6.6}$$

where $\mathbf{M}_{Q\times Q}$, $\mathbf{M}_{P\times Q}$ and $\mathbf{M}_{P\times P}$ are matrices with dimensions indicated by the subscripts, $P + Q = K$, and $\mathbf{0}_{Q\times P}$ is a $Q \times P$ matrix of all zeros. Similarly, the matrix is said to be in *upper-block-triangular form* if the block of zeros is in the lower left corner. If $\mathbf{M}_{P\times Q}$ can be made zero, the matrix is *block-diagonal.*

A matrix representation of a group $\{\mathbf{M}(\mathcal{G}_n)\}$ is *reducible* if there exists a similarity transformation $\mathbf{S}$ such that $\mathbf{S}^{-1}\mathbf{M}(\mathcal{G}_n)\mathbf{S}$ is block-triangular (with the same P and Q) for all n. The representation is said to be *fully reducible* if there is a similarity transformation that transforms all of the matrices into block-diagonal form.

A representation that is not reducible is said to be *irreducible*, and irreducible representations will turn out to be crucial in physical applications of group theory. A 1D representation is irreducible, of course.

Example The 2D representation of $\mathbf{C}_4$ in (6.5) is fully reducible. If we let

$$\mathbf{S} = \frac{1}{\sqrt{2}}\begin{bmatrix} 1 & i \\ i & 1 \end{bmatrix}, \tag{6.7}$$

then the four matrices in (6.5) transform to

$$\mathbf{S}^{-1}\mathbf{M}(\mathcal{E})\,\mathbf{S} = \begin{bmatrix} 1 & 0 \\ 0 & 1 \end{bmatrix}, \qquad \mathbf{S}^{-1}\mathbf{M}(\mathcal{R}_1)\,\mathbf{S} = \begin{bmatrix} i & 0 \\ 0 & -i \end{bmatrix},$$

$$\mathbf{S}^{-1}\mathbf{M}(\mathcal{R}_2)\,\mathbf{S} = \begin{bmatrix} -1 & 0 \\ 0 & -1 \end{bmatrix}, \qquad \mathbf{S}^{-1}\mathbf{M}(\mathcal{R}_3)\,\mathbf{S} = \begin{bmatrix} -i & 0 \\ 0 & i \end{bmatrix}. \tag{6.8}$$

Since each of these matrices is block-diagonal (with 1×1 blocks), we can read off two new 1D representations. The upper-left blocks are, in sequence, (1), (i), (-1), $(-i)$, and this set of scalars satisfies the multiplication table. (In fact, it is precisely the set we used to define the group in the first place.) Similarly, the lower-right blocks in (6.8) give (1), $(-i)$, (-1), (i), and this set also satisfies the multiplication table. Two other 1D representations of $\mathbf{C}_4$ are: (1), (1), (1), (1); and (1), (-1), (1), (-1).

How many irreducible representations are there? There are some rules that let us deduce the number of irreducible representations of a group and the dimensionality of each without actually finding the representations. Derivations of these rules are given in any standard text on group theory (*e.g.*, Hamermesh, 1989).

The first rule is that the number of nonequivalent irreducible representations is also the number of conjugacy classes in the group. If we have partitioned the group into classes, we know immediately how many irreducible representations there are. For example, we saw above that an Abelian group of order N has N classes, so it also has N irreducible representations. In the case of $\mathbf{C}_4$, $N = 4$, and the four irreducible representations are enumerated above just after (6.8).

Another useful rule is an algebraic one: If a group $\mathbf{G}$ of order N has M irreducible representations, and the mth such representation has dimensionality N_m, then it can be shown that (Hamermesh, 1989)

$$\sum_{m=1}^{M} N_m^2 = N\,. \tag{6.9}$$

When applied to an Abelian group, for which $M = N$, this constraint shows that all irreducible representations must be 1D since (6.9) can be satisfied only if $N_m = 1$ for all m.

On the other hand, if the number of classes M is less than the order N, then at least one of the irreducible representations must have a dimensionality greater than 1. For example, we know that $\mathbf{D}_3$ has six elements but only three classes. In this case, (6.9) is sufficient to completely determine the dimensionalities of all three irreducible representations. There is only one combination of three integers such that the sum of their squares is 6, namely $1^2 + 1^2 + 2^2 = 6$, so $\mathbf{D}_3$ must have one 2D irreducible representation and two 1D ones.

6.3.3 Characters

If the set of $K \times K$ matrices $\{\mathbf{M}(\mathcal{G}_n)\}$ constitutes a representation of a group $\mathbf{G}$, then the *character* of $\mathcal{G}_n$ in this representation, denoted $\chi(\mathcal{G}_n)$, is the trace of the corresponding matrix:

$$\chi(\mathcal{G}_n) = \operatorname{tr}\{\mathbf{M}(\mathcal{G}_n)\} = \sum_{k=1}^{K} M_{kk}(\mathcal{G}_n)\,. \tag{6.10}$$

The trace of a square matrix is unchanged by similarity transformations, so equivalent representations have the same sets of characters. For the same reason, elements in the same conjugacy class have the same characters in any given representation.

A matrix representation is irreducible if and only if the sum of the squares of the characters equals the order of the group, *i.e.*,

$$\sum_{n=1}^{N} |\chi(\mathcal{G}_n)|^2 = N\,. \tag{6.11}$$

When it is necessary to distinguish characters arising from different representations, we shall use a superscript in parentheses. Thus $\chi^{(m)}(\mathcal{G}_n)$ is the character of element $\mathcal{G}_n$ in the mth representation. Most texts on group theory include *character tables*, listing the character $\chi^{(m)}(\mathcal{G}_n)$ associated with each group element $\mathcal{G}_n$ for each irreducible representation.

6.3.4 Unitary irreducible representations and orthogonality properties

We now state without proof two classical theorems attributed to Schur and Frobenius. For the proofs, see Schensted (1976), Margenau and Murphy (1956) or Hamermesh (1989).

The first theorem says that any irreducible representation of a group can be transformed into a unitary representation by means of a similarity transformation. That is, it is possible to find a matrix $\mathbf{S}$ such that $\mathbf{S}^{-1}\mathbf{M}(\mathcal{G}_n)\mathbf{S}$ is unitary for all n. This theorem allows us to work with unitary irreducible representations at will. Since similarity transformations do not affect traces, all of the results above on characters still apply if we choose to make the irreducible representations unitary.

The second theorem states that unitary irreducible representations have a powerful orthogonality property. Let $\{\mathbf{M}^{(m)}(\mathcal{G}_n)\}$ and $\{\mathbf{M}^{(m')}(\mathcal{G}_n)\}$ be the unitary irreducible representations m and m', respectively, with dimensionalities N_m and $N_{m'}$. Then it can be shown (Hamermesh, 1989, Chap. 3) that

$$\sum_{n=1}^{N}\left[M_{ij}^{(m)}(\mathcal{G}_n)\right]^* M_{kl}^{(m')}(\mathcal{G}_n) = \frac{N}{N_m}\,\delta_{mm'}\,\delta_{ik}\,\delta_{jl}\,. \tag{6.12}$$

Summing a product of matrix *elements* for unitary irreducible representations over the group elements thus yields zero unless they are exactly corresponding elements from the *same* irreducible representation.

For fixed i, j and m, we can think of the set of numbers $\{M_{ij}^{(m)}(\mathcal{G}_n), n = 1, ..., N\}$ as a vector $\mathbf{m}_{ij}^{(m)}$, so that (6.12) is the scalar product between $\mathbf{m}_{ij}^{(m)}$ and $\mathbf{m}_{kl}^{(m')}$. With this view, it might appear that there are too many orthogonality relations implied by (6.12) since there are at most N orthogonal vectors in an ND space. For each m, there are N_m^2 combinations of i and j and hence that many different vectors $\mathbf{m}_{ij}^{(m)}$, and we must also allow m to vary. From (6.9), however, we see that there are N different vectors $\mathbf{m}_{ij}^{(m)}$, so (6.12) is plausible.

If we let $i = j$ and $k = l$ in (6.12) and sum over i and k, we get an orthogonality relation for the characters:

$$\sum_{n=1}^{N}\left[\chi^{(m)}(\mathcal{G}_n)\right]^* \chi^{(m')}(\mathcal{G}_n) = N\,\delta_{mm'}\,, \tag{6.13}$$

in agreement with (6.11) if $m = m'$.

The orthogonality relations (6.12) and (6.13) pair up matrices or characters associated with the same group element $\mathcal{G}_n$. It is natural to inquire if there is any orthogonality relation involving different elements. For example, is there an orthogonality relation involving $\chi^{(m)}(\mathcal{G}_n)$ and $\chi^{(m)}(\mathcal{G}_k)$? In general, the answer to this question must be no since elements in the same class have the same characters, but there is an important orthogonality relation if we restrict $\mathcal{G}_n$ and $\mathcal{G}_k$ to be in different classes. For an Abelian group, this is no restriction at all since each class consists of one element.

Let L_j be the number of elements in the jth class, and denote the character of all of these elements in the mth unitary irreducible representation by $\chi_j^{(m)}$. Let M denote the number of (nonequivalent) irreducible representations, which is also the number of classes. Hamermesh (1989) proves the following relations:

$$\sum_{j=1}^{M} L_j\left[\chi_j^{(m)}\right]^* \chi_j^{(m')} = N\,\delta_{mm'}\,. \tag{6.14}$$

$$\sum_{m=1}^{M}\left[\chi_j^{(m)}\right]^* \chi_k^{(m)} = \frac{N}{L_j}\,\delta_{jk}\,. \tag{6.15}$$

These matrix orthogonality relations will lead to corresponding orthogonality relations on certain *functions* when we discuss group transformations on Hilbert spaces in Sec. 6.6.

6.4 SOME FINITE GROUPS

We now expand our repertoire of groups, introducing several new ones that are important in physics and image science.

6.4.1 Cyclic groups

A cyclic group is one in which all elements, including the identity, can be generated from one basic element by raising it to various powers. The cyclic group of order N is denoted $\mathbf{C}_N$, and we have already studied the special case $N = 4$. If the basic element is denoted $\mathcal{R}$, then $\mathbf{C}_N$ consists of elements

$$\mathcal{R}_n = \mathcal{R}^n\,, \qquad n = 1, ..., N\,, \qquad \mathcal{R}^N = \mathcal{E}\,. \tag{6.16}$$

Since $\mathcal{R}$ can be interpreted as rotation by an angle of $2\pi/N$, $\mathbf{C}_N$ is also called the *N-fold rotation group.*

It follows from (6.16) that $\mathbf{C}_N$ is Abelian for all N, so it has N classes and N irreducible representations, all 1D. We can label these representations by an index $m = 1, ..., N$. The character of $\mathcal{R}_n$ in the mth irreducible representation is

$$\chi^{(m)}(\mathcal{R}_n) = \exp(-2\pi imn/N)\,, \tag{6.17}$$

which we recognize immediately as the kernel of the discrete Fourier transform (see Sec. 6.8.4 for more on this connection). The orthogonality relation (6.13) becomes

$$\sum_{n=1}^{N} \exp\left[-2\pi i \frac{(m - m')n}{N}\right] = N\,\delta_{mm'}\,, \tag{6.18}$$

which we have encountered in the discussion of the DFT [*cf.* (3.327) and (3.328)].

The cyclic group of order 2 is called the *inversion group.* The elements are frequently denoted $\{\mathcal{E}, \mathcal{I}\}$ with the multiplication table shown in Fig. 6.3. The two elements in this group are the identity $\mathcal{E}$ and the inversion[2] $\mathcal{I}$. The irreducible representations of this group are $\{1, 1\}$ and $\{1, -1\}$.

	$\mathcal{E}$	$\mathcal{I}$
$\mathcal{E}$	$\mathcal{E}$	$\mathcal{I}$
$\mathcal{I}$	$\mathcal{I}$	$\mathcal{E}$

Fig. 6.3 Group multiplication table for the group $\mathbf{C}_2$, known either as the cyclic group of order 2 or the inversion group.

[2]Do not confuse the inversion $\mathcal{I}$ with the identity operator, which we also call $\mathcal{I}$ elsewhere in this book. By convention, the identity operator is denoted by the letter E (for German *Einheit*) in group theory, and we use the script $\mathcal{E}$.

6.4.2 Dihedral groups

Geometrically, the dihedral group $\mathbf{D}_N$ is the set of transformations that leave a regular N-gon (polygon with N sides) unchanged. The example we have used above, $\mathbf{D}_3$, is the symmetry group of an equilateral triangle. This figure is unchanged by a rotation of 120°, so the group includes two rotation operators, but it also includes three mirror reflection operators, one passing through each vertex. The mirror operators are the ones designated as $\mathcal{M}_n$, $n = 1-3$, in Fig. 6.2. The six operators can be divided into three classes, so there are three irreducible representations. Two of the irreducible representations are 1D and one is 2D.

The dihedral group $\mathbf{D}_4$ is the symmetry group of a square. It includes three rotations (90°, 180°, and 270°) as well as four mirror reflections. With the identity operator, this makes a total of 8 elements. This group has five classes, hence five irreducible representations (four 1D and one 2D).

In general, the dihedral group of order $2N$ (which is denoted $\mathbf{D}_N$ here but $\mathbf{D}_{2N}$ in some books), has $\frac{1}{2}(N+3)$ classes if N is odd and $\frac{1}{2}N+3$ classes if N is even. For odd N there are always two 1D irreducible representations, and for even N there are four 1D ones; in both cases, all of the remaining irreducible representations are 2D (James and Liebeck, 1993). The cyclic group $\mathbf{C}_N$ is always a subgroup of $\mathbf{D}_N$.

6.5 CONTINUOUS GROUPS

6.5.1 Basic properties

Consider the limit of the cyclic group $\mathbf{C}_N$ as $N \to \infty$. Since a circle is the limit of a regular polygon as the number of sides tends to ∞, $\mathbf{C}_\infty$ is the group of rotations that leave a circle unchanged. Instead of labeling the rotation operator with a discrete index n in this case, it is convenient to use a continuous label θ, specifying the rotation angle. Thus the group $\mathbf{C}_\infty$ consists of the set of operators $\{\mathcal{R}(\theta), 0 \leq \theta < 2\pi\}$.

Other kinds of continuous groups are also important in various physical applications. A general affine linear transformation of a real number to another real number, for example, is given by $x' = ax + b$. The group of all such transformations is a *two-parameter continuous group*, where a group element is specified by an ordered pair (a, b) and written $\mathcal{G}(a, b)$. Generalizing, we can write an element of a k-parameter continuous group as $\mathcal{G}(\boldsymbol{\theta})$, where $\boldsymbol{\theta}$ is a kD parameter vector.

The requirements for these elements to form a group are directly analogous to those for discrete groups (Hamermesh, 1989):

(a) There must be a parameter value $\boldsymbol{\theta}_0$ such that $\mathcal{G}(\boldsymbol{\theta}_0)$ is the identity element $\mathcal{E}$, *i.e.*, $\mathcal{G}(\boldsymbol{\theta}_0)\,\mathcal{G}(\boldsymbol{\theta}) = \mathcal{G}(\boldsymbol{\theta})\,\mathcal{G}(\boldsymbol{\theta}_0) = \mathcal{G}(\boldsymbol{\theta})$.

(b) For each value $\boldsymbol{\theta}$, there must be a corresponding value $\overline{\boldsymbol{\theta}}$ such that $\mathcal{G}(\boldsymbol{\theta})\,\mathcal{G}(\overline{\boldsymbol{\theta}}) = \mathcal{G}(\overline{\boldsymbol{\theta}})\,\mathcal{G}(\boldsymbol{\theta}) = \mathcal{G}(\boldsymbol{\theta}_0) = \mathcal{E}$.

(c) The product of any two elements must be a member of the group. In other words, for any pair of parameter vectors $\boldsymbol{\theta}_1$ and $\boldsymbol{\theta}_2$, it must be possible to find another vector $\boldsymbol{\theta}_3 = f(\boldsymbol{\theta}_1, \boldsymbol{\theta}_2)$ such that $\mathcal{G}(\boldsymbol{\theta}_1)\,\mathcal{G}(\boldsymbol{\theta}_2) = \mathcal{G}(\boldsymbol{\theta}_3)$.

In point (c), if we require $f(\boldsymbol{\theta}_1, \boldsymbol{\theta}_2)$ to be infinitely differentiable with respect to the components of $\boldsymbol{\theta}_1$ and $\boldsymbol{\theta}_2$, the group is said to be a *Lie group*.

6.5.2 Linear, orthogonal and unitary groups

The *general linear group* or *full linear group* $\mathbf{L}(N)$ on an ND space is the set of all nonsingular $N \times N$ matrices. For $N = 2$, for example, a general linear mapping from $(x, y)^t$ to $(x', y')^t$ is given by

$$\begin{bmatrix} x' \\ y' \end{bmatrix} = \begin{bmatrix} a & b \\ c & d \end{bmatrix} \begin{bmatrix} x \\ y \end{bmatrix}. \tag{6.19}$$

If the matrix elements are real and chosen so that the matrix is invertible, then this transformation is an element of the four-parameter group denoted $\mathbf{L}(2)$. The group operation is 2×2 matrix multiplication, and it is easy to show that $\mathbf{L}(2)$ is a Lie group.

If we add the further condition that the determinant of the transformation matrix be one, $\mathbf{L}(N)$ is known as the *special linear group* of order N, denoted $\mathbf{SL}(N)$. The term *special* in this context indicates simply unit determinant. The group $\mathbf{SL}(2)$ is a 3-parameter continuous group since the condition on the determinant eliminates one of the free parameters.

As discussed in App. A (Sec. A.3.2), an $N \times N$ matrix is called *unitary* if its adjoint equals its inverse. A unitary matrix with real elements is referred to as an *orthogonal* matrix. An orthogonal matrix has determinant ± 1; a unitary matrix can have a complex determinant, but its modulus is necessarily one. The group of all unitary $N \times N$ matrices is called $\mathbf{U}(N)$, and the corresponding group of orthogonal matrices is called $\mathbf{O}(N)$. The group of unitary matrices with determinant $+1$ is $\mathbf{SU}(N)$ (for *special unitary* group), and the group of orthogonal matrices with determinant $+1$ is $\mathbf{SO}(N)$. Since an orthogonal $N \times N$ matrix performs a pure rotation on an $N \times 1$ vector, $\mathbf{SO}(N)$ is the rotation group in N dimensions.

6.5.3 Abelian and non-Abelian Lie groups

Like finite groups, continuous groups are termed Abelian if all elements commute. The 2D rotation group $\mathbf{C}_\infty$ is Abelian since $\mathcal{R}(\theta_1)\mathcal{R}(\theta_2) = \mathcal{R}(\theta_2)\mathcal{R}(\theta_1) = \mathcal{R}(\theta_1 + \theta_2)$. The 3D rotation group $\mathbf{SO}(3)$, on the other hand, is not Abelian since a rotation of θ around the x axis followed by a rotation of ϕ around the z axis does not yield the same point as a rotation of ϕ around the z axis followed by a rotation of θ around the x axis.

Even in 2D, the full symmetry group of a disk is not Abelian. This group includes all of the rotations in $\mathbf{C}_\infty$, but it also includes *improper rotations*, which include inversions. In 2D, a pure rotation can be represented by a 2×2 matrix of real entries with determinant $= 1$. Improper rotations enlarge this group to allow determinant $= -1$. In general, an element of the 2D rotation-inversion group can be represented by (Margenau and Murphy, 1956)

$$\mathbf{M}(\theta, d) = \begin{bmatrix} \cos\theta & \sin\theta \\ -d\sin\theta & d\cos\theta \end{bmatrix}, \tag{6.20}$$

where $d = 1$ for proper rotations and -1 for improper ones. The simplest improper rotation is inversion through the origin, which is described by $\mathbf{M}(\pi, -1)$. It is easy to see that $\mathbf{M}(\theta, 1)$ and $\mathbf{M}(\theta', -1)$ do not commute.

The 1D affine group is another example of a Lie group that is not Abelian (Hamermesh, 1989). An element of this two-parameter group transforms a point x on the real line to another point x' according to

$$x' = \mathcal{S}(a,b)\,x = ax + b\,. \tag{6.21}$$

The group multiplication rule is

$$\mathcal{S}(a',b')\,\mathcal{S}(a,b)x = a'(ax+b) + b' = \mathcal{S}(a'a, a'b+b')\,x\,, \tag{6.22}$$

which is not the same thing as $\mathcal{S}(a,b)\,\mathcal{S}(a',b')x$. The set of pure translation operators $\{\mathcal{S}(1,b)\}$ does, however, form an Abelian subgroup of the 1D affine group. Likewise, the *scale group* $\{\mathcal{S}(a,0)\}$ (also called the *dilation group*) is an Abelian subgroup of the affine group.

Continuous one-parameter Abelian groups such as $\mathbf{C}_\infty$ have an infinite number of 1D irreducible representations, and no other finite-dimensional[3] irreducible representations. As with finite Abelian groups, the characters are identical with the irreducible representations. If we impose the reasonable condition that the representations be single valued, then the set of irreducible representations is denumerably infinite and can be labelled with a single integer m. For $\mathbf{C}_\infty$, the character associated with $\mathcal{R}(\theta)$ in representation m will be denoted $\chi^{(m)}(\theta)$; by analogy to (6.17), it is given by

$$\chi^{(m)}(\theta) = e^{-im\theta}\,, \qquad m = 0, \pm 1, \pm 2, \ldots\,. \tag{6.23}$$

Determination of irreducible representations and characters for non-Abelian Lie groups is much more complicated, leading to discussion of Lie algebras, infinitesimal generators and commutators, all of which are beyond our present needs. An excellent reference on these topics is Hamermesh (1989).

6.6 GROUPS OF OPERATORS ON A HILBERT SPACE

So far we have considered groups of geometrical transformations such as rotations, translations and mirror reflections. These transformations map one point in a Euclidean space to another, say $\mathbf{r} \to \mathbf{r}'$, where $\mathbf{r}$ and $\mathbf{r}'$ are both vectors in $\mathbb{R}^2$. In Sec. 6.6.1 we shall see how to connect these geometrical transformations to transformations of a function $f(\mathbf{r})$ that itself maps $\mathbb{R}^2$ to $\mathbb{R}^1$ or $\mathbb{C}^1$. If the functions in question live in a Hilbert space, we can relate a group of geometrical operators to a group of operators in the Hilbert space. Subspaces of this Hilbert space are discussed in Secs. 6.6.2 and 6.6.3, and some useful orthogonality relations are derived in Sec. 6.6.4.

Other operators on a Hilbert space, including various integral transforms, have been introduced previously, and the mathematics discussed here is equally applicable to these kinds of operators whenever a useful group can be identified. A connection between geometrical transformations and other Hilbert-space operators will be given in Sec. 6.7, and integral transforms will be explored further in Sec. 6.8.

[3]Some one-parameter Lie groups also have infinite-dimensional irreducible representations, though $\mathbf{C}_\infty$ does not.

6.6.1 Geometrical transformations of functions

The connection between geometrical transformations and associated transformations of functions is straightforward. A nonsingular geometrical operator $\mathcal{G}$ maps a point $\mathbf{r}$ to a new point $\mathbf{r}' = \mathcal{G}\mathbf{r}$. A transformed function $f'(\mathbf{r}')$ is defined by assigning it the value at point $\mathbf{r}'$ that the original function had at point $\mathbf{r}$, *i.e.*, $f'(\mathcal{G}\mathbf{r}) = f(\mathbf{r})$. The functional transformation operator $\mathcal{T}$ corresponding to the geometrical operator $\mathcal{G}$ is thus

$$\mathcal{T}f(\mathbf{r}) = f'(\mathbf{r}) = f(\mathcal{G}^{-1}\mathbf{r}) . \tag{6.24}$$

If we have a group of geometrical operators $\{\mathcal{G}_n\}$, we can define a corresponding group of functional operators $\{\mathcal{T}_n\}$ simply by adding subscripts in (6.24). There is an isomorphism between $\{\mathcal{T}_n\}$ and $\{\mathcal{G}_n\}$; the multiplication rule for the set $\{\mathcal{T}_n\}$ is the same as the one for $\{\mathcal{G}_n\}$ since

$$\mathcal{T}_m\mathcal{T}_n f(\mathbf{r}) = f(\mathcal{G}_n^{-1}\mathcal{G}_m^{-1}\mathbf{r}) = f\left[(\mathcal{G}_m\mathcal{G}_n)^{-1}\mathbf{r}\right] . \tag{6.25}$$

To demonstrate that these functional transformation operators are linear, we note that

$$\mathcal{T}_n[\alpha f_1(\mathbf{r}) + \beta f_2(\mathbf{r})] = \alpha f_1(\mathcal{G}_n^{-1}\mathbf{r}) + \beta f_2(\mathcal{G}_n^{-1}\mathbf{r}) = \alpha\mathcal{T}_n f_1(\mathbf{r}) + \beta\mathcal{T}_n f_2(\mathbf{r}) , \tag{6.26}$$

proving the linearity. Moreover, if $\mathcal{G}_n$ is a continuous mapping, then $f(\mathcal{G}_n\mathbf{r})$ is a square-integrable function if $f(\mathbf{r})$ is, so $\mathcal{T}_n$ is a linear operator mapping an $\mathbb{L}_2$ Hilbert space to itself.

6.6.2 Invariant subspaces

A subspace $\mathbb{S}$ of a Hilbert space $\mathbb{U}$ is said to be invariant under a group $\mathbf{T}$ if, for any vector $\mathbf{f}$ in $\mathbb{S}$ and any operator $\mathcal{T}_n$ in $\mathbf{T}$, $\mathcal{T}_n\mathbf{f}$ is also a vector in $\mathbb{S}$. For a concrete mental picture, consider a 2D subspace or plane in Hilbert space. If this plane is an invariant subspace, then application of any group operator to a vector in the plane produces another vector in the plane. In a function space, each vector is a function, and each function in the plane can be written as a linear combination of two basis functions. Saying that this plane is invariant under the group means that a group operator acting on any linear combination of the basis functions can only produce some other linear combination of them.

More generally, K linearly independent (but not necessarily orthogonal) functions $\{u_k(\mathbf{r}), k = 1, ..., K\}$, form a basis for a KD space. Any function $f(\mathbf{r})$ that can be written as

$$f(\mathbf{r}) = \sum_{k=1}^{K} \alpha_k u_k(\mathbf{r}) \tag{6.27}$$

is a vector in this space. Once the basis functions are given, an arbitrary function in the space is specified by a KD column vector of coefficients $\boldsymbol{\alpha}$.

Now consider the effect of the group operator $\mathcal{T}_n$ on $f(\mathbf{r})$. If a KD space containing $f(\mathbf{r})$ is invariant to the group, then the transformed function $\mathcal{T}_n f(\mathbf{r})$ can be written as a linear combination of the basis functions, *i.e.*,

$$\mathcal{T}_n f(\mathbf{r}) = \sum_{k=1}^{K} \alpha_k \mathcal{T}_n u_k(\mathbf{r}) = \sum_{j=1}^{K} \beta_j^{(n)} u_j(\mathbf{r}) , \tag{6.28}$$

where $\{\beta_j^{(n)}\}$ is the set of coefficients for $\mathcal{T}_n f(\mathbf{r})$. Again, we can arrange these coefficients into a column vector $\boldsymbol{\beta}^{(n)}$. Since the basis vectors themselves are in the space, we can also write

$$\mathcal{T}_n u_k(\mathbf{r}) = \sum_{j=1}^{K} M_{jk}(\mathcal{T}_n)\, u_j(\mathbf{r}), \tag{6.29}$$

where $\{M_{jk}(\mathcal{T}_n), j = 1, ..., K\}$ is the appropriate set of expansion coefficients for $\mathcal{T}_n u_k(\mathbf{r})$.

Since the basis functions are linearly independent, (6.28) and (6.29) imply that

$$\beta_j^{(n)} = \sum_{k=1}^{K} M_{jk}(\mathcal{T}_n)\, \alpha_k, \tag{6.30}$$

or, in matrix-vector form,

$$\boldsymbol{\beta}^{(n)} = \mathbf{M}(\mathcal{T}_n)\, \boldsymbol{\alpha}. \tag{6.31}$$

The set of $K \times K$ matrices $\{\mathbf{M}(\mathcal{T}_n)\}$ obeys the group multiplication table of the group of Hilbert-space operators $\mathbf{T}$ and hence forms a KD representation of that group. As noted previously, the multiplication rule for the set $\{\mathcal{T}_n\}$ is the same as the one for $\{\mathcal{G}_n\}$, so the matrices $\{\mathbf{M}(\mathcal{T}_n)\}$ are essentially the matrices $\{\mathbf{M}(\mathcal{G}_n)\}$ introduced in Sec. 6.3, but now associated with a specific invariant subspace of a Hilbert space and a particular basis. The subspace $\mathbb{S}$ is said to be reducible if the matrix representation $\{\mathbf{M}(\mathcal{T}_n)\}$ is reducible; otherwise it is irreducible.

In summary, if the operators of a group $\mathbf{T}$ act on some Hilbert space, and the KD subspace $\mathbb{S}$ of that Hilbert space is invariant under $\mathbf{T}$, then any set of basis functions $\{u_n(\mathbf{r})\}$ for $\mathbb{S}$ can be used to form a KD matrix representation of the group $\mathbf{T}$. The same basis functions can be used to represent any function in the subspace as a KD vector, and the matrices of the group representation allow transformation of these vectors by the group operators.

Construction of invariant subspaces The paragraph above lists some nice features of invariant subspaces, but how do we find such subspaces? The simplest way is to start with an arbitrary function $u(\mathbf{r})$ in the Hilbert space and apply each of the group operators in turn, obtaining the set $\{u_n(\mathbf{r}), n = 1, ..., N\}$, where, with (6.24),

$$u_n(\mathbf{r}) = \mathcal{T}_n u(\mathbf{r}) = u(\mathcal{G}_n^{-1}\mathbf{r}). \tag{6.32}$$

If $\mathcal{T}_1$ is the identity operator $\mathcal{E}$, then $u_1(\mathbf{r})$ is the original function $u(\mathbf{r})$.

These functions are not necessarily orthogonal, but they can be linearly independent if $u(\mathbf{r})$ is sufficiently general. For example, if the group includes rotation by some angle ϕ, we must make sure that the chosen $u(\mathbf{r})$ is not invariant to this rotation. With this caveat, the set $\{u_n(\mathbf{r})\}$ will be assumed to be linearly independent and hence to constitute a basis for an ND subspace of the Hilbert space.

A function in this subspace is one that can be represented in the form

$$f(\mathbf{r}) = \sum_{n=1}^{N} \alpha_n u_n(\mathbf{r}) = \sum_{n=1}^{N} \alpha_n \mathcal{T}_n u(\mathbf{r}). \tag{6.33}$$

The original function $u(\mathbf{r})$ is in the subspace since it corresponds to $\alpha_n = \delta_{n1}$.

To see that the subspace is invariant under the group, operate on $f(\mathbf{r})$ with any operator in the group, yielding

$$\mathcal{T}_k f(\mathbf{r}) = \sum_{n=1}^{N} \alpha_n \mathcal{T}_k \mathcal{T}_n u(\mathbf{r}) . \tag{6.34}$$

By the definition of a group, $\mathcal{T}_k \mathcal{T}_n$ is also a unique element of the group, which we can denote as $\mathcal{T}_j$. As n ranges from 1 to N for any fixed k, $\mathcal{T}_j$ becomes each of the group elements exactly once, so

$$\mathcal{T}_k f(\mathbf{r}) = \sum_{j=1}^{N} \alpha_{n(j,k)} \mathcal{T}_j u(\mathbf{r}) = \sum_{j=1}^{N} \alpha_{n(j,k)} u_j(\mathbf{r}) , \tag{6.35}$$

where $\alpha_{n(j,k)}$ is the coefficient α_n associated with $\mathcal{T}_n = \mathcal{T}_k^{-1}\mathcal{T}_j$ in (6.33). Since $\mathcal{T}_k f(\mathbf{r})$ has the same form as (6.33), it is also in the subspace for all k, so the subspace is indeed invariant under the group.

6.6.3 Irreducible subspaces

The subspace obtained by the procedure just outlined will necessarily be reducible since the dimensionality of the subspace is the order of the group, and the only finite group for which there is an irreducible representation with dimension equal to the order of the group is the trivial group consisting only of the identity [see (6.9)]. To find irreducible subspaces, we could set about to reduce the one specified by (6.33), but there is a shortcut based on character tables. For simplicity, we consider an Abelian group, so that all irreducible representations are 1D.

Irreducible subspaces for Abelian groups Given an arbitrary function $u(\mathbf{r})$ in a Hilbert space and a finite Abelian group of transformations in that space, we define

$$u^{(m)}(\mathbf{r}) = \frac{1}{N} \sum_{n=1}^{N} \left[\chi^{(m)}(\mathcal{T}_n)\right]^* \mathcal{T}_n u(\mathbf{r}) , \tag{6.36}$$

where N is the order of the group and $\chi^{(m)}(\mathcal{T}_n)$ is the character associated with group element $\mathcal{T}_n$ in the mth unitary irreducible representation.

We contend that the function $u^{(m)}(\mathbf{r})$ constructed in this way is the basis for the mth (1D) irreducible representation, so that the only functions in the corresponding 1D subspace of the Hilbert space are scalar multiples of $u^{(m)}(\mathbf{r})$. If this contention is correct, $u^{(m)}(\mathbf{r})$ must transform according to a simple version of (6.29), with $K = 1$ and $M_{jk}(\mathcal{T}_n)$ replaced by the character $\chi^{(m)}(\mathcal{T}_n)$.

To check this contention, we operate on $u^{(m)}(\mathbf{r})$ with any operator in the group, say $\mathcal{T}_k$. By the same argument that led to (6.35), we find

$$\mathcal{T}_k u^{(m)}(\mathbf{r}) = \frac{1}{N} \sum_{j=1}^{N} \left[\chi^{(m)}(\mathcal{T}_k^{-1}\mathcal{T}_j)\right]^* \mathcal{T}_j u(\mathbf{r}) . \tag{6.37}$$

For a 1D irreducible representation, however, the character of a product is the product of the characters, so $\chi^{(m)}(\mathcal{T}_k^{-1}\mathcal{T}_j) = \chi^{(m)}(\mathcal{T}_k^{-1})\,\chi^{(m)}(\mathcal{T}_j)$. Moreover, since we can always work with unitary representations, $\left[\chi^{(m)}(\mathcal{T}_k^{-1})\right]^* = \chi^{(m)}(\mathcal{T}_k)$, and we are left with

$$\mathcal{T}_k u^{(m)}(\mathbf{r}) = \chi^{(m)}(\mathcal{T}_k)\frac{1}{N}\sum_{j=1}^{N}\left[\chi^{(m)}(\mathcal{T}_j)\right]^* \mathcal{T}_j u(\mathbf{r}) = \chi^{(m)}(\mathcal{T}_k)\cdot u^{(m)}(\mathbf{r})\,, \qquad (6.38)$$

in accordance with (6.29). Hence, any $u^{(m)}(\mathbf{r})$ constructed according to (6.36), using characters of a 1D irreducible representation, is invariant under the group; the action of the operator $\mathcal{T}_k$ is simply to multiply $u^{(m)}(\mathbf{r})$ by the corresponding character, and we say that $u^{(m)}(\mathbf{r})$ *transforms according to the* m^{th} *irreducible representation of the group*. Every function in the 1D subspace constructed from $u(\mathbf{r})$ and associated with the mth irreducible representation must be a scalar multiple of $u^{(m)}(\mathbf{r})$.

From the basis functions for all of the irreducible representations, $\{u^{(m)}(\mathbf{r}), m = 1, ..., M\}$, we can recover the original $u(\mathbf{r})$ merely by summing:

$$u(\mathbf{r}) = \sum_{m=1}^{M} u^{(m)}(\mathbf{r}) = \frac{1}{N}\sum_{m=1}^{M}\sum_{n=1}^{N}\left[\chi^{(m)}(\mathcal{T}_n)\right]^* \mathcal{T}_n u(\mathbf{r})\,. \qquad (6.39)$$

This result can be derived from (6.15), but some comment on the notation is needed first. For an Abelian group (the present concern), there is one element per class, so $\chi_j^{(m)} = \chi^{(m)}(\mathcal{T}_j)$ and (6.15) can be rewritten as

$$\frac{1}{N}\sum_{m=1}^{M}\left[\chi^{(m)}(\mathcal{T}_j)\right]^* \chi^{(m)}(\mathcal{T}_{j'}) = \delta_{jj'}\,. \qquad (6.40)$$

Now we simply let $\mathcal{T}_{j'}$ be the identity operator $\mathcal{E}$ and recognize that $\chi^{(m)}(\mathcal{E}) = 1$ for a 1D irreducible representation. Then (6.39) follows readily from (6.36) and (6.40).

Equation (6.39) shows that any function in the Hilbert space can be decomposed into a sum of basis functions of the irreducible representations of a group of operators that act on that Hilbert space, at least if the group is Abelian. For the generalization to the non-Abelian case, the reader is referred to Hamermesh (1989).

Example Consider the cyclic group $\mathbf{C}_4$ of rotations by multiples of $\pi/2$ in the complex plane and denote the function to be decomposed as $u(z) = u(r^{i\theta})$. The transformation operator $\mathcal{T}_n$ in this case corresponds to letting $\theta \to \theta + (n-1)\pi/2$ or, equivalently, letting $z \to i^{n-1}z$. The irreducible representations of $\mathbf{C}_4$ are listed in Sec. 6.3.2, and the characters are identical to these representations since $\mathbf{C}_4$ is Abelian. From the characters and (6.36), we find

$$u^{(1)}(z) = \tfrac{1}{4}\left[1\cdot u(z) + 1\cdot u(iz) + 1\cdot u(-z) + 1\cdot u(-iz)\right]\,;$$

$$u^{(2)}(z) = \tfrac{1}{4}\left[1\cdot u(z) + i\cdot u(iz) - 1\cdot u(-z) - i\cdot u(-iz)\right]\,;$$

$$u^{(3)}(z) = \tfrac{1}{4}\left[1\cdot u(z) - 1\cdot u(iz) + 1\cdot u(-z) - 1\cdot u(-iz)\right]\,;$$

$$u^{(4)}(z) = \tfrac{1}{4}\left[1\cdot u(z) - i\cdot u(iz) - 1\cdot u(-z) + i\cdot u(-iz)\right]\,. \qquad (6.41)$$

Since the sum of these four equations is $u(z)$, (6.39) is satisfied.

Another manifestation of the same mathematics is the group of powers of the Fourier operator. We saw in Sec. 3.3.5 that $\mathcal{F}\mathcal{F}u(x) = u(-x)$, so the square of the Fourier operator is the coordinate-inversion operator which maps $u(x)$ to $u(-x)$. From this observation it is easy to deduce that the group $\{\mathcal{F}^{n-1}, n = 1, ..., 4\}$ is isomorphic to $\mathbf{C}_4$. Application of each operator in turn to an arbitrary function $u(x)$ yields: $u(x)$, $U(x)$, $u(-x)$, $U(-x)$, where $U(x)$ is the Fourier transform of $u(x)$ after the substitution $\xi \to x$. With this interpretation of the operator, the basis functions (6.41) become

$$\begin{aligned}
u^{(1)}(x) &= \tfrac{1}{4}\left[u(x) + U(x) + u(-x) + U(-x)\right] ; \\
u^{(2)}(x) &= \tfrac{1}{4}\left[u(x) + iU(x) - u(-x) - iU(-x)\right] ; \\
u^{(3)}(x) &= \tfrac{1}{4}\left[u(x) - U(x) + u(-x) - U(-x)\right] ; \\
u^{(4)}(x) &= \tfrac{1}{4}\left[u(x) - iU(x) - u(-x) + iU(-x)\right] . \qquad (6.42)
\end{aligned}$$

No matter what we start with for $u(x)$, each of these functions transforms into a constant times the same function under any group operator. In particular, each is a self-Fourier function (Mendlovic *et al.*, 1994).

6.6.4 Orthogonality of basis functions

We are often required to evaluate scalar products such as

$$(f_1(\mathbf{r}), f_2(\mathbf{r})) = \int_\infty d^2r \; f_1^*(\mathbf{r})\, f_2(\mathbf{r}) . \qquad (6.43)$$

It can be shown (Hamermesh, 1989) that this integral must be zero if $f_1(\mathbf{r})$ and $f_2(\mathbf{r})$ are basis functions for two different irreducible representations of some group. The only requirement we need to impose on the group is that it consist of transformations on the Hilbert space in which the scalar product is defined. Moreover, even if $f_1(\mathbf{r})$ and $f_2(\mathbf{r})$ are not basis functions for irreducible representations, we know from the discussion above that they can be written as sums of such basis functions, and the only cross-products in $f_1^*(\mathbf{r})\, f_2(\mathbf{r})$ that contribute to the scalar product are those involving basis functions from the same irreducible representation.

We can go a step further if we choose to make all of the irreducible representations unitary, which we can always do by use of a suitable similarity transformation. Then the orthogonality relations discussed in Sec. 6.3 come into play and lead to a powerful orthogonality relation on the basis functions.

Suppose that the set of functions $\{u_p^{(m)}(\mathbf{r}), p = 1, ..., N_m\}$ transforms according to the mth *unitary* irreducible representation of $\mathbf{T}$, that is, according to (6.29) with unitary matrices. Then it can be shown that (Schensted, 1976; Hamermesh, 1989)

$$\left(u_p^{(m)}(\mathbf{r}), u_{p'}^{(m')}(\mathbf{r})\right) = V\,\delta_{mm'}\,\delta_{pp'} , \qquad (6.44)$$

where the quantity V does not depend on p. In words, (6.44) says that *basis functions of two different unitary irreducible representations* ($m \neq m'$) *or two different rows of the same representation* ($m = m', p \neq p'$) *are orthogonal.* If the basis functions are also normalized, then $V = 1$ in (6.44).

Simple example: Even and odd functions The reader is, no doubt, already familiar with one special case of (6.44). Consider the Hilbert space $\mathbb{L}_2(\mathbb{R})$ and the inversion group of Fig. 6.3. This simple two-element group, when recast as a functional transformation, has elements $\mathcal{T}_E$ and $\mathcal{T}_I$ with the interpretations

$$\mathcal{T}_E f(x) = f(x), \qquad \mathcal{T}_I f(x) = f(-x). \tag{6.45}$$

To construct a basis for a 2D invariant subspace of $\mathbb{L}_2(\mathbb{R})$, we start with an arbitrary function $f(x)$ in that space and, as in (6.32), apply the two group operators, yielding basis vectors $f(x)$ and $f(-x)$. The corresponding group representation is, of course, reducible since the inversion group is Abelian and has only 1D irreducible representations. Since the irreducible representations are $(1,1)$ and $(1,-1)$, the basis functions for the irreducible representations are given from (6.36) as

$$f_e(x) = \tfrac{1}{2}\left[f(x) + f(-x)\right], \qquad f_o(x) = \tfrac{1}{2}\left[f(x) - f(-x)\right], \tag{6.46}$$

where the subscripts e and o stand for *even* and *odd*, respectively.

The original 2D subspace consisted of all functions that could be written as linear combinations of $f(x)$ and $f(-x)$, and it is clear that all of these functions can also be written as linear combinations of $f_e(x)$ and $f_o(x)$. In other words, any function can be written uniquely as the sum of an even and an odd function. However, the new 1D subspaces spanned (separately) by $f_e(x)$ and $f_o(x)$ are also invariant under the inversion group. Explicitly,

$$\mathcal{T}_E f_e(x) = f_e(x), \qquad \mathcal{T}_I f_e(x) = f_e(x); \tag{6.47}$$

$$\mathcal{T}_E f_o(x) = f_o(x), \qquad \mathcal{T}_I f_o(x) = -f_o(x) = const \cdot f_o(x). \tag{6.48}$$

Since $f_e(x)$ and $f_o(x)$ are basis functions for two different unitary irreducible representations of a group of transformations on $\mathbb{L}_2(\mathbb{R})$, it follows from the discussion below (6.43) that their scalar product on $\mathbb{L}_2(\mathbb{R})$ must be zero. Of course, this is nothing more than the familiar statement that the integral of the product of an odd function and an even function over a symmetric interval must vanish, but it serves as a prototype of a much broader class of symmetry relations that can lead to vanishing scalar products.

6.7 QUANTUM MECHANICS AND IMAGE SCIENCE

The main purpose of this section is to convince the reader that a discussion of group theory belongs in a book on image science. Toward this end, we take a brief look at how group theory is useful in quantum mechanics and suggest a formal analogy between quantum mechanics and image science. An excellent general reference for quantum-mechanical applications of group theory is Tinkham (1964).

6.7.1 Smattering of quantum mechanics

A central tenet of quantum mechanics is that all physical observables can be represented by Hermitian operators on a Hilbert space. The only possible outcome of a physical measurement is one of the eigenvalues of the operator. If the system is in an eigenstate of the operator in question, then the result of the measurement will, with probability one, be the corresponding eigenvalue. Otherwise, the probability of obtaining a particular eigenvalue, say λ_n, is the probability of finding the system in the corresponding eigenstate. Thus quantum mechanics is intimately connected with eigenanalysis of Hermitian operators.

The most important Hermitian operator in quantum mechanics is the Hamiltonian $\mathcal{H}$, corresponding to the total energy of a system. It operates on the wavefunction of the system, which for simplicity we assume to be a function of a single 3D spatial variable $\mathbf{r}$. This assumption is valid if we consider only a single electron in a potential field and ignore its spin. With this simplification, the Hilbert space in which $\mathcal{H}$ operates can be taken as $\mathbb{L}_2(\mathbb{R}^3)$. The eigenfunctions of $\mathcal{H}$, vectors in this Hilbert space, are called *stationary states* of the system, and the eigenvalues specify the allowed energy levels in the system. The eigenvalue equation for $\mathcal{H}$, called the time-independent Schrödinger equation, is

$$\mathcal{H}\psi_k(\mathbf{r}) = W_k\psi_k(\mathbf{r}). \tag{6.49}$$

Though the notation does not yet show it, several linearly independent eigenfunctions may correspond to the same eigenvalue W_k, in which case the level is said to be *degenerate*. As we shall see, degeneracies can (and usually do) arise as a result of symmetry properties of the Hamiltonian.

6.7.2 Connection with image science

As discussed in Chap. 1 and developed much more fully in Chap. 7, an imaging system can often be described in terms of a linear operator mapping one Hilbert space, called object space, to another Hilbert space called image space. If the two spaces are not identical, then the operator cannot be Hermitian. This is a pity since Hermitian operators have many nice mathematical properties, as detailed in Chap. 1.

Fortunately, we can always construct a relevant Hermitian operator simply by forming $\mathcal{H}^\dagger\mathcal{H}$, where $\dagger$ denotes the adjoint operator. Eigenanalysis of this new operator, which maps object space to itself, then provides a full system characterization, including sets of orthonormal basis vectors for both object and image space (see Sec. 1.5). This method, called *singular-value decomposition*, is a central theme of this book.

Singular-value decomposition begins with solving the key eigenvalue equation (1.111),

$$\mathcal{H}^\dagger\mathcal{H}u_k(\mathbf{r}) = \mu_k u_k(\mathbf{r}). \tag{6.50}$$

The formal analogy between (6.49) and (6.50) is strong; the most important theoretical task in both quantum mechanics and image science is to solve an eigenvalue equation for a Hermitian operator.

As in quantum mechanics, the eigenfunctions $u_k(\mathbf{r})$ may be degenerate, and again degeneracies usually arise from symmetries; in image science it is symmetries of $\mathcal{H}^\dagger\mathcal{H}$ that play a key role.

6.7.3 Symmetry group of the Hamiltonian

When group theory is applied to quantum mechanics, the discussion usually begins with an enumeration of the operators that commute with the Hamiltonian of the system. That is, we look for a set of operators $\{\mathcal{T}_n\}$ such that

$$\mathcal{T}_j^{-1}\mathcal{H}\mathcal{T}_j = \mathcal{H}\,. \tag{6.51}$$

An equivalent statement is

$$[\mathcal{T}_j, \mathcal{H}] = 0\,, \tag{6.52}$$

where $[\mathcal{T}_j, \mathcal{H}]$ is the *commutator* of the two operators, defined by

$$[\mathcal{T}_j, \mathcal{H}] = \mathcal{T}_j\mathcal{H} - \mathcal{H}\mathcal{T}_j\,. \tag{6.53}$$

The operators $\{\mathcal{T}_j\}$ usually arise from various symmetry properties of $\mathcal{H}$. A simple example is the inversion operator defined by (6.45). If the Hamiltonian itself is symmetric with respect to inversion, $\mathbf{r} \to -\mathbf{r}$, then it does not matter if we apply the inversion to a function and then operate with $\mathcal{H}$ or first operate with $\mathcal{H}$ and then apply the inversion. The inversion operator then commutes with the Hamiltonian. The group of all such symmetry operators is called the *symmetry group of the Hamiltonian*, and that group plus the Hamiltonian itself is called a *complete set of commuting observables.* Note, however, that this latter set does not necessarily form a group since the Hamiltonian may not have an inverse.

An analogous *symmetry group of the imaging system* consists of the set of operators that commute with $\mathcal{H}^\dagger\mathcal{H}$. Thus all of the results obtained by applying group theory to quantum mechanics have immediate applicability to image science.

6.7.4 Symmetry and degeneracy

We now expand on the connection between symmetry and degeneracy. The discussion here is specifically for quantum mechanics, but the corresponding results for image science are obtained by substituting $\mathcal{H}^\dagger\mathcal{H}$ for $\mathcal{H}$ everywhere.

Suppose we have a set of linearly independent eigenfunctions $\psi_k^{(p)}(\mathbf{r})$ of $\mathcal{H}$, satisfying (6.49) with the same eigenvalue W_k, *i.e.*,

$$\mathcal{H}\psi_k^{(p)}(\mathbf{r}) = W_k\psi_k^{(p)}(\mathbf{r})\,, \qquad p = 1, 2, ..., P_k\,. \tag{6.54}$$

This set of eigenfunctions is called a *multiplet* in quantum mechanics, and P_k is called its *multiplicity.* As discussed in Chap. 1, the eigenfunctions can be chosen (via Gram-Schmidt orthogonalization) to be orthonormal.

Now consider the group of transformation operators $\{\mathcal{T}_n, n = 1, ..., N\}$, all of which commute with $\mathcal{H}$. The action of $\mathcal{T}_n$ on an arbitrary function in the relevant Hilbert space is defined by (6.24). Since $\mathcal{H}$ and $\mathcal{T}_n$ commute, we have

$$\mathcal{H}\mathcal{T}_n\psi_k^{(p)}(\mathbf{r}) = \mathcal{T}_n\mathcal{H}\psi_k^{(p)}(\mathbf{r}) = W_k\mathcal{T}_n\psi_k^{(p)}(\mathbf{r})\,. \tag{6.55}$$

Thus the transformed function $\mathcal{T}_n\psi_k^{(p)}(\mathbf{r})$ is also an eigenfunction of $\mathcal{H}$, with the same eigenvalue as for $\psi_k^{(p)}(\mathbf{r})$.

This might be a trivial result since it may happen that $\mathcal{T}_n\psi_k^{(p)}(\mathbf{r})$ is simply a constant times $\psi_k^{(p)}(\mathbf{r})$, which means that the functional form is unchanged by the transformation. In that case, $\psi_k^{(p)}(\mathbf{r})$ is an eigenfunction of both $\mathcal{H}$ and $\mathcal{T}_n$.

On the other hand, $\mathcal{T}_n\psi_k^{(p)}(\mathbf{r})$ might also be a new, linearly independent, function, but it still has to be an eigenfunction of $\mathcal{H}$ with eigenvalue W_k. That means it can be written as a linear combination of functions in the multiplet. The coefficients in this linear superposition can depend on k and $\mathcal{T}_n$, so we write

$$\mathcal{T}_n \psi_k^{(p)}(\mathbf{r}) = \sum_{p'=1}^{P_k} M_{p'p}^{(k)}(\mathcal{T}_n)\, \psi_k^{(p')}(\mathbf{r}) . \tag{6.56}$$

The right-hand side still satisfies the Schrödinger equation with eigenvalue W_k. Moreover, since the transformed function has the general form of (6.29), the set of functions $\{\psi_k^{(p)}(\mathbf{r}), p = 1, ..., P_k\}$ spans an invariant subspace of the symmetry group of the Hamiltonian.

The set of matrices $\{\mathbf{M}^{(k)}(\mathcal{T}_n)\}$ forms a representation of the symmetry group. Since the eigenfunctions in the multiplet are orthonormal, the elements of this matrix are given by[4]

$$M_{p'p}^{(k)}(\mathcal{T}_n) = \left(\psi_k^{(p')}, \mathcal{T}_n \psi_k^{(p)}\right) . \tag{6.57}$$

6.7.5 Reducibility and accidental degeneracy

Is the representation $\{\mathbf{M}^{(k)}(\mathcal{T}_n)\}$ defined by (6.57) reducible? If it is, it means that there is a subset of the functions $\{\psi_k^{(p)}(\mathbf{r})\}$ that transforms among itself under the group transformations $\{\mathcal{T}_n\}$. Functions in this subset are, of course, degenerate with those in its complement since all of the functions $\{\psi_k^{(p)}(\mathbf{r})\}$ are eigenfunctions of the Hamiltonian with the same eigenvalue W_k, but this degeneracy is not mandated by the symmetry of the Hamiltonian. We know that transformed functions $\mathcal{T}_n \psi_k^{(p)}(\mathbf{r})$ are degenerate if $\mathcal{T}_n$ commutes with $\mathcal{H}$, but the converse does not necessarily hold; two degenerate functions are not necessarily related by a group transformation.

Degeneracy unrelated to the elements of the symmetry group is said to be *accidental.* True accidental degeneracy, however, almost never occurs; if we compute eigenvalues to floating-point precision, it would be surprising if two unrelated eigenvalues were exactly the same. What we call an accidental degeneracy is more often due to hidden symmetries, elements of the symmetry group of the Hamiltonian that have not been taken into account. The accident is a result of our state of knowledge.

If there is no accidental degeneracy, the eigenfunctions $\{\psi_k^{(p)}(\mathbf{r})\}$ are degenerate because of symmetry. That is, the eigenfunctions transform among themselves under the group transformations $\{\mathcal{T}_n\}$, and there is no subset that does so. The degenerate eigenfunctions then span an irreducible subspace of the symmetry group of the Hamiltonian, and the matrices in (6.57) form an irreducible matrix representation of that group.

Thus, barring accidental degeneracies, the degenerate eigenfunctions $\{\psi_k^{(p)}(\mathbf{r})\}$ transform according to some irreducible representation of the symmetry group of the Hamiltonian. The multiplicities of the eigenvalues of $\mathcal{H}$ correspond to the dimensionalities of the irreducible representations of its symmetry group. In particular, 1D irreducible representations must be associated with nondegenerate eigenfunctions (again barring accidents).

[4] In Dirac notation, this matrix element would be written $\langle k, p' | \mathcal{T}_n | k, p \rangle$.

6.7.6 Parity

To illustrate the ideas developed above, consider a system with inversion symmetry. Physically, this means that there is no way to distinguish $\mathbf{r}$ from $-\mathbf{r}$; they are equivalent position vectors. Mathematically, inversion symmetry means that the inversion operator $\mathcal{T}_I$ commutes with the Hamiltonian. The 1D definition of $\mathcal{T}_I$ given in (6.45) generalizes, in multiple dimensions, to

$$\mathcal{T}_I f(\mathbf{r}) = f(-\mathbf{r}). \tag{6.58}$$

If there is no other symmetry operator (other than the identity) that also commutes with the Hamiltonian, the symmetry group is the inversion group, with the multiplication table given in Fig. 6.3. Since this group is Abelian, it has only 1D irreducible representations and hence no degeneracies (or, at least, none mandated by symmetry).

Since $\mathcal{H}$ commutes with $\mathcal{T}_I$, a nondegenerate eigenfunction of $\mathcal{H}$ must also be an eigenfunction of $\mathcal{T}_I$. That is, if

$$\mathcal{H}\psi_k(\mathbf{r}) = W_k \psi_k(\mathbf{r}), \tag{6.59}$$

then

$$\mathcal{T}_I \mathcal{H}\psi_k(\mathbf{r}) = \mathcal{H}\mathcal{T}_I \psi_k(\mathbf{r}) = W_k \mathcal{T}_I \psi_k(\mathbf{r}), \tag{6.60}$$

so $\mathcal{T}_I \psi_k(\mathbf{r})$ is also an eigenfunction of $\mathcal{H}$ with eigenvalue W_k.

Since there is no degeneracy, we must have

$$\mathcal{T}_I \psi_k(\mathbf{r}) = C\psi_k(\mathbf{r}), \tag{6.61}$$

where C is a constant.

Another interpretation of (6.61) is that $\psi_k(\mathbf{r})$ is an eigenfunction of $\mathcal{T}_I$ with eigenvalue C. Operating on both sides of (6.61) with $\mathcal{T}_I$ and recognizing that $\mathcal{T}_I^2$ is the identity shows that $C^2 = 1$ or $C = \pm 1$. Thus (6.61) can be rewritten as

$$\mathcal{T}_I \psi_k(\mathbf{r}) = \pm\psi_k(\mathbf{r}). \tag{6.62}$$

Nondegenerate eigenfunctions of an inversion-symmetric Hamiltonian must therefore be either purely even or purely odd. They are said to be functions of definite *parity*, with *even parity* referring to the plus sign and *odd parity* referring to the minus sign in (6.62).

From a group-theoretical perspective, an even eigenfunction transforms according to the irreducible representation $(1, 1)$, while an odd one transforms according to $(1, -1)$.

6.7.7 Rotational symmetry in three dimensions

In classical mechanics and quantum mechanics we often consider central potentials, where the force acting on a particle is a function of only the distance of the particle from some center of attraction. In a coordinate system with origin at that center point, the potential is independent of the angular coordinates. In the language of group theory, the symmetry group of the Hamiltonian is $\mathbf{SO}(3)$.

As we noted in Sec. 6.5.3, $\mathbf{SO}(3)$ is not Abelian. It has a set of irreducible representations usually denoted by an index ℓ, ($\ell = 0, ..., \infty$), and the ℓth irreducible representation has $2\ell+1$ dimensions. This set exhausts the list of odd-dimensional irreducible representations of $\mathbf{SO}(3)$, but there are also even-dimensional ones. The basis functions for the even-dimensional irreducible representations have the apparently unphysical property that $\psi(\phi) \neq \psi(\phi + 2\pi)$, but they must nevertheless be considered for particles with spin (Tinkham, 1964).

The basis functions for the ℓth odd-dimensional irreducible representation are the spherical harmonics $Y_\ell^m(\theta, \phi)$ with m ranging from $-\ell$ to ℓ. (See Sec. 4.1.4 for a definition of the spherical harmonics.) If we ignore spin, the wavefunction for an electron in a central potential is thus specified by three indices, usually taken as n, ℓ and m, and it has the form

$$\psi_{n\ell m}(r, \theta, \phi) = u_{n\ell m}(r)\, Y_\ell^m(\theta, \phi)\,. \tag{6.63}$$

States with the same n and ℓ but different m are necessarily degenerate.

6.8 FUNCTIONS AND TRANSFORMS ON GROUPS

A function is a mapping from some set, called the domain of the function, to the real or complex numbers. A familiar example of the domain is the real line or some portion of it, but we can be much more general in thinking about functions. In this section we shall allow the domain to be the elements of a group. If we have some rule for assigning a real or complex number to each element, we have defined a function on a group. From there it is straightforward to generalize further to functionals or integral transforms on groups. In particular, the familiar concepts of convolution and Fourier transformation can be extended.

We shall introduce these ideas by way of finite groups. Then we shall introduce the important concept of invariant integration, from which we can discuss convolutions and Fourier transforms on infinite groups. Many of the results obtained in Chaps. 3 and 4 will recur and be reinterpreted from a group-theoretic perspective. Then we shall revisit Chap. 5, and especially wavelets, from this new viewpoint.

6.8.1 Functions on a finite group

Consider a finite group $\mathbf{G}$ of order N and denote the elements of this group by $\mathcal{G}_1, \mathcal{G}_2, ..., \mathcal{G}_N$. A function on the group is any rule that associates a unique scalar, denoted $f(\mathcal{G}_n)$, with any element $\mathcal{G}_n$. Since the product of two group elements is also a group element, this same rule gives us a definition of $f(\mathcal{G}_n\mathcal{G}_m)$ for all n and m in $(1, ..., N)$. Since we have not yet imposed any restrictions on either the function or the group structure, *any* sequence of N numbers f_n can be considered a function on any group of order N.

These functions can be embedded in a Hilbert space by defining a scalar product. The definition of the scalar product of two functions $f_1(\mathcal{G}_n)$ and $f_2(\mathcal{G}_n)$ on $\mathbf{G}$ is

$$(f_1, f_2) = \sum_{n=1}^{N} f_1^*(\mathcal{G}_n)\, f_2(\mathcal{G}_n)\,, \tag{6.64}$$

and the norm is defined by

$$||f||^2 = \sum_{n=1}^{N} |f(\mathcal{G}_n)|^2 . \tag{6.65}$$

The space $\mathbb{L}_2(\mathbf{G})$ is the set of all functions on $\mathbf{G}$ with finite norm.

As in Sec. 6.6.1, we can define a group of transformations $\{\mathcal{T}_m\}$ by [*cf.* (6.24)]:

$$\mathcal{T}_m f(\mathcal{G}_n) = f(\mathcal{G}_m^{-1}\mathcal{G}_n) . \tag{6.66}$$

The group $\mathbf{T}$ of transformations is isomorphic to the original group $\mathbf{G}$. That is, if $\mathcal{G}_i\mathcal{G}_j = \mathcal{G}_k$, then $\mathcal{T}_i\mathcal{T}_j = \mathcal{T}_k$.

In ordinary 3D vector analysis, a scalar is a quantity that is independent of coordinate transformations. In the vector space $\mathbb{L}_2(\mathbf{G})$, scalar products are invariant to the transformations defined in (6.66), *i.e.*,

$$(\mathcal{T}_m f_1, \mathcal{T}_m f_2) = \sum_{n=1}^{N} f_1^*(\mathcal{G}_m^{-1}\mathcal{G}_n)\, f_2(\mathcal{G}_m^{-1}\mathcal{G}_n) = \sum_{k=1}^{N} f_1^*(\mathcal{G}_k)\, f_2(\mathcal{G}_k) = (f_1, f_2) , \tag{6.67}$$

where we have used the fact that $\mathcal{G}_m^{-1}\mathcal{G}_n$ is a unique element $\mathcal{G}_k$ of $\mathbf{G}$ (by the definition of a group). Since both k and n run over all group elements, the sum over k is identical to the sum over n.

The invariance demonstrated in (6.67) shows that each operator $\mathcal{T}_m$ is unitary. On a finite-dimensional space, any operator that merely shuffles components is unitary.

Example The considerations above can be illustrated with the rotation group $\mathbf{C}_N$, where the element $\mathcal{G}_n$ can be interpreted as rotation in a plane by an angle $\theta_n = 2\pi n/N$. To define a function on this group, we associate a number $f(\theta_n)$ with each of N angles uniformly distributed around a circle. A transformation of this function by $\mathcal{T}_m$ corresponds to cyclically rotating the numbers by m steps, so that $f(\theta_n) \to f(\theta_n - \theta_m)$. Scalar products are unchanged by this rotation, which just cyclically relabels the numbers.

6.8.2 Extension to infinite groups

The equations above can be extended to encompass denumerably infinite groups just by letting $N \to \infty$. The more interesting case is continuous groups as introduced in Sec. 6.5, where group elements are specified by a continuous parameter $\boldsymbol{\theta}$ (possibly a vector). We shall now denote elements of this group by $\mathcal{G}_{\boldsymbol{\theta}}$ (which is the same thing as $\mathcal{G}(\boldsymbol{\theta})$ in Sec. 6.5) and the group itself as $\mathbf{G}$.

A function on $\mathbf{G}$ can be defined by devising a rule for associating a scalar $f(\mathcal{G}_{\boldsymbol{\theta}})$ with each group element $\mathcal{G}_{\boldsymbol{\theta}}$, or equivalently with each $\boldsymbol{\theta}$. In the latter view, the function could be denoted as $f(\boldsymbol{\theta})$.

A group $\mathbf{T}$ of transformations on functions on $\mathbf{G}$ can be defined by analogy to (6.66) as

$$\mathcal{T}_{\boldsymbol{\theta}'} f(\mathcal{G}_{\boldsymbol{\theta}}) = f(\mathcal{G}_{\boldsymbol{\theta}'}^{-1}\mathcal{G}_{\boldsymbol{\theta}}) . \tag{6.68}$$

As in the case of discrete groups, $\mathbf{T}$ is isomorphic to $\mathbf{G}$.

Example: Translation group A translation in the 2D plane is specified by a 2D vector $\mathbf{r}$, and the set of all possible translations forms an Abelian Lie group. Thus a function on the translation group could be denoted $f(\mathbf{r})$. This function might represent some physical object, for example, the radiant exitance (see Sec. 9.2 for a definition) on a plane at a point defined by translation $\mathbf{r}$ from an origin.

A function on the translation group can itself be translated by the operator $\mathcal{T}_{\mathbf{r}'}$, yielding

$$\mathcal{T}_{\mathbf{r}'} f(\mathbf{r}) = f(\mathbf{r} - \mathbf{r}') . \tag{6.69}$$

With this form, we are not far from being able to discuss convolution; all we need is a definition of integration on a group.

Invariant integration Several of the expressions in Sec. 6.8.1 require summation over group elements. In the continuous case, the sums must be replaced by integrals over $\boldsymbol{\theta}$, but we have many options on how to define this integral. A useful way to resolve the ambiguity is to require an invariance property analogous to the one displayed in (6.67). To achieve this invariance, we define a *left-invariant integration measure* or *left Haar measure* $d\theta_L$ by requiring that

$$\int_{\mathbf{G}} d\theta_L \; f(\mathcal{G}_{\boldsymbol{\theta}'}\mathcal{G}_{\boldsymbol{\theta}}) = \int_{\mathbf{G}} d\theta_L \; f(\mathcal{G}_{\boldsymbol{\theta}}) , \tag{6.70}$$

for all $\mathcal{G}_{\boldsymbol{\theta}'}$ in $\mathbf{G}$. We could also define a *right-invariant integration measure* or *right Haar measure* $d\theta_R$ by requiring that

$$\int_{\mathbf{G}} d\theta_R \; f(\mathcal{G}_{\boldsymbol{\theta}}\mathcal{G}_{\boldsymbol{\theta}'}) = \int_{\mathbf{G}} d\theta_R \; f(\mathcal{G}_{\boldsymbol{\theta}}) , \tag{6.71}$$

but we shall use the left-invariant measure exclusively.

For any particular continuous group, (6.70) and (6.71) are sufficient to determine the invariant measures uniquely. For the translation group, where $\boldsymbol{\theta} = \mathbf{r} = (x, y)$, $d\theta_L = d\theta_R = dxdy$. If the left- and right-invariant measures coincide, as in this example, the group is said to be *unimodular*.

Hilbert space To treat a function $f(\mathcal{G}_{\boldsymbol{\theta}})$ on a continuous group $\mathbf{G}$ as a vector in a Hilbert space, a scalar product is defined by

$$(f_1, f_2) = \int_{\mathbf{G}} d\theta_L \; f_1^*(\mathcal{G}_{\boldsymbol{\theta}}) \, f_2(\mathcal{G}_{\boldsymbol{\theta}}) . \tag{6.72}$$

This scalar product is invariant to the transformations $\mathcal{T}_{\boldsymbol{\theta}'}$ defined in (6.68).

The norm is given by $||f||^2 = (f, f)$, and $\mathbb{L}_2(\mathbf{G})$ is the space of all functions on $\mathbf{G}$ with finite norm.

6.8.3 Convolutions on groups

We encountered several different kinds of convolution (continuous, discrete, Mellin, Laplace) in Chap. 4. Now we shall see that all of them are special cases of a more general concept called *group convolution*.

Given two functions $f(\mathcal{G}_n)$ and $p(\mathcal{G}_n)$ on a discrete group $\mathbf{G}$ of order N (possibly infinite), their group convolution is defined by

$$[p *_G f](\mathcal{G}_m) = \sum_{n=1}^{N} p(\mathcal{G}_n^{-1}\mathcal{G}_m) \, f(\mathcal{G}_n) . \tag{6.73}$$

Similarly, for a continuous group, convolution is defined by

$$[p *_G f](\mathcal{G}_{\boldsymbol{\theta}'}) = \int_{\mathbf{G}} d\theta_L\, p(\mathcal{G}_{\boldsymbol{\theta}}^{-1}\mathcal{G}_{\boldsymbol{\theta}'})\, f(\mathcal{G}_{\boldsymbol{\theta}}) . \tag{6.74}$$

The connection of this expression to ordinary, continuous convolution is immediate. If $\mathbf{G}$ is the translation group, $\boldsymbol{\theta} = \mathbf{r}$, $f(\mathcal{G}_{\boldsymbol{\theta}}) = f(\mathbf{r})$, $p(\mathcal{G}_{\boldsymbol{\theta}}^{-1}\mathcal{G}_{\boldsymbol{\theta}'}) = p(\mathbf{r}' - \mathbf{r})$ and

$$[p *_G f](\mathcal{G}_{\boldsymbol{\theta}'}) = [p * f](\mathbf{r}') = \int_{\infty} d^2r\, p(\mathbf{r}' - \mathbf{r})\, f(\mathbf{r}) , \tag{6.75}$$

which is the usual expression.

For this example, and for Abelian groups in general,

$$[p *_G f](\mathcal{G}_{\boldsymbol{\theta}}) = [f *_G p](\mathcal{G}_{\boldsymbol{\theta}}) , \tag{6.76}$$

but for non-Abelian groups, the order matters.

Mellin convolution The Mellin convolution is defined in (4.85) by

$$[p *_M f](x) = \int_0^{\infty} \frac{da}{a}\, p\left(\frac{x}{a}\right) f(a) . \tag{6.77}$$

We shall now show that this expression is group convolution on the scale group discussed in Sec. 6.5.3; the exercise will also serve to clarify the idea of invariant measures.

A scale transformation is specified by a positive scale factor a, defined so that $p(x/a)$ is a magnified version of $p(x)$; that is, $p(x/a)$ has the same value at $x = ax_0$ as $p(x)$ does at $x = x_0$. Thus the transformation operator $\mathcal{T}_a$ and the scale operator $\mathcal{G}_a$ are defined by

$$\mathcal{T}_a p(x) = p(\mathcal{G}_a^{-1}x) = p(x/a) . \tag{6.78}$$

Now, since the group consists of all scale factors in the range $0 < a < \infty$, any function defined on the positive real line can be regarded as a function on the scale group, and we can write $f(\mathcal{G}_a)$ as $f(a)$; conversely, an arbitrary function $f(x)$ can be interpreted as $f(\mathcal{G}_x^{-1})$. With this notation,

$$p(x/a) = p(\mathcal{G}_a^{-1}x) = p(\mathcal{G}_a^{-1}\mathcal{G}_x) . \tag{6.79}$$

By these manipulations, we have cast the integrand of the Mellin convolution into the same form as for a general group convolution, but we still need to compute the left-invariant measure. By its definition, (6.70), this measure must satisfy

$$\int_0^{\infty} da_L\, f(a/x) = \int_0^{\infty} da_L\, f(a) , \tag{6.80}$$

for all positive x. The change of variables $a' = a/x$ on the left-hand side of (6.80), with subsequent dropping of the prime, yields the right-hand side if $da_L = da/a$. Thus, finally,

$$[p *_M f](x) = \int_0^{\infty} \frac{da}{a}\, p\left(\frac{x}{a}\right) f(a) = \int_{\mathbf{G}} da_L\, p(\mathcal{G}_a^{-1}\mathcal{G}_x)\, f(\mathcal{G}_a) = [p *_G f](\mathcal{G}_x) . \tag{6.81}$$

It will be left as an exercise to reduce the other forms of convolution from Chap. 4 to appropriate group convolutions.

6.8.4 Fourier transforms on groups

We saw in Chap. 4 that each of the convolutions introduced there could be reduced to a simple product by a related integral transform. In particular, the usual continuous convolution becomes a product under ordinary Fourier transformation. Now we shall see why that result is general.

Let $\mathbf{M}^{(m)}(\mathcal{G}_n)$ be the matrix corresponding to $\mathcal{G}_n$ in the mth unitary irreducible representation of group $\mathbf{G}$ (see Sec. 6.3). If $\mathbf{G}$ is discrete (though possibly infinite), then the *group Fourier transform* of a function $f(\mathcal{G}_n)$ on $\mathbf{G}$ is defined by

$$\mathbf{F}_m \equiv \sum_{n=1}^{N} f(\mathcal{G}_n)\,\mathbf{M}^{(m)}(\mathcal{G}_n)\,. \tag{6.82}$$

The corresponding definition for a continuous group is

$$\mathbf{F}_m = \mathcal{F}_{\mathbf{G}}\{f\} \equiv \int_{\mathbf{G}} d\theta_L\; f(\mathcal{G}_{\boldsymbol{\theta}})\,\mathbf{M}^{(m)}(\mathcal{G}_{\boldsymbol{\theta}})\,, \tag{6.83}$$

where the index m is an integer if the group has a countable set of irreducible representations.

Note that $\mathbf{F}_m$ is a matrix of the same dimensions as $\mathbf{M}^{(m)}(\mathcal{G}_n)$. One way to think about $\mathbf{F}_m$ is that it is a *matrix-valued function on representations*. That is, it assigns a matrix to each unitary irreducible representation of $\mathbf{G}$. The index m plays the role of Fourier frequency, but the Fourier transform is not one number for each frequency but, in general, a matrix. To complicate matters even further, different m will yield matrices of different dimensionality.

Most of these complications disappear for Abelian groups, where all irreducible representations are one-dimensional (see Sec. 6.3.2). In that case, $\mathbf{F}_m$ is a scalar F_m and the integrand in (6.83) is a rather conventional integral transform.

Example 1: Finite group The structure of a group Fourier transform can be illustrated by considering the cyclic group $\mathbf{C}_N$, which is a finite-dimensional Abelian group. Since all of the irreducible representations are one-dimensional, the matrices $\mathbf{M}^{(m)}(\mathcal{G}_n)$ reduce to scalars, which are identical to the characters given in (6.17):

$$\mathbf{M}^{(m)}(\mathcal{G}_n) = \exp(-2\pi imn/N)\,. \tag{6.84}$$

It is easy to show that these matrices are unitary and obey the group multiplication rule.

With these representations, (6.82) takes a familiar form,

$$F_m = \sum_{n=1}^{N} f(\mathcal{G}_n)\exp(-2\pi imn/N)\,, \tag{6.85}$$

which is just the discrete Fourier transform of the sequence $f(\mathcal{G}_n)$.

Example 2: Continuous, Abelian group Next consider the translation group in 1D, where the operators are given by a 1D counterpart of (6.69). This group is Abelian, and its irreducible representations are scalars given by $\exp(-2\pi i\xi x)$, where x labels a group element [like n in (6.85)] and ξ labels the representation (like m). The left-invariant measure is just dx, so the Fourier transform is, not surprisingly,

$$F(\xi) = \int_{-\infty}^{\infty} dx\ f(x) \exp(-2\pi i \xi x)\,. \tag{6.86}$$

Example 3: Another continuous, Abelian group We know that the scale group leads to Mellin convolution. What is the corresponding Fourier transform? Once again the group is Abelian, so we seek 1D irreducible representations. We have seen that group elements are specified by nonnegative numbers x, and it can be verified that the following functions form a unitary representation:

$$M^{(\alpha)}(x) = x^{i\alpha}\,, \tag{6.87}$$

where α is a real number labeling the representation. We determined the left-invariant measure in Sec. 6.8.3. Thus the Fourier transform derived from the scale group has the form

$$F(\alpha) = \int_{0}^{\infty} \frac{dx}{x}\ x^{i\alpha} f(x)\,. \tag{6.88}$$

From (4.82) we recognize this integral as the Mellin transform of $f(x)$ with the Mellin variable $s = i\alpha$. (The Mellin transform can be evaluated for arbitrary complex s, but a purely real α is required for $M^{(\alpha)}(x)$ to be a unitary representation.)

Shift theorem One of the most important theorems of ordinary Fourier analysis is the shift theorem, (3.108). We now derive its counterpart for group Fourier analysis.

If $f(\mathcal{G}_{\boldsymbol{\theta}})$ is a function on a continuous group $\mathbf{G}$, then a shifted function, with shift corresponding to $\mathcal{G}_{\boldsymbol{\theta}'}$, can be defined by

$$f_s(\mathcal{G}_{\boldsymbol{\theta}}) = f(\mathcal{G}_{\boldsymbol{\theta}'}^{-1}\mathcal{G}_{\boldsymbol{\theta}})\,. \tag{6.89}$$

From (6.83),

$$\mathcal{F}_{\mathbf{G}}\{f_s\} \equiv \int_{\mathbf{G}} d\theta_L\ f(\mathcal{G}_{\boldsymbol{\theta}'}^{-1}\mathcal{G}_{\boldsymbol{\theta}})\,\mathbf{M}^{(m)}(\mathcal{G}_{\boldsymbol{\theta}})\,. \tag{6.90}$$

The group product of any two group elements is another group element, so we can write

$$\mathcal{G}_{\boldsymbol{\theta}'}^{-1}\mathcal{G}_{\boldsymbol{\theta}} = \mathcal{G}_{\boldsymbol{\theta}''}\,, \tag{6.91}$$

from which we find

$$\mathcal{F}_{\mathbf{G}}\{f_s\} = \int_{\mathbf{G}} d\theta_L\ f(\mathcal{G}_{\boldsymbol{\theta}''})\,\mathbf{M}^{(m)}(\mathcal{G}_{\boldsymbol{\theta}'}\mathcal{G}_{\boldsymbol{\theta}''}) = \mathbf{M}^{(m)}(\mathcal{G}_{\boldsymbol{\theta}'}) \int_{\mathbf{G}} d\theta_L''\ f(\mathcal{G}_{\boldsymbol{\theta}''})\,\mathbf{M}^{(m)}(\mathcal{G}_{\boldsymbol{\theta}''})\,, \tag{6.92}$$

where we have used the invariance of $d\theta_L$ and the fact that the matrices $\mathbf{M}^{(m)}(\mathcal{G}_{\boldsymbol{\theta}})$ obey the group multiplication rule. We recognize the integral in (6.92) as the transform of f, so we have

$$\mathcal{F}_{\mathbf{G}}\{f_s\} = \mathbf{M}^{(m)}(\mathcal{G}_{\boldsymbol{\theta}'})\,\mathcal{F}_{\mathbf{G}}\{f\}\,, \tag{6.93}$$

which is the generalized shift theorem. As an exercise, the reader should show that (6.93) leads to the ordinary shift theorem for the translation group.

Fourier transform of a convolution Next we derive the group counterpart of the convolution theorem, (3.132). Combining (6.74) and (6.83), we have

$$[\mathcal{F}_{\mathrm{G}}\{p *_G f\}]_m = \int_{\mathbf{G}} d\theta'_L \int_{\mathbf{G}} d\theta_L \; p(\mathcal{G}_{\boldsymbol{\theta}}^{-1}\mathcal{G}_{\boldsymbol{\theta}'})\, f(\mathcal{G}_{\boldsymbol{\theta}})\, \mathbf{M}^{(m)}(\mathcal{G}_{\boldsymbol{\theta}'})\,. \tag{6.94}$$

The manipulations are now very similar to the ones used in the derivation of the shift theorem. With $\mathcal{G}_{\boldsymbol{\theta}}^{-1}\mathcal{G}_{\boldsymbol{\theta}'} \equiv \mathcal{G}_{\boldsymbol{\theta}''}$, the invariance of the integration measure and the representational nature of $\mathbf{M}^{(m)}(\mathcal{G}_{\boldsymbol{\theta}})$, we have

$$\begin{aligned}[\mathcal{F}_{\mathrm{G}}\{p *_G f\}]_m &= \left[\int_{\mathbf{G}} d\theta''_L \; p(\mathcal{G}''_{\boldsymbol{\theta}})\, \mathbf{M}^{(m)}(\mathcal{G}''_{\boldsymbol{\theta}})\right] \left[\int_{\mathbf{G}} d\theta_L \; f(\mathcal{G}_{\boldsymbol{\theta}})\, \mathbf{M}^{(m)}(\mathcal{G}_{\boldsymbol{\theta}})\right] \\ &= [\mathcal{F}_{\mathrm{G}}\{p\}\mathcal{F}_{\mathrm{G}}\{f\}]_m\,, \end{aligned} \tag{6.95}$$

as anticipated.

Fourier inversion theorem When we derived various inversion theorems in Chaps. 3 and 4, the procedure was always to multiply a transformed quantity by one of the basis functions, sum or integrate, and appeal to the orthogonality of the basis functions. The situation is more complicated with group Fourier transforms which map scalar-valued functions on a group to matrices of various sizes. The inverse thus has to map a set of matrices back to a scalar function, suggesting that a trace operation might be involved. (The trace of a matrix is a scalar.) Perusal of Sec. 6.3.4 suggests further that the orthogonality relation (6.15) might be useful. This is indeed the case for a finite group, to which we restrict attention here.

To move in the direction of applying (6.15), we multiply (6.82) by $[\mathbf{M}^{(m)}(\mathcal{G}_k)]^\dagger$ and take the trace of the resulting matrix. This procedure yields

$$\mathrm{tr}\left\{\left[\mathbf{M}^{(m)}(\mathcal{G}_k)\right]^\dagger \mathbf{F}_m\right\} = \sum_{n=1}^{N} f(\mathcal{G}_n)\, \chi^{(m)}\{\mathcal{G}_k^{-1}\mathcal{G}_n\}\,, \tag{6.96}$$

where we have appealed to the unitary nature of the representation to set

$$\mathrm{tr}\left\{\left[\mathbf{M}^{(m)}(\mathcal{G}_k)\right]^\dagger \mathbf{M}^{(m)}(\mathcal{G}_n)\right\} = \mathrm{tr}\left\{\mathbf{M}^{(m)}(\mathcal{G}_k^{-1}\mathcal{G}_n)\right\} = \chi^{(m)}(\mathcal{G}_k^{-1}\mathcal{G}_n)\,. \tag{6.97}$$

Guided by (6.15), we now multiply (6.97) by the character for the identity element in the mth irreducible representation, $\chi^{(m)}(\mathcal{E})$, which is just the dimension of the representation, N_m (hence a real number). After summing over representations, we have

$$\sum_m N_m \mathrm{tr}\left\{\left[\mathbf{M}^{(m)}(\mathcal{G}_k)\right]^\dagger \mathbf{F}_m\right\} = \sum_{n=1}^{N} f(\mathcal{G}_n) \sum_m \left[\chi^{(m)}(\mathcal{E})\right]^* \chi^{(m)}(\mathcal{G}_k^{-1}\mathcal{G}_n)\,. \tag{6.98}$$

By (6.15), the sum over m can be nonzero only if $\mathcal{E}$ and $\mathcal{G}_k^{-1}\mathcal{G}_n$ are in the same class. But the identity element $\mathcal{E}$ is always in a class by itself (literally), so we must have $\mathcal{G}_k^{-1}\mathcal{G}_n = \mathcal{E}$ or $\mathcal{G}_k = \mathcal{G}_n$. The sum over m in (6.98) is thus a group representation of a Kronecker delta; from (6.15) with $L_j = 1$ for the identity class,

$$\sum_m \left[\chi^{(m)}(\mathcal{E})\right]^* \chi^{(m)}(\mathcal{G}_k^{-1}\mathcal{G}_n) = N\,\delta_{kn}\,. \tag{6.99}$$

Kronecker performs the remaining sum in (6.98) for us, and we have, finally,

$$f\{\mathcal{G}_k\} = \frac{1}{N}\sum_m N_m \text{tr}\left\{\left[\mathbf{M}^{(m)}(\mathcal{G}_k)\right]^\dagger \mathbf{F}_m\right\}. \tag{6.100}$$

This equation is the inversion theorem for Fourier transforms on a finite group. It holds also for certain continuous groups called *nonunimodular locally compact groups* (Duflo and Moore, 1976) if the order of the group, N, is replaced by the *volume of the group* $V(\mathbf{G})$, defined by

$$V(\mathbf{G}) = \int_{\mathbf{G}} d\theta_L . \tag{6.101}$$

Unfortunately, this volume is not always finite.

6.8.5 Wavelets revisited

Wavelets are related to the affine group, introduced in Sec. 6.5.3. This group brings two new complications to the fore. First, unlike the translation and scale groups used for Fourier and Mellin transforms, respectively, the affine group is not Abelian. Second, its irreducible representations have infinite dimensionality.

To accommodate the second problem, we need to broaden our concept of representation. A matrix representation of dimension K is a set of $K \times K$ matrices that obey the group multiplication rule. Equivalently, as noted in Sec. 6.6.2, the matrices can be regarded as operators on a K-dimensional Hilbert space (usually a subspace of $\mathbb{L}_2$). Thus a representation is a set of *operators* that obey the group multiplication rule. In this sense, the set of operators $\{\mathcal{T}_n\}$ defined in (6.66) is a representation of a finite group, and the set $\{\mathcal{T}_\theta\}$ defined in (6.68) is a representation of an infinite group. Since these operators act on infinite Hilbert spaces, they define representations of infinite dimensionality. Usually such representations are reducible, but in the case of the affine group we are stuck with representations of infinite dimensionality.[5]

There are three infinite-dimensional unitary representations of the affine group, but they are equivalent to one another through a similarity transformation, so there is really only one representation (Sibul, 1995). The representation of the affine group used most commonly, and specifically in wavelet analysis, consists of a set of operators on $\mathbb{L}_2(\mathbb{R})$ defined by

$$\mathcal{T}_{a,b}\, f(x) = \frac{1}{\sqrt{|a|}}\, f\left(\frac{x-b}{a}\right) = \frac{1}{\sqrt{|a|}}\, f\{[\mathcal{S}(a,b)]^{-1}x\}, \tag{6.102}$$

where $\mathcal{S}(a,b)$ is the operator used to define the group in (6.21). The factor of $1/\sqrt{a}$ is required for unitarity.

Since $\mathcal{T}_{a,b}$ is an operator in $\mathbb{L}_2(\mathbb{R})$, its action can be expressed as an integral transform,

$$[\mathcal{T}_{a,b}\,\mathbf{f}]\,(x) = \int_{-\infty}^{\infty} dx'\, t_{a,b}(x,x')\, f(x'), \tag{6.103}$$

[5] There is a trivial exception to this statement. If we fix the translation at 0, we can find one-dimensional representations of the affine group. These representations, which are also representations of Abelian subgroups of the affine group, are of no use in discussing the full affine group.

where $t_{a,b}(x, x')$ is the kernel of $\mathcal{T}_{a,b}$. Comparison of (6.102) and (6.103) shows that we must have

$$t_{a,b}(x, x') = \frac{1}{\sqrt{|a|}}\, \delta\left(x' - \frac{x-b}{a}\right), \tag{6.104}$$

where $\delta(\cdot)$ is the usual 1D Dirac delta function.

In terms of the operator, $\mathcal{T}_{a,b}$, the continuous wavelet transform defined in (5.73) can be written as

$$w(a, b) = \int_{-\infty}^{\infty} dx\; f(x)\, \mathcal{T}_{a,b}\, \psi^*(x)\,. \tag{6.105}$$

The wavelet transform is a function of the parameters a and b specifying an element of the affine group, so it is a function on that group.

Fourier transform on the affine group From the discussion in Sec. 6.8.4, it is natural to examine the group Fourier transform of $w(a, b)$. Recall from the discussion below (6.83) that a group Fourier transformation maps a scalar-valued function on the group to a matrix-valued function on representations. For the affine group, there is only one representation, so there is just one Fourier coefficient like $\mathbf{F}_m$ in (6.83) — the index m is irrelevant! It would seem that we have lost an enormous amount of information by transforming a function of two variables down to a single Fourier coefficient, but the thing that saves us is that this coefficient is an operator in an infinite-dimensional Hilbert space.

Specifically, the Fourier transform of $w(a, b)$ on the affine group is defined, by a modest generalization of (6.83), as

$$\mathcal{F}_{\mathrm{G}}\{w\} \equiv \widetilde{\mathbf{w}} = \int_{-\infty}^{\infty} \frac{da}{a^2} \int_{-\infty}^{\infty} db\; w(a, b)\, \mathcal{T}_{a,b}\,. \tag{6.106}$$

In comparing (6.106) with (6.83), note that $d\theta_L = da\, db/a^2$ and that the $\mathbb{L}_2$ operator $\mathcal{T}_{a,b}$ plays the role of the matrix $\mathbf{M}^{(m)}(\mathcal{G}_\theta)$. If we had chosen to write (6.83) in terms of matrix *elements*, $M_{nn'}^{(m)}(\mathcal{G}_{\boldsymbol{\theta}})$ would have appeared, with the indices n and n' running up to the dimensionality of the representation. Correspondingly, for the infinite-dimensional representation used in (6.106), the indices are the arguments of the kernel $t_{a,b}(x, x')$. Finally, the operator $\widetilde{\mathbf{w}}$ should not be confused with the wavelet transformation operator $\mathcal{W}(a, b)$ used in Chap. 5; the latter maps from $\mathbb{L}_2$ to wavelet space, while the former maps $\mathbb{L}_2$ to itself.

Relation to the inverse wavelet transform The remaining question is how $\widetilde{\mathbf{w}}$ is related back to the original function $f(x)$. The inverse wavelet transform (5.78) provides the answer. From that equation and (6.106), we find

$$f(x) = \frac{1}{C_\psi}\, \widetilde{\mathbf{w}}\, \psi(x)\,. \tag{6.107}$$

Thus the wavelet-transformation operator $\mathcal{W}(a, b)$ maps a function in $\mathbb{L}_2$ to a function in wavelet space, the group Fourier transform maps that function to an *operator* on $\mathbb{L}_2$, and application of that operator to the original mother wavelet recovers the original $\mathbb{L}_2$ function. In shorthand,

$$\frac{1}{C_\psi}\left[\mathcal{F}_{\mathrm{G}} \mathcal{W}(a, b)\mathbf{f}\right] \psi(x) = f(x)\,. \tag{6.108}$$

This result may become a bit more transparent when we realize that the wavelet-transformation operator $\mathcal{W}(a,b)$ corresponds to a scalar product in $\mathbb{L}_2$, so that

$$\mathcal{W}(a,b)\,\mathbf{f} = (\mathcal{T}_{a,b}\psi, \mathbf{f}) = [\mathcal{T}_{a,b}\psi]^\dagger\,\mathbf{f}\,. \tag{6.109}$$

Taking the group Fourier transform and applying the resulting operator to ψ yields

$$\begin{aligned}[\mathcal{F}_{\mathbf{G}}\mathcal{W}(a,b)\,\mathbf{f}]\,\psi &= \int_{\mathbf{G}} d\theta_L\,\left\{[\mathcal{T}_{a,b}\psi]^\dagger\,\mathbf{f}\right\}[\mathcal{T}_{a,b}\psi] \\ &= \int_{\mathbf{G}} d\theta_L\,[\mathcal{T}_{a,b}\psi]\,[\mathcal{T}_{a,b}\psi]^\dagger\,\mathbf{f}\,. \end{aligned} \tag{6.110}$$

Equations (6.108) and (6.110) then state that

$$\frac{1}{C_\psi}\int_{\mathbf{G}} d\theta_L\,[\mathcal{T}_{a,b}\psi]\,[\mathcal{T}_{a,b}\psi]^\dagger = \mathcal{I}\,, \tag{6.111}$$

where $\mathcal{I}$ is the identity operator in $\mathbb{L}_2$. We know independently from (5.80) that this is the correct resolution of the identity for wavelets.

7

Deterministic Descriptions of Imaging Systems

With the background gained from Chapters 1-6, we are now ready to undertake the main task of this book, mathematical analysis of imaging systems. As noted in the Prologue, this analysis must consist of two distinct components, which we can call *deterministic* and *stochastic*. Deterministic analysis is the description of objects, images and the mapping between them without regard to noise or other sources of randomness or uncertainty. Such analysis is the topic of this chapter. Stochastic analysis, which addresses these random aspects, is taken up in the next chapter.

We begin in Sec. 7.1 with a survey of possible deterministic descriptions of objects and images. Since an object or an image can always be regarded as a vector in some Hilbert space, the concepts and tools of linear algebra introduced in Chap. 1 will be essential here, and the reader is assumed to have a good grasp of those topics.

Similarly, an imaging system can always be regarded as a transformation from one Hilbert space to another, and many imaging systems are linear, so the discussion of linear operators in Chap. 1 is also key in this chapter. Specifically, as in Chap. 1, we distinguish three important classes of linear operators: (1) continuous-to-continuous (CC) operators, which map a function of a continuous variable to another function of a (possibly different) continuous variable; (2) continuous-to-discrete (CD) operators, which map functions to finite sets of numbers or discrete vectors; and (3) discrete-to-discrete (DD) operators, which map one discrete vector to another. Each type of operator has an important application in image science, and each is discussed in detail in one major section of this chapter.

Linear CC operators are treated in Sec. 7.2. Not surprisingly, Fourier analysis will be a crucial tool in this section, and the reader will frequently be referred to Chap. 3 for mathematical details.

As mentioned in the prologue, CD mappings are the most natural for describing digital imaging systems since real objects are functions and digital images are discrete vectors. Section 7.3 is a detailed treatment of these mappings, one we shall rely on heavily in later chapters.

On the other hand, we cannot deal directly with functions when we want to represent an object in a computer, so discrete object models and DD operators are also essential in image science. Though such discrete models are often taken for granted in the imaging literature, there are some mathematical subtleties which we shall discuss in Sec. 7.4. This treatment will be of considerable use in Chap. 15 when we discuss inverse problems and image reconstruction.

Finally, in Sec. 7.5, we move beyond linear systems and discuss nonlinear imaging operators.

7.1 OBJECTS AND IMAGES

In the prologue we glimpsed some of the enormous diversity of things that can be regarded as objects to be imaged. There is a similar diversity in the mathematical language that can be used to describe those objects and the images they produce. The goal of this section is to survey these mathematical descriptions and lay the groundwork for discussing imaging systems in the later sections.

7.1.1 Objects and images as functions

An object to be imaged can often be described naturally as a function on a three-dimensional (3D) Euclidean space. A point in this space is described by the Cartesian coordinates (x, y, z) or, more succinctly, by a vector $\mathbf{r}$ from the origin to the point. When an object is regarded as a scalar-valued function in this space, it is denoted $f(\mathbf{r})$ or, equivalently, $f(x, y, z)$. The value of this function is often referred to as the *gray level* of the object. Following common, though loose, terminology, we shall often refer to a scalar-valued function of a 3D vector as a 3D function for short.

The physical meaning of $f(\mathbf{r})$ depends, of course, on the nature of the imaging system. For example, in nuclear medicine, $f(\mathbf{r})$ refers to the density of a radioactive tracer, perhaps measured in radioactive disintegrations per second per cubic centimeter. The goal of an imaging system in nuclear medicine is to produce a map of this density. Similarly, in fluorescence microscopy the object of interest is the fluorescent radiation from the sample. Suitable units could be the number of emitted photons per cm^3 or watts per cubic meter (W/m^3). Alternatively, the object can be described by the density of the fluorophor.

Sometimes it is a useful approximation to regard the object as being confined to a plane. For example, an ordinary camera focused on an object a long distance away (at infinity) does not record any depth information, so the object is effectively planar. As another example, a computed-tomography system that uses a thin fan of radiation receives no information about the object outside the slab defined by the fan, so again the object is essentially planar. In these cases we shall still denote the object by $f(\mathbf{r})$, with the understanding that $\mathbf{r}$ is now a two-dimensional (2D) vector so that $f(\mathbf{r})$ stands for $f(x, y)$. Occasionally 2D and 3D vectors will both occur in the same problem, and in those cases we shall use $\mathfrak{r}$ for the 3D position vector and $\mathbf{r}$ for the 2D one.

In the camera example with a planar object, $f(\mathbf{r})$ could refer to the reflected optical flux per unit area[1] and be measured in $\mathrm{W/m^2}$. In x-ray computed tomography, the quantity of interest is the x-ray attenuation coefficient μ, measured in cm^{-1}, for example. The object $f(\mathbf{r})$ in this case is the two-dimensional distribution of μ.

Other variables Interesting objects are almost always functions of time. They move about from place to place or change their orientation with respect to the imaging system at some rate. The strength of the radiation from the object can also be a function of time. Thus we should be writing $f(\mathbf{r}, t)$ consistently instead of simply $f(\mathbf{r})$. Only if the object function is essentially constant over the imaging time is the static description really justified.

Other variables may also be required. For example, the radiation emitted by or reflected from the object can have a range of energies or wavelengths, and this dependence should be accounted for in the functional argument. Thus we might want to describe an object as $f(\mathbf{r}, t, \lambda)$, where λ is the wavelength of the radiation. Specific functions that incorporate this spectral information are introduced in Chap. 10.

Sometimes the angle of the radiation with respect to the object surface or a symmetry axis of the imaging system is important. In these cases the functional dependence must include one or more angles. Again, specifics as to how this is done are postponed to Chap. 10.

For notational simplicity, we shall usually not list all of the possible variables in the argument of f. Instead we shall write $f(\mathbf{r})$ but, when necessary, interpret $\mathbf{r}$ more broadly than as spatial coordinates. If q variables are required to specify the object, $\mathbf{r}$ will be interpreted as a qD vector. The time variable will be included only when the dynamics of the object play a role in the imaging system.

Restrictions on the value of the function Physical considerations often place restrictions on the values that an object function $f(\mathbf{r})$ can assume. For example, if $f(\mathbf{r})$ represents a concentration of a radiotracer or fluorophor, it cannot be negative, and if it represents a reflectivity it must satisfy $0 \leq f(\mathbf{r}) \leq 1$.

Sometimes it is reasonable to assume that $f(\mathbf{r})$ can have only two values. If we photograph a newspaper, for example, the object is only black and white (even in the half-tone photographs!). Such objects are said to be *binary-valued*, or just binary for short.

Vector-valued functions So far we have considered only scalar-valued functions, but vector-valued functions or *vector fields* arise in some imaging applications. Perhaps the most familiar example is the electric field, which is central to coherent imaging. In this case each of the three Cartesian components of the field is a separate scalar-valued function of space, time and any other relevant variables. While the three components may be interrelated because of the wave equation or other constraints, full specification of the vector field requires all three functions.

Color images can also be regarded as vectors. If a scene is recorded through three separate color filters (say, red, green and blue), then each individual image is

[1]The proper radiometric term for this quantity is *radiant exitance*, but we postpone a complete discussion of radiometric units to Chap. 10.

a scalar field or scalar-valued function of position, and the collection of three images is a 3D vector field.

Sometimes the components of a vector field are entirely different physical quantities. For example, objects encountered in magnetic resonance imaging (MRI) can be described in terms of three spatial maps: (1) spin density, (2) spin-lattice relaxation time and (3) spin-spin relaxation time. We can think of these maps as three separate spatial functions $f_j(\mathbf{r})$, $j = 1, ..., 3$, or as three components of a single vector field $\mathbf{f}(\mathbf{r})$.

This concept can be extended further as needed. If J separate attributes are required to describe an object, and each attribute is a function of q variables, we can denote the object as $\mathbf{f}(\mathbf{r})$, where $\mathbf{r}$ is a qD vector and $\mathbf{f}(\mathbf{r})$ for a fixed $\mathbf{r}$ is a JD vector.

Complex functions In the examples given above, an object is described as a real-valued function, but complex-valued ones are also frequently required. In coherent imaging with ultrasound, microwaves or laser beams, the phase of the radiation plays a crucial role, and a complex notation is mathematically very convenient.

Consider, for example, a holographic imaging system in which an object is illuminated with a monochromatic plane wave of frequency ν. We might choose in this case to define the object by the amplitude of the reflected electric field in some plane near the object. This field is a function of two spatial variables and time, and we can denote it as $e(\mathbf{r}, t)$. This function is real, since it represents an actual physical field, and we can write it as

$$e(\mathbf{r}, t) = A(\mathbf{r}) \cos[\phi(\mathbf{r}) - 2\pi\nu t] , \tag{7.1}$$

where $A(\mathbf{r})$ specifies the strength of the wave and $\phi(\mathbf{r})$ is the phase of its temporal oscillation at point $\mathbf{r}$. Note that we have assumed that both $A(\mathbf{r})$ and $\phi(\mathbf{r})$ are independent of time, which is appropriate if the object is not changing with time in any way and the illumination is constant except for the oscillation at frequency ν.

The real function in (7.1) can be expressed as

$$e(\mathbf{r}, t) = A(\mathbf{r}) \operatorname{Re}\{\exp[i\phi(\mathbf{r}) - 2\pi i\nu t]\} , \tag{7.2}$$

where $\operatorname{Re}\{\cdot\}$ denotes real part. Because the frequency ν is constant and known *a priori*, the factor $\exp(-2\pi i\nu t)$ is an unnecessary part of the specification of the wave. Therefore we can completely define the object in this case by the complex-valued function $f(\mathbf{r})$ given by

$$f(\mathbf{r}) = A(\mathbf{r}) \exp[i\phi(\mathbf{r})] . \tag{7.3}$$

The actual real field can be reconstituted by means of (7.2) if desired.

The complex form will prove to be very useful when we consider the effect of an imaging system. Any linear system operates on the real and imaginary parts of the input separately; the real part of the output is obtained by a linear transformation of the real part of the input, and similarly for the imaginary part. Therefore we can assume, quite unphysically, that the object is a complex wave, compute the image, and then take the real part. We caution the reader, however, that this approach presumes an imaging system that is strictly linear in wave amplitude.

Images as functions So far we have discussed objects, but images can also be treated as functions. Different imaging systems require different choices for the independent variables in these functions.

We need to distinguish variables that affect the response of the detector from ones that the detector *measures*. Photographic film, for example, responds to total energy per unit area to a first approximation. Although the response may depend on angle of arrival of the radiation or the time interval over which it is spread, film does not record these variables. Thus the natural description of an optical image in the film plane of a camera is $g(x_d, y_d)$, where x_d and y_d are spatial coordinates (d denotes *detector*), and the variable g itself denotes the exposure (joules/m^2). If we wish to consider the developed film, the same independent variables can be used but then g will denote transmittance or optical density of the film.

One important variable that affects film response strongly, but is not measured by the film, is the wavelength. To illustrate how such a variable enters into the mathematical description of the image, we can define a spectral response function $R(\lambda)$ so that the effective exposure on the film is given by

$$g(\mathbf{r}_d) = \int_0^\infty d\lambda\ R(\lambda)\, g_\lambda(\mathbf{r}_d, \lambda)\,, \tag{7.4}$$

where $g_\lambda(\mathbf{r}, \lambda)$ is the exposure per unit wavelength on the film. Thus the dependence on wavelength is integrated out, and λ does not appear in g. In order to calculate $g(\mathbf{r}_d)$ we must know $g_\lambda(\mathbf{r}_d, \lambda)$, but the final image is not a function of λ (for black-and-white film). On the other hand, if we consider a detector that does measure wavelength, such as an imaging spectrometer, then the λ variable must be included in the description of g.

7.1.2 Objects and images as infinite-dimensional vectors

As noted in Chap. 1, it is often fruitful to regard objects and images as vectors in linear vector spaces. If the basic description of the object or image is a function of one or more continuous variables, then the space will have an infinite number of dimensions.

In order to decide what vector spaces are most appropriate for imaging applications, we must inquire about the basic mathematical properties of the functions used for describing objects and images. We do not want to impose artificial and physically unrealistic conditions on the functions, but we do want to specify them as completely as possible.

One thing we should not assume is that the functions (especially object functions) are continuous or differentiable. Objects have abrupt edges, so they are not differentiable. Similarly, we do not want to assume that objects are bandlimited (see Sec. 3.5.1) because functions with finite bandwidth also cannot have sharp edges. It may sometimes be useful to approximate *images* as bandlimited, since the edges are blurred by the imaging system, but even there it cannot be an exact description.

One mathematical feature that we can take advantage of is compact support. Objects and images have finite size. An exception to this statement might appear to occur in astronomy — cosmological arguments aside, the universe is infinite. Nevertheless, astronomical objects can usefully be described as functions of finite support. The essentially infinite distances from the earth are no problem since the imaging systems don't measure that distance, so the object is described by angular coordinates which are restricted to 4π steradians. Thus even the universe is, for our purposes, a function with finite support.

Moreover, functions that describe real objects and images do not have infinite values. A function with finite support and no infinite values is necessarily square-integrable and hence a vector in the Hilbert space $\mathbb{L}_2(\mathbf{S})$, where $\mathbf{S}$ is the region of support of the function in terms of its independent variables (not including time).

Explicitly, if an object is described as a bounded function of q variables (or, equivalently, a qD vector $\mathbf{r}$), and $f(\mathbf{r})$ is zero outside some region $\mathbf{S}_f$ in $\mathbb{R}^q$, then it is square-integrable:

$$\int_{\mathbf{S}_f} d^q r \; |f(\mathbf{r})|^2 < \infty \,. \tag{7.5}$$

Thus the object can be regarded as a vector in the Hilbert space $\mathbb{L}_2(\mathbf{S}_f)$. If the object is square-integrable over the infinite domain, it is a vector in $\mathbb{L}_2(\mathbb{R}^q)$. We shall denote an object as $f(\mathbf{r})$ when we want to emphasize its functional dependence, but we shall use $\mathbf{f}$ when we wish to think of it as a vector in Hilbert space.

Similarly, if an image $g(\mathbf{r}_d)$ is defined as a function on $\mathbb{R}^s$ and is zero outside a region $\mathbf{S}_g$, then it corresponds to the vector $\mathbf{g}$ in the Hilbert space $\mathbb{L}_2(\mathbf{S}_g)$, which may be $\mathbb{L}_2(\mathbb{R}^s)$ if no support restriction is required.

If the function includes angles, it can still be regarded as a vector in a Hilbert space. For example, $f(x, y, z, \theta, \phi)$ or $f(\mathbf{r}, \hat{\mathbf{n}})$ is a vector in the Hilbert space with norm defined by

$$||\mathbf{f}||^2 = \int_{\mathbf{S}_f} d^3\mathbf{r} \int_{4\pi} d\Omega \; |f(\mathbf{r}, \hat{\mathbf{n}})|^2 \,, \tag{7.6}$$

where $d\Omega$ is the element of solid angle associated with $\hat{\mathbf{n}}$.

Vector-valued functions A Hilbert-space formalism is also applicable to vector-valued functions. As discussed in Sec. 7.1.1, we can think of a J-component object as J separate spatial functions $f_j(\mathbf{r})$ or as J components of a single vector field $\mathbf{f}(\mathbf{r})$ in $\mathbb{R}^q$. The norm of such a function can be defined by

$$||\mathbf{f}||^2 = \sum_{j=1}^{J} \int_{\mathbf{S}_f} d^q r \; |f_j(\mathbf{r})|^2 \,. \tag{7.7}$$

This norm corresponds to a mixed Hilbert space, the direct product of the JD Euclidean space $\mathbb{E}^J$ and $\mathbb{L}_2(\mathbb{R}^q)$. If the norm in (7.7) is finite, as it will be in real-world applications, the object $\mathbf{f}(\mathbf{r})$ is a vector in the Hilbert space $\mathbb{L}_2(\mathbb{R}^q) \times \mathbb{E}^J$.

Note that boldface $\mathbf{f}$ in this context has two distinct meanings. When we write $\mathbf{f}(\mathbf{r})$, we mean a vector field with a small number of components, but $\mathbf{f}$ without an argument refers to a vector in an infinite-dimensional Hilbert space. To avoid notational complications, we shall consider only scalar-valued functions for the remainder of this chapter, but all results are easily extended to the vector-valued case.

Is $\mathbb{L}_2$ too large? To summarize the considerations above, we shall always assume that object and image functions are square-integrable and hence vectors in an $\mathbb{L}_2$ Hilbert space with appropriate variables and support. No physically obtainable objects or images are ruled out by this assumption. For generality, we shall often denote the object Hilbert space as $\mathbb{U}$ and the image space as $\mathbb{V}$, but $\mathbb{L}_2$ spaces will always be understood. Of course, not all vectors in these spaces correspond to meaningful objects or images. As noted in Sec. 7.1.1, we may have additional restrictions, such as the fact that in many applications objects and images are real and nonnegative. Thus many of the vectors in the $\mathbb{L}_2$ spaces are ruled out for other reasons, but at least we can be assured that $\mathbb{U}$ and $\mathbb{V}$ are large enough to include all physically realizable objects and images, respectively.

Basis vectors The $\mathbb{L}_2$ Hilbert spaces are all *separable*, which means that they are spanned by denumerably infinite sets of basis vectors (see Sec. 1.1.5). Thus any object $f(\mathbf{r})$ can be represented exactly as an infinite series of the form

$$f(\mathbf{r}) = \sum_{n=1}^{\infty} \alpha_n \, \psi_n(\mathbf{r}) \,, \tag{7.8}$$

where $\{\psi_n(\mathbf{r})\}$ is any orthonormal basis for the relevant Hilbert space. If the space is $\mathbb{L}_2(\mathbf{S}_f)$, the coefficients are given by

$$\alpha_n = (\psi_n(\mathbf{r}), f(\mathbf{r})) = \int_{\mathbf{S}_f} d^q r \; \psi_n^*(\mathbf{r}) \, f(\mathbf{r}) \,, \tag{7.9}$$

where $(\psi_n(\mathbf{r}), f(\mathbf{r}))$ denotes a scalar product.

In the more abstract Hilbert-space notation, (7.8) and (7.9) can also be written as

$$\mathbf{f} = \sum_{n=1}^{\infty} \alpha_n \, \boldsymbol{\psi}_n \,, \qquad \alpha_n = (\boldsymbol{\psi}_n, \mathbf{f}) \,. \tag{7.10}$$

Fourier basis functions To illustrate the expansion (7.8), we shall discuss Fourier-series representations, first in one dimension and then in an arbitrary number of dimensions. For background material, see Secs. 3.2.1 and 3.4.6.

As discussed in Sec. 3.2.1, a 1D square-integrable function that vanishes outside $-\frac{1}{2}L < x < \frac{1}{2}L$ can be represented as

$$f(x) = \sum_{k=-\infty}^{\infty} F_k \exp(2\pi i \xi_k x) \, \mathrm{rect}(x/L) \,, \tag{7.11}$$

where ξ_k is a discrete spatial frequency given by $\xi_k = k/L$, and the Fourier coefficients $\{F_k\}$ can be computed from (3.19). Without the rect function, the right-hand side of (7.11) would represent a periodic function consisting of an infinite set of replicas of $f(x)$, but the rect function sets all replicas but one to zero. Thus the Fourier series (7.11) has the same structure[2] as (7.8), with $\psi_k(x) = \exp(2\pi i \, kx/L) \, \mathrm{rect}(x/L)$. These functions form a basis for $\mathbb{L}_2(-\frac{1}{2}L, \frac{1}{2}L)$.

Next we consider a function $f(\mathbf{r})$, where $\mathbf{r}$ is a qD vector. To generalize (7.11) to q dimensions, we need to replace k with a set of q indices. A convenient notation for this set is the multi-index $\mathbf{k}$ (see Sec. 3.4.6), a set of q integers $(k_1, ..., k_q)$ where each integer specifies a component of the qD spatial-frequency

[2]One minor difference between (7.8) and (7.11) is that the index runs from 1 to ∞ in the former and from $-\infty$ to ∞ in the latter, but that is of no great import. A reparameterization, which we won't bother with, could easily solve this problem. If $f(x)$ is real, the negative k values in (7.11) are redundant anyway by dint of (3.45).

vector $\boldsymbol{\rho}_{\mathbf{k}} = (k_1\Delta\rho, ..., k_q\Delta\rho)$. Thus the set of frequency vectors $\{\boldsymbol{\rho}_{\mathbf{k}}\}$ defines an infinite regular lattice of points in the qD frequency space. We shall adopt the convention that summation over the multi-index $\mathbf{k}$ is equivalent to summing over all q of its components:

$$\sum_{\mathbf{k}=-\infty}^{\infty} = \sum_{k_1=-\infty}^{\infty} \sum_{k_2=-\infty}^{\infty} \cdots \sum_{k_q=-\infty}^{\infty} . \tag{7.12}$$

We assume that $f(\mathbf{r})$ vanishes identically unless $\mathbf{r}$ is in a region $\mathbf{S}_f$ of the qD space. This region is assumed to be bounded by a *cube of support*, a region defined by $-\frac{1}{2}L < r_j < \frac{1}{2}L$, $j = 1, ..., q$. For $q = 3$ the cube of support is literally a cube in 3D space, but for $q = 2$ it is a square of side L, while for $q > 3$ it is a hypercube. The support region $\mathbf{S}_f$ can be either the cube of support itself or any smaller region containing the object and lying entirely within the cube.

With this support constraint, an exact Fourier-series representation of $f(\mathbf{r})$ is possible if we choose $\Delta\rho \leq 1/L$ (see Sec. 3.5.4). If this condition is satisfied, then

$$f(\mathbf{r}) = \sum_{\mathbf{k}=-\infty}^{\infty} F_{\mathbf{k}} \exp(2\pi i \boldsymbol{\rho}_{\mathbf{k}} \cdot \mathbf{r})\, S_f(\mathbf{r}) , \tag{7.13}$$

where $\boldsymbol{\rho}_{\mathbf{k}} \cdot \mathbf{r}$ denotes a scalar product of the two qD Euclidean vectors, and $S_f(\mathbf{r})$ is a *support function* that equals 1 for $\mathbf{r}$ in $\mathbf{S}_f$ and 0 otherwise. The support function, like the rect function in (7.11), sets all replicas but one to zero.

A useful way to rewrite (7.13) is [*cf.* (3.279)]

$$f(\mathbf{r}) = \sum_{\mathbf{k}=-\infty}^{\infty} F_{\mathbf{k}}\, \Phi_{\mathbf{k}}(\mathbf{r}) , \tag{7.14}$$

where $\Phi_{\mathbf{k}}(\mathbf{r})$ is a basis function defined by

$$\Phi_{\mathbf{k}}(\mathbf{r}) \equiv \exp(2\pi i \boldsymbol{\rho}_{\mathbf{k}} \cdot \mathbf{r})\, S_f(\mathbf{r}) . \tag{7.15}$$

If the support region $\mathbf{S}_f$ is the cube of support as discussed above, the set $\{\Phi_{\mathbf{k}}(\mathbf{r})\}$ is orthogonal[3] and complete. For some smaller support region lying within the cube of support, the orthogonality is lost but the set is still complete. A few of the basis functions for the case of $q = 3$ and spherical support are shown in Fig. 7.1.

If we do not wish to assume that the object has finite support, we can take the object space to be $\mathbb{L}_2(\mathbb{R}^q)$. Fourier analysis can still be used, of course, but now the object is represented by its qD inverse Fourier transform rather than by a Fourier series. As discussed in Sec. 1.1.6, this approach amounts to using a continuous or nondenumerable basis for $\mathbb{L}_2(\mathbb{R}^q)$, where the basis functions themselves do not lie in the space they span. On the other hand, because $\mathbb{L}_2(\mathbb{R}^q)$ is a separable Hilbert space even without any restriction to finite support, we can also adopt some denumerable basis functions such as Hermite-Gauss functions (see Sec. 4.1.4) or eigenfunctions of some compact, Hermitian operator (see Sec. 1.4.4).

[3] As defined here, the basis functions are not normalized. They could be by inclusion of a factor $1/\sqrt{V}$, where V is the qD volume of the support region. This factor, corresponding to $1/|\det(\mathbf{P})|$ in (3.280), is omitted here for convenience.

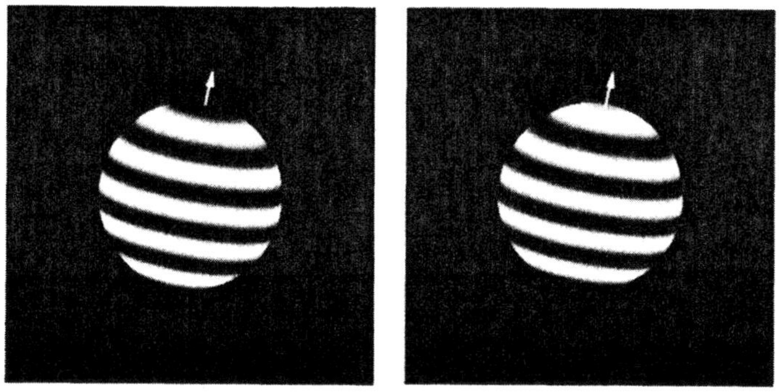

(a) (b)

Fig. 7.1 Illustration of the sine and cosine basis functions $\Phi_{\mathbf{k}}(\mathbf{r})$ for the case of spherical support. (Courtesy of Howard Gifford.)

Spatio-temporal basis functions If the object is explicitly a function of time, $f(\mathbf{r}, t)$, and if it is square-integrable over the spatial support for each time, then it can still be expanded in terms of purely spatial basis functions, but with the coefficients being functions of time; thus (7.8) becomes

$$f(\mathbf{r}, t) = \sum_{n=1}^{\infty} \alpha_n(t)\, \psi_n(\mathbf{r}) . \tag{7.16}$$

If we consider a finite time interval, say $-T < t < T$, then $\alpha_n(t)$ must be square-integrable for any finite object, so it is a vector in $\mathbb{L}_2(-T, T)$, and the coefficients can be expanded in temporal basis functions as

$$\alpha_n(t) = \sum_{j=1}^{\infty} \alpha_{nj}\, \phi_j(t) . \tag{7.17}$$

Combining (7.16) and (7.17), we have

$$f(\mathbf{r}, t) = \sum_{n=1}^{\infty} \sum_{j=1}^{\infty} \alpha_{nj}\, \phi_j(t)\, \psi_n(\mathbf{r}) . \tag{7.18}$$

Thus, under the stated square-integrability conditions, the spatio-temporal function can be expanded in terms of separable[4] basis functions, each a product of a function of time and a function of position.

7.1.3 Objects and images as finite-dimensional vectors

For numerical computations we must represent objects and images as finite sets of numbers. There is a key difference between objects and images in this respect. Objects are actually functions of continuous variables, so a finite discrete representation is necessarily an approximation. Digital images, on the other hand, are

[4]This usage of the term *separable* is distinct from the usage in describing Hilbert spaces; see Sec. 1.1.5.

inherently discrete sets of numbers. Thus a finite object vector *approximates* reality, whereas a finite image vector *is* reality, at least for digital images.

Since finite representations of images are virtually automatic in a digital setting and involve no approximation, we concentrate in this section on finite representations of objects.

Truncation of exact expansions Consider an object $f(\mathbf{r})$, a function of the qD vector $\mathbf{r}$, and assume that it is square-integrable over the support region $\mathbf{S}_f$ in $\mathbb{R}^q$. Then, as in Sec. 7.1.2, $f(\mathbf{r})$ can be regarded as a vector in the Hilbert space $\mathbb{L}_2(\mathbf{S}_f)$, and it can be represented exactly by an infinite series like (7.8).

An approximate representation can be obtained by retaining a finite subset of the terms in this expansion and deleting all others. Often the index n can be chosen so that it ranges monotonically over the subset of terms we want to retain, so the approximation is obtained simply by truncating the series:

$$f_t(\mathbf{r}) = \sum_{n=1}^{N} \alpha_n \, \psi_n(\mathbf{r}) \,, \tag{7.19}$$

where the subscript t stands for truncated and N is the number of terms retained. For example, in a truncated Fourier series, N specifies the highest frequency retained, and in an SVD expansion, N specifies the smallest singular value.

Representation space We can also describe the process of truncating the series in terms of a projection operator $\boldsymbol{\mathcal{P}}_N$, defined such that

$$f_t(\mathbf{r}) = \boldsymbol{\mathcal{P}}_N\{f(\mathbf{r})\} \,. \tag{7.20}$$

As discussed in Sec. 1.3.6, $\boldsymbol{\mathcal{P}}_N$ is idempotent ($\boldsymbol{\mathcal{P}}_N^2 = \boldsymbol{\mathcal{P}}_N$). It projects $f(\mathbf{r})$ from $\mathbb{L}_2(\mathbf{S}_f)$ onto the subspace spanned by $\{\psi_n, n = 1, ..., N\}$. We shall refer to that space as *representation space*. Since the set of all $\psi_n(\mathbf{r})$, $n = 1, ..., \infty$, spans the object space, we can say that representation space approaches object space as $N \to \infty$.

Representation space is a reproducing-kernel Hilbert space (see Sec. 1.8). Any function in the space satisfies

$$f_t(\mathbf{r}) = \int_{\mathbf{S}_f} d^q r' \, p_N(\mathbf{r}, \mathbf{r}') \, f_t(\mathbf{r}') \,, \tag{7.21}$$

where $p_N(\mathbf{r}, \mathbf{r}')$ is the reproducing kernel, given by

$$p_N(\mathbf{r}, \mathbf{r}') = \sum_{n=1}^{N} \psi_n(\mathbf{r}) \, \psi_n^*(\mathbf{r}') \,. \tag{7.22}$$

Since $p_N(\mathbf{r}, \mathbf{r}')$ is also the kernel for the projection operator $\boldsymbol{\mathcal{P}}_N$, (7.21) says simply that once an infinite expansion is truncated, it does not matter if it is truncated again (with the same N).

Truncation as continuous-to-discrete mapping Since $f_t(\mathbf{r})$ is fully described by the set of coefficients $\{\alpha_n, n = 1, ..., N\}$, another way to think about the truncation process is that it is a continuous-to-discrete mapping (see Sec. 1.2.4) from $\mathbb{L}_2(\mathbf{S}_f)$ to a finite-dimensional Euclidean space $\mathbb{E}^N$, where N is the total number of coefficients in the set. Let $\boldsymbol{\alpha}$ denote the vector in $\mathbb{E}^N$ with components $\{\alpha_n\}$. Then we can define a *discretization operator* $\mathcal{D}_\psi$ such that

$$\boldsymbol{\alpha} = \mathcal{D}_\psi\{f(\mathbf{r})\}\,, \tag{7.23}$$

where the components of $\boldsymbol{\alpha}$ are given by (7.9). We shall need several discretization operators in this chapter, so we distinguish them by subscripts indicating the basis functions used.

There is a simple relation between $\mathcal{P}_N$ and $\mathcal{D}_\psi$ if the finite set $\{\psi_n(\mathbf{r}),\ n = 1, ..., N\}$ is orthonormal. From (1.45), we see that[5]

$$\left[\mathcal{D}_\psi^\dagger \boldsymbol{\alpha}\right](\mathbf{r}) = \sum_{n=1}^{N} \alpha_n\, \psi_n(\mathbf{r})\,. \tag{7.24}$$

Hence,

$$\left[\mathcal{D}_\psi^\dagger \mathcal{D}_\psi\, f(\mathbf{r})\right](\mathbf{r}) = f_t(\mathbf{r})\,, \tag{7.25}$$

or

$$\mathcal{D}_\psi^\dagger \mathcal{D}_\psi = \mathcal{P}_N\,. \tag{7.26}$$

Thus, choice of the basis functions $\{\psi_n(\mathbf{r})\}$ and a value of N fully determines the representation space in which $f_t(\mathbf{r})$ lies. Specification of the ND vector of coefficients is equivalent to specifying the components of $f(\mathbf{r})$ in representation space. The space $\mathbb{E}^N$ in which $\boldsymbol{\alpha}$ lies is not identical to representation space, but it is isomorphic to it; each vector $\boldsymbol{\alpha}$ in $\mathbb{E}^N$ is uniquely associated with a vector $\mathbf{f}_t$ in representation space and hence with a function $f_t(\mathbf{r})$. When we need to distinguish $\mathbb{E}^N$ from representation space, we shall refer to the former as *coefficient space*. These spaces are illustrated schematically in Fig. 7.2.

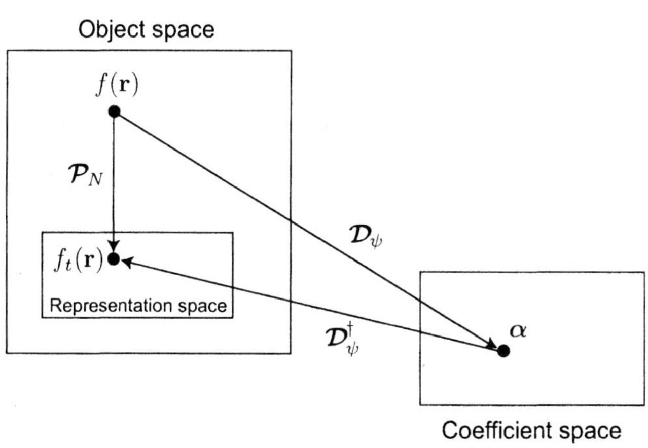

Fig. 7.2 The relationships among object space, representation space (here shown as a subspace of object space) and coefficient space.

[5]The lack of a complex conjugate in (7.24) may be puzzling when that equation is compared with (1.45). In defining $\mathcal{D}_\psi$ to agree with (7.9), we included the conjugate in the forward operator, so we do not need it in the adjoint.

Approximate series expansions Rather than obtaining approximate series representations of $f(\mathbf{r})$ by truncating exact ones, we can also invent the approximations *de novo*. The generic form of a linear approximation to $f(\mathbf{r})$ is

$$f_a(\mathbf{r}) = \sum_{n=1}^{N} \theta_n \, \phi_n(\mathbf{r}) \,, \tag{7.27}$$

where the subscript a stands for approximate. Though superficially similar, (7.27) differs from (7.19) in several important respects:

(a) The functions $\{\phi_n(\mathbf{r})\}$ are not necessarily normalized;

(b) The functions $\{\phi_n(\mathbf{r})\}$ are not necessarily orthogonal;

(c) The coefficients $\{\theta_n\}$ are not necessarily computed as scalar products between $f(\mathbf{r})$ and $\phi_n(\mathbf{r})$;

(d) The function $f_a(\mathbf{r})$ does not necessarily lie in $\mathbb{L}_2(\mathbf{S}_f)$.

Since the functions $\{\phi_n(\mathbf{r})\}$ do not form a basis for the object Hilbert space, we refrain from calling them basis functions (though this term is common in the literature). Instead we call them *expansion functions*. These functions do, however, span some ND space that we can still call representation space; now representation space is simply defined as the space of all functions that can be expressed as linear combinations of $\{\phi_n(\mathbf{r}), n = 1, ..., N\}$. If the expansion functions are chosen to be orthonormal, they form an orthonormal basis for representation space. Moreover, if the expansion functions themselves lie in object space, then representation space is a subspace of object space.

Example—pixels We can illustrate some of the subtleties of (7.27) by considering the simple and familiar *pixel expansion* of a 2D function $f(\mathbf{r})$, where $\mathbf{r} = (x, y)$. We consider a regular lattice of points specified by the multi-index $\mathbf{n}$, a 2D vector with integer components (n_x, n_y). Then a lattice point is denoted $\mathbf{r_n} = (x_\mathbf{n}, y_\mathbf{n})$, where

$$x_\mathbf{n} = n_x \epsilon \,, \qquad y_\mathbf{n} = n_y \epsilon \,, \tag{7.28}$$

and ϵ is the spacing between lattice points. With this notation, we can write the $\mathbf{n}^{th}$ expansion function as

$$\phi_\mathbf{n}(\mathbf{r}) = \text{pix}_\mathbf{n}(\mathbf{r}) \equiv \text{rect}\left(\frac{x - x_\mathbf{n}}{\epsilon}\right) \text{rect}\left(\frac{y - y_\mathbf{n}}{\epsilon}\right) . \tag{7.29}$$

This expansion function is not normalized (though it easily could be), but the set is orthogonal:

$$\int_\infty d^2r \; \text{pix}_\mathbf{n}(\mathbf{r}) \, \text{pix}_\mathbf{m}(\mathbf{r}) = \epsilon^2 \, \delta_\mathbf{nm} \,, \tag{7.30}$$

where the Kronecker delta with vector indices is defined so that $\delta_\mathbf{nm} = 1$ if $n_x = m_x$ and $n_y = m_y$, and it is zero otherwise.

Choosing a set of expansion functions does not finish the job of constructing the approximation $f_a(\mathbf{r})$; we still have to choose the coefficients. One logical way to do so is by analogy to (7.9):

$$\theta_\mathbf{n} = \frac{1}{\epsilon^2} \int_\infty d^2r \; \phi_\mathbf{n}(\mathbf{r}) \, f(\mathbf{r}) \,. \tag{7.31}$$

This choice is, however, not mandated; we might also choose

$$\theta_{\mathbf{n}} = f(\mathbf{r}_{\mathbf{n}}) = \int_{\infty} d^2r\ \delta(\mathbf{r} - \mathbf{r}_{\mathbf{n}})\, f(\mathbf{r})\,. \tag{7.32}$$

Comparing the integral form of (7.32) to (7.31), we see that they can both be written as

$$\theta_{\mathbf{n}} = \int_{\infty} d^2r\ \chi^*_{\mathbf{n}}(\mathbf{r})\, f(\mathbf{r})\,, \tag{7.33}$$

where $\chi_{\mathbf{n}}(\mathbf{r}) = \epsilon^{-2}\,\mathrm{pix}_{\mathbf{n}}(\mathbf{r})$ for (7.31) and $\delta(\mathbf{r} - \mathbf{r}_{\mathbf{n}})$ for (7.32).

Other choices are possible as well. We might, for example, reverse the roles of $\phi_{\mathbf{n}}$ and $\chi_{\mathbf{n}}$, letting $\chi_{\mathbf{n}}(\mathbf{r}) = \mathrm{pix}_{\mathbf{n}}(\mathbf{r})$ and $\phi_{\mathbf{n}}(\mathbf{r}) = \delta(\mathbf{r} - \mathbf{r}_{\mathbf{n}})$, in which case our approximate representation would be

$$f_a(\mathbf{r}) = \sum_{n=1}^{N} \left[\frac{1}{\epsilon^2} \int_{\infty} d^2r'\, \mathrm{pix}_{\mathbf{n}}(\mathbf{r}')\, f(\mathbf{r}') \right] \delta(\mathbf{r} - \mathbf{r}_{\mathbf{n}})\,. \tag{7.34}$$

While this form might be unappealing because the expansion functions are not square-integrable, there is no *a priori* reason to rule it out. Once we begin approximating, we have great flexibility in how we do so.

General operator treatment To formalize what we have learned from this discussion, we restate it in a general operator notation. Construction of a finite linear approximation to a qD function $f(\mathbf{r})$ is a two-step process. First we use a discretization operator $\mathcal{D}_\chi$ to form a vector of coefficients $\boldsymbol{\theta}$ given by

$$\boldsymbol{\theta} = \mathcal{D}_\chi\{f(\mathbf{r})\}\,. \tag{7.35}$$

The operator is defined by a modest generalization of (7.33):

$$\theta_n = [\mathcal{D}_\chi\{f(\mathbf{r})\}]_n = \int_{\infty} d^qr\ \chi^*_n(\mathbf{r})\, f(\mathbf{r})\,, \qquad n = 1, ..., N\,, \tag{7.36}$$

where N is the total number of expansion functions used (and we have returned to a scalar index n). According to the Riesz representation theorem (see Sec. 1.2.1), (7.36) is the general form for a linear functional of $f(\mathbf{r})$. So long as we restrict attention to linear operators, the coefficients $\{\theta_n\}$ can always be written in the form of (7.36) for some choice of $\{\chi_n(\mathbf{r})\}$.

The second step is to use a (potentially different) discretization operator $\mathcal{D}_\phi$ to construct $f_a(\mathbf{r})$. By analogy to (7.24), we rewrite (7.27) as

$$f_a(\mathbf{r}) = \mathcal{D}^\dagger_\phi \boldsymbol{\theta} = \mathcal{D}^\dagger_\phi \mathcal{D}_\chi\{f(\mathbf{r})\} = \sum_{n=1}^{N} \theta_n\, \phi_n(\mathbf{r})\,. \tag{7.37}$$

Construction of a finite approximation to an object function thus requires choice of two sets of functions, $\{\phi_n(\mathbf{r})\}$ and $\{\chi_n(\mathbf{r})\}$, or equivalently, two discretization operators $\mathcal{D}_\phi$ and $\mathcal{D}_\chi$. In much of the literature, the form of one or both of these function sets is simply left unspecified, and the precise relation between an object function and its discrete representation is not spelled out. When it is spelled out, the most common assumption is that $\mathcal{D}_\phi = \mathcal{D}_\chi$, but we shall maintain the distinction for the sake of generality. Taking $\{\phi_n(\mathbf{r})\}$ and $\{\chi_n(\mathbf{r})\}$ to be identical orthonormal function sets means that the coefficients $\{\theta_n\}$ are given by scalar products of the expansion functions and the object. Using distinct function sets means that the coefficients are calculated some other way.

Representation space and projection operators If the expansion functions are square-integrable and have the same support as the objects, then all functions in representation space are also in object space, and hence representation space is a subspace of object space. In that case we can define a projection operator $\mathcal{P}_{rep}$ that projects a general object vector onto representation space as [*cf.* (1.166)]

$$\mathcal{P}_{rep} = \mathcal{D}_\phi^+ \mathcal{D}_\phi \,, \tag{7.38}$$

where $\mathcal{D}_\phi^+$ is the Moore-Penrose pseudoinverse of $\mathcal{D}_\phi$.

We can go a step further if the expansion functions are orthonormal; in that case, we know from Sec. 1.3.6 [*cf.* (1.51)] that

$$\mathcal{P}_{rep} f(\mathbf{r}) = \sum_{n=1}^{N} (\phi_n, \mathbf{f})\, \phi_n(\mathbf{r}) \,. \tag{7.39}$$

In operator form, we can write this relation as

$$\mathcal{P}_{rep} = \mathcal{D}_\phi^\dagger \mathcal{D}_\phi = \sum_{n=1}^{N} \phi_n \phi_n^\dagger \,, \tag{7.40}$$

where the second form makes use of the outer-product notation of (1.57).

Comparison of (7.38) and (7.40) shows that $\mathcal{D}_\phi^\dagger = \mathcal{D}_\phi^+$ if $\{\phi_n(\mathbf{r})\}$ is an orthonormal set. This statement can be proved independently from either the Penrose equations (see Sec. 1.6.1) or the singular-value decomposition of $\mathcal{D}_\phi$ and the results in Sec. 1.6.2.

To summarize, the approximate representation $f_a(\mathbf{r})$, defined generally as $\mathcal{D}_\phi^\dagger \mathcal{D}_\chi f(\mathbf{r})$, is given by $\mathcal{P}_{rep} f(\mathbf{r})$ if $\{\phi_n(\mathbf{r})\}$ and $\{\chi_n(\mathbf{r})\}$ are identical orthonormal sets of square-integrable functions. It also follows from the discussion in Sec. 1.8.2 that representation space is a reproducing-kernel Hilbert space in this case.

7.1.4 Representation accuracy

When constructing approximate representations of object functions, two important questions arise: How much error do we make in any particular representation? How do we choose the best representation? The answers to these questions depend on what we want to do with the representation.

There are several possible applications for finite object representations. One is to simulate images so that we can understand how an imaging system would perform without actually building it. This is the *forward problem*, where we are given an object and want to compute the image it produces. Solving the forward problem on a computer requires a finite object representation that leads to accurate data vectors.

Another application is the *inverse problem*, where we are given a noisy image or data vector and want to determine some information about the object that produced it. If we represent the object by a finite series, the inverse problem is to determine (or estimate) the coefficients in the series. The accuracy of the final estimate depends both on how well we can determine the coefficients from noisy data and how well the series would describe the actual object if the coefficients were known perfectly. We therefore distinguish *representation accuracy* from es-

timation accuracy; the former is discussed in this section, but discussion of the latter is postponed until we take up estimation theory in Chap. 13.

One application where representation accuracy can be assessed, unencumbered by questions of estimation accuracy, is image display. Suppose we have stored a digital representation of an object $f(\mathbf{r})$ as a set of digital values $\{\theta_n\}$ and later want to display the digital data as a luminance pattern on a cathode-ray tube. To do so, we have to convert the digital values to a function like $f_a(\mathbf{r})$, and we are naturally interested in the degree of agreement between the two functions $f(\mathbf{r})$ and $f_a(\mathbf{r})$.

Another way to answer questions about representation accuracy is in terms of objective measures of image quality. We shall argue in Chap. 14 that image quality cannot be defined in the abstract. It must be related to the particular purpose for which the image was acquired, which we refer to as the *task*. It must also be related to just how the task is performed, which we call the *observer*. In Chap. 14, we develop figures of merit for imaging systems for specific tasks and observers and relate these figures of merit to properties of the system and any data-processing algorithms that might be used. In addition, however, the figures of merit are sensitive to the mathematical representation chosen, which is the subject of this section.

One specific task we shall consider in Chap. 14 is the estimation of certain moments of the object, such as its integral over a specified region of interest. Another task is pattern recognition, which begins with estimation of functionals called features. In estimating either moments or features, a prerequisite is an object representation that accurately describes the quantities being estimated.

Object error If our goal is to represent an object as accurately as possible, we can examine the norm of the error between the actual object and any approximate representation of it. This error is defined by

$$\delta f(\mathbf{r}) \equiv f(\mathbf{r}) - f_a(\mathbf{r}) . \tag{7.41}$$

From (7.37) we can also write

$$\delta f(\mathbf{r}) = f(\mathbf{r}) - \mathcal{D}_\phi^\dagger \mathcal{D}_\chi \, f(\mathbf{r}) = \left[\mathbf{I} - \mathcal{D}_\phi^\dagger \mathcal{D}_\chi\right] f(\mathbf{r}) , \tag{7.42}$$

where $\mathbf{I}$ is the identity operator (here simply multiplication by unity).

The error norm is thus given by

$$||\delta f(\mathbf{r})|| = \left|\left| \left[\mathbf{I} - \mathcal{D}_\phi^\dagger \mathcal{D}_\chi\right] f(\mathbf{r}) \right|\right| . \tag{7.43}$$

This norm depends on the choice of the expansion functions and on the particular object function $f(\mathbf{r})$.

It is necessary that $f_a(\mathbf{r})$ be square-integrable for the error norm to be finite. For example, we cannot use a representation like (7.34) where $\phi_n(\mathbf{r}) = \delta(\mathbf{r} - \mathbf{r}_n)$ since that would make $||\delta f(\mathbf{r})||$ infinite. In what follows, we shall therefore usually assume that the functions $\{\phi_n(\mathbf{r})\}$ are square-integrable and hence $f_a(\mathbf{r})$ lies in object space. Representation space will be a subspace of object space unless otherwise stated.

Once we have chosen the set $\{\phi_n(\mathbf{r})\}$, and hence the representation space, the next step is to minimize $||\delta f(\mathbf{r})||$ through choice of the expansion coefficients $\{\theta_n\}$, or equivalently through choice of the set $\{\chi_n(\mathbf{r})\}$. This minimization is essentially least-squares fitting of $f_a(\mathbf{r})$ to $f(\mathbf{r})$. As discussed in Sec. 1.7 and illustrated graphically in Fig. 7.3, the optimal fitting is accomplished by making $f_a(\mathbf{r}) = \mathcal{P}_{rep} \, f(\mathbf{r})$, where $\mathcal{P}_{rep}$ is the projector onto representation space. Since $f_a(\mathbf{r})$ can also be expressed as $\mathcal{D}_\phi^\dagger \mathcal{D}_\chi f(\mathbf{r})$, the optimum (least-squares) choice of $\{\chi_n(\mathbf{r})\}$ implies the operator relation, $\mathcal{D}_\phi^\dagger \mathcal{D}_\chi = \mathcal{P}_{rep} = \mathcal{D}_\phi^+ \mathcal{D}_\phi$. Hence,

$$\delta f(\mathbf{r}) = [\mathbf{I} - \mathcal{P}_{rep}]\, f(\mathbf{r}) = \left[\mathbf{I} - \mathcal{D}_\phi^+ \mathcal{D}_\phi\right] f(\mathbf{r})\,. \tag{7.44}$$

From the idempotency ($\mathcal{P}_{rep}^2 = \mathcal{P}_{rep}$) and Hermiticity ($\mathcal{P}_{rep}^\dagger = \mathcal{P}_{rep}$) of the projector, the minimum achievable error norm is found to be

$$||\delta f(\mathbf{r})||^2 = ||f(\mathbf{r})||^2 - ||\mathcal{P}_{rep}\, f(\mathbf{r})||^2\,, \tag{7.45}$$

where the first norm on the right refers to object space, say $\mathbb{L}_2(\mathbf{R}^q)$, and the second refers to representation space.

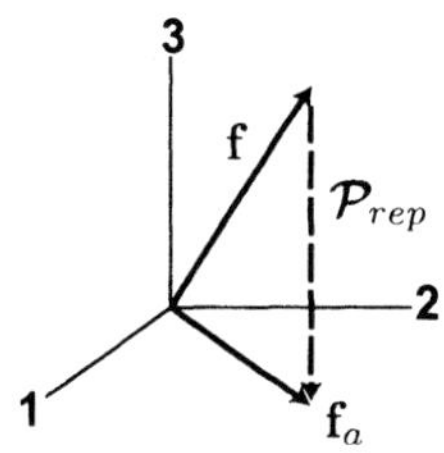

Fig. 7.3 Graphical demonstration that $||\delta \mathbf{f}||$ is minimized by choosing $\mathbf{f}_a = \mathcal{P}_{rep}\mathbf{f}$. Here $\mathbf{f}$ is shown as a vector in a 3D Hilbert space, while its approximate representation $\mathbf{f}_a$ lies in a 2D subspace called representation space, illustrated as the 1-2 plane. Thus $\mathcal{P}_{rep}\mathbf{f}$ is the projection of $\mathbf{f}$ onto the 1-2 plane. Among all vectors $\mathbf{f}_a$ in this plane, the one that minimizes $||\mathbf{f} - \mathbf{f}_a||$ is $\mathbf{f}_a = \mathcal{P}_{rep}\mathbf{f}$.

If $\{\phi_n(\mathbf{r})\}$ is an orthonormal set, then, as shown in Sec. 7.1.3, $\mathcal{D}_\phi^+ = \mathcal{D}_\phi^\dagger$ and

$$\mathcal{P}_{rep} f(\mathbf{r}) = \sum_{n=1}^{N} \theta_n\, \phi_n(\mathbf{r})\,, \qquad \theta_n = (\phi_n(\mathbf{r}), f(\mathbf{r}))\,. \tag{7.46}$$

In this case the norm in representation space is given by

$$\begin{aligned} ||\mathcal{P}_{rep}\, f(\mathbf{r})||^2 &= \int_\infty d^q r \left| \sum_{n=1}^{N} \theta_n\, \phi_n(\mathbf{r}) \right|^2 \\ &= \sum_{n=1}^{N} \theta_n \sum_{n'=1}^{N} \theta_{n'}^* \int_\infty d^q r\; \phi_{n'}^*(\mathbf{r})\, \phi_n(\mathbf{r}) = \sum_{n=1}^{N} |\theta_n|^2\,, \end{aligned} \tag{7.47}$$

and hence

$$||\delta f(\mathbf{r})||^2 = \int_\infty d^q r\; |f(\mathbf{r})|^2 - \sum_{n=1}^{N} |\theta_n|^2 = ||f(\mathbf{r})||^2 - ||\boldsymbol{\theta}||^2\,. \tag{7.48}$$

Thus, so long as we construct a representation by using identical orthonormal sets for $\{\chi_n(\mathbf{r})\}$ and $\{\phi_n(\mathbf{r})\}$, the error norm is simply the squared norm of the function minus the squared norm of its coefficient vector.

If $\{\phi_n(\mathbf{r})\}$ were a complete set on $\mathbb{L}_2(\mathbb{R}^q)$, this error norm would be immediately zero by the Parseval relation (4.6). Similarly, if $f(\mathbf{r})$ were entirely in the representation space, then the error norm would again be zero. Neither of these conditions is likely to prevail in practice since they imply that $f(\mathbf{r})$ can be represented exactly by a finite-dimensional vector.

If the approximate representation is obtained by truncating an exact infinite representation, as in (7.19), then we have

$$||\delta f(\mathbf{r})||^2 = ||f(\mathbf{r})||^2 - \sum_{n=1}^{N} |\alpha_n|^2 = \sum_{n=N+1}^{\infty} |\alpha_n|^2 \tag{7.49}$$

and

$$\lim_{N\to\infty} ||\delta f(\mathbf{r})||^2 = 0\,. \tag{7.50}$$

How rapidly this limit is approached depends on the specific basis functions and the object. For a truncated Fourier series and a differentiable object, for example, the coefficients $F_{\mathbf{k}}$ fall off rapidly as any component of the multi-index $\mathbf{k}$ tends to infinity [see (3.49)].

Ensembles of objects We are rarely interested in the error norm just for one object; we want a representation that performs well for many objects. For this reason it is useful to consider a statistical ensemble of objects and average the error norms over that ensemble. In this view, each object is a sample function of a random process, a topic to be discussed in Chap. 8. For completeness in this section, however, we give expressions for the mean-square representation error (MSRE), averaged over the ensemble of objects.

The MSRE is defined by

$$\text{MSRE} = \langle ||\delta f(\mathbf{r})||^2 \rangle\,, \tag{7.51}$$

where the angle brackets denote an average over the ensemble of objects. For representations derived from a single orthonormal function set, (7.48) holds, and the MSRE is given by

$$\text{MSRE} = \langle ||f(\mathbf{r})||^2 \rangle - \langle ||\boldsymbol{\theta}||^2 \rangle\,. \tag{7.52}$$

Since θ_n is obtained optimally as a scalar product with the expansion functions, $\langle ||\boldsymbol{\theta}||^2 \rangle$ is given by

$$\begin{aligned}\langle ||\boldsymbol{\theta}||^2 \rangle &= \sum_{n=1}^{N} \int_\infty d^q r \int_\infty d^q r'\, \phi_n^*(\mathbf{r}) \langle f^*(\mathbf{r})\, f(\mathbf{r}')\rangle\, \phi_n(\mathbf{r}') \\ &= \sum_{n=1}^{N} \int_\infty d^q r \int_\infty d^q r'\, \phi_n^*(\mathbf{r})\, R(\mathbf{r},\mathbf{r}')\, \phi_n(\mathbf{r}')\,, \end{aligned} \tag{7.53}$$

where $R(\mathbf{r},\mathbf{r}') = \langle f(\mathbf{r})\, f^*(\mathbf{r}')\rangle$ is the autocorrelation function of the random process (see Sec. 8.2). The autocorrelation function can be regarded as the kernel of an integral operator $\boldsymbol{\mathcal{R}}$ called the autocorrelation operator, and we can rewrite (7.53) as a Hermitian form,

$$\langle ||\boldsymbol{\theta}||^2 \rangle = \sum_{n=1}^{N} \boldsymbol{\phi}_n^\dagger \boldsymbol{\mathcal{R}} \boldsymbol{\phi}_n\,. \tag{7.54}$$

To minimize the MSRE, we must maximize the sum in (7.54). The first term, $\boldsymbol{\phi}_1^\dagger \boldsymbol{\mathcal{R}} \boldsymbol{\phi}_1$, is maximized (among vectors with unit norm) by choosing $\boldsymbol{\phi}_1$ to be the eigenvector of $\boldsymbol{\mathcal{R}}$ with the largest eigenvalue. Then the second term, $\boldsymbol{\phi}_2^\dagger \boldsymbol{\mathcal{R}} \boldsymbol{\phi}_2$, is maximized (among vectors with unit norm that are orthogonal to $\boldsymbol{\phi}_1$) by choosing $\boldsymbol{\phi}_1$ to be the eigenvector of $\boldsymbol{\mathcal{R}}$ with the second largest eigenvalue. Continuing in this way, we see that the sum in (7.54) is maximized by choosing the orthonormal vectors $\{\boldsymbol{\phi}_n, n = 1, ..., N\}$ as the eigenvectors of $\boldsymbol{\mathcal{R}}$ corresponding to the N largest eigenvalues. This choice of expansion functions is called the *Karhunen-Loève* (KL) expansion, which we now see is optimal in terms of MSRE when a fixed number of terms is used in the expansion.

With the KL expansion functions, (7.54) simplifies to

$$\langle ||\boldsymbol{\theta}||^2 \rangle = \sum_{n=1}^{N} \lambda_n \boldsymbol{\phi}_n^\dagger \boldsymbol{\phi}_n = \sum_{n=1}^{N} \lambda_n \,, \tag{7.55}$$

where λ_n is the nth eigenvalue of $\boldsymbol{\mathcal{R}}$, and the eigenvalues are all ordered by decreasing value. It will be left as an exercise to show that [*cf.* (7.49)]

$$\text{MSRE} = \sum_{n=N+1}^{\infty} \lambda_n \,. \tag{7.56}$$

In summary, if the objects are sample functions drawn from some ensemble, the KL expansion minimizes the MSRE (for a fixed number of terms), and all we need to know to compute this minimum MSRE is the set of eigenvalues of the covariance operator.

Moment errors In some imaging situations the task is to determine certain linear functionals of the object. As a simple example, in nuclear medicine we might be interested in the total amount of radioactive tracer inside some specified volume of the object. If we knew the actual object $f(\mathbf{r})$, we could simply perform the integral and get the desired number. In reality, we have available some noisy and imperfect data set, so we can only estimate the value. One common approach to this estimation is to adopt a finite representation like (7.19) or (7.27), estimate the coefficients, and then integrate the series term by term. Of course, if the series does not accurately represent the object in the first place, its integral will not accurately reflect the desired integral even if the coefficients are accurately estimated.

To formalize this idea, suppose the linear functional of interest is specified by a *template function* $t(\mathbf{r})$. In the notation of Chap. 2, this functional would be denoted as $\Phi_t\{\mathbf{f}\}$, but we are using Φ for other things in this chapter, so we shall call the functional $\tau(\mathbf{f})$, or just τ for short, and define it by

$$\tau = \int_{\mathbf{S}_f} d^q r \; t(\mathbf{r}) \, f(\mathbf{r}) \,. \tag{7.57}$$

In the nuclear-imaging example mentioned above, $t(\mathbf{r})$ is unity in some region of interest and zero outside, and τ is the integral of the object over the region of interest. More generally, τ is referred to as a moment[6] of the object.

[6]This usage of the word *moment* is more general than, but consistent with, the usage in statistics and mechanics. For example, if $\mathbf{r}$ is a random vector and $f(\mathbf{r})$ is its probability density function

If the object is approximated by the finite series (7.27), then τ is approximated by

$$\tau_a = \int_{\mathbf{S}_f} d^q r\, t(\mathbf{r})\, f_a(\mathbf{r}) = \sum_{n=1}^{N} \theta_n \int_{\mathbf{S}_f} d^q r\, t(\mathbf{r})\, \phi_n(\mathbf{r}) \,. \tag{7.58}$$

If we have obtained a set of estimates $\{\hat{\theta}_n\}$ of the coefficients, then one (not necessarily optimal) way of estimating τ is by

$$\hat{\tau} = \sum_{n=1}^{N} \hat{\theta}_n \int_{\mathbf{S}_f} d^q r\, t(\mathbf{r})\, \phi_n(\mathbf{r}) \,. \tag{7.59}$$

The estimation error is $\tau - \hat{\tau}$, but in this section we are concerned with the *moment error* $\delta\tau$, given by

$$\delta\tau \equiv \tau - \tau_a = \int_{\mathbf{S}_f} d^q r\, t(\mathbf{r})\, \delta f(\mathbf{r}) = \int_{\mathbf{S}_f} d^q r\, t(\mathbf{r}) \big[\mathbf{I} - \mathcal{D}_\phi^\dagger \mathcal{D}_\chi\big] f(\mathbf{r}) \,, \tag{7.60}$$

where we have used (7.41) in the last step. We can go a step further if $\mathcal{D}_\phi = \mathcal{D}_\chi$ and $\{\phi_n(\mathbf{r})\}$ is an orthonormal set. In this case $\mathcal{D}_\phi^\dagger \mathcal{D}_\chi$ is the projector onto representation space, and we have

$$\delta\tau = \int_{\mathbf{S}_f} d^q r\, t(\mathbf{r})\, [\mathbf{I} - \mathcal{P}_{rep}]\, f(\mathbf{r}) = \int_{\mathbf{S}_f} d^q r\, f(\mathbf{r})\, [\mathbf{I} - \mathcal{P}_{rep}]\, t(\mathbf{r}) \,, \tag{7.61}$$

where the second step follows since the projector is Hermitian. Thus τ_a might be a good approximation to τ even if $f_a(\mathbf{r})$ is not a good approximation to $f(\mathbf{r})$, so long as $\mathcal{P}_{rep} t(\mathbf{r})$ is a good approximation to $t(\mathbf{r})$. It is the template rather than the object that has to be accurately approximated if we use the moment error as a figure of merit for the representation. Another way to state this conclusion is that components of $f(\mathbf{r})$ that lie in the orthogonal complement of representation space can greatly influence $\delta f(\mathbf{r})$, but they do not affect $\delta\tau$ at all if $t(\mathbf{r})$ lies entirely in representation space.

An immediate consequence of (7.61) is that $\delta\tau = 0$ if $t(\mathbf{r})$ is any linear combination of the representation functions $\{\phi_n(\mathbf{r})\}$. For example, if we choose to represent an object in pixels, then any moment obtained by integrating over regions defined by an integral number of pixels is exactly represented.

7.1.5 Uniform translates

In many practical situations, the functions $\phi_n(\mathbf{r})$ are chosen as uniform translates of a single function $\phi(\mathbf{r})$, such as a pixel function. In a slightly modified notation, we can write

$$\phi_{\mathbf{n}}(\mathbf{r}) = \phi(\mathbf{r} - \mathbf{r_n}) \,, \tag{7.62}$$

where $\mathbf{n}$ is a multi-index as in Sec. 7.1.3, and the points $\mathbf{r_n}$ form a regular lattice. For pixels in 2D for example, $\mathbf{r_n} = (n_x\epsilon, n_y\epsilon)$ with n_x and n_y being integers.

More generally (see Sec. 3.4.6), we can construct the regular lattice by

$$\mathbf{r_n} = \mathbf{Pn} \,, \tag{7.63}$$

(PDF), then the second moment of the component x can be obtained from (7.57) by letting $t(\mathbf{r}) = x^2$.

where $\mathbf{P}$ is a real, nonsingular $q \times q$ matrix. For the 2D pixel example, $\mathbf{P} = \epsilon\mathbf{I}$, where $\mathbf{I}$ is the 2×2 identity matrix. In q dimensions, the condition $\mathbf{P} = \epsilon\mathbf{I}$ generates a *generalized cubic lattice* where the sides of a unit cell are mutually orthogonal and of equal length.

In this section we shall analyze expansions based on functions of this form, first under the assumption that $\{\phi(\mathbf{r} - \mathbf{r_n})\}$ is an orthonormal set and then more generally. For simplicity, we consider initially a generalized cubic lattice where $\mathbf{P} = \epsilon\mathbf{I}$, but this assumption will also be lifted shortly.

Orthonormal translates Consider an approximation to the object $f(\mathbf{r})$ in the form,

$$f_a(\mathbf{r}) = \sum_{\mathbf{n}=-\infty}^{\infty} \theta_{\mathbf{n}}\, \phi(\mathbf{r} - \mathbf{r_n})\,. \tag{7.64}$$

If $\{\phi(\mathbf{r} - \mathbf{r}_n)\}$ is an orthonormal set, the coefficients are given optimally (in the sense of minimum $||\delta f(\mathbf{r})||$) by

$$\theta_{\mathbf{n}} = \int_\infty d^q r\; f(\mathbf{r})\, \phi^*(\mathbf{r} - \mathbf{r_n})\,. \tag{7.65}$$

If the object and the expansion function $\phi(\mathbf{r})$ both have finite support, then there are a finite number of terms in (7.64) even though the sums on all components of the multi-index run over $(-\infty, \infty)$; the coefficient $\theta_{\mathbf{n}} = 0$ when there is no overlap of $f(\mathbf{r})$ and $\phi^*(\mathbf{r} - \mathbf{r_n})$ in the integrand of (7.65).

With this way of constructing $f_a(\mathbf{r})$, the error norm $||\delta f(\mathbf{r})||^2$ is given, from (7.48), by

$$||\delta f(\mathbf{r})||^2 = \int_\infty d^q r\; |f(\mathbf{r})|^2 - \sum_{\mathbf{n}=-\infty}^{\infty} \left| \int_\infty d^q r\; f(\mathbf{r})\, \phi^*(\mathbf{r} - \mathbf{r_n}) \right|^2 . \tag{7.66}$$

With Parseval's relations (3.225) and (3.226) and the shift theorem (3.238), we have

$$||\delta f(\mathbf{r})||^2 = \int_\infty d^q \rho\; |F(\boldsymbol{\rho})|^2$$

$$- \sum_{\mathbf{n}=-\infty}^{\infty} \int_\infty d^q \rho \int_\infty d^q \rho'\; F(\boldsymbol{\rho})\, F^*(\boldsymbol{\rho}')\, \Phi^*(\boldsymbol{\rho})\, \Phi(\boldsymbol{\rho}') \exp[-2\pi i (\boldsymbol{\rho} - \boldsymbol{\rho}') \cdot \mathbf{r_n}]\,, \tag{7.67}$$

where, as usual, capital letters denote Fourier transforms. Interchanging sum and integral in (7.67) yields

$$||\delta f(\mathbf{r})||^2 = \int_\infty d^q \rho\; |F(\boldsymbol{\rho})|^2$$

$$- \int_\infty d^q \rho \int_\infty d^q \rho'\; F(\boldsymbol{\rho})\, F^*(\boldsymbol{\rho}')\, \Phi^*(\boldsymbol{\rho})\, \Phi(\boldsymbol{\rho}') \sum_{\mathbf{n}=-\infty}^{\infty} \exp[-2\pi i (\boldsymbol{\rho} - \boldsymbol{\rho}') \cdot \mathbf{r_n}$$

$$= \int_\infty d^q \rho\; |F(\boldsymbol{\rho})|^2 - \int_\infty d^q \rho \int_\infty d^q \rho'\; F(\boldsymbol{\rho})\, F^*(\boldsymbol{\rho}')\, \Phi^*(\boldsymbol{\rho})\, \Phi(\boldsymbol{\rho}')\, \mathrm{comb}[\epsilon(\boldsymbol{\rho} - \boldsymbol{\rho}')]\,, \tag{7.68}$$

where we have assumed that $\mathbf{P} = \epsilon\mathbf{I}$. With this assumption, the multidimensional comb function (see Sec. 3.4.6) is given by

$$\mathrm{comb}[\epsilon(\boldsymbol{\rho} - \boldsymbol{\rho}')] = \sum_{\mathbf{n}=-\infty}^{\infty} \delta[\epsilon(\boldsymbol{\rho} - \boldsymbol{\rho}') - \mathbf{n}] = \frac{1}{\epsilon^q} \sum_{\mathbf{n}=-\infty}^{\infty} \delta\!\left(\boldsymbol{\rho} - \boldsymbol{\rho}' - \frac{\mathbf{n}}{\epsilon}\right), \tag{7.69}$$

so we have

$$||\delta f(\mathbf{r})||^2 = \int_\infty d^q\rho\, |F(\boldsymbol{\rho})|^2 - \epsilon^{-q} \sum_{\mathbf{n}=-\infty}^{\infty} \int_\infty d^q\rho\, F(\boldsymbol{\rho})\, F^*\Big(\boldsymbol{\rho} - \frac{\mathbf{n}}{\epsilon}\Big)\, \Phi^*(\boldsymbol{\rho})\, \Phi\Big(\boldsymbol{\rho} - \frac{\mathbf{n}}{\epsilon}\Big)\,. \tag{7.70}$$

The terms in the sum with $\mathbf{n} \neq 0$ result from aliasing. These terms can be neglected if the integrand is small at all $\boldsymbol{\rho}$ for nonzero $\mathbf{n}$, which is a valid assumption if *either* the object $f(\mathbf{r})$ or the expansion function $\phi(\mathbf{r})$ is smooth and nearly bandlimited. For this reason, several authors have advocated using smooth expansion functions; see, for example, Hanson and Wecksung (1985) and Lewitt (1990, 1992).

When aliasing can be neglected, the error norm is

$$||\delta f(\mathbf{r})||^2 = \int_\infty d^q\rho\, |F(\boldsymbol{\rho})|^2 \left[1 - \epsilon^{-q}|\Phi(\boldsymbol{\rho})|^2\right]. \tag{7.71}$$

If $f(\mathbf{r})$ is constant or very slowly varying, then $F(\boldsymbol{\rho})$ is small except near $\boldsymbol{\rho} = 0$. For such functions, aliasing can certainly be neglected, and the error norm is minimized when

$$|\Phi(0)| = \left|\int_\infty d^q r\, \phi(\mathbf{r})\right| = \epsilon^{q/2}\,. \tag{7.72}$$

This condition is satisfied by pixels if they are properly normalized (with a factor of $\epsilon^{-q/2}$) and the rect functions fit together without gaps or overlap.

Non-orthonormal translates We assumed above that the set $\{\phi(\mathbf{r} - \mathbf{r_n})\}$ was orthonormal, but this condition is often violated in practice. For example, it is violated if $\phi(\mathbf{r})$ is nonnegative and the translates are allowed to overlap. We now examine the error norm without the orthonormality assumption, and we also extend the treatment to more general lattices where $\mathbf{P} \neq \epsilon\mathbf{I}$.

Assume that a function $\phi(\mathbf{r})$ has been chosen so that the set $\{\phi(\mathbf{r} - \mathbf{r_n})\}$ can be generated. The only remaining freedom we have is in the choice of the expansion coefficients, or equivalently the functions $\{\chi_\mathbf{n}(\mathbf{r})\}$. We know from the discussion in Sec. 7.1.4 that the object error is minimized if we choose $\{\chi_\mathbf{n}(\mathbf{r})\}$ such that

$$\mathcal{D}_\phi^\dagger \mathcal{D}_\chi = \mathcal{D}_\phi^+ \mathcal{D}_\phi\,. \tag{7.73}$$

The operator $\mathcal{D}_\chi$, a linear mapping from a function of $\mathbf{r}$ to a discrete vector, is given by (7.36), with the functions $\{\chi_\mathbf{n}(\mathbf{r})\}$ yet to be determined. The operator $\mathcal{D}_\phi^\dagger \mathcal{D}_\chi$ is then given by [*cf.* (7.37)]

$$\left[\mathcal{D}_\phi^\dagger \mathcal{D}_\chi \mathbf{f}\right](\mathbf{r}) = \int_\infty d^q r'\, f(\mathbf{r}') \sum_\mathbf{n} \chi_\mathbf{n}^*(\mathbf{r}')\, \phi(\mathbf{r} - \mathbf{r_n})\,. \tag{7.74}$$

Since $\mathcal{D}_\phi^+$ is the Moore-Penrose pseudoinverse of $\mathcal{D}_\phi$, it must satisfy the Penrose equations, (1.130*a-d*). In particular, the fourth Penrose equation requires that $\mathcal{D}_\phi^+ \mathcal{D}_\phi$ be Hermitian, so (7.73) implies that $\mathcal{D}_\phi^\dagger \mathcal{D}_\chi = [\mathcal{D}_\phi^\dagger \mathcal{D}_\chi]^\dagger$, or

$$\sum_\mathbf{n} \chi_n(\mathbf{r})\, \phi^*(\mathbf{r}' - \mathbf{r_n}) = \sum_\mathbf{n} \chi_n^*(\mathbf{r}')\, \phi(\mathbf{r} - \mathbf{r_n})\,. \tag{7.75}$$

This condition holds if we choose

$$\chi_\mathbf{n}(\mathbf{r}) = \sum_\mathbf{k} B_{\mathbf{kn}} \phi(\mathbf{r} - \mathbf{r_k})\,, \tag{7.76}$$

with $B_{\mathbf{kn}}$ being an element (in multi-index notation) of a Hermitian matrix $\mathbf{B}$. In operator form, (7.76) is equivalent to

$$\mathcal{D}_\chi = \mathbf{B}\mathcal{D}_\phi \,. \tag{7.77}$$

If we multiply both sides of (7.73) by $\mathcal{D}_\phi$ and apply the first Penrose equation, (1.130a), we find

$$\mathcal{D}_\phi \mathcal{D}_\phi^\dagger \mathcal{D}_\chi = \mathcal{D}_\phi \,. \tag{7.78}$$

The operator $\mathcal{D}_\phi \mathcal{D}_\phi^\dagger$ is a Hermitian matrix, which we shall call $\mathbf{A}$, with elements given by the overlap integral,

$$A_{\mathbf{mn}} = \left[\mathcal{D}_\phi \mathcal{D}_\phi^\dagger\right]_{\mathbf{mn}} = \int_\infty d^q r \; \phi^*(\mathbf{r} - \mathbf{r_m})\, \phi(\mathbf{r} - \mathbf{r_n}) \,. \tag{7.79}$$

The indices $\mathbf{m}$ and $\mathbf{n}$ on $\mathbf{A_{mn}}$ can, in principle, extend to infinity in each of q directions, so $\mathbf{A}$ is an infinite matrix. In practice, however, $\mathcal{D}_\phi \mathcal{D}_\phi^\dagger$ will be applied to finite vectors, so infinite sums will not be required.

Combining (7.77) and (7.78), we find

$$\mathbf{AB}\mathcal{D}_\phi = \mathcal{D}_\phi \,. \tag{7.80}$$

Multiplying from the right by $\mathcal{D}_\phi^\dagger$ gives

$$\mathbf{ABA} = \mathbf{A} \,. \tag{7.81}$$

This is again the first Penrose equation, so $\mathbf{B}$ is a 1-inverse of $\mathbf{A}$ (see Sec. 1.6.1). Solving (7.81) for $\mathbf{B}$ is equivalent to finding $\mathcal{D}_\chi$ because of (7.77).

In practice, the function $\phi(\mathbf{r})$ will either have compact support or fall off rapidly with $\mathbf{r}$, so there will be very little overlap of the functions in the integrand of (7.79) unless $\mathbf{m}$ is near $\mathbf{n}$. With this assumption, the matrix $\mathbf{A}$ is diagonally dominant. Moreover, if $\mathbf{r_m} = \mathbf{Pm}$, then

$$A_{\mathbf{mn}} = \int_\infty d^q r \; \phi^*(\mathbf{r} - \mathbf{Pm})\, \phi(\mathbf{r} - \mathbf{Pn}) = \int_\infty d^q r' \; \phi^*[\mathbf{r}' - \mathbf{P}(\mathbf{m} - \mathbf{n})]\, \phi(\mathbf{r}') \,. \tag{7.82}$$

Thus $A_{\mathbf{mn}}$ is a function of $\mathbf{m} - \mathbf{n}$. In App. A we noted that a square matrix where each element is determined by the difference in the indices is called a Toeplitz matrix. We can extend the definition of a Toeplitz matrix to include the present case with vector indices, though it is more common in the literature to refer to matrices of this form as *block-Toeplitz*. See Sec. 7.4.4 or Andrews and Hunt (1977) for more discussion of block-Toeplitz matrices.

We can also argue from the discrete shift-invariance that $\mathbf{B}$ must be Toeplitz as well. With these considerations on the structure of $\mathbf{A}$ and $\mathbf{B}$, we can express $\mathbf{ABA}$ in component form as

$$[\mathbf{ABA}]_{\mathbf{mn}} = \sum_{\mathbf{k}=\mathbf{m}-\frac{1}{2}N}^{\mathbf{m}+\frac{1}{2}N-1} \; \sum_{\mathbf{k}'=\mathbf{n}-\frac{1}{2}N}^{\mathbf{n}+\frac{1}{2}N-1} A_{\mathbf{m}-\mathbf{k}} B_{\mathbf{k}-\mathbf{k}'} A_{\mathbf{k}'-\mathbf{n}} \,, \tag{7.83}$$

where the limits are specified component-wise ($e.g.$, $\mathbf{k} = \mathbf{m} - \frac{1}{2}N$ means that $k_i = m_i - \frac{1}{2}N$, $i = 1, ..., q$), and N is any sufficiently large number such that no significant nonzero elements of **A** or **B** are cut off by the limits. With this choice of N, (7.83) is essentially a discrete double convolution, and we do not need to worry about the modulo-N arithmetic usually involved in discrete convolutions (see Sec. 3.6.2). We know from (3.332) that the discrete Fourier transform (DFT) is a natural tool for solving problems involving discrete convolutions, and that is the approach we take here.

Though the DFT is usually defined on the interval $[0, N-1]$ [*cf.* (3.363)], a symmetric definition is more convenient here, so we represent $A_{\mathbf{mn}}$ and $B_{\mathbf{mn}}$ as

$$A_{\mathbf{mn}} = A_{\mathbf{m-n}} = \sum_{\mathbf{j}=-\frac{1}{2}N}^{\frac{1}{2}N-1} a_{\mathbf{j}} \exp[-2\pi i(\mathbf{m}-\mathbf{n}) \cdot \mathbf{j}/N] \, ; \tag{7.84}$$

$$B_{\mathbf{mn}} = B_{\mathbf{m-n}} = \sum_{\mathbf{j}=-\frac{1}{2}N}^{\frac{1}{2}N-1} b_{\mathbf{j}} \exp[-2\pi i(\mathbf{m}-\mathbf{n}) \cdot \mathbf{j}/N] \, , \tag{7.85}$$

where the Fourier coefficients are given by

$$a_{\mathbf{j}} = N^{-q} \sum_{\mathbf{k}=-\frac{1}{2}N}^{\frac{1}{2}N-1} A_{\mathbf{k}} \exp(2\pi i \mathbf{k} \cdot \mathbf{j}/N) \, , \tag{7.86}$$

and similarly for $b_{\mathbf{j}}$. We assume that N is large enough that no significant components are truncated by the limits in either the forward or inverse transform. The individual indices **m** and **n**, however, can still be allowed to run from $-\infty$ to ∞.

With these representations, (7.81) becomes

$$[\mathbf{ABA}]_{\mathbf{mn}} = \sum_{\mathbf{k}=\mathbf{m}-\frac{1}{2}N}^{\mathbf{m}+\frac{1}{2}N-1} \sum_{\mathbf{k}'=\mathbf{n}-\frac{1}{2}N}^{\mathbf{n}+\frac{1}{2}N-1} \sum_{\mathbf{j}=-\frac{1}{2}N}^{\frac{1}{2}N-1} \sum_{\mathbf{j}'=-\frac{1}{2}N}^{\frac{1}{2}N-1} \sum_{\mathbf{j}''=-\frac{1}{2}N}^{\frac{1}{2}N-1} a_{\mathbf{j}} \, b'_{\mathbf{j}} \, a''_{\mathbf{j}}$$

$$\times \exp[-2\pi i(\mathbf{m}-\mathbf{k}) \cdot \mathbf{j}/N] \exp[-2\pi i(\mathbf{k}-\mathbf{k}') \cdot \mathbf{j}'/N] \exp[-2\pi i(\mathbf{k}'-\mathbf{n}) \cdot \mathbf{j}''/N]$$

$$= A_{\mathbf{mn}} = \sum_{\mathbf{j}=-\frac{1}{2}N}^{\frac{1}{2}N-1} a_{\mathbf{j}} \exp[-2\pi i(\mathbf{m}-\mathbf{n}) \cdot \mathbf{j}/N] \, . \tag{7.87}$$

The sum over **k** yields $N^q \, \delta_{\mathbf{jj}'}$ and the one over $\mathbf{k}'$ yields $N^q \, \delta_{\mathbf{j}'\mathbf{j}''}$, so (7.81) becomes

$$N^{2q} \sum_{\mathbf{j}=-\frac{1}{2}N}^{\frac{1}{2}N-1} a_{\mathbf{j}} \, b_{\mathbf{j}} \, a_{\mathbf{j}} \exp[-2\pi i(\mathbf{m}-\mathbf{n}) \cdot \mathbf{j}/N] = \sum_{\mathbf{j}=-\frac{1}{2}N}^{\frac{1}{2}N-1} a_{\mathbf{j}} \exp[-2\pi i(\mathbf{m}-\mathbf{n}) \cdot \mathbf{j}/N] \, , \tag{7.88}$$

which can hold only if

$$N^{2q} a_{\mathbf{j}} \, b_{\mathbf{j}} \, a_{\mathbf{j}} = a_{\mathbf{j}} \, . \tag{7.89}$$

If $a_{\mathbf{j}} = 0$, $b_{\mathbf{j}}$ is arbitrary, but in the spirit of pseudoinverses, we take it to be zero as well (thereby making **B** the Moore-Penrose pseudoinverse of **A**, not just a 1-inverse). Thus we take

$$b_{\mathbf{j}} = \lim_{\eta \to 0^+} \frac{N^{-2q} a_{\mathbf{j}}}{a_{\mathbf{j}}^2 + \eta} \, , \tag{7.90}$$

and the solution for $B_{\mathbf{mn}}$ is

$$B_{\mathbf{mn}} = \lim_{\eta \to 0^+} \sum_{\mathbf{j}=-\frac{1}{2}N}^{\frac{1}{2}N-1} \frac{N^{-2q} a_{\mathbf{j}}}{a_{\mathbf{j}}^2 + \eta} \exp[-2\pi i (\mathbf{m} - \mathbf{n}) \cdot \mathbf{j}/N] \,. \tag{7.91}$$

If the translates are orthonormal, (7.91) reduces to $B_{\mathbf{mn}} = \delta_{\mathbf{mn}}$, which implies that $\chi_{\mathbf{n}}(\mathbf{r}) = \phi(\mathbf{r} - \mathbf{r_n})$ as before. More generally, $\chi_{\mathbf{n}}(\mathbf{r})$ must be constructed from (7.76) and (7.91).

Biorthonormality In the discussions of Gabor expansions and wavelets in Chap. 5, we encountered examples of expansions in non-orthonormal functions where the expansion coefficients could be found by constructing a suitable family of biorthonormal functions [see, *e.g.*, (5.44) and (5.93)]. We shall now show that the family $\{\chi(\mathbf{r} - \mathbf{Pn})\}$ is biorthonormal to $\{\phi(\mathbf{r} - \mathbf{Pn})\}$ with a mild assumption.

Biorthonormality is related to the operator $\mathcal{D}_\chi \mathcal{D}_\phi^\dagger$, which by (7.77) is the same as the matrix $\mathbf{BA}$. This matrix has elements given by

$$\left[\mathcal{D}_\chi \mathcal{D}_\phi^\dagger\right]_{\mathbf{mn}} = [\mathbf{BA}]_{\mathbf{mn}} = \int_\infty d^q r \; \chi^*(\mathbf{r} - \mathbf{Pm}) \, \phi(\mathbf{r} - \mathbf{Pn}) \,. \tag{7.92}$$

This matrix, like the others that occur in this problem, is Toeplitz and diagonally dominant. By analogy to (7.84), we can represent it as

$$\left[\mathcal{D}_\chi \mathcal{D}_\phi^\dagger\right]_{\mathbf{mn}} = \sum_{\mathbf{j}=-\frac{1}{2}N}^{\frac{1}{2}N-1} c_{\mathbf{j}} \exp[-2\pi i (\mathbf{m} - \mathbf{n}) \cdot \mathbf{j}/N] \,. \tag{7.93}$$

By algebra similar to that which led to (7.89), the second Penrose equation, $\mathbf{BAB} = \mathbf{B}$, can be written as

$$N^q c_{\mathbf{j}} \, b_{\mathbf{j}} = b_{\mathbf{j}} \,. \tag{7.94}$$

If the expansion functions are spatially compact, then the Fourier coefficient $a_{\mathbf{j}}$ extends to large $\mathbf{j}$. If we assume that $a_{\mathbf{j}}$ does not vanish identically for any $\mathbf{j}$, then (7.90) shows that $b_{\mathbf{j}}$ is also not zero, and (7.94) requires $c_{\mathbf{j}} = N^{-q}$ for all $\mathbf{j}$. Under this assumption (which is equivalent to replacing the pseudoinverse with a true inverse in the discussion above), we find

$$\left[\mathcal{D}_\chi \mathcal{D}_\phi^\dagger\right]_{\mathbf{mn}} = N^{-q} \sum_{\mathbf{j}=-\frac{1}{2}N}^{\frac{1}{2}N-1} \exp[-2\pi i (\mathbf{m} - \mathbf{n}) \cdot \mathbf{j}/N] = \delta_{\mathbf{mn}} \,, \tag{7.95}$$

which is the anticipated biorthonormality relation.

7.1.6 Other representations

So far in this chapter we have concentrated on linear representations, but nonlinear ones are often useful as well. In this section we briefly survey a few nonlinear representations that occur in the imaging literature.

Nonlinear parametric models The general approximate representation $f_a(\mathbf{r})$ defined in (7.27) is a linear representation since it results from a linear operator acting on $f(\mathbf{r})$. A nonlinear representation can be constructed in the form

$$f_p(\mathbf{r}) = \sum_{n=1}^{N} \Upsilon_n(\mathbf{r}, \boldsymbol{\Theta}_n) , \tag{7.96}$$

where subscript p connotes *parametric* and $\boldsymbol{\Theta}_n$ is the parameter vector. This representation is nonlinear in two senses. First, the function $\Upsilon_n(\mathbf{r}, \boldsymbol{\Theta}_n)$ depends nonlinearly on the free parameters $\boldsymbol{\Theta}_n$, so $f_p(\mathbf{r})$ does as well. Second, the parameters will be determined from $f(\mathbf{r})$ by a nonlinear computation.

The advantage of (7.96) is that it is often possible to use fewer terms than in a pixel representation and to use parameter vectors with few components. As a simple example, consider an astronomical image of a double star, where it is known *a priori* that the object consists of two point sources and the objective of the imaging is to determine the location and strength of each. In that case the object can be described by two terms:

$$f_p(\mathbf{r}) = \sum_{n=1}^{2} A_n \, \delta(\mathbf{r} - \mathbf{r}_n) . \tag{7.97}$$

Now each vector $\boldsymbol{\Theta}_n$ has three components, A_n, x_n and y_n, and the expansion function depends nonlinearly on the latter two. Equation (7.97) is a more compact representation than any linear expansion $f_a(\mathbf{r})$ that we might construct, yet it is capable of being an exact representation for the given problem. With this object representation, the inverse problem is to estimate the six unknown parameters from some data set $\mathbf{g}$.

A simple extension of this example might be appropriate in x-ray astronomy if we know that the objects are point sources but do not know *a priori* how many point sources to expect. In that case the summation limit N in (7.96) can itself be one of the nonlinear parameters to be estimated in the inverse problem.

Shape parameters Sometimes it is valid to regard an object as a superposition of geometrical shapes such as circles, ellipses or polygons. The parameters in the representation can then include things like the center coordinates and semimajor axes of each ellipse or the vertices of each polygon, as well as an amplitude like A_n in (7.97) associated with each term.

Gray levels within a shape Shape alone is an incomplete object specification; functional values (gray levels) within the shape also influence the data. Often we are much more interested in the shape than in the functional values, however, so we can use a much coarser description of the latter than the former. For example, we might approximate a 2D object as

$$f_p(\mathbf{r}) = \left[\sum_{k=0}^{K} \sum_{j=-k}^{k} a_{jk} R_k^j(r) \, e^{ij\theta} \right] S_{\boldsymbol{\Theta}}(\mathbf{r}) , \tag{7.98}$$

where r and θ are the polar coordinates of $\mathbf{r}$ (do not confuse θ and $\boldsymbol{\Theta}$), $R_k^j(r)$ is the Zernike polynomial (see Sec. 4.1.4) and $S_{\boldsymbol{\Theta}}(\mathbf{r})$ is a function that is unity inside some region defined by the parameter vector $\boldsymbol{\Theta}$ and zero outside. The complete parametric description of the object thus consists of the coefficients $\{a_{jk}\}$ and the components of $\boldsymbol{\Theta}$ (*e.g.*, vertex positions). If the gray levels within this region are relatively unimportant to the eventual use of the image, then a small number of Zernike polynomials (small K) can be used. The extreme case is where we assume that the object is uniform within the shape outline, so only the $k = 0$ term is used. Representation of the object in this case requires specifying the shape parameters $\boldsymbol{\Theta}$ and the overall gray level a_{00}.

Geometrical transformations For complex shapes such as the cerebral cortex, it may not be feasible to find an analytic expression, but we may have one or more exemplars that capture the essential features of the shape. Then representation of a particular object can be performed by warping each exemplar to fit to that object, and choosing the best fit if more than one exemplar is available. Once the best fit has been found, the warping parameters and an index identifying the exemplar constitute a nonlinear representation of the object. This method is sometimes called *atlas matching*.

The warping can be an affine transformation of the coordinate system of the form

$$\mathbf{r}' = \mathbf{W}\mathbf{r} + \mathbf{r}_0 , \tag{7.99}$$

where $\mathbf{W}$ and $\mathbf{r}_0$ are a matrix and vector, respectively, to be chosen by the fitting procedure. More complex nonlinear transformations can be also used; for example, the elements of $\mathbf{W}$ and $\mathbf{r}_0$ can be functions of $\mathbf{r}$. Very complex transformations known as *morphing* have been developed for computer graphics.

Adaptive linear expansions In Sec. 7.1.3 we discussed finite representations obtained by truncating infinite, exact representations [see (7.19)]. Such truncations are useful if terms with higher values of the index n convey less information in some sense about the object being represented. An example is the Fourier series representation of (7.11) where the index codes spatial frequency. Truncating the series thus eliminates high frequencies or small structures, with the hazard that these components might be weak but of prime importance for the intended task.

An alternative approach is to start with an exact (and hence infinite) expansion and keep only the N largest terms in order to represent a particular object. This method is commonly used with wavelets. From (5.90) we know that a 1D function can be represented exactly as

$$f(x) = \sum_{j=-\infty}^{\infty} \sum_{k=-\infty}^{\infty} c_{jk} \psi_{jk}(x) , \tag{7.100}$$

where $\psi_{jk}(x)$ is a wavelet. If this expansion is truncated, the resulting approximate representation $f_a(x)$ is linear in both senses: $f_a(x)$ is a linear function of the coefficients c_{jk}, and the coefficients are determined by a linear operation on $f(x)$ as in (5.92). If, on the other hand, we keep all terms for which $|c_{jk}|$ is greater than some threshold c_{th}, then linearity in the latter sense is lost, but perhaps many fewer terms will be needed to get a given representation accuracy.

We can think of the two methods of getting an approximate representation from (7.100) as thresholding on the horizontal axis or on the vertical axis of a wavelet transform; this idea is illustrated in Fig. 7.4. An analysis of the mean-square representation error for thresholding on the vertical axis has been given by Cohen and d'Ales (1995, 1997).

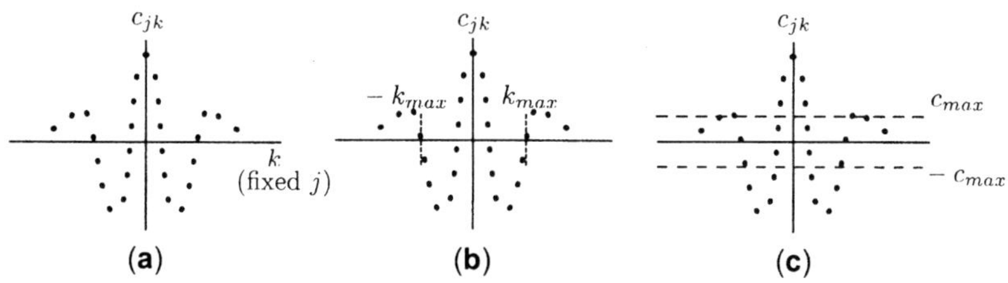

Fig. 7.4 Illustration of two ways of deriving an approximate finite representation from the discrete wavelet transform.

7.2 LINEAR CONTINUOUS-TO-CONTINUOUS SYSTEMS

Having surveyed many different representations of objects and images, we turn now to imaging systems. In this section both the object and the image are described as scalar-valued functions as in Sec. 7.1.1, and we assume that the mapping from object to image is linear. The action of the imaging system is thus described by a linear integral transform (see Sec. 1.2.2). Since the input and output of the system are functions defined on continuous domains, we refer to this system description as the continuous-to-continuous (CC) model. There is, however, no implication that the functions themselves are continuous or even that the system is a continuous mapping as defined in Sec. 1.3.2.

As we shall see in Chap. 9, the linear CC model is valid for optical imaging in two important limits, complete coherence and complete incoherence, and it can be salvaged for partial coherence as well. The CC model also plays a role in indirect imaging. In that case the final image is unlikely to be a function of a continuous variable, since the reconstruction step is usually performed digitally, but the CC model may describe the mapping from an object to an intermediate data set prior to detection and digitization. We shall use the language of direct imaging in this section, but the mathematics is often applicable as well to the first stage in an indirect system.

7.2.1 General shift-variant systems

As in Sec. 7.1.1, we regard the object as a function $f(\mathbf{r})$, where $\mathbf{r}$ is a vector in q dimensions, and the image as a function $g(\mathbf{r}_d)$, where $\mathbf{r}_d$ is a vector in s dimensions. As in Sec. 7.1, the subscript d denotes *detector*, and we shall think of $g(\mathbf{r}_d)$ as a radiation pattern incident on some detector.

The object will be assumed to be supported within a region $\mathbf{S}_f$ in $\mathbb{R}^q$ and square-integrable over that region. Similarly, the image will be assumed to be supported within a region $\mathbf{S}_g$ in $\mathbb{R}^s$ and square-integrable over that region. One or both of these regions might be infinite in a particular problem.

With these assumptions, the imaging system is a mapping from $\mathbb{L}_2(\mathbf{S}_f)$ to $\mathbb{L}_2(\mathbf{S}_g)$, or from object space $\mathbb{U}$ to image space $\mathbb{V}$. If this mapping is linear, the

Riesz representation theorem (see Sec. 1.2.2) tells us that it must have the form

$$g(\mathbf{r}_d) = \int_{\mathbf{S}_f} d^q r \; h(\mathbf{r}_d, \mathbf{r}) \, f(\mathbf{r}) \, . \tag{7.101}$$

In a more abstract notation, we can express this integral as

$$\mathbf{g} = \mathcal{H}\mathbf{f}, \tag{7.102}$$

where $\mathbf{f}$ and $\mathbf{g}$ are Hilbert-space vectors and $\mathcal{H}$ is the linear operator defined by (7.101). We shall also, on occasion, use the notation $[\mathcal{H}f(\mathbf{r})](\mathbf{r}_d)$ or $[\mathcal{H}\mathbf{f}](\mathbf{r}_d)$ to indicate that the function $f(\mathbf{r})$ has been transformed to a function of $\mathbf{r}_d$.

Since we have defined $\mathbb{V}$ without reference to the characteristics of $\mathcal{H}$, it is a larger space than we need to encompass all vectors of the form $\mathbf{g} = \mathcal{H}\mathbf{f}$; the range of $\mathcal{H}$ is a subspace of $\mathbb{V}$. Nevertheless, it is very convenient to choose a familiar $\mathbb{L}_2$ space as image space, and it will prove essential to do so when we consider noise. With noise, the mapping takes the form $\mathbf{g} = \mathcal{H}\mathbf{f} + \mathbf{n}$, where $\mathbf{n}$ is a random vector whose sample space may be all of $\mathbb{V}$.

Basis functions Because the system is linear, we can expand $f(\mathbf{r})$ in any basis we choose, and each basis function will be mapped independently. With the general expansion (7.8), we have

$$g(\mathbf{r}_d) = \sum_{n=1}^{\infty} \alpha_n \int_{\mathbf{S}_f} d^q r \; h(\mathbf{r}_d, \mathbf{r}) \, \psi_n(\mathbf{r}) = \sum_{n=1}^{\infty} \alpha_n [\mathcal{H}\psi_n](\mathbf{r}_d) \, , \tag{7.103}$$

where $\mathcal{H}\psi_n$ is the vector in image space produced by the system in response to vector ψ_n in object space, and $[\mathcal{H}\psi_n](\mathbf{r}_d)$ is the same thing expressed as a function.

Thus, once we know the response of the system to the basis functions, we can obtain the response to an arbitrary object by performing the sum over n. The same conclusion holds for continuous bases (see Sec. 1.1.6), even though the basis functions themselves are not square-integrable.

Point response As discussed in Sec. 2.2.6, Dirac delta functions can be regarded as basis functions for $\mathbb{L}_2$, even though they are not themselves in that space. The sifting property of delta functions can be used to express $f(\mathbf{r})$ as

$$f(\mathbf{r}) = \int_{\mathbf{S}_f} d^q r_0 \; \delta(\mathbf{r} - \mathbf{r}_0) \, f(\mathbf{r}_0) \, . \tag{7.104}$$

When the right-hand side of this equation is substituted for $f(\mathbf{r})$ in (7.101), $f(\mathbf{r}_0)$ is simply a constant and the entire $\mathbf{r}$-dependence is carried by the delta function. Thus we obtain

$$g(\mathbf{r}_d) = \int_{\mathbf{S}_f} d^q r \; h(\mathbf{r}_d, \mathbf{r}) \int_{\mathbf{S}_f} d^q r_0 \; \delta(\mathbf{r} - \mathbf{r}_0) \, f(\mathbf{r}_0) = \int_{\mathbf{S}_f} d^q r_0 \; h(\mathbf{r}_d, \mathbf{r}_0) \, f(\mathbf{r}_0) \, . \tag{7.105}$$

At first glance, the final form here is just a trivial change of variables in (7.101), but it provides an important insight into the meaning of the kernel. Equations (7.103) and (7.104) have the same structure if we think of the integral over $\mathbf{r}_0$ in the latter as equivalent to the sum over n in the former. In this view, $h(\mathbf{r}_d, \mathbf{r}_0)$

plays the same role as $[\mathcal{H}\psi_n](\mathbf{r}_d)$. That is, the kernel $h(\mathbf{r}_d, \mathbf{r}_0)$ is the response of the system at point $\mathbf{r}_d$ in image space to a delta function at point $\mathbf{r}_0$ in object space. Since a delta function is frequently referred to as a *point source* in optics, $h(\mathbf{r}_d, \mathbf{r}_0)$ is called the *point response function* (PRF). Another common term, especially in electrical engineering is impulse response.[7]

Another way to see the interpretation of the kernel as a PRF is to treat the delta function as the limit of a Dirac sequence. If we write

$$\delta(\mathbf{r} - \mathbf{r}_0) = \lim_{k\to\infty} d_k(\mathbf{r} - \mathbf{r}_0), \tag{7.106}$$

where each function $d_k(\mathbf{r} - \mathbf{r}_0)$ is square-integrable and meets the criteria for a Dirac sequence (see Sec. 2.2.2), then we can investigate the effect of the system on each function and pass to the limit. If the operator $\mathcal{H}$ is a continuous mapping (see Secs. 1.3.2 and 2.1.4), then

$$\lim_{k\to\infty} [\mathcal{H} d_k(\mathbf{r} - \mathbf{r}_0)](\mathbf{r}_d) = [\mathcal{H}\,\delta(\mathbf{r} - \mathbf{r}_0)](\mathbf{r}_d) = h(\mathbf{r}_d, \mathbf{r}_0), \tag{7.107}$$

where the last step uses the definition of $\mathcal{H}$ and the sifting property of the delta function. This step can, in fact, be regarded as the definition of $\mathcal{H}\,\delta(\mathbf{r} - \mathbf{r}_0)$, which is otherwise undefined since the delta function is not strictly in the domain of $\mathcal{H}$.

Thus, if the term *point source* is interpreted in the sense of a Dirac sequence as the limit of a very small, very bright source at point $\mathbf{r}_0$, then the kernel $h(\mathbf{r}_d, \mathbf{r}_0)$, regarded as a function of $\mathbf{r}_d$ for a fixed $\mathbf{r}_0$, is the limit of the corresponding image.

The concept of PRF is illustrated in Fig. 7.5 for a 2D imaging system $(q = s = 2)$.

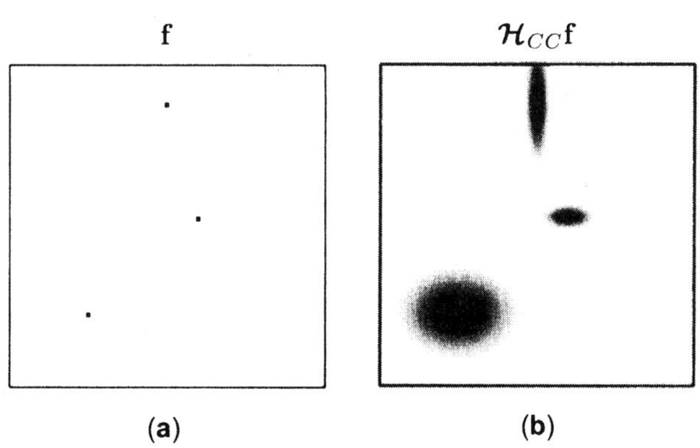

Fig. 7.5 Illustration of the point response function for a 2D CC imaging system. Three different input locations for a point object are shown on the left; the image on the right illustrates the location dependence of the response function.

Resolution measures The full PRF, $h(\mathbf{r}_d, \mathbf{r}_0)$ for all values of $\mathbf{r}_d$ and $\mathbf{r}_0$, is required for a complete specification of a general linear CC system. This specification is a

[7]The reader familiar with the imaging literature will notice that we are avoiding the term *point spread function* here; that term is being reserved for the special case of shift-invariant imaging, to be discussed shortly. Another expression with essentially the same meaning as PRF is *Green's function*, and that terminology will be used in the context of diffraction.

function of $q + s$ variables, in general, or 4 variables for the 2D imaging situation where $q = s = 2$. Since display of even a 4D function is difficult, there is some impetus for finding simpler, though less complete, descriptions of the system. One way to simplify the description is to report some scalar measure of the width of the function rather than the full function. This measure of the width of the PRF is often referred to as the *spatial resolution* of the system. Perversely, however, a system for which the width is *small* is often said to have *high* resolution.

One common width measure is the *full width at half maximum* (FWHM). This idea is best illustrated by assuming first that $q = s = 1$, so $h(\mathbf{r}_d, \mathbf{r}_0)$ becomes $h(x_d, x_0)$, which we shall assume to be real. Let $h_{max}(x_0)$ be the maximum value of $h(x_d, x_0)$ as a function of x_d for fixed x_0, and assume that this maximum occurs at $x_d = x_{d0}$, as illustrated in Fig. 7.6. (The value of x_{d0} depends, of course, on x_0.) If $h(x_d, x_0)$ falls off monotonically away from the peak, it is possible to identify two unique points $x_{d+} > x_{d0}$ and $x_{d-} < x_{d0}$ such that $h(x_{d\pm}, x_0) = \frac{1}{2}h_{max}(x_0)$. A simple measure of the width of the PRF, for a point source at x_0, is then given by

$$\delta_{\mathrm{FWHM}}(x_0) = x_{d+} - x_{d-} \, . \tag{7.108}$$

The FWHM definition can be extended straightforwardly to higher dimensions. If $s = 2$, for example, we can define a 2D vector $\boldsymbol{\delta}_{\mathrm{FWHM}}(\mathbf{r}_0)$ with components $x_{d+} - x_{d-}$ and $y_{d+} - y_{d-}$. If the object is also 2D (*i.e.*, $q = 2$), we can display the resolution by this definition in the form of two separate 2D images, one for each component of $\boldsymbol{\delta}_{\mathrm{FWHM}}(\mathbf{r}_0)$ as a function of the coordinates of the object point, x_0 and y_0. If the PRF is well approximated by a specific functional form such as a Gaussian, then these images fully specify the PRF; otherwise they merely give an indication of general trends.

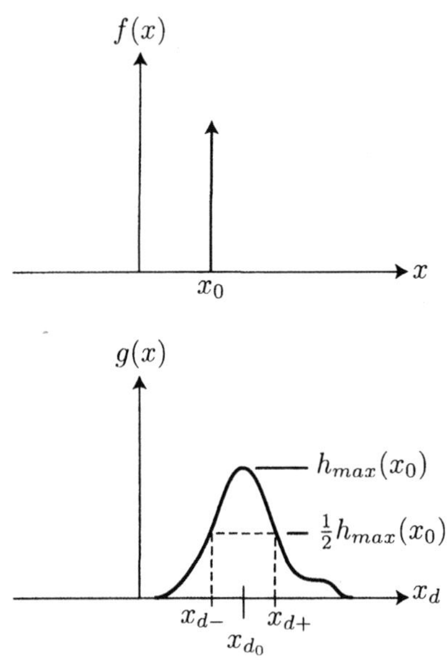

Fig. 7.6 Illustration of the FWHM measure of resolution.

Some authors argue that tails on the PRF are very important, so we should define additional resolution measures such as full width at one-tenth maximum or even one-fiftieth, but these definitions require some strong constraints on the form of the PRF. If, for example, the function has low level ripples, there may be many points where it takes on the values $h_{max}/10$ or $h_{max}/50$. We could take the points closest to the maximum, but the resulting width measure is not very representative of the entire function.

Another way to define a width measure is to normalize the PRF as a probability density function and think of $\mathbf{r}_d$ as a random vector (see Fig. 12.28). The width of the PRF in any direction can then be defined by the variance of the corresponding component of $\mathbf{r}_d$. If we assume that $h(\mathbf{r}_d, \mathbf{r}_0)$ is a real, nonnegative function, the required normalization is

$$p(\mathbf{r}_d, \mathbf{r}_0) = \frac{h(\mathbf{r}_d, \mathbf{r}_0)}{\int_{\mathbf{S}_g} d^s r' \, h(\mathbf{r}', \mathbf{r}_0)} . \tag{7.109}$$

The resolution width in the x direction is then defined by

$$\delta_x^2 = \int_{\mathbf{S}_g} d^2 r_d \; x_d^2 \, p(\mathbf{r}_d, \mathbf{r}_0) - \left[\int_{\mathbf{S}_g} d^2 r_d \; x_d \, p(\mathbf{r}_d, \mathbf{r}_0) \right]^2 , \tag{7.110}$$

and the width in the y direction is defined analogously.

Other definitions of spatial resolution are appropriate in particular problems. In optics, for example, the PRF often has the form of a sinc or besinc function, and it is common to define a resolution width as the distance from the peak to the first zero. This definition is the *Rayleigh criterion* for resolution.

This discussion of resolution has focused on the dependence of the PRF $h(\mathbf{r}_d, \mathbf{r}_0)$ on the image position $\mathbf{r}_d$, with the object point $\mathbf{r}_0$ fixed. It is also useful to consider the dependence of $h(\mathbf{r}_d, \mathbf{r})$ on $\mathbf{r}$ with $\mathbf{r}_d$ fixed. This function of $\mathbf{r}$ can be regarded as the PRF associated with $\mathcal{H}^\dagger$, and summary measures such as δ_{FWHM} can be defined by direct analogy to the ones discussed above. If the system has some magnification associated with it (see Sec. 7.2.7), the numerical values of the resolution measures in object space may be quite different from those in image space.

Flood images and point sensitivities In addition to providing a sharp image of a point, a good imaging system will exhibit good spatial uniformity in two distinct senses. It will produce a uniform image of a uniform object (often called a *flood source*), and its total response to a point source will be independent of where the point is in the field of view.

The first kind of uniformity is measured by the *flood image*, obtained from the basic imaging equation (7.101) with $f(\mathbf{r})$ set to a constant, which we may as well take to be unity. Thus the flood image is given by

$$g_{fld}(\mathbf{r}_d) = \int_{\mathbf{S}_f} d^q r \; h(\mathbf{r}_d, \mathbf{r}) . \tag{7.111}$$

A common goal, at least in direct imaging, is to make $g_{fld}(\mathbf{r}_d)$ constant (independent of $\mathbf{r}_d$). If this condition cannot be achieved by design of the system, it always can be achieved by post-processing; we simply define a normalized image of any object by

$$g_{norm}(\mathbf{r}_d) = \frac{g(\mathbf{r}_d)}{g_{fld}(\mathbf{r}_d)} , \tag{7.112}$$

where $g(\mathbf{r}_d)$ is related to $f(\mathbf{r})$ by (7.101). It follows from the definition that $g_{norm}(\mathbf{r}_d)$ must be constant if $f(\mathbf{r})$ is constant. In addition, large uniform regions of a general object will also be rendered as uniform regions in the image (provided the regions are large compared to the width of the PRF).

Uniformity of response to point sources can be quantified by the *point sensitivity*, defined as

$$s_{pt}(\mathbf{r}_0) = \int_{\mathbf{S}_g} d^s r_d \, h(\mathbf{r}_d, \mathbf{r}_0) \,. \tag{7.113}$$

If $g(\mathbf{r}_d)$ physically represents a mean number of photons per unit area on the detector, then $s_{pt}(\mathbf{r}_0)$ is interpreted as the total mean number of photons collected from a unit point source at $\mathbf{r}_0$.

Constant point sensitivity does not imply constant flood uniformity since the latter depends on where the photons are recorded in the image plane, while the former depends only on the total number of photons recorded. Image distortion, for example, can degrade the flood image without affecting point sensitivity. This point is discussed further in Sec. 7.2.7.

7.2.2 Adjoint operators and SVD

As we saw in Chap. 1, singular-value decomposition (SVD) is a powerful tool for analyzing linear systems. If the system is described by a CC operator $\mathcal{H}$, SVD requires knowledge of the eigenfunctions of $\mathcal{H}^\dagger\mathcal{H}$ and $\mathcal{H}\mathcal{H}^\dagger$, denoted $u_n(\mathbf{r})$ and $v_n(\mathbf{r}_d)$, respectively. In this section we shall discuss the interpretation and uses of these eigenfunctions for CC systems, but first we need to discuss the interpretation of the adjoint operator.

Since $\mathcal{H}$ is a mapping from object space $\mathbb{U}$ to image space $\mathbb{V}$, the adjoint operator $\mathcal{H}^\dagger$ maps from $\mathbb{V}$ to $\mathbb{U}$. In a CC problem, the adjoint converts a function of image coordinates $\mathbf{r}_d$ to a function of object coordinates $\mathbf{r}$. Specifically, for the operator $\mathcal{H}$ defined in (7.102), the adjoint is given (see Sec. 1.3.5) by[8]

$$\left[\mathcal{H}^\dagger \mathbf{g}\right](\mathbf{r}) = \int_{\mathbf{S}_g} d^s r_d \; g(\mathbf{r}_d)\, h^*(\mathbf{r}_d, \mathbf{r}) \,. \tag{7.114}$$

If the PRF is real, the complex conjugate is not needed, but we shall maintain the generality.

Since (7.114) is a linear CC mapping like (7.102), it is again useful to discuss it in terms of a point response function. The adjoint PRF is the response in object space produced by a point source in image space. If that point is at the specific location $\mathbf{r}_{d0}$, then the response is $h^*(\mathbf{r}_{d0}, \mathbf{r})$. Thus there is no need for a separate characterization of the adjoint operator; its PRF is fully determined by the PRF for $\mathcal{H}$ itself.

Figure 7.7 illustrates the adjoint PRF for a 2D system with a nonnegative, real kernel. If we think of $\mathcal{H}$ as a blurring operation in this case, then $\mathcal{H}^\dagger$ is a further blurring, but with $h(\mathbf{r}_d, \mathbf{r})$ replaced by $h(\mathbf{r}, \mathbf{r}_d)$.

[8]Note that we have *not* interchanged the two arguments in $h^*(\mathbf{r}_d, \mathbf{r})$. The interchange stated in (1.44) is implicit in (7.114) since the variable of integration is the first argument, not the second; see the discussion of this point in Sec. 1.3.5.

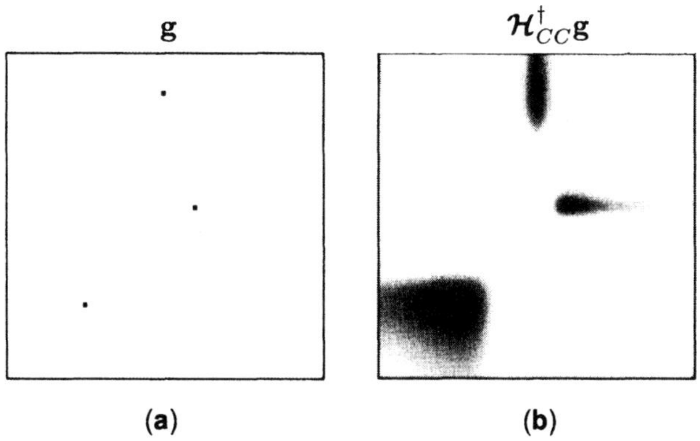

Fig. 7.7 Illustration of the point response function for $\mathcal{H}^\dagger$ for a 2D CC imaging system. As in Fig. 7.5, 3 input points are shown on the left. The resulting image, shown in (b), illustrates that the PRF of the adjoint operator is position-dependent.

The adjoint operation is frequently called *backprojection* (especially in the tomography literature), since each point is, in a sense, projected back into the object space by the operation defined in (7.114). We caution the reader, however, that the word projection in this context does not imply idempotency.

The operator $\mathcal{H}$ is Hermitian if $h(\mathbf{r}, \mathbf{r}_d) = h^*(\mathbf{r}_d, \mathbf{r})$, which cannot happen unless $q = s$. Even if $q = s$, it is rare to find that $\mathcal{H}$ is Hermitian.

Eigenfunctions in object space The operator $\mathcal{H}^\dagger\mathcal{H}$ that forms the starting point for SVD is given by

$$\left[\mathcal{H}^\dagger\mathcal{H}\mathbf{f}\right](\mathbf{r}) = \int_{\mathbf{S}_f} d^q r'\, k(\mathbf{r}, \mathbf{r}')\, f(\mathbf{r}')\,, \tag{7.115}$$

where the kernel $k(\mathbf{r}, \mathbf{r}')$ is

$$k(\mathbf{r}, \mathbf{r}') = \int_{\mathbf{S}_g} d^s r_d\; h^*(\mathbf{r}_d, \mathbf{r})\, h(\mathbf{r}_d, \mathbf{r}')\,. \tag{7.116}$$

This kernel can be interpreted as the PRF for the CC operator $\mathcal{H}^\dagger\mathcal{H}$ (see Fig. 7.8). It is necessarily symmetric, in the sense that $k(\mathbf{r}, \mathbf{r}') = k^*(\mathbf{r}', \mathbf{r})$, so $\mathcal{H}^\dagger\mathcal{H}$ is (not surprisingly) Hermitian.

If $\mathcal{H}^\dagger\mathcal{H}$ is compact (see Sec. 1.3.3), then the eigenvalues form a denumerable set. For a compact operator, the eigenvalue problem is

$$\left[\mathcal{H}^\dagger\mathcal{H}\mathbf{u}_n\right](\mathbf{r}) = \mu_n u_n(\mathbf{r})\,, \qquad n = 1, ..., \infty\,. \tag{7.117}$$

Since $\mathcal{H}^\dagger\mathcal{H}$ is Hermitian, all of the eigenvalues are real.

To clarify the interpretation of the eigenvalue equation, we again consider a 2D system with a nonnegative, real kernel. If an eigenfunction $u_n(\mathbf{r})$ is placed in the object plane of this system, the image $\mathcal{H}\mathbf{u}_n$ is a blurred version of it, and $\mathcal{H}^\dagger\mathcal{H}\mathbf{u}_n$ is blurred further. Since the input is an eigenfunction, however, the result of this double blurring must be the original function simply multiplied by a real number μ_n. It may seem surprising that any function would exist with this property, but from the theory of Hermitian operators, we know there must be an infinite number of them. The eigenfunctions may be complex, however, so they cannot represent real, physical objects. In many imaging situations, the object is constrained to be real and perhaps nonnegative, but no such constraint applies to the eigenfunctions.

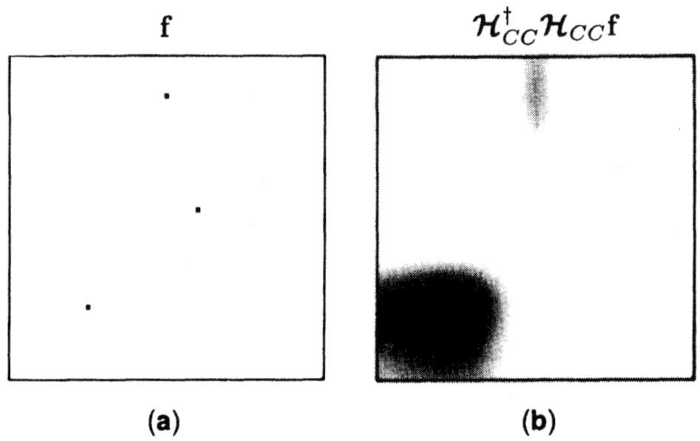

Fig. 7.8 Illustration of the point response function for $\mathcal{H}^{\dagger}\mathcal{H}$ for a CC imaging system.

From the discussion in Sec. 1.5.1, we know that the eigenfunctions can be chosen to form an orthonormal basis in $\mathbb{U}$. The orthonormality is expressed by

$$\int_{\mathbf{S}_f} d^q r \; u_n^*(\mathbf{r}) \, u_m(\mathbf{r}) = \delta_{nm} \tag{7.118}$$

and the completeness by

$$\sum_{n=1}^{\infty} u_n^*(\mathbf{r}) \, u_n(\mathbf{r}') = \delta(\mathbf{r} - \mathbf{r}') \,. \tag{7.119}$$

Because of the completeness, any object can be expanded as

$$f(\mathbf{r}) = \sum_{n=1}^{\infty} \alpha_n u_n(\mathbf{r}) \,. \tag{7.120}$$

From (7.118), the coefficients are given by

$$\alpha_n = \int_{\mathbf{S}_f} d^q r \; u_n^*(\mathbf{r}) \, f(\mathbf{r}) \,. \tag{7.121}$$

Eigenfunctions in image space If we have solved (7.117) and found a $u_n(\mathbf{r})$ corresponding to a nonzero μ_n, then we can construct a function $v_n(\mathbf{r}_d)$ in image space by [*cf.* (1.116) and (1.117)]

$$v_n(\mathbf{r}_d) = \frac{1}{\sqrt{\mu_n}} \left[\mathcal{H}\mathbf{u}_n\right](\mathbf{r}_d) \,. \tag{7.122}$$

This function is easily shown to be a normalized eigenfunction of $\mathcal{H}\mathcal{H}^{\dagger}$. It satisfies

$$\left[\mathcal{H}\mathcal{H}^{\dagger}\mathbf{v}_n\right](\mathbf{r}_d) = \int_{\mathbf{S}_g} d^s r_{d0} \; K(\mathbf{r}_d, \mathbf{r}_{d0}) \, v_n(\mathbf{r}_{d0}) = \mu_n v_n(\mathbf{r}_d) \,, \tag{7.123}$$

where the kernel $K(\mathbf{r}_d, \mathbf{r}_{d0})$ is given by

$$K(\mathbf{r}_d, \mathbf{r}_{d0}) = \int_{\mathbf{S}_f} d^q r \; h^*(\mathbf{r}_{d0}, \mathbf{r}) \, h(\mathbf{r}_d, \mathbf{r}) \, . \tag{7.124}$$

Note that the eigenvalues of $\mathcal{H}\mathcal{H}^\dagger$ are the same as those of $\mathcal{H}^\dagger\mathcal{H}$.

The full set of functions $\{v_n(\mathbf{r}_d)\}$ (not just the ones corresponding to nonzero μ_n) form an orthonormal basis in image space, satisfying relations analogous to (7.118) – (7.120). Any image $g(\mathbf{r}_d)$ can thus be expanded as

$$g(\mathbf{r}_d) = \sum_{n=1}^{\infty} \beta_n v_n(\mathbf{r}_d) \, , \tag{7.125}$$

where the coefficients are given by

$$\beta_n = \int_{\mathbf{S}_g} d^s r_d \; v_n^*(\mathbf{r}_d) \, g(\mathbf{r}_d) \, . \tag{7.126}$$

SVD of a CC operator Having discussed the eigenanalysis of $\mathcal{H}\mathcal{H}^\dagger$ and $\mathcal{H}^\dagger\mathcal{H}$, we are now in a position to treat the SVD of a CC operator. The generic form of the SVD representation of a compact linear operator is given in (1.120), repeated here for convenience:

$$\mathcal{H} = \sum_{n=1}^{R} \sqrt{\mu_n} \, \mathbf{v}_n \mathbf{u}_n^\dagger \, , \tag{7.127}$$

where R is the rank of the operator or the number of nonzero singular values. By the discussion in Sec. 1.4.4, the eigenvalues μ_n cannot be negative, so the singular values $\sqrt{\mu_n}$ are real and nonnegative. Thus we can assume that they are ordered by descending value as in (1.114) and hence the first R terms (where R may be infinite) encompass all nonzero terms in this expansion.

Transcribing (7.127) to the specific case of a CC operator and applying it to an arbitrary object, we obtain

$$\begin{aligned} g(\mathbf{r}_d) = [\mathcal{H}\mathbf{f}](\mathbf{r}_d) &= \sum_{n=1}^{R} \sqrt{\mu_n} \left[\mathbf{v}_n \mathbf{u}_n^\dagger \mathbf{f}\right](\mathbf{r}_d) = \sum_{n=1}^{R} \left[\sqrt{\mu_n} \int_{\mathbf{S}_f} d^q r \; u_n^*(\mathbf{r}) \, f(\mathbf{r})\right] v_n(\mathbf{r}_d) \\ &= \sum_{n=1}^{R} \sqrt{\mu_n} \, \alpha_n v_n(\mathbf{r}_d) \, . \end{aligned} \tag{7.128}$$

That is, given the singular system $\{u_n(\mathbf{r}), v_n(\mathbf{r}), \mu_n\}$, the procedure for computing $[\mathcal{H}\mathbf{f}](\mathbf{r}_d)$ is to form all of the scalar products of $f(\mathbf{r})$ with the object-space singular functions $u_n(\mathbf{r})$, weight these products with the singular values $\sqrt{\mu_n}$ and use the results as coefficients of the image-space singular functions $v_n(\mathbf{r}_d)$. Whenever we use the generic form (7.127) for a CC operator, the detailed implementation of (7.128) will be implied.

Equations (7.125) and (7.128) are two expansions for $g(\mathbf{r}_d)$ in terms of $\{v_n(\mathbf{r}_d)\}$. Since the expansion functions are orthogonal, the sums can be equal only if they are equal term by term, or

$$\beta_n = \sqrt{\mu_n} \, \alpha_n \, . \tag{7.129}$$

Thus SVD expansions reduce the imaging equation $\mathbf{g} = \mathcal{H}\mathbf{f}$ to a simple multiplication.

Null functions A CC operator does not necessarily have null functions. If it does, then the sum over n in (7.128) includes only terms for which $\mu_n \neq 0$. These terms define the *measurement space* $\mathbb{U}_{meas}$ for the operator $\boldsymbol{\mathcal{H}}$, which is a subspace of object space $\mathbb{U}$. The orthogonal complement of this space, spanned by singular functions $u_n(\mathbf{r})$ for which $\mu_n = 0$, is the *null space* $\mathbb{U}_{null}$. A vector in this space is called a null vector and denoted $\mathbf{f}_{null}$, and the corresponding function is called a null function and denoted $f_{null}(\mathbf{r})$.

7.2.3 Shift-invariant systems

Temporal filters In electrical engineering, linear filters are described by a temporal impulse response $p(t, t')$. The voltage on the output of the filter is related to the voltage on the input by

$$v_{out}(t) = \int_{-\infty}^{\infty} dt'\, p(t, t')\, v_{in}(t') . \tag{7.130}$$

As in the spatial case, $p(t, t_0)$ is the filter output when the input is $\delta(t' - t_0)$.

In spite of the analogy of $p(t, t')$ to the PRF $h(\mathbf{r}_d, \mathbf{r})$ introduced above, there are two crucial distinctions (other than dimensionality). First, temporal systems are *causal*, which means that there can be no output before there is an input. Causality imposes the following constraint on the impulse response of any temporal filter:

$$p(t, t') = 0 \qquad \text{if } t < t' . \tag{7.131}$$

Other properties of causal systems have been discussed in Secs. 4.2.1 and 4.2.4. There is no requirement that spatial filters be causal, and they seldom are.

The second distinction applies to temporal filters in which the physical parameters of the filter (capacitances, inductances, etc.) are independent of time. This is the usual case in passive electrical filters, and it implies that $p(t, t')$ is a function of only the time difference $t - t'$. It is convenient to use the same letter to designate the new function of one variable as we used for the function of two variables, and we write

$$p(t, t') = p(t - t') . \tag{7.132}$$

A linear filter satisfying this condition is said to be *linear and shift-invariant* (LSIV), implying that there is no preferred origin in time; if input $v_{in}(t)$ produces output $v_{out}(t)$, then $v_{in}(t - \Delta t)$ will produce $v_{out}(t - \Delta t)$.

Spatial shift-invariance? Unlike temporal filters with constant parameters, spatial linear systems usually do have preferred origins. The optical axis of a rotationally symmetric lens system or the center of the field of view in a tomographic imager establishes a unique point in space. For such systems, the LSIV model is only an approximation, but it is more mathematically tractable than the general CC model, and it may capture the essential features of a real system, especially when deviations from shift-invariance are small or attention can be confined to a small spatial region.

In addition, there are a few linear systems in optics and imaging that can be described accurately as LSIV. One example is a linear scanner, where a detector and imaging optics are moved continuously over an object. In that case, a shift of the object causes a corresponding shift of the image with no other changes.

Another example of a spatial LSIV system is diffraction. In a typical diffraction problem, an open aperture in a planar screen is illuminated with a monochromatic plane wave, and the diffracted field is observed on a plane a distance z from the aperture. As we shall see in detail in Chap. 9, this problem can be formulated as a linear mapping of one 2D function (the field $u_0(\mathbf{r}_a)$ in the aperture) to another 2D function (the field $u_z(\mathbf{r})$ on the observation plane). This mapping is linear, and it is SIV since a joint translation of the aperture and the observation point has no effect; if field $u_0(\mathbf{r}_a)$ produces the diffracted field $u_z(\mathbf{r})$, then $u_0(\mathbf{r}_a - \Delta\mathbf{r})$ produces $u_z(\mathbf{r} - \Delta\mathbf{r})$. Free space has no preferred origin and the wave equations are linear, so diffraction is an LSIV system. We shall exploit this simple observation fully in developing diffraction theory in Chap. 9.

Convolution operator and its adjoint When a spatial system in q dimensions can be described as LSIV, its output is given by

$$g(\mathbf{r}_d) = \int_\infty d^q r \; h(\mathbf{r}_d - \mathbf{r})\, f(\mathbf{r}) \,. \tag{7.133}$$

By a change of variables, this integral can also be written as

$$g(\mathbf{r}_d) = \int_\infty d^q r \; h(\mathbf{r})\, f(\mathbf{r}_d - \mathbf{r}) \,. \tag{7.134}$$

Note that the range of integration is infinite in all variables; LSIV systems cannot have limited fields of view.

From (3.240) we recognize the integral in (7.133) or (7.134) as a convolution, which we can write in shorthand form as

$$g(\mathbf{r}_d) = [h * f]\,(\mathbf{r}_d) = h(\mathbf{r}_d) * f(\mathbf{r}_d) \,. \tag{7.135}$$

The 1D convolution operator was introduced in Sec. 3.3.6, and the multidimensional case was discussed in Sec. 3.4.3. Many properties of convolution were developed in those two sections, and we assume here that the reader is conversant with that material.

The kernel of the convolution operator can again be interpreted as the point response function, but it is now a function of one argument rather than two. We shall use the term *point spread function* or PSF instead of PRF when we want to imply the LSIV case.

The adjoint of the convolution operator can be obtained from (7.114), which now becomes

$$\left[\boldsymbol{\mathcal{H}}^\dagger \mathbf{g}\right](\mathbf{r}) = \int_\infty d^q r_d \; h^*(\mathbf{r}_d - \mathbf{r})\, g(\mathbf{r}_d) \,. \tag{7.136}$$

From (3.242) we recognize this integral as the correlation $[g \star h^*](\mathbf{r})$. Thus the adjoint of convolution with $h(\mathbf{r})$ is correlation with $h^*(\mathbf{r})$. If $h(-\mathbf{r}) = h^*(\mathbf{r})$, then these two operations are identical and $\boldsymbol{\mathcal{H}}$ is Hermitian.

Flood images and point sensitivities As in the general CC case, we can define flood images and point sensitivities for LSIV systems, but they are rather uninteresting. The flood image is given by [*cf.* (7.111)]

$$g_{fld}(\mathbf{r}_d) = \int_\infty d^q r \; h(\mathbf{r}_d - \mathbf{r}) = \int_\infty d^q r' \, h(\mathbf{r}') \,, \tag{7.137}$$

where the last form follows from a simple change of variables. It is evident from this form that $g_{fld}(\mathbf{r}_d)$ is necessarily independent of $\mathbf{r}_d$. Similarly, the point sensitivity is necessarily independent of $\mathbf{r}$. LSIV systems cannot suffer from either of these kinds of nonuniformity (which is another reason why LSIV models do not accurately describe real systems).

7.2.4 Eigenanalysis of LSIV systems

For an LSIV operator, $\mathbf{r}$ and $\mathbf{r}_d$ must have the same dimensionality (q must equal s), since otherwise the algebraic operation $\mathbf{r} - \mathbf{r}_d$ would make no sense. Furthermore, the object and image supports are both infinite ($\mathbf{S}_f = \mathbf{S}_g = \mathbb{R}^q$), so object space $\mathbb{U}$ is identical to image space $\mathbb{V}$. Under these circumstances, it is possible to find eigenfunctions (not just singular functions) and eigenvalues of $\boldsymbol{\mathcal{H}}$.

We do not expect, however, that the eigenvalues will form a denumerable set since the convolution operator is not compact. In Sec. 1.3.3 we introduced the Hilbert-Schmidt condition, (1.33), as a test for compactness. Convolution fails this test, even if the kernel is square-integrable. By a change of variables, the Hilbert-Schmidt integral for a qD convolution can be written

$$\int_\infty d^q r_d \int_\infty d^q r \, |h(\mathbf{r}_d - \mathbf{r})|^2 = \int_\infty d^q r_d \int_\infty d^q r' \, |h(\mathbf{r}')|^2 = \int_\infty d^q r_d \; const = \infty \,, \tag{7.138}$$

so we cannot show by this route that convolution is compact. We must expect that a general convolution operator will have a continuous spectrum.

The eigenvalue problem for an LSIV operator has the form

$$[\boldsymbol{\mathcal{H}}\psi](\mathbf{r}_d) = [\psi * h](\mathbf{r}_d) = \lambda\psi \,, \tag{7.139}$$

where $\psi(\mathbf{r})$ is the eigenfunction and λ is the eigenvalue. From the convolution theorem, (3.132), we can write the Fourier transform of this equation as

$$H(\boldsymbol{\rho})\,\Psi(\boldsymbol{\rho}) = \lambda\Psi(\boldsymbol{\rho}) \,, \tag{7.140}$$

where $H(\boldsymbol{\rho})$ is the Fourier transform of the PSF and $\Psi(\boldsymbol{\rho})$ is the Fourier transform of the eigenfunction.

At first glance it might seem impossible to find a solution to (7.140); we seek a function $\Psi(\boldsymbol{\rho})$ that can be multiplied by some other function, yet retain its functional form. The only function with this property is the Dirac delta function,

$$\Psi(\boldsymbol{\rho}) = \delta(\boldsymbol{\rho} - \boldsymbol{\rho}_0) \,. \tag{7.141}$$

That this function satisfies (7.140) follows from (2.25), according to which

$$H(\boldsymbol{\rho})\,\delta(\boldsymbol{\rho} - \boldsymbol{\rho}_0) = H(\boldsymbol{\rho}_0)\,\delta(\boldsymbol{\rho} - \boldsymbol{\rho}_0) \,. \tag{7.142}$$

Since $H(\boldsymbol{\rho}_0)$ is a constant, it is the eigenvalue λ in (7.140), and the delta function is the required $\Psi(\boldsymbol{\rho})$. Any choice of the constant vector $\boldsymbol{\rho}_0$ works, so $\boldsymbol{\rho}_0$ is the continuous index that distinguishes one eigenfunction from another.

An inverse transform of (7.141) shows that the eigenfunctions, now embellished with a subscript, are given by

$$\psi_{\boldsymbol{\rho}_0}(\mathbf{r}) = \exp(2\pi i \boldsymbol{\rho}_0 \cdot \mathbf{r}) \,, \tag{7.143}$$

and the corresponding eigenvalue is

$$\lambda_{\boldsymbol{\rho}_0} = H(\boldsymbol{\rho}_0) . \tag{7.144}$$

Since these equations hold for any spatial-frequency vector, the subscript on $\boldsymbol{\rho}_0$ is superfluous and will be dropped.

Another way to see that (7.143) and (7.144) satisfy the eigenvalue equation is by direct substitution in (7.139):

$$\left[\mathcal{H}\psi_{\boldsymbol{\rho}}\right](\mathbf{r}_d) = \int_\infty d^q r \; h(\mathbf{r}) \exp[2\pi i \boldsymbol{\rho} \cdot (\mathbf{r}_d - \mathbf{r})]$$

$$= \exp(2\pi i \boldsymbol{\rho} \cdot \mathbf{r}_d) \int_\infty d^q r \; h(\mathbf{r}) \exp(-2\pi i \boldsymbol{\rho} \cdot \mathbf{r}) . \tag{7.145}$$

The last integral is $H(\boldsymbol{\rho})$, but the interesting point for the present discussion is that it is independent of $\mathbf{r}_d$. For the convolution operator, we can thus write

$$[\mathcal{H}\psi_{\boldsymbol{\rho}}](\mathbf{r}_d) = \lambda_{\boldsymbol{\rho}} \psi_{\boldsymbol{\rho}}(\mathbf{r}_d) , \tag{7.146}$$

where the eigenvalue is again $\lambda_{\boldsymbol{\rho}} = H(\boldsymbol{\rho})$.

In summary, the eigenfunctions of an LSIV operator are the Fourier basis functions or complex exponentials, and the eigenvalues are obtained by Fourier-transforming the PSF. Note that the eigenfunctions are not square-integrable, so they are not contained in the space they span.

7.2.5 Singular-value decomposition of LSIV systems

We have just seen that convolution has eigenfunctions and eigenvalues; like any linear operator, it also has singular functions and singular values. In fact, the singular functions are precisely the eigenfunctions. It will be left as an exercise to show that

$$\left[\mathcal{H}^\dagger \mathcal{H}\psi_{\boldsymbol{\rho}}\right](\mathbf{r}) = |\lambda_{\boldsymbol{\rho}}|^2 \psi_{\boldsymbol{\rho}}(\mathbf{r}) = |H(\boldsymbol{\rho})|^2 \psi_{\boldsymbol{\rho}}(\mathbf{r}) , \tag{7.147}$$

where $\mathcal{H}^\dagger$ is given by (7.136). Hence the object-space singular function (or eigenfunction of $\mathcal{H}^\dagger\mathcal{H}$), which we denote by $u_{\boldsymbol{\rho}}(\mathbf{r})$, is also equal in this case to the eigenfunction $\psi_{\boldsymbol{\rho}}(\mathbf{r})$. The corresponding singular value is

$$\sqrt{\mu_{\boldsymbol{\rho}}} = |\lambda_{\boldsymbol{\rho}}| = |H(\boldsymbol{\rho})| . \tag{7.148}$$

As expected, the singular values are all nonnegative.

When $H(\boldsymbol{\rho}) \neq 0$, the singular functions in image space are given by (7.122) as

$$v_{\boldsymbol{\rho}}(\mathbf{r}_d) = \frac{1}{|H(\boldsymbol{\rho})|}[\mathcal{H}\mathbf{u}_\rho](\mathbf{r}_d) = \frac{H(\boldsymbol{\rho})}{|H(\boldsymbol{\rho})|} \exp(2\pi i \boldsymbol{\rho} \cdot \mathbf{r}_d) . \tag{7.149}$$

The factor $H(\boldsymbol{\rho})/|H(\boldsymbol{\rho})|$ is a constant phase factor (± 1 if $H(\boldsymbol{\rho})$ is real); if it is omitted, $v_{\boldsymbol{\rho}}(\mathbf{r}_d)$ is still an eigenfunction of $\mathcal{H}\mathcal{H}^\dagger$, but the factor is needed for consistency in our overall formalism.

The SVD representation of a convolution, applied to an arbitrary object, is given by (7.128) with the sum over n generalized to an integral over $\boldsymbol{\rho}$:

$$g(\mathbf{r}_d) = [\mathcal{H}\mathbf{f}](\mathbf{r}_d) = \int_\infty d^q\rho \, |H(\boldsymbol{\rho})| \left[\mathbf{v}_\rho \mathbf{u}_\rho^\dagger \mathbf{f}\right](\mathbf{r}_d)$$

$$= \int_\infty d^q\rho \, |H(\boldsymbol{\rho})| \left[\int_\infty d^q r \; u_\rho^*(\mathbf{r}) \, f(\mathbf{r})\right] v_\rho(\mathbf{r}_d) \,. \tag{7.150}$$

The integral in square brackets is just the Fourier transform of the object, denoted $F(\boldsymbol{\rho})$, and $v_\rho(\mathbf{r}_d)$ is given by (7.149), so

$$g(\mathbf{r}_d) = \int_\infty d^q\rho \; H(\boldsymbol{\rho}) \, F(\boldsymbol{\rho}) \exp(2\pi i \boldsymbol{\rho} \cdot \mathbf{r}_d) \,. \tag{7.151}$$

Taking Fourier transforms of both sides yields

$$G(\boldsymbol{\rho}) = H(\boldsymbol{\rho}) \, F(\boldsymbol{\rho}) \,, \tag{7.152}$$

where $G(\boldsymbol{\rho}) = \mathcal{F}_q\{g(\mathbf{r}_d)\}$.

This result is the convolution theorem, (3.132), but now it acquires a new interpretation which we can elucidate by writing Fourier expressions beside their SVD counterparts:

$f(\mathbf{r}) = \int_\infty d^q\rho \; F(\boldsymbol{\rho}) \exp(2\pi i \boldsymbol{\rho} \cdot \mathbf{r})$	$\mathbf{f} = \sum_{n=0}^{\infty} \alpha_n \mathbf{u}_n$
$g(\mathbf{r}_d) = \int_\infty d^q\rho \; G(\boldsymbol{\rho}) \exp(2\pi i \boldsymbol{\rho} \cdot \mathbf{r}_d)$	$\mathbf{g} = \sum_{n=0}^{\infty} \beta_n \mathbf{v}_n$
$g(\mathbf{r}_d) = [h * f](\mathbf{r}_d)$	$\mathbf{g} = \mathcal{H}\mathbf{f}$
$G(\boldsymbol{\rho}) = H(\boldsymbol{\rho}) \, F(\boldsymbol{\rho})$	$\beta_n = \sqrt{\mu_n} \, \alpha_n \,.$

Thus the Fourier transform of an object or image gives the coefficients in an SVD representation derived from a convolution operator. The lack of compactness of the convolution means that we have to use integrals rather than sums, but otherwise the familiar SVD results are obtained. In particular, SVD (or Fourier analysis) reduces the convolution operation to a simple multiplication.

7.2.6 Transfer functions

A complex exponential $\exp(2\pi i \boldsymbol{\rho} \cdot \mathbf{r})$ is transferred through an LSIV system unchanged except for multiplication by the eigenvalue $H(\boldsymbol{\rho})$; for this reason $H(\boldsymbol{\rho})$ is often called the *transfer function* of the system.

Of course, the transfer function is also the Fourier transform of the PSF. If the PSF is real, the transfer function must satisfy

$$H(-\boldsymbol{\rho}) = H^*(\boldsymbol{\rho}) \,, \tag{7.153}$$

which shows that the real part of $H(\boldsymbol{\rho})$ is even and the imaginary part is odd (see Sec. 3.2.3). The transfer function is real and even if $h(\mathbf{r})$ is real and even. In any case, (7.153) ensures that $H(0)$ is real if $h(\mathbf{r})$ is real.

As noted in Chap. 3, many books refer to (7.153) as *Hermiticity* or *Hermitian symmetry*, but this terminology is unfortunate since $H(\boldsymbol{\rho})$ does not necessarily relate to a Hermitian operator. As noted in Sec. 7.2.3, the condition that an LSIV operator be Hermitian is that $h(\mathbf{r}) = h^*(-\mathbf{r})$. This condition implies that $H(\boldsymbol{\rho})$ is real for all $\boldsymbol{\rho}$ (which must be the case for a Hermitian operator since its eigenvalues are real). To avoid confusion, we shall refer to (7.153) as *conjugate symmetry*.

For LSIV systems with real PSFs, it is common to define the *optical transfer function* (OTF) by

$$\mathrm{OTF}(\boldsymbol{\rho}) = \frac{H(\boldsymbol{\rho})}{H(0)}. \tag{7.154}$$

Since OTF(0) is unity, OTF($\boldsymbol{\rho}$) gives the strength with which a complex exponential of frequency $\boldsymbol{\rho}$ is transferred through the system relative to the transferred strength of a constant, which can be regarded as an exponential of frequency zero.

Another normalized transfer function is the *modulation transfer function* (MTF) defined by

$$\mathrm{MTF}(\boldsymbol{\rho}) = \frac{|H(\boldsymbol{\rho})|}{H(0)} = |\mathrm{OTF}(\boldsymbol{\rho})|. \tag{7.155}$$

Many systems have the character of a low-pass-filter, which means that $\mathrm{MTF}(\boldsymbol{\rho}) \leq 1$.

To see why $|H(\boldsymbol{\rho})|/H(0)$ is called MTF, consider as the input to the system not a complex exponential but a raised cosine of the form

$$f(\mathbf{r}) = A + B\cos(2\pi\boldsymbol{\rho}\cdot\mathbf{r} - \phi_f), \tag{7.156}$$

where A and B are real, with $A \geq B \geq 0$ and hence $f(\mathbf{r}) \geq 0$. This function has a *modulation* defined by

$$M_f \equiv \frac{f_{max} - f_{min}}{f_{max} + f_{min}} = \frac{B}{A}, \tag{7.157}$$

where f_{max}, the maximum value of $f(\mathbf{r})$, occurs when the cosine $= 1$ and is given by $f_{max} = A + B$. Similarly, $f_{min} = A - B$.

To see how this function is affected by an LSIV system, we decompose it as

$$f(\mathbf{r}) = A + \tfrac{1}{2}B\exp(2\pi i\boldsymbol{\rho}\cdot\mathbf{r} - \phi_f) + \tfrac{1}{2}B\exp(-2\pi i\boldsymbol{\rho}\cdot\mathbf{r} + \phi_f). \tag{7.158}$$

Each of the three terms is now an eigenfunction, so we can write down the image at once:

$$g(\mathbf{r_d}) = AH(0) + \tfrac{1}{2}BH(\boldsymbol{\rho})\exp(2\pi i\boldsymbol{\rho}\cdot\mathbf{r_d} - \phi_f) + \tfrac{1}{2}BH(-\boldsymbol{\rho})\exp(-2\pi i\boldsymbol{\rho}\cdot\mathbf{r_d} + \phi_f). \tag{7.159}$$

Now we assume that the PSF is real, so that (7.153) is satisfied, and write the transfer functions as

$$H(\boldsymbol{\rho}) = |H(\boldsymbol{\rho})|\exp[i\Phi_H(\boldsymbol{\rho})], \qquad H(-\boldsymbol{\rho}) = |H(\boldsymbol{\rho})|\exp[-i\Phi_H(\boldsymbol{\rho})]. \tag{7.160}$$

Some easy algebra shows that

$$g(\mathbf{r_d}) = AH(0) + B|H(\boldsymbol{\rho})|\cos[2\pi\boldsymbol{\rho}\cdot\mathbf{r_d} - \phi_f + \Phi_H(\boldsymbol{\rho})]. \tag{7.161}$$

The modulation of $g(\mathbf{r_d})$, defined analogously to (7.157), is given by

$$M_g = \frac{B|H(\boldsymbol{\rho})|}{AH(0)}. \tag{7.162}$$

Thus the modulation transfer function is aptly named since it is the ratio of output modulation to input modulation:

$$\mathrm{MTF}(\boldsymbol{\rho}) = \frac{M_g}{M_f}\,. \tag{7.163}$$

This computation also provides an interpretation of the phase of $H(\boldsymbol{\rho})$. The cosine is not an eigenfunction unless $\Phi_H(\boldsymbol{\rho}) = 0$, but the change in form as it is transferred through the system is just a phase shift by an amount $\Phi_H(\boldsymbol{\rho})$. For this reason, $\Phi_H(\boldsymbol{\rho})$ is often referred to as the *phase transfer function.*

Fourier transformation as diagonalization Another way of looking at the process of Fourier transformation with an LSIV system is that it diagonalizes the system operator. The basic idea of diagonalization was introduced in Sec. 1.4.5; here we extend the argument specifically to a convolution operator.

Consider first a general linear CC operator as defined in (7.101), with no restriction other than that $\mathbf{r}$ and $\mathbf{r}_d$ have the same dimensionality ($q = s$). The object $f(\mathbf{r})$ can be expressed in terms of its qD inverse Fourier transform as

$$f(\mathbf{r}) = \int_\infty d^q\rho\; F(\boldsymbol{\rho}) \exp(2\pi i \boldsymbol{\rho}\cdot\mathbf{r})\,, \tag{7.164}$$

or in abstract form,

$$\mathbf{f} = \mathcal{F}_q^{-1}\mathbf{F}\,. \tag{7.165}$$

A similar representation holds for the image, and the imaging equation becomes

$$\mathbf{g} = \mathcal{F}_q^{-1}\mathbf{G} = \mathcal{H}\mathcal{F}_q^{-1}\mathbf{F}\,. \tag{7.166}$$

Operating on both sides with $\mathcal{F}_q$ yields

$$\mathbf{G} = \mathcal{F}_q\mathcal{H}\mathcal{F}_q^{-1}\mathbf{F}\,. \tag{7.167}$$

The kernel of the operator $\mathcal{F}_q\mathcal{H}\mathcal{F}_q^{-1}$ is given by

$$\left[\mathcal{F}_q\mathcal{H}\mathcal{F}_q^{-1}\right](\boldsymbol{\rho}_d, \boldsymbol{\rho}) = \int_\infty d^q r_d \int_\infty d^q r\, \exp(-2\pi i \boldsymbol{\rho}_d\cdot\mathbf{r}_d)\, h(\mathbf{r}_d, \mathbf{r}) \exp(2\pi i \boldsymbol{\rho}\cdot\mathbf{r})\,. \tag{7.168}$$

So far this treatment is applicable to any linear CC system with $q = s$, but now we assume shift-invariance, so that $h(\mathbf{r}_d, \mathbf{r}) = h(\mathbf{r}_d - \mathbf{r})$. Then a change of variables and use of (3.217) shows that

$$\left[\mathcal{F}_q\mathcal{H}\mathcal{F}_q^{-1}\right](\boldsymbol{\rho}_d, \boldsymbol{\rho}) = H(\boldsymbol{\rho})\,\delta(\boldsymbol{\rho} - \boldsymbol{\rho}_d)\,. \tag{7.169}$$

If we think of $[\mathcal{F}_q\mathcal{H}\mathcal{F}_q^{-1}](\boldsymbol{\rho}_d, \boldsymbol{\rho})$ as a matrix, with continuous vector indices $\boldsymbol{\rho}$ and $\boldsymbol{\rho}_d$ replacing the usual integer indices, then (7.169) shows that the matrix is diagonal, in the sense that all elements with $\boldsymbol{\rho} \neq \boldsymbol{\rho}_d$ are zero. The diagonal elements are infinite, but the strength of the delta function along the diagonal is the transfer function $H(\boldsymbol{\rho})$. In this sense the Fourier domain is the representation in which the system operator is diagonal.

Diagonal operators, like ordinary diagonal matrices, simplify computations considerably. If we apply the operator $\mathcal{F}_q \mathcal{H} \mathcal{F}_q^{-1}$ to an arbitrary object, expressed in the Fourier domain by $\mathbf{F}$, we quickly find

$$\left[\mathcal{F}_q \mathcal{H} \mathcal{F}_q^{-1} \mathbf{F}\right](\boldsymbol{\rho}_d) = \int_\infty d^q\rho \; H(\boldsymbol{\rho})\, \delta(\boldsymbol{\rho} - \boldsymbol{\rho}_d)\, F(\boldsymbol{\rho}) = H(\boldsymbol{\rho}_d)\, F(\boldsymbol{\rho}_d) \,. \tag{7.170}$$

Again we rediscover that convolution amounts to multiplication by the transfer function in the Fourier domain.

For later reference, we note that *all* LSIV operators are diagonalized by Fourier transformation. In particular, $\mathcal{H}^\dagger \mathcal{H}$ is LSIV if $\mathcal{H}$ is, and it can be shown that

$$\left[\mathcal{F}_q \mathcal{H}^\dagger \mathcal{H} \mathcal{F}_q^{-1}\right](\boldsymbol{\rho}_d, \boldsymbol{\rho}) = |H(\boldsymbol{\rho})|^2\, \delta(\boldsymbol{\rho} - \boldsymbol{\rho}_d) \,. \tag{7.171}$$

7.2.7 Magnifiers

An LSIV system necessarily has unit magnification; if two points on the object are separated by the vector $\Delta\mathbf{r}$, then the two corresponding points on the image must also be separated by $\Delta\mathbf{r}$. Optical systems, however, often function as magnifiers, with the size of the image scaled with respect to the object by a real number m, called the *magnification*. Negative m implies an inverted image, and $|m| < 1$ means that the image is smaller than the object. In this section we examine the effects of magnification from the viewpoint of SVD. Then we generalize the discussion to allow more general affine mappings and space-variant magnification.

Constant magnification An *ideal magnifier* has a PRF given by

$$h(\mathbf{r}_d, \mathbf{r}) = \delta(\mathbf{r}_d - m\mathbf{r}) = |m|^{-q}\, \delta\!\left(\frac{\mathbf{r}_d}{m} - \mathbf{r}\right) , \tag{7.172}$$

so that

$$g(\mathbf{r}_d) = |m|^{-q} f(\mathbf{r}_d/m) \,. \tag{7.173}$$

With the factor of $|m|^{-q}$, the integral of $g(\mathbf{r}_d)$ is equal to the integral of $f(\mathbf{r})$ for all objects, and hence the point sensitivity is unity. This conclusion follows also from the definition of point sensitivity, (7.113), and the sifting property of delta functions.

Next consider a magnifier with blur. If the form of the blur is independent of position in the object, the PRF is similar to (7.172) but with the delta function replaced by a blur function:

$$h(\mathbf{r}_d, \mathbf{r}) = h(\mathbf{r}_d - m\mathbf{r}) \,. \tag{7.174}$$

Though this PRF is not strictly LSIV, a simple redefinition of the image converts it to LSIV form. The input–output relation can be written as

$$g(\mathbf{r}_d) = \int_\infty d^q r \; h(\mathbf{r}_d - m\mathbf{r})\, f(\mathbf{r}) = \int_\infty d^q r \; h[m(\tilde{\mathbf{r}}_d - \mathbf{r})]\, f(\mathbf{r}) \,, \tag{7.175}$$

where $\tilde{\mathbf{r}}_d = \mathbf{r}_d/m$. If we define $\tilde{g}(\tilde{\mathbf{r}}_d) = g(\mathbf{r}_d)$ and $h'(\mathbf{r}) = h(m\mathbf{r})$, then we have

$$\tilde{g}(\tilde{\mathbf{r}}_d) = \int_\infty d^q r \; h'(\tilde{\mathbf{r}}_d - \mathbf{r})\, f(\mathbf{r}) = [h' * f]\,(\tilde{\mathbf{r}}_d) \,. \tag{7.176}$$

Thus, after we refer both the image and the PRF back to the scale of the object, a magnifier with shift-invariant blur is described by a convolution operator.

SVD of a magnifier with blur Another way to salvage the LSIV model with a magnifier is to look at $\mathcal{H}^\dagger\mathcal{H}$ rather than $\mathcal{H}$. From (7.116) and (7.174), the kernel for $\mathcal{H}^\dagger\mathcal{H}$ is given by

$$k(\mathbf{r},\mathbf{r}') = \int_\infty d^q r_d \; h^*(\mathbf{r}_d - m\mathbf{r})\, h(\mathbf{r}_d - m\mathbf{r}')$$

$$= \int_\infty d^q r_0 \; h^*(\mathbf{r}_0)\, h[\mathbf{r}_0 + m(\mathbf{r}-\mathbf{r}')] , \tag{7.177}$$

where $\mathbf{r}_0 = \mathbf{r}_d - m\mathbf{r}$.

By inspection, $k(\mathbf{r},\mathbf{r}')$ is a function of $\mathbf{r}-\mathbf{r}'$, so $\mathcal{H}^\dagger\mathcal{H}$ is LSIV (and hence has unit magnification). Its eigenfunctions $u_{\boldsymbol{\rho}}(\mathbf{r})$, which are also the singular functions of $\mathcal{H}$ in object space, are therefore the complex exponentials $\psi_{\boldsymbol{\rho}}(\mathbf{r})$. With a little algebra it can be shown that

$$\left[\mathcal{H}^\dagger\mathcal{H}\mathbf{u}_{\boldsymbol{\rho}}\right](\mathbf{r}) = \int_\infty d^q r' \, k(\mathbf{r},\mathbf{r}') \exp(2\pi i \boldsymbol{\rho}\cdot\mathbf{r}') = \mu_{\boldsymbol{\rho}} \exp(2\pi i \boldsymbol{\rho}\cdot\mathbf{r}) , \tag{7.178}$$

where the eigenvalues are [*cf.* (7.147)]

$$\mu_{\boldsymbol{\rho}} = |m|^{-q} |H(\boldsymbol{\rho}/m)|^2 . \tag{7.179}$$

The scale factor occurs since the blur function $h(\mathbf{r}_d)$ was originally defined in image space.

When $H(\boldsymbol{\rho}) \neq 0$, the singular functions in image space are given by [*cf.* (7.149)]

$$v_{\boldsymbol{\rho}}(\mathbf{r}_d) = \frac{1}{\sqrt{\mu_{\boldsymbol{\rho}}}}[\mathcal{H}\mathbf{u}_{\boldsymbol{\rho}}](\mathbf{r}_d) = \frac{H(\boldsymbol{\rho}/m)}{|m^{q/2} H(\boldsymbol{\rho}/m)|} \exp(2\pi i \boldsymbol{\rho}\cdot\mathbf{r}_d/m) . \tag{7.180}$$

Thus, with minor insertions of factors of m, magnifiers with shift-invariant blur and constant magnification have the same SVD as LSIV systems.

Affine mapping Next we consider systems where the mapping between a point in the object and the corresponding point in the image is an affine transformation. For simplicity, we neglect blur and take the PRF as [*cf.* (7.172)]

$$h(\mathbf{r}_d,\mathbf{r}) = \delta(\mathbf{r}_d - \mathbf{M}\mathbf{r} - \mathbf{r}_{d0}) = |\det(\mathbf{M})|^{-1}\, \delta(\mathbf{M}^{-1}\mathbf{r}_d - \mathbf{r} - \mathbf{M}^{-1}\mathbf{r}_{d0}) , \tag{7.181}$$

where $\mathbf{M}$ is a nonsingular $q \times q$ matrix and $\det(\mathbf{M})$ is its determinant. With this PRF, the point $\mathbf{r}$ in the object maps to $\mathbf{M}\mathbf{r} + \mathbf{r}_{d0}$ in the image, and the origin in the object ($\mathbf{r} = 0$) maps to the point $\mathbf{r}_d = \mathbf{r}_{d0}$ in the image. If $\mathbf{M} = m\mathbf{I}$ and $\mathbf{r}_{d0} = 0$, then (7.181) reduces to (7.172).

The point sensitivity, defined by (7.113), is a constant for this system since the integral of $h(\mathbf{r}_d,\mathbf{r})$ over $\mathbf{r}_d$ is unity for all $\mathbf{r}$, as we can see by applying the sifting property of delta functions to the first form of (7.181). Similarly, the flood image is also constant, as we can see by integrating the second form over $\mathbf{r}$.

SVD for this system is simple since the kernel for $\mathcal{H}^\dagger\mathcal{H}$ is given by

$$k(\mathbf{r},\mathbf{r}') = \int_\infty d^q r_d \; \delta(\mathbf{r}_d - \mathbf{M}\mathbf{r} - \mathbf{r}_{d0})\, \delta(\mathbf{r}_d - \mathbf{M}\mathbf{r}' - \mathbf{r}_{d0})$$

$$= \delta(\mathbf{M}\mathbf{r} - \mathbf{M}\mathbf{r}') = |\det(\mathbf{M})|^{-1}\, \delta(\mathbf{r}-\mathbf{r}') . \tag{7.182}$$

Thus $\mathcal{H}^\dagger\mathcal{H}$ is a constant times the identity operator. The SVD basis functions can be any orthonormal basis for the object space, including complex exponentials and space-domain delta-functions.

We can modify this discussion to apply to a system with affine mapping and shift-invariant blur. The delta function in (7.181) is replaced by $h(\mathbf{r}_d - \mathbf{M}\mathbf{r} - \mathbf{r}_{d0})$, and $\mathcal{H}^\dagger\mathcal{H}$ is readily shown to be LSIV. Then the SVD analysis of (7.177)–(7.180) holds simply by replacing the scalar magnification m with the matrix $\mathbf{M}$ and $|m|^q$ with $|\det(\mathbf{M})|$.

Shift-variant magnification Now suppose the magnification varies with position in the object field. Again neglecting blur, and setting $\mathbf{r}_{d0}$ to zero for simplicity, we write the PRF as

$$h(\mathbf{r}_d, \mathbf{r}) = \delta\{\mathbf{r}_d - [\mathbf{M}(\mathbf{r})]\,\mathbf{r}\}\ . \tag{7.183}$$

Now point $\mathbf{r}$ in the object maps to $[\mathbf{M}(\mathbf{r})]\mathbf{r}$ in the image.

Since the argument of the delta function vanishes if $\mathbf{r} = [\mathbf{M}(\mathbf{r})]^{-1}\mathbf{r}_d$, we see that $h(\mathbf{r}_d, \mathbf{r}) \propto \delta\{\mathbf{r} - [\mathbf{M}(\mathbf{r})]^{-1}\mathbf{r}_d\}$, but the proportionality factor is a function of $\mathbf{r}$. To compute it rigorously, we would need to apply the multidimensional counterpart of the transformation rule (2.33), which requires computing derivatives of $[\mathbf{M}(\mathbf{r})]\mathbf{r}$ with respect to each of the components of $\mathbf{r}$. A reasonable approximation is available, however, if $\mathbf{M}(\mathbf{r})$ varies slowly with $\mathbf{r}$, in which case

$$h(\mathbf{r}_d,\ \mathbf{r}) \approx \frac{1}{|\det[\mathbf{M}(\mathbf{r})]|}\,\delta\Big\{\mathbf{r} - [\mathbf{M}(\mathbf{r})]^{-1}\,\mathbf{r}_d\Big\}\ . \tag{7.184}$$

For this system, the point sensitivity is still independent of $\mathbf{r}$, as we can see by integrating (7.183) over $\mathbf{r}_d$, but now the flood image is not constant. To compute the flood image, we need to integrate (7.184) over $\mathbf{r}$. For a specific $\mathbf{r}_d$, the argument of the delta function vanishes at $\mathbf{r} = \mathbf{r}_0$, where $\mathbf{r}_0 = [\mathbf{M}(\mathbf{r}_0)]^{-1}\mathbf{r}_d$. Actually finding $\mathbf{r}_0$ may require numerical methods or a specific analytic model for $\mathbf{M}(\mathbf{r})$. If $\mathbf{M}(\mathbf{r})$ is slowly varying in the vicinity of $\mathbf{r}_0$, then the delta function can be integrated, with the result

$$g_{fld}(\mathbf{r}_d) \simeq \frac{1}{|\det[\mathbf{M}(\mathbf{r}_0)]|}\ . \tag{7.185}$$

Thus the flood image at point $\mathbf{r}_d$ is the reciprocal of the determinant of the transformation matrix evaluated at the point $\mathbf{r}_0$ that solves $[\mathbf{M}(\mathbf{r}_0)]\mathbf{r}_0 = \mathbf{r}_d$. The distortion associated with the space-variant mapping results in a nonuniform flood image.

By analogy to (7.182), the kernel of $\mathcal{H}^\dagger\mathcal{H}$ for a non-blurring, shift-variant magnifier is given by

$$\begin{aligned} k(\mathbf{r}, \mathbf{r}') &= \int_\infty d^q r_d\ \delta\{\mathbf{r}_d - [\mathbf{M}(\mathbf{r})]\,\mathbf{r}\}\,\delta\{\mathbf{r}_d - [\mathbf{M}(\mathbf{r}')]\,\mathbf{r}'\} \\ &= \delta\{[\mathbf{M}(\mathbf{r})]\,\mathbf{r} - [\mathbf{M}(\mathbf{r}')]\,\mathbf{r}'\} \simeq \frac{1}{|\det[\mathbf{M}(\mathbf{r})]|}\,\delta(\mathbf{r} - \mathbf{r}')\,, \end{aligned} \tag{7.186}$$

where the last step again assumes $\mathbf{M}(\mathbf{r})$ is slowly varying.

The SVD domain in this case is the space domain since

$$\mathcal{H}^\dagger\mathcal{H}\,\delta(\mathbf{r} - \mathbf{a}) = \frac{1}{|\det[\mathbf{M}(\mathbf{a})]|}\,\delta(\mathbf{r} - \mathbf{a})\,. \tag{7.187}$$

Thus the eigenvectors and eigenvalues of $\mathcal{H}^\dagger\mathcal{H}$ are indexed by a vector $\mathbf{a}$, and we have

$$u_{\mathbf{a}}(\mathbf{r}) = \delta(\mathbf{r}-\mathbf{a})\,, \qquad \mu_{\mathbf{a}} = \frac{1}{|\det[\mathbf{M}(\mathbf{a})]|}\,. \tag{7.188}$$

This result illustrates a general rule: the SVD domain for *any* invertible point-to-point mapping without blur is the space domain. The reader is invited to extend the calculation above to an operator $\mathcal{H}$ with kernel $\delta[\mathbf{r}_d - \mathbf{\Phi}(\mathbf{r})]$, where $\mathbf{\Phi}(\mathbf{r})$ is an arbitrary, invertible, vector-valued function. It will be seen that the eigenvectors of $\mathcal{H}^\dagger\mathcal{H}$ are delta functions; the assumption that $\mathbf{\Phi}(\mathbf{r})$ is slowly varying will be needed only to determine the eigenvalues.

7.2.8 Approximately shift-invariant systems

In this section we examine several other systems where the LSIV model is useful but not exact.

Systems with finite field A real imaging system has a finite field of view (FOV) in object space and records an image of finite size. The image field is the same thing as the support of the image, which we denoted as $\mathbf{S}_g$ in Sec. 7.1.2, but we have to be more careful for the FOV in object space. The object may be known *a priori* to fit within support $\mathbf{S}_f$, but there is no guarantee that the imaging system can respond to all points in this region.

For direct imaging, the main physical effects that limit the fields are finite detector size and vignetting (obscuration of one part of the imaging system by another). The effect of detector size is simply to multiply the image by a function $b_g(\mathbf{r}_d)$, which is 1 for $\mathbf{r}_d$ in $\mathbf{S}_g$ and 0 otherwise. It may be valid to account for vignetting or other field limitations in object space by defining a similar function $b_f(\mathbf{r})$. We shall refer to $b_f(\mathbf{r})$ and $b_g(\mathbf{r}_d)$ as *FOV functions.*

If the blur is shift-invariant except for the FOV functions, the PRF will have the form

$$h(\mathbf{r}_d, \mathbf{r}) = h(\mathbf{r}_d - \mathbf{r})\, b_g(\mathbf{r}_d)\, b_f(\mathbf{r})\,. \tag{7.189}$$

The kernel for $\mathcal{H}^\dagger\mathcal{H}$ is given by [*cf.* (7.116)]

$$k(\mathbf{r}, \mathbf{r}') = b_f(\mathbf{r})\, b_f(\mathbf{r}') \int_\infty d^q r_d\, b_g(\mathbf{r}_d)\, h^*(\mathbf{r}_d - \mathbf{r})\, h(\mathbf{r}_d - \mathbf{r}')\,, \tag{7.190}$$

where we have used the fact that $b_f(\mathbf{r})$ is either 0 or 1, hence it is equal to its square.

Because $k(\mathbf{r}, \mathbf{r}')$ is not a function of $\mathbf{r} - \mathbf{r}'$ alone, the singular functions are not complex exponentials; nevertheless, it is interesting to see what happens if $\mathcal{H}^\dagger\mathcal{H}$ operates on $\psi_{\boldsymbol{\rho}}(\mathbf{r}) = \exp(2\pi i \boldsymbol{\rho}\cdot\mathbf{r})$. With the kernel from (7.190), we have

$$\left[\mathcal{H}^\dagger\mathcal{H}\psi_{\boldsymbol{\rho}}\right](\mathbf{r})$$

$$= b_f(\mathbf{r}) \int_\infty d^q r_d\, b_g(\mathbf{r}_d)\, h^*(\mathbf{r}_d - \mathbf{r}) \int_\infty d^q r'\, b_f(\mathbf{r}')\, h(\mathbf{r}_d - \mathbf{r}') \exp(2\pi i \boldsymbol{\rho}\cdot\mathbf{r}')\,. \tag{7.191}$$

If we represent each of the functions in the double integral by its Fourier transform, we obtain

$$[\mathcal{H}^\dagger\mathcal{H}\psi_\rho](\mathbf{r})$$
$$= b_f(\mathbf{r})\int_\infty d^q r_d \int_\infty d^q r' \int_\infty d^q \rho_d \int_\infty d^q \rho_d' \int_\infty d^q \rho' \int_\infty d^q \rho''\, B_g(\boldsymbol{\rho}_d)\, H^*(\boldsymbol{\rho}_d')\, B_f(\boldsymbol{\rho}')\, H(\boldsymbol{\rho}'')$$
$$\times \exp\{2\pi i\,[\boldsymbol{\rho}\cdot\mathbf{r}' + \boldsymbol{\rho}_d\cdot\mathbf{r}_d - \boldsymbol{\rho}_d'\cdot(\mathbf{r}_d - \mathbf{r}) + \boldsymbol{\rho}'\cdot\mathbf{r}' + \boldsymbol{\rho}''(\mathbf{r}_d - \mathbf{r}')]\}\,. \qquad (7.192)$$

Though this expression may appear frightening, it simplifies readily. By (3.217), the integral over $\mathbf{r}_d$ yields $\delta(\boldsymbol{\rho}_d - \boldsymbol{\rho}_d' + \boldsymbol{\rho}'')$ and the one over $\mathbf{r}'$ yields $\delta(\boldsymbol{\rho} + \boldsymbol{\rho}' - \boldsymbol{\rho}'')$. These two delta functions let us perform two other integrals, and we find

$$[\mathcal{H}^\dagger\mathcal{H}\psi_\rho](\mathbf{r})$$
$$= b_f(\mathbf{r})\int_\infty d^q \rho_d \int_\infty d^q \rho'\, B_g(\boldsymbol{\rho}_d)\, H^*(\boldsymbol{\rho}_d + \boldsymbol{\rho} + \boldsymbol{\rho}')\, B_f(\boldsymbol{\rho}')\, H(\boldsymbol{\rho} + \boldsymbol{\rho}')$$
$$\times \exp[2\pi i(\boldsymbol{\rho} + \boldsymbol{\rho}_d + \boldsymbol{\rho}')\cdot\mathbf{r}]\,. \qquad (7.193)$$

So far this development has been exact, but now we make a reasonable approximation. We assume that the FOV functions $b_f(\mathbf{r})$ and $b_g(\mathbf{r}_d)$ cover a large area so that their Fourier transforms, $B_f(\boldsymbol{\rho}')$ and $B_g(\boldsymbol{\rho}_d)$, respectively, are peaked in a narrow range around the origin. For example, if an FOV function covers a region of width L in each dimension, then its Fourier transform is small when any component of the spatial frequency exceeds about $1/L$. If the transfer function $H(\boldsymbol{\rho})$ is slowly varying on this scale, it may be valid to replace $\boldsymbol{\rho}'$ and $\boldsymbol{\rho}_d$ with zero in the arguments of $H^*(\boldsymbol{\rho}_d + \boldsymbol{\rho} + \boldsymbol{\rho}')$ and $H(\boldsymbol{\rho} + \boldsymbol{\rho}')$. Doing so gives

$$[\mathcal{H}^\dagger\mathcal{H}\psi_\rho](\mathbf{r}) \simeq b_f(\mathbf{r})\,|H(\boldsymbol{\rho})|^2 \int_\infty d^q \rho_d \int_\infty d^q \rho'\, B_g(\boldsymbol{\rho}_d)\, B_f(\boldsymbol{\rho}')\exp[2\pi i(\boldsymbol{\rho} + \boldsymbol{\rho}_d + \boldsymbol{\rho}')\cdot\mathbf{r}]$$
$$= b_f(\mathbf{r})\, b_g(\mathbf{r})\,|H(\boldsymbol{\rho})|^2 \exp(2\pi i\boldsymbol{\rho}\cdot\mathbf{r})\,. \qquad (7.194)$$

We can also define $b_{tot}(\mathbf{r}) = b_f(\mathbf{r})\, b_g(\mathbf{r})$, which is a function that is unity for points inside both fields and zero elsewhere.

Thus the truncated exponential function $b_{tot}(\mathbf{r})\exp(2\pi i\boldsymbol{\rho}\cdot\mathbf{r})$ is a useful approximate singular function for systems described by (7.189); a large but finite field of view does not rule out Fourier analysis so long as the transfer function is relatively slowly varying on a scale given by the reciprocal of the width of the field. Roughly speaking, the errors made with this approximation will be concentrated within about δ of the edge of the field, where δ is one of the resolution measures defined in Sec. 7.2.1.

Weak shift-variance For a shift-variant linear system, the PRF must be a function of two variables. In Sec. 7.2.1 we chose those variables as $\mathbf{r}_d$ and $\mathbf{r}_0$ (where $\mathbf{r}_0$ is the location of the point in the object domain), but if $q = s$ we can equally well choose $\mathbf{r}_d - \mathbf{r}_0$ and $\mathbf{r}_0$, writing

$$h(\mathbf{r}_d, \mathbf{r}_0) = p(\mathbf{r}_d - \mathbf{r}_0; \mathbf{r}_0)\,. \qquad (7.195)$$

The form on the right is particularly useful if $p(\mathbf{r}_d - \mathbf{r}_0; \mathbf{r}_0)$ is a slowly varying function of its second argument.

To clarify this idea, we need to consider two characteristic distances. One is the full width at half maximum of the PRF, $\delta_{\text{FWHM}}(\mathbf{r}_0)$ [see (7.108)]. The other is a *variation distance*, defined loosely as a scalar $\Delta(\mathbf{r}_0)$ such that $p(\mathbf{r}_d - \mathbf{r}_0; \mathbf{r}_0) \simeq p(\mathbf{r}_d - \mathbf{r}_1; \mathbf{r}_1)$ if $|\mathbf{r}_1 - \mathbf{r}_0| < \Delta(\mathbf{r}_0)$. A system is said to be *weakly shift-variant* if $\delta_{\text{FWHM}}(\mathbf{r}_0) \ll \Delta(\mathbf{r}_0)$.

One way to analyze a weakly shift-variant system is via the continuous form of Gabor's signal expansion, introduced in Sec. 5.1.4. Generalizing (5.40) to q dimensions, we can express an object $f(\mathbf{r})$ as

$$f(\mathbf{r}) = \int_\infty d^q\rho \int_\infty d^q r_0 \; F_b(\boldsymbol{\rho}; \mathbf{r}_0)\, b(\mathbf{r} - \mathbf{r}_0) \exp(2\pi i \boldsymbol{\rho} \cdot \mathbf{r}) \,, \tag{7.196}$$

where we have assumed that the window function $b(\mathbf{r})$ is real and has unit norm. With these assumptions, the inverse of (7.196) is the multidimensional local Fourier transform [*cf.* (5.1)]:

$$F_b(\boldsymbol{\rho}; \mathbf{r}_0) = \int_\infty d^q r \; b(\mathbf{r} - \mathbf{r}_0)\, f(\mathbf{r}) \exp(-2\pi i \boldsymbol{\rho} \cdot \mathbf{r}) \,. \tag{7.197}$$

If we apply the operator with kernel (7.195) to (7.196), we obtain

$$g(\mathbf{r}_d) = [\boldsymbol{\mathcal{H}}\mathbf{f}](\mathbf{r}_d)$$

$$= \int_\infty d^q\rho \int_\infty d^q r_0 \; F_b(\boldsymbol{\rho}; \mathbf{r}_0) \int_\infty d^q r \; b(\mathbf{r} - \mathbf{r}_0)\, p(\mathbf{r}_d - \mathbf{r}; \mathbf{r}) \exp(2\pi i \boldsymbol{\rho} \cdot \mathbf{r}) \,. \tag{7.198}$$

For this expression to be useful, we must be able to choose the width δ_b of the window function so that

$$\delta_{\text{FWHM}}(\mathbf{r}_0) \ll \delta_b \ll \Delta(\mathbf{r}_0) \tag{7.199}$$

for all $\mathbf{r}_0$. If this condition is satisfied, $b(\mathbf{r} - \mathbf{r}_0) \simeq b(\mathbf{r}_d - \mathbf{r}_0)$ and $p(\mathbf{r}_d - \mathbf{r}; \mathbf{r}) \simeq p(\mathbf{r}_d - \mathbf{r}; \mathbf{r}_0)$, and we can write

$$g(\mathbf{r}_d) = \int_\infty d^q\rho \int_\infty d^q r_0 \; F_b(\boldsymbol{\rho}; \mathbf{r}_0)\, b(\mathbf{r}_d - \mathbf{r}_0) \int_\infty d^q r \; p(\mathbf{r}_d - \mathbf{r}; \mathbf{r}_0) \exp(2\pi i \boldsymbol{\rho} \cdot \mathbf{r})$$

$$= \int_\infty d^q\rho \int_\infty d^q r_0 \; F_b(\boldsymbol{\rho}; \mathbf{r}_0)\, P(\boldsymbol{\rho}; \mathbf{r}_0)\, b(\mathbf{r}_d - \mathbf{r}_0) \exp(2\pi i \boldsymbol{\rho} \cdot \mathbf{r}_d) \,, \tag{7.200}$$

where $P(\boldsymbol{\rho}; \mathbf{r}_0)$ is the Fourier transform (with respect to $\mathbf{r}_d$) of $p(\mathbf{r}_d; \mathbf{r}_0)$.

The similarity between (7.200) and (7.151) should be noted. The latter equation shows that the output of an LSIV system can be computed by multiplying the Fourier transform of the input by the transfer function and then taking an inverse Fourier transform; the former equation shows that the same thing holds true for weakly shift-variant systems if *local* Fourier transforms are used and the conditions (7.199) can be satisfied.

Under the same conditions, the windowed exponentials $b(\mathbf{r} - \mathbf{r}_0) \exp(2\pi i \boldsymbol{\rho} \cdot \mathbf{r})$ are approximately eigenfunctions (hence also singular functions) of $\boldsymbol{\mathcal{H}}$ and the eigenvalues are $P(\boldsymbol{\rho}; \mathbf{r}_0)$. Though the eigenfunctions and eigenvalues are characterized here by two continuous indices, it follows from the discussion in Sec. 5.1.4 that a countably infinite complete set can be obtained by sampling $\boldsymbol{\rho}$ and $\mathbf{r}_0$ on a Gabor lattice.

7.2.9 Rotationally symmetric systems

A CC system may have rotational symmetry, whether or not it is shift-invariant. A simple example is a rotationally symmetric lens with aberrations. As we shall see in Chap., the aberrations usually spoil the shift-invariance, but the rotational invariance remains and has a strong effect on the allowable forms of $\mathcal{H}$ and $\mathcal{H}^\dagger\mathcal{H}$. In this section we shall discuss the effects of rotational symmetry specifically for systems that map a 2D object to a 2D image. For essential background material, see Chap. 6.

Symmetry group To give a precise definition of rotational symmetry, we need to introduce a Lie group of rotation operators. This group, denoted $\mathbf{C}_\infty$ in Sec. 6.5, can describe either geometric rotations in the plane $\mathbb{R}^2$ or transformations of functions in the Hilbert space $\mathbb{U}$. As in Chap. 6 we make a notational distinction between these two kinds of operators by using $\mathcal{T}$ with various subscripts to denote the functional operator. We denote a geometric rotation through angle ϕ by the operator $\mathcal{R}_\phi$ and the corresponding functional transformation by $\mathcal{T}_\phi$. Specifically, if the 2D vectors $\mathbf{r}$ and $\mathbf{r}'$ are given in polar coordinates by (r, θ) and (r', θ'), respectively, and $\mathbf{r}' = \mathcal{R}_\phi \mathbf{r}$, then $r' = r$ and $\theta' = \theta + \phi$. The functional operator $\mathcal{T}_\phi$ is defined by (6.24) as

$$[\mathcal{T}_\phi \mathbf{f}](\mathbf{r}) = f\left[\mathcal{R}_\phi^{-1}\mathbf{r}\right] = f[\mathcal{R}_{-\phi}\mathbf{r}] = f(r, \theta - \phi), \tag{7.201}$$

where $\mathbf{f}$ is an arbitrary vector in $\mathbb{U}$, corresponding to the function $f(r, \theta)$ in polar coordinates.

As discussed in Sec. 6.7, rotational symmetry means that all rotation operators, $\mathcal{T}_\phi$ for all ϕ, commute with $\mathcal{H}^\dagger\mathcal{H}$, so that

$$\mathcal{H}^\dagger\mathcal{H}\mathcal{T}_\phi = \mathcal{T}_\phi\mathcal{H}^\dagger\mathcal{H}. \tag{7.202}$$

In the language of group theory, this equation says that the symmetry group of the system is $\mathbf{C}_\infty$. In more intuitive terms, it says that we can either rotate the object in its plane by ϕ and then image it and backproject the result into object space, or we can first image and backproject and then rotate, and the result must be the same if the system has rotational symmetry.

Eigenfunctions The eigenfunctions of $\mathcal{H}^\dagger\mathcal{H}$ have been denoted by $u_n(\mathbf{r})$, but in this problem it is natural to express the 2D vector $\mathbf{r}$ in polar coordinates as (r, θ). In addition, it will prove useful to use two indices, j and k, in place of n, so we shall denote the eigenfunctions by $u_{jk}(r, \theta)$, and we shall use $\mathbf{u}_{jk}$ to denote the corresponding Hilbert-space vector. With these notational changes, the eigenvalue equation (7.117) becomes

$$\left[\mathcal{H}^\dagger\mathcal{H}\mathbf{u}_{jk}\right](r, \theta) = \mu_{jk}\, u_{jk}(r, \theta). \tag{7.203}$$

It follows from (7.202) and (7.203) that

$$\mathcal{H}^\dagger\mathcal{H}\mathcal{T}_\phi\mathbf{u}_{jk} = \mathcal{T}_\phi\mathcal{H}^\dagger\mathcal{H}\mathbf{u}_{jk} = \mu_{jk}\mathcal{T}_\phi\mathbf{u}_{jk}. \tag{7.204}$$

Looking at the first and third forms in this equation, we see that $\mathcal{T}_\phi\mathbf{u}_{jk}$ is an eigenvector of $\mathcal{H}^\dagger\mathcal{H}$ with eigenvalue μ_{jk} if $\mathbf{u}_{jk}$ is such an eigenvector (see Sec. 6.7.4). If the eigenvalue μ_{jk} is not degenerate, then $\mathcal{T}_\phi\mathbf{u}_{jk}$ can only be a constant times $\mathbf{u}_{jk}$. In other words, $\mathbf{u}_{jk}$ must be simultaneously an eigenvector of $\mathcal{T}_\phi$ and $\mathcal{H}^\dagger\mathcal{H}$.

To discover the structure of these eigenvectors, we note from (7.201) that $u_{jk}(r,\theta)$ is an eigenvector of $\mathcal{T}_\phi$ if

$$u_{jk}(r,\theta-\phi) = const \cdot u_{jk}(r,\theta)\,. \tag{7.205}$$

The solution to this functional equation is

$$u_{jk}(r,\theta) = u_{jk}(r)\, e^{ik\theta}\,, \tag{7.206}$$

where $u_{jk}(r)$ is an arbitrary function of r.

The constant in (7.205) (the eigenvalue of $\mathcal{T}_\phi$) is given by

$$\chi^{(k)}(\phi) = e^{-ik\phi}\,, \qquad k = 0, \pm1, \pm2, \ldots\,. \tag{7.207}$$

Comparison with (6.23) reveals the reason for the notation chosen: $\chi^{(k)}(\phi)$ is the character associated with $\mathcal{T}_\phi$ in the kth irreducible representation of $\mathbf{C}_\infty$. Since $\mathbf{C}_\infty$ is an Abelian group, all of its irreducible representations are 1D, so all operators in the (infinite) group are represented by 1×1 matrices and act on 1D subspaces of the Hilbert space $\mathbb{U}$. The kth subspace consists of all scalar multiples of $e^{ik\theta}$ (times an arbitrary function of r), and the matrix corresponding to the operator $\mathcal{T}_\phi$ in this representation is the scalar $e^{-ik\phi}$. Since the matrices are 1×1, the character (trace of the matrix) and the eigenvalue of $\mathcal{T}_\phi$ are identical.

The conclusion from this discussion is that the nondegenerate eigenfunctions of $\mathcal{H}^\dagger\mathcal{H}$ for a rotationally symmetric system must have the form,

$$u_{jk}(r,\theta) = u_{jk}(r)\, e^{ik\theta}\,. \tag{7.208}$$

The utility of the double index jk is now apparent: the index k specifies the angular dependence, while different eigenvectors with the same angular dependence are distinguished by j. Since k also indexes the irreducible representation, we can say that $\mathbf{u}_{jk}$ transforms under rotation according to the kth irreducible representation of $\mathbf{C}_\infty$.

A similar argument shows that nondegenerate eigenfunctions of $\mathcal{H}\mathcal{H}^\dagger$ have the form,

$$v_{jk}(r_d,\theta_d) = v_{jk}(r_d)\, e^{ik\theta_d}\,. \tag{7.209}$$

This function also transforms under rotation according to the kth irreducible representation of $\mathbf{C}_\infty$, but now the rotation operator is defined in image space $\mathbb{V}$.

Form of the operators We can use (7.208) along with the spectral decomposition (see Sec. 1.4.5) to discuss the structure of the operators $\mathcal{H}^\dagger\mathcal{H}$ and $\mathcal{H}$. In the present notation, the spectral decomposition of $\mathcal{H}^\dagger\mathcal{H}$ is

$$\mathcal{H}^\dagger\mathcal{H} = \sum_{j,k} \mu_{jk}\, \mathbf{u}_{jk}\, \mathbf{u}_{jk}^\dagger\,. \tag{7.210}$$

In polar coordinates, where $\mathbf{r} = (r,\theta)$ and $\mathbf{r}' = (r',\theta')$, the kernel of this operator is

$$k(\mathbf{r},\mathbf{r}') = k(r,\theta,r',\theta') = \sum_{j,k} \mu_{jk}\, u_{jk}(r)\, u_{jk}^*(r')\, e^{ik(\theta-\theta')}\,. \tag{7.211}$$

By inspection, the kernel is a function of only the three variables, r, r' and $\theta - \theta'$. Angular shift-invariance results from rotational symmetry in the same way that the usual positional shift-invariance results from translational symmetry. The angular shift-invariance can also be stated as

$$k[\mathcal{R}_\phi \mathbf{r}, \mathcal{R}_\phi \mathbf{r}'] = k(\mathbf{r}, \mathbf{r}')\,. \tag{7.212}$$

A similar argument can be used to determine the structure of the operator $\mathcal{H}$. From the SVD expansion (7.127) and the forms (7.208) and (7.209), the kernel for $\mathcal{H}$ is given by [*cf.* (7.127)]

$$h(\mathbf{r}_d, \mathbf{r}) = h(r_d, \theta_d, r, \theta) = \sum_{j,k} \sqrt{\mu_{jk}}\, v_{jk}(r_d)\, u^*_{jk}(r)\, e^{ik(\theta_d - \theta)}\,. \tag{7.213}$$

This kernel is a function of only r_d, r and $\theta_d - \theta$. We shall make use of this structure in Chap. 9 when we discuss aberrations of rotationally symmetric optical systems.

Mirror symmetry and degeneracy We assumed above that $\mathcal{H}^\dagger\mathcal{H}$ has no degeneracies, so each eigenvalue μ_{jk} corresponds uniquely to an eigenvector $\mathbf{u}_{jk}$. This assumption may not hold if the system has other symmetries (see Sec. 6.7.5) so that $\mathbf{C}_\infty$ is just a subgroup of the full symmetry group of the system. An important additional symmetry, which often accompanies $\mathbf{C}_\infty$, is mirror-reflection symmetry. As a geometric transformation in $\mathbb{R}_2$, mirror reflection about the x axis transforms point (x, y) to $(x, -y)$. We denote this operator by $\mathcal{M}_0$, where the subscript indicates that the mirror axis is the line $\phi = 0$, or the x axis. Since $\mathcal{M}_0$ is its own inverse, the corresponding functional transformation, denoted $\mathcal{T}_{\mathcal{M}_0}$, satisfies $[\mathcal{T}_{\mathcal{M}_0}\mathbf{f}](x, y) = f(x, -y)$.

An immediate consequence of mirror symmetry is that the kernel of $\mathcal{H}^\dagger\mathcal{H}$ must satisfy $k[\mathcal{M}_0\mathbf{r}, \mathcal{M}_0\mathbf{r}'] = k(\mathbf{r}, \mathbf{r}')$ [*cf.* (7.212)]. Since mirror reflection about the x axis transforms θ to $-\theta$, it follows (for a system with rotational and mirror symmetry) that $k(r, r', \theta - \theta') = k(r, r', \theta' - \theta)$, where $k(r, r', \theta - \theta')$ is the same function as $k(r, \theta, r', \theta')$ but with the angular shift-invariance displayed explicitly. Thus the kernel can only be an even function of $\theta - \theta'$.

To examine the degeneracies of the eigenvalues, we must first determine the full symmetry group of the system. If a rotationally symmetric $\mathcal{H}^\dagger\mathcal{H}$ is invariant under $\mathcal{T}_{\mathcal{M}_0}$, it is also invariant under all other mirror reflections obtained by rotating the mirror axis by an arbitrary angle. The full symmetry group of the system thus consists of all rotations $\mathcal{T}_\phi$ and all mirror transformations of the form $\mathcal{T}_{\mathcal{M}_\phi} \equiv \mathcal{T}_\phi^{-1}\mathcal{T}_{\mathcal{M}_0}\mathcal{T}_\phi$. This group is isomorphic to the group of geometric operators that leave a uniform disc invariant. It will be denoted $\mathbf{D}_\infty$ since it can be regarded as the limit as $N \to \infty$ of the dihedral group $\mathbf{D}_N$ (the symmetry group of an N-sided regular polygon), which is discussed in Sec. 6.4.2. The group $\mathbf{D}_\infty$ has two 1D irreducible representations and a denumerable infinity of 2D ones. A 2D irreducible representation of the symmetry group of the system leads to two-fold degeneracy of the eigenvectors of $\mathcal{H}^\dagger\mathcal{H}$.

Specifically, $u_{jk}(r, \theta)$ and $[\mathcal{T}_{\mathcal{M}_0}\mathbf{u}_{jk}](r, \theta)$ must be degenerate if $\mathcal{T}_{\mathcal{M}_0}$ commutes with $\mathcal{H}^\dagger\mathcal{H}$. Since the eigenfunctions have the form of (7.206) by rotational symmetry, and mirror reflection about the x axis merely reverses the sign of θ, we have

$$[\mathcal{T}_{\mathcal{M}_0}\mathbf{u}_{jk}]\,(r, \theta) = u_{jk}(r)\, e^{-ik\theta}\,, \tag{7.214}$$

and the eigenfunctions $u_{jk}(r)\exp(ik\theta)$ and $u_{jk}(r)\exp(-ik\theta)$ must be degenerate. If $k = 0$, the two functions are identical, so $k = 0$ corresponds to a 1D irreducible representation. For any other k, however, the two functions are linearly independent and the irreducible representation is 2D.

We can always choose the eigenfunctions to have the form $u_{jk}(r)\exp(ik\theta)$, in spite of the degeneracy. Suppose P linearly independent eigenvectors $\mathbf{u}_{jk}^{(p)}$, $p = 1, ..., P$, have the same eigenvalue μ_{jk}. These vectors define a P-dimensional subspace, and any linear combination of vectors in the space is an eigenvector of $\mathcal{H}^\dagger\mathcal{H}$ with eigenvalue μ_{jk}. Moreover, we know from (7.204) that this subspace is invariant under $\mathbf{C}_\infty$ if $\mathcal{H}^\dagger\mathcal{H}$ is rotationally symmetric. That is, the action of $\mathcal{T}_\phi$ on any vector in the space yields another vector in the space. Following (6.68), we can construct a representation of $\mathbf{C}_\infty$ on this space by forming the $P \times P$ matrices $\mathbf{M}(\mathcal{T}_\phi)$, with elements given by

$$[\mathbf{M}(\mathcal{T}_\phi)]_{p'p} = \left(\mathbf{u}_{jk}^{(p')}, \mathcal{T}_\phi \mathbf{u}_{jk}^{(p)}\right) . \tag{7.215}$$

Since $\mathbf{C}_\infty$ is Abelian, it has only 1D irreducible representations, so the P-dimensional representation defined by (7.215) is necessarily reducible if $P > 1$. We can therefore find a $P \times P$ unitary transformation that will diagonalize each of the P matrices of (7.215), and each diagonal element will be one of the characters defined in (7.207). That is, we can choose the basis of the subspace to consist of eigenvectors of the rotation operators. Since the unitary transformation leaves us within the space of degenerate eigenvectors, the resulting basis vectors are also eigenvectors of $\mathcal{H}^\dagger\mathcal{H}$, and the corresponding eigenfunctions have the form $u_{jk}(r)\exp(ik\theta)$.

Rotational and translational symmetry: The kernels Next we consider systems that are invariant under both rotations and translations. We saw in (7.213) that the kernel of a rotationally symmetric $\mathcal{H}$ is a function of r, r_d and $\theta - \theta_d$, and we know that the kernel (PSF) of a shift-invariant system is a function of $\mathbf{r}_d - \mathbf{r}$. Here we examine the form of the kernel when both symmetries are present.

A system is rotationally and translationally symmetric if $h(\mathbf{r}_d, \mathbf{r}) = p(|\mathbf{r}_d - \mathbf{r}|)$, where $p(\cdot)$ is a scalar-valued function. This form satisfies both conditions in the paragraph above. Any function of $|\mathbf{r}_d - \mathbf{r}|$ is also a function of $\mathbf{r}_d - \mathbf{r}$ as required by translational symmetry and, since $|\mathbf{r}_d - \mathbf{r}| = [r^2 + r_d^2 + 2rr_d\cos(\theta - \theta_d)]^{\frac{1}{2}}$, $p(|\mathbf{r}_d - \mathbf{r}|)$ can be rewritten as a function of r, r_d and $\theta - \theta_d$ as required by rotational symmetry.

Thus any LSIV system where the PSF $p(\mathbf{r})$ is a function of only the magnitude of r is rotationally symmetric, but the converse does not necessarily hold. Since we have defined rotational symmetry by the condition that $\mathcal{T}_\phi\mathcal{H}^\dagger\mathcal{H} = \mathcal{H}^\dagger\mathcal{H}\mathcal{T}_\phi$, it is possible for an LSIV system to have rotational symmetry but for the PSF to depend on θ. A simple example is when $p(\mathbf{r})$ is a function of $|\mathbf{r} - \mathbf{a}|$. The offset $\mathbf{a}$ makes $p(\mathbf{r})$ a function of θ, but $k(\mathbf{r})$, the PSF associated with $\mathcal{H}^\dagger\mathcal{H}$, is the autocorrelation of $p(\mathbf{r})$ and hence insensitive to offset. Thus $k(\mathbf{r}) = k(r)$. The same conclusion emerges in the Fourier domain when we recognize that the transfer function $H(\boldsymbol{\rho})$ has a factor $\exp(2\pi i\boldsymbol{\rho} \cdot \mathbf{a})$ resulting from the offset, but the transfer function associated with $\mathcal{H}^\dagger\mathcal{H}$ is $|H(\boldsymbol{\rho})|^2$. Since the phase factor disappears when we take the squared modulus, we can have $|H(\boldsymbol{\rho})|^2 = |H(\rho)|^2$ even though $H(\boldsymbol{\rho}) \neq H(\rho)$.

We could remove this peculiarity by defining rotational symmetry differently. If object space $\mathbb{U}$ and image space $\mathbb{V}$ are the same, we could require that $\mathcal{H}$ itself, and not just $\mathcal{H}^\dagger\mathcal{H}$, commute with each $\mathcal{T}_\phi$. Then the kernel of $\mathcal{H}$ would have angular shift-invariance and the eigenfunctions of $\mathcal{H}$ would vary as $\exp(ik\theta)$. The advantage of working with $\mathcal{H}^\dagger\mathcal{H}$, however, is that we can define rotational and other symmetries even when $\mathbb{U}$ and $\mathbb{V}$ are different spaces. An example where this feature is useful will be seen in Sec. 7.2.10.

Rotational and translational symmetry: The eigenfunctions We know from Sec. 7.2.4 that the eigenfunctions of $\mathcal{H}^\dagger\mathcal{H}$ for LSIV systems are the complex exponentials, $\exp(2\pi i\boldsymbol{\rho}\cdot\mathbf{r})$, and this statement must still hold true if the system also has rotational symmetry. We have also argued, however, that the eigenfunctions of rotationally symmetric systems should vary as $e^{ik\theta}$. To reconcile these statements, we must recognize that the translation-rotation group, like the affine group, has only infinite-dimensional irreducible representations (see Sec. 6.8.5). That means that all eigenfunctions have infinite degeneracy for an LSIV system with rotational symmetry.

Though this conclusion may be surprising as a mathematical proposition, there is actually a simple physical explanation. We saw in Sec. 7.2.5 that the eigenvalues of $\mathcal{H}^\dagger\mathcal{H}$, now indexed by the continuous vector variable $\boldsymbol{\rho}$, are given by $|H(\boldsymbol{\rho})|^2$, where $H(\boldsymbol{\rho})$ is the transfer function [see (7.147)]. If the system is rotationally symmetric, $|H(\boldsymbol{\rho})|^2 = |H(\rho)|^2$. All points on a circle around the origin in the 2D frequency plane have the same transfer function and hence the same eigenvalue. The subspace associated with a set of degenerate eigenvectors now has infinite dimensionality.

We are free to choose as the basis for this subspace either functions of the form $\exp(2\pi i\boldsymbol{\rho}\cdot\mathbf{r})$ (with $\boldsymbol{\rho}$ constant and the functions indexed by θ_ρ) or functions proportional to $e^{ik\theta}$ for integer k. A function from either set can be expanded in terms of the other set. For example, a Fourier basis function, rewritten in polar coordinates, can be expressed as

$$u_{\boldsymbol{\rho}}(\mathbf{r}) = \exp(2\pi i\boldsymbol{\rho}\cdot\mathbf{r}) = \exp\left[2\pi i\rho\, r\cos(\theta-\theta_\rho)\right]$$

$$= \sum_{k=-\infty}^{\infty} A_k(r,\rho,\theta_\rho)\, e^{ik\theta}\,. \tag{7.216}$$

The coefficients can be found by the usual formula for Fourier-series coefficients, with the result,

$$A_k(r,\rho,\theta_\rho) = \mathrm{J}_k(2\pi\rho\, r)\exp(-2\pi ik\theta_\rho)\,, \tag{7.217}$$

where $\mathrm{J}_k(\cdot)$ is a Bessel function.

7.2.10 Axial systems

Many real-world imaging systems (including our eyes) produce 2D images of 3D objects, so $q = 3$ and $s = 2$. These systems may have a preferred axis, often referred to as the *optic axis*. For a rotationally symmetric system, the axis of rotational symmetry is the optic axis, but rotational symmetry is not the only way to establish a preferred axis. For a pinhole or coded-aperture system, for example, it is natural to take the optic axis to be normal to the plane of the aperture and detector. We shall refer to such imaging systems as *axial systems*, and the optic axis will consistently be taken as the z axis.

It may be a good approximation to model an axial system as LSIV in each plane normal to the axis and as a simple integrator over z. With this model, the system operator is specified by

$$g(\mathbf{r}_d) = [\mathcal{H}\mathbf{f}](\mathbf{r}_d) = \int_\infty d^2r \int_0^\infty dz\, h(\mathbf{r}_d - \mathbf{r}; z)\, f(\mathbf{r}, z)\,, \tag{7.218}$$

where $\mathbf{r}$ and $\mathbf{r}_d$ are both 2D vectors but $(\mathbf{r}, z)$ specifies location in 3D. This model is appropriate for emissive objects that are not opaque to their own radiation (as in fluorescent microscopy). It works also for reflective objects (as in ordinary photography) if structures close to the imaging system do not obscure more distant ones.

By the methods developed in Sec. 1.3.5, the adjoint operator is given by

$$\left[\mathcal{H}^\dagger \mathbf{g}\right](\mathbf{r}, z) = \int_\infty d^2r_d\, h^*(\mathbf{r}_d - \mathbf{r}; z)\, g(\mathbf{r}_d)\,, \tag{7.219}$$

and hence

$$\begin{aligned}\left[\mathcal{H}^\dagger\mathcal{H}\mathbf{f}\right](\mathbf{r}, z) &= \int_\infty d^2r' \int_0^\infty dz'\, f(\mathbf{r}', z') \int_\infty d^2r_d\, h^*(\mathbf{r}_d - \mathbf{r}; z)\, h(\mathbf{r}_d - \mathbf{r}'; z') \\ &= \int_\infty d^2r' \int_0^\infty dz'\, f(\mathbf{r}', z')\, p(\mathbf{r} - \mathbf{r}'; z, z')\,,\end{aligned} \tag{7.220}$$

where $p(\mathbf{r} - \mathbf{r}'; z, z')$ is just the indicated integral over $\mathbf{r}_d$, which is the complex cross-correlation (see Sec. 3.3.6) of $h(\mathbf{r}; z)$ and $h(\mathbf{r}; z')$. The physical picture is that a point source at $(\mathbf{r}', z')$ produces an image value of $p(\mathbf{r} - \mathbf{r}'; z, z')$ at point $(\mathbf{r}, z)$ after projection ($\mathcal{H}$) and backprojection ($\mathcal{H}^\dagger$).

As usual, to perform an SVD we must first find the eigenfunctions of $\mathcal{H}^\dagger\mathcal{H}$. Since the system is LSIV in the lateral dimensions, a 2D Fourier transform is indicated. Denoting the eigenfunctions as

$$u_{\boldsymbol{\rho},j}(\mathbf{r}, z) = \tilde{u}_{\boldsymbol{\rho},j}(z) \exp(2\pi i \boldsymbol{\rho} \cdot \mathbf{r})\,, \tag{7.221}$$

we can write the eigenvalue equation as

$$\begin{aligned}\left[\mathcal{H}^\dagger\mathcal{H}\mathbf{u}_{\boldsymbol{\rho},j}\right](\mathbf{r}, z) &= \int_\infty d^2r' \int_0^\infty dz'\, \tilde{u}_{\boldsymbol{\rho},j}(z') \exp(2\pi i \boldsymbol{\rho} \cdot \mathbf{r}')\, p(\mathbf{r} - \mathbf{r}'; z, z') \\ &= \exp(2\pi i \boldsymbol{\rho} \cdot \mathbf{r}) \int_0^\infty dz'\, \tilde{u}_{\boldsymbol{\rho},j}(z') P(\boldsymbol{\rho}; z, z') \\ &= \exp(2\pi i \boldsymbol{\rho} \cdot \mathbf{r})\, H(-\boldsymbol{\rho}; z) \int_0^\infty dz'\, \tilde{u}_{\boldsymbol{\rho},j}(z')\, H(\boldsymbol{\rho}; z')\,,\end{aligned} \tag{7.222}$$

where $P(\boldsymbol{\rho}; z, z')$ is the 2D Fourier transform of $p(\mathbf{r}; z, z')$, and we have used (3.134) to factor it as $P(\boldsymbol{\rho}; z, z') = H(-\boldsymbol{\rho}; z) H(\boldsymbol{\rho}; z')$.

Because the kernel factors, the integral on the middle line in (7.222) is a rank-one operator and hence has only a single nonzero eigenvalue; by our conventions, we denote that eigenvalue with $j = 1$. The measurement space of $\mathcal{H}^\dagger\mathcal{H}$ is thus spanned by the set $\{u_{\boldsymbol{\rho},1}(\mathbf{r}, z)\}$ for all $\boldsymbol{\rho}$, and there is also a null space spanned by $\{u_{\boldsymbol{\rho},j}(\mathbf{r}, z)\}$ for all $\boldsymbol{\rho}$ and $j > 1$.

We can obtain the eigenfunction and eigenvalue for $j = 1$ by inspection of (7.222). The integral on the last line is a constant, independent of $\mathbf{r}$ and z, and it is therefore the eigenvalue. If this integral is not zero, the only possible spatial dependence of the eigenfunction is

$$u_{\boldsymbol{\rho},1}(\mathbf{r}, z) \propto \exp(2\pi i \boldsymbol{\rho} \cdot \mathbf{r})\, H(-\boldsymbol{\rho}; z)\,. \tag{7.223}$$

The corresponding eigenvalue is

$$\mu_{\boldsymbol{\rho},1} = \int_{S_z} dz' \, |H(\boldsymbol{\rho}; z')|^2\,, \tag{7.224}$$

where the integral is now over the object support in the z direction. (The object support in x and y is infinite since the system is shift-invariant in those directions.) To complete the story, the reader may show that the required normalizing constant in (7.223) is just $1/\sqrt{\mu_{\boldsymbol{\rho},1}}$ and that the singular functions in image space are given by $v_{\boldsymbol{\rho},1}(\mathbf{r}_d) = \exp(2\pi i \boldsymbol{\rho} \cdot \mathbf{r}_d)$.

To summarize, axial systems with lateral shift-invariance have very simple SVDs; all object-space singular functions corresponding to nonzero singular values have the form $\exp(2\pi i \boldsymbol{\rho} \cdot \mathbf{r})\, H(-\boldsymbol{\rho}; z)$. We shall see an example of the usefulness of this result in Sec. 16.2.6.

7.3 LINEAR CONTINUOUS-TO-DISCRETE SYSTEMS

Digital imaging systems view objects defined on a continuous domain and produce finite vectors as their output. A simple example, introduced in the Prologue, is a digital video camera using a charge-coupled device. Each of the M discrete detector elements in this device performs a spatial and temporal integration of the image irradiance.[9] As a result, the image output from this detector in one video frame is simply M numbers $\{g_m, m = 1, ..., M\}$, and we can order these numbers as an $M \times 1$ column vector $\mathbf{g}$.

In the digital-camera example and many other digital imaging systems, it is an excellent approximation to say that each output datum g_m is linearly related to the object, so the correct system description is a linear continuous-to-discrete mapping as introduced in Sec. 1.2.4. It is the goal of this section to describe that mapping in more detail.

7.3.1 System operator

As in Sec. 7.2, we assume that the object is square-integrable and supported within a region $\mathbf{S}_f$ in $\mathbb{R}^q$, so object space $\mathbb{U}$ is $\mathbb{L}_2(\mathbf{S}_f)$. The image vector $\mathbf{g}$ is a finite set of finite numbers, so image space $\mathbb{V}$ is the M-dimensional Euclidean space $\mathbb{E}^M$. Thus the imaging system is a mapping from $\mathbb{L}_2(\mathbf{S}_f)$ to $\mathbb{E}^M$. If this mapping is linear, then, by a generalization of (1.30), it must have the form

$$g_m = \int_{\mathbf{S}_f} d^q r \; h_m(\mathbf{r})\, f(\mathbf{r})\,. \tag{7.225}$$

[9] Irradiance is simply the radiant power per unit area incident on a surface. For more discussion, see Chap. 10.

As usual, we shall express this relation in operator form as

$$\mathbf{g} = \mathcal{H}\mathbf{f}, \tag{7.226}$$

where now $\mathcal{H}$ indicates a linear continuous-to-discrete (CD) mapping.

The index m can have a variety of meanings, depending on the system being considered. A digital camera has a regular array of detectors, which we describe with a single index by lexicographic ordering. Alternatively, we might choose to use a multi-index as in Sec. 7.1.3 and then denote an element of $\mathbf{g}$ as $g_{\mathbf{m}}$. In tomography, for another example, m is a composite index specifying projection angle and location of a particular detector element in an array that measures the projection.

The function $h_m(\mathbf{r})$ specifies how sensitive the mth detector is to radiation originating at point $\mathbf{r}$ in the object. It can be called the *detector sensitivity function* when we wish to consider a single detector and describe how its output varies with location of a point source. As in Sec. 7.2, however, we shall also use the term *point response function* or PRF for $h_m(\mathbf{r})$, especially when we wish to think of $\mathbf{r}$ as fixed and the discrete index m as the variable. If $f(\mathbf{r}) = \delta(\mathbf{r} - \mathbf{r}_0)$, then $g_m = h_m(\mathbf{r}_0)$, so the set $\{h_m(\mathbf{r}_0)\}$ is the digital image of a point at $\mathbf{r}_0$. Fig. 7.9 illustrates the PRF for a CD imaging system.

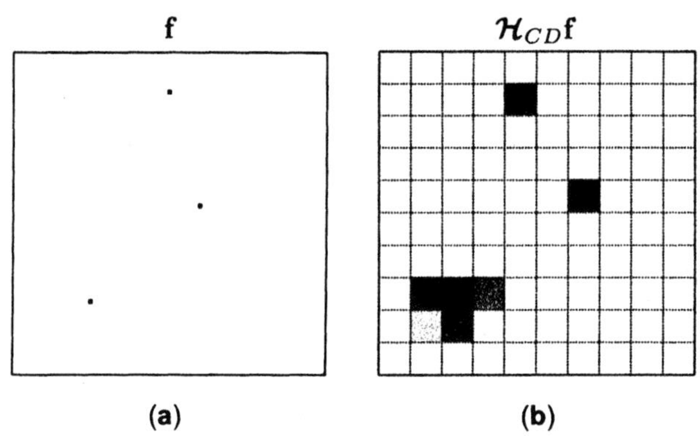

Fig. 7.9 Illustration of the PRF for a CD operator for three input locations.

Continuous imaging followed by sampling In most applications of the CD operator to imaging, we can separate the imaging process into two stages: continuous imaging followed by sampling or discretization. In the digital-camera example, the first stage is accomplished by the imaging lens, and the output of this stage is the continuous irradiance pattern on the detector face. The discretization step arises since the detector integrates the irradiance over each pixel area.

In more mathematical terms, a CC operator $\mathcal{H}_{CC}$ followed by a discretization operator $\mathcal{D}_w$ is equivalent to a composite CD operator:

$$\mathbf{g} = \mathcal{D}_w \mathcal{H}_{CC}\mathbf{f} = \mathcal{H}\mathbf{f}. \tag{7.227}$$

By analogy to (7.33), the general form[10] of $\mathcal{D}_w$ (if it is linear) is

$$[\mathcal{D}_w \mathbf{g}]_m = \int_{\mathbf{S}_g} d^s r_d \, g(\mathbf{r}_d) \, w_m(\mathbf{r}_d) \, . \tag{7.228}$$

The general form of $\mathcal{H}_{CC}$ is given by (7.101), so we can write (7.227) in general as

$$g_m = \int_{\mathbf{S}_g} d^s r_d \, w_m(\mathbf{r}_d) \int_{\mathbf{S}_f} d^q r \, h(\mathbf{r}_d, \mathbf{r}) \, f(\mathbf{r}) \, . \tag{7.229}$$

We see that this form agrees with (7.225) if we set

$$h_m(\mathbf{r}) = \int_{\mathbf{S}_g} d^s r_d \, w_m(\mathbf{r}_d) \, h(\mathbf{r}_d, \mathbf{r}) \, . \tag{7.230}$$

In a digital camera with an ideal detector array, $h(\mathbf{r}_d, \mathbf{r})$ accounts for blur arising from the imaging lens, and $w_m(\mathbf{r})$ is real and equal to a constant whenever $\mathbf{r}$ is within the mth detector pixel. In the terminology of Sec. 3.5.6, $w_m(\mathbf{r})$ is the *sampling aperture.*

Flood uniformity and point sensitivity Flood uniformity and point sensitivity for a CD system can be defined as in Sec. 7.2, but now $\mathbf{g}_{fld}$ is a vector, given by [*cf.* (7.111)]

$$[\mathbf{g}_{fld}]_m = \int_{\mathbf{S}_f} d^q r \, h_m(\mathbf{r}) \, . \tag{7.231}$$

The point sensitivity is, however, still a function, and (7.113) is modified to

$$s_{pt}(\mathbf{r}_0) = \sum_m h_m(\mathbf{r}_0) \, . \tag{7.232}$$

If we decompose the CD operator as a CC operator followed by a discretization operator and use (7.230), the flood image becomes

$$[\mathbf{g}_{fld}]_m = \int_{\mathbf{S}_f} d^q r \int_{\mathbf{S}_g} d^s r_d \, w_m(\mathbf{r}_d) \, h(\mathbf{r}_d, \mathbf{r})$$

$$= \int_{\mathbf{S}_g} d^s r_d \, w_m(\mathbf{r}_d) \, g_{fld}(\mathbf{r}_d) \, , \tag{7.233}$$

where $g_{fld}(\mathbf{r}_d)$ is the flood image for the underlying CC system as defined in (7.111).

We saw in (7.137) that $g_{fld}(\mathbf{r}_d)$ is constant if the CC system is shift-invariant. In that case,

$$[\mathbf{g}_{fld}]_m = const \cdot \int_{\mathbf{S}_g} d^s r_d \, w_m(\mathbf{r}_d) \, . \tag{7.234}$$

The integral accounts for detector nonuniformities; it is a constant if the detectors

[10]To agree strictly with the notation used in Sec. 7.1.3, we would either have to write $w_m^*(\mathbf{r}_d)$ in the integrand of (7.228) or denote the operator by $\mathcal{D}_{w^*}$. This minor notational inconsistency disappears if $w_m(\mathbf{r}_d)$ is real.

are identical, so that $w_m(\mathbf{r}_d) = w(\mathbf{r}_d - \mathbf{r}_{dm})$, where $\mathbf{r}_{dm}$ is the center of the mth detector pixel.

The point sensitivity for a general CD system decomposed as a CC system plus discretization is

$$s_{pt}(\mathbf{r}_0) = \sum_m \int_{\mathbf{S}_g} d^s r_d \; w_m(\mathbf{r}_d)\, h(\mathbf{r}_d, \mathbf{r}_0) = \int_{\mathbf{S}_g} d^s r_d \; \eta(\mathbf{r}_d)\, h(\mathbf{r}_d, \mathbf{r}_0)\,, \qquad (7.235)$$

where

$$\eta(\mathbf{r}_d) = \sum_m w_m(\mathbf{r}_d)\,. \qquad (7.236)$$

This function is proportional to the probability that radiation incident at point $\mathbf{r}_d$ on the detector will contribute to the final image. For an ideal digital camera, $\eta(\mathbf{r}_d)$ is unity whenever $\mathbf{r}_d$ lies in any one of the detector pixels, and it is zero if $\mathbf{r}_d$ lies in a gap between pixels. Because of the blur associated with the CC component, however, gaps in the detector do not necessarily cause $s_{pt}(\mathbf{r}_0)$ to fall to zero.

7.3.2 Adjoint operator and singular-value decomposition

The adjoint of a continuous-to-discrete operator is a discrete-to-continuous operator. That is, if $\mathcal{H}$ maps from the infinite-dimensional object space $\mathbb{U}$ to a finite-dimensional image space $\mathbb{V}$, then $\mathcal{H}^\dagger$ maps from $\mathbb{V}$ to $\mathbb{U}$. Specifically, for the operator $\mathcal{H}$ defined in (7.225), the adjoint is given by [*cf.* (1.45)]

$$\left[\mathcal{H}^\dagger \mathbf{g}\right](\mathbf{r}) = \sum_{m=1}^{M} g_m h_m^*(\mathbf{r})\,. \qquad (7.237)$$

If the PRFs $h_m(\mathbf{r})$ are real, the function $[\mathcal{H}^\dagger \mathbf{g}](\mathbf{r})$ is a linear superposition of the PRFs, with components of $\mathbf{g}$ serving as weights. This operation is illustrated in Fig. 7.10. As in the CC case (see Sec. 7.2.2), this adjoint operator is often referred to as backprojection.

Since the spaces $\mathbb{U}$ and $\mathbb{V}$ have different dimensionalities, a CD operator $\mathcal{H}$ cannot have eigenfunctions. Nevertheless, we can make good use of singular-value decomposition, which is based on eigenanalysis of the Hermitian operators $\mathcal{H}^\dagger\mathcal{H}$ and $\mathcal{H}\mathcal{H}^\dagger$.

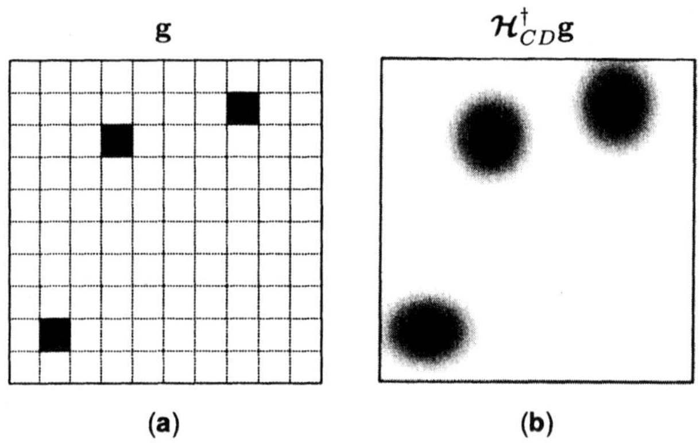

Fig. 7.10 Illustration of the adjoint of a CD operator.

Eigenfunctions in object space The operator $\mathcal{H}^\dagger\mathcal{H}$ maps object space $\mathbb{U}$ to itself. From (7.225) and (7.237), it is given explicitly by

$$\left[\mathcal{H}^\dagger\mathcal{H}\mathbf{f}\right](\mathbf{r}) = \sum_{m=1}^{M} h_m^*(\mathbf{r}) \int_{\mathbf{S}_f} d^q r'\, h_m(\mathbf{r}')\, f(\mathbf{r}')$$

$$= \int_{\mathbf{S}_f} d^q r'\, k(\mathbf{r}, \mathbf{r}')\, f(\mathbf{r}')\,, \tag{7.238}$$

where the kernel is given by

$$k(\mathbf{r}, \mathbf{r}') = \sum_{m=1}^{M} h_m^*(\mathbf{r})\, h_m(\mathbf{r}')\,. \tag{7.239}$$

Comparing this result to the corresponding expression for a CC system, (7.116), we see that the only difference is that the integral over the continuous detector space in the CC case has been replaced by a sum over discrete image components here. Figure 7.11 illustrates the meaning of $k(\mathbf{r}, \mathbf{r}')$.

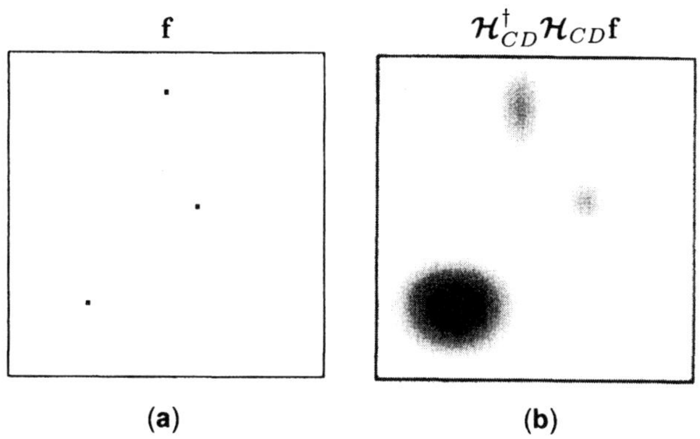

Fig. 7.11 Illustration of the operator $\mathcal{H}^\dagger\mathcal{H}$ when $\mathcal{H}$ is a CD operator.

Compact operators and eigenanalysis The operator $\mathcal{H}^\dagger\mathcal{H}$ is compact (see Sec. 1.3.3) if its kernel satisfies the Hilbert-Schmidt condition [*cf.* (7.138)],

$$\int_{\mathbf{S}_f} d^q r \int_{\mathbf{S}_f} d^q r'\, |k(\mathbf{r}, \mathbf{r}')|^2 < \infty\,. \tag{7.240}$$

This condition is satisfied if $k(\mathbf{r}, \mathbf{r}')$ is bounded and $\mathbf{S}_f$ is finite, as they will be with all physically realizable imaging systems.

We know from Sec. 1.4.4 that a compact Hermitian operator on an infinite-dimensional Hilbert space has a countably infinite set of eigenvectors and real eigenvalues. The eigenvectors of $\mathcal{H}^\dagger\mathcal{H}$ are a set of functions $\{u_n(\mathbf{r})\}$ or, equivalently, a set of vectors $\{\mathbf{u}_n\}$ in the object Hilbert space.

The eigenvalue equation is

$$\mathcal{H}^\dagger\mathcal{H}\mathbf{u}_n = \mu_n \mathbf{u}_n\,, \qquad n = 1, ..., \infty\,, \tag{7.241}$$

which is equivalent to (7.117). The structure of the eigenvalue problem here is exactly the same as in the CC case since in both cases $\mathcal{H}^\dagger\mathcal{H}$ is a Hermitian operator mapping $\mathbb{U}$ to $\mathbb{U}$; the intermediate stop at $\mathbb{V}$ does not invalidate any of the key properties of the eigenfunctions or eigenvalues. In particular, the eigenvalues $\{\mu_n\}$ are real and nonnegative, and $\{\mathbf{u}_n\}$ can be chosen as a complete, orthonormal set in $\mathbb{U}$ (*i.e.*, (7.118) and (7.119) still hold). Thus any object in $\mathbb{U}$ can be expanded as in (7.120) with coefficients given by (7.121):

$$\mathbf{f} = \sum_{n=1}^{\infty} \alpha_n \mathbf{u}_n \, , \qquad \alpha_n = \mathbf{u}_n^\dagger \mathbf{f} = \int_{\mathbf{S}_f} d^q r \; u_n^*(\mathbf{r}) \, f(\mathbf{r}) \, . \tag{7.242}$$

Eigenvectors in image space Though the eigenvalue problems for $\mathcal{H}^\dagger\mathcal{H}$ have the same structure for CC and CD systems, the corresponding problems for $\mathcal{H}\mathcal{H}^\dagger$ are quite different. The operator $\mathcal{H}\mathcal{H}^\dagger$ maps image space $\mathbb{V}$ to itself, and for a CD system $\mathbb{V}$ has a finite dimension.

The action of the operator $\mathcal{H}\mathcal{H}^\dagger$ can be expressed as

$$\begin{aligned}\left[\mathcal{H}\mathcal{H}^\dagger \mathbf{g}\right]_m &= \int_{\mathbf{S}_f} d^q r \; h_m(\mathbf{r}) \sum_{k=1}^{M} g_k h_k^*(\mathbf{r}) \\ &= \sum_{k=1}^{M} \left[\int_{\mathbf{S}_f} d^q r \; h_m(\mathbf{r}) \, h_k^*(\mathbf{r}) \right] g_k \, . \end{aligned} \tag{7.243}$$

In other words, $\mathcal{H}\mathcal{H}^\dagger$ is an $M \times M$ Hermitian matrix with elements given by

$$\left[\mathcal{H}\mathcal{H}^\dagger\right]_{mk} = \int_{\mathbf{S}_f} d^q r \; h_m(\mathbf{r}) \, h_k^*(\mathbf{r}) \, . \tag{7.244}$$

The eigenvalue problem is thus

$$\mathcal{H}\mathcal{H}^\dagger \mathbf{v}_n = \mu_n \mathbf{v}_n \, , \qquad n = 1, ..., M \, , \tag{7.245}$$

where $\mathbf{v}_n$ is an $M \times 1$ column vector. Since $\mathcal{H}\mathcal{H}^\dagger$ is a matrix, this eigenvalue problem can be solved by standard algorithms (Golub and van Loan, 1989; Press *et al.*, 1992) or by various software packages.

From the general discussion in Sec. 1.4.4, we know that the eigenvectors of $\mathcal{H}\mathcal{H}^\dagger$ can be chosen to form a complete, orthonormal set in $\mathbb{V}$, so that

$$\mathbf{v}_k^\dagger \mathbf{v}_n = \delta_{kn} \, ; \tag{7.246}$$

$$\sum_{n=1}^{M} \mathbf{v}_n \mathbf{v}_n^\dagger = \mathbf{I}_M \, , \tag{7.247}$$

where $\mathbf{I}_M$ is the $M \times M$ unit matrix. From these equations it follows that any data vector $\mathbf{g}$ can be expressed as

$$\mathbf{g} = \sum_{n=1}^{M} \beta_n \mathbf{v}_n \, , \qquad \beta_n = \mathbf{v}_n^\dagger \mathbf{g} \, . \tag{7.248}$$

Rank and null space To establish that the eigenvalues for $\mathcal{H}\mathcal{H}^\dagger$ are the same as those for $\mathcal{H}^\dagger\mathcal{H}$, we operate on both sides of (7.245) with $\mathcal{H}^\dagger$ and add some brackets for clarity; we find

$$\mathcal{H}^\dagger\mathcal{H}\left[\mathcal{H}^\dagger \mathbf{v}_n\right] = \mu_n \left[\mathcal{H}^\dagger \mathbf{v}_n\right] , \tag{7.249}$$

so $\mathcal{H}^\dagger \mathbf{v}_n$ is an eigenvector of $\mathcal{H}^\dagger\mathcal{H}$ with eigenvalue μ_n. Similarly, by operating on both sides of (7.241) with $\mathcal{H}$ we can show that $\mathcal{H}\mathbf{u}_n$ is an eigenvector of $\mathcal{H}\mathcal{H}^\dagger$ with eigenvalue μ_n. Thus the two operators $\mathcal{H}\mathcal{H}^\dagger$ and $\mathcal{H}^\dagger\mathcal{H}$ share the same eigenvalue spectrum. But $\mathcal{H}\mathcal{H}^\dagger$ is a finite matrix, so its rank R cannot exceed its dimension M (see Sec. 1.2.3). Moreover, from Sec. 1.4.3 the rank is the number of nonzero eigenvalues. The operator $\mathcal{H}^\dagger\mathcal{H}$ has an infinite number of eigenvalues, since it operates in an infinite-dimensional space, but at most M of them are nonzero.

An important consequence of this discussion is that $\mathcal{H}^\dagger\mathcal{H}$ necessarily has an infinite-dimensional null space if $\mathcal{H}$ itself is a CD operator. If we order the eigenvalues by descending value as in (1.114), so that μ_R is the last non-vanishing eigenvalue, then the measurement space of $\mathcal{H}^\dagger\mathcal{H}$ is the subspace of $\mathbb{U}$ spanned by $\{\mathbf{u}_n, n = 1, ..., R\}$, and the null space is the infinite-dimensional orthogonal complement of the measurement space (see Secs. 1.5.2 and 1.5.3).

Thus an arbitrary object vector $\mathbf{f}$ can be uniquely decomposed into measurement and null functions as follows:

$$\mathbf{f} = \mathbf{f}_{meas} + \mathbf{f}_{null} , \tag{7.250}$$

where

$$\mathbf{f}_{meas} = \mathcal{P}_{meas}\mathbf{f} = \sum_{n=1}^{R} \alpha_n \mathbf{u}_n , \tag{7.251}$$

$$\mathbf{f}_{null} = \mathcal{P}_{null}\mathbf{f} = \sum_{n=R+1}^{\infty} \alpha_n \mathbf{u}_n . \tag{7.252}$$

In these expressions, $\mathcal{P}_{meas}$ and $\mathcal{P}_{null}$ are the projectors onto measurement and null space, respectively, and in both spaces the coefficients α_n are given by $\mathbf{u}_n^\dagger \mathbf{f}$ as in (7.242).

Relation between eigenvectors in object and image space Comparison of (7.249) and (7.241) provides a relation between $\mathbf{v}_n$ and $\mathbf{u}_n$. In fact, the reader's first impulse might be simply to equate $\mathcal{H}^\dagger \mathbf{v}_n$ with $\mathbf{u}_n$, but that would not necessarily produce a properly normalized eigenvector in object space. We can, however, multiply an eigenvector by any constant and it remains an eigenvector with the same eigenvalue. We shall now show that an appropriate choice for the constant is $1/\sqrt{\mu_n}$.

Suppose we have solved the eigenvalue problem for the matrix $\mathcal{H}\mathcal{H}^\dagger$ and found a set of orthonormal eigenvectors $\{\mathbf{v}_n, n = 1, ..., M\}$ and that we form the functions $\{u_n(\mathbf{r})\}$ by

$$\frac{1}{\sqrt{\mu_n}} \left[\mathcal{H}^\dagger \mathbf{v}_n\right](\mathbf{r}) = u_n(\mathbf{r}) , \qquad n = 1, ..., R . \tag{7.253}$$

That is, we backproject each vector $\mathbf{v}_n$ and then divide by $\sqrt{\mu_n}$. By (7.249), the resulting functions are eigenfunctions of $\mathcal{H}^\dagger\mathcal{H}$.

The following manipulations show that the set of eigenfunctions is orthonormal:

$$(\mathbf{u}_n, \mathbf{u}_k) = \frac{1}{\sqrt{\mu_n \mu_k}} (\mathcal{H}^\dagger \mathbf{v}_n, \mathcal{H}^\dagger \mathbf{v}_k) = \frac{1}{\sqrt{\mu_n \mu_k}} (\mathcal{H}\mathcal{H}^\dagger \mathbf{v}_n, \mathbf{v}_k)$$

$$= \frac{\mu_n}{\sqrt{\mu_n \mu_k}} (\mathbf{v}_n, \mathbf{v}_k) = \frac{\mu_n}{\sqrt{\mu_n \mu_k}} \delta_{nk} = \delta_{nk} \,. \qquad (7.254)$$

Similar arguments show that

$$\frac{1}{\sqrt{\mu_n}} \mathcal{H}\mathbf{u}_n = \mathbf{v}_n \,, \qquad n = 1, ..., R. \qquad (7.255)$$

Thus, for $n \leq R$, the eigenvectors of $\mathcal{H}^\dagger\mathcal{H}$ are uniquely determined from those of $\mathcal{H}\mathcal{H}^\dagger$ and vice versa.

SVD and the imaging equation The general form for the SVD of a compact linear operator is (1.120) or (7.127):

$$\mathcal{H} = \sum_{n=1}^{R} \sqrt{\mu_n}\, \mathbf{v}_n \mathbf{u}_n^\dagger \,. \qquad (7.256)$$

For the specific case of a CD operator, the action of this expansion on an arbitrary object is [*cf.* (7.128)]

$$g_m = [\mathcal{H}\mathbf{f}]_m = \sum_{n=1}^{R} \sqrt{\mu_n} \left[\mathbf{v}_n \mathbf{u}_n^\dagger \mathbf{f}\right]_m = \sum_{n=1}^{R} \left[\sqrt{\mu_n} \int_{\mathbf{S}_f} d^q r \; u_n^*(\mathbf{r})\, f(\mathbf{r})\right] v_{nm}$$

$$= \sum_{n=1}^{R} \sqrt{\mu_n}\, \alpha_n v_{nm} \,, \qquad (7.257)$$

where v_{nm} is the mth component of vector $\mathbf{v}_n$. Note especially that only the object components with $n \leq R$ contribute to $\mathbf{g}$; the operator $\mathcal{H}$ has the same infinite-dimensional null space as $\mathcal{H}^\dagger\mathcal{H}$.

Applying the same arguments as used above (7.129) to (7.248) and (7.257), we see that

$$\beta_n = \sqrt{\mu_n}\, \alpha_n \,. \qquad (7.258)$$

Thus SVD again reduces the imaging equation $\mathbf{g} = \mathcal{H}\mathbf{f}$ to a simple multiplication.

7.3.3 Fourier description

The theoretical advantage of SVD is that it reduces the complicated integral operator to a simple multiplication, just as a Fourier transform reduces convolution to multiplication. (In fact, as we saw in Sec. 7.2.4, Fourier transformation *is* SVD for a convolution operator.) The SVD representation is very attractive for this reason, but it has some drawbacks as well, as we shall now enumerate.

The basis functions used in Fourier analysis have a simple analytical form, but it is often impossible to express SVD basis functions analytically. The singular functions corresponding to nonzero singular values can be found numerically by solving the eigenvalue problem for the matrix $\mathcal{H}\mathcal{H}^\dagger$, but this matrix has size $M \times M$, where M is the number of measurements. In practice, even this numerical approach is not tractable for M greater than a few thousand.

Moreover, if one wants to compare two different imaging systems, or even two variants on the same system, it is convenient to use the same basis functions for both. LSIV systems can be compared by SVD since the basis functions for all LSIV systems are plane waves of infinite extent, but with this exception different systems have different SVD basis functions.

Another drawback to SVD analysis is that it is difficult to develop any intuition about the basis functions. With Fourier analysis, the spatial frequency has a simple physical interpretation, and we have no difficulty in visualizing the basis function or the action of an LSIV system on it. It would appear, however, that there is little benefit to using Fourier transforms with systems that are not at least approximately shift-invariant.

In fact, we can develop a fruitful Fourier theory for an arbitrary CD system by employing the Fourier *series* as in (7.14) [see also (3.279)]. A Fourier representation of the system can be obtained by substituting (7.14) into (7.226), yielding

$$\mathbf{g} = \sum_{\mathbf{k}} F_{\mathbf{k}} \mathcal{H}\{\Phi_{\mathbf{k}}(\mathbf{r})\} \equiv \mathbf{\Psi}\mathbf{F}\,, \tag{7.259}$$

where $\mathbf{\Psi}$ is a complex matrix with element $\psi_{m\mathbf{k}}$. In component form, (7.259) is

$$g_m = \sum_{\mathbf{k}} \psi_{m\mathbf{k}} F_{\mathbf{k}}\,. \tag{7.260}$$

Recall from Sec. 7.1.2 that $\mathbf{k}$ is a multi-index, a qD vector with integer components. Since each of the q components of $\mathbf{k}$ ranges over $(-\infty, \infty)$, $\mathbf{F}$ has a q-fold infinity of elements and $\mathbf{\Psi}$ is a matrix with M rows and a q-fold infinity of columns.

The matrix elements of $\mathbf{\Psi}$ are given explicitly by

$$\psi_{m\mathbf{k}} = [\mathcal{H}\Phi_{\mathbf{k}}]_m = \int_\infty d^q r\, \exp(2\pi i \boldsymbol{\rho}_{\mathbf{k}} \cdot \mathbf{r})\, h_m(\mathbf{r})\, S_f(\mathbf{r})\,. \tag{7.261}$$

The integral is the Fourier transform of the product $h_m(\mathbf{r})\, S_f(\mathbf{r})$ evaluated at the specific spatial frequency $\boldsymbol{\rho}_{\mathbf{k}}$. Except for the normalizing constant $1/\sqrt{V}$ [see (7.15)], the integral can also be interpreted as a coefficient in the Fourier-series expansion of the same product. If $h_m(\mathbf{r})$ is real and the multi-index $\mathbf{k}$ is constructed so that $\boldsymbol{\rho}_{-\mathbf{k}} = -\boldsymbol{\rho}_{\mathbf{k}}$, then it follows from (7.261) that

$$\psi_{m(-\mathbf{k})} = \psi^*_{m\mathbf{k}}\,. \tag{7.262}$$

This is the so-called Hermiticity property of Fourier coefficients, (3.227), but $\mathbf{\Psi}$ is definitely not a Hermitian matrix; it cannot be since it is not square.

The complex matrix $\mathbf{\Psi}$ is completely equivalent to the operator $\mathcal{H}$; the latter maps the object $\mathbf{f}$ to the discrete data set $\mathbf{g}$, while the former maps the infinite set of object Fourier coefficients $\mathbf{F}$ to $\mathbf{g}$, but of course $\mathbf{f}$ is uniquely related to $\mathbf{F}$. Since $\mathbf{\Psi}$ has an infinite number of columns, there is no loss of information in using it rather than $\mathcal{H}$.

Fourier crosstalk matrix With the system representation of (7.259), it is natural to investigate the operator $\mathbf{\Psi}^\dagger\mathbf{\Psi}$, where $\mathbf{\Psi}^\dagger$ is the adjoint of the matrix $\mathbf{\Psi}$, *i.e.*,

$\left[\boldsymbol{\Psi}^{\dagger}\right]_{\mathbf{k}m} = \psi^*_{m\mathbf{k}}$. Since $\boldsymbol{\Psi}$ maps a vector of Fourier coefficients to a finite-dimensional vector in image space, $\boldsymbol{\Psi}^{\dagger}$ maps from image space back to the space of Fourier coefficients; in this sense it is another backprojection operator. Thus $\boldsymbol{\Psi}^{\dagger}\boldsymbol{\Psi}$ maps a vector in the space of Fourier coefficients to another vector in the same space.

Since $\boldsymbol{\Psi}^{\dagger}\boldsymbol{\Psi}$ maps an infinite vector to another infinite vector, it can be regarded as a square matrix with an infinite number of rows and columns. For reasons discussed below, this matrix will be referred to as the *Fourier crosstalk matrix*, or simply crosstalk matrix for short. We denote the crosstalk matrix itself as $\mathbf{B}$ (capital beta) and its elements by $\beta_{\mathbf{kk}'}$. The elements are given by

$$\beta_{\mathbf{kk}'} = \sum_{m=1}^{M} \psi^*_{m\mathbf{k}}\psi_{m\mathbf{k}'} = \sum_{m=1}^{M} \left[\mathcal{H}\boldsymbol{\Phi}_{\mathbf{k}}\right]^*_m \left[\mathcal{H}\boldsymbol{\Phi}_{\mathbf{k}'}\right]_m = \left(\mathcal{H}\boldsymbol{\Phi}_{\mathbf{k}}, \mathcal{H}\boldsymbol{\Phi}_{\mathbf{k}'}\right), \tag{7.263}$$

where $(\cdot,\cdot)$ here denotes a scalar product in image space.

From this definition, it follows that the crosstalk matrix is Hermitian:

$$\mathbf{B}^{\dagger} = \mathbf{B} \qquad \text{or} \qquad \beta_{\mathbf{kk}'} = \beta^*_{\mathbf{k}'\mathbf{k}}\,. \tag{7.264}$$

In addition, if $h_m(\mathbf{r})$ is real, it follows from (7.262) that

$$\beta_{(-\mathbf{k})(-\mathbf{k}')} = \beta^*_{\mathbf{kk}'}\,. \tag{7.265}$$

Without any restriction on the form of $\mathcal{H}$ or $\boldsymbol{\Psi}$, $\mathbf{B}$ is positive-semidefinite (see Sec. 1.4.4).

Interpretation of the crosstalk matrix The scalar product in (7.263) is zero if the two M-dimensional vectors $\mathcal{H}\boldsymbol{\Phi}_{\mathbf{k}}$ and $\mathcal{H}\boldsymbol{\Phi}_{\mathbf{k}'}$ are orthogonal to each other. If that is the case, the two Fourier components $\mathbf{k}$ and $\mathbf{k}'$ make linearly independent contributions to the data and can be easily separated by any inversion algorithm. On the other hand, if $\mathcal{H}\boldsymbol{\Phi}_{\mathbf{k}}$ is parallel to $\mathcal{H}\boldsymbol{\Phi}_{\mathbf{k}'}$ in data space, the two Fourier components produce identical data patterns and cannot be separated by any algorithm. Frequencies $\boldsymbol{\rho}_{\mathbf{k}}$ and $\boldsymbol{\rho}_{\mathbf{k}'}$ are fully aliased.

Even if this extreme case does not occur, $\beta_{\mathbf{kk}'}$ with $\mathbf{k} \neq \mathbf{k}'$ is a useful measure of the *degree* of aliasing or crosstalk between two different frequencies. When we discussed aliasing in Sec. 3.5.3, we presented it as a binary concept — either the Nyquist condition is satisfied or it is not. A little reflection reveals the inadequacy of this approach for anything other than regular sampling schemes. Figure 3.8 in Chap. 3 shows how two different cosines are indistinguishable if one knows only their values at a set of evenly spaced points, but all we have to do to make them distinguishable in principle is to displace one of the sampling points slightly.

As a quantitative measure of degree of aliasing, we define the angle $\theta_{\mathbf{kk}'}$ between $\mathcal{H}\boldsymbol{\Phi}_{\mathbf{k}}$ and $\mathcal{H}\boldsymbol{\Phi}_{\mathbf{k}'}$ by (see Sec. 1.1.4)

$$\cos\theta_{\mathbf{kk}'} \equiv \frac{\left(\mathcal{H}\boldsymbol{\Phi}_{\mathbf{k}}, \mathcal{H}\boldsymbol{\Phi}_{\mathbf{k}'}\right)}{||\mathcal{H}\boldsymbol{\Phi}_{\mathbf{k}}|| \cdot ||\mathcal{H}\boldsymbol{\Phi}_{\mathbf{k}'}||} = \frac{\beta_{\mathbf{kk}'}}{\sqrt{\beta_{\mathbf{kk}}\beta_{\mathbf{k}'\mathbf{k}'}}}\,. \tag{7.266}$$

For complete aliasing, $|\cos\theta_{\mathbf{kk}'}| = 1$, and for no aliasing at all, $\cos\theta_{\mathbf{kk}'} = 0$.

The diagonal elements of the crosstalk matrix also have a neat interpretation. If $\mathbf{k} = \mathbf{k}'$, (7.263) becomes

$$\beta_{\mathbf{kk}} = ||\mathcal{H}\boldsymbol{\Phi}_{\mathbf{k}}||^2\,. \tag{7.267}$$

Thus a diagonal element of the crosstalk matrix is a measure of the strength (norm) of the data when the object consists of a single truncated plane wave. It specifies how strongly a particular spatial frequency is transferred through the system to the data. In this sense, the set of diagonal elements $\{\beta_{\mathbf{kk}}\}$ constitutes a kind of transfer function, even though the Fourier-series basis functions are not eigenfunctions of the system.

Eigenanalysis of the crosstalk matrix The eigenvalue equation for $\boldsymbol{\Psi}^\dagger\boldsymbol{\Psi}$ is

$$\boldsymbol{\Psi}^\dagger\boldsymbol{\Psi}\mathbf{U}_n = \mu_n\mathbf{U}_n\,, \tag{7.268}$$

where $\mathbf{U}_n$ has the same structure as $\mathbf{F}$, namely, an infinite vector of Fourier coefficients.

The eigenfunctions of $\boldsymbol{\Psi}^\dagger\boldsymbol{\Psi}$ are related to the eigenfunctions of $\mathcal{H}^\dagger\mathcal{H}$ (which are also the singular functions in object space). If we denote by $\mathcal{F}_s$ the operator that converts a function $f(\mathbf{r})$ to its Fourier-series coefficients, then $\boldsymbol{\Psi}$ and $\mathcal{H}$ are related by

$$\mathcal{H} = \boldsymbol{\Psi}\mathcal{F}_s\,, \tag{7.269}$$

so

$$\mathcal{H}^\dagger\mathcal{H}\mathbf{u}_n = \mathcal{F}_s^\dagger\boldsymbol{\Psi}^\dagger\boldsymbol{\Psi}\mathcal{F}_s\mathbf{u}_n = \mu_n\mathbf{u}_n\,. \tag{7.270}$$

As discussed below (7.15), we can choose the support region so that the basis set $\{\Phi_{\mathbf{k}}(\mathbf{r})\}$ is orthogonal. In that case $\mathcal{F}_s^\dagger = \mathcal{F}_s^{-1}$, and the eigenvalue equation for $\mathcal{H}^\dagger\mathcal{H}$, (7.241), becomes

$$\boldsymbol{\Psi}^\dagger\boldsymbol{\Psi}\mathcal{F}_s\mathbf{u}_n = \mu_n\mathcal{F}_s\mathbf{u}_n\,. \tag{7.271}$$

Comparison with (7.268) shows that $\mathbf{U}_n$ is just the Fourier-series representation of $\mathbf{u}_n$.

7.3.4 Sampled LSIV systems

To gain more insight into the crosstalk matrix, we return to a problem introduced in Sec. 7.3.1. We consider a CD system decomposed into a CC system followed by sampling or discretization, but now we add the additional restriction that the CC system be shift-invariant.

Specifically, we assume that $h(\mathbf{r}_d, \mathbf{r}) = h(\mathbf{r}_d - \mathbf{r})$ and that the discretization functions are translates of a single function, so that $w_m(\mathbf{r}_d) = w(\mathbf{r}_d - \mathbf{r}_{dm})$. Next we make a change of variables and recognize that we can always take $\mathbf{S}_g$ large enough that the limits of integration do not need to be changed. The kernel specified in (7.230) then becomes

$$h_m(\mathbf{r}) \equiv p(\mathbf{r}_{dm} - \mathbf{r}) = \int_{\mathbf{S}_g} d^q r_d\; h(\mathbf{r}_d + \mathbf{r}_{dm} - \mathbf{r})\, w(\mathbf{r}_d)\,. \tag{7.272}$$

Thus $p(\mathbf{r})$, the cross-correlation of $h(\mathbf{r})$ and $w(\mathbf{r})$, serves as an overall shift-invariant PSF prior to point sampling.

With this form for the PRF and (7.261), elements of the system matrix $\boldsymbol{\Psi}$ are given by

$$\psi_{m\mathbf{k}} = \int_\infty d^q r\, \exp(2\pi i \boldsymbol{\rho}_{\mathbf{k}}\cdot\mathbf{r})\, S_f(\mathbf{r})\, p(\mathbf{r}_{dm} - \mathbf{r})\,. \tag{7.273}$$

If $p(\mathbf{r})$ is spatially compact and we choose the region of support large enough, we can replace the support function $S_f(\mathbf{r})$ by unity. Then a change of variables yields

$$\psi_{m\mathbf{k}} = \int_\infty d^q r \, \exp[-2\pi i \boldsymbol{\rho}_\mathbf{k} \cdot (\mathbf{r} - \mathbf{r}_{dm})] p(\mathbf{r}) = \exp(2\pi i \boldsymbol{\rho}_\mathbf{k} \cdot \mathbf{r}_{dm}) \, P(\boldsymbol{\rho}_\mathbf{k}) \,, \qquad (7.274)$$

where $P(\boldsymbol{\rho})$, the Fourier transform of $p(\mathbf{r})$, is the transfer function of the overall LSIV system (including the sampling aperture).

The elements of the crosstalk matrix, defined in (7.263), are now given by

$$\beta_{\mathbf{kk}'} = P(\boldsymbol{\rho}_\mathbf{k}) P^*(\boldsymbol{\rho}_{\mathbf{k}'}) \sum_{m=1}^{M} \exp[2\pi i (\boldsymbol{\rho}_\mathbf{k} - \boldsymbol{\rho}_{\mathbf{k}'}) \cdot \mathbf{r}_{dm}] \,. \qquad (7.275)$$

The diagonal elements,

$$\beta_{\mathbf{kk}} = M |P(\boldsymbol{\rho}_\mathbf{k})|^2 \,, \qquad (7.276)$$

are completely determined by the transfer function $P(\boldsymbol{\rho})$ of the underlying LSIV system, independent of the location of the samples.

It is interesting to compare (7.276) to (7.147), where we saw that the squared modulus of the transfer function, $|H(\boldsymbol{\rho})|^2$, is the eigenvalue of $\mathcal{H}^\dagger \mathcal{H}$ for an LSIV system. Equivalently, $|H(\boldsymbol{\rho})|^2$ is the transfer function of $\mathcal{H}^\dagger \mathcal{H}$, which is also an LSIV system if $\mathcal{H}$ is. In (7.276) we see that the squared modulus of a transfer function (now including a sampling aperture) recurs in the diagonal elements of the crosstalk matrix when the output of an LSIV system is sampled, even though the sampling spoils the shift-invariance.

Aliasing effects can be expressed in terms of $\cos \theta_{\mathbf{kk}'}$, defined in (7.266). In the present problem, we have

$$|\cos \theta_{\mathbf{kk}'}| = \frac{1}{M} \left| \sum_{m=1}^{M} \exp[2\pi i (\boldsymbol{\rho}_\mathbf{k} - \boldsymbol{\rho}_{\mathbf{k}'}) \cdot \mathbf{r}_{dm}] \right| . \qquad (7.277)$$

Note that $|\cos \theta_{\mathbf{kk}'}|$ is determined solely by the location of the sample points $\{\mathbf{r}_{dm}\}$ and is independent of $P(\boldsymbol{\rho})$.

1D Example To understand these results better, consider a 1D problem in which the object $f(x)$ is contained in the interval $-\frac{1}{2}L < x \le \frac{1}{2}L$ and the sample points are spaced by ϵ. It is convenient to replace the index m with another index n, chosen to run symmetrically from $-N$ to N, so that $M = 2N + 1$. Then we can specify the sample points by

$$x_{dn} = n\epsilon \,, \qquad -N \le n \le N \,, \qquad \epsilon = \frac{L}{M} = \frac{L}{2N+1} \,. \qquad (7.278)$$

In 1D, k is an ordinary integer index instead of a multi-index, and the spatial frequency is a scalar ξ_k given by

$$\xi_k = k/L \,, \qquad -\infty < k < \infty \,. \qquad (7.279)$$

Thus (7.277) becomes

$$\cos \theta_{kk'} = \frac{1}{M} \sum_{n=-N}^{N} \exp[2\pi i (\xi_k - \xi_{k'}) \, x_{dn}] = \frac{1}{M} \sum_{n=-N}^{N} \exp[2\pi i (k - k') \, n/M] \,. \qquad (7.280)$$

A geometric progression of this form was discussed in Chap. 2; using (2.49), we find that

$$\cos\theta_{kk'} = \frac{1}{M}\frac{\sin[\pi(k-k')]}{\sin[\pi(k-k')/M]} = \frac{1}{M}\frac{\sin[\pi M\epsilon(\xi_k-\xi_{k'})]}{\sin[\pi\epsilon(\xi_k-\xi_{k'})]}\,. \tag{7.281}$$

This function is plotted in Fig. 7.12, and the crosstalk matrix for this problem is depicted in Fig. 7.13 for the case where $p(x) = \mathrm{rect}(x/\epsilon)$. Both figures show that frequencies ξ_k and $\xi_{k'}$ are fully aliased if $\xi_k - \xi_{k'}$ is an integer multiple of the sampling frequency $1/\epsilon$.

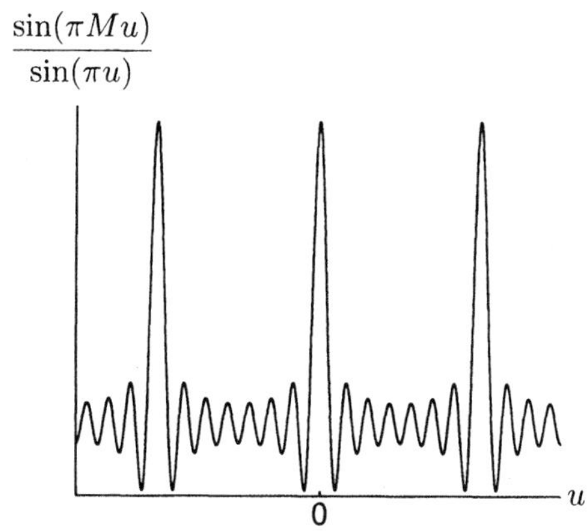

Fig. 7.12 Plot of the function $\sin(\pi Mu)/\sin(\pi u)$.

In the limit of very fine sampling ($\epsilon \to 0$), a finite frequency ξ_k is not aliased with any other finite frequency. In this limit, $\cos\theta_{kk'} \to \delta_{kk'}$, so the crosstalk matrix is diagonal and given by

$$\lim_{\epsilon\to 0}\beta_{kk'} = M|P(\xi_k)|^2\,\delta_{kk'}\,. \tag{7.282}$$

In this limit, the frequencies ξ_k become arbitrarily close together, so ξ_k can be replaced by the continuous variable ξ. The squared modulus of the transfer function of the underlying LSIV system then lies along the diagonal, but multiplied by the number of samples M, which has to go to ∞ as ϵ goes to 0. We have already encountered (7.282); in (7.171) we showed that $\boldsymbol{\mathcal{H}}^\dagger\boldsymbol{\mathcal{H}}$ for an LSIV system is diagonalized by Fourier transformation, and (7.282) reiterates that conclusion.

Another special case of this formalism is when the LSIV system is bandlimited (see Sec. 3.5.1) so that $P(\xi) = 0$ if $|\xi| \geq \xi_{max}$. This means that the crosstalk matrix is zero outside a submatrix of dimension $2\xi_{max}L \times 2\xi_{max}L$. Moreover, if the Nyquist condition (3.291) is satisfied, then there is no aliasing and this submatrix is diagonal. Note that two distinct Nyquist conditions are operative here. We assume that the object is spatially limited to extent L so that we can represent it by a Fourier series with sample spacing $1/L$, and we assume that the LSIV system is bandlimited so that its output can be sampled without aliasing by a discrete detector array. These conditions are not contradictory; we do not need to assume that the object is bandlimited or that the system impulse response is spatially limited.

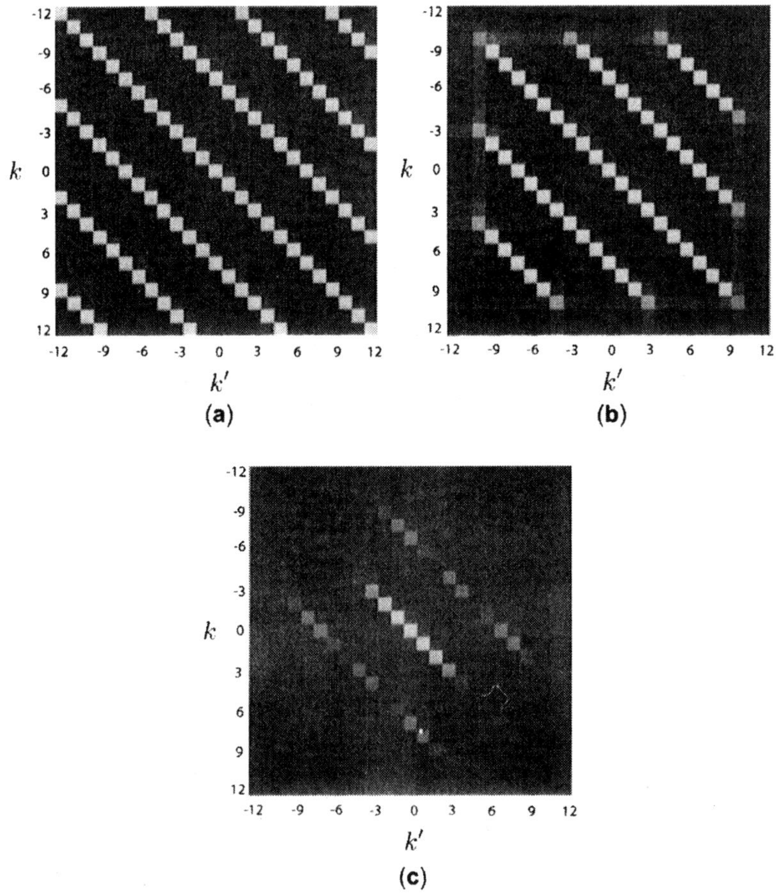

Fig. 7.13 Low-frequency portion of the crosstalk matrix for a 1D CD system with uniformly spaced sampling apertures (Gifford, 1997). (a) Point sampling, no presampling blur. (b) Sampling with rect functions, low-pass presampling blur, Nyquist condition not satisfied. (c) Same as (b) but lower cutoff frequency in presampling blur, so Nyquist is satisfied.

7.3.5 Mixed CC-CD systems

So far we have treated CC and CD systems separately, but sometimes it is useful to mix continuous and discrete variables in the description of the object, the image or both.

One example is a linear system that acquires multiple images of the same object. We can denote the jth image by the vector $\mathbf{g}_j$, with components given by

$$g_j(\mathbf{r}_d) = \int_{\mathbf{S}_f} d^q r \; h_j(\mathbf{r}_d, \mathbf{r})\, f(\mathbf{r})\,. \tag{7.283}$$

In this case, the object is a function of a continuous variable $\mathbf{r}$, but the image is mixed, in the sense that it has both a continuous variable $\mathbf{r}_d$ and a discrete index j.

A system that acquires multiple images of multiple object attributes is mixed in both object and image space. An appropriate linear description would be

$$g_j(\mathbf{r}_d) = \sum_k \int_\infty d^q r \; h_{jk}(\mathbf{r}_d, \mathbf{r}) \, f_k(\mathbf{r}) \,. \tag{7.284}$$

In this case we have a matrix of CC operators.

Color imaging As a practical example of a mixed CC-CD system, consider a color imaging system that views an object $f(\mathbf{r}, \lambda)$, where $\mathbf{r}$ is a 2D vector and λ is the wavelength, through three color filters. Three separate 2D images are produced, each related to the object by

$$g_m(\mathbf{r}_d) = \int_\infty d^2 r \int_0^\infty d\lambda \; f(\mathbf{r}, \lambda) \, h_m(\mathbf{r}_d, \mathbf{r}, \lambda) \,, \tag{7.285}$$

where $m = 1$ indicates the red filter, $m = 2$ green and $m = 3$ blue. This operator, denoted as usual by $\mathcal{H}$, maps a function of three variables (x, y, λ) to three functions of two variables (x_d, y_d) each.

The adjoint operator $\mathcal{H}^\dagger$ maps these data back to a function of $\mathbf{r}$ and λ; its form is

$$\left[\mathcal{H}^\dagger \mathbf{g}\right](\mathbf{r}, \lambda) = \sum_{m=1}^{3} \int_\infty d^2 r_d \; g_m(\mathbf{r}_d) \, h_m^*(\mathbf{r}_d, \mathbf{r}, \lambda) \,. \tag{7.286}$$

It follows that

$$\left[\mathcal{H}^\dagger \mathcal{H} \mathbf{f}\right](\mathbf{r}, \lambda) = \int_\infty d^2 r' \int_0^\infty d\lambda' \, k(\mathbf{r}, \mathbf{r}'; \lambda, \lambda') \, f(\mathbf{r}', \lambda') \,, \tag{7.287}$$

where

$$k(\mathbf{r}, \mathbf{r}'; \lambda, \lambda') = \sum_{m=1}^{3} \int_\infty d^2 r_d \; h_m^*(\mathbf{r}_d, \mathbf{r}, \lambda) \, h_m(\mathbf{r}_d, \mathbf{r}', \lambda') \,. \tag{7.288}$$

This kernel is a combination of the corresponding CC and CD kernels, (7.116) and (7.239), respectively.

The operator $\mathcal{H}\mathcal{H}^\dagger$, which maps the mixed image space to itself, is given by

$$\left[\mathcal{H}\mathcal{H}^\dagger \mathbf{g}\right]_m (\mathbf{r}_d) = \sum_{k=1}^{3} \int_\infty d^2 r_{d0} \; K_{mk}(\mathbf{r}_d, \mathbf{r}_{d0}) \, g_k(\mathbf{r}_{d0}) \,, \tag{7.289}$$

where now the kernel is [*cf.* (7.124)]

$$K_{mk}(\mathbf{r}_d, \mathbf{r}_{d0}) = \int_{\mathbf{S}_f} d^2 r \int_0^\infty d\lambda \; h_m(\mathbf{r}_d, \mathbf{r}, \lambda) \, h_k^*(\mathbf{r}_{d0}, \mathbf{r}, \lambda) \,. \tag{7.290}$$

As with the other system descriptions considered in this chapter, we can now, in principle, find the eigenvectors of $\mathcal{H}^\dagger\mathcal{H}$ and $\mathcal{H}\mathcal{H}^\dagger$ and use them to construct an SVD of the operator $\mathcal{H}$. The singular vectors in image space are functions of $\mathbf{r}_d$ but also require an index m to specify the color band; we denote them by $v_n(\mathbf{r}_d, m)$. These functions must satisfy

$$\left[\mathcal{H}\mathcal{H}^\dagger \mathbf{v}_n\right](\mathbf{r}_d, m) = \mu_n v_n(\mathbf{r}_d, m) \,. \tag{7.291}$$

The singular vectors in object space are functions of $\mathbf{r}$ and λ, so we denote them by $u_n(\mathbf{r}, \lambda)$. For $n \le R$ they can be found by operating on $v_n(\mathbf{r}_d, m)$ with $\mathcal{H}^\dagger$ and renormalizing as in (7.253).

Special case: LSIV color imaging The expressions given above simplify considerably if we assume that the basic imaging system is LSIV and achromatic, so that

$$h_m(\mathbf{r}_d, \mathbf{r}, \lambda) = h(\mathbf{r}_d - \mathbf{r})\, T_m(\lambda)\,, \tag{7.292}$$

where $T_m(\lambda)$ is the transmission of the mth filter. With this form, the kernel for $\mathcal{H}\mathcal{H}^\dagger$, (7.290), becomes

$$K_{mk}(\mathbf{r}_d - \mathbf{r}_{d0}) = \int_\infty d^2r\; h_m(\mathbf{r}_d - \mathbf{r})\, h_k^*(\mathbf{r}_{d0} - \mathbf{r}) \int_0^\infty d\lambda\; T_m(\lambda)\, T_k(\lambda)\,. \tag{7.293}$$

The integral over $\mathbf{r}$ is the complex autocorrelation of the PSF, and the integral over λ expresses the overlap of the filter functions.

Since this kernel is spatially (though not spectrally) shift-invariant, it suggests a spatial Fourier transform. If we denote the 2D Fourier transform of the image-space singular function by $V_{n,\boldsymbol{\rho}}(\boldsymbol{\rho}_d, m)$ and the corresponding Hilbert-space vector as $\mathbf{V}_{n,\boldsymbol{\rho}}$, then the eigenvalue equation, (7.291), becomes

$$\left[\mathcal{F}_2\mathcal{H}\mathcal{H}^\dagger\mathcal{F}_2^{-1}\mathbf{V}_{n,\boldsymbol{\rho}}\right](\boldsymbol{\rho}_d, m) = \mu_n V_{n,\boldsymbol{\rho}}(\boldsymbol{\rho}_d, m)\,. \tag{7.294}$$

Arguments analogous to those leading to (7.169) reveal the structure of the operator $\mathcal{F}_2\mathcal{H}\mathcal{H}^\dagger\mathcal{F}_2^{-1}$. It maps a function of $\boldsymbol{\rho}_d$ and index m to another function of the same form, and its kernel is

$$\left[\mathcal{F}_2\mathcal{H}\mathcal{H}^\dagger\mathcal{F}_2^{-1}\right](\boldsymbol{\rho}_d, \boldsymbol{\rho}_{d0}, m, k) = |H(\boldsymbol{\rho}_d)|^2\, \delta(\boldsymbol{\rho}_d - \boldsymbol{\rho}_{d0})\, A_{mk}\,, \tag{7.295}$$

where

$$A_{mk} = \int_0^\infty d\lambda\; T_m(\lambda)\, T_k(\lambda)\,. \tag{7.296}$$

If we first solve

$$\sum_{k=1}^{3} A_{mk} V_{nk} = \mu_n V_{nm}\,, \tag{7.297}$$

then $V_{n,\boldsymbol{\rho}}(\boldsymbol{\rho}_d, m) = \delta(\boldsymbol{\rho}_d - \boldsymbol{\rho})V_{nm}$. We have thus reduced the complicated mixed CD eigenvalue problem to a 3×3 matrix eigenvalue problem.

This reduction is reminiscent of the one that occurred in Sec. 7.2.10, where we considered axial systems that were LSIV in two of the three dimensions [see especially (7.222)]. In that case as well as the present problem, we were able to make good use of partial shift-invariance.

7.3.6 Discrete-to-continuous systems

Though Sec. 7.3 deals with CD mappings, it is an easy digression to touch on discrete-to-continuous (DC) mappings. As we shall see below, a DC mapping is the adjoint of a CD one.

The most common imaging application of DC mappings is *image display*. For example, if a set of coefficients stored in a computer is converted into a luminance pattern on a cathode-ray tube or other display device, the mapping is DC. Similarly, computer-generated holograms can be regarded as systems for converting discrete vectors into continuous optical fields. Again, the mapping is DC and the goal is display. Linear DC mappings are treated briefly here, and the effect of inevitable display nonlinearities is discussed in Sec. 7.5.

Form of the operator For definiteness, we shall discuss DC mappings specifically in terms of image display. A display system maps a stored set of numbers, described by a vector $\boldsymbol{\theta}$, to a luminance pattern $f_d(\mathbf{r})$, where the subscript stands for display. If the system is linear, the general form of the mapping is

$$f_d(\mathbf{r}) = \sum_{n=1}^{N} d_n(\mathbf{r})\,\theta_n\,. \tag{7.298}$$

By comparison with (7.24), we see that this equation involves the adjoint of a CD operator with kernel $d_n(\mathbf{r})$. It can be written abstractly as

$$\mathbf{f}_d = \mathcal{D}_d^\dagger \boldsymbol{\theta}\,. \tag{7.299}$$

Uniformity A desirable feature of any image display is that it convert a uniform set of coefficients into a uniform displayed function. If all components of $\boldsymbol{\theta}$ are unity, the displayed function is

$$f_d^{(unif)}(\mathbf{r}) = \sum_{n=1}^{N} d_n(\mathbf{r})\,. \tag{7.300}$$

This expression is formally identical to that for the point sensitivity of a CD system, (7.232).

Magnification An image is rarely displayed at the same scale as the original object. For direct imaging, that means that the operator $\mathcal{D}_d^\dagger \mathcal{H}$ functions as a magnifier as discussed in Sec. 7.2.7. An easy way to incorporate an arbitrary magnification into $\mathcal{D}^\dagger$ itself is to replace $d_n(\mathbf{r})$ with $d_n(\mathbf{r}/m)$ in (7.298).

7.4 LINEAR DISCRETE-TO-DISCRETE SYSTEMS

We have argued above that the correct mathematical description for a digital imaging system is a continuous-to-discrete mapping, but when we want to represent such a system in a computer, we must resort to discrete object representations. We learned in Sec. 7.1.3 how to construct such representations, and now we examine how they fit into the analysis of imaging systems.

7.4.1 System matrix

A linear object representation $f_a(\mathbf{r})$ is defined as a linear combination of more or less arbitrarily chosen expansion functions $\{\phi_n(\mathbf{r})\}$. The definition is given in (7.27), repeated here for reference:

$$f_a(\mathbf{r}) = \sum_{n=1}^{N} \theta_n \phi_n(\mathbf{r})\,, \tag{7.301}$$

where the subscript a stands for approximate. The expansion coefficients $\{\theta_n\}$ can be regarded as components of an $N \times 1$ vector $\boldsymbol{\theta}$ in the coefficient space introduced in Fig. 7.2.

Application of an imaging operator $\mathcal{H}$ to this approximate object gives an approximate image vector $\mathbf{g}_a$, defined by

$$\mathbf{g}_a = \mathcal{H}\{f_a(\mathbf{r})\} = \sum_{n=1}^{N} \theta_n \mathcal{H}\{\phi_n(\mathbf{r})\}\,. \tag{7.302}$$

In the literature, the subscript is often omitted on $\mathbf{g}_a$ and the vector $\boldsymbol{\theta}$ is often denoted $\mathbf{f}$, but both of those notations can be misleading. Omitting the subscript on $\mathbf{g}_a$ risks confusing it with an actual, measured data vector, and the use of $\mathbf{f}$ for a vector of expansion coefficients obscures the arbitrariness of the representation; there is no unique finite representation associated with a particular object $f(\mathbf{r})$. Moreover, with our convention of using $\mathbf{f}$ for the infinite-dimensional object vector in Hilbert space, we need another symbol for a finite-dimensional approximation.

If we take $\mathcal{H}$ as a general, linear CD mapping, then the mth component of $\mathbf{g}_a$ is given by

$$g_{am} = \sum_{n=1}^{N} \theta_n \int_{\mathbf{S}_f} d^q r\; h_m(\mathbf{r})\,\phi_n(\mathbf{r})\,, \qquad m = 1, ..., M\,. \tag{7.303}$$

We can write this equation more compactly by defining an $M \times N$ matrix $\mathbf{H}$ with elements H_{mn} given by

$$H_{mn} = \int_{\mathbf{S}_f} d^q r\; h_m(\mathbf{r})\,\phi_n(\mathbf{r})\,, \tag{7.304}$$

so that

$$\mathbf{g}_a = \mathbf{H}\boldsymbol{\theta}\,. \tag{7.305}$$

The element H_{mn} is the response of the mth detector in the actual CD system to an object consisting of a single expansion function with unit weight, *i.e.*, $f(\mathbf{r}) = \phi_n(\mathbf{r})$.

We shall refer to (7.305) as a discrete-to-discrete (or DD) model for an imaging system since both $\mathbf{g}_a$ and $\boldsymbol{\theta}$ are specified by discrete indices, but it should be kept in mind that $\mathbf{g}_a$ is also the result of a CD operator acting on an approximate (continuous but finite-dimensional) object representation, *i.e.*, $\mathbf{H}\boldsymbol{\theta} = \mathcal{H}\mathbf{f}_a$. There is a one-to-one mapping between a vector $\mathbf{f}_a$ in representation space and a vector $\boldsymbol{\theta}$ in coefficient space, so either can be used to compute $\mathbf{g}_a$. As illustrated in Fig. 7.14, $\mathbf{H}$ maps a vector $\boldsymbol{\theta}$ from coefficient space to a vector $\mathbf{g}_a$ in image space, and $\mathcal{H}$ maps the corresponding vector $\mathbf{f}_a$ in representation space to the same vector in image space.

Relation to discretization operators The matrix $\mathbf{H}$ can be expressed in terms of the discretization operators introduced in Sec. 7.1.3. From (7.37), we can write

$$\mathbf{g}_a = \mathcal{H}\mathcal{D}_\phi^\dagger\boldsymbol{\theta} = \mathcal{H}\mathcal{D}_\phi^\dagger\mathcal{D}_\chi\{f(\mathbf{r})\}\,, \tag{7.306}$$

where $\{\phi_n(\mathbf{r})\}$ is the set of expansion functions and $\{\chi_n(\mathbf{r})\}$ is the set of functions used to compute the coefficients in the expansion.

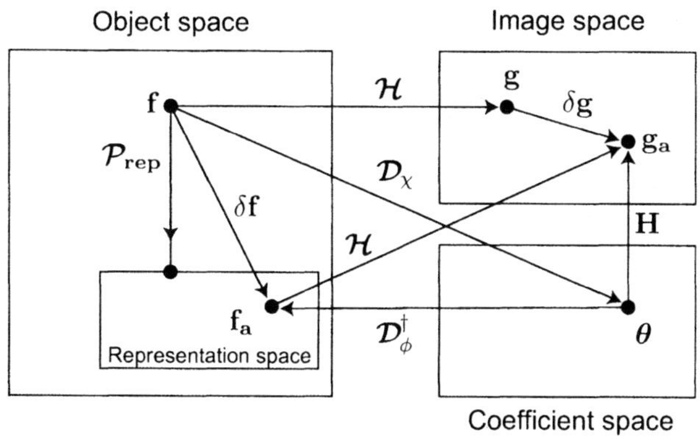

Fig. 7.14 Extension of Fig. 7.2 to include a CD imaging system. Shown are the effects of the actual CD operator $\mathcal{H}$ on an object and its approximate representation as well as the effect of the matrix $\mathbf{H}$ on the vector of coefficients associated with the object representation.

Comparison of (7.306) with (7.305) shows that

$$\mathbf{H} = \mathcal{H}\mathcal{D}_\phi^\dagger \, , \tag{7.307}$$

where, as in (7.35),

$$\boldsymbol{\theta} = \mathcal{D}_\chi\{f(\mathbf{r})\} \, . \tag{7.308}$$

These relations are illustrated in Fig. 7.14.

Measurement and computation of $\mathbf{H}$ The definition of $\mathbf{H}$ immediately suggests a way of measuring its elements if we have an actual digital imaging system in our laboratory. Suppose, for example, that the object is a 2D function and that we have chosen the expansion functions as pixels, so $\phi_n(\mathbf{r}) = \text{pix}_n(\mathbf{r})$ as defined in (7.29). Then we can construct a self-luminous or reflective test object in the form of a single pixel, place it at the nth location in the object space, and record the response of all M detectors. This vector of responses is one column of $\mathbf{H}$, namely, the set of H_{mn} values for the chosen n and all m.

For some kinds of expansion functions, direct measurement of elements of $\mathbf{H}$ is not possible. For example, with a truncated Fourier series, the expansion functions are complex and not realizable as physical objects. In these cases, or when the actual system is not available, we must resort to numerical computation. A straightforward approach is to sample the integrand in (7.304) finely and replace the integral by a sum.

A useful alternative, especially when the dimension q is large, is Monte Carlo integration. If $h_m(\mathbf{r})$ is nonnegative, as it must be in many kinds of imaging, then it can be normalized as a probability density on position $\mathbf{r}$:

$$p_m(\mathbf{r}) \equiv \frac{h_m(\mathbf{r})}{\int_{\mathbf{S}_f} d^q r \; h_m(\mathbf{r})} \, . \tag{7.309}$$

If sample points $\mathbf{r}_j$ can be generated from this density, the elements of $\mathbf{H}$ are given by

$$H_{mn} = \left[\int_{\mathbf{S}_f} d^q r \; h_m(\mathbf{r}) \right] \cdot \lim_{J \to \infty} \frac{1}{J} \sum_{j=1}^{J} \phi_n(\mathbf{r}_j) \, . \tag{7.310}$$

With a finite J, an estimate of H_{mn} is obtained, and it can be shown that this estimate is unbiased. See Gifford (1997) for details of the procedure and error analysis.

Flood uniformity and point sensitivity By analogy to (7.231) and (7.232), flood uniformity and point sensitivity for a DD system model can be defined, respectively, as

$$[\mathbf{g}_{fld}]_m = \sum_{n=1}^{N} H_{mn} \, ; \tag{7.311}$$

$$[\mathbf{s}_{pt}]_n = \sum_{m=1}^{M} H_{mn} \, . \tag{7.312}$$

These two vectors are easily obtained from a calculated or measured system matrix.

We can relate these uniformity measures for the DD model back to the corresponding measures for an actual CD system. Inserting (7.304) into (7.311) yields

$$[\mathbf{g}_{fld}]_m = \sum_{n=1}^{N} \int_\infty d^q r \; h_m(\mathbf{r}) \, \phi_n(\mathbf{r}) = \int_\infty d^q r \; h_m(\mathbf{r}) \sum_{n=1}^{N} \phi_n(\mathbf{r}) \, . \tag{7.313}$$

Comparison with (7.231) shows that the flood image determined from the $\mathbf{H}$ matrix is proportional to the flood image for a CD system provided the sum of expansion functions is a constant independent of $\mathbf{r}$. This condition implies that the expansion functions are adequate to represent a flood object exactly. For example, it is satisfied by pixel functions if they fit together without gaps or overlap. [See the discussion around (7.72).]

The point sensitivity for the DD model is given explicitly by

$$[\mathbf{s}_{pt}]_n = \int_\infty d^q r \; \phi_n(\mathbf{r}) \sum_{m=1}^{M} h_m(\mathbf{r}) = \int_\infty d^q r \; \phi_n(\mathbf{r}) \, s_{pt}(\mathbf{r}) \, , \tag{7.314}$$

where $s_{pt}(\mathbf{r})$ is the CD point sensitivity from (7.232). We see that the DD point sensitivity $\mathbf{s}_{pt}$ is a sampled version of $s_{pt}(\mathbf{r})$ if the latter varies sufficiently slowly that $\phi_n(\mathbf{r})$ can be approximated by $const \cdot \delta(\mathbf{r} - \mathbf{r}_n)$.

7.4.2 Adjoint operator and singular-value decomposition

Unless the number of measurements M is equal to the number of expansion coefficients N, $\mathbf{H}$ is not a square matrix and eigenanalysis of $\mathbf{H}$ is not possible, but SVD will prove to be a useful tool here as in earlier sections of this chapter. After introducing the adjoint of the operator $\mathbf{H}$, we shall discuss the SVD of a DD imaging operator and relate it to the SVD of the CD operator it is intended to model.

Adjoint operator We know from Sec. 1.3.5 that the adjoint of an $M \times N$ matrix is an $N \times M$ matrix obtained by transposing the original matrix (interchanging rows and columns) and taking the complex conjugate of each element. For real matrices, the adjoint is just the transpose. In the context of DD models for imaging systems, the operator $\mathbf{H}^\dagger$ maps a vector in the M-dimensional image space to a vector in the N-dimensional coefficient space, which is isomorphic to representation space.

From (7.307) and a basic property of adjoints (see Sec. 1.3.5), the adjoint of a system matrix $\mathbf{H}$ can be written formally as

$$\mathbf{H}^{\dagger} = \mathcal{D}_{\phi}\mathcal{H}^{\dagger} . \tag{7.315}$$

This operator can be applied to any vector in data space. If it is applied to an actual data vector $\mathbf{g} = \mathcal{H}\mathbf{f}$, where $\mathcal{H}$ is a CD operator, the result is

$$\mathbf{H}^{\dagger}\mathbf{g} = \mathcal{D}_{\phi}\mathcal{H}^{\dagger}\mathcal{H}\mathbf{f} . \tag{7.316}$$

The right-hand side is the same as we would have obtained by backprojecting $\mathbf{g}$ through the actual CD system and then discretizing the result with the functions $\{\phi_n(\mathbf{r})\}$.

On the other hand, if $\mathbf{H}^{\dagger}$ is applied to an approximate data vector $\mathbf{g}_a$ as given by (7.306), the result is

$$\mathbf{H}^{\dagger}\mathbf{g}_a = \mathcal{D}_{\phi}\mathcal{H}^{\dagger}\mathcal{H}\mathcal{D}_{\phi}^{\dagger}\mathcal{D}_{\chi}\mathbf{f} . \tag{7.317}$$

We can simplify this expression if we assume that $\{\chi_n(\mathbf{r})\}$ has been chosen for optimal representational accuracy. From Sec. 7.1.4 we recall that the norm of the object error is minimized if $\mathcal{D}_{\phi}^{\dagger}\mathcal{D}_{\chi} = \mathcal{P}_{rep}$, the projector onto representation space. A simple way to achieve this condition is to take $\{\phi_n(\mathbf{r})\}$ and $\{\chi_n(\mathbf{r})\}$ as identical orthonormal sets. With this assumption, (7.317) becomes

$$\mathbf{H}^{\dagger}\mathbf{g}_a = \mathcal{D}_{\phi}\mathcal{H}^{\dagger}\mathcal{H}\mathcal{D}_{\phi}^{+}\mathcal{D}_{\phi}\mathbf{f} = \mathcal{D}_{\phi}\mathcal{H}^{\dagger}\mathcal{H}\mathcal{P}_{rep}\mathbf{f} . \tag{7.318}$$

The effect of the operator $\mathcal{H}^{\dagger}\mathcal{H}\mathcal{P}_{rep}$ is to eliminate components of $\mathbf{f}$ that do not lie in representation space, then project and backproject what is left through the system (thereby eliminating null functions and weighting measurement functions by eigenvalues of $\mathcal{H}^{\dagger}\mathcal{H}$). The result is a vector (function) in measurement space; a subsequent discretization of this function yields $\mathbf{H}^{\dagger}\mathbf{g}_a$.

Singular-value decomposition As with the other operators discussed in this chapter, we shall construct the SVD of $\mathbf{H}$ by first considering the eigenvalue problems for two Hermitian operators, $\mathbf{H}^{\dagger}\mathbf{H}$ and $\mathbf{H}\mathbf{H}^{\dagger}$.

Since $\mathbf{H}$ is an $M \times N$ matrix, $\mathbf{H}^{\dagger}\mathbf{H}$ is an $N \times N$ matrix mapping one $N \times 1$ vector in coefficient space to another. Its eigenfunctions are $N \times 1$ vectors denoted $\mathbf{u}_n^{(d)}$, where the superscript d, standing for discrete, serves to distinguish these eigenvectors from the Hilbert-space eigenvectors of $\mathcal{H}^{\dagger}\mathcal{H}$, which we have denoted by $\mathbf{u}_n$. The eigenvalue equation for $\mathbf{H}^{\dagger}\mathbf{H}$ is thus

$$\mathbf{H}^{\dagger}\mathbf{H}\mathbf{u}_n^{(d)} = \mu_n^{(d)}\mathbf{u}_n^{(d)} . \tag{7.319}$$

There is no reason to expect the eigenvalues of $\mathbf{H}^{\dagger}\mathbf{H}$, here denoted $\mu_n^{(d)}$, to equal those for $\mathcal{H}^{\dagger}\mathcal{H}$, even if $\mathbf{H}$ is intended to model $\mathcal{H}$, so again we use the superscript.

The set $\{\mathbf{u}_n^{(d)}, n = 1, ..., N\}$ is orthonormal and complete in coefficient space, so any coefficient vector $\boldsymbol{\theta}$ can be expanded as

$$\boldsymbol{\theta} = \sum_{k=1}^{N} a_k \mathbf{u}_k^{(d)} , \qquad a_k = \left[\mathbf{u}_k^{(d)}\right]^{\dagger} \boldsymbol{\theta} , \tag{7.320}$$

and an individual coefficient θ_n is given by

$$\theta_n = \sum_{k=1}^{N} a_k \, u_{kn}^{(d)} , \tag{7.321}$$

where $u_{kn}^{(d)}$ is the nth component of the vector $\mathbf{u}_k^{(d)}$. The set $\{a_k\}$ constitutes a vector $\mathbf{a}$, which is simply the vector $\boldsymbol{\theta}$ in a different basis in coefficient space. Since we always choose $\{\mathbf{u}_n^{(d)}\}$ to be an orthonormal set, $\mathbf{a}$ and $\boldsymbol{\theta}$ differ by a unitary transformation.

The operator $\mathbf{HH}^\dagger$ is an $M \times M$ matrix mapping image space to itself, and its eigenvalue equation is

$$\mathbf{HH}^\dagger \mathbf{v}_m^{(d)} = \mu_m^{(d)} \mathbf{v}_m^{(d)} . \tag{7.322}$$

The set $\{\mathbf{v}_m^{(d)}, m = 1, ..., M\}$ is orthonormal and complete in image space, so any image vector (approximate or real) can be expanded in terms of them. In particular, the approximate image vector $\mathbf{g}_a$ can be written as

$$\mathbf{g}_a = \sum_{m=1}^{M} b_m \mathbf{v}_m^{(d)} , \qquad b_m = \left[\mathbf{v}_m^{(d)}\right]^\dagger \mathbf{g}_a . \tag{7.323}$$

With these preliminaries, we can now express the SVD of $\mathbf{H}$ as

$$\mathbf{H} = \sum_{n=1}^{R} \sqrt{\mu_n^{(d)}} \, \mathbf{v}_n^{(d)} \mathbf{u}_n^{(d)\dagger} , \tag{7.324}$$

where, as discussed in Sec. 1.2.3, the rank R must satisfy the constraint $R \leq \min(M, N)$.

If $R = N$, then $\mathbf{H}$ does not have a null space, but this statement must be interpreted carefully. Since $\mathbf{H}$ operates on coefficient vectors, the lack of a null space means simply that there is no vector $\boldsymbol{\theta}_{null}$ in coefficient space such that $\mathbf{H}\boldsymbol{\theta}_{null} = 0$. It does not imply that the actual CD operator $\boldsymbol{\mathcal{H}}$, which $\mathbf{H}$ is supposed to represent, has no null space; all CD operators have null spaces, even when their approximate matrix representations do not.

Discrete imaging equation in SVD form We saw in (7.129) and (7.258) that SVD reduces an actual imaging equation to a simple multiplication. The same holds true for an approximate imaging equation; (7.305) is equivalent to

$$b_n = \sqrt{\mu_n^{(d)}} \, a_n , \tag{7.325}$$

where a_n and b_n are expansion coefficients for $\boldsymbol{\theta}$ and $\mathbf{g}_a$, respectively, as defined by (7.320) and (7.323).

Relation between discrete and continuous singular vectors Since the matrix $\mathbf{H}$ is intended to be a finite approximation to the CD operator $\boldsymbol{\mathcal{H}}$, we hope that there is some simple relation between the finite-dimensional singular vectors $\{\mathbf{u}_n^{(d)}\}$ appropriate to $\mathbf{H}$ and the infinite-dimensional ones $\{\mathbf{u}_n\}$ appropriate to $\boldsymbol{\mathcal{H}}$. Since $\mathbf{u}_n^{(d)}$ is a discrete vector, the only way it can be related to a function of a continuous variable, such as $\mathbf{u}_n$, is through some discretization operator, and the natural one to consider is the one we have been using all along, $\mathcal{D}_\phi$. We inquire, therefore, whether $\mathcal{D}_\phi \mathbf{u}_n$ might be an eigenvector of the matrix $\mathbf{H}^\dagger\mathbf{H}$.

From (7.307) and (7.315), we can write

$$\mathbf{H}^\dagger\mathbf{H}\mathcal{D}_\phi\mathbf{u}_n = \mathcal{D}_\phi\mathcal{H}^\dagger\mathcal{H}\mathcal{D}_\phi^\dagger\mathcal{D}_\phi\mathbf{u}_n \,. \tag{7.326}$$

If $\{\phi_n\}$ is an orthonormal set, we know from (7.40) that $\mathcal{D}_\phi^\dagger\mathcal{D}_\phi = \mathcal{P}_{rep}$, the projector onto representation space. If $\mathbf{u}_n$ lies in representation space, so that

$$\mathcal{P}_{rep}\mathbf{u}_n = \mathbf{u}_n \,, \tag{7.327}$$

then we have at once that

$$\mathbf{H}^\dagger\mathbf{H}\mathcal{D}_\phi\mathbf{u}_n = \mu_n\mathcal{D}_\phi\mathbf{u}_n \,. \tag{7.328}$$

This equation shows that $\mathcal{D}_\phi\mathbf{u}_n$ is indeed an eigenvector of $\mathbf{H}^\dagger\mathbf{H}$ and that the eigenvalue $\mu_n^{(d)}$ is identical to μ_n, but keep in mind that we have used the assumptions that $\{\phi_n\}$ is an orthonormal set and that $\mathbf{u}_n$ lies in representation space. The latter condition is equivalent to saying that $\mathbf{u}_n$ can be written exactly as a linear superposition of the expansion functions $\{\phi_n\}$.

If pixels are used for the expansion functions, then (7.327) is approximately satisfied for eigenfunctions $\mathbf{u}_n$ that vary slowly on the scale of the pixel. In practice, this often means for large values of μ_n since systems usually image fine details more poorly than coarse ones; thus eigenfunctions that contain fine detail correspond to small eigenvalues.

In Sec. 7.4.3, we shall show how to choose the functions $\{\phi_n\}$ so that representation space is identical to measurement space; in this case all of the eigenvectors of $\mathbf{H}^\dagger\mathbf{H}$ corresponding to nonzero eigenvalues (*i.e.*, the entire measurement space of $\mathbf{H}$) can be found simply by discretizing the measurement-space eigenvectors of $\mathcal{H}^\dagger\mathcal{H}$.

7.4.3 Image errors

In Sec. 7.1.4 we discussed several error norms that could be used to evaluate the degree of agreement between an actual object $\mathbf{f}$ and an approximate representation $\mathbf{f}_a$. These norms were defined in object space, but we can also use an image-space error norm $||\mathcal{H}\,\delta\mathbf{f}||$ as a way of specifying the accuracy of the data produced through an actual CD system by an approximate object representation. Because of the dual interpretation of $\mathbf{g}_a$ as either $\mathcal{H}\mathbf{f}_a$ or $\mathbf{H}\boldsymbol{\theta}$ [see (7.302) and (7.305)], this same norm is also a measure of how accurately the CD system is modeled by a discrete matrix $\mathbf{H}$.

The definition of the image error norm is

$$||\mathcal{H}\,\delta\mathbf{f}|| = ||\mathcal{H}(\mathbf{f}-\mathbf{f}_a)|| = ||\mathbf{g}-\mathbf{g}_a|| \,. \tag{7.329}$$

But we know from (7.306) – (7.308) how $\mathbf{g}_a$ is related to $\mathbf{f}$, so we can write

$$||\mathcal{H}\,\delta\mathbf{f}|| = ||\mathcal{H}\mathbf{f}-\mathbf{H}\boldsymbol{\theta}|| = ||\mathcal{H}(\mathcal{I}-\mathcal{D}_\phi^\dagger\mathcal{D}_\chi)\mathbf{f}|| \,, \tag{7.330}$$

where $\mathcal{I}$ is the identity operator in object space. One immediate conclusion from (7.330) is that only the components of $\delta\mathbf{f}$ within the measurement space of $\mathcal{H}$ contribute to $||\mathcal{H}\,\delta\mathbf{f}||$.

The image error is zero if and only if

$$\mathcal{D}_\phi^\dagger\mathcal{D}_\chi\mathbf{f} = \mathbf{f}_{meas} + \mathbf{f}_0 \,, \tag{7.331}$$

where $\mathbf{f}_0$ is any null vector of $\boldsymbol{\mathcal{H}}$ (not necessarily the null component of $\mathbf{f}$). One way to satisfy this condition is to require that

$$\mathcal{D}_\phi^\dagger \mathcal{D}_\chi = \mathcal{P}_{meas}\,, \tag{7.332}$$

but we could also add any operator whose range lies entirely in the null space of $\boldsymbol{\mathcal{H}}$.

Truncated SVD expansions One way to make $||\boldsymbol{\mathcal{H}}\,\delta\mathbf{f}||$ vanish is to use the singular vectors of $\boldsymbol{\mathcal{H}}$ as the expansion functions, so that

$$\mathbf{f}_a = \sum_{n=1}^{N} \theta_n \mathbf{u}_n\,. \tag{7.333}$$

If we choose $N \geq R$, then $||\boldsymbol{\mathcal{H}}\,\delta\mathbf{f}||$ is identically zero. The first R terms of the SVD expansion are an exact object representation in the sense that the representation produces exactly the same data set as the true object itself. The remaining $N - R$ terms constitute a vector that lies entirely in the null space of $\boldsymbol{\mathcal{H}}$.

If we choose $\boldsymbol{\phi}_n = \mathbf{u}_n$ and $N = R$, then representation space and measurement space are identical and (7.332) holds. Under these same assumptions, the measurement-space eigenvector $\mathbf{u}_n^{(d)}$ of $\mathbf{H}^\dagger\mathbf{H}$ is given by $\mathcal{D}_\phi \mathbf{u}_n$, where $\mathbf{u}_n$ is an eigenvector of $\boldsymbol{\mathcal{H}}^\dagger\boldsymbol{\mathcal{H}}$ with $n \leq R$.

Natural pixels Another way to implement (7.332) is to use the point response functions as the expansion functions. Because we wish to allow the generality of complex kernels, we take the expansion functions as $\{\phi_m(\mathbf{r}), m = 1, ..., M\} = \{h_m^*(\mathbf{r}), m = 1, ..., M\}$, where M is the number of measurements acquired by the CD system we are trying to represent. Buonocore *et al.* (1981) refer to these expansion functions as *natural pixels*.

With natural pixels, the approximate object representation has the functional form,

$$f_a(\mathbf{r}) = \sum_{m=1}^{M} \theta_m h_m^*(\mathbf{r})\,. \tag{7.334}$$

By comparison with (7.237), we see that this function corresponds to a vector $\mathbf{f}_a$ in object space given by

$$\mathbf{f}_a = \boldsymbol{\mathcal{H}}^\dagger \boldsymbol{\theta}\,. \tag{7.335}$$

Since we have chosen the number of coefficients, previously denoted by N, to be equal to the number of measurements M, coefficient space and image space are now identical and $\mathcal{D}_\phi^\dagger = \boldsymbol{\mathcal{H}}^\dagger$.

The natural pixels are not orthonormal, but they span measurement space. Any $\mathbf{f}_a$ of the form (7.335) is necessarily in measurement space since

$$\mathcal{P}_{meas}\mathbf{f}_a = \boldsymbol{\mathcal{H}}^+\boldsymbol{\mathcal{H}}\mathbf{f}_a = \boldsymbol{\mathcal{H}}^+\boldsymbol{\mathcal{H}}\boldsymbol{\mathcal{H}}^\dagger\boldsymbol{\theta} = \boldsymbol{\mathcal{H}}^\dagger\boldsymbol{\theta} = \mathbf{f}_a\,, \tag{7.336}$$

where we have used the pseudoinverse identity (1.153). Moreover, it is always possible to choose the coefficient vector $\boldsymbol{\theta}$ so that $\mathbf{f}_a$ equals $\mathbf{f}_{meas}$. The proper choice is

$$\boldsymbol{\theta} = \left[\boldsymbol{\mathcal{H}}^\dagger\right]^+ \mathbf{f}\,, \tag{7.337}$$

for then, by use of (1.151) and (1.149),

$$\mathbf{f}_a = \mathcal{H}^\dagger \left[\mathcal{H}^\dagger\right]^+ \mathbf{f} = \mathcal{H}^+\mathcal{H}\mathbf{f} = \mathbf{f}_{meas} . \tag{7.338}$$

With coefficients chosen by (7.337), therefore, (7.332) is satisfied and $||\mathcal{H}\,\delta\mathbf{f}|| = 0$.

When $\mathbf{f}_a$ is given by (7.338), we have

$$\mathbf{g} = \mathcal{H}\mathbf{f} = \mathcal{H}\mathbf{f}_a = \mathcal{H}\mathcal{H}^\dagger\left[\mathcal{H}^\dagger\right]^+ \mathbf{f} = \mathcal{H}\mathcal{H}^\dagger\boldsymbol{\theta} . \tag{7.339}$$

The system matrix $\mathbf{H}$ is thus $\mathcal{H}\mathcal{H}^\dagger$ in this case, and its elements are given by the overlap between point response functions (Buonocore *et al.*, 1981):

$$H_{mn} = \left[\mathcal{H}\mathcal{H}^\dagger\right]_{mn} = \int_{\mathbf{S}_f} d^q r \; h_m(\mathbf{r})\, h_n^*(\mathbf{r}) . \tag{7.340}$$

We shall return to natural pixels in Chap. 15 when we discuss inverse problems.

7.4.4 Discrete representations of shift-invariant systems

As discussed in detail in Sec. 7.2.3, a linear shift-invariant CC system is described by a convolution integral. The corresponding DD description models the integral by a matrix-vector multiplication. If we take the expansion functions as uniform translates of a single function and also sample the image on a regular grid, then the matrix has a certain regular structure, which is the topic of this section. We discuss that structure first for 1D imaging and then in an arbitrary number of dimensions.

Matrix description of a 1D LSIV system A 1D linear shift-invariant CC system has a kernel of the form $h(x_d - x)$. As in Sec. 7.3.4, we assume that the image is sampled with detector elements spaced by ϵ. The resulting CD kernel $h_m(x)$ has the form [*cf.* (7.272)],

$$h_m(x) = p(m\epsilon - x) . \tag{7.341}$$

To get a DD model, we also assume that the object is expanded in uniform translates as

$$f(x) = \sum_n \theta_n \phi(x - n\epsilon) , \tag{7.342}$$

where the spacing is the same as in the image.

With these assumptions, the DD system matrix has elements given, from (7.304), by

$$H_{mn} = \int_{-\infty}^{\infty} dx \; p(m\epsilon - x)\, \phi(x - n\epsilon) = \int_{-\infty}^{\infty} dx' \, p(x')\, \phi\left[(m-n)\epsilon - x'\right] , \tag{7.343}$$

where the second integral follows from the change of variables $x' = m\epsilon - x$. By inspection of that integral, we see that the value of each element of $\mathbf{H}$ is a function of only the difference in its indices, so we can write

$$H_{mn} = h_{m-n} . \tag{7.344}$$

As illustrated on the left in Fig. 7.15, each column is identical to the one adjacent to it but shifted by one element. Matrices of this form, called *Toeplitz*, were introduced in App. A and discussed earlier in this chapter in Sec. 7.1.5.

$$\begin{bmatrix} c & b & a & 0 & 0 & 0 & 0 & 0 & 0 \\ d & c & b & a & 0 & 0 & 0 & 0 & 0 \\ e & d & c & b & a & 0 & 0 & 0 & 0 \\ 0 & e & d & c & b & a & 0 & 0 & 0 \\ 0 & 0 & e & d & c & b & a & 0 & 0 \\ 0 & 0 & 0 & e & d & c & b & a & 0 \\ 0 & 0 & 0 & 0 & e & d & c & b & a \\ 0 & 0 & 0 & 0 & 0 & e & d & c & b \\ 0 & 0 & 0 & 0 & 0 & 0 & e & d & c \end{bmatrix} \qquad \begin{bmatrix} c & b & a & 0 & 0 & 0 & 0 & e & d \\ d & c & b & a & 0 & 0 & 0 & 0 & e \\ e & d & c & b & a & 0 & 0 & 0 & 0 \\ 0 & e & d & c & b & a & 0 & 0 & 0 \\ 0 & 0 & e & d & c & b & a & 0 & 0 \\ 0 & 0 & 0 & e & d & c & b & a & 0 \\ 0 & 0 & 0 & 0 & e & d & c & b & a \\ a & 0 & 0 & 0 & 0 & e & d & c & b \\ b & a & 0 & 0 & 0 & 0 & e & d & c \end{bmatrix}$$

Fig. 7.15 Matrix representations of a 1D LSIV system in which the PSF has the form (...0 0 a b c d e 0 0 ...). The object and image vectors are both 9×1. *Left*: Toeplitz matrix; *Right*: Circulant approximation to the Toeplitz matrix.

It is convenient to take the indices m and n to range from 0 to $N-1$. The DD imaging equation (7.305) for a 1D LSIV system represented by a Toeplitz matrix then becomes

$$g_{am} = \sum_{n=0}^{N-1} h_{m-n} \theta_n \, . \tag{7.345}$$

One column of $\mathbf{H}$ can be interpreted as a discrete PSF. A discrete point object is a vector $\boldsymbol{\theta}$ where all elements but one are zero and that one is unity. If the nonzero element is the kth, so that $\theta_n = \delta_{nk}$, then the image elements g_{am} are h_{m-k}. Thus the kth column of $\mathbf{H}$ is the shift-invariant discrete PSF shifted to pixel k.

Truncation issues Even with very fine sampling, (7.345) is a poor representation of an LSIV system in one respect: a convolution integral has infinite limits but the indices m and n have a finite range. If there is no restriction on the form of h_{m-n}, the system is not really shift-invariant; as the PSF is shifted, part of it falls off one end of the column. What we are really modeling is not a true LSIV system but one with a finite field of view and detector size (see Sec. 7.2.8). The DD system of (7.345) corresponds to a CC system with the kernel given in (7.189).

Fortunately, in many imaging applications both the object and the PSF have finite spatial support, so infinite fields of view are not really needed. In particular, for direct imaging the width of the PSF is almost always much less than the size of the object field, so most of the elements in one column of $\mathbf{H}$ are zero. Moreover, for any given object, we can always define the spatial support to extend well beyond the region where the object is nonzero, so the vector $\boldsymbol{\theta}$ will have strings of zeros on both ends. If we choose N and the sample spacing to allow coverage of this extended field, then there is no truncation of nonzero portions of the image by the finite limits in (7.345).

Discrete convolutions and circulant matrices The right-hand side of (7.345) is almost — but not quite — a discrete convolution. When we defined discrete convolutions in (3.329), it was asserted that all indices would be interpreted modulo N. If we apply this convention to the Toeplitz matrix $\mathbf{H}$, then when the PSF is shifted so that part of it falls off the bottom of the matrix, that same part magically reappears on the top as illustrated in the matrix on the right in Fig. 7.15. The matrix elements then satisfy

$$H_{mn} = h_{[m-n]}\,, \tag{7.346}$$

where the square brackets denote modulo-N arithmetic. That is, whenever $m-n$ falls outside the range $[0, N-1]$, an appropriate integer multiple of N is added to bring it back to that range. Matrices with this structure are called *circulant* (Davis, 1979).

The cyclic behavior of circulant matrices has almost no counterpart in real imaging systems,[11] but it is very convenient mathematically. Moreover, if the PSF and the object have finite support and N is chosen large enough, there is no error at all from replacing a Toeplitz matrix by a circulant one. If there is no restriction on the width of the PSF or the size of the field of view, however, the circulant matrix is only an approximation to the Toeplitz one, and significant errors might occur, especially at the edges of an image.

Diagonalization of circulant matrices In Sec. 7.2.4 we showed the key role played by the Fourier transform in the analysis of LSIV systems. A similar role is played by the discrete Fourier transform (DFT) in analyzing discrete models of LSIV systems, but only when the circulant form of the system matrix is used.

To develop this point, we express both $\mathbf{g}_a$ and $\boldsymbol{\theta}$ in terms of their DFTs [*cf.* (3.313)]:

$$\theta_n = \frac{1}{N}\sum_{k=0}^{N-1} A_k \exp(2\pi i n k/N)\,; \tag{7.347}$$

$$g_{am} = \frac{1}{N}\sum_{j=0}^{N-1} G_{aj} \exp(2\pi i m j/N)\,. \tag{7.348}$$

The circulant version of the DD imaging equation now becomes

$$\frac{1}{N}\sum_{j=0}^{N-1} G_{aj} \exp(2\pi i m j/N) = \frac{1}{N}\sum_{n=0}^{N-1}\sum_{k=0}^{N-1} A_k h_{[m-n]} \exp(2\pi i n k/N)\,. \tag{7.349}$$

Following a line parallel to (7.166) - (7.169), we next multiply both sides of (7.349) by $\exp(-2\pi i m p/N)$ and sum over m from 0 to $N-1$. Because of the discrete orthogonality of the complex exponentials, (3.12), this step yields

$$G_{ap} = \frac{1}{N}\sum_{k=0}^{N-1} A_k \sum_{n=0}^{N-1}\sum_{m=0}^{N-1} h_{[m-n]} \exp\left[2\pi i(nk - mp)/N\right]\,. \tag{7.350}$$

Now we can make a change of variables, $\ell = m-n$. It is at this point that the cyclic nature of the kernel comes into play: since both $h_{[m-n]}$ and the exponential factor are cyclic functions, there is no need to change the limits of integration, and we can write

$$G_{ap} = \frac{1}{N}\sum_{k=0}^{N-1} A_k \sum_{n=0}^{N-1} \exp[2\pi i n(k-p)/N] \sum_{\ell=0}^{N-1} h_\ell \exp(-2\pi i \ell p/N)\,. \tag{7.351}$$

[11] For amusement, the reader may try to devise an imaging system in which the cyclic behavior actually occurs. Hint: It is all done with mirrors.

The sum over ℓ is now recognized as the DFT of one column of $\mathbf{H}$, which we shall denote as H_p. The sum over n gives $N\,\delta_{nk}$, and the Kronecker delta allows us to perform the sum over k, so we find

$$G_{ap} = A_p H_p \,. \tag{7.352}$$

Thus discrete Fourier transformation has reduced multiplication by a circulant matrix to a simple product. The discrete transfer function is, not surprisingly, the DFT of the discrete PSF.

We can also view this result in terms of matrix diagonalization. By analogy to (7.167), we can express (7.350) as

$$\mathbf{G}_a = \mathbf{W}_N \mathbf{H} \mathbf{W}_N^{-1} \mathbf{A} \,, \tag{7.353}$$

where $\mathbf{W}_N$ is the matrix operator that performs the DFT [see (3.315)]; the elements of $\mathbf{W}_N$ are $[W_N]_{kn} = \exp(2\pi i k n/N)$. Comparison of (7.353) and (7.352) reveals that

$$\left[\mathbf{W}_N \mathbf{H} \mathbf{W}_N^{-1}\right]_{kp} = H_p\,\delta_{kp} \,, \tag{7.354}$$

which is the discrete counterpart of (7.169). Thus a circulant matrix is diagonalized by the DFT, and the diagonal elements are given by the DFT of one column of the original matrix.

Multidimensional systems The concepts of Toeplitz and circulant matrices can be extended to multidimensional systems by use of multi-indices. Suppose we want to devise a matrix representation for a CC system that maps a qD object to a qD image. If we approximate the object as a weighted sum of uniform translates of a single expansion function as in (7.64), then each term is specified by a multi-index $\mathbf{n}$. Similarly, if we presume that the digital image is obtained by regular sampling of the image $g(\mathbf{r}_d)$, then elements of the discrete image vector can be specified by a multi-index $\mathbf{m}$. An element of the system matrix can then be written as $H_{\mathbf{mn}}$.

In a matrix with scalar indices, the elements H_{mn} for all values of m and one particular n constitute one column of the matrix. With multi-indices, we can still think of the index $\mathbf{n}$ as specifying a column, which is now the collection of all elements with different $m_1, ..., m_q$ for fixed values of $n_1, ..., n_q$.

If the CC system to be represented is LSIV, the matrix $\mathbf{H}$ has a multidimensional Toeplitz structure,

$$H_{\mathbf{mn}} = h_{\mathbf{m}-\mathbf{n}} = \text{function of } (m_1 - n_1, ..., m_q - n_q) \,. \tag{7.355}$$

Like its 1D counterpart, this form shows that each column is identical to the one adjacent to it but shifted by one element.

As in the 1D case, we assume that each component of each multi-index runs from 0 to $N_0 - 1$, so the DD imaging equation (7.305) becomes

$$g_{a\mathbf{m}} = \sum_{\mathbf{n}=0}^{N_0-1} h_{\mathbf{m}-\mathbf{n}} \theta_{\mathbf{n}} \,. \tag{7.356}$$

As in Sec. 7.1.2, summation over a multi-index is equivalent to summing over all q of its components, so here each $n_j (j = 1, ..., q)$ runs from 0 to $N_0 - 1$. The total number of elements in both $\mathbf{g}_a$ and $\boldsymbol{\theta}$ is thus $M = N = N_0^q$.

Multidimensional circulant matrices and DFTs With the same arguments as in the 1D case, the multidimensional Toeplitz matrix can be replaced by a multidimensional circulant one simply by replacing $h_{\mathbf{m}-\mathbf{n}}$ with $h_{[\mathbf{m}-\mathbf{n}]}$, where now the square brackets imply modulo-N arithmetic in each of the elements of the multi-index. Again, this substitution entails no approximation if the object and the PSF have finite support in all dimensions and N_0 is large enough to avoid truncation of the image. If this condition is not satisfied, some error will arise at the edge of the image.

The derivations of (7.352) and (7.354) still hold if all indices are rewritten in boldface type. A qD DFT diagonalizes a qD circulant matrix and hence reduces a discrete qD convolution to simple multiplication.

Display of multidimensional Toeplitz and circulant matrices Though it simplifies the notation, the use of multi-indices makes it tricky to display the matrix $\mathbf{H}$ on a printed page. Another approach is to use lexicographic ordering and specify the matrix by scalar indices. If each component of a multi-index takes on N_0 values, the scalar lexicographic index takes on N_0^q values, so a very large number of elements may have to be displayed, but at least four-dimensional paper is not required. We shall illustrate the procedure for $q = 2$.

A matrix describing a 2D imaging system has two multi-indices and a total of four component indices. To convert such a matrix to one specified by only two indices, let $\mathbf{n} = (n_x, n_y)$ and define the scalar index n by

$$n = n_x + n_y N_0 \,, \qquad n = 0, ..., N-1 \,, \qquad N = N_0^2 \,, \tag{7.357}$$

and similarly for $\mathbf{m}$. Next we define a new matrix of size $N_0^2 \times N_0^2$ with elements equal to $H_{\mathbf{mn}}$ after the conversion (7.357). This new matrix, referred to as the lexicographic matrix, can be displayed as a function of n and m as usual.

For a 2D LSIV system, the lexicographic matrix consists of a set of blocks, each of which is a Toeplitz matrix. Moreover, the blocks themselves have a Toeplitz structure; shifting horizontally by N_0 is equivalent to shifting vertically by N_0. We say that the matrix is *block-Toeplitz with Toeplitz blocks.* Similar terminology applies to lexicographic circulant matrices.

More details on lexicographic matrices can be found in Andrews and Hunt (1977) and Pratt (1991), but we shall generally avoid lexicographic ordering and work mainly with vector indices in this book.

7.5 NONLINEAR SYSTEMS

So far we have discussed only linear imaging systems, but sometimes nonlinear mathematics is either mandated by the physics of the system or convenient for analysis. In this section we survey some of the main nonlinear descriptions that have found use in image science.

7.5.1 Point nonlinearities

One way to classify nonlinear responses is to distinguish *point nonlinearities* from *nonlocal nonlinearities.* Photographic film provides a good way of making this distinction.

Film is a highly nonlinear detector. The output variable (transmission of the developed film or optical density) is a complicated nonlinear function of the input (irradiance or exposure), but it may be a good approximation to say that the output at point $\mathbf{r}_d$ is determined fully by the input at that same point. Under this assumption, we can write the input–output relation of film as

$$g_{out}(\mathbf{r}_d) = \Phi\{g_{in}(\mathbf{r}_d)\}\,, \tag{7.358}$$

where $\Phi\{\cdot\}$ is an ordinary scalar-valued nonlinear function of a scalar argument.

When an equation of this form is valid, we say that the system exhibits a point nonlinearity. On the other hand, if it is necessary to know $g_{in}(\mathbf{r}_d)$ at more than one point in order to compute $g_{out}(\mathbf{r}_d)$ at one point, then the system is said to be nonlocal, and if it is also nonlinear, it is said to exhibit a nonlocal nonlinearity.

In reality, film is indeed nonlocal. Light entering the emulsion diffuses and affects the optical density over a region with a width which we can call the resolution width of the film. If this resolution width is small compared to the width of the PSF associated with the remainder of the system (*e.g.*, the imaging lens), then it may be a useful approximation to treat the film as a point nonlinearity. Alternatively, we can describe the light diffusion as a nonlocal but linear blurring process and then regard the nonlinear step as just the conversion of light to developed photographic grains. In that case (7.358) can be used, with a reinterpreted $g_{in}(\mathbf{r}_d)$, even though blur within the film is not negligible.

Another example of a point nonlinearity occurs in coherent optical imaging. As discussed in Chap. 18, such systems are linear in a complex amplitude $u(\mathbf{r})$ related to the electric field. Thus all of the formalism for linear CC systems developed in Sec. 7.2 is applicable — right up to the detector. Optical detectors, however, do not respond to complex amplitude but rather to the irradiance, which is proportional to the squared modulus of the amplitude, $|u(\mathbf{r}_d)|^2$. Since the irradiance at point $\mathbf{r}_d$ is a nonlinear function of the amplitude at that point and is independent of the amplitude at any other point, the conversion from amplitude to irradiance is a point nonlinearity.

Point nonlinearities in CD systems The concept of point nonlinearities applies to discrete images also. If we think of the CD system as a CC operator $\mathcal{H}_{CC}$ followed by a discretization operator $\mathcal{D}$ as in (7.227), there are two places where the nonlinearity can occur — before or after discretization. In a digital camera with a nonlinear detector array, for example, the output voltage of the mth detector can be written as

$$V_m = \Phi_{det}\{g_m\} = \Phi_{det}\{[\mathcal{D}\mathcal{H}_{CC}\mathbf{f}]_m\}\,, \tag{7.359}$$

where $\Phi_{det}(\cdot)$ is a function describing detector saturation and nonlinearities in the subsequent electronics. In this equation, the image is first integrated over the detector area and then subjected to the nonlinearity. It is a point nonlinearity in a discrete sense because V_m is determined solely by $[\mathcal{D}\mathcal{H}_{CC}\mathbf{f}]_m$ for that same m.

For coherent imaging, on the other hand, it is the irradiance rather than the amplitude that is integrated, so the proper description is

$$g_m = [\mathcal{D}\{\Phi_{sq}[\mathcal{H}_{CC}\mathbf{f}]\}]_m\,, \tag{7.360}$$

where $\Phi_{sq}(u) = |u|^2$. In this case we encounter a point nonlinearity in the continuous sense, before discretization.

Both kinds of nonlinearity may be present simultaneously. If the detector in a coherent imaging system responds nonlinearly, its output voltage is expressed by

$$V_m = \Phi_{det}\left\{[\boldsymbol{\mathcal{D}}[\Phi_{sq}(\boldsymbol{\mathcal{H}}_{CC}\mathbf{f})]\,]_m\right\} . \tag{7.361}$$

This form now involves two nested nonlinearities; both of them are point nonlinearities, but in different senses.

X-ray computed tomography involves a form similar to (7.361). This kind of imaging involves an exponential decay of radiation, and the exponent is a linear CC mapping of the object of interest; the output of this step is then integrated over a detector. As a preprocessing step, the logarithm of the measured data is computed, but this step does not cancel the exponential because of the intervening discretization operator. Instead, the result has the form

$$V_m = \ln\left\{[\boldsymbol{\mathcal{D}}[\exp(\boldsymbol{\mathcal{H}}_{CC}\mathbf{f})]\,]_m\right\} . \tag{7.362}$$

Invertible vs. noninvertible point nonlinearities Often the nonlinear function $\Phi(\cdot)$ will be a monotonic function of its argument, at least over some range of values. In such cases, we can define an inverse function $\Phi^{-1}(\cdot)$ and use it to undo the effects of the nonlinearity. In (7.359), for example, all we require is that an increase in irradiance on the detector always produces an increase in the output voltage, and we can then recover g_m by

$$g_m = \Phi_{det}^{-1}\{V_m\} = [\boldsymbol{\mathcal{D}}\boldsymbol{\mathcal{H}}_{CC}\mathbf{f}]_m . \tag{7.363}$$

In other words, an invertible point nonlinearity does not interfere at all with the nice linear models we have emphasized in this chapter, provided we know the nonlinear response and correct the measured data accordingly.

Conversion of complex amplitude to irradiance is an example of a noninvertible point nonlinearity. We cannot determine the amplitude and phase of a complex number from its squared modulus.

Linearization If the function $\Phi(\cdot)$ is differentiable, it may be possible to approximate a nonlinear system by a linear one for small excursions from some nominal operating point. Consider a nonlinear detector that always receives an input g_m near some mean value $\overline{g}_m$. The output can be written as a Taylor expansion of (7.359):

$$V_m = \Phi_{det}\{\overline{g}_m\} + [(\boldsymbol{\mathcal{D}}\boldsymbol{\mathcal{H}}_{CC}\mathbf{f})_m - \overline{g}_m]\,\Phi'_{det}\{\overline{g}_m\} + \dots , \tag{7.364}$$

where $\Phi'_{det}\{\overline{g}_m\}$ is the derivative of $\Phi_{det}\{\cdot\}$ evaluated at the nominal operating point. If higher terms can be neglected and $\Phi_{det}\{\overline{g}_m\}$ and $\Phi'_{det}\{\overline{g}_m\}$ are known, we can solve (7.364) for $(\boldsymbol{\mathcal{D}}\boldsymbol{\mathcal{H}}_{CC}\mathbf{f})_m$, again salvaging all of our efforts to understand linear systems.

7.5.2 Nonlocal nonlinearities

A general expression covering a wide variety of nonlinear responses is the *Volterra series* (Schetzen, 1980), given by

$$g(\mathbf{r}_d) = \int_{\mathbf{S}_f} d^q r \; f(\mathbf{r}) \, h_1(\mathbf{r}_d; \mathbf{r}) + \int_{\mathbf{S}_f} d^q r \int_{\mathbf{S}_f} d^q r' \; f(\mathbf{r}) \, f(\mathbf{r}') \, h_2(\mathbf{r}_d; \mathbf{r}, \mathbf{r}')$$

$$+ \int_{\mathbf{S}_f} d^q r \int_{\mathbf{S}_f} d^q r' \int_{\mathbf{S}_f} d^q r'' \; f(\mathbf{r}) \, f(\mathbf{r}') \, f(\mathbf{r}'') \, h_3(\mathbf{r}_d; \mathbf{r}, \mathbf{r}', \mathbf{r}'') + \dots . \tag{7.365}$$

If $f(\mathbf{r})$ is complex, additional terms with various combinations of f and f^* are needed for full generality.

Bilinear transforms When only the second term in the Volterra series is present, as in partially coherent imaging (see Chap. 9), we refer to the system as *bilinear*. In this case it is convenient to include a complex conjugate in the definition and write

$$g(\mathbf{r}_d) = \int_{\infty} d^q r \int_{\infty} d^q r' \; f^*(\mathbf{r}) \, f(\mathbf{r}') \, h(\mathbf{r}_d; \mathbf{r}, \mathbf{r}') . \tag{7.366}$$

A more general bilinear transform would be obtained by using two different functions $f_1(\mathbf{r})$ and $f_2(\mathbf{r})$, but this degree of generality is not needed for imaging if we map a single object $f(\mathbf{r})$ to an image.

Other bilinear transforms, such as the Wigner distribution function and the Woodward ambiguity function, were introduced in Chap. 5. Note, however, that both the Wigner and Woodward functions involve single integrals, so the kernel in (7.366) must be a delta function to reproduce those special cases.

A bilinear system in q dimensions can be reformulated as a linear system in $2q$ dimensions (Saleh, 1978; Saleh and Freeman, 1987). Let $\mathbf{R}$ be the $2q$-dimensional vector obtained by concatenating $\mathbf{r}$ and $\mathbf{r}'$, and define an auxiliary function $F(\mathbf{R})$ by

$$F(\mathbf{R}) \equiv f^*(\mathbf{r}) \, f(\mathbf{r}') . \tag{7.367}$$

Then (7.366) becomes

$$g(\mathbf{r}_d) = \int_{\infty} d^{2q} R \; F(\mathbf{R}) \, h(\mathbf{r}_d; \mathbf{R}) , \tag{7.368}$$

where $h(\mathbf{r}_d; \mathbf{R}) \equiv h(\mathbf{r}_d; \mathbf{r}, \mathbf{r}')$. We see that (7.368) has the general form of a linear CC mapping, but now the mapping is from a $2q$D function to an sD one rather than from qD to sD as in (7.101).

7.5.3 Object-dependent system operators

In some imaging systems it is difficult to distinguish the object from the imaging operator. Consider, for example, an optical microscope viewing a thick biological specimen. The light reaching the lens from one layer in the specimen is altered as it passes through other layers (Andrews and Hunt, 1977). If the specimen is transparent, the alteration is in the phase of the light, so the effect is similar to that produced by lens aberrations, which also alter the phase. The medium is both message and messenger.

A general mathematical description for a CC system with an object dependent system operator is

$$g(\mathbf{r}_d) = \int_{\mathbf{S}_f} d^q r \; h(\mathbf{r}_d, \mathbf{r}; \mathbf{f}) \, f(\mathbf{r}) . \tag{7.369}$$

Since the kernel $h(\mathbf{r}_d, \mathbf{r}; \mathbf{f})$ can depend on $f(\mathbf{r})$ at more than one point $\mathbf{r}$, this equation expresses a nonlocal nonlinearity in $\mathbf{f}$.

The Volterra structure emerges if we assume that the kernel depends only weakly on $\mathbf{f}$. To see this, we first decompose the kernel into object-independent and object-dependent parts:

$$h(\mathbf{r}_d, \mathbf{r}; \mathbf{f}) = h_1(\mathbf{r}_d, \mathbf{r}) + \delta h_1(\mathbf{r}_d, \mathbf{r}; \mathbf{f}), \tag{7.370}$$

where $h_1(\mathbf{r}_d, \mathbf{r})$ might be computed by ignoring the effect of the object on the system, and $\delta h_1(\mathbf{r}_d, \mathbf{r}; \mathbf{f})$ is whatever is left over. If we can approximate $\delta h_1(\mathbf{r}_d, \mathbf{r}; \mathbf{f})$ as a linear functional in $\mathbf{f}$, then we can write

$$\delta h_1(\mathbf{r}_d, \mathbf{r}; \mathbf{f}) \simeq \int_{\mathbf{S}_f} d^q r' \, h_2(\mathbf{r}_d; \mathbf{r}, \mathbf{r}') \, f(\mathbf{r}'). \tag{7.371}$$

Substituting (7.370) and (7.371) into (7.369) yields the first two terms in the Volterra expansion, (7.365).

If the approximation of (7.370) is not adequate, h_2 must also depend on $\mathbf{f}$. Then we can write h_2 as a term independent of $\mathbf{f}$ plus a correction term δh_2 which we can approximate as a linear functional in $\mathbf{f}$. This step yields the next term in the Volterra expansion, and repeating the process indefinitely gives the entire series.

Multicomponent objects As noted in Sec. 7.1.1, it is often necessary to consider multiple attributes of an object. If J attributes are needed for a full object description, then the object function $\mathbf{f}(\mathbf{r})$ is a vector field with J components. The object in fluorescence microscopy, for example, is characterized by its emission density $f_1(\mathbf{r})$, but also by a spatially varying absorption coefficient $f_2(\mathbf{r})$ and refractive index $f_3(\mathbf{r})$, so in this case $J = 3$.

These components do not necessarily enter into the imaging equation in the same way. A fluorescent microscope, for example, may respond linearly to the emission density but with a kernel that depends on the absorption coefficient and refractive index. Thus we can write[12]

$$g(\mathbf{r}_d) = \int_{\mathbf{S}_f} d^q r \; h(\mathbf{r}_d, \mathbf{r}; \mathbf{f}_2, \mathbf{f}_3) \, f_1(\mathbf{r}). \tag{7.372}$$

We see that $\mathbf{g}$ is linear in $\mathbf{f}_1$ for fixed $\mathbf{f}_2$ and $\mathbf{f}_3$, but it is nonlinear in the latter two quantities.

We can convert (7.372) to a CD description by integrating $g(\mathbf{r}_d)$ over a detector, as in (7.229), so that

$$g_m = \int_{\mathbf{S}_g} d^s r_d \; w_m(\mathbf{r}_d) \, g(\mathbf{r}_d) = \int_{\mathbf{S}_f} d^q r \; h_m(\mathbf{r}; \mathbf{f}_2, \mathbf{f}_3) \, f_1(\mathbf{r}). \tag{7.373}$$

Nonlinear imaging with known sources In many imaging systems, the radiation source is controllable and known, and the detector may respond linearly to the

[12] The notation here is potentially confusing. We use $f_j(\mathbf{r})$ for one component of the 3D vector field $\mathbf{f}(\mathbf{r})$, but we also use $\mathbf{f}_j$ for $f_j(\mathbf{r})$ when we wish to regard it as a vector in Hilbert space. The kernel in (7.372) is written as $h(\mathbf{r}_d, \mathbf{r}; \mathbf{f}_2, \mathbf{f}_3)$ since it depends on $f_2(\mathbf{r})$ and $f_3(\mathbf{r})$ at all $\mathbf{r}$, not just at a single point.

radiation, but the radiation reaching the detector may be a complicated nonlinear function of the object. An example occurs in optical imaging through a turbid medium, where the light is strongly scattered *en route* from source to detector. The goal of the imaging is to map the distribution of the scatterers. As we shall see in detail in Chap. 10, the propagation of light in a strongly scattering medium can often be described by a diffusion equation of the form,

$$\nabla \cdot [f(\mathbf{r})\, \nabla u(\mathbf{r})] = s(\mathbf{r})\,, \tag{7.374}$$

where $u(\mathbf{r})$ is the photon density in the medium, $s(\mathbf{r})$ is the source distribution and $f(\mathbf{r})$ is the diffusion coefficient, which we denote by $f(\mathbf{r})$ here to emphasize that it is the object we seek to image. A detector placed on the periphery of the medium responds linearly to the photon density, which in turn is a linear functional of the source distribution, but the kernel of that linear mapping depends nonlinearly on $f(\mathbf{r})$. Since many different detectors and source configurations can be used, we write the response of the mth detector to the nth source as

$$g_{mn} = \int_\infty d^q r\ h_m(\mathbf{r};\mathbf{f})\, s_n(\mathbf{r})\,. \tag{7.375}$$

This form is analogous to (7.373) except there is only one object component (the diffusion coefficient) and the source is not considered an attribute of the object.

Parametric models In Sec. 7.1.6, we surveyed a variety of parametric object descriptions. In every case, the end result was an expression for an object function $f(\mathbf{r})$, which could be the input to an imaging system. Even if the system output is a linear functional of $f(\mathbf{r})$, however, it may be nonlinear in the parameters. For the forward problem, this nonlinearity poses no great difficulty. Given a set of parameters describing an object, the image can be computed in two stages: a nonlinear mapping from parameters to $\mathbf{f}$, then a linear mapping from $\mathbf{f}$ to $\mathbf{g}$. The inverse problem, estimating the parameters from $\mathbf{g}$, is more complicated and will be discussed in Chap. 15.

Sometimes a mixed linear/nonlinear parametric representation is useful. An example occurs in magnetoencephalography (MEG), which is the imaging of magnetic sources in the brain. A popular model in that field is that the object consists of a discrete set of magnetic dipoles, each characterized by a dipole moment $\boldsymbol{\mu}_k$ and location $\mathbf{r}_k (k = 1, ..., K)$ (Leahy *et al.*, 1998). The detectors used in MEG respond linearly (though nonlocally) to the dipole strengths, but that response is a complicated nonlinear function of the dipole positions.

To get a mathematical description of an MEG system, we define two new vectors. Since each dipole moment is a 3D vector, we can define a $3K \times 1$ vector $\mathbf{d}$ that contains the information about the strength of each component of each dipole. Similarly, a $3K \times 1$ vector $\mathbf{R}$ contains the information about the location of the dipoles. If M measurements are collected, the imaging equation takes the form

$$\mathbf{g} = [\mathbf{H}(\mathbf{R})]\mathbf{d}\,, \tag{7.376}$$

where $\mathbf{H}(\mathbf{R})$ is an $M \times 3K$ matrix, the elements of which depend nonlinearly on $\mathbf{R}$.

Adaptive imaging systems An *adaptive* system is one where system parameters are modified in response to its input. The prime example in imaging is adaptive optics, where controllable mirrors are often used to compensate for image blurring resulting from atmospheric turbulence.

We can describe adaptive systems as linear CC or CD mappings, where the kernel has one or more controllable parameters. To be specific, we take the CD form and assume that there are J controllable parameters, which we can arrange as a $J \times 1$ vector $\boldsymbol{\Theta}$. The imaging equation is then

$$g_m = \int_{\mathbf{S}_f} d^q r \; h_m(\mathbf{r}; \boldsymbol{\Theta}) \, f(\mathbf{r}) \,. \tag{7.377}$$

For the system to be adaptive, each parameter Θ_j must be determined in some way by the object, but we have access only to the image values. We therefore assume that each Θ_j is a functional of $\mathbf{g}$. In adaptive optics, for example, we might derive some measure of image sharpness and use it in a feedback system to control mirror positions.

If the parameters are controlled linearly, we can write

$$\Theta_j = \Theta_{j0} + \sum_{m=1}^{M} T_{jm} g_m \,, \tag{7.378}$$

where Θ_{j0} is some nominal or preset value that Θ_j assumes when no object is present and $\mathbf{T}$ is a $J \times M$ matrix.

The two simultaneous sets of equations, (7.377) and (7.378), then constitute a nonlinear system, even though each individual equation is linear in $\mathbf{f}$ if other parameters are held fixed.

7.5.4 Postdetection nonlinear operations

So far in this section we have discussed nonlinear imaging systems, where $\mathbf{g}$ is a nonlinear function of $\mathbf{f}$, but often nonlinear operations are performed on an image $\mathbf{g}$ after detection. These operations may have many purposes, including data compression and storage, image reconstruction, image enhancement, and various forms of image analysis, including pattern recognition and estimation of numerical parameters. In this section we take a brief look at some of these topics with the aim of showing how they can be incorporated into the mathematical description of the overall system. Nonlinear estimation and image-reconstruction algorithms are discussed in further detail in Chap. 15.

Digitization and bit planes The output of a detector is often digitized by an analog-to-digital (A/D) converter in order to store it in computer memory. A convenient mathematical description of A/D converters (especially for programmers) is

$$\Phi_{AD}(g_m) = \text{int}(C_1 g_m + C_2) \,, \tag{7.379}$$

where C_1 and C_2 are constants, and $\text{int}(\cdot)$ returns the integer part of its argument [*e.g.*, $\text{int}(2.57) = 2$].

In the language of Sec. 7.5.1, $\Phi_{AD}(\cdot)$ is a noninvertible point nonlinearity. The nonlinear function in this case has a stairstep response with a discrete set of possible output values. It is not invertible since it is not a one-to-one mapping; all inputs in some finite range yield the same output.

If we assume that the output of an A/D converter is always $\leq 2^K - 1$, we can write it as

$$\Phi_{AD}(g_m) = \sum_{k=0}^{K-1} g_{mk} 2^k \, , \qquad g_{mk} = 0 \text{ or } 1 \, . \tag{7.380}$$

The set of all g_{mk} for a chosen level k and all m can be displayed as a binary image, taking on only the values 0 and 1 or black and white. This image is referred to as the kth *bitplane map.*

The conversion of the A/D outputs to a set of bitplane maps is another nonlinear operation. Since any integer $\leq 2^K - 1$, can be expressed as a binary number with K bits, $\Phi_{AD}(g_m)$ can be reconstructed from the full set of K bitplane maps, so the nonlinearity is invertible. On the other hand, conversion from $\Phi_{AD}(g_m)$ to a single bitplane map is noninvertible.

Image compression A simple way of storing an image is to assign one location in computer memory to each value of $\Phi_{AD}(g_m)$. For example, if K in (7.380) is 8, we can use one byte for each m and store values of $\Phi_{AD}(g_m)$ in the range 0 to 255. If more precision is needed, we can use 2 bytes for each m, which makes $K = 16$ and allows storage of values from 0 to 65,535.

In many applications, however, computer storage or data-transmission capability is at a premium, and it is highly desirable to use more efficient digital representations. The generic term for methods that start with one set of digital data and return another set occupying less space is *data compression.* When applied specifically to images, data compression is often called *image compression.*

An image-compression algorithm can be thought of as an imaging system where the input is one digital image and the output is another. In this sense a compression algorithm is a discrete-to-discrete or DD mapping. Since compression is almost always nonlinear, it cannot be described by a matrix as linear DD mappings can.

As with point nonlinearities, it is necessary to distinguish reversible from irreversible compression. A simple example of reversible compression is run-length coding. If an image (or any other digital data set) contains a sequence of N entries all of which have the same value, then it is more efficient to store the value once along with the number of entries N rather than to store the same value N times. It is straightforward to undo this mapping.

Reversible compression is in the realm of computer science rather than image science; the techniques could as well be applied to any digital data. The fact that the data constitute an image is of little import, except perhaps in determining the degree of compression attainable. Irreversible image compression, on the other hand, results in an image that is not mathematically equivalent to its input, so in some sense it is degraded. The methods of image-quality assessment developed later in this book can be used to quantify the degree of degradation for specific tasks and observers.

Specific techniques for image compression are well beyond the scope of this book. For a succinct review, see Dorf (2000), and for many details see Rabbani and Jones (1991).

Nonlinear displays Linear image displays were discussed in Sec. 7.3.6. Real displays, however, exhibit nonlinearities, and their output can be written generally as

$$f_{disp}(\mathbf{r}) = \Phi_{disp}\left\{f_d(\mathbf{r})\right\} \, , \tag{7.381}$$

where $f_d(\mathbf{r})$ is given by (7.298). The display function $\Phi(\cdot)$ is often a monotonic but nonlinear curve. A common form is

$$\Phi_{disp}\{x\} = Ax^{\gamma}, \qquad \gamma > 0, \tag{7.382}$$

where γ is called (not surprisingly) the *gamma* of the display. Higher gammas correspond to more contrast in the displayed image.

In addition to this inherent display nonlinearity, various other point nonlinearities are often applied in software before display in order to manipulate the contrast of an image. When the useful information in an image occupies a limited range of gray levels, for example, we can set upper and lower thresholds and not display values outside this range at all. Values within the range can then be multiplied by a constant to fill up the available range of the display, or more complicated nonlinear mappings such as histogram equalization can be performed. These mappings go under the general term *gray-level transformations.*

Since gray-level transformations are applied to individual image values before display as a function of a continuous variable, the general form of the mapping is obtained by combining (7.298) with a point nonlinearity acting on each θ_n and then inserting the result into (7.381); the result is

$$f_{disp}(\mathbf{r}) = \Phi_{disp}\left\{\sum_{n=1}^{N} d_n(\mathbf{r})\,\Phi_{gl}\{\theta_n\}\right\}, \tag{7.383}$$

where $\Phi_{gl}\{\cdot\}$ is the gray-level mapping.

8

Stochastic Descriptions of Objects and Images

There are many random, unpredictable physical effects that influence the structure of images. The inherent randomness that occurs in photoelectric detection and the noise limits imposed by basic thermodynamics inevitably make images *noisy* or *stochastic* (Greek *stochos*, aim, guess, chance). Additional randomness can arise from a variety of mechanisms in real image detectors. A full description of imaging systems requires analysis of all of these processes. Moreover, any imaging system will be used for a variety of objects, and the randomness of the objects themselves must be taken into account for many purposes.

The natural stochastic description for a digital image is as a finite-dimensional random vector, where each component corresponds to the gray value of a single pixel or to an individual measurement. Objects, on the other hand, are more accurately described as functions of continuous spatial or temporal variables (hence as vectors in an infinite-dimensional vector space); when these functions are stochastic in nature, they are called random processes. In either case, a *stochastic model* is at least a partial description of the statistics of the random vector or process.

Stochastic models have many uses in image science. They are needed for computing simple statistical descriptors such as moments and autocorrelation functions; they allow realistic computer simulation of typical images, and they provide the framework for pattern recognition, image analysis and data compression. In image reconstruction, it is useful to incorporate prior information about the object, and this information is often statistical in nature. Furthermore, as we shall see in detail in Chap. 14, objective assessment of image quality necessarily requires knowledge of the statistical properties of images, and these in turn are sensitive to the statistical properties of the objects being imaged.

It is our objective in this chapter to lay the groundwork for discussing all of these manifestations of randomness. As a starting point, we assume that the reader has a good grasp of the basic concepts of probability and random variables as surveyed in App. C. In Sec. 8.1 we discuss multivariate probability and vector random variables in general terms, though without reference to specific probability laws. Random processes are treated in similar generality in Sec. 8.2. In Sec. 8.3 we discuss an important class of specific probability laws, the Gaussian or normal distributions, as applied to random vectors and random processes. In Sec. 8.4, we introduce a few of the many stochastic models that have been used for random objects, and in Sec. 8.5 we extend the discussion to images (as opposed to objects).

A notable omission in this chapter is any discussion of the Poisson distribution, which plays a crucial role in stochastic modeling of many imaging systems; that omission will be remedied in Chap. 11.

The assistance of Robert F. Wagner in formulating and writing this chapter is gratefully acknowledged.

8.1 RANDOM VECTORS

In Sec. C.2.1 of App. C, a random variable was defined as a function that maps the sample space S of some experiment onto the set of real numbers. That is, each experimental outcome ζ in S is associated with a real scalar $g(\zeta)$. To generalize this idea to a random vector, we need only consider a vector-valued function $\mathbf{g}(\zeta)$.

For example, suppose we want to measure the irradiance of a light beam at some location. We can insert an appropriate photodetector at that location, and the detector output is a scalar random variable. If the beam consists of white light, however, we might want to know the irradiance in each of three color bands. In that case we can use three photodetectors and an arrangement of beamsplitters and filters so that each measurement yields three scalars, which we can regard as components of a three-dimensional (3D) random vector.

Repeated scalar measurements can also be arranged as a vector. If we measure the irradiance at some location K times with a single photodetector, it is often useful to think of the result as a K-dimensional (KD) random vector. Alternatively, we might be interested in the spatial distribution of light in some image plane. We can use an array of M photodetectors and measure the irradiance at M different locations simultaneously, regarding the result as an MD random vector.

Finally, complex scalar random variables can be regarded as 2D vectors. If we measure the amplitude A and phase ϕ of an electromagnetic wave received on an antenna, these quantities can be regarded as two components of a vector. We can also use A and ϕ to compute the real and imaginary parts of a complex number $g = g' + ig''$, and the components g' and g'' are naturally depicted as Cartesian coordinates of a random vector in the complex plane. Equivalently, we can think of g as a complex random scalar if that is convenient. If we measure M complex numbers, we can display the results as either a vector with M complex components or one with $2M$ real components.

It is the goal of this section to establish notation and procedures for dealing with all of these manifestations of random vectors.

8.1.1 Basic concepts

A real MD random vector, denoted $\mathbf{g}$, can be formed from any collection of M real scalar random variables $\{g_m, m = 1, ..., M\}$. For definiteness, the elements will be arranged as a column, so $\mathbf{g}$ is a column vector or $M \times 1$ matrix.

An MD complex random vector $\mathbf{g}$ has components $g_m = g'_m + ig''_m$. It can be represented by the $M \times 1$ column vector of complex random values $(g_1, g_2, ..., g_M)^T$, or as the $2M \times 1$ column vector $(g'_1, g'_2, ..., g'_M, g''_1, g''_2, ..., g''_M)^T$. Hence it is equivalent to think of an MD vector of complex numbers as residing in either $\mathbb{C}^M$ or $\mathbb{R}^{2M}$. Therefore, the treatment in this chapter is often given in terms of real random vectors, with the understanding that the complex case can be obtained by doubling the number of components in the random vector to include both real and imaginary parts as separate elements.

The probability law for a random vector is nothing more than the multivariate probability law for all of its components. Like any other random variable, each component of a random vector is either discrete-valued or continuous-valued. If each component can take on only a finite set of values, or at most a countably infinite set, then we refer to the random vector as discrete-valued. The probability law of a discrete-valued random variable specifies the probability associated with all possible combinations of values for all components. If all of the components of an MD random vector $\mathbf{g}$ are continuous-valued random variables, the full probability law is a multivariate probability density function (PDF) $\mathrm{pr}(g_1, g_2, ..., g_M)$.

The cumulative distribution function for a random vector is defined analogously to that of a scalar random variable [*cf.* (C.26)]:

$$\mathrm{F}(\mathbf{c}) \equiv \Pr(g_1 \leq c_1, g_2 \leq c_2, ..., g_M \leq c_M) , \tag{8.1}$$

where $\mathbf{c}$ is a vector with components $\{c_i\}$.

If $\mathbf{g}$ is a continuous-valued random vector, $\mathrm{F}(\mathbf{c})$ is a continuous function of each c_i. Then, in a generalization of (C.29), the PDF on $\mathbf{g}$ can be defined in terms of partial derivatives of $\mathrm{F}(\mathbf{g})$:

$$\mathrm{pr}(\mathbf{g}) \equiv \frac{\partial^M \mathrm{F}(\mathbf{g})}{\partial g_1 \partial g_2 \cdots \partial g_M} . \tag{8.2}$$

If we integrate (8.2) we retrieve the cumulative distribution function:

$$\mathrm{F}(\mathbf{g}) = \int_{-\infty}^{g_1} dg'_1 \int_{-\infty}^{g_2} dg'_2 \cdots \int_{-\infty}^{g_M} dg'_M \; \mathrm{pr}(\mathbf{g}') . \tag{8.3}$$

A more compact vector notation for (8.3) is

$$\mathrm{F}(\mathbf{g}) = \int_{-\infty}^{\mathbf{g}} d^M g' \; \mathrm{pr}(\mathbf{g}') . \tag{8.4}$$

The corresponding expression for a discrete-valued random vector would involve multiple sums in place of the continuous integrals in (8.4), one for each of the components of $\mathbf{g}$.

Marginal probability densities We are often interested in the statistical behavior of a subset of the components of a random vector regardless of the behavior of the others. The statistical description of a single component g_m of the random vector $\mathbf{g}$ is called the marginal probability density function on g_m. To determine the marginal probability density of g_m, we integrate the joint density of $\mathbf{g}$ over all other components:

$$\mathrm{pr}(g_m) = \int_{-\infty}^{\infty} dg_1 \cdots \int_{-\infty}^{\infty} dg_{m-1} \int_{-\infty}^{\infty} dg_{m+1} \cdots \int_{-\infty}^{\infty} dg_M \; \mathrm{pr}(\mathbf{g}) . \tag{8.5}$$

We can also determine the marginal density of the $(M-1)$-dimensional subvector $\mathbf{g}' = (g_1, g_2, ..., g_{M-1})^t$, which is given by

$$\mathrm{pr}(\mathbf{g}') = \int_{-\infty}^{\infty} dg_M \ \mathrm{pr}(\mathbf{g}) \,. \tag{8.6}$$

Equation (8.5) gives the marginal density of one component of the random vector $\mathbf{g}$; (8.6) gives the marginal density of the random vector $\mathbf{g}'$ formed from all but one component of the random vector $\mathbf{g}$. The marginal density of any other subset of the components of $\mathbf{g}$ is similarly obtained by integrating over all variables not included in the subset.

A simple geometric construction can be used to visualize computation of a marginal. If we compute $\mathrm{pr}(x_0)$ by integrating $\mathrm{pr}(x_0, y)$ over y, we can write that integral as

$$\mathrm{pr}(x_0) = \int_{\infty} dx \int_{\infty} dy \ \mathrm{pr}(x, y) \, \delta(x - x_0) \,. \tag{8.7}$$

The delta function is nonzero on a line parallel to the y-axis, and only values of $\mathrm{pr}(x, y)$ along that line contribute to the integral. The PDF of x_0 is essentially a 1D projection of the 2D joint PDF on (x, y).

Conditional probability densities All of the relations given in Sec. C.4 for joint and conditional probabilities and densities hold for random vectors with minor notational changes. For example, given two random vectors $\mathbf{f}$ and $\mathbf{g}$, Bayes' rule [*cf.* (C.17)] becomes

$$\mathrm{pr}(\mathbf{g}|\mathbf{f}) = \frac{\mathrm{pr}(\mathbf{f}|\mathbf{g}) \, \mathrm{pr}(\mathbf{g})}{\mathrm{pr}(\mathbf{f})} \,. \tag{8.8}$$

Two random vectors $\mathbf{f}$ and $\mathbf{g}$ are *statistically independent* if the value of one of them has no influence on the other, that is, $\mathrm{pr}(\mathbf{f}|\mathbf{g}) = \mathrm{pr}(\mathbf{f})$. When two random vectors are independent, their joint PDF factors:

$$\mathrm{pr}(\mathbf{f}, \mathbf{g}) = \mathrm{pr}(\mathbf{g}) \, \mathrm{pr}(\mathbf{f}) \,. \tag{8.9}$$

It can be shown that the cumulative distribution function of two independent random vectors also factors.

8.1.2 Expectations

Discrete-valued random vectors Expectation values of discrete-valued random vectors are defined by summing over the possible combinations of the components weighted by the corresponding probabilities. Consider, for example, the MD vector $\mathbf{g}$, where each component g_m can take on any of J values x_j, $j = 1, ..., J$. By extension of the discussion in Sec. C.4, the expectation of an arbitrary function of the components is given by

$$\langle h(g_1, g_2, ..., g_M) \rangle$$

$$= \sum_{j_1=1}^{J} \sum_{j_2=1}^{J} \cdots \sum_{j_M=1}^{J} h(x_{j_1}, x_{j_2}, ..., x_{j_M}) \, \mathrm{Pr}(g_1 = x_{j_1}, g_2 = x_{j_2}, ..., g_M = x_{j_M}) \,. \tag{8.10}$$

This notation is cumbersome, but we can shorten it to

$$\langle h(\mathbf{g}) \rangle = \sum_{g_1} \sum_{g_2} \cdots \sum_{g_M} h(\mathbf{g}) \Pr(\mathbf{g}) , \tag{8.11}$$

where it is understood that each sum runs over the possible values of the component. An even more compact notation with the same meaning is

$$\langle h(\mathbf{g}) \rangle = \sum_{\mathbf{g}} h(\mathbf{g}) \Pr(\mathbf{g}) , \tag{8.12}$$

where the sum over a vector index signifies a multiple sum over all components running over all possible values.

As in App. C, we shall use the notations $\langle h(\mathbf{g}) \rangle$ and $\mathrm{E}\{h(\mathbf{g})\}$ interchangeably, and we shall also use an overbar to denote expectation. Thus, $\overline{\mathbf{g}} = \langle \mathbf{g} \rangle = \mathrm{E}\{\mathbf{g}\}$.

Continuous-valued random vectors Given a continuous-valued random vector $\mathbf{g}$, the expectation of an arbitrary function $h(\mathbf{g})$ is written as

$$\langle h(g_1, g_2, ..., g_M) \rangle = \int_{-\infty}^{\infty} dg_1 \int_{-\infty}^{\infty} dg_2 \cdots \int_{-\infty}^{\infty} dg_M \; h(g_1, g_2, ..., g_M) \, \mathrm{pr}(g_1, g_2, ..., g_M) . \tag{8.13}$$

There is no loss of generality in the infinite limits since the density might be zero except on a finite support. In more compact notation, (8.13) becomes

$$\langle h(\mathbf{g}) \rangle = \int_{\infty} d^M g \; h(\mathbf{g}) \, \mathrm{pr}(\mathbf{g}) , \tag{8.14}$$

where the subscript ∞ on the integral sign indicates that it runs over an infinite range for each of the M variables of integration.

We have not specified the nature of the function $h(\mathbf{g})$. It could be a scalar-valued or a vector-valued function of the random vector $\mathbf{g}$. It could even be $\mathbf{g}$ itself, in which case $\langle \mathbf{g} \rangle$ is the *mean vector* $\overline{\mathbf{g}}$. The components of this vector are given by

$$\overline{g}_m = \langle g_m \rangle . \tag{8.15}$$

For complex vectors, the mean is defined separately for real and imaginary parts. Thus $\mathbf{g} = \mathbf{g}' + i\mathbf{g}''$ implies $g_m = g'_m + i g''_m$ and $\overline{\mathbf{g}} = \overline{\mathbf{g}}' + i\overline{\mathbf{g}}''$, which means that $\overline{g}_m = \overline{g}'_m + i\overline{g}''_m$ for all m.

8.1.3 Covariance and correlation matrices

It is often of interest to know whether two different components of a random vector *covary*, that is, whether fluctuations in one are statistically related to fluctuations in the other. To quantify this concept, we define the *covariance matrix* $\mathbf{K}$. For an MD random vector $\mathbf{g}$, $\mathbf{K}$ is an $M \times M$ matrix with elements given by

$$K_{ij} = \left\langle (g_i - \overline{g}_i)(g_j - \overline{g}_j)^* \right\rangle , \tag{8.16}$$

where the asterisk indicates complex conjugate, allowing for the possibility that components of $\mathbf{g}$ might be complex. It is easy to see from this definition that $\mathbf{K}$ is Hermitian, *i.e.*, $K_{ij} = K_{ji}^*$.

For the special case where g_i and g_j are statistically independent, we can write

$$K_{ij} = \langle(g_i - \overline{g}_i)\rangle \left\langle(g_j - \overline{g}_j)^*\right\rangle = 0\,, \qquad i \neq j\,. \tag{8.17}$$

Any random variable covaries with itself. The diagonal elements of the covariance matrix are the variances of the components:

$$K_{jj} = \mathrm{Var}\{g_j\}\,. \tag{8.18}$$

Another way of expressing the covariance matrix is as an *outer product*, as discussed in Sec. 1.3.7. With the notation of (1.53), (8.16) is equivalent to

$$\mathbf{K} = \left\langle(\mathbf{g} - \overline{\mathbf{g}})(\mathbf{g} - \overline{\mathbf{g}})^\dagger\right\rangle = \left\langle\Delta\mathbf{g}\Delta\mathbf{g}^\dagger\right\rangle\,, \tag{8.19}$$

where $\Delta\mathbf{g} \equiv \mathbf{g} - \overline{\mathbf{g}}$.

A related matrix is the *correlation matrix* $\mathbf{R}$, defined as

$$\mathbf{R} = \left\langle\mathbf{g}\,\mathbf{g}^\dagger\right\rangle\,. \tag{8.20}$$

By unscrambling the outer-product notation, we see that $R_{ij} = \left\langle g_i g_j^*\right\rangle$, so $\mathbf{R}$ is the matrix organization of the second moments of the random vector. As a generalization of a well-known relation for two random variables, (C.85), we have

$$\mathbf{K} = \mathbf{R} - \overline{\mathbf{g}}\,\overline{\mathbf{g}}^\dagger\,. \tag{8.21}$$

For zero-mean random vectors, therefore, $\mathbf{R}$ and $\mathbf{K}$ are identical.

When two or more random vectors are involved in the same problem, we shall add appropriate subscripts to $\mathbf{K}$ and $\mathbf{R}$. Thus $\mathbf{R}_\mathbf{g} = \left\langle\mathbf{g}\,\mathbf{g}^\dagger\right\rangle$ and $\mathbf{R}_\mathbf{f} = \left\langle\mathbf{f}\,\mathbf{f}^\dagger\right\rangle$.

Positive-definiteness Every covariance matrix $\mathbf{K}$ is positive-semidefinite, as defined in Sec. A.8. To demonstrate this point, consider an arbitrary quadratic form as in (A.115):

$$\begin{aligned} Q_\mathbf{K}(\mathbf{x}) &= \mathbf{x}^\dagger\mathbf{K}\mathbf{x} = \mathbf{x}^\dagger\left\langle\Delta\mathbf{g}\Delta\mathbf{g}^\dagger\right\rangle\mathbf{x} = \left\langle\mathbf{x}^\dagger\Delta\mathbf{g}\Delta\mathbf{g}^\dagger\mathbf{x}\right\rangle \\ &= \left\langle\left|\mathbf{x}^\dagger\Delta\mathbf{g}\right|^2\right\rangle\,, \end{aligned} \tag{8.22}$$

where $\mathbf{x}$ is a nonrandom vector and we have used elementary properties of scalar products and norms from Chap. 1. Since $|\mathbf{x}^\dagger\Delta\mathbf{g}|^2$ is never negative, its expectation is never negative, so the quadratic form $Q_\mathbf{K}(\mathbf{x})$ is never negative and $\mathbf{K}$ is positive-semidefinite (nonnegative-definite) by definition.

Moreover, it is rare that covariance matrices are not strictly positive-definite. From Sec. A.8 we know that an $M \times M$ positive-semidefinite matrix is positive-definite if its rank R equals its dimension M, and from Sec. A.3 we know that the rank is the number of linearly independent rows or columns. Thus the only way we can have $R < M$ is if at least one of the columns of $\mathbf{K}$ can be written as a linear combination of the other columns. One way in which this can happen is if not all components of $\mathbf{g}$ are measured independently, but instead one component is computed as a weighted sum of the others. Barring such unusual circumstances, however, it is reasonable to assume that $R = M$.

Cross-covariance and cross-correlation The cross-covariance matrix and the cross-correlation matrix for two random vectors $\mathbf{g}$ and $\mathbf{f}$ are defined analogously to (8.19) and (8.20), respectively. They are related by an expression analogous to (8.21):

$$\mathbf{R_{gf}} = \left\langle \mathbf{g}\,\mathbf{f}^\dagger \right\rangle = \mathbf{K_{gf}} + \overline{\mathbf{g}}\,\overline{\mathbf{f}}^\dagger . \tag{8.23}$$

The random vectors $\mathbf{g}$ and $\mathbf{f}$ are said to be *uncorrelated* if their cross-correlation matrix factors as

$$\mathbf{R_{gf}} = \langle \mathbf{g} \rangle \langle \mathbf{f}^\dagger \rangle = \overline{\mathbf{g}}\,\overline{\mathbf{f}}^\dagger , \tag{8.24}$$

or, equivalently, if their cross-covariance matrix is identically zero.

Since the PDF of two independent random vectors separates into the product of their individual PDFs, we have the immediate result that independent random vectors are uncorrelated. No general statement can be made to the converse, but we shall see later in this chapter that uncorrelated normally distributed random vectors are statistically independent.

Two random vectors are said to be *orthogonal* if

$$\mathbf{R_{gf}} = \left\langle \mathbf{g}\,\mathbf{f}^\dagger \right\rangle = 0 . \tag{8.25}$$

Note that this stochastic definition involves the outer product whereas the deterministic definition of orthogonality of two vectors involves the inner product. From (8.23) we see that if the mean of either $\mathbf{g}$ or $\mathbf{f}$ is zero, the cross-correlation and the cross-covariance matrices are equal; in that case orthogonal random vectors are also uncorrelated.

8.1.4 Characteristic functions

The characteristic function $\psi_\mathbf{g}(\boldsymbol{\xi})$ of a random vector can be defined as the natural generalization of the characteristic function of a scalar random variable (see Sec. C.3.3). For a real $M \times 1$ random vector $\mathbf{g}$ (column vector), the characteristic function is defined as

$$\psi_\mathbf{g}(\boldsymbol{\xi}) = \left\langle \exp(-2\pi i \boldsymbol{\xi}^t \mathbf{g}) \right\rangle , \tag{8.26}$$

where $\boldsymbol{\xi}^t$ is a real $1 \times M$ vector[1] (row vector) and hence $\boldsymbol{\xi}^t\mathbf{g}$ is the scalar product of $\mathbf{g}$ and $\boldsymbol{\xi}$.

For the case of a continuous-valued random vector, $\psi_\mathbf{g}(\boldsymbol{\xi})$ can be written as

$$\psi_\mathbf{g}(\boldsymbol{\xi}) = \int_\infty d^M g \; \mathrm{pr}(\mathbf{g}) \exp(-2\pi i \boldsymbol{\xi}^t \mathbf{g}) . \tag{8.27}$$

This integral is the MD Fourier transform of the PDF, so the properties of Fourier transforms from Chap. 3 can be used in its manipulation. In particular, since any PDF is nonnegative and normalized to unity, it is in $\mathbb{L}_1$; thus $\psi_\mathbf{g}(\boldsymbol{\xi})$ is finite for all $\boldsymbol{\xi}$, is continuous everywhere, and vanishes at infinity [see (3.65) and (3.66)]. The PDF on $\mathbf{g}$ is given by the inverse Fourier transform of $\psi_\mathbf{g}(\boldsymbol{\xi})$:

$$\mathrm{pr}(\mathbf{g}) = \int_\infty d^M \xi \; \psi_\mathbf{g}(\boldsymbol{\xi}) \exp(2\pi i \boldsymbol{\xi}^t \mathbf{g}) . \tag{8.28}$$

[1] One should not confuse $\boldsymbol{\xi}$ with the x component of spatial frequency, denoted ξ in other chapters. The vector $\boldsymbol{\xi}$ used here is a frequency in the sense that it is a variable in a Fourier transform, but it is not a spatial frequency.

The characteristic function of a random vector is unique, in that two random vectors have the same characteristic function if and only if they have the same probability distribution. And, as in the univariate case, two random vectors are independent if and only if their joint characteristic function can be written as the product of their marginal characteristic functions.

Moments The characteristic function has great utility not only for deriving PDFs but also as a shortcut to obtaining the moments of a random vector. This property follows from the definition (8.26) by expanding the exponential in a power series before taking the expectation value. This leads to a series of terms involving increasingly higher moments of the random variable $\mathbf{g}$. These moments can be isolated by differentiation of the series and then setting $\boldsymbol{\xi} = \mathbf{0}$, where $\mathbf{0}$ is the vector with all elements equal to zero. Alternatively, one can simply differentiate the characteristic function directly. For example, if we take the gradient we obtain (in the notation of Sec. A.9.2)

$$\frac{\partial \psi_{\mathbf{g}}(\boldsymbol{\xi})}{\partial \boldsymbol{\xi}} = \left\langle (-2\pi i \mathbf{g}) \exp(-2\pi i \boldsymbol{\xi}^t \mathbf{g}) \right\rangle . \tag{8.29}$$

On setting $\boldsymbol{\xi} = \mathbf{0}$, this yields

$$\langle \mathbf{g} \rangle = (-2\pi i)^{-1} \left[\frac{\partial \psi_{\mathbf{g}}(\boldsymbol{\xi})}{\partial \boldsymbol{\xi}} \right]_{\boldsymbol{\xi}=\mathbf{0}} . \tag{8.30}$$

Differentiating twice yields the second moment:

$$\mathbf{R} = \langle \mathbf{g}\,\mathbf{g}^t \rangle = (-2\pi i)^{-2} \left[\frac{\partial^2 \psi_{\mathbf{g}}(\boldsymbol{\xi})}{\partial \boldsymbol{\xi} \partial \boldsymbol{\xi}^t} \right]_{\boldsymbol{\xi}=\mathbf{0}} . \tag{8.31}$$

Higher-order moments can be determined using the following general expression:

$$\mathrm{E}\{g_1^{k_1} g_2^{k_2} ... g_M^{k_M}\} = (-2\pi i)^{k_1+k_2+\cdots+k_M} \left[\frac{\partial^{k_1+k_2+\cdots+k_M} \psi_{\mathbf{g}}(\boldsymbol{\xi})}{\partial \xi_1^{k_1} \partial \xi_2^{k_2} \cdots \partial \xi_M^{k_M}} \right]_{\boldsymbol{\xi}=\mathbf{0}} . \tag{8.32}$$

Complex random vectors The characteristic function of a complex random vector $\mathbf{g}$ can be written

$$\psi_{\mathbf{g}}(\boldsymbol{\xi}) = \left\langle \exp[-2\pi i \,\mathrm{Re}(\boldsymbol{\xi}^\dagger \mathbf{g})] \right\rangle = \left\langle \exp[-\pi i (\boldsymbol{\xi}^\dagger \mathbf{g} + \mathbf{g}^\dagger \boldsymbol{\xi})] \right\rangle$$

$$= \langle \exp\left[-2\pi i (\xi_1' g_1' + \xi_1'' g_1'' + \cdots + \xi_M' g_M' + \xi_M'' g_M'')\right] \rangle , \tag{8.33}$$

where now $\boldsymbol{\xi}$ is an MD complex vector $\boldsymbol{\xi} = \boldsymbol{\xi}' + i\boldsymbol{\xi}''$.

We see that the scalar product in the exponent of (8.33) is the sum of $2M$ real terms, rather than the M terms of (8.26). Another avenue for obtaining this expression is to make use of the fact that complex vectors can be considered to lie in either $\mathbb{C}^M$ or $\mathbb{R}^{2M}$. Thus we could have chosen to represent the MD complex random vector $\mathbf{g}$ by the $2M$D vector of real components $(g_1', g_2', ..., g_M', g_1'', g_2'', ..., g_M'')^t$ and similarly represent $\boldsymbol{\xi}$ by the vector of real components $(\xi_1', \xi_2', ..., \xi_M', \xi_1'', \xi_2'', ..., \xi_M'')^t$. The use of (8.26) with these real vectors would give an expression for the characteristic function identical to (8.33).

The moments of $\mathbf{g}$ can be determined by differentiation of $\psi(\mathbf{g})$ if we mind the rules for differentiation with respect to complex vectors given in Sec. A.9.5. The mean of the random vector $\mathbf{g}$ is found by taking the derivative of $\psi_{\mathbf{g}}(\boldsymbol{\xi})$ with respect to the complex vector $\boldsymbol{\xi}$:

$$\nabla\psi_{\mathbf{g}}(\boldsymbol{\xi}) = \left[\frac{\partial}{\partial\boldsymbol{\xi}'} + i\frac{\partial}{\partial\boldsymbol{\xi}''}\right]\psi_{\mathbf{g}}(\boldsymbol{\xi}) = \left\langle(-2\pi i\mathbf{g})\exp[-\pi i(\boldsymbol{\xi}^\dagger\mathbf{g} + \mathbf{g}^\dagger\boldsymbol{\xi})]\right\rangle, \tag{8.34}$$

where we have made use of (A.159) and (A.160). When we set $\boldsymbol{\xi}$ to zero we find

$$\langle\mathbf{g}\rangle = (-2\pi i)^{-1}\left[\nabla\psi_{\mathbf{g}}(\boldsymbol{\xi})\right]_{\boldsymbol{\xi}=\mathbf{0}}. \tag{8.35}$$

The second moment is found from the generalized Hessian (A.165):

$$\mathbf{R} = \left\langle\mathbf{g}\,\mathbf{g}^\dagger\right\rangle = (-2\pi i)^{-2}\left[\nabla\nabla^\dagger\psi_{\mathbf{g}}(\boldsymbol{\xi})\right]_{\boldsymbol{\xi}=0}. \tag{8.36}$$

Higher-order moments can be derived using successive differentiation, similar to the case of real $\mathbf{g}$.

8.1.5 Transformations of random vectors

Section C.3.1 gives rules for transforming PDFs of scalar random variables. A bivariate extension of these rules is presented in Sec. C.4.5. In this section we extend these rules further so that they apply to random vectors of general dimension. Our treatment is limited to real vectors; the extension to complex vectors can be done by converting the complex vectors to real vectors with double the dimension as described above.

Suppose the random vector $\mathbf{g}$ is related to the random vector $\mathbf{f}$ through the general nonlinear relationship $\mathbf{g} = \mathcal{O}\mathbf{f}$. The mapping from $\mathbf{f}$ to $\mathbf{g}$ is discrete-to-discrete even though the components of the vectors are continuous valued. If we assume that this mapping is differentiable (with respect to the component values) and also one-to-one and onto, then the inverse mapping $\mathbf{f} = \mathcal{O}^{-1}(\mathbf{g})$ exists. The PDF of $\mathbf{g}$ is then obtained from the known PDF of $\mathbf{f}$ by recognizing the equivalence of the probability spaces used to describe random events in terms of either $\mathbf{f}$ or $\mathbf{g}$:

$$\mathrm{pr}_{\mathbf{g}}(\mathbf{g})\,d^N\mathbf{g} = \mathrm{pr}_{\mathbf{f}}(\mathbf{f})\,d^N\mathbf{f}. \tag{8.37}$$

The random vector $\mathbf{g}$ must have the same dimensionality as $\mathbf{f}$ if the mapping from $\mathbf{f}$ to $\mathbf{g}$ is invertible. From (8.37) we obtain

$$\mathrm{pr}_{\mathbf{g}}(\mathbf{g}) = \mathrm{pr}_{\mathbf{f}}(\mathcal{O}^{-1}\mathbf{g})|\det\mathbf{J}|, \tag{8.38}$$

where $\mathbf{J}$ is the Jacobian matrix of partial derivatives relating the components of $\mathbf{f}$ and $\mathbf{g}$ [*cf.* (C.102)]:

$$J_{ij} = \frac{\partial f_i}{\partial g_j}, \tag{8.39}$$

and $|\det\mathbf{J}|$ is the absolute value of its determinant.

Linear transformations If the random vector $\mathbf{g}$ is generated as the output of a linear filter acting on the random vector $\mathbf{f}$, we can characterize the linear transformation by an $M \times N$ matrix $\mathbf{H}$. Then we can write the $M \times 1$ output vector $\mathbf{g}$ in terms of the $N \times 1$ input vector $\mathbf{f}$ as

$$\mathbf{g} = \mathbf{H}\mathbf{f}. \tag{8.40}$$

If $M = N$ and $\mathbf{H}^{-1}$ exists, the PDF of $\mathbf{g}$ can be written in terms of the PDF of $\mathbf{f}$ as a special case of (8.38):

$$\mathrm{pr}_{\mathbf{g}}(\mathbf{g}) = \mathrm{pr}_{\mathbf{f}}(\mathbf{H}^{-1}\mathbf{g})\,|\det \mathbf{H}^{-1}|\,. \tag{8.41}$$

Characteristic function of the transformed vector If $\mathbf{H}$ is not invertible, we cannot use (8.41) to relate the PDF for $\mathbf{g}$ to the PDF for $\mathbf{f}$, but we can relate the corresponding characteristic functions. With (8.40), (8.26) becomes

$$\psi_{\mathbf{g}}(\boldsymbol{\xi}) = \left\langle \exp(-2\pi i \boldsymbol{\xi}^t \mathbf{H}\mathbf{f}) \right\rangle = \left\langle \exp\left[-2\pi i (\mathbf{H}^t \boldsymbol{\xi})^t \mathbf{f}\right] \right\rangle , \tag{8.42}$$

where the last step has used the definition of the adjoint, (1.39). (Since we are considering real matrices here, adjoint is the same as transpose.) Comparison of the last expectation in (8.42) with (8.26) shows that

$$\psi_{\mathbf{g}}(\boldsymbol{\xi}) = \psi_{\mathbf{f}}(\mathbf{H}^t \boldsymbol{\xi})\,, \tag{8.43}$$

so knowledge of $\psi_{\mathbf{f}}$ and $\mathbf{H}$ immediately gives $\psi_{\mathbf{g}}$. As an exercise, the reader can show that (8.43) and (8.38) are equivalent if $\mathbf{H}^{-1}$ exists.

The PDF on $\mathbf{g}$ can in principle be found by taking an inverse MD Fourier transform of (8.43). Formally, we can write

$$\mathrm{pr}(\mathbf{g}) = \int_\infty d^M\xi \; \psi_{\mathbf{f}}(\mathbf{H}^t \boldsymbol{\xi}) \exp(2\pi i \boldsymbol{\xi}^t \mathbf{g})\,, \tag{8.44}$$

but in practice the integral might not be easy. The problem is that we are integrating a function of an ND vector over an MD space.

Alternative approach Another way to derive an expression for the PDF of $\mathbf{g}$, when $\mathbf{g} = \mathbf{H}\mathbf{f}$, is to use the multivariate counterpart of (C.77) to write

$$\mathrm{pr}(\mathbf{g}) = \int_\infty d^N f \; \mathrm{pr}(\mathbf{g}|\mathbf{f})\,\mathrm{pr}(\mathbf{f})\,. \tag{8.45}$$

Here the notation $\mathrm{pr}(\mathbf{g}|\mathbf{f})$ is a bit tricky: $\mathbf{g}$ is *defined* as $\mathbf{H}\mathbf{f}$ (not $\mathbf{H}\mathbf{f} + \mathbf{n}$ here), so once $\mathbf{f}$ is given, $\mathbf{g}$ is no longer random; it is just $\mathbf{H}\mathbf{f}$. Nevertheless, we can still use (8.45) if we let $\mathrm{pr}(\mathbf{g}|\mathbf{f})$ be the MD delta function, $\delta(\mathbf{g} - \mathbf{H}\mathbf{f})$. Then we have

$$\mathrm{pr}(\mathbf{g}) = \int_\infty d^N f \; \delta(\mathbf{g} - \mathbf{H}\mathbf{f})\,\mathrm{pr}(\mathbf{f})\,. \tag{8.46}$$

This form is, in fact, equivalent to (8.42). If we take the MD Fourier transform of both sides of (8.46), we find

$$\psi_{\mathbf{g}}(\boldsymbol{\xi}) = \mathcal{F}_M\{\mathrm{pr}(\mathbf{g})\} = \int_\infty d^M g \int_\infty d^N f \; \delta(\mathbf{g} - \mathbf{H}\mathbf{f})\,\mathrm{pr}(\mathbf{f}) \exp(-2\pi i \boldsymbol{\xi}^t \mathbf{g})\,. \tag{8.47}$$

The delta function allows us to perform the integral over $\mathbf{g}$, and we obtain

$$\psi_{\mathbf{g}}(\boldsymbol{\xi}) = \int_\infty d^N f \; \mathrm{pr}(\mathbf{f}) \exp(-2\pi i \boldsymbol{\xi}^t \mathbf{H}\mathbf{f}) = \left\langle \exp(-2\pi i \boldsymbol{\xi}^t \mathbf{H}\mathbf{f}) \right\rangle . \tag{8.48}$$

This equation is the same as (8.42), and (8.43) follows as before.

Although (8.43) and (8.46) are equivalent, the latter may be easier to interpret geometrically. Suppose $M < N$. Then the integral is over an ND space but the delta function is nonzero only on an MD hyperplane defined by $\mathbf{g} = \mathbf{Hf}$. Only vectors $\mathbf{f}$ that lie on this hyperplane make any contribution to the integral for a particular $\mathbf{g}$. This is similar to the geometric construction we presented for the computation of a marginal in (8.7).

Transformation of the mean and covariance When $\mathbf{g} = \mathbf{Hf}$, all moments of $\mathbf{g}$ can be derived by differentiating (8.43), but often we are interested in just the mean or covariance matrix. From the linearity of the expectation operator, we have immediately for the mean of $\mathbf{g}$,

$$\bar{\mathbf{g}} = \langle \mathbf{g} \rangle = \langle \mathbf{Hf} \rangle = \mathbf{H} \langle \mathbf{f} \rangle = \mathbf{H}\bar{\mathbf{f}}. \tag{8.49}$$

The covariance matrix of $\mathbf{g}$ is found as

$$\mathbf{K_g} = \langle \Delta\mathbf{g}\Delta\mathbf{g}^\dagger \rangle = \langle (\mathbf{Hf} - \mathbf{H}\bar{\mathbf{f}})(\mathbf{Hf} - \mathbf{H}\bar{\mathbf{f}})^\dagger \rangle = \mathbf{H}\left\langle \Delta\mathbf{f}\Delta\mathbf{f}^\dagger \right\rangle \mathbf{H}^\dagger = \mathbf{HK_f H}^\dagger , \tag{8.50}$$

where $\Delta\mathbf{f} \equiv \mathbf{f} - \bar{\mathbf{f}}$. The same results can, of course, also be found from (8.43).

These rules for transforming means and covariance matrices will recur often in this book.

8.1.6 Eigenanalysis of covariance matrices

A covariance matrix is Hermitian, and we saw in Sec. 1.4.4 that eigenvectors and eigenvalues of Hermitian matrices have many nice properties. The eigenvalues are real, and the eigenvectors can be chosen to form a complete, orthonormal set in the domain of the matrix. Expansion of a random vector in eigenvectors of its covariance matrix is a valuable tool in statistical analysis.

Let $\mathbf{K_g}$ be the $M \times M$ covariance matrix for a random vector $\mathbf{g}$. The eigenvalue equation for this matrix is

$$\mathbf{K_g}\boldsymbol{\phi}_m = \mu_m \boldsymbol{\phi}_m , \qquad m = 1, ..., M , \tag{8.51}$$

where $\boldsymbol{\phi}_m$ is an $M \times 1$ eigenvector and μ_m is the corresponding eigenvalue. (Note that the subscript on $\boldsymbol{\phi}_m$ denotes a particular eigenvector, not a component.) Since $\mathbf{K_g}$ is Hermitian, μ_m is real even if $\mathbf{K_g}$ is complex.

We showed above that $\mathbf{K_g}$ is at least positive-semidefinite, so $\mu_m \geq 0$ for all m. For convenience we assume that the eigenvalues are labeled by decreasing value:

$$\mu_1 \geq \mu_2 \geq \cdots \geq \mu_R > 0 , \tag{8.52}$$

where R is the rank. We know from Sec. 1.4.3 that the rank is also the number of nonzero eigenvalues, so μ_R is the smallest nonzero eigenvalue. We also argued above that the rank of $\mathbf{K_g}$ is likely to be the dimension M, in which case there are no nonzero eigenvalues. Then $\mathbf{K_g}$ is positive-definite and hence nonsingular (see Sec. 1.4.3).

We know from Sec. 1.4.4 that the eigenvectors of $\mathbf{K_g}$ can always be chosen as a complete, orthonormal set. The orthonormality can be expressed in inner-product notation as

$$\boldsymbol{\phi}_m^\dagger \boldsymbol{\phi}_n = \delta_{mn} , \tag{8.53}$$

where $\phi_m^\dagger$ is the row vector obtained by transposing the column vector ϕ_m and taking an element-by-element complex conjugate. The completeness of the eigenvectors is expressed by the closure relation,

$$\sum_{m=1}^{M} \phi_m \phi_m^\dagger = \mathbf{I}, \tag{8.54}$$

where $\phi_m \phi_m^\dagger$ is an outer product (see Sec. 1.3.7) and $\mathbf{I}$ is the $M \times M$ unit matrix.

From the discussion in Sec. 1.4.5, we know that the eigenvalue problem (8.51) can also be expressed as

$$\mathbf{K_g \Phi} = \mathbf{\Phi M}, \tag{8.55}$$

where $\mathbf{\Phi}$ is a matrix formed by arraying the column vectors ϕ_m side by side and $\mathbf{M}$ is a diagonal matrix with the mth diagonal element equal to μ_m. (Note that $\mathbf{M}$ is capital μ.) From (8.53) and (8.54), it follows that $\mathbf{\Phi}$ is a unitary matrix, *i.e.*, $\mathbf{\Phi}^{-1} = \mathbf{\Phi}^\dagger$. From this property, we immediately find a useful representation of the covariance matrix [*cf.* (1.85)]:

$$\mathbf{K_g} = \mathbf{\Phi M \Phi}^\dagger. \tag{8.56}$$

This representation can also be expressed in terms of outer products [*cf.* (1.86)] as

$$\mathbf{K_g} = \sum_{m=1}^{M} \mu_m \phi_m \phi_m^\dagger. \tag{8.57}$$

This expression is the *spectral decomposition* of the covariance matrix.

Discrete Karhunen-Loève expansion Since the eigenvectors of a Hermitian operator form a complete, orthonormal set in the relevant space, any $M \times 1$ vector $\mathbf{g}$ can be expressed as

$$\mathbf{g} = \sum_{m=1}^{M} \beta_m \phi_m, \tag{8.58}$$

where the coefficients are given by

$$\beta_m = \phi_m^\dagger \mathbf{g}. \tag{8.59}$$

We can express these relations in matrix-vector form by defining an $M \times 1$ vector $\boldsymbol{\beta}$ with components $\{\beta_m\}$. Then

$$\mathbf{g} = \mathbf{\Phi}\boldsymbol{\beta}, \qquad \boldsymbol{\beta} = \mathbf{\Phi}^\dagger \mathbf{g}. \tag{8.60}$$

These equations are quite general, holding for any $\mathbf{g}$ and any orthonormal basis vectors. If, however, $\mathbf{g}$ is a random vector and the vectors $\{\phi_m\}$ are eigenvectors of its covariance matrix, then the coefficients $\{\beta_m\}$ are uncorrelated random variables. It is easy to demonstrate this point. In component form, the covariance matrix for $\boldsymbol{\beta}$ is given by

$$\langle \Delta\beta_n \Delta\beta_m^* \rangle = \left\langle \left[\phi_n^\dagger \Delta\mathbf{g}\right] \left[\phi_m^\dagger \Delta\mathbf{g}\right]^* \right\rangle = \left\langle \phi_n^\dagger \Delta\mathbf{g} \Delta\mathbf{g}^\dagger \phi_m \right\rangle$$
$$= \phi_n^\dagger \left\langle \Delta\mathbf{g}\Delta\mathbf{g}^\dagger \right\rangle \phi_m = \phi_n^\dagger \mathbf{K_g} \phi_m = \mu_m \phi_n^\dagger \phi_m = \mu_m \, \delta_{nm}, \tag{8.61}$$

where $\Delta\beta_m \equiv \beta_m - \langle\beta_m\rangle$ and we have used the eigenvalue equation (8.51) and the orthonormality of the eigenvectors (8.53). In matrix form, (8.61) reads

$$\mathbf{K}_\beta = \mathbf{\Phi}^\dagger \mathbf{K_g} \mathbf{\Phi} = \mathbf{\Phi}^\dagger \mathbf{\Phi} \mathbf{M} \mathbf{\Phi}^\dagger \mathbf{\Phi} = \mathbf{M}\,, \tag{8.62}$$

where we have used (8.56), (8.60) and the unitarity of $\mathbf{\Phi}$.

Expansion of a random vector in eigenvectors of its covariance matrix is known as *Karhunen-Loève* or KL expansion. The key feature of a KL expansion is that the coefficients are uncorrelated (since $\mathbf{K}_\beta$ is diagonal). A similar expansion for random processes will be presented in Sec. 8.2.7.

The KL expansion enables us immediately to find a useful representation of the inverse of a covariance matrix. Since $\mathbf{\Phi}$ is a unitary matrix, *i.e.*, $\mathbf{\Phi}^{-1} = \mathbf{\Phi}^\dagger$, we can use (8.62) to write the covariance matrix $\mathbf{K_g}$ as

$$\mathbf{K_g} = \mathbf{\Phi} \mathbf{M} \mathbf{\Phi}^\dagger\,. \tag{8.63}$$

The inverse of $\mathbf{K_g}$ is then given by

$$\mathbf{K_g}^{-1} = \mathbf{\Phi} \mathbf{M}^{-1} \mathbf{\Phi}^\dagger\,, \tag{8.64}$$

where $\mathbf{M}^{-1}$ is also diagonal, with the mth diagonal element given by $1/\mu_m$. Thus the same matrix that diagonalizes $\mathbf{K_g}$ also diagonalizes $\mathbf{K_g}^{-1}$.

Whitening As we have just seen, the KL expansion results in a vector $\boldsymbol{\beta}$ with uncorrelated components. It is often useful to go further and force the components all to have the same variance. The concept of a square-root matrix, discussed in Sec. A.8.3, provides us with the tool to accomplish this goal.

By analogy to (A.118), we can define the square root of the covariance matrix of $\mathbf{g}$ by

$$\mathbf{K}_\mathbf{g}^{\frac{1}{2}} = \sum_{m=1}^{M} \sqrt{\mu_m}\, \boldsymbol{\phi}_m \boldsymbol{\phi}_m^\dagger\,. \tag{8.65}$$

If $\mathbf{K_g}$ is nonsingular, as it usually is, we can write the inverse of the square-root matrix as

$$\mathbf{K}_\mathbf{g}^{-\frac{1}{2}} = \sum_{m=1}^{M} \frac{1}{\sqrt{\mu_m}} \boldsymbol{\phi}_m \boldsymbol{\phi}_m^\dagger\,. \tag{8.66}$$

To verify that this is the correct form for the inverse, one can multiply (8.65) by (8.66) and use the orthonormality relation (8.53) to obtain (8.54).

We now define the vector $\mathbf{y}$ by

$$\mathbf{y} = \mathbf{K}_\mathbf{g}^{-\frac{1}{2}} (\mathbf{g} - \overline{\mathbf{g}})\,. \tag{8.67}$$

With this construction $\overline{\mathbf{y}} = \mathbf{0}$, and its covariance matrix is given by

$$\mathbf{K_y} = \langle \mathbf{y}\mathbf{y}^\dagger \rangle = \mathbf{K}_\mathbf{g}^{-\frac{1}{2}} \mathbf{K_g} \mathbf{K}_\mathbf{g}^{-\frac{1}{2}} = \mathbf{I}\,, \tag{8.68}$$

where we have used the definition of the square-root matrix from (A.117) and the fact that covariance matrices are Hermitian, so that $\mathbf{K}_\mathbf{g}^{-\frac{1}{2}}$ is its own adjoint.

Thus the transformation (8.67) always results in a random vector $\mathbf{y}$ such that

$$\langle y_n y_m \rangle = \delta_{nm}\,. \tag{8.69}$$

By analogy to white noise (a topic discussed further in Sec. 8.2.6), this transformation is referred to as *whitening*; it is also called *prewhitening* when it precedes other signal processing. As we shall see in Chap. 13, prewhitening plays a key role in signal-detection theory.

Simultaneous diagonalization We have shown that a Hermitian matrix can always be diagonalized by a unitary transformation. It can be shown that two different Hermitian matrices can be diagonalized by the *same* unitary transformation if and only if they commute. If the different Hermitian matrices do not commute, they can be simultaneously diagonalized by a linear transformation, but the transformation matrix will not be unitary (Fukunaga, 1990). Details of the procedure were given in Sec. 1.4.6.

8.2 RANDOM PROCESSES

8.2.1 Definitions and basic concepts

We now generalize the concept of a random variable further by assigning to every experimental outcome ζ a spatial or temporal function, real or complex, according to some rule (Wentzell, 1981). In the spatial case the function will be denoted $f(\mathbf{r}, \zeta)$, where $\mathbf{r}$ is a position vector, and in the temporal case it will be denoted by $f(t, \zeta)$, where t is the time. We now have a family of functions referred to as a stochastic or random process. The words stochastic and random will be used interchangeably here. If the spatial (or temporal) variable $\mathbf{r}$ (or t) is a continuous one, the family is referred to as a continuous stochastic process; if the variables are taken as discrete, for example by sampling in space or time, the family is referred to as a discrete stochastic process, or a random sequence. A random process or sequence is said to be continuous-valued or discrete-valued according to whether the underlying random variables are continuous- or discrete-valued.

A spatial random process is a function of two variables, $\mathbf{r}$ and ζ. Depending on the context, $f(\mathbf{r}, \zeta)$ can refer to (Papoulis, 1965; Middleton, 1960):

1. The family of spatial functions, referred to as the ensemble; in this case, $\mathbf{r}$ and ζ are variables.

2. A single realization or sample of the spatial functions; in this case, $\mathbf{r}$ is variable and ζ is fixed.

3. The random variable at a single point; in this case, $\mathbf{r}$ is fixed and ζ is variable.

4. A single number; in this case, $\mathbf{r}$ is fixed and ζ is fixed.

The intended interpretation will usually be clear from the context.

Some notational issues require attention here. First, it is frequently cumbersome to carry along the index ζ, so we shall usually refer to the random process simply as $f(\mathbf{r})$. Second, we usually make no notational distinction between the random process *per se*, understood as the ensemble of all possible functions $f(\mathbf{r})$ (interpretation 1), and a specific realization or sample (interpretation 2). This practice is in accord with our conventions on random variables as set out in Sec. C.2.1 of App. C. Occasionally a specific realization will have to be designated, and in those cases we shall either reinstate ζ or use primes, subscripts or other typographical devices. Finally, the variable $\mathbf{r}$ will be understood to be a general q-dimensional position vector unless otherwise stated.

Square-integrable random processes We shall say that a random process lies in some Hilbert space if all sample functions [*i.e.*, $f(\mathbf{r}, \zeta)$ as a function of $\mathbf{r}$ for all ζ] lie in that space. In particular, we shall often be concerned with random processes in $\mathbb{L}_2(\mathbb{R}^q)$ where each sample function is square-integrable.

In many physical problems, $|f(\mathbf{r})|^2$ can be interpreted as an energy density (energy per unit area or volume). For example, that interpretation works when $f(\mathbf{r})$ is an electric field or the amplitude of an acoustic wave. In those cases, the integral of $|f(\mathbf{r})|^2$ is the total energy, and a square-integrable function is one with finite energy. This terminology is used more broadly, and any square-integrable function can be called a finite-energy function without regard to interpretation as a physical energy.

Finite-power random processes For temporal random processes, however, the assumption of finite energy is frequently not warranted. Consider, for example, the thermal noise produced by a resistor. The duration of this noise is completely indefinite. So long as the resistor exists, there will be a fluctuating voltage across it. In this example, ζ designates a particular resistor and $f(t, \zeta)$ is the noise voltage, and there is no reason to assume that the integral of $|f(t, \zeta)|^2$ over $-\infty < t < \infty$ is finite. We could get around this problem by imposing some artificial boundary conditions, *e.g.*, the resistor is manufactured at $t = -T$ and destroyed at $t = T$, but we are not really interested in when the resistor was manufactured.

A more natural approach is just to give up on the restriction to finite energy. The noise voltage across a resistor has finite *power* (energy per unit time). Mathematically, we can state this condition for the random process $f(t, \zeta)$ as

$$0 < \lim_{T \to \infty} \frac{1}{2T} \int_{-T}^{T} dt \, |f(t, \zeta)|^2 < \infty \qquad \text{for all } \zeta \, . \tag{8.70}$$

A random process for which this condition is satisfied will be called a *finite-power random process*. Note that a finite-energy (or $\mathbb{L}_2$) random process cannot simultaneously be a finite-power one because of the left-hand inequality. If the function is in $\mathbb{L}_2$, then the integral is finite as $T \to \infty$, but the factor of $1/2T$ drives the product to zero. It is only when the integral is asymptotically linear in T that (8.70) is satisfied. As we shall see, finite-energy and finite-power random processes require rather different mathematical treatments.

Generalized random processes We shall also have occasion to use random processes constructed with delta functions or other generalized functions. Such constructs are mathematically very convenient, even though no physical process is exactly described by them. We shall refer to random processes where the sample functions are generalized functions as (not surprisingly) *generalized random processes* (Kanwal, 1983). These processes are not in $\mathbb{L}_2$ but instead define a space of tempered distributions (see Chap. 2). If the generalized function in question is a delta function, the generalized random process has neither finite energy nor finite power.

8.2.2 Averages of random processes

We consider here a scalar random process $f(\mathbf{r})$ that is a function of position vector $\mathbf{r}$. The generalization to a vector random process is straightforward using multivariate PDFs. The random process may in principle be either continuous- or discrete-valued, but we shall illustrate the concepts with continuous-valued random processes. The discrete-valued case proceeds via a parallel approach but with sums over discrete values replacing integrals over continuous values.

For fixed $\mathbf{r}$, $f(\mathbf{r})$ is simply a random variable (interpretation 3), and its expectation is defined just as for any other random variable. As before, we use the notations $\mathrm{E}\{\cdot\}$, $\langle\cdot\rangle$ and overbar interchangeably to indicate an expectation, and we can write

$$\mathrm{E}\{f(\mathbf{r})\} = \langle f(\mathbf{r})\rangle = \overline{f}(\mathbf{r}) = \int_{-\infty}^{\infty} df(\mathbf{r})\; f(\mathbf{r})\,\mathrm{pr}[f(\mathbf{r})]\,. \tag{8.71}$$

Computation of this expectation requires only the univariate PDF $\mathrm{pr}[f(\mathbf{r})]$. Note carefully that the integral is over $f(\mathbf{r})$, not $\mathbf{r}$, so $\mathrm{E}\{f(\mathbf{r})\}$ can be (and usually will be) a function of $\mathbf{r}$.

Moments and variance Moments of $f(\mathbf{r})$ are defined easily. For example, the jth moment is given by [*cf.* (C.38)]

$$\left\langle [f(\mathbf{r})]^j \right\rangle = \int_{-\infty}^{\infty} df(\mathbf{r})\; [f(\mathbf{r})]^j\,\mathrm{pr}[f(\mathbf{r})]\,. \tag{8.72}$$

The resultant, $\langle [f(\mathbf{r})]^j\rangle$, can still be a function of $\mathbf{r}$; again, the integral is over $f(\mathbf{r})$, not over $\mathbf{r}$.

Having defined moments, we can also define the variance of a random process. In the general complex case, the variance is given by

$$\mathrm{Var}\{f(\mathbf{r})\} = \mathrm{E}\left\{|f(\mathbf{r})| - |\mathrm{E}\left\{f(\mathbf{r})\right\}|^2\right\} = \mathrm{E}\left\{|f(\mathbf{r})|^2\right\} - |\mathrm{E}\left\{f(\mathbf{r})\right\}|^2$$

$$= \int_{-\infty}^{\infty} df(\mathbf{r})\; |f(\mathbf{r})|^2\,\mathrm{pr}[f(\mathbf{r})] - \left|\int_{-\infty}^{\infty} df(\mathbf{r})\; f(\mathbf{r})\,\mathrm{pr}[f(\mathbf{r})]\right|^2 . \tag{8.73}$$

Note that this definition works equally for finite-energy and finite-power processes. It is possible for a random process to have a finite variance at all points, yet not be square-integrable.

Multiple-point expectations We are often interested in *two-point expectations* or joint second moments of the form $\mathrm{E}\{f(\mathbf{r}_1)f(\mathbf{r}_2)\}$. The usual definitions for joint expectations stand us in good stead here, and we can write

$$\mathrm{E}\{f(\mathbf{r}_1)f(\mathbf{r}_2)\} = \int_{-\infty}^{\infty} df(\mathbf{r}_1) \int_{-\infty}^{\infty} df(\mathbf{r}_2)\; f(\mathbf{r}_1)\,f(\mathbf{r}_2)\,\mathrm{pr}[f(\mathbf{r}_1), f(\mathbf{r}_2)]\,. \tag{8.74}$$

Here, $f(\mathbf{r}_1)$ and $f(\mathbf{r}_2)$ must be regarded as two *distinct* random variables and $\mathrm{pr}[f(\mathbf{r}_1), f(\mathbf{r}_2)]$ is their joint density. Only in very special circumstances will it be possible to write $\mathrm{pr}[f(\mathbf{r}_1), f(\mathbf{r}_2)]$ as $\mathrm{pr}[f(\mathbf{r}_1)]\,\mathrm{pr}[f(\mathbf{r}_2)]$.

A general two-point moment is defined by

$$\mathrm{E}\{[f(\mathbf{r}_1)]^m\,[f(\mathbf{r}_2)]^n\} = \int_{-\infty}^{\infty} df(\mathbf{r}_1) \int_{-\infty}^{\infty} df(\mathbf{r}_2)\; [f(\mathbf{r}_1)]^m\,[f(\mathbf{r}_2)]^n\,\mathrm{pr}[f(\mathbf{r}_1), f(\mathbf{r}_2)]\,. \tag{8.75}$$

Moments involving more points are defined similarly. Any moment involving the K points $\mathbf{r}_1, \mathbf{r}_2, ..., \mathbf{r}_K$ can be computed if $\text{pr}[f(\mathbf{r}_1), f(\mathbf{r}_2), ..., f(\mathbf{r}_K)]$ is known. If this K-fold joint density is known for all values of each of the $\mathbf{r}_k$, the process is said to be *fully characterized* to order K (Snyder and Miller, 1991).

Density of the process Expressing N-fold joint densities using notation of the form $\text{pr}[f(\mathbf{r}_1), f(\mathbf{r}_2), ..., f(\mathbf{r}_N)]$ is cumbersome at best and quite inadequate when we want to define expectations of general functionals $\Phi\{f(\mathbf{r})\}$, which can depend on $f(\mathbf{r})$ at all points $\mathbf{r}$. We now introduce an alternative approach, which works at least for finite-energy random processes (or vectors in $\mathbb{L}_2$). Our objective is to give meaning to an expression like $\text{pr}(\mathbf{f})$, where $\mathbf{f}$ is the Hilbert-space vector corresponding to $f(\mathbf{r})$. We saw in Chap. 1 that $\mathbb{L}_2$ is a *separable* Hilbert space, which means simply that it is spanned by a denumerably infinite set of basis functions. Each sample function of a random process in $\mathbb{L}_2(\mathbb{R}^q)$ can be written as

$$f(\mathbf{r}, \zeta) = \sum_{j=1}^{\infty} \alpha_j(\zeta)\, \psi_j(\mathbf{r}), \tag{8.76}$$

where the set $\{\psi_j(\mathbf{r})\}$ is any orthonormal basis for the space. We can also express this same concept as

$$f(\mathbf{r}) = \lim_{J\to\infty} \sum_{j=1}^{J} \alpha_j \psi_j(\mathbf{r}) . \tag{8.77}$$

We have dropped the index ζ with the understanding that the equation holds for any $f(\mathbf{r}, \zeta)$ so long as the corresponding expansion coefficients $\alpha_j(\zeta)$ are used on the right. The convergence of (8.77) is in the sense of $\mathbb{L}_2(\mathbb{R}^q)$ (see Sec. 3.2.2); if we use the truncated series in place of the original function $f(\mathbf{r})$, the $\mathbb{L}_2$ norm of the error converges to zero as $J \to \infty$. The expansion of the sample function $f(\mathbf{r})$ given in (8.77) is exactly the same form that was used in (7.8) to represent a deterministic object.

Expansion (8.77) provides a convenient way of defining averages involving random processes. Each coefficient α_j is a random variable, and the set of them $\{\alpha_j, j = 1, ..., J\}$ can be regarded as a random vector $\boldsymbol{\alpha}_J$ with J components. In the limit $J \to \infty$, the vector $\boldsymbol{\alpha}_J$ completely defines $f(\mathbf{r})$, and averaging over $f(\mathbf{r})$ is equivalent to averaging over all components of $\boldsymbol{\alpha}$. For finite J, the requisite density can be written as $\text{pr}(\boldsymbol{\alpha}_J)$ or $\text{pr}(\alpha_1, \alpha_2, ..., \alpha_J)$. In the limit,

$$\text{pr}(\boldsymbol{\alpha}) = \lim_{J\to\infty} \text{pr}(\boldsymbol{\alpha}_J) , \tag{8.78}$$

and this density is operationally equivalent to $\text{pr}(\mathbf{f})$.

When $f(\mathbf{r})$ is approximated by the truncated series, any functional $\Phi\{f(\mathbf{r})\}$ is also a function of $\boldsymbol{\alpha}_J$; call it $\Phi_J(\boldsymbol{\alpha}_J)$. If the functional is continuous, in the sense defined in Sec. 1.3.2, then the limit of the functional is the functional of the limit, and we have

$$\Phi\{f(\mathbf{r})\} = \lim_{J\to\infty} \Phi_J(\boldsymbol{\alpha}_J) . \tag{8.79}$$

Moreover, expectation is also a continuous functional, so we can write

$$\text{E}\{\Phi[f(\mathbf{r})]\} = \lim_{J\to\infty} \text{E}\{\Phi_J(\boldsymbol{\alpha}_J)\} = \lim_{J\to\infty} \int_{\infty} d^J\alpha\, \Phi_J(\boldsymbol{\alpha}_J)\, \text{pr}(\boldsymbol{\alpha}_J) . \tag{8.80}$$

For notational convenience, we write this expectation as

$$\mathrm{E}\{\Phi[f(\mathbf{r})]\} = \int_{\mathbb{L}_2} d\mathbf{f}\; \Phi[f(\mathbf{r})]\, \mathrm{pr}(\mathbf{f})\,. \tag{8.81}$$

Here $\mathbf{f}$ is $f(\mathbf{r})$ regarded as a vector in the Hilbert space, and the integral is really a denumerably infinite multiple integral[2] over all basis functions in the space; in other words, (8.81) must be realized operationally by (8.80).

Example: Linear functionals To clarify how (8.81) works in practice, consider a linear functional that depends on $f(\mathbf{r})$ at K points:

$$\Phi\{f(\mathbf{r}_1), ..., f(\mathbf{r}_K)\} \equiv \sum_{k=1}^{K} \beta_k f(\mathbf{r}_k) = \lim_{J\to\infty} \sum_{k=1}^{K} \beta_k \sum_{j=1}^{J} \alpha_j \psi_j(\mathbf{r}_k)\,. \tag{8.82}$$

The random variables here are the coefficients $\{\alpha_j\}$. Using (8.80) and invoking the linearity of the expectation operator, we find

$$\mathrm{E}\{\Phi[f(\mathbf{r}_1), ..., f(\mathbf{r}_K)]\} = \lim_{J\to\infty} \sum_{k=1}^{K} \beta_k \sum_{j=1}^{J} \psi_j(\mathbf{r}_k) \int_\infty d^J\alpha\; \alpha_j\, \mathrm{pr}(\boldsymbol{\alpha}_J)\,. \tag{8.83}$$

In the J-fold multiple integral, we can immediately integrate out all of the variables except α_j. By (C.75), the result is the marginal density on α_j, so

$$\mathrm{E}\{\Phi[f(\mathbf{r}_1), ..., f(\mathbf{r}_K)]\} = \lim_{J\to\infty} \sum_{k=1}^{K} \beta_k \sum_{j=1}^{J} \psi_j(\mathbf{r}_k) \int_{-\infty}^{\infty} d\alpha_j\; \alpha_j\, \mathrm{pr}(\alpha_j)$$

$$= \lim_{J\to\infty} \sum_{k=1}^{K} \beta_k \sum_{j=1}^{J} \psi_j(\mathbf{r}_k)\, \mathrm{E}\{\alpha_j\}\,. \tag{8.84}$$

Thus, for a linear functional of the form (8.82), and by extension any linear functional,

$$\langle \Phi\{\mathbf{f}\}\rangle = \Phi\{\langle \mathbf{f}\rangle\}\,. \tag{8.85}$$

Integrals of random processes An integral of a random process $f(x)$, sometimes called a *stochastic integral*, is another random process, the realizations of which are obtained by integrating corresponding realizations of $f(x)$. For example, the statement

$$g(x) = \int_{-\infty}^{\infty} dx'\, f(x')\, h(x, x') \tag{8.86}$$

means that

$$g(x, \zeta) = \int_{-\infty}^{\infty} dx'\, f(x', \zeta)\, h(x, x') \tag{8.87}$$

[2] We have customarily denoted volume elements by italics rather than boldface, *e.g.*, d^3r rather than $d\mathbf{r}$, on the theory that volume elements are scalars. To preserve a distinction between integrals over Euclidean spaces and ones over Hilbert spaces, however, we use $d\mathbf{f}$ (along with the subscript $\mathbb{L}_2$) to indicate a multiple integral with an infinite number of dimensions.

for all ζ and some fixed kernel $h(x, x')$. A similar definition holds for derivatives of a random process.

Since $g(x)$ is a functional of $f(x')$, its average at any fixed x can be computed by (8.81) as

$$\langle g(x) \rangle = \int_{\mathbb{L}_2} d\mathbf{f} \; g(x) \, \mathrm{pr}(\mathbf{f}) = \int_{\mathbb{L}_2} d\mathbf{f} \int_{-\infty}^{\infty} dx' \, f(x') \, h(x, x') \, \mathrm{pr}(\mathbf{f}) \, . \tag{8.88}$$

It is often useful to interchange the order of these two integrals, but most books gloss over issues of the validity of this step. Middleton (1960) puts it thus: "The condition on the interchangeability of integration and expectation is, *of course*, the existence of the resulting integral" (emphasis added).

When the interchange can be justified, (8.88) becomes

$$\langle g(x) \rangle = \int_{-\infty}^{\infty} dx' \, h(x, x') \int_{\mathbb{L}_2} d\mathbf{f} \; f(x') \, \mathrm{pr}(\mathbf{f}) = \int_{-\infty}^{\infty} dx' \, h(x, x') \, \langle f(x') \rangle \, . \tag{8.89}$$

In other words, the average of a linear integral transform of a random process is the same linear transform of the average of the random process (but only under conditions that we haven't yet stated clearly).

When is the interchange legal? The classical theorem that states when interchange of the order of two integrals is allowed is Fubini's theorem (Lang, 1993). In essence, this theorem tells us that

$$\int_{-\infty}^{\infty} du \int_{-\infty}^{\infty} dv \; k(u, v) = \int_{-\infty}^{\infty} du \left[\int_{-\infty}^{\infty} dv \; k(u, v) \right] = \int_{-\infty}^{\infty} dv \left[\int_{-\infty}^{\infty} du \; k(u, v) \right] \tag{8.90}$$

provided $|k(u, v)|$ is integrable over the product space, here the u-v plane.

There are two difficulties in directly applying Fubini to (8.88). First, we often want to assume that the integrand is in $\mathbb{L}_2$ rather than in $\mathbb{L}_1$, and we know from Sec. 3.3.2 that a function in $\mathbb{L}_2$ need not be in $\mathbb{L}_1$ (the prime example being sinc x). One way around this problem is to consider only random processes where all sample functions are absolutely integrable as well as square-integrable. Another way is to consider a finite interval, say $-\frac{1}{2}X < x \leq \frac{1}{2}X$. This allows use of Fubini with $\mathbb{L}_2$ functions since $\mathbb{L}_2(-\frac{1}{2}X, \frac{1}{2}X)$ is a subspace of $\mathbb{L}_1(-\frac{1}{2}X, \frac{1}{2}X)$.

The second difficulty is that Fubini's theorem can be extended to higher-dimensional multiple integrals, but (8.88) in its most general form requires an infinite nested set of integrals. Fubini's theorem can still be used in this case, but it must be justified with advanced measure-theoretic arguments (Lipster and Shiryayev, 1977). A more elementary argument can be given by using the theory of distributions.

Retreat to distributions Much of the discussion above has hinged on the assumption that the random process lies in a separable Hilbert space. For finite-power processes, we do not have this luxury, and even with $\mathbb{L}_2$ processes, we ran into some problems justifying the interchange of integration and expectation. The solution to these difficulties is the theory of distributions[3] as outlined in Chap. 2. The thing we have

[3] At least three distinctly different meanings attach to the word *distribution* in connection with random processes. A *probability distribution* is, loosely speaking, any probability law, such as the

going for us is that sample functions of a random process may be badly behaved but kernel functions in integral transforms like (8.86) are usually good functions.

Let $t(x)$ denote a good function and $f(x, \zeta)$ be a sample function of a random process. This random process defines a distribution,

$$\Phi_f\{t(x)\} = \int_{-\infty}^{\infty} dx\; t(x)\, f(x, \zeta) \equiv \phi(\zeta)\,. \tag{8.91}$$

Note that $\phi(\zeta)$ is a random variable. It is proved by Kanwal (1983) that this random variable has finite variance if $f(x, \zeta)$ is continuous (in the sense that $f(x + \epsilon, \zeta) \to f(x, \zeta)$ in the limit that $\epsilon \to 0$) and has finite variance at all x. With these mild restrictions, any random process defines a distribution mapping good functions to finite-variance random variables.

By the Schwarz inequality, the finite variance of $\phi(\zeta)$ implies that $\phi(\zeta)$ has finite mean. The expectation $\mathrm{E}\{\phi(\zeta)\}$ is defined conventionally by

$$\mathrm{E}\{\phi(\zeta)\} = \mathrm{E}\{\Phi_f[t(x)]\} = \int_{-\infty}^{\infty} d\phi\; \phi\, \mathrm{pr}(\phi)\,. \tag{8.92}$$

But this is just a linear combination of distributions, which by (2.15) is another distribution. Thus

$$\mathrm{E}\{\Phi_f[t(x)]\} = \Phi_{Ef}\{t(x)\} = \int_{-\infty}^{\infty} dx\; t(x)\, \mathrm{E}\{f(x, \zeta)\}\,, \tag{8.93}$$

where $\Phi_{Ef}\{t(x)\}$ is a distribution defined by using $\mathrm{E}\{f(x, \zeta)\}$ as the generalized function. Equation (8.93) is just what one would obtain by interchanging the expectation operation and the integration over x.

Thus the issue of interchangeability is resolved once we have established that the random process defines a distribution (in the Schwartz sense), and Kanwal did this for us with mild restrictions.

8.2.3 Characteristic functionals

Characteristic functions for scalar random variables were introduced in App. C and extended to random vectors in Sec. 8.1.4. Now we shall extend the concept further to random processes. In a formal sense, the extension is straightforward; all we have to do is to pay attention to the dimensionality of the vectors involved.

As defined in (8.26), the characteristic function of an MD random vector is a function of an MD frequency vector $\boldsymbol{\xi}$. In the case of a random process, each sample function corresponds to a vector $\mathbf{f}$ in an infinite-dimensional Hilbert space, so the frequency vector $\boldsymbol{\xi}$ in (8.26) must be replaced by an infinite-dimensional vector $\mathbf{s}$ in the same Hilbert space as $\mathbf{f}$. That means that $\mathbf{s}$ describes a function $s(\mathbf{r})$, so the characteristic function becomes a characteristic *functional* $\Psi_{\mathbf{f}}\{s(\mathbf{r})\}$ or $\Psi_{\mathbf{f}}(\mathbf{s})$ for short. It is defined by

$$\Psi_{\mathbf{f}}(\mathbf{s}) = \langle \exp[-2\pi i(\mathbf{s}, \mathbf{f})] \rangle\,, \tag{8.94}$$

Poisson distribution. The *distribution function* refers specifically to the cumulative probability distribution function defined in Sec. C.2.3. In the present context the word is used in the Schwartz sense defined in Chap. 2.

where $(\mathbf{s}, \mathbf{f})$ is the usual $\mathbb{L}_2$ scalar product. Note that we use $\Psi(\cdot)$ for characteristic functional and $\psi(\cdot)$ for characteristic function.

The characteristic functional of a random process can be related to the characteristic *function* of any random vector derived from the random process by a linear operator; the calculation is a simple generalization of one performed in Sec. 8.1.5. For example, if $\mathbf{g} = \mathcal{H}\mathbf{f}$, where $\mathcal{H}$ is a continuous-to-discrete (CD) operator as discussed in Secs. 1.2.4 and 7.3, then (8.26) becomes

$$\psi_{\mathbf{g}}(\boldsymbol{\xi}) = \langle \exp[-2\pi i(\boldsymbol{\xi}, \mathcal{H}\mathbf{f})] \rangle = \langle \exp[-2\pi i(\mathcal{H}^{\dagger}\boldsymbol{\xi}, \mathbf{f})] \rangle , \tag{8.95}$$

where the second step follows from the definition of the adjoint, (1.39). Comparison of (8.94) and (8.95) shows that

$$\psi_{\mathbf{g}}(\boldsymbol{\xi}) = \Psi_{\mathbf{f}}(\mathcal{H}^{\dagger}\boldsymbol{\xi}) , \tag{8.96}$$

which is the generalization of (8.43) to random processes.

Thus, if we know the characteristic functional for $\mathbf{f}$, we immediately have the characteristic function for $\mathcal{H}\mathbf{f}$. We shall exploit this relation in Sec. 8.3.5 when we discuss normal random processes.

8.2.4 Correlation analysis

The autocorrelation function $R(\mathbf{r}_1, \mathbf{r}_2)$ of a random process $f(\mathbf{r})$ is defined by

$$R(\mathbf{r}_1, \mathbf{r}_2) = \langle f(\mathbf{r}_1) f^*(\mathbf{r}_2) \rangle , \tag{8.97}$$

which is the two-point expectation defined in (8.74), with the minor modification of the complex conjugate on the second factor [irrelevant if $f(\mathbf{r})$ is real].

The autocovariance function $K(\mathbf{r}_1, \mathbf{r}_2)$ is defined by

$$\begin{aligned} K(\mathbf{r}_1, \mathbf{r}_2) &= \langle [f(\mathbf{r}_1) - \langle f(\mathbf{r}_1) \rangle] \, [f^*(\mathbf{r}_2) - \langle f^*(\mathbf{r}_2) \rangle] \rangle \\ &= R(\mathbf{r}_1, \mathbf{r}_2) - \overline{f}(\mathbf{r}_1) \overline{f}^*(\mathbf{r}_2) . \end{aligned} \tag{8.98}$$

The autocovariance function is thus the two-point moment that is the generalization of the variance; it reduces to the variance when $\mathbf{r}_2 = \mathbf{r}_1 = \mathbf{r}$, *i.e.*,

$$K(\mathbf{r}, \mathbf{r}) = R(\mathbf{r}, \mathbf{r}) - |\overline{f}(\mathbf{r})|^2 = \mathrm{Var}\{f(\mathbf{r})\} \tag{8.99}$$

from (8.73).

When two or more random processes occur in the same problem, their autocorrelation and autocovariance functions will be distinguished with subscripts, *e.g.*, $R_f(\mathbf{r}_1, \mathbf{r}_2)$. It is frequently convenient to define zero-mean random processes such as

$$\Delta f(\mathbf{r}) \equiv f(\mathbf{r}) - \overline{f}(\mathbf{r}) . \tag{8.100}$$

With this definition, $\langle \Delta f(\mathbf{r}) \rangle = 0$ and

$$R_{\Delta f}(\mathbf{r}_1, \mathbf{r}_2) = K_f(\mathbf{r}_1, \mathbf{r}_2) . \tag{8.101}$$

The autocorrelation and autocovariance functions play a fundamental role in the theory of random processes since they specify how far apart two points must be for their fluctuations to be uncorrelated. If $K_f(\mathbf{r}_1, \mathbf{r}_2)$ is zero, the random variables

$f(\mathbf{r}_1)$ and $f(\mathbf{r}_2)$ do not covary; colloquially, they are said to be uncorrelated, though in fact the autocorrelation function $R_f(\mathbf{r}_1, \mathbf{r}_2)$ may be nonzero because of the mean values.

Cross-correlation and cross-covariance functions can also be defined. Consider two functions $f(\mathbf{r})$ and $g(\mathbf{r}')$, where $\mathbf{r}$ and $\mathbf{r}'$ are not necessarily in the same space. The cross-correlation or mutual correlation function is defined by

$$R_{fg}(\mathbf{r}, \mathbf{r}') = \langle f(\mathbf{r})\, g^*(\mathbf{r}') \rangle \,. \tag{8.102}$$

Similarly, the cross-covariance function is

$$K_{fg}(\mathbf{r}, \mathbf{r}') = \langle [f(\mathbf{r}) - \langle f(\mathbf{r}) \rangle]\, [g^*(\mathbf{r}') - \langle g^*(\mathbf{r}') \rangle] \rangle = R_{fg}(\mathbf{r}, \mathbf{r}') - \overline{f}(\mathbf{r})\, \overline{g}^*(\mathbf{r}') \,. \tag{8.103}$$

Two random processes $f(\mathbf{r})$ and $g(\mathbf{r}')$ are said to be uncorrelated if $R_{fg}(\mathbf{r}, \mathbf{r}') = \overline{f}(\mathbf{r})\, \overline{g}(\mathbf{r}')$ for all $\mathbf{r}$ and $\mathbf{r}'$. They are orthogonal if, for all $\mathbf{r}$ and $\mathbf{r}'$, $R_{fg}(\mathbf{r}, \mathbf{r}') = 0$.

Properties of the autocorrelation function From the definition (8.98), we obtain the symmetry property

$$R(\mathbf{r}_1, \mathbf{r}_2) = R^*(\mathbf{r}_2, \mathbf{r}_1) \,. \tag{8.104}$$

In particular, for $\mathbf{r}_1 = \mathbf{r}_2 = \mathbf{r}$, (8.104) shows that $R(\mathbf{r}, \mathbf{r})$ or $\mathrm{Var}\{f(\mathbf{r})\}$ is real.

It follows from the Schwarz inequality that

$$|R(\mathbf{r}_1, \mathbf{r}_2)|^2 \le R(\mathbf{r}_1, \mathbf{r}_1)\, R(\mathbf{r}_2, \mathbf{r}_2) \,. \tag{8.105}$$

It can also be shown (Mandel and Wolf, 1995) that $R(\mathbf{r}_1, \mathbf{r}_2)$ is positive-semidefinite, meaning that [*cf.* (8.22)]

$$\int_\infty d^q r_1\, w^*(\mathbf{r}_1) \int_\infty d^q r_2\; R(\mathbf{r}_1, \mathbf{r}_2)\, w(\mathbf{r}_2) \ge 0 \,, \tag{8.106}$$

for all functions $w(\mathbf{r})$. We shall exploit this property in Sec. 8.2.7 when we discuss the Karhunen-Loève expansion of random processes.

Another way to think about $R(\mathbf{r}_1, \mathbf{r}_2)$ is that it is the kernel of an integral operator $\mathcal{R}$. With this view, the inner integral of (8.106) is recognized as the function $[\mathcal{R}\mathbf{w}](\mathbf{r}_1)$, and the double integral is the scalar $\mathbf{w}^\dagger \mathcal{R} \mathbf{w}$. An autocovariance operator $\mathcal{K}$ can be defined similarly, with the autocovariance function as the kernel.

Temporal stationarity Temporal random processes often have a statistical character that is independent of time, even though any individual realization is a randomly fluctuating function of time. An example is a steady beam of white light, where the electric field fluctuates rapidly, yet there is no preferred origin in time as far as the statistics are concerned. Such processes are said to be *stationary*. Glauber (1965) has phrased it this way: "The term 'stationary' does not mean that nothing is happening. On the contrary, the field is ordinarily oscillating quite rapidly. It means that our knowledge of the field does not change in time."

A temporal random process $f(t)$ is said to be stationary in the strict sense if, for any K, its K-point PDF $\mathrm{pr}[f(t_1), \cdots, f(t_K)]$ is such that

$$\mathrm{pr}[f(t_1), \cdots, f(t_K)] = \mathrm{pr}[f(t_1 + \tau), \cdots, f(t_K + \tau)] \tag{8.107}$$

for any τ. In particular, this requires that the single-point density function be independent of time,

$$\mathrm{pr}[f(t)] = \mathrm{pr}[f(t + \tau)] \,, \tag{8.108}$$

and therefore the mean of the random process is also independent of time,

$$\langle f(t)\rangle = \langle f(t+\tau)\rangle \ . \tag{8.109}$$

Similarly, the two-point density function must be independent of time,

$$\mathrm{pr}[f(t_1), f(t_2)] = \mathrm{pr}[f(t_1+\tau), f(t_2+\tau)]\,, \tag{8.110}$$

and so the autocorrelation function $R(t_1, t_2)$, is also independent of time,

$$R(t_1,t_2) = \langle f(t_1)\, f^*(t_2)\rangle = \langle f(t_1+\tau)\, f^*(t_2+\tau)\rangle \ . \tag{8.111}$$

The only way (8.111) can be satisfied for all t_1 and t_2 is if $R(t_1, t_2)$ is really a function of only $t_1 - t_2$. We shall denote this function by $R(t_1 - t_2)$, but the reader is cautioned that $R(t_1 - t_2)$ is not the same function as $R(t_1, t_2)$; it could not be since the latter has two arguments and the former has only one. With this notation, we have (for a stationary random process),

$$R(t_1,t_2) = R(t_1 - t_2) = R(\Delta t)\,, \tag{8.112}$$

where $\Delta t \equiv t_1 - t_2$. The shift Δt is frequently called the *lag* of the autocorrelation function.

Continuing on in this way, we see that strict stationarity requires that all K-point moments of the process be independent of absolute time. A process is said to be stationary to order M if (8.107) is true only for $K \le M$.

A process is said to be weakly stationary, or stationary in the wide sense, if its expected value does not depend on absolute time t and its autocorrelation depends only on Δt:

$$\langle f(t)\rangle = const\,, \qquad \langle f(t+\Delta t)\, f^*(t)\rangle = R(\Delta t)\,. \tag{8.113}$$

If a process is stationary to second order, then it is wide-sense stationary; however, a wide-sense stationary process is not necessarily stationary to second order because the former involves only the first two moments while the latter involves the entire PDF. One case where we can make a more definitive statement is with normal or Gaussian random processes, to be discussed in Sec. 8.3.5. If a process is normal and stationary in the wide sense, then it is also stationary in the strict sense since the statistical description of a normal process is completely specified once its mean and autocorrelation are specified.

Stationarity is closely connected to the concept of a finite-power random process, introduced in Sec. 8.2.1, but the distinctions should not be overlooked. The finite-power designation applies to individual sample functions of the random process, while stationarity applies to averages. A stationary random process might not have finite power, since it is conceivable (though pathological) that an individual realization could diverge but the average not. Of more practical importance, a process can have finite power yet not be stationary; examples of this situation are discussed below. On the other hand, a nontrivial stationary temporal random process certainly cannot have finite energy.

Properties of the stationary autocorrelation function The general properties of autocorrelations given above specialize in the stationary case as follows: The symmetry property of (8.104) becomes

$$R(\Delta t) = R^*(-\Delta t)\,. \tag{8.114}$$

In particular, for $\Delta t = 0$, (8.114) shows that $R(0)$ is real.

The Schwarz inequality shows that

$$|R(\Delta t)| \leq R(0)\,. \tag{8.115}$$

The condition that $R(\Delta t)$ is positive-semidefinite now means that

$$\int_{-\infty}^{\infty} dt \int_{-\infty}^{\infty} dt'\, w^*(t)\, R(t - t')\, w(t') \geq 0\,, \tag{8.116}$$

for all functions $w(t)$.

Spatial stationarity The spatial counterpart of the wide-sense stationarity condition (8.112) is

$$R(\mathbf{r}_1, \mathbf{r}_2) = R(\mathbf{r}_1 - \mathbf{r}_2) = R(\Delta\mathbf{r})\,, \tag{8.117}$$

where $\Delta\mathbf{r} \equiv \mathbf{r}_1 - \mathbf{r}_2$.

This condition cannot be exactly satisfied[4] by spatial processes representing real objects or images since they have finite support, but it might be a useful mathematical description within a certain boundary. That is, we might be able to assume that $R(\mathbf{r}_1, \mathbf{r}_2) = R(\Delta\mathbf{r})$ provided $\mathbf{r}_1$ and $\mathbf{r}_2$ lie inside the borders of an image. An example would be a piece of x-ray film with a uniform exposure, where the only deviation from stationarity comes from the finite size of the film.

If $f(\mathbf{r})$ vanishes outside the boundary, this kind of stationarity is expressed mathematically by

$$R(\mathbf{r}_1, \mathbf{r}_2) = R(\Delta\mathbf{r})\, W(\mathbf{r}_1)\, W(\mathbf{r}_2)\,, \tag{8.118}$$

where $W(\mathbf{r})$ is a window function that is unity for $\mathbf{r}$ inside the boundary, zero outside.

Quasistationarity In optics and imaging we often encounter spatial random processes whose autocorrelation function can be approximated as a product of two factors — a slowly varying contribution due to slow variations in overall intensity and a short-range function describing correlation between neighboring points. As a simple example, consider a ground glass illuminated nonuniformly with a laser beam. If the statistical character of the ground glass is the same at all points, then we can describe the complex amplitude (see Chap. 9) of the wave emerging from the ground glass by a spatial autocorrelation function of the form,

$$R(\mathbf{r}_1, \mathbf{r}_2) = a(\Delta\mathbf{r})\, b(\mathbf{r}_0)\,, \tag{8.119}$$

where

$$\mathbf{r}_0 = \tfrac{1}{2}(\mathbf{r}_1 + \mathbf{r}_2)\,, \qquad \Delta\mathbf{r} = \mathbf{r}_1 - \mathbf{r}_2\,. \tag{8.120}$$

We shall refer to $\mathbf{r}_0$ as the *center coordinate* (analogous to center of mass) and $\Delta\mathbf{r}$ as the *relative coordinate* or *difference coordinate*. Since the transformation from $(\mathbf{r}_1, \mathbf{r}_2)$ to $(\mathbf{r}_0, \Delta\mathbf{r})$ is unique and invertible (with Jacobian = unity), we always have a choice of which coordinate system to use for any function of two variables, but we

[4]The stationarity condition cannot be exactly satisfied by real temporal processes either. The difference is that we usually do not observe the beginning and end of a temporal process; we almost always observe the boundaries of an object or image.

won't always find that the function can be factored as in (8.119). The factorization is particularly useful if $b(\mathbf{r}_0)$ is slowly varying, in which case the random process is said to be *quasistationary*. If $b(\mathbf{r}_0)$ is a constant and the mean is also constant, the process is wide-sense stationary.

The short-range contribution, $a(\Delta\mathbf{r})$, is usually normalized to be unity at zero shift or lag ($\Delta\mathbf{r} = 0$).

Time averages and ergodicity We have seen that statistical descriptors of a random process, like the mean and autocorrelation function, are determined by averaging over the ensemble of realizations. Knowledge of the ensemble is equivalent to knowledge of the full PDF that describes the random process. However, suppose we are presented with data derived from a single realization of a temporal random process. It is natural to ask how this single data realization might be related to the statistical descriptors of the random process from which it was drawn. The answer to this question rests in the theory of *ergodicity*, a subject that traces its origins to classical statistical mechanics and the works of such luminaries as Maxwell, Boltzmann, Clausius and Gibbs (Ter Haar, 1955).

A random process is said to be *ergodic* if each realization of the process carries the same statistical information as every other realization. The practical ramification of this feature is that when a process is ergodic it becomes possible to derive statistical information about the entire ensemble based on knowledge of a single realization.

In order for a random process to be ergodic, it must first be stationary. The degree of stationarity of the process influences the degree to which the process is ergodic. For example, only wide-sense stationarity is necessary (though not sufficient) for a process to be ergodic in its mean and autocorrelation.

We now present criteria for a random process to be ergodic with respect to its mean and autocorrelation. A more complete development can be found in Papoulis (1965). Let $f(t, \zeta_0)$ denote a particular realization of a random process. Its finite-time average is then given by

$$\langle f(t, \zeta_0) \rangle_T = \frac{1}{T} \int_{-\frac{1}{2}T}^{\frac{1}{2}T} dt \; f(t, \zeta_0) \,, \tag{8.121}$$

where $\langle \, \rangle_T$ denotes a finite-time average over period T. In general this finite-time average is itself a random variable that depends on the particular realization under consideration as well as the interval T.

The time average of the sample function $f(t, \zeta_0)$ is found by taking the limit of (8.121) as $T \to \infty$:

$$\langle f(t, \zeta_0) \rangle_\infty = \lim_{T \to \infty} \frac{1}{T} \int_{-\frac{1}{2}T}^{\frac{1}{2}T} dt \; f(t, \zeta_0) \,. \tag{8.122}$$

The result in (8.122) is independent of time but depends in general on the realization ζ_0. Thus the notational distinction that this average refers to realization ζ_0 must be maintained.

A process is said to be *ergodic in the mean* if the time average of a single realization equals the ensemble average $\langle f(t) \rangle$. We already know that a stationary process has a mean that is independent of time. It can be shown (Papoulis, 1965) that $\langle f(t, \zeta_0)\rangle_T$ approaches this same constant as $T \to \infty$ if and only if

$$\lim_{T\to\infty} \frac{1}{T} \int_{-\frac{1}{2}T}^{\frac{1}{2}T} d\Delta t \; R(\Delta t) = \langle f(t)\rangle^2 , \tag{8.123}$$

where $R(\Delta t)$ is the ensemble autocorrelation function of the stationary random process [*cf.* (8.112)]. In words, (8.123) states that ergodicity in the mean requires the time average of the autocorrelation function of $f(t)$ to be equal to the square of the ensemble mean. When this is true, the variance of the random variable that is the outcome of (8.121) approaches zero as the period T goes to infinity.

As Khinchin (1949) and others have noted, ergodicity in the mean is equivalent to the law of large numbers. In his discussion of ergodicity in statistical mechanics, Khinchin defines an ergodic process as: "On average, a system, whose evolution in time is governed by the equations of motion, remains in different parts of a given manifold of constant energy for fractions of the total time interval which are proportional to the volumes of these parts. Therefore, if we observe any physical quantity associated with a given system over a definite time interval, the arithmetic average of the results of a sufficiently large number of measurements will, as a rule, be close to the (theoretical) statistical average." He goes on to say that it is "hard to prove ergodicity in classical systems and impossible in principle to do so in quantum mechanics."

Multiple-point expectations of one realization of a temporal random process (see Sec. 8.2.2) can also be considered. For example, the finite-time autocorrelation function of one realization with itself is given by

$$R_T(\Delta t, \zeta_0) = \frac{1}{T} \int_{-\frac{1}{2}T}^{\frac{1}{2}T} dt \; f(t + \Delta t, \zeta_0)\, f^*(t, \zeta_0) . \tag{8.124}$$

A random process is said to be *ergodic in autocorrelation* if $R_T(\Delta t, \zeta_0)$ approaches the ensemble quantity $R(\Delta t)$ as $T \to \infty$. We can see that the ensemble average of the sample quantity $R_T(\Delta t, \zeta_0)$ is equal to the ensemble autocorrelation function:

$$\mathrm{E}\{R_T(\Delta t, \zeta_0)\} = \frac{1}{T} \int_{-\frac{1}{2}T}^{\frac{1}{2}T} dt \; \mathrm{E}\{f(t + \Delta t, \zeta_0)\, f^*(t, \zeta_0)\} = R(\Delta t) , \tag{8.125}$$

where the last step follows since $R(\Delta t)$ is independent of the integration time T. It is more difficult to demonstrate that the temporal average of $R_T(\Delta t, \zeta_0)$ approaches $R(\Delta t)$ in the limit as T becomes infinite. While a test for ergodicity of the mean requires knowledge of the ensemble mean and autocorrelation function, Papoulis demonstrates that knowledge of fourth-order moments is required to test for ergodicity of the autocorrelation function.

In general, demonstration of higher levels of ergodicity requires increasing knowledge of the density function that describes the random process. One exception, however, is the special case of the Gaussian random process. We shall see in Sec. 8.3.5 that in that case a straightforward criterion for complete ergodicity can be stated.

Ergodicity comes into play in optics when we consider the output of a detector sensing a rapidly fluctuating optical field. The period of integration in the finite-time average (8.121) is directly analogous to the detector response time. If the field fluctuates rapidly enough that fluctuations in the random process are not evident in the detector output, the random process can be said to be ergodic, and the detector can be assumed to sense an ensemble average.

We have deliberately discussed ergodicity in terms of temporal rather than spatial random processes. Remember that the first condition for ergodicity is that the random process be stationary, but as we stated earlier in this section, the physical boundaries of objects and images make spatial stationarity rarely a plausible assumption. Nevertheless, ergodicity is often assumed in the image-processing community to determine, for example, noise statistics at a single location in an image (an ensemble quantity) based on the characteristics of the fluctuations in a spatial region of that single image.

8.2.5 Spectral analysis

The Fourier transform is an important tool in the analysis of signals in general, and random signals are no exception. The Fourier transform of one sample function of a random process is defined just as for any other function, and all of the properties given in Chap. 3 are applicable. In some cases, particularly finite-power random processes, it may be necessary to consider the sample function as a generalized function and compute its Fourier transform by use of the theory of tempered distributions, but this presents no essential difficulty. With the background on generalized functions presented in Chaps. 2 and 3, we should have no qualms about issues of existence of the transform.

On the other hand, the Fourier transform of a random process is another random process, and we are usually more interested in averages than in properties of individual samples. In particular, with finite-power processes, we often want to know how the average power is distributed as a function of frequency. The branch of stochastic theory that addresses this question is called *spectral analysis*, and a frequency-domain description of the average power is known as a *spectrum*, *power spectrum* or *power spectral density.*

We shall give a brief overview of the historical development of spectral analysis and then give two equivalent definitions of power spectral density. Initially the discussion will consider stationary processes in the time domain, but then we make the transition to the space domain as we see how the theory can be applied to processes that are not exactly stationary.

A brief history of spectra The early history of spectral analysis was motivated by a desire to understand white light (Gouy, 1886; Rayleigh, 1903; Schuster, 1894, 1904, 1906). Gouy's work was based on the Fourier series, while Lord Rayleigh used the newly developed Plancherel ($\mathbb{L}_2$) interpretation of the Fourier transform. Wiener (1930) marvels (though not without a touch of irony) at these forays: "In both cases one is astonished by the skill with which the authors use clumsy and unsuitable tools to obtain the right results, and one is led to admire the unfailing heuristic insight of the true physicist."

Wiener's own pioneering treatise, *Generalized Harmonic Analysis* (Wiener, 1930), was built on the work of Sir Arthur Schuster. Schuster used a windowed or truncated function defined by

$$f_T(t) = f(t)\,\mathrm{rect}(t/T)\,, \tag{8.126}$$

with a Fourier transform defined by

$$F_T(\nu) = \int_{-\frac{1}{2}T}^{\frac{1}{2}T} dt\ f(t) \exp(-2\pi i \nu t)\,. \tag{8.127}$$

Schuster proposed specifying the spectrum of $f(t)$ by the *periodogram*, defined by

$$S_p(\nu) = \lim_{T\to\infty} \frac{1}{T} |F_T(\nu)|^2\,. \tag{8.128}$$

By (3.135), $|F_T(\nu)|^2$ is the Fourier transform of the deterministic autocorrelation *integral* (not to be confused with the statistical autocorrelation function) of $f_T(t)$. Thus (8.128) is equivalent to

$$S_p(\nu) = \lim_{T\to\infty} \frac{1}{T} \mathcal{F}\{[f_T \star f_T^*]\,(x)\}\,, \tag{8.129}$$

where $\mathcal{F}$ is the Fourier operator and, by (3.115),

$$[f_T \star f_T^*]\,(t) = \int_{-\infty}^{\infty} dt'\ f_T(t+t')\, f_T^*(t')\,. \tag{8.130}$$

Wiener's approach was slightly different. He defined

$$R_{W,T}(t) = \frac{1}{T} \int_{-\frac{1}{2}T}^{\frac{1}{2}T} dt'\ f(t+t')\, f^*(t')\,, \tag{8.131}$$

which differs from (8.130) mainly in the fact that the truncation is on the limits rather than on both functions separately; there is also a factor of $1/T$ built into the definition.

The only requirement placed on the function $f(t)$ is that $R_{W,T}(t) < \infty$ for all t, but this turns out to be a very useful mathematical condition (Champeney, 1987). The special case $t = 0$ shows that these functions must be finite-power functions as defined in (8.70). For such functions, Wiener defined a spectrum by

$$S_W(\nu) = \lim_{T\to\infty} \mathcal{F}\{R_{W,T}(t)\}\,. \tag{8.132}$$

Note that neither S_p nor S_W involves any statistical average; both Wiener and Schuster took a functional or deterministic viewpoint and did not invoke ensembles of any kind. Thus their spectra apply to a single realization of the random process, albeit one of infinite length. For any function for which S_W is finite, S_W and S_p are identical (Champeney, 1987).

Convergence issues In practice, one might think that a reasonable approximation of S_p or S_W could be obtained by using a single periodogram of finite length and just omitting the limit $T \to \infty$ in (8.128) or (8.132). It might also be expected that this approximation would get better as T gets larger. In fact, however, the Fourier transform of a single sample function of a random process is a very poor spectral measure.

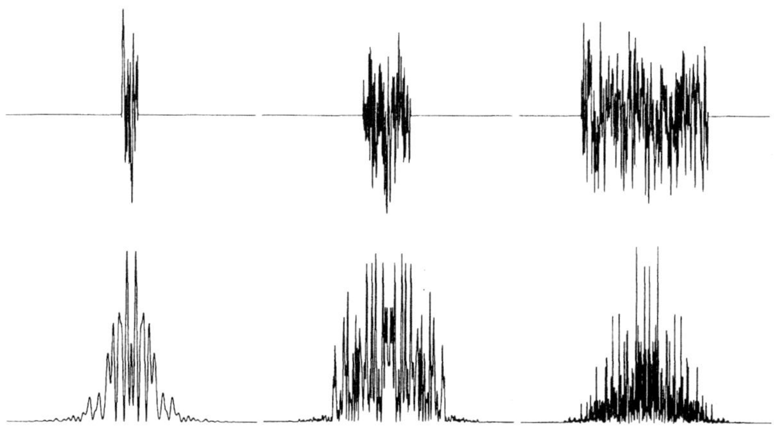

Fig. 8.1 Three sample functions of a random process (top) and their periodograms (bottom). The random process was created by calling a uniform random-number generator independently at each of 1024 sample points, then performing a discrete convolution with a Gaussian to produce a random process with a Gaussian power spectrum. The sample functions were windowed as shown, and the periodograms were computed by discrete Fourier transforms.

This point is illustrated in Fig. 8.1, which shows three sample functions of different length of a stationary random process, along with the corresponding finite-length periodograms. Note that the periodograms do not smoothly approach a limit as $T \to \infty$ but instead oscillate ever more rapidly.

One way to deal with the rapid oscillation is to average the periodogram by convolution with some smooth function. In fact, this average can be built in by windowing the samples with the Fourier transform of the smoothing function. This approach smooths out any fine details that might be present in the spectrum but provides better convergence as T gets large. Some additional approaches to this problem will be discussed briefly in Sec. 8.4.4.

Power spectra as statistical averages Another way to fix the convergence problems associated with S_p and S_W is to use not one but many independent realizations of the random process and to average the resulting periodograms. In the limit of an infinite number of realizations, this approach, pioneered by Khinchin, amounts to incorporating a statistical average in the definition of the spectrum. Khinchin's definition was

$$S_{ac}(\nu) = \mathcal{F}\{R(\Delta t)\} = \int_{-\infty}^{\infty} d\Delta t \ \langle f(t+\Delta t)\, f^*(t)\rangle \exp(-2\pi i \nu \Delta t)\,, \qquad (8.133)$$

where the subscript ac indicates that this version of the spectrum is derived from the autocorrelation function $R(\Delta t)$ of a stationary random process. The spectrum defined this way is well behaved mathematically and universally used. Equation (8.133) is often referred to as the *Wiener-Khinchin theorem*, though it is really a definition rather than a theorem.

Expected periodogram Another way to incorporate an ensemble average into the definition of the spectrum is to take the expectation of the periodogram,

$$S_{ep}(\nu) = \lim_{T\to\infty} \frac{1}{T} \left\langle |F_T(\nu)|^2 \right\rangle . \qquad (8.134)$$

Unlike $S_{ac}(\nu)$, $S_{ep}(\nu)$ is defined for nonstationary as well as stationary random processes, though they have to be finite-power processes for S_{ep} to be nonzero. For stationary processes, however, $S_{ep}(\nu)$ is equivalent to $S_{ac}(\nu)$, as we shall now show.

From the definition of $F_T(\nu)$, we can write

$$S_{ep}(\nu) = \lim_{T\to\infty} \frac{1}{T} \int_{-\infty}^{\infty} dt \int_{-\infty}^{\infty} dt' \, \langle f(t)\, f^*(t') \rangle \, \mathrm{rect}\left(\frac{t}{T}\right) \mathrm{rect}\left(\frac{t'}{T}\right) \exp[2\pi i(t'-t)\nu] \, . \tag{8.135}$$

Now we make the change of variables $(t,t') \to (t, \Delta t)$, where $\Delta t = t - t'$. With the assumption that $\langle f(t)\, f^*(t') \rangle = R(\Delta t)$ and a little algebra, we find

$$S_{ep}(\nu) = \lim_{T\to\infty} \int_{-\infty}^{\infty} d\Delta t \; R(\Delta t) \, \mathrm{tri}\left(\frac{\Delta t}{T}\right) \exp(-2\pi i \nu \Delta t) \, , \tag{8.136}$$

where the function $\mathrm{tri}(\cdot)$ is defined in (3.139).

We can now use the convolution theorem (3.132) along with (3.142) to write

$$S_{ep}(\nu) = \lim_{T\to\infty} S_{ac}(\nu) * T \, \mathrm{sinc}^2(T\nu) \, . \tag{8.137}$$

But we know from (2.87) that $T\,\mathrm{sinc}^2(T\nu)$ is a valid limiting representation of $\delta(\nu)$. From Sec. 3.3.6 we also know that convolution of $S_{ac}(\nu)$ with $\delta(\nu)$ reproduces $S_{ac}(\nu)$ if that function is either a good function (defined in Sec. 2.1.2) or a generalized function of compact support (defined in Sec. 3.3.6). The support can be chosen arbitrarily large, or we can argue as in Sec. 2.3.1 that any generalized function can be approximated arbitrarily closely by a good function.

Thus, with essentially no restrictions beyond stationarity, we have

$$S_{ep}(\nu) = S_{ac}(\nu) \, . \tag{8.138}$$

Because of this equivalence, we shall delete the subscripts henceforth and denote the power spectral density simply by $S(\nu)$. Either definition, (8.133) or (8.134), will be used as convenient.

Spatial power spectra Stationary spatial random processes were discussed in Sec. 8.2.4. If this model is used, the spatial version of the Wiener-Khinchin theorem, (8.133), is

$$S(\boldsymbol{\rho}) = \int_{\infty} d^q \Delta r \; R(\Delta \mathbf{r}) \exp(-2\pi i \boldsymbol{\rho} \cdot \Delta \mathbf{r}) \, . \tag{8.139}$$

Stochastic Wigner distribution function A general way of applying Fourier analysis to nonstationary random processes is to make use of the Wigner distribution function, defined in Sec. 5.2.1. For a spatial random process $f(\mathbf{r})$, we define the stochastic Wigner function by [*cf.* (5.54)]

$$W_f(\mathbf{r}_0, \boldsymbol{\rho}) = \int_{\infty} d^q \Delta r \; \left\langle f(\mathbf{r}_0 + \tfrac{1}{2}\Delta \mathbf{r})\, f^*(\mathbf{r}_0 - \tfrac{1}{2}\Delta \mathbf{r}) \right\rangle \exp(-2\pi i \boldsymbol{\rho} \cdot \Delta \mathbf{r}) \, . \tag{8.140}$$

This expression should be compared to the Wiener-Khinchin theorem for a stationary random process, (8.139), which can be written in symmetrized form as

$$S(\boldsymbol{\rho}) = \int_{\infty} d^q\Delta r \ \langle f(\mathbf{r}+\Delta\mathbf{r})\, f^*(\mathbf{r})\rangle \exp(-2\pi i\boldsymbol{\rho}\cdot\Delta\mathbf{r})$$

$$= \int_{\infty} d^q\Delta r \ \left\langle f(\mathbf{r}_0+\tfrac{1}{2}\Delta\mathbf{r})\, f^*(\mathbf{r}_0-\tfrac{1}{2}\Delta\mathbf{r})\right\rangle \exp(-2\pi i\boldsymbol{\rho}\cdot\Delta\mathbf{r})\,, \tag{8.141}$$

where the second equality follows since the autocorrelation function is independent of shifts of the coordinate system for a stationary process. Thus, if the process is stationary, the stochastic Wigner function is independent of $\mathbf{r}_0$ and is precisely the power spectral density.

For nonstationary processes, however, $W_f(\mathbf{r}_0, \boldsymbol{\rho})$ is a function of $\mathbf{r}_0$ as well as $\boldsymbol{\rho}$; it can be interpreted as the spectral content associated with point $\mathbf{r}_0$. This interpretation is reinforced by examining the quasistationary case. From (8.119) and (8.140) we can write

$$W_f(\mathbf{r}_0, \boldsymbol{\rho}) = b(\mathbf{r}_0) \int_{\infty} d^q\Delta r \ a(\Delta\mathbf{r}) \exp(-2\pi i\boldsymbol{\rho}\cdot\Delta\mathbf{r}) = b(\mathbf{r}_0)\, A(\boldsymbol{\rho})\,. \tag{8.142}$$

Here the Wigner distribution function is just the Fourier transform of the short-range part of the autocorrelation function, modulated by the shift-variant strength of the slowly varying component at $\mathbf{r}_0$.

8.2.6 Linear filtering of random processes

We now derive the autocorrelation function of the output process that results from linear filtering of a given random process. We shall consider stationary and nonstationary random processes and shift-invariant and shift-variant filters.

Nonstationary process, shift-variant filter We first consider the case where a random process $g(\mathbf{r})$ is generated as the output of the transformation of an input random process $f(\mathbf{r})$ by a linear shift-variant filter whose impulse response is denoted $h(\mathbf{r}, \mathbf{r}')$. The output of the filter at positions $\mathbf{r}$ and $\mathbf{r}+\Delta\mathbf{r}$ can be written, respectively, as

$$g(\mathbf{r}) = \int_{\infty} d^q r'\, h(\mathbf{r}, \mathbf{r}')\, f(\mathbf{r}')\,, \tag{8.143}$$

$$g(\mathbf{r}+\Delta\mathbf{r}) = \int_{\infty} d^q r'\, h(\mathbf{r}+\Delta\mathbf{r}, \mathbf{r}')\, f(\mathbf{r}')\,. \tag{8.144}$$

By direct substitution of these expressions into the definition, (8.97), we obtain for the autocorrelation of the output process at positions $\mathbf{r}$ and $\mathbf{r}+\Delta\mathbf{r}$:

$$R_g(\mathbf{r}+\Delta\mathbf{r}, \mathbf{r}) = \langle g(\mathbf{r}+\Delta\mathbf{r})\, g^*(\mathbf{r})\rangle$$

$$= \left\langle \int_{\infty} d^q r'\, h(\mathbf{r}+\Delta\mathbf{r}, \mathbf{r}')\, f(\mathbf{r}') \int_{\infty} d^q r''\, h^*(\mathbf{r}, \mathbf{r}'')\, f^*(\mathbf{r}'') \right\rangle$$

$$= \int_{\infty} d^q r' \int_{\infty} d^q r''\, h(\mathbf{r}+\Delta\mathbf{r}, \mathbf{r}')\, R_f(\mathbf{r}', \mathbf{r}'')\, h^*(\mathbf{r}, \mathbf{r}'')\,. \tag{8.145}$$

The corresponding expression for the autocovariance is

$$K_g(\mathbf{r}+\Delta\mathbf{r},\mathbf{r}) = R_g(\mathbf{r}+\Delta\mathbf{r},\mathbf{r}) - \langle g(\mathbf{r}+\Delta\mathbf{r})\rangle \langle g^*(\mathbf{r})\rangle$$

$$= \int_\infty d^q r' \int_\infty d^q r''\, h(\mathbf{r}+\Delta\mathbf{r},\mathbf{r}')\, K_f(\mathbf{r}',\mathbf{r}'')\, h^*(\mathbf{r},\mathbf{r}'')\,. \tag{8.146}$$

This is the most general form for the autocovariance after linear filtering. It is the continuous analog of the discrete result given in (8.50), as one can see by rewriting it in operator form:

$$\mathcal{K}_{\mathbf{g}} = \mathcal{H}\mathcal{K}_{\mathbf{f}}\mathcal{H}^\dagger\,, \tag{8.147}$$

where $\mathcal{K}_{\mathbf{f}}$ is the autocovariance operator, *i.e.*, the integral operator with kernel $K_{\mathbf{f}}(\mathbf{r},\mathbf{r}')$, and similarly for $\mathcal{K}_{\mathbf{g}}$, while $\mathcal{H}$ describes the filter. There are no restrictions on $\mathcal{H}$ in this equation, except that it must be a linear operator. It even applies to linear CD operators, though in that case the left-hand side is a covariance matrix rather than an autocovariance operator.

Nonstationary process, shift-invariant filter We consider next the case where the random process $g(\mathbf{r})$ is generated as the output of the transformation of a general input random process $f(\mathbf{r})$ by a linear shift-invariant filter with impulse response $h(\mathbf{r})$. The processes $g(\mathbf{r})$ and $f(\mathbf{r})$ are now related by convolution:

$$g(\mathbf{r}) = \int_\infty d^q r'\, h(\mathbf{r}-\mathbf{r}')\, f(\mathbf{r}') = h(\mathbf{r}) * f(\mathbf{r})\,, \tag{8.148}$$

where the notation of Sec. 3.3.6 has been used.

We can obtain the autocorrelation of the output process $g(\mathbf{r})$ from that of the input process $f(\mathbf{r})$ by substituting (8.148) into (8.145):

$$R_g(\mathbf{r}+\Delta\mathbf{r},\mathbf{r}) = \langle g(\mathbf{r}+\Delta\mathbf{r})\, g^*(\mathbf{r})\rangle$$

$$= \left\langle \int_\infty d^q r'\, h(\mathbf{r}+\Delta\mathbf{r}-\mathbf{r}')\, f(\mathbf{r}') \int_\infty d^q r''\, h^*(\mathbf{r}-\mathbf{r}'')\, f^*(\mathbf{r}'') \right\rangle. \tag{8.149}$$

Alternatively, we have

$$R_g(\mathbf{r}+\Delta\mathbf{r},\mathbf{r}) = \left\langle \int_\infty d^q r'\, h(\mathbf{r}')\, f(\mathbf{r}+\Delta\mathbf{r}-\mathbf{r}') \int_\infty d^q r''\, h^*(\mathbf{r}'')\, f^*(\mathbf{r}-\mathbf{r}'') \right\rangle$$

$$= \int_\infty d^q r'\, h(\mathbf{r}') \int_\infty d^q r''\, h^*(\mathbf{r}'')\, R_f(\mathbf{r}+\Delta\mathbf{r}-\mathbf{r}',\mathbf{r}-\mathbf{r}'')\,. \tag{8.150}$$

We can use convolution shorthand to write this equation as

$$R_g(\mathbf{r}+\Delta\mathbf{r},\mathbf{r}) = h(\mathbf{r}+\Delta\mathbf{r}) * R_f(\mathbf{r}+\Delta\mathbf{r},\mathbf{r}) * h^*(\mathbf{r})\,, \tag{8.151}$$

where the notation indicates that the first convolution is evaluated at the position $\mathbf{r}+\Delta\mathbf{r}$ and the second is evaluated at the position $\mathbf{r}$.

Stationary random process, shift-invariant filter For the special case of a stationary input process, the input correlation function in (8.150) can be written solely in terms of the difference vector as

$$R_f(\mathbf{r}+\Delta\mathbf{r}-\mathbf{r}',\mathbf{r}-\mathbf{r}'') = \langle f(\mathbf{r}+\Delta\mathbf{r}-\mathbf{r}')\, f^*(\mathbf{r}-\mathbf{r}'')\rangle = R_f(\Delta\mathbf{r}-\mathbf{r}'+\mathbf{r}'')\,. \quad (8.152)$$

Then (8.150) can be written

$$R_g(\Delta\mathbf{r}) = \int_\infty d^q r'\, h(\mathbf{r}') \int_\infty d^q r''\, h^*(\mathbf{r}'')\, R_f(\Delta\mathbf{r}-\mathbf{r}'+\mathbf{r}'')\,. \quad (8.153)$$

This equation is often written in a shorthand notation as (Papoulis, 1965)

$$R_g(\Delta\mathbf{r}) = \langle g(\mathbf{r}+\Delta\mathbf{r})\, g^*(\mathbf{r})\rangle = h(\Delta\mathbf{r}) * R_f(\Delta\mathbf{r}) * h^*(-\Delta\mathbf{r})\,. \quad (8.154)$$

This notation refers to the fact that the first operation is an ordinary convolution, but the second is actually a correlation. In this shorthand a correlation is written using the convolution notation with a change of sign of the argument. Alternatively, one can use $\star$ to represent the correlation integral:

$$R_g(\Delta\mathbf{r}) = \langle g(\mathbf{r}+\Delta\mathbf{r})\, g^*(\mathbf{r})\rangle = [h * R_f \star h^*]\,(\Delta\mathbf{r})\,. \quad (8.155)$$

Fourier transformation of (8.155) yields the important formula

$$S_g(\boldsymbol{\rho}) = S_f(\boldsymbol{\rho})\, |H(\boldsymbol{\rho})|^2\,, \quad (8.156)$$

where $H(\boldsymbol{\rho}) = \mathcal{F}_q\{h(\mathbf{r})\}$. Thus, when a stationary random process is filtered by a linear shift-invariant filter, the power spectral density on the output of the filter is the power spectral density on the input times the squared modulus of the filter transfer function. This result should be compared to the familiar result for shift-invariant filtering of a deterministic signal. From (3.132) we know that

$$G(\boldsymbol{\rho}) = H(\boldsymbol{\rho})\, F(\boldsymbol{\rho})\,. \quad (8.157)$$

In the context of stationary random processes, (8.157) applies to *individual sample functions* while (8.156) applies to the power spectral densities.

Filtering of delta-correlated processes We are often concerned with random processes where the correlation has such short range that $R_{\Delta f}(\mathbf{r},\mathbf{r}')$ can be approximated by $b(\mathbf{r})\,\delta(\mathbf{r}-\mathbf{r}')$. A prime example, the Poisson random process, will be discussed in detail in Chap. 11. Another example is *white noise*, a stationary process that has a flat power spectrum and hence a delta-function correlation. We now investigate the effect of linear filtering on delta-correlated processes.

With delta correlation, the general space-variant filter equation, (8.144), leads to

$$R_{\Delta g}(\mathbf{r}+\Delta\mathbf{r},\mathbf{r}) = \int_\infty d^q r'\, h(\mathbf{r}+\Delta\mathbf{r},\mathbf{r}') \int_\infty d^q r''\, b(\mathbf{r}')\,\delta(\mathbf{r}'-\mathbf{r}'')\, h^*(\mathbf{r},\mathbf{r}'')$$

$$\int_\infty d^q r'\, h(\mathbf{r}+\Delta\mathbf{r},\mathbf{r}')\, b(\mathbf{r}')\, h^*(\mathbf{r},\mathbf{r}')\,. \quad (8.158)$$

For shift-invariant filters, where $h(\mathbf{r}, \mathbf{r}') = h(\mathbf{r} - \mathbf{r}')$, this equation reduces to

$$R_{\Delta g}(\mathbf{r} + \Delta\mathbf{r}, \mathbf{r}) = \int_\infty d^q r' \, h(\mathbf{r} + \Delta\mathbf{r} - \mathbf{r}') \, b(\mathbf{r}') \, h^*(\mathbf{r} - \mathbf{r}')$$

$$= b(\mathbf{r}) * [h(\mathbf{r} + \Delta\mathbf{r}) \, h^*(\mathbf{r})] \,. \tag{8.159}$$

The shorthand here requires a brief comment. For purposes of the convolution, the function $[h(\mathbf{r} + \Delta\mathbf{r}) \, h^*(\mathbf{r})]$ is to be regarded as a function of $\mathbf{r}$ for fixed $\Delta\mathbf{r}$. As shown by the integral in (8.158), this product function is then convolved with $b(\mathbf{r})$, and the convolution is repeated for different $\Delta\mathbf{r}$ to get the full dependence of the nonstationary autocorrelation $R_{\Delta g}(\mathbf{r} + \Delta\mathbf{r}, \mathbf{r})$.

Even though $f(\mathbf{r})$ is uncorrelated for any finite lag, the filtering results in a correlation on $g(\mathbf{r})$. Suppose $h(\mathbf{r})$ has a width w in each dimension, *i.e.*, $h(\mathbf{r})$ drops to zero if the magnitude of any component of $\mathbf{r}$ exceeds $\frac{1}{2}w$. Then $[h(\mathbf{r} + \Delta\mathbf{r}) \, h^*(\mathbf{r})]$ drops to zero for all $\mathbf{r}$ if the magnitude of any component of $\Delta\mathbf{r}$ exceeds w. The correlation in $g(\mathbf{r})$ thus has a width in $\Delta\mathbf{r}$ determined by the width of the point spread function.

If $b(\mathbf{r})$ is the constant b_0, so that $R_{\Delta f}(\mathbf{r} - \mathbf{r}') = b_0 \, \delta(\mathbf{r} - \mathbf{r}')$, then we are dealing with stationary white noise and a frequency-domain description is appropriate. The power spectral density of $\Delta f(\mathbf{r})$ is just the constant b_0, and by (8.156) that of $\Delta g(\mathbf{r})$ is given by

$$S_{\Delta g}(\boldsymbol{\rho}) = b_0 |H(\boldsymbol{\rho})|^2 \,. \tag{8.160}$$

The corresponding autocorrelation function is obtained by inverse Fourier transformation:

$$R_{\Delta g}(\Delta\mathbf{r}) = b_0 [h \star h^*](\Delta\mathbf{r}) \,. \tag{8.161}$$

Thus the statistical autocorrelation function for filtered white noise is proportional to the deterministic autocorrelation integral of the impulse response.

8.2.7 Eigenanalysis of the autocorrelation operator

In Sec. 8.1.6, we discussed the eigenvectors and eigenvalues of a covariance matrix. In particular, we showed how a random vector could be expanded in a series with uncorrelated coefficients by using eigenvectors of the covariance matrix as basis vectors. This expansion was called the Karhunen-Loève or KL expansion.

In this section we carry out a similar analysis for a random process, substituting the continuous autocovariance or autocorrelation function for the discrete covariance matrix. One result will be a continuous version of the KL expansion — a linear transformation that will render a correlated process uncorrelated for any finite shift.

To maintain parallelism with Sec. 8.1.6, we restrict attention initially to finite-energy random processes (thus ruling out stationarity), but later we extend the analysis to finite-power processes and in particular to wide-sense stationary ones. In that case we shall find that KL expansion is just Fourier analysis.

Autocorrelation operator It is arbitrary whether we develop KL analysis based on the autocorrelation or autocovariance function; from (8.98) we can easily convert between them. We choose the autocorrelation since we shall eventually make contact with the Wiener-Khinchin theorem (8.133) or (8.139), which defines the power spectral density as the Fourier transform of the autocorrelation function.

For a general, nonstationary, spatial random process $f(\mathbf{r})$, where $\mathbf{r}$ is a qD position vector, the autocorrelation function $R(\mathbf{r}, \mathbf{r}')$ is defined by (8.97). For now we restrict attention to square-integrable random processes, so we can regard $R(\mathbf{r}, \mathbf{r}')$ as the kernel of an integral operator $\mathcal{R}$ that maps $\mathbb{L}_2(\mathbb{R}^q)$ to itself. Operating on an arbitrary square-integrable function $t(\mathbf{r})$, the operator $\mathcal{R}$ has the form

$$[\mathcal{R}\mathbf{t}](\mathbf{r}) = \int_\infty d^q r' \, R(\mathbf{r}, \mathbf{r}') \, t(\mathbf{r}') \,. \tag{8.162}$$

Inspection of (8.97) shows that $[R(\mathbf{r}, \mathbf{r}')]^* = R(\mathbf{r}', \mathbf{r})$, so $\mathcal{R}$ is Hermitian (see Sec. 1.3.5).

Moreover, as we shall now show, $\mathcal{R}$ is compact. By the discussion in Sec. 1.3.3, an integral operator is compact if its kernel satisfies the Hilbert-Schmidt condition (1.33), which in the present multidimensional case generalizes to

$$\int_\infty d^q r \int_\infty d^q r' \, |\mathcal{R}(\mathbf{r}, \mathbf{r}')|^2 < \infty \,. \tag{8.163}$$

Denoting this integral by I_{HS} and inserting (8.97), we can rewrite this condition as

$$I_{HS} = \int_\infty d^q r \int_\infty d^q r' \, | \langle f(\mathbf{r}) \, f^*(\mathbf{r}') \rangle |^2 < \infty \,. \tag{8.164}$$

Now, for any random variable x we know from App. C that $| \langle x \rangle |^2 \leq \langle |x|^2 \rangle$. With $x = f(\mathbf{r}') \, f^*(\mathbf{r}')$, this implies that

$$I_{HS} \leq \int_\infty d^q r \int_\infty d^q r' \, \langle | [f(\mathbf{r}) \, f^*(\mathbf{r}')] \, |^2 \rangle \,. \tag{8.165}$$

As discussed in Sec. 8.2.2, we can interchange expectation and integration, yielding

$$I_{HS} \leq \left\langle \int_\infty d^q r \, |f(\mathbf{r})|^2 \int_\infty d^q r' \, |f(\mathbf{r}')|^2 \right\rangle \,. \tag{8.166}$$

Every sample function $f(\mathbf{r})$ is assumed to be square-integrable, so each integral in (8.166) is finite. The output of the expectation operation is therefore finite and $I_{HS} \leq \infty$. Thus we have shown that $\mathcal{R}$ satisfies the Hilbert-Schmidt condition and is therefore compact.

As discussed in Sec. 1.4.4, a compact Hermitian operator has a denumerable set of eigenfunctions and real eigenvalues. Thus $\mathcal{R}$ satisfies an eigenvalue equation of the form

$$\mathcal{R}\phi_n(\mathbf{r}) = \mu_n \phi_n(\mathbf{r}) \,. \tag{8.167}$$

We noted in (8.106) that $\mathcal{R}$ is nonnegative-definite, so $\mu_n \geq 0$. It is convenient to order the eigenvalues by decreasing value:

$$\mu_1 \geq \mu_2 \geq \mu_3 \geq \cdots \geq 0 \,. \tag{8.168}$$

Except in very special cases, none of these eigenvalues will be zero, so $\mathcal{R}$ has infinite rank.

Karhunen-Loève expansions Since the eigenfunctions of a Hermitian operator can be chosen to form an orthonormal basis, any function $f(\mathbf{r})$ in the domain of $\mathcal{R}$, *i.e.*, $\mathbb{L}_2(\mathbb{R}^q)$, can be expanded in the form

$$f(\mathbf{r}) = \sum_{n=1}^{\infty} \alpha_n \phi_n(\mathbf{r}) , \tag{8.169}$$

where the coefficients are given by scalar products of the form

$$\alpha_n = (\phi_n(\mathbf{r}), f(\mathbf{r})) . \tag{8.170}$$

If $f(\mathbf{r})$ is a sample function of a random process, then the coefficients α_n are random variables. If $f(\mathbf{r})$ is drawn from the ensemble described by $\mathcal{R}$, then these coefficients are uncorrelated, as we shall now demonstrate. The cross-correlation of two coefficients, α_n and α_m, is given by

$$\langle \alpha_n \alpha_m^* \rangle = \langle (\phi_n(\mathbf{r}), f(\mathbf{r})) \, (\phi_m(\mathbf{r}'), f(\mathbf{r}'))^* \rangle . \tag{8.171}$$

Writing out the scalar products as integrals and again interchanging integration and expectation, we find

$$\langle \alpha_n \alpha_m^* \rangle = \int_\infty d^q r \int_\infty d^q r' \, \phi_n^*(\mathbf{r}) \, \phi_m(\mathbf{r}') \, \langle f(\mathbf{r}) \, f^*(\mathbf{r}') \rangle . \tag{8.172}$$

By (8.97) and (8.167), we have

$$\langle \alpha_n \alpha_m^* \rangle = \mu_m \int_\infty d^q r \, \phi_n^*(\mathbf{r}) \, \phi_m(\mathbf{r}) , \tag{8.173}$$

and the orthonormality of the eigenfunctions yields, finally,

$$\langle \alpha_n \alpha_m^* \rangle = \mu_n \, \delta_{nm} . \tag{8.174}$$

Thus the expansion in (8.169) generalizes the Karhunen-Loève expansion of random vectors, as discussed in Sec. 8.1.6, to random processes.

Stationary random processes The derivation above of the KL expansion is not directly applicable to stationary random processes since their sample functions are not square-integrable. Hence the autocorrelation operator is not compact and its eigenvalues are not denumerable.

Since the discrete index n on $\phi_n(\mathbf{r})$ and μ_n is no longer appropriate, we shall leave off any index until we discover what to use. The eigenvalue equation for a stationary random process is then

$$\int_\infty d^q r' \, R(\mathbf{r} - \mathbf{r}') \, \phi(\mathbf{r}') = \mu \, \phi(\mathbf{r}) . \tag{8.175}$$

A simple change of variables yields

$$\int_\infty d^q r' \, R(\mathbf{r}') \, \phi(\mathbf{r} - \mathbf{r}') = \mu \, \phi(\mathbf{r}) . \tag{8.176}$$

Direct substitution shows that the solution of this equation is

$$\phi(\mathbf{r}) = \exp(2\pi i \boldsymbol{\rho} \cdot \mathbf{r}), \tag{8.177}$$

$$\mu = \int_\infty d^q r' \, R(\mathbf{r}') \exp(-2\pi i \boldsymbol{\rho} \cdot \mathbf{r}') = \mathcal{F}_q\{R(\mathbf{r})\} = S(\boldsymbol{\rho}), \tag{8.178}$$

where $S(\boldsymbol{\rho})$ is the power spectral density as defined in (8.139). Thus, for a stationary random process, the eigenfunctions of the autocorrelation operator are Fourier basis functions (or plane waves), and the eigenvalues are given by the power spectral density. The problem is mathematically equivalent to singular-value decomposition of a linear, shift-invariant system as discussed in Sec. 7.2.5

The eigenfunctions and eigenvalues are distinguished by a continuous vector index $\boldsymbol{\rho}$ (the spatial frequency), rather than by a discrete index n. Thus we denote the eigenfunction in (8.177) as $\phi_{\boldsymbol{\rho}}(\mathbf{r})$ and the eigenvalue as $\mu_{\boldsymbol{\rho}}$. With this notation, the KL expansion (8.169) becomes

$$f(\mathbf{r}) = \int_\infty d^q \rho \; F(\boldsymbol{\rho}) \, \phi_{\boldsymbol{\rho}}(\mathbf{r}) = \int_\infty d^q \rho \; F(\boldsymbol{\rho}) \exp(2\pi i \boldsymbol{\rho} \cdot \mathbf{r}). \tag{8.179}$$

By analogy with (8.170), the expansion coefficients $F(\boldsymbol{\rho})$ are given by

$$F(\boldsymbol{\rho}) = \int_\infty d^q r \; f(\mathbf{r}) \exp(-2\pi i \boldsymbol{\rho} \cdot \mathbf{r}). \tag{8.180}$$

Formally, (8.179) states that the KL expansion is simply the representation of a sample function of the stationary random process by its inverse Fourier transform, while (8.180) says that the expansion coefficient is the Fourier transform of the sample function. In this sense, KL expansion reduces to Fourier analysis in the stationary case. In a strict mathematical sense, however, this interpretation raises some problems. If $f(\mathbf{r})$ is a sample function from a stationary random process, it must have the same mean value at all points in the infinite domain $\mathbb{R}^q$. Hence it is not square-integrable or absolutely integrable, and the classical Fourier existence and convergence theorems do not apply.

We can fix these problems in one of two ways. One approach is to presume that each sample function is truncated by a window function of finite size, and then let this size go to infinity as in Sec. 8.2.5. A neater approach is simply to regard $f(\mathbf{r})$ as a generalized function related to a tempered distribution. This requires only that the sample function be integrable when multiplied by a good function such as a Gaussian, which is an easy condition to satisfy. From the discussion in Sec. 3.3.4, we know that $F(\boldsymbol{\rho})$ is also a generalized function in that case. For example, if $f(\mathbf{r})$ has a nonzero mean $\overline{f}$ (which must be independent of $\mathbf{r}$ because of the stationarity), then $F(\boldsymbol{\rho})$ must contain a term $\overline{f}\,\delta(\boldsymbol{\rho})$.

From the viewpoint of generalized functions, we can now discuss the correlation properties of the expansion coefficients $F(\boldsymbol{\rho})$. A derivation paralleling the one that led to (8.173) shows that

$$\langle F(\boldsymbol{\rho}) \, F^*(\boldsymbol{\rho}') \rangle = S(\boldsymbol{\rho}) \, \delta(\boldsymbol{\rho} - \boldsymbol{\rho}'). \tag{8.181}$$

Just as in (8.173), the KL expansion coefficients are orthogonal for a stationary random process, but now orthogonality is defined with a Dirac delta rather than a Kronecker delta. Thus Fourier transformation of a stationary random process results in a delta-correlated random process. We shall make use of this result in the next chapter on Poisson random processes.

Another important conclusion from (8.181) is that the second moment $\langle |F(\boldsymbol{\rho})|^2 \rangle$ is infinite for a stationary random process. Since the mean of the Fourier transform, $\langle F(\boldsymbol{\rho})\rangle$, is the same as the Fourier transform of the mean, $\mathcal{F}\{\langle f(\mathbf{r})\rangle\}$, we would not expect $| \langle F(\boldsymbol{\rho})\rangle |^2$ to be infinite (except possibly for $\boldsymbol{\rho} = 0$), so (8.181) implies that the variance of the Fourier transform of a stationary random process is also infinite.

8.2.8 Discrete random processes

As we discussed in Chap. 7, digital images are discrete vectors, and it is often useful to model actual, physical objects as discrete vectors also. When we analyze the stochastic properties of digital images or discrete object models, then, they become random vectors. The general treatment of random vectors from Sec. 8.1 is applicable here, but there is also an additional structure we can exploit. If a random vector $\mathbf{g}$ represents an image and each component of the vector represents a pixel, we are interested above all in the relationship between the values at different pixels. If we shuffled the pixels into a different arrangement, they would not represent the same image.

A similar situation occurs in discussing random temporal signals, where the temporal ordering of the signal values is key. For example, if a random analog waveform $f(t)$ is sampled at regular time points for further digital processing, the sequence of values $\{f(t_n)\}$ constitutes a random vector in which the order of the elements must be maintained.

We shall use the term *discrete random process*[5] to mean a random vector in which crucial information is contained in the temporal or spatial arrangement of the component values. Loosely, a discrete random process is a random vector endowed with a topology. For temporal processes, the term *random sequence* is often used, and some books adopt this term for the spatial case as well.

Discrete stationarity in 1D Suppose the sequence $\{f_n\}$ is obtained by sampling a stationary temporal random process $f(t)$ at regular intervals $t = t_n = n\Delta t$. The sampling could be simple point sampling where $f_n = f(t_n)$, but a more general form is

$$f_n = \int_{-\infty}^{\infty} dt\, f(t)\, s(t_n - t)\,. \tag{8.182}$$

The sampling function $s(t)$ is a delta function for point sampling, but in general it is unrestricted in what follows. Note that (8.182) is in the form of a convolution, so f_n consists of point samples of the random process $[f * s](t)$.

If $f(t)$ is wide-sense stationary, so is $[f * s](t)$. It then follows that the covariance matrix of the samples $\{f_n\}$ satisfies [*cf.* (8.112)]

$$K_{nn'} = k_{n-n'}\,. \tag{8.183}$$

Note that the left-hand side of this equation has two indices but the right-hand side has just one; if there are N elements in the sequence $\{f_n\}$, there are N^2 elements

[5]Note that the elements of the random vector need not be discrete random variables; the term *discrete* here refers to the temporal or spatial variable.

in the matrix $\mathbf{K}$ but only N independent ones. Each row of the matrix is a shifted version of every other row. Matrices with this structure are said to be *Toeplitz.*

Circulant covariance matrices We encountered Toeplitz matrices in a deterministic context in Chap. 7. Specifically, we saw in Sec. 7.4.4 that a considerable mathematical simplification resulted if we could approximate the Toeplitz matrix by a circulant one, where the difference $n - n'$ in (8.183) is interpreted modulo N, with N being the total number of samples. For example, if n and n' run from 0 to 255, then the pairs $(n = 10, n' = 5)$ and $(n = 2, n' = 253)$ have the same value for $n - n'$ modulo 256 and hence the same correlation if $\mathbf{K}$ is a 256×256 circulant matrix. Physically, of course, this makes no sense; elements 5 and 10 of the sequence are close together and might be expected to be correlated, but elements 2 and 253 are widely separated, and there is no reason to believe that they should have the same correlation as elements 5 and 10.

Nevertheless, the circulant approximation to a Toeplitz covariance matrix is often used, just as is the circulant approximation to a discrete convolution [see Sec. 7.4.4, especially (7.344)]. The error might be tolerable if the kernel ($k_{n-n'}$ in the stochastic problem or h_{m-n} in the deterministic problem) is compact and our interest does not extend to the extreme elements in the sequence. Some vigilance is required to be sure that we do not fall into a trap when we assume that a Toeplitz matrix is approximately circulant.

The reason we might want to make this approximation was laid out in Sec. 7.4.4: a circulant matrix is diagonalized by a DFT [see (7.352)]. For the deterministic DD problem considered in Sec. 7.4.4, that meant that the DFT basis was essentially the SVD basis when the system was described by a circulant $\mathbf{H}$ matrix. In the stochastic context of this chapter, the DFT basis is the KL basis when we can use the circulant form for the covariance.

Discrete spatial stationarity Circulant stationarity is even more suspect than continuous stationarity in imaging applications, but for completeness we state the mathematical results explicitly. If we consider an image $\mathbf{g}$ to be a qD discrete random process, then the elements of the image can be denoted by $g_{\mathbf{m}}$, where $\mathbf{m}$ is a qD multi-index as introduced in Sec. 3.4.6. If each component m_i of $\mathbf{m}$ runs from 0 to $M-1$, then circulant stationarity means that $[\mathbf{K_g}]_{\mathbf{mm}'}$ depends on $m_i - m_i'$ modulo M for all i. In that case, as discussed in Sec. 7.4.4, the circulant covariance matrix is diagonalized by a qD DFT, and the basis vectors in this transform comprise the KL basis.

The cyclic character of the covariance matrix becomes less objectionable as the array gets larger if the correlation length is constant. In the limit as $M \to \infty$, the distinction between Toeplitz and circulant vanishes. In that case, the Toeplitz/circulant matrix is diagonalized by the discrete-space Fourier transform (DSFT) introduced in Sec. 3.6.4, and the KL basis vectors form a continuous basis indexed by the spatial-frequency vector $\boldsymbol{\rho}$. To use this basis, however, we must now make two unphysical assumptions: an infinite amount of data and discrete stationarity over an infinite domain.

8.3 NORMAL RANDOM VECTORS AND PROCESSES

Among the many probability laws for continuous random variables, the normal probability law is certainly the most commonly encountered. The fundamental reason for this is that when statistically independent random variables are added together, their sum asymptotically follows the normal distribution. (We shall provide a more rigorous treatment of this principle later in this section.) The second reason for the popularity of the normal law is that, as we shall soon see, its structure leads to straightforward and well-understood manipulations. The third reason follows from the first two: a great collection of practically useful statistical tools develop as elaborations upon the normal probability law.

The normal law is frequently named for C. F. Gauss (1777–1855), whose *Theory of the Combination of Observations* (1823) has earned him this eponymity. We shall use the terms *normal* and *Gaussian* interchangeably.

8.3.1 Probability density functions

For simplicity we consider here only real random variables and vectors, but the complex case is treated in Sec. 8.3.6. The PDF of a real normal random variable g is given (see App. C) by

$$\mathrm{pr}(g) = \left[\frac{1}{2\pi\sigma^2}\right]^{\frac{1}{2}} \exp\left[-\frac{(g-\overline{g})^2}{2\sigma^2}\right], \tag{8.184}$$

where $\overline{g}$ is the mean of the random variable and σ^2 is its variance. To indicate that a random variable g is drawn from a normal distribution with parameters $\overline{g}$ and σ^2, we write $g \sim \mathcal{N}(\overline{g}, \sigma^2)$.

A multivariate normal random vector is a straightforward generalization of the univariate or scalar case. If each component of an MD random vector $\mathbf{g}$ is a normal random variable, the full probability law on $\mathbf{g}$ is a multivariate normal PDF $\mathrm{pr}(\mathbf{g})$, given by

$$\mathrm{pr}(\mathbf{g}) = \left[(2\pi)^M \det(\mathbf{K})\right]^{-1/2} \exp\left[-\tfrac{1}{2}(\mathbf{g}-\overline{\mathbf{g}})^t\, \mathbf{K}^{-1}(\mathbf{g}-\overline{\mathbf{g}})\right], \tag{8.185}$$

where $\overline{\mathbf{g}}$ is the mean vector and $\mathbf{K}$ is the covariance matrix of $\mathbf{g}$ as defined in Sec. 8.1.3. As shown in that section, $\mathbf{K}$ is an $M \times M$, positive-semidefinite Hermitian matrix. The diagonal element K_{mm} of the covariance matrix is the variance of the mth component of $\mathbf{g}$, and the off-diagonal elements of $\mathbf{K}$ are related by $K_{nm} = K_{mn}$ for real vectors. We denote an $M \times 1$ random vector drawn from a multivariate normal distribution with parameters $\overline{\mathbf{g}}$ and $\mathbf{K}$ by $\mathbf{g} \sim \mathcal{N}_M(\overline{\mathbf{g}}, \mathbf{K})$. Its density function is seen from (8.185) to be the exponential of a quadratic form in the random vector.

Diagonalization of the covariance matrix of a Gaussian random vector In Sec. 8.1.6 we showed how the KL expansion of a random vector in terms of the eigenvectors of its covariance matrix results in uncorrelated components. We now revisit the KL expansion procedure for the particular case of Gaussian random vectors. We shall show that, for a multivariate normal, the KL transformation yields a vector with uncorrelated components that are also statistically independent.

From (8.64) we know we can express the inverse of the covariance matrix $\mathbf{K}$ as

$$\mathbf{K}^{-1} = \mathbf{\Phi}\mathbf{M}^{-1}\mathbf{\Phi}^{\dagger}, \tag{8.186}$$

where again $\mathbf{\Phi}$ is the matrix formed from the eigenvectors ϕ_m of $\mathbf{K}$, and $\mathbf{M}$ is a diagonal matrix with the mth diagonal element equal to the eigenvalue μ_m. We can use (8.186) to rewrite the quadratic form of (8.185) as

$$(\mathbf{g}-\overline{\mathbf{g}})^t\mathbf{K}^{-1}(\mathbf{g}-\overline{\mathbf{g}}) = (\mathbf{g}-\overline{\mathbf{g}})^t\mathbf{\Phi}\mathbf{M}^{-1}\mathbf{\Phi}^{\dagger}(\mathbf{g}-\overline{\mathbf{g}})$$

$$= \left[\mathbf{\Phi}^{\dagger}(\mathbf{g}-\overline{\mathbf{g}})\right]^{\dagger}\mathbf{M}^{-1}\left[\mathbf{\Phi}^{\dagger}(\mathbf{g}-\overline{\mathbf{g}})\right], \tag{8.187}$$

where we have used the unitarity of $\mathbf{\Phi}$. We define the random vector $\Delta\boldsymbol{\beta}$ by [*cf.* (8.60)]

$$\Delta\boldsymbol{\beta} = \mathbf{\Phi}^{\dagger}(\mathbf{g}-\overline{\mathbf{g}}). \tag{8.188}$$

Combining (8.187) and (8.188), we obtain

$$\left[\mathbf{\Phi}^{\dagger}(\mathbf{g}-\overline{\mathbf{g}})\right]^{\dagger}\mathbf{M}^{-1}\left[\mathbf{\Phi}^{\dagger}(\mathbf{g}-\overline{\mathbf{g}})\right] = \Delta\boldsymbol{\beta}^{\dagger}\mathbf{M}^{-1}\Delta\boldsymbol{\beta} = \sum_{m=1}^{M}\Delta\beta_m^2/\mu_m. \tag{8.189}$$

From (A.73) in App. A, we know that the determinant of $\mathbf{K}$ is the product of its eigenvalues. Using this fact and (8.189), we can rewrite (8.185) as

$$\mathrm{pr}(\mathbf{g}) = (2\pi)^{-M/2}\left[\prod_{m=1}^{M}\mu_m\right]^{-1/2}\exp\left(-\frac{1}{2}\sum_{m=1}^{M}\frac{\Delta\beta_m^2}{\mu_m}\right)$$

$$= \prod_{m=1}^{M}(2\pi\mu_m)^{-1/2}\exp\left(-\frac{1}{2}\frac{\Delta\beta_m^2}{\mu_m}\right) = \mathrm{pr}(\boldsymbol{\beta}), \tag{8.190}$$

where the last step is valid since the transformation from $\mathbf{g}$ to $\boldsymbol{\beta}$ is unitary and hence the Jacobian is unity.

Thus, when the quadratic form is diagonalized, the Gaussian multivariate PDF can be written as a product of univariate PDFs, which means that the new variables, $\Delta\beta_m$, are statistically independent. While the components of the random vector $\mathbf{g}$ may covary (as represented by the elements of the covariance matrix $\mathbf{K}$), the components of the random vector $\Delta\boldsymbol{\beta}$ are uncorrelated, with diagonal covariance matrix $\mathbf{M}$, and statistically independent. The mean of each component $\Delta\beta_m$ is 0 and its variance is simply μ_m. The product form of the PDF in (8.190) also makes the normalization of the multivariate Gaussian density readily verifiable.

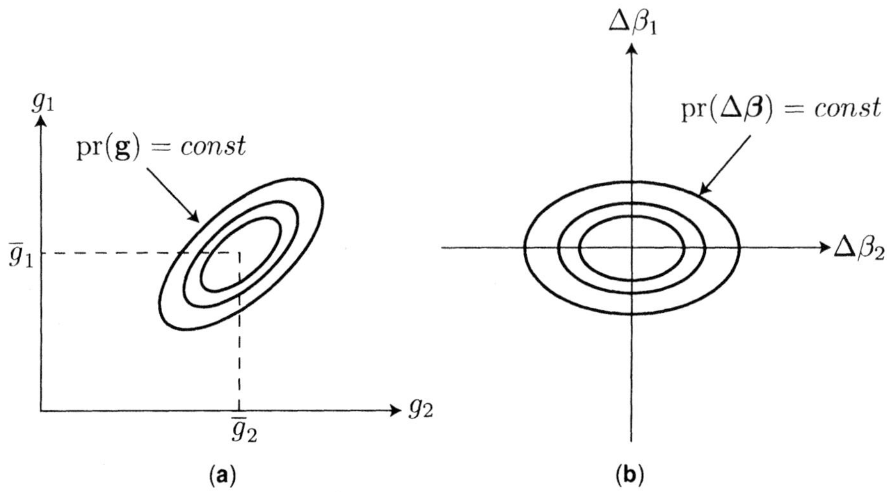

Fig. 8.2 Contours of constant probability density for a multivariate normal, before and after diagonalization.

Figure 8.2 depicts contours of constant probability for the multivariate normal PDF before and after the diagonalization of $\mathbf{K}$. Following the diagonalization operation the surfaces are found to be ellipsoids whose axes have lengths proportional to the square root of the corresponding eigenvalues μ_m. The diagonalization operation rotates the coordinate axes to coincide with the eigenvectors of $\mathbf{K}$.

When does uncorrelated imply independent? We have just seen that a normal random vector with uncorrelated components also has statistically independent components. The converse always holds — statistically independent components must be uncorrelated — but it is *only* the normal law for which uncorrelated components are statistically independent.

8.3.2 Characteristic function

The diagonalized form of the PDF given in Sec. 8.3.1 provides an easy way to derive the characteristic function of a multivariate normal random vector. From (8.188) and the unitarity of $\boldsymbol{\Phi}$, we can write $\mathbf{g}$ as

$$\mathbf{g} = \boldsymbol{\Phi}\Delta\boldsymbol{\beta} + \overline{\mathbf{g}}. \tag{8.191}$$

Thus the characteristic function for $\mathbf{g}$ is given by

$$\psi_{\mathbf{g}}(\boldsymbol{\xi}) = \left\langle \exp\left[-2\pi i\boldsymbol{\xi}^t(\boldsymbol{\Phi}\Delta\boldsymbol{\beta} + \overline{\mathbf{g}})\right]\right\rangle = \exp(-2\pi i\boldsymbol{\xi}^t\overline{\mathbf{g}})\left\langle \exp\left[-2\pi i(\boldsymbol{\Phi}^\dagger\boldsymbol{\xi})^t\Delta\boldsymbol{\beta}\right]\right\rangle, \tag{8.192}$$

where we removed a constant factor from the expectation and used the definition of adjoint, (1.39), to get the last form. Using (8.190) for the PDF and writing out the expectation in detail, we find

$$\psi_{\mathbf{g}}(\boldsymbol{\xi}) = \prod_{m=1}^{M} (2\pi\mu_m)^{-1/2} \exp(-2\pi i \xi_m \overline{g}_m)$$

$$\times \int_{-\infty}^{\infty} d\Delta\beta_m \exp\left(-\frac{1}{2}\frac{\Delta\beta_m^2}{\mu_m}\right) \exp\left[-2\pi i (\boldsymbol{\Phi}^\dagger \boldsymbol{\xi})_m \Delta\beta_m\right] . \tag{8.193}$$

Now we have a product of 1D integrals, each of which is just the Fourier transform of a Gaussian; by (3.180) we have

$$\psi_{\mathbf{g}}(\boldsymbol{\xi}) = \prod_{m=1}^{M} \exp(-2\pi i \xi_m \overline{g}_m) \exp\left[-2\pi^2 \mu_m (\boldsymbol{\Phi}^\dagger \boldsymbol{\xi})_m^2\right] . \tag{8.194}$$

From (8.186) we can see that

$$\sum_{m=1}^{M} \mu_m (\boldsymbol{\Phi}^\dagger \boldsymbol{\xi})_m^2 = \boldsymbol{\xi}^t \mathbf{K} \boldsymbol{\xi} , \tag{8.195}$$

so we have, finally,

$$\psi_{\mathbf{g}}(\boldsymbol{\xi}) = \exp(-2\pi i \boldsymbol{\xi}^t \overline{\mathbf{g}}) \exp\left(-2\pi^2 \boldsymbol{\xi}^t \mathbf{K} \boldsymbol{\xi}\right) . \tag{8.196}$$

For $\overline{\mathbf{g}} = \mathbf{0}$, we obtain $\exp(-2\pi^2 \boldsymbol{\xi}^t \mathbf{K} \boldsymbol{\xi})$, which is easy to remember since it is Gaussian in the Fourier domain with spread inverse to that in the domain of the random variable (*i.e.*, $\mathbf{K}$ occurs in place of $\mathbf{K}^{-1}$). The complete form, (8.196), may then be recalled by invoking the Fourier shift theorem (3.108).

Moments We can use the characteristic function given in (8.196) to determine the moments of a multivariate normal random vector. If we apply (8.30) and (8.31) to (8.196), we obtain $\langle \mathbf{g} \rangle = \overline{\mathbf{g}}$ and $\langle \mathbf{g}\, \mathbf{g}^t \rangle = \mathbf{K} + \overline{\mathbf{g}\mathbf{g}}^t$. If $\overline{\mathbf{g}} = 0$, then $\langle \mathbf{g}\, \mathbf{g}^t \rangle = \mathbf{K}$. By continuing along this path we find that all odd moments of this distribution are zero for $\overline{\mathbf{g}} = 0$, and all even moments are expressible in terms of $\mathbf{K}$.

We shall find that we frequently need fourth moments of the form $\langle g_i g_j g_k g_l \rangle$, where the g_i, etc., are components of a four-dimensional vector $\mathbf{g}$ distributed as $\mathcal{N}_4(\mathbf{0}, \mathbf{K})$. We can obtain the desired result, referred to as the Gaussian moment theorem, by using the rules for differentiation with respect to a real vector given in Sec. A.9.2. We find that

$$\langle g_i g_j g_k g_l \rangle = \left(\frac{\partial^4 \psi_{\mathbf{g}}(\boldsymbol{\xi})}{\partial \xi_l \partial \xi_k \partial \xi_j \partial \xi_i} \right)_{\boldsymbol{\xi}=\mathbf{0}} = K_{ij} K_{kl} + K_{jk} K_{il} + K_{ik} K_{jl} . \tag{8.197}$$

For the case where $i = j = k = l$, we find $\langle g_i^4 \rangle = 3\sigma_i^4$, which is a familiar result for univariate normals given in (C.112).

8.3.3 Marginal densities and linear transformations

In this section we derive various descriptors of the behavior of subsets and transformations of the components of a multivariate Gaussian random vector. We start by analyzing the behavior of a single component, regardless of the behavior of the other components, as described by the marginal PDF. We then discuss the behavior of a random vector obtained from linear transformation of a Gaussian random vector.

According to (8.5), the marginal PDF on component g_i of an MD vector $\mathbf{g}$ is obtained by integrating the multivariate PDF over all g_m except for $m = i$. From the central-ordinate theorem of Fourier analysis, (3.104), we know that integrating a function over $(-\infty, \infty)$ is equivalent to setting the frequency to zero in its Fourier transform. Thus the univariate characteristic function for g_i is related to the multivariate characteristic function for $\mathbf{g}$ by

$$\psi_{g_i}(\xi_i) = \psi_{\mathbf{g}}(0, 0, ..., \xi_i, ..., 0) . \tag{8.198}$$

With (8.196), we have

$$\psi_{g_i}(\xi_i) = \exp(-2\pi i \xi_i \overline{g}_i) \exp(-2\pi^2 K_{ii} \xi_i^2) . \tag{8.199}$$

This is just the characteristic function for a univariate normal with mean $\overline{g}_i$ and variance K_{ii}. Perhaps surprisingly, the form of the marginal on g_i does not depend on K_{im} for $i \neq m$, even though g_i may be correlated with the other components.

Similarly, the bivariate characteristic function for g_i and g_j is given by

$$\begin{aligned} \psi_{g_i, g_j}(\xi_i, \xi_j) &= \psi_{\mathbf{g}}(0, 0, ..., \xi_i, ..., \xi_j, ..., 0) \\ &= \exp\left(-2\pi i \tilde{\boldsymbol{\xi}}^t \tilde{\overline{\mathbf{g}}}\right) \exp\left[-2\pi^2 \tilde{\boldsymbol{\xi}}^t \tilde{\mathbf{K}} \tilde{\boldsymbol{\xi}}\right] , \end{aligned} \tag{8.200}$$

where $\tilde{\boldsymbol{\xi}}^t = (\xi_i, \xi_j)$, $\tilde{\overline{\mathbf{g}}} = (\overline{g}_i, \overline{g}_j)^t$ and

$$\tilde{\mathbf{K}} = \begin{bmatrix} K_{ii} & K_{ij} \\ K_{ij} & K_{jj} \end{bmatrix} . \tag{8.201}$$

Inverse Fourier transformation of (8.200) yields a bivariate normal PDF with the expected mean and covariance. Again, we do not need to know covariance components other than the ones represented in the marginal of interest.

Other linear transformations of normal random vectors Computation of a marginal is equivalent to finding the PDF for the output of a linear transformation of a random vector. For example, the component g_i can be singled out by computing the scalar product of $\mathbf{g}$ with an $1 \times M$ row vector having a one in the ith column and a zero in all others. Similarly, the 2D vector (g_i, g_j) results from applying a $2 \times M$ matrix operator with ones in positions $(1, i)$ and $(2, j)$ and zeros in all other locations. We now compute the PDF for a random vector formed from a general linear transformation.

Consider the random vector $\mathbf{y} = \mathbf{O}\mathbf{g}$, where $\mathbf{y}$ is a $K \times 1$ vector, $\mathbf{O}$ is a real $K \times M$ matrix and $\mathbf{g} \sim \mathcal{N}_M(\overline{\mathbf{g}}, \mathbf{K})$. The characteristic function for $\mathbf{y}$ follows from (8.43) and (8.196):

$$\psi_{\mathbf{y}}(\boldsymbol{\xi}) = \psi_{\mathbf{g}}(\mathbf{O}^t \boldsymbol{\xi}) = \exp(-2\pi i \boldsymbol{\xi}^t \mathbf{O} \overline{\mathbf{g}}) \exp\left(-2\pi^2 \boldsymbol{\xi}^t \mathbf{O} \mathbf{K} \mathbf{O}^t \boldsymbol{\xi}\right) . \tag{8.202}$$

By inspection, then, $\mathbf{y} \sim \mathcal{N}_K(\mathbf{O}\overline{\mathbf{g}}, \mathbf{O}\mathbf{K}\mathbf{O}^t)$. Thus *any* linear transformation of a normal random vector leaves it normal.

In fact, the converse of (8.202) also holds: An $M \times 1$ random vector is normal if and only if its scalar products with all $M \times 1$ vectors are univariate normal (Mardia *et al.*, 1979).

8.3.4 Central-limit theorem

In this section we show that the sum of a large number of random variables tends to be normally distributed. This property, known as the central-limit theorem, is one of the reasons for the prominence of the Gaussian law in probability theory.

We shall introduce the central-limit theorem in stages. Initially we consider *i.i.d.* (independent and identically distributed) scalar random variables, where all moments are finite. These assumptions allow an elementary derivation, though one with restricted validity. Next we discuss the case of i.i.d. random variables where some of the higher moments may be infinite. Then we allow the variables to have different variances and some degree of statistical dependence. Finally we comment briefly on the vector case.

Independent and identically distributed random variables Consider a set of J i.i.d. random variables u_j, $1 \leq j \leq J$, with means $\overline{u}$ and variances σ^2. First we define standardized (zero-mean, unit-variance) random variables by

$$x_j = \frac{u_j - \overline{u}}{\sigma} . \tag{8.203}$$

Then we construct a new random variable z, defined by

$$z = \frac{1}{\sqrt{J}} \sum_{j=1}^{J} x_j . \tag{8.204}$$

Because the variance of a sum of J i.i.d. random variables is J times the individual variances, and the variance of $x_j/\sqrt{J}$ is $1/J$, z has unit variance. Moreover, since z is a sum of zero-mean random variables, it also has zero mean. We want to show that as $J \to \infty$ the PDF on z tends toward a standard normal distribution, from which it follows readily that the sum of the u_j is normal with mean $J\overline{u}$ and variance $J\sigma^2$.

The derivation proceeds most easily with the aid of characteristic functions. We shall denote the characteristic function of x_j as $\psi_x(\xi)$; no index j is needed since the characteristic function has the same form for all of the x_j. If we assume initially that all moments of x_j are finite, we can expand $\psi_x(\xi)$, in a Taylor series:

$$\begin{aligned} \psi_x(\xi) = \langle \exp(-2\pi i \xi x_j) \rangle &= 1 - 2\pi i \xi \langle x_j \rangle - \frac{4\pi^2}{2!} \xi^2 \left\langle x_j^2 \right\rangle + ... \\ &= 1 - \frac{4\pi^2}{2!} \xi^2 + ... , \end{aligned} \tag{8.205}$$

where the second line follows since $\langle x_j \rangle = 0$ and $\left\langle x_j^2 \right\rangle = 1$.

The characteristic function of z is given by

$$\begin{aligned} \psi_z(\xi) = \langle \exp(-2\pi i \xi z) \rangle &= \left\langle \exp\left[-2\pi i \left(\frac{\xi}{\sqrt{J}} \right) \sum_{j=1}^{J} x_j \right] \right\rangle \\ = \prod_{j=1}^{J} \left\langle \exp\left[-2\pi i \left(\frac{\xi}{\sqrt{J}} \right) x_j \right] \right\rangle &= \prod_{j=1}^{J} \psi_x\left(\frac{\xi}{\sqrt{J}} \right) = \left[\psi_x\left(\frac{\xi}{\sqrt{J}} \right) \right]^J , \end{aligned} \tag{8.206}$$

where the independence of the x_j has been invoked on the second line to write the expectation of a product as the product of the expectations, and the fact that the x_j are identically distributed is the key to the last step.

We can now insert the Taylor expansion (8.205) into (8.206), yielding

$$\psi_z(\xi) = \left[1 - \frac{2\pi^2\xi^2}{J} + R_J(\xi)\right]^J , \tag{8.207}$$

where $R_J(\xi)$ is the remainder if the Taylor expansion is truncated with the quadratic term. By Taylor's theorem (Rade and Westgren, 1990), $R_J(\xi)$ tends to zero (for any fixed ξ) at least as fast as $J^{-3/2}$ when $J \to \infty$. Thus, in spite of the Jth power, these higher terms vanish in the limit. The quadratic term must be retained, however, so that

$$\lim_{J\to\infty} \psi_z(\xi) = \lim_{J\to\infty} \left(1 - \frac{2\pi^2\xi^2}{J}\right)^J = \exp(-2\pi^2\xi^2) , \tag{8.208}$$

which is the characteristic function of a standard-normal random variable. It then follows from the celebrated *continuity theorem* of Paul Lévy (see Loève, 1963) that $z \sim \mathcal{N}(0, 1)$.[6]

It is straightforward to go from (8.208) to the probability law for the sum of the original random variables u_j. Defining

$$s_J = \sum_{j=1}^{J} u_j , \tag{8.209}$$

the reader may show that $s_J \sim \mathcal{N}(J\overline{u}, J\sigma^2)$

We have therefore seen that an infinite sum of independent, identically distributed random variables follows a normal distribution, at least when the individual characteristic functions admit of a Taylor expansion. It must be emphasized, however, that the central-limit theorem guarantees normality only asymptotically; it might not be a good approximation for large but finite J. Often the convergence to normality is rapid, requiring as few as perhaps $5-10$ terms, but we should be cautious about finite sums of skewed or otherwise long-tailed PDFs. An extreme example is the case of sums of log-normal distributions, which converge very slowly to the central limit (Barakat, 1976).

Infinite moments There are common PDFs where some of the higher moments are infinite. In Sec. C.5.10, we encountered the Lévy family of distributions, and we noted that the mean was zero but the variance was infinite. A special case of the Lévy distribution is the Cauchy distribution, where $\text{pr}(x) \propto (a^2 + x^2)^{-1}$, a well-known and broadly useful PDF of infinite variance. On the other hand, if we consider $\text{pr}(x) \propto (a^2 + x^2)^{-2}$, then the variance is finite but the fourth moment is infinite. The common feature of these examples is that the characteristic function is not differentiable to all orders and hence cannot be expanded in a Taylor series. Therefore we need to inquire whether it is possible to derive a central-limit theorem.

[6]Thanks to Jack Denny for calling our attention to this theorem.

The key is a theorem proved in Shiryayev (1984). If $\langle |x|^n \rangle$ exists for some $n \geq 1$, then the kth derivative of $\psi_x(\xi)$, denoted $\psi_x^{(k)}(\xi)$, exists for every $k \leq n$, and

$$\psi_x(\xi) = \sum_{k=0}^{n} \frac{(2\pi i \xi)^k}{k!} \langle x^k \rangle + \frac{(2\pi i \xi)^n}{n!} \epsilon_n(\xi) , \tag{8.210}$$

where $|\epsilon_n(\xi)| \leq 3 \langle |x|^n \rangle$ and $\epsilon_n(\xi) \to 0$ as $\xi \to 0$. So long as $\langle |x_j|^3 \rangle$ is finite, this theorem justifies the steps from (8.205) to (8.208), even when the full Taylor expansion for $\psi_x(\xi)$ does not exist.

For the examples given above, $\langle |x_j|^3 \rangle$ is infinite for the Lévy and Cauchy PDFs, so the limiting PDF is not normal; in fact, a sum of any number of Lévy random variables is still a Lévy random variable. For $\text{pr}(x) \propto (a^2 + x^2)^{-2}$, however, $\langle |x_j|^3 \rangle$ is finite and there is a normal central limit.[7]

Independent but not identically distributed random variables Now suppose that the random variables u_j are independent but have different means and variances. Let the mean of u_j be denoted by $\overline{u}_j$ and the variance by σ_j^2, and define

$$x_{jJ} = \frac{u_j - \overline{u}_j}{\sqrt{\sum_{j=1}^{J} \sigma_j^2}} . \tag{8.211}$$

The extra subscript is needed since the denominator depends on J. Note that

$$\langle x_{jJ} \rangle = 0 \qquad \text{and} \qquad \sum_{j=1}^{J} \text{Var}(x_{jJ}) = 1 . \tag{8.212}$$

Now we can define a standardized random variable z by

$$z = \sum_{j=0}^{J} x_{jJ} . \tag{8.213}$$

If the means and variances are independent of j, this definition of z reduces to (8.204).

Shiryayev (1984) discusses various sufficient conditions under which z will tend to a standard normal as $J \to \infty$. They all amount to saying that the variables x_{jJ} are *asymptotically infinitesimal*, in the sense that $\langle x_{jJ}^2 \rangle \to 0$ as $J \to \infty$, or equivalently that, for every ϵ,

$$\Pr(|x_{jJ}| > \epsilon) \to 0 \qquad \text{as} \qquad J \to \infty . \tag{8.214}$$

This condition is plausible in most practical circumstances because of the denominator in (8.211); so long as the variances σ_j^2 do not themselves tend to zero rapidly as j gets large, the sum of the variances will increase as the number of terms increases, so x_{jJ}, which is normalized by this sum, must get smaller in virtually any sense.

Thus the central-limit theorem states that a sum of asymptotically infinitesimal, zero-mean random variables tends to a standard normal, so long as the sum

[7] We thank Dana Clarke for helpful discussions on these examples.

of their variances is normalized to unity (Shiryayev, 1984). From this statement, it is again straightforward to show that the sum of the original variables u_j is also asymptotically normal. Specifically, as $J \to \infty$, s_J becomes distributed as $\mathcal{N}_J[\sum_j \overline{u}_j, \sum_j \mathrm{Var}(u_j)]$.

Sums of dependent random variables Though the central-limit theorem is usually stated for sums of independent random variables, strict independence is not required. For a detailed discussion, see Shiryayev (1984).

Sums of i.i.d. random vectors Central-limit theorems can also be stated for random vectors. We mention here only the simplest case of i.i.d. random vectors where all moments exist.

Let $\mathbf{u}_j$ be an $M \times 1$ random vector with mean $\overline{\mathbf{u}}$ and covariance $\mathbf{K_u}$, both independent of j, and assume that $\mathbf{u}_j$ is independent of $\mathbf{u}_k$ for $j \neq k$. Also let

$$\mathbf{s}_J = \sum_{j=1}^{J} \mathbf{u}_j \,. \tag{8.215}$$

Then, as $J \to \infty$, $\mathbf{s}_J \sim \mathcal{N}_M(J\overline{\mathbf{u}}, J\mathbf{K_u})$. The proof of this statement involves multivariate characteristic functions and the multivariate Taylor expansion (A.179). With this hint, the reader should be able to retrace the steps leading up to (8.208).

8.3.5 Normal random processes

As we shall see in more detail in Sec. 8.4.3, we can sometimes apply the central-limit theorem and argue that the random process representing an object or image is normal. In preparation for that discussion, we examine here some of the mathematical properties of normal random processes. We initially adopt a rather unconventional starting point and define normal random processes in terms of characteristic functionals, but then we shall show that this definition is equivalent to a more common one.

Characteristic functional and linear operators The general form of the characteristic function of a normal random vector is given in (8.196); it can be extended to random processes by use of the characteristic functional, as introduced in Sec. 8.2.3. By analogy to (8.196), we define a real-valued normal random process by requiring that its characteristic functional be given by

$$\Psi_{\mathbf{f}}(\mathbf{s}) = \exp(-2\pi i \mathbf{s}^\dagger \overline{\mathbf{f}}) \exp(-2\pi^2 \mathbf{s}^\dagger \mathcal{K}_{\mathbf{f}} \mathbf{s}) \,, \tag{8.216}$$

where $\mathcal{K}_{\mathbf{f}}$ is the autocovariance operator, *i.e.*, the integral operator with kernel $K_{\mathbf{f}}(\mathbf{r}, \mathbf{r}')$.

From (8.216) and (8.96) we can readily show that all linear functionals of a normal random process are normal. If we let $\mathbf{g} = \mathcal{H}\mathbf{f}$, where $\mathcal{H}$ is a linear CD mapping (see Sec. 7.3) defined by

$$g_m = \int_\infty d^q r \; h_m(\mathbf{r}) \, f(\mathbf{r}) \,, \qquad m = 1, ..., M \,, \tag{8.217}$$

then (8.96) becomes

$$\psi_{\mathbf{g}}(\boldsymbol{\xi}) = \exp(-2\pi i \boldsymbol{\xi}^\dagger \mathcal{H} \overline{\mathbf{f}}) \exp(-2\pi^2 \boldsymbol{\xi}^\dagger \mathcal{H} \mathcal{K}_{\mathbf{f}} \mathcal{H}^\dagger \boldsymbol{\xi}) \,. \tag{8.218}$$

By comparison with (8.196), we see that $\mathbf{g}$ is an MD random vector with mean $\mathcal{H}\bar{\mathbf{f}}$ and covariance $\mathcal{H}\mathcal{K}_{\mathbf{f}}\mathcal{H}^{\dagger}$.

Exactly the same conclusion holds when $\mathcal{H}$ is an integral operator. Linear filtering of a normal random process yields another normal random process. Since normal processes are fully determined by their mean and autocovariance (or autocorrelation) function, the formulas given in Sec. 8.2.6 are all we need for a complete statistical description of the output of a linear filter *if* we know that the input is a normal process.

Multipoint densities and autocovariance functions One way of defining a normal random vector is to require that all of its marginals must be normal (Sec. 8.3.3). Similarly, a normal random process can be defined as one for which all univariate or multivariate marginals are normal. In that approach, a random process $f(\mathbf{r})$ is normal if all M-point PDFs, $\mathrm{pr}[f(\mathbf{r}_1), f(\mathbf{r}_2), ..., f(\mathbf{r}_M)]$ for all M, are normal. We can use (8.218) to show that defining a normal random process by (8.216) is equivalent to requiring that all multipoint densities be normal. Evaluating the random process at the M points $\{\mathbf{r}_m, m = 1, ..., M\}$ is a CD mapping with

$$h_m(\mathbf{r}) = \delta(\mathbf{r} - \mathbf{r}_m)\,. \tag{8.219}$$

Thus $g_m = f(\mathbf{r}_m)$, and it follows at once from (8.218) that $\mathrm{pr}[f(\mathbf{r}_1), f(\mathbf{r}_2), ..., f(\mathbf{r}_M)]$ is an MD normal density. An explicit form for this density can be stated most compactly by defining an $M \times 1$ vector $\mathbf{f}_M$ with m^{th} component given by $f(\mathbf{r}_m)$. For simplicity we assume that $f(\mathbf{r})$ is real. Then the M-point PDF is given by

$$\mathrm{pr}[f(\mathbf{r}_1), f(\mathbf{r}_2), ..., f(\mathbf{r}_M)] = \mathrm{pr}(\mathbf{f}_M)$$

$$= (2\pi)^{-\frac{1}{2}M} |\det \mathbf{K}_M|^{-\frac{1}{2}} \exp\left[-\tfrac{1}{2}(\mathbf{f}_M - \bar{\mathbf{f}}_M)^t\, \mathbf{K}_M^{-1}(\mathbf{f}_M - \bar{\mathbf{f}}_M)\right], \tag{8.220}$$

where $\bar{\mathbf{f}}_M$ is the $M \times 1$ mean vector, with components $\langle f(\mathbf{r}_m)\rangle$, and $\mathbf{K}_M$ is the $M \times M$ covariance matrix, with components given by

$$[\mathbf{K}_M]_{mn} = \langle [f(\mathbf{r}_m) - \langle f(\mathbf{r}_m)\rangle]\,[f(\mathbf{r}_n) - \langle f(\mathbf{r}_n)\rangle]\rangle\,. \tag{8.221}$$

Comparison with (8.98) shows that

$$[\mathbf{K}_M]_{mn} = K_{\mathbf{f}}(\mathbf{r}_m, \mathbf{r}_n)\,. \tag{8.222}$$

Thus the covariance *matrix* in an M-point PDF for a normal random process is fully determined by the autocovariance *function* of the process. Knowledge of this function and $\langle f(\mathbf{r})\rangle$ is therefore sufficient to specify all M-point densities and hence to fully characterize a normal process.

For completeness, we next show that (8.222) also follows from the transformation rule, $\mathbf{K}_{\mathbf{g}} = \mathcal{H}\mathcal{K}_{\mathbf{f}}\mathcal{H}^{\dagger}$. With $\mathbf{K}_{\mathbf{g}} = \mathbf{K}_M$, and the kernel of $\mathcal{H}$ as given by (8.219), we can write

$$[\mathbf{K}_M]_{mn} = \left[\mathcal{H}\mathcal{K}_{\mathbf{f}}\mathcal{H}^{\dagger}\right]_{mn}$$

$$= \int_\infty d^q r \int_\infty d^q r'\, \delta(\mathbf{r} - \mathbf{r}_m) K_{\mathbf{f}}(\mathbf{r}, \mathbf{r}')\, \delta(\mathbf{r}' - \mathbf{r}_n) = K_{\mathbf{f}}(\mathbf{r}_m, \mathbf{r}_n)\,, \tag{8.223}$$

where the last step has used the sifting property of delta functions.

Ergodicity and stationarity Stationarity is defined for normal random processes just as for any other random process. A useful simplification, however, is that we do not have to distinguish wide-sense and narrow-sense stationarity in the normal case. Since the full statistics are inherent in the mean and autocovariance function, wide-sense stationarity (stationary mean and autocovariance) implies narrow-sense or strict stationarity (Papoulis, 1965).

For stationary Gaussian random processes, a straightforward criterion for ergodicity can be stated. Cornfield *et al.* (1982) show that such a process is ergodic if and only if its power spectral density is continuous. From (3.107) and the Wiener-Khinchin theorem (8.133), an equivalent statement is that a stationary Gaussian random process is ergodic if and only if its autocorrelation function vanishes at infinity. Since many physical processes are Gaussian as a result of the central-limit theorem, we can quite often invoke ergodicity on the basis of this theorem.

8.3.6 Complex Gaussian random fields

It is often useful to describe a wave by its complex amplitude. If the wave is regarded as random, perhaps because it has been scattered from a random object, then the wave amplitude $u(\mathbf{r}_1)$ at any point $\mathbf{r}_1$ is a complex random variable. Similarly, the set of amplitudes at K different points, $\{u(\mathbf{r}_k), k = 1, ..., K\}$, is a KD complex vector, and $u(\mathbf{r})$ itself is a complex random process. Moreover, a wave amplitude is usually computed as a diffraction integral or some other linear superposition. If different elements of this superposition are linearly independent random variables, then the central-limit theorem will lead to normal distributions, so we often encounter complex Gaussian random fields.

In one sense, there is nothing new about complex Gaussian random fields; we can describe them with the tools already developed for real Gaussian fields just by considering the real and imaginary parts separately. For example, a $K \times 1$ complex vector can also be written as a $2K \times 1$ real vector, where the first K components are the real parts and the second K are the imaginary parts. The covariance matrix in the first case is a $K \times K$ Hermitian matrix with complex off-diagonal elements, and in the second case it is a $2K \times 2K$ real, symmetric matrix.

Random phase If the complex variables result from random waves, the physics of wave propagation may allow us to impose some additional restrictions, thereby simplifying the mathematics. The phase of a wave relates to the total optical pathlength from a radiation source to the point at which the phase is measured. The natural unit of this pathlength is the wavelength, and typically the paths are very long compared to a wavelength. That means that if we alter the pathlength by a small fraction in absolute terms, it may nevertheless change by several wavelengths, and each change of one wavelength alters the phase by 2π. Now, the pathlength (in units of wavelength) may be random for many reasons: we may consider an ensemble of objects with different positions and different rough surfaces, or we may interpose random phase-altering elements such as diffuse reflectors or ground-glass screens, or we may consider a broad spectrum of wavelengths. The result is that it is frequently an excellent approximation to assume that the phase is completely random.

To state this approximation more mathematically, we denote the wave amplitude (at some unspecified point) by $u = Ae^{i\phi} = x + iy$, where $x = \text{Re}(u)$ and

$y = \text{Im}(u)$ and A is a real number. We do not need to consider phase angles ϕ outside the range $[0, 2\pi)$ since $e^{i\phi}$ is periodic. The phase randomness implies that the PDF on ϕ is constant in this range. The constant can be fixed since the PDF must be normalized to unity, and we can write

$$\text{pr}(\phi) = \frac{1}{2\pi}, \qquad 0 \le \phi < 2\pi. \tag{8.224}$$

We assume that this PDF on ϕ is valid for all A, so $\text{pr}(\phi|A) = \text{pr}(\phi)$, and ϕ and A are statistically independent.

We can use this density to deduce some important properties of u even without specifying the statistics of A. Since the real and imaginary parts of u are given by

$$x = A\cos\phi, \qquad y = A\sin\phi, \tag{8.225}$$

we see that (8.224) implies

$$\langle x\rangle = \langle A\cos\phi\rangle = 0, \qquad \langle y\rangle = \langle A\sin\phi\rangle = 0. \tag{8.226}$$

Thus x and y are both zero-mean, and hence so is the complex u.

The variances of x and y must be equal since

$$\langle x^2\rangle = \langle A^2\cos^2\phi\rangle = \tfrac{1}{2}\langle A^2\rangle, \qquad \langle y^2\rangle = \langle A^2\sin^2\phi\rangle = \tfrac{1}{2}\langle A^2\rangle. \tag{8.227}$$

The marginal PDFs on x and y must also be the same, regardless of the PDF of A, since $\sin\phi$ and $\cos\phi$ have the same PDFs if ϕ is uniform. (As an exercise, the reader can determine what this PDF is.) Moreover, x and y are uncorrelated since

$$\langle xy\rangle = \langle A^2\cos\phi\sin\phi\rangle = 0. \tag{8.228}$$

We can summarize the last two equations in complex form by writing

$$\langle u^2\rangle = \langle u^{*2}\rangle = 0, \qquad \langle uu^*\rangle = \langle u^*u\rangle = \langle A^2\rangle \ne 0. \tag{8.229}$$

Invocation of the central-limit theorem If we now assume that the wave amplitude at any point is the sum of contributions from many independent sources (perhaps points on an illuminated rough surface), then the real and imaginary parts are normal by the central-limit theorem. That means that x and y are not only uncorrelated but also statistically independent; we say that x and y are i.i.d. (independently and identically distributed). Their joint density is given by

$$\text{pr}(x, y) = \frac{1}{2\pi\sigma^2}\exp\left(-\frac{x^2+y^2}{2\sigma^2}\right), \tag{8.230}$$

where σ^2 is the common variance of x and y. Contours of constant PDF in the x-y plane are circles (see Fig. 8.3), so u is referred to as a *circular Gaussian* random variable.

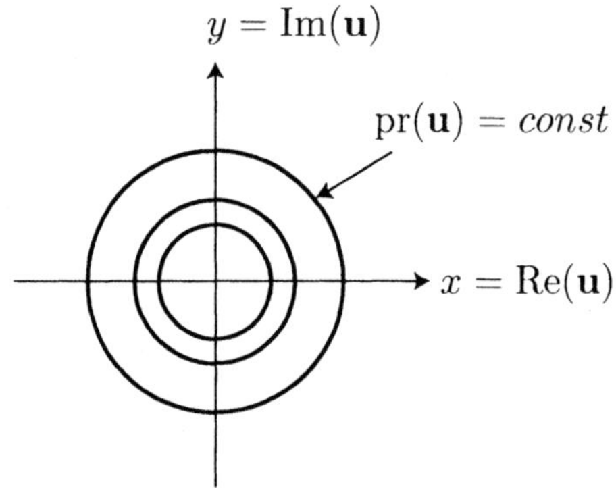

Fig. 8.3 Surfaces of constant probability density for a circular Gaussian random variable.

Other useful PDFs Since $A = \sqrt{x^2 + y^2}$ and $\phi = \tan^{-1}(y/x)$, we can convert $\mathrm{pr}(x, y)$ in (8.230) to $\mathrm{pr}(A, \phi)$ by means of (C.104). The result is the *Rayleigh distribution*, given in (C.140) as

$$\mathrm{pr}(A, \phi) = \frac{A}{2\pi\sigma^2} \exp\left(-\frac{A^2}{2\sigma^2}\right). \tag{8.231}$$

We shall see in Chap. 11 that the irradiance $I = |u|^2$ plays a key role in photodetection and photon counting. If $\mathrm{pr}(x, y)$ is given by (8.230), the PDF on I and ϕ is

$$\mathrm{pr}(I, \phi) = \frac{1}{2\pi\overline{I}} \exp\left(\frac{-I}{\overline{I}}\right), \tag{8.232}$$

where $\overline{I} = 2\sigma^2$. The PDF on I alone, obtained by omitting the 2π in (8.232), is a chi-squared PDF with two degrees of freedom (see Sec. C.5.5). In general, a chi-squared random variable with N degrees of freedom is the sum of the squares of N i.i.d. normal random variables; here $N = 2$ since $I = x^2 + y^2$.

Two-point densities for circular Gaussians Next we examine two-point PDFs involving a complex circular Gaussian random process $u(\mathbf{r})$ at points $\mathbf{r} = \mathbf{r}_1$ and $\mathbf{r} = \mathbf{r}_2$. For notational simplicity, we write $u(\mathbf{r}_1) = u_1 = x_1 + iy_1 = A_1 \exp(i\phi_1)$, and similarly for $u(\mathbf{r}_2)$. One way we could specify the two-point density would be to construct the real 4D vector $\mathbf{U} = (x_1, x_2, y_1, y_2)^t$ and give the 4D PDF for it. If $u(\mathbf{r})$ is to be circular Gaussian, this PDF has to satisfy some constraints. For one thing, if we want u_1 and u_2 to be individual circular Gaussians, the marginals on (x_1, y_1) and (x_2, y_2) must both satisfy (8.230), possibly with different variances. In addition, the joint density on all four variables must be consistent with the autocovariance function of the process,

$$K_{\mathbf{u}}(\mathbf{r}_1, \mathbf{r}_2) = \langle u_1 u_2^* \rangle \equiv k = k' + ik''. \tag{8.233}$$

These conditions lead to

$$\langle x_1 x_2 \rangle = \langle y_1 y_2 \rangle = \tfrac{1}{2}k', \qquad -\langle x_1 y_2 \rangle = \langle y_1 x_2 \rangle = \tfrac{1}{2}k'' \qquad \langle x_1 y_1 \rangle = \langle x_2 y_2 \rangle = 0. \tag{8.234}$$

All of these conditions are satisfied if $\mathbf{U} \sim \mathcal{N}_4(\mathbf{0}, \mathbf{K_U})$, where

$$\mathbf{K_U} = \begin{bmatrix} \sigma_1^2 & \frac{1}{2}k' & 0 & -\frac{1}{2}k'' \\ \frac{1}{2}k' & \sigma_2^2 & \frac{1}{2}k'' & 0 \\ 0 & \frac{1}{2}k'' & \sigma_1^2 & \frac{1}{2}k' \\ -\frac{1}{2}k'' & 0 & \frac{1}{2}k' & \sigma_2^2 \end{bmatrix} . \tag{8.235}$$

The redundancy in the elements of this matrix should be noted. A general 4×4 covariance matrix would have 10 independent elements, but only four real numbers ($\sigma_1^2, \sigma_2^2, k'$ and k'') are required to specify $\mathbf{K}_U$. This redundancy is required in order to represent a circular Gaussian as opposed to a more general complex Gaussian random vector.

Two-dimensional formulation To go from the covariance in (8.235) to the PDF for $\mathbf{U}$ requires inverting $\mathbf{K_U}$ and computing the quadratic form $\mathbf{U}^t\mathbf{K_U}^{-1}\mathbf{U}$. The algebra is not terrible, but a simpler approach, and one that extends more readily to higher dimensions, is to use a 2D complex vector rather than a 4D real one. If we define a 2D vector $\mathbf{u}$ with complex components u_1 and u_2, its covariance matrix is

$$\mathbf{K_u} = \begin{bmatrix} 2\sigma_1^2 & k \\ k^* & 2\sigma_2^2 \end{bmatrix} . \tag{8.236}$$

The inverse covariance, which is what we need in the PDF, is given by

$$\mathbf{K_u}^{-1} = \frac{1}{4\sigma_1^2\sigma_2^2 - |k|^2} \begin{bmatrix} 2\sigma_2^2 & -k \\ -k^* & 2\sigma_1^2 \end{bmatrix} . \tag{8.237}$$

The quadratic form in the PDF is thus

$$\mathbf{u}^\dagger\mathbf{K_u}^{-1}\mathbf{u} = \frac{2\sigma_2^2|u_1|^2 + 2\sigma_1^2|u_2|^2 - ku_1^*u_2 - k^*u_2^*u_1}{4\sigma_1^2\sigma_2^2 - |k|^2} , \tag{8.238}$$

and the corresponding PDF is given by (Neeser and Massey, 1993; Mandel and Wolf, 1995)

$$\mathrm{pr}(\mathbf{u}) = \frac{1}{\pi^2 \det(\mathbf{K_u})} \exp\left(-\mathbf{u}^\dagger\mathbf{K_u}^{-1}\mathbf{u}\right) . \tag{8.239}$$

The reader might have expected a factor of $\frac{1}{2}$ in the exponent and a different normalizing factor [*cf.* (8.185)], but (8.239) is correct as written. One way to make it plausible is to assume there is no correlation, so $k = 0$, so that (8.237) becomes

$$\mathbf{K_u}^{-1} = \begin{bmatrix} \frac{1}{2\sigma_1^2} & 0 \\ 0 & \frac{1}{2\sigma_2^2} \end{bmatrix} . \tag{8.240}$$

Hence, (8.239) becomes

$$\mathrm{pr}(\mathbf{u}) = \frac{1}{4\pi^2\sigma_1^2\sigma_2^2} \exp\left[-\frac{x_1^2 + y_1^2}{2\sigma_1^2} - \frac{x_2^2 + y_2^2}{2\sigma_2^2}\right] , \tag{8.241}$$

which is just what one would get with the 4D real formulation, using (8.235) with $k = 0$ and (8.185). The reader may check that the 2D complex and 4D real formulations also agree when $k \neq 0$. (The 4D determinant must be evaluated by minors.)

Complex Gaussian vectors Most authors use the 2ND real formulation to deal with ND complex random vectors, but there is a significant literature on the complex formulation. The classic text by Doob (1953) discusses the problem, and Wooding (1956) first derived a form like (8.239).

Later authors, however, recognized some surprising features of the complex case (Reed, 1962; Goodman, 1963; Neeser and Massey, 1993). For example, we must revisit the familiar statement that the PDF for a Gaussian random vector is fully determined by its covariance matrix. For a complex random vector, the covariance is defined by $\mathbf{K_u} = \langle(\mathbf{u} - \overline{\mathbf{u}})(\mathbf{u} - \overline{\mathbf{u}})^\dagger\rangle$, but the most general PDF for a Gaussian random vector also involves the *pseudocovariance* $\langle(\mathbf{u} - \overline{\mathbf{u}})(\mathbf{u} - \overline{\mathbf{u}})^t\rangle$, with a transpose in place of the adjoint.

As defined by Neeser and Massey (1993), a complex random vector is said to be *proper* if its pseudocovariance vanishes identically. Any subvector of a proper random vector is proper, but two individually proper random vectors are not necessarily jointly proper. These authors also show that any linear or affine transformation of a proper random vector is another proper random vector, and that a real random vector can be proper if and only if it is a constant.

The condition that the pseudocovariance of a complex vector vanish can be restated in terms of its real and imaginary components. If we write $\mathbf{u} = \mathbf{x} + i\mathbf{y}$, then $\mathbf{u}$ is proper if and only if

$$\langle(\mathbf{x}-\overline{\mathbf{x}})(\mathbf{x}-\overline{\mathbf{x}})^t\rangle = \langle(\mathbf{y}-\overline{\mathbf{y}})(\mathbf{y}-\overline{\mathbf{y}})^t\rangle \quad \text{and} \quad \langle(\mathbf{x}-\overline{\mathbf{x}})(\mathbf{y}-\overline{\mathbf{y}})^t\rangle = -\langle(\mathbf{x}-\overline{\mathbf{x}})(\mathbf{y}-\overline{\mathbf{y}})^t\rangle^t . \tag{8.242}$$

Thus $\mathbf{x}$ and $\mathbf{y}$ must have identical autocovariance matrices, and their cross-covariance matrix must be skew-symmetric.

For optical applications, we are often interested in zero-mean proper Gaussian random vectors and processes, for which the term *circular Gaussian* is commonly used. To be explicit, an ND complex vector $\mathbf{u}$ will be said to obey a circular Gaussian law if all marginals are normal, all components have zero mean and the conditions in (8.242) hold; these conditions can be stated in complex form as

$$\langle u_n u_m\rangle = \langle u_n^* u_m^*\rangle = 0 , \qquad 1 \leq n,\, m \leq N \tag{8.243}$$

and

$$\langle u_n u_m^*\rangle = \langle u_m u_n^*\rangle^* = K_{nm} . \tag{8.244}$$

The intuition behind (8.243) is that u_n can be written as $|u_n| \exp(i\phi_n)$, where ϕ_n is uniformly distributed over $(0, 2\pi)$ but possibly correlated with ϕ_m for $n \neq m$. The expectation $\langle u_n u_m\rangle$ is zero because $\exp[i(\phi_n + \phi_m)]$ takes any value on the unit circle with equal probability. One can think of choosing a ϕ_n from the conditional density $\mathrm{pr}(\phi_n|\phi_m)$ and then choosing ϕ_m from the uniform density; no matter what ϕ_n is chosen in the first step, the second choice means that $\phi_n + \phi_m$ (modulo 2π) is equally likely to be anywhere in $(0, 2\pi)$. On the other hand, $\langle u_n u_m^*\rangle$ depends on $\exp[i(\phi_n - \phi_m)]$, and this average is not zero if ϕ_n and ϕ_m tend to fluctuate together; the second choice tends to undo the first.

The PDF of an ND circular Gaussian random vector is a generalization of (8.239):

$$\mathrm{pr}(\mathbf{u}) = \frac{1}{\pi^N \det(\mathbf{K_u})} \exp\left(-\mathbf{u}^\dagger \mathbf{K}_\mathbf{u}^{-1} \mathbf{u}\right) . \tag{8.245}$$

Thus the only change in going from 2D to ND is the power of π. It is proven in Bellman (1995) that this density is properly normalized, and the reader can check it by considering the basis in which $\mathbf{K_u}$ is diagonal.

The characteristic function for complex random vectors is defined in (8.33); for an ND circular Gaussian it is given by

$$\psi_\mathbf{u}(\boldsymbol{\xi}) = \exp(-\pi^2 \boldsymbol{\xi}^\dagger \mathbf{K_u} \boldsymbol{\xi}) , \tag{8.246}$$

where $\boldsymbol{\xi}$ is an ND complex vector. Note the absence of a factor of 2 in the exponent when compared to the corresponding expression (8.196) for a real Gaussian random vector.

Moments The characteristic function can be used to derive all moments of a random vector. For complex random vectors, the rules for complex differentiation given in Sec. A.9.5 must be used. The reader may use these rules to verify that (8.246) is consistent with the second moments stated in (8.243) and (8.244).

Higher moments are also of interest in many problems. For circular Gaussians, all odd moments vanish, as do all even moments where the number of factors without the complex conjugate is not equal to the number with the conjugate. All other even moments can be expressed in terms of components of the covariance matrix via the *complex Gaussian moment theorem*, first derived by Reed (1962) and discussed by Goodman (1985) in terms of real components and by Osche (2002) in complex form. Osche's statement of the theorem is

$$\langle u_{n_1} u_{n_2} \cdots u_{n_t} u^*_{m_1} u^*_{m_2} \cdots u^*_{m_t} \rangle = \sum_\pi \langle u_{n_1} u^*_{m_{\pi(1)}} \rangle \langle u_{n_2} u^*_{m_{\pi(2)}} \rangle \cdots \langle u_{n_t} u^*_{m_{\pi(t)}} \rangle , \tag{8.247}$$

where $\pi(\,\cdot\,)$ is a permutation of the set of integers $\{1, 2, \cdots, t\}$, and the sum is over all possible permutations. Some useful special cases are:

$$\langle |u_i|^{2n} \rangle = n!\, \langle |u_i|^2 \rangle^n = n!\, \sigma_i^{2n} ; \tag{8.248}$$

$$\langle (u_i u_j^*)^n \rangle = n!\, \langle u_i u_j^* \rangle^n = n!\, K_{ij}^n ; \tag{8.249}$$

$$\langle u_i u_j u_k^* u_\ell^* \rangle = \langle u_i u_k^* \rangle \langle u_j u_\ell^* \rangle + \langle u_j u_k^* \rangle \langle u_i u_\ell^* \rangle = K_{ik} K_{j\ell} + K_{jk} K_{i\ell} . \tag{8.250}$$

This latter equation should be compared to the corresponding real result in (8.197); the complex expression has a sum of two covariances while the real expression has three. We see that $\langle |u_i|^4 \rangle = 2\sigma_i^4$, but for a real, zero-mean, Gaussian random variable, $\langle g_i^4 \rangle = 3\sigma_i^4$. The reader can verify this result by writing $u_i = x_i + iy_i$ and using the real Gaussian moment theorem.

Circular Gaussian random processes A complex random process $u(\mathbf{r})$ will be said to be circular Gaussian if all N-point PDFs are multivariate circular Gaussian random vectors. We can specify this process, as in Sec. 8.3.5, by its characteristic *functional*, given by [*cf.* (8.216)]

$$\Psi_\mathbf{u}(\mathbf{s}) = \exp(-\pi^2 \mathbf{s}^\dagger \boldsymbol{\mathcal{K}}_\mathbf{u} \mathbf{s}) , \tag{8.251}$$

where s is a square-integrable function and $\mathcal{K}_{\mathbf{f}}$ is the autocovariance operator, *i.e.*, the integral operator with kernel $K_{\mathbf{u}}(\mathbf{r}, \mathbf{r}') = \langle u(\mathbf{r})\, u^*(\mathbf{r}') \rangle$. We shall make good use of (8.251) in Chap. 18 when we discuss speckle.

8.4 STOCHASTIC MODELS FOR OBJECTS

We argued in Chap. 7 that an object was best described by a function $f(\mathbf{r})$ (where $\mathbf{r}$ is usually a position vector); now we shall regard this function as a sample function of a random process. The random process is the collection of all possible objects of a given category that might be presented to the imaging system. For example, in computed tomography of the brain, a particular object $f(\mathbf{r})$ is one patient's brain at the time of one imaging procedure, but we can imagine an infinite ensemble of brains from which this one object is drawn. Ideally we would like to specify the full, infinite-dimensional, probability density function (PDF) of the process. As we shall see in Sec. 8.4.1, however, a full PDF is seldom possible, even in principle, and we must make do with less complete models.

The literature on stochastic models in image science is rich and varied, but often the distinction between an object model and an image model is not clear. Many papers claim to address the statistics of images but leave out any consideration of measurement noise or system blur. Moreover, these papers often treat the image as a function of continuous spatial coordinates rather than as a discrete array. Thus they really apply more to objects than to real-world images. On the other hand, if we want to verify our theories by measurements, all we have access to is images, and there is a gap in the current literature on how one can verify stochastic models of objects from observations on noisy, blurred, discrete images.

Another confusing aspect of much of the literature has to do with the meaning of probability. First, there is an unfortunate emphasis on ergodic models where it is assumed, often tacitly, that probabilistic statements can be made for a single object or image. Thus a gray-level histogram of a single image is treated as a probability distribution for pixel values. At best the histogram is an estimate of the probability law for an ensemble of similar images, and then only if ergodicity and hence stationarity are assumed. Except for relatively contrived situations, stationarity is unlikely to hold over the full expanse of an object or image (though local stationarity may be more defensible).

Closely associated with the emphasis on stationarity is the use of loosely defined Fourier measures called *power spectra*. Often this term refers to nothing more than the square modulus of the Fourier transform of a single image. With an assumption of ergodicity this quantity is an estimate of the power spectral density, defined in Sec. 8.2.5 as the Fourier transform of the statistical autocorrelation function. We know from Fig. 8.1, however, that the estimate is poor, and in any case the implicit statistical ensemble is seldom specified, and the underlying stationarity assumption is almost never justified.

Another issue is the conflict between Bayesian and frequentist interpretations of probability, introduced in the Prologue. For many purposes, we want models that emulate reality, in the sense that the model predictions can be verified in principle by measurements on real objects, so we are using a frequentist interpretation of probability. Bayesian interpretations of probability are often useful, however, especially in drawing inferences from images when we have some degree of prior belief about the structure of the object but the frequentist information is incomplete (as it always is). The use of Bayesian priors will be explored further in Chaps. 13 and 15, but the emphasis in this section is descriptive: What can we say about collections of real objects?

In practice, even the very concept of a *real object* must often be expanded. Computer simulations are becoming ever more realistic and ever more essential in image science, and we do not rule out collections of simulations as the ensemble of objects for which we seek a stochastic model.

To state clearly the focus of this section, then, we are considering an ensemble interpretation of probability as applied to objects regarded as sample functions of a random process. The sample function can, in principle, be an actual object $f(\mathbf{r})$, but in practice it may be some approximate representation $f_a(\mathbf{r})$ as introduced in Sec. 7.1.3, and the object can be simulated rather than real.

We begin in Sec. 8.4.1 with a general discussion of just what we mean by the probability density function for an object class and how we might approach the problem experimentally. Included in this section is an introduction to the important concept of independent components.

In Sec. 8.4.2 we revisit the discussion from Sec. 8.2.2 on multipoint densities, but now specifically for objects. Again the focus is on experimental determination of stochastic models.

In Sec. 8.4.3 we do what all statisticians do when problems get difficult: we assume normality. Some implications of the central-limit theorem are discussed, and Gaussian mixture models are introduced. Surprisingly, Gaussian mixture models turn out to account for the highly non-Gaussian character of many filtered images.

In Sec. 8.4.4 we turn to the widely studied but loosely defined topic of texture. For purposes of this section, a texture is regarded as any random field with some degree of stationarity. We discuss here ways of synthesizing sample textures as well as mathematical models for the PDFs.

Sec. 8.4.5 is prelude to the discussion of signal detection in Chap. 13. We make a distinction between signals and backgrounds, and we look at how various assumptions about the signal affect the overall object PDF.

8.4.1 Probability density functions in Hilbert space

To develop a Hilbert-space PDF for objects, we assume that a function $f(\mathbf{r})$ representing a particular object is square-integrable and therefore corresponds to a vector $\mathbf{f}$ in $\mathbb{L}_2(\mathbf{S}_f)$, where $\mathbf{S}_f$ is a support region that will cover all object functions under consideration. Then $\mathbf{f}$ can be expanded as in (8.76):

$$\mathbf{f} = \sum_{n=1}^{\infty} \alpha_n \boldsymbol{\psi}_n , \tag{8.252}$$

where the set $\{\boldsymbol{\psi}_n\}$ is some convenient basis for $\mathbb{L}_2(\mathbf{S}_f)$. The coefficients $\{\alpha_n\}$ are the components of $\mathbf{f}$ in this basis. If the basis is orthonormal, the infinite-dimensional vector of coefficients, denoted $\boldsymbol{\alpha}$, is a unitary transformation of $\mathbf{f}$.

Intuitively, $\mathbf{f}$ corresponds to a single point in the space (or a vector from the origin to the point), and the density $\text{pr}(\mathbf{f})$ is a measure of how these points cluster in the Hilbert space. The density $\text{pr}(\boldsymbol{\alpha})$ describes this same clustering in terms of specific basis vectors $\boldsymbol{\psi}_n$.

A graphical depiction of this clustering is shown in Fig. 8.4. The two axes shown can be construed as any two components $\{\alpha_n, \alpha_m\}$ out of the infinite set.

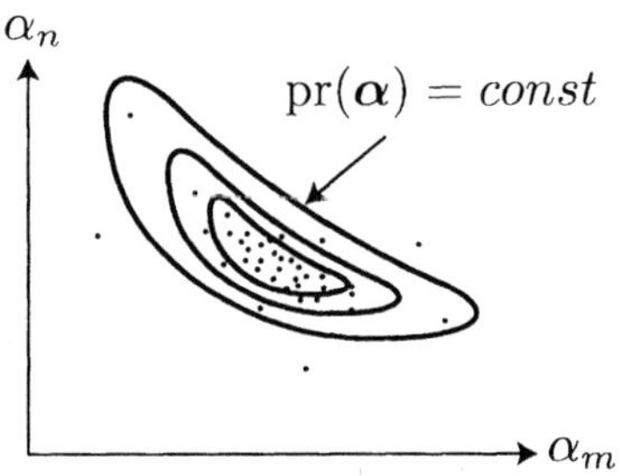

Fig. 8.4 Graphical depiction of the clustering of an object PDF. Two axes out of an infinite-dimensional Hilbert space are shown, and each point corresponds to a different object.

Subspaces We can never hope to know the full PDF in an infinite-dimensional space (and we wouldn't know what to do with it if we had it), but our ultimate goal is always to obtain a PDF $\text{pr}(\mathbf{g})$ for images (see Sec. 8.5). Since the data are insensitive to null functions of the imaging operator $\mathcal{H}$, and all real measurement operators have finite rank R, we can always get by with a finite-dimensional subspace of the object space $\mathbb{U}$. As we know from Sec. 7.4.3, we can use the singular vectors of $\mathcal{H}$ as the expansion functions and truncate the expansion at $n = R$; this truncation produces no error in the data and hence no error in $\text{pr}(\mathbf{g})$.

Another way to restrict the dimensionality is to construct an approximate representation of $\mathbf{f}$, just as we did in Chap. 7, and then consider the PDF of the approximate vector $\mathbf{f}_a$. This procedure can lead to an error in $\text{pr}(\mathbf{g})$, but it will be small if the image error defined in Sec. 7.4.3 is small for all objects in the ensemble. In fact, the image error will be zero if we use natural pixels as the expansion functions (see Sec. 7.4.3).

Experimental determination of the object density We can imagine obtaining information about the object density by examining a large number of typical object functions. There are several ways we could know the object function. For example, we might use a computer program that can simulate sample functions $f(\mathbf{r})$, and for each sample function we could obtain components α_n by computing scalar products with the corresponding basis functions $\psi_n(\mathbf{r})$. (In fact, if a set of components is chosen in advance, the computer program could advantageously generate the sample functions in this basis in the first place.)

Alternatively, we may want to construct a stochastic model useful for one particular imaging system, say a relatively low-resolution, noisy one, but we might have available images from another system with better resolution and less noise. We could then treat the *images* from the better system as good representations of *objects* for the poorer system.

Finally, we might have some physical model, known as a phantom in the medical-imaging literature. If the phantom can be reconfigured into different objects by moving components around in a controllable fashion, it can generate a set of known sample objects.

With any of these sources of sample objects, a histogram estimate of, say, $\text{pr}(\alpha_n, \alpha_m)$ could be obtained by a frequentist interpretation of the PDF. By a multivariate generalization of (C.21), we can write[8]

$$\mathrm{pr}_{(\alpha_n,\alpha_m)}(\alpha_{nk},\alpha_{mk})$$

$$\equiv \lim_{\Delta\alpha\to 0}\frac{1}{(\Delta\alpha)^2}\,\Pr(\alpha_{nk}-\tfrac{1}{2}\Delta\alpha \le \alpha_n < \alpha_{nk}+\tfrac{1}{2}\Delta\alpha,\alpha_{mk}-\tfrac{1}{2}\Delta\alpha \le \alpha_m < \alpha_{mk}+\tfrac{1}{2}\Delta\alpha)\,. \tag{8.253}$$

The histogram estimate is obtained by considering finite bins of width $\Delta\alpha$ (hence omitting the limit) and approximating the probabilities on the right with observed frequencies of occurrence in a finite number of sample objects. Thus we approximate the density as

$$\widehat{\mathrm{pr}}_{(\alpha_n,\alpha_m)}(\alpha_{nk},\alpha_{mk}) \equiv \frac{1}{(\Delta\alpha)^2}\,\frac{J(\alpha_{nk},\alpha_{mk})}{J}\,, \tag{8.254}$$

where $J(\alpha_{nk},\alpha_{mk})$ is the number of times (out of J sample objects) that the computed value of (α_n,α_m) falls in a square of size $(\Delta\alpha)^2$ centered on point $(\alpha_{nk},\alpha_{mk})$. This estimate can, in principle, be extended to an arbitrary number of dimensions.

The problem with this scenario is that the required number of samples may be impractical. As a numerical example, suppose the objects can be adequately specified by 10^4 terms in (8.252), so we are seeking to construct a histogram approximation to a PDF in a ten-thousand-dimensional space. If we choose to use just 10 bins along each axis in the space, then there are $10^{10,000}$ total bins to fill. This is an immense[9] number, and there is no hope of filling the bins with experimental samples. Even with a drastically truncated set of components, $\mathrm{pr}(\boldsymbol{\alpha})$ cannot be interpreted in frequentist terms.

Independent components The number of samples required for a histogram estimate would be much smaller if the components were statistically independent. In that case, for an ND representation, we would have

$$\mathrm{pr}(\boldsymbol{\alpha}) = \prod_{n=1}^{N} \mathrm{pr}(\alpha_n)\,, \tag{8.255}$$

so we would need only a set of N univariate densities instead of an N-dimensional multivariate one.

In contrast to $\mathrm{pr}(\boldsymbol{\alpha})$, the univariate density $\mathrm{pr}(\alpha_n)$ does admit of a frequentist interpretation and a histogram estimate. Suppose, as above, that we have some source of object functions $f(\mathbf{r})$, perhaps a computer simulation code. For each sample function we can evaluate α_n by the usual scalar product, and the histogram estimate of $\mathrm{pr}(\alpha_n)$ is defined by [*cf.* (8.254)]

$$\widehat{\mathrm{pr}}_{\alpha_n}(\alpha_{nk}) = \frac{1}{\Delta\alpha}\,\frac{J_{nk}}{J}\,, \tag{8.256}$$

[8] Recall our notational convention that subscripts on PDFs are deleted where they are redundant with the argument. Thus $\mathrm{pr}(x)$ and $\mathrm{pr}_x(x)$ mean the same thing but the subscript is reinstated on $\mathrm{pr}_x(x_0)$, which means $\mathrm{pr}_x(x)$ evaluated at $x = x_0$.

[9] We use the term *immense* here in its literal sense: incapable of mensuration, immeasurable. Certainly any number exceeding the number of atoms in the universe (of order 10^{80}) qualifies as immense.

where α_{nk} is the specific value of α_n centered on the kth bin, and J_{nk} is the number of times α_n falls in that bin.

The number J_{nk} is a random variable; if the experiment is repeated many times with different sample objects, J_{nk} will be binomially distributed, and the full set of J_{nk} values will be multinomially distributed (see Secs. C.6.1. and 11.2.1). The mean value of J_{nk} will be J times the probability that α_n falls in bin k, or

$$\langle J_{nk} \rangle \approx J \, \mathrm{pr}_{\alpha_n}(\alpha_{nk}) \, \Delta\alpha \, . \tag{8.257}$$

If the number of bins is large, the probability that α_n will fall in one particular bin is small, and any reasonable experiment will use a large value for J, so we are dealing with rare events (see Sec. 11.1.2) where the binomial law on J_{nk} is well approximated by a Poisson.

As a practical example, suppose we want to construct a 100-bin histogram. By the Poisson statistics, a relative error (standard deviation divided by mean) of 10% in the value estimated for the kth bin requires $\langle J_{nk} \rangle = 100$, and a relative error of 1% requires $\langle J_{nk} \rangle = 10^4$. To relate these numbers to the required number of images, we must make some assumptions about the underlying distribution of α_n. If we assume that $\mathrm{pr}(\alpha_n)$ is relatively flat over the range from 0 to α_{max}, then each $\langle J_{nk} \rangle$ is approximately J divided by the number of bins, or $0.01J$ in our example. Thus we require $J = 10^4$ for 10% accuracy and 10^6 for 1% accuracy in a 100-bin histogram. These numbers are large but not immense; they are well within the capabilities of modern computers if the sample objects are simulated. Moreover, each simulated object can be used to evaluate each α_n, so we get the full multivariate PDF for this amount of simulation effort, but only if the components are independent.

Finding the independent components One approach to finding approximately independent components is the Karhunen-Loève (KL) expansion, introduced in Sec. 7.2.4. In Sec. 8.2.7 we showed that the KL expansion yields uncorrelated coefficients, and if we can argue that the process is Gaussian (see Sec. 8.4.3), then uncorrelated implies independent.

To use this argument, we must know the KL expansion. For stationary random processes, as discussed in Sec. 8.2.4, KL expansion is Fourier analysis, but with nonstationary models it can be difficult to determine the autocorrelation function, much less to diagonalize it and find the KL basis. As we shall see in Sec. 8.4.5, some authors argue that wavelet coefficients are approximately uncorrelated for natural scenes, so a wavelet transformation is approximately a KL transformation. Even when this argument can be justified, however, it is still necessary to show that the wavelet coefficients are Gaussian random variables if we want to use (8.255), and we shall present an argument in Sec. 8.4.3 showing why this is *not* the case for a wide class of natural scenes.

When the process is not Gaussian or when we do not know the KL expansion, it may nevertheless be possible to find a transformation that makes the expansion coefficients approximately independent. To make this statement more precise, we need some definition of degree of dependence.

One way to define degree of dependence is in terms of the distance, in some sense, between the multivariate density and the product of its marginals. One distance measure used for this purpose is the *Kullback-Leibler distance*, known also as the *cross-entropy* or *mutual information*. If we consider an $N \times 1$ vector $\boldsymbol{\beta}$ with density $\mathrm{pr}(\boldsymbol{\beta})$, the Kullback-Leibler distance between this density and the product of its marginals is defined by (Comon, 1994)

$$I(\boldsymbol{\beta}) = \int_\infty d^N\beta \ \mathrm{pr}(\boldsymbol{\beta}) \ln \left\{ \frac{\mathrm{pr}(\boldsymbol{\beta})}{\prod_{n=1}^N \mathrm{pr}(\beta_n)} \right\} . \tag{8.258}$$

Note that $I(\boldsymbol{\beta})$ is not a true distance, as defined in Sec. 1.1.2, since it is not symmetric in interchange of $\mathrm{pr}(\boldsymbol{\beta})$ and $\prod_{n=1}^N \mathrm{pr}(\beta_n)$. It does, however, vanish when these two densities are equal, since the argument of the logarithm is unity in that case, and it follows from the convexity of the logarithm that $I(\boldsymbol{\beta}) \geq 0$ (Kendall and Stuart, 1979). Thus independent components can be sought by attempting to find a basis that minimizes $I(\boldsymbol{\beta})$.

Many other measures of degree of dependence are discussed by Comon (1994). In particular, he uses an Edgeworth approximation to argue that independent components will have marginals with large kurtoses, as defined in (C.41). He therefore suggests maximizing the sum of the squares of the marginal kurtoses as a way of finding approximately independent components. We refer the reader to Comon (1994) for a full justification of this approach.

Independent components analysis A structured approach to minimizing some measure of statistical dependence is *independent components analysis* or ICA. ICA is a refinement of *principal components analysis* or PCA, which we shall discuss first.

Though the terms PCA and KL are often used interchangeably in the literature, we make the distinction that PCA is diagonalization of the sample covariance matrix and KL is based on the ensemble covariance. Thus PCA approaches KL analysis as the number of samples goes to infinity.

Suppose we are given J samples of a random vector $\boldsymbol{\alpha}$, denoting the jth sample by $\boldsymbol{\alpha}^{(j)}$. The sample covariance matrix $\widehat{\mathbf{K}}_{\boldsymbol{\alpha}}$ is defined by

$$\widehat{\mathbf{K}}_{\boldsymbol{\alpha}} = \frac{1}{J} \sum_{j=1}^{J} \left[\Delta\boldsymbol{\alpha}^{(j)}\right] \left[\Delta\boldsymbol{\alpha}^{(j)}\right]^\dagger , \tag{8.259}$$

where $\Delta\boldsymbol{\alpha}^{(j)}$ is $\boldsymbol{\alpha}^{(j)}$ minus the sample mean. PCA seeks to find a matrix $\mathbf{M}$ such that the transformed sample vectors,

$$\boldsymbol{\beta}^{(j)} = \mathbf{M}\boldsymbol{\alpha}^{(j)} , \tag{8.260}$$

are uncorrelated and hence the transformed sample covariance matrix $\widehat{\mathbf{K}}_{\boldsymbol{\beta}}$ is diagonal. By retracing the discussion in Sec. 8.1.6 but with $\widehat{\mathbf{K}}$ in place of $\mathbf{K}$, we can see that this diagonalization is accomplished by using the eigenvectors of $\widehat{\mathbf{K}}_{\boldsymbol{\alpha}}$ as the columns of $\mathbf{M}$.

ICA also uses a transformation of the form (8.260), but now the goal is to minimize some measure of statistical dependence as discussed above or in much more detail in Comon (1994) and subsequent literature. Since statistically independent components are necessarily uncorrelated, ICA usually proceeds by first computing the PCA, so that the spectral decomposition of $\widehat{\mathbf{K}}_{\boldsymbol{\alpha}}$ is known, and then applying a prewhitening transformation as in (8.67). At this point we have a set of sample vectors such that the sample covariance matrix is the unit matrix, and all further unitary transformations preserve this property. We thus decompose the matrix $\mathbf{M}$ as

$$\mathbf{M} = \mathbf{U}\widehat{\mathbf{K}}_{\boldsymbol{\alpha}}^{-\frac{1}{2}}, \tag{8.261}$$

where $\mathbf{U}$ is unitary. ICA amounts to choosing $\mathbf{U}$ so as to minimize the chosen measure of statistical dependence.

When ICA is carried out on training sets of natural scenes, the results are quite striking (see Bell and Sejnowski, 1997; Field, 1987; Olshausen and Field, 1996). The columns of $\mathbf{M}$ turn out to be localized, bandpass functions similar to wavelets or to the channels in the human visual system (a topic to be treated in more detail in Chap. 14), suggesting that humans may have evolved in such a way as to process natural scenes through statistically independent channels (see also Barlow, 1989).

One practical implication of the observation that the independent components are localized is that we can treat small pieces of the same object (or image) as independent samples. Bell and Sejnowski (1997), for example, consider 12×12 segments of an image as the samples on which they perform ICA. The resulting ICA filters are smaller than 12 pixels, even though the corresponding PCA filters tend to fill the 12×12 region. The authors note, however, that the restriction to such a small region may be an unrealistic feature of their approach. In addition, pixels themselves are unrealistic if we wish to draw conclusions about "natural scenes."

We shall revisit ICA in the context of texture analysis in Sec. 8.4.4. In that application, ICA is considerably simplified because textures are at least approximately stationary.

8.4.2 Multipoint densities

As we saw in Sec. 8.2.2, another kind of PDF for a random process is a collection of P-point densities of the form $\text{pr}[f(\mathbf{r}_1), f(\mathbf{r}_2), ..., f(\mathbf{r}_P)]$. In principle one needs densities like this for all P to completely characterize the process, but often we must be content with $P = 1$ and 2.

In a sense, multipoint densities are just special cases of the Hilbert-space densities discussed above. If we use delta functions as basis functions for the space (see Sec. 2.2.6), then $f(\mathbf{r}_p)$ is the coefficient α_p associated with basis function $\delta(\mathbf{r} - \mathbf{r}_p)$, and $\text{pr}[f(\mathbf{r}_1), f(\mathbf{r}_2), ..., f(\mathbf{r}_P)]$ is a P-dimensional marginal of a Hilbert-space density. This marginal is, however, a function of P spatial variables, so it is a richer description of the statistics of the random process than $\text{pr}(\alpha_1, \alpha_2, ..., \alpha_P)$ would be with preselected basis functions.

If we have a means of computing $\text{pr}[f(\mathbf{r}_1), f(\mathbf{r}_2), ..., f(\mathbf{r}_P)]$, we can in principle do it for all values of each of the spatial arguments, but a less ambitious goal is to sample the function on a regular spatial grid, making it a discrete random process. If $\mathbf{r}$ is a qD vector and we sample each component to L values, then $\mathbf{f}$ is specified by $N = L^q$ numbers, and the full density is defined in an ND space. In this sampled case, therefore, all of the P-fold multipoint densities can be computed from the ND density on $\mathbf{f}$. Nevertheless, it may be computationally or conceptually simpler to compute the multipoint densities directly rather than as marginals of a high-dimensional multivariate density.

Pointwise evaluation of random functions Before analyzing multipoint densities in more detail, we have to deal with one mathematical subtlety. So far we have assumed only that each sample function $f(\mathbf{r})$ is in an $\mathbb{L}_2$ space, but we noted in Sec. 1.8 that not all functions in $\mathbb{L}_2$ are defined pointwise. If we want an expression like $f(\mathbf{r}_1)$ to be rigorously defined, we must assume that $f(\mathbf{r})$ lies in a reproducing-kernel Hilbert space (RKHS), which might be a subspace of $\mathbb{L}_2$. For imaging purposes, this restriction entails no loss of generality; we saw in Chap. 7 that the imaging operator $\mathcal{H}^\dagger\mathcal{H}$ is a nonnegative-definite Hermitian operator, and we know from Sec. 1.8.2 that such an operator can be used to define an RKHS. Assuming that $f(\mathbf{r})$ lies in that particular RKHS is equivalent to saying that we are discussing the statistics of the measurement component of the object, and that component is necessarily in an RKHS and hence defined pointwise.

The same conclusion can be reached by assuming that we are not interested in the statistics of an actual $f(\mathbf{r})$ but rather those of some linear approximation to it, such as the functions $f_a(\mathbf{r})$ or $f_t(\mathbf{r})$ discussed in Sec. 7.1.3. As we saw there, these functions lie in an RKHS called representation space, so they too can be defined pointwise. For example, we might construct a linear approximation by use of pixel functions, so $f_a(\mathbf{r}_1)$ would refer to the gray level[10] of a pixel centered at $\mathbf{r} = \mathbf{r}_1$.

In what follows we shall use the notation $f(\mathbf{r})$ but always with the implicit assumption that the function corresponds to a vector in an RKHS. Thus we might really mean $f_{meas}(\mathbf{r})$ or $f_a(\mathbf{r})$, but we shall omit subscripts for convenience. As a practical matter, essentially the only thing we rule out with this assumption is that $f(\mathbf{r})$ is white noise or some other generalized, infinite-energy random process.

Single-point PDFs For $P = 1$ and a fixed choice of $\mathbf{r}$, $\text{pr}[f(\mathbf{r})]$ is a univariate PDF for the gray level $f(\mathbf{r})$ at point $\mathbf{r}$. This density can be represented as an ordinary 1D function as in Fig. 8.5. Of course, this function may depend in general on the choice of evaluation point $\mathbf{r}$, so it can also be plotted as a function of the Cartesian coordinates of $\mathbf{r}$, as shown in Fig. 8.6.

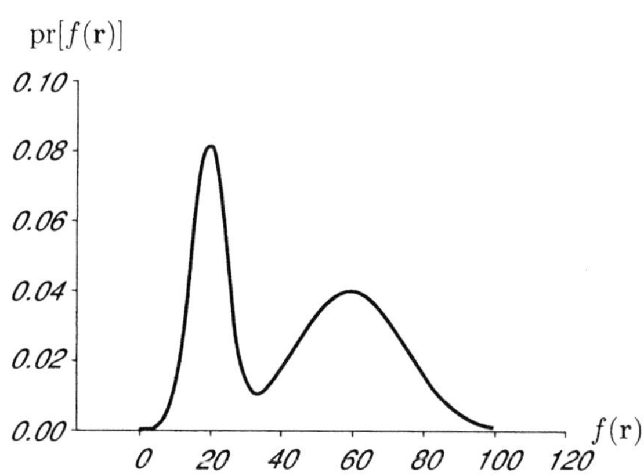

Fig. 8.5 Univariate PDF $\text{pr}[f(\mathbf{r})]$ plotted as a function of $f(\mathbf{r})$ for fixed $\mathbf{r}$.

[10]Even though we are talking about pixels and gray levels here — terms often associated with images — we emphasize that we are discussing *object* models.

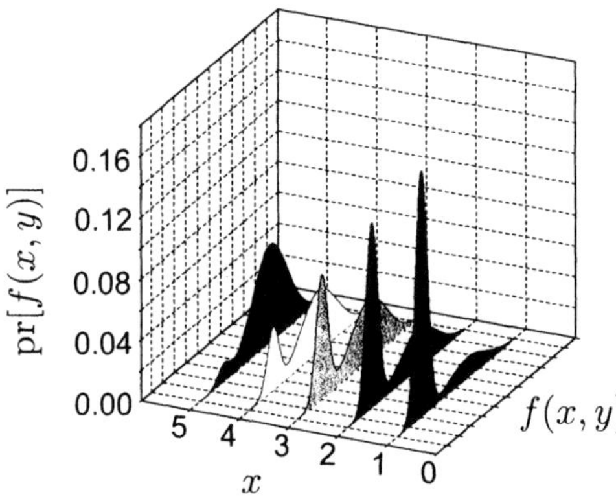

Fig. 8.6 Same PDF as in Fig. 8.5 but now plotted as a function of both $f(\mathbf{r})$ and $\mathbf{r}$.

Since it is univariate, $\mathrm{pr}[f(\mathbf{r})]$ admits of a frequentist interpretation and a histogram estimate. The considerations are essentially the same as for the univariate density $\mathrm{pr}(\alpha_n)$; if we have a source of object functions $f(\mathbf{r})$, such as a computer simulation code, we can evaluate each sample function at any chosen point, say $\mathbf{r} = \mathbf{r}_1$, and define a histogram estimate analogous to (8.256):

$$\widehat{\mathrm{pr}}_{f(\mathbf{r})}[f_k(\mathbf{r}_1)] = \frac{1}{\Delta f} \frac{J\left[f_k(\mathbf{r}_1) - \frac{1}{2}\Delta f \le f(\mathbf{r}_1) < f_k(\mathbf{r}_1) + \frac{1}{2}\Delta f\right]}{J}, \tag{8.262}$$

where the numerator is the number of sample objects for which the value $f(\mathbf{r}_1)$ falls in an interval of width Δf centered on the chosen value $f_k(\mathbf{r}_1)$, and J is the total number of samples. The number of bins in this histogram is just $f_{max}/\Delta f$, where f_{max} is the maximum value of $f(\mathbf{r})$. The kth bin is centered on the point $f_k(\mathbf{r}_1)$ if

$$k = \frac{f_k(\mathbf{r}_1)}{\Delta f}. \tag{8.263}$$

For notational simplicity, we denote the numerator in (8.262) as J_k, which is just the observed number of samples in bin k, but we must keep in mind that the histogram is specific to the point $\mathbf{r}_1$.

The same statistical considerations apply here as in the last section. If the experiment is repeated many times with different sample objects, J_k will be approximately a Poisson random variable. The mean value of J_k will be J times the probability that the gray level will fall in bin k, or

$$\langle J_k \rangle = J\,\mathrm{Pr}\left[f_k(\mathbf{r}_1) - \tfrac{1}{2}\Delta f \le f(\mathbf{r}_1) < f_k(\mathbf{r}_1) + \tfrac{1}{2}\Delta f\right] \approx \mathrm{pr}_{f(\mathbf{r})}[f_k(\mathbf{r}_1)]\,\Delta f\,. \tag{8.264}$$

As in the previous section, we can construct a 100-bin histogram with a relative error of 10% in the value estimated for the kth bin if $\langle J_k \rangle = 100$; a relative error of 1% requires $\langle J_k \rangle = 10^4$. If we assume that $\mathrm{pr}[f(\mathbf{r})]$ is relatively flat over the range from 0 to f_{max}, then we require $J = 10^4$ for 10% accuracy and 10^6 for 1% accuracy in a 100-bin. Again, these numbers are within the capabilities of modern computers if the sample objects are simulated.

One might think that we are far from characterizing the object random process even to order $P = 1$ since we have fixed the evaluation point at $\mathbf{r} = \mathbf{r}_1$ in the discussion above. In fact, however, once we have a source of sample objects $f(\mathbf{r})$, we can evaluate them at as many points as we please, and we can construct histogram estimates of $\mathrm{pr}[f(\mathbf{r})]$ on a grid of spatial points with very little increased effort. A 100×100 grid for a 2D object, for example, requires that we construct 10,000 histograms. If k ranges from 1 to 100 for each sample $\mathbf{r}$ and the observed value of J_k does not exceed 255, then we can store the results in just 1 Megabyte of memory.

As a semantic point, each of the histograms discussed above is a histogram of gray levels; it is not, however, what is usually called a gray-level histogram in the image-processing community. In that community, it is common to compute a histogram of the gray levels at *all points within a single image* for purposes of display manipulation or data compression. The histograms we are discussing here describe the distribution of gray levels *at a single point in an ensemble of images.* Where confusion may result, we shall distinguish between *single-image* histograms and *single-point* or ensemble histograms.

For stationary, ergodic random processes, the single-image histogram can be used in place of the ensemble histogram as an estimator of the single-point PDF, but these two histograms should not be equated in general. The single-image histogram can give a very biased estimate of the PDF if there is even a slight deviation from stationarity across the image. Consider, for example, the common situation where the mean gray level varies slowly across the image; in that case the single-image histogram can be much broader than the ensemble histogram at a fixed point and hence a fixed mean gray level.

Two-point PDFs For fixed $\mathbf{r}_1$ and $\mathbf{r}_2$, the two-point density $\mathrm{pr}[f(\mathbf{r}_1), f(\mathbf{r}_2)]$ is a bivariate density on the two scalar random variables $f(\mathbf{r}_1)$ and $f(\mathbf{r}_2)$. This density can be represented by a 2D plot, where the axes are $f(\mathbf{r}_1)$ and $f(\mathbf{r}_2)$. A full characterization to order $P = 2$ requires evaluation of such bivariate densities for all $\mathbf{r}_1$ and $\mathbf{r}_2$ in $\mathbf{S}_f$.

The two-point density can also be interpreted in frequentist terms, though more sample objects are required than in the single-point case. If we again choose $f_{max}/\Delta f = 100$, then there are 10,000 bins in a histogram representing $\mathrm{pr}[f(\mathbf{r}_1), f(\mathbf{r}_2)]$. A calculation similar to the one above shows that J must be about 10^6 for 10% accuracy and 10^8 for 1% accuracy if the underlying PDF is relatively flat. Moreover, 10^8 such histograms would be required if $\mathbf{r}$ is 2D and both $\mathbf{r}_1$ and $\mathbf{r}_2$ are sampled on 100×100 spatial grids, and 10 GB of storage would be needed to hold the results. In short, full experimental characterization of the random process to order P becomes rapidly more difficult as P increases.

The histogram approximation to the bivariate density $\mathrm{pr}[f(\mathbf{r}_1), f(\mathbf{r}_2)]$ is related to, but not identical to, the *co-occurrence matrix* used in image processing and pattern recognition (Pratt, 1991). The distinction is the same as the one between single-point and single-image histograms. The co-occurrence matrix is a random matrix characteristic of a single image or a smaller region within a single image. It is a histogram of the joint occurrence of binned or quantized gray levels in that image. It is independent of absolute position within the region or image but it does depend on the relative position $\mathbf{r}_2 - \mathbf{r}_1$. The density $\mathrm{pr}[f(\mathbf{r}_1), f(\mathbf{r}_2)]$, on the other hand, is a nonrandom characteristic of the ensemble and a function of two position vectors. A histogram approximation to $\mathrm{pr}[f(\mathbf{r}_1), f(\mathbf{r}_2)]$ is also random since it is formed from a finite number of samples, but this randomness can in principle be reduced arbitrarily by letting the number of samples grow.

If each sample function is drawn from an ergodic random process (see Sec. 8.2.4), then the co-occurrence matrix computed from one sample function is also an estimator of $\mathrm{pr}[f(\mathbf{r}_1), f(\mathbf{r}_2)]$.

Local models If $\mathbf{r}_1$ and $\mathbf{r}_2$ are far apart in an object, $f(\mathbf{r}_1)$ and $f(\mathbf{r}_2)$ might be statistically independent, or nearly so. For example, in a computed-tomography scan of the chest, the gray level at a point in the lungs would be expected to be independent of the gray level at a point in the spine. Two nearby points in the same lung would, however, be expected to be dependent. A stochastic model that takes account of this property is called a *local model.*

To see the structure of a local model, let us first consider two well-separated points. If the gray levels at these two points are statistically independent, the two-point PDFs are determined uniquely from the single-point ones:

$$\mathrm{pr}[f(\mathbf{r}_1), f(\mathbf{r}_2)] = \mathrm{pr}[f(\mathbf{r}_1)]\,\mathrm{pr}[f(\mathbf{r}_2)]\,. \tag{8.265}$$

As discussed in Sec. C.1.6, the independence condition in (8.265) can also be written as

$$\mathrm{pr}[f(\mathbf{r}_1)|f(\mathbf{r}_2)] = \mathrm{pr}[f(\mathbf{r}_1)]\,. \tag{8.266}$$

Now consider a countable set of points, say on a regular lattice in object space. The gray level at some particular point $\mathbf{r}_k$ will often depend on the values at other points $\mathbf{r}_i$ provided they are close to the chosen point $\mathbf{r}_k$, but it could be statistically independent of the values at more distant points. We define the *neighborhood* $\mathcal{N}_k$ of the point $\mathbf{r}_k$ as the set of points close to $\mathbf{r}_k$ in this sense, and we denote the complete set of points in the object support as $\mathcal{S}$. Then a local statistical model is one for which [*cf.* (8.266)]

$$\mathrm{pr}[f(\mathbf{r}_k)|\{f(\mathbf{r}_i), \mathbf{r}_i \in \mathcal{S}, i \neq k\}] = \mathrm{pr}[f(\mathbf{r}_k)|\{f(\mathbf{r}_i), \mathbf{r}_i \in \mathcal{N}_k\}]\,, \tag{8.267}$$

where $\mathbf{r}_i \in \mathcal{N}_k$ is read "point $\mathbf{r}_i$ is an element of the set $\mathcal{N}_k$," or somewhat more colloquially, "$\mathbf{r}_i$ is a neighbor of $\mathbf{r}_k$." As we see from (8.267), the form of the marginal density on $f(\mathbf{r}_k)$ in a local model is determined fully by the values in the neighborhood $\mathcal{N}_k$, and points outside this neighborhood can be neglected for purposes of describing the statistics at $\mathbf{r}_k$.

A local model defined on a discrete lattice as in (8.267) is called a *Markov random field* or MRF. Developed by Besag (1973) and Cross and Jain (1983) for describing textures, MRFs have received considerable attention as Bayesian priors in image reconstruction (see Sec. 15.3.3), but relatively little effort has been expended on establishing their validity as empirical distributions in a frequentist sense. One exception is Herman and Chan (1995), who discussed so-called *image-modeling MRFs* where a sample drawn from the MRF density would have the same neighborhood statistics as the image (object) being modeled.

Regional models and mixture models Often objects can be divided into distinct regions with different statistical properties. In a chest radiograph, for example, the lungs are in more or less the same place for all patients, and the heart is generally situated below the left lung. Before seeing a particular patient's radiograph, we can define a region that is likely to contain lung and another that is likely to contain heart. Of course, this definition is not absolute; a collapsed lung or an enlarged heart, or simply normal variations in patient size and positioning, could mean that the *a priori* region assignment is incorrect. Various strategies are available for refining the region assignments, including image recentering and warping and various segmentation algorithms. None of these methods is perfect, however, and the best we can do is to assess the probability that a particular point is associated with a given region.

If we denote by $\mathcal{S}_i$ the set of points associated with region i, the univariate PDF on the gray level at point $\mathbf{r}$ is given by

$$\mathrm{pr}[f(\mathbf{r})] = \sum_i \mathrm{pr}[f(\mathbf{r})|\mathbf{r} \in \mathcal{S}_i] \Pr(\mathbf{r} \in \mathcal{S}_i) . \tag{8.268}$$

An analogous expression can be given for the two-point PDF:

$$\mathrm{pr}[f(\mathbf{r}_1), f(\mathbf{r}_2)] = \sum_i \sum_k \mathrm{pr}[f(\mathbf{r}_1), f(\mathbf{r}_2)|\mathbf{r}_1 \in \mathcal{S}_i, \mathbf{r}_2 \in \mathcal{S}_k] \Pr(\mathbf{r}_1 \in \mathcal{S}_i, \mathbf{r}_2 \in \mathcal{S}_k) . \tag{8.269}$$

If gray levels in different regions are statistically independent, this equation becomes

$$\begin{aligned} &\mathrm{pr}[f(\mathbf{r}_1), f(\mathbf{r}_2)] \\ &= \sum_i \sum_k (1 - \delta_{ik}) \, \mathrm{pr}[f(\mathbf{r}_1)|\mathbf{r}_1 \in \mathcal{S}_i] \, \mathrm{pr}[f(\mathbf{r}_2)|\mathbf{r}_2 \in \mathcal{S}_k] \Pr(\mathbf{r}_1 \in \mathcal{S}_i, \mathbf{r}_2 \in \mathcal{S}_k) \\ &\quad + \sum_i \mathrm{pr}[f(\mathbf{r}_1), f(\mathbf{r}_2)|\mathbf{r}_1 \in \mathcal{S}_i, \mathbf{r}_2 \in \mathcal{S}_i] \Pr(\mathbf{r}_1 \in \mathcal{S}_i, \mathbf{r}_2 \in \mathcal{S}_i) . \end{aligned} \tag{8.270}$$

Another special case is a *piecewise-constant model* where all points within a given region have the same gray level in each sample function of the random process, though that value (as well as the borders of the region) can vary randomly from one sample function to the next. In that case,

$$\mathrm{pr}[f(\mathbf{r}_1), f(\mathbf{r}_2)|\mathbf{r}_1 \in \mathcal{S}_i, \mathbf{r}_2 \in \mathcal{S}_i] = \delta[f(\mathbf{r}_1) - f(\mathbf{r}_2)] \, \mathrm{pr}[f(\mathbf{r}_1)|\mathbf{r}_1 \in \mathcal{S}_i] . \tag{8.271}$$

The density in (8.268) is an example of a *mixture model* where the random quantity is divided into classes, and the overall density is a weighted sum of the densities for different classes. In (8.268), a class is identified with a spatial region, but other kinds of classes are important in imaging as well. In medical imaging, for example, different disease states are (we hope) described by different PDFs. Similarly, in aerial photography, crops, cities, oceans and forests would require different statistical models.

In such cases, the general form of the object PDF is

$$\mathrm{pr}(\mathbf{f}) = \sum_i \mathrm{pr}[\,\mathbf{f}|\text{class } i\,] \Pr(\text{class } i) . \tag{8.272}$$

The key difference between (8.268) and (8.272) is that the former applies to a univariate density at a specific point $\mathbf{r}$, while the latter is a general statement applying to the entire density of the process.

Specifically, if we represent $\mathbf{f}$ by an $N \times 1$ coefficient vector $\boldsymbol{\alpha}$, the mixture density (8.272) takes the form

$$\mathrm{pr}(\boldsymbol{\alpha}) = \sum_i \mathrm{pr}(\boldsymbol{\alpha}|\text{class } i) \Pr(\text{class } i) , \tag{8.273}$$

and the marginal on a single component of $\boldsymbol{\alpha}$ is

$$\mathrm{pr}(\alpha_n) = \sum_i \mathrm{pr}(\alpha_n|\text{class } i)\,\mathrm{Pr}(\text{class } i)\,. \tag{8.274}$$

8.4.3 Normal models

The basic properties of normal random processes and random vectors were introduced in Sec. 8.3. Here we revisit normal models with the goal of understanding when and how they apply specifically to the statistical description of objects.

When it is possible to use normal models in imaging, a considerable mathematical simplification results. As we saw in Sec. 8.3, the PDF for a normal random vector is fully determined by the mean vector and the covariance matrix. Moreover, any linear transformation of a normal random vector leaves it normal, so a full analysis of the effect of a linear operator requires only that we transform the mean and covariance, using simple formulas developed in Sec. 8.1.5.

These properties of normal random vectors extend readily to normal random processes. The full PDF of any random process is infinite-dimensional, but in the normal case we can take advantage of the fact that any marginal or conditional density derived from a normal PDF, even an infinite-dimensional one, is also normal. Thus if we choose to describe a normal random process by Hilbert-space marginal densities of the form $\mathrm{pr}(\alpha_1, \alpha_2, ..., \alpha_P)$ or by multipoint densities like $\mathrm{pr}[f(\mathbf{r}_1), f(\mathbf{r}_2), ..., f(\mathbf{r}_P)]$, we can be assured that these densities will all be normal.

Central limits To establish the validity of a normal model, we must usually argue that the central-limit theorem applies, as it does when independent random variables or vectors are added together. One way this can happen is when a pixel or voxel representation is used for the object, and subregions of the pixel or voxel are statistically independent. As an example, consider an airborne optical camera viewing a meadow. The camera does not resolve individual blades of grass, and an adequate 2D object representation can use a pixel that covers many blades. It is reasonable to argue that the blades reflect light independently, so the total reflected light in one pixel tends to a normal distribution, at least when we consider only meadows and do not include, say, forests or beaches.

A somewhat more subtle example is nuclear medicine imaging of perfusion patterns in the lungs. In this technique, radioactive albumin particles are injected into a vein and get trapped in the alveoli (the functional units of the lungs where blood becomes oxygenated). The distribution of the trapped tracer is indicative of the perfusion of the lung, and it is this distribution that we regard as the object. Since nuclear medicine systems have very poor spatial resolution compared to the size of alveoli, we can choose a voxel size that contains many alveoli, and the voxel value is the sum of the activities in many alveoli. It is reasonable to presume that these activities are statistically independent, at least when one particular patient is considered. If we were to consider an ensemble of patients, some would have higher perfusion in a particular region than others, and all alveoli in this region would tend to fluctuate together; we avoid this kind of dependence by conditioning the PDF on a particular patient and hence a particular perfusion pattern. In a frequentist sense, this conditional PDF describes the hypothetical distribution that would result from making many different injections of albumin particles into a single patient.

Gaussian mixture models In the two examples just given to justify use of the central-limit theorem, we had to be careful to restrict the ensemble of objects under consideration. In the aerial photography example, we had to consider only meadows and not forests or beaches, and in the nuclear medicine example we had to consider repeated injections into one patient rather than a more realistic ensemble of patients.

To analyze a broader ensemble, we do not necessarily have to abandon the central-limit theorem; instead, we can divide the different objects (or different regions of the same object) into classes and use a mixture density as in (8.272). If we can argue that a normal PDF applies to each component of the mixture, then the resulting model is called a *Gaussian mixture model.*

If $\boldsymbol{\alpha}$ is conditionally multivariate normal for each class, then α_n is conditionally univariate normal, so $\mathrm{pr}(\alpha_n|\text{class } i)$ in (8.274) is fully specified by the conditional mean $\overline{\alpha}_{ni}$ and the conditional variance σ_{ni}^2:

$$\mathrm{pr}(\alpha_n) = \sum_i \frac{1}{\sqrt{2\pi\sigma_{ni}^2}} \exp\left[-\frac{(\alpha_n - \overline{\alpha}_{ni})^2}{2\sigma_{ni}^2}\right] \mathrm{Pr}(\text{class } i)\,. \tag{8.275}$$

If we must use a large number of classes in order to justify the normal law for each class, it might be better to consider a continuum of classes and write

$$\mathrm{pr}(\alpha_n) = \int_{-\infty}^{\infty} d\overline{\alpha}_n \int_0^{\infty} d\sigma_n^2 \; \mathrm{pr}(\overline{\alpha}_n, \sigma_n^2) \frac{1}{\sqrt{2\pi\sigma_n^2}} \exp\left[-\frac{(\alpha_n - \overline{\alpha}_n)^2}{2\sigma_n^2}\right]. \tag{8.276}$$

Similarly, the multivariate density on $\boldsymbol{\alpha}$ for a discrete set of classes is

$$\mathrm{pr}(\boldsymbol{\alpha}) = \sum_i \frac{1}{\sqrt{(2\pi)^N \det(\mathbf{K}_i)}} \exp\left[-\tfrac{1}{2}(\boldsymbol{\alpha} - \overline{\boldsymbol{\alpha}}_i)^t \mathbf{K}_i^{-1} (\boldsymbol{\alpha} - \overline{\boldsymbol{\alpha}}_i)\right] \mathrm{Pr}(\text{class } i)\,, \tag{8.277}$$

where $\overline{\boldsymbol{\alpha}}_i$ and $\mathbf{K}_i$ are, respectively, the mean vector and covariance matrix for $\boldsymbol{\alpha}$ under class i. For a continuum of classes, we can write

$$\mathrm{pr}(\boldsymbol{\alpha}) = \int_\infty d^N\overline{\alpha} \int_\infty d\mathbf{K} \; \frac{1}{\sqrt{(2\pi)^N \det(\mathbf{K}_i)}} \exp\left[-\tfrac{1}{2}(\boldsymbol{\alpha} - \overline{\boldsymbol{\alpha}})^t \mathbf{K}^{-1} (\boldsymbol{\alpha} - \overline{\boldsymbol{\alpha}})\right], \tag{8.278}$$

where $d\mathbf{K}$ is a shorthand for the differential of all components of $\mathbf{K}$.

No matter which of these mixture formulas we use, we do not expect the resulting PDF to be normal. For example, in the simple case of the univariate expression (8.275) with just two classes, we would get a bimodal PDF with one Gaussian peak for each class.

High-pass and band-pass filters There are many circumstances where we either pass an image through a high-pass or band-pass spatial filter or consider an object to consist of a superposition of such components. For example, edges in an image are often detected with some sort of derivative filter, and derivatives suppress the DC component[11] of the image. Other examples of filters with zero DC response include

[11]The common jargon, *DC component*, does not, of course refer to direct current Instead it implies zero spatial frequency, by analogy to the zero temporal frequency of a steady current. In coherent optical processing, the Fourier transform of an object is displayed as an optical amplitude distribution centered on the optical axis of a lens system, and in that case it has been suggested that *DC* stands for *dot in the center*.

wavelets (see Sec. 5.3), channels in the human visual system (see Sec. 14.2) and filters used to extract discrete cosine transforms (except, of course, the DC term in the transform). Continuous objects can be represented by zero-DC components, for example in the Fourier-series basis of (7.13), a wavelet basis or a basis of Gabor functions (see Sec. 5.1.4). As we noted in Sec. 8.4.1, approximately independent components can be obtained by filtering with localized band-pass filters.

In all of these cases, an expansion coefficient is computed by forming a scalar product of the object function with a zero-DC function. For both objects and images, therefore, it is of considerable interest to have a stochastic model for the output of a high-pass filter.

In Sec. 8.3.3 we showed that linear filtering of a Gaussian random process yields a Gaussian random process, so if the input to a filter is Gaussian, the output must be also. It has been observed empirically, however, many images have a decidedly non-Gaussian distribution results after high-pass or band-pass filtering. As seen in the example in Fig. 8.7, the gray-level histograms are typically sharply peaked around zero and display long tails (Heine *et al.*, 1999; Bell and Sejnowski, 1997). In statistical lingo, these histograms have a large kurtosis. As defined in (C.41), the kurtosis for a Gaussian is 3 (though many books subtract off the 3 and make the kurtosis of a Gaussian 0), and gray-level histograms following high-pass filtering often have kurtosis substantially larger than 3. Statistical pedants refer to such distributions as *leptokurtic* (Greek *lepto*, thin or fine); the opposite condition, kurtosis less than that of a Gaussian, is referred to as *platykurtic* (Greek *platys*, broad or flat — behold the platypus!).

Fig. 8.7 *Top*: A typical image before and after high-pass filtering. *Bottom*: Gray-level histogram of the high-pass filtered image (note that the right plot is vs. log frequency of occurrence).

Filtering of Gaussian mixtures Heine *et al.* (1999) offered an explanation for high kurtosis after wavelet filtering, but it made assumptions about scale-invariance that were specific to wavelets. Lam and Goodman (2000) derived the PDF of the coefficients in a discrete cosine transform from a Gaussian mixture model. Clarkson and Barrett (2001) extended that argument and showed that kurtotic distributions were an inevitable consequence of high-pass or band-pass filtering of Gaussian mixtures; we shall sketch here the derivation given by Clarkson and Barrett.

If we think of high-pass filtering as a convolution, then the output is a scalar product of the shifted kernel function with the input. If the kernel contains both positive and negative components, we can suppress the shift variable and write the output for one position of the kernel as

$$z = u - v\,, \tag{8.279}$$

where u arises from the positive part of the filter and v from the negative part. This equation applies whether we think of the input to the filter as a random process or a random vector in a pixel representation. Moreover, it applies also to computation of an expansion coefficient in a representation where the expansion function has positive and negative parts.

We expect u and v to be highly correlated since they come from the same region of the input, but it is reasonable to assume that they have the same mean if the filter has zero DC response. This conclusion follows rigorously if we can assume that all points within the region spanned by the kernel (at a specific shift) have the same mean, and it may also be a good approximation even with a space-variant mean since it requires only that the spatial average of the mean over the positive regions of the kernel equal that over the negative regions. (Consider a difference-of-Gaussians filter, where a positive central peak is surrounded by a negative ring; the means of u and v will be equal if the spatial average of the input mean in the negative ring is the same as the spatial average in the central peak.) Thus we assume

$$\overline{u} = \overline{v}\,; \qquad \overline{z} = 0\,. \tag{8.280}$$

Note that the overbar here implies an ensemble mean; it has nothing to do with spatial averages. We make no assumptions about stationarity or ergodicity, and there is no implication that ensemble averages can be approximated by spatial ones.

Now let us assume that u and v are drawn from a Gaussian mixture. To see the essential results, we assume first that u and v are *conditionally* uncorrelated, for any one component of the mixture, so that the entire correlation between the two variables results from averaging over components in the mixture. Similarly, we assume that u and v have the same conditional variances, so in fact they are conditionally i.i.d. These assumptions may not always be justified, and they will be relaxed below; for now, we write

$$\mathrm{pr}(u, v|\sigma, m) = \frac{1}{2\pi\sigma^2} \exp\left[-\frac{(u-m)^2 + (v-m)^2}{2\sigma^2}\right], \tag{8.281}$$

where m is the common mean of u and v, and σ is the common standard deviation.

The corresponding conditional density on z is given by

$$\mathrm{pr}(z|\sigma, m) = \int_{-\infty}^{\infty} du\ \mathrm{pr}(u, u - z|\sigma, m)\,. \tag{8.282}$$

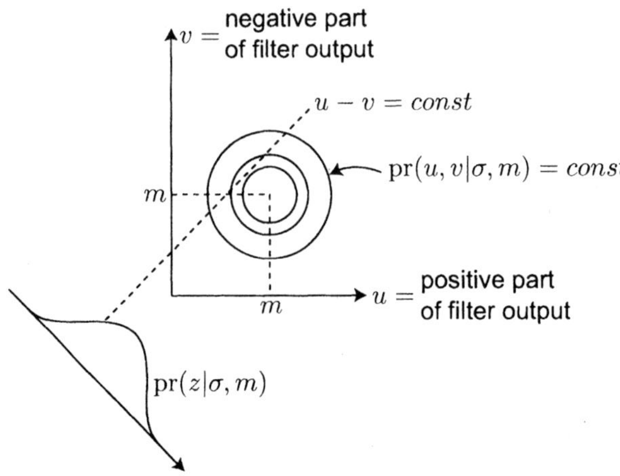

Fig. 8.8 Illustration of the integral encountered in analyzing Gaussian mixture models.

As illustrated in Fig. 8.8, this integral can be interpreted as a 1D projection or Radon transform (see Sec. 4.4) of the 2D function $\mathrm{pr}(u, v|\sigma, m)$. We see graphically that the result is independent of m, and by completing the square we obtain

$$\mathrm{pr}(z|\sigma) = \frac{1}{2\pi\sigma^2} \int_{-\infty}^{\infty} du \, \exp\left[-\frac{u^2 + (u-z)^2}{2\sigma^2}\right] = \frac{1}{2\sigma\sqrt{\pi}} \exp\left(-\frac{z^2}{4\sigma^2}\right). \tag{8.283}$$

Note that we have written this density as conditional on the standard deviation σ rather than the variance σ^2. We are free to choose either, but the standard deviation is convenient when we write out the overall density on z. Since the conditional mean does not influence the statistics of z, the mixture can be specified by a univariate prior on σ, and we find

$$\mathrm{pr}(z) = \frac{1}{2\sqrt{\pi}} \int_0^{\infty} \frac{d\sigma}{\sigma} \exp\left(-\frac{z^2}{4\sigma^2}\right) \mathrm{pr}(\sigma). \tag{8.284}$$

By comparison with (4.85), we recognize (8.284) as a Mellin convolution, and many interesting properties of $\mathrm{pr}(z)$ follow from this observation. Since Mellin transforms convert Mellin convolutions into products, and since Mellin transforms can be interpreted as moments (see Sec. 4.2.2), it follows that moments of z are related simply to moments of σ; from Clarkson and Barrett (2001), the relation is

$$\langle z^k \rangle = \frac{2^k \Gamma\left(\frac{k+1}{2}\right)}{\sqrt{\pi}} \langle \sigma^k \rangle, \tag{8.285}$$

where $\Gamma(\cdot)$ is the gamma function.

From this moment relation and a little algebra, we find

$$\langle z^4 \rangle - 3\langle z^2 \rangle^2 = 12[\langle \sigma^4 \rangle - \langle \sigma^2 \rangle^2]. \tag{8.286}$$

The kurtosis, defined as $\langle z^4 \rangle / \langle z^2 \rangle^2$, is 3 for a Gaussian, so the left-hand side of this expression would be zero for a Gaussian. By the Schwarz inequality, however, the right-hand side is ≥ 0, so z always has a kurtosis greater than or equal to that of a Gaussian, with equality if and only if $\mathrm{pr}(\sigma)$ is a delta function. In short, leptokurtic distributions are inevitable when a Gaussian mixture is filtered with a high-pass or band-pass filter. Moreover, the resulting densities for z often take simple, symmetric forms, quite robust to the detailed assumptions about $\mathrm{pr}(\sigma)$.

Several different analytical forms have been suggested as empirical descriptions of long-tailed densities like those shown in Fig. 8.8. When there is a sharp cusp at the origin, a natural choice is the Laplace or double-exponential density. A family of densities intermediate between Laplace and Gaussian can also be defined with $\mathrm{pr}(z) \propto \exp(-a|z|^p)$, so $p = 1$ is the Laplace density and $p = 2$ is the Gaussian. The parameters p and a can be adjusted to fit empirical densities. Another option is the Lévy family, defined not by the density but by the characteristic function, which has the form $\psi(\xi) = \exp(-b|\xi|^q)$. The corresponding densities cannot be stated as simple analytic functions except when $q = 2$, which is the Gaussian, and $q = 1$, which is the Cauchy density (see Sec. C.5.10). Again, q and b can be treated as adjustable parameters.

Mixtures of correlated Gaussians So far we have considered only a specific Gaussian mixture where u and v were i.i.d. normal, but the result can readily be generalized. Suppose u and v are bivariate normal with a covariance matrix of the form

$$\mathbf{K}_{uv} = \begin{bmatrix} a & b \\ b & c \end{bmatrix}. \tag{8.287}$$

As the reader may show, (8.284) is still valid with this model, only now σ^2 is not a univariate variance but rather $\frac{1}{2}(a + c - 2b)$ (see Clarkson and Barrett, 2001). Thus the initial assumption that u and v are i.i.d. has no essential effect on the conclusions.

Normals and entropy It is not always necessary to invoke the central-limit theorem in order to arrive at a normal probability law. It occurs also in a Bayesian context when one has partial information about a distribution and wishes to complete the description as noncommittally as possible. One way to do this is to use the principle of maximum entropy. A critique of this approach in the context of image reconstruction is given in Sec. 15.3.3, but here we can be content to paraphrase Zhu *et al.* (1998): Entropy is a measure of randomness, and we should choose the density that is as random as possible in all unobserved dimensions and does not attempt to represent information that we do not have.

If we know the mean and variance of a random variable (or mean vector and covariance matrix of a random vector), these moments serve as constraints on the density, and we would like to find the density of maximum entropy consistent with these constraints. We shall carry through the calculation in the univariate case and simply state the multivariate result.

Consider a random variable x and suppose we know that its mean is $\overline{x}$ and its variance is σ^2. According to the principle of maximum entropy, we must choose $\mathrm{pr}(x)$ to maximize $\int_{-\infty}^{\infty} dx\, \mathrm{pr}(x) \ln \mathrm{pr}(x)$, subject to the constraints

$$\int_{-\infty}^{\infty} dx\, \mathrm{pr}(x) = 1\,; \qquad \int_{-\infty}^{\infty} dx\, x\, \mathrm{pr}(x) = \overline{x}\,; \qquad \int_{-\infty}^{\infty} dx\, x^2\, \mathrm{pr}(x) = \sigma^2 + \overline{x}^2\,. \tag{8.288}$$

The maximization can be performed by the method of Lagrange multipliers. We require that the Lagrangian functional,

$$L\{\mathrm{pr}(x)\} \equiv \int_{-\infty}^{\infty} dx\ \mathrm{pr}(x) \ln \mathrm{pr}(x) + \alpha \left[\int_{-\infty}^{\infty} dx\ \mathrm{pr}(x) - 1 \right]$$

$$+\beta \left[\int_{-\infty}^{\infty} dx\ x\ \mathrm{pr}(x) - \overline{x} \right] + \gamma \left[\int_{-\infty}^{\infty} dx\ x^2\ \mathrm{pr}(x) - (\sigma^2 + \overline{x}^2) \right], \tag{8.289}$$

be unchanged by small perturbations of $\mathrm{pr}(x)$. Here, α, β and γ are the Lagrange multipliers, to be fixed by the constraint equations. If we perturb $\mathrm{pr}(x)$ by a small amount $\eta(x)$, and retain only terms linear in the perturbation, we find

$$L\{\mathrm{pr}(x) + \eta(x)\} - L\{\mathrm{pr}(x)\} = \int_{-\infty}^{\infty} dx\ \eta(x) \left\{ 1 + \ln \mathrm{pr}(x) + \alpha + \beta x + \gamma x^2 \right\} = 0\,. \tag{8.290}$$

Since $\eta(x)$ is arbitrary, this equation can hold only if the quantity in braces in the integrand is zero, so $\mathrm{pr}(x)$ must take the form

$$\mathrm{pr}(x) = \exp\left(-1 - \alpha - \beta x - \gamma x^2 \right)\,. \tag{8.291}$$

Both this form and the constraints are satisfied if

$$\mathrm{pr}(x) = \frac{1}{\sqrt{2\pi\sigma^2}} \exp\left[-\frac{(x-\overline{x})^2}{2\sigma^2} \right]\,. \tag{8.292}$$

Thus, if all we know about a random variable is its mean and variance, the maximum-entropy choice for its density is a normal. A similar calculation shows that if all we know about a random vector is its mean vector and covariance matrix, the maximum-entropy density is multivariate normal.

Positivity Appealing though normal distributions may be, they have one serious deficiency in many imaging applications. If the random variable or vector in question is inherently nonnegative, as physical objects often are, then the normal law cannot be strictly correct; it always predicts some finite probability of negative values. We shall now discuss several possible fixes for this problem.

One simple fix is just to consider situations where the standard deviation of the random variable is small compared to its mean; then the probability of getting a negative value is small and can perhaps be neglected without serious error. For the normal law to represent a nonnegative object, in particular, we must consider low-contrast scenes where the variation we are trying to describe is small compared to some spatial-average value. Such situations arise often in medical imaging or other applications involving a faint object on a bright background. They can be particularly useful for local statistical descriptions where the background may vary substantially over the whole scene but relatively little over a region of interest.

Another approach is to use a truncated Gaussian which is not allowed to go negative. Perhaps surprisingly, this is the maximum-entropy choice if we know the mean and variance of a random variable and also know that it is nonnegative. Retracing the calculation above, we see that (8.290) still holds with the simple modification of setting the lower limit of integration to zero, and (8.291) holds without modification. A more substantial modification does occur in (8.292), which can now be written as

$$\mathrm{pr}(x) = N \exp\left[-\frac{(x-x_0)^2}{2v^2}\right] \mathrm{step}(x)\,, \tag{8.293}$$

where N is a normalizing constant, x_0 is *not* the mean and v^2 is *not* the variance. Instead these quantities must be determined by numerically solving constraint equations like (8.288) but with a lower integration limit of 0.

Similarly, if we know the mean and covariance of a nonnegative random vector, a truncated multivariate normal is the maximum-entropy density. Again, however, the known mean and covariance cannot simply be plugged into the standard multivariate normal form.

Log-normals Another solution to the positivity problem is to use log-normals rather than normals. Since a log-normal is a density for a random variable whose log is normal, it is defined for any nonnegative variable, and the density is taken to be zero for negative values of the variable.

The density for a univariate log-normal is given in Sec. C.5.9 of App. C; the corresponding multivariate form is

$$\mathrm{pr}(\mathbf{f}) = \left[\prod_i \frac{1}{\sqrt{2\pi} f_i}\right] \frac{1}{\sqrt{\det(\mathbf{K})}} \exp\left\{-\tfrac{1}{2}\left[\ln(\mathbf{f}) - \boldsymbol{\mu}\right]^t \mathbf{K}^{-1}\left[\ln(\mathbf{f}) - \boldsymbol{\mu}\right]\right\}\,, \tag{8.294}$$

where the logarithm is to be interpreted componentwise, and $\boldsymbol{\mu}$ and $\mathbf{K}$ are the mean vector and covariance matrix of the Gaussian random vector $\ln \mathbf{f}$, not $\mathbf{f}$ itself. The reader may test her understanding of transformations of random variables by showing that the log of $\mathbf{f}$ is indeed a multivariate normal.

Often we can argue on physical grounds that the PDF for an object or image should tend to a log-normal. Consider, for example, transmission x-ray imaging of a thick, inhomogeneous 3D object. The 3D object can be divided into slabs, and the overall transmission of the object is the product of the slab transmissions. If the transmission of each slab is a random process, then the log of the product of the slab transmissions is also a random process, and if the individual slabs are statistically independent, then the log of the product is the sum of logarithms of independent random processes. Thus, regardless of the statistics of the slab log-transmissions, the overall log-transmission tends to a normal by the central-limit theorem, and hence the overall transmission itself tends to a log-normal.

In other situations as well, we can decompose the object function into a product of independent random variables. For example, in nuclear medicine we might inject a radioactive tracer into the blood stream and watch its migration through the circulatory system to some target organ. At each branching of the blood vessels, a tracer molecule can go in one of two directions, and if we consider a point in the vasculature after many branchings, then the number of molecules arriving there is the injected number times a product of a large number of random variables, one for each branch. The central-limit theorem suggests that the log of the tracer concentration at this point is normally distributed, so the concentration itself is log-normal.

There is an essential difference between log-normals and truncated normals as densities for nonnegative random variables or vectors. As the examples above suggest, we can expect the log-normal to be experimentally verifiable, in principle, so it can be interpreted in a frequentist sense. If the variable of interest is a product of many independent random variables, and each experiment results in different values for the individual variables, we can repeat the experiment many times, and the resulting histogram estimate of the density for the product variable will tend to a log-normal, and indeed this distribution is frequently observed experimentally.[12]

There is, in fact, a frequentist rationale for maximum entropy, and it will be sketched in Sec. 15.3.3, but it conceives of the object being constructed by throwing imaginary grains or blobs of gray level; it definitely does not suggest a concrete physical experiment. Thus, even though the truncated normal might be a maximum-entropy density, we should not expect to encounter it as the limit of an experimental histogram. Maximum entropy, as we used it above, is merely a way of going from known moments to a noncommittal PDF.

8.4.4 Texture models

A significant portion of the image-science literature deals with analysis, synthesis, recognition and segmentation of *textures*, defined loosely as spatial random fields with some degree of stationarity. Sometimes the stationarity is periodic, with basic repeating elements such as bricks in a wall or fibers in a woven[13] fabric. Sometimes it is continuous, as with a stucco wall rather than a brick one or the surface of the ocean, where the light reflected from the object can be described as a stationary random process. Sometimes the stationarity is only approximate, in one of the senses discussed in Sec. 8.2.4; the correlation properties might vary slowly, or they might be stationary only within some region boundaries. Sometimes, in fact, the stationarity is purely visual; two regions are said to be the same texture simply because a human observer cannot tell them apart.

Since textures are essentially stationary random processes, Fourier analysis is an important tool for analyzing them. We shall therefore start this section with a discussion of the role of Fourier analysis and power spectral densities, and then we shall briefly discuss methods for estimating power spectra.

Even when stationarity is a good approximation, an autocorrelation function or power spectral density may not capture all of the essential properties of a texture field. It may be necessary to specify also some aspects of the multivariate PDF in order to adequately describe a texture, and we shall describe several means of doing so.

Throughout this section we shall discuss not only methods of characterizing texture as a random process, but also methods for generating sample functions of the random process. An excellent general reference on methods of constructing sample functions with specified correlation properties and marginal distributions is Johnson (1994).

[12]Above we presented an argument that the total amount of tracer in a voxel should tend to a normal, and here we argue that the concentration at a point should be log-normal. These two arguments are not necessarily inconsistent, since a sum of log-normals can converge to a normal, but in fact this convergence is very slow (Barakat, 1976). Which distribution is actually observed is best resolved empirically.

[13]Texture comes from the Latin *texere*, to weave, so a fabric is the prototype of a texture.

Fourier Phase and magnitude Any spatial pattern, whether regarded as a deterministic function or as a sample function of a random process, is completely specified by its Fourier transform. This (continuous or discrete) Fourier transform is complex, but the modulus and phase convey essentially different information about the object. Fourier phase tells you where things are — if the position of an object is shifted, the phase changes but the modulus does not. Fourier modulus, on the other hand, tells you only how strongly different spatial frequencies contribute to the object.

In many cases, Fourier phase is more important than Fourier modulus in conveying the essence of an object. In a famous experiment, Oppenheim and Lim (1981) Fourier-transformed two images, one of the television news anchor Walter Cronkite and one of a clock. They then interchanged the Fourier phases, putting Walter's phase with the clock modulus and vice versa. After inverse transformation, the image with Walter's Fourier phase still looked like Walter, and the one with the phase of a clock looked like a clock.

With textures, on the other hand, the situation can be reversed. In a stationary random process we do not care where things are. One location is as good as another, at least statistically, so Fourier phase is much less important than Fourier modulus. Two stationary random processes with the same modulus but different phases are recognized as sample functions of the same texture. One common way of synthesizing sample textures, therefore, is to generate samples of white noise and pass them through a linear filter.

As an example, Bochud *et al.* (1999b) examined the relative importance of Fourier amplitude and phase in describing coronary angiograms (x-ray images of blood vessels after injection of an x-ray-absorbing material into the blood stream). In agreement with the remarks above, they found that the phase was important for describing the vessel, but not for the random anatomical background against which the vessel was seen. Though the background was not rigorously stationary, they showed that realistic images could be simulated by filtering white noise through a space-variant filter.

Estimation of power spectra or autocorrelation functions of images Suppose we have one or more sample images, and we want to generate additional images with similar texture by filtering white noise. To the extent that the texture is a stationary random process, we need to know the power spectral density or the stationary autocorrelation function. There is a large literature on estimating these quantities from sample images, and we confine ourselves here to a few general observations.

In Sec. 8.2.5 we mentioned — and dismissed — an apparently obvious approach to spectral estimation, the periodogram of a single sample image. Figure 8.1 illustrates the difficulty with this approach. Mallat (1999) refers to periodogram analysis as "naive spectral estimation;" one meaning of *naive* is "lacking information, uninformed" and the periodogram is naive in the sense that it does not incorporate prior information or beliefs into the spectral estimate. We certainly do not believe that the rapid fluctuations seen in Fig. 8.1 are meaningful features of the power spectrum (or if we did, we would need only to repeat the experiment to change our belief system). The situation is very similar to image reconstruction, discussed in much more detail in Chap. 15, where naive attempts at inverse filtering yield large fluctuations in the reconstructed image.

The Bayesian approach to this problem, in both image reconstruction and spectral estimation, is to define a prior probability on the function being estimated and then to seek an estimate consistent with both the data and this prior. In the Bayesian community, a preferred prior is the entropy, and maximum-entropy reconstructions do indeed eliminate the rapid fluctuations and yield smooth estimates. The details of this procedure, in the context of image reconstruction, are given in Sec. 15.3.3.

As we shall also see in Chap. 15, there are many other approaches, referred to collectively as *regularization*, that can be used to suppress fluctuations in reconstructed images, and each of these methods has its analog in spectral estimation. Many of these methods can also be described as Bayesian, but with priors other than entropy (see Sec. 15.3.3); all of them attempt to enforce our prior belief that the function being reconstructed (power spectrum or image) is smooth in some sense.

One way to enforce smoothness in spectral estimation is to model the spectrum as a smooth function with unknown parameters and then to estimate the parameters. For example, we could model the spectrum as a Gaussian and estimate its width, or as a Gaussian times a polynomial and estimate the polynomial coefficients also. One popular model, especially for time-series analysis, is the *autoregressive, moving average* or ARMA model where the spectrum is modeled as a ratio of polynomials (Oppenheim and Schafer, 1989).

Another model is to assume that the power spectrum varies as a power law, $\rho^{-\beta}$, and then to estimate the exponent β. Many images exhibit this behavior in practice (even when there is no reason to assume stationarity), and β is a useful phenomenological descriptor. Physical mechanisms that lead to power-law power spectra in the context of electrical noise are surveyed in Sec. 12.2.3.

Which regularization method is chosen depends on what one wants to do with the spectral estimate. If we want to simulate images that appear realistic to a human observer, we can use one of the psychophysical tests detailed in Sec. 14.2.3 to measure how well the observer can distinguish real texture images from white noise filtered with the estimated spectrum. If the real and simulated images are indistinguishable, it means that the estimated spectrum is good enough for this purpose; on the other hand, if they are readily distinguishable, it may mean that the spectral estimate is poor, or it may mean that the texture is more complicated than just filtered noise.

For many purposes, however, we need more than just visual realism. In texture recognition or discrimination, for example, we need a stochastic model in order to design an optimal discriminant function (see Sec. 13.2.12). If we use a Gaussian model, we need to know the inverse of the covariance matrix, and if we also assume stationarity, that means we need to know the reciprocal of the power spectrum. Even if the spectral estimate accurately represents the actual spectrum for the spatial frequencies where the spectrum is large, it may be a poor estimate in the tails and hence a poor estimate of the reciprocal spectrum. The best spectral estimate in this case is the one that leads to the best discrimination performance for a discriminant function based on the estimated spectrum (but tested on real images — not ones simulated from the estimated spectrum!).

As another example, we shall see in Chaps. 13 and 14 that some important measures of image quality are expressed in terms of the image power spectrum. If we do not know the actual spectrum, we must estimate it, and the adequacy of the spectral estimate must be judged by the accuracy of the corresponding estimates of figures of merit for image quality.

Estimation of power spectra or autocorrelation functions of objects Above we stated our goal as estimation of the power spectral density or autocorrelation function of a set of images. Often, however, what we really want to know is the power spectral density or autocorrelation function of the objects that formed the images.

Suppose we have a set of sample images $\{\mathbf{g}_j, j = 1, ..., J\}$, where the jth image is related to an object $\mathbf{f}_j$ by $\mathbf{g}_j = \mathcal{H}\mathbf{f}_j + \mathbf{n}_j$. We must assume that $\mathbf{f}_j$ is a sample function of a stationary (or at least quasistationary) random process in order to define an object power spectral density $S_f(\boldsymbol{\rho})$, and we need knowledge of $\mathcal{H}$ and of the noise statistics in order to estimate $S_f(\boldsymbol{\rho})$.

As a simple example, suppose the imaging system is well approximated as a convolution (a CC LSIV system in the language of Sec. 7.2.3). Then $\mathbf{g}_j$ is a sample function of a stationary random process, and its power spectrum is denoted by $S_g(\boldsymbol{\rho})$. If we also assume that $\mathbf{n}_j$ is a sample function of a stationary random process, with power spectrum $S_n(\boldsymbol{\rho})$, then use of (8.156) shows that the image power spectrum is given by

$$S_g(\boldsymbol{\rho}) = |H(\boldsymbol{\rho})|^2 \, S_f(\boldsymbol{\rho}) + S_n(\boldsymbol{\rho}) \, . \tag{8.295}$$

The image spectrum $S_g(\boldsymbol{\rho})$ can be estimated by any of the methods suggested above, and the result can be denoted as $\widehat{S}_g(\boldsymbol{\rho})$. If we know the noise spectrum $S_n(\boldsymbol{\rho})$ independently from the physics of the imaging problem, then one reasonable estimate of the object spectrum is

$$\widehat{S}_f(\boldsymbol{\rho}) = \frac{\widehat{S}_g(\boldsymbol{\rho}) - S_n(\boldsymbol{\rho})}{|H(\boldsymbol{\rho})|^2} \, . \tag{8.296}$$

This method gives little information about $S_f(\boldsymbol{\rho})$ at frequencies for which $H(\boldsymbol{\rho})$ is small, and large errors in $S_f(\boldsymbol{\rho})$ can result from small errors in either $S_n(\boldsymbol{\rho})$ or $\widehat{S}_g(\boldsymbol{\rho})$. Moreover, the whole approach depends on modeling the system as CC LSIV and the noise as stationary.

A better approach is to use some parametric description of the object power spectrum, perhaps one that allows quasistationarity, and then to estimate the parameters from the data. This way, the system operator $\mathcal{H}$ can be a general CD mapping and the noise can have an arbitrary covariance matrix $\mathbf{K_n}$, so long as both of these quantities are known. Methods of parameter estimation to be developed in Chap. 13 can then be used to estimate the spectral parameters. Thus a stationary or quasistationary texture field can be imaged through a shift-variant imaging system and have nonstationary noise added to it, yet the parameters describing the spectrum of the texture field can still be estimated.

Gray-level statistics When the correlation properties are not sufficient to characterize a texture, we can also use the single-point PDF $\mathrm{pr}[f(\mathbf{r})]$. For a stationary texture, this density is independent of $\mathbf{r}$, and we might want to generate samples of the texture with this density and some specified autocorrelation function or power spectral density. We shall sketch an iterative algorithm for this purpose.

The algorithm begins by filtering white noise to obtain several samples with the requisite power spectrum. It is probably valid to invoke the central-limit theorem on the filter output since the filter will serve to add up many independent samples of the white noise, so the single-point PDF on the filter output is probably Gaussian, but in any case we can estimate the PDF from the average gray-level histogram of the samples. At this stage we can perform a process known as *histogram equalization*, a pointwise nonlinear transformation that changes the gray-level distribution as described in Sec. C.3.1, and the form of the transformation can be chosen to yield the required PDF. This transformation changes the power spectrum in a complicated way, and it is necessary to estimate the new spectrum from the samples. From the new spectrum, we can devise a new filter to match the current spectrum to the required one, but this changes the PDF so a new histogram-equalization step is needed. The process is then repeated iteratively. Each iteration is a projection onto convex sets, as discussed in detail in Sec. 15.4.5, and convergence can be proven by use of a theorem quoted there. The result is a set of samples that have both the specified power spectrum and the specified single-point PDF.

Texture synthesis with wavelet channels It has been found (Bergen and Adelson, 1991; Chubb and Landy, 1991) that textures that give similar gray-level histograms through a series of wavelet filters appear similar to a human observer. Heeger and Bergen (1995) and Rolland and co-workers (Rolland and Strickland, 1997; Rolland *et al.*, 1998; Rolland, 2000) have used this observation to develop algorithms for synthesizing textures.

The Rolland group uses a digital image of a reference texture and synthesizes additional sample textures of similar visual appearance. The reference texture is decomposed into subbands by means of a discrete wavelet transform (see Sec. 5.3.3). This transform is invertible, so the original reference texture can be recovered by the inverse transform. The stochastic model, however, is that the texture can be characterized by means of gray-level histograms for each subband, basically a histogram estimate of the univariate PDFs for the output of each wavelet filter. In principle, multiple reference images could be used to improve this estimate, but the Rolland algorithm uses just one and implicitly assumes ergodicity.

To synthesize a sample texture, a discrete white noise field is generated, and it is also passed through the same discrete wavelet transform. The histogram of each filter output is computed, just as for the reference texture. A nonlinear point operation is applied in each subband to convert the histograms of the transformed white noise to histograms that match those of the reference texture. An inverse wavelet transform then yields the synthesized texture. The visual correspondence between the reference texture and the synthesized textures is striking, yet all of the synthesized textures are statistically independent since independent noise fields are used.

Multiple filters and maximum entropy The method of Heeger and Bergen permits the synthesis of textures from one or more training images, but it does not give a probability model for the synthesized images. This gap was filled by Zhu *et al.* (1998), whose work can be seen as a combination of wavelet-based texture synthesis and independent components analysis. Rather than restricting attention to some chosen set of wavelets, as in Heeger's method, Zhu *et al.* use a large library of linear filters and compute marginal histograms of the filter outputs for some training set of images (which may consist of just a single image plus an ergodicity assumption). They then use the principle of maximum entropy to construct a multivariate distribution that agrees with the marginals estimated from training data.

The rationale for maximum entropy is the one mentioned in Sec. 8.4.3: maximum-entropy densities are maximally noncommittal and do not attempt to represent information not available empirically. According to Zhu, the maximum-entropy density is the "purest fusion" of the empirical marginals.

Suppose we have a set of linear operators $\boldsymbol{\mathcal{L}}^{(j)}$ in object space, with the output of the jth operator given by $q^{(j)}(\mathbf{r}) = [\boldsymbol{\mathcal{L}}^{(j)}\mathbf{f}](\mathbf{r})$. In the abstract notation of Sec. 8.2.2, the single-point marginal density on the output can be written as

$$\mathrm{pr}\left[q^{(j)}(\mathbf{r})\right] = \int d\mathbf{f}\ \mathrm{pr}\left[q^{(j)}(\mathbf{r})\middle|\mathbf{f}\right]\mathrm{pr}(\mathbf{f})\,. \tag{8.297}$$

But the linear operator is deterministic, so $q^{(j)}(\mathbf{r})$ is known exactly once $\mathbf{f}$ is specified, and we can write

$$\mathrm{pr}\left[q^{(j)}(\mathbf{r})\right] = \int d\mathbf{f}\ \delta\left\{q^{(j)}(\mathbf{r}) - [\boldsymbol{\mathcal{L}}^{(j)}\mathbf{f}](\mathbf{r})\right\}\mathrm{pr}(\mathbf{f})\,, \tag{8.298}$$

where $\delta\{q^{(j)}(\mathbf{r}) - [\boldsymbol{\mathcal{L}}^{(j)}\mathbf{f}](\mathbf{r})\}$ is simply a 1D delta function. Comparing this expression to (4.173), we see that the single-point marginal on the filter output is a Radon-transform projection of the object density $\mathrm{pr}(\mathbf{f})$, where $\mathbf{f}$ here corresponds to the position vector $\mathbf{r}$ in (4.173), and choice of the linear operator here corresponds to the projection direction $\hat{\mathbf{n}}$ in (4.173).

Now suppose we have a set of training "objects" (either good computer simulations or images from a high-resolution, low-noise imaging system as discussed in Sec. 8.4.1) from which we can form a histogram estimate of $\mathrm{pr}\left[q^{(j)}(\mathbf{r})\right]$. If we denote this histogram, defined as in (8.262), by $\widehat{\mathrm{pr}}_{q^{(j)}(\mathbf{r})}\left[q^{(j)}(\mathbf{r})\right]$, then we can pose the maximum-entropy density-estimation problem as

$$-\int d\mathbf{f}\ \mathrm{pr}(\mathbf{f})\ln[\mathrm{pr}(\mathbf{f})] = \max\,, \tag{8.299}$$

subject to the constraints of normalization,

$$\int d\mathbf{f}\ \mathrm{pr}(\mathbf{f}) = 1\,, \tag{8.300}$$

and agreement with the empirical histograms,

$$\widehat{\mathrm{pr}}_{q^{(j)}(\mathbf{r})}(z) = \int d\mathbf{f}\ \delta\left\{z - [\boldsymbol{\mathcal{L}}^{(j)}\mathbf{f}](\mathbf{r})\right\}\mathrm{pr}(\mathbf{f})\,. \tag{8.301}$$

If we assume stationarity, at least over some restricted region, then the histogram should be the same for all positions, and we can drop the argument $\mathbf{r}$ on the subscript, but we still have to satisfy the constraint at all $\mathbf{r}$. In practice, the matching will be done for a discrete set of points $\mathbf{r}_i$, usually on a pixel grid.

This problem can be solved by the method of Lagrange multipliers, just as in (8.288) *ff.*, but now we have an infinite number of constraints! For each operator $\boldsymbol{\mathcal{L}}^{(j)}$, we must satisfy (8.301) for all $\mathbf{r}$ and all z. We thus have a continuum of unknown Lagrange multipliers, which we can express as an unknown function $\Phi^{(j)}\{z\}$. With this view, the general form of the maximum-entropy object density turns out to be [see Zhu *et al.* (1998) for details]

$$\text{pr}(\mathbf{f}) = \frac{1}{Z} \exp\left\{ -\sum_i \sum_j \int dz \; \Phi^{(j)}\{z\}\, \delta\left\{z - \left[\mathcal{L}^{(j)}\mathbf{f}\right](\mathbf{r}_i)\right\}\right\}$$

$$= \frac{1}{Z} \exp\left\{ -\sum_i \sum_j \Phi^{(j)}\left(\left[\mathcal{L}^{(j)}\mathbf{f}\right](\mathbf{r}_i)\right)\right\}, \qquad (8.302)$$

where Z is a normalizing constant.

The problem is not yet solved since we still have to find the functionals $\Phi^{(j)}$ such that the constraints are satisfied. Zhu *et al.* propose an iterative algorithm for this purpose.

One remaining question is how to choose the operators $\mathcal{L}^{(j)}$ in the first place. Since stationarity is probably required to make this whole approach computationally feasible, it is natural to choose the operators as LSIV filters, but another consideration is independence. The maximum-entropy estimate in (8.302) shows that the filter outputs are statistically independent, even if this is not the case in reality. Zhu *et al.* propose use of a large library of filters and an iterative algorithm to select a subset of them that optimize a measure of independence, as in ICA, and Zhang (2001) suggests a Metropolis algorithm.

Parametric descriptions of the marginals The filters chosen in the Zhu approach (or discovered in ICA) are mostly band-pass filters (though Zhu includes a low-pass filter as well). As discussed in Sec. 8.4.3 and illustrated in Fig. 8.7, the outputs of band-pass filters tend to have simple cuspy shapes with long, kurtotic tails. Empirically, we can describe these marginals by simple analytical forms such as Laplacian or Levy densities, with only one or two free parameters per filter output.

This observation suggests an alternative to the Zhu method: instead of trying to choose the functions $\Phi^{(j)}\{z\}$ to match the empirical marginal histograms, we can directly estimate the free parameters in the assumed analytical densities (Kupinski *et al.*, 2003c).

Lumpy backgrounds Another way to generate images (or simulated objects) with specified correlation properties and controllable gray-level statistics is the *lumpy background*, introduced by Rolland and Barrett (1992). In this method, spatial elements, called *lumps* and denoted $l(\mathbf{r})$, are randomly distributed over some area, so the distribution has the form

$$f(\mathbf{r}) = \sum_{n=1}^{N} l(\mathbf{r} - \mathbf{r}_n). \qquad (8.303)$$

A common choice for $l(\mathbf{r})$ is a Gaussian spatial distribution,

$$l(\mathbf{r}) = A \exp\left(-\frac{r^2}{2s^2}\right). \qquad (8.304)$$

The positions $\mathbf{r}_n$ and possibly also the total number of lumps N are random variables.

One important special case is where N is a Poisson random variable; the mathematical tools for analyzing this case will be developed in Sec. 11.3.9, and the characteristic functional for the random process (8.303) will be derived in Sec. 11.3.10.

As we shall see there, $f(\mathbf{r})$ is a stationary random process if the positions $\mathbf{r}_n$ are uniformly distributed over some area, and the statistical autocorrelation function turns out to be just the autocorrelation integral of the lump profile [see (11.140)].

If N is large and the lump positions are statistically independent, the single-point PDF of a lumpy background approaches a Gaussian by the central-limit theorem. In this limit, the details of the lump profile are irrelevant, and the resulting functions are indistinguishable from ones obtained by filtering white, Gaussian noise. If N is small, on the other hand, then the lump profile controls the single-point PDF as well as the correlation properties; for more details, see Sec. 11.3.10.

More general lumpy backgrounds As originally defined by Rolland, the lump profile $l(\mathbf{r})$ in (8.303) is a nonrandom function; the only randomness is in the lump location $\mathbf{r}_n$. To allow more freedom in synthesizing lumpy backgrounds, we can let the lump profile also be random. For example, the amplitude or the width of each blob could vary according to some specified probability law.

One very useful variant of the simple lumpy background is the *clustered lumpy background*, suggested by Bochud *et al.* (1999a), where a cluster of identical blobs forms a *superblob*, and the final model is a superposition of superblobs. With this scheme, (8.303) becomes

$$f(\mathbf{r}) = \sum_{k=1}^{N_s} \sum_{n=1}^{N_k} l_k(\mathbf{r} - \mathbf{r}_{nk} - \mathbf{R}_k) \,, \tag{8.305}$$

where N_s is the number of superblobs, N_k is the number of blobs within the kth superblob, $\mathbf{R}_k$ is the center of the kth superblob, $\mathbf{r}_{nk}$ is the center of the nth blob within the kth superblob, and $l_k(\mathbf{r})$ is the random lump profile associated with that superblob. It is useful to make N_k and N_s Poisson random variables, so we must wait until Chap. 11 to analyze the statistics of (8.305).

Bochud *et al.* chose elongated Gaussians for the lump profiles and used their orientation as the random parameter in $l_k(\mathbf{r})$. With this simple model they were able to synthesize images strikingly similar to clinical mammograms.

Two-point densities As we discussed in Sec. 8.4.2, two-point PDFs of the form $\text{pr}[f(\mathbf{r}_1), f(\mathbf{r}_2)]$ can be an important part of the stochastic description of objects in general, and they are particularly attractive for stationary random processes such as textures. For purposes of stochastic modeling, we can estimate the two-point PDF from the empirical co-occurrence statistics of one or a few images if we assume ergodicity. It was suggested by Julesz (1962) that textures with similar co-occurrence statistics would appear similar, though psychophysical studies have shown that higher-order statistics do have at least some effect on human texture perception (Diaconis and Freedman, 1981).

The use of co-occurrence statistics for synthesis of realistic textures should be distinguished from their use in texture discrimination or segmentation. In the latter application, the goal is to describe the texture pattern within a spatial region by a few features with good discriminatory power, and it is common to reduce pixel values in the region to a co-occurrence matrix and to derive the features from that matrix. There is no need to make any argument about ergodicity or stationarity in this application; if the features are useful in discriminating one region from another or classifying regions, that is justification enough.

Quasistationary textures Most of the discussion above has concentrated on textures as stationary random processes. If exact stationarity is not a good assumption, we may want to model a texture as quasistationary, and in this case the stochastic Wigner distribution function defined in Sec. 8.2.5 is a useful tool. In particular, if the quasistationary form (8.142) is valid, we can estimate the two factors $b(\mathbf{r}_0)$ and $A(\boldsymbol{\rho})$ separately from samples. If we want to generate sample textures with a stochastic Wigner distribution specified by (8.142), we can use a lumpy background with a spatially variable lump density (mean number of lumps per unit area) given by $b(\mathbf{r}_0)$. For more discussion on the statistics of lumpy backgrounds, see Sec. 11.3.10.

Sometimes the pattern we want to synthesize is stationary within prespecified boundaries. For example, we may want to simulate statistically independent sample functions of an abdominal section of the body in order to study image quality in computed tomography (CT). We can start with a good anatomical model, obtained perhaps by manual or automated segmentation of a single reference CT image, and we can identify specific organs such as liver and spleen within this image (Zubal *et al.*, 1994). Then any of the methods described above can be used to characterize the texture within each organ and to generate sample functions consistent with this characterization. These sample functions can then be placed within the specified organ boundaries, and the procedure can be repeated as many times as needed to get a large number of simulated abdomens. These simulations can be regarded as object representations rather than images since the organ boundaries will be sharp and the textures may contain very high spatial frequencies.

Random shapes In addition to simulating random textures within a region, we may wish to make the shape itself random. Simulating a shape usually means adopting some parameterized description of the shape and choosing the parameters. Some simple approaches to describing shapes mathematically were discussed briefly in Sec. 7.1.6. One approach, used for example by Cargill (1989) to describe the human liver, is to specify the distance R from some internal reference point to the boundary as a function of polar angles θ and ϕ. If the surface of the object is smooth, an expansion of $R(\theta, \phi)$ in spherical harmonics can be terminated with relatively few terms ($\sim$ 100 in Cargill's work), and the coefficients in this expansion are the desired parametric representation of the liver. This general approach is applicable to any 3D shape in which a reference point can be found for which $R(\theta, \phi)$ is unique; it is not necessary that the shape be convex, though convexity avoids the necessity of searching for a suitable reference point.

Another general approach, also mentioned briefly in Sec. 7.1.6, is to express the shape as a geometric transformation of a given reference shape. Affine or non-affine transformations can be used, and the parameters of the transformation are then the shape descriptors.

After establishing a parametric description of shape, the next step in shape simulation is to find the PDF on the parameters, for example by analyzing real shapes. One common approach is to compute a sample mean and sample covariance matrix on a set of measured parameters and, in effect, to assume that the PDF is multivariate normal with this mean and covariance. If there are many parameters, it can be advantageous to use principal components analysis or PCA (see Sec. 8.4.1) and retain only components corresponding to a few of the eigenvectors of the sample covariance with the largest eigenvalues. The eigenvectors themselves are sets of shape parameters, and the shapes associated with them are often called *eigenshapes*. It must be kept in mind, however, that these eigenshapes are characteristics of both the particular shape description used and the experimental data set from which the parameters were derived.

However the PDF on the shape parameters is formulated, samples drawn from it can be used to synthesize new shapes consistent with the estimated PDF, and these random shapes can then be used in image-quality studies and many other investigations. For an example of these procedures, see Duta *et al.* (1999), and for general mathematical treatments of statistical shape analysis, see Small (1996), Dryden and Mardia (1998) and Kendall *et al.* (1999).

8.4.5 Signals and backgrounds

In many imaging situations, we do not have equal interest in all parts of the scene. In aerial reconnaissance, for example, we are relatively uninterested in trees and bushes, but we would be extraordinarily interested in a military vehicle that might be hiding in the bushes. Similarly, in an abdominal MRI scan, we have little interest in the myriad features of normal anatomy, but we are much more interested in a small nodule that might turn out to be malignant. As a very general term, we can call an object of interest, that may or may not be present in a given scene, a *signal.* The remainder of the scene can be called *background* or (especially in the radar literature) *clutter.* In Chap. 13 we shall discuss in detail methods of detecting signals, or distinguishing between different signals, but here we introduce the topic by discussing stochastic models for objects with and without signals.

Additive signals Perhaps surprisingly, it entails no loss of generality to decompose an object into a simple sum of signal and background components:

$$f(\mathbf{r}) = f_s(\mathbf{r}) + f_b(\mathbf{r})\,. \tag{8.306}$$

Once we have defined the portion of the object that we regard as signal and denoted it as $f_s(\mathbf{r})$, then the background $f_b(\mathbf{r})$ is just *defined* as $f(\mathbf{r}) - f_s(\mathbf{r})$.

This does not say that $f_b(\mathbf{r})$ is the same as $f(\mathbf{r})$ would be in the absence of the signal, though in fact it may be. In nuclear medicine for example, a tumor is often manifest by an increased uptake of some tumor-seeking radiopharmaceutical, so it is natural to simply add the tumor distribution $f_s(\mathbf{r})$ to the distribution $f_b(\mathbf{r})$ in normal tissue. If both $f_s(\mathbf{r})$ and $f_b(\mathbf{r})$ are sample functions of random processes, then it may be reasonable to take the two processes as statistically independent.

In optical imaging, on the other hand, objects are opaque, so a signal of interest may obscure the background behind it. For purposes of describing the response of an imaging system, the object $f(\mathbf{r})$ is either $f_b(\mathbf{r})$ or $f_s(\mathbf{r})$, not their sum. Nevertheless, we can still use an additive model if the statistical dependence of the two processes is taken into account.

Nonrandom signals The simplest model for a signal on a background is one where the signal function is completely specified and the only randomness is whether or not it is present. This model is often called SKE (signal known exactly). As we shall see in Chap. 13, it is an excellent starting point for discussing signal detection and image quality.

If we adopt the SKE model and assume that the signal is just added to the background rather than obscuring it, then the signal and background are statistically independent. With or without a signal, the PDF on $f(\mathbf{r})$ is fully determined by the PDF on $f_b(\mathbf{r})$ since that is the only random process in the problem. In the absence of a signal, we can write the univariate density on $f(\mathbf{r})$ as

$$\mathrm{pr}[f(\mathbf{r})|\text{signal absent}] = \mathrm{pr}_b[f_b(\mathbf{r})]\,. \tag{8.307}$$

We have added the subscript b to indicate that $\mathrm{pr}_b[f_b(\mathbf{r})]$ is specifically the PDF on the background; the notation is redundant here since the same information is conveyed by the subscript on $f_b(\mathbf{r})$, but its usefulness will become apparent in a moment.

Because of the assumed statistical independence, the form of the PDF for $f_b(\mathbf{r})$ is still the same with a signal present, but to relate it to the PDF on $f(\mathbf{r})$ we must rewrite (8.306) as

$$f_b(\mathbf{r}) = f(\mathbf{r}) - f_s(\mathbf{r})\,. \tag{8.308}$$

We then have

$$\mathrm{pr}[f(\mathbf{r})|\text{signal present}] = \mathrm{pr}_b[f(\mathbf{r}) - f_s(\mathbf{r})]\,. \tag{8.309}$$

Now we see the need for the subscript: the 1D PDF $\mathrm{pr}_b[f_b(\mathbf{r})]$ is merely shifted along the axis by the presence of a nonrandom signal (see Fig. 8.9), and the functional form is unchanged.

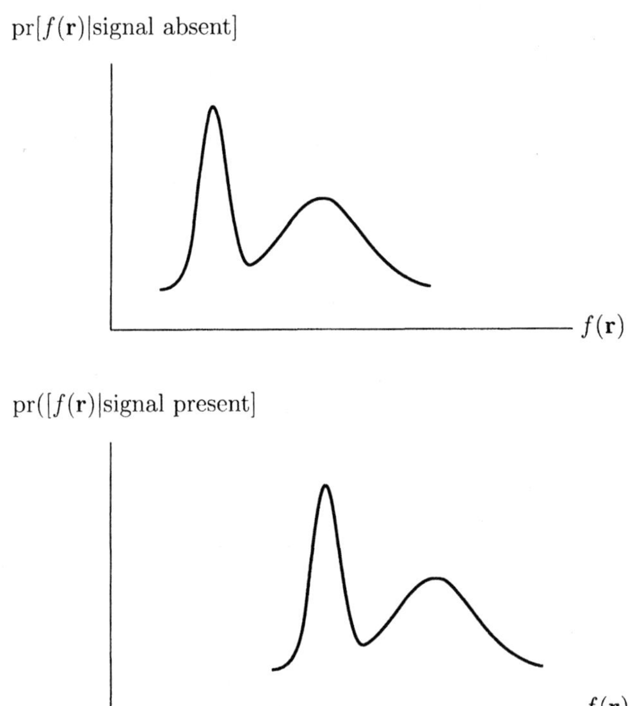

Fig. 8.9 Effect on the univariate PDF of adding a nonrandom signal to a random background.

This discussion has specifically dealt with univariate densities at a single point, but it is easy to extend it to an arbitrary number of points or to general Hilbert-space vectors representing object, signal and background. Abstractly, we can write

$$\text{pr}(\mathbf{f}\,|\text{signal absent}) = \text{pr}_b(\mathbf{f}_b)\,; \tag{8.310}$$

$$\text{pr}(\mathbf{f}\,|\text{signal present}) = \text{pr}_b(\mathbf{f} - \mathbf{f}_s)\,. \tag{8.311}$$

These densities can be interpreted as PDFs on coefficient vectors like $\boldsymbol{\alpha}$ or as multipoint densities. In fact, (8.308) and (8.309) follow from (8.310) and (8.311) just by regarding $f(\mathbf{r})$ as a component of $\mathbf{f}$ (and similarly for $\mathbf{f}_b$ and $\mathbf{f}_s$) and taking marginals on both sides of (8.310) and (8.311). Whatever space we are working in, addition of a nonrandom signal merely shifts the background PDF.

Parametric signal models Sometimes a signal is not known exactly but can be described by a function with a small number of unknown parameters. For example, in nuclear medicine a tumor might be well modeled as a sphere with random location, size and uptake of a radiopharmaceutical. Similarly, in astronomy a pulsar could be modeled as a time-varying point source, where the random parameters are its coordinates in the sky and the amplitude and period of the pulsation.

In these cases we do not need an infinite-dimensional PDF like $\text{pr}(\mathbf{f}_s)$ to describe the signal; if the signal is fully specified by L parameters $\{\theta_{s\ell}, \ell = 1, ..., L\}$, all we need is the L-variate PDF $\text{pr}(\{\theta_{s\ell}\})$. The signal parameters can also be arranged into an $L \times 1$ vector $\boldsymbol{\theta}_s$, so we need the PDF $\text{pr}(\boldsymbol{\theta}_s)$ in order to describe the signal fully.

With a signal described parametrically, the object PDF is given by

$$\text{pr}(\mathbf{f}\,|\text{signal present}) = \int_\infty d^L\theta \; \text{pr}(\mathbf{f}\,|\text{ signal present}, \boldsymbol{\theta}_s)\, \text{pr}(\boldsymbol{\theta}_s)\,. \tag{8.312}$$

The conditional density $\text{pr}(\mathbf{f}\,|\text{ signal present}, \boldsymbol{\theta}_s)$ is just the density for an SKE problem; if we condition on a set of parameters that completely specify the signal, then the signal *is* known exactly. If the signal and background are statistically independent, then this conditional density is given by (8.311), and we have

$$\text{pr}(\mathbf{f}\,|\text{signal present}) = \int_\infty d^L\theta_s \; \text{pr}_b\left[\mathbf{f} - \mathbf{f}_s(\boldsymbol{\theta}_s)\right] \text{pr}(\boldsymbol{\theta}_s)\,. \tag{8.313}$$

The object PDF is now a weighted average of shifted background PDFs.

Obscuring signals If point $\mathbf{r}$ lies within the signal and the signal obscures the background, then $f_b(\mathbf{r})$ can take on only the value zero at this point. Since $f(\mathbf{r})$ is then identical to $f_s(\mathbf{r})$, the univariate density on $f(\mathbf{r})$ for a nonrandom signal is given by

$$\text{pr}[f(\mathbf{r})|\text{signal present at } \mathbf{r}] = \delta[f(\mathbf{r}) - f_s(\mathbf{r})]\,. \tag{8.314}$$

If the signal is absent, or if it is present in the object but not at point $\mathbf{r}$, then (8.307) still holds.

Multipoint densities can be formulated similarly. For example, if a nonrandom signal is present at $\mathbf{r}_1$ but not at $\mathbf{r}_2$, the two-point conditional PDF is

$$\text{pr}[f(\mathbf{r}_1), f(\mathbf{r}_2)|\text{signal present at } \mathbf{r}_1, \text{ absent at } \mathbf{r}_2\,] = \delta[f(\mathbf{r}_1) - f_s(\mathbf{r}_1)]\, \text{pr}_b[f(\mathbf{r}_2)]\,. \tag{8.315}$$

The univariate marginals of (8.315) are consistent with (8.314) and (8.307).

The PDFs specified by (8.314) and (8.315) can be difficult to work with, especially when we extend the discussion to random signals. It is often preferable to work in terms of the expansion coefficients $\{\alpha_n\}$. Consider a nonrandom, obscuring signal with support $\mathbf{S}_s$; that is, the signal obscures the background for all points $\mathbf{r}$ in the region $\mathbf{S}_s$. For an orthonormal basis, the coefficient α_n is given by

$$\alpha_n = \int_{\mathbf{S}_f} d^q r \; \psi_n^*(\mathbf{r}) \, f(\mathbf{r}) \,, \tag{8.316}$$

where $\mathbf{S}_f$ is the overall support of the object. When a signal is present, this integral can be written as

$$\alpha_n = \int_{\mathbf{S}_s} d^q r \; \psi_n^*(\mathbf{r}) \, f_s(\mathbf{r}) + \int_{\mathbf{S}_{sc}} d^q r \; \psi_n^*(\mathbf{r}) \, f_b(\mathbf{r}) \,, \tag{8.317}$$

where $\mathbf{S}_{sc}$ is the complement of $\mathbf{S}_s$, *i.e.*, the set of points in $\mathbf{S}_f$ but not in $\mathbf{S}_s$.

We can think of the first integral in (8.317) as the nth component of an infinite vector $\boldsymbol{\alpha}_s$ describing the signal in the basis $\{\psi_n\}$; since the signal is nonrandom, $\boldsymbol{\alpha}_s$ is nonrandom. The second integral would be the nth expansion coefficient for the background except that we have excluded the region $\mathbf{S}_s$ from the range of integration. Nevertheless, we can think of that integral as the nth component of a random vector which we can denote as $\boldsymbol{\alpha}_b$, and we can write, without approximation,

$$\boldsymbol{\alpha} = \boldsymbol{\alpha}_s + \boldsymbol{\alpha}_b \,, \qquad \text{(signal present)} \,. \tag{8.318}$$

If the signal is absent, then the support of the background is the same as the object support, and we can write

$$\boldsymbol{\alpha} = \boldsymbol{\alpha}_b \,, \qquad \text{(signal absent)} \,, \tag{8.319}$$

where $\boldsymbol{\alpha}_b$ is now computed via integration over all of $\mathbf{S}_f$.

Because of the different regions of integration, the statistics of $\boldsymbol{\alpha}_b$ will, in general, depend on whether or not the signal is present. There is, however, one interesting situation in which we might assume that $\boldsymbol{\alpha}_b$ is independent of the signal. Suppose we have a spatially compact signal but a spatially extended basis function, such as a Fourier basis function (see Sec. 7.1.2). In that case, deletion of a small region may not change the value of the integral very much, so it might be a good approximation to say that $\boldsymbol{\alpha}_b$ is the same with and without the obscuring signal. If that assumption is valid, then we are back to an additive model with a signal-independent background, at least in this basis. If, on the other hand, deletion of the signal support does change the integral significantly, we can still use the additive form (8.318), but we have to use a different PDF on $\boldsymbol{\alpha}_b$ for signal present and signal absent.

8.5 STOCHASTIC MODELS FOR IMAGES

Having just discussed various stochastic models for objects, we turn now to images. In keeping with our emphasis on digital imaging, we consider only CD systems here, and for simplicity we assume they are linear. Our objective will be to characterize an ensemble of such images by its mean vector and covariance matrix and, where possible, a multivariate probability density function.

8.5.1 Linear systems

In the absence of noise, we defined a linear imaging system as one for which the image was a linear functional of the object; with noise, a linear imaging system can be defined as one for which the *average* image, obtained after many repeated images of the same object, is a linear functional of the object. If we denote this mean image by $\overline{\mathbf{g}}(\mathbf{f})$, then for any linear system we can write

$$\overline{\mathbf{g}}(\mathbf{f}) = \boldsymbol{\mathcal{H}}\mathbf{f}\,, \tag{8.320}$$

where $\boldsymbol{\mathcal{H}}$ is a linear operator acting on the object $\mathbf{f}$.

Specifically for the case of digital imaging of an object function, we know from Sec. 7.3.1 that the most general way to write the linear mapping is

$$\overline{g}_m(\mathbf{f}) = \int_{\mathbf{S}_f} d^q r\; h_m(\mathbf{r}) f(\mathbf{r})\,, \qquad m = 1, ..., M\,. \tag{8.321}$$

Except for the overbar and the explicit argument $\mathbf{f}$, this equation is identical to (7.225). We emphasize that the average implied by this overbar is for repeated images of a single object.

To get an expression for the actual random image, we can define an $M \times 1$ noise vector $\mathbf{n}$ by

$$\mathbf{n} \equiv \mathbf{g} - \overline{\mathbf{g}}(\mathbf{f}) = \mathbf{g} - \boldsymbol{\mathcal{H}}\mathbf{f}\,. \tag{8.322}$$

Thus, we have

$$\mathbf{g} = \boldsymbol{\mathcal{H}}\mathbf{f} + \mathbf{n}\,. \tag{8.323}$$

This is the fundamental equation describing noisy, digital imaging of real objects. Now we must understand the statistical properties of $\mathbf{g}$, both for a particular object $\mathbf{f}$ and when a random ensemble of objects is considered.

8.5.2 Conditional statistics for a single object

Conditional density If each component of $\mathbf{g}$ is a continuous random variable, we can denote the conditional probability density function (for a particular object) by $\mathrm{pr}(\mathbf{g}|\mathbf{f})$. If each component of $\mathbf{g}$ can take on only discrete values, we should use the conditional probability $\Pr(\mathbf{g}|\mathbf{f})$, but to avoid considering these two cases in parallel, we shall use the lower-case $\mathrm{pr}(\mathbf{g}|\mathbf{f})$ in both cases, understanding it as a probability density function or probability as needed. Specific forms for $\mathrm{pr}(\mathbf{g}|\mathbf{f})$ will be given later, especially in Chaps. 11 and 12. As we shall see there, independent Poisson models are usually valid when photon-counting detectors are used, and multivariate normal models are valid with most other detectors.

Even without specific models for the detector statistics, we can make some general statements about $\mathrm{pr}(\mathbf{g}|\mathbf{f})$. For one thing, we know that $\mathbf{f}$ affects the data only through the system operator $\boldsymbol{\mathcal{H}}$, so

$$\mathrm{pr}(\mathbf{g}|\mathbf{f}) = \mathrm{pr}(\mathbf{g}|\boldsymbol{\mathcal{H}}\mathbf{f}) = \mathrm{pr}(\mathbf{g}|\boldsymbol{\mathcal{H}}\mathbf{f}_{meas})\,. \tag{8.324}$$

Thus only the measurement component of the object affects the statistics of the image.

Furthermore, for a given $\mathbf{f}$, $\boldsymbol{\mathcal{H}}\mathbf{f}$ is not a random variable, so

$$\mathrm{pr}(\mathbf{g}|\mathbf{f}) = \mathrm{pr}_{\mathbf{n}}(\mathbf{g} - \boldsymbol{\mathcal{H}}\mathbf{f}|\boldsymbol{\mathcal{H}}\mathbf{f})\,, \tag{8.325}$$

where $\mathrm{pr}_{\mathbf{n}}(\mathbf{n}|\overline{\mathbf{g}})$ is the PDF on the noise vector[14] given some mean value for the detector output. If this density is independent of $\overline{\mathbf{g}}$, then we say that the noise is *object-independent* and write

$$\mathrm{pr}(\mathbf{g}|\mathbf{f}) = \mathrm{pr}_{\mathbf{n}}(\mathbf{g} - \mathcal{H}\mathbf{f}) . \tag{8.326}$$

In this case, therefore, the conditional density on $\mathbf{g}$ is just a displaced version of the density on the noise. As we shall see in more detail in later chapters, this object-independent model is often valid for electronic and other excess noise in detectors.

A related approximation is that the noise is object-dependent but *signal-independent*. When we divide an object into signal and background, as in (8.306), it may turn out that the signal is weak compared to the background, and sometimes we can write $\mathrm{pr}(\mathbf{g}|\mathbf{f}) = \mathrm{pr}(\mathbf{g}|\mathbf{f}_b + \mathbf{f}_s) \approx \mathrm{pr}(\mathbf{g}|\mathbf{f}_b)$. This may be a good approximation with photon-counting detectors in low-contrast situations where all components of $\mathcal{H}\mathbf{f}$ are approximately equal.

Another assumption that is often justified in practice is that the components of $\mathbf{n}$ are statistically independent for a fixed object. With discrete arrays of photodiodes, for example, the electronic noise in one element is often statistically independent of noise in all other elements, and we shall see in Chap. 11 that photon-counting detectors viewing a Poisson source almost always yield statistically independent measurements. When this assumption is valid, we have

$$\mathrm{pr}(\mathbf{g}|\mathbf{f}) = \prod_{m=1}^{M} \mathrm{pr}(g_m|\mathbf{f}) . \tag{8.327}$$

Conditional mean and covariance We can also make some general statements about conditional means and covariances. We know already that the conditional mean of $\mathbf{g}$ is

$$\mathrm{E}\{\mathbf{g}|\mathbf{f}\} \equiv \overline{\mathbf{g}} = \mathcal{H}\mathbf{f} , \tag{8.328}$$

from which it follows at once that

$$\mathrm{E}\{\mathbf{n}|\mathbf{f}\} = 0 . \tag{8.329}$$

Thus we can always regard the noise vector as zero-mean.

Since we are conditioning on $\mathbf{f}$ and hence $\mathcal{H}\mathbf{f}$ is not a random variable, the conditional covariance of $\mathbf{g}$ is the same as the covariance of $\mathbf{n}$; notationally, we write

$$\mathbf{K}_{\mathbf{g}|\mathbf{f}} = \mathbf{K}_{\mathbf{n}} , \tag{8.330}$$

but we must allow for the possibility that $\mathbf{K}_{\mathbf{n}}$ depends on $\mathbf{f}$ (in the Poisson case, for example).

8.5.3 Effects of object randomness

Next we examine the image statistics in the case where the object is random. In frequentist terms, we can consider a large number of images, each with a different object drawn from some ensemble. Our knowledge of the object statistics is given by a stochastic model such as those considered in Sec. 8.4.

[14] Recall that we add subscripts to PDFs only when the random variable is not obvious from the argument.

Overall density Formally, we can write the overall probability density as

$$\mathrm{pr}(\mathbf{g}) = \int_{\mathbb{U}} d\mathbf{f} \; \mathrm{pr}(\mathbf{g}|\mathbf{f})\, \mathrm{pr}(\mathbf{f}) \,. \tag{8.331}$$

In principle, this integral runs over the entire infinite-dimensional object space, but from (8.324) we know that only the measurement subspace contributes. The dimensionality of this subspace is R, the rank of the operator $\mathcal{H}$, so really only R components are important. If we expand $\mathbf{f}_{meas}$ in some suitable basis for measurement space as in (7.251), with an $R \times 1$ coefficient vector $\boldsymbol{\alpha}$, then the integral can be written as

$$\mathrm{pr}(\mathbf{g}) = \int_{\infty} d^R\alpha \; \mathrm{pr}(\mathbf{g}|\boldsymbol{\alpha})\, \mathrm{pr}(\boldsymbol{\alpha}) \,. \tag{8.332}$$

Depending on the choice of basis for measurement space, there is some matrix $\mathbf{H}_0$ that exactly maps the coefficients $\boldsymbol{\alpha}$ to $\mathcal{H}\mathbf{f}$ (see Sec. 7.4.3), so we can write

$$\mathrm{pr}(\mathbf{g}) = \int_{\infty} d^R\alpha \; \mathrm{pr}(\mathbf{g}|\mathbf{H}_0\boldsymbol{\alpha})\, \mathrm{pr}(\boldsymbol{\alpha}) \,. \tag{8.333}$$

Derivation of the form of $\mathbf{H}_0$ for the specific case of expansion in natural pixels (Sec. 7.4.3) is an interesting exercise for the reader.

For object-independent noise as in (8.326), (8.333) takes the appealing form,

$$\mathrm{pr}(\mathbf{g}) = \int_{\infty} d^R\alpha \; \mathrm{pr}_{\mathbf{n}}(\mathbf{g} - \mathbf{H}_0\boldsymbol{\alpha})\, \mathrm{pr}(\boldsymbol{\alpha}) \,. \tag{8.334}$$

This equation is not quite a convolution, but nevertheless it can be usefully transformed by Fourier methods. With characteristic functions as defined in (8.27) and some algebra similar to that used in obtaining (8.43), we can show that

$$\psi_{\mathbf{g}}(\boldsymbol{\xi}) = \psi_{\mathbf{n}}(\boldsymbol{\xi})\, \psi_{\boldsymbol{\alpha}}(\mathbf{H}_0^t\boldsymbol{\xi}) \,. \tag{8.335}$$

Increasing the noise level decreases the width of $\psi_{\mathbf{n}}(\boldsymbol{\xi})$ in this Fourier domain, and increasing the degree of object randomness decreases the width of $\psi_{\boldsymbol{\alpha}}(\mathbf{H}_0^t\boldsymbol{\xi})$; either measure decreases the width of $\psi_{\mathbf{g}}(\boldsymbol{\xi})$ and hence increases the spread of $\mathrm{pr}(\mathbf{g})$.

For Poisson noise (8.334) and (8.335) are not valid; instead, (8.334) must be written as[15]

$$\mathrm{Pr}(\mathbf{g}) = \int_{\infty} d^R\alpha \; \mathrm{Pr}(\mathbf{g}|\mathbf{H}_0\boldsymbol{\alpha})\, \mathrm{pr}(\boldsymbol{\alpha}) \,. \tag{8.336}$$

Note that we have written $\mathrm{Pr}(\mathbf{g})$ instead of $\mathrm{pr}(\mathbf{g})$ since Poisson random variables are discrete. The probability (not density) $\mathrm{Pr}(\mathbf{g}|\mathbf{H}_0\boldsymbol{\alpha})$ is just a product of univariate Poisson probabilities, where the mean of g_m is $[\mathbf{H}_0\boldsymbol{\alpha}]_m$.

The transformation of the characteristic function in the Poisson case was derived by Clarkson *et al.* (2002). They show that (8.336) is equivalent to

$$\psi_{\mathbf{g}}(\boldsymbol{\xi}) = \psi_{\boldsymbol{\alpha}}[\mathbf{H}_0^t\, \boldsymbol{\Gamma}(\boldsymbol{\xi})] \,, \tag{8.337}$$

[15]We could also have written $\mathrm{Pr}(\mathbf{g}|\mathbf{H}_0\boldsymbol{\alpha})$ in (8.336) as $\mathrm{Pr}(\mathbf{g}|\boldsymbol{\alpha})$ since $\mathbf{H}_0\boldsymbol{\alpha}$ is fully determined by $\boldsymbol{\alpha}$, but the former version is more useful when we want to write the probability as a product of Poissons; the probability for g_m is specified by a single component of $\mathbf{H}_0\boldsymbol{\alpha}$, but all components of $\boldsymbol{\alpha}$ may be required because of the matrix multiplication.

where $\mathbf{\Gamma}$ is an operator that acts independently on each component of its vector operand; it is defined such that

$$[\mathbf{\Gamma}(\boldsymbol{\xi})]_m = \frac{-1+\exp(-2\pi i\, \xi_m)}{-2\pi i} . \tag{8.338}$$

Clarkson *et al.* (2002) show also that this transformation law applies when $\mathbf{H}_0$ is replaced by a CD operator $\boldsymbol{\mathcal{H}}$ and the full infinite-dimensional vector $\mathbf{f}$ is used in place of the finite-dimensional $\boldsymbol{\alpha}$. In that case the characteristic *function* for $\mathbf{g}$ is related to the characteristic *functional* for $\mathbf{f}$ by

$$\psi_{\mathbf{g}}(\boldsymbol{\xi}) = \Psi_{\mathbf{f}}[\boldsymbol{\mathcal{H}}^\dagger\, \mathbf{\Gamma}(\boldsymbol{\xi})] . \tag{8.339}$$

Overall mean We shall use the notation of (8.331), recognizing that the integral will be realized by expanding the measurement component of $\mathbf{f}$ in some basis and integrating over the coefficients. With this convention, we can write the overall mean image as

$$\mathrm{E}(\mathbf{g}) = \int_\infty d^M g\ \mathbf{g}\, \mathrm{pr}(\mathbf{g}) = \int_\infty d^M g\ \mathbf{g} \int_{\mathbb{U}} d\mathbf{f}\ \mathrm{pr}(\mathbf{g}|\mathbf{f})\, \mathrm{pr}(\mathbf{f}) . \tag{8.340}$$

Shuffling the integrals, we see that

$$\mathrm{E}(\mathbf{g}) = \int_{\mathbb{U}} d\mathbf{f}\ \mathrm{pr}(\mathbf{f}) \int_\infty d^M g\ \mathbf{g}\, \mathrm{pr}(\mathbf{g}|\mathbf{f}) . \tag{8.341}$$

The inner integral is the average of $\mathbf{g}$ with respect to the conditional density, which is precisely what we called $\overline{\mathbf{g}}(\mathbf{f})$ previously, so

$$\mathrm{E}(\mathbf{g}) = \int_{\mathbb{U}} d\mathbf{f}\ \mathrm{pr}(\mathbf{f})\overline{\mathbf{g}}(\mathbf{f}) . \tag{8.342}$$

Another notation that means the same thing is

$$\langle\, \mathbf{g} \,\rangle = \langle \langle \mathbf{g} \rangle_{\mathbf{n}|\mathbf{f}} \rangle_{\mathbf{f}} . \tag{8.343}$$

Yet another notation denotes this overall average as $\overline{\overline{\mathbf{g}}}$, with the double overbar indicating that we have averaged over both the measurement noise and the object variability. This double average can also be seen directly in (8.340) when we recall that $\mathrm{pr}(\mathbf{g}|\mathbf{f})\, \mathrm{pr}(\mathbf{f})$ is also the joint density, $\mathrm{pr}(\mathbf{g}, \mathbf{f})$.

Overall covariance When both measurement noise and object variability are taken into account, the covariance matrix on $\mathbf{g}$ is defined (for real $\mathbf{g}$) by

$$\mathbf{K}_{\mathbf{g}} = \langle\langle\, [\mathbf{g} - \overline{\overline{\mathbf{g}}}]\, [\mathbf{g} - \overline{\overline{\mathbf{g}}}]^t\, \rangle_{\mathbf{n}|\mathbf{f}} \rangle_{\mathbf{f}} . \tag{8.344}$$

Adding and subtracting $\overline{\mathbf{g}}(\mathbf{f})$ in each factor gives

$$\mathbf{K}_{\mathbf{g}} = \langle\langle\, [\mathbf{g} - \overline{\mathbf{g}}(\mathbf{f}) + \overline{\mathbf{g}}(\mathbf{f}) - \overline{\overline{\mathbf{g}}}]\, [\mathbf{g} - \overline{\mathbf{g}}(\mathbf{f}) + \overline{\mathbf{g}}(\mathbf{f}) - \overline{\overline{\mathbf{g}}}]^t\, \rangle_{\mathbf{n}|\mathbf{f}} \rangle_{\mathbf{f}} . \tag{8.345}$$

Noting that $\overline{\mathbf{g}}(\mathbf{f}) - \overline{\overline{\mathbf{g}}}$ does not involve $\mathbf{n}$ (since it has been averaged out) and that $\langle [\mathbf{g} - \overline{\mathbf{g}}(\mathbf{f})] \rangle_{\mathbf{n}|\mathbf{f}} = 0$, we see that

$$\mathbf{K}_{\mathbf{g}} = \langle\langle [\mathbf{g} - \overline{\mathbf{g}}(\mathbf{f})][\mathbf{g} - \overline{\mathbf{g}}(\mathbf{f})]^t \rangle_{\mathbf{n}|\mathbf{f}} \rangle_{\mathbf{f}} + \langle\, [\overline{\mathbf{g}}(\mathbf{f}) - \overline{\overline{\mathbf{g}}}]\, [\overline{\mathbf{g}}(\mathbf{f}) - \overline{\overline{\mathbf{g}}}]^t\, \rangle_{\mathbf{f}} . \tag{8.346}$$

The first term in this expression is just the noise covariance matrix $\mathbf{K_n}$ averaged over $\mathbf{f}$ (though this average is superfluous in the case of object-independent noise); we can denote this term as $\overline{\mathbf{K}}_\mathbf{n}$. The second term has nothing to do with $\mathbf{n}$ but rather reflects the object variability as seen in the mean image; we can denote this term as $\mathbf{K}_{\overline{\mathbf{g}}}$. With this notation, we have

$$\mathbf{K_g} = \overline{\mathbf{K}}_\mathbf{n} + \mathbf{K}_{\overline{\mathbf{g}}} . \tag{8.347}$$

This division of the overall covariance into two terms, one representing the average noise covariance and the other representing the variation in the conditional mean, is exact and does not require any assumptions about the form of either $\mathrm{pr}(\mathbf{g}|\mathbf{f})$ or $\mathrm{pr}(\mathbf{f})$. In particular, it does not require that the noise be object-independent, and it does not require that either the noise or the object be Gaussian.

Other expressions for the object-variability term There are several alternative ways of expressing $\mathbf{K}_{\overline{\mathbf{g}}}$. First, since the object $f(\mathbf{r})$ is a sample function of a random process, we can use the autocovariance operator $\mathcal{K}_\mathbf{f}$, *i.e.*, the integral operator with kernel $K_\mathbf{f}(\mathbf{r}, \mathbf{r}')$. Since $\overline{\mathbf{g}}$ is a linear transformation of $\mathbf{f}$ by (8.320), it follows that [*cf.* (8.50) and (8.145)]

$$\mathbf{K}_{\overline{\mathbf{g}}} = \mathcal{H}\mathcal{K}_\mathbf{f}\mathcal{H}^\dagger . \tag{8.348}$$

Similarly, if we know that $\mathbf{f}_{meas} = \mathbf{H}_0\boldsymbol{\alpha}$ as in (8.333), and if we know the covariance matrix $\mathbf{K}_{\boldsymbol{\alpha}}$, then we have

$$\mathbf{K}_{\overline{\mathbf{g}}} = \mathbf{H}_0\, \mathbf{K}_{\boldsymbol{\alpha}}\, \mathbf{H}_0^t . \tag{8.349}$$

Finally, if we have some approximate object representation as in (7.301) and a system matrix $\mathbf{H}$ as defined in (7.304), and we know a covariance matrix for the coefficients $\boldsymbol{\theta}$, then we can approximate $\mathbf{K}_{\overline{\mathbf{g}}}$ by

$$\mathbf{K}_{\overline{\mathbf{g}}} \approx \mathbf{H}\, \mathbf{K}_{\boldsymbol{\theta}}\, \mathbf{H}^t . \tag{8.350}$$

This approximation will be accurate if the image error defined in (7.329) is small for all objects in the ensemble (and, of course, if $\mathbf{K}_{\boldsymbol{\theta}}$ is accurate).

Sample averages We have written formal expressions for the overall mean and covariance as if we know the densities needed to perform the averages. In practice, we will usually know the conditional density $\mathrm{pr}(\mathbf{g}|\mathbf{f})$, since it follows from the physics of the measurement process; as we have noted, this conditional density will usually be Gaussian or Poisson. The average over objects is much more problematical in practice. In Sec. 8.4 we discussed a variety of statistical models for objects, but we saw that there were many circumstances where we could generate samples of $\mathbf{f}$ but could not develop an analytical expression for $\mathrm{pr}(\mathbf{f})$. In these circumstances we have no choice but to approximate the analytical averages with sample averages; more details on how this is done in practice will be forthcoming in Chap. 14.

8.5.4 Signals and backgrounds in image space

In Sec. 8.4.5, we divided the object into signal and background parts as in (8.306), which we can also write as

$$\mathbf{f} = \mathbf{f}_s + \mathbf{f}_b . \tag{8.351}$$

Now we shall look at how this division affects the image statistics.

Conditional statistics The conditional mean, for a fixed object, is still given by (8.328), but because of the assumed linearity of the operator, we can write separately that

$$\overline{\mathbf{g}}_s = \mathcal{H}\mathbf{f}_s\,, \qquad \overline{\mathbf{g}}_b = \mathcal{H}\mathbf{f}_b\,, \tag{8.352}$$

The conditional covariance is still given by (8.330), but for signal-dependent noise we have to assume in general that the noise covariance matrix depends on both the signal and the background. In many problems, however, we can assume that the signal is weak compared to the background, so $\mathbf{K_n}$ is approximately independent of $\mathbf{f}_s$.

The conditional density is still given by (8.325), which we can now write as

$$\mathrm{pr}(\mathbf{g}|\mathbf{f}) = \mathrm{pr}_{\mathbf{n}}(\mathbf{g} - \mathcal{H}\mathbf{f}_s - \mathcal{H}\mathbf{f}_b|\mathcal{H}\mathbf{f}) \tag{8.353}$$

or, for object-independent noise,

$$\mathrm{pr}(\mathbf{g}|\mathbf{f}) = \mathrm{pr}_{\mathbf{n}}(\mathbf{g} - \mathcal{H}\mathbf{f}_s - \mathcal{H}\mathbf{f}_b)\,. \tag{8.354}$$

For noise that is object-dependent but signal-independent, this expression would become $\mathrm{pr}(\mathbf{g}|\mathbf{f}) = \mathrm{pr}_{\mathbf{n}}(\mathbf{g} - \mathcal{H}\mathbf{f}_s - \mathcal{H}\mathbf{f}_b|\mathcal{H}\mathbf{f}_b)$.

Random background When the background $\mathbf{f}_b$ is random but the signal is not, then the overall probability density function in (8.331) becomes

$$\mathrm{pr}(\mathbf{g}) = \int_{\mathbb{U}} d\mathbf{f}_b\, \mathrm{pr}(\mathbf{g}|\mathbf{f}_b, \mathbf{f}_s)\, \mathrm{pr}(\mathbf{f}_b) \tag{8.355}$$

or, for object-independent noise,

$$\mathrm{pr}(\mathbf{g}) = \int_{\mathbb{U}} d\mathbf{f}_b\, \mathrm{pr}_{\mathbf{n}}(\mathbf{g} - \mathcal{H}\mathbf{f}_b - \mathcal{H}\mathbf{f}_s)\, \mathrm{pr}(\mathbf{f}_b)\,. \tag{8.356}$$

For a nonrandom signal, the overall covariance matrix is almost unchanged from before; from (8.347) and (8.348), we have

$$\mathbf{K_g} = \overline{\mathbf{K}}_{\mathbf{n}} + \mathcal{H}\mathcal{K}_{\mathbf{f}_b}\mathcal{H}^\dagger\,. \tag{8.357}$$

Essentially the only change here is the subscript on $\mathcal{K}$.

Random signals If both signal and background are random but they are statistically independent, the overall density on the data is given by [*cf.* (8.355)]

$$\mathrm{pr}(\mathbf{g}) = \int_{\mathbb{U}} d\mathbf{f}_s \int_{\mathbb{U}} d\mathbf{f}_b\, \mathrm{pr}(\mathbf{g}|\mathbf{f}_b, \mathbf{f}_s)\, \mathrm{pr}(\mathbf{f}_b)\, \mathrm{pr}(\mathbf{f}_s)\,. \tag{8.358}$$

The overall covariance matrix in this case is given by

$$\mathbf{K_g} = \overline{\mathbf{K}}_{\mathbf{n}} + \mathcal{H}\mathcal{K}_{\mathbf{f}_s}\mathcal{H}^\dagger + \mathcal{H}\mathcal{K}_{\mathbf{f}_b}\mathcal{H}^\dagger\,. \tag{8.359}$$

If $\mathbf{f}_s$ and $\mathbf{f}_b$ are not statistically independent, we can write $\mathrm{pr}(\mathbf{f}_b)\,\mathrm{pr}(\mathbf{f}_s) = \mathrm{pr}(\mathbf{f}_b|\mathbf{f}_s)\,\mathrm{pr}(\mathbf{f}_s)$ and do a nested average as in (8.344); the result will be that $\mathcal{K}_{\mathbf{f}_b}$ acquires an overbar indicating that it is to be averaged over signals.

9

Diffraction Theory and Imaging

In Chapter 7 we looked at various ways of describing imaging systems as mathematical operators. Now we begin to connect these operators with physical phenomena and specific imaging systems. In particular, in this chapter we consider the broad category of imaging via wave propagation. The waves in question can be light waves, other electromagnetic waves such as microwaves, sound waves as in medical ultrasound, or even matter waves as in electron microscopy.

To analyze such systems, we need a thorough understanding of wave propagation and diffraction theory, and much of this chapter is devoted to those topics. The story begins with first principles, which for electromagnetic waves is Maxwell's equations. In Sec. 9.1 we use Maxwell's equations to derive the general time-dependent wave equation, and then we specialize to the case of monochromatic or single-frequency waves. Two important special solutions of the wave equation are plane waves and spherical waves, both discussed in Sec. 9.2.

In Sec. 9.3 we develop an important tool called Green's functions for solving the wave equations. Green's functions will prove to be familiar from Chap. 7 since they are the point response functions if we think of the wave equation as a linear system.

Waves passing through a finite aperture or radiating from a finite source propagate in complicated ways, giving rise to beautiful and highly useful patterns. In Secs. 9.4 and 9.5 we develop the essential mathematical techniques needed to calculate these patterns. These techniques are referred to collectively as diffraction theory, though they could equally well be called radiation theory. As we shall see, they also fit nicely into the theory of linear shift-invariant systems developed in Chap. 7.

In Sec. 9.6 we start to apply diffraction theory to imaging. We introduce a mathematical description for an ideal lens and then analyze how it forms an image of a monochromatic point object. Then deviations from ideal behavior, called aberrations, are discussed from a diffraction-theory viewpoint.

The imaging of extended planar objects is discussed in Sec. 9.7. Included in Sec. 9.7 is an introduction to the concept of coherence, needed when the radiation source is random. This development builds on the discussion of random processes in Chap. 8.

The treatment of diffraction and imaging in Secs. 9.4–9.7 is essentially two-dimensional; objects are described as planar apertures or transparencies, and the resulting fields are computed on planes. In Sec. 9.8, however, we extend the discussion to volume objects and fields computed at general points in a 3D space.

9.1 WAVE EQUATIONS

In this section we derive the basic equations that describe wave propagation. The treatment here is for electromagnetic waves, but the resulting theory is valid, with minor modifications, for acoustic waves as well. The key difference is that an electromagnetic wave consists of vector fields (such as the electric and magnetic fields), while an acoustic wave consists of more complicated fields (the stress and strain) called tensors. In practice, however, we usually reduce both the electromagnetic and acoustic wave equations to an equation describing a scalar field, and at that point the mathematical distinction between electromagnetics and acoustics largely disappears. Moreover, the scalar theory is also applicable to electron imaging when the electrons are regarded as waves (Reimer, 1985).

9.1.1 Maxwell's equations

A vector field is a vector, each component of which is a function of spatial position and time. Spatial position is given by a 3D vector which we shall denote by the boldface gothic $\mathfrak{r}$. Later in this chapter, we shall also deal with 2D position vectors, which will be denoted by $\mathbf{r}$. In Cartesian coordinates,

$$\mathfrak{r} = (x, y, z)\,, \qquad \mathbf{r} = (x, y)\,. \tag{9.1}$$

A 3D vector field has three Cartesian components, each a function of $\mathfrak{r}$ and time t. Electromagnetic waves are described by four such vector fields: (1) the electric field $\mathbf{e}(\mathfrak{r}, t)$, (2) the electric flux density or displacement $\mathbf{d}(\mathfrak{r}, t)$, (3) the magnetic field $\mathbf{h}(\mathfrak{r}, t)$ and (4) the magnetic flux density $\mathbf{b}(\mathfrak{r}, t)$. These fields are produced by a current density (current per unit area) $\mathbf{j}(\mathfrak{r}, t)$ and a charge density (charge per unit volume) $q(\mathfrak{r}, t)$. This notation is nonstandard in that it uses lower-case letters where most books use capitals, and in the use of q for charge density instead of the usual ρ, which we want to reserve for spatial frequency. A potential confusion is that q is often used for charge, but here it is charge per unit volume.

The field quantities obey Maxwell's equations, a set of four coupled first-order partial differential equations. In the International System of Units (SI), Maxwell's equations are given by

$$\nabla \times \mathbf{e}(\mathfrak{r}, t) = -\frac{\partial}{\partial t}\mathbf{b}(\mathfrak{r}, t)\,, \tag{9.2a}$$

$$\nabla \times \mathbf{h}(\mathfrak{r}, t) = \mathbf{j}(\mathfrak{r}, t) + \frac{\partial}{\partial t}\mathbf{d}(\mathfrak{r}, t)\,, \tag{9.2b}$$

$$\nabla \cdot \mathbf{d}(\mathfrak{r}, t) = q(\mathfrak{r}, t)\,, \tag{9.2c}$$

$$\nabla \cdot \mathbf{b}(\mathbf{r}, t) = 0\,. \tag{9.2d}$$

In these equations, $\nabla\cdot$ and $\nabla\times$ are the divergence and curl operators, respectively. They can be expressed in terms of the vector operator ∇, specified in Cartesian coordinates as $\left(\frac{\partial}{\partial x}, \frac{\partial}{\partial y}, \frac{\partial}{\partial z}\right)$. The divergence of a vector field such as $\mathbf{b}(\mathbf{r}, t)$ is a scalar field obtained formally by performing a 3D scalar or dot product between ∇ and $\mathbf{b}(\mathbf{r}, t)$, so that

$$\nabla \cdot \mathbf{b}(\mathbf{r}, t) = \frac{\partial b_x(\mathbf{r}, t)}{\partial x} + \frac{\partial b_y(\mathbf{r}, t)}{\partial y} + \frac{\partial b_z(\mathbf{r}, t)}{\partial z}\,. \tag{9.3}$$

The curl is computed similarly except that a vector cross product is used. The curl of a vector field is thus another vector field, with components given by

$$[\nabla \times \mathbf{b}(\mathbf{r}, t)]_i = \frac{\partial b_j(\mathbf{r}, t)}{\partial x_k} - \frac{\partial b_k(\mathbf{r}, t)}{\partial x_j}\,, \tag{9.4}$$

where (i, j, k) is a cyclic permutation of $(1, 2, 3)$.

We shall often consider free space where there are no material media, no currents and no charges. In free space, $q(\mathbf{r}, t) = 0$, $\mathbf{j}(\mathbf{r}, t) = 0$, and the following relations connect $\mathbf{d}(\mathbf{r}, t)$ to $\mathbf{e}(\mathbf{r}, t)$ and $\mathbf{b}(\mathbf{r}, t)$ to $\mathbf{h}(\mathbf{r}, t)$:

$$\mathbf{d}(\mathbf{r}, t) = \epsilon_0 \mathbf{e}(\mathbf{r}, t)\,, \qquad \mathbf{b}(\mathbf{r}, t) = \mu_0 \mathbf{h}(\mathbf{r}, t)\,, \tag{9.5}$$

where ϵ_0 and μ_0 are the permittivity and permeability, respectively, of free space. These equations are called *constitutive relations*. More complicated constitutive relations, to be discussed in Sec. 9.1.3, apply in material media.

In free space, Maxwell's equations can be written entirely in terms of $\mathbf{e}(\mathbf{r}, t)$ and $\mathbf{h}(\mathbf{r}, t)$:

$$\nabla \times \mathbf{e}(\mathbf{r}, t) = -\mu_0 \frac{\partial}{\partial t} \mathbf{h}(\mathbf{r}, t)\,, \tag{9.6a}$$

$$\nabla \times \mathbf{h}(\mathbf{r}, t) = \epsilon_0 \frac{\partial}{\partial t} \mathbf{e}(\mathbf{r}, t)\,, \tag{9.6b}$$

$$\nabla \cdot \mathbf{e}(\mathbf{r}, t) = 0\,, \tag{9.6c}$$

$$\nabla \cdot \mathbf{h}(\mathbf{r}, t) = 0\,. \tag{9.6d}$$

9.1.2 Maxwell's equations in the Fourier domain

By straightforward Fourier analysis, $\mathbf{e}(\mathbf{r}, t)$ can be represented as

$$\mathbf{e}(\mathbf{r}, t) = \int_\infty d^3\sigma \int_{-\infty}^{\infty} d\nu\; \mathbf{E}(\boldsymbol{\sigma}, \nu) \exp[2\pi i(\boldsymbol{\sigma} \cdot \mathbf{r} - \nu t)]\,, \tag{9.7}$$

with the inverse relation

$$\mathbf{E}(\boldsymbol{\sigma}, \nu) = \int_\infty d^3\mathbf{r} \int_{-\infty}^{\infty} dt\; \mathbf{e}(\mathbf{r}, t) \exp[-2\pi i(\boldsymbol{\sigma} \cdot \mathbf{r} - \nu t)]\,. \tag{9.8}$$

Note that we use $\boldsymbol{\sigma}$ for the 3D spatial-frequency vector here, reserving $\boldsymbol{\rho}$ for the 2D frequency later on.

We shall refer to $\mathbf{E}(\boldsymbol{\sigma},\nu)$ as the 4D Fourier transform of $\mathbf{e}(\mathbf{r},t)$ even though it does not conform to the sign conventions for Fourier transforms introduced in Chap. 3. By those conventions, $\mathbf{E}(\boldsymbol{\sigma},\nu)$ is a 3D spatial Fourier transform and a 1D *inverse* Fourier transform of $\mathbf{e}(\mathbf{r},t)$. The convention employed in (9.7) and (9.8) has the advantage that a component with positive ν represents a wave travelling in the $+\boldsymbol{\sigma}$ direction. With the understanding that the usual Fourier sign convention is reversed for temporal transforms, we write

$$\mathbf{E}(\boldsymbol{\sigma},\nu) = \mathcal{F}_4\{\mathbf{e}(\mathbf{r},t)\}\,. \tag{9.9}$$

The Fourier transform of a vector is another vector, obtained by transforming each Cartesian component of the first. In other words, $E_x\{\boldsymbol{\sigma},\nu\} = \mathcal{F}_4\{e_x(\mathbf{r},t)\}$, and similarly for y and z components. Note also that that the vectors being transformed are 3D but the transform operations are 4D; the argument of the 3D vector field $\mathbf{e}(\mathbf{r},t)$ can be regarded as a 4D vector with components (x,y,z,t).

From (3.234) we can establish that

$$\mathcal{F}_4\left\{\tfrac{\partial}{\partial t}\mathbf{e}(\mathbf{r},t)\right\} = -2\pi i\nu\mathbf{E}(\boldsymbol{\sigma},\nu)\,, \tag{9.10}$$

$$\mathcal{F}_4\big\{\nabla\cdot\mathbf{e}(\mathbf{r},t)\big\} = 2\pi i\boldsymbol{\sigma}\cdot\mathbf{E}(\boldsymbol{\sigma},\nu)\,, \tag{9.11}$$

$$\mathcal{F}_4\big\{\nabla\times\mathbf{e}(\mathbf{r},t)\big\} = 2\pi i\boldsymbol{\sigma}\times\mathbf{E}(\boldsymbol{\sigma},\nu)\,. \tag{9.12}$$

With similar Fourier representations for the other fields, Maxwell's equations can be transformed into

$$\boldsymbol{\sigma}\times\mathbf{E}(\boldsymbol{\sigma},\nu) = \nu\mathbf{B}(\boldsymbol{\sigma},\nu)\,, \tag{9.13a}$$

$$2\pi i\boldsymbol{\sigma}\times\mathbf{H}(\boldsymbol{\sigma},\nu) = \mathbf{J}(\boldsymbol{\sigma},\nu) - 2\pi i\nu\mathbf{D}(\boldsymbol{\sigma},\nu)\,, \tag{9.13b}$$

$$2\pi i\boldsymbol{\sigma}\cdot\mathbf{D}(\boldsymbol{\sigma},\nu) = Q(\boldsymbol{\sigma},\nu)\,, \tag{9.13c}$$

$$\boldsymbol{\sigma}\cdot\mathbf{B}(\boldsymbol{\sigma},\nu) = 0\,. \tag{9.13d}$$

Fourier transformation has converted the coupled differential equations into coupled algebraic equations.

Note that it is not legal to set $\mathbf{J}$ and Q to zero and then claim that equations (9.13) are valid in free space. Equations (9.6) are valid in any region of space where $\mathbf{j}$ and q are zero since the differential operators in (9.6) are *local*, relating fields at a point to sources at the *same* point, but Fourier transformation is not local. The Fourier transforms $\mathbf{J}$ and Q would be zero for all $\boldsymbol{\sigma}$ and ν only if $\mathbf{j}$ and q were zero *everywhere*, which is a physically uninteresting case.

SVD of Maxwell's equations In Chap. 7, one of our primary tools for describing linear systems was singular-value decomposition or SVD. The value of SVD is that it reduces complicated linear operators to simple multiplications. Since Maxwell's equations are linear, they should also be amenable to SVD analysis.

In fact, equations (9.13) can be regarded as Maxwell's equations in SVD form. We saw in Sec. 2.4.5 that the vector derivatives that occur in Maxwell's equations can all be expressed in terms of convolutions with corresponding derivatives of delta functions. We also know from Sec. 7.2.3 that Fourier analysis is equivalent to SVD for convolution operators, so the multiplicative forms in (9.13) should not be surprising. A differential equation with constant coefficients is a linear, shift-invariant system.

9.1.3 Material media

Equations (9.6a-d) or (9.13a-d) will form the basis for our discussion of electromagnetic waves, but before proceeding we take a look at how they must be modified if the wave is in a material medium rather than free space. The main thing we have to re-examine is the constitutive relations, (9.5). The electric field and electric displacement in a material medium are related by

$$\mathbf{d}(\mathbf{r}, t) = \epsilon_0 \mathbf{e}(\mathbf{r}, t) + \mathbf{p}(\mathbf{r}, t)\,, \tag{9.14}$$

where $\mathbf{p}(\mathbf{r}, t)$ is the *polarization* of the medium, defined as the electric dipole moment per unit volume averaged over a volume large compared to an atom. A useful physical picture of polarization is that it results from distortion of a charge distribution. A symmetric distribution of charge for which $q(\mathbf{r}, t) = q(-\mathbf{r}, t)$ has no dipole moment and hence no polarization. As an example, an isolated atom in any stationary state (any definite energy level) has no dipole moment because the electronic charge cloud is symmetrically disposed about the nucleus, so a collection of such atoms has no polarization.

Except for certain materials called *ferroelectrics*, there is no permanent polarization; $\mathbf{p}(\mathbf{r}, t)$ is induced by the applied electric field. For example, an electric field exerts a force on the electron cloud around an atom, distorting it and creating a polarization. If the electric field is strong, the polarization can be a nonlinear function of the electric field, but for weak fields the distortion is proportional to the force, so $\mathbf{p}(\mathbf{r}, t)$ is a linear function of $\mathbf{e}(\mathbf{r}, t)$. Except for a few places where we mention nonlinear optics, we shall assume linear media in this sense throughout this book.

The most general linear model would allow $\mathbf{p}(\mathbf{r}, t)$ to be influenced by $\mathbf{e}(\mathbf{r}', t')$ for all $\mathbf{r}'$ and for all $t' \leq t$. The latter restriction is imposed since the response (polarization) at time t has no way of anticipating the stimulus (electric field) at a later time; the response must be causal (see Sec. 7.2.3). Since any Cartesian component of the vector $\mathbf{p}(\mathbf{r}, t)$ can, in general, be affected by any component of $\mathbf{e}(\mathbf{r}, t)$, the most general linear relation between $\mathbf{p}(\mathbf{r}, t)$ and the electric field that induces it has the form

$$\mathbf{p}(\mathbf{r}, t) = \int_{\infty} d^3\mathbf{r}' \int_{-\infty}^{t} dt'\, \mathbf{M}(\mathbf{r}, t; \mathbf{r}', t')\, \mathbf{e}(\mathbf{r}', t')\,, \tag{9.15}$$

where $\mathbf{M}(\mathbf{r}, t; \mathbf{r}', t')$ is a 3×3 response matrix, each element of which is a function of two spatial variables, $\mathbf{r}$ and $\mathbf{r}'$, and two temporal variables, t and t'. If the medium is isotropic, the polarization must be parallel to the electric field by symmetry; there is no preferred direction in the problem except that of the electric field. The mathematical statement of isotropy is that $\mathbf{M}$ is some function $m(\mathbf{r}, t; \mathbf{r}', t')$ times the 3×3 unit matrix. With this form, each Cartesian component of the polarization is affected only by the corresponding component of the electric field.

Two additional assumptions will simplify the formulation further. First, since polarization is the result of the force exerted by an electric field on a charge distribution, we assume that it is local, which means that the polarization at some point $\mathbf{r}$ is influenced only by the field at that same point $\mathbf{r}$. In addition, if the properties of the medium are not themselves time-dependent, the system is temporally shift-invariant (see Sec. 7.2.3). We express these assumptions mathematically by writing $m(\mathbf{r}, t; \mathbf{r}', t') = \delta(\mathbf{r} - \mathbf{r}')\, w(\mathbf{r}, t - t')$. To ensure causality, we assume that

$w(\mathbf{r}, t - t') = 0$ if $t < t'$. With these assumptions, (9.15) becomes

$$\mathbf{p}(\mathbf{r}, t) = \int_{-\infty}^{\infty} dt' \, w(\mathbf{r}, t - t') \, \mathbf{e}(\mathbf{r}, t') \,, \tag{9.16}$$

which is now in the form of a 1D temporal convolution for each spatial position and each Cartesian component. It is necessary to retain the temporal convolution if we wish to consider electric fields with arbitrary temporal dependence.

If the medium is homogeneous, so that its response is independent of position, we can write

$$w(\mathbf{r}, t) = w(t) = \mathcal{F}_1^{-1}\{W(\nu)\} \,. \tag{9.17}$$

With this assumption, a 1D temporal Fourier transform of (9.16) followed by a 3D spatial Fourier transform yields

$$\mathbf{P}(\boldsymbol{\sigma}, \nu) = W(\nu) \, \mathbf{E}(\boldsymbol{\sigma}, \nu) \,, \tag{9.18}$$

where $\mathbf{P}(\boldsymbol{\sigma}, \nu) = \mathcal{F}_4\{\mathbf{p}(\mathbf{r}, t)\}$ and $W(\nu) = \mathcal{F}_1\{w(t)\}$.

If we also assume that $\mathbf{E}(\boldsymbol{\sigma}, \nu)$ is concentrated in a narrow band of frequencies centered around $\nu = \nu_0$, it may sometimes be valid to assume that $\mathbf{P}(\boldsymbol{\sigma}, \nu) \approx W(\nu_0) \, \mathbf{E}(\boldsymbol{\sigma}, \nu)$ for frequencies in this band. In these circumstances, it is customary to define $W(\nu_0) = \epsilon_0 \chi(\nu_0)$, where $\chi(\nu)$ is the *frequency-dependent susceptibility*. With this definition we have

$$\mathbf{P}(\boldsymbol{\sigma}, \nu) \approx \epsilon_0 \chi(\nu_0) \, \mathbf{E}(\boldsymbol{\sigma}, \nu) \,, \qquad \mathbf{p}(\mathbf{r}, t) \approx \epsilon_0 \chi(\nu_0) \, \mathbf{e}(\mathbf{r}, t) \,. \tag{9.19}$$

Plugging this result into (9.14), we find

$$\mathbf{d}(\mathbf{r}, t) = \epsilon_0 [1 + \chi(\nu_0)] \mathbf{e}(\mathbf{r}, t) \equiv \epsilon(\nu_0) \, \mathbf{e}(\mathbf{r}, t) \,. \tag{9.20}$$

Thus, if the long string of assumptions made in this section is valid, all we have to do to use the constitutive relation $\mathbf{d} = \epsilon_0 \mathbf{e}$ in a material medium is to replace ϵ_0 with $\epsilon(\nu_0)$. It is worth recounting the assumptions that go into this result, however. We must assume that the medium is linear, isotropic and homogeneous, that its response is local (which means that $\mathbf{p}(\mathbf{r}, t)$ is determined by $\mathbf{e}(\mathbf{r}, t)$ at that same $\mathbf{r}$), that $\mathbf{e}(\mathbf{r}, t)$ is narrowband or *quasimonochromatic* in the sense that its temporal Fourier transform spans a narrow range of frequencies, and that the response of the medium is approximately independent of the temporal frequency over this range. If all of these assumptions hold, we can use (9.20), but it is incumbent on us to check them in particular problems.

One might think that we would have to go through a similar discussion to find a generalization of the magnetic constitutive relation, $\mathbf{b}(\mathbf{r}, t) = \mu_0 \mathbf{h}(\mathbf{r}, t)$, but in optical problems we have a great simplification. Unlike electric dipoles, magnetic dipoles do not respond at optical frequencies, so magnetization (magnetic dipole moment per unit volume) is independent of the applied field. Thus we can use the free-space magnetic constitutive relation with impunity in a material medium.

The punchline of this section is that we can often get away with (9.6) simply by replacing ϵ_0 with $\epsilon(\nu_0)$. Media for which this approximation is valid will be called *nondispersive*, meaning that $\epsilon(\nu)$ does not vary much over the relevant temporal bandwidth. To simplify the notation we shall use ϵ for $\epsilon(\nu_0)$, but the requirements for narrowband radiation and nondispersive media should be kept in mind.

9.1.4 Time-dependent wave equations

Though Maxwell's equations are a complete description of electromagnetic waves, it is more convenient to work with wave equations, which are second-order partial differential equations.

Consider a region of space where there are no material media, but where there may be charges and currents to generate the field. In this region, the constitutive relations of (9.5) hold. The wave equation for the electric field in this region is obtained by taking the curl of (9.2a) and using the other Maxwell's equations, the constitutive relations and the following identity from vector calculus:

$$\nabla \times (\nabla \times \mathbf{e}) = \nabla(\nabla \cdot \mathbf{e}) - \nabla^2 \mathbf{e}\,, \tag{9.21}$$

where ∇^2 is the Laplacian operator $\frac{\partial^2}{\partial x^2} + \frac{\partial^2}{\partial y^2} + \frac{\partial^2}{\partial z^2}$. The result of these manipulations is

$$\left(\nabla^2 - \mu_0\epsilon_0 \frac{\partial^2}{\partial t^2}\right)\mathbf{e}(\mathbf{r}, t) = \mu_0 \frac{\partial}{\partial t}\mathbf{j}(\mathbf{r}, t) + \frac{1}{\epsilon_0}\nabla q(\mathbf{r}, t)\,. \tag{9.22}$$

The advantage of using a second-order equation is now apparent: (9.22) is an equation for $\mathbf{e}(\mathbf{r}, t)$ alone, uncoupled from the other field quantities. If $\mathbf{j}(\mathbf{r}, t)$ and $q(\mathbf{r}, t)$ are specified, we can solve directly for $\mathbf{e}(\mathbf{r}, t)$ (and we shall do so in Sec. 9.3.4).

Note also that (9.22) is really three equations, one for each of the three Cartesian components of $\mathbf{e}(\mathbf{r}, t)$, but these equations are uncoupled as well. We can solve for $e_x(\mathbf{r}, t)$ knowing only the x component of the right-hand side of (9.22) and without having to solve for the y and z components of $\mathbf{e}(\mathbf{r}, t)$.

Similar wave equations can be obtained for other field quantities of interest. For example, the wave equation for $\mathbf{h}(\mathbf{r}, t)$ has exactly the same differential operator on the left-hand side as in (9.22), but with a different driving term on the right. No matter what field quantity we consider, we arrive at the *time-dependent inhomogeneous scalar wave equation*, the general form of which is (Jackson, 1998; Arfken and Weber, 1995)

$$\left(\nabla^2 - \frac{1}{c_m^2}\frac{\partial^2}{\partial t^2}\right)u(\mathbf{r}, t) = s(\mathbf{r}, t)\,, \tag{9.23}$$

where $u(\mathbf{r}, t)$ is some scalar field and $s(\mathbf{r}, t)$ is a corresponding scalar source term related in some way to charges and currents. The field $u(\mathbf{r}, t)$ may, for example, be a Cartesian component of the electric or magnetic field, or it may be the scalar potential or a Cartesian component of the vector potential. Each choice for $u(\mathbf{r}, t)$ has associated with it an appropriate $s(\mathbf{r}, t)$, the form of which can be derived from Maxwell's equations. If $s(\mathbf{r}, t) = 0$, (9.23) reduces to the *homogeneous time-dependent wave equation*.

The constant c_m in (9.23) has dimensions of speed, and we shall see in Sec. 9.2 that it can be interpreted as the speed of wave propagation in the medium (hence the subscript). Different media and different kinds of waves require different choices for c_m. For electromagnetic waves in vacuum, $c_m^2 = (\mu_0\epsilon_0)^{-1} = c^2$, where c is the speed of light. In a material satisfying the assumptions of Sec. 9.1.3, $c_m^2 = 1/\mu_0\epsilon \equiv c^2/n^2$, where n is the index of refraction of the medium. For homogeneous media, c_m is a constant, but in general it can be a function of position and time.

The scalar wave equation (9.23) applies to other kinds of waves as well. For example, with proper choice of c_m, $u(\mathbf{r},t)$ and $s(\mathbf{r},t)$, it applies to acoustic waves. For the remainder of this chapter, we shall work directly with (9.23) and not worry about the physical interpretation of $u(\mathbf{r},t)$ and $s(\mathbf{r},t)$.

The Fourier transform of the time-dependent scalar wave equation is

$$-4\pi^2\left(\sigma^2 - \frac{\nu^2}{c_m^2}\right)U(\boldsymbol{\sigma},\nu) = S(\boldsymbol{\sigma},\nu)\,, \tag{9.24}$$

where $U(\boldsymbol{\sigma},\nu)$ and $S(\boldsymbol{\sigma},\nu)$ denote 4D Fourier transforms of $u(\mathbf{r},t)$ and $s(\mathbf{r},t)$, respectively. Once again, Fourier transformation has converted a partial differential equation to an algebraic one.

9.1.5 Time-independent wave equations

An important special case of the scalar wave equation arises when the source oscillates at a single frequency, so we can write

$$s(\mathbf{r},t) = s(\mathbf{r})\exp(-2\pi i\nu_0 t)\,. \tag{9.25}$$

Of course, $s(\mathbf{r},t)$ must be real, so the real part of the complex exponential is understood (see Sec. 7.1.1). Because of the linearity of the wave equation, we can use complex notation for source and field and separate real and imaginary parts at the end of the calculation. We allow $s(\mathbf{r})$ to be complex, so the magnitude and phase of the oscillation can vary arbitrarily with position.

The Fourier representation of the source in (9.25) is

$$S(\boldsymbol{\sigma},\nu) = S(\boldsymbol{\sigma})\,\delta(\nu-\nu_0)\,. \tag{9.26}$$

A source that satisfies (9.25), and hence (9.26), is said to be *monochromatic* (the light has a single color). Inserting (9.26) into (9.24) yields

$$-4\pi^2\left(\sigma^2 - \frac{\nu^2}{c_m^2}\right)U(\boldsymbol{\sigma},\nu) = S(\boldsymbol{\sigma})\,\delta(\nu-\nu_0)\,, \tag{9.27}$$

which can be satisfied only if

$$U(\boldsymbol{\sigma},\nu) = U(\boldsymbol{\sigma})\,\delta(\nu-\nu_0)\,, \tag{9.28}$$

or, in the space domain,

$$u(\mathbf{r},t) = u(\mathbf{r})\exp(-2\pi i\nu_0 t)\,. \tag{9.29}$$

In other words, the field must have the same monochromatic time dependence as the source, though possibly with a phase shift since $u(\mathbf{r})$ can be complex. Had we chosen some other time dependence for the source, other than the complex exponential, the time dependence of the field would not have been the same as that of the source. The wave equation describes a temporal linear shift-invariant system, and the complex exponential $\exp(-2\pi i\nu_0 t)$ is an eigenfunction of it.

With a monochromatic source, the Fourier transform of the wave amplitude must satisfy

$$-4\pi^2\left(\sigma^2 - \frac{\nu_0^2}{c_m^2}\right)U(\boldsymbol{\sigma}) = S(\boldsymbol{\sigma})\,. \tag{9.30}$$

Transforming back to the space domain yields

$$\left(\nabla^2 + k^2\right) u(\mathbf{r}) = s(\mathbf{r}) , \tag{9.31}$$

where $k = 2\pi\nu_0/c_m$. This equation is known as the *time-independent scalar wave equation* or the *Helmholtz equation*. If $s(\mathbf{r}) = 0$, (9.31) is called the *homogeneous Helmholtz equation*.

Poisson and Laplace equations If the source is independent of time ($\nu_0 = 0$), then $k = 0$ and the Helmholtz equation reduces to the *Poisson equation*, given by

$$\nabla^2 u(\mathbf{r}) = s(\mathbf{r}) . \tag{9.32}$$

The homogeneous Poisson equation is called the *Laplace equation*. The Poisson and Laplace equations are fundamental in electrostatics and magnetostatics.

9.2 PLANE WAVES AND SPHERICAL WAVES

In later sections we shall determine the general forms of solutions to the time-dependent and time-independent wave equations, but first we examine two important special solutions, plane waves and spherical waves, in order to assemble some notation, terminology and mathematical tools that will be needed later. We consider here only homogeneous, source-free media.

9.2.1 Plane waves

A monochromatic plane wave of wavevector $\mathbf{k}$ and frequency ν_0 has the form

$$u(\mathbf{r}, t) \propto \exp(i\mathbf{k} \cdot \mathbf{r} - 2\pi i \nu_0 t) . \tag{9.33}$$

In the direction parallel to $\mathbf{k}$, this function is periodic with period $2\pi/k$ (where $k = |\mathbf{k}|$), so we can define a period or wavelength by $\lambda = 2\pi/k$.

The 3D wavevector $\mathbf{k}$ can be expressed in terms of its Cartesian components by $\mathbf{k} = (k_x, k_y, k_z)$, but it is convenient to use the 3D spatial frequency $\boldsymbol{\sigma}$ and define

$$\mathbf{k} = 2\pi\boldsymbol{\sigma} = (2\pi\xi, 2\pi\eta, 2\pi\zeta) , \tag{9.34}$$

where $\xi = k_x/2\pi$, etc. In order for this plane wave to be a solution of the homogeneous wave equation, (9.23) with $s(\mathbf{r}, t) = 0$, we require

$$\left(\nabla^2 - \frac{1}{c_m^2}\frac{\partial^2}{\partial t^2}\right) \exp[2\pi i(\boldsymbol{\sigma} \cdot \mathbf{r} - \nu_0 t)]$$

$$= -4\pi^2\left(\xi^2 + \eta^2 + \zeta^2 - \frac{\nu_0^2}{c_m^2}\right) \exp[2\pi i(\boldsymbol{\sigma} \cdot \mathbf{r} - \nu_0 t)] = 0 , \tag{9.35}$$

or

$$\sigma^2 = \xi^2 + \eta^2 + \zeta^2 = \frac{\nu_0^2}{c_m^2} . \tag{9.36}$$

Since $\boldsymbol{\sigma} = 2\pi k$ and $k = 2\pi/\lambda$, we can also write

$$\xi^2 + \eta^2 + \zeta^2 = \frac{1}{\lambda^2} . \tag{9.37}$$

It also follows that the frequency and wavevector must be related by

$$k = \frac{2\pi\nu_0}{c_m} \,. \tag{9.38}$$

An important conclusion from (9.37) is that the three components of $\boldsymbol{\sigma}$ are not independent. If, for example, ξ and η are given (and it is known that we are discussing a simple plane wave of wavelength λ), ζ can be found from (9.37) as

$$\zeta = \pm\sqrt{\frac{1}{\lambda^2} - \xi^2 - \eta^2} \,. \tag{9.39}$$

The only ambiguity is the sign, but if we are dealing with waves propagating generally in the $+z$ direction, we must take the $+$ sign.

Speed of propagation An instructive way to write a monochromatic plane wave is

$$u(\mathbf{r}, t) = \exp[ik(\hat{\boldsymbol{\kappa}} \cdot \mathbf{r} - c_m t)] \,, \tag{9.40}$$

where $\hat{\boldsymbol{\kappa}}$ is a unit vector in the direction of $\mathbf{k}$ (*i.e.*, $\mathbf{k} = k\hat{\boldsymbol{\kappa}}$) and we have used (9.38). This form shows immediately that c_m can indeed be interpreted as the speed of propagation in the medium; the wave has a constant value on the plane $\hat{\boldsymbol{\kappa}} \cdot \mathbf{r} = c_m t$, and this plane moves in direction $\hat{\boldsymbol{\kappa}}$ with speed c_m.

Other mathematical descriptions of plane waves The spatial-frequency components ξ, η and ζ can be related back to the direction of propagation of the plane wave. If α, β and γ denote the direction cosines of $\boldsymbol{\sigma}$ (or equivalently, of $\mathbf{k}$ or $\hat{\boldsymbol{\kappa}}$) with respect to the x, y and z axes, respectively, we have

$$k_x = 2\pi\alpha/\lambda \,, \qquad k_y = 2\pi\beta/\lambda \,, \qquad k_z = 2\pi\gamma/\lambda \,, \tag{9.41}$$

$$\xi = \alpha/\lambda \,, \qquad \eta = \beta/\lambda \,, \qquad \zeta = \gamma/\lambda \,. \tag{9.42}$$

The direction cosines obey the constraint

$$\alpha^2 + \beta^2 + \gamma^2 = 1 \,, \tag{9.43}$$

which is equivalent to (9.37).

In the paraxial approximation, when $\boldsymbol{\sigma}$ is nearly parallel to the z axis and α and β are both near zero, it is convenient to work with angles measured from the z axis. Suppose the projection of $\boldsymbol{\sigma}$ onto the x-z plane makes an angle θ_x with the z axis and the projection onto the y-z plane makes an angle θ_y with the z axis. Then,

$$\sin\theta_x = \frac{\xi}{\sqrt{\xi^2 + \zeta^2}} \,, \qquad \sin\theta_y = \frac{\eta}{\sqrt{\eta^2 + \zeta^2}} \,, \tag{9.44}$$

but if these angles are small, then $\gamma \approx 1$, $\zeta \approx 1/\lambda$, and

$$\sin\theta_x \approx \theta_x \approx \xi\lambda \,, \qquad \sin\theta_y \approx \theta_y \approx \eta\lambda \,. \tag{9.45}$$

Thus, paraxially, ξ and η can be interpreted as angles divided by wavelength, *i.e.*, $\xi = \theta_x/\lambda$, $\eta = \theta_y/\lambda$.

In summary, we can specify a plane wave by giving any of the following pairs of numbers: (k_x, k_y), (ξ, η), (α, β) or (θ_x, θ_y). The missing third component can always be determined by a constraint equation such as (9.37) or (9.43), which must be satisfied if the plane wave is to be a solution to the wave equation.

9.2.2 Spherical waves

A monochromatic spherical wave has the form

$$u(\mathbf{r}, t) = \frac{1}{|\mathbf{r} - \mathbf{r}_0|} \exp(ik|\mathbf{r} - \mathbf{r}_0| - 2\pi i \nu_0 t) \,. \tag{9.46}$$

By inspection, this function is spherically symmetric about the point $\mathbf{r}_0$, and it has a monochromatic time dependence with frequency ν_0. At this stage, k is just a constant, but we shall demonstrate that it must be given by (9.38).

To show that (9.46) is a solution to the wave equation, we need to take its Laplacian. For this purpose, we make a change of variables,

$$\mathbf{R} \equiv \mathbf{r} - \mathbf{r}_0 \,. \tag{9.47}$$

In spherical coordinates centered on $\mathbf{r}_0$, the 3D vector $\mathbf{R}$ has components (R, θ_R, ϕ_R), but $u(\mathbf{r}, t)$ is independent of the angles. Therefore the Laplacian is given by

$$\nabla^2 \left[\frac{e^{ikR}}{R} \right] = \frac{1}{R} \frac{\partial^2}{\partial R^2} \left[R \frac{e^{ikR}}{R} \right] = -k^2 \frac{e^{ikR}}{R} \,, \tag{9.48}$$

provided $R \neq 0$, or $\mathbf{r} \neq \mathbf{r}_0$. The behavior at the point $\mathbf{r} = \mathbf{r}_0$ will be discussed in Sec. 9.3.

From (9.48) and (9.23), we see that the spherical wave of (9.46) satisfies the homogeneous wave equation provided $k = 2\pi\nu_0/c_m$. Exactly this condition was encountered in (9.38) for plane waves, and in fact it is a general condition. We can decompose an arbitrary solution of the homogeneous wave equation into monochromatic plane waves or spherical waves, and for each component we can define a wavelength λ and an associated $k = 2\pi/\lambda$. For propagation in a source-free homogeneous medium, we must always have $k = 2\pi\nu_0/c_m$. If sources are present, however, the condition can be violated since the source imposes its own spatial and temporal dependence, which may not be constrained by $k = 2\pi\nu_0/c_m$.

9.3 GREEN'S FUNCTIONS

Each of the three inhomogeneous wave equations, (9.23), (9.31) and (9.32), has the form $\mathcal{L}\mathbf{u} = \mathrm{s}$, where $\mathcal{L}$ is a linear differential operator, $\mathbf{u}$ is an unknown scalar field[1] and s is the source that generates it. In this section we develop methods to solve such equations.

The basic approach is one we exploited in Chap. 7. We first consider a point source and compute the field produced by it, and then we invoke linear superposition to compute the field due to an arbitrary source. In the imaging literature, the field produced by a point source would be called a point response function (or point spread function in the shift-invariant case), but in electromagnetism it is more commonly called a *Green's function*.[2] As we shall see, the same Green's function

[1] We use boldface for the source and field here since we want to think of them as vectors in a Hilbert space. In the remainder of the chapter, the convention of using non-boldface characters for operators will be followed.

[2] Modern books often drop the possessive and say *Green function* for consistency with *Bessel function*. We make no claim to modernity or consistency.

will allow us to solve problems where there is no source in the region of interest but the field on the boundary of the region is specified.

9.3.1 Differential equations for the Green's functions

We begin with the time-dependent scalar wave equation, (9.23). The equation for the Green's function is obtained simply by replacing the general source $s(\mathbf{r}, t)$ with a point source. The point is an impulse in both space and time, so it is a 4D delta function, $\delta(\mathbf{r} - \mathbf{r}')\,\delta(t - t')$, which is a source that is zero except at position $\mathbf{r}'$ and time t'. The Green's function $p(\mathbf{r}, t; \mathbf{r}', t')$ is the field at space-time point $(\mathbf{r}, t)$ due to the impulsive source at $(\mathbf{r}', t')$. The notation $p(\cdot)$ suggests that a Green's function is really a point response function.

In a homogeneous medium, the Green's function must satisfy the time-dependent wave equation with the impulsive source term:

$$\left[\nabla^2 - \frac{1}{c_m^2}\frac{\partial^2}{\partial t^2}\right] p(\mathbf{r}, t; \mathbf{r}', t') = \delta(\mathbf{r} - \mathbf{r}')\,\delta(t - t') . \tag{9.49}$$

In free space, c_m can be replaced by the vacuum speed of light c, but we shall retain the subscript for generality. Since $c_m = c/n$, the medium is homogeneous if the refractive index is independent of position.

Similarly, the Green's function for the Helmholtz equation must satisfy

$$\left(\nabla^2 + k^2\right) p(\mathbf{r}; \mathbf{r}') = \delta(\mathbf{r} - \mathbf{r}') , \tag{9.50}$$

where $k = 2\pi\nu_0/c_m = 2\pi n\nu_0/c$.

9.3.2 Time-dependent Green's function

To solve any differential equation, we must specify boundary conditions. The effect of physical boundaries will be taken up in Sec. 9.4, but here we seek solutions to (9.49) and (9.50) in a homogeneous medium without boundaries. The solutions for the Green's functions must be causal and decay as the observation point recedes to an infinite distance from the source, but otherwise there are no boundary conditions.

The absence of boundaries makes the problem shift-invariant. For the time-dependent wave equation, this means that $p(\mathbf{r}, t; \mathbf{r}', t')$ is a function only of $\mathbf{r} - \mathbf{r}'$ and $t - t'$. A useful change of variables is thus

$$\mathbf{R} = \mathbf{r} - \mathbf{r}' , \qquad \tau = t - t' , \tag{9.51}$$

so we can write[3]

$$p(\mathbf{r}, t; \mathbf{r}', t') = p(\mathbf{R}, \tau) . \tag{9.52}$$

Similarly, for the Helmholtz equation, we can write $p(\mathbf{r}; \mathbf{r}') = p(\mathbf{R})$.

Explicit forms for the Green's functions can be obtained in the Fourier domain. The 4D Fourier transform of (9.49) is

$$-4\pi^2\left(\sigma^2 - \frac{\nu^2}{c_m^2}\right) P(\boldsymbol{\sigma}, \nu) = 1 , \tag{9.53}$$

[3]Do not confuse $p(\mathbf{R}, \tau)$ with the polarization, which we denoted as $\mathbf{p}(\mathbf{r}, t)$ in Sec. 9.1.3. We shall have no further use for polarization in this chapter.

where $P(\boldsymbol{\sigma}, \nu)$ is the 4D Fourier transform of $p(\mathbf{R}, \tau)$. From (9.53) we get

$$P(\boldsymbol{\sigma}, \nu) = -\frac{1}{4\pi^2} \frac{1}{\left(\sigma^2 - \frac{\nu^2}{c_m^2}\right)} . \tag{9.54}$$

The inverse 4D Fourier transform of $P(\boldsymbol{\sigma}, \nu)$ consists of a 3D spatial part and a 1D temporal part. The spatial part has rotational symmetry, so it is given by (3.250). The temporal part is straightforward, but we must remember the sign convention used in this chapter: the temporal inverse transform has a minus sign in the exponent of the Fourier kernel. We thus have

$$\begin{aligned} p(\mathbf{R}, \tau) &= 4\pi \int_0^\infty \sigma^2 d\sigma \, \mathrm{sinc}(2\sigma R) \int_{-\infty}^\infty d\nu \, \exp(-2\pi i \nu \tau) \, P(\boldsymbol{\sigma}, \nu) \\ &= \frac{c_m^2}{\pi} \int_0^\infty \sigma^2 d\sigma \, \mathrm{sinc}(2\sigma R) \int_{-\infty}^\infty d\nu \, \frac{\exp(-2\pi i \nu \tau)}{(\nu + c_m \sigma)(\nu - c_m \sigma)} , \end{aligned} \tag{9.55}$$

where $R = |\mathbf{R}|$. The integral over ν can be performed by contour integration in the complex-ν plane if we can interpret the singularities on the real axis at $\nu = \pm\sigma c$. As discussed in App. B (Sec. B.3.8), the options are to deform the contour slightly so that it passes over the pole (or, equivalently, displace the pole downward), deform the contour so it passes under the pole, or take the Cauchy principal value. We cannot decide among these three options purely mathematically; instead, we must choose the one that makes physical sense. In particular, in order for the system to be causal, we must impose the condition

$$p(\mathbf{R}, \tau) = 0 \qquad \text{if } \tau < 0 . \tag{9.56}$$

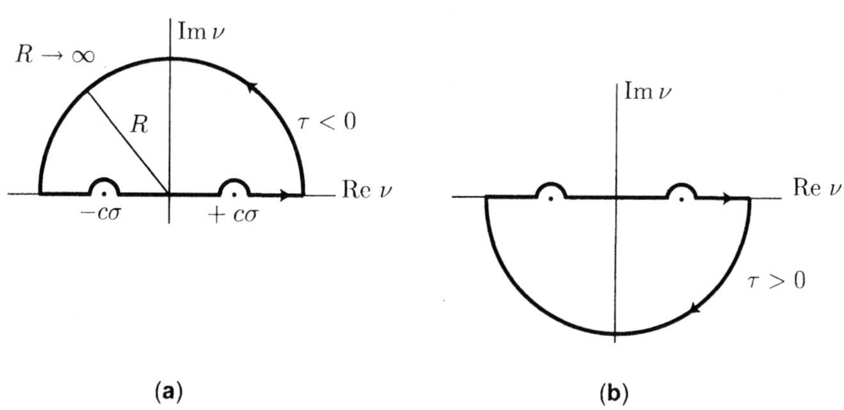

Fig. 9.1 Illustration of the contours needed for calculation of the Green's function for the time-dependent wave equation.

This condition will be satisfied if we indent the contour above the singularities as shown in Fig. 9.1. To see why this contour implies (9.56), note that, for $\tau < 0$, $\exp(-2\pi i \nu \tau)$ vanishes on an infinite semicircle in the upper half-plane, so the contour can be closed by this semicircle without changing the value of the integral. The poles will then lie outside the contour, and (9.56) follows from the Cauchy integral formula. For $\tau > 0$, the contour must be closed by a semicircle in the lower half-plane, so both poles are enclosed and Cauchy's formula yields

$$p(\mathbf{R},\tau) = 2c_m \int_0^{\infty} \sigma d\sigma \ \mathrm{sinc}(2\sigma R)\sin(2\pi c_m \tau\sigma)$$

$$= -\frac{c_m}{8\pi R}\int_{-\infty}^{\infty} d\sigma \ [\exp(2\pi i\sigma R) - \exp(-2\pi i\sigma R)]\,[\exp(2\pi i\sigma c_m\tau) - \exp(-2\pi i\sigma c_m\tau)]\,. \tag{9.57}$$

When we multiply it out, the integrand contains four terms of the form $\exp[2\pi i\sigma(\pm R \pm c_m\tau)]$. By (3.151), a function of this form integrates to $\delta(\pm R \pm c_m\tau)$, but we must recall that both R and τ are positive, so the delta function is nonzero only if we take opposite signs. Also, $\delta(R - c_m\tau) = \delta(-R + c_m\tau) = c_m^{-1}\,\delta(\tau - R/c_m)$, so

$$p(\mathbf{R},\tau) = -\frac{1}{4\pi R}\,\delta\!\left(\tau - \frac{R}{c_m}\right), \tag{9.58}$$

or, with the original variables,

$$p(\mathbf{r},t;\mathbf{r}',t') = -\frac{1}{4\pi|\mathbf{r}-\mathbf{r}'|}\,\delta\!\left(t - t' - \frac{|\mathbf{r}-\mathbf{r}'|}{c_m}\right). \tag{9.59}$$

Note that the delta function here is *one-dimensional*; it describes a function that is zero unless $R = c_m\tau$, which is the equation for a sphere of radius $c\tau$ in 3D space. The Green's function $p(\mathbf{r},t;\mathbf{r}',t')$ is a spherical shell of field, zero except on the sphere of radius $c\tau$ (see Fig. 9.2). The sphere expands outwards at the speed of light in the medium, c_m, so a disturbance originating at some point $\mathbf{r}'$ reaches another point $\mathbf{r}$ in a time given exactly by $|\mathbf{r}-\mathbf{r}'|/c_m$.

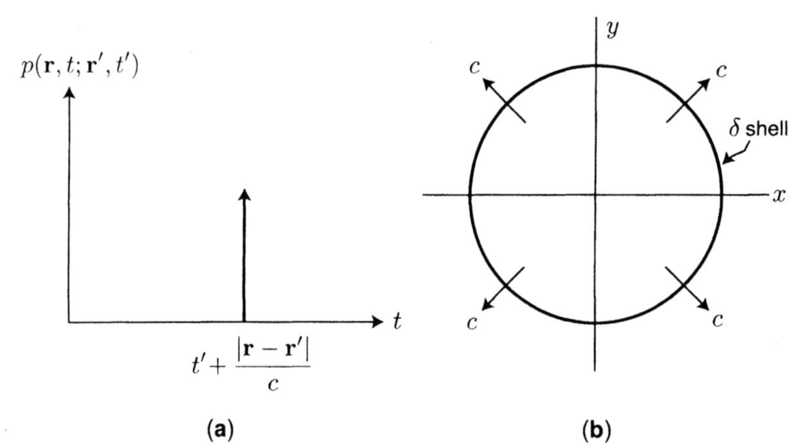

Fig. 9.2 (a) Plot of the Green's function for the time-dependent wave equation as a function of time. (b) Plot of the Green's function for the time-dependent wave equation in the x-y plane.

The Green's function diminishes in amplitude as $1/R$, which is a consequence of conservation of energy. As discussed in Sec. 10.1, the energy flux (power per unit area) is proportional to field squared, and the Green's function is the field produced by a point source. Thus the flux varies as $1/R^2$ while the total area of a sphere of radius R varies as R^2, so the total energy passing through any sphere surrounding the point source is independent of R.

9.3.3 Green's functions for the Helmholtz and Poisson equations

In the absence of physical boundaries, the Helmholtz equation is spatially shift-invariant, so its Green's function can be written as $p(\mathbf{r}, \mathbf{r}') = p(\mathbf{R})$, where again $\mathbf{R} = \mathbf{r} - \mathbf{r}'$. This Green's function can be computed by a Fourier method similar to the one used above, except that there is no temporal transform. The 3D Fourier transform of (9.50) yields

$$P(\boldsymbol{\sigma}) = \frac{1}{-4\pi^2\sigma^2 + k^2} \tag{9.60}$$

and the inverse transform is given by (3.250) as

$$p(\mathbf{R}) = 4\pi \int_0^\infty \sigma^2 d\sigma \; \mathrm{sinc}(2\sigma R) \frac{1}{-4\pi^2\sigma^2 + k^2} \,. \tag{9.61}$$

Again we have poles on the real axis, and again we must appeal to causality for guidance in dealing with them. Depending on which poles are enclosed, contour integration of (9.61) yields terms proportional to $\exp(\pm ikR)$. Either sign would be mathematically acceptable, but we must recall that $p(\mathbf{R})$ is a field and has associated with it a time dependence given by (9.29). Since k was defined as $2\pi\nu_0/c_m$, $\exp(\pm ikR)\exp(-2\pi i\nu_0 t) = \exp[ik(\pm R - c_m t)]$. This exponential represents a *spherical wave*; it is constant on a sphere of radius $R = \pm c_m t$ about the source. Causality dictates that we take the plus sign so that the spherical wave will be expanding outward from the source instead of converging inward toward it. We can ensure that only this sign occurs by proper choice of the contour of integration in (9.61).

Thus the causal Green's function for the Helmholtz equation in a homogeneous medium is given by

$$p(\mathbf{R}) = -\frac{1}{4\pi} \frac{\exp(ikR)}{R} \,. \tag{9.62}$$

This function, illustrated in Fig. 9.3, satisfies

$$\left[\nabla^2 + k^2\right] \frac{\exp(ikR)}{R} = -4\pi\,\delta(\mathbf{R}) \,. \tag{9.63}$$

This result is consistent with the discussion of spherical waves in Sec. 9.2.2, where we showed that $\exp(ikR)/R$ is a solution of the homogeneous wave equation except at the point $\mathbf{r} = 0$; if $\mathbf{R} \neq 0$, $\delta(\mathbf{R}) = 0$ and (9.63) is the homogeneous[4] Helmholtz equation.

The Helmholtz equation reduces to the Poisson equation as $k \to 0$, so the Green's function for the Poisson equation is simply $-1/(4\pi R)$, which satisfies

$$\nabla^2 \frac{1}{R} = -4\pi\,\delta(\mathbf{R}) \,. \tag{9.64}$$

[4] Do not confuse the two meanings of the word *homogeneous*. In the context of linear differential equations, it means that there is no source term (a term independent of the unknown function), but in physics it refers to a medium with all properties independent of position.

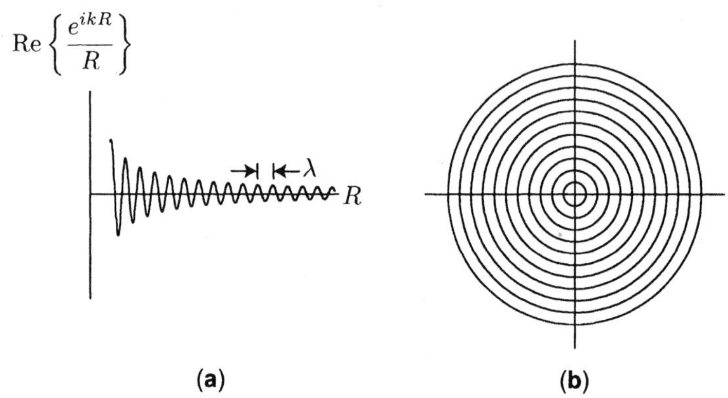

Fig. 9.3 (a) Plot of the Green's function for the Helmholtz equation as a function of radius. (b) Contour plot of the Green's function for the Helmholtz equation in the x-y plane.

9.3.4 Defined-source problems

We now know the Green's functions for the three wave equations in a homogeneous medium. In each case, the Green's function is the field produced by a unit point source. For any other *specified* source distribution, the total field can be obtained by linear superposition. For example, for the time-dependent wave equation, we have

$$u(\mathbf{r},t) = \int_{-\infty}^{\infty} dt' \int_{\infty} d^3\mathbf{r}'\, p(\mathbf{r},t;\mathbf{r}',t')\, s(\mathbf{r}',t')$$
$$= -\frac{1}{4\pi}\int_{-\infty}^{t} dt' \int_{\infty} d^3\mathbf{r}' \frac{1}{|\mathbf{r}-\mathbf{r}'|}\,\delta\!\left(t-t'-\frac{|\mathbf{r}-\mathbf{r}'|}{c_m}\right) s(\mathbf{r}',t') \,. \tag{9.65}$$

We can verify that this field is indeed a solution of the time-dependent wave equation by operating on both sides of the equation with the operator $\nabla^2 - \frac{1}{c_m^2}\frac{\partial^2}{\partial t^2}$, taking the operator under the integrals on the right, and using (9.49).

An instructive alternative form of (9.65) is obtained by using the 1D delta function to perform the 1D integral over t'. The result is

$$u(\mathbf{r},t) = -\frac{1}{4\pi}\int_{\infty} d^3\mathbf{r}' \frac{1}{|\mathbf{r}-\mathbf{r}'|}\, s\!\left(\mathbf{r}',t-\frac{|\mathbf{r}-\mathbf{r}'|}{c_m}\right) . \tag{9.66}$$

This form shows that the field at $(\mathbf{r},t)$ is influenced by the source at point $\mathbf{r}'$ at an earlier time $t-|\mathbf{r}-\mathbf{r}'|/c_m$, called the *retarded time*. The time delay accounts for the finite speed of propagation of the wave from the source point $\mathbf{r}'$ to the observation point $\mathbf{r}$.

Optical path Since we know from Sec. 9.1.4 that $c_m = c/n$, where n is the index of refraction, the retarded time can be written as $t - n|\mathbf{r}-\mathbf{r}'|/c$. The effect of the medium is that the actual physical distance $|\mathbf{r}-\mathbf{r}'|$ between source and observation point is replaced by the distance times the index of refraction. Index-weighted distances, called *optical path lengths*, play a key role in optics, especially in the geometrical-optics approximation. So far, we have treated the index as a constant, so the difference between actual path length and optical path length is just a constant, but in many formulations of optics n is allowed to be a general function of position. Indeed, most of classical optics is derived from *Fermat's principle*, which asserts that light follows a path that minimizes the optical path length, or equivalently the propagation time from source to observation point. We shall have more to say about optical path length in Sec. 9.7.4.

Helmholtz equation The time-independent solution of the Helmholtz equation can be found in several ways. Linear superposition gives at once

$$u(\mathbf{r}) = -\frac{1}{4\pi}\int_\infty d^3\mathbf{r}'\,\frac{\exp(ik|\mathbf{r}-\mathbf{r}'|)}{|\mathbf{r}-\mathbf{r}'|}s(\mathbf{r}') = -\frac{1}{4\pi}\int_\infty d^3\mathbf{r}'\,\frac{\exp(ikR)}{R}s(\mathbf{r}')\,. \tag{9.67}$$

We can verify that this field is a solution of the Helmholtz equation by operating on both sides with $(\nabla^2 + k^2)$ and using (9.63).

An alternative derivation of (9.67) uses (9.66), which is general enough to allow any time dependence, including the monochromatic $\exp(-2\pi i\nu_0 t)$ of (9.25). Evaluating this function at the retarded time, as required by (9.66), yields

$$\exp\left[-2\pi i\nu_0\left(t - \frac{|\mathbf{r}-\mathbf{r}'|}{c_m}\right)\right] = \exp(-2\pi i\nu_0 t)\exp(ik|\mathbf{r}-\mathbf{r}'|)\,. \tag{9.68}$$

Plugging this result into (9.66), we find $u(\mathbf{r},t) = u(\mathbf{r})\exp(-2\pi i\nu_0 t)$, with $u(\mathbf{r})$ given by (9.67). The factor $\exp(ikR)$ in the Helmholtz Green's function thus arises from retardation of the source time dependence for a monochromatic source. The $1/R$ factor is again a consequence of conservation of energy.

The field produced by a static source can be obtained from (9.67) simply by setting $k = 0$ [or $\exp(ikR) = 1$].

Relation to operator theory We can restate the key results of this section in operator form. We have already noted that the inhomogeneous wave equations have the form $\mathcal{L}\mathbf{u} = \mathbf{s}$, where $\mathcal{L}$ is the appropriate linear differential operator. We can also define an integral operator $\mathcal{G}$ whose kernel is the Green's function, and we recall from Chap. 1 [see (1.64)] that the delta function is the kernel of the unit operator $\mathcal{I}$ in the domain of $\mathcal{L}$. Thus (9.49) and (9.50) state that $\mathcal{L}\mathcal{G} = \mathcal{I}$, and hence $\mathcal{G}$ is a right inverse of $\mathcal{L}$ [see (1.37)]. With this right inverse, the solution of the original equation for a specified source is $\mathbf{u} = \mathcal{G}\mathbf{s}$, which is the abstract form of (9.65) or (9.67).

9.3.5 Boundary-value problems

The problems treated in Sec. 9.3.4 require knowledge of the source throughout space; in practice we seldom have such complete knowledge. A more common scenario is that we do not know the details of the source but can describe the field it produces on some surface. The objective is then to calculate the field at points not on this surface. Problems of this type, called *boundary-value problems*, are the subject of this section.

Types of boundary conditions Possible boundary conditions are *Dirichlet* conditions, in which the field is specified on the surface, *Neumann* conditions, in which the normal derivative of the field is specified, and *Cauchy* conditions in which both the field and its normal derivative are specified. Each type can be *homogeneous*,[5] where the specified quantity is zero, or *inhomogeneous*, where it is nonzero. For example, with homogeneous Dirichlet boundary conditions, the field must be zero over the surface.

The surfaces themselves can be either *open* or *closed*. This distinction is intuitive for purely spatial surfaces — a closed surface is one that completely surrounds a finite volume, dividing space uniquely into interior and exterior points such that there is no path connecting an interior point to an exterior one. In discussing the time-dependent wave equation, however, we shall encounter surfaces in 4D where one of the coordinates is time. The situation is best illustrated in 2D, in the x-t plane. The lines $x = const$ for all t or $t = const$ for all x would be open surfaces. Similarly, in 4D, a spatial *volume* at one time is an *open* boundary. To construct a closed surface in the x-t plane, we could consider two open surfaces such as the x axis at time t_1 and the x axis at time t_2 and connect them at $x = \pm\infty$.

The type of equation determines what boundary conditions can be used in its solution. Second-order partial differential equations can be classified as *hyperbolic*, *elliptic* or *parabolic* equations. We do not need to go into the reasons for these designations, but we note that the time-dependent scalar wave equation is hyperbolic while the Helmholtz and Poisson equations are elliptic. An example of a parabolic equation, which we shall not discuss here, is the diffusion equation.

For elliptic equations, such as the Helmholtz equation, there exists a unique, stable solution for Dirichlet or Neumann boundary conditions[6] on a closed surface (Morse and Feshbach, 1953, Chap. 6). Since either Dirichlet or Neumann conditions alone lead to a solution, Cauchy conditions in general are an overspecification. If we were to place independent requirements on both the field and its normal derivative, without requiring that they were consistent with each other, there would be no solution compatible with both conditions and with the differential equation. Of course, any real physical problem leads to *some* values of the field and its derivative on the surface, and if we could specify those actual physical values, a solution would exist, but we usually cannot do so.

The situation is different with a hyperbolic equation such as the time-dependent wave equation. In that case, a unique, stable solution is determined by Cauchy boundary conditions on an open surface (Morse and Feshbach, 1953). To illustrate this statement, consider the homogeneous time-dependent wave equation in one spatial dimension,

$$\left(\frac{\partial^2}{\partial x^2} - \frac{1}{c^2}\frac{\partial^2}{\partial t^2}\right) u(x,t) = 0\,, \tag{9.69}$$

which might describe the free vibrations of a string. If the string has infinite length, a suitable open boundary is the line $t = 0$ between $x = -\infty$ and $x = \infty$. Cauchy boundary conditions on this boundary, *i.e.*, specification of the displacement and velocity of the string at all points for $t = 0$, are sufficient to determine the motion for all later times. On the other hand, if we were to specify displacement and velocity over only a portion of the string, say $a < x < b$, we would not be able to determine the motion at all points at all later times. In fact, we could not even

[5] This is yet another meaning of *homogeneous*; see footnote 4.

[6] For Neumann conditions, there is an additional constraint that the integral of the gradient over the surface must vanish; see Morse and Feshbach (1953), p. 698.

determine the motion in $a < x < b$ at all later times since disturbances outside that region could propagate into it later. Thus Cauchy boundary conditions are sufficient on an infinite open surface, but not on a finite one.[7] Similarly, for the time-dependent wave equation in three spatial dimensions, a spatial volume at one time is an open boundary as noted above. Specification of the field and its time derivative at $t = 0$ at all points in the volume leads to a unique, stable solution for all points at any later time. Another possible open surface is the plane $z = 0$ for all t. Specification of the field and its z derivative on this plane for all x, y and t are acceptable boundary conditions for the time-dependent wave equation. A finite region of the plane $z = 0$ does not suffice in general.

In summary, unique, stable solutions to the Helmholtz and Poisson equations are obtained by specifying either the field or its normal derivative (but not both) on a closed surface. Unique, stable solutions to the time-dependent wave equation result from specifying both the field and its normal derivative on a spatio-temporal open surface, an example of which is the entire spatial volume for one time.

Green's theorem To incorporate boundary conditions into the solution of the Helmholtz equation, we use *Green's theorem*, an important theorem from vector calculus which states that

$$\int_{\mathcal{V}} d^3\mathbf{r}\ [\psi(\mathbf{r})\, \nabla^2\phi(\mathbf{r}) - \phi(\mathbf{r})\, \nabla^2\psi(\mathbf{r})] = \int_{\mathcal{S}} da \left[\psi(\mathbf{r})\, \frac{\partial}{\partial n}\, \phi(\mathbf{r}) - \phi(\mathbf{r})\, \frac{\partial}{\partial n}\, \psi(\mathbf{r})\right], \tag{9.70}$$

where $\psi(\mathbf{r})$ and $\phi(\mathbf{r})$ are scalar fields, $\mathcal{S}$ is a closed surface enclosing volume $\mathcal{V}$, da is an area element on $\mathcal{S}$, and the notation $\frac{\partial}{\partial n}$ means

$$\frac{\partial\psi(\mathbf{r})}{\partial n} = \hat{\mathbf{n}} \cdot \nabla\psi(\mathbf{r}_0)\big|_{\mathcal{S}}, \tag{9.71}$$

where $\hat{\mathbf{n}}$ is the outward unit normal to $\mathcal{S}$. For example, if $\mathcal{S}$ includes the plane $z = 0$ and $\mathcal{V}$ is the right half-space, then $\partial\psi(\mathbf{r})/\partial n = -\partial\psi(\mathbf{r})/\partial z$ on that plane, the minus sign arising since the outward normal points in the $-z$ direction.

To apply Green's theorem to the problem at hand, change the variable of integration in (9.70) from $\mathbf{r}$ to $\mathbf{r}_0$ and let $\phi(\mathbf{r}_0)$ be the field $u(\mathbf{r}_0)$ and $\psi(\mathbf{r}_0)$ be the Green's function $p(\mathbf{r}, \mathbf{r}_0)$. The differential operators ∇ and $\frac{\partial}{\partial n}$ now imply differentiation with respect to $\mathbf{r}_0$ and will accordingly be denoted ∇_0 and $\frac{\partial}{\partial n_0}$, respectively.

The Green's function must satisfy (9.50), which in the present notation becomes

$$\left(\nabla_0^2 + k^2\right) p(\mathbf{r}, \mathbf{r}_0) = \delta(\mathbf{r} - \mathbf{r}_0), \qquad \mathbf{r} \text{ and } \mathbf{r}_0 \text{ in } \mathcal{V}. \tag{9.72}$$

The Green's function is the field at $\mathbf{r}_0$ produced by a point source at $\mathbf{r}$ plus other sources outside $\mathcal{V}$ needed to satisfy the boundary conditions. Suppose, for example, that the field on the boundary is produced by a point source at $\mathbf{r}_1$ outside $\mathcal{V}$. By analogy to (9.50), one might think that we would have to include a term $\delta(\mathbf{r}_0 - \mathbf{r}_1)$ on the right-hand side of (9.72), but we do not need such a term since

[7] A precise statement of the requirements on the boundary uses the concept of *characteristics*, discussed in Morse and Feshbach (1953) or Sokolnikoff and Redheffer (1958), for example. For the 1D wave equation, the characteristics are two families of curves, $x + ct$ = constant and $x - ct$ = constant. Specification of Cauchy conditions on an open surface determines the solution uniquely in a region bounded by those characteristics that intersect the open surface.

$\delta(\mathbf{r}_0 - \mathbf{r}_1)$ is identically zero if $\mathbf{r}_0$ is inside $\mathcal{V}$ and $\mathbf{r}_1$ is outside. Of course, sources outside $\mathcal{V}$ can produce fields inside, so the Green's function is no longer given by (9.62). In fact, it is no longer shift invariant since the absolute position of $\mathbf{r}$ and $\mathbf{r}_0$ with respect to the boundary is important.

Inserting (9.72) into (9.70) and using the Helmholtz equation (9.31) and the sifting property of the delta function, we find

$$u(\mathbf{r}) = \int_{\mathcal{V}} d^3\mathbf{r}_0 \, p(\mathbf{r},\mathbf{r}_0)\, s(\mathbf{r}_0) + \int_{\mathcal{S}} da_0 \left[u(\mathbf{r}_0) \frac{\partial p(\mathbf{r},\mathbf{r}_0)}{\partial n_0} - p(\mathbf{r},\mathbf{r}_0) \frac{\partial u(\mathbf{r}_0)}{\partial n_0} \right] . \quad (9.73)$$

Comparing this result to (9.67), we see that the volume integral has been modified since it now includes only sources inside $\mathcal{V}$, but the new surface integral accounts for sources outside $\mathcal{V}$ through their effect on the boundary conditions.

Equation (9.73) will be our primary tool for calculating fields in a closed volume when we are given knowledge of the fields on the boundary and the sources (if any) inside the volume. In the next section we shall apply this tool to diffraction by an aperture.

9.4 DIFFRACTION BY A PLANAR APERTURE

A common situation in optics and imaging involves an open aperture in an otherwise opaque screen, as shown in Fig. 9.4. It is customary to take the screen as the plane $z = 0$. The standard diffraction problem is to assume some wave incident on the screen from the left and to calculate the resulting field for all points to the right of the screen. Thus the volume $\mathcal{V}$ in Green's theorem is the right half-space. In order to bound this volume with a closed surface $\mathcal{S}$, we can construct a hemisphere of radius R_a centered on the aperture and let $R_a \to \infty$. Equation (9.73) provides a way of determining the field in $\mathcal{V}$ if we know the sources in $\mathcal{V}$ and the boundary conditions on the surface. Here we shall assume that there are no sources in $\mathcal{V}$, so only the surface integral in (9.73) is needed. In Sec. 9.4.1, we argue that the infinite hemispherical surface is irrelevant, so the only important part of $\mathcal{S}$ is the plane $z = 0$. In Sec. 9.4.2, we discuss an approximate way of expressing the field on this plane, and in Sec. 9.4.3 we derive an explicit formula for the field at an arbitrary point in $\mathcal{V}$.

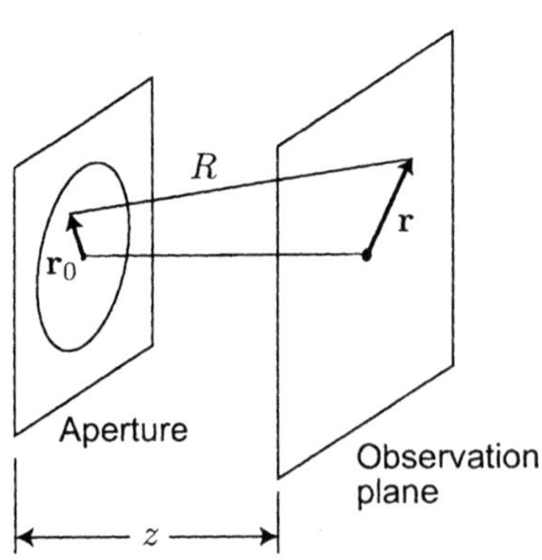

Fig. 9.4 Geometry for diffraction by a planar aperture.

9.4.1 Surface at infinity

For the time-dependent wave equation, we can dispense with the infinite hemisphere. If the field incident on the aperture is turned on at some starting time t_s, rather than existing from $t = -\infty$, then there is no way the boundary condition on the infinite hemisphere can influence the field a finite distance from the aperture at a finite time, and the hemisphere can be safely ignored.

This argument can also be applied, to a good approximation, to a monochromatic source. Strictly speaking, such a source has a time dependence of $\exp(-2\pi i\nu_0 t)$ for all t, but if the source is switched on at a finite time, the field has a time dependence given by $\exp(-2\pi i\nu_0 t)\,\text{step}(t - t_s)$. If $t - t_s$ is finite but sufficiently large that all transient effects from the step function have died out, then the field at a finite distance from the aperture is described adequately by the Helmholtz equation, but the surface at infinity plays no role since the disturbance has not had time to get there.

Next we shall compute how long one must wait for this approximation to be valid. From (9.66), we know that the field at $(\mathbf{r}, t)$ is influenced by the source at point $\mathbf{r}'$ at the earlier time $t - |\mathbf{r} - \mathbf{r}'|/c$. If R_{max} is the maximum distance between the observation point $\mathbf{r}$ and any point on the screen where the field or its derivative is nonzero, then all we have to do is make $t - t_s > R_{max}/c$, and there will be no way to discern when the field was switched on. At later times, the field will be described by the Helmholtz equation, yet the surface at infinity will make no contribution.

Another approach, which we shall not describe here, is to write down formally the effect of the infinite hemisphere on the field at a point in $\mathcal{V}$ and to show that it consists of two integrals which cancel one another as $R_a \to \infty$ as a result of the *Sommerfeld radiation condition*; for details on this approach, see Goodman (1968).

9.4.2 Kirchhoff boundary conditions

Having disposed of the surface at infinity, we turn next to the plane $z = 0^+$. To describe the field in this plane, we introduce the *Kirchhoff boundary condition* or *Kirchhoff approximation*, which assumes that the field in the clear aperture is exactly what it would be if there were no screen. This is a reasonable approximation if the opening is large compared to a wavelength, but we should expect it to break down for small apertures. For large apertures it is also reasonable to assume that the field is zero for any point just to the right of the opaque screen.

For a monochromatic source, all fields have the same time dependence $\exp(-2\pi i\nu_0 t)$, so the Kirchhoff boundary condition can be expressed in terms of the time-independent field amplitude. We refer to the field that would exist in the plane $z = 0$ in the absence of the screen as the incident wave and denote it by $u_{inc}(\mathbf{r})$. The Kirchhoff boundary condition states that the actual field in the plane $z = 0$ is given by

$$u(\mathbf{r})|_{z=0} = \begin{cases} u_{inc}(\mathbf{r}) & \text{if } \mathbf{r} \text{ lies in the clear aperture} \\ 0 & \text{if } \mathbf{r} \text{ is behind the opaque screen}, \end{cases} \tag{9.74}$$

where $\mathbf{r}$ is the 2D vector obtained from $\mathbf{r}$ by setting the z component to zero.

9.4.3 Application of Green's theorem

Using the scalar Kirchhoff boundary condition in (9.73) and assuming there are no sources in $\mathcal{V}$, we find

$$u(\mathbf{r}) = \int_{ap} d^2r_0 \left[u_{inc}(\mathbf{r}_0) \frac{\partial p(\mathbf{r}, \mathbf{r}_0)}{\partial n_0} - p(\mathbf{r}, \mathbf{r}_0) \frac{\partial u_{inc}(\mathbf{r}_0)}{\partial n_0} \right] , \tag{9.75}$$

where the integral is over the clear aperture only.

Several forms of diffraction theory follow from (9.75) depending on what one assumes for the Green's function, $p(\mathbf{r}, \mathbf{r}_0)$ (see Goodman, 1968). We shall not explore all of these options here, but restrict ourselves to the generally accepted *Rayleigh-Sommerfeld diffraction theory.* Rayleigh and Sommerfeld recognized that specifying both $u(\mathbf{r}_0)$ and $\partial u(\mathbf{r}_0)/\partial n_0$ independently on the boundary (Cauchy boundary conditions) would be an overspecification for the Helmholtz equation. To avoid this difficulty, they suggested choosing the Green's function $p(\mathbf{r}, \mathbf{r}_0)$ to be zero on the aperture plane so that $\partial u_{inc}/\partial n_0$ is not needed.

We can construct a Green's function that satisfies this condition by recalling that we are free to add sources in the defining equation for the Green's function, (9.72), so long as they lie outside $\mathcal{V}$. To make $p(\mathbf{r}, \mathbf{r}_0)$ vanish when $\mathbf{r}_0$ lies on the plane $z_0 = 0$, we add a *negative* point source at the point $\mathbf{r}_m = (x, y, -z)$ which is the mirror image of $\mathbf{r}$ in the aperture plane. Thus $p(\mathbf{r}, \mathbf{r}_0)$ is the field at $\mathbf{r}_0$ due to a positive point source at $\mathbf{r}$ and a negative one at $\mathbf{r}_m$. Explicitly,

$$p(\mathbf{r}, \mathbf{r}_0) = -\frac{1}{4\pi} \left[\frac{\exp(ikR)}{R} - \frac{\exp(ikR_m)}{R_m} \right] , \tag{9.76}$$

where

$$R = \sqrt{|\mathbf{r} - \mathbf{r}_0|^2 + (z - z_0)^2} , \qquad R_m = \sqrt{|\mathbf{r} - \mathbf{r}_0|^2 + (-z - z_0)^2} , \tag{9.77}$$

and

$$|\mathbf{r} - \mathbf{r}_0|^2 = (x - x_0)^2 + (y - y_0)^2 . \tag{9.78}$$

Note that $R_m = R$ if $z_0 = 0$.

Even though the Green's function is zero on the aperture plane, its normal derivative is not. Differentiation of (9.76) with respect to z_0 yields

$$\frac{\partial p(\mathbf{r}, \mathbf{r}_0)}{\partial n_0} = -\left.\frac{\partial p(\mathbf{r}, \mathbf{r}_0)}{\partial z_0}\right|_{z_0=0} = -\frac{1}{2\pi} \left(ik - \frac{1}{R} \right) \frac{\exp(ikR)}{R} \cos\theta , \tag{9.79}$$

where θ is the angle between $\mathbf{r} - \mathbf{r}_0$ and the z axis, so that

$$\cos\theta = \frac{z}{\sqrt{|\mathbf{r} - \mathbf{r}_0|^2 + z^2}} . \tag{9.80}$$

Collecting results, we have a general expression for the diffracted field in $\mathcal{V}$:

$$u(\mathbf{r}) = -\frac{1}{2\pi} \int_{ap} d^2r_0 \; u_{inc}(\mathbf{r}_0) \left(ik - \frac{1}{R} \right) \cos\theta \frac{\exp(ikR)}{R} . \tag{9.81}$$

Note that the z component of $\mathbf{r}_0$ is zero, so a point in the aperture is specified equally by the 2D vector $\mathbf{r}_0$ or the 3D vector $\mathbf{r}_0$.

9.4.4 Diffraction as a 2D linear filter

Equation (9.81) can be evaluated for the field at any (x, y, z), so $u(\mathbf{r})$ is a 3D function. If we evaluate it for all (x, y) on a plane of fixed z, however, it is a 2D function which we can denote by $u_z(\mathbf{r})$. Using a similar 2D notation for u_{inc}, we can rewrite (9.81) as

$$u_z(\mathbf{r}) = u(\mathbf{r}) = -\frac{1}{2\pi}\int_\infty d^2r_0\; u_{inc}(\mathbf{r}_0)\, t_{ap}(\mathbf{r}_0)\left(ik - \frac{1}{R}\right)\cos\theta\frac{\exp(ikR)}{R}\,, \tag{9.82}$$

where $t_{ap}(\mathbf{r})$ is the *amplitude transmittance* of the aperture, defined by [*cf.* (9.74)]

$$t_{ap}(\mathbf{r}) \equiv \begin{cases} 1 & \text{if } \mathbf{r} \text{ lies in the clear aperture} \\ 0 & \text{if } \mathbf{r} \text{ is behind the opaque screen}\,. \end{cases} \tag{9.83}$$

Since R and $\cos\theta$ are both functions of $\mathbf{r} - \mathbf{r}_0$, as shown by (9.77) and (9.80), respectively, we recognize (9.82) as a 2D convolution. Thus we can write symbolically

$$u_z(\mathbf{r}) = [u_{inc}(\mathbf{r})\, t_{ap}(\mathbf{r})] * p_z(\mathbf{r})\,, \tag{9.84}$$

where $p_z(\mathbf{r})$ is the 2D point spread function[8] (PSF) for propagation, given by

$$p_z(\mathbf{r}) = -\frac{1}{2\pi}\left(ik - \frac{1}{\sqrt{r^2+z^2}}\right)\frac{z}{\sqrt{r^2+z^2}}\,\frac{\exp\left(ik\sqrt{r^2+z^2}\right)}{\sqrt{r^2+z^2}}\,, \tag{9.85}$$

with $r = |\mathbf{r}|$. As a 2D mapping from the aperture plane to a parallel plane a distance z away, diffraction is a shift-invariant operation. If u_{inc}, t_{ap} and the observation point are all shifted together parallel to the aperture plane, nothing is changed.

Since the product $u_{inc}(\mathbf{r})\, t_{ap}(\mathbf{r})$ is the input to the convolution, it will be useful to write

$$u_{inc}(\mathbf{r})\, t_{ap}(\mathbf{r}) = u_0(\mathbf{r})\,, \tag{9.86}$$

where $u_0(\mathbf{r})$ is to be interpreted as the incident field as modified by the aperture.

As an exercise, the reader can show that $p_z(\mathbf{r}) \to \delta(\mathbf{r})$ as $z \to 0$, so $u_z(\mathbf{r}) \to u_0(\mathbf{r})$, as it should.

9.4.5 Some useful approximations

As it stands, (9.85) is a rather complicated expression from which it is difficult to get any insight. Several approximations are possible, all based on expanding various pieces of $p_z(\mathbf{r})$ in powers of $1/z$ and retaining only the leading terms.

Radiation and paraxial approximations Consider first the factor $ik - (r^2 + z^2)^{-\frac{1}{2}}$ in (9.85). If z is large, we can neglect $(r^2 + z^2)^{-\frac{1}{2}}$ compared to ik. Since $k = 2\pi\nu_0/c = 2\pi/\lambda$, where λ is the wavelength of the radiation, this approximation is valid when $z \gg \lambda$, which it almost always is in optical diffraction problems. This approximation is called the *radiation approximation.*[9] It shows that the radiation

[8] Do not confuse the 2D PSF $p_z(\mathbf{r})$ with the Green's function $p(\mathbf{r}, \mathbf{r}_0)$; by (9.79) the former is the z derivative of the latter.

[9] The radiation approximation is a bit tricky since we are neglecting a real term compared to a pure imaginary one. It can be stated more precisely by noting that $ik - (r^2 + z^2)^{-\frac{1}{2}}$ approaches a complex number with magnitude k and phase $\pi/2$ as $z \to \infty$.

field falls off asymptotically as $1/z$, as required by conservation of energy, but near the aperture it can have a more complicated dependence on z.

With the radiation approximation, (9.82) becomes

$$u_z(\mathbf{r}) = \frac{1}{i\lambda} \int_\infty d^2 r_0 \; u_0(\mathbf{r}_0) \cos\theta \frac{\exp(ikR)}{R} \tag{9.87}$$

and (9.85) becomes

$$p_z(\mathbf{r}) = \frac{1}{i\lambda} \frac{z}{\sqrt{r^2+z^2}} \frac{\exp\left(ik\sqrt{r^2+z^2}\right)}{\sqrt{r^2+z^2}}, \tag{9.88}$$

where we have used $k = 2\pi/\lambda$.

Equation (9.87) is a formal statement of *Huygens' Principle*, enunciated by the Dutch mathematician, astronomer and physicist Christiaan Huygens (1629–1695). Huygens is acknowledged as the founder of the wave theory of light and was one of the first to produce practical lenses, including a nearly perfect achromatic eyepiece. He is even credited in some books (Matteuci, 1970) with the invention of the internal combustion engine. (He proposed gunpowder as the fuel.) Huygens' principle says that every point on a wavefront acts as a source of a secondary spherical wave called a *Huygens' wavelet.* As Huygens knew, the wavelet has a 90° phase shift (relative to the incident wave), and this phase shift is seen in (9.87) as the factor of $1/i$. In (9.87), there is also an angular factor of $\cos\theta$, apparently unknown to Huygens. Since $\cos\theta$ is also the angular dependence of dipole radiation, (9.87) states that the wavefront is equivalent to a fictitious layer of dipoles radiating 90° out of phase with the wave.

The next approximation to consider is the *paraxial approximation*, which is useful if we consider only points that are close to the z-axis (which is assumed to run through the clear aperture). With this approximation, $\cos\theta \approx 1$. With the radiation and paraxial approximations together, (9.85) becomes

$$p_z(\mathbf{r}) \approx \frac{1}{i\lambda z} \exp\left(ik\sqrt{r^2+z^2}\right). \tag{9.89}$$

Note that we have not yet approximated $(r^2+z^2)^{\frac{1}{2}}$ in the exponent. Since k is potentially a very large number, of order 10^5 cm^{-1} at optical wavelengths, we must be very careful in making approximations in the exponent. We shall pursue this point below.

With (9.89), the field in plane z is given by

$$u_z(\mathbf{r}) \approx \frac{1}{i\lambda z} \int_\infty d^2 r_0 \; u_0(\mathbf{r}_0) \exp\left(ik\sqrt{|\mathbf{r}-\mathbf{r}_0|^2+z^2}\right). \tag{9.90}$$

The exponential factor, illustrated in Fig. 9.5, represents a spherical wave (the Huygens' wavelet) emanating from the point $\mathbf{r}_0$ in the plane $z = 0$ and observed at point $\mathbf{r}$ in plane z.

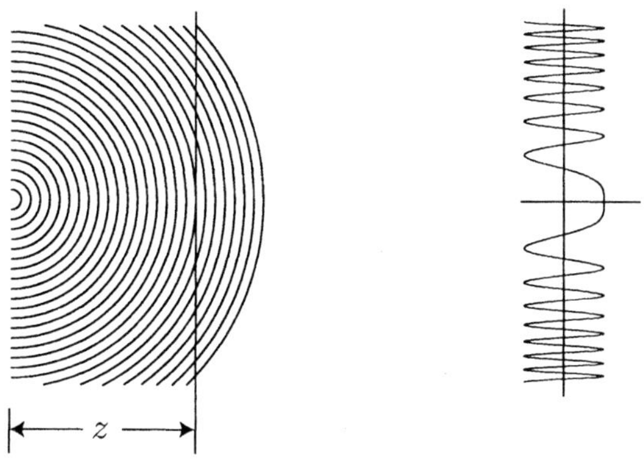

Fig. 9.5 Illustration of the function $\exp(ik\sqrt{|\mathbf{r}-\mathbf{r}_0|^2+z^2})$.

9.4.6 Fresnel diffraction

Next we investigate approximations to the exponential factor in (9.90). A binomial expansion for $z > |\mathbf{r}-\mathbf{r}_0|$ gives

$$R = \sqrt{|\mathbf{r}-\mathbf{r}_0|^2+z^2} = z + \frac{|\mathbf{r}-\mathbf{r}_0|^2}{2z} - \frac{|\mathbf{r}-\mathbf{r}_0|^4}{8z^3} + \cdots , \tag{9.91}$$

so we can write the exponential as

$$\exp\left(ik\sqrt{|\mathbf{r}-\mathbf{r}_0|^2+z^2}\right) = \exp(ikz)\exp\left(ik\frac{|\mathbf{r}-\mathbf{r}_0|^2}{2z}\right)\exp\left(-ik\frac{|\mathbf{r}-\mathbf{r}_0|^4}{8z^3}\right)\cdots . \tag{9.92}$$

One of the *terms* in (9.91) can be neglected only if the corresponding *factor* in (9.92) is approximately unity. It does not suffice if a term in (9.91) is small relative to the previous term; it must be small compared to, say, $\pi/4$ in an absolute sense. The quartic term is negligible if

$$\frac{k|\mathbf{r}-\mathbf{r}_0|^4}{8z^3} \ll \frac{\pi}{4} \qquad \text{or} \qquad |\mathbf{r}-\mathbf{r}_0|^4 \ll \lambda z^3 . \tag{9.93}$$

If z is large enough for this condition to hold, then the quartic term and all higher terms in (9.91) can be neglected, and (9.90) becomes

$$u_z(\mathbf{r}) \approx \frac{\exp(ikz)}{i\lambda z}\int_\infty d^2r_0\; u_0(\mathbf{r}_0)\exp\left(i\pi\frac{|\mathbf{r}-\mathbf{r}_0|^2}{\lambda z}\right) = u_0(\mathbf{r}) * p_z(\mathbf{r}) , \tag{9.94}$$

where now the 2D PSF is given by

$$p_z(\mathbf{r}) \approx \frac{\exp(ikz)}{i\lambda z}\exp\left(i\pi\frac{r^2}{\lambda z}\right) . \tag{9.95}$$

In this form, called the *Fresnel approximation*, the PSF is a constant (independent of $\mathbf{r}$) times the quadratic phase factor $\exp(i\pi r^2/\lambda z)$. Once again, this factor represents a spherical wave as observed on a plane, but now the spherical wavefronts are approximated by paraboloids. To see the full space-time behavior of the waves, it must be recalled that there is an associated time dependence $\exp(-2\pi i\nu_0 t)$. Figure 9.6 displays the behavior of $\exp(i\pi r^2/\lambda z - 2\pi i\nu_0 t)$ as a function of r, z and t.

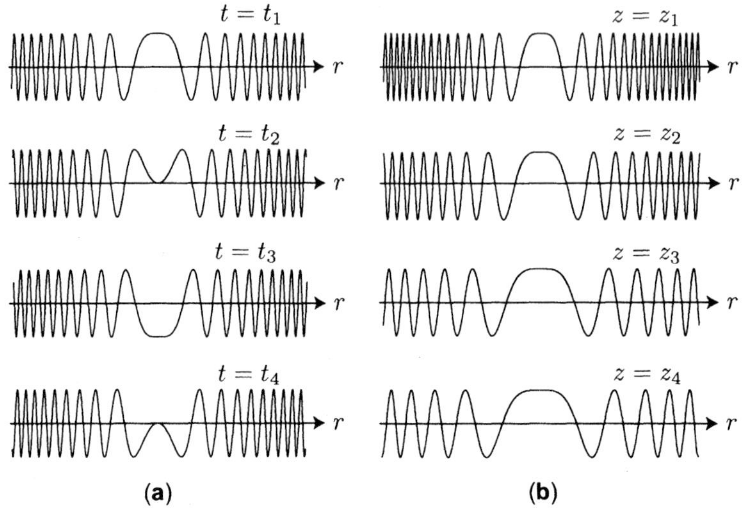

Fig. 9.6 (a) The real part of the time-dependent quadratic phase factor $\exp(i\pi \frac{r^2}{\lambda z} - 2\pi i\nu_0 t)$ as a function of r for different t. (b) The same function plotted as a function of r for different z.

As discussed in Sec. 4.3, a quadratic phase factor is also called a *chirp*, and convolution with a chirp is the same as computing a Fresnel transform. Within the Fresnel approximation, $u_z(\mathbf{r})$ is the Fresnel transform of $u_0(\mathbf{r})$, with the parameter β in (4.122) given by $1/\lambda z$. The properties of Fresnel transforms and chirps given in Sec. 4.3 are very useful in diffraction problems.

In particular, the Fourier implementation of the Fresnel transform discussed in Sec. 4.3 can be used to derive an alternative form for the Fresnel diffraction formula. All we have to do is to extend the 1D argument that led up to (4.138) to 2D. To this end, note that

$$|\mathbf{r} - \mathbf{r}_0|^2 = r^2 + r_0^2 - 2\mathbf{r} \cdot \mathbf{r}_0 \,. \tag{9.96}$$

Substituting this form into (9.94), we find

$$u_z(\mathbf{r}) = \frac{\exp(ikz)}{i\lambda z} \exp\left(i\pi \frac{r^2}{\lambda z}\right) \int_\infty d^2 r_0 \; u_0(\mathbf{r}_0) \exp\left(i\pi \frac{r_0^2}{\lambda z}\right) \exp\left(-2\pi i \frac{\mathbf{r}\cdot\mathbf{r}_0}{\lambda z}\right) . \tag{9.97}$$

This integral can be recognized as the 2D Fourier transform of the product $u_0(\mathbf{r}_0)\exp(i\pi r_0^2/\lambda z)$, but the role of the spatial-frequency vector is played by $\mathbf{r}/\lambda z$. We can thus write

$$u_z(\mathbf{r}) = \frac{\exp(ikz)}{i\lambda z} \exp\left(i\pi \frac{r^2}{\lambda z}\right) \mathcal{F}_2 \left\{ u_0(\mathbf{r}_0) \exp\left(i\pi \frac{r_0^2}{\lambda z}\right) \right\}_{\boldsymbol{\rho}=\mathbf{r}/\lambda z} . \tag{9.98}$$

In spite of the transform, this equation relates the input $u_0(\mathbf{r})$ to the output $u_z(\mathbf{r})$ *in the space domain*; the substitution $\boldsymbol{\rho} = \mathbf{r}/\lambda z$ gets us back from the frequency domain to the space domain, and the Fourier transform is just a convenient way of performing the spatial convolution with the quadratic phase factor (9.95). The resulting expression, (9.98), is mathematically equivalent to (9.94) and is thus an equally valid formula for Fresnel diffraction.

9.4.7 Fraunhofer diffraction

The next approximation takes advantage of the limited support of $t_{ap}(\mathbf{r})$. Suppose the clear aperture fits into a circle of radius a, so that $r_0 < a$ for all $\mathbf{r}_0$ in the range of integration in (9.97). Then, if $z \gg a^2/\lambda$, we can approximate $\exp(i\pi r_0^2/\lambda z)$ by unity inside the Fourier transform in (9.98), and we have

$$u_z(\mathbf{r}) \approx \frac{\exp(ikz)}{i\lambda z} \exp\left(i\pi \frac{r^2}{\lambda z}\right) \mathcal{F}_2\{u_0(\mathbf{r}_0)\}_{\boldsymbol{\rho}=\mathbf{r}/\lambda z}, \qquad z \gg a^2/\lambda. \tag{9.99}$$

Under this *Fraunhofer approximation*, $u_z(\mathbf{r})$ has a factor proportional to the Fourier transform of $u_0(\mathbf{r})$, rescaled by the substitution $\boldsymbol{\rho} = \mathbf{r}/\lambda z$. There is also a factor $\exp(i\pi r^2/\lambda z)$, which is not a constant since it depends on position in the observation plane. This factor can be interpreted as a spherical wave emanating from a point on the z axis, so (9.99) shows that, for z sufficiently large, $u_z(\mathbf{r})$ approaches a simple spherical wave modulated by the rescaled Fourier transform of $u_0(\mathbf{r})$. The region where $z \gg a^2/\lambda$ is called the *Fraunhofer zone* or the *far field.*

Irradiance in the Fraunhofer pattern Colloquially, the term *diffraction pattern* refers to the optical power per unit area incident on a surface, which is called the *irradiance* and denoted $I(\mathbf{r})$. It will be shown in Chap. 10 that, under certain broad assumptions, the irradiance is proportional to $|u(\mathbf{r})|^2$, with the constant of proportionality dependent on the physical interpretation of $u(\mathbf{r})$. In this chapter we dispense with the constants and set $I(\mathbf{r}) = |u(\mathbf{r})|^2$. When the field is random, we shall define the *mean irradiance* $\overline{I}(\mathbf{r})$ as the statistical expectation $\langle |u(\mathbf{r})|^2 \rangle$.

When we compute the irradiance, the leading factor of $\exp(i\pi r^2/\lambda z)$ in (9.99) disappears, and we have

$$|u_z(\mathbf{r})|^2 \approx \frac{1}{\lambda^2 z^2} \left| \int_\infty d^2 r_0 \, \exp(-2\pi i \mathbf{r} \cdot \mathbf{r}_0/\lambda z) \, u_0(\mathbf{r}_0) \right|^2 = \frac{1}{\lambda^2 z^2} \left| U_0\left(\frac{\mathbf{r}}{\lambda z}\right) \right|^2, \tag{9.100}$$

where $U_0(\boldsymbol{\rho})$ is the 2D Fourier transform of $u_0(\mathbf{r})$. In the Fraunhofer approximation, therefore, the diffraction pattern is proportional to the squared modulus of the Fourier transform of the input field. If $u_{inc}(\mathbf{r})$ is a plane wave normally incident on the aperture plane, then $u_0(\mathbf{r})$ is proportional to $t_{ap}(\mathbf{r})$, and we can say that the Fraunhofer pattern is the squared modulus of the Fourier transform of the aperture transmittance.

Fraunhofer diffraction without the paraxial approximation The expressions (9.99) and (9.100) were obtained by first applying the paraxial approximation (or considering observation points near the axis), then taking advantage of the limited support of $t_{ap}(\mathbf{r})$. We can get another useful expression by leaving out the paraxial approximation. We expand R not in powers of $1/z$ as in (9.91) but in powers of $1/|\mathbf{r}|$, obtaining

$$R = |\mathbf{r} - \mathbf{r}_0| = \sqrt{|\mathbf{r}|^2 - 2\mathbf{r} \cdot \mathbf{r}_0 + |\mathbf{r}|_0^2} = |\mathbf{r}| - \frac{\mathbf{r}_0 \cdot \mathbf{r}}{|\mathbf{r}|} + \frac{|\mathbf{r}|_0^2}{2|\mathbf{r}|} + \cdots. \tag{9.101}$$

If the aperture restricts $|\mathbf{r}|_0$ to sufficiently small values that $k|\mathbf{r}|_0^2 \ll |\mathbf{r}|$, then only the first two terms in this series must be retained, and (9.81) becomes

$$u(\mathbf{r}) = \frac{1}{i\lambda} \cos\theta \frac{\exp(ik|\mathbf{r}|)}{|\mathbf{r}|} \int_\infty d^2 r_0 \; u_0(\mathbf{r}_0) \exp\left[-ik\frac{\mathbf{r}_0 \cdot \mathbf{r}}{|\mathbf{r}|}\right]. \tag{9.102}$$

Note that $\cos\theta$ is independent of $\mathbf{r}_0$ in this approximation.

Since $\mathfrak{r}_0 = (\mathbf{r}_0, 0)$ and $\mathfrak{r} = (\mathbf{r}, z)$, the 3D scalar product $\mathfrak{r}_0 \cdot \mathfrak{r}$ is the same as the 2D one, $\mathbf{r}_0 \cdot \mathbf{r}$. Thus we can write the integral in (9.102) as a 2D Fourier transform,

$$u(\mathfrak{r}) = \frac{1}{i\lambda}\cos\theta \frac{\exp(ik|\mathfrak{r}|)}{|\mathfrak{r}|}\int_\infty d^2r_0\; u_0(\mathbf{r}_0)\exp(-2\pi i\mathbf{r}_0\cdot\boldsymbol{\rho}) = \frac{1}{i\lambda}\cos\theta\frac{\exp(ik|\mathfrak{r}|)}{|\mathfrak{r}|}U_0(\boldsymbol{\rho})\,, \tag{9.103}$$

where the 2D spatial frequency vector is now given by

$$\boldsymbol{\rho} = \frac{\mathbf{r}}{\lambda|\mathfrak{r}|}\,. \tag{9.104}$$

The two expressions for $\boldsymbol{\rho}$, $\mathbf{r}/\lambda z$ in (9.99) and $\mathbf{r}/\lambda|\mathfrak{r}|$ in (9.104), agree if $\cos\theta \approx 1$, but the latter is more general.

As in (9.99), there is a factor in (9.103) representing a spherical wave, only now it is an exact spherical wave centered on the origin, not just the Fresnel approximation to such a wave. In both of these equations, however, the interesting factor is the Fourier integral. The amplitude of the field at an observation point is determined by a single component in the 2D Fourier transform of the field in the plane $z = 0$, so Fraunhofer diffraction gives us a way of probing the Fourier domain. Note that the components of the 2D vector $\mathbf{r}/|\mathfrak{r}|$ in (9.104) are direction cosines of the unit vector pointing from the center of the aperture to the observation point, so this direction serves to pick out the Fourier component of interest. Figure 9.7 provides a graphical way of understanding this result.

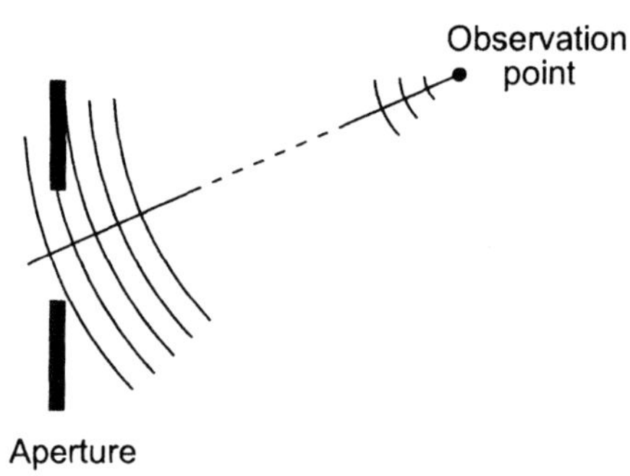

Fig. 9.7 Illustration showing why the Fraunhofer diffraction pattern picks out a single spatial frequency. The Green's function can be interpreted as a spherical wave centered on the observation point, and this spherical wave acts as a weighting function for points in the aperture in the diffraction integral. If the Fraunhofer condition is satisfied, this spherical wave is approximately a plane wave over the aperture.

9.5 DIFFRACTION IN THE FREQUENCY DOMAIN

As we saw in Sec. 9.4.7, Fraunhofer diffraction singles out a particular Fourier component of the field and maps it to a point in the Fraunhofer region, but the role of Fourier analysis in diffraction theory is not limited to the Fraunhofer approximation. Fourier transforms arise naturally in all propagation and diffraction problems because free space (or any homogeneous medium) is a linear shift-invariant system, and plane waves (or Fourier kernels) are eigenfunctions of such systems. When we write a field in terms of its Fourier transform, we are expanding it in the eigenfunction basis, so useful new insights and mathematical formulas can be expected.

In Sec. 9.5.1, we essentially start over and rederive diffraction theory from a Fourier viewpoint. The Fresnel and Fraunhofer approximations are revisited from this new viewpoint in Sec. 9.5.2. In Sec. 9.5.3 we illustrate the usefulness of this formalism by considering beam propagation. Finally, in Sec. 9.5.4 we apply the concept of angular spectrum to obtain the laws governing reflection and refraction of light from an interface between two different material media.

9.5.1 Angular spectrum

As in Sec. 9.4, consider diffraction from a planar aperture in the plane $z = 0$ and assume that the field $u_0(\mathbf{r})$ in this plane can be specified by the Kirchhoff boundary condition. This field can be represented in terms of its 2D Fourier transform $U_0(\boldsymbol{\rho})$ as

$$u_0(\mathbf{r}) = \int_\infty d^2\rho\, U_0(\boldsymbol{\rho}) \exp(2\pi i \boldsymbol{\rho}\cdot\mathbf{r}) = \int_{-\infty}^{\infty} d\xi \int_{-\infty}^{\infty} d\eta\, U_0(\xi,\eta) \exp[2\pi i(\xi x + \eta y)]\,. \tag{9.105}$$

We can interpret $\exp(2\pi i \boldsymbol{\rho}\cdot\mathbf{r})$ as a general plane wave (with the time dependence suppressed) evaluated on the plane $z = 0$, that is,

$$\exp[2\pi i(\xi x + \eta y)] = \exp[2\pi i(\xi x + \eta y + \zeta z)]\,\Big|_{z=0}\,, \tag{9.106}$$

or, in vector form,

$$\exp(2\pi i \boldsymbol{\rho}\cdot\mathbf{r}) = \exp(2\pi i \boldsymbol{\sigma}\cdot\mathbf{r})\Big|_{z=0}\,. \tag{9.107}$$

The integral in (9.105) expresses the wave on $z = 0$ as a superposition of plane waves travelling in different directions. This superposition is known as the *angular spectrum* of the wave. Each plane wave will continue to propagate in the same direction, without coupling to any other plane wave, so all we have to do to get an expression for the field at an arbitrary z is to put back the dependence of the plane wave on z. We can do so since the missing z component of $\boldsymbol{\sigma}$ is determined by (9.39). Taking the + sign in (9.39) since the incident wave propagates in the $+z$ direction, we can construct a solution valid for all z as follows:

$$u_z(\mathbf{r}) = \int_\infty d^2\rho\, U_0(\boldsymbol{\rho}) \exp\left[2\pi i \left(\boldsymbol{\rho}\cdot\mathbf{r} + z\sqrt{\frac{1}{\lambda^2} - \rho^2}\right)\right]$$

$$= \int_{-\infty}^{\infty} d\xi \int_{-\infty}^{\infty} d\eta\, U_0(\xi,\eta) \exp\left[2\pi i \left(\xi x + \eta y + z\sqrt{\frac{1}{\lambda^2} - \xi^2 - \eta^2}\right)\right]\,. \tag{9.108}$$

This form agrees with the boundary condition (9.105) if $z = 0$, and it satisfies the homogeneous wave equation by construction. Since the Dirichlet boundary condition uniquely determines a solution in $\mathcal{V}$, (9.108) must be that solution.

Transfer function Comparing (9.105) and (9.108), we see that the 2D transform of the field in plane z has the form

$$U_z(\boldsymbol{\rho}) = U_0(\boldsymbol{\rho})\, P_z(\boldsymbol{\rho})\,, \tag{9.109}$$

where $P_z(\boldsymbol{\rho})$ is the 2D transfer function for propagation from the plane $z = 0$ to a parallel plane a distance z away. It is given by

$$P_z(\boldsymbol{\rho}) = \exp\left(2\pi i z\sqrt{\frac{1}{\lambda^2} - \rho^2}\right) = \exp\left(2\pi i z\sqrt{\frac{1}{\lambda^2} - \xi^2 - \eta^2}\right). \tag{9.110}$$

Note that $P_z(\boldsymbol{\rho}) \to 1$ as $z \to 0$, which corresponds to the statement that $p_z(\mathbf{r}) \to \delta(\mathbf{r})$.

Another way to derive the transfer function is to Fourier-transform the point spread function. Before any approximations, $p_z(\mathbf{r})$ is given by (9.85), so

$$P_z(\boldsymbol{\rho}) = \mathcal{F}_2\{p_z(\mathbf{r})\} = -\frac{1}{2\pi}\mathcal{F}_2\left\{\left(ik - \frac{1}{R}\right)\frac{\exp(ikR)}{R}\cos\theta\right\}, \tag{9.111}$$

where $R = \sqrt{r^2 + z^2}$, $\cos\theta = z/R$ and $k = 2\pi/\lambda$.

To evaluate the Fourier integral, recall that we obtained the form (9.85) for $p_z(\mathbf{r})$ in the first place by performing a derivative with respect to z in (9.79). Therefore we can write

$$p_z(\mathbf{r}) = -\frac{1}{2\pi}\frac{\partial}{\partial z}\left[\frac{\exp(ik\sqrt{r^2+z^2})}{\sqrt{r^2+z^2}}\right]. \tag{9.112}$$

The expression in brackets is a rotationally symmetric 2D function; taking its Fourier transform by means of (3.248), we find

$$P_z(\boldsymbol{\rho}) = -\frac{\partial}{\partial z}\int_0^\infty r\,dr\ \mathrm{J}_0(2\pi\rho r)\left[\frac{\exp(ik\sqrt{r^2+z^2})}{\sqrt{r^2+z^2}}\right]. \tag{9.113}$$

A change of variables yields

$$P_z(\boldsymbol{\rho}) = -\frac{\partial}{\partial z}\int_z^\infty dt\ \mathrm{J}_0\left(2\pi\rho\sqrt{t^2 - z^2}\right)e^{ikt}. \tag{9.114}$$

This integral is tabulated in a standard source (Gradshteyn and Ryzhik, 1980, p. 736), and we obtain

$$P_z(\boldsymbol{\rho}) = -i\frac{\partial}{\partial z}\frac{\exp(iz\sqrt{k^2 - 4\pi^2\rho^2})}{\sqrt{k^2 - 4\pi^2\rho^2}}, \tag{9.115}$$

from which differentiation and the substitution $k = 2\pi/\lambda$ gives (9.110) again.

Interpretation of the transfer function We see from (9.110) that the transfer function $P_z(\boldsymbol{\rho})$ is a pure phase factor for $\rho < 1/\lambda$. Hence the modulus of the transform is unchanged by propagation:

$$|P_z(\boldsymbol{\rho})| = |P_0(\boldsymbol{\rho})|, \qquad \rho < \frac{1}{\lambda}. \tag{9.116}$$

As required by conservation of energy, no plane waves are either increased or decreased in amplitude as they propagate. All of the wondrous patterns of diffraction are produced by subtle changes of phase among the constituent plane waves in a field.

For $\rho > 1/\lambda$, on the other hand, the complex exponential becomes a real one, and

$$\exp\left(2\pi i z\sqrt{\frac{1}{\lambda^2} - \rho^2}\right) = \exp\left(-2\pi z\sqrt{\rho^2 - \frac{1}{\lambda^2}}\right), \qquad \rho > \frac{1}{\lambda}. \tag{9.117}$$

Waves with $\rho > 1/\lambda$ decay exponentially as they propagate, so they contribute very little to the resulting field if z is more than a few wavelengths. For this reason they are called *evanescent waves.* This exponential decay does not violate conservation of energy since it can be shown that there is no energy flux associated with evanescent waves.

One might wonder whether evanescent waves can even exist. From (9.42), the condition $\rho > 1/\lambda$ implies that the direction cosines α or β have to exceed one, which is unphysical. On the other hand, if we are to represent an arbitrary wave $u_0(\mathbf{r})$ by a Fourier expansion like (9.105), the limits must run over the entire ξ-η plane, not just over a circle of radius $1/\lambda$. In fact, there are several physical ways in which one can create structures with very high spatial frequency in the plane $z = 0$. A simple way is to place a very fine grating with sub-wavelength spacing over the aperture. To correctly describe the field emerging from the aperture requires $\rho > 1/\lambda$ in (9.105), but (9.117) shows that these frequency components die off very quickly as z increases and may safely be ignored in most problems.

As soon as we make the paraxial approximation, we are implicitly ignoring evanescent waves. Paraxial waves propagate nearly parallel to the z axis and hence have $\rho \ll 1/\lambda$ by (9.105).

9.5.2 Fresnel and Fraunhofer approximations

Fresnel To derive the Fresnel approximation for $P_z(\boldsymbol{\rho})$, we assume $\rho \ll 1/\lambda$ and expand the exponent in (9.110). We find

$$P_z(\boldsymbol{\rho}) = \exp\left[2\pi i z\sqrt{\frac{1}{\lambda^2} - \rho^2}\right] \approx \exp(ikz)\exp(-i\pi\lambda z\rho^2). \tag{9.118}$$

This is the expected result since it follows from (3.263) that

$$\mathcal{F}_2^{-1}\{\exp(ikz)\exp(-i\pi\lambda z\rho^2)\} = \frac{\exp(ikz)}{i\lambda z}\exp(i\pi r^2/\lambda z), \tag{9.119}$$

in agreement with (9.95). Thus, in the Fresnel approximation, the angular spectrum propagates according to

$$U_z(\boldsymbol{\rho}) = U_0(\boldsymbol{\rho})\, P_z(\boldsymbol{\rho}) \approx U_0(\boldsymbol{\rho})\exp(ikz)\exp(-i\pi\lambda z\rho^2), \tag{9.120}$$

which is just the 2D Fourier transform of (9.94).

Fraunhofer The general angular-spectrum formula (9.108) expresses the field at a single point $\mathfrak{r}$ or $(\mathbf{r}, z)$ as a superposition of Fourier components $U_0(\boldsymbol{\rho})$, but the Fraunhofer formula (9.104) shows that only a single frequency contributes, namely $\boldsymbol{\rho} = \mathbf{r}/\lambda\mathfrak{r}$. How do we get from an integral over plane-wave components to just one component?

To answer this question, we must recognize that a field $u_0(\mathbf{r})$ contained entirely within an aperture of radius a has a Fourier transform of width $\sim 1/a$ or greater by the Fourier uncertainty relation (5.10). If a is small, $U_0(\boldsymbol{\rho})$ is broad and slowly varying. In (9.108), this broad function is multiplied by the phase factor $\exp i\Phi(\boldsymbol{\rho};\mathbf{r})$, where $\Phi(\boldsymbol{\rho};\mathbf{r}) = 2\pi\boldsymbol{\rho}\cdot\mathbf{r} + 2\pi z\sqrt{\frac{1}{\lambda^2} - \rho^2}$. Regions of the frequency plane where $\Phi(\boldsymbol{\rho};\mathbf{r})$ is a rapidly varying function of $\boldsymbol{\rho}$ do not make much contribution to the integral. Instead, the main contribution comes from the points of *stationary phase* $\boldsymbol{\rho} = \boldsymbol{\rho}_0$, defined by

$$\nabla_\rho \Phi(\boldsymbol{\rho};\mathbf{r}) = 0 \qquad \text{at } \boldsymbol{\rho} = \boldsymbol{\rho}_0 \,. \tag{9.121}$$

Straightforward differentiation shows that

$$\boldsymbol{\rho}_0 = \frac{1}{\lambda}\frac{\mathbf{r}}{\sqrt{r^2+z^2}} = \frac{\mathbf{r}}{\lambda|\mathbf{r}|}\,. \tag{9.122}$$

If $U_0(\boldsymbol{\rho})$ is slowly varying in the vicinity of $\boldsymbol{\rho}_0$, (9.108) becomes

$$u_z(\mathbf{r}) = U_0\left(\frac{\mathbf{r}}{\lambda|\mathbf{r}|}\right)\int_\infty d^2\rho \,\exp\left[2\pi i\left(\boldsymbol{\rho}\cdot\mathbf{r} + z\sqrt{\frac{1}{\lambda^2} - \rho^2}\right)\right]. \tag{9.123}$$

But the integral is now simply the inverse Fourier transform of the transfer function $P_z(\boldsymbol{\rho})$ defined in (9.110), so we have the simple and elegant formula,

$$u_z(\mathbf{r}) = U_0\left(\frac{\mathbf{r}}{\lambda|\mathbf{r}|}\right) p_z(\mathbf{r})\,, \tag{9.124}$$

where the point spread function $p_z(\mathbf{r})$ is given explicitly by (9.85). Thus the wave in the Fraunhofer region is the z-derivative of a spherical wave. The strength of this wave is determined by the Fourier transform of the wave in the aperture evaluated at the single frequency $\boldsymbol{\rho}_0$. This frequency has a simple interpretation; it corresponds to the plane wave with wave vector directed from the center of the aperture towards the observation point, which is just the plane-wave approximation to the spherical wave shown in Fig. 9.7.

9.5.3 Beams

To illustrate the use of the angular spectrum, we examine some optical fields that can be described as *beams* since they propagate predominantly in a single direction.

Gaussian beams Consider a mask with an amplitude transmittance given by

$$t_{mask}(\mathbf{r}) = \exp(-\pi r^2/a^2)\,. \tag{9.125}$$

The constant a specifies the halfwidth of the transmittance function at the $e^{-\pi}$ points.

If this mask is illuminated with a normally incident plane wave of unit amplitude, the field emerging from the mask is also given by $t_{mask}(\mathbf{r})$, and the Fourier transform of that field is

$$U_0(\boldsymbol{\rho}) = a^2 \exp(-\pi a^2 \rho^2)\,. \tag{9.126}$$

Within the Fresnel approximation, the Fourier transform of the field at plane z is given by (9.120) as

$$U_z(\boldsymbol{\rho}) = a^2 \exp(ikz) \exp(-i\pi\lambda z \rho^2 - \pi a^2 \rho^2) . \tag{9.127}$$

Fourier transforms of Gaussians and chirps were discussed in Sec. 3.4.5. To find the field $u_z(\mathbf{r})$ in the present problem, we require the inverse transform of a Gaussian times a chirp. The requisite formula is (3.263); the validity of that formula for complex β is justified simply by deleting the limit in the argument given in the vicinity of (3.181) – (3.185). With a little algebra, we find

$$u_z(\mathbf{r}) = \frac{a^2 \exp(ikz)}{a^2 + i\lambda z} \exp\left(-\pi \frac{r^2}{a^2 + i\lambda z}\right) . \tag{9.128}$$

This form shows that the beam profile is unaltered so long as $a^2 \gg \lambda z$. Eventually, however, for sufficiently large z, it is necessary to consider the beam spread. The general behavior can be seen by rewriting (9.128) as

$$u_z(\mathbf{r}) = \frac{a^2 \exp(ikz)}{a^2 + i\lambda z} \exp\left(i\pi \frac{\lambda z r^2}{a^4 + \lambda^2 z^2}\right) \exp\left(-\pi \frac{a^2 r^2}{a^4 + \lambda^2 z^2}\right) . \tag{9.129}$$

In plane z, the halfwidth of the beam (at the $e^{-\pi}$ point) is given by

$$a_z = \frac{1}{a}\sqrt{a^4 + \lambda^2 z^2} . \tag{9.130}$$

For small z, $a_z \approx a$, while for large z, $a_z \approx \lambda z/a$, as might be anticipated from the Fraunhofer formula (9.99).

The treatment above can be modified for the case where the Gaussian mask is illuminated by a converging spherical wave rather than a plane wave. For an interesting discussion of this case, see Gaskill (1978).

Exact Gaussian-like beams The discussion of Gaussian beams above was based on the Fresnel approximation. Without this approximation, a Gaussian beam does not satisfy the Helmholtz equation, as the reader can verify by direct differentiation. It is possible, however, to construct an exact solution to the Helmholtz equation that has many of the characteristics of Gaussian beams (Landesman and Barrett, 1988). This solution is most readily expressed in oblate spheroidal coordinates, one of the orthogonal coordinates systems in which the Helmholtz equation is separable. Solutions to the separated equations, called oblate spheroidal wavefunctions, have been known for many years; they are discussed in detail by Flammer (1957). Landesman's solution, on the other hand is fundamentally different in that it is not separable. In an appropriate paraxial limit (equivalent to taking a large), it reduces to the simple Gaussian beam, but it remains an exact solution for all values of the parameters.

Moreover, Landesman showed that the Gaussian-like exact solution was just one member of a complete family of solutions; the other members are analogs of Hermite-Gauss beams often used to model laser modes (Kogelnik and Li, 1966; Fox and Li, 1961).

From Gaussian beams to rays Geometric optics deals with thin pencils of light called rays. The theory of Gaussian beams just developed provides one way of conceptualizing rays.

Geometric optics ignores diffraction effects; to turn off diffraction, we can formally take the limit of very short wavelengths. Then (9.129) becomes

$$\lim_{\lambda \to 0} u_z(\mathbf{r}) = \exp\left(-\pi \frac{r^2}{a^2}\right). \tag{9.131}$$

Thus the form of the beam is independent of z in this limit, and we can identify the beam with a geometric ray.

Bessel beams Now suppose we have a mask with an amplitude transmittance given by

$$t_{Bessel}(\mathbf{r}) = \mathrm{J}_0(2\pi\rho_0 r), \tag{9.132}$$

where ρ_0 is a positive real constant and $\mathrm{J}_0(\cdot)$ is the zero-order Bessel function of the first kind. Note that this transmittance, unlike that of the Gaussian mask, can assume negative values. One way to implement such a mask would be to place thin glass plates over the negative regions and adjust the thickness so that the glass shifts the phase of the light by π relative to light that passes through the positive regions.

Again we assume that the mask is illuminated with a normally incident plane wave of unit amplitude, so the field $u_0(\mathbf{r})$ is also given by $t_{mask}(\mathbf{r})$. It can be verified from (3.248) that the Fourier transform of $u_0(\mathbf{r})$ is given by

$$U_0(\boldsymbol{\rho}) = \frac{1}{2\pi\rho_0}\,\delta(\rho - \rho_0). \tag{9.133}$$

Note that the delta function here is 1D; it vanishes except on a ring of radius ρ_0 in the 2D Fourier plane.

In angular-spectrum terms, the ring-delta function selects out plane waves that make a specific angle to the z axis. That is, the wavevectors $\mathbf{k}$ form a cone around the z axis. We could have anticipated this interpretation simply by examining the integral by which we defined the J_0 Bessel function in Chap. 3; from (3.247) with a change of variables, we can write

$$\mathrm{J}_0(2\pi\rho_0 r) = \frac{1}{2\pi}\int_0^{2\pi} d\theta_\rho\, e^{2\pi i \rho_0 r \cos\theta_\rho}. \tag{9.134}$$

If we let $\mathbf{r} = (r, 0)$ and $\boldsymbol{\rho} = (\rho_0, \theta_\rho)$ in polar coordinates, we see from (9.134) that $\mathrm{J}_0(2\pi\rho_0 r)$ is a superposition of plane waves for which ρ_0 is constant and θ_ρ takes on all values in 2π.

To propagate the angular spectrum specified in (9.133), we use (9.109) and (9.110) and obtain

$$U_z(\boldsymbol{\rho}) = \frac{1}{2\pi\rho_0}\,\delta(\rho - \rho_0)\exp\left(2\pi i z\sqrt{\frac{1}{\lambda^2} - \rho^2}\right). \tag{9.135}$$

By (2.25), however, we can replace ρ with ρ_0 in any function that multiplies $\delta(\rho - \rho_0)$, so

$$U_z(\boldsymbol{\rho}) = \frac{1}{2\pi\rho_0}\,\delta(\rho - \rho_0)\exp\left(2\pi i z\sqrt{\frac{1}{\lambda^2} - \rho_0^2}\right) \equiv \frac{e^{i\phi}}{2\pi\rho_0}\,\delta(\rho - \rho_0). \tag{9.136}$$

In other words, except for the constant $e^{i\phi}$, $U_z(\boldsymbol{\rho})$ is identical to $U_0(\boldsymbol{\rho})$, and hence $u_z(\mathbf{r}) = const \cdot u_0(\mathbf{r})$. For this reason, the J_0 Bessel function is sometimes called a *diffraction-free beam*, in spite of the fact that diffraction theory is used to show that its form is unchanged by propagation. A more accurate statement is that it is an eigenfunction of free-space propagation.

More general diffraction-free beams Even the characterization of the Bessel beam as an eigenfunction might be surprising since we have previously noted that plane waves are the eigenfunctions of free space. We have seen that free-space propagation can be regarded as a 2D linear shift-invariant system, so the plane wave $\exp(2\pi i\boldsymbol{\rho}\cdot\mathbf{r})$ is an eigenfunction and the transfer function $P_z(\boldsymbol{\rho})$ is the eigenvalue. The key point, however, is that $P_z(\boldsymbol{\rho})$ depends solely on the magnitude of $\boldsymbol{\rho}$, so free space is a rotationally symmetric LSIV system. As we discussed in Sec. 7.2.9, such systems have infinite degeneracy. *All* plane waves making a specific angle to the z axis have the same eigenvalue (transfer function), and any linear combination of eigenfunctions with the same eigenvalue is another eigenfunction with that eigenvalue.

This observation allows us to synthesize more general beams that are also invariant with propagation. Since the eigenvalues are independent of θ_ρ, we can use any weighting function $A(\theta_\rho)$ that we choose. Hence, the most general form of a so-called diffraction-free beam is (Nieto-Vesperinas, 1991)

$$u_0(\mathbf{r}) = \int_0^{2\pi} d\theta_\rho \, A(\theta_\rho)\, e^{2\pi i\rho_0 r\cos\theta_\rho} . \tag{9.137}$$

The J_0 Bessel function corresponds to $A(\theta_\rho) = \frac{1}{2\pi}$.

Practical issues In the 1980s there was considerable interest in directed-energy applications of high-power lasers. The goal was to apply a large energy per unit area on a distant target, and diffraction spreading was the chief obstacle to achieving this goal. Diffraction-free beams were touted as a way of overcoming this obstacle, but it was immediately recognized that it was not possible to synthesize a true Bessel beam since that would require an optical system with an infinite aperture. Durnin (1987) investigated the diffraction patterns of beams of the form $\mathrm{J}_0(2\pi\rho_0 r)\,\mathrm{cyl}(r/R_a)$ and showed that they maintain a tightly concentrated central peak up to a critical distance determined by the aperture radius R_a.

It does not follow from this observation, however, that it is advantageous to place a Bessel mask over a given lens aperture. From (9.97), the irradiance on the optic axis is proportional to

$$|u_z(0)|^2 = \frac{1}{\lambda z^2}\left|\int_{ap} d^2r_0 \, u_0(\mathbf{r}_0)\exp(i\pi r_0^2/\lambda z)\right|^2 , \tag{9.138}$$

where the integral is over an aperture of radius R_a. We know that $u_0(\mathbf{r}_0) = u_{inc}t_{ap}(\mathbf{r}_0)$, where u_{inc} is the amplitude of a plane wave normally incident on the aperture, and the amplitude transmittance in the aperture must satisfy $|t_{ap}(\mathbf{r}_0)| \leq 1$. Then the irradiance is maximized by choosing $t_{ap}(\mathbf{r}_0) = \exp(-i\pi r_0^2/\lambda z)$. As we shall see in Sec. 9.6.1, this function describes a lens that focuses the beam on the target at distance z; any other function in the aperture can only reduce the irradiance on target.

Bessel beams may be more useful in optical metrology than in directed-energy applications. In surveying, for example, it can be useful to have a fine pencil of light to define a line in space. A Gaussian laser beam may not allow sufficiently precise definition of the center of the pencil, and Durnin's results suggest that a Bessel beam could be advantageous. Even here, however, there are some interesting alternatives, which in fact were well known long before coinage of the term diffraction-free. One such is an axial prism called an *axicon*. This device is simply a conical piece of glass, and it also produces a fine pencil that maintains its form over a substantial distance.

9.5.4 Reflection and refraction of light

The law of reflection of light by a mirror was apparently known to Euclid, but the law of refraction when light passes from one medium to another is of more recent origin. This law is attributed to the Dutch mathematician Willebrord Snell van Royen (1591–1626), but his manuscript on the topic was apparently destroyed by fire; Descartes' *Dioptrique* (1637) is the first extant written account of the law of refraction (Herzberger, 1980).

In this section we shall see how the laws of reflection and refraction can be derived from the angular spectrum. Along the way, we shall acquire some insight into the connection between 2D spatial frequency and 3D direction of propagation.

Consider a plane wave of the form (9.33) incident on a planar interface between two different media, and assume for now that the interface is the plane $z = 0$. The normal to the interface and the incident **k** vector define another plane, called the *plane of incidence*; we assume that the plane of incidence is the x-z plane, so that $k_y = 0$. As in Fig. 9.4, we can draw the z axis horizontal and the interface plane vertical, but now there is no physical aperture in this plane. Assume that the medium to the left of the interface plane is homogeneous and has refractive index n or, equivalently, speed of light $c_m = c/n$. Similarly, the homogeneous medium to the right of the plane is assumed to have refractive index n' and speed of light $c'_m = c/n'$.

The incident field in the plane $z = 0$ is described by the 2D function,

$$u_{inc}(\mathbf{r}, t) = A_{inc} \exp(2\pi i \boldsymbol{\rho} \cdot \mathbf{r} - 2\pi i \nu_0 t) , \tag{9.139}$$

where A_{inc} is a constant amplitude, $\boldsymbol{\rho} = (\xi, 0)$ in the chosen coordinate system, and we have included the time dependence for clarity. We see from (9.139) that $u_{inc}(\mathbf{r}, t)$ oscillates with frequency ν_0 and phase $2\pi\boldsymbol{\rho}\cdot\mathbf{r}$ at point $\mathbf{r}$. It is this temporal oscillation that determines the nature of the reflected and refracted waves emanating from the interface.

Snell's law Let us look first at the wave propagating to the right of the interface. We already know the form of the temporal oscillation at each point in the interface, so we can write

$$u_{tr}(\mathbf{r}, t) = A_{tr} \exp(2\pi i \boldsymbol{\rho} \cdot \mathbf{r} - 2\pi i \nu_0 t) , \tag{9.140}$$

where the subscript tr stands for *transmitted.* Some additional considerations would be needed to determine the amplitude A_{tr}, but we won't need it in what follows. Comparison of (9.139) and (9.140) shows that the incident and transmitted waves have exactly the same dependence on $\mathbf{r}$ and t in the interface plane, and in particular, exactly the same spatial frequency $\boldsymbol{\rho}$. It does not follow, however, that they describe identical waves in the two media; we need to be careful about converting 2D spatial frequency to 3D direction of propagation.

Since $\eta = 0$ in both (9.139) and (9.140), it follows from (9.41) and (9.42) that the y component of the transmitted wavevector $\mathbf{k}_{tr}$ is zero. Another way to state this result is that $\mathbf{k}_{tr}$ lies in the plane of incidence, but we must still determine its angle in this plane.

From (9.44) and (9.39) with $\eta = 0$, we can express the angle θ_{inc} between the incident $\mathbf{k}$ vector and the z axis as[10]

$$\sin\theta_{inc} = \frac{\xi}{\sqrt{\xi^2 + \zeta^2}} = \lambda\xi\,. \tag{9.141}$$

The angle θ_{tr} between the z axis and $\mathbf{k}_{tr}$ is given by a similar expression, but ζ and λ must be replaced by the values appropriate to the medium; by (9.36) and (9.39) we have

$$\zeta' = \sqrt{\frac{1}{\lambda'^2} - \xi^2} = \sqrt{\left(\frac{n'\nu_0}{c}\right)^2 - \xi^2}\,. \tag{9.142}$$

Thus

$$\sin\theta_{tr} = \frac{\xi}{\sqrt{\xi^2 + \zeta'^2}} = \lambda'\xi\,. \tag{9.143}$$

Since ξ is the same in (9.141) and (9.143), and since $\lambda n = \lambda' n'$, we see that

$$n\sin\theta_{inc} = n'\sin\theta_{tr}\,. \tag{9.144}$$

This is the usual scalar formulation of *Snell's law*.

Vector form of Snell's law Equation (9.144) is specific to the coordinate system in which the plane of incidence is the x-z plane and the interface is normal to the z axis. A simple vector form of Snell's law without any reference to specific coordinates is

$$n(\hat{\boldsymbol{\kappa}}_{inc} \times \hat{\mathbf{n}}) = n'(\hat{\boldsymbol{\kappa}}_{tr} \times \hat{\mathbf{n}})\,, \tag{9.145}$$

where $\hat{\mathbf{n}}$ is the unit vector normal to the interface and $\hat{\boldsymbol{\kappa}}_{inc}$ and $\hat{\boldsymbol{\kappa}}_{tr}$ are unit vectors parallel to the incident and transmitted wavevectors, respectively.

A useful alternative form of Snell's law can be found by constructing an orthonormal basis for the plane of incidence. The vectors $\hat{\mathbf{n}}$ and $\hat{\boldsymbol{\kappa}}_{inc}$ are linearly independent (except in the case of normal incidence), so they form a basis for the plane of incidence, but they are not orthonormal. An orthonormal basis, obtained by applying Gram-Schmidt orthogonalization as outlined in Sec. A.4.3 of App. A, consists of $\hat{\mathbf{n}}$ plus an orthogonal unit vector defined by

$$\hat{\mathbf{n}}_\perp = \frac{\hat{\boldsymbol{\kappa}}_{inc} - (\hat{\boldsymbol{\kappa}}_{inc} \cdot \hat{\mathbf{n}})\hat{\mathbf{n}}}{\sqrt{1 - (\hat{\boldsymbol{\kappa}}_{inc} \cdot \hat{\mathbf{n}})^2}}\,. \tag{9.146}$$

The subscript $\perp$ indicates that $\hat{\mathbf{n}}_\perp$ is normal to $\hat{\mathbf{n}}$, which in turn is normal to the interface; hence $\hat{\mathbf{n}}_\perp$ lies in the interface plane, in fact along the intersection of the

[10] Note that θ_{inc} in this equation is the same as θ_x in (9.44); since we have chosen $k_y = 0$ there is no need to consider the projection of $\mathbf{k}$ onto the x-z plane.

interface plane and the plane of incidence. The reader can verify that $\hat{\mathbf{n}} \cdot \hat{\mathbf{n}}_\perp = 0$ and $|\hat{\mathbf{n}}_\perp| = 1$.

In terms of these unit vectors, $\boldsymbol{\rho} = (2\pi)^{-1}(\mathbf{k}_{inc} \cdot \hat{\mathbf{n}}_\perp)\hat{\mathbf{n}}_\perp = (2\pi)^{-1}(\mathbf{k}_{tr} \cdot \hat{\mathbf{n}}_\perp)\hat{\mathbf{n}}_\perp$, and Snell's law is:

$$n(\hat{\boldsymbol{\kappa}}_{inc} \cdot \hat{\mathbf{n}}_\perp) = n'(\hat{\boldsymbol{\kappa}}_{tr} \cdot \hat{\mathbf{n}}_\perp)\,. \tag{9.147}$$

The condition that both $\hat{\boldsymbol{\kappa}}_{inc}$ and $\hat{\boldsymbol{\kappa}}_{tr}$ lie in the plane of incidence is:

$$(\hat{\mathbf{n}}_\perp \times \hat{\mathbf{n}}) \cdot \hat{\boldsymbol{\kappa}}_{inc} = (\hat{\mathbf{n}}_\perp \times \hat{\mathbf{n}}) \cdot \hat{\boldsymbol{\kappa}}_{tr} = 0\,. \tag{9.148}$$

An explicit expression for $\hat{\boldsymbol{\kappa}}_{tr}$ that satisfies both of these conditions is

$$\hat{\boldsymbol{\kappa}}_{tr} = \frac{n}{n'}(\hat{\boldsymbol{\kappa}}_{inc} \cdot \hat{\mathbf{n}}_\perp)\hat{\mathbf{n}}_\perp + \left[\sqrt{1 - \left(\frac{n}{n'}\right)^2 (\hat{\boldsymbol{\kappa}}_{inc} \cdot \hat{\mathbf{n}}_\perp)^2}\,\right]\hat{\mathbf{n}}\,. \tag{9.149}$$

Reflected plane waves In the interface plane, the reflected wave has the same 2D structure as the incident and transmitted waves, namely (9.139) or (9.140). Once again, however, we have to figure out what this means in 3D terms.

We can use the vector form (9.149) for the reflected wave by making two simple modifications. First, since the incident and reflected waves propagate in the same medium, we must set $n' = n$. Second, we must account for the sign ambiguity seen in (9.39) when we convert from 2D spatial frequencies to 3D ones. For the reflected wave we must take the minus sign since the z component of $\mathbf{k}$ or $\hat{\boldsymbol{\kappa}}$ is reversed. Thus, by comparison to (9.149), the unit vector in the direction of the reflected wavevector is given by

$$\begin{aligned}\hat{\boldsymbol{\kappa}}_{refl} &= (\hat{\boldsymbol{\kappa}}_{inc} \cdot \hat{\mathbf{n}}_\perp)\hat{\mathbf{n}}_\perp - \left[\sqrt{1 - (\hat{\boldsymbol{\kappa}}_{inc} \cdot \hat{\mathbf{n}}_\perp)^2}\,\right]\hat{\mathbf{n}} \\ &= (\hat{\boldsymbol{\kappa}}_{inc} \cdot \hat{\mathbf{n}}_\perp)\hat{\mathbf{n}}_\perp - (\hat{\boldsymbol{\kappa}}_{inc} \cdot \hat{\mathbf{n}})\hat{\mathbf{n}}\,.\end{aligned} \tag{9.150}$$

This equation formalizes the usual law of mirror reflection: the angle of reflection equals the angle of incidence.

Reflection and refraction of beams The results obtained above also apply, at least approximately, to collimated beams rather than infinite plane waves. As in Sec. 9.5.3, we can consider a beam with, say, a Gaussian amplitude profile propagating in direction $\hat{\boldsymbol{\kappa}}_{inc}$. For simplicity, we take the Gaussian to be circularly symmetric on the aperture, which means that it is oval on a plane normal to $\hat{\boldsymbol{\kappa}}_{inc}$.

For this beam, the field in the interface plane is given by [*cf.* (9.139)]

$$u_{inc}(\mathbf{r}, t) = A_0 \exp(-\pi r^2/a^2)\exp(2\pi i \boldsymbol{\rho}_0 \cdot \mathbf{r} - 2\pi i \nu_0 t)\,, \tag{9.151}$$

and this form applies also to the transmitted and reflected waves. The effect of the Gaussian is that the incident wave no longer consists of a single spatial frequency (or equivalently, a single wavevector). Instead, the 2D angular spectrum (with the time dependence deleted) is given via (3.237) and (3.262) as

$$U_{inc}(\boldsymbol{\rho}) = A_0\, a^2 \exp(-\pi a^2 |\boldsymbol{\rho} - \boldsymbol{\rho}_0|^2)\,. \tag{9.152}$$

If $a\rho_0 \gg 1$, then this spectrum is concentrated around the frequency $\boldsymbol{\rho}_0$ which corresponds to the average incident wavevector. Under this approximation, we can assume that all wavevectors undergo approximately the same refraction, and Snell's law applies to the beam as well as to a plane wave. The reader is invited to fill in the details.

Reflection and refraction of rays Though we derived the laws of refraction and reflection by considering first plane waves and then beams, we note that the wavelength does not appear explicitly in the final result. For example, (9.145) and (9.150) are written entirely in terms of unit vectors $\hat{\boldsymbol{\kappa}}$ rather than wavevectors $\mathbf{k}$. That means that they are still valid as the wavelength goes to zero. We noted in Sec. 9.5.3 that Gaussian beams in free space behave as geometric rays in this limit. Now we see that geometric rays defined this way obey the same laws of reflection and refraction as plane waves.

9.6 IMAGING OF POINT OBJECTS

We have now laid most of the groundwork needed to analyze imaging systems based on wave propagation, but we must still introduce image-forming elements such as lenses. There are many ways of introducing these elements, including approaches based on Snell's law or Fermat's principle, but the approach taken here is to regard a lens as a device for transforming a wave field.

Several other physical entities are the functional equivalent of lenses, in the sense that they can form images or otherwise alter light beams in a manner analogous to lenses. Included in this category are convex and concave mirrors, holographic optical elements and diffractive optics. In this chapter, however, we shall consider only lenses *per se*.

9.6.1 Ideal thin lens

An *ideal lens* is one that produces an image point that is as compact as it can be according to the laws of physical optics. Hence an ideal lens is also called a *diffraction-limited* lens. Any imperfection in a lens that causes spreading of the light is referred to as an *aberration*.

It is often useful to consider an idealization called the *thin lens*. An operational definition of a thin lens is obtained by constructing two parallel planes as close to the physical lens as possible. In geometrical optics terms, the lens is thin if a ray incident on the first plane undergoes negligible lateral displacement as it goes through the lens to the second plane. In physical-optics terms, a light source to the left of the lens produces an irradiance pattern on the first plane, and the lens is thin if essentially the same irradiance pattern appears on the second plane, no matter what source is used. Since the irradiance is proportional to the squared modulus of the complex field, a thin lens can be defined as one that alters the phase of the field but not its modulus.

Amplitude transmittance In Sec. 9.4 we considered diffraction from an open aperture in the plane $z = 0$. We used the Kirchhoff boundary condition to express the wave in this aperture as the incident field multiplied by a zero-one function $t_{ap}(\mathbf{r})$, where $\mathbf{r}$ is a 2D vector in the plane $z = 0$. We now introduce a similar notation to express the field emerging from a thin lens. We write

$$u_+(\mathbf{r}) = u_-(\mathbf{r})\, t_{lens}(\mathbf{r}) , \tag{9.153}$$

where $u_+(\mathbf{r})$ is the field just after the lens, $u_-(\mathbf{r})$ is the field incident on the lens, and $t_{lens}(\mathbf{r})$ is called the *amplitude transmittance* of the lens. Consistent with the assumption of a thin lens, $u_+(\mathbf{r})$, $u_-(\mathbf{r})$ and $t_{lens}(\mathbf{r})$ are all assumed to be measured in the plane $z = 0$. We assume that the lens is centered on the z-axis, which is called the *optical axis*. If the lens is rotationally symmetric, then the optical axis is also an axis of symmetry; otherwise it is just a convenient line running through the lens.

Equation (9.153) is already an idealization since it assumes that $t_{lens}(\mathbf{r})$ is independent of the form of the incident wave, even though it is well known that the effect of a lens on an incident wave depends on the angle of incidence. We shall return to this point in Sec. 9.6.3.

Since we have defined a thin lens as one that alters only the phase of the incident field, leaving its modulus unchanged, we can write

$$t_{lens}(\mathbf{r}) = \exp[i\Phi_{lens}(\mathbf{r})]\, t_{ap}(\mathbf{r})\,, \tag{9.154}$$

where $t_{ap}(\mathbf{r})$ is unity inside the aperture of the lens and zero outside, and $\Phi_{lens}(\mathbf{r})$ is the phase shift induced at point $\mathbf{r}$ by the lens. Within the lens aperture, $|u_+(\mathbf{r})| = |u_-(\mathbf{r})|$.

To elucidate the meaning of $t_{lens}(\mathbf{r})$, consider the incident field to be a unit-amplitude plane wave travelling parallel to the z axis. Then

$$u_-(\mathbf{r}) = \exp(ikz)|_{z=0} = 1\,, \tag{9.155}$$

where $k = 2\pi/\lambda$, with λ being the wavelength of the light. With (9.155), (9.153) becomes

$$u_+(\mathbf{r}) = t_{lens}(\mathbf{r})\,. \tag{9.156}$$

Thus $t_{lens}(\mathbf{r})$ is the field emerging from a lens when it is illuminated with a plane wave travelling parallel to the z axis.

Amplitude transmittance of an ideal lens An ideal thin lens will convert the incident plane wave into a spherical wave converging toward a point on the optical axis at a distance $z = f$, where f is the *focal length* of the lens. The wavefront is a portion of the *Gaussian sphere*, a sphere of radius f centered on the focal point. For an ideal lens, we thus have

$$t_{lens}(\mathbf{r}) = t_{ideal}(\mathbf{r}) = \exp(-ikR_f)\, t_{ap}(\mathbf{r})\,, \tag{9.157}$$

where $k = 2\pi/\lambda$, and R_f is the distance from point $\mathbf{r}$ in the lens plane to the focal point, given by

$$R_f = \sqrt{r^2 + f^2}\,, \tag{9.158}$$

where $r = |\mathbf{r}|$. Note the minus sign in the exponent in (9.157). Since we use the convention that a wave $u(\mathbf{r})$ has associated with it a time dependence $\exp(-2\pi i\nu_0 t)$, the spatial factor $\exp(-ikR_f)$ represents a *converging* spherical wave. Thus a positive lens (one with a positive focal length) produces a converging spherical wave when illuminated with a plane wave. Conversely, a negative lens produces a diverging spherical wave. A negative lens is also described by (9.157) simply by allowing f to be negative.

Fresnel approximation In the Fresnel approximation, the spherical surface defined by (9.158) is approximated by a paraboloid, so we can write

$$t_{ideal}(\mathbf{r}) \approx \exp\left(-i\pi\frac{r^2}{\lambda f}\right) t_{ap}(\mathbf{r})\,, \tag{9.159}$$

where we have dropped a constant phase factor $\exp(-ikf)$. This factor affects only the overall phase of the field emerging from the lens, which is almost always irrelevant since it can be observed only by interferometric methods. We retain factors that influence the spatial variation of the phase, but we do not need to consider a constant overall phase.

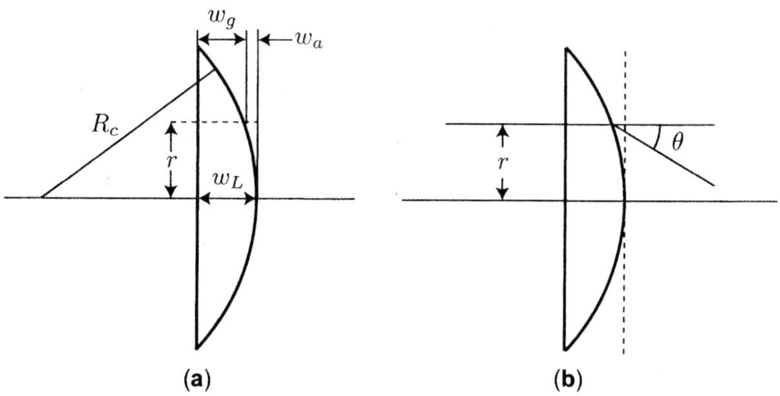

Fig. 9.8 (a) Illustration of a plano-convex lens used for calculation of $t_{lens}(\mathbf{r})$. (b) Field at point $\mathbf{r}$ on the output plane arises from points in a small neighborhood of point $\mathbf{r}$ on the input plane.

Another way to understand (9.159) is to consider specifically a plano-convex lens as illustrated in Fig. 9.8a (Goodman, 1968). We assume that a plane wave is normally incident on the planar side, and we want to compute the field $u_+(\mathbf{r})$ as a function of radius r on a plane tangent to the convex surface as shown. To compute $u_+(\mathbf{r})$ rigorously by diffraction theory, we would have to integrate over the entire planar input surface, but if the lens is thin, the main contribution to $u_+(\mathbf{r})$ at point $\mathbf{r}$ comes from a small neighborhood of the point $\mathbf{r}$ on the input surface (see Fig. 9.8b). Light from each of these points undergoes very nearly the same phase shift in propagating through the lens and air gap, so the sum over all points in the neighborhood undergoes that phase shift. Thus, determining $t_{lens}(\mathbf{r})$ amounts to determining the phase shift *along a horizontal straight line* displaced from the axis by $\mathbf{r}$. We do not need to consider explicitly refraction at the interface since the lens is thin.

The phase shift on this line can be computed from the speeds of light in the two media and the thicknesses of the media. The speed of light in air is essentially c, the speed of light in vacuum, while the speed in glass is c/n, where n is the refractive index. The thickness of air along the line is $w_a = R_c(1 - \cos\theta)$ and the thickness of glass is $w_g = w_L - R_c(1 - \cos\theta)$, where R_c is the radius of curvature of the convex surface and w_L is the overall thickness of the lens (see Fig. 9.8a). The total time $\tau(r)$ required to traverse the horizontal line at radius r is

$$\tau(r) = \frac{w_a}{c} + \frac{nw_g}{c}. \tag{9.160}$$

In time $\tau(r)$, the light traversing the lens undergoes a phase retardation of

$$\Phi_{lens}(r) = -2\pi\nu_0\tau(r) = -\frac{2\pi}{\lambda}(w_a + nw_g), \tag{9.161}$$

where λ is the wavelength of the light in free space. We assume the angle θ (see Fig. 9.8) is small, so that

$$1 - \cos\theta \approx \tfrac{1}{2}\theta^2 \approx \frac{r^2}{2R_c^2} \,. \tag{9.162}$$

With this approximation, (9.161) becomes

$$\Phi_{lens}(r) = -\frac{2\pi}{\lambda}\left[\frac{r^2}{2R_c} + n\left(w_L - \frac{r^2}{2R_c}\right)\right] = \frac{\pi}{\lambda}(n-1)\frac{r^2}{R_c} + const\,. \tag{9.163}$$

An elementary result from geometrical optics (Longhurst, 1973) relates the radius of curvature to the focal length of a plano-convex lens by

$$\frac{1}{f} = \frac{n-1}{R_c} \,, \tag{9.164}$$

so (9.163) agrees with (9.159). A similar analysis applies to double-convex and other kinds of lenses.

9.6.2 Imaging a monochromatic point source

We derived (9.159) by assuming that the incident wave was a plane wave travelling parallel to the z axis. Now consider what happens when the incident wave is produced by a monochromatic point source on the optical axis at $z = -p$. The incident wave measured in the lens plane ($z = 0$) is the diverging spherical wave given by

$$u_-(\mathbf{r}) = A\frac{\exp ikR_p}{R_p} \,, \tag{9.165}$$

where A specifies the strength of the source and

$$R_p = \sqrt{r^2 + p^2} = p + \frac{r^2}{2p} + \ldots \,. \tag{9.166}$$

As in the discussion of the Fresnel approximation in Sec. 9.4.5, we retain the first two terms in this expansion in the exponent of (9.165), but only the first term in the denominator. From (9.153) and (9.159), we find

$$u_+(\mathbf{r}) = A\frac{\exp(ikp)}{p}\exp\left(i\pi\frac{r^2}{\lambda p}\right)\exp\left(-i\pi\frac{r^2}{\lambda f}\right)t_{ap}(\mathbf{r})\,. \tag{9.167}$$

But this equation can be rewritten as

$$u_+(\mathbf{r}) = const \cdot \exp\left(-i\pi\frac{r^2}{\lambda q}\right)t_{ap}(\mathbf{r})\,, \tag{9.168}$$

where q is given by the *imaging condition*,

$$\frac{1}{p} + \frac{1}{q} = \frac{1}{f} \,. \tag{9.169}$$

If $p > f$, then q is a positive number and $u_+(\mathbf{r})$ represents a converging spherical wave, with the spherical wavefronts centered on a point on the optical axis at $z = q$.

Similarly, if $p < f$, then q is negative and $u_+(\mathbf{r})$ represents a diverging spherical wave radiating from a point on the optical axis at $z = -|q|$. In familiar geometrical-optics terms, the image of the point source is always located at $z = q$, and the image is real if q is positive and virtual if q is negative (see Fig. 9.9). The case of an incident plane wave can be recovered by letting $p \to \infty$, in which case $q = f$.

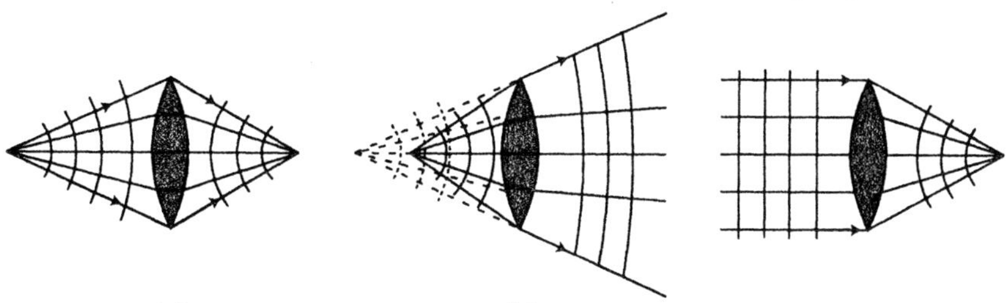

Fig. 9.9 (a) Illustration of a simple lens producing a real image. The arcs are surfaces of constant phase, also called wavefronts, and the straight lines are constructed to be normal to the wavefronts. (b) The same lens producing a virtual image. (c) The same lens imaging a point at infinity.

Field distribution The above discussion is sufficient for determining the center of curvature of the spherical wavefront, also known as the *Gaussian image* of the point source. To determine the field distribution in this image, we must invoke diffraction theory. From (9.98), we can write the field in the image plane ($z = q$) as

$$u_{im}(\mathbf{r}) = \frac{\exp(ikq)}{i\lambda q} \exp\left(i\pi\frac{r^2}{\lambda q}\right)\left[\mathcal{F}_2\left\{\exp\left(i\pi\frac{r'^2}{\lambda q}\right) u_+(\mathbf{r}')\right\}\right]_{\boldsymbol{\rho}=\mathbf{r}/\lambda q} . \tag{9.170}$$

Note that the quadratic phase factor inside the Fourier transform is exactly cancelled by its complex conjugate, which appears in (9.168). We thus find

$$u_{im}(\mathbf{r}) = A\frac{\exp[ik(p+q)]}{i\lambda pq} \exp\left(i\pi\frac{r^2}{\lambda q}\right) T_{ap}\left(\frac{\mathbf{r}}{\lambda q}\right) , \tag{9.171}$$

where $T_{ap}(\boldsymbol{\rho})$ is the 2D Fourier transform of $t_{ap}(\mathbf{r})$.

We are often able to approximate the factor $\exp(i\pi r^2/\lambda q)$ in (9.171) by unity. For example, if $t_{ap}(\mathbf{r})$ is a clear aperture of diameter D, its Fourier transform $T_{ap}(\boldsymbol{\rho})$ has its first zero at $\rho = 1.22/D$ (see (3.260) and Fig. 3.5). Thus the spatial function $T_{ap}(\mathbf{r}/\lambda q)$ has its first zero at $r = 1.22\lambda q/D$. At this radius, the argument of $\exp(i\pi r^2/\lambda q)$ is $i\pi(1.22)^2\lambda q/D^2$, which can also be written as $i\pi(1.22)^2 F^2\lambda/q$, where $F = q/D$ is called the *effective F-number* of the lens.[11] Typically, F is in the range $1-10$ and λ/q is of order 10^{-4} or less, so it is an excellent approximation to set $\exp(i\pi r^2/\lambda q) \approx 1$ for values of r such that $T_{ap}(\mathbf{r}/\lambda q)$ is appreciable.

With this approximation, the field in the plane $z = q$ is given by a scaled version of the Fourier transform of the aperture function:

$$u_{im}(\mathbf{r}) = \frac{A}{\lambda qp} T_{ap}\left(\frac{\mathbf{r}}{\lambda q}\right) , \tag{9.172}$$

where we have dropped constant phase factors (including a factor of i).

[11] The F-number of a lens is usually defined as f/D, which is identical to q/D if $p = \infty$.

Off-axis points The point source considered above was on the optical axis; we now generalize the calculation to a point source at a position specified by 2D vector $\mathbf{r}_0$ in the plane $z = -p$. The field incident on the lens is given by a translated version of (9.165), which in the Fresnel approximation becomes

$$u_-(\mathbf{r}) = \frac{A}{p} \exp\left(i\pi \frac{|\mathbf{r} - \mathbf{r}_0|^2}{\lambda p}\right), \tag{9.173}$$

where we have again dropped an irrelevant constant phase factor. If the lens is ideal, the field emerging from it is

$$u_+(\mathbf{r}) = u_-(\mathbf{r})\, t_{ideal}(\mathbf{r}) = \frac{A}{p} \exp\left(i\pi \frac{|\mathbf{r} - \mathbf{r}_0|^2}{\lambda p}\right) \exp\left(-i\pi \frac{r^2}{\lambda f}\right) t_{ap}(\mathbf{r}). \tag{9.174}$$

Expanding the exponent, we find

$$\begin{aligned} u_+(\mathbf{r}) &= \frac{A}{p} \exp\left[i\pi \frac{(r^2 + r_0^2 - 2\mathbf{r}\cdot\mathbf{r}_0)}{\lambda p}\right] \exp\left(-i\pi \frac{r^2}{\lambda f}\right) t_{ap}(\mathbf{r}) \\ &= \frac{A}{p} \exp\left[i\pi \frac{(r_0^2 - 2\mathbf{r}\cdot\mathbf{r}_0)}{\lambda p}\right] \exp\left(-i\pi \frac{r^2}{\lambda q}\right) t_{ap}(\mathbf{r}), \end{aligned} \tag{9.175}$$

where the last step made use of (9.169). The field in the image plane is obtained by use of (9.170). Again there is a cancellation of quadratic phase factors inside the Fourier transform, and again constant phase factors can be dropped. If the quadratic phase factors outside the transform can be approximated by unity, we are left with

$$\begin{aligned} u_{im}(\mathbf{r}) &= \frac{A}{\lambda q p} \left[\mathcal{F}_2 \left\{\exp\left(-2\pi i \frac{\mathbf{r}'\cdot\mathbf{r}_0}{\lambda p}\right) t_{ap}(\mathbf{r}')\right\}\right]_{\boldsymbol{\rho}=\mathbf{r}/\lambda q} \\ &= \frac{A}{\lambda q p} T_{ap}\left(\frac{\mathbf{r}}{\lambda q} + \frac{\mathbf{r}_0}{\lambda p}\right). \end{aligned} \tag{9.176}$$

This result is most easily compared to (9.172) by rewriting it as

$$u_{im}(\mathbf{r}) = \frac{A}{\lambda q p} T_{ap}\left[\frac{1}{\lambda q}(\mathbf{r} - m\mathbf{r}_0)\right], \tag{9.177}$$

where m is the *lateral magnification*, given just as in geometrical optics by

$$m = -\frac{q}{p}. \tag{9.178}$$

Note that $m = -1$ if $p = q = 2f$.

9.6.3 Transmittance of an aberrated lens

So far we have considered only ideal thin lenses. Such lenses can be described fully by the transmittance of (9.157) or, in the Fresnel approximation, (9.159). In essence, these equations say that an ideal lens produces an ideal spherical wavefront, so the temporal phase of the wave is the same at all points on a spherical cap, *i.e.*, a spherical surface truncated by the lens aperture. An aberrated lens, on the other hand, produces a wavefront that is not exactly spherical. The phase difference between the desired spherical wave and the actual wave emerging from the lens is called the wavefront error and is commonly denoted by $kW(\mathbf{r})$, where $k = 2\pi/\lambda$ as usual. Thus $W(\mathbf{r})$ has units of length and can be interpreted as the spatial distance, along a line parallel to the z axis, between the desired spherical wave and the actual wavefront at vector distance $\mathbf{r}$ from the z axis (see Fig. 9.10). For the moment we assume that $W(\mathbf{r})$ is independent of the wave incident on the lens, but we shall soon remove this restriction.

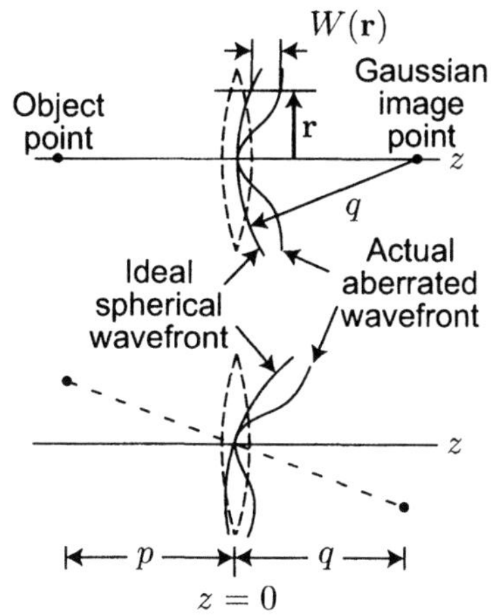

Fig. 9.10 Illustration of an aberrated wavefront.

If we express the phase transformation of the lens by (9.154), then

$$\Phi_{lens}(\mathbf{r}) = \Phi_{ideal}(\mathbf{r}) + kW(\mathbf{r}) \tag{9.179}$$

and

$$t_{lens}(\mathbf{r}) = \exp[i\Phi_{lens}(\mathbf{r})]\, t_{ap}(\mathbf{r}) = \exp[-ikR_f + ikW(\mathbf{r})]\, t_{ap}(\mathbf{r})\,. \tag{9.180}$$

Another way to express this same result is in terms of the *pupil function.* For a thin lens with the aperture stop at the lens, the pupil function is defined by

$$t_{pupil}(\mathbf{r}) = \exp[ikW(\mathbf{r})]\, t_{ap}(\mathbf{r})\,. \tag{9.181}$$

Hence

$$t_{lens}(\mathbf{r}) = \exp(-ikR_f)\, t_{pupil}(\mathbf{r})\,. \tag{9.182}$$

In the Fresnel approximation, we have (apart from a constant phase factor),

$$t_{lens}(\mathbf{r}) \approx \exp\left(-i\pi\frac{r^2}{\lambda f}\right) t_{pupil}(\mathbf{r})\,. \tag{9.183}$$

The pupil function thus describes the aberrations and the overall aperture of the lens, but not the phase transformation characteristic of an ideal lens.

The wavefront error can also depend on the nature of the wave incident on the lens. Suppose that wave is a spherical wave emanating from point $\mathbf{r}_0$ in the plane $z = -p$. In the lens plane ($z = 0$), this wave is expressed by

$$u_-(\mathbf{r}) = A\frac{\exp ikR_p}{R_p}\,, \tag{9.184}$$

where

$$R_p = \sqrt{|\mathbf{r} - \mathbf{r}_0|^2 + p^2} = p + \frac{|\mathbf{r} - \mathbf{r}_0|^2}{2p} + \ldots . \tag{9.185}$$

The wave emerging from the lens is

$$u_+(\mathbf{r}) = \frac{A}{R_p} \exp\left[ikR_p + i\Phi_{lens}(\mathbf{r}; \mathbf{r}_0)\right] t_{ap}(\mathbf{r}) , \tag{9.186}$$

where the dependence of Φ_{lens} on the source location is now shown explicitly. We can again define a wavefront error, now denoted $kW(\mathbf{r}; \mathbf{r}_0)$, by

$$\Phi_{lens}(\mathbf{r}; \mathbf{r}_0) = \Phi_{ideal}(\mathbf{r}) + kW(\mathbf{r}; \mathbf{r}_0) . \tag{9.187}$$

In the Fresnel approximation, this equation becomes

$$\Phi_{lens}(\mathbf{r}; \mathbf{r}_0) = -\pi \frac{r^2}{\lambda f} + kW(\mathbf{r}; \mathbf{r}_0) . \tag{9.188}$$

Without making assumptions about the nature of the lens, this is as far as we can go.

9.6.4 Rotationally symmetric lenses

Almost all lenses used in practice are rotationally symmetric. We learned in Sec. 7.2.9 how to analyze rotationally symmetric linear systems; to apply that theory here, we need to define a suitable system. A convenient approach is to consider the mapping between the 2D field distribution in plane $z = -p$ and the 2D field distribution emerging from the lens. If we denote this mapping by the operator $\mathcal{H}$, we have two choices in defining its symmetry properties. As in Sec. 7.2.9, we could construct the operator $\mathcal{H}^\dagger\mathcal{H}$ and require that $\mathcal{H}^\dagger\mathcal{H}\mathcal{R}_\phi = \mathcal{R}_\phi\mathcal{H}^\dagger\mathcal{H}$, where $\mathcal{R}_\phi$ is the operator for rotation by angle ϕ about the z axis. Alternatively, since the range and domain of $\mathcal{H}$ are both 2D function spaces, we can require that $\mathcal{H}\mathcal{R}_\phi = \mathcal{R}_\phi\mathcal{H}$. The latter condition is the stronger one, but it is satisfied for practical lenses if they are properly centered on the axis.

With this assumption, it follows from the discussion in Sec. 7.2.9 that the kernel of $\mathcal{H}$ can be a function of the lengths of the vectors $\mathbf{r}$ and $\mathbf{r}_0$ and of the angle $\theta - \theta_0$ between them, but it cannot depend on the absolute orientation of $\mathbf{r}$ and $\mathbf{r}_0$. Since the kernel is fully determined by $W(\mathbf{r}; \mathbf{r}_0)$, the same conclusion applies to that function too. Moreover, practical lenses also have mirror symmetry,[12] which implies that $W(\mathbf{r}; \mathbf{r}_0)$ must be an even function of $\theta - \theta_0$ (see Sec. 7.2.9).

Aberration expansion The conclusion from the discussion above is that $W(\mathbf{r}; \mathbf{r}_0)$ is a function of r, r_0 and $\theta - \theta_0$, and it is an even function of the angular difference if the lens has mirror symmetry. Of course, it must also be periodic with period 2π in the angular difference if it is to represent a physically realizable, single-valued function on the plane.

[12] It is possible to concoct systems with rotational symmetry but no mirror symmetry. An image rotator such as a Dove prism, for example, is rotationally symmetric but lacks mirror symmetry.

One way to express a function with these properties is through a power series of the form (Born and Wolf, 1999),

$$W(\mathbf{r};\mathbf{r}_0) = \sum_{\alpha\beta\gamma} C_{\alpha\beta\gamma} r_0^{2\alpha} r^{2\beta} (\mathbf{r}\cdot\mathbf{r}_0)^{\gamma} \,. \tag{9.189}$$

The *order* of a term in this expansion is $\alpha+\beta+\gamma$. A more conventional way to express the same expansion is

$$W(\mathbf{r};\mathbf{r}_0) = \sum_{lmn} w_{lmn} r_0^{l} r^{m} \left[\cos(\theta-\theta_0)\right]^{n} \,, \tag{9.190}$$

where $l = 2\alpha+\gamma$, $m = 2\beta+\gamma$ and $n = \gamma$. The coefficients in either expansion depend on p and the nature of the lens, but they are independent of $\mathbf{r}$ and $\mathbf{r}_0$. In addition, $W(\mathbf{r};\mathbf{r}_0)$ may also depend on the wavelength of the light, but we consider only monochromatic sources here.

The zero-order term ($\alpha = \beta = \gamma = 0$) is uninteresting since it just signifies a constant phase. Of the three possible first-order terms, we need to consider only one. The term $\alpha = 1$, $\beta = \gamma = 0$, can be dismissed since it corresponds to a phase that depends on source location but is constant over the lens aperture. The term $\alpha = \beta = 0$, $\gamma = 1$ cannot occur if the lens surface is a second-order surface, such as a sphere or paraboloid, centered on the optic axis.

The remaining first-order term, $\alpha = \gamma = 0$, $\beta = 1$, corresponding to the coefficient w_{020}, will now be discussed. If only this term is present, (9.188) becomes

$$\Phi_{lens}(\mathbf{r};\mathbf{r}_0) = -\pi\frac{r^2}{\lambda f} + k w_{020} r^2 = -\pi\frac{r^2}{\lambda f'} \,, \tag{9.191}$$

where (since $k = 2\pi/\lambda$)

$$\frac{1}{f'} = \frac{1}{f} - 2w_{020} \,. \tag{9.192}$$

This aberration thus causes the lens to behave as an ideal thin lens with a different focal length. The lens will form a sharp image but not at $z = q$, and the aberration is called *defocus*.

The possible second-order terms are listed below, with their common names and mathematical forms:

Field curvature or Petzval[13] curvature: $w_{220} r_0^2 r^2$;

Distortion: $w_{311} r_0^3 r \cos(\theta-\theta_0) = w_{311} r_0^2 (\mathbf{r}\cdot\mathbf{r}_0)$;

Primary spherical aberration: $w_{040} r^4$;

Coma: $w_{131} r_0 r^3 \cos(\theta-\theta_0) = w_{131} r^2 (\mathbf{r}\cdot\mathbf{r}_0)$;

[13]Born and Wolf (1999) credit the Hungarian mathematician J. Petzval with the earliest investigation of deviations from the Gaussian image formulas. They comment that his manuscript was stolen by thieves, and that much of what we know about his work comes from semi-popular reports.

Astigmatism: $w_{222}r_0^2r^2[\cos(\theta - \theta_0)]^2 = w_{222}(\mathbf{r}\cdot\mathbf{r}_0)^2$.

These second-order terms are known variously as *primary aberrations*, *Seidel aberrations* or (perversely) *third-order aberrations*. The latter designation relates not to the order of a term in the wavefront expansion (9.189) but to the nature of the *ray aberrations*, a topic covered briefly in Sec. 9.6.6. For clear diagrams illustrating these aberrations, see Gaskill (1978) or Hecht (1987).

Note that spherical aberration, like defocus, is independent of $\mathbf{r}_0$, which is the position of the object point in the field of view; for this reason, defocus and spherical aberration are called *field-independent aberrations*. The other Seidel aberrations are field dependent, varying as some power of $\mathbf{r}_0$. For source points on or near the optical axis, the field-dependent aberrations can be ignored.

9.6.5 Field curvature and distortion

If field curvature is the only aberration present, Φ_{lens} takes the form,

$$\Phi_{lens}(\mathbf{r};\mathbf{r}_0) = -\frac{\pi r^2}{\lambda f} + kw_{220}r_0^2r^2 \equiv -\frac{\pi r^2}{\lambda f_{eff}(r_0)}, \tag{9.193}$$

where

$$\frac{1}{f_{eff}(r_0)} = \frac{1}{f} - 2w_{220}r_0^2\,. \tag{9.194}$$

The effective focal length $f_{eff}(r_0)$ thus varies with position of the source point. A point described by the 2D vector $\mathbf{r}_0$ in the plane $z = -p$ forms an image at $z = q(r_0)$, where $q(r_0)$ is given by a slightly modified (9.169):

$$\frac{1}{p} + \frac{1}{q(r_0)} = \frac{1}{f_{eff}(r_0)}\,. \tag{9.195}$$

Since the z coordinate of the image point depends on r_0, the image surface is not a plane, hence the name field curvature.

To understand distortion, we make use of (9.176), replacing T_{ap} by T_{pupil} to allow for the aberration. If distortion is the only aberration present, we have

$$t_{pupil}(\mathbf{r};\mathbf{r}_0) = t_{ap}(\mathbf{r})\exp[ikw_{311}r_0^2(\mathbf{r}\cdot\mathbf{r}_0)]\,. \tag{9.196}$$

Note the second argument in $T_{pupil}(\boldsymbol{\rho};\mathbf{r}_0)$, indicating that the Fourier transform of the pupil function depends on the spatial location of the source point in general.

We see from (9.196) that distortion is described by a linear phase factor in the pupil function (linear in $\mathbf{r}$), so it produces a shift in the Fourier domain. The Fourier transform of the pupil function is obtained via (3.237) as

$$T_{pupil}(\boldsymbol{\rho};\mathbf{r}_0) = T_{ap}\left(\boldsymbol{\rho} - \frac{w_{311}r_0^2\mathbf{r}_0}{\lambda}\right), \tag{9.197}$$

and the field in the image, from (9.176), is given by

$$u_{im}(\mathbf{r}) = const\cdot T_{ap}\left(\frac{\mathbf{r}}{\lambda q} + \frac{\mathbf{r}_0}{\lambda p} - \frac{w_{311}r_0^2\mathbf{r}_0}{\lambda}\right). \tag{9.198}$$

By analogy to (9.177), we can rewrite this result as

$$u_{im}(\mathbf{r}) = const \cdot T_{ap}\left[\frac{1}{\lambda q}(\mathbf{r} - m_{eff}\mathbf{r}_0)\right], \tag{9.199}$$

where the effective lateral magnification m_{eff} is given by

$$m_{eff} = -q\left(\frac{1}{p} - w_{311}r_0^2\right). \tag{9.200}$$

Since the lateral magnification depends on source location, the image is distorted.

9.6.6 Probing the pupil

To understand the other aberrations, it is useful to imagine placing masks over the lens so we can determine where light from different regions of the pupil is directed. The trick is to choose the size of the mask small enough so that the phase of the wave can be approximated by a linear function of position in the pupil, yet large enough that diffraction from the mask is negligible. As we shall see, the mathematics is essentially the same as that developed in Sec. 5.1 for local spectral analysis.

The starting point is (9.170), which gives the field in the image plane for an arbitrary field in the lens plane. The field $u_+(\mathbf{r}')$ in that formula, now interpreted as the field emerging from the mask, is given in the Fresnel approximation by [*cf.* (9.174)]

$$u_+(\mathbf{r}') = const \cdot \exp\left(i\pi\frac{|\mathbf{r}' - \mathbf{r}_0|^2}{\lambda p}\right)\exp\left(-i\pi\frac{r'^2}{\lambda f}\right)\exp[ikW(\mathbf{r}';\mathbf{r}_0)]\,t_{mask}(\mathbf{r}'), \tag{9.201}$$

where t_{mask} is the transmittance of the mask. The Fourier integral in (9.170) now takes the form

$$\left[\mathcal{F}_2\left\{\exp\left(i\pi\frac{r'^2}{\lambda q}\right)u_+(\mathbf{r}')\right\}\right]_{\boldsymbol{\rho}=\mathbf{r}/\lambda q} = \int_{mask} d^2r' \exp[i\Phi(\mathbf{r}';\mathbf{r}_0)]\exp\left(-2\pi\frac{i\mathbf{r}\cdot\mathbf{r}'}{\lambda q}\right), \tag{9.202}$$

where

$$\Phi(\mathbf{r}';\mathbf{r}_0) = \frac{\pi|\mathbf{r}' - \mathbf{r}_0|^2}{\lambda p} - \frac{\pi r'^2}{\lambda f} + kW(\mathbf{r}';\mathbf{r}_0) + \frac{\pi r'^2}{\lambda q}. \tag{9.203}$$

The four terms in (9.203) come from, respectively, the wave incident on the lens, the phase transformation of an ideal lens, the phase distortion due to aberrations and the quadratic phase factor in the Fresnel formula. To recapitulate the meaning of the various 2D position vectors, $\mathbf{r}_0$ is the source position in plane $z = -p$, $\mathbf{r}$ is the observation point in the plane $z = q$, and $\mathbf{r}'$ is a dummy variable of integration.

If we now apply the imaging condition (9.169), all of the terms quadratic in $\mathbf{r}'$ cancel, and we find

$$\Phi(\mathbf{r}';\mathbf{r}_0) = -2\pi\frac{\mathbf{r}'\cdot\mathbf{r}_0}{\lambda p} + kW(\mathbf{r}';\mathbf{r}_0) + \frac{\pi r_0^2}{\lambda q}. \tag{9.204}$$

Except for the last term, which is an irrelevant constant, $\Phi(\mathbf{r}';\mathbf{r}_0)$ now describes a tilted plane wave (with tilt dependent on the object location) plus the wavefront error.

If the mask is a small circular opening of radius b centered at $\mathbf{r}_m$ and $\Phi(\mathbf{r}';\mathbf{r}_0)$ varies sufficiently slowly over this opening, we can express it as a truncated Taylor series of the form

$$\Phi(\mathbf{r}';\mathbf{r}_0) \approx \Phi(\mathbf{r}_m;\mathbf{r}_0) + (\mathbf{r}' - \mathbf{r}_m) \cdot \nabla\Phi(\mathbf{r}_m;\mathbf{r}_0)\,, \tag{9.205}$$

where $\nabla\Phi(\mathbf{r};\mathbf{r}_0)$ denotes the gradient of $\Phi(\mathbf{r};\mathbf{r}_0)$ with respect to the first argument. By analogy to (5.35), we can define

$$\nabla\Phi(\mathbf{r}_m;\mathbf{r}_0) \equiv 2\pi\boldsymbol{\rho}_{loc}(\mathbf{r}_m)\,, \tag{9.206}$$

where $\boldsymbol{\rho}_{loc}(\mathbf{r}_m)$ is a local frequency vector.

The requisite Fourier transform can now be performed by means of (3.237) and (3.259), with the result

$$u_{im}(\mathbf{r}) \propto \left[\mathcal{F}_2\left\{\exp\left(i\pi\frac{r'^2}{\lambda q}\right)u_+(\mathbf{r}')\right\}\right]_{\boldsymbol{\rho}=\mathbf{r}/\lambda q} = \pi b^2 \operatorname{besinc}\left\{2b\left[\frac{\mathbf{r}}{\lambda q} - \frac{\nabla\Phi(\mathbf{r}_m;\mathbf{r}_0)}{2\pi}\right]\right\}. \tag{9.207}$$

In the image plane ($z = q$), the light diffracted from the mask is a besinc function centered at the point $\mathbf{r} = \mathbf{r}_c$ where the argument of the besinc vanishes. Specifically,

$$\mathbf{r}_c = \frac{\lambda q}{2\pi}\nabla\Phi(\mathbf{r}_m;\mathbf{r}_0) = \lambda q\boldsymbol{\rho}_{loc}(\mathbf{r}_m)\,. \tag{9.208}$$

Within the small-angle approximation, this expression is consistent with (9.113). The x component of $\lambda\boldsymbol{\rho}_{loc}(\mathbf{r}_m)$ is the deviation angle with respect to the x axis, and when multiplied by q it gives the corresponding lateral deflection in the x direction (and similarly for y). This deflection arises for two reasons: the source is displaced from the axis and the lens is aberrated.

The width of the besinc function in (9.207), measured from peak to first zero, is $1.22\lambda q/2b$, which is the diffraction-limited spot size for a circular aperture of radius b. If we can choose b relatively large without invalidating (9.205), we can think of the besinc as a small spot, almost a point, and investigate how it moves around as we explore the pupil with the mask.

To see how (9.208) works, consider first an unaberrated lens, where $W(\mathbf{r}';\mathbf{r}_0) = 0$. In that case,

$$\mathbf{r}_c = \frac{\lambda q}{2\pi}\frac{\nabla(-2\pi\mathbf{r}\cdot\mathbf{r}_0)}{\lambda p} = -\frac{q}{p}\mathbf{r}_0\,. \tag{9.209}$$

Thus $\mathbf{r}_c$ is independent of the mask location $\mathbf{r}_m$, so all segments of the lens direct their light to the same spot. This spot is centered at $m\mathbf{r}_0$, where again the lateral magnification m is given by $-q/p$.

If aberrations are present, on the other hand, the field distribution in the plane $z = q$ is centered at

$$\mathbf{r}_c = q\nabla W(\mathbf{r}_m;\mathbf{r}_0) - \frac{q}{p}\mathbf{r}_0\,. \tag{9.210}$$

This equation provides another interpretation of $W(\mathbf{r};\mathbf{r}_0)$. In the paraxial approximation, $\nabla W(\mathbf{r}_m;\mathbf{r}_0)$ is the angle by which the centroid of the light passing through the mask is deflected (see Fig. 9.11). When this angle is multiplied by the distance q, the displacement of $\mathbf{r}_c$ specified by the first term in (9.210) is seen.

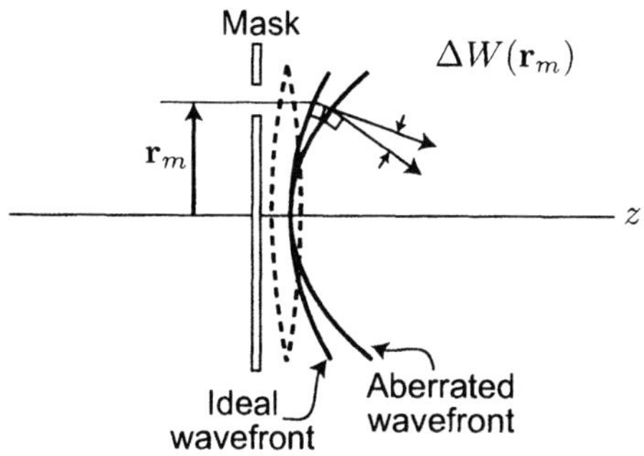

Fig. 9.11 Illustration of an aberrated lens with a mask over the pupil.

9.6.7 Interpretation of the other Seidel aberrations

If spherical aberration is the only aberration present, Φ_{lens} takes the form

$$\Phi_{lens}(\mathbf{r};\mathbf{r}_0) = -\pi\frac{r^2}{\lambda f} + kw_{040}r^4\,. \tag{9.211}$$

Since Φ_{lens} is now independent of $\mathbf{r}_0$, we may as well consider an on-axis point source and take $\mathbf{r}_0 = 0$.

From (9.208), light from a mask at $\mathbf{r}_m$ produces a diffraction pattern centered at

$$\mathbf{r}_c = z\left(\frac{1}{p} - \frac{1}{f} + \frac{1}{z} + 4w_{040}r_m^2\right)\mathbf{r}_m\,. \tag{9.212}$$

Since we are considering an on-axis point source, we would hope that $\mathbf{r}_c = 0$, but this does not occur in the paraxial focal plane $z = q$. Instead, $\mathbf{r}_c = 0$ when

$$\frac{1}{z} = \frac{1}{f} - \frac{1}{p} - 4w_{040}r_m^2\,. \tag{9.213}$$

Since this distance depends on r_m, but not on the orientation of the vector $\mathbf{r}_m$, light passing through different annular zones in the lens focuses at different distances from the lens, as illustrated in Fig. 9.12. If we observe the light in the plane $z = q$ without the mask, the zones near the center of the lens form a sharp focus, but the outer zones are defocused in this plane.

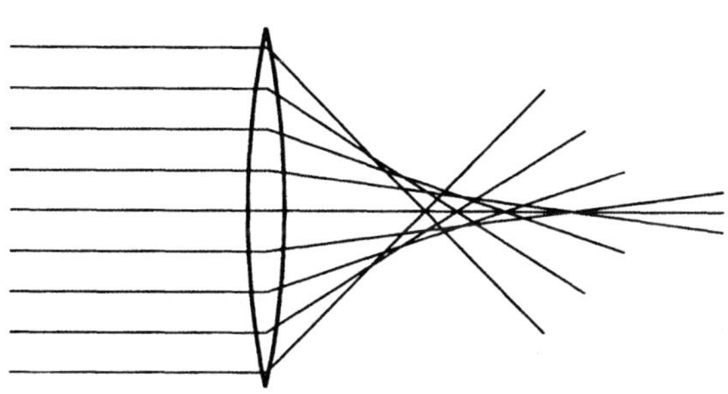

Fig. 9.12 Illustration of spherical aberration.

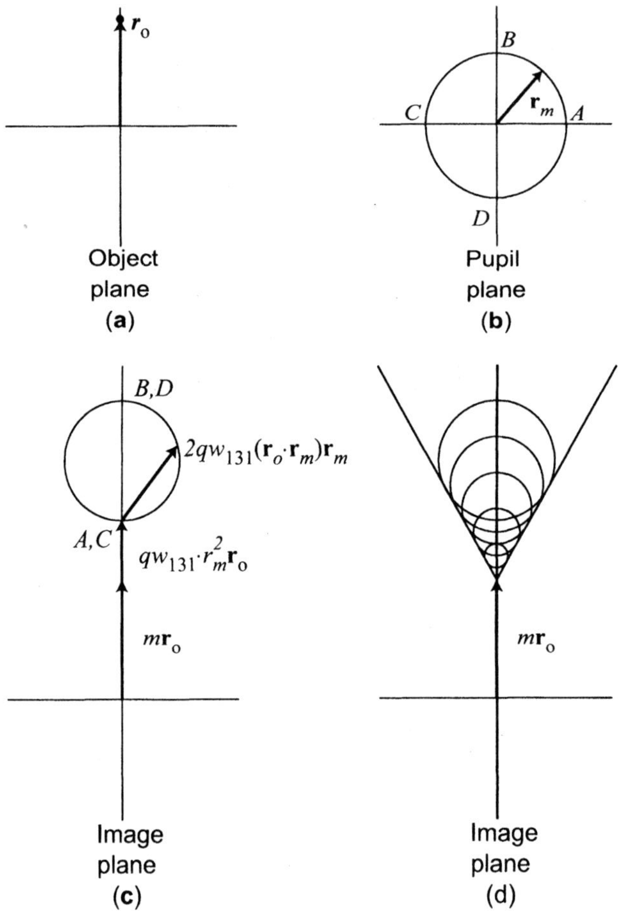

Fig. 9.13 Illustration of coma. (a) A single off-axis point in the object plane. (b) Locus of a probing mask that traverses a circle around the optical axis in the pupil plane. (c) Locus of image points as the mask traverses the pupil-plane path shown in (b). (d) Loci of points in the image plane for several circles of different diameter in the pupil plane.

Next consider coma. If coma is the only aberration present, then

$$\Phi_{lens}(\mathbf{r};\mathbf{r}_0) = -\pi\frac{r^2}{\lambda f} + kw_{131}r^2(\mathbf{r}\cdot\mathbf{r}_0)\,. \tag{9.214}$$

By writing the vectors out in Cartesian coordinates, we can show that

$$\nabla\left[r^2(\mathbf{r}\cdot\mathbf{r}_0)\right] = r^2\mathbf{r}_0 + 2(\mathbf{r}_0\cdot\mathbf{r})\mathbf{r}\,. \tag{9.215}$$

Now the light from the mask at $\mathbf{r}_m$ yields a diffraction pattern centered at $\mathbf{r}_c$ in the plane $z = q$, where

$$\mathbf{r}_c = -\frac{q}{p}\mathbf{r}_0 + qw_{131}\left[r_m^2\mathbf{r}_0 + 2(\mathbf{r}_m\cdot\mathbf{r}_0)\mathbf{r}_m\right]\,. \tag{9.216}$$

The first term is the Gaussian image, $m\mathbf{r}_0$, while the terms proportional to w_{131} specify the lateral displacement of the spot away from the Gaussian image. There are two components to this displacement, one in the direction of $\mathbf{r}_0$ and one in the direction of $\mathbf{r}_m$; both of these components increase linearly with r_0 and quadratically with r_m. If we explore the pupil by rotating the mask in a circle around the optical axis (constant r_m), then the image spot also describes a circle (see Fig. 9.13). Increasing r_m increases the radius of this circle and also shifts its center. If we remove the mask, the resulting superposition of circles of various sizes and center positions resembles a comet, whence the term *coma*.

A similar analysis can be applied to astigmatism. To describe the results of this analysis, we define two orthogonal planes. The *tangential* or *meridional* plane contains the off-axis point and the optical axis (and the Gaussian image point). The *sagittal* plane is perpendicular to the tangential plane and contains the *chief ray* (defined as the ray from the object point that passes through the center of the pupil). Exploring the pupil with a mask shows that the lens has different focal lengths for the two planes. Details can be worked out by the reader as an exercise; since different focal planes are involved, it is necessary to allow both astigmatism and defocus in the pupil function.

Thick lenses and lens systems: Pupil planes and principal planes Our treatment of aberrations to this point has been confined to thin lenses with the stop at the lens. We have made no distinction between the lens plane, entrance and exit pupil planes and front and rear principal planes; all were assumed to be the single plane $z = 0$. Real optical systems, however, consist of lenses with finite thicknesses and usually multiple lens elements, and we can no longer think of the system as confined to a plane. Here we briefly consider how the thin-lens theory must be modified to accommodate more realistic optical systems.

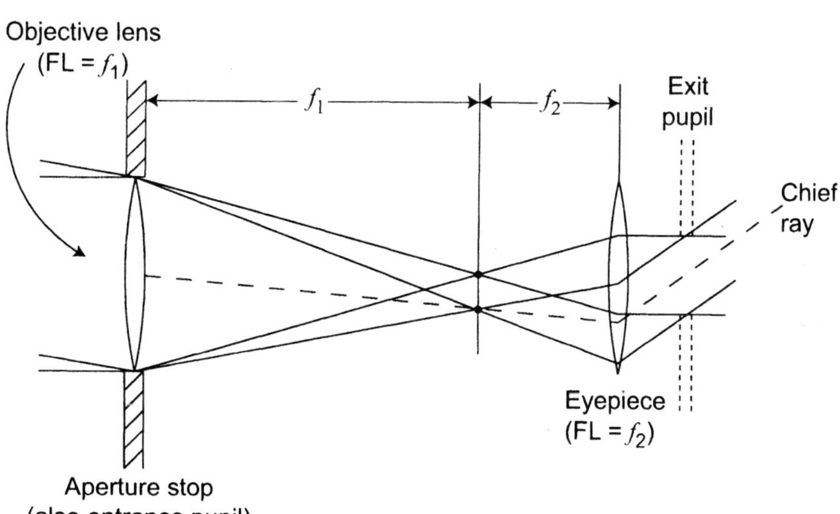

Fig. 9.14 Illustration of the pupil planes of an optical system.

It is usually assumed that there is a single limiting aperture, called the *stop*, in an optical system. The plane where this stop is physically located is called the *stop plane* or *aperture plane*. As shown in Fig. 9.14, the image of the stop on the object side of the system is called the *entrance pupil*, and the image of the stop on the image side is called the *exit pupil.* Thus the entrance pupil, the stop and the exit pupil lie in conjugate planes (*i.e.*, each plane is an image of the other two through the intervening optics).

To compute the field in the image plane, we can write the field in the exit pupil as a spherical wave converging on the Gaussian image point times two factors: a phase factor of the form $\exp[ikW(\mathbf{r})]$ representing the aberrations and a binary-valued factor analogous to $t_{ap}(\mathbf{r})$ representing the stop as imaged to the exit pupil. The pupil function is the product of these two factors as in (9.181), but both $W(\mathbf{r})$ and $t_{ap}(\mathbf{r})$ now refer to the exit pupil rather than the physical aperture. All of the effects of the optical elements are included in the field in the exit pupil (spherical wave times pupil function), and only free-space propagation is needed to get to the image plane. Since the spherical wave converges on the Gaussian image plane, the field in this plane is given by the Fourier transform of the pupil function, and the formalism developed above for thin lenses is still applicable.

The other significant planes in practical optical systems are the front and rear principal planes (see Fig. 9.15). In paraxial optics the principal planes are defined as conjugate planes with unit transverse magnification. Thus a ray that passes the front principal plane at height h passes the rear principal plane at height h also, by definition, so long as h is small (compared to the focal length). For first-order analysis, one can treat a complicated optical system as a thin lens if distances are measured from the principal planes. For example, the basic imaging equation, $p^{-1} + q^{-1} = f^{-1}$, still works and the magnification is still given by $m = -q/p$ so long as p and q are defined from object to front principal plane and from rear principal plane to image, respectively.

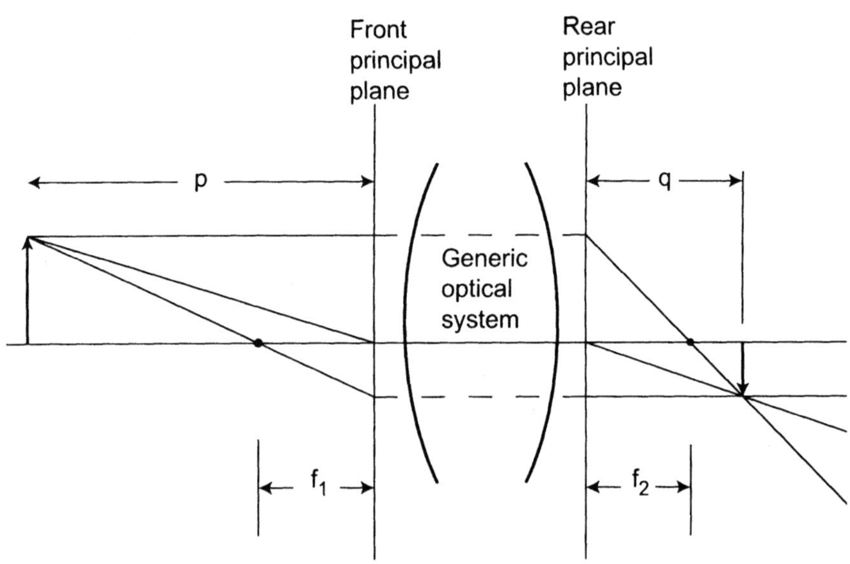

Fig. 9.15 Illustration of the principal planes of an optical system.

Abbe sine condition There is an important condition, known as the Abbe sine condition, that guarantees images free of spherical aberration and coma. In brief, this happy condition occurs when any ray leaving an axial object point at angle θ to the optical axis arrives at the image plane at an angle θ' such that $\sin\theta/\sin\theta'$ is

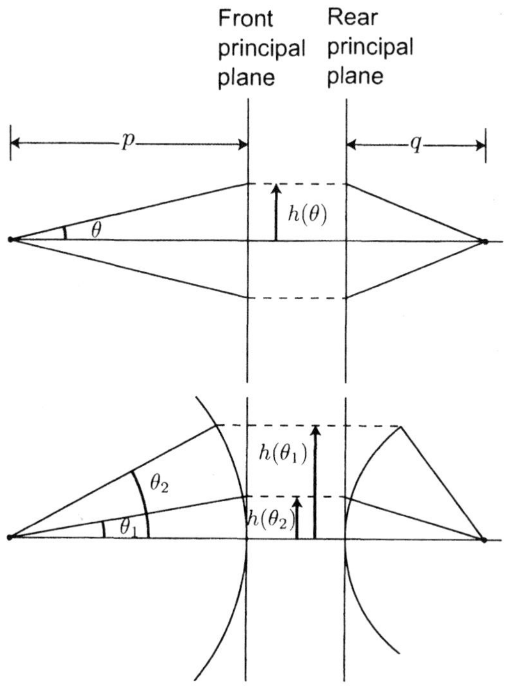

Fig. 9.16 Illustration of the Abbe sine condition.

constant. Remarkably, this condition has to apply only to the rays from a point on the optical axis, even though coma is an off-axis aberration.

There are several ways to understand the Abbe sine condition. A simple argument (Lipson *et al.*, 1995) regards the angles θ and θ' as specifying the directions of plane waves rather than rays. From Sec. 9.2.1 we know that $\sin\theta/\lambda$ is a spatial frequency, so the sine condition says that all spatial frequencies are scaled by the same magnification in going from object to image.

Another viewpoint, illustrated with lucid graphics, is provided by Mansuripur (2002). This treatment starts with the recognition that the front and rear principal planes are conjugate only paraxially, where $\sin\theta \approx \tan\theta \approx \theta$. When this approximation does not hold, it is useful to consider spherical surfaces rather than principal planes. Specifically, for an object point on the optical axis and a distance p from the front principal plane, Mansuripur constructs a sphere of radius p centered on the object point and hence tangent to the principal plane. In a geometrical optics view (see Fig. 9.16), he then defines the height h of a ray by the point where the ray strikes the sphere rather than by where it strikes the principal plane; thus $h(\theta) = \sin\theta/p$. If the image plane is a distance q from the rear focal plane, a similar sphere of radius q is constructed in image space, centered on the Gaussian image point and hence tangent to the rear principal plane. The sine condition is then stated as $\sin\theta/p = \sin\theta'/q$, rather than $\tan\theta/p = \tan\theta'/q$ as would apply if h were the height on the principal planes.

Mansuripur goes on to explain the sine condition in wave-optical terms. He considers an object point a small distance r_0 from the optical axis and shows that the wave emerging from the rear principal plane is a perfect spherical wave converging to the Gaussian image point if the sine condition is satisfied, but not otherwise.

Though the argument is valid only to first order in r_0, this is sufficient to show that coma and spherical aberration vanish.

9.7 IMAGING OF EXTENDED PLANAR OBJECTS

Section 9.6 discussed the images of point objects in considerable detail. In the language of Chap. 7, these images are point response functions[14] (PRFs). To complete the story, we must now show how the PRFs are superimposed to give images of extended objects. In this section we consider specifically planar or 2D objects. Volume objects and 3D imaging are treated in Sec. 9.8.

The reader who has studied Chap. 7 might think that the transition from a point image to the image of an extended object is just a matter of computing an integral, and in one sense that is correct. If we know the object field and the PRF for an arbitrary point in the object, we can simply add up PRFs to get the image; details of this procedure are given in Secs. 9.7.1 – 9.7.3 for various imaging systems.

In many cases, however, we do not — and cannot — know the object field since it is a rapidly oscillating random process. In those cases, we must be content with computing the average irradiance in the image, which is the average of the squared modulus of the field. The averaging process requires that we make some statement about the statistical properties of the source, and coherence theory, the topic of Sec. 9.7.4 provides the language for doing so. The effects of coherence on imaging with quasimonochromatic light will be discussed in detail in Sec. 9.7.5.

Formation of the squared modulus of a complex field is a nonlinear process, but nevertheless, as we shall see in Sec. 9.7.6, there is one important limit (spatial incoherence) in which it is possible to salvage the machinery of linear superposition and define a new PRF for this situation. We shall explore this important case in detail in Secs. 9.7.6 and 9.7.7. In Sec. 9.7.8, however, we shall finally encounter a case — partial spatial coherence — where linear systems theory fails us.

9.7.1 Monochromatic objects and a simple lens

We have discussed in some detail the imaging of a single monochromatic point source with a thin lens. To apply these results to a continuous object, we decompose the object amplitude distribution into delta functions by means of the sifting relation,

$$u_{obj}(\mathbf{r}) = \int_{\infty} d^2r_0 \; u_{obj}(\mathbf{r}_0)\, \delta(\mathbf{r} - \mathbf{r}_0) \,. \tag{9.217}$$

The field in the image plane produced by each delta function was calculated in Sec. 9.6.2, except that we have to get the amplitude A correct. We know from (9.94) that a delta function in plane $z = -p$ produces a spherical wave with amplitude $1/(i\lambda p)$ in the plane $z = 0$. Comparison with (9.173) shows that we should set $A = 1/(i\lambda)$ (though we can again delete the i since constant phase factors are irrelevant). With this substitution, the field in the image plane is given by (9.177), with T_{ap} replaced

[14] As discussed in Chap. 7, we use the term *point response function* or PRF to denote the general shift-variant image of a point, reserving the more common term *point spread function* or PSF specifically for the shift-invariant case.

by T_{pupil} for generality:

$$u_{im}^{(\delta)}(\mathbf{r}) = \frac{1}{\lambda^2 qp} T_{pupil}\left[\frac{1}{\lambda q}(\mathbf{r} - m\mathbf{r}_0); \mathbf{r}_0\right], \tag{9.218}$$

where the superscript (δ) indicates that this is the field due to $\delta(\mathbf{r} - \mathbf{r}_0)$ only.

By linear superposition, the total image-plane field is

$$u_{im}(\mathbf{r}) = \frac{1}{\lambda^2 qp} \int_\infty d^2 r_0 \; u_{obj}(\mathbf{r}_0)\, T_{pupil}\left[\frac{1}{\lambda q}(\mathbf{r} - m\mathbf{r}_0); \mathbf{r}_0\right]. \tag{9.219}$$

This integral describes a magnifier with possibly shift-variant blur; it is not a convolution because of the magnification factor m and also because the pupil function itself depends on $\mathbf{r}_0$ if the lens has off-axis aberrations. Systems of this form were discussed in general terms in Sec. 7.2.7, and the concepts introduced there will now be applied to (9.219).

Reduction to a convolution For the moment, neglect field-dependent aberrations so that $T_{pupil}(\boldsymbol{\rho}; \mathbf{r}_0)$ is independent of $\mathbf{r}_0$ and we can drop the second argument. In that case we can force (9.219) to look like a convolution by defining a rescaled image, the same size as the object, by

$$u_{im}^{(s)}(\mathbf{r}) = m u_{im}(m\mathbf{r}), \tag{9.220}$$

where the superscript s indicates the scaling.

From (9.219), the scaled image field is related to the object field by

$$u_{im}^{(s)}(\mathbf{r}) = \frac{m}{\lambda^2 qp} \int_\infty d^2 r_0 \; u_{obj}(\mathbf{r}_0)\, T_{pupil}\left[\frac{m}{\lambda q}(\mathbf{r} - \mathbf{r}_0)\right] \equiv u_{obj}(\mathbf{r}) * p_{coh}(\mathbf{r}), \tag{9.221}$$

where the PSF $p_{coh}(\mathbf{r})$ is given by

$$p_{coh}(\mathbf{r}) = \frac{m}{\lambda^2 qp} T_{pupil}\left(m \frac{\mathbf{r}}{\lambda q}\right). \tag{9.222}$$

The subscript *coh* stands for *coherent*, a designation that will become clearer in Sec. 9.7.4 when we introduce incoherent fields and objects.

The key conclusion from this discussion is that the coherent PSF is a scaled version of the Fourier transform of the pupil. Note that the scaling factors lead to dimensionally correct expressions. The space-domain pupil function $t_{pupil}(\mathbf{r})$ is dimensionless, so its Fourier transform $T_{pupil}(\boldsymbol{\rho})$ has dimensions of L^2 (from the $d^2 r$ in the Fourier integral). The substitutions in the argument do not alter the dimensions of T_{pupil}, and the factors out front have net dimensions of L^{-4}. Thus p_{coh} has dimensions of L^{-2}, as it must if (9.221) is to be dimensionally correct.

Since $m = -q/p$ we can also write (9.222) as

$$p_{coh}(\mathbf{r}) = \frac{-1}{(\lambda p)^2} T_{pupil}\left(-\frac{\mathbf{r}}{\lambda p}\right). \tag{9.223}$$

This form shows that the PSF, as we have defined it, is really independent of the magnification; we have scaled the image to match the scale of the object, so only the object-to-lens distance p affects the PSF. Recall, however, that we are considering only a simple thin lens with the aperture stop at the lens. For more complicated systems the key distance is from the object to the entrance pupil.

Effect of aberrations If field-dependent aberrations are present, the integral in (9.221) is not a convolution since T_{pupil} is a function of $\mathbf{r}_0$. With field-dependent aberrations, the system is shift variant, and the point response function must depend on two arguments.

To show the field dependence explicitly, we write

$$t_{pupil}(\mathbf{r}; \mathbf{r}_0) = \exp[ikW(\mathbf{r}; \mathbf{r}_0)]\, t_{ap}(\mathbf{r}) . \tag{9.224}$$

The Fourier transform of this function with respect to the $\mathbf{r}$ variable is

$$T_{pupil}(\boldsymbol{\rho}; \mathbf{r}_0) = \mathcal{F}_2\{t_{pupil}(\mathbf{r}; \mathbf{r}_0)\} . \tag{9.225}$$

The counterpart of (9.221) is

$$u_{im}^{(s)}(\mathbf{r}) = \frac{m}{\lambda^2 qp} \int_\infty d^2 r_0\; u_{obj}(\mathbf{r}_0)\, T_{pupil}\left[\frac{m}{\lambda q}(\mathbf{r} - \mathbf{r}_0; \mathbf{r}_0)\right]$$

$$\equiv \int_\infty d^2 r_0\; u_{obj}(\mathbf{r}_0)\, p_{coh}(\mathbf{r} - \mathbf{r}_0; \mathbf{r}_0) , \tag{9.226}$$

where now the *shift-variant* point response function $p_{coh}(\mathbf{r}; \mathbf{r}_0)$ is given by

$$p_{coh}(\mathbf{r}; \mathbf{r}_0) = \frac{m}{\lambda^2 qp} T_{pupil}\left(\frac{m}{\lambda q}\mathbf{r}; \mathbf{r}_0\right) = -\frac{1}{(\lambda p)^2} T_{pupil}\left(-\frac{\mathbf{r}}{\lambda p}; \mathbf{r}_0\right) . \tag{9.227}$$

A useful approximation is to consider a small region of the object plane over which $t_{pupil}(\mathbf{r}, \mathbf{r}_0)$ does not vary much with $\mathbf{r}_0$. If we let $\mathbf{r}_{0c}$ denote the center of this so-called *isoplanatic patch*, we can set

$$p_{coh}(\mathbf{r} - \mathbf{r}_0; \mathbf{r}_0) \approx p_{coh}(\mathbf{r} - \mathbf{r}_0; \mathbf{r}_{0c}) . \tag{9.228}$$

If the object lies entirely within the isoplanatic patch, the image is approximately a convolution. For more discussion of approximately shift-invariant systems, see Sec. 7.2.8.

Coherent transfer function In the shift-invariant case, we can define the coherent transfer function as the Fourier transform of $p_{coh}(\mathbf{r})$:

$$P_{coh}(\boldsymbol{\rho}) = \mathcal{F}_2\{p_{coh}(\mathbf{r})\} = \frac{1}{(\lambda p)^2} \mathcal{F}_2\left\{T_{pupil}\left(-\frac{\mathbf{r}}{\lambda p}\right)\right\} . \tag{9.229}$$

From (3.113) and (3.239),

$$P_{coh}(\boldsymbol{\rho}) = t_{pupil}(\lambda p \boldsymbol{\rho}) . \tag{9.230}$$

An extra factor of -1 occurs in (9.230) since $\mathcal{F}_2\mathcal{F}_2\{f(-\mathbf{r})\} = f(\mathbf{r})$.

Thus *the coherent transfer function is the pupil function itself, suitably scaled.* Object Fourier components for which $\lambda p \boldsymbol{\rho}$ lies outside the lens aperture are not passed by the system.

9.7.2 $4f$ imaging system

Even in the absence of aberrations, a thin lens is not exactly shift invariant; to get convolutions in Sec. 9.7.1, we had to drop quadratic phase factors at several junctures. A somewhat more complicated imaging system that avoids the need for this approximation (but requires some others) is illustrated in Fig. 9.17. We introduce this system here both as an interesting exercise in the application of diffraction theory to imaging and also because it will provide insights into coherence and 3D imaging when we get to those topics.

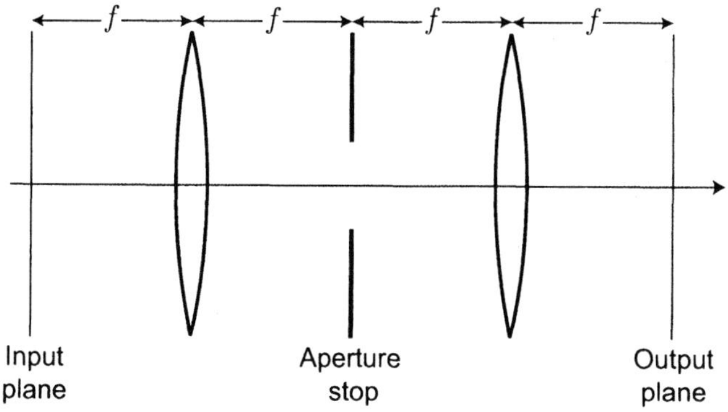

Fig. 9.17 A $4f$ imaging system. The object is a photographic transparency placed in the plane $z = 0$, and the illumination is a monochromatic plane wave normally incident on the transparency.

The focal length of both lenses is f, and the object to be imaged is located a distance f from the first lens. The object plane is called the *front focal plane* of the first lens. The spacing between the lenses is $2f$, and an aperture stop or other mask is placed midway between them, in the back focal plane of the first lens and the front focal plane of the second one. The image is formed in the back focal plane of the second lens, so the total distance from object to image is $4f$, and this configuration of lenses is often referred to as a $4f$ system.

For this analysis, we shall ignore lens aberrations and treat the lenses as thin. We shall assume also that the object is a thin photographic transparency, and that it is illuminated with a monochromatic plane wave propagating along the z axis. The incident wave is thus a constant in the plane of the object, and the wave emerging from the object is this constant (call it A) times the amplitude transmittance of the object transparency. If the plane of the object is $z = 0$, we can write the field emerging from the object as

$$u_{0^+}(\mathbf{r}) = At_{obj}(\mathbf{r}) , \tag{9.231}$$

where the 0^+ subscript indicates the field just to the right of $z = 0$.

As we know from Sec. 9.5.1, this wave can be decomposed into plane-wave components, each of which propagates independently to the lens. To get a qualitative understanding of the action of this imaging system, consider one such plane wave. An ideal lens of focal length f converts an incident plane wave into a spherical wave with radius of curvature f. This wave comes to a focus in the back focal plane, producing a bright spot at a position determined by the direction cosines of the plane wave (see Fig. 9.18). Since these direction cosines correspond to spatial-frequency components by (9.42), the field distribution in the back focal plane of the first lens is the Fourier transform of the object field. For this reason, the back focal plane in this system is often referred to as the *Fourier plane.*

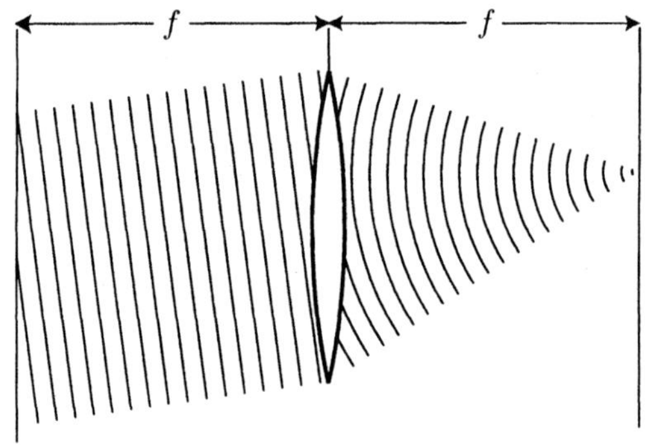

Fig. 9.18 Illustration of one diffracted plane wave emerging from the object and forming a focus in the back focal plane of a lens.

The field in the Fourier plane, incident on the aperture stop, is multiplied by the transmittance of the aperture and then propagates through the second lens to the image plane. Since the second half of the system is identical to the first, it also takes a Fourier transform of the field in its front focal plane. The overall imaging system thus has the following action: it takes the Fourier transform of the input, multiplies it by an aperture stop, and then takes a second Fourier transform.

If the second transform were an inverse Fourier transform, this action would immediately constitute a linear, shift-invariant imaging system, with a transfer function given by the transmittance of the aperture stop. The distinction between forward and inverse transforms, however, is rather trivial. Since $\exp(-2\pi i \boldsymbol{\rho} \cdot \mathbf{r}) = \exp[2\pi i \boldsymbol{\rho} \cdot (-\mathbf{r})]$, a forward transform followed by a coordinate inversion is the same as an inverse transform. In the $4f$ system, the image is inverted (magnification $= -1$), but otherwise it is the expected shift-invariant system.

Fresnel-diffraction analysis Now we look more quantitatively at the $4f$ system, using the framework of Fresnel diffraction theory. If we use the Fresnel diffraction formula (9.98) to propagate the object field to the lens (*i.e.*, to a plane just to the left of $z = f$), the result is

$$u_{f^-}(\mathbf{r}) = \frac{\exp(ikf)}{i\lambda f} \exp\left(i\pi \frac{r^2}{\lambda f}\right) \boldsymbol{\mathcal{F}}_2 \left\{ u_{0^+}(\mathbf{r}_0) \exp\left(i\pi \frac{r_0^2}{\lambda f}\right)\right\}_{\boldsymbol{\rho}=\mathbf{r}/\lambda f} . \tag{9.232}$$

The field emerging from the lens is obtained by multiplying this expression by the lens transmittance, given in the Fresnel approximation by (9.159), so

$$\begin{aligned} u_{f^+}(\mathbf{r}) &= u_{f^-}(\mathbf{r}) \exp\left(-i\pi \frac{r^2}{\lambda f}\right) t_{ap}(\mathbf{r}) \\ &= \frac{\exp(ikf)}{i\lambda f} t_{ap}(\mathbf{r}) \boldsymbol{\mathcal{F}}_2 \left\{ u_{0^+}(\mathbf{r}_0) \exp\left(i\pi \frac{r_0^2}{\lambda f}\right)\right\}_{\boldsymbol{\rho}=\mathbf{r}/\lambda f} . \end{aligned} \tag{9.233}$$

Note that the quadratic phase factor $\exp(i\pi r^2/\lambda f)$ in (9.232) has been cancelled by the quadratic phase factor in the lens transmittance.

Next we assume that the lens diameter D_{lens} is large enough that $t_{ap}(\mathbf{r})$ does not substantially truncate the field incident on it. There are two components to this assumption. It requires that the size of the object transparency be small compared to the lens diameter, but it also requires that t_{obj} not contain fine structures that would diffract light substantially out of the geometric shadow of the object. Roughly speaking, if ρ_{max} is the highest spatial frequency in the object, it will diffract light at an angle given by $\sin\theta_{max} \approx \lambda\rho_{max}$, and the lens aperture is unimportant if $f\tan\theta_{max}$ is less than about $\frac{1}{2}D_{lens}$. This is the major assumption of this analysis: in contrast to the single lens, the lens aperture plays essentially no role in determining the characteristics of a $4f$ system. All of the light is assumed to get through the lens, and the imaging characteristics are set by the stop between the lenses, not by the lens apertures.

If we set $t_{ap}(\mathbf{r}) = 1$, then (9.233) becomes

$$u_{f^+}(\mathbf{r}) = const \cdot \boldsymbol{\mathcal{F}}_2 \left\{ u_{0^+}(\mathbf{r}_0) \exp\left(i\pi \frac{r_0^2}{\lambda f} \right) \right\}_{\boldsymbol{\rho}=\mathbf{r}/\lambda f} . \tag{9.234}$$

A Fourier transformation yields the angular-spectrum representation:

$$U_{f^+}(\boldsymbol{\rho})$$

$$= const \cdot \int_\infty d^2r \, \exp(-2\pi i \boldsymbol{\rho}\cdot\mathbf{r}) \int_\infty d^2r_0 \; u_{0^+}(\mathbf{r}_0) \exp\left(i\pi \frac{r_0^2}{\lambda f} \right) \exp\left(-2\pi i \frac{\mathbf{r}\cdot\mathbf{r}_0}{\lambda f} \right)$$

$$= const \cdot \int_\infty d^2r_0 \; u_{0^+}(\mathbf{r}_0) \exp\left(i\pi \frac{r_0^2}{\lambda f} \right) \delta\left(\boldsymbol{\rho} + \frac{\mathbf{r}_0}{\lambda f} \right)$$

$$= const \cdot u_{0^+}(-\lambda f \boldsymbol{\rho}) \exp(i\pi\lambda f \rho^2) . \tag{9.235}$$

In the plane immediately after the lens, therefore, the Fourier transform of the field is a scaled version of the object field times a quadratic phase factor. To propagate this field to the back focal plane of the lens, we multiply (9.235) by the transfer function for free-space propagation, given in the Fresnel approximation by (9.118); the quadratic phase factors cancel, so

$$U_{2f^-}(\boldsymbol{\rho}) = const \cdot u_{0^+}(-\lambda f \boldsymbol{\rho}) , \tag{9.236}$$

where the subscript $2f^-$ indicates a plane just to the left of $z = 2f$, or immediately before the aperture stop. An inverse Fourier transform of (9.236) yields

$$u_{2f^-}(\mathbf{r}) = const \cdot U_{0^+}\left(\frac{\mathbf{r}}{\lambda f} \right) , \tag{9.237}$$

which confirms the conclusion that we reached qualitatively by considering Fig. 9.18: The space-domain field in the back focal plane is proportional to the Fourier transform of the object field.

Relation to Fraunhofer diffraction The relation of (9.237) to Fraunhofer diffraction theory should not be overlooked. Except for a quadratic phase factor, (9.237) is identical to the Fraunhofer formula (9.99), yet nowhere in this section have we made a Fraunhofer approximation. For (9.99) to be valid, the observation distance z has to be large, but no similar requirement applies to the focal length f.

We can understand this result by realizing that the Fraunhofer approximation becomes exact as the distance z goes to infinity. With the lens in the system, however, a plane at infinity is imaged to $z = f$. Thus (9.237) is a Fraunhofer field imaged to the back focal plane of the first lens.

Propagation to the image plane The field at $z = 2f^+$ is found by multiplying $u_{2f^-}(\mathbf{r})$ by the amplitude transmittance of the aperture stop. If we denote this transmittance by $t_{pupil}(\mathbf{r})$, then

$$u_{2f^+}(\mathbf{r}) = t_{pupil}(\mathbf{r})\, u_{2f^-}(\mathbf{r}) = const \cdot t_{pupil}(\mathbf{r})\, U_{0^+}\left(\frac{\mathbf{r}}{\lambda f}\right). \tag{9.238}$$

We can now use (9.236) to propagate the field from $z = 2f^+$ to the image plane, $z = 4f$; the result is

$$U_{4f}(\boldsymbol{\rho}) = const \cdot u_{2f^+}(-\lambda f \boldsymbol{\rho}) = const \cdot t_{pupil}(-\lambda f \boldsymbol{\rho})\, U_{0^+}(-\boldsymbol{\rho}). \tag{9.239}$$

The constant in this equation turns out just to be the phase factor $\exp(4ikf)$. That the modulus of the constant is unity follows from conservation of energy; we have assumed that the diffracted light all goes through the aperture of the first lens, and if $t_{pupil}(\mathbf{r}) = 1$, there is no light loss at all.

The minus signs in the arguments in (9.239) imply that the image is inverted. As in Sec. 9.7.1, we can define a scaled function that removes the uninteresting magnification factors. In the present case, the scaling amounts to removing a minus sign, so that

$$U_{im}^{(s)}(\boldsymbol{\rho}) \equiv U_{4f}(-\boldsymbol{\rho}) = \exp(4ikf)\, t_{pupil}(\lambda f \boldsymbol{\rho})\, U_{obj}(\boldsymbol{\rho}), \tag{9.240}$$

where $U_{obj}(\boldsymbol{\rho}) = U_{0^+}(-\boldsymbol{\rho})$ is the Fourier transform of the object field but plotted on inverted axes.

The form of (9.240) shows that the $4f$ system is a linear, shift-invariant mapping from the object field to the (scaled) image field. In the frequency domain, this mapping is the simple multiplication shown in (9.239), and in the space domain it is a convolution. By inspection of (9.240), the coherent transfer function for a $4f$ system is

$$P_{coh}(\boldsymbol{\rho}) = \exp(4ikf)\, t_{pupil}(\lambda f \boldsymbol{\rho}). \tag{9.241}$$

If we neglect the constant phase factor $\exp(4ikf)$, (9.241) is essentially the same as the transfer function (9.230) for a simple lens, only now the pupil is interpreted as the transmittance of the aperture stop, and f appears rather than p.

To appreciate the form of (9.241), suppose that the pupil is a square aperture of side L centered on the optic axis, so that $t_{pupil}(\mathbf{r}) = \mathrm{rect}(x/L)\,\mathrm{rect}(y/L)$, where x and y are the Cartesian components of $\mathbf{r}$. Consider an object with an amplitude transmittance of $\exp(2\pi i \xi_0 x)$. This object can be realized as a thin prism, but it can also be regarded as one Fourier component of a more general object. With either interpretation, the field emerging from the object (for a normally incident plane-wave illumination) is a plane wave propagating in the x-z plane at an angle θ_x to the z axis, where, in the small-angle approximation, $\theta_x \approx \xi_0\lambda$.

An elementary function of a lens is to focus a plane wave into a point in the back focal plane. For the plane wave emerging from our test object, this point is located (in the small-angle approximation) at coordinates $(f\theta_x, 0)$. Thus the focal point is at $(\lambda f\xi_0, 0)$, which is precisely the argument of t_{pupil} in (9.241) for the frequency under discussion. If this focal spot lies outside the clear aperture, *i.e.*, if $\lambda f|\xi_0| > \frac{1}{2}L$, then the spot is blocked by the aperture and makes no contribution to the image field. If $\lambda f|\xi_0| < \frac{1}{2}L$, on the other hand, the focused wave travels unimpeded past the focal spot, producing an expanding spherical wave of radius of curvature f incident on the next lens. This lens converts the spherical wave into another tilted plane wave (this time with $\theta_x = -\xi_0\lambda$), which reproduces the object wave. Thus the aperture transmittance directly controls the transfer function of the system.

9.7.3 More complicated lens systems

In geometrical optics, the action of a lens system is described by specifying what it does to rays. Often we consider all of the rays crossing some input plane P_{in} (commonly, but not necessarily, the object plane) and trace them through to some output plane P_{out} (commonly the image plane). If we know what happens to all such rays, we have completely specified the mapping operation from P_{in} to P_{out} so far as geometric optics goes. In this section we shall see that this geometric-optics description also tells us a good deal about the physical-optics properties of the system.

Paraxial ray optics As discussed in Sec. 7.2.10, we often consider systems with a well-defined optical axis (perhaps an axis of rotational symmetry), which we can call the z-axis. For such systems it is natural to construct the planes P_{in} and P_{out} perpendicular to the z-axis. The rays are then described by their x-y coordinates on the plane and their direction cosines relative to the x and y axes. For the output plane P_{out}, we shall denote the coordinates of a particular ray as simply x and y, with the corresponding direction cosines α and β. For P_{in} we shall use x', y', α' and β'.

Elementary geometric optics often considers only *paraxial rays* for which α and β are small, and it restricts the input field of view sufficiently that the system is shift-invariant. It may then be a good approximation to describe the mapping from P_{in} to P_{out} as a simple matrix multiplication. If the system has rotational symmetry, the mapping is *separable*, in the sense that the mapping of x' and α' to x and α is independent of y' and β', and we can write

$$\begin{bmatrix} x \\ \alpha \end{bmatrix} = \begin{bmatrix} A & B \\ C & D \end{bmatrix} \begin{bmatrix} x' \\ \alpha' \end{bmatrix}, \tag{9.242}$$

with a similar expression for the y coordinates.[15]

[15] If the paraxial approximation does not hold, (9.242) can still be used, but the matrix elements must be functions of x' and α'.

For example, if the input and output planes are separated by a distance z in free space, then the $ABCD$ matrix is given by

$$\mathbf{M}_z = \begin{bmatrix} 1 & z \\ 0 & 1 \end{bmatrix}. \tag{9.243}$$

The transformation action of this segment of free space is thus

$$\mathbf{M}_z \begin{bmatrix} x' \\ \alpha' \end{bmatrix} = \begin{bmatrix} x' + \alpha' z \\ \alpha' \end{bmatrix}. \tag{9.244}$$

We see that the angle of the ray remains constant as z is varied but that the ray is displaced parallel to the x-axis by an amount $\alpha' z$.

To describe a thin lens in the paraxial approximation, we take the input and output planes to be immediately adjacent to the lens; thus P_{in} and P_{out} are effectively the same plane since the lens is thin. The matrix that describes a thin lens of focal length f with this choice of planes is

$$\mathbf{M}_{lens} = \begin{bmatrix} 1 & 0 \\ -\frac{1}{f} & 1 \end{bmatrix}, \tag{9.245}$$

and the transformation is

$$\mathbf{M}_{lens} \begin{bmatrix} x' \\ \alpha' \end{bmatrix} = \begin{bmatrix} x' \\ -\frac{x'}{f} + \alpha' \end{bmatrix}. \tag{9.246}$$

Thus the position of the ray is unaltered (which is the operational definition of a thin optical element), but the angle is deviated by an amount proportional to x'.

Concatenation of matrices We can construct matrix representations for more complicated systems from the building blocks $\mathbf{M}_z$ and $\mathbf{M}_{lens}$. For example, if we consider free-space propagation over distance p, followed by a thin lens of focal length f and then propagation over q, where p, q and f are related by the imaging equation (9.169), then the overall matrix is

$$\mathbf{M}_q \mathbf{M}_{lens} \mathbf{M}_p = \begin{bmatrix} 1 & q \\ 0 & 1 \end{bmatrix} \begin{bmatrix} 1 & 0 \\ -\frac{1}{f} & 1 \end{bmatrix} \begin{bmatrix} 1 & p \\ 0 & 1 \end{bmatrix} = \begin{bmatrix} m & 0 \\ -\frac{1}{f} & \frac{1}{m} \end{bmatrix}, \tag{9.247}$$

where m is the magnification, $-q/p$.

The overall transformation from input to output for this example is

$$\begin{bmatrix} m & 0 \\ -\frac{1}{f} & \frac{1}{m} \end{bmatrix} \begin{bmatrix} x' \\ \alpha' \end{bmatrix} = \begin{bmatrix} mx' \\ \frac{\alpha'}{m} - \frac{x'}{f} \end{bmatrix}. \tag{9.248}$$

As expected, $x = mx'$ since m is the magnification. If we define an angular magnification as $m_{ang} = \frac{d\alpha}{d\alpha'}$, then we see that $m_{ang} = m^{-1}$, so systems that magnify the image demagnify the angles. The angular offset $-x'/f$ can be understood by considering the ray with $\alpha' = 0$; this ray must pass through the focal point in image space and hence makes an angle $-x'/f$ with the axis.

As an exercise, the reader may show that the $ABCD$ matrix for the $4f$ system of Fig. 9.17 is just $-\mathbf{I}$, where $\mathbf{I}$ is the 2×2 unit matrix. Thus the lateral and angular magnifications are both -1.

Lens group All matrices considered so far in this section have determinant $= 1$, that is, $AD - BC = 1$. In fact, this is a general rule: whenever an optical system can be described geometrically by a 2×2 $ABCD$ matrix, that matrix must have unit determinant (Stavroudis, 1972). We know from Sec. 6.5.2 that the group of all such matrices is the special linear group $\mathbf{SL}(2)$, which we can now regard as the group of all realizable optical systems that can be described by 2×2 $ABCD$ matrices. (Recall that this description requires the paraxial and shift-invariant approximations and that it applies only to rotationally symmetric or other separable systems where the x and y variables transform independently.)

We know also from Chap. 6 that a set of operators constitutes a group only if the inverse of every operator is in the group. That condition is satisfied for this lens group. For example, the inverse of propagation over distance z is propagation over $-z$, and the inverse of a positive lens of focal length f is a negative lens of focal length $-f$.

Generalized diffraction integral The Fresnel diffraction integral for a rotationally symmetric optical system was first written in terms of matrix optics by Collins (1970) and further explored by Nazarathy and colleagues (Nazarathy and Shamir, 1982a, 1982b) For textbook treatments see Saleh and Teich (1981) and Siegman (1986).

The main result of these discussions is that, when the mapping from P_{in} to P_{out} is described geometrically by a 2×2 $ABCD$ matrix, the field is mapped in the Fresnel approximation according to

$$u_{out}(x, y) = -\frac{i}{B\lambda} \int_{-\infty}^{\infty} dx_0 \int_{-\infty}^{\infty} dy_0 \; u_{in}(x_0, y_0)$$

$$\times \exp\left\{ \frac{i\pi}{\lambda B} \left[A\left(x_0^2 + y_0^2\right) + D\left(x^2 + y^2\right) - 2(xx_0 + yy_0) \right] \right\}, \tag{9.249}$$

or, in our usual vector notation,

$$u_{out}(\mathbf{r}) = -\frac{i}{B\lambda} \int_{\infty} d^2 r_0 \; u_{in}(\mathbf{r}_0) \exp\left[\frac{i\pi}{\lambda B} \left(A r_0^2 + D r^2 - 2\mathbf{r} \cdot \mathbf{r}_0 \right) \right]. \tag{9.250}$$

One special case of this result is already known from earlier in this chapter. For free-space propagation over distance z, $A = 1$, $B = z$, $C = 0$ and $D = 1$, so (9.250) reproduces (9.97) except for the irrelevant constant $\exp(ikz)$.

ABCD matrices for systems without rotational symmetry The use of a 2×2 $ABCD$ matrix presumes that the x and y variables can be transformed independently, as they can for systems with rotational symmetry. More generally, however, we can use a 4×4 matrix and write the transformation as

$$\begin{bmatrix} M_{11} & M_{12} & M_{13} & M_{14} \\ M_{21} & M_{22} & M_{23} & M_{24} \\ M_{31} & M_{32} & M_{33} & M_{34} \\ M_{41} & M_{42} & M_{43} & M_{44} \end{bmatrix} \begin{bmatrix} x' \\ y' \\ \alpha' \\ \beta' \end{bmatrix} = \begin{bmatrix} x \\ y \\ \alpha \\ \beta \end{bmatrix}. \tag{9.251}$$

This expression still assumes shift-invariance, and it applies only to paraxial rays, but it does not require separability in x and y.

It is convenient to define a 2×1 vector $\mathbf{r} = (x, y)^t$ (which is just our usual 2D position vector written in column-vector form) and similarly to define a 2×1 vector of direction cosines by $\boldsymbol{\theta} = (\alpha, \beta)^t$. With this notation, (9.251) can be written in the form

$$\begin{bmatrix} \mathbf{A} & \mathbf{B} \\ \mathbf{C} & \mathbf{D} \end{bmatrix} \begin{bmatrix} \mathbf{r}' \\ \boldsymbol{\theta}' \end{bmatrix} = \begin{bmatrix} \mathbf{r} \\ \boldsymbol{\theta} \end{bmatrix}, \tag{9.252}$$

where $\mathbf{A}$, $\mathbf{B}$, $\mathbf{C}$ and $\mathbf{D}$ are all 2×2 matrices. We shall refer to matrices in this form as **ABCD** matrices rather than $ABCD$ matrices.

Note that (9.252) is equivalent to (9.242) (plus the similar equation for the y direction) if $\mathbf{A} = A\mathbf{I}$, and similarly for $\mathbf{B}$, $\mathbf{C}$ and $\mathbf{D}$.

Symplectic group Stavroudis (1972) shows that the **ABCD** matrix for all realizable optical systems must satisfy the *symplectic condition*,

$$\mathbf{M}^t\mathbf{J}\mathbf{M} = \mathbf{J}, \tag{9.253}$$

where

$$\mathbf{J} = \begin{bmatrix} 0 & 0 & 1 & 0 \\ 0 & 0 & 0 & 1 \\ -1 & 0 & 0 & 0 \\ 0 & -1 & 0 & 0 \end{bmatrix} = \begin{bmatrix} \mathbf{0} & \mathbf{I} \\ -\mathbf{I} & \mathbf{0} \end{bmatrix}, \tag{9.254}$$

with $\mathbf{I}$ being the 2×2 identity matrix and $\mathbf{0}$ being the 2×2 matrix of all zeros. The importance of the symplectic condition is indicated by the fact that Stavroudis calls it the *lens equation*.

The group of all matrices that satisfy the symplectic condition is called the *symplectic group* and denoted $\mathbf{Sp}_2$. All **ABCD** matrices for realizable optical systems are in this group, but the converse does not hold. Since the determinant of a product is the product of the determinants, and since the determinant of $\mathbf{J}$ is 1, it follows from (9.253) that all matrices in the symplectic group have determinant ± 1, and in fact the plus sign is required for optical systems. Moreover, more than one optical system may have the same symplectic matrix.

Diffraction integral further generalized When the full 4×4 **ABCD** matrix is required rather than two 2×2 $ABCD$ matrices, the field propagation from input to output plane is described in the Fresnel approximation by (Siegman, 1986)

$$u_{out}(\mathbf{r}) = -\frac{i}{\lambda\sqrt{\det \mathbf{B}}} \int_\infty d^2r_0 \; u_{in}(\mathbf{r}_0) \exp\left[\frac{i\pi}{\lambda}(\mathbf{r}_0^t\mathbf{B}^{-1}\mathbf{A}\mathbf{r}_0 + \mathbf{r}^t\mathbf{D}\mathbf{B}^{-1}\mathbf{r} - 2\mathbf{r}_0^t\mathbf{B}^{-1}\mathbf{r})\right]. \tag{9.255}$$

Note that (9.255) reduces to (9.250) if $\mathbf{A} = A\mathbf{I}$, and similarly for $\mathbf{B}$, $\mathbf{C}$ and $\mathbf{D}$.

9.7.4 Random fields and coherence

So far in this chapter we have concentrated on perfectly monochromatic radiation, and we have expressed the scalar optical field at a point specified by the 3D vector $\mathbf{r}$ as $\mathrm{Re}\{u(\mathbf{r})\exp(-2\pi i\nu_0 t)\}$. For random but narrowband fields, it is natural to write the field as $\mathrm{Re}\{\tilde{u}(\mathbf{r}, t)\exp(-2\pi i\overline{\nu}t)\}$, where $\overline{\nu}$ is the center frequency of the spectral band and $\tilde{u}(\mathbf{r}, t)$ is a random complex amplitude that varies with spatial position and time. If the field is random but not narrowband, it can be written as the real part of the analytic signal (see Sec. 4.2.4), but there is no loss of generality in writing this signal as $\tilde{u}(\mathbf{r}, t) \exp(-2\pi i \bar{\nu} t)$ for some convenient frequency $\bar{\nu}$. In either case, $\tilde{u}(\mathbf{r}, t)$ is a spatio-temporal random process, and we must specify its statistical properties before we can understand how the randomness affects imaging properties. The basics of random processes were presented in Sec. 8.2.

Statistical description of natural light sources For chaotic light sources, such as incandescent bulbs and fluorescent lights, the complex field is a phasor that is randomly oriented in the complex plane. Since all angles in this plane are equally probable,

$$\langle \tilde{u}(\mathbf{r}, t) \rangle = 0 . \tag{9.256}$$

The angular brackets here denote an ensemble average, but the assumption of ergodicity (see Sec. 8.2.5) allows us to interpret them as time averages as well. If the light were perfectly monochromatic, $\tilde{u}(\mathbf{r}, t)$ would be independent of t and the time average would not be zero, but for chaotic light with a finite spectral bandwidth, the random time-varying phase of $\tilde{u}(\mathbf{r}, t)$ makes the complex field average to zero.

The complex spatio-temporal autocorrelation function of this random process is defined by

$$R_{\tilde{u}}(\mathbf{r}, t; \mathbf{r}', t') \equiv \langle \tilde{u}(\mathbf{r}, t)\, \tilde{u}^*(\mathbf{r}', t') \rangle . \tag{9.257}$$

It is often useful to assume that the process is temporally stationary (though not necessarily spatially stationary), so that $R_u(\mathbf{r}, t; \mathbf{r}', t')$ is a function of $t - t'$ and not t and t' individually. This assumption will be valid if parameters of the light source, such as the temperature of a blackbody radiator, are not themselves functions of time. Under this assumption, we define the *mutual coherence function* $\Gamma(\mathbf{r}, \mathbf{r}', \tau)$ as

$$\Gamma(\mathbf{r}, \mathbf{r}', \tau) \equiv R_{\tilde{u}}(\mathbf{r}, t + \tau; \mathbf{r}', t) = \langle \tilde{u}(\mathbf{r}, t + \tau)\, \tilde{u}^*(\mathbf{r}', t) \rangle . \tag{9.258}$$

A common alternative notation is $\Gamma_{12}(\tau) \equiv \Gamma(\mathbf{r}_1, \mathbf{r}_2, \tau)$, but in imaging applications it is convenient to show the spatial variables explicitly. Note that we define $\Gamma(\mathbf{r}, \mathbf{r}', \tau)$ as the autocorrelation of the envelope $\tilde{u}(\mathbf{r}, t)$ rather than the total complex field $u(\mathbf{r}, t)$; the latter definition is more common in the literature but ours avoids uninteresting factors of $\exp(-2\pi i \bar{\nu} t)$ when we consider narrowband light, and it entails no loss of generality in the broadband case.

Coherence as interferability A normalized version of $\Gamma(\mathbf{r}, \mathbf{r}', \tau)$, called the *complex degree of coherence*, is defined by

$$\gamma(\mathbf{r}, \mathbf{r}', \tau) \equiv \frac{\Gamma(\mathbf{r}, \mathbf{r}', \tau)}{\sqrt{\langle |\tilde{u}(\mathbf{r}, t)|^2 \rangle \langle |\tilde{u}(\mathbf{r}', t)|^2 \rangle}} . \tag{9.259}$$

The complex degree of coherence tells us the degree to which light from the space-time point $(\mathbf{r}, t)$ is capable of interfering with light from point $(\mathbf{r}', t')$. An interferometer is basically a device for bringing together light from these two points and superimposing them at a third space-time point $(\mathbf{r}'', t'')$. When we do so, $|\gamma(\mathbf{r}, \mathbf{r}', t - t')|$ gives the relative strength of the interference term, and the phase tells us about the position of the fringes.

As a simple example, consider the double-pinhole experiment shown in Fig. 9.19. Two identical small pinholes at points $\mathbf{r}$ and $\mathbf{r}'$ are illuminated by a field of arbitrary coherence. The irradiance distribution on an observation screen will show interference fringes if the light arriving there from the two pinholes is capable of interfering.

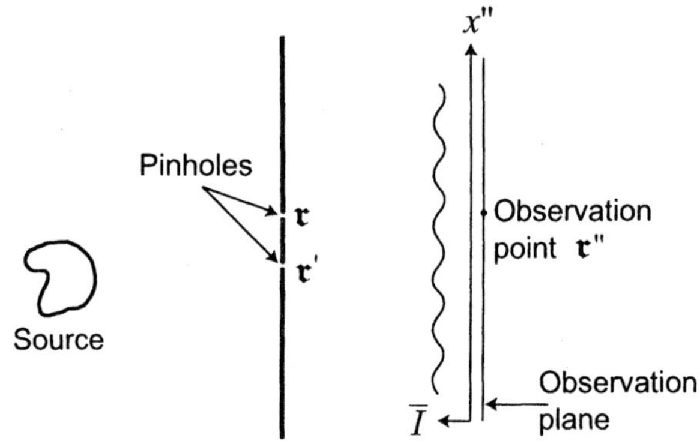

Fig. 9.19 Double-pinhole experiment for measuring the complex degree of coherence.

If we denote the complex field on the screen at $(\mathfrak{r}'', t'')$ resulting from light coming through pinhole 1 as $u_1(\mathfrak{r}'', t'')$, and similarly for light from pinhole 2, then the time-averaged mean irradiance at $\mathfrak{r}''$ is given by

$$\begin{aligned}\overline{I}(\mathbf{r}'') &= \langle |u_1(\mathfrak{r}'', t'') + u_2(\mathfrak{r}'', t'')|^2 \rangle \\ &= \langle |u_1(\mathfrak{r}'', t'')|^2 \rangle + \langle |u_2(\mathfrak{r}'', t'')|^2 \rangle + 2\,\mathrm{Re}\langle u_1(\mathfrak{r}'', t'')\, u_2^*(\mathfrak{r}'', t'') \rangle \,. \end{aligned} \tag{9.260}$$

If we think of the pinholes as effective sources, we know from (9.66) that the field at the observation point is determined by the field at the pinhole at a retarded time, so $u_1(\mathfrak{r}'', t'') \propto u\left(\mathfrak{r}, t'' - \frac{|\mathfrak{r}''-\mathfrak{r}|}{c}\right)$. A similar expression holds for $u_2(\mathfrak{r}'', t'')$, and we can write

$$\frac{\langle u_1(\mathfrak{r}'', t'')\, u_2^*(\mathfrak{r}'', t'')\rangle}{\sqrt{\overline{I}_1(\mathfrak{r}'')\, \overline{I}_2(\mathfrak{r}'')}} = \frac{\left\langle u\left(\mathfrak{r}, t'' - \frac{|\mathfrak{r}''-\mathfrak{r}|}{c}\right) u^*\left(\mathfrak{r}', t'' - \frac{|\mathfrak{r}''-\mathfrak{r}'|}{c}\right)\right\rangle}{\sqrt{\overline{I}(\mathfrak{r})\, \overline{I}(\mathfrak{r}')}}\,, \tag{9.261}$$

where all factors of proportionality have been normalized away. (Note carefully that $\overline{I}(\mathfrak{r})$ is the mean irradiance emerging from the pinhole at $\mathfrak{r}$, while $\overline{I}_1(\mathfrak{r}'') \equiv \langle |u_1(\mathfrak{r}'', t'')|^2 \rangle$ is the mean irradiance that would appear at the observation point if that pinhole alone were open.)

Since $u(\mathfrak{r}, t) = \tilde{u}(\mathfrak{r}, t) \exp(-2\pi i \overline{\nu} t)$, the right-hand side of (9.261) simplifies to

$$\frac{\left\langle u\left(\mathfrak{r}, t'' - \frac{|\mathfrak{r}''-\mathfrak{r}|}{c}\right) u^*\left(\mathfrak{r}', t'' - \frac{|\mathfrak{r}''-\mathfrak{r}'|}{c}\right)\right\rangle}{\sqrt{\overline{I}(\mathfrak{r})\, \overline{I}(\mathfrak{r}')}} = \gamma(\mathfrak{r}, \mathfrak{r}', \tau) \exp(-2\pi i \overline{\nu} \tau)\,, \tag{9.262}$$

where we have invoked temporal stationarity and defined

$$\tau = \frac{1}{c}\,|\mathfrak{r}'' - \mathfrak{r}'| - \frac{1}{c}\,|\mathfrak{r}'' - \mathfrak{r}|\,. \tag{9.263}$$

Thus τ is the difference in propagation times from the two pinholes to the observation point.

If we write $\gamma(\mathbf{r},\mathbf{r}',\tau)$ as $|\gamma(\mathbf{r},\mathbf{r}',\tau)|\exp[i\Phi_\gamma(\mathbf{r},\mathbf{r}',\tau)]$, we obtain

$$\overline{I}(\mathbf{r}'') = \overline{I}_1(\mathbf{r}'') + \overline{I}_2(\mathbf{r}'') + 2|\gamma(\mathbf{r},\mathbf{r}',\tau)|\sqrt{\overline{I}_1(\mathbf{r}'')\,\overline{I}_2(\mathbf{r}'')}\cos[2\pi\overline{\nu}\tau - \Phi_\gamma(\mathbf{r},\mathbf{r}',\tau)]\,. \tag{9.264}$$

Though it is hidden somewhat by our notation, τ is a function of the observation point $\mathbf{r}''$ as well as the pinhole locations $\mathbf{r}$ and $\mathbf{r}'$. For fixed pinholes, the dependence of τ on $\mathbf{r}''$ makes the cosine a spatially oscillating function on the observation plane, *i.e.*, a fringe pattern. The modulation of these fringes is reduced by the factor $|\gamma(\mathbf{r},\mathbf{r}',\tau)|$, and the fringes are shifted by $\Phi_\gamma(\mathbf{r},\mathbf{r}',\tau)$ from what we would observe in the fully coherent case of $\gamma(\mathbf{r},\mathbf{r}',\tau) = 1$. Thus the modulus of the complex degree of coherence tells us the fringe visibility, and the phase gives us fringe position.

We have carried along the arguments in $\gamma(\mathbf{r},\mathbf{r}',\tau)$ to emphasize that it refers to the coherence between the fields at the two pinhole locations. The observation point appears only in that it determines the time delay τ. To probe the dependence of $\gamma(\mathbf{r},\mathbf{r}',\tau)$ on τ, we could move the point $\mathbf{r}$ laterally as shown in Fig. 9.19 or otherwise introduce a time delay between the two paths, but to probe the dependence on the spatial arguments, we have to move the pinholes.

Summary measures of temporal and spatial coherence The state of coherence of a radiation field is fully specified by the complex degree of coherence $\gamma(\mathbf{r},\mathbf{r}',\tau)$, a function of seven variables in all. It is often useful to summarize this complicated function by one or two scalar measures, much as we summarize a point response function by some measure of its width (see Sec. 7.2.1).

If we adopt a measure of width, such as full width at half maximum, then the width of a plot of $\gamma(\mathbf{r},\mathbf{r}',\tau)$ vs. τ will be called the *coherence time* and denoted τ_c. The distance $c\tau_c$, where c is the speed of light, is the *coherence length.* For natural light sources τ_c is approximately the reciprocal of the spectral bandwidth $\Delta\nu$ of the light, so we can control τ_c simply by using narrowband spectral filters to control the bandwidth of the light.

If τ_c is long compared to any relevant time *differences* in a problem, the light is said to be *quasimonochromatic.* In an interferometer, for example, two beams may travel along different paths, requiring different propagation times. If the difference in path lengths is small compared to the coherence length (or, equivalently, the difference in propagation times is small compared to the coherence time), the light can be treated as if it were monochromatic. The implications of this condition for imaging will be taken up in Sec. 9.7.5.

For temporally stationary light, the mutual coherence function at equal time ($\tau = 0$) will be denoted

$$\Gamma(\mathbf{r},\mathbf{r}') \equiv \Gamma(\mathbf{r},\mathbf{r}',0)\,. \tag{9.265}$$

This quantity is referred to variously as the *spatial coherence function*, the *mutual intensity* or the *mutual optical intensity.* Since the word *intensity* has a different meaning in radiometry (see Sec. 10.2), we shall adopt the first of these designations.

The mean irradiance of the field at point $\mathbf{r}$, denoted $\overline{I}(\mathbf{r})$, is related to the spatial coherence function by

$$\overline{I}(\mathbf{r}) = \left\langle|\tilde{u}(\mathbf{r},t)|^2\right\rangle = \Gamma(\mathbf{r},\mathbf{r})\,. \tag{9.266}$$

The complex degree of coherence at $\tau = 0$ is given by

$$\gamma(\mathbf{r}, \mathbf{r}') \equiv \frac{\Gamma(\mathbf{r}, \mathbf{r}')}{\sqrt{\Gamma(\mathbf{r}, \mathbf{r})\,\Gamma(\mathbf{r}', \mathbf{r}')}} = \frac{\Gamma(\mathbf{r}, \mathbf{r}')}{\sqrt{\bar{I}(\mathbf{r})\,\bar{I}(\mathbf{r}')}} . \tag{9.267}$$

From (9.258) and (9.259), we see that

$$\gamma(\mathbf{r}, \mathbf{r}) = 1 . \tag{9.268}$$

Any of the resolution measures discussed in Sec. 7.2.1 can be used to specify the width of $\gamma(\mathbf{r}, \mathbf{r}')$ or $\Gamma(\mathbf{r}, \mathbf{r}')$. Without attempting to be very precise at this stage, we shall denote this width as L_c and refer to it as the *correlation length* (not to be confused with the coherence length defined above).

An incoherent assumption For many natural light sources, $\gamma(\mathbf{r}, \mathbf{r}')$ is a sharply peaked function of $\mathbf{r} - \mathbf{r}'$, and it is convenient mathematically to approximate it with a delta function. When this approximation is valid, we refer to the field as *spatially incoherent*, or *incoherent* for short. Conversely, if $\gamma(\mathbf{r}, \mathbf{r}') \approx 1$ for two points of interest, then the field is said to be spatially coherent between these points.

In many imaging applications, we are interested in the case where $\mathbf{r}$ and $\mathbf{r}'$ lie on a specified plane. If, as usual, we call that plane $z = 0$, then we can characterize the field by the function $\gamma(\mathbf{r}, \mathbf{r}')$, where $\mathbf{r}$ and $\mathbf{r}'$ are 2D vectors. Then the incoherent-field approximation is

$$\gamma(\mathbf{r}, \mathbf{r}') \approx A_c\, \delta(\mathbf{r} - \mathbf{r}') , \tag{9.269}$$

where A_c is a constant with units of area needed for dimensional consistency and proper normalization. We shall refer to A_c as the *coherence area*. Roughly speaking, $A_c \approx L_c^2$, but the exact relation depends on the particular definition of L_c and the functional form of $\gamma(\mathbf{r}, \mathbf{r}')$.

Note that (9.269) does not satisfy (9.268), so it cannot be interpreted pointwise; instead, (9.269) makes sense only when it is used in an integral where the other factors in the integrand are slowly varying in comparison to $\gamma(\mathbf{r}, \mathbf{r}')$.

The closest we can come to justifying (9.269) is with *Lambertian sources* (see Sec. 10.2.1), of which blackbodies are an important example. As we shall see in Sec. 10.2.7, $\gamma(\mathbf{r}, \mathbf{r}') = \sin(k|\mathbf{r} - \mathbf{r}'|)/(k|\mathbf{r} - \mathbf{r}'|)$ for a quasimonochromatic Lambertian. In this expression, $k = 2\pi/\lambda$, so $\gamma(\mathbf{r}, \mathbf{r}')$ has a width (peak to first zero) of $\lambda/2$. If $\lambda/2$ is small compared to the width of the system PSF, then it may be valid to approximate the sinc function with a delta function, but the approximation is not entirely justified with high-resolution imaging systems that can resolve details on the order of a wavelength. For further discussion, see Sec. 9.7.7.

9.7.5 Quasimonochromatic imaging

In Sec. 9.7.4, we presented the vocabulary needed for discussing spatial and temporal coherence, and we introduced two useful single-parameter descriptions needed to speak in qualitative terms about degree of coherence. For spatial coherence, the relevant parameter is the correlation length L_c, and for temporal coherence the parameter is the coherence time τ_c or the coherence length $c\tau_c$. We noted that τ_c is approximately the reciprocal of the spectral bandwidth $\Delta\nu$.

A common situation in imaging is when τ_c is relatively long, in a sense to be defined below, but L_c is very short. When both of these conditions are satisfied for the field emerging from the source, it is common in the literature to speak of *quasimonochromatic, incoherent imaging*, but this terminology is potentially confusing. When the word *incoherent* is used without a clarifying modifier, it usually refers to spatial coherence, and that is the only possible interpretation if the modifier *quasimonochromatic* is included; a nearly monochromatic source has a small spectral bandwidth, hence a large coherence time, and is essentially coherent in a temporal sense whatever its spatial properties.

In this section we shall discuss the effects on imaging systems of using light with a small but finite spectral bandwidth without making any assumptions about spatial coherence. In Sec. 9.7.5, we shall add the assumption that the source is spatially incoherent as well as quasimonochromatic. Later, in Secs. 9.7.6 and 9.7.7, we shall return to polychromatic imaging and partial spatial coherence.

Temporal Fourier analysis In Sec. 9.7.1 we discussed coherent, monochromatic imaging by a simple lens, and in Sec. 9.7.2 we extended the discussion to a $4f$ system. For a nonmonochromatic light source we can perform a temporal Fourier analysis to resolve the object field into monochromatic components, and each component can be analyzed by the formalism we have developed. Linear superposition then yields the total image field.

Recalling from Sec. 9.1.2 that the usual Fourier sign convention is reversed for temporal transforms, we represent a general object field $u_{obj}(\mathbf{r}, t)$ as[16]

$$u_{obj}(\mathbf{r}, t) = \int_{-\infty}^{\infty} d\nu \; U_{obj}(\mathbf{r}, \nu) \exp(-2\pi i \nu t) \,. \tag{9.270}$$

For narrowband light, $U_{obj}(\mathbf{r}, \nu)$ will be appreciable only for ν in the vicinity of some center frequency $\overline{\nu}$.

After rescaling to account for magnification, we can write the image field as

$$u_{im}^{(s)}(\mathbf{r}, t) = \int_{-\infty}^{\infty} d\nu \, \exp(-2\pi i \nu t) \int_{\infty} d^2 r_0 \; U_{obj}(\mathbf{r}_0, \nu) \, p_{coh}(\mathbf{r} - \mathbf{r}_0; \mathbf{r}_0; \nu) \,, \tag{9.271}$$

where we have added an argument ν to the coherent point response function since it can depend on frequency. For all systems we have studied thus far, this PRF has the form [*cf.* (9.227)]

$$p_{coh}(\mathbf{r} - \mathbf{r}_0; \mathbf{r}_0; \nu) = \frac{1}{\lambda^2 q p} \exp[ikw(\mathbf{r}, \mathbf{r}_0)] \left| T_{pupil}\left(\frac{m}{\lambda q}(\mathbf{r} - \mathbf{r}_0); \mathbf{r}_0 \right) \right| \,, \tag{9.272}$$

where $\exp[ikw(\mathbf{r}, \mathbf{r}_0)]$ accounts for the phase of T_{pupil} as well as various phase factors that we somewhat cavalierly dropped in the monochromatic case. For the ideal $4f$ system, for example, $w(\mathbf{r}, \mathbf{r}_0) = 4f$, but for a simple lens $w(\mathbf{r}, \mathbf{r}_0)$ also includes terms quadratic in r and r_0, even if there are no aberrations.

The frequency dependence is hidden in several places in (9.272); the wavenumber k is given by $2\pi\nu/c$ and the wavelength λ is c/ν. (Note that c rather than

[16]Do not confuse $U_{obj}(\mathbf{r}, \nu)$ with $U_{obj}(\boldsymbol{\rho})$ used earlier; the former is a temporal Fourier transform of the object field, the latter a spatial transform.

c_m appears here since we assume that the object and image planes are in air, where c_m is the same as c to an excellent approximation.) In the quasimonochromatic approximation, we replace ν with $\overline{\nu}$ (or equivalently, λ with $\overline{\lambda}$) in the constant in front of the integral and in the scale factor in the argument of $|T_{pupil}|$. In the factor $\exp[ikw(\mathbf{r}, \mathbf{r}_0)]$, however, we must retain the original form since $w(\mathbf{r}, \mathbf{r}_0)$ may vary by several wavelengths as $\mathbf{r}$ and $\mathbf{r}_0$ vary. With (9.271) and (9.272), we then obtain

$$u_{im}^{(s)}(\mathbf{r}, t) = \frac{m}{\overline{\lambda}^2 qp} \int_{-\infty}^{\infty} d\nu \exp(-2\pi i\nu t) \int_{\infty} d^2 r_0 \, U_{obj}(\mathbf{r}_0, \nu)$$
$$\times \exp\left(2\pi i\nu \frac{w(\mathbf{r}, \mathbf{r}_0)}{c}\right) \left| T_{pupil}\left(\frac{m}{\overline{\lambda} q}(\mathbf{r} - \mathbf{r}_0); \mathbf{r}_0\right)\right| . \tag{9.273}$$

The integral over ν is an inverse Fourier transform of $U_{obj}(\mathbf{r}_0, \nu)$ with a shift:

$$\int_{-\infty}^{\infty} d\nu \exp(-2\pi i\nu t)\, U_{obj}(\mathbf{r}_0, \nu) \exp\left(2\pi i\nu \frac{w(\mathbf{r}, \mathbf{r}_0)}{c}\right) = u_{obj}\left(\mathbf{r}_0, t - \frac{w(\mathbf{r}, \mathbf{r}_0)}{c}\right), \tag{9.274}$$

so

$$u_{im}^{(s)}(\mathbf{r}, t) = \frac{m}{\overline{\lambda}^2 qp} \int_{\infty} d^2 r_0 \, u_{obj}\left(\mathbf{r}_0, t - \frac{w(\mathbf{r}, \mathbf{r}_0)}{c}\right) \left| T_{pupil}\left(\frac{m}{\overline{\lambda} q}(\mathbf{r} - \mathbf{r}_0); \mathbf{r}_0\right)\right| . \tag{9.275}$$

As in Sec. 9.7.4, we can write the general object field $u_{obj}(\mathbf{r}_0, t)$ as the product $\tilde{u}_{obj}(\mathbf{r}, t) \exp(2\pi i\overline{\nu} t)$, where the envelope $\tilde{u}(\mathbf{r}, t)$ is slowly varying for narrowband light. With a similar representation for the scaled image field, (9.275) becomes

$$\tilde{u}_{im}^{(s)}(\mathbf{r}, t) = \frac{m}{\overline{\lambda}^2 qp} \int_{\infty} d^2 r_0 \, \tilde{u}_{obj}\left(\mathbf{r}_0, t - \frac{w(\mathbf{r}, \mathbf{r}_0)}{c}\right)$$
$$\times \exp\left[i\overline{k}\, w(\mathbf{r}, \mathbf{r}_0)\right] \left| T_{pupil}\left(\frac{m}{\overline{\lambda} q}(\mathbf{r} - \mathbf{r}_0); \mathbf{r}_0\right)\right| , \tag{9.276}$$

where $\overline{k} = 2\pi\overline{\nu}/c$, and a common factor of $\exp(2\pi i\overline{\nu} t)$ on both sides has been cancelled. From (9.272), we recognize the product of the last two factors in the integrand as the coherent PRF at the mean frequency, yielding finally,

$$\tilde{u}_{im}^{(s)}(\mathbf{r}, t) = \int_{\infty} d^2 r_0 \, \tilde{u}_{obj}\left(\mathbf{r}_0, t - \frac{w(\mathbf{r}, \mathbf{r}_0)}{c}\right) p_{coh}(\mathbf{r} - \mathbf{r}_0; \mathbf{r}_0; \overline{\nu}) . \tag{9.277}$$

Optical path retrod Equations (9.276) and (9.277) are reminiscent of (9.66), where the field at an observation point at time t is related to the source at other points at the retarded time $t - \tau_p$, where τ_p is a propagation time. In (9.66), however, τ_p is the straight-line distance R from source point to observation point divided by the speed of light in the medium, c_m. Since $c_m = c/n$, where n is the refractive index, τ_p is also the optical path nR divided by c.

The simple identification of an optical path in (9.66) came about because the system to which that equation applies was just a homogeneous medium. The discussion leading up to (9.276) and (9.277), however, allowed a rather complicated optical system, with inhomogeneous index distributions (*i.e.*, lenses) interposed between regions of free-space propagation. Nevertheless, we are able to identify an optical path from $\mathbf{r}_0$ to $\mathbf{r}$ without appealing to Fermat's principle or any other construct from geometrical optics. Any time we can write the monochromatic coherent PRF in the form $\exp[2\pi i\nu w(\mathbf{r},\mathbf{r}_0)/c]$ times a slowly varying function of ν, we obtain a retarded-time expression with the phase delay τ_p given by the coefficient of $2\pi i\nu$ in the phase, and we can define $c\tau_p$ as the optical path. In this way we can define optical path in a physical-optics framework without having to say that the light *follows* that path (which, of course, waves are reluctant to do).

How nearly monochromatic is quasi? From (9.277), we see that light emitted from a particular source point $\mathbf{r}_0$ arrives at the image point $\mathbf{r}$ after a time delay $\tau_p(\mathbf{r},\mathbf{r}_0) = w(\mathbf{r},\mathbf{r}_0)/c$. In the fully monochromatic case, this delay did not appear (mainly because we consistently dropped phase factors). The remaining question is: When are we justified in ignoring this delay, or equivalently, when can we drop the phase factors that give rise to the delay?

To answer this question, recall that we always observe the mean image irradiance, given by

$$\overline{I}_{im}^{(s)}(\mathbf{r}) = \left\langle \left[\tilde{u}_{im}^{(s)}(\mathbf{r})\right] \left[\tilde{u}_{im}^{(s)}(\mathbf{r})\right]^* \right\rangle . \tag{9.278}$$

Inserting (9.277) for each appearance of $\tilde{u}_{im}^{(s)}$ (with a different dummy variable for each integral), we obtain

$$\overline{I}_{im}^{(s)}(\mathbf{r}) = \left\langle \int_\infty d^2r'\, \tilde{u}_{obj}[\mathbf{r}',t-\tau_p(\mathbf{r},\mathbf{r}')]\, p_{coh}(\mathbf{r}-\mathbf{r}';\mathbf{r}';\overline{\nu}) \right.$$
$$\left. \cdot \int_\infty d^2r''\, \tilde{u}_{obj}^*[\mathbf{r}'',t-\tau_p(\mathbf{r},\mathbf{r}'')]\, p_{coh}^*(\mathbf{r}-\mathbf{r}'';\mathbf{r}'';\overline{\nu}) \right\rangle . \tag{9.279}$$

Under broad conditions discussed in Sec. 8.2.2, it is valid to interchange the order of integration and statistical averaging, which yields

$$\overline{I}_{im}^{(s)}(\mathbf{r}) = \int_\infty d^2r' \int_\infty d^2r''\, \left\langle \tilde{u}_{obj}[\mathbf{r}',t-\tau_p(\mathbf{r},\mathbf{r}')]\, \tilde{u}_{obj}^*[\mathbf{r}'',t-\tau_p(\mathbf{r},\mathbf{r}'')]\right\rangle$$
$$\times\, p_{coh}(\mathbf{r}-\mathbf{r}';\mathbf{r}';\overline{\nu})\, p_{coh}^*(\mathbf{r}-\mathbf{r}'';\mathbf{r}'';\overline{\nu}) . \tag{9.280}$$

The average is now recognized as the spatio-temporal autocorrelation function of the object field as defined in (9.257). Under the assumption of temporal stationarity, this autocorrelation function reduces to the mutual coherence function defined in (9.258), and we have

$$\left\langle \tilde{u}_{obj}[\mathbf{r}',t-\tau_p(\mathbf{r},\mathbf{r}')]\, \tilde{u}_{obj}^*[\mathbf{r}'',t-\tau_p(\mathbf{r},\mathbf{r}'')]\right\rangle = \Gamma_{obj}[\mathbf{r}',\mathbf{r}'',\Delta\tau(\mathbf{r},\mathbf{r}',\mathbf{r}'')] , \tag{9.281}$$

where $\Delta\tau(\mathbf{r},\mathbf{r}',\mathbf{r}'')$ is the difference in propagation delays to point $\mathbf{r}$ from points $\mathbf{r}'$ and $\mathbf{r}''$, that is,

$$\Delta\tau(\mathbf{r},\mathbf{r}',\mathbf{r}'') = \tau_p(\mathbf{r},\mathbf{r}') - \tau_p(\mathbf{r},\mathbf{r}'') . \tag{9.282}$$

In principle, this difference in delays could be arbitrarily large since $\mathbf{r}$ and $\mathbf{r}'$ range independently over the infinite plane. Remember, however, that we are considering an imaging system, where a design goal is to make the PRF spatially compact. If we achieve this goal to any reasonable degree, then $p_{coh}(\mathbf{r}-\mathbf{r}';\mathbf{r}';\overline{\nu})$ is zero if $\mathbf{r}$ is more than a resolution length from $\mathbf{r}'$, and similarly $p_{coh}(\mathbf{r}-\mathbf{r}'';\mathbf{r}'';\overline{\nu})$ is zero unless $\mathbf{r}$ is near $\mathbf{r}''$ in this sense. One should not jump to the conclusion that the propagation delays themselves are small, since the point referred to as $\mathbf{r}$ is actually quite far from the one referred to as $\mathbf{r}'$, even if the 2D vectors $\mathbf{r}$ and $\mathbf{r}'$ are exactly equal; these vectors refer to physical points in different planes. Nevertheless, the condition that $\mathbf{r}'$ and $\mathbf{r}''$ are both close to $\mathbf{r}$ (in a 2D sense) *does* mean that the two physical points are close together three-dimensionally since $\mathbf{r}'$ and $\mathbf{r}''$ are in the *same* plane, namely, the object plane.

Thus the very fact that we are dealing with an imaging system will tend to make $\Delta\tau(\mathbf{r},\mathbf{r}',\mathbf{r}'')$ small. The essence of the quasimonochromatic approximation is that this difference in delays is so small that we can replace $\Gamma_{obj}[\mathbf{r}',\mathbf{r}'',\Delta\tau(\mathbf{r},\mathbf{r}',\mathbf{r}'')]$ with $\Gamma_{obj}(\mathbf{r}',\mathbf{r}'',0)$, which is the spatial coherence function $\Gamma_{obj}(\mathbf{r}',\mathbf{r}'')$. We can do this with negligible error if $\Delta\tau(\mathbf{r},\mathbf{r}',\mathbf{r}'') \ll \tau_c$, where τ_c is the coherence time, for all points $\mathbf{r}'$ and $\mathbf{r}''$ that contribute to the image at point $\mathbf{r}$. We know from Sec. 9.7.4 that τ_c is approximately $1/\Delta\nu$, where $\Delta\nu$ is the spectral bandwidth of the light, so a sufficiently narrowband source will always allow us to make this approximation.

When the spectral bandwidth is small enough, we can write the image irradiance as

$$\overline{I}_{im}^{(s)}(\mathbf{r}) = \int_\infty d^2r' \int_\infty d^2r'' \, \Gamma_{obj}(\mathbf{r}';\mathbf{r}'')\, p^*_{coh}(\mathbf{r}-\mathbf{r}';\mathbf{r}';\overline{\nu})\, p_{coh}(\mathbf{r}-\mathbf{r}'';\mathbf{r}'';\overline{\nu})$$

$$= \int_\infty d^2r' \int_\infty d^2r'' \, \gamma_{obj}(\mathbf{r}';\mathbf{r}'')\sqrt{\overline{I}_{obj}(\mathbf{r}')\,\overline{I}_{obj}(\mathbf{r}'')}\, p^*_{coh}(\mathbf{r}-\mathbf{r}';\mathbf{r}';\overline{\nu})\, p_{coh}(\mathbf{r}-\mathbf{r}'';\mathbf{r}'';\overline{\nu})\,, \tag{9.283}$$

where $\overline{I}_{obj}(\mathbf{r}') \equiv \langle |\tilde{u}_{obj}(\mathbf{r}')|^2 \rangle$.

It would be tempting to refer to $\overline{I}_{obj}(\mathbf{r}')$ as a mean irradiance, since it is defined analogously to $\overline{I}_{im}(\mathbf{r}')$. Both quantities are proportional to an optical power per unit area, but $\overline{I}_{obj}(\mathbf{r}')$ represents power *leaving* the object surface while $\overline{I}_{im}(\mathbf{r}')$ represents power *arriving* at the detector; for this reason $\overline{I}_{obj}(\mathbf{r}')$ is called the mean *radiant exitance*. For more discussion of these quantities, see Chap. 10.

Implications of the general expression (9.283) will be explored further in Sec. 9.7.7, but now we turn to the special case of spatial incoherence.

9.7.6 Spatially incoherent, quasimonochromatic imaging

In Sec. 9.7.4 we defined the complex degree of coherence $\gamma(\mathbf{r}',\mathbf{r}'')$ as a normalized measure of spatial coherence for quasimonochromatic light. In this section we assume that $\gamma(\mathbf{r}',\mathbf{r}'')$ is sharply peaked compared to the width of the coherent PRF, so that we can use the incoherent-object approximation of (9.269), whereby $\gamma(\mathbf{r}',\mathbf{r}'') = A_c\,\delta(\mathbf{r}'-\mathbf{r}'')$. Inserting this delta function into (9.283) and performing an elementary integral yields

$$\overline{I}_{im}^{(s)}(\mathbf{r}) = A_c \int_\infty d^2r'\, \overline{I}_{obj}(\mathbf{r}')\, |p_{coh}(\mathbf{r}-\mathbf{r}';\mathbf{r}';\overline{\nu})|^2\,. \tag{9.284}$$

If the system is shift-invariant, $p_{coh}(\mathbf{r}-\mathbf{r}';\mathbf{r}';\overline{\nu})$ is a function of $\mathbf{r}-\mathbf{r}'$ only, and we can drop the second spatial argument. Also dropping the frequency argument for simplicity, we write the coherent PSF as just $p_{coh}(\mathbf{r}-\mathbf{r}')$, as we did in the fully monochromatic, shift-invariant case. With this notation, (9.284) reduces to

$$\overline{I}_{im}^{(s)}(\mathbf{r}) = A_c \int_\infty d^2r'\, \overline{I}_{obj}(\mathbf{r}')\, |p_{coh}(\mathbf{r}-\mathbf{r}')|^2\,. \tag{9.285}$$

This is a key result; the mean irradiance in the image is proportional to the mean exitance of the incoherent object convolved with the squared modulus of the coherent PSF. We can thus define an incoherent PSF by

$$p_{incoh}(\mathbf{r}) = A_c|p_{coh}(\mathbf{r})|^2 = \frac{A_c}{(\lambda p)^4}\left|T_{pupil}\left(-\frac{\mathbf{r}}{\lambda p}\right)\right|^2 , \tag{9.286}$$

where we have used (9.223). With this PSF, we can rewrite (9.285) as

$$\overline{I}_{im}^{(s)}(\mathbf{r}) = \overline{I}_{obj}(\mathbf{r}) * p_{incoh}(\mathbf{r}) . \tag{9.287}$$

If aberrations are present, this result, like its coherent counterpart, is valid only if the object is confined to an isoplanatic patch, but (9.284) provides the analogous expression for general shift-variant incoherent imaging. In every case, the incoherent PRF is proportional to the squared modulus of the coherent one if the object is completely spatially incoherent in the sense defined by (9.269).

The important conclusion from this discussion is that linear superposition is still valid in incoherent imaging if we take the input as the radiant exitance of the object and the output as the image-plane irradiance.

Optical transfer function A linear, shift-invariant system can be specified by its transfer function as well as by its PSF. In the case of shift-invariant incoherent imaging, the transfer function is the Fourier transform of the incoherent PSF, given by

$$P_{incoh}(\boldsymbol{\rho}) = \mathcal{F}_2\{p_{incoh}(\mathbf{r})\} = A_c\mathcal{F}_2\{|p_{coh}(\mathbf{r})|^2\} . \tag{9.288}$$

From (9.288), (9.222) and (3.245), we have

$$P_{incoh}(\boldsymbol{\rho}) = \frac{A_c}{(\lambda p)^2}[t_{pupil} \star t^*_{pupil}](\lambda p\boldsymbol{\rho}) , \tag{9.289}$$

where $\star$ denotes correlation. Thus *the incoherent transfer function is a scaled version of the complex autocorrelation of the pupil function.*

Since the support of the autocorrelation of a function is twice as large (in linear dimension) as the support of the function itself, the spatial-frequency cutoff of the incoherent transfer function is twice that of the coherent transfer function.

It is often convenient to normalize the transfer function $P_{incoh}(\boldsymbol{\rho})$ to its value at $\boldsymbol{\rho} = 0$. We define the *optical transfer function* or *OTF* by

$$\text{OTF}(\boldsymbol{\rho}) = \frac{P_{incoh}(\boldsymbol{\rho})}{P_{incoh}(0)} = \frac{[t_{pupil} \star t^*_{pupil}](\lambda p\boldsymbol{\rho})}{[t_{pupil} \star t^*_{pupil}](0)} . \tag{9.290}$$

The modulation transfer function or MTF is the modulus of the OTF. (See Sec. 7.2.6 for a discussion of the significance of the MTF.) For incoherent imaging systems, the MTF is related to the pupil function by

$$\text{MTF}(\boldsymbol{\rho}) = |\text{OTF}(\boldsymbol{\rho})| = \frac{|P_{incoh}(\boldsymbol{\rho})|}{P_{incoh}(0)} . \tag{9.291}$$

Note that it is not necessary to write $|P_{incoh}(0)|$ in the denominator since $P_{incoh}(0)$ is necessarily real. Fig. 9.20 illustrates the PSF and OTF for square and circular apertures.

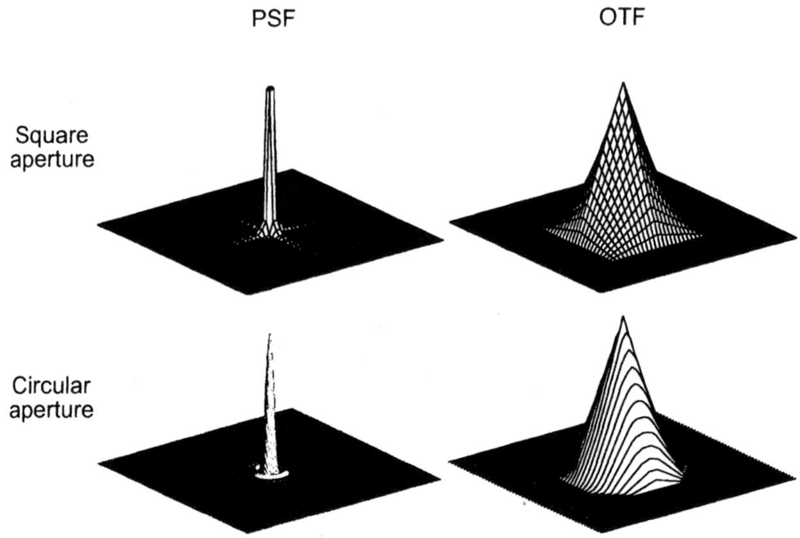

Fig. 9.20 PSF and OTF for incoherent imaging with square and circular pupils.

Incoherent illumination in a 4f system To gain more insight into the statement that the incoherent transfer function is the autocorrelation of the pupil function, we shall revisit the $4f$ system introduced in Sec. 9.7.2. In that section, we assumed that the object transparency was illuminated by a monochromatic plane wave, but we did not say where the plane wave came from. We can create a good approximation to a plane wave by placing a point source in the back focal plane of a condenser lens, as shown in Fig. 9.21a. This setup fits the analysis of Sec. 9.7.2. In the language of coherence theory, the mutual coherence function of the field incident on the object transparency is, to a good approximation, a constant. The system then performs a linear mapping from the object field to the image field, which is what we mean by coherent imaging.

To convert the system into an incoherent one, we can replace the point source by an extended incoherent source as shown in Fig. 9.21b. As seen from Fig. 9.21c, each point on the source produces a plane wave, so the illumination incident on the object transparency consists of many plane waves with random phases. Because different source points radiate independently, the mean irradiance in the image arising from many source points is the sum of the irradiances from the points individually; interference terms average to zero. To compute the total image irradiance, we simply compute it for a single source point and then integrate over the source.

The Fresnel-diffraction analysis presented in Sec. 9.7.2 shows that the field incident on the object is the Fourier transform of the instantaneous field in the object plane, at least to the extent that diffraction from the lens aperture can be neglected. It follows from (9.237) that a point source at $\mathbf{r}_s$ produces a plane wave with amplitude proportional to $\exp(-2\pi i \mathbf{r}\cdot\mathbf{r}_s/\lambda f)$ at point $\mathbf{r}$ in the object plane. This plane wave is multiplied by the object transmittance $t_{obj}(\mathbf{r})$ to produce the field $u_{0+}(\mathbf{r})$ presented to the $4f$ system. The field in the Fourier plane ($z = 2f$) is then proportional to

$$\mathcal{F}_2\left\{\exp\left(-2\pi i\frac{\mathbf{r}\cdot\mathbf{r}_s}{\lambda f}\right)t_{obj}(\mathbf{r})\right\} = T_{obj}\left(\boldsymbol{\rho}+\frac{\mathbf{r}_s}{\lambda f}\right). \tag{9.292}$$

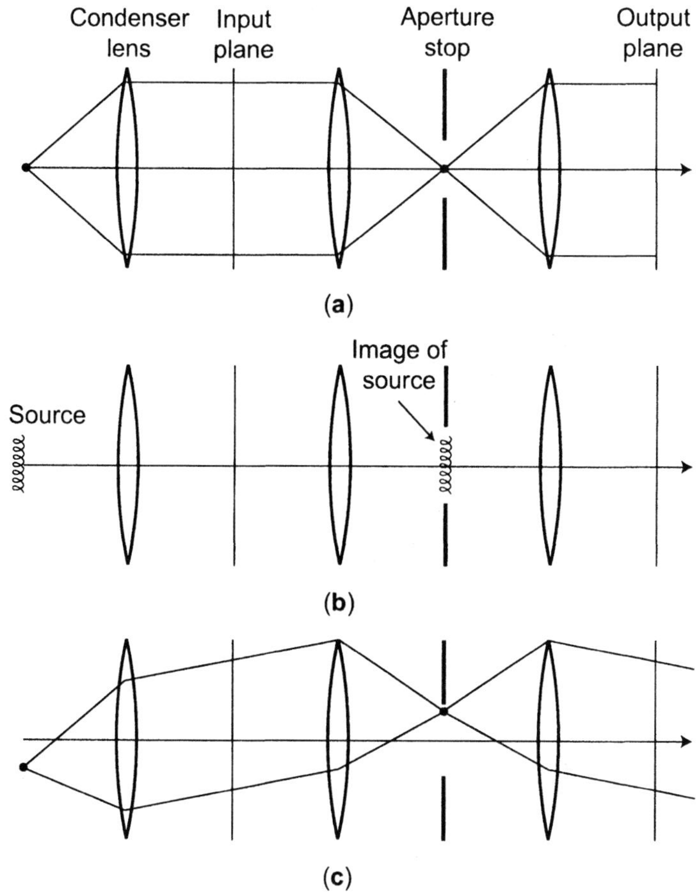

Fig. 9.21 Extension of the diagram of Fig. 9.17, showing the light source explicitly. (a) Illumination with a plane wave created by an on-axis point source and a condenser lens. (b) Illumination with an extended incoherent source. (c) Illumination with a tilted plane wave created by an off-axis point source, which might be one point on the extended source.

Thus the field in the Fourier plane is the Fourier transform of the object shifted laterally by an amount determined by the location of the source point. Mathematically, this result comes from the shift theorem of Fourier analysis, (3.237). Physically, it means that the object Fourier transform is centered on the image of the source point.

Another way to state this conclusion is that the coherent transfer function for a single source point at $\mathbf{r}_s$ is given by

$$P_{coh}(\boldsymbol{\rho}; \mathbf{r}_s) = t_{pupil}(\lambda f \boldsymbol{\rho} + \mathbf{r}_s) . \tag{9.293}$$

A formal way of proceeding from here would be to use this transfer function to filter the object field distribution, take the squared modulus to get the contribution from source point $\mathbf{r}_s$ to the image irradiance and then integrate this result over $\mathbf{r}_s$. As the reader may demonstrate, this procedure would reproduce our previous analysis of incoherent imaging. Perhaps more insight will be obtained, however, if we consider the effect of this procedure on a particular object.

In Sec. 9.7.2 the test object was one with an amplitude transmittance given by $\exp(2\pi i\xi_0 x)$, but this choice will not work in a discussion of incoherent imaging since $|\exp(2\pi i\xi_0 x)|^2 = 1$, hence no spatial modulation can be seen with incoherent illumination. Instead we choose

$$t_{obj}(\mathbf{r}) = 1 + \exp(2\pi i\xi_0 x) = 1 + \exp(2\pi i\boldsymbol{\rho}_0 \cdot \mathbf{r}) , \tag{9.294}$$

where $\boldsymbol{\rho}_0 = (\xi_0, 0)$ for purposes of illustration. An ideal incoherent image of this object would have an image-plane irradiance proportional to

$$|t_{obj}(\mathbf{r})| = 2 + \exp(4\pi i\xi_0 x) + \exp(-4\pi i\xi_0 x) = 2 + 2\cos(2\boldsymbol{\rho}_0 \cdot \mathbf{r}) , \tag{9.295}$$

which is a fringe pattern with 100% modulation and a spatial frequency vector twice that of the amplitude pattern.

The shifted transform in (9.292) is given by

$$T_{obj}\left(\boldsymbol{\rho} + \frac{\mathbf{r}_s}{\lambda f}\right) = \delta\left(\boldsymbol{\rho} + \frac{\mathbf{r}_s}{\lambda f}\right) + \delta\left(\boldsymbol{\rho} - \boldsymbol{\rho}_0 + \frac{\mathbf{r}_s}{\lambda f}\right) . \tag{9.296}$$

Hence, for a point source at $\mathbf{r}_s$, the field distribution in the Fourier plane (immediately before the aperture stop) consists of two spots, one at $\mathbf{r} = -\mathbf{r}_s$ and one at $\mathbf{r} = \boldsymbol{\rho}_0 \lambda f - \mathbf{r}_s$. If only one of these spots passes the stop, it produces an expanding spherical wave which is converted to a plane wave by the final lens. This wave produces a uniform irradiance in the image plane, with no modulation and hence no information about the object.

If both of the spots pass unimpeded through the aperture, however, two plane waves are produced, and the resulting interference pattern has an image irradiance in the ideal form (9.295). Note that this pattern is independent of the source location, so long as both spots get through. The shift $\mathbf{r}_s/\lambda f$ in (9.296) produces a rigid lateral translation of the field pattern in the stop, hence a linear phase factor in its Fourier transform (the image field). This phase factor does not affect the image-plane irradiance.

Now consider an extended incoherent source that is larger than the aperture stop. Each source point produces two spots, and if both get through the aperture they still interfere and produce a fringe pattern; these spots are coherent with each other since they derive from the same source point, even though spots associated with two different source points do not interfere. The overall image irradiance thus contains two contributions, one from source points where both spots get through and one from source points where just one of them does. If we assume that the radiant exitance of the source is spatially uniform, the mean image irradiance has the form

$$\overline{I}_{im}(\mathbf{r}) \propto A_1(\boldsymbol{\rho}_0) + A_2(\boldsymbol{\rho}_0)\left[2 + 2\cos(4\pi\boldsymbol{\rho}_0 \cdot \mathbf{r})\right] , \tag{9.297}$$

where $A_1(\boldsymbol{\rho}_0)$ is the source area such that one spot gets through and $A_2(\boldsymbol{\rho}_0)$ is the source area where both do.

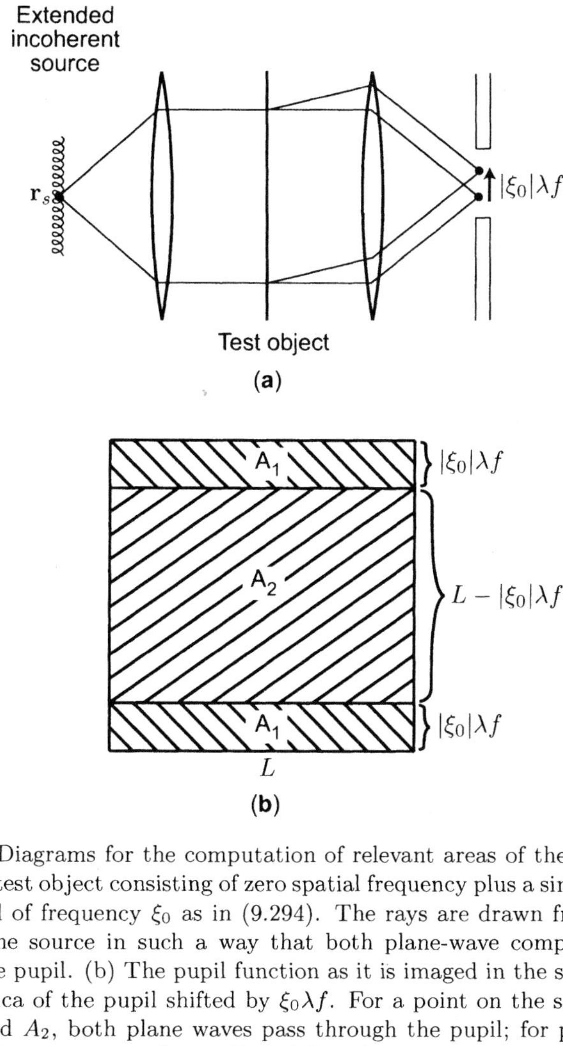

Fig. 9.22 Diagrams for the computation of relevant areas of the source. (a) Image of a test object consisting of zero spatial frequency plus a single complex exponential of frequency ξ_0 as in (9.294). The rays are drawn from a single point on the source in such a way that both plane-wave components pass through the pupil. (b) The pupil function as it is imaged in the source plane, plus a replica of the pupil shifted by $\xi_0 \lambda f$. For a point on the source in the area marked A_2, both plane waves pass through the pupil; for points in A_1 only a single wave passes the pupil, and for other source points no light at all passes.

If we consider a square aperture of side L, these areas are easy to compute, especially for our example where $\boldsymbol{\rho}_0$ is directed along the x axis (parallel to a side of the aperture). From Fig. 9.22, we see that

$$A_1(\boldsymbol{\rho}_0) = 2L|\xi_0|\lambda f\,, \qquad A_2(\boldsymbol{\rho}_0) = L(L - |\xi_0|\lambda f)\,, \qquad (|\xi_0|\lambda f < L)\,. \tag{9.298}$$

The image irradiance now has the form

$$\overline{I}_{im}(\mathbf{r}) \propto 2L^2 + 2L(L - |\xi_0|\lambda f)\cos(4\pi\xi_0 x)\,. \tag{9.299}$$

From (7.157) we see that this pattern has a modulation given by

$$M_{im} = \frac{L - |\xi_0|\lambda f}{L}\,, \qquad (|\xi_0|\lambda f < L)\,. \tag{9.300}$$

Since the object has 100% modulation, M_{im} is also the system MTF (see Sec. 7.2.6). The MTF thus decreases linearly from unity at $\xi_0 = 0$ to zero at $\xi_0 = \pm L/\lambda f$. Precisely this form will be found by autocorrelating the pupil.

As an exercise, the reader may want to repeat this analysis for $t_{obj}(\mathbf{r}) = \cos(2\pi\xi_0 x)$.

9.7.7 Polychromatic, incoherent imaging

So far we have discussed only monochromatic or quasimonochromatic sources, but the results can be extended to broadband (polychromatic) sources. The key physical intuition is that a wave of temporal frequency ν does not interfere with one of frequency ν' if $\nu \neq \nu'$. There are several ways to justify this point. One way is to consider two plane waves of different frequency superimposed on the surface of some optical detector. The total irradiance is then the sum of the individual irradiances of the plane waves plus an interference term that oscillates at frequency $\nu - \nu'$; if this difference frequency is large compared to the reciprocal of the response time of the detector, its time average is zero and the detector output can be computed without consideration of the interference. This viewpoint is developed more fully in Sec. 10.1.5, where we address the question: What do real detectors detect?

A more formal statistical approach is based on properties of stationary random processes. We showed in Sec. 8.2.5 that the Fourier transform of a stationary random process is a delta-correlated process. That is, if $u(t)$ is a sample function of a temporally stationary random process, and we define its Fourier transform (a generalized function) as

$$U(\nu) = \int_{-\infty}^{\infty} dt\ u(t) \exp(-2\pi i\nu t)\,, \tag{9.301}$$

then, by the temporal counterpart of (8.181),

$$\langle U(\nu)\, U^*(\nu')\rangle = S_u(\nu)\,\delta(\nu - \nu')\,. \tag{9.302}$$

In this expression, $S_u(\nu)$ is the *power spectral density*, given by the Wiener-Khinchin theorem (8.133) as

$$S_u(\nu) = \int_{-\infty}^{\infty} d\tau\ \langle u(t+\tau)\, u^*(t)\rangle \exp(-2\pi i\nu\tau)\,. \tag{9.303}$$

To apply (9.302) to imaging, we must include the spatial variables, replacing the temporal random process $u(t)$ by the spatio-temporal one, $\tilde{u}(\mathbf{r}, t)$. The temporal stationarity leads to an expression analogous to (9.302):

$$\left\langle \tilde{U}(\mathbf{r}, \nu)\, \tilde{U}^*(\mathbf{r}', \nu')\right\rangle = W(\mathbf{r}, \mathbf{r}', \nu)\,\delta(\nu - \nu')\,. \tag{9.304}$$

Here, $\tilde{U}(\mathbf{r}, \nu)$ is the temporal Fourier transform of $\tilde{u}(\mathbf{r}, t)$, and $W(\mathbf{r}, \mathbf{r}', \nu)$ is called the *cross-spectral density*; by a generalization of the Wiener-Khinchin theorem, it is given by

$$\begin{aligned} W(\mathbf{r}, \mathbf{r}', \nu) &= \int_{-\infty}^{\infty} d\tau\ \langle \tilde{u}(\mathbf{r}, t+\tau)\, \tilde{u}^*(\mathbf{r}', t)\rangle \exp(-2\pi i\nu\tau) \\ &= \int_{-\infty}^{\infty} d\tau\ \Gamma(\mathbf{r}, \mathbf{r}', \tau) \exp(-2\pi i\nu\tau)\,. \end{aligned} \tag{9.305}$$

We can incorporate the cross-spectral density into our imaging theory by rewriting (9.280) as

$$\overline{I}_{im}^{(s)}(\mathbf{r}) = \int_\infty d^2r' \int_\infty d^2r'' \int_{-\infty}^{\infty} d\nu \int_{-\infty}^{\infty} d\nu' \left\langle \tilde{U}(\mathbf{r}',\nu)\, \tilde{U}^*(\mathbf{r}'',\nu') \right\rangle \exp[2\pi i(\nu-\nu')(t-\tau_p)]$$

$$\times\, p_{coh}^*(\mathbf{r}-\mathbf{r}';\nu)\, p_{coh}(\mathbf{r}-\mathbf{r}'';\nu') \,. \qquad (9.306)$$

As in Sec. 9.7.4, we have added an argument ν to $p_{coh}(\mathbf{r},\nu)$ since the coherent PSF depends on the wavelength. We have also specialized (9.280) to the spatially stationary case for simplicity.

If we now insert (9.304) (adding a subscript obj to denote the object plane) and use the delta function to perform the ν' integral, we obtain

$$\overline{I}_{im}^{(s)}(\mathbf{r}) = \int_\infty d^2r' \int_\infty d^2r'' \int_{-\infty}^{\infty} d\nu\; W_{obj}(\mathbf{r}',\mathbf{r}'',\nu)\, [p_{coh}(\mathbf{r}-\mathbf{r}';\nu)]^*\, p_{coh}(\mathbf{r}-\mathbf{r}'';\nu) \,. \qquad (9.307)$$

This expression is identical to (9.283) except that the cross-spectral density takes the place of the spatial coherence function, and the result is integrated over frequency.

The spatially incoherent limit is obtained by assuming that

$$W_{obj}(\mathbf{r}',\mathbf{r}'',\nu) = A_c \overline{I}_{obj}(\mathbf{r}',\nu)\, \delta(\mathbf{r}'-\mathbf{r}'') \,, \qquad (9.308)$$

where $\overline{I}_{obj}(\mathbf{r}',\nu)$ is called the *spectral radiant exitance* (see Chap. 10). We then have

$$\overline{I}_{im}^{(s)}(\mathbf{r}) = A_c \int_{-\infty}^{\infty} d\nu \int_\infty d^2r'\, \overline{I}_{obj}(\mathbf{r}',\nu)\, |p_{coh}(\mathbf{r}-\mathbf{r}';\nu)|^2 \,. \qquad (9.309)$$

Comparison with (9.285) shows that the polychromatic image is obtained merely by integrating the quasimonochromatic one over frequency. This mathematical result codifies the statement made at the beginning of this section that waves of different frequency do not interfere.

9.7.8 Partially coherent imaging

We have studied 2D imaging in the coherent and incoherent limits. In the completely coherent case, the field emerging from the object is essentially nonrandom, so we just calculated the field in the image plane and showed (basically from the linearity of Maxwell's equations) that the system is a linear mapping from the complex object field to the complex image field. As we have noted (and shall show in more detail in Chap. 10), an image detector responds to the time-averaged squared modulus of the complex field, so the overall imaging system, from object field to detector output, is a nonlinear mapping, but linear systems theory is still sufficient for analyzing a coherent imaging system. The detector is a simple point nonlinearity (see Sec. 7.5.1) applied to the output of a linear system.

With real radiation sources, we argued that the object field is a spatio-temporal random process, and some assumptions were needed to make progress on the analysis. In Sec. 9.7.4 we assumed that the field was a stationary random process temporally and that it was quasimonochromatic. With these assumptions, we argued that it was sufficient to consider the equal-time correlation function $\langle \tilde{u}(\mathbf{r},t)\, \tilde{u}^*(\mathbf{r}',t)\rangle$, which is the spatial coherence function $\Gamma(\mathbf{r}, \mathbf{r}')$. The fully incoherent limit corresponds to assuming further that $\Gamma(\mathbf{r}, \mathbf{r}') \propto \delta(\mathbf{r} - \mathbf{r}')$. With that assumption we were able to show that the system is a linear mapping from the object radiant exitance to the image irradiance. Since detectors often respond linearly to irradiance, we then have a linear description of the overall system, now including the detector.

In this section we look briefly at what happens when we consider a temporally stationary, quasimonochromatic source but do not assume that the correlation function is a spatial delta function. There are two main motivations for studying this case. The first is that actual radiation sources are never delta-correlated. As noted in Sec. 9.7.4, Lambertian sources are correlated over about a wavelength, and many high-resolution imaging systems are sensitive to details on this scale.

The second motivation for avoiding the delta-correlated model is that we often want to cascade linear systems. In imaging, the output of one imaging stage is often the input to a second one. Even if we can regard the input to the first stage as delta-correlated, the output of that stage is correlated over a distance comparable to the width of its PSF, so the assumption fails for analyzing the second stage unless it blurs the image much more than the first one does.

Propagation of spatial coherence Consider a quasimonochromatic but not necessarily spatially incoherent source imaged by an arbitrary system with coherent PRF $p_{coh}(\mathbf{r}, \mathbf{r}_0)$. By a straightforward generalization of the algebra that led to (9.283), we find that the spatial coherence function of the field in the image plane is

$$\Gamma_{im}(\mathbf{r}, \mathbf{r}') = \int_\infty d^2r'' \int_\infty d^2r''' \, \Gamma_{obj}(\mathbf{r}'', \mathbf{r}''') \, p_{coh}(\mathbf{r}, \mathbf{r}'') \, p^*_{coh}(\mathbf{r}', \mathbf{r}''') \, . \tag{9.310}$$

Note that we are not assuming here that the system is shift-invariant, and we are not using scaled variables.

We see from (9.310) that the spatial coherence function is transferred linearly through the system, no matter the functional form of the input. The linear mapping takes place in a 4D space, where one function of two 2D vectors is mapped into another such function. We can then use the output of this mapping, $\Gamma_{im}(\mathbf{r}, \mathbf{r}')$, as the input to a second stage described similarly. If $p_{coh}(\mathbf{r}, \mathbf{r}'')$ is shift invariant in a 2D sense, *i.e.*, $p_{coh}(\mathbf{r}, \mathbf{r}'') = p_{coh}(\mathbf{r} - \mathbf{r}'')$, then the mapping (9.310) is shift invariant in a 4D sense, and cascading of shift-invariant systems involves successive 4D convolutions.

At the end of this process, however, we do not observe $\Gamma_{im}(\mathbf{r}, \mathbf{r}')$; instead, the final detector responds to $\overline{I}_{im}(\mathbf{r})$, which by (9.266) is $\Gamma_{im}(\mathbf{r}, \mathbf{r})$. Moreover, we are rarely interested in the mapping of the spatial coherence function. It is more natural to specify the object in terms of its radiant exitance and the image in terms of irradiance, and this mapping is not linear except in the incoherent limit.

If we know *a priori* the complex degree of coherence in the object, we can cast the mapping from radiant exitance to irradiance into the form of a bilinear transform, as discussed in Sec. 7.5.2. For example, suppose we know that the object is Lambertian so that $\gamma(\mathbf{r}, \mathbf{r}') = \sin(k|\mathbf{r} - \mathbf{r}'|)/(k|\mathbf{r} - \mathbf{r}'|)$ (see Sec. 10.2.7). Then we can write

$$\overline{I}_{im}(\mathbf{r}) = \int_\infty d^2r' \int_\infty d^2r'' \sqrt{\overline{I}_{obj}(\mathbf{r}')}\sqrt{\overline{I}_{obj}(\mathbf{r}'')} \, \gamma_{obj}(\mathbf{r}', \mathbf{r}'') \, [p_{coh}(\mathbf{r}, \mathbf{r}')]^* \, p_{coh}(\mathbf{r}', \mathbf{r}'') \, . \tag{9.311}$$

If we denote $\sqrt{\overline{I}_{obj}(\mathbf{r}')}$ as $f(\mathbf{r}')$ and $\overline{I}_{im}(\mathbf{r})$ as $g(\mathbf{r})$, we see that the form of (9.311) agrees with the general bilinear transform (7.366):

$$g(\mathbf{r}) = \int_\infty d^2r' \int_\infty d^2r''\, f(\mathbf{r}')\, f(\mathbf{r}'')\, h(\mathbf{r}; \mathbf{r}', \mathbf{r}'') . \tag{9.312}$$

Note that $\gamma_{obj}(\mathbf{r}', \mathbf{r}'')$ is now lumped into the description of the system rather than the object. As with the examples discussed in Sec. 7.5.3, it is not altogether obvious what properties of the thing being imaged should be considered object and what properties are part of the system. The key determinant is what we want to map to the output. In the present discussion, we are not interested in mapping $\gamma_{obj}(\mathbf{r}', \mathbf{r}'')$ since we assume it is known *a priori*.

van Cittert-Zernike theorem An important application of (9.310) is when the system described by $p_{coh}(\mathbf{r}, \mathbf{r}')$ isn't really imaging but rather free-space propagation over a distance z. From the discussion of Fresnel diffraction in Sec. 9.4.6, we know that the PSF for this system is the quadratic phase factor given in (9.95). The product of PSFs that appears in (9.310) is then

$$p^*_{coh}(\mathbf{r}, \mathbf{r}'')\, p_{coh}(\mathbf{r}', \mathbf{r}''') = \frac{1}{\lambda^2 z^2} \exp\left(-i\pi \frac{|\mathbf{r} - \mathbf{r}''|^2}{\lambda z}\right) \exp\left(i\pi \frac{|\mathbf{r}' - \mathbf{r}'''|^2}{\lambda z}\right) . \tag{9.313}$$

Now let us assume that the source is incoherent, in the sense discussed in Sec. 9.7.4, so that the source is described by [*cf.* (9.267) and (9.269)]

$$\Gamma(\mathbf{r}'', \mathbf{r}''') \approx A_c\, \delta(\mathbf{r}'' - \mathbf{r}''') \sqrt{\overline{I}_s(\mathbf{r}'')\, \overline{I}_s(\mathbf{r}''')} = A_c\, \delta(\mathbf{r}'' - \mathbf{r}''')\, \overline{I}_s(\mathbf{r}'') , \tag{9.314}$$

where the last step follows from (2.120).

We can now insert (9.313) and (9.314) into (9.310) and use the delta function to perform one integral. The result is the spatial coherence function in a plane at distance z from the source, given by

$$\Gamma_z(\mathbf{r}, \mathbf{r}') = \frac{A_c}{\lambda^2 z^2} \int_\infty d^2r''\, \overline{I}_s(\mathbf{r}'') \exp\left(-i\pi \frac{|\mathbf{r} - \mathbf{r}''|^2}{\lambda z}\right) \exp\left(i\pi \frac{|\mathbf{r}' - \mathbf{r}''|^2}{\lambda z}\right) . \tag{9.315}$$

In the first exponential, $|\mathbf{r} - \mathbf{r}''|^2 = r^2 + r''^2 - 2\mathbf{r} \cdot \mathbf{r}''$ and in the second one, $|\mathbf{r}' - \mathbf{r}''|^2 = r'^2 + r''^2 - 2\mathbf{r}' \cdot \mathbf{r}''$, so

$$\Gamma_z(\mathbf{r}, \mathbf{r}') = \frac{A_c}{\lambda^2 z^2} \exp\left(i\pi \frac{r'^2 - r''^2}{\lambda z}\right) \int_\infty d^2r''\, \overline{I}_s(\mathbf{r}'') \exp\left[-2\pi i \frac{(\mathbf{r} - \mathbf{r}') \cdot \mathbf{r}''}{\lambda z}\right] . \tag{9.316}$$

The integral is recognized as the Fourier transform of the radiant exitance of the source, with the spatial frequency given by the vector distance $\mathbf{r} - \mathbf{r}'$ normalized by λz. If $\overline{I}_s$ is broad and structureless, this transform is peaked near zero frequency, which means that the distance between the points $\mathbf{r}$ and $\mathbf{r}'$ in the image plane must be small for $\Gamma_z(\mathbf{r}, \mathbf{r}')$ to be appreciable. It is then usually a good approximation to replace the remaining quadratic phase factor by unity (though this assumption must be checked in specific cases), yielding, finally,

$$\Gamma_z(\mathbf{r}, \mathbf{r}') = \frac{A_c}{\lambda^2 z^2} \mathcal{F}_2\{\overline{I}_s(\mathbf{r}'')\}|_{\boldsymbol{\rho} = (\mathbf{r} - \mathbf{r}')/\lambda z} . \tag{9.317}$$

In short, the spatial coherence function is the Fourier transform of the source distribution. This result, known as the *van Cittert-Zernike theorem*, depends critically on the assumption that the source is fully incoherent (delta-correlated). It also requires the Fresnel approximation but, in spite of the appearance of a Fourier transform, it does not require the Fraunhofer approximation.

Note that (9.317) shows that the image field is a stationary random process since $\Gamma_z(\mathbf{r},\mathbf{r}')$ is a function of $\mathbf{r}-\mathbf{r}'$ only, but this condition would break down without the paraxial assumptions.

A rule of thumb Consider a uniform, circular, incoherent source of diameter D. The Fourier transform of $\overline{I}(\mathbf{r}'')$ is a besinc function with full width at half maximum approximately $1/D$. That means that $\Gamma_z(\mathbf{r},\mathbf{r}')$ is near unity if $|\mathbf{r}-\mathbf{r}'|$ is less than about $\lambda z/D$, so the correlation length L_c in plane z is about $\lambda z/D$. Another way to state this conclusion is to note that D/z is the angular width $\Delta\theta$ of the source as seen from the observation plane, so $L_c \approx \lambda/\Delta\theta$. We can square both sides and identify L_c^2 with the coherence area $A_c^{(obs)}$ in the observation plane. A rule of thumb worth remembering is then

$$A_c^{(obs)} \approx \frac{\lambda^2}{\Delta\Omega}, \tag{9.318}$$

where, if we neglect factors of order unity, $\Delta\Omega \approx (\Delta\theta)^2$ is the solid angle subtended by the source at the observation plane. The source must be fully incoherent and have a uniform exitance for this rule to hold.

Partial coherence in a 4f system A simple way of generating a partially coherent field is to place a spatially incoherent, quasimonochromatic source with radiant exitance $\overline{I}_s(\mathbf{r}_s)$ in the source plane of a $4f$ system (see Fig. 9.21b). The derivation above of the van Cittert-Zernike theorem is easily modified to permit the propagation of the spatial coherence function through the condenser lens to the object plane; all that is required is to replace z with f and to quit fretting about dropping quadratic phase factors. The field in the object plane is multiplied pointwise by the transmittance $t_{obj}(\mathbf{r})$, and the spatial coherence function of the emerging field is

$$\Gamma_{obj}(\mathbf{r},\mathbf{r}') = \frac{A_c}{\lambda^2 f^2}\, t_{obj}(\mathbf{r})\, t^*_{obj}(\mathbf{r}') \boldsymbol{\mathcal{F}}_2\{\overline{I}_s(\mathbf{r}_s)\}|_{\boldsymbol{\rho}=(\mathbf{r}-\mathbf{r}')/\lambda f}\,. \tag{9.319}$$

We can use (9.310) to propagate to the image plane (primes proliferating in the process), with the result

$$\Gamma_{im}(\mathbf{r},\mathbf{r}') = \frac{A_c}{\lambda^2 f^2}\int_\infty d^2r'' \int_\infty d^2r'''\, t_{obj}(\mathbf{r}'')\, t^*_{obj}(\mathbf{r}''')\, p_{coh}(\mathbf{r},\mathbf{r}'')\, p^*_{coh}(\mathbf{r}',\mathbf{r}''')$$
$$\times \int_\infty d^2r_s\, \overline{I}_s(\mathbf{r}_s) \exp\left[\frac{2\pi i}{\lambda f}(\mathbf{r}''-\mathbf{r}''')\cdot\mathbf{r}_s\right]. \tag{9.320}$$

Since the $4f$ system is shift-invariant to a good approximation, we can replace $p_{coh}(\mathbf{r},\mathbf{r}'')$ by $p_{coh}(\mathbf{r}-\mathbf{r}'')$. Then we recognize the integral over $\mathbf{r}''$ as a windowed Fourier transform [*cf.* (5.1)], where the coherent PSF serves as the window function and $-\mathbf{r}_s/\lambda f$ is the frequency variable. If we denote this transform as

$$g(\mathbf{r}_s;\mathbf{r}) \equiv \int_\infty d^2r''\, t_{obj}(\mathbf{r}'') \exp\left[\frac{2\pi i}{\lambda f}\mathbf{r}''\cdot\mathbf{r}_s\right] p_{coh}(\mathbf{r}-\mathbf{r}''), \tag{9.321}$$

then the image irradiance is given by

$$\overline{I}_{im}(\mathbf{r}) = \Gamma_{im}(\mathbf{r},\mathbf{r}) = \frac{A_c}{\lambda^2 f^2}\int_\infty d^2r_s\, \overline{I}_s(\mathbf{r}_s)\,|g(\mathbf{r}_s;\mathbf{r})|^2\,. \tag{9.322}$$

In Sec. 5.1.3 we referred to the squared modulus of a local Fourier transform as a *local spectrogram*; here we see that the image-plane irradiance for an arbitrary source (hence arbitrary illumination coherence) is a weighted sum of local spectrograms.

The coherent and incoherent limits of (9.322) are instructive. The coherent case corresponds to a point source, so that $\overline{I}_s(\mathbf{r}_s) \propto \delta(\mathbf{r}_s)$. Then the integral over $\mathbf{r}_s$ can be performed, and we find

$$\overline{I}_{im}(\mathbf{r}) \propto |g(0;\mathbf{r})|^2 = |t_{obj} * p_{coh}(\mathbf{r})|^2\,. \tag{9.323}$$

As expected, in the coherent limit we simply convolve the object function with the coherent PSF and take the squared modulus of the result to get the image-plane irradiance.

Complete incoherence, for this system, corresponds to constant $\overline{I}_s(\mathbf{r}_s)$, so

$$\begin{aligned}\overline{I}_{im}(\mathbf{r}) &\propto \int_\infty d^2r_s\; |g(\mathbf{r}_s;\mathbf{r})|^2\\ &= \int_\infty d^2r'' \int_\infty d^2r'''\, t_{obj}(\mathbf{r}'')\, t^*_{obj}(\mathbf{r}''')\, p_{coh}(\mathbf{r}-\mathbf{r}'')\, p^*_{coh}(\mathbf{r}-\mathbf{r}''')\\ &\quad\times \int_\infty d^2r_s\, \exp\left[\frac{2\pi i}{\lambda f}(\mathbf{r}''-\mathbf{r}''')\cdot\mathbf{r}_s\right]\,.\end{aligned} \tag{9.324}$$

The integral over $\mathbf{r}_s$ yields a delta function, with which we can perform the integral on $\mathbf{r}'''$. The result is

$$\overline{I}_{im}(\mathbf{r}) \propto \int_\infty d^2r''\, |t_{obj}(\mathbf{r}'')|^2\, |p_{coh}(\mathbf{r}-\mathbf{r}'')|^2\,, \tag{9.325}$$

which is just what we expect; the system now responds linearly to the object radiant exitance (proportional to $|t_{obj}(\mathbf{r})|^2$), and the incoherent PSF is the squared modulus of the incoherent one.

As an exercise, the reader may show that the result in (9.325) can be obtained even without assuming that $\overline{I}_s(\mathbf{r}_s)$ is constant over all space. It suffices if the source is uniform over the area of the pupil (into which it is imaged). Colloquially, the system is incoherent if the source fills the pupil.

9.8 VOLUME DIFFRACTION AND 3D IMAGING

So far in this chapter we have considered diffraction from planar apertures or imaging systems that map one plane to another. Now we examine scattering from a volume and imaging systems that map one volume to another. Scattering and 3D imaging are closely related, since both boil down to volume diffraction. The difference between scattering and diffraction is merely semantics (see Sec. 10.2.3), and we have seen in this chapter how diffraction theory is the foundation of imaging with waves.

In many scattering problems a spatially compact object is illuminated by some external beam such as a plane wave. The object scatters (diffracts) the incident beam, and one goal is to compute the resulting field at a large distance from the object. Ultimately, we may want to deduce some properties of the object from the scattered field or irradiance, but that is an inverse problem to be solved by methods discussed in Chap. 15; here we concentrate on the forward problem.

Three-dimensional imaging may involve a similar setup, where a volumetric object is illuminated with some known field and the objective is to compute the field at distant points. In this view, the only essential difference between scattering experiments and imaging systems is that lenses or other image-forming elements are interposed in the latter case.

Often, however, we shall be interested in self-luminous volume objects that serve simultaneously as radiation source and object to be imaged. It will simplify the treatment considerably if we can assume that the object does not absorb the radiation. In this case the object is described by a real index of refraction $n(\mathbf{r})$ that depends on position. Some approaches to analyzing imaging systems with absorbing objects will be discussed in Chap. 10, but we ignore absorption here.

As a first step in developing a theory of volume diffraction and imaging, we must find a counterpart to the Kirchhoff boundary condition. There are two common ways to do this — the Born approximation discussed in Sec. 9.8.1 and the Rytov approximation discussed in Sec. 9.8.2. It will turn out that these two approximations lead to basically the same equation, with a slight difference in interpretation. This equation will be applied to scattering in Sec. 9.8.3 and imaging in Sec. 9.8.4.

9.8.1 Born approximation

In a source-free region of a non-absorbing medium, the Helmholtz equation (9.31) can be written as

$$\left(\nabla^2 + n^2 k_0^2\right) u(\mathbf{r}) = 0\,, \tag{9.326}$$

where the wavenumber k, defined in Sec. 9.2.1 by $2\pi\nu_0/c_m$, has been rewritten as $k = nk_0$, where $k_0 \equiv 2\pi\nu_0/c$, and $n = c/c_m$ is the refractive index. This form of the Helmholtz equation is valid even if the index is a function of position, $n = n(\mathbf{r})$.

A simple algebraic manipulation on (9.326) yields

$$\left(\nabla^2 + k_0^2\right) u(\mathbf{r}) = V(\mathbf{r})\, u(\mathbf{r})\,, \tag{9.327}$$

where $V(\mathbf{r})$, which can be called the *scattering potential*, is defined by

$$V(\mathbf{r}) \equiv k_0^2 \left[1 - n^2(\mathbf{r})\right]\,. \tag{9.328}$$

We define $u_{inc}(\mathbf{r})$ as the field that would exist at point $\mathbf{r}$ in the absence of the inhomogeneous index distribution. Specifically, $u_{inc}(\mathbf{r})$ is the solution to

$$\left(\nabla^2 + k_0^2\right) u_{inc}(\mathbf{r}) = 0\,. \tag{9.329}$$

The total field can then be written as

$$u(\mathbf{r}) = u_{inc}(\mathbf{r}) + u_{sc}(\mathbf{r})\,, \tag{9.330}$$

where $u_{sc}(\mathbf{r}) \equiv u(\mathbf{r}) - u_{inc}(\mathbf{r})$ can be interpreted as the field scattered by the index inhomogeneities. With these definitions, (9.327) becomes

$$\left(\nabla^2 + k_0^2\right) u_{sc}(\mathbf{r}) = V(\mathbf{r})\, u(\mathbf{r})\,. \tag{9.331}$$

The right-hand side of this equation is an effective source, though an unknown one since it depends on the total field $u(\mathbf{r})$. Nevertheless, we can use (9.67) to write

$$u_{sc}(\mathbf{r}) = -\frac{1}{4\pi}\int_\infty d^3\mathbf{r}_0 \, \frac{\exp(ik_0|\mathbf{r}-\mathbf{r}_0|)}{|\mathbf{r}-\mathbf{r}_0|} \, V(\mathbf{r}_0)[u_{inc}(\mathbf{r}_0) + u_{sc}(\mathbf{r}_0)] \, . \tag{9.332}$$

This integral equation for $u_{sc}(\mathbf{r})$ is completely equivalent to the Helmholtz equation.

If $n(\mathbf{r})$ is close to one everywhere, then the scattering is weak, and we expect to have $|u_{sc}(\mathbf{r}_0)| \ll |u_{inc}(\mathbf{r}_0)|$ in the region where $V(\mathbf{r}_0)$ is nonzero. Thus it may be a good approximation to neglect $u_{sc}(\mathbf{r})$ inside the integral of (9.332) and write

$$u_{sc}(\mathbf{r}) \approx -\frac{1}{4\pi}\int_\infty d^3\mathbf{r}_0 \, \frac{\exp(ik_0|\mathbf{r}-\mathbf{r}_0|)}{|\mathbf{r}-\mathbf{r}_0|} \, V(\mathbf{r}_0)\, u_{inc}(\mathbf{r}_0) \, . \tag{9.333}$$

This step, known as the *first Born approximation*, is the 3D counterpart of the Kirchhoff approximation introduced in Sec. 9.4.2. If this approximation is valid and the index distribution and the incident field are specified, the scattered field can be found by performing the integral (numerically if necessary).

Born series If the first Born approximation is inadequate, an improved result can be obtained by iteration. If we denote the integral operator defined in (9.332) by $\boldsymbol{\mathcal{L}}$, we can rewrite that equation abstractly as

$$\mathbf{u}_{sc} = \boldsymbol{\mathcal{L}}\mathbf{u} = \boldsymbol{\mathcal{L}}(\mathbf{u}_{inc} + \mathbf{u}_{sc}) \, , \tag{9.334}$$

or

$$(\mathbf{I} - \boldsymbol{\mathcal{L}})\mathbf{u}_{sc} = \boldsymbol{\mathcal{L}}\mathbf{u}_{inc} \, . \tag{9.335}$$

We can obtain a formal solution of this equation by use of the Neumann series introduced in Sec. A.3.4. From (A.59),

$$\mathbf{u}_{sc} = (\mathbf{I} - \boldsymbol{\mathcal{L}})^{-1}\boldsymbol{\mathcal{L}}\mathbf{u}_{inc} = \sum_{j=0}^{\infty} \boldsymbol{\mathcal{L}}^{j+1}\mathbf{u}_{inc} \, . \tag{9.336}$$

The first Born approximation corresponds to retaining only the $j = 0$ term in this expansion. This approximation is equivalent to assuming that the radiation field interacts with the medium just once, producing a scattered field that then propagates as if in free space. The second Born approximation allows the scattered field to be scattered again, and so forth.

9.8.2 Rytov approximation

The first Born approximation is valid when $n(\mathbf{r})$ is close to unity everywhere, so it would hold for a tenuous medium such as fog. A slight reformulation of the theory in Sec. 9.8.1 would allow $n(\mathbf{r})$ to be close to some mean value $\overline{n}$ so long as the variations $n(\mathbf{r}) - \overline{n}$ were small, and in that case it could apply to weak volume phase gratings such as acousto-optic devices where small variations in refractive index are produced by acoustic waves. The Rytov approximation is fundamentally different; it places restrictions on the gradient of the refractive index instead of its value. Large variations are allowed so long as they occur slowly on the scale of the wavelength.

Reformulation of the wave equation To set the stage for the Rytov approximation, we first represent the total field as

$$u(\mathbf{r}) = u_{inc}(\mathbf{r}) \exp[i\Psi(\mathbf{r})] . \tag{9.337}$$

If we allow $\Psi(\mathbf{r})$ to be complex, this representation can be used without approximation in any region where $u_{inc}(\mathbf{r})$ is nonzero. For example, if we take $u_{inc}(\mathbf{r})$ to be a plane wave, the representation will apply in all space. If, however, we illuminate a scattering object with a beam of finite cross-section, then (9.337) applies only in the region that would be covered by the beam in the absence of the scatterer.

Next we substitute (9.337) into the wave equation (9.327). With the identity $\nabla^2 fg = f\nabla^2 g + g\nabla^2 f + 2\nabla f \cdot \nabla g$, we can write

$$\begin{aligned} &\left[\nabla^2 + k_0^2\right] u(\mathbf{r}) \\ &= \left[-|\nabla\Psi(\mathbf{r})|^2 + i\nabla^2\Psi(\mathbf{r})\right] u_{inc}(\mathbf{r}) \exp[i\Psi(\mathbf{r})] + 2i \exp[i\Psi(\mathbf{r})] \nabla u_{inc}(\mathbf{r}) \cdot \nabla\Psi(\mathbf{r}) \\ &= V(\mathbf{r})\, u_{inc}(\mathbf{r}) \exp[i\Psi(\mathbf{r})] , \end{aligned} \tag{9.338}$$

where we have made use of (9.329) to cancel two terms. Using the above identity again, we see that

$$\begin{aligned} \nabla^2[\Psi(\mathbf{r})\, u_{inc}(\mathbf{r})] &= \Psi(\mathbf{r})\, \nabla^2 u_{inc}(\mathbf{r}) + u_{inc}(\mathbf{r})\, \nabla^2\Psi(\mathbf{r}) + 2\nabla\Psi(\mathbf{r}) \cdot \nabla u_{inc}(\mathbf{r}) \\ &= -k_0^2 \Psi(\mathbf{r})\, u_{inc}(\mathbf{r}) + u_{inc}(\mathbf{r})\, \nabla^2\Psi(\mathbf{r}) + 2\nabla\Psi(\mathbf{r}) \cdot \nabla u_{inc}(\mathbf{r}) , \end{aligned} \tag{9.339}$$

where the second line has used (9.329). Thus (9.338) becomes

$$\left(\nabla^2 + k_0^2\right) [\Psi(\mathbf{r})\, u_{inc}(\mathbf{r})] = -i\, V(\mathbf{r})\, u_{inc}(\mathbf{r}) - i|\nabla\Psi(\mathbf{r})|^2 u_{inc}(\mathbf{r}) . \tag{9.340}$$

Now we have an inhomogeneous Helmholtz equation with the unknown $\Psi(\mathbf{r})$ appearing in the source term on the right. As in Sec. 9.8.1, we can use the Green's function to convert the differential equation to an integral equation [*cf.* (9.332)]:

$$\Psi(\mathbf{r})\, u_{inc}(\mathbf{r}) = \frac{i}{4\pi} \int_\infty d^3\mathbf{r}_0\, \frac{\exp(ik_0|\mathbf{r} - \mathbf{r}_0|)}{|\mathbf{r} - \mathbf{r}_0|} \left[V(\mathbf{r}_0) + |\nabla\Psi(\mathbf{r}_0)|^2\right] u_{inc}(\mathbf{r}_0). \tag{9.341}$$

To this point we have made no approximations, and (9.341) is equivalent to the original wave equation (9.326).

Rytov approximation The Rytov approximation consists of neglecting $|\nabla\Psi(\mathbf{r}_0)|^2$ compared to $V(\mathbf{r}_0)$ in the integrand of (9.341). Recall that the scattering potential $V(\mathbf{r}_0)$ is defined in (9.328) with a factor of k_0^2, so it gets larger as the wavelength gets shorter. Thus the Rytov approximation is valid in the geometrical-optics or short-wavelength limit. It does not apply when abrupt changes in index such as air-glass interfaces are present or for scatterers that are small compared to the wavelength. For more discussion of the validity of the Rytov approximation, see Fiddy (1992).

Within the Rytov approximation, the complex phase $\Psi(\mathbf{r})$ is given by

$$\Psi(\mathbf{r}) = \frac{i}{4\pi u_{inc}(\mathbf{r})} \int_\infty d^3\mathbf{r}_0\, \frac{\exp(ik_0|\mathbf{r} - \mathbf{r}_0|)}{|\mathbf{r} - \mathbf{r}_0|}\, V(\mathbf{r}_0)\, u_{inc}(\mathbf{r}_0) . \tag{9.342}$$

Since $u_{inc}(\mathbf{r})$ is known and presumed nonzero, there is no difficulty in dividing through by it. Except for this factor, then, the integral in the Rytov approximation has exactly the same form as in the first Born approximation. The only difference is the interpretation; in the Born approximation, the integral gives the scattered wave directly, while in the Rytov approximation it gives $\Psi(\mathbf{r})u_{inc}(\mathbf{r})$, and the scattered field must be constructed from

$$u_{sc}(\mathbf{r}) = u(\mathbf{r}) - u_{inc}(\mathbf{r}) = u_{inc}(\mathbf{r})\left\{\exp[i\Psi(\mathbf{r})] - 1\right\}. \tag{9.343}$$

The basic problem in scattering or 3D imaging is thus the same within either approximation: evaluate the diffraction integral.

Eikonal equation Closely related to the Rytov approximation is an important relation known as the *eikonal equation.* The eikonal (Greek *eikon*, likeness, image) is a function known also as the *characteristic function* or *point characteristic.* Many theoretical developments of geometrical optics are built on this function.

The starting point for deriving the eikonal equation is a representation of the total field in the form

$$u(\mathbf{r}) = A(\mathbf{r})\exp[ik_0 W(\mathbf{r})], \tag{9.344}$$

where $W(\mathbf{r})$ is the eikonal. The key difference between this form and the Rytov form (9.337) is that $A(\mathbf{r})$ is not the incident field and might be slowly varying.

The Laplacian of (9.344) is

$$\nabla^2 u(\mathbf{r}) = \left[\nabla^2 A(\mathbf{r}) + 2ik_0\nabla W(\mathbf{r})\cdot\nabla A(\mathbf{r}) - k_0^2 A(\mathbf{r})\,|\nabla W(\mathbf{r})|^2\right.$$
$$\left. + ik_0 A(\mathbf{r})\,\nabla^2 W(\mathbf{r})\exp[ik_0 W(\mathbf{r})]\right. . \tag{9.345}$$

In the short-wavelength limit, $k_0 \to \infty$, and we can neglect the term linear in k_0 compared to the quadratic one, so that

$$\nabla^2 u(\mathbf{r}) \to -k_0^2|\nabla W(\mathbf{r})|^2 A(\mathbf{r})\exp[ik_0 W(\mathbf{r})] = -k_0^2|\nabla W(\mathbf{r})|^2 u(\mathbf{r}). \tag{9.346}$$

In this same limit, the Helmholtz equation (9.326) becomes

$$\left[-k_0^2|\nabla W(\mathbf{r})|^2 + n^2 k_0^2\right] u(\mathbf{r}) = 0, \tag{9.347}$$

from which we immediately obtain the eikonal equation,

$$|\nabla W(\mathbf{r})|^2 = [n(\mathbf{r})]^2 . \tag{9.348}$$

This equation is a nonlinear partial differential equation for $W(\mathbf{r})$. Solution is often difficult except for a few textbook problems, but in principle it contains all of geometrical optics. The familiar ray direction in geometric optics is interpreted as $\nabla W(\mathbf{r})$ in this approach; we shall see why in Sec. 10.2.7.

9.8.3 Fraunhofer diffraction from volume objects

In the Fraunhofer approximation, we expand $|\mathbf{r} - \mathbf{r}_0|$ as in (9.101), retaining only the leading term in the denominator of the Green's function but also the term $\mathbf{r}_0 \cdot \mathbf{r}/|\mathbf{r}|$ in the exponent. With this approximation, the scattered field in the first Born approximation, (9.333), can be written as

$$u_{sc}(\mathbf{r}) \approx -\frac{\exp(ik_0|\mathbf{r}|)}{4\pi|\mathbf{r}|}\int_\infty d^3r_0\; V(\mathbf{r}_0)\,u_{inc}(\mathbf{r}_0)\exp\left(-2\pi i\frac{\mathbf{r}_0\cdot\mathbf{r}}{\lambda_0|\mathbf{r}|}\right). \tag{9.349}$$

This result should be compared to its 2D counterpart, (9.103). In both cases, the integral is a Fourier transform, but here it is the 3D transform of the effective field $V(\mathbf{r}_0)\, u_{inc}(\mathbf{r}_0)$, and the 3D spatial frequency that contributes to the scattered field is given by [*cf.* (9.104)]

$$\boldsymbol{\sigma}_{sc} = \frac{\mathbf{r}}{\lambda_0 |\mathbf{r}|} \,. \tag{9.350}$$

Since $\mathbf{r}/|\mathbf{r}|$ is the unit vector $\hat{\mathbf{n}}$ from the origin of coordinates to the observation point, radiation scattered in direction $\hat{\mathbf{n}}$ results from the spatial frequency $\boldsymbol{\sigma}_{sc} = \hat{\mathbf{n}}/\lambda_0$ in the Fourier transform of the effective field.

Ewald sphere An important special case of (9.349) is when the illumination is a plane wave, so that

$$u_{inc}(\mathbf{r}_0) = A \exp(i\mathbf{k}_{inc} \cdot \mathbf{r}_0) \,. \tag{9.351}$$

In order to satisfy the wave equation, the magnitude of $\mathbf{k}_{inc}$ must be $2\pi/\lambda_0$. If we define $\boldsymbol{\sigma}_{inc} = \mathbf{k}_{inc}/2\pi$, then (9.349) becomes

$$u_{sc}(\mathbf{r}) \approx -A \frac{\exp(ik_0|\mathbf{r}|)}{4\pi|\mathbf{r}|} \int_\infty d^3r_0 \; V(\mathbf{r}_0) \exp[-2\pi i(\boldsymbol{\sigma}_{sc} - \boldsymbol{\sigma}_{inc}) \cdot \mathbf{r}_0] \,. \tag{9.352}$$

Now we see explicitly that the scattered field is proportional to the 3D Fourier transform of the scattering potential evaluated at the spatial frequency $\boldsymbol{\sigma}_{sc} - \boldsymbol{\sigma}_{inc}$.

In many scattering experiments, we fix the illumination direction and observe the scattered radiation over a range of points in the far field simultaneously, for example with a viewing screen or a detector array. Since $\boldsymbol{\sigma}_{sc}$ is related to the observation point by (9.350), that means we should consider $\boldsymbol{\sigma}_{sc}$ as a variable but $\boldsymbol{\sigma}_{inc}$ as a constant in (9.352). As shown in Fig. 9.23, $\boldsymbol{\sigma}_{sc} - \boldsymbol{\sigma}_{inc}$ traces out a spherical cap in the 3D Fourier space as the viewing direction $\hat{\mathbf{n}}$ is varied. This sphere, called the *Ewald sphere* in x-ray crystallography, defines the region of the object Fourier transform that contributes to the scattered radiation in this geometry.

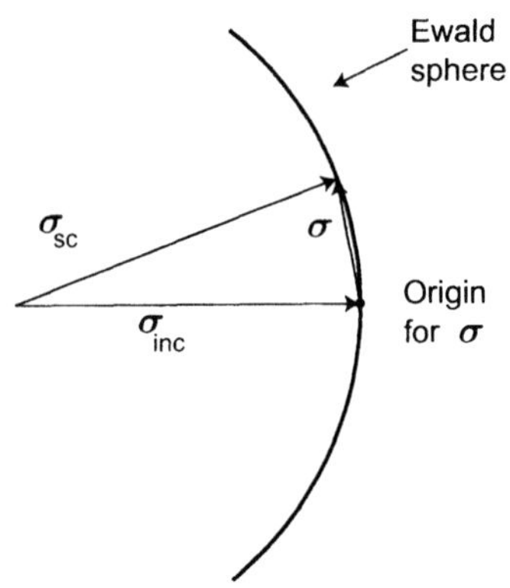

Fig. 9.23 The Ewald sphere.

Crystals are periodic structures, and we know from (3.275) that such structures contain only spatial frequencies that lie on the reciprocal lattice in Fourier space. Thus diffraction occurs in x-ray crystallography only when the Ewald sphere intersects a reciprocal lattice point. This requirement is called the *Bragg condition* or *Bragg matching*.

The Ewald sphere is also an important concept when we are concerned with the inverse problem of reconstructing a scattering potential from measurements in the far field. Since the Ewald sphere is a 2D manifold in a 3D space, we cannot expect to be able to learn very much about the potential with a fixed illumination direction. Some variation of the direction and/or magnitude of $\mathbf{k}_{inc}$ is required.

9.8.4 Coherent 3D imaging

A coherent 3D imaging system is one in which a 3D scattering or reflecting object is illuminated by a coherent monochromatic wave. If the object is transparent and the first Born approximation is applicable, then each point in the object produces a scattered wave proportional to the scattering potential defined in (9.328), and that potential can be regarded as the input to the imaging system. In the Rytov approximation the input can still be regarded as the scattering potential, but there is a more complicated nonlinear relation between this potential and the diffracted field. For a reflecting object, we can consider the system input to be the complex amplitude reflectance defined at every point on the object surface.

In each of these cases, the wave emerging from the object is determined by both the object characteristics and the illumination. This wave is passed through some assembly of lenses, mirrors or other focusing elements to an image space. Though we may eventually observe the wave on a planar image detector in image space, the interest in this section will be on computation of the wave at all points within some volume in this image space. Our goal is thus to analyze the mapping from the object scattering potential or reflectance to the wave in a 3D image space. Since the scattered or reflected field at each object point is a simple product of the illuminating field and an object property, all we really have to do is study the mapping from one 3D field to another.

Coherent imaging with an ideal thin lens Consider a transparent object defined by the scattering potential $V(\mathfrak{r})$ lying entirely in the space $z < 0$. An ideal thin lens is placed in the plane $z = 0$, and we want to find an expression for the resulting field in the space $z > 0$. Methods developed in Sec. 9.6 can be adapted for this purpose.

In the first Born approximation, the field at a point just to the left of the lens can be obtained from (9.333) by setting $\mathfrak{r} = (\mathbf{r}, 0^-)$, where $\mathfrak{r} = (x, y, z)$ and $\mathbf{r} = (x, y)$. If the object is spatially compact and far enough from the lens, we can also make a Fresnel approximation; noting that z_0 is a negative number, we can then write [*cf.* (9.94)]

$$u_-(\mathbf{r}) \approx -\frac{1}{4\pi}\int_\infty d^3\mathfrak{r}_0\, \frac{\exp(-ik_0 z_0)}{z_0} \exp\left(-i\pi\frac{|\mathbf{r}-\mathbf{r}_0|^2}{\lambda_0 z_0}\right) V(\mathfrak{r}_0)\, u_{inc}(\mathfrak{r}_0)\,, \quad (9.353)$$

where we have used $k_0 = 2\pi/\lambda_0$.

The field emerging from the lens (*i.e.*, in the plane $z = 0^+$) is given by (9.153) and (9.159) as

$$u_+(\mathbf{r}) = u_-(\mathbf{r})\, t_{lens}(\mathbf{r})$$

$$= -\frac{1}{4\pi}\exp\left(-i\pi\frac{r^2}{\lambda_0 f}\right) t_{ap}(\mathbf{r}) \int_\infty d^3\mathfrak{r}_0\, \frac{\exp(-ik_0 z_0)}{z_0} \exp\left(-i\pi\frac{|\mathbf{r}-\mathbf{r}_0|^2}{\lambda_0 z_0}\right) V(\mathfrak{r}_0)\, u_{inc}(\mathfrak{r}_0)\,. \quad (9.354)$$

Since we have now specified the field on a plane, we can use ordinary Fresnel diffraction theory to propagate it to an arbitrary point in the space $z > 0$. From (9.97),

$$u(\mathbf{r}) = -\frac{\exp(ik_0 z)}{4\pi i \lambda_0 z} \exp\left(i\pi \frac{r^2}{\lambda_0 z}\right)$$

$$\times \int_\infty d^2r' \exp\left(-i\pi \frac{r'^2}{\lambda_0 f}\right) t_{ap}(\mathbf{r}') \exp\left(i\pi \frac{r'^2}{\lambda_0 z}\right) \exp\left(-2\pi i \frac{\mathbf{r}\cdot\mathbf{r}'}{\lambda_0 z}\right)$$

$$\times \int_\infty d^3\mathbf{r}_0 \, \frac{\exp(-ik_0 z_0)}{z_0} \exp\left(-i\pi \frac{|\mathbf{r}' - \mathbf{r}_0|^2}{\lambda_0 z_0}\right) V(\mathbf{r}_0)\, u_{inc}(\mathbf{r}_0)\,, \qquad (9.355)$$

where $u(\mathbf{r})$ means the same thing as $u_z(\mathbf{r})$.

Point response function To understand the meaning of the complicated expression (9.355), we write

$$u(\mathbf{r}) = \int_\infty d^3\mathbf{r}_0 \; h_{coh}(\mathbf{r}, \mathbf{r}_0)\, V(\mathbf{r}_0)\,, \qquad (9.356)$$

where the PRF is given by

$$h_{coh}(\mathbf{r}, \mathbf{r}_0) = -u_{inc}(\mathbf{r}_0) \frac{\exp[ik_0(z - z_0)]}{4\pi i \lambda_0 z z_0} \exp\left[\frac{i\pi}{\lambda_0}\left(\frac{r^2}{z} - \frac{r_0^2}{z_0}\right)\right]$$

$$\times \int_\infty d^2r' \, t_{ap}(\mathbf{r}') \exp(i\pi\beta r'^2) \exp(-2\pi i \boldsymbol{\rho}\cdot\mathbf{r}')\,, \qquad (9.357)$$

with

$$\beta = \frac{1}{\lambda_0}\left(\frac{1}{z} - \frac{1}{z_0} - \frac{1}{f}\right); \qquad (9.358)$$

$$\boldsymbol{\rho} = \frac{1}{\lambda_0}\left(\frac{\mathbf{r}}{z} - \frac{\mathbf{r}_0}{z_0}\right). \qquad (9.359)$$

Note that β vanishes if plane z_0 is imaged onto plane z since $p = -z_0$ and $q = z$ in that case, and p, q and f are then related by the imaging condition (9.169). The two planes are said to be *conjugate* (one is imaged onto the other) if $\beta = 0$.

When $\beta = 0$, the integral in (9.357) is just the Fourier transform of the aperture transmittance evaluated at the spatial frequency given in (9.359). Since the Fourier transform of a clear aperture is maximum at zero frequency, the maximum value of $h_{coh}(\mathbf{r}, \mathbf{r}_0)$ when $\mathbf{r}$ and $\mathbf{r}_0$ define conjugate planes is determined by setting $\boldsymbol{\rho}$ to zero. This procedure shows that $\mathbf{r} = m_t \mathbf{r}_0$, where the transverse magnification m_t is z/z_0. (Recall that z_0 is negative, so the image is inverted.)

For nonzero β, comparison with (9.97) shows that the integral in (9.357) has the same structure as the Fresnel diffraction pattern of the aperture, but with different constants. In (9.97) β is just $1/\lambda_0 z$, so large values of β correspond to small distances from the aperture, and the Fresnel diffraction pattern is approximately a geometrical shadow of the aperture in this case. Note that β is also proportional to $1/\lambda_0$, so for fixed z (not in a conjugate plane), β increases as the wavelength gets shorter. In the geometric-optics limit ($\lambda_0 \to 0$), the diffraction pattern is exactly the geometric shadow of the aperture except at the focus.

To see where this shadow is located and how big it is, we can use the principle of stationary phase. If we define

$$\Phi(\mathbf{r}') = \pi\beta r'^2 - 2\pi\boldsymbol{\rho}\cdot\mathbf{r}'\,, \qquad (9.360)$$

then the exponential factor in the integrand in (9.357) is rapidly oscillating except in the vicinity of the point where $\nabla\Phi(\mathbf{r}') = 0$, or where $\boldsymbol{\rho} \approx \beta\mathbf{r}'$. For large β, only a small region in the vicinity of $\mathbf{r} = \boldsymbol{\rho}/\beta$ will contribute to the integral. If this point lies in the domain of integration as set by $t_{ap}(\mathbf{r}')$, we can write

$$\int_\infty d^2r'\, t_{ap}(\mathbf{r}') \exp(i\pi\beta r'^2) \exp(-2\pi i\boldsymbol{\rho}\cdot\mathbf{r}')$$

$$\approx t_{ap}(\boldsymbol{\rho}/\beta) \int_\infty d^2r' \exp(i\pi\beta r'^2) \exp(-2\pi i\boldsymbol{\rho}\cdot\mathbf{r}') = \beta^{-1} t_{ap}(\boldsymbol{\rho}/\beta) \exp(-i\pi\rho^2/\beta)\,, \tag{9.361}$$

where the last step follows from (3.263).

The factor $t_{ap}(\boldsymbol{\rho}/\beta)$ in (9.361) is the geometric shadow of the aperture. It is centered at the point where $\boldsymbol{\rho}/\beta = 0$, or $\mathbf{r} = (z/z_0)\mathbf{r}_0$. The size of the shadow is determined by β as well as $\boldsymbol{\rho}$. If we consider a circular lens aperture of radius R_{ap}, then the radius of the shadow R_{shad} is obtained by setting the magnitude of the vector $\boldsymbol{\rho}/\beta$ to R_{ap}; the result is

$$R_{shad} = R_{ap}\left(1 - \frac{z}{z_0} - \frac{z}{f}\right). \tag{9.362}$$

When $z = 0$, $R_{shad} = R_{ap}$ as expected, and R_{shad} decreases linearly with z as we move toward the plane conjugate to z_0. In geometrical-optics terms, the shadow is centered on the chief ray, which in this problem is a line drawn from the point $\mathbf{r}_0$ through the center of the lens and extended into the image space. The 3D PRF is a cone of light centered on the chief ray and converging to the geometrical focus (see Fig. 9.24).

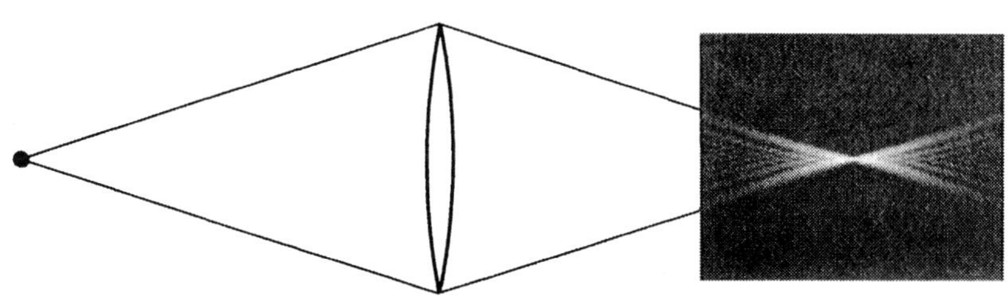

Fig. 9.24 Illustration of the 3D PRF of an ideal lens. The geometric-optics approximation corresponds to the cone shown.

In this large-β or geometric-optics approximation, the full coherent PRF (9.357) becomes (after some messy algebra)

$$h_{coh}(\mathbf{r}, \mathbf{r}_0) \approx -u_{inc}(\mathbf{r}_0)\frac{\exp[ik_0(z - z_0)]}{4\pi i\lambda_0\beta z z_0} t_{ap}(\boldsymbol{\rho}/\beta) \exp\left[-\frac{i\pi}{\lambda_0}\frac{|\mathbf{r} - (q_0/z_0)\mathbf{r}_0|^2}{q_0 - z}\right], \tag{9.363}$$

where q_0 is the distance from the lens plane to the plane where the source point is in focus:

$$\frac{1}{q_0} = \frac{1}{f} + \frac{1}{z_0}. \tag{9.364}$$

The phase factor in (9.363) represents (in the Fresnel approximation) a spherical wave converging toward the geometric focal point. The factor $t_{ap}(\boldsymbol{\rho}/\beta)$ indicates (unphysically) that the spherical wave is truncated abruptly at the edge of the cone shown in Fig. 9.24. The approximate expression (9.363) must break down near the focus since it requires large β. The inset in the figure was computed numerically from (9.357).

10

Energy Transport and Photons

Imaging requires some form of radiation. Objects can emit their own radiation, or they can transmit, reflect or scatter radiation from an external source. The goal of the imaging system is to probe these emission, transmission, reflection or scattering characteristics of the object. The detector in an imaging system responds to the energy in the radiation, so imaging is ultimately about how energy is transported from the object to the detector.

This chapter is about the transport and measurement of radiant energy. In Sec. 10.1 we survey some basic concepts of energy flow specifically for electromagnetic radiation; both classical and quantum-mechanical viewpoints are presented. The concept of irradiance, already alluded to in previous chapters, is elucidated in more detail here and related to detector response, and the concept of a photon is discussed. The reader who is well versed in the basic physics of electromagnetic fields, or one who simply wants to move directly to a more phenomenological description, can skip to Sec. 10.2 without loss of continuity.

In Sec. 10.2 we introduce a variety of quantities that have been used to describe energy flow. Included are the constructs of classical radiometry, as well as a phase-space distribution function more familiar from statistical mechanics.

Section 10.3 deals with the Boltzmann equation, an integro-differential equation that describes radiative energy transport, and in Sec. 10.4 we relate the Boltzmann equation to imaging.

10.1 ELECTROMAGNETIC ENERGY FLOW AND DETECTION

Our main goal in this section is to introduce the word *photon* and show why it is a useful way of thinking about energy transport in imaging systems. We start, however, with a short survey of energy concepts in classical electromagnetic theory.

10.1.1 Energy flow in classical electrodynamics

An important result in electromagnetic theory is *Poynting's theorem.* This theorem, derived by consideration of the work done on a system of charges by an electromagnetic field (see, *e.g.*, Jackson, 1998), states that

$$\nabla \cdot \mathbf{\Pi}(\mathbf{r}, t) + \frac{\partial}{\partial t} U(\mathbf{r}, t) = -\mathbf{j}(\mathbf{r}, t) \cdot \mathbf{e}(\mathbf{r}, t) , \tag{10.1}$$

where $\mathbf{\Pi}(\mathbf{r}, t)$ is the *Poynting vector*, defined by

$$\mathbf{\Pi}(\mathbf{r}, t) \equiv \mathbf{e}(\mathbf{r}, t) \times \mathbf{h}(\mathbf{r}, t) , \tag{10.2}$$

and $U(\mathbf{r}, t)$ is the *energy density*, defined as

$$U(\mathbf{r}, t) \equiv \tfrac{1}{2}\mathbf{e}(\mathbf{r}, t) \cdot \mathbf{d}(\mathbf{r}, t) + \tfrac{1}{2}\mathbf{b}(\mathbf{r}, t) \cdot \mathbf{h}(\mathbf{r}, t) . \tag{10.3}$$

In these equations, $\mathbf{r}$ is a 3D position vector, and the field quantities are defined in Sec. 9.1.1. (The notation is more or less standard, the main exception being that lower-case letters replace the usual capitals for fields.)

The Poynting vector has dimensions of power per unit area (watts/m^2 in SI units), and it provides a measure of energy flow. As the name implies, the energy density has dimensions of energy per unit volume (Joules/m^3); it describes the energy stored in the electric and magnetic fields.

These interpretations are reinforced by considering a region of space where there are no charges, so that the current density $\mathbf{j}(\mathbf{r}, t) = 0$. We denote this region by V and its boundary surface by S. If we integrate both sides of (10.1) over V and use the divergence theorem, we obtain

$$\int_S da\ \hat{\mathbf{n}} \cdot \mathbf{\Pi}(\mathbf{r}, t) = -\frac{\partial}{\partial t} \int_V d^3\mathbf{r}\ U(\mathbf{r}, t) , \tag{10.4}$$

where $\hat{\mathbf{n}}$ is an outwardly directed unit vector normal to S. If $U(\mathbf{r}, t)$ is the energy per unit volume, then the volume integral on the right represents the total energy stored in V. Since no work is done in V if there are no charges there, the surface integral on the left must represent the energy per unit time transported across S. Energy per unit time is power, but in radiometry it is called *energy flux* or *radiant flux*; thus we can interpret $\hat{\mathbf{n}} \cdot \mathbf{\Pi}(\mathbf{r}, t)da$ as the energy flux through the differential area da.

10.1.2 Plane waves

As discussed in Sec. 9.2.1, monochromatic plane waves have the structure $\text{Re}\{A \exp[i(\mathbf{k} \cdot \mathbf{r} - 2\pi\nu_0 t)]\}$, where A is a (possibly complex) amplitude and $\text{Re}\{\cdot\}$ denotes the real part. For plane electromagnetic waves, this structure applies (with different amplitudes) to each Cartesian component of each vector field. For example, we can represent the electric field in a plane wave by

$$\mathbf{e}(\mathbf{r}, t) = \text{Re}\{\mathbf{E} \exp[i(\mathbf{k} \cdot \mathbf{r} - 2\pi\nu_0 t)]\} , \tag{10.5}$$

where Cartesian component E_n (where $n = x, y, z$) of the vector $\mathbf{E}$ is the amplitude associated with the field component $e_n(\mathbf{r}, t)$; similar representations apply to $\mathbf{h}(\mathbf{r}, t)$

and the other fields. We use capital letters for the amplitudes since they are, in effect, in the Fourier domain. Specifically, the 4D Fourier transform of $e_j(\mathbf{r}, t)$ is E_j times a 4D delta function. We shall customarily delete the designator Re with the convention that the real part of all complex quantities is understood.

By Maxwell's equations, the fields in a source-free region must satisfy $\nabla \cdot \mathbf{d} = 0$ and $\nabla \cdot \mathbf{b} = 0$. In terms of the amplitudes, these equations imply that $\mathbf{k} \cdot \mathbf{D} = 0$ and $\mathbf{k} \cdot \mathbf{B} = 0$, so $\mathbf{D}$ and $\mathbf{B}$ are both perpendicular to the wavevector $\mathbf{k}$.

If we assume an isotropic, homogeneous medium, as discussed in Sec. 9.1.3, then the constitutive relations are $\mathbf{b}(\mathbf{r}, t) = \mu_0 \mathbf{h}(\mathbf{r}, t)$ and $\mathbf{d}(\mathbf{r}, t) = \epsilon \mathbf{e}(\mathbf{r}, t)$. These relations imply that $\mathbf{H}$ is parallel to $\mathbf{B}$ and $\mathbf{E}$ is parallel to $\mathbf{D}$, so $\mathbf{E}$ and $\mathbf{H}$ are also perpendicular to $\mathbf{k}$. Since $\nabla \times \mathbf{e} = -\partial \mathbf{b}/\partial t$, we also require

$$i\mathbf{k} \times \mathbf{E} = -2\pi i \nu_0 \mathbf{B} . \tag{10.6}$$

Hence $\mathbf{E}$ and $\mathbf{B}$ are perpendicular to each other as well as to $\mathbf{k}$. It follows, then, that the Poynting vector $\mathbf{\Pi}$ is parallel to $\mathbf{k}$. The energy flow for a plane wave in an isotropic, homogeneous medium is in the direction of the wavevector.

Another consequence of (10.6) is that the amplitudes are related by

$$\frac{|\mathbf{E}|}{|\mathbf{B}|} = \frac{E}{B} = \frac{2\pi\nu_0}{k} = c_m , \tag{10.7}$$

where c_m is the speed of light in the medium, given in Sec. 9.1.4 as $c_m = 1/\sqrt{\mu_0 \epsilon}$, and we have used (9.38). The notation $|\mathbf{E}|$ in (10.7) implies both the norm of the vector and the modulus of the complex quantity:

$$|\mathbf{E}| = \sqrt{\mathbf{E} \cdot \mathbf{E}^*} = \sqrt{\sum_{j=1}^{3} |E_j|^2} . \tag{10.8}$$

Time averages The Poynting vector is a product of factors that oscillate at frequency ν_0, so it consists of a static term and a term that oscillates at $2\nu_0$. As we shall discuss in more detail in Sec. 10.1.5, optical detectors are time-averaging devices, so we are not interested in the rapidly oscillating part, and we must therefore compute the time-averaged Poynting vector. If we choose to average over one cycle of the oscillation, we can employ the *cycle-average theorem*: If $a(t)$ and $b(t)$ are two complex quantities that both vary as $\exp(-2\pi i \nu_0 t)$, then

$$\langle \mathrm{Re}(a)\, \mathrm{Re}(b) \rangle_T = \tfrac{1}{2}\, \mathrm{Re}(ab^*) , \tag{10.9}$$

where $\langle\ \rangle_T$ denotes a time average over one cycle of period $T = 1/\nu_0$. To prove this theorem, one writes $\mathrm{Re}(a) = \frac{1}{2}(a + a^*)$, and similarly for b, and recognizes that $\langle a^2 \rangle_T = \langle b^2 \rangle_T = 0$ and that ab^* and a^*b are independent of time.

Applying (10.9) to the energy density for a plane wave in a homogeneous medium, we find

$$\langle U \rangle_T = \tfrac{1}{4}\epsilon |\mathbf{E}|^2 + \tfrac{1}{4}\mu_0 |\mathbf{H}|^2 = \tfrac{1}{2}\epsilon |\mathbf{E}|^2 , \tag{10.10}$$

where the last form follows from (10.7) and the constitutive relation $\mathbf{B} = \mu_0 \mathbf{H}$.

Similarly, the average Poynting vector for a plane wave is given by

$$\langle \mathbf{\Pi} \rangle_T = \tfrac{1}{2}\, \mathrm{Re}(\mathbf{E} \times \mathbf{H}^*) = \tfrac{1}{2}\epsilon\, c_m |\mathbf{E}|^2 \hat{\boldsymbol{\kappa}} , \tag{10.11}$$

where $\hat{\kappa}$ is a unit vector in the direction of $\mathbf{k}$, and we have used (10.6) and the constitutive relations.

Since $\frac{1}{2}\epsilon|\mathbf{E}|^2$ is the time-averaged energy density, (10.11) has a simple geometric interpretation: If we consider a rectangular volume of dimensions $L_x \times L_y \times L_z$ and take the z axis to be parallel to $\mathbf{k}$, then $\frac{1}{2}\epsilon|\mathbf{E}|^2 L_x L_y L_z$ is the average stored energy. This energy flows through the face of the volume perpendicular to the z axis in a time $\tau = L_z/c_m$, so the average power per unit area through this face is $\frac{1}{2}\epsilon|\mathbf{E}|^2 L_x L_y L_z / L_x L_y \tau$, or simply $\langle \Pi \rangle_T$, the magnitude of the time-averaged Poynting vector.

Terminology and notation The magnitude of the time-averaged Poynting vector is frequently called the *intensity* of the wave, but this term is at variance with the strict radiometric definitions given in Sec. 10.2. In the radiometry community, the term *intensity* (or *radiant intensity*) means power per unit solid angle, while the power per unit area incident on some surface is called *irradiance.*

On the other hand, the magnitude of the time-averaged Poynting vector, though it has units of power per unit area, is not necessarily irradiance; irradiance is a property of the surface as well as the wave. Consider a collimated Gaussian beam propagating along the z axis (see Sec. 9.5.3). The irradiance on a surface normal to the z axis is $\langle|\mathbf{\Pi}|\rangle$, but if we tip the surface so that its normal $\hat{\mathbf{n}}$ makes an angle θ with the axis, then the irradiance is reduced by a factor of $\cos\theta$ since the same amount of power is spread over a larger area, but $\langle|\mathbf{\Pi}|\rangle$ is unchanged. Since $\langle|\mathbf{\Pi}|\rangle$ is the irradiance on a surface normal to the Poynting vector, we shall refer to it as the *normal irradiance* and denote it as I_0.

We caution the reader that our notation for irradiance is not standard in the radiometry literature, where irradiance is denoted by E, from the French *éclairement.* The letter E gets used for many other purposes (Big Bird, 1969). We use E (in script form) for the photon energy and for the Fourier transform of the electric field, and the photography community uses it for exposure, which is irradiance times exposure time. To avoid wasting E on an application where it has no mnemonic value (except to Francophones), we denote the irradiance by I.

10.1.3 Photons

So far our discussion of energy transport in an electromagnetic field has been cast in terms of classical fields and energies, but it is very useful to picture energy transport in terms of radiation quanta called *photons.* A complete understanding of photons, and more generally quantum states of the radiation field, requires the use of quantum electrodynamics (QED). We present here a brief introduction to QED, including a simplified treatment of field quantization, photon-number states and photodetection.

For this section, the reader is presumed to be familiar with the basic principles of quantum mechanics, including the concept of a state vector. Dirac notation is used freely.

Modes A useful starting point for QED is the classical Fourier series. If we consider a cube of side L in free space (where L can eventually be allowed to approach infinity), the real electric field in this region can be expressed as [*cf.* (10.5)]

$$\mathbf{e}(\mathbf{r},t) = \tfrac{1}{2}\sum_j \gamma_j [E_j \exp(i\mathbf{k}_j \cdot \mathbf{r} - i\omega_j t) + E_j^* \exp(-i\mathbf{k}_j \cdot \mathbf{r} + i\omega_j t)] . \tag{10.12}$$

Each term in (10.12), called a *mode*, is characterized by its wavevector $\mathbf{k}_j$, polarization direction[1] γ_j and radian frequency $\omega_j = 2\pi\nu_j$. The vectors $\mathbf{k}_j$ fall on the *reciprocal lattice* (see Sec. 3.4.6), which in this problem is a cubic lattice of spacing $2\pi/L$. For each $\mathbf{k}_j$ there are two orthogonal directions for the electric field, hence two unit vectors γ_j. The radian frequency associated with mode j is $\omega_j = ck_j$, where c is the speed of light and $k_j = |\mathbf{k}_j|$.

It will prove convenient to define time-dependent amplitudes $a_j(t)$ by

$$E_j \exp(-i\omega_j t) = 2iN_j a_j(t) , \qquad N_j = \sqrt{\frac{\hbar\omega_j}{\epsilon_0 L^3}} , \tag{10.13}$$

where $\hbar$ is Planck's constant divided by 2π and N_j is a constant with dimensions of electric field. With these definitions,

$$\mathbf{e}(\mathbf{r},t) = i\sum_j \gamma_j N_j [\exp(i\mathbf{k}_j \cdot \mathbf{r})\, a_j(t) - \exp(-i\mathbf{k}_j \cdot \mathbf{r})\, a_j^*(t)] . \tag{10.14}$$

A similar expansion can be used for the magnetic field, and the coefficients in the two expansions can be related to each other by (10.6). When the resulting expansions are plugged into (10.3), the energy density in the field is found to be

$$U = \frac{1}{L^3}\sum_j \tfrac{1}{2}\hbar\omega_j [a_j^* a_j + a_j a_j^*] . \tag{10.15}$$

Details of this calculation can be found in Cohen-Tannoudji *et al.* (1989).

The expression in (10.15) can be cast into a more familiar form by defining two other amplitudes:

$$q_j = \sqrt{\frac{\hbar}{2\omega_j}}\,(a_j^* + a_j) , \qquad p_j = i\sqrt{\frac{\hbar\omega_j}{2}}\,(a_j^* - a_j) . \tag{10.16}$$

We can also define a total energy or *Hamiltonian* $H = L^3 U$; in terms of the new variables q_j and p_j, the Hamiltonian is given by

$$H = \tfrac{1}{2}\sum_j \left[p_j^2 + \omega_j^2 q_j^2\right] . \tag{10.17}$$

Both p_j and q_j are functions of time. Specifically, if E_j is a complex constant $|E_j|\exp(i\phi_j)$, then

$$\omega_j q_j = \sqrt{\frac{\hbar\omega_j}{2}}\,\frac{|E_j|}{N_j}\cos(\omega_j t - \phi_j) , \tag{10.18a}$$

[1] The vector γ_j is a unit vector but we leave off the usual caret since we shall shortly need that ornament for another purpose. Also, the word *polarization* is ambiguous; here it refers to the direction of the electric field, not to dipole moment per unit volume.

$$p_j = \sqrt{\frac{\hbar\omega_j}{2}} \frac{|E_j|}{N_j} \sin(\omega_j t - \phi_j). \tag{10.18b}$$

Since the amplitudes of these two oscillations are the same, a plot of p_j vs. $\omega_j q_j$ traces out a circle during one period of the oscillation. The energy, on the other hand, remains constant.

Each term in the sum in (10.17) has the same form as for a classical particle of unit mass executing simple harmonic motion in a quadratic potential. If we think of p_j as the momentum of the oscillating particle and q_j as its displacement, the kinetic energy is given by $\frac{1}{2}p_j^2$ and the potential energy is given by $\frac{1}{2}\omega_j^2 q_j^2$.

The formal analogy to a mechanical harmonic oscillator is useful, but we should not lose sight of the fact that we are merely choosing different representations for the amplitude of the electric field in each mode. Choice of p_j and q_j leads to a real representation, while choice of a_j and a_j^* leads to a complex representation. In all cases, the final expression for the field $\mathbf{e}(\mathbf{r}, t)$ is real.

Field quantization A simple heuristic way to get a quantized field theory is just to replace the field amplitudes by quantum-mechanical operators. More sophisticated approaches based on the Hamiltonian or Lagrangian formulation of continuum mechanics can also be invoked (Cohen-Tannoudji *et al.*, 1989), but in the end they lead to the same results, so we shall jump immediately to the heuristic method.[2]

We shall denote quantum-mechanical operators with a caret. Thus $\hat{H}$ is the total Hamiltonian operator, $\hat{P}_j$ is the momentum operator associated with mode j, and $\hat{Q}_j$ is the corresponding position operator. For a single mode, these operators act in the Hilbert space $\mathbb{L}_2(\mathbb{R})$. In Dirac notation, a general vector in this space will be denoted $|\psi\rangle$, analogous to the abstract vector $\mathbf{f}$ we have used in previous chapters to represent a square-integrable function $f(x)$. Since $\hat{H}$, $\hat{P}_j$ and $\hat{Q}_j$ all represent physical observables, we know from the basic axioms of quantum mechanics that they are Hermitian operators. See Sec. 1.4.4 for important background information on Hermitian operators.

In spite of the mechanical terminology, $\hat{Q}_j$ and $\hat{P}_j$ represent field amplitudes, but we shall assume (heuristically) that they obey the same rules as for mechanical position and momentum operators. Specifically, $\hat{Q}_j$ and $\hat{P}_j$ do not commute; instead they satisfy

$$\left[\hat{Q}_j, \hat{P}_j\right] \equiv \hat{Q}_j\hat{P}_j - \hat{P}_j\hat{Q}_j = i\hbar, \tag{10.19}$$

where the square brackets denote a *commutator*. Any operator commutes with itself, so $[\hat{Q}_j, \hat{Q}_j] = [\hat{P}_j, \hat{P}_j] = 0$. Moreover, operators associated with different field modes commute, so $[\hat{Q}_j, \hat{P}_{j'}] = 0$ if $j \neq j'$.

Two useful non-Hermitian operators are the operator equivalents of a_j and a_j^*, which we shall call $\hat{a}_j$ and $\hat{a}_j^\dagger$, respectively. The dagger (†) denotes an adjoint, and quantum-mechanical adjoints are defined just as in classical linear algebra (see Sec. 1.3.5). For reasons that will emerge below, $\hat{a}_j$ is called an *annihilation operator* and $\hat{a}_j^\dagger$ is called a *creation operator*. From (10.19) and the operator counterparts

[2]One meaning of *heuristic* is that it refers to a method of teaching or discovery based on experimentation or trial and error. That meaning applies here since we cannot know if the procedure of replacing amplitudes by operators will have any physical meaning until we perform many experiments. To date, all such experiments have been spectacularly successful.

of (10.16), we can show that

$$\left[\hat{a}_j, \hat{a}_{j'}^\dagger\right] = \delta_{jj'} \,. \tag{10.20}$$

In terms of the creation and annihilation operators, the Hamiltonian can be written as

$$\hat{H} = \sum_j \tfrac{1}{2}\hbar\omega_j(\hat{a}_j^\dagger\hat{a}_j + \hat{a}_j\hat{a}_j) = \sum_j \hbar\omega_j(\hat{a}_j^\dagger\hat{a}_j + \tfrac{1}{2})\,, \tag{10.21}$$

where the equivalence of the two forms follows from (10.20). In either form, we can identify a single term as the partial Hamiltonian for the jth mode and denote it as $\hat{H}_j$.

The operator for the total electric field is given by [*cf.* (10.14)]

$$\hat{\mathbf{e}}(\mathbf{r}, t) = i\sum_j \gamma_j N_j \left[\exp(i\mathbf{k}_j \cdot \mathbf{r})\,\hat{a}_j - \exp(-i\mathbf{k}_j \cdot \mathbf{r})\,\hat{a}_j^\dagger\right] . \tag{10.22}$$

In quantum optics it is often useful to label the individual terms in the sum (10.22) according to whether they involve positive or negative frequencies. We know from (10.13) that the classical amplitude $a_j(t)$ varies as $\exp(-i\omega_j t)$ for positive ω_j, and in fact the same is true for the operator $\hat{a}_j$. We can therefore refer to $i\gamma_j N_j \exp(i\mathbf{k}_j\cdot\mathbf{r})\hat{a}_j$ as the *positive-frequency part* of the field in mode j and denote it as $\hat{\mathbf{e}}_j^+(\mathbf{r}, t)$, and similarly for the negative-frequency part.[3] Thus we write

$$\hat{\mathbf{e}}(\mathbf{r}, t) = \hat{\mathbf{e}}^+(\mathbf{r}, t) + \hat{\mathbf{e}}^-(\mathbf{r}, t) = \sum_j \left[\hat{\mathbf{e}}_j^+(\mathbf{r}, t) + \hat{\mathbf{e}}_j^-(\mathbf{r}, t)\right] . \tag{10.23}$$

Note that $\hat{\mathbf{e}}^+$ contains all the annihilation operators and $\hat{\mathbf{e}}^-$ contains all the creation operators. The usefulness of this decomposition in the analysis of optical detectors will be seen in Sec. 10.1.4.

Quantum harmonic oscillator The properties of quantum-mechanical harmonic oscillators are discussed in detail in virtually every book on quantum mechanics; some key points are listed here for reference.

An important difference between classical and quantum-mechanical harmonic oscillators is in the allowed energies. The total energy of a classical oscillator of resonant frequency ν_j can be any positive number since the amplitude can have any value. A quantum oscillator with the same resonant frequency, however, can take on only the energies $\mathcal{E}_{jn_j}$ which are eigenvalues of the Hamiltonian. An eigenvector of $\hat{H}_j$ will be denoted $|n_j\rangle$, and the eigenvalue equation is

$$\hat{H}_j|n_j\rangle = \mathcal{E}_{jn_j}|n_j\rangle \,. \tag{10.24}$$

For the harmonic oscillator, this equation will have a solution if and only if

$$\mathcal{E}_{jn_j} = (n_j + \tfrac{1}{2})\hbar\omega_j\,, \qquad n_j = 0, 1, \ldots . \tag{10.25}$$

Thus the nonnegative integer n_j specifies the number of quanta of excitation of the oscillator, and each quantum has energy ω_j. If the oscillator describes the jth mode

[3]The reader should not confuse the plus superscript with a pseudoinverse.

of the radiation field, we say that the mode contains n_j photons. The minimum energy $\frac{1}{2}\omega_j$ is called the *zero-point energy*; in quantum mechanics each mode of the radiation field contains some energy even if no photons are present.

Since $\hat{H}_j = \hbar\omega_j(\hat{a}_j^\dagger \hat{a}_j + \frac{1}{2})$ by (10.21), and since any vector is an eigenvector of an operator that simply multiplies the vector by a constant, we can see that $|n_j\rangle$ is also an eigenvector of $\hat{a}_j^\dagger \hat{a}_j$:

$$\hat{a}_j^\dagger \hat{a}_j |n_j\rangle = n_j |n_j\rangle\,. \tag{10.26}$$

As a result of this equation, $\hat{a}_j^\dagger \hat{a}_j$ is called the *number operator*; its eigenvalue is the number of photons in the state described by vector $|n_j\rangle$. This state is known variously as a *number state*, *Fock state*, *energy eigenstate* or *stationary state*.

Though $|n_j\rangle$ is an eigenstate of $\hat{a}_j^\dagger \hat{a}_j$, it is not an eigenstate of $\hat{a}_j^\dagger$ or $\hat{a}_j$ separately. Instead, these operators have the following effect on number states:

$$\hat{a}_j^\dagger |n_j\rangle = \sqrt{n_j + 1}\,|n_j + 1\rangle\,, \tag{10.27a}$$

$$\hat{a}_j |n_j\rangle = \sqrt{n_j}\,|n_j - 1\rangle\,. \tag{10.27b}$$

Thus the creation operator acting on a number state increases the number of photons by one and the annihilation operator decreases the number by one—whence the names.

Other quantum states So far we have discussed only number states. A more general state of the jth mode is a linear superposition of number states, expressed by

$$|\psi_j\rangle = \sum_{n_j=0}^{\infty} c_{jn_j} |n_j\rangle\,. \tag{10.28}$$

The coefficients in this expansion are the *probability amplitudes* for the individual states. In terms of these (generally complex) amplitudes, the actual probability of observing an eigenenergy $\mathcal{E}_{jn_j}$ is $|c_{jn_j}|^2$. According to the basic tenets of quantum mechanics, only an eigenenergy can ever be observed, even in this superposition state.

As an example, the so-called *coherent states* (Glauber, 1963) are the best quantum approximation to a classical nonrandom wave. They have probabilities given by the Poisson distribution,

$$|c_{jn_j}|^2 = \exp\left(-|\alpha_j|^2\right) \frac{|\alpha_j|^{2n}}{n!}\,, \tag{10.29}$$

where α_j is a complex number designating a particular coherent state for mode j. Thus the probability of observing n photons in mode j obeys a Poisson law with mean $|\alpha_j|^2$ if the mode is excited to a coherent state. We shall discuss coherent states in more detail in Chap. 11 when we analyze photon-counting experiments.

Many other quantum states of the field are possible, some with no classical analog (see Sec. 11.5), but all can be expressed as linear superpositions of states with definite numbers of photons. Regardless of the amplitudes c_{jn_j}, however, the spatio-temporal character of the radiation in a single mode is that of a monochromatic plane wave.

Multimode states and localized photons Strictly speaking, photons are defined as quanta of excitation of a single mode of the radiation field, but colloquially we often speak as though photons were localized bundles of energy. Since all excitations of a single mode are monochromatic plane waves, a photon must strictly be uniformly distributed over the cavity, and the only way we can construct localized excitations is by considering multiple modes.

A multimode number state can be denoted $|\{n_j\}\rangle$; it is simultaneously an eigenstate of all of the single-mode number operators, *i.e.*,

$$\hat{a}_i^\dagger \hat{a}_i | \{n_j\}\rangle = n_i |\{n_j\}\rangle \,, \tag{10.30}$$

where $\{n_j\}$ denotes an infinite set of nonnegative integers. The state $|\{n_j\}\rangle$ is also an eigenstate of the total number operator $\sum_i \hat{a}_i^\dagger \hat{a}_i$, so the total number of photons in all modes is precisely $\sum_j n_j$. A special case of $|\{n_j\}\rangle$ is the state with one photon in mode i and zero in all other modes; we shall denote this state as $|1_i, \{0\}_{j\neq i}\rangle$.

A straightforward approach to constructing a localized one-photon state uses an analogy to classical wave packets (Mandel and Wolf, 1995). Suppose we define a multimode state $|\psi\rangle$ by

$$|\psi\rangle = \sum_i c_i |1_i, \{0\}_{j\neq i}\rangle \,. \tag{10.31}$$

No matter what we choose for the coefficients c_i, this state is an eigenstate of the total number operator with eigenvalue 1:

$$\left[\sum_m \hat{a}_m^\dagger \hat{a}_m\right] |\psi\rangle = \sum_i c_i \sum_m \hat{a}_m^\dagger \hat{a}_m |1_i, \{0\}_{j\neq i}\rangle = \sum_i c_i |1_i, \{0\}_{j\neq i}\rangle = 1 \cdot |\psi\rangle \,. \tag{10.32}$$

As an example, we can choose c_i such that the only wavevectors included in the sum are those with $|\mathbf{k}_i| \simeq \omega_0/c$, thus creating a quasimonochromatic state, and we can further apply a Gaussian weighting on directions around some chosen vector $\mathbf{k}_0$. More precisely, we can set $c_i = 0$ if $|k_i - \omega_0/c| > \delta$ and, within this range,

$$c_i = C \exp\left[-\frac{|\mathbf{k}_i - \mathbf{k}_0|^2}{2\sigma^2}\right] \,, \tag{10.33}$$

where C is a suitable normalizing constant. With these conditions, $|\psi\rangle$ represents a Gaussian beam (see Sec. 9.5.3) with mean wavevector $\mathbf{k}_0$ and some spread around this direction, but all spatially dependent operators such as $\hat{\mathbf{e}}(\mathbf{r}, t)$ are confined laterally (in a sense that will be clarified in Sec. 10.1.4) in the plane perpendicular to $\mathbf{k}_0$, just as a classical Gaussian beam would be. We can say that the state is localized to this beam, yet it contains precisely one photon.

Another way to associate photons with particular positions is through an operator corresponding to the energy density:

$$\hat{U}(\mathbf{r}, t) \equiv \epsilon_0 \hat{\mathbf{e}}^-(\mathbf{r}, t) \cdot \hat{\mathbf{e}}^+(\mathbf{r}, t) \,. \tag{10.34}$$

Like the classical energy density, this operator has dimensions of energy per unit volume. To get the total energy, we can integrate $\hat{U}(\mathbf{r}, t)$ over the cavity volume and make use of the orthogonality of the cavity modes along with (10.22) and (10.23); the result is

$$\int_{cav} d^3\mathbf{r}\ \hat{U}(\mathbf{r}, t) = \sum_j \hbar\omega_j \hat{a}_j^\dagger \hat{a}_j \,, \tag{10.35}$$

which is the same as the Hamiltonian operator except that there are no zero-point terms.

Integration of $\hat{U}(\mathbf{r},t)$ over a smaller volume yields an operator related to the energy in that volume. For quasimonochromatic radiation of frequency ω_0, the operator

$$\frac{1}{\hbar\omega_0}\int_V d^3\mathbf{r}\;\hat{U}(\mathbf{r},t) \tag{10.36}$$

corresponds to a local photon number, and its expectation in any quantum state can be interpreted as the average number of photons in the volume V for that state.

Similarly, for quasimonochromatic radiation with all wavevectors nearly parallel, we can define an operator $\hat{I}_p(\mathbf{r},t)$ by

$$\hat{I}_p(\mathbf{r},t) = \frac{c}{\omega_0}\hat{U}(\mathbf{r},t) = \frac{c\epsilon_0}{\omega_0}\hat{\mathbf{e}}^-(\mathbf{r},t)\cdot\hat{\mathbf{e}}^+(\mathbf{r},t)\,. \tag{10.37}$$

In the quantum-optics literature, $\hat{I}_p(\mathbf{r},t)$ is called the *photon intensity operator*, but in our terminology $\omega_0\hat{I}_p(\mathbf{r},t)$ corresponds to the classical normal irradiance I_0, or magnitude of the Poynting vector.

Though the operators $\hat{I}_p(\mathbf{r},t)$ and $\hat{U}(\mathbf{r},t)$ have an appealing analogy to the corresponding classical quantities, they must be used with caution. For example, $\hat{I}_p(\mathbf{r},t)$ and $\hat{I}_p(\mathbf{r}',t')$ do not commute in general, and theories making use of $\hat{I}_p(\mathbf{r},t)$ are often restricted in the kinds of fields for which they apply. Many books simply dismiss the issue by saying that it is impossible to define a strict photon-position operator. Nevertheless, as we shall see in Sec. 10.1.4, the basic operator $\hat{\mathbf{e}}^-(\mathbf{r},t)\cdot\hat{\mathbf{e}}^+(\mathbf{r},t)$ used in the definition of $\hat{I}_p(\mathbf{r},t)$ and $\hat{U}(\mathbf{r},t)$ is of fundamental importance in describing photon absorption by a small detector located at point $\mathbf{r}$, and in this sense at least we can speak of localized photons.

Mandel and Wolf (1995) give a detailed discussion of issues associated with photon localization. They conclude that localized photons can be defined, at least approximately, provided we do not try to localize them to a scale comparable to or smaller than a wavelength. The problem was also analyzed in detail by Bialynicki-Birula (1998) who concluded that localization with exponential falloff in energy density and photodetection rates was possible.

These results provide some impetus for the intuitive notion of a photon as a fuzzy blob, but there are some well-known situations where this intuition can fail us. In an interferometer, for example, any attempt to localize the photon to one arm will necessarily destroy the fringes. We shall revisit the issue of photon localization in Sec. 10.1.4 after developing some background on photodetection.

10.1.4 Physics of photodetection

There are many kinds of detectors of electromagnetic radiation (Dereniak and Crowe, 1984; Kingston, 1995). They are often classified as either thermal or photoelectric, but in fact even thermal detectors start by absorbing the radiation through a photoelectric interaction. The distinction is whether the photoelectron is observed directly as a current or it produces heat, which in turn alters some other property of the material such as its conductivity. Basically, all detectors of electromagnetic radiation are photoelectric.

In this section we briefly discuss the photoelectric interaction process, first by use of QED and then from a semiclassical perspective in which the atom is treated quantum mechanically but the field is assumed to be classical. Excellent references for both approaches include Loudon (1973), Sargent *et al.* (1974), Meystre and Sargent (1990), Cohen-Tannoudji *et al.* (1989) and Mandel and Wolf (1995).

Quantum-mechanical perspective Electromagnetic fields interact with matter predominantly through electric-dipole interactions. Considered as a point charge $-e$, an electron bound to an atom has a classical electrical dipole moment given by $-e\mathbf{r}$, where $\mathbf{r}$ is the position of the electron. The energy of the electron in an external electric field $\mathbf{e}(\mathbf{r},t)$ is $e\mathbf{r}\cdot\mathbf{e}(\mathbf{r},t)$. To get the corresponding quantum-mechanical interaction Hamiltonian, we replace the classical field by the operator defined in (10.23) and the classical electron position by the operator $\hat{\mathbf{r}}$. The $\mathbf{r}$ in the argument of $\mathbf{e}(\mathbf{r},t)$, on the other hand, specifies the center-of-mass position of the atom, which we do not usually quantize. To distinguish these two positions, we simply omit the caret on the atom's center-of-mass position and denote it as $\mathbf{r}$, which we regard as a classical parameter. Then the interaction Hamiltonian has the form

$$\hat{H}_{int} = e\hat{\mathbf{r}}\cdot\hat{\mathbf{e}}(\mathbf{r},t) = e\hat{\mathbf{r}}\cdot\left[\hat{\mathbf{e}}^{+}(\mathbf{r},t)+\hat{\mathbf{e}}^{-}(\mathbf{r},t)\right]. \tag{10.38}$$

We now want to use this Hamiltonian to compute the transition rate $R_{i\to f}$ for transitions from some specified initial state $|i\rangle$ to any final state $|f\rangle$. A well-known result, called *Fermi's Golden Rule*, states that this rate is given by

$$R_{i\to f} = \frac{2\pi}{\hbar}\sum_{f}|\langle f|\hat{H}_{int}|i\rangle|^{2}\,\delta(\mathcal{E}_f-\mathcal{E}_i)\,, \tag{10.39}$$

where in principle the sum is over all possible final states, but the delta function picks out only transitions for which the total energy (atom plus field) in the initial state, denoted $\mathcal{E}_i$, is the same as in the final state, $\mathcal{E}_f$. For photodetection problems we are interested only in final states where the electron is free and can be detected somehow, so the sum is over free-electron states. Moreover, if we assume that the transition is from the ground state of the atom to some excited state, it must involve absorption of energy from the field, so only photon annihilation operators contribute. These operators are contained in the term $e\hat{\mathbf{r}}\cdot\hat{\mathbf{e}}^{+}$ in the interaction Hamiltonian, and matrix elements of $e\hat{\mathbf{r}}\cdot\hat{\mathbf{e}}^{-}$ can be dropped.

For simplicity, we now assume that the light is linearly polarized, which means that all modes present have the same polarization unit vector γ_j and the index j is superfluous. Then the important term in $\hat{H}_{int}$ is the product of two scalar operators, $e\gamma\cdot\hat{\mathbf{r}}$ and a scalar operator $\hat{e}^{+}$. The first of these factors involves only the atomic operators and the second involves only the field operators; therefore the interaction matrix element factors as

$$\langle f|\hat{H}_{int}|i\rangle = \langle f_{at}|e\gamma\cdot\hat{\mathbf{r}}|i_{at}\rangle\langle f_{rad}|\hat{e}^{+}|i_{rad}\rangle\,, \tag{10.40}$$

where $|i_{rad}\rangle$ and $|f_{rad}\rangle$ are the initial and final states of the radiation field and $|i_{at}\rangle$ and $|f_{at}\rangle$ are those of the atom. The summation over final states factors similarly, and we can write

$$R_{i\to f} = \frac{2\pi}{\hbar}\sum_{f_{rad}}|\langle f_{rad}|\hat{e}^{+}|i_{rad}\rangle|^{2}\sum_{f_{at}}|\langle f_{at}|e\gamma\cdot\hat{\mathbf{r}}|i_{at}\rangle|^{2}\,\delta(\mathcal{E}_f-\mathcal{E}_i)\,. \tag{10.41}$$

For quasimonochromatic radiation of radian frequency ω_0, the initial and final energies of the field will differ by $\hbar\omega_0$ since $e\gamma \cdot \hat{\mathfrak{r}}$ involves only annihilation operators. Then the delta function in (10.41) can be written as $\delta(\mathcal{E}_{f,at} - \mathcal{E}_{i,at} - \hbar\omega_0)$, which does not depend explicitly on the states of the field. Thus the sum over final atomic states can be performed independently of the one over final field states.

To proceed in detail, we would convert the sum over atomic states in (10.41) into an integral over $\mathcal{E}_{f,at}$ with an appropriate density of states. The essential points can be seen, however, simply by defining an atomic property $C(\omega_0)$ by

$$C(\omega_0) \equiv \frac{2\pi}{\hbar} \sum_{f_{at}} |\langle f_{at}|e\gamma \cdot \hat{\mathfrak{r}}|i_{at}\rangle|^2 \, \delta(\mathcal{E}_{f,at} - \mathcal{E}_{i,at} - \hbar\omega_0) \,. \tag{10.42}$$

With this definition, we have

$$R_{i\to f} = C(\omega_0) \sum_{f_{rad}} |\langle f_{rad}|\hat{e}^+|i_{rad}\rangle|^2 \,. \tag{10.43}$$

A further simplification follows by realizing that $\hat{e}^-$ is the adjoint of $\hat{e}^+$, so

$$R_{i\to f} = C(\omega_0) \sum_{f_{rad}} \langle i_{rad}|\hat{e}^-|f_{rad}\rangle\langle f_{rad}|\hat{e}^+|i_{rad}\rangle = C(\omega_0)\langle i_{rad}|\hat{e}^-\hat{e}^+|i_{rad}\rangle \,, \tag{10.44}$$

where we have used the closure relation (see Sec. 1.3.7),

$$\sum_{f_{rad}} |f_{rad}\rangle\langle f_{rad}| = \hat{I}_{rad} \,, \tag{10.45}$$

with $\hat{I}_{rad}$ being the unit operator in the state space of the field (not to be confused with the photon intensity operator). The final form for the transition rate can be written most simply as

$$R_{i\to f} = C(\omega_0)\langle \hat{e}^-\hat{e}^+\rangle \,, \tag{10.46}$$

where $\langle \; \rangle$ denotes a quantum-mechanical expectation value, in this case an expectation of the field operator $\hat{e}^-\hat{e}^+$ in the initial state of the field. We encountered this same operator in Sec. 10.1.3 and saw that it could be identified with the square of the classical field. Moreover, within a constant, $\hat{e}^-\hat{e}^+$ is the same as the photon intensity operator or normal irradiance [see (10.37)]. Thus (10.46) shows that the photoelectric detection rate is proportional to the expectation of the square of the field or to the expectation of the normal irradiance.

Semiclassical perspective Perhaps surprisingly, photodetection can also be analyzed successfully (in most cases) with a semiclassical model in which the atom is treated quantum mechanically but the field is assumed to be classical, obeying Maxwell's equations. For a nonrandom, quasimonochromatic classical field $\mathbf{e}(\mathfrak{r}, t)$, the interaction Hamiltonian has the same form as in (10.38) but with no caret on $\mathbf{e}$. The sum over final states now includes only the atomic states (since the field is not quantized), and (10.46) becomes

$$R_{i\to f} = C(\omega_0)\langle [e(\mathfrak{r}, t)]^2\rangle_T \,, \tag{10.47}$$

where now the averaging is just a time average over one cycle.

The important conclusion here is that photoelectric transition rates are proportional to the average of the square of the electric field. Basically the same conclusion was reached quantum-mechanically, but there the square of the field is an operator, and the average must be interpreted as the expectation in a quantum state describing the incident light. Semiclassically, the square of the field has its natural interpretation, and the average is over time. In both cases, the important quantity is a mean-square field.

The great utility of the semiclassical result is that one can do completely classical calculations of electrical fields and, at the end, simply compute a classical mean-square field. This quantity then gives directly the rate of photoelectric transitions.

In almost all cases of importance in imaging, the semiclassical procedure will give essentially the same result as a fully quantum-electrodynamical calculation, but with much less effort. If one works very hard, it is possible to devise experiments in which the semiclassical and quantum approaches predict different results, and in those cases QED invariably turns out to be correct. We shall pursue these issues further in Chap. 11.

Who needs localized photons? The discussion above shows that calculation of photon-counting rates, either semiclassically or quantum-electrodynamically, boils down to computation of a mean-square field. There is never any need to think of photons falling on a detector. Indeed, the quantum purist would insist that such a statement is meaningless since photons are single-mode quanta and hence extend throughout all space. The localized event, in this view, is the photoelectric interaction, which involves the collapse of the wavefunction; the wavefunction itself is not localized, and does not have to be.

In practice, however, computation of a mean-square field may be very difficult. Consider, for example, a small source of short-wavelength radiation such as x rays or gamma rays and a small detector some distance away. We know from practical experience that if we place an opaque obstacle, such as a piece of lead, between the source and detector, we get no photocounts, but it is not straightforward to reach this obvious conclusion by computing a mean-square field. We might, for example, model the source as an ideal point, say it emits a spherical wave and compute the field classically behind the obstacle at the detector location. If this field is zero, there will be no counts on the semiclassical detection model. The problem is that the conclusion — no counts — is far more general than the point-source calculation would suggest. It holds for any spatial and temporal distribution of the source and any state of coherence, so long as the line of sight is blocked and diffraction around the obstacle can be neglected. We may not know these details of the source, and we probably do not care. All we know, or need to know, is that the line of sight is blocked.

In this problem, and many others of practical importance in imaging, it serves our computational needs better to think of the source as emitting localized wavepackets, which we shall call photons for want of a better term, and to think of these packets as travelling in a straight line from the source to the detector. If all relevant dimensions in the problem are large compared to a wavelength, then we can ignore diffraction in the classical perspective, and we can ignore the mathematical issues, raised in Sec. 10.1.3, associated with defining localized photon states.

This approach has much in common with the usual description of electron transport in semiconductors. Elementary solid-state physics tells us that an electron wavefunction in an ideal crystal is a modulated plane wave (called a *Bloch function*), so the electron, like the photon, extends throughout the medium. To think clearly about diodes and transistors, however, it is essential to assign a degree of localization to the electron and to imagine a fuzzy bundle of charge being swept through a P-N junction. The bundle is formed, as in (10.31), by superimposing pure electron wavefunctions with different wavevectors. So long as all states in the superposition respond to external forces in essentially the same way, it is legitimate to think of the localized packet as responding this way.

Another way to think about the issue of localization is to inquire why the plane-wave descriptions of electrons or photons are used in the first place. Why, for example, did we start in (10.12) with a Fourier series or plane-wave expansion for the field? The answer was given in Chap. 7, where we emphasized the role of Fourier analysis in analyzing linear shift-invariant (LSIV) systems. An electric field in free space, or in a cavity that will be allowed to become infinite, is governed by linear equations with no preferred origin of coordinates, so these equations describe an LSIV system. Similarly, the Schrödinger equation for an electron in an ideal crystal has discrete translational symmetry. In both cases, the eigenfunctions (normal modes) are essentially plane waves.

When we break the translational symmetry, for example with a P-N junction or an opaque obstacle, the normal modes are no longer plane waves, and we probably cannot calculate exactly what they are. Instead, in the localized approach, we use the plane waves as basis functions to construct approximate normal modes in much the same way as we used local Fourier transforms to analyze weakly shift-variant systems in Sec. 7.2.8. So long as we do not attempt to localize too finely and we can neglect interference effects, no significant error results.

10.1.5 What do real detectors detect?

The results in Sec. 10.1.4 were obtained by considering an isolated single atom as the detector. Real detectors consist of many atoms, and the processes that occur after the initial photoelectric interaction can be quite complicated.

One complication is that the response of the detector to a photon will usually depend to some degree on where in the detector material the photoelectron is produced. For example, photoelectrons generated at different depths within the photocathode of a photomultiplier will have different probabilities of escaping the cathode and starting a cascade of secondary electrons. Similarly, in a photodiode, there is a region of high electric field (the *depletion region*, for readers familiar with P-N junctions), and photoelectrons produced in this region are much more likely to contribute to the external current than those produced outside the region.

In addition, the squared electric field itself can vary over the volume of the detector, and it can depend in a complicated way on the time t. Moreover, the response produced by radiation of frequency ω is a complicated function of ω; this latter issue is discussed at the end of this section, but initially we consider quasimonochromatic radiation.

The mean detector output for quasimonochromatic radiation can be defined as

$$g_{out}(\omega_0) = n_{at} C(\omega_0) \int_0^{\tau} dt \int_{det} d^3\mathbf{r}\ S(\mathbf{r}) \left\langle |\mathbf{e}(\mathbf{r}, t)|^2 \right\rangle , \tag{10.48}$$

where n_{at} is the number of absorbing atoms per unit volume in the detector, $\mathbf{e}(\mathbf{r}, t)$ is the electric field at point $\mathbf{r}$ inside the detector at time t, τ is the exposure time, and $S(\mathbf{r})$ is the probability that an interaction at point $\mathbf{r}$ will contribute to the detector output. Thus $g_{out}(\omega_0)$ is the mean number of contributing interactions within the detector volume during the exposure time τ.

Note that (10.48) works equally well from a classical or quantum-mechanical perspective, depending on how we interpret the field and the average. Various special cases of this formula will now be discussed.

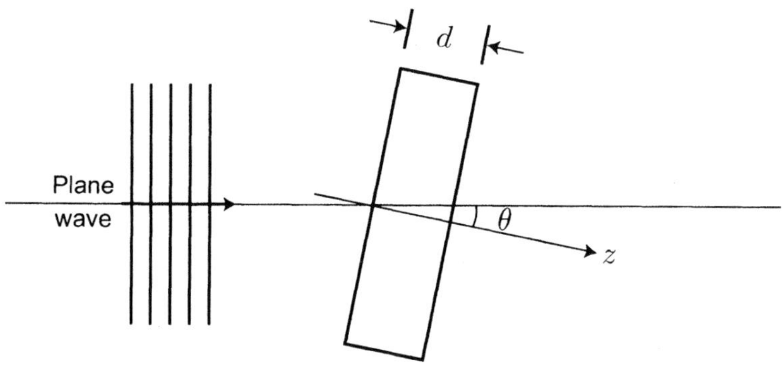

Fig. 10.1 Illustration of a photodetector illuminated with a plane wave. Note that the coordinate system is fixed to the detector.

Response to a plane wave Consider a photodetector of dimensions $L \times L \times d$ illuminated with a monochromatic plane wave tipped at angle θ as shown in Fig. 10.1. For simplicity, we imagine that the detector is immersed in an index-matching fluid so we do not have to consider Snell's law. We assume for now that $S(\mathbf{r})$ is a constant and focus attention on the $\mathbf{r}$ dependence of the electric field.

Since the radiation is absorbed, the wave is exponentially attenuated in the detector, but because of the tip angle the beam has traversed a thickness $z/\cos\theta$ when it is at depth z in the detector (see Fig. 10.1). Thus, in a coordinate system fixed to the detector, (10.48) becomes

$$\begin{aligned} g_{out}(\omega_0) &= n_{at} C(\omega_0) |E_0|^2 \tau \int_{-\frac{1}{2}L}^{\frac{1}{2}L} dx \int_{-\frac{1}{2}L}^{\frac{1}{2}L} dy \int_0^d dz \exp\left(-\frac{\mu z}{\cos\theta}\right) \\ &= n_{at} C(\omega_0) |E_0|^2 \tau L^2 \frac{\cos\theta}{\mu} \left[1 - \exp\left(-\frac{\mu d}{\cos\theta}\right)\right], \end{aligned} \tag{10.49}$$

where μ is the attenuation coefficient of the light in the detector material and E_0 is the field amplitude just inside the detector. Note that E_0 is nonrandom for now, so $\langle\cdot\rangle$ is not required.

An important limit of (10.49) is when $\mu d \gg 1$. In the optical wavelength range, μ is about 10^6 m^{-1} in typical semiconductor detector materials, so this limit prevails when d is at least a few μm. In this case,[4]

$$g_{out}(\omega_0) = n_{at} C(\omega_0) |E_0|^2 \tau \, L^2 \frac{\cos\theta}{\mu}. \tag{10.50}$$

[4]The factor of $1/\mu$ in (10.50) may be puzzling since it seems to imply that we get a smaller response by using a more absorbing material, even though all of the energy is absorbed when $\mu d \gg 1$, but in fact $C(\omega_0)$ is also proportional to μ. This issue is discussed further below under the heading Broadband polychromatic radiation.

Comparing (10.50) with (10.4) and (10.11) and noting that $\cos\theta = \hat{\boldsymbol{\kappa}} \cdot \hat{\mathbf{n}}$, we see that

$$g_{out}(\omega_0) \propto \int_S da\ \hat{\mathbf{n}} \cdot \mathbf{\Pi}(\mathbf{r}, t)\,. \tag{10.51}$$

The integral in this limit is the energy flux (total power transmitted across the detector face) or, equivalently, the integral of the irradiance over the detector face. Because of the factor of τ, the detector senses the total energy transmitted across the face in the measurement time.

In the limit of an optically thin detector, however, where $\mu d/\cos\theta \ll 1$, (10.49) becomes

$$g_{out}(\omega_0) = n_{at} C(\omega_0) |E_0|^2 \tau L^2 d\,. \tag{10.52}$$

In this case the detector senses the square of the field, not the energy flux. The power intercepted by the detector varies as $\cos\theta$ but the fraction absorbed is $\mu d/\cos\theta$, so the total absorbed power is independent of θ. An optically thick detector absorbs all of the energy of a beam and gives a response proportional to the energy flux, while an optically thin detector measures the square of the field. For normal incidence, the difference is only in the constant of proportionality, but thin and thick detectors behave differently as the angle of incidence is changed.

Photocathodes In a popular type of photomultiplier, light is incident on one side of the photocathode and electrons are emitted from the opposite side. These photocathodes cannot be made optically thick since then the electrons would have to traverse a large thickness of the material, and their probability of escape would be reduced. To develop a simplified model of such cathodes, assume that the electrons are exponentially attenuated with an attenuation coefficient μ_e, so that an electron produced at depth z has a probability $\exp[-\mu_e(d-z)]$ of escaping. This probability is the factor $S(\mathbf{r})$ in (10.48), so (10.49) is modified to

$$\begin{aligned} g_{out}(\omega_0) &= n_{at} C(\omega_0) |E_0|^2 \tau \int_{-\frac{1}{2}L}^{\frac{1}{2}L} dx \int_{-\frac{1}{2}L}^{\frac{1}{2}L} dy \int_0^d dz \exp\left[-\frac{\mu z}{\cos\theta} - \mu_e(d-z)\right] \\ &= n_{at} C(\omega_0) |E_0|^2 \tau L^2 e^{-\mu_e d} \left[\frac{\mu}{\cos\theta} - \mu_e\right]^{-1} \left\{1 - \exp\left[-\left(\frac{\mu}{\cos\theta} - \mu_e\right) d\right]\right\}. \end{aligned} \tag{10.53}$$

This expression approaches zero as either $d \to 0$ or $d \to \infty$, so some intermediate thickness must be chosen. Then the detector is neither optically thin nor optically thick, so it measures neither field squared nor energy flux.

Multiple plane waves Now consider two plane waves of the same frequency and phase, with parallel electric field amplitudes $\mathbf{E}_1$ and $\mathbf{E}_2$. For simplicity we again assume that the detector is immersed in an index-matching fluid so there is no need to worry about reflection or refraction effects. We assume also that $\mu d \gg 1$ and that $S(\mathbf{r})$ is constant.

With these assumptions, the total amplitude $\mathbf{E}$ at an arbitrary point $\mathbf{r}$ is

$$E^2 = E_1^2 + E_2^2 + 2E_1E_2 \cos[(\mathbf{k}_2 - \mathbf{k}_1) \cdot \mathbf{r}]\,, \tag{10.54}$$

and (10.48) becomes

$$g_{out}(\omega_0) = n_{at}C(\omega_0)\frac{\tau L^2}{\mu}\left[|E_1|^2 + |E_2|^2\right] + \frac{\tau}{\mu}2E_1E_2\int_{-\frac{1}{2}L}^{\frac{1}{2}L} dx \int_{-\frac{1}{2}L}^{\frac{1}{2}L} dy\ \cos[(\mathbf{k}_2 - \mathbf{k}_1)\cdot\mathbf{r}]$$

$$= n_{at}C(\omega_0)\frac{\tau L^2}{\mu}\left[|E_1|^2 + |E_2|^2 + 2E_1E_2\,\mathrm{sinc}\,L\Delta\xi\,\mathrm{sinc}\,L\Delta\eta\right] , \qquad (10.55)$$

where $\Delta\xi$ is the x component of $(\mathbf{k}_2 - \mathbf{k}_1)/2\pi$ and $\Delta\eta$ is the y component. The sinc functions are small if $L\Delta\xi$ or $L\Delta\eta$ is greater than about one. From the discussion in Sec. 9.2.1, we can relate these conditions to the angle $\Delta\theta$ between the two waves. From (9.45), we can see that $L\Delta\xi > 1$ or $L\Delta\eta > 1$ is equivalent to $L\Delta\theta/\lambda > 1$, and the term proportional to E_1E_2 can be neglected if this condition is satisfied. Since L/λ is usually $\gg 1$ with practical detectors, this condition can be satisfied even if $\Delta\theta$ is quite small. Physically, the interference term does not contribute to the mean detector response if there are one or more fringes across the width of the detector. Under these conditions, the detector again measures total flux, proportional to $|E_1|^2 + |E_2|^2$. The fact that the waves can interfere with each other does not affect the detector reading unless the detector is smaller than a fringe.

This argument extends to any number of plane waves and hence to general noncollimated radiation. We can ignore interference fringes that are not spatially resolved by the detector.

Narrowband polychromatic radiation Now consider two plane waves of slightly different frequencies but parallel $\mathbf{E}$. At time t there is an instantaneous phase difference between the waves given by $\Delta\phi = 2\pi\Delta\nu t$, where $\Delta\nu$ is the difference in frequencies. For simplicity assume that the two wave vectors are parallel to each other and to the z axis. Then the interesting part of (10.48) is the time integral, and we find

$$g_{out}(\omega_0) = n_{at}C(\omega_0)\frac{L^2}{\mu}\int_0^\tau dt\ \left\{E_1^2 + E_2^2 + 2E_1E_2\cos(2\pi\Delta\nu t)\right\} . \qquad (10.56)$$

The cosine term integrates to approximately zero if $\Delta\nu\tau \gg 1$, a condition that is quite easy to satisfy. For example, the frequency difference $\Delta\nu$ between the two sodium D lines is about 5×10^{11} Hz, and between two adjacent modes in a typical helium-neon laser, $\Delta\nu$ is about 5×10^7 Hz. With a 1 μsec integration time, the cosine would make no significant contribution to the response in either of these cases.

The conclusion is that the response is proportional to the total flux in the two beams, and once again the interference effects can be safely ignored. Instantaneous fringes are indeed produced, but many fringes move through the detector volume in the measurement time τ, so they have no measurable effect. As in the spatial case, we can extend the argument to many different frequencies. Any two frequencies ν and ν' for which $(\nu - \nu')\tau \gg 1$ are detected independently.

Broadband polychromatic radiation To extend the discussion above to broadband radiation, one obvious factor that must be considered is $C(\omega)$, defined in (10.42). Most detectors have a threshold photon energy below which there is no response, simply because there are no energy-conserving transitions. In semiconductor detectors, this minimum energy is the bandgap energy, and in photomultipliers, it is the workfunction of the photocathode. Thus $C(\omega) = 0$ for $\omega < \omega_{thr}$, where $\hbar\omega_{thr}$ is the minimum or threshold energy.

For $\omega > \omega_{thr}$, $C(\omega)$ has two effects: it relates the photoelectric transition rate to field squared [see (10.46) and (10.47)], and it controls the absorption coefficient μ of the radiation in the detector material. Suppose we increase $C(\omega)$, either by changing ω or by changing the material properties. Then the transition rate is increased and more energy per unit time is extracted from the beam as it propagates, and μ is increased proportionally. Thus the factor $C(\omega)/\mu$ in (10.50) is a constant, independent of both frequency and material properties.

The only residual effect of $C(\omega)$ is that it determines the field distribution within the detector through the factor $\exp[-\mu z/\cos\theta]$ in (10.49) or (10.53). For optically thick detectors, this factor influences the response only when $S(\mathbf{r})$ varies over the detector volume, as it does in (10.53) but not in (10.49).

One other issue that we need to discuss with broadband radiation is the precise meaning of the word response. As defined in (10.48), the response has units of total observable photoelectric transitions during the exposure time. In many detectors, each interaction yields either zero or one observable electron in an external circuit, with probability $S(\mathbf{r})$. In that case the total charge produced by narrowband radiation in time τ is just $eg_{out}(\omega_0)$, where e is the charge on the electron. To get the corresponding expression for broadband radiation, we must integrate against a radiation spectrum defined in terms of the photon flux per unit energy; precise ways of defining such spectra are given in Sec. 10.2. To the extent that $g_{out}(\omega)$ is a slowly varying function of ω, the detector output is then proportional to the total number of photons with energy above $\hbar\omega_{thr}$.

Some detectors, on the other hand, have a response that depends on the energy of each photon, not just on the number of photons absorbed. Consider, for example, an x-ray scintillation detector where the initial photoelectric interaction produces a high-energy electron, but the electron energy is quickly converted into optical photons. Each photoelectron in this case can produce many optical photons, and the mean number of optical photons is approximately proportional to the energy of the original x ray.

A similar situation occurs with certain infrared detectors in which the initial photoelectric interaction excites a free electron from an impurity atom. If the photon energy is larger than the binding energy of the impurity electron, then the photoelectron will have sufficient energy to excite lattice vibrations called *phonons*. Since lattice vibrations are basically the same thing as thermal energy, the photoelectron heats the medium, and the amount of heating is proportional to the photon energy. If we observe the heating with a temperature sensor or through its effect on the conductivity of the medium, then the observed response is related to the total absorbed energy, not the total number of photons.

When the response to an individual photon is proportional to the energy of the photon, as in the x-ray and infrared examples, then the response must be integrated against a spectrum defined in terms of energy flux per unit energy to get the total response to broadband radiation. When the response to an individual photon is approximately independent of its energy (at least above the threshold), as in photodiodes and photomultipliers, then the relevant spectrum involves the photon flux per unit energy.

In the remainder of this chapter we shall learn how these fluxes are specified precisely and how they move about in an imaging system.

10.2 RADIOMETRIC QUANTITIES AND UNITS

Classical radiometry describes the energy in a radiation field but ignores many other characteristics, including coherence, the wave nature of light, the fact that light is a vector field and all aspects of quantum electrodynamics. The main justification for this seemingly egregious oversimplification is the point made at the end of Sec. 10.1: practical detectors often respond rather simply to the energy deposited in them. Hence the most important thing we need to know is how much energy gets to the detector.

In this section we focus on ways of describing the energy content of a radiation field as a function of spatial and angular coordinates. The culmination of the discussion will be the definition of a quantity called the *distribution function* from which many other radiometric quantities can be derived. In defining this function, we use the language of photons, regarded as localized particles, but there is nothing quantum mechanical about the treatment. In Sec. 10.2.7, however, we shall return to the foundations of radiometry and explore further how the quantities defined in this section relate to both classical and quantum electrodynamics.

10.2.1 Self-luminous surface objects

We begin with self-luminous or emissive objects, but we need to distinguish surface emitters, considered here, from volume emitters, to be considered in Sec. 10.2.2. A surface emitter is one where the radiation originates on or very near the surface, and structures within the volume of the object do not substantially influence the radiation received by the imaging system. In a volume emitter, by contrast, the internal structure does influence the radiation received by the imaging system, and it is usually the goal of the system to image this internal, three-dimensional, structure.

If a surface emitter is planar, we can choose the coordinate system so that the surface lies in the plane $z = 0$. Hence, any measure of the strength of the emitted radiation will be a 2D function $f(x, y)$. Small deviations from a planar surface can often be accommodated by writing the z coordinate of a point on the surface as a function of x and y, so that the object function becomes $f(x, y, z(x, y))$, which is still a 2D function. This quasiplanar description is useful only in situations such as that illustrated in Fig. 10.2a where, from the vantage point of the imaging system, two coordinates are sufficient to determine uniquely a point on the surface; more convoluted surfaces as in Fig. 10.2b cannot be described fully this way. Most of the optics literature implicitly assumes a planar or quasiplanar surface object, and we shall also adopt that assumption for the remainder of Sec. 10.2.1. Points in the plane $z = 0$ will be denoted by a 2D vector $\mathbf{r} = (x, y)$.

Basic radiometric quantities We still have to specify just what physical entity is described by the function $f(\mathbf{r})$. Intuitively, $f(\mathbf{r})$ corresponds to the strength of the object, but we need a more precise definition. To arrive at such a definition, we begin with the concept of *radiant energy*.

Radiant energy Q is simply the total energy emitted by a source (over some specified time). The SI units of radiant energy are joules, abbreviated J. In many imaging applications, radiant energy per unit area is an important quantity. In photography, this quantity is called the *exposure*, and in radiological imaging it is called the *energy fluence*. We shall use the latter terminology and define the energy fluence Ψ by

$$\Psi = \frac{\partial Q}{\partial A}, \tag{10.57}$$

where A denotes area. SI units of Ψ are J/m^2.

There is still an ambiguity in this definition since it depends on how the area is specified. In radiology, fluence is defined by using a small sphere of cross-sectional area ΔA and dividing the energy intercepted by this sphere by ΔA. Though this device has some advantages when considering energy flow in a volume, we do not adopt it here; instead the area should be regarded as a small portion of a surface, so the fluence can, in general, depend on the orientation of the surface.

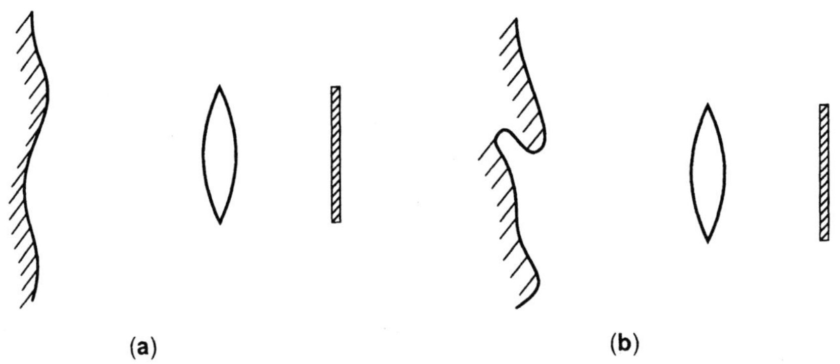

Fig. 10.2 (a) Illustration of a quasiplanar self-luminous object and an imaging system for viewing it. (b) Illustration of a more convoluted surface that, for the imaging system shown, cannot be described as quasiplanar.

The radiant flux Φ was already introduced in Sec. 10.1.1; it is the power, or rate of emission of radiant energy, with SI units of J/sec or watts, W. Thus

$$\Phi = \frac{\partial Q}{\partial t}. \tag{10.58}$$

A simple and broadly applicable description of a surface emitter is the *radiant exitance* M, defined as the radiant flux per unit area emitted by a surface, or

$$M = \frac{\partial \Phi}{\partial A} = \frac{\partial \Psi}{\partial t} = \frac{\partial^2 Q}{\partial t \partial A}. \tag{10.59}$$

SI units of radiant exitance are W/m^2. If the object is time varying, the energy emission can be expressed by $M(\mathbf{r}, t)$.

A radiometric quantity closely related to radiant exitance is the irradiance I, which we discussed in some detail in Sec. 10.1. In the notation of the present section, $I = \partial\Phi/\partial A$, which is (10.59) except that the flux in question is incident on a surface rather than emitted by it. Thus irradiance is flux per unit area in an image and exitance is flux per unit area emitted from a self-luminous object.

Directional properties of the energy flux can be expressed by the *radiant intensity*[5] Υ, defined as the radiant flux per unit solid angle, or

$$\Upsilon = \frac{\partial \Phi}{\partial \Omega} = \frac{\partial^2 Q}{\partial t \partial \Omega}. \tag{10.60}$$

[5]The letter Υ (upsilon) for radiant intensity is definitely not standard in the literature, but perhaps it suggests an angular spreading.

Solid angle is measured in steradians (ster), so the SI units of radiant intensity are W/ster. (Strictly speaking, angles are dimensionless, but it is handy to carry along the units of radians or steradians in order to check the consistency of various formulas.)

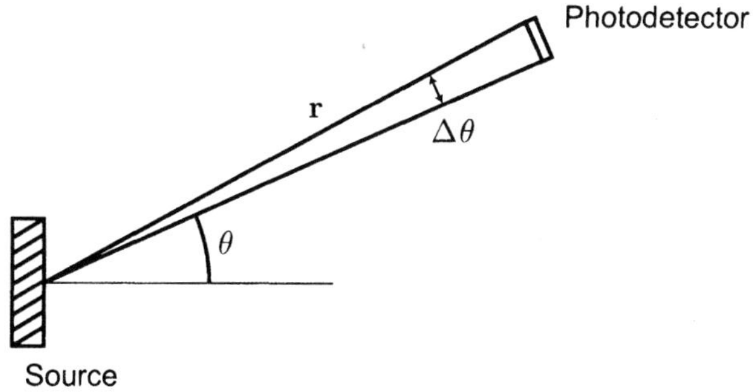

Fig. 10.3 Apparatus for measuring the radiant intensity of a small source. The solid angle $\Delta\Omega$, determined by the distance $\mathfrak{r}$ and the detector area, is constant for all detector positions.

Radiant intensity is particularly useful when a source is observed from distances large compared to its size. Then the source is effectively a point, but it does not necessarily radiate uniformly in all directions. A practical measurement setup for observing the radiant intensity is shown in Fig. 10.3, where the directionality of the source is observed by swinging a detector on an arc about the source. As we shall see in Sec. 10.2.7, this directionality is related to the correlation properties of the source considered as a random process, so the intensity Υ cannot, in general, be determined from knowledge of the exitance M.

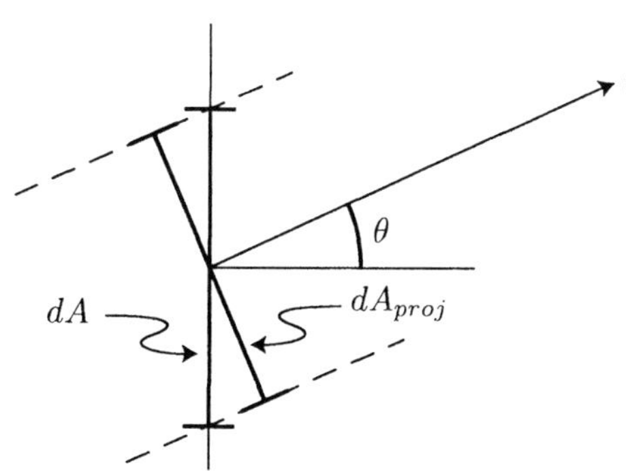

Fig. 10.4 Geometry for interpreting the concept of radiance. An infinitesimal area dA on a surface emits power $d\Phi$ into a differential solid angle $d\Omega$ in a direction making an angle θ to the surface normal. The radiance is the power per unit solid angle per unit *projected* area, where the projected area $dA_{proj} = \cos\theta\, dA$.

Radiance If we wish to specify both the spatial and angular dependence of energy flux, we can use the *radiance* L, which gives the radiant flux per unit area per unit solid angle. SI units of L are $\mathrm{W/(m^2 \cdot ster)}$. One quirk in the definition of radiance is that the area involved is not the actual area of an element of the source but rather that area projected onto the direction of the flux as shown in Fig. 10.4. An element of projected area dA_{proj} is related to an element of actual area by

$$dA_{proj} = \cos\theta\, dA = \hat{\mathbf{s}} \cdot d\mathbf{A}\,, \tag{10.61}$$

where θ is the angle between the direction of interest $\hat{\mathbf{s}}$ and the surface normal, and $d\mathbf{A}$ is a vector surface element with magnitude dA and orientation parallel to the surface normal. The definition of radiance is thus

$$L = \frac{\partial^2 \Phi}{\partial\Omega \partial A_{proj}} = \frac{1}{\cos\theta} \frac{\partial^2 \Phi}{\partial\Omega \partial A}\,. \tag{10.62}$$

Radiance can depend on position on the surface $\mathbf{r}$ and direction of the flux as specified by the unit vector $\hat{\mathbf{s}}$, so it will be denoted $L(\mathbf{r}, \hat{\mathbf{s}})$. A time argument will also be included when needed.

Radiance is equally applicable to flux incident on a surface or emitted from it; all that changes is the direction $\hat{\mathbf{s}}$. The irradiance/exitance dichotomy does not arise.

Lambertian surfaces One advantage of using projected area in the definition of radiance is that many natural surfaces then have a radiance that is approximately independent of θ. A surface emitter or reflector with L completely independent of direction $\hat{\mathbf{s}}$ is called a *Lambertian* surface. Another common term applied to a Lambertian emitter is *isotropic* (Chandrasekhar, 1960), implying that the radiance is independent of direction. We shall explore this designation further in Sec. 10.2.6.

A semantic confusion often arises because a Lambertian surface is said to obey *Lambert's cosine law of emission.* This statement does not imply a cosine dependence of radiance; rather, it refers to the radiant intensity. Comparison of (10.60) and (10.62) shows that a Lambertian surface has a radiant intensity that varies as $\cos\theta$.

For a Lambertian surface emitter, the relation between exitance and radiance is simple. Integrating $L\cos\theta$ (where the cosine comes from converting projected area to actual area) over 2π ster, we find (for a Lambertian)

$$M = \int_{2\pi} L\cos\theta\, d\Omega = \int_0^{2\pi} d\phi \int_0^{\pi/2} \sin\theta\, d\theta\, L\cos\theta = \pi L\,, \tag{10.63}$$

where θ and ϕ are the polar angles of the unit vector $\hat{\mathbf{s}}$, with θ measured from the surface normal.

Spectral dependence To specify the radiation completely, it may also be necessary to give its spectral distribution. If the source emits a range of wavelengths λ, each of the radiometric quantities defined above can be generalized to a function of λ. For example, the *spectral exitance* M_λ, defined by

$$M_\lambda = \frac{\partial^2 \Phi}{\partial A \partial \lambda}\,, \tag{10.64}$$

gives the emitted power per unit area per unit wavelength. If λ is measured in nanometers (nm), units of M_λ are $\mathrm{W/(m^2 \cdot nm)}$. Similarly, *spectral radiance* L_λ has units of units of $\mathrm{W/(m^2 \cdot ster \cdot nm)}$, being defined by

$$L_\lambda = \frac{1}{\cos\theta} \frac{\partial^3 \Phi}{\partial\Omega \partial A \partial\lambda}\,. \tag{10.65}$$

Spectral quantities can also be specified in energy or frequency units. For quantized electromagnetic radiation in some medium, the energy per photon is $\mathcal{E} = h\nu = hc_m/\lambda$, where h is Planck's constant, ν is the frequency of the radiation and c_m is the speed of light at frequency ν in the medium. Spectral radiance in frequency units (*i.e.*, radiance per unit frequency) is given by

$$L_\nu = \frac{1}{\cos\theta}\frac{\partial^3\Phi}{\partial\Omega\partial A\partial\nu} = L_\lambda\left|\frac{d\lambda}{d\nu}\right| = L_\lambda\frac{c_m}{\nu^2}\,, \tag{10.66}$$

with SI units $\mathrm{W/(m^2\cdot ster\cdot Hz)}$. In the older literature, *e.g.*, Chandrasekhar (1960), L_ν is called the *specific intensity* or often just *intensity*, but this risks confusion with the radiant intensity Υ or even with the irradiance, which is called intensity in many parts of the optics literature.

Spectral radiance per unit energy is defined by

$$L_\mathcal{E} = \frac{1}{\cos\theta}\frac{\partial^3\Phi}{\partial\Omega\partial A\partial\mathcal{E}} = L_\nu\left|\frac{d\nu}{d\mathcal{E}}\right| = \frac{1}{h}L_\nu\,. \tag{10.67}$$

Units of $L_\mathcal{E}$ depend, of course, on the units chosen for $\mathcal{E}$. The SI units would be joules, but photon energy is virtually never expressed in joules. If electron volts (eV) are used, units of $L_\mathcal{E}$ are $\mathrm{W/(m^2\cdot ster\cdot eV)}$.

The spectral descriptions are more complete than the original functions. From spectral exitance, for example, we can compute exitance via

$$M = \int_0^\infty d\lambda\; M_\lambda\,, \tag{10.68}$$

but without additional knowledge we cannot go in the opposite direction.

Photon radiometry Sometimes it is more convenient to describe the radiation in terms of photon flux rather than energy flux. If we consider a monochromatic source emitting photons of energy $\mathcal{E}$ (see Secs. 10.1.3. and 10.1.4), the photon flux Φ_p is related to the energy flux Φ by

$$\Phi_p = \frac{\Phi}{\mathcal{E}}\,. \tag{10.69}$$

Units of photon flux are photons per second or simply sec^{-1}. Similarly, photon exitance and photon radiance are defined by $M_p = M/\mathcal{E}$ and $L_p = L/\mathcal{E}$, with SI units $\mathrm{sec^{-1}m^{-2}}$ and $\mathrm{sec^{-1}m^{-2}\cdot ster^{-1}}$, respectively. If the radiation consists of particles, such as neutrons, the word 'particle' can be substituted for 'photon' throughout.

Spectral counterparts of these photon quantities can also be defined. For example, the *spectral photon radiance* (per unit wavelength) is given by

$$L_{p,\lambda} = \frac{\lambda}{hc\cos\theta}\frac{\partial^3\Phi}{\partial\Omega\partial A\partial\lambda}\,. \tag{10.70}$$

Spectral photon radiance is closely related to *brightness* (sometimes called *brilliance*) as used in the laser and synchrotron communities. Papers in those fields may omit the factor of $1/\cos\theta$ in defining brightness, and they may mean spectral photon radiance per unit energy rather than per unit wavelength depending on the application. We prefer to avoid the term brightness.

All of these source-strength functions can depend on the 2D position vector $\mathbf{r}$ and on the time t. Radiance and spectral radiance also depend on direction $\hat{\mathbf{s}}$, and the spectral quantities depend on wavelength or energy. The most complete description we have discussed so far is thus $L_\lambda(\mathbf{r}, \hat{\mathbf{s}}, \lambda, t)$. If needed, further arguments for state of polarization of the radiation can be added.

It is not necessary to specify all of these attributes of the source in all circumstances. We need only include those attributes that influence the response of the imaging system. For example, if the system is insensitive to wavelength, the spectral functions are not needed.

10.2.2 Self-luminous volume objects

Next we consider objects where the radiant energy is produced throughout a 3D volume. Position in the volume is specified by a 3D vector $\mathfrak{r}$ with components (x, y, z). Initially we assume that the object is transparent to its own radiation; the effects of internal absorption and scattering will be taken up later.

As with surface emitters, the discussion begins with radiant energy Q or radiant flux Φ. The volumetric counterpart of radiant exitance is the emitted radiant flux per unit volume, denoted S and defined by

$$S = \frac{\partial \Phi}{\partial V}\,, \tag{10.71}$$

with units of W/m^3. There is no generally accepted term for S; we shall call it simply the *emission density*.

If the radiation consists of photons of energy $\mathcal{E}$, the emitted photon flux per unit volume, denoted S_p, is given by

$$S_p = \frac{S}{\mathcal{E}} = \frac{1}{\mathcal{E}}\frac{\partial \Phi}{\partial V}\,. \tag{10.72}$$

We shall refer to S_p as the *photon emission density*. Units are photons/(sec $\cdot$ m^3) or equivalently sec^{-1}m^{-3}.

If the volume emitter is a radioactive source, a conventional descriptor is its *activity*, which is the total number of radioactive decays per second. SI units of activity are *Becquerels*,[6] abbreviated Bq, and 1 Bq = 1 sec^{-1}. The number of decays per second per unit volume of the source is its *specific activity*. If every radioactive decay resulted in exactly one photon, specific activity would be identical to S_p, but that is not usually the case.

Directional and spectral dependence To specify the directional dependence of the emitted radiation, we need to know the emitted flux per unit volume per unit solid angle (or, equivalently, emitted radiance per unit path length). We shall denote this quantity as Ξ and define it by

$$\Xi = \frac{\partial^2 \Phi}{\partial \Omega \partial V}\,. \tag{10.73}$$

[6] In the older literature activity was specified in *curies*, abbreviated Ci and defined as the activity of one gram of radium-226. The SI system abolished the curie, the only common unit named for a woman, and replaced it with one named for Henri Becquerel, Marie Curie's mentor.

Again, there is no accepted name for this quantity, though a few books use the obscure and nondescriptive term *sterisent* (Spiro *et al.*, 1965). We shall call Ξ simply the *source distribution.*

Many natural volume emitters are isotropic, emitting radiation uniformly into 4π ster. For such sources, $\partial/\partial\Omega$ can be replaced by $1/4\pi$ wherever it occurs, so

$$\Xi = \frac{1}{4\pi}\frac{\partial\Phi}{\partial V} = \frac{S}{4\pi}\,. \tag{10.74}$$

If the source emits radiation with a range of wavelengths or energies, spectral source distributions Ξ_λ and $\Xi_\mathcal{E}$ can be defined by

$$\Xi_\lambda = \frac{\partial^3\Phi}{\partial\Omega\partial V\partial\lambda}\,, \qquad \Xi_\mathcal{E} = \frac{\partial^3\Phi}{\partial\Omega\partial V\partial\mathcal{E}}\,. \tag{10.75}$$

Photon counterparts of these functions are given by

$$\Xi_{p,\lambda} = \left(\frac{\lambda}{hc_m}\right)\Xi_\lambda\,, \qquad \Xi_{p,\mathcal{E}} = \frac{1}{\mathcal{E}}\,\Xi_\mathcal{E}\,. \tag{10.76}$$

10.2.3 Surface reflection and scattering

Objects to be imaged need not generate their own radiation; they can also reflect or scatter radiation from external sources. Though etymologically distinct,[7] the terms reflection and scattering as used in optics are virtually synonymous, both referring to a redirection of incident radiation. In practice, scattering is more likely to be used for interaction of radiation with small particles within a volume, while reflection is used for surfaces.

The directionality of the redirected radiation is not the crucial determinant of which word is used. Surface reflection can be either *specular* or *diffuse.* Specular reflection (from Latin *speculum*, mirror) occurs when the surface is smooth on the scale of the wavelength of the radiation. In that case a highly directional beam of radiation remains highly directional, and the angle of reflection (measured from the surface normal) equals the angle of incidence. If the surface is rough, however, each element of the surface redirects radiation in a different direction, a phenomenon that can be called either diffuse reflection or surface scattering.

In fact, both specular and diffuse scattering are just manifestations of diffraction. As we saw in Sec. 9.5.4, the diffracted radiation from different surface elements on a smooth surface adds coherently to produce a highly directional reflected/scattered beam. In this view surface roughness imparts a random phase shift to the radiation, destroying its directionality.

Throughout this section we shall assume that the reflective object is planar and lies in the plane $z = 0$. Small deviations from the plane are easily accommodated if the surface can be approximated as locally planar. Very convoluted surfaces as in Fig. 10.2b are more difficult to analyze since one portion of the surface can obstruct either the radiation incident on or reflected from another portion. Analysis of such obstructions is central to computer graphics, image understanding and scene analysis, but will not be treated further here.

[7] Reflect comes from Latin *flectere*, to bend, and scatter comes from Middle English *scatere* (related to the Dutch *schateren*), to burst out laughing.

Reflectance and BRDF The officially sanctioned measure of strength of reflection from a surface is the *reflectance*, defined fundamentally as the ratio of reflected radiant flux to incident radiant flux (Palmer, 1995). We shall call this quantity R_{tot}. For imaging purposes, however, R_{tot} conveys little information since it refers to the total flux and not its spatial variation. A more useful quantity is the position-dependent reflectance $R(\mathbf{r})$, defined as the ratio of radiant exitance to irradiance. The two definitions coincide for a uniform surface where $R(\mathbf{r})$ is independent of position, but such objects are rather uninteresting from an imaging perspective.

We can quantify the diffusive properties of a reflective surface by use of the *bidirectional reflectivity distribution function* or BRDF, defined as the ratio of reflected radiance to *radiant incidence* (Palmer, 1995; Nicodemus, 1963*a*, *b*, 1973). Radiant incidence, with units of $\mathrm{W/m^2}$, is the irradiance of a highly collimated beam travelling in direction $\hat{\mathbf{s}}$, and hence BRDF specifies the reflected radiance produced by such a beam.

By its definition, BRDF has units ster^{-1}, suggesting that it can be used in an integral over solid angle. In fact, BRDF is the kernel in an integral transform relating reflected radiance to incident radiance. Since BRDF can depend on position $\mathbf{r}$ on the surface as well as the two directions, we write

$$L_{refl}(\mathbf{r},\hat{\mathbf{s}}) = \int_{2\pi} d\Omega' \, \mathrm{BRDF}(\mathbf{r},\hat{\mathbf{s}},\hat{\mathbf{s}}') \, L_{inc}(\mathbf{r},\hat{\mathbf{s}}') \cos\theta' \,, \tag{10.77}$$

where the angular integral covers the hemisphere of unit vectors $\hat{\mathbf{s}}'$ directed towards the surface, and $\hat{\mathbf{s}}$ lies in the opposite hemisphere. The factor of $\cos\theta'$ in the integrand is necessary since BRDF is defined in terms of radiant incidence, which depends on the actual surface area illuminated by a beam, not the projected area.

For a Lambertian surface, the reflected radiance must be independent of both $\hat{\mathbf{s}}$ and the details of incident radiance; only the total irradiance matters. Conservation of energy requires (for a Lambertian) that

$$L_{refl}(\mathbf{r},\hat{\mathbf{s}}) = \frac{1}{\pi} R(\mathbf{r}) \, I(\mathbf{r}) \,, \tag{10.78}$$

where $I(\mathbf{r})$ is the surface irradiance, $R(\mathbf{r})I(\mathbf{r})$ is thus the radiant exitance, and the factor $1/\pi$ converts exitance to radiance according to (10.63). The irradiance $I(\mathbf{r})$ is obtained by integrating $L_{inc}\cos\theta$ over solid angle, so

$$I(\mathbf{r}) = \int_{2\pi} d\Omega' \, L_{inc}(\mathbf{r},\hat{\mathbf{s}}') \cos\theta' \,. \tag{10.79}$$

Inserting (10.79) into (10.78) and comparing the result to (10.77) shows that the BRDF of a Lambertian surface is

$$\mathrm{BRDF}_{Lamb}(\mathbf{r},\hat{\mathbf{s}},\hat{\mathbf{s}}') = \frac{1}{\pi} R(\mathbf{r}) \,. \tag{10.80}$$

A *perfect Lambertian* surface is one that absorbs no energy, so that $R(\mathbf{r}) = 1$ and $\mathrm{BRDF} = 1/\pi$.

Spectral dependence None of the reflectance descriptors used so far takes any account of the spectral distribution of the radiation. If there is no wavelength shift on reflection, we can still use (10.77) but with a spectral BRDF defined as the ratio of reflected spectral radiance to spectral radiant incidence. This new $\mathrm{BRDF}(\mathbf{r}, \hat{\mathbf{s}}, \hat{\mathbf{s}}', \lambda)$ still has units of ster^{-1}. The original $\mathrm{BRDF}(\mathbf{r}, \hat{\mathbf{s}}, \hat{\mathbf{s}}')$ is not obtained by integrating $\mathrm{BRDF}(\mathbf{r}, \hat{\mathbf{s}}, \hat{\mathbf{s}}', \lambda)$ over λ. Instead,

$$\mathrm{BRDF}(\mathbf{r}, \hat{\mathbf{s}}, \hat{\mathbf{s}}') = \frac{\int_0^\infty L_\lambda(\mathbf{r}, \hat{\mathbf{s}}, \lambda)\, d\lambda}{\int_0^\infty I_\lambda(\mathbf{r}, \hat{\mathbf{s}}', \lambda)\, d\lambda}, \tag{10.81}$$

where $I_\lambda(\mathbf{r}, \hat{\mathbf{s}}, \lambda)$ is the spectral radiant incidence of a beam in direction $\hat{\mathbf{s}}$.

If inelastic processes such as Raman or Compton scattering can occur (see Sec. 10.2.5), BRDF has to be generalized to a function of $\mathbf{r}$, $\hat{\mathbf{s}}$, $\hat{\mathbf{s}}'$, λ and λ', with units of $\mathrm{ster}^{-1}\mathrm{nm}^{-1}$ if λ is measured in nm. Equation (10.77) then generalizes to

$$L_{\lambda,refl}(\mathbf{r}, \hat{\mathbf{s}}, \lambda) = \int_0^\infty d\lambda' \int_{2\pi} d\Omega'\, \mathrm{BRDF}(\mathbf{r}, \hat{\mathbf{s}}, \hat{\mathbf{s}}', \lambda, \lambda')\, L_{\lambda,inc}(\mathbf{r}, \hat{\mathbf{s}}', \lambda') \cos\theta' . \tag{10.82}$$

10.2.4 Transmissive objects

For transmissive objects, we must distinguish between thin objects and thick ones. A thin object is one in which there is little lateral spread of the radiation as it passes through the object. If a transmissive object is a slab of thickness L with flat, parallel faces, it can be placed so that light is incident on it in the plane $z = 0$ and emerges from the plane $z = L$. Position on both the input face and the output face can be specified by the 2D vector $\mathbf{r} = (x, y)$. The object is considered thin if the exitance M at point $\mathbf{r}$ in the exit plane is determined to a good approximation by the irradiance I at that same 2D point in the entrance plane. By direct analogy with the reflectance, the *transmittance* of such an object is a dimensionless quantity defined by

$$T(\mathbf{r}) = \frac{M(\mathbf{r})}{I(\mathbf{r})} . \tag{10.83}$$

Another common measure of transmission, especially for photographic film, is the *optical density* $D(\mathbf{r})$, defined by

$$D(\mathbf{r}) = -\log_{10}[T(\mathbf{r})] . \tag{10.84}$$

With either T or D, we can express the radiant exitance from a thin object as a simple multiplicative factor times the irradiance. For a thick object, by contrast, an integral is required, so we write

$$M(\mathbf{r}) = \int_\infty d^2r\, T(\mathbf{r}, \mathbf{r}')\, I(\mathbf{r}') , \tag{10.85}$$

where $T(\mathbf{r}, \mathbf{r}')$ has SI units of m^{-2}.

Directional properties Transmission, like reflection, can be either diffuse or specular. A general descriptor of the directional properties of thin transmissive objects is the *bidirectional transmission distribution function* or BTDF, defined by analogy to BRDF as the ratio of transmitted radiance L_{trans} to radiant incidence. The counterpart of (10.77) is

$$L_{trans}(\mathbf{r}, \hat{\mathbf{s}}) = \int_{2\pi} d\Omega'\, \mathrm{BTDF}(\mathbf{r}, \hat{\mathbf{s}}, \hat{\mathbf{s}}')\, L_{inc}(\mathbf{r}, \hat{\mathbf{s}}') \cos\theta' . \tag{10.86}$$

For thick transmissive objects, we can construct two parallel reference planes $z = 0$ and $z = L$ and specify two coordinates and two angles on each plane. The general linear input-output relation between these planes is

$$L_{trans}(\mathbf{r}, \hat{\mathbf{s}}) = \int_{\infty} d^2r' \int_{2\pi} d\Omega' \, T(\mathbf{r}, \hat{\mathbf{s}}; \mathbf{r}', \hat{\mathbf{s}}') \, L_{inc}(\mathbf{r}', \hat{\mathbf{s}}') \,, \qquad (10.87)$$

where L_{inc} is measured in the plane $z = 0$, L_{trans} is measured in the plane $z = L$, and $T(\mathbf{r}, \hat{\mathbf{s}}; \mathbf{r}', \hat{\mathbf{s}}')$ is a generalized transmittance function.

10.2.5 Cross sections

BRDF and BTDF describe absorption and scattering on a macroscopic scale, but often it is useful to relate these functions to elementary interaction processes on a microscopic scale. These processes are conveniently discussed in terms of *cross sections*, quantities with dimensions of area that specify the relative strengths of various physical interactions. As we shall see, cross sections can be defined for both absorption and scattering.

Scattering Scattering can be either *elastic* or *inelastic*. In terms of photons, elastic scattering is when the scattered photon has the same energy as the incident photon, while inelastic scattering involves a change in energy (usually a loss).

A classical view of elastic scattering is that an incident electric field of frequency ν induces dipole oscillations in the scattering object, and the dipoles then radiate at the same frequency. When this kind of scattering occurs in an atom or molecule, it is referred to as *Thomson* or *Rayleigh scattering*, and if it occurs in a larger object, such as a dielectric sphere, it is called *Mie scattering*.

Another form of elastic scattering is *resonant fluorescence* where an atom absorbs a photon of energy $h\nu$ from the field by a resonant transition and then emits a photon of the same energy by spontaneous emission.

Inelastic scattering involves a transfer of energy to some other particle or elementary excitation. In *Compton scattering*, important for x rays and gamma rays, the incident photon transfers part of its energy (and momentum) to an electron, producing a scattered photon of lower energy in a different direction.

Raman scattering is similar to Compton scattering except that the energy is transferred to a vibrational mode in a molecule or solid. One difference in practice between the Compton and Raman effects is that the vibrational modes have relatively low energy and hence have appreciable thermal excitation at room temperature. In addition to imparting energy to the molecule or solid, the photon can also absorb energy from these thermal vibrations, so the scattered photon can have either higher or lower energy than the incident photon. *Brillouin scattering* is a form of Raman scattering where the energy exchange is with vibrational waves called *acoustic phonons* in a solid.

Definitions of cross sections Scattering cross sections are defined for discrete entities such as atoms, molecules or dielectric spheres. In the radar literature, cross sections are even defined for airplanes and missiles. In all of these cases, a beam of radiation incident on the entity yields a certain scattered flux, and the scattering cross section

σ_{sc} is a measure of flux per unit irradiance. Specifically, for elastic processes,

$$\sigma_{sc} = \frac{\Phi}{I_0} = \frac{\text{scattered flux}}{\text{normal irradiance}} . \qquad (10.88)$$

The term *normal irradiance* was introduced and explained in Sec. 10.1.2.

For inelastic scattering, we must recognize that the incident and scattered photons have different energies. We can define the scattering cross section by

$$\sigma_{sc} = \frac{\Phi_p}{I_{0,p}} = \frac{\text{scattered photon flux}}{\text{normal photon irradiance}} , \qquad (10.89)$$

where normal photon irradiance is normal irradiance divided by the *incident* photon energy, while scattered photon flux is scattered flux divided by the *scattered* photon energy.

A cross section for absorption can be defined similarly, except that it makes little sense to talk about absorbed flux; after the absorption, there is no flux. Instead we back up to a basic definition, (10.58), and relate the rate of energy absorption by the atom (which, by conservation of energy, is also the rate of energy loss from the field) to the normal irradiance. We saw in (10.46) and (10.47) that the transition rate is linear in field squared or normal irradiance, so we can write

$$\frac{\partial Q_{atom}}{\partial t} = \sigma_{abs} I_0 , \qquad (10.90)$$

where σ_{abs} is the *absorption cross section.* Thus the basic definition is

$$\sigma_{abs} = \frac{\partial Q_{atom}/\partial t}{I_0} = \frac{\text{rate of energy absorption}}{\text{normal irradiance}} . \qquad (10.91)$$

As the term implies, a cross section is dimensionally an area. Since flux is measured in watts and irradiance in watts per m^2, SI units of cross section are m^2. A common unit for cross section is the *barn*, defined as $10^{-24} m^2$. The implication of this designation is that 1 barn is a very large cross section, a statement that is true in nuclear physics but not in optics. If an optical beam induces a resonant transition between energy levels in an atom, the cross section is[8] $\lambda_0^2/2\pi$, where λ_0 is the resonant wavelength (Loudon, 1973). In the visible region, λ_0 is around 500 nm, so optical absorption cross sections are of order $4 \times 10^{-14} m^2$.

If we think in terms of photons, the cross section has a simple geometric interpretation. The normal photon irradiance is the mean number of incident photons per second per unit area (where area is measured in a plane normal to the beam direction), and the mean number of photons per second scattered or absorbed is given by the average number per second passing through area σ.

[8] This formula holds when the incident radiation is exactly resonant and the only energy broadening of the transition results from the natural lifetime. It may be surprising that no dipole moment, oscillator strength or other quantity related to strength of the interaction between the atom and the radiation field enters into the cross section. Basically, this comes about since the linewidth (reciprocal of the lifetime) and the integral of the absorption over the line both scale as oscillator strength; thus the peak value exactly on resonance is independent of oscillator strength.

Differential scattering cross section Often we need to know not only the total scattered flux but also its angular distribution. For this purpose we use the *differential scattering cross section* $\partial\sigma_{sc}/\partial\Omega$, defined as the ratio of scattered photon intensity to the normal photon irradiance, so

$$\Upsilon_p = \frac{\partial\sigma_{sc}}{\partial\Omega} I_{0,p} , \tag{10.92}$$

where the photon intensity Υ_p is the photon flux per unit solid angle.

For elastic scattering, we can multiply both sides of this equation by the common photon energy and get

$$\Upsilon = \frac{\partial\Phi}{\partial\Omega} = \frac{\partial\sigma_{sc}}{\partial\Omega} I_0 . \tag{10.93}$$

This expression does not work for inelastic scattering where I_0 refers to a stream of photons of one energy, and Υ refers to a stream of photons of a different (usually lower) energy.

It is evident from the definition that the solid angle Ω that appears in $\partial\sigma_{sc}/\partial\Omega$ refers to the direction of the scattered radiation, but the differential cross section can also depend on the incident direction. In fact, unless we take heroic measures, the scattering centers will have random orientation. Averaged over orientations, $\partial\sigma_{sc}/\partial\Omega$ can be a function of only the deflection angle θ, defined by

$$\cos\theta = \hat{\mathbf{s}} \cdot \hat{\mathbf{s}}' , \tag{10.94}$$

where $\hat{\mathbf{s}}'$ is the direction of the incident radiation and $\hat{\mathbf{s}}$ is the direction of the scattered radiation. Therefore we can fully specify the directional character of the scattering by fixing the incident direction, say along the z axis in polar coordinates, and plotting the radiant intensity as a function of the polar angle θ. We shall denote this function as $[\partial\sigma_{sc}/\partial\Omega](\theta)$.

Differential scattering cross sections often depend strongly on the energy of the incident radiation. To describe this dependence, we need the *differential scattering cross section per unit energy*, $\partial^2\sigma/\partial\Omega\partial\mathcal{E}$, where $\mathcal{E}$ here refers to the incident energy.

10.2.6 Distribution function

Radiometric quantities such as exitance, irradiance and radiant intensity are defined in two-dimensional terms, but often we need to consider in more detail the three-dimensional structure of an object being imaged. For example, if an imaging system is viewing a self-luminous object and the emitted radiation can be scattered or absorbed within the object, knowledge of the emission density is not sufficient to compute the image. Regions of the object that contain no emitter can scatter radiation and act as secondary sources, and radiation from the main source can be absorbed and not get out of the object. Similarly, if we want to compute the BTDF or the generalized transmittance of a transmissive object, we must account for the internal scattering and absorption.

To describe these effects more fully, we can use the *phase-space distribution function* (or simply *distribution function* for short). This function is denoted $w(\mathbf{r}, \hat{\mathbf{s}}, \mathcal{E}, t)$ and defined by

$$w = \frac{1}{\mathcal{E}} \frac{\partial^3 Q}{\partial V \partial\Omega \partial\mathcal{E}} . \tag{10.95}$$

In terms of localized photons, $w(\mathbf{r}, \hat{\mathbf{s}}, \mathcal{E}, t)\, \Delta V \Delta\Omega \Delta\mathcal{E}$ can be interpreted loosely as the number of photons contained in volume ΔV centered on point $\mathbf{r}$, travelling in solid angle $\Delta\Omega$ about direction $\hat{\mathbf{s}}$, and having energies between $\mathcal{E}$ and $\mathcal{E} + \Delta\mathcal{E}$ at time t. All of the other radiometric quantities can be derived from the distribution function.

Units of w are $m^{-3}(\text{ster})^{-1}(\text{eV})^{-1}$ if energies are expressed in eV. If the emission density and the absorption and scattering properties of the medium are known, the distribution function can be found (inside and outside the medium) by solving the *Boltzmann transport equation*, as discussed in detail in Sec. 10.3.

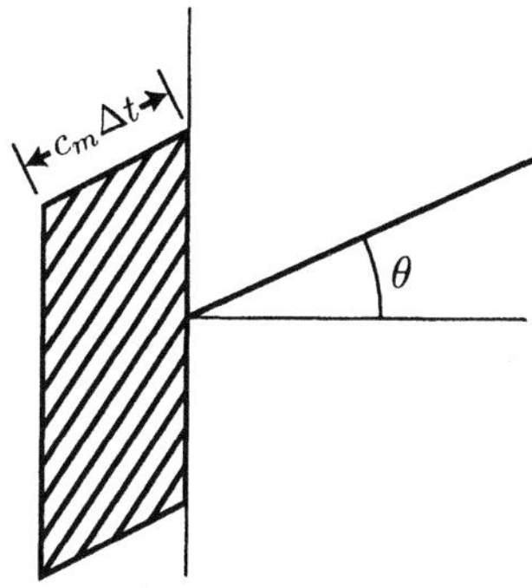

Fig. 10.5 Geometry for computing the radiance at an arbitrary plane when the distribution function $w(\mathbf{r}, \hat{\mathbf{s}}, \mathcal{E}, t)$ is known. Photons within the shaded parallelepiped and travelling in the direction $\hat{\mathbf{s}}$ will pass through the area element ΔA in time Δt.

Relation to radiance The distribution function w differs from the other radiometric quantities defined above in that it is a derivative of the radiant energy rather than of the radiant flux. In other words, w is not defined on a per-unit-time basis. There is, however, a simple relation between the distribution function and the flux-based quantities such as radiance. To deduce this relation, consider Fig. 10.5. A small area ΔA on an arbitrary plane inside or outside the object forms one face of a parallelepiped. The sides of the parallelepiped are parallel to $\hat{\mathbf{s}}$, and the length of these sides is $c_m \Delta t$ as shown, where c_m is the speed of light in the medium (possibly a function of $\mathbf{r}$) and Δt is some small time interval. During this interval, all photons within the parallelepiped and travelling in a small solid angle $\Delta\Omega$ around direction $\hat{\mathbf{s}}$ will pass through the area ΔA. By the definition of w, the number of such photons is $w \Delta V \Delta\Omega \Delta\mathcal{E}$. Here ΔV is the volume of the parallelepiped, given by

$$\Delta V = c_m \Delta t \Delta A \cos\theta = c_m \Delta t \Delta A_{proj}\,, \tag{10.96}$$

where θ is the angle between $\hat{\mathbf{s}}$ and the surface normal, and we have used (10.62) to relate area to projected area. The photon flux through the face ΔA in solid angle $\Delta\Omega$ and energy range ΔE is thus

$$\Delta\Phi_p = w \Delta V \Delta\Omega \Delta\mathcal{E} / \Delta t = c_m w \Delta\Omega \Delta\mathcal{E} \Delta A_{proj}\,. \tag{10.97}$$

From the definition of spectral photon radiance (per unit energy), we now have

$$L_{p,\mathcal{E}}(\mathbf{r}, \hat{\mathbf{s}}, \mathcal{E}, t) = \frac{\Delta\Phi_p}{\Delta\Omega \Delta\mathcal{E} \Delta A_{proj}} = c_m(\mathbf{r})\, w(\mathbf{r}, \hat{\mathbf{s}}, \mathcal{E}, t)\,. \tag{10.98}$$

Thus the distribution function times the photon speed is identically the spectral photon radiance.

One subtlety here is that radiance, originally defined for a 2D emitter, is now related to the distribution function, which has a 3D vector $\mathfrak{r}$ in its argument. The 2D vector $\mathbf{r}$ in L specifies a point on a plane passing through the point $\mathfrak{r}$ in the 3D volume. The above derivation shows that it is quite irrelevant just how the plane is chosen; the flux per unit projected area is independent of the orientation of the surface through which the flux passes. Radiance is a property of the position $\mathfrak{r}$ in the volume and the direction of the radiation.

The directional dependence of the radiance is determined entirely by the directional properties of $w(\mathfrak{r}, \hat{\mathbf{s}}, \mathcal{E}, t)$. If $w(\mathfrak{r}, \hat{\mathbf{s}}, \mathcal{E}, t)$ is independent of $\hat{\mathbf{s}}$, the radiance on any plane passing through point $\mathfrak{r}$ is independent of $\hat{\mathbf{s}}$ also. This observation explains why Lambertian surfaces are so common; all that is required to obtain a Lambertian is a physical mechanism that thoroughly randomizes the photon directions. Strong scattering as in a thick layer of paint or opal glass is one such mechanism. Also, a blackbody cavity in thermal equilibrium necessarily has $w(\mathfrak{r}, \hat{\mathbf{s}}, \mathcal{E}, t)$ independent of $\hat{\mathbf{s}}$, so both of these situations give rise to a radiance that is independent of direction. For this reason, *isotropic* and *Lambertian* are often used synonymously in radiometric parlance.

10.2.7 Radiance in physical optics and quantum optics

As presented above, the distribution function rests heavily on the corpuscular view of light; neither the classical vector-wave characteristics nor the quantum properties of light were considered. For completeness, we now look briefly at the meaning of the distribution function and radiance in classical scalar wave theory, electromagnetic theory and quantum electrodynamics.

Walther's generalized radiance In a pioneering paper, Adriaan Walther (1968) suggested a connection between radiance and the correlation structure of a scalar optical field. His general approach starts with the spatio-temporal autocorrelation function of the field (see Sec. 9.7.4), performs a temporal Fourier transform to get a correlation function involving the spatial variables and the temporal frequency ν, and then uses this quantity to define radiance. We can see the essential features of the problem, however, by assuming quasimonochromatic light at the outset.

For quasimonochromatic light of wavelength λ, Walther (1968) defined the *generalized spectral radiance* at a point $\mathfrak{r}'$ on plane P by

$$L_\nu(\mathfrak{r}', \hat{\mathbf{s}}, \nu) = \frac{\cos\theta}{\lambda^2} \int_P d^2r'' \, \langle u(\mathfrak{r}' + \tfrac{1}{2}\mathfrak{r}'')\, u^*(\mathfrak{r}' - \tfrac{1}{2}\mathfrak{r}'') \rangle \exp(-ik\,\hat{\mathbf{s}} \cdot \mathfrak{r}'') , \qquad (10.99)$$

where θ is the angle between $\hat{\mathbf{s}}$ and the normal to the plane, $k = 2\pi/\lambda = 2\pi\nu/c_m$, $u(\mathfrak{r})$ is any suitable scalar field (see Chap. 9) and $\mathfrak{r}'$ and $\mathfrak{r}''$ are 3D position vectors (but confined to the plane). If we take plane P to be $z = 0$, then $\mathfrak{r}' = (x', y', 0)$, $\mathfrak{r}'' = (x'', y'', 0)$ and $d^2r'' = dx''dy''$.

As Walther noted, this definition is closely related to the Wigner distribution function. In fact, the integral in (10.99) is identical to the expectation value of the Wigner distribution function of the field [*cf.* (5.54)]. This expectation, called the *stochastic Wigner distribution function*, was introduced in Sec. 8.2.5. The connection between radiance and the Wigner distribution was explored in detail by Bastiaans (1978, 1979*a*, 1979*b*).

Generalized radiance and wavefronts There are many situations where it is convenient to represent a monochromatic optical field in the form

$$u(\mathbf{r}) = A(\mathbf{r})\exp[ik\,W(\mathbf{r})]\,, \tag{10.100}$$

where $W(\mathbf{r})$ is real and $A(\mathbf{r})$ might be slowly varying compared to the exponential factor. In an imaging context, the function $W(\mathbf{r})$ is called the eikonal (see Sec. 9.8.2), and the surface $W(\mathbf{r})$ = constant is called the wavefront.

Plane waves and spherical waves fit the description in (10.100), as do many of the coherent point response functions discussed in Sec. 9.6. In Sec. 9.8.2 we derived the eikonal equation from (10.100) with the assumption that k is very large (or λ very small). We shall now investigate the generalized radiance with (10.100).

If we assume that the wave is perfectly monochromatic and nonrandom, then the expectation in (10.99) is not needed, and we have

$$L_\nu(\mathbf{r}',\hat{\mathbf{s}},\nu)$$
$$= \frac{\cos\theta}{\lambda^2}\int_P d^2r''\,A(\mathbf{r}'+\tfrac{1}{2}\mathbf{r}'')\,A^*(\mathbf{r}'-\tfrac{1}{2}\mathbf{r}'')\exp\{ik[W(\mathbf{r}'+\tfrac{1}{2}\mathbf{r}'')-W(\mathbf{r}'-\tfrac{1}{2}\mathbf{r}'')-\hat{\mathbf{s}}\cdot\mathbf{r}'']\}\,. \tag{10.101}$$

If $W(\mathbf{r})$ is sufficiently slowly varying in the vicinity of the point $\mathbf{r}'$, then we can expand $W(\mathbf{r}' \pm \frac{1}{2}\mathbf{r}'')$ in a Taylor series about this point and retain only terms up to second order in $\mathbf{r}''$ (*i.e.*, make the Fresnel approximation). The constant and second-order terms cancel, and we have

$$W(\mathbf{r}'+\tfrac{1}{2}\mathbf{r}'') - W(\mathbf{r}'-\tfrac{1}{2}\mathbf{r}'') \simeq \mathbf{r}''\cdot\nabla W(\mathbf{r}')\,. \tag{10.102}$$

If, in addition, k is large, then the exponential factor is rapidly varying, and only the vicinity of the point $\mathbf{r}''=0$ contributes significantly to the integral. To a good approximation, then,

$$L_\nu(\mathbf{r}',\hat{\mathbf{s}},\nu) = \frac{\cos\theta}{\lambda^2}|A(\mathbf{r}')|^2\int_P d^2r''\exp\{ik[\mathbf{r}''\cdot\nabla W(\mathbf{r}') - \hat{\mathbf{s}}\cdot\mathbf{r}'']\}\,. \tag{10.103}$$

To evaluate this integral, we must relate the 3D vectors in the integrand to their 2D counterparts needed for the integration. Taking the plane of integration as $z=0$ as usual, we write $\mathbf{r}$ as $(\mathbf{r}'',0) = (x'',y'',0)$. Similarly, if $\hat{\mathbf{s}} = (\alpha,\beta,\gamma)$, then we can define a 2D vector $\mathbf{s}_\perp$ as (α,β). Since $\hat{\mathbf{s}}$ is a unit vector, $\alpha^2+\beta^2+\gamma^2=1$, but $\alpha^2+\beta^2\neq 1$ in general, and $\mathbf{s}_\perp$ is not a unit vector. Since $\nabla = (\frac{\partial}{\partial x},\frac{\partial}{\partial y},\frac{\partial}{\partial z})$, we define a transverse gradient operator by $\nabla_\perp = (\frac{\partial}{\partial x},\frac{\partial}{\partial y})$. Finally, we denote $W(\mathbf{r})$ on the plane $z=0$ as $W_0(\mathbf{r})$.

With this notation and (3.217), we have

$$\begin{aligned} L_\nu(\mathbf{r}',\hat{\mathbf{s}},\nu) &= \frac{\cos\theta}{\lambda^2}|A(\mathbf{r}')|^2\int_\infty d^2r''\exp\{ik[\mathbf{r}''\cdot\nabla_\perp W_0(\mathbf{r}') - \mathbf{s}_\perp\cdot\mathbf{r}'']\} \\ &= \frac{\cos\theta}{\lambda^2}|A(\mathbf{r}')|^2\,\delta\left\{\frac{k}{2\pi}[\nabla_\perp W_0(\mathbf{r}') - \mathbf{s}_\perp]\right\}, \end{aligned} \tag{10.104}$$

where the delta function is 2D. Since $k = 2\pi/\lambda$, the scaling property of the delta function, (2.1201), allows us to write

$$L_\nu(\mathbf{r}', \hat{\mathbf{s}}, \nu) = \cos\theta\, |A(\mathbf{r}')|^2\, \delta[\nabla_\perp W_0(\mathbf{r}') - \mathbf{s}_\perp] \,. \tag{10.105}$$

The delta function is zero except where $\nabla_\perp W_0(\mathbf{r}') = \mathbf{s}_\perp$. If we consider the short-wavelength limit ($k \to \infty$), however, we know from the eikonal equation (9.348) that $|\nabla W(\mathbf{r}')| = 1$ in free space, so $\nabla W(\mathbf{r}')$ is a 3D unit vector and $\nabla_\perp W_0(\mathbf{r}')$ is its component in plane P. Thus the 2D vector relation $\nabla_\perp W_0(\mathbf{r}') = \mathbf{s}_\perp$ implies the 3D one, $\nabla W(\mathbf{r}') = \hat{\mathbf{s}}$. It is an interesting exercise in delta functions to show that $\cos\theta\, \delta[\nabla_\perp W_0(\mathbf{r}') - \mathbf{s}_\perp] = \delta[\nabla W(\mathbf{r}') - \hat{\mathbf{s}}]$, where the latter is the angular delta function discussed in Sec. 2.4.7.

With this observation, we have, finally,

$$L_\nu(\mathbf{r}', \hat{\mathbf{s}}, \nu) = |A(\mathbf{r}')|^2\, \delta[\nabla W(\mathbf{r}') - \hat{\mathbf{s}}] \,, \tag{10.106}$$

so the radiance is zero (in the short-wavelength limit) except in the direction normal to the wavefront. Note that (10.106) is a pure vector relation, with no reference to the plane P remaining. This is as it should be, since radiance is defined at a point. The quantity $|A(\mathbf{r}')|^2$ is then the normal irradiance (a property of the field alone) and not the irradiance (which depends also on the plane chosen).

Radiant intensity and generalized radiance To further motivate Walther's definition, we shall now compute the radiant intensity from the generalized radiance and then compare the result to what would be obtained from scalar diffraction theory. A more complete treatment of this problem is given by Marchand and Wolf (1974*a*).

It follows from (10.60) and (10.62) that the radiant intensity is related to radiance by

$$\Upsilon(\hat{\mathbf{s}}) = \cos\theta \int_\infty d^2r'\, L(\mathbf{r}', \hat{\mathbf{s}}) \,, \tag{10.107}$$

where θ is the angle between $\hat{\mathbf{s}}$ and the normal to the plane of integration. We noted earlier that radiant intensity is particularly useful when a source is observed from distances large compared to its size as in Fig. 10.3. In that case, $\hat{\mathbf{s}}$ is the unit vector from anywhere on the source to a small, distant detector.

If the radiance is given by (10.99), then the intensity is

$$\Upsilon(\hat{\mathbf{s}}) = \frac{\cos^2\theta}{\lambda^2} \int_\infty d^2r' \int_\infty d^2r''\, \langle u(\mathbf{r}' + \tfrac{1}{2}\mathbf{r}'')\, u^*(\mathbf{r}' - \tfrac{1}{2}\mathbf{r}'')\rangle \exp(-ik\,\hat{\mathbf{s}}\cdot\mathbf{r}'') \,. \tag{10.108}$$

The radiant flux on the detector in Fig. 10.3 is $\Upsilon(\hat{\mathbf{s}})$ times the solid angle subtended by the detector, or $\Upsilon(\hat{\mathbf{s}})A_d/\mathfrak{r}^2$, where A_d is the detector area and $\mathfrak{r}$ is its distance from the small source. Thus the detector response will be determined by $\Upsilon(\hat{\mathbf{s}})$, which in turn is determined by the stochastic Wigner distribution function of the field.

The experimental arrangement in Fig. 10.3 was analyzed in Sec. 9.4.7 by Fraunhofer diffraction theory, and we shall now show how that theory also predicts (10.108). If we identify $\langle |u(\mathbf{r})|^2\rangle$ as the irradiance on the detector, then the radiant flux $\Upsilon(\hat{\mathbf{s}})A_d/\mathfrak{r}^2$ is equal to $\langle |u(\mathbf{r})|^2\rangle A_d$. By use of (9.102) we can write

$$\Upsilon(\hat{\mathbf{s}}) = \mathfrak{r}^2 \langle |u(\mathfrak{r})|^2 \rangle$$
$$= \frac{\cos^2\theta}{\lambda^2} \int_\infty d^2r_0 \int_\infty d^2r_1 \, \langle u(\mathfrak{r}_0)\, u^*(\mathfrak{r}_1)\rangle \exp\left[-ik\,\frac{(\mathfrak{r}_0 - \mathfrak{r}_1)\cdot \mathfrak{r}}{\mathfrak{r}}\right]. \quad (10.109)$$

A change of variables and the recognition that $\mathfrak{r}/\mathfrak{r} = \hat{\mathbf{s}}$ then reproduces (10.108).

Lambertians We know from Sec. 10.2.1 that a Lambertian source has a radiant intensity that varies as $\cos\theta$, but (10.108) has a leading factor of $\cos^2\theta$, so the integral in that equation must somehow provide a factor of $1/\cos\theta$ for a Lambertian.

As noted by Walther (1968), the correlation function of a quasimonochromatic Lambertian source must have the form

$$\langle u(\mathfrak{r} + \tfrac{1}{2}\mathfrak{r}')\, u^*(\mathfrak{r} - \tfrac{1}{2}\mathfrak{r}')\rangle \propto \operatorname{sinc}(k\mathfrak{r}'), \quad (10.110)$$

whenever $\mathfrak{r} + \frac{1}{2}\mathfrak{r}'$ and $\mathfrak{r} - \frac{1}{2}\mathfrak{r}'$ both lie within the source region. Since $k = 2\pi/\lambda$, the sinc function in (10.110) has a width of approximately λ. If the source region is large compared to λ, so that we need not fret much over limits, then the integral in (10.108) is approximately

$$\int_\infty d^2r \int_\infty d^2r' \, \langle u(\mathfrak{r} + \tfrac{1}{2}\mathfrak{r}')\, u^*(\mathfrak{r} - \tfrac{1}{2}\mathfrak{r}')\rangle \exp(-ik\,\hat{\mathbf{s}}\cdot\mathfrak{r}')$$
$$\simeq A_s \int_\infty d^2r' \, \frac{\sin(k\mathfrak{r}')}{k\mathfrak{r}'} \exp(-ik\,\hat{\mathbf{s}}\cdot\mathfrak{r}'), \quad (10.111)$$

where A_s is the area of the source.

Since the integral is over the plane $z = 0$, we can replace $\mathfrak{r}'$ with its 2D counterpart $\mathbf{r}'$ and write the scalar product $\hat{\mathbf{s}}\cdot\mathfrak{r}'$ as $\mathbf{s}_\perp\cdot\mathbf{r}'$, where $\mathbf{s}_\perp$ is the projection of $\hat{\mathbf{s}}$ onto the plane. Specifically, if θ and ϕ are the spherical polar coordinates of $\hat{\mathbf{s}}$ (with the z axis being the polar axis), then $\mathbf{s}_\perp = (\sin\theta\cos\phi, \sin\theta\sin\phi)$. Hence $\mathbf{s}_\perp$ is not a unit vector.

With these notational changes, the integral is the 2D Fourier transform of a rotationally symmetric function, so it can be written with the aid of (3.248) as

$$\int_\infty d^2r' \, \frac{\sin(k\mathfrak{r}')}{k\mathfrak{r}'} \exp(-ik\,\hat{\mathbf{s}}\cdot\mathfrak{r}') = 2\pi \int_0^\infty r'dr' \, \frac{\sin(kr')}{kr'} J_0(kr's_\perp), \quad (10.112)$$

where $s_\perp = |\mathbf{s}_\perp| = \sin\theta$. A tabulated integral (6.671(7) in Gradshteyn and Ryzhik, 1980) then shows that

$$\int_0^\infty r'dr' \, \frac{\sin(kr')}{kr'} J_0(kr's_\perp) = \frac{1}{k^2\sqrt{1 - s_\perp^2}} = \frac{1}{k^2\cos\theta}. \quad (10.113)$$

Thus the sinc-function correlation of (10.110) indeed implies a Lambertian character. Sarfatt (1963) showed that (10.110) is applicable to blackbodies, so blackbodies are Lambertian (see also Mandel and Wolf, 1976).

In fact, the sinc correlation is equivalent to saying that the radiation is completely randomized in direction. As an exercise, the reader can start with the mode expansion (10.12) and assume that the amplitudes are uncorrelated random variables and that every direction $\hat{\boldsymbol{\kappa}}_j$ is equally probable. If it is assumed further that the source is observed through a narrowband filter so that only a small range of values for k_j contribute, then the sinc form (10.110) will be obtained.

Propagation through paraxial optical systems In Sec. 9.7.3 we discussed the Fresnel propagation of the scalar wave field through optical systems that can be described in geometrical optics by a 2×2 or 4×4 matrix. The key results cited there are the generalized Fresnel diffraction integral (9.255) and its special case, (9.250), which holds for systems with rotational symmetry.

Friberg (1991) computed the generalized radiance associated with the output field as given in (9.250), and the theory can be extended to make use of (9.255). The result is most neatly stated if we use 2D vectors as the arguments of the radiance and write

$$L_\nu(\mathbf{r}, \hat{\mathbf{s}}, \nu) = L\left(\begin{bmatrix} \mathbf{r} \\ \mathbf{s}_\perp \end{bmatrix}\right) , \tag{10.114}$$

where we have dropped the ν argument and subscript, but a spectral radiance is still implied. With this notation, we can now append subscripts to distinguish the radiances associated with input and output fields for some optical system described by an **ABCD** matrix denoted **M**. As Friberg showed, these radiances are related by

$$L_{out}\left(\begin{bmatrix} \mathbf{r} \\ \mathbf{s}_\perp \end{bmatrix}\right) = L_{in}\left(\mathbf{M}\begin{bmatrix} \mathbf{r} \\ \mathbf{s}_\perp \end{bmatrix}\right) . \tag{10.115}$$

Thus the radiance is constant along the ray defined by geometrical optics. The same conclusion was reached by Walther (1978) by a stationary-phase approximation, valid asymptotically in the limit of zero wavelength. Friberg's derivation does not require this limit (though it does use the Fresnel approximation), and it is valid for any state of coherence of the fields. Moreover, since it holds for all ν, (10.115) works for the overall radiance as well as the spectral radiance.

Some mathematical issues As noted in Sec. 5.2.1, the Wigner distribution function can take on negative values. Radiance, however, is defined in (10.62) in terms of nonnegative quantities, so the negative values of the generalized radiance are unphysical.

On the other hand, we know from (5.57) and (5.59) that an integral of the Wigner distribution function over either variable is nonnegative. In a radiometric context, the integral of $L(\mathbf{r}, \hat{\mathbf{s}}) \cos\theta$ over angles in a hemisphere is the radiant exitance (or irradiance), and the integral of the same quantity over area is radiant intensity. Fortunately, both of these integrals can be shown to be nonnegative, so the negative values of the generalized radiance do not affect the more directly measurable radiometric quantities (Marchand and Wolf, 1974*b*). In addition, Mandel and Wolf (1995) demonstrate that the generalized radiance itself turns out to be nonnegative for *quasistationary*[9] sources, as defined in Sec. 8.2.4. This category includes thermal sources, diffuse reflectors and many other sources with short-range correlations.

Just as a bivariate probability density function (PDF) is not determined uniquely by its two marginals, neither is the radiance fully determined by the exitance and radiant intensity. As a result, several different definitions of generalized radiance have been proposed. Walther himself noted some conceptual problems with the Wigner definition and proposed an alternative (Walther, 1973; Baltes *et*

[9]The term *quasihomogeneous*, often used in the literature on radiometry and coherence, is synonymous with quasistationary, at least when spatial stationarity is being discussed.

al., 1978; Marchand and Wolf, 1974*b*). Since all of these definitions lead to the same expressions for exitance and intensity, and none avoids the negativity problem inherent in Walther's original approach, there is little reason to choose an alternative. Moreover, for quasistationary sources, the various definitions are equivalent (Mandel and Wolf, 1995).

Radiometry and Maxwell's equations Emil Wolf and his co-workers have developed an extensive and rigorous electromagnetic theory of radiative transfer in free space. (See Wolf, 1976; Wolf, 1978; Zubairy and Wolf, 1977; and Mandel and Wolf, 1995.) Fante (1981) extended the theory to inhomogeneous dielectric media and discussed the connection between Maxwell's equations and classical radiometry. Fante concludes that the generalized radiance, though it can go negative, satisfies the other postulates of classical radiative transfer if the source is either quasistationary or highly directional; the fluctuations in index of refraction are small and nearly stationary, and the longitudinal components of the field (components of $\mathbf{E}$ parallel to $\mathbf{k}$) are small.

Quantum electrodynamics A fully quantum-mechanical definition of radiance can be given by an extension of ideas introduced in Sec. 10.1.3. A straightforward generalization of (10.99) is

$$L_\nu(\mathbf{r}, \hat{\mathrm{s}}, \nu) = \frac{\cos\theta}{\lambda^2} \int_\infty d^2r' \, \langle \hat{\mathrm{e}}^-(\mathbf{r} + \tfrac{1}{2}\mathbf{r}') \cdot \hat{\mathrm{e}}^+(\mathbf{r}' - \tfrac{1}{2}\mathbf{r}') \rangle \exp(-ik\,\hat{\mathbf{s}} \cdot \mathbf{r}') \,, \quad (10.116)$$

where the operators are defined below (10.22) and the angle brackets now denote a quantum-mechanical expectation. The other radiometric quantities can then be computed just as in the classical case. *E.g.*, radiant intensity is given by (10.107).

In most imaging applications, neither the classical electromagnetic approach nor the fully quantum-mechanical approach adds much to the more intuitive concept of photon, which we shall use throughout the remainder of this chapter.

10.3 BOLTZMANN TRANSPORT EQUATION

In Sec. 10.2 we introduced a menagerie of radiometric quantities, including the distribution function from which all of the other quantities can be derived. Now we shall derive an important equation that governs the spatio-temporal behavior of the distribution function.

This equation is known by various names in the literature. In statistical mechanics and neutron-transport theory, it is called the *Boltzmann equation*, and we shall adopt that terminology here. In optics, it is often called the *radiative transport equation*, or simply, the *transport equation*. In the quantum-optics literature, one often encounters the term *Fokker-Planck equation*, which describes a wide variety of equations for various distribution functions (Risken, 1984). For example, if we focus on the distribution of photon directions and energies, then we need the distribution function that is obtained from $w(\mathbf{r}, \hat{\mathrm{s}}, \mathcal{E}, t)$ by integrating out the spatial dependence. The resulting reduced distribution function satisfies a certain Fokker-Planck equation. In this sense, the Boltzmann equation is the most general Fokker-Planck equation. Finally, one form of the *Liouville equation* is the Boltzmann equation without absorption or scattering terms.

We shall stick with the name Boltzmann equation in honor of the seminal contributions of Ludwig Boltzmann (1844–1906) in mechanics, statistics, radiation and thermodynamics. Boltzmann's scientific vision had a large impact not only on these many fields but also on his own mental health, leading eventually to his tragic suicide; for a fascinating biography, see Cercignani (1998) .

The Boltzmann equation has many applications in imaging. In medical imaging it can be used to analyze the distribution of x rays in chest radiography and computed tomography, gamma rays in nuclear medicine, or infrared photons in breast imaging. In optical imaging it can be used to discuss scattering by the atmosphere or ocean.

Though a useful tool in all of these areas, the Boltzmann equation is not a complete description since it ignores the wave aspects of the radiation. When applied to optics, the Boltzmann equation treats light as if it were made up of localized photons. Interference and diffraction effects are not accounted for, and the photons are assumed to travel in straight lines in homogeneous media. In this sense, transport theory and the Boltzmann equation make the same approximations as in geometric optics.

10.3.1 Derivation of the Boltzmann equation

The Boltzmann equation is an equation for the time derivative of $w(\mathbf{r}, \hat{\mathbf{s}}, \mathcal{E}, t)$, or w for short. This derivative has contributions from the physical processes of absorption, emission, propagation and scattering of radiation, so we have

$$\frac{dw}{dt} = \left[\frac{\partial w}{\partial t}\right]_{abs} + \left[\frac{\partial w}{\partial t}\right]_{em} + \left[\frac{\partial w}{\partial t}\right]_{prop} + \left[\frac{\partial w}{\partial t}\right]_{sc}, \tag{10.117}$$

where the subscripts have the obvious meanings. We examine each of these terms in succession.

Absorption Suppose we have ΔN identical atoms in a small volume ΔV in a medium where the distribution function is w. If we consider only photons travelling in a small solid angle $\Delta\Omega$ around direction $\hat{\mathbf{s}}$ and having energies in a narrow range $(\mathcal{E}, \mathcal{E} + \Delta\mathcal{E})$, then the irradiance on the atoms in a plane normal to $\hat{\mathbf{s}}$ is $I_0 = c_m \mathcal{E} w \Delta\mathcal{E}\Delta\Omega$. We assume that the atoms absorb radiation independently of each other.

From (10.91), the total energy absorbed by this group of atoms in time Δt is $\Delta N \sigma_{abs} I_0 \Delta t = c_m \mathcal{E} \Delta N \sigma_{abs} w \Delta\mathcal{E}\Delta\Omega\Delta t$. From the definition of w, (10.95), the corresponding change in w is given by

$$\Delta w = -\frac{c_m \mathcal{E} \Delta N \sigma_{abs} w \Delta\mathcal{E}\Delta\Omega\Delta t}{\mathcal{E}\Delta\mathcal{E}\Delta V \Delta\Omega} = -\frac{\Delta N c_m \sigma_{abs} w \Delta t}{\Delta V}. \tag{10.118}$$

If we divide through by Δt and let all small quantities tend to zero, we find

$$\left[\frac{\partial w}{\partial t}\right]_{abs} = -c_m \mu_{abs} w, \tag{10.119}$$

where μ_{abs} is the *absorption coefficient*, defined by

$$\mu_{abs} = n_{abs} \sigma_{abs}, \tag{10.120}$$

and $n_{abs} = \Delta N/\Delta V$ is the number of absorbing atoms per unit volume. Since n_{abs} has units of reciprocal volume and σ_{abs} is an area, μ_{abs} has units of reciprocal length. The interpretation of that length will be seen in Sec. 10.3.3.

If there are several different kinds of atoms or other absorbing particles interspersed in the same volume, then we define

$$\mu_{abs} = \sum_j n_{abs,j}\, \sigma_{abs,j}\,, \tag{10.121}$$

where subscript j identifies a particular species. With this definition, (10.119) still holds if each species absorbs radiation independently of the others, a valid assumption for x-ray or gamma-ray absorption and for optical absorption by isolated impurities in solids.

Note that both μ_{abs} and c_m can depend on wavelength or energy of the radiation, position in the medium and time. In addition, c_m depends on direction of propagation in anisotropic media. (Anisotropic absorption can also occur, but is rare.)

Emission To determine the effect of emission of radiation on w, we need the quantities $\Xi_{p,\mathcal{E}}$ and $S_{p,\mathcal{E}}$ defined in Sec. 10.2.2. It follows immediately from these definitions, along with the definition of w in (10.95), that

$$\left[\frac{\partial w}{\partial t}\right]_{em} = \Xi_{p,\mathcal{E}} = \frac{1}{4\pi} S_{p,\mathcal{E}}\,, \tag{10.122}$$

where the second form holds for an isotropic emitter. Recall that $S_{p,\mathcal{E}}(\mathbf{r},\mathcal{E})\Delta V \Delta\mathcal{E}$ is the total number of photons per second emitted from volume ΔV in energy range $\Delta\mathcal{E}$, while $\Xi_{p,\mathcal{E}}(\mathbf{r},\hat{\mathbf{s}},\mathcal{E})\Delta V\Delta\mathcal{E}\Delta\Omega$ is the total number of photons per second emitted from volume ΔV in energy range $\Delta\mathcal{E}$ and solid angle $\Delta\Omega$ about direction $\hat{\mathbf{s}}$. The dimensions in (10.122) are consistent if the factor of $1/4\pi$ is assigned dimensions of inverse steradians.

If the emission is monoenergetic at photon energy $\mathcal{E}_0$, we have

$$\left[\frac{\partial w}{\partial t}\right]_{em} = \Xi_p\, \delta(\mathcal{E} - \mathcal{E}_0) = \frac{1}{4\pi} S_p\, \delta(\mathcal{E} - \mathcal{E}_0)\,. \tag{10.123}$$

Again, the second form is only for an isotropic emitter; it applies, for example, to a radioactive source.

Propagation in a homogeneous medium In a homogeneous medium where the speed of light c_m is a constant, it is straightforward to compute the propagation term in the Boltzmann equation. In a short time interval Δt, photons at point $\mathbf{r}$ travelling in direction $\hat{\mathbf{s}}$ move to point $\mathbf{r} + c_m\Delta t\,\hat{\mathbf{s}}$, so

$$w(\mathbf{r},\hat{\mathbf{s}},\mathcal{E},t) = w(\mathbf{r} + c_m\Delta t\,\hat{\mathbf{s}},\hat{\mathbf{s}},\mathcal{E},t+\Delta t)\,. \tag{10.124}$$

Expanding the right-hand side in a Taylor series yields

$$w(\mathbf{r},\hat{\mathbf{s}},\mathcal{E},t) = w(\mathbf{r},\hat{\mathbf{s}},\mathcal{E},t) + c_m\Delta t\,\hat{\mathbf{s}}\cdot\nabla w(\mathbf{r},\hat{\mathbf{s}},\mathcal{E},t) + \Delta t\left[\frac{\partial}{\partial t} w(\mathbf{r},\hat{\mathbf{s}},\mathcal{E},t)\right]_{prop}\,. \tag{10.125}$$

The time derivative in the last term is the desired $[\partial w/\partial t]_{prop}$, so we have

$$\left[\frac{\partial w}{\partial t}\right]_{prop} = -c_m \hat{\mathbf{s}} \cdot \nabla w \,. \tag{10.126}$$

Propagation in inhomogeneous media The form of the propagation term in (10.126) assumes that the speed of light in the medium, c_m, is a constant, but there are many situations where we need to consider a variable speed of light. We saw in Sec. 9.1.4 that $c_m = c/n$, where n is the index of refraction of the medium, and in general we should write the index as a function of position $n(\mathbf{r})$. We need to distinguish three distinct kinds of spatial variations of the refractive index: slow variations, point variations and abrupt changes at interfaces.

The index variations can be considered slow if $n(\mathbf{r})$ is approximately the same at two points one wavelength apart. Mathematically, the condition is $k|\nabla n(\mathbf{r})| \ll 1$, where, as usual, $k = 2\pi/\lambda$. Slow variations may be introduced deliberately as in gradient-index (GRIN) optics in order to focus the light, or they may be random as in propagation of light through the atmosphere. The effect of such variations is that the light follows a complicated curved path, which can be determined by solving the eikonal equation (9.348) or by including an additional term in the Boltzmann equation (Ferwerda, 1999).

Point variations are ones for which $n(\mathbf{r})$ is a constant except over a discrete set of regions which are all small compared to a wavelength. For example, small (subwavelength) bubbles in glass or dust particles in the atmosphere might be well modeled as point variations. The effect of point variations is to scatter the radiation, so these effects are best described by a scattering cross section (see Sec. 10.2.5) rather than in terms of a position-dependent refractive index. Point scatterers can be treated by including a scattering term in the Boltzmann equation.

An important kind of inhomogeneity occurs at the interface between two different homogeneous media, for example at a glass-air interface on the surface of a lens. We do not need a separate term in the Boltzmann equation to account for such interfaces since we can solve the equation separately in each homogeneous medium and then match boundary conditions to get the full solution. Interfaces are discussed in more detail in Sec. 10.3.2.

Scattering Scattering has two effects on the distribution function. Consider photons in a small volume element ΔV travelling in a small solid angle $\Delta\Omega$ and having energies in a narrow range $(\mathcal{E}, \mathcal{E} + \Delta\mathcal{E})$. The mean number of photons in this group is $w\Delta\Omega\Delta\mathcal{E}\Delta V$. Scattering processes occurring in the volume element can either increase or decrease the number of photons in this group. The decrease comes about because photons in the group can change direction, energy or both as a result of scattering. On the other hand, photons not in the group under consideration can scatter *into* the angular range $\Delta\Omega$ and the energy band $\Delta\mathcal{E}$.

Scattering out of the group is described by exactly the same mathematics as in the absorption case; as far as removal from the group is concerned, there is no distinction between absorption and scattering. Thus we can write at once, by analogy to (10.119),

$$\left[\frac{\partial w}{\partial t}\right]_{out} = -c_m \mu_{sc} w \,, \tag{10.127}$$

where μ_{sc}, called the *scattering coefficient* or *linear attenuation coefficient for scattering*, is defined by

$$\mu_{sc} = n_{sc}\sigma_{sc} , \tag{10.128}$$

and n_{sc} is the number of scatterers per unit volume. As in the absorption term, μ_{sc} can depend on position, time, photon energy and possibly direction $\hat{\mathbf{s}}$.

Scattering into the group under consideration is more complicated. It has to involve integrals over energy and direction since photons of any energy or direction can, in principle, scatter into the group. On the other hand, no integral over position or time is needed since the scattering processes occur at a definite location and definite time. Thus we are looking for an integral transform that connects $w(\mathbf{r}, \hat{\mathbf{s}}, \mathcal{E}, t)$ to $w(\mathbf{r}, \hat{\mathbf{s}}', \mathcal{E}', t)$ for all other $\hat{\mathbf{s}}'$ and $\mathcal{E}'$.

At this point an important distinction between optics and statistical mechanics arises. In the latter field, the dominant scattering occurs between molecules, so the probability of scattering depends on a product of the form $w(\mathbf{r}, \hat{\mathbf{s}}, \mathcal{E}, t)w(\mathbf{r}, \hat{\mathbf{s}}', \mathcal{E}', t)$, which makes the scattering term in the Boltzmann equation a nonlinear function of w. In optics and radiology, however, the distribution of scattering centers has nothing to do with w; the scattering characteristics are determined by electrons, atoms and molecules in the medium, while w relates to the distribution of photons. Since photons do not scatter off other *photons*, the Boltzmann equation is linear.[10]

The general form of the term that describes scattering into the group of interest is thus

$$\left[\frac{\partial}{\partial t} w(\mathbf{r}, \hat{\mathbf{s}}, \mathcal{E}, t)\right]_{in} = \int_{4\pi} d\Omega' \int_0^\infty d\mathcal{E}'\, K(\hat{\mathbf{s}}, \mathcal{E}; \hat{\mathbf{s}}', \mathcal{E}'|\mathbf{r}, t)\, w(\mathbf{r}, \hat{\mathbf{s}}', \mathcal{E}', t) . \tag{10.129}$$

The kernel $K(\hat{\mathbf{s}}, \mathcal{E}; \hat{\mathbf{s}}', \mathcal{E}'|\mathbf{r}, t)$ can depend on both the initial and final energy and direction, and it can also depend on position $\mathbf{r}$ in the medium and, for time-varying media, on the time t. Since it is reasonable to assume that the scatterers are randomly oriented, the dependence on $\hat{\mathbf{s}}$ and $\hat{\mathbf{s}}'$ is through the scalar product $\hat{\mathbf{s}} \cdot \hat{\mathbf{s}}'$.

If we think of the scattering centers as mathematical points, then the kernel $K(\hat{\mathbf{s}}, \mathcal{E}; \hat{\mathbf{s}}', \mathcal{E}'|\mathbf{r}, t)$ should be written as a sum of spatial delta functions with weights dependent on the other variables. This kind of description, called a *point process*, is discussed in detail in Chap. 11. In practice, however, we would rarely have enough knowledge of the medium to specify the location of these points. Usually the best we can do is specify an average density of scatterers $n_{sc}(\mathbf{r})$, defined such that $n_{sc}(\mathbf{r})\Delta V$ is the mean number of scattering centers in a small (but not infinitesimal) volume ΔV centered on $\mathbf{r}$.

We can assume that each scatterer is described by a differential scattering cross section per unit energy $\partial^2\sigma_{sc}/\partial\Omega\partial\mathcal{E}$ as discussed in Sec. 10.2.5. Then, by arguments similar to those used earlier (really little more than dimensional analysis), we find that

$$K(\hat{\mathbf{s}}, \mathcal{E}; \hat{\mathbf{s}}', \mathcal{E}'|\mathbf{r}, t) = c_m n_{sc} \frac{\partial^2\sigma_{sc}}{\partial\Omega\partial\mathcal{E}} . \tag{10.130}$$

[10]Nonlinear optical processes do, of course, occur, but only at high intensities and in carefully controlled experimental geometries. Unless one sets out to do a nonlinear optical experiment, it is probably safe to ignore nonlinear effects.

To simplify the notation, we denote the operator defined by (10.129) and (10.130) as $\mathcal{K}$ and write

$$\left[\frac{\partial w}{\partial t}\right]_{scat} = \left[\frac{\partial w}{\partial t}\right]_{in} + \left[\frac{\partial w}{\partial t}\right]_{out} = \mathcal{K}w - c_m \mu_{sc} w \,. \tag{10.131}$$

Complete equation Collecting together (10.119), (10.122), (10.126) and (10.131), we get the following form for the Boltzmann equation:

$$\frac{dw}{dt} = -c_m \mu_{tot} w + \Xi_{p,\mathcal{E}} + \mathcal{K}w - c_m \hat{\mathbf{s}} \cdot \nabla w \,, \tag{10.132}$$

where μ_{tot} is the total attenuation coefficient, given by

$$\mu_{tot} = \mu_{sc} + \mu_{abs} = n_{sc}\sigma_{sc} + n_{abs}\sigma_{abs} \,. \tag{10.133}$$

In (10.132), w, $\Xi_{p,\mathcal{E}}$ and μ_{tot} can depend, in general, on $\mathbf{r}$, $\mathcal{E}$, $\hat{\mathbf{s}}$ and t. Similarly, the kernel of the operator $\mathcal{K}$ can depend on $\mathbf{r}$ and t, and it couples w at one $\hat{\mathbf{s}}$ and $\mathcal{E}$, in general, to w at all other $\hat{\mathbf{s}}$ and $\mathcal{E}$. On the other hand, our derivations considered the speed of light c_m just to be a constant; point variations in refractive index are included in the scattering term, but we postpone any discussion of slow variations in index or interfaces between media with different indices.

10.3.2 Steady-state solutions in non-absorbing media

In many problems we are interested in steady-state solutions to the Boltzmann equation, where $dw/dt = 0$. For example, in medical applications with either light or x rays, the transit time of photons across the body is only a few nanoseconds. If the radiation source is independent of time, or at least slowly varying on the nanosecond scale, then the distribution function reaches its steady-state value very quickly. Only when the source or the medium is varying rapidly do we need the full time-dependent Boltzmann equation. When it is necessary to include the arguments, we shall denote a steady-state solution by $w(\mathbf{r}, \hat{\mathbf{s}}, \mathcal{E})$, without a time argument.

Emission and propagation terms In a medium with no absorption or scattering, the Boltzmann equation consists of just the source term and the propagation term; the steady-state equation is then

$$c_m \hat{\mathbf{s}} \cdot \nabla w = \Xi_{p,\mathcal{E}} \,. \tag{10.134}$$

To solve this equation, we choose a Cartesian coordinate system such that the z axis is parallel to $\hat{\mathbf{s}}$. Then $\hat{\mathbf{s}} \cdot \nabla = \frac{\partial}{\partial z}$, and (10.134) reduces to an ordinary differential equation in z, which integrates to

$$w(x, y, z, \hat{\mathbf{s}}, \mathcal{E}) = \frac{1}{c_m} \int_{-\infty}^{z} dz' \, \Xi_{p,\mathcal{E}}(x, y, z', \hat{\mathbf{s}}, \mathcal{E}) \,. \tag{10.135}$$

No constant of integration is needed if there is no radiation source other than $\Xi_{p,\mathcal{E}}$.

Defining $\ell = z - z'$ and reverting to a general vector notation, we have

$$w(\mathbf{r}, \hat{\mathbf{s}}, \mathcal{E}) = \frac{1}{c_m} \int_{0}^{\infty} d\ell \, \Xi_{p,\mathcal{E}}(\mathbf{r} - \hat{\mathbf{s}}\ell, \hat{\mathbf{s}}, \mathcal{E}) \,. \tag{10.136}$$

The interpretation of this equation is that $w(\mathbf{r}, \hat{\mathbf{s}}, \mathcal{E})$ is obtained by integrating the source distribution along a line parallel to $\hat{\mathbf{s}}$ and passing through the point $\mathbf{r}$. Only photons originating along this line can contribute to the distribution function with the specified $\mathbf{r}$ and $\hat{\mathbf{s}}$, and of those originating along the line, only those travelling in direction $\hat{\mathbf{s}}$ contribute. Only positive values of ℓ are needed since photons originating at a point $\mathbf{r} + \hat{\mathbf{s}}|\ell|$ propagate away from the observation point $\mathbf{r}$.

Equation (10.136) defines an integral transform, known as the *x-ray transform*,[11] which maps a function of $\mathbf{r}$ and $\hat{\mathbf{s}}$ to another function of $\mathbf{r}$ and $\hat{\mathbf{s}}$. We shall denote this operator as $\boldsymbol{\mathcal{X}}$ and write (10.136) symbolically as

$$w = \boldsymbol{\mathcal{X}} \Xi_{p,\mathcal{E}} \,. \tag{10.137}$$

Note that we are now including the factor $1/c_m$ in the definition of $\boldsymbol{\mathcal{X}}$, so the operator is dimensionally a time. This factor will be dropped in Chap. 16 when we discuss actual x-ray applications of the x-ray transform.

Optical case Many problems in optics involve propagation of radiation in a homogeneous, source-free medium with no absorption or scattering. In that case, the steady-state Boltzmann equation is just

$$\hat{\mathbf{s}} \cdot \nabla w = 0 \,. \tag{10.138}$$

This form implies that w is constant in direction $\hat{\mathbf{s}}$.

Recall from (10.98) that $L_{p,\mathcal{E}}(\mathbf{r}, \hat{\mathbf{s}}, \mathcal{E}, t) = c_m(\mathbf{r})\, w(\mathbf{r}, \hat{\mathbf{s}}, \mathcal{E}, t)$. Since c_m is a constant in a homogeneous medium, (10.138) shows that $L_{p,\mathcal{E}}(\mathbf{r}, \hat{\mathbf{s}}, \mathcal{E}, t)$ is also constant along $\hat{\mathbf{s}}$. This conclusion is true for all $\mathcal{E}$, and we can drop the t dependence in steady-state situations, so the total radiance $L(\mathbf{r}, \hat{\mathbf{s}})$ is constant. We can move an arbitrary distance in direction $\hat{\mathbf{s}}$ and $L(\mathbf{r}, \hat{\mathbf{s}})$ does not change. In the optics literature, this result is often capsulized by saying that radiance is constant along the ray.

Constant radiance, however, does not imply constant irradiance on a detector or other surface. To illustrate this point, we consider a uniform volume radiator in the form of a disc of diameter D and thickness a (see Fig. 10.6). At a point along the positive z axis (which is the axis of the disc), the spectral photon radiance in the z direction is given by (10.98) and (10.136) as

$$L_{p,\mathcal{E}}(\mathbf{r}, \hat{\mathbf{z}}) = \frac{a}{4\pi} S_0 \,, \tag{10.139}$$

where the factor of a results from the line integral through the disc and S_0 is the value of $S_{p,\mathcal{E}}(\mathbf{r})$ inside the disc. Thus $L_{p,\mathcal{E}}(\mathbf{r}, \hat{\mathbf{z}})$ remains constant as we move along the z axis, in direction $\hat{\mathbf{z}}$. If we consider a general direction $\hat{\mathbf{s}}$, however, $L_{p,\mathcal{E}}(\mathbf{r}, \hat{\mathbf{s}})$ will be zero unless we can draw a line from point $\mathbf{r}$ backwards in the direction $-\hat{\mathbf{s}}$ to the source. If this line misses the source, the line integral in (10.136) is zero. Thus, the radiance at point $\mathbf{r}$ is zero except for a range of directions $\hat{\mathbf{s}}$ defined by the angular subtense of the source at the point.

[11] The name of this transform comes about since it was originally defined in the radiology literature, but there is nothing in the mathematics specific to x rays. Moreover, some authors in radiology would call (10.136) the *cone-beam transform* in general, reserving the term *x-ray transform* for the special case where all rays are parallel. For more details on cone-beam tomography, see Chap. 17.

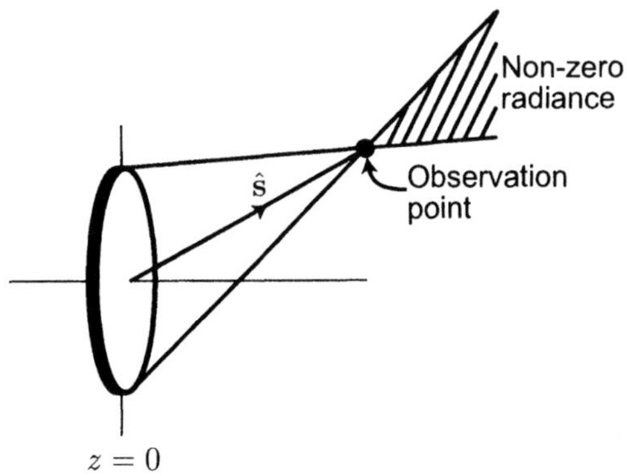

Fig. 10.6 Diagram for illustrating the relation between radiance and irradiance.

The spectral photon irradiance is obtained by integrating the corresponding radiance over angles [*cf.* (10.79)]:

$$I_{p,\mathcal{E}} = \int_{2\pi} d\Omega \; L_{p,\mathcal{E}}(\mathbf{r}, \hat{\mathbf{s}}) \cos\theta \,. \tag{10.140}$$

If the point $\mathbf{r}$ is on the z axis and $z \gg D$, then the source subtends a solid angle of $\Delta\Omega = \pi D^2/4z^2$, and the radiance is given by (10.139) for all $\hat{\mathbf{s}}$ in this range. Since $\cos\theta \simeq 1$ with these assumptions, we have

$$I_{p,\mathcal{E}} = \frac{D^2}{16z^2} \, a S_0 \,. \tag{10.141}$$

Now we see the familiar inverse-square law; the irradiance varies inversely as the square of the distance from the source, so long as that distance is sufficiently large that all cosine factors can be approximated by unity. The radiance, on the other hand, is power per unit (projected) area per unit solid angle, so it varies as $I/\Delta\Omega$, and both I and $\Delta\Omega$ vary as $1/z^2$, leaving the radiance independent of z.

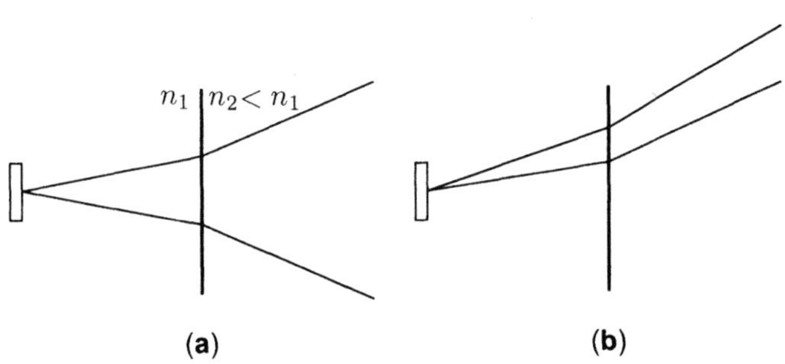

Fig. 10.7 Diagram illustrating the change of radiance at a smooth interface. (a) Normal incidence. (b) Oblique incidence.

Interfaces From the discussion of Snell's law in Chap. 9, we know what happens to light at a smooth interface between media with different refractive indices. In particular, (9.148) gives an explicit expression for the direction $\hat{\mathbf{s}}_{tr}$ of the transmitted light when the incident direction $\hat{\mathbf{s}}_{inc}$ and the indices are known, and (9.149) is a similar expression for the reflected direction. Now we need to determine what happens to the radiance across this interface.

The interface plane is defined as $z = 0$, and the index of refraction is n for $z < 0$ and n' for $z > 0$. As in Sec. 9.2.1, the projection of $\hat{\mathbf{s}}_{inc}$ onto the x-z plane makes an angle θ_x with the z axis and the projection onto the y-z plane makes an angle θ_y with the z axis. Similarly, $\hat{\mathbf{s}}_{tr}$ is defined by angles θ'_x and θ'_y. Snell's law shows that

$$n \sin\theta_x = n' \sin\theta'_x \,; \qquad n \sin\theta_y = n' \sin\theta'_y \,. \tag{10.142}$$

As seen in Fig. 10.7, Snell's law causes the light to diverge at the interface upon entering a medium with smaller index. Differentiating (10.142), we find

$$n\Delta\theta_x \cos\theta_x = n'\Delta\theta'_x \cos\theta'_x \,; \qquad n\Delta\theta_y \cos\theta_y = n'\Delta\theta'_y \cos\theta'_y \,. \tag{10.143}$$

At normal incidence, all of the cosines are one, so $n^2\Delta\Omega = n'^2\Delta\Omega'$, where $\Delta\Omega = \Delta\theta_x\Delta\theta_y$ and $\Delta\Omega' = \Delta\theta'_x\Delta\theta'_y$. In this case, the total flux incident on a small area ΔA on the interface is given by $L\Delta A\Delta\Omega$. By conservation of energy, exactly the same flux must emerge from the area, part of it in the reflected beam and part in the transmitted beam. Therefore, at normal incidence, we must have

$$L_{inc}\Delta\Omega = L_{refl}\Delta\Omega + L_{tr}\Delta\Omega' \,. \tag{10.144}$$

If we denote by R the fraction of the energy reflected and by $T \equiv 1 - R$ the fraction transmitted, then the transmitted radiance must satisfy

$$\frac{1}{n'^2} L_{tr} = \frac{T}{n^2} L_{inc} \,. \tag{10.145}$$

For a glass-air interface, $R \simeq 0.04$ and $T \simeq 0.96$, and it is common to approximate T as unity. With that approximation, (10.145) shows that L/n^2 is conserved across the interface. To be explicit, L here refers to $L(\mathbf{r}, \hat{\mathbf{n}})$, where $\hat{\mathbf{n}}$ is normal to the interface and $\mathbf{r}$ is any point on the interface.

The calculation is similar for oblique incidence, but we must remember that radiance is defined in terms of projected area. If we consider $\hat{\mathbf{s}}_{inc}$ in the x-z plane, then we can still take $\cos\theta_y \simeq \cos\theta'_y \simeq 1$, but we must retain the cosines in the x direction. Now the projected area is $\Delta A \cos\theta_x$ and the total flux incident on ΔA is given by $L\Delta A\Delta\Omega \cos\theta_x$. The solid angle $\Delta\Omega$ is still $\Delta\theta_x\Delta\theta_y$, and a little algebra shows that (10.145) still holds. Explicitly,

$$\frac{1}{n'^2} L_{tr}(\mathbf{r}, \hat{\mathbf{s}}_{tr}) = \frac{T}{n^2} L_{inc}(\mathbf{r}, \hat{\mathbf{s}}_{inc}) \,, \tag{10.146}$$

where $\hat{\mathbf{s}}_{tr}$ is related to $\hat{\mathbf{s}}_{inc}$ and $\hat{\mathbf{n}}$ by (9.148). Up to reflection losses, then, L/n^2 is conserved along a ray direction defined by Snell's law.

10.3.3 Steady-state solutions in absorbing media

Now we can add absorption to the treatment above. With the absorption term and a general volume source, the steady-state Boltzmann equation is

$$\hat{\mathbf{s}} \cdot \nabla w = \frac{1}{c_m} \Xi_{p,\mathcal{E}} - \mu_{abs} w \,. \tag{10.147}$$

In a Cartesian coordinate system with the z axis parallel to $\hat{\mathbf{s}}$, (10.147) is again an ordinary differential equation in z, as it was in Sec. 10.3.2, but now the coefficients are not constant. In particular, μ_{abs} is a function of z, in general, so we incorporate an integrating factor by defining

$$\tilde{w}(z) = w(x, y, z, \hat{\mathbf{s}}, \mathcal{E}) \exp\left[\int_{-\infty}^{z} dz'' \, \mu_{abs}(x, y, z'', \mathcal{E})\right] , \tag{10.148}$$

where we show all of the arguments on the right for clarity but focus on the z dependence on the left. Of course, $\tilde{w}$ is still a function of the other variables.

With this substitution, the equation for $\tilde{w}$ is

$$\frac{\partial \tilde{w}}{\partial z} = \frac{1}{c_m} \Xi_{p,\mathcal{E}}(z) \exp\left[\int_{-\infty}^{z} dz'' \, \mu_{abs}(z'')\right] . \tag{10.149}$$

The solution for $\tilde{w}$ is obtained by simple integration, and then the original w is given by

$$\begin{aligned} w &= \exp\left[-\int_{-\infty}^{z} dz'' \, \mu_{abs}(z'')\right] \frac{1}{c_m} \int_{-\infty}^{z} dz' \, \Xi_{p,\mathcal{E}}(z') \exp\left[\int_{-\infty}^{z'} dz'' \, \mu_{abs}(z'')\right] \\ &= \frac{1}{c_m} \int_{-\infty}^{z} dz' \, \Xi_{p,\mathcal{E}}(z') \exp\left[-\int_{z'}^{z} dz'' \, \mu_{abs}(z'')\right] . \end{aligned} \tag{10.150}$$

In a vector notation analogous to (10.136) (and with all of the variables reinstated), we can write

$$w(\mathbf{r}, \hat{\mathbf{s}}, \mathcal{E}) = \frac{1}{c_m} \int_0^{\infty} d\ell \, \Xi_{p,\mathcal{E}}(\mathbf{r} - \hat{\mathbf{s}}\ell, \hat{\mathbf{s}}, \mathcal{E}) \exp\left[-\int_0^{\ell} d\ell' \, \mu_{abs}(\mathbf{r} - \hat{\mathbf{s}}\ell', \mathcal{E})\right] . \tag{10.151}$$

As in (10.136), $w(\mathbf{r}, \hat{\mathbf{s}}, \mathcal{E})$ is still found by integrating the source distribution along a line parallel to $\hat{\mathbf{s}}$ and passing through the point $\mathbf{r}$, but more distant points along this line contribute less because of the exponential attenuation factor.

We can now see the interpretation of μ_{abs}; if it is independent of $\mathbf{r}$, then radiation traversing a distance ℓ is attenuated by a factor of $\exp(-\mu_{abs}\ell)$, so μ_{abs} is the reciprocal of the distance required for attenuation by $1/e$. If μ_{abs} depends on position, however, the attenuation is determined by a line integral of μ_{abs}.

The integral in (10.151) defines the *attenuated x-ray transform* $\boldsymbol{\mathcal{X}}_\mu$. Like the x-ray transform, $\boldsymbol{\mathcal{X}}_\mu$ maps a function of $\mathbf{r}$ and $\hat{\mathbf{s}}$ to another function of $\mathbf{r}$ and $\hat{\mathbf{s}}$. In terms of this operator, (10.151) is

$$w = \boldsymbol{\mathcal{X}}_\mu \Xi_{p,\mathcal{E}} . \tag{10.152}$$

As we shall see in Chap. 16, this transform is fundamental to emission-imaging modalities such as nuclear medicine and fluorescence microscopy in which the object is the radiation source but there is significant self-absorption of the radiation in the object.

Point sources In many kinds of imaging, including transmission microscopy and radiography with x rays, we are interested in the attenuating properties of the medium rather than the properties of the source. In these cases we can choose the form of the source in order to facilitate probing of the attenuation distribution. A common choice is a very small source, approximating a mathematical point. In radiography, for example, x rays are generated by focusing a beam of electrons on a metal target, and it is desirable to make the focal spot as small as possible.

An ideal point source at point $\mathbf{r}_0$ is described by

$$\Xi_{p,\mathcal{E}}(\mathbf{r}, \hat{\mathbf{s}}, \mathcal{E}) = A\,\delta(\mathbf{r} - \mathbf{r}_0)\,. \tag{10.153}$$

The strength A of the delta function can, in general, depend on energy $\mathcal{E}$, but to focus on the spatial structure we treat A as a constant and drop the $\mathcal{E}$ argument in μ_{abs}.

When we insert (10.153) into (10.151), the result is

$$w(\mathbf{r}, \hat{\mathbf{s}}, \mathcal{E}) = \frac{A}{c_m} \int_0^\infty d\ell\; \delta(\mathbf{r} - \mathbf{r}_0 - \hat{\mathbf{s}}\ell\,) \exp\left[- \int_0^\ell d\ell'\, \mu_{abs}(\mathbf{r} - \hat{\mathbf{s}}\ell') \right]. \tag{10.154}$$

The argument of the delta function vanishes when $\mathbf{r} - \mathbf{r}_0 = \hat{\mathbf{s}}\ell$, which requires that

$$\hat{\mathbf{s}} = \frac{\mathbf{r} - \mathbf{r}_0}{|\mathbf{r} - \mathbf{r}_0|} \tag{10.155}$$

and

$$\ell = |\mathbf{r} - \mathbf{r}_0|\,. \tag{10.156}$$

If we substitute (10.156) for ℓ in the upper limit of the integral over ℓ' in (10.154), we can write

$$w(\mathbf{r}, \hat{\mathbf{s}}, \mathcal{E}) = \frac{A}{c_m} \exp\left[- \int_0^{|\mathbf{r}-\mathbf{r}_0|} d\ell'\, \mu_{abs}(\mathbf{r} - \hat{\mathbf{s}}\ell') \right] \int_0^\infty d\ell\; \delta(\mathbf{r} - \mathbf{r}_0 - \hat{\mathbf{s}}\ell\,)\,. \tag{10.157}$$

The remaining integral is a representation of an angular delta function, enforcing the condition (10.155). Before continuing with the discussion of (10.157), we must derive a new representation of such delta functions.

Digression: Angular delta functions An angular delta function was defined in Sec. 2.4.7 by requiring that

$$\int_{4\pi} d\Omega_s\; \delta(\hat{\mathbf{s}} - \hat{\mathbf{s}}_0)\, t(\hat{\mathbf{s}}) = t(\hat{\mathbf{s}}_0)\,, \tag{10.158}$$

where $d\Omega_s$ is the element of solid angle associated with $\hat{\mathbf{s}}$, and $t(\hat{\mathbf{s}})$ is a test function in the angular variables.

With this notation, we shall now show that

$$|\mathbf{r} - \mathbf{r}_0|^2 \int_0^\infty d\ell\; \delta(\mathbf{r} - \mathbf{r}_0 - \hat{\mathbf{s}}\ell\,) = \delta\left(\hat{\mathbf{s}} - \frac{\mathbf{r} - \mathbf{r}_0}{|\mathbf{r} - \mathbf{r}_0|} \right). \tag{10.159}$$

To demonstrate that this equation is correct, we multiply both sides by an arbitrary angular test function $t(\hat{\mathbf{s}})$ and integrate over solid angle. The integral on the left-hand side becomes

$$|\mathbf{r} - \mathbf{r}_0|^2 \int_{4\pi} d\Omega_s\; t(\hat{\mathbf{s}}) \int_0^\infty d\ell\; \delta(\mathbf{r} - \mathbf{r}_0 - \hat{\mathbf{s}}\ell\,) = \int_{4\pi} d\Omega_s \int_0^\infty \ell^2\, d\ell\; t(\hat{\mathbf{s}})\, \delta(\mathbf{r} - \mathbf{r}_0 - \hat{\mathbf{s}}\ell\,)\,, \tag{10.160}$$

where we have moved $|\mathbf{r}-\mathbf{r}_0|^2$ inside the integral and replaced it with ℓ^2 by use of (10.156). Now we can define a 3D vector $\mathbf{r}_s = \hat{\mathbf{s}}\ell$ and recognize that $d^3\mathbf{r}_s = \ell^2 d\ell\, d\Omega_s$. This substitution puts the integral into the correct form for applying the familiar 3D sifting property, and we obtain

$$|\mathbf{r}-\mathbf{r}_0|^2 \int_{4\pi} d\Omega_s\, t(\hat{\mathbf{s}}) \int_0^\infty d\ell\, \delta(\mathbf{r}-\mathbf{r}_0-\hat{\mathbf{s}}\ell) = \int_\infty d^3\mathbf{r}_s\, t(\hat{\mathbf{s}})\, \delta(\mathbf{r}-\mathbf{r}_0-\mathbf{r}_s) = t\left(\frac{\mathbf{r}-\mathbf{r}_0}{|\mathbf{r}-\mathbf{r}_0|}\right). \tag{10.161}$$

Exactly the same result would be obtained with the right-hand side of (10.159) and the angular sifting property (10.158), so (10.161) proves the validity of (10.159).

Return With (10.159), (10.157) becomes

$$w(\mathbf{r}, \hat{\mathbf{s}}, \mathcal{E}) = \frac{A}{c_m} \exp\left[-\int_0^{|\mathbf{r}-\mathbf{r}_0|} d\ell'\, \mu_{abs}(\mathbf{r}-\hat{\mathbf{s}}\ell')\right] \frac{\delta\left(\hat{\mathbf{s}} - \frac{\mathbf{r}-\mathbf{r}_0}{|\mathbf{r}-\mathbf{r}_0|}\right)}{|\mathbf{r}-\mathbf{r}_0|^2}. \tag{10.162}$$

The inverse-square factor might surprise the reader, but it is consistent with our previous conclusions. In the absence of attenuation, we showed that the radiance (and hence w) was constant along the direction $\hat{\mathbf{s}}$, with no inverse-square factor. From (10.162), we see only that w is infinite for all distances when $\hat{\mathbf{s}}$ is directed precisely away from the source, and it is zero for all distances for any other $\hat{\mathbf{s}}$. The inverse-square factor is evident only when one performs an angular integral as in (10.140) to get irradiance. This integral can be performed via (10.158), and the resulting spectral photon irradiance (measured at point $\mathbf{r}$ on a surface normal to the line of sight to the source) is

$$I_{p,\mathcal{E}} = \frac{A}{|\mathbf{r}-\mathbf{r}_0|^2} \exp\left[-\int_0^{|\mathbf{r}-\mathbf{r}_0|} d\ell'\, \mu_{abs}\left(\mathbf{r} - \frac{\mathbf{r}-\mathbf{r}_0}{|\mathbf{r}-\mathbf{r}_0|}\ell'\right)\right]. \tag{10.163}$$

As expected, the irradiance diminishes as the inverse square of the distance from the source, and it is further attenuated by an exponential factor involving the integral of the total attenuation coefficient along the path.

For fixed locations of the source and observation point, (10.163) shows that the log of the irradiance is linearly related to the line integral of the attenuation coefficient. We shall have much more to say about this relation in Chap. 16 when we discuss x-ray imaging.

10.3.4 Scattering effects

Now we consider the full steady-state Boltzmann equation with the scattering term:

$$\hat{\mathbf{s}} \cdot \nabla w = \frac{1}{c_m}\Xi_{p,\mathcal{E}} - \mu_{tot} w + \frac{1}{c_m}\mathcal{K}w. \tag{10.164}$$

Because of the scattering term, this equation cannot be solved by simple integration, but an integral along $\hat{\mathbf{s}}$ is nevertheless useful. It transforms (10.164) to

$$w(\mathbf{r}, \hat{\mathbf{s}}, \mathcal{E}) = \frac{1}{c_m} \int_0^\infty d\ell \; \Xi_{p,\mathcal{E}}(\mathbf{r} - \hat{\mathbf{s}}\ell, \hat{\mathbf{s}}, \mathcal{E}) \exp\left[- \int_0^\ell d\ell' \; \mu_{tot}(\mathbf{r} - \hat{\mathbf{s}}\ell') \right]$$

$$+ \frac{1}{c_m} \int_0^\infty d\ell \; [\mathcal{K}w](\mathbf{r} - \hat{\mathbf{s}}\ell, \hat{\mathbf{s}}, \mathcal{E}) \exp\left[- \int_0^\ell d\ell' \; \mu_{tot}(\mathbf{r} - \hat{\mathbf{s}}\ell') \right] . \tag{10.165}$$

This equation is not a solution since the unknown w still appears in the second integral. With the operators defined above, (10.165) can be written as

$$w = \mathcal{X}_\mu \Xi_{p,\mathcal{E}} + \mathcal{X}_\mu \mathcal{K} w \,, \tag{10.166}$$

or

$$[\mathbf{I} - \mathcal{X}_\mu \mathcal{K}] \, w = \mathcal{X}_\mu \Xi_{p,\mathcal{E}} \,, \tag{10.167}$$

where $\mathbf{I}$ is the identity operator. Since $\mathcal{X}_\mu$ and $\mathcal{K}$ are integral operators, (10.167) is really an integral equation for w.

A formal solution of (10.167) is given by the Neumann series (introduced in App. A and discussed in more detail in Chap. 1). From (A.59) and (10.167), we can write

$$w = [\mathbf{I} - \mathcal{X}_\mu \mathcal{K}]^{-1} \, \mathcal{X}_\mu \Xi_{p,\mathcal{E}}$$

$$= \mathcal{X}_\mu \Xi_{p,\mathcal{E}} + \mathcal{X}_\mu \mathcal{K} \mathcal{X}_\mu \Xi_{p,\mathcal{E}} + \mathcal{X}_\mu \mathcal{K} \mathcal{X}_\mu \mathcal{K} \mathcal{X}_\mu \Xi_{p,\mathcal{E}} + \dots . \tag{10.168}$$

Convergence conditions for this series are discussed in Sec. 1.7.6.

The series in (10.168) can be interpreted as the attenuated x-ray transform of an effective source distribution Ξ_{eff}, *i.e.*,

$$w = \mathcal{X}_\mu \Xi_{eff} \,, \tag{10.169}$$

where

$$\Xi_{eff} = \Xi_{p,\mathcal{E}} + \mathcal{K} \mathcal{X}_\mu \Xi_{p,\mathcal{E}} + \mathcal{K} \mathcal{X}_\mu \mathcal{K} \mathcal{X}_\mu \Xi_{p,\mathcal{E}} + \dots . \tag{10.170}$$

Successive terms in this series represent successively more scattering; photons that have scattered n times contribute the term $[\mathcal{K}\mathcal{X}_\mu]^n \Xi_{p,\mathcal{E}}$ to Ξ_{eff}.

A useful tool for dealing with scatter problems is an expansion in spherical harmonics, as discussed immediately below. Other useful techniques are treated in Secs. 10.3.6 and 10.3.7.

10.3.5 Spherical harmonics

Most scatter problems have an important symmetry that we can exploit. For randomly oriented scatterers, as noted below (10.129), the scatter kernel $K(\hat{\mathbf{s}}, \mathcal{E}; \hat{\mathbf{s}}', \mathcal{E}' | \mathbf{r}, t)$ depends on $\hat{\mathbf{s}}$ and $\hat{\mathbf{s}}'$ only through the scalar product $\hat{\mathbf{s}} \cdot \hat{\mathbf{s}}'$. That means that the operator $\mathcal{K}$ is invariant to arbitrary rotations of the coordinate system used to express the directions $\hat{\mathbf{s}}$ and $\hat{\mathbf{s}}'$. In the language of group theory (see Sec. 6.7), the symmetry group of $\mathcal{K}$ is $\mathbf{SO}(3)$. As we discussed in Sec. 6.7.7, the spherical harmonics are basis functions for irreducible representations of $\mathbf{SO}(3)$. Since $\mathcal{K}$ is invariant to this group, the scatter kernel will take a simple diagonal form in this basis.

Transformation of the distribution function We can expand the angular dependence of the distribution function in spherical harmonics as

$$w(\mathbf{r}, \hat{\mathbf{s}}, \mathcal{E}, t) = \sum_{\ell=0}^{\infty} \sum_{m=-\ell}^{\ell} W_{\ell m}(\mathbf{r}, \mathcal{E}, t)\, Y_{\ell m}(\hat{\mathbf{s}})\,, \tag{10.171}$$

where $Y_{\ell m}(\hat{\mathbf{s}})$ means the same thing as $Y_{\ell m}(\theta, \phi)$. Note carefully that θ and ϕ are the spherical coordinates of $\hat{\mathbf{s}}$, not $\mathbf{r}$.

From the orthogonality relation (4.37), the coefficients are given by

$$W_{\ell m}(\mathbf{r}, \mathcal{E}, t) = \int_{4\pi} d\Omega\; Y^*_{\ell m}(\hat{\mathbf{s}})\, w(\mathbf{r}, \hat{\mathbf{s}}, \mathcal{E}, t)\,. \tag{10.172}$$

This equation defines a CD operator $\boldsymbol{\mathcal{Y}}$ that maps a function of direction $\hat{\mathbf{s}}$ (or equivalently, polar angles θ and ϕ) to its spherical-harmonic coefficients, *i.e.*,

$$[\boldsymbol{\mathcal{Y}} w(\mathbf{r}, \hat{\mathbf{s}}, \mathcal{E}, t)]_{\ell m} = W_{\ell m}(\mathbf{r}, \mathcal{E}, t)\,. \tag{10.173}$$

If we consider the coefficients to be components of an infinite-dimensional vector $\mathbf{W}$, then we can express this operator relation even more abstractly as

$$\boldsymbol{\mathcal{Y}}\mathbf{w} = \mathbf{W}\,, \tag{10.174}$$

where the boldface is used since we now want to think of $\mathbf{w}$ as a vector in a Hilbert space. Since the spherical harmonics are orthonormal [see (4.37)], the operator $\boldsymbol{\mathcal{Y}}$ is unitary.

Transformation of the source Like the distribution function w, the source distribution $\Xi_{p,\mathcal{E}}$ is a function of $\mathbf{r}$, $\hat{\mathbf{s}}$, $\mathcal{E}$ and t, and its angular dependence (if any) can also be transformed to spherical harmonics. That is, we can define an infinite vector $\boldsymbol{\xi} = \boldsymbol{\mathcal{Y}}\Xi_{p,\mathcal{E}}$, with components $\xi_{\ell m}$ given by

$$\xi_{\ell m}(\mathbf{r}, \mathcal{E}, t) = \int_{4\pi} d\Omega\; Y_{\ell m}(\hat{\mathbf{s}})\, \Xi_{p,\mathcal{E}}(\mathbf{r}, \hat{\mathbf{s}}, \mathcal{E}, t)\,. \tag{10.175}$$

For an isotropic source this vector has just one nonzero element (ξ_{00}), but it is useful to be more general.

Steady-state Boltzmann equation We now assume steady-state conditions, so that the source and distribution function are independent of time. To get the steady-state Boltzmann equation in spherical-harmonic form, we make the substitutions $\mathbf{w} = \boldsymbol{\mathcal{Y}}^{-1}\mathbf{W}$ and $\Xi_{p,\mathcal{E}} = \boldsymbol{\mathcal{Y}}^{-1}\boldsymbol{\xi}$ in (10.132) and then operate from the left with $\boldsymbol{\mathcal{Y}}$, obtaining

$$\mu_{tot}\mathbf{W} - \frac{1}{c_m}\boldsymbol{\xi} - \frac{1}{c_m}\boldsymbol{\mathcal{Y}}\boldsymbol{\mathcal{K}}\boldsymbol{\mathcal{Y}}^{-1}\mathbf{W} + \boldsymbol{\mathcal{Y}}(\hat{\mathbf{s}} \cdot \nabla)\,\boldsymbol{\mathcal{Y}}^{-1}\mathbf{W}\,. \tag{10.176}$$

This equation, where the unknown is the infinite vector $\mathbf{W}$, is equivalent to the steady-state Boltzmann equation. To solve it, we need to learn how to compute the scatter term $\boldsymbol{\mathcal{Y}}\boldsymbol{\mathcal{K}}\boldsymbol{\mathcal{Y}}^{-1}\mathbf{W}$ and the propagation term $\boldsymbol{\mathcal{Y}}(\hat{\mathbf{s}} \cdot \nabla)\,\boldsymbol{\mathcal{Y}}^{-1}\mathbf{W}$. It will turn out that the scatter term is diagonal, but the propagation term is complicated. We shall discuss both terms below.

Another way of expressing the steady-state Boltzmann equation is by the integral equation (10.166), and it is useful to transform this equation to spherical-harmonic form also. By applying $\mathcal{Y}$ to (10.166) and inserting $\mathcal{Y}^{-1}\mathcal{Y}$ judiciously, we find

$$\mathbf{W} = \mathcal{Y}\mathcal{X}_\mu\mathcal{Y}^{-1}\left[\boldsymbol{\xi} + \mathcal{Y}\mathcal{K}\mathcal{Y}^{-1}\mathbf{W}\right] . \tag{10.177}$$

The scatter term $\mathcal{Y}(\hat{\mathbf{s}}\cdot\nabla)\mathcal{Y}^{-1}\mathbf{W}$ from (10.176) appears here as an effective source term. The overall source is then operated on by the attenuated x-ray transform, expressed in the spherical-harmonic representation as $\mathcal{Y}\mathcal{X}_\mu\mathcal{Y}^{-1}$. Ways of implementing this operator are discussed below.

Transformation of the scatter operator We look first at the scattering term $\mathcal{Y}\mathcal{K}\mathcal{Y}^{-1}\mathbf{W}$ which appears in (10.176) and (10.177). In component form, this term is

$$\left[\mathcal{Y}\mathcal{K}\mathcal{Y}^{-1}\mathbf{W}\right]_{\ell m} = \sum_{\ell'=0}^{\infty}\sum_{m'=-\ell'}^{\ell'}\left[\mathcal{Y}\mathcal{K}\mathcal{Y}^{-1}\right]_{\ell m,\ell' m'} W_{\ell' m'} . \tag{10.178}$$

Now we make use of the fact that the scatter kernel is a function of $\hat{\mathbf{s}}\cdot\hat{\mathbf{s}}'$, which is just the cosine of the angle between the incoming and scattered photons. As we noted in Sec. 4.1.4, Legendre polynomials are particularly useful for expanding functions of the cosine of an angle, so we expand the scatter kernel as

$$K(\hat{\mathbf{s}}\cdot\hat{\mathbf{s}}') = \sum_{\ell''=0}^{\infty} k_{\ell''} P_{\ell''}(\hat{\mathbf{s}}\cdot\hat{\mathbf{s}}') = \sum_{\ell''=0}^{\infty} k_{\ell''}\frac{4\pi}{2\ell''+1}\sum_{m''=-\ell''}^{\ell''} Y_{\ell'' m''}(\hat{\mathbf{s}})\, Y^*_{\ell'' m''}(\hat{\mathbf{s}}') , \tag{10.179}$$

where we have used the addition theorem (4.38) to get the second form. The notation here may be somewhat misleading since $k_{\ell''}$ looks like a constant but can be substantially more complicated. For elastic scattering in an inhomogeneous medium, $k_{\ell''}$ is a function of $\mathbf{r}$, and for inelastic scattering it is an integral operator with respect to the energy variable; it transforms a function of $\mathcal{E}'$ to a function of $\mathcal{E}$.

With this expansion, the matrix elements in (10.178) are given by

$$\begin{aligned}\left[\mathcal{Y}\mathcal{K}\mathcal{Y}^{-1}\right]_{\ell m,\ell' m'} & \\ = \sum_{\ell''=0}^{\infty} k_{\ell''}\frac{4\pi}{2\ell''+1}\sum_{m''=-\ell''}^{\ell''}&\int_{4\pi} d\Omega\, Y_{\ell'' m''}(\hat{\mathbf{s}})\, Y^*_{\ell m}(\hat{\mathbf{s}})\int_{4\pi} d\Omega'\, Y^*_{\ell'' m''}(\hat{\mathbf{s}}')\, Y_{\ell' m'}(\hat{\mathbf{s}}') \\ &= \frac{4\pi}{2\ell+1} k_\ell\, \delta_{\ell\ell'}\, \delta_{mm'} ,\end{aligned} \tag{10.180}$$

where the final form has made use of the orthogonality of the spherical harmonics, (4.37).

It follows from (10.180) that the spherical harmonics are eigenfunctions of the angular part of $\mathcal{K}$. (The full operator $\mathcal{K}$ still requires an integral over energy if the scattering is inelastic.) As a practical matter, (10.180) shows that the angular integral in $\mathcal{K}$ can be implemented in the spherical-harmonic domain, where it corresponds to multiplication by a diagonal matrix.

For notational ease, we define

$$\mathcal{D} = \mathcal{Y}\mathcal{K}\mathcal{Y}^{-1} , \tag{10.181}$$

where the letter $\mathcal{D}$ suggests that the operator is diagonal with respect to the spherical-harmonic indices. For inelastic scattering, $\mathcal{D}$ is still an operator on the energy variables, and the matrix elements can depend on both position and energy.

Propagation term Unfortunately, the spherical-harmonic expansion does not diagonalize the full Boltzmann equation. In particular, the propagation term in (10.176) is not diagonalized.

In detail, this term takes the form

$$[\mathcal{Y}(\hat{\mathbf{s}}\cdot\nabla)\,\mathcal{Y}^{-1}\mathbf{W}]_{\ell m}(\mathbf{r},\mathcal{E}) = \int_{4\pi} d\Omega\; Y^*_{\ell m}(\hat{\mathbf{s}})\,\hat{\mathbf{s}}\cdot\nabla w(\mathbf{r},\hat{\mathbf{s}},\mathcal{E})$$

$$= \sum_{\ell' m'}\int_{4\pi} d\Omega\; Y^*_{\ell m}(\hat{\mathbf{s}})\,Y_{\ell' m'}(\hat{\mathbf{s}})\,\hat{\mathbf{s}}\cdot\nabla W_{\ell' m'}(\mathbf{r},\mathcal{E})\,. \tag{10.182}$$

Since $\hat{\mathbf{s}} = (\sin\theta\cos\phi, \sin\theta\sin\phi, \cos\theta)$ in Cartesian coordinates, we can also write

$$[\mathcal{Y}(\hat{\mathbf{s}}\cdot\nabla)\,\mathcal{Y}^{-1}\mathbf{W}]_{\ell m}(\mathbf{r},\mathcal{E}) = \sum_{\ell' m'}\frac{\partial W_{\ell' m'}(\mathbf{r},\mathcal{E})}{\partial x}\int_{4\pi} d\Omega\; Y^*_{\ell m}(\hat{\mathbf{s}})\,Y_{\ell' m'}(\hat{\mathbf{s}})\sin\theta\cos\phi$$

$$+\sum_{\ell' m'}\frac{\partial W_{\ell' m'}(\mathbf{r},\mathcal{E})}{\partial y}\int_{4\pi} d\Omega\; Y^*_{\ell m}(\hat{\mathbf{s}})\,Y_{\ell' m'}(\hat{\mathbf{s}})\sin\theta\sin\phi$$

$$+\sum_{\ell' m'}\frac{\partial W_{\ell' m'}(\mathbf{r},\mathcal{E})}{\partial z}\int_{4\pi} d\Omega\; Y^*_{\ell m}(\hat{\mathbf{s}})\,Y_{\ell' m'}(\hat{\mathbf{s}})\cos\theta\,, \tag{10.183}$$

where we have used (10.171).

This relation can be written more compactly if we define an infinite-dimensional, vector-valued matrix $\mathbf{C}$, where each element is a 3D vector given by

$$\mathbf{C}_{\ell m,\ell' m'} = \int_{4\pi} d\Omega\; Y^*_{\ell m}(\hat{\mathbf{s}})\,Y_{\ell' m'}(\hat{\mathbf{s}})\,\hat{\mathbf{s}}\,. \tag{10.184}$$

In terms of $\mathbf{C}$, (10.182) is

$$\left[\frac{\partial \mathbf{W}}{\partial t}\right]_{prop} = -c_m\mathbf{C}\cdot\nabla\mathbf{W}\,. \tag{10.185}$$

Full equation With the forms derived above for the scattering and propagation terms, the steady-state Boltzmann equation in the spherical-harmonic representation, (10.176), becomes

$$\mathbf{C}\cdot\nabla\mathbf{W} = \frac{1}{c_m}\boldsymbol{\xi} - \mu_{tot}\mathbf{W} - \frac{1}{c_m}\mathcal{D}\mathbf{W}\,, \tag{10.186}$$

Attenuated x-ray transform Next we look at the operator $\mathcal{Y}\mathcal{X}_\mu\mathcal{Y}^{-1}$ that appears in (10.177). This operator maps an infinite vector in the spherical-harmonic space to another such infinite vector, but the mapping is not local.

To see the nonlocal character, consider a general vector $\mathbf{V}(\mathbf{r})$ with components $V_{\ell m}(\mathbf{r})$. If we let $\mathcal{Y}^{-1}\{\mathbf{V}(\mathbf{r})\} = v(\mathbf{r},\hat{\mathbf{s}})$, then the equation

$$\mathbf{U} = \mathcal{Y}\mathcal{X}_\mu\mathcal{Y}^{-1}\mathbf{V} \tag{10.187}$$

means that

$$U_{\ell m}(\mathbf{r}) = [\mathcal{Y}\mathcal{X}_\mu \mathbf{v}]_{\ell m}(\mathbf{r})$$

$$= \frac{1}{c_m}\int_{4\pi} d\Omega\, Y^*_{\ell m}(\hat{\mathbf{s}}) \int_0^\infty dt\; v(\mathbf{r}-\hat{\mathbf{s}}t, \hat{\mathbf{s}}) \exp\left[-\int_0^t dt'\, \mu_{tot}(\mathbf{r}-\hat{\mathbf{s}}t')\right]. \qquad (10.188)$$

From (10.171) and the unitarity of $\mathcal{Y}$, the function $v(\mathbf{r}, \hat{\mathbf{s}})$ is related to the components of $\mathbf{V}(\mathbf{r})$ by

$$v(\mathbf{r}, \hat{\mathbf{s}}) = \sum_{\ell' m'} V_{\ell' m'}(\mathbf{r})\, Y_{\ell' m'}(\hat{\mathbf{s}}). \qquad (10.189)$$

By a change of variables,

$$v(\mathbf{r}-\hat{\mathbf{s}}t, \hat{\mathbf{s}}) = \sum_{\ell' m'} V_{\ell' m'}(\mathbf{r}-\hat{\mathbf{s}}t)\, Y_{\ell' m'}(\hat{\mathbf{s}}), \qquad (10.190)$$

so (10.188) becomes

$$U_{\ell m}(\mathbf{r}) =$$

$$\frac{1}{c_m}\sum_{\ell' m'}\int_{4\pi} d\Omega\, Y^*_{\ell m}(\hat{\mathbf{s}}) \int_0^\infty dt\; V_{\ell' m'}(\mathbf{r}-\hat{\mathbf{s}}t)\, Y_{\ell' m'}(\hat{\mathbf{s}}) \exp\left[-\int_0^t dt'\, \mu_{tot}(\mathbf{r}-\hat{\mathbf{s}}t')\right]. \qquad (10.191)$$

If we define $\mathbf{r}' = \mathbf{r} - \hat{\mathbf{s}}t$ and recognize that $t^2 dt\, d\Omega = d^3\mathbf{r}'$ and $t = |\mathbf{r}-\mathbf{r}'|$, then

$$U_{\ell m}(\mathbf{r}) = \frac{1}{c_m}\sum_{\ell' m'}\int_\infty d^3\mathbf{r}'\, \frac{Y^*_{\ell m}\left(\frac{\mathbf{r}-\mathbf{r}'}{|\mathbf{r}-\mathbf{r}'|}\right) Y_{\ell' m'}\left(\frac{\mathbf{r}-\mathbf{r}'}{|\mathbf{r}-\mathbf{r}'|}\right)}{|\mathbf{r}-\mathbf{r}'|^2}\, V_{\ell' m'}(\mathbf{r}')$$

$$\times \exp\left[-\int_0^{|\mathbf{r}-\mathbf{r}'|} dt'\, \mu_{tot}\left(\mathbf{r} - t'\frac{\mathbf{r}-\mathbf{r}'}{|\mathbf{r}-\mathbf{r}'|}\right)\right]. \qquad (10.192)$$

This equation shows in detail the action of the operator $\mathcal{Y}\mathcal{X}_\mu\mathcal{Y}^{-1}$.

We can simplify the notation by defining

$$\mathcal{A} = \mathcal{Y}\mathcal{X}_\mu\mathcal{Y}^{-1}. \qquad (10.193)$$

The letter $\mathcal{A}$ here indicates that the operator is the spherical-harmonic transform of the attenuated x-ray transform. From (10.192) we see that $\mathcal{A}$ is a combination of a matrix operator and an integral operator,

$$[\mathcal{A}\mathbf{V}]_{\ell m}(\mathbf{r}) = \sum_{\ell' m'}\int_\infty d^3\mathbf{r}'\, A_{\ell m,\ell' m'}(\mathbf{r}, \mathbf{r}')\, V_{\ell' m'}(\mathbf{r}'), \qquad (10.194)$$

with an element/kernel given by

$$A_{\ell m,\ell' m'}(\mathbf{r}, \mathbf{r}') = \frac{1}{c_m}\frac{1}{|\mathbf{r}-\mathbf{r}'|^2} Y^*_{\ell m}\left(\frac{\mathbf{r}-\mathbf{r}'}{|\mathbf{r}-\mathbf{r}'|}\right) Y_{\ell' m'}\left(\frac{\mathbf{r}-\mathbf{r}'}{|\mathbf{r}-\mathbf{r}'|}\right)$$

$$\times \exp\left[-\int_0^{|\mathbf{r}-\mathbf{r}'|} dt'\, \mu_{tot}\left(\mathbf{r} - t'\frac{\mathbf{r}-\mathbf{r}'}{|\mathbf{r}-\mathbf{r}'|}\right)\right]. \qquad (10.195)$$

If the attenuation coefficient is a constant, independent of position, then the exponential factor becomes $\exp[-\mu_{tot}|\mathbf{r}-\mathbf{r}'|]$, and in this case the spatial part of the operator is shift-invariant.

Scattered distribution in spherical harmonics We now know the effect of each of the operators in (10.177), and we can use that equation to construct an equation for the scattered distribution in spherical harmonics. In the present notation, (10.177) reads

$$\mathbf{W} = \boldsymbol{\mathcal{A}}[\boldsymbol{\xi} + \boldsymbol{\mathcal{D}}\mathbf{W}], \tag{10.196}$$

where $\boldsymbol{\mathcal{A}}$ is the attenuated x-ray transform in spherical-harmonic space, and the operator $\boldsymbol{\mathcal{D}}$, which is a diagonal matrix for elastic scattering, is in general the transform of the scatter operator $\boldsymbol{\mathcal{K}}$.

It is useful to divide $\mathbf{W}$ into contributions arising from primary (unscattered) and scattered radiation:

$$\mathbf{W} = \mathbf{W}^{pri} + \mathbf{W}^{sc}, \tag{10.197}$$

where

$$\mathbf{W}^{pri} \equiv \boldsymbol{\mathcal{A}}\boldsymbol{\xi}. \tag{10.198}$$

With this division, (10.196) leads to

$$\mathbf{W}^{sc} = \boldsymbol{\mathcal{A}}\boldsymbol{\mathcal{D}}[\boldsymbol{\mathcal{A}}\boldsymbol{\xi} + \mathbf{W}^{sc}], \tag{10.199}$$

or

$$[\mathbf{I} - \boldsymbol{\mathcal{A}}\boldsymbol{\mathcal{D}}]\,\mathbf{W}^{sc} = \boldsymbol{\mathcal{A}}\boldsymbol{\mathcal{D}}\boldsymbol{\mathcal{A}}\xi. \tag{10.200}$$

The right-hand side of this equation represents an effective source distribution produced by the attenuated primary photons plus singly scattered photons. Nevertheless, because of the operator on the left, the equation still accounts for multiple scatters of all orders. No approximations have yet been made.

Solution methods A formal solution to (10.200) is

$$\mathbf{W}^{sc} = [\mathbf{I} - \boldsymbol{\mathcal{A}}\boldsymbol{\mathcal{D}}]^{-1}\boldsymbol{\mathcal{A}}\boldsymbol{\mathcal{D}}\boldsymbol{\mathcal{A}}\boldsymbol{\xi}, \tag{10.201}$$

and a numerical solution can, in principle, be implemented with a Neumann series like (10.168). As discussed below, three practical ways to solve (10.200) are (1) the weak-scattering approximation, (2) the diffusion approximation (Ishimaru, 1978) and (3) trickle-down theory (Barrett *et al.*, 1998d).

The weak-scattering limit of (10.201), which amounts to neglecting multiple scatters, is

$$\mathbf{W}^{sc} \simeq \boldsymbol{\mathcal{A}}\boldsymbol{\mathcal{D}}\boldsymbol{\mathcal{A}}\boldsymbol{\xi}. \tag{10.202}$$

This approximation is valid if the dimensions of the scattering medium are all small compared to $1/\mu_{sc}$, so that a photon will probably escape the medium before scattering more than once.

The diffusion approximation, discussed in Sec. 10.3.6, is the opposite extreme from the weak-scattering approximation; it is appropriate with strong elastic scattering where the dimensions of the medium are all much greater than $1/\mu_{sc}$.

The trickle-down approach is useful with inelastic scattering, such as Compton scattering, where the photon loses energy in each scatter event. In that case, the distribution at one energy is influenced by the distribution at all higher energies but not that at lower energies. This approach is developed in Sec. 10.3.7.

10.3.6 Elastic scattering and diffusion

Elastic scattering is the dominant scattering mechanism for optical or infrared photons propagating in a turbid medium such as breast tissue or sea water. Even Compton scattering of x rays and gamma rays is approximately elastic if the photon energy is very small compared to the rest-mass energy of the electron (511 keV).

If many elastic scattering events can occur before a photon is absorbed or escapes the medium, the effect of elastic scattering is to thoroughly randomize the photon directions. In that case only a few low-order terms in the spherical-harmonic expansion of the distribution function are needed to describe the scattered radiation. As we shall see, the result of this approximation is to reduce the Boltzmann equation to the diffusion equation.

Since elastic scattering does not change the photon energy, we can consider each energy group individually, or equivalently just drop the energy argument and denote the distribution function as $w(\mathbf{r}, \hat{\mathbf{s}})$.

Diffusion approximation As in (10.197), we split the distribution function into contributions arising from primary and scattered radiation,

$$w(\mathbf{r}, \hat{\mathbf{s}}) = w^{pri}(\mathbf{r}, \hat{\mathbf{s}}) + w^{sc}(\mathbf{r}, \hat{\mathbf{s}}) \,, \tag{10.203}$$

where $w^{pri} = \mathcal{Y}^{-1}\mathbf{W}^{pri}$ and $w^{sc} = \mathcal{Y}^{-1}\mathbf{W}^{sc}$. The Boltzmann equation now splits into two equations,

$$\hat{\mathbf{s}} \cdot \nabla w^{pri} = \frac{1}{c_m}\xi_{p,\mathcal{E}} - \mu_{tot}\, w^{pri} \,; \tag{10.204}$$

$$\hat{\mathbf{s}} \cdot \nabla w^{sc} = -\mu_{tot}\, w^{sc} + \frac{1}{c_m}\mathcal{K}(w^{sc} + w^{pri}) \,. \tag{10.205}$$

These equations are coupled since both w^{pri} and w^{sc} appear in the last term of (10.205).

The diffusion approximation consists of retaining only the terms corresponding to $\ell = 0$ and $\ell = 1$ in the spherical-harmonic expansion of w^{sc}, thus writing

$$w^{sc}(\mathbf{r}, \hat{\mathbf{s}}) \simeq W_{00}^{sc}(\mathbf{r})\, Y_{00}(\hat{\mathbf{s}}) + \sum_{m=-1}^{1} W_{1m}^{sc}(\mathbf{r})\, Y_{1m}(\hat{\mathbf{s}}) \,. \tag{10.206}$$

Since $Y_{00}(\hat{\mathbf{s}})$ is the constant $1/\sqrt{4\pi}$, the first term represents an isotropic distribution of photons. Specifically, with the use of (10.172),

$$W_{00}^{sc}(\mathbf{r})\, Y_{00}(\hat{\mathbf{s}}) = \frac{1}{4\pi}\int_{4\pi} d\Omega'\, w^{sc}(\mathbf{r}, \hat{\mathbf{s}}') \,. \tag{10.207}$$

From the definition of the distribution function, the integral in (10.207) can be interpreted as a spatial density of scattered photons, which we denote as $u^{sc}(\mathbf{r})$ and define by

$$u^{sc}(\mathbf{r}) \equiv \int_{4\pi} d\Omega'\, w^{sc}(\mathbf{r}, \hat{\mathbf{s}}') \,. \tag{10.208}$$

To see the meaning of the second term in (10.206), we use (10.172) again to write

$$\sum_{m=-1}^{1} W_{1m}^{sc}(\mathbf{r})\,Y_{1m}(\hat{\mathbf{s}}) = \sum_{m=-1}^{1} Y_{1m}(\hat{\mathbf{s}}) \int_{4\pi} d\Omega'\, Y_{1m}^{*}(\hat{\mathbf{s}}')\, w^{sc}(\mathbf{r},\hat{\mathbf{s}}')$$

$$= \frac{3}{4\pi}\int_{4\pi} d\Omega'\, P_1(\hat{\mathbf{s}}\cdot\hat{\mathbf{s}}')\, w^{sc}(\mathbf{r},\hat{\mathbf{s}}') = \frac{3}{4\pi}\hat{\mathbf{s}}\cdot\int_{4\pi} d\Omega'\,\hat{\mathbf{s}}' w^{sc}(\mathbf{r},\hat{\mathbf{s}}')\,, \qquad (10.209)$$

where the second line has used the addition theorem (4.38) and the fact that $P_1(x) = x$. We now define a vector quantity $\mathbf{J}^{sc}(\mathbf{r})$ by

$$\mathbf{J}^{sc}(\mathbf{r}) = c_m \int_{4\pi} d\Omega'\,\hat{\mathbf{s}}' w^{sc}(\mathbf{r},\hat{\mathbf{s}}')\,. \qquad (10.210)$$

If $w^{sc}(\mathbf{r},\hat{\mathbf{s}}')$ is isotropic, then the integral vanishes since the integrand is an odd function of $\hat{\mathbf{s}}$. Thus $\mathbf{J}^{sc}(\mathbf{r})$ measures the magnitude and direction of the residual anisotropy. Moreover, since $c_m\hat{\mathbf{s}}$ is the photon velocity, $\mathbf{J}^{sc}(\mathbf{r})$ can be thought of as the average of the velocity of the photons at point $\mathbf{r}$ times their density. Units of $\mathbf{J}^{sc}$ are photons per second per unit area per unit energy, but for elastic scattering we ignore the energy dependence (we could integrate over energy) and think of $\mathbf{J}^{sc}$ as photons per second per unit area. Then $\mathbf{J}^{sc}(\mathbf{r})\cdot\hat{\mathbf{n}}$ gives the rate at which photons traverse unit area on a surface normal to $\hat{\mathbf{n}}$, so we can think of $\mathbf{J}^{sc}$ as a *photon current density*. In terms of the radiometric quantities introduced in Sec. 10.2, $\mathbf{J}^{sc}(\mathbf{r})\cdot\hat{\mathbf{n}}$ is the spectral photon irradiance.

Putting the pieces together, we see that

$$w^{sc}(\mathbf{r},\hat{\mathbf{s}}) \simeq \frac{1}{4\pi}u^{sc}(\mathbf{r}) + \frac{3}{4\pi c_m}\hat{\mathbf{s}}\cdot\mathbf{J}^{sc}(\mathbf{r}) \qquad (10.211)$$

in the diffusion approximation. This equation involves both u^{sc} and $\mathbf{J}^{sc}$, but these quantities are not independent. As we shall now show, we can eliminate $\mathbf{J}^{sc}$ and derive the diffusion equation, a partial differential equation for u^{sc} alone. The derivation given here follows Ishimaru (1978).

If we integrate both sides of (10.205) over $\hat{\mathbf{s}}$ and use (10.208), the result is

$$\int_{4\pi} d\Omega\;\hat{\mathbf{s}}\cdot\nabla w^{sc} = -\mu_{tot}u^{sc} + \frac{1}{c_m}\int_{4\pi} d\Omega\;\mathcal{K}(w^{sc}+w^{pri})\,. \qquad (10.212)$$

Since $\hat{\mathbf{s}}$ is a constant so far as the ∇ operator is concerned,

$$\hat{\mathbf{s}}\cdot\nabla w^{sc} = \nabla\cdot(\hat{\mathbf{s}}w^{sc})\,, \qquad (10.213)$$

which, with (10.210), shows that the left-hand side of (10.212) is $\nabla\cdot\mathbf{J}^{sc}$.

The term in (10.212) involving $\mathcal{K}$ also simplifies considerably. Since the kernel of $\mathcal{K}$ for elastic scattering and randomly oriented scatterers is just $K(\hat{\mathbf{s}}\cdot\hat{\mathbf{s}}'|\mathbf{r})$, we have

$$\int_{4\pi} d\Omega\;\mathcal{K}w^{sc} = \int_{4\pi} d\Omega \int_{4\pi} d\Omega'\, K(\hat{\mathbf{s}}\cdot\hat{\mathbf{s}}'|\mathbf{r})\, w^{sc}(\mathbf{r},\hat{\mathbf{s}}')$$

$$= c_m\mu_{sc}(\mathbf{r})\int_{4\pi} d\Omega'\, w^{sc}(\mathbf{r},\hat{\mathbf{s}}') = c_m\mu_{sc}(\mathbf{r})\,u^{sc}(\mathbf{r})\,, \qquad (10.214)$$

where we have used (10.128) and an integral of (10.130) to relate the scatter kernel to the attenuation coefficient. A similar result holds for the term involving $\mathcal{K}w^{pri}$, and (10.212) becomes simply

$$\nabla \cdot \mathbf{J}^{sc} = -c_m \mu_{tot} u^{sc} + c_m \mu_{sc}(u^{sc} + u^{pri}) . \tag{10.215}$$

This equation connects the two unknown quantities u^{sc} and $\mathbf{J}^{sc}$. To get another equation in these unknowns, we substitute (10.211) into (10.205) to yield

$$\frac{1}{4\pi}\hat{\mathbf{s}} \cdot \nabla u^{sc} + \frac{3}{4\pi c_m}\hat{\mathbf{s}} \cdot \nabla(\hat{\mathbf{s}} \cdot \mathbf{J}^{sc})$$

$$= -\mu_{tot}\left(\frac{1}{4\pi}u^{sc} + \frac{3}{4\pi c_m}\hat{\mathbf{s}} \cdot \mathbf{J}^{sc}\right) + \frac{1}{c_m}\mathcal{K}\left(w^{pri} + \frac{1}{4\pi}u^{sc} + \frac{3}{4\pi c_m}\hat{\mathbf{s}} \cdot \mathbf{J}^{sc}\right) . \tag{10.216}$$

The goal now is to find simultaneous solutions for u^{sc} and $\mathbf{J}^{sc}$ satisfying (10.215) and (10.216). The general procedure will be to multiply both sides of (10.216) by $\hat{\mathbf{s}}$, integrate over $\hat{\mathbf{s}}$, use all of the symmetries we can, and plug the result back into (10.215).

The following identities (Dorn, 1997; Ishimaru, 1978) are useful:

(*a*) $\int_{4\pi} d\Omega = 4\pi$;

(*b*) $\int_{4\pi} d\Omega \; \hat{\mathbf{s}} = 0$;

(*c*) $\int_{4\pi} d\Omega \; \hat{\mathbf{s}} \cdot \mathbf{A} = 0$;

(*d*) $\int_{4\pi} d\Omega \; \hat{\mathbf{s}} \, (\hat{\mathbf{s}} \cdot \mathbf{A}) = \frac{4\pi}{3}\mathbf{A}$;

(*e*) $\int_{4\pi} d\Omega \; (\hat{\mathbf{s}} \cdot \mathbf{A}) \, (\hat{\mathbf{s}} \cdot \mathbf{B}) = \frac{4\pi}{3}\mathbf{A} \cdot \mathbf{B}$;

(*f*) $\int_{4\pi} d\Omega \; \hat{\mathbf{s}} \, (\hat{\mathbf{s}} \cdot \mathbf{A}) \, (\hat{\mathbf{s}} \cdot \mathbf{B}) = 0$.

In these relations, $\mathbf{A}$ and $\mathbf{B}$ are any vectors that are independent of $\hat{\mathbf{s}}$, including the vector operator ∇. The relations with 0 on the right-hand side follow immediately from symmetry since the integrands are odd functions of $\hat{\mathbf{s}}$. The others can be derived by writing out the integrands in component form.

Anisotropic scattering Let us look first at the term $\mathcal{K}(\hat{\mathbf{s}} \cdot \mathbf{J}^{sc})$ in (10.216). This term can be simplified by taking advantage of the form of the kernel of $\mathcal{K}$ as in (10.214); by using polar coordinates with $\hat{\mathbf{s}}$ as the polar axis, we can show that

$$\mathcal{K}(\hat{\mathbf{s}} \cdot \mathbf{J}^{sc}) = \int_{4\pi} d\Omega \; K(\hat{\mathbf{s}} \cdot \hat{\mathbf{s}}')\, \hat{\mathbf{s}}' \cdot \mathbf{J}^{sc} = \hat{\mathbf{s}} \cdot \mathbf{J}^{sc} \int_{4\pi} d\Omega \; K(\hat{\mathbf{s}} \cdot \hat{\mathbf{s}}')\, \hat{\mathbf{s}} \cdot \hat{\mathbf{s}}' \equiv c_m \, \mu_{sc} \gamma \, \hat{\mathbf{s}} \cdot \mathbf{J}^{sc} , \tag{10.217}$$

where γ (often called g in the literature) is defined by

$$\gamma \equiv \frac{\int_{4\pi} d\Omega \; K(\hat{\mathbf{s}} \cdot \hat{\mathbf{s}}')\, \hat{\mathbf{s}} \cdot \hat{\mathbf{s}}'}{\int_{4\pi} d\Omega \; K(\hat{\mathbf{s}} \cdot \hat{\mathbf{s}}')} = \langle \cos\theta_{sc} \rangle , \tag{10.218}$$

where $\cos\theta_{sc} = \hat{\mathbf{s}}\cdot\hat{\mathbf{s}}'$ is the cosine of the scattering angle. Note that the denominator here equals $c_m\mu_{sc}$ as in (10.214).

The integral in the numerator of (10.218) is a measure of the difference in cross section for forward and backward scattering. In many situations, including optical scattering from small (subwavelength) particles and Compton scattering of low-energy x rays, forward and backward scattering are almost equally probable and $\gamma \approx 0$ For optical scattering in biological tissue, however, it turns out that $\gamma \approx 0.8 - 0.9$, so the scattering is forward-peaked.

Messy manipulations, minimally mentioned As advertised, we now multiply both sides of (10.216) by $\hat{\mathbf{s}}$ and integrate over $\hat{\mathbf{s}}$. The result is

$$\tfrac{1}{3}\nabla u^{sc} + \frac{3}{4\pi c_m}\int_{4\pi} d\Omega\; (\hat{\mathbf{s}}\cdot\nabla)\,(\hat{\mathbf{s}}\cdot\mathbf{J}^{sc})\,\hat{\mathbf{s}}$$

$$= -\frac{\mu_{tot}}{c_m}\int_{4\pi} d\Omega\;\hat{\mathbf{s}}\,u^{sc} - \frac{\mu_{abs} + (1-\gamma)\mu_{sc}}{c_m}\,\mathbf{J}^{sc} + \frac{1}{c_m}\int_{4\pi} d\Omega\;\hat{\mathbf{s}}\,\mathcal{K}\left(w^{pri} + \frac{1}{4\pi}u^{sc}\right), \tag{10.219}$$

where we have used identity (d) above.

Several of the integrals in (10.219) vanish by symmetry. The integral $\int_{4\pi} d\Omega\;\hat{\mathbf{s}}u^{sc}$ vanishes by identity (b) since u^{sc} is independent of $\hat{\mathbf{s}}$. Similarly, $\int_{4\pi} d\Omega\;\hat{\mathbf{s}}\,\mathcal{K}u^{sc}$ vanishes, again by identity (b), since $\mathcal{K}u^{sc}$ is also independent of $\hat{\mathbf{s}}$. Finally, the first integral in (10.218) vanishes by identity (f) with $\mathbf{A} = \nabla$. We are left with

$$\tfrac{1}{3}\nabla u^{sc} = -\frac{\mu_{abs} + (1-\gamma)\mu_{sc}}{c_m}\,\mathbf{J}^{sc} + \frac{1}{c_m}\int_{4\pi} d\Omega\;\hat{\mathbf{s}}\,\mathcal{K}w^{pri}\,. \tag{10.220}$$

Fick's law and diffusion coefficient In a very strongly scattering medium, the primary flux may make a small contribution to w, so we can approximate w^{pri} by 0. In that case, (10.220) expresses *Fick's law* (Fick, 1855), which says that the photon current is proportional to the gradient of the photon density:

$$\mathbf{J}^{sc} = -D\nabla u^{sc}\,, \tag{10.221}$$

where D is the *diffusion coefficient*,

$$D \equiv \frac{c_m}{3[\mu_{abs} + (1-\gamma)\mu_{sc}]} \equiv \frac{c_m}{3\mu_{tr}}\,. \tag{10.222}$$

The quantity $\mu_{tr} \equiv \mu_{abs} + (1-\gamma)\mu_{sc}$ is an effective total attenuation coefficient for transport problems; it reduces to μ_{tot} for isotropic scattering. Though we have suppressed the arguments in our shorthand notation, D and μ_{tr} depend in general on position $\mathbf{r}$ and energy $\mathcal{E}$.

Slightly different forms for D are also found in the literature. If $\gamma \approx 0$, we see that $D \approx c_m/(3\mu_{tot})$, a form often used for scattering from subwavelength particles. In the context of neutron transport, Weinberg and Wigner (1958) give $D = c_m\mu_{sc}/(3\mu_{tot}^2)$, but diffusion theory depends on the assumption that the particle undergoes many scatter events before it is absorbed, so $\mu_{abs} \ll \mu_{sc}$, $\mu_{sc} \approx \mu_{tot}$, and $c_m\mu_{sc}/(3\mu_{tot}^2) \approx c_m/(3\mu_{tot})$.

More messy manipulations Now we return to (10.220), without approximating the last term by zero. If we multiply that equation by $3D$, take the divergence of both sides, solve for $\nabla \cdot \mathbf{J}^{sc}$ and substitute the result into (10.215), we find

$$\nabla \cdot (D\nabla u^{sc}) - c_m \mu_{abs} u^{sc} = -c_m \mu_{sc} u^{pri} + \frac{3}{c_m} \nabla \cdot \left[D \int_{4\pi} d\Omega \; \hat{\mathbf{s}} \, \mathcal{K} w^{pri} \right] . \quad (10.223)$$

If μ_{sc} and μ_{abs} are independent of position, this formula reduces to

$$(\nabla^2 - \kappa^2) u^{sc} = -3\mu_{tr}\mu_{sc} u^{pri} + \frac{3}{c_m} \nabla \cdot \int_{4\pi} d\Omega \; \hat{\mathbf{s}} \, \mathcal{K} w^{pri} , \quad (10.224)$$

where

$$\kappa^2 \equiv 3\mu_{tr}\mu_{abs} = 3[\mu_{abs} + (1 - \gamma)\mu_{sc}]\mu_{abs} . \quad (10.225)$$

Discussion The right-hand side of (10.224) is a source term that can be found by solving the scatter-free Boltzmann equation (10.204), using methods developed in Sec. 10.3.3. With that source, u^{sc} can be found by solving (10.224), $\mathbf{J}^{sc}$ can be found from (10.220), and the scatter distribution function in the diffusion approximation follows from (10.211).

An interesting result of this analysis is that κ is $\sqrt{3}$ times the geometric mean of μ_{abs} and μ_{tr}. If there is no absorption, $\kappa = 0$ in spite of the strong attenuation due to scattering. In that case the diffusion equation reduces to the Poisson equation discussed in Sec. 9.1.5. With absorption, the diffusion equation resembles the Helmholtz equation, but $\nabla^2 + k^2$ is replaced by $\nabla^2 - \kappa^2$, where k and κ are both real. Thus the Helmholtz equation admits of wave-like solutions proportional to $\exp(i\mathbf{k} \cdot \mathbf{r})$, but the time-independent diffusion equation has only damped solutions.

10.3.7 Inelastic (Compton) scattering

Compton scattering is often the dominant interaction mechanism for low-energy x rays or gamma rays, especially in media with low atomic number such as air, water or human tissue. In Compton scattering, the incident photon interacts with an electron in the medium. The photon changes direction and the electron recoils, gaining some energy and momentum from the photon in the process.

If we assume that the electron is initially at rest, its energy after the interaction must equal the loss in energy of the photon. By conservation of energy and momentum, it can be shown (see, for example, Krane, 1983) that the scattered photon has an energy $\mathcal{E}$ given by

$$\frac{1}{\mathcal{E}} = \frac{1}{\mathcal{E}_0} + \frac{1}{mc^2}(1 - \cos\theta_s) , \quad (10.226)$$

where θ_s is the scattering angle, $\mathcal{E}_0$ is the energy of the incident photon, m is the rest mass of the electron and c is the speed of light. Thus mc^2 is the rest-mass energy of the electron, numerically equal to 511 keV. For example, 45° scattering of a 140 keV photon gives a scattered photon of energy 129.6 keV, 90° scattering gives 109.9 keV and 180° scattering gives 90.4 keV.

Form of the scattering kernel To get an explicit form for the kernel of $\mathcal{K}$ with Compton scattering, we make use of the differential scattering cross section defined in Sec. 10.2.5. If the scatterers (electrons in the Compton problem) act independently, the scattering kernel must be proportional to the density of scatterers, n_{sc}, times $\partial\sigma_{sc}/\partial\Omega$. Moreover, since it is reasonable to assume that the scatterers are randomly oriented, the dependence on $\hat{\mathbf{s}}$ and $\hat{\mathbf{s}}'$ is through the scalar product $\hat{\mathbf{s}} \cdot \hat{\mathbf{s}}'$, which is the same as $\cos\theta_s$ in (10.226).

The dependence of the kernel on energy must take account of the conservation rule (10.226) and must therefore involve a delta function. Since (10.226) shows that the initial energy $\mathcal{E}'$ is determined if the final energy $\mathcal{E}$ and $\cos\theta_s$ are specified, we have the choice of including a delta function of the form $\delta[\mathcal{E} - \gamma_1(\mathcal{E}', \cos\theta_s)]$, $\delta[\mathcal{E}' - \gamma_2(\mathcal{E}, \cos\theta_s)]$ or $\delta[\cos\theta_s - \gamma_3(\mathcal{E}, \mathcal{E}')]$, where the functions $\gamma_j(\cdot)$ are determined by solving (10.226). These delta functions all impose the constraint (10.226), but they differ from each other by Jacobians. As we shall demonstrate in a moment, for the first delta function listed, the kernel has the structure,

$$K(\hat{\mathbf{s}}, \mathcal{E}; \hat{\mathbf{s}}', \mathcal{E}'|\mathbf{r}) = cn_{sc}\frac{\partial\sigma_{sc}}{\partial\Omega}\,\delta\left\{\mathcal{E} - \left[\frac{1}{\mathcal{E}'} + \frac{1}{mc^2}(1 - \cos\theta_s)\right]^{-1}\right\}. \tag{10.227}$$

By the usual rule for transforming delta functions, an equivalent form is

$$K(\hat{\mathbf{s}}, \mathcal{E}; \hat{\mathbf{s}}', \mathcal{E}'|\mathbf{r}) = cn_{sc}\frac{mc^2}{\mathcal{E}^2}\frac{\partial\sigma_{sc}}{\partial\Omega}\,\delta\left[\cos\theta_s - 1 + mc^2\left(\frac{1}{\mathcal{E}} - \frac{1}{\mathcal{E}}'\right)\right]. \tag{10.228}$$

The first form is useful when we wish to use the delta function to perform an integral over energy, while the second form is used in angular integrals.

To show that (10.227) is correct, consider a collimated beam of monoenergetic photons for which

$$w(\mathbf{r}, \hat{\mathbf{s}}', \mathcal{E}') = A\,\delta(\hat{\mathbf{s}}' - \hat{\mathbf{s}}_0)\,\delta(\mathcal{E}' - \mathcal{E}_0)\,, \tag{10.229}$$

where A is a constant and $\delta(\hat{\mathbf{s}}' - \hat{\mathbf{s}}_0)$ is an angular delta function, with sifting property given by (10.158). As discussed in Chap. 2, the sifting property can be extended to situations where $t(\hat{\mathbf{s}})$ is a generalized function rather than a test function or a good function by noting that a generalized function can be approximated arbitrarily closely by a sequence of good functions.

With these considerations, (10.227) and (10.229) yield

$$[\mathcal{K}w]\,(\mathbf{r}, \hat{\mathbf{s}}, \mathcal{E}, t) = cAn_{sc}\frac{\partial\sigma_{sc}}{\partial\Omega}\,\delta\left\{\mathcal{E} - \left[\frac{1}{\mathcal{E}_0} + \frac{1}{mc^2}(1 - \cos\theta_{s0})\right]^{-1}\right\}, \tag{10.230}$$

where $\cos\theta_{s0} = \hat{\mathbf{s}} \cdot \hat{\mathbf{s}}_0$. If $\partial\sigma_{sc}/\partial\Omega$ depends on energy, it must be evaluated at energy $\mathcal{E}_0$ as a result of the second delta function in (10.229).

The delta function in (10.230) shows that the photons have the correct energy, depending on the scatter angle θ_{s0} from the original beam direction. The right-hand side of (10.230) is also consistent with the definitions of $\partial\sigma_{sc}/\partial\Omega$ and w; from the definition of w, the incident photon irradiance is cA, so $cA\partial\sigma_{sc}/\partial\Omega$ is the scattered intensity per scatterer. The total scattered intensity can also be obtained by integrating $\mathcal{K}w$ over a volume ΔV and over all energies, so it is given by $cAN_{sc}\partial\sigma_{sc}/\partial\Omega$, where $N_{sc} = n_{sc}\Delta V$ is the total number of scatterers in ΔV. Since this result is just what we would have obtained directly from the definition of $\partial\sigma_{sc}/\partial\Omega$ without going through the integrals, it verifies that the form (10.227) is correct.

Trickle-down theory Since the photon loses energy on each scattering event, the distribution function at energy $\mathcal{E}$ is influenced by higher energies, $\mathcal{E}' > \mathcal{E}$, but it is insensitive to lower energies. If we start with a monoenergetic source, as we usually do in nuclear medicine, we can solve the Boltzmann equation one energy at a time.

To show the trickle-down of photon energy explicitly, we use discrete energy bins and define

$$w_k(\mathbf{r}, \hat{\mathbf{s}}) = w(\mathbf{r}, \hat{\mathbf{s}}, \mathcal{E}_0 - k\Delta\mathcal{E}) , \qquad k = 0, ..., k_{max} , \tag{10.231}$$

where $\Delta\mathcal{E} \equiv \mathcal{E}_0/k_{max}$ is the width of the energy bin.

The steady-state Boltzmann equation now takes the form

$$-c\mu_{tot} w_k + \Xi_k + \Delta\mathcal{E} \sum_{j=0}^{k-1} \boldsymbol{\mathcal{K}}_{kj} w_j - c\,\hat{\mathbf{s}} \cdot \nabla w_k = 0 , \tag{10.232}$$

where $\Xi_k(\mathbf{r})$ is the source distribution for the kth energy bin, and $\boldsymbol{\mathcal{K}}_{kj}$ is the angular part of the operator $\boldsymbol{\mathcal{K}}$ with its kernel sampled at $\mathcal{E} = \mathcal{E}_0 - k\Delta\mathcal{E}$ and $\mathcal{E}' = \mathcal{E}_0 - j\Delta\mathcal{E}$, *i.e.*,

$$[\boldsymbol{\mathcal{K}}_{kj} w_j]\,(\mathbf{r}, \hat{\mathbf{s}}) = \int_{4\pi} d\Omega'\, K(\hat{\mathbf{s}}, \mathcal{E}_0 - k\Delta\mathcal{E}; \hat{\mathbf{s}}', \mathcal{E}_0 - j\Delta\mathcal{E}|\mathbf{r})\, w_j(\mathbf{r}, \hat{\mathbf{s}}') . \tag{10.233}$$

Note that the kernel still depends on position $\mathbf{r}$ and that there is still an implicit delta function in it. The trickle-down aspect of (10.232) is contained in the summation limits; with discrete energy bins as we have defined them, the requirement $\mathcal{E}' > \mathcal{E}$ translates to $j < k$.

Consider a monoenergetic source, in which case

$$\Xi_k(\mathbf{r}) = \frac{f(\mathbf{r})}{4\pi\Delta\mathcal{E}}\, \delta_{k0} . \tag{10.234}$$

We have already obtained the solution to (10.232) for $k = 0$, which is the only bin with a real source in it. For this bin, the scattering term makes no contribution, and (10.152) shows that

$$w_0 = \boldsymbol{\mathcal{X}}_\mu \Xi_0 . \tag{10.235}$$

Next look at $k = 1$, where there is no true source term but the summation over j in (10.232) contains the single term $\boldsymbol{\mathcal{K}}_{10} w_0$, which acts as an effective source of photons of energy $\mathcal{E}_0 - \Delta\mathcal{E}$. We then have

$$w_1 = \boldsymbol{\mathcal{X}}_\mu \boldsymbol{\mathcal{K}}_{10} w_0 = \boldsymbol{\mathcal{X}}_\mu \boldsymbol{\mathcal{K}}_{10} \boldsymbol{\mathcal{X}}_\mu \Xi_0 . \tag{10.236}$$

In the next bin, $k = 2$, we have two effective source terms since photons of energy $\mathcal{E}_0 - 2\Delta\mathcal{E}$ can be generated by scattering photons of energy $\mathcal{E}_0$ or $\mathcal{E}_0 - \Delta\mathcal{E}$. Thus,

$$w_2 = \boldsymbol{\mathcal{X}}_\mu \left[\boldsymbol{\mathcal{K}}_{20} w_0 + \boldsymbol{\mathcal{K}}_{21} w_1\right] . \tag{10.237}$$

In general, we can compute w_k by the iteration rule,

$$w_k = \boldsymbol{\mathcal{X}}_\mu \sum_{j=0}^{k-1} \boldsymbol{\mathcal{K}}_{kj} w_j . \tag{10.238}$$

Note that this series is fundamentally different from the Neumann series used in (10.168), where each term represents the contribution at all energies from a particular order of scattering. In (10.238), by contrast, each term gives the contribution at a particular energy from all orders of scatter.

10.4 TRANSPORT THEORY AND IMAGING

Transport theory gives us a way of determining the spatial, spectral and angular distribution of energy in any region of space, but we still need to relate this information to what we measure in an imaging system. In this section we treat the imaging system as a continuous-to-discrete (CD) mapping from the source strength, which is a function of spatial position $\mathfrak{r}$, to a set of discrete measurements. The Boltzmann equation maps the source strength to the distribution function, so we must now learn how to map the distribution function to the mean detector outputs.

In Sec. 10.4.1, we begin the discussion of the Boltzmann equation in imaging by deriving a general equation applicable to all linear CD imaging systems in which diffraction and polarization effects can be ignored. As we shall see, all such systems measure linear functionals of the distribution function (or, equivalently, of the radiance). Examples of the use of this equation are given in Secs. 10.4.1, where we discuss pinhole imaging or x rays or gamma rays, and 10.4.3, where we treat optical imaging of a planar source.

Some practical computational issues that arise when the Boltzmann equation is applied to imaging are discussed in Secs. 10.4.4 and 10.4.5. We show in Sec. 10.4.4 how the adjoint of the Boltzmann operator can be put to good use. In Sec. 10.4.5 we introduce the important computational technique of Monte Carlo integration and show how it can be applied to solving the Boltzmann equation.

10.4.1 General imaging equation

To get a general expression for $\overline{g}_m$, we need only two simple assumptions. First, we assume that that the detector responds linearly to the energy incident on it. Second, we assume that the radiation source and the detector are spatially separated so that we can set up a reference plane P somewhere between them. Then, if we know the distribution function on this plane we can compute the mean output[12] of each detector element as a linear functional of w. By the Riesz representation theorem, this functional must have the form,

$$\overline{g}_m = \int_P d^2r \int_0^\infty d\mathcal{E} \int_{2\pi} d\Omega \int_0^\tau dt \; d_m(\mathfrak{r}, \hat{\mathbf{s}}, \mathcal{E}, t)\, w(\mathfrak{r}, \hat{\mathbf{s}}, \mathcal{E}, t)\,, \tag{10.239}$$

where τ is the exposure time and $\mathfrak{r}$ is a general 3D vector, but the spatial integral is over the two variables needed to specify points in the plane. If we take the plane to be $z = 0$, then $\mathfrak{r} = (x, y, z)$ but $d^2r = dxdy$. We shall frequently jump back and forth between 2D and 3D notation in this section, with the convention that $\mathfrak{r}$ and $\mathbf{r}$ specify the same point.[13] Note that the integral over solid angle in (10.239) covers only 2π ster since photons on the plane but directed away from the detector do not contribute to the output. Usually the properties of the detector system will be independent of time, so we can drop the t argument in d_m. If we consider only steady-state sources, w is independent of t also, so the t-integral simply gives a

[12]To be specific, the mean output $\overline{g}_m$ considered in this section is the expectation of g_m conditioned on a particular source. The averaging implied by the overbar is thus over measurement noise, not over object randomness.

[13]A different convention was used in Chap. 9, where $\mathbf{r}$ could refer to the same x-y coordinates in two different planes.

factor of τ. The function $d_m(\mathbf{r}, \hat{\mathbf{s}}, \mathcal{E})$ in (10.239) is called the *detector response function* since it specifies how the mth detector element responds to photons at point $\mathbf{r}$ travelling in direction $\hat{\mathbf{s}}$ and having energy $\mathcal{E}$. The detector system must necessarily include imaging elements such as lenses, collimators or pinholes so that information about the source distribution can be captured in the image, and the effect of these elements is contained in d_m. Specific examples will be given in Secs. 10.4.2 and 10.4.3.

Imaging equation in the absence of scatter In operator form, (10.239) can be written as

$$\overline{\mathbf{g}} = \boldsymbol{\mathcal{M}}\mathbf{w}\,, \tag{10.240}$$

where $\boldsymbol{\mathcal{M}}$ is called the *measurement operator.* If there is no scatter and the source is isotropic, then $\mathbf{w}$ is given by (10.152), so

$$\overline{\mathbf{g}} = \frac{1}{4\pi}\boldsymbol{\mathcal{M}}\boldsymbol{\mathcal{X}}_\mu \mathbf{S}_{p,\mathcal{E}}\,. \tag{10.241}$$

We would like to put this expression into the same form as the standard imaging equation in Chap. 7, $\overline{\mathbf{g}} = \boldsymbol{\mathcal{H}}\mathbf{f}$, but to do so we must be more precise about what $\mathbf{f}$ means. If we consider emission imaging, where the object to be imaged is the radiation source, and assume that the emission is isotropic, we might think that $\mathbf{f}$ would be synonymous with $\mathbf{S}_{p,\mathcal{E}}$. Usually, however, we are more interested in the spatial properties of the source than in the spectral properties, so we want $\mathbf{f}$ to correspond to a function of $\mathbf{r}$ alone. On the other hand, the response of a detector system is inevitably a function of $\mathcal{E}$, so we need the full $S_{p,\mathcal{E}}(\mathbf{r}, \mathcal{E})$ to compute $\overline{\mathbf{g}}$.

In some problems it is valid to assume that the spectral properties of the source are independent of position. This assumption holds, for example, in nuclear medicine if a single isotope is used and in fluorescence microscopy with a single fluorophore. On the other hand, almost any other emissive or reflective optical source would have a spectrum dependent on position, so a full description of the source is necessarily a spatio-spectral function.

When the spectral properties of the source are independent of position, we can write

$$S_{p,\mathcal{E}}(\mathbf{r}, \mathcal{E}) = f(\mathbf{r})\, N(\mathcal{E})\,, \tag{10.242}$$

where $N(\mathcal{E})$ is a normalized source spectrum, defined such that

$$\int_0^\infty d\mathcal{E}\, N(\mathcal{E}) = 1\,. \tag{10.243}$$

With this definition, $f(\mathbf{r})$ is identical to the photon emission density S_p. Units of f are thus photons/(sec $\cdot$m^3).

Since $S_{p,\mathcal{E}}$ is linearly related to $f(\mathbf{r})$, it must be possible to find a linear operator $\boldsymbol{\mathcal{H}}$ such that

$$\overline{\mathbf{g}} = \boldsymbol{\mathcal{H}}\mathbf{f}\,, \tag{10.244}$$

or, as an integral,

$$\overline{g}_m = \int_\infty d^3r'\, f(\mathbf{r}')\, h_m(\mathbf{r}') = [\boldsymbol{\mathcal{H}}\mathbf{f}]_m\,. \tag{10.245}$$

The remaining problem is to determine the kernel $h_m(\mathbf{r}')$.

Explicit form of the kernel The overall mapping from $\mathbf{f}$ to $\overline{\mathbf{g}}$ is found by substituting (10.151) and (10.242) into (10.239); the result is

$$\overline{g}_m = \frac{T}{4\pi c_m} \int_P d^2r \int_0^\infty d\mathcal{E}\, N(\mathcal{E}) \int_{2\pi} d\Omega\, d_m(\mathbf{r}, \hat{\mathbf{s}}, \mathcal{E})$$

$$\times \int_0^\infty d\ell\, f(\mathbf{r} - \hat{\mathbf{s}}\ell) \exp\left[-\int_0^\ell d\ell'\, \mu_{abs}(\mathbf{r} - \hat{\mathbf{s}}\ell') \right]. \tag{10.246}$$

To identify the kernel $h_m(\mathbf{r}')$, we use a manipulation similar to the one that led to (10.192): we define $\mathbf{r}' = \mathbf{r} - \hat{\mathbf{s}}\ell$ and recognize that $\ell^2 d\ell d\Omega = d^3\mathbf{r}'$ and $\ell = |\mathbf{r} - \mathbf{r}'|$. Then (10.246) becomes

$$\overline{g}_m = \frac{T}{4\pi c_m} \int_\infty d^3\mathbf{r}'\, f(\mathbf{r}') \int_P d^2r\, \frac{d_m(\mathbf{r}, \hat{\mathbf{s}})}{|\mathbf{r} - \mathbf{r}'|^2} \exp\left\{ -\int_0^{|\mathbf{r}-\mathbf{r}'|} d\ell'\, \mu_{abs}\left[\mathbf{r} - \left(\frac{\mathbf{r} - \mathbf{r}'}{|\mathbf{r} - \mathbf{r}'|} \right) \ell' \right] \right\}, \tag{10.247}$$

where

$$d_m(\mathbf{r}, \hat{\mathbf{s}}) = \int_0^\infty d\mathcal{E}\, d_m(\mathbf{r}, \hat{\mathbf{s}}, \mathcal{E})\, N(\mathcal{E}). \tag{10.248}$$

By comparing (10.245) and (10.247), we see that

$$h_m(\mathbf{r}') = \frac{T}{4\pi c_m} \int_P d^2r\, \frac{d_m(\mathbf{r}, \hat{\mathbf{s}})}{|\mathbf{r} - \mathbf{r}'|^2} \exp\left\{ -\int_0^{|\mathbf{r}-\mathbf{r}'|} d\ell'\, \mu_{abs}\left[\mathbf{r} - \left(\frac{\mathbf{r} - \mathbf{r}'}{|\mathbf{r} - \mathbf{r}'|} \right) \ell' \right] \right\}. \tag{10.249}$$

The factor $1/|\mathbf{r} - \mathbf{r}'|^2$ arose from the change of variables $\mathbf{r}' = \mathbf{r} - \hat{\mathbf{s}}\ell$, but it has an important physical interpretation; at the source point $\mathbf{r}$, an area element d^2r subtends a solid angle $\cos\theta d^2r/|\mathbf{r} - \mathbf{r}'|^2$, where θ is the angle between $\hat{\mathbf{s}}$ and the normal to plane P. As we shall see in more detail in Sec. 10.4.2, the cosine is hidden in $d_m(\mathbf{r}, \hat{\mathbf{s}})$, but the inverse-square factor appears explicitly.

Imaging equation with weak scatter If we substitute the Neumann series (10.168) into (10.240), we have

$$\overline{\mathbf{g}} = \mathcal{M}[\mathbf{I} - \mathcal{X}_\mu \mathcal{K}]^{-1} \mathcal{X}_\mu \Xi_{p,\mathcal{E}} = \mathcal{M}\mathcal{X}_\mu \sum_{j=0}^\infty (\mathcal{K}\mathcal{X}_\mu)^j \Xi_{p,\mathcal{E}}. \tag{10.250}$$

If the object being imaged is the source distribution and we assume monoenergetic, isotropic emission, we can write $\Xi_{p,\mathcal{E}}(\mathbf{r}, \mathcal{E}) = f(\mathbf{r})\, \delta(\mathcal{E} - \mathcal{E}_0)/4\pi$.

The single-scatter term ($j = 1$) in (10.250) has the same structure as the no-scatter term except that $\mathcal{M}\mathcal{X}_\mu$ operates on $\mathcal{K}\mathcal{X}_\mu\mathbf{S}$ rather than on $\mathbf{S}$ directly. We can therefore compute the single-scatter kernel $h_m^{(1)}(\mathbf{r})$ in two steps, considering first the operator $\mathcal{K}\mathcal{X}_\mu$ and then $\mathcal{M}\mathcal{X}_\mu$. From (10.129) and (10.151), the first of these operators has the form,

$$[\mathcal{K}\mathcal{X}_\mu \mathbf{S}](\mathbf{r}, \hat{\mathbf{s}}, \mathcal{E}) = \int_\infty d^3r'\, K\left(\hat{\mathbf{s}}, \mathcal{E}; \frac{\mathbf{r} - \mathbf{r}'}{|\mathbf{r} - \mathbf{r}'|}, \mathcal{E}_0 \middle| \mathbf{r} \right)$$

$$\times \frac{f(\mathbf{r}')}{4\pi c|\mathbf{r} - \mathbf{r}'|^2} \exp\left\{ -\int_0^{|\mathbf{r}-\mathbf{r}'|} d\ell'\, \mu_{tot}\left[\mathbf{r} - \left(\frac{\mathbf{r} - \mathbf{r}'}{|\mathbf{r} - \mathbf{r}'|} \right) \ell' \right] \right\}, \tag{10.251}$$

where we have used the delta function in $\Xi_{p,\mathcal{E}}(\mathbf{r}, \mathcal{E})$ to good end, and we have also made a change of variables, $\mathbf{r}' = \mathbf{r} - \hat{\mathbf{s}}'\ell'$.

The interpretation of (10.251) is straightforward. Photons of energy $\mathcal{E}_0$ originate at $\mathbf{r}'$ and travel to $\mathbf{r}$, diminishing in number per unit area because of the inverse-square factor and the attenuation factor. At $\mathbf{r}$, they scatter into direction $\hat{\mathbf{s}}$ with energy $\mathcal{E}$.

Now we can apply the operator $\boldsymbol{\mathcal{M}}\boldsymbol{\mathcal{X}}_\mu$ to propagate these photons to the plane P where they are measured by the imaging system. The operator $\boldsymbol{\mathcal{X}}_\mu$ has the effect of replacing $\mathbf{r}$ with $\mathbf{r} - \hat{\mathbf{s}}\ell$ everywhere, inserting another exponential factor and a $1/c$, and integrating over ℓ. The measurement operator $\boldsymbol{\mathcal{M}}$ is implemented by multiplication by d_m and integration over the plane P, energy $\mathcal{E}$ and solid angle Ω. The overall kernel is thus

$$h_m^{(1)}(\mathbf{r}') = \tau \int_P d^2r \int_0^\infty d\mathcal{E} \int_{2\pi} d\Omega \, d_m(\mathbf{r}, \hat{\mathbf{s}}, \mathcal{E}) \int_0^\infty d\ell \, K\left(\hat{\mathbf{s}}, \mathcal{E}; \frac{\mathbf{r} - \hat{\mathbf{s}}\ell - \mathbf{r}'}{|\mathbf{r} - \hat{\mathbf{s}}\ell - \mathbf{r}'|}, \mathcal{E}_0 \middle| \mathbf{r} - \hat{\mathbf{s}}\ell\right)$$
$$\times \frac{1}{4\pi c^2 |\mathbf{r} - \hat{\mathbf{s}}\ell - \mathbf{r}'|^2} \exp\left\{ - \int_0^{|\mathbf{r} - \hat{\mathbf{s}}\ell - \mathbf{r}'|} d\ell' \mu_{tot} \left[\mathbf{r} - \hat{\mathbf{s}}\ell - \left(\frac{\mathbf{r} - \hat{\mathbf{s}}\ell - \mathbf{r}'}{|\mathbf{r} - \hat{\mathbf{s}}\ell - \mathbf{r}'|} \right) \ell' \right] \right\}$$
$$\times \exp\left[- \int_0^\ell d\ell' \, \mu_{tot}(\mathbf{r} - \hat{\mathbf{s}}\ell') \right] . \tag{10.252}$$

When integrated against a source distribution, this complicated expression gives the contribution to $\overline{g}_m$ of photons that have scattered exactly once. If patience suffices, higher-order terms can be computed similarly.

10.4.2 Pinhole imaging

To illustrate the formalism developed in Sec. 10.4.1, we consider a circular pinhole imaging system of the kind that might be used to image a gamma-ray source in nuclear medicine. For short-wavelength radiation such as gamma rays, diffraction can be neglected and transport theory should give accurate answers.

The imaging geometry is illustrated in Fig. 10.8. We consider a circular pinhole of diameter D_{ph} in a thin sheet of highly absorbing material such as lead. The origin of coordinates is taken as the center of the pinhole, and the aperture plane is $z = 0$. The detectors are assumed to lie in the plane $z = -s$, hence parallel to the aperture plate and a distance s away. Two different choices for the reference plane P will be discussed below.

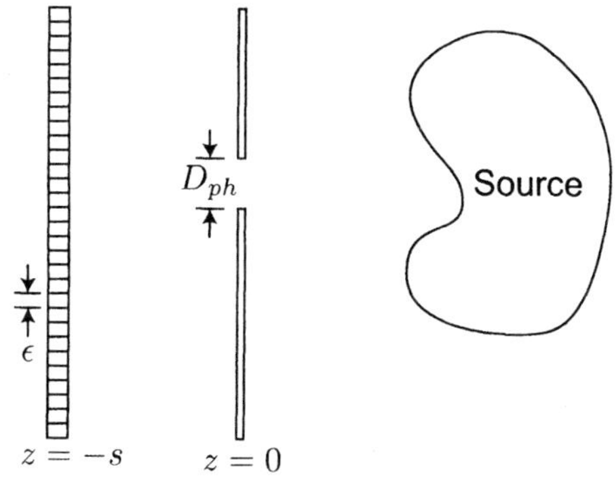

Fig. 10.8 Pinhole imager viewing a volume source.

Source model The source is assumed to be time-independent, monoenergetic and isotropic, so that

$$\Xi_{p,\mathcal{E}}(\mathbf{r},\hat{\mathbf{s}},\mathcal{E}) = \frac{1}{4\pi} f(\mathbf{r})\,\delta(\mathcal{E}-\mathcal{E}_0)\,. \tag{10.253}$$

For simplicity, we assume that there is no absorption or scattering in the object itself, but we may still need to consider the absorption term in the Boltzmann equation, depending on where the reference plane P is with respect to the aperture plate.

Detector model The detector system is assumed to be a regular array of square detector elements of size $\epsilon \times \epsilon$, with the mth element centered at position $\mathbf{r}_m$ in the plane $z = -s$. We assume that $\epsilon \ll D_{ph}$ and that the detectors respond uniformly to photons that strike them. That is, the mth detector element measures an integral of the spectral photon irradiance in the detector plane:

$$\overline{g}_m = \tau \int_0^\infty d\mathcal{E}\,\eta(\mathcal{E}) \int_\infty d^2r \;\mathrm{rect}\left[\frac{\mathbf{r}-\mathbf{r}_m}{\epsilon}\right] I_{p,\mathcal{E}}(\mathbf{r},\mathcal{E})\,, \tag{10.254}$$

where $\eta(\mathcal{E})$, called the *quantum efficiency* of the detector, is the probability that a photon of energy $\mathcal{E}$ incident on the detector will, in fact, be detected.

The spectral photon irradiance is related to the spectral photon radiance by (10.140), and the radiance is related in turn to the distribution function by a factor of c_m [*cf.* (10.98)], so

$$\overline{g}_m = \tau c_m \int_0^\infty d\mathcal{E}\,\eta(\mathcal{E}) \int_\infty d^2r \;\mathrm{rect}\left[\frac{\mathbf{r}-\mathbf{r}_m}{\epsilon}\right] \int_{2\pi} d\Omega\; w(\mathbf{r},\hat{\mathbf{s}})\,\hat{\mathbf{n}}\cdot\hat{\mathbf{s}}\,, \tag{10.255}$$

where $\hat{\mathbf{n}}$ is the unit vector normal to the detector plane (hence parallel to the z axis).

Reference plane coincident with detector plane Suppose initially that the reference plane coincides with the detector plane. Comparison of (10.255) with (10.239) shows immediately that

$$d_m(\mathbf{r},\hat{\mathbf{s}},\mathcal{E}) = \tau\eta(\mathcal{E})\,c_m\,\mathrm{rect}\left[\frac{\mathbf{r}-\mathbf{r}_m}{\epsilon}\right]\hat{\mathbf{n}}\cdot\hat{\mathbf{s}}\,, \tag{10.256}$$

where $\mathbf{r}$ and $\mathbf{r}_m$ are the 2D vectors in the image plane specifying points $\mathbf{r}$ and $\mathbf{r}_m$, respectively. Since the detector plane is $z = -s$, $\mathbf{r}_m = (\mathbf{r}_m, -s)$.

For small detectors such that $w(\mathbf{r},\hat{\mathbf{s}}) \simeq w(\mathbf{r}_m,\hat{\mathbf{s}})$ within the mth detector, we can write

$$\overline{g}_m \simeq \epsilon^2 \tau c_m \int_0^\infty d\mathcal{E}\,\eta(\mathcal{E}) \int_{2\pi} d\Omega\; w(\mathbf{r}_m,\hat{\mathbf{s}},\mathcal{E})\,\hat{\mathbf{n}}\cdot\hat{\mathbf{s}}\,. \tag{10.257}$$

Now we can make use of (10.151) along with the source model (10.253) to obtain

$$\overline{g}_m = \frac{\epsilon^2\tau\eta(\mathcal{E}_0)}{4\pi} \int_{2\pi} d\Omega \int_0^\infty d\ell\; f(\mathbf{r}_m - \hat{\mathbf{s}}\ell)\,\hat{\mathbf{n}}\cdot\hat{\mathbf{s}}\,\exp\left[-\int_0^\ell d\ell'\,\mu_{abs}(\mathbf{r}_m - \hat{\mathbf{s}}\ell')\right]. \tag{10.258}$$

As we have done several times in this chapter, we make a change of variables by defining $\mathbf{r} = \mathbf{r}_m - \hat{\mathbf{s}}\ell$, noting that the minus sign reverses the direction of the hemisphere of integration. Whenever $f(\mathbf{r}) = 0$ between the reference plane and the detector, we can extend the integral to the full sphere. In this case the reference plane is the detector plane, so (10.258) becomes

$$\overline{g}_m = \int_\infty d^3\mathbf{r}\; f(\mathbf{r})\, h_m(\mathbf{r}) \,, \tag{10.259}$$

where

$$h_m(\mathbf{r}) = \frac{\epsilon^2 \tau \eta(\mathcal{E}_0)}{4\pi} \frac{\hat{\mathbf{n}} \cdot (\mathbf{r}_m - \mathbf{r})}{|\mathbf{r}_m - \mathbf{r}|^3} \exp\left[- \int_0^{|\mathbf{r}_m - \mathbf{r}|} d\ell' \, \mu_{abs} \left(\mathbf{r}_m - \frac{\mathbf{r}_m - \mathbf{r}}{|\mathbf{r}_m - \mathbf{r}|} \ell' \right) \right] . \tag{10.260}$$

Since we are neglecting attenuation in the object, μ_{abs} refers only to the aperture. If we assume that μ_{abs} is very high for the aperture material, the exponential factor is nearly zero if the line $\mathbf{r}_m - \hat{\mathbf{s}}\ell$ passes through the absorbing portion of the aperture plate. Thus, for a small, point-like detector element, $h_m(\mathbf{r})$ is nearly zero except over a cone-shaped region of object space as shown in Fig. 10.9. If the detector element cannot be approximated by a point, the cone has fuzzy edges.

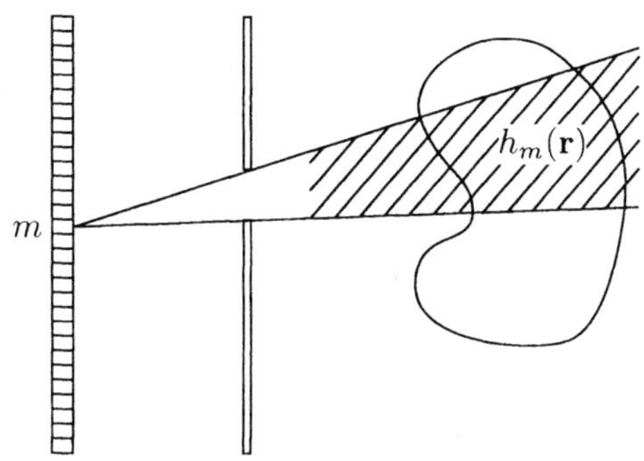

Fig. 10.9 Illustration of the detector response function for a pinhole imager.

The weighting within this conical region is just what we would calculate by elementary geometrical considerations. If we consider a unit-strength isotropic emitter at point $\mathbf{r}_0$, then this source emits, on average, one photon per second or τ photons during the integration time. The mth detector element subtends a solid angle Ω_m given by

$$\Omega_m = \frac{\epsilon^2}{|\mathbf{r}_m - \mathbf{r}_0|^2} \hat{\mathbf{n}} \cdot \frac{\mathbf{r}_m - \mathbf{r}_0}{|\mathbf{r}_m - \mathbf{r}_0|} \,, \tag{10.261}$$

where the scalar product is simply the cosine of the angle between the detector normal and the line of sight from $\mathbf{r}_0$ to the detector. If this line of sight passes through the open part of the pinhole, then the exponential is unity, otherwise it is near zero. The interpretation of (10.260) is then straightforward: the mean response of the mth detector to a point equals the mean number of photons emitted by that point times the fractional solid angle $\Omega_m/4\pi$ times the probability $\eta(\mathcal{E}_0)$ that a photon reaching the detector will be detected times a 0–1 function indicating whether the photon is blocked by the aperture.

Reference plane coincident with aperture plane The reference plane P need not coincide with the detector plane; it can be placed anywhere between the source and the detector. If it is immediately after the aperture plane, then $d_m(\mathbf{r}, \hat{\mathbf{s}}, \mathcal{E})$ is near zero unless $\mathbf{r}$ lies in the open portion of the pinhole, so d_m includes a factor of $\mathrm{cyl}(r/D_{ph})$. The cylinder function is defined in (3.257).

For a point within the pinhole, however, the mth detector element responds only to radiation directed from this point to the detector. If the detector size ϵ is small, as we have assumed above, then this angular selectivity can be described by a delta function of the form

$$\epsilon^2 \frac{\cos\theta}{|\mathbf{r}_m - \mathbf{r}|^2} \delta\left(\hat{\mathbf{s}} - \frac{\mathbf{r}_m - \mathbf{r}}{|\mathbf{r}_m - \mathbf{r}|^2}\right) = \epsilon^2 \frac{\hat{\mathbf{n}} \cdot (\mathbf{r}_m - \mathbf{r})}{|\mathbf{r}_m - \mathbf{r}|^3} \delta\left(\hat{\mathbf{s}} - \frac{\mathbf{r}_m - \mathbf{r}}{|\mathbf{r}_m - \mathbf{r}|^2}\right), \tag{10.262}$$

where $\epsilon^2 \cos\theta$ is the area of the detector element projected onto the direction from point $\mathbf{r}$ in the open aperture to the detector. The unit vector here looks much like the one in (10.261), but there is an essential difference: the vector $\mathbf{r}$ here specifies a point in the aperture plane, not a point in the object, and the weight of the delta function is independent of location in the object.

The spectral photon radiance at point $\mathbf{r}$ in the aperture plane is $c_m w$, and we recall from Sec. 10.2.1 that radiance is radiant flux per unit solid angle per unit *projected* area [see (10.62)]. To get the total photon flux from an area element d^2r into the direction defined by (10.262), we must multiply $c_m w$ by the projected area $\cos\theta d^2r$. With respect to the line from $\mathbf{r}$ to $\mathbf{r}_m$, both the aperture plane and the detector plane are tipped by an angle θ, so we need two cosine factors. In addition, the response is proportional to the quantum efficiency $\eta(\mathcal{E})$ and the exposure time τ, so we now have

$$d_m(\mathbf{r}, \hat{\mathbf{s}}, \mathcal{E}) = \epsilon^2 c_m \tau \eta(\mathcal{E}) \frac{\cos^2\theta}{|\mathbf{r}_m - \mathbf{r}|^2} \mathrm{cyl}\left(\frac{r}{D_{ph}}\right) \delta\left(\hat{\mathbf{s}} - \frac{\mathbf{r}_m - \mathbf{r}}{|\mathbf{r}_m - \mathbf{r}|^2}\right). \tag{10.263}$$

With this response function, the detector output is given by

$$\overline{g}_m = \epsilon^2 c_m \tau \int_0^\infty d\mathcal{E}\, \eta(\mathcal{E}) \int_\infty d^2r \frac{\cos^2\theta}{|\mathbf{r}_m - \mathbf{r}|^2} \mathrm{cyl}\left(\frac{r}{D_{ph}}\right) w\left(\mathbf{r}, \frac{\mathbf{r}_m - \mathbf{r}}{|\mathbf{r}_m - \mathbf{r}|}, \mathcal{E}\right). \tag{10.264}$$

With (10.151) and (10.253), we find

$$\overline{g}_m = \frac{\epsilon^2 \tau \eta(\mathcal{E}_0)}{4\pi} \int_0^\infty d\ell \int_\infty d^2r \frac{\cos^2\theta}{|\mathbf{r}_m - \mathbf{r}|^2} \mathrm{cyl}\left(\frac{r}{D_{ph}}\right) f\left(\mathbf{r} - \ell \frac{\mathbf{r}_m - \mathbf{r}}{|\mathbf{r}_m - \mathbf{r}|}\right). \tag{10.265}$$

There is now no exponential factor since we are assuming that the object is non-absorbing. Nevertheless, the integral is over the conical region depicted in Fig. 10.9 since the cylinder function selects out lines in that region. Moreover, $d^2r \cos\theta / |\mathbf{r}_m - \mathbf{r}|^2$ is the solid angle subtended by d^2r from the detector, so we can replace that expression by $d\Omega$, and (10.258) is then recovered.

Thus the two choices for reference plane lead to the same expression for $\overline{g}_m$ and hence for $h_m(\mathbf{r})$.

10.4.3 Optical imaging of a planar source

In the last section we discussed pinhole imaging of a volume source, such as a radioisotope distribution in nuclear medicine. In optics, however, we are more often concerned with surface emitters, and we can use lenses and mirrors rather than simple apertures such as pinholes. In this section we shall see how the general imaging equation (10.239) can be applied in this case.

Source description Consider a source confined to the plane $z = 0$. If the source is independent of time and if its spectral and directional properties are independent of position, then we can express the source distribution as

$$\Xi_{p,\mathcal{E}}(\mathbf{r}, \hat{\mathbf{s}}, \mathcal{E}) = A(\hat{\mathbf{s}})\, N(\mathcal{E})\, f(\mathbf{r})\, \delta(z)\,, \tag{10.266}$$

where $A(\hat{\mathbf{s}})$ describes the angular dependence, $f(\mathbf{r})$ describes the dependence on position in the plane, and $N(\mathcal{E})$ gives the spectral dependence. We shall generalize this source description below and allow different points on the object plane to have different spectral characteristics.

When we discussed isotropic volume sources, the angular factor $A(\hat{\mathbf{s}})$ was assumed to be a constant [*cf.* (10.242)], but in the case of Lambertian surface emitters, it must contain a cosine factor. To see why, let us use (10.136) to compute the distribution function w in the plane $z = 0^+$, just to the right of the source. Since $c_m w$ is the spectral photon radiance, and since a Lambertian source has constant radiance in the source plane, this calculation must give w independent of $\hat{\mathbf{s}}$. With a general $A(\hat{\mathbf{s}})$, however, (10.136) shows that

$$c_m w(\mathbf{r}, z, \hat{\mathbf{s}}, \mathcal{E}) = A(\hat{\mathbf{s}})\, N(\mathcal{E}) \int_0^\infty d\ell \,.f(x - s_x\ell\,, y - s_y\ell\,)\, \delta(z - s_z\ell\,)\,, \tag{10.267}$$

where $\hat{\mathbf{s}} = (s_x, s_y, s_z)$ and of course $f(x, y) = f(\mathbf{r})$. With the help of (2.28), we find

$$c_m w(\mathbf{r}, z, \hat{\mathbf{s}}, \mathcal{E}) = A(\hat{\mathbf{s}})\, N(\mathcal{E})\, f\left(x - \frac{s_x}{s_z} z,\; y - \frac{s_y}{s_z} z\right) \frac{1}{s_z}\,. \tag{10.268}$$

In order for this expression to be independent of $\hat{\mathbf{s}}$ as $z \to 0^+$, we must have $A(\hat{\mathbf{s}}) \propto s_z$, which is just the cosine of the angle between $\hat{\mathbf{s}}$ and the z-axis (or surface normal).

To summarize and generalize, a planar Lambertian source lying in the plane $\mathbf{r} \cdot \hat{\mathbf{n}} = p$ is described by the source distribution[14]

$$\Xi_{p,\mathcal{E}}(\mathbf{r}, \hat{\mathbf{s}}, \mathcal{E}) = (\hat{\mathbf{s}} \cdot \hat{\mathbf{n}})\, N(\mathcal{E})\, f(\mathbf{r})\, \delta(p - \mathbf{r} \cdot \hat{\mathbf{n}})\,, \tag{10.269}$$

provided the spatial and spectral properties are independent. The spectral photon radiance immediately adjacent to this source is given by

$$L_{p,\mathcal{E}}(\mathbf{r}, \hat{\mathbf{s}}, \mathcal{E}, t) = c_m w(\mathbf{r}, \hat{\mathbf{s}}, \mathcal{E}, t) = N(\mathcal{E})\, f(\mathbf{r})\,. \tag{10.270}$$

Imaging with a simple lens Consider a planar Lambertian source lying in the plane $z = -p$, a thin lens of focal length f and diameter D_{lens} in the plane $z = 0$, and a discrete detector array in the plane $z = q$, where p, q and f are related by the imaging equation $p^{-1} + q^{-1} = f^{-1}$. In terms of paraxial geometrical optics, this system is described by the ABCD matrix (9.247):

$$\mathbf{M} = \begin{bmatrix} m & 0 \\ -\frac{1}{f} & \frac{1}{m} \end{bmatrix}, \tag{10.271}$$

where $m = -p/q$.

[14] The reader should not lose sight of our convention that $\mathbf{r}$ and $\mathbf{r}$ refer to the same point on the plane.

We can use this matrix in (10.115) to determine the radiance in the detector plane when the input radiance is specified by (10.270). If there were no aperture on the lens, the result would be $N(\mathcal{E})f(m\mathbf{r})$, but the radiance in the detector plane cannot be independent of angle if only a finite range of ray angles emerges from the lens; radiance is constant along a ray only if that ray is not interrupted by an aperture. If we neglect diffraction from the aperture, we can write

$$L_{p,\mathcal{E}}(\mathbf{r},\hat{\mathbf{s}},\mathcal{E},t) = N(\mathcal{E})\, f(m\mathbf{r})\, I(\mathbf{r},\hat{\mathbf{s}})\,, \qquad (z=q)\,, \tag{10.272}$$

where $I(\mathbf{r},\hat{\mathbf{s}})$ is an indicator function that takes on the value 1 when a ray from point $\mathbf{r}$ in the detector plane extended backward along $\hat{\mathbf{s}}$ passes through the lens aperture, and $I(\mathbf{r},\hat{\mathbf{s}})$ is 0 otherwise.

Detector response The simplest model for the response of a detector is that it integrates the irradiance over its active area and is independent of incidence angle or photon energy. With this model, the output of the jth detector has the form [*cf.* (10.257)]

$$\overline{g}_j \propto \int_0^\infty d\mathcal{E}\, N(\mathcal{E}) \int_P d^2r\; f(m\mathbf{r})\, \mathrm{rect}\left[\frac{\mathbf{r}-\mathbf{r}_j}{\epsilon}\right] \int_{2\pi} d\Omega\; I(\mathbf{r},\hat{\mathbf{s}})\, \hat{\mathbf{n}}\cdot\hat{\mathbf{s}}\,, \tag{10.273}$$

where plane P coincides with the detector.

A common approximation at this stage is to consider a lens with a large F-number so that $q \gg D_{lens}$ and to assume that the pixel size ϵ is much smaller than D_{lens}. Then the factor $\hat{\mathbf{n}}\cdot\hat{\mathbf{s}}$ can be approximated by the cosine of the angle θ_j from the jth detector pixel to the center of the lens and removed from the integral. Similarly, the indicator function $I(\mathbf{r},\hat{\mathbf{s}})$ can be approximated by $I(\mathbf{r}_j,\hat{\mathbf{s}})$. The integral over $d\Omega$ is then just the solid angle subtended by the lens from the point $\mathbf{r}_j$; within the approximation we are making, this solid angle is given explicitly by

$$\Omega_{lens}(\mathbf{r}_j) \equiv \int_{2\pi} d\Omega\; I(\mathbf{r}_j,\hat{\mathbf{s}}) \simeq \frac{\pi D_{lens}^2}{4R_j^2}\cos\theta_j = \frac{\pi D_{lens}^2}{4q^2}\cos^3\theta_j\,, \tag{10.274}$$

where $R_j \equiv q/\cos\theta_j$ is the distance from the detector element to the center of the lens. Thus we have

$$\overline{g}_j \propto \cos^4\theta_j \int_P d^2r\; f(m\mathbf{r})\, \mathrm{rect}\left[\frac{\mathbf{r}-\mathbf{r}_j}{\epsilon}\right]. \tag{10.275}$$

With the current approximations, therefore, the detector output is the integral of the magnified object weighted with a factor $\cos^4\theta_j$. (One cosine factor comes from converting projected area to area on the detector, and the other three factors come from the solid angle of the lens.) This result shows that the image of a uniform object (the flood image discussed in Sec. 7.2.1) will not be uniform.

To reiterate the approximations made in deriving (10.275), we assumed that the system could be described by a simple ABCD matrix and that diffraction from the lens aperture could be neglected; that the lens F-number was large and that the detector element was small; that the source was Lambertian with spectral properties independent of position; and that the detector simply integrated the irradiance without regard for spectral or angular properties of the radiation. It is straightforward to avoid these approximations and to allow more complicated detector models, but numerical evaluation will usually be required.[15]

Spatio-spectral sources The source description of (10.266) is seldom valid in optics. It might hold in fluorescence microscopy with a single fluorophore, but most emissive or reflective objects have different spectral properties at different locations. For such objects, a better description for a planar source is

$$\Xi_{p,\mathcal{E}}(\mathbf{r}, \hat{\mathbf{s}}, \mathcal{E}) = A(\hat{\mathbf{s}})\, f(\mathbf{r}, \mathcal{E})\, \delta(z)\,. \tag{10.276}$$

This form allows a general spatio-spectral function but it still assumes that the angular properties are independent of location and energy; for example, this model would hold for Lambertian objects.

Without further assumptions, we would now have to consider $f(\mathbf{r}, \mathcal{E})$ as a 3D object function, even though it is only 2D spatially. If we can assume, however, that the spectral response of the detector is independent of position and angle, then a simplification results. Suppose the detector response can be written as

$$d_j(\mathbf{r}, \hat{\mathbf{s}}, \mathcal{E}) \propto D(\mathcal{E})\, \mathrm{rect}\left[\frac{\mathbf{r} - \mathbf{r}_j}{\epsilon}\right], \tag{10.277}$$

where $D(\mathcal{E})$ is a normalized spectral response function. Then retracing the steps leading up to (10.275) shows that

$$\overline{g}_j \propto \cos^4 \theta_j \int_P d^2r\ f_{eff}(m\mathbf{r})\, \mathrm{rect}\left[\frac{\mathbf{r} - \mathbf{r}_j}{\epsilon}\right], \tag{10.278}$$

where $f_{eff}(\mathbf{r})$ is an effective 2D object function defined by

$$f_{eff}(\mathbf{r}) = \int_0^\infty d\mathcal{E}\, D(\mathcal{E})\, f(\mathbf{r}, \mathcal{E})\,. \tag{10.279}$$

Thus the conventional view of the lens system as a 2D-to-2D mapping is preserved with this detector model. If neither the object nor the detector has spectral properties independent of position, however, we have to use a 3D-to-2D mapping, with the third dimension being photon energy or wavelength.

10.4.4 Adjoint methods

So far we have derived a general imaging equation and applied it to pinhole imaging and optical imaging. In these simple problems, we were able to obtain useful results completely analytically. In more complicated problems, however, numerical methods will be needed. This section and the next introduce important practical techniques for the numerical computation of images. In particular, here we show the role of adjoint operators.

[15]The reader may well question the consistency of these approximations. Since we used an ABCD matrix derived in the last chapter under paraxial conditions, should we not have set $\cos\theta_j$ to unity throughout? The answer might be yes for a simple lens, but for well-designed lenses with low distortion and low field curvature, the ABCD matrix of (10.271) will be valid at much larger angles than ones for which $\cos^4\theta_j \simeq 1$.

Linear CD systems To introduce adjoint methods, let us reexamine the discussion in Sec. 7.2.1 on linear CD systems that can be decomposed into a CC system followed by a detector array modeled as a linear CD mapping. From (7.227) and (7.229), we know that the mean image produced by such a system can be described by

$$\overline{\overline{\mathbf{g}}} = \mathcal{D}_a \mathcal{H}_{CC} \mathbf{f} \equiv \mathcal{H}\mathbf{f}, \tag{10.280}$$

or in component form as

$$\overline{g}_m = \int_{\mathbf{S}_g} d^s r_d \; a_m(\mathbf{r}_d) \int_{\mathbf{S}_f} d^q r \; h(\mathbf{r}_d, \mathbf{r}) \, f(\mathbf{r}) \,, \tag{10.281}$$

where $a_m(\mathbf{r}_d)$ is the response function of the mth detector. This response function can, for example, take the value 1 inside the aperture of the detector and 0 outside. Recall that $\mathbf{r}$ is a position vector of dimensionality q in the object domain, and $\mathbf{r}_d$ is a position vector of dimensionality s on the detector plane, before sampling by the discrete detector array.

This last equation can be rewritten in two equivalent forms,

$$\overline{g}_m = (\mathbf{a}_m, \mathcal{H}_{CC}\mathbf{f}) = (\mathcal{H}_{CC}^{\dagger}\,\mathbf{a}_m, \mathbf{f}) \,, \tag{10.282}$$

where $\mathbf{a}_m$ is the Hilbert-space vector corresponding to the function $a_m(\mathbf{r}_d)$. The first scalar product is thus in the continuous data space (variable $\mathbf{r}_d$), and the second scalar product is in the continuous object space (variable $\mathbf{r}$). In the first version we map the object through $\mathcal{H}_{CC}$ onto the detector response function, and in the second version we map the detector response function through $\mathcal{H}_{CC}^{\dagger}$ onto the object.

Though the two versions of (10.282) are mathematically equivalent, the second or adjoint version may have considerable computational benefit in simulation studies. To implement the first version for a particular object, we have to transform the object through the operator $\mathcal{H}_{CC}$; if this operator is represented by a matrix with K sample points for each dimension, then altogether it is a $K^s \times K^q$ matrix, and for realistic choices of K, this size may be completely impractical. In the second version, we have a matrix of the same size but need only apply it to a compact vector, most elements of which are zero. For example, it often suffices to use a 3×3 or 5×5 array of sampling points across a 2D detector element, and in that case $a_m(\mathbf{r}_d)$ is represented by just 9 or 25 nonzero samples. The operation $\mathcal{H}_{CC}^{\dagger}\mathbf{a}_m$ thus becomes feasible, and the result is represented as K^q points in the discretized object space. Many of these points may be near zero, but the scalar product with any given object vector requires at most K^q multiplies. In addition, the calculation of $\mathcal{H}_{CC}^{\dagger}\mathbf{a}_m$ does not have to be repeated for each new object.

In some problems $\mathcal{H}_{CC}^{\dagger}\mathbf{a}_m$ can be computed completely analytically without any matrix representations. For example, in diffraction-limited optical systems, $\mathcal{H}_{CC}^{\dagger}\mathbf{a}_m$ is determined by the diffraction pattern of the detector aperture, which can be expressed analytically in terms of Fresnel integrals for Fresnel diffraction (see Sec. 9.4.6) or as the Fourier transform of $a_m(\mathbf{r}_d)$ for Fraunhofer diffraction (see Sec. 9.4.7).[16] When such analytic calculations are possible, discretization is needed

[16]For incoherent optical systems, one must take the squared modulus of the amplitude of these diffraction patterns, but the system is nevertheless linear when the input and output are stated in terms of irradiance.

only for the final scalar product in the second version of (10.282). The first version, on the other hand, is nearly useless for analytic purposes since the object is not expressed analytically in the first place, and even if it was, its analytic diffraction pattern would almost never be calculable.

Adjoint methods for transport calculations Adjoint methods are also useful for the Boltzmann equation. To illustrate, consider the steady-state Boltzmann equation (10.164) for photons, which can be written in simplified form as

$$v\,\hat{\mathbf{s}}\cdot\nabla w + v\mu_{tot}w - \mathcal{K}w = \Xi\,, \tag{10.283}$$

where v is the speed of light in the medium (denoted c_m earlier in this chapter, but we now need the subscript m for something else) and Ξ is the spectral photon distribution function of the source (denoted $\Xi_{p,\mathcal{E}}$ earlier), $\mathcal{K}$ is the scattering operator and w denotes the distribution function $w(\mathbf{r},\hat{\mathbf{s}},\mathcal{E})$. In general, Ξ depends on direction $\hat{\mathbf{s}}$, photon energy $\mathcal{E}$ and 3D position $\mathbf{r}$, but it may be valid to assume that the directional and energy dependences are independent of position, and then the source description is reduced to a purely spatial function like our familiar $f(\mathbf{r})$ (see Sec. 10.4.2).

We can also regard the distribution function as a vector in a Hilbert space of square-integrable functions of 3D position, direction and energy. We shall call this new Hilbert space *distribution space* and denote it as $\mathbb{D}$, and the distribution function itself, regarded as a vector in this space, will be denoted as $\mathbf{w}$. We can then express the Boltzmann equation in operator form as

$$\mathcal{B}\mathbf{w} = \Xi\,, \tag{10.284}$$

where $\mathcal{B}$ is an operator given by

$$\mathcal{B} = v\,\hat{\mathbf{s}}\cdot\nabla + v\mu_{tot} - \mathcal{K}\,. \tag{10.285}$$

In the absence of scattering, $\mathcal{B}$ is a first-order partial differential operator, and in free space where both the absorption and scattering vanish, $\mathcal{B} = v\,\hat{\mathbf{s}}\cdot\nabla$. In all cases, $\mathcal{B}$ is a local operator; the scattering operator $\mathcal{K}$ mixes angles and energies but maps a distribution function at one spatial location to another distribution function at the *same* location.

If there are no radiation sources outside some volume V, the operator $\mathcal{B}$ has a left inverse, which simply means that the distribution function $\mathbf{w}$ in V is fully determined by the source Ξ within V. Denoting this left inverse as $\mathcal{L}$, we can write

$$\mathbf{w} = \mathcal{L}\Xi\,. \tag{10.286}$$

When the source is describable by its spatial dependence alone, we can go a step further and write

$$\mathbf{w} = \mathcal{L}_f\mathbf{f}\,. \tag{10.287}$$

Expressions for $\mathcal{L}$ and $\mathcal{L}_f$ are given for various cases in Sec. 10.3; with scattering it is difficult to express them analytically except by an infinite series such as (10.168).

Of course, the distribution function $\mathbf{w}$ is not an image, but in many cases a (noise-free) digital image consists of a set of linear functionals of $\mathbf{w}$ (see Sec. 10.4.1). Specifically, if the detector is linear and its response function for the mth

measurement is denoted $d_m(\mathbf{r}, \hat{\mathbf{s}}, \mathcal{E})$, then the mean of the measurement is given under steady-state conditions by [*cf.* (10.239)]

$$\overline{g}_m = \tau \int_P d^2r \int_0^\infty d\mathcal{E} \int_{2\pi} d\Omega \, d_m(\mathbf{r}, \hat{\mathbf{s}}, \mathcal{E}) \, w(\mathbf{r}, \hat{\mathbf{s}}, \mathcal{E}) \,, \tag{10.288}$$

where τ is the exposure time. As we discussed in Sec. 10.4.1, $\mathbf{r}$ is a general 3D vector, but the spatial integral is over the two variables needed to specify a reference plane P somewhere between the source and detector. If we denote that plane as $z = 0$ and assume that only photons with $s_z \geq 0$ are directed towards the detector, we can redefine the detector response function in 3D by letting

$$p_m(\mathbf{r}, \hat{\mathbf{s}}, \mathcal{E}) \equiv \tau \, d_m(\mathbf{r}, \hat{\mathbf{s}}, \mathcal{E}) \, \delta(z) \, \text{step}(s_z) \,, \tag{10.289}$$

and (10.288) becomes

$$\overline{g}_m = \int_V d^3\mathbf{r} \int_0^\infty d\mathcal{E} \int_{4\pi} d\Omega \, p_m(\mathbf{r}, \hat{\mathbf{s}}, \mathcal{E}) \, w(\mathbf{r}, \hat{\mathbf{s}}, \mathcal{E}) \,. \tag{10.290}$$

We can regard this expression as a scalar product[17]:

$$\overline{g}_m = (\mathbf{p}_m, \mathbf{w})_{\mathfrak{D}} \,. \tag{10.291}$$

We can also define adjoints of the operators $\boldsymbol{\mathcal{B}}$, $\boldsymbol{\mathcal{L}}$ and $\boldsymbol{\mathcal{L}}_f$; note that the first two of these operators map $\mathfrak{D}$ to itself, but the third maps $\mathbb{L}_2(\mathfrak{R}^q)$ to $\mathfrak{D}$. The usefulness of $\boldsymbol{\mathcal{L}}^\dagger$ and $\boldsymbol{\mathcal{L}}_f^\dagger$ is that we can write, analogously to (10.282),

$$\overline{g}_m = (\boldsymbol{\mathcal{L}}^\dagger \mathbf{p}_m, \boldsymbol{\Xi})_{\mathfrak{D}} = (\boldsymbol{\mathcal{L}}_f^\dagger \mathbf{p}_m, \mathbf{f})_{\mathbb{L}_2(\mathfrak{R}^q)} \,. \tag{10.292}$$

The latter form in particular is useful computationally since $\boldsymbol{\mathcal{L}}_f^\dagger$ maps the detector response from a 6D space to a 2D or 3D one. Also, as in the discussion of adjoint methods above, the difficult calculation $\boldsymbol{\mathcal{L}}_f^\dagger \mathbf{p}_m$ does not need to be repeated for each $\mathbf{f}$, and in some problems (usually without scatter) it can be performed analytically.

Adjoint Boltzmann equation The adjoint of $\boldsymbol{\mathcal{B}}$ plays a rather different role than the adjoints of $\boldsymbol{\mathcal{L}}$ and $\boldsymbol{\mathcal{L}}_f$. It is not used with solutions of the Boltzmann equation itself, but rather with solutions of the *adjoint Boltzmann equation*:

$$\boldsymbol{\mathcal{B}}^\dagger \widetilde{\mathbf{w}}_m = \mathbf{p}_m \,. \tag{10.293}$$

In this equation, $\widetilde{\mathbf{w}}_m$ is a function in $\mathfrak{D}$ but not the photon distribution produced by the actual source. To understand its significance, let us temporarily ignore scatter so that $\boldsymbol{\mathcal{B}}$ becomes a differential operator,

$$\boldsymbol{\mathcal{B}} = v \, \hat{\mathbf{s}} \cdot \nabla + v\mu_{abs} \,. \tag{10.294}$$

Adjoints of differential operators were discussed in Sec. 4.1.3. Integration by parts [*cf.* (4.16)] shows that

$$\boldsymbol{\mathcal{B}}^\dagger = -v \, \hat{\mathbf{s}} \cdot \nabla + v\mu_{abs} \,, \tag{10.295}$$

[17] A technical difficulty here is that $p_m(\mathbf{r}, \hat{\mathbf{s}}, \mathcal{E})$ is not a square-integrable function because of the factor $\delta(z)$, but this problem can be handled by a limiting argument. The problem would not arise in the first place if we considered the finite thickness of the detector.

provided the boundary terms vanish.[18] Thus the forward operator $\mathcal{B}$ describes particles moving at speed v in direction $\hat{\mathbf{s}}$, while the adjoint $\mathcal{B}^\dagger$ describes particles moving at the same speed in direction $-\hat{\mathbf{s}}$. The adjoint distribution function $\widetilde{\mathbf{w}}_m$ can be viewed as the distribution function in the object domain that would result if the detector response function $\mathbf{p}_m$ were a source of these backward-travelling photons.

Without scatter, the ordinary Boltzmann equation is solved by the attenuated x-ray transform (10.151). For the adjoint Boltzmann equation without scatter, (10.151) is again the solution provided we replace the actual source Ξ with the effective source p_m and also let $\hat{\mathbf{s}} \to -\hat{\mathbf{s}}$.

When the scatter term is reinstated, the adjoint operator becomes

$$\mathcal{B}^\dagger = -v\,\hat{\mathbf{s}} \cdot \nabla + v\mu_{tot} - \mathcal{K}^\dagger \, . \tag{10.296}$$

The kernel for the forward scatter operator $\mathcal{K}$ is defined by (10.129), and the reader should be able to construct the kernel for $\mathcal{K}^\dagger$ by taking a little care with the primes on the energy variables. Actually solving the adjoint Boltzmann equation with scatter usually requires Monte Carlo methods, to be discussed in Sec. 10.4.5.

If we have solved the adjoint Boltzmann equation, the mean image can be computed by combining (10.291) and (10.293) as follows:

$$\overline{g}_m = (\mathbf{p}_m, \mathbf{w}) = \left(\mathcal{B}^\dagger \widetilde{\mathbf{w}}_m, \mathbf{w}\right) = (\widetilde{\mathbf{w}}_m, \mathcal{B}\mathbf{w}) = (\widetilde{\mathbf{w}}_m, \Xi) \, , \tag{10.297}$$

where the last step has used (10.284). Thus the scalar product between the adjoint distribution function and the source is the same as that between the actual physical distribution function and the detector response, namely the desired mean data $\overline{g}_m$.

10.4.5 Monte Carlo methods

The term *stochastic simulation* refers to a broad class of methods in which some quantity is estimated by performing random experiments, either physically or in a computer, in such a way that the mean value of the experimental outcome is the quantity of interest. The quantities being estimated are often naturally interpretable as probabilities, but many applications of stochastic simulation have no relation to probability except as a computational tool.

Historically, stochastic simulation substantially predates computers. For example, the eighteenth-century French polymath Georges-Louis Leclerc De Buffon[19] (1707–1768) posed the question: If a needle of length L is thrown randomly onto a grid of parallel lines of spacing L, what is the probability that the needle will overlap one of the lines? The answer turns out to be $2/\pi$, and throughout the nineteenth and twentieth century many people used either experiments with real needles or simulations as a way of estimating π.

Similarly, W. S. Gosset ("Student") used numerical experiments to verify his analytic form for the celebrated t distribution. Lord Rayleigh and A. N. Kolmogorov showed the connection between random walks and differential equations.

[18] In fact, the boundary terms vanish if it is possible, as we have assumed, to set up a reference plane completely separating the source and detector; see Lewis and Miller (1984).

[19] It was also Buffon who in 1748 suggested what we now call Fresnel lenses. Fresnel's contribution was applying them to the construction of lighthouse lenses (*Encyclopedia Brittanica*, 2001).

The real power of stochastic simulation, however, emerged with the advent of computers, and the application that sparked widespread interest was simulation of neutron transport in connection with the Manhattan Project during World War II. Key figures in this development included John von Neumann, Edward Teller, Nicolas Metropolis, Herman Kahn and Stanislaw Ulam.

When one person is singled out as the originator of this method, it is usually Stanislaw Ulam (1909–1984). Ulam obtained his Ph.D. in 1933 from the Polytechnic Institute in Lemberg, Poland (now Lviv, Ukraine), where he studied under Banach and drank coffee at the Scottish Cafe. In 1935 he was invited by von Neumann to spend a few months at the Institute for Advanced Studies of Princeton, and in 1940 he emigrated permanently to the U.S. In 1943 he received a second invitation from von Neumann, this time to participate in secret wartime research. It was apparently Ulam who suggested the colorful term *Monte Carlo* for the general class of methods that relies on computer-generated random numbers. Von Neumann, on the other hand, remained skeptical; in 1951 he said, "Anyone who considers arithmetical methods of producing random digits is, of course, in a state of sin."

An important practical advantage of Monte Carlo methods in general is that they are inherently well suited to parallel computation (with either many peasants throwing needles or many computers computing particle trajectories). Since the events are independent, the processing speed grows linearly with the number of processing units.

We shall introduce Monte Carlo methods here with a brief discussion of their use in performing definite integrals, but then we shall return to their use in transport calculations and image simulation.

Monte Carlo integration Consider an integral of the form

$$I = \int_a^b dx\, f(x)\, p(x)\,, \tag{10.298}$$

where $p(x)$ is a PDF on (a, b) [*i.e.*, $p(x) \geq 0$ and $\int_a^b dx\, p(x) = 1$]. Thus I is the expectation of $f(x)$ with respect to this PDF. If we can draw a set of samples $\{x_i\}$ from $p(x)$, say by one of the methods discussed in Sec. C.7, then we can construct a random point process $u(x)$ defined by

$$u(x) = \frac{1}{N} \sum_{i=1}^{N} \delta(x - x_i)\,. \tag{10.299}$$

We can then estimate I by

$$\hat{I}_N \equiv \int_a^b dx\, f(x)\, u(x) = \frac{1}{N} \sum_{i=1}^{N} f(x_i)\,. \tag{10.300}$$

This viewpoint of a Monte Carlo estimate as a scalar product with a random point process will prove useful below.

A full treatment of random point processes will be given in Chap. 11, but we anticipate a few simple results here. As we shall show in (11.81), $\langle u(x) \rangle = p(x)$, so it follows that $\langle \hat{I}_N \rangle = I$. It can also be shown that

$$\lim_{N \to \infty} \hat{I}_N = I\,, \qquad \mathrm{Var}\{\hat{I}_N\} \propto \frac{1}{N}\,. \tag{10.301}$$

Thus the estimate of I is unbiased and consistent.

The basic idea of Monte Carlo integration extends readily to multidimensional integrals, and in fact the Monte Carlo approach becomes increasingly advantageous as the dimension increases. If we want to perform a qD integral by sampling on a regular grid with a total of N function evaluations, then we have $N^{\frac{1}{q}}$ samples in each dimension, and the precision of the result (expressed as RMS error) varies as $N^{-\frac{1}{2q}}$. For Monte Carlo estimation, again with N function evaluations, the variance of the estimate varies as $1/N$ for all q, so the precision (standard deviation of the estimate) varies as $N^{-\frac{1}{2}}$. For large q, convergence is therefore much faster with Monte Carlo.

Monte Carlo transport calculations For simulating imaging systems (as opposed to bombs), the essence of a Monte Carlo transport calculation is simply to track photons in a computer from a source towards a detector. The initial photon directions and the locations and outcomes of all scattering and absorption events are assigned by calling suitable random-number generators, mimicking the actual physical processes. Each photon is tracked until it is either absorbed, reaches the reference plane P between the source and the detector, or escapes from the system altogether. We repeat the procedure for a total of N initial photons and denote the number that make it to P as N_P.

At this stage, we know the photon coordinates $\{\mathbf{r}_i, \hat{\mathbf{s}}_i, \mathcal{E}_i, i = 1, ..., N_P\}$, and we can define a random point process analogous to (10.299) by

$$u(\mathbf{r}, \hat{\mathbf{s}}, \mathcal{E}) = \frac{1}{N_P} \sum_{i=1}^{N_P} \delta(\mathbf{r} - \mathbf{r}_i)\, \delta(\hat{\mathbf{s}} - \hat{\mathbf{s}}_i)\, \delta(\mathcal{E} - \mathcal{E}_i)\,. \tag{10.302}$$

The mean measurement $\overline{g}_m$ is defined in (10.291), and it can be estimated by analogy to (10.300) as

$$\hat{\overline{g}}_m = \frac{\overline{N}_{em} N_P}{N_T} (\mathbf{p}_m, \mathbf{u})_{\mathbb{D}} = \frac{\overline{N}_{em}}{N} \sum_{i=1}^{N_P} d_m(\mathbf{r}_i, \hat{\mathbf{s}}_i, \mathcal{E}_i)\,, \tag{10.303}$$

where N is the number of initial photons used in the Monte Carlo simulation and $\overline{N}_{em}$ is the mean number of photons emitted during the exposure time when an actual physical source is used. (In most Monte Carlo problems, $\overline{N}_{em}$ can be computed directly from the original object description and does not have to be estimated.) The reader may demonstrate her virtuosity with point processes by showing that this estimator is unbiased, at least to the extent that the Monte Carlo simulation accurately models the real physical processes that would occur with the physical source.

Mechanics of Monte Carlo We shall illustrate some of the practical aspects of Monte Carlo simulation by describing how to simulate the image of a self-luminous object that both absorbs and scatters its own radiation. This description is motivated by single-photon emission computed tomography or SPECT (see Chap. 17), but it is applicable also to fluorescence microscopy, solar imaging and studies of flames, for example.

The first step is to choose the emission point for the photon. If the object is specified in a voxel representation, one approach is to step systematically through the voxels and to let each voxel emit a random number of photons, Poisson-distributed about a mean proportional to the coefficient θ_n associated with that voxel. This approach amounts to assuming that the object is the mean of a Poisson random process (see Sec. 11.3).

When the object is specified geometrically, for example in terms of geometric shapes, a variant on the rejection method of Sec. C.7.1 can be used. First the object is normalized by defining $f_0(\mathbf{r}) \equiv f(\mathbf{r})/\max\{f(\mathbf{r})\}$. Then a point $\mathbf{r}$ is chosen randomly from a PDF that is uniform over the object support, and a photon is assumed to be emitted from this point if $f_0(\mathbf{r}) > t$, where t is a uniform random number on (0, 1). This method amounts to assuming that the object is the mean of a random process but not necessarily a Poisson one.

Having chosen the emission point, the next step is to launch a photon from that point in a random direction, taking care to get the proper angular distribution. For example, if the source is an isotropic emitter, we must ensure that all emission directions are equally probable. If the direction is specified by the usual polar angles θ and ϕ, it is not correct to choose values for these angles by calling a uniform random-number generator. The problem is that the differential solid angle is given by $d\Omega = \sin\theta\, d\theta\, d\phi$, so a uniform distribution of θ and ϕ does not correspond to a constant number of photons per unit solid angle (or constant radiant intensity). A simple fix is to choose $\cos\theta$ from a uniform distribution over $(-1, 1)$, then to compute θ itself by taking an arccosine. Since $d\Omega = d(\cos\theta)\, d\phi$, this procedure gives a source radiant intensity independent of direction. On the other hand, if we wish to simulate a Lambertian surface emitter, we must make the radiant intensity vary as $\cos\theta$, where this angle is measured from the surface normal (see Sec. 10.2.1). We can do this by drawing $\cos\theta$ from $\mathrm{pr}(\cos\theta) = \cos\theta$, $(-1 < \cos\theta < 1)$.

The next step is to decide whether the photon escapes from the object without interacting, or instead undergoes an absorption or scattering process at some point. For this purpose it is necessary to specify the spatial distribution of the total attenuation coefficient $\mu_{tot}(\mathbf{r}, \mathcal{E})$. We shall use the medical terminology here and refer to this distribution as the *body*, thereby distinguishing it from the *object*, which refers to the distribution of the radiation source. The simplest assumption is that $\mu_{tot}(\mathbf{r}, \mathcal{E})$ is the constant μ_{tot}, independent of position and energy within the boundaries of a convex body. Under these assumptions, the probability of the photon escaping from the body without interacting is given by $P_{esc} = \exp\left[-\mu_{tot} L(\mathbf{r}, \hat{\mathbf{s}})\right]$, where $L(\mathbf{r}, \hat{\mathbf{s}})$ is the total length of attenuating medium between point $\mathbf{r}$ and the detector in direction $\hat{\mathbf{s}}$. If the attenuation is not uniform, numerical integration must be used to compute P_{esc}. To decide whether the particular photon being simulated escapes, we can draw a random number t_1 uniformly from $(0, 1)$; if $t_1 < P_{esc}$, the photon escapes.

If the photon does not escape, we must decide where it interacts, whether the interaction is absorption or scattering, and in the latter case the direction of the scattered photon. For x rays or gamma rays, we must also decide whether the scattering is elastic or inelastic (Compton). Each of these decisions can be made by drawing suitably distributed random numbers. As an example, to decide where the interaction takes place, we must draw a random number from the density $\mathrm{pr}(\ell)$ on interaction at distance ℓ from the source. For uniform attenuation, source location $\mathbf{r}$ and propagation direction $\hat{\mathbf{s}}$, this density is given by

$$\mathrm{pr}_\ell(\ell) = \frac{\mu_{tot} \exp(-\mu_{tot}\ell)}{1 - \exp\left[-\mu_{tot} L(\mathbf{r}, \hat{\mathbf{s}})\right]}, \qquad 0 \le \ell \le L(\mathbf{r}, \hat{\mathbf{s}}). \tag{10.304}$$

The corresponding cumulative distribution function is

$$F_\ell(\ell) = \int_0^\ell d\ell' \, \mathrm{pr}_\ell(\ell') = \frac{1 - \exp(-\mu_{tot}\ell)}{1 - \exp[-\mu_{tot}L(\mathbf{r}, \hat{\mathbf{s}})]} . \tag{10.305}$$

Following the procedure of Sec. C.7.2, we draw a random variable t_2 uniformly from $(0, 1)$ and solve the equation $t_2 = F_\ell(\ell)$; the result is

$$\ell = -\frac{1}{\mu_{tot}} \ln\{1 - t_2 + t_2 \exp[-\mu_{tot}L(\mathbf{r}, \hat{\mathbf{s}})]\} . \tag{10.306}$$

The variables generated this way will follow the PDF of (10.304). A similar procedure can be used for nonuniform attenuation, but the equation $t_2 = F_\ell(\ell)$ must be solved numerically in that case.

To decide whether the interaction after traversing distance ℓ is absorption or scattering, we draw another random number t_3 uniformly from (0, 1) and choose the absorption interaction if $t_3 < \mu_{pe}/\mu_{tot}$. If the interaction is an absorption, the photon is terminated, but if it is scattering it is necessary to choose a direction for the scattered photon. This step is accomplished by the methods of Sec. C.7.2 along with knowledge of the differential scattering cross section (see Sec. 10.2.5). For Compton scattering, the energy of the scattered photon is then determined from the scattering angle and (10.226). The process is repeated until the photon either undergoes an absorption or escapes the body.

When the photon escapes the body, we have several options, depending on how we choose the reference plane separating the source and detector. As we saw by example in Sec. 10.4.2, we can choose this plane such that the photons do not encounter any apertures or image-forming elements before reaching the plane, or we can choose it immediately adjacent to the detector, or anywhere in between. If there are no obstacles between the source and the reference plane, a photon escaping the body can be propagated along a straight line until it strikes the reference plane. If, on the other hand, we choose the reference plane after some image-forming elements (lenses, pinholes, collimators, ...), then we can continue tracing the photon through these elements to the plane. In either case, if we know the response function $d_m(\mathbf{r}, \hat{\mathbf{s}}, \mathcal{E})$ analytically, we can use the photon coordinates on P to compute the contribution of that photon to $\overline{g}_m$. If we do not know the response analytically, we can continue the Monte Carlo simulation inside the detector. This entire process is repeated for many photons until the desired accuracy in the estimate is obtained.

Adjoint Monte Carlo and forced detection A major drawback of Monte Carlo methods as described so far is that many simulated photons — possibly a great majority — will be absorbed either in the body or in apertures outside the body and never reach the detector. Far more photons must be launched than are eventually incorporated in the simulated data set, and there will be large Poisson fluctuations in the estimated image.

One way around this problem is the *adjoint Monte Carlo* method. A Monte Carlo simulation begins with a source distribution and transports it to a detector; adjoint Monte Carlo begins with a detector response function and transports it back to all source points; thus it implements the adjoint Boltzmann equation. As a practical matter, most Monte Carlo simulation codes can be run either in the normal or adjoint mode; it does not matter to the code what one considers as source and what as detector.

Another technique for more efficiently using the simulated photons is *forced detection*. Like many other variance-reduction methods described in the literature, forced detection does not regard photons as physically indivisible but rather assigns weights to the various fates that can befall them. In straightforward Monte Carlo as described above, a scattered photon can go in an arbitrary direction, and only a small portion of these directions will lead it through the image-forming elements to the detector. In forced detection, the probability that the photon will head towards the detector and escape the body from that scattering point is computed, and then this photon is assumed to contribute to the point process on plane P weighted with this probability. Then, however, the photon is allowed to scatter in an arbitrary direction, and a similar weighted contribution is computed from the next scattering site. Thus many more terms appear in the point process, yet the weights keep the estimate of $\overline{g}_m$ unbiased.

We must note, however, that these methods are intended for accurate and efficient estimation of mean images, not for generating sample images with realistic statistical properties. Thus variance-reduction techniques are profitable in estimating the mean image $\overline{\mathbf{g}}$, but they should be avoided if the objective is to simulate $\mathbf{g}$ itself.

11

Poisson Statistics and Photon Counting

This chapter deals with *photon-counting detectors*, though this term is somewhat of a misnomer since what such detectors count is actually photoelectrons or photoelectric interactions, not photons. As we discussed in Chap. 10, it is usually not necessary to consider a quantized description of the radiation field at all. Photon-counting experiments can often be analyzed successfully with a semiclassical theory in which matter is treated quantum mechanically but electromagnetic radiation is assumed to obey Maxwell's equations.

Nevertheless, as we argued in Sec. 10.1, it can be very useful, both conceptually and computationally, to think of electromagnetic radiation as transported by localized excitations which, for want of a better term, we call photons. This viewpoint can be justified through quantum electrodynamics if we define localized (multi-mode) states of the radiation field and localized operators that correspond to classical concepts like irradiance and energy density. With the background from Chap. 10, therefore, we adopt the language of photons throughout most of this chapter, returning to quantum electrodynamics only in the last section.

In photon counting and many other problems, the most important probability law is the Poisson distribution. The eminent statistician Sir Ronald Fisher made this point forcefully, saying that, "Among discontinuous distributions, the Poisson series is of first importance" (Haight, 1967). There are two main reasons for this preeminence. First, the Poisson distribution arises as an almost inevitable consequence of statistical independence in counting problems. This statement is made more precise in Sec. 11.1.1 by stating three fundamental postulates from which the Poisson law can be derived. Somewhat colloquially, these postulates say merely that we are counting statistically independent events. It is not much of a stretch to say that if events are independent, the number of them in any fixed time interval must be a Poisson random variable. Conversely, when a counting distribution is determined not to be Poisson, we can usually discover a reason why the events are not independent. One of the authors of this book goes so far as to tell his students that Poisson is French for independent.

The second reason for the importance of the Poisson distribution is that it also occurs in spite of statistical dependence when the events in question are rare, in a sense to be discussed below. This result, sometimes called the *law of small numbers*, was derived by S. D. Poisson in 1837 in the course of research on the probability of judgments in civil and criminal trials.[1]

When electromagnetic radiation falls on a photon-counting detector, the resulting pattern of photoelectric interactions is a spatio-temporal random process. That is, each sample function depends on both spatial and temporal variables. Which of these variables we are interested in depends on what the detector measures. In Sec. 11.1, we assume that the detector reports only the total number of accumulated counts, so the concern is with a single random variable, which under many circumstances will turn out to be Poisson distributed. In Sec. 11.2 we discuss discrete arrays of detectors where the outputs can be considered as random vectors, often described by the multivariate Poisson law. These two sections provide the basis for discussing digital imaging systems that are noise-free except for the inevitable limits imposed by counting statistics. We refer to such systems as *photon-limited* or *quantum-noise-limited*; they are the imaging counterparts of electronic systems that are limited by *shot noise*, which arises from the discrete nature of electrons.

Section 11.3 moves on to random point processes, which can be used to describe detector outputs more fully than by total counts. If the temporal waveform of the detector output is observed without regard to the spatial coordinates of the interactions, then the proper description is a temporal random process. On the other hand, if a detector (such as film) records spatial locations without regard to time of interaction, then the relevant random process is a spatial one. In both cases, we shall encounter Poisson random processes if the postulates are satisfied.

Section 11.4 deals with devices such as photomultipliers and image intensifiers where one input event initiates a random number of output events. Since many output events arise from each input event, the independence is lost and the output random process is usually not Poisson. Nevertheless, the principles developed earlier in the chapter for Poisson processes will be put to good use.

In Sec. 11.5 we look at photon counting from a quantum-mechanical perspective. Here, too, we shall see that the Poisson distribution plays a fundamental role, and that for certain kinds of light a Poisson distribution of photocounts will be observed. In fact, in most cases the classical results of the previous sections will turn out to be valid also in quantum-mechanical terms. We shall see, however, that there are also some purely quantum-mechanical forms of light for which the classical results do not hold.

Excellent general references on Poisson statistics include Haight (1967), Johnson and Kotz (1969), Feller (1968) and the Encyclopedia of Statistical Sciences (Kotz *et al.*, 1982). An exhaustive bibliography up to 1966 is given by Haight (1967). Discussions relevant to optics and imaging are given in Metz (1969), Barrett and Swindell (1981, 1996), Goodman (1985), Frieden (1983) and Snyder and Miller (1991). A comprehensive book that covers many of the topics in this chapter is Saleh (1978).

[1]While the Poisson probability law is alive and well today, perhaps the more important part of that research — the stochastic nature of judicial proceedings — is seldom acknowledged.

11.1 POISSON RANDOM VARIABLES

In this section we focus on individual Poisson random variables, but we make use of Poisson random processes as a way of elucidating the properties of the variables. We look first at two basic principles, independence and rarity, from which the Poisson probability law can be derived.

11.1.1 Poisson and independence

The fundamental postulates that lead to the Poisson probability law are discussed by many authors, for example Davenport and Root (1958), Barrett and Swindell (1981, 1996) and Goodman (1985). Our treatment will rely heavily on these previous works.

We could state the postulates in abstract form, but it will help to develop mental imagery if we consider a specific experimental setup. Imagine a beam of radiation falling on a detector and producing photoelectrons. Each photoelectron will produce an impulse of current in an external circuit, and we assume that the circuit includes a counter that will tally the total number N of impulses produced in some time interval of duration T. The probability of N photoelectrons in this interval is denoted $\Pr(N \text{ in } T)$. Since each photoelectron causes a current impulse and thus a change in counter reading, $\Pr(N \text{ in } T)$ can also be interpreted as the probability law for the number of current impulses in time T or for the change in counter reading during time T. In the jargon of nuclear physics and radiology, each impulse is referred to as a *count*, and N is then the number of counts in time T. If we further assume that each photoelectron is produced by absorption of a single photon (see Sec. 10.1.4), then N is also the number of detected photons. More generally, of course, N is simply the number of events of an unspecified nature in time T, and we shall refer to the events here as counts.

Poisson postulates The postulates that will lead to the Poisson law are:

(a) The number of counts in any time interval $t_1 < t \leq t_2$ is statistically independent of the number in any other *nonoverlapping* interval $t_3 < t \leq t_4$, where $t_1 < t_2 \leq t_3 < t_4$.

(b) If we consider a very small time interval of duration ΔT, the probability of a single count in this interval approaches some constant a times ΔT, *i.e.*, up through terms linear in ΔT,

$$\Pr(1 \text{ in } \Delta T) = a\Delta T . \tag{11.1}$$

(c) The probability of more than one count in a vanishingly small interval ΔT is zero. Thus, again through terms linear in ΔT,

$$\Pr(1 \text{ in } \Delta T) + \Pr(0 \text{ in } \Delta T) = 1 . \tag{11.2}$$

Discussion of the postulates Postulate (a) is a clear statement of the independence of the counts; no matter how many counts are observed in (t_1, t_2), that number has no influence on the number observed in (t_3, t_4). Postulate (c) also requires that the counts be independent; if, for example, counts always came in pairs, postulate (c) would not hold since then $\Pr(2 \text{ in } \Delta T)$ would be linear in ΔT and $\Pr(1 \text{ in } \Delta T)$ would be zero.

Postulate (b) goes beyond statistical independence. As stated here, it also requires that $\Pr(1 \text{ in } \Delta T)$ be independent of the absolute time; if the small interval ΔT is defined as $T < t \leq T + \Delta T$, the probability depends only on ΔT and is independent of T. As we shall see in more detail in Sec. 11.2, this condition implies that we are dealing with a stationary random process. A more general form of postulate (b) would allow $\Pr(1 \text{ in } \Delta T)$ to have the form $a(t)\Delta T$, in which case we would be dealing with a nonstationary random process. Later in this section we shall consider the case where $a(t)$ is a nonrandom function of time, and in Secs. 11.1.4 and 11.3.7 we shall allow $a(t)$ to be random. For now we treat a as a constant and hence concentrate on the stationary case. In this case the parameter a is the average rate of counts. From (11.1) it can be seen that a has dimensions of reciprocal time since probabilities are dimensionless.

Bernoulli trials and independence Another familiar situation involving independent events is Bernoulli trials, such as successive flips of a coin. (See Sec. C.6.1 in App. C.) The common assumption in this problem is that the trials are statistically independent, but the result is the binomial law, not the Poisson. It is easy to see that our postulates are not satisfied for Bernoulli trials. If we think of coins being flipped at a regular pace, say once per second, then the numbers of heads in two different time intervals are not statistically independent. If we consider a half-second interval containing a flip, then the number of heads in the next half second is fully determined; it can only be zero since no flip takes place. Moreover, the number of heads in N flips (where N is nonrandom) is not statistically independent of the number of tails. Since the total is N, the number of tails is fully determined by the number of heads. Thus, even though the individual flips are independent, Bernoulli trials have a degree of statistical regularity not consistent with the postulates and not characteristic of Poisson processes.

Derivation of the Poisson law As a first step toward computing $\Pr(N \text{ in } T)$, we consider an interval of duration $T + \Delta T$ and compute the probability of getting *no* counts in this interval. This probability is denoted $\Pr(0 \text{ in } T + \Delta T)$. The only way we can find no counts in $(0, T + \Delta T)$ is to find none in $(0, T)$ and also none in the subsequent interval of $(T, T + \Delta T)$. Since the intervals are statistically independent, we have

$$\Pr(0 \text{ in } T + \Delta T) = \Pr(0 \text{ in } T) \Pr(0 \text{ in } \Delta T) = \Pr(0 \text{ in } T)\,(1 - a\Delta T)\,, \tag{11.3}$$

where the last step makes use of (11.1) and (11.2). Some simple algebra gives

$$\Pr(0 \text{ in } T + \Delta T) - \Pr(0 \text{ in } T) = -a\Delta T \Pr(0 \text{ in } T)\,. \tag{11.4}$$

Dividing through by ΔT and passing to the limit, we find

$$\frac{d}{dT} \Pr(0 \text{ in } T) = -a \Pr(0 \text{ in } T)\,. \tag{11.5}$$

The solution of this elementary differential equation, with the boundary condition $\Pr(0 \text{ in } 0) = 1$, is

$$\Pr(0 \text{ in } T) = \exp(-aT)\,. \tag{11.6}$$

Thus the probability of getting no counts decays exponentially with the length of the interval.

Next consider the event where exactly one count occurs in $T + \Delta T$. There are two mutually exclusive ways in which this can happen: one count in the first interval of T and none in the subsequent interval of ΔT; or none in T and one in ΔT. The probability of one count in the overall interval is thus

$$\Pr(1 \text{ in } T + \Delta T) = \Pr(1 \text{ in } T)\Pr(0 \text{ in } \Delta T) + \Pr(0 \text{ in } T)\Pr(1 \text{ in } \Delta T)\,. \tag{11.7}$$

With (11.1), (11.2), (11.6) and a smidgen of algebra, we find

$$\Pr(1 \text{ in } T + \Delta T) - \Pr(1 \text{ in } T) = -a\Delta T\Pr(1 \text{ in } T) + a\Delta T\exp(-aT)\,. \tag{11.8}$$

Again dividing through by ΔT and passing to the limit, we find

$$\frac{d}{dT}\Pr(1 \text{ in } T) = -a\Pr(1 \text{ in } T) + a\exp(-aT)\,. \tag{11.9}$$

With the boundary condition $\Pr(1 \text{ in } 0) = 0$, the solution to this differential equation is

$$\Pr(1 \text{ in } T) = aT\exp(-aT)\,. \tag{11.10}$$

Now consider the general situation. What is the probability of getting exactly N counts in $T + \Delta T$? Again, there are two ways in which this event can occur: N counts in T followed by 0 in ΔT; or $N - 1$ in T followed by 1 in ΔT. By the third postulate, there is vanishingly small probability as ($\Delta T \to 0$) of getting more than one count in ΔT. Since the two outcomes that lead to N counts in $T + \Delta T$ are mutually exclusive, and since the counts in the two time intervals are statistically independent, we have

$$\begin{aligned}\Pr(N \text{ in } T + \Delta T) &= \Pr(N \text{ in } T)\Pr(0 \text{ in } \Delta T) + \Pr(N-1 \text{ in } T)\Pr(1 \text{ in } \Delta T) \\ &= \Pr(N \text{ in } T)(1 - a\Delta T) + \Pr(N-1 \text{ in } T)a\Delta T\,.\end{aligned} \tag{11.11}$$

If we presume $\Pr(N-1 \text{ in } T)$ is known, we obtain the recursion relation

$$\frac{d}{dT}\Pr(N \text{ in } T) = -a\Pr(N \text{ in } T) + a\Pr(N-1 \text{ in } T)\,. \tag{11.12}$$

It can be verified by substitution (or induction) that the solution is

$$\Pr(N \text{ in } T) = \frac{(aT)^N}{N!}\exp(-aT)\,. \tag{11.13}$$

Equations (11.6) and (11.10) are special cases of (11.13) for $N = 0$ and 1, respectively. (Note that $0! \equiv 1$.)

From the discussion in Sec. C.6.2 of App. C, (11.13) is recognized as the Poisson probability law with parameter aT. As shown in the appendix, the parameter is also the mean of a Poisson distribution. That the mean number of counts in time T is aT should come as no surprise; from (11.1) a is the mean rate, and mean rate times total time is the mean total number of counts.

Time-dependent rate There are many circumstances in which the number of counts in a small interval $(t, t + \Delta T)$ depends on t as well as ΔT. One simple example is radioactive decay, where the mean rate $a(t)$ decays as $2^{-t/\tau}$, where τ is the half-life. A fundamentally different example is an unstable light source, where $a(t)$ fluctuates randomly. The random case is considered briefly in Sec. 11.1.4 and in more detail in Sec. 11.3.7, but we investigate here the case where $a(t)$ is a prescribed function of time, such as an exponential decay. The derivation parallels the one just given for $a =$ constant, but now a more elaborate notation is required. The probability of getting N counts in a time interval $(t, t + T)$ will be denoted $\Pr[N \text{ in } (t, t + T)]$.

If $a(t)$ is a nonrandom function of time, (11.3) becomes

$$\begin{aligned}\Pr[0 \text{ in } (t, t + T + \Delta T)] &= \Pr[0 \text{ in } (t, t + T)] \Pr[0 \text{ in } (t + T, t + T + \Delta T)] \\ &= \Pr[0 \text{ in } (t, t + T)] \left[1 - a(t + T)\Delta T\right]. \end{aligned} \tag{11.14}$$

Thus the differential equation (11.5) becomes

$$\frac{d}{dT} \Pr[0 \text{ in } (t, t + T)] = -a(t + T) \Pr[0 \text{ in } (t, t + T)], \tag{11.15}$$

with the solution

$$\Pr[0 \text{ in } (t, t + T)] = \exp\left[-\int_t^{t+T} dt'\, a(t')\right]. \tag{11.16}$$

To verify that (11.16) solves (11.15), one uses Leibniz' rule, according to which

$$\frac{d}{dT} \int_t^{t+T} dt'\, a(t') = a(t + T). \tag{11.17}$$

Equation (11.16) differs from (11.6) in that the product of the mean rate and the elapsed time has been replaced by the time-integral of the rate. If $a(t)$ is constant, (11.6) is recovered easily. The remainder of the derivation follows the derivation for the stationary case ($a =$ constant) with similar modifications. The final result is

$$\Pr[N \text{ in } (t, t + T)] = \frac{\overline{N}^N}{N!} \exp(-\overline{N}), \tag{11.18}$$

where now $\overline{N}$ is a function of time given by

$$\overline{N} = \int_t^{t+T} dt'\, a(t'). \tag{11.19}$$

The key point is that a Poisson law for the number of counts is obtained even if the mean rate is a function of time; all that is required is that the expected number of counts be computed by integrating the rate. We emphasize, however, that this conclusion requires that the mean rate be nonrandom; it does not hold if $a(t)$ fluctuates randomly.

11.1.2 Poisson and rarity

Another route to the Poisson law — indeed, the one used by Poisson himself — is to take the limit of a binomial law. Consider a set of M Bernoulli trials (coin flips, say), where the probability of success (heads) is p. We know that the total number N of successes in M trials is given by the binomial law, (C.161). We denote this probability as $\Pr(N|M,p)$, and we recall from (C.163) that the mean number of successes, $\overline{N}$, is given by Mp.

Now let M become very large and p become very small in such a way that Mp (or $\overline{N}$) remains constant. Since the probability of success is getting vanishingly small, success is a rare event. In the context of coin flipping, very few of the flips (on average, $\overline{N}/M$) result in heads.

With $p = \overline{N}/M$ and some algebra, the binomial law can be written as

$$\Pr(N|M,p) = \frac{M!}{(M-N)!N!}p^N(1-p)^{M-N}$$

$$= \left[\frac{M(M-1)\cdots(M-N+1)}{M^N}\left(1-\frac{\overline{N}}{M}\right)^{-N}\right]\left(1-\frac{\overline{N}}{M}\right)^M\frac{\overline{N}^N}{N!}. \tag{11.20}$$

As $M \to \infty$, the factor in square brackets tends to 1. Also, since $(1-\overline{N}/M)^M$ limits to $\exp(-\overline{N})$, the remaining factors limit to the Poisson law, *i.e.*,

$$\lim_{M\to\infty}\Pr(N|M,p) = \frac{(\overline{N})^N}{N!}\exp(-\overline{N}), \qquad (\overline{N} = Mp = const). \tag{11.21}$$

This formula gives the correct mean $\overline{N}$ by construction, and it also gives the correct variance. From App. C, we know that the variance of a binomial law is $Mp(1-p)$, which here is $\overline{N}(1-p)$. In the limit $p \to 0$, this variance approaches $\overline{N}$, as it must for a Poisson.

Another way to see that the binomial limits to a Poisson is to show that the Poisson postulates are satisfied in the limit. Consider a sequence of coin flips at a regular pace, one every ΔT seconds. The total time over which the flips occur is held constant at T, so the number of flips $M = T/\Delta T$ goes to infinity as $\Delta T \to 0$. As before, the probability of success is taken as $p = \overline{N}/M$, where $\overline{N}$ is constant.

Since the flips are independent, the number in any time interval is statistically independent of the number in any nonoverlapping interval, providing we consider only intervals given by integer multiples of ΔT. In the limit as $\Delta T \to 0$, this restriction is irrelevant and the first postulate holds. Similarly, since $p = \overline{N}/M = \overline{N}\Delta T/T$, the second postulate holds with $a = \overline{N}/T$. Finally, the third postulate holds easily since there is only one flip in any interval of length ΔT. Thus all three postulates are satisfied in the limit, and the Poisson law is inevitable.

11.1.3 Binomial selection of a Poisson

In most discussions of the binomial law, it is implicitly assumed that the number of trials M is a predetermined constant. In many physical situations, however, M is random. An important example in optics is radiation detection by an inefficient photon-counting detector. We analyze this problem here using the intuitive picture where discrete photons fall on the detector and produce photoelectrons with some probability; in Sec. 11.5 we shall revisit the problem from the viewpoint of quantum electrodynamics.

Suppose M photons are incident on the detector in time T and that each has a probability η (called the *quantum efficiency*) of producing a photoelectron. If the photoelectric interactions are statistically independent, then the *conditional* probability $\Pr(N|M)$ for getting N photoelectrons is a binomial; the problem is formally equivalent to Bernoulli trials with η being the probability of success (detection). The margninal probability $\Pr(N)$, however, is given by

$$\Pr(N) = \sum_{M=N}^{\infty} \Pr(N|M) \Pr(M) . \tag{11.22}$$

Note the lower limit of the sum; getting N photoelectrons requires that there be at least N photons.

Now suppose that the photons satisfy the three postulates so that M is a Poisson random variable. With $\Pr(N|M)$ being binomial, we have

$$\Pr(N) = \sum_{M=N}^{\infty} \binom{M}{N} \eta^N (1-\eta)^{M-N} \exp(-\overline{M}) \frac{\overline{M}^M}{M!} . \tag{11.23}$$

A change of variables, $K = M - N$, and a little algebra shows that (Barrett and Swindell, 1981, 1996)

$$\Pr(N) = \exp(-\eta\overline{M}) \frac{(\eta\overline{M})^N}{N!} . \tag{11.24}$$

Thus N obeys a Poisson law with mean $\overline{N} = \eta\overline{M}$. This result should come as no surprise; if the postulates are satisfied for M, they are also satisfied for N. No statistical dependence is introduced by the selection process.

This fundamental result is known as *binomial selection theorem*; we restate it for emphasis as:

> *Binomial selection of a Poisson yields a Poisson, and the mean of the output of the selection process is the mean of the input times the binomial probability of success.*

An interesting corollary of this theorem is that the number of times that the photon is *not* detected is also a Poisson random variable. Moreover, the number of nondetections is statistically independent of the number of detections. These statements will come as a surprise to someone used to thinking about Bernoulli trials in terms of coin flipping. If a coin with probability of heads equal to η is flipped exactly M times, the number of heads N_h is binomial with parameter η, the number of tails N_t is binomial with parameter $1 - \eta$, and N_h and N_t are not statistically independent since $N_h + N_t = M$. The situation is entirely different when M itself is a Poisson random variable. Then N_h is Poisson with parameter $\eta\overline{M}$, N_t is Poisson with parameter $(1-\eta)\overline{M}$, and N_t and N_h are statistically independent. This peculiar result is obtained *only* if M is Poisson-distributed. Haight (1967) states it in formal terms as follows: In a sequence of Bernoulli trials where the number of attempts is random, the number of successes is independent of the number of failures if and only if the number of attempts is a Poisson random variable.

Cascaded binomial selection Many situations in imaging can be described mathematically as a cascade of binomial-selection stages. Consider, for example, the optical system depicted in Fig. 11.1. A light bulb emits photons randomly in all directions, some of the photons are collected with a lens, pass through a neutral-density filter, and illuminate the photocathode of a photomultiplier tube. We assume that the photons are fully independent, in the sense that neither the direction nor the time of emission of one photon has any influence on the properties of any other photon, and hence that the photons satisfy the Poisson postulates. The number of photons M emitted in some observation time T is then a Poisson random variable with mean $\overline{M}$.

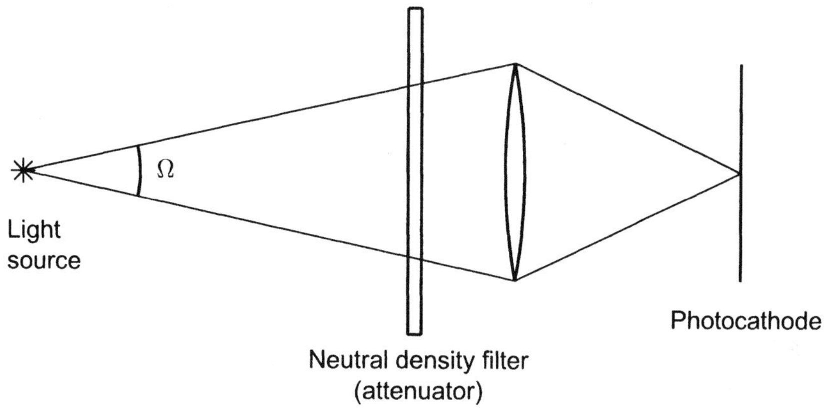

Fig. 11.1 An optical system to illustrate the idea of binomial selection of a Poisson distribution.

Since the photons are independent in all respects, including direction, collection of photons by the lens can be treated as a binomial selection. If the lens subtends a solid angle Ω from the source and the emission is isotropic, on average a fraction $\Omega/4\pi$ of the photons pass through the lens. The binomial probability of success p in this case is $\Omega/4\pi$, and the mean number of photons passed by the lens is $\overline{M}\Omega/4\pi$. Since we know that the binomial selection of a Poisson yields a Poisson, and that a Poisson is fully characterized by its mean, we now have the full probability law for the number of photons passed by the lens: it is a Poisson with mean $\overline{M}\Omega/4\pi$.

The neutral-density filter is another binomial selection. Each photon is either passed by the filter or it is not. If the average transmittance of the filter is τ, then the probability of success (in getting through) for an individual photon is also τ, and the mean number passed by the filter is $\overline{M}\tau\Omega/4\pi$. Again we have a binomial selection of a Poisson, which yields a Poisson, so the probability law on the number of photons passing the filter is Poisson with mean $\overline{M}\tau\Omega/4\pi$.

Next we come to the photocathode, which we assume to have quantum efficiency η. That is, each photon can independently produce a photoelectron with probability η. This is another binomial selection, and again the input to the selection is a Poisson, so we know at once that the probability law on the number of photoelectrons is Poisson with mean $\overline{M}\tau\eta\Omega/4\pi$.

We can extend this kind of reasoning indefinitely. If we know that at some stage in a system we have photons (or particles of any nature) that satisfy the Poisson postulates and that subsequent stages in the chain are well modeled as binomial selections, we need only compute the *average* number of photons passing through, and the binomial-selection theorem will then demonstrate that the probability law on this number is Poisson with the computed mean. Since the Poisson law is so ubiquitous, this is a very powerful means of analyzing the statistical properties of imaging systems.

11.1.4 Doubly stochastic Poisson random variables

A random variable for which a parameter of the probability law is itself random is said to be *doubly stochastic.* In this section we discuss Poisson random variables where the Poisson mean is random. Such variables are called *doubly stochastic Poisson random variables.* The closely related topic of doubly stochastic Poisson random *processes* is discussed in Secs. 11.3.6 and 11.3.7.

To understand the usefulness of doubly stochastic Poisson random variables, suppose that a large supply of light bulbs is available for use with a photon-counting detector. Assume that each bulb produces a steady light output and that the photons satisfy the Poisson postulates. Then, as we have seen, the counting distribution is Poisson so long as only that light bulb is used. In an experiment where the number of counts N in a fixed interval T is observed, N is a Poisson random variable. That means that if we repeat the experiment a large number of times (always with the same light bulb), a histogram of the observed values of N would approach $\Pr(N|\overline{N})$, where $\overline{N}$ is related to the brightness of the particular light bulb. A different bulb produces a different amount of light, however, so this is only a conditional probability law, conditional on the particular bulb.

On the other hand, if we used a different light bulb for each measurement, then the histogram of observed counts would be broadened because of the variation in bulb output (see Fig. 11.2). To describe this situation fully, we now need to account for two kinds of randomness, the Poisson randomness in photon counting and the manufacturing randomness in bulb output. To do so, we allow $\overline{N}$ to be a random variable.

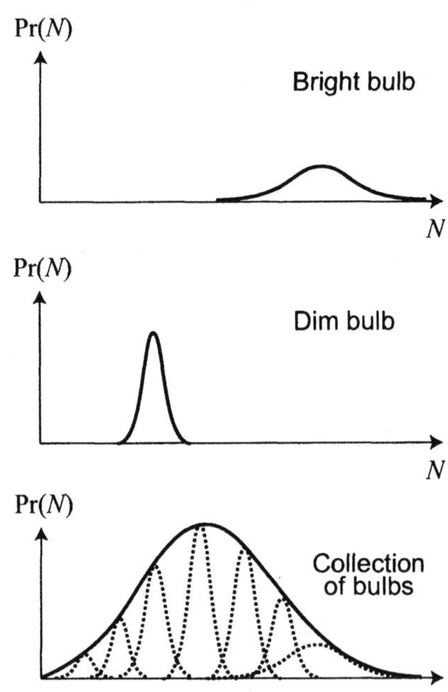

Fig. 11.2 Illustration of the broadening of Pr(N) that occurs when the Poisson rate is a random variable.

Another situation in which $\overline{N}$ is random is when only one light source is used but the mean count rate $a(t)$ is random. For any particular realization of $a(t)$, the calculation in Sec. 11.1.1 holds and the probability law on N is a Poisson with mean $\overline{N}$ given by the integral in (11.19). If $a(t)$ is random, however, $\overline{N}$ is also necessarily random. In Sec. 11.3.7 we develop the tools needed to discuss this case fully, but we shall now consider what happens if we simply let $\overline{N}$ be a random *variable* (ultimately derived from some random *process*), with an unspecified probability law.

Poisson transform Though N is a discrete random variable with only integer values, its mean $\overline{N}$ can take on any value in $(0, \infty)$. Thus, when we consider $\overline{N}$ to be random, it must be described by a probability density function $\text{pr}(\overline{N})$. The usual rule for computing a marginal from a joint density, (C.75), then leads to

$$\Pr(N) = \int_0^\infty d\overline{N}\, \Pr(N|\overline{N})\, \text{pr}(\overline{N}) = \frac{1}{N!}\int_0^\infty d\overline{N}\, \overline{N}^N \exp(-\overline{N})\, \text{pr}(\overline{N})\,. \qquad (11.25)$$

This expression is called the *Poisson transform* of $\text{pr}(\overline{N})$. Many properties of Poisson transforms and a table of examples are given in (Saleh, 1978). Of note is the fact that the Poisson transform of an exponential is a Bose-Einstein; the relevance of this result to optics will be explored in Sec. 11.5.3.

Though it is necessary to carry out the integral in the Poisson transform if one wants the full probability law on N, the mean and variance of N can be obtained more simply. The mean is given by

$$\langle N\rangle = \sum_{N=0}^{\infty} N \Pr(N) = \sum_{N=0}^{\infty} \frac{N}{N!}\int_0^\infty d\overline{N}\, \overline{N}^N \exp(-\overline{N})\, \text{pr}(\overline{N})\,. \qquad (11.26)$$

Interchanging order of summation and integration, we find

$$\langle N\rangle = \int_0^\infty d\overline{N} \left[\sum_{N=0}^{\infty} \frac{N}{N!}\overline{N}^N \exp(-\overline{N})\right] \text{pr}(\overline{N})\,. \qquad (11.27)$$

The quantity in square brackets is recognized as the mean of the Poisson distribution with parameter $\overline{N}$, and we know from App. C that this mean is just the parameter. Therefore,

$$\langle N\rangle = \int_0^\infty d\overline{N}\, \overline{N}\, \text{pr}(\overline{N}) \equiv \overline{\overline{N}}\,. \qquad (11.28)$$

The double-overbar notation indicates that two separate averages are being performed, one with respect to the probability $\Pr(N|\overline{N})$ and the other with respect to the probability density function $\text{pr}(\overline{N})$.

Computation of the second moment of N proceeds similarly. After the interchange of order of summation and integration, we have

$$\langle N^2\rangle = \int_0^\infty d\overline{N} \left[\sum_{N=0}^{\infty} \frac{N^2}{N!}\overline{N}^N \exp(-\overline{N})\right] \text{pr}(\overline{N}) = \int_0^\infty d\overline{N} \left(\overline{N} + \overline{N}^2\right) \text{pr}(\overline{N})\,, \qquad (11.29)$$

where we have used the result from (C.168) for the second moment of a Poisson.

We can express (11.29) in a neater form by recognizing that, for any random variable x,

$$\langle x^2\rangle = \text{Var}(x) + \langle x\rangle^2\,. \qquad (11.30)$$

For the problem at hand, x is $\overline{N}$, so

$$\langle N^2 \rangle = \overline{\overline{N}} + \text{Var}(\overline{N}) + \overline{\overline{N}}^2 . \tag{11.31}$$

Finally,

$$\text{Var}(N) = \langle N^2 \rangle - \overline{\overline{N}}^2 = \overline{\overline{N}} + \text{Var}(\overline{N}) . \tag{11.32}$$

The first term on the right is just the Poisson variance appropriate to the average value of the mean, while the second term, often called the *excess variance*, is the result of randomness in the Poisson mean.

Another way to derive (11.32) is to apply a general result from Sec. C.4.4 of Appendix C. In the notation used there, $\text{Var}_{\overline{N}}\text{E}\{N|\overline{N}\} = \text{Var}\{\overline{N}\}$ and $\text{E}_{\overline{N}}\text{Var}\{N|\overline{N}\} = \overline{\overline{N}}$, so (11.32) is equivalent to (C.84).

One trivial case of this formalism is when $\overline{N}$ is not really random. In that case, $\text{pr}(\overline{N}) = \delta(\overline{N} - \overline{N}_0)$ and $\text{Var}(\overline{N}) = 0$. Then $\Pr(N)$ is just a Poisson of mean $\overline{N}_0$, so $\text{Var}(N) = \overline{N}_0$ and the excess variance vanishes.

For any $\text{pr}(\overline{N})$ other than a delta function, however, the excess variance is positive, and the variance associated with $\Pr(N)$ is greater than that associated with $\Pr(N|\overline{N})$. Within the confines of the theory as developed so far, therefore, the *minimum* variance of any counting process is its mean value, and this minimum is achieved only when the mean itself is nonrandom, *i.e.*, for a Poisson random variable. Any randomness of the mean will lead to an increase in the variance over the Poisson value. Curiously, this statement breaks down in the world of quantum optics, where sub-Poisson statistics are indeed possible. This point will be discussed further in Sec. 11.5.

Binomial selection from a non-Poisson source We now revisit the binomial-selection theorem for the case where the source does not obey Poisson statistics (Barrett and Swindell, 1981, 1996). The probability law on N is given by (11.22), where $\Pr(N|M)$ is a binomial with probability of success η and $\Pr(M)$ is unspecified. With this problem statement and some properties of binomial distributions, the mean of N is given by

$$\langle N \rangle = \sum_{N=0}^{\infty} \sum_{M=N}^{\infty} N \Pr(N|M) \Pr(M) = \sum_{M=0}^{\infty} \sum_{N=0}^{M} N \Pr(N|M) \Pr(M)$$

$$= \sum_{M=0}^{\infty} M\eta \Pr(M) = \overline{M}\eta , \tag{11.33}$$

and the second moment is given by

$$\langle N^2 \rangle = \sum_{M=0}^{\infty} \sum_{N=0}^{M} N^2 \Pr(N|M) \Pr(M)$$

$$= \sum_{M=0}^{\infty} \left[M\eta(1-\eta) + M^2\eta^2\right] \Pr(M) = \overline{M}\eta(1-\eta) + \eta^2 \left[\text{Var}(M) + \overline{M}^2\right] , \tag{11.34}$$

where we have used (11.30). Finally, the variance of N is

$$\text{Var}(N) = \langle N^2 \rangle - \overline{N}^2 = \eta(1-\eta)\overline{M} + \eta^2 \, \text{Var}(M) . \tag{11.35}$$

There are several interesting features of (11.35). First, if $\text{Var}(M) = 0$ so that M is not really random, (11.35) shows that $\text{Var}(N) = \eta(1-\eta)M$, just as we would expect from the binomial law. Second, if η is small, (11.35) can be approximated as

$$\text{Var}(N) \approx \overline{N}\,, \tag{11.36}$$

just as we would expect for a Poisson random variable. For η small enough, the excess variance is negligible.

Additional insight can be obtained by using $\overline{N} = \eta\overline{M}$ and rewriting (11.35) as (Barrett and Swindell, 1981, 1996)

$$\text{Var}(N) - \overline{N} = \eta^2\left[\text{Var}(M) - \overline{M}\right]. \tag{11.37}$$

This form shows that a sufficient condition for having $\text{Var}(N) = \overline{N}$ is that $\text{Var}(M) = \overline{M}$. This condition is, of course, satisfied if M is Poisson, in which case (11.37) is in accord with the binomial selection theorem. We see also from (11.37) that $\text{Var}(N) \to \overline{N}$ as $\eta \to 0$, regardless of the statistics of M. Once again, rarity begets Poissonicity.

Equations (11.35) and (11.37) are general results for binomial selection where the input to the selection process, M, has arbitrary statistics. If M is specifically a doubly stochastic Poisson random variable, its mean is denoted $\overline{\overline{M}}$ and its variance is given by [*cf.* (11.32)]

$$\text{Var}(M) = \langle M^2\rangle - \overline{\overline{M}}^2 = \overline{\overline{M}} + \text{Var}(\overline{M}). \tag{11.38}$$

Inserting these results into (11.35), we obtain

$$\text{Var}(N) = \eta(1-\eta)\overline{\overline{M}} + \eta^2\left[\overline{\overline{M}} + \text{Var}(\overline{M})\right] = \eta\overline{\overline{M}} + \eta^2\,\text{Var}(\overline{M}). \tag{11.39}$$

Now the factor $\eta(1-\eta)$ usually seen in binomial variances does not appear explicitly. Instead, we see that the Poisson part of the variance scales as η while the excess variance scales as η^2. Again, small efficiency reduces the excess variance relative to the Poisson part.

11.2 POISSON RANDOM VECTORS

Many photon-counting detectors have a discrete output array, where the output value in each element is the number of counts that occurred within the area of the element during the exposure time. For simplicity, we can think of an array of discrete detector elements, each with its own electronic system and counter. Suppose that there are J elements and let the number of photons detected by element j be denoted g_j, $(j = 1, ..., J)$. We wish to find a probability law for the set $\{g_j\}$. In this section we shall call this probability $\Pr(\{g_j\})$, but later it will be convenient to consider the set $\{g_j\}$ as components of a J-dimensional vector $\mathbf{g}$, so the probability can be denoted more compactly as $\Pr(\mathbf{g})$.

11.2.1 Multivariate Poisson statistics

It is easy to find $\Pr(\{g_j\})$ for a Poisson source since it emits photons independently. Suppose that detector element j has probability P_j of detecting a photon from the source, and that the mean number emitted by the source during time T is the nonrandom quantity $\overline{M}$. Then, by the binomial-selection theorem (Sec. 11.1.3), g_j is a Poisson with mean $\overline{g}_j = P_j\overline{M}$.

Since the photons arriving at different elements are independent, the multivariate probability law $\Pr(\{g_j\})$ on the set of all counts $\{g_j, j = 1, ..., J\}$ is a product of Poissons of the form

$$\Pr(\{g_j\}) = \prod_{j=1}^{J} \exp(-P_j\overline{M}) \frac{(P_j\overline{M})^{g_j}}{g_j!} = \prod_{j=1}^{J} \exp(-\overline{g}_j) \frac{(\overline{g}_j)^{g_j}}{g_j!} . \tag{11.40}$$

Because of this product form, counts in different elements are uncorrelated. Also, the variance of the counts in one element is equal to its mean, so we can write the covariance matrix elements as

$$K_{jk} = \langle \Delta g_j \Delta g_k \rangle = \overline{g}_j \, \delta_{jk} , \tag{11.41}$$

where $\Delta g_j = g_j - \overline{g}_j$. This relation will prove very useful in later chapters where we discuss statistical properties of images.

Multinomial statistics Another way of obtaining (11.40) is by generalizing the discussion in Sec. 11.1.3 on binomial selection of a Poisson. Suppose that the total number of photons detected by an array is N and that element j has probability α_j of getting any particular photon. If the photons are independent, the conditional multivariate probability law on g_j for fixed N is the multinomial (see App. C):

$$\Pr(\{g_j\}|N) = N! \prod_{j=1}^{J} \frac{(\alpha_j)^{g_j}}{g_j!} , \tag{11.42}$$

where

$$\sum_{j=1}^{J} g_j = N , \qquad \sum_{j=1}^{J} \alpha_j = 1 . \tag{11.43}$$

Note that (11.42) cannot be factored into independent probabilities because of the factor $N!$.

If N is itself a random variable, we have

$$\Pr(\{g_j\}) = \sum_{N=0}^{\infty} \Pr(\{g_j\}|N) \Pr(N) = \sum_{N=0}^{\infty} \Pr(N)\, N! \prod_{j=1}^{J} \frac{(\alpha_j)^{g_j}}{g_j!} \, \delta\left(N, \sum_{j=1}^{J} g_j \right) , \tag{11.44}$$

where $\delta(\cdot,\cdot)$ is another notation for the Kronecker delta. If $\Pr(N)$ is a Poisson of mean $\overline{N}$, we have

$$\Pr(\{g_j\}) = \sum_{N=0}^{\infty} \exp(-\overline{N}) \, (\overline{N})^N \prod_{j=1}^{J} \frac{(\alpha_j)^{g_j}}{g_j!} \, \delta\left(N, \sum_{j=1}^{J} g_j \right) . \tag{11.45}$$

Since $\overline{N} = \sum_j \overline{g}_j$, we can rewrite this expression as

$$\Pr(\{g_j\}) = \sum_{N=0}^{\infty} \prod_{j=1}^{J} \exp(-\overline{g}_j) \frac{(\overline{N}\alpha_j)^{g_j}}{g_j!} \, \delta\left(N, \sum_{j=1}^{J} g_j \right) . \tag{11.46}$$

The Kronecker delta allows us to perform the sum, with the result:

$$\Pr(\{g_j\}) = \prod_{j=1}^{J} \exp(-\alpha_j \overline{N}) \frac{(\alpha_j \overline{N})^{g_j}}{g_j!} . \tag{11.47}$$

We have thus generalized the binomial selection theorem:

> *Multinomial selection of a univariate Poisson yields a multivariate Poisson, and the mean of each cell after multinomial selection is the mean of the input times the probability of the event going to that cell.*

Equation (11.47) has the same form as (11.40), but it does not seem to be identical; $\alpha_j \overline{N}$ appears in (11.47) while $P_j \overline{M}$ appears in (11.40). (Recall that $\overline{N}$ is the total number of *detected* photons and α_j is the probability that one of them goes in detector j, while $\overline{M}$ is the total number of photons *emitted* by the source and P_j is the probability that an emitted photon is detected in detector j.) The distinction between $\alpha_j \overline{N}$ and $P_j \overline{M}$ is, however, illusory. Of the M photons emitted by the source, a fraction β on average is detected *somewhere* in the array. (As discussed in Sec. 11.1.3, β is the product of the solid-angle factor $\Omega/4\pi$ and the quantum efficiency η if the photons are emitted isotropically and independently, but we can just leave β as a general parameter, interpreted as the overall probability of detection.) Thus $\overline{N} = \beta \overline{M}$.

Moreover, under the same independence assumptions, N is a binomial selection from M, so $\Pr(N)$ is a Poisson of mean $\overline{N} = \beta \overline{M}$ if $\Pr(M)$ is a Poisson of mean $\overline{M}$. The counts g_j in any single detector element are obtained by another *binomial* selection with probability of success P_j, so the univariate probability $\Pr(g_j)$ is a Poisson of mean $P_j \overline{N} = \beta P_j \overline{M}$. Finally, we can argue from independence that the multivariate law $\Pr(\{g_j\})$ is just the product of the individual probabilities $\Pr(g_j)$ as given in (11.40) *or* (11.47); the two are equivalent since $\alpha_j \overline{N} = P_j \overline{M}$, with $P_j = \beta \alpha_j$.

Preset counts vs. preset time We used the multinomial law above as a mathematical device to derive the multivariate Poisson, but there are circumstances where the multinomial itself is the correct multivariate distribution.

In many event-counting systems, data can be acquired either for a given time or until a given number of events is reached. In the nuclear-medicine literature, these two acquisition modes are referred to as preset time and preset counts, respectively, and we shall adopt that terminology here also. The key distinction is that the total number of events N is a random variable for preset time but a fixed number for preset counts.

For preset counts, $\Pr(\{g_j\}|N)$ as given in (11.42) is directly the multivariate probability law for the set $\{g_j\}$. Even though the multinomial is derived on the assumption that the *events* are independent, the *numbers of counts* in different elements are not independent because of the constraint on the total number.

Multinomials and rarity Consider a preset-counts acquisition by a detector array where the number of elements J is large and the counts are spread fairly evenly over the elements, so that α_j is small for all j. Small α_j means that the event of a count going into the jth detector is rare. Since we almost always collect a large total number of counts N, the arguments of Sec. 11.1.2 apply, and the univariate law on any g_j is well approximated by a Poisson of mean $N\alpha_j$.

It is a little trickier to see how the multinomial limits to a multivariate Poisson. The problem is that the total number of counts is fixed, so any one of the g_j is fully determined by the sum of all of the others. Thus the variance of one of the g_j conditional on all of the others is zero.

Let us arbitrarily (and without loss of generality) single out the detector with index $j = J$ as the one with zero conditional variance. Then, with algebra similar to that in (11.20), (11.42) becomes

$$\Pr(\{g_j\}|N) = \delta\left(g_J, N - \sum_{j=1}^{J-1} g_j\right) \prod_{j=1}^{J-1} \exp(-\alpha_j N) \frac{(\alpha_j N)^{g_j}}{g_j!} . \tag{11.48}$$

If we leave out one detector, the probability law on the rest of them is multivariate Poisson in this approximation. Even for the one we leave out, it is easy to show that the mean equals the variance:

$$\langle g_J \rangle = N\alpha_J , \qquad \mathrm{Var}(g_J) = \langle g_J \rangle = N\alpha_J . \tag{11.49}$$

Moreover, to the same approximation, the covariance matrix is still given by the Poisson covariance, (11.41).

Once again, we see two routes to the Poisson: independence and rarity. In a preset-time acquisition mode, the counts are independent and the Poisson results from binomial-selection arguments. In a preset-counts mode, strict independence is lost but the Poisson is an excellent approximation if counts in any one detector are rare compared to total counts.

11.2.2 Doubly stochastic multivariate statistics

The lack of correlation and the product form in (11.40) are consequences of the Poisson (independent) nature of the source. They do not apply when the source is random. There are two distinct kinds of randomness that could invalidate those results: either the overall source strength could change from measurement to measurement, or the spatial configuration of the source could change. The first circumstance could occur in passive imaging where a static object is illuminated with a fluctuating external source; then the effective source strength is a constant spatial distribution of reflectance or scattering amplitude times a random illumination. The second circumstance could occur if an ensemble of objects is imaged with a steady source; the statistical properties of interest then include both counting statistics and object statistics.

Variable source strength Assume first that the source fluctuates only in overall strength, and that the spatial distribution of radiation on the detector array is constant. This means that the probabilities P_j are not random variables, and the only random variable describing the source is the strength $\overline{M}$.

Under these assumptions, the joint probability $\Pr(\{g_j\})$ is obtained simply by averaging (11.40) over the source fluctuations, *i.e.*,

$$\begin{aligned}\Pr(\{g_j\}) &= \int_0^\infty d\overline{M}\ \mathrm{pr}(\overline{M}) \Pr(\{g_j\}\,|\overline{M}) \\ &= \int_0^\infty d\overline{M}\ \mathrm{pr}(\overline{M}) \prod_{j=1}^{J} \exp(-P_j\overline{M}) \frac{(P_j\overline{M})^{g_j}}{g_j!} .\end{aligned} \tag{11.50}$$

Any marginal $\Pr(g_i)$ can be recovered by summing (11.50) over all g_j for $j \neq i$. All of the sums are one by the normalization of the Poisson probability, and we are left with

$$\Pr(g_i) = \int_0^\infty d\overline{M}\ \mathrm{pr}(\overline{M}) \exp(-P_i\overline{M}) \frac{(P_i\overline{M})^{g_i}}{g_i!} . \tag{11.51}$$

This equation is a Poisson transform, just as in our original discussion of doubly stochastic Poisson random variables [*cf.* (11.25)]. Indeed, we could have obtained it just by ignoring all other elements.

Of more interest is a pairwise marginal like $\Pr(g_i, g_k)$ for $i \neq k$. Summing (11.50) over all of the g_j except $j = i$ and $j = k$, we find

$$\Pr(g_i, g_k) = \int_0^\infty d\overline{M}\ \mathrm{pr}(\overline{M}) \exp(-P_i\overline{M}) \frac{(P_i\overline{M})^{g_i}}{g_i!} \exp(-P_k\overline{M}) \frac{(P_k\overline{M})^{g_k}}{g_k!} . \tag{11.52}$$

Various moments can be obtained from this probability via methods used previously. It will be left as an exercise to show that

$$\langle g_i \rangle = P_i\overline{\overline{M}} ; \tag{11.53}$$

$$\langle g_i g_k \rangle = P_k\overline{\overline{M}}\,\delta_{ik} + P_i P_k \left[\mathrm{Var}(\overline{M}) + \overline{\overline{M}}^2 \right] ; \tag{11.54}$$

$$[\mathbf{K_g}]_{ik} = \langle \Delta g_i \Delta g_k \rangle = P_k\overline{\overline{M}}\,\delta_{ik} + P_i P_k \,\mathrm{Var}(\overline{M}) , \tag{11.55}$$

where $\Delta g_i = g_i - \langle g_i \rangle$. Another useful form for the covariance is obtained by use of (11.32) for $\mathrm{Var}(\overline{M})$ (see Barrett and Swindell, 1981, Sec. 3.4, 1996), which leads to

$$[\mathbf{K_g}]_{ik} = P_k\overline{\overline{M}}\,\delta_{ik} + P_i P_k \left[\mathrm{Var}(M) - \overline{\overline{M}} \right] . \tag{11.56}$$

Equation (11.55) shows that the off-diagonal elements of $\mathbf{K_g}$ vanish if $\overline{M}$ is nonrandom, which means that the individual counts g_i and g_k are Poisson. Equation (11.56) shows that the off-diagonal elements also vanish if $\mathrm{Var}(M) = \overline{\overline{M}}$; substituting M for N in (11.32), we see that this condition can occur if and only if the density on $\overline{M}$ is a delta function, again implying that the individual counts are Poisson. In addition, both (11.55) and (11.56) show that the off-diagonal elements are small if $P_i P_k$ is small so that detection events are rare (compared to emission events).

Variable source distribution Next we consider what happens if the spatial configuration of the source is random. Different realizations of the source produce different values of the mean counts $\{\overline{g}_j\}$. A full description of the source randomness now requires a specification of the multivariate density $\mathrm{pr}(\{\overline{g}_j\})$, not just the univariate $\mathrm{pr}(\overline{M})$. Note that variable source strength, treated above, is a special case of variable source distribution where all of the $\overline{g}_j$ covary together as a result of the common factor $\overline{M}$. Here we analyze the more general case.

The joint probability $\Pr(\{g_j\})$ is obtained by averaging (11.40) over the set $\{\overline{g}_j\}$, and (11.50) generalizes to

$$\Pr(\{g_j\}) = \int_0^\infty d\overline{g}_1 \int_0^\infty d\overline{g}_2 \cdots \int_0^\infty d\overline{g}_J \; \mathrm{pr}(\{\overline{g}_j\}) \Pr(\{g_j\}|\{\overline{g}_j\})$$

$$= \int_0^\infty d\overline{g}_1 \int_0^\infty d\overline{g}_2 \cdots \int_0^\infty d\overline{g}_J \; \mathrm{pr}(\{\overline{g}_j\}) \prod_{j=1}^{J} \exp(-\overline{g}_j) \frac{(\overline{g}_j)^{g_j}}{g_j!} \,. \tag{11.57}$$

This expression appears formidable, but it is actually fairly easy to obtain means and covariances from it. Showing all of the steps for pedagogical purposes, we find the mean of g_k via

$$\langle g_k \rangle = \sum_{g_1=0}^{\infty} \cdots \sum_{g_k=0}^{\infty} \cdots \sum_{g_J=0}^{\infty} g_k \Pr(\{g_j\})$$

$$= \int_0^\infty d\overline{g}_1 \int_0^\infty d\overline{g}_2 \cdots \int_0^\infty d\overline{g}_J \; \mathrm{pr}(\{\overline{g}_j\}) \sum_{g_1=0}^{\infty} \cdots \sum_{g_k=0}^{\infty} \cdots \sum_{g_J=0}^{\infty} g_k \prod_{j=1}^{J} \exp(-\overline{g}_j) \frac{(\overline{g}_j)^{g_j}}{g_j!}$$

$$= \int_0^\infty d\overline{g}_1 \int_0^\infty d\overline{g}_2 \cdots \int_0^\infty d\overline{g}_J \; \mathrm{pr}(\{\overline{g}_j\}) \sum_{g_k=0}^{\infty} g_k \exp(-\overline{g}_k) \frac{(\overline{g}_k)^{g_k}}{g_k!}$$

$$= \int_0^\infty d\overline{g}_k \; \mathrm{pr}(\overline{g}_k)\, \overline{g}_k = \overline{\overline{g}}_k \,. \tag{11.58}$$

Along the way, we have used the normalization of the Poisson law (from line 2 to line 3) and the definition of marginal densities (line 3 to line 4). The final result, of course, could have been written down virtually by inspection.

Similar manipulations yield the second moment. Under Poisson statistics alone, $\langle g_i g_k \rangle = \overline{g}_i \, \delta_{ik} + \overline{g}_i \overline{g}_k$, but with the additional randomness in the means, we have

$$\langle g_i g_k \rangle = \int_0^\infty d\overline{g}_i \int_0^\infty d\overline{g}_k \; \mathrm{pr}(\overline{g}_i, \overline{g}_k) \left[\overline{g}_i \, \delta_{ik} + \overline{g}_i \overline{g}_k\right] = \overline{\overline{g}}_i \, \delta_{ik} + [\mathbf{K}_{\overline{\mathbf{g}}}]_{ik} + \overline{\overline{g}}_i \overline{\overline{g}}_k \,, \tag{11.59}$$

where $\mathbf{K}_{\overline{\mathbf{g}}}$ is the covariance matrix for the means,

$$[\mathbf{K}_{\overline{\mathbf{g}}}]_{ik} = \left\langle \left[\overline{g}_i - \overline{\overline{g}}_i\right] \left[\overline{g}_k - \overline{\overline{g}}_k\right] \right\rangle \,. \tag{11.60}$$

The final covariance matrix for the counts themselves is

$$[\mathbf{K}_{\mathbf{g}}]_{ik} = \left\langle \left[g_i - \overline{\overline{g}}_i\right] \left[g_k - \overline{\overline{g}}_k\right] \right\rangle = \overline{\overline{g}}_i \, \delta_{ik} + [\mathbf{K}_{\overline{\mathbf{g}}}]_{ik} \,. \tag{11.61}$$

Several limits and special cases of this result are of interest. First, if none of the means is random, we are back to Poisson statistics, and the covariance matrix has the diagonal form given in (11.41). Second, if the randomness in the means comes from randomness in overall source strength without any variation in source configuration, then all of the means covary together, $[\mathbf{K}_{\overline{\mathbf{g}}}]_{ik}$ is $P_i P_k \,\mathrm{Var}(\overline{M})$, and (11.55) is recovered.

Finally, as in the case of randomness in overall source strength [*cf.* (11.55)], the two terms in (11.61) have different dependencies on average number of counts. The first term $\overline{\overline{g}}_i \, \delta_{ik}$ varies as the mean number of counts, which means it is linear in source strength, detector efficiency or counting time. The second term $[\mathbf{K}_{\overline{\mathbf{g}}}]_{ik}$ is quadratic in all of these quantities, so it becomes relatively more important as mean counts increase. Conversely, when there are relatively few counts, the second term can be neglected and the covariance matrix $\mathbf{K_g}$ is diagonal in spite of randomness in the individual means $\overline{g}_j$. Thus Poisson sources always produce uncorrelated, Poisson-distributed counts, and non-Poisson sources produce approximately uncorrelated and approximately Poisson-distributed counts in inefficient detectors.

11.3 RANDOM POINT PROCESSES

A random point process is a random process for which each sample function is localized to a set of spatial or temporal points (Snyder and Miller, 1991). A sample function of a point process is thus a sum of delta functions in some number of dimensions. The amplitude and arguments of the delta functions can be random variables, and the number of terms in the sum can be random as well.

In this section we develop the mathematical tools necessary to describe the first- and second-order statistics of point processes. As in Sec. 11.1, we cast the discussion in terms of radiation falling on a photon-counting detector since this is the most common manifestation of point processes in optics and imaging.

11.3.1 Temporal point processes

Consider first a detector that produces a current pulse for each photoelectric interaction, and assume that this current pulse is independent of the spatial coordinates of the event. If the properties of the detector and its associated electronic circuitry are time-invariant (though the radiation source need not be temporally stationary), the current waveform can be expressed as

$$i(t) = \sum_{n=1}^{N} i_0(t - t_n)\,, \tag{11.62}$$

where $i_0(t)$ is the current pulse produced by a count at $t = 0$, t_n is the time of occurrence of the nth count $(0 < t_n \leq T)$, and N is the total number of counts in $(0, T)$.

This current waveform can be written as

$$i(t) = z(t) * i_0(t)\,, \tag{11.63}$$

where $z(t)$ is a random point process defined by

$$z(t) = \sum_{n=1}^{N} \delta(t - t_n)\,, \qquad (0 < t \leq T)\,. \tag{11.64}$$

Since the delta function is a generalized function, $z(t)$ is a generalized random process. Once the statistical properties of $z(t)$ are understood, those of $i(t)$ can be derived by using the methods developed in Sec. 8.2 for filtered random processes.

Invocation of the Poisson postulates A full specification of the statistics of $z(t)$ requires knowledge of the probability laws for each of the random variables involved.

In (11.64) there are $N+1$ random variables, namely each of the t_n and N itself. We require, therefore, the joint probability law on N and the arrival times $\{t_n\}$, which we can write as $\text{pr}(\{t_n\}|N)\,\text{Pr}(N)$. It is easiest to specify this probability law in the case of a Poisson point process where the postulates of Sec. 11.1.1 are satisfied, and that is what we do here. Deviations from the Poisson model are considered in Sec. 11.3.6 and 11.3.7.

First, note that the index n labels the individual counts but does not specify the order of arrival. The counts are assumed to be indistinguishable, so $\text{pr}(\{t_n\}|N)$ must be invariant to permutations of the labels. Second, under the Poisson postulates the counts are assumed to be statistically independent. With these assumptions, the variables $\{t_n\}$ are *i.i.d.* (independent and identically distributed) and we have

$$\text{pr}(\{t_n\}\,|N) = \prod_{n=1}^{N} \text{pr}(t_n)\,. \tag{11.65}$$

Note that $\text{pr}(t_n)$ cannot depend on the total number N if the counts are to be statistically independent.

We can relate $\text{pr}(t_n)$ to the mean arrival rate $a(t)$ introduced in Sec. 11.1.1. If we allow $a(t)$ to be a nonrandom function of time, postulate (b) states that the probability of getting one count in $(t, t+\Delta T)$ is given by

$$\text{Pr}[1 \text{ in } (t, t+\Delta T)] = a(t)\Delta T\,, \qquad (\Delta T \to 0)\,. \tag{11.66}$$

This same probability is also the probability that *one* of the t_n lies in this time interval. The event that t_n lies in the interval $(t, t+\Delta T)$ and the event that $t_m (m \neq n)$ lies in the same interval are mutually exclusive by postulate (c) as $\Delta T \to 0$. Thus the probability that either t_n or t_m is in the interval is the sum of the individual probabilities, or twice the probability that specifically t_n is in the interval. By extension, if exactly N counts occur in (0, T) and both t and $t+\Delta T$ lie in this interval, then the probability that one count occurs in $(t, t+\Delta T)$ is given by

$$\text{Pr}[1 \text{ in } (t, t+\Delta T)|N] = N\,\text{Pr}(t < t_n \le t+\Delta T) = N\int_t^{t+\Delta T} dt_n\ \text{pr}(t_n) \approx N\Delta T\,\text{pr}(t)\,. \tag{11.67}$$

The approximation in the last step becomes exact as $\Delta T \to 0$.

The probability $\text{Pr}[1 \text{ in } (t, t+\Delta T)]$ is obtained by averaging over N:

$$\text{Pr}[1 \text{ in } (t, t+\Delta T)] = \sum_{N=0}^{\infty} \text{Pr}(N)\,\text{Pr}[1 \text{ in } (t, t+\Delta T)|N] = \overline{N}\Delta T\,\text{pr}(t)\,. \tag{11.68}$$

Comparison of (11.68) with (11.66) shows that

$$\overline{N}\,\text{pr}(t_n) = a(t_n)\,. \tag{11.69}$$

Thus $a(t)$, originally introduced in Sec. 11.1 as a mean arrival rate at time t, now takes on a second meaning. When evaluated at t_n, it is the constant $\overline{N}$ times the probability density for occurrence of a count at $t = t_n$. (Note that both $\text{pr}(t_n)$ and $a(t_n)$ have dimensions of reciprocal time.)

Since we know from (11.19) that $\overline{N}$ is the integral of $a(t)$, we also have

$$\text{pr}(t_n) = \frac{a(t_n)}{\int_0^T dt\ a(t)}\,, \tag{11.70}$$

which shows immediately that

$$\int_0^T dt_n \, \mathrm{pr}(t_n) = 1 \,. \tag{11.71}$$

Thus the probability density on arrival times is just the mean arrival rate properly normalized.

When do the postulates break down? The essence of the Poisson postulates is the independence of the events. Thus the statistical independence expressed in (11.65) is required by the postulates. We shall now consider several physical circumstances in which independence is not a valid assumption.

One such circumstance is detector saturation, manifested as dead time or loss of efficiency at high counting rates. If one photon temporarily paralyzes the detector and there is a significant probability of another photon arriving before it recovers, the probability of detection of the second photon is dependent on the presence of the first.

Statistical independence also fails in random multiplication processes where one primary event gives rise to a random number of secondary events. In a scintillation detector, for example, a single gamma-ray photon produces a large number of optical photons, and these secondary events are not statistically independent since they arise from the same gamma-ray photon. This situation will be discussed further in Sec. 11.4.

Statistical independence also breaks down if $a(t)$ is itself random, in which case we refer to $z(t)$ as a *doubly stochastic Poisson random process*. The effect of such processes on the statistics of the total number of counts was discussed in Sec. 11.1.4, and the effect on the mean and autocorrelation of $z(t)$ will be discussed below in Sec. 11.3.7.

The postulates are intimately related not only to the probability density on the arrival times but also to the probability law on the total number of counts. The postulates cannot be satisfied unless N is a random variable, and specifically a Poisson random variable. With photon-counting detectors, it is always possible to remove the randomness in N; all we have to do is to count for a preset number of counts rather than for a preset time. When exactly N_0 counts have been accumulated, the accumulation is terminated. Under these conditions, $\mathrm{Pr}(N)$ is the Kronecker delta δ_{N,N_0}, but this form of $\mathrm{Pr}(N)$ is incompatible with postulate (a). Suppose that $N_0 - 1$ counts occur in an interval $(0, t')$ and the total number of counts is constrained to be N_0. Then the number of counts in (t', T) is fully determined, in contradiction to postulate (a); it can only be one.

In summary, the Poisson postulates can be satisfied only if the arrival times are statistically independent and identically distributed as in (11.65), the probability density on each arrival time is a normalized version of the mean arrival rate as in (11.70), and the total number of counts N is a Poisson random variable.

11.3.2 Spatial point processes

The spatial counterpart of the temporal point process $z(t)$ is $g(\mathbf{r})$, defined by

$$g(\mathbf{r}) = \sum_{n=1}^{N} \delta(\mathbf{r} - \mathbf{r}_n) \,, \tag{11.72}$$

where $\mathbf{r}$ is a spatial position vector in q dimensions. For example, with $q = 2$, $g(\mathbf{r})$ could describe the pattern of photon interactions on a piece of film.

As in the temporal case, there are $N+1$ random variables in $g(\mathbf{r})$, namely the N positions $\mathbf{r}_n$ and N itself. A full specification of the statistics of $g(\mathbf{r})$ requires the joint probability law $\text{pr}(\{\mathbf{r}_n\}|N)\,\text{Pr}(N)$.

Spatial Poisson postulates and their consequences As in the temporal case, there are certain postulates that will lead to $g(\mathbf{r})$ being a Poisson random process. For definiteness we take $q = 2$ and speak in terms of areas, but the extension to 3D volumes or regions in spaces of other dimensionalities is straightforward. We consider an overall area A and an exposure time T during which counts occur at points $\{\mathbf{r}_n\}$ contained in A. The spatial counterparts of the postulates given for the temporal case in Sec. 11.1.1 are:

(a) The number of counts in any area A_1 is statistically independent of the number in any other nonoverlapping area A_2, where A_1 and A_2 are subareas of A.

(b) If we consider a very small area ΔA contained in A and centered on point $\mathbf{r}$, the probability of a single count in this area during an observation time approaches a deterministic function $b(\mathbf{r})$ times ΔA, *i.e.*, up through terms linear in ΔA,

$$\text{Pr}(1 \text{ in } \Delta A) = b(\mathbf{r})\,\Delta A\,. \tag{11.73}$$

(c) The probability of more than one count in a vanishingly small area ΔA is zero. Thus, again through terms linear in ΔA,

$$\text{Pr}(1 \text{ in } \Delta A) + \text{Pr}(0 \text{ in } \Delta A) = 1\,. \tag{11.74}$$

Note that we have allowed $\text{Pr}(1 \text{ in } \Delta A)$ to be a function of position, but that $b(\mathbf{r})$ in (11.73) is a fixed function and not yet a random process.

By arguments analogous to those of the temporal case, these postulates can be satisfied only if

$$\text{pr}(\{\mathbf{r}_n\}|N) = \prod_{n=1}^{N} \text{pr}(\mathbf{r}_n)\,; \tag{11.75}$$

$$\text{pr}(\mathbf{r}_n) = \frac{b(\mathbf{r}_n)}{\int_A d^2r\; b(\mathbf{r})}\,; \tag{11.76}$$

$$\text{Pr}(N) = \frac{\overline{N}^N}{N!}\exp(-\overline{N})\,; \tag{11.77}$$

$$\overline{N} = \int_A d^2r\; b(\mathbf{r})\,. \tag{11.78}$$

As in the temporal case, we see that $b(\mathbf{r})$ has a dual interpretation. It is the *count density* or mean number of counts per unit area, and after normalization as in (11.76), it is also the probability density on the position of any individual count. In the language of radiometry (see Chap. 10), $b(\mathbf{r})$ is the mean *photon fluence* if $g(\mathbf{r})$ represents a photon distribution.

The spatial Poisson postulates can break down for reasons similar to those discussed above in the temporal case. In particular, they do not hold for random multiplication processes where two or more correlated secondary events are produced by one primary event. This issue is discussed further in Sec. 11.4.

11.3.3 Mean and autocorrelation of point processes

Since sample functions of a point process are generalized functions (sums of delta functions) with no finite values other than zero, a probability density function does not have much meaning. As with many other random processes, the most we can do is compute the first- and second-order statistics of the process, or its mean and autocorrelation function. We carry out this calculation here for the spatial point process $g(\mathbf{r})$; simple changes in notation will yield the corresponding results for the temporal case.

Mean of a general spatial point process To compute the expectation value of $g(\mathbf{r})$, we must average over the random variables $\{\mathbf{r}_n, n = 1, ..., N\}$ and N itself. Equations (11.75) – (11.78) specify the probability laws of these variables for Poisson processes, but we shall delay for a while invoking these equations. Instead, we first derive a general expression that will prove useful later.

The joint probability law for $\{\mathbf{r}_n\}$ and N can be written as $\text{pr}(\{\mathbf{r}_n\}|N)\,\text{Pr}(N)$, and the general formula of interest is obtained by first taking the conditional expectation of $g(\mathbf{r})$ for fixed N and then averaging over N. With the definition of $g(\mathbf{r})$ from (11.72), the first step requires evaluation of

$$\text{E}\{g(\mathbf{r})|N\} = \int_A d^2r_1 \int_A d^2r_2 \cdots \int_A d^2r_N \ \text{pr}(\{\mathbf{r}_n\}|N) \sum_{n=1}^{N} \delta(\mathbf{r} - \mathbf{r}_n)\,. \tag{11.79}$$

For any particular n, all of the integrals over $\mathbf{r}_j$ except the one with $j = n$ can be performed by use of (C.75). What remains after performing these $N - 1$ integrals is the marginal density for $\mathbf{r}_n$, so

$$\text{E}\{g(\mathbf{r})|N\} = \sum_{n=1}^{N} \int_A d^2r_n \ \text{pr}(\mathbf{r}_n|N)\,\delta(\mathbf{r} - \mathbf{r}_n)\,. \tag{11.80}$$

With the sifting property of delta functions, we have

$$\text{E}\{g(\mathbf{r})|N\} = \sum_{n=1}^{N} \text{pr}(\mathbf{r}|N) = N\,\text{pr}(\mathbf{r}|N)\,. \tag{11.81}$$

The notation here is a bit tricky; $\text{pr}(\mathbf{r}|N)$ must be interpreted as $\text{pr}(\mathbf{r}_n|N)$ after the substitution $\mathbf{r}_n = \mathbf{r}$. The random variable is $\mathbf{r}_n$, and $\mathbf{r}$ is a deterministic position vector. A further average over N yields

$$\text{E}\{g(\mathbf{r})\} = \langle N\,\text{pr}(\mathbf{r}|N)\rangle_N = \sum_{N=0}^{\infty} \text{Pr}(N)\,N\,\text{pr}(\mathbf{r}|N)\,. \tag{11.82}$$

Mean of a Poisson point process The expression (11.82) holds for any point process; we have not used any specific characteristics of Poisson point processes. For the Poisson case, $\text{pr}(\mathbf{r}|N)$ is independent of N, and we can use (11.76) to write

$$\text{pr}(\mathbf{r}) = \frac{b(\mathbf{r})}{\int_A d^2r'\,b(\mathbf{r}')}\,. \tag{11.83}$$

With (11.78) we have, finally,

$$\text{E}\{g(\mathbf{r})\} = \overline{N}\,\text{pr}(\mathbf{r}) = b(\mathbf{r})\,. \tag{11.84}$$

This result provides a third interpretation of $b(\mathbf{r})$ for Poisson processes. We have already seen that it is both the mean number of counts per unit area and, when normalized, the probability density on the position of any individual count. We now see from (11.84) that it is also the expectation value of the random process $g(\mathbf{r})$. All three of these interpretations are dimensionally consistent since counts per unit area, $\mathrm{pr}(\mathbf{r}_j)$ and $g(\mathbf{r})$ all have dimensions of reciprocal area (see Sec. 2.4.6 for a discussion of dimensions of delta functions).

Autocorrelation function of a general point process We consider next the autocorrelation function for the random point process $g(\mathbf{r})$. Initially we avoid use of the Poisson model in order to obtain a general result; then we specialize to the Poisson case.

The nonstationary autocorrelation function of $g(\mathbf{r})$ is defined by

$$R_{\mathbf{g}}(\mathbf{r},\mathbf{r}') = \mathrm{E}\{g(\mathbf{r})\,g(\mathbf{r}')\}\,. \tag{11.85}$$

Again we do the calculation in two stages, first over the set $\{\mathbf{r}_n\}$ for fixed N, then over N. The first stage requires computation of

$$\mathrm{E}\{g(\mathbf{r})\,g(\mathbf{r}')|N\} = \int_A d^2r_1 \int_A d^2r_2 \cdots \int_A d^2r_N \;\mathrm{pr}(\{\mathbf{r}_n\}|N) \sum_{n=1}^{N} \delta(\mathbf{r}-\mathbf{r}_n) \sum_{j=1}^{N} \delta(\mathbf{r}'-\mathbf{r}_j)\,. \tag{11.86}$$

The double sum over j and n has N^2 terms, N of which have $j = n$ and $N^2 - N$ of which have $j \neq n$. Consider first a term with $j = n$. All of the integrals over $\mathbf{r}_l$ except the one with $l = j = n$ can be performed by (C.75), resulting in the marginal density for $\mathbf{r}_n$. Then, with a property of delta functions given in (2.78) and (2.79), we have

$$\int_A d^2r_1 \int_A d^2r_2 \cdots \int_A d^2r_N \;\mathrm{pr}(\{\mathbf{r}_n\}|N)\,\delta(\mathbf{r}-\mathbf{r}_n)\,\delta(\mathbf{r}'-\mathbf{r}_n)$$

$$= \int_A d^2r_n \;\mathrm{pr}(\mathbf{r}_n|N)\,\delta(\mathbf{r}-\mathbf{r}_n)\,\delta(\mathbf{r}'-\mathbf{r}_n) = \mathrm{pr}(\mathbf{r}|N)\,\delta(\mathbf{r}-\mathbf{r}')\,, \tag{11.87}$$

where again $\mathrm{pr}(\mathbf{r}|N)$ is to be interpreted as $\mathrm{pr}(\mathbf{r}_n|N)$ evaluated at $\mathbf{r}_n = \mathbf{r}$. Note that (11.87) is independent of n. There are N terms with $j = n$, and all of them have the same expectation. Thus the contribution to $\mathrm{E}\{g(\mathbf{r})\,g(\mathbf{r}')|N\}$ from all terms with $j = n$ is

$$[\mathrm{E}\{g(\mathbf{r})\,g(\mathbf{r}')|N\}]_{j=n} = N\,\mathrm{pr}(\mathbf{r}|N)\,\delta(\mathbf{r}-\mathbf{r}')\,. \tag{11.88}$$

Next consider the case $j \neq n$. Now all but two of the integrals can be performed by (C.75), and we obtain

$$\int_A d^2r_1 \int_A d^2r_2 \cdots \int_A d^2r_N \;\mathrm{pr}(\{\mathbf{r}_n\}|N)\,\delta(\mathbf{r}-\mathbf{r}_n)\,\delta(\mathbf{r}'-\mathbf{r}_j)$$

$$= \int_A d^2r_n \int_A d^2r_j \;\mathrm{pr}(\mathbf{r}_n,\mathbf{r}_j|N)\,\delta(\mathbf{r}-\mathbf{r}_n)\,\delta(\mathbf{r}'-\mathbf{r}_j) = \mathrm{pr}(\mathbf{r},\mathbf{r}'|N)\,, \tag{11.89}$$

where $\mathrm{pr}(\mathbf{r},\mathbf{r}'|N)$ is the joint density $\mathrm{pr}(\mathbf{r}_n,\mathbf{r}_j|N)$ evaluated at $\mathbf{r}_n = \mathbf{r}$ and $\mathbf{r}_j = \mathbf{r}'$.

Since (11.89) is independent of j and n, the contribution to $\mathrm{E}\{g(\mathbf{r})\,g(\mathbf{r}')|N\}$ from the $N^2 - N$ terms with $j \neq n$ is

$$[\mathrm{E}\{g(\mathbf{r})\,g(\mathbf{r}')|N\}]_{j\neq n} = (N^2 - N)\,\mathrm{pr}(\mathbf{r},\mathbf{r}'|N)\,. \tag{11.90}$$

The conditional autocorrelation function $\{g(\mathbf{r})g(\mathbf{r}')|N\}$ is obtained by adding (11.88) and (11.90). A subsequent average over N yields the general result,

$$R_{\mathbf{g}}(\mathbf{r},\mathbf{r}') = \mathrm{E}\{g(\mathbf{r})\,g(\mathbf{r}')\} = \langle N\,\mathrm{pr}(\mathbf{r}|N)\rangle_N\,\delta(\mathbf{r}-\mathbf{r}') + \big\langle \big[N^2 - N\big]\,\mathrm{pr}(\mathbf{r},\mathbf{r}'|N)\big\rangle_N\,. \tag{11.91}$$

The autocovariance function for $g(\mathbf{r})$ is given by

$$K_{\mathbf{g}}(\mathbf{r},\mathbf{r}') = \langle \Delta g(\mathbf{r})\,\Delta g(\mathbf{r}')\rangle = R_{\mathbf{g}}(\mathbf{r},\mathbf{r}') - \mathrm{E}\{g(\mathbf{r})\}\,\mathrm{E}\{g(\mathbf{r}')\}$$

$$= \langle N\,\mathrm{pr}(\mathbf{r}|N)\rangle_N\,\delta(\mathbf{r}-\mathbf{r}') + \big\langle \big[N^2 - N\big]\,\mathrm{pr}(\mathbf{r},\mathbf{r}'|N)\big\rangle_N - \langle N\,\mathrm{pr}(\mathbf{r}|N)\rangle_N\,\langle N\,\mathrm{pr}(\mathbf{r}'|N)\rangle_N\,, \tag{11.92}$$

where $\Delta g(\mathbf{r}) = g(\mathbf{r}) - \langle g(\mathbf{r})\rangle$.

Autocorrelation function of a Poisson point process We now specialize the general results (11.91) and (11.92) to the case of a Poisson process. Under that model, $\mathrm{pr}(\mathbf{r},\mathbf{r}'|N) = \mathrm{pr}(\mathbf{r})\,\mathrm{pr}(\mathbf{r}')$ (independent of N) and $\mathrm{pr}(\mathbf{r})$ is given by (11.83). Moreover, for Poisson random variables, $\langle N^2 - N\rangle_N = \overline{N}^2$, and $\overline{N}$ is given by the denominator of (11.83). Thus,

$$R_{\mathbf{g}}(\mathbf{r},\mathbf{r}') = b(\mathbf{r})\,\delta(\mathbf{r}-\mathbf{r}') + b(\mathbf{r})\,b(\mathbf{r}')\,. \tag{11.93}$$

For the autocovariance function in the Poisson case, the second and third terms in the last form of (11.92) cancel, and we obtain

$$K_{\mathbf{g}}(\mathbf{r},\mathbf{r}') = b(\mathbf{r})\,\delta(\mathbf{r}-\mathbf{r}')\,. \tag{11.94}$$

Note that the variance, $K_{\mathbf{g}}(\mathbf{r},\mathbf{r})$, is infinite; it could hardly be otherwise for a random process that takes on only the values zero and infinity. In Sec. 11.3.9 we shall see that this is not a worry in practice since a filtered Poisson random process always has finite variance.

Temporal Poisson processes We can easily transcribe the results of this section to the temporal case just by replacing $g(\mathbf{r})$ with $z(t)$ and $b(\mathbf{r})$ with $a(t)$. For a nonrandom but possibly time-varying rate $a(t)$, the mean of $z(t)$ is [*cf.* (11.84)]

$$\mathrm{E}\{z(t)\} = a(t)\,. \tag{11.95}$$

Similarly, the autocorrelation function is [*cf.* (11.93)]

$$R_z(t,t') = a(t)\,\delta(t-t') + a(t)\,a(t')\,. \tag{11.96}$$

The autocovariance function for $z(t)$ is [*cf.* (11.94)]

$$K_{\mathbf{z}}(t,t') = a(t)\,\delta(t-t')\,. \tag{11.97}$$

We emphasize that these results hold only if the Poisson postulates are satisfied, which means that $a(t)$ is nonrandom and N is a Poisson random variable.

11.3.4 Relation between Poisson random vectors and processes

We have so far discussed Poisson random vectors and random processes separately. In this section we explore the connections between them.

From process to vector In Sec. 11.2.1 we discussed multivariate Poisson statistics, and especially the multivariate probability law (11.40). That law can also be derived another way, starting from what we know about Poisson random processes. Suppose we have a 2D detector array with uniformly spaced pixels of area ϵ^2. Assume that the pattern of photons incident on the detector is described by the 2D Poisson random process $g(\mathbf{r})$ and that every photon incident on the area of the jth pixel is detected and assigned to that pixel. Then the (random) number of counts in the jth pixel is given by

$$g_j = \int_j d^2r \; g(\mathbf{r}) \,, \tag{11.98}$$

where the integral is over the area of the jth pixel. The mean number of counts in this pixel is given by

$$\overline{g}_j = \int_j d^2r \; b(\mathbf{r}) \,. \tag{11.99}$$

The variance of g_j is given by

$$\mathrm{Var}(g_j) = \mathrm{E}\left\{[g_j - \overline{g}_j]^2\right\} = \int_j d^2r \int_j d^2r' \, \langle \Delta g(\mathbf{r}) \, \Delta g(\mathbf{r}') \rangle$$

$$= \int_j d^2r \int_j d^2r' \, K_{\mathbf{g}}(\mathbf{r},\mathbf{r}') = \int_j d^2r \int_j d^2r' \, b(\mathbf{r}) \, \delta(\mathbf{r}-\mathbf{r}') = \int_j d^2r \; b(\mathbf{r}) = \overline{g}_j \,, \tag{11.100}$$

where $\Delta g(\mathbf{r}) = g(\mathbf{r}) - \langle g(\mathbf{r}) \rangle = g(\mathbf{r}) - b(\mathbf{r})$.

The covariance of the counts in two different pixels is computed similarly:

$$\mathrm{E}\{(g_j - \overline{g}_j)(g_k - \overline{g}_k)\} = \int_j d^2r \int_k d^2r' \, \langle \Delta g(\mathbf{r}) \, \Delta g(\mathbf{r}') \rangle$$

$$= \int_j d^2r \int_k d^2r' \, K_{\mathbf{g}}(\mathbf{r},\mathbf{r}') = \int_j d^2r \int_k d^2r' \, b(\mathbf{r}) \, \delta(\mathbf{r}-\mathbf{r}') \,. \tag{11.101}$$

Now, however, the double integral must be zero since $\mathbf{r}$ cannot equal $\mathbf{r}'$ if the former lies in the jth pixel and the latter in the kth. We can combine the cases $j = k$ and $j \neq k$ by writing

$$\mathrm{E}\{(g_j - \overline{g}_j)(g_k - \overline{g}_k)\} = \overline{g}_j \, \delta_{jk} \,. \tag{11.102}$$

This result is in accord with (11.41).

We can also justify the multivariate Poisson form in (11.40). Again the key is the binomial selection theorem, with the probability of a photon being detected in the jth pixel given by

$$P_j = \frac{\int_j d^2r \; b(\mathbf{r})}{\int_{det} d^2r \; b(\mathbf{r})} \,, \tag{11.103}$$

where the integral in the denominator is over the whole detector. Since binomial selection of a Poisson yields a Poisson, and since the events are independent, (11.40) follows.

From vector to process Having derived the probability law (11.40) from the properties of Poisson random processes, it is instructive to go in the other direction and derive the autocorrelation function of the random processes from the probability law (Metz, 1969). Since (11.40) implies (11.41), we start with the latter equation, rewriting it as

$$\mathrm{E}\left[\frac{\Delta g_j}{\epsilon^2}\frac{\Delta g_k}{\epsilon^2}\right] = \frac{\overline{g}_j}{\epsilon^2}\frac{\delta_{jk}}{\epsilon^2}, \tag{11.104}$$

where $\Delta g_k = g_k - \overline{g}_k$. Since ϵ^2 is the area of a pixel, $\overline{g}_j/\epsilon^2$ is the mean number of counts per unit area in the jth pixel. Similarly, g_j/ϵ^2 is the actual random number per unit area. If we pass to the limit $\epsilon \to 0$, then $g_j/\epsilon^2 \to g(\mathbf{r}_j)$ and $\overline{g}_j/\epsilon^2 \to b(\mathbf{r}_j)$. Furthermore,

$$\lim_{\epsilon\to 0}\frac{\delta_{jk}}{\epsilon^2} = \delta(\mathbf{r}_j - \mathbf{r}_k). \tag{11.105}$$

This result can be proved by integrating both sides over $\mathbf{r}_j$ (or $\mathbf{r}_k$) and representing the integral on the left by a Riemann sum. Collecting results, we now have

$$\lim_{\epsilon\to 0}\mathrm{E}\left\{\frac{\Delta g_j}{\epsilon^2}\frac{\Delta g_k}{\epsilon^2}\right\} = \mathrm{E}\{\Delta g(\mathbf{r}_j)\,\Delta g(\mathbf{r}_k)\} = b(\mathbf{r}_j)\,\delta(\mathbf{r}_j - \mathbf{r}_k), \tag{11.106}$$

which is identical to (11.94).

11.3.5 Karhunen-Loève analysis of Poisson processes

In Sec. 8.2.7 we introduced the concept of Karhunen-Loève analysis, which is just eigenanalysis of the autocovariance operator. What are the eigenfunctions of the temporal autocovariance operator with kernel $K_{\mathbf{z}}(t,t')$ as given in (11.97)? The eigenvalue problem is

$$\int_{-\infty}^{\infty} dt'\, a(t)\,\delta(t-t')\,\phi(t') = \mu\phi(t). \tag{11.107}$$

The integral can be performed by means of the sifting property of delta functions, so the eigenvalue problem can also be stated as

$$a(t)\,\phi(t) = \mu\phi(t). \tag{11.108}$$

Since μ cannot be a function of t, we require a function which, when multiplied by an arbitrary $a(t)$, will yield a constant times the original function. The required function is

$$\phi(t) = \delta(t - t_0), \tag{11.109}$$

since, by (2.25),

$$a(t)\,\delta(t-t_0) = a(t_0)\,\delta(t-t_0) = const\cdot\delta(t-t_0). \tag{11.110}$$

Thus the eigenfunctions are the delta functions $\delta(t-t_0)$, indexed by the continuous variable t_0, and the eigenvalues are just the rates $a(t_0)$. The continuous index arises because this particular correlation operator is not compact (see Sec. 8.2.7). The fact that the eigenfunctions are delta functions means that the Karhunen-Loève domain is the original time domain in this problem. An exactly parallel analysis holds for spatial Poisson random processes.

We are now up to four interpretations of the rate $a(t)$. It is (1) the mean number of counts per unit time; (2) when normalized, the probability density on the time of arrival of any individual count; (3) the expectation value of the random process $z(t)$, and now (4) an eigenvalue of the autocovariance operator. Just as the statistics of a Poisson random variable are fully determined by its mean, so too are the statistics of a Poisson random *process* fully determined by its mean *rate*. Similar conclusions hold for spatial Poisson processes, and the same set of four interpretations applies to the spatial density $b(\mathbf{r})$.

The reader may have noticed an anomaly. In Sec. 8.2.7 we learned that the Karhunen-Loève domain was the frequency domain for stationary random processes. Here we see that it is the time (or space) domain for Poisson random processes, which might in fact be stationary. How do we reconcile these apparently contradictory statements?

In order for $z(t)$ to be stationary, $a(t)$ must be constant. In that case, the stationary autocovariance function $K_{\mathbf{z}}(t)$ is $a\,\delta(t)$, so the autocovariance operator is a multiple of the unit operator. If we transform this operator to the frequency domain, we find

$$\Gamma(\nu_1, \nu_2) = \int_{-\infty}^{\infty} dt_1 \int_{-\infty}^{\infty} dt_2 \; a\,\delta(t_1 - t_2) \exp[-2\pi i(\nu_1 \cdot t_1 - \nu_2 \cdot t_2)] = a\,\delta(\nu_1 - \nu_2)\,. \tag{11.111}$$

Thus the operator has exactly the same form in the frequency and time domains; unitary transformations leave the unit operator unchanged. It is therefore not surprising that both domains are Karhunen-Loève.

11.3.6 Doubly stochastic spatial Poisson random processes

In Sec. 11.1.4 we studied the effect of randomness in the rate $a(t)$ on the statistics of the total counts N, and in Sec. 11.2.2 we examined similar effects for random vectors, but so far in Sec. 11.3 $a(t)$ and its spatial counterpart $b(\mathbf{r})$ have been considered nonrandom. We now extend the treatment in Sec. 11.1.4 and compute the mean and autocorrelation function of a doubly stochastic spatial Poisson random process. The corresponding temporal case will be studied in Sec. 11.3.7.

Figure 11.3 illustrates the basic idea of a doubly stochastic Poisson process in 1D. Like a Poisson point process, each sample function $g(x)$ of this process is still a sum of delta functions, but now the density of the delta functions is controlled by the random process $b(x)$.

From (11.84) we know that the mean of $g(\mathbf{r})$ is $b(\mathbf{r})$ if that function is deterministic. We could equally well regard $b(\mathbf{r})$ as a sample function of a random process, however, and (11.84) would still be valid for a specific $b(\mathbf{r})$. The mean of a doubly stochastic Poisson random process is obtained simply by averaging (11.84) over all realizations of the process $b(\mathbf{r})$. All we need is some notation.

We now designate the kind of average we have been using in this chapter, where $b(\mathbf{r})$ is fixed, as a conditional average $\mathrm{E}\{\cdot|\mathbf{b}\}$, where $\mathbf{b}$ is the Hilbert-space vector corresponding to $b(\mathbf{r})$. With this notation, (11.84) can be written as

$$\mathrm{E}\{g(\mathbf{r})|\mathbf{b}\} = b(\mathbf{r})\,. \tag{11.112}$$

A further average over realizations of $b(\mathbf{r})$ will be denoted by $\mathrm{E}_{\mathbf{b}}\{\cdot\}$, so

$$\mathrm{E}_{\mathbf{b}}\{\mathrm{E}\{g(\mathbf{r})|\mathbf{b}\}\} = \mathrm{E}_{\mathbf{b}}\{b(\mathbf{r})\} \equiv \overline{b}(\mathbf{r})\,. \tag{11.113}$$

If $b(\mathbf{r})$ is a stationary random process, then $\overline{b}(\mathbf{r})$ is a constant.

Now consider the same averaging process for the autocorrelation function. Averaging (11.93) over $b(\mathbf{r})$ yields

$$\mathrm{E}_{\mathbf{b}}\{\mathrm{E}\{g(\mathbf{r})\,g(\mathbf{r}')|\mathbf{b}\}\} = R_{\mathbf{g}}(\mathbf{r},\mathbf{r}') = \overline{b}(\mathbf{r})\,\delta(\mathbf{r}-\mathbf{r}') + R_{\mathbf{b}}(\mathbf{r},\mathbf{r}')\,, \tag{11.114}$$

where $R_{\mathbf{b}}(\mathbf{r},\mathbf{r}')$ is the autocorrelation function of $b(\mathbf{r})$, defined by

$$R_{\mathbf{b}}(\mathbf{r},\mathbf{r}') = \mathrm{E}_{\mathbf{b}}\{b(\mathbf{r})\,b(\mathbf{r}')\}\,. \tag{11.115}$$

The corresponding autocovariance function is

$$\begin{aligned} K_{\mathbf{g}}(\mathbf{r},\mathbf{r}') = R_{\mathbf{g}}(\mathbf{r},\mathbf{r}') - \overline{b}(\mathbf{r})\,\overline{b}(\mathbf{r}') &= \overline{b}(\mathbf{r})\,\delta(\mathbf{r}-\mathbf{r}') + R_{\mathbf{b}}(\mathbf{r},\mathbf{r}') - \overline{b}(\mathbf{r})\,\overline{b}(\mathbf{r}') \\ &= \overline{b}(\mathbf{r})\,\delta(\mathbf{r}-\mathbf{r}') + K_{\mathbf{b}}(\mathbf{r},\mathbf{r}')\,. \end{aligned} \tag{11.116}$$

Thus the autocovariance consists of two terms, a delta-correlated term $\overline{b}(\mathbf{r})\,\delta(\mathbf{r}-\mathbf{r}')$ which represents the average Poisson random process and a term $K_{\mathbf{b}}(\mathbf{r},\mathbf{r}')$ which is just the autocovariance of the rate process $b(\mathbf{r})$. This result should be compared to (11.32) where the variance of a doubly stochastic Poisson random *variable* is similarly decomposed. Such decompositions will turn out to be very important when we consider the statistics of images where the fluctuations result from both Poisson noise and object variability.

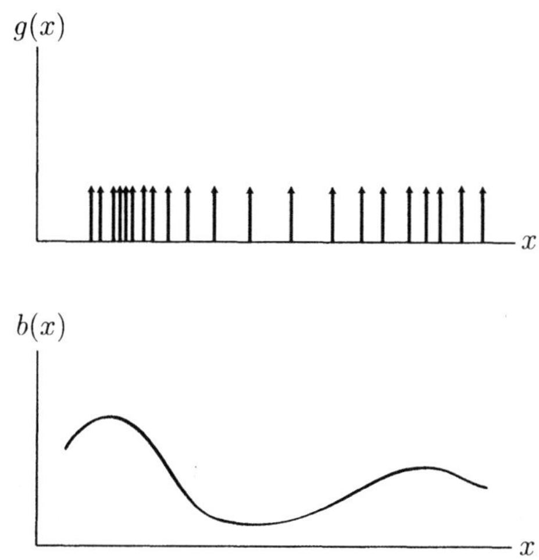

Fig. 11.3 *Top*: Sample function from the doubly stochastic process $g(x)$. *Bottom*: Corresponding sample function of the rate process $b(x)$.

11.3.7 Doubly stochastic temporal Poisson random processes

Next we examine the temporal counterparts of (11.113)–(11.116). For a doubly stochastic temporal random process where the rate is the random process $a(t)$, the overall mean of $z(t)$, averaged over both the Poisson process and the rate process, is given by

$$\langle z(t)\rangle = \overline{a}(t)\,, \tag{11.117}$$

and the overall autocorrelation function is [*cf.* (11.114)]

$$R_{\mathbf{z}}(t,t+\tau) = \langle z(t)\,z(t+\tau)\rangle = \overline{a}(t)\,\delta(\tau) + R_{\mathbf{a}}(t,t+\tau)\,, \tag{11.118}$$

where $R_{\mathbf{a}}(t, t+\tau)$ is the temporal autocorrelation function of the rate process $a(t)$.

The autocovariance function for $z(t)$ is

$$K_{\mathbf{z}}(t, t+\tau) = \langle [z(t) - \overline{a}(t)] \, [z(t+\tau) - \overline{a}(t+\tau)] \rangle = \overline{a}(t) \, \delta(\tau) + K_{\mathbf{a}}(t, t+\tau) \,, \quad (11.119)$$

where $K_{\mathbf{a}}(t, t+\tau)$ is the autocovariance function of the rate process.

Statistics of the total count In Sec. 11.1.4 we considered a doubly stochastic random variable where the mean number of counts $\overline{N}$ was itself a random variable with mean $\overline{\overline{N}}$ and variance $\mathrm{Var}(\overline{N})$. In (11.32) we related the overall variance of N to these parameters. With what we have learned about doubly stochastic random processes, we are now in a position to relate $\overline{\overline{N}}$ and $\mathrm{Var}(\overline{N})$ back to the statistics of $a(t)$.

We can immediately relate the (random) total number of counts N to $z(t)$ simply by integrating. From (11.64),

$$\int_0^T dt \; z(t) = \sum_{n=1}^{N} \int_0^T dt \; \delta(t - t_n) = \sum_{n=1}^{N} 1 = N \,. \quad (11.120)$$

The average of N over the Poisson process, conditional on $a(t)$, is

$$\overline{N} = \mathrm{E}\{N|\mathbf{a}\} = \int_0^T dt \; a(t) \,, \quad (11.121)$$

where we have used (11.95). A second average $\mathrm{E}_{\mathbf{a}}\{\cdot\}$ over the rate process yields

$$\overline{\overline{N}} = \mathrm{E}_{\mathbf{a}}\{\mathrm{E}\{N|\mathbf{a}\}\} = \int_0^T dt \; \overline{a}(t) \,. \quad (11.122)$$

If $a(t)$ is a stationary random process, $\overline{a}(t)$ is a constant and we find

$$\overline{\overline{N}} = \overline{a} T \,. \quad (11.123)$$

The overall variance of N is obtained by a similar double-averaging process. The definition is

$$\mathrm{Var}(N) = \mathrm{E}_{\mathbf{a}}\{\mathrm{E}\{N^2|\mathbf{a}\}\} - \overline{\overline{N}}^2 \,. \quad (11.124)$$

With a bit of algebra, we find

$$\mathrm{Var}(N) = \overline{\overline{N}} + \int_0^T dt \int_0^T dt' \; K_{\mathbf{a}}(t, t') \,. \quad (11.125)$$

This equation has important physical consequences, most easily seen in the stationary case. If $K_{\mathbf{a}}(t, t') = K_{\mathbf{a}}(t - t')$, it is convenient to transform to center coordinate $t_0 = \frac{1}{2}(t + t')$ and difference coordinate $\Delta t = t' - t$ as in Sec. 8.2.4. If $a(t)$ fluctuates rapidly compared to the measurement time T, then $K_{\mathbf{a}}(\Delta t)$ is a sharply peaked function of Δt, and the limits on the Δt integral can be extended to infinity. Using $K_{\mathbf{a}}(\Delta t) = K_{\mathbf{a}}(-\Delta t)$, we can show that

$$\mathrm{Var}(N) \approx \overline{\overline{N}} + T \int_{-\infty}^{\infty} d\Delta t \; K_{\mathbf{a}}(\Delta t) \,. \quad (11.126)$$

We can define a correlation time $\tau_{\mathbf{a}}$ for the process $a(t)$ by

$$\tau_{\mathbf{a}} = \frac{\int_{-\infty}^{\infty} d\Delta t \; K_{\mathbf{a}}(\Delta t)}{K_{\mathbf{a}}(0)} = \frac{\int_{-\infty}^{\infty} d\Delta t \; K_{\mathbf{a}}(\Delta t)}{\mathrm{Var}(a)} . \tag{11.127}$$

With (11.123) and (11.127), Var(N) is given by

$$\mathrm{Var}(N) \approx \overline{\overline{N}} + \overline{\overline{N}}^2 \, \frac{\tau_{\mathbf{a}}}{T} \, \frac{\mathrm{Var}(a)}{\overline{a}^2} . \tag{11.128}$$

This equation reveals several situations under which N would behave as a Poisson random variable, at least in the sense that $\mathrm{Var}(N) \approx \overline{N}$. First, if $a(t)$ is nonrandom, $\mathrm{Var}(a) = 0$ and $\mathrm{Var}(N) = \overline{\overline{N}} = \overline{N}$. Second, if $\overline{\overline{N}}$ is very small compared to one (events are rare on the time scale T), then $\overline{\overline{N}}^2 \ll \overline{\overline{N}}$ and again $\mathrm{Var}(N) \approx \overline{N}$. Finally, in the limit of short correlation time or long measurement time, *i.e.*, $\tau_{\mathbf{a}}/T \to 0$, then $\mathrm{Var}(N) \to \overline{\overline{N}}$ regardless of the statistics of $a(t)$. This last condition comes into play with white light, where $\tau_{\mathbf{a}}$ is of order 10^{-15} sec and $\mathrm{Var}(a)/\overline{a}^2 = 1$.

11.3.8 Point processes in other domains

We have discussed spatial and temporal point processes separately, but of course it is possible that both spatial and temporal aspects are important in the same problem. In that case we must be concerned with spatio-temporal point processes, with sample functions of the form,

$$g(\mathbf{r}, t) = \sum_{n=1}^{N} \delta(t - t_n) \, \delta(\mathbf{r} - \mathbf{r}_n) . \tag{11.129}$$

Now the rate process can depend on both space and time, so we denote it as $b(\mathbf{r}, t)$. We shall say that the events are fully statistically independent if $\mathbf{r}_j$ and $\mathbf{r}_k$ $(j \neq k)$ are independent, t_j and t_k are independent, and $\mathbf{r}_j$ is independent of t_j. Full statistical independence will occur only if $b(\mathbf{r}, t)$ is nonrandom.

For the fully independent case, both the spatial and temporal Poisson postulates are satisfied, and it is easy to show that

$$\langle g(\mathbf{r}, t) \rangle = b(\mathbf{r}, t) ; \tag{11.130}$$

$$K_{\mathbf{g}}(\mathbf{r}, \mathbf{r}'; t, t') \equiv \langle g(\mathbf{r}, t) \, g(\mathbf{r}', t') \rangle - \langle g(\mathbf{r}, t) \rangle \, \langle g(\mathbf{r}', t') \rangle = b(\mathbf{r}, t) \, \delta(\mathbf{r} - \mathbf{r}') \, \delta(t - t') . \tag{11.131}$$

For the general doubly stochastic case where $b(\mathbf{r}, t)$ is a spatio-temporal random process, (11.131) becomes

$$K_{\mathbf{g}}(\mathbf{r}, \mathbf{r}'; t, t') = \overline{b}(\mathbf{r}, t) \, \delta(\mathbf{r} - \mathbf{r}') \, \delta(t - t') + K_{\mathbf{b}}(\mathbf{r}, \mathbf{r}'; t, t') . \tag{11.132}$$

Other point processes are also important in imaging applications. For example, in Chap. 10 we introduced the concept of radiance L in a deterministic sense, but it can also be regarded as the mean of a 4D point process. We can define

$$g(\mathbf{r}, \hat{\mathbf{n}}) = \sum_j \delta(\mathbf{r} - \mathbf{r}_j) \, \delta(\hat{\mathbf{n}} - \hat{\mathbf{n}}_j) , \tag{11.133}$$

where the second delta function is an angular one, satisfying the sifting property (2.155). The mean of this process is the mean number of photons per unit area per unit solid angle, which we called the photon radiance in Chap. 10. Thus,

$$\langle g(\mathbf{r}, \hat{\mathbf{n}}) \rangle = L_{phot}(\mathbf{r}, \hat{\mathbf{n}}) . \tag{11.134}$$

If $g(\mathbf{r}, \hat{\mathbf{n}})$ is a Poisson point process, then its statistics are fully determined by its mean and hence by the photon radiance. This is a useful result since there are many ways of determining the radiance in imaging systems. For example, if we consider film illuminated with collimated light, then the output radiance is proportional to the bidirectional transmittance distribution function or BTDF (see Sec. 10.2.4).

One last kind of point process worth mentioning because of its radiological applications is the spatio-spectral process,

$$g(\mathbf{r}, \mathcal{E}) = \sum_j \delta(\mathbf{r} - \mathbf{r}_j)\, \delta(\mathcal{E} - \mathcal{E}_j) , \tag{11.135}$$

where $\mathcal{E}$ denotes a photon energy. As we shall see in Chap. 12, certain types of gamma-ray cameras estimate position and energy of each gamma ray, and the output of such cameras can be described by (11.135).

11.3.9 Filtered point processes

The general topic of filtered random processes was introduced in Sec. 8.2.6. We saw there that each sample function of the random process goes through a linear filter just as any deterministic function would. If the random process $g(\mathbf{r})$ is the input to a general shift-variant filter with impulse response $h(\mathbf{r}, \mathbf{r}')$, the output random process is given by

$$g_o(\mathbf{r}) = \int_\infty d^q r'\, h(\mathbf{r}, \mathbf{r}')\, g(\mathbf{r}') . \tag{11.136}$$

For a point process, where $g(\mathbf{r})$ is given by (11.72), the filter output becomes

$$g_o(\mathbf{r}) = \int_\infty d^q r'\, h(\mathbf{r}, \mathbf{r}') \sum_{n=1}^{N} \delta(\mathbf{r}' - \mathbf{r}_n) = \sum_{n=1}^{N} h(\mathbf{r}, \mathbf{r}_n) , \tag{11.137}$$

so the output is just a sum of randomly displaced impulse responses. If the filter is shift-invariant, we have

$$g_o(\mathbf{r}) = \sum_{n=1}^{N} h(\mathbf{r} - \mathbf{r}_n) , \tag{11.138}$$

which is a sum of identical, randomly displaced replicas of $h(\mathbf{r})$. The expressions in (11.137) and (11.138) are illustrated in Fig. 11.4.

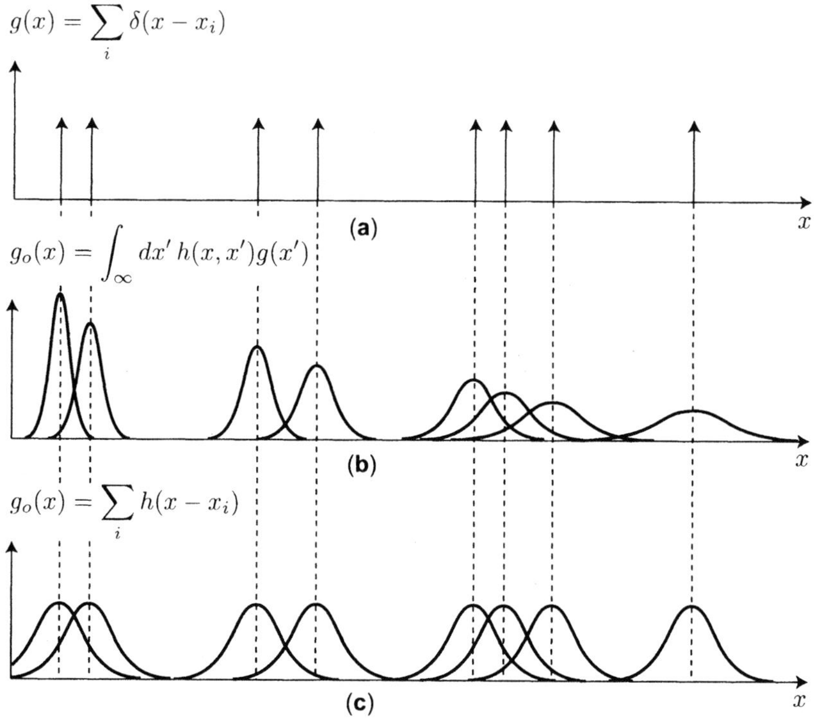

Fig. 11.4 Illustration in one dimension of the output of a linear filter when the input is a random point process. (a) Input; (b) Output of shift-variant filter; (b) Output of shift-invariant filter.

Mean value of the filter output Calculation of the expectation of $g_o(\mathbf{r})$ is straightforward. For the case of a general shift-variant filter and a doubly stochastic Poisson process, we have

$$\langle g_o(\mathbf{r})\rangle = \int_\infty d^q r'\, h(\mathbf{r}, \mathbf{r}')\, \langle g(\mathbf{r}')\rangle = \int_\infty d^q r'\, h(\mathbf{r}, \mathbf{r}')\, \overline{b}(\mathbf{r}')\,. \tag{11.139}$$

Various special cases are easily derived from this general result. If we consider a Poisson point process rather than a doubly stochastic one, all we have to do is delete the overbar on $\overline{b}(\mathbf{r}')$. If the filter is shift-invariant, $h(\mathbf{r}, \mathbf{r}')$ can be replaced with $h(\mathbf{r} - \mathbf{r}')$. And if the input process is stationary, $\langle g(\mathbf{r})'\rangle$ is a constant and can be removed from the integral.

The autocorrelation is a bit more complicated; it will be discussed separately for Poisson processes and doubly stochastic ones.

Autocovariance of filtered Poisson point processes In Sec. 8.2.6 we investigated filtering of delta-correlated processes. By (11.94), a Poisson point process $g(\mathbf{r})$ has a delta-function autocovariance. Equivalently, the zero-mean process $\Delta g(\mathbf{r}) = g(\mathbf{r}) - \overline{g}(\mathbf{r})$ has a delta-function autocorrelation, and the results in Sec. 8.2.6 can be applied immediately. From (11.94) we have

$$K_{g_o}(\mathbf{r}, \mathbf{r} + \Delta\mathbf{r}) = R_{\Delta g_o}(\mathbf{r}, \mathbf{r} + \Delta\mathbf{r}) = \int_\infty d^q r'\, h(\mathbf{r}, \mathbf{r}')\, b(\mathbf{r}')\, h^*(\mathbf{r} + \Delta\mathbf{r}, \mathbf{r}')\,. \tag{11.140}$$

For shift-invariant filters, this equation reduces to

$$K_{g_o}(\mathbf{r}, \mathbf{r} + \Delta\mathbf{r}) = \int_\infty d^q r' \, h(\mathbf{r} - \mathbf{r}') \, \bar{b}(\mathbf{r}') \, h^*(\mathbf{r} + \Delta\mathbf{r} - \mathbf{r}')$$
$$= \bar{b}(\mathbf{r}) * [h(\mathbf{r}) \, h^*(\mathbf{r} + \Delta\mathbf{r})] \,. \qquad (11.141)$$

Thus, in order to compute the autocovariance of a filtered Poisson point process, all we have to do is convolve the mean of the input, $\bar{b}(\mathbf{r})$, with the function $[h(\mathbf{r}) \, h^*(\mathbf{r} + \Delta\mathbf{r})]$. As discussed in Sec. 8.2.6, this induces a correlation width determined by the impulse response.

The variance can be easily derived from the autocovariance. For a shift-variant filter with a Poisson process on the input, we have

$$\text{Var}\{g_o(\mathbf{r})\} = K_{g_o}(\mathbf{r}, \mathbf{r}) = \int_\infty d^q r' \, |h(\mathbf{r}, \mathbf{r}')|^2 \, \bar{b}(\mathbf{r}') \,, \qquad (11.142)$$

and when the filter is shift-invariant,

$$\text{Var}\{g_o(\mathbf{r})\} = \bar{b}(\mathbf{r}) * |h(\mathbf{r})|^2 \,. \qquad (11.143)$$

In the literature on shot noise (*e.g.*, Davenport and Root, 1958), (11.143) is called *Campbell's theorem*.

It is interesting to compare (11.142) and (11.143) to the corresponding expressions for the mean of $g_o(\mathbf{r})$, as obtained from (11.139):

$$\text{E}\{g_o(\mathbf{r})\} = \int_\infty d^q r' \, h(\mathbf{r}, \mathbf{r}') \, \bar{b}(\mathbf{r}') \,, \qquad \text{(shift-variant)} \,; \qquad (11.144)$$

$$\text{E}\{g_o(\mathbf{r})\} = \bar{b}(\mathbf{r}) * h(\mathbf{r}) \,, \qquad \text{(shift-invariant)} \,. \qquad (11.145)$$

Thus the expression for the variance of $g_o(\mathbf{r})$ has the same form as for the mean, save only that the squared modulus of the filter kernel appears in place of the filter kernel itself. We emphasize, however, that this neat result holds only when the input is a Poisson (delta-correlated) random process.

Another important result follows from (11.142) and (11.143). We saw in (11.94) that the variance of a Poisson process was infinite, but (11.142) shows that the variance on the filter output is finite any time either $|h(\mathbf{r}, \mathbf{r}')|^2$ or $\bar{b}(\mathbf{r}')$ has finite spatial support. To be mathematically precise, we should also require that neither $|h(\mathbf{r}, \mathbf{r}')|^2$ nor $\bar{b}(\mathbf{r}')$ can go to infinity anywhere, but physically that doesn't happen anyway. For any real-world filter, a filtered Poisson random process will have finite variance.

For a stationary random process, $\bar{b}(\mathbf{r})$ is a constant $\bar{b}_0$, and (11.142) shows that the variance is

$$\text{Var}\{g_o(\mathbf{r})\} = K_{g_o}(\mathbf{r}, \mathbf{r}) = \bar{b}_0 \int_\infty d^q r' \, |h(\mathbf{r}, \mathbf{r}')|^2 \,. \qquad (11.146)$$

Note that this variance can still be a function of $\mathbf{r}$ if the filter is shift-variant. For a shift-invariant filter, however,

$$\text{Var}\{g_o(\mathbf{r})\} = \bar{b}_0 \int_\infty d^q r' \, |h(\mathbf{r} - \mathbf{r}')|^2 = \bar{b}_0 \int_\infty d^q r' \, |h(\mathbf{r}')|^2 = const \,. \qquad (11.147)$$

If a stationary random process is filtered with a shift-invariant filter, then neither the input nor the filter imposes any preferred origin and hence the output is also stationary.

Autocovariance of filtered doubly stochastic Poisson processes In Sec. 11.3.6 we derived the autocovariance of a doubly stochastic Poisson random process. These expressions can be substituted into the formulas from Sec. 8.2.6 to find the effects of filtering. We consider explicitly the case of a nonstationary process but a shift-invariant filter; the computation for a shift-variant filter is similar but messier.

From (8.146) and (8.147), we find

$$K_{g_o}(\mathbf{r}, \mathbf{r} + \Delta\mathbf{r})$$

$$= \int_\infty d^q r' \, h(\mathbf{r}') \int_\infty d^q r'' \, h^*(\mathbf{r}'') \left[\bar{b}(\mathbf{r} - \mathbf{r}') \, \delta(\mathbf{r}' + \Delta\mathbf{r} - \mathbf{r}'') + K_\mathbf{b}(\mathbf{r} - \mathbf{r}', \mathbf{r} + \Delta\mathbf{r} - \mathbf{r}'') \right]$$

$$= \bar{b}(\mathbf{r}) * [h(\mathbf{r}) \, h^*(\mathbf{r} + \Delta\mathbf{r})] + \int_\infty d^q r' \, h(\mathbf{r}') \int_\infty d^q r'' \, h^*(\mathbf{r}'') \, K_\mathbf{b}(\mathbf{r} - \mathbf{r}', \mathbf{r} + \Delta\mathbf{r} - \mathbf{r}'')$$

$$= \bar{b}(\mathbf{r}) * [h(\mathbf{r}) \, h^*(\mathbf{r} + \Delta\mathbf{r})] + \left[\mathcal{H} \mathcal{K}_\mathbf{b} \mathcal{H}^\dagger \right] (\mathbf{r}, \mathbf{r} + \Delta\mathbf{r}), \qquad (11.148)$$

where $\mathcal{H}$ is the integral operator corresponding to convolution with $h(\mathbf{r})$, $\mathcal{H}^\dagger$ is its adjoint (see Sec. 1.3.5) and $\mathcal{K}_\mathbf{b}$ is the integral operator for which the kernel is the autocovariance function of $b(\mathbf{r})$ (see Sec. 8.2.7).

The first term in (11.148) is identical to (11.141) except for the overbar on $\bar{b}(\mathbf{r})$; it represents the average Poisson random process as seen through the filter. The second term is just what we would get by filtering the process $b(\mathbf{r})$ without regard to the fact that it is the mean of a Poisson process [*cf.* (11.129)].

11.3.10 Characteristic functionals of filtered point processes

The discussion above of filtered point processes is incomplete since it does not give the full PDF of the output process. This PDF would have several applications. For example, it would describe the output of a photon-counting detector with high spatial resolution but post-detection smoothing. Also, as we saw in Sec. 8.4.4, a filtered Poisson process called a lumpy background is a useful model for background inhomogeneities in objects.

It is not possible to get the PDF in its full infinite-dimensional generality, but we can compute the characteristic functional, from which all other statistical properties can be derived (see Sec. 8.2.3). We shall carry out this computation first for an unfiltered Poisson random process, and then we shall consider the effects of filtering and show how various marginal PDFs can be derived. We shall also make contact with Sec. 8.4.4 and discuss the statistics of lumpy-background texture models.

As defined in (8.94), the characteristic functional involves an $\mathbb{L}_2$ scalar product between the random process and the function $s(\mathbf{r})$ in the argument of the functional. To apply this definition to a Poisson process, we must either take $s(\mathbf{r})$ as a test function or treat $\delta(\mathbf{r} - \mathbf{r}_n)$ as the limit of a sequence of $\mathbb{L}_2$ functions; with either approach, we can substitute the definition of the point process from (11.72) into (8.94) and obtain

$$\Psi_\mathbf{g}(\mathbf{s}) = \left\langle \exp\left[-2\pi i \int_A d^2 r \; s(\mathbf{r}) \sum_{n=1}^N \delta(\mathbf{r} - \mathbf{r}_n) \right] \right\rangle = \left\langle \exp\left[-2\pi i \sum_{n=1}^N s(\mathbf{r}_n) \right] \right\rangle . \qquad (11.149)$$

The expectation is performed in two steps, first over the set of random variables $\{\mathbf{r}_n\}$ for fixed N, then over N. By using Poisson assumptions and the methods of Sec. 11.3.3 [or by peeking at the derivation below (11.160)], the reader can show that

$$\Psi_{\mathbf{g}}(\mathbf{s}) = \exp\left\{-\overline{N} + \int_A d^2r_n\ b(\mathbf{r}_n)\, e^{-2\pi i\, s(\mathbf{r}_n)}\right\}, \tag{11.150}$$

where $b(\mathbf{r})$ is the usual photon fluence (which must be nonrandom for this expression to hold). For a stationary Poisson random process where $b(\mathbf{r})$ is constant over an area A, the fluence is $\overline{N}/A$, and we have

$$\Psi_{\mathbf{g}}(\mathbf{s}) = \exp\left\{-\overline{N}\left[1 + \frac{1}{A}\int_A d^2r_n\ e^{-2\pi i\, s(\mathbf{r}_n)}\right]\right\}. \tag{11.151}$$

Effect of filtering Though similar in form to the characteristic function of a Poisson random variable [see (C.171)], the expressions in (11.150) and (11.151) are not very useful as they stand. In particular, we know that the variance and higher moments of a Poisson random process are infinite, so there is little point in trying to compute them from the characteristic functionals. We can, however, go immediately from (11.150) to the characteristic functional for a filtered Poisson process.

If we consider a general nonstationary Poisson random process and a general linear filter $\boldsymbol{\mathcal{H}}$ as defined in (11.136), then from (11.150) and (8.96) we see that

$$\Psi_{\mathbf{g}_o}(\mathbf{s}) = \Psi_{\mathbf{g}}(\boldsymbol{\mathcal{H}}^\dagger\mathbf{s}) = \exp\left\{-\overline{N} + \int_A d^2r_n\ b(\mathbf{r}_n)\exp\left[-2\pi i\int_A d^2r'\ s(\mathbf{r}')\,h(\mathbf{r}',\mathbf{r}_n)\right]\right\}. \tag{11.152}$$

This expression, the full characteristic function for the output process, is equivalent to the full infinite-dimensional density of the process. To relate it to familiar single-point or multiple-point characteristic functions and densities, we need to make additional linear transformations.

For example, to get the univariate characteristic function on the scalar random variable $g_o(\mathbf{r})$ for some fixed point $\mathbf{r} = \mathbf{R}$, we define a sampling operator $\boldsymbol{\mathcal{S}}_{\mathbf{R}}$ (see Sec. 3.5) such that $\boldsymbol{\mathcal{S}}_{\mathbf{R}}\mathbf{g}_o = g_o(\mathbf{R})$. The kernel of this operator is $\delta(\mathbf{r}-\mathbf{R})$, and the characteristic function (not functional this time) is given by (8.96) as

$$\psi_{g_o(\mathbf{R})}(\xi) = \Psi_{\mathbf{g}_o}(\boldsymbol{\mathcal{S}}_{\mathbf{R}}^\dagger\xi). \tag{11.153}$$

Since $\boldsymbol{\mathcal{S}}_{\mathbf{R}}^\dagger\xi = \xi\,\delta(\mathbf{r}-\mathbf{R})$, we see from (11.152) that

$$\psi_{g_o(\mathbf{R})}(\xi) = \exp\left\{-\overline{N} + \int_A d^2r_n\ b(\mathbf{r}_n)\exp[-2\pi i\xi\, h(\mathbf{R},\mathbf{r}_n)]\right\}. \tag{11.154}$$

To instill some confidence in this formalism, we can use (11.154) to compute the mean and variance of $g_o(\mathbf{R})$. With (C.55) and a little algebra, we find that

$$\langle g_o(\mathbf{R})\rangle = \int_A d^2r\ h(\mathbf{R},\mathbf{r})\,b(\mathbf{r}), \qquad \mathrm{Var}\,\{g_o(\mathbf{R})\} = \int_A d^2r\ |h(\mathbf{R},\mathbf{r})|^2\,b(\mathbf{r}), \tag{11.155}$$

in agreement with (11.139) and (11.142).

Single-point density The single-point (univariate) PDF on $g_o(\mathbf{R})$ can be obtained by performing an inverse Fourier transform on (11.154), but usually this computation will have to be performed numerically. The Fourier integral is one-dimensional (with ξ as the variable of integration), but each point in the integrand requires another two-dimensional integral (the one over $\mathbf{r}_n$), so the overall computational effort is equivalent to a three-dimensional numerical integration — no great problem on modern computers.

To gain some insight into the problem without doing numerical integrals, let us consider a stationary problem with constant fluence b_0 and a shift-invariant filter such that $h(\mathbf{R}, \mathbf{r}_n) = \text{rect}\left[(\mathbf{R} - \mathbf{r}_n)/\epsilon\right]$. This filter function is either 0 or 1, so the inner integral in (11.154) simplifies to

$$\int_A d^2r_n \; b(\mathbf{r}_n) \exp[-2\pi i\xi \, h(\mathbf{R}, \mathbf{r}_n)] = \left[A - \epsilon^2 + \epsilon^2 \exp(-2\pi i\xi)\right] b_0 \, , \tag{11.156}$$

provided $\mathbf{R}$ is not within ϵ of the border. Since $b_0 A = \overline{N}$, (11.154) becomes

$$\psi_{g_o(\mathbf{R})}(\xi) = \exp\left\{-\epsilon^2 b_0[1 - \exp(-2\pi i\xi)]\right\} . \tag{11.157}$$

This expression will be recognized as the characteristic function for a Poisson, (C.171), with mean given by $\epsilon^2 b_0$.

The corresponding single-point PDF is essentially also a Poisson, but now as a density rather than a probability; specifically [*cf.* (C.24)],

$$\text{pr}[g_o(\mathbf{R})] = \sum_{N=0}^{\infty} \frac{(\epsilon^2 b_0)^N}{N!} \exp(-\epsilon^2 b_0) \, \delta[g_o(\mathbf{R}) - N] . \tag{11.158}$$

The delta functions at integer values arise since (11.157) is periodic with period 1.

The characteristic functional can, of course, be used also to compute various multivariate densities, such as $\text{pr}[g_o(\mathbf{R}_1), g_o(\mathbf{R}_2)]$. In that case it is necessary to define a sampling operator that maps the function $g_o(\mathbf{r})$ to a 2D vector, the components of which are the scalars $g_o(\mathbf{R}_1)$ and $g_o(\mathbf{R}_2)$. Because of the filtering operation, the bivariate density will not factor into two univariate densities if $\mathbf{R}_1$ is within a filter width of $\mathbf{R}_2$.

Generalized lumpy backgrounds In Sec. 8.4.4 we discussed the concept of a lumpy background as a model for random texture fields. The simplest definition was given in (8.303) as

$$f(\mathbf{r}) = \sum_{n=1}^{N} l(\mathbf{r} - \mathbf{r}_n) , \tag{11.159}$$

where $l(\mathbf{r})$ is the lump profile, taken as nonrandom in the initial formulation by Rolland and Barrett (1992). If the lump positions are independent and N is a Poisson random variable, then this $f(\mathbf{r})$ is a filtered Poisson random process, and its statistics can be analyzed as discussed above. We also mentioned in Sec. 8.4.4, however, that it can be very useful to take the lump profile as random. We shall now extend the theory of filtered random processes to allow for random lump profile (or filter function). Our tool will again be the characteristic functional.

We consider a random process $f(\mathbf{r})$ that is composed of a sum of N statistically independent and identically distributed realizations of some other random process

$l(\mathbf{r})$, where N is a Poisson random variable. Denoting the nth realization by $l_n(\mathbf{r})$, we write

$$f(\mathbf{r}) = \sum_{n=1}^{N} l_n(\mathbf{r}) \,. \tag{11.160}$$

The original lumpy background of (11.159) fits this mold if the only randomness is in the location, but (11.160) is much more general.

To find the characteristic functional, we start with the basic definition (8.94) and write [*cf.* (11.149)]

$$\Psi_{\mathbf{f}}(\mathbf{s}) = \left\langle \exp\left[-2\pi i \sum_{n=1}^{N} \int_A d^2r \; s(\mathbf{r})\, l_n(\mathbf{r})\right] \right\rangle . \tag{11.161}$$

As usual in this chapter, we perform the expectation in two steps, first over the randomness of each member of the set $\{l_n(\mathbf{r}), n = 1, ...N\}$ for fixed N, then over N itself. We therefore have

$$\Psi_{\mathbf{f}}(\mathbf{s}) = \left\langle \left\langle \prod_{n=1}^{N} \exp\left[-2\pi i \int_A d^2r \; s(\mathbf{r})\, l_n(\mathbf{r})\right] \right\rangle_{\{l_n(\mathbf{r})\}|N} \right\rangle_N . \tag{11.162}$$

Since the realizations are independent, the expectation of the product is the product of the expectations, and each factor is the same since the realizations are identically distributed; hence

$$\Psi_{\mathbf{f}}(\mathbf{s}) = \left\langle \prod_{n=1}^{N} \left\langle \exp\left[-2\pi i \int_A d^2r \; s(\mathbf{r})\, l_n(\mathbf{r})\right] \right\rangle_{\{l_n(\mathbf{r})\}|N} \right\rangle_N = \left\langle [\Psi_{\mathbf{l}}(\mathbf{s})]^N \right\rangle_N , \tag{11.163}$$

where $\Psi_{\mathbf{l}}(\mathbf{s})$ is the characteristic functional of the individual random process $l_n(\mathbf{r})$.

To perform the remaining expectation, we temporarily define N_0 by

$$N_0 = \overline{N}\Psi_{\mathbf{l}}(\mathbf{s}) \,, \tag{11.164}$$

so that

$$\Psi_{\mathbf{f}}(\mathbf{s}) = \sum_{N=0}^{\infty} \exp(-\overline{N}) \frac{\overline{N}^N}{N!} [\Psi_{\mathbf{l}}(\mathbf{s})]^N = \exp(-\overline{N} + N_0) \sum_{N=0}^{\infty} \exp(-N_0) \frac{N_0^N}{N!} \,. \tag{11.165}$$

The sum is unity (since Poisson probabilities are normalized), and we have, finally,

$$\Psi_{\mathbf{f}}(\mathbf{s}) = \exp(-\overline{N} + N_0) = \exp[-\overline{N} + \overline{N}\Psi_{\mathbf{l}}(\mathbf{s})] \,. \tag{11.166}$$

From this characteristic functional, we can in principle compute any desired statistical properties of this generalized lumpy background, just as we discussed above for the simple filtered Poisson process. Moreover, various special cases can be treated by making different assumptions about the randomness inherent in $l(\mathbf{r})$. The reader should, for example, be able to go from (11.166) to (11.154).

Another useful exercise is to let $\overline{N}$ get large and show that (11.166) approaches the characteristic functional for a normal random process, (8.216). Thus, no matter the statistics of $l(\mathbf{r})$, a random process created by adding a large number of i.i.d. realizations of it approaches normality.

Characteristic functionals of other kinds of point processes have been derived by Ramirez-Perez and Serfling (2001). Statistical properties of other kinds of lumpy backgrounds can be obtained by modifying their results to include the filtering operation as above.

11.3.11 Spectral properties of point processes

As we have seen in Chap. 8, Fourier analysis is often a useful tool for studying random processes. In this section we use the Fourier domain to study Poisson point processes and their relatives.

Fourier transform of a sample function In Chap. 3 we discussed the Fourier properties of generalized functions at some length. We can apply these methods directly to random point processes. For example, the Fourier transform of a sample function of the qD point process $g(\mathbf{r})$, as defined in (11.72), is

$$G(\boldsymbol{\rho}) = \mathcal{F}_q\{g(\mathbf{r})\} = \sum_{n=1}^{N} \exp(-2\pi i \boldsymbol{\rho} \cdot \mathbf{r}_n) . \tag{11.167}$$

Thus $G(\boldsymbol{\rho})$ is a sum of a random number of plane waves with random *frequencies* $\mathbf{r}_n$.

Power spectral density of stationary Poisson processes The power spectral density (see Sec. 8.2.5) is an important way of specifying the frequency content of a stationary random process. By the Wiener-Khinchin theorem, the power spectral density $S_{\mathbf{g}}(\boldsymbol{\rho})$ is the Fourier transform of the autocorrelation function $R_{\mathbf{g}}(\Delta\mathbf{r})$. For a stationary Poisson point process, $R_{\mathbf{g}}(\Delta\mathbf{r})$ is obtained from (11.93) by setting $b(\mathbf{r}) = b_0 =$ constant and $\Delta\mathbf{r} = \mathbf{r}' - \mathbf{r}$. We then obtain for the spectral density,

$$S_{\mathbf{g}}(\boldsymbol{\rho}) = \mathcal{F}_q\{b_0\, \delta(\Delta\mathbf{r}) + b_0^2\} = b_0 + b_0^2\, \delta(\boldsymbol{\rho}) . \tag{11.168}$$

The term $b_0^2\, \delta(\boldsymbol{\rho})$ is a result of the nonzero mean of $g(\mathbf{r})$. If we define a zero-mean process $\Delta g(\mathbf{r}) = g(\mathbf{r}) - b_0$, its power spectral density is

$$S_{\Delta g}(\boldsymbol{\rho}) = \mathcal{F}_q\{b_0\, \delta(\Delta\mathbf{r})\} = b_0 . \tag{11.169}$$

Thus a stationary Poisson random process is *white*; its power spectral density is a constant, namely the mean photon fluence.

Filtered Poisson processes Suppose a stationary Poisson process is passed through a filter with shift-invariant impulse response $h(\mathbf{r})$, and hence transfer function $H(\boldsymbol{\rho})$. Let the output of the filter be denoted $g_o(\mathbf{r})$ and define $\Delta g_o(\mathbf{r}) = g_o(\mathbf{r}) - \langle g_o(\mathbf{r})\rangle$. Then, by (8.156), the power spectral density of $\Delta g(\mathbf{r})$ is given by

$$S_{\Delta g_o}(\boldsymbol{\rho}) = b_0 |H(\boldsymbol{\rho})|^2 . \tag{11.170}$$

Thus the output spectrum is a direct measure of $|H(\boldsymbol{\rho})|^2$, but of course only because we know that the input is white.

If the poisson process is not stationary, we cannot use the power spectral density to describe its frequency content, but we can use the stochastic Wigner distribution function defined in Sec. 8.2.5. As an exercise, the reader can show that $W_{\Delta g_o}(\mathbf{r}, \boldsymbol{\rho}) = b(\mathbf{r}) * W_h(\mathbf{r}, \boldsymbol{\rho})$ and that this expression reduces to (11.170) when the fluence is constant.

Stationary doubly stochastic processes The autocovariance function of a doubly stochastic Poisson process is given by (11.116). The process is (wide-sense) stationary if $\overline{b}(\mathbf{r}) = \overline{b} =$ constant and $K_{\mathbf{b}}(\mathbf{r}, \mathbf{r}') = K_{\mathbf{b}}(\mathbf{r} - \mathbf{r}')$. In that case, the power spectral density of $\Delta g(\mathbf{r})$ is given by

$$S_{\Delta g}(\boldsymbol{\rho}) = \mathcal{F}_q\{K_{\mathbf{g}}(\Delta \mathbf{r})\} = \bar{b} + S_{\Delta b}(\boldsymbol{\rho}) . \tag{11.171}$$

In contrast to (11.169), this spectrum is not white; in addition to the constant term $\bar{b}$, there is also a term $S_{\Delta b}(\boldsymbol{\rho})$ arising from the rate process. The presence of these two terms can, perhaps, be appreciated from the sample functions shown in Fig. 11.3. These images were constructed with a rate process having a low-pass character, and the low spatial frequencies are evident in the images if the fine structure arising from the white-noise component is ignored. The same low frequencies are also seen in the sample functions of $b(\mathbf{r})$ shown in the figure.

11.4 RANDOM AMPLIFICATION

Many detectors and other imaging devices include a built-in gain mechanism. Examples include photomultipliers, avalanche photodiodes, fluorescent screens, image intensifiers and photographic film. In this section we develop the basic statistical tools for the analysis of such devices.

The basic principles of random amplification are introduced in Sec. 11.4.1 in the context of single-element detectors. Such devices are analyzed again in Sec. 11.4.2 in order to introduce the useful mathematical technique of generating functions. In Sec. 11.4.3 we study the spatial and temporal dependence of amplified point processes, and the same topics are treated in the frequency domain in Sec. 11.4.4. These sections build heavily on the discussion of random point processes in Sec. 11.3. Finally, in Sec. 11.4.5, we apply the formalism of generating functions to imaging arrays with gain.

11.4.1 Random amplification in single-element detectors

We begin by considering a simple nonimaging detector such as a photomultiplier or avalanche photodiode in which the input is photons and the output is electrons. All we need to know about the mechanism of the detector for now is that it puts out some random number of electrons for each input photon it absorbs. Since the same mathematics will apply to detectors with other inputs (*e.g.*, electrons or ions) and/or other outputs (especially photons), we refer to the input events as *primaries* and the output events as *secondaries*.

If N primaries are absorbed in time T and the nth primary produces k_n secondaries, then the total number of secondaries is

$$K = \sum_{n=1}^{N} k_n . \tag{11.172}$$

There are $N+1$ random variables in this problem: each of the k_n and N itself. The objective of the problem is to determine the statistics of K from those of N and the set $\{k_n\}$.

Conditional probabilities The statistics of k_n are governed by a probability law $\Pr(k_n = k)$. Since the primaries are indistinguishable, the gain mechanism must be the same for all primaries. Hence $\Pr(k_n = k)$ has the same form for all n, and we shall denote it as $\gamma(k)$. If $N = 1$, this same probability law also governs the total number of output secondaries K, *i.e.*,

$$\Pr(K|N = 1) = \gamma(K)\,. \tag{11.173}$$

The conditional mean and variance of K are given by

$$\mathrm{E}\{K|N = 1\} = \sum_{K=0}^{\infty} K \Pr(K|N = 1) = \sum_{k=0}^{\infty} k\,\gamma(k) = m_1\,; \tag{11.174}$$

$$\mathrm{Var}\{K|N = 1\} = \sum_{k=0}^{\infty} k^2\gamma(k) - m_1^2 = m_2 - m_1^2\,, \tag{11.175}$$

where m_j is the jth moment of $\gamma(k)$. The first moment m_1 is the average gain $\overline{k}_n$, *i.e.*, the mean number of output secondaries per input primary.

If we assume that all input primaries are amplified independently, then the mean and variance for N primaries are given by

$$\mathrm{E}\{K|N\} = Nm_1\,, \tag{11.176}$$

$$\mathrm{Var}\{K|N\} = N(m_2 - m_1^2)\,. \tag{11.177}$$

From these results it follows that the conditional second moment of K is

$$\mathrm{E}\left\{K^2|N\right\} = N(m_2 - m_1^2) + N^2m_1^2\,. \tag{11.178}$$

The overall first and second moments of K can now be calculated by averaging over N:

$$\mathrm{E}\{K\} = \overline{N}m_1\,, \tag{11.179}$$

$$\mathrm{E}\{K^2\} = \overline{N}(m_2 - m_1^2) + \left[\mathrm{Var}(N) + \overline{N}^2\right] m_1^2\,. \tag{11.180}$$

The overall variance of K is not obtained just by replacing N with $\overline{N}$ in (11.177). Instead, we must write

$$\mathrm{Var}(K) = \mathrm{E}\{K^2\} - [\mathrm{E}\{K\}]^2 = \overline{N}(m_2 - m_1^2) + m_1^2\,\mathrm{Var}(N)\,. \tag{11.181}$$

Since $m_2 - m_1^2$ is the variance of k_n (for one input primary) and m_1 is the corresponding mean, we can also write

$$\mathrm{Var}(K) = \overline{N}\,\mathrm{Var}(k_n) + \overline{k}_n^2\,\mathrm{Var}(N)\,. \tag{11.182}$$

This result is often referred to as the *Burgess variance theorem* (Burgess, 1959), but it has been rederived many times in the literature; see Shockley and Pierce (1938), Mandel (1959) and Zweig (1965).

For a Poisson input, $\mathrm{Var}(N) = \overline{N}$, so (11.182) becomes

$$\mathrm{Var}(K) = \overline{N}\left[\mathrm{Var}(k_n) + \overline{k}_n^2\right] = \overline{N}m_2\,, \tag{11.183}$$

showing that in this case the variance of the output is the mean number of input primaries times the second moment of the gain distribution. We can also express this result in terms of an output signal-to-noise ratio, defined as the mean of K divided by its standard deviation:

$$\mathrm{SNR}_{\mathrm{out}} \equiv \frac{\overline{K}}{\sqrt{\mathrm{Var}(K)}} = \overline{N}^{\frac{1}{2}} \frac{m_1}{\sqrt{m_2}} . \tag{11.184}$$

For a Poisson random variable, the SNR is the square root of the mean, so we also have

$$[\mathrm{SNR}_{\mathrm{out}}]^2 = [\mathrm{SNR}_{\mathrm{in}}]^2 \frac{m_1^2}{m_2} . \tag{11.185}$$

By the Schwarz inequality, $m_1^2 \leq m_2$ for any probability law, so $\mathrm{SNR}_{\mathrm{out}}$ is always less than or equal to $\mathrm{SNR}_{\mathrm{in}}$. The gain mechanism gives us a stronger signal, but it cannot improve the SNR beyond that inherent in the Poisson statistics of the input. In the radiology literature (see *e.g.*, Barrett and Swindell, 1981, 1996), the ratio m_1^2/m_2 is known as the *Swank factor* since R. K. Swank (1973) derived it in the course of analyzing fluorescent screens for x-ray imaging.

If $m_2 = m_1^2$, then the Swank factor is unity, the variance of k_n is zero and the gain mechanism is noise-free, so $\mathrm{SNR}_{\mathrm{out}} = \mathrm{SNR}_{\mathrm{in}}$ for a Poisson input N. This does not mean, however, that the output K is also Poisson. Instead, from (11.179) and (11.183), $\mathrm{Var}(K) = m_1\overline{K} = \overline{k}_n^2\overline{N}$, so the variance is increased in absolute terms by the square of the gain, while the mean is increased only linearly. Rarity begets Poissonicity, but amplification (even noise-free amplification) inevitably destroys it. Since the secondaries come in bursts, they are not independent.

11.4.2 Random amplification and generating functions

Discussions of random amplification in the literature often make use of probability-generating functions (or generating functions for short). In order to give the reader the necessary background to follow this literature, we shall now rederive all of the results in Sec. 11.4.1 with generating functions. There is no loss of continuity in skipping this section and going directly to Sec. 11.4.3, but the material here will be needed in Sec. 11.4.5 when we discuss random amplification in arrays.

Sums of a random number of random variables By its definition (11.172), K is a sum of a random number of integer-valued random variables. A natural tool for dealing with integer-valued random variables is the generating function $\Phi(\zeta)$ defined in Sec. C.3.3. Here we have three separate integer-valued random variables, K, k_n and N, so we distinguish their generating functions with appropriate subscripts. If the individual k_n are i.i.d., there is a compact expression for the generating function $\Phi_K(\zeta)$ (see Sec. C.3.3) for the total output K in terms of the generating function $\Phi_N(\zeta)$ for the random number of terms N and the (common) generating function $\Phi_k(\zeta)$ for each of the k_n.

The generating function for K is defined by

$$\Phi_K(\zeta) = \langle \zeta^K \rangle = \sum_{K=0}^{\infty} \zeta^K \Pr(K) , \tag{11.186}$$

and $\Phi_k(\zeta)$ and $\Phi_N(\zeta)$ are defined similarly. With the usual probability calculus, we can write

$$\Phi_K(\zeta) = \sum_{N=0}^{\infty} \sum_{K=0}^{\infty} \zeta^K \Pr(K|N) \Pr(N) . \qquad (11.187)$$

The sum over K is the *conditional* generating function for K given a fixed N; in light of the assumed independence of the k_n, it can be written as

$$\sum_{K=0}^{\infty} \zeta^K \Pr(K|N) = \mathrm{E}\{\zeta^K|N\} = \prod_{n=1}^{N} \mathrm{E}\{\zeta^{k_n}\} = \prod_{n=1}^{N} \Phi_k(\zeta) = [\Phi_k(\zeta)]^N . \qquad (11.188)$$

Plugging this result back into (11.187), we find

$$\Phi_K(\zeta) = \sum_{N=0}^{\infty} \Pr(N) [\Phi_k(\zeta)]^N . \qquad (11.189)$$

But this expression has the same form as $\Phi_N(\zeta)$ [*cf.* (11.186)] except that $\Phi_k(\zeta)$ appears in place of ζ; we can therefore write

$$\Phi_K(\zeta) = \Phi_N[\Phi_k(\zeta)] . \qquad (11.190)$$

Hence the generating function for the total number of secondaries K is a compound generating function (function of a function) involving the separate generating functions for number of primaries N and number of secondaries per primary k_n. If there are many stages of gain, the compounding can be repeated for each stage (Lombard and Martin, 1961).

One limit in which we can verify (11.190) is when N is not really random. Then $\Pr(N) = \delta_{NN_0}$, $\Phi_N(\zeta) = \zeta^{N_0}$ and $\Phi_K(\zeta) = [\Phi_k(\zeta)]^{N_0}$, just as we would expect for the sum of a *fixed* number of independent random variables.

Moments From (C.63) we know that $\Phi_K(\zeta)$ is essentially the z-transform of the probability law for K, so we can in principle go from (11.190) to $\Pr(K)$, though it requires a model for $\Pr(k)$ or $\Phi_k(\zeta)$. As we shall see in later chapters, however, many practical figures of merit for image quality require only the first- and second-order statistics, so we turn now to computation of moments of K. The reader who is interested in $\Pr(K)$ is referred to Saleh (1978) and Saleh and Teich (1982).

To compute moments of K, we must compute derivatives of the compound function $\Phi_K(\zeta)$ (see Sec. C.3.3). For the first derivative, we have

$$\frac{d}{d\zeta}[\Phi_K(\zeta)] = \frac{d}{d\zeta}\{\Phi_N[\Phi_k(\zeta)]\} = \frac{d}{d\Phi_k}[\Phi_N(\Phi_k)]\frac{d}{d\zeta}[\Phi_k(\zeta)] = \Phi_N'(\Phi_k)\,\Phi_k'(\zeta) , \qquad (11.191)$$

where prime denotes the derivative of a function with respect to its argument. For the second derivative, we obtain

$$\frac{d^2}{d\zeta^2}[\Phi_K(\zeta)] = \Phi_N'(\Phi_k)\,\Phi_k''(\zeta) + \Phi_N''(\Phi_k)\,[\Phi_k'(\zeta)]^2 . \qquad (11.192)$$

Note the asymmetry in this result; the functional compounding takes place in a specific order.

We can now use the results in (11.191) and (11.192) to obtain the first two moments of K. From (C.65) we know how to compute factorial moments c_k from derivatives of the generating function. The first factorial moment c_1 is simply the mean m_1, and the second factorial moment is related to the mean and variance by (C.43). Thus we can write

$$\langle N\rangle = \Phi_N'(1)\,, \qquad \mathrm{Var}(N) = \Phi_N''(1) - [\Phi_N'(1)]^2 + \Phi_N'(1)\,. \tag{11.193}$$

Similarly, the mean and variance of each of k_n is given by

$$\langle k_n\rangle = \Phi_k'(1)\,, \qquad \mathrm{Var}(k_n) = \Phi_k''(1) - [\Phi_k'(1)]^2 + \Phi_k'(1)\,. \tag{11.194}$$

For the total output K, we obtain

$$\langle K\rangle = \Phi_N'[\Phi_k(1)]\,\Phi_k'(1) = \Phi_N'(1)\,\Phi_k'(1) = \langle N\rangle\langle k_n\rangle\,, \tag{11.195}$$

$$\mathrm{Var}(K) = \Phi_N'(1)\,\Phi_k''(1) + \Phi_N''(1)\,[\Phi_k'(1)]^2 - [\Phi_N'(1)\Phi_k'(1)]^2 + \Phi_N'(1)\,\Phi_k'(1)\,. \tag{11.196}$$

The variance of the compound process can be expressed in terms of the means and variances of the component processes by adding and subtracting $\Phi_N'(1)[\Phi_k'(1)]^2$ on the righthand side of (11.196):

$$\mathrm{Var}(K) = \Phi_N'(1)\left\{\Phi_k''(1) - [\Phi_k'(1)]^2 + \Phi_k'(1)\right\}$$
$$+ [\Phi_k'(1)]^2\left\{\Phi_N''(1) - [\Phi_N'(1)]^2 + \Phi_N'(1)\right\}. \tag{11.197}$$

With the use of (11.193) and (11.194), we now have

$$\mathrm{Var}(K) = \langle N\rangle\,\mathrm{Var}(k_n) + \langle k_n\rangle^2\,\mathrm{Var}(N)\,, \tag{11.198}$$

in agreement with (11.182).

These results have been used in the literature to analyze nonimaging detectors with gain, most notably photomultipliers (Lombard and Martin, 1961). In Sec. 11.4.5 we shall learn how to extend the method to imaging arrays.

11.4.3 Random amplification of point processes

In Sec. 11.4.1 we considered the statistics of the total number of output secondaries in a detector with gain, but we did not look at the random output process itself. If we describe each secondary as a delta function in time, then a sample function of the temporal random process on the detector output can be written as

$$y(t) = \sum_{n=1}^{N}\sum_{k=1}^{k_n}\delta(t - t_{nk}) = \sum_{n=1}^{N}\sum_{k=1}^{k_n}\delta(t - t_n - \Delta t_{nk})\,, \tag{11.199}$$

where t_n is the time at which the nth primary is absorbed, t_{nk} is the time at which the kth secondary (of those resulting from the nth primary) is produced, and Δt_{nk} is the random time delay from absorption of the nth primary to production of its kth secondary. The total number of output secondaries $K = \sum_{n=1}^{N} k_n$.

If the spatial distribution is of interest, a sample function of the amplified point process is given by

$$y(\mathbf{r}) = \sum_{n=1}^{N}\sum_{k=1}^{k_n} \delta(\mathbf{r} - \mathbf{r}_{nk}) = \sum_{n=1}^{N}\sum_{k=1}^{k_n} \delta(\mathbf{r} - \mathbf{R}_n - \Delta\mathbf{r}_{nk}), \tag{11.200}$$

where $\mathbf{R}_n$ is the location at which the nth primary is absorbed, $\mathbf{r}_{nk}$ is the location for the kth secondary produced by that primary, and $\Delta\mathbf{r}_{nk}$ is the random displacement. The only essential difference in the spatial and temporal cases is that Δt_{nk} must be positive since the secondary cannot be produced before the primary is absorbed, while the components of $\Delta\mathbf{r}_{nk}$ can have either sign.

We shall apply the procedures of Sec. 11.3.3 to calculate the mean and autocorrelation function of $y(\mathbf{r})$, and the results can be readily transcribed for the temporal case. For definiteness, we assume that $\mathbf{r}$, $\mathbf{R}_n$ and $\Delta\mathbf{r}_{nk}$ are 2D vectors, but that restriction too is easily lifted.

Specification of the probability laws The random quantities in $y(\mathbf{r})$ are the sets $\{\Delta\mathbf{r}_{nk}\}$, $\{\mathbf{R}_n\}$ and $\{k_n\}$ plus the random number of primaries N and, for a doubly stochastic source, the primary fluence $b(\mathbf{r})$. In order to decide how to average over all of these quantities, we must know how each affects the probability laws on the others.

Consider first the displacements $\{\Delta\mathbf{r}_{nk}\}$ of the secondaries from the primary positions. The secondaries from different primaries are generated independently since the devices in question have no memory from one primary to the next. Thus $\Delta\mathbf{r}_{nk}$ must be independent of $\Delta\mathbf{r}_{n'k'}$ for $n \neq n'$. Moreover, even for the same primary ($n = n'$), it is reasonable to assume that $\Delta\mathbf{r}_{nk}$ is independent of $\Delta\mathbf{r}_{nk'}$ for $k \neq k'$ unless there is some specific mechanism (perhaps space charge) coupling the secondaries. We can thus assume that the multivariate density on $\{\Delta\mathbf{r}_{nk}\}$ is a product of univariate densities on each of the $\Delta\mathbf{r}_{nk}$. Similarly, the *conditional* multivariate density $\mathrm{pr}(\{\mathbf{r}_{nk}\}|\mathbf{R}_n)$ is a product of conditional univariate densities $\mathrm{pr}(\mathbf{r}_{nk}|\mathbf{R}_n)$. Notice, however, that the statistical independence of the $\mathbf{r}_{nk}$ is lost when we consider the randomness in $\mathbf{R}_n$; then $\mathbf{r}_{nk}$ and $\mathbf{r}_{nk'}$ $(k \neq k')$ are not independent because of the common term $\mathbf{R}_n$. It remains quite reasonable to assume that $\Delta\mathbf{r}_{nk}$ and $\Delta\mathbf{r}_{nk'}(k \neq k')$ are independent even with random $\mathbf{R}_n$, and this is one advantage of working with $\Delta\mathbf{r}_{nk}$ rather than $\mathbf{r}_{nk}$.

In principle the univariate density on $\Delta\mathbf{r}_{nk}$ could depend on N, k_n or $b(\mathbf{r})$, but such dependence is ruled out if we consider only linear detectors where the mean spatial pattern on the output is independent of the input fluence. (With this assumption we ignore an effect called *blooming*, where an image is blurred more at high fluences than at lower ones.) On the other hand, we should allow the spatial pattern of secondaries to depend on the position of the primary interaction, since the blur in the amplification process is shift-variant in many detectors. Thus the density on $\Delta\mathbf{r}_{nk}$ can depend on $\mathbf{R}_n$ (though not on $\mathbf{R}_{n'}$ for $(n' \neq n)$, and the relevant univariate density describing the displacement is denoted $\mathrm{pr}_{\Delta\mathbf{r}}(\Delta\mathbf{r}_{nk}|\mathbf{R}_n)$.

The specific form of $\mathrm{pr}_{\Delta\mathbf{r}}(\Delta\mathbf{r}_{nk}|\mathbf{R}_n)$ can be deduced by an argument similar to the one used to obtain (11.70) or (11.76). Let $p_d(\mathbf{r}, \mathbf{R})$ be the shift-variant point spread function of the gain mechanism, defined as the mean number of secondaries per unit area at $\mathbf{r}$ from one primary absorbed at $\mathbf{R}$. By analogy to (11.66), the probability of one secondary falling in a vanishingly small area $\Delta A(\mathbf{r})$ centered at

$\mathbf{r}$ is given by

$$\Pr(1 \text{ sec in } \Delta A(\mathbf{r})|1 \text{ pri at } \mathbf{R}) = p_d(\mathbf{r}, \mathbf{R})\, \Delta A(\mathbf{r})\,. \tag{11.201}$$

Another way of computing the probability of one secondary in $\Delta A(\mathbf{r})$ is to consider a specific secondary, say the kth secondary produced by the nth primary. If exactly k_n indistinguishable secondaries are produced by the nth primary, the probability that one of them falls in $\Delta A(\mathbf{r})$ is k_n times the probability that specifically the kth secondary falls in that area. Thus

$$\begin{aligned} &\Pr(1 \text{ sec in } \Delta A(\mathbf{r})|1 \text{ pri at } \mathbf{R}_n, k_n \text{ secs produced}) \\ &= k_n \Pr(\mathbf{r}_{nk} \text{ in } \Delta A(\mathbf{r})|1 \text{ pri at } \mathbf{R}_n, k_n \text{ secs produced}) \\ &\quad = k_n \Delta A(\mathbf{r})\, \mathrm{pr}_{\Delta\mathbf{r}}(\Delta\mathbf{r}_{nk}|\mathbf{R}_n)|_{\Delta\mathbf{r}_{nk}=\mathbf{r}-\mathbf{R}_n}\,, \end{aligned} \tag{11.202}$$

where $\mathrm{pr}_{\Delta\mathbf{r}}(\Delta\mathbf{r}_{nk}|\mathbf{R}_n)$ is the probability density function on $\Delta\mathbf{r}_{nk}$ given that the kth secondary was produced by the nth primary at $\mathbf{R}_n$. Averaging over k_n yields

$$\Pr(1 \text{ sec in } \Delta A(\mathbf{r})|1 \text{ pri at } \mathbf{R}_n) = \overline{k}_n(\mathbf{R}_n)\, \Delta A(\mathbf{r})\, \mathrm{pr}_{\Delta\mathbf{r}}(\Delta\mathbf{r}_{nk}|\mathbf{R}_n)|_{\Delta\mathbf{r}_{nk}=\mathbf{r}-\mathbf{R}_n}\,, \tag{11.203}$$

where

$$\overline{k}_n(\mathbf{R}_n) = \int_\infty d^2r\; p_d(\mathbf{r}, \mathbf{R}_n)\,. \tag{11.204}$$

Comparison of (11.201) and (11.203) shows that

$$\mathrm{pr}_{\Delta\mathbf{r}}(\Delta\mathbf{r}_{nk}|\mathbf{R}_n) = \left[\overline{k}_n(\mathbf{R}_n)\right]^{-1} p_d(\Delta\mathbf{r}_{nk} + \mathbf{R}_n, \mathbf{R}_n)\,. \tag{11.205}$$

We shall also need the probability density function on $\mathbf{R}_n$. For a given photon fluence function $b(\mathbf{r})$ (or Hilbert-space vector $\mathbf{b}$), this density is given by (11.76), written here as

$$\mathrm{pr}_{pri}(\mathbf{R}_n|\mathbf{b}) = \frac{b(\mathbf{R}_n)}{\overline{N}(\mathbf{b})}\,, \tag{11.206}$$

where $\overline{N}(\mathbf{b})$ is the mean total number of counts for fixed $\mathbf{b}$.

We still need probability laws on N and $\{k_n\}$. Since the mechanisms of generating primaries and secondaries are unrelated, and since the secondaries produced by one primary have no effect on those produced by another primary, it is reasonable to assume that

$$\Pr(N, \{k_n\}|\mathbf{b}) = \Pr(N|\mathbf{b}) \Pr(\{k_n\}|N) = \Pr(N|\mathbf{b}) \prod_{n=1}^{N} \Pr(k_n)\,. \tag{11.207}$$

As we saw in Sec. 11.1.1, a fixed $\mathbf{b}$ implies a Poisson law for N, but by treating the probability on N as conditional on $\mathbf{b}$, we can later average over $\mathbf{b}$ and get a more general doubly stochastic model.

To summarize, averaging any function of the point process $y(\mathbf{r})$ requires the following steps, in sequence:

(a) Average over displacements $\{\Delta\mathbf{r}_{nk}\}$ for fixed $\mathbf{R}_n$ with density (11.205);

(b) Average over number of secondaries k_n for fixed $\mathbf{R}_n$;

(c) Average over $\mathbf{R}_n$ for fixed $\mathbf{b}$ with density (11.206);

(d) Average over total number of primaries N for fixed $\mathbf{b}$;

(e) Average over $\mathbf{b}$ if the primaries are doubly stochastic.

Calculation of the mean With the recipe given above, calculation of the mean of $y(\mathbf{r})$ is straightforward. Step (a), the average over $\{\Delta\mathbf{r}_{nk}\}$, yields

$$\langle y(\mathbf{r})\rangle_{\{\Delta\mathbf{r}_{nk}\}} = \sum_{n=1}^{N}\sum_{k=1}^{k_n}\int_\infty d^2\Delta r_{nk}\ \delta(\mathbf{r}-\mathbf{R}_n-\Delta\mathbf{r}_{nk})\left[\overline{k}_n(\mathbf{R}_n)\right]^{-1} p_d(\Delta\mathbf{r}_{nk}+\mathbf{R}_n,\mathbf{R}_n)$$

$$= \sum_{n=1}^{N}\sum_{k=1}^{k_n}\left[\overline{k}_n(\mathbf{R}_n)\right]^{-1} p_d(\mathbf{r},\mathbf{R}_n) = \sum_{n=1}^{N} k_n\left[\overline{k}_n(\mathbf{R}_n)\right]^{-1} p_d(\mathbf{r},\mathbf{R}_n)\,, \qquad (11.208)$$

where the last form follows from the observation that the summand is independent of k and that there are k_n terms in the sum over k. Step (b), the average over all k_n for fixed $\mathbf{R}_n$ simply replaces k_n by $\overline{k}(\mathbf{R}_n)$, so

$$\langle\langle y(\mathbf{r})\rangle_{\{\Delta\mathbf{r}_{nk}\}}\rangle_{\{k_n\}} = \sum_{n=1}^{N} p_d(\mathbf{r},\mathbf{R}_n)\,. \qquad (11.209)$$

Step (c) involves integration against the density $\mathrm{pr}_{pri}(\mathbf{R}_n|\mathbf{b})$ from (11.206), yielding

$$\langle\langle\langle y(\mathbf{r})\rangle_{\{\Delta\mathbf{r}_{nk}\}}\rangle_{\{k_n\}}\rangle_{\{\mathbf{R}_n\}} = \sum_{n=1}^{N}\left[\overline{N}(\mathbf{b})\right]^{-1}\int_\infty d^2R_n\ p_d(\mathbf{r},\mathbf{R}_n)\, b(\mathbf{R}_n)$$

$$= N\left[\overline{N}(\mathbf{b})\right]^{-1}\int_\infty d^2R\ p_d(\mathbf{r},\mathbf{R})\, b(\mathbf{R})\,, \qquad (11.210)$$

where the last line follows since all N terms in the sum are independent of n once the dummy variable of integration $\mathbf{R}_n$ is renamed $\mathbf{R}$.

Steps (d) and (e) are now easy. An average over N conditional on $\mathbf{b}$ replaces N with $\overline{N}(\mathbf{b})$, so

$$\mathrm{E}\{y(\mathbf{r})|\mathbf{b}\} = \int_\infty d^2R\ p_d(\mathbf{r},\mathbf{R})\, b(\mathbf{R})\,. \qquad (11.211)$$

Step (e), the average over $\mathbf{b}$, replaces $b(\mathbf{R})$ with $\overline{b}(\mathbf{R})$, yielding, finally,

$$\langle y(\mathbf{r})\rangle = \int_\infty d^2R\ p_d(\mathbf{r},\mathbf{R})\,\overline{b}(\mathbf{R}) = \int_\infty d^2R\ \mathrm{pr}_{\Delta\mathbf{r}}(\mathbf{r}-\mathbf{R}|\mathbf{R})\,\overline{k}_n(\mathbf{R})\,\overline{b}(\mathbf{R})\,, \quad (11.212)$$

where the second form has incorporated (11.205).

For later convenience, we now define a linear operator $\boldsymbol{\mathcal{H}}_1$ such that

$$[\boldsymbol{\mathcal{H}}_1\mathbf{b}]\,(\mathbf{r}) = \int_\infty d^2R\ p_d(\mathbf{r},\mathbf{R})\, b(\mathbf{R})\,, \qquad (11.213)$$

where $\mathbf{b}$ is the Hilbert-space vector corresponding to the function $b(\mathbf{R})$. The operator $\boldsymbol{\mathcal{H}}_1$ maps a function of $\mathbf{R}$ to a function of $\mathbf{r}$. Comparison with (11.212) shows that the kernel of the operator is $\mathrm{pr}_{\Delta\mathbf{r}}(\mathbf{r}-\mathbf{R}|\mathbf{R})\,\overline{k}_n(\mathbf{R})$. With this notation, we have

$$\langle y(\mathbf{r})\rangle = \left[\boldsymbol{\mathcal{H}}_1\overline{\mathbf{b}}\right](\mathbf{r})\,. \qquad (11.214)$$

Discussion The expressions for the mean in (11.212) may be anticlimactic; the first form, at least, could have been written down at once by someone conversant with shift-variant imaging systems (see Sec. 7.2.1) but ignorant of the statistics of point processes. The second form is also fairly obvious since it says merely that the input fluence pattern is first multiplied by the space-variant gain and then subjected to a space-variant blurring process.

Several limits of (11.212) are worth examining. If the system is really shift-invariant, then $p_d(\mathbf{r}, \mathbf{R}) = p_d(\mathbf{r} - \mathbf{R})$ and $\langle y(\mathbf{r}) \rangle$ is just the convolution of the mean input fluence $\overline{b}(\mathbf{R})$ with the point spread function. If the system is shift-variant but $\overline{b}(\mathbf{R})$ is slowly varying compared to the blur width, then $\overline{b}(\mathbf{R})$ can be approximated by $\overline{b}(\mathbf{r})$ and removed from the integral. The integral can then be evaluated via (11.205), giving

$$\langle y(\mathbf{r}) \rangle \approx \overline{b}(\mathbf{r})\, \overline{k}_n(\mathbf{r}) \,, \tag{11.215}$$

which is just the mean input fluence times the position-dependent gain. If the image amplifier is ideal in the sense that the blur width is negligible and the gain is independent of position, then $\langle y(\mathbf{r}) \rangle$ is the mean input fluence times a constant gain factor.

The final average over $\mathbf{b}$ may not be appropriate in all cases. Often we want to know the statistics of the image for *one* object, which implies one fluence pattern $b(\mathbf{r})$, so the conditional expectation $\mathrm{E}\{y(\mathbf{r})|\mathbf{b}\}$ is the important quantity; it is recovered from (11.212) just by deleting the overbar.

Calculation of the autocorrelation function The autocorrelation function of $y(\mathbf{r})$ is defined by

$$R_{\mathbf{y}}(\mathbf{r}, \mathbf{r}') = \langle y(\mathbf{r})\, y(\mathbf{r}') \rangle = \left\langle \sum_{n=1}^{N} \sum_{k=1}^{k_n} \delta(\mathbf{r} - \mathbf{R}_n - \Delta \mathbf{r}_{nk}) \sum_{n'=1}^{N} \sum_{k'=1}^{k_{n'}} \delta(\mathbf{r}' - \mathbf{R}_{n'} - \Delta \mathbf{r}_{n'k'}) \right\rangle , \tag{11.216}$$

where the angle brackets imply the five-step recipe outlined above. Executing the recipe is a bit tedious because of the four sums. As in Sec. 11.3.3, various special cases must be considered separately.

In the double sum over n and n', there are N terms with $n = n'$ (each term being itself a double sum over k and k') and $N^2 - N$ terms with $n \neq n'$. For $n = n'$, there are k_n terms with $k = k'$ in the double sum over k and k', and there are $k_n^2 - k_n$ terms with $k \neq k'$. For $n \neq n'$, it is irrelevant whether $k = k'$, so there are three cases to consider.

Case 1: $n = n'$ and $k = k'$

The calculation in this case parallels the derivation of (11.88), with the additional step of averaging over $\mathbf{R}_n$. From steps (a)–(d) in the recipe, the total contribution from all terms with $n = n'$ and $k = k'$ is

$$[\mathrm{E}\{y(\mathbf{r})\, y(\mathbf{r}')|\mathbf{b}\}]_{n=n',\, k=k'} = \left[\int_\infty d^2R\; p_d(\mathbf{r}, \mathbf{R})\, b(\mathbf{R}) \right] \delta(\mathbf{r} - \mathbf{r}') = [\boldsymbol{\mathcal{H}}_1 \mathbf{b}]\,(\mathbf{r})\, \delta(\mathbf{r} - \mathbf{r}') \,. \tag{11.217}$$

Step (e) in this case just replaces $b(\mathbf{R})$ with $\overline{b}(\mathbf{R})$, but we shall postpone this step for a while.

Case 2: $n = n'$ and $k \neq k'$

Now step (a) in the recipe requires averaging $\delta(\mathbf{r}-\mathbf{R}_n-\Delta\mathbf{r}_{nk})\,\delta(\mathbf{r}'-\mathbf{R}_n-\Delta\mathbf{r}_{nk'})$ over $\Delta\mathbf{r}_{nk}$ and $\Delta\mathbf{r}_{nk'}$ for fixed $\mathbf{R}_n$. Since $\Delta\mathbf{r}_{nk}$ and $\Delta\mathbf{r}_{nk'}$ are independent, this average is $\mathrm{pr}_{\Delta\mathbf{r}}(\mathbf{r}-\mathbf{R}_n|\mathbf{R}_n)\,\mathrm{pr}_{\Delta\mathbf{r}}(\mathbf{r}'-\mathbf{R}_n|\mathbf{R}_n)$. Each of the $k_n^2-k_n$ terms with $k \neq k'$ gives this same form, so the result of averaging over displacements and summing over k and k' for fixed n is

$$\sum_{k=1}^{k_n}\sum_{k'=1}^{k_n}(1-\delta_{kk'})\mathrm{E}_{\{\Delta\mathbf{r}_{nk}\}}\{\delta(\mathbf{r}-\mathbf{R}_n-\Delta\mathbf{r}_{nk})\,\delta(\mathbf{r}'-\mathbf{R}_n-\Delta\mathbf{r}_{nk'})\,|\mathbf{R}_n\}$$
$$= (k_n^2-k_n)\,\mathrm{pr}_{\Delta\mathbf{r}}(\mathbf{r}-\mathbf{R}_n|\mathbf{R}_n)\,\mathrm{pr}_{\Delta\mathbf{r}}(\mathbf{r}'-\mathbf{R}_n|\mathbf{R}_n)\,. \qquad (11.218)$$

Note that the factor $k_n^2-k_n$ eliminates primary interactions for which $k_n = 0$ or 1; there must be at least two secondaries for this expression to be nonzero since we are considering $k \neq k'$.

The average over k_n required in step (b) yields

$$\mathrm{E}\{k_n^2-k_n|\mathbf{R}_n\} = \mathrm{Var}(k_n|\mathbf{R}_n) + [\mathrm{E}\{k_n|\mathbf{R}_n\}]^2 - \mathrm{E}\{k_n|\mathbf{R}_n\} \equiv s(\mathbf{R}_n)\,, \qquad (11.219)$$

where we have introduced $s(\mathbf{R}_n)$ as a shorthand for the indicated combination of conditional mean and variance. The mnemonic is that $s(\mathbf{R})$ is related to the statistical spread of the gain mechanism at point $\mathbf{R}$.

Steps (c) and (d) now give

$$[\mathrm{E}\{y(\mathbf{r})\,y(\mathbf{r}')|\mathbf{b}\}]_{n=n',k\neq k'} = \int_\infty d^2R\;\mathrm{pr}_{\Delta\mathbf{r}}(\mathbf{r}-\mathbf{R})\,\mathrm{pr}_{\Delta\mathbf{r}}(\mathbf{r}'-\mathbf{R}|\mathbf{R})\,b(\mathbf{R})\,s(\mathbf{R})\,. \qquad (11.220)$$

We can express this term more compactly by defining a linear operator $\mathcal{H}_2$ that maps a function of $\mathbf{R}$ to a function of $\mathbf{r}$ and $\mathbf{r}'$. The definition is

$$[\mathcal{H}_2\mathbf{b}](\mathbf{r},\mathbf{r}') = \int_\infty d^2R\;\mathrm{pr}_{\Delta\mathbf{r}}(\mathbf{r}-\mathbf{R}|\mathbf{R})\,\mathrm{pr}_{\Delta\mathbf{r}}(\mathbf{r}'-\mathbf{R}|\mathbf{R})\,b(\mathbf{R})\,s(\mathbf{R})\,. \qquad (11.221)$$

Case 3: $n \neq n'$

If $n \neq n'$, $\Delta\mathbf{r}_{nk}$ and $\Delta\mathbf{r}_{n'k'}$ are independent, and there is no correlation between $\mathbf{R}_n$ and $\mathbf{R}_{n'}$ except possibly that induced by randomness in the fluence $b(\mathbf{r})$. For fixed fluence, then,

$$[\mathrm{E}\,\{y(\mathbf{r})\,y(\mathbf{r}')|\mathbf{b}\}]_{n\neq n'}$$
$$= \mathrm{E}\{N^2-N|\mathbf{b}\}\,\mathrm{E}\{k_n\,\delta(\mathbf{r}-\mathbf{R}_n-\Delta\mathbf{r}_{nk})|\mathbf{b}\}\,\mathrm{E}\{k_{n'}\,\delta(\mathbf{r}'-\mathbf{R}_{n'}-\Delta\mathbf{r}_{n'k})|\mathbf{b}\}\,. \qquad (11.222)$$

Since N is a Poisson random variable if the fluence is fixed, we know that

$$\mathrm{E}\{N^2-N|\mathbf{b}\} = \mathrm{Var}(N|\mathbf{b}) + [\mathrm{E}\{N|\mathbf{b}\}]^2 - \mathrm{E}\{N|\mathbf{b}\} = [\mathrm{E}\{N|\mathbf{b}\}]^2 = \left[\overline{N}(\mathbf{b})\right]^2\,. \qquad (11.223)$$

The remaining two expectations in (11.222) were calculated in (11.211), and we have

$$[\mathrm{E}\{y(\mathbf{r})\,y(\mathbf{r}')|\mathbf{b}\}]_{n\neq n'} = \int_\infty d^2R\;p_d(\mathbf{r},\mathbf{R})\,b(\mathbf{R})\int_\infty d^2R'\,p_d(\mathbf{r}',\mathbf{R}')\,b(\mathbf{R}')$$
$$= [\mathcal{H}_1\mathbf{b}](\mathbf{r})\,[\mathcal{H}_1\mathbf{b}](\mathbf{r}')\,, \qquad (11.224)$$

where $\mathcal{H}_1$ is the operator defined in (11.213).

Conditional autocorrelation and autocovariance Adding (11.217), (11.220) and (11.224), we find

$$R_y(\mathbf{r},\mathbf{r}'|\mathbf{b}) = \mathrm{E}\{y(\mathbf{r})\,y(\mathbf{r}')|\mathbf{b}\} = [\mathcal{H}_1\mathbf{b}](\mathbf{r})\,\delta(\mathbf{r}-\mathbf{r}') + [\mathcal{H}_2\mathbf{b}](\mathbf{r},\mathbf{r}') + [\mathcal{H}_1\mathbf{b}](\mathbf{r})\,[\mathcal{H}_1\mathbf{b}](\mathbf{r}')\,. \tag{11.225}$$

This expression is the conditional autocorrelation function (for fixed $\mathbf{b}$) of the amplified point process $y(\mathbf{r})$. The corresponding conditional autocovariance function is obtained by subtracting off the product of the conditional means; comparison of (11.225) and (11.211) shows that the term to be subtracted off is just the third term in (11.225), so the conditional autocovariance is

$$K_{\mathbf{y}}(\mathbf{r},\mathbf{r}'|\mathbf{b}) = \mathrm{E}\{y(\mathbf{r})y(\mathbf{r}')|\mathbf{b}\} - \mathrm{E}\{y(\mathbf{r})|\mathbf{b}\}\mathrm{E}\{y(\mathbf{r}')|\mathbf{b}\}$$

$$= \left[\int_\infty d^2R\; p_d(\mathbf{r},\mathbf{R})\,b(\mathbf{R})\right]\delta(\mathbf{r}-\mathbf{r}') + \int_\infty d^2R\; \mathrm{pr}_{\Delta\mathbf{r}}(\mathbf{r}-\mathbf{R}|\mathbf{R})\,\mathrm{pr}_{\Delta\mathbf{r}}(\mathbf{r}'-\mathbf{R}|\mathbf{R})\,b(\mathbf{R})\,s(\mathbf{R})$$

$$= [\mathcal{H}_1\mathbf{b}](\mathbf{r})\,\delta(\mathbf{r}-\mathbf{r}') + [\mathcal{H}_2\mathbf{b}](\mathbf{r},\mathbf{r}')\,. \tag{11.226}$$

This expression contains a delta-correlated part, which comes from the fact that each sample function of the amplified random process is a sum of delta functions, and it also has a term with a finite correlation range. The factor $\mathrm{pr}_{\Delta\mathbf{r}}(\mathbf{r}-\mathbf{R}|\mathbf{R})\,\mathrm{pr}_{\Delta\mathbf{r}}(\mathbf{r}'-\mathbf{R}|\mathbf{R})$ in the kernel of $\mathcal{H}_2$ drops to zero when $|\mathbf{r}-\mathbf{r}'|$ is approximately equal to the width of $\mathrm{pr}_{\Delta\mathbf{r}}(\mathbf{r}|\mathbf{R})$.

The similarity of the second term in $K_{\mathbf{y}}(\mathbf{r},\mathbf{r}'|\mathbf{b})$ to (11.140) should be noted (see also the discussion in Sec. 8.2.6). The randomness in $\Delta\mathbf{r}_{nk}$ induces a correlation in $y(\mathbf{r})$ of the same form as that induced by filtering. A key difference, however, is that there is no delta-correlated term in the autocorrelation or autocovariance function of a filtered point process. An amplified point process is still a point process, but a filtered point process is not. Instead, each input delta function produces a shifted replica of the filter spread function, so there is no longer a delta-correlated component.

A possible simplifying assumption at this point would be that the gain process obeys Poisson statistics, so that $\mathrm{Var}(k_n|\mathbf{R}_n) = \overline{k}_n(\mathbf{R}_n)$. In fact, a Poisson law for detectors with gain is rarely valid (see Sec. 11.4.2), but if it is, then $s(\mathbf{R})$ is just the square of the gain $\mathrm{E}\{k_n|\mathbf{R}_n\}$. In that case, the second term in (11.226) is identical to (11.140), with the effective filter spread function $h(\mathbf{r},\mathbf{R})$ given by the product of the gain and the probability density function $\mathrm{pr}_{\Delta\mathbf{r}}(\mathbf{r}-\mathbf{R}|\mathbf{R})$ that controls the blur.

Amplification without blur Another interesting limit is where the gain process has no significant blur associated with it, so that $\mathrm{pr}_{\Delta\mathbf{r}}(\mathbf{r}-\mathbf{R}|\mathbf{R})$ is well approximated by $\delta(\mathbf{r}-\mathbf{R})$. Then,

$$[\mathcal{H}_2\mathbf{b}](\mathbf{r},\mathbf{r}') \approx \int_\infty d^2R\; \delta(\mathbf{r}-\mathbf{R})\,\delta(\mathbf{r}'-\mathbf{R})\,b(\mathbf{R})\,s(\mathbf{R}) = b(\mathbf{r})\,s(\mathbf{r})\,\delta(\mathbf{r}-\mathbf{r}')\,. \tag{11.227}$$

Under this approximation, both terms are delta-correlated, and the conditional autocovariance is given by

$$K_{\mathbf{y}}(\mathbf{r},\mathbf{r}'|\mathbf{b}) = \overline{k}_n(\mathbf{r})\,b(\mathbf{r})\,\delta(\mathbf{r}-\mathbf{r}') + s(\mathbf{r})\,b(\mathbf{r})\,\delta(\mathbf{r}-\mathbf{r}')$$

$$= \left\{\mathrm{Var}(k_n|\mathbf{r}) + \overline{k}_n^2(\mathbf{r})\right\} b(\mathbf{r})\,\delta(\mathbf{r}-\mathbf{r}')\,, \tag{11.228}$$

where we have used (11.219). We recall from (11.94) that the autocovariance of a Poisson random process is just $b(\mathbf{r})\,\delta(\mathbf{r}-\mathbf{r}')$, and by computing the autocovariance of $y(\mathbf{r})$ conditional on $\mathbf{b}$, we are, in effect, assuming that the input to the amplifier is a Poisson random process. For an amplifier without blur, (11.228) shows that the autocovariance is increased by a factor of $\text{Var}(k_n|\mathbf{r}) + \overline{k}_n^2(\mathbf{r})$. For a noise-free, blur-free amplifier, the factor would be $\overline{k}_n^2(\mathbf{r})$, but noise in the amplification process imparts a random amplitude to the impulses.

Since $\text{Var}(k_n|\mathbf{r}) + \overline{k}_n^2(\mathbf{r})$ is just the second moment m_2 of the number of secondaries, the autocovariance is increased by m_2/m_1^2, which is the reciprocal of the Swank factor discussed in Sec. 11.4.1. Here the factor refers to the relative strength of an autocorrelation function rather than a variance, so it can be a function of position.

Mislocation without gain Some imaging detectors do not amplify the point process but rather estimate (with some error) the location of each impulse; an example is the Anger scintillation camera to be introduced in Sec. 12.3. In such devices, k_n is identically one but $\Delta\mathbf{r}_{n1}$ is still a random variable. The derivation just given still applies if we set $\text{Var}(k_n) = 0$ and $\overline{k}_n(\mathbf{R}) = 1$. Then $s(\mathbf{R}) = 0$ and the conditional autocovariance function (11.226) becomes

$$K_{\mathbf{y}}(\mathbf{r},\mathbf{r}'|\mathbf{b}) = \left[\int_\infty d^2R\; \text{pr}_{\Delta\mathbf{r}}(\mathbf{r}-\mathbf{R}|\mathbf{R})\, b(\mathbf{R})\right] \delta(\mathbf{r}-\mathbf{r}')\,, \tag{11.229}$$

which is just what we would have for a Poisson random process and an ideal detector, except that the mean fluence pattern is blurred with the spread function $\text{pr}_{\Delta\mathbf{r}}(\mathbf{r}|\mathbf{R})$. With no gain and no randomness in $\mathbf{b}$, only the delta-correlated term appears.

Random fluence If the fluence pattern is random, the overall autocorrelation function is obtained by averaging (11.225) over $\mathbf{b}$. The first two terms, both linear in $\mathbf{b}$, are easily averaged just by adding an overbar. The final term becomes

$$\begin{aligned}\text{E}_{\mathbf{b}}\{[\mathcal{H}_1\mathbf{b}](\mathbf{r})\,[\mathcal{H}_1\mathbf{b}](\mathbf{r}')\} &= \int_\infty d^2R\; p_d(\mathbf{r},\mathbf{R}) \int_\infty d^2R'\, p_d(\mathbf{r}',\mathbf{R}')\, \text{E}_{\mathbf{b}}\{b(\mathbf{R})\,b(\mathbf{R}')\} \\ &= \int_\infty d^2R\; p_d(\mathbf{r},\mathbf{R}) \int_\infty d^2R'\, p_d(\mathbf{r}',\mathbf{R}')\, R_{\mathbf{b}}(\mathbf{R},\mathbf{R}')\,,\end{aligned} \tag{11.230}$$

where $R_{\mathbf{b}}(\mathbf{r},\mathbf{r}')$ is the autocorrelation function of $b(\mathbf{r})$. The integral in (11.230) can be interpreted as the autocorrelation of $b(\mathbf{r})$ as transformed by $\mathcal{H}_1$; we denote it as $[\mathcal{H}_1 R_{\mathbf{b}} \mathcal{H}_1^\dagger](\mathbf{r},\mathbf{r}')$. The final expression for R_y is

$$R_y(\mathbf{r},\mathbf{r}') = \text{E}\{y(\mathbf{r})\,y(\mathbf{r}')\} = \left[\mathcal{H}_1\overline{\mathbf{b}}\right](\mathbf{r})\,\delta(\mathbf{r}-\mathbf{r}') + \left[\mathcal{H}_2\overline{\mathbf{b}}\right](\mathbf{r},\mathbf{r}') + \left[\mathcal{H}_1 R_{\mathbf{b}} \mathcal{H}_1^\dagger\right](\mathbf{r},\mathbf{r}')\,. \tag{11.231}$$

To get the autocovariance, we must subtract the product of the means, now including the average over $\mathbf{b}$. The result is

$$\begin{aligned}K_{\mathbf{y}}(\mathbf{r},\mathbf{r}') &= \text{E}\{y(\mathbf{r})\,y(\mathbf{r}')\} - \text{E}\{y(\mathbf{r})\}\,\text{E}\{y(\mathbf{r}')\} \\ &= \left[\mathcal{H}_1\overline{\mathbf{b}}\right](\mathbf{r})\,\delta(\mathbf{r}-\mathbf{r}') + \left[\mathcal{H}_2\overline{\mathbf{b}}\right](\mathbf{r},\mathbf{r}') + \left[\mathcal{H}_1 \mathbf{R}_{\mathbf{b}} \mathcal{H}_1^\dagger\right](\mathbf{r},\mathbf{r}') - \left[\mathcal{H}_1\overline{\mathbf{b}}\right](\mathbf{r})\,\left[\mathcal{H}_1\overline{\mathbf{b}}\right](\mathbf{r}')\,.\end{aligned} \tag{11.232}$$

The third and fourth terms comprise the autocovariance function $K_{\mathbf{b}}(\mathbf{r},\mathbf{r}')$ transformed by the operator $\mathcal{H}_1$, so we can write, finally,

$$K_{\mathbf{y}}(\mathbf{r},\mathbf{r}') = \left[\mathcal{H}_1\overline{\mathbf{b}}\right](\mathbf{r})\,\delta(\mathbf{r}-\mathbf{r}') + \left[\mathcal{H}_2\overline{\mathbf{b}}\right](\mathbf{r},\mathbf{r}') + \left[\mathcal{H}_1 K_{\mathbf{b}}\mathcal{H}_1^\dagger\right](\mathbf{r},\mathbf{r}')\,. \tag{11.233}$$

As in the case of nonrandom fluence, the delta-correlated term arises because the amplified point process is still a sum of delta functions. The second term arises from the amplification process, and the third is the contribution from the doubly stochastic nature of the source, as in (11.116).

11.4.4 Spectral analysis

As we saw in Sec. 11.2, stationary random processes are usefully described in the frequency domain by their power spectral densities. There are, however, many reasons why the amplified point process $y(\mathbf{r})$ might not be stationary. We shall enumerate these reasons here along with the assumptions needed for stationarity; then we shall compute the power spectral density of $y(\mathbf{r})$ under these assumptions.

For a random process to be stationary in the wide sense, its mean must be constant and its autocorrelation function must depend only on $\mathbf{r}-\mathbf{r}'$. From (11.212), the mean is constant if the blurring process $\mathrm{pr}_{\Delta\mathbf{r}}(\mathbf{r}-\mathbf{R}|\mathbf{R})$ is shift-invariant, meaning that it can be written as $\mathrm{pr}_{\Delta\mathbf{r}}(\mathbf{r}-\mathbf{R})$, and the mean gain $\overline{k}_n(\mathbf{R})$ and mean input fluence $\overline{b}(\mathbf{R})$ are independent of position $\mathbf{R}$. One way to achieve a constant input fluence is to illuminate the detector with a uniform flood source of radiation. In that case, $b(\mathbf{R})$ is nonrandom and independent of $\mathbf{R}$.

One additional condition for stationarity that is impossible to satisfy in practice is that the detector and the input fluence must both extend to infinity so that the limits of integration in (11.212) are infinite. As discussed in Sec. 8.2.4, boundary effects require a modified definition of stationarity. For present purposes we simply ignore the boundaries and use infinite limits.

If all of these conditions — constant gain and fluence, shift-invariant blurring and infinite field of view — are satisfied, (11.212) is the convolution of $\mathrm{pr}_{\Delta\mathbf{r}}(\mathbf{r})$ with a constant, which gives a constant.

To achieve stationarity for the conditional autocovariance (11.226), we must impose the additional condition that $s(\mathbf{R})$ is constant, which requires that the variance of the amplification process as well as its mean be constant. If we set $s(\mathbf{R}) = s_0$ and $b(\mathbf{R}) = b_0$ and make the change of variables $\mathbf{R}' = \mathbf{r}-\mathbf{R}$, we can rewrite the second term in (11.226) as

$$\begin{aligned}\left[\mathcal{H}_2\mathbf{b}\right](\mathbf{r},\mathbf{r}') &= s_0 b_0 \int_\infty d^2R'\,\mathrm{pr}_{\Delta\mathbf{r}}(\mathbf{R}')\,\mathrm{pr}_{\Delta\mathbf{r}}(\mathbf{r}'-\mathbf{r}+\mathbf{R}') \\ &= s_0 b_0 \left[\mathrm{pr}_{\Delta\mathbf{r}} \star \mathrm{pr}_{\Delta\mathbf{r}}\right](\mathbf{r}'-\mathbf{r})\,,\end{aligned} \tag{11.234}$$

where $\star$ denotes the spatial (not statistical) autocorrelation function as defined in (3.115). The conditional autocovariance under the accumulated assumptions is thus

$$K_{\mathbf{y}}(\mathbf{r},\mathbf{r}'|\mathbf{b}) = \overline{k}b_0\,\delta(\mathbf{r}-\mathbf{r}') + s_0 b_0 \left[\mathrm{pr}_{\Delta\mathbf{r}} \star \mathrm{pr}_{\Delta\mathbf{r}}\right](\mathbf{r}-\mathbf{r}')\,, \tag{11.235}$$

where $\overline{k}$ is the constant mean gain, and we have used the normalization of $\mathrm{pr}_{\Delta\mathbf{r}}(\mathbf{r})$.

Since this expression is manifestly a function of only $\mathbf{r}-\mathbf{r}'$, it describes a wide-sense stationary random process. The conditional power spectral density of the zero-mean process $\Delta y(\mathbf{r})$ is the Fourier transform of the autocovariance function $K_{\mathbf{y}}$ (see Sec. 8.2.5). An easy transform leads to

$$S_{\Delta \mathbf{y}}(\boldsymbol{\rho}|\mathbf{b}) = \overline{k} b_0 + s_0 b_0 |\psi_{\Delta \mathbf{r}}(\boldsymbol{\rho})|^2 , \tag{11.236}$$

where $\psi_{\Delta \mathbf{r}}(\boldsymbol{\rho})$ is the Fourier transform of $\mathrm{pr}_{\Delta \mathbf{r}}(\mathbf{r})$ (*i.e.*, the characteristic function for $\Delta \mathbf{r}$), and we have used (3.245).

The first term in (11.236) represents white noise with power spectral density $\overline{k} b_0$ equal to the mean number of secondaries per unit area. Since $\mathrm{pr}_{\Delta \mathbf{r}}(\mathbf{r})$ is the point spread function of the amplifier, $\psi_{\Delta \mathbf{r}}(\boldsymbol{\rho})$ is its transfer function, and the second term in (11.236) is proportional to the squared modulus of the transfer function. This term has a low-pass characteristic since $\mathrm{pr}_{\Delta \mathbf{r}}(\mathbf{r})$ is a blurring filter.

If the fluence pattern is a random process, then the autocovariance function has another term as seen in (11.233). In order for $y(\mathbf{r})$ to be stationary, $b(\mathbf{r})$ must be stationary as well. Under that assumption, the third term in (11.233) is just a double convolution, and we obtain

$$S_{\Delta \mathbf{y}}(\boldsymbol{\rho}) = \overline{k}\,\overline{b} + \left[s_0 \overline{b} + \overline{k}^2 S_{\Delta \mathbf{b}}(\boldsymbol{\rho}) \right] |\psi_{\Delta \mathbf{r}}(\boldsymbol{\rho})|^2 . \tag{11.237}$$

Now we have a superposition of three distinctly different spectral functions: a white-noise term $\overline{k}\,\overline{b}$, a term $s_0 \overline{b} |\psi_{\Delta \mathbf{r}}(\boldsymbol{\rho})|^2$ proportional to the squared modulus of the transfer function, and a term $\overline{k}^2 S_{\Delta \mathbf{b}}(\boldsymbol{\rho}) |\psi_{\Delta \mathbf{r}}(\rho)|^2$. The first of these arises since $y(\mathbf{r})$ is still a sum of impulses, the second arises from the spread of the gain mechanism, and the third represents the fluence spectrum as amplified and filtered by the transfer function.

11.4.5 Random amplification in arrays

If a detector with gain is to be used as part of a digital imaging system, the output must be binned into discrete pixels. The binned output is then a random vector rather than a random process, and a second-order statistical analysis must give the mean vector and the covariance matrix. There are three basic ways in which we can compute these quantities, corresponding to three different descriptions of imaging systems. As we saw in Chap. 7, an imaging system can be modeled as a continuous-to-continuous (CC), continuous-to-discrete (CD) or discrete-to-discrete (DD) mapping. The discussion in Secs. 11.3.3 and 11.3.4 was based on a stochastic CC model, where both input and output of the detector are random processes. However, we showed in Sec. 11.3.4 how to convert mean and autocovariance function of a random process to mean vector and covariance matrix by integrating over pixels. Applying this procedure to (11.226), we obtain immediately the conditional (fixed-fluence) covariance matrix,

$$[\mathbf{K}_{\mathbf{y}}(\mathbf{b})]_{mm'} = \delta_{mm'} \int_m d^2 r \, [\boldsymbol{\mathcal{H}}_1 \mathbf{b}](\mathbf{r}) + \int_\infty d^2 R \; p_m(\mathbf{R}) \, p_{m'}(\mathbf{R}) \, b(\mathbf{R}) \, s(\mathbf{R}) , \tag{11.238}$$

where

$$p_m(\mathbf{R}) = \int_m d^2 r \; \mathrm{pr}_{\Delta r}(\mathbf{r} - \mathbf{R}|\mathbf{R}) . \tag{11.239}$$

The second approach is to model the detector from the outset as a CD mapping and to modify the derivation given in Sec. 11.4.3 so that it uses probabilities rather than probability density functions. For example, instead of the conditional density $\mathrm{pr}(\mathbf{r}_{nk}|\mathbf{R}_n)$, we would consider the probability that a secondary produced by a primary at the continuous position $\mathbf{R}$ will fall in the jth output pixel. This approach will also lead to (11.238).

The third approach is to imagine that the position of the primary interaction is described not by a continuous variable $\mathbf{R}$, but by another discrete pixel index. In this approach we are interested in the probability that a secondary will fall in pixel j given that it was produced by a primary in pixel i. Few real detectors are accurately described by this DD model, just as few real imaging systems are accurately described by deterministic DD models. Nevertheless, such models are deeply entrenched in the literature, and it is worthwhile learning how to analyze them. Furthermore, the stochastic DD model affords a chance to gain more fluency with the useful technique of generating functions. For these reasons, we sketch here an analysis of amplifying detectors based on multivariate generating functions and a DD model, an approach used by Rabbani *et al.* (1987).

Multivariate generating functions We denote the random number of primaries absorbed on the jth input pixel by g_j, $j = 1, ..., J$, and the random number of secondaries on the mth output pixel by y_m, $m = 1, ..., M$. The set $\{g_j\}$ can be regarded as the components of a $J \times 1$ vector $\mathbf{g}$, while the set $\{y_m\}$ forms an $M \times 1$ vector $\mathbf{y}$. Note that the number of input pixels need not be the same as the number of output pixels, and in fact J can be allowed to go to infinity to get results for the CD case.

The multivariate generating function (more specifically, the probability-generating function) is a generalization of the univariate generating function discussed in Sec. C.3.3. For the output vector $\mathbf{y}$, the generating function is defined by

$$\Phi_{\mathbf{y}}(\zeta_1, \zeta_2, ..., \zeta_M) = \Phi_{\mathbf{y}}(\boldsymbol{\zeta}) = \mathrm{E}_{\mathbf{y}}\left\{\prod_{m=1}^{M} (\zeta_m)^{y_m}\right\}, \tag{11.240}$$

where $\boldsymbol{\zeta}$ is an $M \times 1$ vector with components $\{\zeta_m\}$. The generating function for the input vector $\mathbf{g}$ is defined similarly:

$$\Phi_{\mathbf{g}}(\xi_1, \xi_2, ..., \xi_J) = \Phi_{\mathbf{g}}(\boldsymbol{\xi}) = \mathrm{E}_{\mathbf{g}}\left\{\prod_{j=1}^{J} (\xi_j)^{g_j}\right\}, \tag{11.241}$$

where $\boldsymbol{\xi}$ is a $J \times 1$ vector.

It will also be useful to consider various conditional generating functions, defined like (11.240) or (11.241) but with conditional expectations. Consider first the conditional generating function $\Phi_{\mathbf{y}}(\boldsymbol{\zeta}|1 \text{ pri in } j, 1 \text{ sec})$ where one primary is absorbed in the jth input pixel and exactly one secondary is produced. Since there is only one secondary, the sample space for each of the y_m is just $(0, 1)$. The conditional probability that $y_m = 1$, given that one secondary was produced in input pixel j, will be denoted p_{jm}. If a particular y_m (say the one with $m = n$) equals one, all others must be zero since there is only one secondary. When that event occurs, the M-fold product in (11.240) consists of one factor given by $(\zeta_n)^1 = \zeta_n$ and $M - 1$ factors of $(\zeta_n)^0 = 1$. This event has probability p_{jn}, and it is mutually exclusive of the similar events for other m values, so the corresponding probabilities add. The conditional generating function is thus

$$\Phi_{\mathbf{y}}(\boldsymbol{\zeta}|1 \text{ pri in } j, 1 \text{ sec}) = \sum_{m=1}^{M} p_{jm}\zeta_m \,. \tag{11.242}$$

The righthand side of (11.242) has the form of a scalar product in an M-dimensional space, so we define an $M \times 1$ vector $\mathbf{p}_j$ with components p_{jm} and write

$$\Phi_{\mathbf{y}}(\boldsymbol{\zeta}|1 \text{ pri in } j, 1 \text{ sec}) = \mathbf{p}_j^t\boldsymbol{\zeta} \,. \tag{11.243}$$

We have assumed throughout that secondaries are independent, so if exactly k_j secondaries are produced at input pixel j, they are distributed independently among the output pixels. We thus have

$$\Phi_{\mathbf{y}}(\boldsymbol{\zeta}|1 \text{ pri in } j, k_j \text{ secs}) = (\mathbf{p}_j^t\boldsymbol{\zeta})^{k_j} \,. \tag{11.244}$$

Averaging over k_j yields

$$\Phi_{\mathbf{y}}(\boldsymbol{\zeta}|1 \text{ pri in } j) = \sum_{k_j=0}^{\infty} (\mathbf{p}_j^t\boldsymbol{\zeta})^{k_j} \Pr(k_j) = \Phi_{k_j}(\mathbf{p}_j^t\boldsymbol{\zeta}) \,, \tag{11.245}$$

where we have used the definition of the generating function, (11.186).

We can also take advantage of the fact that secondaries produced by different primaries are independent. If one primary is absorbed in pixel i and another in pixel j, the conditional expectations factor, and we find

$$\Phi_{\mathbf{y}}(\boldsymbol{\zeta}|1 \text{ pri in } i \text{ and } 1 \text{ pri in } j) = \Phi_{k_i}(\mathbf{p}_i^t\boldsymbol{\zeta})\, \Phi_{k_j}(\mathbf{p}_j^t\boldsymbol{\zeta}) \,. \tag{11.246}$$

By extension, given a specific realization of the random primary vector $\mathbf{g}$ (each component of which is an integer), we have

$$\Phi_{\mathbf{y}}(\boldsymbol{\zeta}|\mathbf{g}) = \prod_{j=1}^{J} \left[\Phi_{k_j}(\mathbf{p}_j^t\boldsymbol{\zeta})\right]^{g_j} \,. \tag{11.247}$$

Averaging over $\mathbf{g}$ yields

$$\Phi_{\mathbf{y}}(\boldsymbol{\zeta}) = \mathrm{E}_{\mathbf{g}}\{\Phi_{\mathbf{y}}(\boldsymbol{\zeta}|\mathbf{g})\} = \mathrm{E}_{\mathbf{g}}\left\{\prod_{j=1}^{J} \left[\Phi_{k_j}(\mathbf{p}_j^t\boldsymbol{\zeta})\right]^{g_j}\right\} \,. \tag{11.248}$$

Comparison with (11.241) shows that

$$\Phi_{\mathbf{y}}(\boldsymbol{\zeta}) = \Phi_{\mathbf{g}}\left[\Phi_{k_1}(\mathbf{p}_1^t\boldsymbol{\zeta}), \Phi_{k_2}(\mathbf{p}_2^t\boldsymbol{\zeta}), ..., \Phi_{k_J}(\mathbf{p}_J^t\boldsymbol{\zeta})\right] \,. \tag{11.249}$$

Mean and covariance We can derive the mean vector and covariance matrix for $\mathbf{y}$ by differentiating (11.249) in a manner similar to (11.195) – (11.198). Chain-rule differentiation gives

$$\frac{\partial \Phi_{\mathbf{y}}(\boldsymbol{\zeta})}{\partial \zeta_m} = \sum_{j=1}^{J} \frac{\partial \Phi_{\mathbf{g}}\left[\Phi_{k_1}(\mathbf{p}_1^t\boldsymbol{\zeta}), ..., \Phi_{k_J}(\mathbf{p}_J^t\boldsymbol{\zeta})\right]}{\partial \Phi_{k_j}(\mathbf{p}_j^t\boldsymbol{\zeta})} \frac{\partial \Phi_{k_j}(\mathbf{p}_j^t\boldsymbol{\zeta})}{\partial \mathbf{p}_j^t\boldsymbol{\zeta}} \frac{\partial \mathbf{p}_j^t\boldsymbol{\zeta}}{\partial \zeta_m} \,. \tag{11.250}$$

The mean of y_m is found by evaluating (11.250) with $\zeta_j = 1$ for all j (or $\boldsymbol{\zeta} = \mathbf{1}$, where $\mathbf{1}$ is a vector with all elements equal to one). The last factor in the summand, $\partial \mathbf{p}_j^t \boldsymbol{\zeta} / \partial \zeta_m$, is simply p_{jm}, independent of $\boldsymbol{\zeta}$, so no evaluation is needed. The middle factor evaluates to

$$\left.\frac{\partial \Phi_{k_j}(\mathbf{p}_j^t \boldsymbol{\zeta})}{\partial \mathbf{p}_j^t \boldsymbol{\zeta}}\right|_{\boldsymbol{\zeta}=\mathbf{1}} = \left.\frac{\partial \Phi_{k_j}(\nu)}{\partial \nu}\right|_{\nu=\sum_m p_{jm}=1} = \overline{k}_j\,, \tag{11.251}$$

where the middle step makes use of the normalization of probability, from which it follows that setting $\boldsymbol{\zeta} = \mathbf{1}$ is the same thing as setting $\mathbf{p}_j^t \boldsymbol{\zeta} = 1$.

The first factor in (11.251) evaluates to

$$\left.\frac{\partial \Phi_{\mathbf{g}}\left[\Phi_{k_1}(\mathbf{p}_1^t \boldsymbol{\zeta}), ..., \Phi_{k_J}(\mathbf{p}_J^t \boldsymbol{\zeta})\right]}{\partial \Phi_{k_j}(\mathbf{p}_j^t \boldsymbol{\zeta})}\right|_{\boldsymbol{\zeta}=\mathbf{1}} = \left.\frac{\partial \Phi_{\mathbf{g}}(\xi_1, ..., \xi_J)}{\partial \xi_j}\right|_{\boldsymbol{\xi}=\mathbf{1}} = \overline{g}_j\,, \tag{11.252}$$

where in this step we have used the fact that $\Phi_{k_j}(\mathbf{p}_j^t \mathbf{1}) = \Phi_{k_j}(1) = 1$ since the zeroth moment of any random variable is unity.

Collecting results, we find

$$\langle y_m \rangle = \sum_{j=1}^{J} \overline{g}_j \overline{k}_j p_{jm}\,. \tag{11.253}$$

A similar analysis shows that

$$\left.\frac{\partial^2 \Phi_{\mathbf{y}}(\boldsymbol{\zeta})}{\partial \zeta_m \partial \zeta_n}\right|_{\boldsymbol{\zeta}=\mathbf{1}} = \langle y_m y_n \rangle$$

$$= \sum_{j=1}^{J} \overline{g}_j \overline{k}_j p_{jm} \sum_{l=1}^{J} \overline{g}_l \overline{k}_l p_{lm} + \sum_{j=1}^{J} \overline{g}_j p_{jm} p_{jn} \langle k_j(k_j - 1)\rangle\,, \qquad (m \neq n)\,. \tag{11.254}$$

For the special case $m = n$ we find

$$\left.\frac{\partial^2 \Phi_{\mathbf{y}}(\boldsymbol{\zeta})}{\partial \zeta_m^2}\right|_{\boldsymbol{\zeta}=\mathbf{1}} = \langle y_m(y_m - 1)\rangle$$

$$= \sum_{j=1}^{J} \overline{g}_j \overline{k}_j p_{jm} \sum_{l=1}^{J} \overline{g}_l \overline{k}_l p_{lm} + \sum_{j=1}^{J} \overline{g}_j p_{jm}^2 \langle k_j(k_j - 1)\rangle\,. \tag{11.255}$$

From these expressions, the covariance matrix of $\mathbf{y}$ (for fixed fluence) is found to be

$$\left[\mathbf{K}_{\mathbf{y}}(\mathbf{b})\right]_{mn} = \left[\langle y_m y_n \rangle - \langle y_m \rangle \langle y_n \rangle\right](1 - \delta_{mn}) + \left[\langle y_m(y_m - 1)\rangle + \langle y_m \rangle - \langle y_m \rangle^2\right]\delta_{mn}$$

$$= \sum_{j=1}^{J} \overline{g}_j p_{jm} p_{jn} \langle k_j(k_j - 1)\rangle + \left[\sum_{j=1}^{J} \overline{g}_j p_{jm} \overline{k}_j\right]\delta_{mn}\,. \tag{11.256}$$

This expression is the discrete counterpart of (11.226). The latter can be derived from (11.256) by sprinkling in factors of $1/\epsilon^2$, where ϵ is the pixel width, and passing to the limit $\epsilon \to 0$ [*cf.* (11.106)]. In carrying out this procedure, note that $\langle k_j(k_j - 1)\rangle$ corresponds to $s(\mathbf{R})$, which appears in (11.226) in the guise of $\mathcal{H}_2$ [see (11.219) and (11.221)].

The covariance matrix in the CD case can also be derived by letting $J \to \infty$ in (11.256) but keeping the output pixel size constant. In this case the Kronecker delta function remains instead of limiting to a Dirac delta as in (11.105). This procedure will reproduce (11.238).

11.5 QUANTUM MECHANICS OF PHOTON COUNTING

In this section we examine the role of quantum electrodynamics (QED) in the analysis of photon-counting experiments. Much of the groundwork for this discussion was laid in Chap. 10, where we introduced the concepts of modes of the radiation field, number operators and photons, along with the elements of photodetection theory. The reader is presumed to be familiar with these ideas and with the basic principles of quantum mechanics, including the concept of a state vector and its relation to a wavefunction. Dirac notation is used freely, and quantum-mechanical operators are denoted with a caret.

11.5.1 Coherent states

Since this chapter is mainly about Poisson statistics, we begin our discussion of QED and photon counting by considering a set of quantum states where the Poisson distribution is of central importance. These states, introduced briefly in Sec. 10.1.3, are called variously *coherent states*, *canonical coherent states*, *Glauber states* or *minimum uncertainty states*. They are states of a harmonic oscillator, but we saw in Chap. 10 that a single mode of the radiation field is in fact described by a harmonic oscillator.

The fundamental reference on coherent states is Glauber (1963), and excellent discussions are found in Meystre and Sargent (1990), Mandel and Wolf (1995) and Cohen-Tannoudji *et al.* (1989). Here we summarize, without derivation, the main properties needed to better understand the role of Poisson statistics in imaging. Along the way, we shall note some interesting connections with local Fourier transforms, wavelets and other mathematical concepts introduced earlier in this book.

Eigenstates of the annihilation operator We saw in Sec. 10.1.3 that the electric field operator for the jth mode of the radiation field is [*cf.* (10.22)]

$$\hat{\mathbf{e}}_j(\mathbf{r},t) = i\boldsymbol{\gamma}_j N_j \left[\exp(i\mathbf{k}_j\cdot\mathbf{r})\,\hat{a}_j - \exp(-i\mathbf{k}_j\cdot\mathbf{r})\,\hat{a}_j^\dagger\right] , \tag{11.257}$$

where $\boldsymbol{\gamma}_j$ is a unit vector in the direction of the field, N_j is a constant defined in (10.13), and $\hat{a}_j^\dagger$ and $\hat{a}_j$ are, respectively, the creation and annihilation operators. As shown in (10.27), the creation operator increases the number of photons in the field by one and the annihilation operator decreases the number by one.

In Sec. 10.1.3 we discussed mainly the number states, eigenstates of the Hamiltonian. Another very useful set of states consists of the eigenstates of the annihilation operator. The annihilation operator is especially important since (as stated more precisely in Sec. 10.1.4) photodetection takes place by annihilating photons.

We now drop the index j designating a particular mode of the field. Then the annihilation operator is denoted by $\hat{a}$, and its eigenstate $|\alpha\rangle$ is defined by

$$\hat{a}|\alpha\rangle = \alpha|\alpha\rangle\,. \tag{11.258}$$

Since $\hat{a}$ is not Hermitian, its eigenvalue α need not be real, and in fact it can take on any value in the complex plane.

The state $|\alpha\rangle$ can be expressed in terms of number states as

$$|\alpha\rangle = \sum_{n=0}^{\infty} |n\rangle\langle n|\alpha\rangle\,, \tag{11.259}$$

where (Glauber, 1963)

$$\langle n|\alpha\rangle = \exp(-\tfrac{1}{2}|\alpha|^2)\,\frac{\alpha^n}{\sqrt{n!}}\,. \tag{11.260}$$

As noted in Sec. 10.1.3, the probability that the state $|\alpha\rangle$ contains n photons is given by

$$|c_n|^2 = |\langle n|\alpha\rangle|^2 = \exp(-|\alpha|^2)\,\frac{|\alpha|^{2n}}{n!}\,, \tag{11.261}$$

which is a Poisson distribution with mean $|\alpha|^2$.

Orthogonality and completeness The states $|\alpha\rangle$ are not orthogonal; instead they satisfy (Glauber, 1963)

$$|\langle\alpha|\alpha'\rangle|^2 = \exp(-|\alpha-\alpha'|^2)\,. \tag{11.262}$$

In spite of the lack of orthogonality, the coherent states form a complete set in terms of which any state of the mode can be expanded. The set is overcomplete in the sense that some subset of it would also span the state space; the state space is an $\mathbb{L}_2$ space and can therefore be spanned by a denumerable basis (see Sec. 1.1.5).

The closure relation (resolution of the unit operator) is

$$\hat{I} = \frac{1}{\pi}\int_\infty d^2\alpha\;|\alpha\rangle\langle\alpha|\,, \tag{11.263}$$

where $d^2\alpha$ is an area element in the complex α plane. The factor of $1/\pi$ is a measure of overcompleteness of the states; for a complete, orthonormal set, the sum of outer products is $\hat{I}$ rather than $\pi\hat{I}$. Equation (11.263) is reminiscent of closure relations for other overcomplete sets or extended representations introduced in Chap. 5; see, in particular, (5.39) and (5.85). We shall explore this connection further below.

One use of the closure relation is to express a number state as

$$|n\rangle = \frac{1}{\pi}\int_\infty d^2\alpha\;|\alpha\rangle\langle\alpha|n\rangle\,, \tag{11.264}$$

where $\langle\alpha|n\rangle$ is the complex conjugate of (11.260).

Displacement operator As discussed by Glauber (1963), the coherent state $|\alpha\rangle$ can be generated by the *Weyl displacement operator* $\hat{D}(\alpha)$ acting on the harmonic-oscillator ground state (or vacuum state):

$$|\alpha\rangle = \hat{D}(\alpha)|0\rangle\,. \tag{11.265}$$

Explicitly, $\hat{D}(\alpha)$ is given by

$$\hat{D}(\alpha) = \exp\left(\alpha\hat{a}^\dagger - \alpha^*\hat{a}\right) = \exp(-\tfrac{1}{2}|\alpha|^2)\exp(\alpha\hat{a}^\dagger)\,. \tag{11.266}$$

The exponential operator is defined in the same way as a matrix exponential and obeys the same manipulation rules (see Sec. A.7.1). Note that $\hat{D}(\alpha)$ reduces to the unit operator when $\alpha \to 0$. Thus the vacuum state corresponds to both $n = 0$ and $\alpha = 0$, and the ambiguity of the notation $|0\rangle$ causes no problem.

and we also define real scalars q and p by

$$q = \sqrt{\frac{\hbar}{2\omega}}\,(\alpha^* + \alpha)\,, \qquad p = i\sqrt{\frac{\hbar\omega}{2}}\,(\alpha^* - \alpha)\,, \tag{11.268}$$

respectively. With these definitions,

$$\hat{D}(\alpha) = \exp\left[\frac{i}{\hbar}\left(p\hat{Q} - q\hat{P}\right)\right] \equiv \hat{D}(q,p)\,. \tag{11.269}$$

To discover the reason for the term *displacement operator*, we use the coordinate representation where

$$\hat{Q}|q'\rangle = q'|q'\rangle\,. \tag{11.270}$$

In this representation, the matrix elements of $\hat{Q}$ are given by (Messiah, 1961)

$$\langle q'|\hat{Q}|q''\rangle = q'\delta(q' - q'')\,, \tag{11.271}$$

and those of $\hat{P}$ are given by

$$\langle q'|\hat{P}|q''\rangle = -i\hbar\frac{\partial}{\partial q'}\,\delta(q' - q'')\,. \tag{11.272}$$

With (11.270), it follows that

$$\langle q'|\hat{D}(q,p)|q''\rangle = \exp\left(\frac{i}{\hbar}pq' - q\frac{\partial}{\partial q'}\right)\delta(q' - q'')\,. \tag{11.273}$$

In the coordinate representation, the stationary-state wavefunctions $\psi_n(q') = \langle q'|n\rangle$ are Hermite-Gauss functions, and in particular the vacuum state is represented by (Messiah, 1961)

$$\langle q'|0\rangle = \left(\frac{\omega}{\pi\hbar}\right)^{\frac{1}{4}} \exp\left(-\frac{\omega}{2\hbar}q'^2\right)\,. \tag{11.274}$$

Thus

$$\langle q'|\alpha\rangle = \int dq''\,\langle q'|\hat{D}(p,q)|q''\rangle\langle q''|0\rangle = \left(\frac{\omega}{\pi\hbar}\right)^{\frac{1}{4}} \exp\left(\frac{i}{\hbar}pq' - q\frac{\partial}{\partial q'}\right)\exp\left(-\frac{\omega}{2\hbar}q'^2\right)\,. \tag{11.275}$$

We recognize $\exp(-q\frac{\partial}{\partial q'})$ as the displacement operator defined in Sec. A.10.1 of App. A. From (A.177) and (11.275), it follows that

$$\langle q'|\alpha\rangle = \int dq''\,\langle q'|\hat{D}(p,q)|q''\rangle\langle q''|0\rangle = \left(\frac{\omega}{\pi\hbar}\right)^{\frac{1}{4}} \exp\left(\frac{i}{\hbar}pq'\right)\exp\left[-\frac{\omega}{2\hbar}(q' - q)^2\right]\,. \tag{11.276}$$

Thus the effect of $\hat{D}(q,p)$ on the vacuum wavefunction $\langle q'|0\rangle$ is to displace it from the origin by an amount q and to multiply it by the linear phase factor $\exp(ipq'/\hbar)$.

Uncertainty We showed in Sec. 5.1.2 that the uncertainty product $\sigma_q\sigma_p$ is minimized when the wavefunction is a Gaussian. Thus the uncertainty is minimal for a harmonic oscillator when it is in its ground state. The same uncertainty product is, however, obtained for all coherent states. The phase factor in (11.276) vanishes when we compute the probability density $|\langle q'|\alpha\rangle|^2$, and a shift of a probability density function does not affect its variance. In fact, the phase factor is merely a shift in the momentum wavefunction, which is the Fourier transform of $\langle q'|\alpha\rangle$, and the squared modulus of the momentum wavefunction is the probability density function on the momentum. Hence both σ_q and σ_p are independent of α by the observation that variance is independent of shift.

Thus the coherent states comprise a family of states, all of which have the smallest possible uncertainty. Recall also that the coherent states are quantum states of a single mode of the radiation field and that the mode is a plane wave. A perfectly monochromatic classical plane wave would have constant amplitude, but in quantum mechanics the real and imaginary parts of the amplitude are operators subject to the uncertainty relation. The coherent state is the best quantum-mechanical counterpart of a classical monochromatic plane wave in the sense of minimizing that uncertainty.

Local Fourier transforms In Sec. 5.1 we defined a local Fourier transform by windowing a function with, say, a Gaussian, multiplying it by a linear phase factor (or Fourier kernel), and then integrating. The local Fourier transform (5.1) is then a function of the shift of the window function and the spatial frequency, which is essentially the slope of the linear phase. Equivalently, the local Fourier transform is a scalar product with a kernel of the form $b(x - x_0)\exp(2\pi i\xi x)$. Exactly the same operations—shifting and multiplying by a linear phase factor—occur in (11.276), so we can think of the coherent-state wavefunction as a kernel in a local Fourier transform.

This analogy sheds some light on the orthogonality and completeness properties of the coherent states. We saw in Sec. 5.1 that the local Fourier transform was always invertible without any requirement that the kernels be orthogonal. As seen in (5.40), inversion of the local Fourier transform requires integration over both the shift variable and the spatial frequency. The closure relation (5.39) is, in fact, identical to (11.263) since the integral over the complex α plane can be expressed in terms of integrals over p and q. The reader should fill in the details of this assertion as an exercise.

Coherent states and group theory In Chap. 5 we saw the similarities between the local Fourier transform and the wavelet transform, and in Chap. 6 we investigated the relation between wavelets and group theory. There is also an intimate connection between coherent states and group theory (Klauder, 1985; Daubechies, 1992).

As we saw in Sec. 6.8.5, wavelets are derived from the affine group, a Lie group that involves shifting and scaling a function. Coherent states and the local Fourier transform are related to the *Weyl-Heisenberg group*, a group of functional transformations of the form, $f(x) \to \exp(2\pi i\xi x)\, f(x - x_0)$. This group is sometimes referred to as the *translate-modulate* group while the affine group is the *translate-scale* group.

There are many similarities between the Weyl-Heisenberg and affine groups. Both are multiparameter Lie groups, both are non-Abelian and both have only infinite-dimensional irreducible representations (see Sec. 6.8.5). A minor difference is that the affine group has two parameters, the shift and the scale, but the Weyl-Heisenberg group actually has three; we must include multiplication by a constant phase factor $e^{i\phi}$ in order for the functional transformations to form a group. An element of the group is then specified by x_0, ξ and ϕ (or q, p and ϕ in quantum language).

Klauder (1985) has shown how *any* Lie group can be used to define generalized coherent states. Consider a Lie group **G** represented by a set of unitary operators $\{\hat{T}_\theta\}$ on some Hilbert space. (In Chap. 6 such an operator would have been denoted by $\boldsymbol{T}_\theta$, but here we use the caret to suggest a quantum-mechanical operator.) A set of vectors in the Hilbert space can be defined by

$$|\theta\rangle = \hat{T}_\theta|0\rangle\,, \tag{11.277}$$

where $|0\rangle$ is an arbitrary reference state, called the *fiducial state*. Klauder calls $|\theta\rangle$ a generalized coherent state if it satisfies two conditions:

Continuity: The vector $|\theta\rangle$ is a strongly continuous function of the label θ.

Completeness: There exists a positive measure $\delta\theta$ on the Hilbert space such that the unit operator can be expressed as

$$\hat{I} = \int \delta\theta\ |\theta\rangle\langle\theta|\,. \tag{11.278}$$

These conditions are satisfied by Glauber's coherent states (which Klauder refers to as the *canonical coherent states*); in that case, the fiducial state is the vacuum state of a harmonic oscillator, and the resolution of unity is given in (11.263).

Note that the continuity condition rules out discrete orthogonal vectors or δ-normalized continuum orthogonal vectors such as delta functions and plane waves. Moreover, the vectors $|\theta\rangle$ are not orthogonal in general, so $|\theta\rangle$ cannot be an eigenstate of a Hermitian operator.

11.5.2 Density operators

According to the basic axioms of quantum mechanics, any physical observable Ω is represented by a Hermitian operator $\hat{\Omega}$, and the expectation value of the observable in any pure quantum state $|\psi\rangle$ is

$$\langle\hat{\Omega}\rangle = \langle\psi|\hat{\Omega}|\psi\rangle\,. \tag{11.279}$$

This expectation can be computed in any convenient basis. For example, in the basis formed by the number states,

$$\langle\hat{\Omega}\rangle = \sum_{n,m}\langle\psi|n\rangle\langle n|\hat{\Omega}|m\rangle\langle m|\psi\rangle\,, \tag{11.280}$$

and in the coordinate basis,

$$\langle\hat{\Omega}\rangle = \int_\infty dq' \int_\infty dq''\ \langle\psi|q'\rangle\langle q'|\hat{\Omega}|q''\rangle\langle q''|\psi\rangle\,. \tag{11.281}$$

It is useful to define an operator, called the *density operator*, by

$$\hat{\rho} = |\psi\rangle\langle\psi|\,. \tag{11.282}$$

The matrix elements of this operator in the number-state representation are given by

$$\rho_{mn} = \langle m|\psi\rangle\langle\psi|n\rangle\,, \tag{11.283}$$

and the matrix with these elements is called the *density matrix*.

In terms of the density operator, (11.280) becomes

$$\langle\hat{\Omega}\rangle = \sum_m \left[\sum_n \rho_{mn}\langle n|\hat{\Omega}|m\rangle\right] = \mathrm{tr}\left\{\hat{\rho}\,\hat{\Omega}\right\}\,, \tag{11.284}$$

where $\mathrm{tr}\{\cdot\}$ denotes the trace or sum of diagonal elements.

Since the trace is invariant to unitary transformations, the final form of (11.284) applies also in the coordinate representation where

$$\langle q'|\hat{\rho}|q''\rangle = \langle q'|\psi\rangle\,\langle\psi|q''\rangle = \psi(q')\psi^*(q'')\,, \tag{11.285}$$

and $\psi(q')$ is the usual Schrödinger wavefunction. In this representation, the counterpart of (11.284) is

$$\langle\hat{\Omega}\rangle = \int_\infty dq' \int_\infty dq''\,\langle q'|\hat{\rho}|q''\rangle\langle q''|\hat{\Omega}|q'\rangle = \int_\infty dq'\,\langle q'|\hat{\rho}\,\hat{\Omega}|q'\rangle\,, \tag{11.286}$$

where the last step follows from the closure relation,

$$\int_\infty dq''\,|q''\rangle\langle q''| = \hat{I}\,. \tag{11.287}$$

The single integral in the last form of (11.286) can be interpreted as a trace, but with the discrete indices replaced with continuous variables and the sum replaced by an integral.

Mixed or doubly stochastic states As we have just seen, the density operator, in any representation, is sufficient for computing the expectation value of any physical observable; knowledge of $\hat{\rho}$ is equivalent to knowledge of the wavefunction or state vector. Often, however, we do not know that the system is in a particular state. Perhaps we can say only that it is in state $|\psi_k\rangle$ with probability $\Pr(k)$. In the quantum-mechanics literature, such a system is often said to be in a *mixed state*, though this may be misleading since we are really just saying that we do not know what state it is in. In a frequentist interpretation of probability, the system is in state $|\psi_k\rangle$ with relative frequency $\Pr(k)$ (over repeated experiments), and in an ensemble interpretation a fraction $\Pr(k)$ of the systems are described by state vector $|\psi_k\rangle$. In either case, $\Pr(k)$ is a classical probability satisfying the Kolmogorov axioms (see Sec. C.1.4), not a quantum-mechanical probability amplitude.

In a mixed state, the expectation must be computed by a double averaging process: first the quantum-mechanical expectation for a particular state, then a classical averaging over states. The resulting expression is

$$\langle\hat{\Omega}\rangle = \sum_k \Pr(k)\,\langle\psi_k|\hat{\Omega}|\psi_k\rangle\,. \tag{11.288}$$

In Sec. 11.1.4, we defined a doubly stochastic random variable as one for which a parameter of the probability law is itself random. That definition applies here if we think of the index k as the random parameter. Thus the averaging process in (11.288) is the same sort of doubly stochastic averaging we have used often in this chapter.

We can still use the concept of a density operator if we now define it as

$$\hat{\rho} = \sum_k \Pr(k) |\psi_k\rangle\langle\psi_k| \,. \tag{11.289}$$

The density matrix in the number-state representation (or any other denumerable basis) is given by

$$\rho_{nm} = \langle n|\rho|m\rangle = \sum_k \Pr(k)\, \langle n|\psi_k\rangle\langle\psi_k|m\rangle \,, \tag{11.290}$$

and we can still write the overall, doubly averaged, expectation of $\hat{\Omega}$ as

$$\langle\hat{\Omega}\rangle = \operatorname{tr}\left\{\hat{\rho}\hat{\Omega}\right\} . \tag{11.291}$$

Density operator in the coherent-state representation The density operator can be expressed in terms of coherent states as (Glauber, 1963)

$$\hat{\rho} = \int d^2\alpha \; P(\alpha)|\alpha\rangle\langle\alpha| \,. \tag{11.292}$$

The function $P(\alpha)$ appears to play the same role as $\Pr(k)$ in (11.290), but it cannot be interpreted as a classical probability or probability density function. In (11.290), the states $|\psi_k\rangle$ are distinct (orthogonal), and the system is in state k with probability $\Pr(k)$. If the system is in a particular $|\psi_k\rangle$, say $k = k_0$, with probability one, then by the Kolmogorov axioms (see Sec. C.1.4), its probability of being in some other state must be zero. Because of the lack of orthogonality, however, the states $|\alpha\rangle$ are not distinct. If the system is described exactly by $|\psi\rangle = |\alpha_0\rangle$, then its probability $|\langle\alpha_1|\psi\rangle|^2$ of being in some other coherent state $|\alpha_1\rangle$ is not zero. This problem causes $P(\alpha)$ to have some properties that are not allowed in classical probability theory. As we shall see in some examples below, $P(\alpha)$ can go negative, and it can be more singular than a delta function; that is, it does not integrate to a finite value and hence it cannot be normalized as a probability density function.

Nevertheless, $P(\alpha)$ is quite useful for computing expectation values of functions of the operators $\hat{a}$ and $\hat{a}^\dagger$. Suppose we have a classical complex random variable $z = x + iy$, and we want to compute the expectation of some function $f(x, y)$. Since x and y can be expressed in terms of z and z^*, we can define $f(x, y) = f'(z^*, z)$ and write the expectation as

$$\langle f'(z^*, z)\rangle = \int_\infty d^2z \; \mathrm{pr}(z)\, f'(z^*, z) \,, \tag{11.293}$$

where the notation $\mathrm{pr}(z)$ is used as a shorthand for the joint density of z and z^*. One might think that a similar formula would hold for an operator function $f(\hat{a}^\dagger, \hat{a})$, but $\hat{a}^\dagger$ and $\hat{a}$ do not commute, so any expectation must depend on the order in which the operators are written.

One important way of ordering operators is *normal ordering*, where all creation operators appear to the left of all annihilation operators; an example is $[a^\dagger]^3 a^2$.

Another possibility is *antinormal ordering* where the annihilation operators appear to the left, as in $a^2[a^\dagger]^3$. There are also several ways of defining a *symmetrical ordering* (Tatarskii, 1983). Since the commutation relations are known, it is straightforward to convert from one ordering to another. Normal ordering is important in optics since, as we saw in Sec. 10.1.4, the photodetection rate is proportional to an expectation value of a normally ordered operator.

The function $P(\alpha)$ plays the role of a probability for normally ordered operators. The quantum-mechanical counterpart of (11.293) is

$$\langle \{f(\hat{a}^\dagger, \hat{a})\}_{nor} \rangle = \int_\infty d^2\alpha \; P(\alpha)\, f(\alpha^*, \alpha)\,, \tag{11.294}$$

where $\{\cdot\}_{nor}$ indicates a normal ordering. Since it appears where a true probability density function would appear in this formula, $P(\alpha)$ is called a *quasiprobability.*

Other quasiprobabilities Equations analogous to (11.294) can be given for other ordering schemes. The quasiprobability for antinormal ordering is usually denoted $Q(\alpha)$. If $P(\alpha)$ is known, $Q(\alpha)$ can be found by convolving it with a Gaussian of the form $\exp(-|\alpha|^2)$ (Cohen-Tannoudji *et al.*, 1989). As a result of the smoothing action of this convolution, $Q(\alpha)$ is much better behaved than $P(\alpha)$; it is never negative and it can always be normalized as a probability density function.

For a certain kind of symmetric ordering (called *Weyl ordering*), the appropriate quasiprobability is the Wigner distribution function or WDF (Tatarskii, 1983). The WDF is also a smoothed version of $P(\alpha)$, but the convolution is with $\exp(-\frac{1}{2}|\alpha|^2)$ rather than $\exp(-|\alpha|^2)$. The WDF is intermediate between P and Q in terms of its mathematical behavior; it can go negative but nevertheless it obeys the normalization rules of a true joint probability density function.

From $\mathbf{P(\alpha)}$ to the density matrix In terms of $P(\alpha)$, the number-state matrix elements of the density operator are given by

$$\rho_{nm} = \langle n|\hat{\rho}|m\rangle = \int_\infty d^2\alpha \; P(\alpha)\, \langle n|\alpha\rangle\langle\alpha|m\rangle\,, \tag{11.295}$$

where $\langle n|\alpha\rangle$ is given by (11.260). The diagonal element ρ_{nn} is the probability of having n photons in the mode; it is given by

$$\rho_{nn} = \langle n|\hat{\rho}|n\rangle = \frac{1}{n!}\int_\infty d^2\alpha \; |\alpha|^{2n} \exp(-|\alpha|^2)\, P(\alpha)\,. \tag{11.296}$$

This integral is formally identical to the Poisson transform defined in (11.25), and it might appear that we have just rederived the classical result. Since $P(\alpha)$ does not behave like a true probability density function, however, some strange and distinctly nonclassical behavior can occur. To illustrate, let us look at a few forms for $P(\alpha)$ and the resulting ρ_{nn}.

Example 1: Coherent state Consider first the coherent state $|\alpha_0\rangle$. For this state, $P(\alpha) = \delta(\alpha - \alpha_0)$, and ρ_{nn} is given from (11.296) as

$$\rho_{nn} = \frac{1}{n!}\int_\infty d^2\alpha \; |\alpha|^{2n} \exp(-|\alpha|^2)\, \delta(\alpha - \alpha_0) = \frac{|\alpha_0|^{2n}}{n!} \exp(-|\alpha_0|^2)\,, \tag{11.297}$$

which is a Poisson probability of mean $|\alpha_0|^2$. One consequence is that the variance of the number of photons is equal to its mean.

Recall that in Sec. 11.1 we considered a source where the rate of photon emission was constant and the photons were independent, and we showed that the probability law on the number of photons was Poisson. Here we see that a coherent state satisfies the assumptions of that calculation. A classical steady wave gives rise to a Poisson distribution of counts, and the coherent state is the best quantum-mechanical approximation to a classical steady wave.

Example 2: Thermal equilibrium If the system is in thermal equilibrium, $P(\alpha)$ is a Gaussian centered at the origin (Glauber, 1963):

$$P(\alpha) = \frac{1}{\pi\langle n\rangle}\exp\left[-\frac{|\alpha|^2}{\langle n\rangle}\right]. \tag{11.298}$$

This $P(\alpha)$ can be interpreted as a classical probability density function. In fact, it is just the circular Gaussian density discussed in Sec. 8.3.6. The real and imaginary parts of α are independent and identically distributed.

The diagonal elements of the density matrix are now given by

$$\rho_{nn} = \frac{1}{\pi\langle n\rangle n!}\int_\infty d^2\alpha\ |\alpha|^{2n}\exp(-|\alpha|^2)\exp\left[-\frac{|\alpha|^2}{\langle n\rangle}\right]. \tag{11.299}$$

The integral is easily performed in polar coordinates. If we set $\alpha = Re^{i\theta}$, then $d^2\alpha = R\,dR\,d\theta$, and the θ integral yields a factor of 2π. A further change of variables, $u = R^2 = |\alpha|^2$, converts the integral to

$$\rho_{nn} = \frac{1}{\langle n\rangle}\int_0^\infty du\ \frac{u^n}{n!}e^{-u}\exp\left[-\frac{u}{\langle n\rangle}\right]. \tag{11.300}$$

Recalling (11.25), we recognize this form as the Poisson transform of an exponential probability law. From the definition of the gamma or factorial function and some algebra, we find (Saleh, 1978)

$$\rho_{nn} = \frac{\langle n\rangle^n}{[\langle n\rangle + 1]^{n+1}}, \tag{11.301}$$

which can also be written as

$$\rho_{nn} = \frac{1}{\langle n\rangle + 1}\exp\left[-n\ln\left(\frac{\langle n\rangle + 1}{\langle n\rangle}\right)\right]. \tag{11.302}$$

This is the *Bose-Einstein* distribution of photons in a single mode.

This Bose-Einstein distribution should not be confused with another function often given the same designation. The latter function specifies the variation of $\langle n\rangle$ with the frequency ω of the mode and the absolute temperature T:

$$\langle n\rangle = \frac{1}{\exp\left(\frac{\hbar\omega}{k_B T}\right) - 1}, \tag{11.303}$$

where k_B is Boltzmann's constant. The Bose-Einstein distribution in (11.301) gives the *probability* distribution of photons, but (11.303) gives their distribution over *frequency*; it would better be called a spectrum, but the term distribution is almost universal for it. Both (11.301) and (11.303) are needed for a complete statistical description of the statistics of photons in multiple modes, but we concentrate here on a single mode, where the statistics in thermal equilibrium are fully determined by the single number $\langle n \rangle$.

The (single-mode) Bose-Einstein distribution ρ_{nn} is plotted in Fig. 11.5. As seen from (11.302), ρ_{nn} is a simple exponential sampled at integer values; the most probable value of n in thermal equilibrium is always zero.

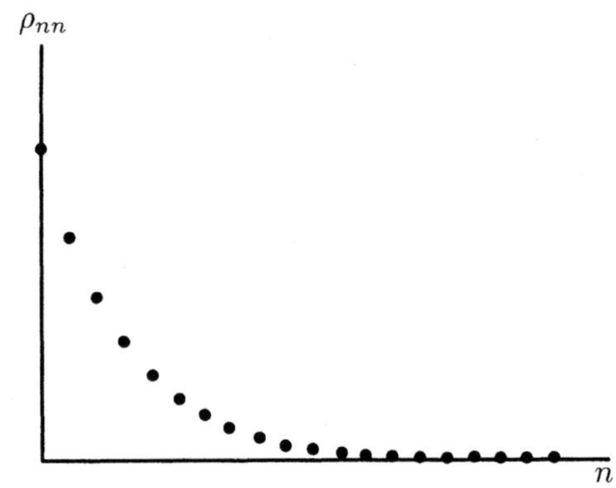

Fig. 11.5 Plot of the Bose-Einstein probability distribution of photons in a single mode, as given by the expression for ρ_{nn} in (11.301).

The variance associated with this distribution is given by (C.187) as

$$\mathrm{Var}(n) = \langle n \rangle + \langle n \rangle^2 . \tag{11.304}$$

This value is necessarily greater than $\langle n \rangle$ (unless $\langle n \rangle = 0$), so there is an excess variance (see Sec. 11.1.4), beyond the variance associated with a Poisson distribution of the same mean.

Example 3: Number state Finally, consider a particular number state $|N\rangle$. The density operator for this state is $\hat{\rho} = |N\rangle\langle N|$, and the matrix elements of this operator in the number-state representation are given by

$$\rho_{mn} = \langle m|N\rangle\langle N|n\rangle = \delta_{Nm}\,\delta_{Nn} . \tag{11.305}$$

Hence the probability of finding n photons is

$$\rho_{nn} = \delta_{Nn} , \tag{11.306}$$

so the state contains exactly N photons.

Since n can take on only the single value N, the mean of n is N and the variance of n is zero in the number state $|N\rangle$. Since 0 is less than N, the variance of the number of photons in this state is less than the value predicted for a Poisson with the same mean, and the statistics of n are said to be *sub-Poisson*. The classical expression (11.32) says that sub-Poisson statistics cannot occur, so the number state is an example of nonclassical light.

The density operator for a number state can also be expressed in the coherent-state representation. The relevant $P(\alpha)$ is given by Glauber (1963) as

$$P(\alpha) = \frac{1}{\pi}(-1)^N e^{|\alpha|^2} \delta^{(N)}(|\alpha|^2)\,, \tag{11.307}$$

where $\delta^{(N)}(\cdot)$ denotes the Nth derivative of a delta function as defined in Sec. 2.2.4. This $P(\alpha)$ is unbounded and takes on negative values, which is another indication that we are dealing with nonclassical light. It is a nontrivial exercise in delta functions to go from (11.307) to (11.306).

11.5.3 Counting statistics

So far we have discussed the statistics of photons in a single mode of the radiation field, but the number of photons is not directly observable. All we can ever do is use some photoelectric detector and study the statistics of the number of photoelectrons.

For a single mode, Scully and Lamb (1969) did a quantum mechanical calculation of the probability of counting k photoelectrons in a certain time τ. They considered first the probability of getting one photoelectron in τ when the mode contained exactly one photon, so the state was $|1\rangle$. Denoting this probability as p, they then showed that the probability of getting k photoelectrons in τ is given by

$$\Pr(k) = \sum_{n=k}^{\infty} \binom{n}{k} p^k (1-p)^{n-k} \rho_{nn}\,. \tag{11.308}$$

This result is exactly like the classical expression (11.23), but now we know that ρ_{nn} is given by (11.296), so

$$\Pr(k) = \sum_{n=k}^{\infty} \binom{n}{k} p^k (1-p)^{n-k} \int_\infty d^2\alpha\; P(\alpha) \exp(-|\alpha|^2)\, \frac{|\alpha|^{2n}}{n!}\,. \tag{11.309}$$

The same algebra as used in deriving the Poisson transform, (11.25), shows that

$$\Pr(k) = \frac{1}{k!} \int_\infty d^2\alpha\; P(\alpha) \exp(-p|\alpha|^2)\, (p|\alpha|^2)^k\,. \tag{11.310}$$

This expression looks just like (11.296) except that the number of photons n is replaced with the number of photoelectrons k and $|\alpha|^2$ is replaced with $p|\alpha|^2$. Thus the photoelectron statistics reflect the photon statistics as modified by the properties of the detector, including its quantum efficiency η and the exposure time τ. To understand better the nature of this modification, we shall revisit the examples used in Sec. 11.4.2.

Example 1: Coherent state For the coherent state $|\alpha_0\rangle$, $P(\alpha) = \delta(\alpha - \alpha_0)$, and (11.310) becomes

$$\begin{aligned}\Pr(k) &= \frac{1}{k!} \int_\infty d^2\alpha\; \delta(\alpha-\alpha_0) \exp(-p|\alpha|^2)\, (p|\alpha|^2)^k \\ &= \exp(-p|\alpha_0|^2)\, \frac{(p|\alpha_0|^2)^k}{k!}\,. \end{aligned} \tag{11.311}$$

This is a Poisson probability law, so the photoelectrons always obey Poisson statistics if the incident light is in a coherent state.

The mean number of photoelectrons is $p|\alpha_0|^2$, and p was defined as the probability of one photoelectron in the observation time if there is exactly one photon in the mode. In the coherent state $|\alpha_0\rangle$, the mean number of photons in the mode is $|\alpha_0|^2$ [see (11.297)], so we see that the mean number of photoelectrons is just the mean number of photons times the probability of one photoelectron from one photon. In other words, we have recovered the binomial selection theorem (11.24), but now in a quantum-mechanical context. The coherent state is the correct quantum-mechanical description of a source that produces a Poisson-distributed photon stream on a detector.

Example 2: Thermal state For a single mode in thermal equilibrium, $P(\alpha)$ is given by (11.298), and the distribution of photoelectrons is given by

$$\Pr(k) = \frac{1}{\pi \langle n \rangle k!} \int_{\infty} d^2\alpha \, \exp\left[-\frac{|\alpha|^2}{\langle n \rangle}\right] \exp(-p|\alpha|^2)\,(p|\alpha|^2)^k \,. \tag{11.312}$$

This integral has the same form as (11.299), and similar manipulations show that [*cf.* (11.301)]

$$\Pr(k) = \frac{\langle k \rangle^k}{[\langle k \rangle + 1]^{k+1}} \,, \tag{11.313}$$

where $\langle k \rangle = p\langle n \rangle$. Thus the distribution of photoelectrons is also a Bose-Einstein, and it follows that

$$\mathrm{Var}(k) = \langle k \rangle + \langle k \rangle^2 \,. \tag{11.314}$$

These results might lead the unwary to think that it would be rather easy to observe a Bose-Einstein distribution of photocounts in the laboratory, but we must remember that we are considering only a single mode of the radiation field here. Since the modes are characterized by their frequency and the direction of their wavevector, (11.313) and (11.314) apply only to highly directional and essentially monochromatic thermal light. To get the directionality in the laboratory, we might place a small thermal source at the rear focal plane of a lens, and to get a nearly monochromatic source we might use a narrowband optical filter. The resulting light would become more directional as the source size got smaller and more nearly monochromatic as the bandwidth of the filter got smaller, but both of these measures result in loss of light (unless we somehow let the temperature of the thermal source get correspondingly larger). In practice, the thermal light we can produce in the laboratory is not well described by a single mode of the radiation field. To analyze the photoelectron distribution with practical thermal sources, we must consider multiple modes.

Actually, we have already analyzed the case of nonmonochromatic thermal light, though we did not use that language. In Sec. 11.3.7, we considered a doubly stochastic temporal random process where the rate is the random process $a(t)$. If modes of different frequencies and random phases are present, the rate of photoemission is a temporal random process, and the variance of the number of counts is given by (11.128) rather than (11.314). (Note that N in (11.128) is the same thing as k in this section.) We can regard (11.314) as the (hypothetical) photocount variance for a single-mode thermal source, while (11.128) describes the actual multimode variance for a practical source. Though derived classically in Sec. 11.3.7,

(11.128) would also result from a multimode generalization of the calculation by Scully and Lamb.

We can get a rough idea of the differences between (11.128) and (11.314) by considering thermal light that is somehow perfectly directional but which has a finite bandwidth $\Delta\nu$ imposed by a narrowband filter. The rate process $a(t)$ in Sec. 11.3.7 has correlation time τ_a of approximately $1/\Delta\nu$. Achievable filter bandwidths in the optical region are of order 10^{11} Hz, and a typical observation time T might be 1 μsec, so the factor τ_a/T in (11.128) is of order 10^{-5}, and the factor $\mathrm{Var}(a)/\overline{a}^2$ is unity for a thermal source since $\overline{a}$ is exponentially distributed Sec. C.5.3). Thus the excess variance is reduced by a factor of about 10^{-5}. Any deviation from perfect directionality would reduce the excess variance still further (Goodman, 1985). In other words, it is very difficult to discern any deviation between the photocount variance produced by a practical thermal source and that produced by a coherent state.

Example 3: Number state Now consider our prime example of nonclassical light, the single-mode number state. Substituting (11.306) into (11.308) yields

$$\Pr(k) = \binom{N}{k} p^k (1-p)^{N-k}\,, \qquad k \leq N\,. \tag{11.315}$$

This is the binomial law (C.161). From (C.163), the mean of k is pN and the variance is

$$\mathrm{Var}(k) = pN - p^2 N = \langle k\rangle - p\langle k\rangle\,. \tag{11.316}$$

The excess variance $-p\langle k\rangle$ is negative, again reflecting the nonclassical nature of the light, but we note that $\mathrm{Var}(k) \to \langle k\rangle$ as $p \to 0$. As we saw several times in Sec. 11.1, rare events tend to a Poisson law. In the present context, rare means that the probability of getting one photocount when there is one photon in the mode is small.

12

Noise in Detectors

The physical principles underlying photodetection were introduced in Secs. 10.1.4 and 10.1.5, but there we discussed only the mean response. In Chap. 11 we developed the tools needed to discuss fluctuations about this mean, and in this chapter we use these tools to analyze the noise properties of a variety of practical photodetectors.

We begin the discussion in Sec. 12.1 with a specific class of detectors called photodiodes. These devices are important in their own right, but they also serve as a useful pedagogical device for discussing the discrete nature of photoelectric interactions and the consequent fundamental limits to optical detection. In Sec. 12.2 we discuss a variety of other noise sources that afflict practical optical detectors.

Finally, Sec. 12.3 addresses detectors for x rays and gamma rays, with particular attention to photon-counting detectors. The interesting new feature of these devices is that we can do more than detect the presence or absence of a photon interaction; we can also estimate the energy of the photon, the depth in the detector at which it interacts and other attributes.

12.1 PHOTON NOISE AND SHOT NOISE IN PHOTODIODES

The terms *Poisson noise*, *shot noise*, and *photon noise* occur frequently in the literature on radiation detectors, and they are often used interchangeably. The simplest description of shot noise is that it is the noise associated with the random arrival of discrete electrons, and photon noise is often similarly ascribed to the random arrival of discrete photons. As noted in Chaps. 10 and 11, however, we seldom have to consider the quantum properties of the electromagnetic field, so it is not really relevant whether light consists of photons. On the other hand, photoelectric interactions are discrete events, and the electrons produced in these events result in shot noise just as any other free electrons do. Thus so-called photon noise is just shot noise with photoelectrons.

The term Poisson noise is somewhat more problematic since certain conditions, discussed in detail in Chap. 11, must be satisfied for photoelectrons to obey Poisson statistics. In this section we shall generally assume that these conditions are satisfied, so shot noise (or photon noise) will be well described by Poisson random processes.

In Section 12.1.1 we discuss a simple device called a vacuum photodiode. Though rarely used in practice, the vacuum photodiode provides a convenient way of understanding the role of Poisson random processes in photodetection.

Section 12.1.2 is a digression from the main theme of Sec. 12.1. It is included to ensure that readers are familiar with some basic concepts and terminology of semiconductor physics. Many readers will be able to skip this section with impunity. In particular, Sec. 12.1.2 deals entirely with average quantities such as mean carrier concentrations and conductivities; fluctuations about these means and their effects on observable noise are taken up later, beginning in Sec. 12.1.3.

Section 12.1.3 discusses P-N junctions as detectors of electromagnetic radiation, especially in the visible and infrared portions of the spectrum. The treatment parallels the discussion of vacuum photodiodes in Sec. 12.1.1, with considerable emphasis on the role of Poisson statistics.

12.1.1 Vacuum photodiodes

A vacuum photodiode is illustrated in Fig. 12.1. To fix the geometry, assume that the light is propagating in the $+z$ direction, that the photocathode lies in the plane $z = 0$, and that the anode lies in $z = L_z$. Assume also that the lateral dimensions L_x and L_y are large compared to L_z.

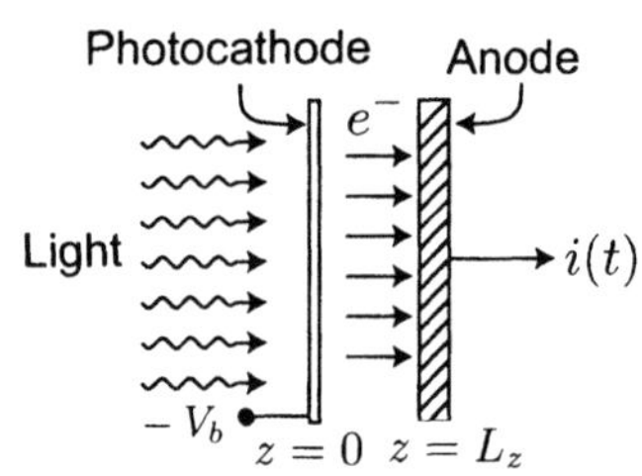

Fig. 12.1 Basic geometry for a vacuum photodiode.

The light falls on the photocathode and liberates photoelectrons into the vacuum gap with a quantum efficiency η. The electrons are accelerated across the vacuum gap by an applied potential V_b, producing a current in the anode circuit. In addition, electrons may also be liberated from the photocathode by thermal excitation; to distinguish these two mechanisms, we refer to current arising from thermal excitation as *dark current* and that arising from the incident radiation as *photocurrent.* Once liberated, however, all electrons produce the same effect in the output current.

To understand the noise properties of the anode current, we must first determine the form of the current pulse $i_0(t)$ produced by a single electron emitted from the photocathode at time $t = 0$. We assume that the electron has very little kinetic energy when it leaves the photocathode, but the electric field $E_0 = -V_b/L_z$ in the vacuum gap accelerates it towards the anode. By Newton's second law, the

z-component of the electron velocity at time t is given by

$$v_z(t) = \frac{eV_b}{mL_z}t\,, \tag{12.1}$$

where $-e$ and m are, respectively, the charge and mass of the electron.

The work done by the field on the electron when it moves a distance dz is $dW = -eE_0dz$, and the power delivered to the electron is $P(t) = dW/dt = -eE_0v_z(t) = +eV_bv_z(t)/L_z$. By conservation of energy, this power must be delivered to the photodiode by the external circuit, so the power flow in the circuit is $-P(t)$. If the circuit maintains the potential across the diode at V_b (*e.g.*, with an operational amplifier, as discussed below), then the current during the time the electron is moving across the gap is

$$i_0(t) = -\frac{P(t)}{V_b} = -\frac{e}{L_z}v_z(t) = -\frac{e^2V_b}{mL_z^2}t\,, \qquad 0 \le t < T_{tr}\,, \tag{12.2}$$

where T_{tr} is the transit time, given by

$$T_{tr} = L_z\sqrt{\frac{2m}{eV_b}}\,. \tag{12.3}$$

Note that it is not necessary for the electron actually to reach the anode to induce a current in the external circuit. We see from (12.2) that this current has the form of a triangular pulse, with the current beginning as soon as the electron is liberated from the cathode and then increasing in magnitude linearly with time as the electron accelerates.

Delta-function approximation With practical values for V_b and L_z, it often turns out that the induced current pulse is very narrow compared to the response time of the circuit. For example, if $V_b = 10$ Volts and $L_z = 1$ mm, then $T_{tr} \simeq 10^{-9}$ sec. If the circuit bandwidth is small compared to 1 GHz, we can approximate the triangular pulse by a delta function. As an exercise, the reader should show that the approximation takes the form

$$i_0(t) \simeq -e\,\delta(t)\,. \tag{12.4}$$

That the coefficient is $-e$ should not be surprising; the time integral of the current is the charge on the electron. The units are also worthy of note: current is measured in amperes or Coulombs per second, which is dimensionally consistent with the right-hand side of (12.4) since e is measured in Coulombs and a temporal delta function has units of sec^{-1} (see Sec. 2.4.3).

RC filtering Suppose the photodiode output is filtered by a simple RC circuit as shown in Fig. 12.2a or the slightly more complicated circuit of Fig. 12.2b. In both cases, the input to the filter is the current from the photodiode, and the output of the filter is the voltage drop across the RC network. The circuit in Fig. 12.2b has the advantage that the voltage on the output plate of the photodiode is held very near zero potential. Because an operational amplifier has very high gain, the voltage across its input terminals remains very small; it is often said that the input port of an operational amplifier is a *virtual ground*. Thus the potential difference

between the two plates of the diode remains essentially at the fixed potential V_b regardless of how much current flows. In the circuit of Fig. 12.2a, on the other hand, the potential difference between the plates is $V_b - V_{out}(t)$, so there can be a change in the field inside the diode if V_{out} is significant compared to V_b. We shall neglect this so-called *debias* effect here, either by using Fig. 12.2b or by keeping $V_{out}(t) \ll V_b$.

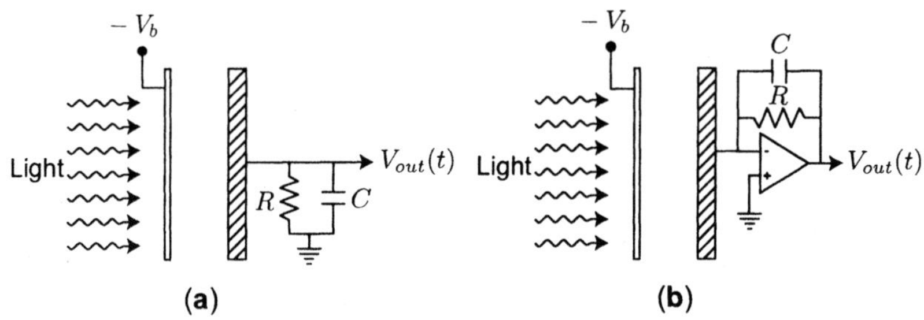

Fig. 12.2 Two circuits for filtering the output of a vacuum photodiode. In (a), the output current flows through a simple parallel RC circuit, and the voltage across this circuit is the filter output. In (b), an operational amplifier with very high gain is used to maintain the anode of the photodiode at ground potential, and the output voltage of the amplifier is the filter output.

With this assumption, the current from the photodiode is equal to the current through the parallel RC combination, which is the sum of the current through the capacitor and the current through the resistor. From elementary circuit theory,

$$i(t) = C\frac{dV_{out}(t)}{dt} + \frac{V_{out}(t)}{R}. \tag{12.5}$$

If we let $i(t)$ be the impulsive current (12.4) produced by a single electron, then the corresponding $V_{out}(t)$ is the impulse response of the filter, denoted $h(t)$, which must then satisfy

$$C\frac{dh(t)}{dt} + \frac{h(t)}{R} = -e\,\delta(t). \tag{12.6}$$

The solution of this differential equation is

$$h(t) = -\frac{e}{C}\exp\left(-\frac{t}{RC}\right)\text{step}(t). \tag{12.7}$$

Verification of this solution requires some properties of delta functions, including (2.73) and (2.25).

Filtering a photoelectron stream So far we have computed the filter output for a single photoelectron. Now suppose the photocathode is illuminated with steady radiation satisfying the Poisson postulates. From the discussion in Sec. 11.1, we know that the photoelectrons are statistically independent in this case, and we assume that the dark-current electrons are also independent (of each other and of the photoelectrons). Then, with approximation (12.4), the total output current from the photodiode has the form [see also (11.62)]

$$i(t) = \sum_{n=1}^{N} i_0(t - t_n) = -ez(t), \tag{12.8}$$

where $z(t)$ is a stationary Poisson process of rate a. With this input, the filter output is a stationary random process given by

$$V_{out}(t) = h(t) * z(t)\,. \tag{12.9}$$

Note that the input to the filter is $z(t)$, not $i(t)$; we have included the constant $-e$ in the definition of $h(t)$. Thus $h(t)$ has dimensions of voltage, and the dimensions of $z(t)$ (namely $\sec^{-1}$) cancel those of the asterisk (the time integral in the convolution) in (12.9).

The statistical properties of filtered Poisson point processes were discussed in Sec. (11.3.9). From the temporal counterpart of (11.139), we know that the mean of $V_{out}(t)$ is

$$\langle V_{out}(t)\rangle = h(t) * \langle z(t)\rangle = h(t) * a = -\frac{ea}{C}\int_0^\infty dt'\,\exp(-t'/RC) = -eaR\,. \tag{12.10}$$

This result is really just a statement of Ohm's law since $-ea$ is the mean current. Note that the mean filter output is proportional to the rate of the input process; for this reason an RC filter is sometimes called a *ratemeter.*

The variance of the output is given by the temporal counterpart of (11.143),

$$\mathrm{Var}\{V_{out}(t)\} = [h(t)]^2 * \langle z(t)\rangle = \frac{e^2 a}{C^2}\int_0^\infty dt'\,\exp(-2t'/RC) = \frac{e^2 aR}{2C}\,. \tag{12.11}$$

The ratio of the square of the mean voltage to its variance is

$$\frac{[\langle V_{out}(t)\rangle]^2}{\mathrm{Var}\{V_{out}(t)\}} = 2aRC\,. \tag{12.12}$$

One way to interpret this ratio is to compare it to the corresponding expression for a perfect counter which observes a Poisson source for a time T. For this counter, we know from Chap. 11 that the number of counts N in time T is Poisson and the variance is equal to the mean; specifically, $\mathrm{Var}(N) = \overline{N} = aT$. Thus the ratio of squared mean to variance is also aT for the counter. The expression in (12.12) shows that the filter output has the same ratio if $T = 2RC$; the effective averaging time of the filter is thus $2RC$.

Once one knows that the effective averaging time T_{eff} is $2RC$, the expressions above for the mean and variance of $V_{out}(t)$ can be deduced from Poisson statistics. Suppose N electrons occur in time T_{eff}. These electrons carry a charge $-Ne$, so the effective current is $-Ne/T_{eff}$ and the resulting voltage drop across the load resistor is $-RNe/T_{eff}$. Thus the factor $-Re/T_{eff}$ converts number of electrons to voltage. The mean voltage is obtained by replacing the random number N by its mean aT_{eff}, reproducing (12.10). The variance of the voltage is the variance in N times the *square* of the conversion factor $-Re/T_{eff}$. Since $\mathrm{Var}(N) = aT_{eff}$, (12.11) follows readily.

Power spectral density and effective noise bandwidth Since we are assuming that the photoelectron stream is a stationary temporal random process, we can also discuss the noise properties in the frequency domain. If we subtract off the mean of all random processes, the input to our filter is the random process $\Delta z(t)$, and we have seen in Sec. 11.3.11 that its power spectral density is the constant rate a, so the input is white noise. The power spectral density of the current is

$$S_{\Delta i}(\nu) = e^2 a = e|\langle i(t)\rangle| \,. \tag{12.13}$$

By the temporal counterpart of (11.170), the power spectral density on the output of the filter is

$$S_{\Delta V_{out}}(\nu) = a|H(\nu)|^2 \,, \tag{12.14}$$

where $H(\nu)$ is the Fourier transform of the impulse response $h(t)$. Hence this power spectral density has the same functional form as the squared modulus of the filter transfer function. A straightforward transform of (12.7) shows that, for the problem at hand,

$$|H(\nu)|^2 = \frac{e^2 R^2}{1 + (2\pi\nu RC)^2} \,. \tag{12.15}$$

The variance of $V_{out}(t)$ is the autocovariance function at zero lag, or

$$\begin{aligned}\mathrm{Var}\{V_{out}(t)\} = R_{\Delta V_{out}}(0) &= \int_{-\infty}^{\infty} d\nu \; S_{\Delta V_{out}}(\nu) \\ &= \int_{-\infty}^{\infty} d\nu \; \frac{e^2 R^2 a}{1 + (2\pi\nu RC)^2} \,, \end{aligned} \tag{12.16}$$

where we have used the Wiener-Khinchin theorem (8.133). The integral in (12.16) is elementary, and we find

$$\mathrm{Var}\{V_{out}(t)\} = \frac{e^2 a R}{2C} \,, \tag{12.17}$$

in agreement with (12.11).

Another way of viewing this result is to define an *effective noise bandwidth* B by

$$B \equiv \frac{1}{|H(0)|^2} \int_0^{\infty} d\nu \; |H(\nu)|^2 \,. \tag{12.18}$$

Note that the integral runs from 0 to ∞ in accord with the common practice in electrical engineering of defining bandwidth over positive frequencies only.

For our RC filter,

$$B = \int_0^{\infty} d\nu \; \frac{1}{1 + (2\pi\nu RC)^2} = \frac{1}{4RC} \,, \tag{12.19}$$

so (12.17) can be written as

$$\mathrm{Var}\{V_{out}(t)\} = 2Be^2 a R^2 = 2BeR^2 |\langle i(t)\rangle| \,, \tag{12.20}$$

which is a familiar formula in the literature on electronic shot noise. The appearance of $2B$ rather than just B often confuses students, but in fact $2B$ is the total bandwidth, including negative frequencies. Note that $2B$ is the reciprocal of the effective averaging time $2RC$.

SNR and DQE The rate a that appears in (12.10) and (12.11) is the total rate of emission of photoelectrons from the photocathode. If we consider the radiation to be made up of photons and assume that each incident photon produces a single photoelectron with probability η (the quantum efficiency), then a can be written as

$$a = \eta a_{phot} + a_{dark} , \tag{12.21}$$

where a_{phot} is the rate of arrival of photons and a_{dark} is the rate associated with dark current or thermionic emission. The first term, ηa_{phot}, can be considered a *signal* since it represents the mean response of the detector to the incident radiation. From (12.10), the mean signal in the output voltage is then $eR\eta a_{phot}$. We can thus define a *signal-to-noise ratio*[1] or SNR by

$$\mathrm{SNR}_{out}^2 = \frac{(eR\eta a_{phot})^2}{\mathrm{Var}\{V_{out}(t)\}} = 2RC\frac{(\eta a_{phot})^2}{\eta a_{phot} + a_{dark}} . \tag{12.22}$$

The ideal detector has $\eta = 1$ and $a_{dark} = 0$. We can express how closely the real detector approaches the ideal by defining a *detective quantum efficiency* or DQE as

$$\mathrm{DQE} \equiv \left(\frac{\mathrm{SNR}_{out}}{\mathrm{SNR}_{in}}\right)^2 , \tag{12.23}$$

where SNR_{in} is the signal-to-noise ratio of the input photon stream as seen by an ideal detector and the same RC filter. Thus $\mathrm{SNR}_{in} = (2RCa_{phot})^{\frac{1}{2}}$, and

$$\mathrm{DQE} = \frac{\eta^2 a_{phot}}{\eta a_{phot} + a_{dark}} . \tag{12.24}$$

One way to think about DQE is that it would require N/DQE photons in some time interval to achieve the same SNR_{out} with a real detector as would be obtained with N photons and an ideal detector. Since $\eta \leq 1$ and $a_{dark} \geq 0$, DQE is necessarily ≤ 1; real detectors always require more photons for the same SNR than an ideal detector.

Note that the detective quantum efficiency reduces to the quantum efficiency η if there is no thermionic emission or other noise sources. Moreover, DQE for this problem is a function of the photon arrival rate a_{phot}; it is possible to make $\mathrm{DQE} \simeq \eta$ just by using a bright light source so that $a_{phot} \gg a_{dark}$.

12.1.2 Basics of semiconductor detectors

Vacuum photodiodes are seldom used in practice today, having been replaced by semiconductor photodiodes. The fundamental mechanisms of photodetection with semiconductor devices are surveyed briefly in this section; the reader with a basic understanding of semiconductor physics can skip to Sec. 12.1.3 without loss of continuity.

[1]The usage of the term SNR given here is common in electrical engineering, but a distinctly different meaning is emerging in image science. In objective assessment of image quality (one of the main themes of this book), SNR is a measure of performance of some specific task. The ratio of the mean of a random variable to its standard deviation, on the other hand, does not relate directly to tasks. Perhaps the best interpretation of this SNR is that its reciprocal is the average fluctuation in units of the mean.

For detection of visible light, the most common semiconductor material is crystalline silicon (Si). Pure silicon is a poor conductor (actually an insulator at 0 K) because all of its outer electrons are used up in covalent bonds. Each Si atom is bound covalently to four neighbors in the crystalline lattice, and each bond requires two electrons, one from each of the neighboring atoms. Except for a few electrons released from the bonds by thermal excitation, there are no free electrons to participate in conduction.

The same phenomenon can also be explained another way. In a crystalline solid, electrons are confined to specific *energy bands*, consisting of the quantum states allowed by quantum mechanics. According to the Pauli exclusion principle, each state can be occupied by only zero or one electron. In Si there are two energy bands of interest, the *valence band* and the *conduction band*, and there is a forbidden region or *band gap* between them. At low temperature, the valence band is fully occupied (exactly one electron per state) and the conduction band is empty. In this condition there is no electrical conduction in the valence band since there are no available states for electrons to be accelerated into, and there is no conduction in the conduction band because there are no electrons there.

Doping In order to provide charge carriers for conduction, trace amounts of impurity elements are added to the Si in a process called *doping*. There are two kinds of dopants, called *donors* and *acceptors*. For doping of Si, donors are elements such as arsenic (As) or antimony (Sb) from Group V of the periodic table, and acceptors are elements such as gallium (Ga) or indium (In) from Group III. Since Si is in Group IV, it requires four valence electrons to complete its chemical bonds. When an element from Group V is substituted for Si, there is one electron not needed for bonding and hence only loosely bound to the impurity atom. At room temperature there is a high probability that this extra electron will be thermally excited and free to move around in the Si lattice; the impurity has donated an electron to the conduction band. The doped Si is said to be *N-type* (N for negative) because of the excess free electrons.

When an element from Group III is substituted for Si, the three available valence electrons are not sufficient to complete all of the bonds. At room temperature, there is a high probability that one of the electrons in the valence band will be incorporated into a bond to the impurity, thereby creating a vacancy, called a *hole*, in the valence band. Under an applied electric field, a valence-band electron can be accelerated into the state vacated when the hole was formed, filling that state but creating a vacancy in another state. If the field is in the $+z$ direction, the actual valence band electron moves in the $-z$ direction but the hole appears to move in $+z$, thus behaving like a mobile positive charge carrier. The acceptor impurity, by accepting an electron, has in effect donated a hole to the valence band. The doped Si in this case is said to be *P-type* because of the excess free holes.

Optical absorption in semiconductors In addition to thermal excitation, holes and electrons can also be created in semiconductors by optical excitation. The physics of this process was discussed in Sec. 10.1.4 where we saw that a transition from an initial state of energy $\mathcal{E}_i$ to a final state of energy $\mathcal{E}_f$ can be induced by light of frequency ν if $h\nu = \mathcal{E}_f - \mathcal{E}_i$. If $h\nu$ is less than the bandgap energy $\mathcal{E}_g$, this condition cannot be satisfied for an initial state in the valence band and a final state in the conduction band, and no transitions between these two bands are induced.

Moreover, transitions between two states within the valence band generally do not occur since both states are filled, and transitions within the conduction band do not occur since both states are empty. For $h\nu < \mathcal{E}_g$, therefore, there is very little absorption of light.

For $h\nu$ substantially greater than $\mathcal{E}_g$, however, there are many possible combinations of filled valence-band initial states and empty conduction-band final states, and the light is strongly absorbed by producing these transitions. As a practical matter, the attenuation coefficient for light of energy $h\nu > \mathcal{E}_g$ is about 10^4 cm^{-1} or 1 μm^{-1}. Thus a photon travels a path length of only about 1 μm before absorption. When an absorption event occurs, it creates one electron in the conduction band and one hole in the valence band; both contribute to the electrical conductivity.

Conductivity in semiconductors Consider a homogeneous semiconductor with an electron concentration[2] of n electrons/cm^3 and a hole concentration of p holes/cm^3. A uniform electric field E_0 in the $+z$ direction will exert a force in the same direction on the holes and in the opposite direction on the electrons. In addition, however, the carriers experience forces from lattice imperfections, from lattice vibrations or *phonons* and from interactions with other free carriers. All of these interactions retard the acceleration of the carriers by the field, and the carriers quickly reach a *terminal velocity*, much as a sky diver falling from an airplane reaches a terminal velocity at which the force of gravity is balanced by the forces from collisions with air molecules. To a good approximation, the terminal velocity is proportional to the field, and the constant of proportionality, called the *mobility*, is denoted by μ. Adding subscripts e for electrons and h for holes, we can write the terminal velocities as

$$\mathbf{v}_e = -\mu_e \mathbf{E}_0 \,, \qquad \mathbf{v}_h = \mu_h \mathbf{E}_0 \,. \tag{12.25}$$

More discussion of mobility and its physical origins will be found in Sec. 12.2.1.

The mean current density (Amperes/cm^2) is the carrier concentration (cm^{-3}) times the charge per carrier (Coul) times the mean velocity (cm/sec); it is given, for electrons and holes, respectively, by

$$\mathbf{J}_e = -ne\mathbf{v}_e \,, \qquad \mathbf{J}_h = pe\mathbf{v}_h \,. \tag{12.26}$$

Note that these two currents are in the same direction since both velocity and charge have opposite signs for electrons and holes.

The total mean current density $\mathbf{J}$, is thus given by

$$\mathbf{J} = \mathbf{J}_e + \mathbf{J}_h \equiv \sigma \mathbf{E}_0 \,, \tag{12.27}$$

where σ, called the *conductivity*, is given by

$$\sigma = e(n\mu_e + p\mu_h) \,. \tag{12.28}$$

Practical units of mobility are cm^2/V-sec and those of conductivity are (Ohm-cm)$^{-1}$, where 1 Ohm = 1 Volt/Ampere.

[2] Consistent use of SI units would require concentrations to be measured in m^{-3}, but the semiconductor literature normally uses hybrid units with lengths in cm.

Photoconductivity A simple way to make a photodetector is to shine light on a homogeneous semiconductor and observe the change in conductivity. Two experimental setups are shown in Fig. 12.3. In both geometries, the light is travelling in the $+z$ direction, and for simplicity we assume that the detector material is thin in this direction so that photoelectric interactions occur uniformly within the material. Suppose the incident photon irradiance[3] is I_p photons/(cm^2 sec) and a fraction η of the photons are absorbed. (For an optically thin detector, $\eta = \alpha L_z$, where α is the absorption coefficient for the light.) If we assume that each absorption event results in the production of one electron-hole pair, then the light causes the mean carrier densities to increase at the rates

$$\left[\frac{\partial n}{\partial t}\right]_{light} = \left[\frac{\partial p}{\partial t}\right]_{light} = \frac{1}{L_z}\eta I_p \,. \tag{12.29}$$

The carrier concentrations can also change because of recombination and trapping. In a recombination event, a conduction-band electron fills a valence-band hole, with the energy being converted to photons or phonons. Trapping occurs when an electron or hole is attracted to an impurity or lattice defect and the binding energy is such that thermal fluctuations are unlikely to free it over the time scale of interest in some measurement. The traps correspond to localized energy states somewhere near the middle of the forbidden gap. (They are not themselves forbidden since an impurity breaks the translational symmetry, which is what gives rise to the band structure in the first place.)

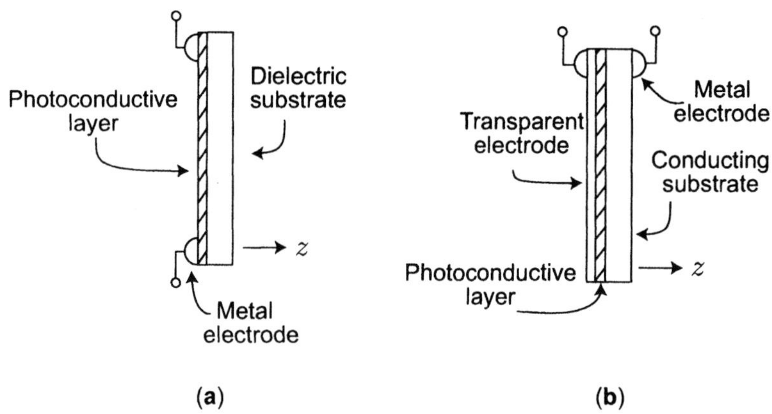

Fig. 12.3 Two configurations for observing photoconductivity. In (a) the light impinges directly on a face of the photoconductive material, and the conductivity is measured by observing a current that flows perpendicular to the direction of the light flux. In (b), the light passes through a transparent electrode before striking the photoconductive material, and the current flow is parallel to the light flux.

Trapping and recombination are actually quite complicated, but a simple phenomenological model will serve our purposes here. We assume that the carrier

[3]Note that we use capital I for irradiance, and to avoid confusion we therefore use lower-case i for current. One might, of course, wonder why either I or i is used for current. It apparently goes back to Georg Simon Ohm, who often referred to the 'intensity of the current' through a load. Thus there was once the same degree of confusion about the word *intensity* in electrical engineering as there is today in optics!

concentrations n and p relax towards their thermal equilibrium values, n_0 and p_0 respectively, according to

$$\left[\frac{\partial n}{\partial t}\right]_{tr} = -\frac{n - n_0}{\tau_e}, \qquad \left[\frac{\partial p}{\partial t}\right]_{tr} = -\frac{p - p_0}{\tau_h}, \tag{12.30}$$

where τ_e and τ_h are characteristic time constants for electrons and holes, respectively. These time constants are called *lifetimes* since an excess carrier lasts an average time τ before it recombines or is trapped. It is usually a good approximation to assume that τ_e and τ_h are independent of any applied electric field.

In the steady state, the electron concentration is given by

$$\left[\frac{\partial n}{\partial t}\right]_{light} + \left[\frac{\partial n}{\partial t}\right]_{tr} = 0, \tag{12.31}$$

or

$$n = n_0 + \frac{\tau_e}{L_z}\eta I_p, \tag{12.32}$$

and similarly for holes. The mean steady-state conductivity is then found from (12.28) to be

$$\sigma = \sigma_0 + \frac{\tau_e\mu_e + \tau_h\mu_h}{L_z}\eta e I_p, \tag{12.33}$$

where σ_0 is the dark conductivity. This formula shows that the change in conductivity is linearly related to the mean photon irradiance. Real photoconductors may exhibit nonlinearities since the lifetimes and mobilities may depend on the carrier concentrations (Bube, 1992).

P-N junctions A P-N junction is formed when P-type and N-type materials are brought together as shown in Fig. 12.4. Both pieces are initially electrically neutral; the N-type material has a density n of free electrons, but since these all came from donor atoms, there is an equal density $N_d = n$ of ionized (hence positively charged) donors. Similarly, the P-type material has an initial density p of free holes and an equal density $N_a = p$ of negatively charged ionized acceptors.

When the two materials are joined as shown in Fig. 12.4b, the free electrons from the N-type material diffuse into the P-type material, leaving behind the immobile positive donor ions. Once in the P-type material, the electrons encounter many holes and quickly recombine. Similarly, the free holes from the P-type material diffuse into the N-type material and recombine, leaving behind the immobile negative acceptor ions. In the steady state, there is a region devoid of free carriers but containing a layer of fixed negative charges on the P side and fixed positive charges on the N side (see Fig. 12.4b). The field due to this double layer retards further diffusion of the free carriers. The region without any free carriers is called the *depletion region.* The potential as a function of position across the depletion region is illustrated in Fig. 12.4c and the associated electric field is shown in Fig. 12.4d.

In the absence of any applied bias voltage, there is no net current flowing across a P-N junction, but there are four individual current components, as illustrated in Fig. 12.5. First, there are a few holes generated by thermal fluctuations in the N-type material, where they are referred to as *minority carriers.* If one of these holes diffuses into the depletion region, it feels an accelerating field that sweeps it across the junction. (The field direction is such that it retards the diffusion of electrons from N to P, hence it must accelerate holes in that direction.) Similarly, electrons generated thermally on the P side can diffuse into the depletion region and be swept into the N-type material. We refer to these two current components as *hole generation current* and *electron generation current*, respectively. Note that they are in the same direction, so there is a nonzero net generation current.

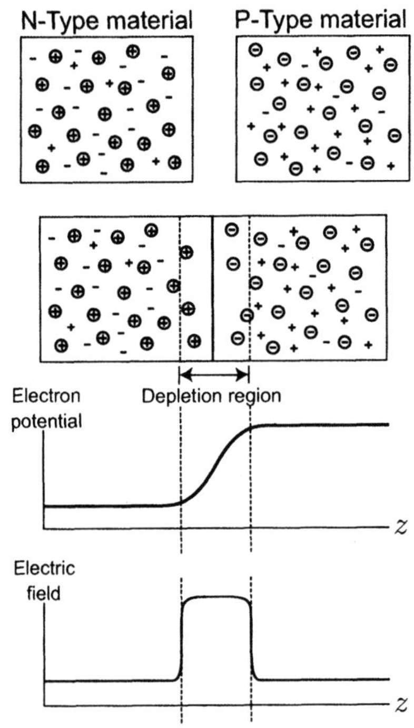

Fig. 12.4 Diagrams to illustrate the formation of a P-N junction. (a) N-type and P-type semiconductor materials before they are joined. In the N-type material, circles with plus signs denote ionized donor atoms and minus signs denote the free electrons. In the P-type material, circles with minus signs denote ionized acceptor atoms and plus signs denote the free holes. Note that a few free holes are created on the N side and a few free electrons are created on the P side by thermal excitation across the band gap. (b) After the two materials are joined, free holes diffuse into the N side and free electrons diffuse into the P side, where they recombine. A depletion region is created by the ionized dopant atoms left behind, and equilibrium is reached when the electric field from these ions retards further diffusion. (c) Illustration of the potential energy of an electron as it tries to move through the depletion region. Note that the direction of the potential is such that it opposes passage of an electron from N side to P side. (d) Electric field corresponding to the potential of (c) in the absence of any applied voltage. Note that a positive field accelerates a hole to the right and an electron to the left.

The two other current components arise from carriers that happen to be energetic enough to diffuse across the potential barrier and still reach the other side. There are copious electrons on the N side, but if they diffuse into the depletion region, they are repelled by the negative charges associated with acceptor ions, and they are dragged back by the positive charges of the donor ions. The electrons have a thermal distribution of velocities, however, and a tiny fraction of them will have sufficient velocity in the $+z$ direction to get over the potential barrier. When they do, they enter the undepleted portion of the P-type material and recombine. For this reason, the resulting current component is called the electron *recombination current.*

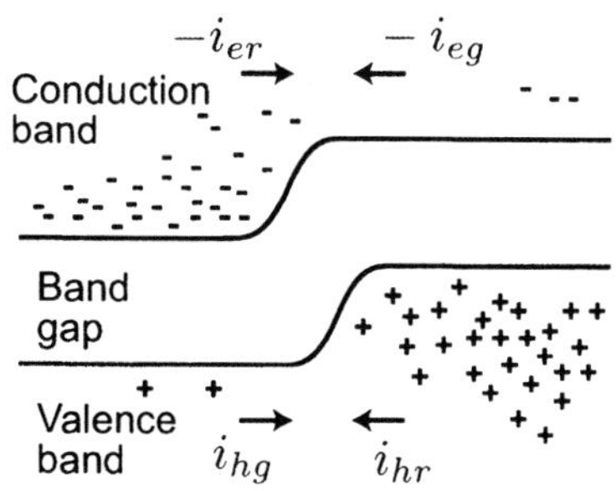

Fig. 12.5 Energy bands in a P-N junction at zero bias, showing the free electrons and free holes. On the N side, there are many free electrons but only a few energetic enough to surmount the potential barrier; those that do constitute the electron recombination current i_{er}, an electron flow from left to right on the figure. On the P side, there are a few thermally generated free electrons, but they are readily swept across the barrier from right to left, producing the electron generation current i_{eg}. Similar considerations apply to holes, though the diagram must be turned upside down in order to depict a hole potential. Thus the few free holes on the left are readily swept across the barrier to form the hole generation current i_{hg}.

In thermal equilibrium with no applied external bias, the electron recombination current must be exactly equal and opposite to the electron generation current. This cancellation is required by the thermodynamic *principle of detailed balance,* but it is also plausible on physical grounds. Recall that there are many electrons on the N side but very few of them are sufficiently energetic to surmount the potential barrier. On the other hand, there are very few electrons on the P side, but essentially all of those that reach the depletion region are swept across. The product of the number of carriers times the probability of crossing the barrier must be the same for the two current components to maintain thermodynamic equilibrium. Similar considerations show that there is a hole recombination current to exactly balance the hole generation current at zero bias.

Effect of bias Now suppose that a bias voltage V_b is applied to the P-N junction as shown in Fig. 12.6a, with the positive terminal attached to the P side. Now the potential barrier is reduced in height by an amount V_b, and a larger fraction of holes on the P side and electrons on the N side can surmount the barrier; thus both recombination currents are increased. The generation currents are essentially unchanged since they arise from minority carriers (holes on the N side or electrons on the P side) that diffuse into the depletion region and get swept across by the field; even the reduced field is adequate for this purpose. With the recombination currents increased and the generation currents unchanged by the applied voltage, there is now a net current in the conventional direction from positive terminal to negative. The current increases rapidly with bias voltage in this direction, which is called *forward bias.*

If the direction of the bias is reversed and the positive terminal is attached to the N side as shown in Fig. 12.6b, then the height of the potential barrier is increased. Fewer majority carriers (electrons on the N side, holes on the P side) are energetic enough to surmount the barrier in this case, and the recombination currents are reduced. In the limit of a large reverse bias, the only currents remaining are the small generation currents due to minority carriers.

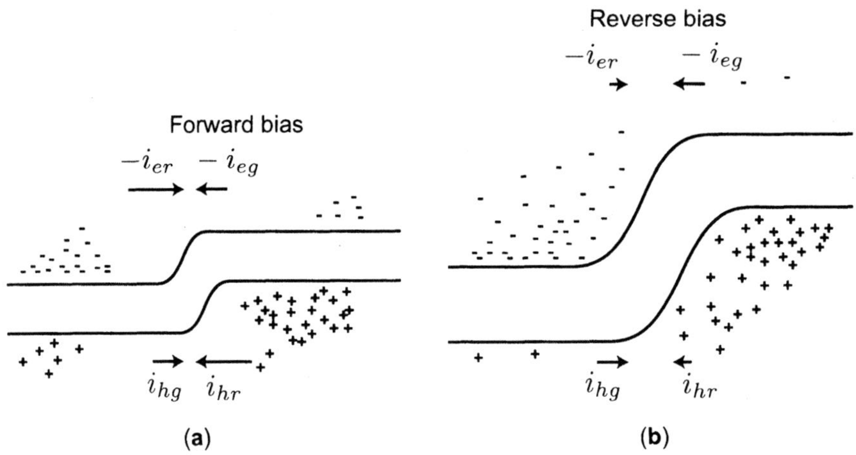

Fig. 12.6 Effect of a bias voltage on the energy bands of Fig. 12.5. (a) Forward bias; (b) reverse bias. Note that the two generation currents are relatively unaffected by the bias, but the two recombination currents are increased by forward bias and reduced by reverse bias.

These considerations are summarized in the *current-voltage characteristic curve* illustrated in Fig. 12.7. The difference in behavior in the forward- and reverse-bias conditions makes the junction into a rectifying device called a *diode*. Mathematically, the current i through an ideal P-N junction when a voltage V_b is applied is given by

$$i = i_s \left[\exp\left(\frac{eV_b}{k_B T} \right) - 1 \right] , \tag{12.34}$$

where k_B is Boltzmann's constant, T is the absolute temperature and i_s is just the sum of the two generation currents. When V_b has a large negative value, $i = -i_s$, so i_s is called the *reverse-bias saturation current*.

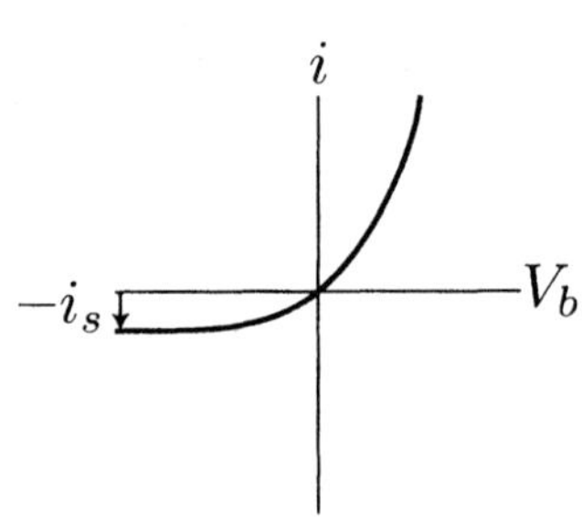

Fig. 12.7 Current-voltage characteristic of a P-N junction without illumination.

Effect of light If we want to use a P-N junction as a photodetector, we have to get light into the depletion region. Light absorbed in an undepleted region changes the conductivity of that region but does not greatly affect the total current since most of the voltage drop is across the depletion region. Since light with $h\nu > \mathcal{E}_g$ penetrates only about 1 μm or so, we must minimize the amount of material between the external surface of the semiconductor and the depletion region. A suitable geometry is shown in Fig. 12.8. The light traverses a transparent conducting layer and penetrates into the semiconductor. The junction is designed so that there is very little undepleted material (< 1 μm) before the light enters the depletion region. The thickness of the depletion region is substantially greater than 1 μm, so essentially all of the light is absorbed in the depletion region.

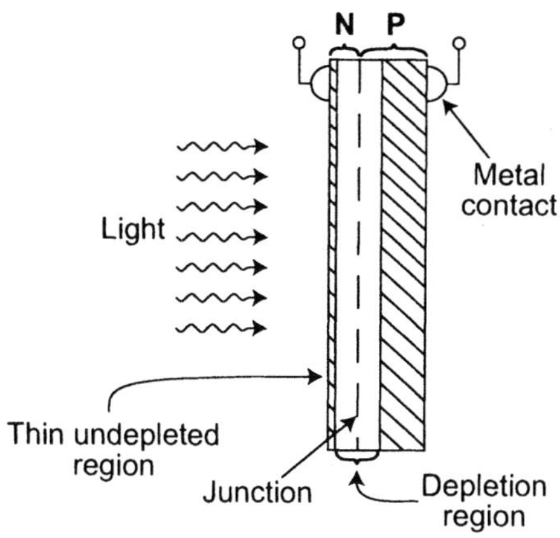

Fig. 12.8 Geometry of a photodiode.

When a photon is absorbed in the depletion region, producing a hole-electron pair, both carriers are accelerated by the field, holes towards the P side and electrons towards the N side. One might think that each absorbed photon would contribute a total charge of $2e$ to the external circuit, so the mean current would be $-2e\eta I_p A$ (where A is the entrance area of the detector, I_p is the photon irradiance and η is the fraction of light absorbed in the depletion region), but the factor of 2 in this expression is incorrect. We shall see why in Sec. 12.1.3 when we analyze the P-N junction photodetector in more detail. For now, suffice it to say that the mean current due to the light is $-e\eta I_p A$.

Because we now have another current source — photogeneration of electron-hole pairs in the depletion region — the current-voltage characteristic (12.34) is modified to

$$i = -e\eta I_p A + i_s \left[\exp\left(\frac{eV_b}{k_B T}\right) - 1\right] . \tag{12.35}$$

This characteristic is illustrated in Fig. 12.9. Note that the sign of the photocurrent is such that it effectively increases the reverse-bias saturation current. As discussed earlier in this subsection, this current arises from thermally generated minority carriers that are swept across the depletion region, and the light provides additional carriers to be swept across.

One very useful limit of (12.35) is when a large reverse bias is applied, so that V_b is large and negative. Then

$$i = -e\eta I_p A - i_s \,. \tag{12.36}$$

The current flowing in the circuit in this case is a direct linear (affine) measure of the photon irradiance, independent of the bias voltage.

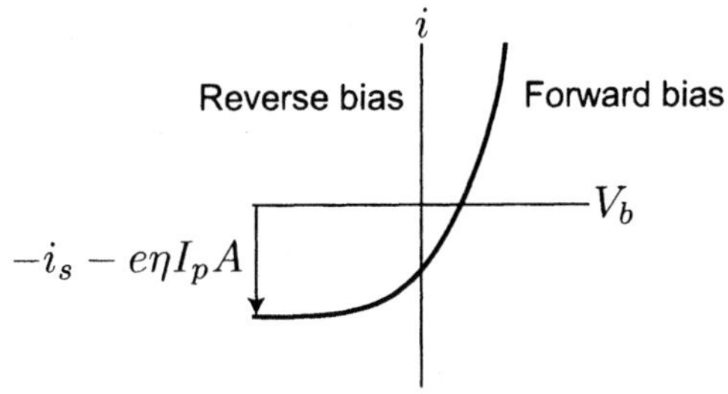

Fig. 12.9 Current-voltage characteristic of a photodiode with illumination.

12.1.3 Shot noise in semiconductor photodiodes

The discussion in Sec. 12.1.1 of shot noise in vacuum photodiodes is generally applicable to semiconductor photodiodes, but there are some important differences between the two kinds of devices. In the vacuum diode, only one kind of carrier is produced, and it is always produced at the same location, the photocathode. In the semiconductor, holes and electrons are produced, and the position of the photoelectric interaction is a random variable. Finally, electrons are accelerated across the gap in the vacuum diode, but both carriers quickly reach a terminal velocity in a semiconductor.

We can analyze these effects with a simple model. Suppose that light is propagating in the $+z$ direction and that the depletion region extends from $z = 0$ on the P side to $z = L_z$ on the N side. Assume also that the lateral dimensions (L_x and L_y) of the P-N junction are large compared to L_z. If an electron-hole pair is produced at the interaction point $z = z_{int}$ in the depletion region, the hole must travel a distance z_{int} and the electron must travel $L_z - z_{int}$ to reach undepleted material. Since the undepleted material is a good conductor, it may be considered part of the external circuit, and we need to consider the effect of carrier motion only within the depletion region.

For simplicity, we assume that the field in the depletion region has a constant value (which may be a good approximation at large reverse bias). In that case, the hole moves at constant speed $v_h = \mu_h E_0$ for a time $T_h = z_{int}/(\mu_h E_0)$, and the electron moves at speed $v_e = \mu_e E_0$ for a time $T_e = (L_z - z_{int})/(\mu_e E_0)$. A conservation-of-energy argument similar to the one in Sec. 12.1.1 then shows that the hole and electron currents in the external circuit (for a single photoelectric interaction) are, respectively,

$$i_{0h}(t) = -\frac{eE_0\mu_h}{L_z}\,, \qquad 0 \le t < \frac{z_{int}}{\mu_h E_0}\,; \tag{12.37}$$

$$i_{0e}(t) = -\frac{eE_0\mu_e}{L_z}\,, \qquad 0 \le t < \frac{L_z - z_{int}}{\mu_e E_0}\,. \tag{12.38}$$

Comparison of these expressions with (12.2) shows that the semiconductor diode gives two rect-function pulses rather than the single triangular one obtained with the vacuum diode. Both pulses are negative since the electrons move towards the load resistor in the external circuit and the holes move away.

At zero bias, the field E_0 is approximately given by the bandgap potential difference (*i.e.*, $\mathcal{E}_g$ expressed in eV) divided by L_z. Since $\mathcal{E}_g$ is around 1 eV and L_z is typically a few microns, E_0 is of order 10^3-10^4 V/cm. Typical mobilities are a few thousand cm^2/V-sec, so the carrier speeds in the depletion region are of order 10^7 cm/sec. Therefore the resulting pulse widths in (12.37) and (12.38) are of order 10^{-11} sec for zero bias, and even less for reverse bias. If these widths are short compared to the reciprocal bandwidth of the circuit, then we can approximate the rects by delta functions. Taking care to preserve normalization, we write

$$i_{0h}(t) \simeq -\frac{z_{int}}{L_z}\, e\,\delta(t)\,, \qquad i_{0e}(t) \simeq -\frac{L_z - z_{int}}{L_z}\, e\,\delta(t)\,. \tag{12.39}$$

One interpretation of this result is that each carrier induces a total charge proportional to the distance it travels and independent of the field and mobilities.

Of course, we do not observe these pulses separately. The total current flowing in the circuit for one electron-hole pair is

$$i_0(t) = i_{0h}(t) + i_{0e}(t) = -e\,\delta(t)\,. \tag{12.40}$$

Thus the total induced charge per electron-hole pair is e, not $2e$.

Since (12.40) is identical to (12.4), the remainder of the discussion in Sec. 12.1.1 is still applicable to semiconductor diodes. Note especially that $i_0(t)$ is independent of the interaction position z_{int}, so this random variable does not affect the statistics of the photocurrent under the present model. The key assumptions that caused z_{int} to cancel out were that both carriers eventually reach the undepleted region (no trapping or recombination in the depletion region) and that the lateral dimensions of the diode are large ($L_x, L_y \gg L_z$). We shall come back to the effects of trapping in the depletion region below.

Mean photocurrent As in Sec. 12.1.1, we now assume that the radiation is temporally stationary and that the Poisson postulates are satisfied. The rate of photoelectric generation of electron-hole pairs is ηa_{phot}, where η is the fraction of photons absorbed in the depletion region and a_{phot} is the rate of arrival of photons,

$$a_{phot} = A I_p\,. \tag{12.41}$$

Here, $A = L_x L_y$ is the cross-sectional area of the diode and I_p is the (nonrandom) photon irradiance.

The total photocurrent $i_p(t)$ is now a Poisson random process of the form (12.8), and from (11.95) its mean is

$$\langle i_p(t)\rangle = -e\eta a_{phot} = -e\eta A I_p\,. \tag{12.42}$$

Total current There is a dark current in semiconductor photodiodes analogous to the thermionic emission from the photocathode in a vacuum diode, but the physics is rather different. As we saw in Sec. 12.1.2, there are actually four components to the dark current in a P-N junction: electron and hole generation currents and electron and hole recombination currents. The sum of the means of these currents is zero at zero bias in the dark, but each current component carries its own noise, and the noises do not cancel even if the means do.

To the extent that each electron or hole flowing in the diode produces a current pulse short compared to the reciprocal of the system bandwidth, we can express the total current as a sum of point processes,

$$i_{tot}(t) = -ez_p(t) - ez_{eg}(t) - ez_{hg}(t) + ez_{er}(t) + ez_{hr}(t)\,, \tag{12.43}$$

where the subscripts have the following meanings: $p \Rightarrow$ *photoelectric*, $eg \Rightarrow$ *electron generation*, $hg \Rightarrow$ *hole generation*, $er \Rightarrow$ *electron recombination*, $hr \Rightarrow$ *hole recombination*. Each random process is a sum of delta functions as in (12.8), and each has its own rate a with an appropriate subscript. All of the rates are positive by definition, so the signs in (12.43) indicate the directions of the corresponding currents. The rates for the generation and recombination components depend on temperature and on the bias voltage across the diode, but in thermal equilibrium at zero bias, $a_{eg} = a_{er}$ and $a_{hg} = a_{hr}$.

A great simplification arises if we can assume that each of the five random processes in (12.43) is statistically independent of the other four, but to make this assumption we must neglect several effects. The first is Coulomb interaction of one charge carrier with another. The basic argument that validates this assumption is that the field in the depletion region is rather strong, of order 10^4 V/cm, so the motion of each carrier is determined by this field and not by the weaker fields of the other carriers. We must also assume that the rate for one current is independent of the magnitudes of the other currents. Effects that could invalidate this assumption include heating of the material at high current and changes in bias arising from the voltage drop across the external circuit. The latter effect is minimized by use of an operational amplifier as in Fig. 12.2b. Finally, fluctuations in ambient temperature or bias voltage can produce correlations among the current components. Good engineering practice will usually control these factors adequately, however, and there is seldom a serious objection to assuming statistical independence of the component currents.

An additional useful assumption is that each of the five random processes is stationary. From the discussion in Sec. 11.3.1, we know the conditions under which $z_p(t)$ is a stationary Poisson process; basically, the photon irradiance must be nonrandom and independent of time, and all photoelectric events must produce identical output pulses. Similarly, the four thermal processes are stationary if the temperature is constant since the rates are determined by the temperature. Moreover, all carriers entering the depletion region produce identical current pulses if there is no trapping or recombination in the depletion region. Under these conditions, $z_{eg}(t)$, $z_{hg}(t)$, $z_{er}(t)$ and $z_{hr}(t)$ are all stationary Poisson random processes.

With these assumptions, the overall current is a stationary random point process, though not exactly a Poisson point process because of the factors of $\pm e$. Instead, from the discussion in Sec. 11.3.3, we can see that the mean current is

$$\langle i_{tot}(t)\rangle = -ea_p - ea_{eg} - ea_{hg} + ea_{er} + ea_{hr}\,, \tag{12.44}$$

and the stationary autocovariance function is [*cf.* (11.97)]

$$K_{i_{tot}}(\Delta t) = \left[e^2a_p + e^2a_{eg} + e^2a_{hg} + e^2a_{er} + e^2a_{hr}\right]\delta(\Delta t) = e^2a_{tot}\,\delta(\Delta t)\,, \tag{12.45}$$

where a_{tot} is the total rate of electrons crossing the junction due to all processes, with no regard to the direction of the currents. Because the factors $\pm e$ are squared, the covariances add even when the mean currents tend to cancel.

The power spectral density of the total current is the Fourier transform of (12.45), or

$$S_{i_{tot}}(\nu) = e^2 a_{tot} , \tag{12.46}$$

so the current noise is white (within our approximation of short transit time).

***RC* filtering** As with vacuum photodiodes, we now investigate the effect of passing $i_{tot}(t)$ through an RC filter. Repeating the calculations in Sec. 12.1.1 with due attention to the signs of various terms, we find [*cf.* (12.10) and (12.11)]

$$\langle V_{out}(t)\rangle = R[-ea_p - ea_{eg} - ea_{hg} + ea_{er} + ea_{hr}] , \tag{12.47}$$

$$\text{Var}\{V_{out}(t)\} = \frac{R}{2C}\left[e^2 a_p + e^2 a_{eg} + e^2 a_{hg} + e^2 a_{er} + e^2 a_{hr}\right] . \tag{12.48}$$

To simplify these expressions, we consider a large reverse bias so that the recombination currents are zero (no majority carriers can surmount the potential barrier) and the reverse-bias saturation current i_s is given by $ea_{eg} + ea_{hg}$. Since $a_p = \eta A I_p$, we can now write the mean and variance of the filter output as

$$\langle V_{out}(t)\rangle = -Re\eta A I_p - Ri_s , \tag{12.49}$$

$$\text{Var}\{V_{out}(t)\} = \frac{R}{2C}\left[e^2 \eta A I_p + e i_s\right] . \tag{12.50}$$

The mean photocurrent is $-e\eta A I_p$, and $-i_s$ is the mean dark current, so (12.49) is Ohm's law and (12.50) is the shot noise associated with the total mean current [*cf.* (12.11)].

Trapping So far we have assumed that any carrier that enters the depletion region is swept across. One consequence of this assumption is that all photoelectric events in the depletion region produce identical output current pulses, so the noise is determined solely by Poisson statistics. In real photodiodes, however, there is always some probability that an electron or hole will be trapped at a defect in the depletion region. When this occurs, the induced current pulse in the external circuit is smaller than for a carrier that makes it all the way across. Since trapping is a random phenomenon, the noise is increased in the presence of trapping.

With trapping, there are three new random variables to consider for each photoelectric event: the random depth of interaction z_{int}, the random distance travelled by the electron and the random distance travelled by the hole. Suppose a hole and an electron are generated at time $t = 0$ and at depth $z = z_{int}$, that the electron moves in the $+z$ direction by a distance d_e (where $0 < d_e \leq L_z - z_{int}$), and that the hole moves in the $-z$ direction a distance d_h (where $0 < d_h \leq z_{int}$). (Hence d_e and d_h are both positive numbers.) If we can assume, as we did above, that the transit times are small compared to the reciprocal bandwidth of the circuit, the current induced in the external circuit can be written as

$$i_0(t) = i_{0h}(t) + i_{0e}(t) = -e\left[\frac{d_h}{L_z} + \frac{d_e}{L_z}\right]\delta(t) \equiv -e\beta\,\delta(t) , \tag{12.51}$$

where $0 < \beta \leq 1$. Note that $\beta = 1$ if there is no trapping since then $d_e = L_z - z_{int}$ and $d_h = z_{int}$.

Equation (12.51) may seem to imply that the induced charge need not be an integral multiple of e, in violation of charge quantization, but two points should be kept in mind. First, charge is quantized but current is not; an electron can move an arbitrary distance and hence induce an arbitrary current. Second, even when the total current is integrated on a capacitor to get a voltage, it is the total charge, arising from many current pulses of the form (12.51) that is quantized, not the charge from each pulse independently. Quantization of charge on a capacitor plate, called *Coulomb blockade*, has indeed been observed experimentally, but it is of little practical concern in electronic devices.

A first-principles treatment from this point would deduce the probability density function for β from knowledge of the physics of photon absorption and carrier trapping, but for present purposes it will suffice to assume that the mean $\overline{\beta}$ and variance σ_β^2 of the random amplitude β are known. It is not necessary to know how the statistics of β depend on those of z_{int}, d_e and d_h.

Suppose the photoelectric events comprise a stationary Poisson point process of rate a_p, with the nth event occurring at time t_n $(0 < t_n \leq T)$ but inducing a delta-function current pulse with strength $-e\beta_n$. Then the current is not a Poisson process because the pulses are not identical. We assume, as usual, that all events are statistically indistinguishable and independent, so the mean and variance of β_n are the same for all events, and β_n is statistically independent of β_m for $n \neq m$. The total photocurrent is then

$$i_p(t) = -e \sum_{n=1}^{N} \beta_n \, \delta(t - t_n) \,. \tag{12.52}$$

There are now $2N + 1$ random variables: the set $\{t_n\}$, the set $\{\beta_n\}$ and N itself. It will serve as an excellent test of the reader's comprehension of random point processes to retrace the derivation in Sec. 11.3.3 and show that the mean and autocorrelation function of the photocurrent are given by

$$\langle i_p(t) \rangle = -ea_p\overline{\beta} \,, \tag{12.53}$$

$$\langle i_p(t) \, i_p(t') \rangle = e^2 a_p \left(\sigma_\beta^2 + \overline{\beta}^2 \right) \delta(t - t') + e^2 a_p^2 \overline{\beta}^2 \,. \tag{12.54}$$

If there is no trapping, $\overline{\beta} = 1$ and $\sigma_\beta^2 = 0$, and we are back to the usual results for stationary Poisson random processes.

With trapping, however, the mean and variance on the output of the RC filter are both modified. Neglecting generation and recombination currents for simplicity, we find [*cf.* (12.47) and (12.48)]

$$\langle V_{out}(t) \rangle = -ea_p R\overline{\beta} \,, \tag{12.55}$$

$$\text{Var}\{V_{out}(t)\} = \frac{e^2 a_p R}{2C} \left(\sigma_\beta^2 + \overline{\beta}^2 \right) . \tag{12.56}$$

The detective quantum efficiency in this case is given by

$$\text{DQE} = \frac{\overline{\beta}^2 \eta}{\sigma_\beta^2 + \overline{\beta}^2} \,, \tag{12.57}$$

where η is the usual quantum efficiency (the probability of a photon producing a photoelectric event). Note that DQE $\to \eta$ as $\sigma_\beta^2 \to 0$, regardless of $\overline{\beta}$; the absolute size of the current pulse does not affect DQE (if there are no other noise sources), but its randomness does.

12.2 OTHER NOISE MECHANISMS

So far we have focused on different manifestations of shot noise or photon noise, but in real detectors there are other noise mechanisms as well. In this section we shall discuss the most important ones, including thermal or Johnson noise, generation-recombination noise arising from random fluctuations in the number of free carriers in a semiconductor, a somewhat mysterious process known as $1/f$ noise, and a kind of noise called kTC noise, which is peculiar to gated integrators.

12.2.1 Thermal noise

The thermal motion of charge carriers in any conductor leads inevitably to fluctuations in the current. This phenomenon was reported independently in 1928 by J. B. Johnson and H. Nyquist in two papers with nearly identical titles in the same issue of Physical Review (Johnson, 1928; Nyquist, 1928). The terms *Johnson noise* and *Nyquist noise* are both used in the literature, with perhaps a 75% chance that Johnson will be given sole credit. We shall use the broad term *thermal noise* to describe Johnson/Nyquist noise and any other fluctuations that would disappear at absolute zero temperature.

There are many ways of looking at thermal noise, including the equipartition principle, the fluctuation-dissipation theorem and stochastic differential equations. We shall touch briefly on these various approaches, emphasizing concepts useful in analyzing imaging systems.

Thermodynamic probabilities and equipartition A basic result of statistical mechanics is that the probability of occurrence of any state j in thermal equilibrium is given by

$$\Pr(j) = \frac{1}{Z} \exp\left(-\frac{\mathcal{E}_j}{k_B T}\right), \tag{12.58}$$

where $\mathcal{E}_j$ is the energy of the state. The normalizing constant Z, called the *partition function*, is usually written as

$$Z = \sum_j \exp\left(-\frac{\mathcal{E}_j}{k_B T}\right), \tag{12.59}$$

where the sum is over all possible states of the system. In many cases, there is a continuum of states, so the probability becomes a probability density function and the discrete sum over the index j must be replaced by an appropriate integral.

For a system consisting of N free particles of mass m, the state j is specified by stating the velocity $\mathbf{v}_n$ for each particle. The energy is the sum of the kinetic energies of the particles:

$$\mathcal{E} = \sum_{k=1}^{N} \tfrac{1}{2} m v_k^2 = \sum_{k=1}^{N} \tfrac{1}{2} m \left(v_{kx}^2 + v_{ky}^2 + v_{kz}^2\right), \tag{12.60}$$

where v_{kx} is the x-component of $\mathbf{v}_k$, etc.

An immediate consequence of (12.58) and (12.60) is that all components of all velocities are normally distributed with zero mean. For example, the marginal probability density function on v_{kz} is

$$\mathrm{pr}(v_{kz}) = \sqrt{\frac{m}{2\pi k_B T}} \exp\left(-\frac{\frac{1}{2}mv_{kz}^2}{k_B T}\right). \tag{12.61}$$

Since current in a circuit is linearly related to the velocity of charge carriers, it follows that thermal noise currents must be normally distributed as well.

The average energy associated with any velocity component (say v_{kz}) is

$$\left\langle \tfrac{1}{2}mv_{kz}^2 \right\rangle = \tfrac{1}{2}k_B T. \tag{12.62}$$

This is quite a general result, known as the *equipartition principle*: Each quadratic contribution to the energy has an average value of $\frac{1}{2}k_B T$, and the associated dynamical variable (here v_{kz}) is normally distributed.

Basic equations for thermal noise A simple way to get from the equipartition principle to a practical formula for thermal noise is to consider a resistor R in parallel with a capacitor C. The energy stored in the capacitor is $\frac{1}{2}CV^2$, where V is the voltage across the parallel combination. This voltage is the only dynamical variable necessary to specify the energy of the system. By equipartition, we must have

$$\tfrac{1}{2}C\left\langle V^2 \right\rangle = \tfrac{1}{2}k_B T, \tag{12.63}$$

or

$$\left\langle V^2 \right\rangle = \frac{k_B T}{C}. \tag{12.64}$$

But we know from (12.19) that the effective noise bandwidth B is given by $(4RC)^{-1}$, so we can also write

$$\left\langle V^2 \right\rangle = 4R\,k_B T B. \tag{12.65}$$

This fundamental equation shows that the noise variance is proportional to the resistance, to the absolute temperature and to the circuit bandwidth. An alternative — and more general — derivation of (12.65) will be given below after we state the fluctuation-dissipation theorem.

One way to interpret (12.65) is in terms of equivalent circuits, as shown in Fig. 12.10. The actual noisy resistor can be modeled as a noise free resistor in series with a zero-mean random voltage source that produces an RMS noise of $\sqrt{4R\,k_B T B}$. By Thevenin's theorem an alternative equivalent circuit is a noise-free resistor in parallel with a zero-mean random current source that produces an RMS current of $\sqrt{4k_B T B/R}$. The latter circuit is convenient for comparing the thermal noise current to the photocurrent from a photodiode.

Another interpretation of (12.65) is in terms of *available noise power*. Maximum power is transferred from a source to a load if the resistance of the load equals the internal resistance of the source. Thus a noisy resistor R in parallel with another resistor of resistance R' will transfer the maximum possible power to the second resistor if $R = R'$ (and of course the second resistor will transfer the same amount of power back to the first if they are in thermal equilibrium). The noise voltage from the first resistor is then divided evenly between itself and the second resistor, and the voltage that appears across the load is $V/2$, where the variance of V is still given by (12.65). The average power transferred to a matched load is

$$P_{\text{max}} = \frac{\langle V^2 \rangle}{4R} = k_B T B\,. \tag{12.66}$$

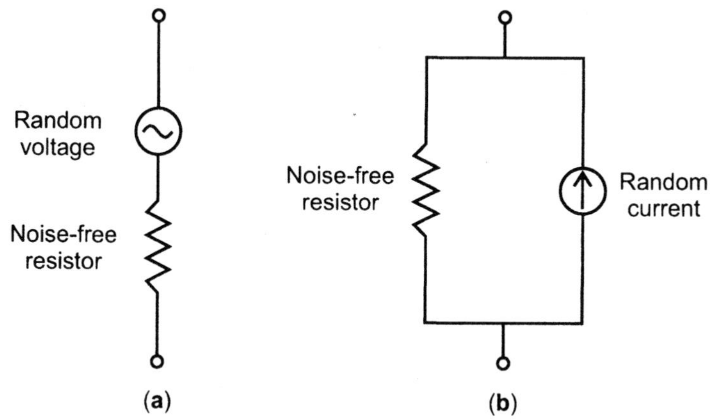

Fig. 12.10 Equivalent circuits for a resistor with thermal noise. *Left*: Series equivalent circuit with a random voltage source. *Right*: Parallel equivalent circuit with a random current source.

Thus the available power is $k_B T B$, and the available power per unit bandwidth is $k_B T$. Since $k_B T$ is an energy (with SI units of Joules) and B has units of Hz or sec^{-1}, $k_B T B$ has units of Joules/sec or Watts. Numerically, at room temperature (300 K) and a bandwidth of 1 MHz, $k_B T B = 4 \times 10^{-15}$ Watts, a seemingly tiny number but nevertheless often the dominant noise in practical electronic circuits and photodetectors.

Finally, we can also interpret (12.65) in terms of the power spectral density of the noise, $S_V(\nu)$. Since the frequency ν does not appear in (12.65), we must be dealing with a white-noise process. If we assume that

$$S_V(\nu) = 2R\, k_B T\,, \tag{12.67}$$

then (12.65) follows since the variance is the autocovariance function at zero shift and, for a zero-mean process, the autocovariance function is the inverse Fourier transform of the power spectral density. Thus (12.67) requires that

$$\langle V^2 \rangle = \int_{-B}^{B} d\nu\; S_V(\nu) = 4R\, k_B T B\,, \tag{12.68}$$

in agreement with (12.65).

The expression in (12.67) can be regarded as the power spectral density of the voltage source in the equivalent circuit of Fig. 12.10a. The corresponding spectral density for the current source in Fig. 12.10b is, by Thevenin's theorem,

$$S_i(\nu) = \frac{2k_B T}{R}\,. \tag{12.69}$$

Thermal noise in P-N junctions We have already analyzed the noise properties of generation and recombination currents in P-N junctions. In Sec. 12.1.3 we considered the shot noise from these currents and showed in (12.46) that the power spectral density of the current was white. Since generation and recombination currents arise from thermal fluctuations, however, one might wonder how they relate to thermal noise. We shall now show that in thermal equilibrium (no bias or illumination), they are in fact the same thing.

A P-N junction can be considered to be a nonlinear resistance. For a linear resistor, the resistance R is V/i, and for a nonlinear element an effective differential resistance R_d can be defined as $\partial V/\partial i$. We can find R_d for a P-N junction in the dark at zero bias by differentiating the current-voltage characteristic (12.34):

$$\frac{1}{R_d} \equiv \left[\frac{\partial i}{\partial V_b}\right]_{V_b=0} = \frac{e i_s}{k_B T}\,. \tag{12.70}$$

Since the thermal fluctuations are always so small that the current-voltage characteristic is locally linear, we can compute the thermal noise associated with R_d just as we would from a linear resistor. In particular, we know from (12.69) that the power spectral density for the thermal-noise current is $2k_BT/R$, and if we replace R with R_d we find

$$S_{\Delta i}(\nu) = \frac{2k_BT}{R_d} = 2ei_s\,. \tag{12.71}$$

There is no net current across an unilluminated P-N junction at zero bias, but there are four component currents: electron and hole generation and recombination currents. As discussed in Sec. 12.1.2, the two generation currents persist at reverse bias, and i_s is given by $ea_{eg} + ea_{hg}$. Thus the positive quantity $a_{eg} + a_{hg} = i_s/e$. At zero bias there are two other contributions to the total rate, namely a_{er} and a_{hr}, and the total rate a_{tot} is $2i_s/e$. The power spectral density of the shot noise associated with this total rate is then $e^2 a_{tot} = 2ei_s$, which is identical to (12.71).

The shot noise due to thermally generated generation and recombination currents is thus the same thing as thermal noise in an unilluminated P-N junction at zero bias. At large reverse bias, on the other hand, the recombination currents are zero and the rate of carrier transport across the junction is reduced from $2i_s/e$ to i_s/e, so the noise power spectral density (and variance after filtering) are also reduced a factor of 2. The differential resistance R_d is very high at large reverse bias, so the thermal contribution to $S_{\Delta i}(\nu)$ is negligible, but the shot-noise associated with the two generation currents remains. The system is no longer in thermal equilibrium at reverse bias, so there is no reason to expect that the thermal expressions will account for all of the noise.

Fluctuation-dissipation theorem The basic principle that dictates the inevitability of thermal noise is the *fluctuation-dissipation theorem*, which was stated precisely and given a quantum-mechanical basis by Callen and Welton (1951). A succinct derivation is given by Mandel and Wolf (1995), and a more detailed treatment is given by Kogan (1996). Here we merely state the key results without derivation.

Consider a temporally shift-invariant system in which some dynamical variable $y(t)$ responds linearly to a stimulus $x(t)$. For a monochromatic stimulus expressed as the real part of $X(\nu)\exp(-2\pi i\nu t)$, the response has a similar form with amplitude $Y(\nu)$ given by

$$Y(\nu) = H(\nu)\,X(\nu)\,, \tag{12.72}$$

where $H(\nu)$ is the transfer function. In addition, assume that the variables are defined so that a time average of the product $x(t)\,y(t)$ is the power delivered from the stimulus to the system. For example, $x(t)$ could be a current across a resistor and $y(t)$ the resulting voltage, or $x(t)$ could be the force exerted by an electric field and $y(t)$ the velocity of a charged particle. Note, however, that we cannot take $x(t)$ as the electric field and $y(t)$ as the resulting displacement of a charged particle; even though these variables are linearly related, the product $x(t)\,y(t)$ is not a power, dimensionally or physically.

With the proper choice of variables and a monochromatic stimulus, the power transfer is given by (see Sec. 10.1.2)

$$P = \tfrac{1}{2}\,\mathrm{Re}\{Y(\nu)\,X^*(\nu)\} = \tfrac{1}{2}|X(\nu)|^2\,\mathrm{Re}\,H(\nu)\,. \tag{12.73}$$

Thus the real part of the transfer function controls the power dissipation.

The fluctuation-dissipation theorem says that the dynamical variable $y(t)$ will undergo thermal fluctuations $\Delta y(t)$ with a power spectral density $S_{\Delta y}(\nu)$ proportional to $\mathrm{Re}\,H(\nu)$. The full quantum-mechanical calculation (Mandel and Wolf, 1995) shows that

$$S_{\Delta y}(\nu) = 2h\nu\left[\frac{1}{2} + \frac{1}{e^{h\nu/k_B T} - 1}\right]\mathrm{Re}\,H(\nu)\,. \tag{12.74}$$

The reader should verify the dimensional consistency of this equation.

The factor in square brackets in (12.74) is the mean number of quanta associated with a mode of the radiation field in thermal equilibrium, and $h\nu$ times this factor is the mean energy. As $T \to 0$, the mean energy approaches the zero-point value $\frac{1}{2}h\nu$. For practical problems with ordinary electrical circuits or photodetectors, it is almost always valid[4] to say that $h\nu \ll k_B T$. With this approximation, the mean number of quanta in the mode is very large, so quantum-mechanical effects become negligible and (12.74) takes a simple form (independent of Planck's constant h):

$$S_{\Delta y}(\nu) = 2k_B T\,\mathrm{Re}\,H(\nu)\,. \tag{12.75}$$

Now we see that fluctuations in $y(t)$ have a power spectral density controlled entirely by the dissipative part of the transfer function, with the constant of proportionality just given by $2k_B T$.

As an application of this theorem, let the dissipative system be an ideal resistor where the input $x(t)$ is the current through the resistor and the output $y(t)$ is the voltage $V(t)$ across it. Then $H(\nu)$ is the constant R for all ν, and the power spectral density of the voltage fluctuations is

$$S_{\Delta V}(\nu) = 2R\,k_B T\,, \tag{12.76}$$

in agreement with (12.67).

Stochastic differential equation The fluctuation-dissipation theorem tells us how large the thermal fluctuations in a dissipative system must be, but it sheds no light

[4] As a useful rule of thumb, an energy $\mathcal{E} = 1$ eV corresponds to a temperature $\mathcal{E}/k_B = 11,600$ K and to a frequency $\mathcal{E}/h = 2.4 \times 10^{14}$ Hz.

on how they arise in the first place. We shall now show how both the fluctuations and the dissipation can be derived from a common differential equation.

Consider a single free electron in a solid. Forces are exerted on the electron by an applied (possibly time-dependent) electric field $\mathbf{E}_0(t)$ in the $+z$ direction and by thermal motion of atoms and other electrons in the material. The equation of motion for the z component of the electron velocity is (Reif, 1965; Riskin, 1984)

$$m\frac{dv_z(t)}{dt} = -eE_0(t) + F_z(t)\,, \tag{12.77}$$

where $F_z(t)$ is the z-component of the total fluctuating force on the electron. Since $F_z(t)$ is random, (12.77) is a *stochastic differential equation.*

The main thing we need to know about $F_z(t)$ is that it fluctuates very rapidly. That is, $F_z(t)$ is a sample function of a temporal random process with a very short correlation time, of order 10^{-13} sec in semiconductors, so it is short compared to any time interval over which we can observe the system in the laboratory. In particular, the correlation time is short compared to the lifetime τ introduced in Sec. 12.1.2. The lifetime is the mean time before trapping or recombination, and it is typically milliseconds or microseconds for electrons in semiconductors. We shall neglect trapping and recombination altogether in this section but reintroduce them in Sec. 12.2.2.

We can write the velocity component $v_z(t)$ as

$$v_z(t) = \overline{v}_z(t) + \Delta v_z(t)\,, \tag{12.78}$$

where the overbar indicates a time average over an interval long compared to the correlation time of the force but short compared to the time dependence of E_0. Similarly, we can write the random force as

$$F_z(t) = \overline{F}_z(t) + \Delta F_z(t)\,. \tag{12.79}$$

One might expect $\overline{F}_z(t)$ to be zero because forces acting in the $+z$ direction have the same probability as forces acting in $-z$. That would be true for an electron at rest, but the average motion $\overline{v}_z(t)$ breaks the symmetry. To first order in $\overline{v}_z(t)$ we can write

$$\overline{F}_z(t) = -\alpha\overline{v}_z(t)\,, \tag{12.80}$$

where $\alpha > 0$ since the average force must be in the direction of decelerating the moving electron.

A time average of (12.77) yields

$$m\frac{d\overline{v}_z(t)}{dt} = -eE_0(t) - \alpha\overline{v}_z(t)\,. \tag{12.81}$$

For constant field, the steady-state solution of this equation is $\overline{v}_z = -eE_0/\alpha$. If we identify $\overline{v}_z$ with the average electron velocity in (12.25), we see that e/α is the electron mobility μ_e. Thus, after reinserting the fluctuating terms, we can rewrite (12.77) as

$$\frac{dv_z(t)}{dt} + \frac{e}{m\mu_e}v_z(t) = -\frac{e}{m}E_0(t) + \frac{1}{m}\Delta F_z(t)\,. \tag{12.82}$$

Equation (12.82) is known as the *Langevin equation.*

Solution of the Langevin equation For notational simplicity, we define $e/(m\mu_e)$ as γ and $\Delta F_z(t)/m$ as the force per unit mass, denoted $\Delta f_z(t)$. Then the Langevin equation in the absence of an applied field is

$$\frac{dv_z(t)}{dt} + \gamma v_z(t) = \Delta f_z(t) . \qquad (12.83)$$

If one considers a specific sample function $\Delta f_z(t)$ of the fluctuating force, then this is an ordinary inhomogeneous differential equation with the solution

$$v_z(t) = v_z(0)e^{-\gamma t} + \int_0^t dt'\, e^{-\gamma(t-t')}\Delta f_z(t') . \qquad (12.84)$$

Of course, $v_z(t)$ is also a sample function of a random process, and we can now study its mean and autocorrelation function. The mean is immediately zero since there is no applied field and the $+z$ and $-z$ directions are indistinguishable. To find the autocorrelation function of $v_z(t)$, we take advantage of the fact that $\Delta f_z(t)$ fluctuates very rapidly (with zero mean), so that its autocorrelation function can be written as

$$\langle \Delta f_z(t')\, \Delta f_z(t'')\rangle = C\,\delta(t'-t'') , \qquad (12.85)$$

where C will be determined below. With this assumption, the autocorrelation function for $v_z(t)$ becomes

$$\begin{aligned}\langle v_z(t_1)\, v_z(t_2)\rangle &= v_z^2(0)\, e^{-\gamma(t_1+t_2)} + \int_0^{t_1} dt' \int_0^{t_2} dt''\, e^{-\gamma(t_1+t_2-t'-t'')} C\,\delta(t'-t'') \\ &= v_z^2(0)\, e^{-\gamma(t_1+t_2)} + C\int_0^{\min[t_1,t_2]} dt'\, e^{-\gamma(t_1+t_2-2t')} \\ &= v_z^2(0)\, e^{-\gamma(t_1+t_2)} + \frac{C}{2\gamma} e^{-\gamma(t_1+t_2)}\left[e^{2\gamma\min[t_1,t_2]} - 1\right] \\ &= v_z^2(0)\, e^{-\gamma(t_1+t_2)} + \frac{C}{2\gamma}\left[e^{-\gamma|t_1-t_2|} - e^{-\gamma(t_1+t_2)}\right] . \end{aligned} \qquad (12.86)$$

In typical semiconductors at room temperature, μ_e is about 1000 $\text{cm}^2/\text{V-sec}$, and m must be interpreted as the effective mass, typically 0.1 times the actual electron mass; with these numbers, γ $(= e/m\mu_e)$ is of order $10^{13}\,\text{sec}^{-1}$. Therefore, we can neglect the exponential factors involving $t_1 + t_2$, so that

$$\langle v_z(t_1)\, v_z(t_2)\rangle \simeq \frac{C}{2\gamma} e^{-\gamma|t_1-t_2|} . \qquad (12.87)$$

This equation shows that the correlation time of the random process $v_z(t)$, defined as the value of $|t_1 - t_2|$ for which the autocorrelation drops by a factor of $1/e$, is $1/\gamma$. We can express the correlation time as

$$\tau_{sc} = \frac{1}{\gamma} , \qquad (12.88)$$

where the notation τ_{sc} is used since this correlation time can be interpreted as the mean time between scattering events.

We can now determine the constant C by appealing to the equipartition principle. Since $\frac{1}{2}m\left\langle v_z^2(t)\right\rangle = \frac{1}{2}k_BT$, we must have

$$\langle v_z(t_1)\, v_z(t_2)\rangle = \frac{k_BT}{m}\, e^{-\gamma|t_1-t_2|}\,. \tag{12.89}$$

If we are not concerned with time scales as small as τ_{sc}, we can use

$$\lim_{\gamma\to\infty} \frac{\gamma}{2}\, e^{-\gamma|\Delta t|} = \delta(\Delta t) \tag{12.90}$$

to obtain

$$\langle v_z(t_1)\, v_z(t_2)\rangle = \frac{2k_BT\mu_e}{e}\, \delta(t_1 - t_2)\,. \tag{12.91}$$

Relation to current noise Consider a homogeneous semiconductor of dimensions $L_x \times L_y \times L_z$ $(L_x, L_y \gg L_z)$ with electrodes on the faces $z = 0$ and $z = L_z$ connected to an external circuit. We know from (12.2) that a moving electron with velocity $\mathbf{v}(t)$ induces a current in the circuit given by

$$i_0(t) = -\frac{e}{L_z}\, v_z(t)\,. \tag{12.92}$$

It follows from (12.91) and (12.92) that the autocorrelation function of this current is

$$\langle i_0(t)\, i_0(t+\Delta t)\rangle = \frac{2k_BTe\mu_e}{L_z^2}\, \delta(\Delta t)\,. \tag{12.93}$$

Now suppose the semiconductor has a density of n electrons per unit volume, or a total of $N = nL_xL_yL_z$ electrons. The total current is still a zero-mean random process. Since the fluctuations in the motions of different electrons are uncorrelated, the autocorrelation function of the total current is

$$\langle i(t)\, i(t+\Delta t)\rangle = N\langle i_0(t)\, i_0(t+\Delta t)\rangle = \frac{2nAk_BTe\mu_e}{L_z}\, \delta(\Delta t)\,, \tag{12.94}$$

where $A = L_xL_y$.

We can rewrite (12.94) in terms of the resistance of the specimen, which, with (12.28), is given by

$$R = \frac{L_z}{\sigma A} = \frac{L_z}{en\mu_e A}\,. \tag{12.95}$$

Thus the autocorrelation function for the current is

$$\langle i(t)\, i(t+\Delta t)\rangle = \frac{2k_BT}{R}\, \delta(\Delta t) \tag{12.96}$$

and the corresponding power spectral density is

$$S_i(\nu) = \frac{2k_BT}{R}\,, \tag{12.97}$$

in agreement with (12.69).

This expression shows that the noise power is independent of frequency, but recall that we approximated the autocorrelation function by a delta function in (12.91). The reader is invited to show that the more general result is

$$S_i(\nu) = \frac{2k_BT}{R}\, \frac{1}{1 + (2\pi\nu\tau_{sc})^2}\,. \tag{12.98}$$

The white-noise expression (12.97) is thus valid if $2\pi\nu\tau_{sc} \ll 1$, which is easily satisfied in all practical electronic circuits, but the total power, integrated over all frequencies is finite.

Another straightforward generalization is to consider an applied field in the derivation above. Then the mean velocity is not zero, but the autocorrelation function of the velocities is essentially unchanged; thermal velocities are so much larger than drift velocities that the drift can be neglected in the autocorrelation. Thus, even though thermal noise can be described microscopically as current noise or shot noise, it does not depend on the mean current.

Diffusion and the Einstein relation We have seen that random thermal forces on an electron account for both the finite mobility and the thermal noise. As we shall now demonstrate, they also account for Brownian motion and diffusion.

Consider an electron (or any other particle) whose z coordinate at time $t = 0$ is denoted $z(0)$. As a result of the random thermal forces, it will be at some other coordinate $z(t)$ at time t. The mean-squared distance moved in the $+z$ direction is given by

$$\left\langle [z(t) - z(0)]^2 \right\rangle = \left\langle \left[\int_0^t dt_1 \; v_z(t_1) \right]^2 \right\rangle . \tag{12.99}$$

If we write the squared integral as a product of two identical integrals and interchange order of integration and expectation,[5] we obtain

$$\left\langle [z(t) - z(0)]^2 \right\rangle = \int_0^t dt_1 \int_0^t dt_2 \; \langle v_z(t_1) \, v_z(t_2) \rangle . \tag{12.100}$$

Inserting the delta-function form (12.91) for the autocorrelation of $v_z(t)$ and performing two easy integrals, we find

$$\left\langle [z(t) - z(0)]^2 \right\rangle = \frac{2k_B T \mu_e}{e} t . \tag{12.101}$$

We have thus rediscovered a well-known result from the theory of Brownian motion and other diffusion processes: the RMS displacement grows as the square-root of time. The *diffusion constant* D is defined such that (Reif, 1965)

$$\left\langle [z(t) - z(0)]^2 \right\rangle = 2Dt . \tag{12.102}$$

We see at once that

$$D = \frac{k_B T}{e} \mu_e , \tag{12.103}$$

which is the celebrated *Einstein relation* linking the diffusion constant to the mobility.

[5] The validity of this step is discussed in Sec. 8.2.2, but the theorem cited there specifically rules out white-noise processes. To justify the interchange we have to say that it is done before letting $\gamma \to \infty$.

12.2.2 Generation-recombination noise

Generation-recombination noise or *GR noise* is the noise associated with thermal fluctuations in carrier density in a semiconductor. Since the conductivity is proportional to the carrier density, GR noise can be thought of as fluctuations in resistance of a specimen. If an electric field is applied to the specimen, the resistance fluctuations lead to current fluctuations, so this mechanism is an important noise source in photoconductors.

GR noise should not be confused with the shot noise due to generation and recombination currents in P-N junctions. In Sec. 12.1.3 we analyzed the latter noise in terms of shot noise, and then in Sec. 12.2.1 we saw that it could also be interpreted as thermal noise, at least at zero bias. What we are calling GR noise in this section occurs in homogeneous semiconductors (as opposed to junctions), and it is fundamentally the result of resistance fluctuations rather than shot noise.

Simple model for GR noise Consider again a piece of semiconductor with dimensions $L_x \times L_y \times L_z$ ($L_x, L_y \gg L_z$) with electrodes on the faces $z = 0$ and $z = L_z$. If a bias voltage V_b is applied to the electrodes, the charge carriers experience a field $E_0 = V_b/L_z$ in the $-z$ direction. For simplicity we consider only electrons.

Each *free* electron moves with constant velocity $\mu_e E_0$ in the $+z$ direction, corresponding to a current $-e\mu_e E_0/L_z$, but the electrons do not remain free so the current is not constant. If an electron recombines or is trapped, its current immediately ceases. If the electron is thermally excited out of a trap at a later time, or a new free electron is generated by thermal excitation from the valence band, then this electron is very quickly accelerated to its terminal velocity (on a time scale of picoseconds), and the current resumes. The resulting current waveform due to a single electron is illustrated in Fig. 12.11; this random process is often referred to as a *random telegraph wave*, and its statistics directly control the statistics of GR noise.

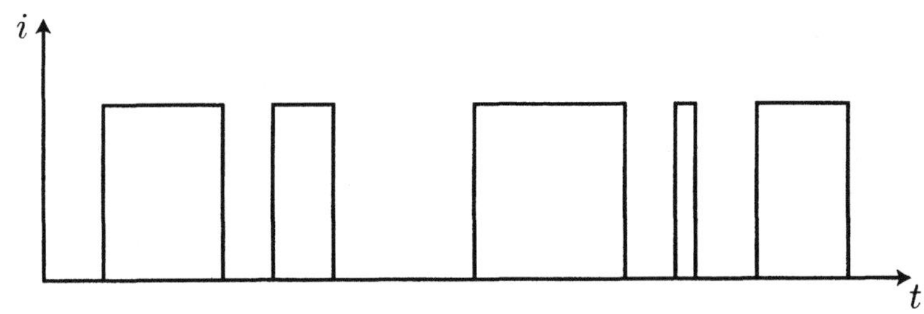

Fig. 12.11 Typical waveform of a single electron that is randomly trapped.

Statistics of the random telegraph wave Consider a single electron and define a random process $y(t)$ to have the value 1 when the electron is free and zero when it is trapped. The mean of this process is

$$\langle y(t)\rangle = \Pr[y(t) = 1] = \Pr[\text{electron is free}] \equiv p_1 . \tag{12.104}$$

The autocorrelation function (for $\Delta t > 0$) is

$$\begin{aligned}\langle y(t)\, y(t+\Delta t)\rangle &= \Pr[y(t) = 1 \text{ and } y(t+\Delta t) = 1] \\ &= \Pr[y(t+\Delta t) = 1 | y(t) = 1]\, p_1 .\end{aligned} \tag{12.105}$$

If p_1 is independent of time, the process is stationary.

We can further decompose the event that $y(t + \Delta t) = 1$ given that $y(t) = 1$ into two mutually exclusive possibilities: either no transitions between the free and trapped states occur in the interval Δt, or one or more transitions occur. If no transitions occur, then $\Pr[y(t + \Delta t) = 1|y(t) = 1] = 1$. If one or more transitions occur, then it is reasonable to assume that the electron loses all memory of its previous state, and $\Pr[y(t + \Delta t) = 1|y(t) = 1] = \Pr[y(t + \Delta t) = 1] = p_1$. Thus

$$\langle y(t)\, y(t + \Delta t)\rangle = p_1\{\Pr(\text{no trans in } \Delta t) + p_1[1 - \Pr(\text{no trans in } \Delta t)]\}\,, \tag{12.106}$$

where $\Pr(\text{no trans in } \Delta t)$ is the probability that the electron makes no transitions and hence remains free for a time Δt. If the transitions are independent events satisfying the Poisson postulates, this probability is given by [*cf.* 11.6]

$$\Pr(\text{no trans in } \Delta t) = e^{-\Delta t/\tau}\,, \tag{12.107}$$

where τ is the lifetime of the electron in the free state.

The discussion above was for $\Delta t > 0$, but an autocorrelation function must be symmetrical, so

$$\begin{aligned}\langle y(t)\, y(t + \Delta t)\rangle &= p_1\left[e^{-|\Delta t|/\tau} + p_1\left(1 - e^{-|\Delta t|/\tau}\right)\right] \\ &= (p_1 - p_1^2)\, e^{-|\Delta t|/\tau} + p_1^2\,.\end{aligned} \tag{12.108}$$

Defining the zero-mean process $\Delta y(t) = y(t) - \langle y(t)\rangle$, we see that the autocovariance of $y(t)$ is given by

$$K_y(\Delta t) = \langle \Delta y(t)\, \Delta y(t + \Delta t)\rangle = (p_1 - p_1^2)\, e^{-|\Delta t|/\tau}\,. \tag{12.109}$$

From telegraph wave to GR noise Now assume there are N_{tot} electrons in the volume and that the generation and recombination processes act independently on different electrons. Under this assumption, the mean and autocovariance of the current are given by

$$\langle i(t)\rangle = -N_{tot}\frac{e\mu_e E_0}{L_z}\, p_1\,; \tag{12.110}$$

$$K_i(\Delta t) = \langle \Delta i(t)\, \Delta i(t + \Delta t)\rangle = N_{tot}\left[\frac{e\mu_e E_0}{L_z}\right]^2 (p_1 - p_1^2) e^{-|\Delta t|/\tau}\,. \tag{12.111}$$

But $N_{tot}p_1$ is the mean number of free electrons, denoted $\overline{N}$ and given in terms of the mean density n of free electrons by

$$\overline{N} = N_{tot}\, p_1 = nAL_z\,, \tag{12.112}$$

where $A = L_x L_y$. The mean resistance of the specimen, now denoted $\overline{R}$, is given by (12.95), and it follows from (12.110) and (12.112) that the mean current obeys Ohm's law,

$$\langle i(t)\rangle = V_0/\overline{R}\,, \tag{12.113}$$

where $V_0 = E_0 L_z$ is the applied voltage.

The autocovariance can also be expressed in terms of V_0 and $\overline{R}$. Since the number of free carriers is much less than the total number of carriers in a semiconductor (or else it wouldn't be *semi*), $p_1 \ll 1$ and we can neglect the term proportional to p_1^2 in (12.111). A little algebra then yields

$$K_i(\Delta t) = nA\frac{(e\mu_e E_0)^2}{L_z} e^{-|\Delta t|/\tau} = \frac{\langle i(t)\rangle^2}{\overline{N}} e^{-|\Delta t|/\tau} = \frac{V_0^2}{\overline{N}\,\overline{R}^2} e^{-|\Delta t|/\tau}\,. \tag{12.114}$$

Power spectral density The power spectral density of the GR current fluctuations is obtained by Fourier transforming (12.114), with the result,

$$S_{\Delta i}(\nu) = \frac{V_0^2}{\overline{N}\,\overline{R}^2} \frac{2\tau}{1 + (2\pi\nu\tau)^2} . \tag{12.115}$$

A useful alternative form is

$$S_{\Delta i}(\nu) = \overline{N} \left[\frac{e\mu_e E_0}{L_z} \right]^2 \frac{2\tau}{1 + (2\pi\nu\tau)^2} . \tag{12.116}$$

Though one of these expressions has $\overline{N}$ in the denominator and one has it in the numerator, they are nonetheless equivalent since $\overline{R} \propto 1/\overline{N}$.

This power spectral density is similar in form to the thermal-noise expression in (12.98), but there are two key differences. First, the relevant relaxation time for GR noise is the lifetime, which is microseconds to milliseconds, while the relaxation time for thermal noise is τ_{sc}, which is of order 10^{-13} sec. For frequencies of importance with ordinary laboratory electronics, $2\pi\nu\tau_{sc} \ll 1$ but $2\pi\nu\tau$ may be comparable to or greater than one, so the rolloff in (12.115) should not be neglected. In fact, measurement of the power spectral density of GR noise is a useful technique for determining the lifetime.

Second, we see that the power spectrum of the GR noise current in a homogeneous semiconductor is proportional to the square of the mean current and hence to the square of the applied voltage. The power spectral density of thermal noise, on the other hand, is independent of applied voltage (basically because thermal velocities are large compared to drift velocities).

Variance and its interpretation In discussing shot noise and thermal noise, we assumed that the relevant fluctuations were very rapid so that the power spectral density was essentially constant. To that approximation, the variance of the current was then infinite; to get back to a finite variance, we had to invoke an additional filtering step with finite bandwidth. With GR noise, as we have noted, the fluctuations tend to be much slower, so the power spectral density in (12.115) will not usually be well approximated by a constant. Thus the variance of the GR noise current is finite even without a filter.

Specifically, from (12.114) the variance is given by

$$\mathrm{Var}\{i(t)\} = K_i(0) = \frac{V_0^2}{\overline{N}\,\overline{R}^2} . \tag{12.117}$$

The ratio of the square of the mean to the variance is thus

$$\mathrm{SNR}_i^2 \equiv \frac{\langle i(t) \rangle^2}{\mathrm{Var}\{i(t)\}} = \overline{N} . \tag{12.118}$$

This expression has a simple interpretation. The mean number of free electrons is $\overline{N}$, but the electrons are generated independently and thus obey Poisson statistics. The variance in the number is also $\overline{N}$, as is the ratio of the mean number squared to the variance. Since $i(t) = V_0/R(t)$ and $R(t)$ is proportional to $1/N(t)$, $\Delta i(t)/\langle i(t) \rangle = -\Delta R(t)/\overline{R} = \Delta N(t)/\overline{N}$ and (12.118) follows. Thus GR noise is basically the result of the Poisson fluctuations in the number of free carriers, leading to fluctuations in resistance which are converted to fluctuations in current if a constant voltage is applied.

GR noise in photoconductors As we saw in Sec. 12.1.2, a photoconductor is essentially a resistor whose resistance depends on illumination. The resistance fluctuations of GR noise are therefore an important limitation to the ability to detect weak illumination with photoconductors. In addition, a photoinduced carrier has a random lifetime, and this randomness is a further limitation to detection performance. This is also a form of GR noise, even though it involves photogeneration rather than thermal generation (Kingston, 1995). We shall now analyze a photoconductor where both of these effects are present.

The starting point for the discussion will be a slight modification of (12.116). If we write the total mean number of free electrons $\overline{N}$ as the sum of a thermal component $\overline{N}_{th}$ and a component $\overline{N}_p$ from photoexcitation, we have

$$S_{\Delta i}(\nu) = \left(\overline{N}_{th} + \overline{N}_p\right) \left[\frac{e\mu_e E_0}{L_z}\right]^2 \frac{2\tau}{1 + (2\pi\nu\tau)^2} \,. \tag{12.119}$$

Similarly, the mean current from (12.110) becomes

$$\langle i(t) \rangle = -\left(\overline{N}_{th} + \overline{N}_p\right) \frac{e\mu_e E_0}{L_z} \,. \tag{12.120}$$

Now consider explicitly the geometry of Fig. 12.3b, where both the illumination and the mean current flow are in the $+z$ direction. Also, assume for simplicity that the hole mobility and/or lifetime are small so that the photoconductivity is mediated solely by electrons. From (12.32) we obtain

$$\overline{N}_p = \tau A \eta I_p \,. \tag{12.121}$$

If we recognize that $L_z/(\mu_e E_0)$ is the transit time T_{tr} across the detector, we can write the mean photocurrent as

$$\langle i_p(t) \rangle = -\frac{\tau}{T_{tr}} eA\eta I_p \,. \tag{12.122}$$

Since $A\eta I_p$ is the mean number of electrons per second generated by the illumination, we see that each electron contributes a current equivalent to that produced by a charge $-e\tau/T_{tr}$. The ratio of lifetime to transit time,

$$G \equiv \frac{\tau}{T_{tr}} \,, \tag{12.123}$$

is called the *photoconductive gain.*

In many practical detectors, G is greater than 1, which requires a little explanation. One might think that the life of an electron would end when it reached the anode, as indeed it does in a photodiode. In a linear photoconductor with ohmic (non-rectifying) contacts, however, a new electron is injected at the cathode as soon as one disappears at the anode. Depending on τ and T_{tr}, this process can be repeated many times (Bube, 1992).

From (12.119), the low-frequency power-spectral density of the current can now be written as

$$S_{\Delta i}(0) = 2\tau \overline{N}_{th} \left[\frac{e\mu_e E_0}{L_z}\right]^2 + 2Ge\langle i_p \rangle \,. \tag{12.124}$$

The first term is what we had before for the GR noise associated with thermally generated carriers. The second term, however, looks more like shot noise. Comparison with (12.13) shows that this term can be interpreted (except for a factor of 2 which we explain below) as the shot noise associated with carriers of charge Ge (Kingston, 1995). Note, however, that both G and $\langle i_p \rangle$ are proportional to the applied voltage, so the second term in (12.124) still varies as voltage squared.

The extra factor of 2 in (12.124) comes from the randomness in the time an electron lives before recombination or trapping, an effect we have already analyzed in discussing trapping in P-N junctions. We saw in (12.56) that the noise variance (or power spectral density) is increased by a factor of $\sigma_\beta^2 + \overline{\beta}^2$ if each carrier contributes a random charge βe to the output current. In the present problem, β is proportional to the free time T, which is exponentially distributed:

$$\text{pr}(T) = \frac{1}{\tau} e^{-T/\tau}, \qquad (T \geq 0). \tag{12.125}$$

From the mean and variance of this density, as given in Sec. C.5.3, it follows that $(\sigma_T^2 + \overline{T}^2) = 2\tau^2$, accounting for the extra 2 in (12.124).

12.2.3 $1/f$ noise

In 1925 J. B. Johnson (later to discover thermal noise) was studying noise in vacuum tubes with thermionic cathodes. He observed the shot noise associated with the thermionic emission and saw that it had a flat temporal power spectrum, but he also discovered a component of the noise that had increased power at lower frequencies, with the power spectral density varying approximately as the reciprocal of the temporal frequency (Johnson, 1925; Kogan, 1996). A year later, Schottky (who had earlier discovered shot noise) proffered an explanation in terms of slow, random changes in the emission from the thermionic cathode, and he dubbed the effect *flicker noise* (Schottky, 1926). The flickering consists of slow, large-amplitude variations (see Fig. 12.12) that might today be called *drift* or *trending*.

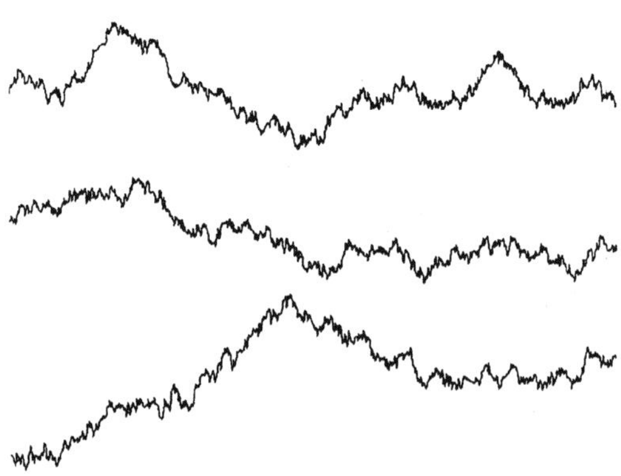

Fig. 12.12 Three typical waveforms of $1/f$ noise.

Since 1925 it has been found that virtually any current-carrying device has such excess low-frequency noise. Moreover, very similar behavior has been found in a wide variety of other fields, including seasonal temperatures, average rainfall, vehicular traffic flow, potentials across nerve membranes, loudness and pitch of music and even stock-market indices (Keshner, 1982). The most common term for this phenomenon, in all of these fields, is *1/f noise*, even though the power spectrum seldom varies exactly as $1/f$ (and even though it is common to use ν or ω rather than f for temporal frequency).

A comprehensive treatment of electronic $1/f$ noise in solids is given by Kogan (1996), and broad general reviews are given by Mandelbrot (1999), West and Shlesinger (1990), Gardner (1978), Keshner (1982) and Schroeder (1991). Another excellent source with a different bent is Lukyanchikova (1996); this book views noise processes in semiconductors not so much as a limitation on device performance but rather as a tool for investigating the fundamental physics and measuring key parameters of semiconductor materials and devices.

Power spectral density and autocorrelation function The most striking manifestation of $1/f$ noise is that the power spectral density of some temporal random process $x(t)$ varies as

$$S_x(\nu) \propto \frac{1}{|\nu|^\beta}, \tag{12.126}$$

where β is a positive number, usually in the range $1 \le \beta < 2$. Since the variance of a stationary, zero-mean random variable is given by

$$\text{Var}(x) = \int_{-\infty}^{\infty} d\nu\ S_x(\nu), \tag{12.127}$$

we must have infinite variance if (12.126) holds for all ν. For this reason, many workers have assumed that there must be some low-frequency cutoff, but all attempts to date to find one have failed. For example, Caloyannides (1974) studied the noise in certain transistors down to 5×10^{-7} Hz, or 1 cycle in 3 weeks,[6] and found no deviation from (12.126).

The power spectral density in (12.126) implies that the temporal correlations persist over very long times. One might think that the form of the autocorrelation would be given by (3.168), which gives the Fourier transform of $|\nu|^{-\beta}$, but in fact the function $|\nu|^{-\beta}$ in (12.126) is not the same thing as the generalized function $|\nu|^{-\beta}$ defined by Lighthill and discussed in Sec. 2.3.3 [see (2.96)]. As we saw there, the generalized function must have a strong negative singularity at the origin, while power spectral densities are never negative. We can, however, directly transform the power spectral density (at least for $\beta < 1$) and get the autocorrelation function. Since a power spectral density is an even function, we can write

$$R(\Delta t) = 2\int_0^{\infty} d\nu\ S_x(\nu)\cos(2\pi\nu\Delta t). \tag{12.128}$$

With (12.126) and the change of variables $u = 2\pi\nu\Delta t$, we find

$$R(\Delta t) \propto |\Delta t|^{\beta-1}\int_0^{\infty} du\ u^{-\beta}\cos u. \tag{12.129}$$

The integral converges for $0 < \beta < 1$, but the important point for this discussion is that it is just some constant,[7] independent of Δt, so we see that $R(\Delta t) \propto |\Delta t|^{1-\beta}$,

[6] The number of seconds in a year is approximately $\pi \times 10^7$.

[7] The reader who really must know what this constant is can consult Gradshteyn and Ryzhik (1980), formula 3.761.9.

which approaches $|\Delta t|^0$ or constant as $\beta \to 1$. A $1/f$ process is therefore one where the distant past strongly influences the present.

Observed variance If (12.126) is valid for all frequencies, the ensemble-average variance is infinite, but it does not follow that an experimenter would ever measure an infinite variance. Suppose a single sample function of duration T of the random process $x(t)$ is available, say for $-\frac{1}{2}T < t \leq \frac{1}{2}T$. The experimenter might define a sample mean m and variance s^2 by

$$m = \frac{1}{T} \int_{-\frac{1}{2}T}^{\frac{1}{2}T} dt\; x(t)\,, \tag{12.130}$$

$$s^2 = \frac{1}{T} \int_{-\frac{1}{2}T}^{\frac{1}{2}T} dt\; [x(t) - m]^2\,. \tag{12.131}$$

Now suppose that $x(t)$ is zero-mean and stationary, with the power spectral density given by (12.126). What are the expected values of m and s^2?

To find the expected value of m, we must interchange the order of statistical averaging and integration over t. This interchange is legal since the integral of the absolute value of the integrand is finite before and after taking the expectation (see Sec. 8.2.2). It is finite before expectation since experimental values of $x(t)$ must be finite and the range of integration is finite, and after taking the expectation inside the integral, the result is in fact zero:

$$\langle m \rangle = \frac{1}{T} \int_{-\frac{1}{2}T}^{\frac{1}{2}T} dt\; \langle x(t) \rangle = 0\,. \tag{12.132}$$

The expectation of s^2 is more delicate. The integral in (12.131) is still finite before taking the average, but the ensemble average of $x^2(t)$ is infinite if (12.126) holds for all ν. Surprisingly, however, $\langle s^2 \rangle$ is finite. To see why, express the sample function $x(t)$ in terms of its Fourier transform as

$$x(t) = \int_{-\infty}^{\infty} d\nu\; X(\nu) \exp(2\pi i \nu t)\,. \tag{12.133}$$

Then the sample mean is given by

$$m = \frac{1}{T} \int_{-\frac{1}{2}T}^{\frac{1}{2}T} dt \int_{-\infty}^{\infty} d\nu\; X(\nu) \exp(2\pi i \nu t) = \int_{-\infty}^{\infty} d\nu\; X(\nu)\, \mathrm{sinc}(\nu T)\,, \tag{12.134}$$

and the sample variance is

$$\begin{aligned} s^2 &= \frac{1}{T} \int_{-\frac{1}{2}T}^{\frac{1}{2}T} dt \left| \int_{-\infty}^{\infty} d\nu\; X(\nu)\, [\exp(2\pi i \nu t) - \mathrm{sinc}(\nu T)] \right|^2 \\ &= \frac{1}{T} \int_{-\frac{1}{2}T}^{\frac{1}{2}T} dt \int_{-\infty}^{\infty} d\nu \int_{-\infty}^{\infty} d\nu'\; X(\nu)\, X^*(\nu') \\ &\quad \times [\exp(2\pi i \nu t) - \mathrm{sinc}(\nu T)][\exp(-2\pi i \nu' t) - \mathrm{sinc}(\nu' T)]\,. \end{aligned} \tag{12.135}$$

Now we want to take the expectation inside not one but three integrals, and two of them are on the infinite line. To legalize this step, suppose that $S_x(\nu) \propto |\nu|^{-\beta}$ only for $|\nu| > \nu_1$, where ν_1 will be allowed to approach zero. All integrals then remain finite, and we have

$$\langle s^2 \rangle = \frac{1}{T} \int_{-\frac{1}{2}T}^{\frac{1}{2}T} dt \int_{-\infty}^{\infty} d\nu \int_{-\infty}^{\infty} d\nu' \, \langle X(\nu)\, X^*(\nu') \rangle$$

$$\times \left[\exp(2\pi i \nu t) - \mathrm{sinc}(\nu T)\right]\left[\exp(-2\pi i \nu' t) - \mathrm{sinc}(\nu' T)\right]. \tag{12.136}$$

We know from the temporal counterpart of (8.181), however, that

$$\langle X(\nu)\, X^*(\nu') \rangle = S_x(\nu)\, \delta(\nu - \nu') \tag{12.137}$$

for any stationary random process, so

$$\langle s^2 \rangle = \frac{1}{T} \int_{-\frac{1}{2}T}^{\frac{1}{2}T} dt \int_{-\infty}^{\infty} d\nu \; S_x(\nu) |\exp(2\pi i \nu t) - \mathrm{sinc}(\nu T)|^2 . \tag{12.138}$$

A little algebra then yields

$$\langle s^2 \rangle = \int_{-\infty}^{\infty} d\nu \; S_x(\nu) \left[1 - \mathrm{sinc}^2(\nu T)\right]. \tag{12.139}$$

The factor $[1 - \mathrm{sinc}^2(\nu T)]$ suppresses low frequencies, $|\nu| < T^{-1}$, so the integral in (12.139) no longer diverges, even if $\nu_1 \to 0$. Specifically, if $S_x(\nu) = C/|\nu|^\beta$ for all ν, then

$$\langle s^2 \rangle = C T^{\beta - 1} \int_{-\infty}^{\infty} \frac{dy}{|y|^\beta} \left[1 - \mathrm{sinc}^2(y)\right], \tag{12.140}$$

where $y \equiv \nu T$. The integral is finite for $1 < \beta < 2$, so the expectation of s^2 increases with increasing measurement time in this range. Arbitrarily large T implies arbitrarily large experimental variances. On the other hand, for any finite T, s^2 is an infinitely biased estimate of the ensemble variance.

Other experimental observations For linear electrical devices, such as resistors and photoconductors, it is usually found that the power spectral density of the fluctuating voltage across the device is proportional to the square of the mean voltage and inversely proportional to the mean number of free carriers $\overline{N}$ (Voss and Clarke, 1976). This observation prompted Hooge (1969) to suggest the following empirical relation:

$$S_{\Delta V}(\nu) = \frac{\alpha_H \overline{V}^2}{\overline{N} |\nu|}, \tag{12.141}$$

where α_H is often called the *Hooge constant*, though it varies quite a bit from one material to another (Kogan, 1996).

Since $\Delta i(t) = \Delta V(t)/\overline{R}$, the Hooge relation can also be expressed as

$$S_{\Delta i}(\nu) = \frac{\alpha_H \overline{V}^2}{\overline{N}\, \overline{R}^2 |\nu|}. \tag{12.142}$$

The factor $\overline{V}^2/\overline{N}\,\overline{R}^2$ is found also in the GR power spectrum of (12.115), where it occurs because the basic noise mechanism is random fluctuations in resistance. The

Hooge relation thus suggests strongly that $1/f$ noise is also the result of random resistance variations.

The single-point amplitude statistics of $1/f$ noise have also been studied extensively. Occasionally deviations from Gaussian behavior are observed in electronic $1/f$ noise, but usually the probability density function $\mathrm{pr}[V(t)]$ is remarkably close to a univariate normal (Kogan, 1996).

Control of 1/f noise Even if the variance of $1/f$ noise is not actually infinite, it is at least very large, and the only way to control it in practice is to make measuring systems insensitive to low frequencies. For electronic systems, a simple coupling capacitor may be effective in suppressing $1/f$ noise. More sophisticated high-pass filters or active systems such as automatic level controls can also be used. In optical systems, a common remedy is to modulate the light by periodically interrupting it with a shutter. (A periodically opened shutter is often called a *chopper*.) Then the electronics of the detector system does not need to pass low frequencies, and most of the $1/f$ spectrum can be filtered out.

Distribution of relaxation times The first attempts at explaining $1/f$ noise mathematically were in the 1930s (Bernamont, 1937; Surdin, 1939). These early workers took the view that $1/f$ noise was basically GR noise but with a distribution of relaxation times. If we start with (12.115) for the GR spectrum but regard τ as a random variable with probability density function $\mathrm{pr}_\tau(\tau)$, we can write

$$S_{\Delta i}(\nu) = \frac{\overline{V}^2}{\overline{N}\,\overline{R}^2} \int_0^\infty d\tau \; \mathrm{pr}_\tau(\tau) \frac{2\tau}{1+(2\pi\nu\tau)^2} \,. \tag{12.143}$$

A convenient assumption, made with essentially no justification in the 1930s, is that τ takes on values in some broad range (τ_1, τ_2) and that its density in this range is given by

$$\mathrm{pr}_\tau(\tau) \propto \frac{1}{\tau}\,, \qquad \tau_1 \le \tau < \tau_2 \,. \tag{12.144}$$

With this assumption, we find

$$S_{\Delta i}(\nu) \propto \int_{\tau_1}^{\tau_2} d\tau \, \frac{1}{1+(2\pi\nu\tau)^2} = \frac{1}{2\pi|\nu|}\left[\tan^{-1}(2\pi|\nu|\tau_2) - \tan^{-1}(2\pi|\nu|\tau_1)\right] . \tag{12.145}$$

Over a broad range of frequencies, this expression can be approximated as

$$\frac{1}{2\pi|\nu|}\left[\tan^{-1}(2\pi|\nu|\tau_2) - \tan^{-1}(2\pi|\nu|\tau_1)\right] \simeq \frac{1}{4|\nu|}\,, \qquad \frac{1}{\tau_2} \ll 2\pi|\nu| \ll \frac{1}{\tau_1} \,. \tag{12.146}$$

Thus this distribution of relaxation times explains the $1/|\nu|$ behavior over a limited range, but of course it still remains to explain the distribution of relaxation times. To account for the experimental observations, the range from τ_1 to τ_2 must cover many decades.

Thermal activation The next step was taken by van der Ziel (1950). He recognized that trapping is a thermally activated process obeying an Arrhenius relation of the form [*cf.* (12.58)]

$$\tau = \tau_0 \exp\left(\frac{\mathcal{E}}{k_B T}\right) , \tag{12.147}$$

where $\mathcal{E}$ is the binding energy of a trapped electron and τ_0 is a characteristic of the material and the trap, assumed to be a constant. In semiconductors, $\mathcal{E}$ is of order 0.1 – 1.0 eV. Thus a distribution in τ can result from a distribution in $\mathcal{E}$, but because of the exponential a relatively narrow range in $\mathcal{E}$ can lead to a very large range in τ.

If the probability density function of $\mathcal{E}$, denoted $\text{pr}_\mathcal{E}(\mathcal{E})$, is assumed to be known, the corresponding distribution on τ can be found from (C.45); the result is

$$\text{pr}_\tau(\tau) = \frac{\text{pr}_\mathcal{E}(\mathcal{E})}{|d\tau/d\mathcal{E}|} = \frac{k_BT}{\tau}\,\text{pr}_\mathcal{E}[k_BT\log(\tau/\tau_0)]\,. \tag{12.148}$$

Now we see that $\text{pr}_\tau(\tau)$ varies as $1/\tau$ so long as $\text{pr}_\mathcal{E}(\mathcal{E})$ is approximately constant. Moreover, since τ appears in the argument of $\text{pr}_\mathcal{E}(\mathcal{E})$ only logarithmically, $\text{pr}_\mathcal{E}(\mathcal{E})$ can vary substantially with $\mathcal{E}$ without causing large deviations from the $1/\tau$ behavior.

Some numbers should prove instructive. Suppose $\tau_0 = 10^{-11}$ sec, which is typical for semiconductors, and suppose $\mathcal{E}$ is distributed more or less uniformly over 0.2-0.8 eV. Since $k_BT = 0.025$ eV at room temperature, (12.147) shows that τ varies from 3×10^{-8} sec to about 800 sec. Thus a modest range of $\mathcal{E}$ values can lead to an enormous range of τ values if the process is thermally activated, and (12.148) shows that $\text{pr}_\tau(\tau) \propto 1/\tau$ over this range. With the numbers in this example, the approximation in (12.146) is valid from milliHertz to gigaHertz frequencies, and the spectrum follows $1/\nu$ over this range.

A caveat It is important to note that the derivation of a $1/f$ spectrum from a distribution of relaxation times implicitly assumes that the medium is inhomogeneous. In a homogeneous medium, multiple relaxation mechanisms would combine to give an overall effective relaxation rate

$$\frac{1}{\tau_{eff}} = \int_0^\infty d\tau\,\frac{\text{pr}(\tau)}{\tau}\,. \tag{12.149}$$

Then (12.107) would still hold with τ replaced by τ_{eff}, and the characteristic Lorentzian GR spectrum of (12.116) would be found in spite of the distribution of τ. It is only when the medium can be divided into independent subregions, each characterized by a certain τ and hence a certain GR spectrum, that we should expect multiple relaxation mechanisms to lead to a $1/f$ spectrum.

Relation to log-normals A rather different view of $1/f$ noise and other power-law phenomena has been championed by Montroll, West and Shlesinger in various combinations (*e.g.*, Montroll and Shlesinger, 1982; West and Shlesinger, 1989, 1990). They note that a wide variety of physical processes are cascaded random selections, where the overall probability of a success is the product of probabilities of success on individual tasks. They cite the probability of publishing a scientific paper, the probability of a water droplet making it all the way to the end of the Nile and the probability of an air molecule reaching a particular alveolus in the lung. In all of these cases, they argue, some final random variable should follow a log-normal probability law.

The basic premise is simple: If one has a product of independent random variables,

$$X = \prod_{n=1}^{N} x_n\,, \tag{12.150}$$

then the log of the product must be a sum of independent random variables,

$$Y = \ln X = \sum_{n=1}^{N} \ln x_n \,. \tag{12.151}$$

By the central-limit theorem (see Sec. 8.3.4), Y tends to a normal distribution as N increases, so X tends to a log-normal. Specifically, the density on X takes the form (see Sec. C.5.9)

$$\mathrm{pr}(X) \simeq \frac{1}{X} \frac{1}{\sqrt{2\pi\sigma_Y^2}} \exp\left[-\frac{\left(\ln X - \overline{Y}\right)^2}{2\pi\sigma_Y^2} \right] . \tag{12.152}$$

Shlesinger and his colleagues note that this density mimics a $1/X$ over a range that can be large if σ_Y^2 is large, and they offer this observation as an explanation of the ubiquitous appearance of $1/X$ probability densities. To make the connection to electrical $1/f$ noise, they identify X as the relaxation time τ and then appeal to the same reasoning as in (12.144) – (12.146). They do not, however, identify the component variables x_n, nor do they explain why the variance should be large or how the log-normal can approximate a $1/\tau$ behavior over the many decades required to match observed noise spectra. At best, this work is an intriguing suggestion without obvious applicability to electrical noise.

Nonlinear dynamics Another class of explanations for $1/f$ noise centers on chaos and nonlinear dynamics. A readable, semi-popular survey of these approaches is Schroeder (1991).

In a classic paper, Bak *et al.* (1988) argue that many dissipative dynamical systems evolve naturally toward a critical state with no characteristic time or length scale. In this view, $1/f$ noise is not noise at all but rather reflects the intrinsic dynamics of self-organized critical systems.

The prolific scientific output of Benoit Mandelbrot on these themes is collected in Mandelbrot (1999).

Maybe it isn't stationary Another view on $1/f$ noise is that it doesn't have a $1/f$ power spectral density at all — in fact, it doesn't have any power spectral density because it isn't a stationary random process. Mandelbrot (1967) suggested that the paradox of infinite variance could be avoided by treating $1/f$ noise as a nonstationary random process with a time-dependent variance.

Keshner (1982) developed this theme further and gave several intriguing examples, all of which were based on the assumption that the process had a definite beginning at some point in the distant past. If we call the beginning point $t = 0$ and we observe the process from $t = t_1$ to $t = t_2$, where $t_2 - t_1 \gg t_1$, then we might not be able to discern that the process was nonstationary. If we then assumed that the observed segment was a sample function of a stationary random process, we could estimate a power spectral density even though the actual process did not possess one. Under broad assumptions, Keshner's models yield apparent power spectra of $1/f$ form.

There have been several experimental attempts to detect nonstationarity in electronic $1/f$ noise by making repeated measurements on the same device over long periods of time, but uniformly they fail to find evidence for the kinds of mechanism postulated by Keshner (Kogan, 1996). For example, Stoisek and Wolf (1976) prepared resistors by ion implantation, and no manifestation of nonstationarity was found for 2.5 years after the birth of the devices. Similarly, Tandon and Bilger (1976) observed a semiconductor device called a stabilitron and concluded that it remained essentially invariant for 4.5 years.

Other mechanisms A wide variety of other mechanisms have been postulated for electronic $1/f$ noise, including temperature fluctuations, surface trapping, random scattering of carriers and hopping of carriers from one site to another (van der Ziel, 1988). Each of these mechanisms is capable of limited success in explaining the behavior of certain devices, but none appears to reach the stature of a universal mechanism for this virtually universal phenomenon. Indeed, many authors assert that no single mechanism can ever account for the myriad experimental manifestations of $1/f$ noise.

12.2.4 Noise in gated integrators

In Secs. 12.1.1 and 12.1.3, we considered the effect of an RC filter on the output current from a photodiode. This circuit can be thought of as a *leaky integrator*: the charge on the capacitor builds up by integrating the current but leaks off through the resistor. As we saw in Sec. 12.1.1, the effective averaging time is $2RC$, though the averaging occurs with an exponential temporal weighting.

An alternative to the leaky RC integrator is the *gated integrator* shown in Fig. 12.13. At time $t = 0$ the electronic switch is closed briefly, shorting out the capacitor and setting its voltage to approximately zero. Then the switch is opened and the capacitor begins to integrate the current, so the mean output voltage at time t is

$$\overline{V}_{out}(t) = \frac{1}{C} \int_0^t dt' \, \overline{i}(t') \,. \tag{12.153}$$

In practice, the device is usually allowed to integrate for a fixed time t_0, after which the output voltage is sampled and stored, and then the voltage is reset and the cycle is repeated.

Gated integrators form the basis for a number of important image detectors. It is possible, for example, to replicate the circuit of Fig. 12.13 many times on a silicon integrated-circuit chip and to bond it to a photoconductor made of some other material. Such devices are known as hybrid focal-plane arrays (*hybrid* because two materials are involved, *focal-plane* because they are often placed in the image plane of an optical system). Since the different cells on the device function independently, the analysis presented below for single-element gated integrators is directly applicable to the arrays.

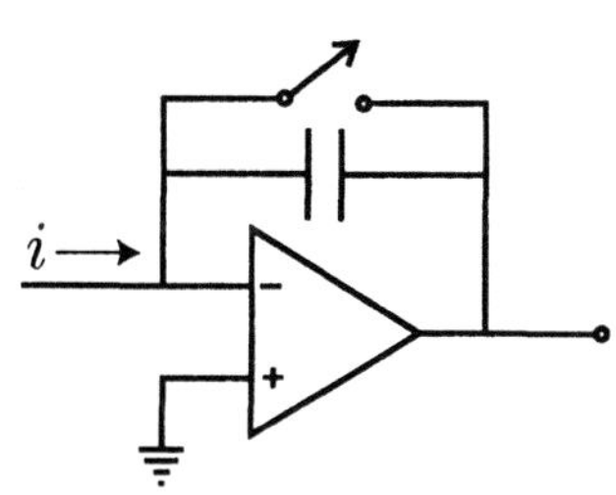

Fig. 12.13 Schematic of a gated integrator.

Charge-coupled devices (CCDs) and various other video detectors can also be regarded as arrays of gated integrators. In a CCD, charge is accumulated for a fixed period in each unit cell, and then the charge signals of all cells in one row of the array are transferred to a shift-register and stepped to an output line. The charge in each cell is set approximately to zero by the transfer operation, so the basic operation of a CCD is that of a gated integrator.

Similarly, in television camera tubes such as vidicons, charge is accumulated on a photoconductive surface and read out periodically by a scanning electron beam. The beam effectively sets the charge to zero, so a vidicon too is a kind of gated integrator. In this case, however, there are no discrete cells and hence no way to identify discrete elementary circuits like Fig. 12.13.

kTC noise The mean output voltage from a single gated integrator has the form (12.153). The corresponding equation for the actual random output is

$$V_{out}(t) = V_{out}(0^+) + \frac{1}{C}\int_0^t dt'\, i(t')\,, \tag{12.154}$$

where $V_{out}(0^+)$ is the voltage immediately after the switch is opened, and $i(t)$ is the total random current (dark current, photocurrent and any noise currents that might be present).

One might think that $V_{out}(0^+)$ should be zero, but a small random voltage is required by the equipartition principle. The mean of this voltage is zero, and the variance is given by (12.64) as

$$\mathrm{Var}\{V_{out}(0^+)\} = \frac{k_B T}{C}\,. \tag{12.155}$$

Since the charge on the capacitor at $t = 0^+$ is $Q(0^+) = CV_{out}(0^+)$, the charge variance corresponding to (12.155) is

$$\mathrm{Var}\{Q(0^+)\} = C^2\,\mathrm{Var}\{V_{out}(0^+)\} = k_B TC\,. \tag{12.156}$$

Because of this expression, the randomness in $Q(0^+)$ is called *kTC noise*, and the same term (rather than kT-over-C noise) is applied to the randomness in $V_{out}(0^+)$.

Though the equipartition principle gives the right answer for the kTC noise variance, there are some puzzling aspects to it. First, the kTC noise is of thermal origin, since it vanishes if $T \to 0$, but it does not seem to be in accord with the fluctuation-dissipation theorem. Ideal capacitors are purely reactive elements and do not dissipate any power, so why should there be fluctuations associated with them? Moreover, when the capacitor is shorted out by the switch, the voltage across it should be zero, so why shouldn't it remain zero when the switch is opened?

To answer these questions, it is necessary to look in more detail at the characteristics of the switch. A switch is a device that changes its resistance from a very low value to a very high value when it is opened. For typical electronic switches, the closed resistance is $1 - 10$ ohms and the open resistance is perhaps $10^{10} - 10^{11}$ ohms. This resistance is in parallel with the capacitance C, and we know from (12.19) that the bandwidth B of this combination is $(4RC)^{-1}$. The integrating capacitors used in gated integrators are quite small in order to get a large voltage for a small amount of charge. If we take $C = 0.1$ pF, then B is of order $25 - 250$ Hz when the switch is open and $250 - 2500$ GHz when it is closed.

In the closed state, then, the voltage is not zero because the switch is not a perfect short circuit; the voltage fluctuates at hundreds of GHz, and its variance is given by (12.155). When the switch is open, (12.155) still applies, but now the voltage fluctuates very slowly. Indeed, if the integrator is well designed, there should be negligible discharging of the capacitor during one integration period, so this period sets the time scale for the voltage fluctuations. Opening the switch does not change the noise variance but instead freezes in a value that will remain constant for the integration period. During the next integration period, a different value will be frozen in. Since the fluctuations are so rapid in the closed state, the values during successive integration periods are statistically independent.

Correlated double sampling There is a simple way to suppress kTC noise. Suppose the output voltage is sampled twice per integration cycle, once just after a reset ($t = 0^+$) and once just before the next reset ($t = t_0^-$). Both of these voltages can be stored either digitally or on additional capacitors. Since there is very little decay of the initial voltage during the integration period, the first of these voltages is $V_1 = V_{out}(0^+)$ and the second, V_2, is given by (12.154). If we take the difference, $V_2 - V_1$, the kTC noise disappears and only the desired integrated current remains. We can effectively remove the kTC noise because it is so highly correlated between the two sample points.

Dark current The gated integrator, like any integrator used with a photoconductor, must integrate dark current as well as photocurrent. If we integrate for a time T, the mean output voltage from dark current is $-(e/C)a_{dark}T$ and the variance, under the usual Poisson model, is $(e/C)^2 a_{dark}T$. Since kTC noise and dark current are statistically independent, this variance component is just added to (12.155).

12.2.5 Arrays of noisy photodetectors

So far we have considered single photodetectors, but in imaging applications we usually have imaging detectors, often in the form of regular arrays. In that case the detector performance cannot be summarized by a single number such as DQE; instead the characteristics must be stated as a function of spatial position or spatial frequency.

As a simple example, consider a $J \times J$ array of contiguous gated integrators with center-to-center spacing of ϵ and area ϵ^2. Thus $J = L/\epsilon$, where L is the width of the array, and the total number of detectors is $M = J^2$. The easiest way to characterize the noise properties of this array is in terms of the mean vector and covariance matrix for its output. We have already discussed this problem in detail in Sec. 11.2 for the case where each element is an ideal photon counter, and now we need to extend the discussion to include the additional noise sources introduced above.

To do so, we write the output of the mth detector as a sum of two random variables,

$$g_m = g_m^{(phot)} + \delta g_m \,, \tag{12.157}$$

where $g_m^{(phot)}$ results from photon absorption and δg_m is the contribution from dark current, Johnson noise and other noise sources. In many real detector arrays, the elements are electrically and optically isolated, so photons absorbed in one element have no influence on neighboring elements and any excess noise is statistically independent from element to element. With this assumption the elements of $\delta\mathbf{g}$ are independent, and the covariance matrix of this term is diagonal. If we further assume that the detector elements are identical, then

$$\mathbf{K}_{\delta\mathbf{g}} = \sigma_{exc}^2 \mathbf{I}\,, \tag{12.158}$$

where $\mathbf{I}$ is the $M \times M$ unit matrix and σ_{exc}^2 is the variance of the excess noise in each element.

Moreover, with the exception of trapping noise discussed in Sec. 12.1.3, all of the excess noise sources presented so far in this chapter are independent of the photon flux, so $\delta\mathbf{g}$ is independent of $\mathbf{g}^{(phot)}$. With this restriction, the overall covariance matrix is given by

$$\mathbf{K}_{\mathbf{g}} = \mathbf{K}_{\mathbf{g}}^{(phot)} + \mathbf{K}_{\delta\mathbf{g}}\,. \tag{12.159}$$

The first term, $\mathbf{K}_{\mathbf{g}}^{(phot)}$, is just what we computed in Sec. 11.2. We know from that discussion that $\mathbf{K}_{\mathbf{g}}^{(phot)}$ is diagonal for a Poisson source, but it can be non-diagonal if we consider random fluence or other non-Poisson effects.

Frequency-domain descriptions The covariance matrix is a complete description of the second-order statistics of a random vector. When the random vector is produced by a regular detector array, however, many authors apparently feel an impulse to invoke a Fourier description. We know from Chaps. 7 and 8 that Fourier descriptions are useful for signals and systems that exhibit translational invariance, and an array of identical detectors looks almost the same when it is shifted by one element, so one might expect a Fourier transform to do something useful; there are several reasons why this expectation might be forlorn.

First, the apparent translational invariance is discrete rather than continuous, so a discrete Fourier transform must be used instead of a Fourier integral. Second, the array is not really invariant to a shift by an integer number of elements since it has finite extent; the only way to get any kind of invariance is to assume that the array wraps around cyclically when shifted (see Sec. 8.2.8). Third, even if we assume that the array itself has this unphysical cyclic property, it does not follow that the noise covariance does; we saw in Chap. 11 that a nonuniform photon fluence leads to nonstationary noise. Finally, to get something like an SNR or DQE in the frequency domain, we must pick a frequency of interest, and a discrete DFT contains only a discrete set of frequencies even though the actual continuous fluence pattern incident on the detector is not so constrained.

We shall return to the topic of DQE in the spatial-frequency domain in Sec. 13.2.9, and we shall deal specifically with some of the problems associated with discrete spatial frequencies in Sec. 16.1 when we discuss digital radiography.

12.3 X-RAY AND GAMMA-RAY DETECTORS

Though x rays and gamma rays are electromagnetic radiation like visible light, the crucial difference is that the energy per photon is much higher. While a photoelectric interaction of visible light in a semiconductor detector, for example, produces a single electron-hole pair, thousands of such pairs are produced in each x-ray or gamma-ray interaction.

A more subtle issue concerns the rate of arrival of photons. In almost all gamma-ray imaging systems in nuclear medicine or gamma-ray astronomy, the average interval between photons is large compared to the resolving time of the electronics, so it is possible to count individual photons. In medical radiography, on the other hand, a relatively short exposure is made with a high flux of x rays, so the individual photons are usually not temporally resolved. When talking about x rays and gamma rays, therefore, we need to distinguish *photon-counting detectors* from *integrating detectors*, the latter term implying that only the integrated effect of many photons is observed.

In Sec. 12.3.1 we review the basic physics of the interaction of high-energy photons with a detector material; readers conversant with these processes can skip to the next section without loss of continuity.

In Sec. 12.3.2 we analyze single-element, photon-counting semiconductor detectors in detail, building on the discussion of optical semiconductor detectors in Sec. 12.1.2.

The discussion of imaging detectors begins in Sec. 12.3.3 with an analysis of photon-counting semiconductor detector arrays. This discussion continues in Sec. 12.3.4, where we show how techniques from estimation theory can be used to get better information about the position and energy of gamma-ray photons. This section and much of the remainder of the chapter presumes some acquaintance with maximum-likelihood estimation, a topic to be treated in detail in Sec. 13.3.4.

Section 12.3.5 discusses photon-counting scintillation detectors, especially the Anger camera, which has been the mainstay of nuclear medicine since the 1950s, and Sec. 12.3.6 treats estimation of photon energy and interaction position in these devices. Then in Sec. 12.3.7 we discuss the statistical properties of images formed from these position and energy estimates.

In Sec. 12.3.8 we discuss integrating detectors of the kind that might be used in digital radiography. This treatment builds heavily on the theory of amplified random processes from Sec. 11.4.

Finally, in Sec. 12.3.9 we discuss the effects that arise specifically in x-ray and gamma-ray detectors when the energy from one incident photon is deposited at two distinct locations due to K x-ray emission or Compton scattering. This section also builds on basic principles developed in Chap. 11, and it should serve to deepen the reader's understanding of random point processes.

12.3.1 Interaction mechanisms

A full understanding of the detection of x rays and gamma rays must take into account the mechanisms by which they interact with the detector material. For the energy range of interest in medical applications, there are two main processes for the initial interaction: *photoelectric absorption* and *Compton scattering*. There is also an elastic scattering mechanism at relatively low energies called *Rayleigh scattering*, and at high energies (above 1 MeV) gamma rays can produce positron-electron pairs, but we shall ignore both of these effects in this discussion.

Many of the interesting properties of x-ray and gamma-ray detectors arise from secondary interactions that take place after the initial photoelectric or Compton event. For example, a high-energy photoelectron or Compton electron can produce many electron-hole pairs in a semiconductor, or it can excite luminescent centers in a scintillator. We then need to understand how these secondary excitations are distributed spatially and how they result in a detector output. Moreover, the initial interaction events will usually result in secondary high-energy photons as well as electrons, and these photons can also produce electron-hole pairs or excite luminescent centers.

We shall discuss these processes in this section to the extent needed to understand the characteristics of image detectors. Some background material relevant to this section has already been presented in Chap. 10. In particular, the reader is assumed to be familiar with the physics of photoelectric interactions (Sec. 10.1.4) and Compton scattering (Sec. 10.3.7) and with the concept of a cross section (Sec. 10.2.5).

Photoelectric interactions A photoelectric interaction is one in which a photon is absorbed by an electron, the photon disappearing completely in the process and the electron acquiring some kinetic energy. These interactions cannot occur with free electrons since momentum would not be conserved in that case. A photon of energy $\mathcal{E}_0$ has momentum $\mathcal{E}_0/c$, and a free electron at rest has zero momentum. If the photon energy is transferred to the electron, it will have kinetic energy $\mathcal{E}_{kin} = \frac{1}{2}mv^2 = \mathcal{E}_0$, where v is the speed of the electron after the interaction and m is its mass. Since momentum $p = mv$ for an electron, it will then have momentum $\sqrt{2\mathcal{E}_0 m}$ rather than $\mathcal{E}_0/c$, so we cannot simultaneously conserve energy and momentum in photoelectric absorption by a free electron.

If the electron is bound to an atom, however, then any excess momentum can be transferred to motion of the atom. The atom is much heavier than an electron, so when it carries away momentum Δp, it acquires only a very small energy $\Delta p^2/2M_a$, where M_a is the mass of the atom. To a good approximation, therefore, we can neglect the energy transferred to the atom and write the kinetic energy of the photoelectron as

$$\mathcal{E}_{kin} = \mathcal{E}_0 - \mathcal{E}_b \, , \tag{12.160}$$

where $\mathcal{E}_b$ is the initial binding energy of the electron to the atom.

The photon $\mathcal{E}_0$ is typically in the range 10 – 100 keV for x rays used in diagnostic radiology and up to 500 keV or so for the gamma rays used in nuclear medicine. The binding energy $\mathcal{E}_b$ depends on the type of atom and the specific energy level. Photoelectric interactions have higher cross sections for more tightly bound electrons, so the dominant interaction is usually with the innermost electron shell, called the *K shell* in x-ray parlance. As an example, iodine is a constituent of scintillation materials such as NaI or CsI, and the K-shell binding energy of iodine is about 35 keV. Thus if a photon of energy 140 keV (a common energy in nuclear medicine) is absorbed by photoelectric interaction with a K-shell electron in iodine, the electron gains a kinetic energy of about 105 keV.

Compton interactions As discussed in Sec. 10.3.7, Compton scattering is the process by which a photon scatters inelastically from a free (or loosely bound) electron. Suppose the electron is initially at rest and a photon of initial energy $\mathcal{E}_0$ scatters through an angle θ_s. Energy and momentum are conserved if the photon energy $\mathcal{E}$ after interaction is given by (10.226) and the electron acquires an energy $\mathcal{E}_0 - \mathcal{E}$.

Numerically, we saw in Sec. 10.3.7 that 45^0 scattering of a 140 keV photon gives a scattered photon of energy 129.6 keV, 90^0 scattering gives 109.9 keV and 180^0 scattering gives 90.4 keV. The corresponding energies for the Compton electrons are 11.4 keV, 30.1 keV and 49.6 keV, respectively.

Secondary photons Both photoelectric interactions and Compton interactions produce high-energy electrons, and Compton scattering also leaves a high-energy scattered photon. In fact, secondary photons are also produced in photoelectric interactions. Let us suppose, as in the example above, that a photon of energy $\mathcal{E}_0$ greater than 35 keV is absorbed by a K-shell electron in iodine. This process leaves a vacancy in the K shell that can be filled by a transition of an electron from some higher shell into the K shell. The most probable occurrence is that the transition is from the next higher shell, called the L shell (see Fig. 12.14). This transition results in the emission of an x-ray photon of energy corresponding to the difference in energies of the two shells. This photon, called a K_α x ray, has an energy of about 28 keV in iodine. If the K-shell vacancy is filled by a transition from the next higher level, the M shell, then the photon is called a K_β x ray and has energy about 30 keV.

Of course, the process does not stop there. The transition from, say, the L shell to the K shell leaves a vacancy in the L shell. That vacancy can be filled by a transition from the M shell, yielding a photon of energy around 2 keV, called an L_α x ray. Often we ignore such low energy photons since they are reabsorbed very near the point of emission, but the K x rays can play a significant role in detector performance.

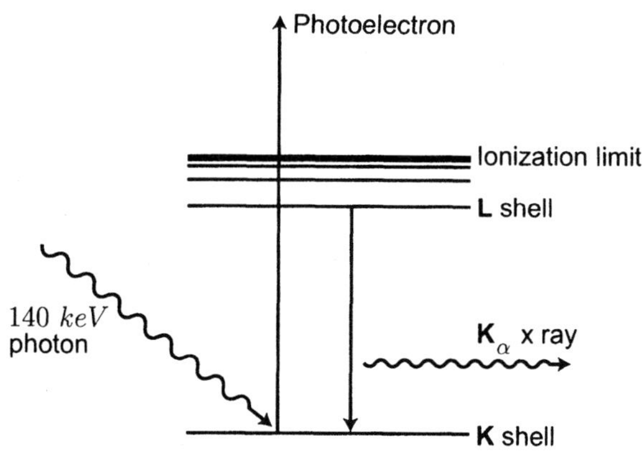

Fig. 12.14 Processes involved in production of a K x ray following absorption of a 140 keV photon in iodine.

Two things can happen to a K x ray when it is produced after a photoelectric interaction: it can interact within the material or it can escape from the volume of the material without interacting. If it interacts within the material, it can do so by either a Compton or photoelectric interaction. In either case, another energetic electron will be produced and another photon. Similarly, if the initial interaction is Compton scattering, the scattered photon can either escape the material or interact within it, and in the latter case the interaction can be either Compton or photoelectric. Accounting for all events in this complex cascade is difficult; usually the best approach is Monte Carlo simulation.

High-energy electrons As we have just seen, an initial interaction, either Compton or photoelectric, results in one or more energetic electrons in the material. If the material is to be used as a detector, we must sense the effects of these electrons. How this is done depends on the type of detector. In a semiconductor detector, some of the energy of the high-energy electrons is converted to electron-hole pairs that can eventually be sensed on external electrodes. In a scintillation detector, some of the energy of the high-energy electrons is converted to light which we can detect with ordinary optical detectors. In either case, the characteristics of the detector depend on the fraction of the energy converted to a useful form, its spatial distribution, and its statistics.

The major interaction mechanisms for a high-energy electron in a solid are scattering by other free electrons, emission or absorption of phonons (lattice vibrations), and ionization of the lattice via creation of electron-hole pairs. Since detector materials are typically good insulators, there are very few free electrons, and we can concentrate on the two competing processes of electron–phonon interactions and lattice ionization.

We do not need to study these processes in detail here, but a useful mental image is that the electron travels a tortuous path from the point of interaction, generating phonons and electron-hole pairs in its wake. The energy loss along this path is approximately uniform, so a relatively uniform cloud of low-energy electrons and holes is produced. The size of this cloud is roughly 20 μm for a 100 keV electron. In a scintillator, there is another step in which the low-energy electrons are trapped in luminescent centers and eventually recombine with holes and emit optical photons.

12.3.2 Photon-counting semiconductor detectors

Having sketched the process of production of electron-hole pairs following an initial interaction of an x ray or gamma ray, we can now begin to develop more detailed statistical descriptions of the detector output, taking into account charge transport and induction of measurable signals in an external circuit. The basic principles we need for this purpose were developed in Secs. 12.1.2 and 12.1.3 in the context of optical semiconductor detectors, but there are some new complications when x rays and gamma rays are involved.

First, x-ray and gamma-ray detectors must be thick enough to absorb the radiation; it is common to use thicknesses from 1 mm to 1 cm or more, as opposed to $1-10$ μm which will suffice for optical detectors. A result of the increased thickness is that trapping effects may be much more severe in semiconductor x-ray and gamma-ray detectors than in optical ones. Moreover, it is quite common for one type of charge carrier (usually holes) to be trapped much more strongly than the other. As we shall see, the combined effect of the random depth of interaction and the strong trapping plays a key role in the detector statistics.

In a semiconductor photodiode for optical use, the material initially has fairly low resistivity, but a thin, high-resistivity depletion region is formed by a P-N junction, and it is this region that is sensitive to light. In a well-designed gamma-ray detector, on the other hand, the whole volume has high resistivity and hence is sensitive to photons. It is perhaps more accurate to refer to gamma-ray detectors as semi-insulators rather than semiconductors.

Finally, because we can detect individual photons, we may ask for more information from each one. Often we want to know not only the position of a photon but also its energy. For example, in nuclear medicine we usually use a radioisotope that emits only a single photon energy, but photons may undergo Compton scatter in the patient's body and arrive at the detector with a reduced energy; it is important to be able to recognize and reject these scattered photons since they convey relatively little useful information about the object. In some applications — notably positron emission tomography — we also want to estimate the time of arrival of a photon to high precision.

Problem statement Throughout this section we shall consider a single-element (nonimaging) detector in the same configuration as in Sec. 12.1.3, a slab of dimensions $L_x \times L_y \times L_z$, with $L_x, L_y \gg L_z$. Even though L_z is small compared to the lateral dimensions, we shall assume that it is large compared to the dimensions of the charge cloud produced by a gamma-ray absorption; typical sizes might be a charge cloud of 20 μm, a slab thickness of 1 – 2 mm and lateral dimensions of 1 cm. The electrodes are on the faces $z = 0$ and $z = L_z$, and the photons are incident in the $+z$ direction. The material has homogeneous resistivity (no P-N junction), so a potential difference V_0 between the electrodes establishes a uniform field $E_0 = V_0/L_z$ in the material. Holes drift toward the cathode and electrons toward the anode, with each carrier having some probability of being trapped *en route*.

As we know from Sec. 12.1.3, a moving charge carrier will induce a current pulse in the external circuit even if it is trapped before reaching the electrode. In common practice, this current pulse is integrated with a leaky integrator having a time constant that is large compared to the duration of the current pulse. If the time interval between pulses is large compared to this time constant, then each absorbed gamma ray produces a distinct output pulse, and the amplitude of the pulse is proportional to the total induced charge for that photon. The distribution of output signals for a large number of incident photons is called the *pulse-height spectrum*. Except for normalization, the pulse-height spectrum is the univariate probability density function for the charge induced by a single absorbed gamma ray.

The random variables contributing to the pulse-height spectrum are the depth of interaction z_{int}, the initial number of electron-hole pairs N_{eh}, and the distances travelled by each electron and hole under the influence of the field. In addition, we need to know the probability that the initial interaction is photoelectric vs. Compton, the probability that a K x ray or Compton-scattered photon will escape the detector volume, and if not, the probability density on where it will be reabsorbed. From knowledge of the statistics of these variables, we would first like to compute the pulse-height spectrum. Then, from knowledge of the spectrum, we would like to set some criterion for acceptance of photons and to compute the statistics on the number accepted.

The initial interaction Consider a beam of gamma rays travelling in the $+z$ direction, entering the detector at $z = 0$. The probability density function that an incident gamma ray interacts at $z = z_{int}$ is

$$\mathrm{pr}(z_{int}) = \alpha_{tot} \exp[-\alpha_{tot} z_{int}] \,, \tag{12.161}$$

where α_{tot} is the total attenuation coefficient[8] including contributions from the Compton and photoelectric effects:

$$\alpha_{tot} = \alpha_C + \alpha_{pe} \,. \tag{12.162}$$

[8]In Chap. 10, attenuation coefficients were denoted by μ, but we use α here to avoid confusion with mobility.

Since we are considering only two types of interaction, the probability that the initial interaction is photoelectric is α_{pe}/α_{tot} and the probability that it is Compton is α_C/α_{tot}.

The probability that an interaction occurs at all in a detector of thickness L_z is

$$\Pr(\text{int in } L_z) = \int_0^{L_z} dz_{int}\, \text{pr}(z_{int}) = 1 - e^{\alpha_{tot} L_z} . \tag{12.163}$$

If we consider only the gamma rays that undergo an interaction in the detector, the probability density function on their interaction depths is

$$\text{pr}(z_{int}|\text{int in } L_z) = \frac{\text{pr}(z_{int})}{\Pr(\text{int in } L_z)} = \frac{\alpha_{tot}\exp(-\alpha_{tot} z_{int})}{1 - \exp(-\alpha_{tot} L_z)} . \tag{12.164}$$

Probability law for N_{eh} As noted above, generation of electron-hole pairs by a high-energy electron is a competition between phonon emission and lattice ionization. The minimum energy required to generate a pair is the bandgap energy $\mathcal{E}_g$, so an electron of kinetic energy $\mathcal{E}_{kin}$ could in principle produce $\mathcal{E}_{kin}/\mathcal{E}_g$ pairs, but because of the competing process of phonon generation, a smaller number will be produced on average. We can define the mean energy expended per electron-hole pair, $\mathcal{E}_{eh}$, such that the mean number of pairs is

$$\overline{N}_{eh} = \frac{\mathcal{E}_{kin}}{\mathcal{E}_{eh}} . \tag{12.165}$$

The distribution of N_{eh} about its mean is interesting. If there were no competition from phonon generation or other mechanisms, conservation of energy would require that $N_{eh} = \mathcal{E}_{kin}/\mathcal{E}_g$, which is not a random number at all. At the opposite extreme, if the phonon interaction were so strong that only a small fraction of the electron energy went into creation of electron-hole pairs, the pairs would be generated approximately independently, and N_{eh} would therefore be a Poisson random variable (see the discussion of Poisson and rarity in Sec. 11.1.2).

Thus the variance of N_{eh} should fall between 0, for a material where the electron-phonon interaction is very weak, and $\overline{N}_{eh}$, for a material where the interaction is strong. This problem was first studied by Ugo Fano (1947), and we define the *Fano factor* F as

$$F = \frac{\text{Var}(N_{eh})}{\overline{N}_{eh}} . \tag{12.166}$$

A value of $F < 1$ implies sub-Poisson behavior, and that is indeed observed experimentally. In Si and Ge, measurements of F are in the range 0.07 to 0.15.

The full probability law $\Pr(N_{eh})$ is difficult to compute but fortunately not very important. If N_{eh} were a Poisson random variable with mean of order 10^4, we would not hesitate to approximate it by a Gaussian; with the actual sub-Poisson character, we can do the same, writing

$$\Pr(N_{eh}) \simeq \frac{1}{\sqrt{2\pi F\overline{N}_{eh}}} \exp\left[-\frac{(N_{eh} - \overline{N}_{eh})^2}{2F\overline{N}_{eh}}\right] . \tag{12.167}$$

To make contact with Sec. 11.3.1, we note that $\Pr(N_{eh})$ is what we called $\Pr(K|N=1)$ or $\gamma(K)$ there. The moments m_1 and m_2 are now $\overline{N}_{eh}$ and $F\overline{N}_{eh} + \overline{N}_{eh}^2$, respectively.

Drift and diffusion If a photon of energy $\mathcal{E}_0$ is absorbed photoelectrically, then N_{eh} electron-hole pairs are generated in a compact cloud around the interaction point, and $\overline{N}_{eh} = (\mathcal{E}_0 - \mathcal{E}_b)/\mathcal{E}_{eh}$. Additional pairs may be generated at another location by reabsorption of a K x ray, but for now we focus on this initial charge cloud. There are three forces acting on the charges in the cloud: Coulomb forces between charges, the applied electric field, and random forces from interactions of the charges with lattice vibrations.

The Coulomb forces are initially small since there is a cloud of electrons interspersed with an equal and oppositely charged cloud of holes, but as the two clouds separate, self-repulsion may come into play. Nevertheless, we shall neglect this effect here since the resultant field is usually small compared to the applied field (Marks, 2000). The total field is then approximately the applied field V_0/L_z.

Interactions of the charge carriers with phonons have two effects. As discussed in Sec. 12.1.2, one effect is to create a viscous drag so that the carriers quickly reach a terminal or drift velocity given by the mobility times the field. The other effect is that the carriers acquire a random, zero-mean thermal velocity superimposed on the drift velocity. The effect of this thermal motion is that the carriers can diffuse radially outward from the (moving) center of the cloud, so the electrons comprise a fuzzy ball of charge moving uniformly towards the anode and increasing in radius as it moves. The holes form a similar fuzzy ball moving at a different speed toward the cathode.

The initial radius of the fuzzy ball is approximately the range of the photoelectron, of order 20 μm for a 100 keV electron in a typical detector material. To see the effects of diffusion, we neglect this initial radius and suppose that a pointlike distribution of charge is created at $t = 0$. With this assumption, we know from (12.102) that the radius of the distribution at time t is $\sqrt{6Dt}$, where D is the diffusion coefficient.

To estimate how large the ball can get, we need to know D and the maximum time available for diffusion; an important link between these two parameters is provided by the Einstein relation, (12.103). The largest distance either carrier can drift is L_z, so the maximum drift time is

$$t_{\max} = \frac{L_z}{\mu E_0} = \frac{L_z^2}{\mu V_0}, \tag{12.168}$$

where μ is either μ_e or μ_h. With the Einstein relation and a little algebra, we have

$$\sqrt{6Dt_{\max}} = L_z \sqrt{\frac{6k_B T}{eV_0}}. \tag{12.169}$$

Note that no specific material properties enter into this result. Materials with higher mobility have smaller $t_{\max}$ but larger diffusion coefficients, so the radius of the charge cloud after drifting the full thickness of the detector is independent of the mobility.

As a numerical example, if we consider a 1 mm thick detector with 100 Volts bias at room temperature, then $\sqrt{6Dt_{\max}} = 38$ μm. Since this number is small compared to typical detector dimensions, we shall ignore the diffusion for the remainder of this section.

Trapping As the carriers drift, they may be captured by localized defects known as traps (see Sec. 12.1.2). The trapped carriers will eventually be released (detrapped) by thermal excitation, but this process usually takes a very long time compared to the duration of the current pulse generated by a single photon, so we shall neglect detrapping here; once a carrier has been trapped, it disappears (at least from our theory).

We know from (12.30) that hole trapping occurs at a constant rate determined by the hole lifetime τ_h, and similarly for electrons, so the mean number of carriers diminishes exponentially with time. Since the carriers are drifting at a constant velocity, the mean number also diminishes exponentially along the drift path, and the probability of an individual carrier surviving for a distance d without being trapped decays exponentially with d.

Specifically, the probability of a hole not being trapped after traveling a distance d_h is

$$\Pr(\text{no trapping in } d_h) = e^{-d_h/\lambda_h}\,, \tag{12.170}$$

where λ_h is the *hole drift length*, given by

$$\lambda_h = \mu_h E_0 \tau_h\,. \tag{12.171}$$

Similar relations exist for electrons, of course, but we shall continue to discuss holes for definiteness.

Next we need an expression for the probability density function for a hole being trapped at distance d_h from the interaction point. Note that we are taking d_h as a positive number even though the hole is travelling in the $-z$ direction, so the maximum value for d_h is the interaction depth z_{int}. For $d_h < z_{int}$, the probability density on d_h is given by analogy to (12.161) as $\lambda_h^{-1}\exp(-d_h/\lambda_h)$, but we also have to consider what happens when the hole strikes the cathode.

If the cathode contact is ohmic, then it is possible that a compensating hole will be injected at the anode when the initial hole hits the cathode, and then a photoconductive gain will ensue [see (12.123)]. More commonly, however, the cathode contact will be such that there is a very high probability that the hole will be trapped (and eventually recombine) when it hits the cathode. In that case, the overall PDF for trapping at d_h must include a delta function at $d_h = z_{int}$. The weight of this delta function is just the probability that the hole makes it to the cathode, $\exp(-z_{int}/\lambda_h)$, so the desired PDF is[9]

$$\begin{aligned}\mathrm{pr}(d_h) &= e^{-z_{int}/\lambda_h}\,\delta(d_h - z_{int}) + \frac{1}{\lambda_h}\,e^{-d_h/\lambda_h}\\ &= e^{-d_h/\lambda_h}\left[\delta(d_h - z_{int}) + \frac{1}{\lambda_h}\right].\end{aligned} \tag{12.172}$$

The mean and variance of d_h (conditional on the interaction depth z_{int}) can be computed from this density as

$$E\{d_h|z_{int}\} = \lambda_h\left[1 - e^{-z_{int}/\lambda_h}\right], \tag{12.173}$$

$$\mathrm{Var}\{d_h|z_{int}\} = \lambda_h^2 - 2\lambda_h z_{int}\,e^{-z_{int}/\lambda_h} - \lambda_h^2\,e^{-2z_{int}/\lambda_h}\,. \tag{12.174}$$

[9]The normalization of this density is correct on the interval $0 \le d_h \le z_{int} + \epsilon$, where ϵ is a vanishingly small positive quantity. Alternatively, we could multiply the second term by a rect function that is nonzero on the interval $0 \le d_h \le z_{int}$, and then the normalization would work on $0 \le d_h < \infty$.

The corresponding results for electrons are obtained by changing subscript h to e throughout and changing z_{int} to $L_z - z_{int}$.

Conditional PDF of the induced charge By integrating the expression for $i_0(t)$ in (12.51), we see that the random charge induced in the circuit by a single electron-hole pair (say the jth) is

$$Q_j = -\frac{e}{L_z}(d_{hj} + d_{ej}), \tag{12.175}$$

where the electron has travelled a distance d_{ej} and the hole d_{hj}. An assemblage of N_{eh} pairs generated by a single gamma-ray interaction at depth z_{int} thus induces a pulse with total charge Q given by

$$Q = -\frac{e}{L_z}\sum_{j=1}^{N_{eh}}(d_{hj} + d_{ej}). \tag{12.176}$$

We know the probability laws for N_{eh}, z_{int}, d_{hj} and d_{ej}, so we are ready to study the statistics of Q (that is, to compute the pulse-height spectrum). As a first step, we shall compute the conditional PDF of Q for fixed z_{int} and N_{eh}.

Recall that N_{eh} is a large number, of order $10^4 - 10^5$, so Q is a sum of a large number of random variables. Since the trapping events are independent, the central-limit theorem (see Sec. 8.3.4) tells us that the conditional PDF is

$$\text{pr}(Q|z_{int}, N_{eh}) = \frac{1}{\sqrt{2\pi\sigma_Q^2}} \exp\left[-\frac{\left[Q - \overline{Q}(z_{int}, N_{eh})\right]^2}{2\,\text{Var}\{Q|z_{int}, N_{eh}\}}\right], \tag{12.177}$$

where the conditional mean and variance are

$$\overline{Q}(z_{int}, N_{eh}) = -\frac{e}{L_z}N_{eh}[E\{d_h|z_{int}\} + E\{d_e|z_{int}\}], \tag{12.178}$$

$$\text{Var}\{Q|z_{int}, N_{eh}\} = \left(\frac{e}{L_z}\right)^2 N_{eh}[\text{Var}\{d_h|z_{int}\} + \text{Var}\{d_e|z_{int}\}]. \tag{12.179}$$

With (12.173) and the similar expression for electrons, we can also express the conditional mean of the induced charge via the *Hecht relation*,

$$\overline{Q}(z_{int}, N_{eh}) = -\frac{e}{L_z}N_{eh}\left\{\lambda_h\left[1 - e^{-z_{int}/\lambda_h}\right] + \lambda_e\left[1 - e^{-(L_z - z_{int})/\lambda_e}\right]\right\}. \tag{12.180}$$

Some limits of this expression are instructive. If λ_e and λ_h are both large compared to L_z, as they usually are in Si or Ge detectors, then an expansion of the exponentials shows that

$$\overline{Q}(z_{int}, N_{eh}) \to -\frac{e}{L_z}N_{eh}\left\{\lambda_h\frac{z_{int}}{\lambda_h} + \lambda_e\frac{L_z - z_{int}}{\lambda_e}\right\} = -eN_{eh}. \tag{12.181}$$

In this limit, therefore, trapping is negligible and the full charge of N_{eh} electron-hole pairs appears in the external circuit.

The opposite limit is where both carriers are heavily trapped, so that λ_e and λ_h both approach zero. In this case,

$$\overline{Q}(z_{int}, N_{eh}) \to -eN_{eh}\frac{\lambda_h + \lambda_e}{L_z}. \tag{12.182}$$

Much less charge is induced in this case, but both (12.181) and (12.182) show that the induced charge is independent of z_{int}. Unfortunately, that conclusion does not hold in the intermediate cases.

Suppose, for example, that holes are heavily trapped and electrons are not, so that $\lambda_h \ll L_z$ and $\lambda_e \gg L_z$ (which is often a realistic assumption in compound semiconductors such as CdTe). In the limit as $\lambda_h \to 0$,

$$\overline{Q}(z_{int}, N_{eh}) \to -eN_{eh}\frac{L_z - z_{int}}{L_z}\,. \tag{12.183}$$

In this case, the average induced charge depends linearly on the random depth of interaction; as we shall see shortly, this randomness smears out the pulse-height spectrum.

Effect of carrier-generation statistics The desired overall PDF for Q is given by

$$\text{pr}(Q) = \int_0^{L_z} dz_{int}\ \text{pr}(z_{int})\,\text{pr}(Q|z_{int})$$

$$= \int_0^{L_z} dz_{int}\ \text{pr}(z_{int}) \sum_{N_{eh}=0}^{\infty} \text{Pr}(N_{eh})\,\text{pr}(Q|z_{int}, N_{eh})\,. \tag{12.184}$$

We have approximated $\text{pr}(Q|z_{int}, N_{eh})$ by a normal in (12.177), and in (12.167) we also approximated $\text{Pr}(N_{eh})$ by a normal. Since N_{eh} is very large, we can replace the sum over N_{eh} with an integral, and then we have a convolution[10] of two normals, which is another normal. To specify $\text{pr}(Q|z_{int})$ in this approximation, therefore, we need only compute the mean and variance of Q conditional on z_{int} alone.

The mean of Q conditioned on z_{int} is obtained by averaging (12.178) over N_{eh}, with the result

$$E\{Q|z_{int}\} = -\frac{e}{L_z}\overline{N}_{eh}[E\{d_h|z_{int}\} + E\{d_e|z_{int}\}] \equiv -\frac{e}{L_z}\overline{N}_{eh}\overline{d}_{eh}\,, \tag{12.185}$$

where $\overline{d}_{eh}$ is the mean total distance travelled by a single hole and electron. (Recall that d_e and d_h are both positive numbers, even though the carriers travel in opposite directions.)

The procedure for computing $\text{Var}(Q|z_{int})$ is first to compute the second moment, then to subtract off the square of the relevant mean. A similar procedure was used several times in Sec. 11.4 for discussing random amplification, but that problem should not be confused with the present one. In Sec. 11.4, the double randomness came about since a random number of primaries each generated a random number of identical secondaries. Here we are concerned with a single primary, and the secondaries are not identical since they induce random amounts of charge in the output.

[10]The reader who wishes to fill in the details of this calculation will discover that it is necessary to replace N_{eh} with $\overline{N}_{eh}$ in (12.179), but this step is easily justified since $\text{Pr}(N_{eh})$ is sharply peaked.

From (12.178) and (12.179), the second moment of interest is

$$E\{Q^2|z_{int}\} = \frac{e^2}{L_z^2}\left\{\langle N_{eh}\rangle[\mathrm{Var}\{d_h|z_{int}\} + \mathrm{Var}\{d_e|z_{int}\}] + \langle N_{eh}^2\rangle\, \overline{d}_{eh}^2\right\}$$

$$= \frac{e^2}{L_z^2}\left\{\overline{N}_{eh}[\mathrm{Var}\{d_h|z_{int}\} + \mathrm{Var}\{d_e|z_{int}\}] + \left[F\overline{N}_{eh} + \overline{N}_{eh}^2\right]\overline{d}_{eh}^2\right\}, \qquad (12.186)$$

where we have used (12.166). The desired variance is now

$$\mathrm{Var}\{Q|z_{int}\} = \frac{e^2}{L_z^2}\overline{N}_{eh}\left[\mathrm{Var}\{d_h|z_{int}\} + \mathrm{Var}\{d_e|z_{int}\} + F\overline{d}_{eh}^2\right]. \qquad (12.187)$$

With these expressions for the mean and variance, the overall PDF $\mathrm{pr}(Q)$ is given by

$$\mathrm{pr}(Q) = \int_0^{L_z} dz_{int}\, \mathrm{pr}(z_{int})\frac{1}{\sqrt{2\pi\,\mathrm{Var}\{Q|z_{int}\}}}\exp\left[-\frac{[Q - E\{Q|z_{int}\}]^2}{2\,\mathrm{Var}\{Q|z_{int}\}}\right]. \qquad (12.188)$$

All of the pieces needed for a numerical computation of $\mathrm{pr}(Q)$ with arbitrary trapping are in place; the key equations are (12.164), (12.173), (12.174), (12.185), (12.187) and (12.188), and only a 1D integral is needed. This procedure can be used to study the pulse-height spectrum as a function of material parameters, bias voltage and photon energy, but more insight can be obtained by developing approximate analytic expressions. We shall do so for various assumptions about the degree of trapping.

Pulse-height spectrum when trapping is negligible As we saw in (12.181), $\overline{Q}$ is independent of z_{int} in the limit that trapping is negligible for both carriers, as in Si and Ge. The variances of the drift lengths, $\mathrm{Var}\{d_h|z_{int}\}$ and $\mathrm{Var}\{d_e|z_{int}\}$, also go to zero in this limit, as one can see physically by noting that the drift length for holes approaches its maximum value of z_{int} and the one for electrons approaches $L_z - z_{int}$, and both of these quantities are nonrandom for fixed z_{int}. The same conclusion follows more formally from (12.174) and the corresponding expression for d_e by expanding the exponentials and letting λ_h and λ_e get large.

Thus, when trapping is negligible, the overall mean and variance, from (12.185) and (12.187) respectively, become

$$\overline{Q} = -e\overline{N}_{eh}, \qquad (12.189)$$

$$\mathrm{Var}\{Q|z_{int}\} = e^2 F\overline{N}_{eh}. \qquad (12.190)$$

We then have $\mathrm{pr}(Q|z_{int}) = \mathrm{pr}(Q)$, so this quantity can be removed from the integral in (12.188), and the integral is then unity. To an excellent approximation, the PDF $\mathrm{pr}(Q)$ for negligible trapping is thus a normal with the mean and variance specified in (12.189) and (12.190).

We can define a signal-to-noise ratio for Q as

$$\mathrm{SNR}_Q = \frac{|\overline{Q}|}{\sqrt{\mathrm{Var}(Q)}} = \frac{\sqrt{\overline{N}_{eh}}}{\sqrt{F}}, \qquad (12.191)$$

and the sub-Poisson character of the electron-hole generation process can be seen in the denominator.

Photopeaks and energy resolution If we recall from (12.165) that $\overline{N}_{eh} = \mathcal{E}_{kin}/\mathcal{E}_{eh}$, then we see from (12.189) that (without trapping) the mean induced charge in the external circuit is proportional to the mean kinetic energy of the high-energy electron. For photoelectric events, we also know from (12.160) how the kinetic energy depends on the photon energy, but that equation does not tell the whole story. A photoelectric interaction with a K-shell electron may be followed by emission of a K x ray, and that photon may or may not be reabsorbed in the detector material; if it is, additional electron-hole pairs are created and add to the induced charge Q. The maximum Q, over all possible sequelae of a photoelectric absorption, occurs when the K x ray and all subsequent photons are reabsorbed within the detector. Then the total kinetic energy released is the photon energy $\mathcal{E}_0$. Such events contribute a distinct peak, called the *photopeak*, to the pulse-height spectrum. Similarly, events where the K x ray escapes, contribute a lower peak called the *K-escape peak* (see Fig. 12.15), and peaks corresponding to the escape of other photons can often be identified as well.

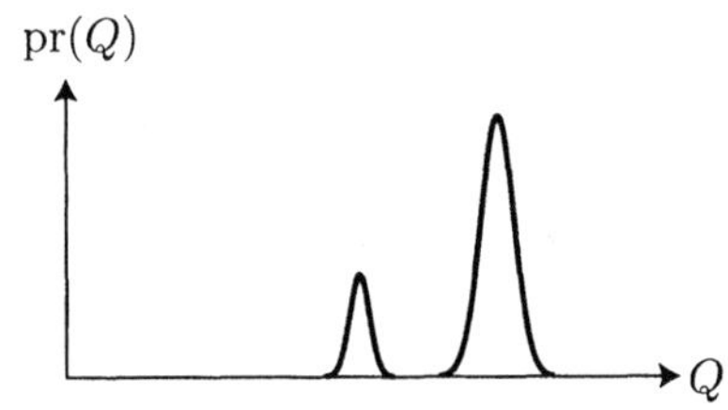

Fig. 12.15 Schematic pulse-height spectrum showing K-escape peak. The lower-energy peak is centered at a pulse height corresponding to the initial gamma-ray energy minus the K x-ray energy.

Since $\overline{Q} = -e\overline{N}_{eh} = -e\mathcal{E}_0/\mathcal{E}_{eh}$ for photopeak events in a good detector material, it is natural to use the actual induced charge as an estimate of the photon energy $\mathcal{E}_0$. We can define an estimated energy by

$$\hat{\mathcal{E}} \equiv -\mathcal{E}_{eh}\frac{Q}{e}, \tag{12.192}$$

and then the pulse-height spectrum can be reinterpreted as a PDF on the new random variable $\hat{\mathcal{E}}$. For all events, $\hat{\mathcal{E}}$ is an unbiased estimator of $\mathcal{E}_{kin}$, and for photopeak events it is an unbiased estimator of $\mathcal{E}_0$. For this reason, a pulse-height spectrum is often called (quite misleadingly) an *energy spectrum*. At best, it is the distribution of *estimated* energies, and the estimate is useful only when trapping is negligible and escape peaks are ignored. We can also reinterpret SNR_Q, defined above, as an SNR for $\hat{\mathcal{E}}$; since the constant $-e/\mathcal{E}_{eh}$ affects numerator and denominator in the same way, we have

$$\frac{\langle\hat{\mathcal{E}}\rangle}{\sqrt{\text{Var}(\hat{\mathcal{E}})}} = \text{SNR}_Q = \frac{\sqrt{\overline{N}_{eh}}}{\sqrt{F}}. \tag{12.193}$$

For photopeak events where $\langle\hat{\mathcal{E}}\rangle = \mathcal{E}_0$, we can write

$$\frac{\sqrt{\text{Var}\{\hat{\mathcal{E}}\}}}{\mathcal{E}_0} = \sqrt{F}\sqrt{\frac{\mathcal{E}_{eh}}{\mathcal{E}_0}}. \tag{12.194}$$

Thus the precision of the energy estimate is better for higher-energy photons and for detector materials with smaller Fano factor and smaller average energy per electron-hole pair.

It is almost universal in the literature to quote the energy resolution as the full-width at half maximum (FWHM) of the photopeak divided by the center position of the photopeak. This ratio is usually denoted $\Delta\mathcal{E}/\mathcal{E}$ (ignoring the fact that the $\mathcal{E}$ in question is an estimate, not an actual energy). As we have seen, Q and hence $\mathcal{E}$ are normally distributed in the absence of trapping, and the FWHM of a normal is 2.35 times its standard deviation, so $\Delta\mathcal{E}/\mathcal{E}$ is 2.35 times the expression in (12.194).

Strong trapping of both carriers If both carriers are heavily trapped, so that λ_h and λ_e approach zero, we see from (12.182) that $\overline{Q}$ is again independent[11] of z_{int}, and we have

$$\overline{Q} = -e\frac{\lambda_h + \lambda_e}{L_z}\overline{N}_{eh}\,. \tag{12.195}$$

Similarly, we see from (12.174) that $\mathrm{Var}\{d_h|z_{int}\} \to \lambda_h^2$ and $\mathrm{Var}\{d_e|z_{int}\} \to \lambda_e^2$ in this limit, so the variance of Q from (12.187) becomes

$$\mathrm{Var}\{Q\} = \frac{e^2}{L_z^2}\overline{N}_{eh}\left[\lambda_h^2 + \lambda_e^2 + F(\lambda_h + \lambda_e)^2\right]\,. \tag{12.196}$$

The SNR on Q is now

$$\mathrm{SNR}_Q = \sqrt{\overline{N}_{eh}}\,\frac{\lambda_h + \lambda_e}{\sqrt{\lambda_h^2 + \lambda_e^2 + F(\lambda_h + \lambda_e)^2}}\,. \tag{12.197}$$

Thus the beneficial effects of the Fano factor, evidenced in (12.193), are reduced with strong trapping. If $F = 0$, SNR_Q is infinite for no trapping but finite with strong trapping. If $\lambda_h = \lambda_e$, for example, SNR_Q approaches $\sqrt{2\overline{N}_{eh}}$ as $F \to 0$.

Carrier-generation statistics negligible In the two limits of no trapping and very strong trapping of both carriers, $\mathrm{pr}(Q|z_{int})$ is independent of the interaction depth z_{int}, but in the intermediate cases the spread of the pulse-height spectrum can be dominated by the randomness in z_{int}. In these cases, we can see the basic shape of the spectrum by neglecting the variance of N_{eh} and approximating $\mathrm{pr}(Q|z_{int})$ with a delta function in (12.188). This approximation becomes increasingly valid at higher photon energies since the SNR for Q conditional on z_{int} varies as $\sqrt{\overline{N}_{eh}}$, so the relative width of $\mathrm{pr}(Q|z_{int})$ decreases as energy (and hence $\overline{N}_{eh}$) increases. In addition, low-energy photons may be absorbed near the surface of the detector, but higher-energy photons are more penetrating so $\mathrm{pr}(z_{int})$ is broader, and hence the width of $\mathrm{pr}(Q|z_{int})$ plays less of a role as energy increases.

With this approximation, (12.188) becomes

$$\mathrm{pr}(Q) = \int_0^{L_z} dz_{int}\ \mathrm{pr}_{z_{int}}(z_{int})\,\delta[Q - \overline{Q}(z_{int})]\,, \tag{12.198}$$

[11]The limit here is a little tricky. For λ_h small but finite, we have to exclude events where z_{int} is less than about λ_h (*i.e.*, interactions very near the cathode), and for λ_e small but finite, we have to exclude interactions near the anode.

where $\text{pr}_{z_{int}}(z_{int})$ means the same thing as $\text{pr}(z_{int})$. (The reason for the notational change will become apparent shortly.) To perform the integral, we can use (2.33) to write

$$\delta[Q - \overline{Q}(z_{int})] = \frac{\delta[z_{int} - \tilde{z}(Q)]}{\left|\frac{\partial Q(z_{int})}{\partial z_{int}}\right|_{z_{int}=\tilde{z}(Q)}}, \quad (12.199)$$

where $\tilde{z}(Q)$ is found by solving $\overline{Q}(\tilde{z}) = Q$. Then we have

$$\text{pr}(Q) = \frac{\text{pr}_{z_{int}}[\tilde{z}(Q)]}{\left|\frac{\partial Q(z_{int})}{\partial z_{int}}\right|_{z_{int}=\tilde{z}(Q)}}. \quad (12.200)$$

Since $\text{pr}_{z_{int}}[\tilde{z}(Q)]$ is simply $\text{pr}_{z_{int}}(z_{int})$ evaluated at the specific point $z_{int} = \tilde{z}(Q)$, the pulse-height spectrum in the current approximation is just a remapped version of the PDF on the depth of interaction. Each random depth corresponds to a specific (nonrandom, we assume) amount of charge in the external circuit, so the number of events producing charge between Q and $Q + \Delta Q$ is the same as the number of gamma-ray photons interacting between $\tilde{z}(Q)$ and $\tilde{z}(Q + \Delta Q)$. Some typical pulse-height spectra are shown in Figs. 12.16 – 12.18.

Electronic noise The analysis above is not complete since it does not include electronic noise in the circuit that reads out the detector signal. As we know from Sec. 12.2, electronic noise is usually a Gaussian random process, but we need to understand how it affects a measured pulse height.

For definiteness, consider an operational amplifier with a parallel RC circuit in the feedback loop, and assume that the amplifier noise can be modeled as a random voltage source in series with the input as shown in Fig. 12.19. A separate low-pass filter is used to limit the overall bandwidth, and the fluctuating output voltage of this filter is denoted $v_f(t)$. For purposes of noise analysis, we can neglect the DC voltage applied to the detector and the current pulse that results when a gamma ray is absorbed. The problem is thus to compute the variance of $v_f(t)$ in terms of the properties of the noise source.

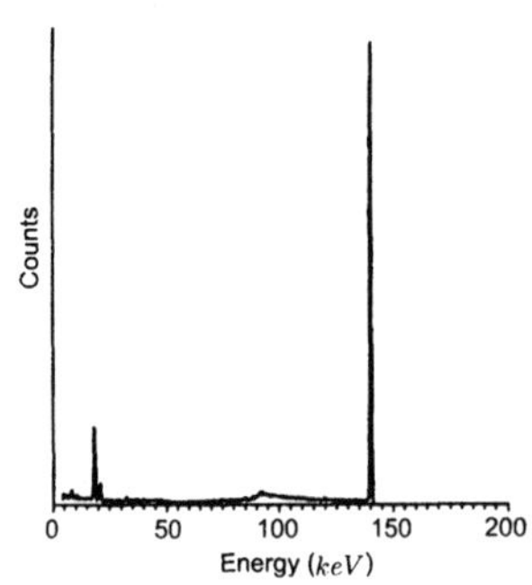

Fig. 12.16 Pulse-height spectrum for a high-quality Germanium detector at cryogenic temperature (≃100K). The gamma-ray source is ^{99m}Tc, which emits photons at energies 18, 20, 21 and 140 keV. The weak response near 90 keV arises from 140 keV photons that are Compton scattered in the detector material, with the scattered photon escaping from the detector. Trapping is negligible in Ge, and the K-escape peaks are not seen since K x rays in Ge have an energy near 10 keV and are therefore reabsorbed without much probability of escape. Note that pulse heights have been converted to energy units for this plot by use of (12.192).

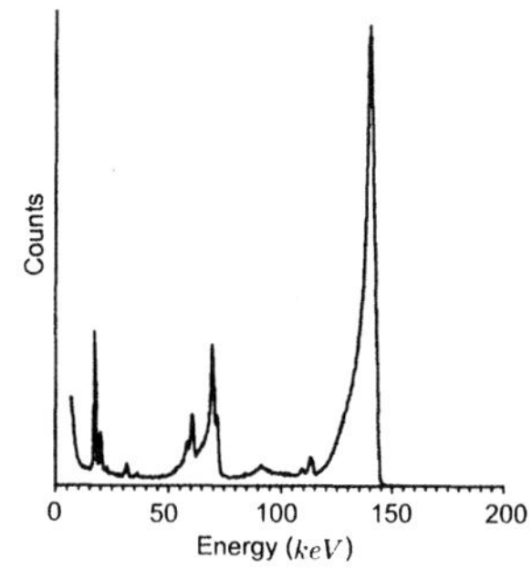

Fig. 12.17 Pulse-height spectrum for a high-quality Mercuric Iodide (HgI_2) detector at room temperature. As in Fig. 12.16, the gamma-ray source is ^{99m}Tc. Moderate trapping is evidenced by the tails on the low-energy side of each peak. The peaks around 110 – 115 keV correspond to escape of iodine K x rays, and the peaks around 6 – 75 keV correspond to escape of Hg K_α and K_β photons.

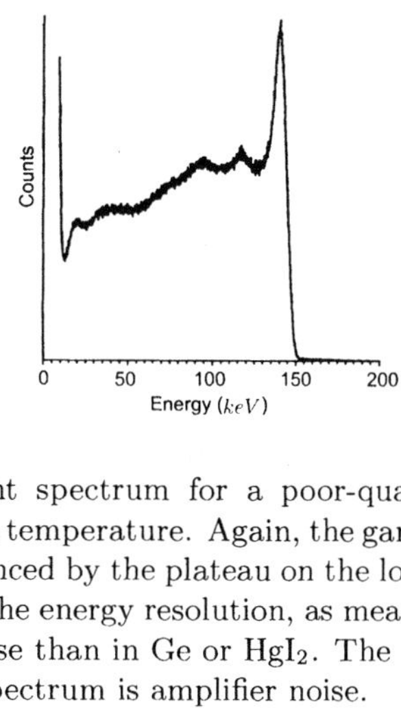

Fig. 12.18 Pulse-height spectrum for a poor-quality Cadmium Telluride (CdTe) detector at room temperature. Again, the gamma-ray source is ^{99m}Tc. Severe trapping is evidenced by the plateau on the low-energy side of the photopeak. Note also that the energy resolution, as measured by the width of the photopeak, is much worse than in Ge or HgI_2. The rapid rise at the extreme low-energy end of the spectrum is amplifier noise.

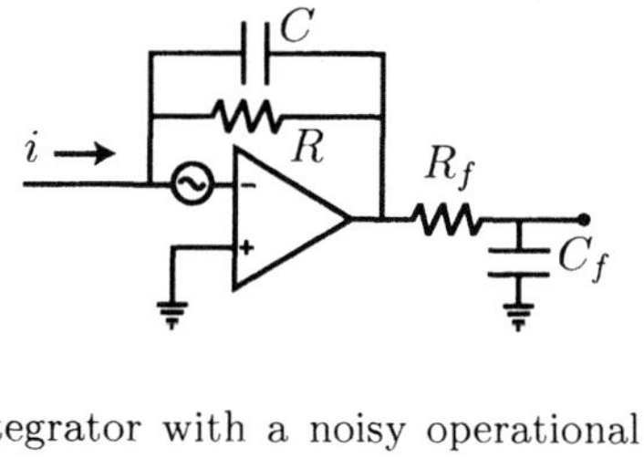

Fig. 12.19 Leaky integrator with a noisy operational amplifier and added low-pass filter.

As in Sec. 12.1.1, we can assume that the operational amplifier has very high gain and hence maintains the voltage across its input terminals very near zero. Thus the noise voltage $v_n(t)$ is also the voltage across the detector capacitance, and the fluctuating current through the detector is $C_{det} dv_n/dt$. Since the amplifier has very high input impedance, no current can flow into its input terminals, and the current through the detector must be equal and opposite to the current in the feedback loop. This condition can be expressed as [*cf.* (12.5)]

$$C_{det}\frac{dv_n(t)}{dt} = \frac{v_{out}(t) - v_n(t)}{R} + C\frac{d}{dt}[v_{out}(t) - v_n(t)]\,, \tag{12.201}$$

where $v_{out}(t)$ is the voltage at the output of the amplifier. A Fourier transform and a bit of algebra yield

$$V_{out}(\nu) = V_n(\nu)\frac{1 + 2\pi i\nu RC_{tot}}{1 + 2\pi i\nu RC}\,, \tag{12.202}$$

where $C_{tot} = C + C_{det}$. If the detector had negligible capacitance, this equation shows that the amplifier would function as a voltage follower with $V_{out}(\nu) = V_n(\nu)$, but with real detectors there is a frequency-dependent gain factor that approaches C_{tot}/C as ν gets large. Even though the amplifier with RC feedback functions as a low-pass filter for the current pulse produced by a gamma ray, it does not have a low-pass effect on the noise. The transfer function relating the noise voltage to the amplifier output voltage is given by the ratio in (12.202).

To proceed, we must specify the power spectral density of the noise. For simplicity, we ignore $1/f$ noise and assume that all other sources of electronic noise can be lumped together into a white-noise spectrum $S_n(\nu) = S_n(0) =$ constant. With this assumption and (8.156), the power spectral density for $v_{out}(t)$ is

$$S_{out}(\nu) = S_n(0)\cdot\left|\frac{1 + 2\pi i\nu RC_{tot}}{1 + 2\pi i\nu RC}\right|^2 = S_n(0)\frac{1 + (2\pi\nu RC_{tot})^2}{1 + (2\pi\nu RC)^2}\,. \tag{12.203}$$

This function is plotted in Fig. 12.20. Note that $S_{out}(\nu)$ approaches $S_n(0)[C_{tot}/C]^2$ as $\nu \to \infty$, so the power spectral density is not integrable and the variance of $v_{out}(t)$ is infinite.

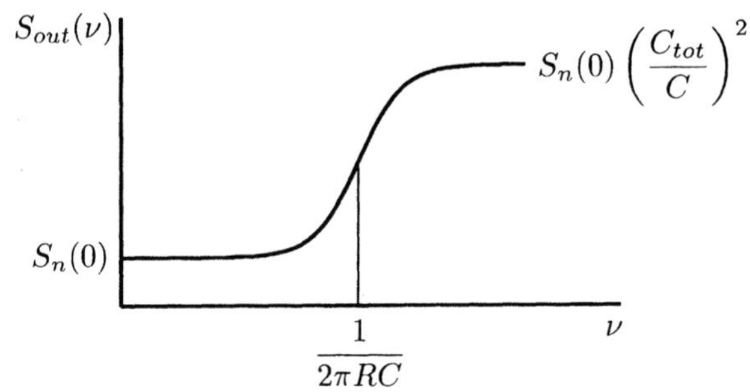

Fig. 12.20 Power spectral density of noise on the output of the operational amplifier of Fig. 12.18. [Plot of (12.203)]

To get a finite variance, we must include the low-pass filter as in Fig. 12.19. The power spectral density for the filter output voltage $v_f(t)$ is obtained by multiplying (12.203) by $|H_f(\nu)|^2$, where $H_f(\nu)$ is the filter transfer function. With (8.156) and (12.203), the variance of $v_f(t)$ is given by

$$\text{Var}\{v_f(t)\} = S_n(0)\int_{-\infty}^{\infty} dv\,\frac{1 + (2\pi\nu RC_{tot})^2}{1 + (2\pi\nu RC)^2}|H_f(\nu)|^2\,. \tag{12.204}$$

The integral simplifies if we assume that the filter has the same low-pass characteristic as the amplifier ($R_f = R$, $C_f = C$) and that $H_f(0) = 1$. Then

$$\text{Var}\{v_f(t)\} = S_n(0)\int_{-\infty}^{\infty} d\nu\,\frac{1 + (2\pi\nu RC_{tot})^2}{[1 + (2\pi\nu RC)^2]^2} = S_n(0)\frac{1}{4RC}\left(1 + \frac{C_{tot}^2}{C^2}\right)\,, \tag{12.205}$$

where we have used Gradshteyn and Ryzhik (1980), formula 3.241.5.

Dark current The variance of $v_f(t)$ as given in (12.205) arises from the amplifier noise alone. An additional variance component comes from the shot noise of the dark current through the detector. This component was analyzed in Sec. 12.1.1, and we do not need to repeat the discussion here, but one key point should be noted: The dark-current originates outside the feedback loop of the amplifier, so it is subject to the filtering action of both the integrating amplifier and the low-pass filter. Unlike the amplifier noise, the dark current does not produce white noise at the amplifier output.

Photopeak variance and SNR In Sec. 12.1.1, we analyzed an operational amplifier with a parallel RC circuit in the feedback loop. By a slight generalization of (12.7), we know that an input current pulse of total charge Q produces an output voltage pulse of height Q/C if the pulse duration is very short compared to RC. If Q is random, the pulse height is also random, with variance given by $C^{-2}\,\mathrm{Var}\{Q\}$. When the output pulse from the amplifier is fed through the low-pass filter, however, a further change in the pulse height occurs. As illustrated in Fig. 12.21, the pulse height is reduced by a factor of e^{-1} (where e is the base of the natural logarithms, not the charge on the electron) and the peak is shifted to $t = RC$ (where $t = 0$ is the time of the gamma-ray interaction). The contribution of randomness in Q to the variance of $v_f(t)$ at $t = RC$ is thus $(eC)^{-2}\,\mathrm{Var}\{Q\}$.

If the amplifier noise (but not the dark current) is included, the total variance of the output voltage of the filter is obtained by adding the variances of the independent components:

$$\mathrm{Var}\{v_f(RC)\} = \frac{1}{e^2C^2}\,\mathrm{Var}\{Q\} + S_n(0)\frac{1}{4RC}\left(1 + \frac{C_{tot}^2}{C^2}\right). \tag{12.206}$$

It is tempting to identify the variance of $v_f(RC)$ as the variance in the height of the pulse out of the filter, but that step requires some justification. The difficulty is that there are several ways to measure pulse height. Some pulse-height analyzers become active after the pulse exceeds some threshold, and then they find the next maximum of the voltage and call it the height. With this kind of system, the probability density function for the heights relates to the density of maxima of the random process, a notoriously difficult problem (see Middleton, 1996, Chap. 9, and Rice, 1948). An alternative approach is to trigger a delay generator on the leading edge of the pulse and to take a voltage sample after a fixed delay approximately equal to RC. In this approach, the sample need not occur at a maximum, but we regard the voltage at that time as an estimate of the pulse height anyway. Then (12.206) can be directly interpreted as the variance in the (estimated) pulse height. We shall assume this latter approach in what follows.

The SNR on the pulse heights is defined as

$$[\mathrm{SNR}_{ph}]^2 = \frac{(\overline{Q}/eC)^2}{\mathrm{Var}\{v_f(RC)\}}. \tag{12.207}$$

With (12.206), we have

$$[\mathrm{SNR}_{ph}]^2 = \frac{\overline{Q}^2}{\mathrm{Var}\{Q\} + S_n(0)\frac{e^2}{4RC}(C^2 + C_{tot}^2)}. \tag{12.208}$$

One implication of this expression is that there is an optimum choice for the capacitance C; regarded as a function of C, (12.208) has a maximum when $C = C_{det}/\sqrt{2}$. Other implications are also apparent. To optimize SNR_{ph}, we should use low-noise amplifiers, detectors with low capacitance and large values of RC. The latter measure, however, reduces the count rate that can be accommodated without pulse overlap.

Overall pulse-height spectrum Implicit in most of the discussion above is the assumption that the initial gamma-ray interaction transfers a definite energy $\mathcal{E}_{kin}$ to the detector material, and for the most part we have considered photopeak events where $\mathcal{E}_{kin} = \mathcal{E}_0 - \mathcal{E}_b$. In practice, however, $\mathcal{E}_0$ is a random variable, and $\mathcal{E}_{kin}$ is also a random variable even when conditioned on $\mathcal{E}_0$. In nuclear medicine, for example, gamma rays may scatter in the patient's body and lose a random amount of energy, so the energy $\mathcal{E}_0$ of a photon striking the detector is random. Moreover, in any detector there is an additional randomness arising from escape of secondary photons, so $\mathcal{E}_{kin}$ is random as well. Under these circumstances, the pulse-height spectrum is quite complicated (see Figs. 12.16 – 12.18), and it would be very misleading to characterize it with a single number like an SNR. What we need now is an expression for the overall probability density on the pulse heights, accounting for randomness in $\mathcal{E}_{kin}$ and $\mathcal{E}_0$ as well as all of the random detector effects treated so far.

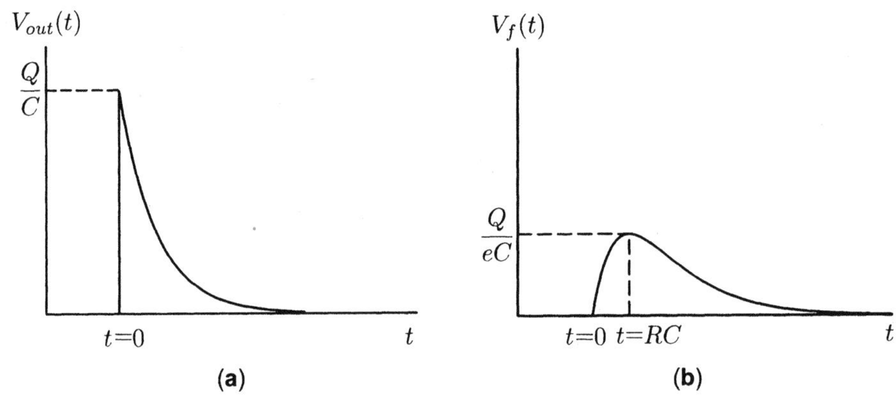

Fig. 12.21 Pulse outputs from (a) the operational amplifier and (b) the low-pass filter of Fig. 12.19.

A good starting point is (12.188), which we can now view as a conditional density and relabel as $\text{pr}(Q|\mathcal{E}_{kin})$. Then we can write[13]

$$\text{pr}(Q) = \int_0^\infty d\mathcal{E}_{kin}\ \text{pr}(Q|\mathcal{E}_{kin})\,\text{pr}(\mathcal{E}_{kin})$$

$$= \int_0^\infty d\mathcal{E}_{kin} \int_0^\infty d\mathcal{E}_0\ \text{pr}(Q|\mathcal{E}_{kin})\,\text{pr}(\mathcal{E}_{kin}|\mathcal{E}_0)\,\text{pr}(\mathcal{E}_0)\,. \qquad (12.209)$$

[13]Equation (12.209) leaves out one potentially significant effect. When an x ray or scattered photon is reabsorbed within the detector, the PDF on the total induced charge Q depends not only on the initial interaction depth z_{int}, which was already accounted for in (12.188), but also on the depth at which the secondary photon is absorbed. (Lateral position of the absorption is unimportant in the slab geometry.)

Usually the only feasible way to evaluate this expression is Monte Carlo simulation, but two separate simulations are required: a simulation of scatter and escape in the detector material to estimate $\mathrm{pr}(\mathcal{E}_{kin}|\mathcal{E}_0)$, and a simulation of scatter and absorption in the patient's body and in any collimating structures to estimate $\mathrm{pr}(\mathcal{E}_0)$.

After Monte Carlo evaluation of $\mathrm{pr}(Q)$, we still need to transfer the charge signal through the integrating amplifier and low-pass filter to get the pulse height. With the assumptions listed above, this is merely an amplitude scaling by a factor of $1/(eC)$, but we also need to add the electronic noise. This latter step is just convolution with a zero-mean Gaussian with variance given by (12.205).

Windowing and counting statistics What do we do with a pulse-height spectrum in gamma-ray imaging systems? Ideally, we would use the entire spectrum from every detector and attempt to extract as much information about the object as possible. In common practice, however, a simple binary decision is made on each pulse: it is either accepted into the image or rejected. The usual approach is to accept an event only if its pulse height lies in some preset range of values around the photopeak. This range is usually called an *energy window*, but that is a distinct misnomer; it is a window on pulse heights rather than energies of incident photons. With complicated spectra such as Figs. 12.16 – 12.18, the relation between pulse height and energy is tenuous to say the least.

The fraction of photons absorbed by the detector that are accepted by the window depends in a complicated way on the actual distribution of incident energies, all of the noise sources considered above, and the escape processes. If we know the pulse-height spectrum, we can compute the probability of an incident photon being accepted by the window, denoted $P^{(acc)}$, by integrating the spectrum over the window range. Knowing $P^{(acc)}$, we can discuss the statistics of the number of detected photons.

Consider first the case where the detector observes a Poisson source for a fixed time τ, and assume that a fraction α of the emitted photons strike the detector and a fraction η of those are absorbed. By the discussion in Sec. 11.1.3, we know that binomial selection of a Poisson random process yields a Poisson random variable; in the present problem we have three successive binomial selections, so the number of recorded counts is a Poisson random variable with mean $f_0 \tau \alpha \eta P^{(acc)}$, where f_0 is the mean number of photons per second emitted by the source. In effect, the quantum efficiency of the detector is reduced by the factor $P^{(acc)}$.

Other results from Chap. 11 are also applicable with this factor of $P^{(acc)}$. If we have a system of M detectors, we have to add subscripts to the quantities defined above to specify a particular detector. If we count until a preset total number of counts N_{tot} is recorded, then the number recorded in the mth detector follows a binomial law with mean $N_{tot} \alpha_m \eta_m P_m^{(acc)}$, and the multivariate distribution on the counts in all detectors is multinomial as in (11.42). If the mean number of counts in each detector is small, the multinomial limits to a multivariate Poisson as in (11.48).

Finally, all of the discussion of doubly stochastic processes in Sec. 11.2.2 is immediately applicable with the additional factor of $P_m^{(acc)}$ in each quantum efficiency.

12.3.3 Semiconductor detector arrays

So far we have discussed single-element, nonimaging detectors for x rays or gamma rays. With the slab geometry, no information is obtained about the interaction position of a photon except that it lies within the slab, and we have assumed that the lateral dimensions L_x and L_y are large, so only rudimentary spatial information is obtained. More complete information could be obtained by using an array of single-element detectors — a lot of separate slabs, each with its own electronics — but this would be expensive because of cost of fabricating many small detectors and the amount of electronics required. Moreover, the resolution would still be limited by the conditions on the slab geometry. In this section we shall discuss photon counting semiconductor detectors in which spatial resolution is obtained by using a slab detector with many separate electrodes rather than one continuous one.

The objective of this section is to develop mathematical and statistical descriptions of the output signals from such multi-electrode semiconductor detectors. In Sec. 12.3.4 we shall learn how to use the signals to form an image.

Electrode geometries and readouts The basic slab detector can be made into a detector array just by adding more electrodes. An approach dating back to the 1960s is to place a set of strip electrodes on one side of the slab and a set of orthogonal strips on the other side as shown in Fig. 12.22. When a high-energy photon is absorbed, a charge is induced primarily on one strip on each side, so the event is localized to approximately the area of overlap of the orthogonal strips. The index of the active strip on one side can specify the x or row index in a pixel matrix and the index of the active strip on the other side can specify the y or column index, so this device is called a *row-by-column detector*. If J strips are used on each side, a $J \times J$ pixel array is defined, but only $2J$ channels of electronics are needed (plus logic circuitry to detect the strongest signal on each side).

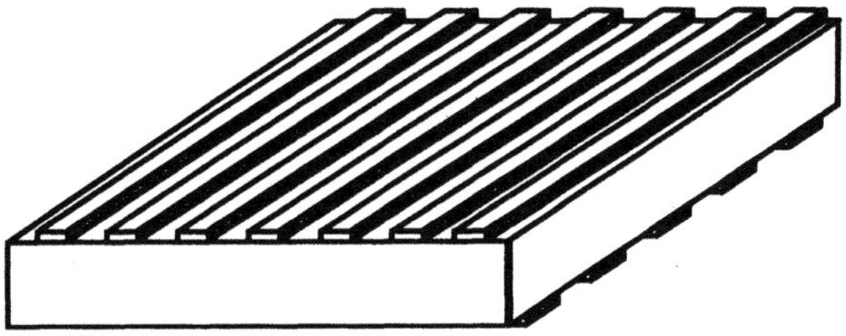

Fig. 12.22 Strip detector, with orthogonal strip electrodes on the two sides.

One drawback of row-by-column readout is increased detector capacitance compared to an array of single-element detectors with the same spatial resolution. If each strip has width ϵ and length $J\epsilon$, then the capacitance on the input to each amplifier is proportional to the strip area $J\epsilon^2$ while the resolution area is ϵ^2. As we saw in (12.208), increased detector capacitance results in poorer pulse-height resolution.

Another useful geometry, shown in Fig. 12.23, consists of a slab detector with a continuous metal electrode on one side and a set of small square electrodes on the other side. The small electrodes define pixels in the image array, so this geometry is referred to as *pixellated*.[13] Now each pixel electrode needs its own electronics chain, but fortunately they can all be fabricated on a single silicon chip in an *application-specific integrated circuit* or *ASIC*. The ASIC can then be bonded to the semiconductor detector slab as shown in Fig. 12.23. The resulting assembly is often called a *hybrid array* because two different semiconductor materials are used, one to absorb the gamma rays and provide charge signals and one to read out the signals and pass them on to additional off-chip electronics.

The readout ASIC may use gated integrators, as shown in Fig. 12.13 and discussed in Sec. 12.2.4, with one integrator provided for each pixel electrode. In operation, the voltage across the integrating capacitor is set to zero, and the system then integrates the incoming charge for a fixed time T. This charge includes both leakage current and the induced charge following a gamma-ray absorption. At the end of the integration time, the voltage is sampled and passed to the external electronics. This transfer operation can be done serially on all pixels during the next integration period, so different pixel signals can be temporally multiplexed to a single output line via electronic switches.

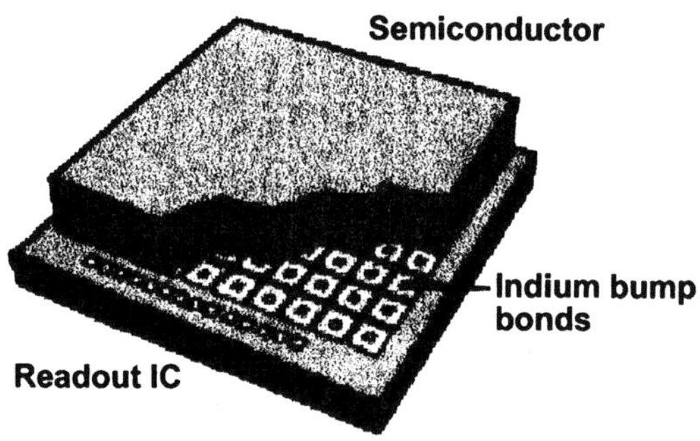

Fig. 12.23 Schematic of a hybrid semiconductor detector with a readout integrated circuit (IC). A continuous metal electrode is deposited on the upper side of the semiconductor slab, and small pixel electrodes are produced photolithographically on the underside. The semiconductor material is cold-welded to the readout IC with indium bumps.

If two or more gamma-ray photons are absorbed in the slab during time T, then the integrated charge is the sum of the leakage charge plus the charges due to all gamma rays. In many applications, however, the rate of arrival of the gamma rays is such that it is unlikely that more than one photon will be absorbed in time T. When we can make this assumption, the hybrid array of gated integrators functions as a photon-counting detector except for an occasional double hit, which we can usually reject by windowing.

Charge spreading and induction In both row-by-column and pixellated detectors, more than one output electrode may receive charge signals from a single gamma-ray interaction. As we saw in Sec. 12.3.2, the initial charge cloud produced by

[13]In commenting on the origin of the word *pixel*, William Safire (Arizona Daily Star, Apr. 3, 1995) cites two unrelated but similar words: *pixilated*—bemused, fey, whimsical (from pixie); and *pixilate* (which Safire suggests we should spell *pixelate*), referring to a photographic technique to make cinematography look like animation by deleting frames. We intend neither of these usages here.

a gamma-ray interaction has a finite size, and it grows as the charges drift as a result of Coulomb interactions and diffusion. These effects, collectively known as *charge spreading*, were not an issue in slab detectors with lateral dimensions large compared to the thickness, but pixellated or row-by-column detectors may have electrodes as small as 50 μm, so the charge cloud might spread out over several electrodes. Moreover, even without spread, the charge cloud can induce a charge on several electrodes. We need either to minimize these effects or to understand them and use the signals from multiple electrodes intelligently.

One way to minimize charge spreading is to cut grooves around the electrodes, but this approach increases fabrication costs and makes the device more difficult to analyze. In what follows we shall assume that there are no grooves and that the detector is a slab of homogeneous semiconductor with a regular array of pixel electrodes on one side.

We shall also assume that each electrode is connected to a gated integrator, for example on a readout ASIC, and that the integrator holds the electrode at ground potential. We shall neglect the small gaps between the pixels, so essentially the entire pixellated surface of the slab ($z = L_z$) is at ground potential. The continuous electrode on the other side of the slab ($z = 0$) will be the cathode and held at potential $-V_b$. Thus the field inside the slab is still uniform, so all of the calculations on slab detectors above still apply to the total induced charge and the total current through all pixel electrodes, but we now want to compute the signal on each electrode. The mathematical tools needed for this purpose were developed in Chap. 9; specifically, we shall make use of the Green's function for the Poisson equation (Sec. 9.3.3) and Green's theorem (Sec. 9.3.5).

Suppose a gamma ray is absorbed at an arbitrary point $\mathbf{r}_{int}$ in the semiconductor. As in Sec. 12.3.2, the result of the interaction is a cloud of holes and a cloud of electrons, and the overall charge density[15] can be written as

$$q(\mathbf{r}, t) = q_e(\mathbf{r}, t) + q_h(\mathbf{r}, t) \,, \tag{12.210}$$

where the first term comes from the electrons and the second from holes. Both terms include a moving cloud of charge as well as any trapped charge.

This time-varying charge density produces a time-varying potential $\phi(\mathbf{r}, t)$, which is difficult to compute in full generality. A useful approximation, justified by Eskin *et al.* (1999), is that the charge density varies sufficiently slowly that the potential can be computed by the equations of electrostatics rather than the full Maxwell's equations. With this *quasistatic* assumption, the potential must satisfy Poisson's equation,

$$\nabla^2 \phi(\mathbf{r}, t) = -\frac{1}{\epsilon_s} \, q(\mathbf{r}, t) \,, \tag{12.211}$$

where ϵ_s is the permittivity of the semiconductor. The potential must also satisfy the inhomogeneous Dirichlet boundary conditions $\phi(\mathbf{r}, t) = -V_b$ for $z = 0$ and $\phi(\mathbf{r}, t) = 0$ for $z = L_z$. We shall assume that L_x and L_y are very large compared to L_z, so the lateral boundaries of the slab are not important.

[15] Recall from Sec. 9.1.1 that we use q for charge density instead of the usual ρ, reserving the latter for spatial frequency. Do not confuse charge density $q(\mathbf{r}, t)$ with total charge $Q(t)$.

To solve (12.211) subject to these boundary conditions, we need a Green's function $p(\mathbf{r}, \mathbf{r}_0)$ that satisfies [*cf.* (9.72)]

$$\nabla_0^2 p(\mathbf{r}, \mathbf{r}_0) = \delta(\mathbf{r} - \mathbf{r}_0)\,, \qquad \mathbf{r} \text{ and } \mathbf{r}_0 \text{ in } \mathcal{V}\,, \tag{12.212}$$

where $\mathcal{V}$ is the volume of the slab. We also require the Green's function to satisfy homogeneous Dirichlet boundary conditions, $p(\mathbf{r}, \mathbf{r}_0) = 0$ if $\mathbf{r}_0$ is in $\mathcal{V}$ and $\mathbf{r}$ lies on the surface $z = 0$ or $z = L_z$ (or vice versa, $\mathbf{r}$ in $\mathcal{V}$ and $\mathbf{r}_0$ on the surface). Physically, $p(\mathbf{r}, \mathbf{r}_0)$ is the potential at point $\mathbf{r}$ due to a unit point source at $\mathbf{r}_0$ in $\mathcal{V}$ plus whatever configuration of sources outside $\mathcal{V}$ is needed to enforce the boundary conditions. Explicit forms for the Green's function will be derived below.

Once we know the Green's function, we can express $\phi(\mathbf{r}, t)$ via (9.73) as

$$\phi(\mathbf{r}, t) = -\frac{1}{\epsilon_s} \int_{\mathcal{V}} d^3\mathbf{r}_0\; p(\mathbf{r}, \mathbf{r}_0)\, q(\mathbf{r}_0, t) - V_b \int_{\mathcal{S}} dx_0\, dy_0\; \frac{\partial p(\mathbf{r}, \mathbf{r}_0)}{\partial z_0}\,, \tag{12.213}$$

where $\mathcal{S}$ is now just the surface $z_0 = 0$. (The potential vanishes on $z_0 = L_z$, and we are neglecting the lateral boundaries.)

Next, we need to compute the induced charge on each pixel electrode. For this purpose, we construct a Gaussian pillbox, shown in Fig. 12.24, straddling the inner surface of the mth pixel electrode. The field inside the electrode is zero if the electrode is an ideal conductor, and the normal component (in the $-z$ direction) of the field on the surface of the pillbox inside the semiconductor is $\partial\phi(\mathbf{r}, t)/\partial z$. By Gauss's theorem, the enclosed charge is

$$Q_m(t) = \epsilon_s \int_m da\; \frac{\partial \phi(\mathbf{r}, t)}{\partial z} = -\int_m da \int_{\mathcal{V}} d^3\mathbf{r}_0\; \frac{\partial p(\mathbf{r}, \mathbf{r}_0)}{\partial z}\, q(\mathbf{r}_0, t) + const\,, \tag{12.214}$$

where the area integral is over the inner surface of the mth pixel (in the plane $z = L_z$), and $da = dx\, dy$.

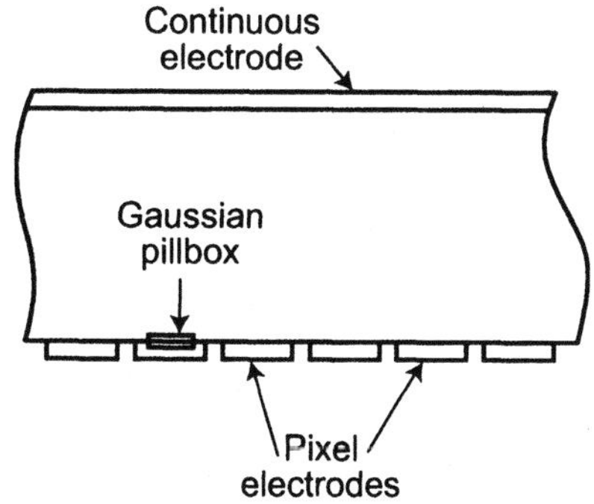

Fig. 12.24 Gaussian pillbox used to compute induced charge.

The physical interpretation of (12.214) is that a surface charge is induced with just the right distribution to maintain the electrode at ground potential, and the total charge induced on the electrode is the area integral of the surface charge density. The constant term in (12.214) [arising from the second integral in (12.213)] is the charge needed to maintain the DC field in the absence of gamma rays. Our interest is in the charge induced by $q(\mathbf{r}_0, t)$, so we shall drop the constant term.

The time derivative of the charge pulse is the current, which is integrated by the gated integrator. The output voltage of the mth gated integrator at $t = T$ is given from (12.153) as

$$V_m(T) = \frac{1}{C}\int_0^T dt'\, i(t') = \frac{1}{C}Q_m(T)\,, \tag{12.215}$$

where we have neglected leakage current (for now) and assumed that $Q_m(0) = 0$. With (12.214), we can write (12.215) in the continuous-to-discrete form,

$$V_m(T) = -\frac{1}{C}\int_{\mathcal{V}} d^3\mathbf{r}_0\, \Phi_m(\mathbf{r}_0)\, q(\mathbf{r}_0, T)\,, \tag{12.216}$$

where the dimensionless function $\Phi_m(\mathbf{r}_0)$ is given by

$$\Phi_m(\mathbf{r}_0) = \int_m da\, \frac{\partial p(\mathbf{r}, \mathbf{r}_0)}{\partial z}\Big|_{z=L_z}\,. \tag{12.217}$$

In some of the literature on gamma-ray detectors, $\Phi_m(\mathbf{r})$ is called a *weighting potential*, but it isn't really a potential, so we call it simply the *weighting function*.[15]

We see from (12.216) that the voltage at the end of the integration period depends on only the final charge distribution $q(\mathbf{r}_0, T)$. Since T is usually long compared to the transit time of charge across the detector, this distribution consists of static charge that was trapped either *en route* to the electrode or at the electrode. (Recall that we are not considering photoconductive gain, so any charge that makes it to the electrode can be considered to be trapped there.)

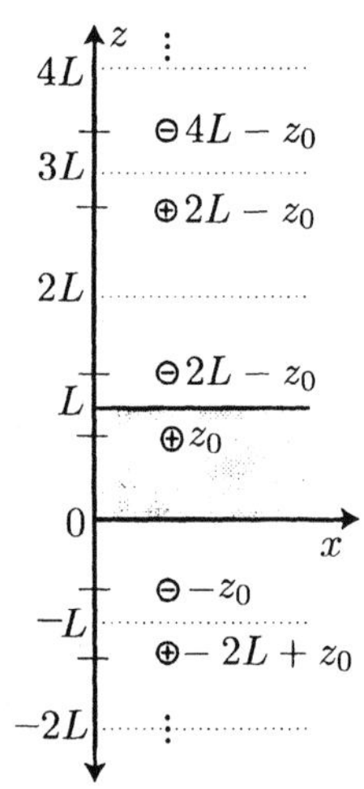

Fig. 12.25 Infinite sequence of pairs of image charges needed to satisfy boundary conditions.

[15]The term *Ramo's theorem* is sometimes used for an equation similar to (12.216). This theorem was proved by S. Ramo (1939) after being introduced by Shockley (1938). Neither Ramo nor Shockley envisioned the kinds of integrating devices under consideration here, and their theorem is essentially the time derivative of (12.216).

Image charges and the Green's function The method of images was introduced in Sec. 9.4.3 as a way of constructing a Green's function satisfying Dirichlet boundary conditions. By (12.212), the Green's function is the potential at point $\mathbf{r}$ in $\mathcal{V}$ due to a point charge at point $\mathbf{r}_0$ in $\mathcal{V}$, but this charge produces a nonzero potential on the boundaries, so additional charges outside $\mathcal{V}$ are needed. As in Sec. 9.4.3, we could cancel the potential on $z = 0$ by placing a compensating negative charge at the mirror-image point $\mathbf{r}_{0m}$ outside $\mathcal{V}$, where $\mathbf{r}_{0m} = (x_0, y_0, -z_0)$ if $\mathbf{r}_0 = (x_0, y_0, z_0)$, but then we would have a nonzero potential on $z = L_z$ from these two charges. Two additional charges would correct this problem and force the potential on $z = L_z$ to zero, but that would then make the potential on $z = 0$ nonzero. To force the Green's function to be zero on both surfaces simultaneously, we need an infinite sequence of pairs of image charges as shown in Fig. 12.25. The Green's function can then be written as (Barrett *et al.*, 1995; Eskin *et al.*, 1999)

$$p(\mathbf{r}, \mathbf{r}_0)$$

$$= \sum_{k=-\infty}^{\infty} \left[\frac{1}{[(z - 2kL_z - z_0)^2 + |\mathbf{r} - \mathbf{r}_0|^2]^{\frac{1}{2}}} - \frac{1}{[(z - 2kL_z + z_0)^2 + |\mathbf{r} - \mathbf{r}_0|^2]^{\frac{1}{2}}} \right], \tag{12.218}$$

where $\mathbf{r} = (x, y)$ and $\mathbf{r}_0 = (x_0, y_0)$.

Alternative form of the Green's function There is another way of deriving the Green's function that will not only give a useful alternative form, but will also serve to illustrate some important mathematical concepts introduced earlier in the book.

The basic idea is to expand the Green's function in eigenfunctions of the Laplacian operator with the relevant Dirichlet boundary conditions. Since we are ignoring the lateral boundaries, this operator is shift invariant in x and y, and the eigenfunctions are complex exponentials of the form $\exp[2\pi i(\xi x + \eta y)]$ (see Sec. 7.2.4). With any real values of ξ and η, these functions are eigenfunctions of ∇^2, so the operator has a continuous spectrum (see Sec. 1.4.5) as far as its x-y dependence is concerned.

In the $+z$ direction, we could use sines and cosines as eigenfunctions [see (4.24)], but the boundary conditions now restrict our choices for the spatial frequency, and the spectrum is discrete. The function $\sin(\pi nz/L_z)$ is an eigenfunction of ∇^2 and vanishes at $z = 0$ and $z = L_z$ if n is an integer, but the corresponding cosine does not satisfy the boundary conditions, and no other spatial frequencies can be used in the sine.

Thus the eigenfunctions we need are

$$u_{n\xi\eta}(\mathbf{r}) = \sqrt{\frac{2}{L_z}} \exp[2\pi i(\xi x + \eta y)] \sin(\pi nz/L_z). \tag{12.219}$$

These functions are orthonormal on the slab, satisfying

$$(\mathbf{u}_{n\xi\eta}, \mathbf{u}_{n'\xi'\eta'}) = \int_{-\infty}^{\infty} dx \int_{-\infty}^{\infty} dy \int_0^{L_z} dz \; u^*_{n\xi\eta}(\mathbf{r})\, u_{n'\xi'\eta'}(\mathbf{r}) = \delta_{nn'}\, \delta(\xi - \xi')\, \delta(\eta - \eta'). \tag{12.220}$$

Any function that is square-integrable on the slab and vanishes for $z = 0$ and L_z can be expanded in these functions.

Acting on the eigenfunctions with ∇^2 shows that the corresponding eigenvalues are

$$\lambda_{n\xi\eta} = -\left(\frac{n\pi}{L_z}\right)^2 - 4\pi^2\rho^2 \,, \tag{12.221}$$

where $\rho^2 = \xi^2 + \eta^2$. The eigenvalues are real since ∇^2 is Hermitian.

The spectral decomposition of the Laplacian (with the pertinent boundary conditions) is [*cf.* (1.86)]

$$\nabla^2 = \sum_{n=1}^{\infty} \int_{-\infty}^{\infty} d\xi \int_{-\infty}^{\infty} d\eta \; \lambda_{n\xi\eta} \mathbf{u}_{n\xi\eta} \mathbf{u}_{n\xi\eta}^{\dagger} \,, \tag{12.222}$$

and the inverse operator is given formally by [*cf.* (1.87)]

$$\nabla^{-2} = \sum_{n=1}^{\infty} \int_{-\infty}^{\infty} d\xi \int_{-\infty}^{\infty} d\eta \; \frac{1}{\lambda_{n\xi\eta}} \mathbf{u}_{n\xi\eta} \mathbf{u}_{n\xi\eta}^{\dagger} \,. \tag{12.223}$$

The inverse Laplacian is an integral operator whose kernel is the Green's function, which can therefore be written as

$$p(\mathbf{r}, \mathbf{r}_0) = \sum_{n=1}^{\infty} \int_{-\infty}^{\infty} d\xi \int_{-\infty}^{\infty} d\eta \; \frac{1}{\lambda_{n\xi\eta}} u_{n\xi\eta}(\mathbf{r})\, u_{n\xi\eta}^{*}(\mathbf{r}_0) \,. \tag{12.224}$$

Note the limits on the sum; the eigenfunctions vanish identically for $n = 0$, and terms with negative n are redundant with those for positive n. Therefore we are never dividing by zero in (12.224), and there is no worry about the existence of the inverse.

To get an explicit form for the Green's function, we note that the double integral is an inverse Fourier transform of a rotationally symmetric 2D function, which we can reduce to a single integral by (3.248). By use of (12.219), (12.221), (12.224) and formula 6.532.4 in Gradshteyn and Ryzhik (1980), we then find (Eskin *et al.*, 1999)

$$p(\mathbf{r}, \mathbf{r}_0) = -\frac{1}{\pi L_z} \sum_{n=1}^{\infty} K_0\left(\frac{\pi n |\mathbf{r} - \mathbf{r}_0|}{L_z}\right) \sin\left(\frac{\pi n z}{L_z}\right) \sin\left(\frac{\pi n z_0}{L_z}\right) , \tag{12.225}$$

where $K_0(\cdot)$ is the zero-order modified Bessel function. The reader may show the equivalence of this form to (12.218) by use of the Poisson summation formula, (3.197). Both forms are numerically tractable (Eskin, 1997).

Computation of the weighting function We now need to differentiate the Green's function with respect to z and integrate over a pixel area to get $\Phi_m(\mathbf{r})$ as defined in (12.217). This step is most easily accomplished with the eigenfunction expansion of (12.224) where the z-dependence is exhibited in a simple form. We need the derivative only on the pixel surface, $z = L_z$, and we find from (12.219) that

$$\frac{\partial}{\partial z} u_{n\xi\eta}(\mathbf{r})|_{z=L_z} = \sqrt{\frac{2}{L_z}} (-1)^n \frac{\pi n}{L_z} \exp[2\pi i(\xi x + \eta y)] \,. \tag{12.226}$$

If we define a 2D function $w_m(\mathbf{r})$ to be unity on the surface of the mth pixel and zero elsewhere, and let $W_m(\boldsymbol{\rho})$ be its 2D Fourier transform, we can combine

(12.217), (12.221) and (12.224) to obtain

$$\Phi_m(\mathbf{r}_0) = \frac{1}{\pi}\sum_{n=1}^{\infty}(-1)^n n \sin\left(\frac{n\pi z_0}{L_z}\right)\int_{\infty} d^2\rho \, \frac{W_m(\boldsymbol{\rho})}{n^2+(2L_z\rho)^2}\exp(2\pi i\boldsymbol{\rho}\cdot\mathbf{r}_0)\,. \quad (12.227)$$

We can interchange sum and integral and then perform the sum by means of formula 1.445.4 in Gradshteyn and Ryzhik (1980); the result is

$$\Phi_m(\mathbf{r}_0) = -\frac{1}{2}\int_{\infty} d^2\rho \, \frac{\sinh(2\pi z_0\rho)}{\sinh(2\pi L_z\rho)} W_m(\boldsymbol{\rho})\exp(2\pi i\boldsymbol{\rho}\cdot\mathbf{r}_0)\,. \quad (12.228)$$

For $z_0 = L_z$, the ratio of sinh functions is unity, and the integral is proportional to the 2D inverse Fourier transform of $W_m(\boldsymbol{\rho})$. Thus $\Phi_m(\mathbf{r}_0) \propto w_m(\mathbf{r}_0)$, and the weighting function is the pixel function itself in the pixel plane. For $z_0 = 0$, $\sinh(2\pi z_0\rho) = 0$, so the weighting function vanishes identically on the cathode plane.

To get the weighting function for other planes, we recognize that the integral in (12.228) is the 2D inverse Fourier transform of a product, which is a 2D convolution given by

$$\Phi_m(\mathbf{r}_0) = [w_m * s_{z_0}](\mathbf{r}_0)\,, \quad (12.229)$$

where, with the help of (3.248),

$$s_{z_0}(\mathbf{r}_0) = -\pi\int_0^{\infty}\rho d\rho \, \frac{\sinh(2\pi z_0\rho)}{\sinh(2\pi L_z\rho)} J_0(2\pi\rho r_0)\,. \quad (12.230)$$

Note the curious mixture of coordinates here: $\mathbf{r}_0$ is the 3D position vector comprised of the 2D vector $\mathbf{r}_0$ and the distance z_0, and $r_0 = |\mathbf{r}_0|$. The 2D convolution with a rotationally symmetric, z-dependent kernel yields a 3D function.

In general, numerical methods must be used to evaluate the convolution, but some insight can be gained by examining two limits.

The slab limit If the width ϵ is large compared to L_z, then $W_m(\boldsymbol{\rho}) \simeq \delta(\boldsymbol{\rho})$, and (12.228) yields

$$\Phi_m(\mathbf{r}_0) = -\frac{1}{2}\lim_{\rho\to 0}\frac{\sinh(2\pi z_0\rho)}{\sinh(2\pi L_z\rho)} = -\frac{z_0}{2L_z}\,. \quad (12.231)$$

Thus the pixel response to charge near the surface $z_0 = 0$ is very small, and the largest magnitude response is at $z_0 = L_z$. Recall, however, that the weighting function applies to the charge distribution at $t = T$, when presumably all of the charge drift has finished. Electrons that make it to the pixel electrode (the anode) have the full effect, and holes that make it to the cathode have zero effect. As explained in Sec. 12.1.3, an electron-hole pair induces an anode charge of $-e$, not $-2e$, in the absence of trapping.

The small-pixel limit To obtain high sensitivity in a gamma-ray detector, we want to make L_z large, and to obtain high spatial resolution, we want to make the pixel size ϵ small, so it often turns out that the aspect ratio L_z/ϵ is large compared to 1. For the energies used in nuclear medicine and gamma-ray astronomy, L_z is typically in the range 1 – 10 mm, and pixels can be as small as 0.05 mm.

The function $W_m(\boldsymbol{\rho})$ extends to frequencies of order $1/\epsilon$, and when $L_z \gg \epsilon$, we can use the asymptotic limit $2\pi L_z\rho \gg 1$ over most of this range. To see why this is useful, note that

$$S_{z_0}(\rho) \equiv \frac{\sinh(2\pi z_0 \rho)}{\sinh(2\pi L_z \rho)} = e^{-2\pi(L_z - z_0)\rho}\left[\frac{1 - e^{-4\pi z_0 \rho}}{1 - e^{-4\pi L_z \rho}}\right]. \tag{12.232}$$

When $z_0\rho$ and $L_z\rho$ are both large, the factor in square brackets is near unity and $S_{z_0}(\rho) \simeq \exp[-2\pi(L_z - z_0)\rho]$. The greatest error in this approximation occurs near $\rho = 0$, where $S_{z_0}(\rho) \simeq (z_0/L_z)\exp[-2\pi(L_z - z_0)\rho]$, but even this error is small for z_0 near the pixel plane $z = L_z$.

Thus a reasonable approximation to (12.230) is

$$s_{z_0}(\mathbf{r}_0) = -\pi \int_0^\infty \rho d\rho \, \exp[-2\pi(L_z - z_0)\rho] J_0(2\pi\rho r_0) = \frac{-(L_z - z_0)}{4\pi\left[(L_z - z_0)^2 + (r_0)^2\right]^{3/2}}, \tag{12.233}$$

where we have used Gradshteyn and Ryzhik (1980), formula 6.623.2. In the limit that $L_z - z_0 \to 0$, $s_{z_0}(\mathbf{r}_0) \to \delta(\mathbf{r}_0)$, so the weighting function exactly in the pixel plane is again proportional to the pixel function.

For finite values of $L_z - z_0$, $s_{z_0}(\mathbf{r}_0)$ has a spatial width of about $L_z - z_0$, so we can regard it as approximately a delta function whenever it is integrated against a substantially broader function. In particular,

$$\Phi_m(\mathbf{r}_0) \simeq w_m(\mathbf{r}_0) \qquad \text{for} \qquad L_z - z_0 \ll \epsilon. \tag{12.234}$$

Thus the weighting function is independent of z_0 for points within about ϵ of the pixel.

For more distant points, such that $L_z - z_0 \gg \epsilon$, $s_{z_0}(\mathbf{r}_0)$ is broad compared to $w_m(\mathbf{r}_0)$, and we can do the convolution by regarding the latter as the delta function this time. If $w_m(\mathbf{r})$ is centered on $\mathbf{r} = \mathbf{r}_m$, we obtain

$$\Phi_m(\mathbf{r}_0) \simeq \epsilon^2 s_{z_0}(\mathbf{r}_0 - \mathbf{r}_m) \qquad \text{for} \qquad L_z - z_0 \gg \epsilon. \tag{12.235}$$

By inspection of (12.233), we see that the weighting function now falls off as the inverse square of the z distance from the pixel in this limit.

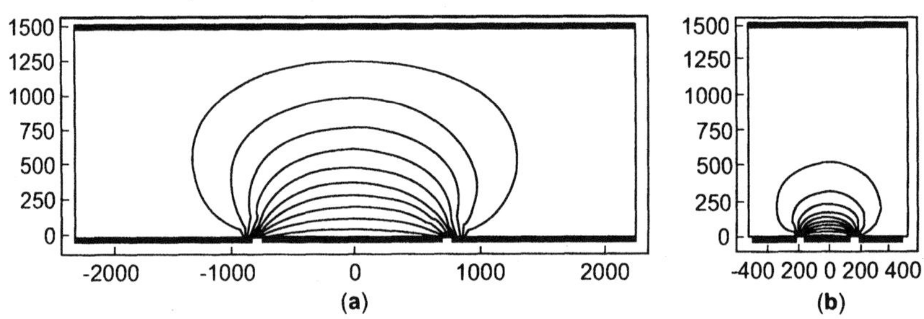

Fig. 12.26 Behavior of the weighting function in the small-pixel limit. (Courtesy of Josh Eskin.)

The behavior of the weighting function for small and large pixels is shown in Fig. 12.26. The key point is that the pixel is sensitive only to charge that comes within about one pixel width of the electrode; charge trapped outside this sensitive zone is irrelevant (Barrett *et al.*, 1995; Eskin *et al.*, 1999). If the pixel is the anode, as we have assumed, then holes travel away from the sensitive zone, and hole trapping is far less important than it is with large electrodes. As a result of this *small-pixel effect*, the trapping plateau in a pulse-height spectrum is greatly reduced, and the photopeak fraction is increased.

Mean signals Now that we know the weighting function, we can use it in (12.216) to compute the mean signal from each pixel under various assumptions about charge spreading and trapping. We just need to replace the charge density $q(\mathbf{r}_0, T)$ by its average over all possible paths for the carriers, which we denote by $\overline{q}(\mathbf{r}_0, T)$. Though the notation does not show it, the average here is conditional on both $\mathbf{r}_{int}$ and N_{eh}.

For example, if we neglect carrier diffusion and the finite size of the initial charge cloud, but consider arbitrary trapping of both holes and electrons, then the carriers travel along straight lines, undergoing exponential trapping as they go. After time T (assumed to be much longer than the drift times), the mean electron charge density $\overline{q}_e(\mathbf{r}_0, T)$ is a line of trapped charge extending from the interaction point to the pixel plane (which we assume is the anode) plus a delta function to account for the electrons that make it to the pixel. Specifically,

$$\overline{q}_e(\mathbf{r}_0, T) = -eN_{eh}\,\delta(\mathbf{r}_0 - \mathbf{r}_{int})$$
$$\times \left[\frac{1}{\lambda_e} \exp\left(-\frac{z_0 - z_{int}}{\lambda_e} \right) \text{step}(z_0 - z_{int}) + \exp\left(-\frac{L_z - z_{int}}{\lambda_e} \right) \delta(z_0 - L_z) \right], \tag{12.236}$$

where $\mathbf{r}_0 = (\mathbf{r}_0, z_0)$. A similar expression gives the charge density for holes:

$$\overline{q}_h(\mathbf{r}_0, T) = eN_{eh}\,\delta(\mathbf{r}_0 - \mathbf{r}_{int})$$
$$\times \left[\frac{1}{\lambda_h} \exp\left(-\frac{z_{int} - z_0}{\lambda_h} \right) \text{step}(z_{int} - z_0) + \exp\left(-\frac{z_{int}}{\lambda_e} \right) \delta(z_0) \right]. \tag{12.237}$$

Note that these densities of trapped charge are assumed to be proportional to N_{eh}, which is a valid assumption if each trap has only a small probability of being filled. Most practical detectors satisfy this condition.

Now we want to use these charge densities to average the voltage $V_m(T)$ defined in (12.216). For notational simplicity, we shall drop the argument and shorten $V_m(T)$ to V_m. Then we can insert (12.236) and (12.237) into (12.216) and perform the integrals for which we have delta functions, yielding

$$E\{V_m|\mathbf{r}_{int}\} = -\frac{e}{C}\overline{N}_{eh} \left[\exp\left(-\frac{z_{int}}{\lambda_e} \right) \delta_{mm_{int}} \right.$$
$$+ \frac{1}{\lambda_e} \int_{z_{int}}^{L_z} dz_0\; \Phi_m(\mathbf{r}_{int}, z_0) \exp\left(-\frac{z_0 - z_{int}}{\lambda_e} \right)$$
$$\left. - \frac{1}{\lambda_h} \int_0^{z_{int}} dz_0\; \Phi_m(\mathbf{r}_{int}, z_0) \exp\left(-\frac{z_{int} - z_0}{\lambda_h} \right) \right], \tag{12.238}$$

where m_{int} denotes the pixel nearest to the interaction (*i.e.*, the one for which $w_{m_{int}}(\mathbf{r}_{int}) = 1$). The notation $\Phi_m(\mathbf{r}_0, z_0)$ means the same thing as $\Phi_m(\mathbf{r}_0)$, and we have used the fact that $\Phi_m(\mathbf{r}_{int}, 0) = 0$. Note also that the expectation has now included an average over N_{eh}.

Several limits of this expression are of interest. First, if there is very little trapping ($\lambda_e, \lambda_h \gg L_z$), then $E\{V_m|\mathbf{r}_{int}\} \simeq -(e/C)\overline{N}_{eh}\, \delta_{mm_{int}}$; all of the electrons arrive at a single pixel ($m = m_{int}$) and induce the full charge there. Second, for strong trapping of both carriers ($\lambda_e, \lambda_h \ll L_z$), $E\{V_m|\mathbf{r}_{int}\} \simeq -(e/C)\overline{N}_{eh}\Phi_m(\mathbf{r}_{int}, z_{int})$; in contrast to slab detectors, pixellated detectors exhibit a strong dependence on depth of interaction in the strong-trapping limit.

Finally, many materials of interest exhibit good electron transport ($\lambda_e \gg L_z$) but poor hole transport ($\lambda_h \ll L_z$); in these materials,

$$E\{V_m|\mathbf{r}_{int}\} \simeq -\frac{e}{C}\overline{N}_{eh}[\,\delta_{mm_{int}} - \Phi_m(\mathbf{r}_{int}, z_{int})]\,. \tag{12.239}$$

Except for interactions that occur within about ϵ of the pixel, $\Phi_m(\mathbf{r}_{int}, z_{int})$ is small compared to one, so the second term makes little contribution to the signal on pixel m_{int}. Its contribution to adjacent pixels may, however, be significant. Hole trapping produces a small positive signal on an anode pixel, while electrons produce a negative one; if no electrons reach a pixel for which $m \neq m_{int}$, only the positive signal is observed. Since this signal depends strongly on z_{int}, it can be used to estimate the depth of interaction (Marks, 2000).

It is not difficult to generalize this calculation to include charge spreading and nonlocal charge deposition, but Monte Carlo methods are usually required to get the relevant charge densities. For examples and many details, see Eskin (1997) and Marks (2000).

Covariance of the output signals The conditional covariance matrix on the output voltages for a specified interaction position $\mathbf{r}_{int}$ is defined as

$$[\mathbf{K}_V(\mathbf{r}_{int})]_{mm'} \equiv E\{\Delta V_m \Delta V_{m'}|\mathbf{r}_{int}\}\,, \tag{12.240}$$

where $\Delta V_m \equiv V_m - \overline{V}_m(\mathbf{r}_{int})$, and $\overline{V}_m(\mathbf{r}_{int}) \equiv E\{V_m|\mathbf{r}_{int}\}$. The major stochastic effects that influence this matrix are generation of electron-hole pairs, trapping, dark current and electronic noise.

Some of these effects contribute only to the diagonal terms of this matrix (the variance), and some are correlated from pixel to pixel and hence contribute to the off-diagonal elements as well. In particular, dark current and electronic noise are certainly statistically independent from pixel to pixel. In fact, since the pixels are at least nominally identical, a good model for these noise sources is an i.i.d. Gaussian distribution. Carrier generation and trapping may lead to significant correlations.

We shall see, however, that the off-diagonal terms in $\mathbf{K}_V(\mathbf{r}_{int})$ will often vanish if N_{eh} is a Poisson random variable (so that the Fano factor is unity). This result should be expected from Chap. 11 where the word Poisson was taken to be almost synonymous with independent (and hence uncorrelated). On the other hand, in the discussion in Sec. 11.4 on integrating detectors with gain, we did not find that the correlations vanished when the gain process was Poisson. In Sec. 11.4, the variable k_n, the number of secondaries produced by the nth primary, corresponds to what we call N_{eh} here, but there are correlations, evident in (11.226) and (11.238), even when the variance of k_n equals its mean. [From (11.219) we see that $\text{Var}(k_n) = \overline{k}_n$ does not correspond to $s(\mathbf{R}) = 0$.]

The resolution of this apparent paradox is that N_{eh} is only *conditionally* Poisson, since it is computed for a single interaction event. When a Poisson number of interaction events occur over the integration time, the total number of secondaries is no longer Poisson [*cf.* (11.198)], and the correlations evident in (11.226) and (11.238) must occur. The reader should not confuse $\mathbf{K}_V(\mathbf{r}_{int})$, which applies to a single interaction event at the specified location $\mathbf{r}_{int}$, with the covariance matrix $\mathbf{K}_\mathbf{y}$ of an integrating detector with gain as given in (11.238).

Carrier-generation statistics with no trapping In the absence of trapping, all electrons make it to the pixel plane (anode), and all holes make it to the cathode. Since the weighting function $\Phi_m(\mathbf{r}_0)$ is zero for the cathode plane, we can ignore the holes and just account for the distribution of electrons among the pixels. This distribution is affected by nonlocal charge deposition and charge spreading, but for present purposes all we need to know is that there is some probability $\beta_m(\mathbf{r}_{int})$ that an electron generated at $\mathbf{r}_{int}$ arrives at the mth pixel. In the absence of trapping, $\sum_m \beta_m(\mathbf{r}_{int}) = 1$. If N_m is the actual number that arrive at the pixel, then $\sum_m N_m = N_{eh}$, and the conditional mean of N_m is $N_{eh}\beta_m(\mathbf{r}_{int})$.

If we neglect Coulomb interactions between electrons, then each electron will choose a pixel independently of the others, and the conditional probability law (for fixed N_{eh} and $\mathbf{r}_{int}$) will be the multinomial, (C.164). For $m \neq m'$, we can write

$$\langle N_m N_{m'} \rangle \equiv E\{N_m N_{m'}|\mathbf{r}_{int}\} = \sum_{N_{eh}=0}^{\infty} \Pr(N_{eh})\, E\{N_m N_{m'}|N_{eh}, \mathbf{r}_{int}\}$$

$$= \beta_m(\mathbf{r}_{int})\, \beta_{m'}(\mathbf{r}_{int}) \sum_{N_{eh}=0}^{\infty} \Pr(N_{eh})\, N_{eh}(N_{eh}-1)$$

$$= \beta_m(\mathbf{r}_{int})\, \beta_{m'}(\mathbf{r}_{int})[\mathrm{Var}\{N_{eh}\} + \overline{N}_{eh}^2 - \overline{N}_{eh}] \,. \tag{12.241}$$

Since $\mathrm{Var}\{N_{eh}\} = F\overline{N}_{eh}$, where F is the Fano factor, we quickly find that

$$\langle \Delta N_m \Delta N_{m'} \rangle = \overline{N}_{eh}(F-1)\, \beta_m(\mathbf{r}_{int})\, \beta_{m'}(\mathbf{r}_{int}) \,, \qquad (m \neq m') \,, \tag{12.242}$$

where $\Delta N_m = N_m - E\{N_m|\mathbf{r}_{int}\}$. By a similar procedure, the variance of N_m is found to be

$$\mathrm{Var}\{N_m\} = \overline{N}_{eh}\beta_m(\mathbf{r}_{int}) + \overline{N}_{eh}(F-1)\, \beta_m^2(\mathbf{r}_{int}) \,. \tag{12.243}$$

We can combine these last two equations into the covariance matrix (Barrett and Swindell, 1981, 1996),

$$[\mathbf{K}_N(\mathbf{r}_{int})]_{mm'} \equiv \langle \Delta N_m \Delta N_{m'} \rangle = \overline{N}_{eh}\beta_m(\mathbf{r}_{int})\, \delta_{mm'} + \overline{N}_{eh}(F-1)\, \beta_m(\mathbf{r}_{int})\, \beta_{m'}(\mathbf{r}_{int}) \,. \tag{12.244}$$

We see that N_m and $N_{m'}$ are uncorrelated if N_{eh} is a Poisson random variable where $F = 1$; this result is expected from the discussion in Sec. 11.2.1 on multinomial selection of a Poisson. We know from earlier discussions in this chapter, however, that N_{eh} is not Poisson, and F may be much less than unity. With such sub-Poisson statistics, there is a negative correlation between N_m and $N_{m'}$; in the limit where $F = 0$ and only two pixels receive charge, a fluctuation in N_m must be accompanied by an equal and opposite fluctuation in $N_{m'}$ since their sum is fixed.

If we consider only the statistics of N_m and neglect electronic noise and dark current, the output voltage of the mth integrator is $-eN_m/C$, so the covariance matrix for the output voltages is

$$[\mathbf{K}_V(\mathfrak{r}_{int})]_{mm'} = \frac{e^2}{C^2}\beta_m(\mathfrak{r}_{int})\,\overline{N}_{eh}\,\delta_{mm'} + \frac{e^2}{C^2}(F-1)\,\overline{N}_{eh}\beta_m(\mathfrak{r}_{int})\,\beta_{m'}(\mathfrak{r}_{int})\,. \tag{12.245}$$

We reiterate that (12.245) refers to the signals after one integration period in which exactly one gamma-ray interaction occurs, that it is conditional on the interaction location, and that it applies only in the absence of trapping.

Random distribution of trapped charge With trapping, we cannot use the multinomial arguments because we are not just counting particles; as we have seen, electrons and holes induce signals even without reaching the electrodes. The new stochastic effect is the random spatial distribution of trapped charge after time T.

Since the trapped charges are single electrons or holes, the charge density is a spatial point process, and we can use the properties of such processes as developed in Sec. 11.3.3. Specifically, the electron charge density can be written as

$$q_e(\mathfrak{r},T) = -e\sum_{j=1}^{N_{eh}}\delta(\mathfrak{r}-\mathfrak{r}_j) \equiv -eg_e(\mathfrak{r})\,, \tag{12.246}$$

with a similar expression for holes. Since the number of carriers N_{eh} is not a Poisson random variable, however, we are not dealing with a Poisson point process. Instead, we can start with the general expression (11.92) for the autocovariance of an arbitrary point process.

To adapt (11.92) to the present discussion, we must first distinguish the 3D vector $\mathfrak{r}$ from the 2D vector $\mathbf{r}$. Also, we shall need separate expressions for electrons and holes, so we add subscripts. Finally, we recognize the dependence of all quantities on the interaction position $\mathfrak{r}_{int}$. With these notational changes, the probability density $\mathrm{pr}(\mathbf{r}|N)$ that appears in (11.92) can be replaced (for electrons) by $\mathrm{pr}_e(\mathfrak{r}|\mathfrak{r}_{int}, N_{eh})$, which is the spatial density of trapped electrons after time T. We include in this density the number of electrons that make it to the electrode as well as the number trapped in the bulk, so the total number is just N_{eh}.

If we assume, as we did in (12.236) and (12.237), that each trap has a small probability of being filled, then the trapping events are independent, and the spatial density of trapped electrons is given, by analogy to (11.83), as

$$\mathrm{pr}_e(\mathfrak{r}|\mathfrak{r}_{int}, N_{eh}) = \frac{\overline{q}_e(\mathfrak{r},T)}{\int_V d^3\mathfrak{r}'\;\overline{q}_e(\mathfrak{r}',T)} = -\frac{1}{eN_{eh}}\overline{q}_e(\mathfrak{r},T)\,, \tag{12.247}$$

and similarly for holes. Recall from the discussion above (12.236) that $\overline{q}_e(\mathfrak{r},T)$ is conditional on both $\mathfrak{r}_{int}$ and N_{eh}, so these dependences are hidden on the right-hand side of (12.247). With (12.236), however, $\overline{q}_e(\mathfrak{r},T)$ is linearly related to N_{eh}, and the resulting spatial density $\mathrm{pr}_e(\mathfrak{r}|\mathfrak{r}_{int})$ is independent of N_{eh}. This conclusion is again a consequence of the assumption that each trap has a small probability of being filled.

By this same assumption, the joint density $\mathrm{pr}(\mathbf{r},\mathbf{r}'|N)$ that appears in (11.92) factors into the product of the two marginals, and (11.92) becomes

$$K_{g_e}(\mathbf{r},\mathbf{r}') = \overline{N}_{eh}\,\mathrm{pr}_e(\mathfrak{r}|\mathfrak{r}_{int})\,\delta(\mathbf{r}-\mathbf{r}') + (F-1)\overline{N}_{eh}\,\mathrm{pr}_e(\mathfrak{r}|\mathfrak{r}_{int})\,\mathrm{pr}_e(\mathfrak{r}'|\mathfrak{r}_{int})\,, \tag{12.248}$$

where we have used the definition of the Fano factor, (12.166). Not unexpectedly, we see that the random process $g_e(\mathfrak{r})$ is delta-correlated for the Poisson case where $F = 1$.

Covariance of the output signals with trapping To compute the covariance matrix on the signals out of the gated integrators, we can use (12.248) along with results from Sec. 8.2.6 on filtering of random processes. The filtering action is described by (12.216), which has the general form of a continuous-to-discrete (CD) mapping from the point process $q(\mathbf{r}_0, T)$ to the random vector of voltages. With (12.216), we can write the covariance matrix of the voltages as [*cf.* (8.147)]

$$[\mathbf{K}_V(\mathbf{r}_{int})]_{mm'} = \left(\frac{e}{C}\right)^2 \int_\infty d^3\mathbf{r} \int_\infty d^3\mathbf{r}'\ \Phi_m(\mathbf{r})\,[K_{g_e}(\mathbf{r},\mathbf{r}') + K_{g_h}(\mathbf{r},\mathbf{r}')]\,\Phi_{m'}(\mathbf{r}')\,. \tag{12.249}$$

There are no cross-covariance terms since the random processes for holes and electrons are uncorrelated with each other.

Inserting (12.248) and its hole counterpart into (12.249), we get the following generalization of (12.245):

$$\begin{aligned}[\mathbf{K}_V(\mathbf{r}_{int})]_{mm'} &= \left(\frac{e}{C}\right)^2 \overline{N}_{eh} \int_\infty d^3\mathbf{r}\ [\mathrm{pr}_e(\mathbf{r}|\mathbf{r}_{int}) + \mathrm{pr}_h(\mathbf{r}|\mathbf{r}_{int})]\,\Phi_m(\mathbf{r})\,\Phi_{m'}(\mathbf{r}) \\ &+ \left(\frac{e}{C}\right)^2 (F-1)\,\overline{N}_{eh} \int_\infty d^3\mathbf{r}\ \mathrm{pr}_e(\mathbf{r}|\mathbf{r}_{int})\,\Phi_m(\mathbf{r}) \int_\infty d^3\mathbf{r}'\ \mathrm{pr}_e(\mathbf{r}'|\mathbf{r}_{int})\,\Phi_{m'}(\mathbf{r}') \\ &+ \left(\frac{e}{C}\right)^2 (F-1)\,\overline{N}_{eh} \int_\infty d^3\mathbf{r}\ \mathrm{pr}_h(\mathbf{r}|\mathbf{r}_{int})\,\Phi_m(\mathbf{r}) \int_\infty d^3\mathbf{r}'\,\mathrm{pr}_h(\mathbf{r}'|\mathbf{r}_{int})\,\Phi_{m'}(\mathbf{r}')\,.\end{aligned} \tag{12.250}$$

This expression reduces to (12.245) in the absence of trapping, where $\mathrm{pr}_e(\mathbf{r}|\mathbf{r}_{int}) \propto \delta(z - L_z)$ and $\mathrm{pr}_h(\mathbf{r}|\mathbf{r}_{int}) \propto \delta(z)$.

One interesting observation about (12.250) concerns the Poisson limit, $F \to 1$. We are accustomed to Poisson processes leading to uncorrelated measurements, but that is not necessarily the case here. The first integral in (12.250) can be nonzero for $m \neq m'$ if $\Phi_m(\mathbf{r})$ and $\Phi_{m'}(\mathbf{r})$ overlap. A particular trapped charge in the bulk can contribute to the signals on different pixels, so there is a positive correlation. In the absence of trapping, however, the delta function $\delta(z - L_z)$ in the electron distribution reduces the volume integral to an integral over the pixel surface where $\Phi_m(\mathbf{r})$ is the pixel function; then $\Phi_m(\mathbf{r})$ and $\Phi_{m'}(\mathbf{r})$ cannot overlap if $m \neq m'$ and hence there is no correlation.

The non-Poisson parts of (12.250) (the second and third terms) give negative correlations for the usual case where $F < 1$. These correlations do not require that $\Phi_m(\mathbf{r})$ and $\Phi_{m'}(\mathbf{r})$ overlap, but they do require that the charge distribution overlap with both functions. Thus the correlations are stronger for interactions farther from the pixel plane, and they arise only from charge trapped in the bulk. Charge spreading by diffusion enhances these off-diagonal terms in $\mathbf{K}_V(\mathbf{r}_{int})$.

Correlations induced by random depth of interaction So far we have concentrated on the conditional covariance matrix $\mathbf{K}_V(\mathbf{r}_{int})$ for fixed interaction location, as defined in (12.240). As we shall see in Sec. 12.3.5, this is the relevant matrix when we wish to estimate all three coordinates of $\mathbf{r}_{int}$ from the observed signals, but often we want to estimate just the two lateral coordinates x_{int} and y_{int}. In that case we shall need the covariance $\mathbf{K}_V(\mathbf{r}_{int})$ conditional on only the 2D vector $\mathbf{r}_{int}$; it is defined by

$$\left[\mathbf{K}_V(\mathbf{r}_{int})\right]_{mm'} \equiv E\left\{\left[V_m - \overline{V}_m(\mathbf{r}_{int})\right]\left[V_{m'} - \overline{V}_{m'}(\mathbf{r}_{int})\right]\right\}\,, \tag{12.251}$$

where $\overline{V}_m(\mathbf{r}_{int}) \equiv E\{V_m|\mathbf{r}_{int}\}$. By adding and subtracting $\overline{V}_m(\mathbf{r}_{int})$ to each factor in the expectation and doing a little algebra, we find

$$\mathbf{K}_V(\mathbf{r}_{int}) = \langle \mathbf{K}_V(\mathbf{r}_{int})\rangle_{z_{int}} + \mathbf{K}_{\overline{V}}(\mathbf{r}_{int}) , \tag{12.252}$$

where

$$[\mathbf{K}_{\overline{V}}(\mathbf{r}_{int})]_{mm'} = \left\langle \left[\overline{V}_m(\mathbf{r}_{int}) - \overline{V}_m(\mathbf{r}_{int})\right] \left[\overline{V}_{m'}(\mathbf{r}_{int}) - \overline{V}_{m'}(\mathbf{r}_{int})\right] \right\rangle_{z_{int}} . \tag{12.253}$$

The second term in (12.252) arises because the random depth of interaction affects the distribution of charge trapped in the bulk, and this distribution is sensed by two pixels m and m', especially if they are adjacent. The resulting correlation can be either positive or negative. It is positive when the photon energy is low and L_z/ϵ is large, so that most photons are absorbed a long distance from the pixel plane; then the photons that happen to be absorbed nearer to the pixels contribute more to both $\overline{V}_m(\mathbf{r}_{int})$ and $\overline{V}_{m'}(\mathbf{r}_{int})$. For higher photon energies and smaller L_z/ϵ, more photons are absorbed within ϵ of the pixels where $\Phi_m(\mathbf{r})$ and $\Phi_{m'}(\mathbf{r})$ have little overlap, and the positive correlation is reduced or may become negative. For no trapping, of course, depth of interaction is irrelevant, and $[\mathbf{K}_{\overline{V}}(\mathbf{r}_{int})]_{mm'} = 0$ for $m \neq m'$.

12.3.4 Position and energy estimation with semiconductor detectors

The analysis in Sec. 12.3.3 provides us with a statistical description of the electrode signals in a photon-counting semiconductor gamma-ray detector array, but we still need to form an image from these signals. Two distinct approaches are possible: we can either devise some algorithm based on the pixel signals from a single event to assign that event to an image bin; or we can use all signals from all events to estimate the fluence pattern. The first approach, called *event estimation*, is by far the more common, and it will be the focus of this section. The second approach, *fluence estimation*, requires knowledge of image reconstruction algorithms, so it will be postponed to Chap. 15. For application of fluence estimation to semiconductor detector arrays, see Marks (2000).

In event estimation, the x and y coordinates of the interaction position (*i.e.*, the 2D vector $\mathbf{r}_{int}$) can be estimated and used to assign the event to a 2D array of image bins, or the depth z_{int} can also be estimated, leading to a 3D array of image bins. In either case, the image bins are referred to as pixels, but they need not be the same size as the pixel electrodes on the detector array. As we shall see, it is possible to obtain spatial resolution better than the size of the electrodes.

As discussed in Sec. 12.3.2, we also usually want to estimate the energy of the event. The energy estimate can be used to create another index into the image array, but more commonly it is used in some decision algorithm for accepting an event into the spatial image array.

The mathematical framework underlying this process of image formation is statistical decision theory, as presented in Chap. 13. This theory includes the theory of parameter estimation, obviously relevant for both position and energy estimation, and classification theory, applicable to the decision to accept or reject an event. We presume here that the reader is conversant with that theory, especially as developed in Sec. 13.3. In particular, we shall use concepts of bias and variance of an estimator freely, and we shall soon make use of maximum-likelihood estimation.

The hottest pixel The simplest algorithm for assigning events to image pixels is just to identify the electrode with the largest signal in the electrode array and to assign the event to the corresponding pixel in an image array of the same size. There is then a 1:1 correspondence between electrodes and image pixels.

This procedure is the obvious one to use in the absence of trapping since then only one electrode gets charge from the gamma-ray interaction; all others merely integrate the dark current. If σ^2 is the variance of the signals due to dark current and electronic noise, and if the interaction produces a mean signal that is at least, say, 3σ or 4σ, then there is essentially zero probability that the wrong pixel will be identified this way. With trapping, however, several electrodes may receive charge from a single gamma-ray interaction, and there may be a significant probability of an incorrect assignment.

In addition to identifying the interaction pixel, we also want to estimate the energy of the incident photon and to accept or reject it based on some preset window (see Sec. 12.3.2). In the absence of trapping, the obvious estimate of energy is the observed signal on the interaction pixel minus an estimate of the contribution from dark current, but with trapping it is not so obvious how to proceed. To use just the signal from one electrode to estimate energy is to ignore useful information from other electrodes.

Linear and quasilinear estimates One simple way to use information from pixels other than the one with the largest signal is to form linear combinations of the signals in some neighborhood around that pixel. For example, if we have identified pixel m_0 as the one with the largest signal, we can define a neighborhood $\mathcal{N}(m_0)$ consisting of a square array of pixels centered on m_0. With this reduced data set, a possible estimate of $\mathcal{E}_0$ is

$$\hat{\mathcal{E}} = A \sum_{m \text{ in } \mathcal{N}(m_0)} \left[V_m - \overline{V}_m^{(dark)} \right] , \tag{12.254}$$

where A is a constant and $\overline{V}_m^{(dark)}$ is the average of V_m over many integration periods without gamma-ray illumination (a quantity we can easily measure). Alternatively, if we assume that exactly one interaction has occurred in a single integration period, then we can use the entire set of electrode signals for that period. In either case, except for the fact that we have subtracted off the mean dark current, $\mathcal{E}$ is a *linear estimator* of $\mathcal{E}_{kin}$. We can choose the constant A to minimize the bias in this estimate. With charge spreading but no trapping, for example, $\hat{\mathcal{E}}$ is an unbiased estimator of the deposited energy $\mathcal{E}_{kin}$ (or $\mathcal{E}_0$ if we consider only photopeak events) if $A = -\mathcal{E}_{eh}C/e$ [*cf.* (12.192)]. With this choice, the mean of $\hat{\mathcal{E}}$ is $\mathcal{E}_{kin}$ since the mean of the total generated electron charge is given by $-e\mathcal{E}_{kin}/\mathcal{E}_{eh}$, and this charge induces a total mean voltage of Q_e/C. A linear estimator of the x component of the 2D interaction position has a form analogous to (12.254),

$$\hat{x} = \sum_m B_{xm} \left[V_m - \overline{V}_m^{(dark)} \right] , \tag{12.255}$$

and similarly for the y component. Again, the coefficients B_{xm} and B_{ym} could be chosen to minimize bias.

One immediate difficulty with (12.255), however, is that the position estimate depends on $\mathcal{E}_{kin}$ since the mean of all terms in the sum is proportional to the deposited energy. To avoid this problem, we can define a *quasilinear estimator* as

$$\hat{\mathbf{r}} = \frac{1}{\hat{\mathcal{E}}} \sum_m \mathbf{B}_m \left[V_m - \overline{V}_m^{(dark)} \right] , \tag{12.256}$$

where $\mathbf{B}_m = (B_{xm}, B_{ym})$. This estimator is not linear in the signals $\{V_m\}$ since the denominator $\hat{\mathcal{E}}$ is itself a linear function of the signals. Instead, the form is

$$\hat{\mathbf{r}} = \frac{\sum_m \mathbf{w}_m \left[V_m - \overline{V}_m^{(dark)} \right]}{\sum_m \left[V_m - \overline{V}_m^{(dark)} \right]} , \tag{12.257}$$

where $\mathbf{w}_m = \mathbf{B}_m / A$.

We can determine the weighting coefficients by making use of the lateral shift-invariance of the slab. If the lateral dimensions L_x and L_y are large compared to L_z and ϵ, then the mean of $V_m - \overline{V}_m^{(dark)}$ (averaged over all random effects including depth of interaction) is some function of $\mathbf{r}_m - \mathbf{r}_{int}$, where $\mathbf{r}_m$ is the 2D location of the mth electrode. Calling this function $f(\mathbf{r}_m - \mathbf{r}_{int})$, we can write

$$E\{\hat{\mathbf{r}} | \mathbf{r}_{int}\} \simeq \frac{\sum_m \mathbf{w}_m f(\mathbf{r}_m - \mathbf{r}_{int})}{\sum_m f(\mathbf{r}_m - \mathbf{r}_{int})} . \tag{12.258}$$

If we now simply take $\mathbf{w}_m$ as the pixel position $\mathbf{r}_m$, and if we assume that the pixels are small compared to the width of $f(\mathbf{r}_m - \mathbf{r}_{int})$, we find

$$E\{\hat{\mathbf{r}} | \mathbf{r}_{int}\} = \frac{\sum_m \mathbf{r}_m f(\mathbf{r}_m - \mathbf{r}_{int})}{\sum_m f(\mathbf{r}_m - \mathbf{r}_{int})} \simeq \frac{\int_\infty d^2 r_m \; \mathbf{r}_m f(\mathbf{r}_m - \mathbf{r}_{int})}{\int_\infty d^2 r_m \; f(\mathbf{r}_m - \mathbf{r}_{int})} . \tag{12.259}$$

Next we make the obvious change of variables $\mathbf{r}' = \mathbf{r}_m - \mathbf{r}_{int}$. Since left and right are indistinguishable in this problem, we must have $f(\mathbf{r}') = f(-\mathbf{r}')$, and we find readily that

$$E\{\hat{\mathbf{r}} | \mathbf{r}_{int}\} = \mathbf{r}_{int} . \tag{12.260}$$

Thus the simple expedient of choosing the weights in a quasilinear estimator as the pixel coordinates leads to an unbiased estimator of $\mathbf{r}_{int}$ if we can assume that the charge spread is large compared to the pixel width and we can neglect effects from the lateral boundaries. This approach was originally proposed by Hal Anger (1958) in the context of scintillation cameras (to be discussed in Sec. 12.3.5), but it is also applicable to semiconductor arrays.

Problems with linear estimators In spite of the easy implementation, linear estimators have some deficiencies. One way to see them is to reconsider the neighborhood $\mathcal{N}(m_0)$ introduced in (12.254). If we assume that exactly one interaction has occurred in the integration period, then the neighborhood can be as large as desired, up to the size of the array, without changing the choice of weighting coefficients. The mean of $V_m - \overline{V}_m^{(dark)}$ is nonzero only for a few pixels surrounding m_0, so including additional pixels does not affect the mean of $\hat{\mathcal{E}}$ or $\hat{\mathbf{r}}$, and we deduced the weighting coefficients solely by consideration of the mean values. Intuitively, however, we expect bad things to happen when we make the neighborhood unnecessarily large since we are adding in signals that convey no useful information, yet which are corrupted with noise. In the position estimator, moreover, this effect is exacerbated because we are weighting the signals proportionally to the pixel position, so a signal from a pixel far from the interaction site could receive a large weight and hence a large noise amplification. To minimize these noise problems, we need estimators that account for the variances and covariances of the signals, not just their means. Statistical estimation theory, the topic of Sec. 13.3, tells us how to find such estimators under different assumptions about our knowledge of the noise in the signals and the distribution of the parameters being estimated. In particular, we show in Sec. 13.3.6 that maximum-likelihood (ML) estimators have certain desirable noise characteristics, so we shall now discuss the application of ML estimation to the problem at hand.

Maximum-likelihood estimation of position and energy An ML estimate of any parameter $\boldsymbol{\theta}$ from a data vector $\mathbf{g}$ requires knowledge of the conditional probability $\mathrm{pr}(\mathbf{g}|\boldsymbol{\theta})$, which is called the likelihood when it is regarded as a function of $\boldsymbol{\theta}$ for fixed $\mathbf{g}$. In the present problem, $\boldsymbol{\theta}$ includes the interaction position $\mathbf{r}_{int}$ or $\mathbf{\mathfrak{r}}_{int}$ and the energy $\mathcal{E}_0$ of the gamma-ray photon. The data vector $\mathbf{g}$ is now the set of pixel signals $\{V_m\}$ for m either in some neighborhood of the hottest pixel or in the entire array. We can denote the signal set as an $M \times 1$ vector $\mathbf{V}$, where M is either the size of the neighborhood or the size of the array, so the ML estimation problem requires that we maximize $\mathrm{pr}(\mathbf{g}|\mathbf{r}_{int}, \mathcal{E}_0)$ or $\mathrm{pr}(\mathbf{g}|\mathbf{\mathfrak{r}}_{int}, \mathcal{E}_0)$, depending on whether we want to know the interaction position in 2D or 3D. The 2D and 3D problems are rather different, so we shall discuss them separately.

ML estimation of position in 3D and energy The PDF on $\mathbf{V}$ is fully determined by $\mathcal{E}_0$ and $\mathbf{\mathfrak{r}}_{int}$ so long as the entire photon energy is deposited locally at the interaction point, which means we are ignoring K escape and Compton scatter. The only remaining random effects are the random generation and propagation of charge carriers. Since the electrode signals result from many statistically independent electrons and holes, we can appeal to the central-limit theorem (Sec. 8.3.4) to assert that the likelihood is multivariate normal. In Sec. 12.3.3, we developed expressions for the mean of $\mathbf{V}$ and its covariance matrix for fixed $\mathbf{\mathfrak{r}}_{int}$ and $\mathcal{E}_0$, so we can write the likelihood for 3D position estimation as

$$\mathrm{pr}(\mathbf{V}|\mathbf{\mathfrak{r}}_{int}, \mathcal{E}_0) = (2\pi)^{-\frac{1}{2}M}[\det \mathbf{K}_V(\mathbf{\mathfrak{r}}_{int}, \mathcal{E}_0)]^{-\frac{1}{2}}$$

$$\times \exp\left\{-\tfrac{1}{2}[\mathbf{V} - \overline{\mathbf{V}}(\mathbf{\mathfrak{r}}_{int}, \mathcal{E}_0)]^t\, \mathbf{K}_V^{-1}(\mathbf{\mathfrak{r}}_{int}, \mathcal{E}_0)[\mathbf{V} - \overline{\mathbf{V}}(\mathbf{\mathfrak{r}}_{int}, \mathcal{E}_0)]\right\}. \qquad (12.261)$$

ML estimation requires maximization of $\mathrm{pr}(\mathbf{V}|\mathbf{\mathfrak{r}}_{int}, \mathcal{E}_0)$ with respect to the unknown parameters $\mathbf{\mathfrak{r}}_{int}$ and $\mathcal{E}_0$. Equivalently, we can maximize the log-likelihood, requiring that

$$\log[\mathrm{pr}(\mathbf{V}|\mathbf{\mathfrak{r}}_{int}, \mathcal{E}_0)] = \max \text{ at } \mathbf{\mathfrak{r}}_{int} = \hat{\mathbf{\mathfrak{r}}}_{ML} \text{ and } \mathcal{E}_0 = \hat{\mathcal{E}}_{ML}\,. \qquad (12.262)$$

Explicitly,

$$\log[\mathrm{pr}(\mathbf{V}|\mathbf{\mathfrak{r}}_{int}, \mathcal{E}_0)]$$

$$= -\tfrac{1}{2}\log\{\det[\mathbf{K}_V(\mathbf{\mathfrak{r}}_{int}, \mathcal{E}_0)]\} - \tfrac{1}{2}[\mathbf{V} - \overline{\mathbf{V}}(\mathbf{\mathfrak{r}}_{int}, \mathcal{E}_0)]^t\, \mathbf{K}_V^{-1}(\mathbf{\mathfrak{r}}_{int}, \mathcal{E}_0)[\mathbf{V} - \overline{\mathbf{V}}(\mathbf{\mathfrak{r}}_{int}, \mathcal{E}_0)]\,. \qquad (12.263)$$

Often we can assume that $\log\{\det[\mathbf{K}_V(\mathbf{\mathfrak{r}}_{int}, \mathcal{E}_0)]\}$ is a slowly varying function of its arguments and treat it as approximately a constant. In that case, maximizing the log-likelihood is equivalent (because of the minus sign) to minimizing the quadratic form in (12.263):

$$[\mathbf{V}-\overline{\mathbf{V}}(\mathbf{r}_{int},\mathcal{E}_0)]^t\,\mathbf{K}_V^{-1}(\mathbf{r}_{int},\mathcal{E}_0)[\mathbf{V}-\overline{\mathbf{V}}(\mathbf{r}_{int},\mathcal{E}_0)] = \min \text{ at } \mathbf{r}_{int} = \hat{\mathbf{r}}_{ML} \text{ and } \mathcal{E}_0 = \hat{\mathcal{E}}_{ML}\,. \tag{12.264}$$

Since $\overline{\mathbf{V}}(\mathbf{r}_{int},\mathcal{E}_0)$ and $\mathbf{K}_V^{-1}(\mathbf{r}_{int},\mathcal{E}_0)$ are nonlinear functions of their arguments, (12.264) states a nonlinear least-squares problem. Many methods of solution exist, but all amount to searching systematically through the 4D space defined by $\mathcal{E}_0$ and the three components of $\mathbf{r}_{int}$. The linear estimators discussed above are useful starting points for the search.

ML estimation in 2D If we want to know only the 2D interaction position $\mathbf{r}_{int}$, then the depth of interaction z_{int} is a *nuisance parameter*, a topic to be discussed in Sec. 13.3.8. As we shall see in that section, there are three general ways of dealing with a nuisance parameter: we can estimate it along with the parameters of interest; we can assign it some value *a priori*, or we can marginalize the likelihood over the nuisance parameter and then do ML estimation solely on the parameters of interest. Under very general assumptions, the marginalization approach is optimal, but the other two approaches are often computationally easier.

We have already seen how to estimate the nuisance parameter z_{int} when $\mathbf{r}_{int}$ and $\mathcal{E}_0$ are the parameters of interest. We simply solve the nonlinear least-squares problem (12.264) and discard the estimate of z_{int}. The main drawback to this approach is that the search is over a 4D space when we are interested in only a 3D one.

If we want to assign a typical value to z_{int}, one obvious choice is the mean $\overline{z}_{int}$ as computed from (12.164); it varies from α_{tot}^{-1} to $\frac{1}{2}L_z$ depending on the value of $\alpha_{tot}L_z$. With the substitution $\mathbf{r}_{int} \rightarrow (\mathbf{r}_{int}, \overline{z}_{int})$, the nonlinear least-squares problem in (12.264) is again solved by a 3D search.

The final approach is true 2D ML position estimation, where the likelihood is

$$\mathrm{pr}(\mathbf{V}|\mathbf{r}_{int},\mathcal{E}_0) = \int_0^{L_z} dz_{int}\; \mathrm{pr}(\mathbf{V}|\mathbf{r}_{int},\mathcal{E}_0)\,\mathrm{pr}(z_{int})\,, \tag{12.265}$$

where $\mathrm{pr}(z_{int})$ is given in (12.164). We can no longer argue that $\mathrm{pr}(\mathbf{V}|\mathbf{r}_{int},\mathcal{E}_0)$ is normal; instead, it is essentially a multivariate pulse-height spectrum. The marginal $\mathrm{pr}(V_m|\mathbf{r}_{int},\mathcal{E}_0)$ is precisely the pulse-height spectrum that would be observed on the mth electrode if the detector were illuminated at normal incidence with photons of energy $\mathcal{E}_0$ at the point $\mathbf{r}_{int}$, and we know from inspection of Figs. 12.16 – 12.18 that pulse-height spectra are far from normal.

ML estimation is not the same as nonlinear least-squares in this case, but we can still maximize (12.265) by means of a 3D search. Evaluation of the likelihood at each step in the search requires numerical integration over z_{int}, but in practice this integral might be approximated by a sum with relatively few terms.

12.3.5 Scintillation cameras

All detectors for x rays and gamma rays operate by converting an absorbed photon into charge. In semiconductor detectors, this charge is sensed directly, but in scintillation detectors the charge is converted to light, and the light is then sensed by optical detectors such as photomultipliers or photodiodes. If multiple optical detectors are used to provide spatial information, and if their temporal response is sufficient to resolve individual pulses from each absorbed gamma ray, then the detector is called a *scintillation camera.*

The basic geometry of a scintillation camera is often quite similar to that of a semiconductor array: a slab crystal absorbs a high-energy photon and produces light, the light spreads out as it propagates, and many optical detectors receive light from an absorption event. The analogy should be clear — light in a scintillation camera plays the role of charge, and optical detectors substitute for electrodes.

In one respect, however, scintillation detectors are simpler than semiconductor detectors: an optical photon causes no response on an optical detector until it actually reaches the detector surface, while a moving charge induces an output signal without reaching the electrode. Optical photons can be absorbed, which is analogous to charge trapping, but the absorbed photons have no effect on the output signals.

Anger camera The most common detector in nuclear medicine is the *Anger camera* (Anger, 1958), illustrated in Fig. 12.27. It usually consists of a large single crystal of sodium iodide (NaI) doped with thallium (Tl), an optical window and an array of photomultiplier tubes (PMTs). In Anger's original design, seven PMTs were used in a hexagonal configuration, but modern cameras use up to 100 PMTs and cover a field of view of up to a half meter. Many different configurations of PMTs have been tried.

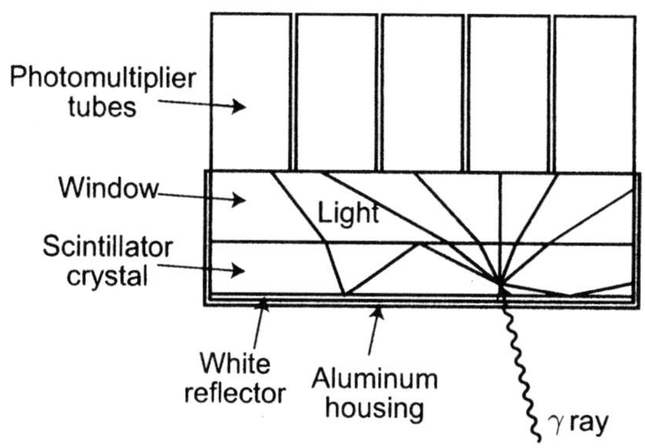

Fig. 12.27 Schematic of an Anger scintillation camera.

In NaI(Tl), an optical photon of energy around 3 eV is produced for every 30 eV of gamma-ray energy on average, so a 150 keV gamma-ray photon produces about 5000 optical photons. The optical window allows the optical photons to spread out over several PMTs, and typically the PMT nearest the interaction point might receive only 500 – 1000 photons. The photons that reach the photocathode of a particular PMT produce photoelectrons with some quantum efficiency η, usually below 0.3. The PMT amplifies this weak pulse of photoelectrons by about a factor of 10^6, and an easily measurable output current pulse is produced. All PMTs that receive light produce pulses simultaneously, and the heights of the pulses are measured and used to estimate the interaction position and energy of the gamma-ray photon. The accuracy of these estimates is determined by the statistics of the PMT signals.

The two major noise sources that affect a PMT signal are the random number of photoelectrons produced at each photocathode and the randomness in the gain process. Because the output of the PMT is large, noise in the subsequent electronics can usually be neglected.

Dark current is also negligible, for two reasons. First, the scintillation light from NaI(Tl) is in the blue and near-ultraviolet portion of the spectrum, so there is no need for the PMT to respond to longer wavelengths. Therefore, a photocathode with a large work function can be used, and few thermal electrons can overcome the potential barrier at the photocathode surface. Second, even if there were a dark current, only the electrons during a gamma-ray pulse would have any effect; thermal electrons between pulses would not trigger the pulse circuitry.

Photoelectron statistics If N_m photoelectrons are produced at the mth photocathode by a single gamma-ray interaction, we want to compute the mean $\overline{N}_m$ and the covariance matrix $\mathbf{K}_N$. Both of these quantities depend on $\mathbf{r}_{int}$, the 3D position of the interaction point in the scintillation crystal. In addition, both depend on the energy deposited in the crystal and its spatial distribution; for simplicity, we shall consider only photopeak events and neglect nonlocal charge deposition, so the photon energy $\mathcal{E}_0$ is assumed to be deposited at $\mathbf{r}_{int}$.

We denote the mean number of optical photons as $\overline{N}_{opt}(\mathcal{E}_0)$ (independent of $\mathbf{r}_{int}$ for a homogeneous material) and the average fraction of those that reach the mth photocathode as $\beta_m(\mathbf{r}_{int})$. To a first approximation, $\beta_m(\mathbf{r}_{int})$ is $\Omega_m/4\pi$, where Ω_m is the solid angle subtended by the photocathode from point $\mathbf{r}_{int}$, but a practical Anger camera includes various reflecting surfaces to increase the light collection, so actual computation of $\beta_m(\mathbf{r}_{int})$ is complicated. We shall simply treat $\beta_m(\mathbf{r}_{int})$ as a known function here. The mean number of photoelectrons produced in the mth photocathode, $\overline{N}_m(\mathbf{r}_{int}, \mathcal{E}_0)$, is then the mean number of optical photons that reach the cathode times the quantum efficiency η, or $\overline{N}_{opt}(\mathcal{E}_0)\,\eta\beta_m(\mathbf{r}_{int})$.

We have already done most of the work necessary to determine the covariance matrix $\mathbf{K}_N(\mathbf{r}_{int})$. In Sec. 12.3.3 we considered the analogous problem for semiconductor detectors without trapping, and we computed the covariance matrix (12.244) for the number of electrons that reach each pixel. To do so, we had to neglect Coulomb interactions between electrons, so that each electron would choose a pixel independently of the others. Photons do not have Coulomb interactions in the first place, so this assumption is rigorously satisfied in scintillation detectors.

All of the steps leading up to (12.244) are then valid for the scintillation camera as long as we replace $\beta_m(\mathbf{r}_{int})$ with $\eta\beta_m(\mathbf{r}_{int})$, and the resulting covariance matrix is[17]

$$[\mathbf{K}_N(\mathbf{r}_{int}, \mathcal{E}_0)]_{mm'} = \overline{N}_{opt}(\mathcal{E}_0)\,\eta\beta_m(\mathbf{r}_{int})\,\delta_{mm'} + (F-1)\,\overline{N}_{opt}(\mathcal{E}_0)\,\eta^2\beta_m(\mathbf{r}_{int})\,\beta_{m'}(\mathbf{r}_{int}), \tag{12.266}$$

where we have assumed that η is the same for all PMTs.

For several reasons, the off-diagonal elements of $\mathbf{K}_N(\mathbf{r}_{int}, \mathcal{E}_0)$ are less important here than in semiconductor detectors. First, because of the extra conversion

[17]The reader may worry about the requirement in the derivation of (12.244) that the β_m factors sum to unity, which is what led to the use of the multinomial law. This requirement is not met in scintillation cameras because light may be absorbed and because $\eta \neq 1$. We can, however, remove the restriction by supposing that there is an unobserved light sink in which a fraction β_0 of the optical photons are collected, where $\beta_0 = 1 - \sum_{m=1}^{M} \eta\beta_m$. The conditional multinomial law then holds if all $M+1$ locations (the M actual detectors and the light sink) are considered, but only the actual detectors are included in the $M \times M$ covariance matrix.

step (charge to light), scintillators are less efficient than semiconductors, and we argued in Sec. 12.3.2 that lower efficiency implies a probability law on N_{opt} that is more nearly Poisson and hence a Fano factor nearer to unity. For $F = 1$, the off-diagonal part of $\mathbf{K}_N(\mathbf{r}_{int}, \mathcal{E}_0)$ vanishes, basically because the distribution of optical photons among the PMTs is a multinomial selection of a Poisson (see Sec. 11.2.1).

The off-diagonal elements are also relatively unimportant, even if $F \neq 1$, because the quantum efficiency η and the collection efficiency β_m are both small. From (12.266) we see that the off-diagonal terms scale as η^2 while the diagonal ones scale as η, so there is a factor of η reducing the off-diagonal elements relative to the diagonal ones. Similarly, the off-diagonal terms involve the product $\beta_m(\mathbf{r}_{int})\,\beta_{m'}(\mathbf{r}_{int})$ while the diagonal terms are linear in $\beta_m(\mathbf{r}_{int})$. As we noted above, η is about 0.3 and $\beta_m(\mathbf{r}_{int})$ is about $0.1 - 0.2$ for the PMT nearest the interaction point, and much less for more distant ones. Thus the event that a photoelectron is produced on a particular photocathode is rare compared to the emission of optical photons at the interaction point, and we know from Chap. 11 that rarity implies Poisson implies independent.

An excellent approximation to (12.266) is therefore

$$[\mathbf{K}_N(\mathbf{r}_{int}, \mathcal{E}_0)]_{mm'} = \overline{N}_m(\mathbf{r}_{int}, \mathcal{E}_0)\,\delta_{mm'}\,, \tag{12.267}$$

where $\overline{N}_m(\mathbf{r}_{int}, \mathcal{E}_0) = \overline{N}_{opt}(\mathcal{E}_0)\eta\beta_m(\mathbf{r}_{int})$.

Another approximation that is usually justified in Anger cameras is that $\mathbf{K}_N(\mathbf{r}_{int}, \mathcal{E}_0)$ and $\overline{N}_m(\mathbf{r}_{int}, \mathcal{E}_0)$ are independent of the depth of interaction z_{int}. A practical camera will have a reflector on the entrance surface $z = 0$, and the amount of light reaching a particular photocathode is relatively independent of how far from the reflector the light is produced. Without the reflector, the subtended solid angle would depend on z_{int}, but the reflected light compensates for this effect. Thus we can often use (12.267) with $\overline{N}_m(\mathbf{r}_{int}, \mathcal{E}_0)$ replaced by $\overline{N}_m(\mathbf{r}_{int}, \mathcal{E}_0)$.

Statistics of the PMT outputs Knowing the statistics of N_m, we next want to study the statistics of K_m, the number of electrons produced on the output of the mth PMT in a single event. The theory needed for this purpose was developed in Sec. 11.4.1.

Computation of the mean of K_m is little more than a matter of definition. The gain G of the PMT is defined as the average number of output electrons per electron emitted from the photocathode. It is reasonable to assume that all electrons are amplified independently, so the PMT is a linear detector, and the mean total number of photoelectrons in the mth PMT for a single event is given by[18]

$$\overline{K}_m(\mathbf{r}_{int}, \mathcal{E}_0) = G\overline{N}_m(\mathbf{r}_{int}, \mathcal{E}_0)\,. \tag{12.268}$$

In the notation of Sec. 11.4.1, G is the same as m_1, the first moment of the gain distribution (normalized such that $m_0 = 1$). If we know m_1 and the corresponding second moment m_2, we can express the variance of K_m by the Burgess variance theorem, (11.182). Moreover, if N_m is approximately Poisson, as we have argued above, then we can use (11.183) and write

$$\mathrm{Var}\{K_m\} = \overline{N}_m m_2 = G\overline{K}_m \frac{m_2}{G^2} = G\overline{K}_m \frac{m_2}{m_1^2}\,. \tag{12.269}$$

[18]Do not confuse K_m or $\overline{K}_m$ with a covariance matrix, which we denote $\mathbf{K}$ with some subscript. We use K here as the number of electrons on the PMT output for consistency with Sec. 11.4.1.

As noted in Sec. 11.4.1, the ratio m_1^2/m_2 is called the Swank factor, and we shall denote it here as s. For typical PMTs, s is about $0.8-0.9$.

The PMTs act independently, so there is no correlation in their outputs if there is none in their inputs. Thus, if the input covariance is given by (12.267), the output covariance is

$$[\mathbf{K}_k(\mathbf{r}_{int})]_{mm'} = \frac{G}{s}\overline{K}_m(\mathbf{r}_{int}, \mathcal{E}_0)\,\delta_{mm'}\,. \tag{12.270}$$

Finally, if the output pulse is integrated with a simple RC integrator with time constant long compared to the pulse duration, then the mean pulse height (for photopeak events) is

$$\overline{V}_m(\mathbf{r}_{int}, \mathcal{E}_0) = \frac{e}{C}\overline{K}_m(\mathbf{r}_{int}, \mathcal{E}_0) = \frac{e}{C}G\overline{N}_{opt}(\mathcal{E}_0)\,\eta\beta_m(\mathbf{r}_{int})\,, \tag{12.271}$$

and the covariance matrix on the pulse heights is [*cf.* (12.245)]

$$[\mathbf{K}_V(\mathbf{r}_{int}, \mathcal{E}_0)]_{mm'} = \frac{e^2}{C^2}\frac{G}{s}\overline{K}_m(\mathbf{r}_{int}, \mathcal{E}_0)\,\delta_{mm'} = \frac{e}{C}\frac{G}{s}\overline{V}_m(\mathbf{r}_{int}, \mathcal{E}_0)\,\delta_{mm'}\,. \tag{12.272}$$

Next, we use these results in estimating the interaction position and the energy.

12.3.6 Position and energy estimation with scintillation cameras

In Sec. 12.3.4 we discussed position and energy estimation in the context of semiconductor detectors, but in fact most of the methods introduced there were originally developed for scintillation detectors. A quasilinear estimator was pioneered by Anger in the mid-1950s, and it has been widely emulated and embellished since. ML position estimation was first suggested for scintillation cameras by Gray and Macovski (1976), and it was implemented in practical cameras at the University of Michigan (Clinthorne *et al.*, 1987) and the University of Arizona (Milster *et al.* 1984, 1985, 1990). As in the semiconductor case, we prefer the ML approach because of its optimal bias and variance properties.

Statistical model To do ML estimation, we need a probability law, and it is much easier to find one for scintillation cameras than for semiconductors since there are fewer random processes. In semiconductors, as we saw, we have to account for random trapping, which does not arise in scintillators, and dark current, which is usually negligible for the PMTs commonly used in scintillation cameras. In both cases, depth of interaction is random, but the effect is exacerbated in semiconductors because of the trapping and ameliorated in scintillators by the reflecting surfaces.

The dominant stochastic effect in a scintillation camera is fluctuations in the number of photoelectrons. Because the output voltage is the sum of independent contributions from each photoelectron, the central-limit theorem applies, and we can assume that the likelihood $\mathrm{pr}(\mathbf{V}|\mathbf{r}_{int}, \mathcal{E}_0)$ is a multivariate normal with mean and covariance given by (12.271) and (12.272), respectively. Several simplifications of these expressions will serve us well.

First, since the signals are only weakly dependent on the depth of interaction z_{int}, we can replace $\mathbf{r}_{int}$ with $\mathbf{r}_{int}$. Second, as with semiconductors, the mean signals from scintillators are linear in $\mathcal{E}_0$ to a good approximation, so (12.271) has the form

$$\overline{V}_m(\mathbf{r}_{int}, \mathcal{E}_0) = \mathcal{E}_0 f_m(\mathbf{r}_{int})\,. \tag{12.273}$$

The function $f_m(\mathbf{r}_{int})$ is just some constant times $\beta_m(\mathbf{r}_{int})$; it can be determined by calibration measurements or by modeling of the light propagation in the camera.

The third simplification is the diagonal form of the covariance in (12.272), which we can rewrite as

$$[\mathbf{K}_V(\mathbf{r}_{int}, \mathcal{E}_0)]_{mm'} = A\mathcal{E}_0 f_m(\mathbf{r}_{int})\,\delta_{mm'}\,, \tag{12.274}$$

where A is a constant that will turn out not to be very important.

Log-likelihood We argued above that $\text{pr}(\mathbf{V}|\mathbf{r}_{int}, \mathcal{E}_0)$ is a multivariate normal, and we now have the simplified expressions (12.273) and (12.274) for the mean and covariance, respectively. Since the covariance is diagonal and none of the elements is identically zero, it is easy to compute the inverse covariance needed in the multivariate normal PDF. Also, the determinant of a diagonal matrix, which is needed in the normalizing factor of a multivariate normal, is just the product of the diagonal elements. Thus the log-likelihood for estimation of 2D position and energy takes the form [*cf.* (12.263)]

$$\begin{aligned}
&\log[\text{pr}(\mathbf{V}|\mathbf{r}_{int}, \mathcal{E}_0)] \\
&= -\tfrac{1}{2}\log\{2\pi \det[\mathbf{K}_V(\mathbf{r}_{int}, \mathcal{E}_0)]\} - \tfrac{1}{2}[\mathbf{V} - \overline{\mathbf{V}}(\mathbf{r}_{int}, \mathcal{E}_0)]^t \mathbf{K}_V^{-1}(\mathbf{r}_{int}, \mathcal{E}_0)[\mathbf{V} - \overline{\mathbf{V}}(\mathbf{r}_{int}, \mathcal{E}_0)] \\
&= -\frac{1}{2}\sum_{m=1}^{M} \log[2\pi A\mathcal{E}_0 f_m(\mathbf{r}_{int})] - \frac{1}{2}\sum_{m=1}^{M} \frac{[V_m - \mathcal{E}_0 f_m(\mathbf{r}_{int})]^2}{A\mathcal{E}_0 f_m(\mathbf{r}_{int})}\,.
\end{aligned} \tag{12.275}$$

ML estimation of 2D position and energy The ML estimates of position and energy are found by searching for the maximum of the log-likelihood over the three parameters x_{int}, y_{int} and $\mathcal{E}_0$. If we neglect the slow dependence of the log-determinant on these parameters, as we did in Sec. 12.3.4, then the maximum of the log likelihood occurs when the quadratic form in the exponent is minimized. Thus we have the nonlinear least-squares problem [*cf.* (12.264)]:

$$\sum_{m=1}^{M} \frac{[V_m - \mathcal{E}_0 f_m(\mathbf{r}_{int})]^2}{\mathcal{E}_0 f_m(\mathbf{r}_{int})} = \min \text{ at } \mathbf{r}_{int} = \hat{\mathbf{r}}_{ML} \text{ and } \mathcal{E}_0 = \hat{\mathcal{E}}_{ML}\,. \tag{12.276}$$

Note that the constant A has disappeared; the same minimum will be found for all A.

The similarity in form between (12.276) and (12.264) should not conceal the fact that the latter was for 3D position estimation in semiconductor detectors while the former is for 2D estimation in scintillation cameras. We got to (12.276) by assuming that the random depth of interaction in a scintillation camera had no appreciable effect on the mean PMT signals. In a semiconductor material with trapping, on the other hand, depth-of-interaction effects dominate the pulse-height spectrum (see Sec. 12.3.2), so the only way we could get to a simple least-squares problem was to estimate the depth along with the lateral coordinates.

12.3.7 Imaging characteristics of photon-counting detectors

From the position and energy estimates, derived in Sec. 12.3.4 for semiconductor detectors and 12.3.6 for scintillation cameras, we need to form an image. The most common way to do so is to apply some window test, as discussed in Sec. 12.3.2, and then to bin the position estimates for the events that pass the test into a pixel array. Intuitively, there is some loss of information in this operation, both because photons that do not pass the window test are simply rejected, even though they might convey useful information about the object being imaged, and because the finite bin width adds further uncertainty to the position estimates.

In principle, this information loss could be avoided by storing the raw position and energy estimates or, equivalently, forming from them a random point process. In this section we shall first discuss the properties of this point process, building on the theory developed in Chap. 11, and then we shall discuss the effects of energy windowing and spatial binning.

This treatment is largely motivated by gamma-ray detectors as used in nuclear medicine, but there are position-sensitive optical detectors that fit into a similar framework (except that no energy estimation is involved).

Point processes Suppose we have observed a set of J events and estimated that the jth event occurred at the 2D interaction position $\hat{\mathbf{r}}_j$ and that it deposited energy $\hat{\mathcal{E}}_j$ there. We can then construct the spatio-spectral point process (see Sec. 11.3.8)

$$g_{det}(\hat{\mathbf{r}}, \hat{\mathcal{E}}) \equiv \sum_{j=1}^{J} \delta(\hat{\mathbf{r}} - \hat{\mathbf{r}}_j)\, \delta(\hat{\mathcal{E}} - \hat{\mathcal{E}}_j)\,, \tag{12.277}$$

where the subscript *det* indicates that we are dealing with the detected events. We shall refer to $g_{det}(\hat{\mathbf{r}}, \hat{\mathcal{E}})$ as the *detected image*, with the understanding that the detection process includes estimation of position and energy.

The statistics of the detected image follow from the theory of Poisson random processes developed in Sec. 11.3. In that section we discussed the conditions needed for a point process to be a Poisson process. In essence, these conditions amount to saying that the events are independent. We saw that randomness in the source configuration or source strength could spoil the independence, but let us assume here that the source is nonrandom, so the incident photons are independent and satisfy the conditions for a Poisson point process. In that case $g_{det}(\hat{\mathbf{r}}, \hat{\mathcal{E}})$ is also a Poisson point process since the actions of detecting the interaction events and estimating their position and energy do not introduce any dependence or otherwise invalidate the Poisson model. As we know from Sec. 11.1.3, detection is a binomial selection process, which preserves the Poisson character, and the estimation step is a random displacement of each point, which also preserves the Poisson character [see Sec. 11.4.3, especially (11.229)].

Mean detected image The mean of $g_{det}(\hat{\mathbf{r}}, \hat{\mathcal{E}})$ is the *spatio-spectral fluence* (or *spectral photon fluence*) $b_{det}(\hat{\mathbf{r}}, \hat{\mathcal{E}})$, defined as the mean number of photons per unit area per unit energy in the detected image. Thus

$$\langle g_{det}(\hat{\mathbf{r}}, \hat{\mathcal{E}}) \rangle \equiv b_{det}(\hat{\mathbf{r}}, \hat{\mathcal{E}})\,. \tag{12.278}$$

By an extension of the discussion of spatial Poisson processes in Sec. 11.3.2, however, we know that $b_{det}(\hat{\mathbf{r}}, \hat{\mathcal{E}})$ also has another interpretation: after proper normalization, it is the probability density on the position of any individual count. Specifically, the PDF for recording a count at $\hat{\mathbf{r}}$ and $\hat{\mathcal{E}}$ is [*cf.* (11.76)]

$$\mathrm{pr}(\hat{\mathbf{r}}, \hat{\mathcal{E}}) = \frac{b_{det}(\hat{\mathbf{r}}, \hat{\mathcal{E}})}{\int_\infty d^2r \int_0^\infty d\mathcal{E}\; b_{det}(\mathbf{r}, \mathcal{E})} = \frac{1}{\bar{J}}\, b_{det}(\hat{\mathbf{r}}, \hat{\mathcal{E}})\,, \tag{12.279}$$

where $\overline{J}$ is the mean total number of detected counts.

We can now rewrite (12.278) as

$$\langle g_{det}(\hat{\mathbf{r}}, \hat{\mathcal{E}}) \rangle = \overline{J}\, \mathrm{pr}(\hat{\mathbf{r}}, \hat{\mathcal{E}}) = \overline{J} \int_{\infty} d^2r \int_0^{\infty} d\mathcal{E}\ \mathrm{pr}(\hat{\mathbf{r}}, \hat{\mathcal{E}}|\mathbf{r}, \mathcal{E})\, \mathrm{pr}(\mathbf{r}, \mathcal{E}) \,, \tag{12.280}$$

where $\mathrm{pr}(\hat{\mathbf{r}}, \hat{\mathcal{E}}|\mathbf{r}, \mathcal{E})$ is the PDF for obtaining estimates $(\hat{\mathbf{r}}, \hat{\mathcal{E}})$ when the event actually occurred at $(\mathbf{r}, \mathcal{E})$; in other words, it is the spatio-spectral point response function.

To put this result into a more familiar form, suppose that the incident photons are described by a spectral fluence $b(\mathbf{r}, \mathcal{E})$ and that the detector has a quantum efficiency of η for all $\mathbf{r}$ and $\mathcal{E}$. Suppose also that the photons strike the detector at normal incidence so that we do not need to distinguish the lateral coordinates of the interaction point, $\mathbf{r}_{int}$, from the 2D vector that appears in the fluence. Then, by analogy to (12.279), we see that

$$\mathrm{pr}(\mathbf{r}, \mathcal{E}) = \frac{\eta}{\overline{J}}\, b(\mathbf{r}, \mathcal{E}) \,, \tag{12.281}$$

where $\overline{J}/\eta$ is the mean number of incident photons. Combining (12.280) and (12.281), we find

$$\langle g_{det}(\hat{\mathbf{r}}, \hat{\mathcal{E}}) \rangle = \eta \int_{\infty} d^2r \int_0^{\infty} d\mathcal{E}\ \mathrm{pr}(\hat{\mathbf{r}}, \hat{\mathcal{E}}|\mathbf{r}, \mathcal{E})\, b(\mathbf{r}, \mathcal{E}) \,. \tag{12.282}$$

Unsurprisingly, the mean detected image is the incident spectral fluence blurred by a spatio-spectral point response function and multiplied by the quantum efficiency.

We had, in fact, obtained essentially this same result earlier in Sec. 11.4.3, where we considered gain processes where the secondaries are randomly displaced from the primary photons. A special case considered in that section was mislocation without gain, where each primary produces exactly one secondary; this is just what happens in position and energy estimation, where one absorbed photon results in one point in the detected image.

Bias and variance; distortion and spatial resolution Perhaps the only new insight in (12.282) is that the point response function is also the probability density function associated with the estimation procedure. In fact, we can relate this function to the bias and variance of the estimate as defined in Sec. 13.3.1. To see this relation more clearly, let us ignore the energy estimation and write the PRF as $\mathrm{pr}(\hat{\mathbf{r}}|\mathbf{r})$. From this density we can compute the bias and variance of the position estimate via (13.276) and (13.279).

The bias of the position estimate is a vector defined by

$$\mathbf{B}(\mathbf{r}) \equiv E\{\hat{\mathbf{r}}|\mathbf{r}\} - \mathbf{r} \,, \tag{12.283}$$

where the conditional expectation is computed with the density $\mathrm{pr}(\hat{\mathbf{r}}|\mathbf{r})$. Note that this bias can, in general, depend on the true position $\mathbf{r}$.

Since $\hat{\mathbf{r}}$ is a 2D random vector, its second-order statistics are specified by a 2×2 covariance matrix, but we focus here on the two diagonal elements, the variances of $\hat{x}$ and $\hat{y}$. The variance of the x estimate is defined by

$$\mathrm{Var}\{\hat{x}|\mathbf{r}\} \equiv E\{\hat{x}^2|\mathbf{r}\} - [E\{\hat{x}|\mathbf{r}\}]^2 \,, \tag{12.284}$$

and similarly for the y estimate. Note especially that the variance of $\hat{x}$ measures the spread around the mean estimate, not around the actual x.

The bias and variance of the 2D position estimates can be measured by using a thin beam of gamma rays, all of which strike the detector at normal incidence very near point $\mathbf{r}$. After a large number of gamma rays from this source have been detected, the resulting image (see Fig. 12.28) is a good representation of the PDF on $\hat{\mathbf{r}}$ conditional on $\mathbf{r}$. As the number of collected photons approaches infinity, the sample mean of $\hat{x} - x$ approaches the bias component $B_x(\mathbf{r})$, and similarly for the y component. The sample variances also approach the ensemble variances defined in (12.284). The off-diagonal terms manifest themselves as a tilt of the distribution.

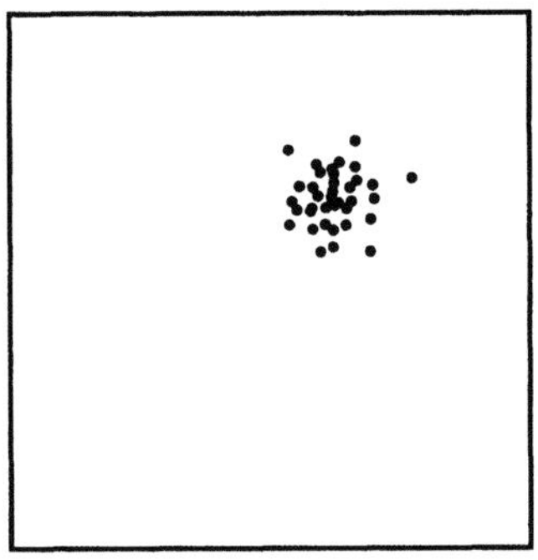

Fig. 12.28 Image of many gamma rays, all incident at the same position. As the number of gamma rays approaches infinity, this image approaches the conditional PDF $\text{pr}(\mathbf{r}|\hat{\mathbf{r}})$.

We can relate the biases and variances back to the general descriptions of shift-variant imaging systems introduced in Sec. 7.2. The bias measures the distortion of the image detector, and the variances measure the spatial resolution (see Sec. 7.2.1). If we repeated the measurement described above with a uniformly spaced grid of true interaction positions $\mathbf{r}$, the position estimates would cluster around a distorted grid, and both the displacement of the estimates from the true grid (distortion) and the spread of the estimates (variance) could vary over the surface of the detector, in general.

Contributions to bias and variance Bias and variance in any estimate can arise from several sources, including modeling error, inaccurate or incomplete system calibration, statistical noise in the data or suboptimal estimators.

In the present problem, an example of modeling error is neglecting the depth of interaction. When we assume that the mean and covariance of the PMT signals are independent of z_{int}, we make an error that can affect both the bias and the variance. The magnitude of this error can be determined only by doing more careful modeling.

System calibration comes into our problem since we need to know the functions $f_m(\mathbf{r}_{int})$ defined in (12.273). To calibrate the camera, we can measure these functions for a finite grid of interaction points and interpolate between grid points, but errors of measurement and interpolation again affect bias and variance.

Finally, even with full knowledge of the system, there may be bias and variance associated with the estimator. As discussed more fully in Sec. 13.3.5, there is a minimum variance, called the Cramér-Rao bound, that can be attained by any estimator operating on a specific data set. An estimator that achieves this lower bound is said to be *efficient*. An efficient estimator may not exist for many problems, but if one does exist, it is the ML estimator.

Moreover, the ML estimator is always asymptotically unbiased and asymptotically efficient. This statement is usually applied to situations where N independent measurements are made for the same value of the unknown parameter, and the term asymptotic implies the limit as $N \to \infty$. In a scintillation camera, we get only a single light flash for each $\mathbf{r}_{int}$ and $\mathcal{E}$ we want to estimate, but nevertheless an asymptotic argument can be applied. Specifically, if the number of optical photons gets very large, the ML estimate is again asymptotically unbiased and efficient, which means that it will give better spatial resolution and less distortion than any other estimator in this limit.

Autocovariance of the detected image Now let us assume that we have accounted for the modeling and calibration errors and the estimator performance, so that we know the PDF $\mathrm{pr}(\hat{\mathbf{r}}, \hat{\mathcal{E}}|\mathbf{r}, \mathcal{E})$. Then, for a nonrandom incident spectral fluence, we can compute the mean detected image, or equivalently the detected spectral fluence $b_{det}(\hat{\mathbf{r}}, \hat{\mathcal{E}})$, via (12.282). We argued above that the detected image is a Poisson random process, so we can use a generalization of (11.94) to express its autocovariance function as

$$K_{g_{det}}(\hat{\mathbf{r}}, \hat{\mathbf{r}}', \hat{\mathcal{E}}, \hat{\mathcal{E}}') = b_{det}(\hat{\mathbf{r}}, \hat{\mathcal{E}})\, \delta(\hat{\mathbf{r}} - \hat{\mathbf{r}}')\, \delta(\hat{\mathcal{E}} - \hat{\mathcal{E}}') . \tag{12.285}$$

Binning For storage or display of the image, we must bin the events into some digital matrix. We can, for example, apply an energy window and accept only those events for which $\hat{\mathcal{E}}$ falls in the window and then bin the remaining events into spatial pixels. Thus if $\hat{\mathbf{r}}$ falls in the region of the mth pixel, it contributes one count to that pixel in the final image. Alternatively, we can use multiple energy bins and make one spatial image for each energy.

In both of these cases, we can compute the statistics of the discrete image from what we know about the random process $g_{det}(\hat{\mathbf{r}}, \hat{\mathcal{E}})$. Suppose the mth pixel includes only those events for which $\mathcal{E}_m - \frac{1}{2}\Delta\mathcal{E} < \hat{\mathcal{E}} \leq \mathcal{E}_m + \frac{1}{2}\Delta\mathcal{E}$, $x_m - \frac{1}{2}\epsilon < \hat{x} \leq x_m + \frac{1}{2}\epsilon$ and $y_m - \frac{1}{2}\epsilon < \hat{y} \leq y_m + \frac{1}{2}\epsilon$. Then the number of counts g_m in this pixel is obtained by integrating $g_{det}(\hat{\mathbf{r}}, \hat{\mathcal{E}})$, as defined in (12.277), over this region:

$$g_m = \int_{\mathcal{E}_m - \frac{1}{2}\Delta\mathcal{E}}^{\mathcal{E}_m + \frac{1}{2}\Delta\mathcal{E}} d\hat{\mathcal{E}} \int_{x_m - \frac{1}{2}\epsilon}^{x_m + \frac{1}{2}\epsilon} d\hat{x} \int_{y_m - \frac{1}{2}\epsilon}^{y_m + \frac{1}{2}\epsilon} d\hat{y}\; g_{det}(\hat{\mathbf{r}}, \hat{\mathcal{E}}) . \tag{12.286}$$

From (12.282), the mean of g_m is

$$\begin{aligned}
\overline{g}_m &= \int_{\mathcal{E}_m - \frac{1}{2}\Delta\mathcal{E}}^{\mathcal{E}_m + \frac{1}{2}\Delta\mathcal{E}} d\hat{\mathcal{E}} \int_{x_m - \frac{1}{2}\epsilon}^{x_m + \frac{1}{2}\epsilon} d\hat{x} \int_{y_m - \frac{1}{2}\epsilon}^{y_m + \frac{1}{2}\epsilon} d\hat{y}\; \overline{g}_{det}(\hat{\mathbf{r}}, \hat{\mathcal{E}}) \\
&= \eta \int_{\mathcal{E}_m - \frac{1}{2}\Delta\mathcal{E}}^{\mathcal{E}_m + \frac{1}{2}\Delta\mathcal{E}} d\mathcal{E} \int_{x_m - \frac{1}{2}\epsilon}^{x_m + \frac{1}{2}\epsilon} d\hat{x} \int_{y_m - \frac{1}{2}\epsilon}^{y_m + \frac{1}{2}\epsilon} d\hat{y} \int_{\infty} d^2r \int_0^{\infty} d\mathcal{E}\; \mathrm{pr}(\hat{\mathbf{r}}, \hat{\mathcal{E}}|\mathbf{r}, \mathcal{E})\, b(\mathbf{r}, \mathcal{E}) \\
&\equiv \int_{\infty} d^2r \int_0^{\infty} d\mathcal{E}\, h_m(\mathbf{r}, \mathcal{E})\, b(\mathbf{r}, \mathcal{E}) ,
\end{aligned} \tag{12.287}$$

where $h_m(\mathbf{r}, \mathcal{E})$ is the overall kernel for the CD mapping from $b(\mathbf{r}, \mathcal{E})$ to the discrete vector $\mathbf{g}$. This kernel includes all of the complicated effects that go on in the detector material, position and energy estimation and binning into pixels and energy bins of finite size. At the end, we have our standard linear CD mapping formula.

Autocovariance of the discrete image Since we have argued that $g_{det}(\hat{\mathbf{r}}, \hat{\mathcal{E}})$ is a Poisson random process (for nonrandom fluence), counts in different spatial or energy bins must be independent Poisson random variables. Thus we have at once that

$$[\mathbf{K_g}]_{mm'} = \overline{g}_m \, \delta_{mm'} \,. \qquad (12.288)$$

The indices m and m' can refer to two different spatial pixels in the same energy image or to different energy images; so long as any event can go into only one bin and the fluence is nonrandom, the counts in different bins are independent.

We shall make considerable use of (12.288) in Chap. 14 when we discuss image quality and in Chap. 15 when we discuss inverse problems.

12.3.8 Integrating detectors

So far in this section we have discussed mainly photon-counting detectors, but both semiconductor and scintillator arrays can be operated in an integrating mode where no attempt is made to identify individual gamma rays or x rays. None of the discussion above on position and energy estimation is applicable in this case, and we must go back to the theory developed in Sec. 11.4 to determine the statistical properties of the images.

Scintillator-photodiode arrays For definiteness, we shall consider a scintillation detector with an array of photodiodes for the readout, but semiconductor detectors can be treated similarly. The geometry will be a slab of scintillation material with a regular array of photodiodes of width ϵ on the face $z = L_z$. The specific application of most interest for these detectors is transmission radiography using x rays with a broad spectrum of energies; for a more detailed discussion of this application, see Sec. 16.1.

Though the detector being considered here is qualitatively similar to a scintillation camera, two important differences must be noted. First, x-ray detectors often use fluorescent screens consisting of scintillating grains held together with a partially transparent binder. Scattering from binder-grain interfaces causes the light to diffuse significantly as it propagates from the interaction point to the photodetectors, and light can be absorbed in both the binder and the grains. As a result, depth of interaction is much more important here than it is with the single-crystal scintillators used in scintillation cameras.

Secondly, scintillation cameras are used mainly in nuclear medicine where the gamma-ray source is usually monoenergetic. In medical transmission radiography, on the other hand, the x rays are generated by allowing energetic electrons to strike a metal target, and most of the x rays are *Bremsstrahlung* (German: braking radiation) generated when the electrons are decelerated in the metal. This process produces a broad spectrum of x-ray energies extending from near zero up to the energy of the electron.

The light diffusion and the broad energy spectrum contribute to the variability in the amount of light reaching the photodetectors. Our goal in this section is to show how these effects can be incorporated into the derivation given in Sec. 11.4 of the mean vector and covariance matrix for the output signals after a finite integration time during which more than one x ray can be absorbed. The procedure will be first to compute the mean and autocovariance function of the output random process as in Sec. 11.4.3 and then to convert to the mean and covariance matrix of the discrete random vector as in Sec. 11.4.5. The reader may wish to review these sections before continuing.

Interaction parameters and the output random process In Sec. 11.4.3, the interaction position of the nth primary event was specified by the 2D vector $\mathbf{R}_n$, but for x rays interacting in a thick detector we must also specify the depth of interaction z_n and the energy deposited in the interaction, $\mathcal{E}_n$. We can incorporate these new parameters formally simply by regarding $\mathbf{R}_n$ as a 4D vector with components $(x_n, y_n, z_n, \mathcal{E}_n)$ or $(\mathbf{r}_n, z_n, \mathcal{E}_n)$.

The optical photons created in this interaction can propagate to the plane $z = L_z$, and there they define a 2D random point process $y(\mathbf{r})$ given by [*cf.* (11.200)]

$$y(\mathbf{r}) = \sum_{n=1}^{N_x} \sum_{k=1}^{k_n} \delta(\mathbf{r} - \mathbf{r}_{nk}), \tag{12.289}$$

where N_x is the number of absorbed x-ray photons, $\mathbf{r}_{nk}$ is the 2D position of the kth optical photon produced by the nth absorbed x ray, and k_n is the number of optical photons reaching $z = L_z$ for the nth absorption. The random variables in (12.289) are the sets $\{\mathbf{r}_{nk}\}$ and $\{k_n\}$ as well as N_x.

It is convenient to define a 2D displacement vector $\Delta\mathbf{r}_{nk}$ such that $\mathbf{r}_{nk} = \mathbf{r}_n + \Delta\mathbf{r}_{nk}$, and then we can write

$$y(\mathbf{r}) = \sum_{n=1}^{N_x} \sum_{k=1}^{k_n} \delta(\mathbf{r} - \mathbf{r}_n - \Delta\mathbf{r}_{nk}). \tag{12.290}$$

In contrast to (11.200), $\mathbf{r}_n$ rather than $\mathbf{R}_n$ appears here. (It would not make sense to subtract a 4D vector from a 2D one.)

Our task now is to compute the mean and autocovariance function of $y(\mathbf{r})$ by methods introduced in Sec. 11.4.3. This will turn out to be relatively easy since we took care in that section not to assume shift invariance. (Even if the fluorescent screen is laterally shift-invariant, it cannot have this property with respect to z_n or $\mathcal{E}_n$.)

Probability density for the interaction parameters To incorporate the new random variables z_n and $\mathcal{E}_n$ into the derivation of Sec. 11.4.3, we need a probability density function $\mathrm{pr}(\mathbf{R}_n)$, which we can write as

$$\mathrm{pr}(\mathbf{R}_n) = \mathrm{pr}(\mathbf{r}_n, z_n, \mathcal{E}_n) = \mathrm{pr}(z_n|\mathcal{E}_n, \mathbf{r}_n)\, \mathrm{pr}(\mathcal{E}_n|\mathbf{r}_n)\, \mathrm{pr}(\mathbf{r}_n). \tag{12.291}$$

To proceed, we need to make some assumptions about the x-ray beam and its interactions. The simplest form of $\mathrm{pr}(\mathbf{R}_n)$ will arise if we assume that all of the x rays are travelling parallel to the z axis, that all of the x-ray interactions are photoelectric and that all of the photon energy is deposited at the interaction site. From the discussion in Sec. 12.3.1, we know that even a photoelectric interaction can result in a secondary photon (a K x ray), but for present purposes we assume that it is reabsorbed a negligible distance from the initial interaction. The effects of nonlocal charge deposition will be discussed in more detail in Sec. 12.3.9.

With these restrictive assumptions, we can relate $\mathrm{pr}(\mathbf{r}_n)$ to the x-ray fluence $b_0(\mathbf{r})$ on the entrance face $z = 0$ by (11.206):

$$\mathrm{pr}(\mathbf{r}_n) = \frac{b_0(\mathbf{r}_n)}{\int_A d^2r\, b_0(\mathbf{r})}, \tag{12.292}$$

where $A = L_x L_y$ is the area of the detector.

In (12.291), $\mathrm{pr}(\mathcal{E}_n|\mathbf{r}_n)$ is the energy spectrum of the incident gamma rays, which is presumed to be known, and $\mathrm{pr}(z_n|\mathcal{E}_n, \mathbf{r}_n)$ is the PDF on interaction depths. If the properties of the fluorescent screen are independent of x and y, $\mathrm{pr}(z_n|\mathcal{E}_n, \mathbf{r}_n)$ is independent of $\mathbf{r}_n$ and given by (12.161) as

$$\mathrm{pr}(z_n|\mathcal{E}_n) = \alpha(\mathcal{E}_n) \exp[-\alpha(\mathcal{E}_n) z_n], \tag{12.293}$$

where $\alpha(\mathcal{E}_n)$ is the energy-dependent photoelectric absorption coefficient.

With all of the accumulated assumptions, the desired PDF is thus

$$\mathrm{pr}(\mathbf{R}_n) = \alpha(\mathcal{E}_n) \exp[-\alpha(\mathcal{E}_n) z_n]\, \mathrm{pr}(\mathcal{E}_n|\mathbf{r}_n) \frac{b_0(\mathbf{r}_n)}{\int_A d^2r\, b_0(\mathbf{r})}. \tag{12.294}$$

To remove the assumption that all of the x rays are travelling in the $+z$ direction, we must specify the angular distribution of the x rays on $z = 0$, for example by stating the photon radiance on that plane; this point will be pursued further in Sec. 16.1.

Spatio-spectral fluence Another way of looking at the distribution of interaction sites and energies is that they define a 4D spatio-spectral point process [*cf.* (11.135)]

$$g_{ss}(\mathbf{R}) \equiv \sum_{n=1}^{N_x} \delta(\mathbf{r} - \mathbf{r}_n)\, \delta(z - z_n)\, \delta(\mathcal{E} - \mathcal{E}_n) = \sum_{n=1}^{N_x} \delta(\mathbf{R} - \mathbf{R}_n). \tag{12.295}$$

Since we are neglecting nonlocal energy deposition (Compton scattering and K x rays), each of the interactions is produced by a different x-ray photon, and these photons are independent unless we consider random fluence, so $g(\mathbf{R})$ is a 4D Poisson point process. We can define the mean of this Poisson process as the *spatio-spectral fluence* $b_{ss}(\mathbf{R})$, and we know from the discussion in Sec. 11.3 that $\mathrm{pr}(\mathbf{R}_n)$ is just a normalized version of that fluence. Specifically, by a slight generalization of (11.83),

$$\mathrm{pr}(\mathbf{R}_n) = \frac{b_{ss}(\mathbf{R}_n)}{\int_D d^4R\, b_{ss}(\mathbf{R})}, \tag{12.296}$$

where $\int_D d^4R$ implies integration of x and y over the lateral dimensions of the detector, integration over z from 0 to L_z and integration over $\mathcal{E}$ from 0 to the maximum energy in the spectrum.

We can also work backwards and regard (12.296) as a definition of $b_{ss}(\mathbf{R}_n)$ if $\mathrm{pr}(\mathbf{R}_n)$ is known, for example from (12.294). In either case, if we know that $g_{ss}(\mathbf{R})$ is a spatio-spectral Poisson process, its statistics are fully specified by $b_{ss}(\mathbf{R})$.

Averaging over $\mathbf{R}_n$ It is now straightforward to modify the treatment in Sec. 11.4.3 with the new density $\mathrm{pr}(\mathbf{R}_n)$ from (12.294) or (12.296). The modification comes at step (c) of the five-step averaging procedure summarized below (11.207), and mainly it is a matter of replacing $b(\mathbf{r})$ with $b_{ss}(\mathbf{R})$ at various places.

A slight subtlety arises with the function $p_d(\mathbf{r}, \mathbf{R})$, defined above (11.201) as the shift-variant spread function of the gain mechanism. Had we succumbed in Chap. 11 to the temptation of writing this function as $p_d(\mathbf{r} - \mathbf{R})$, we would not be able to use subsequent results in the present context where $\mathbf{r}$ is 2D and $\mathbf{R}$ is 4D, but with the shift-variant notation there is no problem. In particular, $p_d(\mathbf{r}, \mathbf{R})$ is now defined as mean number of optical photons per unit area at point $\mathbf{r}$ in the plane $z = 0$ when the x-ray interaction is at the 4D point $\mathbf{R}$.

With this reinterpretation of $p_d(\mathbf{r}, \mathbf{R})$, the definition of the operator $\mathcal{H}_1$ in (11.213) is modified to

$$[\mathcal{H}_1 \mathbf{b}_{ss}](\mathbf{r}) = \int_D d^4R \; p_d(\mathbf{r}, \mathbf{R}) \, b_{ss}(\mathbf{R}) \,. \tag{12.297}$$

Similarly, the definition of $\mathcal{H}_2$ in (11.221) becomes

$$[\mathcal{H}_2 \mathbf{b}_{ss}](\mathbf{r}, \mathbf{r}') = \int_D d^4R_n \; \mathrm{pr}_{\Delta\mathbf{r}}(\mathbf{r} - \mathbf{r}_n|\mathbf{R}_n) \, \mathrm{pr}_{\Delta\mathbf{r}}(\mathbf{r}' - \mathbf{r}_n|\mathbf{R}_n) \, b_{ss}(\mathbf{R}_n) \, s(\mathbf{R}_n) \,, \tag{12.298}$$

where $s(\mathbf{R}_n)$ is still given by (11.219).

As the reader should verify, no further changes are required in Sec. 11.4.3, and the conditional autocovariance for the optical-photon random process $y(\mathbf{r})$ is given from (11.226) as

$$K_y(\mathbf{r}, \mathbf{r}'|\mathbf{b}_{ss}) = [\mathcal{H}_1 \mathbf{b}_{ss}](\mathbf{r}) \, \delta(\mathbf{r} - \mathbf{r}') + [\mathcal{H}_2 \mathbf{b}_{ss}](\mathbf{r}, \mathbf{r}') \,. \tag{12.299}$$

This autocovariance is conditional on a specific spatio-spectral fluence $b_{ss}(\mathbf{R})$, which is just the average distribution of interaction points within the detector as a function of 3D position and energy deposition. For x rays at normal incidence, $b_{ss}(\mathbf{R})$ is given by (12.296) and (12.294), but for more complicated x-ray beams a numerical calculation might be required. Nevertheless, (12.299) gives the autocovariance of any nonrandom x-ray source. It would not apply if the source strength fluctuated or if different absorbing objects were placed in the x-ray beam on different repetitions of the experiment.

Discrete photodetectors Now consider an array of photodiodes on the surface $z = L_z$ of the scintillator. If we assume that each photodiode is connected to a gated integrator, the output voltage of the mth photodiode after the integration time T is given by (12.215), where the charge $Q_m(T)$ that appears in that equation results from both dark current and photocurrent. The photocurrent contribution to $V_m(T)$ is related to $y(\mathbf{r})$ by

$$V_m^{ph}(T) = -\frac{e\eta_m}{C_m} \int_m d^2r \; y(\mathbf{r}) \,, \tag{12.300}$$

where C_m is the integrating capacitor for photodiode m, η_m is its quantum efficiency, and the integral is over the area of that diode. We carry along the subscript m on η_m and C_m since these parameters may not be well controlled in the semiconductor manufacturing process. We can define an effective gain for the mth photodiode as

$$\Gamma_m = -\frac{e\eta_m}{C_m} \,. \tag{12.301}$$

This gain is the ratio of the output voltage from the photodiode to the number of optical photons incident on it.

Since (12.300) is a linear continuous-to-discrete mapping, we can obtain the covariance matrix of the output voltages by applying (8.147) to (12.299). All other noise sources, including dark current, kTC noise and electronic noise, can be lumped into a variance term σ_m^2 in V_m. These noise sources are uncorrelated, so they contribute only to the diagonal elements of the covariance matrix. The covariance is thus given by

$$[\mathbf{K}_{\mathbf{V}}(\mathbf{b}_{ss})]_{mm'}$$

$$= \left[\Gamma_m^2 \int_m d^2r\, [\mathcal{H}_1 \mathbf{b}_{ss}](\mathbf{r}) + \sigma_m^2\right] \delta_{mm'} + \Gamma_m \Gamma_{m'} \int_m d^2r \int_{m'} d^2r'\, [\mathcal{H}_2 \mathbf{b}_{ss}](\mathbf{r}, \mathbf{r}') . \tag{12.302}$$

The diagonal elements come from both the excess noise and the fact that $y(\mathbf{r})$ is a point process, and the off-diagonal terms come from the fact that a single x ray can contribute to the output of more than one photodiode.

Gain and offset correction In addition to the gain variations described by Γ_m, there can also be offset voltages arising from dark current or electronic offsets. It is common practice with photodiode arrays to correct the voltages for these variations, defining the final output data by

$$g_m = \frac{V_m - \overline{V}_m^{dark}}{\Gamma_m} , \tag{12.303}$$

where $\overline{V}_m^{dark}$ is the mean of V_m in the absence of x rays.

The covariance matrix for the data vector $\mathbf{g}$ is given by

$$[\mathbf{K}_{\mathbf{g}}(\mathbf{b}_{ss})]_{mm'} = \left[\int_m d^2r\, [\mathcal{H}_1 \mathbf{b}_{ss}](\mathbf{r}) + \frac{\sigma_m^2}{\Gamma_m^2}\right] \delta_{mm'} + \int_m d^2r \int_{m'} d^2r'\, [\mathcal{H}_2 \mathbf{b}_{ss}](\mathbf{r}, \mathbf{r}') . \tag{12.304}$$

Note especially that the effect of gain variations has not disappeared. Photodiodes with small gain have relatively more noise after the gain correction if there is any noise source that is independent of the optical flux.

Stationarity? We have by now accumulated a long list of conditions that must be satisfied if we want to consider the noise in $\mathbf{g}$ to be stationary in any sense. First, we know from Sec. 11.4.4 that the optical-photon point process $y(\mathbf{r})$ is stationary only if the 2D x-ray fluence is constant (immediately ruling out any interesting images) and the gain process is independent of lateral position. In the present context, the latter condition requires that the amount of light produced by an x ray and the spread of that light in propagating to the photodiode plane must be independent of x and y. There are many different kinds of material inhomogeneities that could invalidate this assumption. For example, many scintillators use a dopant to provide the light, and stationarity of $y(\mathbf{r})$ requires uniform doping density.

In addition, we know from Sec. 12.2.5 that stationarity is useful in a finite, discrete array only if it is cyclic. If we specify the photodiodes by the 2D multi-index $\mathbf{j}$, cyclic stationarity requires that $[\mathbf{K}]_{\mathbf{jj}}'$ depend on $\mathbf{j} - \mathbf{j}'$ modulo $\mathbf{J}$. This unphysical wrap-around effect never occurs in practice, but it might have a negligible effect on subsequent uses of the covariance matrix if the array is large. Or it might not (Pineda *et al.*, 2001).

Finally, we have just encountered a new reason for the stationarity assumption to fail — the inevitable gain variations in practical arrays. Even after correction, we see from (12.304) that the excess noise variance will depend on m (or $\mathbf{j}$ if we choose to use a multi-index). To see why this effect negates any possible advantage of a Fourier description of the data, consider the data vector $\mathbf{g}$ in the absence of any x rays, so that

$$\left[\mathbf{K}_{\mathbf{g}}^{dark}\right]_{\mathbf{jj}'} = \frac{\sigma_{\mathbf{j}}^2}{\Gamma_{\mathbf{j}}^2}\,\delta_{\mathbf{jj}'} \,. \tag{12.305}$$

We now take a discrete Fourier transform (DFT) of $\mathbf{g}$, using the multi-index notation. If both components of $\mathbf{j}$ are assumed to run from 0 to $J-1$, we can write

$$G_{\mathbf{k}} = \sum_{\mathbf{j}=0}^{J-1} g_{\mathbf{j}} \exp\left[-2\pi i \frac{\mathbf{j}\cdot\mathbf{k}}{J}\right] . \tag{12.306}$$

The covariance matrix is then transformed to

$$\begin{aligned}\left[\mathbf{K}_{\mathbf{G}}^{dark}\right]_{\mathbf{kk}'} &= \sum_{\mathbf{j}=0}^{J-1}\sum_{\mathbf{j}'=0}^{J-1} \left[\mathbf{K}_{\mathbf{g}}^{dark}\right]_{\mathbf{jj}'} \exp\left[-2\pi i \frac{\mathbf{j}\cdot\mathbf{k}-\mathbf{j}'\cdot\mathbf{k}'}{J}\right] \\ &= \sum_{\mathbf{j}=0}^{J-1} \frac{\sigma_{\mathbf{j}}^2}{\Gamma_{\mathbf{j}}^2} \exp\left[-2\pi i \frac{\mathbf{j}\cdot(\mathbf{k}-\mathbf{k}')}{J}\right] .\end{aligned} \tag{12.307}$$

This is as far as we can go in general. If we could assume that $\sigma_{\mathbf{j}}^2/\Gamma_{\mathbf{j}}^2$ was constant, we could perform the sum and find that $\mathbf{K}_{\mathbf{G}}^{dark}$ was diagonal, but in real detectors there will be off-diagonal terms. The natural (Karhunen-Loève) domain for describing the excess noise is the original data domain, where $\mathbf{K}_{\mathbf{g}}^{dark}$ is diagonal; the covariance matrix for this noise component gets more complicated, not less, after the DFT.

We thus have a quandary: The dark noise is diagonal in the original data domain, but if we make all of the other assumptions needed for stationarity, the x-ray part of the covariance is diagonalized in the Fourier domain. To diagonalize the overall covariance, we would have to use a Karhunen-Loève transformation (see Sec. 8.1.6), but it would be just a numerical device; no general theory or new insights would emerge.

The resolution of this quandary will be presented in Sec. 16.1. As we shall see there, we can use the noise characterizations derived in this chapter to compute meaningful, task-based figures of merit for image quality for x-ray detector arrays without needing the Fourier domain.

12.3.9 K x rays and Compton scattering

Throughout most of Sec. 12.3, we have ignored the secondary photons that are created in the initial x-ray or gamma-ray interaction. We know from the qualitative discussion in Sec. 12.3.1 that an initial photoelectric interaction can produce a K x ray that might then be reabsorbed at another point in the detector material. Similarly, if the initial interaction is Compton scattering, the scattered photon can also be reabsorbed in the detector. These reabsorbed photons can deposit energy at points distant from the initial interaction, creating light or charge there that can be sensed on the exit face of the detector.

In this section we shall discuss the effect of these new energy deposition points with integrating x-ray detectors. The main tool for this purpose will be the 4D spatio-spectral point process defined in (12.295), only now it will be necessary to distinguish the primary (initial) interaction parameters from those for the secondary interactions. We thus rewrite (12.295) as

$$g_{ss}(\mathbf{R}) = g_{ss}^{pri}(\mathbf{R}) + g_{ss}^{sec}(\mathbf{R}) = \sum_{n=1}^{N_x} \delta(\mathbf{R} - \mathbf{R}_n^{pri}) + \sum_{n=1}^{N_{sec}} \delta(\mathbf{R} - \mathbf{R}_n^{sec}), \tag{12.308}$$

where N_x is the number of absorbed x rays and N_{sec} is the number of secondary photon absorptions in the detector (which can be less than N_x since the secondary photons can escape from the detector).

Spatio-spectral fluences We can define spatio-spectral fluences as the means of the component random processes:

$$b_{ss}^{pri}(\mathbf{R}) = \left\langle g_{ss}^{pri}(\mathbf{R}) \right\rangle, \qquad b_{ss}^{sec}(\mathbf{R}) = \left\langle g_{ss}^{sec}(\mathbf{R}) \right\rangle. \tag{12.309}$$

Though it may be complicated in practice, there is no difficulty in principle in computing these fluences from knowledge of the incident x-ray flux and properties of the detector material. Henceforth we shall assume that the fluences are known.

If the primary fluence is nonrandom, $g_{ss}^{pri}(\mathbf{R})$ is a Poisson random process, and its autocovariance function is given by

$$K^{pri}(\mathbf{R}, \mathbf{R}') = b_{ss}^{pri}(\mathbf{R})\, \delta(\mathbf{R} - \mathbf{R}'). \tag{12.310}$$

Moreover, by the binomial-selection theorem of Sec. 11.1.3, N_{sec} is a Poisson random variable if N_x is. Once the binomial selection is made, the secondary random process results from a 4D displacement from the primary process, and we know from the discussion around (11.229) that the Poisson character is preserved. Thus we also have

$$K^{sec}(\mathbf{R}, \mathbf{R}') = b_{ss}^{sec}(\mathbf{R})\, \delta(\mathbf{R} - \mathbf{R}'). \tag{12.311}$$

It does not follow, however, that the overall interaction pattern $g_{ss}(\mathbf{R})$ is a Poisson random process; it cannot be since $\mathbf{R}_n^{sec}$ and $\mathbf{R}_n^{pri}$ both arise from the nth absorbed x ray and hence are not independent.

Cross-correlation To compute the overall autocovariance, we need to study the cross-correlation function,

$$\left\langle g_{ss}^{pri}(\mathbf{R}) g_{ss}^{sec}(\mathbf{R}') \right\rangle = \left\langle \sum_{n=1}^{N_x} \delta(\mathbf{R} - \mathbf{R}_n^{pri}) \sum_{n'=1}^{N_{sec}} \delta(\mathbf{R}' - \mathbf{R}_{n'}^{sec}) \right\rangle. \tag{12.312}$$

The average will be performed by methods developed in Sec. 11.3.3. As in that section, we must consider separately terms for which $n = n'$ and those for which $n \neq n'$.

If $n = n'$, then

$$\langle \delta(\mathbf{R} - \mathbf{R}_n^{pri})\, \delta(\mathbf{R}' - \mathbf{R}_n^{sec}) \rangle$$

$$= \int_D d^4\mathbf{R}_n^{pri} \int_D d^4\mathbf{R}_n^{sec}\ \mathrm{pr}(\mathbf{R}_n^{sec}|\mathbf{R}_n^{pri})\, \mathrm{pr}(\mathbf{R}_n^{pri})\, \delta(\mathbf{R} - \mathbf{R}_n^{pri})\, \delta(\mathbf{R}' - \mathbf{R}_n^{sec})$$

$$= \mathrm{pr}_{p\to s}(\mathbf{R}'|\mathbf{R})\, \mathrm{pr}_{pri}(\mathbf{R}), \tag{12.313}$$

where $\mathrm{pr}_{p\to s}(\mathbf{R}_n^{sec}|\mathbf{R}_n^{pri})$ means the same thing as $\mathrm{pr}(\mathbf{R}_n^{sec}|\mathbf{R}_n^{pri})$, but the new notation is required when we substitute $\mathbf{R}$ for $\mathbf{R}_n^{pri}$ and $\mathbf{R}'$ for $\mathbf{R}_n^{sec}$ as dictated by the delta functions.

For $n \neq n'$, we have

$$\langle \delta(\mathbf{R} - \mathbf{R}_n^{pri})\, \delta(\mathbf{R}' - \mathbf{R}_{n'}^{sec}) \rangle$$

$$= \int_D d^4\mathbf{R}_n^{pri}\ \mathrm{pr}(\mathbf{R}_n^{pri})\, \delta(\mathbf{R} - \mathbf{R}_n^{pri}) \int_D d^4\mathbf{R}_{n'}^{sec}\ \mathrm{pr}(\mathbf{R}_{n'}^{sec})\, \delta(\mathbf{R} - \mathbf{R}_{n'}^{sec})$$

$$= \mathrm{pr}_{pri}(\mathbf{R})\, \mathrm{pr}_{sec}(\mathbf{R}'). \tag{12.314}$$

Since $g_{ss}^{pri}(\mathbf{R})$ and $g_{ss}^{sec}(\mathbf{R}')$ are individually Poisson processes (albeit correlated with each other), $\mathrm{pr}_{pri}(\mathbf{R})$ and $\mathrm{pr}_{sec}(\mathbf{R}')$ are just normalized versions of the respective fluences:

$$\mathrm{pr}_{pri}(\mathbf{R}) = \overline{N}_x^{-1} b_{ss}^{pri}(\mathbf{R}), \qquad \mathrm{pr}_{sec}(\mathbf{R}') = \overline{N}_{sec}^{-1} b_{ss}^{sec}(\mathbf{R}'). \tag{12.315}$$

Since each of the secondary interactions derives from exactly one primary interaction, there are N_{sec} terms for which $n = n'$ and $N_x N_{sec} - N_{sec}$ terms with $n \neq n'$. Thus

$$\langle g_{ss}^{pri}(\mathbf{R})\, g_{ss}^{sec}(\mathbf{R}') \rangle$$

$$= \overline{N}_{sec}\Big[\mathrm{pr}_{p\to s}(\mathbf{R}'|\mathbf{R})\, \overline{N}_x^{-1} b_{ss}^{pri}(\mathbf{R})\Big] + (\langle N_x N_{sec}\rangle - \overline{N}_{sec})\left[\overline{N}_x^{-1} b_{ss}^{pri}(\mathbf{R})\, \overline{N}_{sec}^{-1} b_{ss}^{sec}(\mathbf{R}')\right]. \tag{12.316}$$

The only remaining expectation to perform is

$$\langle N_x N_{sec}\rangle = \sum_{N_x=0}^{\infty} \Pr(N_x) \sum_{N_x=0}^{\infty} \Pr(N_{sec}|N_x)\, N_x N_{sec}. \tag{12.317}$$

The conditional probability $\Pr(N_{sec}|N_x)$ is a binomial for which the probability of success can be denoted $p_{p\to s}$, which is the probability that an absorbed primary x ray will produce an absorbed secondary somewhere in the detector. The average of N_{sec} with respect to $\Pr(N_{sec}|N_x)$ is $p_{p\to s}N_x$. Since $\Pr(N_x)$ is a Poisson, we have

$$\langle N_x N_{sec}\rangle = \sum_{N_x=0}^{\infty} \Pr(N_x)\, p_{p\to s} N_x^2 = p_{p\to s}\left(\overline{N}_x + \overline{N}_x^2\right). \tag{12.318}$$

Inserting this result into (12.317) and doing a little algebra, we find

$$\langle g_{ss}^{pri}(\mathbf{R})\, g_{ss}^{sec}(\mathbf{R}')\rangle = p_{p\to s}\, \mathrm{pr}_{p\to s}(\mathbf{R}'|\mathbf{R})\, b_{ss}^{pri}(\mathbf{R}) + b_{ss}^{pri}(\mathbf{R})\, b_{ss}^{sec}(\mathbf{R}'). \tag{12.319}$$

Overall 4D autocovariance function The autocovariance of $g_{ss}(\mathbf{R})$ (conditional on the specified fluences) is defined by

$$K_{g_{ss}}(\mathbf{R}, \mathbf{R}'|\mathbf{b}_{ss})$$
$$= \left\langle [g_{ss}^{pri}(\mathbf{R}) + g_{ss}^{sec}(\mathbf{R})][g_{ss}^{pri}(\mathbf{R}') + g_{ss}^{sec}(\mathbf{R}')] \right\rangle$$
$$- \; [\overline{g}_{ss}^{pri}(\mathbf{R}) + \overline{g}_{ss}^{sec}(\mathbf{R})][\overline{g}_{ss}^{pri}(\mathbf{R}') + \overline{g}_{ss}^{sec}(\mathbf{R}')] \,. \tag{12.320}$$

With (12.309) – (12.311) and (12.319), we see that

$$K_{g_{ss}}(\mathbf{R}, \mathbf{R}'|\mathbf{b}_{ss})$$
$$= [b_{ss}^{pri}(\mathbf{R}) + b_{ss}^{sec}(\mathbf{R})] \, \delta(\mathbf{R} - \mathbf{R}')$$
$$+ \; p_{p \to s}[\mathrm{pr}_{p \to s}(\mathbf{R}'|\mathbf{R}) \, b_{ss}^{pri}(\mathbf{R}) + \mathrm{pr}_{p \to s}(\mathbf{R}|\mathbf{R}') \, b_{ss}^{pri}(\mathbf{R}')] \,. \tag{12.321}$$

We see that the 4D autocovariance contains the expected delta-correlated part proportional to the total fluence plus another part with a correlation structure determined by $\mathrm{pr}_{p \to s}(\mathbf{R}'|\mathbf{R})$. This function is the probability density for a secondary interaction at $\mathbf{R}'$ given that the primary interaction was at $\mathbf{R}$ (and given that the secondary photon does not escape).

The dependence of $\mathrm{pr}_{p \to s}(\mathbf{R}'|\mathbf{R})$ on the spatial parts of $\mathbf{R}$ and $\mathbf{R}'$ (the first three components of the 4D vectors) can be computed from knowledge of the attenuation coefficient and the geometry of the detector material. If the detector were infinitely thick, the spatial correlation would extend over a range in $|\mathbf{r}' - \mathbf{r}|$ approximately equal to the reciprocal of the attenuation coefficient, but of course the correlation length cannot exceed L_z in the $+z$ direction. We can envision $\mathrm{pr}_{p \to s}(\mathbf{R}'|\mathbf{R})$ as a fuzzy ball centered on the primary interaction site and truncated by the detector boundaries.

The fourth component of $\mathbf{R}'$ is the energy deposited in the secondary interaction. If we assume that the secondary photon gives up all of its remaining energy in the secondary interaction, then conservation of energy requires that $\mathcal{E}' + \mathcal{E} = \mathcal{E}_0$ (where $\mathcal{E}_0$ is the initial energy of the primary photon), so $\mathrm{pr}_{p \to s}(\mathbf{R}'|\mathbf{R}) \propto \delta(\mathcal{E}' + \mathcal{E} - \mathcal{E}_0)$.

Discrete readouts The 4D autocovariance is not directly observable, but it determines the correlation properties of discrete readout signals in both semiconductor and scintillation detectors. Consider, for example, a scintillation detector with an array of photodiodes as analyzed in Sec. 12.3.8. In that section the energy deposition was described by a Poisson point process, so the only pixel-to-pixel correlation was that induced by the spreading of the optical photons *en route* to the photodiodes. With the secondary interactions, there is another source of correlation since optical photons are generated not only at the primary site, but also at a nearby secondary site. Even if there were no spread of the optical photons, there would be correlations between different pixels.

The implications of these correlations for x-ray imaging will be discussed in Sec. 16.1.

13

Statistical Decision Theory

In the preceding chapters we have considered various descriptions for the deterministic image-formation process (Chaps. 7, 9 and 10) as well as the possible sources of randomness in the resulting images (Chaps. 8, 11 and 12). Armed with this background, we are now able to consider the weighty topic of image quality and imaging system evaluation.

It is our premise that the quality of an imaging system is *defined* by how well inferences about an underlying scene can be made using its image as input. More importantly, image quality must be assessed on the basis of *average* performance of some inference task by some observer or decision-maker. Image quality is thus a statistical concept; statistical decision theory is the key to the mathematical characterization of image quality.

This chapter discusses the kinds of inferences that might be of interest to various imaging communities, how these tasks are approached by predefined decision-makers or observers, and figures of merit for quantifying the performance of these observers. In Chap. 14 we shall discuss many of the practical issues that arise in the application of the mathematical tools provided in this chapter to the problem of the objective assessment of imaging systems. In particular, issues related to estimation of image statistics and figures of merit from a finite set of sample images are presented in Chap. 14. Throughout this chapter we assume full knowledge of the ensemble statistics of the data necessary to calculate the figures of merit under discussion.

13.1 BASIC CONCEPTS

We begin with a discussion of the kinds of inferences or decisions that can be derived from the output of an imaging system. In addition to the image itself, inputs to the decision-making process can include statistical and deterministic models of the imaging system (combined into a likelihood function), prior information about the objects, and a judgment as to the costs of incorrect decisions and the benefits of correct ones. The discussion of priors and costs will inevitably take us back to the conflict between Bayesian and frequentist approaches to inference. Our general approach to this issue, sketched in the Prologue, will be taken up in more mathematical detail here.

13.1.1 Kinds of decisions

As described in the Prologue, we divide statistical inference into two types: *classification* and *parameter estimation*. Any time there are a finite number of possible outcomes, we refer to the inference as classification. In the imaging literature, the terms *pattern recognition*, *signal detection*, *discrimination*, *discriminant analysis*, *differential diagnosis*, *segmentation*, and *hypothesis testing* fall under this category. Parameter estimation, on the other hand, can be regarded as the limit of hypothesis testing when the number of hypotheses becomes infinite (usually uncountably infinite). Then the premise is that we are interested in extracting one or more numerical parameters from the data, and these parameters can take on any value in some specified range. The medical literature sometimes calls this task *quantitation*.

We treat both kinds of inference within this chapter for, as we shall demonstrate, there are strong connections between the approaches to classification and estimation problems provided by statistical decision theory, as demonstrated in Fig. 13.1. For both inferences the first step is the development of a model for the objects under investigation. For this we make use of the mathematical models for the deterministic aspects of objects described in Chap. 7 and the tools provided in Chap. 8 for characterization of the variability of the objects under consideration. The next step is the generation of a model for the acquired data. Here we must apply whatever knowledge we have of the image-formation process, including the deterministic mapping of an object through the imaging system as well as the noise sources of the system.

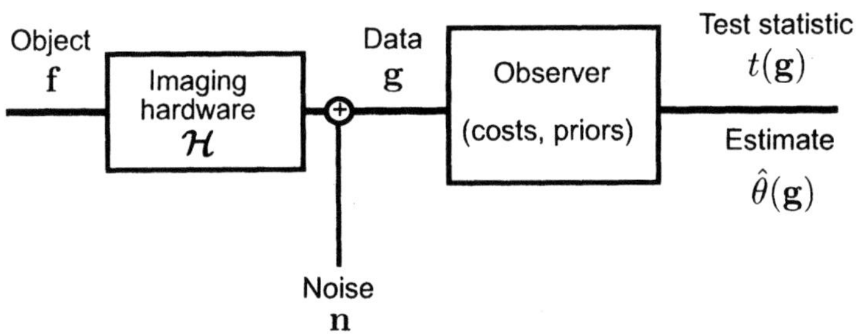

Fig. 13.1 Flow chart showing steps toward the extraction of a test statistic or a parameter estimate.

We call the means by which the task gets done, or the strategy, the observer or decision-maker. For example, the observer in a lesion-detection task on real clinical images might be a radiologist. Alternatively, much research effort is being expended on the development of computer algorithms for such tasks. A special observer known as the *ideal observer* or *Bayesian observer* is defined as the observer that utilizes all statistical information available regarding the task to maximize task performance as measured by Bayes risk or some other related measure of performance. Thus the performance of the Bayesian observer provides an upper bound against which all other observers can be compared.

Estimation tasks using images as inputs are most often performed by computer algorithms. These algorithms can run the gamut from very *ad hoc* procedures to ones based on optimality criteria involving the bias and/or variance of the resulting estimates.

Returning to Fig. 13.1, we see that one or more operations are applied to the data by the observer to facilitate the inference. If the task is classification, the output of these operations is a *test statistic*. If the task is estimation, the output is the estimate itself. These quantities are random, because they are based on operations on random data. Thus both classification and estimation tasks demand the same kind of information from the practitioner, and both result in the computation of a random quantity at the output to the inference process. As we shall show, the assessment of system performance for both kinds of tasks also has a similar construction.

Classification In classification, the output of an imaging system is observed, and the observer must infer the class membership of the object at the input to the imaging system. The observer usually has some information about the possible objects being imaged, the way in which objects are distorted by the image-acquisition process, and the sources of randomness in the data. In radar, the classification task might be the determination of the presence or absence of a target in the radar signal. In medical applications the task might be tumor detection, or it might be the classification of a detected tumor into a particular class, *e.g.*, malignant or benign. The discrimination of tanks from trucks in aerial photographs is a common military application of statistical decision theory.

The classification problem was studied as early as the middle of the 18th century, when Thomas Bayes published his theory for the testing of hypotheses by statistical inference (Bayes, 1764). The advent of radar and communications technology in the mid-twentieth century rekindled interest in Bayes' theories for statistical decision making (Van Meter and Middleton, 1954; Peterson, 1954). More recently, statistical decision theory has been applied to the evaluation of medical imaging systems (Swets, 1979; Swets and Pickett, 1982; Wagner and Brown, 1985; Barrett, 1990; ICRU Report, 1996) and has contributed to the study of human perception (Tanner and Swets, 1954; Swets, 1964; Lusted, 1968; Burgess *et al.*, 1981; Swensson and Judy, 1981).

A classification problem is categorized according to the number of hypotheses to be distinguished, the nature of the hypotheses, the structure of the data, and the statistics of the signal and noise. Classification tasks can be as simple as the detection of a known (nonrandom) signal in a known background, or as complex as the discrimination of multiple classes comprised of fully random objects described by different probability density functions.

Classification tasks with just two underlying hypotheses or classes from which the data might be drawn are known as *binary* decision problems. Determining whether a reconnaissance image contains a tank is an example of a binary decision problem, because the classification is into one of two possible alternatives — tank or no-tank. As we shall see, the same theory can be extended to the multiple-decision or L-class task, where the data are to be assigned to one of L possible hypotheses.

The L hypotheses may correspond to L different signals, all in the same location (Burgess and Ghandeharian, 1984b; Eckstein and Abbey, 2001), or one signal in one of L possible locations (Goodenough, 1975; Starr *et al.*, 1975), or a more general set of L hypotheses.

A common example of an L-class task is the pattern-recognition task of character recognition. The character to be classified might be one of a finite number of letters of known font, so that the signals are nonrandom. Alternatively, the characters might have some randomness in one or more parameters; for example, perhaps they might be of known font but with some randomness in scale. At the other extreme, the characters to be classified might be freehand letters, so that the objects have a great deal of randomness associated with them. In Sec. 13.3 we shall show in mathematical detail how increasing levels of randomness in the classes to be discriminated are incorporated in signal-detection theory.

Estimation Classical estimation begins with the assumption that we are given data $\mathbf{g}$ from some known probability law $\mathrm{pr}(\mathbf{g}|\theta)$ and the task is to estimate the scalar parameter θ. As an example, a photon-counting probe might be used to determine the activity in a draining lymph node. Poisson statistics govern the number of detected counts in this case. Since the Poisson law has just one free parameter, the mean count rate, measurements over equal counting intervals will allow the estimation of this rate for the node under test. As a second example, suppose we use a photoconductor to detect light transmitted by an object. If the current at the output of the detector suffers from Johnson noise fluctuations, the randomness of the current can be modeled by a Gaussian random process. While a Gaussian probability density function (PDF) is generally parameterized by a mean and variance, we might make the assumption that the noise variance is independent of the signal, and that this variance has already been characterized for this detector. An estimate of the mean of the Gaussian PDF can then be formed based on measurements at the output of the detector.

The estimation of a single parameter from a measurement, like the example of the photon-counting detector above, is *scalar parameter estimation. Vector parameter estimation* involves the determination of more than one parameter, which we arrange in a parameter vector $\boldsymbol{\theta}$. If we had to estimate both the mean and variance of the noise in the Gaussian example above, these two parameters would constitute the vector $\boldsymbol{\theta}$. The task of image reconstruction, a subject we treat in detail in Chap. 15, may also be thought of as a kind of vector parameter estimation.

Determination of the single "best" estimate of one or more parameters is known as *point estimation*. In point estimation no information is provided about the uncertainty in the estimated value. In contrast, *interval* or *region* estimation results in a set of values the parameters might reasonably take on (Casella and Berger, 1990). An interval estimate is not as precise as a point estimate, but is intended to provide confidence that the true value of the parameter is within the estimated interval. The result of an interval estimation procedure for a scalar quantity is straightforward to present in graphical or numerical terms. However, due to the inherent difficulty of representing the output of a region estimation procedure for a large set of numbers, region estimation is not commonly applied to vector estimation problems.

In some circumstances the task is the estimation of the entire probability density function of a random variable from sample values. One approach to this so-called *density estimation* problem is to parameterize a well-known density function kernel and seek estimates for the underlying parameters, *e.g.*, the mean and variance of a Gaussian PDF. The most common nonparametric method of density estimation is *kernel estimation*, in which a smoothing kernel such as a Gaussian is associated with each observed point in the sample space (Silverman, 1986).

In a sense, image reconstruction is density estimation. We saw in Chap. 11 that an object, when properly normalized, can often be regarded as the PDF for photon emissions, so to estimate the object is to estimate this PDF. In practice, we may adopt some approximate representation such as a voxel expansion for the object, and in that case reconstruction is parameter estimation, albeit with a large number of parameters. As we shall see in Sec. 15.2.2, however, it is also possible to estimate the underlying object function — a density — without discretization. With this exception, we shall say nothing more about density estimation.

Hybrid estimation-classification tasks We are often interested in forming an estimate of one or more parameters that will then be used as input to subsequent decision-making operations. Examples abound of quantities that are extracted from medical images with the end goal being the classification of the tissue or organ as normal or diseased. For instance, a series of nuclear medicine images might be obtained to quantify organ time-activity dynamics, which can be used in the classification of disease in such organs as the liver, heart, or kidney. Alternatively, estimation may follow detection, as is the case when a decision is made in favor of a particular hypothesis, and an estimate is then made of a parameter underlying the objects known to be present under the selected hypothesis. For example, a tumor may first be detected, and then its size is estimated to determine the stage of the disease. When estimation and classification tasks are combined, the overall procedure is termed a *hybrid* estimation-classification task.

Tests of the null hypothesis Classification can also be called hypothesis testing. For some readers this term will conjure up the standard form of significance testing taught in most statistics classes and found in much of the biomedical sciences literature. In this approach, a null hypothesis H_0 is formulated; for example, the parameter θ is a random variable distributed as $\mathcal{N}(0, \sigma^2)$. The complement of the null hypothesis, denoted H_1, is called the alternative to H_0. Often the null hypothesis is that one data set is drawn from the same density function as another; the alternative to this hypothesis is that the data sets are different (or, one product is better!). Once the hypotheses are formulated, observed values are tested to see whether the null hypothesis can be accepted or rejected.

Consider the typical scenario in which data are collected from two sets of biological samples, one which underwent a treatment program and one which did not. Now the investigators want to determine whether the mean value of the data from each set is different to see if the treatment program has an effect. A test of the null hypothesis that the data have the same means, assuming Gaussian PDFs with equal variance, is performed. Sample means $\overline{x}_1$ and $\overline{x}_2$ are determined from the data and used to test whether $\overline{x}_2 - \overline{x}_1$ is statistically significantly different from 0. Based on standard tables, the probability of obtaining the observed difference in the means is computed. If this probability (the so-called p-value) is deemed small enough, at the .01 or .05 level, say, these investigators would claim they have grounds for rejecting the null hypothesis.

There are a number of serious flaws in this approach to assessing the separability of two data sets. First of all, the simple point hypothesis can virtually always be shown to be wrong. Consider two normal distributions $\mathcal{N}(1,10)$ and $\mathcal{N}(1.01,10)$. A single sample would have almost no value for distinguishing the two distributions. A physician would say that a diagnostic test based on this variable was worthless, and a physicist would say the two distributions were not significantly different since the difference in means is very small compared to the standard deviation. A determined statistician, on the other hand, could draw not one but many thousand samples and discern that the difference in means is significant at the .01 level. The same statement could be made if the distributions were $\mathcal{N}(1,1)$ and $\mathcal{N}(2,1)$ and many fewer samples were used. The point is that so-called statistical significance is a function of the diligence of the experimenter as well as the size of the deviation from the null hypothesis relative to the spread in measurements of a sample. A statistician who reports only a p-value has ignored perhaps the most important aspect of the data: the difference in estimated means divided by the estimated average standard deviation.

A second serious flaw in this approach is that the alternative to the null hypothesis is vague and meaningless. The alternative to the hypothesis that the data have equal means is that the means are different. How different? It does not matter in this framework.

A further objection to this sort of hypothesis testing is that the determination of statistical significance allows the investigator to reject the null hypothesis, when in fact it might be that the model is what is really in need of being rejected. One can easily concoct an illustration where two data sets would result in a rejected null hypothesis for Gaussian PDFs with equal means when in actuality the Gaussian PDF assumption that was made was inappropriate. In the end, it can be quite difficult to know whether it is the model or the hypothesis that should be rejected. For all these reasons we reject the usage of this methodology for the assessment of imaging systems, recommending instead the task-based approach of statistical decision theory with clearly specified alternatives.

13.1.2 Inputs to the process

The data As shown in Fig. 13.1, estimation and classification involve the computation of estimates, in the former, or test statistics, in the latter, based on the data at the output of an imaging system. For a digital imaging system the data consist of the set of M measurements $\{g_m\}$, which might be the set of pixel values or gray levels from a direct imaging system, or the raw measurements from a tomographic imaging system. As usual, these data can be thought of as a point in an M-dimensional observation space, denoted by the $M \times 1$ column vector $\mathbf{g}$. Each of the elements of $\mathbf{g}$ is a random variable.

The data are the result of an image-formation process whereby a continuous object $f(\mathbf{r})$ is mapped to the data set. The mapping can be represented quite generally by

$$\mathbf{g} = \mathcal{H}\mathbf{f} + \mathbf{n}, \tag{13.1}$$

where $\mathcal{H}$ is some appropriate imaging operator, as discussed in Chap. 7. Note that we have written the continuous object as a vector $\mathbf{f}$ in the Hilbert space of square-integrable functions; we emphasize that this notation does not limit our treatment to discrete objects.

The M-dimensional vector $\mathbf{n}$ in (13.1) represents the noise in the data set. The fact that the noise is represented as additive does not restrict us to additive noise conditions. It is understood that the noise is the difference between the expected data set in the absence of noise and the actual data set. That is,

$$\mathbf{n} \equiv \mathbf{g} - \langle \mathbf{g} \rangle_{\mathbf{n}} = \mathbf{g} - \boldsymbol{\mathcal{H}}\mathbf{f}, \tag{13.2}$$

where the angle brackets subscripted by $\mathbf{n}$ denote a statistical average over the noise for an ensemble of data sets acquired with a particular object $\mathbf{f}$.

Raw images are just one form of input data upon which an observer might base decisions. Decisions might also be based on features extracted from images. These features might be numerified qualitative observations (*e.g.*, spicularity of a lesion on a scale of 1 to 10), or features derived from an image by computer algorithms.

It follows from (13.1) that a model of the deterministic characteristics of the imaging system, specified by $\boldsymbol{\mathcal{H}}$, is required for drawing inferences based on images. To whatever extent is possible, we want our system model to be accurate so that inference errors are solely the result of the limiting variability in the data, and not due to modeling errors.

Conditional probability model In order to draw an inference from random data, it is vital that a model for the conditional probability density function on the data be constructed and brought to bear on the classification or estimation problem by the decision-maker. Classification strategies hinge on how the data are distributed given a particular underlying object or hypothesis; this information is captured by the conditional density function $\text{pr}(\mathbf{g}|\mathbf{f})$. Alternatively, estimation involves the determination of the value of some feature θ of the object given the data in hand. The strategy for estimating this feature will depend to a large degree on the effect this feature has on the data set. We write the probability density function for the data conditioned on a particular value of this feature by $\text{pr}(\mathbf{g}|\theta)$. The conditional probability densities $\text{pr}(\mathbf{g}|\mathbf{f})$ and $\text{pr}(\mathbf{g}|\theta)$ are referred to as *likelihood* functions since they tell how likely it is that a given data set is obtained when some underlying state of nature is true.

Imperfect or incomplete knowledge of the likelihood function results in additional classification or estimation errors beyond what would occur as a result of the limiting uncertainty in the data due to measurement noise. In particular, it may be hard to determine $\text{pr}(\mathbf{g}|\boldsymbol{\theta})$ in an estimation problem if the vector $\boldsymbol{\theta}$ has low dimensionality. For example, many possible data sets could be obtained from objects with the same tumor volume, so that it would be quite difficult to determine the probabilistic relationship between tumor volume and the data. If, on the other hand, $\boldsymbol{\theta}$ has high dimensionality, we can be more hopeful of making sense of $\text{pr}(\mathbf{g}|\boldsymbol{\theta})$. For example, if $\boldsymbol{\theta}$ is a vector of expansion coefficients used to represent the object, we might be able to approximate the mapping of (13.1) by writing $\mathbf{g} = \boldsymbol{\mathcal{H}}\mathbf{f} + \mathbf{n} \approx \mathbf{H}\boldsymbol{\theta} + \mathbf{n}$ (for a linear imaging system). In that case we can presume that $\text{pr}(\mathbf{g}|\boldsymbol{\theta}) \approx \text{pr}(\mathbf{g}|\mathbf{f})$. This density is something we can compute by making use of the techniques presented in Chaps. 8, 11, and 12.

In the Prologue we describe our philosophy regarding the interpretation of $\text{pr}(\mathbf{g}|\mathbf{f})$, which we believe is naturally interpreted in terms of relative frequencies and hence can, in principle, be obtained by repeated experimental observation. In most cases, however, the basic physics of the measurement process tells us the form of $\text{pr}(\mathbf{g}|\mathbf{f})$, in a frequentist sense, and we do not have to actually do the experiments. For example, in a photon-counting situation, we know that the statistics will be Poisson under broad conditions discussed in Chap. 11.

Priors When the object has some randomness, the density function for the data depends on the probability density function on the object as well as the measurement noise. Consider a classification problem for which there exists a distribution of possible objects for each hypothesis H_j. We write the conditional density function on the data under H_j as[1]

$$\mathrm{pr}(\mathbf{g}|H_j) = \int_{\mathbb{U}} d\mathbf{f}\; \mathrm{pr}(\mathbf{g}|\mathbf{f})\, \mathrm{pr}(\mathbf{f}\,|H_j)\,, \tag{13.3}$$

where the conditional probability $\mathrm{pr}(\mathbf{f}\,|H_j)$ is the object prior. From Chap. 8 we know that it is difficult to be precise about the meaning of $\mathrm{pr}(\mathbf{f}\,|H_j)$ for realistic problems since $\mathbf{f}$ is in principle infinite-dimensional. As discussed in Sec. 8.4, we can avoid this problem by using a finite-dimensional object representation such as (7.27) with an $N \times 1$ vector of coefficients $\boldsymbol{\alpha}$. In that case (13.3) becomes

$$\mathrm{pr}(\mathbf{g}|H_j) = \int_{\infty} d^N\alpha\; \mathrm{pr}(\mathbf{g}|\boldsymbol{\alpha})\, \mathrm{pr}(\boldsymbol{\alpha}|H_j)\,. \tag{13.4}$$

Now we need only an N-dimensional prior on $\boldsymbol{\alpha}$ instead of the infinite-dimensional prior on **f**. As discussed in Sec. 8.4.1, it is sometimes possible to evaluate the integral in (13.4) through simulation methods.

Analogously, in estimation a Bayesian decision-maker assumes that $\boldsymbol{\theta}$ is itself a random variable with a probability density function given by $\mathrm{pr}(\boldsymbol{\theta})$. As we shall see, the Bayesian's estimation strategy recasts the information contained in this prior density in light of the acquired data.

As discussed in the Prologue, Bayesians and frequentists have vastly differing views on the relevance and interpretation of object priors. Methods for deriving and interpreting prior object information are introduced in the Prologue; a number of models for object variability are presented in Chap. 8.

Prevalence in classification tasks From a Bayesian point of view a final bit of information is needed to perform a classification task. For classification a Bayesian would need to know the prior probability or *prevalence* of the underlying classes to which the data set must be assigned, denoted $\Pr(H_j) = P_j$. For example, information regarding troop and equipment movements will influence the reading of newly acquired aerial reconnaissance photographs by an expert military observer. Likewise, the prevalence of a disease will influence a radiologist's decision process when reading a set of films where the task is to determine if that disease is present or absent. There are many ways of determining prevalence, all typically frequentist in approach. Prevalence might be estimated from nationwide statistics, local community statistics, data gathered from a particular medical practice so that the demographics of that small patient population are represented, or it might even take into account patient-specific information such as age, weight, or family risk factors.

[1]The integral here is, in principle, over the entire object Hilbert space $\mathbb{U}$. See Sec. 8.2.2 and especially (8.81) for details.

Costs and risks Costs can be associated with making correct and incorrect decisions, and a decision strategy can be designed to minimize these costs. Similarly, costs can be associated with estimation errors, and an estimator can be chosen to minimize these. The philosophy of the decision-maker greatly influences the manner by which costs are utilized in the design of an inference strategy.

For estimating a scalar parameter, we write the cost of an estimation error as $C(\hat{\theta}, \theta)$, which is a function of both the underlying parameter, θ, and its estimated value, $\hat{\theta}$. The squared distance $(\hat{\theta} - \theta)^2$ between the estimate and the actual value of the parameter is a common cost function. The *risk*, or average cost, can be defined in three ways, depending on the functional dependence we are interested in probing.

Table 13.1 Definitions of Risk

Average Cost	Functional Dependence
$\left\langle C(\hat{\theta}, \theta) \right\rangle_{\mathbf{g}\mid\theta}$	Function of θ
$\left\langle C(\hat{\theta}, \theta) \right\rangle_{\theta\mid\mathbf{g}}$	Function of $\mathbf{g}$
$\left\langle \left\langle C(\hat{\theta}, \theta) \right\rangle_{\mathbf{g}\mid\theta} \right\rangle_{\theta} = \left\langle \left\langle C(\hat{\theta}, \theta) \right\rangle_{\theta\mid\mathbf{g}} \right\rangle_{\mathbf{g}}$	Pure scalar

The first definition of risk in Table 13.1 is essentially frequentist, in that it averages over many realizations of the data for a particular underlying parameter θ. In contrast, a Bayesian considers only the single data set in hand to be relevant. Thus the second row in the table above, which expresses the average cost of the estimate for a particular data set $\mathbf{g}$, is a Bayesian measure of risk. The left-most form in the third row defines the average cost found by first averaging over all data sets conditioned on a particular parameter θ, followed by averaging over the ensemble of possible values of θ. An equivalent value is obtained by first computing the average cost of the estimate for a fixed data set and then averaging over the ensemble of possible data vectors. The resulting risk is a scalar that summarizes the overall performance of the estimator in the presence of measurement noise as well as randomness of the underlying object. This measure is commonly (but confusingly) referred to as *Bayes risk*. Hard-nosed Bayesians would resist the average over $\mathbf{g}$.

A similar hierarchy exists when considering classification errors. Let C_{ij} be the cost associated with making decision D_i (in favor of hypothesis H_i) when hypothesis H_j is actually true. Then $C_{ij} = C(D_i, H_j)$ is the classification analog of the estimation cost function $C(\hat{\theta}, \theta)$. In general, a cost can be assigned to a correct decision as well as an incorrect decision (a positive cost indicates a penalty). If there are L total classes, the average cost of making decision D_i is:

$$\overline{C}(D_i) = \sum_{j=1}^{L} C_{ij} \Pr(H_j \mid D_i) \,. \tag{13.5}$$

The average cost, over all possible decisions, when hypothesis H_j is true is given by

$$\overline{C}(H_j) = \sum_{i=1}^{L} C_{ij} \Pr(D_i|H_j) \, . \tag{13.6}$$

The overall average cost of a decision, also called the Bayes risk, is given by

$$\begin{aligned}\overline{C} &= \sum_{i=1}^{L}\sum_{j=1}^{L} C_{ij} \Pr(H_j, D_i) \\ &= \sum_{i=1}^{L}\sum_{j=1}^{L} C_{ij} \Pr(H_j|D_i) \Pr(D_i) = \sum_{i=1}^{L}\sum_{j=1}^{L} C_{ij} \Pr(D_i|H_j) \Pr(H_j) \\ &= \sum_{i=1}^{L} \overline{C}(D_i) \Pr(D_i) = \sum_{j=1}^{L} \overline{C}(H_j) \Pr(H_j) \, . \end{aligned} \tag{13.7}$$

Note the parallel between the three kinds of averages represented by (13.5) – (13.7) and the three kinds of average cost functions for estimation presented in Table 13.1. We can average over the decision (estimate), the underlying truth state (underlying true parameter value), or both.

13.2 CLASSIFICATION TASKS

In this section we present a formal theory for the detection and classification of objects from data distorted by noise. We begin in Sec. 13.2.1 with the concept of classification as a partitioning of the data space. The specific case of a binary decision task is considered in Sec. 13.2.2, with methods for summarizing binary classification performance given in Secs. 13.2.3 – 13.2.5. Once we have defined figures of merit for task performance, we can make use of these tools in Secs. 13.2.6 and 13.2.7 to determine optimal strategies for classification. The optimal classifier for known signals on known backgrounds in Gaussian noise is treated in Sec. 13.2.8, and the non-Gaussian case is the subject of Sec. 13.2.9. Section 13.2.10 considers situations in which the signal is random; the generalization to random backgrounds is given in Sec. 13.2.11. Finally, we consider the performance of the best linear observer in Sec. 13.2.12. All these sections treat the data as discrete random vectors; continuous data sets are treated in Sec. 13.2.13. Readers already conversant with the approach to statistical decision theory found in such classic texts as Van Trees (1968), Whalen (1971), and Melsa and Cohn (1978) should be able to skim the material in Secs. 13.2.1 – 13.2.4.

13.2.1 Partitioning the data space

In a classification problem, an observer makes use of a data vector $\mathbf{g}$ to infer which of the underlying classes or hypotheses was the source of the detected data. That is, given all the inputs to the process the observer makes decision D_i, deciding in favor of hypothesis H_i. We shall impose two restrictions on the manner by which the observer makes this decision. First, we allow no randomness in the decision rule; repeated observations of the same data $\mathbf{g}$ must result in the same decision D_i. We also assume that every observation results in a decision; we shall not allow for the possibility of an equivocal test. With these assumptions, classification is equivalent to partitioning the observation space into distinct (nonoverlapping) volumes. Combined, these regions in observation space must contain all possible observation points. (See App. C for a presentation of the basics of set theory needed to understand the concept of an observation space and its underlying probability structure.)

An example data space is shown in Fig. 13.2 with nonoverlapping partitions labeled Γ_i to indicate that a data vector in that region results in decision D_i. Regardless of the topology of the decision boundaries, the union of the partitions covers all data space $\mathbb{V}$:

$$\bigcup_i \Gamma_i = \mathbb{V}, \tag{13.8a}$$

where

$$\Gamma_i \bigcap \Gamma_j = \emptyset \qquad \text{for } i \neq j. \tag{13.8b}$$

For a binary or two-class problem, the process of assigning the data to one of the decision regions is accomplished by computation of a scalar *test statistic*, t. The test statistic is related to the data through a *discriminant function* $T(\mathbf{g}) = t$. The discriminant function can be either a linear or nonlinear functional of the random data. The observer assigns a given data set to a particular decision region by comparing t to a *decision threshold* or cutoff t_c. The partition lines in Fig. 13.2 are thus contours of constant value of t.

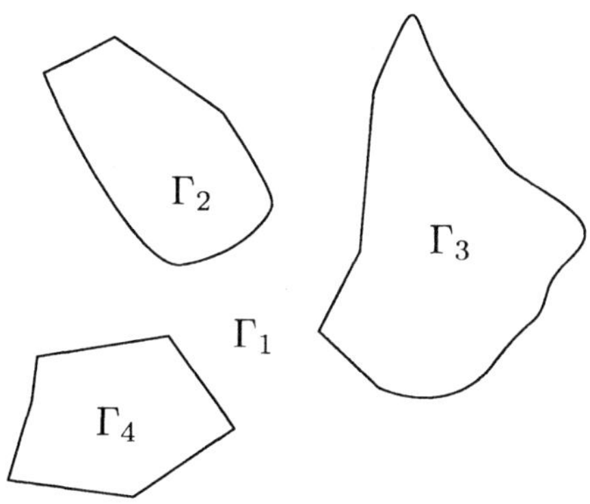

Fig. 13.2 Example of a partitioned data space showing four decision regions.

In an L class problem with $L > 2$, we need multiple discriminant functions and some partitioning rule. For example, we could compute a set of functions $\{T_\ell(\mathbf{g}), \ell = 1, ..., L\}$ and assign a particular $\mathbf{g}$ to region Γ_j if $T_j(\mathbf{g}) > T_k(\mathbf{g})$ for all $j \neq k$.

Partitioning with hyperplanes Consider the special case in which $t = T(\mathbf{g})$ is a linear function of the data. That is,

$$T(\mathbf{g}) = \mathbf{w}^t \mathbf{g}, \tag{13.9}$$

where $\mathbf{w}$ is an $M \times 1$ vector and $\mathbf{w}^t\mathbf{g}$ is the scalar product of $\mathbf{w}$ and $\mathbf{g}$. The decision boundary is an isocontour of this function, which is a hyperplane in an

M-dimensional space.[2] For a binary classification problem, the data space can be partitioned with a single such hyperplane, though the use of several hyperplanes to allow a region Γ_i to consist of disjoint subregions is not precluded. If there are L classes, at least $L-1$ hyperplanes are needed.

General discriminants and their density functions More generally, the test statistic is a nonlinear function of the data, in which case the decision boundary can deviate greatly from a hyperplane. For example, when the test statistic depends quadratically on the data, the decision boundaries shown in Fig. 13.2 would be quadratic in shape.

Whatever its functional dependence on the data, the test statistic t is a random variable through its dependence on $\mathbf{g}$. The probability density function on t depends on the underlying hypothesis:

$$\mathrm{pr}(t|H_j) = \int_\infty d^M g \; \mathrm{pr}(t|\mathbf{g})\,\mathrm{pr}(\mathbf{g}|H_j) = \int_\infty d^M g \; \delta[t - T(\mathbf{g})]\,\mathrm{pr}(\mathbf{g}|H_j)\,, \tag{13.10}$$

where we have represented the deterministic function $T(\mathbf{g}) = t$ as a probabilistic mapping.

The ellipses in Figure 13.3 represent the PDFs $\mathrm{pr}(\mathbf{g}|H_1)$ and $\mathrm{pr}(\mathbf{g}|H_2)$ in a 2D subspace for the binary decision problem. Overlapping univariate PDFs representing $\mathrm{pr}(g_1|H_1)$ and $\mathrm{pr}(g_1|H_2)$ are plotted along the x axis. Similarly, overlapping univariate PDFs representing $\mathrm{pr}(g_2|H_1)$ and $\mathrm{pr}(g_2|H_2)$ are plotted along the y axis. A linear discriminant is also shown in the figure; the line perpendicular to the discriminant function is the t-axis. When $t \geq t_c$, H_2 is chosen; when $t < t_c$, H_1 is chosen. By varying t_c we move the line in Fig. 13.3 perpendicular to itself. In the general L-class problem, changes in decision boundaries cause the partitions in Fig. 13.2 to shrink or grow.

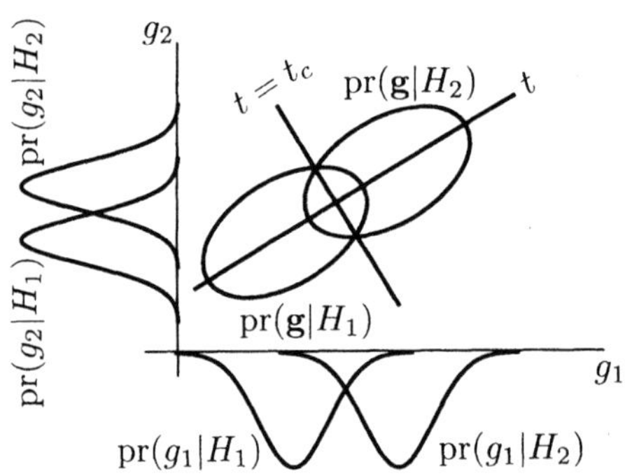

Fig. 13.3 Illustration of a linear discriminant for a 2-pixel (feature) problem. The ellipsoids indicate isocontours of constant joint probability density on the data values for each class. The line labeled $t = t_c$ is the linear discriminant for one threshold setting. Also shown are the marginal PDFs for each data component.

[2] We have assumed that both $\mathbf{g}$ and $\mathbf{w}$ are real vectors in (13.9), resulting in a real test statistic t. In the case of complex data, the form of the linear discriminant is generalized to $t = \mathrm{Re}[\mathbf{w}^\dagger \mathbf{g}]$, with $\mathbf{w}$ possibly also complex, such that t is a real scalar which is then compared to a threshold.

13.2.2 Binary decision outcomes

We again restrict our discussion to the binary decision problem depicted in Fig. 13.3, in which the object belongs to one of two classes or hypotheses: H_1 and H_2. In the detection case the hypotheses correspond to signal present or signal absent, but the hypotheses more generally represent signal family 1 vs. signal family 2. We assume that the decision made by the observer is also binary: D_1 denotes a decision that hypothesis H_1 is true, and D_2 similarly for H_2.

The observer's decision is based on the data **g**, which constitute only incomplete clues to the underlying object because they are obtained through some imaging system and contaminated by noise. Because this is the case, whatever the decision rule adopted, the decisions to which it leads cannot always be correct. Instead, as shown in Table 13.2, four scenarios exist for each experimental observation.

Table 13.2 Decision outcomes

1.	True positive (TP):	H_2 is true; observer decides H_2 is true.
2.	False positive (FP):	H_1 is true; observer decides H_2 is true.
3.	False negative (FN):	H_2 is true; observer decides H_1 is true.
4.	True negative (TN):	H_1 is true; observer decides H_1 is true.

Two of the above alternatives result in the observer correctly determining the underlying hypothesis, but we also see that two types of errors can be made. If the problem is to decide whether signal is present or absent, and the observer says a signal is present when in fact it is not, a type I error is made. In radar terminology this is called a *false alarm*, while in the medical literature it is called a *false positive*. When the signal is present, but the observer chooses the noise-only alternative, we say a *miss* or *false negative* has occurred. This is known as a Type II error.

The four scenarios for the binary detection problem can be represented by the 2×2 decision table shown in Table 13.3.

Table 13.3 Decision states

	H_2: Signal present	H_1: Signal absent
Decide H_2 D_2	True positive Hit Correct detection	False positive False alarm Type I error
Decide H_1 D_1	False negative Miss Type II error	True negative "good no call"

Sensitivity and specificity Let N be the total number of decisions made by an observer. Further, let N_{TP} denote the number of true positive decisions made by the observer. Similar notation can be used for the number of times a decision is made corresponding to the other three scenarios represented in the table, so that $N = N_{TP} + N_{FP} + N_{TN} + N_{FN}$. The observed fractions of correct and incorrect decisions under each truth state are random quantities; in the limit of an infinite number of trials they yield the actual true- and false-positive fractions:

$$\mathrm{TPF} = \Pr(D_2|H_2) = \left\langle \frac{N_{TP}}{N_{TP} + N_{FN}} \right\rangle = \lim_{N\to\infty} \left[\frac{\text{Number of true positive decisions}}{\text{Number of actually positive cases}} \right], \tag{13.11a}$$

$$\mathrm{TNF} = \Pr(D_1|H_1) = \left\langle \frac{N_{TN}}{N_{TN} + N_{FP}} \right\rangle = \lim_{N\to\infty} \left[\frac{\text{Number of true negative decisions}}{\text{Number of actually negative cases}} \right], \tag{13.11b}$$

$$\mathrm{FPF} = \Pr(D_2|H_1) = \left\langle \frac{N_{FP}}{N_{TN} + N_{FP}} \right\rangle = 1 - \mathrm{TNF}\,, \tag{13.11c}$$

$$\mathrm{FNF} = \Pr(D_1|H_2) = \left\langle \frac{N_{FN}}{N_{TP} + N_{FN}} \right\rangle = 1 - \mathrm{TPF}\,. \tag{13.11d}$$

By definition each of these fractions has value in the range from zero to one.

We see from (13.11*c*) and (13.11*d*) that the observer's performance is fully specified by two of the fractions, *e.g.*, TPF and FPF. In the medical literature, the TPF is referred to as the *sensitivity*, since it is an indication of the sensitivity of the test to the presence of an abnormality. The TNF is commonly referred to as the *specificity*, because a test with low specificity is one where there are many false or meaningless positive decisions.

We can compute the fraction of true and false decisions for each truth state (see Table 13.2) given knowledge of the probability density functions on t, which we presume we have via (13.10). As we see in Fig. 13.4, these fractions are the areas under the appropriate PDF on t for a given threshold t_c:

$$\mathrm{TPF} = \Pr(t \geq t_c|H_2) = \int_{t_c}^{\infty} dt\ \mathrm{pr}(t|H_2)\,, \tag{13.12a}$$

$$\mathrm{FPF} = \Pr(t \geq t_c|H_1) = \int_{t_c}^{\infty} dt\ \mathrm{pr}(t|H_1)\,, \tag{13.12b}$$

$$\mathrm{TNF} = 1 - \mathrm{FPF} = \int_{-\infty}^{t_c} dt\ \mathrm{pr}(t|H_1)\,, \tag{13.12c}$$

$$\mathrm{FNF} = 1 - \mathrm{TPF} = \int_{-\infty}^{t_c} dt\ \mathrm{pr}(t|H_2)\,. \tag{13.12d}$$

13.2.3 The ROC curve

We see from (13.12*a*) and (13.12*b*) that the TPF and FPF depend on the decision threshold or criterion t_c. A decision-maker is said to be conservative when a strict criterion is used that results in relatively few positive decisions, some of which are correct and some of which are false. On the other hand, a lax criterion yields a higher number of positive decisions, both true and false. By varying the decision threshold, a plot showing the relationship between the TPF and the FPF can be generated. This plot, known as a *receiver operating characteristic curve*, or an ROC curve, is a method of portraying test performance that is becoming increasingly popular in the medical community. The ROC curve is an extremely useful tool because, as we shall see, it summarizes the difficulty of the task, the performance of the decision strategy, and the quality of the data for enabling the observer to perform the specified task. Additionally, the ROC curve and figures of merit derived from it are independent of prevalence, the main complaint lodged against the use of accuracy (see Sec. 13.2.4) as a measure of classification performance.

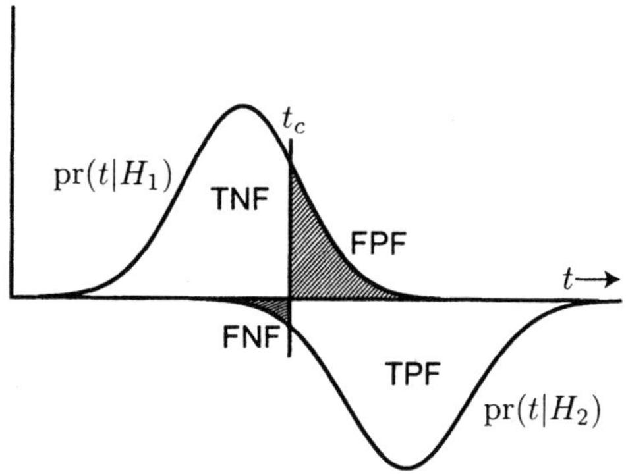

Fig. 13.4 The probability density functions for the test statistic under the two hypotheses and one setting of the threshold, with the resulting TPF, FPF, TNF, FNF.

A sample ROC curve is shown in Fig. 13.5. There are three points on the ROC curve labeled A, B, and C. Point A corresponds to a very strict criterion level, so that the observer has very few false-positive responses, but few true-positive responses as well. As the observer's threshold is relaxed, the number of true-positive and false-positive responses increases. The point indicated by the letter B on the graph is a moderate criterion level, and point C corresponds to a very lenient criterion. At point C the observer often says signal is present in the images, resulting in many true-positive calls, but many false-positive calls, too.

Properties of ROC curves are covered thoroughly in the three-volume series by Van Trees (1968). The application of ROC techniques to nuclear medical imaging and radiography has been discussed by Lusted (1971), Metz *et al.* (1973), and Anderson *et al.* (1973). Two early tutorial papers on ROC analysis in medical imaging were published by Metz (1978) and Turner (1978). Good general overviews have been presented by Swets (1979), Swets and Pickett (1982), Swets (1988), and Metz (1999).

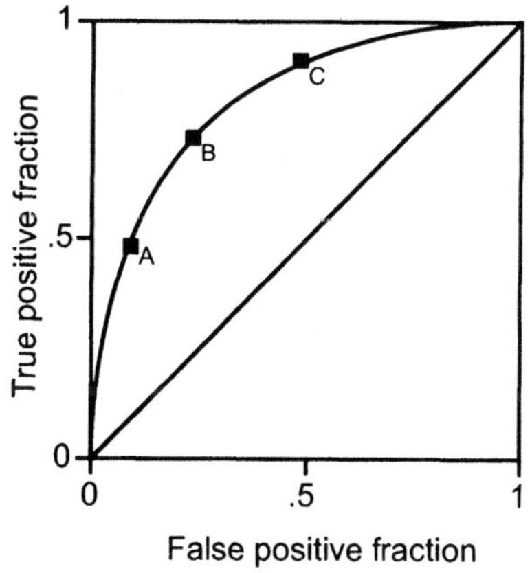

Fig. 13.5 A sample ROC curve with three threshold levels identified.

13.2.4 Performance measures for binary tasks

A variety of measures exist for summarizing task performance for a given imaging system and observer. More importantly, meaningful summary measures of task performance allow the quantitative comparison of imaging systems and observers.

Accuracy It is tempting to summarize the performance of a classification task by its *accuracy*, which is the fraction of decisions that are correct:

$$\text{accuracy} = \lim_{N\to\infty} \frac{N_{TP} + N_{TN}}{N} . \tag{13.13}$$

As mentioned briefly in Sec. 13.2.3, the problem with accuracy as a figure of merit is that it is highly dependent on the prevalence of the underlying hypotheses. Consider a physician involved in screening patients for a disease so rare it affects only 0.1% of the population $[\Pr(H_2) = 0.001]$. The physician can perform with 99.9% accuracy simply by calling all patients normal, all the while miscalling every case where disease is present, a clearly worthless strategy. Accuracy can make a bad test look good when the prevalences are unbalanced.

A similar misrepresentation of test performance occurs for any figure of merit that involves only one of the probabilities of error. Beware of studies that claim excellent performance based on high values for sensitivity or specificity alone. A test with high sensitivity could be achieved by calling all patients positive for disease, but this strategy would yield abysmal specificity.

Positive and negative predictive value Two other measures for summarizing the performance of a diagnostic test or classification task are sometimes found in the literature. These are the *positive predictive value* (PPV) and *negative predictive value* (NPV) of the test. They are defined by:

$$\text{PPV} = \Pr(H_2|D_2) = \lim_{N\to\infty} \frac{N_{TP}}{N_{TP} + N_{FP}} , \tag{13.14a}$$

and

$$\text{NPV} = \Pr(H_1|D_1) = \lim_{N\to\infty} \frac{N_{TN}}{N_{TN} + N_{FN}} . \tag{13.14b}$$

These measures have a decidedly Bayesian flavor, in that they tell us how likely an underlying condition is given that the test decides in favor of that condition.

Cost and utility We already described in Sec. 13.1 how we can characterize a decision-making strategy by associating costs with making correct and incorrect decisions. Three kinds of average classification cost are defined in (13.5) – (13.7). Any one of these average costs can be used as a summary measure of performance, though the choice among them and the assignment of costs is inherently arbitrary.

The *utility* of a classification outcome has been defined as a measure of the desirability of the outcome relative to other outcomes (Patton and Woolfenden, 1989). In medical applications the overall diagnostic utility of a test can be evaluated as an expectation of the utility of the test over the population of patients and outcomes. The cost effectiveness of a test is determined by the utility of the test and its cost (Fryback and Thornbury, 1991). Early models for analyzing the cost effectiveness of classification decisions were presented by McNeil and Adelstein (1976), Weinstein and Fineberg (1980), and Swets and Pickett (1982). Cost-effectiveness analysis of medical imaging exams, including the determination of decision costs and the valuation of life quality with and without various medical interventions, is an area of active research (Gold *et al.*, 1996; Russell *et al.*, 1996).

TPF-FPF pairs The ROC curve suggests a number of summary measures of classification performance. The first is the TPF at a specified FPF, which is no more than a presentation of two of the elements in Table 13.2. A detection strategy that maximizes this performance measure is known as the *Neyman-Pearson criterion* (Sec. 13.2.6).

When we can make the argument that t is normally distributed, then the TPF and FPF can be derived from (13.12) to give

$$\text{TPF} = \frac{1}{2}\left[1 - \text{erf}\left(\frac{t_c - \bar{t}_2}{\sqrt{2\sigma_2^2}}\right)\right], \qquad (13.15a)$$

$$\text{FPF} = \frac{1}{2}\left[1 - \text{erf}\left(\frac{t_c - \bar{t}_1}{\sqrt{2\sigma_1^2}}\right)\right], \qquad (13.15b)$$

at each threshold level t_c. Here $\langle t \rangle_j = \bar{t}_j$ denotes the mean of the test statistic under hypothesis H_j, and σ_j^2 is the variance of the test statistic under hypothesis j, defined by

$$\sigma_j^2 = \left\langle \left(t - \bar{t}_j\right)^2 \right\rangle_j . \qquad (13.16)$$

Throughout this chapter the notation $\langle \cdot \rangle_j$ indicates an average over the data when hypothesis j is true, or equivalently, an average over the density of t given that H_j is true. The *error function*, erf(z), is defined by (C.115)

$$\text{erf}(z) = \frac{2}{\sqrt{\pi}} \int_0^z dy \exp\left[-y^2\right] . \qquad (13.17)$$

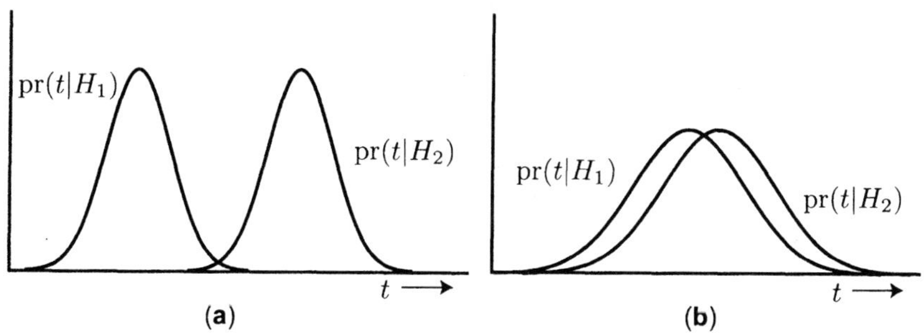

Fig. 13.6 Plots of $\text{pr}(t|H_1)$ and $\text{pr}(t|H_2)$ for two cases. In (a) the PDFs exhibit little overlap, resulting in relatively few decision errors. In (b) the PDFs are virtually identical and the TPF and FPF are approximately equal for each value of t_c.

Figure 13.6 contains plots of $\text{pr}(t|H_1)$ and $\text{pr}(t|H_2)$ for two cases. In Fig. 13.6a the PDFs exhibit little overlap, enabling the observer to choose a threshold that separates the two classes quite well. Figure 13.6b shows the case where the PDFs are virtually identical. In this case, the TPF and FPF are approximately equal for each value of t_c, and the observer using this test statistic has difficulty separating the classes. Plots of the TPF vs. FPF for these cases are shown in Fig. 13.7. Two PDFs with no overlap (perfect class separation) yield an ROC curve formed by the left and top borders of the square, while the 45° or chance line results when $\text{pr}(t|H_1) = \text{pr}(t|H_2)$ for all t.

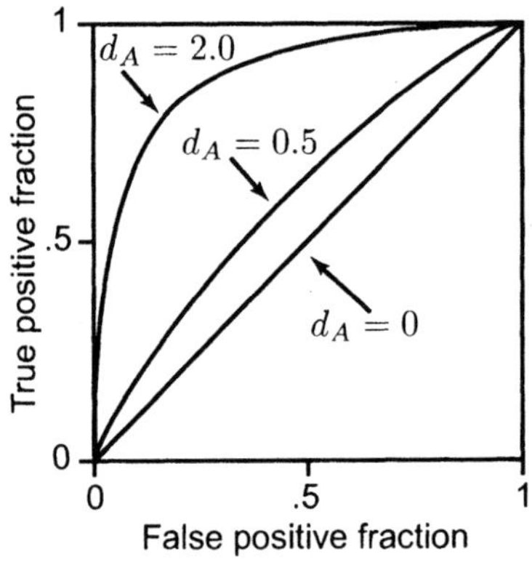

Fig. 13.7 Plots of the TPF vs. FPF for the PDFs depicted in Fig. 13.6, as well as the chance line ($d_A = 0$). The ROC curves are symmetric about the negative diagonal.

AUC Another figure of merit is the area under the ROC curve, AUC, defined as

$$\text{AUC} = \int_0^1 d\text{FPF TPF(FPF)}. \tag{13.18}$$

Since both the FPF and the TPF range from 0 to 1, the area under the ROC curve also ranges from 0 to 1. The ROC curves shown in Fig. 13.7 are symmetric about the negative diagonal. For this special case, higher values of AUC indicate higher true positive fraction for any given false positive fraction; intuitively one can see that classification systems with higher AUC are then preferable. On the other hand, ROC curves need not be symmetric, as Fig. 13.8 illustrates. The ROC curves in Fig. 13.8 have the same area, which is the *average* TPF over all FPF. Thus the AUC may be the same for two imaging systems, even though one may be superior within some range of FPF values.

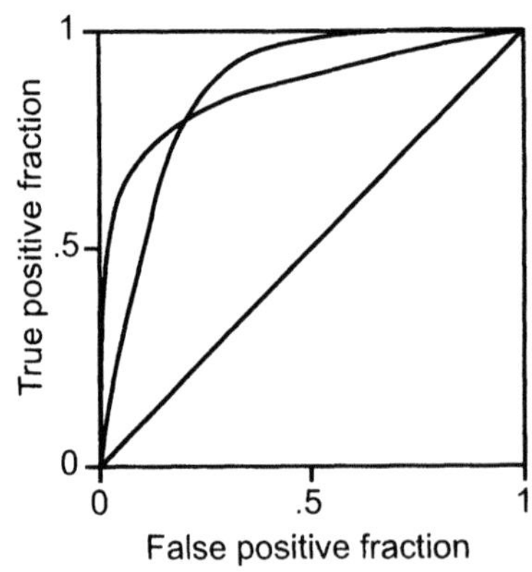

Fig. 13.8 Two crossing ROC curves with the same area.

SNR$_t$ The degree of overlap of the density functions of the test statistic determines the separability of the classes in the general classification problem; in the detection problem this overlap determines the detectability of the signal. The AUC is one measure of this overlap. Another is the signal-to-noise ratio associated with t:

$$\text{SNR}_t = \frac{\langle t\rangle_2 - \langle t\rangle_1}{\sqrt{\frac{1}{2}\sigma_1^2 + \frac{1}{2}\sigma_2^2}} . \tag{13.19}$$

SNR$_t$ should be used with caution if the test statistic is not normally distributed under both hypotheses. In particular, when the test statistic has a highly skewed PDF, the variance is not a good measure of the spread of the decision variable, and this figure of merit is not useful.

SNR under assumptions of normality When the test statistic is normally distributed under both hypotheses, the area under the ROC curve can be derived from SNR$_t$ through the following relationship:

$$\text{AUC} = \tfrac{1}{2} + \tfrac{1}{2}\,\text{erf}\left(\frac{\text{SNR}_t}{2}\right) , \tag{13.20}$$

where erf(z) is defined in (13.17). A proof of this relationship is given in Sec. 13.2.5.

When the variance of a normally-distributed test statistic is the same under each hypothesis, SNR$_t$ is granted the special name d' (Simpson and Fitter, 1973; Swets, 1979). In this particular case, the ROC curve bows upward toward the left-hand corner of the ROC plot and is symmetric about the negative diagonal. The value of d' is related monotonically to the distance between the ROC curve and the positive diagonal. Figure 13.7 contains two equal-variance ROC curves with different nonzero values of d'. When d' is zero, pr$(t|H_1)$ is equal to pr$(t|H_2)$, and the resulting ROC curve is the chance line. When d' is infinite, there is perfect separation between pr$(t|H_2)$ and pr$(t|H_1)$. Consequently, d' values range from 0 to ∞ and classification performance increases monotonically with d'.

If the test statistic is normal with unequal variance under the two hypotheses, the ROC curve will be asymmetric, like those shown in Fig. 13.8. The shape of the curve in that case is determined by the ratio of the variances under the two hypotheses (Green and Swets, 1966). In this circumstance, neither SNR$_t$ nor AUC is sufficient to specify the system completely. Of course, we can still find an area under the ROC curve, but two skewed ROC curves may have the same area, as shown in the figure. We need another way of deciding which imaging system yields the best performance when this happens. Metz and Kronman (1980) have discussed several possible approaches. One option is to choose the system with the highest partial area under the ROC curve, given as a modified version of (13.18) involving only the portion of ROC-space thought to be most critical for the system under consideration (McClish, 1989). Another is to compare the ROC curves by considering the utility associated with their optimal operating points (Halpern *et al.*, 1996).

The estimate of AUC given in (13.20) can be grossly in error for test statistics that are far from normally distributed. In such cases, it is advisable to determine AUC directly from (13.18).

Detectability index d_A When AUC is known, it can be used to compute an effective signal-to-noise ratio simply by inverting (13.20). The resulting figure of merit is denoted d_A and calculated via (ICRU Report 54, 1996)

$$d_A = 2\,\mathrm{erf}^{-1}[2(\mathrm{AUC}) - 1]\,. \tag{13.21}$$

The quantity is invariant to any monotonic transformation on the test statistic used to obtain AUC.

Most experimental ROC curves from psychophysical studies of human performance can be modeled quite well with the assumption that the test statistic is normal under each hypothesis, although with unequal variance, otherwise referred to as a *binormal* model for the test statistic (Swets, 1986). In that case d_A equals the difference in means divided by the square root of the average variance under each hypothesis as given in (13.19).

Many studies of human-observer performance measure AUC directly through the use of a forced-choice technique, a method described theoretically in Sec. 13.2.5 and revisited in Sec. 14.2. Such a measurement method gives an estimate of AUC without determining the shape of the ROC curve, so no information is obtained regarding the relevance of the Gaussian model for the underlying PDFs of the observer's test statistic. Even so, the measured AUC is commonly converted to d_A for comparison with other studies (see Sec. 14.2.3).

13.2.5 Computation of AUC

We now consider various ways of computing AUC, depending on what knowledge we have of the statistics of the problem. Our treatment follows Barrett *et al.* (1998b), though each of the methods has a considerable prior literature.

Discriminant function with known probability law Consider the case where we have complete statistical descriptions $\mathrm{pr}(t|H_1)$ and $\mathrm{pr}(t|H_2)$ for the discriminant function t. First let us rewrite the expression for the area under the ROC curve in (13.18) to explicitly show the role of the threshold t_c:

$$\mathrm{AUC} = \int_0^1 d\mathrm{FPF}(t_c)\;\mathrm{TPF}(t_c)\,. \tag{13.22}$$

For notational convenience, we define the shorthand $\mathrm{pr}(t|H_j) \equiv p_j(t)$. Since FPF is a monotonic function of t_c, we can change the variable of integration from $\mathrm{FPF}(t_c)$ to t_c, yielding

$$\mathrm{AUC} = -\int_{-\infty}^{\infty} dt_c\;\mathrm{TPF}(t_c)\,\frac{d}{dt_c}\mathrm{FPF}(t_c)\,, \tag{13.23}$$

where the minus sign arises since $\mathrm{FPF}(t_c) \to 1$ as $t_c \to -\infty$.

From (13.12*b*) and Leibniz' rule, we have

$$\frac{d}{dt_c}\mathrm{FPF}(t_c) = -p_1(t_c)\,, \tag{13.24}$$

so

$$\mathrm{AUC} = \int_{-\infty}^{\infty} dt_c\;p_1(t_c)\int_{t_c}^{\infty} dy\;p_2(y)\,. \tag{13.25}$$

We can rewrite this expression in a variety of ways. One is to recognize that the cumulative probability distribution function (Sec. C.2.3) of t under H_2 is given by

$$\mathrm{F}_2(t_c) \equiv \Pr(t < t_c|H_2) = \int_{-\infty}^{t_c} dy\;p_2(y) = 1 - \int_{t_c}^{\infty} dy\;p_2(y)\,. \tag{13.26}$$

Thus

$$\text{AUC} = 1 - \int_{-\infty}^{\infty} dt_c \, p_1(t_c)\, \text{F}_2(t_c)\,. \tag{13.27}$$

We can obtain another form for AUC by using the step function to rewrite (13.25) as

$$\text{AUC} = \int_{-\infty}^{\infty} dt_c \int_{-\infty}^{\infty} dy \; p_1(t_c)\, p_2(y)\, \text{step}(y - t_c)\,. \tag{13.28}$$

With the change of variables $x = y - t_c$, we obtain

$$\text{AUC} = \int_{-\infty}^{\infty} dt_c \int_{-\infty}^{\infty} dx \; p_1(t_c)\, p_2(x + t_c)\, \text{step}(x) = \int_0^{\infty} dx \; [p_1 \star p_2](x)\,, \tag{13.29}$$

where $\star$ denotes a 1D correlation integral [*cf.* (3.115)]. Computation of AUC by this formula thus requires cross-correlating p_1 and p_2 and then integrating the result over the half line from 0 to ∞.

The Fourier transform of the step function given in (3.163) allows us to write

$$\text{step}(x) = \frac{1}{2} + \frac{1}{2\pi i}\mathcal{P}\int_{-\infty}^{\infty} \frac{d\xi}{\xi} \exp(2\pi i \xi x)\,, \tag{13.30}$$

where $\mathcal{P}$ indicates that the singular integral must be interpreted as a Cauchy principal value (see Sec. B.3.9). Then (13.29) becomes

$$\begin{aligned}\text{AUC} &= \frac{1}{2} + \frac{1}{2\pi i}\mathcal{P}\int_{-\infty}^{\infty} \frac{d\xi}{\xi} \int_{-\infty}^{\infty} dt_c \int_{-\infty}^{\infty} dy \; p_1(t_c)\, p_2(y) \exp[2\pi i \xi (y - t_c)] \\ &= \frac{1}{2} + \frac{1}{2\pi i}\mathcal{P}\int_{-\infty}^{\infty} \frac{d\xi}{\xi}\, \psi_{t1}(\xi)\, \psi_{t2}^*(\xi)\,, \end{aligned} \tag{13.31}$$

where $\psi_{tj}(\xi) = \langle \exp(-2\pi i \xi t) | H_j \rangle$ is the characteristic function for t under hypothesis H_j [see (C.53)].

AUC with normal probability law When we know that the functions $\text{pr}_1(t)$ and $\text{pr}_2(t)$ are both univariate Gaussians, the characteristic function for t under hypothesis H_j is given by

$$\psi_{tj}(\xi) = \exp(-2\pi i \bar{t}_j \xi - 2\pi^2 \sigma_j^2 \xi^2)\,. \tag{13.32}$$

Then the AUC of (13.31) becomes

$$\begin{aligned}\text{AUC} &= \frac{1}{2} + \frac{1}{2\pi i}\mathcal{P}\int_{-\infty}^{\infty} \frac{d\xi}{\xi} \exp\left[-2\pi i (\bar{t}_1 - \bar{t}_2)\,\xi - 2\pi^2 (\sigma_1^2 + \sigma_2^2)\,\xi^2\right] \\ &= \frac{1}{2} + \frac{1}{2\pi}\mathcal{P}\int_{-\infty}^{\infty} \frac{d\xi}{\xi} \sin[2\pi(\bar{t}_2 - \bar{t}_1)\,\xi] \exp\left[-2\pi^2 (\sigma_1^2 + \sigma_2^2)\,\xi^2\right] \\ &= \frac{1}{2} + (\bar{t}_2 - \bar{t}_1)\int_{-\infty}^{\infty} d\xi \; \text{sinc}[2(\bar{t}_2 - \bar{t}_1)\,\xi]\; \text{gaus}\left[\sqrt{2\pi(\sigma_1^2 + \sigma_2^2)}\,\xi\right]\,, \end{aligned} \tag{13.33}$$

where gaus(·) is defined in (3.173), and we have dropped the principal value in the third line because the integrand is now well behaved at the origin. By Parseval's theorem we have

$$\begin{aligned}\text{AUC} &= \frac{1}{2} + \frac{1}{2\sqrt{2\pi(\sigma_1^2+\sigma_2^2)}}\int_{-\infty}^{\infty} dx\ \text{rect}\left[\frac{x}{2(\bar{t}_2-\bar{t}_1)}\right]\text{gaus}\left[\frac{x}{\sqrt{2\pi(\sigma_1^2+\sigma_2^2)}}\right] \\ &= \frac{1}{2} + \frac{1}{\sqrt{2\pi(\sigma_1^2+\sigma_2^2)}}\int_0^{(\bar{t}_2-\bar{t}_1)} dx\ \exp\left[-\frac{x^2}{2(\sigma_1^2+\sigma_2^2)}\right]. \end{aligned} \tag{13.34}$$

A change of variables then yields the error-function relationship of (13.20).

Discriminant function with unknown probability law In the last sections we assumed that the densities $p_j(t)$ were known, and we derived general expressions for AUC that are independent of the specific form of the densities and for the specific case of a normal probability law. When t is a complicated function of $\mathbf{g}$, we may not know its densities or even its moments. We may, however, know $\text{pr}(\mathbf{g}|H_j)$ from the basic physics of the image-formation process and knowledge of the signal and background statistics. We shall now develop expressions for AUC in terms of integrals over $\mathbf{g}$ rather than ones over t.

If we again define $\text{pr}(t|H_j) \equiv p_j(t)$ and also define the shorthand $q_j(\mathbf{g}) \equiv \text{pr}(\mathbf{g}|H_j)$, (13.10) becomes

$$p_j(t) = \int_\infty d^M g\ q_j(\mathbf{g})\,\delta[t - T(\mathbf{g})]\,. \tag{13.35}$$

The 1D delta function defines an $(M-1)$-dimensional surface in the M-dimensional data space; all points on this surface have $T(\mathbf{g}) = t$ and hence contribute to the probability density on t at the same t.

From (13.28) and (13.35), we have

$$\text{AUC} = \int_{-\infty}^{\infty} dt_c \int_{-\infty}^{\infty} dy \int_\infty d^M g\ q_1(\mathbf{g})\,\delta[t_c - T(\mathbf{g})] \int_\infty d^M g'\ q_2(\mathbf{g}')\,\delta[y - T(\mathbf{g}')]\,\text{step}(y - t_c)\,. \tag{13.36}$$

The delta functions allow us to perform the integrals over t_c and y, with the result

$$\text{AUC} = \int_\infty d^M g \int_\infty d^M g'\ q_1(\mathbf{g})\,q_2(\mathbf{g}')\,\text{step}[T(\mathbf{g}') - T(\mathbf{g})]\,. \tag{13.37}$$

Note that if we replace $T(\mathbf{g})$ with $T'(\mathbf{g}) = h[T(\mathbf{g})]$, where $h(x)$ is a monotonically increasing function, then the step function remains unchanged: if $\text{step}[T(\mathbf{g}') - T(\mathbf{g})] = 1$ for some set of values of $\mathbf{g}$ and $\mathbf{g}'$, then $\text{step}[T'(\mathbf{g}') - T'(\mathbf{g})] = 1$ for precisely this same set. Thus AUC is unchanged by a monotonic point transformation of the data.

If we express the step function in terms of its Fourier transform, as given in (3.163), we obtain

$$\begin{aligned}\text{AUC} &= \frac{1}{2} + \frac{1}{2\pi i}\mathcal{P}\int_{-\infty}^{\infty}\frac{d\xi}{\xi}\int_\infty d^M g \int_\infty d^M g'\ q_1(\mathbf{g})\,q_2(\mathbf{g}')\exp\{2\pi i\xi\,[T(\mathbf{g}') - T(\mathbf{g})]\} \\ &= \frac{1}{2} + \frac{1}{2\pi i}\mathcal{P}\int_{-\infty}^{\infty}\frac{d\xi}{\xi}\,\langle\exp[-2\pi i\xi T(\mathbf{g})]\rangle_1\,\langle\exp[2\pi i\xi T(\mathbf{g}')]\rangle_2 \\ &= \frac{1}{2} + \frac{1}{2\pi i}\mathcal{P}\int_{-\infty}^{\infty}\frac{d\xi}{\xi}\,\psi_{t1}(\xi)\,\psi_{t2}^*(\xi)\,, \end{aligned} \tag{13.38}$$

where we have again made use of the definition of the characteristic function (C.53) in the last line. The final form of (13.38) is identical to (13.31); AUC can be obtained by computing expectations over either the probability density on $\mathbf{g}$ or the one on t.

2AFC interpretations The area under the ROC curve has another interesting interpretation. Consider an experiment where an observer is presented with two data sets $\mathbf{g}$ and $\mathbf{g}'$ simultaneously, where $\mathbf{g}$ is drawn from $\text{pr}(\mathbf{g}|H_1)$ and $\mathbf{g}'$ is drawn from $\text{pr}(\mathbf{g}|H_2)$. The observer's task is to choose the image from class 2. This experimental paradigm, called the *two-alternative forced-choice* method, or 2AFC procedure, is often used in studies of human performance (see Sec. 14.2.3).

To make a decision, the observer computes two test statistics $T(\mathbf{g})$ and $T(\mathbf{g}')$, and the data vector that gives the higher value is assigned to H_2. This assignment is correct if $T(\mathbf{g}') > T(\mathbf{g})$. Thus the probability of a correct decision is

$$\Pr(\text{correct}) = \Pr[T(\mathbf{g}') > T(\mathbf{g})] = \int_\infty d^M g \int_\infty d^M g' \, q_1(\mathbf{g}) \, q_2(\mathbf{g}') \, \text{step}[T(\mathbf{g}') - T(\mathbf{g})] \,, \tag{13.39}$$

which, by (13.37), is AUC. For any test statistic, the probability of a correct decision in a 2AFC experiment is the AUC for that observer.

A similar interpretation applies to (13.28). If we denote the test statistics by $t = T(\mathbf{g})$ and $y = T(\mathbf{g}')$, the 2AFC decision is correct if $t > y$, and (13.28) gives the probability of this event.

Linear discriminants When the test statistic is linear in the data, as in (13.9), (13.37) becomes

$$\text{AUC}_{lin} = \int_\infty d^M g \int_\infty d^M g' \, q_1(\mathbf{g}) \, q_2(\mathbf{g}') \, \text{step}[\mathbf{w}^t(\mathbf{g}' - \mathbf{g})] \,. \tag{13.40}$$

The change of variables $\mathbf{g}'' = \mathbf{g}' - \mathbf{g}$ yields

$$\begin{aligned} \text{AUC}_{lin} &= \int_\infty d^M g \int_\infty d^M g'' \, q_1(\mathbf{g}) \, q_2(\mathbf{g} + \mathbf{g}'') \, \text{step}(\mathbf{w}^t \mathbf{g}'') \\ &= \int_\infty d^M g'' \, [q_1 \star q_2](\mathbf{g}'') \, \text{step}(\mathbf{w}^t \mathbf{g}'') \,, \end{aligned} \tag{13.41}$$

where $[q_1 \star q_2](\mathbf{g}'')$ denotes a multidimensional cross-correlation integral with shift $\mathbf{g}''$. This equation shows that AUC_{lin} can be found by cross-correlating q_1 and q_2 and then integrating the result over the half-space $\mathbf{w}^t \mathbf{g} > 0$.

The similarity in form between (13.29) and (13.41) should be noted; (13.29) holds for an arbitrary discriminant function (but requires the probability densities for that function), while (13.41) holds specifically for a linear discriminant and requires knowledge of the data densities.

With a linear discriminant, we can also relate AUC to the multivariate characteristic functions for $\mathbf{g}$, defined by (see Sec. 8.1.4)

$$\psi_{\mathbf{g}j}(\boldsymbol{\rho}) \equiv \int_\infty d^M g \, q_j(\mathbf{g}) \exp(-2\pi i \boldsymbol{\rho}^t \mathbf{g}) \,, \tag{13.42}$$

where $\boldsymbol{\rho}$ is the M-dimensional frequency vector conjugate to the data vector $\mathbf{g}$. We can relate the characteristic function for the scalar test statistic t to the characteristic function for the MD data vector $\mathbf{g}$ via

$$\psi_{tj}(\xi) = \int_\infty dt\; p_j(t) \exp(-2\pi i \xi t) = \int_\infty dt \int_\infty d^M g\; \delta[t - T(\mathbf{g})] \operatorname{pr}(\mathbf{g}|H_j) \exp(-2\pi i \xi t)$$

$$= \int_\infty d^M g\; \operatorname{pr}(\mathbf{g}|H_j) \exp(-2\pi i \xi \mathbf{w}^t \mathbf{g}) = \psi_{\mathbf{g}j}(\mathbf{w}\xi)\,, \tag{13.43}$$

where we have used (13.19), (13.35), and the sifting property of the delta function. Thus, in the case of a linear discriminant, we see that the characteristic function of the test statistic is determined by the characteristic function of the data along a line through the origin and parallel to $\mathbf{w}$ in the MD Fourier space. For linear discriminants, (13.38) becomes

$$\text{AUC}_{lin} = \frac{1}{2} + \frac{1}{2\pi i} \mathcal{P} \int_{-\infty}^{\infty} \frac{d\xi}{\xi}\; \psi_{\mathbf{g}1}(\mathbf{w}\xi)\, \psi_{\mathbf{g}2}^*(\mathbf{w}\xi)\,. \tag{13.44}$$

Another way of interpreting this result is to say that all we need to calculate AUC_{lin} are integrals of the data densities q_1 and q_2 over $(M-1)$D hyperplanes normal to $\mathbf{w}$. By the MD central-slice theorem (4.178), integral of an MD density over these hyperplanes gives information about the Fourier transform of the density along the line specified in (13.43). With nonlinear discriminants, the central-slice theorem does not provide any assistance; we need integrals over $(M-1)$D surfaces defined by constant $T(\mathbf{g})$ rather than integrals over hyperplanes.

In practice, we can often simplify the computation of AUC_{lin} greatly by realizing that a linear discriminant $\mathbf{w}^t\mathbf{g}$ is univariate normal if $\mathbf{g}$ is multivariate normal. Even if $\mathbf{g}$ follows some other probability law, the output of a linear discriminant acting on the data can often be assumed to approximate a univariate normal law by the central-limit theorem (see Sec. 8.3.4). Then we can use the expression given in (13.19) for SNR_t with confidence. Thus it is almost always safe to compute AUC from SNR_t for a linear discriminant, though it is useful to check the assumption by plotting histograms of t.

Linear discriminants with independent additive noise We can be more specific about the AUC of a linear discriminant if we can assume that the noise is additive and independent of the signal. In that case, we can write:

$$\operatorname{pr}(\mathbf{g}|H_2) = \int_\infty d^M s\; \operatorname{pr}(\mathbf{g}|H_2, \mathbf{s})\; \operatorname{pr}(\mathbf{s}) = \int_\infty d^M s\; \operatorname{pr}(\mathbf{g} - \mathbf{s}|H_1)\; \operatorname{pr}(\mathbf{s})\,. \tag{13.45}$$

Since this is an MD convolution, its Fourier transform yields

$$\psi_{\mathbf{g}2}(\boldsymbol{\rho}) = \psi_{\mathbf{g}1}(\boldsymbol{\rho})\, \psi_{\mathbf{s}}(\boldsymbol{\rho})\,, \tag{13.46}$$

where $\psi_{\mathbf{s}}(\boldsymbol{\rho})$ is the characteristic function of the signal. Thus (13.44) becomes

$$\text{AUC}_{lin} = \frac{1}{2} + \frac{1}{2\pi i} \mathcal{P} \int_{-\infty}^{\infty} \frac{d\xi}{\xi}\; \psi_{\mathbf{s}}^*(\mathbf{w}\xi)\; |\psi_{g1}(\mathbf{w}\xi)|^2\,. \tag{13.47}$$

Note that $\psi(\boldsymbol{\rho}) = \psi^*(-\boldsymbol{\rho})$ for any characteristic function since it is the Fourier transform of a real function. Thus $|\psi_{\mathbf{g}1}(\mathbf{w}\xi)|^2$ is an even function of ξ, and the integral in (13.47) vanishes unless $\psi_s(\mathbf{w}\xi)$ has an odd component.

One situation where we can readily identify the odd component of $\psi_{\mathbf{s}}(\mathbf{w}\xi)$ is when the signal is nonrandom and known exactly. In that case,

$$\mathrm{pr}(\mathbf{s}) = \delta(\mathbf{s} - \mathbf{s}_0)\,, \qquad \psi_{\mathbf{s}}(\boldsymbol{\rho}) = \exp(-2\pi i \mathbf{s}_0^t \boldsymbol{\rho})\,. \tag{13.48}$$

Only the odd component of $\psi_{\mathbf{s}}(\boldsymbol{\rho})$ contributes to the integral in (13.47), so

$$\mathrm{AUC}_{lin} = \frac{1}{2} + \frac{1}{2\pi}\int_{-\infty}^{\infty} d\xi \, \frac{\sin(2\pi \mathbf{s}_0^t \mathbf{w}\xi)}{\xi} \, |\psi_{\mathbf{g}1}(\mathbf{w}\xi)|^2\,. \tag{13.49}$$

The principal-value indicator $\mathcal{P}$ is no longer necessary, and the expression is patently real.

As $\mathbf{s}_0^t\mathbf{w} \to \infty$, the factor $\sin(2\pi \mathbf{s}_0^t \mathbf{w}\xi)/\xi$ approaches $\pi\delta(\xi)$ [*cf.* (2.44)], so $\mathrm{AUC}_{lin} \to 1$; strong, nonrandom signals can be perfectly detected by any linear discriminant not orthogonal to the signal. If $\mathbf{s}_0^t\mathbf{w}$ is small, on the other hand, then

$$\mathrm{AUC}_{lin} \approx \frac{1}{2} + \mathbf{s}_0^t\mathbf{w} \int_{-\infty}^{\infty} d\xi \, |\psi_{\mathbf{g}1}(\mathbf{w}\xi)|^2\,. \tag{13.50}$$

This expression shows, not surprisingly, that $\mathrm{AUC}_{lin} \to \frac{1}{2}$ as $\mathbf{s}_0^t\mathbf{w} \to 0$, which occurs either for weak signals or for a linear discriminant orthogonal to the signal.

13.2.6 The likelihood ratio and the ideal observer

We have defined AUC and related figures of merit for classification, and these metrics provide a basis for the optimization of classification performance. We shall now explore several optimization strategies and show that all optimization roads lead to the ideal or Bayesian observer.

Minimum average cost One possible decision strategy is to formulate a test statistic that minimizes the overall average cost or Bayes risk of making a decision. The average cost given in (13.7) can be written in the binary case as

$$\begin{aligned}\overline{C} &= C_{22}\,\mathrm{Pr}(D_2|H_2)\,\mathrm{Pr}(H_2) + C_{12}\,\mathrm{Pr}(D_1|H_2)\,\mathrm{Pr}(H_2)\\ &\quad + C_{21}\,\mathrm{Pr}(D_2|H_1)\,\mathrm{Pr}(H_1) + C_{11}\,\mathrm{Pr}(D_1|H_1)\,\mathrm{Pr}(H_1)\,. \end{aligned}\tag{13.51}$$

The probability of making decision D_i when hypothesis H_j is true can be written as an integral over the M-dimensional data space:

$$\mathrm{Pr}(D_i|H_j) = \int_{\Gamma_i} d^M g \; \mathrm{pr}(\mathbf{g}|H_j)\,, \tag{13.52}$$

where Γ_i defines the observation region over which a decision is made in favor of H_i (see Fig. 13.2). The average cost can thus be written in terms of integrals over the regions Γ_1 and Γ_2:

$$\begin{aligned}\overline{C} &= C_{22}\,\mathrm{Pr}(H_2)\int_{\Gamma_2} d^M g \; \mathrm{pr}(\mathbf{g}|H_2) + C_{12}\,\mathrm{Pr}(H_2)\int_{\Gamma_1} d^M g \; \mathrm{pr}(\mathbf{g}|H_2)\\ &\quad + C_{21}\,\mathrm{Pr}(H_1)\int_{\Gamma_2} d^M g \; \mathrm{pr}(\mathbf{g}|H_1) + C_{11}\,\mathrm{Pr}(H_1)\int_{\Gamma_1} d^M g \; \mathrm{pr}(\mathbf{g}|H_1)\,. \end{aligned}\tag{13.53}$$

To minimize the overall cost is to choose an optimal division between the regions Γ_1 and Γ_2.

We can use the fact that

$$\Pr(D_2|H_2) + \Pr(D_1|H_2) = 1\,, \tag{13.54}$$

and that the regions are nonoverlapping and complete to say

$$\int_{\Gamma_2} d^M g \; \mathrm{pr}(\mathbf{g}|H_2) + \int_{\Gamma_1} d^M g \; \mathrm{pr}(\mathbf{g}|H_2) = 1\,. \tag{13.55}$$

We can then rewrite the average cost as a function of region Γ_2 only:

$$\overline{C} = C_{12}\Pr(H_2) + C_{11}\Pr(H_1)$$

$$+ \int_{\Gamma_2} d^M g \; [(C_{21} - C_{11})\Pr(H_1)\,\mathrm{pr}(\mathbf{g}|H_1) - (C_{12} - C_{22})\Pr(H_2)\,\mathrm{pr}(\mathbf{g}|H_2)]\,. \tag{13.56}$$

The first two terms in the cost function are constants, leaving the integral to be minimized by choice of the region Γ_2. This is done by choosing to include in the region only that portion of the observation space for which the integrand is negative. Thus the region Γ_2, where H_2 is chosen, is the region for which

$$(C_{12} - C_{22})\,\mathrm{pr}(\mathbf{g}|H_2)\Pr(H_2) > (C_{21} - C_{11})\,\mathrm{pr}(\mathbf{g}|H_1)\Pr(H_1)\,, \tag{13.57}$$

where we have assumed that the cost of making an error is greater than the cost of a correct decision, so $(C_{12} - C_{22})$ and $(C_{21} - C_{11}) > 0$, and we made use of the knowledge that all PDFs are nonnegative. The decision rule is then to choose H_2 when

$$\frac{\mathrm{pr}(\mathbf{g}|H_2)}{\mathrm{pr}(\mathbf{g}|H_1)} > \frac{(C_{21} - C_{11})\Pr(H_1)}{(C_{12} - C_{22})\Pr(H_2)}\,. \tag{13.58}$$

Otherwise, choose H_1.

The quantity $\mathrm{pr}(\mathbf{g}|H_2)/\,\mathrm{pr}(\mathbf{g}|H_1)$ is called the *likelihood ratio* and is often written $\Lambda(\mathbf{g})$. It is a scalar random variable that depends on the data vector $\mathbf{g}$. In terms of $\Lambda(\mathbf{g})$, the decision rule becomes

$$\Lambda(\mathbf{g}) \underset{D_1}{\overset{D_2}{\gtrless}} \frac{(C_{21} - C_{11})\Pr(H_1)}{(C_{12} - C_{22})\Pr(H_2)}\,. \tag{13.59}$$

This inequality is to be read "decide hypothesis H_2 true whenever the greater-than sign holds; decide hypothesis H_1 when the less-than sign holds."

We have found the region Γ_2 that minimizes the average cost of (13.51). The resulting minimum average cost is called the Bayes risk, and this decision criterion is called the Bayes criterion. A detector that uses this criterion to do signal detection is called a Bayesian detector.

The overall average cost is determined by repeated decision-making trials. Thus the performance of the Bayesian detector that minimizes this value is determined using frequentist methods.[3] Similarly, AUC and all measures derived from it are found by keeping score in frequentist fashion.

[3] Recall from Sec. 13.1 that a hard-nosed Bayesian would not use the Bayes criterion. Strict Bayesians resist the last averaging step, relying instead on the cost determined from the posterior as represented in the second row of Table 13.1.

Minimum-error detector The costs C_{ij} can be quite difficult to know. Many times they are determined in some *ad hoc* manner. As a result, it is often desirable to find a decision rule that minimizes the average probability of error, which is equivalent to maximizing the AUC. The probability of error in the binary decision problem is given by

$$P_e = \Pr(D_1|H_2)\Pr(H_2) + \Pr(D_2|H_1)\Pr(H_1)\,. \tag{13.60}$$

Comparing (13.60) to (13.51), we discover that minimizing the probability of error is equivalent to minimizing the average cost, provided no costs are associated with correct decisions, and each kind of error is assigned an equal cost. From (13.59) we ascertain that the minimum-error decision rule is written

$$\Lambda(\mathbf{g}) \underset{D_1}{\overset{D_2}{\gtrless}} \frac{\Pr(H_1)}{\Pr(H_2)}\,. \tag{13.61}$$

This decision rule tells us to choose H_2 when

$$\frac{\mathrm{pr}(\mathbf{g}|H_2)}{\mathrm{pr}(\mathbf{g}|H_1)} > \frac{\Pr(H_1)}{\Pr(H_2)}\,, \tag{13.62}$$

or when

$$\mathrm{pr}(\mathbf{g}|H_2)\Pr(H_2) > \mathrm{pr}(\mathbf{g}|H_1)\Pr(H_1)\,, \tag{13.63}$$

since $\Pr(H_j)$ and $\mathrm{pr}(\mathbf{g}|H_j)$ are always positive.

We can use Bayes' theorem,

$$\mathrm{pr}(\mathbf{g}|H_j)\Pr(H_j) = \Pr(H_j|\mathbf{g})\,\mathrm{pr}(\mathbf{g})\,, \tag{13.64}$$

to rewrite the condition for choosing H_2 as

$$\Pr(H_2|\mathbf{g})\,\mathrm{pr}(\mathbf{g}) > \Pr(H_1|\mathbf{g})\,\mathrm{pr}(\mathbf{g}) \tag{13.65}$$

or

$$\Pr(H_2|\mathbf{g}) > \Pr(H_1|\mathbf{g})\,. \tag{13.66}$$

The decision rule is to choose H_2 when the *a posteriori* probability of H_2 given the data vector $\mathbf{g}$ is greater than the *a posteriori* probability of H_1 given $\mathbf{g}$. For this reason the minimum-error criterion is also called the *maximum a posteriori probability*, or *MAP*, criterion.

Neyman-Pearson criterion In certain applications we may wish to constrain the false-positive fraction to be less than or equal to some level α. The Neyman-Pearson test seeks the maximum true-positive fraction given this constraint. We now derive the decision strategy that yields this outcome.

We want to impose the constraint that FPF $= \alpha' \leq \alpha$ while maximizing the TPF. We can solve this problem using Lagrange multipliers. The Lagrangian function we want to maximize is

$$\begin{aligned} F &= \mathrm{TPF} + \gamma[\alpha' - \mathrm{FPF}] \\ &= \int_{\Gamma_2} d^M g\, [\mathrm{pr}(\mathbf{g}|H_2) - \gamma\,\mathrm{pr}(\mathbf{g}|H_1)] + \gamma\alpha'\,. \end{aligned} \tag{13.67}$$

The problem is thus one of choosing the discriminant function such that the region Γ_2 results in maximum F, similar to the approach we used earlier to find the decision rule that gives minimum average cost. For positive γ, we obtain maximum F if we include all $\mathbf{g}$ in Γ_2 for which $\mathrm{pr}(\mathbf{g}|H_2) > \gamma\, \mathrm{pr}(\mathbf{g}|H_1)$. In other words, the Neyman-Pearson decision boundaries are obtained by assigning to Γ_2 all data for which

$$\frac{\mathrm{pr}(\mathbf{g}|H_2)}{\mathrm{pr}(\mathbf{g}|H_1)} > \gamma . \tag{13.68}$$

We see that a constraint on the likelihood ratio results. When $\gamma < 0$, a likelihood-ratio constraint also results, of opposite sign in that instance. Since the probability density functions are continuous, the probability that these quantities are equal is zero, and thus we have neglected this case.

Lastly, we need a solution that satisfies the constraint FPF $= \alpha' \leq \alpha$. We can rewrite the decision rule of (13.68) in terms of the likelihood ratio as

$$\Lambda(\mathbf{g}) \underset{D_1}{\overset{D_2}{\gtrless}} \gamma , \tag{13.69}$$

whereby we see that the Lagrange multiplier plays the role of a threshold for the scalar decision variable $\Lambda(\mathbf{g})$. The FPF can thus be expressed in terms of $\Lambda(\mathbf{g})$ as

$$\mathrm{FPF} = \int_{\gamma}^{\infty} d\Lambda\ \mathrm{pr}[\Lambda(\mathbf{g})|H_1] = \alpha' . \tag{13.70}$$

Note that decreasing γ increases the region for which we decide in favor of H_2, which thereby increases the TPF (and, of course, the FPF). Since our goal is to maximize the TPF while keeping FPF $\leq \alpha$, we continue to decrease γ until we run into the $\alpha' = \alpha$ condition (we are assuming the FPF to be a continuous function of γ). Thus the maximum TPF is obtained when the FPF equals its upper bound, α.

The particular value of α in the above derivation is arbitrary. Thus, the decision rule given by (13.69) yields maximum TPF at every predetermined FPF, and this is equivalent to a decision rule that yields maximum AUC by virtue of (13.18).

Maximum-likelihood criterion There is one last decision criterion important to us. This is the decision rule that results when the observer has no information about the *a priori* probabilities of the hypotheses, $\Pr(H_2)$ and $\Pr(H_1)$, which in medical imaging correspond to the probabilities of disease. When this is the case the observer has no reason not to assume they are equivalent (there is no information to the contrary, so this is the least prejudiced assumption). The decision rule under this assumption becomes

$$\Lambda(\mathbf{g}) \underset{D_1}{\overset{D_2}{\gtrless}} 1 . \tag{13.71}$$

We call (13.71) a *maximum-likelihood* criterion because we choose the hypothesis which results in the greatest probability or likelihood of the data vector given that hypothesis. This is the decision strategy that results when the observer has the least amount of information about the decision problem.

The ideal observer We have described four decision strategies, all of the form

$$\text{Choose } H_2 \text{ if } \Lambda(\mathbf{g}) > \Lambda_c , \tag{13.72}$$

where $\Lambda(\mathbf{g})$ is the likelihood ratio and Λ_c is the threshold determined by the particular objective of the test. Note that the data-dependent part of the decision rule in each case is the same. Thus all four decision strategies described above yield the same ROC curve. The observers differ only in their choice of threshold, the operating point along the ROC curve.

A test that can be written in the form of (13.72) is termed a *likelihood-ratio test*. Any observer that performs a likelihood-ratio test properly is referred to as an *ideal observer*.

What is required to perform a likelihood-ratio test properly? Looking back at our four test strategies, we see that the ideal observer must have no internal noise mechanism that would further corrupt the data. It must know the threshold and maintain it exactly at that level on each trial. Most importantly, it must have all the information necessary to formulate $\text{pr}(\mathbf{g}|H_j)$, including descriptions of the objects to be classified and complete information regarding the measurement process and the noise statistics.

In the previous subsections we have seen how particular choices of threshold result in minimum-Bayes risk or minimum-error performance for the ideal observer. These are just two operating points on the same ROC curve. These operating points are determined by the particular values of the priors and decision costs that enter into the calculation of the threshold. Since the ideal observer is optimal for every choice of costs C_{ij} in (13.59), it is optimal at all operating points. In other words, the ideal observer is that observer that achieves maximum TPF for any specified FPF. It follows that the ideal observer achieves maximum AUC of all observers. AUC is a widely-used figure of merit for summarizing ideal-observer performance among scientists studying image quality, since figures of merit that require the specification of the operating point depend on the costs and priors assigned by the various users of the system and are therefore fairly arbitrary.

Monotonic transformations and performance of the ideal observer Since monotonic transformations of the discriminant function do not affect decision outcomes, an observer that uses a monotonically transformed version of the likelihood ratio is also called an ideal observer. In particular, when the data statistics are normal, the log-likelihood ratio, denoted $\lambda(\mathbf{g})$ or just λ, is often found to be an easier test statistic to work with. Since the decision outcomes are unchanged, $\text{AUC}_\lambda = \text{AUC}_\Lambda$. Therefore, d_A, as defined in (13.21), is also invariant to whether the likelihood ratio or the log-likelihood ratio is used as the test statistic. On the other hand, the SNR computed from (13.19) is very much dependent on the statistical properties of the decision variable. Thus SNR_λ may be quite different from SNR_Λ. An SNR computed from (13.19) is useful only when the decision variable is approximately Gaussian; otherwise it can be very misleading as a summary measure of observer performance.

In the same spirit, image processing does not affect the ideal observer's AUC, so long as the processing operation is invertible in the sense that the original data before processing are recoverable. When it is not invertible, the processing may reduce the ideal observer's AUC. Alternatively, post-processing cannot improve ideal-observer performance. By definition the ideal observer operates optimally on the data to achieve the maximum AUC for the specified classification task, so if any algorithm is useful, she will use it.

To see the invariance of the ideal observer to post-processing, let $\mathbf{y} = \mathcal{B}\mathbf{g}$ represent the output of an invertible processing operation applied to the data. Since the data are random, $\mathbf{y}$ is also a random vector in the range of $\mathcal{B}$. The PDF on $\mathbf{y}$ is given by

$$\mathrm{pr}_{\mathbf{y}}(\mathbf{y}) = \mathrm{pr}_{\mathbf{y}}(\mathcal{B}\mathbf{g}) = \frac{\mathrm{pr}_{\mathbf{g}}(\mathbf{g})}{|J|}, \tag{13.73}$$

where J is the Jacobian of the transformation [see (C.102)], assumed to be nonzero for all $\mathbf{g}$.

An ideal observer who has only $\mathbf{y}$ to perform the classification task forms the likelihood ratio in $\mathbf{y}$-space:

$$\Lambda(\mathbf{y}) = \frac{\mathrm{pr}_{\mathbf{y}}(\mathbf{y}|H_2)}{\mathrm{pr}_{\mathbf{y}}(\mathbf{y}|H_1)} = \frac{\mathrm{pr}_{\mathbf{g}}(\mathbf{g}|H_2)}{\mathrm{pr}_{\mathbf{g}}(\mathbf{g}|H_1)} = \Lambda(\mathbf{g}) \tag{13.74}$$

since the Jacobians cancel. Thus the performance of an ideal observer given only $\mathbf{y}$ is identical to the performance of an ideal observer with access to the original data $\mathbf{g}$. This conclusion is true whether $\mathcal{B}$ is linear or nonlinear and whether $\mathbf{g}$ and $\mathbf{y}$ are continuous or discrete; all we require is that the dimensionality of $\mathbf{g}$ equal the dimensionality of $\mathbf{y}$ and the transformation from $\mathbf{g}$ to $\mathbf{y}$ be invertible.

13.2.7 Statistical properties of the likelihood ratio

Knowledge of the PDFs on Λ or λ is required to calculate the TPF and FPF at each threshold and thus compute AUC. We shall now show that the particular form of the ideal observer's discriminant function leads to some interesting and useful relationships between its moments under each hypothesis.

The likelihood ratio is the ratio of the two densities $q_2(\mathbf{g})$ and $q_1(\mathbf{g})$. These same densities are the ones needed to compute moments of Λ under the two hypotheses. The general $(k+1)^{th}$ moment of Λ under H_1 is given by

$$\left\langle \Lambda^{k+1} \right\rangle_1 = \int_\infty d^M\!g\; q_1(\mathbf{g}) \left[\frac{q_2(\mathbf{g})}{q_1(\mathbf{g})}\right]^{k+1} = \int_\infty d^M\!g\; q_2(\mathbf{g}) \left[\frac{q_2(\mathbf{g})}{q_1(\mathbf{g})}\right]^{k} = \left\langle \Lambda^{k} \right\rangle_2 . \tag{13.75}$$

This relationship is true for *any* task, regardless of the form of the PDF on Λ, simply by virtue of the special form of the ideal discriminant function.

It follows immediately from (13.75) that the mean of Λ under H_1 is always 1, since

$$\langle \Lambda \rangle_1 = \int_\infty d^M\!g\; q_1(\mathbf{g}) \frac{q_2(\mathbf{g})}{q_1(\mathbf{g})} = \int_\infty d^M\!g\; q_2(\mathbf{g}) = 1 . \tag{13.76}$$

The variance of Λ under H_1 is given by

$$\mathrm{Var}(\Lambda|H_1) = \langle \Lambda^2 \rangle_1 - \overline{\Lambda}_1^2 = \overline{\Lambda}_2 - 1 . \tag{13.77}$$

Since $\Lambda = e^\lambda$, we can use (13.75) to write a similar relationship for the moments of λ:

$$\left\langle e^{(k+1)\lambda} \right\rangle_1 = \left\langle e^{k\lambda} \right\rangle_2 . \tag{13.78}$$

This relationship holds for all k, including complex values, so long as the expectations exist.

From the definition of the moment-generating function (C.56), we see that the moment-generating function of λ evaluated at $k = \beta + 1$ under H_1 is the moment-generating function of λ evaluated at $k = \beta$ under H_2:

$$M_1(\beta + 1) = M_2(\beta)\,, \tag{13.79}$$

where β is an arbitrary complex number. This property has been shown by Swensson and Green (1977) to be unique to log-likelihood ratios. It is left as an exercise for the reader to show how $M_1(\beta)$ can be used to generate moments of both λ and Λ under both hypotheses.

A corresponding relation for the characteristic functions for λ can be found from (13.79) and (C.53) to be

$$\psi_{\lambda 1}\left(\xi + \frac{i}{2\pi}\right) = \psi_{\lambda 2}(\xi)\,. \tag{13.80}$$

The densities $p_1(\lambda)$ and $p_2(\lambda)$ can then be related by an inverse Fourier transformation of (13.80):

$$\begin{aligned} p_2(\lambda) = \mathcal{F}^{-1}\{\psi_{\lambda 2}(\xi)\} &= \int_{-\infty}^{\infty} d\xi\ \psi_{\lambda 1}\left(\xi + \frac{i}{2\pi}\right)\exp(2\pi i\xi\lambda) \\ &= e^{\lambda}\int_{-\infty+\frac{i}{2\pi}}^{\infty+\frac{i}{2\pi}} dz\ \psi_{\lambda 1}(z)\exp(2\pi i z\lambda)\,, \end{aligned} \tag{13.81}$$

where $z = \xi + i/(2\pi)$. If $\psi_{\lambda 1}(z)$ is analytic in the strip $0 \le \text{Im}(z) \le 1/(2\pi)$, we can shift the contour to obtain (see Barrett *et al.* 1998b, App. A)

$$p_2(\lambda) = e^{\lambda}\int_{-\infty}^{\infty} dz\ \psi_{\lambda 1}(z)\exp(2\pi i z\lambda) = e^{\lambda}p_1(\lambda)\,. \tag{13.82}$$

The shift is allowed as long as $\langle\Lambda\rangle_2$ is finite.

We can now use (13.82) to find a relation between the densities for Λ under the two hypotheses. From (C.45) we know the transformation of the probability density functions is given by

$$p_j(\lambda) = \frac{\text{pr}(\Lambda|H_j)}{|d\lambda/d\Lambda|}\,, \tag{13.83}$$

where the Jacobian $|d\lambda/d\Lambda|$ is the same under H_1 and H_2. Then we have

$$\text{pr}(\Lambda|H_2) = e^{\lambda}\,\text{pr}(\Lambda|H_1) = \Lambda\,\text{pr}(\Lambda|H_1)\,. \tag{13.84}$$

Another way of writing (13.84) is

$$\frac{\text{pr}(\Lambda|H_2)}{\text{pr}(\Lambda|H_1)} = \Lambda\,. \tag{13.85}$$

Green and Swets (1966) describe this result by stating, "To paraphrase Gertrude Stein, the likelihood ratio of the likelihood ratio is the likelihood ratio." The revelation of (13.85) is that all the information necessary for discriminating between H_2 and H_1 is contained in Λ. In statistical terminology, the likelihood ratio is said to be a *sufficient statistic* for the task.

Likelihood-generating function We have considered various ways of computing AUC, depending on the knowledge we have of the form of the probability law on the data or the test statistic. In this section we introduce the concept of the likelihood-generating function, which will provide us with another avenue for obtaining approximations for AUC.

The relationship between the density functions given in (13.82) tells us that one function is sufficient to specify both $p_2(\lambda)$ and $p_1(\lambda)$. In terms of this arbitrary function, which we shall call $f(\lambda)$, we have

$$p_2(\lambda) = e^{\lambda/2} f(\lambda) \qquad \text{and} \qquad p_1(\lambda) = e^{-\lambda/2} f(\lambda) . \tag{13.86}$$

The characteristic function of λ can be written in terms of $f(\lambda)$:

$$\psi_{\lambda 2}(\xi) = F\left(\xi + \frac{i}{4\pi}\right) \qquad \text{and} \qquad \psi_{\lambda 1}(\xi) = F\left(\xi - \frac{i}{4\pi}\right) . \tag{13.87}$$

The moment-generating functions on λ can be rewritten in terms of this function:

$$M_2(\beta) = F_L\left(\beta + \tfrac{1}{2}\right) \qquad \text{and} \qquad M_1(\beta) = F_L\left(\beta - \tfrac{1}{2}\right) , \tag{13.88}$$

where $F_L(\beta)$ is the two-sided Laplace transform of $f(\lambda)$ [see (4.77) and (C.59)].

The normalization of the moment-generating function requires that $M_j(0) = 1$, so that $F_L(\pm\frac{1}{2}) = 1$. Normalization of the characteristic function requires $\psi_{\lambda j}(0) = 1$, so that $F(\pm i/4\pi) = 1$. We can enforce these constraints by rewriting the characteristic function and moment-generating function in terms of new functions $A(\xi)$ and $G(\beta)$:

$$F(\xi) = \exp\left[\left(\xi + \frac{i}{4\pi}\right)\left(\xi - \frac{i}{4\pi}\right) A(\xi)\right] \tag{13.89}$$

and

$$F_L(\beta) = \exp\left[\left(\beta + \tfrac{1}{2}\right)\left(\beta - \tfrac{1}{2}\right) G(\beta)\right] , \tag{13.90}$$

where $G(\beta)$ is the *likelihood-generating function.* It is straightforward to show the following relationship between these new functions:[4]

$$A(\xi) = -4\pi^2 G(2\pi i \xi) . \tag{13.91}$$

In terms of $G(\beta)$ the moment-generating functions under each hypothesis are

$$M_2(\beta) = \exp\left[\beta(\beta+1)\, G\left(\beta + \tfrac{1}{2}\right)\right] \quad \text{and} \quad M_1(\beta) = \exp\left[\beta(\beta-1)\, G\left(\beta - \tfrac{1}{2}\right)\right] . \tag{13.92}$$

The characteristic function for λ under each hypothesis is given in terms of the likelihood-generating function by

$$\psi_{\lambda 2}(\xi) = \exp\left[-4\pi^2 \xi \left(\xi + \tfrac{i}{2\pi}\right) G\left(2\pi i \xi - \tfrac{1}{2}\right)\right] \qquad \text{and}$$

$$\psi_{\lambda 1}(\xi) = \exp\left[-4\pi^2 \xi \left(\xi - \tfrac{i}{2\pi}\right) G\left(2\pi i \xi + \tfrac{1}{2}\right)\right] . \tag{13.93}$$

[4]The difference in sign between the argument in this expression and that found in (5.7) of Barrett *et al.* (1998b) is the result of the difference between their definition of the Laplace transform and the one given in (4.77).

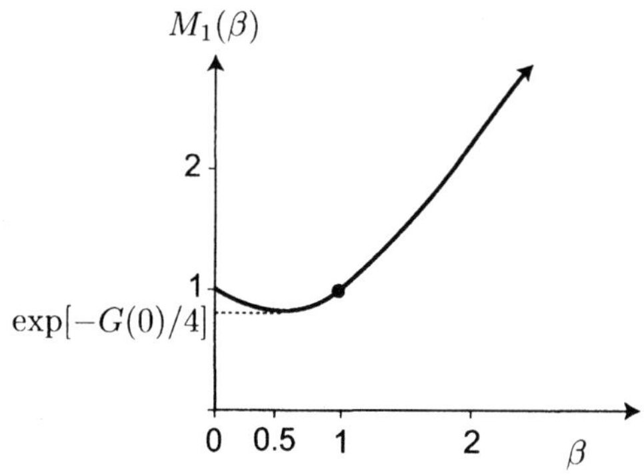

Fig. 13.9 The function $M_1(\beta)$, which passes through unity at $\beta = 0$ and $\beta = 1$.

Figure 13.9 contains a plot of $M_1(\beta)$ as a function of β for real β. From (13.92) we know the value of the function at $\beta = \frac{1}{2}$ defines $G(0)$.

We can make use of Marcinkiewicz's theorem, presented in Shiryayev (1984), to draw a conclusion regarding the form of $G(\beta)$. By this theorem, when a characteristic function has the form $\exp(P)$, where P is a polynomial, the order of the polynomial cannot exceed 2. Thus $G(\beta)$ cannot be a polynomial (other than a polynomial of order 0, where $G(\beta)$ is a constant).

Since all moments of the likelihood ratio are determined by $G(\beta)$, the likelihood-generating function determines the statistics of the likelihood ratio under both hypotheses. We can apply (C.54) to (13.92) to determine the moments of λ in terms of $G(\beta)$. We leave it to the reader to show that

$$\overline{\lambda}_2 = G\left(\tfrac{1}{2}\right) \qquad \text{and} \qquad \overline{\lambda}_1 = -G\left(-\tfrac{1}{2}\right),$$

$$\mathrm{Var}(\lambda|H_2) = 2\left[G\left(\tfrac{1}{2}\right) + G'\left(\tfrac{1}{2}\right)\right] \qquad \text{and} \qquad \mathrm{Var}(\lambda|H_1) = 2\left[G\left(-\tfrac{1}{2}\right) - G'\left(-\tfrac{1}{2}\right)\right]. \tag{13.94}$$

With these moments we can write the SNR for the log-likelihood ratio in terms of $G(\beta)$:

$$\mathrm{SNR}_\lambda^2 = \frac{\left[G\left(\frac{1}{2}\right) + G\left(-\frac{1}{2}\right)\right]^2}{G\left(\frac{1}{2}\right) + G\left(-\frac{1}{2}\right) + G'\left(\frac{1}{2}\right) - G'\left(-\frac{1}{2}\right)}. \tag{13.95}$$

For a Gaussian test statistic we can obtain AUC from (13.95) via (13.20).

Note that (13.95) takes on a particularly simple form if $G'(1/2) \approx G'(-1/2)$. When we can make this assumption,

$$\mathrm{SNR}_\lambda^2 = G\left(-\tfrac{1}{2}\right) + G\left(\tfrac{1}{2}\right) \approx 2G(0). \tag{13.96}$$

Clarkson and Barrett (2000) have found that values of ideal-observer AUC derived from $2G(0)$ are more accurate than those derived from the SNR given in (13.19) for non-Gaussian noise models, including Poisson, and one- and two-sided exponential models.

Setting $\beta = \frac{1}{2}$ in (13.92) gives $G(0) = -4\ln M_1(\frac{1}{2})$. Thus the SNR of (13.96) can be computed directly from the probability density functions on the data according to

$$G(0) = -4 \ln M_1\left(\tfrac{1}{2}\right) = -4 \ln \langle \Lambda^{\frac{1}{2}}(\mathbf{g}) \rangle_1$$

$$= -4 \ln \left\{ \int d^M g \left[\frac{q_2(\mathbf{g})}{q_1(\mathbf{g})} \right]^{\frac{1}{2}} q_1(\mathbf{g}) \right\} = -4 \ln \left\{ \int d^M g \, [q_1(\mathbf{g})\, q_2(\mathbf{g})]^{\frac{1}{2}} \right\} \equiv 4 d_B \,, \tag{13.97}$$

where d_B is the *Bhattacharyya distance* between the probability densities on the data under the two hypotheses (Bhattacharyya, 1943).

A complete description of the performance of any observer is contained in the general expression for AUC of (13.38), which holds for any test statistic. For the special case of the ideal observer, we can rewrite this expression in terms of the characteristic function for the log-likelihood ratio under H_1:

$$\mathrm{AUC} = \frac{1}{2} + \frac{1}{2\pi i} \mathcal{P} \int_{-\infty}^{\infty} \frac{d\xi}{\xi} \, \psi_{\lambda 1}(\xi)\, \psi_{\lambda 1}\left(-\xi + \frac{i}{2\pi} \right) , \tag{13.98}$$

where we have made use of (13.80) and the Hermiticity of the characteristic function. By moving the contour of integration, (13.98) becomes (Barrett *et al.*, 1998b)

$$\mathrm{AUC} = 1 + \frac{1}{2\pi i} \int_{-\infty}^{\infty} \frac{d\xi}{\xi + \frac{i}{4\pi}} \, \psi_{\lambda 1}\left(\xi + \frac{i}{4\pi} \right) \psi_{\lambda 1}\left(-\xi + \frac{i}{4\pi} \right) . \tag{13.99}$$

We can write (13.99) in terms of $F(\xi)$ using (13.87):

$$\mathrm{AUC} = 1 + \frac{1}{2\pi i} \int_{-\infty}^{\infty} \frac{d\xi}{\xi + \frac{i}{4\pi}} \, F(\xi)\, F(-\xi) . \tag{13.100}$$

Substituting the expression for $F(\xi)$ given in (13.89) and doing some algebra yields

$$\mathrm{AUC} = 1 + \frac{1}{2\pi i} \int_{-\infty}^{\infty} \frac{d\xi}{\xi + \frac{i}{4\pi}} \, \exp\left\{ \left(\xi^2 + \frac{1}{16\pi^2} \right) 2\,\mathrm{Re}[A(\xi)] \right\} , \tag{13.101}$$

where we have made use of the fact that $A(-\xi) + A(\xi) = 2\,\mathrm{Re}[A(\xi)]$. The relationship between $A(\xi)$ and $G(\xi)$ given in (13.91) leads finally to

$$\mathrm{AUC} = 1 + \frac{1}{2\pi i} \int_{-\infty}^{\infty} \frac{d\xi}{\xi + \frac{i}{4\pi}} \, \exp\left\{ -8\pi^2 \left(\xi^2 + \frac{1}{16\pi^2} \right) 2\,\mathrm{Re}[G(2\pi i \xi)] \right\} . \tag{13.102}$$

Thus the values of the likelihood-generating function along the imaginary axis determine AUC.

Bounds on AUC We have seen that the single value $G(0)$ can be used to derive AUC with the approximation above (13.96). It has also been shown that $G(0)$ establishes a lower bound on AUC with no approximation (Barrett *et al.*, 1998b). Additional bounds on AUC have been derived that may help to bracket ideal-observer performance for tasks where exact calculation of AUC is difficult (Shapiro, 1999; Clarkson, 2002). An active area of research continues to be the development of approximations and bounds for AUC for use in system optimization for more realistic noise and object models.

13.2.8 Ideal observer with Gaussian statistics

We now consider the decision problem of discriminating between two nonrandom signals in additive Gaussian noise. This is the so-called SKE/BKE (signal-known-exactly/background-known-exactly) problem. While this level of knowledge about the object is unrealistic in practice, it provides a useful starting place for understanding the theory of signal detection and the calculation of ideal decision strategies.

Let the first hypothesis, H_1, denote that a single, nonrandom object $\mathbf{f}_1$ is present at the input. Under the second hypothesis, H_2, a different, but nonrandom, object $\mathbf{f}_2$ is present. Under each hypothesis we then have:

$$H_1 : \mathbf{g} = \boldsymbol{\mathcal{H}}\mathbf{f}_1 + \mathbf{n} = \mathbf{s}_1 + \mathbf{n}$$

$$H_2 : \mathbf{g} = \boldsymbol{\mathcal{H}}\mathbf{f}_2 + \mathbf{n} = \mathbf{s}_2 + \mathbf{n}\,, \tag{13.103}$$

where the signal $\mathbf{s}_j$ is the MD data-space vector resulting from the imaging operator acting on the object-space vector $\mathbf{f}_j$. These signals are assumed to be real and known (though it is not known which of the two is present), and they are nonrandom since the two possible objects are nonrandom.

We shall now derive expressions for the ideal-observer test statistic and associated figures of merit for this problem for both correlated and uncorrelated Gaussian noise. Non-Gaussian noise is treated in Sec. 13.2.9.

Independent, identically distributed Gaussian noise As a simple example, assume that the noise corrupting the data is zero-mean, independent, identically distributed (i.i.d.) Gaussian with variance σ^2 for each data component. The conditional probability density function for each component is then

$$\mathrm{pr}(g_m|H_j) = \left(\frac{1}{2\pi\sigma^2}\right)^{\frac{1}{2}} \exp\left[-\frac{(g_m - s_{jm})^2}{2\sigma^2}\right]. \tag{13.104}$$

The noise samples are statistically independent, which tells us that the conditional PDF on the data vector $\mathbf{g}$ is simply the product of the densities of the components. Thus the PDF on the data vector under the jth hypothesis can be written as follows:

$$\mathrm{pr}(\mathbf{g}|H_j) = \left(\frac{1}{2\pi\sigma^2}\right)^{\frac{M}{2}} \prod_{m=1}^{M} \exp\left[-\frac{(g_m - s_{jm})^2}{2\sigma^2}\right]. \tag{13.105}$$

Using this expression for the conditional PDF on the data vector and the definition of the likelihood ratio below (13.58), we can write $\Lambda(\mathbf{g})$ as

$$\Lambda(\mathbf{g}) = \frac{\mathrm{pr}(\mathbf{g}|H_2)}{\mathrm{pr}(\mathbf{g}|H_1)} = \frac{\prod_{m=1}^{M} \sqrt{2\pi\sigma^2} \exp\left[-\frac{(g_m - s_{2m})^2}{2\sigma^2}\right]}{\prod_{m=1}^{M} \sqrt{2\pi\sigma^2} \exp\left[-\frac{(g_m - s_{1m})^2}{2\sigma^2}\right]}. \tag{13.106}$$

The ideal decision rule is to evaluate this expression and compare it to a threshold Λ_c. We can simplify this expression by first cancelling the common terms in the numerator and denominator to get

$$\Lambda(\mathbf{g}) = \prod_{m=1}^{M} \exp\left[\frac{(s_{2m} - s_{1m})g_m}{\sigma^2} - \frac{s_{2m}^2 - s_{1m}^2}{2\sigma^2}\right] \mathop{\gtrless}_{D_1}^{D_2} \Lambda_c\,. \tag{13.107}$$

Since the natural logarithm is a monotonic function, we can construct an equivalent test by taking the log of both sides of this expression to give

$$\ln \Lambda(\mathbf{g}) = \lambda(\mathbf{g}) = \sum_{m=1}^{M} \frac{(s_{2m} - s_{1m}) g_m}{\sigma^2} - \frac{s_{2m}^2 - s_{1m}^2}{2\sigma^2} \underset{D_1}{\overset{D_2}{\gtrless}} \ln \Lambda_c = \lambda_c \,. \tag{13.108}$$

The threshold is the logarithm of Λ_c, which we have rewritten as λ_c. We can go one step further and combine the constant term relating the nonrandom difference in the signal energies into the threshold, giving the new decision rule:

$$\sum_{m=1}^{M} (s_{2m} - s_{1m}) g_m \underset{D_1}{\overset{D_2}{\gtrless}} \lambda_c' \,. \tag{13.109}$$

This expression gives the well-known "matched-filter" discriminator in which the expected signal is used as a filter that is correlated[5] with the data. Since the signal is known exactly, the filtering operation is performed with the data and the filter perfectly aligned; no relative shift between filter and data is implied. Matched filtering has long been known as the optimal strategy for the detection of known signals buried in Gaussian noise (North, 1943; Van Vleck and Middleton, 1946; Zadeh and Ragazzini, 1952).

We can rewrite (13.109) using vector notation as

$$\Delta \mathbf{s}^t \mathbf{g} \underset{D_1}{\overset{D_2}{\gtrless}} \lambda_c' \,, \tag{13.110}$$

where $\Delta \mathbf{s}$ is the expected difference signal $\mathbf{s}_2 - \mathbf{s}_1$, and we have assumed that the data and expected signals are real.

Correlated Gaussian noise The covariance matrix under the jth hypothesis is defined to be

$$\mathbf{K}_j = \langle (\mathbf{g} - \mathbf{s}_j)(\mathbf{g} - \mathbf{s}_j)^t | H_j \rangle = \langle \mathbf{n}\mathbf{n}^t \rangle \,, \tag{13.111}$$

where the angle brackets denote an average over all possible noise realizations for a given hypothesis with known signal $\mathbf{s}_j$, and we have again assumed that the noise and signals are real. We see immediately from (13.111) that $\mathbf{K}_1 = \mathbf{K}_2$, because the noise is independent of the signal.[6] We can therefore drop the hypothesis-specific subscript on the covariance matrix in the discussion that follows and refer to it as $\mathbf{K}_\mathbf{n}$, where the subscript $\mathbf{n}$ refers to the noise.

The probability density function on $\mathbf{g}$ under the jth hypothesis is the multivariate Gaussian [*cf.* (8.185)]

$$\mathrm{pr}(\mathbf{g}|H_j) = \left[(2\pi)^M \det(\mathbf{K}_\mathbf{n})\right]^{-\frac{1}{2}} \exp\left[-\tfrac{1}{2}(\mathbf{g} - \mathbf{s}_j)^t \, \mathbf{K}_\mathbf{n}^{-1} (\mathbf{g} - \mathbf{s}_j)\right] \,, \tag{13.112}$$

[5] Though the term "correlation filter" is commonly used for a matched filter, we note that (13.109) is just a scalar product, not a correlation. The confusion arises since scanning matched filters are often used with signals whose form is known but whose spatial or temporal location is not. In that case, the matched filter is not optimal, even with i.i.d. Gaussian noise. For more details see Sec. 13.2.10.

[6] Two random variables that have the same variance, or two random vectors that have the same covariance, are said to be *homoscedastic*. The word derives from the ancient Greek verb *skedanumi* ($\sigma\kappa\epsilon\delta\alpha\nu o\upsilon\mu\iota$), meaning "scatter." A related word is diaskedanumi, which can mean "throw the troubles out, entertain, have fun." In many statistical problems, homoscedasticity is more fun than heteroscedasticity. (Thanks to Mary Ruddick for assistance with ancient Greek etymology.)

where $\det(\mathbf{K_n})$ is the determinant of the matrix $\mathbf{K_n}$.

The likelihood ratio test follows:

$$\Lambda(\mathbf{g}) = \frac{\exp\left[-\frac{1}{2}(\mathbf{g}-\mathbf{s}_2)^t\, \mathbf{K_n}^{-1}(\mathbf{g}-\mathbf{s}_2)\right]}{\exp\left[-\frac{1}{2}(\mathbf{g}-\mathbf{s}_1)^t\, \mathbf{K_n}^{-1}(\mathbf{g}-\mathbf{s}_1)\right]} \underset{D_1}{\overset{D_2}{\gtrless}} \Lambda_c\,. \tag{13.113}$$

The log-likelihood ratio is then

$$\begin{aligned}\lambda(\mathbf{g}) &= \tfrac{1}{2}(\mathbf{g}-\mathbf{s}_1)^t\, \mathbf{K_n}^{-1}(\mathbf{g}-\mathbf{s}_1) - \tfrac{1}{2}(\mathbf{g}-\mathbf{s}_2)^t\, \mathbf{K_n}^{-1}(\mathbf{g}-\mathbf{s}_2)\\ &= \tfrac{1}{2}\left[\mathbf{g}^t\mathbf{K_n}^{-1}(\mathbf{s}_2-\mathbf{s}_1) + (\mathbf{s}_2-\mathbf{s}_1)^t\mathbf{K_n}^{-1}\mathbf{g} + \mathbf{s}_1^t\mathbf{K_n}^{-1}\mathbf{s}_1 - \mathbf{s}_2^t\mathbf{K_n}^{-1}\mathbf{s}_2\right]\,. \end{aligned}\tag{13.114}$$

The ideal observer evaluates this expression and compares it to the threshold λ_c to make its decision. We can form an equivalent test by incorporating the last two terms into the threshold, since they are scalars independent of the data, and make use of the fact that the covariance matrix is symmetric, to obtain this form for the ideal decision strategy:

$$\Delta\mathbf{s}^t\mathbf{K_n}^{-1}\mathbf{g} = \lambda'(\mathbf{g}) \underset{D_1}{\overset{D_2}{\gtrless}} \lambda'_c\,. \tag{13.115}$$

Equation (13.115) describes the operations an ideal observer would perform to decide which hypothesis was responsible for the received data. First the data vector is filtered by the inverse of the covariance matrix. The output of this operation is then correlated with the difference signal, $\Delta\mathbf{s}$, a matched-filter operation like that found in (13.110). The correlation function is evaluated at zero shift because the ideal observer makes use of the *a priori* information that the signal is at a known location in the object; we can assume without loss of generality that the signal location is the origin.

SNR and AUC in the Gaussian case The expression for the log-likelihood ratio in (13.115) was derived under the assumption that the data are multivariate normal with $\mathbf{K}_1$ and $\mathbf{K}_2$ equal. The decision variable is a linear transformation of the data, and since linear transformations of Gaussian random variables are Gaussian random variables, we conclude that $\lambda'(\mathbf{g})$ must also be Gaussian. Thus we can confidently use the expression for SNR given in (13.19) as a measure of the test performance. We shall now determine the SNR for this case, dropping the prime on the decision variable to avoid clutter in the equations that follow.

To derive SNR_λ, we must first find the mean of $\lambda(\mathbf{g})$ under each hypothesis:

$$\langle\lambda(\mathbf{g})|H_j\rangle = \langle\Delta\mathbf{s}^t\mathbf{K_n}^{-1}\mathbf{g}|H_j\rangle = \Delta\mathbf{s}^t\mathbf{K_n}^{-1}\mathbf{s}_j\,. \tag{13.116}$$

The variance of $\lambda(\mathbf{g})$ under the jth hypothesis is

$$\begin{aligned}\sigma_\lambda^2 &= \langle[\lambda(\mathbf{g}) - \langle\lambda(\mathbf{g})|H_j\rangle]^2|H_j\rangle = \langle\lambda^2(\mathbf{g})|H_j\rangle - \langle\lambda(\mathbf{g})|H_j\rangle^2\\ &= \langle\Delta\mathbf{s}^t\mathbf{K_n}^{-1}\mathbf{g}\mathbf{g}^t\mathbf{K_n}^{-1}\Delta\mathbf{s}|H_j\rangle - \Delta\mathbf{s}\mathbf{K_n}^{-1}\mathbf{s}_j\mathbf{s}_j^t\mathbf{K_n}^{-1}\Delta\mathbf{s}\\ &= \Delta\mathbf{s}^t\mathbf{K_n}^{-1}\Delta\mathbf{s}\,,\end{aligned}\tag{13.117}$$

which is independent of the hypothesis, as we would expect.

Now we can put all the pieces together to write down the expression for SNR_λ^2:

$$\text{SNR}_\lambda^2 = \frac{\left[\Delta\mathbf{s}^t\mathbf{K_n}^{-1}\mathbf{s}_2 - \Delta\mathbf{s}^t\mathbf{K_n}^{-1}\mathbf{s}_1\right]^2}{\sigma_\lambda^2} = \frac{\left[\Delta\mathbf{s}^t\mathbf{K_n}^{-1}\Delta\mathbf{s}\right]^2}{\Delta\mathbf{s}^t\mathbf{K_n}^{-1}\Delta\mathbf{s}} = \Delta\mathbf{s}^t\mathbf{K_n}^{-1}\Delta\mathbf{s}\,. \tag{13.118}$$

As we should expect, the performance of the observer depends only on the difference in the two signals and the noise covariance.

In the special case where the noise in the data is uncorrelated, the covariance matrix can be written $[K_{\mathbf{n}}]_{mm'} = \sigma_m^2 \, \delta_{mm'}$ and the SNR becomes

$$\mathrm{SNR}_\lambda^2 = \sum_{m=1}^{M} \frac{[\Delta \mathrm{s}_m]^2}{\sigma_m^2} . \tag{13.119}$$

If, in addition, the noise variance is uniform ($\sigma_m = \sigma$ for all m), this expression further simplifies to

$$\mathrm{SNR}_\lambda^2 = \frac{1}{\sigma^2} \sum_{m=1}^{M} [\Delta \mathrm{s}_m]^2 = \frac{||\Delta \mathbf{s}||^2}{\sigma^2} . \tag{13.120}$$

In this case the form of the signal does not enter into the SNR expression; all that matters is the norm of the difference signal relative to the noise variance.

Likelihood-generating function for Gaussian data The SNR given in (13.118) can be used to determine AUC exactly via (13.20) in this Gaussian case. Moreover, the approximation based on $G(0)$ given in (13.96) gives precisely the same SNR values. These approximations are all exact for Gaussian data.

For Gaussian data the log-likelihood ratio is Gaussian because of its linear relationship with $\mathbf{g}$. Thus its characteristic function takes the form given in (C.116) for scalar Gaussian random variables (although with a constant phase factor determined by the mean of λ under each hypothesis). When we revisit the general forms for the characteristic function for λ given in (13.93), we see that $G(\beta)$ must therefore be independent of β. Thus the likelihood-generating function is a constant in the Gaussian case. All the information needed to describe the performance of the ideal observer is contained in one number, $2G(0)$.

KL formulation for SKE/BKE tasks Equivalent expressions for the SNRs given in the previous section can be written in terms of the Karhunen-Loève expansion of the signal and data vectors. From (8.58) we can write the data as

$$\mathbf{g} = \sum_{m=1}^{M} \beta_m \boldsymbol{\phi}_m = \boldsymbol{\Phi}\boldsymbol{\beta} , \tag{13.121}$$

where, by (8.59), the vector of coefficients is given by $\boldsymbol{\beta} = \boldsymbol{\Phi}^\dagger \mathbf{g}$. For any set of expansion vectors the $\boldsymbol{\beta}$ would be random variables through their linear relationship to $\mathbf{g}$. In the special case where we use a KL expansion, so that the $\boldsymbol{\phi}_m$ are the eigenvectors of the covariance matrix of the data, we know by (8.62) that the coefficients are uncorrelated random variables. Thus we can write the covariance matrix for the coefficients as $\mathbf{K}_\beta = \mathbf{M}$, with diagonal elements μ_m.

We can similarly define the mean difference in the data under each hypothesis by

$$\Delta \mathbf{s} = \Delta \overline{\mathbf{g}} = \boldsymbol{\Phi} \Delta \overline{\boldsymbol{\beta}} . \tag{13.122}$$

We can use (13.122) and (8.64) to rewrite (13.118) as

$$\mathrm{SNR}_\lambda^2 = \Delta\mathbf{s}^t\mathbf{K}_\mathbf{n}^{-1}\Delta\mathbf{s} = \left[\mathbf{\Phi}\Delta\overline{\boldsymbol{\beta}}\right]^t\left[\mathbf{\Phi}\mathbf{M}^{-1}\mathbf{\Phi}^\dagger\right]\left[\mathbf{\Phi}\Delta\overline{\boldsymbol{\beta}}\right]$$

$$= \sum_{m=1}^{M}\frac{\left[\Delta\overline{\beta}_m\right]^2}{\mu_m} = \sum_{m=1}^{M}\frac{|(\boldsymbol{\phi}_m, \Delta\mathbf{s})|^2}{\mu_m}. \tag{13.123}$$

The last form is analogous to the result we found in (13.119) for data contaminated by uncorrelated Gaussian noise. The KL domain is defined to be the domain where the random vectors are uncorrelated, so the SNR in that domain always has the form of (13.119).

Prewhitening matched filter If the noise covariance matrix $\mathbf{K_n}$ is nonsingular, we can make use of (8.66) and (8.67) to define a new random variable $\mathbf{z}$ in terms of the square-root matrix $\mathbf{K}_\mathbf{n}^{-\frac{1}{2}}$:

$$\mathbf{z} = \mathbf{K}_\mathbf{n}^{-\frac{1}{2}}\mathbf{g}. \tag{13.124}$$

Similarly,

$$\overline{\mathbf{z}}_j = \mathbf{K}_\mathbf{n}^{-\frac{1}{2}}\overline{\mathbf{g}}_j, \tag{13.125}$$

so that

$$\Delta\overline{\mathbf{z}} = \mathbf{K}_\mathbf{n}^{-\frac{1}{2}}\Delta\overline{\mathbf{g}} = \mathbf{K}_\mathbf{n}^{-\frac{1}{2}}\Delta\mathbf{s}, \tag{13.126}$$

and the test statistic of (13.115) becomes

$$\lambda = \Delta\mathbf{s}^t\mathbf{K}_\mathbf{n}^{-1}\mathbf{g} = \left[\mathbf{K}_\mathbf{n}^{-\frac{1}{2}}\Delta\mathbf{s}\right]^t\left[\mathbf{K}_\mathbf{n}^{-\frac{1}{2}}\mathbf{g}\right] = \Delta\overline{\mathbf{z}}^t\mathbf{z}. \tag{13.127}$$

The covariance of $\mathbf{z}$ is given by

$$\mathbf{K_z} = \left\langle\left[\mathbf{K}_\mathbf{n}^{-\frac{1}{2}}\Delta\mathbf{g}\right]\left[\mathbf{K}_\mathbf{n}^{-\frac{1}{2}}\Delta\mathbf{g}\right]^t\right\rangle = \mathbf{K}_\mathbf{n}^{-\frac{1}{2}}\mathbf{K}_\mathbf{n}\mathbf{K}_\mathbf{n}^{-\frac{1}{2}} = \mathbf{I}. \tag{13.128}$$

We see that $\mathbf{z}$ is an uncorrelated Gaussian random variable with unit variance in each element. The process represented by (13.124) is called *prewhitening*; it yields a Gaussian random variable whose correlation matrix has a flat, or white, eigenspectrum. From (13.127) we see that the ideal decision strategy is to first prewhiten the data and then perform a filtering operation with a prewhitened version of the expected difference signal. This observer is therefore referred to as the *prewhitening matched filter* or PWMF.

The simple form for the covariance of $\mathbf{z}$ leads to a particularly simple form for the SNR given in (13.118) when it is written in terms of $\mathbf{z}$:

$$\mathrm{SNR}_\lambda^2 = ||\Delta\overline{\mathbf{z}}||^2. \tag{13.129}$$

13.2.9 Ideal observer with non-Gaussian data

So far we have restricted our attention to the SKE/BKE classification task in additive Gaussian noise. We now consider the SKE/BKE classification problem for other statistical descriptions of the measurement noise that might arise in imaging.

Poisson noise In many photon-limited imaging situations, the probability law on the data is Poisson (see Chap. 11). If the task is the discrimination between two exactly-specified signals in a radiological image with only Poisson noise, the conditional PDF of the data under hypothesis H_j is given by [*cf.* (11.40)]

$$\text{pr}(\mathbf{g}|H_j) = \prod_{m=1}^{M} \frac{e^{-\overline{g}_{jm}} [\overline{g}_{jm}]^{g_m}}{(g_m)!}, \tag{13.130}$$

where $\overline{g}_{jm}$ is the mean of the mth element of the data vector when hypothesis H_j is true. When we form the likelihood ratio using (13.130) for the two hypotheses, we find the ideal observer's test statistic to be (Helstrom, 1964)

$$\lambda(\mathbf{g}) = \sum_{m=1}^{M} g_m \ln \frac{\overline{g}_{2m}}{\overline{g}_{1m}}. \tag{13.131}$$

We see that in the Poisson case the optimum operation on the data is again a linear filtering operation, only now the filter is a nonlinear functional of the expected signals under the two hypotheses.

We can use (13.131) to determine the signal-to-noise ratio associated with the test statistic, according to (13.19) (Cunningham *et al.*, 1976; Wagner *et al.*, 1981)

$$\text{SNR}^2 = \frac{\left[\sum_{m=1}^{M} (\overline{g}_{2m} - \overline{g}_{1m}) \ln\left(\frac{\overline{g}_{2m}}{\overline{g}_{1m}}\right)\right]^2}{\frac{1}{2}\sum_{m=1}^{M} (\overline{g}_{2m} + \overline{g}_{1m}) \ln^2\left(\frac{\overline{g}_{2m}}{\overline{g}_{1m}}\right)}. \tag{13.132}$$

The derivation of (13.132) makes use of a definition for SNR that is relevant whenever the test statistic is a Gaussian random variable. While the data are not assumed here to be Gaussian, the fact that λ as defined in (13.131) is a weighted sum of random variables implies that the assumption of Gaussianity is often valid owing to the central-limit theorem.

It is illustrative to consider the behavior of the ideal observer when certain pixels in the image are known to have expected values equal to zero for one hypothesis and nonzero for the other. From (13.131) we see that the test statistic becomes infinite when the data in those elements are nonzero. The SNR is infinite as well, as (13.132) confirms. This example demonstrates the tremendous power of such complete prior information regarding the expected values in a particular pixel in the image, information that is possible only for contrived problems in which there is no uncertainty in the expected images under each hypothesis.

An approximation to the SNR for the Poisson noise case can be obtained by utilizing the SNR expression for the multivariate Gaussian case given in (13.118). Poisson noise results in a diagonal covariance matrix in which the mean of each element specifies the variance:

$$[\mathbf{K_n}]_{mm'} = \sigma_m^2 \, \delta_{mm'} = \overline{g}_m \, \delta_{mm'}, \tag{13.133}$$

where we have assumed that the signal is of sufficiently low contrast that the variance in each detector element is independent of hypothesis. Substituting this form for $\mathbf{K_n}$ into (13.118) yields

$$\text{SNR}^2 = \sum_{m=1}^{M} \frac{(\overline{g}_{2m} - \overline{g}_{1m})^2}{\overline{g}_m} = \sum_{m=1}^{M} \frac{\Delta\overline{g}_m^2}{\overline{g}_m}, \tag{13.134}$$

similar to (13.119). To make use of (13.118) in deriving (13.134) we are assuming that the number of counts in each detector is greater than about 10, so that a Gaussian approximation to the Poisson law on the data can be made (Barrett and Swindell, 1981, 1996).

Perhaps surprisingly, even in the absence of a large number of counts per detector element, we can obtain the same expression for the SNR given in (13.134), provided we again assume that the signal is weak. To see this, consider the task of detecting a low-contrast signal, in which case we can write the expected data under hypothesis H_2 as $\overline{g}_{2m} = \overline{g}_{1m} + s_m$, with $s_m \ll \overline{g}_{1m}$ for all m. Then a Taylor expansion of the logarithms in (13.132) through terms linear in the signal yields

$$\text{SNR}_\lambda^2 \approx \sum_{m=1}^{M} \frac{s_m^2}{\overline{g}_m}, \tag{13.135}$$

where $\overline{g}_m$ can be either $\overline{g}_{1m}$ or $\overline{g}_{2m}$ to this approximation.

Exponential noise Speckle noise, found in coherent imaging applications involving ultrasound or laser sources, is characterized by an exponential probability density function (see Chap. 18 for more detail). If the task is the detection of a known signal in speckle noise, where the signal affects only the mean, the conditional PDF on the data is written [*cf.* (C.118)]

$$\text{pr}(\mathbf{g}|H_j) = \prod_{m=1}^{M} \frac{1}{\overline{g}_{jm}} \exp\left[-\frac{g_m}{\overline{g}_{jm}}\right]. \tag{13.136}$$

This expression is valid in the idealized case in which the detectors are small relative to a speckle cell and separated by a distance greater than a speckle cell, such that the data elements are uncorrelated.

The log-likelihood ratio for this case can be shown to be

$$\lambda(\mathbf{g}) = \sum_{m=1}^{M} \left(\frac{\Delta\overline{g}_m}{\overline{g}_{1m}\overline{g}_{2m}}\right) g_m, \tag{13.137}$$

where we have dropped all terms that are independent of the data. We see that the ideal observer again performs a matched filter on the detected data, only now the filter is the expected difference signal normalized by the square of the geometric mean.

We can find SNR_λ by direct calculation of the relevant expectations. The mean of λ under hypothesis H_j is

$$\langle\lambda(\mathbf{g})\rangle_j = \sum_{m=1}^{M} \left(\frac{\Delta\overline{g}_m}{\overline{g}_{1m}\overline{g}_{2m}}\right) \overline{g}_{jm}. \tag{13.138}$$

The variance of λ under the jth hypothesis is

$$\sigma_j^2 = \sum_{m=1}^{M} \left(\frac{\Delta\overline{g}_m}{\overline{g}_{1m}\overline{g}_{2m}}\right)^2 (\overline{g}_{jm})^2, \tag{13.139}$$

since we have assumed there are no correlations between the data components. The SNR is thus

$$\mathrm{SNR}_\lambda^2 = \frac{\left[\sum_m \frac{\Delta\overline{g}_m^2}{\overline{g}_{1m}\overline{g}_{2m}}\right]^2}{\frac{1}{2}\sum_m \frac{\Delta\overline{g}_m^2\left(\overline{g}_{2m}^2+\overline{g}_{1m}^2\right)}{\overline{g}_{1m}^2\overline{g}_{2m}^2}}. \tag{13.140}$$

An interesting limit is the one in which $\overline{g}_{1m} \approx \overline{g}_{2m} = \overline{g}_m$. Then (13.140) becomes

$$\mathrm{SNR}_\lambda^2 = \sum_m \frac{\Delta\overline{g}_m^2}{\overline{g}_m^2}. \tag{13.141}$$

There is an interesting difference between this expression and that obtained in the low-contrast limit for Poisson noise [see (13.135)]. For exponential noise there is no SNR increase to be found by increasing the exposure, because this affects both the numerator and the denominator equally. This is in contrast to the behavior found for the low-contrast Poisson case, where an increase in exposure yields a linear improvement in SNR^2 with count density.

Log-normal In some circumstances the data can be modeled as log-normally distributed. For example, log-normal statistics have been shown to describe the statistics of images that have been reconstructed from projections using nonlinear algorithms (Barrett *et al.*, 1994; see also Sec. 15.4.7).

Independent log-normal data are described by

$$\mathrm{pr}(\mathbf{g}|H_j) = \prod_m \frac{1}{\sqrt{2\pi}\,\sigma_{jm} g_m} \exp\left[-\frac{(\ln g_m - \ln A_m - \overline{x}_{jm})^2}{2\sigma_{jm}^2}\right], \tag{13.142}$$

assuming the density for the random variable g_m is determined by the two parameters $\overline{x}_{jm}$ and σ_{jm} under hypothesis j (see Sec. C.5.9).

The log-likelihood ratio can be shown to be

$$\lambda(\mathbf{g}) = \sum_m \ln g_m \left[\frac{\ln A_m + \overline{x}_{2m}}{\sigma_{2m}^2} - \frac{\ln A_m + \overline{x}_{1m}}{\sigma_{1m}^2}\right] + \sum_m (\ln g_m)^2 \left[\frac{1}{2\sigma_{1m}^2} - \frac{1}{2\sigma_{2m}^2}\right], \tag{13.143}$$

where we have again combined all terms independent of the data into the threshold. If we can assume $\sigma_{1m} \approx \sigma_{2m}$, the optimal decision variable is linear in the logarithm of each data element. In this case the log-likelihood ratio takes the form of a matched filter operation acting on the elements of $\ln \mathbf{g}$, which is intuitively sensible since the log operation recovers Gaussian data.

13.2.10 Signal variability and the ideal observer

In the previous section we assumed that the signals to be classified were known exactly, so that the only limitation to task performance by the ideal observer was the randomness in the data resulting from measurement noise. Let us now consider the case where the task is the detection of a signal which is in some way random. Under the two hypotheses the data are now given by

$$H_1 : \mathbf{g} = \boldsymbol{\mathcal{H}}\mathbf{f}_1 + \mathbf{n} = \mathbf{b} + \mathbf{n} \tag{13.144a}$$

and

$$H_2 : \mathbf{g} = \mathcal{H}\mathbf{f}_2 + \mathbf{n} = \mathbf{s} + \mathbf{b} + \mathbf{n}\,. \tag{13.144b}$$

In this model, when the signal is absent the data are the sum of two components: the nonrandom (but not necessarily uniform) background $\mathbf{b}$ and the measurement noise $\mathbf{n}$. The background $\mathbf{b}$ is the image that would be obtained in the signal-absent case in the limit of an infinite exposure time. The signal present under H_2 is whatever represents the distinguishing image feature(s) between class 1 and class 2. We assume here that the presence of the signal does not occlude or alter the background.

Both $\mathbf{b}$ and $\mathbf{s}$ are defined in the data space $\mathbb{V}$. They are the result of some unspecified deterministic mapping $\mathcal{H}$ that maps each object, either background-only or signal-plus-background, to the data domain. For example, the signal might be the image of a nodule to be detected in a detection task. The background is whatever is left that makes up the image of the object, for example, the ribs and other "non-nodule" structures in a chest film.

One approach to the problem of signal variability is to let $\mathbf{s}$ be characterized by randomness in P of its defining parameters (Sec. 8.4). For example, $\mathbf{s}$ might have known shape, but unknown amplitude or location or the signal might be a sinusoidal pattern of unknown frequency and phase. We can define a P-dimensional vector of random parameters $\boldsymbol{\theta}$ with probability density $\mathrm{pr}_{\boldsymbol{\theta}}(\boldsymbol{\theta})$. Then

$$H_2 : \mathbf{g} = \mathbf{s}(\boldsymbol{\theta}) + \mathbf{b} + \mathbf{n}\,. \tag{13.145}$$

We assume that the statistics of $\mathbf{n}$ and $\boldsymbol{\theta}$ are independent.

The ideal observer uses the likelihood ratio as a test statistic. To compute this statistic we must first determine the conditional PDF of the data under each hypothesis. The PDF of the data under the signal-absent condition is given by

$$\mathrm{pr}(\mathbf{g}|H_1) = \mathrm{pr}_{\mathbf{n}}(\mathbf{g} - \mathbf{b})\,, \tag{13.146}$$

which is the noise probability density function centered on the background $\mathbf{b}$. The PDF of the data under the signal-present condition is a weighted average of the noise probability density function shifted to each signal-plus-background combination, with the weighting given by the probability of each signal as described by $\mathrm{pr}_{\boldsymbol{\theta}}(\boldsymbol{\theta})$:

$$\mathrm{pr}(\mathbf{g}|H_2) = \int_\infty d^P\theta\; \mathrm{pr}_{\mathbf{n}}[\mathbf{g} - \mathbf{b} - \mathbf{s}(\boldsymbol{\theta})|\boldsymbol{\theta}]\, \mathrm{pr}_{\boldsymbol{\theta}}(\boldsymbol{\theta}) = \langle \mathrm{pr}(\mathbf{g}|H_2, \boldsymbol{\theta})\rangle_{\boldsymbol{\theta}}\,. \tag{13.147}$$

The likelihood ratio is thus given by

$$\Lambda(\mathbf{g}) = \frac{\mathrm{pr}(\mathbf{g}|H_2)}{\mathrm{pr}(\mathbf{g}|H_1)} = \frac{\langle \mathrm{pr}(\mathbf{g}|H_2, \boldsymbol{\theta})\rangle_{\boldsymbol{\theta}}}{\mathrm{pr}(\mathbf{g}|H_1)} = \left\langle \frac{\mathrm{pr}(\mathbf{g}|H_2, \boldsymbol{\theta})}{\mathrm{pr}(\mathbf{g}|H_1)} \right\rangle_{\boldsymbol{\theta}} = \langle \Lambda_{\mathrm{SKE}}(\mathbf{g}, \boldsymbol{\theta})\rangle_{\boldsymbol{\theta}} \tag{13.148}$$

in the case where the signal depends on a random parameter vector.

Consider the case of independent, identically distributed Gaussian noise, so that $\mathbf{K_n} = \sigma^2\mathbf{I}$. Then

$$\begin{aligned}\Lambda_{\mathrm{SKE}}(\mathbf{g}, \boldsymbol{\theta}) &= \frac{\left(\frac{1}{2\pi\sigma^2}\right)^{\frac{M}{2}} \exp\left\{-\frac{1}{2}[\mathbf{g} - \mathbf{b} - \mathbf{s}(\boldsymbol{\theta})]^t\, \mathbf{K}_{\mathbf{n}}^{-1}[\mathbf{g} - \mathbf{b} - \mathbf{s}(\boldsymbol{\theta})]\right\}}{\left(\frac{1}{2\pi\sigma^2}\right)^{\frac{M}{2}} \exp\left[-\frac{1}{2}(\mathbf{g} - \mathbf{b})^t\, \mathbf{K}_{\mathbf{n}}^{-1}(\mathbf{g} - \mathbf{b})\right]} \\ &= \exp\left\{\frac{1}{\sigma^2}[\mathbf{s}(\boldsymbol{\theta})]^t(\mathbf{g} - \mathbf{b}) - \frac{1}{2\sigma^2}||\mathbf{s}(\boldsymbol{\theta})||^2\right\}\,.\end{aligned} \tag{13.149}$$

When we average (13.149) over the distribution of random parameters we obtain

$$\Lambda(\mathbf{g}) = \int_\infty d^P\theta \; \mathrm{pr}_{\boldsymbol\theta}(\boldsymbol\theta) \exp\left\{\tfrac{1}{\sigma^2}[\mathbf{s}(\boldsymbol\theta)]^t(\mathbf{g}-\mathbf{b}) - \tfrac{1}{2\sigma^2}||\mathbf{s}(\boldsymbol\theta)||^2\right\}. \tag{13.150}$$

We see from (13.150) that the decision function is no longer a linear function of the data, even in this i.i.d. Gaussian-noise case, and the log-likelihood function is not linear in the data either. In general, when the signal is random, the ideal observer performs a nonlinear operation on the data.

An exception to this general rule can be found for the case of weak signals or large noise. Suppose the energy of the signal is independent of $\boldsymbol\theta$, so that $||\mathbf{s}(\boldsymbol\theta)||^2 = ||\mathbf{s}_0||^2$. This holds, for example, if the random parameter is location and the signal is large with respect to the detector elements. Now let the signal be sufficiently weak that it satisfies

$$\frac{1}{\sigma^2}[\mathbf{s}(\boldsymbol\theta)]^t(\mathbf{g}-\mathbf{b}) \ll 1. \tag{13.151}$$

In this case we can rewrite (13.150) as

$$\begin{aligned}\Lambda(\mathbf{g}) &= \exp\left\{-\tfrac{1}{2\sigma^2}||\mathbf{s}_0||^2\right\} \int_\infty d^P\theta \; \mathrm{pr}_{\boldsymbol\theta}(\boldsymbol\theta)\left\{1 + \tfrac{1}{\sigma^2}\left[\mathbf{s}(\boldsymbol\theta)\right]^t(\mathbf{g}-\mathbf{b})\right\} \\ &= \exp\left\{-\tfrac{1}{2\sigma^2}||\mathbf{s}_0||^2\right\}\left\{1 + \tfrac{1}{\sigma^2}\overline{\mathbf{s}}^t(\mathbf{g}-\mathbf{b})\right\},\end{aligned} \tag{13.152}$$

where $\overline{\mathbf{s}} = \langle \mathbf{s}(\boldsymbol\theta)\rangle_{\boldsymbol\theta}$ is the average signal. When we incorporate all additive and multiplicative constants into the threshold, we find the ideal observer's decision function to be

$$\Lambda(\mathbf{g}) = \overline{\mathbf{s}}^t(\mathbf{g}-\mathbf{b}). \tag{13.153}$$

The strategy is a matched filter with the expected average signal, once the known background has been subtracted from the data. Note that this matched filter yields the likelihood ratio, not the log-likelihood ratio as found in (13.110) and (13.115).

Example: Weak sinusoid of random frequency If the task is the detection of a weak grating pattern of uncertain frequency, the signal contribution to the mth detector element is given by

$$[\mathbf{s}(\theta)]_m = \cos(2\pi i\theta x_m) \tag{13.154}$$

when the pattern has frequency θ. If the PDF on θ is a Gaussian centered at frequency θ_0 and with variance σ_θ^2, the average signal is given by

$$\begin{aligned}\overline{s}_m &= \left(\frac{1}{2\pi\sigma_\theta^2}\right)^{\frac{1}{2}} \int_{-\infty}^{\infty} d\theta \; \cos(2\pi i\theta x_m)\exp\left[-\frac{(\theta-\theta_0)^2}{2\sigma_\theta^2}\right] \\ &= \cos(2\pi\theta_0 x_m)\exp\left[-2\pi^2 x_m^2 \sigma_\theta^2\right],\end{aligned} \tag{13.155}$$

which has the form of a Gabor function centered at frequency θ_0, with width dictated by σ_θ. It is this average signal that the ideal observer would use as a template in the matched filter expression of (13.153). As we shall discuss in greater detail in Chap. 14, the visual system has been shown to have frequency-selective filters or channels of the form given in (13.155). It has been suggested that the evolution of such channels was motivated by the need for a mechanism for detecting weak signals of unknown scale, leading to Gabor-like channels in the visual system.

Example: Location uncertainty Consider the problem of detecting a signal with unknown location in uncorrelated Gaussian measurement noise. The task is to determine whether the data do or do not contain the signal, without needing to determine its location. Under H_1 the signal is absent, and the conditional PDF of the data (13.146) is equivalent to (13.105), where the nonrandom signal component in that expression is now given by $\mathbf{b}$. Under H_2 the underlying continuous signal $s(\mathbf{r})$ is parameterized by a random location vector $\mathbf{r_s}$ such that when the signal is present at location $\mathbf{r_s}$, its contribution to the mth data element is

$$s_m(\mathbf{r_s}) = \int_m d^2r \; s(\mathbf{r} - \mathbf{r_s}) \,, \tag{13.156}$$

where the subscripted integral denotes an integration over the sensitive area of the mth detector element. Using (13.105) and (13.147), the PDF of the data under H_2 is given by

$$\mathrm{pr}(\mathbf{g}|H_2) = \left(\frac{1}{2\pi\sigma^2}\right)^{\frac{M}{2}} \int_\infty d^2r_\mathbf{s} \; \mathrm{pr}_\mathbf{r}(\mathbf{r_s}) \exp\left\{-\frac{1}{2\sigma^2}\sum_{m=1}^{M}[g_m - s_m(\mathbf{r_s}) - b_m]^2\right\} . \tag{13.157}$$

The likelihood ratio is thus given by

$$\Lambda(\mathbf{g}) \propto \int_\infty d^2r_\mathbf{s} \; \mathrm{pr}_\mathbf{r}(\mathbf{r_s}) \exp\left\{\frac{1}{\sigma^2}\sum_{m=1}^{M}[g_m - b_m]s_m(\mathbf{r_s})\right\} , \tag{13.158}$$

where we have dropped all factors that are independent of the data. We see from this expression that the ideal decision strategy in the location-uncertain problem is to subtract the known background contribution from each pixel value, correlate with the signal for a particular location $\mathbf{r_s}$, and exponentiate. The observer performs this set of operations for each possible signal shift and averages with respect to the prior density on signal locations.

The generalization of (13.158) to correlated Gaussian noise is straightforward, leading to

$$\Lambda(\mathbf{g}) \propto \int_\infty d^2r_\mathbf{s} \; \mathrm{pr}_\mathbf{r}(\mathbf{r_s}) \exp\left[(\mathbf{g} - \mathbf{b})^t \mathbf{K_n}^{-1} \mathbf{s}(\mathbf{r_s})\right] . \tag{13.159}$$

It is interesting to consider the weak-signal case in the presence of location uncertainty. If the signal is equally likely to be located anywhere in the data, the average signal is a uniform gray level. For i. i. d. Gaussian noise we see from (13.153) or (13.159) that ideal observer uses this template to detect the signal, meaning it simply computes the sum of the data elements to determine the total number of counts collected and compares this to a threshold.

The decision strategy represented by (13.158) was explored by Nolte and Jaarsma (1967), who also considered several approximations to this expression. They showed that the integral is dominated by the point of maximum correlation with the signal, provided the signal is equally likely at all locations [uniform $\mathrm{pr}(\boldsymbol{\theta})$] and the noise variance is small. The ideal observer is approximated quite well by a cross-correlator that compares the maximum output to a threshold in this case.

More recently, Brown *et al.* (1995) revisited the location-uncertainty problem and showed that the likelihood ratio has a very non-Gaussian probability density function, rendering the SNR expression of (13.19) invalid as a measure of performance. Subsequently, Barrett *et al.* (1998a) showed that this problem is resolved if one makes use of the log-likelihood ratio as the ideal observer's test statistic. They also demonstrated that the likelihood-generating function can be used to determine the ideal observer's SNR for this task in terms of $G(0)$.

One caution should be observed in making use of (13.158), in that we have dropped a data-independent term in the exponent that is proportional to s_m^2, the energy of the signal in each pixel. This quantity is independent of $\mathbf{r_s}$ provided the signal is large relative to the detector pixels and always within the field of view; otherwise this quantity becomes random with signal position. When this is the case, another component of uncertainty must be incorporated into the ideal observer's decision rule.

We have considered the ideal-observer strategy in the location-uncertain problem in which the task is to decide whether signal is present or absent without providing an indication of signal position. When the location of the signal must also be reported, the strategies discovered above are no longer optimal. The detection of a signal in one of many possible signal locations can be considered to be an L-class task, with hypotheses connected with each of the location alternatives (potentially an infinite number of them). The ideal observer would then determine the signal location with the greatest likelihood and compare this to the likelihood that signal is absent in all locations, choosing to report the signal as present in location $\mathbf{r_s}$ or absent, depending on the observer's threshold.

Example: Location and scale uncertainty A signal with random location $\mathbf{r_s}$ and random scale parameter θ can be represented by

$$[\mathbf{s}(\theta, \mathbf{r}_s)]_m = \int_m d^2r \left(\frac{1}{\theta^2}\right) s_m \left(\frac{\mathbf{r} - \mathbf{r}_s}{\theta}\right) . \tag{13.160}$$

Scaling the signal amplitude by $1/\theta^2$ ensures that all signals have the same integrated value regardless of their size. The likelihood ratio in this case is given by

$$\Lambda(\mathbf{g}) \propto \int_\infty d^2 r_\mathbf{s} \int_0^\infty d\theta \; \mathrm{pr}_\theta(\theta)\, \mathrm{pr}_\mathbf{r}(\mathbf{r_s}) \exp\left[(\mathbf{g} - \mathbf{b})^t \mathbf{K_n}^{-1} \mathbf{s}(\theta, \mathbf{r_s})\right] . \tag{13.161}$$

We see that in this case the ideal observer performs the following steps:

1. Subtract the known background from the image.
2. Prewhiten with the noise covariance matrix $\mathbf{K_n}^{-\frac{1}{2}}$.
3. Compute the inner product with the shifted, scaled, prewhitened signal.
4. Exponentiate.
5. Average over all shifts and scales.
6. Compare to a threshold.

Unequal variance and the quadratic discriminant Another model for variability in the object is to consider a signal with known mean $\overline{\mathbf{s}}$, and covariance given by $\mathbf{K_s}$. This might be a reasonable model for a signal with known position, shape, and scale and random amplitude described by a Gaussian PDF. The likelihood ratio for the detection of such a signal is [*cf.* (13.113)]

$$\Lambda = \frac{\exp\left[-\frac{1}{2}(\mathbf{g}-\overline{\mathbf{s}})^t \mathbf{K}_2^{-1}(\mathbf{g}-\overline{\mathbf{s}})\right]}{\exp\left[-\frac{1}{2}\mathbf{g}^t\mathbf{K}_1^{-1}\mathbf{g}\right]}, \tag{13.162}$$

where we have dropped the normalizing constants out front, since they are independent of the data and will therefore only affect the operating point of the observer and not the form of the discriminant function. We have also assumed the signal-absent image has zero mean; a nonzero mean simply requires the addition of a background term as in previous examples. The covariance matrix on the data is now different under the two hypotheses so we have used subscripts to label them; $\mathbf{K}_1 = \mathbf{K}_\mathbf{n}$ and $\mathbf{K}_2 = \mathbf{K}_\mathbf{s} + \mathbf{K}_\mathbf{n}$. Note that we are again describing the signal as a vector in the data space.

The log-likelihood ratio test can be shown to be

$$\lambda(\mathbf{g}) = -\tfrac{1}{2}\mathbf{g}^t\left(\mathbf{K}_2^{-1} - \mathbf{K}_1^{-1}\right)\mathbf{g} + \overline{\mathbf{s}}^t\mathbf{K}_2^{-1}\mathbf{g} \underset{D_1}{\overset{D_2}{\gtrless}} \lambda_c\,, \tag{13.163}$$

where we have incorporated all data-independent terms into the threshold. We see that the inequality of the data covariance under the hypotheses results in a decision function that is no longer linear in the data. The discriminant contains a term that is quadratic in the data in addition to the prewhitening matched filter term.

The L-class problem Another approach to signal variability is to consider the task to be the classification of the data into L distinct classes. For example, the classes might represent L different signal shapes or locations. One of the L classes might be a signal-absent alternative, but this is not required. For each random data set, the observer chooses in favor of D_ℓ, indicating the decision that hypothesis H_ℓ is true,[7] without equivocation or randomness in the decision process.

The ideal observer is defined as that observer who makes optimal use of the available information to perform this task. Not surprisingly, the optimal strategy in this L-class paradigm is to choose the hypothesis H_ℓ associated with a likelihood $\text{pr}(\mathbf{g}|H_\ell)$ that is higher than the likelihood of the data conditioned on all competing hypotheses (Melsa and Cohn, 1978). More generally, the ideal observer decides in favor of the hypothesis with greatest utility, which can be defined as minimization of Bayes risk if the costs and prevalences are known, or as maximum likelihood if this information is not to be used.

While the optimal strategy is straightforward (in principle) in the L-class problem, the full ROC curve is an $(L^2 - L - 1)$-parameter hypersurface in an $(L^2 - L)$-dimensional probability space. Thus an analog to the area under the ROC curve is harder to determine. AUC is an appealing summary measure of binary classification performance because it is a scalar figure of merit that is independent of prevalence and threshold; these attributes are difficult to capture in L-class problems. As an alternative metric, some investigators have made use of the percent of correct responses in an L-alternative forced-choice experiment (Goodenough, 1975; Burgess

[7]This is in contrast to the binary task of the previous section, in which the observer chose between two hypotheses, signal present (with random parameters) versus signal absent.

and Ghandeharian, 1984b). Another approach is to compute the original Bayes risk, recognizing that this figure of merit depends on prevalence and threshold. In Chap. 14 we shall discuss the experimental use of L-alternative tasks for measurement of human performance.

13.2.11 Background variability and the ideal observer

Up to this point we have assumed that the background is a known, nonrandom quantity. In most instances this is an unrealistic model for the imaging task. We now remove the restriction that the background be nonrandom and consider the effect this has on the ideal observer's decision function.

Under the signal-absent hypothesis, the presence of a random background results in a data vector with a probability density given by

$$\mathrm{pr}(\mathbf{g}|H_1) = \int_\infty d^M b \; \mathrm{pr}(\mathbf{g}|H_1, \mathbf{b}) \, \mathrm{pr}_\mathbf{b}(\mathbf{b}) \,. \tag{13.164}$$

Similarly, the signal-present density is given by

$$\mathrm{pr}(\mathbf{g}|H_2) = \int_\infty d^M b \; \mathrm{pr}(\mathbf{g}|H_2, \mathbf{b}) \, \mathrm{pr}_\mathbf{b}(\mathbf{b}) \,. \tag{13.165}$$

Note that we have assumed that the background density is the same under both hypotheses.

The likelihood ratio now becomes

$$\Lambda(\mathbf{g}) = \frac{\int_\infty d^M b \; \mathrm{pr}(\mathbf{g}|H_2, \mathbf{b}) \, \mathrm{pr}_\mathbf{b}(\mathbf{b})}{\int_\infty d^M b' \; \mathrm{pr}(\mathbf{g}|H_1, \mathbf{b}') \, \mathrm{pr}_\mathbf{b}(\mathbf{b}')} = \frac{\langle \mathrm{pr}(\mathbf{g}|H_2, \mathbf{b}) \rangle_\mathbf{b}}{\langle \mathrm{pr}(\mathbf{g}|H_1, \mathbf{b}) \rangle_\mathbf{b}} \,. \tag{13.166}$$

The quantities $\mathrm{pr}(\mathbf{g}|H_j, \mathbf{b})$ in this expression are usually easy to compute from the physics of the measurement noise, at least when the signal is nonrandom. In some problems we will also be able to perform the average over $\mathbf{b}$ either analytically or numerically by drawing samples as discussed in Sec. 8.4, but in other problems this will prove difficult.

To get a useful alternative form for the likelihood ratio, we multiply and divide the integrand in the numerator by $\mathrm{pr}(\mathbf{g}|H_1, \mathbf{b})$ and regroup terms,[8] yielding

$$\Lambda(\mathbf{g}) = \int_\infty d^M b \, \frac{\mathrm{pr}(\mathbf{g}|H_2, \mathbf{b})}{\mathrm{pr}(\mathbf{g}|H_1, \mathbf{b})} \left[\frac{\mathrm{pr}(\mathbf{g}|H_1, \mathbf{b}) \, \mathrm{pr}_\mathbf{b}(\mathbf{b})}{\int_\infty d^M b' \, \mathrm{pr}(\mathbf{g}|H_1, \mathbf{b}') \, \mathrm{pr}_\mathbf{b}(\mathbf{b}')} \right] . \tag{13.167}$$

The factor in square brackets is recognized as the posterior density on the background after observation of $\mathbf{g}$ under the no-signal hypothesis:

$$\frac{\mathrm{pr}(\mathbf{g}|H_1, \mathbf{b}) \, \mathrm{pr}_\mathbf{b}(\mathbf{b})}{\int_\infty d^M b' \, \mathrm{pr}(\mathbf{g}|H_1, \mathbf{b}') \, \mathrm{pr}_\mathbf{b}(\mathbf{b}')} = \frac{\mathrm{pr}(\mathbf{g}|H_1, \mathbf{b}) \, \mathrm{pr}_\mathbf{b}(\mathbf{b})}{\mathrm{pr}(\mathbf{g}|H_1)} = \mathrm{pr}(\mathbf{b}|\mathbf{g}, H_1) \,. \tag{13.168}$$

Thus we can write the likelihood ratio as

$$\Lambda(\mathbf{g}) = \langle \Lambda_{\mathrm{BKE}}(\mathbf{g}, \mathbf{b}) \rangle_{\mathbf{b}|\mathbf{g}, H_1} \,, \tag{13.169}$$

[8]The authors thank Brandon D. Gallas for suggesting this approach and Hongbin Zhang for demonstrating its practicality.

where the subscript BKE indicates *background known exactly*, and

$$\Lambda_{\text{BKE}}(\mathbf{g}, \mathbf{b}) \equiv \frac{\text{pr}(\mathbf{g}|H_2, \mathbf{b})}{\text{pr}(\mathbf{g}|H_1, \mathbf{b})} . \tag{13.170}$$

The interpretation is that $\Lambda_{\text{BKE}}(\mathbf{g}, \mathbf{b})$ is the likelihood that we would have if we knew the background exactly; since we don't, we must average over the backgrounds, but with the posterior density $\text{pr}(\mathbf{b}|H_1, \mathbf{g})$ rather than the prior $\text{pr}(\mathbf{b})$ as in (13.166). Note also that (13.166) uses separate averages of numerator and denominator, while (13.169) averages the ratio.

For nonrandom signals, $\Lambda_{\text{BKE}}(\mathbf{g}, \mathbf{b})$ is easy to compute since numerator and denominator are just noise densities for a specified object. The average over the posterior in (13.169) can often be performed by Markov-chain Monte Carlo methods, especially in simulation studies. (See Secs. 14.3.3 and 15.4.8 for more on Markov-chain Monte Carlo methods.)

When the signal has parameter uncertainty in addition to randomness in the background, the signal-present density is given by

$$\text{pr}(\mathbf{g}|H_2) = \int_\infty d^P\theta \int_\infty d^M b \; \text{pr}_{\mathbf{n}}[\mathbf{g} - \mathbf{b} - \mathbf{s}(\boldsymbol{\theta})|\mathbf{b}] \; \text{pr}_{\mathbf{b}}(\mathbf{b}) \, \text{pr}_{\boldsymbol{\theta}}(\boldsymbol{\theta}) . \tag{13.171}$$

Now the likelihood ratio is given by

$$\Lambda(\mathbf{g}) = \langle \Lambda_{\text{SKE}}(\mathbf{g}, \boldsymbol{\theta}) \rangle_{\boldsymbol{\theta}} , \tag{13.172}$$

where the SKE likelihood ratio must first be computed by taking the background variability into account for each possible signal, as is done in (13.169). Then an average over all possible signals is performed.

Gaussian random backgrounds Several models for random backgrounds were presented in Sec. 8.4. One useful model for a statistically defined background is a multivariate Gaussian with mean $\overline{\mathbf{b}}$ and covariance matrix $\mathbf{K}_{\mathbf{b}}$ in the image domain.[9] If the noise covariance matrix is also Gaussian, the data are described by a form similar to that given in (13.112):

$$\text{pr}(\mathbf{g}|H_1) = \left[(2\pi)^M \det(\mathbf{K}_{\mathbf{g}})\right]^{\frac{1}{2}} \exp\left[-\tfrac{1}{2}(\mathbf{g} - \overline{\mathbf{b}})^t \, \mathbf{K}_{\mathbf{g}}^{-1} (\mathbf{g} - \overline{\mathbf{b}})\right] , \tag{13.173}$$

where the overall data covariance matrix $\mathbf{K}_{\mathbf{g}} = \mathbf{K}_{\mathbf{b}} + \mathbf{K}_{\mathbf{n}}$. Similarly, if the signal is known exactly, then

$$\text{pr}(\mathbf{g}|H_2) = \left[(2\pi)^M \det(\mathbf{K}_{\mathbf{g}})\right]^{\frac{1}{2}} \exp\left[-\tfrac{1}{2}(\mathbf{g} - \overline{\mathbf{b}} - \mathbf{s})^t \, \mathbf{K}_{\mathbf{g}}^{-1} (\mathbf{g} - \overline{\mathbf{b}} - \mathbf{s})\right] . \tag{13.174}$$

When the signal is nonrandom the data covariance matrix is the same under the two hypotheses and the likelihood ratio can be determined in the same fashion by which (13.115) was derived; the ideal observer is again a prewhitening matched filter, this time applied to the image once the mean background has been subtracted.

[9]The process of integrating a 2D stationary Gaussian process over a finite detector element is one way to obtain the Gaussian random background vector model in data space, although other forms for the background in object space can also yield a Gaussian random vector when mapped through an imaging system by virtue of the central-limit theorem.

Random signals on Gaussian backgrounds It is straightforward to incorporate a Gaussian random background into the previous variable-signal examples in which the signal is a function of one or more random parameters. The randomness in the signal results in a data covariance matrix that is no longer the same under the two hypotheses. For this reason the ideal observer's discriminant function is no longer linear in the data. Barrett and Abbey (1997) give specific forms for the likelihood ratio for the cases of random signal location and scale on a Gaussian background.

Random signals on non-Gaussian random backgrounds Non-Gaussian statistical forms for a random background can be incorporated into the general framework of (13.164) and (13.165). All that is required to make use of these expressions is an analytical form for $\text{pr}(\mathbf{g}|H_1) = \text{pr}_{\mathbf{b+n}}(\mathbf{g}|H_1)$. The likelihood ratio can be written again as in (13.148), with the randomness in the background implicitly included in the randomness in $\mathbf{g}$. The likelihood ratio is thus written

$$\Lambda(\mathbf{g}) = \frac{\langle \text{pr}(\mathbf{g}|H_2, \boldsymbol{\theta})\rangle_{\boldsymbol{\theta}}}{\text{pr}(\mathbf{g}|H_1)} = \left\langle \frac{\text{pr}_{\mathbf{b+n}}[\mathbf{g} - \mathbf{s}(\boldsymbol{\theta})]}{\text{pr}_{\mathbf{b+n}}(\mathbf{g})} \right\rangle_{\boldsymbol{\theta}} = \left\langle \Lambda_{\text{BKS}}(\mathbf{g}, \boldsymbol{\theta}) \right\rangle_{\boldsymbol{\theta}}, \tag{13.175}$$

where the subscript BKS indicates *background known statistically*. This expression tells us that for a general random background model the ideal detection strategy is to

1. Compute the signal-absent density function on the data for the given background model, $\text{pr}(\mathbf{g}|H_1)$.

2. Shift by $\mathbf{s}(\boldsymbol{\theta})$ for particular $\boldsymbol{\theta}$.

3. Compute $\Lambda_{\text{BKS}}(\mathbf{g}, \boldsymbol{\theta})$, the conditional likelihood ratio for that $\boldsymbol{\theta}$.

4. Average over all $\boldsymbol{\theta}$.

5. Compare to a threshold.

Many useful models for non-Gaussian backgrounds are described in Chap. 8, which presents a variety of analytical models for random background as well as suggestions for their simulation.

13.2.12 The optimal linear discriminant

We began this chapter with a discussion of figures of merit for discriminant functions of arbitrary form. We then considered the particular form and behavior of the ideal observer, which requires full knowledge of the density function of the data for each hypothesis. In this section we consider discriminant functions that are optimal amongst all discriminant functions constrained to be linear in the data. As we shall see, linear discriminant functions are easy to compute, their performance is easy to summarize, and far less information regarding the data statistics is needed along the way.

Linear discriminants have the general form [*cf.* (13.9)]

$$T(\mathbf{g}) = \mathbf{w}^t\mathbf{g}, \tag{13.176}$$

where $\mathbf{g}$ and $\mathbf{w}$ are assumed to be real. We define the optimal linear discriminant as the $T(\mathbf{g})$ that maximizes a certain measure of class separability, to be discussed below.

In the binary classification problem, a measure of separability is the SNR defined in (13.19). As we shall demonstrate below, the linear discriminant that maximizes this measure takes the form

$$\mathbf{w}_{opt\ lin} = \mathbf{K}_{\mathbf{g}}^{-1}\Delta\overline{\mathbf{g}} \tag{13.177}$$

when the data have equal covariance $\mathbf{K}_{\mathbf{g}}$ under each hypothesis. The resulting SNR is then given by

$$\mathrm{SNR}^2_{opt\ lin} = \Delta\overline{\mathbf{g}}^t\mathbf{K}_{\mathbf{g}}^{-1}\Delta\overline{\mathbf{g}} = \mathrm{tr}\left[\mathbf{K}_{\mathbf{g}}^{-1}\Delta\overline{\mathbf{g}}\Delta\overline{\mathbf{g}}^t\right], \tag{13.178}$$

where $\Delta\overline{\mathbf{g}}$ is the average difference in the data under the two hypotheses, averaged over all sources of variability, and $\mathrm{tr}[\,\cdot\,]$ denotes the trace of the matrix.

An expression very similar to (13.178) (albeit with sample means and covariances instead of population quantities) was first given by the American statistician Harold Hotelling (Hotelling, 1931). By extension, $\mathbf{w}_{opt\ lin}$ has been called the *Hotelling discriminant*, though in fact it is the population equivalent of the familiar Fisher linear discriminant, introduced five years after Hotelling's 1931 paper (Fisher, 1936).[10] Similarly, an observer who implements the optimal linear discriminant has been called the *Hotelling observer*. We shall adopt that terminology here and replace the subscript *opt lin* with *Hot* henceforth.

Relation between Hotelling and ideal observers We have already seen in Sec. 13.2.8 that the optimal discriminant *is* linear in the data whenever the data are Gaussian distributed with the same covariance matrix under the two hypotheses, as is the case when the signal is known exactly and the background is either known exactly or random but Gaussian under both hypotheses. The ideal observer's SNR is then [*cf.* (13.118)]

$$\mathrm{SNR}^2_\lambda = \Delta\mathbf{s}^t\mathbf{K}_{\mathbf{g}}{}^{-1}\Delta\mathbf{s} = \mathrm{tr}\left[\mathbf{K}_{\mathbf{g}}{}^{-1}\Delta\mathbf{s}\Delta\mathbf{s}^t\right]. \tag{13.179}$$

The significant difference between the Hotelling SNR of (13.178) and this expression is that the former allows for signal variability through its reference to the mean data vector $\Delta\overline{\mathbf{g}}$. The ideal observer's SNR takes the form given in (13.179) only when the signals to be discriminated are known exactly; hence it is written in terms of $\Delta\mathbf{s}$.

When the data are Gaussian, with equal covariance for the classes, the Hotelling observer is equal to the ideal observer for the task. When the data are not Gaussian, due to signal variability or non-Gaussian noise or both, the ideal observer can be (usually is) nonlinear in the data. Calculation of the likelihood ratio requires knowledge of the full probability density functions for the data; this requirement can be a major impediment to calculation of the ideal discriminant function for more realistic tasks. The advantage of the Hotelling approach is that it requires knowledge of only the first- and second-order statistics of the data. In essence, the Hotelling approach models the data as Gaussian, regardless of the data's true

[10] Hotelling is credited with establishing one of the first statistics departments in the United States at the University of North Carolina in 1946. Before World War II, many statisticians, including Fisher, had worked in eugenics departments.

statistics. Moreover, as we shall discuss in Chap. 14, the Hotelling observer has been found to be a useful predictor of human performance for a variety of discrimination tasks. Another advantage to the Hotelling formalism is that, as we shall see below, it readily extends to the L-class problem.

Demonstration of optimality We now demonstrate that the Hotelling template of (13.177) is indeed optimal in an SNR sense. To do this, we must show that the Hotelling observer achieves equal or better SNR than that achieved by an arbitrary template $\mathbf{w}$. An arbitrary template would yield an SNR according to (13.19) of

$$\mathrm{SNR}^2_{\mathbf{w}} = \frac{(\mathbf{w}^t \Delta\overline{\mathbf{g}})^2}{\mathbf{w}^t \mathbf{K}_{\mathbf{g}} \mathbf{w}} . \tag{13.180}$$

Thus we must show that

$$\frac{(\mathbf{w}^t \Delta\overline{\mathbf{g}})^2}{\mathbf{w}^t \mathbf{K}_{\mathbf{g}} \mathbf{w}} \le \Delta\overline{\mathbf{g}}^t \mathbf{K}_{\mathbf{g}}^{-1} \Delta\overline{\mathbf{g}} \tag{13.181}$$

or

$$(\mathbf{w}^t \Delta\overline{\mathbf{g}})^2 \le (\mathbf{w}^t \mathbf{K}_{\mathbf{g}} \mathbf{w})(\Delta\overline{\mathbf{g}}^t \mathbf{K}_{\mathbf{g}}^{-1} \Delta\overline{\mathbf{g}}) , \tag{13.182}$$

where we have used the positive-definiteness of the denominator in the first line to move it to the right-hand side in the second line.

We can insert an identity in the form of the product of the square root of the covariance matrix and its inverse, and then make use of the triangle inequality, to write the left-hand side as

$$(\mathbf{w}^t \Delta\overline{\mathbf{g}})^2 = \left[\mathbf{w}^t \mathbf{K}_{\mathbf{g}}^{1/2} \mathbf{K}_{\mathbf{g}}^{-1/2} \Delta\overline{\mathbf{g}}\right]^2 \le ||\mathbf{w}^t \mathbf{K}_{\mathbf{g}}^{1/2}||^2 \, ||\mathbf{K}_{\mathbf{g}}^{-1/2} \Delta\overline{\mathbf{g}}||^2 . \tag{13.183}$$

Now we write the norms of the vectors as inner products and we have

$$(\mathbf{w}^t \Delta\overline{\mathbf{g}})^2 \le (\mathbf{K}_{\mathbf{g}}^{1/2}\mathbf{w})^t (\mathbf{K}_{\mathbf{g}}^{1/2}\mathbf{w}) \, (\mathbf{K}_{\mathbf{g}}^{-1/2}\Delta\overline{\mathbf{g}})^t (\mathbf{K}_{\mathbf{g}}^{-1/2}\Delta\overline{\mathbf{g}}) = (\mathbf{w}^t \mathbf{K}_{\mathbf{g}} \mathbf{w}) \, (\Delta\overline{\mathbf{g}}^t \mathbf{K}_{\mathbf{g}}^{-1} \Delta\overline{\mathbf{g}}) . \tag{13.184}$$

This is what we set out to show in (13.182). Thus the template given in (13.177) indeed achieves the maximum SNR of all linear observers when $\mathbf{K}_1 = \mathbf{K}_2 = \mathbf{K}_{\mathbf{g}}$.

It is left to the reader to show that a template given by $\mathbf{w} = [\frac{1}{2}(\mathbf{K}_1 + \mathbf{K}_2)]^{-1} \Delta\overline{\mathbf{g}}$ achieves maximum SNR in the binary classification problem when $\mathbf{K}_1 \neq \mathbf{K}_2$. As we shall soon demonstrate, this is the Hotelling template for this case.

Hotelling trace The beauty of the Hotelling formalism is that it is readily extended to the L-class problem, where $L > 2$. To describe the performance of the optimal linear discriminant in the L-class problem, we first define two scatter matrices. The *interclass scatter matrix*,

$$\mathbf{S}_1 = \tfrac{1}{L} \sum_{\ell=1}^{L} (\overline{\mathbf{g}}_\ell - \overline{\overline{\mathbf{g}}})(\overline{\mathbf{g}}_\ell - \overline{\overline{\mathbf{g}}})^t , \tag{13.185}$$

describes the average distance between the means of the distributions of the data under each hypothesis from the overall mean $\overline{\overline{\mathbf{g}}} = (1/L) \sum_\ell \overline{g}_\ell$. The rank of $\mathbf{S}_1$ is $L - 1$, owing to the relationship of the class means to the overall mean. In the 2-class problem, the interclass scatter matrix reduces to:

$$\mathbf{S}_1 = \tfrac{1}{4} \Delta\overline{\mathbf{g}} \, \Delta\overline{\mathbf{g}}^t . \tag{13.186}$$

The *intraclass scatter matrix*, $\mathbf{S}_2$, is given by

$$\mathbf{S}_2 = \tfrac{1}{L}\sum_{\ell=1}^{L}\langle(\mathbf{g} - \overline{\mathbf{g}}_\ell)(\mathbf{g} - \overline{\mathbf{g}}_\ell)^t\rangle_\ell = \tfrac{1}{L}\sum_{\ell=1}^{L}\mathbf{K}_{\mathbf{g}|\ell}\,, \tag{13.187}$$

and describes the average covariance matrix of the data, found by averaging the covariance matrices of the L classes.

We assume knowledge of the ensemble mean and covariance of the data under each hypothesis in writing (13.185) and (13.187). The Hotelling observer is the linear observer that makes use of this knowledge to achieve maximum discrimination performance of all linear observers. The generalized measure of class separability in the L-class problem is referred to as the *Hotelling trace* and often given the label J:

$$J = \mathrm{tr}\left[\mathbf{S}_2^{-1}\mathbf{S}_1\right]\,. \tag{13.188}$$

For a binary classification task, this reduces to

$$J = \tfrac{1}{4}\Delta\overline{\mathbf{g}}^t\mathbf{S}_2^{-1}\Delta\overline{\mathbf{g}}^t = \tfrac{1}{4}\mathrm{SNR}_{Hot}^2 \tag{13.189}$$

by (13.186).

Note that the scatter matrices of (13.186) and (13.187) often are written as sums of class contributions weighted by class prevalence, even in literature from the authors. We have deliberately defined the scatter matrices here with equally weighted contributions from the L classes so that the Hotelling observer's discriminant function is independent of prevalence. With this formulation the Hotelling observer achieves maximum SNR and maximum J.

Finding the optimal linear discriminant Given the scatter matrices above, we wish to determine the linear discriminant in the L-class problem that achieves maximum J. We follow the derivation given by Fukunaga (1990).

The key to finding the optimal linear discriminant is the knowledge that we can simultaneously diagonalize two noncommuting Hermitian matrices by the process outlined in Sec. 1.4.6. The simultaneous diagonalization of $\mathbf{S}_{2\mathbf{g}}$ and $\mathbf{S}_{1\mathbf{g}}$ is written [*cf.* (1.100)]

$$\mathbf{W}^\dagger\mathbf{S}_{2\mathbf{g}}\mathbf{W} = \mathbf{I} \qquad \text{and} \qquad \mathbf{W}^\dagger\mathbf{S}_{1\mathbf{g}}\mathbf{W} = \mathbf{D}\,, \tag{13.190}$$

where $\mathbf{D}$ is diagonal. Note that this transformation whitens the intraclass scatter matrix. It is shown in Sec. 1.4.6 that an equivalent eigenvalue problem is given by [*cf.* (1.110)]

$$\mathbf{S}_{2\mathbf{g}}^{-1}\mathbf{S}_{1\mathbf{g}}\mathbf{W} = \mathbf{W}\mathbf{D}\,. \tag{13.191}$$

Thus $\mathbf{W}$ is the matrix of eigenvectors of the product $\mathbf{S}_{2\mathbf{g}}^{-1}\mathbf{S}_{1\mathbf{g}}$, and $\mathbf{D}$ is the diagonal matrix of eigenvalues. We shall call the M eigenvalues $\{\mu_m\}$ and the eigenvectors $\{\mathbf{w}_m\}$.

Recall that the trace of a matrix is equal to the sum of its eigenvalues. Thus the Hotelling trace for the data has these equivalent forms:

$$J_\mathbf{g} = \mathrm{tr}[\mathbf{S}_{2\mathbf{g}}^{-1}\mathbf{S}_{1\mathbf{g}}] = \mathrm{tr}[\mathbf{D}] = \sum_{m=1}^{M}\mu_m\,. \tag{13.192}$$

The mth eigenvalue measures the separability associated with a projection of the data along the direction in feature space defined by $\mathbf{w}_m$. Since the rank of $\mathbf{S}_{1\mathbf{g}}$ is

$(L-1)$, where L is the number of classes, the ranks of $\mathbf{S}_{2\mathbf{g}}^{-1}\mathbf{S}_{1\mathbf{g}}$ and $\mathbf{D}$ are also $(L-1)$. There are therefore only $(L-1)$ nonzero eigenvalues in the sum; all the separability in the data is carried by these eigenvalues. Projection of the data onto the subspace spanned by the eigenvectors of $\mathbf{S}_{2\mathbf{g}}^{-1}\mathbf{S}_{1\mathbf{g}}$ results in no loss of discriminability.

The optimal linear discriminant is thus found by solving the eigenvalue equation of (13.191). The Hotelling observer classifies the data using the feature vector

$$\mathbf{t} = \mathbf{W}^t\mathbf{g}\,, \tag{13.193}$$

where $\mathbf{W}$ is the matrix whose columns are the eigenvectors of $\mathbf{S}_{2\mathbf{g}}^{-1}\mathbf{S}_{1\mathbf{g}}$. We shall label the mth column vector $\mathbf{w}_m$, corresponding to the eigenvalue μ_m. Then

$$t_m = \mathbf{w}_m^t\mathbf{g}\,. \tag{13.194}$$

In the binary classification problem the separability inherent in an M-dimensional data set is preserved in a 1D feature space, and (13.191) reduces to

$$\mathbf{S}_{2\mathbf{g}}^{-1}\mathbf{S}_{1\mathbf{g}}\mathbf{w}_{Hot} = \mu\mathbf{w}_{Hot}\,. \tag{13.195}$$

The reader can show that this equation is satisfied when

$$\mathbf{w}_{Hot} = \mathbf{S}_{2\mathbf{g}}^{-1}\Delta\overline{\mathbf{g}}\,, \tag{13.196}$$

which is a generalized form for the optimal feature vector of (13.177) when the covariances under the hypotheses are unequal. The binary discriminant is thus given by

$$t = \mathbf{w}_{Hot}^t\mathbf{g}\,, \tag{13.197}$$

and the resulting Hotelling trace is given by μ.

Fukunaga (1990) also addresses the question of the best linear transformation that yields an ND feature vector, where $N < (L-1)$. It can be shown that this *inefficient* transformation, so named because it does not preserve the separability in the original data, is again obtained by solving the eigenvector problem of (13.191), only now the transformation matrix is composed of the N eigenvectors with the largest eigenvalues as its columns.

From features to classification In the binary decision problem the connection between feature extraction and classification is straightforward. The Hotelling observer forms the single test statistic of (13.197) and compares this value to a threshold to decide between hypothesis 1 and 2. The test statistic t can be thought of as a single feature derived from a given image; that feature is used to classify the image.

The relationship between feature extraction and classification in the L-class task is more complex. Comparing each of the $L-1$ features to a threshold is equivalent to using $L-1$ hyperplanes to partition the feature space, which can lead to regions of ambiguous class assignment.

A number of options are available for avoiding the problem of ambiguous areas in the decision space (Duda *et al.*, 2001). We shall consider one option, inspired by the approach taken by the ideal observer in the L-class problem as described in Sec. 13.2.10. Recall that the ideal observer selects the hypothesis associated with the greatest likelihood of the data. Analogously, one option for the Hotelling observer's strategy is to choose the hypothesis that gives the largest probability of obtaining the data, only now under Gaussian assumptions for each $\mathrm{pr}(\mathbf{g}|H_\ell)$.

Under the Gaussian assumption, $\mathrm{pr}(\mathbf{t}|H_\ell)$ is a multivariate Gaussian PDF as given in (13.112), with $\overline{\mathbf{t}}_\ell = \mathbf{W}^t\overline{\mathbf{g}}_\ell$ and $\mathbf{K}_{\mathbf{t}|\ell} = \mathbf{W}^t\mathbf{K}_{\mathbf{g}|\ell}\mathbf{W}$. The Hotelling observer chooses the hypothesis that gives maximum

$$\overline{\mathbf{t}}_\ell^t \mathbf{K}_{\mathbf{t}|\ell}^{-1} \mathbf{t} = \overline{\mathbf{g}}_\ell^t \mathbf{W} \left[\mathbf{W}^t \mathbf{K}_{\mathbf{g}|\ell} \mathbf{W}\right]^{-1} \mathbf{W}^t \mathbf{g} . \tag{13.198}$$

When $\mathbf{K}_{\mathbf{g}|\ell}$ is approximately independent of ℓ, then by (13.190)

$$\left[\mathbf{W}^t \mathbf{K}_{\mathbf{g}} \mathbf{W}\right]^{-1} = \mathbf{I} . \tag{13.199}$$

With (13.199) we can rewrite the Hotelling observer's decision strategy of (13.198) as

$$\text{Choose } H_\ell \text{ that gives max} \quad [\overline{\mathbf{g}}_\ell^t \mathbf{W}][\mathbf{W}^t \mathbf{g}] = \overline{\mathbf{t}}_\ell^t \mathbf{t} . \tag{13.200}$$

This is a matched filter applied to the feature vector; no prewhitening is required because the covariance of the feature vectors is already white by (13.199).

Thus, in the L-class classification problem, the Hotelling observer forms the matched filter output for each possible signal in the reduced-dimensionality feature space and chooses the alternative that gives the maximum value. No ambiguity in the decision space arises.

It should be noted that the Hotelling trace has limitations in describing the performance of the Hotelling observer in the L-class problem. For example, it fails to distinguish between situations in which one class is perfectly separated from alternative classes that highly overlap *vs.* situations in which all L classes are highly separable. Performance measures in higher dimensions are required to summarize the Hotelling observer for the L-class problem; this is an area of active research.[11]

Post-processing and feature extraction We found in Sec. 13.2.6 that digital image processing does not affect the ideal observer, so long as the processing operation is invertible in the sense that the original digital data before processing is recoverable. The same can be said for linear post-processing and the performance of the Hotelling observer. Any nonsingular $M \times M$ linear transformation $\mathbf{A}$ preserves the the separability in the data because if

$$\mathbf{y} = \mathbf{A}^t \mathbf{g} , \tag{13.201}$$

then

$$J_{\mathbf{y}} = \mathrm{tr}[(\mathbf{A}^t \mathbf{S}_{2\mathbf{g}} \mathbf{A})^{-1} \mathbf{A}^t \mathbf{S}_{1\mathbf{g}} \mathbf{A}] = \mathrm{tr}[\mathbf{S}_{2\mathbf{g}} \mathbf{S}_{1\mathbf{g}}] = J_{\mathbf{g}} . \tag{13.202}$$

Similarly, a nonsingular linear transformation can be applied to the features $\mathbf{t}$ of (13.197); the inherent discriminability of the data is preserved by these features, and an invertible transformation applied to $\mathbf{t}$ will still yield $J_{\mathbf{t}} = J_{\mathbf{g}}$ by the logic of (13.202).

We have seen that the dimensionality of the data can be reduced and yet maintain the separability inherent in the data. Suppose instead that the post-processing resulted in an increase in dimensionality. For example, linear image

[11] We thank Xin He and Eric Frey for helpful comments on this point.

reconstruction algorithms process the raw data to yield a number of features N (pixel values) that can be greater than M. We now show under what conditions the resulting features preserve the separability of the data.

Let the post-processing operation again be written as (13.201), with $\mathbf{A}$ now an $M \times N$ post-processing matrix, $N > M$. Then

$$J_{\mathbf{y}} = \mathrm{tr}[(\mathbf{A}^t\mathbf{S}_{2\mathbf{g}}\mathbf{A})^{-1}\mathbf{A}^t\mathbf{S}_{1\mathbf{g}}\mathbf{A}] = \mathrm{tr}[\mathbf{A}(\mathbf{A}^t\mathbf{S}_{2\mathbf{g}}\mathbf{A})^{-1}\mathbf{A}^t\mathbf{S}_{1\mathbf{g}}]\,. \tag{13.203}$$

However, we cannot equate $J_{\mathbf{y}}$ with $J_{\mathbf{g}}$ as we did in (13.202) because in the current example $\mathbf{A}^{-1}$ does not exist.

Now form the $M \times M$ matrix $\mathbf{L} = \mathbf{A}\mathbf{A}^t$. If we assume that $\mathbf{A}$ has a left inverse (that is, L is nonsingular; see Sec. 1.7.2), we can insert the identity $\mathbf{S}_{2\mathbf{g}}^{-1}\mathbf{L}^{-1}\mathbf{L}\mathbf{S}_{2\mathbf{g}}$ inside the trace in (13.203) to find

$$\begin{aligned} J_{\mathbf{y}} &= \mathrm{tr}[\mathbf{S}_{2\mathbf{g}}^{-1}\mathbf{L}^{-1}\mathbf{L}\mathbf{S}_{2\mathbf{g}}\mathbf{A}(\mathbf{A}^t\mathbf{S}_{2\mathbf{g}}\mathbf{A})^{-1}\mathbf{A}^t\mathbf{S}_{1\mathbf{g}}] \\ &= \mathrm{tr}[\mathbf{S}_{2\mathbf{g}}^{-1}\mathbf{L}^{-1}\mathbf{A}\mathbf{A}^t\mathbf{S}_{2\mathbf{g}}\mathbf{A}(\mathbf{A}^t\mathbf{S}_{2\mathbf{g}}\mathbf{A})^{-1}\mathbf{A}^t\mathbf{S}_{1\mathbf{g}}] = \mathrm{tr}[\mathbf{S}_{2\mathbf{g}}^{-1}\mathbf{L}^{-1}\mathbf{A}\mathbf{A}^t\mathbf{S}_{1\mathbf{g}}] \\ &= \mathrm{tr}[\mathbf{S}_{2\mathbf{g}}^{-1}\mathbf{S}_{1\mathbf{g}}] = J_{\mathbf{g}}\,. \end{aligned} \tag{13.204}$$

We have thus demonstrated that the features preserve the inherent discriminability in the original data so long as a left inverse of the post-processing matrix exists.

The effect associated with nonlinear processing is more subtle. Appropriately chosen nonlinear processing or feature extraction can improve Hotelling observer performance by rendering the data more separable by a linear discriminant. For example, consider the case of a binary detection task in which the signal can have either a positive amplitude a or negative amplitude $-a$ with equal probability. Since the mean data vector under each hypothesis is zero, a linear observer would yield an SNR of zero for this problem. The reader can show that, for additive multivariate Gaussian noise, use of (13.112) and (13.148) to compute the log-likelihood ratio yields the following optimal strategy:

$$\lambda = \mathbf{g}^t\mathbf{K}_{\mathbf{n}}^{-1}\mathbf{g}\,. \tag{13.205}$$

When the noise is independent and identically distributed, the ideal observer's decision variable reduces to

$$\lambda = \sum_{m=1}^{M} g_m^2\,. \tag{13.206}$$

We see that the ideal observer uses a decision variable that is quadratic in the data. Taking inspiration from the ideal observer, the application of a point-wise absolute-value operator to each data element, $\mathbf{y} = |\mathbf{g}|$, improves the performance of the Hotelling observer in that $\mathrm{SNR}_{\mathbf{y}}^2 > \mathrm{SNR}_{\mathbf{g}}^2 = 0$.

More generally, for any task where the ideal observer's decision strategy is nonlinear, it is possible to improve a linear observer's AUC by nonlinear preprocessing of the data. As demonstrated in the previous example, the optimal nonlinear processing takes the form of computing the likelihood ratio. Designing post-processing to improve the Hotelling trace, however, is trickier business, requiring an improvement in the difference in class means relative to their average variance, as quantified by (13.178). This point is particularly relevant to the performance of human observers. There are some tasks for which the human and Hotelling SNRs are well below the detectability achievable by an ideal observer [computed from AUC via (13.21)]. In such circumstances, preprocessing of the data may improve human observer performance by better matching the data to the abilities of the human visual system. This has implications for image processing and computer-aided diagnosis.

Signal variability and the Hotelling observer The Hotelling approach allows arbitrary variability in the data resulting from random signals and/or random backgrounds. If the task is discrimination between two classes with statistically defined signals, the Hotelling template is given by

$$\mathbf{w} = \mathbf{S}_2^{-1}\Delta\overline{\mathbf{g}} = \mathbf{S}_2^{-1}\Delta\overline{\mathbf{s}}\,, \tag{13.207}$$

requiring the mean of the variable signal under each hypothesis to calculate $\Delta\overline{\mathbf{s}}$. This linear discriminant may work very well in the presence of signal variability, especially for compact signals (Eckstein and Abbey, 2001).

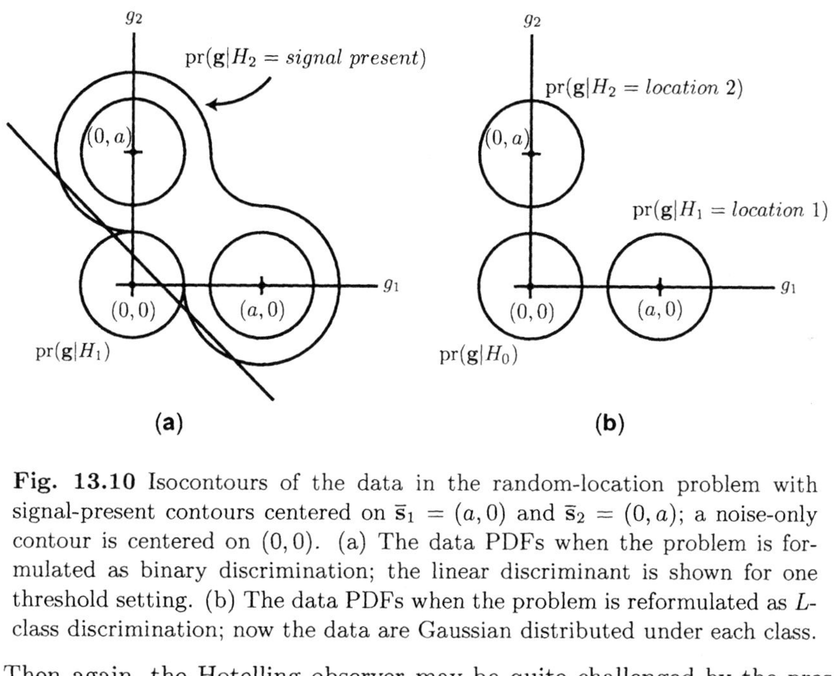

Fig. 13.10 Isocontours of the data in the random-location problem with signal-present contours centered on $\overline{\mathbf{s}}_1 = (a, 0)$ and $\overline{\mathbf{s}}_2 = (0, a)$; a noise-only contour is centered on $(0, 0)$. (a) The data PDFs when the problem is formulated as binary discrimination; the linear discriminant is shown for one threshold setting. (b) The data PDFs when the problem is reformulated as L-class discrimination; now the data are Gaussian distributed under each class.

Then again, the Hotelling observer may be quite challenged by the presence of signal uncertainty, particularly when it leads to highly non-Gaussian data. We shall illustrate this challenge by considering the problem of detection of a signal with large location uncertainty. Let H_1 denote the absence of a signal and H_2 denote the signal-present hypothesis, with signal of known shape and amplitude in any one of L nonoverlapping locations on the detector. The task of the observer is to decide whether or not a signal is present. Figure 13.10 is a simplistic representation of the data PDFs for this task in a two-dimensional subspace. Under H_1, the data are centered on the origin with circular isocontours of constant probability density. Under H_2 the data are distributed equally into L regions of data space, two of which are shown in the plot. The circular isocontours show the scatter about these two locations due to the additive Gaussian noise. The overall density function of the data in the signal-present case is highly non-Gaussian and multimodal. The ideal observer decision variable for this problem, quite nonlinear in the data, is given in (13.150). The Hotelling observer attempts to separate the data from the two hypotheses using features derived from the eigenanalysis of (13.191). It has been shown that in the limit of a very large number of locations, the linear discriminant is not an effective classifier for this task (Brown *et al.*, 1995). However, this binary detection example can be reformulated as a $(L+1)$-hypothesis classification task, with H_0: signal absent and H_ℓ: signal present at location ℓ. Then a set of linear discriminants will separate the data quite nicely.

Note that a similar reformulation can be imposed on the bipolar-signal example discussed above in the context of post-processing. Splitting the signal-present hypothesis into two, where one is that the signal is present with amplitude $+a$ and the other is that the signal is present with amplitude $-a$, yields improved performance over the 2-class problem.

More generally, situations in which signal uncertainty produces a data space populated by multiple approximately Gaussian clouds, such as occurs in the detection of a signal in one of L orthogonal locations, or the detection of one of L orthogonal signal profiles (*e.g.*, sinusoids of different phase, or multiple Hadamard signals) often yield improved Hotelling performance when the number of candidate hypotheses is increased. In these cases the signal is defined in terms of a vector of random parameters $\boldsymbol{\theta}$ as in (13.145), where $\text{pr}_{\boldsymbol{\theta}}(\boldsymbol{\theta})$ takes on the special form of a finite set of delta functions. Increasing the number of hypotheses reduces the contribution from signal variability to the intraclass scatter matrix, leaving measurement noise (which is often well modeled by a Gaussian PDF) as the dominant source of variability in the data. The data are thus characterized by localized Gaussian distributions such that a set of linear discriminants can be used to partition the space.

Not all decision problems with signal-parameter uncertainty may be reformulated to suit the Hotelling observer so readily. When the signal uncertainty does not take the form of a finite set of delta functions, the Hotelling observer's efficiency relative to the ideal observer can decrease dramatically, and no amount of fiddling with the number of hypotheses will help. For example, discrimination of texture differences is often cited as a decision task that requires nonlinear strategies. The choice of observer (Hotelling or ideal) used to design or evaluate an imaging system will be driven by the information available regarding the task and the observer for which the system is being optimized.

Detectability maps Another approach to tasks with randomness in the signal is to compute the optimum linear test statistic for the SKE case as a function of the random parameters:

$$\mathbf{w}(\boldsymbol{\theta}) = [\mathbf{S}_2(\boldsymbol{\theta})]^{-1}\Delta\mathbf{s}(\boldsymbol{\theta}) \,. \tag{13.208}$$

The SNR for the Hotelling observer for each value of the random parameter vector is thus given by

$$\text{SNR}^2_{Hot}(\boldsymbol{\theta}) = \frac{\{[\mathbf{w}(\boldsymbol{\theta})]^t\,\Delta\mathbf{s}(\boldsymbol{\theta})\}^2}{[\mathbf{w}(\boldsymbol{\theta})]^t\,\mathbf{S}_2(\boldsymbol{\theta})\,\mathbf{w}(\boldsymbol{\theta})} = [\mathbf{w}(\boldsymbol{\theta})]^t\mathbf{s}(\boldsymbol{\theta}) \,. \tag{13.209}$$

Using this expression, detectability maps showing sensitivity of SNR to $\boldsymbol{\theta}$, *e.g.*, location, can be made (Pineda *et al.*, 2000). The beauty of (13.208) is that in many cases the linear observer is the ideal observer for fixed signal parameters, because the randomness in the data from measurement noise (and even background variability in some cases) is well approximated by a Gaussian.

Signal known exactly, but variable The SNR of (13.209) gives the Hotelling observer's performance for a particular value of the random parameter $\boldsymbol{\theta}$. A summary measure of observer performance can be obtained by averaging this expression over $\boldsymbol{\theta}$. System optimization can thus be pursued with this overall measure of performance as the figure of merit, so that the system is "best" for the set of all possible signals. We refer to this assessment method as *signal known exactly, but variable (SKEV)*.

Detection in a random background We found previously that a Gaussian random background results in an ideal decision strategy that is linear in the data. The ideal observer and the Hotelling observer are thus equivalent for this special case. Other statistical models for random backgrounds may yield nonlinear optimal discriminant functions; the exact form of the optimal discriminant given in (13.175) may not even be calculable in many cases. We are not without resources, though, because calculation of the Hotelling SNR requires only knowledge of the first- and second-order statistics of the data, and often we have knowledge of these.

We can characterize a general random background in terms of its mean contribution to detector element m, denoted $\bar{b}_m$ and its covariance matrix

$$[\mathbf{K_b}]_{ij} = \langle (b_i - \bar{b}_i)(b_j - \bar{b}_j) \rangle_{\mathbf{b}} \,. \tag{13.210}$$

In the weak-signal approximation, the covariance of the data is the same under each hypothesis and equal to

$$[\mathbf{K_g}]_{ij} = \langle \langle (g_i - \bar{b}_i)(g_j - \bar{b}_j) \rangle_{\mathbf{n}|\mathbf{b}} \rangle_{\mathbf{b}} = \bar{b}_i \, \delta_{ij} + [\mathbf{K_b}]_{ij} \,, \tag{13.211}$$

where we have assumed Poisson noise in writing the last result. The Hotelling observer's discriminant is given by (13.208) (presuming some randomness in the signal) with $\mathbf{S}_2$ equal to $\mathbf{K_g}$ of (13.211).

A uniform background model with a random level can be considered as the limit of a random background model in which the mean background is the same in all detector elements, and the background variation is completely correlated across the detector. Let $\bar{b}$ be the average level of the uniform background, with variance σ_b^2. The covariance matrix for the data in this case is given by

$$[\mathbf{K_g}]_{ij} = \bar{b}\, \delta_{ij} + \sigma_b^2$$

or

$$\mathbf{K_g} = \bar{b}\mathbf{I} + \mathbf{u}\mathbf{u}^t \,, \tag{13.212}$$

where $\mathbf{u}$ is an $M \times 1$ column vector in which each element is equal to σ_b. The inverse of $\mathbf{K_g}$ can be determined via (A.55):

$$\mathbf{K_g^{-1}} = (\bar{b}\,)^{-1}\mathbf{I} - \frac{\mathbf{u}\mathbf{u}^t}{\bar{b}^2 + M\bar{b}\sigma_b^2} \,. \tag{13.213}$$

The Hotelling observer's test statistic for discriminating between two known signals on a flat background of random level is then given by:

$$t = \Delta \mathbf{s}^t \mathbf{K_g^{-1}} \mathbf{g} = \sum_{n=1}^{M} \sum_{m=1}^{M} \Delta s_n \left[\frac{\delta_{nm}}{\bar{b}} - \frac{[\mathbf{u}\mathbf{u}^t]_{nm}}{\bar{b}^2 + M\bar{b}\sigma_b^2} \right] g_m$$

$$= (\bar{b})^{-1} \sum_{n=1}^{M} \Delta s_n \left[g_n - \frac{\sigma_b^2 \sum_m g_m}{\bar{b} + M\sigma_b^2} \right] = (\bar{b})^{-1} \sum_{n=1}^{M} \Delta s_n \left[g_n - \frac{M\sigma_b^2 \hat{b}}{\bar{b} + M\sigma_b^2} \right], \quad (13.214)$$

where $\hat{b} = \frac{1}{M} \sum_m g_m$ is an estimate of the background level formed from the data set in hand. Note that as $\sigma_b^2 \to 0$ the Hotelling observer approaches the background-known-exactly matched filter result. Alternatively, for large σ_b^2, the Hotelling test statistic approaches

$$t = \frac{1}{\bar{b}} \sum_{n=1}^{M} \Delta s_n (g_n - \hat{b}) . \quad (13.215)$$

In this case the Hotelling observer estimates the background from the data prior to applying a matched filter.

The Hotelling observer's performance in the random-background-level task is summarized by

$$\mathrm{SNR}_{Hot}^2 = \Delta \mathbf{s}^t \mathbf{K}_\mathbf{g}^{-1} \Delta \mathbf{s} = \sum_{n=1}^{M} \sum_{m=1}^{M} \Delta s_n \left[\frac{\delta_{nm}}{\bar{b}} - \frac{[\mathbf{u}\mathbf{u}^t]_{nm}}{\bar{b}^2 + M\bar{b}\sigma_b^2} \right] \Delta s_m$$

$$= \frac{||\Delta \mathbf{s}||^2}{\bar{b}} - \frac{[\Delta \mathbf{s}^t \mathbf{u}]^2}{\bar{b}^2 + M\bar{b}\sigma_b^2} . \quad (13.216)$$

The Hotelling observer's SNR in the presence of a more general background is found by an extension to the approach that led to (13.216); in particular, a more general background model would give a more general expression for the data covariance of (13.212). Of note are models for statistically defined backgrounds with known first- and second-order statistics such as the lumpy and clustered-lumpy backgrounds (described in Sec. 8.4.4) that allow for task-based assessment through exact computation of Hotelling-observer performance even though the ideal observer's strategy is unknown. Figure 13.11 shows the radial profile of the Hotelling template for classification of a known Gaussian signal on a Gaussian lumpy background of known correlation length. The Hotelling observer attempts to match the signal in the expected location, while subtracting off an estimate of the local surrounding background.

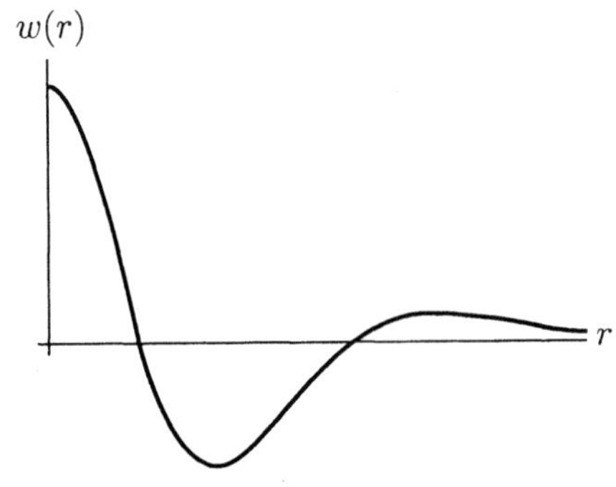

Fig. 13.11 Radial profile of the Hotelling template for detection of a Gaussian signal on a lumpy background.

Practical issues related to the use of random backgrounds in system evaluation will be discussed further in Chap. 14.

Non-Gaussian noise and the Hotelling observer The Hotelling observer results in a suboptimal AUC whenever the statistics of the data are non-Gaussian. Even when the statistics on the data are Gaussian, the Hotelling observer achieves suboptimal AUC for situations where data covariance matrices are unequal across hypotheses; the ideal-observer's discriminant is a nonlinear functional of the data in that case.

In Sec. 13.2.9 we determined the ideal observer's decision strategy for several examples involving non-Gaussian data.[11] While the ideal observer's test statistic can be highly nonlinear in the data for non-Gaussian statistics, recall that certain non-Gaussian statistics yield a linear ideal decision surface. We found this to be the case for Poisson noise (13.131) as well as exponential noise (13.137). Nevertheless, the ideal observer's linear strategy and that of the Hotelling observer can differ because the Hotelling observer uses only first- and second-order statistics to describe the data and not the complete joint density function.

As an example, consider the problem of detection in speckle noise, for which the log-likelihood ratio gives a linear filtering operation with the following template [*cf.* (13.137)]:

$$[\mathbf{w}_{ideal}]_m = \frac{\Delta\overline{g}_m}{\overline{g}_{1m}\,\overline{g}_{2m}} \,. \tag{13.217}$$

The Hotelling observer makes use of the fact that the mean in the mth detector element under hypothesis j is $\overline{g}_{jm}$ and its variance is $\overline{g}^2_{jm}$ to yield

$$[\mathbf{w}_{Hot}]_m = [\mathbf{S}_2^{-1}\Delta\overline{\mathbf{g}}]_m = \frac{\Delta\overline{g}_m}{\frac{1}{2}(\overline{g}^2_{1m} + \overline{g}^2_{2m})} \,. \tag{13.218}$$

The numerators are the same, but the ideal observer divides by the geometric mean of the variances while the Hotelling observer divides by the arithmetic mean. The resulting decision surface for each observer is a hyperplane in data space, although with different orientation. As a result, the Hotelling observer attains maximum SNR, while the ideal observer attains maximum AUC. The observer with the higher AUC is preferred, of course, since AUC is a more accurate measure of observer performance for arbitrary data statistics.

A similar set of calculations can be performed to compare the Hotelling observer to the ideal observer for other non-Gaussian data distributions. In particular, since Poisson noise is well approximated by a Gaussian PDF when the number of counts per detector element is greater than about ten, the Hotelling observer is asymptotically optimal in both AUC and SNR in that limit. Both the ideal observer and the Hotelling observer are linear in the data for Poisson noise; the orientation of their decision surfaces converges as the count rate increases.

AUC-optimal linear discriminants As we have just seen, there are some cases where the log-likelihood is a linear discriminant but not equivalent to the Hotelling discriminant. This raises two questions: 1) In general, what linear discriminant is optimal in the sense of maximizing AUC? 2) When is the log-likelihood a linear discriminant? We shall address the former question here and the latter below.

[11]Note that these strategies do not necessarily yield maximum SNR as computed from (13.19). Rather, they are optimal in the attributes relevant to the ideal observer: maximum AUC and thus maximum d_A, minimum decision error, and minimum Bayes risk.

Our starting point for finding the linear discriminant that maximizes AUC is (13.47), which gives the AUC for an arbitrary linear discriminant when the noise is additive and independent of the signal. As that equation shows, only the odd part of the signal characteristic function $\psi_{\mathbf{s}}^*(\mathbf{w}\xi)$ contributes to AUC_{lin}. Since $\psi_{\mathbf{s}}^*(\mathbf{w}\xi) = \psi_{\mathbf{s}}(-\mathbf{w}\xi)$, the odd part of $\psi_{\mathbf{s}}^*(\mathbf{w}\xi)$ is i times its imaginary part, and the AUC for a linear discriminant is optimized by

$$\mathbf{w}_{opt} = \underset{\mathbf{w}}{\mathrm{argmax}} \left\{ \int_{-\infty}^{\infty} \frac{d\xi}{\xi} \left|\psi_{\mathbf{g}1}(\mathbf{w}\xi)\right|^2 \mathrm{Im}\left[\psi_{\mathbf{s}}^*(\mathbf{w}\xi)\right] \right\}. \tag{13.219}$$

For SKE problems [*cf.* (13.49) and (13.50)],

$$\begin{aligned}\mathbf{w}_{opt} &= \underset{\mathbf{w}}{\mathrm{argmax}} \left\{ \int_{-\infty}^{\infty} \frac{d\xi}{\xi} \left|\psi_{\mathbf{g}1}(\mathbf{w}\xi)\right|^2 \sin(2\pi \mathbf{s}_0^t \mathbf{w}\xi) \right\} \\ &= \underset{\mathbf{w}}{\mathrm{argmax}} \left\{ \mathbf{s}_0^t \mathbf{w} \int_{-\infty}^{\infty} d\xi \left|\psi_{\mathbf{g}1}(\mathbf{w}\xi)\right|^2 \mathrm{sinc}(2\mathbf{s}_0^t \mathbf{w}\xi) \right\} \\ &\approx \underset{\mathbf{w}}{\mathrm{argmax}} \left\{ \mathbf{s}_0^t \mathbf{w} \int_{-\infty}^{\infty} d\xi \left|\psi_{\mathbf{g}1}(\mathbf{w}\xi)\right|^2 \right\},\end{aligned} \tag{13.220}$$

where the last form holds for weak signals. Note that the magnitude of $\mathbf{w}$ does not affect the quantity being maximized; if $\mathbf{w} \to \alpha\mathbf{w}$, the integrals in the last two forms scale as $1/\alpha$, $\mathbf{s}_0^t\mathbf{w}$ scales as α, and the quantity in curly brackets remains constant.

As applications of these results, the reader can rederive the linear discriminants for uncorrelated Poisson or exponential noise, as well as the prewhitening matched filter for correlated Gaussian noise.

When is the log-likelihood ratio linear? Now we address the second question posed above: When is the AUC-optimal linear discriminant in fact the log-likelihood ratio? We require that

$$\lambda(\mathbf{g}) = \ln\left[\frac{\mathrm{pr}(\mathbf{g}|H_2)}{\mathrm{pr}(\mathbf{g}|H_1)}\right] = \mathbf{w}^t\mathbf{g} + c\,, \tag{13.221}$$

where c is a constant that can be lumped into the decision threshold. This condition implies that

$$\mathrm{pr}(\mathbf{g}|H_j) = A_j\, d(\mathbf{g}) \exp(\mathbf{b}_j^t\mathbf{g})\,, \tag{13.222}$$

where A_j and $\mathbf{b}_j$ are independent of the data $\mathbf{g}$, and $d(\mathbf{g})$ is independent of the hypothesis j. The constants $\mathbf{b}_j$ are related to the discriminant function by

$$\mathbf{w} = \mathbf{b}_2 - \mathbf{b}_1\,. \tag{13.223}$$

As a point of terminology, (13.222) is a *multivariate linear exponential-type distribution*. The general form for such a PDF is (Kotz *et al.*, 2000)

$$\mathrm{pr}(\mathbf{g}|\boldsymbol{\theta}) = d(\mathbf{g}) \exp\left[\mathbf{b}^t\mathbf{g} - q(\boldsymbol{\theta})\right]. \tag{13.224}$$

Rather than a continuous parameter $\boldsymbol{\theta}$, (13.222) has a discrete index, but (13.222) and (13.224) are essentially the same form. The reader can show that (13.222) holds for several SKE problems we have encountered: multivariate Gaussian, Poisson and exponential PDFs, all of which are exponential families.

We can recast (13.222) in terms of the characteristic functions:

$$\psi_{\mathbf{g}j}(\boldsymbol{\xi}) = A_j \int_\infty d^M g\, d(\mathbf{g}) \exp\left[(\mathbf{b}_j - 2\pi i\boldsymbol{\xi})^t\, \mathbf{g}\right]. \tag{13.225}$$

A little algebra shows that

$$\psi_{\mathbf{g}2}(\boldsymbol{\xi}) = \frac{A_2}{A_1}\, \psi_{\mathbf{g}1}\left(\boldsymbol{\xi} + \frac{i}{2\pi}\mathbf{w}\right). \tag{13.226}$$

Note that $\mathbf{b}_j$ and hence $\mathbf{w}$ have to be real since PDFs are real; thus, in order for the log-likelihood ratio to be a linear discriminant, $\psi_{\mathbf{g}2}(\boldsymbol{\xi})$ must be proportional to $\psi_{\mathbf{g}1}(\boldsymbol{\xi})$ with each component shifted along the imaginary axis; the shift is immediately the ideal-observer discriminant function.

13.2.13 Detectability in continuous data

The previous sections assumed that the data were a discrete set of M measurements, relevant to a digital imaging system. We can conceive of two scenarios in which we might be willing to analyze an imaging system using a continuous framework. The first would be the case where the data are indeed continuous, as would be found when the detection system is film-based. The second scenario would be when we have a sufficiently large number of discrete detector elements that we are tempted to consider the limit of an infinite number of infinitely fine samples. We shall treat the true continuous detection system here. Our treatment is inspired by that of Helstrom (1995).

SKE/BKE tasks in Gaussian noise Consider first the problem of binary classification in Gaussian noise, and assume that the signal and background are both exactly known. We write the continuous data set as a linear functional of the known object under each hypothesis:

$$H_j : g(\mathbf{r}) = \int_\infty d^q r'\, h(\mathbf{r}, \mathbf{r}')\, f_j(\mathbf{r}') + n(\mathbf{r}) \equiv s_j(\mathbf{r}) + n(\mathbf{r}). \tag{13.227}$$

Because the noise is a zero-mean Gaussian random process, $E[g(\mathbf{r})|H_j] = s_j(\mathbf{r})$. The noise is described by the continuous autocovariance function $K_\mathbf{n}$, which is assumed to be independent of the underlying hypothesis:

$$K_\mathbf{n}(\mathbf{r}, \mathbf{r}') = \langle n(\mathbf{r})\, n^*(\mathbf{r}')\rangle. \tag{13.228}$$

From Sec. 8.2.7 we know that the noise autocovariance operator $\mathcal{K}_\mathbf{n}$, whose kernel is the noise autocovariance function $K_\mathbf{n}(\mathbf{r}, \mathbf{r}')$, is a compact Hermitian operator provided $\mathcal{K}_\mathbf{n}$ is Hilbert-Schmidt. The conditions for compactness are further detailed in Sec. 1.3.3; from that discussion we know that if the random variable $\mathbf{g}$ has finite support, then $\mathcal{K}_\mathbf{n}$ is Hilbert-Schmidt and hence has a denumerable eigenfunction expansion. The eigenfunctions of $\mathcal{K}_\mathbf{n}$ can be used as an infinite series of orthonormal expansion functions for $g(\mathbf{r})$, which is written analogously to (13.121) as

$$\mathbf{g} = \sum_{m=1}^{\infty} \beta_m \boldsymbol{\phi}_m. \tag{13.229}$$

Using these expansion functions as a basis for the data $g(\mathbf{r})$ allows us to follow the same steps that led to the SNR given in (13.123), giving

$$\mathrm{SNR}_\lambda^2 = \sum_{m=1}^{\infty} \frac{\left[\Delta\overline{\beta}_m\right]^2}{\sigma_m^2} = \sum_{m=1}^{\infty} \frac{|(\phi_m, \Delta\mathbf{s})|^2}{\mu_m}. \tag{13.230}$$

The infinite sums in (13.230) cannot diverge so long as the discrimination task cannot be done perfectly.

Stationarity Let us now assume that the autocovariance function is wide-sense stationary: $K_\mathbf{n}(\mathbf{r}, \mathbf{r}') = K_\mathbf{n}(\mathbf{r} - \mathbf{r}')$. Note that to make this assumption we are abandoning the requirement stated in the previous paragraph that the data have finite support. The data are thus no longer square-integrable, the Hilbert-Schmidt condition is gone, and the eigenvalues are no longer denumerable. In Chap. 8 we showed that the eigenfunctions of the autocorrelation operator of a stationary random process are the Fourier expansion functions. The eigenfunctions of $\mathcal{K}_\mathbf{n}$ are thus

$$\phi_{\boldsymbol{\rho}}(\mathbf{r}) = \exp(2\pi i \boldsymbol{\rho} \cdot \mathbf{r}), \tag{13.231}$$

where $\boldsymbol{\rho}$ is the continuous vector spatial-frequency index. With these expansion functions we can write the data as

$$g(\mathbf{r}) = \int_\infty d^q\rho \; G(\boldsymbol{\rho})\, \phi_{\boldsymbol{\rho}}(\mathbf{r}) = \int_\infty d^q\rho \; G(\boldsymbol{\rho}) \exp(2\pi i \boldsymbol{\rho} \cdot \mathbf{r}). \tag{13.232}$$

The inner product of the Fourier expansion functions and the expected difference signal is given by

$$(\phi_{\boldsymbol{\rho}}, \Delta\mathbf{s}) = \int_\infty d^q r \; \Delta\overline{g}(\mathbf{r}) \exp(-2\pi i \boldsymbol{\rho} \cdot \mathbf{r}) = \Delta\overline{G}(\boldsymbol{\rho}); \tag{13.233}$$

and thus

$$|(\phi_{\boldsymbol{\rho}}, \Delta\mathbf{s})|^2 = |\Delta\overline{G}(\boldsymbol{\rho})|^2. \tag{13.234}$$

From (8.181) we know that the autocovariance function in the Fourier basis becomes

$$\left\langle [\mathcal{F}_2\{n(\mathbf{r})\}]\, [\mathcal{F}_2\{n(\mathbf{r}')\}]^\dagger \right\rangle = [\mathcal{F}_2 \mathcal{K}_\mathbf{n} \mathcal{F}_2^\dagger](\boldsymbol{\rho}, \boldsymbol{\rho}') = W_\mathbf{n}(\boldsymbol{\rho})\, \delta(\boldsymbol{\rho} - \boldsymbol{\rho}'), \tag{13.235}$$

where $W_\mathbf{n}(\boldsymbol{\rho})$ is the power spectral density of the noise. The noise autocorrelation function is diagonalized by the Fourier operator, so that the components are delta-correlated in that domain.

We now have all the pieces in place for writing the ideal observer's SNR by analogy with (13.230). The numerator is given in (13.234). By (8.178), the eigenvalues in the denominator are given by the power spectral density $W_\mathbf{n}(\boldsymbol{\rho})$. Thus, for the continuous, stationary case the SNR is given by

$$\mathrm{SNR}_\lambda^2 = \int_\infty d^q\rho \; \frac{|\Delta\overline{G}(\boldsymbol{\rho})|^2}{W_\mathbf{n}(\boldsymbol{\rho})}. \tag{13.236}$$

For a linear, shift-invariant imaging system, we can rewrite the expected difference in the data as (see Sec. 7.2.6)

$$|\Delta\overline{G}(\boldsymbol{\rho})|^2 = |H(\boldsymbol{\rho})\, \Delta F(\boldsymbol{\rho})|^2, \tag{13.237}$$

where $H(\boldsymbol{\rho})$ is the transfer function of the system and $\Delta F(\boldsymbol{\rho})$ is the difference object, $f_2(\mathbf{r}) - f_1(\mathbf{r})$, in the Fourier domain. By (7.154) and (7.155) we can factor $H(\boldsymbol{\rho})$ into two components and thus rewrite the SNR of (13.236) as

$$\mathrm{SNR}_\lambda^2 = \int_\infty d^q\rho \, |\Delta F(\boldsymbol{\rho})|^2 \left\{ \frac{|H(0)|^2 \, \mathrm{MTF}^2(\boldsymbol{\rho})}{W_{\mathbf{n}}(\boldsymbol{\rho})} \right\}, \tag{13.238}$$

where $\mathrm{MTF}(\boldsymbol{\rho})$ is the modulation transfer function of the imaging system. The integrand of (13.238) is the product of two factors, one specifying the objects to be discriminated and the other (in curly brackets) characterizing the performance of the imaging system. We shall have more to say about (13.238) after we consider the effects of quantum noise.

Quantum noise: The weak-signal limit Just as we did in the SKE/BKE task for Gaussian noise, we can consider the limit of an infinite number of infinitely fine samples to reach the limit of a continuous data set for Poisson noise. In this limit, the SNR of (13.135) becomes

$$\mathrm{SNR}_\lambda^2 \approx \int_\infty d^2r \, \frac{s^2(\mathbf{r})}{\overline{g}(\mathbf{r})} = \int_\infty d^2r \, \frac{s^2(\mathbf{r})}{b(\mathbf{r})} \tag{13.239}$$

in the 2D case, where $b(\mathbf{r})$ is the background fluence that describes the mean background image at each location [*cf.* (11.84) and (11.94)].

The SNR of (13.239) is the continuous formulation of the SNR of (13.135), which was derived for the case of a weak signal. Note that for the current problem — the limit of infinitely fine detector elements — it cannot be the case that each mean data component $\overline{g}_m$ is much greater than 1. Thus the weak-signal limit must be achieved by requiring that the signal have low amplitude relative to the background. Even so, the SNR may still be large since it is computed as an integral over a large area.

Poisson noise plus additive Gaussian noise Suppose the data have an additional noise contribution that can be modeled as an additive Gaussian random process, e.g., the excess noise due to dark current or noisy amplification. Then the data covariance is given by [*cf.* (12.159)]

$$K_{\mathbf{g}}(\mathbf{r}, \mathbf{r}') = K^{phot}(\mathbf{r}, \mathbf{r}') + K^{exc}(\mathbf{r}, \mathbf{r}') = b(\mathbf{r}) \, \delta(\mathbf{r} - \mathbf{r}') + K^{exc}(\mathbf{r}, \mathbf{r}'), \tag{13.240}$$

where the first term is the delta-correlated photon noise and the second term is the excess Gaussian noise. For this form to be valid, the excess noise must have the same units as the Poisson random process, namely fluence or inverse area. For example, suppose there is a noisy gain mechanism that senses the Poisson process and produces a random output voltage. If the mean conversion gain (from fluence to voltage) is denoted G_{conv}, then the voltage must be divided by G_{conv} before calculating $K^{exc}(\mathbf{r}, \mathbf{r}')$ in (13.240). Thus $K^{exc}(\mathbf{r}, \mathbf{r}') = (G_{conv})^{-2} K_V^{exc}(\mathbf{r}, \mathbf{r}')$, where $K_V^{exc}(\mathbf{r}, \mathbf{r}')$ is the autocovariance function of the voltage random process. We say that the excess noise has been *referred back to the input.*

With the noise described by (13.240), we seek to derive a Fourier-domain extension to (13.239) to describe the SNR for detection of a weak signal. To do so, we again need to impose the assumptions of a linear, shift-invariant imaging system and wide-sense stationary noise. The Fourier-domain counterpart of the numerator is then $|H(\boldsymbol{\rho})F(\boldsymbol{\rho})|^2$, where $F(\boldsymbol{\rho})$ is the Fourier transform of the weak signal to be detected. Noise stationarity requires that the background be flat, that is, $b(\mathbf{r}) = b_0$. Then the mean image and the variance in the data due to the quantum fluctuations are independent of position (in this weak-signal case). The Fourier-domain noise description under the present model then becomes $W_{\mathbf{g}}(\boldsymbol{\rho}) = b_0 + W_{exc}(\boldsymbol{\rho})$.

We can rewrite (13.239) for the current problem in the Fourier domain as

$$\mathrm{SNR}_\lambda^2 = \int_\infty d^2\rho \, \frac{|H(\boldsymbol{\rho})\, F(\boldsymbol{\rho})|^2}{W_{\mathbf{g}}(\boldsymbol{\rho})} = \int_\infty d^2\rho \, |F(\boldsymbol{\rho})|^2 \left\{ \frac{|H(0)|^2\, \mathrm{MTF}^2(\boldsymbol{\rho})}{W_{\mathbf{g}}(\boldsymbol{\rho})} \right\} . \qquad (13.241)$$

We define the relative object $\Delta f_{rel}(\mathbf{r}) = s(\mathbf{r})/b_0$, so that $F(\boldsymbol{\rho}) = b_0[\Delta F_{rel}(\boldsymbol{\rho})]$. Then

$$\begin{aligned} \mathrm{SNR}_\lambda^2 &= |H(0)|^2 \int_\infty d^2\rho \, |\Delta F_{rel}(\boldsymbol{\rho})|^2 \left\{ \frac{b_0^2\, \mathrm{MTF}^2(\boldsymbol{\rho})}{W_{\mathbf{g}}(\boldsymbol{\rho})} \right\} \\ &\equiv |H(0)|^2 \int_\infty d^2\rho \, |\Delta F_{rel}(\boldsymbol{\rho})|^2\, \mathrm{NEQ}(\boldsymbol{\rho}) , \end{aligned} \qquad (13.242)$$

where

$$\mathrm{NEQ}(\boldsymbol{\rho}) = \frac{b_0^2\, \mathrm{MTF}^2(\boldsymbol{\rho})}{W_{\mathbf{g}}(\boldsymbol{\rho})} = \frac{b_0^2\, \mathrm{MTF}^2(\boldsymbol{\rho})}{b_0 + W_{exc}(\boldsymbol{\rho})} . \qquad (13.243)$$

The expression in (13.243) is referred to as the *noise equivalent quanta*, or NEQ of the system. Its name originates in the historical interpretation of NEQ as the equivalent number of input quanta per unit area required by an ideal imaging system to give the same SNR achieved by an actual system whose output is an image degraded by measurement noise and blur. Shaw (1963) proposed the frequency decomposition of system performance measures of (13.243) in response to the problems that were being encountered at the time when evaluating detector performance using finite apertures. Dainty and Shaw (1974) explored the NEQ concept in great detail for photographic processes.

NEQ and signal-detection theory The ideal-observer SNR of (13.242) is an elegant factorization of hardware characteristics, bundled into NEQ, and the signal to be detected, expressed in terms of its Fourier spectrum. Thus $\mathrm{NEQ}(\boldsymbol{\rho})$ can be interpreted as the weight applied to each frequency component of the signal in computing the ideal observer's SNR for an LSIV imaging system on an SKE/BKE task when there is quantum noise and possibly also some excess noise mechanism. $\mathrm{NEQ}(\boldsymbol{\rho})$ is a function of the system MTF, which describes the resolution properties of the imaging system in terms of how well each exponential eigenfunction of an LSIV system is transferred by the system; the noise power spectrum $W_{\mathbf{g}}(\boldsymbol{\rho})$, which characterizes the noise variance in the data as a function of spatial frequency, and finally, the fluence of the uniform input exposure level, b_0, also referred to as the operating point of interest. Since the MTF is dimensionless, $\mathrm{NEQ}(\boldsymbol{\rho})$ has dimensions of fluence (quanta per unit area).

Wagner and his colleagues were the first to show this connection between NEQ and ideal-observer SNR (Wagner, 1978; Wagner *et al.*, 1979). Soon after, Wagner and Brown (1985) developed expressions for NEQ as a function of spatial frequency for a variety of medical imaging modalities, including radiography, CT, PET, and MRI.

It would appear from (13.242) that system optimization could be performed via maximization of NEQ. However, very restrictive assumptions regarding the deterministic and stochastic properties of the imaging system were required to derive this SNR form, as well as strong assumptions regarding the task. As Wagner and Brown point out in their 1985 paper, their treatment is strictly valid only when the signal is low contrast and the noise is additive and Gaussian. As we shall see, more complicated tasks, or imaging systems in which the assumptions of shift invariance and stationarity are violated, rarely yield such an elegant factorization for SNR as found in (13.242).

Relationship between NEQ and DQE The *detective quantum efficiency*, or DQE, was first encountered in Chap. 12 as a measure of detector performance (without any discussion of tasks or observers) for the case of a single photodiode [*cf.* (12.24)]. The nonimaging DQE of (12.24) is in keeping with the basic definition of DQE given by Rose (1948), who compared the noise level of an actual radiation detector with that of an ideal one. Rose applied this concept to the determination of the efficiency of the eye as a photoreceptor.

Dainty and Shaw (1974) explored the Rose concept of DQE in great detail. One of the ways in which they interpreted DQE was as a ratio of squared SNRs, so that the DQE described the transfer of SNR in photographic processes. From there it was a simple step to relate the squared input and output SNRs to the ratio of the ideal and effective number of counts, and Dainty and Shaw's NEQ was born.

Consider the DQE we obtain when we form the ratio of the frequency-dependent NEQ of (13.243) with the actual number of quanta (per unit area) at the input.

$$\mathrm{DQE}(\boldsymbol{\rho}) = \frac{\mathrm{NEQ}(\boldsymbol{\rho})}{b_0} = \left[\frac{b_0}{b_0 + W_{exc}(\boldsymbol{\rho})}\right] \mathrm{MTF}^2(\boldsymbol{\rho}) . \qquad (13.244)$$

We see that this frequency-dependent DQE is dimensionless, as we would expect. Note also the similarity between the DQE expressions of Chap. 12 and that obtained here in the context of the performance of the ideal observer on a specific imaging task. The denominator is again the sum of two terms representing the photon and excess noise contributions. Furthermore, (13.244) has the limiting behavior we would expect for a perfect system: as the excess noise and blur approach zero, DQE $\rightarrow$ 1.

However, the expression for DQE given in (13.244) is limited in that it is suitable for characterization of an imaging system only when NEQ($\boldsymbol{\rho}$) is an appropriate measure of system performance. Specifically, this definition of DQE is not appropriate when the system is shift variant, or the noise is nonstationary, or the task is not SKE/BKE.

In order for the DQE concept to be applicable to more realistic systems and tasks, we must return to its original definition in terms of a ratio of squared detector SNRs. To generalize this definition, we define a task-dependent DQE as the squared SNR at the output of an imaging system or component divided by the squared SNR at the input, for a given task and observer. This ratio describes the degree to which the SNR has been degraded by the system, in terms of observer performance. Thus this definition is relevant to the usefulness of the system for its intended task, as Wagner and Brown advocated, while going beyond the limitations of NEQ($\boldsymbol{\rho}$). Since the numerator and denominator are SNRs, and not in units of quanta, we call this ratio simply the detection efficiency, η. As we shall see in Chap.

14, a similar efficiency measure is often used to describe the performance of the human observer relative to a model observer for a given task.

We shall now relate (13.244) to a ratio of observer SNRs. The squared SNR of an ideal system can be obtained by considering (13.242) in the limit of a perfect MTF and no excess noise:

$$[\mathrm{SNR}_{\lambda,\,\text{ideal system}}]^2 \equiv |H(0)|^2\, b_0 \int_\infty d^2\rho\, |\Delta F_{rel}(\boldsymbol{\rho})|^2 , \tag{13.245}$$

where we use subscripts to label the observer as well as the system. The efficiency of the system in presenting information to the ideal observer is then

$$\eta = \frac{[\mathrm{SNR}_\lambda]^2}{[\mathrm{SNR}_{\lambda,\,\text{ideal system}}]^2} = \frac{|H(0)|^2 \int_\infty d^2\rho\, |\Delta F_{rel}(\boldsymbol{\rho})|^2\, \mathrm{NEQ}(\boldsymbol{\rho})}{|H(0)|^2\, b_0 \int_\infty d^2\rho\, |\Delta F_{rel}(\boldsymbol{\rho})|^2}$$
$$= \frac{\int_\infty d^2\rho\, |\Delta F(\boldsymbol{\rho})|^2\, \mathrm{DQE}(\boldsymbol{\rho})}{\int_\infty d^2\rho\, |\Delta F(\boldsymbol{\rho})|^2} . \tag{13.246}$$

To interpret these results, consider the detection of a cosinusoidal signal of the form

$$\Delta f(\mathbf{r}) = \frac{A}{L} \operatorname{rect}\left(\frac{x}{L}\right) \operatorname{rect}\left(\frac{y}{L}\right) \cos(2\pi \boldsymbol{\rho}_0 \cdot \mathbf{r}) . \tag{13.247}$$

It follows from (2.86) and (3.255) that

$$\lim_{L\to\infty} |\Delta F(\boldsymbol{\rho})|^2 = \tfrac{1}{2} A^2 [\delta(\boldsymbol{\rho} - \boldsymbol{\rho}_0) + \delta(\boldsymbol{\rho} + \boldsymbol{\rho}_0)] . \tag{13.248}$$

We note that $\mathrm{NEQ}(\boldsymbol{\rho})$ depends only on the MTF and the noise power spectrum, both of which are even functions of $\boldsymbol{\rho}$. Hence $\mathrm{DQE}(\boldsymbol{\rho})$ is also even, and in the limit $L \to \infty$, (13.246) becomes

$$\eta = \mathrm{DQE}(\boldsymbol{\rho}_0) . \tag{13.249}$$

For this highly specialized signal, the efficiency of the system is directly its DQE evaluated at the frequency of the signal. For more realistic signals, $\mathrm{DQE}(\boldsymbol{\rho})$ is best interpreted as the weighting factor giving the contribution of each frequency component of $|\Delta F(\boldsymbol{\rho})|^2$ to the efficiency η. We reiterate, however, that the whole formalism rests on the assumptions of continuous data, a linear, shift-invariant imaging system, and stationary noise. The concept of detective quantum efficiency and its application to digital radiography are treated further in Sec. 16.1.6.

Random backgrounds and generalized NEQ The concept of NEQ was presented in (13.243) in the context of detection SNRs for classification tasks limited by stationary noise processes and for which the imaging system was assumed to be linear and shift-invariant (LSIV). Both the signals to be classified and the background were assumed to be known exactly. The NEQ concept can be generalized to the detection of known signals on Gaussian random backgrounds (Barrett *et al.*, 1989; Barrett *et al.*, 1995); to do this we again assume the imaging system to be LSIV and the random background to be stationary in the wide sense.

In the analysis of NEQ for the SKE problem we modeled the data as having two sources of randomness, one due to the quantum fluctuations in the incoming photons and the other resulting from an excess noise source [see (13.240)]. To generalize that derivation to incorporate a Gaussian random background, we add another term to the autocovariance function for the data:

$$K_{\mathbf{g}}(\mathbf{r},\mathbf{r}') = K^{phot}(\mathbf{r},\mathbf{r}') + K^{exc}(\mathbf{r},\mathbf{r}') + K^{bg}(\mathbf{r},\mathbf{r}') , \tag{13.250}$$

where the third term is the contribution to the covariance of the data from the random background. For a particular background, the data have a covariance given by (13.240). In the limit of no excess noise, and for a fixed background, the data are Poisson.

Now let us make the assumption that the random background can be described by a 2D stationary Gaussian random process in the object domain with power spectral density $W_{\mathbf{f}_b}(\boldsymbol{\rho})$. We assume that the mean image averaged over all random backgrounds is independent of location and denoted by $\bar{b}_0$. When we also make the assumption that the excess noise contributions are stationary, the Fourier domain expression for (13.250) is written [*cf.* above (13.241)]:

$$W_{\mathbf{g}}(\boldsymbol{\rho}) = \bar{b}_0 + W_{exc}(\boldsymbol{\rho}) + |H(0)|^2\,\mathrm{MTF}^2(\boldsymbol{\rho})\,W_{\mathbf{f}_b}(\boldsymbol{\rho})\,. \tag{13.251}$$

The factors describing the imaging system transfer characteristics impact only the background power spectral density; they represent the mapping of the background randomness from object space to image space.

Given the Fourier description for the data covariance of (13.251), we can write the SNR for the random-background detection problem analogously to (13.242):

$$\mathrm{SNR}_\lambda^2 = |H(0)|^2 \int d^2\rho\, |\Delta F_{rel}(\boldsymbol{\rho})|^2 \left\{ \frac{\bar{b}_0^{-2}\mathrm{MTF}^2(\boldsymbol{\rho})}{\bar{b}_0 + W_{exc}(\boldsymbol{\rho}) + |H(0)|^2\,\mathrm{MTF}^2(\boldsymbol{\rho})\,W_{\mathbf{f}_b}(\boldsymbol{\rho})} \right\}. \tag{13.252}$$

We identify the quantity in brackets as the generalized NEQ:

$$\mathrm{GNEQ}(\boldsymbol{\rho}) = \frac{\bar{b}_0^{-2}\mathrm{MTF}^2(\boldsymbol{\rho})}{\bar{b}_0 + W_{exc}(\boldsymbol{\rho}) + |H(0)|^2\,\mathrm{MTF}^2(\boldsymbol{\rho})\,W_{\mathbf{f}_b}(\boldsymbol{\rho})}\,. \tag{13.253}$$

The important attribute of this result is that, even when the imaging system is LSIV and the background process is Gaussian and stationary, we find that object variability results in an SNR that couples together task and hardware contributions in a complicated fashion. Simple measures of the properties of the imaging system alone cannot be reported as measures of system performance across all tasks.

A more meaningful approach would be to a) compute the SNR of a specified observer at the input to the system, taking into account the background randomness as well as the quantum fluctuations; b) determine the same observer's SNR at the output of the system; and c) compute the ratio of those squared measures, the detection efficiency η. This quantity describes the efficiency of the imaging system in transferring the information to the observer for performing the specified task.

Hotelling SNR in the continuous limit The SNR given in (13.178) is maximum over all possible linear decision strategies without making any assumptions regarding the stationarity of the data statistics. For continuous data (13.178) becomes

$$\mathrm{SNR}_{Hot}^2 = \int d^2r \int d^2r'\,\Delta\bar{g}(\mathbf{r})\,S_2^{(-1)}(\mathbf{r},\mathbf{r}')\,\Delta\bar{g}(\mathbf{r}')\,, \tag{13.254}$$

where $S_2(\mathbf{r},\mathbf{r}')$ is the average of the data autocovariance functions, and $S_2^{(-1)}(\mathbf{r},\mathbf{r}')$ satisfies

$$\int d^2r\; S_2^{(-1)}(\mathbf{r}'',\mathbf{r})\,S_2(\mathbf{r},\mathbf{r}') = \delta(\mathbf{r}''-\mathbf{r}')\,. \tag{13.255}$$

The superscript -1 is in parentheses here to indicate that $S_2^{(-1)}(\mathbf{r}'', \mathbf{r})$ is the kernel of the operator $\mathbf{S}_2^{-1}$, not the reciprocal of the function $S_2(\mathbf{r}'', \mathbf{r})$.

If we assume that the autocovariance functions are stationary, we can follow the derivation that led to the ideal observer's SNR in the Fourier domain given in (13.242). The Hotelling observer's SNR is given by

$$\text{SNR}^2_{Hot} = \int_\infty d^2\rho \, \frac{|\Delta\overline{G}(\boldsymbol{\rho})|^2}{W_\mathbf{g}(\boldsymbol{\rho})} \,, \tag{13.256}$$

where $\Delta\overline{G}(\boldsymbol{\rho})$ is the Fourier transform of the average signal and $W_\mathbf{g}(\boldsymbol{\rho})$ is the power spectral density determined from the stationary average autocovariance function. While the SNR given in (13.242) holds for the ideal observer in the case of stationary Gaussian noise and signal known exactly, (13.256) holds more generally for the Hotelling observer. The signal may be random and the data statistics may be non-Gaussian. All that is required is that the data covariance be stationary under each hypothesis, and that the mean signal under each hypothesis be known.

Quasistationary noise We now consider the form of the SNR that results when the task is the detection of a known signal on a quasistationary background. We shall continue to assume that the data are continuous in order to make use of the Fourier descriptors developed in Sec. 8.2.5 for quasistationary random processes.

We assume the signal is spatially compact and localized at $\mathbf{r}_0$, so that $\Delta\overline{g}(\mathbf{r}) = 0$ if $|\mathbf{r} - \mathbf{r}_0| > R$. As in (8.120), we can make use of the following coordinate transformation:

$$\overline{\mathbf{r}} = \tfrac{1}{2}(\mathbf{r} + \mathbf{r}') \,, \qquad \Delta\mathbf{r} = \mathbf{r} - \mathbf{r}' \tag{13.257}$$

so that

$$\mathbf{r} = \overline{\mathbf{r}} + \tfrac{1}{2}\Delta\mathbf{r} \,, \qquad \mathbf{r}' = \overline{\mathbf{r}} - \tfrac{1}{2}\Delta\mathbf{r} \,. \tag{13.258}$$

Note that the Jacobian of this transformation is unity.

We assume the autocovariance function of the background $K_\mathbf{g}(\mathbf{r}, \mathbf{r}')$ is the same under each hypothesis. When we rewrite the autocovariance function using the transformation of (13.258) we obtain

$$K_\mathbf{g}(\mathbf{r}, \mathbf{r}') = K_\mathbf{g}(\overline{\mathbf{r}} + \tfrac{1}{2}\Delta\mathbf{r}, \overline{\mathbf{r}} - \tfrac{1}{2}\Delta\mathbf{r}) \equiv \widetilde{K}_\mathbf{g}(\overline{\mathbf{r}}, \Delta\mathbf{r}) \,. \tag{13.259}$$

We can similarly transform $K_\mathbf{g}^{-1}(\mathbf{r}, \mathbf{r}')$ or any other function of $\mathbf{r}$ and $\mathbf{r}'$. Thus the SNR of (13.254) becomes

$$\text{SNR}^2_{Hot} = \int d^2\overline{r} \int d^2\,\Delta r \; \Delta\overline{g}(\overline{\mathbf{r}} + \tfrac{1}{2}\Delta\mathbf{r}) \, \widetilde{K}_\mathbf{g}^{(-1)}(\overline{\mathbf{r}}, \Delta\mathbf{r}) \, \Delta\overline{g}(\overline{\mathbf{r}} - \tfrac{1}{2}\Delta\mathbf{r}) \,, \tag{13.260}$$

where $\widetilde{K}_\mathbf{g}^{(-1)}(\overline{\mathbf{r}}, \Delta\mathbf{r})$ is the kernel of the operator $\widetilde{\mathbf{K}}_\mathbf{g}^{-1}$. We have made a coordinate transformation, but no approximations thus far.

The compact support of the signal implies that $\Delta\overline{g}(\overline{\mathbf{r}} + \frac{1}{2}\Delta\mathbf{r}) \, \Delta\overline{g}(\overline{\mathbf{r}} - \frac{1}{2}\Delta\mathbf{r}) = 0$ unless $|\overline{\mathbf{r}} - \mathbf{r}_0| < 2R$. If we can assume that $\widetilde{K}_\mathbf{g}^{(-1)}(\overline{\mathbf{r}}, \Delta\mathbf{r})$ is a slowly varying function of the $\bar{\mathbf{r}}$ variable, so that it is approximately constant over this range, then $\widetilde{K}_{\mathbf{g}}^{(-1)}(\bar{\mathbf{r}}, \Delta\mathbf{r}) \approx \widetilde{K}_{\mathbf{g}}^{(-1)}(\mathbf{r}_0, \Delta\mathbf{r})$. Then the SNR becomes

$$\mathrm{SNR}^2_{Hot}(\mathbf{r}_0) \approx \int d^2\Delta r\ \widetilde{K}_{\mathbf{g}}^{(-1)}(\mathbf{r}_0, \Delta\mathbf{r}) \int d^2\bar{r}\ \Delta\overline{g}(\bar{\mathbf{r}} + \tfrac{1}{2}\Delta\mathbf{r})\, \Delta\overline{g}(\bar{\mathbf{r}} - \tfrac{1}{2}\Delta\mathbf{r})$$

$$= \int d^2\Delta r\ \widetilde{K}_{\mathbf{g}}^{(-1)}(\mathbf{r}_0, \Delta\mathbf{r})\, [\Delta\overline{g} \star \Delta\overline{g}](\Delta\mathbf{r})\,, \qquad (13.261)$$

where $[\Delta\overline{g} \star \Delta\overline{g}]$ is the spatial autocorrelation function of the known signal, and the argument indicates that the SNR depends on the signal location $\mathbf{r}_0$.

The SNR expression above is ripe for transformation to the Fourier domain. We know that $\mathcal{F}_2\{\Delta\overline{g}(\mathbf{r})\} = \Delta\overline{G}(\boldsymbol{\rho})$ and $\mathcal{F}_2\{[\Delta\overline{g} \star \Delta\overline{g}](\Delta\mathbf{r})\} = |\Delta\overline{G}(\boldsymbol{\rho})|^2$. We define $\mathcal{F}_2\{\widetilde{K}_{\mathbf{g}}^{(-1)}(\mathbf{r}_0, \Delta\mathbf{r})\} = C(\mathbf{r}_0, \boldsymbol{\rho})$. By Parseval's rule (13.261) can be rewritten as

$$\mathrm{SNR}^2_{Hot}(\mathbf{r}_0) = \int d^2\rho\ C(\mathbf{r}_0, \boldsymbol{\rho}) |\Delta\overline{G}(\boldsymbol{\rho})|^2\,. \qquad (13.262)$$

To better understand (13.262), we must determine the relationship between $C(\mathbf{r}_0, \boldsymbol{\rho})$ and the autocovariance of the quasistationary noise, $\widetilde{K}_{\mathbf{g}}(\bar{\mathbf{r}}, \Delta\mathbf{r})$. To ascertain this relationship, we rewrite the defining equation (13.255) for the inverse of $K_{\mathbf{g}}$ with the approximation $\frac{1}{2}(\mathbf{r} + \mathbf{r}') \approx \frac{1}{2}(\mathbf{r} + \mathbf{r}'') \approx \mathbf{r}_0$, yielding

$$\int d^2r\ K_{\mathbf{g}}^{(-1)}(\mathbf{r}'', \mathbf{r})\, K_{\mathbf{g}}(\mathbf{r}, \mathbf{r}') \approx \int d^2r\ \widetilde{K}_{\mathbf{g}}^{(-1)}(\mathbf{r}_0, \mathbf{r}'' - \mathbf{r})\, \widetilde{K}_{\mathbf{g}}(\mathbf{r}_0, \mathbf{r} - \mathbf{r}') = \delta(\mathbf{r}'' - \mathbf{r}')\,. \qquad (13.263)$$

With the change of variables[12] $\Delta\mathbf{r} = \mathbf{r} - \mathbf{r}'$, this relation becomes

$$\int d^2\Delta r\ \widetilde{K}_{\mathbf{g}}^{(-1)}(\mathbf{r}_0, \mathbf{r}'' - \Delta\mathbf{r} - \mathbf{r}')\, \widetilde{K}_{\mathbf{g}}(\mathbf{r}_0, \Delta\mathbf{r}) = \delta(\mathbf{r}'' - \mathbf{r}')\,. \qquad (13.264)$$

Since (13.264) is a convolution integral, taking the Fourier transform of both sides produces

$$\mathcal{F}_2\{\widetilde{K}_{\mathbf{g}}^{(-1)}(\mathbf{r}_0, \Delta\mathbf{r})\}\, \mathcal{F}_2\{\widetilde{K}_{\mathbf{g}}(\mathbf{r}_0, \Delta\mathbf{r})\} = \mathcal{F}_2\{\delta(\Delta\mathbf{r})\} = 1$$

or

$$\mathcal{F}_2\{\widetilde{K}_{\mathbf{g}}^{(-1)}(\mathbf{r}_0, \Delta\mathbf{r})\} = C(\mathbf{r}_0, \boldsymbol{\rho}) = \frac{1}{\mathcal{F}_2\{\widetilde{K}_{\mathbf{g}}(\mathbf{r}_0, \Delta\mathbf{r})\}} = \frac{1}{W_{\Delta\mathbf{g}}(\mathbf{r}_0, \boldsymbol{\rho})}\,, \qquad (13.265)$$

where $W_{\Delta\mathbf{g}}(\mathbf{r}_0, \boldsymbol{\rho})$ is the stochastic Wigner distribution function [*cf.* (5.54) and (8.140)] for the zero-mean process $\Delta g(\mathbf{r})$.

Thus the SNR in (13.262) in the quasistationary approximation is

$$\mathrm{SNR}^2_{Hot}(\mathbf{r}_0) = \int d^2\rho\ \frac{|\Delta\overline{G}(\boldsymbol{\rho})|^2}{W_{\Delta\mathbf{g}}(\mathbf{r}_0, \boldsymbol{\rho})}\,. \qquad (13.266)$$

This expression is strikingly similar to the one found in (13.256), only now we have the Wigner distribution function in place of the power spectral density of the stationary noise, $W_{\mathbf{n}}(\boldsymbol{\rho})$. For stationary noise, (13.266) reduces to (13.256).

[12]Note that $\mathbf{r}'$ is simply a constant here, independent of the variable of integration.

Factorable quasistationarity and prewhitening As we saw in Sec. 8.2.4, an autocorrelation or autocovariance function can often be factored into a slowly varying contribution arising from variations in overall intensity and a short-range function describing correlation between neighboring points. As in (8.119), we can write

$$K_{\mathbf{g}}(\mathbf{r}, \mathbf{r}') = a(\Delta \mathbf{r})\, b(\overline{\mathbf{r}}) . \tag{13.267}$$

For example, this formula holds for Poisson random processes with $a(\Delta \mathbf{r}) = \delta(\Delta \mathbf{r})$ and $b(\overline{\mathbf{r}})$ being the fluence [see (11.94)]. It also has use for describing the field autocovariance in coherent imaging systems [see (18.115)].

From (8.142) we know that the stochastic Wigner distribution function corresponding to (13.267) is

$$W_{\Delta \mathbf{g}}(\overline{\mathbf{r}}, \boldsymbol{\rho}) = A(\boldsymbol{\rho})\, b(\overline{\mathbf{r}}) . \tag{13.268}$$

If $b(\overline{\mathbf{r}})$ is slowly varying so that it is approximately equal to $b(\mathbf{r}_0)$ over the signal support, then (13.266) becomes

$$\mathrm{SNR}^2_{Hot}(\mathbf{r}_0) = \frac{1}{b(\mathbf{r}_0)} \int d^2\rho\, \frac{|\Delta \overline{G}(\boldsymbol{\rho})|^2}{A(\boldsymbol{\rho})} . \tag{13.269}$$

The factor $1/b(\mathbf{r}_0)$ may be surprising, but in fact $\Delta \overline{G}(\boldsymbol{\rho})$ will scale as $b(\mathbf{r}_0)$ in most problems, so the overall result will be that $\mathrm{SNR}^2_{Hot}(\mathbf{r}_0) \propto b(\mathbf{r}_0)$. This proportionality can be seen explicitly in the expression for the ideal observer's SNR given in (13.242).

It is interesting to relate these results to prewhitening. Let the output of a linear processing operation on the data be

$$y(\mathbf{r}) = \int d^2 r'\, o(\mathbf{r}, \mathbf{r}')\, g(\mathbf{r}') . \tag{13.270}$$

This operation can legitimately be referred to as prewhitening if $\langle \Delta y(\mathbf{r}_1) \Delta y(\mathbf{r}_2) \rangle = \delta(\mathbf{r}_1 - \mathbf{r}_2)$. We leave it to the reader to demonstrate that this property is achieved for the covariance model of (13.267) if

$$o(\mathbf{r}, \mathbf{r}') = \frac{1}{\sqrt{b(\mathbf{r}')}} \int d^2\rho\, \frac{1}{\sqrt{A(\boldsymbol{\rho})}} \exp[2\pi i \boldsymbol{\rho} \cdot (\mathbf{r} - \mathbf{r}')] . \tag{13.271}$$

Note that $y(\mathbf{r})$ is simply $g(\mathbf{r})/\sqrt{b(\mathbf{r})}$ if $a(\Delta \mathbf{r}) = \delta(\Delta \mathbf{r})$ and hence $A(\boldsymbol{\rho}) = 1$.

From (13.129) we know that the SNR after prewhitening is given by

$$\mathrm{SNR}^2_{pw} = \int d^2 r\, |\Delta \overline{y}(\mathbf{r})|^2 , \tag{13.272}$$

and the reader can show that this expression is equivalent to (13.269) under the assumption of (13.267). Thus (13.269) is the SNR^2 for a *locally prewhitening matched filter*. The more general expression for the locally prewhitened SNR^2 when the model of (13.267) is not valid is (13.266).

Observer efficiency We defined DQE as a measure of the efficiency of a detector in providing information to the ideal observer for performing a specified task. The DQE concept can be applied equally well to observer models, as a means of describing the relative performance of two model observers on the same task using the same data. Specifically, the efficiency of the Hotelling observer relative to the ideal observer is written

$$\eta_{Hot} = \frac{\mathrm{SNR}^2_{Hot}}{d^2_{A,ideal}} . \tag{13.273}$$

The figure of merit for the ideal observer to be used here is d_A, derived from AUC via (13.21), because d_A summarizes the area under the ROC curve, while the SNR of (13.19) might not. We aren't too interested in computing η_{Hot} for SKE/BKE tasks in Gaussian noise, since we know the efficiency must be unity for that circumstance. What is interesting is the quantification of the difference in performance between the observers for more complex tasks, particularly ones where (13.19) would not be a good summary measure of the ideal observer's performance. Hence the use of d_A.

In Chap. 14 we shall make use of similar definitions of observer efficiency in terms of human performance relative to that of various model observers.

13.3 ESTIMATION THEORY

In Sec. 13.1 we presented a joint description of classification and estimation tasks, emphasizing their connections. As described in detail in Sec. 13.2, classification is partitioning of data space, assigning labels to regions. In contrast, estimation is assigning numbers or vectors to points in data space. We now take up the subject of estimation in detail.

As we saw in Sec. 13.1.1, estimation problems can be categorized by the particular quantity being estimated. In *point estimation*, also called *parameter estimation*, we are given a data vector $\mathbf{g}$ and a parametric form for the conditional density $\mathrm{pr}(\mathbf{g}|\boldsymbol{\theta})$. Our job is to form an estimate of all elements of the vector $\boldsymbol{\theta}$. A second class of problems is one in which two complementary vectors $\boldsymbol{\theta}$ and $\boldsymbol{\theta}_n$ determine the data. We seek to estimate the P-dimensional vector $\boldsymbol{\theta}$; the components of $\boldsymbol{\theta}_n$ are termed *nuisance parameters*. This problem is therefore referred to as *estimation with nuisance parameters*. Most texts on estimation deal only with the problem of point estimation, with little mention of nuisance parameters. We shall first consider parameter estimation without regard for nuisance parameters here as well, later turning to the topic of nuisance parameters and means of dealing with them.

We begin in Sec. 13.3.1 with a discussion of the basic concepts of estimation, including the essential ingredients of the estimation process and figures of merit for parameter estimation. Common terms that are defined and discussed in that section are *bias* and *mean-square error* or *MSE*. As we shall see, there is considerable subtlety in defining these terms; indeed, in many practical problems in imaging, no satisfactory definition exists. In Sec. 13.3.2 we digress from our development of estimation methods to detail some of the difficulties that arise in trying to apply MSE to imaging.

In Sec. 13.3.3 we return to our main thread and discuss Bayesian estimation, in which the underlying parameters are assumed random and characterized by a prior probability density. The powerful method of maximum-likelihood (ML) estimation and properties of ML estimators are presented in Secs. 13.3.4 – 13.3.6. Other classical estimators are discussed briefly in Sec. 13.3.7. In Sec. 13.3.8, we present a definition of nuisance parameters and discuss ways of dealing with them in pure estimation problems. In Sec. 13.3.9 we treat hybrid classification-estimation tasks, again from the perspective of nuisance parameters.

13.3.1 Basic concepts

The first ingredient in an estimation problem is the PD vector of parameters $\boldsymbol{\theta}$ we seek to estimate. In an imaging context, $\boldsymbol{\theta}$ may be a low-dimensional vector that parameterizes object size, shape, location, amplitude, etc. Alternatively, we may be interested in the estimation of a figure of merit for the objective evaluation of an imaging system. For example, we may be given a limited set of images for use in the estimation of the area under the ROC curve or related measures of imaging system performance, a problem we consider in more detail in Chap. 14. These applications are to be contrasted with the situation in which a large number of parameters are to be estimated, for example, a large set of expansion coefficients used to represent the object. Chap. 15 is devoted to the complex topic of estimation of coefficients of approximate object expansions. With one exception, we limit the presentation in this chapter to estimation problems in which a small number of parameters are to be determined. The exception is the treatment of some issues that arise in the use of mean-square error as a figure of merit in digital imaging; that presentation sets the stage for a related discussion in Chap. 14.

The prior probability density, $\text{pr}(\boldsymbol{\theta})$, describes the underlying randomness in the parameter(s). This density is analogous to $\Pr(H_j)$ in the classification problem. As described in the Prologue, Bayesian estimation makes use of such a prior in formulating an estimation procedure; a Bayesian considers the underlying parameter vector to be random. This is in contrast to classical estimation, in which the underlying $\boldsymbol{\theta}$ is assumed to be fixed for a given data set.

Another essential element, as we emphasized in Sec. 13.1.2, is knowledge of the mapping from parameter space to the measurement (data) space. This mapping is the probability law on the data conditioned on the true parameter $\boldsymbol{\theta}$, written $\text{pr}(\mathbf{g}|\boldsymbol{\theta})$. In imaging, this density function is determined by the (possibly nonlinear) relationship between $\boldsymbol{\theta}$ and the object, the deterministic mapping from object space to data space, as well as the noise characteristics of the imaging system.

Finally, estimation requires a rule, or procedure, for mapping from the data space to the estimate $\hat{\boldsymbol{\theta}}(\mathbf{g})$. We assume that the estimation rule is deterministic, so that the same data vector always yields the same estimate. We made an analogous assumption in Sec. 13.2, stating that the decision function always yields the same decision given the same data as input. We shall often drop the explicit dependence of the estimate on the data, writing simply $\hat{\boldsymbol{\theta}}$.

Many estimation procedures are formulated as the minimization of some cost function $C(\hat{\boldsymbol{\theta}}, \boldsymbol{\theta})$ that describes the penalty associated with reporting $\hat{\boldsymbol{\theta}}$ when the true parameter is equal to $\boldsymbol{\theta}$. This cost function is analogous to the classification cost C_{ij} of making decision D_i when hypothesis H_j is true.

There are a variety of methods for characterizing the performance of an estimator. We begin here with a discussion of costs and risks, followed by definitions of the bias and variance of a scalar estimate, and then we shall generalize these expressions to vector estimation. Other relevant aspects, such as whether the estimate is consistent, efficient, or sufficient, will be taken up in Sec. 13.3.6.

Costs and risks Table 13.1 gives three definitions of risk using three different kinds of average cost. Let us consider the assumptions required for the computation of the various definitions of risk. The first row gives a frequentist definition of risk, which we shall call $\overline{C}(\boldsymbol{\theta})$ since it is a function of $\boldsymbol{\theta}$. The computation of the average requires knowledge of the prior probability density on the data conditioned on the underlying parameter, $\text{pr}(\mathbf{g}|\boldsymbol{\theta})$. This averaging operation is analogous to the average performed in (13.6) over all decision outcomes given an underlying true hypothesis state.

The second row of Table 13.1 defines average cost as a function of a particular $\mathbf{g}$. This Bayesian definition requires knowledge of the posterior probability on $\boldsymbol{\theta}$, that is, $\text{pr}(\boldsymbol{\theta}|\mathbf{g})$. The Bayesian regards $\boldsymbol{\theta}$ as random, but has no concept of an ensemble of $\mathbf{g}$ vectors; the data are fixed. We call this cost function $\overline{C}(\mathbf{g})$ since it is a function of $\mathbf{g}$. This cost is analogous to the cost computed in (13.5), where the average is performed over all underlying classes for a given decision.

The third row of Table 13.1 defines the overall average cost, found by averaging over the ensemble of possible data vectors $\{\mathbf{g}\}$ for a particular $\boldsymbol{\theta}$, followed by an average over the distribution of underlying parameters $\text{pr}_{\boldsymbol{\theta}}(\boldsymbol{\theta})$. The double average results in a scalar cost $\overline{C}$ that is no longer a function of either $\boldsymbol{\theta}$ or $\mathbf{g}$. This cost is analogous to the overall average cost of a classification decision computed in (13.7).

In summary, estimates are statistical quantities, hence their evaluation requires statistical methods. We conceive the data as being derived from some noisy measurement system, so that the data would be random given repeated trials of the measurement procedure on the same object, with additional randomness in the case of underlying parameter variability. We shall thus treat two problems in parallel, considering the performance of an estimator acting on noisy data when the underlying parameter is fixed, as well as the extension to the case where there is randomness in the underlying parameter.

Bias Given a data vector $\mathbf{g}$, let $\hat{\theta}$ denote an estimate of some fixed scalar parameter θ underlying the data. Because $\hat{\theta}$ is a function of noisy data, it is a random variable with a distribution (see Fig. 13.12) that depends on the true value of the underlying parameter. If we know $\text{pr}(\mathbf{g}|\theta)$ and the estimation rule $\hat{\theta}(\mathbf{g})$ we can write the mean of $\hat{\theta}$ as

$$\overline{\hat{\theta}} = \left\langle \hat{\theta}(\mathbf{g}) \right\rangle_{\mathbf{g}|\theta} = \int d^M g \ \text{pr}(\mathbf{g}|\theta)\, \hat{\theta}(\mathbf{g}) \,, \tag{13.274}$$

where the subscript $\mathbf{g}|\theta$ in the middle form indicates that the average is over all sources of randomness in the data when the parameter has fixed value equal to θ. We therefore call $\overline{\hat{\theta}}$ the conditional mean of $\hat{\theta}$. If we know the conditional PDF of $\hat{\theta}$ itself, we can write

$$\overline{\hat{\theta}} = \int d\hat{\theta} \ \text{pr}(\hat{\theta}|\theta)\, \hat{\theta} \,, \tag{13.275}$$

where now we no longer need an explicit form for $\hat{\theta}(\mathbf{g})$. On the other hand, we need to know $\text{pr}(\hat{\boldsymbol{\theta}}|\boldsymbol{\theta})$ in this formulation. Since $\hat{\boldsymbol{\theta}}$ is often a nonlinear or even an implicit function of $\mathbf{g}$, it can be challenging to determine $\text{pr}(\hat{\boldsymbol{\theta}}|\boldsymbol{\theta})$.

Note that the mean of the estimated value, $\overline{\hat{\theta}}$, may not equal the true value θ. In this case a systematic error, or *bias*, exists in the estimation procedure. We denote the conditional bias by $b(\theta)$, where

$$b(\theta) = \bar{\hat{\theta}} - \theta\,. \tag{13.276}$$

The overline indicates an average over all realizations of the data, given the true value θ. An *unbiased* estimate is one for which $b(\theta) = 0$ for all θ.

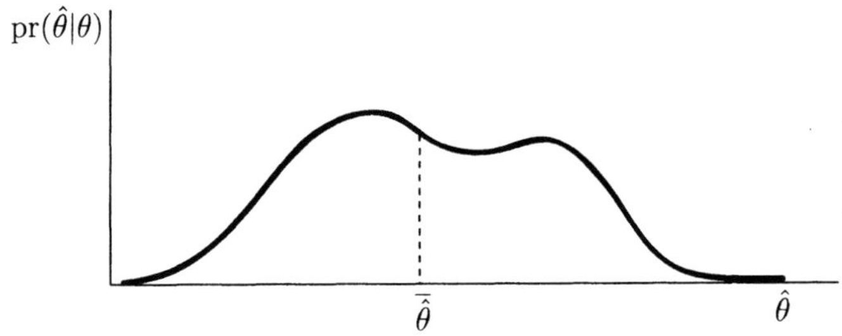

Fig. 13.12 Distribution of $\hat{\theta}$ conditioned on θ.

Estimability It is not always possible to find an unbiased estimate. It often happens that even noise-free data are not sufficient to determine θ unambiguously. As a simple example, suppose that θ is the integral of some object over a region $\mathbf{S}$ but that the data available consist of an image that covers only the smaller region $\mathbf{S}'$ contained in $\mathbf{S}$. One could still make an estimate of θ, but there would be no satisfactory way of determining the bias $b(\theta)$. The estimate $\hat{\theta}(\mathbf{g})$ is fully determined by $\mathbf{g}$, which in turn is determined by the part of the object in $\mathbf{S}'$, so the mean of $\hat{\theta}$ is well defined but insensitive to the part of the object outside $\mathbf{S}'$. The true value θ, however, depends on the exterior part of the object also, so different objects will give different θ but the same mean estimate, hence different $b(\theta)$. The bias might be zero for some particular object, but that is not of much use; we would like to have an estimator that is unbiased for all values of the parameter. A stopped clock is an unbiased estimator of the time twice a day.

A parameter is said to be be *estimable* or *identifiable* with respect to some data set if there exists an unbiased estimator of it for all true values of the underlying parameter. Some books impose the additional restriction that there must be a *linear* unbiased estimator, but we use the broader definition of estimability.

An alternative approach to defining estimability is in terms of the likelihood $\mathrm{pr}(\mathbf{g}|\theta)$. A parameter is estimable if different values of the parameter lead to different likelihoods. Putting it the other way around, if the statement $\mathrm{pr}(\mathbf{g}|\theta_1) = \mathrm{pr}(\mathbf{g}|\theta_2)$ does not imply that $\theta_1 = \theta_2$, then θ is not estimable.

For linear measurement systems, estimability is closely linked to null functions. Consider our familiar imaging equation, $\mathbf{g} = \mathcal{H}\mathbf{f} + \mathbf{n}$. The conditional density $\mathrm{pr}(\mathbf{g}|\mathbf{f})$ depends on $\mathbf{f}$ only through $\mathcal{H}\mathbf{f}$, so if there are two objects $\mathbf{f}_1$ and $\mathbf{f}_2$ such that $\mathcal{H}\mathbf{f}_1 = \mathcal{H}\mathbf{f}_2$ but $\theta(\mathbf{f}_1) \neq \theta(\mathbf{f}_2)$, then $\theta(\mathbf{f})$ is not estimable. The two objects differ by a null function, since $\mathcal{H}(\mathbf{f}_1 - \mathbf{f}_2) = 0$, and it is fundamentally the existence of null functions that causes some parameters not to be estimable.

Estimability is critical in digital imaging, especially in indirect methods such as computed tomography, since it determines what one can, in principle, determine about the object from a particular data set. Most importantly, it turns out that the integral of an object over small pixels or voxels is almost never an estimable parameter, so the bias in pixel values is not very useful in saying how well an imaging system performs. We shall be more specific about this point below in Sec. 13.3.2, and we shall discuss it in detail in Chap. 15 when we consider image reconstruction algorithms.

Ensemble-average bias One way of dealing with the issue of estimability is to average the bias over all possible true values of the parameter, defining an average bias by

$$\overline{\overline{\hat{\theta}}} = \int d\theta \ \mathrm{pr}(\theta) \int d^M g \ \mathrm{pr}(\mathbf{g}|\theta)\, \hat{\theta}(\mathbf{g}) \,, \tag{13.277}$$

or equivalently

$$\overline{b} = \langle b(\theta) \rangle_\theta = \overline{\overline{\hat{\theta}}} - \overline{\theta} \,. \tag{13.278}$$

The average bias $\overline{b}$ can be zero even for a biased estimator if positive and negative biases cancel out.

Actually computing the average bias requires a probability density on θ. For our example above of data limited to a subregion, we would need a probability density function for the integral of the object over the region of $\mathbf{S}$ that is not contained in the measured region $\mathbf{S}'$. Note that we do not need a density on the object itself, just its integral.

Modeling errors and bias Bias may be the result of the estimation procedure itself, as we shall see below in the discussion of the use of prior information to form an estimate. Bias can also be the result of an incorrect model for the mapping from the parameter space to the data space. When incorrect assumptions are made regarding the likelihood $\mathrm{pr}(\mathbf{g}|\theta)$, a nonzero bias can be expected. Theoretical and simulation studies sometimes skirt this issue by using the same incorrect likelihood in the computation of both $\hat{\theta}$ *and* in the evaluation of the bias.

Variance Another common measure of estimator performance is the variance, which describes the fluctuations in the estimate of θ that would be obtained over a repeated number of trials:

$$\mathrm{Var}(\theta) = \sigma^2_{\hat{\theta}} = \left\langle |\hat{\theta}(\mathbf{g}) - \overline{\hat{\theta}}|^2 \right\rangle_{\mathbf{g}|\theta} = \int d^M g \ \mathrm{pr}(\mathbf{g}|\theta)\, |\hat{\theta} - \overline{\hat{\theta}}|^2 \,. \tag{13.279}$$

The variance describes the average squared deviations of the estimate from the mean of the estimate, not from the true value.

The bias and variance of an estimator are closely related to the *accuracy* and *precision* of a measurement, since a measurement is essentially an estimate of some physical quantity. The accuracy of a measurement is the closeness of the average measured result to the true value, which is specified by the bias. The precision of a measurement is the reproducibility of multiple measurements, which is specified by the variance.

Mean-square error The bias and variance defined in (13.276) and (13.279) are conditioned on a particular value for the parameter θ. The *mean-square error*, or *MSE*, of an estimate is the overall fluctuation in the estimate, conditioned on a particular value for θ:

$$\text{MSE}(\theta) = \left\langle |\hat{\theta} - \theta|^2 \right\rangle_{\mathbf{g}|\theta} . \tag{13.280}$$

Note that the MSE is a distance from the true θ, while the variance is a measure of the spread relative to the mean estimate $\bar{\hat{\theta}}$. For an unbiased estimate the variance and the MSE are identical.

If the parameter θ has some randomness associated with it, the *ensemble mean-square error*, or *EMSE*, is found by taking an additional average over the parameter:

$$\text{EMSE} = \left\langle \left\langle |\hat{\theta} - \theta|^2 \right\rangle_{\mathbf{g}|\theta} \right\rangle_{\theta} . \tag{13.281}$$

Thus computation of the EMSE requires knowledge of the probability density function $\text{pr}(\theta)$. The prior probability density function required to compute the average over θ could be an actual sampling density or a subjective Bayesian prior.

Vector generalizations The generalization of the discussion of performance measures to vector estimates is straightforward. A P-dimensional parameter vector $\boldsymbol{\theta}$ has an estimate $\hat{\boldsymbol{\theta}}$ with mean given by

$$\bar{\hat{\boldsymbol{\theta}}} = \int d^M g \; \text{pr}(\mathbf{g}|\boldsymbol{\theta}) \, \hat{\boldsymbol{\theta}}(\mathbf{g}) = \int d^P \hat{\theta} \; \text{pr}(\hat{\boldsymbol{\theta}}|\boldsymbol{\theta}) \, \hat{\boldsymbol{\theta}} . \tag{13.282}$$

The first integral requires knowledge of the data density $\text{pr}(\mathbf{g}|\boldsymbol{\theta})$ and the explicit form of the mapping $\hat{\boldsymbol{\theta}}(\mathbf{g})$, while the second form assumes we know the conditional PDF of $\hat{\boldsymbol{\theta}}$ itself.

The bias $\mathbf{b}(\boldsymbol{\theta})$ is now a vector quantity:

$$\mathbf{b}(\boldsymbol{\theta}) \equiv \bar{\hat{\boldsymbol{\theta}}} - \boldsymbol{\theta} \equiv \int_\infty d^M g \left[\hat{\boldsymbol{\theta}}(\mathbf{g}) - \boldsymbol{\theta} \right] \text{pr}(\mathbf{g}|\boldsymbol{\theta}) = \int_\infty d^P \theta \left[\hat{\boldsymbol{\theta}} - \boldsymbol{\theta} \right] \text{pr}(\hat{\boldsymbol{\theta}}|\boldsymbol{\theta}) . \tag{13.283}$$

The average bias is now written $\bar{\mathbf{b}} = \langle \mathbf{b}(\boldsymbol{\theta}) \rangle_{\boldsymbol{\theta}}$.

If we denote the mean of the pth element of the random vector $\hat{\boldsymbol{\theta}}$ by $\langle \hat{\theta} \rangle_p = \bar{\hat{\theta}}_p$, the variance of the pth element is given by

$$\text{Var}(\hat{\theta}_p) \equiv \left\langle \left[\hat{\theta}_p - \langle \hat{\theta}_p \rangle \right] \left[\hat{\theta}_p - \langle \hat{\theta}_p \rangle \right]^* \right\rangle_{\mathbf{g}|\boldsymbol{\theta}}$$

$$= \int_\infty d^M g \; |\hat{\theta}_p(\mathbf{g}) - \langle \hat{\theta}_p(\mathbf{g}) \rangle|^2 \, \text{pr}(\mathbf{g}|\boldsymbol{\theta}) = \int_\infty d^P \theta \; |\hat{\theta}_p - \langle \hat{\theta}_p \rangle|^2 \, \text{pr}(\hat{\boldsymbol{\theta}}|\boldsymbol{\theta}) , \tag{13.284}$$

and the full covariance matrix is written:

$$\mathbf{K}_{\hat{\boldsymbol{\theta}}} = \left\langle \left(\hat{\boldsymbol{\theta}} - \bar{\hat{\boldsymbol{\theta}}} \right) \left(\hat{\boldsymbol{\theta}} - \bar{\hat{\boldsymbol{\theta}}} \right)^\dagger \right\rangle = \langle \Delta \hat{\boldsymbol{\theta}} \Delta \hat{\boldsymbol{\theta}}^\dagger \rangle . \tag{13.285}$$

The MSE in the vector case is:

$$\text{MSE} = \left\langle ||\hat{\boldsymbol{\theta}} - \boldsymbol{\theta}||^2 \right\rangle_{\mathbf{g}|\boldsymbol{\theta}} = \int_\infty d^M g \; ||\hat{\boldsymbol{\theta}}(\mathbf{g}) - \boldsymbol{\theta}||^2 \, \text{pr}(\mathbf{g}|\boldsymbol{\theta}) = \int_\infty d^P \theta \; ||\hat{\boldsymbol{\theta}} - \boldsymbol{\theta}||^2 \, \text{pr}(\hat{\boldsymbol{\theta}}|\boldsymbol{\theta}) . \tag{13.286}$$

The reader can show that the MSE can be written equivalently in terms of the covariance matrix and the bias vector as:

$$\text{MSE} = \text{tr}\left[\mathbf{K}_{\hat{\boldsymbol{\theta}}}\right] + \text{tr}\left[\mathbf{b}\mathbf{b}^{\dagger}\right] . \tag{13.287}$$

The mean-square error is thus the sum of two contributions: one from the bias and the other resulting from the variance about the mean estimate.

The EMSE is obtained by averaging the MSE over all $\boldsymbol{\theta}$:

$$\text{EMSE} = \left\langle \left\langle ||\hat{\boldsymbol{\theta}} - \boldsymbol{\theta}||^2 \right\rangle_{\mathbf{g}|\boldsymbol{\theta}} \right\rangle_{\boldsymbol{\theta}} = \text{tr}\left[\overline{\mathbf{K}}_{\hat{\boldsymbol{\theta}}}\right] + \text{tr}\left\langle \mathbf{b}\mathbf{b}^{\dagger}\right\rangle . \tag{13.288}$$

As noted below (13.281), the PDF used to perform the average over $\boldsymbol{\theta}$ could be an actual sampling density or a subjective Bayesian prior. As described in the Prologue and Chap. 8, it becomes increasingly difficult to determine $\text{pr}(\boldsymbol{\theta})$ through sampling as the dimensionality P increases. In such cases, it is common to assume some subjective Bayesian prior, examples of which are explored further in Chap. 15.

13.3.2 MSE in digital imaging

In this section we discuss some technical issues in using MSE in digital imaging, and specifically in image reconstruction from indirect data. The reader wanting to learn the basics of estimation without necessarily applying them to imaging may skip ahead to Sec. 13.3.3.

It is common in digital imaging to specify how well an imaging system is functioning by specifying the MSE between object and image. The implicit assumption is that the purpose of the imaging system is to reproduce the object, so the best image must be the one that looks the most like the object. Our goal here is to examine that approach critically; alternative approaches will be offered in Chap. 14.

The commensurability problem An immediate problem in trying to define an MSE for digital images is that the object and image are in different spaces; the object is a function of continuous variables and the image is a discrete set of numbers. To compute a difference between the two, we must either make the object discrete or the image continuous.

As we saw in detail in Chap. 7, a digital data set is described by a continuous-to-discrete (CD) mapping of the form

$$g_m = \int_{S_f} d^q r \; f(\mathbf{r})\, h_m(\mathbf{r}) + n_m \tag{13.289}$$

or, in operator form,

$$\mathbf{g} = \mathcal{H}\mathbf{f} + \mathbf{n}\,, \tag{13.290}$$

where $\mathbf{g}$ is the $M \times 1$ data vector, $\mathbf{n}$ the zero-mean $M \times 1$ noise vector, and $\mathbf{f}$ is the infinite-dimensional Hilbert-space vector corresponding to the function $f(\mathbf{r})$. Though we shall use the notations $f(\mathbf{r})$ and $\mathbf{f}$ interchangeably, as we have in previous chapters, we emphasize here that the object is a function; no discretization is implied by the notation $\mathbf{f}$.

For direct-imaging systems, $\mathbf{g}$ may be the final digital image, but often some additional processing is applied. For indirect imaging systems, a reconstruction step is always required. We can treat these cases together by denoting the final digital image by the $N \times 1$ vector $\hat{\boldsymbol{\theta}}$, where N may be equal to M, but could also in general be different. If the digital processing is linear, we can write

$$\hat{\boldsymbol{\theta}} = \mathbf{A}\mathbf{g} = \mathbf{A}\boldsymbol{\mathcal{H}}\mathbf{f} + \mathbf{A}\mathbf{n}\,, \tag{13.291}$$

where $\mathbf{A}$ is an $N \times M$ matrix. For direct imaging without additional processing, $N = M$, and $\mathbf{A}$ is the unit matrix. Specific forms for $\mathbf{A}$ for indirect imaging will be discussed in Chap. 15.

Our problem now is to compute an MSE between the $N \times 1$ vector $\hat{\boldsymbol{\theta}}$ and the original object function $f(\mathbf{r})$. Building on the formalism in Chap. 7, we shall present three possible solutions to this problem.

Discrete error norm One way to compare $\hat{\boldsymbol{\theta}}$ to $f(\mathbf{r})$ is to discretize $f(\mathbf{r})$ to a vector of the same dimension as $\hat{\boldsymbol{\theta}}$. If this discretization is a linear mapping, its general form is given by (7.35) as

$$\boldsymbol{\theta} = \boldsymbol{\mathcal{D}}_\chi \mathbf{f}\,, \tag{13.292}$$

where the discretization operator $\boldsymbol{\mathcal{D}}_\chi$ is defined explicitly in (7.36).

We can now define the discrepancy between this discretized version of the object and the output of the digital imaging system as

$$\delta\boldsymbol{\theta} \equiv \hat{\boldsymbol{\theta}} - \boldsymbol{\theta}\,. \tag{13.293}$$

With (13.291), the norm of this error is given by

$$||\delta\boldsymbol{\theta}||^2 = ||(\mathbf{A}\boldsymbol{\mathcal{H}} - \boldsymbol{\mathcal{D}}_\chi)\mathbf{f} + \mathbf{A}\mathbf{n}\,||^2\,, \tag{13.294}$$

where the norm is in the ND Euclidean space $\mathbb{E}^N$. The numerical value of this norm depends, of course, on the choice of the discretization functions $\{\chi_n(\mathbf{r})\}$ as well as on the system operator $\boldsymbol{\mathcal{H}}$, the noise $\mathbf{n}$, and the processing matrix $\mathbf{A}$.

Continuous error norm The second option is to make a continuous version of the digital image. If this operation is a linear mapping, its general form is given from (7.37) as

$$\hat{f}(\mathbf{r}) = \left[\boldsymbol{\mathcal{D}}_\phi^\dagger \hat{\boldsymbol{\theta}}\right](\mathbf{r}) = \sum_{n=1}^{N} \hat{\theta}_n\, \phi_n(\mathbf{r})\,. \tag{13.295}$$

The functions $\{\phi_n(\mathbf{r})\}$ thus serve as interpolating functions.

We can define the discrepancy between the two functions $f(\mathbf{r})$ and $\hat{f}(\mathbf{r})$ as

$$\delta f(\mathbf{r}) \equiv \hat{f}(\mathbf{r}) - f(\mathbf{r})\,. \tag{13.296}$$

The norm of this error, now defined in $\mathbb{L}_2(\mathbb{R}^q)$, is given by

$$||\delta\mathbf{f}||^2 = ||(\boldsymbol{\mathcal{D}}_\phi^\dagger \mathbf{A}\boldsymbol{\mathcal{H}} - \mathbf{I})\mathbf{f} + \boldsymbol{\mathcal{D}}_\phi^\dagger \mathbf{A}\mathbf{n}||^2\,. \tag{13.297}$$

The simulation solution The third solution to the commensurability problem is to ignore it and do a simulation! This approach, favored in most of the imaging literature, starts by discretizing some assumed continuous object $f(\mathbf{r})$, yielding an $N \times 1$ vector $\boldsymbol{\theta}_a$, defined by

$$\boldsymbol{\theta}_a = \boldsymbol{\mathcal{D}}_{\psi}\mathbf{f} \tag{13.298}$$

for some set of discretization functions $\{\psi_n\}$.

This vector is now regarded as the "true" object. It is imaged through a matrix $\mathbf{H}$, as described in Sec. 7.4.1, and the resulting approximate data vector is given by [*cf.* (7.305)]

$$\mathbf{g}_a = \mathbf{H}\boldsymbol{\theta}_a + \mathbf{n}\,. \tag{13.299}$$

The noise term should really be written as $\boldsymbol{\epsilon}$ instead of $\mathbf{n}$ since we use the former notation to include both true measurement noise and modeling error, but this distinction is virtually always ignored in simulation studies. Customarily, a zero-mean noise vector $\mathbf{n}$ is generated in the computer and added to $\mathbf{H}\boldsymbol{\theta}$ to give $\mathbf{g}_a$. This $\mathbf{g}_a$ is then processed just as a real data vector would be, yielding a digital image $\hat{\boldsymbol{\theta}}_a$.

Now, however, there is no commensurability problem since the true object is regarded as the discrete vector $\boldsymbol{\theta}_a$. The error norm is given by

$$||\delta\boldsymbol{\theta}||^2_{sim} \equiv ||\hat{\boldsymbol{\theta}}_a - \boldsymbol{\theta}_a||^2_{sim} = ||(\mathbf{A}\mathbf{H}\boldsymbol{\mathcal{D}}_{\psi} - \boldsymbol{\mathcal{D}}_{\psi})\mathbf{f} + \mathbf{A}\mathbf{n}||^2\,. \tag{13.300}$$

What do we mean by mean? To convert an error norm to a mean-square error, we must specify what kind of averaging is implicit in the word *mean*. There are three options: First, we can do a spatial average over a single image for a single object and a single realization of the noise. In spite of the terminology used, the MSE that results in this case is a random variable. Second, we can choose to average the error norm over all realizations of the noise for a single object. If we regard the object as nonrandom, the resulting MSE is not a random variable, but it does depend on the specific object chosen. Third, we can average over the noise and also over some ensemble of objects. This approach thus requires some knowledge of the object statistics, the topic of Sec. 8.4.

The spatial average is just the norm divided by the number of voxels in the discrete cases or the area of the object support in the continuous case, so the MSE is proportional to the error norm if we choose this kind of averaging. Explicitly, if we do only the spatial averaging, the continuous MSE is defined by

$$\text{MSE}_{cont} \equiv \frac{1}{A}||\delta\mathbf{f}||^2 = \frac{1}{A}\int_{S_f} d^q r\; |\delta f(\mathbf{r})|^2\,, \tag{13.301}$$

where A is the area of the object support if $q = 2$ (or the volume if $q = 3$).

For the discrete error norm, the corresponding MSE is defined by

$$\text{MSE}_{disc} \equiv \frac{1}{N}\sum_{n=1}^{N} |\hat{\theta}_n - \theta_n|^2\,. \tag{13.302}$$

When we include the other kinds of average, we shall append additional subscripts. For example,

$$\text{MSE}_{cont,\mathbf{n},\mathbf{f}} \equiv \frac{1}{A}\Big\langle\Big\langle ||\delta\mathbf{f}||^2\Big\rangle_{\mathbf{n}|\mathbf{f}}\Big\rangle_{\mathbf{f}}\,. \tag{13.303}$$

With three ways of defining the error norm and three ways of averaging it, we have nine definitions of MSE. Some of them are common in the literature. For example, MSE_{sim}, with only the spatial average, probably appears in the majority of papers on image processing. The Wiener-Helstrom estimator (Wiener, 1942; Helstrom, 1967) minimizes $\text{MSE}_{cont,\mathbf{n},\mathbf{f}}$, though usually under the assumption that $f(\mathbf{r})$ is a sample function of a stationary random process, and also that the imaging system is so finely sampled that the distinction between continuous and discrete MSEs can be ignored. When the Wiener-Helstrom estimator is modified to use discrete models, it minimizes $\text{MSE}_{disc,\mathbf{n},\mathbf{f}}$. The best linear unbiased estimator or BLUE, to be discussed in Sec. 13.3.7, minimizes $\text{MSE}_{disc,\mathbf{n}}$ in two steps: first it is designed to be unbiased — which implicitly assumes that all components of $\boldsymbol{\theta}$ are estimable — and then it minimizes the MSE, which for an unbiased estimator is just the average variance.

Note that the issue of estimability does not arise if the MSE includes an average over the object class. While it is true that some objects in the class will almost invariably contain null functions, any MSE with an average over $\mathbf{f}$ will tell us how close we come, on average, to reconstructing an object from the class.

This point is largely moot, however, for two reasons: we never have enough information about the object class to do a believable average over $\mathbf{f}$, and even if we did, MSE has nothing to do with the intended use of the image. At best, $\text{MSE}_{cont,\mathbf{n},\mathbf{f}}$ and $\text{MSE}_{disc,\mathbf{n},\mathbf{f}}$ say something about how well we can represent a class of objects, but nothing about how well we can distinguish between two different classes. The details that are important to the intended use of the image may make a very small contribution to the quadratic content (energy), yet accurate reproduction of these details may be essential.

Another objection to MSE is that any value computed will depend on an arbitrary choice of the functions used in the discretization or interpolation steps. Sometimes these choices can make large changes in the numerical values and can even change the rank ordering of the systems supposedly being evaluated.

Effect of null functions To understand better the role of null functions when no average over $\mathbf{f}$ is involved, let us look more closely at the mathematical forms of $\text{MSE}_{cont,\mathbf{n}}$, $\text{MSE}_{disc,\mathbf{n}}$ and $\text{MSE}_{sim,\mathbf{n}}$. With some algebra, the expressions we wish to compare can be rewritten as

$$\text{MSE}_{cont,\mathbf{n}} = \frac{1}{A}||(\mathcal{D}_\phi^\dagger \mathbf{A}\mathcal{H} - \mathbf{I})\mathbf{f}||^2 + \frac{1}{A}\text{tr}\{\mathbf{AKA}^\dagger \mathcal{D}_\phi \mathcal{D}_\phi^\dagger\}\,; \tag{13.304}$$

$$\text{MSE}_{disc,\mathbf{n}} = \frac{1}{N}||(\mathbf{A}\mathcal{H} - \mathcal{D}_\chi)\mathbf{f}||^2 + \frac{1}{N}\text{tr}\{\mathbf{AKA}^\dagger\}\,; \tag{13.305}$$

$$\text{MSE}_{sim,\mathbf{n}} = \frac{1}{N}||(\mathbf{AH}\mathcal{D}_\psi - \mathcal{D}_\psi)\mathbf{f}||^2 + \frac{1}{N}\text{tr}\{\mathbf{AKA}^\dagger\}\,, \tag{13.306}$$

where $\mathbf{K}$ is the covariance matrix of the noise and $\text{tr}\{\cdot\}$ denotes the trace operation, or sum of the diagonal elements[13] of the matrix.

[13]Note that the noise contribution to the MSE in (13.304) can also be written as $\text{tr}\{\mathcal{D}_\phi^\dagger \mathbf{AKA}^\dagger \mathcal{D}_\phi\}$ by using the cyclic property of the trace, (A.96), but in this form the trace would have to be interpreted as an integral rather than a sum.

To see how null functions influence the continuous MSE, we can decompose the object function as in (1.127):

$$f(\mathbf{r}) = f_{meas}(\mathbf{r}) + f_{null}(\mathbf{r}) , \tag{13.307}$$

where $\mathcal{H}\mathbf{f}_{null} = 0$. With this decomposition, we can rewrite (13.304) as

$$\text{MSE}_{cont,\mathbf{n}} = \frac{1}{A}||(\mathcal{D}_\phi^\dagger \mathbf{A}\mathcal{H} - \mathbf{I})\mathbf{f}_{meas} - \mathbf{f}_{null}||^2 + \frac{1}{A}\text{tr}\{\mathcal{D}_\phi^\dagger \mathbf{A}\mathbf{K}\mathbf{A}^\dagger \mathcal{D}_\phi\} . \tag{13.308}$$

By the triangle inequality, the first term has a minimum value of $A^{-1}||\mathbf{f}_{null}||^2$.

We could simply accept this offset and rate systems by how close they came to the minimum if all systems of interest had the same null space (optical systems with the same aperture but different aberrations, for example), but modern imaging ranges over many different systems with different null spaces. Even different processing algorithms on the same data can differ in the way they attempt to recover null functions. Hence $||\mathbf{f}_{null}||^2$ is of critical importance if we want to compare systems on the basis of $\text{MSE}_{cont,\mathbf{n}}$. Since this MSE uses just a single object, it depends strongly on the null functions of that object. The nefarious (or unwitting) investigator wanting to show that System A was better than system B could, for example, construct an object as a linear superposition of natural pixels (see Sec. 7.4.3) for system A, thereby removing the contribution $||\mathbf{f}_{null}||^2$ from the continuous MSE.

The situation is more complicated in $\text{MSE}_{disc,\mathbf{n}}$ since both $\mathcal{H}$ and $\mathcal{D}_\psi$ have null spaces. If these null spaces were identical, then null functions would play no role at all in $\text{MSE}_{disc,\mathbf{n}}$. Furthermore, if we chose an object such that $\mathcal{D}_\chi\mathbf{f}$ lay entirely in the measurement space of $\mathcal{H}$, then again null components of the object would play no role. Even for complicated objects with lots of fine detail not resolved by the system, we can still make $\text{MSE}_{disc,\mathbf{n}}$ small just by choosing the matrix $\mathbf{A}$ to reconstruct on a coarse grid and choosing $\{\chi_n(\mathbf{r})\}$ to discretize the object on the same grid. In that case, $\mathbf{A}\mathcal{H}\mathbf{f}$ could easily be a good approximation to the coarsely discretized object $\mathcal{D}_\chi\mathbf{f}$.

The main feature of $\text{MSE}_{sim,\mathbf{n}}$ is that it can be completely insensitive to null functions for all choices of the object. Since both $\mathbf{AH}\mathcal{D}_\psi$ and $\mathcal{D}_\psi$ erase the components of $f(\mathbf{r})$ in the null space of $\mathcal{D}_\psi$, this null space is always irrelevant for any version of MSE_{sim}. Moreover, the entire first term in (13.308) vanishes when $\mathbf{A} = \mathbf{H}^{-1}$ or $\mathbf{AH}\mathcal{D}_\psi = \mathcal{D}_\psi$; since one or both of these conditions often hold in simulations, all we can ever hope to learn about the system from MSE measures in such studies is how much the noise is amplified by the processing.

13.3.3 Bayesian estimation

Bayesian estimation is the determination of an estimate of a random $\boldsymbol{\theta}$ through minimization of the Bayes risk. Knowledge of $\text{pr}(\boldsymbol{\theta})$ is assumed, and a cost function $C(\hat{\boldsymbol{\theta}}, \boldsymbol{\theta})$ must be specified. Choosing a cost function is equivalent to choosing a particular tradeoff between bias and variance. In the following sections we shall determine the form of the Bayesian estimator for several commonly chosen cost functions. In each case we begin with the scalar estimation problem. Where possible, the generalization to the vector problem is then given.

Quadratic cost functions and the MMSE estimator The EMSE is a common form for the cost measure $C(\hat{\theta}, \theta)$ to be minimized in the design of an estimation procedure.

Since the EMSE is quadratic in $(\hat{\theta}-\theta)$, it is referred to as a quadratic cost function. We shall now derive the form of the estimator that minimizes this cost function, starting with the scalar case.

The EMSE in the scalar case, (13.281), can be written as

$$\text{EMSE} = \left\langle \left(\hat{\theta}-\theta\right)^2 \right\rangle_{\mathbf{g},\theta} = \int d^M g \; \text{pr}(\mathbf{g}) \int d\theta \; \text{pr}(\theta|\mathbf{g}) \left(\hat{\theta}-\theta\right)^2$$

$$= \int d^M g \; \text{pr}(\mathbf{g}) \int d\theta \; \text{pr}(\theta|\mathbf{g}) \left(\hat{\theta}^2 - 2\theta\hat{\theta} + \theta^2\right)$$

$$= \int d^M g \; \text{pr}(\mathbf{g}) \left[\hat{\theta}^2 - 2\hat{\theta}\int d\theta \; \theta \, \text{pr}(\theta|\mathbf{g}) + \int d\theta \; \theta^2 \, \text{pr}(\theta|\mathbf{g})\right]. \tag{13.309}$$

We know that pr($\mathbf{g}$) is always nonnegative, so the EMSE is minimized by setting

$$\frac{\partial}{\partial\hat{\theta}}\left[\hat{\theta}^2 - 2\hat{\theta}\int d\theta \; \theta \, \text{pr}(\theta|\mathbf{g}) + \int d\theta \; \theta^2 \, \text{pr}(\theta|\mathbf{g})\right] = 0. \tag{13.310}$$

The solution is given by

$$2\hat{\theta} - 2\int d\theta \; \theta \, \text{pr}(\theta|\mathbf{g}) = 0, \tag{13.311a}$$

or

$$\hat{\theta}_{\text{MMSE}} = \int d\theta \; \theta \, \text{pr}(\theta|\mathbf{g}) = \langle\theta\rangle_{\theta|\mathbf{g}}, \tag{13.311b}$$

where the subscript MMSE stands for *minimum mean-squared-error*. A better term might be MEMSE, or *minimum ensemble mean-squared-error* since the average over θ is required. Note, however, that the average over $\mathbf{g}$ is not required; exactly the same estimate would be obtained with a quadratic cost function averaged over the prior alone, so $\hat{\theta}_{\text{MMSE}}$ is a true Bayesian estimate.

The density function in the integrand of (13.311b) is the posterior probability of θ conditioned on the measured data vector. We see that $\hat{\theta}_{\text{MMSE}}$ is the mean of θ with respect to this density, or the *posterior mean*.

We can use Bayes' rule to write (13.311b) as

$$\hat{\theta}_{\text{MMSE}} = \int d\theta \; \theta \left[\frac{\text{pr}(\mathbf{g}|\theta)\,\text{pr}(\theta)}{\text{pr}(\mathbf{g})}\right] = \frac{\int d\theta \; \text{pr}(\mathbf{g}|\theta)\,\text{pr}(\theta)\,\theta}{\int d\theta \; \text{pr}(\mathbf{g}|\theta)\,\text{pr}(\theta)}. \tag{13.312}$$

Example: MMSE estimation of the rate of a Poisson process Suppose we have a radioactive sample of unknown concentration and we wish to estimate the rate parameter a of the Poisson process. We further assume that a follows an exponential distribution (C.118): $\text{pr}_a(a) = (1/\beta)\exp(-a/\beta)$. We denote the integer number of counts detected in the sampling interval by n. Then $\text{pr}(n|a)$ is given by the well-known Poisson probability law of (C.165):

$$\text{pr}(n|a) = \frac{e^{-a}a^n}{n!}. \tag{13.313}$$

We find the MMSE estimate for a by solving (13.312):

$$\hat{a}_{\text{MMSE}} = \frac{\int_0^\infty da \; e^{-a[1+(1/\beta)]}\, a^{(n+1)}}{\int_0^\infty da \; e^{-a[1+(1/\beta)]}\, a^n}. \tag{13.314}$$

A tabulated integral (Gradshteyn and Ryzhik 3.351(3.), 1980) can be used to determine the numerator and denominator, giving

$$\hat{a}_{\mathrm{MMSE}} = \frac{n+1}{1+\frac{1}{\beta}}. \tag{13.315}$$

We shall return to this example in the next sections, to demonstrate how differences in the choice of prior or cost function alter the form of the resulting estimate.

Posterior mean and symmetric cost functions It can be shown that any symmetric, convex-upward cost function gives the same form for the optimal estimator we have derived using a quadratic cost function, provided the posterior density is symmetric as well. Furthermore, if the cost function is symmetric and nondecreasing, like the one shown in Fig. 13.13, the optimal estimator is again the posterior mean if the posterior density is a symmetric unimodal function that satisfies the following property (Van Trees, 1968):

$$\lim_{\theta\to\infty} C(\theta)\, \mathrm{pr}_{\theta|\mathbf{g}}(\theta|\mathbf{g}) = 0. \tag{13.316}$$

There are many cost functions and posterior PDFs that satisfy (13.316), leading to optimal estimators that are equivalent to $\hat{\theta}_{\mathrm{MMSE}}$. We shall have occasion to refer to this property in later sections.

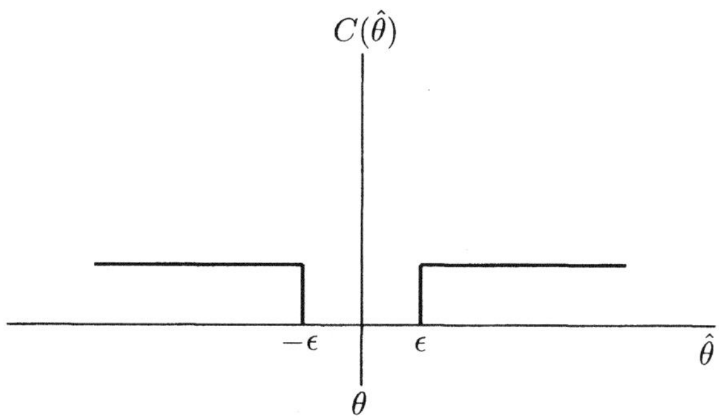

Fig. 13.13 An example of a symmetric, nondecreasing cost function, in particular, a cost function that is uniform outside a region of width 2ϵ.

Vector generalization The vector generalization of the MMSE is a straightforward extension of (13.311). The vector estimate that minimizes (13.288) is given by

$$\hat{\boldsymbol{\theta}}_{\mathrm{MMSE}} = \int d^P\theta\; \boldsymbol{\theta}\, \mathrm{pr}(\boldsymbol{\theta}|\mathbf{g}). \tag{13.317}$$

The conditional mean of the MMSE estimate is given by

$$\overline{\hat{\boldsymbol{\theta}}}_{\mathrm{MMSE}}(\boldsymbol{\theta}) = \left\langle \hat{\boldsymbol{\theta}}_{\mathrm{MMSE}} \right\rangle_{\mathbf{g}|\boldsymbol{\theta}} = \int d^P\theta'\; \boldsymbol{\theta}' \int d^M g\; \mathrm{pr}(\boldsymbol{\theta}'|\mathbf{g})\, \mathrm{pr}(\mathbf{g}|\boldsymbol{\theta}), \tag{13.318}$$

for all values of the underlying parameter $\boldsymbol{\theta}$. The estimator is biased, but the average bias [see the discussion below (13.283)] is zero.

Linear cost functions A linear cost function is one that takes the form $C(\hat{\theta}, \theta) = c|\hat{\theta} - \theta|$. A plot of this cost function is shown in Fig. 13.14. The estimator that minimizes this cost function is found by solving

$$\frac{\partial}{\partial \hat{\theta}} \int d^M g \; \mathrm{pr}(\mathbf{g}) \int d\theta \; \mathrm{pr}(\theta|\mathbf{g}) \, c|\hat{\theta} - \theta| = 0 \,. \tag{13.319}$$

We again solve for the value of $\hat{\theta}$ that minimizes the inner integral. We can rewrite the absolute value to give

$$\frac{\partial}{\partial \hat{\theta}} \int_{-\infty}^{\infty} d\theta \; \mathrm{pr}(\theta|\mathbf{g}) \left(|\hat{\theta} - \theta|\right)$$

$$= \frac{\partial}{\partial \hat{\theta}} \left[\int_{-\infty}^{\hat{\theta}} d\theta \; \mathrm{pr}(\theta|\mathbf{g}) \, (\hat{\theta} - \theta) + \int_{\hat{\theta}}^{\infty} d\theta \; \mathrm{pr}(\theta|\mathbf{g}) \, (\theta - \hat{\theta}) \right] = 0 \tag{13.320}$$

as the expression to be solved. When we use Leibniz' rule for taking the partial inside the integrals, we find

$$\int_{-\infty}^{\hat{\theta}} d\theta \; \mathrm{pr}(\theta|\mathbf{g}) - \int_{\hat{\theta}}^{\infty} d\theta \; \mathrm{pr}(\theta|\mathbf{g}) = 0 \tag{13.321a}$$

or

$$\int_{-\infty}^{\hat{\theta}} d\theta \; \mathrm{pr}(\theta|\mathbf{g}) = \int_{\hat{\theta}}^{\infty} d\theta \; \mathrm{pr}(\theta|\mathbf{g}) \,. \tag{13.321b}$$

This equality holds when $\hat{\theta}$ is equal to the median of the posterior density, so we say that $\hat{\theta}_{\mathrm{lin}}$ is equal to the *posterior median.*

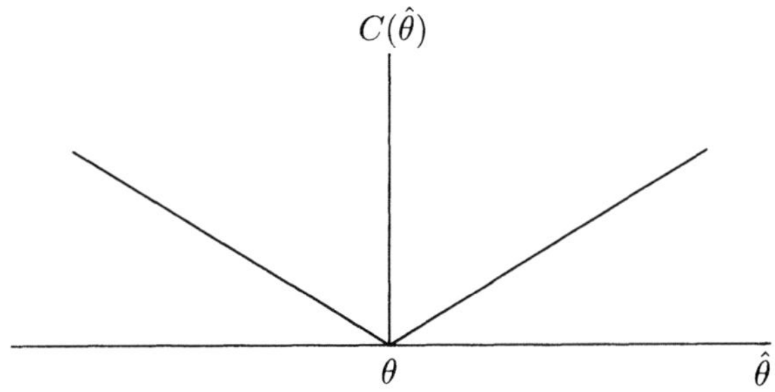

Fig. 13.14 Illustration of a linear cost function for a scalar parameter.

The optimal estimator for a linear cost function in the vector case does not lead to such an easily interpretable form. However, given the fact that the linear cost function is a symmetric, nondecreasing function, $\hat{\boldsymbol{\theta}}_{\mathrm{lin}} = \hat{\boldsymbol{\theta}}_{\mathrm{MMSE}}$ for all unimodal posterior PDFs that satisfy (13.316).

Uniform cost functions and MAP estimation Sometimes the cost is considered negligible if smaller than some tolerance ϵ, and all estimator errors beyond that tolerance are regarded as equally costly. In the scalar case, the form of such a cost function is given by

$$C(\hat{\theta}, \theta) = C(\hat{\theta} - \theta) = 1 - \mathrm{rect}\left(\frac{\hat{\theta} - \theta}{2\epsilon}\right) . \tag{13.322}$$

Figure 13.13 shows a plot of this cost function, which is referred to as a *uniform* cost function.

We seek the estimator that minimizes the cost function of (13.322), averaged over all θ and $\mathbf{g}$. This average cost, or risk is written

$$R = 1 - \int d\theta \, \mathrm{pr}_\theta(\theta) \int d^M g \, \mathrm{pr}(\mathbf{g}|\theta) \, \mathrm{rect}\left(\frac{\hat{\theta} - \theta}{2\epsilon}\right) . \tag{13.323}$$

We assume that ϵ is sufficiently small that $\mathrm{pr}_\theta(\theta) \approx \mathrm{pr}_\theta(\hat{\theta})$ and $\mathrm{pr}(\mathbf{g}|\theta) \approx \mathrm{pr}(\mathbf{g}|\hat{\theta})$ in the integral. Then

$$\begin{aligned} R &= 1 - \int d^M g \, \mathrm{pr}(\mathbf{g}|\hat{\theta}) \int d\theta \, \mathrm{pr}_\theta(\hat{\theta}) \, \mathrm{rect}\left(\frac{\hat{\theta} - \theta}{2\epsilon}\right) \\ &= 1 - 2\epsilon \int d^M g \, \mathrm{pr}(\mathbf{g}|\hat{\theta}) \, \mathrm{pr}_\theta(\hat{\theta}) . \end{aligned} \tag{13.324}$$

This expression is minimized if at each $\mathbf{g}$ we choose $\hat{\theta}(\mathbf{g})$ such that $\mathrm{pr}(\mathbf{g}|\hat{\theta}) \, \mathrm{pr}_\theta(\hat{\theta})$ is maximized with respect to $\hat{\theta}$. That is,

$$\hat{\theta}_{\mathrm{Unif}} = \underset{\theta}{\mathrm{argmax}} \; \mathrm{pr}(\mathbf{g}|\theta) \, \mathrm{pr}(\theta) . \tag{13.325}$$

Bayes' rule lets us rewrite (13.325) as

$$\hat{\theta}_{\mathrm{Unif}} = \underset{\theta}{\mathrm{argmax}} \; \mathrm{pr}(\theta|\mathbf{g}) \, \mathrm{pr}(\mathbf{g}) . \tag{13.326}$$

Since $\mathrm{pr}(\mathbf{g})$ is independent of θ, we have

$$\hat{\theta}_{\mathrm{Unif}} = \underset{\theta}{\mathrm{argmax}} \; \mathrm{pr}(\theta|\mathbf{g}) . \tag{13.327}$$

The quantity $\mathrm{pr}(\theta|\mathbf{g})$ describes the posterior probability of θ after the data $\mathbf{g}$ are obtained; thus the estimation rule of (13.327) is known as *maximum a posteriori*, or MAP, estimation. Equivalently, the MAP estimate is the mode of the posterior, or the *posterior mode*. We shall therefore refer to the resulting estimate as $\hat{\theta}_{\mathrm{MAP}}$.

Interestingly, MAP estimation is often what is meant in the literature on Bayesian estimation. However, as we have just seen, MAP estimation is just a special case of Bayesian estimation in which the particular form of the cost function is (13.322).

Some authors present MAP estimation as the procedure to follow in the absence of a cost function (Whalen, 1971). This is really a special case of the uniform cost function of (13.322) in which ϵ is taken to be small, which was the assumption we made below (13.323).

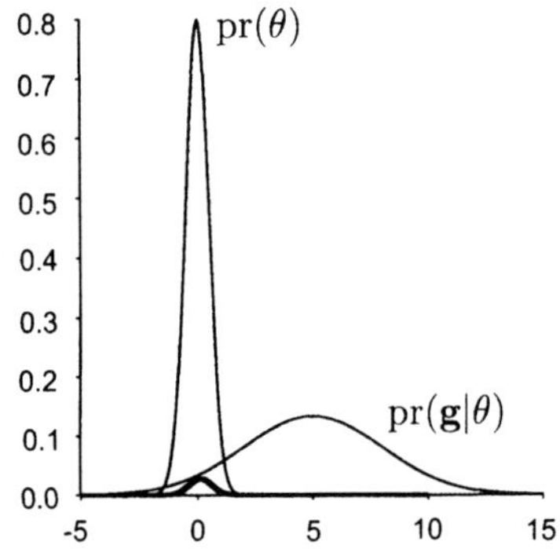

Fig. 13.15 Graph of the prior $\mathrm{pr}(\theta)$ and the conditional density function $\mathrm{pr}(\mathbf{g}|\theta)$; the product of these, referred to as the weighted likelihood function, is the dark curve.

Figure 13.15 shows a graph of the prior $\mathrm{pr}(\theta)$ and the likelihood $\mathrm{pr}(\mathbf{g}|\theta)$ that appear in (13.325). Their product is shown in bold line width in the figure. The prior serves as a scalar weight on the likelihood function, playing an increasing role as the width of $\mathrm{pr}(\theta)$ narrows, as happens when there is less prior uncertainty in the value of θ. For this reason, MAP estimation is also referred to as weighted-likelihood estimation. The MAP estimate is located where the weighted likelihood reaches a maximum, as shown in the figure.

An equivalent expression for the MAP estimation rule of (13.326) is

$$\hat{\theta}_{\mathrm{MAP}} = \underset{\theta}{\mathrm{argmax}} \; \{\ln[\mathrm{pr}(\mathbf{g}|\theta)] + \ln[\mathrm{pr}(\theta)]\} \, . \tag{13.328}$$

The monotonicity of the logarithm function means it does not affect the location of the maximum in the estimation procedure.

Another way of interpreting MAP estimation is that $\mathrm{pr}(\theta)$ characterizes the prior uncertainty in the parameter, which is often subjective in nature. Then $\mathrm{pr}(\theta|\mathbf{g})$ is the (presumably reduced) uncertainty *after* data are collected, hence the term posterior. We now revisit the problem of estimating the rate of a Poisson process to demonstrate this relationship between the estimate, the data, and the prior.

Example: MAP estimation of the rate of a Poisson process We again wish to estimate the rate parameter a of a Poisson process, where a is assumed to follow an exponential distribution as in the previous example. We are given an MD data vector $\mathbf{g} = \{n_m\}$ of independent samples of the Poisson process, so that

$$\mathrm{pr}(\mathbf{g}|a) = \prod_{m=1}^{M} \frac{e^{-a} a^{n_m}}{n_m!} \, . \tag{13.329}$$

The MAP estimate is found by solving

$$0 = \frac{\partial}{\partial a} \ln[\mathrm{pr}(\mathbf{g}|a)] + \frac{\partial}{\partial a} \ln[\mathrm{pr}(a)] = \sum_{m=1}^{M} \left(-1 + \frac{n_m}{a}\right) - \frac{1}{\beta} \, , \tag{13.330}$$

or

$$\hat{a}_{\text{MAP}} = \left[\frac{1}{M + (1/\beta)}\right] \sum_{m=1}^{M} n_m \,. \tag{13.331}$$

We see that the estimate approaches the sample mean when M is large. In the case of a limited number of samples, however, the estimate is influenced by the value of β, especially for low numbers of counts per sample, where $\hat{a}_{\text{MAP}} \propto \beta$. Note that in the case of a single sample (13.331) reduces to $\hat{a}_{\text{MAP}} = n/[1 + (1/\beta)]$. The MAP estimate is not equal to the MMSE estimate [*cf.* (13.315)] because the posterior is not symmetric, thanks to the exponential prior.

Vector generalization The vector generalization of (13.325) and (13.327) is straightforward:

$$\hat{\boldsymbol{\theta}}_{\text{MAP}} = \underset{\boldsymbol{\theta}}{\text{argmax}} \; \text{pr}(\boldsymbol{\theta}|\mathbf{g}) = \underset{\boldsymbol{\theta}}{\text{argmax}} \; \text{pr}(\mathbf{g}|\boldsymbol{\theta})\,\text{pr}(\boldsymbol{\theta})\,. \tag{13.332}$$

Similarly, a vector form for (13.328) is written

$$\hat{\boldsymbol{\theta}}_{\text{MAP}} = \underset{\boldsymbol{\theta}}{\text{argmax}} \; \{\ln[\text{pr}(\mathbf{g}|\boldsymbol{\theta})] + \ln[\text{pr}(\boldsymbol{\theta})]\}\,. \tag{13.333}$$

The uniform cost function is a symmetric, nondecreasing function; hence, $\hat{\boldsymbol{\theta}}_{\text{MAP}} = \hat{\boldsymbol{\theta}}_{\text{MMSE}}$ for all unimodal forms for $\text{pr}(\mathbf{g}|\boldsymbol{\theta})$ that satisfy (13.316). A Gaussian form for $\text{pr}(\boldsymbol{\theta}|\mathbf{g})$ satisfies this property. In a series of examples, we shall derive the form of the MAP estimator for multivariate Gaussian data and Gaussian priors on the underlying parameters, keeping in mind that the same estimator satisfies the MMSE criterion.

Example: Gaussian parameter uncertainty and Gaussian noise Consider a data vector composed of the sum of $\mathbf{s}(\boldsymbol{\theta})$, which denotes a signal parameterized by $\boldsymbol{\theta}$, a known background $\mathbf{b}$, and Gaussian noise, so that $\mathbf{g} = \mathbf{s}(\boldsymbol{\theta}) + \mathbf{b} + \mathbf{n}$, with $\mathbf{n} \sim \mathcal{N}_M(\mathbf{0}, \mathbf{K_n})$ [*cf.* (13.145)].[14] The parameter vector might contain signal location, scale, etc. (See Sec. 8.4 for a review of parametric signal descriptions.) The conditional PDF of the data is given by [*cf.* (13.174)]

$$\text{pr}(\mathbf{g}|\boldsymbol{\theta}) = (2\pi)^{-M/2}[\det(\mathbf{K_n})]^{-1/2} \exp\left\{-\tfrac{1}{2}[\mathbf{g} - \mathbf{b} - \mathbf{s}(\boldsymbol{\theta})]^t \mathbf{K_n}^{-1}[\mathbf{g} - \mathbf{b} - \mathbf{s}(\boldsymbol{\theta})]\right\}\,. \tag{13.334}$$

Suppose the parameter vector is normally distributed according to $\mathcal{N}_P(\overline{\boldsymbol{\theta}}, \mathbf{K}_{\boldsymbol{\theta}})$, or

$$\text{pr}(\boldsymbol{\theta}) = (2\pi)^{-P/2}[\det(\mathbf{K}_{\boldsymbol{\theta}})]^{-1/2} \exp\left[-\tfrac{1}{2}(\boldsymbol{\theta} - \overline{\boldsymbol{\theta}})^t \mathbf{K}_{\boldsymbol{\theta}}^{-1}(\boldsymbol{\theta} - \overline{\boldsymbol{\theta}})\right]\,. \tag{13.335}$$

By (13.333), the MAP estimate is found by minimizing

$$[\mathbf{g} - \overline{\mathbf{g}}(\boldsymbol{\theta})]^t \mathbf{K_n}^{-1}[\mathbf{g} - \overline{\mathbf{g}}(\boldsymbol{\theta})] + (\boldsymbol{\theta} - \overline{\boldsymbol{\theta}})^t \mathbf{K}_{\boldsymbol{\theta}}^{-1}(\boldsymbol{\theta} - \overline{\boldsymbol{\theta}}) \tag{13.336}$$

with respect to $\boldsymbol{\theta}$. The first term tells us to minimize the difference between the data in hand and the data we would expect to obtain given the parameter vector; hence it

[14]The reader should not confuse the background $\mathbf{b}$ with the bias vector discussed earlier.

is referred to as the data-agreement term. The second term, which comes from the prior, drives the solution toward agreement with the mean parameter vector. The prior serves to control the contribution of the measurement noise to the estimate; hence it is often referred to as a regularizer. The MAP solution strikes a balance between these two terms, giving a solution that matches the data more closely as the prior broadens, and in contrast, giving an estimate that is increasingly biased toward the prior as the noise in the data increases. We shall have much more to say about regularizers in Chap. 15.

To obtain a more explicit form for the MAP estimate of the parameters in (13.336), an explicit form for the expected data is required. In the next examples we find the form of the MAP estimate when we have a signal of random amplitude, random background level, or both.

Example: Amplitude uncertainty Suppose the random parameter is the amplitude A of a known signal $\mathbf{s}_0$, whose other characteristics are all nonrandom and known. Furthermore, we assume $A \sim \mathcal{N}(\overline{A}, \sigma_A^2)$ and all $n_m \sim \mathcal{N}(0, \sigma_n^2)$. Then $\mathbf{g}(A) = A\,\mathbf{s}_0 + \mathbf{b} + \mathbf{n}$ and $\overline{\mathbf{g}} = \overline{A}\,\mathbf{s}_0 + \mathbf{b}$. An explicit solution to (13.336) is found by solving

$$\frac{\partial}{\partial A}\left\{\frac{1}{\sigma_n^2}||\mathbf{g} - A\,\mathbf{s}_0 - \mathbf{b}||^2 + \frac{(A - \overline{A})^2}{\sigma_A^2}\right\} = 0\,, \tag{13.337}$$

leading to

$$\hat{A}_{\text{MAP}} = \left[||\mathbf{s}_0||^2 + \frac{\sigma_n^2}{\sigma_A^2}\right]^{-1}\left[(\mathbf{g} - \mathbf{b})^t\mathbf{s}_0 + \left(\frac{\sigma_n^2}{\sigma_A^2}\right)\overline{A}\right] = C_A\left[(\mathbf{g} - \mathbf{b})^t\mathbf{s}_0 + \overline{A}'\right]\,, \tag{13.338}$$

where $\overline{A}'$ is the average amplitude scaled by the ratio of the noise to amplitude variances. In this simple Gaussian case the MAP estimate is linear in the data. It is determined by subtracting the known background from the data, matched filtering with the signal, adding in the scaled mean signal amplitude, and rescaling again. We can see from (13.338) how the relative strengths of the noise and parameter variances influence the MAP estimate.

Recall that the SKE classification problem in Gaussian noise leads to an optimal discriminant that is linear in the data [*cf.* (13.110) and (13.115)]; similarly we find here that a signal of Gaussian amplitude embedded in Gaussian noise is optimally estimated in a MAP sense through linear estimation.

Example: Random background level Now consider the problem in which we have a signal of nonrandom, known amplitude A on a random background $\mathbf{b}$. The background has uniform level $b_m = b$ for all m, with b distributed according to $\mathcal{N}(\overline{b}, \sigma_b^2)$. The MAP estimate for the background level is found by solving

$$\frac{\partial}{\partial b}\left\{\frac{1}{\sigma_n^2}||\mathbf{g} - A\,\mathbf{s}_0 - \mathbf{b}||^2 + \frac{(b - \overline{b})^2}{\sigma_b^2}\right\} = 0\,, \tag{13.339}$$

which gives

$$\hat{b}_{\text{MAP}} = \left(M + \frac{\sigma_n^2}{\sigma_b^2}\right)^{-1}\left[\sum_m(\mathbf{g} - A\mathbf{s}_0)_m + \left(\frac{\sigma_n^2}{\sigma_b^2}\right)\overline{b}\right] \equiv C_b\left[\sum_m(\mathbf{g} - A\mathbf{s}_0)_m + \overline{b}'\right]\,, \tag{13.340}$$

where $\overline{b}'$ is the average background level multiplied by the ratio of noise to background variances. The MAP estimation procedure starts with the expected value for the background level, scales it by the ratio of the noise and background variances, does a correction that depends on the sum of the differences in each detector element between the actual data and what is expected given the signal, and finally rescales. Like the amplitude estimate of (13.338), the background estimate is also linear in the data.

Note that the M in (13.340) is the dimensionality of the data vector, or the number of detector elements. When this number is much greater than σ_n^2/σ_b^2, (13.340) reduces to

$$\hat{b}_{\mathrm{MAP}} \approx \frac{1}{M}\sum_m (\mathbf{g} - A\mathbf{s})_m . \tag{13.341}$$

In this case the optimal strategy is to first subtract off the known signal and then find the average data value. In most instances we would expect M to be quite large relative to σ_n^2/σ_b^2; otherwise the noise variance would have to be very large relative to the fluctuations in the random background level. We would not use (13.341) only when the prior knowledge is very strong, meaning σ_b^2 is so small that even when it is multiplied by M it is still comparable to σ_n^2. In general, the availability of M independent measurements for use in estimating the single number b renders the prior knowledge useless relative to the quality of the data as M gets large.

Example: Random amplitude and background level Now let us take up the multiparameter estimation problem in which both the signal amplitude and background level are random. If we make the assumption that A and b are independent, the MAP solution is found by simultaneously solving (13.337) and (13.339) for the two unknown parameters. The solutions are given by the same expressions for $\hat{A}_{\mathrm{MAP}}$ and $\hat{\mathbf{b}}_{\mathrm{MAP}}$ we gave in (13.338) and (13.340), only now the parameter fixed in each solution is replaced by a MAP estimate. Thus

$$\hat{A}_{\mathrm{MAP}} = C_A \left[(\mathbf{g} - \hat{\mathbf{b}}_{\mathrm{MAP}})^t \mathbf{s} + \overline{A}'\right] \tag{13.342a}$$

and

$$\hat{b}_{\mathrm{MAP}} = C_b \left[\sum_m (\mathbf{g} - \hat{A}_{\mathrm{MAP}}\mathbf{s})_m + \overline{b}'\right] . \tag{13.342b}$$

The reader can show that the MAP estimates are now given by

$$\hat{A}_{\mathrm{MAP}} = \left[1 - C_A C_b \left(s_{tot}\right)^2\right]^{-1} C_A \left[\mathbf{g}^t\mathbf{s} + \overline{A}' - C_b(g_{tot} + \overline{b}')\, s_{tot}\right] \tag{13.343a}$$

and

$$\hat{b}_{\mathrm{MAP}} = \left[1 - C_A C_b \left(s_{tot}\right)^2\right]^{-1} C_b \left[g_{tot} + \overline{b}' - C_A(\mathbf{g}^t\mathbf{s} + \overline{A}')\, s_{tot}\right] , \tag{13.343b}$$

where $g_{tot} = \sum_m g_m$ and $s_{tot} = \sum_m s_m$ are the sums of the data and signal components, respectively. Note that, while the data appear twice in each of the estimators given in (13.343), the estimates are still linear in the data. It can be shown that the optimal estimator (in a minimum mean-square error sense) is always linear if the object and the noise obey Gaussian statistics, regardless of the dimensionality of the problem (Van Trees, 1968). The resulting parameter estimates are Gaussian distributed and therefore easy to characterize in terms of bias, variance, and so on.

In contrast, we found in Sec. 13.2.11 that the detection of a random Gaussian signal on a random Gaussian background gives a discriminant function that is nonlinear in the data, owing to the fact that the data covariance matrices cannot be the same under the two hypotheses (with the exception of the low-contrast limit). The decision variable in that problem is not Gaussian distributed, unlike the estimates we just derived.[15]

Adding in the imaging operator In the above examples, the signal and background components of $\mathbf{g}$ were expressed in the data domain without concern for how these entities came to be in that space. We now reconsider the MAP estimation problem when the data are derived from a continuous-to-discrete imaging system. Our goal is to see the role the imaging system plays in the form of the estimator in that case.

Consider a linear imaging model, $\mathbf{g} = \boldsymbol{\mathcal{H}}\mathbf{f}(\boldsymbol{\theta}) + \mathbf{n}$, with $\mathbf{n} \sim \mathcal{N}_M(\mathbf{0}, \mathbf{K_n})$, so that the conditional PDF on the data is given by

$$\text{pr}[\mathbf{g}|\mathbf{f}(\boldsymbol{\theta})] = \text{pr}(\mathbf{g}|\boldsymbol{\theta})$$

$$= (2\pi)^{-M/2}[\det(\mathbf{K_n})]^{-1/2} \exp\left\{-\tfrac{1}{2}[\mathbf{g} - \boldsymbol{\mathcal{H}}\mathbf{f}(\boldsymbol{\theta})]^t\, \mathbf{K_n}^{-1}[\mathbf{g} - \boldsymbol{\mathcal{H}}\mathbf{f}(\boldsymbol{\theta})]\right\}. \quad (13.344)$$

The parameter could be signal size, amplitude, location, etc. We again assume that the parameter vector is normally distributed according to (13.335). By (13.333), the MAP estimate is found by minimizing

$$[\mathbf{g} - \boldsymbol{\mathcal{H}}\mathbf{f}(\boldsymbol{\theta})]^t \mathbf{K_n}^{-1}[\mathbf{g} - \boldsymbol{\mathcal{H}}\mathbf{f}(\boldsymbol{\theta})] + (\boldsymbol{\theta} - \overline{\boldsymbol{\theta}})^t \mathbf{K}_{\boldsymbol{\theta}}^{-1}(\boldsymbol{\theta} - \overline{\boldsymbol{\theta}}) \quad (13.345)$$

with respect to $\boldsymbol{\theta}$.

Let us return to the problem of signal-amplitude uncertainty, so that the object is $f(\mathbf{r}) = Af_s(\mathbf{r}) + f_b(\mathbf{r})$. We shall assume that the imaging operator $\boldsymbol{\mathcal{H}}$ is linear and therefore object-independent. We further assume that $\boldsymbol{\mathcal{H}}$ is a CD operator, with $\overline{g}_m = (\boldsymbol{\mathcal{H}}\mathbf{f})_m$. Not surprisingly, when the noise is again assumed to be i.i.d. Gaussian, the MAP estimate of the signal amplitude is given by [*cf.* (13.338)]

$$\hat{A}_{\text{MAP}} = \left[||\boldsymbol{\mathcal{H}}\mathbf{f}_s||^2 + \frac{\sigma_n^2}{\sigma_A^2}\right]^{-1} \left[(\mathbf{g} - \boldsymbol{\mathcal{H}}\mathbf{f}_b)^t \boldsymbol{\mathcal{H}}\mathbf{f}_s + \left(\frac{\sigma_n^2}{\sigma_A^2}\right)\overline{A}\right]. \quad (13.346)$$

More generally, when the noise is correlated, we find

$$\hat{A}_{\text{MAP}} = \left[(\boldsymbol{\mathcal{H}}\mathbf{f}_s)^t \mathbf{K_n}^{-1} \boldsymbol{\mathcal{H}}\mathbf{f}_s + \frac{1}{\sigma_A^2}\right]^{-1} \left[(\mathbf{g} - \boldsymbol{\mathcal{H}}\mathbf{f}_b)^t \mathbf{K_n}^{-1} \boldsymbol{\mathcal{H}}\mathbf{f}_s + \left(\frac{\overline{A}}{\sigma_A^2}\right)\right]. \quad (13.347)$$

Other MAP problems and priors Problems involving other types of parameter uncertainty, such as location, scale, or frequency, using different forms for the PDFs in (13.333), can be solved to determine the resulting MAP estimators by following the steps outlined in the previous examples. We have considered Gaussian PDFs for the data and parameters in the examples we have presented for their ease of

[15] While strictly speaking the decision variable is non-Gaussian, we argued that it is approximately Gaussian in some cases by way of the central-limit theorem.

manipulation in making the concepts more concrete. However, we recognize the artificiality of this assumption in many problems. A Gaussian model for the amplitude of a signal implies that the amplitude can take on negative values. Perhaps we can justify this model by claiming we are interested in objects that have either positive or negative contrast relative to the average background. However, allowing the background to also have the possibility of negative values is less appealing on physical grounds. At best, we need to assume that $\overline{b}$ is large enough, and σ_b^2 sufficiently small that the probability of a negative background is negligible. Alternative prior models for random objects that have been suggested for imaging applications are presented in Chap. 8.

MAP estimation is applicable when the parameters to be estimated are random as well as nonrandom. For random parameters, the prior $\text{pr}(\boldsymbol{\theta})$ describes the distribution of parameter values that would be observed over repeated samples. When the parameters are assumed to be nonrandom, $\text{pr}(\boldsymbol{\theta})$ is a description of our belief that the nonrandom parameters take on any particular value. The prior serves to bias the estimate toward our expectation of the underlying parameter, before collecting data.

In the absence of prior information, or in deference to the quality of the data, one might eschew the use of a prior altogether. This is one avenue leading to the approach referred to as maximum-likelihood estimation, a method we now consider.

13.3.4 Maximum-likelihood estimation

Maximum-likelihood or *ML* estimation uses the following rule to determine the underlying parameters:

$$\hat{\boldsymbol{\theta}}_{\text{ML}} \equiv \underset{\boldsymbol{\theta}}{\text{argmax}} \;\; \text{pr}(\mathbf{g}|\boldsymbol{\theta}) \,. \tag{13.348}$$

This procedure can be written equivalently as [*cf.* (13.333)]

$$\hat{\boldsymbol{\theta}}_{\text{ML}} = \underset{\boldsymbol{\theta}}{\text{argmax}} \;\; \ln[\text{pr}(\mathbf{g}|\boldsymbol{\theta})] \,. \tag{13.349}$$

As stated in the previous section, ML estimation can be considered as a limit to MAP estimation when the prior $\text{pr}(\boldsymbol{\theta})$ is sufficiently broad that $\text{pr}(\mathbf{g}|\boldsymbol{\theta})\,\text{pr}(\boldsymbol{\theta})$ is dominated by $\text{pr}(\mathbf{g}|\boldsymbol{\theta})$. However, ML estimation is much more than a limiting form of MAP estimation. As we shall see, ML estimation is a powerful procedure in its own right, owing to the unique properties of ML estimates.

Example: Poisson data revisited Consider once again the estimation of the rate parameter of a Poisson process. The conditional PDF of the data has the form given in (13.329). The ML estimate of a is obtained by solving

$$0 = \frac{\partial}{\partial a} \ln \text{pr}(\mathbf{g}|a) = \sum_{m=1}^{M} \frac{\partial}{\partial a} \left(-a + n_m \ln a + const\right) , \tag{13.350}$$

which leads to

$$\hat{a}_{\text{ML}} = \frac{1}{M} \sum_{m=1}^{M} n_m \,. \tag{13.351}$$

The ML estimate is the sample mean, which can be compared to the MAP and MMSE results found earlier.

Example: Mean of a Gaussian process Suppose we are given a data vector $\mathbf{g}$ that contains M i.i.d. samples of a Gaussian process with mean μ and variance σ^2. Our goal is to form an ML estimate of the underlying mean μ.

The conditional PDF of the data is given by

$$\text{pr}(\mathbf{g}|\mu) = \prod_{m=1}^{M} \left(\frac{1}{2\pi\sigma^2}\right)^{\frac{1}{2}} \exp\left[-\frac{1}{2}\frac{(g_m - \mu)^2}{\sigma^2}\right]. \tag{13.352}$$

The estimate is found by solving (13.349), which in this case becomes

$$0 = \frac{\partial}{\partial\mu} \ln \text{pr}(\mathbf{g}|\mu) = \sum_{m=1}^{M} \frac{\partial}{\partial\mu}\left[C - \frac{1}{2}\frac{(g_m - \mu)^2}{\sigma^2}\right] = \sum_{m=1}^{M} \frac{(g_m - \mu)}{\sigma^2}, \tag{13.353}$$

leading to

$$\hat{\mu}_{\text{ML}} = \frac{1}{M}\sum_{m=1}^{M} g_m. \tag{13.354}$$

We see that the ML estimate is the sample mean, which is the most likely approach someone would use without any knowledge of statistics! This is a common finding (it seems to us) in ML estimation.

Since the estimate of (13.354) is a linear function of the Gaussian data, it too is Gaussian distributed. Note also that $\overline{\hat{\mu}}_{\text{ML}} = \mu$, so the estimate is unbiased. It is left to the reader to show that

$$\sigma_{\hat{\mu}}^2 = \langle(\hat{\mu} - \overline{\hat{\mu}})^2\rangle = \frac{\sigma^2}{M}. \tag{13.355}$$

The greater the number of samples used to form the estimate, the smaller the variance in the result. When a single sample is used, the variance in the ML estimate is the variance in the sample.

Multivariate ML estimation in the Gaussian case Suppose we are given data contaminated by Gaussian noise so that the likelihood of the data takes the form given in (13.334). When the noise is correlated, the ML estimate can be determined from (13.349) to be

$$\hat{\boldsymbol{\theta}}_{\text{ML}} = \underset{\boldsymbol{\theta}}{\text{argmin}}\ [\mathbf{g} - \overline{\mathbf{g}}(\boldsymbol{\theta})]^t \mathbf{K}_{\mathbf{n}}^{-1} [\mathbf{g} - \overline{\mathbf{g}}(\boldsymbol{\theta})], \tag{13.356}$$

which has the form of a weighted least-squares procedure. The presence of the inverse of the noise covariance matrix serves to weight the low-noise components preferentially in the formulation of the ML estimate. When the noise is uncorrelated, the ML estimate simplifies to

$$\hat{\boldsymbol{\theta}}_{\text{ML}} = \underset{\boldsymbol{\theta}}{\text{argmin}}\ ||\mathbf{g} - \overline{\mathbf{g}}(\boldsymbol{\theta})||^2. \tag{13.357}$$

This expression is the well-known form of a least-squares estimator [*cf.* (1.191)]. We simply find the parameter vector that results in an expected data vector that most closely matches the data in hand in a mean-squared-error sense.

13.3.5 Likelihood and Fisher information

In all of the cost functions described in Sec. 13.3.1 we found that a central role is played by $\text{pr}(\mathbf{g}|\boldsymbol{\theta})$. In ML estimation it is *the* quantity being maximized, but even in MAP estimation its role has some level importance that depends on the relative weight of the prior. The quantity $\text{pr}(\mathbf{g}|\boldsymbol{\theta})$ means different things to different people. The frequentist interpretation is that it is a function of $\mathbf{g}$ for fixed $\boldsymbol{\theta}$. Repeated observations can be used to determine the nature of this quantity for a given underlying object (parameter vector). Alternatively, $\text{pr}(\mathbf{g}|\boldsymbol{\theta})$ can be viewed as a function of $\boldsymbol{\theta}$ for fixed $\mathbf{g}$. In this viewpoint $\text{pr}(\mathbf{g}|\boldsymbol{\theta})$ is a measure of the likelihood of any $\boldsymbol{\theta}$ once the data are in hand. In this section we explore the likelihood function and functions of the likelihood in greater depth.

Score The *score* is a random vector that tells us how sensitive the likelihood is to changes in the parameters:

$$\mathbf{s}(\mathbf{g}) = \frac{\frac{\partial}{\partial \boldsymbol{\theta}} \text{pr}(\mathbf{g}|\boldsymbol{\theta})}{\text{pr}(\mathbf{g}|\boldsymbol{\theta})} = \frac{\partial}{\partial \boldsymbol{\theta}} \ln[\text{pr}(\mathbf{g}|\boldsymbol{\theta})] \,. \tag{13.358}$$

In words, the score is the gradient of the log-likelihood. Since the score is a function of the log-likelihood, which is a random variable through its dependence on $\mathbf{g}$, the score is also random. Note that $\langle \mathbf{s} \rangle_{\mathbf{g}|\boldsymbol{\theta}} = 0$, where $\langle \cdot \rangle_{\mathbf{g}|\boldsymbol{\theta}}$ denotes an average with respect to $\text{pr}(\mathbf{g}|\boldsymbol{\theta})$:

$$\begin{aligned} \langle \mathbf{s} \rangle_{\mathbf{g}|\boldsymbol{\theta}} &= \int_\infty d^M g \; \text{pr}(\mathbf{g}|\boldsymbol{\theta}) \frac{\frac{\partial}{\partial \boldsymbol{\theta}} \text{pr}(\mathbf{g}|\boldsymbol{\theta})}{\text{pr}(\mathbf{g}|\boldsymbol{\theta})} \\ &= \frac{\partial}{\partial \boldsymbol{\theta}} \int_\infty d^M g \; \text{pr}(\mathbf{g}|\boldsymbol{\theta}) = \frac{\partial}{\partial \boldsymbol{\theta}} (1) = 0 \,. \end{aligned} \tag{13.359}$$

Thus the score is a zero-mean random vector.

Our interest in the score and its properties stems from its close relationship to maximum-likelihood estimation. As the gradient of the log-likelihood, $\mathbf{s}(\mathbf{g}, \boldsymbol{\theta}) = 0$ when $\boldsymbol{\theta} = \hat{\boldsymbol{\theta}}(\mathbf{g})$. The process of finding the ML estimate is thus equivalent to determining the point in parameter space where all components of the score vanish.[16] The score is useful more generally, though, as we shall see next.

Fisher information and performance bounds The covariance matrix of the score is called the *Fisher information matrix*:

$$\mathbf{F} = \langle \mathbf{s}\mathbf{s}^t \rangle_{\mathbf{g}|\boldsymbol{\theta}} \,. \tag{13.360}$$

The components of $\mathbf{F}$ are given by

$$\begin{aligned} F_{jk} &= \left\langle \left[\frac{\partial}{\partial \theta_j} \ln \text{pr}(\mathbf{g}|\boldsymbol{\theta}) \right] \left[\frac{\partial}{\partial \theta_k} \ln \text{pr}(\mathbf{g}|\boldsymbol{\theta}) \right] \right\rangle_{\mathbf{g}|\boldsymbol{\theta}} \\ &= \int_\infty d^M g \; \text{pr}(\mathbf{g}|\boldsymbol{\theta}) \left[\frac{1}{\text{pr}(\mathbf{g}|\boldsymbol{\theta})} \frac{\partial}{\partial \theta_j} \text{pr}(\mathbf{g}|\boldsymbol{\theta}) \right] \left[\frac{1}{\text{pr}(\mathbf{g}|\boldsymbol{\theta})} \frac{\partial}{\partial \theta_k} \text{pr}(\mathbf{g}|\boldsymbol{\theta}) \right] . \end{aligned} \tag{13.361}$$

[16] We are restricting our attention here to unconstrained ML estimation. Constraints such as positivity could lead to solutions away from the $\mathbf{s} = 0$ point.

We can rewrite (13.361) as

$$F_{jk} = \int_\infty d^M g \, \frac{1}{\text{pr}(\mathbf{g}|\boldsymbol{\theta})} \left[\frac{\partial}{\partial \theta_j} \text{pr}(\mathbf{g}|\boldsymbol{\theta})\right] \left[\frac{\partial}{\partial \theta_k} \text{pr}(\mathbf{g}|\boldsymbol{\theta})\right]$$

$$= - \int_\infty d^M g \, \text{pr}(\mathbf{g}|\boldsymbol{\theta}) \frac{\partial}{\partial \theta_j} \left[\frac{1}{\text{pr}(\mathbf{g}|\boldsymbol{\theta})} \frac{\partial}{\partial \theta_k} \text{pr}(\mathbf{g}|\boldsymbol{\theta})\right]$$

$$= - \int_\infty d^M g \, \text{pr}(\mathbf{g}|\boldsymbol{\theta}) \frac{\partial^2}{\partial \theta_j \partial \theta_k} \ln \text{pr}(\mathbf{g}|\boldsymbol{\theta}) = - \left\langle \frac{\partial^2}{\partial \theta_j \partial \theta_k} \ln \text{pr}(\mathbf{g}|\boldsymbol{\theta}) \right\rangle_{\mathbf{g}|\boldsymbol{\theta}} . \quad (13.362)$$

Thus F_{jk} is the second derivative of the average log-likelihood, evaluated at $\boldsymbol{\theta}$. One interpretation, then, is that the components of the Fisher information matrix describe the average degree of curvature of the log-likelihood, where the average is over all data sets given the underlying parameter vector $\boldsymbol{\theta}$.

Consider an arbitrary (not necessarily ML) estimate $\hat{\boldsymbol{\theta}}(\mathbf{g})$. We assume the estimate is unbiased; if we define $\mathbf{a} = \hat{\boldsymbol{\theta}}(\mathbf{g}) - \boldsymbol{\theta}$ to be the estimation error, then $\langle \mathbf{a} \rangle_{\mathbf{g}|\boldsymbol{\theta}} = 0$ and $\langle \mathbf{a}\mathbf{a}^t \rangle = \mathbf{K}_{\hat{\boldsymbol{\theta}}}$. We can build a new random vector $\mathbf{x}$ from the components of $\mathbf{a}$ and $\mathbf{s}$:

$$\mathbf{x} = \begin{pmatrix} \mathbf{a} \\ \mathbf{s} \end{pmatrix} \qquad \langle \mathbf{x}\mathbf{x}^t \rangle = \begin{pmatrix} \langle \mathbf{a}\mathbf{a}^t \rangle & \langle \mathbf{a}\mathbf{s}^t \rangle \\ \langle \mathbf{s}\mathbf{a}^t \rangle & \langle \mathbf{s}\mathbf{s}^t \rangle \end{pmatrix} , \quad (13.363)$$

where all averages are over the data conditioned on the true parameter $\boldsymbol{\theta}$. We already know the diagonal components of the covariance matrix $\mathbf{x}$; we now show that $\langle \mathbf{s}\mathbf{a}^t \rangle = \langle \mathbf{a}\mathbf{s}^t \rangle = \mathbf{I}$:

$$\langle \mathbf{s}\mathbf{a}^t \rangle_{ij} = \langle s_i(\hat{\theta}_j - \theta_j) \rangle = \langle s_i \hat{\theta}_j \rangle - \langle s_i \rangle \theta_j = \langle s_i \hat{\theta}_j \rangle$$

$$= \int d^M g \, \text{pr}(\mathbf{g}|\boldsymbol{\theta}) \left[\frac{\partial}{\partial \theta_i} \ln \text{pr}(\mathbf{g}|\boldsymbol{\theta})\right] \hat{\theta}_j(\mathbf{g})$$

$$= \int d^M g \, \hat{\theta}_j(\mathbf{g}) \frac{\partial}{\partial \theta_i} \text{pr}(\mathbf{g}|\boldsymbol{\theta}) = \frac{\partial \langle \hat{\theta}_j \rangle}{\partial \theta_i} = \frac{\partial \theta_j}{\partial \theta_i} = \delta_{ij} \, . \quad (13.364)$$

Putting this result together with our knowledge of the covariance on $\mathbf{a}$ and $\mathbf{s}$ gives

$$\langle \mathbf{x}\mathbf{x}^t \rangle = \mathbf{K}_{\mathbf{x}} = \begin{pmatrix} \mathbf{K}_{\hat{\boldsymbol{\theta}}} & \mathbf{I} \\ \mathbf{I} & \mathbf{F} \end{pmatrix} . \quad (13.365)$$

We know that the determinant of any covariance matrix must be greater than or equal to zero; therefore $\mathbf{K}_{\mathbf{x}}$, $\mathbf{K}_{\hat{\boldsymbol{\theta}}}$, and $\mathbf{F}$ are all positive-semidefinite. By definition, a positive-semidefinite matrix satisfies $\mathbf{u}^t \mathbf{K}_{\mathbf{x}} \mathbf{u} \geq 0$ for all nonzero vectors $\mathbf{u}$ (see Sec. A.8.1). The same condition holds for quadratic forms involving a nonzero matrix $\mathbf{U}$ in place of the vector $\mathbf{u}$ (Harville, 1997): $\mathbf{U}^t \mathbf{K}_{\mathbf{x}} \mathbf{U} \geq 0$ when $\mathbf{K}_{\mathbf{x}}$ is positive-semidefinite. We now use this property to prove that $\mathbf{K}_{\hat{\boldsymbol{\theta}}} \geq \mathbf{F}^{-1}$.

We shall assume that there are P features to be estimated, so that $\mathbf{K}_{\hat{\boldsymbol{\theta}}}$ and $\mathbf{F}$ are both $P \times P$ matrices. Let $\mathbf{U}$ be a $2P \times 1$ matrix with

$$\mathbf{U} = \begin{pmatrix} \mathbf{u}_1 \\ \mathbf{u}_2 \end{pmatrix} , \quad (13.366)$$

where the vectors $\mathbf{u}_1$ and $\mathbf{u}_2$ are both $P\times 1$ column vectors. The quadratic condition then becomes

$$(\mathbf{u}_1{}^t\,\mathbf{u}_2^t)\begin{pmatrix}\mathbf{K}_{\hat{\boldsymbol{\theta}}} & \mathbf{I}\\ \mathbf{I} & \mathbf{F}\end{pmatrix}\begin{pmatrix}\mathbf{u}_1\\ \mathbf{u}_2\end{pmatrix} = \mathbf{u}_1^t\mathbf{K}_{\hat{\boldsymbol{\theta}}}\mathbf{u}_1 + \mathbf{u}_1^t\mathbf{u}_2 + \mathbf{u}_2^t\mathbf{u}_1 + \mathbf{u}_2^t\mathbf{F}\mathbf{u}_2 \geq 0\,. \quad (13.367)$$

For any nonzero $\mathbf{u}_1$, we can let $\mathbf{u}_2 = -\mathbf{F}^{-1}\mathbf{u}_1$. Then (13.367) becomes[17]

$$\mathbf{u}_1^t\mathbf{K}_{\hat{\boldsymbol{\theta}}}\mathbf{u}_1 - \mathbf{u}_1^t\mathbf{F}^{-1}\mathbf{u}_1 - \mathbf{u}_1^t\mathbf{F}^{-1}\mathbf{u}_1 + \mathbf{u}_1^t\mathbf{F}^{-1}\mathbf{u}_1 \geq 0\,, \quad (13.368)$$

or

$$\mathbf{u}_1^t\mathbf{K}_{\hat{\boldsymbol{\theta}}}\mathbf{u}_1 \geq \mathbf{u}_1^t\mathbf{F}^{-1}\mathbf{u}_1\,. \quad (13.369)$$

This inequality is proof that $\mathbf{K}_{\hat{\boldsymbol{\theta}}} - \mathbf{F}^{-1}$ is positive-semidefinite (see Sec. A.11.2). The *Loewner convention* can be used to indicate that (13.369) holds by writing simply

$$\mathbf{K}_{\hat{\boldsymbol{\theta}}} \geq \mathbf{F}^{-1}\,. \quad (13.370)$$

Consider the particular case where we let $\mathbf{u}_1 = \mathbf{e}_n$ be the column vector $\{0, ..., 1, ..., 0\}^t$ with n^{th} component equal to 1 and all other components equal to zero. In this case (13.370) reduces to

$$[\mathbf{K}_{\hat{\boldsymbol{\theta}}}]_{nn} = \mathrm{Var}\{\hat{\theta}_n\} \geq \left[\mathbf{F}^{-1}\right]_{nn}\,. \quad (13.371)$$

This inequality is the well-known *Cramér-Rao lower bound* on the variance of an estimator, after H. Cramér (1946) and C. R. Rao (1945). The unbiased estimate of the nth parameter has a variance that must be at least as large as the nth diagonal element of the inverse of the Fisher information matrix.

For a scalar parameter (13.371) reduces to

$$\mathrm{Var}\{\hat{\theta}\} \geq \frac{1}{\left\langle\left[\frac{\partial}{\partial\theta}\ln \mathrm{pr}(\mathbf{g}|\theta)\right]^2\right\rangle}\,. \quad (13.372)$$

We can extend the CR bound to biased estimators by allowing the mean of $\mathbf{a}$ to be nonzero in the preceding derivation. In this case (13.364) generalizes to

$$\langle\mathbf{s}\mathbf{a}^t\rangle_{ij} = \frac{\partial\langle\hat{\theta}_j\rangle}{\partial\theta_i} = \frac{\partial}{\partial\theta_i}\left[(\langle\hat{\theta}_j\rangle - \theta_j) + \theta_j\right] = \frac{\partial b_j}{\partial\theta_i} + \delta_{ij}\,, \quad (13.373)$$

where b_j is a component of the bias vector as defined in (13.276), making $\partial b_j/\partial\theta_i$ the bias gradient. In vector form (13.373) becomes

$$\langle\mathbf{s}\mathbf{a}^t\rangle = \nabla_{\boldsymbol{\theta}}\mathbf{b} + \mathbf{I}\,. \quad (13.374)$$

Now (13.365) becomes

$$\langle\mathbf{x}\mathbf{x}^t\rangle = \mathbf{K}_{\mathbf{x}} = \begin{pmatrix}\mathbf{K}_{\hat{\boldsymbol{\theta}}} & \nabla_{\boldsymbol{\theta}}\mathbf{b} + \mathbf{I}\\ (\nabla_{\boldsymbol{\theta}}\mathbf{b} + \mathbf{I})^t & \mathbf{F}\end{pmatrix}, \quad (13.375)$$

[17] We are assuming here that $\mathbf{F}^{-1}$ exists; Stoica and Marzetta (2001) provide an alternative theory for situations for which the Fisher matrix is singular.

and

$$\det\left[\mathbf{K}_{\hat{\theta}}\mathbf{F} - (\nabla_{\theta}\mathbf{b} + \mathbf{I})(\nabla_{\theta}\mathbf{b} + \mathbf{I})^t\right] \geq 0\,,$$

or

$$\mathbf{K}_{\hat{\theta}} \geq (\nabla_{\theta}\mathbf{b} + \mathbf{I})\mathbf{F}^{-1}(\nabla_{\theta}\mathbf{b} + \mathbf{I})^t\,. \tag{13.376}$$

This inequality gives the lower bound on the variance achievable by *any* estimator. To add to our intuition, let us see what form (13.376) takes in the case of a scalar parameter. We find that

$$\text{Var}\{\hat{\theta}\} \geq \frac{\left(\frac{db(\theta)}{d\theta} + 1\right)^2}{\left\langle\left[\frac{\partial}{\partial\theta}\ln \text{pr}(\mathbf{g}|\theta)\right]^2\right\rangle}\,. \tag{13.377}$$

It is interesting to compare this expression to (13.372). The bias of an estimator alters the lower bound on the variance by an amount that depends on the bias gradient. If the bias is independent of the parameter, no impact on the variance is felt. But, if the bias strongly depends on the parameter, the bound can change dramatically. Note that bias can also decrease the variance if the bias gradient is negative. Consider the scalar estimation problem where we choose an estimation strategy of always setting $\hat{\theta}$ to 3. Then $\partial b/\partial\theta = -1$ and the variance is zero. Thus bias does not uniformly increase variance. In all cases, though, the mean-squared error in $\hat{\theta}$ is inversely proportional to the average of the squared gradient of the log-likelihood.

The derivations of (13.371) and (13.376) rely on the assumption that the Fisher information matrix is nonsingular. Stoica and Marzetta (2001) have shown that a singular Fisher information matrix typically leads to estimates with infinite variance.

An unbiased estimator that achieves the CR bound is called *efficient*. As we shall prove below, when an efficient estimator exists, the ML estimator is efficient.

13.3.6 Properties of ML estimators

Maximum likelihood estimators offer multiple advantages for problems that can be categorized as well-posed. These are problems in which the number of parameters to be estimated is low relative to the amount of data available to form the estimate. Furthermore, the data are known to be influenced by the parameters in such a manner that we can unambiguously determine $\boldsymbol{\theta}$ from the data in the noise-free case. Such are the problems we shall focus on in this section. We defer until Chap. 15 the contrasting situation we find ourselves in when attempting to reconstruct a continuous object from a discrete set of measurements, where there is a significant null space that confounds the use of ML methods.

In the early 1900s, R.A. Fisher wrote a series of papers in which he considered various criteria for the evaluation of estimator performance (Fisher, 1922, 1925, 1934, 1935). Chief among these criteria were: consistency, efficiency, and sufficiency.[19] In this classical (non-Bayesian) approach, an estimate is assumed to be a

[19]While Fisher derived these criteria, and then proceeded to evaluate ML methods based upon them, it is interesting to note that the early history of the maximum-likelihood principle dates back to the late 18th and early 19th centuries, with contributions by Lagrange, Bernoulli, Gauss, and Laplace (Edwards, 1974).

random variable, through its dependence on the random data. The performance of the estimator is assessed using its sampling distribution, that is, the distribution of estimates obtained over repeated trials for the same underlying true parameter. For example, the bias (13.276) and the variance (13.279) are computed using the sampling distribution. Since these measures can be difficult to compute in some cases, an alternative approach is to consider bounds on estimator performance. This is the motivation for the development of the CR bound on variance presented in the previous section. While the CR bound was developed without regard to the form of the estimator, we shall see that it is particularly relevant to ML estimation.

Before we return to the topic of the CR bound on variance and its relationship to Fisher's criteria, we shall establish one of the properties of ML estimators that makes them exceedingly useful. Recall that up to this point we have restricted our attention to the direct estimation of a parameter vector $\boldsymbol{\theta}$. In what follows we shall expand our treatment to include estimation of *any* arbitrary function of the unknown $\boldsymbol{\theta}$. This extension is made eminently doable thanks to a powerful theorem in ML estimation.

ML estimation of a function of a parameter Let the true value of a parameter be $\boldsymbol{\theta}$, and the function we want to estimate be $\tau(\boldsymbol{\theta})$. If $\boldsymbol{\Theta} = \tau(\boldsymbol{\theta})$, then (Scharf, 1991)

$$\hat{\boldsymbol{\Theta}}_{\text{ML}} = \tau(\hat{\boldsymbol{\theta}}_{\text{ML}})\,. \tag{13.378}$$

In words, the maximum-likelihood estimate of a function is the function of the maximum-likelihood estimate. This property of the ML estimator has been termed *invariance* (Tan and Drossos, 1975; Scharf, 1991). By this principle, all the discussion that follows on ML estimators applies to both direct parameter estimation and estimation of the function of an unknown parameter.

Efficiency According to (13.371), the variance of any unbiased estimator must exceed some minimum value greater than zero. If the bound is achieved, so that the equality in (13.371) holds, the estimator is said to be *efficient*. We now show that if the CR bound is attainable, it will be attained by an ML estimator. Without loss of generality, we consider the scalar case.

We can rewrite (13.372) as

$$\text{Var}\{\hat{\theta} - \theta\}\left\langle \left[\frac{\partial}{\partial\theta}\ln \text{pr}(\mathbf{g}|\theta)\right]^2 \right\rangle \geq 1$$

or

$$\left\{\int_\infty d^M g\ \text{pr}(\mathbf{g}|\boldsymbol{\theta})\,[\hat{\theta}(\mathbf{g}) - \theta]^2\right\}\left\{\int_\infty d^M g\ \text{pr}(\mathbf{g}|\boldsymbol{\theta})\left[\frac{\partial}{\partial\theta}\ln \text{pr}(\mathbf{g}|\theta)\right]^2\right\} \geq 1\,. \tag{13.379}$$

This expression is a restatement of the Schwarz inequality. The equality holds if and only if

$$\frac{\partial \ln \text{pr}(\mathbf{g}|\theta)}{\partial\theta} = \alpha(\theta)\,[\hat{\theta}(\mathbf{g}) - \theta]\,, \tag{13.380}$$

where $\alpha(\theta)$ is a constant that depends on θ. Also, (13.380) must hold for all θ and $\mathbf{g}$.

The ML estimate is defined by the likelihood equation, (13.349), which we can rewrite as

$$\left.\frac{\partial \ln \text{pr}(\mathbf{g}|\boldsymbol{\theta})}{\partial\boldsymbol{\theta}}\right|_{\boldsymbol{\theta}=\hat{\boldsymbol{\theta}}_{\text{ML}}} = 0\,. \tag{13.381}$$

Combining (13.380) and (13.381) gives

$$\alpha(\theta)[\hat{\theta}(\mathbf{g}) - \theta]\Big|_{\boldsymbol{\theta}=\hat{\boldsymbol{\theta}}_{\mathrm{ML}}} = 0\,. \tag{13.382}$$

The data-dependent solution to (13.382) is to require that $\hat{\theta}(\mathbf{g}) = \hat{\theta}_{\mathrm{ML}}$. Thus, if an efficient estimator exists, the ML estimator is efficient.

Sufficiency and uniqueness We found in Sec. 13.2 that the likelihood ratio is a sufficient statistic for hypothesis testing. It contains all the necessary information for performing the classification task. The concept of a sufficient statistic extends quite naturally to estimation tasks. In estimation, a sufficient statistic is one that captures all the essential features in the data necessary for optimal performance of a given estimation task. The maximum-likelihood estimator is a sufficient statistic for estimation; it makes optimal use of the information in the data. No other estimator can yield more information.

A necessary and sufficient condition for $\hat{\theta}$ to be a sufficient estimate is that the likelihood function must be factorable into the product

$$\mathrm{pr}(\mathbf{g}|\theta) = \mathrm{pr}(\hat{\theta}|\theta)\, f(\mathbf{g})\,, \tag{13.383}$$

where $f(\mathbf{g})$ is independent of θ. We leave it to the reader to show that this can be done for the case of Gaussian samples of unknown mean.

A sufficient estimator may exist even when an efficient estimator does not; efficiency is a stricter criteria than sufficiency. A sufficient estimator is unique; we can use a function of the sufficient statistic because it will also be sufficient, and we can choose it such that the estimate may be consistent (defined below) and unbiased.

In general there can be many solutions to (13.349) that give equal likelihood. This is especially the case in high-dimensional problems, where the null space can be very large. In Chap. 15 we explore the application of ML estimation to the high-dimensional problem of image reconstruction; as we shall see there, prior information or smoothness conditions can be introduced to address the nonuniqueness of the ML estimate.

Asymptotic properties Asymptotic properties of an estimate describe the behavior of the estimate as the number of observations approaches infinity. Suppose an estimate based on the data vector $\mathbf{g}$ is denoted $\hat{\boldsymbol{\theta}}_M(\mathbf{g})$, where the data are M independent observations or samples. The estimate is *conditionally consistent* if, for any arbitrarily small, positive ϵ and α, there exists an N such that

$$\Pr\left[||\hat{\boldsymbol{\theta}}_M(\mathbf{g}) - \boldsymbol{\theta}|| < \epsilon|\boldsymbol{\theta}\right] > 1 - \alpha \tag{13.384}$$

for all $M > N$. An estimate is *unconditionally consistent* when (13.384) is true for all $\boldsymbol{\theta}$. This property is called *stochastic convergence*, or *convergence in probability*. Cramér proved that, under reasonably general conditions, an ML estimate is consistent. The density function for a consistent estimate becomes increasingly narrow about the value of the underlying parameter as the number of samples increase. For example, we found that the ML estimate of the variance of a Gaussian process [*cf.* (13.354)] is proportional to $1/M$.

An example of an estimate that is not consistent is the sample power spectrum. For a single record, an increase in the number of samples (an increase in the record length) does not increase the probability that the estimate of the power spectrum is within ϵ of the true power spectrum. For an example, see Fig. 8.1.

While we might know that an estimate is consistent, that knowledge tells us nothing regarding its behavior for a finite set of observations. Consistency is a property that applies to the estimate as $M \to \infty$.

The relationship between bias and consistency is not as strong as it appears. A consistent estimator is not necessarily unbiased, and an unbiased estimate is not necessarily consistent. However, a consistent estimator whose asymptotic distribution has finite mean must be asymptotically unbiased.

We have established that the ML estimate is efficient, if an efficient estimator exists. ML estimates achieve the equality sign in the bound given by (13.371) as the number of samples goes to infinity; that is, ML estimates are asymptotically efficient.

When estimates satisfy (13.371), they are said to demonstrate *minimum variance.* Moreover, in the limit of a large number of samples, Fisher showed that an efficient estimate is one whose sampling distribution approaches a minimum-variance Gaussian (Fisher, 1922). Thus an ML estimate is asymptotically Gaussian distributed with minimum variance. It can be shown that, for a large class of consistent statistics, the sampling distribution of the ML estimate is approximately Gaussian as the number of observations increases because of the central-limit theorem (Cramér, 1946). The asymptotic Gaussian nature of ML estimates provides tremendous ease in understanding their properties, even for those ML estimates that are not efficient.

In summary, ML estimates are asymptotically unbiased, efficient, normally distributed and consistent.

Other bounds When ML estimates are not normally distributed, the Cramér-Rao bound may not be a good predictor of the variance of the estimates. For example, in problems where the data are nonlinearly related to the parameter and the noise is high, Müller *et al.* (1995) have shown that the CR bound is not a good measure of the variance of the estimates. Abbey *et al.* (1998) demonstrated how a non-Gaussian density on the estimates can be approximated starting with a Taylor series expansion about the true parameter vector. The isocontours of the resulting approximate densities were shown to capture the skewed distributions of the estimates quite well. This approach was subsequently applied to the estimation of parameters from phase-shifting interferometer/ellipsometer data by Rogala (1999).

Alternatives to the Cramér-Rao bound have been explored for the characterization of estimates. The *Bhattacharyya bound* (Bhattacharyya, 1946, 1947, 1948) has received significant attention, but it can be challenging to compute. While the Cramér-Rao bound involves the calculation of second-order partial derivatives of the likelihood, the Bhattacharyya bound involves higher partial derivatives (Van Trees, 1968).

Another alternative is the *Barankin bound* (Barankin, 1949), which has the advantage that it gives a greatest lower bound and it does not require that the underlying PDFs be differentiable. Barankin demonstrated that an unbiased estimator that achieves the Barankin bound must exist if an unbiased estimator exists. McAulay and Hofstetter (1971) evaluated the performance of the Barankin bound for estimation of vector parameters in additive white Gaussian noise; while the Barankin bound reduced to the Cramér-Rao bound when the SNR was high, large differences between these bounds were exhibited in other regimes. Kijewski *et al.* (1992) compared CR and Barankin bounds on estimates of lesion parameters in simulated nuclear medicine images for the purpose of collimator optimization. This approach is limited since the Barankin bound is infinite if no unbiased estimator exists (Müller, 1995; Abbey and Denny, 1996). The difficulty in practice is thus in knowing whether an unbiased estimator exists.

If no efficient estimate exists, an unbiased estimator with lower variance might exist. The difficulty comes in knowing how to find it. A significant advantage of ML estimation is that it is straightforward.

13.3.7 Other classical estimators

We have described Bayesian estimation in general and determined the resulting estimators for a variety of cost functions. We then explored ML estimation in considerable detail. The full Bayesian approach requires that the cost function as well as the prior probability of the random parameters and their likelihood function be completely specified. The ML approach does not require a prior, but it still requires full knowledge of the likelihood function. In this section we consider some other estimation methods and their properties.

Linear estimators Linear estimation is always much easier mathematically than general, nonlinear estimation. Sometimes it is even optimal. As we shall see, there is a parallel between linear classification strategies and linear estimators. In this section we shall present basic properties of some well-known linear estimation strategies. We shall limit the discussion here to basic theory, without reference to specific applications. In particular, while image reconstruction is frequently formulated as a linear estimation problem, the high dimensionality of this application requires an understanding of the specific role played by null functions of the imaging system in the assessment of the estimator. In Chap. 15 we discuss linear and nonlinear image reconstruction algorithms in great detail. Thus we shall defer to that chapter further discussion of linear image reconstruction.

A linear estimator is one that takes the general form $\hat{\boldsymbol{\theta}} = \mathbf{W}^t\mathbf{g}$, where $\mathbf{g}$ is an $M \times 1$ data vector and $\mathbf{W}$ is a $M \times P$ estimator matrix. The P columns of $\mathbf{W}$ represent templates that each yield a parameter estimate. For example, a region-of-interest (ROI) estimator is realized by a column vector $\mathbf{w}$ defined on the space of voxels, where the elements corresponding to voxels in the region of interest take on the value 1, and voxels outside this region are taken as 0.

For a linear estimator, we can write the ensemble mean-square error as

$$\text{EMSE} = \langle ||\hat{\boldsymbol{\theta}} - \boldsymbol{\theta}||^2\rangle_{\mathbf{g},\boldsymbol{\theta}} = \langle (\mathbf{W}^t\mathbf{g} - \boldsymbol{\theta})^t(\mathbf{W}^t\mathbf{g} - \boldsymbol{\theta})\rangle_{\mathbf{g},\boldsymbol{\theta}}\,, \tag{13.385}$$

and we can use calculus of variations to find the $\mathbf{W}$ that gives minimum EMSE. It will be convenient to assume that $\boldsymbol{\theta}$ and $\mathbf{g}$ are zero-mean random vectors, but this assumption does not limit the generality of the treatment; if the means are not zero, we can always define new vectors with zero means by subtracting off the means, and we can reinsert the means in the final result.

Let $\mathbf{W}' = \mathbf{W} + \epsilon\mathbf{V}$. The optimal $\mathbf{W}$ is found by solving

$$\frac{\partial(\text{EMSE})}{\partial\epsilon}\Big|_{\epsilon=0} = 0\,. \tag{13.386}$$

This leads to the requirement that

$$\langle(\mathbf{V}^t\mathbf{g})^t(\mathbf{W}^t\mathbf{g} - \boldsymbol{\theta})\rangle_{\mathbf{g},\boldsymbol{\theta}} = 0\,. \tag{13.387}$$

This expression is a condition on the optimal $\mathbf{W}$ for all possible $\mathbf{V}$, including the case $\mathbf{V} = \mathbf{W}$. In that case

$$\langle(\mathbf{W}^t\mathbf{g})^t(\mathbf{W}^t\mathbf{g} - \boldsymbol{\theta})\rangle_{\mathbf{g},\boldsymbol{\theta}} = 0\,, \tag{13.388}$$

which says that the error $\mathbf{W}^t\mathbf{g} - \boldsymbol{\theta}$ must be orthogonal to the estimate $\hat{\boldsymbol{\theta}} = \mathbf{W}^t\mathbf{g}$, when averaged over all sources of randomness. When this condition is satisfied, the minimum mean-square error is achieved.

We have derived a condition for the EMSE to reach a minimum. Now we shall derive a form of the estimator $\mathbf{W}$ that achieves this performance. We can rewrite (13.387) as

$$\left\langle\langle\text{tr}[(\mathbf{W}^t\mathbf{g} - \boldsymbol{\theta})(\mathbf{V}^t\mathbf{g})^t]\rangle_{\mathbf{g}|\boldsymbol{\theta}}\right\rangle_{\boldsymbol{\theta}} = \text{tr}\left[\mathbf{V}\left\langle\langle(\mathbf{W}^t\mathbf{g} - \boldsymbol{\theta})\mathbf{g}^t\rangle_{\mathbf{g}|\boldsymbol{\theta}}\right\rangle_{\boldsymbol{\theta}}\right]$$
$$= \text{tr}\left[\mathbf{V}\left\langle\mathbf{W}^t\mathbf{K}_{\mathbf{g}|\boldsymbol{\theta}} - \boldsymbol{\theta}\overline{\mathbf{g}}_{\boldsymbol{\theta}}^t\right\rangle_{\boldsymbol{\theta}}\right] = 0\,, \tag{13.389}$$

where $\overline{\mathbf{g}}_{\boldsymbol{\theta}}$ denotes the average data vector conditioned on the underlying parameter $\boldsymbol{\theta}$ and $\mathbf{K}_{\mathbf{g}|\boldsymbol{\theta}}$ is the data covariance matrix when the underlying parameter is $\boldsymbol{\theta}$. (Recall that we are assuming zero-mean vectors, so $\mathbf{K}_{\mathbf{g}|\boldsymbol{\theta}} \equiv \langle\mathbf{g}\mathbf{g}^t\rangle_{\mathbf{g}|\boldsymbol{\theta}}$.)

The last step we must take to find the form of the minimum-EMSE estimator is to average the conditional quantities in (13.389) over the prior probability density on $\boldsymbol{\theta}$, obtaining

$$\text{tr}\left[\mathbf{V}(\mathbf{W}^t\mathbf{K}_{\mathbf{g}} - \mathbf{K}_{\boldsymbol{\theta},\overline{\mathbf{g}}})\right] = 0\,, \tag{13.390}$$

where $\mathbf{K}_{\mathbf{g}}$ is the overall covariance of the data (including the average over $\boldsymbol{\theta}$) and $\mathbf{K}_{\boldsymbol{\theta},\overline{\mathbf{g}}}$ is the cross-covariance of $\boldsymbol{\theta}$ and $\overline{\mathbf{g}}_{\boldsymbol{\theta}}$. Note that $\mathbf{K}_{\boldsymbol{\theta},\overline{\mathbf{g}}}$ is a $P \times M$ matrix and asymmetric even if $P = M$. Since (13.390) must hold for all $\mathbf{V}$, the linear estimator $\mathbf{W}$ that achieves minimum mean-square error is given by

$$\mathbf{W}^t = \mathbf{K}_{\boldsymbol{\theta},\overline{\mathbf{g}}}\mathbf{K}_{\mathbf{g}}^{-1}\,. \tag{13.391}$$

Reinserting the means, we obtain the final form for the minimum-EMSE linear estimator:

$$\hat{\boldsymbol{\theta}} = \overline{\boldsymbol{\theta}} + \mathbf{K}_{\boldsymbol{\theta},\overline{\mathbf{g}}}\mathbf{K}_{\mathbf{g}}^{-1}\,[\mathbf{g} - \overline{\overline{\mathbf{g}}}]\,, \tag{13.392}$$

where $\overline{\boldsymbol{\theta}}$ is the prior mean of the parameter vector and $\overline{\overline{\mathbf{g}}} \equiv \langle\langle\mathbf{g}\rangle_{\mathbf{g}|\boldsymbol{\theta}}\rangle_{\boldsymbol{\theta}}$. The corresponding optimal EMSE is

$$\text{EMSE} = \text{tr}\,[\mathbf{K}_{\boldsymbol{\theta}}] - \text{tr}\left[\mathbf{K}_{\boldsymbol{\theta},\overline{\mathbf{g}}}\mathbf{K}_{\mathbf{g}}^{-1}\mathbf{K}_{\boldsymbol{\theta},\overline{\mathbf{g}}}^t\right]\,. \tag{13.393}$$

An estimator of the form (13.392) is sometimes referred to as the Wiener estimator (Wiener, 1942) or the Wiener-Helstrom estimator (Helstrom, 1967), though both terms are more commonly encountered for applications where the data are assumed to be continuous and stationary. We have shown that the Wiener estimator is the linear estimator that achieves minimum EMSE. If $\mathbf{g}$ and $\boldsymbol{\theta}$ are jointly Gaussian, the Wiener estimator achieves minimum EMSE of all estimators. (In this special case the Wiener estimator is the posterior mean given in (13.312).)

Recall from Sec. 13.2 that for classification problems where knowledge of the full joint probability on the data is not available, we often can compute the first- and second-order statistics of the data. Then we can determine the performance of the Hotelling observer for the task. If the data are truly Gaussian distributed, the Hotelling observer is the overall optimal observer in an SNR sense. Analogously, when our knowledge is limited to the first- and second-order statistics of the data, we can determine the optimal linear estimator, that is, the Wiener estimator. If the data are truly Gaussian, this estimator achieves minimum EMSE of all estimators.

Best linear unbiased estimators A *best linear unbiased estimator*, or *BLUE*, achieves the minimum conditional variance among all unbiased estimators constrained to be linear. Unlike the Wiener estimator, a BLUE takes no account of object variability. Rather, it minimizes the conditional variance of $\hat{\theta}$ for a given $\mathbf{f}$; it minimizes MSE, not EMSE. This estimator is also referred to as the Gauss-Markov estimator. When the data covariance is large, the Wiener estimator approaches the Gauss-Markov estimator.

Most texts on estimation derive mathematical expressions for Gauss-Markov estimates for problems involving data linearly related to the underlying parameter according to the model $\mathbf{g} = \mathbf{H}\boldsymbol{\theta} + \mathbf{n}$. We have deliberately avoided examples of this form because they are not directly applicable to estimation of object parameters such as tumor size or activity. However, this data model is the basis for many image reconstruction approaches, as we discuss in great detail in Chap. 15.

Uniformly minimum-variance unbiased estimates A minimum-variance estimate is one that achieves a variance less than that of any other estimate. *Uniformly minimum-variance unbiased estimates*, or *UMVU* estimates, are unbiased and have minimum variance among all unbiased estimates, hence the name. UMVU estimates are usually unique. They may or may not be linear. Rade and Westergren (1990) suggest two approaches for the determination of an UMVU estimate. The first is to find an unbiased estimator that is a function of a complete sufficient statistic. Such an estimate is automatically the minimum variance estimator. As Cox and Hinkley (1994) put it, any function of a complete sufficient statistic is the unique minimum-variance estimator of its expectation. The second approach is to find an unbiased estimate $\hat{\theta}$ and solve for its expectation conditioned on the sufficient statistic.

Rade and Westergren provide a table of parameters and their UMVU estimates, along with the variance of the estimate, for a variety of parameter distributions.

13.3.8 Nuisance parameters

Thus far, our treatment of estimation theory has assumed that we are given data $\mathbf{g}$ from some known probability law $\text{pr}(\mathbf{g}|\boldsymbol{\theta})$ and that we want to estimate a parameter $\boldsymbol{\theta}$ that fully determines the PDF. We have ignored the fact that this assumption is often not valid. Here we consider the role of *nuisance parameters*, which we are not interested in estimating but which must be stated in order for the PDF on the data to be fully specified.

Definition of nuisance parameters Suppose that the PDF on $\mathbf{g}$ is fully determined by the PD vector of parameters of interest $\boldsymbol{\theta}$ and some complementary LD vector of parameters $\boldsymbol{\theta}_n$, called a nuisance vector. We define a nuisance parameter as any parameter that does not influence the overall cost, or Bayes risk (see Sec. 13.3.1).

There are basically four ways we might deal with nuisance parameters:

(1) Replace $\boldsymbol{\theta}_n$ with some typical value $\boldsymbol{\theta}_0$, so that $\mathrm{pr}(\mathbf{g}|\boldsymbol{\theta}) \approx \mathrm{pr}(\mathbf{g}|\boldsymbol{\theta}, \boldsymbol{\theta}_{n0})$.

(2) Ignore the problem and assume a form for $\mathrm{pr}(\mathbf{g}|\boldsymbol{\theta})$.

(3) Estimate $\boldsymbol{\theta}_n$.

(4) Assume some prior and marginalize over $\boldsymbol{\theta}_n$.

We shall consider each of these approaches in turn.

The first option, replacing $\boldsymbol{\theta}_n$ with some typical value $\boldsymbol{\theta}_{n0}$, amounts to taking $\mathrm{pr}(\boldsymbol{\theta}_n)$ as a delta function — a very strong prior! This approach sounds extreme but in fact is often used in imaging with little or no justification. For example, in SPECT (single-photon emission computed tomography, treated in detail in Chap. 17), scatter correction is often performed using an assumed scatter coefficient at each location; this is a nuisance parameter when the activity at each location is to be estimated.

The second approach often taken is to ignore the problem altogether and assume some form for $\mathrm{pr}(\mathbf{g}|\boldsymbol{\theta})$. This approach sounds even more extreme, but it describes the approach being taken when image reconstruction is done using a discrete object model and the modeling errors are ignored.

The third approach is to estimate both $\boldsymbol{\theta}$ and $\boldsymbol{\theta}_n$, in hopes that doing so will improve the estimate of $\boldsymbol{\theta}$. Note that many times $\boldsymbol{\theta}_n$ is infinite-dimensional, so that this approach is not realizable. Suppose, however, that it is possible to estimate $\hat{\boldsymbol{\theta}}_n$. We can show how having to estimate the nuisance parameters affects our ability to estimate the parameters of interest, in terms of the variance of the estimate.

From (13.371) we know that the variance on an unbiased estimate of any parameter θ_p is bounded according to $\mathrm{Var}\{\hat{\theta}_p\} \geq [\mathbf{F}^{-1}]_{pp}$, where $\mathbf{F}$ is the $P \times P$ Fisher information matrix for $\boldsymbol{\theta}$. If only 1 of the P parameters is unknown and in need of estimation, then the bound on that estimate is given by (13.372). That is, $\mathrm{Var}\{\hat{\theta}_p\} \geq 1/\mathbf{F}_{pp}$ bounds the estimate on the pth parameter when all the other parameters, presumed for this discussion to be nuisance parameters, are known. We can use the extended Cauchy-Schwarz inequality of (A.192) to write

$$|\mathbf{a}^t\mathbf{b}|^2 \leq (\mathbf{a}^t\mathbf{F}\mathbf{a})(\mathbf{b}^t\mathbf{F}^{-1}\mathbf{b}), \tag{13.394}$$

since $\mathbf{F}$ is positive-definite. If we let $\mathbf{a} = \mathbf{b} = \mathbf{e}_p$, the column vector $\{0, ..., 1, ..., 0\}^t$ with the pth component equal to 1 and all other components equal to zero, we obtain

$$|\mathbf{e}_p^t\mathbf{e}_p|^2 \leq (\mathbf{F}_{pp})(\mathbf{F}^{-1})_{pp}\,. \tag{13.395}$$

The left side is equal to 1, giving

$$\frac{1}{(\mathbf{F}_{pp})} \leq (\mathbf{F}^{-1})_{pp} \,. \tag{13.396}$$

Thus the bound on the variance for the parameter estimated in the absence of nuisance parameters is lower than the bound that is active in the presence of nuisance parameters.

Now let us consider the fourth option, which is to assume some prior $\mathrm{pr}(\boldsymbol{\theta}_n)$ on the nuisance parameters and form the marginalized likelihood:[19]

$$\mathrm{pr}(\mathbf{g}|\boldsymbol{\theta}) = \int_\infty d\boldsymbol{\theta}_n \; \mathrm{pr}(\mathbf{g}|\boldsymbol{\theta}, \boldsymbol{\theta}_n) \, \mathrm{pr}(\boldsymbol{\theta}_n|\boldsymbol{\theta}) \,, \tag{13.397}$$

where now all P elements of $\boldsymbol{\theta}$ are unknown parameters we want to estimate. The risk is a function of only these parameters, and is given by

$$\begin{aligned}
\langle C(\boldsymbol{\theta}, \hat{\boldsymbol{\theta}}) \rangle &= \int d\boldsymbol{\theta} \int d\boldsymbol{\theta}_n \int d\hat{\boldsymbol{\theta}} \int d\hat{\boldsymbol{\theta}}_n \; C(\boldsymbol{\theta}, \hat{\boldsymbol{\theta}}) \, \mathrm{pr}(\boldsymbol{\theta}, \hat{\boldsymbol{\theta}}, \boldsymbol{\theta}_n, \hat{\boldsymbol{\theta}}_n) \\
&= \int d\boldsymbol{\theta} \int d\boldsymbol{\theta}_n \int d\hat{\boldsymbol{\theta}} \int d\hat{\boldsymbol{\theta}}_n \; C(\boldsymbol{\theta}, \hat{\boldsymbol{\theta}}) \, \mathrm{pr}(\hat{\boldsymbol{\theta}}, \hat{\boldsymbol{\theta}}_n|\boldsymbol{\theta}, \boldsymbol{\theta}_n) \, \mathrm{pr}(\boldsymbol{\theta}_n|\boldsymbol{\theta}) \, \mathrm{pr}(\boldsymbol{\theta}) \\
&= \int d\boldsymbol{\theta} \int d\boldsymbol{\theta}_n \int d\hat{\boldsymbol{\theta}} \int d\hat{\boldsymbol{\theta}}_n \; C(\boldsymbol{\theta}, \hat{\boldsymbol{\theta}}) \int d^M g \; \mathrm{pr}(\hat{\boldsymbol{\theta}}, \hat{\boldsymbol{\theta}}_n|\mathbf{g}) \, \mathrm{pr}(\mathbf{g}|\boldsymbol{\theta}, \boldsymbol{\theta}_n) \, \mathrm{pr}(\boldsymbol{\theta}_n|\boldsymbol{\theta}) \, \mathrm{pr}(\boldsymbol{\theta}) \,,
\end{aligned} \tag{13.398}$$

where the last line explicitly shows that the cost is averaged over all possible realizations of the data and the parameters.

We can rearrange the order of integration to obtain:

$$\begin{aligned}
&\langle C(\boldsymbol{\theta}, \hat{\boldsymbol{\theta}}) \rangle \\
&= \int d\boldsymbol{\theta} \; \mathrm{pr}(\boldsymbol{\theta}) \int d\hat{\boldsymbol{\theta}} \; C(\boldsymbol{\theta}, \hat{\boldsymbol{\theta}}) \int d\hat{\boldsymbol{\theta}}_n \int d^M g \; \mathrm{pr}(\hat{\boldsymbol{\theta}}, \hat{\boldsymbol{\theta}}_n|\mathbf{g}) \int d\boldsymbol{\theta}_n \; \mathrm{pr}(\mathbf{g}|\boldsymbol{\theta}, \boldsymbol{\theta}_n) \, \mathrm{pr}(\boldsymbol{\theta}_n|\boldsymbol{\theta}) \\
&= \int d\boldsymbol{\theta} \; \mathrm{pr}(\boldsymbol{\theta}) \int d\hat{\boldsymbol{\theta}} \; C(\boldsymbol{\theta}, \hat{\boldsymbol{\theta}}) \int d\hat{\boldsymbol{\theta}}_n \int d^M g \; \mathrm{pr}(\hat{\boldsymbol{\theta}}, \hat{\boldsymbol{\theta}}_n|\mathbf{g}) \, \mathrm{pr}(\mathbf{g}|\boldsymbol{\theta}) \,.
\end{aligned} \tag{13.399}$$

We could stop right here since only the marginalized likelihood $\mathrm{pr}(\mathbf{g}|\boldsymbol{\theta})$ appears. The estimation rule cannot depend separately on the original joint likelihood.

To continue, however, note that the estimators are functions of the data,

$$\hat{\boldsymbol{\theta}} = f(\mathbf{g}) \qquad \text{and} \qquad \hat{\boldsymbol{\theta}}_n = f_n(\mathbf{g}) \,, \tag{13.400}$$

so

$$\mathrm{pr}(\hat{\boldsymbol{\theta}}, \hat{\boldsymbol{\theta}}_n|\mathbf{g}) = \delta\Big[\hat{\boldsymbol{\theta}} - f(\mathbf{g})\Big] \; \delta\Big[\hat{\boldsymbol{\theta}}_n - f_n(\mathbf{g})\Big] \,. \tag{13.401}$$

Substituting (13.401) into (13.399), we find

$$\langle C(\boldsymbol{\theta}, \hat{\boldsymbol{\theta}}) \rangle = \int d^M g \int d\boldsymbol{\theta} \; C[\boldsymbol{\theta}, f(\mathbf{g})] \, \mathrm{pr}(\mathbf{g}|\boldsymbol{\theta}) \, \mathrm{pr}(\boldsymbol{\theta}) \,. \tag{13.402}$$

[19] We thank Dan Marks for posing the problem and formulating the solution we present here.

Now we will stop, because to go on would be to rederive familiar results from Sec. 13.3.3. Whatever estimation rule we would obtain from that section still holds, dependent only on the forms of the prior $\text{pr}(\boldsymbol{\theta})$ and cost function, as if $\boldsymbol{\theta}_n$ had never existed! The optimal strategy in the presence of nuisance parameters is to marginalize over them rather than estimate them. The key step that led us to this conclusion was assuming a cost function that was independent of $\boldsymbol{\theta}_n$.

13.3.9 Hybrid detection/estimation tasks

In Sec. 13.2.10 we considered the problem of detection of signals with one or more random parameters $\boldsymbol{\theta}$. We derived the optimal classification strategy that minimizes Bayes risk. We found that the optimal decision strategy, given knowledge of the prior probability densities $\text{pr}_{\boldsymbol{\theta}}(\boldsymbol{\theta})$ of the parameters, involved marginalizing over $\boldsymbol{\theta}$ [see (13.147)] to compute the likelihood of the data in the signal-present case. We presented examples where signal scale, frequency, or location were random. These were nuisance parameters; the observer does not report an estimate of their value, and no penalty is associated with errors in their estimate. In the language used here, the cost is just a 2×2 matrix, not a function of the signal parameter. Thus, similar to what we found above in the estimation case, a cost function that is independent of the nuisance parameter results in an optimal strategy involving marginalizing over the nuisance parameters.

Generalized likelihood-ratio detection Suppose under H_2 that a signal with random parameters $\boldsymbol{\theta}_{\text{s}}$ is present, along with a background with random parameters $\boldsymbol{\theta}_{\text{b}}$. Under H_1 only the random background is present. If the prior probability density on these random parameter vectors is unknown, a common approach is to form the generalized likelihood ratio

$$\Lambda(\mathbf{g}) = \frac{\max_{\boldsymbol{\theta}_{\text{s}},\boldsymbol{\theta}_{\text{b}}}[\text{pr}(\mathbf{g}|\boldsymbol{\theta}_{\text{s}},\boldsymbol{\theta}_{\text{b}},H_2)]}{\max_{\boldsymbol{\theta}_{\text{b}}}[\text{pr}(\mathbf{g}|\boldsymbol{\theta}_{\text{b}},H_1)]} . \tag{13.403}$$

The interpretation of this detection strategy is that the observer forms maximum-likelihood estimates of the unknown parameters $\boldsymbol{\theta}_{\text{s}}$ and $\boldsymbol{\theta}_{\text{b}}$ under H_2, forms a maximum-likelihood estimate of $\boldsymbol{\theta}_{\text{b}}$ under H_1, and chooses the hypothesis with greater overall likelihood. This approach is known as generalized likelihood-ratio detection. No estimates of the parameters are reported; a simple binary classification is performed.

The reasonableness of the generalized likelihood-ratio approach does not imply optimality in a minimum-decision-error or minimum-cost sense. In fact, as we shall now show, when the cost function depends on the underlying parameters, the result is a decision strategy that does not involve computation of the maximum-likelihood estimate under each hypothesis.

Parameter-dependent cost functions Suppose the task is again the detection of a random signal on a random background and that the classification cost function depends on the random parameters and their estimates. For example, there is a penalty for a false location in the random-location problem, or the penalty for missing a lesion increases with lesion size. We can construct a cost function that involves both the parameter estimates and the decision and determine the optimal classification strategy given that cost function.

If $\boldsymbol{\theta}_s$ is a random parameter vector associated with the signal and $\boldsymbol{\theta}_b$ is one associated with the background, the elements of the cost matrix are:

$C_{22}(\boldsymbol{\theta}_s, \hat{\boldsymbol{\theta}}_s, \boldsymbol{\theta}_b, \hat{\boldsymbol{\theta}}_b)$ = cost of deciding H_2 and giving estimates $\hat{\boldsymbol{\theta}}_s$ and $\hat{\boldsymbol{\theta}}_b$ when H_2 is true and the actual values of the parameters are $\boldsymbol{\theta}_s$ and $\boldsymbol{\theta}_b$.

$C_{21}(\hat{\boldsymbol{\theta}}_s, \boldsymbol{\theta}_b, \hat{\boldsymbol{\theta}}_b)$ = cost of deciding H_2 and giving estimates $\hat{\boldsymbol{\theta}}_s$ and $\hat{\boldsymbol{\theta}}_b$ when H_1 is true and the actual value of the background parameter vector is $\boldsymbol{\theta}_b$. (In this case, $\boldsymbol{\theta}_s$ does not exist, so it has no actual value.)

$C_{12}(\boldsymbol{\theta}_s, \boldsymbol{\theta}_b, \hat{\boldsymbol{\theta}}_b)$ = cost of deciding H_1 and giving estimate $\hat{\boldsymbol{\theta}}_b$ for the background when H_2 is true and the actual values of the parameters are $\boldsymbol{\theta}_s$ and $\boldsymbol{\theta}_b$. (Since the decision is signal-absent, no $\hat{\boldsymbol{\theta}}_s$ is reported.)

$C_{11}(\boldsymbol{\theta}_b, \hat{\boldsymbol{\theta}}_b)$ = cost of deciding H_1 and giving estimate $\hat{\boldsymbol{\theta}}_b$ when H_1 is true and the actual background parameter vector is $\boldsymbol{\theta}_b$. (In this case $\boldsymbol{\theta}_s$ does not exist, and no value of $\hat{\boldsymbol{\theta}}_s$ is reported.)

A reasonable step at this point would be to drop the dependence of all costs on $\boldsymbol{\theta}_b$ and $\hat{\boldsymbol{\theta}}_b$ since there rarely is a cost associated with an inaccurate background estimate. (The background parameters are nuisance parameters.) With this assumption, the elements of the cost matrix are

$C_{22}(\boldsymbol{\theta}_s, \hat{\boldsymbol{\theta}}_s)$ = cost of deciding H_2 and obtaining estimate $\hat{\boldsymbol{\theta}}_s$ when H_2 is true.

$C_{21}(\hat{\boldsymbol{\theta}}_s)$ = cost of deciding H_2 and giving estimate $\hat{\boldsymbol{\theta}}_s$ when H_1 is true.

$C_{12}(\boldsymbol{\theta}_s)$ = cost of deciding H_1 when H_2 is true.

C_{11} = constant = cost of deciding H_1 when H_1 is true.

The cost C_{21} is dependent on the estimate $\hat{\boldsymbol{\theta}}_s$, allowing for the possibility that it might be more risky to falsely detect a lesion at some locations, or that the cost of estimating that the pseudo-lesion is large is different from estimating it to be small. Similarly, C_{12} is a function of the underlying signal parameters, recognizing that the cost of missing the lesion might be a function of lesion size or location.

By analogy to (13.398), the risk is given by

$$\langle C \rangle = \sum_{i=1}^{2} \sum_{j=1}^{2} \int d\boldsymbol{\theta}_s \int d\hat{\boldsymbol{\theta}}_s \, C_{ij}(\boldsymbol{\theta}_s, \hat{\boldsymbol{\theta}}_s) \, \mathrm{pr}(D_i, H_j, \boldsymbol{\theta}_s, \hat{\boldsymbol{\theta}}_s)$$

$$= \sum_{i=1}^{2} \sum_{j=1}^{2} \int d\boldsymbol{\theta}_s \int d\hat{\boldsymbol{\theta}}_s \, C_{ij}(\boldsymbol{\theta}_s, \hat{\boldsymbol{\theta}}_s) \, \mathrm{pr}(D_i, \hat{\boldsymbol{\theta}}_s | H_j, \boldsymbol{\theta}_s) \, \mathrm{pr}(H_j, \boldsymbol{\theta}_s)$$

$$= \int d\boldsymbol{\theta}_s \int d\hat{\boldsymbol{\theta}}_s \, C_{22}(\boldsymbol{\theta}_s, \hat{\boldsymbol{\theta}}_s) \, \mathrm{pr}(D_2, \hat{\boldsymbol{\theta}}_s | H_2, \boldsymbol{\theta}_s) \, \mathrm{pr}(\boldsymbol{\theta}_s | H_2) \, \mathrm{Pr}(H_2)$$

$$+ \int d\hat{\boldsymbol{\theta}}_s \, C_{21}(\hat{\boldsymbol{\theta}}_s) \, \mathrm{pr}(D_2, \hat{\boldsymbol{\theta}}_s | H_1) \, \mathrm{Pr}(H_1)$$

$$+ \int d\boldsymbol{\theta}_s \, C_{12}(\boldsymbol{\theta}_s) \int d\hat{\boldsymbol{\theta}}_s \, \mathrm{pr}(D_1, \hat{\boldsymbol{\theta}}_s | H_2, \boldsymbol{\theta}_s) \, \mathrm{pr}(\boldsymbol{\theta}_s | H_2) \, \mathrm{Pr}(H_2)$$

$$+ C_{11} \int d\hat{\boldsymbol{\theta}}_s \, \mathrm{pr}(D_1, \hat{\boldsymbol{\theta}}_s | H_1) \, \mathrm{Pr}(H_1), \tag{13.404}$$

where D_i is the decision and H_j is the true state.

When we insert the dependence on $\mathbf{g}$ we obtain

$$\langle C\rangle = \int d^M g \int d\boldsymbol{\theta}_s \int d\hat{\boldsymbol{\theta}}_s \; C_{22}(\boldsymbol{\theta}_s, \hat{\boldsymbol{\theta}}_s)\,\mathrm{pr}(D_2, \hat{\boldsymbol{\theta}}_s|\mathbf{g})\,\mathrm{pr}(\mathbf{g}|H_2, \boldsymbol{\theta}_s)\,\mathrm{pr}(\boldsymbol{\theta}_s|H_2)\,\mathrm{Pr}(H_2)$$
$$+ \int d^M g \int d\hat{\boldsymbol{\theta}}_s \; C_{21}(\hat{\boldsymbol{\theta}}_s)\,\mathrm{pr}(D_2, \hat{\boldsymbol{\theta}}_s|\mathbf{g})\,\mathrm{pr}(\mathbf{g}|H_1)\,\mathrm{Pr}(H_1)$$
$$+ \int d^M g \int d\boldsymbol{\theta}_s \; C_{12}(\boldsymbol{\theta}_s) \int d\hat{\boldsymbol{\theta}}_s \; \mathrm{pr}(D_1, \hat{\boldsymbol{\theta}}_s|\mathbf{g})\,\mathrm{pr}(\mathbf{g}|H_2, \boldsymbol{\theta}_s)\,\mathrm{pr}(\boldsymbol{\theta}_s|H_2)\,\mathrm{Pr}(H_2)$$
$$+ C_{11} \int d^M g \int d\hat{\boldsymbol{\theta}}_s \; \mathrm{pr}(D_1, \hat{\boldsymbol{\theta}}_s|\mathbf{g})\,\mathrm{pr}(\mathbf{g}|H_1)\,\mathrm{Pr}(H_1)\,. \tag{13.405}$$

Using a delta function representation of the estimator, as in (13.401), gives

$$\mathrm{pr}(D_i, \hat{\boldsymbol{\theta}}_s|\mathbf{g}) = \mathrm{pr}(D_i|\hat{\boldsymbol{\theta}}_s, \mathbf{g})\,\mathrm{pr}(\hat{\boldsymbol{\theta}}_s|\mathbf{g}) = \mathrm{pr}(D_i|\mathbf{g})\,\mathrm{pr}(\hat{\boldsymbol{\theta}}_s|\mathbf{g}) = \mathrm{pr}(D_i|\mathbf{g})\,\delta[\hat{\boldsymbol{\theta}}_s - f(\mathbf{g})]\,. \tag{13.406}$$

Then

$$\langle C\rangle = \int d^M g \int d\boldsymbol{\theta}_s \; C_{22}(\boldsymbol{\theta}_s, f(\mathbf{g}))\,\mathrm{pr}(D_2|\mathbf{g})\,\mathrm{pr}(\mathbf{g}|H_2, \boldsymbol{\theta}_s)\,\mathrm{pr}(\boldsymbol{\theta}_s|H_2)\,\mathrm{Pr}(H_2)$$
$$+ \int d^M g \; C_{21}(f(\mathbf{g}))\,\mathrm{pr}(D_2|\mathbf{g})\,\mathrm{pr}(\mathbf{g}|H_1)\,\mathrm{Pr}(H_1)$$
$$+ \int d^M g \int d\boldsymbol{\theta}_s \; C_{12}(\boldsymbol{\theta}_s)\,\mathrm{pr}(D_1|\mathbf{g})\,\mathrm{pr}(\mathbf{g}|H_2, \boldsymbol{\theta}_s)\,\mathrm{pr}(\boldsymbol{\theta}_s|H_2)\,\mathrm{Pr}(H_2)$$
$$+ C_{11} \int d^M g \; \mathrm{pr}(D_1|\mathbf{g})\,\mathrm{pr}(\mathbf{g}|H_1)\,\mathrm{Pr}(H_1)\,. \tag{13.407}$$

When we combine the integrals over $\mathbf{g}$ and $\boldsymbol{\theta}$ we find:

$$\langle C\rangle = \int d^M g \Big\{ \int d\boldsymbol{\theta}_s \Big[[C_{22}(\boldsymbol{\theta}_s, f(\mathbf{g}))\,\mathrm{pr}(D_2|\mathbf{g})$$
$$+ C_{12}(\boldsymbol{\theta}_s)\,\mathrm{pr}(D_1|\mathbf{g})]\,\mathrm{pr}(\mathbf{g}|H_2, \boldsymbol{\theta}_s)\,\mathrm{pr}(\boldsymbol{\theta}_s|H_2)\,\mathrm{Pr}(H_2)\Big]$$
$$+ [C_{21}(f(\mathbf{g}))\,\mathrm{pr}(D_2|\mathbf{g}) + C_{11}\,\mathrm{pr}(D_1|\mathbf{g})]\,\mathrm{pr}(\mathbf{g}|H_1)\,\mathrm{Pr}(H_1)\Big\}\,. \tag{13.408}$$

Let Γ_i denote the region of $\mathbf{g}$-space for which the decision is D_i. Since $\mathrm{pr}(D_i|\mathbf{g}) = 1$ if $\mathbf{g}$ lies in Γ_i and zero otherwise, we have

$$\langle C\rangle = \int_{\Gamma_2} d^M g \Big\{ \int d\boldsymbol{\theta}_s \; [C_{22}(\boldsymbol{\theta}_s, f(\mathbf{g}))\,\mathrm{pr}(\mathbf{g}|H_2, \boldsymbol{\theta}_s)\,\mathrm{pr}(\boldsymbol{\theta}_s|H_2)\,\mathrm{Pr}(H_2)]$$
$$+ C_{21}(f(\mathbf{g}))\,\mathrm{pr}(\mathbf{g}|H_1)\;\mathrm{Pr}(H_1)\Big\}$$
$$+ \int_{\Gamma_1} d^M g \Big\{ \int d\boldsymbol{\theta}_s \; [C_{12}(\boldsymbol{\theta}_s)\,\mathrm{pr}(\mathbf{g}|H_2, \boldsymbol{\theta}_s)\,\mathrm{pr}(\boldsymbol{\theta}_s|H_2)\,\mathrm{Pr}(H_2)]$$
$$+ C_{11}\,\mathrm{pr}(\mathbf{g}|H_1)\;\mathrm{Pr}(H_1)\Big\}\,. \tag{13.409}$$

We now have a double optimization problem: choose Γ_2 and $f(\mathbf{g})$ to minimize $\langle C \rangle$. There is no need to choose Γ_1 separately since Γ_1 and Γ_2 comprise all of $\mathbf{g}$-space.

To make further headway, we must assume specific forms for the cost functions in (13.409). We shall adopt a MAP cost function for $C_{22}[\boldsymbol{\theta}_s, f(\mathbf{g})]$:

$$C_{22}[\boldsymbol{\theta}_s, f(\mathbf{g})] = C_{22} \times \left\{1 - \text{rect}\left[\frac{\boldsymbol{\theta}_s - f(\mathbf{g})}{\epsilon}\right]\right\}, \tag{13.410}$$

where C_{22} without an argument is the cost of a true-positive classification with a tolerance in the error of the signal-parameter estimate specified by ϵ. To simplify the example we shall assume that the cost of rendering a false-negative decision is independent of $\boldsymbol{\theta}_s$, and similarly, the cost associated with the erroneous signal parameter estimates in the absence of a signal does not depend on the estimate:

$$C_{12}(\boldsymbol{\theta}_s) = C_{12} \tag{13.411}$$

and

$$C_{21}[f(\mathbf{g})] = C_{21}\,. \tag{13.412}$$

With ϵ tending to zero, (13.409) becomes

$$\begin{aligned}\langle C \rangle = \int_{\Gamma_2} d^M g \, \Big\{ & C_{22}\big[\text{pr}(\mathbf{g}|H_2)\Pr(H_2) - \epsilon\, \text{pr}(\mathbf{g}|H_2, \hat{\boldsymbol{\theta}}_s)\, \text{pr}(\hat{\boldsymbol{\theta}}_s|H_2)\Pr(H_2)\big] \\ & + C_{21}\, \text{pr}(\mathbf{g}|H_1)\Pr(H_1)\Big\} \\ + \int_{\Gamma_1} d^M g \, & \left\{C_{12}\, \text{pr}(\mathbf{g}|H_2)\Pr(H_2) + C_{11}\, \text{pr}(\mathbf{g}|H_1)\Pr(H_1)\right\},\end{aligned} \tag{13.413}$$

where we have substituted $\hat{\boldsymbol{\theta}}_s$ for $f(\mathbf{g})$.

We can write the integrals over Γ_1 in terms of integrals over Γ_2 to find

$$\begin{aligned}\langle C \rangle = {} & C_{12}\Pr(H_2) + C_{11}\Pr(H_1) \\ & + \int_{\Gamma_2} d^M g \, \left\{(C_{22} - C_{12})\, \text{pr}(\mathbf{g}|H_2)\Pr(H_2) + (C_{21} - C_{11})\, \text{pr}(\mathbf{g}|H_1)\Pr(H_1)\right. \\ & \left. - \epsilon C_{22}\, \text{pr}(\mathbf{g}|H_2, \hat{\boldsymbol{\theta}}_s)\, \text{pr}(\hat{\boldsymbol{\theta}}_s|H_2)\, \Pr(H_2)\right\}.\end{aligned} \tag{13.414}$$

By Bayes rule we can write the last term in (13.414) as

$$\epsilon C_{22}\, \text{pr}(\mathbf{g}|H_2, \hat{\boldsymbol{\theta}}_s)\, \text{pr}(\hat{\boldsymbol{\theta}}_s|H_2)\Pr(H_2) = \epsilon C_{22}\, \text{pr}(\hat{\boldsymbol{\theta}}_s|\mathbf{g}, H_2)\, \text{pr}(\mathbf{g}|H_2)\, \Pr(H_2)\,, \tag{13.415}$$

and the average cost becomes, finally,

$$\begin{aligned}\langle C \rangle = {} & C_{12}\Pr(H_2) + C_{11}\Pr(H_1) \\ & + \int_{\Gamma_2} d^M g \, \left\{(C_{22} - C_{12})\, \text{pr}(\mathbf{g}|H_2)\Pr(H_2) + (C_{21} - C_{11})\Pr(H_1)\, \text{pr}(\mathbf{g}|H_1)\right. \\ & \left. - \epsilon C_{22}\, \text{pr}(\hat{\boldsymbol{\theta}}_s|\mathbf{g}, H_2)\, \text{pr}(\mathbf{g}|H_2)\Pr(H_2)\right\}.\end{aligned} \tag{13.416}$$

If $\epsilon \to 0$, we must choose Γ_2 so that

$$(C_{11} - C_{21})\Pr(H_1)\, \text{pr}(\mathbf{g}|H_1) > (C_{22} - C_{12})\, \text{pr}(\mathbf{g}|H_2)\Pr(H_2)\,. \tag{13.417}$$

That is, we make decision D_2 if

$$\frac{\mathrm{pr}(\mathbf{g}|H_2)}{\mathrm{pr}(\mathbf{g}|H_1)} > \frac{(C_{21} - C_{11})\,\mathrm{Pr}(H_1)}{(C_{12} - C_{22})\,\mathrm{Pr}(H_2)}, \tag{13.418}$$

which is the usual decision strategy for the ideal observer in the absence of parameter uncertainty. The term proportional to ϵ in (13.416) has the following interpretation: choose $\hat{\boldsymbol{\theta}}_s$ such that $\mathrm{pr}(\hat{\boldsymbol{\theta}}_s|\mathbf{g}, H_2)$ is maximal (*i.e.*, the expected MAP estimator).

In summary, with the stated assumptions about costs, the strategy is to do the hypothesis test with the marginalized likelihood $\mathrm{pr}(\mathbf{g}|H_2)$, followed by MAP estimation of the underlying parameter if D_2 is made. Contrary to popular belief, one should not use the maximum of $\mathrm{pr}(\mathbf{g}|\boldsymbol{\theta}, H_2)$ in the hypothesis test.

Bayesians and frequentists In both pure estimation problems (Sec. 13.3.8) and hybrid detection/estimation problems (this section), we have found that the optimal strategy for handling nuisance parameters is to marginalize rather than to estimate them. This finding begs the question of the appropriate prior to utilize. To a frequentist, the prior should be a sampling prior, verifiable by experiment; an example is the exponential absorption law for photons. To a Bayesian, the prior could incorporate prior beliefs. Indeed, a Bayesian would say that the problem of nuisance parameters is an example of a fundamental dilemma that arises in any inference problem: we *never* have enough empirical information to solve the problem at hand, and we must always bring in prior belief. Our reply, which we shall expound more fully in Chaps. 14 and 15, is that, yes, one can use prior beliefs in estimation problems in imaging, but the final measure of the efficacy of the belief is a long-run, frequentist measure of task performance.

14
Image Quality

In this chapter we consider the many practical issues one must wrestle with in the objective evaluation of imaging systems. Unlike Chap. 13, where knowledge of the relevant population statistics of the image classes is assumed, the emphasis here is on the practical issues that come to the fore when only a finite sample of images is available for determining the image statistics or the observer's performance, or both.

We begin in Sec. 14.1 with a description of various approaches to the assessment of image quality, including methods based on preference assessments, fidelity measures, and information-theoretic approaches. Then, in Sec. 14.1.5, we introduce the key elements that are required for the approach we advocate: the method must be objective, task-based, and account for the statistical properties of the relevant images and observers.

Properties of the human visual system and the determination of classification performance by human observers is the subject of Sec. 14.2, including the conduct of psychophysical experiments and the estimation of summary statistics for human performance. In Sec. 14.3 we turn to the subject of model or algorithmic observers for classification and estimation tasks. The approaches presented in Secs. 14.2 and 14.3 may make use of actual data sets derived from real imaging systems or, more often in research investigations, simulated images. Methods for image simulation are discussed in Sec. 14.4. As emphasized in that section, accurate models of the properties of the object and the physics of the image acquisition system are required if simulated images are to lead to accurate assessments of system performance.

14.1 SURVEY OF APPROACHES

14.1.1 Subjective assessment

The simplest approach to the assessment of image quality is to rely on a viewer's subjective assessment regarding how good an image looks. This approach can be as crass as the presentation of just a single pair of images, one processed by algorithm A and the other processed by contender B, with the developer of algorithm A drawing sweeping conclusions regarding the merits of A over B. A panel of experts might be used to make a stronger case regarding the merits of one algorithm over another, but here again the panel's decision is based on subjective preference rather than objective, task-based performance. There may be a place for beauty contests in the evaluation of imagery, such as when an individual selects a home-entertainment video system. We would argue that even then, most buyers base their subjective preference of one system over others by viewing a range of images; buyers usually take into account technical data across competing systems as well.

In an effort toward putting subjective preference methods on more solid footing, Zetzsche and Hauske (1989) developed a model based on the visual system with the goal of predicting subjective ratings of image quality. If this goal were met, the authors reasoned that they could determine image quality without the need for building physical prototypes of display devices. The predictions of the model were found to have correlations with mean subjective ratings ranging from 0.74 to 0.95 for images in which various artifacts were present.

Methods based on multidimensional scaling (MDS) have been applied to the analysis of subjective image quality ratings (Ahumada and Null, 1993). MDS methods incorporate various approaches for collecting numerical rating from multiple observers given the task of rating the quality of a set of images. Images can be presented in pairs, with the observer given the task of selecting the one with higher quality, or a set of images can be rank-ordered by quality. Normalizations can be done to account for differences in how observers scale the rating values; Thurstone scaling is a procedure that allows observers to use a rating scale nonlinearly (Torgerson, 1958). Once the rating data are in hand, MDS enables the dimensions of image quality to be extracted (Farrell *et al.*, 1991). Standard software packages are available for performing MDS. The difficulty with the MDS approach is that the labeling of the extracted dimensions, in terms of physical characteristics of the images or the image acquisition system, is left to the investigator (Shepard *et al.*, 1972). Moreover, the connection between an observer's rating of the quality of an image and the usefulness of the image for a specified task is never made.

Structured preference assessments formalize the subjective approach through the use of trained observers who perform a prescribed set of analyses. The well-known National Imagery Interpretability Rating Scale (NIIRS) system, which uses an interpretability rating scale for analyzing military reconnaissance images, is an example of a structured-preference approach. The NIIRS system was developed under the leadership of the U.S. Imagery Resolution Assessments and Reporting Standards (IRARS) Committee in the early 1970s. The first NIIRS system evaluated the visibility of military objects in images acquired in the visible spectrum. Later, the NIIRS system was extended to incorporate objects like buildings, roads, railroads and bridges, enabling the evaluation of images without military objects.

The NIIRS system is now able to handle data outside the visible spectrum, including thermal, radar, and multispectral imagery.

Models have been developed for predicting NIIRS ratings just as models have been developed for predicting subjective preference ratings. Given a set of input variables that can include the scene contrast, scene illumination, and imaging system characteristics, the models generate measures of image quality that can be related to the NIIRS scale. Another approach to the estimation of an image quality metric that correlates with the NIIRS scale is based on the power spectrum of the image to be rated, indicating that the measure is heavily influenced by the noise properties of the image.

The NIIRS approach is almost exclusively used for military applications; NIIRS refers to the value of an image for "intelligence purposes," rather than image quality *per se*. Furthermore, the NIIRS approach is not amenable to the analysis of the variation in true- and false-positive fractions [TPF and FPF, defined in (13.11)] of image interpretations as a function of the reader's mindset (see Fig. 13.5). Some argue that a preference-based approach is appropriate whenever the task is not well defined. We have not encountered an example where there truly is no specific task. There may be several tasks, in which case the system could be evaluated for each.

We regard preference assessments as useful in go/no-go decisions, giving information on the adequacy of images for further, more rigorous, testing. Indeed, rank-order studies of image quality have been proposed as a formal approach to determining whether the cost of a large-scale objective study is justified (Gur *et al.*, 1997; Rockette *et al.*, 1997; Good *et al.*, 1999; Towers *et al.*, 2000). These studies can make use of highly trained observers and specified tasks; their drawback is that they identify trends without providing an absolute measure of image quality. Statistical methods for planning and analyzing rank-order experiments have been introduced (Rockette *et al.*, 2001).

14.1.2 Fidelity measures

A common approach to image assessment is to assume that the goal in imaging is to reproduce a likeness of the object, leading to the conclusion that the best imaging system is the one that gives the smallest discrepancy between object and image. The most common measure of fidelity is the *mean-square error* (MSE) between object and image; some flavor of MSE is quoted in the majority of papers on image processing or image reconstruction. As we saw in Sec. 13.3.2, however, there are some arbitrary choices to be made in defining MSE, and different choices can lead to quite different conclusions about the quality of an imaging system or processing algorithm.

Problems with fidelity measures MSE and any other fidelity measure will be sensitive to many different properties of an image. If we rotate an image slightly with respect to the object or change the magnification, for example, we can produce a large discrepancy between the object and the image, even if they would otherwise be identical. Similarly, image distortion, such as barrel or pincushion effects, can lead to a large MSE. Finally, gray-scale errors such as nonlinear mapping of the image intensity or even an error in overall brightness can contribute heavily to any fidelity measure.

In many cases, these image modifications are trivial in the sense that they do not degrade the information we want to extract from the images. For example, a radiologist can interpret a chest radiograph just as well if it is rotated by a few degrees on a light box or displayed at a different magnification on a computer monitor.[1] MSE or other measures of fidelity would show that the rotated or scaled image was a poor representation of the object, but the user might not even notice the discrepancy.

Sometimes, however, apparently trivial modifications of an image are important. A cartographer wanting to derive accurate distances from an aerial photograph, for example, would worry a great deal about the magnification, and an astronomer wanting to track the angle between the two members of a binary star would worry about the rotation angle. In designing a lens system for photolithography, distortion might be critical, though for portrait photography it would be imperceptible. Even in these cases, however, a fidelity measure such as MSE is too blunt an instrument to say anything meaningful about the usefulness of an image.

Why not MSE? As delineated in Sec. 13.3.2, there are many arbitrary choices to be made in defining an MSE. For digital images, we must decide whether to discretize the assumed object for comparison with the digital output, to interpolate the digital image in order to get a continuous function to compare to the real object, or just to do a simulation and hope that the results will mean something. For each of these options, we must select a set of functions for discretization or interpolation, and we must select either a single object or a class of objects for comparison in some sense to the images. If the object contains null functions of either the system operator or the discretization operator, as any real object will, then any MSE will be very sensitive to the choice of object.

MSE measures can be very sensitive to relatively trivial image modifications such as magnification, rotation and gray-scale mappings, but they may be completely insensitive to small details that we really want to capture in the image. Furthermore, MSE measures make no distinction between blur and noise. It is easy to construct two very different images, one with high noise but good sharpness and a blurred one with low noise that have the same MSE. The main objection to any MSE metric, however, is that it has nothing to do with the intended use of the image..

14.1.3 JND models

There exists a school of thought in the field of image evaluation that the goal of an image processing or compression algorithm is to create an image that is perceptually equivalent to the original. This school measures image degradation in units of just-noticeable differences, or JNDs, between the original image and its processed counterpart. One JND unit corresponds to a fixed probability, say 50 or 75 percent, that an observer would detect the difference between two images or image regions (Lubin, 1993).

[1]There is anecdotal evidence that sometimes such image modifications can even aid the observer by changing the appearance of an image such that a previously-missed signal becomes visible.

The JND approach to image quality is rooted in the threshold theory of vision. Threshold theory states that signal detection occurs when a signal's perceptibility exceeds an observer's threshold; signal detection is a yes or no event. Furthermore, by the Weber-Fechner law, discussed in more detail in Sec. 14.2.1, the threshold for detecting an extended signal increases proportionally with background intensity. In the early days of vision science, much effort was expended on the measurement of the detection thresholds of various signals on different backgrounds. In the JND approach to image quality, the "signal" is a difference in a pair of images; if that difference is below threshold, the images are of equal quality.

All JND models are based on a model of the human visual system with the intent of predicting human performance in the ranking of image quality or the detection of image differences. The simplest approach is to weight image differences using a function that models the sensitivity of the human visual system to spatial frequency, referred to as a contrast sensitivity function (Daly, 1993). The JND model of Carlson and Cohen (1980) decomposes the input images into frequency bands. After the contents of the bands are processed nonlinearly, the outputs are compared to determine where image differences as seen through this simple model of the visual system are greatest. This model has been used to predict the detectability of edges and artifacts. Barten has also presented a model of the visual system that has been used to predict image quality (Barten, 1992, 1993). The Barten JND model utilizes a single integral over spatial frequencies rather than a decomposition into frequency bands, making use of an average contrast sensitivity function of the visual system. The Barten model has been shown to predict subjective image quality for several simple tasks and is the centerpiece of a recent National Electrical Manufacturers Association standard on display quality (NEMA, 2001).

More complex mechanistic models of the visual system have been developed for use in the prediction of visually perceptible differences in gray scale, color, and video imagery (Hultgren, 1990; Lubin, 1993; Daly, 1993). The models can account for such observation factors as viewing distance and light level (pupil diameter). The most comprehensive models include a nonlinearity representing the visual system's nonlinear response to luminance, a contrast sensitivity function, a bank of spatial-frequency and orientation-sensitive filters, and models of the chromatic and temporal properties of the visual system. The output is a JND map of the image differences, quantified per pixel, field, frame, or sequence.

One argument for the use of a JND metric is that the approach implies the matching of the processing algorithm with the visual system, similar to the way in which the information in a color television signal is matched to the human; because color resolution in the visual system is less than gray-scale resolution, the National Television Standards Commission (NTSC) represents color information more sparsely than luminance information.

Advocates of the JND approach argue that it is objective, it correlates with subjective assessments of image quality, and it predicts a large body of human data for both detection and discrimination tasks without the need to fit any free model parameters. The tasks have included disk detection, sine grating detection, checkerboard detection and edge-sharpness discrimination. The task can utilize real objects on real backgrounds; a recent comparison of image quality for the task of microcalcification detection in mammographic images showed a high correlation between JND measures of image quality and human observer performance (Krupinski *et al.*, 2003). Commercial JND-based image evaluation packages are readily available.

JND measures suffer from some of the same problems we have enumerated for fidelity measures, including the lack of distinction between blur and noise and the questionable definition of task. Both fidelity and JND measures quantify some form of image discrepancy: fidelity measures give all image differences equal importance, while JND measures weigh image differences according to their predicted manifestation at the output of the visual system. In order to calculate perceptual image differences, the JND approach requires twinned-noise image pairs, that is, two images in which the noise realization in each is the same. This paradigm is significantly different from the one underlying statistical decision theory, in which each image represents an independent sample from the signal, background, and noise distributions. It is not clear how the JND approach can be extended beyond simulated targets to real images with real signals because it is not possible to acquire real images that are identical except for the presence or absence of some target. An active area of current research is the usefulness of the JND approach for predicting the quality of an imaging system given random signals on random backgrounds in images with unpaired noise realizations.

Nevertheless, the JND community has much to offer the field of objective assessment of image quality. For example, we shall see that model observers play a significant role in the objective assessment of image quality; the sophisticated models of the visual system developed by the JND community may be of use in the development of predictive models of human task performance for more realistic tasks.

14.1.4 Information-theoretic assessment

In 1948, Claude Shannon published his now-famous theory of communication, in which he defined the information content of a message as a measure of the degree to which it is unexpected.[2] Shannon defined the information content of a single message state n as $I(n) = \log[1/\Pr(n)]$, where $\Pr(n)$ is the prior probability of occurrence of the nth message. Messages with high probability carry little information; high information content is associated with messages that are least expected. By this definition, the mean information content of a message is

$$\overline{I} = \sum_{n=1}^{N} \Pr(n)\, I(n) = \sum_{n=1}^{N} \Pr(n) \log\left[\frac{1}{\Pr(n)}\right] = -\sum_{n=1}^{N} \Pr(n) \log[\Pr(n)]\,, \qquad (14.1)$$

which becomes

$$\overline{I} = \sum_{n=1}^{N} \frac{1}{n} \log[n] = -\sum_{n=1}^{N} \frac{1}{n} \log\left[\frac{1}{n}\right] \qquad (14.2)$$

when the messages are equally likely.

Shannon's model for a communications system was a nonimaging system comprised of a single source (the message), an encoder, a communications channel that transmitted the message, and a decoder. The purpose of the communications system was to provide the user with a reproduction of the message. Designers of

[2]In his book on the relationship between information theory and thermodynamic entropy, Brillouin (1956) points out that the theory developed by Shannon came to light earlier in Szilard's discussion of the Maxwell demon (1929). (We thank B. R. Frieden for this historical note.)

encoders, decoders, and transmitters were seeking to ensure that the user received the message that was sent. Not surprisingly, systems whose goal was to reproduce a transmitted message were most often evaluated using fidelity measures.

There is a large literature on the application of information theory to the evaluation of imaging systems. Fellgett and Linfoot (1955) and Linfoot (1955) considered a simplified model of an optical system in which the source is divided into small discrete elements, each capable of a finite number of discrete brightness levels. The information content of the values of the elements can then be defined in terms of their degree of unexpectedness. That is, the information carried by a particular object $\mathbf{f}$ is given by

$$I(\mathbf{f}) = \sum_{n=1}^{N} \mathrm{pr}(f_n) \log \left[\frac{1}{\mathrm{pr}(f_n)} \right] = - \sum_{n=1}^{N} \mathrm{pr}(f_n) \log \left[\mathrm{pr}(f_n) \right] , \qquad (14.3)$$

where f_n is the brightness of the nth object element. In this simple model the object values are assumed to be independent, and we see that the entropy of the set of values becomes the measure of information content [*cf.* (15.158)].

Fellgett and Linfoot generalized this simple model to allow for a continuous distribution of object values and a division of object space into isoplanatic patches. With these additions to the model, Fourier methods can be used to describe the transfer characteristics of the imaging system. Felgett and Linfoot considered the assessment of an optical system for two tasks: the formation of an image that is similar to the object, and the production of an image that carries the most information about the object without regard to a specific inference or interpretation process. Assessment by similarity leads to fidelity measures; the same issues raised in the previous section on fidelity measures then apply, and Felgett and Linfoot point out many of these shortcomings as well. Thus Felgett and Linfoot turn to assessment by information content. Using the object's information measure as a starting point, an imaging system's ability to transfer information is computed and maximized. However, their resulting figure of merit is independent of the statistics of the object set and the measurement noise (film type, in those days). This is seen as a positive result by these authors, because it allows for optimization of optical systems without regard to the statistical properties of the object and the measurements, and no specific task must be considered.

More modern works have followed the approach of Fellgett and Linfoot, emphasizing the information rate of an imaging system (Huck *et al.*, 1997) and its correlation with the visual quality of the resulting images, where visual quality is measured in terms of image sharpness, clarity, and fidelity. Of course, all of these measures encounter the commensurability problem discussed in Sec. 13.3.2. Moreover, these measures are not uniquely related to the performance of a specified observer on a particular task.

Dainty and Shaw (1974) and Shaw (1978) related the information theory of Shannon to their noise-equivalent quanta (NEQ) approach to image assessment. According to these authors, an actual imaging system that degrades the information content of the input is associated with an NEQ relative to the real exposure quanta. As described in Sec. 13.2.13, this theory assumes a linear shift-invariant imaging system and stationary noise, leading to a Fourier-domain framework for describing the detection SNR as a function of spatial frequency. Spatial frequencies correspond to Shannon's channels in this approach (Wagner and Brown, 1985).

A broader view of information-theoretic image formation and assessment exists (O'Sullivan *et al.*, 1998). Object representation is achieved by combinations (not necessarily linear) of basis functions that may or may not be orthogonal; this approach does not automatically assume the object space is decomposed into pixels. The objects may be known exactly or random. The imaging system may be deterministic (low noise, nonrandom) or may be stochastic, and may be direct or indirect. This view of information-theoretic image formation is consistent with the framework shown in Fig. 7.14 for the imaging process. Moreover, in this treatment the task is more generally cast to include measures of optimality for detection, recognition (classification), parameter estimation, and scene estimation (image reconstruction). When the task is detection or classification, the overall performance of a system is measured by the performance of the recognition or detection function; performance measures for detection and recognition tasks include such familiar measures from Chap. 13 as the probability of detection and the probability of a false alarm. Optimal estimation for random objects is achieved using the familiar *maximum a posteriori* (MAP) procedure derived in Chap. 13 when a prior for the object exists; without a prior, maximum-likelihood methods result and are characterized by the Fisher information matrix and the Cramér-Rao bound.

Thus we see that the information-theoretic approach, when presented in this broad manner, is akin to the statistical-decision-theoretic approach presented in Chap. 13. In the information-theoretic approach, all performance metrics quantify the information provided by the measurements and the likelihood function plays a fundamental role in all cases. Similarly, we found in Chap. 13 that the likelihood ratio is central to all measures of task performance that characterize optimal decision/estimation strategies in statistical decision theory. The information-theoretic approach postulates that the user knows "everything except the decision" (O'Sullivan *et al.*, 1998). In other words, an ideal observer is assumed. Information measures are therefore useful for the assessment of raw data, but they are not necessarily good predictors of human performance. This point is particularly relevant to the use of information criteria in deriving optimal reconstruction algorithms. There is no guarantee that the resulting images are optimal when assessed in terms of human performance.

14.1.5 Objective assessment of image quality

For an image-assessment method to be acceptable, it must objectively quantify the usefulness of the images for performing a given task. Task-based measures of image quality have been advocated for many decades, starting with Harris (1964), and including Hanson (1977), Wagner (1978), Judy *et al.* (1981) and Myers *et al.* (1986). The resulting figure of merit must be computable and scalar, so that it can be used unambiguously in the optimization of imaging systems and the assessment of observer performance. Methods based on statistical decision theory satisfy these requirements.

Four key elements are essential in the objective assessment of image quality (Barrett, 1990):

1. Specification of a task;
2. Description of the object class(es) and imaging process, leading to a description of the data;

3. Delineation of the observer;

4. Figure of merit.

Let's consider each of these elements in more detail.

The task In Chap. 13 we considered two kinds of tasks in some detail. One kind of task is the detection of an object in the presence of a background or clutter. The object might have one or more random parameters and the background may or may not be random. A related task is the classification of an image into one of a finite number of alternative classes. A second type of task is the estimation of parameters describing the object or background or both. Chap. 13 gives many examples of detection, classification, and estimation tasks.

We have seen that many of the approaches described in earlier sections define the task as the reproduction of a single object. While object reproduction might be construed as an estimation task, there are several important differences between estimation and object reproduction. First, defining the task as object reproduction leads to the problem of commensurability delineated in Sec. 13.3.2: objects and images live in different spaces. No imaging system can exactly reproduce a continuous object. How then, to choose among systems that all fall short of this impossible goal? In addition, when the stated task is object reproduction, an assumption is being made that all object locations/elements/parameters are equally important; this is not the case in real situations. Finally, no imaging system will be utilized for a single object, so the task definition should encompass the use of the system over the expected range of objects.

Properties of objects and images From the preceding discussion we know that the evaluation of an imaging system should take into account the physical and statistical properties of the set of objects to be imaged. In a classification task, the objects are categorized into a finite set of classes. For example, the evaluation of mammographic imaging systems for the task of breast lesion detection requires the characterization of normal breast tissues and breast lesions in terms of the full probability density function of the objects under each class. While this is an impossible task, tremendous progress is being made toward the characterization of the mean and low-order joint densities of real tissues using ultra-high-resolution projection imaging and autoradiography, among other methods (Hoeschen *et al.*, 2000).

Another method for creating and characterizing a set of objects is through the use of simulations. The use of numerical algorithms to generate random objects gives the investigator the ability to characterize the deterministic and stochastic properties of the objects. Modern simulations are becoming increasingly realistic. Investigators have added simulated targets to real images (creating so-called hybrid images) with sufficient realism that in some cases human observers were unable to discriminate the artificial targets from real ones (Revesz *et al.*, 1974; Eckstein and Whiting, 1996). The future will bring even greater flexibility and realism to simulated images, with the entire anatomy and physiology of a human being modeled on a fine scale as a starting point toward the creation of simulated, highly realistic imagery of normal and abnormal states. Nonmedical imaging applications are following the same trajectory; in astronomy, acoustical imaging, radar, and so on, simulations of objects and imaging systems are vastly improving and leading to new abilities to generate realistic data sets for image evaluation. Image simulation methods are described in some detail in Sec. 14.4.

The observer Given a task and a set of objects, the next requirement for the assessment of image quality is an observer or strategy for performing the task. The observer might be a human, such as a radiologist or an expert photointerpreter. Models of human observers can be used to predict human performance. Model observers make it possible to optimize imaging systems without the need for lengthy human-observer studies at every design stage. Human observers and their models are relevant to the assessment of images to be displayed for human consumption. For example, the assessment of display devices, reconstruction algorithms, and all manner of image-processing routines are evaluated appropriately using human observers or their surrogates.

The ideal observer is defined in Chap. 13 as the observer that makes optimal use of all available information to perform the specified task. Having no need for image reconstruction, the ideal observer is appropriate for the evaluation of the quality of the raw data for classification tasks.[3] Thus the ideal observer is the observer of choice for the assessment of imaging hardware. As detailed in Chap. 13, the ideal observer requires the complete PDF of the data under each hypothesis. In cases where this information is not available, the Hotelling observer can be a useful alternative, requiring only the first- and second-order statistics of the data.

The figure of merit Having specified the task, the objects, and the observer, all that is needed is some way of telling how well the observer performs. For classification tasks, useful figures of merit include the area under the receiver operating characteristic (ROC) curve (AUC), partial ROC areas, sensitivity/specificity pairs, the percent of correct decisions (PC), and the classification signal-to-noise ratio, or SNR. Those readers unfamiliar with the theory of ROC curves are referred to Chap. 13 for background material necessary for understanding the terminology here.

Possible figures of merit for estimation tasks include bias, variance, mean-square error (MSE), and ensemble mean-square error (EMSE). The MSE summarizes the performance of an estimation algorithm in determining the estimable parameters of a single object averaged over multiple data sets. In contrast, EMSE describes estimation performance averaged over both measurement noise and a distribution of objects, allowing for nonestimable parameters. Estimators can also be evaluated using bounds on their performance, the most notable being the Cramér-Rao bound for maximum-likelihood estimators. In Sec. 13.3 the reader can find a lengthier treatment of performance measures for estimation tasks.

Returning to the set of requirements listed at the beginning of this section, we can see that each of the methods described in the previous sections lacks one or more of these key elements. For example, JND methods measure image quality using a distance between two scenes without specification of a task or an object class. Thus, in the remainder of this chapter we shall rely on the approach to the objective assessment of image quality outlined in this section.

[3] Wagner, Brown, and Pastel suggested the division of imaging systems into detection and display components for assessment purposes as early as 1979.

14.2 HUMAN OBSERVERS AND CLASSIFICATION TASKS

A wide variety of imaging applications make use of a human as the observer or expert reader. The task is almost always classification, because humans are not as adept as machine algorithms at the absolute quantitation of parameters using images as input. The purpose of this section is to chronicle what is known regarding the perception of form by the human visual system, how we measure human performance on classification tasks, and what we have learned regarding human performance for various classification tasks. We shall focus on the perception of pattern and form, with the goal of connecting this to an understanding of human performance on single, static images. The extension to tasks involving temporal information, color, or stereo are beyond our scope, although in many cases the generalizations required to include this kind of information will be suggested.

14.2.1 Methods for investigating the visual system

Centuries ago, the human eye was assumed to work as a simple camera. This view was espoused by the famous astronomer Johannes Kepler as early as 1604. Not long after, Réné Descartes' famous treatise, *La Dioptrique* (1637), described an experiment in which an eye from an ox was used to "view" the image formed on the retina, which had been scraped away to make the eye translucent. The discovery that the image formed by the eye's lens was inverted was a source of much confusion, since none of us has the experience of seeing the world upside down. Since that time we have come to realize that we do not directly "see" the retinal image; what we perceive is a processed and interpreted version of the image formed at the back of the eye. The retina and the visual components of the eye-brain system are complex entities that have been the subject of amazing discovery since the time of Kepler.

The images formed by the eye's lens onto the retina stimulate the approximately 130 million photoreceptors we know as the rods and cones. These units stimulate bipolar cells that lead to the ganglion cells, whose axons form the optic nerve. The axons of the optic nerve terminate in the lateral geniculate nucleus (LGN) of the thalamus. The cells of the LGN relay signals to a region of the striate cortex called the primary visual cortex. The activity of a cortical cell is thus the result of millions of retinal inputs. Within the visual cortex further signal processing and feature extraction occurs, leading to our visual perception of the world around us.

Early discoveries of the visual system were anatomical, as described so graphically by Descartes. Anatomical studies tell us the spatial sampling of the rods and cones, the number of fibers making up the optic nerve, and the location of their termini. We need other means of determining how these entities function and interrelate.

One means of elucidating the functional properties of the elements of the visual system is through electrophysiological studies in animals. These studies involve the placement of electrodes into single cells in the visual pathway and the subsequent measurement of the cell's response to visual stimuli. In 1940, Hartline became the first to insert electrodes into a single ganglion cell in a vertebrate (a frog) and record axon potentials, following his earlier experiments in the horseshoe crab (1934). Hartline's work was the precursor to the acclaimed work of Hubel and

Wiesel (1962), who shared the Nobel Prize for their pioneering study of the visual system of the cat. Hubel and Wiesel studied the response of single cortical cells to visual patterns of specific orientation and location (bars, edges, and spots) and found that the cells demonstrate orientation selectivity and binocularity. They soon reported similar findings in monkeys (1968).

Many electrophysiological investigations in animal models have followed in the giant footsteps of Hartline, Hubel and Wiesel. For such studies to be relevant to the human visual system, the animal's characteristics must be able to be extrapolated to the human. Since the visual systems of all vertebrates are similar, these measurements provide especially valuable information regarding the behavior of the human visual system.

The functioning of the visual system can also be studied using *psychophysics*, the measurement of the reactions of observers to visual scenes and the development of quantitative relationships between response data and physical characteristics of the input images. The physical characteristics of the images include quantities such as the display luminance, the noise and resolution properties of the images, as well as parameters that specify the target and background. Observer performance is measured in terms of indices such as the area under the ROC curve or the percentage of correct detection or localization responses. Thus psychophysical experiments determine external measures of the visual-system function. Methods for the conduction of psychophysical studies using human observers are presented in Sec. 14.2.3.[4]

Modern imaging methods have brought new tools to the study of the function of the visual system. Using functional imaging methods such as functional magnetic resonance imaging (fMRI) and positron emission tomography (PET), investigators are determining areas of the brain involved in the performance of visual tasks. Imaging provides a noninvasive alternative to electrophysiological techniques with the ability to map both spatial and temporal response to stimuli.

In what follows we shall describe the more salient features of the visual system that are relevant to understanding human performance on classification tasks using images as inputs. These characteristics play a key role in the development of predictive models of the human observer.

Receptive fields A *receptive field* is an area on the retina that gives excitation or inhibition of a neuron's activity upon changes in illumination. Receptive fields can be defined for ganglion, geniculate, and cortical cells. The receptive field is evidence of a many-to-one relationship between photoreceptors in a region of the retina and the neural cell. In fact, there are about 1000 cortical neurons per retinal cone for visual information processing (Kronauer and Zeevi, 1985).

Receptive fields for the ganglia can be organized into two broad classes: those that have plain receptive fields, and those that have complex receptive fields. Plain receptive fields have a center-surround structure. When a spot of light illuminates their center, an increase in firing rate occurs (excitation); light on the surround region decreases the rate (inhibition). Diffuse light that illuminates both regions gives a cancellation of the signal, resulting in no response. Simple cells are often

[4]While it might be expected that psychophysics is exclusively applied to the study of human observers, psychophysical experiments using trained animals have been performed to elucidate properties of the cat and monkey visual system.

referred to as Kuffler cells, after the early investigator who mapped their behavior in the cat (Kuffler, 1953).

Complex cells, not surprisingly, are tuned to more complex retinal patterns, such as gratings. Enroth-Cugell and Robson (1966) were the first to measure the response of single ganglion cells to cosine gratings in the cat. Both even and odd receptive fields exist, giving the cat visual sensitivity to gratings and edges. At one time it was thought that complex cells were in series with simple cells, but we now know that simple and complex cells act in parallel. For some tasks the response of complex cells occurs earlier than that of simple cells, and for other tasks the opposite is the case (Hoffman and Stone, 1971).

The LGN and cortical neurons are also associated with receptive fields at the retina. Cortical cells have been found that are tuned to edges, lines, movement of lines and gratings at certain orientations, speeds and accelerations, and even angles between lines. While the responses of ganglion, LGN, and cortical neurons to stimuli have significant similarities, there are interesting differences across them as well. Maffei and Fiorentini (1973) compared the responses of these neurons to cosine gratings in the cat and found that the stages respond to different ranges of spatial frequencies. Moreover, the spatial frequency selectivity becomes narrower from the retina to the LGN to the cortex. DeValois *et al.* (1982) also found in the macaque that the sharpest tuning occurs in the cortex.

Lateral inhibition The output of a receptor's ganglion cell is not only impacted by multiple retinal inputs, but also by the behavior of nearby neurons. In studies of the horseshoe crab, Hartline and Ratliff (1957) were the first to show that the output of a ganglion cell can be inhibited when a nearby neuron is excited. This effect is referred to as *lateral inhibition.* The opposite can also occur, in which case the effect is termed *lateral summation.* Lateral inhibition and summation demonstrate one role of the synapse of the bipolar cells that communicate with the ganglion cells, enabling ganglion cells to interact.

Contrast sensitivity function By measuring the electrophysiological response of animals to patterns, the structure and function of multiple receptive fields have been elucidated. In humans, the ability to detect patterns is measured via psychophysics, giving a global response to the pattern rather than a response localized to a single neuron. The *contrast sensitivity function* (CSF) describes the overall sensitivity of the visual system to sinusoidal patterns as a function of pattern frequency. High sensitivity signifies that the pattern can be seen with little contrast; low sensitivity implies that a large contrast is required for the observer to detect the pattern. A great many psychophysical studies have been conducted to determine the sensitivity of human observers to grating patterns, starting with DePalma and Lowry in 1962. Robson (1966) measured both spatial and temporal CSFs in humans in the 1960s. Campbell and Robson (1968) measured the contrast sensitivity to single sinusoidal gratings over a broad range of spatial frequencies at fixed background luminances. By determining the just-visible contrast of sine-wave targets, these authors found that the CSF follows a band-pass shape with a pronounced maximum at 2 to 4 cycles per degree, falling off at both low and high spatial frequency. An idealized CSF is shown in Fig. 14.1.

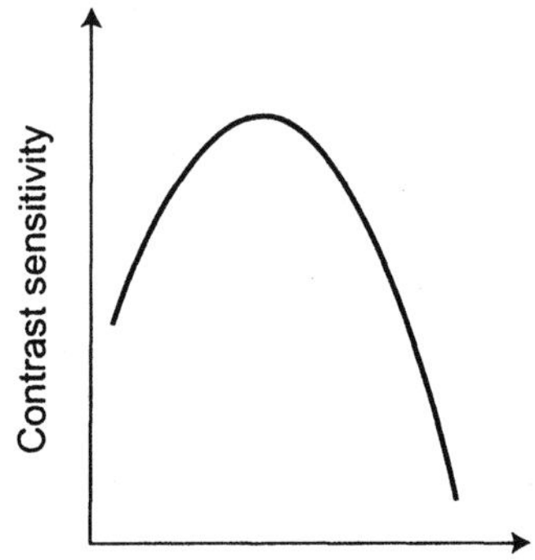

Fig. 14.1 Idealized example of a contrast sensitivity function.

We now know that the visual system is extraordinarily adaptive; the CSF is dependent on mean luminance level, noise level, color, accommodation, eccentricity, and image size (Kelly, 1977). While color CSFs have a shape similar to that shown in Fig. 14.1, high-frequency color patterns are less detectable than luminance patterns of the same frequency (Cornsweet, 1970). There is also significant variation in contrast sensitivity functions, as well as other parameters of the visual system, across human observers (Ginsburg *et al.*, 1982; Owsley *et al.*, 1983). Ginsburg and Evans (1984) measured the CSFs of a large population of observers and found that the peak value is dependent on the individual.

The CSF is often depicted as the envelope of multiple narrow spatial-frequency-selective responses internal to the visual system. This view stems from evidence that the bandwidth determined via psychophysical study in animals can be much greater than that determined via electrophysiological experimentation. For example, individual neurons in the cat have been found to respond to a narrower range of frequencies (Movshon *et al.*, 1978) than what is found externally via behavioral studies (Blake *et al.*, 1974).

Lateral inhibition reduces sensitivity to signals with large extent. That the CSF is very low at low frequency is consistent with the inhibitory behavior of receptive fields at low frequencies. Several studies have established that the human observer is unable to efficiently integrate information beyond a certain spatial extent (Blackwell, 1946; Burgess *et al.*, 1979; Boff *et al.*, 1986).

Masking Research has shown that the presence of one pattern can make another pattern less visible to an observer. This property is known as *masking*. The opposite of masking is *facilitation*, defined as the improved detection of a pattern in the presence of another. The pattern to be detected is referred to as the *signal*; the additional pattern is referred to as the *mask*. The mask is usually supra-threshold, meaning its contrast is above that required for detection. When the mask contrast becomes sufficiently low, the signal threshold is identical to the the signal threshold in the presence of a uniform background; that is, the signal threshold is what is expected based on the observer's CSF and no masking occurs.

Periodic patterns such as gratings or sinusoids have been shown to mask patterns with similar orientation or spatial frequency (Legge and Foley, 1980; Phillips and Wilson, 1984). This effect is known as *phase-coherent* masking. Another experimental paradigm is to use noise fields of different bandwidths as masks; the effect is then called *phase-incoherent* masking (Pollehn and Roehrig, 1970; Pelli, 1981; Thomas, 1985). The presence of an aperiodic pattern such as an edge or a gradient can also mask a nearby feature (Fiorentini *et al.*, 1955). Masking demonstrates orientational selectivity as well as frequency-dependent behavior (Campbell and Kulikowski, 1966).

Diffuse light can mask signals. For this reason radiologists are trained to read images in a darkened viewing area after they have adapted to the ambient light level, to better detect low-contrast signals. Scattering in the lens and cornea of the eye can also mask low-contrast signals. This problem is known to worsen with age.

Channels *Channels* are independent processors tuned to different narrow ranges of spatial or temporal frequency. Channels were first hypothesized as visual scientists pondered data from studies using compound-frequency patterns such as sawtooth and rectangular gratings (Campbell and Robson, 1968). These data seemed to indicate that detection of the pattern occurs only when the most detectable component reaches its own threshold, independent of the presence of the other frequency components. Sachs *et al.* (1971) then carried out experiments using compound gratings consisting of just two frequency components. Whenever the second component differed in frequency from the first by more than a certain ratio, the data were consistent with the hypothesis that the two frequency components were being detected independently. Moreover, when two components with frequencies related by an even larger ratio were combined, the grating was no more detectable when the two components were phased so that their peaks added than when their peaks subtracted (Graham and Nachmias, 1971). The investigators concluded that different spatial-frequency components were detected by independent processors tuned to different narrow ranges of spatial frequencies. Detection of a stimulus occurs whenever the activity in one of these processors rises above a threshold. These processors were referred to as channels. Channels can be thought of as mosaics of receptive fields (Sachs *et al.*, 1971).

Many scientists have worked to corroborate the presence of frequency-selective channels in the visual system (Mostafavi and Sakrison, 1976) and to determine their properties in finer detail (Halter, 1976). The electrophysiological recordings of Hubel and Wiesel (1962) are construed by many as the first evidence for channels. Adaptation and masking experiments support the hypothesis that the channels are medium-bandwidth mechanisms (Blakemore and Campbell, 1969; Stromeyer and Julesz, 1972; Stromeyer and Klein, 1975; Legge and Foley, 1980). Narrow-bandwidth channels are suggested by the results of frequency-discrimination tasks (Campbell *et al.*, 1970). The entirety of the data suggests the presence of approximately octave bandwidth spatial-frequency channels over the entire visible range.

There is ample evidence, starting with the work of Hubel and Wiesel (1962), that the visual system also contains orientation-selective channels. DeValois *et al.* (1982) investigated simple cells in the macaque and found them to have an angular resolution of $\pm 20°$. These data are quite similar to the estimates of orientation selectivity in humans obtained using masking experiments (Campbell and Kulikowski, 1966; Phillips and Wilson, 1984). There are also channels tuned to object motion that have direction selectivity (Tolhurst, 1973), with a temporal two-octave bandwidth (Tolhurst, 1975; Watson and Robson, 1981).

Internal noise Human observers are noisy measurement devices. Thus, even if the images presented to a human observer were noise-free, the output of the human would have some variability. While it requires only one optical photon to excite a rod, the number necessary for "seeing" is larger (Hecht *et al.*, 1942). Barlow was the first to suggest that this discrepancy is the result of an internal noise mechanism (1956).

Burgess *et al.* (1981) compared human SKE (signal-known-exactly) detection performance in white noise to an ideal detector with an added internal noise contribution. While this modification to the ideal-observer model improved the model's agreement with the human data, it was suggested that some form of observer sampling inefficiency was also needed for the model to match the slope of the human data vs. noise spectral density. The authors further suggested that perhaps the observer noise might be a function of image noise. Data from subsequent classification experiments have borne out the suggestion that the visual system has two internal noise components (Burgess and Colborne, 1988). The first component is an additive noise term that is independent of the image luminance. This noise component may be the result of neural noise (Tolhurst *et al.*, 1983), as well as fluctuations in the observer's decision criterion (Eckstein *et al.*, 1997). The second component is an induced, or image-dependent, component. The induced internal noise has been shown to be proportional to the variance of the image noise (Burgess and Colborne, 1988).

Weber-Fechner law As stated earlier, diffuse light can mask low-contrast signals. As a result, objects on bright backgrounds are harder to detect than objects on dark ones (Cornsweet, 1970). The Weber-Fechner law states that the relative contrast of an object, given by $(L_{max} - L_{min})/L_{mean} = \Delta L/L$, is equal to a constant for a given probability of detection. By this law, the detection of a difference in luminance depends on the baseline, so that relative luminance is important, rather than absolute differences.

Evidence of behavior following the Weber-Fechner Law has been interpreted as a local gain mechanism or a saturating nonlinearity in the visual system, coupled with internal noise (Shapley and Enroth-Cugell, 1985). This law also plays a significant role in the approach used by many investigators in choosing the calibration method for their soft-copy display (Blume and Hemminger, 1997). Many investigators choose to use a *perceptually linearized* display, in which the output luminance at each digital driving level is set so that the step sizes between gray levels is higher at higher absolute luminance levels (Pizer, 1981).

Psychometric functions A *psychometric function* is a plot of the probability of a signal being detected as a function of signal contrast. For a signal of contrast c, the probability of detection is usually fit by a sigmoidal function of the form (Nachmias, 1981)

$$\Pr(D_2|c) = 1 - \exp[-(c/\alpha)^{\beta}], \tag{14.4}$$

where D_2 indicates that the observer chose in favor of the signal being present, β is a slope parameter, and α shifts the function relative to the signal contrast. Many experiments have been found to indicate approximately equal slope parameters (Mayer and Tyler, 1986).

14.2.2 Modified ideal-observer models

Given the vast array of anatomical, electrophysiological and psychophysical data now available to us, many researchers have worked to develop models for all or portions of the visual system. Some models are highly specialized, with the minimum number of components required to demonstrate the model's ability to predict data obtained in a narrow range of psychophysical experiments. Other models are extraordinarily complex, incorporating foveal sampling, a hierarchy of neural stages, and higher-level signal processing and decision making in an effort to replicate the entire visual system. We shall focus on models that have been developed for the specific purpose of objective evaluation of imaging systems for classification tasks.

In Chap. 13, the ideal observer was introduced as the optimal decision maker for classification tasks as determined by statistical decision theory. The ideal observer sets the upper bar for classification performance. Statistical-decision-theoretic models of the human observer thus use the ideal-observer model as a starting point. We do not need a model with millions of photoreceptors and receptive fields, so long as the model predicts human data on a range of tasks that are useful for image assessment. In fact, a simpler model facilitates imaging system evaluation and optimization over high-dimensional optimization spaces.

The modified-ideal-observer approach to modeling human performance is this: begin with the concept of the ideal observer; compare performance predictions with human performance on actual classification tasks; modify the model to better predict human performance. Modifications to the model should be grounded in the known features of the visual system described in the previous section.

We therefore require a rigorous basis for comparing observer performance. For this, we return to the concept of observer efficiency.

Observer efficiency In Chap. 12 we introduced the concept of detective quantum efficiency as a measure of the SNR transfer characteristics of a detector [*cf.* (12.23)]. In Chap. 13 we extended this concept to describe the efficiency of the Hotelling observer relative to the ideal observer [*cf.* (13.273)]. Analogously, we can define the statistical efficiency of the human observer relative to the ideal observer as

$$\eta_{human} = \frac{\mathrm{SNR}^2_{human}}{\mathrm{SNR}^2_{ideal}} . \tag{14.5}$$

The relative efficiency of any two observers can be similarly defined.[5]

When human and ideal performance are comparable, the efficiency approaches one and we conclude that the human observer is able to make almost complete use of the information in the data to perform the visual task. For efficiencies much less than one, we can conclude that the human observer is inefficient at extracting the relevant information in the image for performing the task. When this occurs, we look for features of the human visual system that might be the basis for the human observer's reduced performance.

[5]Some authors have defined observer efficiency as the ratio of SNRs required by the observers to perform the task. In this school, human efficiency equals the SNR required by the ideal observer divided by the SNR required by the human, where SNR is a physical quantity such as contrast; smaller SNRs denote better performance. We prefer the definition given in (14.5), where SNR quantifies task performance and high SNR is good!

Classification in uncorrelated noise As described in Sec. 13.2.13, the definition of observer efficiency given in (14.5) comes from the basic definition of DQE first given by Albert Rose (1948) as a means of comparing the noise level of an actual radiation detector with that of an ideal one. Rose compared the performance of the eye to an ideal picture pickup device and determined that the minimum contrast c_{min} required for detecting a uniform object on a flat background with quantum noise satisfies

$$c_{min}^2 NA = k\,, \tag{14.6}$$

where N is the photon density of the uniform background, A is the area of the object and k is a constant dependent on the observer; from experiments on human subjects, Rose determined that k is in the range of 3 to 7. A lower value of k implies a lower c_{min} and hence a more efficient observer.

Recall from Sec. 13.2.8 that the ideal observer takes on a special form when the task is the discrimination of two nonrandom signals in additive Gaussian noise. In this case the ideal observer is equivalent to a prewhitening matched filter (PWMF), which reduces to a simple matched filter when the noise is white. Lawson (1971) demonstrated that the Rose model of (14.6) is a special case of the PWMF for a pillbox signal in Poisson noise of sufficient count rate that the Poisson statistics can be approximated by Gaussian statistics.

The calculation of the ideal observer's SNR is straightforward for SKE/BKE (signal-known-exactly/background-known-exactly) tasks in Gaussian noise and can be done analytically. For this reason the first comparisons of human performance to ideal-observer performance were achieved in SKE/BKE tasks in white, or uncorrelated, Gaussian noise. Burgess *et al.* (1981) found human observers to be highly efficient (η of 0.5 to 0.8) for SKE/BKE detection and discrimination tasks in white noise. Human performance is well predicted by an ideal observer that positions a template over the location of the expected signal and performs a linear summation of the output. The fact that the efficiency is less than one can be explained by internal noise (Burgess and Colborne, 1988).

When the signal extent becomes sufficiently large, human detection efficiency in white noise declines (Burgess *et al.*, 1979). In effect, there is a spatial limit to the human's ability to perform the template-matching operation. We might have expected this from the shape of the CSF of the visual system. Other investigators have found that the human is unable to efficiently process "DC" information (Ratliff, 1965; Van Nes and Bouman, 1967). For this reason some investigators proposed that the PWMF model be modified by adding an "eye filter" (Loo *et al.*, 1984; Burgess, 1994).

Correlated noise Many experiments have been performed to investigate the impact of correlated noise on human discrimination performance (Judy, 1981; Guignard, 1982; Burgess, 1985b; Myers *et al.*, 1985; Blackwell, 1998). Of particular interest in the early 1980s was the character of the noise in computed tomography (CT) images and its impact on human perception. Raw CT data sets have Poisson noise, which is uncorrelated. When CT images are reconstructed from the raw data using the method of filtered backprojection (see Sec. 4.4.3), a filter with a ramp shape in the frequency domain is used, and the resulting images have a ramp-shaped power spectrum at low spatial frequency. Early on, Wagner (1978) hypothesized that human observers would be inefficient when faced with this noise-correlation structure, and suggested that a non-prewhitening matched filter model might be a good predictor of human performance. Soon after, several studies found that human efficiency relative to the ideal observer is about 20% in CT noise, much less than the efficiencies found in white noise (Judy *et al.*, 1981; Burgess *et al.*, 1985b). Myers *et al.* (1985) investigated human performance for a family of noise power spectra of the form ρ^n, for $n = 1, 2, 3, 4$, where ρ is spatial frequency. Thus $n = 1$ corresponds to the CT case. These studies showed that human efficiency falls rapidly as n increases from 1 to 4.

A natural conclusion to draw from the reduced efficiency of the human observer in tasks limited by correlated noise is that the human observer is indeed unable to perform the prewhitening operation. For this reason the human observer was modeled by some investigators as a matched filter without the prewhitening operation. The efficiency of the human relative to this so-called non-prewhitening matched filter (NPWMF) was shown to be around 50% (Judy and Swensson, 1985), with the difference again explainable by internal noise.

Since the NPWMF equals the PWMF in white noise, the NPWMF model predicts human performance in both correlated and uncorrelated noise. Furthermore, by combining an eye filter and an internal noise mechanism with the NPWMF, an even larger body of human psychophysical data can be explained (Ishida *et al.*, 1984; Loo *et al.*, 1985; Ohara *et al.*, 1986; Giger and Doi, 1987, deBelder *et al.*, 1971; and Wolf, 1980). This observer is often called the NPWE in the literature, to denote the addition of an eye filter to the non-prewhitening matched filter; we shall use this same shorthand below.

The NPWE models the spatial-frequency response of the visual system with a single spatial-frequency filter. Given the experimental evidence that the human visual system has multiple narrow spatial-frequency channels, a preferred approach to modifying the ideal observer is to incorporate this recognized characteristic of the visual system.

Adding channels to the ideal observer The model of the human visual system as a matched filter is effectively a model with an infinite number of channels. Yet there is substantial evidence that the visual system processes images through a finite number of finite-width channels. Myers and Barrett (1987) introduced a handicapped ideal observer, constrained to process scenes through frequency-selective channels, and demonstrated that this modified Bayesian observer ably predicted human performance in correlated noise. They found that this model was robust to the choice of a channel width parameter. By requiring the lowest-frequency channel to have a finite turn-on frequency, this model also predicts the inefficient performance of human observers on tasks that have significant DC content.

Myers and Barrett found the performance predictions of the channelized ideal observer and the NPWMF to be indistinguishable for the problems they studied (stationary Gaussian noise, signal known exactly). They argued in favor of the channelized ideal observer because this model is consistent with a known mechanism of the visual system. Moreover, as we shall see in the following sections, this model has been found to be predictive of human performance over a much broader range of signal detection and discrimination tasks.

Random backgrounds The tasks described in the previous section were ones in which the background was known exactly; the only variation in the data was due to measurement noise. We now consider tasks in which the data are random due to both background variability as well as measurement noise. The noise in the data is therefore said to have two components.

In Sec. 8.4 we described several approaches for generating random backgrounds, and in Chap. 13 we discussed model observers for tasks in which the background is random and known only in a statistical sense. Several investigators have made use of these methods to study the performance of human observers in random backgrounds and compare the results to model-observer predictions. Rolland and Barrett (1992) generated lumpy backgrounds by randomly superimposing Gaussian blobs on a uniform background according to the procedure described in Sec. 8.4.4. For the task of detecting Gaussian signals of known size and location on the lumpy backgrounds, Rolland and Barrett compared human performance to the performance of the Hotelling or optimal linear observer defined in Sec. 13.2.12 as well as the NPWMF. Rolland and Barrett found that the Hotelling observer was a good predictor of the human performance data. The performance of the NPWMF was not able to predict human performance over the range of system parameters investigated in the study.

Yao and Barrett (1992) combined the background model of Rolland and Barrett with power-law noise of the type investigated by Myers *et al.* (1987) and found that a channelized Hotelling observer was a good predictor of all the human data acquired in these experiments (Barrett *et al.*, 1993). Burgess *et al.* (1994, 1997, 1999) studied human performance in random lumpy backgrounds generated by filtering a Gaussian field. Their results were consistent with the findings of Rolland and Yao: a Hotelling observer constrained to process the frequency-selective channels is able to predict the data over the range of experimental parameters describing the signals and backgrounds. A NPWMF is not predictive, even when modified to include an eye filter. More recent experiments in power-law backgrounds generated by filtering a Gaussian random process were less conclusive; the most predictive model depended on the signal profile in a study by Burgess (2001).

Several studies have been performed to compare human performance to model observers using real images as backgrounds. In a study using backgrounds drawn from real x-ray coronary angiograms, Eckstein *et al.* (1999) found the channelized Hotelling model to be predictive of human performance in detecting simulated abnormalities. Bochud *et al.* (1995, 1999a, 1999b) studied human performance using simulated nodules in mammographic and angiographic backgrounds and compared their results to a non-prewhitening observer with and without an eye filter (a single channel). They found that, owing to the nonstationarity of the images, the models must be allowed to adapt to the statistics of the local background around the signal in order to better predict human performance. Interestingly, the data of Bochud *et al.* (1999b) suggest that the clinical backgrounds have higher-order statistical properties used by the human observer, although not by the Hotelling observer. Similarly, Caelli and Moraglia (1986) showed that a cross-correlator does not predict human performance when the background is a natural scene.

Signals of large spatial extent The inability of human observers to efficiently detect signals of large spatial extent described in Sec. 14.2.1 has direct ramifications on the task-based assessment of the quality of images derived from systems with significant artifact content. For example, the effective point response function (PRF) for images reconstructed from limited-angle tomographic data can be quite noncompact, yielding long-range streak artifacts. The images of compact objects are thus quite extended and human efficiency for detecting such objects suffers a penalty (Wagner *et al.*, 1992; Myers *et al.*, 1993). These studies found that human performance is modeled quite well by an observer that performs only linear operations on the images. These studies involved signals at random locations, leading to location-dependent artifacts; the ideal observer is nonlinear in this case if the problem is cast as a binary signal-detection problem.

A long-tailed PRF can also arise when veiling glare is present in a display device or gamma rays penetrate the collimator in gamma-ray imaging. Rolland *et al.* (1989) has shown that human classification performance is inefficient for images formed by a system with a long-tailed PRF, consistent with the earlier literature on the inefficient spatial integration properties of the human. Rolland found that human performance is improved by linear filtering designed to narrow the overall system PRF, even though the ideal observer performance is unchanged by image processing (Sec. 13.2.6), as long as it is invertible.

Texture perception In some special circumstances the human can detect signals of large spatial extent quite efficiently. An example is the detection of a known grid of bright lattice points on a noisy background (Wagner *et al.*, 1990a). Another example is the detection of mirror symmetry patterns of dots (Barlow, 1978; Barlow and Reeves, 1979) buried in a background of random dots (Glass patterns). These results can be explained by an observer who uses the strategy of performing a series of local template-matching operations, skirting the need for integration over a large area (Wagner *et al.*, 1989).

A particular form of extended signal is a pattern of a different texture than the texture of the background in which the signal is embedded. In tasks where such an extended signal is to be detected, human efficiency can be extremely low. For example, the detection of a regular grid or lattice of objects, where some randomization of the object locations is involved, results in low human efficiency (Wagner *et al.*, 1990a). Similarly, the detection of random dot patterns (Maloney *et al.*, 1987; Tapiovaara, 1990) and the detection of diffuse liver disease (Garra *et al.*, 1989) can also be low-efficiency tasks. While many investigators have considered human performance in texture discrimination tasks (Julesz, 1981), these studies are rarely placed in the context of ideal-observer performance. Much more work is needed to understand human performance in textured tasks on an absolute scale.

Nonlinear tasks While the channelized Hotelling observer has been found to predict human performance over a wide range of experimental paradigms, that observer is constrained to perform linear operations on the data. In addition, as the previous section describes, there are ample examples of psychophysical studies showing low human efficiency relative to the ideal observer for nonlinear tasks. The question then arises, can the human do nonlinear operations?

There are many examples of tasks for which human efficiency is fairly high even though the optimal strategy is nonlinear. One example is the task of noise variance discrimination, wherein observers are asked to determine which of two scenes has higher pixel variance. The optimal discrimination strategy is quadratic in the data as seen in (13.163). In unpublished studies, we found that humans were able to perform this task quite efficiently. Does this mean the humans are able to do the computations of (13.163)? Maybe not. It can be shown (Wagner *et al.*, 1990b) that a combination of linear and logic operations can approximate this ideal nonlinear strategy quite efficiently.

Similarly, the detection of 1 of M orthogonal signals in white noise is optimally performed with a nonlinear strategy [see (13.159)]. However, Nolte and Jaarsma (1967) showed that a series of linear operations, followed by the nonlinear operation of selecting the filter with the maximum output, approximates the ideal nonlinear strategy well over much of the signal parameter space of interest (the range of contrasts of use for psychophysical study). Other investigators have also shown that the "maximum-of" detector gives performance predictions very close to those of the optimal observer in the SNR ranges of experimental interest (Pelli, 1985; Wagner, 1990b).

Burgess and colleagues (Burgess and Ghandeharian, 1984a, 1984b; Burgess, 1985a) measured human efficiency in studies with signal uncertainty in white noise. To approximate ideal-observer performance, they computed the performance of an observer that compared the maximum of a series of matched-filter outputs to a threshold, following the theory of Nolte and Jaarsma (1967). Human observer performance was well predicted by this model observer, with an efficiency around 50%. Judy *et al.* (1997) found little degradation in human performance for the detection of sharp-edged disks and Gaussian signals when the disk diameter or Gaussian width was variable, relative to the SKE task.

Since selecting the maximum of a set of outputs from linear filters is a nonlinear or logical operation, we call this model a *linear+logic observer*. The closeness of the optimal observer to the linear+logic model may preclude one model being rejected in favor of the other using psychophysical data.

The field of neural networks sheds some light on the similarity of these models. A fully connected neural network can be shown to approximate ideal-observer performance. The neural network applies a series of filters in the form of weights to each input value, followed by a sigmoidal nonlinearity—a neural network is a linear+logic observer. As described earlier, there is good evidence that neural responses in the human visual system can be represented by a set of linear filtering operations followed by thresholding, and neural networks represent this scheme. Thus, when the human observer performs nonlinear tasks efficiently, we must be cautious before concluding the human can perform the optimal higher-order (in the data) nonlinear operations. It may well be that the human is smartly performing a series of linear operations, followed by a threshold nonlinearity, to obtain near-optimal performance.

Optimal processing of channelized data As demonstrated in Fig. 14.2, the addition of a channel mechanism to a model observer can be visualized by adding a block to the processing steps shown in Fig. 13.1. In the figure we suppose there are P channels, each represented by the column vector $\mathbf{u}_p$. The output of the channels is given by

$$\mathbf{v} = \mathbf{U}^t \mathbf{g}, \tag{14.7}$$

where $\mathbf{U}$ is the $M \times P$ matrix whose columns are the channel profiles $\mathbf{u}_p$, and $\mathbf{v}$ is the $P \times 1$ vector of channel outputs. The $\mathbf{u}_p$ represent the channel profiles, which we assume to be real. Processing the images through channels reduces the dimensionality of the data set from M to P. Possible choices for channel profiles are presented below. In some applications, P can be as small as 3.

Each channel output v_p is a random variable, made random by measurement noise and object variability—whatever sources of randomness are in the raw data. The probability density of each channel output is obtained using the methods for transforming random vectors presented in Sec. 8.1.5. Life is usually simpler, though, because in most models each channel output is the sum of multiple data values; the central-limit theorem tells us that the resulting random variable tends to be Gaussian distributed in that case.

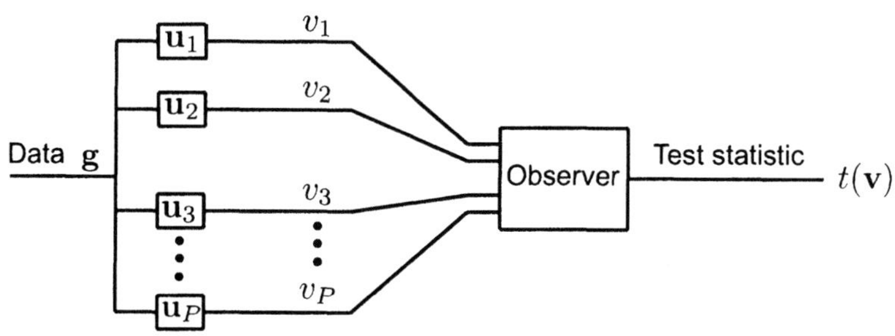

Fig. 14.2 Block diagram of a channelized observer.

Given the $\{v_p\}$, a strategy must be defined by which the channel outputs are combined to arrive at a decision that a stimulus either is or is not present. Possible options include adding the responses of the channels or using only the channel with the maximum response (Graham and Nachmias, 1971). Alternatively, the channel outputs may be combined via *probability summation* (Pirenne, 1943). Pirenne conjectured that binocular vision yields lower detection thresholds than monocular vision because the probabilities of detection from the left and right eyes are independent, and signals are detected if they are detected by the right eye, the left eye, or both. Formally, he suggested that the probability of detecting a signal using both eyes is

$$\Pr(D|L+R) = 1 - [1 - \Pr(D|L)][1 - \Pr(D|R)] \, , \tag{14.8}$$

where $\Pr(D|L)$ and $\Pr(D|R)$ are the probabilities of detecting the stimulus with the left and right eyes, respectively. While probability summation has been rejected as an explanation for the relative performance of binocular to monocular vision, it is encountered in some vision-system models as a means of combining the outputs of parallel channels (Daly, 1993). Combinations of differences in channels at each location/pixel have also been suggested (Lubin, 1993; Lloyd and Beaton, 1990; Zetzsche and Hauske, 1989). In some channel models, the sigmoidal form of (14.4) is imposed on the outputs of the frequency- and orientation-selective filters at each location (Legge and Foley, 1980) before the decision-making step.

Optimal methods for combining channel outputs The human observer can also be modeled as a quasi-ideal observer, that is, an observer who is constrained to process visual scenery through channels, but who is otherwise optimal in how the channel outputs are used to perform the task. If the human is modeled as a channelized ideal obsever, the model will achieve maximal AUC among all observers constrained to process data through the visual channels. A channelized ideal observer forms the likelihood ratio of the channel outputs under each hypothesis, giving

$$\Lambda(\mathbf{v}) = \frac{\text{pr}(\mathbf{v}|H_2)}{\text{pr}(\mathbf{v}|H_1)} . \tag{14.9}$$

The model observer's decision strategy is to compare $\Lambda(\mathbf{v})$ to a threshold, choosing H_2 when $\Lambda(\mathbf{v})$ is greater than this value, and H_1 otherwise. As detailed in Chap. 13, the ROC curve and related performance measures for the channelized ideal observer can be determined using (14.9) as a starting point.

Alternatively, a channelized Hotelling observer (CHO) model might be invoked, thereby assuming that the human observer forms an optimal linear combination of the channel outputs. As described in Sec. 13.2.12, there is a well-established theory for determining the optimal linear combination of the channel outputs and the resulting CHO figure of merit using the statistical properties of the channel outputs. For a binary discrimination task, the Hotelling observer's template in the channel space is given by

$$\mathbf{w}_{Hot,\mathbf{v}} = \mathbf{S}_{2\mathbf{v}}^{-1} \Delta \mathbf{v} , \tag{14.10}$$

where $\mathbf{S}_{2\mathbf{v}}$ is the $P \times P$ intraclass scatter matrix of the channel outputs [*cf.* (13.187)] and $\Delta\mathbf{v}$ is the expected difference in the channel outputs under each hypothesis.

The separability of the data in channel space is written in terms of the interclass and intraclass scatter matrices for $\mathbf{v}$:

$$J_{\mathbf{v}} = \text{tr}[\mathbf{S}_{2\mathbf{v}}^{-1} \mathbf{S}_{1\mathbf{v}}] = \text{tr}[(\mathbf{U}^\dagger \mathbf{S}_{2\mathbf{g}} \mathbf{U})^{-1} (\mathbf{U}^\dagger \mathbf{S}_{1\mathbf{g}} \mathbf{U})] , \tag{14.11}$$

where $\mathbf{S}_{1\mathbf{v}}$ is the interclass scatter matrix of the channel outputs [*cf.* (13.186)]. While (14.7) has the form of the linear transformation given in (14.10), including a dimensionality reduction, these expressions differ significantly because transformation using the matrix of visual channel functions $\mathbf{U}$ may result in the separability of the channel outputs being less than the separability of the data, while the operation of (14.10) generates a test statistic that preserves the separability in the channel outputs.

When the channel outputs are Gaussian random variables with equal covariance under the hypotheses, the channelized Hotelling observer and the channelized ideal observer are equivalent. In Sec. 14.3 we shall discuss methods for computing performance measures for channelized model observers.

Channel choices The nature of the signal and the background play a significant role in determining an appropriate choice for the channel profiles $\{\mathbf{u}_p\}$ and the way they are imposed on the data. For example, in an SKE task the channels are centered at the known signal location and the sum represented by (14.7) is done. If, on the other hand, the signal can be located at N multiple orthogonal locations, the channels could be centered at each location to give an $N \times P$ vector of outputs for decision-making purposes. When the signals and background are rotationally symmetric, the channels do not require any angular dependence; orientation-dependent signals and backgrounds require channels with orientation-selective responses.

The channelized ideal-observer model of Myers and Barrett (1987) incorporated radially concentric channels to predict human performance in correlated noise. The model's predictive ability was found to be insensitive to channel width and low-frequency turn-on parameters for the tasks considered in that work. The simplicity of this channel structure was possible because the task was the detection of radially symmetric signals at known locations. Because the signals were low contrast and the image noise was a filtered Gaussian random process, the model was equivalent to a channelized Hotelling observer.

More complex tasks involving asymmetric signals at varying locations may require more complex channel models. A variety of approaches for representation of channel mechanisms have been pursued by the developers of models of the visual system; these approaches can be incorporated into a CHO framework. Models based on Gabor functions (Daugman, 1988; Lloyd and Beaton, 1990; Watson, 1987) and wavelets (Daugman, 1985; Mallat, 1989; Marcelja, 1980; Watson, 1983) can be made to have both spatial and location specificity. Other options include ratio-of-Gaussian channels (Zetzsche and Hauske, 1989) and difference-of-Gaussian (DOG) models (Wilson and Bergen, 1979). Difference-of-mesa (DOM) filters can be used to model radial-frequency filters as well (Daly, 1993). "Mesa" is Spanish for table; a difference of two mesa functions gives a filter with a flat passband, a transition region, and a flat no-pass band. To give radial-frequency filters orientation selectivity, they can be multiplied by a set of functions tuned to orientation. For example, Daly uses what he calls *fan* filters to model the orientation-selective response. The product of the DOM and fan filters are termed cortex filters.

Once a selection has been made of the functions to be used to create a family of channels tuned to an array of orientations and frequencies, the next question is the number of such channels to include in the model. Many studies have indicated that only a fairly small number of channels is required for adequate modeling of human data. Myers and Barrett (1987) found good agreement between human data and CHO predictions with about 6 radial channels. The Daly visual-difference predictor model, designed to predict human performance for JND tasks (Sec. 14.1.3), uses only 6 DOM filters, combined with as few as 6 fan filters (30 degrees each), leading to 31 cortex filters in all [(# of fans) $\times$ (# of DOMs -1) + 1], since the lowest frequency filter has no orientation specificity). Wilson and Gelb (1984) also suggested the use of 6 spatial-frequency selective DOG filters, each with a range of orientations. There seems to be reasonable consensus that only about 6 channels are needed to cover frequency space; adding about 6 orientation-specific channels to each frequency-selective filter gives a complete model.

Internal noise To achieve even better matching between CHO and human data, the internal noise of the visual system must also be addressed. One way to account for internal noise is to scale the detectability of the human observer to that of the model observer (Burgess *et al.*, 1997; Burgess, 1999), giving $\text{SNR}_{human} = \kappa \text{SNR}_{model}$, where κ represents the impact of internal noise on detectability. From (14.5) it can be seen that the scaling factor is related to observer efficiency according to $\eta = \kappa^2$.

Another approach is to add noise injectors to the channel model, as shown in Fig. 14.3, giving a modified definition of the channel outputs of (14.7):

$$v_p = \mathbf{u}_p^\dagger \mathbf{g} + \epsilon_p \,, \tag{14.12}$$

where ϵ_p is an additive Gaussian noise contribution in channel p (Legge and Foley, 1980).

For a channelized linear observer, (14.12) is equivalent to adding noise to the decision variable (Abbey and Bochud, 2000). In particular, the channelized Hotelling observer forms the scalar test statistic t according to

$$t = \mathbf{w}_{Hot}^t \mathbf{v} = \mathbf{w}_{Hot}^t [\mathbf{U}^t \mathbf{g} + \boldsymbol{\epsilon}] = \tilde{v} + \tilde{\epsilon} \,, \tag{14.13}$$

where the tildes represent the transformation to the decision-variable space. Both $\tilde{v}$ and $\tilde{\epsilon}$ are scalar random variables, and both can usually be assumed to be Gaussian. The point of (14.13) is that a single Gaussian-noise injector at the decision variable level can model additive internal noise.

Observer uncertainty acts as a source of internal noise. It might be expected that observer uncertainty about parameters of the signal would result in a signal-dependent noise contribution, or a channel-dependent noise contribution. For example, one might hypothesize that the magnitude of a matched-filter's template uncertainty would depend on actual signal size, which would result in signal-dependent internal noise. However, Eckstein *et al.* (2000a) note that signal uncertainty is most often modeled as statistically independent of the signal.

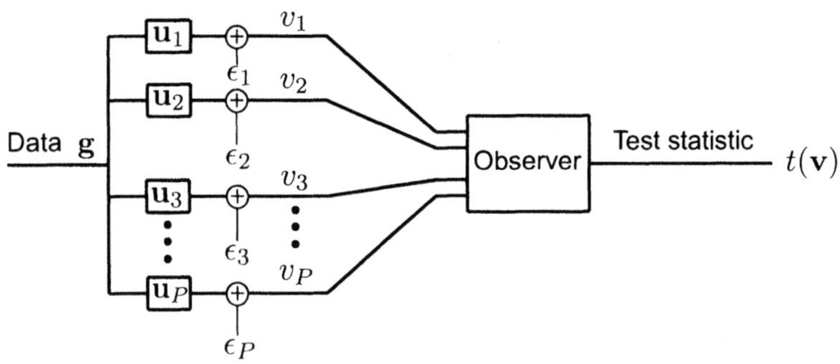

Fig. 14.3 Block diagram of a channelized observer including an internal noise mechanism.

Observer efficiency revisited The efficiency of the human observer relative to any model observer can be defined via (14.5). Choices for the denominator include the full ideal observer, the PWMF in SKE/BKE tasks in Gaussian noise, the NPWMF in SKE tasks, and the channelized Hotelling observer. A growing volume of psychophysical data, combined with model-observer calculations, is establishing the channelized Hotelling observer as an excellent predictor of human classification performance over a wide range of experimental paradigms, including ones in which the PWMF and NPWMF models are far less able to predict human performance. For this reason the remainder of this section will emphasize human efficiency relative to the CHO and review some of the many additional studies that have demonstrated the predictive capacity of the CHO model.

CHO success stories Many of the first comparisons of Hotelling and human performance involved the assessment of image quality in nuclear medicine. Fiete *et al.* (1987) investigated human performance in detecting simulated lesions in simulated liver scans and found excellent correlation with the Hotelling observer. Cargill (1989) used a more elaborate simulation for nuclear medicine, involving the detection of abnormalities in simulated images of a computer-generated 3D model of the liver with several possible disease states. She found excellent correlation between human performance and Hotelling predictions of image quality for 9 different collimator designs.

The CHO has also been shown to correlate well with human performance in the assessment of acquisition systems and reconstruction algorithms in tomographic imaging. Abbey and Barrett (1995) found good agreement between the human and

CHO across a range of linear iterative reconstruction algorithm parameters. Gifford *et al.* (1999, 2000a) used human and model observers to evaluate the impact of detector-response compensation on tumor detection in SPECT.

Acceptable levels of lossy image compression have been hotly debated for years. When the compressed images will be used by human observers, the evaluation of the compression algorithms must involve the assessment of the impact of the compression on human performance. Observer models can play a significant role in the evaluation of the large number of potential compression algorithms and the many parameters defined by each, provided the model observer predicts human performance. Eckstein *et al.* (1999) found the CHO and the NPWE to correlate well with human observer performance in the evaluation of image-compression algorithm settings. Based on this fact these investigators used a model observer to optimize the quantization parameters of the JPEG algorithm (Eckstein *et al.*, 2000b); the optimized parameter settings were then validated by psychophysical determination of improved human performance.

Human observers are known to be adaptive to noise level and image content, among other things. The CHO is also adaptive, with a decision strategy that changes when the signal or noise characteristics of the images are altered. Rolland and Barrett (1992) demonstrated that the adaptation of the human observer can be predicted by the CHO. In nuclear medicine, increasing exposure time shifts the dominant source of variability in the data from quantum noise toward the contribution from object variability. Rolland showed that human detection performance improves as exposure time increases, providing evidence of the human's ability to incorporate improved quantum statistics into its decision strategy. Similarly, the Hotelling observer's performance increases with increasing exposure. The correlation between the CHO predictions and the human data was extremely high, over several decades of observer performance. Conversely, the NPWMF strategy is not adaptive; the NPWMF applies a template determined by the difference of the signals under each hypothesis without regard for the character of the background. This observer's performance saturates as exposure time increases, failing to predict the performance of the human observer. No nonadaptive model could possibly predict human performance in this study.

A number of studies have extended the body of knowledge regarding CHO performance in random backgrounds. In addition to the work involving lumpy backgrounds of Rolland, Yao and Burgess discussed previously, it has been shown that the CHO correlates well with human performance in the presence of anatomical backgrounds (Eckstein and Whiting, 1995). Abbey and Barrett (2001) measured human-observer performance in several SKE tasks to investigate the effects of regularization and object variability in tomographic image reconstructions. Across a range of experiments that investigated parameters determining the signal profile, exposure time, and data covariance, the channelized-Hotelling observer was most able to predict the array of human data.

Abbey *et al.* (1999) give an elegant theoretical derivation of an unbiased procedure for determining the template of a linear observer for a detection task. The only inputs to the procedure are the images presented to the observer on each trial and the observer's decision as to which image was deemed "signal-present." The procedure requires the means and covariances of the data under each hypothesis. While the template-estimation procedure is applicable to any linear observer, Abbey *et al.* made use of the procedure for estimating the templates of human observers and comparing them to the templates of model observers. Edwards *et al.* (2000) extended the template-estimation procedure to the case where the noise is a mixture distribution of Gaussians. Recently, Abbey and Eckstein (2001) suggested the use of Bayesian template-estimation methods; the reduction in variance obtained through these methods may outweigh the small bias that also results. These template-estimation methods are pointing the way toward a better understanding of the human-observer's decision strategy. Perhaps in the future they may even find use in the development of improved methods for computer-aided diagnosis (CAD).

14.2.3 Psychophysical methods for image evaluation

Psychophysical methods are used to measure human-observer performance and assess diagnostic accuracy. In this section we shall review the history of the development of ROC methodology as a tool for understanding the visual system and assessing imaging technologies. We shall then describe methods for the conduct of psychophysical studies.[6]

Early applications of ROC analysis ROC techniques were initially developed during World War II for analyzing the performance of radar systems for detecting aircraft. One of the earliest applications of psychophysical methods to a medical application was the work of Garland (1949), who investigated the diagnostic accuracy of roentgenographic and photofluorographic techniques and presented some of the earliest evidence of reader error and variability. The cross-fertilization that brought ROC methods to visual science was greatly facilitated when W. P. Tanner, a graduate student in psychology at the University of Michigan, was assigned a desk in the office of T. G. Birdsall, one of the early pioneers of ROC methods (Cohn, 1993). In 1954, Tanner and J. A. Swets, also of the University of Michigan, published a seminal paper in which statistical decision theory was first applied to the study of visual performance, even including a section entitled, "A new theory of visual detection." This paper demonstrated that the core principles of statistical decision theory were applicable to observer performance. Most notably, the mathematical model of Fig. 13.4 is applicable to human decision variables, and human observers can control their decision criterion and manipulate it in response to information regarding the prior probability of each hypothesis and the decision costs. The paper presented data collected by yes-no and forced-choice experiments and showed them to be consistent.

It took some time for the perception community to relinquish the theory of an absolute detection threshold for "seeing." In 1963, Nachmias and Steinman published an ROC study meant to determine whether humans have a decision criterion that could be altered by directives from the investigator. The paper concluded that the data supported the variable-criterion hypothesis, but did not rule out the absolute-threshold theory entirely. Finally, in 1969, Kratz published an analysis of the Nachmias and Steinman data that concluded that the absolute-threshold theory could be rejected.

[6]We gratefully acknowledge the presentation materials made available to us by Charles Metz for use in writing this section.

While the variable-threshold theory was being established, Swets and his colleagues were working with great gusto at extending the use of statistical decision theory to the study of decision processes in perception (Swets *et al.*, 1961; Swets, 1964; Green and Swets, 1966). Another mathematical psychologist at the University of Michigan, D. D. Dorfman, and his colleague E. Alf, Jr. published a maximum-likelihood method for estimating ROC curve parameters and determining confidence intervals (1968, 1969). In the same timeframe, L. Lusted became the first investigator to apply ROC methods to medicine in general and medical imaging in particular (1968, 1971). Also, in 1960, the First Freiburg Conference on the Neurophysiology and Psychophysics of the Visual System was held, creating a forum for the movement toward combining and correlating information about the visual system derived from electrophysiological investigation with that derived using ROC methods (Jung and Kornhumber, 1961). This was a time of tremendous growth in methodology and accumulation of data in visual science.

The next decade saw a shift in the center of the ROC universe from the University of Michigan to the University of Chicago, where ROC analysis was applied to a variety of problems in medical imaging. Metz *et al.* demonstrated the relationship between ROC analysis and Shannon's information theory (1973) and published a tutorial on the basic principles of ROC analysis for a medical imaging audience (1978). Goodenough (1975) made use of an L-alternative forced-choice paradigm. Starr *et al.* (1975) investigated the detectability of low-contrast disks and spheres on uniform backgrounds in radiography. The early work of Starr *et al.* was one of the first of a set of studies that together demonstrated the limitations of using single measures of imaging system performance like resolution as a measure of image quality. It was also one of the first investigations of the effect of search-region size on ROC curves.

In the 1980s, the advent of relatively inexpensive, fast computers enabled the development and dissemination of free software for curve-fitting of ROC data, making ROC analysis much more widely utilized for image evaluation.[7] Software for statistical testing also became available. There is now a wide variety of free packages available for the analysis of data acquired under a variety of experimental paradigms and providing an assortment of possible model fits, as well as the statistical comparison of results across imaging systems, observers, and tasks.

The last decade has seen continued development of numerical tools for the analysis of ROC data, the generalization of ROC methods to more complex and clinically relevant tasks, and a significant increase in the utilization of ROC-based methods for studies of image evaluation and observer performance. In the next sections we shall describe in more detail how ROC experiments are designed and performed and the methods for data analysis that are available to investigators today. Our purpose is not to provide a complete "how-to" manual, but rather to give an idea of the many options available to the investigator and the relevant literature where more specific experimental and analytical tools can be found.

The yes-no experiment A single point on an ROC curve can be determined for a given observer on a given binary-classification task using a simple "yes-no" experiment. In each experiment, an observer is presented with a set of images one at a

[7]Some twenty years later, the number of registered users of the free Metz software package is close to 4000!

time, and the observer responds either "yes – the signal is present" or "no – the signal is absent." (More generally, "yes – class 2 is true" or "no – class 2 is not true.") By tabulating the fraction of true and false responses at the end of the experiment, a single point on the ROC curve is determined. By instructing the observer to use a different mindset on each of a set of yes-no experiments, a set of points on the ROC curve is found, as depicted in Fig. 13.5. The finer the curve desired, the more yes-no experiments that must be performed.

Rating-scale approach Swets *et al.* (1961) showed that a single rating-scale experiment gives equivalent ROC estimates to that obtained via the inefficient process of repeated yes-no experiments. The rating-scale approach involves the presentation of single images to the observer at a time, with each image presentation referred to as a "trial." The data collected on each trial is the observer's certainty that the image belongs to class 2. Table 14.1 gives an example rating scale.

There are many variations on this theme. At one extreme, class 2 can be defined by the presence of an exactly-specified object at an exact location, while at the other extreme it can encompass the presence of any pathology of any kind, with the range of object variability in the middle ground. The number of rating levels can be as few as 5, although 6 or 7 is more commonly encountered, or the experiment can use a continuous rating scale. The use of a continuous rating scale, first advocated by Rockette *et al.* (1992) and validated by King *et al.* (1993), allows for finer distinctions of certainty levels by the observer, and a smaller chance of degenerate data (where the cells of the rating scale are not fairly evenly distributed with responses) in the analysis stage (Wagner *et al.*, 2001). However, some investigators shy away from this method because of a concern that diligent observers will find it difficult to report their rating so finely, and the concern that the intra-observer variability will be increased (the likelihood that the observer will rate the same case at the same level on two independent trials will be infinitely small).

A current controversy is the use of *action scales* like the BI-RADS scale (ACR, 1998) for classification of mammographic images. Action scales incorporate patient management as well as the reader's level of suspicion. Some investigators have recommended that a pure probability-of-disease rating be acquired in addition to an action rating to avoid the bias that can occur when using an action rating alone for ROC purposes.

Table 14.1 Example rating scale

Rating	Description of certainty level
1	Object is definitely a member of class 1
2	Object is likely to belong to class 1
3	Object is equally likely to belong to class 1 or class 2
4	Object is likely to belong to class 2
5	Object is definitely a member of class 2

Relationship to contrast-detail (CD) diagrams An early paradigm for image assessment was the *contrast-detail* approach. In this method the observer is shown an image containing multiple signals with a range of contrasts and sizes. The observer reports the smallest detectable signal at each contrast. A plot of the detection-contrast versus size (detail) is then generated. When a set of CD diagrams are plotted as a function of exposure or dose, it is termed a CDD diagram (Cohen *et al.*, 1981).

There are several difficulties with the CD-diagram approach. One is that the approach is subjective because it does not control for the observer's variable decision criterion; different observers can be lax or strict in their judgment and even the same observer's criterion for "seeing" the signal can vary. Also, there is no ability to correct for "wishful thinking" on the part of the observer, and without signal-absent locations there is no ability to determine the trade-off with false-positive responses. Thus, while the CD-diagram approach is routinely used as a quality-assurance protocol in many imaging applications, it is not recommended as a quantitative tool in the assessment of imaging systems unless the aforementioned concerns are addressed in the study.

The CD-diagram can be of use when machine observers are used in place of humans. Then the observer's threshold can be set to a fixed level, and the algorithm can be forced to evaluate both signal-present and signal-absent locations. Chakraborty and Eckert (1995) have developed a procedure for the machine evaluation of phantom images for use in the evaluation of image quality.

Forced-choice experiments We first encountered the forced-choice (FC) experimental paradigm in Sec. 13.2.5. In a forced-choice experiment, an observer is forced to make a decision in favor of one of the alternative hypotheses. In the binary-classification task, a pair of images is presented to the observer, either at the same time or sequentially, one from class 1 and the other from class 2. The order/placement of the images is randomized, and usually there is no restriction on viewing time. The observer must decide which alternative belongs to class 2. As derived in Sec. 13.2.5, the percentage of correct responses in a two-alternative forced choice (2AFC) experiment equals the area under the ROC curve. We shall have more to say on this when we discuss the analysis of forced-choice data in Sec. 14.2.4.

The generalization of the FC paradigm to the L-alternative task requires the observer to state which of L-alternative signals is present in a signal-present image, or which of L regions contains a specified signal, for example.

Generalized ROC methods In a classic ROC experiment designed to evaluate an SKE binary classification task, the observer records a single rating on each trial describing his or her certainty that the image belongs to class 1 or class 2 (see Table 14.1). The application of this experimental paradigm to more realistic problems in which the signal is not known exactly is problematic. When there are multiple possible signals, or multiple locations, a single probability score does not capture all the data available from the observer. Most notably, the observer may indicate a high certainty that a signal is present in a signal-present image, but in fact the observer may have missed the true signal and be responding to a noise-only location that is perceived to be signal. Without requiring the observer to provide location data along with the probability rankings, there is no ability to correct for this effect.

An alternative is to require the observer to point to the signal that is detected on an image, and rate the probability that it is there. A *localization ROC* (LROC) curve is a plot of the actually positive images detected with the lesion correctly localized vs. the fraction of actually negative images falsely called positive (Swensson, 1996). The x axis of an LROC curve is thus the same as in a conventional ROC plot. On each image there is either a single signal at an unknown location or there is no lesion. An example LROC curve is shown in Fig. 14.4.

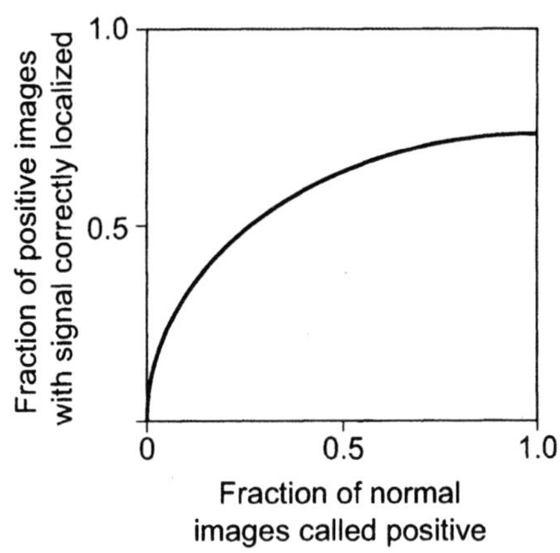

Fig. 14.4 An example LROC curve.

Free-response ROC curves (FROC) were introduced by Bunch *et al.* (1978) to enable the detection-and-localization analysis of images with an arbitrary number of signals. An FROC curve is a plot of the fraction of lesions detected vs. the average number of false-positive detections per image. FROC curves are often used in the assessment of CAD algorithms, where the number of false positives can be high. An example FROC curve is shown in Fig. 14.5

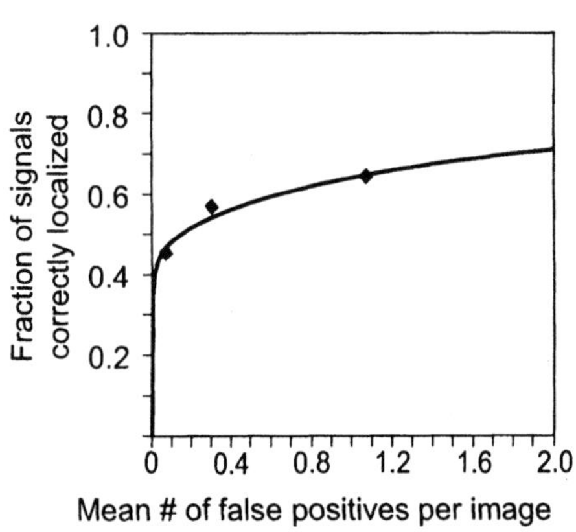

Fig. 14.5 An example FROC curve.

The *alternative free-response ROC* or AFROC curve is a plot of the fraction of lesions detected vs. the fraction of actually negative images falsely called positive (Chakraborty and Winter, 1990). An actually negative image is included in the fraction of those called positive whenever one or more false-positive locations are identified on it. The x axis of an AFROC curve is similar to the x axis of ROC and LROC curves, only now the ability to mark more than one location on an image is allowed. An example AFROC curve is shown in Fig. 14.6.

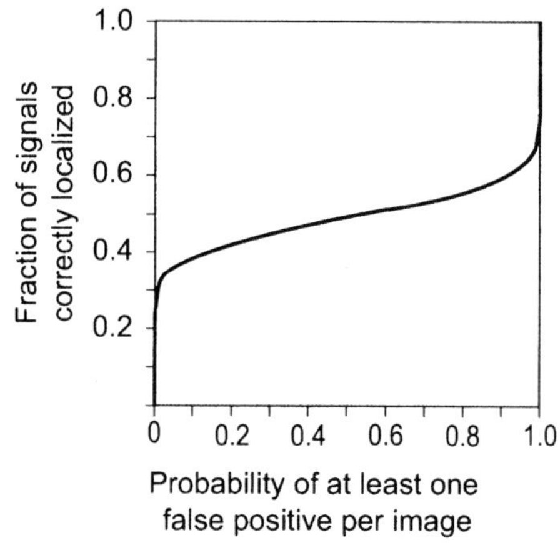

Fig. 14.6 An example AFROC curve.

14.2.4 Estimation of figures of merit

Once the experimental procedure has been run and the observer-response data have been collected, the question arises as to how best to analyze the data. In this section we shall describe methods for the analysis of ROC and related data, and the estimation of figures of merit for summarizing image quality for classification tasks.

Analysis of conventional ROC data The foundation for the analysis of ROC-like data is the analysis of conventional rating data. The simplest approach to the generation of an ROC curve from rating data is to determine the number of true- and false-positive responses associated with each rating level. For a rating scale with N levels, this will give a graph with $N-1$ TPF-FPF pairs, plus the $(0,0)$ and $(1,1)$ anchors for the plot. An *empirical* ROC curve is obtained by "connecting the dots" to generate a stairstep plot consisting of vertical and horizontal line segments obtained by adding true and positive responses to the curve as a threshold is swept across the response data. For continuous rating data, the AUC estimate obtained by integrating the area under an empirical ROC curve is a Wilcoxon statistic (Bamber, 1975).

When the number of rating levels is small, the empirical ROC curve will be jagged, but a fitting approach can yield a smooth estimate of the curve. However, the assumptions of conventional least-squares curve fitting are invalid, owing to the joint dependence of the ratings on the observer's mindset (Metz, 1986a). Hence a maximum-likelihood estimation procedure must be used instead. Dorfman and Alf (1968) published the first ML solution to the analysis of rating data.

The most widely used ML method assumes that the data underlying the ratings take on a parametric form with adjustable parameters under each hypothesis. Fig. 13.4 helps to make this concept more concrete: the fitting procedure assumes the distributions for the decision variable conditioned on each hypothesis take on a particular form, most commonly a Gaussian, and the goal of the estimation procedure is to estimate the parameters of the two distributions given the rating data. The assumption that the two distributions are Gaussians is the so-called *binormal model* (Swets *et al.*, 1961). The binormal model does not limit the decision-variable data to Gaussian distributions; all that is required is that the data obtained under each hypothesis be transformable to Gaussian distributed random variables by the same unknown transformation.

The binormal model yields an estimated ROC curve with a straight-line plot on double-probability paper; the axes are given by the normal deviates z_{TPF} and z_{FPF} [*cf.* (13.15)]. The conventional ML procedure estimates the slope and intercept of this line, which can be related to the difference in means and the ratio of the variances of the two underlying distributions (Dorfman and Alf, 1969). Methods for obtaining ML binormal fits to ROC curves are also available for continuous rating data (Metz *et al.*, 1998b).

A large number of experiments have demonstrated the validity of the binormal model (Swets, 1986; Metz, 2000). Hanley (1988) has shown that ROC curves obtained from ML parameter fits to a variety of non-Gaussian distribution models, including binomial, Poisson, gamma, χ^2, rectangular, and triangular forms, are indistinguishable from the Gaussian-based curve, provided the number of data samples is large. Nevertheless, the standard binormal model can result in fitted ROC curves that cross the chance line and have a slope that does not decrease monotonically as the FPF increases (Berbaum *et al.*, 1990). These so-called "improper" ROC curves can be obtained from the standard binormal model when the number of cases is low, the data scale is discrete, or the operating points are not well distributed (degenerate data). To avoid this outcome, the "proper" ROC analysis was introduced by Dorfman *et al.* (1996). Proper ROC curves are constrained from crossing the chance line. Proper models based on bigamma distributions (Dorfman *et al.*, 1996) and binormal distributions (Pan and Metz, 1997; Metz and Pan, 1999) have been investigated. The proper binormal model transforms the data by forming the likelihood ratio associated with the two underlying normal distributions; the result is an ROC curve with a monotonically decreasing slope. An objection to this procedure is that the calculation of the likelihood ratio is not something that the actual observer under test is hypothesized to do. Rather, it is an additional transformation applied to the observer data and thus may not be representative of the observer to which the ROC curve applies.

An alternative fitting approach for ROC rating data is the "contaminated" binormal model (Dorfman *et al.*, 2000a). This model assumes the distribution of decision variables is the bimodal sum of two Gaussians under the signal-present alternative (Dorfman and Berbaum, 2000b). The model has been found to be useful in the analysis of data with small false-positive fractions and to give results very similar to those of the standard binormal fitting procedure for nondegenerate data (Dorfman and Berbaum, 2000c).

Analysis of forced-choice data We described the 2AFC experiment mathematically in Sec. 13.2.5 as one in which the observer is presented with two images $\mathbf{g}$ and $\mathbf{g}'$, where $\mathbf{g}$ is drawn from $\mathrm{pr}(\mathbf{g}|H_1)$ and $\mathbf{g}'$ is drawn from $\mathrm{pr}(\mathbf{g}|H_2)$. The images are presented simultaneously in different spatial locations or separately in time at the same location. The assignment of the two underlying sources of images to the two presentation locations is randomized. The observer's task is to choose the image from class 2. To make the decision, the observer computes two test statistics $T(\mathbf{g})$ and $T(\mathbf{g}')$, and the data vector that gives the higher value is assigned to H_2. This assignment is correct if $T(\mathbf{g}') > T(\mathbf{g})$. By (13.39), the probability of a correct decision on any trial is AUC. Viewed this way, AUC is a criterion-free parameter-free distribution-independent figure of merit for a classification task (Massof and Emmel, 1987).

To estimate AUC from a forced-choice experiment, the percentage of correct decisions over a large number of trials is determined. To keep score of the number of correct responses, let n_i take on the value 1 for a correct response on trial i and 0 for an incorrect response. Mathematically, $n_i = \text{step}[T(\mathbf{g}_i') > T(\mathbf{g}_i)]$, where the subscript i denotes the ith trial. Thus n_i is a Bernoulli random variable (see Sec. C.6.1). Over N trials, the AUC estimate is the proportion of correct responses (PC):

$$\widehat{\text{AUC}} = \text{PC} = \frac{1}{N}\sum_{i=1}^{N} n_i \,. \tag{14.14}$$

If we assume the response variables are independent from trial to trial, (14.14) is the sum of N i.i.d. Bernoulli random variables. From (C.159) we know that the summand must be a binomial random variable with parameters N and the true AUC. It is well known that (14.14) is an unbiased ML estimate of AUC.

The early work of Tanner and Swets (1954) demonstrated the consistency of data collected in yes-no and forced-choice experiments. While an FC experiment yields an estimate of AUC, it has the disadvantage of not providing any information regarding the shape of the underlying ROC curve. Burgess (1985b) compared ROC and FC experimental methods and concluded that ROC methods make more efficient use of the available images, giving AUC estimates with lower variance, while FC methods make more efficient use of observer time. In an effort to make more efficient use of the available images, some experimenters use a multiple-pass paradigm in which different images from each hypothesis are paired for presentation in each pass. It can be shown that the full ROC curve can be obtained in the limit of every image being paired with every other. Note that the response variables are no longer independent Bernoulli random variables in this case.

Analysis of generalized-ROC experiments The primary advantage of the generalized-ROC approaches described above is their applicability to tasks in which signal uncertainty, usually location uncertainty, plays a key role. Many advances have been made in the last decade toward the development of robust procedures for the analysis of generalized ROC data from LROC, FROC, and AFROC experiments.

Maximum-likelihood methods have been introduced for the fitting of LROC data (Swensson, 1996). An ROC curve can be obtained from LROC data. Swensson (1996) gives the following relationship between the area under the ROC curve and the area under the LROC curve: $A_{\text{LROC}} = 0.5(\text{AUC} + 1)$.

ML analysis tools for FROC data have been introduced by Chakraborty (1989). The procedure assumes that the underlying signal and noise distributions are Gaussians and the number of false-positive responses per image follows a Poisson distribution. The assumptions underlying the analysis of FROC and AFROC are detailed and their validity argued thoroughly in a recent book chapter by Chakraborty (2000), who also suggests that the FROC analysis gives estimates of system performance with greater statistical power than those obtained using conventional ROC analysis.

While the use of localization brings a significant degree of reality to the task, compared to the classical ROC experimental design, there is also the added requirement for deciding what region around a signal will be considered a "true-positive" response in the analysis. The choice for this tolerance is arbitrary; yet it has ramifications on the results of the data analysis. There is yet no consensus on the appropriate localization tolerance to use in the analysis of LROC, FROC, and AFROC experiments.

There are few resources for the analysis of more complex experiments involving multiple hypotheses. Kijewski *et al.* (1989) developed an analysis procedure for determining the parameters specifying ROC curves between all pairs of classes in an L-class problem given ratings of the multiple alternatives. Still needed is a practical method for the analysis of multi-alternative tasks using ROC analysis.

Summary measures Once a satisfactory fit to the ROC rating data has been obtained, summary measures of performance can be derived. Global measures of system performance include the AUC, the detectability measure d_A obtained from AUC via (13.21), or the parameters of the binormal model. If it is known that a certain operating point on the ROC curve is more significant for the intended use of the system than others, a local measure of performance might be reported at that operating point; that is, the TPF at a given FPF or the FPF at a given TPF might be reported. Partial area measures giving either the area to the right or below the ROC curve from a specified operating point give a regional measure of performance (McClish, 1989). Jiang *et al.* (1996) provided an extension to partial-area index analyses for systems with high AUC. Finally, if sufficient information is available regarding the cost and benefit of decisions is known, these can be reported at the optimal operating point, or a full cost/benefit curve can be reported and summarized. There are many open questions regarding the best approach to summarizing performance. The AUC is the most widely used figure of merit today.

The ML theorem of (13.378) enables us to say something about the ML estimates of other performance measures based on the ML estimate of AUC. For example, an ML estimate of the observer SNR can be derived from $\widehat{\text{AUC}}_{\text{ML}}$ by inverting (13.20). An ML estimate of the observer's squared SNR can be obtained by similar reasoning and used in an ML estimation of observer efficiency.

When more than one observer has participated in an ROC study, there are two options for deriving an overall figure of merit. The first is to derive estimates of the binormal model parameters for each observer and average the parameters. The second is to pool the rating data and then perform the ML estimation procedure. Metz (1986b) has discussed the advantages and disadvantages of these alternatives. When multiple observers are used in the evaluation of multiple imaging systems, correlations in the data result. Tools for the analysis of multiple-reader, multiple-case experiments are discussed below.

Error analysis and the comparison of imaging performance When a measure of imaging performance is obtained, it is natural to ask how large the error bars are about that estimate. Moreover, when two imaging systems are being compared, we seek methods for determining the significance of the difference between figures of merit for the competing systems. In Sec. 13.1.1 we discussed a number of drawbacks to the use of statistical tests of the null hypothesis. These same drawbacks are equally applicable to tests of the null hypothesis using estimated figures of merit for imaging systems.

In his 1920s work on estimation, R. A. Fisher (see Sec. 13.3.6) set the stage for randomized clinical trials by discussing randomized experiments in agricultural research. The analysis of independent imaging modalities was firmly established in the late 1970s, when the National Cancer Institute funded a contract to J. A. Swets and R. A. Pickett of Bolt, Beranek and Newman to develop methods for the assessment of diagnostic technologies. The outcome was a landmark text that presented computer code for the analysis of ROC data, including an analysis of the error in the the estimate of AUC for a single modality (1982).

Soon after, Metz and Kronman (1980) and Hanley and McNeil (1982) proposed methods for the comparison of ROC curves for which the data were assumed to be independent. In 1983, Hanley and McNeil extended their work to the situation where the data were obtained from the same set of patients. In 1984, Metz *et al.* provided a method for analyzing differences between ROC curves measured from correlated data. Differences could be given in terms of the difference in AUC, the TPF at a specified FPF, or the parameters of the standard binormal model. Non-parametric methods for comparing the areas under correlated ROC curves based on Wilcoxon statistics have been presented by DeLong *et al.* (1988) and Campbell *et al.* (1988). These early methods for estimating the uncertainty in AUC and comparing ROC curves took into account the variability in the data resulting from the measurement noise and the object variability sampled by the finite set of cases but did not describe or compensate for the contribution from observer variability.

Observer variability is a complex, multivariate phenomenon that was understood in principle as early as the text by Swets and Pickett (1982), which contains two chapters on the subject. Observers respond differently to different cases, and even the same observer's responses are not 100% correlated across repeated readings of the same data set. As we have seen, a reader's response depends on the latent decision criterion, but it also depends on the observer's training, experience, age, fatigue, and other factors. Readers have different skill levels, and some readers are better at some case sets or modalities than others. An excellent example of reader variability due to differences in decision criteria is contained in data published by Elmore *et al.* (1994) and the subsequent commentary by D'Orsi and Swets (1995). Beam *et al.* (1996) have published the largest study to date demonstrating radiologist variability in skill level and decision criterion in the case of mammographic interpretations.

The first practical multivariate method for the analysis of the variance in AUC estimates for correlated tests with the assumption that both observers and images (readers and cases in the medical literature) are random effects was the multi-reader, multi-case (MRMC) method of Dorfman, Berbaum and Metz (Dorfman *et al.*, 1992), now commonly referred to as the DBM MRMC method. The method makes use of a jackknife procedure to generate multiple estimates of AUC, each derived by leaving out one of the observations and analyzing those that remain. The results of each leave-one-out procedure are termed *pseudovalues.* An analysis of the variance in the pseudovalues gives an estimate of the variance in the estimate of AUC. By analyzing the statistics of pseudovalues, the contribution to the variance in the estimate of AUC from the cases or the readers can be obtained.

The DBM MRMC method was first developed for the analysis of discrete rating data. Roe and Metz (1997a, 1997b) further developed and validated the DBM method and made software freely available for either continuous or discrete rating data. An alternative, nonparametric method for analyzing the components of variance in ROC studies based on bootstrapping has been suggested by Beiden *et al.* (2000a). Gifford *et al.* (2001) have recently simulated the application of the DBM method to LROC studies and found it to be useful for studies with low numbers of readers and cases.

In some countries, double reading of certain clinical images is the standard, as a method for reducing the number of incorrect interpretations and reducing reader variability. In the U.S., several commercial CAD systems are now available for use as a second reader to the radiologist. The analysis of adjunctive systems requires careful consideration of the appropriate assumptions regarding the variability of the readers (for example, no threshold variability for a computer) and the means for combining their interpretations. The overall performance of the system will be dependent on these considerations. A method for analyzing the improvements in the accuracy of imaging studies derived by repeated observations was suggested by Metz and Shen (1992). Beiden *et al.* (2001a, 2001b) have presented a nonparametric estimate of the components of variance of AUC for comparing two modalities with different variance structures, for example, where one modality involves a CAD adjunctive device and the other does not.

The original MRMC method required every reader to interpret every image in each modality. Recently, the statistical analysis of "partially paired" data sets has been presented (Zhou and Gatsonis, 1996; Metz *et al.*, 1998a).

Ordinal regression Standard ROC methodology reports the performance of a particular observer on a particular task given a specified imaging system. The dependence of the performance measure on a parameter describing the object (size or amplitude, say) or observer (age or number of years of training) would require a series of studies across the range of the parameter of interest. Given the time and cost required for a single psychophysical study, the notion of performing repeated studies of this sort is daunting.

Tosteson and Begg (1988) proposed the use of ordinal-regression techniques for combining studies of multiple object and observer characteristics in a single study. Toledano and Gatsonis (1995, 1996, 1999) have further developed the method and provided extensions for handling incomplete data. The use of ordinal-regression methods in the optimization of imaging system parameters using realistic models for the imaging process deserves greater attention.

Sources of bias We have described methods for analyzing the uncertainty in estimated measures of system performance without mention of possible sources of error in the estimated mean system performance. There are many sources of bias that can creep into the evaluation of an imaging system (Begg and Greenes, 1983). Probably the most significant is the ground-truth problem, which we shall address again in Sec. 14.4.5. It is difficult in real imaging applications to know the true status of an object, be it an enemy aircraft in a reconnaissance image, a stellar object in astronomy, or an unknown feature in a medical image. Knowledge of ground truth can require expensive verification procedures like long-term follow-up, biopsy, or imaging using an alternative system. Thus, in order to know the truth status required for scoring observer responses in an ROC study, the investigator might design the study with an absence of subtle objects or confounding cases (Rockette *et al.*, 1995, 1998). Without these cases in the study, the results of the study will not describe the performance of the system on these kinds of cases. Similarly, bias can also result from the skill of the observers involved in the study. In the design of the study, the investigator should carefully consider whether to use experts vs. nonexperts and the extent that they represent the intended use of the system.

In summary, the estimated AUC is a joint description of the performance of the imaging system and the population of images and observers used in the study.

Field tests vs. stress tests A *field test* samples the objects and observers as they are expected to be sampled in routine use. A *stress test* limits the objects or the observers (or both) in order to "challenge" the performance of systems where differences are expected. Studies over subpopulations of observers or objects can potentially enable significant differences in system performance to be demonstrated for those subpopulations. For example, it may be that expert and nonexpert radiologists utilize the output of a CAD algorithm differently, but a study that averages over the two sets of readers would possibly miss this important finding. In another example, the fraction of women with dense/heterogeneous breasts is small; a study comparing film-screen to digital mammography using a broad sample of patients might not uncover a significant advantage of one system over the other for that subpopulation of women.

As described in Sec. 13.1.1, the diligent investigator can always increase the number of cases in a study until a statistically significant result is obtained. However, a judicious selection of the cases used in the study can sometimes reduce the number of images required to show significant differences in system performance, by taking into account known differences in the physical performance characteristics of the imaging systems under comparison.

Summary of process Although there are many open areas of research, methods based on ROC analysis are still the best approach for evaluating classification tasks performed by human and model observers. Before beginning a psychophysical investigation, a few questions should be considered. The first is the nature of the classification task — will standard ROC methods suffice, or is a generalized method that incorporates localization/search needed? Consideration should be given to the need for realism and the adequacy of the information that will be gained, the available methods for data analysis as well as methods for statistical analysis.

Careful consideration should be given to sampling issues for images and readers, recognizing the impact these will have on the conclusions that can be drawn from the study.

The specific viewing conditions should be considered, including the degree of observer adaptation, the display settings, the observation distance, and so on. The human-machine interface is critical; small numbers of observer mistakes due to a poor interface can impact the data appreciably.

It is recommended that a block design be used to avoid image-order effects. Observers can read a subset of the images representing one imaging system, then another, then back to the first, until the entire set under all conditions has been read. Randomize the ordering across readers. Do not expect observation sessions to last more than about an hour, or fatigue can degrade observer performance.

Pilot studies can be used to determine the imaging conditions that will yield the best study power and highest efficiency of observer effort. Staircase methods have been described for determining the object contrast that will give high statistical power (Watson and Pelli, 1983; Watson and Fitzhugh, 1990). These methods adjust the object contrast iteratively over a sequence of 2AFC trials to find the signal contrast that yields a SNR_{human} of ~ 0.75 by decreasing the signal amplitude when the observer is correct, and increasing it after decision errors. Once the contrast has been determined that corresponds to that level of performance, the final study can make use of a fixed signal at that contrast — the method of constant stimulus — to give a more precise measure of system performance for that stimulus. Alternatively, the method of ordinal regression allows the evaluation of system performance across a range of stimuli.

Once the data are collected, they can be analyzed by the chosen fitting method. Free software packages are readily available on the worldwide web for this purpose. Then the figure of merit can be estimated along with its confidence interval. Tests for the differences between estimates of figures of merit are also included in several of the freeware packages.

14.3 MODEL OBSERVERS

Model observers serve many purposes. They can be used as tools in the study of the human visual system; by comparing the results of psychophysical studies to model-observer performance measures, researchers gain insights into human perception that can lead to improved models of the human visual system. Such studies give information regarding what tasks the human performs well and what image characteristics impact the human most significantly, potentially leading to improved imaging system designs for generating images for human interpretation. Model observers can also act in place of humans, or in concert with them, in which case we refer to the model as a computer-aided diagnosis (CAD) system.

Above and beyond the use of model observers as tools for understanding the visual system, model observers are an extremely valuable tool in the objective assessment of image quality. Model observers that operate on raw images or detected data enable the objective evaluation and optimization of image acquisition systems. Model observers designed to operate on reconstructed or processed images are useful for the assessment of image-processing algorithms without lengthy human-observer experiments. Because our emphasis in this text is on the design and evaluation of imaging systems, and not on the development of a better understanding of human perception, we shall focus on the use of model observers for the purpose of OAIQ — the objective assessment of image quality — in what follows.

Many of the same statistical methods used to evaluate imaging systems with human observers are applicable to the evaluation and comparison of model observers. The goal of this section is to present methods for determining the performance of a model observer for a given study. As we shall see, the particular model observer chosen and the method of determining its figure of merit will depend on the task as well as the extent to which we know or can characterize the statistics of the data.

We shall begin in Sec. 14.3.1 with a brief review of selected model observers for classification tasks and the requirements for determining each model observer's performance. We shall then discuss how one chooses a model observer for system evaluation based on classification performance. Sec. 14.3.2 deals with the particular issues involved in the determination of classification performance by linear model observers. The determination of performance measures for ideal observers is the subject of Sec. 14.3.3. Finally, in Sec. 14.3.4, we shall discuss the use of estimation tasks in the objective assessment of image quality.

14.3.1 General considerations

Structure of observer models All model observers used in the objective assessment of imaging systems have a similar structure, illustrated in Fig. 13.1. As described in Sec. 13.2, for classification tasks every model observer computes a scalar test statistic t of the form

$$t = T(\mathbf{g}), \tag{14.15}$$

where $\mathbf{g}$ might be either the raw data or a processed image and $T(\mathbf{g})$ is the observer's discriminant function. A decision is made in favor of hypothesis H_2 if t is greater than some threshold; otherwise H_1 is selected. By determining the number of images classified correctly for all threshold settings, an ROC curve can be generated.

The performance of the model observer can then be summarized using some metric related to the ROC curve. The area under the ROC curve (13.18) and the detectability d_A derived from AUC via (13.21) are common figures of merit. Alternatively, the SNR associated with the test statistic (13.19) can be determined from the first- and second-order statistics of t as a measure of the separability of the data from the two classes. The SNR and the detectability are the same when the test statistic is Gaussian under the two classes.

Categories of observer models for classification Observers can be classified according to whether $T(\mathbf{g})$ is optimal or suboptimal and whether it is a linear or nonlinear function of $\mathbf{g}$. By definition, optimal observers are the best possible in some sense. The Bayesian or ideal observer makes optimal use of all available information in the data and any additional nonimage information to achieve the highest AUC attainable. The ideal observer's test statistic is the likelihood ratio. In general, the ideal observer's discriminant function is a nonlinear function of the input data.

The Hotelling observer is the ideal linear observer; this observer's discriminant function is optimal in the sense that it achieves maximum SNR amongst all linear observers. The AUC-optimal linear observer is another privileged linear observer; as the name suggests, this observer employs the linear discriminant that achieves the highest possible AUC of all linear discriminants for the task.

Because of the large dimensionality of modern images, it may be necessary to make use of "efficient" features, or channels, that preserve the information in the data while enabling the determination of observer performance. Sec. 13.2.12 describes a method for deriving information-preserving linear features from an analysis of the known first- and second-order statistics of the data. Thus we can design a channelized Hotelling observer (CHO) such that it is still the optimal linear observer in spite of the reduced dimensionality.

The objective assessment of image quality may involve suboptimal model observers, particularly when the goal is to predict human performance. From Sec. 14.2 we know that the human observer has been modelled as an observer that processes images through frequency-selective and orientation-selective channels. The channelized Hotelling observer has been shown to be a useful predictor of the human observer for a variety of tasks, where the channels in this case are not efficient, but are instead chosen to predict human performance. Alternatively, more mechanistic models of the human visual system might be employed as surrogates for the human. These "anthropomorphic" models can incorporate highly nonlinear building blocks such as adaptive gain and contrast nonlinearity. Such models reduce the dimensionality of the data and incur an information loss as well.

Table 14.2 summarizes the types of model observers that can be employed in the objective assessment of image quality.

Table 14.2 Classification of observer models used in OAIQ

	Optimal	Suboptimal
Nonlinear	Ideal observer	Nonlinear model of human
Linear	Hotelling (max-SNR) CHO (efficient) AUC-optimal linear	CHO (visual channels, internal noise)

Computation vs. estimation As noted in Sec. 14.2, the goal of a psychophysical experiment is the *estimation* of human performance from a finite sample of images. This is in contrast to the methods presented in Chap. 13, which addressed the *computation* of ensemble performance measures for model observers. In this chapter we are concerned with the issues that arise when limited data are available for the estimation of observer performance. As we shall see, we might estimate the model observer's decision function from finite samples, and use that function to estimate the model's performance from the same or another set of finite data. Alternatively, a finite data set might be utilized to estimate the statistics of the data under competing hypotheses, with this information then used to estimate a figure of merit for the model observer's performance directly.

Why OAIQ is easier than pattern recognition While the objective assessment of image quality has striking similarities to classical pattern recognition, the two problems are significantly different. Whenever we evaluate an imaging system we do so in terms of a particular task and a specific observer performing the task; thus we have considerable prior information regarding the objects to be classified and the discriminant function to be utilized. In many circumstances we can make use of a signal-known-exactly task, where the background might be simulated or might be a real clinical background. In contrast to most pattern recognition problems, we also have tremendous knowledge of the physics and statistics of the imaging system under evaluation that we can exploit to simulate noise-free training images. Thus the mean data under each hypothesis is fairly easily determined. Furthermore, the noise PDF $\text{pr}(\mathbf{g}|\mathbf{f})$ is usually known from the physics; hence the noise covariance matrix is also known. With this information we are well positioned for determining a linear observer's discriminant function. Note also that we can avoid the gold-standard problem to be discussed in Sec. 14.4.5 by using simulated images; then we always know the underlying truth status of each image.

While we might estimate the model observer's template and evaluate the model observer's performance from finite data, the feature-extraction step is not ad hoc. It is dictated by the statistics of the data. If the purpose of the study is the prediction of human performance, the features are further dictated by physiology — a channel model representing the visual system is then used as well.

In OAIQ the amount of prior information we bring to bear on the problem is tremendous relative to various approaches found in pattern recognition and data mining, where the statistics of the data may be completely unknown, the features are unspecified, and even the number of classes is uncertain. Moreover, OAIQ often makes use of simulated images, so there is no limitation to the number of images available, and there is no issue about their true classification.

Basic equations describing the ideal observer As derived in Sec. 13.2.6, the ideal observer achieves maximum AUC, maximum TPF at any FPF, and minimum Bayes risk. The ideal observer's test statistic is the likelihood ratio, given by

$$\Lambda(\mathbf{g}) \equiv \frac{\text{pr}(\mathbf{g}|H_2)}{\text{pr}(\mathbf{g}|H_1)} . \tag{14.16}$$

To classify a data set, the ideal observer compares $\Lambda(\mathbf{g})$ to a threshold.

Alternatively, the ideal observer forms the log-likelihood ratio, given by

$$\lambda(\mathbf{g}) \equiv \ln[\Lambda(\mathbf{g})] = \ln\left[\frac{\text{pr}(\mathbf{g}|H_2)}{\text{pr}(\mathbf{g}|H_1)}\right] , \tag{14.17}$$

which is then compared to a threshold to classify an image. Because the log-likelihood ratio is a monotonic function of the likelihood ratio, the AUC of the ideal observer is unchanged by this transformation.

Data needed for ideal-observer studies We see from (14.16) or (14.17) that the computation of the ideal observer's performance requires full knowledge of the probability density function for the data under the competing hypotheses. In general, these are high-dimensional functions, describing the full joint statistical behavior of M data values. There are well-known examples for which the ideal-observer's performance is calculable, most notably the SKE case in Gaussian noise (Sec. 13.2.8) and some non-Gaussian noise models as well (Sec. 13.2.9). However, for random signals and backgrounds, (see Secs. 13.2.10 and 13.2.11), the ideal observer's decision variable takes the form of an integral of huge dimensionality over the posterior density of the data conditioned on known signals and backgrounds. In Sec. 14.3.3 we shall consider various techniques for estimation of the ideal observer's performance.

Basic equations describing linear observers We may not be able to evaluate ideal-observer performance because of the computational complexity or because we simply do not have the statistical information required to use those tools. Or, we may not want to estimate the performance of the ideal observer because the goal of the assessment process is the prediction of human, rather than ideal, performance. For these reasons the assessment effort may focus on the estimation of the performance of linear model observers.

In a binary classification problem, an arbitrary linear discriminant computes a scalar test statistic t from the $M \times 1$ data vector $\mathbf{g}$ using a transformation of the form

$$t = \mathbf{w}^t\mathbf{g} , \tag{14.18}$$

where $\mathbf{w}$ is an $M \times 1$ template. The observer classifies each data set by comparing the value of t to a threshold. The statistics of t determine the performance of the observer, as measured by AUC or SNR_t. When t is Gaussian-distributed, AUC and SNR_t are related according to (13.20). Given that the linear observer's test statistic is a linear weighted sum of many random variables, the Gaussian assumption for the PDF of t is usually valid as a result of the central-limit theorem.

Optimal linear (Hotelling) observer When the ensemble mean and covariance for $\mathbf{g}$ are known, the observer that maximizes SNR can be derived according to the procedure presented in Sec. 13.2.12. By (13.177) the Hotelling observer's template is known to be

$$\mathbf{w}_{Hot} = \mathbf{K}_{\mathbf{g}}^{-1}\Delta\overline{\mathbf{g}}, \tag{14.19}$$

where $\mathbf{K}_{\mathbf{g}}$ is the ensemble data covariance, assumed to be the same under each hypothesis, and $\Delta\overline{\mathbf{g}}$ is the difference in the mean data vector under the two hypotheses. The assumption of equal data covariance under each hypothesis is a reasonable approximation for weak signals, even though the signals may be random under each hypothesis. The subscript $\mathbf{g}$ on the covariance matrix, which refers to the raw data, is required because we shall later encounter covariance matrices that describe channel outputs, which will be subscripted accordingly. It can be seen that the Hotelling test statistic is the output of a prewhitening matched filter operation that attempts to compensate for all contributions to the correlations in the data.

The performance of the Hotelling observer is given by [*cf.* (13.178)]

$$\mathrm{SNR}_{Hot}^2 = \Delta\overline{\mathbf{g}}^t\mathbf{K}_{\mathbf{g}}^{-1}\Delta\overline{\mathbf{g}} = \mathrm{tr}\left[\mathbf{K}_{\mathbf{g}}^{-1}\Delta\overline{\mathbf{g}}\,\Delta\overline{\mathbf{g}}^t\right]. \tag{14.20}$$

In the SKE detection problem this expression simplifies to

$$\mathrm{SNR}_{Hot}^2 = \mathbf{s}^t\mathbf{K}_{\mathbf{g}}^{-1}\mathbf{s}, \tag{14.21}$$

where we denote the signal to be detected by $\mathbf{s}$ in the data domain. Note that the data covariance matrix is assumed to be the same under each hypothesis in (14.20) and (14.21) because the contributions from background variations and measurement noise are assumed to dominate contributions from signal variability in the random-signal case.

The Hotelling observer achieves maximum SNR of all linear observers. An alternative approach is to determine the template $\mathbf{w}$ that gives maximum AUC of all linear observers. In Sec. 13.2.12 we presented the problem of classification in Poisson noise as an example for which the ideal observer is linear (without actually imposing a linearity requirement); this is the observer that achieves maximum AUC as discussed in the previous section. However the ideal observer is not the linear observer that achieves maximum SNR for this task. There is a much smaller literature on the AUC-optimal linear observer relative to the large literature on the Hotelling or max-SNR observer. In the case of a normally distributed test statistic, these two observers coincide.

Data needed for Hotelling-observer studies We see from (14.20) and (14.21) that computation of the performance of the Hotelling observer requires knowledge of the ensemble first- and second-order statistics of the data under each hypothesis. When information regarding the mean and covariance of $\mathbf{g}$ is unavailable, we must resort to procedures for estimating the performance of the optimal linear observer from samples.

The difference in the class means under each hypothesis, $\Delta\overline{\mathbf{g}}$, is an $M \times 1$ vector, where each element $\Delta\overline{g}_m = \overline{g}_{2m} - \overline{g}_{1m}$ is the difference in the average value in the mth pixel in the image or data set under the two hypotheses. Its estimate can be obtained by determining the sample mean from sets of images known to be from each class; the behavior of the sample mean as an estimator is well-understood. Moreover, in many studies the signal is simulated and nonrandom, so that (14.21) is relevant and no estimation of the mean is required. Thus the determination of the mean data under each hypothesis is not a major stumbling block in most applications.

The most daunting issue in imaging applications is the determination of an estimate of $\mathbf{K_g}$, which we shall denote $\widehat{\mathbf{K}}_\mathbf{g}$. A natural inclination is to assume that $\widehat{\mathbf{K}}_\mathbf{g}$ is the sample covariance matrix, but the reader is cautioned against acting on this impulse. If the number of image samples, N_s, is less than the number of pixels in each image, M will be singular and noninvertible. This option therefore requires $N_s \geq M$.

Consider the number of elements of a covariance matrix to be estimated in typical imaging scenarios. A flat-panel digital x-ray imager can have 1024 × 1024 elements. A SPECT system with a 128 × 128 detector that collects data over 64 projection angles has the same number of elements. Thus these systems have a data vector with $\sim 10^6$ data elements, so $\mathbf{K_g}$ is a $10^6 \times 10^6$ matrix with about a trillion elements. The symmetry of this matrix allows us to reduce the number of elements to be estimated by about a factor of 2, but a half trillion is still a large number.

The linear discriminant based on sample means and covariances for the pixels in the raw data set is the approach commonly referred to as the Fisher discriminant. Because the number of values to be estimated to form the sample covariance matrix is almost always far greater than the number of samples available for the estimation procedure, the Fisher discriminant is rarely a useful estimate of the optimal linear discriminant in imaging applications.

We shall discuss several alternative approaches to the estimation of the Hotelling observer's performance in Sec. 14.3.2.

Basic equations describing channelized linear observers Any linear channel model can be represented by a matrix-vector multiplication like the one given in (14.7):

$$\mathbf{v} = \mathbf{U}^t\mathbf{g}, \tag{14.22}$$

where $\mathbf{U}$ is an $M \times P$ matrix whose columns are the channel profiles $\mathbf{u}_p$, and $\mathbf{v}$ is the $P \times 1$ vector of channel outputs. The $\mathbf{u}_p$ represent the channel profiles, which we have assumed to be real. Each channel output v_p is a number.

While both (14.7) and (14.22) represent a reduction of the dimensionality of the data set, the critical difference is that we are free to choose the channel profiles in (14.22) to suit our purpose. The channels could be designed to be efficient, giving minimal loss of detectability and thereby providing an estimate of the separability inherent in the data. Alternatively, the channels could be designed to estimate the separability of the data after processing through visual-system channels to predict human performance, which may or may not be efficient depending on the task. A number of possible channel profiles used in the vision literature are described in Sec. 14.2.

The performance of a channelized observer is given by the SNR associated with the channel outputs under each hypothesis:

$$\mathrm{SNR}_{\mathbf{v}}^2 = \Delta\overline{\mathbf{v}}^t \mathbf{K}_{\mathbf{v}}^{-1} \Delta\overline{\mathbf{v}} = \Delta\overline{\mathbf{g}}^t \mathbf{U}[\mathbf{U}^t \mathbf{K}_{\mathbf{g}} \mathbf{U}]^{-1} \mathbf{U}^t \Delta\overline{\mathbf{g}} \,. \tag{14.23}$$

Data needed for channelized-observer studies The information required to evaluate the performance of a channelized observer is the first- and second-order statistics of the data *as seen through the channels.* We see immediately from (14.23) that the channel covariance matrix to be inverted is much smaller than the data covariance matrix. If the number of channels is P, then $\mathbf{K}_{\mathbf{v}}$ is a $P \times P$ matrix, where P can be as small as 3 to 6. Even if P is 30 to 50, the matrix to be inverted is still a reasonably manageable size.

The second advantage to the use of a channelized model is the flexibility we have in choosing the channel profiles. As we shall see, prior knowledge of the characteristics of the signal and background can suggest particular forms for efficient channels. Alternatively, the channels can be chosen to model human performance. Given the nontrivial time required to perform psychophysical evaluations, the ability to evaluate a set of imaging system parameters using a model that predicts human performance can offer significant advantages.

Which model observer? The question of which model observer to employ is answered by the objective of the evaluation study. If the goal is to evaluate or optimize the hardware of the data acquisition system, then the ideal observer is the model of choice. Optimization with this observer will result in a system with the maximum information in the raw data in the sense of being able to perform the specified task. If it is not possible to compute ideal-observer performance because the calculation of the likelihood of the data under each hypothesis is not tractable, then the ideal linear observer is a useful alternative for use in hardware evaluation and optimization.

When the task is the evaluation of image-processing algorithms, ideal observers are of no use, because they are invariant to invertible image processing (see Sec. 13.2.6). Image processing algorithms, reconstruction methods and display devices exist for presenting images to human observers; thus the appropriate model should be one that predicts human performance. The model might be a highly detailed, mechanistic model of the visual system or a simpler linear channel model like the CHO.

In the next subsections we discuss in more detail each of these model observers and methods for estimating their classification performance.

14.3.2 Linear observers

In this subsection we shall present a number of approaches for determining the performance of linear model observers from finite data sets. We shall first consider the Hotelling observer that makes use of the raw data and describe several methods for estimating the SNR of this observer. As suggested by (14.20) and the discussion that followed, the estimation of this Hotelling observer's SNR must involve some method for dealing with (or circumventing) the need to estimate the inverse of $\mathbf{K}_{\mathbf{g}}$. Once we have exhausted our list of possible approaches to this problem, we shall explore methods that invoke dimensionality-reducing linear channels.

In many instances, image quality can be ascertained through a classification task involving nonrandom signals that are added to real or simulated backgrounds. Thus we shall first assume that the problem is the detection of a known signal, the so-called SKE problem, while allowing for a random background. In this case there is no need to estimate $\Delta\overline{\mathbf{g}}$ in (14.20); it is known, and our goal is to find an estimate of the SNR given in (14.21). This objective is only hampered by the fact that $\mathbf{K_g}$ is unknown. Subsequently, we shall consider methods for estimating linear-observer performance for random signals.

We shall then briefly discuss the characteristics of the estimated figures of merit. Finally, the subsection concludes with a short discussion of methods for determining the AUC-optimal linear observer. Throughout this subsection we make the assumption that the truth status of each image sample is known; methods for dealing with the no-gold-standard problem are presented in Sec. 14.4.5.

Nonrandom signals We consider the object to be the sum of a known signal and a random background according to the decomposition introduced in (8.306):

$$\mathbf{f} = \mathbf{f}_s + \mathbf{f}_b \,. \tag{14.24}$$

In the detection task, $\mathbf{f}_s$ is zero under H_1. The backgrounds are assumed to be random and drawn from the same ensemble under each hypothesis.

From (14.24), the mean data for a fixed object and a linear imaging operator $\mathcal{H}$ can be written as a linear superposition of signal and background [*cf.* (8.352)]

$$\overline{\mathbf{g}}(\mathbf{f}) = \mathcal{H}\mathbf{f}_s + \mathcal{H}\mathbf{f}_b \equiv \mathbf{s} + \mathbf{b}\,, \tag{14.25}$$

where $\mathbf{b}$ is the image of the particular background realization.

Without signal variability, the covariance matrix $\mathbf{K_g}$ describes the randomness in the data due to background variability and measurement noise. It can be written formally in terms of an expectation of the covariance of the data about the mean taken first over the noise for a single background, followed by an average over all backgrounds:

$$\mathbf{K_g} = \langle\langle(\mathbf{g} - \overline{\mathbf{g}})(\mathbf{g} - \overline{\mathbf{g}})^t\rangle_{\mathbf{n}|\mathbf{b}}\rangle_{\mathbf{b}} \,. \tag{14.26}$$

In the absence of object variability the data covariance matrix reduces to the noise covariance matrix, an entity that is usually known or computable through our knowledge of the image-formation process. Nonrandom backgrounds can be very useful in the validation of software intended to simulate realistic noise properties of an imaging system. However, the objective evaluation of imaging systems in the absence of object variability can yield misleading conclusions; thus image evaluation should employ a random background model if at all possible. Sec. 14.4 describes a number of approaches for simulation of random objects and images.

We have cautioned against the use of sampling methods to directly estimate the sample covariance matrix, and the use of exactly-specified backgrounds in the objective assessment of image quality. How, then, to simplify the calculation of Hotelling SNR in the presence of a random background? One assumption that is often made is that the background is stationary.

Stationarity? A stationarity assumption is attractive because the covariance matrix is then diagonalized by an appropriate Fourier transformation. For example, we know from Sec. 7.4.4 that a circulant covariance matrix that satisfies $K_{\mathbf{mm}'} = K_{[\mathbf{m}-\mathbf{m}']_M}$ (where the subscript indicates modulo-M arithmetic in both components of the multi-index) is diagonalized by a discrete Fourier transform. And from Sec. 8.2.8 we know that an infinite covariance matrix that satisfies $K_{\mathbf{mm}'} = K_{\mathbf{m}-\mathbf{m}'}$ for all $\mathbf{m}$ and $\mathbf{m}'$ is diagonalized by a discrete-space Fourier transform. Following diagonalization by Fourier methods, $\mathbf{K}_{\mathbf{g}}^{-1}$ can be found by taking the reciprocal of each diagonal element.

While Fourier methods based on stationarity assumptions may seem attractive, this approach is fraught with problems. Real covariance matrices are neither infinite nor circulant. The assumption that $\mathbf{K}_{\mathbf{g}}$ is circulant implies digital wrap-around, meaning the statistical correlation of two pixel values representing adjacent detector elements is assumed to be equal to the correlation of two elements on opposite sides of the detector, or even in different projections. In an investigation of image quality in digital radiography, Pineda and Barrett (2001) have shown that stationarity assumptions can give misleading results.

Local stationarity Requiring stationarity of any sort over the whole image field is not only unrealistic, it is also unnecessary if our goal is to compute the SNR of a spatially localized lesion. Since (14.21) is the norm of the vector $\mathbf{K}_{\mathbf{g}}^{-1/2}\mathbf{s}$, we can compute it by summing over only those pixels for which the vector is substantially different from zero. Typically, in direct imaging systems, those pixel elements correspond to a restricted region in data space. If so, we can express the SNR in terms of the Wigner distribution function computed over this region, as discussed in Sec. 13.2.13.

For indirect imaging systems, a spatially localized lesion can contribute to a very nonlocalized set of detector elements. In this situation it is unlikely that an assumption of approximate stationarity would hold over the entire region for which $\mathbf{K}_{\mathbf{g}}^{-1/2}\mathbf{s}$ is significantly greater than zero. Thus for tomographic systems it is necessary to perform a reconstruction first to restore the local nature of the signal to be detected and allow the use of methods that invoke an assumption of approximate stationarity. The argument of the previous paragraph holds if we let $\mathbf{g}$ be the reconstruction and $\mathbf{s}$ be the reconstructed signal.

If $\mathbf{K}_{\mathbf{g}}$ were diagonal (in the multi-indices[8]), the region where approximate stationarity is required would be the same as the subset of pixels for which $\mathbf{s}$ is nonzero, but a nondiagonal covariance means that some elements of $\mathbf{K}_{\mathbf{g}}^{-1/2}\mathbf{s}$ are nonzero even if the corresponding elements of $\mathbf{s}$ are zero. Moreover, the range of the correlations is only a rough guide to selecting the correct subset of pixels; the matrix $\mathbf{K}_{\mathbf{g}}^{-1/2}$ can occupy a substantially larger band around the diagonal than $\mathbf{K}_{\mathbf{g}}$.

We do not know the width of this band if we cannot compute $\mathbf{K}_{\mathbf{g}}^{-1/2}$, but we can proceed experimentally. If we start with a measured covariance matrix, or one computed on a realistic nonstationary model, we can select an $L \times L$ subset of it centered on the signal location. Calling this matrix $\mathbf{K}_L$, we can compute $\mathbf{s}^t\mathbf{K}_L^{-1}\mathbf{s}$, which would be an estimate of the Hotelling SNR without any stationarity assumption if we were given only this subset of the data. We can then vary L and observe the behavior of this SNR; when it no longer changes, we can assume that we have

[8]See Sec. 8.2.8 for a discussion of discrete random processes and diagonality in multi-index notation.

found the band containing the nonzero elements of $\mathbf{K_g}^{-1/2}$, and we can compare the resulting SNR to that computed with the Wigner distribution function. If agreement is good, we can use the Wigner expression to compute SNR for a variety of signals and all positions in the field and to define local NEQ and DQE as functions of spatial frequency and signal location (see Sec. 13.2.13). This approach may result in a number of nonzero elements in need of estimation that is small enough that the finite number of image samples can support their estimation.

Decomposition of the covariance matrix Another approach is to make use of our knowledge of the physics of the imaging process, which often gives us powerful information regarding the distribution of data for a fixed object. Statistically speaking, we often know $\text{pr}(\mathbf{g}|\mathbf{f})$, from which we can determine the conditional mean $\overline{\mathbf{g}}(\mathbf{f})$ and the conditional covariance $\mathbf{K_{n|f}}$.

Key to making use of this prior information is a decomposition of the overall data covariance given in Sec. 8.5.3; we know from (8.347) that $\mathbf{K_g}$ is the sum of two terms, written

$$\mathbf{K_g} = \langle\langle[\mathbf{g} - \overline{\mathbf{g}}(\mathbf{f})]\,[\mathbf{g} - \overline{\mathbf{g}}(\mathbf{f})]^t\rangle_{\mathbf{n|f}}\rangle_{\mathbf{f}} + \left\langle \left[\overline{\mathbf{g}}(\mathbf{f}) - \overline{\overline{\mathbf{g}}}\right]\left[\overline{\mathbf{g}}(\mathbf{f}) - \overline{\overline{\mathbf{g}}}\right]^t \right\rangle_{\mathbf{f}}$$

$$= \langle\mathbf{K_{n|f}}\rangle_{\mathbf{f}} + \mathbf{K_{\overline{g}}} \equiv \overline{\mathbf{K}}_{\mathbf{n}} + \mathbf{K_{\overline{g}}}\,, \tag{14.27}$$

where $\overline{\mathbf{K}}_{\mathbf{n}}$ represents the noise covariance averaged over all objects. While both $\overline{\mathbf{K}}_{\mathbf{n}}$ and $\mathbf{K_{\overline{g}}}$ are influenced by object variability, we emphasize that they are covariances for vectors in data space.

When the signal is random but statistically independent of the background, we can write the covariance matrix for $\overline{\mathbf{g}}$ as [see (8.359)]

$$\mathbf{K_{\overline{g}}} = \left\langle \left[\overline{\mathbf{g}}(\mathbf{f}) - \overline{\overline{\mathbf{g}}}\right]\left[\overline{\mathbf{g}}(\mathbf{f}) - \overline{\overline{\mathbf{g}}}\right]^t \right\rangle_{\mathbf{f}} = \mathcal{H}\mathcal{K}_{\mathbf{f}}\mathcal{H}^\dagger$$

$$= \mathcal{H}\mathcal{K}_{\mathbf{f_s}}\mathcal{H}^\dagger + \mathcal{H}\mathcal{K}_{\mathbf{f_b}}\mathcal{H}^\dagger \equiv \mathbf{K_s} + \mathbf{K_b}\,, \tag{14.28}$$

where $\mathbf{K_s}$ and $\mathbf{K_b}$ are the covariance of the data about the conditional mean resulting from signal and object variability, respectively. When the signal is nonrandom (14.28) simplifies to $\mathbf{K_{\overline{g}}} = \mathbf{K_b}$. Even in the random-signal case this simplification can be relevant; if the signal is of sufficiently low contrast, then $\mathbf{K_{\overline{g}}} \approx \mathbf{K_b}$ because the contribution due to the random background dominates.

Much of what follows on estimation of linear-observer performance is based on the decomposition of (14.27). Though we shall often use the approximation that $\mathbf{K_{\overline{g}}} \approx \mathbf{K_b}$, we note that (14.27) itself is exact; it requires no Gaussian assumptions regarding either the objects or the noise, and it does not assume that the noise is object-independent. Alternative forms for the object-variability term that make use of alternative ways of expressing the autocovariance of the object in object space are given in Sec. 8.5.3.

Role of the measurement noise To be more explicit about $\overline{\mathbf{K}}_{\mathbf{n}}$, we need to distinguish direct from indirect imaging and object-dependent from object-independent noise.

The simplest case is direct imaging with additive Gaussian measurement noise. As discussed in detail in Chap. 12, electronic noise in different detector elements is usually statistically independent and hence uncorrelated. If every detector element has the same noise variance σ^2, which is independent of the object $\mathbf{f}$, then

$$\overline{\mathbf{K}}_{\mathbf{n}} = \mathbf{K_{n|f}} = \sigma^2\mathbf{I}\,. \tag{14.29}$$

Thus $\overline{\mathbf{K}}_{\mathbf{n}}$ is a multiple of the unit matrix and hence full rank.

The situation is only slightly more complicated with Poisson noise. Since Poisson measurements are conditionally statistically independent with variance equal to the mean, we can write

$$\left[\mathbf{K}_{\mathbf{n}|\mathbf{f}}\right]_{mm'} = \overline{g}_m(\mathbf{f})\,\delta_{mm'} = [\mathcal{H}\mathbf{f}]_m\,\delta_{mm'}\,, \tag{14.30}$$

where the last form is for a linear digital imaging system characterized by the CD operator $\mathcal{H}$. Averaging over object variability is now straightforward:

$$\overline{\mathbf{K}}_{\mathbf{n}} = \langle \overline{g}_m(\mathbf{f})\rangle_{\mathbf{f}}\,\delta_{mm'} = \overline{\overline{g}}_m\,\delta_{mm'} = \left[\mathcal{H}\overline{\mathbf{f}}\right]_m\delta_{mm'}\,. \tag{14.31}$$

Thus the average noise covariance matrix is diagonal in spite of the object variability, though of course the overall covariance matrix $\mathbf{K}_{\mathbf{g}}$ is not diagonal.

It is not immediately obvious, however, that $\overline{\mathbf{K}}_{\mathbf{n}}$ is full rank. Indeed, the conditional noise covariance $\mathbf{K}_{\mathbf{n}|\mathbf{f}}$ is not full rank if any of the $\overline{g}_m$ is zero. Similarly, $\overline{\mathbf{K}}_{\mathbf{n}}$ is not full rank if any of the $\overline{\overline{g}}_m$ is zero, but this turns out to be of much less concern; the only way a particular $\overline{\overline{g}}_m$ could be zero is if the mth detector element never receives radiation for any object in the ensemble, and in that case we might as well delete that detector element from the data set. Thus we can always assume that $\overline{\mathbf{K}}_{\mathbf{n}}$ is full rank for direct imaging, even with Poisson noise.

For indirect imaging the measurement noise is modified by the reconstruction algorithm. This issue will be discussed at length in the next chapter, but for now we note that analytic expressions for $\mathbf{K}_{\mathbf{n}|\mathbf{f}}$ can be developed, where $\mathbf{n}$ refers to the noise in an image reconstructed by a linear algorithm from either Gaussian or Poisson data (see Sec. 15.4.2). For nonlinear algorithms, analytic covariances are generally not possible, but practical computational methods are available for determining $\mathbf{K}_{\mathbf{n}|\mathbf{f}}$ numerically; for details, see Sec. 15.4.7. These numerical expressions can then be averaged over $\mathbf{f}$ to obtain $\overline{\mathbf{K}}_{\mathbf{n}}$.

Sample averages In Sec. 8.4 we discussed a variety of statistical models for objects and found that there were many circumstances where we could generate samples of $\mathbf{f}$; more discussion of methods for simulating random objects is also given in Sec. 14.4. However, it is usually not possible to determine $\mathrm{pr}(\mathbf{f})$ from samples, and we almost always have to resort to the use of sample averages to determine the statistical properties of the data resulting from random objects.

Consider again the case of nonrandom signals (or where the signal is random but of low contrast), so that $\mathbf{K}_{\overline{\mathbf{g}}} = \mathbf{K}_{\mathbf{b}}$, and we want to estimate this covariance matrix. From Sec. 13.2.12 we know that the data covariance resulting from a general random background is given by

$$[\mathbf{K}_{\mathbf{b}}]_{mm'} = \left\langle (b_m - \overline{b}_m)(b_{m'} - \overline{b}_{m'})\right\rangle_{\mathbf{b}}\,, \tag{14.32}$$

where $\overline{b}_m$ is the mean contribution of the random background to detector element m. This expression describes the fluctuations in the data that would be observed over a large set of simulated or real noise-free images.

One approach to finding an estimate of $\mathbf{K}_{\mathbf{b}}$ is to use a theoretical object model such as a lumpy background (Sec. 8.4) for which the autocovariance function can be specified. This function is then mapped through the blur associated with the imaging system to produce the covariance matrix $\mathbf{K}_{\mathbf{b}}$. If we choose some functional form (*e.g.*, fractal) for the autocorrelation, we can use sample images to estimate any unknown parameters in the function.

Another approach is to acquire a set of low-noise images and estimate the covariance matrix for the background (in data space) from them. If we do not want to make any assumptions about the form of the autocovariance, we can simply form the sample covariance matrix as a low-rank approximation to the desired ensemble covariance. The samples might be simulated noise-free backgrounds, or they might be experimental background images with low but nonzero noise, obtained with image-averaging or high-dose techniques. Methods for simulating noise-free backgrounds are discussed in Sec. 14.4.

Suppose we have a set of sample background images $\{\mathbf{g}_j, j = 1, ..., N_s\}$, which are either noise-free (simulated) or for which the noise is negligible compared to the effects of object variability (perhaps because the images were acquired with a long exposure time). We can array each of these images as $M \times 1$ column vectors. We can then subtract the sample mean from each image to form the set $\{\delta\mathbf{g}_j, j = 1, ..., N_s\}$, and the covariance matrix $\mathbf{K}_{\overline{\mathbf{g}}}$ can be estimated by

$$\widehat{\mathbf{K}}_{\overline{\mathbf{g}}} = \mathbf{W}\mathbf{W}^t\,, \tag{14.33}$$

where $\mathbf{W}$ is the $M \times N_s$ matrix with columns given by sample images:

$$\mathbf{W} = \frac{1}{\sqrt{N_s}}\left[\delta\mathbf{g}_1,\, \delta\mathbf{g}_2,\, ...,\delta\mathbf{g}_{N_s}\right]. \tag{14.34}$$

The sample covariance matrix of (14.33) is equally applicable when the set of sample images contains random signals of unknown statistical description as well as random backgrounds.

Once the background covariance matrix is estimated, the noise contribution can be determined (if it is not already known) to yield the full data covariance matrix. For example, we can make use of (14.31) to write the covariance of the data in the weak-signal approximation under Poisson measurement noise as

$$\begin{aligned}[\widehat{\mathbf{K}}_{\mathbf{g}}]_{mm'} &= \langle\langle(g_m - \hat{\bar{b}}_m)(g_{m'} - \hat{\bar{b}}_{m'})\rangle_{\mathbf{n}|\mathbf{b}}\rangle_{\mathbf{b}} \\ &= \hat{\bar{b}}_m\,\delta_{mm'} + [\widehat{\mathbf{K}}_{\mathbf{b}}]_{mm'}\,. \end{aligned} \tag{14.35}$$

The first term in the last line is $\widehat{\overline{\mathbf{K}}}_{\mathbf{n}}$, which in this case is a diagonal matrix with elements given by sample averages of the mean background. The second term $\widehat{\mathbf{K}}_{\mathbf{b}}$ is an estimate of the covariance $\mathbf{K}_{\overline{\mathbf{g}}}$ due to the random backgrounds.

Other non-Poisson forms of object-dependent measurement noise can be simulated to generate noisy images once the simulation of random objects and noise-free data sets is achieved satisfactorily. These images can be used to determine the first- and second-order statistics of the data necessary to determine the SNR of the linear observer, using methods described below.

Matrix-inversion tools Once we are assured that we have a covariance matrix with full rank, the next step is to compute the SNR. Given the size of $\widehat{\mathbf{K}}_{\mathbf{g}}$, direct inversion of the estimated covariance matrix to estimate the detectability via (14.20) or (14.21) is not feasible. We shall consider the following alternative approaches, none of which assumes stationarity in any sense:

1. Iterative computation of the template;
2. Neumann series;
3. Matrix-inversion lemma.

Iterative computation When the observer's template is known, it can be applied to a set of sample images (for which the ground truth is known) to yield a set of test statistics under each class that can be used to compute the observer's AUC or SNR. The optimal linear observer has a template given in (14.19); thus it would appear that the determination of $\mathbf{w}_{Hot}$ also requires the inversion of $\mathbf{K_g}$. Not so! Fiete *et al.* (1987) suggested that the Hotelling template could be calculated iteratively.

Finding the template amounts to solving the equation $\mathbf{K_g w} = \mathbf{s}$, where the signal $\mathbf{s}$ is assumed known and $\mathbf{K_g}$ is either known or estimated. This equation is analogous to the imaging equation $\mathbf{Hf} = \mathbf{g}$, where the unknown template takes the place of the unknown object and the covariance matrix plays the role of the imaging system. However, the covariance matrix is square, making it invertible in principle, unlike the system operator in most imaging problems.

The solution can be found by any of the iterative methods enumerated in Chap. 1 or by the regularized methods to be discussed in Chap. 15. One possible solution is given by the Landweber algorithm (1.231), which gives the following template estimates at each iteration:

$$\hat{\mathbf{w}}_{n+1} = \hat{\mathbf{w}}_n + \alpha[\widehat{\overline{\mathbf{K}}}_\mathbf{n}]^{-1}[\mathbf{s} - \widehat{\mathbf{K}}_\mathbf{g}\hat{\mathbf{w}}_n] , \tag{14.36}$$

where n denotes the iteration number and we have made use of the knowledge that the noise contribution to the covariance matrix is full rank. The beauty of this iterative approach is that no inversion of the full $\widehat{\mathbf{K}}_\mathbf{g}$ is required.

Once the template has been estimated, the SNR can be found by applying the template to a set of sample images, determining the mean and variance of the resulting scalar test statistic under each hypothesis, and computing the observer's performance via (13.19). Alternatively, we can directly estimate SNR^2 by (14.21) as $\mathbf{s}^t\hat{\mathbf{w}}$.

Neumann series The covariance matrix may not be diagonal in real situations, but it may be nearly diagonal (at least with the multi-index convention). For example, as we shall see in Chap. 16, for direct imaging applications using x rays the nondiagonal contributions to the data covariance are due to correlations in the object statistics and physical processes like escape of K x rays from the phosphor. When these contributions are not very long-range, the Neumann series approach can be advantageous.

To see why the near-diagonal character of $\mathbf{K_g}$ is useful, suppose initially that

$$\mathbf{K_g} = \sigma^2\mathbf{I} + \mathbf{A} = \sigma^2\left[\mathbf{I} + \frac{1}{\sigma^2}\mathbf{A}\right] , \tag{14.37}$$

where $\mathbf{A}$ describes the off-diagonal elements. Then we can use the Neumann series (A.59) to write the inverse covariance as

$$\mathbf{K}_\mathbf{g}^{-1} = \frac{1}{\sigma^2}\sum_{j=0}^{\infty}\left[-\frac{1}{\sigma^2}\mathbf{A}\right]^j = \frac{1}{\sigma^2}\mathbf{I} - \frac{1}{\sigma^4}\mathbf{A} + \frac{1}{\sigma^6}\mathbf{A}^2 - \dots . \tag{14.38}$$

The Hotelling SNR then becomes

$$\mathrm{SNR}_{Hot}^2 = \mathbf{s}^t\mathbf{K}_\mathbf{g}^{-1}\mathbf{s} = \frac{||\mathbf{s}||^2}{\sigma^2} - \frac{\mathbf{s}^t\mathbf{A}\mathbf{s}}{\sigma^4} + \frac{\mathbf{s}^t\mathbf{A}^2\mathbf{s}}{\sigma^6} - \dots . \tag{14.39}$$

Formally, the Neumann series will converge if $||\mathbf{A}||/\sigma^2 < 1$, but that requirement is too stringent for our purposes since it takes no account of the nature of the signal. By the ratio test, the series in (14.39) will converge if

$$\frac{\mathbf{s}^t\mathbf{A}^{n+1}\mathbf{s}}{\sigma^2\mathbf{s}^t\mathbf{A}^n\mathbf{s}} < 1 \tag{14.40}$$

for all n, and it may still converge (because of the alternating signs) even if (14.40) is violated. In practice, convergence will be rapid if the correlations are weak and short-range and the signal is spatially compact.

More generally, we can always decompose $\mathbf{K}_\mathbf{g}$ into a diagonal part $\mathbf{D}$ plus a matrix $\mathbf{A}$ with only off-diagonal terms. Assuming convergence, we then have

$$\mathbf{K}_\mathbf{g} = \mathbf{D} + \mathbf{A} = \mathbf{D}\left[\mathbf{I} + \mathbf{D}^{-1}\mathbf{A}\right] ; \tag{14.41}$$

$$\mathbf{K}_\mathbf{g}^{-1} = \left[\sum_{j=0}^{\infty}\left[-\mathbf{D}^{-1}\mathbf{A}\right]^j\right]\mathbf{D}^{-1} ; \tag{14.42}$$

$$\mathrm{SNR}^2 = \mathbf{s}^t\mathbf{K}_\mathbf{g}^{-1}\mathbf{s} = \mathbf{s}^t\mathbf{D}^{-1}\mathbf{s} - \mathbf{s}^t\mathbf{D}^{-1}\mathbf{A}\mathbf{D}^{-1}\mathbf{s} + \mathbf{s}^t\mathbf{D}^{-1}\mathbf{A}\mathbf{D}^{-1}\mathbf{A}\mathbf{D}^{-1}\mathbf{s} - \dots . \tag{14.43}$$

The first term in this expansion, $\mathbf{s}^t\mathbf{D}^{-1}\mathbf{s}$, is what we computed above when we assumed there were no off-diagonal terms, and the remaining terms are the corrections arising from correlations induced by the detector. If these correlations are sufficiently weak, we may be able to truncate the series after a few terms, making the calculation of SNR easy.

The banded character of the covariance is especially useful if we are trying to detect a spatially compact signal. At the extreme, suppose $\mathbf{s}$ is confined to a single detector element, say $\mathbf{m} = \mathbf{n}$. Then SNR^2 is simply $s_\mathbf{n}^2\,[\mathbf{K}_\mathbf{g}^{-1}]_{\mathbf{nn}}$, and the first correction term in (14.43) becomes

$$\mathbf{s}^t\,\mathbf{D}^{-1}\mathbf{A}\mathbf{D}^{-1}\mathbf{s} = \sum_\mathbf{j}\sum_\mathbf{k} s_\mathbf{n}\,[\mathbf{D}^{-1}]_{\mathbf{nj}}\,[\mathbf{A}]_{\mathbf{jk}}\,[\mathbf{D}^{-1}]_{\mathbf{kn}}\,s_\mathbf{n} = \frac{s_\mathbf{n}^2}{D_\mathbf{nn}^2}A_\mathbf{nn} = 0 \tag{14.44}$$

since the diagonal elements of $\mathbf{A}$ are zero by definition.

The next term in the series is also simplified if we consider a signal confined to a single pixel:

$$\mathbf{s}^t\mathbf{D}^{-1}\mathbf{A}\mathbf{D}^{-1}\mathbf{A}\mathbf{D}^{-1}\mathbf{s} = \frac{s_\mathbf{n}^2}{D_\mathbf{nn}^2}[\mathbf{A}\mathbf{D}^{-1}\mathbf{A}]_\mathbf{nn} = \frac{s_\mathbf{n}^2}{D_\mathbf{nn}^2}\sum_\mathbf{k}\frac{[A_\mathbf{nk}]^2}{D_\mathbf{kk}} . \tag{14.45}$$

If we say that $\mathbf{A}_\mathbf{nk} \simeq 0$ when $|\mathbf{n} - \mathbf{k}|\epsilon > \delta$, then the number of terms we have to sum is of order $[\delta/\epsilon]^2$, which could be quite small. Moreover, if the elements of $\mathbf{A}$ are small compared to $\mathbf{D}_\mathbf{nn}$, then the correction terms are small and the series converges rapidly.

If the signal covers P pixels, the number of computations required is increased by a factor of P^2, and a convergence condition analogous to (14.40) must be satisfied.

Matrix-inversion lemma Suppose we want to invert an overall covariance matrix of the form

$$\mathbf{K_g} = \overline{\mathbf{K}}_\mathbf{n} + \widehat{\mathbf{K}}_{\overline{\mathbf{g}}} = \overline{\mathbf{K}}_\mathbf{n} + \mathbf{W}\mathbf{W}^t , \tag{14.46}$$

where we have assumed that $\widehat{\mathbf{K}}_{\overline{\mathbf{g}}}$ is given by (14.33). For electronic or Poisson noise $\overline{\mathbf{K}}_\mathbf{n}$ will be diagonal, but in some applications correlations will be introduced by the detector and $\overline{\mathbf{K}}_\mathbf{n}$ will be a nearly diagonal, banded matrix (see, for example, the discussion of x-ray detectors in Sec. 12.3.8).

By the matrix-inversion lemma (A.56*a*), we see that

$$\left[\overline{\mathbf{K}}_\mathbf{n} + \mathbf{W}\mathbf{W}^t\right]^{-1} = \overline{\mathbf{K}}_\mathbf{n}^{-1} - \overline{\mathbf{K}}_\mathbf{n}^{-1}\mathbf{W}\left[\mathbf{I} + \mathbf{W}^t\overline{\mathbf{K}}_\mathbf{n}^{-1}\mathbf{W}\right]^{-1}\mathbf{W}^t\overline{\mathbf{K}}_\mathbf{n}^{-1} . \tag{14.47}$$

The advantage of this form is that $[\mathbf{I} + \mathbf{W}^t\overline{\mathbf{K}}_\mathbf{n}^{-1}\mathbf{W}]$ is an $N_s \times N_s$ matrix, where N_s is a few hundred in practice, rather than an $M \times M$ matrix, where M may be 10^6. Moreover, since $\mathbf{W}^t\mathbf{K}_n^{-1}\mathbf{W}$ is positive-semidefinite, the inverse of the $N_s \times N_s$ matrix will always exist. Thus, if $\overline{\mathbf{K}}_\mathbf{n}$ can be inverted, either trivially because it is diagonal or by use of a rapidly convergent Neumann series, then it becomes feasible to add the sample covariance representing object variability.[9]

The matrix-inversion lemma reduces the size of the required inverse from $M \times M$ to $N_s \times N_s$ but it is not a dimensionality-reduction method in the sense that it does not entail potential information loss.

Dimensionality reduction using efficient channels The previous approaches depend upon writing the data covariance matrix as the sum of a full-rank, near-diagonal component representing the measurement noise and a low-rank contribution obtained from samples. When we do not have access to low-noise or noise-free samples from which to estimate the second term, an alternative approach is to make use of efficient channels that allow us to estimate the Hotelling observer's SNR in a lower-dimensional space.

In Sec. 13.2.12 we showed that a limited set of features, when properly chosen, preserves the information in the data in terms of yielding the same SNR for a linear observer. We found that an eigenanalysis of the inter- and intra-class scatter matrices results in full preservation of the separability using only $(L - 1)$ features for optimal linear discrimination between L classes. In the binary classification problem, a single feature is all that is needed—quite a dimensionality reduction.

The requirement for finding that single privileged feature is that complete knowledge of the scatter matrices is available in order to do the eigenanalysis. Without such complete knowledge, we must judiciously apply whatever prior information we have regarding the signals to be discriminated and the background statistics to find channels that reduce the dimensionality of the problem with limited loss of detectability.

For a particular set of channel profiles, the Hotelling formalism can be applied in the channel space to determine the $\mathbf{w_v}$, which is the vector of optimal channel weights. By analogy with (14.19), we write the template in the channel space as

$$\mathbf{w_v} = \mathbf{K}_\mathbf{v}^{-1}\Delta\overline{\mathbf{v}} , \tag{14.48}$$

[9]This idea was suggested to us by Brandon D. Gallas (see Barrett *et al.*, 2001).

where $\Delta\overline{\mathbf{v}}$ is the difference in channel outputs under the two hypotheses,

$$\Delta\overline{\mathbf{v}} = \mathbf{U}^t \Delta\overline{\mathbf{g}}, \tag{14.49}$$

and $\mathbf{K_v}$ is the $P \times P$ covariance matrix of the channel outputs:

$$\mathbf{K_v} = \mathbf{U}^t \mathbf{K_g} \mathbf{U}. \tag{14.50}$$

The SNR on the channel outputs is given by

$$\text{SNR}_\mathbf{v}^2 = \Delta\overline{\mathbf{v}}^t \mathbf{K}_\mathbf{v}^{-1} \Delta\overline{\mathbf{v}}. \tag{14.51}$$

From Sec. 13.2.12 we know that efficient features are ones that preserve the separability of the data in a space of reduced dimensionality, achieving $\text{SNR}_\mathbf{v}^2 = \text{SNR}_\mathbf{g}^2$.

Laguerre-Gauss channels Consider the example of a detection task in which the detected signal is approximately radially symmetric, centrally peaked and smooth, and situated at a known location on a stationary background with a correlation that has no preferred orientation. With these assumptions it can be expected that the ideal linear template will be centered at the known position of the signal, rotationally symmetric and smooth before discretization to match the CD nature of the imaging system.[10] Laguerre-Gauss channel profiles have been proposed by Barrett *et al.* (1998c) for this task because they form a basis on the space of rotationally-symmetric square-integrable functions in $\mathbb{R}^2$.

The Laguerre polynomials are defined in (4.57) as

$$L_p(x) = \sum_{k=0}^{p} (-1)^p \binom{p}{k} \frac{x^k}{k!}. \tag{14.52}$$

The orthogonality relation for these polynomials is given by (4.58):

$$\int_0^\infty dx\; e^{-x} L_p(x)\, L_{p'}(x) = \delta_{pp'}. \tag{14.53}$$

We can transform this relationship to a two-dimensional form with the change of variables $x = 2\pi r^2/a_u^2$, where r is the radial distance and a_u plays the role of a scaling factor, giving

$$\frac{1}{2\pi}\int_0^{2\pi} d\theta \int_0^\infty \frac{4\pi r\, dr}{a_u^2} \exp\left(\frac{-2\pi r^2}{a_u^2}\right) L_p\left(\frac{2\pi r^2}{a_u^2}\right) L_{p'}\left(\frac{2\pi r^2}{a_u^2}\right) = \delta_{pp'}. \tag{14.54}$$

We see that the exponential factor of (14.53) has been transformed to a Gaussian factor in (14.54). From this equation we can define the Laguerre-Gauss (LG) functions as

$$u_p(r|a_u) = \frac{\sqrt{2}}{a_u} \exp\left(\frac{-\pi r^2}{a_u^2}\right) L_p\left(\frac{2\pi r^2}{a_u^2}\right), \tag{14.55}$$

where the $\{u_p\}$ are orthogonal (without weighting factors) over $\mathbb{R}^2$ by (14.54).

Figure 14.7 shows radial dependencies of the first, third, and ninth LG functions, as well as their 2D forms. In order to apply these continuous functions to a

[10] Of course, a signal defined on a square pixel grid cannot be exactly rotationally symmetric, but we can ignore this problem if the template covers many pixels.

discrete data set, the functions must be sampled on the same grid used to discretize the data.

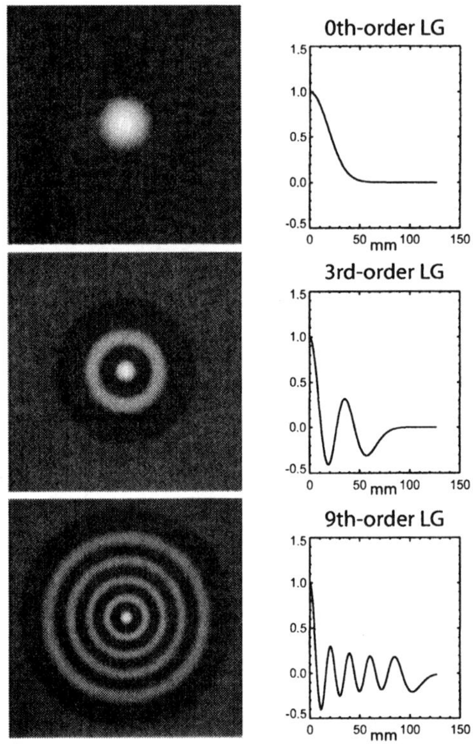

Fig. 14.7 The first, third and ninth Laguerre-Gauss functions: *Left*: 2D functions; *Right*: Radial forms. (Courtesy of Brandon Gallas.)

Because the LG functions form a basis for radially-symmetric functions in 2D, they can be used to exactly represent any rotationally symmetric function $f(r)$ by

$$f(r) = \frac{\sqrt{2}}{a_u} \exp\left(\frac{-\pi r^2}{a_u^2}\right) \sum_{p=0}^{\infty} \alpha_p \, L_p\left(\frac{2\pi r^2}{a_u^2}\right), \tag{14.56}$$

where

$$\alpha_p = \int d^2r \; u_p(r) \, f(r) \, . \tag{14.57}$$

Knowledge of the signal and background can be used to choose a_u and estimate the coefficients α_p for a finite set of channels. Alternatively, a range of values for a_u can be investigated, with the number of channels increased until the detectability reaches a maximum over a_u and P. This approach has been investigated extensively by Gallas and Barrett (2003) for an SKE task on a lumpy background with widely varying statistical parameters. These authors found excellent agreement between the channelized linear observer's performance and the ideal linear observer's performance with a small number of channels $(5-30)$. The number of channels needed was found to depend on the complexity of the background statistics.

Channel models for predicting human performance While the previous paragraphs specifically address the considerations that come to the fore when using channels to estimate the performance of the optimal linear observer, the advantages offered by the dimensionality reduction of the channelized-Hotelling approach are common to all linear channel models. As described in Sec. 14.2.2, a variety of linear channel models have been proposed for use in the prediction of human performance. All such models have the similar characteristic that they result in a calculable figure of merit for model-observer performance based on dimensionality reduction.

All channels designed to model the human have another similarity: because the human visual system is insensitive to broad, structureless regions, channel models designed to predict human performance have zero response at zero spatial frequency. As stated in Sec. 4.1.4, Laguerre-Gauss functions are eigenfunctions of the 2D rotationally symmetric Fourier operator. Thus the LG channel profiles in the Fourier domain have the same form as the space-domain channels shown in Fig. 14.7. The LG channels are peaked at $\rho = 0$ in the Fourier domain, just as they are peaked at $r = 0$ in the space domain. The LG channels are therefore not recommended for use in modeling human performance.

The body of literature providing the range of applicability of the various candidate channel models for predicting human performance continues to grow. Whenever a given model is utilized, it is important to validate the performance predictions with psychophysical studies involving human observers if the task or the statistics of the data sets are outside the range of experimental conditions for which the model has previously been shown to be predictive of human performance.

Random signals We have described a variety of methods for estimating the Hotelling observer's performance for SKE tasks. Random signals present an additional level of complexity (and realism). Even so, the generalization of the Hotelling approach to random signals is often straightforward. In particular, when the signals are low contrast, we have already stressed that the data covariance is approximately equal to its composition in the SKE case. In that case the only new question that arises is the estimation of the mean data vector that appears in (14.20).

Estimation of the mean data vector If the task is the detection of a random signal, and there is no prior information regarding the signal distribution, it is straightforward to estimate the sample means for the two classes and subtract them to determine $\Delta\hat{\overline{\mathbf{g}}}$. The sample mean is the maximum-likelihood estimate of the true mean. The number of values to be estimated is the number of nonzero elements in the difference $(\hat{\overline{\mathbf{g}}}_2 - \hat{\overline{\mathbf{g}}}_1)$, which is determined by the extent of the signal as seen through the imaging system.

Prior information can be brought to bear on the estimation of the mean difference vector in a number of ways. If the signal is compact and there is prior information regarding its location, this information can be used to limit the number of values to be estimated to those within a certain region of the image. In the case of random signals of a specified shape, prior information regarding the signal's form can be used to reduce the number of parameters to be estimated to a small set, for example, signal amplitude, width, or location. Furthermore, prior information regarding the underlying distributions of the random parameters can be used to form Bayesian estimation procedures according to the theory presented in Chap. 13.

It should be noted that Hotelling SNR may be a poor indicator of system performance with large signal variability, as discussed in Sec. 13.2.12. If a signal can be located anywhere within a wide field of view, the signal averaged over location is a broad, structureless function and the detectability of the Hotelling observer, or any linear observer, becomes very small. One way around this problem is to replace the original two-alternative detection problem with an $(L+1)$-alternative problem where the signal can be at one of L nonoverlapping locations. The simple detection decision can then be made by choosing the location for which the response of the Hotelling observer is maximum, but we also get information on lesion location this way. Another possibility is to allow signal location to be a parameter in the SNR and compute a detectability map as described next.

Signal known exactly, but variable Let us assume that the signal varies randomly but is known to the observer on each trial (the only uncertainty being whether it is present). This task is sometimes referred to as the signal-known-exactly-but-variable, or SKEV, task (Eckstein and Abbey, 2001; Eckstein *et al.*, 2002). Let the randomness in the signal be captured by a random parameter vector $\boldsymbol{\theta}$. For each value of $\boldsymbol{\theta}$, the optimum linear test statistic is given by [*cf.* (13.208)]

$$\hat{\mathbf{w}}(\boldsymbol{\theta}) = [\widehat{\mathbf{K}}_{\mathbf{g}}(\boldsymbol{\theta})]^{-1}\mathbf{s}(\boldsymbol{\theta}) , \tag{14.58}$$

where the estimate of $\mathbf{K_g}$ and its inverse must be determined using the methods described above. In particular, the method of template estimation given above may be used to estimate (14.58) without the need for finding an inverse of $\mathbf{K_g}$ in some cases.

The Hotelling SNR can be estimated for each value of the random parameter, following (13.209):

$$\widehat{\mathrm{SNR}}^2_{Hot}(\boldsymbol{\theta}) = \frac{\{[\hat{\mathbf{w}}(\boldsymbol{\theta})]^t\mathbf{s}(\boldsymbol{\theta})\}^2}{[\hat{\mathbf{w}}(\boldsymbol{\theta})]^t\widehat{\mathbf{K}}_{\mathbf{g}}(\boldsymbol{\theta})\,\hat{\mathbf{w}}(\boldsymbol{\theta})} = [\hat{\mathbf{w}}(\boldsymbol{\theta})]^t\mathbf{s}(\boldsymbol{\theta}) , \tag{14.59}$$

where the second form follows from (14.58).

A summary measure of observer performance can be obtained by averaging (14.59) over $\boldsymbol{\theta}$ if $\mathrm{pr}(\boldsymbol{\theta})$ is known. Alternatively, a detectability map, which plots the SNR^2 as a function of $\boldsymbol{\theta}$, can be presented. Eckstein *et al.* (2002) have found that the optimal parameters for image compression are the same when evaluated using either an SKE or an SKEV paradigm.

AUC and the linear discriminant Thus far we have concentrated on the estimation of the SNR for the Hotelling observer. As discussed in Chap. 13, the Hotelling observer gives maximal SNR and maximal AUC when the data are Gaussian distributed. For non-Gaussian data, the Hotelling observer may not give the best AUC that can be achieved by a linear observer. It is therefore of interest to consider the behavior of AUC for an arbitrary linear discriminant and investigate methods for maximizing this alternative, and arguably superior, figure of merit.

It was shown in (13.44) that

$$\mathrm{AUC}_{lin} = \frac{1}{2} + \frac{1}{2\pi i}\mathcal{P}\int_{-\infty}^{\infty}\frac{d\xi}{\xi}\,\psi_{\mathbf{g}1}(\mathbf{w}\xi)\,\psi^*_{\mathbf{g}2}(\mathbf{w}\xi) , \tag{14.60}$$

where $\mathbf{w}$ is the arbitrary $M\times 1$ template of (14.18) that generates the test statistic t from each data vector $\mathbf{g}$, and $\psi_{\mathbf{g}j}(\cdot)$ is the characteristic function for the data under hypothesis j. Limiting forms of (14.60) for nonrandom signals and for weak signals are given in Sec. 13.2.5. Note that (14.60) is just a 1D integral—only one line through the multivariate characteristic function under each hypothesis is needed once $\mathbf{w}$ is specified.

This formula for AUC is useful when we have analytic forms for the characteristic functions of the data under the two hypotheses. In background-known-exactly (BKE) problems, we might know the characteristic functions directly from the data statistics, but if the background is random we have to first characterize the object statistics and then propagate them into the data domain as discussed in Sec. 8.5.3. If the object is regarded as a continuous function, we need first to obtain an analytic expression for its characteristic functional, then apply (8.335) or (8.339) to obtain the characteristic functions for the data. For example, lumpy and clustered lumpy backgrounds were introduced in Sec. 8.4.4, and their characteristic functionals were derived in Sec. 11.3.10. Additional examples of analytic characteristic functions will be given in Chap. 18.

When the needed characteristic functions are available, an iterative search can be used to maximize the AUC given by (14.60); useful search algorithms are discussed in Sec. 15.4.3. Since the integral is one-dimensional, this search is not particularly computationally expensive.

A major advantage of the approach suggested by (14.60) is that no matrix inversion is required, unlike the determination of the full Hotelling SNR. While an iterative approach can be used to determine the Hotelling SNR when the noise contribution to the covariance matrix is known, it works by searching for the optimum linear template and indirectly obtaining the SNR. An iterative solution for (14.60) directly yields AUC.

The linear discriminant obtained by searching for the $\mathbf{w}$ that maximizes AUC may differ from the Hotelling observer, as discussed in Sec. 13.2.12. When this occurs, the linear discriminant that gives higher AUC is to be preferred whenever our goal is the linear approximation to the ideal observer.

Errors in estimates of SNR for linear observers It is natural to ask how close the estimated SNR is to the true SNR that would have been obtained with full knowledge of the ensemble statistics of the data. That is, we would like to know the bias and variance of the estimate. In this context, bias and variance refer to the first-and second-order statistics of the estimate when different finite sets of images are used. There are several methods, briefly surveyed below, to estimate the magnitude of the bias and variance from this source.

As with any real-world estimation problem, however, there can also be a systematic bias arising from invalid assumptions or modeling errors, and this kind of bias is much more difficult to assess. With computer-generated images, a major source of systematic bias is unrealistic or oversimplified simulation; with real images, a major problem is uncertainty in the true diagnosis. Both of these issues are discussed in Sec. 14.4; here we focus on statistical errors.

It is straightforward to estimate the variance of estimates of SNR or AUC when simulated images are used; all that is needed is to repeat the simulation several times with independent sets of images and compute the sample variance of the values obtained. More sophisticated resampling methods (see below) can also be used, but their only advantage is a saving in computer time, seldom a primary concern these days. In fact, with simulated images the variance and the statistical bias can be made arbitrarily small simply by running the computer long enough. If real images are used, however, the number of images might be quite limited, and it becomes more critical to estimate the error associated with an estimate of SNR^2.

Errors in direct estimation of SNR in channel space One situation in which we can give not only the bias and variance but indeed the full probability density function of the estimated SNR^2 is when dimensionality reduction is performed with efficient or anthropomorphic channels and the resulting channel outputs are normally distributed. In that case we can estimate SNR^2 by

$$\widehat{\mathrm{SNR}^2} \equiv [\widehat{\Delta\overline{\mathbf{v}}}]^t\, \widehat{\mathbf{K}}_{\mathbf{v}}^{-1} [\widehat{\Delta\overline{\mathbf{v}}}]\,, \qquad (14.61)$$

where the hats here denote estimates obtained by sample averages; it is assumed that the number of sample images is larger than the number of channels so that the sample covariance matrix is invertible.

The estimator defined in (14.61) is precisely the one studied by Hotelling in his classic 1931 paper, and it is often referred to as *Hotelling's* T^2 *statistic.* The PDF of T^2 is closely related to the F distribution; for details see Hotelling (1931) or Anderson (1971). The general behavior of the estimate is illustrated in Fig. 14.8, where it is seen that the estimate is highly biased unless the number of sample images is much larger than the number of channels (and of course it is not even defined if the number of sample images is less than the number of channels).

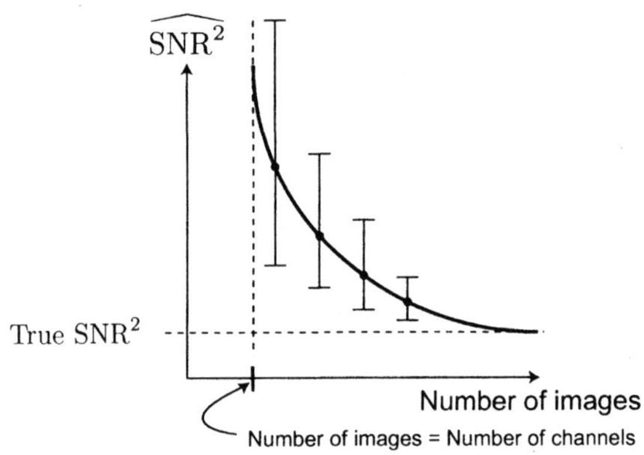

Fig. 14.8 Schematic behavior of the Hotelling T^2 estimate of SNR^2 as defined in (14.61). The dashed horizontal line indicates the true value of SNR^2 on the channel outputs, and the solid curve shows the mean of the estimate. The error bars are indicative of the variance. We thank Andy Alexander for suggesting this kind of plot.

The basic problem with the Hotelling T^2 estimate is that it makes no use of prior information about the quantity being estimated, namely the SNR^2 on the channel outputs. One key piece of prior information in many cases is knowledge of the mean difference signal in data space, $\overline{\mathbf{g}}$, from which we can determine the mean difference signal in channel space, $\Delta\overline{\mathbf{v}}$, by (14.49). If we regard $\Delta\overline{\mathbf{v}}$ as known and nonrandom, we can define a better estimate of SNR^2 by

$$\widehat{\mathrm{SNR}^2} \equiv [\Delta\overline{\mathbf{v}}]^t\, \widehat{\mathbf{K}}_{\mathbf{v}}^{-1} [\Delta\overline{\mathbf{v}}]\,. \qquad (14.62)$$

Another key piece of prior information is the covariance decomposition (14.27). If we regard $\overline{\mathbf{K}}_{\mathbf{n}}$ as known and nonrandom and use (14.51), the estimate in (14.62) is modified to

$$\widehat{\mathrm{SNR}^2} \equiv [\Delta\overline{\mathbf{v}}]^t \left[\mathbf{U}^t \left(\overline{\mathbf{K}}_{\mathbf{n}} + \widehat{\mathbf{K}}_{\overline{\mathbf{g}}}\right) \mathbf{U}\right]^{-1} [\Delta\overline{\mathbf{v}}] . \tag{14.63}$$

Note that the hat, denoting sample estimates, now appears over only $\widehat{\mathbf{K}}_{\overline{\mathbf{g}}}$, so only that term contributes to the bias and variance of the estimate of SNR^2.

The statistical properties of (14.62) and (14.63) have not yet been derived, but they should offer substantially smaller bias and variance than the T^2 estimate of (14.61) simply because they use more prior information. All of these estimates, however, assume that the channel outputs are normally distributed; it is advisable to plot experimental histograms to check this assumption.

Training and testing An alternative to direct estimation of SNR^2 is first to estimate the template $\mathbf{w}$ and then to apply it to a set of sample images. When only a single, finite set of images is available, the experimenter must use the set of images for two purposes: training the observer (choosing the number of channels, their weights, and any parameters that characterize the channel profiles); and testing the observer (estimating its performance). This is the so-called "training-testing" paradigm. The training-testing label applies even without dimensionality-reducing feature extraction. When we estimate the template from samples by any method, we are training the observer.

There are two common ways to train and test an observer with a single set of sample images. The first option is to split the data into two independent sets, one set to be used to train the observer and the other to be used for testing the observer. The split does not need to be into subsets of equal size. This approach is sometimes referred to as the holdout method. A related method is the use of $N_s - 1$ images to train the observer, with the final sample used to test the observer. This method is known as the round-robin approach; by repeating the training/testing sequence N_s times, keeping score of the observer's decision variable each time, an estimate of the observer's performance is obtained over the entire data set. However, the round-robin method does not yield a single observer, but rather, each held-out image is tested on a different observer.

Gallas (2003) investigated various resampling approaches for determining the bias and variance of the performance estimate for the channelized linear observer trained and tested using variations on the hold-out method. Using a very large set of independent estimates of observer performance (the beauty of Monte Carlo image simulation), Gallas was able to determine the true performance of the channelized observer and thus calculate the bias as well as the variance of the finite-sample methods.

The second training-testing option is the resubstitution method, where the observer is trained and tested on the same set of images. The use of a single set of images to estimate the observer's template, followed by an estimation procedure that applies that template to the data to determine the first- and second-order statistics of t under each hypothesis to derive an SNR, will give an optimistic result (Wagner *et al.*, 1997). The resulting estimates of observer performance correspond to the results obtained via (14.61) and illustrated in Fig. 14.8.

14.3.3 Ideal observers

We learned in Chap. 13 that the ideal observer for binary classification tasks is one that bases its decision on the likelihood ratio. Many properties of the likelihood ratio and its logarithm, and of performance metrics derived from them, were given in Sec. 13.2. In this section we review a variety of approaches to using these often abstract mathematical concepts in the practical assessment of image quality.

Analogies with the Hotelling problem The basic challenge in computing the test statistic for both the Hotelling and the ideal observer is dimensionality. For the Hotelling observer, we need to construct and invert a huge covariance matrix; for the ideal observer, we need to form huge-dimensional multivariate probability density functions and take ratios of them. In neither case are brute-force methods likely to be fruitful; in both cases we must make use of prior information about the task and imaging system in order to make progress.

An important piece of prior information for the Hotelling problem is the conditional covariance $\mathbf{K}_{\mathbf{n}|\mathbf{f}}$, which is known from the physics of the measurement process. For example, $\mathbf{K}_{\mathbf{n}|\mathbf{f}}$ for raw, unprocessed data and Gaussian noise is given in (14.29), and for Poisson noise it is given by (14.30). The analogous prior information for the ideal observer is the conditional PDF $\text{pr}(\mathbf{g}|\mathbf{f})$, which is again known from the physics. Before processing, $\text{pr}(\mathbf{g}|\mathbf{f})$ is often multivariate Gaussian or multivariate Poisson, and in both cases the multivariate PDF can often be written as a product of univariate PDFs. The effect of processing is discussed in Secs. 15.2.6, 15.4.2 and 15.4.7.

In both Hotelling and ideal-observer studies, it is necessary to choose the object model carefully, allowing enough complexity and variability to capture the essence of real objects, yet retaining adequate mathematical tractability. In both cases, object models such as the lumpy and clustered lumpy backgrounds introduced in Sec. 8.4 are very useful.

The signal model, too, can be chosen to facilitate the computation. In particular, nonrandom signals are very attractive, though it remains an open question how well conclusions from SKE studies can be applied to more realistic tasks.

Decomposition of the PDFs The likelihood ratio is the ratio of two PDFs, each of which can be written somewhat abstractly as

$$\text{pr}(\mathbf{g}|H_j) = \int d\mathbf{f}\ \text{pr}(\mathbf{g}|\mathbf{f})\ \text{pr}(\mathbf{f}|H_j)\,, \qquad (j = 1, 2)\,. \tag{14.64}$$

The notation $\text{pr}(\mathbf{f})$ is explained in Sec. 8.2.2 [see especially (8.78) and (8.81)]. In brief, it denotes the density on the full (potentially infinite) set of parameters needed to specify the object as a random process $f(\mathbf{r})$.

The density on the data can also be written as

$$\text{pr}(\mathbf{g}|H_j) = \langle \text{pr}(\mathbf{g}|\mathbf{f}) \rangle_{\mathbf{f}|H_j}\ . \tag{14.65}$$

Numerous alternative forms of $\text{pr}(\mathbf{g}|H_j)$, with various assumptions about the object and the noise, are given in Sec. 8.5.4.

Thus, in order to determine the densities needed in the likelihood ratio, we need both the conditional density on the data for a given object, $\text{pr}(\mathbf{g}|\mathbf{f})$, and the densities $\text{pr}(\mathbf{f}|H_j)$ on the object under the two hypotheses. Note that $\text{pr}(\mathbf{g}|\mathbf{f})$ does not depend directly on the hypothesis H_j; specifying the object specifies the mean of $\mathbf{g}$, and that in turn specifies the full density in most cases.[11] Note also that we do not refer to $\mathrm{pr}(\mathbf{g}|\mathbf{f})$ as a likelihood since it is never our goal to estimate $\mathbf{f}$; it is not the goal in this section since we are discussing a classification problem, and it is not even the goal in image reconstruction (see Chap. 15).

Conditional PDFs To be more specific about $\mathrm{pr}(\mathbf{g}|\mathbf{f})$, we need to distinguish direct from indirect imaging and object-dependent from object-independent noise, just as we did in Sec. 14.3.2 when we discussed $\mathbf{K}_{\mathbf{n}|\mathbf{f}}$ [see (14.29) and (14.30)].

Consider first the case of direct imaging with a detector array limited by Gaussian electronic noise. If we assume that all elements in the array are identical and that each generates its own noise independently of the other elements, then the probability density function of $\mathbf{n}$ is

$$\mathrm{pr}_{\mathbf{n}}(\mathbf{n}) = (2\pi\sigma^2)^{-M/2} \prod_{m=1}^{M} \exp\left(-\frac{n_m^2}{2\sigma^2}\right) . \tag{14.66}$$

Since the electronic noise is independent of the mean detector output, the conditional density on the data is just a shifted version of the noise density:

$$\mathrm{pr}(\mathbf{g}|\mathbf{f}) = \mathrm{pr}_{\mathbf{n}}[\mathbf{g} - \overline{\mathbf{g}}(\mathbf{f})] = (2\pi\sigma^2)^{-M/2} \prod_{m=1}^{M} \exp\left\{-\frac{[g_m - \overline{g}_m(\mathbf{f})]^2}{2\sigma^2}\right\} . \tag{14.67}$$

For linear systems we can go a step further and write $\overline{g}_m(\mathbf{f}) = [\mathcal{H}\mathbf{f}]_m$.

Similarly, with raw Poisson measurements we have

$$\mathrm{pr}(\mathbf{g}|\mathbf{f}) = \prod_{m=1}^{M} \exp\left[-\overline{g}_m(\mathbf{f})\right] \frac{[\overline{g}_m(\mathbf{f})]^{g_m}}{g_m!} . \tag{14.68}$$

Thus in both of these cases the multivariate density is a product of univariate densities.

The situation is more complicated if we regard $\mathbf{g}$ as the output of some data-processing or image-reconstruction step. Linear processing leaves Gaussian data Gaussian but introduces correlations. Nevertheless, it is straightforward to write down a multivariate expression for $\mathrm{pr}(\mathbf{g}|\mathbf{f})$ since we know how to compute mean vectors and covariance matrices after linear operations, and a multivariate normal is fully specified by its mean and covariance. There is no simple way of expressing $\mathrm{pr}(\mathbf{g}|\mathbf{f})$ after linear processing of Poisson data, but it may be valid to approximate it with a suitably correlated multivariate normal (see Sec. 15.2.6).

Noise on the output of iterative reconstruction algorithms is discussed in Secs. 15.4.2 and 15.4.7. If the algorithm is nonlinear and enforces a positivity constraint, then the noise cannot be Gaussian since negative values cannot occur. Specifically, with multiplicative algorithms such as MLEM (maximum-likelihood expectation-maximization), it often happens that the PDF on the reconstructed image is approximately a correlated log-normal (Wilson *et al.*, 1994; Barrett *et al.*, 1994).

[11]An exception to this statement will be given in Sec. 18.6.4 where we discuss speckle. As we shall see there, in some speckle problems the variance of the data is different for the signal-present and signal-absent hypotheses.

To summarize, with raw, unprocessed data, $\mathrm{pr}(\mathbf{g}|\mathbf{f})$ usually has a simple analytic form (independent Gaussian or Poisson). With processing, the elements of $\mathbf{g}$ are no longer statistically independent, but it is usually possible to give at least an approximate form for the conditional density. In what follows we shall assume throughout that $\mathrm{pr}(\mathbf{g}|\mathbf{f})$ is known analytically.

As a notational point, we see from (14.66) and (14.67) that the conditional density on unprocessed data $\mathbf{g}$ is completely determined by its mean with both the Gaussian and Poisson noise models, so $\mathrm{pr}(\mathbf{g}|\mathbf{f}) = \mathrm{pr}[\mathbf{g}|\overline{\mathbf{g}}(\mathbf{f})]$. The same is true after processing; if we know $\overline{\mathbf{g}}$, we can specify the density on $\mathbf{g}$, and we know $\overline{\mathbf{g}}$ if we know $\mathbf{f}$. We shall therefore write $\mathrm{pr}(\mathbf{g}|\mathbf{f})$ and $\mathrm{pr}(\mathbf{g}|\overline{\mathbf{g}})$ interchangeably, depending on which conditional variable we wish to emphasize.

Object statistics Statistical properties of objects were the subject of Sec. 8.4. The viewpoint adopted there regards the object as a random process for which each sample function is a vector in a Hilbert space. Since we are concerned only with the measurement component of the object, the Hilbert space of interest has a finite but huge dimensionality. We saw a few cases where the object statistics could be specified analytically, for example as a Gaussian random process or a Gaussian mixture, but in most cases analytic models are either unavailable or unrealistic. The two main options in those cases are to reduce the dimensionality of the statistical description of the object or to use a constructive model that allows us to simulate sample objects even if we cannot specify their statistics.

Dimensionality reduction rests on the assumption that somehow the essential features of a complicated random process can be captured with a relatively small number of parameters. As discussed in Sec. 8.4.1, approaches to finding this low-dimensional representation include principal components analysis (PCA) and independent components analysis (ICA). When ICA is applied to images, it is found that the independent components are the outputs of bandpass filters similar to wavelets or the channels in the human visual system; indeed, some have speculated that our visual system has evolved to extract approximately statistically independent components of natural scenes, thereby permitting efficient transformation of information to the brain.

We postulate that there exist similar low-dimensional representations of objects, as opposed to images, and that they again involve bandpass filters or channels. We know from the discussion in Sec. 8.4.3 that the univariate PDFs on the channel outputs have a long-tailed, kurtotic form. Sometimes they are described empirically in terms of the Lévy family, defined not by the density but by the characteristic function, which has the form $\psi(\xi) = \exp(-b|\xi|^q)$. If the channels are chosen so that the outputs are approximately statistically independent, the multivariate object statistics are described by a finite product of characteristic functions of this form.

The constructive models that have received the most attention in image-quality studies are lumpy and clustered lumpy backgrounds. As defined in (8.303), a sample function of a lumpy background is specified exactly by stating the lump positions $\{\mathbf{r}_n\}$ as well as the number of lumps N. The statistical properties are fully specified by giving the probability laws for $\mathbf{r}_n$ and N.

Alternatively, for many constructive models, the object statistics can be specified by giving an analytic form for the characteristic functional associated with the random field. This concept was introduced in Sec. 8.2.3, and the specific forms for lumpy and clustered lumpy backgrounds were calculated in Sec. 11.3.9. Other constructive models that can be used to synthesize texture fields, and for which analytic characteristic functionals are available, will be introduced in Chap. 18.

To summarize, random objects may be specified by huge-dimensional PDFs, by lower-dimensional PDFs on channel outputs, by rules that let us construct sample functions, and/or by characteristic functionals. In what follows we shall see how each of these descriptions aids us in the computation of ideal-observer performance.

From object domain to data domain If we have either a statistical or a constructive specification of a random object, the next step is to transform it into the data domain. For constructive models, this step is straightforward in principle; one generates the random object and uses it to simulate the random image. Simulation methods are discussed in Sec. 14.4.

To discuss transformation of the PDF, we need to distinguish linear from nonlinear imaging systems. A general rule for nonlinear transformation of bivariate PDFs is given in (C.104), but it does not extend usefully to high-dimensional multivariate problems since the Jacobian cannot be evaluated. So far as the authors can see, there is no hope of transforming an object PDF through a nonlinear imaging system.

The transformation rules for linear systems are most easily expressed in terms of characteristic functions and functionals. If $\overline{\mathbf{g}}(\mathbf{f}) = \boldsymbol{\mathcal{H}}\mathbf{f}$, with $\boldsymbol{\mathcal{H}}$ a linear CD operator, then we know from (8.96) that the characteristic function for the random vector $\overline{\mathbf{g}}$ under hypothesis H_j is given by

$$\psi_{\overline{\mathbf{g}}|H_j}(\boldsymbol{\xi}) = \Psi_{\mathbf{f}|H_j}(\boldsymbol{\mathcal{H}}^\dagger \boldsymbol{\xi}), \tag{14.69}$$

where $\boldsymbol{\xi}$ is an $M \times 1$ vector, $\psi_{\overline{\mathbf{g}}|H_j}(\boldsymbol{\xi})$ is the characteristic function for $\overline{\mathbf{g}}$, and $\Psi_{\mathbf{f}|H_j}(\boldsymbol{\sigma})$ is the characteristic functional of the object $\mathbf{f}$, with $\boldsymbol{\sigma}$ being a vector in the same Hilbert space as $\mathbf{f}$, *e.g.*, $\boldsymbol{\sigma}$ corresponds to a function $\sigma(\mathbf{r})$.

If we write $\mathbf{g} = \boldsymbol{\mathcal{H}}\mathbf{f} + \mathbf{n}$ and assume that $\mathbf{n}$ is object-independent, then we know from (8.335) that

$$\psi_{\mathbf{g}|H_j}(\boldsymbol{\xi}) = \psi_{\mathbf{n}}(\boldsymbol{\xi})\, \psi_{\overline{\mathbf{g}}|H_j}(\boldsymbol{\xi}) = \psi_{\mathbf{n}}(\boldsymbol{\xi})\, \Psi_{\mathbf{f}|H_j}(\boldsymbol{\mathcal{H}}^\dagger \boldsymbol{\xi}). \tag{14.70}$$

For Poisson noise, (8.339) tells us that

$$\psi_{\mathbf{g}|H_j}(\boldsymbol{\xi}) = \Psi_{\mathbf{f}|H_j}[\boldsymbol{\mathcal{H}}^\dagger\, \boldsymbol{\Gamma}(\boldsymbol{\xi})], \tag{14.71}$$

where

$$[\boldsymbol{\Gamma}(\boldsymbol{\xi})]_m = \frac{-1 + \exp(-2\pi i\, \xi_m)}{-2\pi i}. \tag{14.72}$$

Expressions for the data PDFs can be obtained by performing an inverse MD Fourier transform on each expression for $\psi_{\mathbf{g}|H_j}(\boldsymbol{\xi})$.

For signal-known-exactly tasks, it follows from the Fourier shift theorem that

$$\psi_{\mathbf{g}|H_2}(\boldsymbol{\xi}) = \exp(-2\pi i \boldsymbol{\xi}^t \mathbf{s})\, \psi_{\mathbf{g}|H_1}(\boldsymbol{\xi}), \tag{14.73}$$

where $\mathbf{s}$ is the nonrandom signal in data space. Thus it suffices to know the no-signal or background-only characteristic function in this case.

Estimation of object statistics In many cases we can express the object statistics in parametric form. For example, with stationary lumpy backgrounds the width of a single lump and the number of lumps per unit area (or volume) fully describe the random process. Similarly, if the object statistics are specified in terms of the outputs of bandpass channels, we could assume that the univariate characteristic function for the nth channel has the Lévy form $\psi_n(\xi) = \exp(-b_n|\xi|^{q_n})$; if we assume further that the channel outputs are statistically independent, then the multivariate object statistics are specified by the sets $\{b_n\}$ and $\{q_n\}$.

We can use the freedom in choosing these parameters to create a wide variety of random object fields. Moreover, if we can estimate the parameters from a training set of real images, we can tailor the object description to a particular physical situation. The problem is that the training set will consist of *images*, and we want to find the parameters for describing *objects*, in spite of the blur and noise associated with whatever imaging system was used to form the images.

A way of estimating the object parameters from blurred, noisy images was devised by Kupinski *et al.* (2003a). They assumed that the object characteristic functional under the no-signal hypothesis was known except for some parameter vector $\boldsymbol{\alpha}$, so it could be written as $\Psi_{\mathbf{f}|H_1}(\mathbf{s};\boldsymbol{\alpha})$. The corresponding characteristic function in the data domain could then be obtained by one of the transformation rules given above; for example, (14.71) applies with Poisson noise, and

$$\psi_{\mathbf{g}|H_1}(\boldsymbol{\xi};\boldsymbol{\alpha}) = \Psi_{\mathbf{f}|H_1}\left[\boldsymbol{\mathcal{H}}^\dagger\boldsymbol{\Gamma}(\boldsymbol{\xi});\boldsymbol{\alpha}\right] . \tag{14.74}$$

Given a set of signal-absent training images $\{\mathbf{g}_n, n = 1, ..., N_s\}$, Kupinski *et al.* formed the *empirical characteristic function* for the data, which is basically a Monte Carlo estimate of $\psi_{\mathbf{g}|H_1}(\boldsymbol{\xi};\boldsymbol{\alpha})$, defined by

$$\widehat{\psi}(\boldsymbol{\xi}) \equiv \frac{1}{N_s}\sum_{n=1}^{N_s} \exp(-2\pi i\boldsymbol{\xi}^t\mathbf{g}_n) . \tag{14.75}$$

The estimation procedure was then basically minimization of the norm of the difference between the known $\widehat{\psi}(\boldsymbol{\xi})$ and the known analytic form $\psi_{\mathbf{g}|H_1}(\boldsymbol{\xi};\boldsymbol{\alpha})$ from (14.74), minimization being carried out by varying $\boldsymbol{\alpha}$. In practice a weighted least-squares norm was used, taking advantage of the fact that all characteristic functions are unity at $\boldsymbol{\xi} = 0$, so the variance of the estimate $\widehat{\psi}(\boldsymbol{\xi})$ is zero at that point. Moreover, a set of channels was applied to each $\mathbf{g}_n$ to reduce the dimensionality and ease the computational burden. For details, see Kupinski *et al.* (2003a).

The beauty of this procedure is that it gives a statistical description of the underlying objects, independent of the imaging system. Thus, even though a particular imaging system, say one described by an operator $\boldsymbol{\mathcal{H}}_0$, was used to obtain the training images, the characteristic function for another system, described by a general $\boldsymbol{\mathcal{H}}$, can be found from (14.74) once $\boldsymbol{\alpha}$ has been estimated. If we can devise a way of computing ideal-observer performance from this information, we can in principle vary $\boldsymbol{\mathcal{H}}$ and optimize the imaging system for the class of objects from which the training set was drawn.

Estimation of the likelihood ratio In an ideal-observer study, the basic quantity to be calculated is the likelihood ratio, defined by

$$\Lambda(\mathbf{g}) = \frac{\mathrm{pr}(\mathbf{g}|H_2)}{\mathrm{pr}(\mathbf{g}|H_1)} = \frac{\int d\mathbf{f}\ \mathrm{pr}(\mathbf{g}|\mathbf{f})\ \mathrm{pr}(\mathbf{f}|H_2)}{\int d\mathbf{f}\ \mathrm{pr}(\mathbf{g}|\mathbf{f})\ \mathrm{pr}(\mathbf{f}|H_1)} . \tag{14.76}$$

The integrals here are over a potentially infinite-dimensional Hilbert space, but they can be reduced to M dimensions (where M is the number of measurements) by observing that $\text{pr}(\mathbf{g}|\mathbf{f}) = \text{pr}[\mathbf{g}|\overline{\mathbf{g}}(\mathbf{f})]$. If we use (8.351) to decompose the object into background and signal parts,

$$\mathbf{f} = \mathbf{f}_b + \mathbf{f}_s\,, \tag{14.77}$$

and (for a linear system) transform the background and signal into data space as

$$\overline{\mathbf{g}} \equiv \mathbf{b} + \mathbf{s}\,, \qquad \mathbf{b} \equiv \boldsymbol{\mathcal{H}}\mathbf{f}_b\,, \qquad \mathbf{s} \equiv \boldsymbol{\mathcal{H}}\mathbf{f}_s\,, \tag{14.78}$$

then we can write the likelihood ratio as [*cf.* (13.166)]

$$\Lambda(\mathbf{g}) = \frac{\int_\infty d^M b\ \text{pr}(\mathbf{g}|H_2, \mathbf{b})\,\text{pr}(\mathbf{b})}{\int_\infty d^M b\ \text{pr}(\mathbf{g}|H_1, \mathbf{b})\,\text{pr}(\mathbf{b})}\,. \tag{14.79}$$

A useful alternative form of the likelihood ratio is given by (13.169) and (13.170) as

$$\Lambda(\mathbf{g}) = \langle \Lambda_{\text{BKE}}(\mathbf{g}, \mathbf{b}) \rangle_{\mathbf{b}|\mathbf{g},H_1}\,, \tag{14.80}$$

where the subscript BKE indicates background-known-exactly, and

$$\Lambda_{\text{BKE}}(\mathbf{g}, \mathbf{b}) \equiv \frac{\text{pr}(\mathbf{g}|H_2, \mathbf{b})}{\text{pr}(\mathbf{g}|H_1, \mathbf{b})}\,. \tag{14.81}$$

The advantage of this form is that $\Lambda_{\text{BKE}}(\mathbf{g}, \mathbf{b})$ is easy to calculate. In fact, for nonrandom signals it is just the ratio of two conditional densities like (14.67) or (14.68). Note, however, that the required average in (14.80) is with respect to the posterior density on the background, $\text{pr}(\mathbf{b}|\mathbf{g}, H_1)$; we shall learn shortly how to do this average by Monte Carlo methods.

One way of evaluating the performance of the ideal observer on a signal-detection task is to generate sets of signal-present and signal-absent sample images, estimate the likelihood ratio of each image and form an ROC curve. Methods discussed in Sec. 14.2.4 can then be used to estimate the area under the curve or ideal-observer AUC.

A useful surrogate for ideal-observer AUC is the likelihood-generating function evaluated at the origin. We know from (13.97) that this quantity is given by

$$G(0) = -4\ln\left\{\int d^M g\ [\text{pr}(\mathbf{g}|H_1)\ \text{pr}(\mathbf{g}|H_2)]^{\frac{1}{2}}\right\}. \tag{14.82}$$

We can use $G(0)$ to estimate AUC by [*cf.* (13.20) and (13.96)]

$$\text{AUC} \approx \tfrac{1}{2} + \tfrac{1}{2}\,\text{erf}\left(\sqrt{\frac{G(0)}{2}}\right). \tag{14.83}$$

If the log-likelihood ratio is normally distributed or $G(0)$ is large (which means that AUC approaches 1), then this result is exact. Clarkson and Barrett (2000) have found it to be an excellent approximation in a variety of cases with practical values of AUC.

Monte Carlo methods Perusal of expressions such as (14.79) or (14.82) shows that computation of ideal-observer performance in nontrivial cases requires evaluation of huge-dimensional integrals. In Sec. 10.4.5 we introduced the concept of Monte Carlo simulation and commented that it was useful in numerical evaluation of multidimensional integrals. As we shall show, Monte Carlo integration is a very valuable tool in ideal-observer evaluations, but in fact we need to move beyond the simple Monte Carlo methods of Sec. 10.4.5 to the more sophisticated and powerful approach of Markov-chain Monte Carlo (MCMC). Book-length treatments of MCMC are given by Robert and Casella (1999) and Gilks *et al.* (1996). We begin here, however, with simple Monte Carlo integration to illustrate the principles and problems.

To evaluate the numerator or denominator in the likelihood ratio as given in (14.76), we must in principle integrate over an infinite-dimensional space, though we could also use (14.79) to reduce it to M dimensions (which is of little consolation if M is of order 10^6). If, however, we can simulate a set of objects $\{\mathbf{f}_n, n = 1, ..., N_s\}$, then we can approximate those integrals by [*cf.* (10.300)]

$$\mathrm{pr}(\mathbf{g}|H_j) = \int d\mathbf{f}\ \mathrm{pr}(\mathbf{g}|\mathbf{f})\ \mathrm{pr}(\mathbf{f}|H_j) \approx \frac{1}{N_s}\sum_{n=1}^{N_s} \mathrm{pr}(\mathbf{g}|\mathbf{f}_n)\,, \tag{14.84}$$

where the sample must be drawn from $\mathrm{pr}(\mathbf{f}|H_j)$. That is, if H_2 denotes signal-present and H_1 denotes signal-absent, the simulations must include the signal and background for $j = 2$ but only the background for $j = 1$.

Recall that $\mathrm{pr}(\mathbf{g}|\mathbf{f}_n)$ in (14.84) is a known function, for example given by (14.67) or (14.68). In essence, the Monte Carlo integration associates this known function with every sample point $\mathcal{H}\mathbf{f}_n$ in the data space. The method is thus reminiscent of *kernel estimation*, a technique often used to estimate probability densities from a discrete set of samples. The key difference is that choosing the kernel in kernel estimation is a black art. The kernel must be broad enough to fill in the gaps between samples, yet not so broad as to smooth out essential details in the density being estimated. No such issue arises with (14.84); the form of the kernel is dictated by the physics of the problem, and its width is dictated by the noise level.

This is not to say that (14.84) is a panacea. The kernel $\mathrm{pr}(\mathbf{g}|\mathbf{f}_n)$ falls off rapidly as $\mathcal{H}\mathbf{f}_n$ gets farther from the particular $\mathbf{g}$ for which $\Lambda(\mathbf{g})$ is being calculated. If the noise level is small, most randomly chosen $\mathcal{H}\mathbf{f}_n$ will be so far from $\mathbf{g}$ that $\mathrm{pr}(\mathbf{g}|\mathbf{f}_n)$ will be zero to computer precision, and few of the samples will make any contribution to the sum in (14.84). Even though the sum will asymptotically approach $\mathrm{pr}(\mathbf{g}|H_j)$ as N_s goes to infinity, and the estimator is unbiased for all N_s, the variance can be huge for practical finite values of N_s. The problem gets worse as M gets larger or as the noise level gets smaller.

One way to ameliorate this problem in some cases is by *importance sampling*. Suppose we know an analytic form for $\mathrm{pr}(\mathbf{f}|H_j)$, say as a Gaussian mixture or in terms of independent components. Then we are free to rewrite the data density as

$$\mathrm{pr}(\mathbf{g}|H_j) = \int d\mathbf{f}\ \frac{\mathrm{pr}(\mathbf{g}|\mathbf{f})\ \mathrm{pr}(\mathbf{f}|H_j)}{q(\mathbf{f})} q(\mathbf{f})\,, \tag{14.85}$$

where $q(\mathbf{f})$ is a probability density function (*i.e.*, a nonnegative function normalized to unity) with a support large enough that dividing by zero does not become an issue. We can then approximate the data density as

$$\mathrm{pr}(\mathbf{g}|H_j) \approx \frac{1}{N_s} \sum_{n=1}^{N_s} \frac{\mathrm{pr}(\mathbf{g}|\mathbf{f}_n)\,\mathrm{pr}(\mathbf{f}_n|H_j)}{q(\mathbf{f}_n)}, \tag{14.86}$$

where now the samples are drawn from $q(\mathbf{f})$. For this modification to be useful, we must choose $q(\mathbf{f})$ so that the samples in data space, $\mathcal{H}\mathbf{f}_n$, are clustered near the actual $\mathbf{g}$. In simulation studies, we can do this by taking advantage of the knowledge of how we produced $\mathbf{g}$ in the first place. If we did so by simulating some particular object $\mathbf{f}_0$, then we know what this object was and can use this knowledge in computing the likelihood ratio. For example, if we describe objects by their independent components, with expansion coefficients $\{\alpha_k\}$, then the initial object $\mathbf{f}_0$ is described by $\{\alpha_{k0}\}$, and the importance sampler can generate random objects by random perturbations about $\{\alpha_{k0}\}$. So long as the perturbations are large enough to adequately sample the integrand, the sum in (14.86) is still an unbiased estimator of $\mathrm{pr}(\mathbf{g}|H_j)$, and the variance is greatly reduced by using the prior knowledge of the point about which to take samples. A related approach, suggested by Zhang *et al.* (2001a), is to draw the samples from $\mathrm{pr}(\mathbf{g}|\mathbf{f})$, renormalized as a density on $\mathbf{f}$.

Markov-chain Monte Carlo Direct Monte Carlo integration as sketched above has limited applicability because of the need for an analytic form for $\mathrm{pr}(\mathbf{f}|H_j)$ in the importance sampler. A more general technique is MCMC, which will be discussed in the context of image reconstruction in Sec. 15.4.8. As we shall see there, the essence of MCMC is to propose random perturbations in the vector that is the variable of integration, and to accept or reject the proposed perturbations with a carefully chosen rule such that the sequence of accepted perturbations forms a Markov chain, and the equilibrium PDF for the chain is precisely the one from which we wish to sample.

For ideal-observer studies, MCMC is particularly applicable to the expression for the likelihood ratio given in (14.80). A Monte Carlo implementation of this formula is

$$\Lambda(\mathbf{g}) \approx \frac{1}{N_s} \sum_{n=1}^{N_s} \frac{\mathrm{pr}(\mathbf{g}|H_2, \mathbf{b}_n)}{\mathrm{pr}(\mathbf{g}|H_1, \mathbf{b}_n)}, \tag{14.87}$$

where the samples $\mathbf{b}_n$ are drawn from the posterior $\mathrm{pr}(\mathbf{b}|\mathbf{g}, H_1)$.

To sample from the posterior, we can use a Metropolis-Hastings algorithm, which we shall discuss in more detail in Sec. 15.4.8. As applied to the present problem, the basic idea is to generate a sequence of samples of the background $\mathbf{b}$ in such a way that the samples are drawn from some target density $\pi(\mathbf{b})$ such as the posterior $\mathrm{pr}(\mathbf{b}|\mathbf{g})$. If the current background in the sequence is $\mathbf{b}^{(k)}$, a new trial background $\mathbf{b}'$ is generated from a proposal density $q(\mathbf{b}'|\mathbf{b}^{(k)})$, which can depend on the current state. The probability of accepting this proposed change is [*cf.* (15.328)]

$$\Pr(acc) = \min\left\{1, \frac{\pi(\mathbf{b}')\,q(\mathbf{b}^{(k)}|\mathbf{b}')}{\pi(\mathbf{b}^{(k)})\,q(\mathbf{b}'|\mathbf{b}^{(k)})}\right\}. \tag{14.88}$$

If the change is accepted, we set $\mathbf{b}^{(k+1)} = \mathbf{b}'$; otherwise $\mathbf{b}^{(k+1)} = \mathbf{b}^{(k)}$. By a detailed-balance argument (see Sec. 15.4.8), it can be shown that the equilibrium distribution is indeed $\pi(\mathbf{b})$. Note that only ratios of target densities are required; if $\pi(\mathbf{b})$ is the posterior, then we can write it as $\mathrm{pr}(\mathbf{b}|\mathbf{g}) \propto \mathrm{pr}(\mathbf{g}|\mathbf{b})\,\mathrm{pr}(\mathbf{b})$. The constant of proportionality cancels out in (14.88), and we do not need to know the normalization of the posterior. We do, however, need to know the ratios $\mathrm{pr}(\mathbf{b}')/\mathrm{pr}(\mathbf{b}^{(k)})$.

Kupinski *et al.* (2003b) showed how this approach could be applied to likelihood-ratio calculations with a lumpy background model. Two types of perturbations to the background were allowed: changes in the location of a particular lump and changes in the number of lumps. This procedure was used to compare three rather stylized pinhole imaging systems in terms of ideal-observer AUC for an SKE task. By running the Markov chain multiple times, the variance in the estimate of the AUC was estimated. In subsequent work, Park *et al.* (2003) extended this method to random signals.

Channelized ideal observer We have mentioned low-dimensional representations of object statistics, but we can also consider dimensionality reduction in data space as a way of facilitating ideal-observer studies. Though dimensionality reduction would also be called feature extraction in pattern recognition, we have several advantages in assessment of image quality that we do not have in pattern recognition. As we discussed in the context of the channelized Hotelling observer in Sec. 14.3.2, we can consider SKE tasks where all details of the signal are known, we can construct backgrounds with known statistical properties, and we can simulate noise-free samples with these statistics.

Armed with this information, we can construct so-called efficient channels in such a way that the performance of the Hotelling observer operating on the channel outputs is a good approximation to that of the true Hotelling observer operating on the original data $\mathbf{g}$. For example, if we consider a rotationally symmetric signal in a known location in a statistically isotropic background, we can use rotationally symmetric channels defined by Laguerre-Gauss functions. Gallas and Barrett (2003) demonstrated that only 5–10 such channels were needed for good estimates of Hotelling-observer performance; we anticipate that a similar result will hold for ideal observers when we have a similar amount of prior information, but this hypothesis has not yet been confirmed.

Suppose we have a set of linear channels that we believe, based on our knowledge of the classification task, might be efficient with respect to the ideal observer. To check this possibility, we need to compute the likelihood ratio on the channel outputs for a training set of signal-present and signal-absent images and then create an ROC curve. Recent work by Subok Park and Matthew Kupinski offers some possible ways of computing the likelihood ratio. Though this work is unpublished at this writing, we sketch the main ideas here with the permission of the originators.

The approaches suggested by Park and Kupinski apply to situations where we have analytic expressions for the characteristic functions of the data $\mathbf{g}$ but no PDFs, yet still want to compute a likelihood ratio (LR). The basic idea is to reduce the dimensionality of the data by use of a set of P channels and then attempt to compute the LR on the channel outputs rather than on $\mathbf{g}$ itself.

As Park formulated the problem, the PD channel output vector is given by

$$\mathbf{v} = \mathbf{Tg} = \mathbf{T}\left\{\boldsymbol{\mathcal{H}}\mathbf{f} + \mathbf{n}\right\}, \tag{14.89}$$

and the characteristic function for $\mathbf{v}$ under hypothesis j (in the case of Poisson noise) is given by an extension of (14.71) as

$$\psi_{\mathbf{v}|H_j}(\boldsymbol{\omega}) = \Psi_{\mathbf{f}|H_j}\left[\boldsymbol{\mathcal{H}}^\dagger \boldsymbol{\Gamma}\left(\mathbf{T}^\dagger\boldsymbol{\omega}\right)\right], \tag{14.90}$$

where $\boldsymbol{\omega}$ is a $P\times 1$ vector. It is assumed that P is relatively small and that $\psi_{\mathbf{v}|H_j}(\boldsymbol{\omega})$ can be computed analytically. The likelihood ratio for a given $\mathbf{v}$ is[12]

$$\Lambda(\mathbf{v}) = \frac{\int d^P\omega \ \psi_{\mathbf{v}|H_2}(\boldsymbol{\omega}) \exp(2\pi i\boldsymbol{\omega}^\dagger \mathbf{v})}{\int d^P\omega \ \psi_{\mathbf{v}|H_1}(\boldsymbol{\omega}) \exp(2\pi i\boldsymbol{\omega}^\dagger \mathbf{v})} . \tag{14.91}$$

Since P is small, Park proposed doing these integrals as FFTs. Kupinski refined the idea by suggesting Monte Carlo integration with importance sampling:

$$\int d^P\omega \ \psi_{\mathbf{v}|H_j}(\boldsymbol{\omega}) \exp(2\pi i\boldsymbol{\omega}^\dagger \mathbf{v}) = \int d^P\omega \ \frac{\mathrm{pr}(\boldsymbol{\omega})}{\mathrm{pr}(\boldsymbol{\omega})} \psi_{\mathbf{v}|H_j}(\boldsymbol{\omega}) \exp(2\pi i\boldsymbol{\omega}^\dagger \mathbf{v})$$

$$\approx \frac{1}{N}\sum_{n=1}^{N} \frac{\psi_{\mathbf{v}|H_j}(\boldsymbol{\omega}_n) \exp(2\pi i\boldsymbol{\omega}_n^\dagger \mathbf{v})}{\mathrm{pr}(\boldsymbol{\omega}_n)} , \tag{14.92}$$

where the samples $\boldsymbol{\omega}_n$ are drawn from $\mathrm{pr}(\boldsymbol{\omega})$. Kupinski also suggested using a few sample images to determine the mean and covariance of $\mathbf{v}$ and then constructing a PD Gaussian with these estimated parameters to use as $\mathrm{pr}(\boldsymbol{\omega})$.

Much further work is needed to validate this approach and explore possible choices for the channels, but if efficient linear channels in the ideal-observer sense exist, it opens up many new avenues for evaluating imaging systems with the ideal observer and classification tasks.

Nonlinear features Several nonlinear approaches to dimensionality reduction have been suggested by Hongbin Zhang. Zhang *et al.* (2001a) discusses features derived from the ideal observer and based on $\Lambda_{\mathrm{BKE}}(\mathbf{g}, \mathbf{b})$ as defined in (14.81). Rather than attempt to average this expression over the posterior on the backgrounds, as required by (14.80), Zhang reasoned that a useful set of features for an SKE classification task could be defined as

$$\theta_p = \Lambda_{\mathrm{BKE}}(\mathbf{g}, \hat{\mathbf{b}}_p) , \qquad (p = 1, ..., P) , \tag{14.93}$$

where $\hat{\mathbf{b}}_p$ is some estimate of the background at the known signal location. Specifically, he argued that the background was likely to be slowly varying compared to a small signal, so he suggested that $\hat{\mathbf{b}}_p$ be taken as a smoothed version of $\mathbf{g}$, with different p corresponding to different widths of the smoothing filter. The resulting values of θ_p would not immediately be the ideal-observer discriminant function, but Zhang suggested that an artificial neural network might find a good approximation to the likelihood ratio in the PD space.

In related work, Zhang also suggested using a set of wavelets centered on the known signal location, followed by a nonlinear point transformation on each wavelet coefficient (Zhang *et al.*, 2001b). He suggested an iterative algorithm to train the nonlinear transformation so that the outputs would follow a PD multivariate normal law. Then, when a new image is passed through the same transformation, the likelihood ratio can readily be calculated (see Sec. 13.2.8).

[12]Even though the channels are linear, this likelihood ratio will usually be a nonlinear functional of the channelized data $\mathbf{v}$; it should not be confused with the AUC-optimal linear observer introduced in Sec. 13.2.12.

Checking the results We have sketched a number of approximate methods for estimating AUC for the ideal observer. How do we know if the results are correct? That is, how can we estimate the bias and variance of an estimate of ideal-observer AUC?

Variance in the estimate comes from two sources. First, as with any observer study, there is a variance arising from the random selection of images, or *cases* in medical parlance. Second, whenever the likelihood ratio is evaluated by using Monte Carlo or Markov-chain Monte Carlo methods to average over backgrounds, there is a variance associated with the random selection of backgrounds. This kind of variance is analogous to internal noise in the human observer; if the Monte Carlo calculation is repeated with the same image but a different random-number seed, it will not return the same value for the likelihood ratio.

Both kinds of variance can be estimated in simulation studies just by repeating the study many times, with different sets of images or with the same images but different random-number seeds. Alternatively, variance can be analyzed with MRMC methods as discussed in Sec. 14.2.4 or by using resampling methods as discussed in Sec. 14.3.2.

Bias is much more difficult to assess since we do not know what systematic errors we might be making in the likelihood-ratio calculation. To study the bias, Clarkson *et al.* (2003) proposed a set of consistency checks that must be satisfied in an ROC study if the test statistic is indeed a likelihood ratio. For example, we know from (13.85) that

$$\frac{\mathrm{pr}(\Lambda|H_2)}{\mathrm{pr}(\Lambda|H_1)} = \Lambda\,, \tag{14.94}$$

and in fact this relation holds if and only if Λ is a likelihood ratio (Clarkson and Barrett, 2000). It follows from (14.94) that

$$2\,(1-\mathrm{AUC}_\Lambda) = \int_0^\infty d\Lambda_t\,\left[\mathrm{FPF}(\Lambda_t)\right]^2\,, \tag{14.95}$$

where $\mathrm{FPF}(\Lambda_t)$ is the false-positive fraction for threshold Λ_t. Again, this relation holds if and only if Λ is a likelihood ratio (Clarkson *et al.*, 2003).

Other useful relations are derived from moment-generating functions and from the likelihood-generating function. The moment-generating function for the log-likelihood ratio λ under hypothesis H_j is defined by (C.56) as

$$M_j(\beta) \equiv \langle \exp(\beta\lambda)\rangle_{\mathbf{g}|H_j} = \left\langle \Lambda^\beta \right\rangle_{\mathbf{g}|H_j}\,. \tag{14.96}$$

From (13.79) we know that $M_j(\beta)$, must satisfy

$$M_1(\beta+1) = M_2(\beta)\,, \tag{14.97}$$

from which it follows that $M_1(1) = 1$. Moreover, a plot of $M_1(\beta)$ vs. β must be concave upward and pass through the points (0,1) and (1,1) as in Fig. 13.9. Once again, these properties are unique to the ideal observer.

Finally, a number of investigators have derived inequalities relating AUC to the likelihood-generating function (Clarkson, 2002; Clarkson and Barrett, 2000; Shapiro, 1999; Burnashev, 1998). One example is

$$\tfrac{1}{2}G(0) \le -\ln[2(1-\mathrm{AUC}_\Lambda)] \le \tfrac{1}{2}G(0) + \sqrt{G(0) - \tfrac{1}{8}G''(0)}\,, \tag{14.98}$$

where $G(\beta)$ is the likelihood-generating function and primes denote derivatives.

If we have a set of simulated or real sample images and a way of estimating the likelihood ratio for each, we can check the validity of these relationships. For example, we can use no-signal images to estimate $M_1(\beta)$ directly from its definition (14.96) and see if it indeed passes through $(1, 1)$. Similar numerical methods can be devised for each of the relations that must be satisfied for the ideal observer.

If the relations are not verified, we must look for some error in our calculation of the likelihood ratio. If they are satisfied, we can have confidence that the test statistic we are calculating is *some* likelihood ratio, though not necessarily the likelihood ratio we think we are calculating, namely the one applicable to the image data. Clarkson *et al.* (2003) admit the possibility that the algorithm is finding a good estimate of some other likelihood ratio, but say they have a "natural tendency to regard (that) possibility as unlikely."

One case where it is quite likely, however, is when linear or nonlinear features have been extracted from the original image data for dimensionality reduction. Then the Markov chain or other algorithm applied to the features may indeed give a good estimate of the likelihood ratio on the features, and all of the consistency checks mentioned above will be passed, but there is no guarantee that this likelihood ratio will give the same performance as one calculated on the original image data; the consistency checks do not ensure that the features preserve the information content of the images.

14.3.4 Estimation tasks

Compared to the large literature on model observers for detection and classification tasks in image-quality assessment, much less attention has been given to computational methods for estimation tasks, and there is much less agreement about what one should be computing in the first place. Of course, there is a huge body of work on estimation of pixel values in image processing and reconstruction, but we have argued in Sec. 13.3.2 that there is no meaningful way of relating accuracy of the pixel values to image quality. We shall discuss image reconstruction further in the next chapter, but for now we concentrate on estimation problems other than reconstruction.

Thus, by "estimation task" we mean estimation of one or a few parameters characteristic of the object being imaged and (unlike pixel values) of direct relevance to the purpose for which the image was obtained. Our goal here is to survey some of the computational methods that can be used for assessment of performance on such tasks.

Since the parameter being estimated is determined by the object being imaged, we write it as $\mathbf{\Theta}(\mathbf{f})$. Boldface is used since the parameter will often be a vector, though almost always a low-dimensional one; when we intend a scalar parameter, we shall denote it as $\Theta(\mathbf{f})$. Upper case is used for $\mathbf{\Theta}(\mathbf{f})$ since we use $\boldsymbol{\theta}$ for several other things, including expansion coefficients (*e.g.*, pixel coefficients) in approximate object representations like (7.27), and that is definitely not what we mean here.

We emphasize here that we are viewing the parameters to be estimated as characteristics of the object. This is in contrast to the view of Sec. 13.3 where we were concerned with parameters characterizing the probability density function of the data. The relationship between the two viewpoints is subtle, yet critical for assessing image quality on the basis of estimation tasks; we shall return to it at several points below.

Dichotomies Two useful dichotomies for the parameters are linear vs. nonlinear and estimable vs. nonestimable. The imaging systems that deliver the data from which the estimates are derived can also be categorized as linear or nonlinear. The estimators themselves can be linear or nonlinear functionals of the data, and they can be either biased or unbiased.

We encountered linear parameters in Sec. 7.1.4 when we discussed moment errors. In brief, a linear parameter is a linear functional of the object. If the components of $\boldsymbol{\Theta}(\mathbf{f})$ are derived linearly from the object, we know from (7.33) that they can be written as

$$\Theta_n(\mathbf{f}) = \int_\infty d^q r \; \chi_n^*(\mathbf{r}) \, f(\mathbf{r}) = \boldsymbol{\chi}_n^\dagger \mathbf{f} \, . \tag{14.99}$$

Equations of this form will be used in Chap. 15 for discussing image reconstruction, but here we should think of the components Θ_n merely as weighted integrals of the object. If the weighting function $\chi_n(\mathbf{r})$ is constant over some spatial region, we refer to $\Theta_n(\mathbf{f})$ as a region-of-interest integral, and its estimate $\hat{\Theta}_n(\mathbf{g})$ as a region-of-interest estimator. Of course, the estimate depends on the data $\mathbf{g}$ while the parameter itself depends on $\mathbf{f}$ but not on $\mathbf{g}$.

An important class of nonlinear parameters occurs in *mensuration tasks*, where the goal is to measure some physical dimension of a portion of the object. Examples include the area of an agricultural field in aerial photography, volume of the left ventricle in cardiology, and distance to a target in radar.

As we saw in Sec. 13.3.1, a parameter is said to be *estimable* or *identifiable* with respect to some data set if there is an estimator of it that is unbiased for all true values of the parameter. In terms of the likelihood $\mathrm{pr}(\mathbf{g}|\Theta)$, a parameter is estimable if different values of the parameter lead to different likelihoods.

The imaging system that acquires the data $\mathbf{g}$ may be linear or nonlinear as defined in Chaps. 1 and 7. The distinction rests on the form of the *mean* data; the system is linear if $\overline{\mathbf{g}}$ is a linear functional of $\mathbf{f}$. We denote a general linear system by the operator $\boldsymbol{\mathcal{H}}$, so $\overline{\mathbf{g}} = \boldsymbol{\mathcal{H}}\mathbf{f}$.

For a linear system, we can be more precise about estimability, because in that case we can divide object space $\mathbb{U}$ into subspaces called measurement space and null space, and any object can be uniquely decomposed as

$$\mathbf{f} = \mathbf{f}_{meas} + \mathbf{f}_{null} \, . \tag{14.100}$$

We know from the discussion in Sec. 14.3.3 that the probability density function on the data in most cases is fully determined by the mean data, so for a linear system we have

$$\mathrm{pr}(\mathbf{g}|\mathbf{f}) = \mathrm{pr}[\mathbf{g}|\overline{\mathbf{g}}(\mathbf{f})] = \mathrm{pr}(\mathbf{g}|\boldsymbol{\mathcal{H}}\mathbf{f}) = \mathrm{pr}(\mathbf{g}|\mathbf{f}_{meas}) \, . \tag{14.101}$$

A general definition of estimability in this case is that $\boldsymbol{\Theta}(\mathbf{f})$ is estimable if and only if $\boldsymbol{\Theta}(\mathbf{f}) = \boldsymbol{\Theta}(\mathbf{f}_{meas})$ for all $\mathbf{f}$. If this condition is met, then a change in $\mathbf{f}_{meas}$ leads to a different $\boldsymbol{\Theta}(\mathbf{f})$ and a different likelihood $\mathrm{pr}(\mathbf{g}|\boldsymbol{\Theta})$. Another definition of estimability is that $\boldsymbol{\Theta}(\mathbf{f})$ is estimable if and only if $\mathrm{pr}(\mathbf{g}|\boldsymbol{\Theta}_1) = \mathrm{pr}(\mathbf{g}|\boldsymbol{\Theta}_2)$ implies that $\boldsymbol{\Theta}_1 = \boldsymbol{\Theta}_2$.

We can go a step further for a linear parameter. We can decompose the templates $\chi_n(\mathbf{r})$ into measurement and null components, and the nth component of the parameter vector can be written as

$$\Theta_n(\mathbf{f}) = \boldsymbol{\chi}_{n,meas}^{\dagger}\mathbf{f}_{meas} + \boldsymbol{\chi}_{n,null}^{\dagger}\mathbf{f}_{null} \,. \tag{14.102}$$

Then $\boldsymbol{\Theta}(\mathbf{f}) = \boldsymbol{\Theta}(\mathbf{f}_{meas})$ for all $\mathbf{f}$ if and only if $\boldsymbol{\chi}_{n,null} = 0$ for all n. Otherwise a change in $\mathbf{f}_{null}$ would give a different value of the parameter but the same mean data and hence the same likelihood. Note that it is not necessary that the system have no null space, just that the templates have no components in that space. Since null components tend to involve high spatial frequencies, linear parameters derived from large, blobby templates are more likely to be estimable than ones derived from small or highly structured templates. In particular, as we shall discuss in more detail in the next chapter, integrals of the object over small pixels are almost never estimable.

The final dichotomies involve the estimator itself, which can be linear or nonlinear and biased or unbiased. Linear estimators were discussed briefly in Sec. 13.3, but considerable emphasis was placed there on maximum-likelihood (ML) estimators. Like the likelihood ratio used in ideal-observer classification problems, ML estimators are usually nonlinear functionals of the data. An exception in both cases occurs with Gaussian data. For Gaussian data with equal covariances under the two hypotheses, the ideal observer computes a test statistic (the log-likelihood ratio) that is linear in the data, and for Gaussian data and any linear parameter, the ML estimator is also linear in the data. In most interesting cases, however, neither the log-likelihood ratio nor the ML estimator is linear.

If the parameter is estimable, there exists an unbiased estimator, but we may not know it, or we may choose not to use it; Bayesian estimation, for example, deliberately introduces a bias toward the prior. Thus we must distinguish biased from unbiased estimators even for estimable parameters.

Performance metrics: MSE and EMSE From the discussion in Sec. 13.3.1, a natural choice for a figure of merit is the mean-square error or MSE, defined for a scalar parameter in (13.280) and for a vector in (13.286) or (13.287).

For estimable parameters, MSE has much to recommend it. It can be computed for any chosen object and estimator, it takes into account both bias and variance, and it is a scalar that can be used for system optimization. One drawback is that MSE is defined by averaging the error with respect to the density $\text{pr}(\mathbf{g}|\boldsymbol{\Theta})$, so it will depend on the true value of $\boldsymbol{\Theta}$ in general. One solution to this problem is simply to plot MSE($\boldsymbol{\Theta}$) vs. $\boldsymbol{\Theta}$, much in the same manner that one can plot SKE detectability as a function of signal location or other parameters [see (13.209)].

With nonestimable parameters, MSE is more problematical. Since null components of the object influence $\boldsymbol{\Theta}(\mathbf{f})$ but not $\overline{\mathbf{g}}(\mathbf{f})$ in that case, many different objects can give the same mean data but different true values of $\boldsymbol{\Theta}$, and it is quite arbitrary which true value one associates with a given data set. Indeed, if there are no other constraints, it is usually possible to find an object so that *any* estimator of a nonestimable parameter is unbiased; whether that object is one that would ever be encountered is another matter. As we shall see in Sec. 15.1.4, positivity constraints limit the magnitude of null functions and alleviate issues of estimability, but they don't eliminate them.

Perhaps the best solution to defining a scalar figure of merit for estimates of nonestimable parameters[13] is to use the *ensemble mean-square error* or *EMSE* defined in (13.281) for scalars or (13.288) for vectors. The vector definition can be rewritten for our purposes as

$$\text{EMSE} = \left\langle \left\langle ||\hat{\boldsymbol{\Theta}} - \boldsymbol{\Theta}||^2 \right\rangle_{\mathbf{g}|\boldsymbol{\Theta}} \right\rangle_{\boldsymbol{\Theta}} = \left\langle \left\langle ||\hat{\boldsymbol{\Theta}} - \boldsymbol{\Theta}(\mathbf{f})||^2 \right\rangle_{\mathbf{g}|\mathbf{f}} \right\rangle_{\mathbf{f}} . \tag{14.103}$$

In the last form, the average is over some ensemble of objects. For any particular object in the ensemble, a bias and hence an MSE can be defined, and the ensemble-average MSE is the quadratic error norm specific to the imaging system, the estimator *and* the chosen ensemble. Note that the use of an average over objects in the figure of merit does not imply that this same information was used in the estimator. The quantity $\hat{\boldsymbol{\Theta}}(\mathbf{g})$ might have been obtained by Bayesian methods, but it might also be an ML estimate or some other one that eschews prior information.

The question that remains is what ensemble to use in the averaging. The Bayesian answer would be to average over the prior, and indeed to use that same prior in the estimation process in order the minimize the EMSE. To a pragmatist, there are several difficulties with this approach. First, in practice we might not have enough verifiable prior information (as opposed to subjective or noninformative priors) that we would be willing to build it into the inference process. In practice, the only computationally tractable priors for Bayesian estimation might be some noninformative prior like entropy or simple analytic expressions such as conjugate priors[14] or the regularizing functions to be discussed in Sec. 15.3.3. Even if we were willing to use one of these analytic priors to do the estimation, there is no reason to think that samples drawn from it would bear any relation to the true distribution of $\boldsymbol{\Theta}(\mathbf{f})$ or $\mathbf{f}$, so it would be hard to have any confidence (belief) in the MSE computed from that prior.

What pragmatists can do well, however, is to perform realistic simulations (*i.e.*, ones consistent with a belief system honed in the field, laboratory or clinic), and these simulations can be used to compute sample approximations to the EMSE defined in (14.103). Specifically, if a set of sample objects $\{\mathbf{f}_n, n = 1, ..., N_s\}$ is generated, then we can approximate the EMSE by

$$\widehat{\text{EMSE}} = \frac{1}{N_s} \sum_{n=1}^{N_s} \left\langle ||\hat{\boldsymbol{\Theta}} - \boldsymbol{\Theta}(\mathbf{f}_n)||^2 \right\rangle_{\mathbf{g}|\mathbf{f}_n} . \tag{14.104}$$

The remaining average can be performed either analytically or by additional Monte Carlo simulations of $\mathbf{g}$ for a fixed $\mathbf{f}_n$.

Why ML? And how? As we saw in Secs. 13.3.4 – 13.3.6, ML estimators have many desirable properties. We know that ML estimators are efficient (*i.e.*, they achieve

[13] In spite of the terminology, nonestimable parameters can indeed be estimated. An estimate is merely a number associated with a data set. To be perverse, one could associate the number 3 with *any* data set. Then an estimate would be given for all $\mathbf{g}$ no matter whether the parameter was estimable, and in fact the variance of the estimate would be zero. The bias would, however, be completely meaningless.

[14] A conjugate prior is one chosen purely for mathematical convenience, to make the posterior have the same mathematical form as the prior. Unless one believes that nature is constructed for the convenience of statisticians, there is no reason to ascribe any degree of belief to conjugate priors.

the minimum possible variance as given by the Cramér-Rao bound) if any efficient estimator exists. Also, ML estimators are asymptotically efficient, asymptotically unbiased and asymptotically normally distributed. In the statistics literature, "asymptotic" refers to accumulating N i.i.d. data sets and letting $N \to \infty$, but it can have a broader meaning. All of the nice asymptotic properties of ML estimators apply if the variance of additive Gaussian noise goes to zero or if the number of counts in a photon-limited measurement gets large. Thus there is considerable motivation for using ML estimators, especially if we can get into one of these asymptotic regimes.

It is not obvious how we can perform ML estimation in general, since we seldom know the likelihood $\text{pr}(\mathbf{g}|\boldsymbol{\Theta})$ directly. Instead, as discussed in Sec. 14.3.3, we usually know the conditional density $\text{pr}(\mathbf{g}|\mathbf{f})$ or $\text{pr}[\mathbf{g}|\overline{\mathbf{g}}(\mathbf{f})]$; for direct imaging and Gaussian and Poisson noise, they are given by (14.67) and (14.68), respectively.

The general relation between the conditional densities on the data and the likelihood can be expressed either as an integral over the object space or an integral over data space:

$$\text{pr}(\mathbf{g}|\boldsymbol{\Theta}) = \int d\mathbf{f}\, \text{pr}(\mathbf{g}|\mathbf{f})\, \text{pr}(\mathbf{f}|\boldsymbol{\Theta}) = \int d^M\overline{g}\, \text{pr}(\mathbf{g}|\overline{\mathbf{g}})\, \text{pr}(\overline{\mathbf{g}}|\boldsymbol{\Theta}) . \tag{14.105}$$

These forms are equivalent whenever the conditional probability on the data is determined solely by its mean, which is the case with our usual Gaussian or Poisson noise models, with or without post-acquisition data processing (see Sec. 14.3.3).

One situation where we can easily go from these conditional densities to the likelihood is in the estimation counterpart of the SKE/BKE problem. Suppose we decompose the object into background and signal as in (14.77), and we assume that the signal is known to be present but that it is characterized by some unknown parameter vector $\boldsymbol{\Theta}$. For a linear system, we can write the mean data for background and signal, respectively, as

$$\mathbf{b} \equiv \boldsymbol{\mathcal{H}}\mathbf{f}_b , \qquad \mathbf{s}(\boldsymbol{\Theta}) \equiv \boldsymbol{\mathcal{H}}\mathbf{f}_s(\boldsymbol{\Theta}) . \tag{14.106}$$

For example, in medical imaging $\mathbf{f}_s(\boldsymbol{\Theta})$ might describe a spherical tumor with unknown center coordinates, gray level and diameter. In military reconnaissance, it might refer to a tank with unknown coordinates and heading.

If the background is known exactly and the signal is known except for these parameters, then

$$\text{pr}(\overline{\mathbf{g}}|\boldsymbol{\Theta}) = \delta[\overline{\mathbf{g}} - \mathbf{b} - \mathbf{s}(\boldsymbol{\Theta})] , \tag{14.107}$$

and the likelihood becomes

$$\text{pr}(\mathbf{g}|\boldsymbol{\Theta}) - \text{pr}(\mathbf{g}|\overline{\mathbf{g}})\Big|_{\overline{\mathbf{g}} = \mathbf{b} + \mathbf{s}(\Theta)} . \tag{14.108}$$

Explicit expressions for the likelihood ratio in the case of direct imaging can be found by substituting $\overline{\mathbf{g}} = \mathbf{b} + \mathbf{s}(\boldsymbol{\Theta})$ into (14.67) or (14.68). Since the number of parameters is small, there is no difficulty in maximizing the likelihood numerically.

Random backgrounds Just as in the signal-detection problem, the BKE assumption in estimation is oversimplified and can be misleading. It is much more realistic to consider random, cluttered backgrounds when we want to estimate signal parameters. We can regard the background components as a set of nuisance parameters, in the sense that they do not enter into the overall cost or Bayes risk associated with the estimation problem. As we learned in Sec. 13.3.8, the optimal strategy for this problem is to marginalize over the nuisance parameters, at least if we have a believable way of generating or approximating the prior density or drawing realistic samples. The likelihood is then given by

$$\mathrm{pr}(\mathbf{g}|\mathbf{\Theta}) = \int d^M b \; \mathrm{pr}(\mathbf{g}|\mathbf{\Theta}, \mathbf{b}) \; \mathrm{pr}(\mathbf{b}) \,, \tag{14.109}$$

where $\mathrm{pr}(\mathbf{g}|\mathbf{\Theta}, \mathbf{b})$ is to be computed from (14.108). This form is quite similar to the likelihood expressions encountered in Sec. 14.3.3 [*cf.* (14.84) – (14.86)], and similar Monte Carlo and Markov-chain Monte Carlo methods can be devised to evaluate it (Kupinski *et al.*, 2003c). As in the detection case, direct sampling of backgrounds from $\mathrm{pr}(\mathbf{b})$ is unlikely to work well since a randomly chosen $\mathbf{b}$ will probably lead to a vanishingly small $\mathrm{pr}(\mathbf{g}|\mathbf{\Theta}, \mathbf{b})$, but importance sampling can be used as in (14.85). If an analytic form is known for $\mathrm{pr}(\mathbf{b})$, samples $\mathbf{b}_n$ can also be drawn from the BKE likelihood $\mathrm{pr}(\mathbf{g}|\mathbf{\Theta}, \mathbf{b})$, renormalized as a density on $\mathbf{b}$, and the likelihood estimate is proportional to $\frac{1}{N}\sum_{n=1}^{N} \mathrm{pr}(\mathbf{b}_n)$.

For a detailed survey of Monte Carlo methods in ML estimation, see Geyer and Thompson (1992).

PDFs of the estimates Monte Carlo methods can also be used to study the distribution of the estimates themselves. If we simulate multiple data sets with the same true value, say $\mathbf{\Theta} = \mathbf{\Theta}_0$, and compute $\hat{\mathbf{\Theta}}$ for each, then we have, in effect, drawn samples from $\mathrm{pr}(\hat{\mathbf{\Theta}}|\mathbf{\Theta}_0)$. From these samples we can estimate the bias, variance, MSE and any other figure of merit we might devise.

In many problems, it is also possible to compute $\mathrm{pr}(\hat{\mathbf{\Theta}}|\mathbf{\Theta}_0)$ directly. Building on earlier work by Müller *et al.* (1990, 1995), Abbey *et al.* (1998) developed a method for approximating the density of maximum-likelihood and MAP estimates under a Gaussian noise model. They showed that the method was directly applicable to estimating parameters such as tumor volume from medical images, and they found that the predicted analytic PDFs were in good agreement with Monte Carlo simulation.

Rogala and Barrett (1997, 1998a, b, c) applied Abbey's method to a combination interferometer/ellipsometer where the goal was to estimate surface height and the real and imaginary parts of the refractive index at all points on a metal surface. Again, the analytic results were confirmed by Monte Carlo simulation.

Cramér-Rao bounds Rather than using the performance of a particular estimator as a figure of merit, it is also possible to use various performance bounds that might be easier to compute. In particular, the Cramér-Rao bound, introduced in Sec. 13.3.5, sets a lower limit to the variance of an unbiased estimator. For an unbiased estimator, the Cramér-Rao bound is given in (13.371) or (13.372), and for a biased estimator, the appropriate forms are (13.376) and (13.377). Both the biased and unbiased form are derived from the Fisher information matrix.

Kupinski *et al.* (2003c) developed MCMC methods to estimate the Fisher information matrix for the problem of estimating the position, width and amplitude of a Gaussian signal in a lumpy background. They did not assume that the signal was always present, so their treatment applied to a hybrid detection/estimation problem, but the figure of merit was based only on the estimation performance, marginalized over the probability of detection.

Approaches based on the Cramér-Rao bound are attractive, but they have their limitations. For one thing, if more than one parameter is to be estimated, it is not clear how to combine the individual bounds into a single scalar figure of merit that can be used for system optimization. Second, in many problems no efficient estimator exists, and it is not clear in practice how far actual variance will be from the bound. Similarly, it is often the case that no unbiased estimator exists, so use of the unbiased form of the bound can be misleading; the biased form (13.376) is less useful since it requires knowledge of the bias gradient (derivative of the bias with respect to the parameter). Considerable work has been done at the University of Michigan on variance bounds in which the norm of the bias gradient is constrained, though mostly in the context of estimation of pixel values (Gorman and Hero, 1990; Hero and Fessler, 1994; Hero *et al.*, 1996).

14.4 SOURCES OF IMAGES

Simulated images play an important role in the practical assessment of image quality. They can be used to get a subjective impression of the effects of changing parameters of the imaging system, and they can serve as input for objective studies with either model observers or humans. If the simulations are realistic, they may even be preferable to real images since there is no question about the true state of the object. Most importantly, simulations can be used to assess imaging systems that do not exist, so they are essential to any program of systematic optimization.

Realistic simulations involve computer implementations of the object, the image-formation process and the detector, and they must accurately reflect both the deterministic and stochastic aspects of each of these components. The art of good simulation is thus necessarily specific to both the imaging system and the use to which the simulation will be put. Nevertheless, it is our goal in this section to give some general guidelines on the simulation process, with reference to specific systems only as examples. We shall refer to the methods for representing deterministic and random objects given in Chaps. 7 and 8, along with material on the simulation of image formation provided in Chap. 10.

In Secs. 14.4.1 and 14.4.2 we survey methods for deterministic and stochastic simulation of objects, and in Secs. 14.4.3 and 14.4.4 we treat deterministic and stochastic simulation of image formation. Finally, in Sec. 14.4.5 we discuss the gold-standard problem that arises when using real images instead of simulated ones.

14.4.1 Deterministic simulation of objects

In Sec. 7.1 we emphasized that real objects are functions, but we also acknowledged that numerical computations require approximate discrete representations. In all fields of image science, there is considerable emphasis on linear representations, and we know from (7.27) that the general form of such a representation is

$$f_a(\mathbf{r}) = \sum_{n=1}^{N} \theta_n \phi_n(\mathbf{r}) . \tag{14.110}$$

Thus object simulation involves two steps: choosing the expansion functions $\phi_n(\mathbf{r})$ and choosing the coefficients θ_n.

It is all but universal in simulation studies to choose the expansion functions as pixels or voxels, for two reasons. First, if we humans are simply inventing the simulated objects, it is easiest for us to think in terms of spatial variables. Pixels and voxels are discretizations of our natural visual domain, and it would be much harder for us to think in terms of, say, Fourier basis functions. Second, as we shall see below, we may also want to use images from some high-resolution imaging system as objects for another system of lower resolution. Since the first system is designed to present data to humans, it is likely to provide us with digital data in a pixel or voxel representation. Thus we have a ready-made discrete simulation if we stick with those expansion functions.

When the goal of the simulation is to evaluate imaging systems, it is not so much the simulated objects as the resulting simulated images that interest us. Our goal is to use the discrete representations of objects and systems to produce images that are as near as possible to those that would be obtained with actual continuous objects and continuous-to-discrete systems (see Sec. 7.4.3). That means that we should take N in (14.110) as large as possible. The only cost to increasing the number of pixels and voxels, in most cases, is increased computational time, and that commodity continues to plummet in price. In particular, we do not need to worry about whether the resulting system matrix is highly non-square and hence leads to an underdetermined inverse problem. In this section we are concerned only with accurate simulation of the forward problem; issues associated with choice of representation in inverse problems are discussed in detail in the next chapter.

Geometric objects The easiest way to get started in object simulation is to use superpositions of simple geometric shapes (circles, squares, ellipses...). In a pixel representation, the coefficients θ_n are assumed to have the same value for all pixels within one elemental shape, but generally different values within different elements. If the number of pixels N is large, we need not worry too much about pixels that straddle the border between elements. By using a range of sizes for the elemental shapes, we can get a simulated object that has small structures to challenge the spatial resolution of a simulated imaging system and large uniform structures with which to study system uniformity, radiometric accuracy and noise properties.

Such seemingly naive simulated objects have proven particularly valuable in tomographic imaging. They are known as *mathematical phantoms* in that field, and some have been so durable that they are commonly referred to by the name(s) of the investigators who devised them. Thus we have the Shepp-Logan phantom (an arrangement of ellipses somewhat resembling a 2D cross-section of the human brain; Shepp and Logan, 1974) and the Defrise phantom (a 3D set of thin parallel disks meant to challenge certain cone-beam tomographic systems; Defrise and Clack, 1994).

Geometrical shapes can also be manipulated to mimic much more complicated objects. For example, Tsui *et al.* (1993) devised a 3D representation of the human torso that includes a static model of the heart, and Pretorius *et al.* (1997) extended the work to a beating heart. This so-called MCAT (mathematical cardiac torso) phantom has become a *de facto* standard in simulation of nuclear-medicine cardiac studies.

The mathematical theory that treats efficient ways of representing and manipulating geometrical forms within the computer is called *computational geometry*. Two useful textbooks in this emerging field are O'Rourke (1998) at an undergraduate level and the more comprehensive graduate-level text by Preparata and Shamos (1985).

Digitized real objects Useful though these geometric objects may be, they do not capture the complexity of object variation from pixel to pixel within a given geometric element, and for this reason they may not give accurate results when used for the objective assessment of image quality. One way around this difficulty is stochastic simulation, discussed in Sec. 14.4.2, but another approach is use of real image data.

As mentioned above, we might have access to high-resolution images of objects that we also wish to image with a lower-resolution system. Often the high-resolution system will measure fundamentally different parameters of the object, or it might be that the higher-resolution system is more expensive or more invasive than the system under development. Under these circumstances, the higher-resolution system might not be one we would use in practice, but we can nevertheless use the images it produces to guide the development of the new system.

An example of considerable interest for medical imaging is the Visible Human Project. In this project a human cadaver was imaged with computed tomography at high spatial resolution and high (but irrelevant) radiation dose. High-resolution magnetic resonance imaging was also performed, and then literal tomograms[15] were obtained by slicing the cadaver into thin layers and photographing each.

The CT images obtained in this project have higher resolution and lower noise than any obtainable with living patients, so they can serve directly as objects for simulation studies of new CT systems. The MRI images are less useful for this purpose since the object in MRI is specified in a complicated way by three distinct scalar fields, the spin density and two relaxation times (see Prologue and Sec. 7.1.1). Any particular image represents some nonlinear combination of these three components and cannot be used to simulate objects for imaging systems that respond to other combinations. The optical images are useful mainly because they accurately delineate borders of the organs, so they provide an alternative to the stylized geometric shapes discussed above. The actual gray levels (or colors) do not, however, correspond to anything that would be seen with any real medical imaging system.

For many further details on the Visible Human images and their applications, the reader may consult the proceedings of conferences that have been held on the project (Banvard, 2000).

Similarly, Zubal *et al.* (1994) at Yale have developed torso and brain phantoms by starting with high-resolution CT images and painstakingly labelling different anatomical regions by hand. To simulate objects in lower-resolution nuclear-medicine simulations, these labelled regions can be assigned different gray levels, corresponding to uptakes of some radiopharmaceutical of interest.

Computer graphics Perhaps the greatest impetus to progress in image simulation today is computer games and the closely related field of virtual reality. Since

[15]Greek $\tau o \mu o \sigma$ = slice.

the everyday reality we see around us consists mainly of surface reflections from opaque objects, virtual reality and computer graphics are particularly useful for simulating such objects. Useful books in this area include works by Neelamkavil (1987), Anand (1993), Sillion and Puech (1994), Glassner (1995) and Rogers (1998). Graphics-related journals and magazines include: IEEE Transactions on Visualization and Computer Graphics, IEEE Computer Graphics and Applications and IEEE Multimedia; ACM Transactions on Modeling and Computer Simulation and ACM Transactions on Graphics, and Computer Vision, Graphics and Image Processing (CVGIP).

14.4.2 Stochastic simulation of objects

In Sec. 8.4 we discussed a wide variety of statistical models for objects. Each of these models provides a PDF that at least partially describes the random variation in objects, and stochastic simulation of objects amounts to drawing sample functions (or vectors) from those PDFs. Often the first thing we need to simulate is the overall shape of the object or of key components in the object; see Chap. 8 for a brief discussion of the statistical description of shape. Then we need to add in a random texture.

Random textures Methods of generating samples of texture fields with specified statistics were discussed in Sec. 8.4.4, and the literature on computer graphics can provide additional approaches. A common approach in image simulation is to assume that the texture is stationary within the boundaries of a single geometric element of the simulated object.

How accurately the texture needs to be simulated depends critically on the purpose of the simulation. If the task used to assess image quality is detection of a low-contrast lesion in a medical image, then, as we have noted earlier in this chapter, the texture results from anatomical variations that may, in fact, constitute the main noise source limiting task performance, so accurate modeling is essential. On the other hand, if measurement noise is high or if the task is estimation or mensuration, then task performance might be relatively insensitive to fine details of the object structure.

We urge the reader to be skeptical of simulations that omit texture modeling, particularly if the goal of the simulations is to provide input for image reconstruction. As we shall see in the next chapter, any reconstruction algorithm involves a choice of how much fine detail to attempt to reconstruct. Often this choice is made on the basis of claimed prior information, and the most common such claim amounts to saying that the object contains little or no fine detail. At the extreme, it may be asserted that the object is piecewise constant within boundaries of regions such as organs. Of course, it is easily possible to simulate objects and hence tomographic data consistent with this assertion, but the simulations then provide essentially no information about how the algorithms would perform on tasks that are sensitive to fine details.

Random signals In signal-detection studies it is useful to think of the object as a superposition of signal and background (see Sec. 8.4.5), and the signal component might be particularly amenable to simulation. In medical imaging, for example, a common task is tumor detection, and it might suffice to model the tumor as a small sphere or ellipsoid of low contrast. When the task is discrimination between types of tumors or between benign and malignant lesions, however, it may be necessary to include other features such as spicules (needle-like protrusions from the body of the tumor), but these too can be incorporated in realistic simulations.

Simulated signals may be superimposed on simulated or real backgrounds. As we saw in Sec. 8.4.5, the signal can sometimes be regarded as simply added to background, and in those cases we can maintain separate files of simulated signals and real or simulated backgrounds, adding them together in various combinations as needed. Moreover, when we are dealing with linear systems, we can choose to add the images rather than the objects (see Sec. 8.5.4). This makes it possible to add simulated signals to actual images of normal (signal-absent) objects as seen through real imaging systems. Since normal images are much easier to acquire and verify than abnormal ones, this approach can be very beneficial in avoiding the gold-standard problem.

14.4.3 Deterministic simulation of image formation

Linear systems Once we have a discrete object representation, it is generally straightforward to compute its mean image through a linear imaging system by matrix multiplication. The only real difficulty is in formulating the matrix, and that problem is specific to the imaging modality. We shall give an example of how to construct the matrix for emission computed tomography in Sec. 17.2.6.

We emphasize again, however, that it is important in simulation studies to sample the object finely, especially in image-reconstruction problems. If the reconstruction algorithm assumes that the object consists of voxels of a certain size, and the data are generated on precisely this same assumption, then a false consistency may result. When the same matrix is used in data simulation and reconstruction, and the resulting images are good in some sense, all that has been proved is that the matrix is nonsingular; no useful conclusions can be drawn about the true CD system or about real data. Simulation studies that use the same matrix for both a forward problem and its inverse problem should be regarded with strong suspicion.

Sparseness of the H matrix Though the $\mathbf{H}$ matrix used in simulation may be huge, it is often very sparse, with most of its elements equal to or very near zero. In direct-imaging systems, for example, the point response function $h_m(\mathbf{r})$ will tend to be highly concentrated; for any chosen source point $\mathbf{r}$, only a small subset of the detector pixels will receive radiation. (Indeed, this is essentially a definition of direct imaging.) When such systems are represented by matrices, the same thing holds: for any chosen n, H_{mn} is nonzero for only a small subset of m. Put another way, each column of $\mathbf{H}$ is mostly zero, no matter how many columns we choose to use. The zero elements need not be stored, and of course there is never any point in multiplying by zero.

Indirect-imaging systems may also result in sparse matrices. In tomography, for example, radiation is received from points along or near a thin pencil through the object (at least when scatter is neglected). Conversely, for any chosen object point and any projection direction, only a small subset of the detector elements receive radiation. In this case elements in one column of $\mathbf{H}$ are indexed by both the detector index and the projection direction, but nevertheless only a small fraction of the elements in each column are nonzero. For more discussion of this point in the context of emission computed tomography, see Sec. 17.2.6.

Shift-invariance Another structure that we might consider using is shift-invariance. As we have discussed in Sec. 7.2.3, it may be reasonable to describe certain CC systems with shift-invariant point spread functions. When the output of such a system is sampled with a regular detector array, and the object is represented by a regular grid of the same spacing, it is tempting to say that the system exhibits discrete shift-invariance and hence that the images are convolutions that can be computed efficiently with fast Fourier transforms or FFTs.

There are several problems with this approach. The first is that discrete convolutions (with N samples in 1D) are described by modulo-N arithmetic (see Sec. 3.6.2). The result is an entirely unphysical wrap-around such that images that disappear from one edge of the detector as the object point is shifted magically reappear on the other side. Various stratagems can be employed to minimize this effect, but their adequacy is seldom verified.

The second problem is that any real CC system must have some departures from strict shift-invariance. In a lens system with aberrations, for example, the form of the PSF varies with field angle. This problem is incompatible with any convolutional description. It can be minimized by restricting the object field and/or the image field, but then wrap-around effects may become more significant.

Finally, a great hazard of working with FFTs and discrete convolutions is that it entices the user to choose the number of samples in object space to be the same as the number of samples in image space. As we stressed above, accurate simulation of CD systems requires fine sampling of the object. The FFT approach to simulation requires sacrificing accuracy for speed; this tradeoff becomes increasingly difficult to justify as computers get faster, and it is especially questionable when only one or a few images are to be simulated.

These warnings do not imply that we should ignore approximate shift-invariance when constructing an $\mathbf{H}$ matrix or performing simulation studies. If neighboring columns of $\mathbf{H}$ are nearly equal but for a shift, we can take advantage of this structure and reduce the computation time needed to find the matrix and the memory needed to store it. For an example in the context of emission computed tomography, see Sec. 17.2.6.

Deterministic transport calculations In principle, the Boltzmann transport equation, discussed in detail in Secs. 10.3 and 10.4, allows us to compute the image of any object where the radiation can be considered particle-like, which for electromagnetic radiation means that interference and diffraction, polarization and quantum-mechanical effects such as squeezing can be neglected. To oversimplify, the domain of the Boltzmann equation is the same as that of geometric optics.

14.4.4 Stochastic simulation of image formation

Stochastic simulation was introduced in Sec. 10.4.5 as a broad class of methods in which some quantity is estimated by performing random experiments, either physically or in a computer. These methods can be applied to the generation of samples of noisy data for use in psychophysical experiments and model observer calculations.

Detectors and image noise So far we have discussed ways of computing or estimating the mean data $\{\overline{g}_m\}$, but for many purposes we need to simulate the actual noisy data $\{g_m\}$, so we must also be able to simulate the noise contributions $\{n_m\}$.

For simple noise processes, we can just call an appropriate random-number generator to generate a noisy image. For example, many detector arrays are dominated by electronic noise, which we know from the discussion in Chap. 12 to be usually well described by Gaussian probability laws. Moreover, we can often argue from physical grounds that the noise in different detector elements is statistically independent, so the noise can be simulated by calling an independent Gaussian random-number generator at each element. Similarly, if Poisson noise dominates, we can first calculate the mean number of counts at each element by deterministic methods and then call a Poisson random-number generator with this mean.

Some detectors generate excess noise as a result of a random amplification process (see Secs. 11.4), and the detector therefore introduces noise correlations. To accurately simulate a noisy image in this case, we must draw random vectors from a multivariate PDF, or we must simulate the amplification process itself.

In summary, for the results of an evaluation study to be valid, it is crucial that realistic object models and accurate models of the imaging system be employed. Particularly when the investigation involves an imaging system or a task for which model and human observer data have not been compared before, performance estimates based on simulated data sets and model observers should be verified using real data and human observers.

14.4.5 Gold standards

Conventional ROC analysis requires knowledge of the truth status of the images in order to score observer responses as correct or incorrect. Thus standard ROC methods are not directly applicable when the truth status of the images is unknown. The requirement that independent truth status be known can lead to case-selection bias that can favor one system over another.

Even when a method for establishing the truth status of the images exists, giving a so-called "gold standard," the method is more often a bronze standard rather than gold. New modalities are often evaluated with an older technology as the gold standard, even though the new modality may allow for the detection of subtle objects missed when using images from the older device. In medical applications, biopsy proof is the gold standard, but even biopsy is not perfect. Biopsy needles can miss their mark, and pathologists have been shown to make mistakes as well. Pathologist is another name for a human observer performing a classification task, so the process should be amenable to objective evaluation based on task performance. But what would be the gold standard?

Given the need to keep score of the observer's performance using a specified figure of merit, it is clear that simulations offer an added advantage — they solve the *ground truth* problem. For simulated images, the truth state of the objects are known because this information is in the hand of the investigator.

In this section we shall describe the effect of inaccurate gold standards on ROC methods and present approaches to the evaluation of imaging systems in the absence of ground truth. As we shall see, methods for the assessment of imaging systems in the absence of ground truth exist, but the uncertainty in the estimate of the system's performance is much larger than what is achieved when ground truth is known.

Truth by expert panel One approach to the establishment of truth is the use of an expert panel of observers. This method raises a multitude of questions and concerns regarding the number of experts to be used, how they will be chosen, and how their responses will be combined to establish "truth." Revesz *et al.* (1983) showed that the ranking of 3 systems could be made to favor any one of the 3, depending on the way in which the expert opinions were used to determine truth.

Mixture-distribution analysis Mixture-distribution analysis is based on the assumption that experts are likely to be correct when they agree. Kundel and Polansky (1997) suggested the use of a mixture-distribution analysis as an alternative to ROC methods when ground truth is not available. The method is based on dichotomizing the images into groups on the basis of the extent of a set of observers' agreement on them. It is assumed that the image groups represent different levels of case difficulty; *i.e.*, lower agreement indicates harder cases while higher agreement indicates easy cases. The number of groups is arbitrary. Thus the underlying model is a mixture distribution with a user-defined number of groups. Given the observer's ratings, an expectation-maximization (EM) method can be used to estimate the proportion of images in each group and the probability of truth given a certain level of agreement. Having estimated the truth status, the reader ratings can be used to determine the ROC curve.

Kundel and Polansky (1997) have compared mixture-distribution analysis to the results of an ROC analysis, where the ROC method used a separate expert panel to determine truth. Both methods gave similar estimates for the percentage of correct diagnoses for the task of image interpretation in chest radiography. Kundel and Polansky (1998) have shown that the results are fairly robust to the number of groups used in the model. The method may be especially useful in the evaluation of CAD algorithms; Kundel *et al.* (2001) recently demonstrated the use of the mixture-distribution approach for the evaluation of CAD in mammography. Recent emphasis on lung cancer screening programs using high-resolution CT raises the spectre of a very large number of potential lesions in the images for each patient; biopsy proof is simply not viable. Mixture-distribution analysis may be useful for the assessment of adjunctive CAD algorithms for this application.

See Polansky (2000) for a tutorial on the mixture-distribution method and other agreement-based approaches.

ROC analysis without truth of diagnosis It is not possible to perform an ROC evaluation of a single imaging system in the absence of ground truth because the problem is underdetermined. However, if each object has associated with it ratings from images obtained on two or more modalities, Henkelman *et al.* (1990) demonstrated that an EM algorithm can be used to estimate the class prevalences and the model parameters of a mixture distribution for the underlying objects.

The EM model makes the assumption that there are two underlying distributions for the decision variables, one for each class, and the distributions are correlated by an unknown amount. The dimensionality of each distribution is equal to the number of modalities under test. The EM algorithm estimates the relative proportion of each distribution (the prevalences), the locations of the observer's thresholds corresponding to each rating level along the decision axis for each modality, and the parameters specifying the distributions. For example, the use of a 5-point rating scale for 2 medical imaging modalities involves the estimation of 10 category boundaries, one disease prevalence and, in the case of a bivariate normal model for the distributions, the difference in means of the two distributions, their widths in two dimensions, and their correlations. Thus, given a sufficient number of images and observers, the estimation problem becomes tractable.

In a commentary on the Henkelman approach, Begg and Metz (1990) point out that the method breaks down if the imaging systems have low AUC (Henkelman *et al.* restricted their investigations to systems with an AUC ≥ 0.92). Begg and Metz suggest that each system must have an AUC of 0.80 or better for this technique to be applicable.

The work of Henkelman *et al.* has been extended by Beiden *et al.* (2000b), who performed Monte Carlo simulations to determine the uncertainties in the EM estimates of AUC obtained in the absence of ground truth. These authors found that many more patients were required in the truth-unknown case to yield estimates of AUC with standard deviations of those determined in the truth-known case, for the particular choice of true underlying distributions they investigated.

More investigation is required to better understand the usefulness of the EM approach to the no-gold-standard problem in ROC analysis in order to better understand the impact of the forms of the underlying distributions, the number of samples, the number of observers, the model assumptions made in the EM algorithm, and so on. Henkelman *et al.* suggest that the estimation problem may become better conditioned as the number of imaging modalities increases. More research is required to investigate this issue as well. Nonimaging diagnostic tests, including pathology readings, might also be included as additional modalities along with one or more imaging tests in the EM procedure.

Evaluation of estimation performance without a gold standard The issue of ground truth arises also in evaluating imaging systems on the basis of estimation tasks. For example, cardiac ejection fraction (the fraction of the blood expelled on each beat) can be measured by many different methods, including SPECT, planar nuclear medicine, MRI, ultrasound, CT and biplanar projection x rays. Each of these methods has significant errors, and none is universally accepted (except by its practitioners) as the "gold standard." When a new method is developed, it is customary (perhaps even mandatory) to publish a plot of ejection fractions obtained by the new method against ones obtained on the same patients with some older method. Ideally such plots would show a high correlation, with regression slopes near one and intercepts near zero. It is not uncommmon, however to find slopes around 0.6-0.8 and intercepts around 0.2-0.3. Something is wrong with one or both methods, but there appears to be no way of telling which without a gold standard.

It would be desirable to regress the estimates obtained from each modality against the true value of the parameter rather than against another estimate, and in fact it is possible to do so if each patient is studied on each of two or more modalities (Hoppin *et al.*, 2002; Kupinski *et al.*, 2002). The basic assumption is that there exists a linear relation between the mean value of the estimates and the true value for each patient (though nonlinear relations can also be used). If θ_{pm} is the estimate obtained from patient p on modality m, and Θ_p is the true value for that patient, the assumed relation is

$$\theta_{pm} = a_m \Theta_p + b_m + n_{pm} , \tag{14.111}$$

where n_{pm} is a zero-mean random variable. For simplicity, Hoppin *et al.* assumed that n_{pm} was normally distributed, but this does not appear to be critical. It was also assumed that n_{pm} was statistically independent of $n_{p'm'}$ for $p \neq p'$ or $m \neq m'$, and that the random variables for different patients but the same modality had the same variance. If P patients are each studied on M modalities, there are a total of PM measurements and $3M$ unknowns, namely the M values of a_m and b_m as well as the variances of n_{pm} for each m.

The basic idea is to estimate the $3M$ unknowns from the PM measurements by maximum-likelihood methods. With the assumptions made about n_{pm} it is straightforward to write down a probability density function on the measurements conditional on the unknown parameters and on the true values Θ_p, but of course we don't know these true values. Therefore Hoppin *et al.* assumed that the Θ_p were drawn independently from some parametric density $\text{pr}(\Theta_p|\boldsymbol{\alpha})$, where $\boldsymbol{\alpha}$ is a vector of unknown parameters describing the density. For example, since ejection fraction is defined on 0-1, a natural choice for $\text{pr}(\Theta_p|\boldsymbol{\alpha})$ is a beta distribution, which has two free parameters. These two parameters are of course unknown, so they are simply added to the list of parameters to be estimated. For example, with three modalities and the beta distribution, there are a total of $3M + 2 = 11$ unknowns, but if 100 patients are studied, there are 300 measurements.

This method has been well validated in simulation studies, and it has been placed on a firm theoretical footing by calculation of the Fisher information matrix. Not only does it give accurate estimates for the desired regression parameters, it also gives good values for the nuisance parameters contained in $\boldsymbol{\alpha}$.

15

Inverse Problems

In an imaging context, a *forward problem* is to determine the image produced by a given object. One might think that the corresponding *inverse problem* would be to determine the object that produced a given image. Were that so, this chapter would be quite short. As we have emphasized in previous chapters, imaging systems have null functions, and infinitely many different objects can produce the same image. It is virtually never possible to determine *the* object. A less ambitious goal is just to learn something about the object, perhaps to characterize it with a finite set of parameters in an approximate object description. It is in this sense that we shall understand the term *inverse problem*. Excellent general discussions of inverse problems are given by Sabatier (1987, 1991), Bertero (1989), Pike and Bertero (1992), Engl *et al.* (1996), Kirsch (1996), Glasko (1988) and Bertero and Boccacci (1998).

Various terms are used for particular inverse problems. The term *image reconstruction* can be used quite generally for almost any inverse problem in imaging, but it is usually used in a tomographic context, meaning reconstruction from projection data. An *inverse-source* problem (Baltes *et al.*, 1978) is one where the unknown object is a radiation source, and an *inverse scattering* problem (Fiddy, 1992) is one where the unknown is a distribution of scatterers, such as variations in refractive index. When the object is reconstructed from a blurred and noisy direct image, the terms *deblurring* and *deconvolution* are often used (even when the forward problem is not a convolution). Closely associated with deconvolution is *system identification*, which refers to a situation in which a signal from a known source propagates through some linear system, and the problem is to characterize the system. The even more difficult problem of *blind deconvolution* arises when neither the input nor the system response is known, and it is nevertheless desired to characterize the input, the system or both.

Many inverse problems can also be cast as *estimation* problems, especially when the statistics of the data are taken into account. In this view, the goal is usually to estimate pixel values or other parameters of the object. Many of the principles of estimation theory, introduced in Chap. 13, will recur in this chapter.

In Sec. 15.1, we give an overview of several basic issues in image reconstruction, including the key concept of estimability. In Sec. 15.2 we discuss various ways of deriving linear operators that can be applied to image data to give a reconstructed image in one step. In Sec. 15.3 we formulate solutions to inverse problems as minimization of some functional of the solution and the data. One component of this functional is a measure of the discrepancy between the actual measured data and the data that would be produced by some approximate object representation. Minimization of this term alone would force the solution to be consistent with the data, but strict agreement with noisy data will inevitably yield noisy and unsatisfactory images, so some means of backing off from strict data agreement is required. The general term for noise control is *regularization*, and in Sec. 15.3 we impose regularization by adding another term to the objective functional. In Sec. 15.4, some specific iterative algorithms for finding this minimum are discussed, and the statistical properties of the resulting images are derived.

15.1 BASIC CONCEPTS

Several basic concepts that run through this chapter are introduced here. In Sec. 15.1.1 we present a useful taxonomy of inverse problems. In Sec. 15.1.2 we recognize that most inverse problems are approached by adopting a discrete representation of the object, but a number of often overlooked problems arise when we do so. One such problem is that the coefficients in this representation may not be uniquely determined even by noise-free data. This leads to a discussion of *estimability* in Sec. 15.1.3. In Sec. 15.1.4 we begin to examine some of the implications of the fact that many physical objects cannot assume negative values. Finally, in Sec. 15.1.5 we present some remarks on applying the general principles of objective, task-based assessment of image quality, as introduced in Chap. 14, to inverse problems.

15.1.1 Classifications of inverse problems

Classification by data type We have previously classified forward problems as CC, CD and DD, depending on whether the object and the image data were discrete vectors (D) or functions of a continuous variable (C). Usually nature dictates that the object is continuous in this sense, and the nature of digital image acquisition dictates that the image is discrete, so the CD model is the natural one. When we want to include a reconstruction step, however, we can choose to produce either a function or a finite vector. If we want to think of the data-acquisition system plus the reconstruction algorithm as one big imaging system, we must expand our classification scheme, adding another C or D to specify whether the output of the reconstruction is continuous or discrete. Some interesting cases are:

CCC: This designation refers to a CC system for the forward problem followed by a CC mapping to the reconstruction. Hence it applies to an integral equation and its analytic solution.

CCD: Again we have a CC system for the forward problem, but now with a reconstruction algorithm that produces a discrete vector. This designation would apply to numerical inversion of an integral transform.

CDD: This is the usual case in image reconstruction, where discrete data are reconstructed on a discrete grid.

CDDC: This designation could apply to image reconstruction on a discrete grid followed by display as a continuous luminance pattern.

CDC: This refers to direct reconstruction of functions without adopting a discrete object model (see Sec. 15.2.2).

DDD: Pure simulation.

DDDC: Simulation plus display.

Linear vs. nonlinear reconstructions We can further distinguish linear and nonlinear operations in both the forward and inverse step. Linear and nonlinear forward mappings were discussed in Chap. 7, so we concentrate here on the inverse mapping.

If we denote the reconstruction by $\hat{\mathbf{f}}$, whether it be continuous or discrete, and if the reconstruction can be performed with a linear operator $\mathcal{O}_{lin}$, we can write

$$\hat{\mathbf{f}} = \mathcal{O}_{lin}\mathbf{g}\,. \tag{15.1}$$

For a discrete reconstruction from discrete data (CDD or DDD), the operator $\mathcal{O}_{lin}$ is a matrix; for a continuous reconstruction from discrete data, it is a DC operator as discussed in Sec. 7.3.6.

If the operations involved in reconstruction of $\hat{\mathbf{f}}$ from $\mathbf{g}$ are nonlinear, we can write generically

$$\hat{\mathbf{f}} = \mathcal{O}_{nl}\mathbf{g}\,, \tag{15.2}$$

where $\mathcal{O}_{nl}$ is some nonlinear operator. Nonlinear reconstruction operators can be applied to data obtained from either linear or nonlinear imaging systems.

One example where it is possible to state a simple operational form for $\mathcal{O}_{nl}$ is when point nonlinearities are applied following an otherwise linear reconstruction. In the notation of Sec. 7.5, we can write

$$\hat{\mathbf{f}} = \Phi\{\mathcal{O}_{lin}\mathbf{g}\}\,, \tag{15.3}$$

where $\Phi\{\cdot\}$ is a nonlinear functional applied pointwise. A simple example is where $\Phi\{x\} = x\,\mathrm{step}(x)$, so that the effect of Φ is to clip negative values.

Implicit and iterative solutions Equations (15.1) and (15.2) suggest direct, one-step reconstructions, where some operator is applied once to the data to get a solution. Often, however, the desired $\hat{\mathbf{f}}$ is found by minimizing some scalar-valued functional $Q(\mathbf{f}, \mathbf{g})$ that depends on the object and the data. Various terms for $Q(\mathbf{f}, \mathbf{g})$ can be found in the literature, including *objective function* (or functional), *merit function*, *penalty function*, *cost function* and *energy* (the latter designation arising from analogies to statistical mechanics).

We encountered one example of an implicit reconstruction procedure in Chap. 1 when we discussed least-squares reconstruction in a DDD context. As we saw in (1.191), the solution in that case can be written as

$$\hat{\mathbf{f}} = \underset{\mathbf{f}}{\operatorname{argmin}}\; Q(\mathbf{f}, \mathbf{g})\,, \tag{15.4}$$

where, in the least-squares case, $Q(\mathbf{f}, \mathbf{g}) = ||\mathbf{g} - \mathbf{H}\hat{\mathbf{f}}||^2$. The argmin notation means that $\hat{\mathbf{f}}$ is the vector $\mathbf{f}$ for which $Q(\mathbf{f}, \mathbf{g})$ is minimum. Since this minimization is performed for a given $\mathbf{g}$, the resulting $\hat{\mathbf{f}}$ is a function of $\mathbf{g}$. In the least-squares case, the $\hat{\mathbf{f}}$ found by minimizing Q is a linear (or at least affine) function of $\mathbf{g}$ [*cf.* (1.195) or (1.200)], but more generally $\hat{\mathbf{f}}$ is a nonlinear function of $\mathbf{g}$.

Implicit formulations often lead to iterative algorithms for finding the $\hat{\mathbf{f}}$ that minimizes the functional. In these algorithms, successive estimates $\hat{\mathbf{f}}^{(k)}$ are generated according to a recursion rule with the general form,

$$\hat{\mathbf{f}}^{(k+1)} = \mathcal{O}^{(k)}\left\{\hat{\mathbf{f}}^{(k)}, \mathbf{g}\right\}, \tag{15.5}$$

where $\mathcal{O}^{(k)}\{\hat{\mathbf{f}}^{(k)}, \mathbf{g}\}$ is some operator with two operands, so that its output at each step depends (often nonlinearly) on both the previous estimate $\hat{\mathbf{f}}^{(k)}$ and the original data $\mathbf{g}$. The superscript on $\mathcal{O}$ indicates that the operator itself can change as the iteration proceeds. Often the recursion rule (15.5) will be chosen so that the argmin solution of (15.4) will coincide with $\hat{\mathbf{f}}^{(\infty)}$.

15.1.2 Discretization dilemma

In Chap. 7 we discussed in detail various approximate object representations. The general form of a linear approximation to an object function was given in (7.27) as

$$f_a(\mathbf{r}) = \sum_{n=1}^{N} \theta_n \phi_n(\mathbf{r}), \tag{15.6}$$

where the subscript a denotes *approximate*, and $\{\phi_n(\mathbf{r}), n = 1, ..., N\}$ is any convenient set of expansion functions. In a more compact operator notation, (15.6) becomes

$$\mathbf{f}_a = \mathcal{D}_\phi^\dagger \boldsymbol{\theta}, \tag{15.7}$$

where $\mathcal{D}_\phi$ is a CD discretization operator and $\mathcal{D}_\phi^\dagger$ is its adjoint (hence a DC operator).

If the coefficients $\{\theta_n\}$ in (15.6) are derived linearly from the object, we know from (7.33) that they can be written as

$$\theta_n = \int_\infty d^q r \; \chi_n^*(\mathbf{r}) \, f(\mathbf{r}), \tag{15.8}$$

or in operator form as

$$\boldsymbol{\theta} = \mathcal{D}_\chi \mathbf{f}. \tag{15.9}$$

Mapping a discrete object representation through a CD system In Sec. 7.3 we saw how object functions map through a linear CD system to form discrete data. If the CD system acquires M noisy measurements, the discrete data vector $\mathbf{g}$ is an $M \times 1$ random vector given by

$$\mathbf{g} = \mathcal{H}\mathbf{f} + \mathbf{n}, \tag{15.10}$$

where $\mathcal{H}$ is the linear CD operator defined in Sec. 7.3.1 and $\mathbf{n}$ is an $M \times 1$ noise vector. A simple mathematical tautology allows us to write

$$\mathbf{g} = \mathcal{H}\mathbf{f}_a + \mathcal{H}\mathbf{f} - \mathcal{H}\mathbf{f}_a + \mathbf{n} \equiv \mathbf{H}\boldsymbol{\theta} + \boldsymbol{\epsilon}, \tag{15.11}$$

where the overall error $\boldsymbol{\epsilon}$ (modeling error plus noise) is given by

$$\boldsymbol{\epsilon} = \mathcal{H}\mathbf{f} - \mathcal{H}\mathbf{f}_a + \mathbf{n}, \tag{15.12}$$

and the system matrix $\mathbf{H}$ is given in operator form by (7.307):

$$\mathbf{H} \equiv \mathcal{H}\mathcal{D}_{\phi}^{\dagger}. \tag{15.13}$$

More specifically, the elements of $\mathbf{H}$ are given by (7.304):

$$H_{mn} = \int_{\mathbf{S}_f} d^q r \; h_m(\mathbf{r})\, \phi_n(\mathbf{r}). \tag{15.14}$$

Thus the $(mn)^{th}$ element is the nth expansion function as imaged onto the mth detector element.

Considerations on choosing a discretization scheme The linear discretization problem boils down to selecting two sets of functions, $\{\phi_n(\mathbf{r})\}$ and $\{\chi_n(\mathbf{r})\}$, or equivalently, two sets of Hilbert-space vectors $\{\boldsymbol{\phi}_n\}$ and $\{\boldsymbol{\chi}_n\}$. These vectors affect the accuracy with which $f_a(\mathbf{r})$ approximates the actual object $f(\mathbf{r})$, and they also affect the form and dimensions of the system matrix $\mathbf{H}$ as well as the size and nature of the error vector $\boldsymbol{\epsilon}$. Finally, through (15.8), the set $\{\boldsymbol{\chi}_n\}$ affects the meaning of the parameters $\{\theta_n\}$ that we want to determine.

There are several ways we can approach the problem of choosing $\{\boldsymbol{\phi}_n\}$ and $\{\boldsymbol{\chi}_n\}$. We might say at the outset that we are interested in certain functionals of the object, such as pixel values, and select the functions $\{\boldsymbol{\chi}_n\}$ accordingly. That would leave us free to choose $\{\boldsymbol{\phi}_n\}$ by some other criterion, such as minimizing the data-space modeling error $\mathcal{H}\mathbf{f} - \mathcal{H}\mathbf{f}_a$. Unfortunately, as we shall discover in Sec. 15.1.3, most functionals that we might choose to estimate do not admit of an unambiguous estimate. That is, we cannot determine them from the data even in the absence of noise.

Alternatively, we might want to construct as accurate a representation of the object as possible for a specified number of terms in the expansion (15.6). In Secs. 7.1.4 and 7.1.5, we learned how to choose $\{\boldsymbol{\chi}_n\}$ for minimum object error once $\{\boldsymbol{\phi}_n\}$ was specified. In particular, if $\{\boldsymbol{\phi}_n\}$ is an orthonormal set, then $\{\boldsymbol{\chi}_n\}$ should be chosen to be the same orthonormal set if representational accuracy is our concern.

The choice of $\{\boldsymbol{\phi}_n\}$ itself might be dictated by statistical considerations. If we consider a statistical ensemble of objects, then the ensemble-average representational error is minimized by using the N eigenfunctions of the object covariance matrix corresponding to the N largest eigenvalues as $\{\boldsymbol{\phi}_n\}$. As noted in Sec. 7.1.4, this representation is called the Karhunen-Loève or KL expansion. One problem is that we do not usually have enough information about the ensemble to be able to compute the eigenvectors needed in a KL expansion.

No matter how we choose $\{\boldsymbol{\phi}_n\}$, representational accuracy can be improved by increasing the number of terms N in the representation. The rank of the matrix $\mathbf{H}$, however, cannot exceed the rank R of the CD operator $\mathcal{H}$, which in turn cannot exceed the number of measurements M. Whenever N exceeds R, therefore, the problem of finding $\boldsymbol{\theta}$ when given $\mathbf{g}$ is underdetermined. Moreover, the nature of the effective noise term is unknown, except that $\boldsymbol{\epsilon} \to \mathbf{n}$ as $N \to \infty$ with any sensible choice of expansion functions.

We are thus faced with a conundrum: If we use an accurate object model (large N), we cannot possibly find all the coefficients, and if we use a less accurate model ($N \leq R$), we make unknown modeling errors before even starting to estimate coefficients, and the approximate object $\mathbf{f}_a$ may not resemble the actual object $\mathbf{f}$, even if the coefficients can be determined exactly. It is the view of the authors that the only satisfactory way to resolve this problem is via task-based assessment of image quality, as introduced in Chap. 14 and discussed further in Sec. 15.1.5.

15.1.3 Estimability

If $N < R$ in a linear expansion like (15.6), the number of unknowns is less than the rank of the system operator, but it is still not evident that we can estimate the coefficients $\{\theta_n\}$ from the data $\mathbf{g}$, even in the absence of noise. The coefficients may not be *estimable parameters*. The concept of estimability was briefly introduced in Sec. 13.3; we shall revisit the subject here from the viewpoint of image reconstruction.

Estimability of a single linear parameter Consider a scalar parameter θ defined by the linear functional,

$$\theta = \int_{\infty} d^q r \; \chi^*(\mathbf{r}) \, f(\mathbf{r}) \,. \tag{15.15}$$

For a mental image, think of θ as the integral of the object over a region defined by a 0-1 function $\chi(\mathbf{r})$, but the mathematics will be more general. If we are given a noise-free data vector $\mathbf{g} = \mathcal{H}\mathbf{f}$, where $\mathcal{H}$ is a linear CD operator, we would like to know whether we can determine θ uniquely from $\mathbf{g}$. An equivalent question is whether we can find an unbiased estimate of θ when zero-mean noise is present.

To answer these questions, note that we can write θ as a scalar product in object space:

$$\theta = (\boldsymbol{\chi}, \mathbf{f}) \,. \tag{15.16}$$

If the system operator $\mathcal{H}$ is linear, we can define two orthogonal subspaces of object space, called measurement space and null space, with the latter defined as the space of all vectors $\mathbf{a}_{null}$ such that $\mathcal{H}\mathbf{a}_{null} = 0$. The vectors $\mathbf{f}$ and $\boldsymbol{\chi}$ can be uniquely decomposed as

$$\mathbf{f} = \mathbf{f}_{meas} + \mathbf{f}_{null} \,, \tag{15.17}$$

$$\boldsymbol{\chi} = \boldsymbol{\chi}_{meas} + \boldsymbol{\chi}_{null} \,. \tag{15.18}$$

Since these two subspaces are orthogonal, we can write

$$\theta = (\boldsymbol{\chi}_{meas}, \mathbf{f}_{meas}) + (\boldsymbol{\chi}_{null}, \mathbf{f}_{null}) \,. \tag{15.19}$$

Since the data vector is insensitive to null components, the first term represents what one can learn about θ from noise-free data, and the second term is the component of θ that cannot cannot be measured with the system in question. This term is zero if *either* $\mathbf{f}_{null} = 0$ *or* $\boldsymbol{\chi}_{null} = 0$. The first condition is often satisfied in simulation studies but seldom in reality; we have no control over the object our imaging system is pointed at, so we cannot assert that $\mathbf{f}_{null} = 0$. We can, however, choose the function $\chi(\mathbf{r})$ defining θ, so we can make the error zero by choosing it so that $\boldsymbol{\chi}_{null} = 0$.

If $\boldsymbol{\chi}_{null} = 0$, the associated parameter θ is said to be *estimable* or *identifiable*. If θ is not estimable, there is an inevitable error of unknown magnitude arising from the second term in (15.19). Objects differing by null functions will give the same data, and hence the same value for any estimate derived from the data, even though they might have vastly different true values for θ.

There are various equivalent mathematical statements of the estimability condition. A linear parameter $\theta = (\boldsymbol{\chi}, \mathbf{f})$ is estimable if and only if

$$\boldsymbol{\chi}_{null} = 0\,, \tag{15.20}$$

or

$$\mathcal{P}_{meas}\boldsymbol{\chi} = \boldsymbol{\chi}\,, \tag{15.21}$$

where $\mathcal{P}_{meas}$ is the orthogonal projector onto measurement space. As we saw in Chap. 1, this projector is given by $\mathcal{H}^+\mathcal{H}$, where $\mathcal{H}^+$ is the Moore-Penrose pseudoinverse of the CD operator $\mathcal{H}$. Thus we can also state the estimability condition as

$$\mathcal{H}^+\mathcal{H}\boldsymbol{\chi} = \boldsymbol{\chi}\,. \tag{15.22}$$

Finally, we can also say that θ is estimable if there exists a set of coefficients B_m such that

$$\chi(\mathbf{r}) = \sum_{m=1}^{M} B_m h_m(\mathbf{r})\,. \tag{15.23}$$

This equation implies estimability since measurement space is spanned by the point response functions $\{h_m(\mathbf{r})\}$, so a function in the form of (15.23) lies entirely in measurement space and cannot have null components.

Since null components of an imaging system usually contain high spatial frequencies, a large blobby template will be more likely to lead to an estimable parameter than a small or structured one. In particular, integrals of the object over small pixels are unlikely to be estimable.

Nonlinear parameters All of these conditions apply to linear functionals, which are the subject of this section, but for completeness we also give an estimability condition that is applicable to nonlinear parameters (see also Sec. 13.3.1). If $\theta(\mathbf{f})$ is an arbitrary (possibly nonlinear) functional of $\mathbf{f}$, we say that it is estimable if

$$\theta(\mathbf{f}) = \theta(\mathbf{f}_{meas})\,. \tag{15.24}$$

That is, $\theta(\mathbf{f})$ is completely determined by the measurement component of $\mathbf{f}$, so the null space is irrelevant. It is straightforward to show that (15.24) is equivalent to (15.20) if θ is linear.

Vector of linear parameters So far, we have considered a single parameter θ, but now we extend the discussion to the vector $\boldsymbol{\theta}$ defined by (15.9). The estimability conditions (15.21) – (15.23) now generalize to

$$\mathcal{P}_{meas}\boldsymbol{\chi}_n = \boldsymbol{\chi}_n\,, \qquad n = 1, ..., N\,, \tag{15.25}$$

$$\mathcal{H}^+\mathcal{H}\boldsymbol{\chi}_n = \boldsymbol{\chi}_n\,, \qquad n = 1, ..., N\,, \tag{15.26}$$

$$\chi_n(\mathbf{r}) = \sum_{m=1}^{M} B_{nm} h_m(\mathbf{r})\,, \qquad n = 1, ..., N\,. \tag{15.27}$$

In terms of the discretization operator $\mathcal{D}_\chi$, (15.26) can also be written as

$$\mathcal{D}_\chi \mathcal{H}^+\mathcal{H} = \mathcal{D}_\chi\,, \tag{15.28}$$

and (15.27) becomes

$$\mathcal{D}_\chi = \mathbf{B}\mathcal{H}\,, \tag{15.29}$$

where $\mathbf{B}$ is the $N \times M$ matrix with elements B_{nm}. By comparison of (15.28) and (15.29), $\mathbf{B}$ is given explicitly by

$$\mathbf{B} = \mathcal{D}_\chi \mathcal{H}^+\,. \tag{15.30}$$

Since the dimensionality of measurement space is the rank R of the system operator, it is not possible to satisfy these conditions for N linearly independent parameters unless $N \le R$. Conversely, a set of N estimable parameters must be linearly dependent if $N > R$.

Natural pixels in inverse problems It follows from (15.27) that all of the θ_n are estimable if $B_{mn} = \delta_{mn}$ for $n \le N \le M$, so that $\chi_n(\mathbf{r}) = h_n(\mathbf{r})$. If we take $N = M$ and assume that $h_n(\mathbf{r})$ is real, this means that $\mathcal{D}_\chi = \mathcal{H}$ and $\mathbf{B} = \mathcal{H}\mathcal{H}^+$, and (15.29) in this case is just the Penrose equation, $\mathcal{H} = \mathcal{H}\mathcal{H}^+\mathcal{H}$. Thus, if we use the point response functions to define our parameters, we need not worry about null functions or estimability. Buonocore *et al.* (1981) refer to the $h_n(\mathbf{r})$ as *natural pixels*.

We encountered natural pixels in Sec. 7.4.3, where we showed that the modeling error $\mathcal{H}\mathbf{f} - \mathcal{H}\mathbf{f}_a$ could be made zero if we chose the set $\{\phi_n(\mathbf{r})\}$ to coincide with the natural pixels. Now we are using natural pixels for a different purpose, to ensure estimability. These two goals are not contradictory; a set of estimable pixels with zero modeling error can be achieved in two distinct ways:

$$\mathcal{D}_\chi = \mathcal{H}\,, \qquad \mathcal{D}_\phi^\dagger = \mathcal{H}^+, \qquad \mathbf{H} = \mathcal{H}\mathcal{H}^+\,, \tag{15.31}$$

or

$$\mathcal{D}_\phi = \mathcal{H}\,, \qquad \mathcal{D}_\chi = \mathcal{H}^{\dagger +}, \qquad \mathbf{H} = \mathcal{H}\mathcal{H}^\dagger\,. \tag{15.32}$$

The reader should prove that both of these options imply estimability and that both allow us to write $\mathcal{H}\mathbf{f} = \mathbf{H}\boldsymbol{\theta}$ without approximation.

With (15.31), the parameters $\{\theta_n\}$ are easy to interpret, since they are just scalar products with the natural pixels, but the matrix $\mathbf{H}$ may be hard to compute since it requires a pseudoinverse of a CD operator. With (15.32), on the other hand, the elements of $\mathbf{H}$ are given by [*cf.* (7.340)]

$$H_{mn} = \left[\mathcal{H}\mathcal{H}^\dagger\right]_{mn} = \int_{\mathbf{S}_f} d^q r\; h_m(\mathbf{r})\, h_n^*(\mathbf{r})\,. \tag{15.33}$$

Thus each H_{mn} is determined by the overlap between two natural pixels, so it is relatively easy to compute since no pseudoinverse is needed, but it is not so clear what the parameters mean. Even if $h_n(\mathbf{r})$ is nonnegative everywhere, some or all of the kernels associated with $\mathcal{H}^{\dagger +}$ will have negative values, and some of the resulting integrals $\hat{\theta}_n$ may go negative as well (even for nonnegative objects).

Subspace constraints and estimability We have always taken object space to be some $\mathbb{L}_2$ space, but we have acknowledged that not all vectors in this space should be construed as realizable objects (see Sec. 7.1.2). As we shall discuss below, we may know that real objects cannot be negative, and this constraint will cause us to revise our view of estimability. Even without a positivity constraint, however, we may be able to restrict our attention to some subset of the vectors in object space.

For example, if an object is described by a spatio-temporal function $f(\mathbf{r}, t)$, we may know on physical grounds that the rate of temporal change is bandlimited, in which case we could restrict attention to a Paley-Wiener subspace (see Sec. 3.5.1). Similarly, if $f(\mathbf{r}, t)$ represents the response to some stimulus applied at $t = 0$, then we would know that the function is causal, so $f(\mathbf{r}, t) = 0$ for $t \leq 0$, and this also limits the object to a subspace of the spatio-temporal $\mathbb{L}_2$ space.

It is more difficult to find examples of purely spatial functions that are limited *a priori* to subspaces. Of course, it is often necessary to approximate the object by a function in a subspace as in (15.6), but this is not the same thing as saying that all realizable objects lie in this space. In particular, virtually all objects we might want to image have sharp edges, so we cannot assume they are spatially bandlimited.

On the rare occasions where we do know with certainty that the object is confined to some subspace, we can define a projector onto that subspace as $\mathcal{P}_{sub}$ and write $\mathbf{f} = \mathcal{P}_{sub}\mathbf{f}$. As an exercise, the reader may show that the estimability condition in this case becomes $\mathcal{P}_{sub}\mathcal{P}_{null}\boldsymbol{\chi} = 0$, which could be much easier to satisfy than the original condition.

Bayesian view of estimability As discussed in the Prologue and in Sec. 8.4, we may be able to make statements about the prior PDF on the object, $\text{pr}(\mathbf{f})$. Partial information about this density is called a stochastic model. Sometimes there is a frequentist justification for the stochastic model, and sometime it is a purely Bayesian statement of beliefs. In either case, we may be able to use the model to say something about the magnitude of the error term in (15.19). For example, if a Bayesian believes that the objects of interest contain no null functions for some particular imaging system, then there is no need for an estimability condition; $(\boldsymbol{\chi}_{null}, \mathbf{f}_{null})$ is zero if *either* $\boldsymbol{\chi}_{null}$ or $\mathbf{f}_{null}$ is zero. An unfortunate corollary of this belief is that it implies that it is not necessary to build better imaging systems. If the system at hand captures all of the information about the objects, why attempt to capture more? In particular, if the objects have no fine details beyond those in some assumed prior, why try to build a system with better spatial resolution?

A Bayesian could also use the assumed prior to compute the average error $(\boldsymbol{\chi}_{null}, \langle\mathbf{f}_{null}\rangle)$. So long as this quantity is negligible, the Bayesian could assume that the parameter $(\boldsymbol{\chi}, \mathbf{f})$ is estimable. The only hazard is in trying to convince someone with a different belief system that the assumptions are valid. The skeptical non-Bayesian would need only to collect a few sample objects and compute the scalar product to check the model.

15.1.4 Positivity

As we discussed in Chap. 7, many physical objects are constrained by their nature to be nonnegative functions. For example, the object being imaged in nuclear medicine or fluorescence microscopy is the concentration of a tracer, and concentrations by definition cannot be negative. Similarly, in incoherent optical imaging, the object is a radiant exitance or a transmittance, neither of which can be negative. As we shall see, many reconstruction algorithms enforce this physical reality and yield only nonnegative images. Such algorithms are said to enforce a *positivity constraint*, though strictly speaking it should be *nonnegativity* since most of them do admit zero values. We shall often use a common but loose parlance and speak of positivity when an object or image is restricted to be greater than *or equal* to zero.

In Chaps. 1 and 7, we viewed a linear imaging system as an operator $\mathcal{H}$ that maps from a Hilbert space $\mathbb{U}$ to a Hilbert space $\mathbb{V}$. As we have noted many times, the object space is divided into two subspaces, called measurement space and null space, and the data space is divided into consistency space and inconsistency space. In this section we discuss the implications of positivity for these spaces. Later, in the context of specific algorithms, we shall consider the effects of a positivity constraint on reconstructed images.

Positivity and null functions Consider an object function satisfying $f(\mathbf{r}) \geq 0$ for all $\mathbf{r}$. Like any other vector in $\mathbb{U}$, this object can be decomposed uniquely into measurement and null components. It is not true, however, that $f_{meas}(\mathbf{r})$ and $f_{null}(\mathbf{r})$ are nonnegative; only their sum is so constrained. In fact, it is almost always the case that the null component of a nonnegative object will contain negative values. The reason for this statement can be seen by considering a Fourier decomposition of the object (perhaps a Fourier series like (7.13) if the object has finite support). Almost any imaging system is capable of correctly responding to the low-frequency terms, so the null components will consist solely of higher frequencies. If the zero-frequency term is in measurement space, all null functions must satisfy

$$\int_{\mathbf{S}_f} d^q r \; f_{null}(\mathbf{r}) = 0 \,, \tag{15.34}$$

which implies that $f_{null}(\mathbf{r})$ must have both positive and negative values.

Estimability revisited As we saw above, estimability is closely related to null functions. Suppose $\mathbf{f}_1$ and $\mathbf{f}_2$ are two objects that differ by a null function, so that $\mathcal{H}\mathbf{f}_1 = \mathcal{H}\mathbf{f}_2$. A linear parameter θ defined as in (15.16) can, in general, take on very different values for these two objects. We can define $\theta_1 = (\boldsymbol{\chi}, \mathbf{f}_1)$ and $\theta_2 = (\boldsymbol{\chi}, \mathbf{f}_2)$. If $\boldsymbol{\chi}$ has a null component, then $\theta_2 - \theta_1$ can be arbitrarily large, and any estimate derived from the data can have an arbitrarily large bias. To avoid this problem, we must choose $\boldsymbol{\chi}$ to be free of null functions.

The situation changes, however, if we consider only nonnegative objects. In that case, there are various ways to set bounds on null functions, which in turn lead to bounds on the bias of θ. Full details can be found in Clarkson and Barrett (1997, 1998a), and only a simple special case will be treated here.

We consider the usual linear CD system for which, in the absence of noise,

$$g_m = \int_{\mathbf{S}_f} d^q r \; h_m(\mathbf{r}) \, f(\mathbf{r}) \,, \tag{15.35}$$

and we define the point sensitivity function $s(\mathbf{r})$ as in (7.232) by

$$s(\mathbf{r}) = \sum_{m=1}^{M} h_m(\mathbf{r}) \,. \tag{15.36}$$

Suppose first that $s(\mathbf{r})$ is a constant s_0, independent of $\mathbf{r}$. Then, if $f_1(\mathbf{r})$ and $f_2(\mathbf{r})$ produce the same data, we have

$$\sum_{m=1}^{M} g_m = s_0 \int_{\mathbf{S}_f} d^q r \; f_1(\mathbf{r}) = s_0 \int_{\mathbf{S}_f} d^q r \; f_2(\mathbf{r}) \,. \tag{15.37}$$

Since the objects are nonnegative, we can add absolute-value signs inside the integrals, obtaining

$$\int_{\mathbf{S}_f} d^q r \; |f_1(\mathbf{r})| = \int_{\mathbf{S}_f} d^q r \; |f_2(\mathbf{r})| \,. \tag{15.38}$$

Thus two nonnegative objects that give the same data must have the same $\mathbb{L}_1$ norm if $s(\mathbf{r})$ is a constant. Since the difference $\mathbf{f}_2 - \mathbf{f}_1$ is a null function $\mathbf{f}_{null}$, it follows from the triangle inequality (see Sec. 1.1.2) that

$$\int_{\mathbf{S}_f} d^q r \; |f_{null}(\mathbf{r})| \leq 2 \int_{\mathbf{S}_f} d^q r \; |f_j(\mathbf{r})| \,, \tag{15.39}$$

where j can be either 1 or 2.

A more general expression that does not require $s(\mathbf{r})$ to be constant was derived by Clarkson and Barrett (1997); they found that

$$\int_{\mathbf{S}_f} d^q r \; |f_{null}(\mathbf{r})| \leq \left(1 + \frac{s_{max}}{s_{min}}\right) \int_{\mathbf{S}_f} d^q r \; |f_j(\mathbf{r})| \,, \tag{15.40}$$

where s_{max} and s_{min} are, respectively, the maximum and minimum values of $s(\mathbf{r})$ over the object support.

To relate this result to the parameter θ, note that

$$\theta_2 - \theta_1 = (\boldsymbol{\chi}, \mathbf{f}_2 - \mathbf{f}_1) = (\boldsymbol{\chi}, \mathbf{f}_{null}) = (\boldsymbol{\chi}_{null}, \mathbf{f}_{null}) \,, \tag{15.41}$$

where the last step follows since the measurement and null spaces are orthogonal. Thus

$$|\theta_2 - \theta_1| = |(\boldsymbol{\chi}_{null}, \mathbf{f}_{null})| = \left| \int_{\mathbf{S}_f} d^q r \; \chi_{null}(\mathbf{r}) \, f_{null}(\mathbf{r}) \right|$$

$$\leq \int_{\mathbf{S}_f} d^q r \; |\chi_{null}(\mathbf{r}) \, f_{null}(\mathbf{r})| \leq \max_{\mathbf{r}} \left\{ |\chi_{null}(\mathbf{r})| \right\} \int_{\mathbf{S}_f} d^q r \; |f_{null}(\mathbf{r})| \,. \tag{15.42}$$

With (15.40), we have, finally,

$$|\theta_2 - \theta_1| \leq \max_{\mathbf{r}} \left\{ |\chi_{null}(\mathbf{r})| \right\} \left(1 + \frac{s_{max}}{s_{min}}\right) \int_{\mathbf{S}_f} d^q r \; |f_j(\mathbf{r})| \,. \tag{15.43}$$

If $\chi(\mathbf{r})$ has no null component, the right-hand side is zero and we are back to the usual definition of estimability: Two objects that give the same data give the same value for the parameter θ. If $\chi(\mathbf{r})$ does have a null component, however, the positivity constraint gives us a bound on how large the difference between θ_1 and θ_2 can be. This bound gets smaller as the null component of $\chi(\mathbf{r})$ gets smaller (in peak value), as the system sensitivity gets more uniform and as the object itself gets weaker (and hence θ gets smaller).

Positivity and data consistency We have defined consistency space $\mathbb{V}_{con}$ as the range of $\boldsymbol{\mathcal{H}}$, which is a subset of the overall data space $\mathbb{V}$. Every vector in consistency space can be realized as the (noise-free) image through the operator $\boldsymbol{\mathcal{H}}$ of some object in the domain $\mathbb{U}$ of $\boldsymbol{\mathcal{H}}$. Fundamentally, this domain is constrained only by the definition of the Hilbert space $\mathbb{U}$; for example, it might consist of all square-integrable functions of some specified support. Many of the vectors in this domain, however, may not be physically realizable objects. For example, we may know that physical objects are nonnegative. We can define a *positive consistency set* $\mathbf{V}_+$ as the set of all noise-free images of nonnegative objects. Thus $\mathbf{V}_+$ is a subset of $\mathbb{V}_{con}$. Note that it is not correct to say that $\mathbf{V}_+$ is a sub*space* of $\mathbb{V}_{con}$ since it does not satisfy the requirements for a linear vector space (see Sec. 1.1.1). In particular, multiplying an image of a positive object by a negative number would remove it from the set.

Moment cone Another name for $\mathbf{V}_+$ is *moment cone.* To see how this name arises, consider first a unit point source $\delta(\mathbf{r}-\mathbf{r}_0)$ (and ignore the fact that it is not square-integrable). The image of this source is a point in $\mathbf{V}_+$. To be precise, for this object and a CD model of the imaging system as in (15.35),

$$g_m = [\boldsymbol{\mathcal{H}}\,\delta(\mathbf{r}-\mathbf{r}_0)]_m = h_m(\mathbf{r}_0)\,. \tag{15.44}$$

If we vary $\mathbf{r}_0$, we generate a set of points in data space. For example, if we scan object space in raster-like[1] fashion, we generate a line of points in data space as shown in Fig. 15.1a.

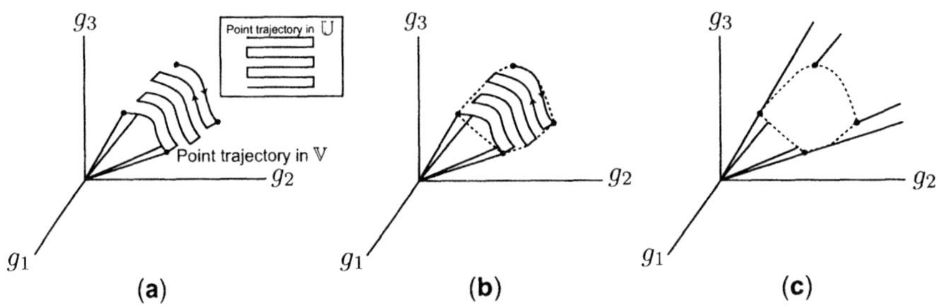

Fig. 15.1 Construction of the moment cone. Three of the M dimensions of data space are shown, and the object is a 2D delta function $\delta(\mathbf{r}-\mathbf{r}_0)$. The line shown is traced out as the point $\mathbf{r}_0$ is scanned back and forth over the object space. As drawn, all three detectors receive radiation for all $\mathbf{r}_0$.

Now add a second point source and renormalize the object to unit strength. The new object, $\beta\,\delta(\mathbf{r}-\mathbf{r}_0) + (1-\beta)\,\delta(\mathbf{r}-\mathbf{r}_1)$ with $0 \le \beta \le 1$, is nonnegative everywhere, so its image must lie in $\mathbf{V}_+$. This object is a convex combination of the two point objects $\delta(\mathbf{r}-\mathbf{r}_0)$ and $\delta(\mathbf{r}-\mathbf{r}_1)$, and since the system is linear, the resulting point in data space is the same convex combination, $\beta\boldsymbol{\mathcal{H}}\,\delta(\mathbf{r}-\mathbf{r}_0)+(1-\beta)\,\boldsymbol{\mathcal{H}}\,\delta(\mathbf{r}-\mathbf{r}_1)$. (Recall the definition of a convex set of points: If $\mathbf{g}_0$ and $\mathbf{g}_1$ are both in the set, then so is $\beta\mathbf{g}_0 + (1-\beta)\mathbf{g}_1$ for $0 \le \beta \le 1$.) Thus we can include in $\mathbf{V}_+$ all image points within the convex set generated this way with all $\mathbf{r}_0$, $\mathbf{r}_1$ and β (see Fig. 15.1b). We shall call this set $\mathbf{V}_+^{(2)}$ since it is the set of image points generated by unit-strength, nonnegative objects consisting of pairs of point sources.

[1]The word raster comes from the Latin *rastrum*, rake, so it suggests a set of lines all moving in the same direction, as in a TV raster. The back-and-forth pattern in Fig. 15.1 is called a *boustrophedonic* scan, referring to an ancient method of writing in which the lines run in alternating directions. The root *bou* occurs also in bovine, and the boustrophedon is the pattern of an ox plowing a field.

If we add a third point source and renormalize the object to have an integral of unity, the new image point will also lie in $\mathbf{V}_+^{(2)}$ since it is just a convex combination of the image of the third point and the image obtained when the first two are present simultaneously; both of these images are in the convex set $\mathbf{V}_+^{(2)}$, and by the definition of convexity the image of the new three-point object must lie in it also. We can extend this argument to an arbitrary number of point sources and hence to a general positive object of unit integral; the image of all of these objects lies in $\mathbf{V}_+^{(2)}$.

To remove the restriction that the objects must integrate to one, consider what happens when an object is scaled, $\mathbf{f} \to \alpha\mathbf{f}$, with $\alpha \geq 0$ so that a nonnegative object remains nonnegative. Since the system is linear, the image is scaled similarly, $\mathbf{g} \to \alpha\mathbf{g}$. Thus, when we consider all α from 0 to ∞, each point in the convex set discussed above becomes a line extending from the origin through the original point to infinity (see Fig. 15.1b). Carrying out this extension for all points in $\mathbf{V}_+^{(2)}$, we generate a cone of points as shown in Fig. 15.1c. This cone is the positive consistency set $\mathbf{V}_+$. It is known as the *moment cone* since each data component is a weighted integral or moment of the object function.

In summary, the moment cone, a convex cone with vertex at the origin in data space, is the set of all noise-free images of nonnegative objects. It is a subset of consistency space or the range of $\mathcal{H}$. Moreover, if each $h_m(\mathbf{r})$ is nonnegative for all $\mathbf{r}$, the moment cone is a subset of the *positive orthant* of data space, *i.e.*, the set of all $\mathbf{g}$ such that $g_m \geq 0$ for all m.

15.1.5 Choosing the best algorithm

We have already seen that there are many choices to be made in formulating and solving an inverse problem. If we adopt a discrete model, we must directly or indirectly select the sets $\{\phi_n\}$ and $\{\chi_n\}$. If we use an implicit formulation, we must choose the objective functional, and if we use an iterative algorithm, we must select the iteration rule and the number of iterations. We must choose whether or not to enforce positivity. In addition, many algorithms have various other free parameters, called things like regularization parameters, acceleration parameters or hyperparameters.

There is a large literature on choosing these parameters and optimizing a reconstruction algorithm, and we shall discuss several aspects of the problem in the remainder of this chapter. In our view, however, any meaningful discussion of image quality must ultimately relate back to the intended use of the image, that is, to the task and the observer. The general principles enunciated in Chap. 14 on assessment of image quality apply equally well to the assessment and optimization of reconstruction algorithms.

For classification tasks, the observer is most often a human, and the reconstruction algorithm serves only to transduce the original data into a form where it is useful to a human observer. Under these conditions, assessment and optimization of an algorithm must necessarily take into account the properties and limitations of human visual perception. For classification tasks to be performed by human observers, image quality can be measured *only* by psychophysical studies or mathematical model observers that predict the outcome of such studies.

The ideal observer is generally useless for evaluating algorithms since its performance is usually invariant to algorithm. The performance of the ideal observer on a detection task can be evaluated on the raw data $\mathbf{g}$, before any algorithm, and the algorithm cannot improve the performance (see Sec. 13.2.7). Virtually by definition, if the algorithm were useful, it would be incorporated into the detection strategy employed by the ideal observer; otherwise, that observer would not be ideal.

A similar conclusion holds for the Hotelling observer. It was shown in Sec. 13.2.12 that the Hotelling trace is invariant to any invertible linear algorithm, *i.e.*, one where it is possible to go backwards from the reconstruction to the original data. Thus neither the ideal nor the Hotelling observer tells us very much about the algorithm if the task is classification and the end user will be a human; their only use in evaluating algorithms is to make sure that no information is lost.

The situation is rather different for estimation tasks, which are rarely performed by humans. Instead, some kind of image-analysis program is used to extract quantitative information from images, *i.e.*, to estimate some parameter of the object. If we denote the parameter of interest by $\mathbf{\Theta}(\mathbf{f})$, where the capitalization of theta serves to distinguish it from the vector of coefficients, then the output of the analysis program is $\widehat{\mathbf{\Theta}}$. The input to the program could be the raw image data $\mathbf{g}$, in which case the output would be some (possibly nonlinear) function $\widehat{\mathbf{\Theta}}_{raw}(\mathbf{g})$. Often, however, it is easier to perform the analysis using a reconstructed image as the data. In that case, two successive estimation steps are needed, one to obtain the input to the image-analysis program and one carried out in that program. The final estimate would be written as $\widehat{\mathbf{\Theta}}_{recon}(\hat{\boldsymbol{\theta}})$, and its accuracy could, as usual, be evaluated in terms of the bias, variance or mean-squared error (MSE). For more on how to compute MSE, see Sec. 14.3.4. A unique specification of bias and MSE presupposes, of course, that $\mathbf{\Theta}(\mathbf{f})$ is estimable, which pixel values rarely are. Some of the problems in defining an MSE on pixel values are discussed in Sec. 13.3.2.

Since the accuracy of $\widehat{\mathbf{\Theta}}(\hat{\boldsymbol{\theta}})$ depends on the noise in the data, on the reconstruction algorithm and on the final estimation procedure, the MSE can be used to assess the quality of any combination of these steps. If we consider a fixed imaging system and reconstruction algorithm, for example, the MSE of $\widehat{\mathbf{\Theta}}(\hat{\boldsymbol{\theta}})$ would be interpreted, as it traditionally is in the statistics literature, as the performance of the final estimate. For assessing reconstruction algorithms in terms of estimation tasks, however, we can consider the imaging system and the image-analysis program as fixed, and then the MSE is a figure of merit for the reconstruction.

No matter which task and observer we pick, task performance depends on the statistical properties of the data. For tasks performed on reconstructed images, that means that we need to know the statistical properties of the images, and much of this chapter is devoted to this goal. In particular, for classification tasks and linear observers, we know from Sec. 13.2.12 that the performance can be computed from the mean vector and the covariance matrix, so we concentrate in what follows on calculation of these quantities for reconstructed images.

15.2 LINEAR RECONSTRUCTION OPERATORS

We shall now discuss various linear operators that can be applied to a data set to yield a reconstructed image in one step. Since the operators are linear, the resulting images will seldom satisfy the positivity constraint.

Section 15.2.1 deals with the CDD problem of finding the coefficients in an approximate object expansion from discrete data. In Sec. 15.2.2 we look at various ways of getting a continuous estimate either from the discrete coefficients or directly; thus this section deals with CDC inverse problems.

In Sec. 15.2.3 we consider a broad class of imaging systems that can be called *Fourier samplers* since the data consist of discrete samples of the object Fourier transform. These systems will provide a concrete example of the formalism developed in Sec. 15.2.2.

In Sec. 15.2.4 we begin to examine the relation between CC operators and CD operators. Specifically, we consider situations where a CC operator with a known inverse might be a reasonable description of an imaging system in the limit of very fine sampling in the data space. We then investigate the actual CDD inverse problem using the known solution to the CCC problem as a starting point.

In Sec. 15.2.5, we examine CCC problems in which the adjoint operator $\mathcal{H}^\dagger$ has a null space. Such operators arise frequently in tomographic problems, and their properties have implications for practical (*i.e.*, CDD) tomographic reconstruction algorithms.

Finally, the effect of noise on estimates obtained with linear reconstruction operators is discussed in Sec. 15.2.6.

15.2.1 Matrix operators for estimation of expansion coefficients

In this section we view the reconstruction process as estimation of the coefficients $\{\theta_n\}$ in an approximate object expansion. Since there are N coefficients and M measurements, we seek an $N \times M$ reconstruction matrix that will yield an estimate of $\boldsymbol{\theta}$ in a single matrix-vector multiply. If the coefficients are estimable parameters (see Sec. 15.1.3), it is reasonable to require that the matrix give the correct values in the absence of noise; that condition will be the starting point for this discussion, and it will lead to pseudoinverse matrices. As the reader might expect from Chap. 1, pseudoinverse solutions have poor noise performance, but we postpone until Sec. 15.2.6 a discussion of ways of dealing with this problem.

Let us suppose that the parameters $\{\theta_n\}$ are estimable and hence that (15.27) holds. The matrix $\mathbf{B}$ defined in that equation and given explicitly in (15.30) is, in fact, just the kind of reconstruction matrix we are seeking. We can define estimates of θ_n by

$$\hat{\theta}_n \equiv \sum_{m=1}^{M} B_{nm} g_m \,. \tag{15.45}$$

In the absence of noise,

$$\begin{aligned}\hat{\theta}_n &= \sum_{m=1}^{M} B_{nm} \int_\infty d^q r \; h_m(\mathbf{r}) \, f(\mathbf{r}) = \int_\infty d^q r \left[\sum_{m=1}^{M} B_{nm} h_m(\mathbf{r}) \right] f(\mathbf{r}) \\ &= \int_\infty d^q r \; \chi_n(\mathbf{r}) \, f(\mathbf{r}) = \theta_n \,, \end{aligned} \tag{15.46}$$

where the last line has used (15.8). Hence the matrix $\mathbf{B}$ is the reconstruction operator, yielding $\hat{\boldsymbol{\theta}} = \boldsymbol{\theta}$ exactly in the absence of noise. With zero-mean additive noise, $\hat{\boldsymbol{\theta}}$ is an unbiased estimate, where $\langle \hat{\theta}_n \rangle = \theta_n$ for all n. These conclusions require, however, that $\boldsymbol{\theta}$ be estimable.

We can restate the results of the last paragraph more succinctly in operator form. With (15.30), (15.10), (15.28) and the assumption that $\mathbf{n} = 0$,

$$\mathbf{Bg} = \mathcal{D}_\chi \mathcal{H}^+ \mathbf{g} = \mathcal{D}_\chi \mathcal{H}^+ \mathcal{H} \mathbf{f} = \mathcal{D}_\chi \mathbf{f} = \boldsymbol{\theta}\,. \tag{15.47}$$

Thus we can get $\boldsymbol{\theta}$ by first applying the pseudoinverse operator $\mathcal{H}^+$ to $\mathbf{g}$, then discretizing the result with $\mathcal{D}_\chi$, but only if $\boldsymbol{\theta}$ is estimable and there is no noise.

Bias for nonestimable parameters What happens if $\boldsymbol{\theta}$ is not estimable, but we pretend it is and apply $\mathbf{B}$ anyway? Then we get (still without noise)

$$\mathbf{Bg} = \mathcal{D}_\chi \mathcal{H}^+ \mathcal{H} \mathbf{f} = \mathcal{D}_\chi \mathbf{f}_{meas} \neq \boldsymbol{\theta}\,. \tag{15.48}$$

For some particular object, we can determine the error (bias) in this procedure. The error norm is

$$||\mathbf{Bg} - \boldsymbol{\theta}|| = ||\mathcal{D}_\chi (\mathbf{f}_{meas} - \mathbf{f})|| = ||\mathcal{D}_\chi \mathbf{f}_{null}||\,. \tag{15.49}$$

To evaluate this expression exactly, we have to know $\mathbf{f}_{null}$, but it is interesting to try to set a bound analogous to (15.43) that involves only χ_{null} and not $\mathbf{f}_{null}$. For a scalar parameter θ, $\mathcal{D}_\chi = \boldsymbol{\chi}^\dagger$, and

$$|\boldsymbol{\chi}^\dagger \mathcal{H}^+ \mathbf{g} - \theta| = |\boldsymbol{\chi}^\dagger \mathbf{f}_{null}|\,. \tag{15.50}$$

Since null and measurement spaces are orthogonal, we can write the scalar product as $\boldsymbol{\chi}^\dagger \mathbf{f}_{null} = \boldsymbol{\chi}_{null}^\dagger \mathbf{f}_{null} = \boldsymbol{\chi}_{null}^\dagger \mathbf{f}$. Then we have

$$|\boldsymbol{\chi}^\dagger \mathcal{H}^+ \mathbf{g} - \theta| = \left| \int_{\mathbf{S}_f} d^q r\ \chi_{null}(\mathbf{r})\, f(\mathbf{r}) \right|$$

$$\leq \int_{\mathbf{S}_f} d^q r\ |\chi_{null}(\mathbf{r})\, f(\mathbf{r})| \leq \max_{\mathbf{r}} \{\, |\chi_{null}(\mathbf{r})|\, \} \int_{\mathbf{S}_f} d^q r\ |f(\mathbf{r})|\,. \tag{15.51}$$

Note that we are now computing the error between θ for a particular object and its pseudoinverse estimate, not between the θ values for two objects that give the same data as in (15.43); this difference accounts for the absence of the factor involving s_{max}/s_{min} here.

A tighter bound can be stated if $\chi_{null}(\mathbf{r})$ has compact support, at least approximately. If we can define a region $\mathbf{S}(\chi_{null})$ in object space such that $\chi_{null}(\mathbf{r}) \simeq 0$ for $\mathbf{r}$ outside this region, then the integrals in (15.51) are restricted to this region, and we find

$$|\boldsymbol{\chi}^\dagger \mathcal{H}^+ \mathbf{g} - \theta| \leq \max_{\mathbf{r}} \{\, |\chi_{null}(\mathbf{r})|\, \} \int_{\mathbf{S}(\chi_{null})} d^q r\ |f(\mathbf{r})|\,. \tag{15.52}$$

Note that the integral here is less than or equal to the one in (15.51), so the bound is tighter.

Matrix pseudoinverses In practice, it may be difficult to implement the pseudoinverse of the CD operator $\mathcal{H}$, as required in (15.48), and we should not have to since the objective is to go from one vector ($\mathbf{g}$) to another ($\hat{\boldsymbol{\theta}}$). It may be much more convenient to compute the matrix $\mathbf{H}$ and try to solve the problem $\mathbf{g} = \mathbf{H}\boldsymbol{\theta} + \boldsymbol{\epsilon}$ directly, without worrying about the meaning of $\boldsymbol{\theta}$, but what do we get this way? It seems to be tacitly assumed in much of the literature that some sort of matrix pseudoinversion will suffice to recover $\mathcal{D}_\phi\mathbf{f}$, at least from noise-free data. For example, in a pixel basis one might hope that a matrix constructed from pixels will suffice to recover integrals of the object over pixels. It is rarely the case that this hope will be fulfilled.

Suppose we choose some arbitrary set of expansion functions $\{\phi_n\}$ and define $\mathbf{H} = \mathcal{H}\mathcal{D}_\phi^\dagger$. Even in the absence of measurement noise, all we can say in general about the pseudoinverse solution is that

$$\mathbf{H}^+\mathbf{g} = \left[\mathcal{H}\mathcal{D}_\phi^\dagger\right]^+ \mathcal{H}\mathbf{f}\,. \tag{15.53}$$

To the extent that this is an estimate of some parameter $\boldsymbol{\theta}$, it must be the one defined by $\mathcal{D}_\chi = [\mathcal{H}\mathcal{D}_\phi^\dagger]^+\mathcal{H}$, and there is no evident way to simplify this expression. In particular, it is not true in general that $[\mathcal{H}\mathcal{D}_\phi^\dagger]^+$ is the same as $\mathcal{D}_\phi^{\dagger +}\mathcal{H}^+$.

There is, however, one circumstance under which $\mathbf{H}^+\mathbf{g}$ becomes more transparent. Suppose $\{\phi_n\}$ is an orthonormal basis for measurement space, for example the first R singular vectors of $\mathcal{H}$ (see Sec. 7.4.3). In that case, $\mathcal{D}_\phi^\dagger = \mathcal{D}_\phi^+$ (by the orthonormality) and $\mathcal{D}_\phi^+\mathcal{D}_\phi = \mathcal{H}^+\mathcal{H} = \mathcal{P}_{meas}$, the projector onto measurement space. Then we can apply a theorem quoted in Sec. 1.6.3 for the pseudoinverse of a product:

$$(\mathbf{XY})^+ = (\mathbf{X}^+\mathbf{XY})^+(\mathbf{XYY}^+)^+\,. \tag{15.54}$$

With this theorem and the assumption that $\{\phi_n\}$ is an orthonormal basis for measurement space, we can rewrite (15.53) as

$$\mathbf{H}^+\mathbf{g} = \left[\mathcal{H}\mathcal{D}_\phi^+\right]^+ \mathcal{H}\mathbf{f} = \mathcal{D}_\phi\mathbf{f}\,. \tag{15.55}$$

In this special case, therefore, application of a matrix pseudoinverse to noise-free data yields scalar products of the object with the same expansion functions used to form $\mathbf{H}$. These scalar products are necessarily estimable parameters since the functions $\{\chi_n(\mathbf{r})\}$, which are the same as $\{\phi_n(\mathbf{r})\}$ in this case, lie entirely in measurement space.

Natural pixels and matrix pseudoinverses We can also use the natural pixels as expansion functions, which is how they were introduced by Buonocore *et al.* (1981). Since natural pixels span measurement space, we might expect them to simplify the calculations in much the same way as the singular functions of $\mathcal{H}$; we give up the orthonormality but gain an advantage since the expansion functions are given up front and it is easy to compute $\mathbf{H}$.

If we take $\mathcal{D}_\phi = \mathcal{H}$ as in (15.32), the system matrix $\mathbf{H} = \mathcal{H}\mathcal{H}^\dagger$. In the absence of noise, application of $\mathbf{H}^+$ to $\mathbf{g}$ yields

$$\hat{\boldsymbol{\theta}} = \mathbf{H}^+\mathbf{g} = \left[\mathcal{H}\mathcal{H}^\dagger\right]^+ \mathcal{H}\mathbf{f} = \mathcal{H}^{\dagger +}\mathbf{f}\,, \tag{15.56}$$

where the last step follows from (1.148) with $\mathcal{H}$ and $\mathcal{H}^\dagger$ interchanged. The last form in (15.56) will be recognized as the $\boldsymbol{\theta}$ defined by the operator $\mathcal{D}_\chi$ given in (15.32). Thus application of the matrix pseudoinverse now yields the same result as a CD pseudoinverse followed by discretization as in (15.47).

Because $\mathcal{D}_\chi$ involves a pseudoinverse in this case, it is not easy to interpret the resulting parameters $\hat{\theta}_n$; in particular, they are not scalar products with natural pixels. We shall return to the interpretation of (15.56) in Sec. 15.2.2.

15.2.2 Reconstruction of functions from discrete data

Since the object of interest in most imaging situations is a function $f(\mathbf{r})$, one possible goal of image reconstruction would be to produce another function $\hat{f}(\mathbf{r})$ that resembles $f(\mathbf{r})$ in some way. If we have obtained estimates $\{\hat{\theta}_n\}$ of the expansion coefficients, a straightforward way to construct such an estimate is [*cf.* (15.6)]

$$\hat{f}(\mathbf{r}) = \sum_{n=1}^{N} \hat{\theta}_n \phi_n(\mathbf{r}), \tag{15.57}$$

or, in operator notation,

$$\hat{\mathbf{f}} = \mathcal{D}_\phi^\dagger \hat{\boldsymbol{\theta}}. \tag{15.58}$$

There is, however, no rule that says we have to use the functions $\{\phi_n(\mathbf{r})\}$ in this step; we could use some entirely different set $\{\psi_n(\mathbf{r})\}$ and define

$$\hat{\mathbf{f}} = \mathcal{D}_\psi^\dagger \hat{\boldsymbol{\theta}}. \tag{15.59}$$

For example, if we display a set of estimates $\{\hat{\theta}_n\}$ on a computer screen, the functions $\{\psi_n(\mathbf{r})\}$ might be uniform square pixels even if pixels played no role in defining $\{\theta_n\}$ in the first place.

Moreover, it may not be necessary to perform the intermediate step of estimating $\boldsymbol{\theta}$ at all. If $\hat{\boldsymbol{\theta}}$ is a linear function of $\mathbf{g}$, then the $\hat{\mathbf{f}}$ defined in (15.59) is also a linear function of $\mathbf{g}$, so we can write

$$\hat{\mathbf{f}} = \mathcal{O}\mathbf{g}, \tag{15.60}$$

where $\mathcal{O}$ is a linear DC operator. Denoting the kernel of this operator as $o_m(\mathbf{r})$, we can write

$$\hat{f}(\mathbf{r}) = \sum_{m=1}^{M} g_m o_m(\mathbf{r}). \tag{15.61}$$

In this view, the reconstruction operator consists of the set of functions $\{o_m(\mathbf{r})\}$.

Backus-Gilbert method Backus and Gilbert (1968) proposed a method for obtaining pseudoinverses of CD operators without adopting any discrete object representation. They assumed that $\mathcal{H}\mathcal{H}^\dagger$ was invertible, or that $R = M$, but this assumption is not always warranted. Therefore the treatment here is based on the identity (1.149), which allows us to write

$$\mathcal{H}^+\mathbf{g} = \mathcal{H}^\dagger \left[\mathcal{H}\mathcal{H}^\dagger\right]^+ \mathbf{g}. \tag{15.62}$$

Since $\mathcal{H}\mathcal{H}^\dagger$ is a matrix, its pseudoinverse can be computed by SVD (discussed later in this section) or by iterative methods (discussed in Secs. 1.7 and 15.4). Then, since $\mathcal{H}^\dagger$ is a DC operator, a reconstructed image is given by

$$\hat{f}(\mathbf{r}) = \left[\mathcal{H}^+\mathbf{g}\right](\mathbf{r}) = \sum_{n=1}^{M} \left[(\mathcal{H}\mathcal{H}^\dagger)^+\mathbf{g}\right]_n h_n^*(\mathbf{r}) = \sum_{m=1}^{M}\sum_{n=1}^{M} \left[(\mathcal{H}\mathcal{H}^\dagger)^+\right]_{nm} g_m h_n^*(\mathbf{r}) . \tag{15.63}$$

The last expression has the form of (15.61) with

$$o_m(\mathbf{r}) = \sum_{n=1}^{M} \left[(\mathcal{H}\mathcal{H}^\dagger)^+\right]_{nm} h_n^*(\mathbf{r}) . \tag{15.64}$$

Thus a continuous reconstructed image can be obtained either by superimposing the functions $o_m(\mathbf{r})$ with data values g_m as weights or by superimposing the point response function $h_n(\mathbf{r})$ with pseudoinverse values $[(\mathcal{H}\mathcal{H}^\dagger)^+\mathbf{g}]_n$ as weights.

Relation of Backus-Gilbert to natural pixels There is no need to select a discrete object representation in the Backus-Gilbert method, but it can be interpreted after the fact in terms of natural pixels. In fact, (15.63) is just the natural-pixel expansion formed by using the estimated coefficients $\{\hat{\theta}_n\}$ specified in (15.56). If we take the expansion functions $\{\phi_n(\mathbf{r})\}$ to coincide with the point response functions $\{h_n(\mathbf{r})\}$ (or their complex conjugates if the PRFs are complex), then $\mathcal{D}_\phi = \mathcal{H}$. The adjoint operator $\mathcal{D}_\phi^\dagger$ is then the same as $\mathcal{H}^\dagger$, which is often referred to as *back-projection*, especially in a tomographic context; it serves to project a vector in the M-dimensional data space back into the infinite-dimensional object space (see Sec. 7.3.2).

If we define $\mathbf{H}$ as $\mathcal{H}\mathcal{H}^\dagger$ and apply its pseudoinverse to $\mathbf{g}$, we obtain

$$\hat{\mathbf{f}} = \mathcal{D}_\phi^\dagger\hat{\boldsymbol{\theta}} = \mathcal{H}^\dagger\hat{\boldsymbol{\theta}} = \mathcal{H}^\dagger\left[\mathcal{H}\mathcal{H}^\dagger\right]^+\mathbf{g} , \tag{15.65}$$

which is identical to the Backus-Gilbert result. To gain more insight into this result, suppose there is no noise so that $\mathbf{g} = \mathcal{H}\mathbf{f}$. Then we have

$$\hat{\mathbf{f}} = \mathcal{H}^\dagger\left[\mathcal{H}\mathcal{H}^\dagger\right]^+\mathcal{H}\mathbf{f} = \mathcal{H}^+\mathcal{H}\mathbf{f} = \mathbf{f}_{meas} , \tag{15.66}$$

where we have used (1.149) and (1.166). Thus either the natural-pixel estimate $\hat{\boldsymbol{\theta}}$ when backprojected into object space or the direct Backus-Gilbert estimate reduces in the no-noise limit to $\mathbf{f}_{meas}$, which is all we can ever expect from the given data.

SVD methods The Backus-Gilbert method can also be formulated in terms of singular-value decomposition (Bertero *et al.*, 1985, 1988). For background on SVD of CD operators, see Secs. 1.6.2 and 7.3.2.

From (1.131) we know that

$$\mathcal{H}^+ = \sum_{n=1}^{R} \frac{1}{\sqrt{\mu_n}}\mathbf{u}_n\mathbf{v}_n^\dagger , \tag{15.67}$$

where $\mathbf{v}_n$ is an $M \times 1$ eigenvector of the matrix $\mathcal{H}\mathcal{H}^\dagger$, $\mathbf{u}_n$ is an eigenfunction of the CC operator $\mathcal{H}^\dagger\mathcal{H}$, and μ_n is their common eigenvalue. If we can find the eigenvectors of $\mathcal{H}\mathcal{H}^\dagger$ by numerical means, the eigenfunctions of $\mathcal{H}^\dagger\mathcal{H}$ corresponding to nonzero eigenvalues can be found by (7.253):

$$\frac{1}{\sqrt{\mu_n}}\left[\mathcal{H}^\dagger \mathbf{v}_n\right](\mathbf{r}) = u_n(\mathbf{r}), \qquad n = 1, ..., R. \tag{15.68}$$

Thus

$$\hat{f}(\mathbf{r}) \equiv \left[\mathcal{H}^+\mathbf{g}\right](\mathbf{r}) = \sum_{n=1}^{R} \frac{1}{\sqrt{\mu_n}}(\mathbf{v}_n^\dagger \mathbf{g})\, u_n(\mathbf{r}). \tag{15.69}$$

Since $\mathbf{v}_n^\dagger \mathbf{g}$ is just the coefficient β_n in an SVD expansion of $\mathbf{g}$, we see that we can get a pseudoinverse reconstruction by superimposing the eigenfunctions with weight $\beta_n/\sqrt{\mu_n}$.

Another way of looking at this result is to recall the SVD expansion for the original object $f(\mathbf{r})$:

$$f(\mathbf{r}) = \sum_{n=1}^{\infty} \alpha_n u_n(\mathbf{r}). \tag{15.70}$$

To put (15.69) in a similar format, we write

$$\hat{f}(\mathbf{r}) = \sum_{n=1}^{R} \hat{\alpha}_n u_n(\mathbf{r}), \tag{15.71}$$

where $\hat{\alpha}_n$ is an estimate of the coefficient α_n given by

$$\hat{\alpha}_n = \frac{\beta_n}{\sqrt{\mu_n}}, \qquad n \leq R. \tag{15.72}$$

This estimate is plausible since we know from (1.209) that the imaging equation, $\mathbf{g} = \mathcal{H}\mathbf{f} + \mathbf{n}$, in the SVD domain takes the form,

$$\beta_n = \sqrt{\mu_n}\alpha_n + \gamma_n, \tag{15.73}$$

where γ_n is a coefficient in the SVD expansion of $\mathbf{n}$. Thus

$$\hat{\alpha}_n = \alpha_n + \frac{\gamma_n}{\sqrt{\mu_n}}, \qquad n \leq R, \tag{15.74}$$

so $\hat{\alpha}_n = \alpha_n$ for $n \leq R$ in the absence of noise. That means, once again, that $\hat{\mathbf{f}} = \mathbf{f}_{meas}$ in the no-noise limit. With noise, however, the second term in (15.74) can be quite large since μ_n will often be very small for n near R. Methods for dealing with this problem are discussed in Sec. 15.3.

15.2.3 Reconstruction from Fourier samples

Many imaging systems, including various optical interferometers and most magnetic resonance imagers, directly measure Fourier components of the object at discrete spatial frequencies. In this section we analyze such systems as a way of illustrating the ideas introduced in Sec. 15.2.2.

We assume throughout this section that the object $f(\mathbf{r})$ has finite support $\mathbf{S}_f$. If we define a support function $s(\mathbf{r})$ (not to be confused with the sensitivity function used earlier) that takes the value 1 for points $\mathbf{r}$ inside $\mathbf{S}_f$ and zero for points outside, then $f(\mathbf{r})s(\mathbf{r}) = f(\mathbf{r})$. We now define a *Fourier sampler* as a CD system where

$$[\boldsymbol{\mathcal{H}}\mathbf{f}]_m = \int_\infty d^q r \; s(\mathbf{r})\, f(\mathbf{r}) \exp[-2\pi i \boldsymbol{\rho}_m \cdot \mathbf{r}] \,, \tag{15.75}$$

where $\{\boldsymbol{\rho}_m, m = 1, ..., M\}$ is a set of discrete frequencies at which the system takes measurements. There is no implication that these frequencies fall on a regular grid.

The kernel of $\boldsymbol{\mathcal{H}}$ is given by

$$h_m(\mathbf{r}) = s(\mathbf{r}) \exp[-2\pi i \boldsymbol{\rho}_m \cdot \mathbf{r}] \,. \tag{15.76}$$

The adjoint is given by

$$\left[\boldsymbol{\mathcal{H}}^\dagger \mathbf{g}\right](\mathbf{r}) = \sum_{m=1}^{M} g_m \, s(\mathbf{r}) \exp[2\pi i \boldsymbol{\rho}_m \cdot \mathbf{r}] \,. \tag{15.77}$$

The matrix $\boldsymbol{\mathcal{H}}\boldsymbol{\mathcal{H}}^\dagger$ has elements

$$\left[\boldsymbol{\mathcal{H}}\boldsymbol{\mathcal{H}}^\dagger\right]_{mk} = \int_\infty d^q r \; s(\mathbf{r}) \exp[-2\pi i (\boldsymbol{\rho}_m - \boldsymbol{\rho}_k) \cdot \mathbf{r}] = S(\boldsymbol{\rho}_m - \boldsymbol{\rho}_k) \,, \tag{15.78}$$

where $S(\boldsymbol{\rho}) = \mathcal{F}_q\{s(\mathbf{r})\}$. If we define a matrix $\mathbf{S}$ (not to be confused with the support set $\mathbf{S}_f$) with elements $S_{mk} = S(\boldsymbol{\rho}_m - \boldsymbol{\rho}_k)$, then

$$\boldsymbol{\mathcal{H}}\boldsymbol{\mathcal{H}}^\dagger = \mathbf{S} \,. \tag{15.79}$$

Since the Fourier transform of the support function is usually known analytically, it is straightforward to compute this matrix.

Pseudoinverse Depending on the choice of the sample frequencies $\{\boldsymbol{\rho}_m\}$, the matrix $\mathbf{S}$ may be invertible, but in any case its pseudoinverse exists and can presumably be computed with no great difficulty. Then the pseudoinverse of $\boldsymbol{\mathcal{H}}$ can be calculated from (1.149), which in the present problem becomes

$$\boldsymbol{\mathcal{H}}^+ = \boldsymbol{\mathcal{H}}^\dagger (\boldsymbol{\mathcal{H}}\boldsymbol{\mathcal{H}}^\dagger)^+ = \boldsymbol{\mathcal{H}}^\dagger \mathbf{S}^+ \,. \tag{15.80}$$

A pseudoinverse reconstruction from a noise-free data vector $\mathbf{g}$ is thus

$$\left[\boldsymbol{\mathcal{H}}^+ \mathbf{g}\right](\mathbf{r}) = \sum_{m=1}^{M} \left[\mathbf{S}^+ \mathbf{g}\right]_m s(\mathbf{r}) \exp\left[2\pi i \boldsymbol{\rho}_m \cdot \mathbf{r}\right] \,. \tag{15.81}$$

We see that a superposition of plane waves, weighted by the values $[\mathbf{S}^+\mathbf{g}]_m$ and truncated by the support function, gives $[\boldsymbol{\mathcal{H}}^+\mathbf{g}](\mathbf{r})$ at all points $\mathbf{r}$.

Van de Walle *et al.* (2001) have used this approach to reconstruct magnetic resonance images from Fourier samples on irregular grids.

Nyquist sampling Let us examine the solution (15.81) in the case where the Fourier samples fall on a regular grid and satisfy the Nyquist condition. (Recall from Sec. 3.5.4 that the Nyquist condition for sampling in the Fourier domain depends on the *spatial* support of the object; there is no need for the object to be bandlimited.)

If the support is a cube of side L and if the sampled frequencies fall on a cubic lattice of spacing $1/L$, then $S(\boldsymbol{\rho})$ is a qD sinc function, and $S(\boldsymbol{\rho}_m - \boldsymbol{\rho}_k)$ vanishes unless $m = k$, so we can write

$$\left[\mathcal{H}\mathcal{H}^\dagger\right]_{mk} = S_{mk} = L^q\,\delta_{mk}\,. \tag{15.82}$$

In this case, $\mathbf{S}$ is proportional to the $M \times M$ unit matrix, so (15.81) becomes

$$\left[\mathcal{H}^+\mathbf{g}\right](\mathbf{r}) = \frac{1}{L^q}\sum_{m=1}^{M} g_m\, s(\mathbf{r})\exp[2\pi i\boldsymbol{\rho}_m \cdot \mathbf{r}]\,. \tag{15.83}$$

Now all we have to do is superimpose truncated plane waves weighted by data values. If we are content to observe the resulting function on a regular grid of points, the final formula is precisely a multidimensional discrete Fourier transform as introduced in Sec. 3.6.6.

15.2.4 Discretization of analytic inverses

In many inverse problems, we would be able to perform the inverse if only we sampled finely enough. An example is computed tomography, where the data are samples of the Radon transform of a function. The inverse Radon transform was derived in Sec. 4.4, but it requires that we know the data for a continuous set of projection angles and as a function of a continuous variable at each angle; instead we know only a discrete set of data values.

Another example occurs when a system performs a linear shift-invariant operation on the object. We know that the system operator, in a CC sense, is then a convolution: $g(\mathbf{r}) = f(\mathbf{r}) * h(\mathbf{r})$. If we knew the full function $g(\mathbf{r})$, we could perform a Fourier transform, obtaining $G(\boldsymbol{\rho}) = H(\boldsymbol{\rho})\,F(\boldsymbol{\rho})$, and then divide through by the transfer function $H(\boldsymbol{\rho})$ to get $F(\boldsymbol{\rho})$, at least for frequencies where $H(\boldsymbol{\rho}) \neq 0$. An inverse Fourier transform would then recover $f(\mathbf{r})$ or a good approximation to it. As in the Radon example, however, we need to consider what happens when we have available only a finite set of points from the convolution output and when we cannot neglect noise.

Consider the CC problem $\mathbf{y} = \boldsymbol{\mathcal{L}}\mathbf{f}$, where $\boldsymbol{\mathcal{L}}$ is a nonsingular linear operator with a known inverse. Instead of observing the function $\mathbf{y}$, however, we observe a discrete, noisy data vector $\mathbf{g}$ given by

$$\mathbf{g} = C\boldsymbol{\mathcal{D}}_w\mathbf{y} + \mathbf{n}\,, \tag{15.84}$$

where C is a constant related to the sensitivity of the system and the exposure time, and $\boldsymbol{\mathcal{D}}_w$ is some appropriate discretization operator. In computed tomography, for example, the operator $\boldsymbol{\mathcal{D}}_w$ samples the projection angles and perhaps integrates over finite detector apertures. The discretization functions associated with $\boldsymbol{\mathcal{D}}_w$ are thus delta functions in angle and rect functions in the detector plane.

Since we have only a finite-dimensional data vector, we cannot determine the infinite-dimensional object; instead we attempt to estimate some vector related to $\mathbf{f}$ by

$$\boldsymbol{\theta} = \boldsymbol{\mathcal{D}}_\chi\mathbf{f}\,. \tag{15.85}$$

The discretization functions associated with $\boldsymbol{\mathcal{D}}_\chi$ might, for example, be pixels.

Though we do not know $\mathbf{y}$, we can nevertheless write $\boldsymbol{\theta}$ as

$$\boldsymbol{\theta} = \boldsymbol{\mathcal{D}}_\chi\boldsymbol{\mathcal{L}}^{-1}\mathbf{y}\,. \tag{15.86}$$

We now insert the operator $\mathcal{D}_w^\dagger \mathcal{D}_w$ between $\mathcal{L}^{-1}$ and $\mathbf{y}$ and promptly subtract off the error we make in doing so; the result is

$$\boldsymbol{\theta} = \mathcal{D}_\chi \mathcal{L}^{-1} \mathcal{D}_w^\dagger \mathcal{D}_w \mathbf{y} + \mathcal{D}_\chi \mathcal{L}^{-1} \left[\mathbf{I} - \mathcal{D}_w^\dagger \mathcal{D}_w \right] \mathbf{y} . \tag{15.87}$$

We can now define a matrix $\mathbf{O}$ by

$$\mathbf{O} = \mathcal{D}_\chi \mathcal{L}^{-1} \mathcal{D}_w^\dagger . \tag{15.88}$$

This matrix is a discretization of the (known) CC inverse operator $\mathcal{L}^{-1}$. Note that the discretization functions in the continuous data domain are generally different from those used in the object domain.

With this matrix, (15.87) becomes

$$\boldsymbol{\theta} = \mathbf{O}\mathcal{D}_w \mathbf{y} + \mathcal{D}_\chi \mathcal{L}^{-1} \left[\mathbf{I} - \mathcal{D}_w^\dagger \mathcal{D}_w \right] \mathbf{y} . \tag{15.89}$$

Comparison of the first term in this equation with (15.84) suggests that we define an estimate of $\boldsymbol{\theta}$ by

$$\hat{\boldsymbol{\theta}} = \frac{1}{C} \mathbf{O}\mathbf{g} . \tag{15.90}$$

Straightforward algebra shows that

$$\hat{\boldsymbol{\theta}} = \mathbf{O}\mathcal{D}_w \mathbf{y} + \frac{1}{C}\mathbf{O}\mathbf{n} = \boldsymbol{\theta} + \frac{1}{C}\mathbf{O}\mathbf{n} - \mathcal{D}_\chi \mathcal{L}^{-1} [\mathbf{I} - \mathcal{D}_w^\dagger \mathcal{D}_w] \mathcal{L}\mathbf{f} . \tag{15.91}$$

This reconstruction procedure thus gives the correct answer $\boldsymbol{\theta}$ plus a noise term plus an error term related to the discretization.

To understand the error term, recall from Sec. 7.1.3 that $\mathcal{D}_w^\dagger \mathcal{D}_w = \mathcal{D}_w^+ \mathcal{D}_w$ for orthonormal discretization functions. The operator $\mathcal{D}_w^\dagger \mathcal{D}_w$ is thus the projector onto the space spanned by the discretization functions [*cf.* (7.38)], and $\mathbf{I} - \mathcal{D}_w^\dagger \mathcal{D}_w$ is the projector onto its orthogonal complement. The error term is approximately zero if $\mathcal{D}_w^\dagger \mathcal{D}_w$ is a good approximation to the unit operator in the continuous data domain.

We must emphasize, however, that (15.91) is based on the data model (15.84); if that model is inaccurate, additional errors will arise. It is common to assume that data are described by simple discretization of an idealized CC operator, neglecting physical effects such as scattered radiation or detector blur. These effects must be taken into account when computing the overall error associated with some reconstruction matrix $\mathbf{O}$. In Chap. 17 we shall discuss both types of error in the context of emission computed tomography.

15.2.5 More on analytic inverses

In Sec. 15.2.4, we assumed that the inverse of the CC operator $\mathcal{L}$ existed. Strictly speaking, all we really needed in that section was that the *left inverse* existed (see Sec. 1.3.4). In other words, we required that $\mathcal{L}$ had no null functions so that we could retrieve $\mathbf{f}$ uniquely from $\mathcal{L}\mathbf{f}$. In many cases of interest in imaging, especially in tomography, this condition is satisfied, but the adjoint operator $\mathcal{L}^\dagger$ does have null functions. In the language of Sec. 1.5.2, there is a nontrivial *inconsistency space*. For such operators, as we shall now show, any continuous noise-free data must satisfy certain conditions called *consistency conditions*. As a consequence, the left inverse is not unique.

Consistency conditions Let $\tilde{v}_n(\mathbf{r}_d)$ denote a null function of $\mathcal{L}^\dagger$, and let $\tilde{\mathbf{v}}_n$ be the corresponding vector in the Hilbert space $\mathbb{V}$, so that

$$\mathcal{L}^\dagger \tilde{\mathbf{v}}_n = \mathbf{0}, \tag{15.92}$$

where $\mathbf{0}$ is a vector of zero length in object space $\mathbb{U}$. If we take the scalar product of this zero vector with any object $\mathbf{f}$, the result will be the scalar 0:

$$(\mathcal{L}^\dagger \tilde{\mathbf{v}}_n, \mathbf{f})_{\mathbb{U}} = 0. \tag{15.93}$$

By the definition of the adjoint [see (1.39)],

$$(\tilde{\mathbf{v}}_n, \mathcal{L}\mathbf{f})_{\mathbb{V}} = 0. \tag{15.94}$$

That is, if $\mathbf{g} = \mathcal{L}\mathbf{f}$ is a noise-free data vector, its scalar product with each of the $\tilde{\mathbf{v}}_n$ must vanish. Because of noise or other measurement errors, not all data vectors will satisfy these consistency conditions.

Varieties of left inverses It is common to preprocess a data vector before performing an inversion. If $\mathcal{A}$ is a linear operator that maps data space $\mathbb{V}$ to itself, the preprocessed data vector has the form

$$\mathbf{g}' = \mathcal{A}\mathbf{g}. \tag{15.95}$$

Applying $\mathcal{L}^\dagger$ to the modified data vector $\mathbf{g}'$ yields

$$\mathcal{L}^\dagger \mathbf{g}' = \mathcal{L}^\dagger \mathcal{A}\mathbf{g} = \mathcal{L}^\dagger \mathcal{A}\mathcal{L}\mathbf{f}, \tag{15.96}$$

where the latter form applies only in the noise-free case. Note that $\mathcal{L}^\dagger \mathcal{A}\mathcal{L}\mathbf{f}$ is in object space.

If we restrict $\mathcal{A}$ so that $\mathcal{L}^\dagger \mathcal{A}\mathcal{L}$ is an invertible operator, then we can invert (15.96) to get

$$\mathbf{f} = \left[\mathcal{L}^\dagger \mathcal{A}\mathcal{L}\right]^{-1} \mathcal{L}^\dagger \mathcal{A}\mathbf{g}. \tag{15.97}$$

Thus any choice of $\mathcal{A}$ satisfying the stated condition generates a left inverse. The reader may show that all of these inverses are identical if $\mathcal{L}^\dagger$ has no null space.

Filtering before backprojection One interesting choice for $\mathcal{A}$ is

$$\mathcal{A} = \left[\mathcal{L}\mathcal{L}^\dagger\right]^+. \tag{15.98}$$

With this choice, (15.96) becomes

$$\mathcal{L}^\dagger \mathcal{A}\mathbf{g} = \mathcal{L}^\dagger \left[\mathcal{L}\mathcal{L}^\dagger\right]^+ \mathbf{g} = \mathcal{L}^\dagger \left[\mathcal{L}\mathcal{L}^\dagger\right]^+ \mathcal{L}\mathbf{f}, \tag{15.99}$$

where the latter form is valid for noise-free data. By (1.149) we have

$$\mathcal{L}^\dagger \mathcal{A}\mathbf{g} = \mathcal{L}^+ \mathcal{L}\mathbf{f}, \tag{15.100}$$

which, without any assumptions on $\mathcal{L}$, is the measurement component of $\mathbf{f}$. If $\mathcal{L}$ has no null functions, then

$$\mathcal{L}^\dagger \mathcal{A}\mathbf{g} = \mathbf{f}. \tag{15.101}$$

Thus no further filtering is required in object space when this $\mathcal{A}$ is used; simple application of $\mathcal{L}^\dagger$ to $\mathcal{A}\mathbf{g}$ yields $\mathbf{f}$.

As discussed in Sec. 4.4, $\mathcal{L}^\dagger$ is referred to in the tomographic literature as *back-projection*, and (15.101) is an abstract formulation of the *filtered-backprojection* algorithm, with the data-space filter function given by (15.98). Explicit forms of this algorithm were given in Chap. 4.

The similarity between (15.99) and (15.62) should be noted; filtered backprojection is essentially the same thing as the Backus-Gilbert method, though the latter designation is usually restricted to reconstruction from discrete data, so of course the object is not recovered exactly.

15.2.6 Noise with linear reconstruction operators

All of the linear reconstruction operators discussed above were designed to extract the maximum possible information from a data set in the absence of noise. In the real world, of course, data are always corrupted by noise, and most of the art of image reconstruction is in developing ways to control that noise. In this section we begin to discuss the effects of noise on reconstructed images; the theme will continue throughout the chapter.

Discrete linear reconstructions Consider a linear CDD reconstruction problem in which an $N \times 1$ vector $\boldsymbol{\theta}$ of expansion coefficients is estimated by

$$\hat{\boldsymbol{\theta}} = \mathbf{O}\mathbf{g}\,, \tag{15.102}$$

where $\mathbf{O}$ is an $N \times M$ matrix and $\mathbf{g}$ is an $M \times 1$ data vector. Our objective is to derive the statistical properties of $\hat{\boldsymbol{\theta}}$ from those of $\mathbf{g}$.

For linear imaging systems, we know from (15.10) and (15.11) that $\mathbf{g}$ can be expressed in two equivalent ways:

$$\mathbf{g} = \mathcal{H}\mathbf{f} + \mathbf{n} = \mathbf{H}\boldsymbol{\theta} + \boldsymbol{\epsilon}\,, \tag{15.103}$$

where $\mathbf{n}$ describes the random measurement noise and $\boldsymbol{\epsilon}$ describes both measurement noise and modeling error. For present purposes the first of these forms is most useful since the properties of $\mathbf{n}$ can be specified more easily than those of $\boldsymbol{\epsilon}$. In particular, $\mathbf{n}$ is by definition a zero-mean random vector since $\mathcal{H}\mathbf{f}$ is the mean of $\mathbf{g}$ for a particular object function $\mathbf{f}$. The vector $\boldsymbol{\epsilon}$, on the other hand, depends in a complicated way on the object and the system model.

Uncorrelated Gaussian noise In Chap. 12 we discussed a variety of noise mechanisms in radiation detectors. Many of these mechanisms, especially those discussed in Sec. 12.2, are well described by normal (Gaussian) statistics. If we have a discrete detector array dominated by one of these mechanisms, and if we assume that all elements in the array are identical and that each generates its own noise independently of the other elements, then a good description of the probability density function of $\mathbf{n}$ may be

$$\mathrm{pr}(\mathbf{n}) = (2\pi\sigma^2)^{-M/2} \prod_{m=1}^{M} \exp\left[-\frac{n_m^2}{2\sigma^2}\right]. \tag{15.104}$$

This form assumes that the mean of each component n_m is zero and that the variance of each is the same constant σ^2. Thus the covariance matrix of $\mathbf{n}$ is

$$\mathbf{K_n} = \sigma^2 \mathbf{I}, \tag{15.105}$$

where $\mathbf{I}$ is the $M \times M$ identity matrix. Using a notation introduced in Chap. 8, we say that $\mathbf{n} \sim \mathcal{N}(0, \sigma^2 \mathbf{I})$. In words, the noise is i.i.d. (independent, identically distributed) normal. The data vector itself is also independent normal but not zero mean; instead, $\mathbf{g} \sim \mathcal{N}(\mathcal{H}\mathbf{f}, \sigma^2\mathbf{I})$.

We know from Sec. 8.3.3 that a linear transformation of a normal random vector yields another normal random vector. From (8.49) and (8.50), the mean and covariance of $\hat{\boldsymbol{\theta}}$ (both conditional on a specific $\mathbf{f}$) are given by

$$\mathrm{E}\{\hat{\boldsymbol{\theta}}|\mathbf{f}\} = \mathbf{O}\mathrm{E}\{\mathbf{g}|\mathbf{f}\} = \mathbf{O}\mathcal{H}\mathbf{f}\,, \tag{15.106}$$

$$\mathbf{K}_{\hat{\boldsymbol{\theta}}|\mathbf{f}} = \mathbf{O}\mathbf{K_n}\mathbf{O}^\dagger = \sigma^2 \mathbf{O}\mathbf{O}^\dagger\,. \tag{15.107}$$

Thus $\hat{\boldsymbol{\theta}} \sim \mathcal{N}(\mathbf{O}\mathcal{H}\mathbf{f}, \sigma^2\mathbf{O}\mathbf{O}^\dagger)$. To be precise, this normal distribution describes the conditional density $\mathrm{pr}(\hat{\boldsymbol{\theta}}|\mathbf{f})$. The mean of this density depends on the actual object and the actual CD system operator $\mathcal{H}$, and it is generally not true that $\mathcal{H}\mathbf{f}$ equals $\mathbf{H}\boldsymbol{\theta}$, so we have not computed $\mathrm{pr}(\hat{\boldsymbol{\theta}}|\boldsymbol{\theta})$. On the other hand, for fixed $\mathbf{f}$ the modeling error is nonrandom, so $\mathbf{K_n} = \mathbf{K}_{\boldsymbol{\epsilon}}$, and (15.107) is correct no matter how bad our system model is (so long as the noise model is correct).

In component form, the covariance matrix is

$$\left[\mathbf{K}_{\hat{\boldsymbol{\theta}}|\mathbf{f}}\right]_{nn'} = \mathrm{E}\{\Delta\hat{\theta}_n \Delta\hat{\theta}_{n'}|\mathbf{f}\} = \sigma^2 \sum_{m=1}^{M} O_{nm} O^*_{n'm}\,, \tag{15.108}$$

where $\Delta\hat{\theta}_n = \hat{\theta}_n - \mathrm{E}\{\hat{\theta}_n|\mathbf{f}\}$. The conditional variance of $\hat{\theta}_n$ then takes the simple form

$$\mathrm{Var}\left\{\hat{\theta}_n|\mathbf{f}\right\} = \sigma^2 \left[\mathbf{O}\mathbf{O}^\dagger\right]_{nn} = \sigma^2 \sum_{m=1}^{M} |O_{nm}|^2\,. \tag{15.109}$$

Noise amplification—an example As an example of the formalism just developed, suppose that $\mathbf{O}$ is the pseudoinverse of $\mathbf{H}$, which we can express in SVD form as

$$\mathbf{O} = \mathbf{H}^+ = \sum_{j=1}^{R} \frac{1}{\sqrt{\mu_j}} \mathbf{u}_j \mathbf{v}_j^\dagger\,, \tag{15.110}$$

where $\mathbf{u}_j$ and $\mathbf{v}_j$ are the singular vectors associated with $\mathbf{H}$. With this form, (15.107) becomes

$$\mathbf{K}_{\hat{\boldsymbol{\theta}}|\mathbf{f}} = \sigma^2\mathbf{O}\mathbf{O}^\dagger = \sigma^2 \sum_{j=1}^{R} \frac{1}{\sqrt{\mu_j}} \mathbf{u}_j \mathbf{v}_j^\dagger \sum_{k=1}^{R} \frac{1}{\sqrt{\mu_k}} \mathbf{v}_k \mathbf{u}_k^\dagger\,. \tag{15.111}$$

Since $\mathbf{v}_j^\dagger \mathbf{v}_k = \delta_{jk}$ by the orthonormality of the singular vectors, we have

$$\mathbf{K}_{\hat{\boldsymbol{\theta}}|\mathbf{f}} = \sigma^2 \sum_{j=1}^{R} \frac{1}{\mu_j} \mathbf{u}_j \mathbf{u}_j^\dagger\,. \tag{15.112}$$

The conditional variance of the component $\hat{\theta}_n$ is given by

$$\text{Var}\left\{\hat{\theta}_n|\mathbf{f}\right\} = \left[\mathbf{K}_{\hat{\boldsymbol{\theta}}|\mathbf{f}}\right]_{nn} = \sum_{j=1}^{R} \frac{\sigma^2}{\mu_j}|u_{jn}|^2 \, , \tag{15.113}$$

where u_{jn} is the nth component of $\mathbf{u}_j$. Since μ_j decreases (often rapidly) as j increases, the variance is amplified because of division by small singular values.

Noise control in the SVD domain If we implement the pseudoinverse in the SVD domain, we can control the noise either by truncating the sum over j at some value less than R or by modifying the reconstruction operator in some way to avoid division by small μ_j. One possibility is to define

$$\mathbf{O} = \sum_{j=1}^{R} \frac{\sqrt{\mu_j}}{\mu_j + \eta} \mathbf{u}_j \mathbf{v}_j^{\dagger} \, , \tag{15.114}$$

where η is a positive constant. With this approach, the variance of $\hat{\theta}_n$ becomes

$$\text{Var}\left\{\hat{\theta}_n\right\} = \left[\mathbf{K}_{\hat{\boldsymbol{\theta}}}\right]_{nn} = \sum_{j=1}^{R} \frac{\sigma^2 \mu_j}{(\mu_j + \eta)^2}|u_{jn}|^2 \, . \tag{15.115}$$

This option reduces the noise since the denominator is not allowed to approach zero; larger values for η lead to smaller variances. (Recall from Chap. 1 that μ_j is real and greater than zero for $j \leq R$.)

The use of a large η suppresses the singular components corresponding to small μ_j. These components usually correspond to fine details in the image since most imaging systems function essentially as low-pass filters, showing smaller response to higher spatial frequencies. Thus use of a large η will limit the spatial resolution in the reconstructed image. As with virtually any inverse problem, there is a parameter that sets the tradeoff between noise and resolution. (It would be a mistake, however, to assert that η sets the tradeoff between bias and variance; as we have seen in Sec. 15.1, bias is usually not well defined, and setting $\eta = 0$ seldom produces an unbiased estimate of anything.)

Noise control by smoothing Another way to control the noise with any reconstruction operator $\mathbf{O}$ is to smooth the image after reconstruction. A Hermitian smoothing operator in the discrete reconstruction space is an $N \times N$ matrix $\mathbf{S}$ satisfying $\mathbf{S}^{\dagger} = \mathbf{S}$. For simplicity, assume that $\mathbf{S}$ commutes with $\mathbf{H}^{\dagger}\mathbf{H}$; in that case, as we saw in Chap. 6, $\mathbf{S}$ is diagonal in the basis formed by the right singular vectors of $\mathbf{H}$, and we can write

$$\mathbf{S} = \sum_{j=1}^{N} s_j \mathbf{u}_j \mathbf{u}_j^{\dagger} \, . \tag{15.116}$$

For $\mathbf{S}$ to serve as a smoothing filter, we should take s_j to be positive and to decrease with increasing j.

If we apply this operator after the reconstruction operator $\mathbf{O}$, we get

$$\hat{\boldsymbol{\theta}}' = \mathbf{S}\hat{\boldsymbol{\theta}} = \mathbf{SOg} \, . \tag{15.117}$$

For the example used above, where $\mathbf{O} = \mathbf{H}^{+}$, we have

$$\mathbf{SO} = \sum_{j=1}^{R} \frac{s_j}{\sqrt{\mu_j}} \mathbf{u}_j \mathbf{v}_j^{\dagger} \,. \tag{15.118}$$

The rolloff of s_j with increasing j can be used to counteract the noise amplification associated with small μ_j, but of course it also limits the fine detail in the image.

An advantage of post-reconstruction smoothing is that we can implement it without knowing the SVD of $\mathbf{H}$ or $\mathbf{O}$. Any Hermitian operator, including simple discrete convolution with a nonnegative blur function, will serve to limit the noise amplification. Conversely, any modification of the SVD pseudoinverse that controls the noise can be interpreted as an equivalent smoothing operator. For example, (15.114) and (15.118) are identical if we take $s_j = \mu_j/(\mu_j + \eta)$.

Apodization A common name for a smoothing filter, especially when it is incorporated in the reconstruction operator, is *apodizing function*. In zoology, *apodal* means without feet, so to apodize is to cut off the feet. In the early development of radar, it was observed that some cutoff in the temporal frequency response was needed to control noise, but that a sharp cutoff led to sidelobes (feet) on the impulse response. Since these sidelobes could be confused with weak targets adjacent to a strong one, a smooth rolloff of the frequency response was found to be preferable to an abrupt cutoff. Much research has gone into studying the effects of different apodizing functions on target detection, and many of the filters developed for radar have found use in tomographic image reconstruction as well.

Poisson noise The discussion above requires some modification if the noise is Poisson rather than i.i.d. normal. As discussed in detail in Sec. 11.2, it is often an excellent model for an array of photon-counting detectors to assume that the components of $\mathbf{g}$ are independent Poisson, so that the covariance matrix is given by

$$\left[\mathbf{K}_{\mathbf{g}|\mathbf{f}}\right]_{mm'} = [\mathcal{H}\mathbf{f}]_m \, \delta_{mm'} \,. \tag{15.119}$$

With this covariance for the data, the covariance and variance in the reconstruction are [*cf.* (15.108) and (15.109)]

$$\left[\mathbf{K}_{\hat{\boldsymbol{\theta}}|\mathbf{f}}\right]_{nn'} = \sum_{m=1}^{M} [\mathcal{H}\mathbf{f}]_m \, O_{nm} O_{n'm}^{*} \,, \tag{15.120}$$

$$\mathrm{Var}\left\{\hat{\theta}_n|\mathbf{f}\right\} = \sum_{m=1}^{M} [\mathcal{H}\mathbf{f}]_m \, |O_{nm}|^2 \,. \tag{15.121}$$

With Poisson noise, we cannot say rigorously that $\hat{\boldsymbol{\theta}}$ is normally distributed, but a normal distribution is usually an excellent approximation, for two reasons. First, if $[\mathcal{H}\mathbf{f}]_m$ is large (greater than 10 or so), the Poisson can be well approximated by a Gaussian with variance equal mean. In that case, the reconstruction is a linear transformation of a normal random vector, yielding another normal random vector. Second, even if some of the $[\mathcal{H}\mathbf{f}]_m$ are small, $\hat{\boldsymbol{\theta}}$ may still be approximately normal as the result of the central-limit theorem. A linear reconstruction is equivalent to forming a weighted sum of the data values. The weighting changes the mean and variance, but as we discussed in Sec. 8.3.4 any sum of independent random variables, not necessarily identically distributed ones, tends to a normal under broad conditions [see (8.211) *ff.* and Shiryayev (1984)]. Since Poisson random variables are inherently independent, we should expect to get a good approximation to a normal when the reconstruction operator serves to add many Poisson data values with some weights. Moreover, strict independence is also not required for the central-limit theorem (Shiryayev, 1984), so a normal distribution for $\hat{\boldsymbol{\theta}}$ can result from linear reconstruction with almost any data statistics.

Noise kernel and point response The variance expression for Poisson noise, (15.121), can be rewritten as

$$\text{Var}\left\{\hat{\theta}_n|\mathbf{f}\right\} = \int_{\mathbf{S}_f} d^q r\ \aleph_n(\mathbf{r})\, f(\mathbf{r})\,, \tag{15.122}$$

where

$$\aleph_n(\mathbf{r}) \equiv \sum_{m=1}^{M} |O_{nm}|^2\, h_m(\mathbf{r})\,. \tag{15.123}$$

Because of the Poisson character of the noise and the linearity of the processing algorithm, there is a linear CD mapping from the object $f(\mathbf{r})$ to the variance in the reconstructed discrete image. The kernel of this mapping, $\aleph_n(\mathbf{r})$, is called the *noise kernel* (Barrett and Swindell, 1981, 1996). We emphasize that this linear mapping from object to variance holds only for Poisson noise where the noise covariance is given by (15.119). From (15.108), for example, we see that the variance in the reconstruction is independent of the object for the noise model of (15.105).

Equations (15.122) and (15.123) should be compared to similar expressions for the mean:

$$\bar{\hat{\theta}}_n = \int_{\mathbf{S}_f} d^q r\ p_n(\mathbf{r})\, f(\mathbf{r})\,, \tag{15.124}$$

where

$$p_n(\mathbf{r}) \equiv \sum_{m=1}^{M} O_{nm} h_m(\mathbf{r})\,. \tag{15.125}$$

Thus $p_n(\mathbf{r})$ is the overall CD point response function (PRF) mapping the object function through the CD system and the DD processing algorithm to the mean discrete image.

The formal difference between the noise kernel and the PRF is that $|O_{nm}|^2$ appears in the former and O_{nm} in the latter. In more practical terms, O_{nm} can have both positive and negative values and hence the filter can serve a sharpening function. In the noise kernel, on the other hand, $|O_{nm}|^2$ is confined to nonnegative values, so the variance distribution will be a blurred and discretized version of the object. We shall explore the nature of the noise kernel further in Sec. 17.3 in the context of emission computed tomography.

15.3 IMPLICIT ESTIMATES

As we noted in Sec. 15.1.1, image reconstruction is often framed in terms of minimization of a scalar-valued objective functional. The resulting images are often referred to as *implicit estimates* or *implicit reconstructions* since there is no explicit formula for generating them. In Sec. 15.4 we shall discuss iterative algorithms for finding implicit estimates, but in this section we study some of their properties without regard to any particular algorithm.

15.3.1 Functional minimization

As given in (15.4), the objective functional depends on the data vector $\mathbf{g}$ and the object $\mathbf{f}$, where the latter notation implies a vector in an infinite-dimensional Hilbert space. As we shall see in Sec. 15.3.5, it is indeed possible to carry out this minimization and find a continuous estimate $\hat{\mathbf{f}}$ without ever adopting a discrete representation, but it is much more common in the literature to first choose a discretization. and then minimize with respect to the free parameters in that representation. With a linear representation like (15.6), then, the functional is redefined so that it depends on the coefficient vector $\boldsymbol{\theta}$ rather than the function $\mathbf{f}$.

An important consideration in choosing the objective functional is that it should encourage agreement between the actual measured data vector $\mathbf{g}$ and the approximate data vector $\mathbf{g}_a$ generated when the exact system operator operates on an approximate object representation. For a linear system, we know from (7.302) and (7.305) that $\mathbf{g}_a = \boldsymbol{\mathcal{H}}\mathbf{f}_a = \mathbf{H}\boldsymbol{\theta}$, where $\mathbf{H}$ is a matrix. A key component of the objective functional is thus some measure of the distance in data space between $\mathbf{g}$ and $\mathbf{H}\boldsymbol{\theta}$, so we are led to define a *data-agreement functional* $Q_{data}(\mathbf{g}, \mathbf{H}\boldsymbol{\theta})$. In least-squares methods, this functional is just the squared Euclidean distance $||\mathbf{g} - \mathbf{H}\boldsymbol{\theta}||^2$, but many other choices are also possible. We shall survey some of them in Sec. 15.3.2.

Another key component of the functional is some means of controlling noise amplification. As we saw in Secs. 1.7.5 and 15.2.6, forcing exact agreement with the data is equivalent to dividing by small singular values, and any noise or other error in the data is thus multiplied by large numbers. One way of controlling this noise is to add another functional, called the *regularizing functional* to $Q_{data}(\mathbf{g}, \mathbf{H}\boldsymbol{\theta})$. Most often the regularizing functional depends only on $\boldsymbol{\theta}$, so we write

$$Q(\boldsymbol{\theta}, \mathbf{g}) = Q_{data}(\mathbf{g}, \mathbf{H}\boldsymbol{\theta}) + \eta Q_{reg}(\boldsymbol{\theta})\,. \tag{15.126}$$

The parameter η adjusts the relative weights of the two functionals and hence serves to control the tradeoff between data agreement and noise amplification. Several common forms for $Q_{reg}(\boldsymbol{\theta})$ will be discussed in Sec. 15.3.3.

Given the data-agreement and regularizing functionals, there are four distinct ways in which they can be used. The most common is simply to minimize $Q(\boldsymbol{\theta}, \mathbf{g})$, so that the estimate of $\boldsymbol{\theta}$ is given by

$$\hat{\boldsymbol{\theta}} = \underset{\boldsymbol{\theta}}{\operatorname{argmin}} \left[Q_{data}(\mathbf{g}, \mathbf{H}\boldsymbol{\theta}) + \eta Q_{reg}(\boldsymbol{\theta})\right]. \tag{15.127}$$

If $Q(\boldsymbol{\theta}, \mathbf{g})$ is a strictly convex function of $\boldsymbol{\theta}$ for fixed $\mathbf{g}$, then the estimate $\hat{\boldsymbol{\theta}}$ defined by (15.127) will be unique. It will, however, be a random vector since different realizations of $\mathbf{g}$ will yield different $\hat{\boldsymbol{\theta}}$.

The other three approaches use Q_{data} and Q_{reg} separately rather than in the sum $Q(\boldsymbol{\theta}, \mathbf{g})$. For example, we could minimize $Q_{data}(\mathbf{g}, \mathbf{H}\boldsymbol{\theta})$ subject to the constraint that $Q_{reg}(\boldsymbol{\theta})$ is less than some preset constant δ. Alternatively, we could minimize $Q_{reg}(\boldsymbol{\theta})$ subject to a similar constraint on $Q_{data}(\mathbf{g}, \mathbf{H}\boldsymbol{\theta})$. Finally, we could minimize $Q_{data}(\mathbf{g}, \mathbf{H}\boldsymbol{\theta})$ subject to the constraint that $\boldsymbol{\theta}$ belong to some subset of its possible values. Given the freedom in choice of $Q_{data}(\mathbf{g}, \mathbf{H}\boldsymbol{\theta})$, $Q_{reg}(\boldsymbol{\theta})$ and the parameter η, there does not seem to be any particular advantage to these alternatives (other than possibly fitting into unexplored niches in the literature), so we shall confine our attention here to estimates of the form (15.127).

Positivity As discussed in Sec. 15.1.4, we often know on physical grounds that the object cannot be negative, and we may want to incorporate this information into our reconstruction. The easiest way to do so is to first choose our expansion functions $\phi_n(\mathbf{r})$ to be nonnegative and then to allow only nonnegative coefficients, $\theta_n \geq 0$ for all n. We can state the latter condition in vector form as $\boldsymbol{\theta} \geq 0$. With this constraint, we can restate the minimization principle as

$$\hat{\boldsymbol{\theta}} = \underset{\boldsymbol{\theta} \geq 0}{\operatorname{argmin}} \left[Q_{data}(\mathbf{g}, \mathbf{H}\boldsymbol{\theta}) + \eta Q_{reg}(\boldsymbol{\theta})\right] . \tag{15.128}$$

This simple modification may require substantial changes in the algorithm used to find the minimum, a topic taken up in Sec. 15.4.

Inversions from nonlinear data As written, (15.127) and (15.128) are applicable only to linear systems for which the system operator $\boldsymbol{\mathcal{H}}$ can be approximated by a matrix, but they are easily modified to apply to any of the nonlinear systems discussed in Sec. 7.5. Let us write the noise-free mapping from object to image data as

$$\mathbf{g} = \boldsymbol{\mathcal{N}}\{\mathbf{f}\} , \tag{15.129}$$

where $\boldsymbol{\mathcal{N}}$ is some nonlinear operator[2]. For example, in electrical impedance tomography, $\boldsymbol{\mathcal{N}}$ represents the mapping by way of the Poisson equation from the object impedance to the voltages at a set of surface points for specified current sources.

If we use (15.7) to construct an approximate object representation, then the data-agreement functional can be chosen to enforce agreement between $\mathbf{g}$ and the approximate data vector $\mathbf{g}_a$ given by

$$\mathbf{g}_a = \boldsymbol{\mathcal{N}}\{\mathbf{f}_a\} = \boldsymbol{\mathcal{N}}\{\boldsymbol{\mathcal{D}}_\phi^\dagger \boldsymbol{\theta}\} . \tag{15.130}$$

Thus (15.127) becomes

$$\hat{\boldsymbol{\theta}} = \underset{\boldsymbol{\theta}}{\operatorname{argmin}} \left[Q_{data}(\mathbf{g}, \boldsymbol{\mathcal{N}}\{\boldsymbol{\mathcal{D}}_\phi^\dagger \boldsymbol{\theta}\}) + \eta Q_{reg}(\boldsymbol{\theta})\right] . \tag{15.131}$$

The same functional form for $Q_{data}(\mathbf{g}, \cdot)$ can be used for the linear and nonlinear cases, though it may be more work to compute it if the system is nonlinear. For example, a least-squares form is commonly used with nonlinear mappings, but computation of $\boldsymbol{\mathcal{N}}\{\boldsymbol{\mathcal{D}}_\phi^\dagger \boldsymbol{\theta}\}$ for a single $\boldsymbol{\theta}$ may require numerical solution of a differential equation.

Bayesian interpretation The estimate defined in (15.127) or (15.131) can be interpreted as a Bayesian MAP estimate as introduced in Sec. 13.3.3. We saw there that the MAP estimate maximizes the log posterior, or

[2]The boldface $\boldsymbol{\mathcal{N}}$ should not be confused with the symbol $\mathcal{N}$ used to signify a normal distribution.

$$\hat{\boldsymbol{\theta}}_{MAP} = \underset{\boldsymbol{\theta}}{\text{argmax}} \ \{\ln[\text{pr}(\mathbf{g}|\boldsymbol{\theta})] + \ln[\text{pr}(\boldsymbol{\theta})]\} \, . \tag{15.132}$$

The first term agrees with (15.127) if we ignore issues of estimability and modeling error and write the likelihood as

$$\text{pr}(\mathbf{g}|\boldsymbol{\theta}) = \frac{1}{Z_1} \exp[-Q_{data}(\mathbf{g}, \mathbf{H}\boldsymbol{\theta})] \, , \tag{15.133}$$

where Z_1 is a normalizing constant required to make $\text{pr}(\mathbf{g}|\boldsymbol{\theta})$ a probability density function (or probability if $\mathbf{g}$ is discrete). The log-likelihood can now be written as

$$\ln[\text{pr}(\mathbf{g}|\boldsymbol{\theta})] = -Q_{data}(\mathbf{g}, \mathbf{H}\boldsymbol{\theta}) - \ln(Z_1) \, . \tag{15.134}$$

Note that $\boldsymbol{\theta}$ appears here only as $\mathbf{H}\boldsymbol{\theta}$ since $\boldsymbol{\theta}$ can influence the likelihood only through its effect on the data vector.

Similarly, we can write an arbitrary prior as

$$\text{pr}(\boldsymbol{\theta}) = \frac{1}{Z_2} \exp[-\eta Q_{reg}(\boldsymbol{\theta})] \, , \tag{15.135}$$

so that

$$\ln[\text{pr}(\boldsymbol{\theta})] = -\eta Q_{reg}(\boldsymbol{\theta}) - \ln(Z_2) \, . \tag{15.136}$$

The minus signs in (15.134) and (15.136) convert argmax to argmin, and the logs of the normalizing constants do not affect the argmin. Thus minimization of any functional of the form (15.127) or (15.131) can be interpreted as maximizing a log-posterior. This observation reinforces the Bayesian conceit that any reasonable estimation procedure can be interpreted as a Bayesian method with respect to some prior.

It is not possible, however, to interpret (15.135) in frequentist terms since it depends on the free parameter η, which is often called a *hyperparameter*. If $\text{pr}(\boldsymbol{\theta})$ is to represent the frequency of occurrence of $\boldsymbol{\theta}$, its form and any parameters in it must be determined (according to the frequentist) by direct observations of the random vector $\boldsymbol{\theta}$. The parameter η, on the other hand, is set by the user of an inversion algorithm as a means of controlling noise, and its value can (and usually does) vary from image to image. A true frequentist prior distribution on the object would be independent of noise in any particular image. A true Bayesian would consider this point irrelevant and assert that one can choose any prior that reflects one's degree of belief, which may be influenced by noise, whim or any other factor. Key among these factors is often mathematical simplicity since the Bayesian, like the frequentist faced with the same inversion problem, must eventually find the minimum of a complicated functional. One believes in what one can compute.

To reiterate a point from Sec. 15.1.5, the authors of this book believe in objective, task-based assessment of image quality, so their prior is one that maximizes task performance. Instead of being established prior to data collection, the prior/regularizer in this pragmatist view is determined essentially by what one wants to do with the data after acquisition.

Bayesians and modeling In Sec. 15.3.2, we shall give specific mathematical forms for $\text{pr}(\mathbf{g}|\boldsymbol{\theta})$, but in fact we rarely know this density. The difficulty is that the data statistics are not determined by $\boldsymbol{\theta}$ alone. As we emphasized in Chap. 7 and reiterated in Sec. 15.1.2, the mean of $\mathbf{g}$ is given by two terms, $\mathbf{H}\boldsymbol{\theta} + \mathcal{H}\,\delta\mathbf{f}$. In image reconstruction we often assume that we can ignore the second term, but it is worthwhile to try to articulate just what prior knowledge we are assuming when we do so.

Let $\mathbf{f} = \mathbf{f}_a + \mathbf{f}_\perp$, where $\mathbf{f}_\perp$ is the component of $\mathbf{f}$ in the (infinite-dimensional) orthogonal complement of representation space. Then

$$\Pr(\mathbf{g}|\boldsymbol{\theta}) = \Pr(\mathbf{g}|\mathbf{f}_a) = \int_{\mathbb{U}_\perp} d\mathbf{f}_\perp \, \Pr(\mathbf{g}|\mathbf{f}_a, \mathbf{f}_\perp)\,\mathrm{pr}(\mathbf{f}_\perp)\,. \tag{15.137}$$

Hence, setting $\Pr(\mathbf{g}|\boldsymbol{\theta}) = \Pr(\mathbf{g}|\mathbf{f})$ is equivalent to adopting the prior that $\mathbf{f}_\perp$ is zero. With this assumption, $\overline{\mathbf{g}} = \mathbf{H}\boldsymbol{\theta}$, and simple forms for the density $\Pr(\mathbf{g}|\boldsymbol{\theta})$ follow.

As with many other Bayesian priors, this one should more accurately be called a prior desire rather than prior knowledge. We desire a solution where the modeling error can be neglected (since otherwise we cannot solve the problem). Again, the usefulness of this desire can be ascertained by task-based assessment methods.

15.3.2 Data-agreement functionals

In this section we survey a variety of data-agreement functionals that have been used in the literature and comment on how each is related to a likelihood model.

Agreement in the least-squares sense The simplest data-agreement functional is the $\mathbb{L}_2$ norm of the difference between $\mathbf{g}$ and $\mathbf{H}\boldsymbol{\theta}$. As discussed in detail in Sec. 1.7, this choice leads to least-squares solutions of the set of linear equations $\mathbf{g} = \mathbf{H}\boldsymbol{\theta}$.

As a log-likelihood, the least-squares functional stems from an independent Gaussian noise model. Suppose we write

$$\mathbf{g} = \mathbf{H}\boldsymbol{\theta} + \boldsymbol{\epsilon} \tag{15.138}$$

and (ignoring modeling errors) assume that $\boldsymbol{\epsilon}$ is an $M \times 1$ multivariate normal random vector with zero mean and covariance matrix $\sigma^2\mathbf{I}$. Then the likelihood is given by

$$\mathrm{pr}(\mathbf{g}|\boldsymbol{\theta}) = (2\pi\sigma^2)^{-\frac{1}{2}M} \exp\left[-\frac{1}{2\sigma^2}\sum_{m=1}^{M}\left[g_m - (\mathbf{H}\boldsymbol{\theta})_m\right]^2\right]. \tag{15.139}$$

The corresponding log-likelihood is

$$\ln[\mathrm{pr}(\mathbf{g}|\boldsymbol{\theta})] = -\tfrac{1}{2}M\ln(2\pi\sigma^2) - \frac{1}{2\sigma^2}\sum_{m=1}^{M}\left[g_m - (\mathbf{H}\boldsymbol{\theta})_m\right]^2. \tag{15.140}$$

From (15.134), we have

$$Q_{data}(\mathbf{g}, \mathbf{H}\boldsymbol{\theta}) = \frac{1}{2\sigma^2}\sum_{m=1}^{M}\left[g_m - (\mathbf{H}\boldsymbol{\theta})_m\right]^2. \tag{15.141}$$

The multiplicative constant $1/(2\sigma^2)$ does not affect the position of the minimum as a function of $\boldsymbol{\theta}$, so we may as well write

$$Q_{data}(\mathbf{g}, \mathbf{H}\boldsymbol{\theta}) = \sum_{m=1}^{M}\left[g_m - (\mathbf{H}\boldsymbol{\theta})_m\right]^2 = ||\mathbf{g} - \mathbf{H}\boldsymbol{\theta}||^2, \tag{15.142}$$

which is the usual least-squares form.

Weighted least-squares In least-squares problems we may know that some measurements are more reliable than others, so we weight them more heavily. For example, suppose the components of $\boldsymbol{\epsilon}$ are independent normals with different variances. Then the covariance matrix of $\boldsymbol{\epsilon}$ is given by

$$[\mathbf{K}_{\boldsymbol{\epsilon}}]_{mm'} = \sigma_m^2 \, \delta_{mm'} \,, \tag{15.143}$$

and an argument similar to one just given shows that

$$Q_{data}(\mathbf{g}, \mathbf{H}\boldsymbol{\theta}) = \sum_{m=1}^{M} \frac{[g_m - (\mathbf{H}\boldsymbol{\theta})_m]^2}{\sigma_m^2} \,. \tag{15.144}$$

More generally, when $\boldsymbol{\epsilon}$ is a real-valued normal random vector with covariance $\mathbf{K}_{\boldsymbol{\epsilon}}$, we find

$$Q_{data}(\mathbf{g}, \mathbf{H}\boldsymbol{\theta}) = (\mathbf{g} - \mathbf{H}\boldsymbol{\theta})^t \mathbf{K}_{\boldsymbol{\epsilon}}^{-1} (\mathbf{g} - \mathbf{H}\boldsymbol{\theta}) \,. \tag{15.145}$$

This is still an $\mathbb{L}_2$ norm, but now one defined with respect to the weight $\mathbf{K}_{\boldsymbol{\epsilon}}^{-1}$.

An interesting way to rewrite (15.145) is

$$Q_{data}(\mathbf{g}, \mathbf{H}\boldsymbol{\theta}) = \left[\mathbf{K}_{\boldsymbol{\epsilon}}^{-\frac{1}{2}} (\mathbf{g} - \mathbf{H}\boldsymbol{\theta})\right]^t \left[\mathbf{K}_{\boldsymbol{\epsilon}}^{-\frac{1}{2}} (\mathbf{g} - \mathbf{H}\boldsymbol{\theta})\right] = ||\mathbf{K}_{\boldsymbol{\epsilon}}^{-\frac{1}{2}} (\mathbf{g} - \mathbf{H}\boldsymbol{\theta})||^2 \,, \tag{15.146}$$

where $\mathbf{K}_{\boldsymbol{\epsilon}}^{-\frac{1}{2}}$ is the prewhitening matrix discussed in Secs. 8.1.6 and 13.2.8. Thus the data-agreement functional with any normal model for the noise is the $\mathbb{L}_2$ norm of the prewhitened residual vector.

Gaussian approximation to Poisson likelihood One important situation where (15.144) is used is when the data vector is Poisson but the mean number of counts per measurement is large, say greater than about 10. In that case, it is an excellent approximation to ignore the discrete nature of the data and consider each g_m to be a normal random variable with conditional mean $(\mathbf{H}\boldsymbol{\theta})_m$ (Barrett and Swindell, 1981, 1996). As we emphasized repeatedly in Chap. 11, Poisson random variables are inherently independent, so the covariance of (15.143) applies with $\sigma_m^2 = (\boldsymbol{\mathcal{H}}\mathbf{f})_m \simeq (\mathbf{H}\boldsymbol{\theta})_m$, and (15.144) becomes

$$Q_{data}(\mathbf{g}, \mathbf{H}\boldsymbol{\theta}) = \sum_{m=1}^{M} \frac{[g_m - (\mathbf{H}\boldsymbol{\theta})_m]^2}{(\mathbf{H}\boldsymbol{\theta})_m} \,. \tag{15.147}$$

This functional is no longer quadratic in $\boldsymbol{\theta}$, so it does not, strictly speaking, correspond to a least-squares problem. However, to the same degree of approximation that allowed us to replace the Poisson with a Gaussian in the first place, we can replace $(\mathbf{H}\boldsymbol{\theta})_m$ with the observed g_m in the denominator, yielding

$$Q_{data}(\mathbf{g}, \mathbf{H}\boldsymbol{\theta}) = \sum_{m=1}^{M} \frac{[g_m - (\mathbf{H}\boldsymbol{\theta})_m]^2}{g_m} \,, \tag{15.148}$$

which is now quadratic in $\boldsymbol{\theta}$.

Even though g_m will only rarely go to zero if $(\mathbf{H}\boldsymbol{\theta})_m \gg 1$, as we assumed in deriving (15.148), it is nevertheless good programming practice to avoid division by zero, either by a conditional statement or by adding some small constant to the denominator.

Kullback-Leibler distance Of course, it is not necessary to approximate a Poisson by a Gaussian. For independent Poisson data, the log-likelihood is given exactly by

$$\ln[\text{pr}(\mathbf{g}|\boldsymbol{\theta})] = \ln \prod_{m=1}^{M} \left\{ \exp[-(\mathbf{H}\boldsymbol{\theta})_m] \frac{[(\mathbf{H}\boldsymbol{\theta})_m]^{g_m}}{g_m!} \right\}$$

$$= \sum_{m=1}^{M} \{-(\mathbf{H}\boldsymbol{\theta})_m + g_m \ln[(\mathbf{H}\boldsymbol{\theta})_m] - \ln(g_m!)\} . \tag{15.149}$$

Since $\ln(g_m!)$ is a constant that does not affect the minimization, we can write

$$Q_{data}(\mathbf{g}, \mathbf{H}\boldsymbol{\theta}) = \sum_{m=1}^{M} \{(\mathbf{H}\boldsymbol{\theta})_m - g_m \ln[(\mathbf{H}\boldsymbol{\theta})_m]\} . \tag{15.150}$$

This form is closely related to the *Kullback-Leibler distance*, defined by[3]

$$D_{\text{KL}}(\mathbf{g}, \mathbf{g}_0) \equiv \sum_{m=1}^{M} \left\{ (g_{0m} - g_m + g_m \ln\left[\frac{g_m}{g_{0m}}\right] \right\} . \tag{15.151}$$

Thus

$$Q_{data}(\mathbf{g}, \mathbf{H}\boldsymbol{\theta}) = D_{\text{KL}}(\mathbf{g}, \mathbf{H}\boldsymbol{\theta}) + const. \tag{15.152}$$

The Kullback-Leibler distance is also known as Csiszár's *I-divergence* or the *cross-entropy*. It is not truly a distance metric, as defined in Sec. 1.1.2, since $D_{\text{KL}}(\mathbf{g}, \mathbf{g}_0) \neq D_{\text{KL}}(\mathbf{g}_0, \mathbf{g})$. Instead, it is an example of a *generalized distance* or *Bregman distance*. The Kullback-Leibler distance will form the basis for the *expectation-maximization* algorithm to be discussed in Sec. 15.4.6.

15.3.3 Regularizing functionals

This section surveys some choices for the regularizing functional and discusses how each is related to a Bayesian prior.

Tikhonov regularization We encountered one example of a regularizing functional in Chap. 1, though we did not call it that. In Sec. 1.7.5 we introduced the minimum-norm, least-squares (MNLS) solution of $\mathbf{g} = \mathbf{H}\boldsymbol{\theta}$ as the unique least-squares solution with no null functions. Alternatively, we could have defined it as [*cf.* (1.191)]

$$\hat{\boldsymbol{\theta}}_{\text{MNLS}} = \lim_{\eta \to 0^+} \underset{\boldsymbol{\theta}}{\text{argmin}} \left\{ ||\mathbf{g} - \mathbf{H}\boldsymbol{\theta}||^2 + \eta||\boldsymbol{\theta}||^2 \right\} . \tag{15.153}$$

As an exercise, the reader should show that

$$\hat{\boldsymbol{\theta}}_{\text{MNLS}} = \mathbf{H}^+\mathbf{g} , \tag{15.154}$$

just as we found in Sec. 1.7.5.

[3]The sharp-eyed reader might note that a slightly different definition of the Kullback-Leibler distance was given in (8.258), but in that case it was a distance between two probability density functions. If we required g_m and g_{0m} to sum to unity, then (15.151) would reduce to (8.258).

We also showed in Sec. 1.7.5, however, that the MNLS solution is very noisy since we are forced, in the limit $\eta \to 0$, to divide by very small singular values. An interesting *ad hoc* solution to this problem is simply to forgo the limit and write

$$\hat{\boldsymbol{\theta}} = \underset{\boldsymbol{\theta}}{\text{argmin}} \left\{ ||\mathbf{g} - \mathbf{H}\boldsymbol{\theta}||^2 + \eta||\boldsymbol{\theta}||^2 \right\} . \tag{15.155}$$

We have dropped the subscript MNLS here since this $\hat{\boldsymbol{\theta}}$ is not a least-squares solution. For nonzero η, the regularizing term $\eta||\boldsymbol{\theta}||^2$ drags the estimate away from the least-squares point, suppressing noise in the process.

In fact, we have already analyzed the noise in this problem. An extension of the discussion in Sec. 1.7.5 shows that (15.155) has the solution,

$$\hat{\boldsymbol{\theta}} = (\mathbf{H}^t\mathbf{H} + \eta\mathbf{I})^{-1}\mathbf{H}^t\mathbf{g} . \tag{15.156}$$

By using SVD relations derived in Chap. 1, we can show that this reconstruction operator is exactly the one stated in the SVD domain in (15.114). Thus the modification of the denominator in (15.114) to avoid division by zero is also accomplished implicitly by this regularizer.

The functional $||\boldsymbol{\theta}||^2$ is widely known as the *Tikhonov regularizer* since its properties have been studied in many contexts by A. N. Tikhonov. Tikhonov's original work was published in the Soviet Union during World War II (Tikhonov, 1943), but more recent accounts are given by Tikhonov and Arsenin (1979) and Morosov (1993).

Tikhonov and Bayes The Tikhonov regularizer can be interpreted as a Bayesian prior. From (15.135) we have

$$\text{pr}(\boldsymbol{\theta}) = \frac{1}{Z_2} \exp\left[-\eta||\boldsymbol{\theta}||^2\right] , \tag{15.157}$$

which implies that the components of $\boldsymbol{\theta}$ are independent, zero mean normals, each with variance $1/(2\eta)$.

In frequentist terms, this result is quite puzzling. Suppose, for example, that the coefficients $\{\theta_n\}$ represent pixel values in a discrete object representation. There is no reason to expect *a priori* that these pixels would be normally distributed, and even if they were, they would hardly be zero-mean or uncorrelated. The zero mean implies that negative values are as probable as positive ones, even though in most situations θ_n is constrained to be nonnegative by its physical meaning. An *ad hoc* fix for this problem is to assume that (15.157) applies only when $\theta_n \geq 0$ for all n, with the probability of negative values being assigned zero prior probability.

This does not remove all of the objections to (15.157), however, since different components are still uncorrelated. Correlations between different points are what separate meaningful objects and images from snow on a television set tuned to an inactive channel, so an uncorrelated prior model is *a priori* wrong, at least in a frequentist sense. Nevertheless, the Tikhonov prior/regularizer can lead to quite useful reconstructed images, so it cannot be ruled out *a priori* by a pragmatist.

Entropy Another regularizer, for which many Bayesians show great fondness, is the entropy, defined by

$$Q_{reg}(\boldsymbol{\theta}) = \sum_{n=1}^{N} \theta_n \ln(\theta_n) . \tag{15.158}$$

The corresponding prior is

$$\mathrm{pr}(\boldsymbol{\theta}) = \frac{1}{Z} \exp\left[-\eta \sum_{n=1}^{N} \theta_n \ln(\theta_n)\right], \qquad \theta_n \geq 0, \tag{15.159}$$

where Z is a normalizing constant. Since this density is a product of factors that each involve only a single θ_n, the *a priori* assumption is again that the components are statistically independent. The density is defined only for positive values since the logarithm is imaginary if its argument is negative, but we shall see in Sec. 15.3.4 that this condition is satisfied almost automatically for any implicit estimate based on an entropy regularizer.

A more general form of entropy prior is

$$\mathrm{pr}(\boldsymbol{\theta}) = \frac{1}{Z} \exp\left[-\eta \sum_{n=1}^{N} \theta_n \ln\left(\frac{\theta_n}{m_n}\right)\right], \tag{15.160}$$

where m_n is a component of a vector $\mathbf{m}$ known as the *model*. The model is not the mean of the prior distribution, but instead is more closely related to the mode (most probable value). The reader may show that the peak of $\mathrm{pr}(\boldsymbol{\theta})$ occurs when $\theta_n = m_n/e$ for all n. As $\eta \to \infty$, $\mathrm{pr}(\boldsymbol{\theta}) \to \delta(\boldsymbol{\theta} - \mathbf{m}/e)$, so in this limit the mean equals the mode, but for finite η the density is skewed and there is no simple relationship between the model and the prior mean.

In practice, the most common model is a uniform field, $m_n = C$ for all n. This prior pulls the reconstruction towards the point where each $\theta_n = C/e$. If we really knew *a priori* that θ_n was likely to be close to some value θ_0, we would therefore set $m_n = e\theta_0$, but usually the only purpose of the flat model is to smooth the image, so the actual value of C is of little import.

Why entropy? Many Bayesians argue fervently that the use of an entropy prior is virtually a moral imperative. For example, Gull and Skilling (1984) assert that "the maximum entropy method is ... the *only consistent way* of combining different data into a single positive image" [emphasis added]. Several different lines of argument are used to justify this fervor.

One common Bayesian argument postulates a set of axioms which, they say, any tenable prior must satisfy, and then they show that the axioms can be satisfied only by entropy (Shore and Johnson, 1990; Tikochinsky *et al.*, 1984; Skilling, 1988, Csiszár, 1991). Upon examination, however, the axioms themselves seem to be untenable since they require that the components of $\boldsymbol{\theta}$ be statistically independent *a priori*. As we noted above, statistically independent densities describe random noise fields, not meaningful objects and images. The Bayesian reply is that if we know something about correlations, we should construct the prior that has maximum entropy subject to the constraint of having a prescribed covariance matrix. This approach leads to correlated normal models and Wiener filters (see Sec. 13.3.7), topics that otherwise receive little attention in the Bayesian imaging literature.

The entropy distribution can also be derived on combinatorial grounds, as in statistical mechanics. If we suppose (for some reason) that all objects are constructed by randomly throwing indistinguishable elements of brightness called *grains* into pixels, then the most probable object is the one that can be constructed in the greatest number of ways.

Let ϵ be the grain size and suppose that pixel n contains k_n grains, so that $\theta_n = \epsilon k_n$. Let p_n be the probability that a grain thrown at random falls in pixel n, and let $K = \sum_n k_n$ be the total number of grains thrown. For fixed grain size, the probability law on $\boldsymbol{\theta}$ is the same as the probability law on the vector $\mathbf{k}$ with components k_n. With the assumption that the grains are independent, that probability is the multinomial [*cf.* (C.164)],

$$\text{pr}(\mathbf{k}) = K! \prod_{n=1}^{N} \frac{p_n^{k_n}}{k_n!} . \tag{15.161}$$

If ϵ is small, then each of the k_n is large and we can use Stirling's approximation,

$$k! \approx \sqrt{2\pi k}\, k^k e^{-k} , \tag{15.162}$$

to show that

$$\begin{aligned} \ln[\text{pr}(\mathbf{k})] &= -\sum_{n=1}^{N} k_n \ln\left(\frac{k_n}{p_n}\right) + const \\ &= -\frac{1}{\epsilon}\sum_{n=1}^{N} \theta_n \ln\left(\frac{\theta_n}{m_n}\right) + const . \end{aligned} \tag{15.163}$$

This form agrees with (15.159) with η given by $1/\epsilon$ and the model components m_n given by p_n/ϵ. From this observation it is argued that $\text{pr}(\boldsymbol{\theta})$ must have the entropy form, even when $\boldsymbol{\theta}$ takes on a continuous range of values.

Finally, one way in which Bayesians attempt to appeal to the broader imaging community is to compare images reconstructed with an entropy prior to ones using, say, a Tikhonov regularizer; they then assert that the ones based on entropy are obviously superior on some subjective grounds. One problem with this comparison is that the Tikhonov images are almost always produced without a positivity constraint, while the maximum-entropy algorithms build in this physical reality. When constrained Tikhonov images are compared with maximum-entropy images, it is very hard even to tell them apart, much less to say which is subjectively superior. Moreover, when task performance is compared, even the unconstrained Tikhonov images seem to be as useful to a human observer as the entropy images (Gooley and Barrett, 1992).

Since none of these arguments for entropy says anything about its relation to task performance, they imply that the same prior and regularizing parameter should be used for all tasks and all observers. We have noted that the prior is irrelevant for an ideal observer and classification tasks, so it cannot be argued that entropy is in any sense optimal in that case. For the human observer, there is substantial psychophysical evidence that detection performance is optimized with different degrees of regularization for different background structures and signals, neither of which enter into the rationale for the entropy prior, so again it is difficult to argue that entropy is optimal. In short, since the entropy approach takes no account of task or observer, we fail to see why it should be accorded a privileged position in image-reconstruction problems; it is simply one more tool to be used and evaluated as any other method would be.

Nonlocal regularizers In practice, the Tikhonov and entropy regularizers usually produce smooth images, simply because they suppress the recovery of singular vectors that correspond to small singular values, which are the components that convey the fine detail about an object. Even though the Bayesian priors are uncorrelated, they induce strong correlations in a reconstructed image. But if we know that adjacent points in an object are likely to be correlated, we can also build that knowledge directly into the regularizer or prior. Haynor (1997) argues that the most important function of a prior is to specify neighborhood relations since any decent imaging system will get the large-scale features correct without prior information.

There are many ways to build neighborhood information into a regularizer. For example, if we know that neighboring pixels are unlikely to have significantly different values, we can define the regularizer to penalize deviations between a pixel value and some average of its neighboring pixels. For real $\boldsymbol{\theta}$, one possible form is

$$Q_{reg}(\boldsymbol{\theta}) = \sum_{n=1}^{N} \left[\theta_n - (\mathbf{S}\boldsymbol{\theta})_n\right]^2 , \tag{15.164}$$

where $\mathbf{S}$ is a local smoothing operator and hence $(\mathbf{S}\boldsymbol{\theta})_n$ is a weighted average of θ_n and its neighbors. A common choice in 2D imaging is to average over a 3×3 or 5×5 neighborhood, with weights chosen such that $\theta_n - (\mathbf{S}\boldsymbol{\theta})_n = 0$ if the components of $\boldsymbol{\theta}$ are constant in the neighborhood. If the object is constant in this neighborhood, there is no penalty.

General quadratic regularizer The regularizer defined in (15.164) is a quadratic functional of $\boldsymbol{\theta}$. It can be rewritten as

$$Q_{reg}(\boldsymbol{\theta}) = ||(\mathbf{I}-\mathbf{S})\boldsymbol{\theta}||^2 = \boldsymbol{\theta}^t(\mathbf{I}-\mathbf{S})^\dagger(\mathbf{I}-\mathbf{S})\boldsymbol{\theta} . \tag{15.165}$$

The matrix $(\mathbf{I}-\mathbf{S})^\dagger(\mathbf{I}-\mathbf{S})$ is positive-semidefinite and Hermitian, as any matrix of the form $\mathbf{A}^\dagger\mathbf{A}$ is. This observation suggests that we define the general quadratic regularizer as

$$Q_{reg}(\boldsymbol{\theta}) = \boldsymbol{\theta}^t\mathbf{C}\boldsymbol{\theta} , \tag{15.166}$$

where $\mathbf{C}$ is an arbitrary positive-semidefinite Hermitian matrix. We can define the square-root of $\mathbf{C}$ as in Sec. A.8.3 and write

$$Q_{reg}(\boldsymbol{\theta}) = ||\mathbf{C}^{\frac{1}{2}}\boldsymbol{\theta}||^2 . \tag{15.167}$$

If we choose $\mathbf{C}$ to be positive-definite rather than just positive-semidefinite, then the $\hat{\boldsymbol{\theta}}$ defined by (15.127) or (15.128) will be unique in spite of null functions in $\mathbf{H}$.

The Tikhonov regularizer is a special case of (15.167) with $\mathbf{C}=\mathbf{I}$. The entropy regularizer, however, does not fit this form since it is not quadratic in $\boldsymbol{\theta}$.

In Bayesian terms, the general quadratic regularizer amounts to choosing a zero-mean multivariate normal as the prior, with the inverse covariance matrix given by $\mathbf{K}_\theta^{-1} = 2\eta\mathbf{C}$. If we wanted to include a nonzero prior mean $\overline{\boldsymbol{\theta}}$, we would write

$$Q_{reg}(\boldsymbol{\theta}) = (\boldsymbol{\theta}-\overline{\boldsymbol{\theta}})^t\,\mathbf{C}(\boldsymbol{\theta}-\overline{\boldsymbol{\theta}}) . \tag{15.168}$$

Gradient norms One common quadratic approach to regularization is to penalize large values of the image gradient. When dealing with digital images, the word gradient must be understood in a discrete sense, but it is easiest to explain the form of the regularizer in continuous terms and then worry about discretization later. Therefore, consider an object $f(\mathbf{r})$, where for definiteness $\mathbf{r} = (x,y)$ is a

2D vector. The object is scalar-valued, and its gradient is a 2D vector field with components

$$(\nabla f)_x = \frac{\partial f(\mathbf{r})}{\partial x}, \qquad (\nabla f)_y = \frac{\partial f(\mathbf{r})}{\partial y}. \tag{15.169}$$

The norm of this vector field, defined in (7.7), is given by

$$||\nabla f||^2 = \int_{\mathbf{S}_f} d^2r \left\{ \left[\frac{\partial f(\mathbf{r})}{\partial x}\right]^2 + \left[\frac{\partial f(\mathbf{r})}{\partial y}\right]^2 \right\}. \tag{15.170}$$

Note that the norm here includes the sum of the squares of the components of the vector field at each position as well as an integral over position.

One might be tempted to compare $||\nabla f||^2$ with (15.167) and to identify $\mathbf{C}^{\frac{1}{2}}$ as a discrete gradient, but there are several problems with this approach. First, $\mathbf{C}^{\frac{1}{2}}$ acts on an $N \times 1$ vector and produces another $N \times 1$ vector, each such vector being the digital counterpart of a scalar field. The continuous gradient operator, on the other hand, acts on a scalar field and produces two scalar fields, $(\nabla f)_x$ and $(\nabla f)_y$. Moreover, the gradient operator is neither Hermitian nor positive-semidefinite, and both of these characteristics are required for $\mathbf{C}$ if we are to define its square root.

To make (15.170) look like (15.167), we must identify $\mathbf{C}$ not with the gradient but with the square root of the Laplacian operator. The Laplacian of $f(\mathbf{r})$ is defined by

$$\nabla^2 f(\mathbf{r}) = \frac{\partial^2 f(\mathbf{r})}{\partial x^2} + \frac{\partial^2 f(\mathbf{r})}{\partial y^2}. \tag{15.171}$$

The 2D Fourier transform of $\nabla^2 f(\mathbf{r})$ is given by (3.236) as

$$\mathcal{F}_2\{\nabla^2 f(\mathbf{r})\} = -4\pi^2\rho^2 F(\boldsymbol{\rho}), \tag{15.172}$$

where $\boldsymbol{\rho} = (\xi, \eta)$ is the spatial frequency vector and hence $\rho^2 = \xi^2 + \eta^2$. From this form it can be shown that the CC operator $\mathcal{C} \equiv -\nabla^2$ is a positive-definite Hermitian operator. Then we can define $\mathcal{C}^{\frac{1}{2}}$ by

$$\mathcal{F}_2\{\mathcal{C}^{\frac{1}{2}} f(\mathbf{r})\} = 2\pi\rho F(\boldsymbol{\rho}), \tag{15.173}$$

and we find from Parseval's theorem that

$$||(\nabla^2)^{\frac{1}{2}} f||^2 = 4\pi^2 \int_\infty d^2\rho\, \rho^2\, |F(\boldsymbol{\rho})|^2. \tag{15.174}$$

If we apply Parseval's theorem to (15.170), we get the same Fourier-domain integral, so $||(\nabla^2)^{\frac{1}{2}} f||^2 = ||\nabla f||^2$.

In the frequency domain, this regularizer grows as ρ^2 and hence suppresses high spatial frequencies, giving smooth images. An even stronger bias against high frequencies would be obtained by use of $||\nabla^2 f||^2$ as the regularizer; in that case the penalty grows as ρ^4.

Discretization of these continuous norms is straightforward. For example, a 2D discrete Laplacian can be realized by discrete convolution with a 3×3 kernel such as

$$\frac{1}{12}\begin{bmatrix} -1 & -2 & -1 \\ -2 & 12 & -2 \\ -1 & -2 & -1 \end{bmatrix}. \tag{15.175}$$

Nonquadratic regularizers A problem with quadratic regularizers is that they encourage smoothness even where the object itself may not be smooth, for example at a boundary between organs in medical imaging. We would like to smooth out small variations, which probably result from noise, but retain larger ones, which are more likely to be true edges.

One way to construct an edge-preserving regularizer is to use a nonlinear function of a scalar argument, $\Phi\{x\}$, and write

$$Q_{reg}(\boldsymbol{\theta}) = \sum_{n=1}^{N} \Phi\{[\mathbf{C}^{\frac{1}{2}}\boldsymbol{\theta}]_n\}, \tag{15.176}$$

where $\mathbf{C}^{\frac{1}{2}}$ is a positive-definite operator chosen, as above, to sense differences between a pixel's value and some average of its neighbors. If $\Phi\{x\} = x^2$, we are back to the general quadratic regularizer, but if we choose $\Phi\{x\}$ to grow less rapidly than x^2 for large x, then large variations will be penalized proportionally less than with the quadratic. For example, we can let $\Phi\{x\} = |x|^2$ for $|x| \leq 1$ and $\Phi\{x\} = 2|x| - 1$ for $|x| > 1$.

Another approach to edge-preserving, nonquadratic priors is *Markov random fields*, introduced in Sec. 8.4.4. In a pixel representation, the general form of the regularizer associated with a Markov random field is

$$Q_{reg}(\boldsymbol{\theta}) = \sum_{n=1}^{N} \sum_{j \in \mathcal{N}_n} U(\theta_n, \theta_j), \tag{15.177}$$

where $\mathcal{N}_n$ is a set of pixels in some suitably defined neighborhood of pixel n, and $U(\theta_n, \theta_j)$ is called the *potential.*

One commonly used potential has the form

$$U(\theta_n, \theta_j) = |\theta_n - \theta_j|^p, \tag{15.178}$$

where p is usually chosen to be in the range $1 < p \leq 2$. The corresponding prior density is then

$$\mathrm{pr}(\boldsymbol{\theta}) \propto \prod_{n=1}^{N} \prod_{j \in \mathcal{N}_n} \exp[-\eta|\theta_n - \theta_j|^p]. \tag{15.179}$$

Densities of this form have a cusp at the origin and high kurtosis (long tails) compared to a Gaussian. We saw in Sec. 8.4.3 that a similar behavior was often observed experimentally and could be explained very broadly in terms of Gaussian mixtures. Thus the prior of (15.179) has some frequentist justification, in that it exhibits characteristics seen empirically in collections of images.

Data-dependent and space-variant regularizers We have denoted the regularizing functional as $Q_{reg}(\boldsymbol{\theta})$, but there is no reason why it cannot depend on $\mathbf{g}$ also. One might object that it can then no longer be interpreted as a prior, but we lost that feature anyway when we allowed the regularizing parameter η to be determined from the data. In a sense, it is mandatory for the regularizing term to depend on the data since the two terms in (15.126) are defined in different spaces. The data agreement term is a norm or some other distance measure in data space while the regularizing term is defined in the reconstruction space. If we scale the matrix $\mathbf{H}$

by a constant, such that $\mathbf{H}' = C\mathbf{H}$ and $\mathbf{g}' = C\mathbf{g}$, then we change the ratio of the two terms and hence the degree of regularization. To compensate for this effect, a user of a regularizing functional would adjust η in accordance with the scaling of $\mathbf{H}$.

This global adjustment alone may not achieve the desired result, however, if the system sensitivity [defined in (7.232) or (15.36)] is a strong function of position. The value of η that achieves a desired balance between data agreement and smoothness at one point will lead to either oversmoothing or undersmoothing at other points in the reconstruction. With pixels or other local representations, it is straightforward to make η a function of position, thereby achieving a nonlocal regularization.

An alternative approach suggested by Fessler (1994) is to make the regularizing functional depend on the data $\mathbf{g}$ as well as the reconstruction $\boldsymbol{\theta}$. Fessler has demonstrated that this approach can yield a spatial resolution independent of position even when the system sensitivity varies with position. (See also Fessler and Rogers, 1996, and Stayman and Fessler, 2000).

15.3.4 Effects of positivity

If there are no constraints on the value of $\boldsymbol{\theta}$, and if we assume that $Q(\boldsymbol{\theta}, \mathbf{g})$ is everywhere differentiable with respect to its first argument, then the implicit estimate must satisfy

$$\frac{\partial}{\partial \theta_n} Q(\boldsymbol{\theta}, \mathbf{g}) = 0\,, \tag{15.180}$$

for all n at $\boldsymbol{\theta} = \hat{\boldsymbol{\theta}}$. If we allow only nonnegative solutions, however, the minimum may not occur at a point of zero derivative; it can also occur when one or more of the components θ_n are zero, so long as the derivative is positive at this point (see Fig. 15.2). We can allow for this possibility by requiring that

$$\theta_n \frac{\partial}{\partial \theta_n} Q(\boldsymbol{\theta}, \mathbf{g}) = 0 \qquad \text{and} \qquad \frac{\partial}{\partial \theta_n} Q(\boldsymbol{\theta}, \mathbf{g}) \geq 0\,, \tag{15.181}$$

for all n at $\boldsymbol{\theta} = \hat{\boldsymbol{\theta}}$. The first condition says that the solution can occur when either θ_n or the derivative is zero, and the second requires that the derivative be nonnegative if the solution occurs at $\theta_n = 0$, since otherwise there would be a smaller value with $\theta_n > 0$.

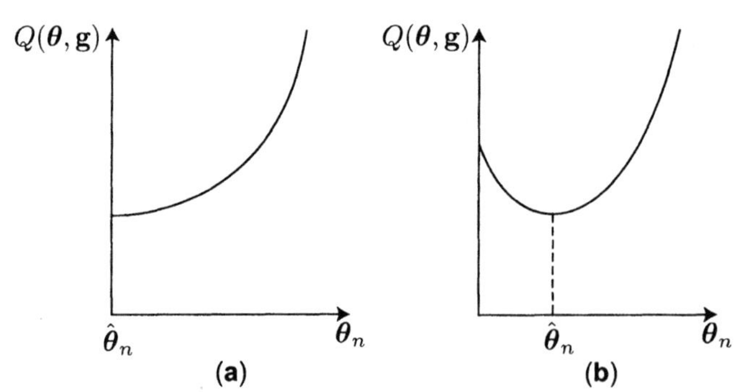

Fig. 15.2 Illustration of the Karush-Kuhn-Tucker conditions.

These conditions are often ascribed to Kuhn and Tucker[4] (1951), but in the simple form given here they were published earlier in a master's thesis by Karush (1939), so we shall refer to (15.181) as the *Karush-Kuhn-Tucker* or *KKT* conditions. More general Kuhn-Tucker conditions, some of which do not require the existence of the derivative, are discussed in detail in Mangasarian (1994).

Tikhonov and KKT To illustrate the effects of the KKT conditions, consider a least-squares data-agreement term and a Tikhonov regularizer. The derivative we need was discussed in Sec. 1.7.4, where we showed that

$$\frac{\partial}{\partial\theta_n}Q(\boldsymbol{\theta},\mathbf{g}) = \frac{\partial}{\partial\theta_n}\left[||\mathbf{g}-\mathbf{H}\boldsymbol{\theta}||^2 + \eta||\boldsymbol{\theta}||^2\right]$$

$$= -2[\mathbf{H}^t(\mathbf{g}-\mathbf{H}\boldsymbol{\theta})]_n + 2\eta\theta_n\,. \qquad (15.182)$$

Since $\eta > 0$, the KKT conditions require that

$$\hat{\theta}_n\left[\mathbf{H}^t(\mathbf{g}-\mathbf{H}\hat{\boldsymbol{\theta}})\right]_n - \eta\hat{\theta}_n = 0 \qquad \text{and} \qquad \hat{\theta}_n \geq \eta^{-1}\left[\mathbf{H}^t(\mathbf{g}-\mathbf{H}\hat{\boldsymbol{\theta}})\right]_n. \qquad (15.183)$$

If we define a vector $\hat{\mathbf{a}}$ in data space by

$$\hat{\mathbf{a}} = \frac{1}{\eta}(\mathbf{g}-\mathbf{H}\hat{\boldsymbol{\theta}})\,, \qquad (15.184)$$

then (15.183) can be rewritten as

$$\hat{\theta}_n\left[(\mathbf{H}^t\hat{\mathbf{a}})_n - \hat{\theta}_n\right] = 0 \qquad \text{and} \qquad \hat{\theta}_n \geq (\mathbf{H}^t\hat{\mathbf{a}})_n\,. \qquad (15.185)$$

It is useful to define a nonlinear operator $\mathbf{P}_+$ that has the effect of clipping off all negative values. Applied to a vector $\mathbf{b}$, $\mathbf{P}_+$ is defined by

$$[\mathbf{P}_+\mathbf{b}]_m = b_m\,\text{step}(b_m) = \begin{cases} b_m & \text{if } b_m \geq 0 \\ 0 & \text{if } b_m < 0 \end{cases}. \qquad (15.186)$$

With this operator, (15.185) is formally solved by

$$\hat{\boldsymbol{\theta}} = \mathbf{P}_+\mathbf{H}^t\hat{\mathbf{a}}\,, \qquad (15.187)$$

where $\hat{\mathbf{a}}$ must satisfy

$$\mathbf{H}\mathbf{P}_+\mathbf{H}^t\hat{\mathbf{a}} + \eta\hat{\mathbf{a}} = \mathbf{g}\,. \qquad (15.188)$$

Of course (15.187) is not really a solution to our problem since it depends on the vector $\hat{\mathbf{a}}$, which must be found by solving (15.188) by some iterative algorithm. It does show, however, that the solution must have the form of the clipped backprojection of some vector in data space.

[4]It was something of an accident that the topologist Albert W. Tucker (1905 – 1955) got into mathematical programming. He happened to be free to give George Dantzig a ride to the Princeton train station after Dantzig had made an unsuccessful trip to Princeton to try to interest John von Neumann in the new field, and Dantzig ended up recruiting Tucker rather than von Neumann (S. B. Maurer, *SIAM News*, July 1995).

Relation to the unconstrained Tikhonov problem It is interesting to relate (15.187) and (15.188) back to the corresponding equations for the same problem without the positivity constraint. If we simply delete the clipping operator, (15.188) becomes

$$\left[\mathbf{H}\mathbf{H}^t + \eta\mathbf{I}\right]\hat{\mathbf{a}} = \mathbf{g}\,. \tag{15.189}$$

Since $\eta > 0$, $\mathbf{H}\mathbf{H}^t + \eta\mathbf{I}$ is a positive-definite (hence invertible) operator, and we can write

$$\hat{\mathbf{a}} = \left[\mathbf{H}\mathbf{H}^t + \eta\mathbf{I}\right]^{-1}\mathbf{g}\,. \tag{15.190}$$

Thus (15.187) becomes

$$\hat{\boldsymbol{\theta}} = \mathbf{H}^t\left[\mathbf{H}\mathbf{H}^t + \eta\mathbf{I}\right]^{-1}\mathbf{g} = \left[\mathbf{H}^t\mathbf{H} + \eta\mathbf{I}\right]^{-1}\mathbf{H}^t\mathbf{g}\,, \tag{15.191}$$

where the equivalence of these two forms can be demonstrated with SVD expressions derived in Chap. 1. The second form agrees with (15.156).

Entropy and KKT The positivity constraint enters in a somewhat different way with the entropy regularizer. Consider the implicit estimate defined by

$$\hat{\boldsymbol{\theta}} = \underset{\boldsymbol{\theta} \geq \mathbf{0}}{\operatorname{argmin}} \left[||\mathbf{g} - \mathbf{H}\boldsymbol{\theta}||^2 + \eta \sum_{n=1}^{N} \theta_n \ln\left(\frac{\theta_n}{m_n}\right) \right]. \tag{15.192}$$

Now the required derivative is [*cf.* (15.182)]

$$\frac{\partial}{\partial \theta_n} Q(\boldsymbol{\theta}, \mathbf{g}) = -2\left[\mathbf{H}^t(\mathbf{g} - \mathbf{H}\boldsymbol{\theta})\right]_n + \eta \ln\left(\frac{\theta_n}{m_n}\right) + \eta\,. \tag{15.193}$$

This derivative vanishes if

$$\theta_n = \frac{m_n}{e}\exp\left(\frac{2}{\eta}\left[\mathbf{H}^t(\mathbf{g} - \mathbf{H}\boldsymbol{\theta})\right]_n\right), \tag{15.194}$$

so this equation must be satisfied at the minimum if that minimum does not occur when $\theta_n = 0$. Note that $\theta_n \to m_n/e$ as $\eta \to \infty$, in which case the prior dominates the data-agreement term. For small η, on the other hand, the solution must have g_m close to $(\mathbf{H}\boldsymbol{\theta})_m$, emphasizing the data-agreement term.

In fact, the solution cannot be at $\theta_n = 0$ since the derivative approaches $-\infty$ at this point and the second KKT condition cannot be satisfied there. Since entropy is a convex regularizer, there must be at least one minimum, and it must occur where (15.194) is satisfied. Moreover, (15.194) is automatically consistent with the positivity constraint (if $m_n > 0$) since the exponential lies between 0 and infinity, never reaching either extreme for finite argument. We do not need to include a clipping operator like $\mathbf{P}_+$ since there are never any negative values to clip.

Some in the Bayesian community regard the inability to get a zero value as a virtue. Csiszár (1991) notes that "use of an entropy regularizer/prior ensures that a nonnegative quantity is never inferred to be 0 when the available information *permits* it to be positive, which is generally considered to be desirable" [emphasis added]. Even from a Bayesian perspective, this feature seems to be at odds with any reasonable prior. We often have the prior knowledge that many pixels in the object representation will be zero. For example, a point outside the patient's body in medical imaging must be zero, though we do not know beforehand where these points are located in the scene. A Bayesian could take this obvious fact into account by including a delta function at the origin in the prior density, but we know of no case where this has been done.

15.3.5 Reconstruction without discretization

So far in this section we have discussed minimization of a functional involving the vector $\boldsymbol{\theta}$ of coefficients in an approximate discrete object representation. As we shall show in this section, however, it is also possible to formulate the implicit reconstruction problem in continuous terms without ever adopting a discrete representation. The results will be analogous to what we obtained with the Backus-Gilbert method in Sec. 15.2.2, but with an implicit formulation and a positivity constraint. The treatment here owes a great deal to the authors' interactions with Eric Clarkson.

If we denote a continuous estimate as $\hat{f}(\mathbf{r})$, with the corresponding vector in an infinite-dimensional Hilbert space denoted by $\hat{\mathbf{f}}$, then we can define the estimate by [*cf.* (15.128)]

$$\hat{\mathbf{f}} = \underset{\mathbf{f} \geq 0}{\text{argmin}} \left[Q_{data}(\mathbf{g}, \boldsymbol{\mathcal{H}}\mathbf{f}) + \eta Q_{reg}(\mathbf{f}) \right] . \tag{15.195}$$

Here, the constraint $\mathbf{f} \geq 0$ means $f(\mathbf{r}) \geq 0$ for all $\mathbf{r}$ in $\mathbf{S_f}$, and $\boldsymbol{\mathcal{H}}$ is a CD operator.

To find the minimum of this functional, we introduce the concept of a *Fréchet derivative* (Stakgold, 1979). An ordinary gradient of a scalar-valued function with a vector argument is a vector of the same dimensionality as the argument. The same is true of the Fréchet derivative of a scalar-valued functional, even though the argument is infinite-dimensional. The definition is also familiar; if the vector $\mathbf{f}$ is perturbed to $\mathbf{f} + \epsilon\mathbf{u}$, where $\mathbf{u}$ is some vector in the Hilbert space, then we can define the Fréchet derivative $\boldsymbol{\delta}_f\psi$ of the scalar functional $\psi(\mathbf{f})$ via

$$\lim_{\epsilon \to 0} \frac{\psi(\mathbf{f} + \epsilon\mathbf{u}) - \psi(\mathbf{f})}{\epsilon} = (\boldsymbol{\delta}_f\psi, \mathbf{u}) . \tag{15.196}$$

If $\mathbf{u}$ is a basis vector in the space, the right-hand side can be viewed as a component of the vector $\boldsymbol{\delta}_f\psi$. With only a slight loss of generality, we can restrict attention to reproducing-kernel Hilbert spaces where functions can be evaluated pointwise (see Sec. 1.8). Then if we let $\mathbf{u}$ correspond to the evaluation function for the Hilbert space [analogous to $\delta(\mathbf{r} - \mathbf{r}_0)$], the scalar product on the right in (15.196) is the function $[\boldsymbol{\delta}_f\psi](\mathbf{r})$ evaluated at $\mathbf{r} = \mathbf{r}_0$. The Fréchet derivative of a scalar functional is thus a function.

Two examples of (15.196) will prove useful. For the continuous Tikhonov regularizer,

$$\psi(\mathbf{f}) = \int_{\mathbf{S_f}} d^2r \, [f(\mathbf{r})]^2 , \tag{15.197}$$

we can write the Fréchet derivative as a vector in Hilbert space,

$$\boldsymbol{\delta}_f\psi = 2\,\mathbf{f} , \tag{15.198}$$

or equivalently as a function,

$$[\boldsymbol{\delta}_f\psi]\,(\mathbf{r}) = 2f(\mathbf{r}) . \tag{15.199}$$

Similarly, for the continuous entropy regularizer,

$$\psi(\mathbf{f}) = \int_{\mathbf{S_f}} d^2r \, f(\mathbf{r}) \ln[f(\mathbf{r})] , \tag{15.200}$$

we find

$$[\boldsymbol{\delta}_f\psi]\,(\mathbf{r}) = \ln f(\mathbf{r}) + 1 . \tag{15.201}$$

In both of these examples we have assumed that $f(\mathbf{r})$ is real.

A somewhat more complicated example comes from a least-squares data-agreement term. If we consider

$$\psi(\mathbf{f}) = ||\mathbf{g} - \mathcal{H}\mathbf{f}||^2 = \sum_{m=1}^{M} \left[g_m - \int_{\mathbf{S_f}} d^2r'\, h_m(\mathbf{r}')\, f(\mathbf{r}') \right]^2 , \tag{15.202}$$

with all quantities being real, then we can show that

$$[\boldsymbol{\delta}_f \psi](\mathbf{r}) = -2 \sum_{m=1}^{M} h_m(\mathbf{r}) \left[g_m - \int_{\mathbf{S_f}} d^2r'\, h_m(\mathbf{r}')\, f(\mathbf{r}') \right] . \tag{15.203}$$

To get a more abstract operator form, we can use the definition of the adjoint of a CD operator from (7.237) and write

$$\boldsymbol{\delta}_f \psi = -2\mathcal{H}^\dagger [\mathbf{g} - \mathcal{H}\mathbf{f}] . \tag{15.204}$$

Continuous Tikhonov reconstruction Consider the regularized least-squares problem with the data-agreement functional (15.202) and the continuous Tikhonov regularizer (15.197). From results above, we see that the Fréchet derivative is a modest generalization of (15.182):

$$\boldsymbol{\delta}_f Q(\mathbf{f}, \mathbf{g}) = -2\big[\mathcal{H}^\dagger (\mathbf{g} - \mathcal{H}\mathbf{f})\big] + 2\eta\, \mathbf{f} . \tag{15.205}$$

Similarly, (15.183) becomes

$$\hat{f}(\mathbf{r})\big[\mathcal{H}^\dagger (\mathbf{g} - \mathcal{H}\hat{\mathbf{f}})\big](\mathbf{r}) - \eta \hat{f}(\mathbf{r}) = 0 ; \qquad \hat{f}(\mathbf{r}) \geq \eta^{-1}\big[\mathcal{H}^\dagger (\mathbf{g} - \mathcal{H}\hat{\mathbf{f}})\big](\mathbf{r}) . \tag{15.206}$$

Just as (15.183) had to hold for all n at the solution point, so too must these continuous conditions hold for all $\mathbf{r}$. They thus comprise a set of infinite-dimensional KKT conditions.[5]

From this point it is just a matter of notational changes to get a set of equations for the continuous estimate. Specifically, (15.187) becomes

$$\hat{f}(\mathbf{r}) = \big[\mathbf{P}_+ \mathcal{H}^\dagger \hat{\mathbf{a}}\big](\mathbf{r}) , \tag{15.207}$$

where $\hat{\mathbf{a}}$ is still a finite-dimensional vector in data space and satisfies [*cf.* (15.188)]

$$\mathcal{H}\mathbf{P}_+ \mathcal{H}^\dagger \hat{\mathbf{a}} + \eta \hat{\mathbf{a}} = \mathbf{g} . \tag{15.208}$$

From (15.207) we see that $\hat{f}(\mathbf{r})$ is simply a clipped backprojection of $\hat{\mathbf{a}}$, or equivalently a clipped superposition of natural pixels with weights $\{a_m\}$. Thus the reconstruction problem boils down to finding the M components of $\hat{\mathbf{a}}$ by solving (15.208). One approach would be a fixed-point iteration (see Sec. 15.4.4). In practice, the nonlinear operator $\mathcal{H}\mathbf{P}_+\mathcal{H}^\dagger$ could be implemented by replacing the integral over object space by a finely sampled sum, but this sampling would not imply that a discrete object representation had been adopted.

[5]These conditions were originally derived by Eric Clarkson but have not been published elsewhere at this writing.

Other regularized reconstructions By arguments parallel to those leading to (15.194), we can show that the continuous maximum-entropy solution must satisfy

$$\hat{f}(\mathbf{r}) = \frac{m(\mathbf{r})}{e} \exp\left\{\frac{2}{\eta}\left[\mathcal{H}^\dagger(\mathbf{g} - \mathcal{H}\mathbf{f})\right](\mathbf{r})\right\} . \tag{15.209}$$

As in the discrete case, any solution to this equation automatically satisfies the positivity constraint.

From (15.207) and (15.209), we see that the solution must have the form of a nonlinear point operator acting on a superposition of natural pixels, at least for least-squares problems with the Tikhonov or entropy regularizers. The same is true for a wide variety of other continuous implicit estimates.[6]

Night skies The purpose of regularization in any inverse problem is to reduce noise variance, but with implicit estimation of functions, a completely new class of solutions arises if we do not regularize. In particular, we can have solutions that consist of a finite number of delta functions:

$$\hat{f}(\mathbf{r}) = \sum_{k=1}^{K} \alpha_k \, \delta(\mathbf{r} - \mathbf{r}_k) , \qquad \alpha_k > 0 , \qquad K \le M . \tag{15.210}$$

Since this set of delta functions might represent stars in the sky, solutions of this form can be called *night-sky reconstructions*. Needless to say, a night-sky reconstruction will usually not resemble the actual object, so solutions like (15.210) are infinitely biased with respect to almost any real object, and they have infinite variance. Nevertheless, they can always occur if we do not regularize or discretize.

To understand the existence of night-sky reconstructions, consider the moment cone, introduced in Sec. 15.1.4, and suppose first that $\mathbf{g}$ lies in its interior. Then we can express $\mathbf{g}$ as a positive linear combination of points on the boundary of the moment cone. Each of these points is the image of a unit delta function in object space, so it is always possible to find an $\hat{\mathbf{f}}$ in the form of (15.210) that will give $\mathcal{H}\hat{\mathbf{f}} \equiv \mathbf{g}$. Any distance measure $Q_{data}(\mathbf{g}, \mathcal{H}\mathbf{f})$ will be minimized when $\mathcal{H}\mathbf{f} = \mathbf{g}$, so the night-sky estimate will necessarily minimize the distance. Since $\mathcal{H}$ has null functions, it is also possible that there will be other objects $\mathbf{f}$ such that $\mathcal{H}\mathbf{f} \equiv \mathbf{g}$, yielding the same value of zero for the data-agreement functional. Some or all of these objects may not be night skies; which object is found depends on the specific minimization algorithm and its initial estimate.

Because of noise, $\mathbf{g}$ may not lie in the moment cone. In fact, in many cases it is overwhelmingly likely that $\mathbf{g}$ will not lie in the moment cone, because the moment cone is a lower-dimensional manifold in the data space. If the DC operator $\mathcal{H}^\dagger$ has null functions, then as we saw in Sec. 15.2.5 there are consistency conditions in data space, and different components of noise-free data are not linearly independent. If there are K consistency conditions and M components to the data vector, then consistency space is an $(M-K)$-dimensional manifold in data space. Since the moment cone is a subset of consistency space, it also has just $M - K$ dimensions. A noisy data vector, on the other hand, does not have the linear dependence, so it will lie at a general point in data space, with a vanishingly small probability of hitting the cone.

[6]Unpublished results of Eric Clarkson.

If a noisy $\mathbf{g}$ lies outside in the moment cone, and if the data-agreement functional is strictly convex, then there must be a unique point $\mathbf{g}_0$ on the surface of the moment cone such that $Q_{data}(\mathbf{g}, \mathcal{H}\mathbf{f})$ is minimized (though not zero). Since $\mathbf{g}_0$ is on the surface of the moment cone, it is either the image of a single point object or a positive combination of at most M point objects. Thus the equation $\mathcal{H}\mathbf{f} = \mathbf{g}_0$ can always be solved exactly by a night-sky object.

Moreover, if $\mathbf{g}$ is outside the moment cone, the night-sky object will be unique in the absence of exact consistency conditions. Consider a CD imaging system with three detectors viewing a 1D object function $f(x)$. As shown in Fig. 15.3, moving a delta function along the x-axis in object space traces out a curve C in data space, and the moment cone is formed by taking the convex hull and drawing a line from the origin through each point in the hull to infinity. As the figure is drawn, the cone is a 3D manifold, so there are no consistency conditions. Nevertheless, two examples of data vectors, $\mathbf{g}$ and $\mathbf{g}'$, outside the cone are shown. The point on the cone nearest to $\mathbf{g}$ is $\mathbf{g}_0$ and the one nearest to $\mathbf{g}'$ is $\mathbf{g}'_0$. As shown, $\mathbf{g}'_0$ lies on the original curve C, so it is uniquely the image of a single delta function. By contrast, $\mathbf{g}_0$ lies on a flat (two-dimensional) face created when the convex hull was formed. This face touches C at just two points, each of which is the image of a delta function, so $\mathbf{g}_0$ is uniquely the image of an object consisting of two delta functions. The only way in which a smooth object could produce $\mathbf{g}_0$ would be if the curve C itself had a face, which would be the image of delta functions $\delta(\mathbf{r} - \mathbf{r}_0)$ for a continuous range of $\mathbf{r}_0$; then any point in the face could be the image of an integral superposition of these delta functions, not just a finite sum.

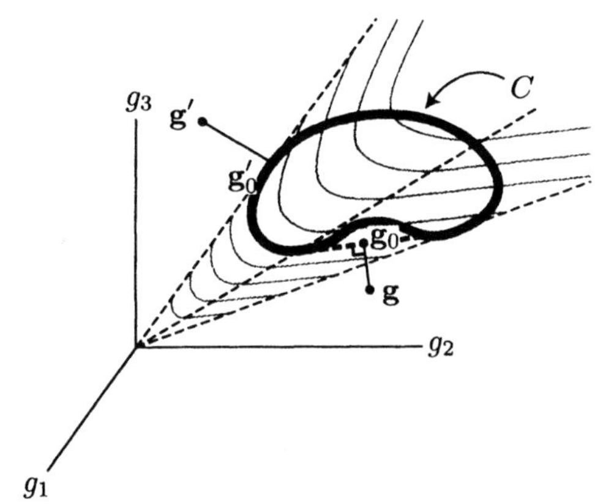

Fig. 15.3 The moment cone for a linear system with three detectors viewing a 1D object $f(x)$.

In summary, there is always a nonnegative night-sky object that will minimize the unregularized data-agreement functional. If there are consistency conditions, the data vector $\mathbf{g}$ will usually lie outside the moment cone, and the night sky object will usually be unique. Thus night skies are almost inevitable in unregularized positive reconstruction without discretization. Similar features occur in DD problems, as discussed by Byrne (1993, 1995).

15.3.6 Resolution and noise in implicit estimates

Sometimes the minimum of an objective function $Q(\boldsymbol{\theta}, \mathbf{g})$ occurs at a point where it is differentiable with respect to both its arguments. This is the case, for example, with the least-squares data-agreement term and any popular regularizer if there are no constraints on the solution, and even a positivity constraint may not spoil the differentiability, as we saw in the case of the entropy regularizer in Sec. 15.3.4. In this section we shall explore some properties of the argmin estimate when $Q(\boldsymbol{\theta}, \mathbf{g})$ is differentiable at the minimum with respect to both arguments.

Implicit derivatives If x and y are real scalars, the equation $f(x, y) = 0$ defines y as an implicit function of x. The total differential of f is

$$df = \frac{\partial f}{\partial x} dx + \frac{\partial f}{\partial y} dy = 0\,, \tag{15.211}$$

from which we see that

$$\frac{dy}{dx} = -\frac{\partial f / \partial x}{\partial f / \partial y}\,. \tag{15.212}$$

Now suppose $\mathbf{x}$ is a real $M \times 1$ column vector and $\mathbf{y}$ is a real $N \times 1$ column vector. To define $\mathbf{y}$ as an implicit function of $\mathbf{x}$, we need N linearly independent equations of the form $f_n(\mathbf{x}, \mathbf{y}) = 0$. Then, by the chain rule,

$$df_n = \frac{\partial f_n}{\partial \mathbf{x}^t} d\mathbf{x} + \frac{\partial f_n}{\partial \mathbf{y}^t} d\mathbf{y} = 0\,, \tag{15.213}$$

where $\partial f_n / \partial \mathbf{x}^t$ is a row vector of partial derivatives. (For more discussion of the vector notation used here, see Sec. A.9.2.) If we consider a vector $d\mathbf{x}$ where all components except dx_j are zero, we see that

$$\frac{\partial f_n}{\partial x_j} = -\sum_k \frac{\partial f_n}{\partial y_k} \frac{\partial y_k}{\partial x_j}\,. \tag{15.214}$$

This result can be written in matrix-vector form by defining a matrix $\partial \mathbf{y} / \partial \mathbf{x}^t$ with jk^{th} component given by $\partial y_k / \partial x_j$, so that

$$\frac{\partial f_n}{\partial \mathbf{x}} = -\frac{\partial \mathbf{y}}{\partial \mathbf{x}^t} \frac{\partial f_n}{\partial \mathbf{y}}\,. \tag{15.215}$$

Application to implicit estimates Suppose that the minimum of $Q(\boldsymbol{\theta}, \mathbf{g})$ occurs at a point $\boldsymbol{\theta} = \hat{\boldsymbol{\theta}}(\mathbf{g})$ where $\partial Q(\boldsymbol{\theta}, \mathbf{g}) / \partial \theta_n = 0$ for $n = 1, ..., N$. Then we can apply (15.215) with $f_n = \partial Q(\hat{\boldsymbol{\theta}}, \mathbf{g}) / \partial \hat{\theta}_n$, $\mathbf{y} = \hat{\boldsymbol{\theta}}$ and $\mathbf{x} = \mathbf{g}$, yielding

$$\frac{\partial}{\partial \mathbf{g}} \left[\frac{\partial Q(\hat{\boldsymbol{\theta}}, \mathbf{g})}{\partial \hat{\theta}_n} \right] = -\frac{\partial \hat{\boldsymbol{\theta}}}{\partial \mathbf{g}^t} \frac{\partial}{\partial \hat{\boldsymbol{\theta}}} \left[\frac{\partial Q(\hat{\boldsymbol{\theta}}, \mathbf{g})}{\partial \hat{\theta}_n} \right], \tag{15.216}$$

or, in full matrix form,

$$\frac{\partial^2 Q(\hat{\boldsymbol{\theta}}, \mathbf{g})}{\partial \mathbf{g} \partial \hat{\boldsymbol{\theta}}^t} = -\frac{\partial \hat{\boldsymbol{\theta}}}{\partial \mathbf{g}^t} \frac{\partial^2 Q(\hat{\boldsymbol{\theta}}, \mathbf{g})}{\partial \hat{\boldsymbol{\theta}} \partial \hat{\boldsymbol{\theta}}^t}\,. \tag{15.217}$$

The second derivative on the right is the Hessian matrix (see Sec. A.9.2), which will be invertible whenever the objective function has a unique minimum. Multiplying (15.217) from the right with the inverse Hessian and taking a transpose yields the multivariate counterpart of (15.212):

$$\frac{\partial \hat{\boldsymbol{\theta}}}{\partial \mathbf{g}} = -\left[\frac{\partial^2 Q(\hat{\boldsymbol{\theta}}, \mathbf{g})}{\partial \hat{\boldsymbol{\theta}} \partial \hat{\boldsymbol{\theta}}^t}\right]^{-1} \frac{\partial^2 Q(\hat{\boldsymbol{\theta}}, \mathbf{g})}{\partial \hat{\boldsymbol{\theta}} \partial \mathbf{g}^t}. \tag{15.218}$$

We can rewrite (15.218) as

$$d\hat{\boldsymbol{\theta}} = \left[\mathbf{A}(\hat{\boldsymbol{\theta}}(\mathbf{g}), \mathbf{g})\right] d\mathbf{g}, \tag{15.219}$$

where $\mathbf{A}(\hat{\boldsymbol{\theta}}, \mathbf{g})$ is the $N \times M$ matrix defined by the right-hand side of (15.218):

$$\mathbf{A}(\hat{\boldsymbol{\theta}}, \mathbf{g}) = -\left[\frac{\partial^2 Q(\hat{\boldsymbol{\theta}}, \mathbf{g})}{\partial \hat{\boldsymbol{\theta}} \partial \hat{\boldsymbol{\theta}}^t}\right]^{-1} \frac{\partial^2 Q(\hat{\boldsymbol{\theta}}, \mathbf{g})}{\partial \hat{\boldsymbol{\theta}} \partial \mathbf{g}^t}. \tag{15.220}$$

As it stands, (15.219) is a nonlinear mapping since $\mathbf{A}(\hat{\boldsymbol{\theta}}, \mathbf{g})$ depends on $\hat{\boldsymbol{\theta}}$. To linearize it, let $\overline{\mathbf{g}}$ be the mean data vector and let $\hat{\boldsymbol{\theta}}(\mathbf{g})$ be the argmin estimate obtained when $\mathbf{g} = \overline{\mathbf{g}} \equiv \boldsymbol{\mathcal{H}}\mathbf{f}$. [Note that $\hat{\boldsymbol{\theta}}(\overline{\mathbf{g}})$ is not necessarily the mean of $\hat{\boldsymbol{\theta}}$.] For small deviations of the data vector from its mean, we can write

$$\Delta\hat{\boldsymbol{\theta}} = \left[\mathbf{A}(\hat{\boldsymbol{\theta}}(\boldsymbol{\mathcal{H}}\mathbf{f}), \boldsymbol{\mathcal{H}}\mathbf{f})\right] \Delta\mathbf{g}. \tag{15.221}$$

This formula, derived by Fessler (1996), is now a linearized mapping from $\Delta\mathbf{g}$ to $\Delta\hat{\boldsymbol{\theta}}$, though the mapping operator depends on the actual object $\mathbf{f}$. A change in $\mathbf{g}$ can result from either a change in the object or from noise, so the formula can be used to discuss the resolution and noise properties of implicit estimates.

Resolution Now suppose $\Delta\mathbf{g}$ results from adding a weak delta function $\epsilon\,\delta(\mathbf{r} - \mathbf{r}_0)$ to the object, where ϵ is small enough that $\Delta g_m \ll g_m$ for all m. For our usual CD model, $\Delta g_m = \epsilon h_m(\mathbf{r}_0)$. The point response function for the overall system (CD operator plus implicit estimator) is the $N \times 1$ vector $\mathbf{p}(\mathbf{r}_0)$ defined as the image of this point object divided by ϵ; it is given by

$$\mathbf{p}(\mathbf{r}_0) = \left[\mathbf{A}(\hat{\boldsymbol{\theta}}(\boldsymbol{\mathcal{H}}\mathbf{f}), \boldsymbol{\mathcal{H}}\mathbf{f})\right] \mathbf{h}(\mathbf{r}_0), \tag{15.222}$$

where $\mathbf{h}(\mathbf{r}_0)$ is a vector in data space with m^{th} component given by $h_m(\mathbf{r}_0)$.

In practice, this PRF might be evaluated with the aid of a DD model for the imaging system. If we neglect null functions and modeling errors, then $\boldsymbol{\mathcal{H}}\mathbf{f} \simeq \mathbf{H}\boldsymbol{\theta}$. If $\boldsymbol{\theta}$ is perturbed by adding $\epsilon\,\delta_{nn_0}$, representing a change of ϵ in the element $n = n_0$, then the PRF in $\hat{\boldsymbol{\theta}}$ is

$$\mathbf{p}(n_0) = \left[\mathbf{A}(\hat{\boldsymbol{\theta}}(\mathbf{H}\boldsymbol{\theta}), \mathbf{H}\boldsymbol{\theta})\right] \mathbf{h}(n_0), \tag{15.223}$$

where $\mathbf{h}(n_0)$ is an $M \times 1$ vector with the mth element given by H_{mn_0}. If $\mathbf{h}(n_0)$ is a good approximation to $\mathbf{h}(\mathbf{r}_0)$, then the PRF of (15.223) could accurately represent (15.222), but the assumptions should be checked in particular cases. An additional possible simplification is that the noise-free reconstruction $\hat{\boldsymbol{\theta}}(\mathbf{g})$ accurately reproduces $\boldsymbol{\theta}$, in which case $\mathbf{A}(\hat{\boldsymbol{\theta}}(\mathbf{H}\boldsymbol{\theta}), \mathbf{H}\boldsymbol{\theta}) \simeq \mathbf{A}(\boldsymbol{\theta}, \mathbf{H}\boldsymbol{\theta})$, but again this step should be taken with caution.

Variance and covariance in the estimate The linearized mapping of (15.221) allows us to use the formalism of Sec. 15.2.6 to discuss the statistical properties of $\hat{\boldsymbol{\theta}}$. From (15.107) and (15.221), the conditional covariance matrix for the reconstruction, given a specified object, is

$$\mathbf{K}_{\hat{\theta}|\mathbf{f}} = \mathbf{A}\mathbf{K}_{\mathbf{g}}\mathbf{A}^{\dagger}\,, \tag{15.224}$$

where $\mathbf{A} = \mathbf{A}(\hat{\boldsymbol{\theta}}(\boldsymbol{\mathcal{H}}\mathbf{f}), \boldsymbol{\mathcal{H}}\mathbf{f})$, but it might be approximated as in (15.223) if a DD model is used.

As discussed in Secs. 12.2 and 15.2.6, it is frequently a good model to assume that $\mathbf{g} \sim \mathcal{N}(\boldsymbol{\mathcal{H}}\mathbf{f}, \sigma^2\mathbf{I})$, and in that case the conditional covariance matrix, in component form, is given by

$$\left[\mathbf{K}_{\hat{\theta}|\mathbf{f}}\right]_{nn'} = \sigma^2 \sum_{m=1}^{M} A_{nm} A^*_{n'm}\,, \tag{15.225}$$

and the corresponding conditional variance is

$$\mathrm{Var}\left\{\hat{\theta}_n|\mathbf{f}\right\} = \sigma^2\left[\mathbf{A}\mathbf{A}^{\dagger}\right]_{nn} = \sigma^2 \sum_{m=1}^{M} |A_{nm}|^2\,. \tag{15.226}$$

These expressions are very similar to (15.108) and (15.109), respectively, but there is a key difference. Since $\mathbf{A}$ is a function of $\mathbf{f}$ (or approximately, $\boldsymbol{\theta}$), the noise in the reconstruction depends on the object and hence on position in the image.

The term *signal-dependent noise* is often encountered in the literature, and it can be applied here if the signal is construed to be the object itself. By the terminology introduced in Chap. 8, however, *signal* refers to a part of the object that we might be interested in detecting. If the signal in this sense is a small perturbation to the object, then it may have little influence on the statistics of $\hat{\boldsymbol{\theta}}$, so we would say that the noise is object-dependent but signal-independent.

This same argument applies even in the case of Poisson noise, which is often called signal-dependent. In that case the conditional covariance and variance of $\hat{\boldsymbol{\theta}}$ are given by (15.120) and (15.121) (with $\mathbf{O}$ replaced by $\mathbf{A}$). Then the noise depends on position in the image for two reasons: the data covariance depends on the object, and the implicit reconstruction procedure also introduces an object dependence. Nevertheless, if we think of a signal as a small perturbation to an object, then the noise in the reconstruction is signal-independent but object-dependent.

Example 1: Tikhonov regularization and least-squares Within the linearized approximation, computation of resolution and noise in the reconstruction requires only that we be able to compute the $N \times M$ matrix $\mathbf{A}$ defined in (15.220). We shall now show how to do that for two special cases.

First, consider a least-squares data-agreement term and a Tikhonov regularizer. The first derivative of the objective function with respect to $\boldsymbol{\theta}$ is given by (15.182), and the Hessian matrix is obtained by differentiating (15.182) again; the result is

$$\frac{\partial^2}{\partial\theta_n \partial\theta_m} Q(\boldsymbol{\theta}, \mathbf{g}) = 2\left[\mathbf{H}^t\mathbf{H}\right]_{nm} + 2\eta\, \delta_{nm}\,. \tag{15.227}$$

The requisite cross-derivative is

$$\frac{\partial^2}{\partial\theta_n \partial g_m} Q(\boldsymbol{\theta}, \mathbf{g}) = 2H_{mn}\,. \tag{15.228}$$

Thus (15.220) takes the familiar form,

$$\mathbf{A}(\hat{\boldsymbol{\theta}}, \mathbf{g}) = \left[\mathbf{H}^t\mathbf{H} + \eta\mathbf{I}\right]^{-1}\mathbf{H}^t\,, \tag{15.229}$$

which is identical to the regularized least-squares operator derived in Sec. 1.7.5 and reproduced in (15.156). Since this operator is independent of $\hat{\boldsymbol{\theta}}$ and $\mathbf{g}$, the linearization is exact in this case.

Example 2: Entropy regularization For a least-squares data-agreement functional and the entropy regularizer of (15.158), the Hessian is

$$\frac{\partial^2}{\partial\theta_n\partial\theta_m}Q(\boldsymbol{\theta}, \mathbf{g}) = 2\left[\mathbf{H}^t\mathbf{H}\right]_{nm} + 2\left(\frac{\eta}{\theta_n}\right)\delta_{nm}\,. \tag{15.230}$$

The cross-derivative is still given by (15.228), so

$$\mathbf{A}(\hat{\boldsymbol{\theta}}, \mathbf{g}) = \left[\mathbf{H}^t\mathbf{H} + \eta\mathbf{D}(\hat{\boldsymbol{\theta}})\right]^{-1}\mathbf{H}^t\,, \tag{15.231}$$

where

$$\left[\mathbf{D}(\hat{\boldsymbol{\theta}})\right]_{nn'} = \frac{1}{\hat{\theta}_n}\,\delta_{nn'}\,. \tag{15.232}$$

Now the matrix $\mathbf{A}$ depends on $\hat{\boldsymbol{\theta}}$, so different mappings apply at different points in the image. Since the regularizing parameter is divided by $\hat{\theta}_n$, the qualitative effect is to reduce the variance in regions of the image where $\hat{\theta}_n$ is small.

15.4 ITERATIVE ALGORITHMS

Once we have chosen an objective functional to minimize, we next need an algorithm to find the minimum. In one sense, this choice is more a matter of computer science than image science. If the functional is strictly convex, then the minimum is unique and all algorithms should obtain the same image if run to convergence. The only issues are which algorithm will find the minimum most efficiently and with the least computing resources, and those questions are peripheral to the goal of this book.

In practice, however, iterative algorithms may not be run to convergence. In that case, the resulting image depends on the algorithm, the initial estimate and the stopping rule (as well as the data and the model, of course). One focus in this section, therefore, will be on the properties of images produced by iterative algorithms as a function of iteration number.

We shall distinguish linear from nonlinear algorithms, depending on whether the update rule gives the next estimate as a linear (or at least affine) functional of the previous estimate. Linear algorithms are, of course, easier to analyze, but nonlinear algorithms may offer considerable advantages. For example, imposition of a positivity constraint at each iteration requires nonlinearity, but it may be effective in controlling noise and artifacts in the image.

We begin in Sec. 15.4.1 with a survey of linear iterative algorithms, building on the discussion of iteration computation of the pseudoinverse in Sec. 1.7.6. Noise properties of linear iterative algorithms are derived in Sec. 15.4.2.

In Sec. 15.4.3 we give a rapid survey of optimization methods. Some emphasis is placed on conjugate-gradient methods, but mainly for pedagogical purposes; as we shall see, conjugate gradients are closely related to two main themes of this book, pseudoinverses and prewhitening.

In Sec. 15.4.4, we shall see how to introduce nonlinear constraints into iterative algorithms, and we shall revisit the KKT conditions in the context of a specific class of nonlinear algorithms, fixed-point iterations. Another important general class of algorithms is projections onto convex sets, surveyed briefly in Sec. 15.4.5.

The popular expectation-maximization (EM) algorithm is introduced in Sec. 15.4.6, and its relation to Poisson likelihood is discussed, and in Sec. 15.4.7, we discuss noise propagation in nonlinear iterative algorithms, using the EM algorithm as an example.

Finally, in Sec. 15.4.8 we consider iterative algorithms in which the update step is random. Starting with an algorithm called *simulated annealing*, we discuss a general class of algorithms called *Markov-chain Monte Carlo*. Applications of this algorithm to image-quality assessment have already been mentioned in Sec. 14.3.3 and 14.3.4.

15.4.1 Linear iterative algorithms

In Sec. 1.7.6 we introduced several iterative methods for minimizing a quadratic data-agreement functional; included there were the Landweber, Gauss-Seidel and Jacobi methods. For these algorithms and many others, the iteration rule has the general form [*cf.* (1.236)]

$$\hat{\boldsymbol{\theta}}^{(k+1)} = \hat{\boldsymbol{\theta}}^{(k)} + \mathbf{B}\left[\mathbf{g} - \mathbf{H}\hat{\boldsymbol{\theta}}^{(k)}\right] , \tag{15.233}$$

where $\mathbf{B}$ is an $N \times M$ matrix. Since the correction term is added to the previous estimate, (15.233) is often referred to as an *additive algorithm*. It is also called a *linear algorithm* since only linear operations are performed on the kth estimate to obtain the correction term; the relation between $\hat{\boldsymbol{\theta}}^{(k+1)}$ and $\hat{\boldsymbol{\theta}}^{(k)}$ is, however, affine rather than strictly linear.

The choices to be made in using this kind of algorithm are the matrix $\mathbf{B}$, the starting estimate $\hat{\boldsymbol{\theta}}^{(0)}$ and the number of iterations K. For example, a modified Landweber algorithm can be defined by taking $\mathbf{B} = \alpha\mathbf{H}^t$, where α is called an *acceleration parameter* since it controls the rate of convergence. Common choices for $\hat{\boldsymbol{\theta}}^{(0)}$ are $\mathbf{H}^t\mathbf{g}$ or a uniform field (*i.e.*, $\hat{\theta}_n^{(0)} = c$ for all n).

Conditions under which algorithms like (15.233) converge were discussed in Sec. 1.7.6. For present purposes, let us simply assume that the algorithm converges and see what it gives when it does. Convergence means that $\hat{\boldsymbol{\theta}}^{(k+1)} - \hat{\boldsymbol{\theta}}^{(k)} \to 0$, or $\mathbf{B}[\mathbf{g} - \mathbf{H}\hat{\boldsymbol{\theta}}^{(k)}] \to 0$. If the rank of $\mathbf{B}$ is the data dimension M, then $\mathbf{B}$ has no null functions in data space, so convergence can occur only if $\mathbf{g} = \mathbf{H}\hat{\boldsymbol{\theta}}^{(k)}$. Thus the algorithm attempts to enforce agreement between the actual noisy data $\mathbf{g}$ and the estimated data $\mathbf{H}\hat{\boldsymbol{\theta}}$; as we have seen several times, strict agreement with noisy data results in a very noisy image, so this algorithm is almost never run to convergence.

A more general linear algorithm has the form

$$\hat{\boldsymbol{\theta}}^{(k+1)} = \hat{\boldsymbol{\theta}}^{(k)} - \eta\mathbf{C}\hat{\boldsymbol{\theta}}^{(k)} + \mathbf{B}\left[\mathbf{g} - \mathbf{H}\hat{\boldsymbol{\theta}}^{(k)}\right] , \tag{15.234}$$

where $\mathbf{C}$ is an $N \times N$ matrix. Now the algorithm no longer forces $\mathbf{g} = \mathbf{H}\hat{\boldsymbol{\theta}}^{(k)}$ but instead converges (if it does so at all) to a $\hat{\boldsymbol{\theta}}^{(k)}$ such that

$$\eta \mathbf{C}\hat{\boldsymbol{\theta}}^{(k)} = \mathbf{B}\left[\mathbf{g} - \mathbf{H}\hat{\boldsymbol{\theta}}^{(k)}\right] . \tag{15.235}$$

Equivalently, the algorithm converges to

$$\hat{\boldsymbol{\theta}}^{(k)} = (\mathbf{BH} + \eta\mathbf{C})^{-1}\mathbf{Bg}\,, \tag{15.236}$$

provided the indicated inverse exists. If $\mathbf{C} = \mathbf{I}$ and $\mathbf{B} = \mathbf{H}^t$, then we are back to the regularized least-squares solution of (15.156). In any case, $\mathbf{C}$ controls the form of the regularization and η controls the amount.

The algorithm of (15.234) can be written as

$$\hat{\boldsymbol{\theta}}^{(k+1)} = \boldsymbol{\Omega}\hat{\boldsymbol{\theta}}^{(k)} + \mathbf{Bg}\,, \tag{15.237}$$

where

$$\boldsymbol{\Omega} = \mathbf{I} - \eta\mathbf{C} - \mathbf{BH}\,. \tag{15.238}$$

Since (15.237) has the form of a general linear (more precisely, affine) mapping, any linear iterative algorithm with a fixed update rule can be put in this form with proper choice of $\boldsymbol{\Omega}$ and $\mathbf{B}$.

The only possible remaining generalization is to let the update rule vary with iteration number. For example, the regularizing constant η or the accelerating parameter α could vary with k. The general form of the algorithm then becomes

$$\hat{\boldsymbol{\theta}}^{(k+1)} = \boldsymbol{\Omega}^{(k)}\hat{\boldsymbol{\theta}}^{(k)} + \mathbf{B}^{(k)}\mathbf{g}\,. \tag{15.239}$$

This form can encompass algorithms, called *row-action methods*, that use only a subset of the data at each iteration. We have already encountered one important row-action method in Chap. 1, where we introduced the Gauss-Seidel or ART method in the context of pseudoinverse calculations. For a detailed survey of row-action methods, see Censor (1981).

15.4.2 Noise propagation in linear algorithms

Since (15.239) describes the most general linear algorithm, we shall now use this equation to discuss the evolution of the covariance matrix with iteration number, a process sometimes referred to as *noise propagation*.

The first step is to find an expression for the evolution of the mean vector. If we take a conditional average of both sides of (15.239) for a fixed object $\mathbf{f}$, the result is

$$\langle\hat{\boldsymbol{\theta}}^{(k+1)}\rangle_{\mathbf{g}|\mathbf{f}} = \boldsymbol{\Omega}^{(k)}\langle\hat{\boldsymbol{\theta}}^{(k)}\rangle_{\mathbf{g}|\mathbf{f}} + \mathbf{B}^{(k)}\boldsymbol{\mathcal{H}}\mathbf{f}\,, \tag{15.240}$$

where, as discussed in Sec. 15.2.6, $\langle\mathbf{g}\rangle_{\mathbf{g}|\mathbf{f}} = \boldsymbol{\mathcal{H}}\mathbf{f}$. In practice (or at least in simulation studies), this mean vector can be found as a function of k simply by running the iterative algorithm on a noise-free data set.

Subtraction of (15.240) from (15.239) yields

$$\Delta\hat{\boldsymbol{\theta}}^{(k+1)} = \boldsymbol{\Omega}^{(k)}\Delta\hat{\boldsymbol{\theta}}^{(k)} + \mathbf{B}^{(k)}\mathbf{n}\,, \tag{15.241}$$

where $\mathbf{n} \equiv \mathbf{g} - \boldsymbol{\mathcal{H}}\mathbf{f}$, and $\Delta\hat{\boldsymbol{\theta}}^{(k)}$ indicates the deviation of the estimate from its mean.

If the initial estimate $\hat{\boldsymbol{\theta}}^{(0)}$ is nonrandom, the deviation after the first iteration is

$$\Delta\hat{\boldsymbol{\theta}}^{(1)} = \boldsymbol{\Omega}^{(0)}\Delta\hat{\boldsymbol{\theta}}^{(0)} + \mathbf{B}^{(0)}\mathbf{n} = \mathbf{B}^{(0)}\mathbf{n}\,. \tag{15.242}$$

The next two iterations yield

$$\Delta\hat{\boldsymbol{\theta}}^{(2)} = \boldsymbol{\Omega}^{(1)}\Delta\hat{\boldsymbol{\theta}}^{(1)} + \mathbf{B}^{(1)}\mathbf{n} = \boldsymbol{\Omega}^{(1)}\mathbf{B}^{(0)}\mathbf{n} + \mathbf{B}^{(1)}\mathbf{n}. \tag{15.243}$$

$$\Delta\hat{\boldsymbol{\theta}}^{(3)} = \boldsymbol{\Omega}^{(2)}\Delta\hat{\boldsymbol{\theta}}^{(2)} + \mathbf{B}^{(2)}\mathbf{n} = \boldsymbol{\Omega}^{(2)}\mathbf{B}^{(1)}\mathbf{n} + \boldsymbol{\Omega}^{(2)}\boldsymbol{\Omega}^{(1)}\mathbf{B}^{(0)}\mathbf{n} + \mathbf{B}^{(2)}\mathbf{n}. \tag{15.244}$$

By extension (or induction), the general form is

$$\Delta\hat{\boldsymbol{\theta}}^{(k)} = \mathbf{U}^{(k)}\mathbf{n}, \tag{15.245}$$

where

$$\mathbf{U}^{(k+1)} = \mathbf{B}^{(k)} + \boldsymbol{\Omega}^{(k)}\mathbf{U}^{(k)}. \tag{15.246}$$

With the initial condition $\mathbf{U}^{(1)} = \mathbf{B}^{(0)}$, (15.246) constitutes an iterative algorithm for the matrix $\mathbf{U}^{(k)}$ that relates the deviation $\Delta\hat{\boldsymbol{\theta}}^{(k)}$ back to the noise vector $\mathbf{n}$. Even with large images, the algorithm is feasible in practice since it requires only a single matrix-matrix multiply and a matrix addition at each step.

If we have computed $\mathbf{U}^{(k)}$ by (15.246), we can immediately determine the conditional covariance matrix for $\hat{\boldsymbol{\theta}}^{(k)}$. From (8.50) we know how to transform a covariance matrix through a linear transformation, so we have

$$\mathbf{K}_{\hat{\boldsymbol{\theta}}^{(k)}|\mathbf{f}} = \mathbf{U}^{(k)}\mathbf{K}_{\mathbf{n}}\mathbf{U}^{(k)\dagger}. \tag{15.247}$$

For i.i.d. noise, [*cf.* (15.107)], we have

$$\mathbf{K}_{\hat{\boldsymbol{\theta}}^{(k)}|\mathbf{f}} = \sigma^2\,\mathbf{U}^{(k)}\,\mathbf{U}^{(k)\dagger}. \tag{15.248}$$

The covariance expression (15.248) must be modified if the data noise is Poisson. By comparison to (15.120), we see that

$$\left[\mathbf{K}_{\hat{\boldsymbol{\theta}}^{(k)}|\mathbf{f}}\right]_{nn'} = \sum_{m=1}^{M} [\boldsymbol{\mathcal{H}}\mathbf{f}]_m\, U_{nm}^{(k)} \left[U_{n'm}^{(k)}\right]^*. \tag{15.249}$$

Conditional PDFs and optimization of algorithms If the noise in the data is normally distributed, then any linear transformation leaves it normal, so (15.240) and (15.247) are sufficient to specify fully the statistics of the linear reconstruction at each iteration. For Poisson noise, the reconstruction is not exactly normal, but as we argued in Sec. 15.2.6, the normal law will often be an excellent approximation. Thus in either case we have good knowledge of the conditional statistics for any linear algorithm and number of iterations. This statistical description can be used in conjunction with any of the task-based figures of merit developed in Chap. 14 to optimize the number of iterations and other free parameters of the algorithm.

15.4.3 Search algorithms for functional minimization

As we saw in Sec. 15.3, image reconstruction is often formulated as minimization of some functional $Q(\boldsymbol{\theta}, \mathbf{g})$. A host of methods for performing this minimization can be found in the literature on optimization, and only a brief overview will be given here. For more details, see Golub and van Loan (1989), Scales (1985), Pierre (1986) or Gill *et al.* (1981).

A general approach to functional minimization is to choose a search direction in the reconstruction space, take a step in that direction, and then repeat the process iteratively. The iteration rule for algorithms in this class is thus

$$\hat{\boldsymbol{\theta}}^{(k+1)} = \hat{\boldsymbol{\theta}}^{(k)} + h^{(k)}\mathbf{s}^{(k)}, \qquad (15.250)$$

where the vector $\mathbf{s}^{(k)}$ specifies the search direction for the kth iteration, and $h^{(k)}$ is the corresponding step size. Each algorithm is specified by the initial estimate, a procedure for setting the sequence of directions $\{\mathbf{s}^{(k)}\}$ and a rule for determining the step sizes $\{h^{(k)}\}$.

The most common initial estimates are a uniform field (all components equal) and the discrete backprojection $\mathbf{H}^t\mathbf{g}$. If enough iterations are used and $Q(\boldsymbol{\theta}, \mathbf{g})$ has a unique minimum, the final solution does not depend on the initialization, but with a finite number of iterations the solution is biased towards the initial estimate, which can thus be regarded as a kind of regularization or prior knowledge.

Common choices for $\{\mathbf{s}^{(k)}\}$ lead to algorithms known as *iterative coordinate descent*, *steepest descent* and *conjugate gradient*, all of which will be summarized below, but first we look at ways of choosing the step size.

One-dimensional minimization algorithms Since the objective of the search is to find a minimum of $Q(\boldsymbol{\theta}, \mathbf{g})$, it is usual (though not mandatory) to choose $h^{(k)}$ at each step so that $Q(\hat{\boldsymbol{\theta}}^{(k+1)}, \mathbf{g}) \leq Q(\hat{\boldsymbol{\theta}}^{(k)}, \mathbf{g})$. Colloquially, the step is in a downhill direction. For example, we could choose a step size h_0 and take each $h^{(k)}$ to be either h_0 or $-h_0$, depending on which direction is downhill. If neither h_0 nor $-h_0$ results in a reduction in the functional, no change is made. If few changes are made in the course of many iterations, we can change h_0, for example by halving it, and repeat the process.

Rather than just moving in the direction of the minimum, one can also attempt to come as close as possible to the minimum along each search direction before moving on to a new direction. Many methods for finding this minimum are available. The brute-force method is just to take many small steps in the downhill direction until you start uphill again. More sophisticated methods such as the *golden-section search* (Pierre, 1986) vary the step size in an attempt to bracket the minimum.

Other methods such as *Newton-Raphson* (Pierre, 1986; Scales, 1985) attempt to estimate the location of the minimum from the gradient and/or Hessian (curvature) of the functional. All of these methods approximate the functional at its current estimate by means of a multivariate Taylor series as discussed in Sec. A.10.2. From (15.250) and (A.181), we can write

$$Q(\hat{\boldsymbol{\theta}}^{(k+1)}, \mathbf{g}) = Q(\hat{\boldsymbol{\theta}}^{(k)}, \mathbf{g}) + h^{(k)}\mathbf{s}^{(k)t}\nabla Q^{(k)} + \tfrac{1}{2}\left[h^{(k)}\right]^2\mathbf{s}^{(k)t}\mathbf{A}^{(k)}\mathbf{s}^{(k)} + \ldots, \qquad (15.251)$$

where $\nabla Q^{(k)}$ and $\mathbf{A}^{(k)}$ are, respectively, the gradient vector and Hessian matrix evaluated at $\boldsymbol{\theta} = \hat{\boldsymbol{\theta}}^{(k)}$; in the notation of App. A,

$$\nabla Q^{(k)} = \left[\frac{\partial}{\partial \boldsymbol{\theta}} Q(\boldsymbol{\theta}, \mathbf{g})\right]_{\boldsymbol{\theta}=\hat{\boldsymbol{\theta}}^{(k)}}, \qquad \mathbf{A}^{(k)} = \left[\frac{\partial^2}{\partial \boldsymbol{\theta} \partial \boldsymbol{\theta}^t} Q(\boldsymbol{\theta}, \mathbf{g})\right]_{\boldsymbol{\theta}=\hat{\boldsymbol{\theta}}^{(k)}}. \qquad (15.252)$$

If we truncate the Taylor series at the quadratic term, the minimum of $Q(\hat{\boldsymbol{\theta}}^{(k+1)}, \mathbf{g})$ in direction $\mathbf{s}^{(k)}$ occurs when

$$h^{(k)} = -\frac{\mathbf{s}^{(k)t}\nabla Q^{(k)}}{\mathbf{s}^{(k)t}\mathbf{A}^{(k)}\mathbf{s}^{(k)}} \,. \tag{15.253}$$

This is the univariate Newton-Raphson formula, and in the local quadratic approximation to the functional it provides a way to reach the minimum in direction $\mathbf{s}^{(k)}$ in a single step. If the functional is not locally quadratic, then (15.253) may still be a reasonable choice when the gradient and Hessian can be computed. It is easy to compute the gradient and Hessian for some functionals, such as the Tikhonov-regularized least-squares one, but difficult or impossible for others, so (15.253) and its variants are not universally applicable.

Iterative coordinate descent When we adopt an approximate object representation $f_a(\mathbf{r})$ in the form of (15.6), the functions $\phi_n(\mathbf{r})$ are the basis vectors in the representation space, and the coefficients θ_n specify the components of the vector $\mathbf{f}_a$ in this basis. Thus, if we vary one coefficient at a time, we are sequentially moving along the different coordinate axes in representation space. Algorithms that use the coordinate axes as the search directions and that reduce the functional at each step are known generically as *iterative coordinate descent.* Their advantage is that no computation is needed to determine the search directions, and in particular no knowledge of the gradient or Hessian is required for this purpose.

If iteration k varies the component n_k of the estimate, then the update has the form

$$\left[\hat{\boldsymbol{\theta}}^{(k+1)}\right]_{n_k} = \left[\hat{\boldsymbol{\theta}}^{(k)}\right]_{n_k} + h^{(k)}\,, \tag{15.254}$$

and all other components are left unchanged at this iteration. If the estimate has N components, the index n_k should cycle through all N of them in the course of N iterations. The sequence of components $\{n_k\}$ is called the *control sequence.* It can be some natural order such as lexicographic ordering of pixels, or it can assure that adjacent pixels are not altered in immediate succession, or the order can even be chosen randomly.

Any of the one-dimensional minimization techniques mentioned above can be used to determine the step size $h^{(k)}$. For example, the step can be fixed at each iteration as $\pm h_0$, whichever is downhill. Alternatively, we can make a fixed fractional change in the component at each step, so that

$$\left[\hat{\boldsymbol{\theta}}^{(k+1)}\right]_{n_k} = \left(1 \pm \alpha^{(k)}\right)\left[\hat{\boldsymbol{\theta}}^{(k)}\right]_{n_k}\,, \tag{15.255}$$

with $0 < \alpha^{(k)} < 1$ and the sign chosen so that the step is downhill. Again, if neither sign results in a downhill step, then no change is made, and the size of $\alpha^{(k)}$ can be reduced when this happens for many components. Both of these coordinate-descent algorithms are conceptually simple and easy to code, requiring only the ability to compute the functional rather than its gradient or Hessian; both will find the minimum if it is unique and computational time is not a limitation.

If we actually find the minimum in one coordinate direction (for example by Newton-Raphson) before moving on to the next direction, the algorithm is called *iterative conditional modes.* This term arises from the interpretation of $Q(\boldsymbol{\theta}, \mathbf{g})$ as the negative logarithm of a posterior (see Sec. 15.3.1), so that minimizing it is equivalent to maximizing the probability density, $\mathrm{pr}(\boldsymbol{\theta}|\mathbf{g})$. Similarly, minimizing $Q(\boldsymbol{\theta}, \mathbf{g})$ along the nth coordinate axis is the same as maximizing $\mathrm{pr}(\theta_n|\mathbf{g}, \{\theta_j, j \neq n\})$. Since the peak of a probability density function is called its *mode*, all coordinate-descent methods that minimize the functional individually for each direction iteratively determine this conditional mode.

Steepest descent The coordinate-descent algorithms are usually slow since most directions chosen do not point very close to the minimum. Another general approach is try to choose the direction optimally at each iteration. For example, one might choose to go downhill as steeply as possible. Since the direction in which a scalar functional changes most rapidly is the direction of its gradient, the new search direction in this approach is taken as the negative of the local gradient. This method is called *steepest descent* or *gradient descent.* Typically it is combined with one of the search methods mentioned above so the actual minimum in the gradient direction is found or estimated.

Steepest descent is particularly simple when it is applied to a functional that is either exactly or approximately quadratic. Dropping the dependence on $\mathbf{g}$ for notational convenience, we consider a general quadratic functional of the form

$$Q(\boldsymbol{\theta}) = \tfrac{1}{2}\boldsymbol{\theta}^t\mathbf{A}\boldsymbol{\theta} - \mathbf{b}^t\boldsymbol{\theta}\,. \tag{15.256}$$

This form could represent a least-squares problem with a quadratic regularizer, or it could be the local quadratic approximation to a more general functional as in (15.251).

The gradient of $Q(\boldsymbol{\theta})$ is given, from (A.126) and (A.127), as

$$\nabla Q(\boldsymbol{\theta}) = \mathbf{A}\boldsymbol{\theta} - \mathbf{b}\,. \tag{15.257}$$

Finding the minimum of $Q(\boldsymbol{\theta})$ requires setting the gradient to zero, which we see is equivalent to solving the set of linear equations $\mathbf{A}\boldsymbol{\theta} = \mathbf{b}$. If $\mathbf{A}$ is positive-definite (hence nonsingular), then $Q(\boldsymbol{\theta})$ has a unique minimum at $\boldsymbol{\theta} = \mathbf{A}^{-1}\mathbf{b}$, but in large problems it is not feasible to compute this inverse even though it exists. In nonsingular quadratic problems, therefore, steepest-descent is a way of finding $\mathbf{A}^{-1}\mathbf{b}$ without knowing $\mathbf{A}^{-1}$.

If we have a current estimate $\hat{\boldsymbol{\theta}}^{(k)}$, the next search direction is parallel to $-\nabla Q(\hat{\boldsymbol{\theta}}^{(k)})$, which we called $-\nabla Q^{(k)}$ earlier. Thus

$$\nabla Q^{(k)} = \mathbf{A}\hat{\boldsymbol{\theta}}^{(k)} - \mathbf{b}\,, \tag{15.258}$$

which is just the residual vector $\mathbf{r}^{(k)}$ specifying the difference between $\mathbf{A}\boldsymbol{\theta}$ and $\mathbf{b}$ at the current estimate. The descent direction is the negative of the residual, and finding the overall minimum of $Q(\boldsymbol{\theta})$ is equivalent to driving the residual to zero.

Conjugacy Though steepest descent sounds appealing, it may converge slowly when the functional has long regions (canyons) where it varies slowly in one direction but rapidly in another (see Fig. 15.4). An alternative approach which may converge more rapidly is the *conjugate-gradient* method, described below.

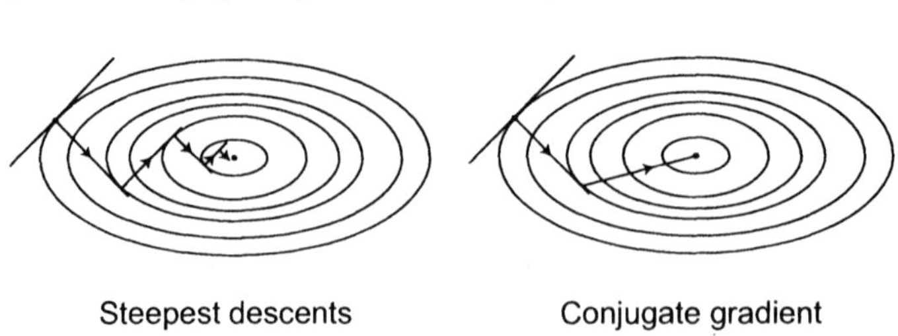

Fig. 15.4 (a) Illustration of a situation where steepest descent converges very slowly. (b) Convergence of the conjugate-gradient algorithm for the same situation.

To explain the conjugate-gradient method, we must first explain the concept of *conjugacy*. Let $\mathbf{A}$ be a real $N \times N$ Hermitian matrix of rank R. Two real $N \times 1$ vectors $\mathbf{p}_j$ and $\mathbf{p}_k$ are said to be $\mathbf{A}$-conjugate (or conjugate with respect to $\mathbf{A}$) if

$$\mathbf{p}_j^t \mathbf{A} \mathbf{p}_k = d_j \, \delta_{jk} \, . \tag{15.259}$$

One possible choice for the set $\{\mathbf{p}_j, j = 1, ..., R\}$ is the set of eigenvectors of $\mathbf{A}$, but there are many other possibilities. As shown in Sec. 1.4.4, the eigenvectors are mutually orthogonal (or can be chosen to be via Gram-Schmidt if there are degeneracies). Conjugate vectors can be shown to be linearly independent (Scales, 1985), but they are not necessarily orthogonal.

Conjugacy is formally related to the idea of prewhitening, introduced in a statistical context in Sec. 8.1.6. Here we do not need any particular statistical interpretation; we simply define[7]

$$\tilde{\mathbf{p}}_j = \mathbf{A}^{\frac{1}{2}} \mathbf{p}_j \, , \tag{15.260}$$

where $\mathbf{A}^{\frac{1}{2}}$ is the square root of the Hermitian matrix $\mathbf{A}$ (see Sec. A.8.3). With this transformation and the definition of the adjoint in (1.39), the conjugacy condition (15.259) can be written as

$$\mathbf{p}_j^t \mathbf{A} \mathbf{p}_k = (\mathbf{A}^{\frac{1}{2}} \mathbf{p}_j)^t (\mathbf{A}^{\frac{1}{2}} \mathbf{p}_k) = \tilde{\mathbf{p}}_j^t \tilde{\mathbf{p}}_k = d_j \, \delta_{jk} \, . \tag{15.261}$$

Thus the prewhitened vectors $\{\mathbf{A}^{\frac{1}{2}} \mathbf{p}_j, j = 1, ..., R\}$ form an orthogonal set, and they can be normalized to yield the orthonormal set, $\{d_j^{-\frac{1}{2}} \mathbf{A}^{\frac{1}{2}} \mathbf{p}_j, j = 1, ..., R\}$.

Conjugate vectors and minimization of a quadratic Let us return to the quadratic problem of (15.256) and assume for simplicity that $\mathbf{A}$ is Hermitian and positive-definite. As applied to this problem, conjugate-gradient algorithms all have the general iteration rule

$$\hat{\boldsymbol{\theta}}^{(k+1)} = \hat{\boldsymbol{\theta}}^{(k)} + h^{(k)} \mathbf{p}_k \, , \tag{15.262}$$

where each $\mathbf{p}_k$ is conjugate (with respect to $\mathbf{A}$) to all previous $\mathbf{p}_j$, $j < k$. For an exact linear search, $h^{(k)}$ is chosen by (15.253) to minimize $Q(\hat{\boldsymbol{\theta}}^{(k+1)})$:

$$h^{(k)} = -\frac{\mathbf{p}_k^t \nabla Q^{(k)}}{d_k} \, . \tag{15.263}$$

The advantage of choosing conjugate vectors to define the search directions can be seen by applying the prewhitening transformation $\mathbf{A}^{\frac{1}{2}}$ as in (15.260). If we define

$$\tilde{\boldsymbol{\theta}} = \mathbf{A}^{\frac{1}{2}} \boldsymbol{\theta} \qquad \text{and} \qquad \tilde{\mathbf{b}} = \mathbf{A}^{-\frac{1}{2}} \mathbf{b} \, , \tag{15.264}$$

then the functional to be minimized becomes

$$\tilde{Q}(\tilde{\boldsymbol{\theta}}) = \tilde{\mathbf{b}}^t \tilde{\boldsymbol{\theta}} + \tfrac{1}{2} \tilde{\boldsymbol{\theta}}^t \tilde{\boldsymbol{\theta}} \, . \tag{15.265}$$

[7]When $Q(\boldsymbol{\theta}, \mathbf{g})$ is the negative of a multivariate Gaussian log-likelihood, then $\mathbf{A} = \mathbf{K}^{-1}$, where $\mathbf{K}$ is the covariance matrix. In that case, (15.249) is identical to what we called prewhitening in Secs. 8.1.6 and 13.2.18.

Thus the transformation converts the isocontours of the functional into N-dimensional spheres in the prewhitened space. It follows from (15.261) that any algorithm of the form (15.262) will make a succession of minimizations along mutually orthogonal directions in the prewhitened space if the vectors $\{\mathbf{p}_k\}$ satisfy the conjugacy condition (15.259). This search will reach the overall minimum in at most N steps; zig-zagging along the canyon as in Fig. 15.4 does not occur after prewhitening. Of course, a steepest-descent algorithm would reach the minimum in one step after prewhitening, but we would need to know the operator $\mathbf{A}^{\frac{1}{2}}$ in order to find the steepest-descent direction. For large problems we must assume that $\mathbf{A}^{\frac{1}{2}}$ and $\mathbf{A}^{-1}$ are not accessible.

Generating conjugate vectors To generate a sequence of conjugate vectors, we can in principle generate a sequence of orthogonal vectors in the prewhitened space and then apply the operator $\mathbf{A}^{-\frac{1}{2}}$; fortunately, as we shall see, we can do this without actually knowing $\mathbf{A}^{-\frac{1}{2}}$.

To set notation, let $\tilde{\mathbb{U}}^{(k)}$ be the kD space spanned by $\{\tilde{\mathbf{p}}_j, 0 \leq j \leq k-1\}$, and let $\mathbb{U}^{(k)}$ be the kD space spanned by $\{\mathbf{p}_j, 0 \leq j \leq k-1\}$. Since $\mathbf{A}^{\frac{1}{2}}$ is square and nonsingular, there is a 1:1 correspondence between these two spaces.

If we know the vectors $\{\tilde{\mathbf{p}}_j, 0 \leq j \leq k-1\}$, we can generate a new vector $\tilde{\mathbf{p}}_k$ orthogonal to all of them by choosing a vector $\tilde{\mathbf{q}}_k$, which is arbitrary except that it cannot lie entirely in $\tilde{\mathbb{U}}^{(k)}$, and projecting it onto the orthogonal complement of $\tilde{\mathbb{U}}^{(k)}$, denoted $\tilde{\mathbb{U}}_{\perp}^{(k)}$. Since the set $\{d_j^{-\frac{1}{2}}\mathbf{A}^{\frac{1}{2}}\mathbf{p}_j,\ 0 \leq j \leq k-1\}$ is an orthonormal basis for $\tilde{\mathbb{U}}^{(k)}$, the projector onto $\tilde{\mathbb{U}}_{\perp}^{(k)}$ has the form [*cf.* (1.60)]

$$\tilde{\mathbf{p}}_k \equiv \tilde{\mathcal{P}}_{\perp}^{(k)}\hat{\mathbf{q}}_k = \left[\mathbf{I} - \sum_{j=0}^{k-1}\frac{1}{d_j}\tilde{\mathbf{p}}_j\tilde{\mathbf{p}}_j^t\right]\tilde{\mathbf{q}}_k$$

$$= \tilde{\mathbf{q}}_k - \sum_{j=0}^{k-1}\frac{1}{d_j}(\mathbf{A}^{\frac{1}{2}}\mathbf{p}_j)(\mathbf{A}^{\frac{1}{2}}\mathbf{p}_j)^t\tilde{\mathbf{q}}_k\,. \tag{15.266}$$

Letting $\mathbf{q}_k = \mathbf{A}^{-\frac{1}{2}}\tilde{\mathbf{q}}_k$ and multiplying (15.266) by $\mathbf{A}^{-\frac{1}{2}}$, we find

$$\mathbf{p}_k = \mathbf{q}_k - \sum_{j=0}^{k-1}\frac{\mathbf{p}_j^t\mathbf{A}\mathbf{q}_k}{d_j}\mathbf{p}_j\,. \tag{15.267}$$

Note that $\mathbf{A}^{\frac{1}{2}}$ no longer appears in this expression.

To show explicitly that this $\mathbf{p}_k$ is conjugate to the previous $\mathbf{p}_i (i < k)$, we take the scalar product of $\mathbf{p}_k$ with $\mathbf{A}\mathbf{p}_i$, yielding

$$(\mathbf{p}_k, \mathbf{A}\mathbf{p}_i) = \mathbf{q}_k^t\mathbf{A}\mathbf{p}_i - \sum_{j=0}^{k-1}\frac{\mathbf{p}_j^t\mathbf{A}\mathbf{q}_k}{d_j}(\mathbf{p}_j^t\mathbf{A}\mathbf{p}_i) = \mathbf{q}_k^t\mathbf{A}\mathbf{p}_i - \mathbf{q}_k^t\mathbf{A}\mathbf{p}_i = 0\,, \tag{15.268}$$

where we have used the conjugacy condition to set $\mathbf{p}_j^t\mathbf{A}\mathbf{p}_i$ to zero for $j \neq i$ and $i, j \leq k-1$. Thus, with any choice of $\mathbf{q}_k$, $\mathbf{p}_k$ as defined by (15.267) can be used as the next conjugate vector, provided only that $\mathbf{p}_k$ is nonzero.

Figure 15.5 provides a geometrical interpretation of (15.267). Since each $\mathbf{p}_k$ is built from $\mathbf{q}_k$ and the previous $\mathbf{p}_i$, the space $\mathbb{U}^{(k)}$ is spanned either by

$\{\mathbf{p}_j, 0 \leq j \leq k-1\}$ or by $\{\mathbf{q}_j, 0 \leq j \leq k-1\}$. The figure is drawn for $N = 3$ and $k = 2$, so $\mathbb{U}^{(k)}$ is an ordinary 2D plane, but in general it is a hyperplane. The vector $\mathbf{p}_k$ is conjugate (not orthogonal) to this hyperplane.

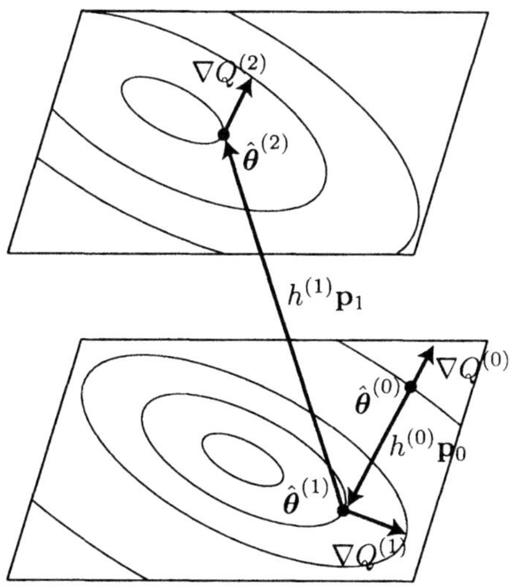

Fig. 15.5 Illustration of the conjugate-gradient algorithm in 3D.

When the sequence of vectors $\{\mathbf{p}_k\}$ is used with the exact linear search algorithm of (15.262) and (15.263), the residual $\mathbf{r}^{(k)}$ is systematically reduced. Recalling that the residual $\mathbf{r}^{(k)}$ is the same thing as the gradient $\nabla Q^{(k)}$, we see from (15.258) and (15.262) that

$$\mathbf{r}^{(k+1)} = \mathbf{r}^{(k)} + h^{(k)}\mathbf{A}\mathbf{p}_k \,. \tag{15.269}$$

In fact, $\mathbf{r}^{(k+1)}$ turns out to be orthogonal to $\mathbb{U}^{(k)}$. We know that $\mathbf{A}\mathbf{p}_k$ is orthogonal to all previous $\mathbf{p}_j$ and hence orthogonal to the subspace $\mathbb{U}^{(k)}$, and a proof by induction can be used to show that $\mathbf{r}^{(k)}$ is orthogonal to $\mathbb{U}^{(k)}$ (see Scales, 1985).

Hestenes-Stiefel algorithm The algorithm implied by (15.267) is not an efficient way of generating conjugate vectors since each term in the sum requires a matrix-vector multiplication, and the number of terms increases with k. Hestenes and Stiefel (1952) realized that a recursive algorithm requiring only one matrix-vector multiplication for each new conjugate vector could be obtained by choosing $\mathbf{q}_k$ to be specifically the steepest-descent direction:

$$\mathbf{q}_k = -\nabla Q^{(k)} = -\mathbf{r}^{(k)} \,. \tag{15.270}$$

The algorithm is initialized by performing the initial search in the steepest-descent direction:

$$\mathbf{p}_0 = -\mathbf{r}^{(0)} \,. \tag{15.271}$$

With these choices and (15.267), the kth conjugate vector is given by

$$\mathbf{p}_k = -\left[\mathbf{I} - \sum_{j=0}^{k-1} \frac{1}{d_j}\mathbf{p}_j(\mathbf{A}\mathbf{p}_j)^t\right]\mathbf{r}^{(k)} = -\mathbf{r}^{(k)} + \sum_{j=0}^{k-1} \frac{(\mathbf{A}\mathbf{p}_j)^t\mathbf{r}^{(k)}}{d_j}\mathbf{p}_j \,. \tag{15.272}$$

The key observation in the Hestenes-Stiefel approach is that

$$(\mathbf{A}\mathbf{p}_j)^t \mathbf{r}^{(k)} = 0 \qquad \text{if } j < k-1\,, \tag{15.273}$$

which follows from (15.269) since $\mathbf{A}\mathbf{p}_j \propto \mathbf{r}^{(j)} - \mathbf{r}^{(j-1)}$, a vector in $\mathbb{U}^{(j)}$, and we have already noted that successive residuals are orthogonal. Thus only one term in the sum in (15.272) survives, and we have

$$\mathbf{p}_k = -\mathbf{r}^{(k)} + \beta_k \mathbf{p}_{k-1}\,, \tag{15.274}$$

where

$$\beta_k = \frac{(\mathbf{A}\mathbf{p}_{k-1})^t \mathbf{r}^{(k)}}{d_{k-1}}\,. \tag{15.275}$$

Hence computation of $\mathbf{p}_k$ requires knowledge of only the current residual and the immediately preceding conjugate vector, $\mathbf{p}_{k-1}$.

Quadratic termination An often-cited advantage of conjugate-gradient algorithms is their *quadratic-termination property*: If the functional is strictly quadratic, the true minimum will be reached in at most N steps. This property follows from (15.269); the residual is reduced component-by-component at each iteration and must be zero after N iterations.

As applied to image reconstruction, however, where N can be of order 10^5 or 10^6, this property is not very useful. We can virtually never afford the computing resources to run this many iterations of any algorithm. Typically, reconstruction algorithms are run for only 10–100 iterations.

Conjugacy and pseudoinverses Conjugate vectors are also related to pseudoinverses of $\mathbf{A}$ (Hestenes, 1975). Suppose $\mathbf{A}$ is an $N \times N$ matrix of rank R. If we have a set of conjugate vectors, we can form the pseudoinverse of $\mathbf{A}$ via

$$\mathbf{A}^+ = \sum_{j=1}^{R} \frac{1}{d_j} \mathbf{p}_j \mathbf{p}_j^t\,. \tag{15.276}$$

It follows from (15.259) that $\mathbf{A}^+$ constructed this way is a (1,2)-pseudoinverse (as defined in Sec. 1.6.1), and Hestenes shows that it is a Moore-Penrose pseudoinverse if and only if all of the $\mathbf{p}_j$ are orthogonal to the null space of $\mathbf{A}$. Thus any algorithm for generating a sequence of conjugate vectors also generates a pseudoinverse of $\mathbf{A}$.

Moreover, if we have an arbitrary $N \times K$ matrix $\mathbf{M}$ of rank R and wish to generate its pseudoinverse, we can simply form the $K \times K$ square matrix $\mathbf{A} = \mathbf{M}^t\mathbf{M}$ and generate the $\mathbf{A}$-conjugate $K \times 1$ vectors $\{\mathbf{p}_j, j = 1, ..., R\}$. The pseudoinverse of $\mathbf{M}$ is then given by (Hestenes, 1975)

$$\mathbf{M}^+ = \sum_{j=1}^{R} \frac{1}{d_j} \mathbf{p}_j (\mathbf{M}\mathbf{p})_j^t\,. \tag{15.277}$$

This result can be compared to (1.131); the structure is the same but we do not need to find the singular vectors and singular values in order to use (15.277).

15.4.4 Nonlinear constraints and fixed-point iterations

Linear algorithms or search methods can always be used to minimize the functional $Q(\boldsymbol{\theta}, \mathbf{g})$ if there are no constraints on the solution, but often we want to impose some nonlinear constraints, the most important of which is that the solution cannot go negative. In this section we first discuss some simple modifications of linear algorithms to incorporate such constraints. Then we introduce a broad class of algorithms that will often converge to a solution satisfying the constraints.

Modified linear algorithms As a starting point, consider the form of linear algorithm given in (15.233), which is general enough to encompass the Jabobi, Gauss-Seidel and Landweber algorithms, among others. As we discussed in Sec. 15.4.1, when this algorithm converges it ensures that $\mathbf{g} = \mathbf{H}\hat{\boldsymbol{\theta}}^{(k)}$, but it places no other constraints on the solution. It can be very useful to modify the algorithm so as to enforce prior knowledge such as positivity or finite support.

One way to enforce a positivity constraint is to clip off negative values at each iteration. Then (15.233) is modified to

$$\hat{\boldsymbol{\theta}}^{(k+1)} = \mathbf{P}_{+}\{\hat{\boldsymbol{\theta}}^{(k)} + \mathbf{B}[\mathbf{g} - \mathbf{H}\hat{\boldsymbol{\theta}}^{(k)}]\}, \tag{15.278}$$

where $\mathbf{P}_{+}$ is the positivity operator defined in (15.186). At each iteration the estimate is nonnegative, so if the algorithm converges it must do so by finding a nonnegative solution such that $\mathbf{B}[\mathbf{g} - \mathbf{H}\hat{\boldsymbol{\theta}}^{(k)}] = 0$. If $\mathbf{B}$ has no null functions in data space, that means that $\mathbf{g} = \mathbf{H}\hat{\boldsymbol{\theta}}^{(k)}$.

Other constraint operators can be used in addition to or in place of the operator $\mathbf{P}_{+}$ in (15.278). For example, if we are working in a pixel representation and we know that the object is zero outside a support region $\mathbf{S}$, we can use an operator $\mathbf{P}_{\mathbf{S}}$ that sets pixel values to zero outside this region.

A general framework for successively enforcing constraints such as positivity and support is the method of projections onto convex sets, to be discussed in Sec. 15.4.5.

Fixed-point iterations As we saw in Sec. 15.3.4, it is often possible to work from the KKT conditions to an implicit equation that must be satisfied when the functional $Q(\boldsymbol{\theta}, \mathbf{g})$ is minimized; an example is given in (15.194). The general form of this implicit equation is

$$\hat{\boldsymbol{\theta}} = \mathbf{T}\{\hat{\boldsymbol{\theta}}, \mathbf{g}\}, \tag{15.279}$$

where $\mathbf{T}\{\hat{\boldsymbol{\theta}}, \mathbf{g}\}$ is some vector-valued (and usually nonlinear) functional of its arguments.

To convert (15.279) to an iterative algorithm, we can define a sequence of estimates $\hat{\boldsymbol{\theta}}^{(k)}$ by a *fixed-point iteration* of the form

$$\hat{\boldsymbol{\theta}}^{(k+1)} = \mathbf{T}\{\hat{\boldsymbol{\theta}}^{(k)}, \mathbf{g}\}. \tag{15.280}$$

It is not obvious that this sequence will converge, but if it does the convergent point must satisfy (15.279); this point is called the fixed point of the nonlinear equation.

A sufficient condition for the convergence of a fixed-point iteration is provided by the *contraction-mapping theorem* generally attributed to Banach (see Stakgold, 1979; Kreysig, 1978). For fixed $\mathbf{g}$, the mapping $\mathbf{T}\{\boldsymbol{\theta}, \mathbf{g}\}$ is called a *contraction* if

$$||\mathbf{T}\{\boldsymbol{\theta}_1, \mathbf{g}\} - \mathbf{T}\{\boldsymbol{\theta}_2, \mathbf{g}\}||^2 < ||\boldsymbol{\theta}_1 - \boldsymbol{\theta}_2||^2, \tag{15.281}$$

for any vectors $\boldsymbol{\theta}_1$ and $\boldsymbol{\theta}_2$. That is, the effect of the mapping is to bring two arbitrary points closer together. The contraction-mapping theorem states that the algorithm (15.280) will converge to the fixed point for all starting points if $\mathbf{T}\{\boldsymbol{\theta}, \mathbf{g}\}$ is a contraction. A graphical way of understanding this theorem is given in Fig. 15.6. Weaker versions of the theorem that require the operator to be a contraction only within some local region are given by Kreysig (1978).

An example of a fixed-point iteration is (15.278). For the special case where $\mathbf{B} = \mathbf{H}^\dagger$, it can be shown (details are left to the reader) that the mapping in that algorithm is a contraction provided $\mu_1 < 2$, where μ_1 is the largest eigenvalue of $\mathbf{H}^\dagger\mathbf{H}$. This is precisely the condition derived in Sec. 1.7.6 for the convergence of the linear Landweber algorithm, and now we see that it is sufficient for the nonlinear version as well.

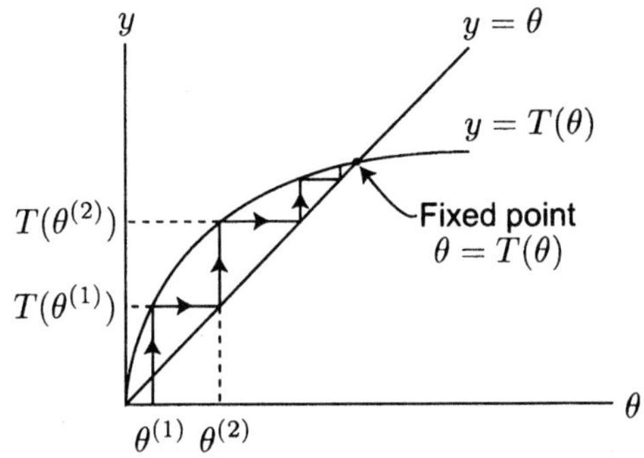

Fig. 15.6 Illustration of a fixed-point algorithm in 1D.

15.4.5 Projections onto convex sets

As we have seen, image reconstruction often seeks to find an image that satisfies certain constraints such as positivity, support, smoothness or agreement with the data. Many of these constraints can be formulated as convex sets (defined more precisely below), and a satisfactory image can be regarded as one that satisfies all of the constraints and hence lies in the intersection of these sets. The *method of projections* (MOP) and more specifically *projections onto convex sets* (POCS) is a very general tool for finding such a satisfactory image.

MOP was originated by John von Neumann (1950) who considered projections onto closed linear manifolds. The extension to convex sets was made by Bregman (1965), and the potential of POCS for signal processing was enunciated by Youla (1978). Applications to imaging were discussed by Youla and Webb (1982). Levi and Stark (1987) extended MOP to nonconvex projections, yielding the *method of generalized projections.* Excellent reviews are given by Sezan (1992), Combettes (1993) and Stark and Sezan (1994).

Convexity A convex set is defined as follows: If $\boldsymbol{\theta}$ and $\boldsymbol{\theta}'$ are members of the set, so is $\alpha\boldsymbol{\theta} + (1-\alpha)\boldsymbol{\theta}'$ for $0 \leq \alpha \leq 1$. The geometric interpretation of this condition is that all points along the line joining $\boldsymbol{\theta}$ and $\boldsymbol{\theta}'$ are members of the set if $\boldsymbol{\theta}$ and $\boldsymbol{\theta}'$ are members (see Fig. 15.7). The set of all nonnegative objects is a convex set, as is the set of all objects with a specified support region.

The operators $\mathbf{P}_+$ and $\mathbf{P}_\mathbf{S}$ defined in Sec. 15.4.4 are projectors onto convex sets. Acting on a $\boldsymbol{\theta}$ that is not a member of the set, they produce the member of the set that is nearest to $\boldsymbol{\theta}$ in the sense of minimizing the $\mathbb{L}_2$ norm $||\mathbf{P}\boldsymbol{\theta} - \boldsymbol{\theta}||$. Projectors onto convex sets are not necessarily projectors onto subspaces as discussed in Sec. 1.3.6. The set of vectors with nonnegative components, for example, is not a linear vector space.

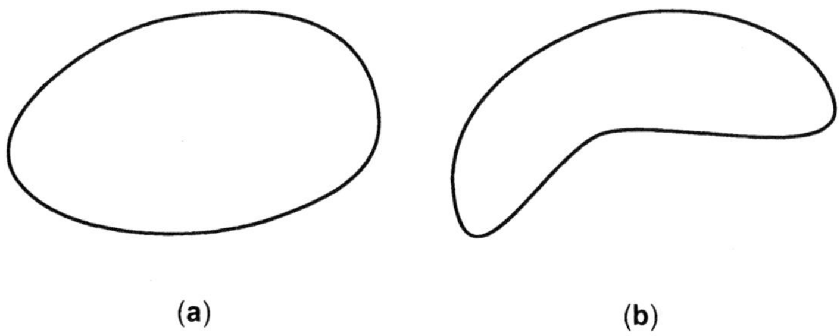

Fig. 15.7 Examples of convex and nonconvex sets.

Besides positivity and support, many other convex sets have been proposed in the imaging literature (Stark and Sezan, 1994). For example, if the object is defined as a transmittance, it can take on values only between 0 and 1, and the upper limit as well as the lower limit defines a convex set. Smoothness can be enforced by setting an upper bound to the norm of some derivative of the image.

Data agreement Agreement with the data also defines a convex set. Strict agreement implies the set of all $\boldsymbol{\theta}$ such that $\mathbf{H}\boldsymbol{\theta} \equiv \mathbf{g}$, so members of the set differ by null functions. Since linear combinations of null functions are null functions, it follows readily that this set is convex.

To discover the form for the projector onto this set, recall from (1.188) that the general exact solution to $\mathbf{H}\boldsymbol{\theta} = \mathbf{g}$ is

$$\boldsymbol{\theta} = \mathbf{H}^+\mathbf{g} + \left[\mathbf{I}_N - \mathbf{H}^+\mathbf{H}\right]\mathbf{y}\,, \tag{15.282}$$

where $\mathbf{y}$ is an arbitrary $N \times 1$ vector and $[\mathbf{I}_N - \mathbf{H}^+\mathbf{H}]\mathbf{y}$ is the component of $\mathbf{y}$ in the null space (if any) of the $M \times N$ matrix $\mathbf{H}$. Projection of an arbitrary $\hat{\boldsymbol{\theta}}$ onto the set of functions of this form is accomplished by replacing the measurement component of $\hat{\boldsymbol{\theta}}$ with $\mathbf{H}^+\mathbf{g}$ and leaving the null component unchanged; in this way we find the vector in the convex set that is closest (in a Euclidean sense) to the original $\hat{\boldsymbol{\theta}}$. Explicitly, if we denote the projector onto the set defined by $\mathbf{H}\boldsymbol{\theta} = \mathbf{g}$ as $\mathbf{P}_\mathbf{g}$, then

$$\mathbf{P}_\mathbf{g}\{\hat{\boldsymbol{\theta}}\} = \mathbf{H}^+\mathbf{g} + \left[\mathbf{I}_N - \mathbf{H}^+\mathbf{H}\right]\hat{\boldsymbol{\theta}}\,. \tag{15.283}$$

We shall illustrate the use of this equation below when we discuss the Gerchberg-Papoulis algorithm.

We can also define the set of all $\boldsymbol{\theta}$ such that $||\mathbf{g} - \mathbf{H}\boldsymbol{\theta}|| \le \epsilon$. To see that this set is convex, note that

$$||\mathbf{g} - \alpha\mathbf{H}\boldsymbol{\theta} - (1-\alpha)\mathbf{H}\boldsymbol{\theta}'|| = ||\alpha(\mathbf{g} - \mathbf{H}\boldsymbol{\theta}) - (1-\alpha)(\mathbf{g} - \mathbf{H}\boldsymbol{\theta}')||$$

$$\le \alpha||\mathbf{g} - \mathbf{H}\boldsymbol{\theta}|| + (1-\alpha)||\mathbf{g} - \mathbf{H}\boldsymbol{\theta}'|| \le \alpha\epsilon + (1-\alpha)\epsilon = \epsilon\,, \tag{15.284}$$

where we have used some properties of the norm, including the triangle inequality (see Sec. 1.1.2).

Another convex set related to data agreement is the set of all $\boldsymbol{\theta}$ such that $(\mathbf{H}\boldsymbol{\theta})_m = g_m$ for a particular m. Derivation of the projectors for these latter two sets will be left as an exercise.

Fundamental theorem of POCS Suppose we have J different convex sets $\mathbf{C}_j$, $j = 1, ..., J$, with corresponding projectors $\mathbf{P}_j, j = 1, ..., J$. Suppose also that the intersection of these sets is not empty. If we apply each of the projectors sequentially (in any order), so that the update rule is

$$\hat{\boldsymbol{\theta}}^{(k+1)} = \mathbf{P}_J \mathbf{P}_{J-1} \cdots \mathbf{P}_2 \mathbf{P}_1 \hat{\boldsymbol{\theta}}^{(k)}, \tag{15.285}$$

then the iteration will converge, regardless of the initialization, to a point in the intersection of the sets (Youla, 1978). This result is illustrated geometrically in Fig. 15.8.

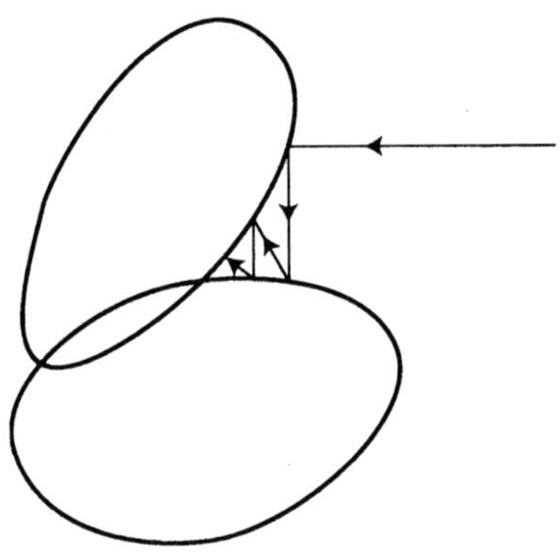

Fig. 15.8 Illustration of successive projections onto convex sets.

Corresponding to the projector $\mathbf{P}_j$, there is a *relaxed* projector defined by

$$\mathbf{P}_j^\lambda = \mathbf{I} + \lambda_j (\mathbf{P}_j - \mathbf{I}), \tag{15.286}$$

where $\lambda_j \geq 0$ and $\mathbf{I}$ is the identity operator. The iteration rule for the relaxed projectors is

$$\hat{\boldsymbol{\theta}}^{(k+1)} = \mathbf{P}_J^\lambda \mathbf{P}_{J-1}^\lambda \cdots \mathbf{P}_2^\lambda \mathbf{P}_1^\lambda \hat{\boldsymbol{\theta}}^{(k)}, \tag{15.287}$$

and this iteration can be shown (Youla, 1978) to converge to a point in the intersection of the sets provided $0 < \lambda_j < 2$ for all j.

Example of POCS: Gerchberg-Papoulis algorithm Suppose we have acquired image data with a diffraction-limited lens and a discrete detector array such as a CCD (charge-coupled device). We know from the discussion in Sec. 9.7 that this system is approximately shift-invariant and bandlimited, and we know how to compute its presampling transfer function $P(\boldsymbol{\rho})$. If the detector array samples finely enough, simple inverse filtering will allow us to estimate accurately the object Fourier components for $\rho < \rho_c$, where ρ_c is the cutoff frequency and $\rho = |\boldsymbol{\rho}|$. Can we also recover components with $\rho > \rho_c$, which lie in the null space of the imaging system? Doing so is often referred to as *superresolution*.

One approach that has had a modicum of success in achieving superresolution is a form of POCS known as the *Gerchberg-Papoulis algorithm* (Gerchberg, 1974;

Papoulis, 1975). It successively enforces data agreement and positivity and support constraints in the hope that the prior information infused by the latter two constraints will help to fill in the null space.

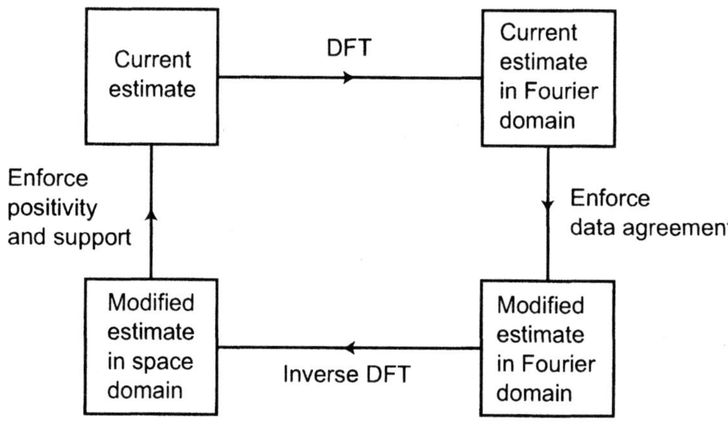

Fig. 15.9 Illustration of the Gerchberg-Papoulis algorithm.

In practice, the Gerchberg-Papoulis algorithm uses a DFT (discrete Fourier transform) of $\boldsymbol{\theta}$ and assumes that the components are known (after inverse filtering) within the system passband. The algorithm, illustrated in Fig. 15.9, is specified by

$$\hat{\boldsymbol{\theta}}^{(k+1)} = \mathbf{P_g P_+ P_S} \hat{\boldsymbol{\theta}}^{(k)}, \tag{15.288}$$

where the three projection operators have been defined above. Agreement with the data is enforced in the discrete Fourier domain by replacing the Fourier components within the band with their known values but retaining the current estimates for out-of-band components. Roughly speaking, out-of-band Fourier components are identified with null functions, and Fourier-domain inverse filtering is essentially the pseudoinverse for shift-invariant systems, so this projection operation is an implementation of (15.283). Then the estimate is transformed back to the (discrete) space domain so that the positivity and support constraints can easily be enforced. The algorithm will converge to a $\hat{\boldsymbol{\theta}}$ satisfying all three constraints if such a solution exists.

Gerchberg-Papoulis and Fourier crosstalk The discussion above glossed over the CD nature of the imaging system and the distinctions between continuous and discrete Fourier transforms. It was also based implicitly on a finite-dimensional object representation with coefficients specified by the $N \times 1$ vector $\boldsymbol{\theta}$. We shall now sketch a treatment based on the infinite-dimensional Fourier-series representation of the object and the Fourier crosstalk matrix. Essential background information will be found in Sec. 7.3.3.

Consider a spacelimited object but a bandlimited imaging system consisting again of a lens and a detector array, and assume that the system is linear and shift-invariant (LSIV) before sampling. The system thus maps a 2D object $f(\mathbf{r})$ into a 2D irradiance distribution on the detector plane.

If the noise is negligible (as it must be for superresolution to work), then the data can be expressed as [*cf.* (7.260)]

$$g_m = \sum_{\mathbf{k}} \psi_{m\mathbf{k}} F_{\mathbf{k}} \,, \tag{15.289}$$

where $F_{\mathbf{k}}$ is the doubly infinite vector of object Fourier coefficients. In vector form, (15.289) is $\mathbf{g} = \mathbf{\Psi F}$. We know from (7.274) that

$$\psi_{m\mathbf{k}} = \exp(2\pi i \boldsymbol{\rho}_{\mathbf{k}} \cdot \mathbf{r}_{dm}) \, P(\boldsymbol{\rho}_{\mathbf{k}}) \,, \tag{15.290}$$

where $\mathbf{r}_{dm}$ is the location of the mth detector element and $P(\boldsymbol{\rho})$ is the presampling transfer function of the overall LSIV system (including the finite size of the detector elements). Since the system is bandlimited, $P(\boldsymbol{\rho}) = 0$ if ρ is greater than some cutoff frequency ρ_c.

It often happens that a useful first step in an inverse problem is to apply an adjoint operator to get back to the object domain. If the object is specified by its Fourier coefficients, the relevant adjoint is $\mathbf{\Psi}^\dagger$; applying $\mathbf{\Psi}^\dagger$ to $\mathbf{g}$ yields

$$\mathbf{\Psi}^\dagger \mathbf{g} = \mathbf{\Psi}^\dagger \mathbf{\Psi F} = \mathbf{BF} \,, \tag{15.291}$$

where $\mathbf{B}$ is the Fourier crosstalk matrix. As discussed in Sec. 7.3, $\mathbf{B}$ is an infinite matrix, but in the present problem all of its elements are zero except within a finite-dimensional submatrix. Moreover, if the pitch of the detector array satisfies the Nyquist condition[8] for sampling the irradiance pattern incident on the detector, this submatrix is strictly diagonal, and its elements are known from (7.276) to be

$$\beta_{\mathbf{kk}} = M |P(\boldsymbol{\rho}_{\mathbf{k}})|^2 \,, \tag{15.292}$$

where M is the total number of detector elements.

With these assumptions, the pseudoinverse of $\mathbf{\Psi}$ consists of division by the nonzero diagonal elements. If we assume that $P(\boldsymbol{\rho}_{\mathbf{k}})$ is not exactly zero at any in-band sampling frequency $\boldsymbol{\rho}_k$, the in-band Fourier coefficients are given by

$$F_{\mathbf{k}} = \mathbf{\Psi}^+ \mathbf{g} = \frac{\left[\mathbf{\Psi}^\dagger \mathbf{g}\right]_{\mathbf{k}}}{\beta_{\mathbf{kk}}} = \frac{P^*(\boldsymbol{\rho}_{\mathbf{k}}) \sum_m g_m \exp(-2\pi i \boldsymbol{\rho}_{\mathbf{k}} \cdot \mathbf{r}_{dm})}{M |P(\boldsymbol{\rho}_{\mathbf{k}})|^2} \,, \qquad |\boldsymbol{\rho}_{\mathbf{k}}| < \rho_c \,. \tag{15.293}$$

To get an infinite-dimensional form of Gerchberg-Papoulis, the projector $\mathbf{P}_{\mathbf{g}}$ defined in (15.283) is replaced by the infinite-dimensional operator $\mathcal{P}_{\mathbf{g}}$ that projects a function $\hat{\mathbf{f}}$ onto the subset of $\mathbb{U}$ for which $\mathcal{H}\mathbf{f} = \mathbf{g}$ or equivalently, $\mathbf{\Psi F} = \mathbf{g}$. By analogy to (15.283), this operator is formally

$$\mathcal{P}_{\mathbf{g}}\{\hat{\mathbf{F}}\} = \mathbf{\Psi}^+ \mathbf{g} + \left[\mathcal{I} - \mathbf{\Psi}^+ \mathbf{\Psi}\right] \hat{\mathbf{F}} \,. \tag{15.294}$$

Operationally, this projector can be implemented at each POCS iteration by replacing in-band Fourier coefficients with the known values from (15.293) and leaving

[8] There are two distinct Nyquist conditions in this problem, one for the object and one for the system, and both can be exactly satisfied simultaneously. The finite spatial support of the object sets the required sampling in the Fourier domain (see Sec. 3.5.4), and the finite bandwidth of the system before sampling sets the required pitch of the CCD to avoid aliasing. We do not need to assume that the object is bandlimited or that the irradiance pattern on the CCD plane is spatially limited.

the out-of-band ones alone.

Positivity and support constraints cannot be implemented readily on Fourier coefficients, so transformation back to the space domain is still required. The exact transformation rule is

$$\hat{f}(\mathbf{r}) = \sum_{\mathbf{k}} \hat{F}_{\mathbf{k}} \exp(2\pi i \boldsymbol{\rho}_{\mathbf{k}} \cdot \mathbf{r}) , \tag{15.295}$$

where the sum is, in principle, infinite in two dimensions. In practice, of course, a finite sum must be used, with the range determined by the amount of superresolution desired. To transform back to Fourier coefficients, the exact equation is

$$\hat{\mathbf{F}}_{\mathbf{k}} = \frac{1}{L^2} \int_{\mathbf{S_f}} d^2r \; \hat{f}(\mathbf{r}) \exp(-2\pi i \boldsymbol{\rho}_{\mathbf{k}} \cdot \mathbf{r}) , \tag{15.296}$$

where the object support $\mathbf{S_f}$ is assumed to be a square of side L. In practice the integral would be implemented as a sum, most likely as a DFT (though one with many more than M elements).

15.4.6 MLEM algorithm

An important iterative technique is the *maximum-likelihood expectation-maximization* or *MLEM* algorithm, so called because it can be derived by alternating expectation (E) and maximization (M) steps, and because it maximizes the likelihood for a Poisson data model. MLEM has been rediscovered several times. To the authors' knowledge, the earliest paper to present the algorithm was by Metz and Pizer (1971) at the second international conference on Information Processing in Medical Imaging (IPMI). Unfortunately, the untimely death of the conference organizer, Eberhard Jahns, led to the promised Proceedings of IPMI II never appearing and thus the Metz and Pizer paper never being published.

In the optics literature, the MLEM algorithm was presented independently by Richardson (1972) and Lucy (1974), and it is still referred to often as the Richardson-Lucy algorithm. The paper that ignited widespread interest for tomographic applications was by Shepp and Vardi (1982). Another important early contribution to the tomographic literature was Lange and Carson (1984).

MLEM is but one example of a broad class of algorithms that alternate expectation and maximization steps. A rigorous treatment of these more general EM algorithms was given in an important paper by Dempster, Laird and Rubin (1977), and an excellent monograph on the subject is McLachlan and Krishnan (1997). In this section, however, we consider only the MLEM algorithm.

MLEM as a multiplicative algorithm The iteration rule for the basic MLEM algorithm is

$$\hat{\theta}_n^{(k+1)} = \hat{\theta}_n^{(k)} \frac{1}{s_n} \sum_{m=0}^{M} \frac{g_m}{(\mathbf{H}\hat{\boldsymbol{\theta}}^{(k)})_m} H_{mn} , \tag{15.297}$$

where s_n is the nth component of the point sensitivity vector, defined in (7.312) as

$$s_n = \sum_{m=0}^{M} H_{mn} . \tag{15.298}$$

To interpret s_n, consider the usual voxel description of a source where θ_n is the mean number of photons emitted from the nth voxel and H_{mn} is the probability that a photon from voxel n is detected in detector m. Then $s_n\theta_n$ is the mean number of photons from that voxel detected by all detectors, and s_n is the probability that a photon emitted from voxel n is detected somewhere. We need not worry about dividing by zero in (15.297) since voxels with $s_n = 0$ have zero probability of ever contributing to the data and should not be included in the representation in the first place.

Unlike the linear algorithms discussed in Sec. 15.4.1 and the modified linear algorithms discussed in Sec. 15.4.4, MLEM is a *multiplicative algorithm* where an estimate is modified by multiplying it by a correction factor rather than adding a correction term. Other important examples of multiplicative algorithms include multiplicative ART or MART (Gordon *et al.*, 1970) and its variants (Byrne, 1993; 1995) and the SAGE (space-alternating generalized EM) algorithms (Fessler and Hero, 1994), but we shall not discuss any of these methods further.

The MLEM algorithm preserves positivity; that is, if the initial estimate $\hat{\boldsymbol{\theta}}^{(0)}$ is nonnegative, and if all elements of $\mathbf{g}$ and $\mathbf{H}$ are nonnegative, then all subsequent iterations remain nonnegative since we always multiply by a nonnegative factor. By the same token, however, a component of the estimate will seldom be driven exactly to zero; if H_{mn} is nonzero for *any* m for which g_m is nonzero, then the correction factor for $\hat{\theta}_n^{(k)}$ will not be zero. In this respect, the MLEM algorithm is like maximum entropy in that it tends to drive the estimate towards zero but never quite gets there. The exception would be if $g_m = 0$ for all detectors for which $H_{mn} \neq 0$ for a given n; in that case $\hat{\theta}_n^{(k)}$ would be immediately set to zero.

If we know the support of the object *a priori*, on the other hand, we can set the elements of the estimate (in a voxel representation) outside the support to zero in the initial estimate and they will remain zero for all subsequent iterations.

Finally, note that the algorithm strives for agreement between the actual data and the image of the estimate. If $(\mathbf{H}\hat{\boldsymbol{\theta}}^{(k)})_m = g_m$ for all m, then the correction factor is unity and no further change in the estimate occurs. Of course, it may not be possible to find an estimate such that $(\mathbf{H}\hat{\boldsymbol{\theta}}^{(k)})_m = g_m$ for all m, and in that case it turns out (as we shall see below) that the algorithm converges to an estimate that minimizes the Kullback-Leibler distance (see Sec. 15.3.2) $D_{\mathrm{KL}}(\mathbf{g}, \mathbf{H}\hat{\boldsymbol{\theta}})$ between the data and the image of the estimate.

Poisson likelihood As presented so far, the MLEM algorithm is just a convenient way of finding an estimate that agrees as well as possible (in the Kullback-Leibler sense) with the data. It has no particular relation to the statistics of the data and in fact will work with many different kinds of data. We know from Sec. 15.3.2, however, that the Kullback-Leibler distance is closely related to the log-likelihood for Poisson data, and we shall now explore this relation further.

If we consider an MD Poisson random vector $\mathbf{g}$ with mean $\mathbf{H}\boldsymbol{\theta}$, then

$$\Pr(\mathbf{g}|\boldsymbol{\theta}) = \prod_{m=1}^{M} \exp[-(\mathbf{H}\boldsymbol{\theta})_m] \frac{[(\mathbf{H}\boldsymbol{\theta})_m]^{g_m}}{g_m!}, \tag{15.299}$$

and $\Pr(\mathbf{g}|\boldsymbol{\theta})$ is the likelihood of $\boldsymbol{\theta}$ for a given $\mathbf{g}$. One must view this equation with caution, however, since we are free to choose any representation we like for the object, with any number of components N. It is only when the finite object representation is an adequate representation of the data, in the sense that $\mathbf{H}\boldsymbol{\theta}$ is a good approximation to $\boldsymbol{\mathcal{H}}\mathbf{f}$, that $\Pr(\mathbf{g}|\boldsymbol{\theta})$ is really the likelihood of $\boldsymbol{\theta}$. [See (15.137) and the associated discussion.]

With this caveat, the logarithm of the likelihood is given by (15.149), repeated here for convenience:

$$\ln[\Pr(\mathbf{g}|\boldsymbol{\theta})] = \sum_{m=1}^{M} \{-(\mathbf{H}\boldsymbol{\theta})_m + g_m \ln[(\mathbf{H}\boldsymbol{\theta})_m] - \ln(g_m!)\} . \tag{15.300}$$

An extremum of this function occurs at a point where the derivative with respect to all components vanishes:

$$\frac{\partial}{\partial \theta_j} \ln[\Pr(\mathbf{g}|\boldsymbol{\theta})] = \sum_{m=1}^{M} \left\{ -H_{mj} + \frac{g_m}{(\mathbf{H}\boldsymbol{\theta})_m} H_{mj} \right\} = 0 , \qquad j = 1, ..., N . \tag{15.301}$$

To see whether the extremum is a minimum or a maximum, we take another derivative:

$$\frac{\partial^2}{\partial \theta_j \partial \theta_k} \ln[\Pr(\mathbf{g}|\boldsymbol{\theta})] = \sum_{m=1}^{M} \left\{ -\frac{g_m}{[(\mathbf{H}\boldsymbol{\theta})_m]^2} H_{mj} H_{mk} \right\} . \tag{15.302}$$

All components of $\mathbf{g}$ and $\mathbf{H}\boldsymbol{\theta}$ must be nonnegative for the Poisson law to be applicable; a negative number of counts makes no sense. Moreover, all elements of $\mathbf{H}$ must be nonnegative, since otherwise a negative component of $\mathbf{H}\boldsymbol{\theta}$ could occur for some nonnegative $\boldsymbol{\theta}$. Thus the second derivative is negative everywhere (*i.e.*, the log-likelihood is concave), and any extremum must be a maximum. Maximizing the likelihood is thus equivalent to solving the implicit equation (15.301) for $\boldsymbol{\theta}$. Moreover, from the discussion in Sec. 15.3.2, it is also equivalent to minimizing the Kullback-Leibler distance $D_{\mathrm{KL}}(\mathbf{g}, \mathbf{H}\boldsymbol{\theta})$.

MLEM as a fixed-point iteration We can rewrite (15.301) (with the dummy index j changed to n) as

$$\frac{1}{s_n} \sum_{m=1}^{M} \frac{g_m}{(\mathbf{H}\boldsymbol{\theta})_m} H_{mn} = 1 , \tag{15.303}$$

where s_n is defined in (15.298). We now multiply both sides of (15.303) by θ_n, yielding

$$\theta_n = \theta_n \frac{1}{s_n} \sum_{m=1}^{M} \frac{g_m}{(\mathbf{H}\boldsymbol{\theta})_m} H_{mn} . \tag{15.304}$$

To get an iterative algorithm, we replace $\boldsymbol{\theta}$ by a succession of estimates $\hat{\boldsymbol{\theta}}^{(k)}$ and use the fixed-point iteration procedure introduced in Sec. 15.4.4; the result is the MLEM algorithm:

$$\hat{\theta}_n^{(k+1)} = \hat{\theta}_n^{(k)} \frac{1}{s_n} \sum_{m=1}^{M} \frac{g_m}{(\mathbf{H}\hat{\boldsymbol{\theta}}^{(k)})_m} H_{mn} . \tag{15.305}$$

If this algorithm converges, it must find a maximum of the log-likelihood (and hence of the likelihood itself).

Convergence and stopping rules The mapping defined by (15.305) is not a contraction and hence does not necessarily converge to a fixed point independent of the initial estimate. It can, however, be shown to converge in the sense that the likelihood increases monotonically at each step (Dempster *et al.*, 1977). Of course, just because the algorithm approaches a specified likelihood (the maximum value) does not mean it approaches a specified image. The likelihood is a function of $\mathbf{H}\boldsymbol{\theta}$, not $\boldsymbol{\theta}$ alone. If $\mathbf{H}$ has null functions, many different $\boldsymbol{\theta}$ can give the same $\mathbf{H}\boldsymbol{\theta}$ and hence the same likelihood; which one is obtained by the algorithm depends on the null components of the initial estimate.

Moreover, maximum likelihood is seldom a desirable end point in image reconstruction. As we have stressed repeatedly, forcing agreement with noisy data (in any sense) results in noisy images. In practice, running the MLEM algorithm for a large number of iterations usually results in a virtually useless image, often one consisting of a few bright pixels like the night-sky reconstructions discussed in Sec. 15.3.5. (For an example, see Fig. 17.9.)

The most common way of avoiding these problems is just to stop the algorithm before it gives a poor image in some sense. In this case, the image is not a maximum-likelihood estimate, and it depends on the number of iterations and the initial estimate. The choice of stopping point is usually made purely subjectively, though various statistical stopping rules have been proposed (see, for example, Llacer and Veklerov, 1989) as a means of avoiding excessive noise amplification. The stopping point should ideally be chosen to optimize some objective measure of image quality, such as the ability of a human observer to detect an abnormality (see Sec. 14.2), but in practice it is usually done without regard to task.

15.4.7 Noise propagation in nonlinear algorithms

We have already discussed the noise properties of reconstructed images in several cases in this chapter. In Sec. 15.2.6 we treated the effect of a linear reconstruction operator on noise in the data, and in Sec. 15.3.6 we considered implicit estimates and found that we could get useful analytical forms for the covariance without specifying the algorithm for actually finding the estimate. Then, in Sec. 15.4.2 we studied noise propagation through linear iterative algorithms. Now we shall show how that analysis needs to be modified for nonlinear iterative algorithms. The main difference will turn out to be that constant matrices are replaced by ones that depend on the current estimate.

Differentiable update rules All of the iterative algorithms considered in this chapter have the general form,

$$\hat{\boldsymbol{\theta}}^{(k+1)} = \mathbf{D}^{(k)}\{\hat{\boldsymbol{\theta}}^{(k)}, \mathbf{g}\}\,, \tag{15.306}$$

where $\mathbf{D}^{(k)}\{\cdot\,,\,\cdot\}$ is a vector-valued functional of its two arguments. The fixed-point iteration of (15.280) is immediately in this form, and POCS, MLEM and conjugate-gradient fit as well.

It will be very useful to assume that $\mathbf{D}^{(k)}\{\cdot\,,\,\cdot\}$ is differentiable with respect to both arguments. That assumption is justified by inspection for MLEM, conjugate-gradient and many other algorithms, but it may not hold for POCS or other algorithms that employ a clipping operator such as $\mathbf{P}_+$ as defined in (15.186). To get around this difficulty, we can redefine $\mathbf{P}_+\{x\}$ as the limit of a differentiable functional, for example,

$$\mathbf{P}_+\{x\} = x\,\mathrm{step}(x) = \lim_{\beta\to\infty} \frac{x}{1-\exp(-\beta x)}\,. \tag{15.307}$$

Basic propagation equations Assuming the update operator $\mathbf{D}^{(k)}\{\hat{\boldsymbol{\theta}}^{(k)}, \mathbf{g}\}$ is differentiable, we want to expand it in a Taylor series and retain only terms linear in small perturbations of both arguments. For the perturbation of the second argument, we have two options: we can write either $\mathbf{g} = \mathbf{H}\boldsymbol{\theta} + \boldsymbol{\epsilon}$ or $\mathbf{g} = \boldsymbol{\mathcal{H}}\mathbf{f} + \mathbf{n}$, and therefore we can consider either $\boldsymbol{\epsilon}$ or $\mathbf{n}$ to be the perturbation. As we did in earlier noise analysis in this chapter, we shall choose $\mathbf{n}$ because we know a great deal more about its statistics: it has zero mean by definition, and its covariance can usually be determined from the physics of the imaging process.

Similarly, we have some options about how we expand the first argument. A natural choice would seem to be to consider small perturbations about the true $\boldsymbol{\theta}$, but that runs afoul of what we have learned in this chapter about estimability. We know that many different objects can give the same data and hence the same sequence of estimates $\{\hat{\boldsymbol{\theta}}^{(k)}\}$, yet they may have quite different true $\boldsymbol{\theta}$; a small perturbation about one of them might be a large perturbation with respect to another. Another option would be to expand about the final estimate, $\hat{\boldsymbol{\theta}}^{(\infty)}$ or some approximation to it, but this is not wise either; the algorithm may not converge at all (which is precisely why we need to know the statistics as a function of iteration number), or it may converge to a solution that is substantially different from $\hat{\boldsymbol{\theta}}^{(k)}$ at earlier iterations, so $\hat{\boldsymbol{\theta}}^{(\infty)}$ is a poor choice for the center of expansion.

A better choice is to expand $\hat{\boldsymbol{\theta}}^{(k)}$ about its mean *at each iteration* (Barrett *et al.*, 1994). We thus use a different Taylor series for each k, and so long as $\mathbf{n}$ is small enough compared to $\boldsymbol{\mathcal{H}}\mathbf{f}$, we can be confident that truncation of the series with linear terms is valid.

To simplify the notation, we denote the conditional mean of $\hat{\boldsymbol{\theta}}^{(k)}$ (for specified $\mathbf{f}$) as $\mathbf{a}^{(k)}$ and write

$$\hat{\boldsymbol{\theta}}^{(k)} = \mathbf{a}^{(k)} + \Delta\hat{\boldsymbol{\theta}}^{(k)} . \tag{15.308}$$

The linearized Taylor expansion is then

$$\hat{\boldsymbol{\theta}}^{(k+1)} = \mathbf{D}^{(k)}\{\hat{\boldsymbol{\theta}}^{(k)}, \mathbf{g}\} \simeq \mathbf{D}^{(k)}\{\hat{\mathbf{a}}^{(k)}, \boldsymbol{\mathcal{H}}\mathbf{f}\} + \boldsymbol{\Omega}^{(k)}\Delta\hat{\boldsymbol{\theta}}^{(k)} + \mathbf{B}^{(k)}\mathbf{n} , \tag{15.309}$$

where $\boldsymbol{\Omega}^{(k)}$ is an $N \times N$ matrix with components

$$[\boldsymbol{\Omega}^{(k)}]_{nn'} = \frac{\partial}{\partial\hat{\theta}_{n'}} D_n^{(k)}\{\hat{\boldsymbol{\theta}}^{(k)}, \boldsymbol{\mathcal{H}}\mathbf{f}\} , \tag{15.310}$$

and $\mathbf{B}^{(k)}$ is an $N \times M$ matrix with components

$$[\mathbf{B}^{(k)}]_{nm} = \left[\frac{\partial}{\partial g_m} D_n^{(k)}\{\hat{\boldsymbol{\theta}}^{(k)}, \mathbf{g}\} \right]_{\mathbf{g}=\boldsymbol{\mathcal{H}}\mathbf{f}} . \tag{15.311}$$

Within this linearized formulation, therefore, the mean estimate evolves as

$$\hat{\mathbf{a}}^{(k+1)} = \mathbf{D}^{(k)}\{\hat{\mathbf{a}}^{(k)}, \boldsymbol{\mathcal{H}}\mathbf{f}\} \tag{15.312}$$

and the deviation from the mean evolves as

$$\Delta\hat{\boldsymbol{\theta}}^{(k+1)} = \boldsymbol{\Omega}^{(k)}\Delta\hat{\boldsymbol{\theta}}^{(k)} + \mathbf{B}^{(k)}\mathbf{n} . \tag{15.313}$$

These results should be compared to (15.240) and (15.241), which were derived for linear iterative algorithms. Like (15.240), (15.312) shows that the sequence of mean values $\{\hat{\mathbf{a}}^{(k)}\}$ can be generated by running the algorithm on noise-free data, $\mathbf{g} = \mathcal{H}\mathbf{f}$, which can be simulated or obtained experimentally by frame-averaging or by using a long exposure time (in the case of Poisson noise).

The equations for propagation of the deviations through linear and nonlinear algorithms, (15.241) and (15.313) respectively, have identical forms, but the matrices $\boldsymbol{\Omega}^{(k)}$ and $\mathbf{B}^{(k)}$ have somewhat different meanings. They are independent of $\mathbf{g}$ and $\boldsymbol{\theta}^{(k)}$ in the linear case, and they are also independent of k unless the update rule itself varies with iteration number. In the nonlinear case, by contrast, the matrices depend on the current estimate $\boldsymbol{\theta}^{(k)}$, so they must vary with k.

Since (15.241) and (15.313) have the same form, the remainder of the derivation in Sec. 15.4.2 still holds, and in particular the expressions (15.247) – (15.249) for the covariance matrix are still valid. As noted in Sec. 15.4.2, computation of the covariance requires running a recursive algorithm based on (15.246) to compute $\mathbf{U}^{(k)}$, but this turns out to be computationally feasible even for large images and nonlinear algorithms.

Application to the MLEM algorithm Wilson *et al.* (1994) applied a recursive scheme analogous to (15.246) to the MLEM algorithm, (15.305). Because MLEM is a multiplicative algorithm, they took the logarithm of both sides of (15.305) and obtained an update rule for $\hat{y}_n^{(k)} \equiv \ln \hat{\theta}_n^{(k)}$; they then recursively computed the matrix $\mathbf{U}^{(k)}$ relevant to the covariance matrix for $\hat{\mathbf{y}}^{(k)}$.

Because the update rule for $\hat{\mathbf{y}}^{(k)}$ was nearly linear, it could be argued from the central-limit theorem that $\hat{\mathbf{y}}^{(k)}$ should be a multivariate-normal random vector, so its mean and covariance as computed by the recursion should fully specify its statistics. Moreover, if $\hat{\mathbf{y}}^{(k)}$ is normal, it implies that $\hat{\boldsymbol{\theta}}^{(k)}$ is multivariate log-normal. As a consequence, the univariate PDF on each pixel value is also a log-normal, exhibiting the characteristic long tails of that density. Finally, it was predicted that a map of the pixel variance would resemble the mean image, with low variance in regions of low object brightness. These predictions were borne out to high accuracy in a detailed Monte Carlo study performed by Wilson and in later studies by Wang and Gindi (1997) and Soares *et al.* (1998).

15.4.8 Stochastic algorithms

All of the iterative algorithms discussed so far will converge to the minimum of the functional if there is only one such minimum. Sometimes, however, we are interested in functionals with multiple local minima, such that small changes in any direction about a local minimum will increase the value, yet there may be a distant point with a lower value. If an iterative algorithm is initialized with an estimate that is near one of the local minima, straightforward minimization (say by gradient descent or conjugate gradient) may draw it towards this point even if a distant minimum would result in a lower value.

In image reconstruction, multiple minima can occur with Bayesian priors based on *mixture models*, where it is assumed that the object can be drawn from one of two or more classes, with each class having its own prior density. Thus we write

$$\mathrm{pr}(\boldsymbol{\theta}) = \sum_{j=1}^{J} \mathrm{pr}(\boldsymbol{\theta}|C_j)\,\mathrm{Pr}(C_j)\,, \tag{15.314}$$

where $\text{pr}(\boldsymbol{\theta}|C_j)$ is the density associated with class C_j, and $\Pr(C_j)$ is the prior probability that the object came from that class. A mixture prior is often multimodal (exhibiting multiple maxima), and the posterior can also be multimodal if the likelihood is not too sharply peaked. Since the objective functional $Q(\boldsymbol{\theta}, \mathbf{g})$ can be interpreted as the negative logarithm of the posterior (see Sec. 15.3.1), it can have multiple local minima.

Another way in which local minima can occur is with parametric object descriptions (see Sec. 7.1.6) where the mean data vector is a nonlinear functional of the parameters. For example, if we describe an astronomical object as a set of N stars at locations $\{\mathbf{r}_n, n = 1, ...N\}$ and brightnesses $\{b_n, n = 1, ...N\}$, we may want to estimate the N locations and brightnesses as well as N itself; for one choice of N, there is a minimum of the objective functional with respect to the locations and brightnesses, but a different N might give a lower minimum.

Yet another way in which local minima can arise is when the gray levels are quantized. In the extreme where only black or white pixels are allowed, it often happens that an estimate will be reached where changing the state of any one pixel will increase the objective functional, but some entirely new configuration may have a lower value.

This section describes several methods for dealing with local minima, with emphasis on the *simulated-annealing* algorithm. Simulated annealing was developed initially by Metropolis *et al.* (1953) in the context of statistical mechanics, but its general applicability to optimization problems was later recognized by Kirkpatrick *et al.* (1983, 1984). Following quickly after Kirkpatrick, the method was applied to imaging by W. E. Smith *et al.* (1983), Geman and Geman (1984) and Geman and McClure (1985). For a detailed treatment of simulated annealing in a variety of problems, see van Laarhoeven and Aarts (1987).

As we shall see, simulated annealing is a special case of a very powerful technique known as *Markov-chain Monte Carlo* or *MCMC*. We have already alluded to MCMC in Chap. 14 as a tool for computing averages needed in observer-performance studies, and more detail on this use is also given in this section. Excellent books on MCMC include Robert and Casella (1999) and Gilks *et al.* (1996).

The approach taken here will be first to provide some background on statistical mechanics, then specifically describe the simulated-annealing algorithm and its use in both statistical mechanics and image reconstruction. Finally, we shall describe the more general construct of MCMC and briefly mention some of its applications in imaging.

Basic results from statistical mechanics Consider a gas of N particles, where the state of the nth particle is described by its 3D position $\mathbf{r}_n$ and 3D velocity $\mathbf{v}_n$. The complete state of the system is thus described by $6N$ coordinates, and the system can be represented as a point in a $6N$-dimensional phase space. The total energy of the system can thus be written as $\mathcal{E}(\{\mathbf{r}_n\}, \{\mathbf{v}_n\})$ (where the brackets denote sets with N members). Since the positions and velocities are random, the energy fluctuates about some mean $\overline{\mathcal{E}}$; we assume that the system is coupled to a heat bath that fixes the temperature, so the mean energy is a constant.

For bookkeeping purposes, we sample the phase space on a regular $6N$-dimensional grid,[9] so the system has a finite (but immense) number of possible states. In the jth state, $\mathbf{r}_n = \mathbf{r}_{nj}$, $\mathbf{v}_n = \mathbf{v}_{nj}$ and $\mathcal{E}_j = \mathcal{E}(\{\mathbf{r}_{nj}\}, \{\mathbf{v}_{nj}\})$. A fundamental result of statistical mechanics (Reif, 1965) is that the probability of occurrence of this state in thermal equilibrium is given by

$$\Pr(j) = \frac{1}{Z}\exp\left(-\frac{\mathcal{E}_j}{k_B T}\right) = \frac{1}{Z}\exp\left(-\beta\mathcal{E}_j\right) , \tag{15.315}$$

where T is the absolute temperature, k_B is Boltzmann's constant, and

$$\beta \equiv \frac{1}{k_B T} . \tag{15.316}$$

The normalizing constant Z, called the *partition function*, is given by

$$Z = \sum_j \exp\left(-\beta\mathcal{E}_j\right) , \tag{15.317}$$

where the sum is over all possible states of the system.

All important thermodynamic quantities can be expressed in terms of derivatives of the partition function with respect to β. For example, as the reader may show, the mean energy is given by

$$\overline{\mathcal{E}} = -\frac{1}{Z}\frac{\partial Z}{\partial \beta} = -\frac{\partial \ln Z}{\partial \beta} , \tag{15.318}$$

and the thermodynamic entropy (the quantity that appears in the second law of thermodynamics) is given by

$$S \equiv -k_B \sum_j \Pr(j)\ln \Pr(j) = k_B \ln Z + \frac{\overline{\mathcal{E}}}{T} . \tag{15.319}$$

We know from Sec. 15.3.3 that $-\sum_j \Pr(j)\ln\Pr(j)$ is (within constants) the logarithm of the number of ways[10] that the state can be constructed from indistinguishable molecules [*cf.* (15.163)]; the mythical grains used to justify entropy priors in image reconstruction correspond to actual molecules in statistical mechanics.

The equation of state is the relationship between mean pressure $\overline{p}$, volume V and temperature T. In terms of the partition function,

$$\overline{p} = \frac{1}{\beta}\frac{\partial \ln Z}{\partial V} . \tag{15.320}$$

[9] In classical statistical mechanics, the phase-space sampling interval is arbitrary and can eventually be allowed to go to zero, so that sums become integrals. In quantum statistical mechanics, on the other hand, there is a natural interval set by the uncertainty principle. For more discussion, see Reif (1965).

[10] The number of ways the state can be constructed is frequently denoted W, and the equation $S = k\log W$ appears on Ludwig Boltzmann's gravestone. For a photograph, see Cercignani (1998).

Metropolis algorithm in statistical mechanics Since the partition function is very difficult to calculate for complex systems, Metropolis *et al.* (1953) proposed estimation of the equation of state by Monte Carlo sampling. If we can generate many random samples of the configurations $\{\mathbf{r}_n, \mathbf{v}_n, n = 1, ..., N\}$, then properties of the system can be estimated by replacing ensemble averages by sample averages. In Sec. 10.4.5, we discussed ways of producing the samples by generating and tracing individual particles, but this method does not correspond to thermal equilibrium where the probability of occurrence of different states must obey (15.315). The problem addressed by the Metropolis paper was to generate configurations $\{\mathbf{r}_n, \mathbf{v}_n\}$ from this probability law.

The Metropolis algorithm starts with a system in some initial state i with particle configuration $\{\mathbf{r}_{ni}, \mathbf{v}_{ni}\}$ and energy $\mathcal{E}_i$. It then proposes a change to the configuration by altering one or more $\mathbf{r}_n$ or $\mathbf{v}_n$, so the proposed new state j has configuration $\{\mathbf{r}_{nj}, \mathbf{v}_{nj}\}$ and energy $\mathcal{E}_j$. If $\mathcal{E}_j < \mathcal{E}_i$, the proposed change is accepted and the system makes a transition to the new configuration. If $\mathcal{E}_j > \mathcal{E}_i$, however, the change is not necessarily rejected; instead it is accepted with a probability

$$\Pr(i \to j) = \exp[-\beta\,(\mathcal{E}_j - \mathcal{E}_i)]\,, \qquad \mathcal{E}_j > \mathcal{E}_i\,. \tag{15.321}$$

In practice, the decision to accept or reject the proposed change is made by drawing a random number t uniformly distributed on $(0, 1)$ and accepting the change if $\exp[-\beta\,(\mathcal{E}_j - \mathcal{E}_i)] > t$. Since $\beta > 0$, this condition is always satisfied for any t on $(0, 1)$ if $\mathcal{E}_j < \mathcal{E}_i$, so we can write

$$\Pr(i \to j) = \min\{1, \exp[-\beta(\mathcal{E}_j - \mathcal{E}_i)]\}\,. \tag{15.322}$$

This process is then repeated iteratively, generating many successive configurations. Though we won't attempt to prove it, eventually an equilibrium is reached. Equilibrium means that any two states i and j satisfy the condition of *detailed balance*, so that

$$\Pr(i)\Pr(i \to j) = \Pr(j)\Pr(j \to i)\,. \tag{15.323}$$

A more verbose way to state this condition is in terms of joint probabilities:

$$\Pr(\text{state } i \text{ at iteration } k, \text{ state } j \text{ at iteration } k+1)$$

$$= \Pr(\text{state } j \text{ at iteration } k, \text{ state } i \text{ at iteration } k+1)\,. \tag{15.324}$$

Were this condition not satisfied, the relative probabilities of the two states would change with iteration number and the system would not be in equilibrium. (Note that one cycle through the process described above is counted as one iteration, whether or not the proposed change is accepted; thus iteration number is discretized time, and equilibrium means that all probabilities are independent of time.)

The beauty of the Metropolis algorithm is that the basic relation of statistical mechanics, (15.315), is satisfied at equilibrium. To see this, consider two states i and j for which $\mathcal{E}_j > \mathcal{E}_i$. Then (15.323) and (15.322) require that

$$\Pr(i)\exp[-\beta\,(\mathcal{E}_j - \mathcal{E}_i)] = \Pr(j)\cdot 1\,, \tag{15.325}$$

in accord with (15.315).

Simulated annealing in optimization problems As described so far, the Metropolis algorithm is concerned with systems in thermal equilibrium (in spite of its origins in the Manhattan project). It is also possible, however, to gradually reduce the temperature T (increase β) so that the system evolves through a series of near-equilibrium states. In this fashion properties such as specific heat and thermal expansion can be studied.

In materials science, a gradual reduction in temperature is called *annealing*; its opposite, an abrupt reduction in temperature, is called *quenching*. Suppose, for example, that one wants to make a sample of silicon for use in manufacturing semiconductor devices. Basic solid-state physics tells us that the lowest-energy configuration of silicon atoms at $T = 0$ is a perfect crystal with discrete translational symmetry. At room temperature the equilibrium configuration involves small fluctuations (phonons) around the ideal lattice, but it is nevertheless a good approximation to consider silicon in thermal equilibrium to be nearly a regular crystal lattice with a high degree of translational order. If excess electrons or holes are somehow produced in such a crystal, they exhibit very high mobility (see Sec. 12.1.2) and lead to excellent device properties.

On the other hand, the equilibrium state of silicon at very high temperature is a liquid with little translational order and poor electrical properties. When the liquid is cooled below the melting point, the silicon atoms are not initially arranged into a regular lattice, but instead inherit the disorder of the liquid. If the solid is held at a temperature just below the melting point, the atoms gradually approach an equilibrium where the atoms fluctuate around the ideal lattice points, and if the temperature is gradually reduced, so too are the fluctuations. If the temperature reduction is too rapid, however, the atoms get frozen into non-equilibrium positions, and the deviations from ideal translational symmetry degrade the electrical properties at room temperature. Thus physical annealing is a way of assuring that the material finds its way to its lowest-energy state.

Kirkpatrick *et al.* (1983, 1984) proposed using the Metropolis algorithm as a way of simulating the physical annealing process in problems having nothing to do with thermodynamics. Specifically, he was interested in the design of integrated circuits consisting of many electrical gates and a complex system of interconnections. The locations of the gates on the silicon crystal were the main design variables, and he defined a cost function involving both performance and actual cost of manufacture. The objective was to choose the gate location so as to minimize the cost. Thus the analogy to thermodynamics: gate locations correspond to coordinates in a gas, and the cost function corresponds to an energy.

Kirkpatrick began with some initial configuration of gates, just as Metropolis began with an initial configuration of gas molecules, and he proposed a change in the configuration at each step. The change was accepted or rejected according to (15.322) with the cost function in place of the physical energy. Since the probability of accepting a transition to a higher-cost design depends on β (or equivalently T), choice of this free parameter is critical. If β is large (T small), then there is very little chance of getting out of a local minimum, and if β is small there are large fluctuations around a minimum and the algorithm explores many high-cost designs. The solution is to increase β gradually, always remaining near equilibrium. Then, as illustrated qualitatively in Fig. 15.10, the equilibrium probability density function approaches a delta function centered on the true optimal configuration.

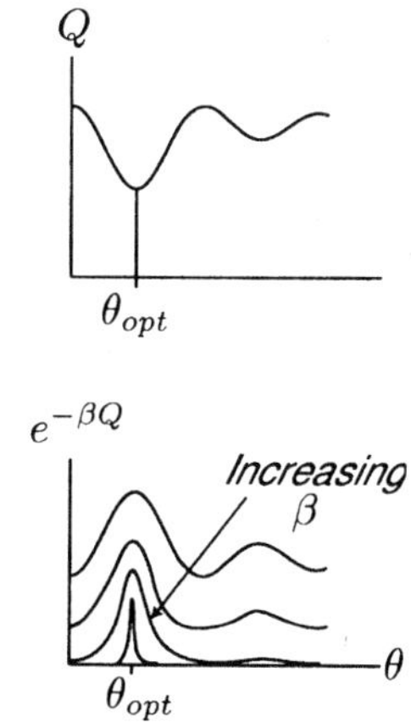

Fig. 15.10 Illustration of a cost function in 1D and the corresponding equilibrium probability density function at three different temperatures.

The art of simulated annealing Many questions arise in actually implementing simulated annealing for optimization problems, and many relevant theorems have been derived. In practice, however, simulated annealing remains more art than science.

The first concern is how rapidly the temperature should be reduced. The true minimum will be attained so long as the temperature is proportional to the reciprocal of the logarithm of the iteration number, but this is an extraordinarily slow approach to the optimum, and to the authors' knowledge never used. Instead, various *ad hoc* annealing schedules are used in an effort to trade off computational time and final cost (another optimization problem). One useful technique is to monitor the cost function and change the temperature when it seems to be fluctuating about some constant value.

Another open question is how the proposed changes to the system should be chosen. In one sense, this choice doesn't matter; so long as all possible states of the system can eventually be reached, the equilibrium distribution will be (15.315). Practically, however, the approach to equilibrium following a temperature change can be slow if only small changes are proposed.

Simulated annealing in image reconstruction Since we have presented image reconstruction as an optimization problem, it is natural to consider simulated annealing whenever there might be multiple minima. In image reconstruction, the state of the system is specified by the estimated coefficient vector $\hat{\boldsymbol{\theta}}$. At the kth iteration, a change to the state is proposed in which one or more pixel values are altered in some way, so that the proposed new configuration is given by

$$\hat{\boldsymbol{\theta}}^{(k+1)} = \hat{\boldsymbol{\theta}}^{(k)} + \mathbf{h}^{(k)} . \tag{15.326}$$

By analogy to (15.322), this proposed change is accepted with probability

$$\Pr(acc) = \min\left\{1, \exp\left[-\beta\left(Q\left(\hat{\boldsymbol{\theta}}^{(k+1)}\right) - Q\left(\hat{\boldsymbol{\theta}}^{(k)}\right)\right)\right]\right\} . \tag{15.327}$$

As in the Kirkpatrick application, the temperature can be gradually lowered in order to entice the reconstruction into the true minimum of $Q(\hat{\boldsymbol{\theta}})$.

As with any other image-reconstruction algorithm, there are free parameters to set and choices to be made. The strategy for choosing the proposed changes $\mathbf{h}^{(k)}$ must be specified, the annealing schedule must be chosen, and of course the cost function itself must be selected. In principle, all of these choices should be made in such a way as to optimize observer performance, though in practice this nested optimization would be very difficult to carry out.

Metropolis-Hastings algorithms Hastings (1970) proposed a generalization of the Metropolis algorithm. In a general notation, let us assume that the goal is to generate a sequence of samples of a parameter $\boldsymbol{\theta}$ from a density $\pi(\boldsymbol{\theta})$ and that the current value of the parameter is $\boldsymbol{\theta}^{(k)}$. A new trial parameter $\boldsymbol{\theta}'$ is generated from a density $q(\boldsymbol{\theta}'|\boldsymbol{\theta}^{(k)})$, which can depend on the current state. The probability of accepting this proposed change is[11]

$$\Pr(acc) = \min\left\{1, \frac{\pi(\boldsymbol{\theta}')\, q(\boldsymbol{\theta}^{(k)}|\boldsymbol{\theta}')}{\pi(\boldsymbol{\theta}^{(k)})\, q(\boldsymbol{\theta}'|\boldsymbol{\theta}^{(k)})}\right\}. \tag{15.328}$$

If the change is accepted, we set $\boldsymbol{\theta}^{(k+1)} = \boldsymbol{\theta}'$; otherwise $\boldsymbol{\theta}^{(k+1)} = \boldsymbol{\theta}^{(k)}$. By a detailed-balance argument similar to the one used in (15.323), the equilibrium distribution is indeed $\pi(\boldsymbol{\theta})$; details are left as an exercise.

Comparison of (15.328) and (15.323) shows that the original Metropolis algorithm is a Hastings algorithm with a symmetric proposal density, $q(\boldsymbol{\theta}'|\boldsymbol{\theta}) = q(\boldsymbol{\theta}|\boldsymbol{\theta}')$. In terms of proposed changes $\mathbf{h}^{(k)}$ as in (15.326), the proposal density is symmetric if $+\mathbf{h}^{(k)}$ and $-\mathbf{h}^{(k)}$ are equally likely. When this condition is satisfied, (15.328) becomes

$$\Pr(acc) = \min\left\{1, \frac{\pi(\boldsymbol{\theta}')}{\pi(\boldsymbol{\theta}^{(k)})}\right\}, \tag{15.329}$$

which is the general form of a Metropolis algorithm. The specific form (15.322) is recovered by choosing $\pi(\boldsymbol{\theta}) \propto \exp[-\beta\mathcal{E}(\boldsymbol{\theta})]$ as in (15.315). Note that the partition function Z is not required since it cancels out in (15.328) and (15.329); Metropolis and Metropolis-Hastings algorithms can be used for drawing samples from densities that we cannot normalize.

Another special case of interest is the *single-component Metropolis-Hastings algorithm* in which only a single component of the vector is altered at a time. In an imaging context, for example, the candidate image differs from the current image at only a single pixel.

To describe the single-component algorithm, we need some additional notation. Let θ_n denote the nth component of the $N \times 1$ vector $\boldsymbol{\theta}$, and let $\boldsymbol{\theta}_{-n}$ be the $(N-1)\times 1$ vector obtained by deleting θ_n from $\boldsymbol{\theta}$. Thus the set $\{\theta_n, \boldsymbol{\theta}_{-n}\}$ is the same thing as $\boldsymbol{\theta}$.

With this notation, proposal of a new vector $\boldsymbol{\theta}'$ is equivalent to proposing a new value for θ_n, and we write the proposal density as $q_n(\theta_n', |\boldsymbol{\theta})$, where the subscript on q_n indicates that different proposal densities can be used for different components.

[11]We use $\pi(\boldsymbol{\theta})$ and $q(\boldsymbol{\theta}'|\boldsymbol{\theta}^{(k)})$ here instead of our usual notations, $\mathrm{pr}(\boldsymbol{\theta})$ and $\mathrm{pr}(\boldsymbol{\theta}'|\boldsymbol{\theta}^{(k)})$, respectively, since we mean specific functional forms, not generic densities. To see what would happen without this notational distinction between $\pi(\cdot)$ and $q(\cdot|\cdot)$, see the discussion below of the Gibbs sampler.

Thus (15.328) becomes

$$\Pr(acc) = \min\left\{1, \frac{\pi(\boldsymbol{\theta}')\, q_n(\theta_n^{(k)}|\boldsymbol{\theta}')}{\pi(\boldsymbol{\theta}^{(k)})\, q_n(\theta_n'|\boldsymbol{\theta}^{(k)})}\right\} . \tag{15.330}$$

An advantage of this algorithm is that the samples are drawn from a univariate density, so any of the methods discussed in Sec. C.7 can be used.

A succinct tutorial on Metropolis-Hastings algorithms is given by Chib and Greenberg (1995), and a collection of practical papers on the subject is found in Gilks *et al.* (1996).

Gibbs sampler In a general Metropolis-Hastings algorithm, the proposal density $q(\boldsymbol{\theta}'|\boldsymbol{\theta})$ need not have anything to do with the density $\pi(\boldsymbol{\theta})$ from which we wish to draw samples. All that is required is that it be possible to generate any $\boldsymbol{\theta}$ in the sample space by a sequence of proposals. In a *Gibbs sampler*, on the other hand, the proposal density is a conditional density derived from $\pi(\boldsymbol{\theta})$. Specifically,

$$q_n(\theta_n'|\boldsymbol{\theta}) = \pi(\theta_n'|\boldsymbol{\theta}_{-n}) , \tag{15.331}$$

where $\pi(\theta_n|\boldsymbol{\theta}_{-n})$ is the *full conditional*, defined by [*cf.* (C.76)]

$$\pi(\theta_n|\boldsymbol{\theta}_{-n}) = \frac{\pi(\boldsymbol{\theta})}{\int d\theta_n \; \pi(\boldsymbol{\theta})} . \tag{15.332}$$

Note carefully that the quantity that appears on the right in (15.331), namely $\pi(\theta_n'|\boldsymbol{\theta}_{-n})$, is simply $\pi(\theta_n|\boldsymbol{\theta}_{-n})$ evaluated at $\theta_n = \theta_n'$; this substitution causes no notational problems since $\pi(\cdot)$, unlike $\mathrm{pr}(\cdot)$, denotes a specific functional form.

If we use (15.331) and (15.332), the ratio that appears in (15.330) becomes

$$\frac{\pi(\boldsymbol{\theta}')\, q_n(\theta_n^{(k)}|\boldsymbol{\theta}')}{\pi(\boldsymbol{\theta}^{(k)})\, q_n(\theta_n'|\boldsymbol{\theta}^{(k)})} = \frac{\pi(\boldsymbol{\theta}')\, \pi(\boldsymbol{\theta}^{(k)})/\int d\theta_n^{(k)} \; \pi(\boldsymbol{\theta}^{(k)})}{\pi(\boldsymbol{\theta}^{(k)})\, \pi(\boldsymbol{\theta}')/\int d\theta_n' \; \pi(\boldsymbol{\theta}')} = 1 , \tag{15.333}$$

where the integrals cancel since $\boldsymbol{\theta}'$ and $\boldsymbol{\theta}^{(k)}$ are identical except for the nth component, which is integrated out. Thus the probability of acceptance is given by

$$\Pr(acc) = \min\{1, 1\} = 1 , \tag{15.334}$$

and no proposal is ever rejected in a Gibbs sampler.

In summary, Gibbs samplers always draw univariate samples from the full conditionals. They are very efficient since no proposal is ever rejected, but they require that the conditionals be known. For a survey of methods of drawing the samples, see Gilks *et al.* (1996), Chap. 5. For a tutorial on Gibbs samplers, see Casella and George (1992).

Markov chains A *Markov chain* is a sequence of random vectors $\{\boldsymbol{\theta}^{(k)}\}$ in which the probability of occurrence of one value depends only on the immediate previous value in the chain, *i.e.*, $\mathrm{pr}(\boldsymbol{\theta}^{(k+1)}|\boldsymbol{\theta}^{(k)}, \boldsymbol{\theta}^{(k-1)}, \boldsymbol{\theta}^{(k-2)}, \dots) = \mathrm{pr}(\boldsymbol{\theta}^{(k+1)}|\boldsymbol{\theta}^{(k)})$. All of the stochastic algorithms described above generate Markov chains, and the general term for these methods is *Markov-chain Monte Carlo*. If the proposal density and acceptance rule are independent of iteration number k, the Markov chain is *stationary*. Thus the basic Metropolis-Hastings and Gibbs algorithms generate stationary chains, but simulated annealing is at best quasistationary since the temperature parameter changes as the iteration proceeds.

Bayesian applications MCMC methods can in principle be used for drawing samples from any PDF, but for Bayesian applications the important PDF is the posterior. Indeed, a good working definition of a Bayesian is that it is someone who draws inferences only from posteriors.

In the context of Bayesian image analysis, we can identify four distinct uses of MCMC. The first is to find a MAP estimate where the posterior $\mathrm{pr}(\mathbf{g}|\boldsymbol{\theta})$ is maximized with respect to $\boldsymbol{\theta}$. As noted above, this is an optimization problem that can be solved by simulated annealing or a more general Metropolis-Hastings algorithm in the limit as $T \to 0$.

The second use of MCMC is to derive other point estimates from the posterior. For example, we noted in Sec. 13.3.3 that a posterior-mean estimator minimizes a quadratic loss function. Except for the case of Gaussian noise and a Gaussian object model, where the posterior mean leads to the Wiener filter, this estimator is rarely used in image analysis, largely because of computational difficulties. MCMC offers a potential way of overcoming these difficulties.

The third Bayesian application of MCMC is to study uncertainties in estimates. Uncertainties in the Bayesian world are based on the posterior, so posterior variance and covariance are regarded as measures of accuracy of estimates. It must be noted, however, that Bayesian priors reflect belief rather than any sampling properties of real-world objects, so posterior variance can be reduced arbitrarily just by adopting a stringent prior—if the prior is sharply peaked, so too is the posterior. Many Bayesians avoid this trap by using neutral or noninformative priors such as entropy, but they then face computational difficulties in actually determining the posterior variance.

Finally, MCMC can be used to study the prior itself. It is rarely the case that samples drawn from the prior used in image reconstruction bear any resemblance to real objects, but we might hope that they capture certain local features such as the joint density of neighboring pixels (Herman and Chan, 1995). MCMC applied to the prior offers a way of determining if this hope is justified.

For an excellent treatment of Bayesian applications of MCMC, see Besag *et al.* (1995)—and don't overlook the 25 pages of lively comment following the article.

16

Planar Imaging with X Rays and Gamma Rays

This chapter is the first of four that provide detailed case studies of selected imaging modalities. The goal of these applications chapters is to illustrate how the mathematical techniques developed in earlier chapters can be used in the analysis and optimization of real-world imaging systems. The applications have been chosen to illustrate specific mathematical methods, with the hope that these case studies will provide templates for the reader to use in extending the analysis to other imaging realms.

Specifically, this chapter covers two distinct modalities in medical radiological imaging. Both modalities are direct imaging, in the sense that the measured data are immediately the desired image without the need for image reconstruction or further processing. Both produce 2D projection images of 3D objects, and for this reason they are referred to as *planar imaging*. Tomographic methods, where 3D objects yield 3D images, are treated in the next chapter.

The first modality considered here, in Sec. 16.1, is digital radiography, where discrete arrays of detectors are used to sense the distribution of x rays transmitted through a patient's body. We chose to include this application because of high current interest in the medical community, but also because it is an excellent vehicle to elucidate many aspects of imaging introduced earlier. Digital radiography is fundamentally a continuous-to-discrete mapping, but only approximately a linear one; thus it provides an object lesson on how to apply the linear deterministic descriptions of Chap. 7 as well as an illustration of their limitations. Image formation in radiography is most rigorously described by the Boltzmann transport equation of Chap. 10, and the relation of that approach to linear methods is described in this chapter.

In stochastic terms as well, digital radiography provides concrete illustrations of many subtle points. X-ray images suffer from Poisson noise, so the whole theoretical structure developed in Chap. 11 comes into play, but the detectors themselves have excess, non-Poisson, noise as treated in Chap. 12.

Finally, digital radiography provides a excellent application of our general approach to objective assessment of image quality. Image quality in radiology is closely linked to the radiation dose given to the patient; Poisson noise can be reduced by using more intense x-ray beams, but the potential biological damage cannot be ignored. Therefore it is crucial to provide quantitative assessments of task performance and to identify the system components that limit that performance. The methodology of Chaps. 13 and 14 will stand us in good stead in this endeavor.

The second modality considered here, in Sec. 16.2, is planar nuclear medicine. In brief, nuclear medicine is the use of radioactive tracers to follow some physiological pathway or to identify tumors or other pathologies. The tracers can have exquisite biological specificity, targeting specific cell-surface receptors or disease-specific antigens. Indeed, a whole new approach to biology and medicine, referred to as molecular imaging, is now emerging, and nuclear imaging is one of the key technologies making it possible.

In physical terms, a key distinction between radiography and nuclear medicine is that the radiation source is outside the body in the former, inside the body in the latter. In radiography the object is the distribution of x-ray attenuation coefficient, while in nuclear medicine the object is the radiation source itself. Thus radiography is *transmission imaging* while nuclear medicine is *emission imaging*. Both, however, use high-energy photons; there is no essential distinction between the x rays used in radiography and the gamma rays used in nuclear medicine except that x rays result from electronic transitions and gamma rays come from nuclear transitions during radioactive decay.

Another difference between the two modalities is the method of image formation. In transmission radiography the image is produced by simple shadow casting, but in nuclear medicine some image-forming element is required. Since gamma rays are not appreciably refracted or reflected by matter, image formation must rely on absorption. Options include pinholes, multibore collimators and coded apertures, but we concentrate here on collimators.

Many of the same mathematical themes adduced in Sec. 16.1 will recur in Sec. 16.2 in the context of gamma-ray emission imaging, but some differences are worth noting. The Boltzmann transport equation is again the rigorous means of describing deterministic image formation in nuclear medicine, but the scattering term is more important than in transmission radiography. CD imaging models are again required, but linearity is much more justifiable in nuclear medicine than in radiography. Both x-ray and gamma-ray detectors respond approximately linearly to the flux incident on them, and that flux is linearly related to the object in the case of emission imaging. In transmission radiography, on the other hand, the flux is nonlinearly related to the object, which is the distribution of x-ray attenuation coefficient.

Stochastically, Poisson noise is paramount in nuclear medicine, and it is still closely linked to the radiation dose given to the patient. In nuclear medicine, however, there is an important trade-off between noise and spatial resolution; coarser collimators collect more photons but produce blurrier images. Moreover, the detectors used in nuclear medicine, unlike those used in digital radiography, are essentially ideal photon counters, so non-Poisson noise seldom arises in gamma-ray images. Instead, photon-counting gamma cameras require an estimation step to relate the observable output signals to position and energy of the individual gamma-ray photon. These differences will lead to rather different assessment methodologies and design considerations when we discuss image quality in nuclear medicine at the end of Sec. 16.2.

16.1 DIGITAL RADIOGRAPHY

From Roentgen's discovery of x rays at the end of the 19th century through the end of the 20th century, x-ray imaging was predominantly based on fluorescent screens and photographic film. At the beginning of the 21st century, however, film-based radiography is giving way to fully electronic systems. Farsighted prognosticators (*e.g.*, Capp, 1981) who have been heralding the all-digital radiology department for decades may soon be proven correct.

Many technological advances have contributed to this change, including computers to process the images, large disks to store them, and high-resolution image displays to reproduce fine details. The enabling technology, however, has been large-area digital x-ray detectors with spatial resolution comparable to that of film-screen systems.

Before launching into the mathematical analysis of digital radiography systems, we present a qualitative discussion of the underlying physical principles that impact the imaging performance. In Sec. 16.1.1 we give a brief account of x-ray sources and some considerations on the object being imaged. In Sec. 16.1.2 we describe the main detectors used for digital radiography, and in Sec. 16.1.3, we discuss an important practical issue, scattered radiation.

In Sec. 16.1.4, we begin the mathematical analysis, starting with deterministic properties. One goal here is to see what approximations are needed if we wish to use linear systems theory. In Sec. 16.1.5, we treat the stochastic properties of digital radiographs in detail. Sections 16.1.6 and 16.1.7 bring together the deterministic and stochastic properties and show how they influence image quality for detection and estimation tasks, respectively.

Essential background material for this discussion is found in Chaps. 10 and 12, especially Secs. 10.3, 10.4 and 12.3. An excellent general reference is the three-volume SPIE Handbook of Medical Imaging.

16.1.1 The source and the object

X-ray sources Though radioisotope sources and synchrotrons have been used for x-ray imaging, by far the most common source today is the one used by Roentgen himself, a vacuum tube in which high-energy electrons bombard a metal anode and create x rays. Two kinds of x rays are produced, Bremsstrahlung (German for *braking radiation*) and characteristic x rays. Bremsstrahlung arises from the deceleration of the electrons in the anode material and has a continuous energy spectrum extending up to the electron energy. The characteristic radiation is produced when the electron collides with an atom in the anode and creates a vacancy in an inner electron shell; this vacancy is then filled by an electron dropping into it from a higher energy level, liberating energy in the form of an x ray. X rays produced in this way by transitions between discrete energy levels have a discrete spectrum characteristic of the anode material. As discussed in Sec. 12.3.1, characteristic x rays are also produced in x-ray detectors, though there the initial inner-shell vacancy is produced by photoelectric absorption rather than electron impact.

As a numerical example, suppose the anode is tungsten (W), for which the K-shell binding energy is about 70 keV, and that the applied potential across the x-ray tube is 100 kV. Since the electrons are produced thermally, they have very low energy when they exit the cathode, but they are accelerated toward the anode and have an energy of 100 keV when they arrive there. This energy is sufficient to create a vacancy in the K shell, and a transition from the L shell to the K shell will then produce an x ray of about 59 keV. Vacancies in the L shell itself result in x rays around 8–9 keV. The 100 keV electrons also produce Bremsstrahlung photons with energies ranging up to 100 keV. Thus the overall energy spectrum, shown in Fig. 16.1, consists of discrete lines characteristic of W plus a continuous Bremsstrahlung spectrum depending mainly on the electron energy (or accelerating voltage).

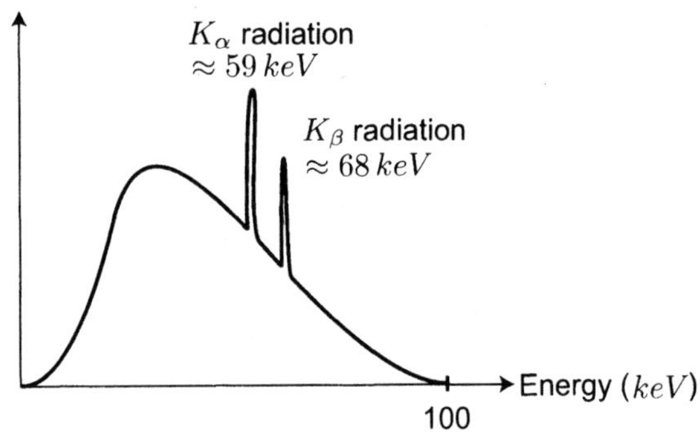

Fig. 16.1 Energy spectrum for a tungsten-target x-ray tube operated at 100 kV.

To get sharp images, it is necessary to focus the electrons tightly so that the x-ray source approximates a point source, but it is also desirable to use a large current in order to get a large x-ray flux. These two conditions are contradictory since forcing a large amount of current into a small area can result in strong local heating of the anode, possibly even to the melting point. One way to minimize this problem is to use a tilted anode as shown in Fig. 16.2 and to focus the electrons into something approximating a line rather than a point; viewed end-on, the line looks like a small point, but from other directions it has a greater extent. Another

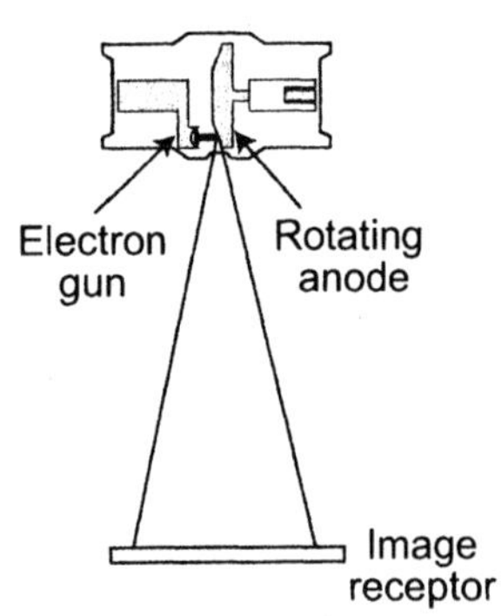

Fig. 16.2 Illustration of an x-ray imaging system using an x-ray tube with a tilted anode. The anode may rotate to spread the heat load over a larger area.

common measure to allow greater current and hence greater x-ray flux is to rotate the anode, spreading the heat over a much larger area. As we shall see in Sec. 16.1.4, tilting the anode leads to a strongly shift-variant blur, but the rotation has no effect on the blur because the electron focal spot does not move.

Objects At the energies used in radiography, x-ray wavelengths are very small compared to resolvable image details; for example, 60 keV x rays have a wavelength of about 0.2 Angstrom or 0.02 nm, and digital radiography systems have spatial resolutions around 50–100 μm, so diffraction in the object is negligible on this scale. Moreover, the refractive index of all materials for x rays is very close to unity, so refractive effects are negligible as well. The physical effects that we do have to consider are photoelectric absorption and Compton scattering. (See Sec. 12.3.1 for a brief review of the x-ray physics.) The object is thus described by the distribution of x-ray absorption and scattering coefficients, and the propagation of x rays through the object is well described by the Boltzmann transport equation (Chap. 10).

A transmission image of a scattering and absorbing object, obtained with a point-like x-ray source, consists of two components. First, an x-ray photon may travel without scattering along a straight line from the source point to the detector, but the probability of this happening depends on the total x-ray attenuation (absorption plus scattering) along the line. The unscattered photons thus produce a 2D shadow image of the 3D object on the detector. The shadow is sharp if the x-ray source is small.

The second component of the image consists of photons that have undergone one or more Compton-scattering events in the object. These photons form a diffuse background that reduces the contrast of the shadow image and increases the noise level.

Motion of the object is also an issue in medical imaging. Beating of the heart, respiration in the lung or peristalsis in the esophagus during an exposure time can blur the images of these organs, and patient fidgeting can also be problematic. Breath-holding and various Torquemadian restraint devices will reduce motion blur, but the best solution is to shorten the exposure time; unfortunately this requires an increased current through the x-ray tube if the same total number of photons are to be detected.

Radiation dose In any imaging application, we need to know how the object affects the image, but in medical radiography we also need to know how the imaging system affects the object. X-ray photons that are absorbed or scattered in a patient's body deliver energy to it and can cause biological damage. A full analysis of the system should therefore include computation of the *radiation dose* (absorbed energy per unit mass) along with some assessment of the biological hazards from the dose. Part of the problem of system design is to choose the total x-ray flux and the energy spectrum in such a way as to minimize patient dose while maximizing the image information (as measured by task performance).

16.1.2 X-ray detection

X rays are detected either by converting them to light and detecting the light or by converting them to electrical charge and detecting the charge. Detectors operating on the first principle include film-screen systems, x-ray image intensifiers, scintillator-photodiode arrays, systems in which a lens or mirror couples light from a scintillator to an optical detector, and photostimulable phosphors. Systems that avoid the conversion to light include semiconductor detector arrays and x-ray-sensitive television cameras such as vidicons.

All of these detectors involve an amplification process: one x-ray photon produces a large number of secondary particles, either optical photons or electrons and holes. The randomness in this amplification process, discussed in general terms in Sec. 11.4 and specifically for x-ray detectors in Sec. 12.3, contributes to the image statistics and the resulting image quality. Whatever detector is used, it is always the goal to have the final image quality limited only by the inevitable Poisson statistics of the x-ray beam. If this happy condition is achieved, we say that the detector is *quantum limited.*

In the remainder of this section, we shall sketch the operation of each of these detectors and comment on the major contributions to blur and image statistics.

Film-screen detectors In a film-screen detector, shown in Fig. 16.3, the fluorescent screen absorbs an x-ray photon and produces hundreds of optical photons, which then diffuse out of the screen and expose the film placed in contact with the screen. It is also possible for x rays to be photoelectrically absorbed in the film itself, in which case the exposure is produced by the high-energy photoelectron, but this process is much less probable than absorption in the screen since the screen is usually much thicker than the photographic emulsion and has a higher atomic number.

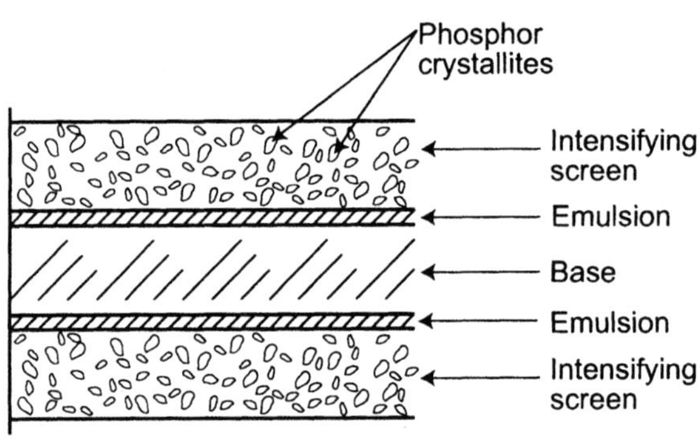

Fig. 16.3 Film-screen detector.

Since thin photographic emulsions have very high spatial resolution, blur in a film-screen detector is dominated by the diffusion of the light from the interaction point to the emulsion. The screen usually consists of relatively small (1–10 μm) fluorescent crystallites or grains in a transparent or translucent binder, so the diffusion of light can be controlled to a degree by varying the grain size and the optical properties of the binder. When these properties are optimized, the only way to get better resolution is to use a thinner screen, but this increases image noise by reducing the number of x-ray photons absorbed. For this reason practical x-ray detectors often use two screens and films with emulsions on both sides, thereby doubling the total screen thickness for a given spatial resolution.

Major statistical limitations in a film-screen detector arise from the random x-ray absorption process, random conversion of the x-ray energy to light, and random conversion of light to latent (developable) photographic grains. We refer to these noise sources, respectively, as quantum noise, amplification noise and film-grain noise. For an excellent treatise on these topics, see Dainty and Shaw (1974).

The developed x-ray film can, of course, be digitized with a scanning microdensitometer, in which case the overall system — film-screen, development process and microdensitometer — could be considered as a digital radiography system, but we shall understand that term in a narrower sense and not consider film-based systems any further in this chapter.

X-ray image intensifiers In an x-ray image intensifier, illustrated in Fig. 16.4, the initial x-ray absorption again takes place in a fluorescent screen, but now the light from the screen impinges on a photocathode rather than a photographic emulsion. Electrons emerging from the photocathode are accelerated and focused by an electrode structure serving as an electron lens, and they then impinge on a second fluorescent screen known as the output phosphor. The physical processes in the output phosphor are essentially the same as in the first fluorescent screen except that there the x-ray photon produces a high-energy photoelectron that excites the luminescence, while in the output phosphor the high-energy electron impinges from the outside.

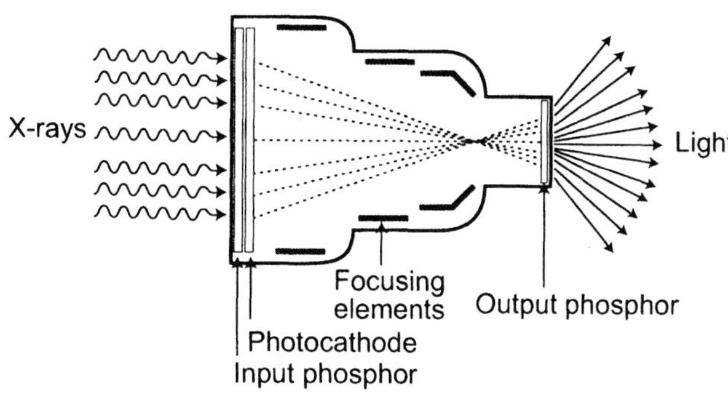

Fig. 16.4 An x-ray image intensifier.

The x-ray image intensifier thus involves two separate amplification steps. Each x-ray photon produces many photoelectrons in the photocathode, and each photoelectron is accelerated to high energy and can produce many optical photons in the output phosphor. The overall gain — optical photons per x-ray photon — is quite high, so there are many options for reading out the final optical image. One important method, known as fluoroscopy or ciné-radiography, records the output with a movie camera so that dynamic images of moving objects can be captured.

There are several sources of blurring in an image intensifier. As with a film-screen system, the light diffuses in the input screen before reaching the photocathode. Then the photoelectrons are not focused perfectly by the electron lens, and the diffusion of light in the output phosphor increases the blur still further. Sometimes the input screen consists of columnar structures such as long, thin crystallites of CsI or hollow tubes loaded with a polycrystalline scintillator material. If these columns are oriented perpendicular to the photocathode, they restrict the lateral diffusion of the light and reduce the blur in the input screen. Also, the output phosphor can be made relatively thin since it must absorb only electrons, not x-ray photons. Under these conditions, the blur may be dominated by the electron optics.

An insidious source of blur in many image intensifiers is the thin glass window that separates the input screen from the photocathode. Light that strikes this window near normal incidence passes through with relatively little reflection loss, but at greater incidence angles the reflection coefficient increases and the light might be reflected several times before finally emerging from the glass. These multiple reflections produce long tails on the point spread function. Even though the tails have fairly low amplitude, they can greatly reduce the contrast of a small detail in a large background since the convolution of the tails with the background can be large. A common term for this phenomenon is *veiling glare.*

Factors influencing the image statistics in an image-intensifier system include the statistics of light production in both the input screen and the output phosphor, the quantum efficiency of the photocathode and, of course, the Poisson statistics of the x-ray flux. Further noise can, in principle, be contributed by the movie camera or other optical device used to record the output image, but noise from this source is usually negligible because of the high gain of the intensifier.

Fluorescent screens with optically coupled readouts Because of the cost and complexity of image intensifiers and the blur in the electron optics, there is an incentive to eliminate the optical intensification and couple the light from a fluorescent screen directly to a CCD camera or other optical readout with a lens, focusing mirror or fiber optics.

The basic problem with this approach is that optical readouts have a small area, while the fluorescent screen must have a large area to cover the desired field of view. If a lens or mirror is used to image the large screen onto the small readout, the optical magnification M must be small. For a simple lens, we know from Sec. 9.6.2 that $M = -q/p$, where p is the distance from the object to the lens and q is the distance from the lens to the image plane. To make M small and still satisfy the imaging equation (9.169), we must take $q \approx f$ and $p \approx -f/M$ (where f is the focal length of the lens). The lens-to-screen distance p must increase as M gets smaller (larger demagnification). As a result, the lens subtends a smaller solid angle at the screen and hence collects fewer of the optical photons emitted by it. A similar conclusion holds for more complicated lenses, mirror systems and even fiber optics; large demagnifications always mean inefficient photon collection.

In fact, the photon collection efficiency in these systems may be so poor that less than one optical photon is collected per absorbed x-ray photon. Since photon collection is a rare event, the statistics of the collected photons approaches Poisson (see Sec. 11.4.1) as M gets small, and in this limit the image noise is much larger, relative to the mean image, than when the Poisson statistics of the x-ray beam dominate. Moreover, even if we succeed in getting the collection efficiency up to the point where more than one optical photon is collected per absorbed x-ray photon, the optical image on the readout is still weak, and noise in the readout itself is more important than it would be with an image intensifier.

Scintillator-photodiode arrays One way to improve the optical collection efficiency is to make a large-area readout such as a photodiode array and place it in contact with the fluorescent screen, thereby producing the electronic counterpart of a film-screen detector. One of the more promising configurations at this writing is a CsI screen and an amorphous-Si photodiode array. These photodiodes are noisier than ones made of crystalline Si, but they are much easier to fabricate into large arrays.

The amplifiers for each photodiode can also be fabricated in the same amorphous-Si structure.

With any kind of integrated-circuit detector array, there is an inevitable problem with variations in gain of the amplifiers. In addition, there can be variations across the array in the mean dark signal because of variations in dark current through the photodiode and offset voltages in the amplifiers. If left uncorrected, these gain and offset variations would show up as an objectionable structure, called *fixed-pattern noise*, superimposed on the desired x-ray image. Fixed-pattern noise can be removed by measuring the gains and offsets of each element in the array and applying the appropriate correction during readout, but one must then be cognizant of the effect of the correction on the image statistics.

Semiconductor detector arrays In any x-ray detector with optical readout, the x-ray photon first produces a high-energy photoelectron (or Compton electron) which then generates a large number of electron-hole pairs. These electrons and holes recombine at a luminescent center and produce optical photons, which are collected with some efficiency in an optical detector where they again produce electron-hole pairs. In a semiconductor detector array, the initial electron-hole pairs produced by the photoelectron are sensed directly with an electrode array.

Detectors based on this principle were treated in detail in Sec. 12.3. As discussed there, trapping of the charge carriers before they reach the electrodes contributes in a complicated way to both the image statistics and the blur.

Other readout mechanisms have also been proposed for semiconductor detectors. One approach is to use a semiconductor with no electrode on one side and to charge this surface with an electron beam or corona discharge. The charge will leak off as a result of the finite conductivity of the material, but absorbed x rays will create free carriers and increase the conductivity, so the remaining charge after some exposure time will be related to the x-ray exposure. This charge can be sensed with a moving electrode scanned over the free surface or with a scanning electron beam as in a vidicon. It is also possible to read out the charge pattern by converting it into a pattern of toner particles as in the Xerox copying process.

Photostimulable phosphors Though trapping is generally deleterious in semiconductor detectors, it is essential in x-ray detectors based on *photostimulable phosphors*. In these materials an absorbed x ray produces many hole-electron pairs, just as in a semiconductor detector, but they are trapped in impurity levels deep within the bandgap. Because the binding energy of these traps is so large, thermal detrapping essentially does not occur, and the spatial pattern of trapped charge is a latent image of the pattern of absorbed x rays. To read out this pattern, the phosphor is scanned with a focused laser beam, usually red light. The energy of a red photon is sufficient to excite a trapped carrier to the conduction band, from which there is some probability that it will recombine and produce an optical photon with approximately the bandgap energy, which makes it blue or ultraviolet. A simple optical detector such as a photomultiplier can then be used to detect this shorter-wavelength light, with a spectral filter to reject the red excitation light. No imaging optics are needed on the collection channel since the emitted light must have come from the point illuminated by the scanning red beam.

Digital imaging systems based on photostimulable phosphors are known commercially as *computed radiography* or CR systems. This is an unfortunate misnomer since the computer plays very little role in this kind of imaging. Of course, the image is stored digitally, possibly manipulated in the computer and then displayed on a CRT screen, but the image-formation process is purely analog. Unlike computed tomography, where no image at all could be formed without computation, computed radiography could have been invented before the computer.

The main sources of blur in CR systems are the finite focal spot of the scanning beam and diffusion of the red light into the photostimulable phosphor. Image statistics are again determined by amplification processes; we must consider the random number of trapped carriers per x-ray photon, the probability of detrapping under red excitation and the probability that a short-wavelength optical photon will be registered in the collection detector. These processes are complicated but understandable within the framework presented in Sec. 11.4.

16.1.3 Scattered radiation

So far we have considered only unscattered radiation, but in many practical situations the majority of x-ray photons arriving at the detector plane have been scattered at least once in the object. These scattered photons reduce the image contrast and increase the noise level. Methods for computing the scatter distribution function were discussed in detail in Chap. 10, so here we confine the discussion to methods for minimization of the scatter component of radiographic images. The photon flux at the detector plane is characterized by the distribution function $w(\mathbf{r}, \hat{\mathbf{s}}, \mathcal{E}, t)$, and we shall discuss in turn how each of the variables in this function can potentially be used to distinguish scattered from unscattered photons.

Temporal gating The time variable is of interest only if the source is pulsed rapidly compared to the transit time of photons across the object and if the detector has similar time resolution; neither of these conditions is satisfied in systems designed for medical radiography. Since the speed of light in American units is one foot per nanosecond, and typical clinical objects are a foot or so across, we would need a subnanosecond source and detector. Such rapid pulses of x rays can be created with pulsed beams of electrons or pulsed lasers, but the technology required is too expensive for the medical market, and it would be of no use without a detector with nanosecond resolution.

An individual x-ray detector can indeed have the requisite response time, but it requires a fast amplifier and electronic gate or sample-and-hold circuit, which would then have to be duplicated for each element in a detector array. There is no reason in principle why this could not be done with an array of gated integrators as discussed in Sec. 12.2.4, but the detector arrays being developed at this writing have much slower response.

The other option to consider is a rapid shutter that admits photons to the detector surface for only a brief time. Optical detectors can be gated in this way by use of electro-optic shutters, but no comparable shutter exists for x rays. An x-ray image intensifier could, in principle, be gated either by putting an optical shutter between the output phosphor and the final optical readout or by gating the electron-amplification stage, but the response time would always be limited by the input fluorescent screen. Current scintillators have an optical rise time of at least a few nanoseconds and decay times up to a microsecond or more, so it is unlikely that subnanosecond response could be obtained by gating an image intensifier.

Energy discrimination The energy variable is potentially useful for discriminating against scattered radiation since photons lose energy upon Compton scattering (see Sec. 10.3.7). As with the time variable, however, energy is useful only if the source has a narrow spread in energies and the detector has good energy resolution. As seen in Fig. 16.1, commonly used x-ray sources have both a discrete line component and a continuous spectrum. With judicious choice of operating voltage and filtration in the x-ray beam, the relative strength of the discrete component can be enhanced, but some continuum background will always remain with electron-beam tubes. Heavier charged particles such as protons or heavy ions can also be used to excite x rays, and they produce a higher ratio of characteristic radiation to Bremsstrahlung, but the expense is prohibitive for clinical use. Finally, synchrotrons can produce quite monoenergetic x-ray beams, but again the issue is expense.

The energy resolution of semiconductor and scintillation detectors was discussed in Sec. 12.3. Since energy discrimination depends on the ability to analyze an individual photon absorption event, energy information is useless if more than one event occurs in a detector element during one readout period. Moreover, as discussed in Sec. 12.3, an event can produce signals in neighboring readout channels as a result of light spread in scintillators or charge trapping in semiconductors, so the requirement is really that there be no more than one event per readout period in some cluster of detector elements. Furenlid *et al.* (2000) have analyzed the probability of this happening as a function of input flux rate, detector element size and readout time. As a rule of thumb, the flux should be kept to less than 0.1 x-ray photons per cluster per readout period.

Even if a monoenergetic source could be used and the input flux could be kept low enough to resolve individual photons, it is still not evident that energy discrimination would be useful in digital radiography. The problem is that the fractional energy loss upon Compton scattering, $\Delta\mathcal{E}/\mathcal{E}_0$, is small at energies much less than mc^2, where m is the mass of the electron [see (10.226)]. Since $mc^2 = 511$ keV and diagnostic radiology mainly uses photons in the 10–100 keV range, the energy loss is small, and excellent energy resolution would be required to distinguish a scattered photon from an unscattered one.

While photon-by-photon energy discrimination is probably not useful for scatter rejection, it may still be advantageous to modify the energy-dependence of the source function $\Xi_{p,\mathcal{E}}(\mathbf{r}, \hat{\mathbf{s}}, \mathcal{E}, t)$ and the detector response $d_{\mathbf{m}}(\mathbf{r}, \hat{\mathbf{s}}, \mathcal{E})$. For example, if the x rays are produced by bombarding a W target, we can place a tantalum (Ta) sheet in front of the x-ray tube. Since Ta has a K absorption edge of 67.4 keV, it has a relatively low absorption for W characteristic x rays of energy around 58 keV, and the relative strength of the line spectrum will be increased. Then, if we use Gd_2O_2S (gadolinium oxysulfide) as the detector material, the detector will be very efficient for the unscattered W characteristic photons, but relatively inefficient for scattered radiation that falls below the Gd K edge at 50 keV. The thicknesses of the Ta sheet and the Gd_2O_2S then become parameters to be optimized in terms of task performance.

Angular discrimination The angular distributions of scattered and unscattered radiation are quite different. All of the unscattered photons travel along straight lines from the focal spot to the detector, so the radiance at any point on the detector is nearly an angular delta function. The scattered photons, on the other hand, have a very broad angular distribution, and they can be rejected by making $d_{\mathbf{m}}(\mathbf{r}, \hat{\mathbf{s}}, \mathcal{E})$ a sharp function of $\hat{\mathbf{s}}$.

One way to do this is to to use an anti-scatter collimator as shown in Fig. 16.5. The bores in the collimator point to the focal spot, so they do not impede any unscattered photons that hit the open areas, but they have a narrow acceptance angle and reject most of the scattered photons. Sometimes the collimator is moved during the exposure to wash out moiré effects between the collimator and the detector array.

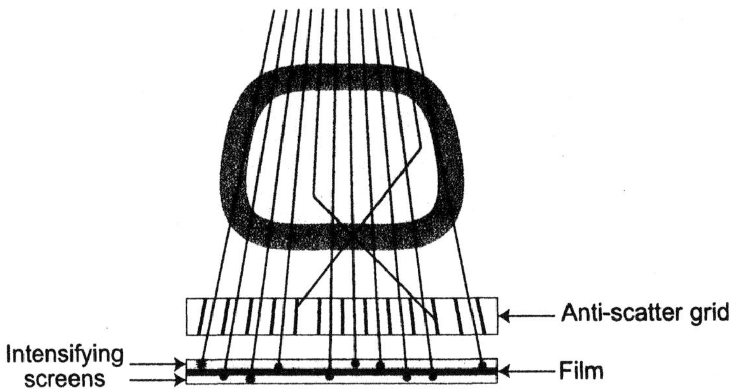

Fig. 16.5 Illustration of the use of an anti-scatter grid in a film-screen x-ray imaging system.

A related approach is to confine the x-ray beam to a thin slab with a slot collimator placed between the x-ray source and the object as shown in Fig. 16.6. With a similar slot collimator in front of the detector, unscattered photons pass unimpeded from source to detector, but scattered photons are likely to scatter out of the slab and miss the detector. Even if they scatter back into the slab, they are likely to be out of the angular acceptance of the detector collimator. The desired field of view is then covered by scanning the entire structure — source, detector and two collimators — in the direction perpendicular to the slab. A side benefit of this geometry is that the required detector area is much less than the area of the field of view; even 1D detector arrays can be used.

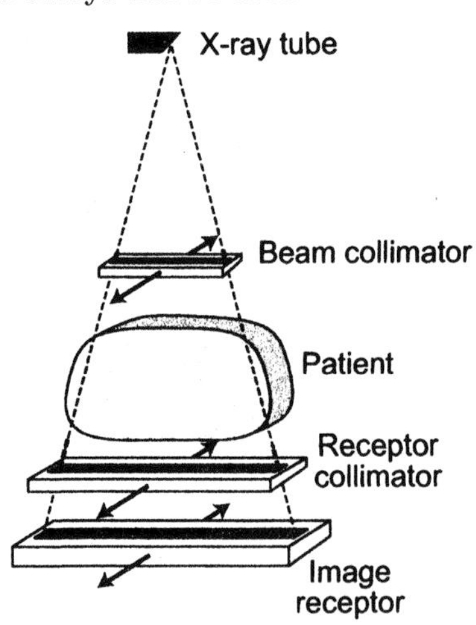

Fig. 16.6 Slot-scanning system.

Spatial filtering The final variable we can use for discriminating against scattered radiation is spatial position. There is nothing useful we can do to modify the dependence of $d_{\mathbf{m}}(\mathbf{r}, \hat{\mathbf{s}}, \mathcal{E})$ on $\mathbf{r}$ since scattered and unscattered photons occupy the same area on the detector face, but we can operate on the recorded data as a function of the spatial index m. That is, we can spatially filter the digital image.

One approach to spatial filtering is to recognize that scattered radiation contains no fine details, hence consists mainly of low spatial frequencies, so a high-pass filter implemented in the discrete Fourier domain should preferentially exclude scattered radiation and preserve the interesting image details. On average, this statement is true, but scattered radiation also has spatially uncorrelated fluctuations or noise associated with it, and this noise will pass through the filter. Moreover, the relative contribution of primary and scattered radiation will vary over the image, so shift-invariant filtering may not be optimal.

Another approach, which sounds different but turns out to be equivalent to Fourier-domain filtering, is unsharp masking. In this approach, a local estimate of the low-frequency background is made by averaging the image over a region and then subtracting the average from the original image. The result is equivalent to convolving the image with a digital filter having a central positive core one pixel wide and a small negative value over the averaging region.

A more sophisticated approach is to estimate the amount of scatter locally and subtract off just that component of the background rather than all low-frequency components. This estimate could be obtained from an approximate solution of the Boltzmann equation or through a Monte Carlo simulation. In both cases a greatly simplified model of the object, such as a uniform cylinder, might give an adequate estimate of the scatter image. If there are unknown parameters such as the diameter of the cylinder, they could be estimated from the measured image data.

16.1.4 Deterministic properties of shadow images

As noted above, propagation of x-ray photons through an object can be described by the Boltzmann transport equation (10.132), with terms accounting for the x-ray source, photon propagation, Compton scattering and photoelectric absorption. In this section we shall ignore the scattering term and compute the mean image produced by unscattered photons.

Since the x-ray source is pulsed in digital radiography, it is necessary in general to use the full time-dependent Boltzmann equation, but we can usually assume that the pulse width is long compared to the propagation time of the x rays across the object. For simplicity, we shall also assume that the object is independent of time over the pulse duration. With these assumptions, we can use the steady-state equation even with a time-dependent source, and (10.132) takes the form

$$-c\mu_{tot}(\mathbf{r}, \mathcal{E})\, w(\mathbf{r}, \hat{\mathbf{s}}, \mathcal{E}, t) + \Xi_{p,\mathcal{E}}(\mathbf{r}, \hat{\mathbf{s}}, \mathcal{E}, t) + [\mathcal{K}\mathbf{w}](\mathbf{r}, \hat{\mathbf{s}}, \mathcal{E}, t) - c\hat{\mathbf{s}} \cdot \nabla w(\mathbf{r}, \hat{\mathbf{s}}, \mathcal{E}, t) = 0\,, \tag{16.1}$$

where c is the speed of light (recall that the refractive index is near one for x rays, so $c_m \approx c$), and the other quantities are defined in Chap. 10.

If a solution to (16.1) has been found for a given object and source-detector configuration, the next step is to compute the mean output of each detector. It will be convenient to designate the detector elements by the 2D multi-index $\mathbf{m} = (m_x, m_y)$, where the components take on integer values. If the detector is linear, then the mean output of the $\mathbf{m}^{th}$ detector is given by (10.239) as

$$\overline{g}_{\mathbf{m}} = \int_P d^2r \int_0^\infty d\mathcal{E} \int_{2\pi} d\Omega \int_0^\tau dt\; d_{\mathbf{m}}(\mathbf{r}, \hat{\mathbf{s}}, \mathcal{E})\, w(\mathbf{r}, \hat{\mathbf{s}}, \mathcal{E}, t)\,, \tag{16.2}$$

where τ is the exposure time, $d_{\mathbf{m}}(\mathbf{r}, \hat{\mathbf{s}}, \mathcal{E})$ is the detector response function (assumed independent of time), and the spatial integral is over a plane immediately adjacent to the detector. Recall from Chap. 10 that we use the 2D vector $\mathbf{r}$ and the 3D vector $\mathbf{r}$ to refer to the same spatial point. Thus, if we take the plane P to be $z = 0$, then $\mathbf{r} = (x, y, 0)$, $\mathbf{r} = (x, y)$ and $d^2r = dx\, dy$. Note also that we have added an overbar to $g_{\mathbf{m}}$ to emphasize that the Boltzmann equation gives the mean photon distribution and hence the mean detector output.

The unscattered image If we consider only the unscattered photons, the Boltzmann equation takes the form of (10.147):

$$\hat{\mathbf{s}} \cdot \nabla w = \frac{1}{c}\Xi_{p,\mathcal{E}} - \mu_{tot} w\,. \tag{16.3}$$

We know from Sec. 10.3.3 that the general solution to this equation is the attenuated x-ray transform of the source distribution. An explicit solution was given in that section for an ideal point source, and an ideal detector model was discussed in Sec. 10.4.2; in this section we extend the discussion to more realistic source and detector models.

To describe the source, we assume that the anode of the x-ray tube lies in the plane defined by $\mathbf{r} \cdot \hat{\mathbf{n}}_a = p_a$ (where subscript a indicates *anode*), $\hat{\mathbf{n}}_a$ is a unit vector normal to the anode and p_a specifies the location of the anode along the line defined by this unit vector. We assume also that the temporal and spectral dependences of the source function are independent of the spatial and angular dependences, so we can write

$$\Xi_{p,\mathcal{E}}(\mathbf{r}, \hat{\mathbf{s}}, \mathcal{E}, t) = A(t)\, N(\mathcal{E})\, (\hat{\mathbf{n}}_a \cdot \hat{\mathbf{s}})\, L_p(\mathbf{r}, \hat{\mathbf{s}})\, \delta(p_a - \mathbf{r} \cdot \hat{\mathbf{n}}_a)\,, \tag{16.4}$$

where $N(\mathcal{E})$ is a normalized spectral function satisfying $\int_0^\infty N(\mathcal{E})\, d\mathcal{E} = 1$, and $A(t)\, L_p(\mathbf{r}, \hat{\mathbf{s}})$ is the photon radiance of the source. The factor $(\hat{\mathbf{n}}_a \cdot \hat{\mathbf{s}})$ is discussed in Sec. 10.4.3 [see (10.269)].

With this source model and (10.151), the distribution function is given by

$$w(\mathbf{r}, \hat{\mathbf{s}}, \mathcal{E}, t) = \frac{1}{c} A(t)\, N(\mathcal{E})$$

$$\times \int_0^\infty d\ell\; (\hat{\mathbf{n}}_a \cdot \hat{\mathbf{s}})\, L_p(\mathbf{r} - \hat{\mathbf{s}}\ell, \hat{\mathbf{s}})\, \delta[p_a - (\mathbf{r} - \hat{\mathbf{s}}\ell) \cdot \hat{\mathbf{n}}_a] \exp\left[-\int_0^\ell d\ell'\, \mu_{tot}(\mathbf{r} - \hat{\mathbf{s}}\ell', \mathcal{E})\right]. \tag{16.5}$$

The argument of the delta function vanishes when

$$\ell = \frac{\hat{\mathbf{n}}_a \cdot \mathbf{r} - p_a}{\hat{\mathbf{n}}_a \cdot \hat{\mathbf{s}}} \equiv \ell_0\,, \tag{16.6}$$

which occurs when a ray extended along $-\hat{\mathbf{s}}$ from point $\mathbf{r}$ intersects the anode plane. The delta function transforms via (2.28) to $(\hat{\mathbf{n}}_a \cdot \hat{\mathbf{s}})\, \delta[p_a - (\mathbf{r} - \hat{\mathbf{s}}\ell) \cdot \hat{\mathbf{n}}_a] = \delta(\ell - \ell_0)$, so (16.5) integrates to

$$w(\mathbf{r}, \hat{\mathbf{s}}, \mathcal{E}, t) = \frac{1}{c} A(t)\, N(\mathcal{E})\, L_p(\mathbf{r} - \hat{\mathbf{s}}\ell_0, \hat{\mathbf{s}}) \exp\left[-\int_0^{\ell_0} d\ell'\, \mu_{tot}(\mathbf{r} - \hat{\mathbf{s}}\ell', \mathcal{E})\right]. \tag{16.7}$$

The interpretation of this equation is straightforward when one recalls from Sec. 10.2 that c times the distribution function is the spectral photon radiance, and that radiance is conserved along rays in free space; thus (16.7) says that the radiance at point $\mathbf{r}$ and direction $\hat{\mathbf{s}}$ is the source radiance along the same ray but attenuated by passage through the object. Note that the factor $(\hat{\mathbf{n}}_a \cdot \hat{\mathbf{s}})$ has cancelled out; we do not need to know the tilt angle of the anode explicitly if we know the source radiance.

With this distribution function and (16.2), the mean output of the $\mathbf{m}^{th}$ detector is

$$\overline{g}_{\mathbf{m}} = \frac{1}{c}\int_{-\infty}^{\infty} dt\ A(t) \int_0^{\infty} d\mathcal{E}\ N(\mathcal{E})$$

$$\times \int_P d^2r \int_{2\pi} d\Omega\ d_{\mathbf{m}}(\mathbf{r}, \hat{\mathbf{s}}, \mathcal{E})\, L_p(\mathbf{r} - \hat{\mathbf{s}}\ell_0, \hat{\mathbf{s}}) \exp\left[-\int_0^{\ell_0} d\ell'\ \mu_{tot}(\mathbf{r} - \hat{\mathbf{s}}\ell', \mathcal{E})\right] . \quad (16.8)$$

This is the general nonlinear mapping from the object, described by the function $\mu_{tot}(\mathbf{r}, \mathcal{E})$, to the digital data.

To simplify this equation a bit, we can assume that the spectral response of the detector is the same for all $\mathbf{m}$ and is independent of direction and position of the radiation, so we can write

$$d_{\mathbf{m}}(\mathbf{r}, \hat{\mathbf{s}}, \mathcal{E}) = d_{\mathbf{m}}(\mathbf{r}, \hat{\mathbf{s}})\, S_d(\mathcal{E})\,, \quad (16.9)$$

where $S_d(\mathcal{E})$ is the spectral response function. If we also assume that the x-ray attenuation coefficient is approximately independent of energy, we have

$$\overline{g}_{\mathbf{m}} = C \int_P d^2r \int_{2\pi} d\Omega\ d_{\mathbf{m}}(\mathbf{r}, \hat{\mathbf{s}})\, L_p(\mathbf{r} - \hat{\mathbf{s}}\ell_0, \hat{\mathbf{s}}) \exp\left[-\int_0^{\ell_0} d\ell'\ \mu_{tot}(\mathbf{r} - \hat{\mathbf{s}}\ell')\right], \quad (16.10)$$

where

$$C \equiv \frac{1}{c}\int_{-\infty}^{\infty} dt\ A(t) \int_0^{\infty} d\mathcal{E}\ N(\mathcal{E})\, S_d(\mathcal{E})\,. \quad (16.11)$$

The object is now described by the purely spatial function $\mu_{tot}(\mathbf{r})$, and (16.10) is the mapping of this object to the mean data.

Linearization In spite of the simplifications, (16.10) is still highly nonlinear. To secure the blessings of linear systems theory, we might consider expanding the exponential in (16.10) and retaining only linear terms in $\mu_{tot}(\mathbf{r})$, but this approximation is rarely valid; the x-ray attenuation coefficient in soft tissue is about 0.2 cm^{-1} for energies used in diagnostic radiology, so 10 cm of tissue corresponds to an attenuation of e^{-2}.

A better way to linearize is to consider a thick object to be made up of thin slabs parallel to the detector and to look at the image of one slab at a time. To do this, we adopt a coordinate system in which the detector lies in the plane $z = 0$ and the x rays propagate generally in the $+z$ direction. In this plane, then, $\mathbf{r} = (x, y, 0) \equiv (\mathbf{r}, 0)$, and $d_{\mathbf{m}}(\mathbf{r}, \hat{\mathbf{s}})$ becomes $d_{\mathbf{m}}(\mathbf{r}, \hat{\mathbf{s}})$. Also, if $\hat{\mathbf{s}} = (s_x, s_y, s_z)$, we define the 2D vector $\mathbf{s}_\perp = (s_x, s_y)$. Note that $\mathbf{s}_\perp$ is not a unit vector and that $s_z = \cos\theta$, where θ is the angle between $\hat{\mathbf{s}}$ and the z axis. With these definitions, the argument of μ_{tot} in (16.10) is $\mathbf{r} - \hat{\mathbf{s}}\ell' = (\mathbf{r} - \mathbf{s}_\perp\ell', -s_z\ell')$.

We can now divide the object into slabs of thickness Δz and approximate the integral in the exponential factor of (16.10) with a Riemann sum:

$$\exp\left[-\int_0^{\ell_0} d\ell'\, \mu_{tot}(\mathbf{r} - \hat{\mathbf{s}}\ell')\right] \approx \exp\left[-\Delta\ell \sum_{j=1}^{J} \mu_{tot}(\mathbf{r} - \hat{\mathbf{s}}j\Delta\ell)\right]$$

$$= \prod_{j=1}^{J} \exp\left[-\frac{\Delta z}{s_z}\, \mu_{tot}\left(\mathbf{r} + \mathbf{s}_\perp \frac{z_j}{s_z}, -z_j\right)\right], \tag{16.12}$$

where $J = \ell_0/\Delta\ell$, $z_j = -js_z\Delta\ell$ and $\Delta z/s_z = \Delta\ell$. Each factor in this product can be expressed in terms of the transmission of the slab for x rays travelling in the $\hat{\mathbf{s}}$ direction. Since this transmission depends on the angle θ between the slab normal and $\hat{\mathbf{s}}$, we write it as

$$t_j^{(\theta)}(\mathbf{r}) \equiv \exp\left[-\frac{\Delta z}{\cos\theta}\, \mu_{tot}(\mathbf{r}, -z_j)\right], \tag{16.13}$$

and (16.10) becomes

$$\overline{g}_{\mathbf{m}} = C \int_P d^2r \int_{2\pi} d\Omega\, d_{\mathbf{m}}(\mathbf{r}, \hat{\mathbf{s}})\, L_p(\mathbf{r} - \hat{\mathbf{s}}\ell_0, \hat{\mathbf{s}}) \left[\prod_{j=1}^{J} t_j^{(\theta)}\left(\mathbf{r} + \mathbf{s}_\perp \frac{z_j}{s_z}\right)\right]. \tag{16.14}$$

If we are interested in one particular slab, say $j = k$, we can factor out the transmittance of that slab and write

$$\overline{g}_{\mathbf{m}} = C \int_P d^2r \int_{2\pi} d\Omega\, d_{\mathbf{m}}(\mathbf{r}, \hat{\mathbf{s}})\, L_p(\mathbf{r} - \hat{\mathbf{s}}\ell_0, \hat{\mathbf{s}}) \left[\prod_{j\neq k}^{J} t_j^{(\theta)}\left(\mathbf{r} + \mathbf{s}_\perp \frac{z_j}{s_z}\right)\right] t_k^{(\theta)}\left(\mathbf{r} + \mathbf{s}_\perp \frac{z_k}{s_z}\right). \tag{16.15}$$

This equation is now a linear mapping from $t_k^{(\theta)}(\mathbf{r})$ to $\overline{g}_m$, but it is is highly shift-variant since the effective source distribution depends on all of the other $t_j^{(\theta)}(\mathbf{r})$. In fact, (16.15) is a good example of an object-dependent system function as introduced in Sec. 7.5.3.

Note that the linearization has been obtained without expanding the exponential; $\overline{g}_{\mathbf{m}}$ is a linear functional of the transmittance of the slab of interest, though it is still a nonlinear functional of the attenuation coefficient in the slab. If we choose Δz small enough, of course, we can expand the exponential and retain only the constant and linear terms. Then we find that $\overline{g}_{\mathbf{m}}$ is an affine functional of μ_{tot} in the slab.

Another way to attempt to linearize the imaging equation is to take its logarithm. It is not obvious that this will work since we cannot take the logarithm under the integral sign in (16.10), but it often leads to a good linear approximation. We shall discuss this option in more detail in Sec. 16.1.7 in the context of estimation tasks.

Spatial resolution in a single slab To better understand (16.15) and the spatial-resolution limitations in digital radiography, let us consider an object consisting of just one slab. In fact, radiographic systems are often evaluated with test patterns consisting of fine openings in a thin slab of lead, tungsten or other highly absorbing material. This test pattern then mimics the transmittance of a single slab out of a thick object, but at much higher contrast and without the confusion caused by overlying layers.

With an object consisting of a single slab in the plane $z = -z_0$ and specified by the transmittance $t_{obj}^{(\theta)}(\mathbf{r})$, (16.15) becomes

$$\overline{g}_{\mathbf{m}} = C \int_P d^2r \int_{2\pi} d\Omega \; d_{\mathbf{m}}(\mathbf{r}, \hat{\mathbf{s}}) \, L_p(\mathbf{r} - \hat{\mathbf{s}}\ell_0, \hat{\mathbf{s}}) \, t_{obj}^{(\theta)}\left(\mathbf{r} + \mathbf{s}_\perp \frac{z_0}{s_z}\right) . \tag{16.16}$$

This expression is now strictly linear in the object transmittance $t^{(\theta)}(\mathbf{r})$; the system function no longer depends on the object in any way, though the measured object property depends on the system since the transmittance depends on ray angle θ in general. (The transmittance is approximately independent of θ for a test pattern consisting of openings in a thin plate if the size of the openings is large compared to the thickness, but continuous slabs of attenuating material such as tissue will always exhibit some angular dependence.)

To simplify the integrals in (16.16), we can take advantage of the small size of the detector elements in practical digital radiography systems and the small solid angle subtended by practical focal spots. Even though the integral over solid angle allows a range of 2π ster in general, the spatial dependences (the first arguments) of the functions $d_{\mathbf{m}}$ and L_p restrict the range to rays close to $\hat{\mathbf{s}}_{\mathbf{m}}$, the unit vector directed from the center of the focal spot to the center of the $\mathbf{m}^{th}$ detector element. If the angular dependences (the second arguments) of $d_{\mathbf{m}}$ and L_p are weak over this range, then $d_{\mathbf{m}}(\mathbf{r}, \hat{\mathbf{s}}) \approx d_{\mathbf{m}}(\mathbf{r}, \hat{\mathbf{s}}_{\mathbf{m}})$ and $L_p(\mathbf{r} - \hat{\mathbf{s}}\ell_0, \hat{\mathbf{s}}) \approx L_p(\mathbf{r} - \hat{\mathbf{s}}\ell_0, \hat{\mathbf{s}}_{\mathbf{m}})$ in (16.16).

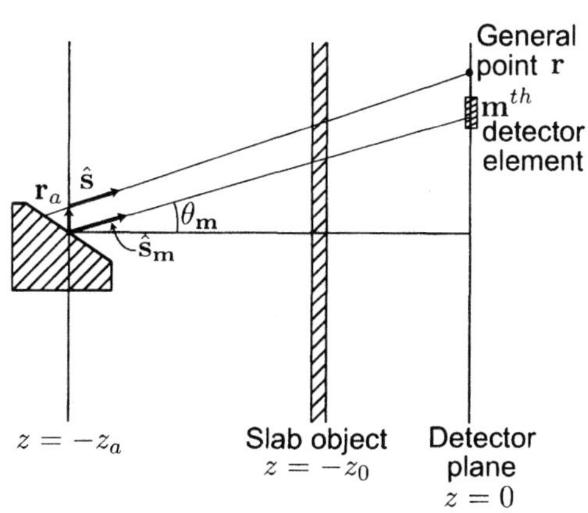

Fig. 16.7 Geometry for imaging a thin slab object with a tilted-anode x-ray source.

We can now erect a plane normal to the z axis and passing through the focal spot. We denote this plane by $z = -z_a$ and let transverse position in the plane be specified by the 2D vector $\mathbf{r}_a$ (see Fig. 16.7). We know that ℓ_0 is the distance along direction $-\hat{\mathbf{s}}_{\mathbf{m}}$ from a general point $\mathbf{r}$ in the detector plane ($z = 0$) to a point on the anode plane, and we can approximate it with ℓ_{0m}, the distance from the center of the $\mathbf{m}^{th}$ detector element to the center of the focal spot. Noting also that $z_a \approx -\ell_{0m} \cos\theta_{\mathbf{m}}$, where $\theta_{\mathbf{m}}$ is the angle between the z axis and the line from the focal spot to the $\mathbf{m}^{th}$ detector, we can write the element of solid angle as

$$d\Omega = \frac{\cos\theta_{\mathbf{m}}}{\ell_{0m}^2}\, d^2 r_a = \frac{\cos^3\theta_{\mathbf{m}}}{z_a^2}\, d^2 r_a \,. \tag{16.17}$$

Moreover, $\mathbf{r}_a = \ell_0\,\mathbf{s}_\perp \approx \ell_{0m}\,\mathbf{s}_\perp$. Thus, since $\mathfrak{r} = (\mathbf{r}, 0)$ in plane P, we can rewrite the 3D spatial argument of L_p as

$$\mathfrak{r} - \hat{\mathbf{s}}\ell_0 = (\mathbf{r} - \mathbf{r}_a, -z_a)\,. \tag{16.18}$$

With these approximations and substitutions, (16.16) becomes

$$\overline{g}_{\mathbf{m}} = \frac{C\cos^3\theta_{\mathbf{m}}}{z_a^2} \int_\infty d^2 r \int_\infty d^2 r_a\, d_{\mathbf{m}}(\mathbf{r}, \hat{\mathbf{s}}_{\mathbf{m}})\, L_p(\mathbf{r} - \mathbf{r}_a, -z_a, \hat{\mathbf{s}}_{\mathbf{m}})\, t_{obj}^{(\theta_{\mathbf{m}})}\left(\mathbf{r} - \mathbf{r}_a \frac{z_0}{z_a}\right). \tag{16.19}$$

Note that we have extended the ranges of integration to the infinite planes, assured that the functions $d_{\mathbf{m}}$ and L_p will limit the integrands to small portions of the planes.

Magnification To help interpret (16.19), consider a point source of x rays and a small point-like detector element. Without loss of generality, we can put the point source at the 2D origin of coordinates in the plane $z = -z_a$, so $L_p(\mathbf{r}', -z_a, \hat{\mathbf{s}}_{\mathbf{m}}) \propto \delta(\mathbf{r}')$ and $d_{\mathbf{m}}(\mathbf{r}, \hat{\mathbf{s}}_{\mathbf{m}}) \propto \delta(\mathbf{r} - \mathbf{r}_{\mathbf{m}})$, where $\mathbf{r}_{\mathbf{m}}$ is the 2D location of the $\mathbf{m}^{th}$ detector in the plane $z = 0$. Then (16.19) becomes

$$\overline{g}_{\mathbf{m}} \propto \int_\infty d^2 r \int_\infty d^2 r_a\, \delta(\mathbf{r} - \mathbf{r}_{\mathbf{m}})\, \delta(\mathbf{r} - \mathbf{r}_a)\, t_{obj}^{(\theta_{\mathbf{m}})}\left(\mathbf{r} - \mathbf{r}_a \frac{z_0}{z_a}\right) = t_{obj}^{(\theta_{\mathbf{m}})}\left[\mathbf{r}_{\mathbf{m}}\left(1 - \frac{z_0}{z_a}\right)\right]. \tag{16.20}$$

Thus the image, as a function of the detector coordinate $\mathbf{r}_{\mathbf{m}}$, is a magnified version of the object transmittance. The magnification M is

$$M = \left(1 - \frac{z_0}{z_a}\right)^{-1} = \frac{z_a}{z_a - z_0}\,, \tag{16.21}$$

which is the source-to-detector distance divided by the source-to-object distance. If the object is in contact with the detector ($z_0 = 0$), then $M = 1$, but if the object is close to the source there is a large geometric magnification. Since z_a and z_0 are both negative numbers with our conventions, M is always positive (there is no inversion), and in fact $M \geq 1$.

Point response function If we allow a finite extent for the source and detector element, then the image of a slab object is not just a magnified version of the object, but it can always be cast into our usual form of a linear CD mapping. If we define the object as $f(\mathbf{r}) \equiv t_{obj}^{(\theta)}(\mathbf{r})$, then by a change of variables we can write

$$\overline{g}_{\mathbf{m}} = \int_\infty d^2 r'\, h_{\mathbf{m}}(\mathbf{r}')\, f(\mathbf{r}')\,, \tag{16.22}$$

where

$$h_{\mathbf{m}}(\mathbf{r}') = \frac{C\cos^3\theta_{\mathbf{m}}}{z_a^2} \int_\infty d^2 r_a\, d_{\mathbf{m}}\left(\mathbf{r}' + \mathbf{r}_a \frac{M-1}{M}, \hat{\mathbf{s}}_{\mathbf{m}}\right) L_p\left(\mathbf{r}' - \mathbf{r}_a \frac{1}{M}, -z_a, \hat{\mathbf{s}}_{\mathbf{m}}\right). \tag{16.23}$$

In essence, this kernel is the cross-correlation of the detector response function and the spatial part of the source function, though these two functions appear with different magnifications in the integrand.

Two limits of (16.23) are of interest. First, if the object is in contact with the detector so that $z_0 = 0$ and $M = 1$, then $d_{\mathbf{m}}$ is independent of $\mathbf{r}_a$ and can be removed from the integral, and the integral of L_p becomes independent of $\mathbf{r}'$ by a change of variables. In that limit, therefore $h_{\mathbf{m}}(\mathbf{r}') \propto d_{\mathbf{m}}(\mathbf{r}', \hat{\mathbf{s}}_{\mathbf{m}})$. In other words, the size of the source is of no importance for contact printing, and the PRF is simply the detector function.

The opposite limit is when the object is placed close to the source and a long distance from the detector. The magnification is large in this case, so L_p is a broad function of $\mathbf{r}_a$, and the width of the detector response contributes little to the width of $h_{\mathbf{m}}$.

16.1.5 Stochastic properties

Many different random phenomena can affect the statistical properties of digital radiographic images. A convenient organization is to consider separately noise mechanisms that produce no correlations among the elements of $\mathbf{g}$, ones that produce short-range correlations and ones that produce long-range correlations.

As we know from Chap. 11, the x-ray photons produce uncorrelated noise if we can argue that they are described by Poisson statistics; we make that argument below in the context of x-ray production by electron bombardment. Other sources of uncorrelated noise include various kinds of electronic noise such as amplifier noise, shot noise due to dark current and kTC noise (see Sec. 12.2.4).

The gain processes in scintillation and semiconductor x-ray detectors produce both uncorrelated and correlated noise components. There is an uncorrelated component since the secondary particles (optical photons or charge carriers) form a point process, but there is a correlation since the secondaries produced by a single primary x ray are not independent. In the limit of a very large number of secondaries per primary, the gain mechanism is just a deterministic blur, so it produces a correlation over a scale approximately equal to the blur width as determined by the spread of the secondaries before detection. In practical detectors, this blur is only a few pixels, so the correlations associated with the gain process are short range.

Not all blur mechanisms produce correlation, however. If we consider an optically coupled scintillation detector with poor collection efficiency, for example, so that less than one optical photon is collected per x-ray photon, the rarity of secondary collection converts the statistics to Poisson and destroys the correlation in spite of the blur. As another example, blurring before detection by the finite size of the focal spot leaves the x-ray photons independent and does not introduce any correlation.

Another important source of short-range correlations is production of secondary x-ray photons in the detector by Compton scattering or K x-ray emission. These secondaries produce a correlation over a scale related to the mean distance they travel before reabsorption in the detector (see Sec. 12.3.9).

Long-range correlations can arise from blur mechanisms with long tails on their point response functions, such as veiling glare in image intensifiers. They can also arise, however, when the object being imaged is considered to be random, so that the image is doubly stochastic (see Sec. 11.3.6). Almost by definition, interesting objects have long-range correlations, so this component of the image covariance is usually long-range.

All of these random phenomena have been discussed in previous chapters; in this section we shall try to bring the pieces together and present a comprehensive statistical model for digital radiographic images.

Poisson statistics The starting point for discussion of the statistical properties of digital radiographs is the statistics of the incident x-ray beam. We would like to assume that the number of photons detected during the exposure time is a Poisson random variable and that the pattern of photoelectric interactions in the detector is a Poisson random process, but we should look critically at the conditions that must be satisfied for the Poisson models to hold.

As we saw in Chap. 11, Poisson statistics can arise from two different principles: independence and rarity. X rays are produced by electron bombardment of an anode, and the number of emitted x-ray photons will obey Poisson statistics if the electrons act independently, but that requires that the mean current be constant; as illustrated qualitatively in Fig. 11.2, a random current makes the overall strength of the x-ray source random and the photon statistics non-Poisson.

The effect of a random source strength on image statistics for an array of ideal photon-counting detectors was discussed in Sec. 11.2.2. In particular, (11.56) shows that the covariance matrix of the data consists of two terms, a diagonal matrix accounting for the Poisson statistics and a nondiagonal one arising from the source fluctuations:

$$[\mathbf{K_g}]_{ik} = P_k \overline{\overline{M}}\, \delta_{ik} + P_i P_k \left[\mathrm{Var}(M) - \overline{\overline{M}} \right] , \tag{16.24}$$

where, in digital radiography, $\overline{\overline{M}}$ is the mean number of x-ray photons emitted by the tube, and the average is over both the random photon-generation process and fluctuations in the tube current; P_k is the probability that an emitted photon is detected in the kth detector. Physically, if the current in the x-ray tube is high on a particular exposure, all signals in a detector array tend to fluctuate high together, and conversely if the current fluctuates low.

We argued in Sec. 11.2.2 that the nondiagonal term was likely to be small if the probability of an x-ray photon being detected in a given detector element was small, but this argument is hard to sustain in digital radiography, as a numerical example will show.

Typical x-ray images are taken with integrated currents of ~10–100 mA-s, or 10^{17}–10^{18} electrons incident on the anode of the x-ray tube during the exposure time. Only about 1% of the electrons produce x rays, and only a small fraction of the emitted x rays reach an individual detector element. If we consider a $100\mu\mathrm{m} \times 100\mu\mathrm{m}$ detector element 1 m from a source, it collects $\sim 10^{-9}$ of the emitted photons if there is no intervening object; attenuation in the object reduces the number detected even further, say by a factor of 100–1000. With all of these factors, we might get $\sim 10^4$ detected photons in each detector element. even though $\overline{\overline{M}} \sim 10^{15}$–$10^{16}$. Thus photon detection in this element certainly sounds like a rare event, but a flux of 10^4 photons has an associated standard deviation that is 1% of the mean; the nondiagonal term in the covariance is negligible only if the current is stable to much better than 1%, which is very difficult to accomplish in practice.

One might question whether this effect is of any practical concern. Since only a single image is collected for a given patient, a small change in the current for that image is equivalent to a small change in exposure time, the effect of which would be hard to discern. The only way to observe the nondiagonal term in the covariance matrix (11.56) would be to do repeated measurements on the same patient and compare the results.

In the spirit of this book, however, we pose the question differently: Does the departure from Poisson statistics and the resulting nondiagonal term in the covariance have any effect on task performance? We shall return to this point in the context of specific tasks in Secs. 16.2.5 and 16.2.6, but for now we make two remarks. First, an overall fluctuation in the strength of the x-ray beam arriving at the detector could have arisen from a spatially uniform change in transmittance through the object, so the second term in (16.24) can be lumped into object variability. Second, the nondiagonal term in the covariance is a rank-one matrix, and we will see that it has little effect on the performance of detection or estimation tasks. In terms of task performance, we shall show that we can treat the x-ray tube as a Poisson source, and we make that assumption for the remainder of this section.

Poisson random processes If we neglect the current variations, the total number of x-ray photons incident on the detector plane is well approximated by a Poisson random variable, and if we assume that the spatial position, angle of arrival and energy of one photon are independent of the properties of all other photons, then the pattern of photoelectric interactions in the detector is a Poisson random process. The particular Poisson process we should consider depends on what detector model we want to use, but in all cases the statistics of a Poisson random process are fully determined by its mean.

The simplest model of the detector is that the $\mathbf{m}^{th}$ element responds with some quantum efficiency η to each x-ray photon that hits it, so the mean output $\overline{g}_{\mathbf{m}}$ is just η times the integral of the x-ray fluence $b(\mathbf{r})$ across the area of that element. There are no correlations from element to element with this model, so the image statistics are fully determined by the incident fluence, a 2D function.

In more realistic x-ray detectors, as discussed in Sec. 12.3.8, the important Poisson process is the spatio-spectral random process $g_{ss}(\mathbf{R})$, defined in (12.295), which specifies the spatial distribution of photoelectric interactions in the volume of the detector and the energy deposited in each interaction. In the argument of this process, $\mathbf{R}$ is a 4D vector consisting of the 3D interaction position $\mathfrak{r}$ and energy $\mathcal{E}$.

All statistical properties of $g_{ss}(\mathbf{R})$ are determined by its mean, the spatio-spectral fluence $b_{ss}(\mathbf{R})$. To calculate this mean, we can first solve the Boltzmann equation for the distribution function $w_{det}(\mathfrak{r}, \hat{\mathbf{s}}, \mathcal{E}, t)$ inside the detector, subject to a boundary condition obtained by solving the Boltzmann equation outside the detector. For example, if the detector is adequately modeled as a homogeneous slab and the only interaction process considered is photoelectric absorption, then the Boltzmann equation consists of just the propagation and absorption terms. The distribution function inside the detector is then given by [*cf.* (10.148)]

$$w_{det}(\mathfrak{r}, \hat{\mathbf{s}}, \mathcal{E}, t) = w(\mathfrak{r}_P - \hat{\mathbf{s}}|\mathfrak{r} - \mathfrak{r}_P|, \hat{\mathbf{s}}, \mathcal{E}, t) \exp\left(-\alpha_{pe}|\mathfrak{r} - \mathfrak{r}_P|\right), \tag{16.25}$$

where $w(\mathfrak{r}_P, \hat{\mathbf{s}}, \mathcal{E}, t)$ is the distribution function on plane P (the entrance face of the detector) and α_{pe} is the photoelectric attenuation coefficient[1] in the detector.)

[1]We use α for attenuation coefficients in the detector material for consistency with Chap. 12 and to distinguish them from attenuation coefficients in the object, which we are denoting as μ.

It then follows from the definitions of w_{det} and b_{ss} that

$$b_{ss}(\mathbf{R}) = b_{ss}(\mathbf{r}, \mathcal{E}) = c \int_0^{\mathcal{T}} dt \int_{4\pi} d\Omega \; \alpha_{pe} w_{det}(\mathbf{r}, \hat{\mathbf{s}}, \mathcal{E}, t) \,. \tag{16.26}$$

Statistics of the detector output For a given object, the spatio-spectral fluence can be determined (albeit by a complicated nonlinear equation), and the mean vector and covariance matrix for the digital images can be found if we specify the detector readout mechanism. For example, for a scintillation detector read out by an array of photodiodes and gated integrators, the conditional covariance matrix for a specified object is given by (12.302), which we repeat here for convenience with a slight change in notation:

$$[\mathbf{K_g}(\mathbf{b}_{ss})]_{\mathbf{mm'}}$$

$$= \left\{ \Gamma_{\mathbf{m}}^2 \int_{\mathbf{m}} d^2r \; [\mathcal{H}_1 \mathbf{b}_{ss}](\mathbf{r}) + \sigma_{\mathbf{m}}^2 \right\} \delta_{\mathbf{mm'}} + \Gamma_{\mathbf{m}} \Gamma_{\mathbf{m'}} \int_{\mathbf{m}} d^2r \int_{\mathbf{m'}} d^2r' \; [\mathcal{H}_2 \mathbf{b}_{ss}](\mathbf{r}, \mathbf{r}') \,, \tag{16.27}$$

where $\Gamma_{\mathbf{m}}$ is the gain of the $\mathbf{m}^{th}$ photodiode as defined in (12.301), and $\sigma_{\mathbf{m}}^2$ is the variance of its excess noise (electronic, dark current and kTC). The operators $\mathcal{H}_1$ and $\mathcal{H}_2$, originally introduced in Sec. 11.4.3, are given explicitly in (12.297) and (12.298), respectively, which we repeat here as

$$[\mathcal{H}_1 \mathbf{b}_{ss}](\mathbf{r}) = \int_D d^4R_n \; p_d(\mathbf{r}, \mathbf{R}_n) \, b_{ss}(\mathbf{R}_n) \,; \tag{16.28}$$

$$[\mathcal{H}_2 \mathbf{b}_{ss}](\mathbf{r}, \mathbf{r}') = \int_D d^4R_n \; \mathrm{pr}_{\Delta \mathbf{r}}(\mathbf{r} - \mathbf{r}_n | \mathbf{R}_n) \, \mathrm{pr}_{\Delta \mathbf{r}}(\mathbf{r}' - \mathbf{r}_n | \mathbf{R}_n) \, b_{ss}(\mathbf{R}_n) \, s(\mathbf{R}_n) \,, \tag{16.29}$$

where $p_d(\mathbf{r}, \mathbf{R}_n)$ is the mean number of secondaries per unit area on the output plane if the nth x-ray interaction is at the 4D point $\mathbf{R}_n = (\mathbf{r}_n, z_n, \mathcal{E}_n)$; $\mathrm{pr}_{\Delta \mathbf{r}}(\mathbf{r} - \mathbf{r}_n | \mathbf{R}_n)$ is the probability density function for the lateral displacement of the secondaries about the interaction position as they propagate to the output plane, and $s(\mathbf{R}_n)$ is defined in (11.219) as

$$s(\mathbf{R}_n) = \mathrm{E}\{k_n^2 - k_n | \mathbf{R}_n\} = \mathrm{Var}\{k_n | \mathbf{R}_n\} + \left[\mathrm{E}\{k_n | \mathbf{R}_n\}\right]^2 - \mathrm{E}\{k_n | \mathbf{R}_n\} \,, \tag{16.30}$$

where k_n is the number of secondaries produced by the nth primary. The 4D integrals in (16.28) and (16.29) run over the volume of the detector and all energies from 0 to ∞. Since the x rays are presumed to be indistinguishable, the integrals are independent of the index n.

The operator $\mathcal{H}_1$ serves also to express the mean output. From (11.214) as generalized to the spatio-spectral case, we have that

$$\overline{g}_{\mathbf{m}} = \Gamma_{\mathbf{m}} \int_{\mathbf{m}} d^2r \; [\mathcal{H}_1 \mathbf{b}_{ss}](\mathbf{r}) \,. \tag{16.31}$$

Expressions very similar to (16.27) and (16.31) apply to semiconductor detectors with an array of pixel electrodes.

To make contact with Sec. 16.1.4, note that $\mathbf{b}_{ss}$ is a linear functional of its boundary condition, specified by the distribution function w on the entrance face, and (16.31) shows that $\overline{g}_{\mathbf{m}}$ is a linear functional of $\mathbf{b}_{ss}$, so $\overline{g}_{\mathbf{m}}$ is linear in w. From these nested functionals we can derive the measurement function $d_{\mathbf{m}}(\mathbf{r}, \hat{\mathbf{s}}, \mathcal{E})$; the reader is invited to do so for the detector model of (16.25).

If we consider only unscattered photons, then (16.31) is equivalent to (16.8) or (16.10), depending on what we want to assume about the detector response, but the spatio-spectral fluence must, in general, include contributions from both scattered and unscattered photons. Actually computing the input quantity $\mathbf{b}_{ss}$ in (16.27) and (16.31) is thus a daunting task, requiring solution of the Boltzmann equation with a scatter term in the region between the source and the detector and then solving the Boltzmann equation again in the detector material.

One might worry that scatter in the object would introduce correlation in the data since scattered photons could enter the detector far from normal incidence, even if the unscattered ones are nearly normal. Fortunately, this complication does not occur since $\mathbf{b}_{ss}$ is the mean of a Poisson random process if the x rays obey Poisson statistics. Each impulse in the spatio-spectral random process produces an independent response, and the only correlation is the result of the correlated secondary particles (*e.g.*, optical photons) reaching different detectors.

On the other hand, the spatio-spectral distribution can influence the form of the correlation function. For example, if the incident photons are mainly of low energy and therefore absorbed near the entrance face of the detector, then secondary optical photons in a scintillation detector can spread out as they propagate through the full thickness of the detector and therefore contribute to several adjacent photodiodes. X-ray photons of higher energy, however, may be absorbed much closer to the photodiode plane and therefore spread less.

Random objects If we consider random objects, then the spatio-spectral fluence becomes a doubly stochastic random process and $\mathbf{g}$ becomes a doubly stochastic random vector. By generalizing (11.233) and integrating over detector areas, we find

$$[\mathbf{K_g}]_{\mathbf{mm}'} = \left[\Gamma_{\mathbf{m}}^2 \int_{\mathbf{m}} d^2r\, [\mathcal{H}_1 \overline{\mathbf{b}}_{ss}](\mathbf{r}) + \sigma_{\mathbf{m}}^2\right] \delta_{\mathbf{mm}'}$$

$$+ \Gamma_{\mathbf{m}}\Gamma_{\mathbf{m}'} \int_{\mathbf{m}} d^2r \int_{\mathbf{m}'} d^2r' \left\{[\mathcal{H}_2 \overline{\mathbf{b}}_{ss}](\mathbf{r}, \mathbf{r}') + \left[\mathcal{H}_1 \mathbf{K}_{\mathbf{b}_{ss}} \mathcal{H}_1^\dagger\right](\mathbf{r}, \mathbf{r}')\right\}, \qquad (16.32)$$

where $\mathbf{K}_{\mathbf{b}_{ss}}$ is the covariance for the random spatio-spectral fluence, which must in general be calculated from the properties of the random object by use of the nonlinear equation (16.7). Note that the object randomness has the same qualitative effect as blur in the detector, introducing off-diagonal terms in the data covariance matrix.

Nonlocal charge deposition The effect of reabsorption of Compton-scattered or K x-ray photons on the image statistics was discussed in Sec. 12.3.9. The key result there was the autocovariance function (12.321) for the spatio-spectral random process $g_{ss}(\mathbf{R})$. That formula states that

$$K_{g_{ss}}(\mathbf{R}, \mathbf{R}'|\mathbf{b}_{ss}) = [b_{ss}^{pri}(\mathbf{R}) + b_{ss}^{\sec}(\mathbf{R})]\, \delta(\mathbf{R} - \mathbf{R}')$$

$$+ p_{p\to s}[pr_{p\to s}(\mathbf{R}'|\mathbf{R})\, b_{ss}^{pri}(\mathbf{R}) + pr_{p\to s}(\mathbf{R}|\mathbf{R}')\, b_{ss}^{pri}(\mathbf{R}')], \qquad (16.33)$$

where $p_{p\to s}$ is the probability that a primary x-ray photon will produce a secondary event somewhere in the detector, and $pr_{p\to s}(\mathbf{R}|\mathbf{R}')$ is the probability density function for a secondary x ray to be absorbed at the 4D point $\mathbf{R}$ given a primary x-ray interaction at $\mathbf{R}'$, or vice versa. The delta-correlated terms represent the independent Poisson-noise contributions of the primary and secondary x rays, and the next two terms represent their correlation.

Overall covariance To summarize this section, the physical effects that contribute to the noise in digital radiography are not all statistically independent, but it is nevertheless possible to write the overall covariance matrix in the form:

$$\mathbf{K_g} = \mathbf{K_g}^{(elec)} + \mathbf{K_g}^{(x)} + \mathbf{K_g}^{(gain)} + \mathbf{K_g}^{(Kx)} + \mathbf{K_g}^{(obj)}\,, \tag{16.34}$$

where the terms represent, respectively, the electronic noise; the Poisson statistics of the x rays as reflected through the gain mechanism; the excess noise of the gain mechanism; the effect of reabsorbed Compton-scattered and K x rays, and the effect of object randomness

The first two terms in (16.34) are diagonal matrices, even in the presence of noisy gain, object randomness or secondary x-ray events. As we have noted, the electronic noise has a covariance of the form

$$\left[\mathbf{K_g}^{(elec)}\right]_{\mathbf{mm}'} = \sigma_{\mathbf{m}}^2\, \delta_{\mathbf{mm}'}\,, \tag{16.35}$$

which is independent of the x-ray exposure, the object and the gain mechanism in the detector.

On the other hand, the second term, $\mathbf{K_g}^{(x)}$, does depend on the x-ray exposure, the object and the gains. It can be read off explicitly as the term involving $\boldsymbol{\mathcal{H}}_1$ in (16.27) or (16.32), but with (16.31) it can also be summarized as

$$\left[\mathbf{K_g}^{(x)}\right]_{\mathbf{mm}'} = \Gamma_{\mathbf{m}}\, \overline{g}_{\mathbf{m}}\, \delta_{\mathbf{mm}'}\,. \tag{16.36}$$

The covariance $\mathbf{K_g}^{(gain)}$ refers specifically to the term involving $\boldsymbol{\mathcal{H}}_2$ in (16.27) or (16.32). It can be thought of as the correlated part of the gain process, but in fact it may also be a diagonal matrix, or nearly so. In a scintillator-photodiode detector, for example, the random process $y(\mathbf{r})$ describing optical photons on the photodiode plane has a correlated term involving $\boldsymbol{\mathcal{H}}_2$ [see (12.299)], but the range of this correlation is approximately the range over which the optical photons spread in propagating to the output plane, which in turn is approximately the detector thickness. This range might be small compared to the size ϵ of a photodiode, and if so, then $\mathbf{K_g}^{(gain)}$ will be diagonal even though $y(\mathbf{r})$ is not delta-correlated. If the spread of the optical photons is comparable to ϵ, then a correlation will be induced in $\mathbf{g}$ between adjacent photodiodes but not between more distant ones.

The situation is only slightly more complicated in semiconductor detectors. As discussed in Sec. 12.3, the charge carriers not only spread out as they propagate to the electrode plane, but they can also be trapped *en route*. Trapped carriers can induce correlated charges on neighboring electrodes, but there is no significant correlation from this effect on two electrodes that are separated by several times the detector thickness.

Our use of the multi-index $\mathbf{m}$ to denote the detector elements makes it very simple to characterize the short-range correlations associated with $\mathbf{K_g}^{(gain)}$. We can say that

$$\left[\mathbf{K_g}^{(gain)}\right]_{\mathbf{mm}'} \approx 0 \qquad \text{if} \qquad \epsilon\,|\mathbf{m}-\mathbf{m}'| > \delta_{gain}\,, \tag{16.37}$$

where δ_{gain} is the correlation length of the secondary process $y(\mathbf{r})$. If we had not used multi-indices, it would have been more complicated to state which elements of $\mathbf{K}_\mathbf{g}^{(gain)}$ were zero; with multi-indices, we can think of this covariance matrix as confined to a narrow band around the diagonal, and we shall refer to it as a *banded matrix.*

The covariance $\mathbf{K}_\mathbf{g}^{(Kx)}$ is the covariance associated with reabsorption of K x rays or Compton-scattered photons. If they don't escape from the detector material, the secondary x rays have a pathlength of order $1/\alpha_{tot}$, where α_{tot} is the total attenuation coefficient for the secondary photons in the detector material, and twice this pathlength is a reasonable estimate for the correlation range for this term. We can thus say, roughly, that

$$\left[\mathbf{K}_\mathbf{g}^{(Kx)}\right]_{\mathbf{mm}'} \approx 0 \qquad \text{if} \qquad \epsilon\, \alpha_{tot}|\mathbf{m} - \mathbf{m}'| > 2 \,. \tag{16.38}$$

Often this condition will lead to the conclusion that $\mathbf{K}_\mathbf{g}^{(Kx)}$ is confined to a band one or two elements wide around the diagonal (in the multi-index notation).

Finally, $\mathbf{K}_\mathbf{g}^{(obj)}$ in (16.34) refers to the term involving $\mathbf{K}_{\mathbf{b}_{ss}}$ in (16.32). A random object creates a random spatio-spectral fluence, which is then transformed through the amplification process to the output data. A key point about this term is that it varies quadratically with the x-ray exposure (since $\mathcal{H}_1$ varies linearly with exposure), so at large exposures it will be the dominant noise contribution. The range of the correlations associated with this term might be quite large since object structures can have large scales.

Estimating the covariance In order to make use of the covariance in image-quality assessment, we must be able to evaluate or estimate each of the five terms. Possible methods include model-based theoretical calculation, theory augmented by measurement, Monte Carlo simulation, and collection of sample images.

To evaluate the expressions for $\mathbf{K}^{(x)}$ and $\mathbf{K}^{(gain)}$, we need to know the first two moments of the random number of secondaries as well as how the secondaries are distributed on the readout plane of the detector. These unknowns can be determined from theoretical models or Monte Carlo simulation, but simple measurements with tightly collimated x-ray beams can be used also.

Similarly, $\mathbf{K}^{(elec)}$ can be computed theoretically from knowledge of the circuit design and standard electronic simulation software. If we know that this component matrix is diagonal as in (16.35), then the only unknowns are the variances, and we can also get those by direct measurement with no x-ray beam.

The contribution $\mathbf{K}^{(Kx)}$ is probably best determined by Monte Carlo simulation. Alternatively, the combination $\mathbf{K}^{(elec)} + \mathbf{K}^{(x)} + \mathbf{K}^{(gain)} + \mathbf{K}^{(Kx)}$ can be measured directly by using a uniform x-ray fluence and acquiring a large number of image frames. All of these terms are diagonal or banded around the diagonal, so only a very small subset of all possible elements in the covariance matrix needs to be computed; elements far from the diagonal are zero *a priori* by (16.37) and (16.38) and do not need to be computed or measured. Once the diagonal and near-diagonal terms are established, we have a full-rank estimate of the sum of the first four terms in (16.34). We denote that estimate as $\hat{\mathbf{K}}_\mathbf{g}^{(noise)}$.

The most difficult term is $\mathbf{K}_\mathbf{g}^{(obj)}$, which is simply neglected in many detectability studies. Any covariance matrix can be estimated from samples, but if we want the resulting sample covariance matrix to represent $\mathbf{K}_\mathbf{g}^{(obj)}$, the other noise sources must be negligible. To achieve this goal, we can take advantage of the fact that $\mathbf{K}_{\mathbf{g}}^{(obj)}$ is the only term that varies quadratically with the x-ray exposure. If clinical (or animal or cadaver) images can be taken at high enough exposure, the resulting sample covariance matrix on the images is directly an estimate of $\mathbf{K}_{\mathbf{g}}^{(obj)}$, denoted $\hat{\mathbf{K}}_{\mathbf{g}}^{(obj)}$. Alternatively, several images of each object can be taken at lower exposures, and the sample covariance matrix can be analyzed element by element to tease out $\hat{\mathbf{K}}_{\mathbf{g}}^{(obj)}$ separately.

Another useful approach is to simulate realistic objects and then simulate their images without the other noise sources being present. Methods of simulation were discussed in Sec. 8.4. In the context of digital mammography we mention particularly the clustered lumpy background of Bochud *et al.* (1999a), which gives quite realistic simulated mammograms.

Whether actual images or simulated ones are used to estimate $\mathbf{K}_{\mathbf{g}}^{(obj)}$, it is important to note that the overall covariance estimate $\hat{\mathbf{K}}_{\mathbf{g}}^{(noise)} + \hat{\mathbf{K}}_{\mathbf{g}}^{(obj)}$ will have full rank even if the number of sample images is small. The great advantage we have in this application is that $\mathbf{K}_{\mathbf{g}}^{(elec)}$ and $\mathbf{K}_{\mathbf{g}}^{(x)}$ are diagonal (hence full rank) and do not have to be estimated from samples.

16.1.6 Image quality: Detection tasks

We shall begin the discussion of image quality in digital radiography with the simplest of detection tasks, SKE/BKE (signal known exactly, background known exactly). As we saw in Sec. 13.2.12, the Hotelling observer is ideal if the noise is Gaussian and independent of the signal. We shall argue below that these conditions prevail, to a good approximation, and the Hotelling SNR is the relevant figure of merit for this task in digital radiography.

Computation of this SNR requires knowledge of the mean signal in the data and the covariance matrix, along with some method for performing the inversion of the covariance. We shall show how to carry out this computation first for an an x-ray detector that has no element-to-element noise correlation and then with short-range correlations of the sort that can arise from reabsorption of secondary x-ray photons or spread of optical photons. Then we shall allow random backgrounds and signals, and in Sec. 16.1.7 we shall discuss estimation tasks.

The signal To derive an SNR for an SKE/BKE task, we must first define precisely what we mean by the signal. In medical radiography, the signal is the result of a local change in the x-ray attenuation coefficient in the patient's body. Such local changes, known generically as *lesions*, may result from tumors, cysts, blockages in blood-vessels, etc., and it is of great clinical importance to be able to detect them.

A change in x-ray attenuation in the object will, in general, cause rather complicated changes in the detector output. First, it will change the mean signal from the unscattered radiation since it changes the mean number of unscattered x-ray photons reaching the detector as well as their spatial distribution. Second, it will also change the mean distribution of scattered radiation. And finally, since the noise properties of the image all depend on the x-ray distribution function on the detector face, any change in x-ray attenuation coefficient will change the noise covariance.

To make the problem of computing the SNR tractable, we assume that the lesion makes only a small change in the unscattered x-ray distribution function on the detector face and therefore only a small change in the mean detector outputs. Moreover, we shall neglect any changes in the scattered distribution on the basis that small changes in scatter properties of the object result in changes in the distribution function spread out over the entire detector surface and hence of very low contrast.

If we specify the lesion by a change in attenuation coefficient $\Delta\mu(\mathbf{r}, \mathcal{E})$, and if we ignore any changes in the scattered radiation, then the change in $\overline{g}_{\mathbf{m}}$ can be computed from (16.8). If we make a few simplifying assumptions about the detector, as indicated in (16.9), and if we also assume that the attenuation coefficients of the lesion and the background object are independent of $\mathcal{E}$, we can use the simpler form (16.10). We can then define the mean signal as

$$s_{\mathbf{m}} \equiv \Delta\overline{g}_{\mathbf{m}} = C \int_P d^2r \int_{2\pi} d\Omega \; d_{\mathbf{m}}(\mathbf{r}, \hat{\mathbf{s}}) \, L_p(\mathbf{r} - \hat{\mathbf{s}}\ell_0, \hat{\mathbf{s}})$$
$$\times \left\{ \exp\left[-\int_0^{\ell_0} d\ell' \, [\mu_0(\mathbf{r} - \hat{\mathbf{s}}\ell') + \Delta\mu(\mathbf{r} - \hat{\mathbf{s}}\ell')] \right] - \exp\left[-\int_0^{\ell_0} d\ell' \, \mu_0(\mathbf{r} - \hat{\mathbf{s}}\ell') \right] \right\}, \tag{16.39}$$

where $\mu_0(\mathbf{r})$ is the total attenuation coefficient of the object in the absence of the lesion.

A useful alternative way to write (16.39) is

$$s_{\mathbf{m}} = C \int_P d^2r \int_{2\pi} d\Omega \; d_{\mathbf{m}}(\mathbf{r}, \hat{\mathbf{s}}) \, L_p(\mathbf{r} - \hat{\mathbf{s}}\ell_0, \hat{\mathbf{s}})$$
$$\times \left\{ \exp\left[-\int_0^{\ell_0} d\ell'' \, \Delta\mu(\mathbf{r} - \hat{\mathbf{s}}\ell'') \right] - 1 \right\} \exp\left[-\int_0^{\ell_0} d\ell' \, \mu_0(\mathbf{r} - \hat{\mathbf{s}}\ell') \right]. \tag{16.40}$$

The factor in large curly brackets can be interpreted as $t_{les} - 1$, where t_{les} is the relative transmission of the lesion (possibly greater than one if the lesion exhibits reduced attenuation compared to the background object). The signal is thus a line-integral projection image of $t_{les} - 1$ modulated by the projection of the background object and blurred by the finite focal spot and detector response function. The modulation is similar to what we saw in (16.15) when we divided the object into slabs and considered the effect of all other slabs on the slab of interest. Here the slab of interest is defined by the lesion, and there is no requirement that the slab be physically thin.

If the lesion has sufficiently low contrast that $\int_0^{\ell_0} d\ell'' \, \mu_0(\mathbf{r} - \hat{\mathbf{s}}\ell'') \ll 1$ for all $\mathbf{r}$ and $\hat{\mathbf{s}}$, then we can expand the first exponential in (16.40) and write

$$s_{\mathbf{m}} \approx -C \int_P d^2r \int_{2\pi} d\Omega \; d_{\mathbf{m}}(\mathbf{r}, \hat{\mathbf{s}}) \, L_p(\mathbf{r} - \hat{\mathbf{s}}\ell_0, \hat{\mathbf{s}}) \left[\int_0^{\ell_0} d\ell'' \, \Delta\mu(\mathbf{r} - \hat{\mathbf{s}}\ell'') \right]$$
$$\times \exp\left[-\int_0^{\ell_0} d\ell' \, \mu_0(\mathbf{r} - \hat{\mathbf{s}}\ell') \right]. \tag{16.41}$$

With this approximation, the signal is linear in the attenuation coefficient of the lesion rather than its transmission minus one.

SKE/BKE detectability for detectors with uncorrelated noise Consider a detector in which the gain process entails a negligible blur and in which reabsorption of Compton and K x rays can be neglected. Then the first three covariance terms in (16.34) are diagonal matrices, the fourth is neglected and the fifth is absent for a BKE task. The noise sources we retain in this model are all well described as Gaussian. Electronic noise is certainly Gaussian, as discussed in Chap. 12, and the terms associated with the x-ray flux are Gaussian if the number of absorbed x-ray photons per detector element is larger than 10 or so. The term associated with object variability can be decidedly non-Gaussian (see Sec. 8.4), but we are not considering that term here, and we are also neglecting signal variability. Thus an overall Gaussian noise model is accurate, and the Hotelling SNR is the relevant figure of merit.

We know from Sec. 13.2.12 how to compute the Hotelling SNR for an SKE/BKE task with signal-independent noise, and we have argued above that the noise is independent of the signal if the latter results from a small change in the x-ray attenuation coefficient of the object. We know also from (13.123) that the Hotelling SNR is particularly simple if we can choose a data representation, called the Karhunen-Loève or KL domain, where the data covariance matrix is diagonal. The KL domain is the original data domain (the detector pixel domain) under the present assumptions, and we have at once from (13.119) that[2]

$$\mathrm{SNR}^2 = \sum_{\mathbf{m}=1}^{M} \frac{s_{\mathbf{m}}^2}{\mathrm{Var}(g_{\mathbf{m}})} \,. \tag{16.42}$$

The only remaining problem is to compute $\mathrm{Var}(g_{\mathbf{m}})$. We know from (16.27) that, for the present noise model,

$$\mathrm{Var}(g_{\mathbf{m}}) = \sigma_{\mathbf{m}}^2 + \Gamma_{\mathbf{m}}^2 \int_{\mathbf{m}} d^2r\ [\mathcal{H}_1 \mathbf{b}_{ss}](\mathbf{r}) + \Gamma_{\mathbf{m}}^2 \int_{\mathbf{m}} d^2r \int_{\mathbf{m}} d^2r'\ [\mathcal{H}_2 \mathbf{b}_{ss}](\mathbf{r}, \mathbf{r}') \,. \tag{16.43}$$

To evaluate this expression, we need to know the quantities $p_d(\mathbf{r}, \mathbf{R}_n)$ and $\mathrm{pr}_{\Delta\mathbf{r}}(\mathbf{r} - \mathbf{r}_n | \mathbf{R}_n)$ that appear in the definitions of $\mathcal{H}_1$ and $\mathcal{H}_2$. If there is no blur in the gain process, then $\mathrm{pr}_{\Delta\mathbf{r}}(\mathbf{r} - \mathbf{r}_n | \mathbf{R}_n)$ is the 2D delta function $\delta(\mathbf{r} - \mathbf{r}_n)$, and

$$p_d(\mathbf{r}, \mathbf{R}_n) = \mathrm{E}\{k_n | \mathbf{R}_n\}\, \delta(\mathbf{r} - \mathbf{r}_n) \,, \tag{16.44}$$

where k_n is the number of secondaries produced in the nth x-ray interaction. With a little algebra, we then find

$$\mathrm{Var}(g_{\mathbf{m}}) = \sigma_{\mathbf{m}}^2 + \Gamma_{\mathbf{m}}^2 \int_{\mathbf{m}} d^2r_n \int_0^{L_z} dz_n \int_0^{\infty} d\mathcal{E}\ b_{ss}(\mathbf{r}_n, z_n, \mathcal{E}_n)\, \mathrm{E}\{k_n^2 | \mathbf{R}_n\} \,, \tag{16.45}$$

where L_z is the thickness of the detector.

If the gain process were noise-free, then $\mathrm{E}\{k_n^2\}$ would be the constant $\overline{k}_n^2$, independent of $\mathbf{R}_n$, and we could remove it from the integral. The product $\Gamma_{\mathbf{m}}^2 \overline{k}_n^2$ is then the square of the overall gain, including (for scintillation detectors) the production of optical photons, the photodiode and any electronic gain. Therefore, in

[2]The summation convention used here is the one introduced in Sec. 7.1.2: letting $\mathbf{m}$ go from 1 to M means letting each component of the vector run over this range, so (16.42) applies to an $M \times M$ detector array, with M^2 total elements.

this unrealistic case, (16.45) just says that the variance of g_m is the mean number of absorbed x-ray photons times the gain squared, plus the variance of the electronic noise.

With realistic models for the amplification noise, even without blur, the variance of g_m is increased by the random variation of the number of secondaries with depth of interaction and deposited energy. We encountered this situation in Sec. 11.4.1 where we initially discussed random amplification in single-element detectors, and it should be no surprise that the same mathematics recurs here since an array without blur or correlation is just a collection of independent single-element detectors. From (11.183) we know that the variance of the number of secondaries is the mean number of primary x-ray absorptions times the second moment of the gain distribution, and (16.45) generalizes that statement to include random depth of interaction and energy per event. Moreover, if we factor out $\overline{k}_n^2$, we can rewrite (16.45) as

$$\mathrm{Var}(g_{\mathbf{m}}) = \sigma_{\mathbf{m}}^2 + \Gamma_{\mathbf{m}}^2 \overline{k}_n^2 \int_{\mathbf{m}} d^2 r_n \int_0^{L_z} dz_n \int_0^{\infty} d\mathcal{E}\; b_{ss}(\mathbf{r}_n, z_n, \mathcal{E}_n) \left[\frac{\mathrm{E}\{k_n^2 | \mathbf{R}_n\}}{\overline{k}_n^2} \right], \tag{16.46}$$

and the factor in square brackets can be interpreted as the reciprocal of a Swank factor, as introduced in Sec. 11.4.1.

To summarize, the Hotelling SNR for this case is

$$\mathrm{SNR}^2 = \sum_{\mathbf{m}=1}^{M} \frac{s_{\mathbf{m}}^2}{\sigma_{\mathbf{m}}^2 + \Gamma_{\mathbf{m}}^2 \int_{\mathbf{m}} d^2 r_n \int_0^{L_z} dz_n \int_0^{\infty} d\mathcal{E}\; b_{ss}(\mathbf{r}_n, z_n, \mathcal{E}_n)\, \mathrm{E}\{k_n^2 | \mathbf{R}_n\}} . \tag{16.47}$$

For sufficiently small x-ray fluences, $\sigma_{\mathbf{m}}^2$ will be larger than the second term in the denominator, and the detectability will be determined solely by the signal strength and the electronic noise. A similar limit can occur if the gain factor $\Gamma_{\mathbf{m}} \overline{k}_n$ gets small, as it might, for example, in an optically coupled detector with large demagnification. For sufficiently large fluences, on the other hand, the SNR becomes independent of the gain factor $\Gamma_{\mathbf{m}} \overline{k}_n$ since $s_{\mathbf{m}}$ is also linear in this factor.

The dependence of the SNR on the background object is contained in the spatio-spectral fluence $b_{ss}(\mathbf{r}_n, z_n, \mathcal{E}_n)$. The gross effect is that thicker objects absorb more x rays and reduce the value of both $s_{\mathbf{m}}^2$ and the second term in the variance. Since $s_{\mathbf{m}}^2$ is quadratic in b_{ss}, the net effect is the expected one: thicker objects lead to fewer detected x rays and a smaller SNR for the same signal. In addition, thicker objects usually produce more scatter, increasing that component of $b_{ss}(\mathbf{r}_n, z_n, \mathcal{E}_n)$ and reducing the SNR further.

Effect of correlations on SKE/BKE tasks When we consider nondiagonal covariance matrices, we can no longer simply write down the Hotelling SNR. Instead we must somehow compute or estimate $\mathbf{s}^t \mathbf{K}^{-1} \mathbf{s}$. Direct inversion loses its appeal when one contemplates the size of $\mathbf{K}$ in digital radiography; for a $1,000 \times 1,000$ detector, $\mathbf{K}$ is $1,000,000 \times 1,000,000$! Therefore we must look for methods that bring in prior information and thereby reduce the computational burden; several ways of doing so were developed in Sec. 14.3.2. As we saw there, useful prior information can include signal location (in an SKE problem), smoothness and symmetry of the Hotelling template, and knowledge of the structure of the covariance matrix.

For SKE/BKE problems in digital radiology, one key piece of prior information is that the noise correlations have short range. Since the background is being treated as nonrandom for now, we are considering the first four terms in the overall covariance (16.34). We argued in Sec. 16.1.5 that these covariance components are likely to be nearly diagonal (at least with the multi-index convention), and we should take advantage of that fact.

Two methods from Sec. 14.3.2 that are particularly useful for a nearly diagonal covariance matrix are iterative estimation of the Hotelling template and Neumann-series expansion for the SNR. Estimation of the template by, say, the Landweber algorithm works for any covariance matrix, but it is much more efficient for nearly diagonal ones since a large fraction of the elements are zero.

The Neumann series makes use of the decomposition of the covariance $\mathbf{K}$ into a diagonal part $\mathbf{D}$ plus a matrix $\mathbf{A}$ with only off-diagonal terms. Conditions for convergence of the series are discussed in Sec. 14.3.2. If the series converges, we know from (14.43) that

$$\mathrm{SNR}^2 = \mathbf{s}^t\mathbf{K}^{-1}\mathbf{s} = \mathbf{s}^t\mathbf{D}^{-1}\mathbf{s} - \mathbf{s}^t\mathbf{D}^{-1}\mathbf{A}\mathbf{D}^{-1}\mathbf{s} + \mathbf{s}^t\mathbf{D}^{-1}\mathbf{A}\mathbf{D}^{-1}\mathbf{A}\mathbf{D}^{-1}\mathbf{s} + \dots \,. \tag{16.48}$$

The first term in this expansion, $\mathbf{s}^t\mathbf{D}^{-1}\mathbf{s}$, is what we would get if there were no off-diagonal terms, and the remaining terms are the corrections arising from correlations induced by the detector. If these correlations are sufficiently weak, we may be able to truncate the series after a few terms.

If we keep only the first two terms in this expansion, we can see that the effect of the off-diagonal terms is to reduce the SNR; the expression $\mathbf{s}^t\mathbf{D}^{-1}\mathbf{A}\mathbf{D}^{-1}\mathbf{s}$ is nonnegative and it appears with a negative sign in (16.48). In fact, this is a general result: for fixed signal $\mathbf{s}$ and fixed diagonal part $\mathbf{A}$, addition of off-diagonal terms can only reduce the SNR. One should not conclude, however, that radiographic detectors should never have correlations. Using a thicker detector, for example, increases the correlations associated with $\mathbf{K}_{\mathbf{g}}^{(gain)}$ and $\mathbf{K}_{\mathbf{g}}^{(Kx)}$, but it also increases the mean number of detected photons; detecting more photons increases SNR, but increasing blur could reduce SNR. In short, changing detector characteristics can have complicated effects on the SNR, perhaps increasing it overall for some tasks and decreasing it for others (Pineda and Barrett, 2004a, 2004b).

Random signals and backgrounds To be more realistic, we need to move away from SKE/BKE tasks and consider random signals and backgrounds.

Random signals do not pose any special difficulties with the Hotelling observer. We know from Sec. 13.2.12 that we merely have to replace the known signal in an SKE task with the average signal. As we discussed in that section, this approach is likely to work well for variations in signal size or shape, but randomness in signal location can greatly degrade the performance of any linear discriminant.

Random backgrounds arising from spatial inhomogeneity in the object can be treated by methods developed in Sec. 14.3.2. When we add in the object-variability part of the covariance, we inevitably introduce longer-range correlations; meaningful radiographic objects are almost always correlated over distances of many detector pixels. Therefore we cannot assume that the overall covariance (or an estimate of it) is nearly diagonal. Of the methods developed in Sec. 14.3.2, the one that is most appealing in this case is iterative estimation of the Hotelling template. Use of the Woodbury matrix-inversion lemma as in (14.47) greatly reduces the size of the matrices involved. Another viable option is to use Laguerre-Gauss or wavelet filters to reduce the dimensionality.

Source fluctuations One extreme case of random backgrounds with long-range correlations is when there are fluctuations in the strength of the x-ray source. In this case, the number of x-ray photons impinging on the detector is not Poisson, even for a nonrandom object, but the signals from all detector elements fluctuate together, and the range of the correlations is the whole width of the detector array. Mathematically, as we saw in Sec. 16.1.5 [*cf.* (16.24)], the effect of source fluctuations is to add a rank-one matrix to the covariance matrix associated with Poisson statistics.

We analyzed a similar problem in Sec. 13.2.12. For a spatially uniform background fluence of mean $\bar{b}$ and variance σ_b^2 and an array of ideal photon counters, we found that the SKE discriminability for a difference signal $\Delta\mathbf{s}$ was given in (13.216) by

$$\mathrm{SNR}^2 = \frac{||\Delta\mathbf{s}||^2}{\bar{b}} - \frac{[\Delta\mathbf{s}^t\mathbf{u}]^2}{\bar{b}^2 + M\bar{b}\sigma_b^2}, \tag{16.49}$$

where M is the total number of detector elements and $\mathbf{u}$ is an $M \times 1$ vector with all elements equal to σ_b. Note that both $\Delta\mathbf{s}$ and $\bar{b}$ are dimensionless numbers, representing detected photons per pixel; for this discussion we are neglecting the subtleties of real x-ray detectors and just considering the SNR as limited by Poisson noise and current fluctuations in the x-ray tube.

As a numerical example, suppose we have $\bar{b} \sim 10^4$ photons per pixel and $M \sim 10^6$ pixels in the array, and that the tube current fluctuates by 1% so that $\sigma_b \sim 100$. Then $M\bar{b}\sigma_b^2$ is 10^6 times larger than $\bar{b}^2$. With any similar assumptions about the number of pixels and number of photons per pixel, we will always have $M\bar{b}\sigma_b^2 \gg \bar{b}^2$, and we can write

$$\mathrm{SNR}^2 = \frac{\sum_{m=1}^M \Delta s_m^2}{\bar{b}} - \frac{1}{M\bar{b}}\left[\sum_{m=1}^M \Delta s_m\right]^2, \tag{16.50}$$

The first term in this expression is the detectability as limited only by photon-counting statistics, and the second term represents the reduction in detectability arising from source fluctuations. The key point is the factor of $1/M$; the ideal observer can look over the whole array and estimate the background on a particular x-ray exposure with high precision. For any practical number of elements in the array, the effect of source fluctuations is completely negligible for this task and observer.

Human observers, too, are quite insensitive to fluctuations in the source strength. We know from Sec. 14.2 that human observers are very insensitive to low spatial frequencies, and that is what we have in this problem since all elements in the array fluctuate together as the source strength varies. By contrast, the Poisson term corresponds to completely uncorrelated fluctuations in different detector elements, hence high spatial frequencies that do go through the human visual system.

Finally, we note that the second term in (16.50) is absent if $\sum \Delta s_m \equiv 0$. This condition holds for the Rayleigh discrimination task, to be discussed in Sec. 16.2.5 and illustrated in Fig. 16.15. In general, the Rayleigh task (or any task where the difference signal sums to zero) is a good way of assessing SKE detectability without having to worry about source fluctuations.

Why not Fourier? By far the most common way of attempting to estimate an SNR in digital radiography is to assume stationarity and then assume implicitly that the covariance matrix is diagonalized by a discrete Fourier transform. Both parts of this assumption are usually wrong: digital radiographs are not really stationary in any sense, and even if some degree of stationarity can be justified, the covariance may not be diagonalized by a DFT.

The first problem is that the Poisson noise is never stationary for interesting radiographic images. An interesting image, virtually by definition, is one where the x-ray fluence conveys interesting information and hence varies spatially. The x rays constitute a random point process, and the mean and autocovariance function of this process are functions of position on the detector. For a stationary random process, by contrast, the mean would have to be constant, and the autocovariance function could depend only on the relative location of the two points to which it refers, not the absolute position. Thus the very fact that we are imaging an actual object, as opposed to just a uniform x-ray beam, immediately invalidates stationarity, even if we regard that object as nonrandom. Moreover, the simple fact that radiographs have finite size also spoils strict stationarity; the autocovariance between two image points cannot be invariant when one or both of the points falls off the detector.

The stationarity assumption runs into several additional difficulties for digital radiography. The sampling of the x-ray beam by an array of detector elements converts the autocovariance function to a covariance matrix, and we need to assume that this matrix is stationary in some sense. One sense of stationarity in a discrete image is that the covariance matrix is *Toeplitz*, so that the covariance between the signals from two detector elements depends on their relative position in the array, not on absolute location. For an ideal detector that simply samples a continuous random process, the covariance matrix is Toeplitz if the random process is stationary, in the sense that the autocovariance function depends only on relative position. If we ignore the anatomical noise and consider only the Poisson noise of the x rays, then the covariance matrix is diagonal, and if we further assume that the x-ray fluence does not depend on position across the array, then the covariance matrix is a multiple of the unit matrix, hence Toeplitz.

As we know from the discussion above, digital x-ray detectors involve a gain mechanism where each x-ray photon is converted to many secondary optical photons or charge carriers, which are then detected. This gain process is random and hence a source of noise. Moreover, there is electronic noise in the amplifiers associated with each detector element, and additional noise arises from production and possible escape of Compton-scattered or K x rays. All of these effects must be stationary for the overall system to be stationary.

The Poisson and electronic contributions to the covariance are diagonal, so they are Toeplitz only if they are multiples of the unit matrix. That means that not only must the x-ray fluence be constant, but also the electronic noise variance must be the same in every element. The random gain mechanism and Compton and K x rays introduce short-range correlations between neighboring detector elements, and these effects must also be be invariant to absolute location for stationarity to hold. Again, this can happen only if the x-ray fluence is spatially uniform, but it also requires that the detector itself be uniform from element to element, which real detectors never are.

Finally, the inevitable departures from stationarity in the anatomy induce a departure from Toeplitz character in the covariance term $\mathbf{K}_{\mathbf{g}}^{(obj)}$. There is no reason at all to regard anatomy as stationary in any global sense, though local stationarity may sometimes describe the local texture (see Sec. 8.4).

The Toeplitz assumption, questionable though it may be, is still not strong enough for Fourier methods to be useful. To diagonalize a matrix by a discrete Fourier transform (DFT), we must assume that it is *circulant*, not Toeplitz. For a circulant covariance matrix, two elements at opposite sides of the array have the same covariance as two adjacent elements in the center. This "digital wrap-around" has no physical basis, but it is what we must assume if we want to diagonalize the matrix with DFTs. Fourier aficionados skirt the Toeplitz-vs.-circulant issue by arguing that the correlations are short range, but this statement is not true for anatomical noise. Moreover, even if one considers only BKE problems without anatomical noise and does a simulation in which all of the other requisite assumptions are valid, substantial errors can still be made in the computation of SNR by approximating a Toeplitz matrix by a circulant one (Pineda *et al.*, 2003).

NEQ and DQE In spite of these difficulties, Fourier methods are firmly entrenched in the digital radiography community, and it is common, even mandatory in some circles, to express the performance of digital x-ray detectors in terms of a frequency-dependent NEQ (noise-equivalent quanta) or DQE (detective quantum efficiency). We defined DQE as a ratio of squared SNRs for single-element detectors in Sec. 12.1.1, and the concept was extended to continuous shift-invariant systems with stationary noise in Sec. 13.2.13. For such systems and SKE/BKE tasks, we found that the ideal-observer SNR is given approximately[3] by [*cf.* (13.242)]

$$\mathrm{SNR}_\lambda^2 = |H(0)|^2 \int_\infty d^2\rho\, |\Delta F_{rel}(\boldsymbol{\rho})|^2\, \mathrm{NEQ}(\boldsymbol{\rho}), \tag{16.51}$$

where $H(0)$ is the system transfer function at zero spatial frequency, $\Delta F_{rel}(\boldsymbol{\rho})$ is the difference signal expressed as a fractional or relative change in the Fourier transform of the nonrandom object, and $\mathrm{NEQ}(\boldsymbol{\rho})$ is defined in (13.243) as

$$\mathrm{NEQ}(\boldsymbol{\rho}) = \frac{b_0^2 \mathrm{MTF}^2(\boldsymbol{\rho})}{W_{\mathbf{g}}(\boldsymbol{\rho})} = \frac{b_0^2 \mathrm{MTF}^2(\boldsymbol{\rho})}{b_0 + W_{exc}(\boldsymbol{\rho})}. \tag{16.52}$$

Here, $W_{\mathbf{g}}(\boldsymbol{\rho})$ is the noise power spectrum (NPS) of the data (treated as a stationary random process), b_0 is the Poisson contribution to this power spectrum from a nonrandom photon stream, and $W_{exc}(\boldsymbol{\rho})$ is the excess noise arising from amplification processes or electronics, but again treated as stationary random processes. The frequency-dependent DQE is just $\mathrm{NEQ}(\boldsymbol{\rho})$ normalized by the photon fluence [*cf.* (13.244)]:

$$\mathrm{DQE}(\boldsymbol{\rho}) = \frac{\mathrm{NEQ}(\boldsymbol{\rho})}{b_0} = \left[\frac{b_0}{b_0 + W_{exc}(\boldsymbol{\rho})}\right] \mathrm{MTF}^2(\boldsymbol{\rho}). \tag{16.53}$$

One challenge in applying the NEQ and DQE concepts to digital radiography — or any other digital imaging system — is knowing how to interpret the NPS

[3] As discussed in Sec. 13.2.12, the expression in (16.49) is the correct SNR for the log-likelihood ratio if the noise is Gaussian, but it is approximately correct if the noise is Poisson and the fluence is large and/or the signal is weak. It is the correct expression for the Hotelling observer whenever the task is SKE/BKE, the system is CC and shift-invariant, and the noise is stationary, without regard to whether the noise is exactly or approximately Gaussian.

when the data are discrete. It is always tacitly assumed that Fourier transforms are to be replaced with discrete Fourier transforms and hence the integral in (16.51) is replaced by a sum. Most methods then simulate or measure x-ray images of a uniform object, perform the DFT and interpret the variance of the DFT values as an NPS.

This interpretation is analogous to the situation with Fourier *transforms* (not DFTs) of continuous random processes; we know from Sec. 8.2.7 that Fourier transformation is also Karhunen-Loève transformation for stationary random processes, and we saw in (8.181) that

$$\langle F(\boldsymbol{\rho})\, F^*(\boldsymbol{\rho}')\rangle = S(\boldsymbol{\rho})\, \delta(\boldsymbol{\rho} - \boldsymbol{\rho}')\,, \tag{16.54}$$

where $S(\boldsymbol{\rho})$ is the NPS of a random process $f(\mathbf{r})$, whose Fourier transform is $F(\boldsymbol{\rho})$. Practitioners of digital DQE analysis implicitly *assume* that DFTs are Karhunen-Loève transformations for discrete images, and they *define* a discrete NPS as

$$\langle F_{\mathbf{n}} F_{\mathbf{n}}^*\rangle \equiv S_{\mathbf{n}}\,, \tag{16.55}$$

where $F_{\mathbf{n}}$ is the DFT value associated with the discrete spatial frequency $\boldsymbol{\rho}_{\mathbf{n}}$.

A complete analogy with the continuous result in (16.54) would require, in addition to (16.55), that

$$\langle F_{\mathbf{n}} F_{\mathbf{n}'}^*\rangle \equiv S_{\mathbf{n}}\, \delta_{\mathbf{n}\mathbf{n}'}\,, \tag{16.56}$$

but this additional requirement is seldom checked. A simple example will reveal why. Suppose that we have an x-ray detector array with electronic and Poisson noise but no other noise sources. We know from (16.35) and (16.36) that the covariance matrix in this case is

$$[\mathbf{K}_{\mathbf{g}}]_{\mathbf{m}\mathbf{m}'} = \sigma_{\mathbf{m}}^2\, \delta_{\mathbf{m}\mathbf{m}'} + \Gamma_{\mathbf{m}}\, \overline{g}_{\mathbf{m}}\, \delta_{\mathbf{m}\mathbf{m}'}\,. \tag{16.57}$$

Thus the covariance matrix is diagonal without any need for Fourier or any other transformation, and the ideal-observer SNR for a SKE/BKE task can be written down immediately from (16.42). In fact, performing a DFT will serve only to *undiagonalize* the covariance unless $\sigma_{\mathbf{m}}^2$, $\Gamma_{\mathbf{m}}$ and $\overline{g}_{\mathbf{m}}$ are all independent of the pixel index $\mathbf{m}$, in which case the covariance matrix is a multiple of the unit matrix, so the DFT and all other unitary transforms are simply irrelevant. It is a useful exercise to compute the exact Hotelling SNR for this problem and to compare it to some version of SNR based on (16.56). As a side benefit, this computation will demonstrate why it is *not* useful to normalize the gain variations.

Other difficulties with the Fourier methods have already been mentioned and need not be reiterated, but we can take a broader look. The basic assumption underlying *all* uses of Fourier transformations for computing detection-based SNRs is that the same transform diagonalizes both the deterministic imaging operator and the noise covariance matrix. Since the imaging operator is fundamentally a CD mapping and the image is fundamentally a discrete array of finite extent, no single transform can perform both functions.

Some recommendations regarding Fourier analysis One can, of course, regard digital NPS, NEQ or DQE as empirical characterizations of digital x-ray detectors, much the same as quantum efficiency, uniformity, pixel size or even cost. These are things we want to know before buying a detector, and they should be reported in a standardized way for ease of comparison. To the extent that detective quantum efficiency is to be associated with detection tasks, however, some additional considerations arise. We offer the following recommendations:

- Check the off-diagonal elements. If measured or simulated images are available, one can estimate the covariances of the DFT components as well as the variances. A full row of the DFT covariance matrix can be computed for no more effort than computing the diagonal elements, the so-called digital NPS.

- Determine whether the off-diagonal elements are significant for detection tasks. For some methods of doing this, see Pineda and Barrett (2004a, 2004b) and Gallas and Barrett (2003).

- Check other measures of stationarity. Stationarity requires that the mean image, the noise variance and the pixel-to-pixel covariances should all be independent of the absolute position in the array. These items can be studied empirically.

- Check boundary effects. Since the distinction between Toeplitz and circulant matrices disappears as the array size goes to infinity, it is advisable to treat array size as an experimental variable. It is difficult to make the array larger, but straightforward to mask the array and use fewer pixels.

- If only local stationarity applies, use local Fourier methods. As we have noted in Chaps. 13 and 14, the stochastic Wigner distribution describes the frequency content of the noise as a function of location and can be related to variations of SKE-detection SNRs with position.

- Check the result against other methods for computing SNR. In Chap. 14 we surveyed methods for computing SNRs for ideal, Hotelling and channelized Hotelling observers. The researcher who doubts the validity of Fourier-based methods can check it against one of these methods.

- Consider the use of more realistic tasks. At best, Fourier methods apply only to SKE/BKE tasks in noise that is stationary in some sense. There are numerous examples in the literature where such idealized tasks can give misleading conclusions, even when the figure of merit is computed correctly. As facility is gained with other methods, the researcher will be able to explore the limits of this task choice.

Detectability by a human observer Methods for measuring and predicting lesion detectability by a human observer were discussed extensively in Chap. 14, and there is relatively little to add that is specific to digital radiography. Human detection performance can be assessed by psychophysical methods and analyzed in terms of ROC curves, even for complex random signals and background, so long as the task is posed as a binary decision. For simple signals and backgrounds, at least, the outcome of these psychophysical experiments can be predicted with observer models such as the channelized Hotelling observer.

One factor that is prevalent in radiographic images, and that can degrade the performance of a human observer, is the low contrast of the lesion images. The Hotelling and channelized Hotelling observers are sensitive to this contrast since low contrast means low SNR, but it would make no difference to them if the signal and noise were both reduced by the same factor. For the human, on the other hand, there is an additional randomness, which we described in Sec. 14.2.1 as internal noise. If the signal and the noise in the image are reduced in the same proportion, the internal noise becomes more significant and human performance degrades relative to that of models without internal noise. To make the models predict this behavior, we must include internal noise in them as discussed in Sec. 14.2.2.

To minimize the degradation of human performance, we must take care to display the images at adequate contrast. Interactive contrast manipulations and automatic ones such as histogram equalization can be very useful in this regard. The qualitative guidelines given in Sec. 14.2.3 for the conduct of psychophysical studies are equally applicable to the display of real images in real clinical settings: the dominant image noise, whether it be the anatomical background or the measurement noise, should be readily apparent to the observer, and the lesion should be readily detectable if it were displayed without image noise.

Another aspect of the human observer that is not accounted for in our models, yet which is important in making real radiological diagnoses, is the human search strategy. If the lesion is randomly located within the image field, a physician will have some prior knowledge on anatomical and physiological grounds of where to look for it and will search this field in a way that attempts to optimize detection performance for a limited total viewing time. Radiologists differ greatly in their ability to carry out this optimization, and the relation of their performance to our objective measures of image quality has not been elucidated at this writing.

16.1.7 Image quality: Estimation tasks

Though the task in radiology is usually detection or classification, there are some circumstances where the task is estimation. For example, a radiographic contrast agent might be injected into the left ventricle of the heart with the goal of estimating the volume of the chamber or the volume of blood expelled on each beat. The interest might also be in whether the diameter of a tumor is decreasing in response to therapy, or in the degree of stenosis (narrowing) of a blood vessel.

Performance on an estimation task is specified in terms of the bias and variance of the estimator, but we have stressed in Chaps. 13 and 15 that bias is not well defined if the parameter is not estimable (see Secs. 13.3.1 and 15.1.3). The linear tests of estimability stated in Sec. 15.1.3 are not applicable here since the data are not linearly related to the object distribution, and in many cases the parameter of interest is not linearly related to the object either. We shall therefore consider various situations where estimability holds, at least approximately, in spite of the nonlinearity.

Estimability in the absence of blur If there were no blur from the detector or the focal spot, then certain line integrals of the object attenuation coefficient would be estimable. Suppose

$$d_{\mathbf{m}}(\mathbf{r}, \hat{\mathbf{s}})\, L_p(\mathbf{r} - \hat{\mathbf{s}}\ell_0) \propto \delta(\mathbf{r} - \mathbf{r_m})\, \delta(\hat{\mathbf{s}} - \hat{\mathbf{s}}_{\mathbf{m}}) \,. \tag{16.58}$$

Then (16.10) would become

$$\overline{g}_{\mathbf{m}} \propto \exp\left[-\int_0^{\ell_0} d\ell'\, \mu_{tot}(\mathbf{r_m} - \hat{\mathbf{s}}_{\mathbf{m}}\ell') \right] , \tag{16.59}$$

and the simple operation of taking a logarithm would recover the line integral of μ_{tot} along a line determined by the positions of the detector and focal spot.

Thus, in the absence of blur, a parameter of the form

$$\Theta = \int_\infty d^3\mathbf{r}\; \mu_{tot}(\mathbf{r})\, \chi(\mathbf{r}) \tag{16.60}$$

is estimable[4] if $\chi(\mathbf{r})$ can be written as a linear superposition of line delta functions corresponding to the measured projections. If only a single image is taken from a single focal-spot location, then these line deltas form a cone, and $\chi(\mathbf{r})$ must be a weighted sum of the rays in this cone. In computed tomography, however, many different source locations are used, and a much wider variety of parameters is at least approximately estimable.

Subtraction imaging A common technique in digital radiography is *subtraction imaging* in which two images are recorded, one before and one after administration of a contrast agent that locally increases the x-ray attenuation coefficient. The most important applications of this method are to visualize blood vessels or the cardiac chambers.

To isolate the structure of interest, the logarithm of each image is computed pixel-by-pixel, and the second logarithmic image is subtracted from the first. When there is no blur, the difference of the logarithmic images is directly related to the change in line integrals of the attenuation coefficient. Denoting the elements of the first image by $g_{\mathbf{m}}$ and those of the second by $g'_{\mathbf{m}}$ and ignoring noise, we see from (16.59) that

$$\ln g'_{\mathbf{m}} - \ln g_{\mathbf{m}} = \int_0^{\ell_0} d\ell'\, \Delta\mu_{tot}(\mathbf{r_m} - \hat{\mathbf{s}}_{\mathbf{m}}\ell') \,, \tag{16.61}$$

where $\Delta\mu_{tot}(\mathbf{r})$ is the change in attenuation coefficient between the two images. This difference image is thus a projection of the change in μ from the point source to the detector plane.

Often the parameter of interest in subtraction imaging is the total amount of contrast agent in a volume of interest, such as the left ventricle. If one assumes that the agent is mixed thoroughly with the blood, then the total amount of the agent is proportional to the blood volume. One way to attempt to estimate this parameter

[4]Note that the estimate is a nonlinear function of the data because of the logarithm. The statistics literature usually says that a parameter is estimable (identifiable) if there exists a linear unbiased estimator for all values of the parameter, but here we must consider nonlinear estimators and a more general definition of estimability.

is to sum both sides of (16.61) over $\mathbf{m}$. If we assume that the detector spacing is sufficiently small that we can replace the sum by an integral, we have

$$\sum_{\mathbf{m}} [\ln g'_{\mathbf{m}} - \ln g_{\mathbf{m}}] \approx \frac{1}{\epsilon^2} \int_P d^2 r_{\mathbf{m}} \int_0^{\ell_0} d\ell' \, \Delta\mu_{tot}(\mathbf{r}_{\mathbf{m}} - \hat{\mathbf{s}}_{\mathbf{m}} \ell') \,, \tag{16.62}$$

where ϵ^2 is the area of a detector element. We next make the change of variables $\mathbf{r}' = (\mathbf{r}_{\mathbf{m}} - \hat{\mathbf{s}}_{\mathbf{m}} \ell')$. The volume element is

$$d^3 \mathbf{r}' = \frac{|\mathbf{r}' - \mathbf{r}_a|^2}{z_a^2} \cos^3 \theta_{\mathbf{m}} \, d^2 r_{\mathbf{m}} \, d\ell' \,. \tag{16.63}$$

If we incorporate the factor $\cos^3 \theta_{\mathbf{m}}$ into the sum, (16.62) becomes

$$\sum_{\mathbf{m}} \cos^3 \theta_{\mathbf{m}} [\ln g'_{\mathbf{m}} - \ln g_{\mathbf{m}}] \approx \frac{z_a^2}{\epsilon^2} \int_V d^3 \mathbf{r}' \frac{1}{|\mathbf{r}' - \mathbf{r}_a|^2} \Delta\mu_{tot}(\mathbf{r}') \,, \tag{16.64}$$

where V is a conical volume defined by the detector elements chosen and the source location. Thus, even without noise, blur or finite detector sampling, the integrated change in μ is still not determined. All we can get is the weighted integral indicated in (16.64); no choice of weighting factors in the sum will remove the weighting in the integral since we cannot synthesize a constant from rays emanating from a single point. Even if we assume that $\Delta\mu$ is spatially compact, so that $|\mathbf{r}' - \mathbf{r}_a|^2$ is nearly constant over the volume of interest, we seldom know accurately how far this volume is from the detector, so we do not know $|\mathbf{r}' - \mathbf{r}_a|^2$ and hence cannot estimate the integral of $\Delta\mu$. The only recourse is to place the source at a large distance from the object and detector so that $|\mathbf{r}' - \mathbf{r}_a|^2 \approx \mathbf{r}_a^2$. Absent this approximation, even the simplest of parameters, an integral over the object, turns out not to be estimable from a single radiograph.

Effects of blur We now return to the more realistic situation where the image is blurred by the detector response and the focal spot, but we shall continue to ignore noise for a while longer. If we take a logarithm of (16.10) in an attempt to linearize, we obtain

$$\ln g_{\mathbf{m}} = \ln C + \ln \int_P d^2 r \int_{2\pi} d\Omega \; d_{\mathbf{m}}(\mathbf{r}, \hat{\mathbf{s}}) L_p(\mathbf{r} - \hat{\mathbf{s}}\ell_0, \hat{\mathbf{s}}) \exp\left[- \int_0^{\ell_0} d\ell' \, \mu_{tot}(\mathbf{r} - \hat{\mathbf{s}}\ell') \right] . \tag{16.65}$$

Since we cannot take the logarithm inside the integral, the second term on the right is not a linear transformation of the line integral of μ_{tot}, but a linear approximation might be valid. Two such approximations will now be described.

The exponential is a function of $\mathbf{r}$ and $\hat{\mathbf{s}}$, which can be written as $\exp[f(\mathbf{r}, \hat{\mathbf{s}})]$. If this function varies sufficiently slowly compared to $d_{\mathbf{m}}(\mathbf{r}, \hat{\mathbf{s}}) \, L_p(\mathbf{r} - \hat{\mathbf{s}}\ell_0, \hat{\mathbf{s}})$, it can be replaced by the constant $\exp[f(\mathbf{r}_{\mathbf{m}}, \hat{\mathbf{s}}_{\mathbf{m}})]$ and pulled out of the integral, and application of the logarithm then yields a linear functional of μ_{tot}.

Instead of just replacing the exponential by a constant, a better approximation can be obtained by expanding it in a Taylor series and retaining the constant and linear terms. To simplify the notation temporarily, consider the expression

$$\ln[g_m] = \ln \int_\infty dx \; p_m(x) \exp[f(x)] \,, \tag{16.66}$$

where $p_m(x)$ is peaked around $x = x_m$ and normalized such that $\int_\infty dx\, p(x) = 1$. We can write the exponential as

$$\exp[f(x)] = \exp[f(x_m) + f(x) - f(x_m)] = \exp[f(x_m)]\exp[f(x) - f(x_m)]$$
$$= \exp[f(x_m)]\left\{1 + \, f(x) - f(x_m) \, + \ldots\right\}. \tag{16.67}$$

If $\exp[f(x)]$ varies slowly in the vicinity of x_m, we can drop the unstated terms, all of which are nonlinear in $f(x)$. Inserting the truncated version of (16.67) into (16.66), we find

$$\ln[g_m] \approx \ln\left\{\exp[f(x_m)]\int_\infty dx\; p_m(x)[1 + \; f(x) - f(x_m)]\right\}$$
$$= f(x_m) + \ln\left\{1 - f(x_m) + \int_\infty dx\; p_m(x)\, f(x)\right\}, \tag{16.68}$$

where we have used the normalization of $p(x)$. If we now use $\ln(1+\epsilon) \approx \epsilon$, we see that

$$\ln[g_m] \approx \int_\infty dx\; p_m(x)\, f(x). \tag{16.69}$$

Up through terms linear in $f(x)$, therefore, the logarithm commutes with blurring, and a linear functional results. Note that this kind of linearization is rather different from the ones we discussed in Sec. 16.1.4. We do not require here that $\exp[f(x)]$ be near one or have small total variation; it suffices if it varies slowly over the range of x where $p(x)$ is nonzero.

Applying a similar argument to (16.65), we see that

$$\ln g_{\mathbf{m}} \approx A_{\mathbf{m}} - \int_P d^2r \int_{2\pi} d\Omega\; d_{\mathbf{m}}(\mathbf{r}, \hat{\mathbf{s}})\, L_p(\mathbf{r} - \hat{\mathbf{s}}\ell_0, \hat{\mathbf{s}}) \left[\int_0^{\ell_0} d\ell'\; \mu_{tot}(\mathbf{r} - \hat{\mathbf{s}}\ell')\right], \tag{16.70}$$

where $A_{\mathbf{m}}$ is a constant. We can define $y_{\mathbf{m}} \equiv A_{\mathbf{m}} - \ln g_{\mathbf{m}}$, and within the stated approximations $y_{\mathbf{m}}$ is a blurred version of the line integral of μ_{tot}.

To see the implications of this result for estimability, we apply a change of variables similar to the one that led to (16.63) and obtain

$$y_{\mathbf{m}} \approx \int_\infty d^3\mathbf{r}'\, \chi_{\mathbf{m}}(\mathbf{r}')\, \mu_{tot}(\mathbf{r}'). \tag{16.71}$$

It will be left as an exercise to determine the mathematical form of $\chi_{\mathbf{m}}(\mathbf{r})$, but physically it describes a blurred ray extending from the focal spot to the $\mathbf{m}^{th}$ detector. The line integral of μ_{tot} is not itself an estimable parameter, but the integral of μ_{tot} over this blurred ray is approximately estimable, and so is any linear parameter formed from a linear combination of the functions $\{\chi_{\mathbf{m}}(\mathbf{r})\}$. This observation is critical to the success of computed tomography, but it is seldom stated explicitly.

Effects of noise So far we have neglected noise in the discussion of estimation tasks, but we shall now remedy this defect. The actual measured data from the $\mathbf{m}^{th}$ detector can be written, as usual, as

$$g_{\mathbf{m}} = \overline{g}_{\mathbf{m}} + n_{\mathbf{m}}. \tag{16.72}$$

If the noise level is low enough, $y_\mathbf{m}$ can be approximated as

$$y_\mathbf{m} \approx \int_\infty d^3\mathbf{r}\, \chi_\mathbf{m}(\mathbf{r}')\, \mu_{tot}(\mathbf{r}') + \frac{n_\mathbf{m}}{\overline{g}_\mathbf{m}} \,. \tag{16.73}$$

To the extent that the modeling error attributable to neglecting terms nonlinear in μ_{tot} is negligible, the mean of $n_\mathbf{m}$ is zero and $y_\mathbf{m}$ is an unbiased estimator of the indicated integral. The variance of this estimate is given by

$$\mathrm{Var}\{y_\mathbf{m}\} = \frac{\mathrm{Var}\{n_\mathbf{m}\}}{\overline{g}^2_\mathbf{m}} \,. \tag{16.74}$$

If the only noise in the data were the Poisson noise of the x rays, then we would have $\mathrm{Var}\{n_\mathbf{m}\} = \overline{g}_\mathbf{m}$ and $\mathrm{Var}\{y_\mathbf{m}\} = 1/\overline{g}_\mathbf{m}$. If the dominant noise is independent of the x-ray flux, as it would be for dark current or electronic noise, then $\mathrm{Var}\{y_\mathbf{m}\} \propto 1/\overline{g}^2_\mathbf{m}$. In both cases, increasing the mean number of detected photons decreases the variance in the estimate of the integral.

Effect of source fluctuations In Sec. 16.1.5 we pointed out that fluctuations in the current in an x-ray tube can lead to an additional term in the data covariance matrix [see (16.24)]. The elements in this extra part of the covariance might be comparable to or even larger than those that we have been discussing, but the matrix has rank one, and we showed in Sec. 16.1.6 that its contribution to SKE detectability was negligible [see (16.50)]. Now we need to see what effect it has on estimation tasks.

Suppose the goal is to estimate the average x-ray transmittance of an object over some large area, so that issues of estimability do not arise. If the area covers many pixels, a reasonable estimator is just the sum of the pixel values over the region of interest. Consider the limit of large x-ray flux so that the only limitation is variations in source strength. Since all pixels fluctuate together as the source strength varies, a change of $x\%$ in current through the x-ray tube leads to an $x\%$ change in the estimated transmittance.

If this number is too large, we need to take some auxiliary measurement in order to normalize for the source strength. A separate detector in the x-ray beam can be used, or the imaging detector can be used by looking at x rays that miss the object. Since the auxiliary detector can have large area, Poisson noise in the estimate of source strength should be small. The situation is analogous to the SKE detection problem discussed in 16.1.6 where the ideal observer could average over the entire array and estimate and correct for the source fluctuations.

Similar considerations will apply to other detection or estimation problems. In short, even though fluctuations in tube current lead to large variance and distinctly non-Poisson statistics, they can almost always be overcome by making good use of the measured data or by acquiring auxiliary data for normalization. In terms of task performance, we can almost always assume that the x-ray tube is a Poisson source.

16.2 PLANAR IMAGING IN NUCLEAR MEDICINE

The second direct-imaging method we discuss in this chapter is gamma-ray imaging of radioactive sources, especially for diagnosis in clinical nuclear medicine. We begin in Sec. 16.2.1 with a qualitative overview of the modality and the basic issues affecting image quality. In Secs. 16.2.2 and 16.2.3 we analyze the image-formation and detection processes deterministically. Stochastic considerations are added in Sec. 16.2.4, and objective measures of image quality are developed in Secs. 16.2.5 and 16.2.6.

For an excellent, succinct introduction to clinical aspects of diagnostic nuclear medicine, see Alazraki (1988), and for a not-so-succinct survey of instrumentation issues, see Barrett and Swindell (1981, 1996).

16.2.1 Basic issues

The tracer principle In biology and medicine, a *tracer* is a fluorescent or radioactive marker that can be attached to a biologically important molecule without altering its biological properties. When the tracer is subsequently detected, its location and strength convey information about the function of the tagged molecule.

The advantage of radioactive tracers over fluorescent ones is that gamma rays are more penetrating than optical photons. Therefore gamma-ray imaging can be used to study physiological function deep within a patient's body. For this reason, nuclear medicine is often called *functional imaging*; it gives information about the function of the body as opposed to the anatomical or morphological information provided by transmission radiography or ultrasound. In addition, since the tracer acts at a molecular level, nuclear medicine is a form of *molecular medicine*.

The earliest use of radioactive tracers in biomedical research was by the Hungarian chemist George Charles de Hevesy. Working with Ernest Rutherford in 1911, de Hevesy was staying at a boarding house in Manchester. The guests strongly suspected that the landlady was serving them leftovers, but she insisted she was not. One night, de Hevesy surreptitiously sprinkled an "isotopic indicator," as he called it, into some leftover beef pie, and several days later he was able to detect the tracer in a soufflé (de Hevesy, 1962; Patton, 2000). In 1943, de Hevesy received the Nobel Prize in Chemistry, but this initial experiment was not cited.

Radioisotopes The radioisotopes used in nuclear medicine can emit gamma rays, beta particles (electrons) or positrons (anti-electrons). Beta particles have very short range in tissue, a few millimeters. They have found some use for imaging superficial structures or during surgery where a detector can be placed in the body, but they are not useful for imaging deeper structures with detectors outside the body. Positrons also have short range, but they can annihilate by interacting with electrons, producing high-energy gamma rays that can be detected outside the body in a technique called positron emission tomography or PET. In this chapter we consider only isotopes that emit low-energy gamma rays, typically in the range 100–400 keV.

Most tracers used in nuclear medicine are labelled with an isotope of technetium, $^{99\mathrm{m}}\mathrm{Tc}$ (where m stands for *metastable*). This isotope has a half-life of 6 hours and emits a single gamma ray of energy 140 keV. We shall use this energy when we give numerical examples in this section.

Technetium, especially in the form of the pertechnetate ion TcO_4^-, can be used to label a wide variety of pharmaceuticals, some with exquisite biological specificity. As an example, a tracer called sestamibi has an affinity to mitochondria in cells. Since the mitochondria are the cell's power plants, they are found in abundance in hard-working cells as in the heart muscle (myocardium) or in rapidly dividing cells as in tumors. Thus sestamibi has found use in studying myocardial perfusion and in detecting breast cancer.

Attenuation and scatter Photons of energy 140 keV have an attenuation coefficient of about 0.14 cm^{-1} in soft tissue, so they travel about 7 cm on average before interacting with the tissue. The predominant interaction is Compton scattering, in which the photon changes its direction by scattering from a free electron, transferring some of its energy to the electron in the process. See Secs. 10.3.7 and 12.3.1 for more discussion of the physics of Compton scattering.

At 140 keV, about 1% of the interactions in soft tissue are photoelectric absorption rather than Compton scattering. In photoelectric absorption, as discussed in Sec. 12.3.1, the gamma-ray photon disappears, transferring all of its energy to a photoelectron.[5] The relative probability of photoelectric absorption increases rapidly as the photon energy decreases, so after multiple Compton scatters, the photon is likely to be absorbed if it has not escaped the body. The primary mathematical tools for accounting for all of these possible events are Monte Carlo simulation and the Boltzmann transport equation (Sec. 10.3).

Radiation dose to the patient Both Compton scattering and photoelectric absorption result in a high-energy electron in the patient's body. This electron is absorbed in the body very near the interaction point, possibly causing local biological damage.

Though the basic interaction mechanisms are the same for x rays and gamma rays, there is an important difference between transmission radiography and emission imaging in terms of radiation dose. In transmission imaging, dose is delivered to the patient while the x-ray source is turned on, but dose delivery ceases instantaneously when the source is turned off. In emission imaging, the source cannot be turned off; it is injected into the body and delivers dose continuously until it decays or is biologically excreted. In transmission imaging, therefore, an increased exposure time or a stronger source will give more dose to the patient, but it will also result in more collected photons and hence better image quality. In emission imaging, on the other hand, the dose to the patient depends on the quantity of radiotracer used, but it has nothing to do with exposure time.

Usually the allowed exposure time in nuclear medicine is limited to a few minutes by patient motion or economic considerations, and relatively few gamma-ray photons can be collected in this time with allowable radiation dose. Typical planar nuclear medicine images consist of only 100,000 to 1,000,000 detected photons; the resulting Poisson noise is the main limitation to image quality, as we shall see in more detail in Secs. 16.2.5 and 16.2.6.

Collimators For all practical purposes, gamma rays are not reflected or refracted by matter, so all image-forming apertures in nuclear medicine must work by selective absorption of the photons. The two major alternatives are pinhole apertures and

[5]As discussed in Sec. 12.3.1, the energy imparted to the photoelectron is the gamma-ray energy minus the electron binding energy [see (12.160)], but in soft tissue this binding energy is less than 1 keV and can easily be neglected. For the same reason, we do not have to consider emission of K x rays in the body, though they are important in the detector material as we saw in Sec. 12.3.

multibore collimators. Pinholes were discussed in Sec. 10.4.2 as an illustration of the Boltzmann equation in imaging, and collimators will be analyzed below.

Collimators are basically just slabs of lead or other highly absorbing material with multiple bores through which gamma rays can pass. The bores may be parallel to one another, in which case we have a *parallel-hole collimator* (see Fig. 16.8), or they may be slanted in some manner.

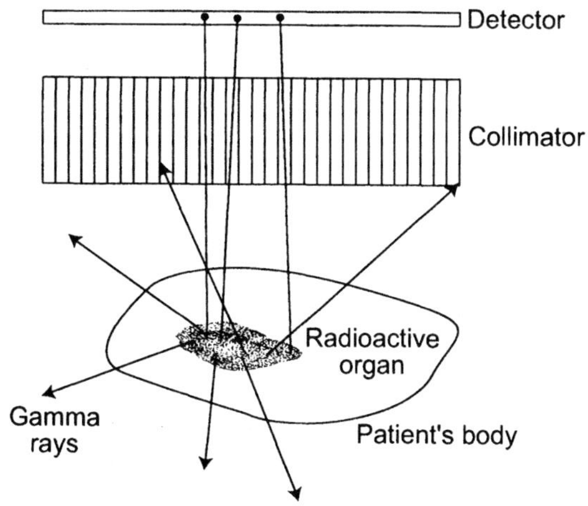

Fig. 16.8 Illustration of a parallel-hole collimator.

For a parallel-hole collimator, key design parameters are the shape and size of each bore, the thickness of the septa between bores and the overall thickness or length of each bore. To a first approximation, only photons that pass down the bore without hitting any absorbing material can reach the detector. Making the bore size larger or the bore length smaller increases the number of photons that get through but reduces the spatial resolution, so these parameters provide a way of controlling the tradeoff between resolution and noise. The septal thickness and the collimator material are chosen to minimize penetration of gamma rays through nominally opaque portions of the collimator.

Typical bore diameters are 1–2 mm for 140 keV radiation, and typical bore lengths are 2–4 cm. With these numbers only about 10^{-4} of the emitted photons pass through the collimator.

Detectors Almost all commercial nuclear-medicine systems at this writing use an Anger scintillation camera as the detector. This detector is analyzed in Secs. 12.3.5 and 12.3.6, and the statistical properties of images obtained with it are discussed in Sec. 12.3.7.

There are several salient points to recall from Chap. 12 about the Anger camera. It counts individual gamma-ray photons and hence suffers from photon noise, but it has no other significant noise source. It provides continuous estimates of the 2D position and energy of each absorbed gamma-ray photon. Each photon is accepted into the final image based on its estimated energy, which is an indication of whether or not the photon has undergone a scattering event in the patient's body. This scatter-rejection technique is not perfect, however, and some scattered photons will contribute to the image.

Even with the position and energy estimation and rejection of some events, the continuous image can still be described as a sample function of a Poisson random process. If the continuous image is binned into a pixel array for storage, processing or display, the resulting discrete image is a Poisson random vector. It is only when the effects of object randomness are considered that the Poisson character is lost.

In spite of the widespread use of anger cameras, there is considerable interest in discrete detector arrays such as semiconductor devices or scintillator-photodiode arrays, discussed in Secs. 12.3.3 and 12.3.4. Major advantages of discrete arrays are that the element size can be smaller than the spatial resolution of the Anger camera and (at least for the semiconductor devices) the energy resolution can be better, improving the ability to reject scattered radiation.

16.2.2 Image formation

The mathematical tools needed to analyze gamma-ray imaging systems were developed in Chap. 10. Since gamma rays have very short wavelength (less than 10^{-9} cm at 140 keV), they behave as particles, and their transport is well described by the Boltzmann equation as derived in Sec. 10.3.

Specifically, if we assume that scattered radiation is rejected by the detector and hence is effectively absorbed, the steady-state Boltzmann equation is given by (10.147) as

$$\hat{\mathbf{s}} \cdot \nabla w = \frac{1}{c} \Xi_{p,\mathcal{E}} - \mu_{tot} w \,, \tag{16.75}$$

where c is the speed of light, μ_{tot} is the total attenuation coefficient (arising mainly from Compton scattering), $w \equiv w(\mathfrak{r}, \hat{\mathbf{s}}, \mathcal{E})$ is the distribution function and $\Xi_{p,\mathcal{E}} \equiv \Xi_{p,\mathcal{E}}(\mathfrak{r}, \hat{\mathbf{s}}, \mathcal{E})$ is the source distribution. For nuclear imaging with a monoenergetic radioisotope, the source has the form (10.253):

$$\Xi_{p,\mathcal{E}}(\mathfrak{r}, \hat{\mathbf{s}}, \mathcal{E}) = \frac{1}{4\pi} f(\mathfrak{r})\, \delta(\mathcal{E} - \mathcal{E}_0) \,. \tag{16.76}$$

The function $f(\mathfrak{r})$ describes the spatial distribution of the radiotracer and hence conveys the functional information that we want to extract. We can interpret $f(\mathfrak{r})$ as the mean rate of photon emission per unit volume; *i.e.*, $f(\mathfrak{r}) d^3\mathfrak{r}$ is the mean number of photons per second emitted in all directions from a volume element $d^3\mathfrak{r}$ centered at point $\mathfrak{r}$. This density is assumed to be independent of time over the exposure time for an image.

As in Sec. 10.4.2, we set up a reference plane P somewhere between the object and the detector. The photon distribution function on that plane is denoted $w(\mathbf{r}, \hat{\mathbf{s}}, \mathcal{E})$. Since we are assuming for now that scattered photons are rejected, all photons of interest have the original energy $\mathcal{E}_0$, and we must have

$$w(\mathbf{r}, \hat{\mathbf{s}}, \mathcal{E}) = w_0(\mathbf{r}, \hat{\mathbf{s}})\, \delta(\mathcal{E} - \mathcal{E}_0) \,. \tag{16.77}$$

The factor $w_0(\mathbf{r}, \hat{\mathbf{s}})$ is given from (16.76) and (10.151) as

$$w_0(\mathbf{r}, \hat{\mathbf{s}}) = \frac{1}{4\pi c} \int_0^\infty d\ell \; f(\mathfrak{r} - \hat{\mathbf{s}}\ell) \exp\left[- \int_0^\ell d\ell' \; \mu_{tot}(\mathfrak{r} - \hat{\mathbf{s}}\ell', \mathcal{E}_0) \right] , \tag{16.78}$$

where the 2D vector $\mathbf{r}$ and the 3D vector $\mathfrak{r}$ denote the same point on plane P. The photon distribution at this point is thus given by a line integral of the source distribution weighted with an attenuation factor.

Mathematical description of the collimator Given an expression for the photon distribution on a plane P, the next step is to analyze the effect of the collimator. The first comprehensive analysis of properties of collimators was performed by Robert Beck (Beck, 1964a, 1964b, 1968a, 1968b). More recent treatments can be found in Barrett and Swindell (1981, 1996), Gunter (1996) and Tsui *et al.* (1996).

A convenient place to put plane P is *behind* the collimator, in which case $w_0(\mathbf{r}, \hat{\mathbf{s}})$ is the distribution in that plane when the collimator is not present. The actual distribution on P with the collimator in place has the form

$$w_c(\mathbf{r}, \hat{\mathbf{s}}) = w_0(\mathbf{r}, \hat{\mathbf{s}})\, T(\mathbf{r}, \hat{\mathbf{s}}) \,, \tag{16.79}$$

where $T(\mathbf{r}, \hat{\mathbf{s}})$ is the transmission of the collimator for photons travelling in direction $\hat{\mathbf{s}}$ and striking plane P at point $\mathbf{r}$.

To obtain an expression for $T(\mathbf{r}, \hat{\mathbf{s}})$, we need a geometrical model for the collimator. We consider a parallel-hole collimator with bores on a regular grid indexed by the 2D vector index $\mathbf{n}$, and we let $\mathbf{r_n}$ denote the center of the $\mathbf{n}^{th}$ bore. We define $\beta(\mathbf{r} - \mathbf{r_n})$ as a function that is unity within the open area of that bore and zero otherwise.

If we neglect penetration through the septa, $T(\mathbf{r}, \hat{\mathbf{s}})$ is unity only if the point where the photon exits the collimator and the point where it enters both lie within the same bore. The exit point is just $\mathbf{r}$ since the reference plane is the exit plane, but the entrance point is found by tracing backward from the exit point along direction $-\hat{\mathbf{s}}$. The 2D vector specifying this point is $\mathbf{r} - \mathbf{s}_\perp \frac{L_b}{s_z}$, where $\mathbf{s}_\perp = (s_x, s_y)$ if $\hat{\mathbf{s}} = (s_x, s_y, s_z)$ and plane P is $z = 0$. Thus we have

$$T(\mathbf{r}, \hat{\mathbf{s}}) = \sum_{\mathbf{n}} \beta(\mathbf{r} - \mathbf{r_n})\, \beta\left(\mathbf{r} - \mathbf{s}_\perp \frac{L_b}{s_z} - \mathbf{r_n}\right) , \tag{16.80}$$

where the sum is over all bores in the collimator.

Photon distribution on the collimator exit plane Combining (16.79) and (16.78), we arrive at the CC mapping from $f(\mathbf{r})$ to $w_c(\mathbf{r}, \hat{\mathbf{s}})$:

$$w_c(\mathbf{r}, \hat{\mathbf{s}}) = \frac{1}{4\pi c} T(\mathbf{r}, \hat{\mathbf{s}}) \int_0^\infty d\ell\; f(\mathbf{r} - \hat{\mathbf{s}}\ell\,) \exp\left[- \int_0^\ell d\ell'\, \mu_{tot}(\mathbf{r} - \hat{\mathbf{s}}\ell', \mathcal{E}_0) \right] . \tag{16.81}$$

We can use this expression to compute the photon irradiance (mean number of photons per unit area) in plane P. In (10.140) we saw that the spectral photon irradiance $I_{p,\mathcal{E}}$ (photons per unit area per unit energy) is obtained from the spectral photon radiance $L_{p,\mathcal{E}}$ (photons per unit *projected* area per steradian per unit energy) by multiplying by a cosine factor (to convert projected area to true area) and then integrating over solid angle. We also know from (10.98) that $L_{p,\mathcal{E}}$ is c times the distribution function w_c. In the present problem the cosine factor is near unity, and the spectral dependence is simple because $L_{p,\mathcal{E}}(\mathbf{r}, \mathcal{E}) = L_p(\mathbf{r})\, \delta(\mathcal{E} - \mathcal{E}_0)$, so we can also write $I_{p,\mathcal{E}}(\mathbf{r}, \mathcal{E}) = I_p(\mathbf{r})\, \delta(\mathcal{E} - \mathcal{E}_0)$. Thus we have

$$\begin{aligned} &I_p(\mathbf{r}) \\ &= c \int_{2\pi} d\Omega\; w_c(\mathbf{r}, \hat{\mathbf{s}}) = \frac{1}{4\pi} \int_{2\pi} d\Omega\; T(\mathbf{r}, \hat{\mathbf{s}}) \int_0^\infty d\ell\; f(\mathbf{r} - \hat{\mathbf{s}}\ell\,) \exp\left[- \int_0^\ell d\ell'\, \mu_{tot}(\mathbf{r} - \hat{\mathbf{s}}\ell', \mathcal{E}_0) \right] . \end{aligned} \tag{16.82}$$

To interpret this expression, we make a change of variables that proved useful several times in Chap. 10; we define $\mathbf{r}' = \mathbf{r} - \hat{\mathbf{s}}\ell$ and recognize that $\ell^2\, d\ell\, d\Omega = d^3\mathbf{r}'$ and $\ell = |\mathbf{r} - \mathbf{r}'|$, yielding

$$I_p(\mathbf{r})$$

$$= \frac{1}{4\pi}\int_\infty \frac{d^3\mathbf{r}'}{|\mathbf{r}-\mathbf{r}'|^2}\, T\left(\mathbf{r}, \frac{\mathbf{r}-\mathbf{r}'}{z'}\right) f(\mathbf{r}') \exp\left[-\int_0^{|\mathbf{r}-\mathbf{r}'|} d\ell'\, \mu_{tot}\left(\mathbf{r} - \frac{\mathbf{r}-\mathbf{r}'}{|\mathbf{r}-\mathbf{r}'|}\ell', \mathcal{E}_0\right)\right]. \tag{16.83}$$

Since we are concerned with the irradiance on the plane $z = 0$, the vectors $\mathbf{r} = (x, y, 0)$ and $\mathbf{r} = (x, y)$ specify the same point. The vector $\mathbf{r}'$ is not confined to $z' = 0$, but we can write it as $\mathbf{r}' = (x', y', z') \equiv (\mathbf{r}', z')$; then, with (16.80) and a little algebra, we have

$$I_p(\mathbf{r}) = \frac{1}{4\pi}\sum_{\mathbf{n}} \beta(\mathbf{r} - \mathbf{r_n}) \int_{L_b}^\infty \frac{dz'}{z'^2} \int_\infty d^2r'\, \beta\left[\left(1 - \frac{L_b}{z'}\right)\mathbf{r} + \frac{L_b}{z'}\mathbf{r}' - \mathbf{r_n}\right] f(\mathbf{r}', z')$$

$$\times \exp\left[-\int_0^{|\mathbf{r}-\mathbf{r}'|} d\ell'\, \mu_{tot}\left(\mathbf{r} - \frac{\mathbf{r}-\mathbf{r}'}{|\mathbf{r}-\mathbf{r}'|}\ell', \mathcal{E}_0\right)\right], \tag{16.84}$$

where we have replaced $1/|\mathbf{r} - \mathbf{r}'|^2$ with $1/z'^2$ since the collimator accepts only photons that travel almost parallel to the z axis. In the z' integral, we have set the lower limit to L_b on the assumption that there is no radioactive material inside the collimator bores (though this condition presumes some minimal care in laboratory practice).

The transformation in (16.84) is a CC mapping from the 3D function $f(\mathbf{r}', z')$ to the 2D function $I_p(\mathbf{r})$; it has the general form

$$I_p(\mathbf{r}) = \int_{L_b}^\infty dz' \int_\infty d^2r'\, h(\mathbf{r}; \mathbf{r}', z')\, f(\mathbf{r}', z'), \tag{16.85}$$

where the kernel $h(\mathbf{r}; \mathbf{r}', z')\, f(\mathbf{r}', z')$ is the point response function (PRF) giving the response at point $\mathbf{r}$ in the collimator exit plane to a point source at point $(\mathbf{r}', z')$ in the 3D space. The remainder of this section is devoted to studying this PRF.

Point response function in air For a point source in air, where $\mu_{tot} \approx 0$, the PRF is given by

$$h(\mathbf{r}; \mathbf{r}', z') = \frac{1}{4\pi z'^2}\sum_{\mathbf{n}} \beta(\mathbf{r} - \mathbf{r_n})\, \beta\left[\left(1 - \frac{L_b}{z'}\right)\mathbf{r} + \frac{L_b}{z'}\mathbf{r}' - \mathbf{r_n}\right]. \tag{16.86}$$

This PRF is highly shift-variant, depending in a complicated way on the lateral source coordinates $\mathbf{r}'$, the longitudinal position of the source z', and the position on the collimator exit face $\mathbf{r}$. Some examples are shown in Fig. 16.9.

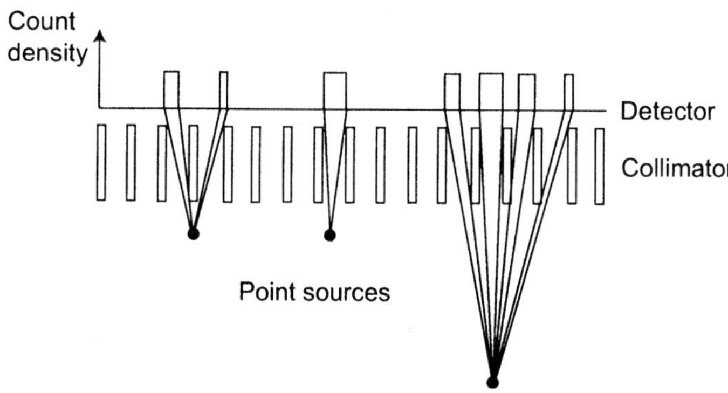

Fig. 16.9 Illustration of photon paths through the collimator for different source positions.

When the point source is on the collimator entrance face, we see graphically from Fig. 16.9 that the PRF as a function of $\mathbf{r}$ is just the transmission function of a single bore, so long as the source point lies over a bore; the response is zero if the source is hidden behind a septum. We reach this same conclusion analytically by setting $z' = L_b$ in (16.86), so that the summand becomes $\beta(\mathbf{r} - \mathbf{r_n})\beta(\mathbf{r}' - \mathbf{r_n})$. The second factor is nonzero only when the source lies within bore $\mathbf{n}$, and the first factor requires that the output point $\mathbf{r}$ lie within that same bore.

For z' substantially greater than L_b, however, photons can reach the exit plane through several different bores, as shown graphically in Fig. 16.9. To understand how this happens analytically, consider circular bores of diameter D_b, so that $\beta(\mathbf{r}) = \text{cyl}(\mathbf{r}/D_b)$. For simplicity, consider a point source centered laterally over the origin, so that $\mathbf{r}' = 0$ but $z' > L_b$. Then the summand of (16.86) can be rewritten as

$$\beta(\mathbf{r} - \mathbf{r_n})\,\beta\left[\left(1 - \frac{L_b}{z'}\right)\mathbf{r} + \frac{L_b}{z'}\mathbf{r}' - \mathbf{r_n}\right]$$

$$= \text{cyl}\left[\frac{1}{D_b}(\mathbf{r} - \mathbf{r_n})\right]\text{cyl}\left\{\frac{1}{D_b}\left[\left(1 - \frac{L_b}{z'}\right)\mathbf{r} - \mathbf{r_n}\right]\right\}. \tag{16.87}$$

The first factor is nonzero when $\mathbf{r}$ lies within a circle of diameter D_b centered on $\mathbf{r_n}$. The second factor, however, is nonzero when $\mathbf{r}$ lies in a circle of diameter $D_b z'/(z' - L_b)$ centered on $z'\mathbf{r_n}/(z' - L_b)$. The complicated PRF seen in Fig. 16.9 is the overlap of these two cylinder functions summed over bores.

Another way to think about $h(\mathbf{r}; \mathbf{r}', z')$ is that it is the kernel of the adjoint mapping (or backprojection) from the 2D collimator exit face to the 3D object volume. In that view, we pick an $\mathbf{r}$ and study $h(\mathbf{r}; \mathbf{r}', z')$ as a function of $\mathbf{r}'$ and z'. The first factor in (16.87) requires that only a single bore, or a single index $\mathbf{n}$, can contribute to $h(\mathbf{r}; \mathbf{r}', z')$. The second factor defines a cone in the 3D space centered on this bore, as shown in Fig. 16.10. On plane z', this kernel is constant for points $\mathbf{r}'$ within a circle of diameter $D_b z'/L_b$ centered on the origin.

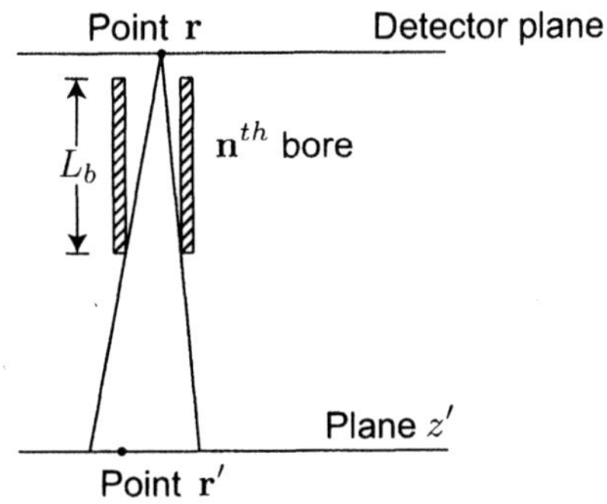

Fig. 16.10 Illustration of the kernel of the adjoint mapping from the collimator exit plane to the 3D object space.

Average over collimator shifts The lateral shift variance of the PRF comes about since it matters just where the point source is with respect to the center of a collimator bore. Anger (1964) suggested that a more meaningful PRF could be defined by averaging over all positions of the collimator. Since shifting the collimator laterally is equivalent to shifting the object and detector together, the averaging blurs out the fine structure of the collimator but does not blur the image.

Following Barrett and Swindell (1981, 1996), we compute the average PRF by adding a 2D vector $\mathbf{R}$ to each bore position $\mathbf{r_n}$, integrating the PRF over all $\mathbf{R}$ within a disc of radius R_{max}, and dividing by the area of that disc. With the form of PRF given in (16.86), the average PRF is given by

$$\overline{h}(\mathbf{r};\mathbf{r}',z')$$
$$= \frac{1}{4\pi z'^2}\sum_{\mathbf{n}}\frac{1}{\pi R_{max}^2}\int_{disc} d^2R\ \beta(\mathbf{r}-\mathbf{r_n}-\mathbf{R})\,\beta\left[\left(1-\frac{L_b}{z'}\right)\mathbf{r}+\frac{L_b}{z'}\mathbf{r}'-\mathbf{r_n}-\mathbf{R}\right]. \tag{16.88}$$

Making the change of variables $\mathbf{R}'=\mathbf{r}-\mathbf{r_n}-\mathbf{R}$ and letting R_{max} get large so we don't have to worry about limits of integration, we find

$$\overline{h}(\mathbf{r};\mathbf{r}',z') = \frac{1}{4\pi z'^2}\sum_{\mathbf{n}}\frac{1}{\pi R_{max}^2}\int_{\infty} d^2R\ \beta(\mathbf{R}')\,\beta\left[\frac{L_b}{z'}(\mathbf{r}'-\mathbf{r})+\mathbf{R}'\right]. \tag{16.89}$$

The integral is recognized as the autocorrelation integral of the bore function (see Sec. 3.4.3); since this autocorrelation is independent of $\mathbf{n}$, it can be taken out of the sum, and we have

$$\overline{h}(\mathbf{r};\mathbf{r}',z') = \frac{1}{4\pi z'^2}\frac{N(R_{max})}{\pi R_{max}^2}[\beta\star\beta]\left[\frac{L_b}{z'}(\mathbf{r}-\mathbf{r}')\right], \tag{16.90}$$

where $\star$ denotes the autocorrelation integral as defined in (3.115), and $N(R_{max})$ is the number of bores within the disc. If the collimator bores are arranged on a regular lattice with each bore lying in a unit cell of area A_{cell}, then $N(R_{max}) = \pi R_{max}^2/A_{cell}$. The bore itself has an area A_{bore} (equal to $\pi D_b^2/4$ for circular bores), and we can define the *packing fraction* $\alpha_{pf} \equiv A_{bore}/A_{cell}$, so that $N(R_{max}) = \pi R_{max}^2\alpha_{pf}/A_{bore}$, and we obtain, finally,

$$\overline{h}(\mathbf{r}-\mathbf{r}',z') = \frac{1}{4\pi z'^2}\frac{\alpha_{pf}}{A_{bore}}[\beta\star\beta]\left[\frac{L_b}{z'}(\mathbf{r}-\mathbf{r}')\right]. \tag{16.91}$$

Note that we have rewritten the average PRF as $\overline{h}(\mathbf{r}-\mathbf{r}', z')$ since the averaging process has removed the shift-variance that came from the collimator structure. The system is now an axial shift-invariant system of the kind discussed in Sec. 7.2.10, where the mapping from 3D object to 2D image is a convolution in the lateral variables x' and y' and an integration over depth z'.

Flood image and point sensitivity We know from Sec. 7.2.1 that the uniformity of a CC imaging system can be specified by its flood image and point-source sensitivity. The flood uniformity can be measured in practice by placing a uniform planar source parallel to the collimator face and recording an image. In the present discussion, the image is specified by the photon irradiance in plane P, and we denote the flood image as $I_{fld}(\mathbf{r})$. Mathematically, if $f(\mathbf{r}', z') = \delta(z' - z_0)$, then [*cf.* (7.111)]

$$I_{fld}(\mathbf{r}) = \int_{L_b}^{\infty} dz' \int_{\infty} d^2r'\, h(\mathbf{r};\mathbf{r}', z')$$

$$= \frac{1}{4\pi z_0^2} \sum_{\mathbf{n}} \beta(\mathbf{r}-\mathbf{r_n}) \int_{\infty} d^2r'\, \beta\left[\left(1-\frac{L_b}{z_0}\right)\mathbf{r} + \frac{L_b}{z_0}\mathbf{r}' - \mathbf{r_n}\right]. \qquad (16.92)$$

The integral can be performed if we again assume circular bores of diameter D_b. Since we know from Fig. 16.10 that the integrand is unity within a circle of diameter $D_b z_0/L_b$ and zero otherwise, we find that

$$I_{fld}(\mathbf{r}) = \frac{D_b^2}{16L_b^2} \sum_{\mathbf{n}} \beta(\mathbf{r}-\mathbf{r_n}). \qquad (16.93)$$

Note that this result is independent of z_0; you cannot tell how far you are from a perfectly uniform source. The dependence on $\mathbf{r}$ is just the collimator bore pattern.

The point sensitivity is defined generally in (7.113), and in the present problem it is given by

$$s_{pt}(\mathbf{r}', z') \equiv \int_{\infty} d^2r\; h(\mathbf{r};\mathbf{r}', z')$$

$$= \frac{1}{4\pi z'^2} \sum_{\mathbf{n}} \int_{\infty} d^2r\; \beta(\mathbf{r}-\mathbf{r_n})\, \beta\left[\left(1-\frac{L_b}{z'}\right)\mathbf{r} + \frac{L_b}{z'}\mathbf{r}' - \mathbf{r_n}\right]. \qquad (16.94)$$

The integral is straightforward for circular bores and a point source on the collimator entrance face ($z' = L_b$), in which case we find

$$s_{pt}(\mathbf{r}', L_b) = \frac{D_b^2}{16L_b^2} \sum_{\mathbf{n}} \beta(\mathbf{r}'-\mathbf{r_n}). \qquad (16.95)$$

Note that $D_b^2/(16L_b^2)$ can also be written as $\frac{1}{4\pi}(\pi D_b^2/4L_b^2)$; the factor in parentheses in this form is the solid angle subtended by a bore on the exit face of the collimator from a point within that bore on the entrance face, and 4π is the solid angle subtended by a sphere centered on this point. Thus $D_b^2/(16L_b^2)$ is the fraction of the photons emitted from the point that pass through the collimator on average. The sum is unity unless the point is behind an opaque septum, in which case the sum is zero.

Computation of the point sensitivity is more difficult for $z_0 \neq L_b$, but an interesting insight can be obtained by considering $z' \gg L_b$. In that case, photons from a point at $(\mathbf{r}', z')$ can pass through many different bores of the collimator (see Fig. 16.9), and it may be a good approximation to replace the sum over $\mathbf{n}$ by an integral:

$$\sum_{\mathbf{n}} \to \frac{1}{A_{cell}} \int_\infty d^2 r_{\mathbf{n}} = \frac{\alpha_{pf}}{A_{bore}} \int_\infty d^2 r_{\mathbf{n}} \,. \tag{16.96}$$

With this approximation and a change of variables, we find

$$s_{pt}(\mathbf{r}', z') = \frac{1}{4\pi z'^2} \frac{\alpha_{pf}}{A_{bore}} \int_\infty d^2 r \, [\beta \star \beta] \left[\frac{L_b}{z'} (\mathbf{r} - \mathbf{r}') \right] , \tag{16.97}$$

which we could also have obtained immediately from the average PSF of (16.91). The remaining integral can be performed with the help of (3.135) and the central-ordinate theorem, (3.229); the result is

$$s_{pt}(\mathbf{r}', z') = \frac{1}{4\pi L_b^2} \alpha_{pf} A_{bore} = \frac{\alpha_{pf} D_b^2}{16 L_b^2} , \tag{16.98}$$

where the last form is specifically for circular bores. Thus, when $z' \gg L_b$ so that many bores can be seen from one source point, the point sensitivity is independent of source position in both $\mathbf{r}'$ and z'. In addition, this sensitivity is essentially the same as that given in (16.91) even though that expression was for the opposite limit of $z' = L_b$; averaging (16.91) over $\mathbf{r}'$ just introduces a factor of α_{pf} since that fraction of source positions fall in open collimator bores. In this average sense, therefore, the point sensitivity is approximately independent of z' for the full range of depths. As we shall see in Chap. 17, this observation is important in SPECT imaging.

Attenuation in the object The sensitivity is by no means independent of source depth when attenuation is considered, and the PRF becomes difficult to analyze in full generality in this case. Fortunately, it is often valid to assume that the attenuation coefficient is either constant within the object or at least a slowly varying function of position.

When $\mu_{tot}(\mathbf{r}', z', \mathcal{E}_0)$ varies slowly with $\mathbf{r}'$ (perpendicular to the collimator bores), the exponential factor in (16.84) can be approximated as

$$\exp\left[- \int_0^{|\mathbf{r}-\mathbf{r}'|} d\ell' \, \mu_{tot} \left(\mathbf{r} - \frac{\mathbf{r} - \mathbf{r}'}{|\mathbf{r} - \mathbf{r}'|} \ell', \mathcal{E}_0 \right) \right] \approx \exp\left[- \int_0^{z'} d\ell' \, \mu_{tot}(\mathbf{r}, \ell', \mathcal{E}_0) \right] . \tag{16.99}$$

Moreover, if μ_{tot} is a constant, this factor becomes $\exp[-\mu_{tot} L(\mathbf{r}', z')]$, where $L(\mathbf{r}', z')$ is the total path length through the attenuating material from point $(\mathbf{r}', z')$ to the collimator face.

It is straightforward to carry these attenuation factors along in an analysis of the forward problem in gamma-ray imaging, but, as we shall see in Sec. 17.2, it is very tricky to compensate for them in the inverse problem, SPECT.

Other effects The discussion to this point has left out several effects that can be important in practical nuclear medicine systems. We have not yet explicitly included Compton scatter in the formalism above; though its effects are implicitly contained in the total attenuation μ_{tot}, we have not considered what happens to the scattered photons and how they might affect the image. In addition, we considered only an idealized collimator, neglecting septal penetration and scattering and x-ray generation within the bore. All of these effects can, in principle, be analyzed with the Boltzmann equation.

Scatter within the object can be analyzed as in Sec. 10.3.4, and the scattered radiation can be treated as an additional emissive source to be imaged by the collimator. Because of the energy loss on Compton scattering, this secondary source is no longer monoenergetic, so the attenuation factors may be different for scattered and unscattered photons, and the detector response will definitely be different.

To include septal penetration in the collimator description, we require a double sum over bore indices $\mathbf{n}$ and $\mathbf{n}'$ in the transmission expression (16.80), and we have to compute the amount of absorbing material interposed between the exit point and the entrance point. The formal expression is not difficult to obtain but it must usually be evaluated numerically.

To include scatter in the collimator, we can use the *bidirectional transmission distribution function* or BTDF, as defined in (10.86), in place of the transmission $T(\mathbf{r}, \hat{\mathbf{s}})$. In fact, we could adopt the BTDF as the general description of the collimator, and in the limit of no scatter or x-ray generation it would be given by $\text{BTDF}(\mathbf{r}, \hat{\mathbf{s}}, \hat{\mathbf{s}}') = T(\mathbf{r}, \hat{\mathbf{s}})\, \delta(\hat{\mathbf{s}} - \hat{\mathbf{s}}')$, with $T(\mathbf{r}, \hat{\mathbf{s}})$ given by (16.80).

16.2.3 The detector

Next we consider how a detector responds to the photons emerging from the collimator. For simplicity we assume that the detector entrance face exactly coincides with the collimator exit face (plane P), though in practice there is always a small gap. We assume also that the photons are normally incident on the detector since they have passed down a bore of a parallel-hole collimator; thus we can adequately specify the photon distribution by the irradiance on P rather than the full distribution function. We consider separately the cases of discrete detector arrays and Anger scintillation cameras.

Discrete detector array Consider first discrete detector elements, such as the photon-counting semiconductor detectors analyzed in Sec. 12.3.2. Suppose each element absorbs a fraction $\eta(\mathcal{E})$ of the incident photons of energy $\mathcal{E}$ and that an absorbed photon has a probability $P_{acc}(\mathcal{E})$ of being accepted by the energy window and hence contributing to the image; both $\eta(\mathcal{E})$ and $P_{acc}(\mathcal{E})$ are assumed for simplicity to be independent of where the photon strikes the detector face. Thus the mean number of photons accepted in the $\mathbf{m}^{th}$ detector in some exposure time τ is

$$\overline{g}_{\mathbf{m}} = \tau \int_{\mathbf{m}} d^2r \int_0^\infty d\mathcal{E}\; P_{acc}(\mathcal{E})\, \eta(\mathcal{E})\, I_p(\mathbf{r}, \mathcal{E})\,, \tag{16.100}$$

where the spatial integral is over the face of the detector. Note that $\tau I_p(\mathbf{r}, \mathcal{E})$ is the spectral photon fluence on the detector face.

If we consider only the unscattered photons, $I_p(\mathbf{r}, \mathcal{E}) = I_p(\mathbf{r})\, \delta(\mathcal{E} - \mathcal{E}_0)$, and

$$\overline{g}_{\mathbf{m}} = \tau \eta(\mathcal{E}_0)\, P_{acc}(\mathcal{E}_0) \int_{\mathbf{m}} d^2r\; I_p(\mathbf{r})\,. \tag{16.101}$$

Next we insert (16.85) into (16.101), yielding

$$\overline{g}_{\mathbf{m}} = \tau\eta(\mathcal{E}_0)\,P_{acc}(\mathcal{E}_0)\int_{\mathbf{m}} d^2r \int_{L_b}^{\infty} dz' \int_{\infty} d^2r'\,h(\mathbf{r};\mathbf{r}',z')\,f(\mathbf{r}',z')\,. \tag{16.102}$$

We can put this result in our standard CD form,

$$\overline{g}_{\mathbf{m}} = \int_{\infty} d^3\mathfrak{r}'\,h_{\mathbf{m}}(\mathfrak{r}')\,f(\mathfrak{r}')\,, \tag{16.103}$$

by noting that $\mathfrak{r}' = (\mathbf{r}', z')$ and defining

$$h_{\mathbf{m}}(\mathfrak{r}') = \tau\eta(\mathcal{E}_0)\,P_{acc}(\mathcal{E}_0)\int_{\mathbf{m}} d^2r\ h(\mathbf{r};\mathfrak{r}')\,. \tag{16.104}$$

With $h(\mathbf{r};\mathfrak{r}')$ given by (16.86), this equation describes the overall mapping from the emissive object to the detector output when attenuation and scatter can be neglected. We shall put these effects back into the formalism shortly, but for now we concentrate on the effects of the collimator and the detector.

Small detector elements To perform the integral in (16.104), we assume initially that one detector element covers exactly one collimator bore, so that we can renumber the collimator bores with the detector index $\mathbf{m}$ and drop the sum over bores. Neglecting attenuation, we can use (16.86) and write

$$h_{\mathbf{m}}(\mathfrak{r}') = \eta\tau P_{acc}(\mathcal{E}_0)\frac{1}{4\pi z'^2}\int_{\infty} d^2r\ \beta(\mathbf{r}-\mathbf{r}_{\mathbf{m}})\,\beta\!\left[\left(1-\frac{L_b}{z'}\right)\mathbf{r} + \frac{L_b}{z'}\mathbf{r}' - \mathbf{r}_{\mathbf{m}}\right], \tag{16.105}$$

where we have been able to extend the limits of integration to ∞ because of the first factor in the integrand.

To visualize this kernel, recall from Fig. 16.10 that the integrand for fixed $\mathbf{r}$ defines a cone in the 3D $\mathfrak{r}'$ space; the integral in (16.105) merely adds a lot of slightly tilted cones, fuzzing the edges as shown in Fig. 16.11.

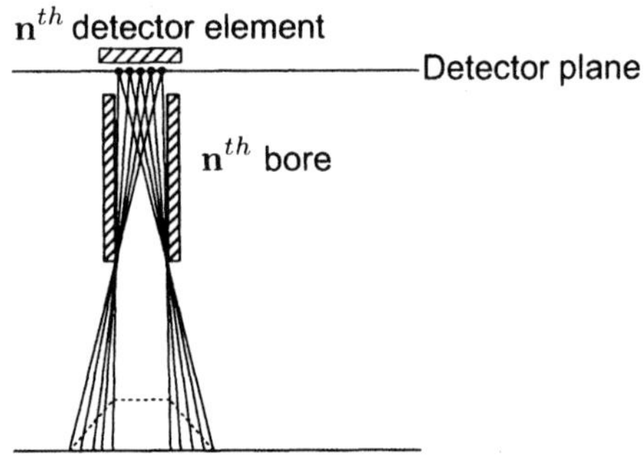

Fig. 16.11 Illustration of the kernel of the adjoint mapping from a discrete detector element to the 3D object space when the element covers exactly one bore

Larger detector elements Now suppose that each detector element covers many collimator bores. Let $S_{\mathbf{m}}$ denote the set of collimator indices $\mathbf{n}$ such that the open bore of collimator $\mathbf{n}$ falls entirely within the area of detector $\mathbf{m}$; assume that no bores fall partially on a detector element. Then (16.105) is modified to

$$h_{\mathbf{m}}(\mathbf{r}') = \tau\eta(\mathcal{E}_0)\, P_{acc}(\mathcal{E}_0)\frac{1}{4\pi z'^2}\sum_{\mathbf{n}\in S_{\mathbf{m}}}\int_\infty d^2r\ \beta(\mathbf{r}-\mathbf{r_n})\,\beta\left[\left(1-\frac{L_b}{z'}\right)\mathbf{r}+\frac{L_b}{z'}\mathbf{r}'-\mathbf{r_n}\right]. \tag{16.106}$$

This kernel is again a superposition of cones, as shown in Fig. 16.12, but now cones associated with different bores are displaced laterally. Thus the resulting kernel after integrating over each bore and summing over bores is more nearly a cylinder with fuzzy edges. The spatial resolution, as measured by the width of the intersection of the kernel with a plane of constant z', is larger because the detector element is larger, but it is less dependent on z', which may be an advantage in some applications. At the least, it will simplify the mathematics in SPECT imaging (see Sec. 17.1).

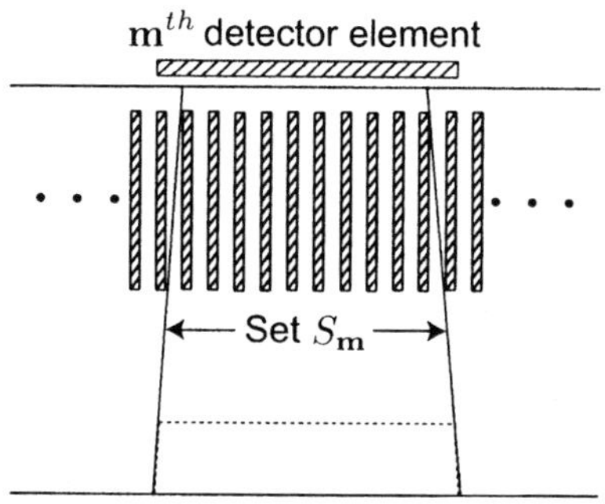

Fig. 16.12 Illustration of the kernel for the adjoint mapping when one discrete detector element covers several collimator bores.

Anger camera An Anger camera does not merely integrate the spectral photon irradiance over a fixed area. Instead, as discussed in Sec. 12.3.6, it estimates the position of each photon, and (in modern cameras, at least) it assigns the photon to one bin in a digital matrix based on the estimated coordinates. We can now use the index $\mathbf{m}$ to refer to a bin in the final digital image rather than a discrete detector element.

We can describe the position-estimation process with a conditional probability density function $\mathrm{pr}(\hat{\mathbf{r}}|\mathbf{r})$, where $\mathbf{r}$ is the actual position where the photon strikes the camera face and $\hat{\mathbf{r}}$ is the estimated position. With the position-estimation step included, $\overline{g}_{\mathbf{m}}$ is given by

$$\begin{aligned}\overline{g}_{\mathbf{m}} &= \tau\eta(\mathcal{E}_0)\, P_{acc}(\mathcal{E}_0)\int_{\mathbf{m}} d^2\hat{r}\int_\infty d^2r\ \mathrm{pr}(\hat{\mathbf{r}}|\mathbf{r})\, I_p(\mathbf{r}) \\ &= \tau\eta(\mathcal{E}_0)\, P_{acc}(\mathcal{E}_0)\int_{\mathbf{m}} d^2\hat{r}\int_\infty d^2r\ \mathrm{pr}(\hat{\mathbf{r}}|\mathbf{r})\int_\infty d^3\mathbf{r}'\, h(\mathbf{r};\mathbf{r}')\, f(\mathbf{r}').\end{aligned} \tag{16.107}$$

The overall kernel for the CD mapping, in the absence of scatter and attenuation, is thus

$$h_{\mathbf{m}}(\mathbf{r}') = \tau\eta(\mathcal{E}_0)\, P_{acc}(\mathcal{E}_0)\int_{\mathbf{m}} d^2\hat{r}\int_\infty d^2r\ \mathrm{pr}(\hat{\mathbf{r}}|\mathbf{r})\, h(\mathbf{r};\mathbf{r}'), \tag{16.108}$$

where $h(\mathbf{r};\mathbf{r}')$ is given explicitly by (16.86). This kernel accounts for the projection of the 3D object onto a 2D plane, collimator blur, blur introduced by position estimation and integration over a bin. Of these effects, only the collimator blur depends on depth in the object, z'; blurring by position estimation and discrete binning occur after the 3D-to-2D projection.

Scatter and attenuation To include scattered radiation in the formalism, we must consider the spectral photon irradiance $I_{p,\mathcal{E}}$ (or equivalently, the spatio-spectral fluence $b(\mathbf{r},\mathcal{E}) \equiv \tau I_{p,\mathcal{E}}$) incident on the detector. The first form of (16.107) generalizes to [*cf.* (16.100)]

$$\overline{g}_{\mathbf{m}} = \tau \int_{\mathbf{m}} d^2\hat{r} \int_0^\infty d\hat{\mathcal{E}}\; P_{acc}(\hat{\mathcal{E}}) \int_\infty d^2 r \int_0^\infty d\mathcal{E}\; \mathrm{pr}(\hat{\mathbf{r}},\hat{\mathcal{E}}|\mathbf{r},\mathcal{E})\,\eta(\mathcal{E})\,I_{p,\mathcal{E}}(\mathbf{r},\mathcal{E})\,, \tag{16.109}$$

where $P_{acc}(\hat{\mathcal{E}})$ is unity if $\hat{\mathcal{E}}$ lies in the energy window and zero otherwise. (By contrast, $P_{acc}(\mathcal{E})$ in (16.100) can take on a continuous range of values since it depends on the actual energy rather than the estimated one.)

The spectral photon irradiance $I_{p,\mathcal{E}}$ can be found from the spectral photon radiance by integrating $L_{p,\mathcal{E}}\cos\theta$ over solid angle, and $L_{p,\mathcal{E}}$ in turn can be found by solving the Boltzmann equation. In this calculation, it may be a useful approximation to neglect photons that have undergone more than one scatter event on the grounds that multiply scattered photons are likely to have lost enough energy that they are rejected by the energy window. This approximation will break down for large objects or cameras with poor energy resolution.

The Boltzmann equation is linear, so (16.109) is an overall linear mapping from the source $f(\mathbf{r})$ to the data value $\overline{g}_{\mathbf{m}}$. Computation of the kernel for this mapping without the position and energy estimation is treated in Sec. 10.4.1, and the integrals over $\hat{\mathcal{E}}$ and $\hat{\mathbf{r}}$ in (16.109) account for the estimation steps. It is a useful exercise to derive an explicit expression for the kernel in the single-scatter approximation with position and energy estimation.

16.2.4 Stochastic properties

We already have a comprehensive account of the stochastic properties of gamma-ray images from the discussions in Chaps. 11 and 12. From Chap. 11 we know that Poisson statistics arise whenever we are counting independent events. From a nonrandom source, gamma rays are emitted independently, in full accord with the postulates presented in Sec. 11.1.1. Position and energy estimation and energy windowing do not introduce any dependence, and the postulates still apply after these operations (see Secs. 12.3.6 and 12.3.7). It follows, then, that the number of gamma rays emitted in some time interval is a Poisson random variable, the pattern of photons incident on any plane is a Poisson random process, and a discrete detected image is a Poisson random vector.

In this section we shall apply these general principles specifically to the case of planar nuclear medicine, but we shall be looking ahead to Chap. 17, where we discuss tomographic gamma-ray imaging; the stochastic models developed here will find their full use in that reconstruction problem. Three distinct ways of representing the random data — list, random process and image histogram — will be discussed, and the statistical properties of each representation will be presented.

List mode The first data representation to discuss is just a list of the raw data values. In the nuclear-medicine literature, this form of data storage is known as *list mode.* There is considerable freedom in choosing what information to include in the list.

For a scintillation camera, the rawest data are the photomultiplier signals for each absorbed events. If we wanted to preserve this information in its entirety, we could just construct a list containing all of these values. Suppose that J photons are absorbed in the scintillation crystal of an Anger camera, and that the jth photon produces a peak voltage on the kth photomultiplier of V_k, $(k = 1, ...K)$. We can store these values in a list J elements long, with each entry in the list consisting of K numbers. We might also add other information to each entry, such as the time of arrival of the photon or an identifier specifying which of several cameras the photon struck. With this form of storage, we have preserved all of the information in the raw PMT data, but we must do substantial further processing to convert the list into an image.

A less raw approach is to first estimate position and energy of each photon and then to store these estimated values in a list. If the jth photon is estimated to have a 2D position $\hat{\mathbf{r}}_j$ and an energy $\hat{\mathcal{E}}_j$, then each entry in the list consists of three numbers, $\hat{x}_j$, $\hat{y}_j$ and $\hat{\mathcal{E}}_j$, plus time and possibly other identifiers.

To be general about the statistics of list-mode data, we can define an *attribute vector* $\mathbf{A}_j$ for the jth photon. For example, if we store raw PMT signals, the attribute vector has K elements, and for position and energy estimates it has three elements. Nonrandom parameters such as time might be included in the list but are not considered to be part of the attribute vector; instead they can be lumped into a parameter vector $\boldsymbol{\alpha}_j$, and the full list is $\{\mathbf{A}_j, \boldsymbol{\alpha}_j, j = 1, ..., J\}$.

Statistical independence Whenever the Poisson postulates of Sec. 11.1.1 are satisfied, the attribute vectors for different photons are statistically independent. This is not a universally valid assumption, however, since it ignores some effects that can occur in a detector at high count rates. As we noted in Sec. 11.1.1, if one photon temporarily paralyzes the detector and there is a significant probability of another photon arriving before it recovers, the probability of detection of the second photon is dependent on the presence of the first. By the same token, the measured attributes of the second photon can depend on those of the first. For example, PMT signals can have tails extending for a microsecond or so after the peak. If the second photon occurs within a microsecond of the first, the observed signal will be the second signal plus the tail of the first (see Fig. 16.13); the peak signal will be altered, as will any position or energy estimates derived from the peak signals, and the attributes will not be independent.

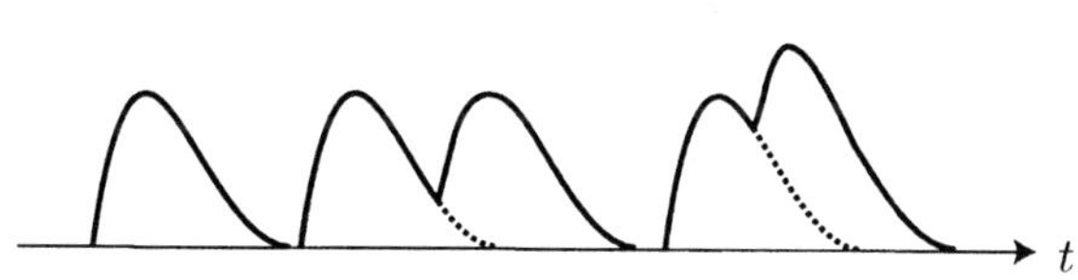

Fig. 16.13 Typical photomultiplier signals showing pulse pileup.

If we neglect these high-count-rate effects, then we can assume that attribute vectors for different photons are statistically independent, and we can write

$$\mathrm{pr}(\{\mathbf{A}_j\}|\mathbf{f}) = \prod_{j=1}^{J} \mathrm{pr}(\mathbf{A}_j|\mathbf{f}) \,. \tag{16.110}$$

Preset time vs. preset count As we discussed in Sec. 11.2.1, there are two different ways of acquiring data in nuclear medicine; we can collect photons for a preset time or until a preset number of counts is accumulated. If we collect for a preset time τ, the number of items in the list is random, and J therefore becomes part of the data. The overall data distribution is then[6]

$$\mathrm{pr}(\{\mathbf{A}_j\}, J|\mathbf{f}) = \mathrm{pr}(\{\mathbf{A}_j\}|J, \mathbf{f})\, \mathrm{Pr}(J|\mathbf{f}) = \exp\left[-\overline{J}(\mathbf{f})\right] \frac{\left[\overline{J}(\mathbf{f})\right]^J}{J!} \prod_{j=1}^{J} \mathrm{pr}(\mathbf{A}_j|\mathbf{f}) \,, \tag{16.110}$$

where the last form recognizes that J is a Poisson random variable with mean dependent on the object (and, of course, on the exposure time).

If we collect exactly J counts, the image itself contains no information on the absolute strength of the source, but we might also record the time τ_J needed to collect the counts. In that case, the data set includes τ_J as well as the attribute list, and we can write the overall PDF as

$$\mathrm{pr}(\{\mathbf{A}_j\}, \tau_J|\mathbf{f}) = \mathrm{pr}(\{\mathbf{A}_j\}|\mathbf{f})\, \mathrm{pr}(\tau_J|\mathbf{f}) = \mathrm{pr}(\tau_J|\mathbf{f}) \prod_{j=1}^{J} \mathrm{pr}(\mathbf{A}_j|\mathbf{f}) \,. \tag{16.111}$$

The time needed to collect J counts from a Poisson source is a random variable, but for large J its density can be approximated by a delta function $\delta[\tau_J - J/a(\mathbf{f})]$, where $a(\mathbf{f})$ is the overall object-dependent count rate.

To complete these statistical descriptions, we need a way of computing the attribute density $\mathrm{pr}(\mathbf{A}_j|\mathbf{f})$; we shall return to this problem shortly.

Random process For theoretical analysis, it is convenient to represent the items in a list in terms of a random process. For example, if the attribute vectors consist of position and energy estimates, we can use them to construct a spatio-spectral random process as in (12.277):

$$g_{det}(\hat{\mathbf{r}}, \hat{\mathcal{E}}) \equiv \sum_{j=1}^{J} \delta(\hat{\mathbf{r}} - \hat{\mathbf{r}}_j)\, \delta(\hat{\mathcal{E}} - \hat{\mathcal{E}}_j) \,, \tag{16.112}$$

where $\hat{\mathbf{r}}$ is the estimated 2D position on the camera face and $\hat{\mathcal{E}}$ is the estimated energy.

All properties of a Poisson random process are fully determined by its fluence. For the spatio-spectral random process $g_{det}(\hat{\mathbf{r}}, \hat{\mathcal{E}})$, the relevant fluence is given by (12.282) as [*cf.* (16.109)]:

$$b_{det}(\hat{\mathbf{r}}, \hat{\mathcal{E}}) = \tau \int_{\infty} d^2r \int_0^{\infty} d\mathcal{E}\; \mathrm{pr}(\hat{\mathbf{r}}, \hat{\mathcal{E}}|\mathbf{r}, \mathcal{E})\, \eta(\mathcal{E})\, I_{p,\mathcal{E}}(\mathbf{r}, \mathcal{E}) \,. \tag{16.113}$$

[6] Recall that we use the lower-case $\mathrm{pr}(\cdot)$ when we want to specify jointly a probability for a discrete random variable and a PDF for a continuous one.

This fluence is directly the mean of the random process, and it also specifies the autocovariance function in position and energy by (12.285):

$$K_{g_{det}}(\hat{\mathbf{r}},\hat{\mathbf{r}}',\hat{\mathcal{E}},\hat{\mathcal{E}}') = b_{det}(\hat{\mathbf{r}},\hat{\mathcal{E}})\,\delta(\hat{\mathbf{r}}-\hat{\mathbf{r}}')\,\delta(\hat{\mathcal{E}}-\hat{\mathcal{E}}')\,. \tag{16.114}$$

PDF for the attribute vector We know from Secs. 11.3.2 and 12.3.7 that the spatio-spectral fluence can be interpreted, after proper normalization, as the probability density function on the estimated position and energy of any individual count. Thus, if the attributes for an item in the list are estimated position and energy, the PDF needed in (16.110) and (16.111) is [*cf.* (12.281)]

$$\text{pr}(\hat{\mathbf{r}},\hat{\mathcal{E}}|\mathbf{f}) = \frac{1}{\overline{J}(\mathbf{f})} b_{det}(\hat{\mathbf{r}},\hat{\mathcal{E}})\,. \tag{16.115}$$

Image histograms In gamma-ray imaging with a scintillation camera, the final image is usually obtained from the position and energy estimates by accepting photons whose estimated energies lie in some energy window and binning or histogramming the spatial positions as in (12.286). The integral of a Poisson random process over any area is a Poisson random variable, so we know the statistics of the discrete data vector at once. In particular, the covariance matrix is given by (12.288) as

$$[\mathbf{K_g}]_{\mathbf{mm}'} = \overline{g}_{\mathbf{m}}\,\delta_{\mathbf{mm}'}\,. \tag{16.116}$$

If each $\overline{g}_{\mathbf{m}}$ is large, say greater than about 10, it is a good approximation to consider the data to be normally distributed with this covariance.

Noise due to scatter Since it was couched entirely in terms of the spatio-spectral fluence, this discussion on stochastic properties of gamma-ray images has made no distinction between scattered and unscattered photons. The scattering does not introduce any dependence among different photons and does not affect the Poisson character of the spatio-spectral random process or the random vector that results after binning. Each accepted photon contributes one count to the digital image; the energy loss influences the probability of acceptance but not the contribution of the photon to the image if it is accepted.

Object randomness Object randomness does lead to non-Poisson statistics in nuclear medicine. When the object varies, the spatio-spectral fluence varies also, and $g_{det}(\hat{\mathbf{r}},\hat{\mathcal{E}})$ becomes a doubly stochastic Poisson random process as discussed in Sec. 11.3.6. Similarly, the discrete image after energy windowing and spatial binning becomes a doubly stochastic Poisson random vector (see Sec. 11.2.2).

16.2.5 Image quality: Classification tasks

Planar nuclear medicine affords a good opportunity to discuss basic issues in image quality with relatively simple noise and system models. Unlike the discussion of image quality in digital radiography in Sec. 16.1.6, we need not worry about non-Poisson noise (other than object variability), and the system is strictly linear so we do not need contorted discussions of linearization.

In this section we shall discuss image quality in planar nuclear medicine for detection and discrimination tasks. We shall begin with SKE/BKE tasks, using them to discuss the importance of detector resolution. When we get to collimator resolution, however, we shall see that it is hazardous to rely too much on SKE/BKE tasks. At that point we shall turn to discrimination tasks and random backgrounds.

SKE/BKE tasks and the ideal observer SKE/BKE detection tasks in pure Poisson noise were analyzed in Sec. 13.2.9. We know from (13.131) that the log-likelihood ratio for this case is given by

$$\lambda = \sum_{\mathbf{m}=1}^{M} g_{\mathbf{m}} \ln \frac{\overline{g}_{2\mathbf{m}}}{\overline{g}_{1\mathbf{m}}}, \tag{16.117}$$

and the ideal-observer SNR is given by (13.132) as

$$\mathrm{SNR}_\lambda^2 = \frac{\left[\sum_{\mathbf{m}=1}^{M}(\overline{g}_{2\mathbf{m}} - \overline{g}_{1\mathbf{m}}) \ln\left(\frac{\overline{g}_{2\mathbf{m}}}{\overline{g}_{1\mathbf{m}}}\right)\right]^2}{\frac{1}{2}\sum_{\mathbf{m}=1}^{M}(\overline{g}_{2\mathbf{m}} + \overline{g}_{1\mathbf{m}}) \ln^2\left(\frac{\overline{g}_{2\mathbf{m}}}{\overline{g}_{1\mathbf{m}}}\right)}. \tag{16.118}$$

Now suppose that the signal to be detected makes a small contribution to the mean data, so that we can write $\overline{g}_{2\mathbf{m}} = \overline{g}_{1\mathbf{m}} + s_{\mathbf{m}}$, with $s_{\mathbf{m}} \ll \overline{g}_{j\mathbf{m}}$ for $j = 1, 2$ and all $\mathbf{m}$. As we saw in (13.135), a Taylor expansion of the logarithms through terms linear in the signal yields

$$\mathrm{SNR}_\lambda^2 \approx \sum_{\mathbf{m}=1}^{M} \frac{s_{\mathbf{m}}^2}{\overline{g}_{\mathbf{m}}}, \tag{16.119}$$

where $\overline{g}_{\mathbf{m}}$ can be either $\overline{g}_{1\mathbf{m}}$ or $\overline{g}_{2\mathbf{m}}$ to this approximation.

An alternative way to arrive at this expression is to assume that the mean number of counts in each bin is large enough that we can approximate the Poisson law with a Gaussian ($\overline{g}_{\mathbf{m}} > 10$ will suffice). In that case the Hotelling observer is approximately ideal, and the Hotelling SNR for a weak signal is given by

$$\mathrm{SNR}_{Hot}^2 = \mathbf{s}^t \mathbf{K}_{\mathbf{g}}^{-1} \mathbf{s}. \tag{16.120}$$

With the help of (16.116), this general expression again reduces to (16.119).

A useful approximation to (16.119) is obtained by considering a spatially compact signal and a slowly varying object, such that $\overline{g}_{\mathbf{m}}$ is approximately the same for all pixels where the signal $s_{\mathbf{m}}$ is nonzero. If the signal is centered on pixel $\mathbf{m}_0$, we then have

$$\mathrm{SNR}_\lambda^2 \approx \frac{1}{\overline{g}_{\mathbf{m}_0}} \sum_{\mathbf{m}=1}^{M} s_{\mathbf{m}}^2, \tag{16.121}$$

and the sum is recognized as the $\mathfrak{L}_2$ norm of the detected signal. Anything that reduces this norm, reduces the detectability of the signal in this approximation. Moreover, anything that increases $\overline{g}_{\mathbf{m}_0}$ without also increasing the sum also decreases the detectability; an immediate implication is that background activity overlapping the signal in the 2D projection must be deleterious.

As written, (16.119) is simpler than the corresponding expression for digital radiography, (16.47), because the covariance matrix is simpler, but it hides some essential factors that contribute to detection performance. We cannot see immediately, for example, how the spatial and energy resolution of the detector or the design of the collimator affect SNR^2. We shall now consider each of these effects in turn for the weak-signal SKE/BKE problem.

Effect of position estimation In an Anger camera, the bins in the digital matrix can be made arbitrarily small, but there is still a resolution limitation arising from the position-estimation step, and this limitation can affect the detectability.

To isolate this effect, suppose that the bins in the digital image are so small that a sum over bins can be approximated as an integral. Suppose also that the estimation blur is shift-invariant, so that $\text{pr}(\hat{\mathbf{r}}|\mathbf{r}) = p_{est}(\hat{\mathbf{r}} - \mathbf{r})$. Then we can write (16.121) as

$$\text{SNR}_\lambda^2 = \frac{1}{b_0} \int_\infty d^2\hat{r} \left[\int_\infty d^2r \; p_{est}(\hat{\mathbf{r}} - \mathbf{r}) \, \Delta b(\mathbf{r}) \right]^2 . \tag{16.122}$$

The integral over $\mathbf{r}$ is recognized as a convolution, so we can use the convolution theorem (3.243) and Parseval's theorem (3.226) to write

$$\text{SNR}_\lambda^2 = \frac{1}{b_0} \int_\infty d^2\rho \, |P_{est}(\boldsymbol{\rho}) \, \Delta B(\boldsymbol{\rho})|^2 . \tag{16.123}$$

Since $p_{est}(\hat{\mathbf{r}} - \mathbf{r})$ is a properly normalized probability density function, it follows from the central-ordinate theorem (3.229) that $P_{est}(0) = 1$. Moreover, since $p_{est}(\hat{\mathbf{r}} - \mathbf{r})$ is nonnegative, it follows from the first inequality in (3.65) that $|P_{est}(\boldsymbol{\rho})| \leq 1$ for all $\boldsymbol{\rho}$. Therefore,

$$\text{SNR}_\lambda^2 \leq \frac{1}{b_0} \int_\infty d^2\rho \, |\Delta B(\boldsymbol{\rho})|^2 . \tag{16.124}$$

The right-hand side is what one would get with no estimation error. Thus the finite intrinsic resolution of the Anger camera will reduce the ideal-observer detectability unless the blur is negligible compared to the width of the pre-detection difference signal.

Effect of the collimator Another source of blur is the collimator; if we increase the bore diameter or reduce the bore length, the mean image incident on the detector plane is a more blurred representation of the 2D projection of the 3D object. Unlike the blur due to detector resolution or post-detection processing, however, collimator blur has a countervailing advantage: collimators with larger bore diameter or smaller bore length collect more photons and hence reduce the Poisson noise relative to the mean. It is not obvious how the collimator parameters should be chosen, but in keeping with our general philosophy, the collimator should be designed to optimize task performance. As we shall now see, this requirement leads to a surprising result if the task is SKE/BKE detection.

Since the collimator blur depends on distance from the collimator, it is convenient for this discussion to consider a planar object parallel to the face of the collimator. Thus we take $f(\mathbf{r}, z) = f(\mathbf{r}) \, \delta(z - z_0)$ in the absence of a signal and $f(\mathbf{r}, z) = [f(\mathbf{r}) + \Delta f(\mathbf{r})] \, \delta(z - z_0)$ when a signal is present. We assume for simplicity that $z_0 \gg L_b$ so that radiation can reach the detector plane through many bores. In that case, we can approximate the sum over bores with an integral as in (16.96), and the point response function is well approximated by the autocorrelation of the bore function as in (16.91). Recall, however, that this PRF maps the object $f(\mathbf{r})$ to the photon irradiance $I_p(\mathbf{r})$ *incident* on the detector plane P; the corresponding fluence for the *detected* photons requires some constants. Neglecting scatter and attenuation, we obtain

$$b(\mathbf{r}) = C \int_\infty d^2r' \, [\beta \star \beta] \left[\frac{L_b}{z_0} (\mathbf{r} - \mathbf{r}') \right] f(\mathbf{r}', z_0) , \tag{16.125}$$

where

$$C = \frac{1}{4\pi z_0^2} \frac{\alpha_{pf}}{A_{bore}} \tau\eta(\mathcal{E}_0)\, P_{acc}(\mathcal{E}_0)\,. \tag{16.126}$$

Note that the system is now shift-invariant since we have considered a planar object sufficiently far away that the fine structure of the collimator bores is not important. The same conclusion would hold for any z_0 if we averaged the PSF over collimator shifts.

To focus on the effect of the collimator, we assume an ideal detector and compute the detectability associated with the continuous image on plane P. If we consider a weak, spatially compact signal centered at $\mathbf{r} = \mathbf{r}_0$ and assume that the background object is slowly varying over the support of the signal, the expression we must evaluate is [*cf.* (16.121)]

$$\mathrm{SNR}_\lambda^2 = \frac{1}{b_0}||\Delta \mathbf{b}||^2\,. \tag{16.127}$$

The denominator in (16.127) is given by

$$b_0 = C\, f(\mathbf{r}_0) \int_\infty d^2r'\, [\beta \star \beta] \left[\frac{L_b}{z_0}(\mathbf{r} - \mathbf{r}')\right], \tag{16.128}$$

and the numerator is given by

$$||\Delta \mathbf{b}||^2 = C^2 \int_\infty d^2r \left\{ \int_\infty d^2r'\, [\beta \star \beta] \left[\frac{L_b}{z_0}(\mathbf{r} - \mathbf{r}')\right] \Delta f(\mathbf{r}') \right\}^2 . \tag{16.129}$$

Various Fourier theorems can be applied to simplify this expression; details are left as an exercise, but hints can be found in Barrett and Swindell (1981, 1996). The result is

$$\mathrm{SNR}_\lambda^2 = \frac{\tau\eta_{tot}}{f(\mathbf{r}_0)} \int_\infty d^2\rho \left| M_{coll}\left(\frac{z_0}{L_b}\boldsymbol{\rho}\right) \Delta F(\boldsymbol{\rho}) \right|^2 , \tag{16.130}$$

where η_{tot} is the total efficiency of the collimator and detector, defined by

$$\eta_{tot} = \frac{\alpha_{pf} A_{bore}}{4\pi L_b^2}\, \eta(\mathcal{E}_0)\, P_{acc}(\mathcal{E}_0)\,, \tag{16.131}$$

and $M_{coll}(\boldsymbol{\rho})$ is the collimator MTF, given by

$$M_{coll}(\boldsymbol{\rho}) = \left[\frac{2\mathrm{J}_1(\pi D_b \rho)}{\pi D_b \rho}\right]^2 . \tag{16.132}$$

Two limits It is instructive to look at the behavior of the SNR for this problem in the limits of large and small collimator bores. In the limit as $D_b \to 0$, we can evaluate (16.130) by noting that the collimator MTF is a broad function compared to the signal transform $\Delta F(\boldsymbol{\rho})$. Since $M_{coll}(0) = 1$, we can then write

$$\mathrm{SNR}_\lambda^2 \to \frac{\tau\eta_{tot}}{f(\mathbf{r}_0)} \int_\infty d^2\rho\, |\Delta F(\boldsymbol{\rho})|^2 = \frac{\tau\eta_{tot}||\Delta \mathbf{f}||^2}{f(\mathbf{r}_0)}, \tag{16.133}$$

where the last step follows from Parseval's theorem. In this limit, therefore, SNR_λ^2 grows linearly with the efficiency and the exposure time, and it is directly proportional to the squared $\mathfrak{L}_2$ norm of the signal in the object domain.

In the opposite limit, the signal transform $\Delta F(\boldsymbol{\rho})$ is large compared to $M_{coll}(z_0\boldsymbol{\rho}/L_b)$, and we have

$$\mathrm{SNR}_\lambda^2 \to \frac{\tau\eta_{tot}}{f(\mathbf{r}_0)}|\Delta F(0)|^2 \int_\infty d^2\rho \left| M_{coll}\left(\frac{z_0}{L_b}\boldsymbol{\rho}\right)\right|^2$$

$$= \frac{\tau\eta_{tot}}{f(\mathbf{r}_0)}\left|\int_\infty d^2r\ \Delta f(\mathbf{r})\right|^2 \int_\infty d^2\rho \left| M_{coll}\left(\frac{z_0}{L_b}\boldsymbol{\rho}\right)\right|^2 , \qquad (16.134)$$

where now the last step has used the central-ordinate theorem. A change of variables shows that the integral over $\boldsymbol{\rho}$ varies as $1/D_b^2$, cancelling the factor of D_b^2 in η_{tot} and yielding an SNR that is independent of D_b.

The factor $|\int_\infty d^2r\ \Delta f(\mathbf{r})|^2$ is also of interest. No longer does the $\mathfrak{L}_2$ norm of $\Delta f(\mathbf{r})$ appear in the SNR; instead we now have the integrated activity in the difference object. If the difference object is nonnegative, then this integral is the $\mathfrak{L}_1$ norm, but there is no absolute value in the integral, and there is nothing that requires $\Delta f(\mathbf{r}) \geq 0$.

What collimator should be used? Knowing the limiting behavior of SNR_λ, we can now sketch its dependence on bore diameter as in Fig. 16.14. The main conclusion from this plot is that the optimum collimator is no collimator at all! The bore diameter should be as large as possible in order to collect as many photons as possible, and there is no advantage in this task to getting good spatial resolution. The explanation for this counterintuitive result is that the task is so tightly specified that it is not necessary to resolve fine details; every detail of the two possible objects is known in advance, and it is required only to decide which object (background or signal plus background) is present. When the two objects differ by a known amount in the total number of photons they emit, the best strategy for deciding between them is just to reduce as much as possible the uncertainty in estimating this number, which we can do by collecting as many photons as possible.

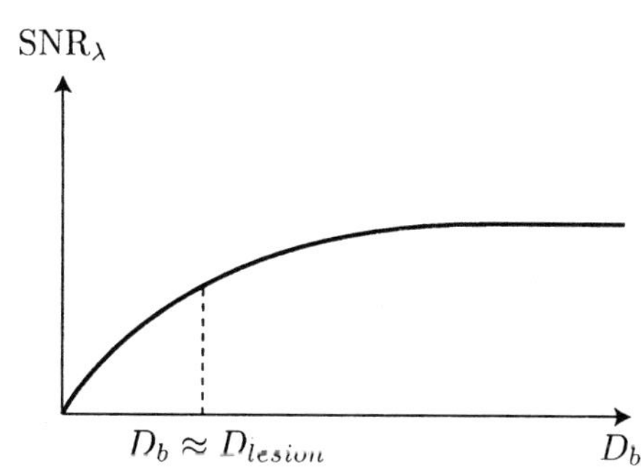

Fig. 16.14 Dependence of the ideal-observer SNR for detection of a disc signal on diameter of the collimator bore.

Zero-DC tasks Since $\int_\infty d^2r\ \Delta f(\mathbf{r})$ is the zero-frequency or DC Fourier component of the difference object, we can refer to tasks where this component vanishes as *zero-DC tasks.* We see from (16.134) that the detectability goes to zero in the limit of large collimator blur for a zero-DC task; task performance requires more than just estimating the total activity in this case.

The first use of a zero-DC task in image-quality assessment was by Harris (1964), who suggested the *Rayleigh task* illustrated in Fig. 16.15. The goal in this task is to discriminate between a single Gaussian blob and two separated blobs of half the amplitude, so that the two signals have the same $\int_\infty d^2r\ f(\mathbf{r})$, and hence $\int_\infty d^2r\ \Delta f(\mathbf{r}) = 0$. Harris suggested using the minimum value of the separation at some specified level of discrimination performance as a measure of spatial resolution.

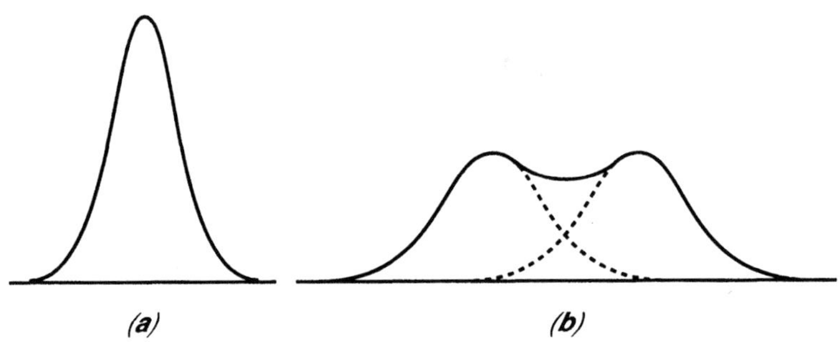

Fig. 16.15 Illustration of the Rayleigh task, where the object is either two Gaussian blobs or one Gaussian with the same integrated activity.

Wagner *et al.* (1981) used the Rayleigh task to evaluate nuclear-medicine imaging systems, but they still found paradoxical results. In particular, they found that coded apertures had no resolution advantage, in the Harris sense, over very large pinholes that collected the same number of photons. Coded-aperture images can be decoded to give a spatial resolution, as defined by the width of the point response function, that is much better than that of a pinhole of equal collection efficiency. Therefore the conclusion of Wagner *et al.* is equivalent to saying that spatial resolution in this conventional sense is not important for the stylized SKE/BKE task they considered. To get a more realistic assessment of the system, we must consider more realistic tasks.

Random signals and backgrounds As discussed in Sec. 13.2.11, one way to make the task more realistic is to consider detection on a random background. The first step in this direction in emission imaging was taken by Tsui *et al.* (1978, 1983), who considered detection of a spherical tumor of known size and location in a spatially uniform background of unknown level. One observer model they considered closely approximated the ideal Bayesian strategy for this task [see (13.215)]. The observer estimated the random background level by integrating the counts over an annular region surrounding the tumor and compared that value with the integrated counts over the signal location. With this model, the optimum aperture size was approximately equal to the size of the lesion to be detected, and increasing the aperture size beyond this point resulted in reduced detectability in spite of increased counts. When the background was assumed to be known, however, the optimum aperture size increased to infinity, in accord with the ideal-observer analysis presented in Fig. 16.14.

A somewhat more realistic background, the lumpy background described in Secs. 8.4.4 and 13.2.12, was used by Myers *et al.* (1990) and Rolland and Barrett (1992) to discuss aperture optimization in emission imaging. It was found that spatial resolution in the conventional sense (width of the point response function) was required to distinguish the signal from the background inhomogeneities. As in the Tsui work, the optimum aperture size was found to be approximately equal to the size of the signal.

Still more realistic random backgrounds, and random signals, were used by Fiete *et al.* (1987) and later by White (summarized in Barrett *et al.*, 1992). Using a 3D mathematical liver model developed by Cargill (1989), these authors studied both human and Hotelling detection performance as a function of collimator bore diameter and bore length. Two key findings were that a long bore and small bore diameter optimized Hotelling performance, in spite of the relatively low collection efficiency, and that Hotelling performance was an accurate predictor of human performance in this case.

Psychophysical studies Psychophysical studies can be used without correlation with model observers to evaluate components of nuclear medicine systems. One example is the work of Buvat *et al.* (2001), who used a physical breast phantom and human ROC studies to compare collimators in scintimammography. Breast tumors were represented by hot spheres. It was demonstrated that ultrahigh-resolution collimators were advantageous for this task in spite of reduced collection efficiency.

Detection tasks and list-mode data So far we have assumed that the detection or discrimination tasks were performed by an observer with access to image histograms, but there may be some loss of information in going from the raw list-mode data to the binned histograms. For SKE/BKE discrimination tasks, the SNR for an ideal observer operating on list-mode data was computed by Barrett *et al.* (1997b). For preset time, they showed that

$$\mathrm{SNR}_\lambda^2 = \frac{\left[\int_\infty d^2\hat{r} \int_0^\infty d\hat{\mathcal{E}}\ (b_2 - b_1)\ln\left(\frac{b_2}{b_1}\right)\right]^2}{\frac{1}{2}\int_\infty d^2\hat{r}\int_0^\infty d\hat{\mathcal{E}}\ (b_2 + b_1)\ln^2\left(\frac{b_2}{b_1}\right)}, \tag{16.135}$$

where we have simplified the notation by writing b_j for $b_{det,j}(\hat{r}, \hat{\mathcal{E}})$; with the j subscript distinguishing hypothesis 1 from hypothesis 2.

If we consider weak signals, so that $b_2 = b_1 + \Delta b$, with $\Delta b \ll b_j$ for $j = 1, 2$, then we find

$$\mathrm{SNR}_\lambda^2 \approx \int_\infty d^2\hat{r} \int_0^\infty d\hat{\mathcal{E}}\ \frac{(\Delta b)^2}{b_j}. \tag{16.136}$$

Comparing (16.135) and (16.136) with their binned counterparts, (16.118) and (16.119), respectively, we see that the main difference is that the sum over bins has been replaced by an integral over attribute space. Thus (16.135) and (16.136) are the limits of the binned expressions as the bin widths go to zero. Recall, however, that the binned expressions were derived on the assumption that there was only one energy bin; either $\hat{\mathcal{E}}$ fell in the energy window and the event was accepted, or it did not and the event was rejected. In the list-mode expressions, there is no windowing and every event contributes, in principle. Even scattered photons are useful to the extent that Δb is nonzero.

Linear discriminants and random processes If we wish to move away from SKE/BKE tasks and consider random signals or backgrounds with list-mode data, we run into difficulties computing the ideal-observer detectability. In that case, it is again useful to consider the ideal linear or Hotelling observer. This observer is not defined as a linear functional of the attributes in the list, but rather as a linear discriminant acting on the corresponding random process. For the spatio-spectral process given in (16.112), the linear discriminant function has the form [*cf.* (13.9)]

$$T(\mathbf{g}_{det}) = \mathbf{w}^\dagger \mathbf{g}_{det} = \int_\infty d^2\hat{r} \int_0^\infty d\hat{\mathcal{E}}\; w_{det}(\hat{\mathbf{r}}, \hat{\mathcal{E}})\, g_{det}(\hat{\mathbf{r}}, \hat{\mathcal{E}})\,. \tag{16.137}$$

To compute the SNR for this discriminant, we need to know the mean and autocorrelation function for the random process. For the important special case of a doubly stochastic spatial Poisson random process, these expressions are found in Sec. 11.3.6, and they extend readily to the spatio-spectral case.

16.2.6 Image quality: Estimation tasks

Planar nuclear medicine is poorly suited to estimation tasks because of the 3D-to-2D mapping. If, for example, we wanted to estimate the activity in the left ventricle of the heart, counts from the ventricle would be confused with counts from tissues that overlap the ventricle in a single 2D projection. We encountered a similar situation in the context of digital radiography in Sec. 16.1.7, but there we avoided the issue of overlapping tissues by considering only changes in x-ray attenuation coefficient following injection of a contrast agent. The main problems that arose in that case were the system nonlinearity and the variation of system sensitivity with depth. In planar nuclear medicine, these two system problems do not come up, but we do not usually have the possibility of doing subtraction imaging, and we must therefore deal with the overlapping activity.

This fundamental difficulty can be solved only by tomographic imaging, which is the subject of Chap. 17, but it is instructive to analyze the problem further in the context of planar imaging. This discussion will illustrate the concept of estimability, introduced in Secs. 13.3.1 and 15.1.3, and it will provide some insights into the role of prior information in estimation problems.

Bias and estimability Suppose the task is to estimate the activity in a spherical region of interest (ROI) of known location in the body. As usual, we would like to assess performance on this task by bias and variance of the estimate, but we have emphasized that bias is well defined only for estimable parameters. If the parameter in question is the activity in an ROI, then the estimability condition is that the template defining the ROI be a linear combination of system sensitivity functions [see (15.23)]. For a parallel-hole collimator, the sensitivity functions are long, thin cones parallel to the axis of the bores (the z axis). The spherical template cannot be synthesized as a linear combination of such functions, so any estimate derived from a single 2D projection will necessarily have a bias of unknown magnitude arising from the overlapping tissue.

To discuss this problem quantitatively, we need to decompose the template into null and measurement components. To see the essential features of the calculation, let us first ignore blur and attenuation. The kernel for the mapping from the 3D object to the continuous 2D projection is obtained from (16.91) by letting $D_b \to 0$; a factor of $(z'/L_b)^2$ arises because of the scale factor in the argument of the autocorrelation, and we see that

$$\overline{h}(\mathbf{r} - \mathbf{r}', z') \to \frac{\alpha_{pf}}{4\pi L_b^2}\, \delta(\mathbf{r} - \mathbf{r}')\,. \tag{16.138}$$

There is no longer any dependence on z'; the system sensitivity is independent of depth, so the projection is a simple line integral in the z' direction.

We learned in Sec. 7.2.10 how to compute the SVD for axial systems, which blur laterally and integrate along the z axis. For shift-invariant blur, we saw in (7.223) that the object-space singular functions corresponding to nonzero singular values are 2D plane waves $\exp(2\pi i\boldsymbol{\rho}\cdot\mathbf{r})$ modulated by the depth-dependent transfer function $H(-\boldsymbol{\rho}, z)$. In the present problem, there is no lateral blur and no dependence on z, so $H(-\boldsymbol{\rho}, z)$ is independent of $\boldsymbol{\rho}$ and constant over the region of support in the z direction; if we take that support as $0 < z \leq L_z$, we get

$$u_{\boldsymbol{\rho},j}(\mathbf{r}, z) = \frac{1}{\sqrt{L_z}}\exp(2\pi i\boldsymbol{\rho}\cdot\mathbf{r})\,\mathrm{rect}\left(\frac{z-\frac{1}{2}L_z}{L_z}\right), \qquad (j=1). \tag{16.139}$$

There is also an infinite set of functions $\{u_{\boldsymbol{\rho},j}(\mathbf{r}, z)\}$ with $j > 1$, but they are in the null space; the measurement space is spanned by $\{u_{\boldsymbol{\rho},j}(\mathbf{r}, z)\}$ for all $\boldsymbol{\rho}$ and $j = 1$.

With these singular functions, the projection of the ROI template onto measurement space is [*cf.* (1.165)]

$$\chi_{meas}(\mathbf{r}, z) = [\boldsymbol{\mathcal{P}}_{meas}\boldsymbol{\chi}](\mathbf{r}, z) = \int_\infty d^2\rho\; \mathbf{u}_{\boldsymbol{\rho},1}\mathbf{u}^\dagger_{\boldsymbol{\rho},1}\boldsymbol{\chi}$$

$$= \frac{1}{L_z}\int_\infty d^2\rho\int_\infty d^2r'\int_0^{L_z} dz'\exp[2\pi i\boldsymbol{\rho}\cdot(\mathbf{r}-\mathbf{r}')]\,\chi(\mathbf{r}', z') = \frac{1}{L_z}\int_0^{L_z} dz'\,\chi(\mathbf{r}, z'), \tag{16.140}$$

where we have used the completeness relation (3.218) in the last step. Thus, not surprisingly, the measurement component of the template is just the template averaged over depth.

One way to restate this result is to use a Fourier series in z for $\chi(\mathbf{r}, z)$:

$$\chi(\mathbf{r}, z) = \sum_{n=-\infty}^{\infty} \chi_n(\mathbf{r})\exp(2\pi i n z/L_z). \tag{16.141}$$

As one might expect from the central-slice theorem (4.150), the measurement component is just the zero-frequency term:

$$\chi_{meas}(\mathbf{r}, z) = \chi_0(\mathbf{r}). \tag{16.142}$$

The null component contains all other frequencies; it is given by

$$\chi_{null}(\mathbf{r}, z) - \chi(\mathbf{r}, z) - \frac{1}{L_z}\int_0^{L_z} dz'\,\chi(\mathbf{r}, z'). \tag{16.143}$$

Integrating this expression over $0 < z \leq L_z$ yields zero, so it is indeed a null function for the system model of (16.138).

With uniform attenuation but no blur, the measurement-space singular functions are given by (16.139) with an extra factor of $\exp(-\mu z)$ and a modified normalization. In that case,

$$\chi_{meas}(\mathbf{r}, z) = \frac{\mu\exp(-\mu z)}{1-\exp(-\mu L_z)}\int_0^{L_z} dz'\,\chi(\mathbf{r}, z')\exp(-\mu z'). \tag{16.144}$$

The reader may fill in the missing steps.

Knowing the decomposition of the template into measurement and null components, we can now begin to make some statements about bias in an ROI estimate. There are several ways to proceed. We can compute the bias (or bounds on the bias) for particular objects as in Sec. 15.1.4, or we can assume some prior knowledge about the class of possible objects. Prior knowledge can be either a statistical model or a deterministic model with unknown parameters. We shall sketch each of these approaches and then discuss objective evaluation of biased estimates.

Pseudoinverse estimators We shall first consider pseudoinverse estimators as discussed in Sec. 15.2.1 and see what we can say about bias for particular objects. Since the parameters of interest and the estimators are linear, we do not have to know anything about the noise in the data to compute the bias, but we note that if the noise is i.i.d. normal, the pseudoinverse estimators are also maximum likelihood.

The general form of a pseudoinverse estimator is given in (15.47). For estimation of a scalar parameter $\Theta \equiv \boldsymbol{\chi}^\dagger \mathbf{f}$, that equation reduces to

$$\widehat{\Theta} = \boldsymbol{\chi}^\dagger \boldsymbol{\mathcal{H}}^+ \mathbf{g} . \tag{16.145}$$

This estimate is unbiased if and only if Θ is estimable, which it isn't in the present discussion.

We can construct $\boldsymbol{\mathcal{H}}^+$ from what we know about the SVD of laterally shift-invariant axial systems (see Sec. 7.2.10). It follows from (7.223), (7.224), (1.116) and (1.121) that

$$[\boldsymbol{\mathcal{H}}^+ \mathbf{g}](\mathbf{r}, z) = \operatorname{rect}\left(\frac{z - \frac{1}{2}L_z}{L_z}\right) \int_\infty d^2\rho \, \frac{H^*(\boldsymbol{\rho}, z)}{\int_0^{L_z} dz' \, |H(\boldsymbol{\rho}, z')|^2} G(\boldsymbol{\rho}) \exp(2\pi i \boldsymbol{\rho} \cdot \mathbf{r}) . \tag{16.146}$$

If the transfer function is independent of z (as it would be for the intrinsic resolution of an Anger camera), this equation expresses a lateral inverse filter[7] followed by backprojection in z. For depth-dependent blur, for example due to the collimator, the pseudoinverse is *not* a separate inverse filter at each z.

To go from $\boldsymbol{\mathcal{H}}^+ \mathbf{g}$ to the desired $\widehat{\Theta}$, we must take the scalar product with $\chi(\mathbf{r}, z)$; by Parseval's theorem, the result is

$$\widehat{\Theta} = \int_0^{L_z} dz \int_\infty d^2\rho \, \frac{H^*(\boldsymbol{\rho}, z)}{\int_0^{L_z} dz' \, |H(\boldsymbol{\rho}, z')|^2} X^*(\boldsymbol{\rho}, z) \, G(\boldsymbol{\rho}) , \tag{16.147}$$

where $X(\boldsymbol{\rho}, z)$ is the 2D Fourier transform of $\chi(\mathbf{r}, z)$. (Note that X is capital χ.)

In the absence of blur and attenuation, $H(\boldsymbol{\rho}, z) = \alpha_{pf}/(4\pi L_b^2)$, so (16.147) simplifies to

$$\begin{aligned} \widehat{\Theta} &= \frac{4\pi L_b^2}{\alpha_{pf}} \int_\infty d^2\rho \left[\frac{1}{L_z} \int_0^{L_z} dz \; X^*(\boldsymbol{\rho}, z) \right] G(\boldsymbol{\rho}) \\ &= \frac{4\pi L_b^2}{\alpha_{pf}} \int_\infty d^2 r_d \left[\frac{1}{L_z} \int_0^{L_z} dz \; \chi(r_d, z) \right] g(r_d) , \end{aligned} \tag{16.148}$$

[7]Note that $H^*(\boldsymbol{\rho})/|H(\boldsymbol{\rho})|^2 = 1/H(\boldsymbol{\rho})$.

where the last step has again invoked Parseval. In this case, therefore, the pseudoinverse estimate is obtained essentially by taking the scalar product of the data with the projection of the ROI template.

To determine a bound on the bias in the pseudoinverse estimate for some particular object, we can use (15.52). We shall illustrate the procedure by again neglecting attenuation and lateral blur. In that case, it follows from (16.143) that $\max_{\mathbf{r}} |\chi_{null}(\mathbf{r}, z)| = 1$, so the difference between the true Θ and its pseudoinverse estimate is bounded by

$$|\Theta - \widehat{\Theta}| \leq \int_{\mathbf{S}(\chi_{null})} d^3\mathbf{r} \, |f(\mathbf{r})| \, . \tag{16.149}$$

Since there is no lateral blur, the support of $\boldsymbol{\chi}_{null}$ is a cylinder parallel to the z axis and encompassing the ROI; object activity outside this cylinder is irrelevant to estimation of the activity in the ROI. Even with this restriction, however, the bound in (16.149) is not very useful. It says merely that the bias in estimating an integral of the object over an ROI cannot exceed the integral of the object over a cylinder encompassing the ROI. The worst case is when all of the object activity is outside the ROI.

Model-based background subtraction To get a better estimate of Θ in this grossly underdetermined problem, we must have better prior knowledge of the object. One form of prior information is a model for the object with a small number of free parameters.

As an example, consider a spherical ROI immersed in a uniform slab of known thickness but unknown background activity, so there are just two parameters of interest: the integrated activity in the ROI and the activity per unit volume in the background. We assume also, for now, that there is no blur from the detector or collimator, no attenuation and no scattered radiation.

With this model, we can define an unbiased estimator of the ROI activity, in spite of the lack of estimability, since we are not allowing much freedom in the object being imaged. Though the object can have arbitrary structure over the ROI, its behavior outside the region is specified by one number, the activity per unit volume. Recall the basic argument for estimability from Sec. 15.1.3. If a parameter θ is defined by a template $\chi(\mathbf{r})$ as in (15.15), it can be written via (15.19) as $\theta = (\boldsymbol{\chi}_{meas}, \mathbf{f}_{meas}) + (\boldsymbol{\chi}_{null}, \mathbf{f}_{null})$. The first term represents what we can learn about θ from noise-free data, and the second term is zero if either $\mathbf{f}_{null} = 0$ or $\boldsymbol{\chi}_{null} = 0$. We argued in Chap. 15 that we cannot be guaranteed that $\mathbf{f}_{null} = 0$, so therefore we must require $\boldsymbol{\chi}_{null} = 0$; this requirement is unnecessary here since we are, in essence, assuming that the portion of the object outside the ROI contains no null components.

In the absence of blur and attenuation, unbiased estimators of the two unknown parameters can be obtained rather simply. Both parameters contribute to the counts in region A of the image (see Fig. 16.16), but only the background activity contributes to region B. Knowing the geometry, we can estimate the background level from the pixels in B, and we can then construct a new image with no background contribution; summing the pixels in region A for this corrected image then gives an estimate of the activity in the ROI. In terms of estimability, the ROI can be taken as a cylinder in the corrected image rather than a sphere, and the template defining the cylinder is well approximated as a linear combination of the sensitivity functions, so ROI activity is estimable within this model.

Attenuation can spoil estimability, even in this greatly oversimplified model, since it causes the system sensitivity to vary with depth. In the digital radiography problem discussed in Sec. 16.1.7, the integrated change in x-ray attenuation coefficient over a region of interest was not estimable, even in the absence of blur, because the $1/R^2$ factor seen in (16.64) caused the system sensitivity to depend on depth. For emission-imaging with a parallel-hole collimator, we know from (16.95) that the system sensitivity in air is approximately independent of depth, but attenuation causes a depth dependence. The sensitivity functions all contain an exponential factor, and even a cylinder cannot be synthesized by superimposing them. In order to ignore this effect, we must add more assumptions to our already overburdened model. If we assume that the depth of the spherical ROI is known and the diameter of the sphere is small compared to the reciprocal of the attenuation coefficient, we can correct for attenuation.

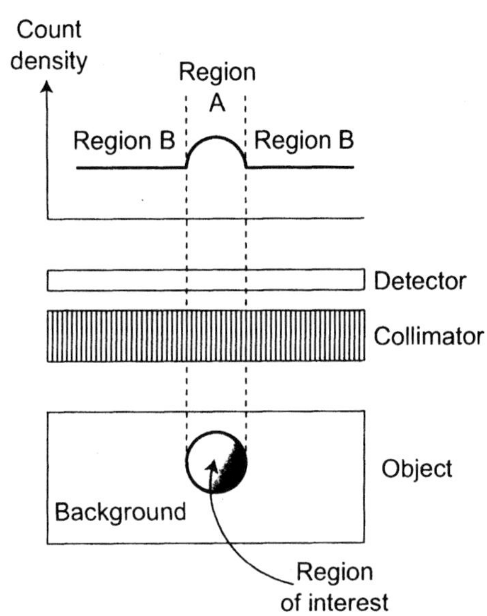

Fig. 16.16 Slab geometry for an idealized estimation problem.

An attempt at being Bayesian Another form of prior information is statistical. A Bayesian assumes that a prior PDF on the object is known and uses that information to minimize some risk function. We shall now pursue this approach in the context of ROI estimation with a quadratic risk.

As with most Bayesian analyses, we choose our prior knowledge not as something we really know but as something that will lead to a tractable solution. In statistics, *tractable* is almost synonymous with *Gaussian*, so we begin by assuming Gaussian statistics for the object and the measurement noise. Since we are regarding the data as continuous here, the noise $n(\mathbf{r}_d)$ is a 2D Gaussian random process and the object $f(\mathbf{r}, z)$ is a 3D one. The object random process is fully specified by the mean vector $\overline{f}(\mathbf{r}, z)$ and the autocovariance function $K_{\mathbf{f}}(\mathbf{r}, \mathbf{r}'; z, z')$, corresponding to the autocovariance operator $\mathcal{K}_{\mathbf{f}}$. The noise is, by definition, zero mean, and its autocovariance function is denoted $K_{\mathbf{n}}(\mathbf{r}_d, \mathbf{r}'_d)$, with the corresponding operator being $\mathcal{K}_{\mathbf{n}}$.

We wish to use this prior model to minimize the mean-square error (MSE), averaged over both data realizations and the prior. The minimum-MSE estimate is given in terms of the generalized Wiener filter (see Sec. 13.3.7) as

$$\widehat{\Theta}_{MMSE} = \boldsymbol{\chi}^\dagger \hat{\mathbf{f}}_{MMSE} = \boldsymbol{\chi}^\dagger \mathcal{K}_{\mathbf{f}} \mathcal{H}^\dagger \left[\mathcal{H} \mathcal{K}_{\mathbf{f}} \mathcal{H}^\dagger + \mathcal{K}_{\mathbf{n}} \right]^{-1} (\mathbf{g} - \overline{\mathbf{g}}) + \boldsymbol{\chi}^\dagger \overline{\mathbf{f}}. \tag{16.150}$$

Since we have assumed that the imaging system is laterally shift-invariant, the only tractable way to proceed is to assume that the object and noise are laterally stationary. Thus we assume that

$$K_{\mathbf{f}}(\mathbf{r}, \mathbf{r}'; z, z') = K_{\mathbf{f}}(\mathbf{r} - \mathbf{r}'; z, z'), \qquad K_{\mathbf{n}}(\mathbf{r}_d, \mathbf{r}'_d) = K_{\mathbf{n}}(\mathbf{r}_d - \mathbf{r}'_d). \tag{16.151}$$

These assumptions are plausible since the template confines our interest in (16.150) to a limited spatial region where the statistics might not vary too much.

Stationarity in z would be much more problematical. We are assuming a slab geometry where the object is contained in $0 < z \leq L_z$, so we definitely cannot assume strict stationarity. A common ruse in this situation is to assume cyclic stationarity, as discussed in Sec. 8.2.8, so that a DFT can be used, but there is no physical justification for this assumption either.

Fortunately, there is no need to assume anything about the z and z' dependence of $K_{\mathbf{f}}(\mathbf{r} - \mathbf{r}'; z, z')$ if we are considering only LSIV axial systems. If we represent this function by its 2D Fourier transform as

$$K_{\mathbf{f}}(\mathbf{r} - \mathbf{r}'; z, z') = \int_\infty d^2\rho \; S_{\mathbf{f}}(\boldsymbol{\rho}; z, z') \exp[2\pi i \boldsymbol{\rho} \cdot (\mathbf{r} - \mathbf{r}')], \tag{16.152}$$

similarly represent $\mathcal{H}$ and $\mathcal{K}_{\mathbf{n}}$ as 2D inverse Fourier transforms, and do a considerable amount of algebra, we find [*cf.* (16.147)]

$$\widehat{\Theta} = \overline{\overline{\Theta}} + \int_0^{L_z} dz \int_\infty d^2\rho \; X^*(\boldsymbol{\rho}, z)\, W(\boldsymbol{\rho}, z)\, [G(\boldsymbol{\rho}) - \overline{G}(\boldsymbol{\rho})], \tag{16.153}$$

where $\overline{\overline{\Theta}} = \boldsymbol{\chi}^\dagger \overline{\mathbf{f}}$ is the prior mean of Θ, $\overline{G}(\boldsymbol{\rho})$ is the Fourier transform of $[\mathcal{H}\overline{\mathbf{f}}](\mathbf{r}_d)$ (*i.e.*, the prior mean of the data in the Fourier domain), and $W(\boldsymbol{\rho}, z)$ is a Wiener-like filter defined by

$$W(\boldsymbol{\rho}, z) = \frac{\int_0^{L_z} dz' \, S_{\mathbf{f}}(\boldsymbol{\rho}; z, z')\, H^*(\boldsymbol{\rho}, z')}{S_{\mathbf{n}}(\boldsymbol{\rho}) + \int_0^{L_z} dz' \int_0^{L_z} dz'' \, H(\boldsymbol{\rho}, z')\, S_{\mathbf{f}}(\boldsymbol{\rho}; z', z'')\, H^*(\boldsymbol{\rho}, z'')}. \tag{16.154}$$

To understand the behavior of this filter, consider again the case of no blur or attenuation, and suppose that the object has long-range lateral correlations so that $S_{\mathbf{f}}(\boldsymbol{\rho}; z, z')$ is a sharply peaked function of $\boldsymbol{\rho}$. By contrast, $S_{\mathbf{n}}(\boldsymbol{\rho})$ is a constant for uncorrelated noise, so the denominator has a large value near $\boldsymbol{\rho} = 0$, and it rapidly approaches a smaller constant value as ρ increases. We saw just this behavior in Sec. 13.2.12 when we discussed the form of the Hotelling observer for detection in a lumpy background. As shown in Fig. 13.11, the space-domain counterpart of this frequency-domain behavior is a positive peak at the origin and long negative tails. The function of the tails is to estimate and subtract off the background, just as with the Hotelling template. In other words, this Bayesian estimate is simply a more formal way of arriving at the same sort of background subtraction we discussed qualitatively above.

Performance evaluation In both the background-subtraction and Bayesian approaches, we had to make many simplifying assumptions in order to derive an estimate. If we really believed these assumptions, we could use them to compute figures of merit for image quality.

With the model used for background subtraction, the estimator is unbiased and the only randomness arises from the Poisson noise, so we can readily compute the variance in the estimate and use it as the figure of merit. Unfortunately this model is far from reality. We oversimplified the system (neglecting attenuation and blur) and we oversimplified the object description (assuming uniform activity outside the ROI).

With the Bayesian estimator, the objective was to minimize the posterior mean-square error, and a pure Bayesian would then use this MSE as the figure of merit. The difficulty in this approach is that we had to choose the prior for mathematical tractability, not because it accurately represented the randomness in real objects. Indeed, any object prior simple enough for incorporation in a Bayesian estimator is almost certainly unrealistic, and posterior MSE based on this prior is only a measure of self-consistency of our assumptions, not something indicative of performance in practice.

There is, however, nothing that says the estimate has to be evaluated on the basis of the same model under which it was derived. For any estimator, we can simulate a large number of realistic 3D models for the objects, compute the corresponding images, and apply the estimator to each. Since we know the true ROI activity for each simulated image, we can compute bias, variance and hence MSE in a frequentist sense. This MSE is then a joint figure of merit for the estimator and the imaging system.

Estimation of signal parameters So far we have been discussing estimation of parameters of the object, but we know from Sec. 8.4.5 that it is often useful to divide the object into signal and background components; loosely speaking, the signal is the component of the object in which we are most interested. Normally we talk abut detection of signals, but there are many situations where we know the signal is there and just want to estimate parameters associated with it (see Sec. 14.3.4). For example, we may want to estimate the volume of a tumor to see if a patient is responding to therapy.

In a series of papers, Müller, Moore, Kijewski and co-workers have used such estimation tasks to evaluate and optimize collimators for planar nuclear medicine (Müller *et al.*, 1986; Müller *et al.*, 1990; Moore *et al.*, 1995). They use relatively simple models for the background and signal, but reach important conclusions regarding the tradeoffs between resolution and sensitivity as bore length and diameter and septal thickness are varied. Their modeling of the physical characteristics of the collimator is quite detailed, taking into account septal penetration and scattering and K x-ray production in the collimator material as well as the effects of multiple energy emissions from some isotopes. Though much further work is needed, the approach suggested by these workers is a model for the use of estimation tasks in system optimization.

17

Single-Photon Emission Computed Tomography

In the last chapter we considered two direct-imaging systems using high-energy radiation; in this chapter we discuss an indirect method called *single-photon emission computed tomography*, or *SPECT*. The modifier *single-photon* distinguishes it from *positron emission tomography*, or *PET*, in which two photons are generated simultaneously and the projection data are formed by coincident detection of both photons. In SPECT, a single gamma ray is emitted for each nuclear disintegration, and some sort of collimator is needed to form a projection image.

As in planar nuclear medicine, the object in SPECT is a self-luminous volume source. Unlike the planar case, however, the goal in SPECT is to provide a 3D map of the source, not a projection of it onto a 2D plane. We shall discuss SPECT in detail here because it is an important modality in its own right, but also because it serves as a vehicle for introducing many topics that are important in other kinds of tomography.

As with all digital imaging, SPECT systems are most accurately described as CD mappings from a continuous object function to a discrete data set. Much of the literature on SPECT and other forms of tomography, however, is based on CC formulations. This literature ignores data sampling (at least initially) and makes various other simplifying assumptions in order to find a linear integral transform, which we shall call the *forward tomographic transform*, that maps a continuous object to continuous mean data. Depending on the geometric model and assumptions made, this forward transform might be the 2D or 3D Radon transform (Sec. 4.4.5), the x-ray transform (Sec. 10.3.2), the cone-beam transform (Sec. 17.1.5), or attenuated versions of any of these.

The objective in this approach is to find another linear integral transform that, under various idealized assumptions, will allow recovery of the object function from the continuous data. We shall call this transform the *inverse tomographic transform*. This inverse transform is then discretized in some manner to get a discrete linear algorithm, which is applied to real, noisy, discrete data rather than to the idealized, noise-free, continuous data assumed in the derivation. This approach works well, in the sense of giving useful images, in a surprisingly large number of cases.

There are two motivations for this emphasis on linear transforms and algorithms. First, there is the basic mathematical understanding that can be obtained from linear transforms. If we can see, based even on an idealized model, what information is contained in our data and how to extract it, we can perhaps use this understanding to improve the design of the imaging system and the algorithm. Second, there is a computational and economic motivation. We can use nonlinear iterative reconstruction algorithms (see Sec. 15.4) to account for factors such as blur, sampling and noise that are necessarily left out of the integral transforms, but iterative algorithms require more computation than one-step linear transforms. Computed tomography came of age at a time when computational power was very limited, and there was a strong motivation for noniterative linear algorithms. Of course, this computational power has grown exponentially, but so too has the ability of practical tomographic systems to acquire data. With 3D data acquisition and ever finer detector resolution, the demands on iterative algorithms have grown apace, and the motivation for linear algorithms remains. It will be interesting to see whether linear transforms or iterative algorithms win out in the decades to come.[1] (The authors of this book believe that the choice must be based on objective measures of image quality as well as on economics.)

Though this chapter is essentially about an inverse problem, we take seriously the maxim: If you want to solve an inverse problem, concentrate on the forward problem. Accordingly, Sec. 17.1 is devoted to analysis of the forward problem for several different data-acquisition geometries. The inverse problem is treated in Sec. 17.2, and issues of noise and image quality are in Sec. 17.3.

17.1 FORWARD PROBLEMS

Methods for the deterministic analysis of imaging systems were developed in Chaps. 6, 7 and 10, and in this section we apply those methods to the mathematical description of the forward problem in SPECT. Our goals are to provide concrete applications of the mathematics from earlier chapters and to lay the ground work for discussing image reconstruction and image quality later in this chapter.

Sections 17.1.1 – 17.1.4 are all devoted to one popular SPECT geometry, where a gamma camera and parallel-hole collimator are rotated around the object. The first three of these sections are based on CD models where a discrete set of projection angles is used, and the projection data at each angle are measured on a discrete array. In particular, Sec. 17.1.1 develops a formal operator theory so that we can see how the system operators $\boldsymbol{\mathcal{H}}$, $\boldsymbol{\mathcal{H}}^\dagger$ and $\boldsymbol{\mathcal{H}}^\dagger\boldsymbol{\mathcal{H}}$ for SPECT are related to their counterparts for planar nuclear medicine as treated in Sec. 16.2. Then, in Sec. 17.1.2, we consider specifically equally spaced angles and see some of the implications of

[1]The tension between linear transforms and iterative algorithms is seen historically as well. Some would argue that tomography could not have existed without the fundamental work of Radon, Bracewell and Cormack on integral transforms, but in fact it was an iterative algorithm implemented by Godfrey Hounsfield that was used in the first commercial CT scanner, and Cormack and Hounsfield shared the Nobel prize for tomography.

group theory; acquaintance with Chap. 6 is essential for this discussion, but the section can be skipped without loss of continuity.

In Sec. 17.1.3 we bring in a different mathematical tool and show how Fourier analysis, in the form of the Fourier crosstalk matrix introduced in Sec. 7.3.3, can be used to analyze practical SPECT systems. The interesting point here is that we can make good use of Fourier analysis without assuming shift-invariance.

In Sec. 17.1.4, we continue the discussion of SPECT systems with parallel-hole collimators, but now from a CC perspective. Our goal here is to relate the actual data-acquisition process to the commonly used idealization, the 2D Radon transform.

Continuing on the theme of forward problems, Sec. 17.1.5 considers 3D CC formulations applicable to pinhole imaging and focused collimators. The goal in this section is to relate the continuous data to the 3D Radon transform (Sec. 4.4.5) and the x-ray or cone-beam transform (Sec. 10.3.2), and also to relate these two apparently different transforms to each other. The effects of attenuation on the forward tomographic operators are discussed in Sec. 17.1.6.

17.1.1 CD formulations for parallel-beam SPECT

The simplest way to acquire data for SPECT is to use a parallel-hole collimator with any suitable gamma-ray detector, such as an Anger camera, and to rotate the collimator-detector assembly around the patient as shown in Fig. 17.1. We already know a great deal about this system from the discussion in Sec. 16.2 of planar nuclear medicine with a parallel-hole collimator, so our main goal here is to see how these planar results apply to rotating-camera SPECT. We shall discuss this problem in the language of CD mappings, where the detector array is a discrete set of elements and the data are acquired at discrete projection angles, but the same formalism will prove useful in a CC context in Sec. 17.2.1. Only the forward problem is discussed here; image reconstruction for this geometry is treated in Secs. 17.2.2 and 17.2.5.

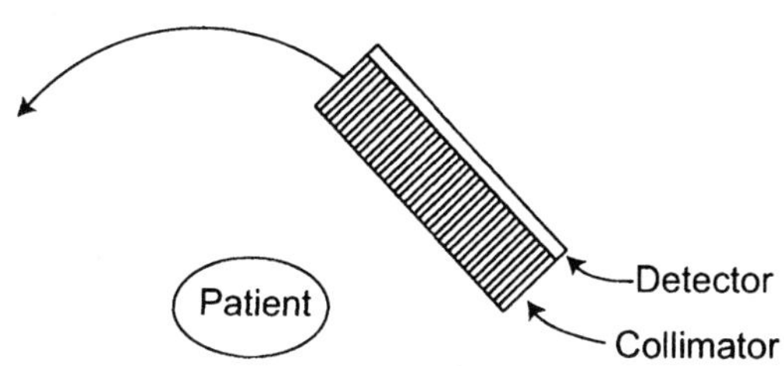

Fig. 17.1 Configuration for acquiring SPECT data with a rotating gamma camera and a parallel-hole collimator. The collimator and camera are rotated as a unit around the patient.

Notation and conventions The formalism developed in Sec. 16.2 for planar imaging with parallel-hole collimators is applicable to parallel-beam SPECT with minor modifications. First, we need to add an index to denote the projection angle. One way to do so is just to consider the vector index on $\overline{g}_{\mathbf{m}}$ to have three components, two to index position on the detector as before and now a third, say j, to indicate projection angle. To avoid confusion with previous notation, we shall use gothic bold $\mathfrak{m}$ to denote the 3D index and retain $\mathbf{m}$ as 2D; thus $\mathfrak{m} = (m_x, m_y, j) = (\mathbf{m}, j)$.

We also have to be more careful about the specification of the object support now than we were in Sec. 16.2. There we considered the object support to extend to $\pm\infty$ in x and y and from 0 to ∞ in z (where z is normal to the collimator face), or we considered a slab of infinite lateral extent but finite thickness in z. Neither of these options is viable here—it is hard to rotate a camera around an infinite slab. We could consider a cube, but it is useful for the object support to have the same projection for all angles. Therefore we define the region $\mathbf{S}_f$ as a cylinder whose axis coincides with the axis of rotation of the camera, and we define a support function $S_f(\mathbf{r})$ to be unity within this cylinder and zero outside. There is no loss of generality with this definition so long as all objects of interest fit entirely in the support.

With the cylindrical support, the projection operator at angle ϕ_j, denoted $\mathcal{H}_j$, appears to be just a rotated version of the operator for $\phi = 0$, which we denote $\mathcal{H}_0$. This would be true if there were no attenuation, or if the attenuation coefficient were uniform within the cylinder of support, but with a general attenuation distribution, $\mathcal{H}_j$ depends in a more complicated way on the projection angle. We shall come back to the subject of attenuation in Sec. 17.1.6, but for now we simply ignore it.

The forward operator With the cylindrical support and no attenuation, rotating the detector and collimator by ϕ_j is equivalent to rotating the object by $-\phi_j$. Either the object function or the kernel function is transformed according to (7.198), and the mean of the $\mathbf{m}^{th}$ measurement can be written in two equivalent forms as

$$\overline{g}_{\mathbf{m}} = \int_{\mathbf{S}_f} d^3r\; h_{\mathbf{m}}(\mathbf{r})\, f(\mathbf{R}_j\mathbf{r}) = \int_{\mathbf{S}_f} d^3r\; h_{\mathbf{m}}(\mathbf{R}_j^{-1}\mathbf{r})\, f(\mathbf{r})\,, \tag{17.1}$$

where $\mathbf{R}_j$ is a 3×3 rotation matrix,[2] corresponding to rotation of the detector about the x axis by an angle ϕ_j, and $h_{\mathbf{m}}(\mathbf{r})$ (with a 2D subscript) is the kernel for $\phi_j = 0$. When convenient, we shall also denote $\overline{g}_{\mathbf{m}}$ by $\overline{g}_{\mathbf{m}j}$, and if we wish to consider a fixed j and all $\mathbf{m}$, we can define a vector $\overline{\mathbf{g}}_j$ with M^2 elements.

We can also express (17.1) in terms of functional operators defined as in Secs. 6.6.1 and 7.2.9. We let $\mathcal{T}_j$ be the functional transformation corresponding to the geometric rotation $\mathbf{R}_j$, so that

$$\mathcal{T}_j\, t(\mathbf{r}) = t(\mathbf{R}_j^{-1}\mathbf{r}) \tag{17.2}$$

for an arbitrary function $t(\mathbf{r})$. Then, since $\mathcal{T}_j$ is unitary,

$$f(\mathbf{R}_j\mathbf{r}) = \mathcal{T}_j^{\dagger} f(\mathbf{r})\,, \tag{17.3}$$

and (17.1) becomes

$$\overline{\mathbf{g}}_j = \mathcal{H}_0\mathcal{T}_j^{\dagger}\mathbf{f}. \tag{17.4}$$

Thus the operator for projection with the camera at angle ϕ_j is

$$\mathcal{H}_j = \mathcal{H}_0\mathcal{T}_j^{\dagger}\,. \tag{17.5}$$

Note that it is not correct to write $\overline{\mathbf{g}}$ as $\sum_j \overline{\mathbf{g}}_j$ or $\mathcal{H}$ as $\sum_j \mathcal{H}_j$; the vector $\overline{\mathbf{g}}_j$ has M^2 elements for an $M \times M$ detector array, but $\overline{\mathbf{g}}$ has JM^2 elements if there are J projection angles; the individual projections are stored separately, not summed. Mathematically speaking, the range of $\mathcal{H}$ is the direct sum of the ranges of the $\mathcal{H}_j$.

[2]In Chaps. 6 and 7, rotation matrices were denoted by $\mathcal{R}$, but we reserve that symbol in this chapter for the Radon transform.

Adjoint operators Next we consider the adjoint or backprojection operator. From (17.5) and property (c) of Sec. 1.3.5, the adjoint operator for a single projection is given formally by

$$\mathcal{H}_j^\dagger = \mathcal{T}_j \mathcal{H}_0^\dagger \,. \tag{17.6}$$

The meaning of this operator was discussed in Sec. 16.2.2. If we think of $\mathcal{H}_j$ as simple line-integral projection of the 3D object onto a 2D detector, then backprojection amounts to smearing the 2D projection back into the 3D volume defined by the support function. If there is blurring associated with the collimator or detector, then there is an additional blurring in the backprojection step.

When all projections are considered, the adjoint operator is given by

$$\mathcal{H}^\dagger = \sum_{j=0}^{J-1} \mathcal{T}_j \mathcal{H}_0^\dagger \,, \tag{17.7}$$

or in detail as[3]

$$\left[\mathcal{H}^\dagger \mathbf{g}\right](\mathbf{r}) = \sum_{j=0}^{J-1} \sum_{\mathbf{m}=1}^{M} h_{\mathbf{m}}(\mathbf{R}_j^{-1}\mathbf{r})\, g_{\mathbf{m}} \,. \tag{17.8}$$

Now we do have the sum over j; all projections contribute to the backprojected image at each $\mathbf{r}$.

Projection-backprojection operator We can now write down expressions for the projection-backprojection operator $\mathcal{H}^\dagger\mathcal{H}$, which we know is fundamental to both SVD and inverse problems. Formally,

$$\mathcal{H}^\dagger\mathcal{H} = \sum_{j=0}^{J-1} \mathcal{H}_j^\dagger \mathcal{H}_j = \sum_{j=0}^{J-1} \mathcal{T}_j \mathcal{H}_0^\dagger \mathcal{H}_0 \mathcal{T}_j^\dagger \,. \tag{17.9}$$

Explicitly, the kernel of the CC operator $\mathcal{H}^\dagger\mathcal{H}$ is given by (7.238) and (17.1) as

$$k(\mathbf{r},\mathbf{r}') = \sum_{j=0}^{J-1} \sum_{\mathbf{m}=1}^{M} h_{\mathbf{m}}(\mathbf{R}_j^{-1}\mathbf{r})\, h_{\mathbf{m}}(\mathbf{R}_j^{-1}\mathbf{r}') = \sum_{j=0}^{J-1} k_0(\mathbf{R}_j^{-1}\mathbf{r}, \mathbf{R}_j^{-1}\mathbf{r}') \,, \tag{17.10}$$

where $k_0(\mathbf{r},\mathbf{r}')$ is the kernel for $\mathcal{H}_0^\dagger\mathcal{H}_0$.

Weighted Hilbert spaces In the analysis above we have taken both $\mathbb{U}$ and $\mathbb{V}$ as simple Euclidean spaces, but in tomography it is often advantageous to use weighted Hilbert spaces. As in (1.12), we can define the object space by means of the weighted scalar product,

$$(\mathbf{f}_1,\mathbf{f}_2)_{\mathbb{U}} = \int_{\mathbf{S}_f} d^3r\, W(\mathbf{r})\, f_1^*(\mathbf{r})\, f_2(\mathbf{r}) \,, \tag{17.11}$$

where $W(\mathbf{r})$ is an arbitrary real, nonnegative function; in the present problem it is useful to assume that $W(\mathbf{r})$ is invariant to rotations about the x axis. Similarly,

[3]Recall that the sum over the 2D multi-index $\mathbf{m}$ means that both components m_x and m_y run from 1 to M, so we are considering an $M \times M$ detector array stepped to J angles.

the finite-dimensional image space can be defined via

$$(\mathbf{g}_1, \mathbf{g}_2)_{\mathbb{V}} = \sum_{j=0}^{J-1} \sum_{\mathbf{m}=1}^{M} w_{\mathbf{m}} g_{1\mathbf{m}}^* g_{2\mathbf{m}} , \tag{17.12}$$

where $\{w_{\mathbf{m}}\}$ is a set of real, nonnegative weights. It is convenient to choose these weights to be independent of the angular index j.

Using the definition of the adjoint from Sec. 1.3.5, the reader may show that (17.8) now becomes

$$\left[\mathcal{H}^\dagger \mathbf{g}\right](\mathbf{r}) = \frac{1}{W(\mathbf{r})} \sum_{j=0}^{J-1} \sum_{\mathbf{m}=1}^{M} w_{\mathbf{m}} h_{\mathbf{m}}(\mathbf{R}_j^{-1}\mathbf{r})\, g_{\mathbf{m}} . \tag{17.13}$$

17.1.2 Equally spaced angles

So far we have not placed any restrictions on the J projection angles, but now we assume that they are equally spaced around 360°. We therefore let the angular index j run from 0 to $J-1$ and set $\phi_j = j\Delta\phi \equiv 2\pi j/J$. Our goal is to derive an SVD for this system; group theory (see Chap. 6) will prove useful in this endeavor.

Symmetry considerations For J equally spaced angles, the matrices $\{\mathbf{R}_j\}$ form a representation of the group $\mathbf{C}_J$ (see Sec. 6.4.1). If the support functions are invariant under any rotation about the axis, as we have assumed above, then the transformation operators $\{\mathcal{T}_j\}$ also form a representation of $\mathbf{C}_J$, and each of these operators commutes with $\mathcal{H}^\dagger\mathcal{H}$ (see Sec. 6.7.3).

Mirror symmetries may also be present. If the detector array is invariant to reflection in a mirror plane passing through the axis of rotation, then the full symmetry group is the dihedral group $\mathbf{D}_J$, discussed in Sec. 6.4.2. The dihedral groups are significantly more complicated than the rotation groups since they are not Abelian. To avoid this complication, we can assume that the detector array is offset laterally by a fraction of a pixel so that the mirror symmetry is broken.

If there are no other symmetries, then, in the language of Sec. 6.7.3, the symmetry group of the system is the Abelian group $\mathbf{C}_J$. This simple statement allows us to say a great deal about the eigenfunctions of $\mathcal{H}^\dagger\mathcal{H}$, which are also the object-space singular functions in an SVD. We know from Secs. 6.7.5 and 7.2.9 that the eigenfunctions are nondegenerate (*i.e.*, all eigenvalues are distinct). Moreover, each eigenfunction transforms under rotation according to a specific irreducible representation of $\mathbf{C}_J$. Since all of these irreducible representations are 1D, that means that each eigenfunction of $\mathcal{H}^\dagger\mathcal{H}$ is also an eigenfunction of every $\mathcal{T}_j$.

These properties suggest that we denote the eigenfunctions with two indices n and k, where k specifies the irreducible representation, and write the eigenvalue equation for $\mathcal{H}^\dagger\mathcal{H}$ as

$$\mathcal{H}^\dagger\mathcal{H}\mathbf{u}_{nk} = \mu_{nk}\mathbf{u}_{nk} . \tag{17.14}$$

The eigenvalue equation for $\mathcal{T}_j$ is

$$\mathcal{T}_j\mathbf{u}_{nk} = \chi_j^{(k)}\mathbf{u}_{nk} . \tag{17.15}$$

We have written the eigenvalue here as $\chi_j^{(k)}$, which is also the character (see Sec. 6.3.3) for $\mathcal{T}_j$ in the kth irreducible representation; since the representation is 1D, matrix, character and eigenvalue are identical. Specifically, we know from (6.17) that

$$\chi_j^{(k)} = \exp(-2\pi ikj/J)\,. \tag{17.16}$$

Singular-value decomposition Suppose we have solved the eigenvalue problem for a fixed projection angle, say $\phi_j = 0$. That is, we know the solutions to

$$\mathcal{H}_0^\dagger \mathcal{H}_0 \mathbf{u}_n = \mu_n \mathbf{u}_n\,. \tag{17.17}$$

We would like to use these single-view eigenfunctions to construct eigenfunctions of $\mathcal{H}^\dagger\mathcal{H}$ as given by (17.9).

As we have already noted, each of the eigenfunctions must transform under rotation according to a specific irreducible representation of $\mathbf{C}_J$, and we saw in Sec. 6.6.3 that we can construct functions with the desired transformation properties by starting with an arbitrary function and projecting it onto the kth irreducible representation. From (6.36), we know that the form of this projection is

$$u_{nk}(\mathbf{r}) = \frac{1}{J}\sum_{j=0}^{J-1}\left[\chi_j^{(k)}\right]^* \mathcal{T}_j\, q_n(\mathbf{r}) = \frac{1}{J}\sum_{j=0}^{J-1}\exp\left(\frac{2\pi ikj}{J}\right)\mathcal{T}_j\, q_n(\mathbf{r}) \equiv \mathcal{P}_k\, q_n(\mathbf{r})\,, \tag{17.18}$$

where $\mathcal{P}_k$ is the projection operator and $q_n(\mathbf{r})$ is yet to be determined. The reader can verify that this $u_{nk}(\mathbf{r})$ satisfies (17.15) as required, and also that $\mathcal{P}_k$ is idempotent and Hermitian as required of all projection operators (see Sec. 1.3.6).

If we use (17.9) along with (17.18) and the spectral representation for $\mathcal{H}_0^\dagger\mathcal{H}_0$, the eigenvalue equation (17.14) becomes

$$\begin{aligned}\mathcal{H}^\dagger\mathcal{H}\mathbf{u}_{nk} &= \frac{1}{J}\sum_{n'=1}^{R}\sum_{j'=0}^{J-1}\sum_{j=0}^{J-1}\mu_{n'}\exp\left(\frac{2\pi ikj}{J}\right)\mathcal{T}_{j'}\mathbf{u}_{n'}\mathbf{u}_{n'}^\dagger\mathcal{T}_{j'}^\dagger\mathcal{T}_j\,\mathbf{q}_n \\ &= \mu_{nk}\frac{1}{J}\sum_{j=0}^{J-1}\exp\left(\frac{2\pi ikj}{J}\right)\mathcal{T}_j\,\mathbf{q}_n\,,\end{aligned} \tag{17.19}$$

where R is the rank of $\mathcal{H}_0^\dagger\mathcal{H}_0$. If we make the change of variables $\ell = [j - j']$ (where the square brackets denote modulus-J arithmetic) and recognize that $\mathcal{T}_{j'}^\dagger\mathcal{T}_j = \mathcal{T}_\ell$, we can write the left-hand side of (17.19) as

$$\begin{aligned}\mathcal{H}^\dagger\mathcal{H}\mathbf{u}_{nk} &= \frac{1}{J}\sum_{n'=1}^{R}\sum_{j'=0}^{J-1}\sum_{\ell=0}^{J-1}\mu_{n'}\exp\left[\frac{2\pi ik(\ell+j')}{J}\right]\mathcal{T}_{j'}\mathbf{u}_{n'}\mathbf{u}_{n'}^\dagger\mathcal{T}_\ell\,\mathbf{q}_n \\ &= J\sum_{n'=1}^{R}\mu_{n'}[\mathbf{u}_{n'}^\dagger\mathcal{P}_k\,\mathbf{q}_n]\mathcal{P}_k\mathbf{u}_{n'}\,.\end{aligned} \tag{17.20}$$

Thus the eigenvalue equation becomes

$$J\sum_{n'=1}^{R}\mu_{n'}[\mathbf{u}_{n'}^\dagger\mathcal{P}_k\,\mathbf{q}_n]\mathcal{P}_k\mathbf{u}_{n'} = \mu_{nk}\mathcal{P}_k\,\mathbf{q}_n\,, \tag{17.21}$$

and the objective is to find $\mathbf{q}_n$. In most cases numerical methods are necessary, though we shall encounter a situation in Sec. 17.2.1 where a fully analytical solution is possible. To gain further insight at this point, however, we shall describe an approximate analytic solution and argue qualitatively that the approximation is reasonable.

Approximate SVD A useful first approximation is $\mathbf{q}_n \propto \mathbf{u}_n$. This choice is based on numerical studies of parallel-hole collimators and multiple-pinhole systems[4] that indicate that, for all ℓ,

$$\mathbf{u}_{n'}^\dagger \boldsymbol{\mathcal{T}}_\ell \mathbf{u}_n \approx 0 \qquad \text{if } n \neq n' . \tag{17.22}$$

If $\ell = 0$, then $\mathbf{u}_{n'}^\dagger \boldsymbol{\mathcal{T}}_\ell \mathbf{u}_n \equiv 0$ since $\boldsymbol{\mathcal{T}}_0$ is the unit operator and $\mathbf{u}_n$ and $\mathbf{u}_{n'}$ are both eigenvectors of the Hermitian operator $\boldsymbol{\mathcal{H}}_0^\dagger \boldsymbol{\mathcal{H}}_0$, hence orthogonal if $n \neq n'$. For $\ell \neq 0$, $\mathbf{u}_{n'}$ and $\boldsymbol{\mathcal{T}}_\ell \mathbf{u}_n$ may not be exactly orthogonal, but, as we shall see in an example below, we expect their scalar product to be small.

If (17.22) is satisfied, it follows that $\mathbf{u}_{n'}^\dagger \boldsymbol{\mathcal{P}}_k \mathbf{u}_n$ is approximately zero for $n \neq n'$, so we can dispense with the sum and set $n' = n$ in (17.21). By inspection, we then have

$$\mu_{nk} = J\mu_n \mathbf{u}_n^\dagger \boldsymbol{\mathcal{P}}_k \mathbf{u}_n = \mu_n \sum_{\ell=0}^{J-1} \exp\left(\frac{2\pi i k \ell}{J}\right) \mathbf{u}_n^\dagger \boldsymbol{\mathcal{T}}_\ell \mathbf{u}_n \tag{17.23}$$

and

$$\mathbf{u}_{nk} = N_{nk} \boldsymbol{\mathcal{P}}_k \mathbf{u}_n = \frac{N_{nk}}{J} \sum_{j=0}^{J-1} \exp\left(\frac{2\pi i k j}{J}\right) \boldsymbol{\mathcal{T}}_j \mathbf{u}_n , \tag{17.24}$$

where N_{nk} is a normalizing constant. Solution of the single-view eigenvalue problem thus gives us the solution of the J-view problem under $\mathbf{C}_J$ symmetry in this approximation. As exercises, the reader may show that (a) μ_{nk} is real; (b) $\mathbf{u}_{nk}$ and $\mathbf{u}_{n'k'}$ are orthogonal unless $n = n'$ and $k = k'$; and (c) (17.23) and (17.24) reproduce (17.9).

How good is the approximation? The essential step that led to (17.23) and (17.24) is the assumption that $\boldsymbol{\mathcal{P}}_k \mathbf{u}_n$ is at least approximately orthogonal to $\mathbf{u}_{n'}$ (which follows from the stronger assumption that $\boldsymbol{\mathcal{T}}_\ell \mathbf{u}_n$ is approximately orthogonal to $\mathbf{u}_{n'}$). To check these assumptions, we shall derive the single-view eigenfunctions for a simple model of the imaging system. The model we choose is one treated briefly in Sec. 16.2.3, where we considered a large detector element used with a fine collimator bore, so that the collimator blur was negligible and the system response function was a thin cylindrical ray. The key point of this model is that there is no overlap of the system response functions in a single view.

With this assumption, the eigenfunction $\mathbf{u}_n$ is just the response function itself. To demonstrate this point, we first note that the response functions are indexed by detector elements, so we can replace the index n by the 2D multi-index $\mathbf{m}$ that we customarily use in detector space. From (7.238) with $h_\mathbf{m}(\mathbf{r})$ real, we then have

$$\left[\boldsymbol{\mathcal{H}}_0^\dagger \boldsymbol{\mathcal{H}}_0\right] h_\mathbf{m}(\mathbf{r}) = \sum_{\mathbf{m}'=1}^{M} h_{\mathbf{m}'}(\mathbf{r}) \int_{\mathbf{S}_f} d^3r'\, h_{\mathbf{m}'}(\mathbf{r}')\, h_\mathbf{m}(\mathbf{r}') . \tag{17.25}$$

[4]Numerical and analytic SVDs of pinhole systems can be found in Aarsvold (1993), and some of the results can be found in Barrett *et al.* (1991). Aarsvold has also done numerical studies of collimator systems but they are unpublished at this writing.

But if there is no overlap of the response functions, the integral is zero unless $\mathbf{m} = \mathbf{m}'$, so

$$\left[\mathcal{H}_0^\dagger \mathcal{H}_0\right] h_{\mathbf{m}}(\mathbf{r}) = \mu_{\mathbf{m}} h_{\mathbf{m}}(\mathbf{r}), \tag{17.26}$$

where

$$\mu_{\mathbf{m}} = \int_{\mathbf{S}_f} d^3r' \, [h_{\mathbf{m}}(\mathbf{r}')]^2 \,. \tag{17.27}$$

The normalized single-view eigenfunctions are given by

$$u_{\mathbf{m}}(\mathbf{r}) = \frac{h_{\mathbf{m}}(\mathbf{r})}{\sqrt{\int_{\mathbf{S}_f} d^3r \, h_{\mathbf{m}}^2(\mathbf{r})}} \,. \tag{17.28}$$

Thus, as expected, the single-view eigenfunctions are proportional to the response functions, provided only that the response functions are orthogonal. In particular, attenuation does not invalidate (17.28). As an exercise, the reader may show that (17.27) and (17.28) lead to the correct expression for $\mathcal{H}_0^\dagger \mathcal{H}_0$.

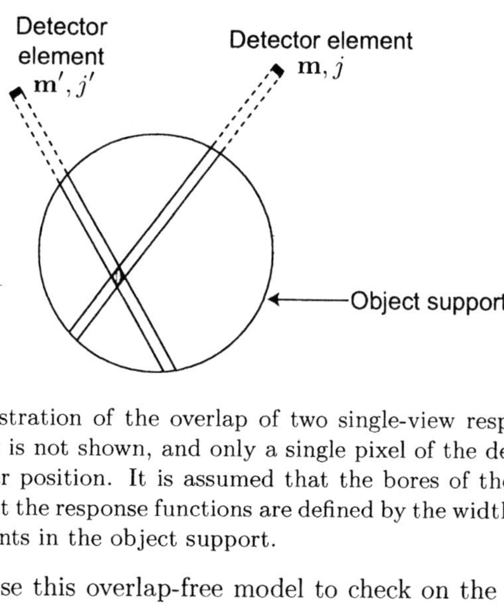

Fig. 17.2 Illustration of the overlap of two single-view response functions. The collimator is not shown, and only a single pixel of the detector is shown at each angular position. It is assumed that the bores of the collimator are very fine so that the response functions are defined by the width of the detector pixel at all points in the object support.

We can now use this overlap-free model to check on the accuracy of (17.22). With (17.28), we can write

$$\mathbf{u}_{\mathbf{m}}^\dagger \mathcal{T}_\ell \mathbf{u}_{\mathbf{m}'} = \frac{\int_{\mathbf{S}_f} d^3r \, h_{\mathbf{m}}(\mathbf{r}) \, [\mathcal{T}_\ell \mathbf{h}_{\mathbf{m}'}] \, (\mathbf{r})}{\sqrt{\int_{\mathbf{S}_f} d^3r \, h_{\mathbf{m}}^2(\mathbf{r}) \int_{\mathbf{S}_f} d^3r \, h_{\mathbf{m}'}^2(\mathbf{r})}} \,. \tag{17.29}$$

For $\ell \neq 0$, the integrand in the numerator is nonzero only where the single-view response functions $h_{\mathbf{m}}(\mathbf{r})$ and $[\mathcal{T}_\ell \mathbf{h}_{\mathbf{m}'}] \, (\mathbf{r})$ overlap. The overlap is identically zero if $\mathbf{m}$ and $\mathbf{m}'$ refer to different x-planes (*i.e.*, $m_x \neq m_x'$), where the x-axis is the axis of rotation. For $m_x' = m_x$, the overlap is the shaded region of Fig. 17.2. For simplicity, we consider a square bore of side ϵ, and we see that

$$\mathbf{u}_{\mathbf{m}}^\dagger \mathcal{T}_\ell \mathbf{u}_{\mathbf{m}'} \approx \frac{\epsilon}{\sqrt{L_{\mathbf{m}} L_{\mathbf{m}'}} \, |\sin(2\pi\ell/J)|} \,, \qquad \ell \neq 0 \,, \qquad m_x' = m_x \,, \tag{17.30}$$

where $L_{\mathbf{m}}$ is the length of the intersection of the response function $h_{\mathbf{m}}(\mathbf{r})$ and the object support, and we have assumed that $\epsilon/|\sin(2\pi\ell/J)|$ is less than either $L_{\mathbf{m}}$ or $L_{\mathbf{m}'}$. The important point to note about (17.30) is that it goes to zero as $\epsilon \to 0$ (for fixed J).

To summarize, $\mathbf{u}_{\mathbf{m}}^{\dagger}\mathcal{T}_{\ell}\mathbf{u}_{\mathbf{m}'}$ is always identically zero if $\ell = 0$ and $\mathbf{m} \neq \mathbf{m}'$. If there is no overlap of the single-view response functions, then the eigenfunctions are the response functions, and in that case $\mathbf{u}_{\mathbf{m}}^{\dagger}\mathcal{T}_{\ell}\mathbf{u}_{\mathbf{m}'} = 0$ if $m_x' \neq m_x$ (for rotation about the x-axis). If the single-view response functions do not overlap, the only case where $\mathbf{u}_{\mathbf{m}}^{\dagger}\mathcal{T}_{\ell}\mathbf{u}_{\mathbf{m}'}$ is nonzero is when $\ell \neq 0$, $m_x' = m_x$ and there is overlap of the rays from different views; even in that case $\mathbf{u}_{\mathbf{m}}^{\dagger}\mathcal{T}_{\ell}\mathbf{u}_{\mathbf{m}'}$ approaches zero when $\epsilon \to 0$.

We shall revisit this approximation in Sec. 17.2.1 and give a condition under which it is exact.

17.1.3 Fourier analysis in the CD formulation

Even though the system under consideration is not shift-invariant, we can nevertheless profitably use Fourier methods to describe it. We know from Sec. 7.3.3 that a Fourier *series* is a useful description for a function of compact support and that it leads to the Fourier crosstalk matrix as an exact description of the CD imaging system.

From (7.261) we know that the first step in determining the crosstalk matrix is to compute the response in the data to a single Fourier component, denoted by the vector index[5] $\mathbf{k}$. In our present notation, this response is given by [*cf.* (17.1)]

$$\psi_{\mathbf{mk}} = \int_{\mathbf{S}_f} d^3r\; h_{\mathbf{m}}(\mathbf{r}) \exp(2\pi i \boldsymbol{\rho}_{\mathbf{k}} \cdot \mathbf{R}_j \mathbf{r}) = \int_{\mathbf{S}_f} d^3r\; h_{\mathbf{m}}(\mathbf{r}) \exp(2\pi i \mathbf{R}_j^{-1} \boldsymbol{\rho}_{\mathbf{k}} \cdot \mathbf{r}), \quad (17.31)$$

where the last form is valid since $\mathbf{R}_j^{\dagger} = \mathbf{R}_j^{-1}$. Note that we now have a total of six indices on $\psi_{\mathbf{mk}}$. In the data domain, m_x and m_y specify location on the 2D detector and j specifies projection angle; recall that $\mathbf{m} = (m_x, m_y, j)$. In the object domain, the three components of $\mathbf{k}$ are needed to specify the 3D spatial frequency.

To compute $\psi_{\mathbf{mk}}$, we can use the same simple model as above where the response function $h_{\mathbf{m}}(\mathbf{r})$ is a thin cylinder of cross-sectional area ϵ^2, with its axis normal to the detector plane at location $\mathbf{m}$. For this model and the coordinates we are using (with the origin on the axis of rotation),

$$\psi_{\mathbf{mk}} \approx \epsilon^2 \exp\left[2\pi i \left(\mathbf{R}_j^{-1}\boldsymbol{\rho}_{\mathbf{k}}\right)_{\perp} \cdot \mathbf{r}_{\mathbf{m}}\right] \int_{-L_{\mathbf{m}}/2}^{L_{\mathbf{m}}/2} dz\, \exp\left[2\pi i \left(\mathbf{R}_j^{-1}\boldsymbol{\rho}_{\mathbf{k}}\right)_z z\right]$$

$$= \epsilon^2 L_{\mathbf{m}} \exp\left[2\pi i \left(\mathbf{R}_j^{-1}\boldsymbol{\rho}_{\mathbf{k}}\right)_{\perp} \cdot \mathbf{r}_{\mathbf{m}}\right] \operatorname{sinc}\left[L_{\mathbf{m}} \left(\mathbf{R}_j^{-1}\boldsymbol{\rho}_{\mathbf{k}}\right)_z\right], \quad (17.32)$$

where subscript $\perp$ denotes the projection of the 3D vector onto the x-y plane.

We know from (7.263) that the crosstalk matrix $\mathbf{B}$ has elements given by

$$\beta_{\mathbf{kk}'} = \sum_{\mathbf{m}=1}^{M} \sum_{j=0}^{J-1} \psi_{\mathbf{mk}}^{*} \psi_{\mathbf{mk}'} \,. \quad (17.33)$$

[5] The reader should not confuse the index $\mathbf{k}$ used here with the k used above; both denote Fourier series coefficients, but in different senses.

The diagonal elements, which tell us how strongly particular Fourier components are transferred into the data, are given by

$$\beta_{\mathbf{kk}} = \epsilon^4 \sum_{\mathbf{m}=1}^{M} \sum_{j=0}^{J-1} L_{\mathbf{m}}^2 \operatorname{sinc}^2 \left[L_{\mathbf{m}} \left(\mathbf{R}_j^{-1} \boldsymbol{\rho}_{\mathbf{k}} \right)_z \right] . \qquad (17.34)$$

This equation is the CD counterpart of the central-slice theorem (see Sec. 4.4.2). It says that the only Fourier components that contribute to the data, for each j, are the ones for which the crests of the wave are nearly parallel to the z axis, so that $\left(\mathbf{R}_j^{-1}\boldsymbol{\rho}_{\mathbf{k}}\right)_z$ is near zero .

The sum over $\mathbf{m}$ in (17.34) doesn't do much since the summand depends only weakly on that index. For the off-diagonal elements, however, the sum over detector locations is critical. These elements are given by

$$\beta_{\mathbf{kk}'} = \epsilon^4 \sum_{\mathbf{m}=1}^{M} \sum_{j=0}^{J-1} L_{\mathbf{m}}^2 \operatorname{sinc} \left[L_{\mathbf{m}} \left(\mathbf{R}_j^{-1} \boldsymbol{\rho}_{\mathbf{k}} \right)_z \right] \operatorname{sinc} \left[L_{\mathbf{m}} \left(\mathbf{R}_j^{-1} \boldsymbol{\rho}_{\mathbf{k}'} \right)_z \right]$$

$$\times \exp \left[-2\pi i \left(\mathbf{R}_j^{-1} \boldsymbol{\rho}_{\mathbf{k}} - \mathbf{R}_j^{-1} \boldsymbol{\rho}_{\mathbf{k}'} \right)_\perp \cdot \mathbf{r}_{\mathbf{m}} \right] . \qquad (17.35)$$

Now the summand varies rapidly with the detector index $\mathbf{m}$, and the sum exhibits complicated aliasing behavior, as we shall see below.

Numerical study of the crosstalk matrix In Fig. 17.3 we show a sequence of crosstalk matrices, generated by D. W. Wilson for various SPECT systems. The figure is a plot of $\beta_{\mathbf{kk}'}$ as the gray level, with k_x and k'_x as the axes. All of the systems used a detector with 64 elements in one dimension, and it acquired 64 projection at equally spaced angles. The first figure (upper left) shows a nearly ideal system with a parallel-hole collimator but no detector blur, no attenuation and no scatter. In this limit the crosstalk matrix is nearly diagonal, indicating that $\mathcal{H}^\dagger\mathcal{H}$ is nearly shift invariant and that image reconstruction should be very easy. Then we progressively add these various degrading effects, moving steadily away from the diagonal condition.

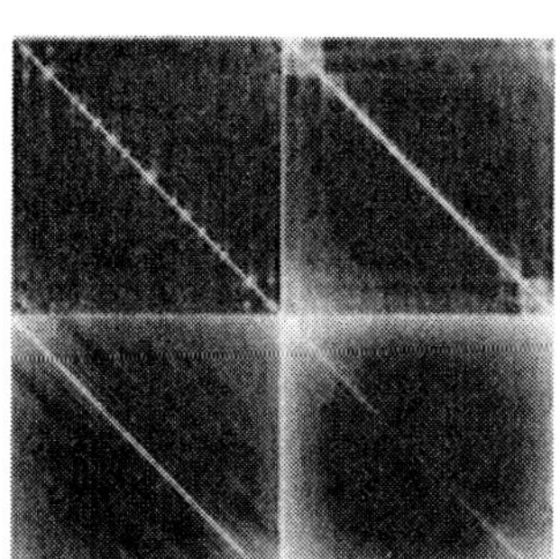

Fig. 17.3 Crosstalk matrices for various SPECT systems, courtesy of D. W. Wilson. *Upper left*: near-ideal system with no attenuation or scatter in the patient. *Upper right*: near-ideal system with attenuation. *Lower left*: same system with attenuation and detector blur. *Lower right*: same system with attenuation, detector blur and scatter.

17.1.4 2D Radon transform and parallel-beam SPECT

There are two ways we can pass to the limit of continuous data in SPECT or other kinds of tomography. First, we can consider very fine sampling by the collimator and detector, so that the discrete detector positions $\mathbf{r_m}$ are replaced by the continuous 2D vector $\mathbf{r}_d$ on the detector plane. Second, we can consider a very large number of projection angles, so that $J \to \infty$ and the discrete projection angle ϕ_j is replaced by an arbitrary angle ϕ. If we take both of these limits, the mean data can be denoted $\overline{g}(\mathbf{r}_d, \phi)$.

It is also common in the literature to consider only one of the two continuous limits, treating $\mathbf{r}_d$ as continuous but still considering a discrete set of angles. From an engineering perspective, this approach is really backwards: if we want to collect more angles in a rotating-camera system, we can just do so, without any physical or engineering limitation except that more storage space is required for the data. If we want more detector elements, however, it is a major engineering effort, and if we want finer collimators, not only must we be able to fabricate them, we must also accept the inevitable loss of photon collection efficiency. Nevertheless, the hybrid model of discrete angles and continuous detectors has been well studied, and like the fully CC model, it often leads to useful linear algorithms.

Just formally passing to the limit of fine sampling is not sufficient if we want to find a simple forward transform, and hence to have a chance of finding the inverse transform. We must also ignore many physical effects such as detector and collimator blur, attenuation and scatter. In this section we shall specify in detail what assumptions and approximations are needed if we want to reduce the parallel-beam SPECT problem to the 2D Radon transform.

2D Radon transform As discussed in Sec. 4.4.1, the 2D Radon transform consists of 1D line-integral projections of a 2D function. Specifically, the definition is given in (4.139) as

$$\lambda(p, \phi) = [\boldsymbol{\mathcal{R}}_2 \mathbf{f}](p, \phi) = \int_\infty d^2r\, f(\mathbf{r})\, \delta(p - \mathbf{r} \cdot \hat{\mathbf{n}}) , \tag{17.36}$$

or simply as $\boldsymbol{\lambda} = \boldsymbol{\mathcal{R}}_2 \mathbf{f}$, where $\boldsymbol{\mathcal{R}}_2$ is the 2D Radon-transform operator. Recall that $p = \mathbf{r} \cdot \hat{\mathbf{n}}$ is the equation of a line normal to the 2D unit vector $\hat{\mathbf{n}}$ and a distance p from the origin (see Fig. 2.7) and that the origin coincides with the center of rotation.

From irradiance to Radon We shall now build on the discussion in Sec. 16.2.2 to show that the 2D Radon transform is applicable to rotating-camera SPECT if we ignore attenuation in the object and blur from the detector and collimator. A useful starting point for making the connection is (16.82); if we set $\mu_{tot} = 0$ in that equation, we can write the photon irradiance on the detector face as

$$I_p(\mathbf{r}) = \frac{1}{4\pi} \int_{2\pi} d\Omega\, T(\mathbf{r}, \hat{\mathbf{s}}) \int_0^\infty d\ell\, f(\mathbf{r} - \hat{\mathbf{s}}\ell) . \tag{17.37}$$

In this expression, $T(\mathbf{r}, \hat{\mathbf{s}})$ is the transmission of the collimator, defined by (16.80) as

$$T(\mathbf{r}, \hat{\mathbf{s}}) = \sum_{\mathbf{n}} \beta(\mathbf{r} - \mathbf{r_n})\, \beta\left(\mathbf{r} - \mathbf{s}_\perp \frac{L_b}{s_z} - \mathbf{r_n}\right) , \tag{17.38}$$

where the sum is over all bores in the collimator, $\beta(\mathbf{r}-\mathbf{r_n})$ is a 2D function that is unity within the open area of the $\mathbf{n}^{th}$ bore and zero otherwise, and $\mathbf{s}_\perp = (s_x, s_y)$ if $\hat{\mathbf{s}} = (s_x, s_y, s_z)$ and the detector plane is normal to the z axis.

We pass to the limit of a very fine collimator by letting

$$\beta(\mathbf{r}-\mathbf{r_n}) \to A_{bore}\,\delta(\mathbf{r}-\mathbf{r_n})\,, \qquad \sum_n \cdots \to \frac{\alpha_{pf}}{A_{bore}}\int_\infty d^2r_n \cdots, \tag{17.39}$$

where A_{bore} is the open area of the collimator bore and α_{pf} is the packing fraction as defined above (16.91). In this limit, we have

$$T(\mathbf{r},\hat{\mathbf{s}}) \to \alpha_{pf}A_{bore}\int_\infty d^2r_n\;\delta(\mathbf{r}-\mathbf{r_n})\,\delta\!\left(\mathbf{r}-\mathbf{s}_\perp\frac{L_b}{s_z}-\mathbf{r_n}\right) = \frac{\alpha_{pf}A_{bore}}{L_b^2}\,\delta(\mathbf{s}_\perp)\,, \tag{17.40}$$

since $s_z = 1$ if $\mathbf{s}_\perp = 0$. The factor $\delta(\mathbf{s}_\perp)$ indicates that the collimator accepts only the photons that travel normal to the detector face, in a direction that we can denote by $\hat{\mathbf{s}}_d$ (d for detector). As a nontrivial exercise in delta functions, the reader may show that

$$\int_{2\pi} d\Omega\;\delta(\mathbf{s}_\perp) = 1\,, \tag{17.41}$$

so (17.37) becomes

$$I_p(\mathbf{r}) \to \frac{\alpha_{pf}A_{bore}}{4\pi L_b^2}\int_0^\infty d\ell\; f(\mathbf{r}-\hat{\mathbf{s}}_d\ell)\,. \tag{17.42}$$

The factor in front of the integral is recognized from (16.98) as the point-source sensitivity, which is independent of distance from the collimator face with a parallel-hole collimator; it is this fact that makes the Radon transform useful for describing projection data obtained with such collimators.

If there is no detector blur, then the detector measures the photon irradiance directly, but we are not interested in the full 2D photon distribution in this discussion. Instead, we consider a thin stripe of height ϵ_x (where the camera rotation is about the x axis). Photons detected in this stripe originate from a thin slab parallel to the y-z plane. If we denote the coordinates in the detector plane as (x_d, y_d), we can choose the origin so that the center of this slab is at $x_d = 0$. Then we can define a linear photon density $g_\phi(y_d)$ at projection angle ϕ by

$$g_\phi(y_d) \equiv \int_{-\epsilon_x/2}^{\epsilon_x/2} dx_d\; I_p(x_d, y_d) \to \epsilon_x I_p(0, y_d)\,, \tag{17.43}$$

where the last form holds in the limit as $\epsilon_x \to 0$.

To finish forcing the data to look like a 2D Radon transform, we define a 2D function $f_0(y,z) = f(0,y,z)$ and we let $\mathbf{r}$ now be (y,z). Then we see that

$$g_\phi(y_d) = \frac{\alpha_{pf}A_{bore}\,\epsilon_x}{4\pi L_b^2}\int_\infty d^2r\; f(\mathbf{r})\,\delta(y_d - \mathbf{r}\cdot\hat{\mathbf{n}})\,, \tag{17.44}$$

where $\hat{\mathbf{n}}$ is a 2D unit vector in the y-z plane perpendicular to $\hat{\mathbf{s}}_d$. Thus, finally, we have something that looks like (17.36).

To summarize the assumptions and approximations needed to get (17.44), we had to pass to the limit of vanishingly small collimator bores, we had to neglect detector blur and sampling, and we had to consider a thin stripe of the photon irradiance on the detector plane. We also implicitly ignored septal penetration in the collimator since (17.38) was derived earlier on the assumption that all photons striking the collimator material were absorbed.

There are many obvious difficulties in these assumptions. Very fine collimators have very little efficiency, as indicated by the factor of A_{bore} in (17.44), and if we chose to make a collimator with fine bores, we would run into problems of septal penetration, invalidating (17.38). Moreover, selecting out a thin stripe of height ϵ_x reduces the photon count further. Finally, we know from the discussions in Chap. 12 that gamma-ray detectors such as the Anger scintillation camera have fundamental limitations on their spatial resolution.

17.1.5 3D transforms and cone-beam SPECT

One consequence of the collimator and detector blur in SPECT is that we cannot really treat the problem as 2D, even with a parallel-hole collimator. When we integrate over a stripe of the data, we are necessarily including contributions from portions of the object outside the corresponding slab in object space. For realistic systems, there is no good substitute for a fully 3D treatment.

For a fully 3D analysis of tomographic systems in CC terms, we have two candidates for the idealized tomographic transform: the 3D Radon transform and the x-ray transform. In this section we shall discuss each of them and show how they relate to practical acquisition geometries. Then we shall show a way in which they relate to each other.

3D Radon transform The 3D Radon transform is very similar in form to the 2D version, (17.36). The definition of the 3D Radon transform is (4.173), but for clarity we modify the notation slightly and write it here as

$$\lambda(p, \hat{\mathbf{n}}) = [\mathcal{R}_3\mathbf{f}](p, \hat{\mathbf{n}}) = \int_\infty d^3\mathbf{r}\, f(\mathbf{r})\, \delta(p - \mathbf{r} \cdot \hat{\mathbf{n}})\,, \tag{17.45}$$

where $\mathbf{r}$ and $\hat{\mathbf{n}}$ are 3D vectors (the latter a unit vector). In pure operator form, $\boldsymbol{\lambda} = \mathcal{R}_3\mathbf{f}$.

In spite of the mathematical similarity of (17.36) and (17.45), they describe rather different physical systems. The equation $p = \mathbf{r} \cdot \mathbf{n}$ describes a line in 2D, but $p = \mathbf{r} \cdot \hat{\mathbf{n}}$ is the equation of a *plane* in 3D. Thus the 2D Radon transform is a useful description for systems where the data consist of integrals of the object over thin tubes, approximating lines, while the 3D Radon transform is useful when the data consist of integrals of the object over thin slabs, approximating planes.

One way to obtain data related to the 3D Radon transform is to acquire 2D line-integral-like projections with a parallel-hole collimator and then to sum over stripes in the 2D image. This is what we did in (17.43) to get to the 2D Radon transform, but in that case the stripe of integration was perpendicular to the rotation axis (x), and the corresponding slab of integration in object space reduced to the plane $x = 0$ as ϵ_x vanished. If we consider more general stripes, we can obtain integrals over other planes. For example, if we integrate the 2D projection at angle $\phi = 0$ over stripes at all angles in the 2D data plane, we obtain the 3D Radon transform for all $\hat{\mathbf{n}}$ lying in the x-y plane. If we then repeat the process for all projection angles, we get the full 3D Radon transform of the object (neglecting attenuation and blur).

A more direct way of acquiring the 3D Radon transform (and collecting more photons in the process) is to use a slat collimator, as shown in Fig. 17.4. This "collimator" actually collimates in only one direction, allowing photons travelling near a plane defined by one of the openings to pass through unimpeded. Thus the detector need not have any spatial resolution in the direction parallel to the slats, and something approximating planar integrals of the object are obtained immediately. It is an interesting exercise in radiometry to work out the precise relation between object and image in this case.

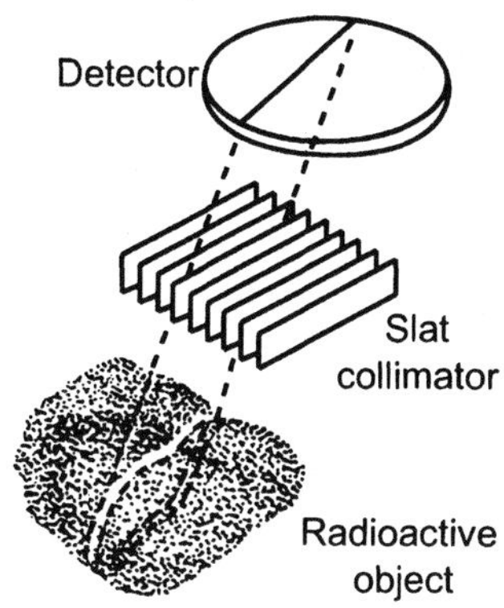

Fig. 17.4 Slat collimator viewing a 3D radioactive object.

3D x-ray transform The x-ray transform, denoted by the operator $\boldsymbol{\mathcal{X}}$, was defined in Chap. 10 by (10.136) and (10.137). For simplicity, we drop the factor $1/c_m$ used in Chap. 10 and define

$$[\boldsymbol{\mathcal{X}}\mathbf{f}](\mathbf{r},\hat{\mathbf{s}}) = \int_0^\infty d\ell\ f(\mathbf{r}-\hat{\mathbf{s}}\ell)\,, \tag{17.46}$$

where $\hat{\mathbf{s}}$ is a 3D unit vector. (Note that we use $\hat{\mathbf{n}}$ or $\hat{\mathfrak{n}}$ for a unit vector normal to a line or plane, but $\hat{\mathbf{s}}$ for a unit vector in a ray direction.)

It is also useful to recast (17.46) into our standard form for a 3D integral transform. By the change of variables $\mathbf{r}' = \mathbf{r} - \hat{\mathbf{s}}\ell$ and some algebra, or more directly by use of the identity (10.159), we can write

$$[\boldsymbol{\mathcal{X}}\mathbf{f}](\mathbf{r},\hat{\mathbf{s}}) = \int_\infty d^3r'\, f(\mathbf{r}')\, \frac{\delta\left(\hat{\mathbf{s}} - \frac{\mathbf{r}-\mathbf{r}'}{|\mathbf{r}-\mathbf{r}'|}\right)}{|\mathbf{r}-\mathbf{r}'|^2}\,. \tag{17.47}$$

In this form, $\boldsymbol{\mathcal{X}}$ appears to map a function of three variables to a function of five variables (x, y, z and two angles), but in practice some subset of these variables will be addressed in any particular measurement geometry.

Like the 2D Radon transform, the x-ray transform involves line integrals. Within the approximations that led to (17.42), each measurement taken with a parallel-hole collimator is one sample of the 3D x-ray transform of the object (as well as one sample of the 2D Radon transform). The difference is that the 2D Radon transform involves lines in the plane of a 2D object, while the 3D x-ray transform involves lines through a volume object. The 3D Radon transform, on the other hand, appears rather different since it involves integrals over planes in the object, but in fact there is an important relation between these two 3D transforms, as we shall see below.

Cone-beam geometries So far the 2D projection systems we have discussed have all used parallel-hole collimators rotated in a circle around the object. An alternative data-acquisition scheme is to acquire each 2D data set with a pinhole rather than a collimator (see Sec. 10.4.2), and to attempt to acquire enough data for 3D reconstruction by moving the pinhole-detector assembly through a sequence of positions. The locus of all positions of the pinhole is called the *trajectory* or *vertex path*, and data are usually acquired at a discrete set of positions along the trajectory (though continuous acquisition is also possible). Many authors argue that a simple circular trajectory is inadequate for 3D reconstruction and advocate more complicated trajectories such as a spiral or a circle plus a line segment.

SPECT data can also be acquired with focused collimators. With both pinholes and focused collimators, the rays involved in each 2D projection of the 3D object pass approximately through a point and hence form a cone. A similar situation occurs in x-ray computed tomography where the rays all emanate from a point (the focal spot on the x-ray tube). The term *cone-beam tomography* is used to describe all of these situations, and the point of the cone is often called the *vertex*.

The idealized CC description of the data in cone-beam tomography in all of these geometries is the 3D x-ray transform (17.46) when we interpret the variable $\mathbf{r}$ as the vertex position $\mathbf{r}_v$ and $\hat{\mathbf{s}}$ as a unit vector along the line from the vertex to a position on the detector.[6] Since location of the vertex along the trajectory can be specified by a single parameter (distance along the trajectory, say), the 3D x-ray transform maps a function of three variables to another function of three variables in this case.

Many authors refer to (17.46) as the *cone-beam transform*, reserving the term x-ray transform for the special case where the unit vectors $\hat{\mathbf{s}}$ are all parallel. Physically, this would be appropriate if the vertex point $\mathbf{r}_v$ were at an infinite distance from the object support. We prefer not to make this distinction since we often want to consider situations where the object support can be unbounded.

Relation to the Boltzmann transport equation Fundamentally, it is the law of conservation of radiance that makes the x-ray transform so widely applicable. In Sec. 10.3.2 we used the Boltzmann transport equation to show that radiance was conserved at all points along a geometric ray in a non-attenuating medium. By properly defining the source function, we showed in (10.136) that the radiance ($c_m w$) was just a line integral of the source distribution.

This result is exact and holds at all points in space for a non-attenuating medium, but it takes no account of the effects of the detector and image-forming aperture. For a pinhole of finite diameter, a conical set of rays through the object reaches each point on the detector, as illustrated in Fig. 10.9. The *irradiance* at that point is obtained from the *radiance* by integrating it over a range of angles defined by the pinhole (see Sec. 10.4.2). Thus the irradiance at a point is an an-

[6] A minor difference between the pinhole and focused-collimator geometries concerns the direction of $\hat{\mathbf{s}}$. In (17.46), $\mathbf{r}_v - \hat{\mathbf{s}}\ell$ describes a point displaced from the vertex by an amount ℓ in the direction $-\hat{\mathbf{s}}$. If we think of $\hat{\mathbf{s}}$ as pointing generally to the right, that means that the source is to the left of the vertex. Since the detector is to the right of the vertex for pinhole imaging, (17.46) is correct in that case. With a focused collimator, on the other hand, the source is to the right of the vertex, and $\mathbf{r}_v - \hat{\mathbf{s}}\ell$ would describe a point to the left, hence never passing through the source. We can fix that problem either by regarding $-\hat{\mathbf{s}}$ as the ray direction or by changing $\mathbf{r}_v - \hat{\mathbf{s}}\ell$ to $\mathbf{r}_v + \hat{\mathbf{s}}\ell$ in (17.46).

gular integral of line integrals through the object. This cone must not, however, be confused with the cone in cone-beam tomography. Even an ideal point detector *integrates* the radiance over a cone of rays defined by the pinhole, but it is assumed in cone-beam tomography that we can *measure separately* the radiance associated with each ray. The only way to reconcile these views is to assume that the pinhole diameter goes to zero, so that each point on an ideal detector is uniquely associated with a single ray.

Relation between the 3D Radon and x-ray transforms Most approaches to reconstruction from cone-beam data start with a mathematical relation between the 3D Radon and x-ray transforms. That there should be such a relation is perhaps not surprising since we can integrate a single 2D cone-beam projection over a stripe in data space as shown in Fig. 17.5, and when we do, we see that the integral is sensitive to source points on a slab, approximating a plane through the object.

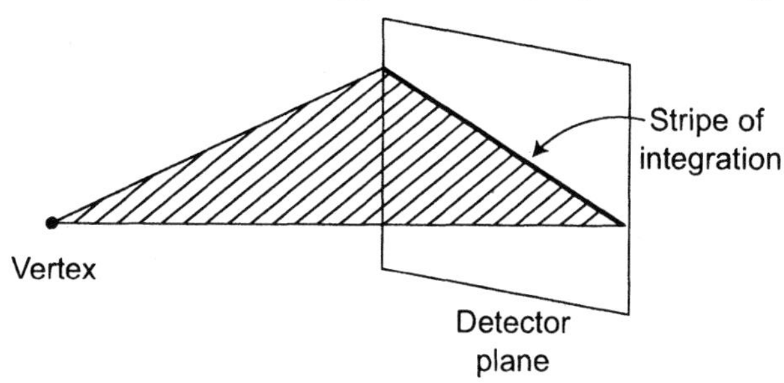

Fig. 17.5 Integration of a single 2D cone-beam projection over a stripe in data space.

One might think that we have, in fact, estimated one sample of the 3D Radon transform in this way, just as we did when we integrated over a stripe in a parallel-beam projection. The key difference, however, is that there is a Jacobian of the form $1/|\mathbf{r}-\mathbf{r}'|^2$ in the integral over the plane in the cone-beam case [see (17.47)], but no such factor with parallel projections. We encountered a similar Jacobian in (16.63) when discussing another form of cone-beam imaging, transmission radiography, and we noted there that we can get an unweighted integral over the object only if the vertex is far away and/or the object is small, so that the cone-beam rays through the object are approximately parallel. Since the vertex is usually placed close to the object in SPECT in order to collect many photons, this is seldom a useful approximation.

Nevertheless, it is possible to recover the 3D Radon transform of a function from knowledge of an appropriate set of values of the x-ray transform. In fact, many different formulas for making this conversion appear in the literature; for an excellent survey, see Natterer and Wübbeling (2001). As this reference emphasizes, all of these formulas can be derived from a single formula originally obtained by Hamaker *et al.*, (1980). In our notation, Hamaker's result can be written as

$$\int_{4\pi} d\Omega_{\hat{s}}\, [\boldsymbol{\mathcal{X}}\mathbf{f}](\mathbf{r},\hat{\mathbf{s}})\, h(\hat{\mathbf{s}}\cdot\hat{\mathbf{n}}) = \int_{-\infty}^{\infty} dp\, [\boldsymbol{\mathcal{R}}_3\mathbf{f}](p,\hat{\mathbf{n}})\, h(p-\mathbf{r}\cdot\hat{\mathbf{n}})\,, \tag{17.48}$$

where $h(\,\cdot\,)$ is any function of one variable that is homogeneous of degree -2, *i.e.*,

$$h(\alpha x) = \frac{1}{\alpha^2}\, h(x)\,. \tag{17.49}$$

To derive (17.48), we substitute (17.46) into the left-hand side and use our favorite change of variables, $\mathbf{r}' = \mathbf{r} - \hat{\mathbf{s}}\ell$, yielding

$$\int_{4\pi} d\Omega_{\hat{s}}\, [\mathcal{X}\mathbf{f}](\mathbf{r},\hat{\mathbf{s}})\, h(\hat{\mathbf{s}}\cdot\hat{\mathbf{n}}) = \int_{\infty} d^3r'\, f(\mathbf{r}')\, \frac{1}{|\mathbf{r}-\mathbf{r}'|^2}\, h\left(\frac{\mathbf{r}-\mathbf{r}'}{|\mathbf{r}-\mathbf{r}'|}\cdot\hat{\mathbf{n}}\right), \tag{17.50}$$

which, with the use of (17.45) and (17.49), gives the right-hand side of (17.48).

A derivative formula An important special case of Hamaker's formula is where $h(x)$ is the derivative of a delta function, $h(x) = \delta'(x)$. This function satisfies the homogeneity condition (17.49) for $\alpha > 0$, which is all we need. With this choice and (2.57), (17.48) becomes

$$\begin{aligned}\int_{4\pi} d\Omega_{\hat{s}}\, [\mathcal{X}\mathbf{f}](\mathbf{r},\hat{\mathbf{s}})\, \delta'(\hat{\mathbf{s}}\cdot\hat{\mathbf{n}}) &= \int_{-\infty}^{\infty} dp\, [\mathcal{R}_3\mathbf{f}](p,\hat{\mathbf{n}})\, \delta'(p-\mathbf{r}\cdot\hat{\mathbf{n}}) \\ &= -\left\{\frac{\partial}{\partial p}[\mathcal{R}_3\mathbf{f}](p,\hat{\mathbf{n}})\right\}_{p=\mathbf{r}\cdot\hat{\mathbf{n}}}.\end{aligned} \tag{17.51}$$

Geometrically, all three forms in (17.51) involve the neighborhood of the same plane, namely the plane $p = \mathbf{r}\cdot\hat{\mathbf{n}}$. Since $[\mathcal{X}\mathbf{f}](\mathbf{r},\hat{\mathbf{s}})$ is measured only for $\mathbf{r} = \mathbf{r}_v$, where $\mathbf{r}_v$ is one of the vertex positions, we must have $p = \mathbf{r}_v\cdot\hat{\mathbf{n}}$, so the equation of the plane of interest can also be written as $(\mathbf{r}-\mathbf{r}_v)\cdot\hat{\mathbf{n}} = 0$. The derivative on the right-hand side in (17.51) can be computed just from values of $\mathcal{R}_3\mathbf{f}$ arbitrarily close to this plane. The delta derivative on the left-hand side then requires that $\hat{\mathbf{s}}\cdot\hat{\mathbf{n}} = 0$, which means that the only rays of interest are ones that travel in the plane from $\mathbf{r}_v$ toward the detector.

The derivative of a delta function in the middle form of (17.51) is essentially the basis function for the dipole-sheet transform, introduced in Sec. 4.4.5.

17.1.6 Attenuation

Our discussion of ECT so far has consistently neglected attenuation of the radiation in the object, but in practice attenuation can lead to serious image degradation. Ways of compensating for attenuation during reconstruction are discussed in Sec. 17.2.4, but here we collect some mathematical properties of the forward transforms.

Attenuated x-ray transform The attenuated x-ray transform was originally defined in (10.151), but we write it slightly differently here. We drop the factor of $1/c_m$, which was handy in the radiometry discussion but superfluous here, and we drop the subscript on the total attenuation coefficient μ_{tot}. Thus we define

$$[\mathcal{X}_\mu\mathbf{f}]\,(\mathbf{r},\hat{\mathbf{s}}) = \int_0^{\infty} d\ell\; f(\mathbf{r}-\hat{\mathbf{s}}\ell)\exp\left[-\int_0^{\ell} d\ell'\,\mu(\mathbf{r}-\hat{\mathbf{s}}\ell')\right]. \tag{17.52}$$

With the same manipulation as in (17.47), we can also write

$$[\mathcal{X}_\mu\mathbf{f}]\,(\mathbf{r},\hat{\mathbf{s}}) = \int_{\infty} d^3r'\, f(\mathbf{r}')\,\frac{\delta\left(\hat{\mathbf{s}}-\frac{\mathbf{r}-\mathbf{r}'}{|\mathbf{r}-\mathbf{r}'|}\right)}{|\mathbf{r}-\mathbf{r}'|^2}\exp\left[-\int_0^{|\mathbf{r}-\mathbf{r}'|} d\ell'\,\mu(\mathbf{r}-\hat{\mathbf{s}}\ell')\right]. \tag{17.53}$$

Thus, if we let $\mathbf{r}$ be the vertex position $\mathbf{r}_v$, radiation originating at point $\mathbf{r}'$ and travelling toward the vertex is attenuated by a factor determined by the line integral of μ along the line from $\mathbf{r}'$ to $\mathbf{r}_v$.

Attenuated 2D Radon transform Since the 2D Radon transform is really the same thing as the x-ray transform with rays confined to a plane, we can regard (17.52) as also defining the 2D attenuated Radon transform if we simply treat $\mathbf{r}$ and $\hat{\mathbf{s}}$ as 2D vectors. To make the transform look like (17.36), however, we write

$$\lambda_\mu(p,\phi) = [\mathcal{R}_{2,\mu}\mathbf{f}](p,\phi) = \int_\infty d^2r\ f(\mathbf{r})\,\delta(p - \mathbf{r}\cdot\hat{\mathbf{n}})\exp\left[-\int_0^\infty d\ell'\,\mu(\mathbf{r} + \hat{\mathbf{n}}_\perp \ell')\right], \tag{17.54}$$

where $\hat{\mathbf{n}}_\perp$ is normal to $\hat{\mathbf{n}}$ and in the 2D plane where $f(\mathbf{r})$ is defined (see Fig. 17.6). Since the delta function selects out a line perpendicular to $\hat{\mathbf{n}}$, it follows that $\hat{\mathbf{n}}_\perp$ is the ray direction. Specifically, if $\hat{\mathbf{n}}$ makes an angle ϕ with the y axis, then $\hat{\mathbf{n}}_\perp$ makes an angle ϕ with the z axis in the 3D coordinate system where the axis of rotation is the x axis. Now the line integral runs from the source point $\mathbf{r}$ to infinity, but that really means to the boundary of the attenuating medium since $\mu(\mathbf{r})$ is zero beyond that point.

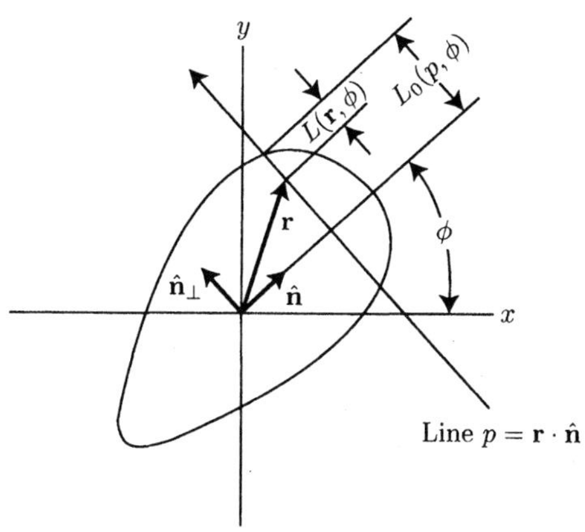

Fig. 17.6 Geometry used in discussing the attenuated 2D Radon transform.

Exponential 2D Radon transform Sometimes it is useful to assume that the attenuation coefficient is constant within the object of interest. For example, in SPECT imaging of the abdomen, the attenuating material is mostly soft tissue. Differences in attenuation among different soft tissues are small, so little error is made by replacing the actual attenuation distribution $\mu(\mathbf{r})$ with a constant μ. Then the attenuation factor in (17.54) becomes

$$\exp\left[-\int_0^\infty d\ell'\,\mu(\mathbf{r} + \hat{\mathbf{n}}_\perp \ell')\right] \approx \exp\left[-\mu L(\mathbf{r},\phi)\right], \tag{17.55}$$

where $L(\mathbf{r},\phi)$ is the total length of attenuating medium between point $\mathbf{r}$ and the detector in direction $\hat{\mathbf{n}}_\perp$. If we assume that the boundary of the attenuating medium is convex, we can decompose this length further as (see Fig. 17.6)

$$L(\mathbf{r},\phi) = L_0(p,\phi) - \mathbf{r}\cdot\hat{\mathbf{n}}_\perp, \tag{17.56}$$

where $L_0(p,\phi)$ is the distance to the boundary of the medium in direction $\hat{\mathbf{n}}_\perp$ from a point on the line running through the center of rotation. Since $L_0(p,\phi)$ is independent of $\mathbf{r}$, it can be taken out of the integral, and the attenuated 2D Radon transform becomes

$$\lambda_\mu(p,\phi) = \exp[-\mu L_0(p,\phi)] \int_\infty d^2r \; f(\mathbf{r})\,\delta(p - \mathbf{r}\cdot\hat{\mathbf{n}})\exp(\mu\mathbf{r}\cdot\hat{\mathbf{n}}_\perp)\,. \tag{17.57}$$

The first exponential factor is known if the object contour is specified, so we can move it to the other side of the equation and define a modified projection,

$$g_\mu(p,\phi) \equiv \exp[\mu L_0(p,\phi)]\,\lambda_\mu(p,\phi) = \int_\infty d^2r \; f(\mathbf{r})\,\delta(p - \mathbf{r}\cdot\hat{\mathbf{n}})\exp(\mu\mathbf{r}\cdot\hat{\mathbf{n}}_\perp)\,. \tag{17.58}$$

The right-hand side defines the *2D exponential Radon transform* $\mathcal{R}_{2e,\mu}$. The reader is cautioned that this transform is often seen with a minus sign in the exponent, but that comes about when $\hat{\mathbf{n}}_\perp$ is taken to be antiparallel to the ray direction instead of parallel.

Why not a 3D attenuated Radon? Since the 3D Radon transform involves integrals of the activity distribution $f(\mathfrak{r})$ over planes and the attenuation factors involve integrals of the attenuation distribution $\mu(\mathfrak{r})$ along lines, there is no unique way to define an attenuated 3D Radon transform. We noted in Sec. 17.1.5 that we could get 3D Radon data by starting with line-integral projections taken with a parallel-hole collimator and then integrating over a stripe in the 2D data space; in this case the attenuation factors would be determined by lines of sight through the collimator bores. Many different combinations of projection angles and stripes of integration would, however, integrate the activity over the same plane, but with different attenuation factors. For example, if we want to know the integral of $f(\mathfrak{r})$ over the plane $x = 0$, we can get it by integrating the photon irradiance $I_p(x_d, y_d)$ over $-\frac{1}{2} < x_d \le \frac{1}{2}$ and over all y_d in the detector plane [*cf.* (17.43)]. This procedure gives the same answer for all projection angles ϕ in the absence of attenuation, but the attenuation factors are different for different ϕ.

Similarly, if we consider other detection geometries such as the slat collimator of Fig. 17.4, or if we derive the 3D Radon data by mathematical manipulation of cone-beam data as suggested by (17.51), still other attenuation factors arise. In short, as soon as we combine planar integrals of the activity with line integrals of the attenuation, then many different versions of the attenuated transform can occur; very few of them have been explored in the literature.

17.2 INVERSE PROBLEMS

Having catalogued the various formulations of the forward problem in SPECT, we can now discuss inverse problems. The discussion will range from purely theoretical inversion formulas based on idealized models of the forward problem, through practical issues of how the idealized expressions are discretized for computation, to comprehensive iterative algorithms that make essentially no assumptions about the data-acquisition process.

We begin in Secs. 17.2.1 – 17.2.3 by discussing the SVD of some specific CC forward operators. The operator formalism from Sec. 17.1.1 and the group-theoretic considerations from Sec. 17.1.2 recur in this discussion. In particular, Sec. 17.2.1 derives SVDs for the 2D Radon operator, first with infinite object support and continuous angular sampling, then with finite support and a finite number of projection angles. In Sec. 17.2.2, the SVD results are used to derive analytic inverses of the full 2D Radon transform and pseudoinverses in the case of finite angular sampling. In Sec. 17.2.3, we discuss inversion formulas for the 3D x-ray transform, and in Sec. 17.2.4 we consider transforms with attenuation factors.

In Secs. 17.2.5 and 17.2.6, we turn, at last, to practical reconstruction algorithms that can be applied to discrete data. Linear algorithms obtained by discretizing analytic inverses are discussed in Sec. 17.2.5, and ways of forming system matrices for iterative nonlinear algorithms are discussed in Sec. 17.2.6. The algorithms themselves have previously been treated in Sec. 15.4.

17.2.1 SVD of the 2D Radon transform

Study of the properties of tomographic transforms is sometimes referred to as *integral geometry*. The ultimate goal of research in this field is the full singular-value decomposition of each transform, from which one can discuss inversion formulas, null functions, data-consistency conditions and ill-posedness. This goal represents a rich source of papers since there are many different transforms, many specific geometries for each transform (*e.g.*, many trajectories in cone-beam tomography) and many ways to define scalar products and hence Hilbert spaces.

In this section we look in some detail at SVD of the 2D Radon transform and its close relatives; in the following section we shall use these SVDs to discuss inverses and pseudoinverses in various 2D problems.

Radon transform on the infinite plane With what we have learned in previous chapters, it is straightforward to give a singular-value decomposition of the 2D Radon transform, so long as we do not impose any constraints on the object support. That is, we use the Hilbert spaces defined as in Sec. 4.4.1, where the object is square-integrable over the infinite plane[7] and the scalar product in data space is defined by the measure $dp\,d\phi$; neither space involves any weighting function. We shall discuss this SVD problem first and then address finite object support, finite angular sampling and weighted Hilbert spaces.

With infinite support and no weights, we know from (4.167) that the operator $\mathcal{R}_2^\dagger \mathcal{R}_2$ is a 2D convolution:

$$\left[\mathcal{R}_2^\dagger \mathcal{R}_2 \mathbf{f}\right](\mathbf{r}) = \int_\infty d^2r' \, \frac{f(\mathbf{r}')}{|\mathbf{r}-\mathbf{r}'|} \, . \tag{17.59}$$

We know also from Sec. 7.2.4 that the eigenfunctions of convolution operators are complex exponentials and that the eigenvalues are given by the transfer function; see (7.140) and (7.141). For the 2D Radon transform, the relevant eigenvalue equation is

$$\mathcal{R}_2^\dagger \mathcal{R}_2 \mathbf{u}_\rho = \mu_\rho \mathbf{u}_\rho \, . \tag{17.60}$$

[7]Use of this Hilbert space does not imply that any actual object has infinite extent; objects with finite support and finite values are, of course, square-integrable over the plane, hence members of the space. The fact that some members of the space are not realizable objects should not concern us. *Any* Hilbert space we choose will contain functions that cannot be objects, if for no other reason than that objects in tomography are nonnegative.

From (7.140), the Hilbert-space eigenvector $\mathbf{u}_{\boldsymbol{\rho}}$ corresponds to the eigenfunction

$$u_{\boldsymbol{\rho}}(\mathbf{r}) = \exp(2\pi i \boldsymbol{\rho} \cdot \mathbf{r}) , \tag{17.61}$$

where $\boldsymbol{\rho}$ is a 2D spatial-frequency vector. The operator is not compact (see Sec. 1.3.3), so a continuous index $\boldsymbol{\rho}$ is needed, and the eigenfunctions span the space $\mathbb{L}_2(\mathbb{R}^2)$ but are not themselves square-integrable; see Sec. 1.4.4 for further discussion of these subtleties.

From (7.141) and (4.169), the continuous eigenvalue spectrum is given by

$$\mu_{\boldsymbol{\rho}} = \mathcal{F}_2 \left\{ \frac{1}{r} \right\} = \frac{1}{|\rho|} , \tag{17.62}$$

where the absolute-value signs are needed if we allow ρ to be signed, so that $\rho = \pm|\boldsymbol{\rho}|$. Eigenfunctions of the Hermitian operator $\mathcal{R}_2^\dagger \mathcal{R}_2$ are the same as the object-space singular functions of $\mathcal{R}_2$, but we still need to determine the image-space singular functions to complete the SVD.

To get the image-space singular functions corresponding to nonzero singular values, we can use (7.122), which in the present case is written

$$v_{\boldsymbol{\rho}}(p, \phi) = \frac{1}{\sqrt{\mu_{\boldsymbol{\rho}}}} \left[\mathcal{R}_2 \mathbf{u}_{\boldsymbol{\rho}} \right] (p, \phi) . \tag{17.63}$$

The reader with some degree of virtuosity in delta functions can show that

$$v_{\boldsymbol{\rho}}(p, \phi) = \sqrt{|\rho|} \exp(2\pi i p \rho) \, \delta(\boldsymbol{\rho} \cdot \hat{\mathbf{n}}_\perp) = \frac{1}{\sqrt{|\rho|}} \exp(2\pi i p \rho) \, \delta(\hat{\boldsymbol{\rho}} \cdot \hat{\mathbf{n}}_\perp) , \tag{17.64}$$

where $\hat{\boldsymbol{\rho}}$ is the unit vector parallel to $\boldsymbol{\rho}$. A similar exercise shows that both sets of singular functions are orthonormal, in the sense that

$$\mathbf{u}_{\boldsymbol{\rho}}^\dagger \mathbf{u}_{\boldsymbol{\rho}'} = \delta(\boldsymbol{\rho} - \boldsymbol{\rho}') , \qquad \mathbf{v}_{\boldsymbol{\rho}}^\dagger \mathbf{v}_{\boldsymbol{\rho}'} = \delta(\boldsymbol{\rho} - \boldsymbol{\rho}') , \tag{17.65}$$

and that both sets of functions are complete,[8] so that

$$\int_\infty d^2\rho \; u_{\boldsymbol{\rho}}(\mathbf{r}) \, u_{\boldsymbol{\rho}}^*(\mathbf{r}') = \delta(\mathbf{r} - \mathbf{r}') , \qquad \int_\infty d^2\rho \; v_{\boldsymbol{\rho}}(p, \phi) \, v_{\boldsymbol{\rho}}^*(p', \phi') = \delta(p - p') \, \delta(\phi - \phi') . \tag{17.66}$$

Another interesting exercise is to show that (17.64) – (17.66) are dimensionally correct.

Note that the delta function in (17.64) is 1D, since its argument is a scalar product. For fixed $\boldsymbol{\rho}$, this delta function shows that the image-space singular function is zero unless $\boldsymbol{\rho}$ is normal to $\hat{\mathbf{n}}_\perp$, which means that the line of integration is parallel to the crests of the function $\exp(2\pi i \boldsymbol{\rho} \cdot \mathbf{r})$. If we integrate in any other direction, the line of integration will cut across positive and negative portions of the function, giving a zero integral.

Since both sets of singular functions are complete, neither $\mathcal{R}_2$ nor $\mathcal{R}_2^\dagger$ has null functions for the present choice of Hilbert spaces. We shall soon see, however, that $\mathcal{R}_2$ does have null functions if we consider a finite set of angles, and $\mathcal{R}_2^\dagger$ has null functions if we impose a finite object support.

[8] Completeness of the image-space functions requires that ρ be allowed to take on positive and negative values.

Finite set of angles Suppose we acquire J 2D projections at the angles $\{\phi_j, j = 0, ..., J-1\}$ (not necessarily equally spaced), so that we have a mixed CC/CD problem: the projection data are functions of the continuous p variable and the discrete angular index j. Then we can define a new Radon operator $\boldsymbol{\mathcal{R}}_{2J}$ by

$$[\boldsymbol{\mathcal{R}}_{2J}\mathbf{f}](p,\phi_j) = \int_\infty d^2r\; f(\mathbf{r})\,\delta(p - \mathbf{r}\cdot\hat{\mathbf{n}}_j) \qquad j = 0, ..., J-1\,, \tag{17.67}$$

where $\hat{\mathbf{n}}_j$ makes angle ϕ_j with the x axis.

The projection-backprojection operator is given by

$$\left[\boldsymbol{\mathcal{R}}_{2J}^\dagger\boldsymbol{\mathcal{R}}_{2J}\mathbf{f}\right](\mathbf{r}) = \sum_{j=0}^{J-1}\int_{-\infty}^{\infty} dp \int_\infty d^2r'\, f(\mathbf{r}')\,\delta(p-\mathbf{r}\cdot\hat{\mathbf{n}}_j)\,\delta(p-\mathbf{r}'\cdot\hat{\mathbf{n}}_j)\,. \tag{17.68}$$

Using one of the delta functions to perform the p integral, we obtain

$$\left[\boldsymbol{\mathcal{R}}_{2J}^\dagger\boldsymbol{\mathcal{R}}_{2J}\mathbf{f}\right](\mathbf{r}) = \int_\infty d^2r'\, f(\mathbf{r}')\left\{\sum_{j=0}^{J-1}\delta[(\mathbf{r}-\mathbf{r}')\cdot\hat{\mathbf{n}}_j]\right\}. \tag{17.69}$$

The key point here is that the kernel (the quantity in large curly brackets) is a function of $\mathbf{r}-\mathbf{r}'$, so this operator is still a convolution. The PSF is a radial spoke pattern that limits to a $1/r$ function as $J \to \infty$.

Since $\boldsymbol{\mathcal{R}}_{2J}^\dagger\boldsymbol{\mathcal{R}}_{2J}$ is a convolution, its eigenfunctions are still given by the complex exponentials (17.61), and its eigenvalues are given by the Fourier transform of the PSF as

$$\mu_{\boldsymbol{\rho}} = \boldsymbol{\mathcal{F}}_2\left\{\sum_{j=0}^{J-1}\delta(\mathbf{r}\cdot\hat{\mathbf{n}}_j)\right\} = \sum_{j=0}^{J-1}\int_\infty d^2r\,\exp(-2\pi i\boldsymbol{\rho}\cdot\mathbf{r})\,\delta(\mathbf{r}\cdot\hat{\mathbf{n}}_j)\,, \tag{17.70}$$

where the delta function is 1D since its argument is a scalar. The integral can be evaluated in a Cartesian coordinate system where $\mathbf{r}\cdot\hat{\mathbf{n}}_j = x$, and the result is

$$\mu_{\boldsymbol{\rho}} = \sum_{j=0}^{J-1}\delta(\boldsymbol{\rho}\cdot\hat{\mathbf{n}}_{\perp j}) = \frac{1}{\rho}\sum_{j=0}^{J-1}\delta[\sin(\theta_\rho - \phi_j)]\,, \tag{17.71}$$

where $\hat{\mathbf{n}}_{\perp j}$ is the unit vector normal to $\hat{\mathbf{n}}_j$. The delta function vanishes unless $\boldsymbol{\rho}$ is perpendicular to one of the directions $\hat{\mathbf{n}}_{\perp j}$, hence parallel to one of the $\hat{\mathbf{n}}_j$. As we would expect from the central-slice theorem (4.150), the eigenvalues $\mu_{\boldsymbol{\rho}}$ are zero except for values of $\boldsymbol{\rho}$ along a set of radial spokes in the 2D frequency plane. Frequencies not on these spokes correspond to null functions of $\boldsymbol{\mathcal{R}}_{2J}$.

As a check on this formalism, we can compute the spectral representation of the projection-backprojection operator, which is given abstractly by

$$\boldsymbol{\mathcal{R}}_{2J}^\dagger\boldsymbol{\mathcal{R}}_{2J} = \int_\infty d^2\rho\;\mu_{\boldsymbol{\rho}}\mathbf{u}_{\boldsymbol{\rho}}\mathbf{u}_{\boldsymbol{\rho}}^\dagger = \sum_{j=0}^{J-1}\int_\infty d^2\rho\;\delta(\boldsymbol{\rho}\cdot\hat{\mathbf{n}}_{\perp j})\,\mathbf{u}_{\boldsymbol{\rho}}\mathbf{u}_{\boldsymbol{\rho}}^\dagger\,. \tag{17.72}$$

Explicitly,

$$\left[\boldsymbol{\mathcal{R}}_{2J}^\dagger\boldsymbol{\mathcal{R}}_{2J}\mathbf{f}\right](\mathbf{r}) = \sum_{j=0}^{J-1}\int_\infty d^2\rho\;\delta(\boldsymbol{\rho}\cdot\hat{\mathbf{n}}_{\perp j})\exp(2\pi i\mathbf{r}\cdot\boldsymbol{\rho})\int_\infty d^2r'\exp(-2\pi i\mathbf{r}'\cdot\boldsymbol{\rho})\,f(\mathbf{r}')\,. \tag{17.73}$$

The integral over $\boldsymbol{\rho}$ can be performed in Cartesian coordinates with axes defined by $\hat{\mathbf{n}}_j$ and $\hat{\mathbf{n}}_{\perp j}$, and we obtain

$$\left[\mathcal{R}_{2J}^{\dagger}\mathcal{R}_{2J}\mathbf{f}\right](\mathbf{r}) = \sum_{j=0}^{J-1}\int_{\infty} d^2r'\,\delta[(\mathbf{r}-\mathbf{r}')\cdot\hat{\mathbf{n}}_j]\,f(\mathbf{r}')\,, \tag{17.74}$$

in accord with (17.69).

To complete the SVD of $\mathcal{R}_{2J}$, we need the image-space eigenfunctions. These functions are given by (17.63), but there is a complication since $\mu_{\boldsymbol{\rho}}$ involves a delta function, and we don't know how to take the square root of a delta function. We might solve this problem with limiting representations, but we can avoid it altogether by going directly to the SVD representation of $\mathcal{R}_{2J}$. From (1.120) (as extended to continuous indices) and (17.63),

$$\mathcal{R}_{2J} = \int_{\infty} d^2\rho\,\sqrt{\mu_{\boldsymbol{\rho}}}\,\mathbf{v}_{\boldsymbol{\rho}}\mathbf{u}_{\boldsymbol{\rho}}^{\dagger} = \int_{\infty} d^2\rho\,\sqrt{\mu_{\boldsymbol{\rho}}}\left[\frac{1}{\sqrt{\mu_{\boldsymbol{\rho}}}}\mathcal{R}_{2J}\mathbf{u}_{\boldsymbol{\rho}}\right]\mathbf{u}_{\boldsymbol{\rho}}^{\dagger} = \int_{\infty} d^2\rho\,[\mathcal{R}_{2J}\mathbf{u}_{\boldsymbol{\rho}}]\,\mathbf{u}_{\boldsymbol{\rho}}^{\dagger}\,. \tag{17.75}$$

Now the awkward square-root factor has cancelled,[9] and we need only compute $\mathcal{R}_{2J}\mathbf{u}_{\boldsymbol{\rho}}$, which is given by

$$[\mathcal{R}_{2J}\mathbf{u}_{\boldsymbol{\rho}}](p,\phi_j) = \int_{\infty} d^2r\,\delta(p-\mathbf{r}\cdot\hat{\mathbf{n}}_j)\exp(2\pi i\boldsymbol{\rho}\cdot\mathbf{r}) = \exp(2\pi i\rho p)\,\delta(\boldsymbol{\rho}\cdot\hat{\mathbf{n}}_{j\perp})\,. \tag{17.76}$$

The reader should verify that (17.75) gives the right answer when applied to an arbitrary $f(\mathbf{r})$.

Finite support and finite angular sampling Next we consider the case of finite object support and J equally spaced angles around 2π ($\mathbf{C}_J$ symmetry). We laid much of the groundwork for this problem in Secs. 17.1.1 and 17.1.2, and we now apply that treatment to the 2D Radon transform.

Specifically, the operator $\mathcal{H}$ used in those sections is now to be identified with $\mathcal{R}_{2J}$, and the single-view projection operator $\mathcal{H}_0$ is defined by

$$\lambda_0(p) = [\mathcal{H}_0\mathbf{f}](p) = \int_{\mathbf{S}_f} d^2r\,f(\mathbf{r})\,\delta(p-\mathbf{r}\cdot\hat{\mathbf{n}}_0)\,, \tag{17.77}$$

where $\hat{\mathbf{n}}_0$ is parallel to the x axis if $\mathbf{r}$ is chosen to lie in the x-y plane.

As in Sec. 17.1.2, we suppose that we have solved the eigenvalue problem for $\mathcal{H}_0$, so we know the solutions to

$$\mathcal{H}_0^{\dagger}\mathcal{H}_0\mathbf{u}_n = \mu_n\mathbf{u}_n\,. \tag{17.78}$$

We would like to use these single-view eigenfunctions to solve the eigenvalue problem for $\mathcal{H}^{\dagger}\mathcal{H}$, which is now the same as $\mathcal{R}_{2J}^{\dagger}\mathcal{R}_{2J}$. The solution is given by (17.23) and (17.24), but these results were based on the approximation in (17.22) that $\mathbf{u}_{n'}^{\dagger}\mathcal{T}_{\ell}\mathbf{u}_n \approx 0$ if $n \neq n'$. We shall now show how this approximation can be made exact in the Radon problem.

[9] Equivalently, we could have derived the final form of (17.75) immediately by writing $\mathcal{R}_{2J} = \mathcal{R}_{2J}\mathbf{I}$ and using the completeness of the object-space eigenfunctions to express the unit operator $\mathbf{I}$.

Weighted Hilbert spaces Weighted Hilbert spaces, which we first mentioned in a CD context in Sec. 17.1.1, are a great help in finding an SVD of $\mathcal{R}_{2J}$. If the object support $\mathbf{S}_f$ is the disk of radius R, a general weighted scalar product is defined similarly to (17.11) as

$$(\mathbf{f}_1, \mathbf{f}_2)_{\mathbb{U}} = \int_{\mathbf{S}_f} d^2r\; W(r)\, f_1^*(\mathbf{r})\, f_2(\mathbf{r})\,, \tag{17.79}$$

where $W(r) > 0$ for all $r < R$.

For a finite set of angles $\{\phi_j, j = 0, ..., J-1\}$, the scalar product in 2D Radon space can be defined by

$$(\boldsymbol{\lambda}_1, \boldsymbol{\lambda}_2)_{\mathbb{V}} = \sum_{j=0}^{J-1} \int_{-R}^{R} dp\; w(p)\, \lambda_1^*(p, \phi_j)\, \lambda_2(p, \phi_j)\,. \tag{17.80}$$

(Note that we have maintained the rotational symmetry by making the weighting functions independent of angle.) With these scalar products, the single-view adjoint operator is given by [*cf.* (17.13)]

$$\left[\mathcal{H}_0^\dagger \boldsymbol{\lambda}_0\right](\mathbf{r}) = \frac{1}{W(\mathbf{r})} \int_{-R}^{R} dp\; w(p)\, \lambda_0(p)\, \delta(p - \mathbf{r}\cdot\hat{\mathbf{n}}_0) = \frac{w(\mathbf{r}\cdot\hat{\mathbf{n}}_0)}{W(\mathbf{r})} \lambda_0(\mathbf{r}\cdot\hat{\mathbf{n}}_0)\,. \tag{17.81}$$

Davison and Grunbaum (1981) suggest that the weights be chosen such that

$$\frac{1}{w(p)} = \int_{\mathbf{S}_f} d^2r\; \frac{1}{W(r)}\, \delta(p - \mathbf{r}\cdot\hat{\mathbf{n}}) \tag{17.82}$$

for all $\hat{\mathbf{n}}$. The beauty of this condition is that it turns the single-view operator $\mathcal{H}_0\mathcal{H}_0^\dagger$ into the unit operator, as we can see by calculating the kernel:

$$\left[\mathcal{H}_0\mathcal{H}_0^\dagger\right](p, p') = \int_{\mathbf{S}_f} d^2r\; \frac{w(p')}{W(r)}\, \delta(p - \mathbf{r}\cdot\hat{\mathbf{n}}_0)\, \delta(p' - \mathbf{r}\cdot\hat{\mathbf{n}}_0)$$

$$= \delta(p - p')w(p') \int_{\mathbf{S}_f} d^2r\; \frac{1}{W(r)}\, \delta(p - \mathbf{r}\cdot\hat{\mathbf{n}}_0) = \delta(p - p')\,. \tag{17.83}$$

Thus *any* function of p is an eigenfunction of $\mathcal{H}_0\mathcal{H}_0^\dagger$, and if we want the full set of single-view image-space eigenfunctions, we need only pick a complete set of functions that are orthonormal on $-R < p < R$ with respect to the weight $w(p)$, *i.e.*,

$$\int_{-R}^{R} dp\; w(p)\, v_n^*(p)\, v_{n'}(p) = \delta_{nn'}\,. \tag{17.84}$$

An important example, treated by Logan and Shepp (1975) and Hamaker and Solmon (1978), is $W(r) = 1$ over the support disk. In this case, $w(p) = \left[2\sqrt{R^2 - p^2}\,\right]^{-1}$, a weighting function that we encountered in connection with Chebyshev polynomials in Sec. 4.1.4 [see (4.60) and (4.61)].

A commuting family Weights satisfying (17.82) have a second salutary effect; in addition to converting $\mathcal{H}_0\mathcal{H}_0^\dagger$ to the unit operator, they simplify the entire family of operators of the form $\mathcal{H}_0\mathcal{T}_\ell\mathcal{H}_0^\dagger$. With these weights, Davison and Grunbaum have shown that each member of this family commutes with every other member. As we shall now show, this is precisely what we need in order to make the approximate SVD expressions from Sec. 17.1.2 exact.

If every member of the family commutes with every other member, then it is possible to find a set of vectors $\mathbf{v}_n$ in data space that are simultaneously eigenvectors for each member of the family, including $\mathcal{H}_0\mathcal{H}_0^\dagger$. The eigenvalue equation is

$$\mathcal{H}_0\mathcal{T}_\ell\mathcal{H}_0^\dagger\mathbf{v}_n = \mu_n^{(\ell)}\mathbf{v}_n \,. \tag{17.85}$$

Note that $\mu_n^{(0)}$ is what we called μ_n above, and note also that there is no index ℓ on $\mathbf{v}_n$. We shall verify (17.85) for one choice of weights below, but for now we simply assume it is true.

From Sec. 1.5.1 we know that the eigenvectors of $\mathcal{H}_0^\dagger\mathcal{H}_0$ with nonzero singular values are given by

$$\mathbf{u}_n = \frac{1}{\sqrt{\mu_n^{(0)}}}\mathcal{H}_0^\dagger\mathbf{v}_n \,, \tag{17.86}$$

but $\mu_n^{(0)} = 1$ here since $\mathcal{H}_0\mathcal{H}_0^\dagger$ is the unit operator in the single-view data space. Thus

$$\mathbf{u}_n^\dagger\mathcal{T}_\ell\mathbf{u}_{n'} = \left[\mathcal{H}_0^\dagger\mathbf{v}_n\right]^\dagger \mathcal{T}_\ell\mathcal{H}_0^\dagger\mathbf{v}_{n'} \,. \tag{17.87}$$

By an elementary property of the adjoint (see Sec. 1.3.5) and (17.85), we have

$$\mathbf{u}_n^\dagger\mathcal{T}_\ell\mathbf{u}_{n'} = \mathbf{v}_n^\dagger\mathcal{H}_0\mathcal{T}_\ell\mathcal{H}_0^\dagger\mathbf{v}_{n'} = \mu_n^{(\ell)}\mathbf{v}_n^\dagger\mathbf{v}_{n'} = \mu_n^{(\ell)}\,\delta_{nn'} \,, \tag{17.88}$$

which is just the condition we needed in Sec. 17.1.2 to derive (17.23) and (17.24).

Computation of eigenfunctions Next we need to find the eigenfunctions that satisfy (17.85). For simplicity, we take $W(\mathbf{r}) = 1$ on the object support, so that $w(p) = \left[2\sqrt{R^2-p^2}\,\right]^{-1}$. With this choice of weights, one might think we should take $v_n(p)$ to be a Chebyshev polynomial of the first kind [see (4.60)], but there is a less-obvious choice that is needed if we want to satisfy (17.85) for all ℓ. Following Davison and Grunbaum, we take

$$v_n(p) = \frac{2}{R\sqrt{\pi}}\sqrt{R^2-p^2}\;U_n\!\left(\frac{p}{R}\right) , \tag{17.89}$$

where $U_n(p/R)$ is the Chebyshev polynomial of the *second* kind. With (4.61) and a change of variables, we can see that these functions are orthonormal with respect to the weight $\left[2\sqrt{R^2-p^2}\,\right]^{-1}$ on $(-R,R)$; in essence, we have artificially introduced a factor of $\sqrt{R^2-p^2}$ in each of the functions, and one of these factors cancels the weight, leaving an orthogonality integral with weight $\sqrt{R^2-p^2}$ rather than its reciprocal.

To verify that these eigenfunctions satisfy (17.85), we shall apply each of the three operators in that expression in turn. The first of these operators is $\mathcal{H}_0^\dagger$, which is defined by (17.81). Since $\mu_n^{(0)} = 1$ in (17.86), application of this operator gives the single-view, object-space eigenfunctions $u_n(\mathbf{r})$; thus

$$u_n(\mathbf{r}) = \left[\mathcal{H}_0^\dagger \mathbf{v}_n\right](\mathbf{r}) = \frac{1}{R\sqrt{\pi}}\, U_n\!\left(\frac{\mathbf{r}\cdot\hat{\mathbf{n}}_0}{R}\right). \tag{17.90}$$

The rotation operator $\mathcal{T}_\ell$ simply replaces $\hat{\mathbf{n}}_0$ with $\hat{\mathbf{n}}_\ell$, so the left-hand side of (17.85) becomes

$$\left[\mathcal{H}_0 \mathcal{T}_\ell \mathcal{H}_0^\dagger \mathbf{v}_n\right](p) = \frac{1}{R\sqrt{\pi}} \int_{\mathbf{S}} d^2r\ \delta(p - \mathbf{r}\cdot\hat{\mathbf{n}}_0)\, U_n\!\left(\frac{\mathbf{r}\cdot\hat{\mathbf{n}}_\ell}{R}\right). \tag{17.91}$$

To perform the integral, we take $\mathbf{r}$ in the x-y plane and choose the x axis parallel to $\hat{\mathbf{n}}_0$, so that $\mathbf{r}\cdot\hat{\mathbf{n}}_0 = x$ and $\mathbf{r}\cdot\hat{\mathbf{n}}_\ell = x\cos\phi_\ell + y\sin\phi_\ell$. The delta function takes care of one integral, and we see that

$$\left[\mathcal{H}_0 \mathcal{T}_\ell \mathcal{H}_0^\dagger \mathbf{v}_n\right](p) = \frac{1}{R\sqrt{\pi}} \int_{-\sqrt{R^2-p^2}}^{\sqrt{R^2-p^2}} dy\ U_n\!\left(\frac{p\cos\phi_\ell + y\sin\phi_\ell}{R}\right). \tag{17.92}$$

Next we define $\cos\theta_p \equiv p/R$ and

$$\alpha \equiv \frac{p\cos\phi_\ell + y\sin\phi_\ell}{R}, \tag{17.93}$$

so that

$$\left[\mathcal{H}_0 \mathcal{T}_\ell \mathcal{H}_0^\dagger \mathbf{v}_n\right](p) = \frac{1}{\sqrt{\pi}}\,\frac{1}{\sin\phi_\ell} \int_{\cos(\phi_\ell-\theta_p)}^{\cos(\phi_\ell+\theta_p)} d\alpha\ U_n(\alpha)\,. \tag{17.94}$$

With the definition of the Chebyshev polynomial from (4.63) and some trigonometry, we obtain

$$\left[\mathcal{H}_0 \mathcal{T}_\ell \mathcal{H}_0^\dagger \mathbf{v}_n\right](p) = \frac{\sin[(n+1)\phi_\ell]}{(n+1)\sin\phi_\ell}\, v_n(p)\,. \tag{17.95}$$

This result confirms (17.85) and shows that

$$\mu_n^{(\ell)} = \frac{\sin[(n+1)\phi_\ell]}{(n+1)\sin\phi_\ell}\,. \tag{17.96}$$

The limit $\phi_\ell \to 0$ shows that $\mu_n^{(0)} = 1$ as expected.

The eigenvalue spectrum Now we can put the pieces together and compute the eigenvalues and eigenfunctions needed in the full SVD of $\mathcal{R}_{2J}$ with circular support and the chosen weights.

The eigenvalue spectrum is found from (17.23), which contains the factor $\mathbf{u}_n^\dagger \mathcal{T}_\ell \mathbf{u}_n$. From (17.88), this factor is just $\mu_n^{(\ell)}$ in the present problem, so we have

$$\mu_{nk} = \sum_{\ell=0}^{J-1} \exp\left(\frac{2\pi i k\ell}{J}\right) \frac{\sin[2\pi(n+1)\,\ell/J]}{(n+1)\sin(2\pi\ell/J)}\,. \tag{17.97}$$

This sum can be performed numerically by an inverse DFT, but special attention must be given to the points where the denominator vanishes ($\ell = 0$ for any J and also $\ell = J/2$ for J even). The eigenvalue $\mu_n^{(\ell)}$ limits to 1 at these points.

Davison and Grunbaum (1981) point out that μ_{nk} can also be expressed as

$$\mu_{nk} = \frac{J}{n+1} \sum_{\ell=0}^{n} \delta[n-k, 2\ell]_J \,, \tag{17.98}$$

where the delta function is a modulo-J Kronecker function, *i.e.*, it equals one when the modulo-J difference of the two arguments is zero, and zero otherwise. This form can be verified by taking the DFT of both sides and summing a geometric series.

Several features of the spectrum can be seen without numerics. First, as noted in Sec. 17.1.2, all μ_{nk} must be real and nonnegative, though some can be zero. Next, if $k = 0$, then the Fourier kernel is always 1, and if $n = 0$, the ratio of sines is also 1, so μ_{00} is equal to the number of projections J.

We can also get an analytic result if n and k are nonzero but $J > n + k$. As ℓ ranges from n to 0 in (17.98), the difference of arguments $n - k - 2\ell$ ranges from $-(n + k)$ to $n - k$. If $J > n + k$, the negative values can never be brought back to zero by addition of a multiple of J, so the modulo-J condition never comes into play. We need only consider $n - k \geq 0$, and the Kronecker delta takes the value unity exactly once in the range of summation if $n - k$ is even, and never if $n - k$ is odd. Thus, for $J > n + k$,

$$\mu_{nk} = \begin{cases} \frac{J}{n+1} & \text{if} \quad n \geq k\,, \quad n-k \text{ even} \\ 0 & \text{if} \quad n \geq k\,, \quad n-k \text{ odd} \\ 0 & \text{if} \quad n < k \end{cases} . \tag{17.99}$$

This form is useful in regularized pseudoinverses where the sums over n and k are truncated so that $n + k < J$. It will also prove useful in Sec. 17.2.2 when we discuss the limit $J \to \infty$.

At the opposite extreme, if $n \gg J$ and J is odd, then only the $\ell = 0$ term contributes much to the sum [*cf.* (2.47)], and this term limits to 1 as $\frac{n}{J} \to \infty$. Thus $\mu_{nk} \to 1$ in this limit for all k. If J is even, then there is also a contribution from $\ell = J/2$, so $\mu_{nk} \to 2$ as $\frac{n}{J} \to \infty$. (The 2 comes in since the list of angles is redundant for J even.)

Eigenfunctions We get the multiple-view, object-space eigenfunctions from (17.24) and (17.90) as

$$u_{nk}(\mathbf{r}) = \frac{1}{R\sqrt{\pi}} \frac{N_{nk}}{J} \sum_{j=0}^{J-1} \exp\left(\frac{2\pi i k j}{J}\right) U_n\left(\frac{\mathbf{r}\cdot\hat{\mathbf{n}}_j}{R}\right) . \tag{17.100}$$

Calculation of the normalizing constant N_{nk} follows the same lines that led to (17.97); when $\mu_{nk} \neq 0$, the result is

$$N_{nk} = \sqrt{\frac{J}{\mu_{nk}}} . \tag{17.101}$$

Each multiple-view, image-space eigenvector $\mathbf{v}_{nk}$ comprises a set of projections of the corresponding object-space eigenvector $\mathbf{u}_{nk}$; the jth member of the set is given by

$$v_{nkj}(p) = \sqrt{\frac{1}{\mu_{nk}}}\left[\mathcal{H}_0\mathcal{T}_j^\dagger \mathbf{u}_{nk}\right](p) = \sqrt{\frac{1}{\mu_{nk}}}\int_{\mathbf{S}_f} d^2r\ \delta(p-\mathbf{r}\cdot\hat{\mathbf{n}}_j)\, u_{nk}(\mathbf{r})\,, \qquad (\mu_{nk}\neq 0)\,. \tag{17.102}$$

By use of (17.24), (17.85) (with the recognition that $\mathcal{H}_0^\dagger \mathbf{v}_n = \mathbf{u}_n$) and a change of variables, we find

$$v_{nkj}(p) = \frac{1}{\sqrt{J}}\exp(2\pi ikj/J)\, v_n(p)\,. \tag{17.103}$$

This remarkably simple result is actually mandated by group theory. Since the projection data are stored separately rather than summed (see Sec. 17.1.1), the effect of a rotation is to permute the projections cyclically, and any function that transforms according to the kth irreducible representation must have the form of (17.103).

Null functions of the adjoint With continuous angular sampling, the forward operator $\mathcal{R}_2$ is nonsingular, but the adjoint may nevertheless have null functions. The procedure used in (17.63) gives only the image-space singular functions corresponding to nonzero singular values. To complete the analysis, we must also find the null functions of the adjoint. This task is sometimes referred to as *characterizing the range* of the operator, since the null space of the adjoint (inconsistency space) is the orthogonal complement of the range of the forward operator (consistency space) as discussed in Sec. 1.5.2 (see especially Fig. 1.5).

Group theory provides us some assistance in finding null functions of the adjoint operator (Clarkson and Barrett, 1998b). From representation theory, it is known (Naimark and Stern, 1982) that any subspace of $\mathbb{V}$ invariant under all rotations and dilations must have basis functions of the form

$$\tilde{v}_{mk} \propto p^m \exp(ik\phi)\,. \tag{17.104}$$

Applying the adjoint operator to these functions, we see that

$$\left[\mathcal{R}_2^\dagger p^m \exp(ik\phi)\right](\mathbf{r}) = \int_{-\infty}^{\infty} dp \int_0^{\pi} d\phi\ p^m \exp(ik\phi)\,\delta(p-\mathbf{r}\cdot\hat{\mathbf{n}}) = \int_0^{\pi} d\phi\ (\mathbf{r}\cdot\hat{\mathbf{n}})^m \exp(ik\phi)\,. \tag{17.105}$$

The integral is invariant to the orientation of $\mathbf{r}$ in the x-y plane, so we may as well take $\mathbf{r}$ to be parallel to the x axis (from which ϕ is measured). Then we have

$$\int_0^{\pi} d\phi\ (\mathbf{r}\cdot\hat{\mathbf{n}})^m \exp(ik\phi) = r^m \int_0^{\pi} d\phi\ (\cos\phi)^m \exp(ik\phi)\,. \tag{17.106}$$

This integral vanishes if $m+|k|$ is even and $0 \le m < |k|$, so functions $\tilde{v}_{mk}(p,\phi)$ with m and k satisfying this condition are null functions of $\mathcal{R}_2^\dagger$.

The existence of these null functions does not contradict the earlier claim that the image-space eigenfunctions given in (17.64) form a complete set if we define the Hilbert spaces without an object support. Functions proportional to p^m are not square-integrable over the infinite line, so are not part of the Hilbert space $\mathbb{V}$ if we impose no support limitation. If we limit the object support to the disk of radius R, then the projection is restricted to $-R < p < R$, so p^m becomes integrable. The math leading to (17.106) is still the same (so long as $r < R$) and the functions $\tilde{v}_{mk}(p,\phi)$ are still null functions, but now part of the data space.

Consistency conditions We saw in Sec. 15.2.5 that any null function of the adjoint leads to a data-consistency condition. In the present problem, we have

$$\left(\boldsymbol{\mathcal{R}}_2^\dagger \tilde{\mathbf{v}}_{mk}, \mathbf{f}\right) = 0 = \left(\tilde{\mathbf{v}}_{mk}, \boldsymbol{\mathcal{R}}_2 \mathbf{f}\right), \tag{17.107}$$

provided $m + |k|$ is even and $0 \le m < |k|$. Thus any vector in consistency space (*i.e.*, one that can be written as $\boldsymbol{\mathcal{R}}_2\mathbf{f}$ for some $\mathbf{f}$) must be orthogonal to $p^m \exp(ik\phi)$ under the stated conditions on m and k. Noisy data will generally not lie entirely in consistency space.

For finite support and equally spaced angles ($\mathbf{C}_J$ symmetry), the consistency conditions are derived by Clarkson and Barrett (1998b) using group-theoretic considerations. They show that the null functions of the adjoint $\boldsymbol{\mathcal{R}}_{2J}^\dagger$ are $\tilde{v}_{mk}(p, \phi_j) \propto p^m \exp(ik\phi_j)$, but now the restrictions on m and k must be stated differently for J even or odd; see Clarkson and Barrett for the details. Again, any consistent data vector must be orthogonal to these functions.

17.2.2 Inverses and pseudoinverses in 2D

Some of the 2D transforms treated in Sec. 17.2.1 are singular and some are nonsingular. Our goal in this section is to show how SVD methods can be used to find inverses of the nonsingular transforms and pseudoinverses of the singular ones.

Inverse 2D Radon transform From Sec. 4.4, we already know various forms of the inverse 2D Radon transform with infinite object support, but we shall now show how some of them are related to the SVD expressions above. We begin by rederiving (4.155).

Since $\boldsymbol{\mathcal{R}}_2$ is nonsingular, its inverse is the same as its pseudoinverse; one operator form is found from (1.131) with the discrete index k replaced by the continuous vector index $\boldsymbol{\rho}$:

$$\boldsymbol{\mathcal{R}}_2^{-1} = \int_\infty d^2\rho \, \frac{1}{\sqrt{\mu_\rho}} \mathbf{u}_\rho \mathbf{v}_\rho^\dagger \,. \tag{17.108}$$

With (17.61) – (17.64), we obtain the explicit expression,

$$f(\mathbf{r}) = \left[\boldsymbol{\mathcal{R}}_2^{-1}\boldsymbol{\lambda}\right](\mathbf{r})$$

$$= \int_{-\infty}^{\infty} |\rho| \, d\rho \int_0^\pi d\theta_\rho \, \exp(2\pi i \boldsymbol{\rho}\cdot\mathbf{r}) \int_0^\pi d\phi \, \delta(\boldsymbol{\rho}\cdot\hat{\mathbf{n}}_\perp) \int_{-\infty}^{\infty} dp \, \exp(-2\pi i p\rho) \, \lambda(p,\phi) \,. \tag{17.109}$$

The integral over p yields the 1D Fourier transform of the projection, $\Lambda(\rho, \phi)$, and the integral over $d\theta_\rho$ can be performed by using (4.166), resulting in

$$f(\mathbf{r}) = \int_0^\pi d\phi \int_{-\infty}^{\infty} |\rho| \, d\rho \, \exp(2\pi i \mathbf{r}\cdot\hat{\mathbf{n}}\rho) \, \Lambda(\rho,\phi) \,. \tag{17.110}$$

This expression is equivalent to (4.155).

Filtered backprojection We can also express the inverse 2D Radon transform in the filtered-backprojection form of (4.161). To derive this form from the SVD, we write the inverse Radon operator as

$$\boldsymbol{\mathcal{R}}_2^{-1} = \boldsymbol{\mathcal{R}}_2^\dagger \left(\boldsymbol{\mathcal{R}}_2 \boldsymbol{\mathcal{R}}_2^\dagger\right)^{-1} . \tag{17.111}$$

In this expression, $\mathcal{R}_2^\dagger$ is the backprojection operator, and $(\mathcal{R}_2\mathcal{R}_2^\dagger)^{-1}$ is the filter applied to the projections before backprojection.

The filtering operator is given in SVD form as

$$\left(\mathcal{R}_2\mathcal{R}_2^\dagger\right)^{-1} = \int_\infty d^2\rho \, \frac{1}{\mu_\rho}\mathbf{v}_\rho\mathbf{v}_\rho^\dagger \,. \tag{17.112}$$

The kernel of this operator is

$$\left[\left(\mathcal{R}_2\mathcal{R}_2^\dagger\right)^{-1}\right](p,\phi;p'\phi') = \int_\infty d^2\rho \, \frac{1}{\mu_\rho} v_\rho(p,\phi)\, v_\rho^*(p',\phi') \,. \tag{17.113}$$

With (17.62) and (17.64), we see that

$$\left[\left(\mathcal{R}_2\mathcal{R}_2^\dagger\right)^{-1}\right](p,\phi;p'\phi')$$

$$= \int_{-\infty}^{\infty} |\rho|\, d\rho \, \exp[2\pi i(p-p')\rho] \int_0^\pi d\theta_\rho \, \delta\left[\cos(\phi-\theta_\rho)\right] \delta\left[\cos(\phi'-\theta_\rho)\right] \,. \tag{17.114}$$

Performing the angular integral as in (4.166) and using (3.169) for the integral over ρ, we get

$$\left[\left(\mathcal{R}_2\mathcal{R}_2^\dagger\right)^{-1}\right](p,\phi;p'\phi') = -\frac{1}{2\pi^2}\frac{1}{(p-p')^2}\,\delta(\phi-\phi') \,, \tag{17.115}$$

where $1/p^2$ is the strange generalized function discussed in detail in Sec. 2.3.3. Thus the filtered projection is given by

$$\left[\left(\mathcal{R}_2\mathcal{R}_2^\dagger\right)^{-1}\boldsymbol{\lambda}\right](p,\phi) = -\frac{1}{2\pi^2}\int_{-\infty}^{\infty} dp' \, \frac{\lambda(p',\phi)}{(p-p')^2} \,, \tag{17.116}$$

and application of the backprojection operator $\mathcal{R}_2^\dagger$ then yields (4.161).

Apodization Though the inverse 2D Radon transform exists, it involves dividing by singular values that tend to zero as $\rho \to \infty$, leading to a large noise amplification for image components associated with small singular values. As we know from Sec. 15.2.6, this amplification can be controlled — at the expense of spatial resolution — by regularization in the SVD domain [see (15.114)]. Such regularization is also called *apodization*, especially in a tomographic context.

Since the singular values in the 2D Radon transform with infinite object support are indexed by the spatial frequency $\boldsymbol{\rho}$, apodization amounts to weighting each component with a function that tends to 0 as the frequency gets large. For example, (17.110) can be modified to

$$\hat{f}(\mathbf{r}) = \int_0^\pi d\phi \int_{-\infty}^{\infty} |\rho|\, d\rho \, \exp(2\pi i\mathbf{r}\cdot\hat{\mathbf{n}}\rho)\, A(\rho)\, \Lambda(\rho,\phi) \,, \tag{17.117}$$

where the apodizing function $A(\rho)$ is typically unity at $\rho = 0$. Note that this expression is no longer an exact inverse, hence the hat on $\hat{f}(\mathbf{r})$.

Pseudoinverses with finite angular sampling? In Sec. 17.2.1, we derived the SVD for a 2D Radon transform with infinite object support and J angular samples, but it is not obvious whether a pseudoinverse exists in this case. The difficulty is that the eigenvalues $\mu_{\boldsymbol{\rho}}$ are angular delta functions. They vanish unless $\boldsymbol{\rho}$ is parallel to one of the projection directions $\hat{\mathbf{n}}_j$, but in a sense they are infinite along these directions. This problem prevented us from defining properly normalized singular functions in the image space; an expression like (17.63) does not work since the eigenvalues involve delta functions. This same problem comes up again in trying to define a pseudoinverse, which necessarily involves dividing by $\mu_{\boldsymbol{\rho}}$ or its square root.

There is no theorem that says a pseudoinverse has to exist in this problem. We noted in Sec. 1.6.1 that a pseudoinverse exists for any bounded linear operator with closed range, and in particular when the range is a finite-dimensional space. The operator $\mathcal{R}_{2J}$ has an infinite-dimensional range, however, and with infinite support $\mathcal{R}_{2J}^\dagger$ is unbounded; backprojection of any function into the infinite plane yields a function of infinite $\mathbb{L}_2$ norm. (Of course, both of these problems apply also to the full 2D Radon operator $\mathcal{R}$, and in that case there is not only a pseudoinverse but even a true inverse.)

If a Moore-Penrose pseudoinverse of $\mathcal{R}_{2J}$ exists, it must also be a 1-inverse and satisfy the first Penrose equation, $\mathcal{R}_{2J}\mathcal{R}_{2J}^+\mathcal{R}_{2J} = \mathcal{R}_{2J}$ (see Sec. 1.6.1). To check this equation, consider first the operator $\mathcal{R}_{2J}^+\mathcal{R}_{2J}$, which is the projector onto the measurement space of $\mathcal{R}_{2J}$. Measurement space in this problem is defined by the radial spokes parallel to $\hat{\mathbf{n}}_k$ in the 2D Fourier plane, so the projection operator must have the form,

$$\begin{aligned}\left[\mathcal{R}_{2J}^+\mathcal{R}_{2J}\mathbf{f}\right](\mathbf{r}) &= \sum_{k=0}^{J-1}\int_\infty d^2\rho\; C(\rho)\,\delta(\boldsymbol{\rho}\cdot\hat{\mathbf{n}}_{\perp k})\exp(2\pi i\boldsymbol{\rho}\cdot\mathbf{r})\int_\infty d^2r'\exp(-2\pi i\boldsymbol{\rho}\cdot\mathbf{r}')\,f(\mathbf{r}') \\ &= \sum_{k=0}^{J-1}\int_0^\infty d\rho\; C(\rho)\exp(2\pi i\rho\,\hat{\mathbf{n}}_k\cdot\mathbf{r})\,F(\rho\,\hat{\mathbf{n}}_k)\,. \end{aligned} \tag{17.118}$$

Applying $\mathcal{R}_{2J}$ to both sides, we obtain

$$\left[\mathcal{R}_{2J}\mathcal{R}_{2J}^+\mathcal{R}_{2J}\mathbf{f}\right](p,\phi_j) = \sum_{k=0}^{J-1}\int_\infty d^2r\;\delta(p-\hat{\mathbf{n}}_j\cdot\mathbf{r})\int_0^\infty d\rho\; C(\rho)\exp(2\pi i\rho\,\hat{\mathbf{n}}_k\cdot\mathbf{r})\,F(\rho\,\hat{\mathbf{n}}_k)\,. \tag{17.119}$$

Next we interchange the order of integration,[10] so that

$$\left[\mathcal{R}_{2J}\mathcal{R}_{2J}^+\mathcal{R}_{2J}\mathbf{f}\right](p,\phi_j) = \sum_{k=0}^{J-1}\int_0^\infty d\rho\; C(\rho)\exp(2\pi i\rho\,\hat{\mathbf{n}}_k\cdot\mathbf{r})\,\delta(\rho\,\hat{\mathbf{n}}_k\cdot\hat{\mathbf{n}}_{\perp j})F(\rho\,\hat{\mathbf{n}}_k)\,. \tag{17.120}$$

What we would like to get this way is

$$\begin{aligned}\left[\mathcal{R}_{2J}\mathbf{f}\right](p,\phi_j) &= \int_\infty d^2\rho\,\exp(2\pi i\rho\hat{\mathbf{n}}_j\cdot\mathbf{r})\,\delta(\boldsymbol{\rho}\cdot\hat{\mathbf{n}}_{\perp j})\,F(\rho\,\hat{\mathbf{n}}_j) \\ &= \int_0^\infty d\rho\,\exp(2\pi i\rho\,\hat{\mathbf{n}}_j\cdot\mathbf{r})\,F(\rho\,\hat{\mathbf{n}}_j)\,. \end{aligned} \tag{17.121}$$

[10]If this step is not legal, then the projection operator did not exist in the first place.

To make (17.120) the same as (17.121), we would have to make $C(\rho)\,\delta(\rho\,\hat{\mathbf{n}}_k \cdot \hat{\mathbf{n}}_{\perp j})$ be the Kronecker delta δ_{kj}, which isn't possible. Hence no 1-inverse of $\mathcal{R}_{2J}$ exists.

It is instructive to repeat this exercise with $\mathcal{R}_2$ instead of $\mathcal{R}_{2J}$ and to find $C(\rho)$ in that case.

Finite support, finite angular sampling The issue of existence of a pseudoinverse disappears if we consider a finite object support. In Sec. 17.2.1, we derived an SVD for J angles equally spaced around 2π, objects supported on a disc of radius R, and weighted Hilbert spaces with $W(r) = 1$ on the object support and $w(p) = \left[2\sqrt{R^2 - p^2}\,\right]^{-1}$ in the projection data. With these choices, the singular functions in object and image space are given by (17.100) and (17.103), respectively, and the singular values are found by taking the square root of (17.98). Using these pieces in (1.138) (again extended to continuous indices), we can write the pseudoinverse as

$$\left[\mathcal{R}_{2J}^{+}\boldsymbol{\lambda}\right](\mathbf{r})$$

$$= \lim_{\eta\to 0^+} \frac{1}{\pi R^2} \sum_{n=0}^{\infty} \sum_{k=0}^{J-1} \frac{1}{\mu_{nk} + \eta} \left[\frac{1}{J} \sum_{j=0}^{J-1} \exp(2\pi i k j/J)\, U_n\!\left(\frac{\mathbf{r}\cdot\hat{\mathbf{n}}_j}{R}\right)\right] \int_{-R}^{R} dp\; U_n\!\left(\frac{p}{R}\right) \lambda_k(p)\,, \tag{17.122}$$

where

$$\lambda_k(p) \equiv \sum_{j'=0}^{J-1} \exp(-2\pi i k j'/J)\, \lambda(p, \phi_{j'})\,. \tag{17.123}$$

Thus SVD inversion in this problem begins by performing a DFT in the angular variable to yield a function of the index k, then using a Chebyshev polynomial of the second kind to transform from the resulting function of p to a function of order number n. Filtering takes place in this n-k domain, and transformation back to the object domain is performed by multiplying by a Chebyshev polynomial with $\mathbf{r}\cdot\hat{\mathbf{n}}_j$ in its argument, doing an inverse DFT on the j index, and summing over n and k.

Limit of infinite angular sampling More insight into (17.122) can be obtained by considering the limit of very fine angular sampling ($J \to \infty$). In this limit, the eigenvalues are given by (17.99), and we can write

$$\left[\mathcal{R}_{2J}^{+}\boldsymbol{\lambda}\right](\mathbf{r})$$

$$= \frac{1}{\pi R^2} \sum_{n=0}^{\infty} \sum_{k=0}^{J-1} \frac{n+1}{J} \left[\frac{1}{J} \sum_{j=0}^{J-1} \exp\left(2\pi i k j/J\right) U_n\!\left(\frac{\mathbf{r}\cdot\hat{\mathbf{n}}_j}{R}\right)\right] \int_{-R}^{R} dp\; U_n\!\left(\frac{p}{R}\right) \lambda_k(p)\,. \tag{17.124}$$

One might think it would be necessary to modify the summation limits here since, according to (17.99), the sum includes only the terms for which $n \geq k$ and $n - k$ is even, but in fact no modification is needed; in the absence of noise, the consistency condition (17.107) causes the integral over p to vanish for the terms we would like to leave out. (Recall that $U_n(p/R)$ is a polynomial of degree n, containing terms in p^n, p^{n-2}, and so on.)

We can now reinsert (17.123) and do the sum over k, yielding $J\delta_{jj'}$, and we can use this delta function to do the sum over j'. We then take advantage of the limit

$J \to \infty$, where the discrete angle ϕ_j (or $2\pi j/J$) becomes the continuous variable ϕ, to replace the sum over j with an integral: $\sum_{j=0}^{J-1} \to \frac{J}{2\pi}\int_0^{2\pi} d\phi$. Thus

$$[\mathcal{R}_{2J}^{+}\boldsymbol{\lambda}](\mathbf{r}) = \frac{1}{\pi R^2}\sum_{n=0}^{\infty}(n+1)\frac{1}{2\pi}\int_0^{2\pi} d\phi\, U_n\left(\frac{\mathbf{r}\cdot\hat{\mathbf{n}}}{R}\right)\int_{-\infty}^{\infty} dp\, U_n\left(\frac{p}{R}\right)\lambda(p,\phi)\,. \tag{17.125}$$

To put this expression into a more familiar form, we write

$$[\mathcal{R}_{2J}^{+}\boldsymbol{\lambda}](\mathbf{r}) = \int_0^{2\pi} d\phi \int_{-\infty}^{\infty} dp\, h(\mathbf{r}\cdot\hat{\mathbf{n}}, p)\,\lambda(p,\phi)\,, \tag{17.126}$$

where

$$h(\mathbf{r}\cdot\hat{\mathbf{n}}, p) \equiv \frac{1}{\pi R^2}\frac{1}{2\pi}\sum_{n=0}^{\infty}(n+1)\,U_n\left(\frac{\mathbf{r}\cdot\hat{\mathbf{n}}}{R}\right)U_n\left(\frac{p}{R}\right)\,. \tag{17.127}$$

Comparing these expressions to (4.159) and (4.160), we see that (17.126) is the shift-variant, finite-support counterpart of filtered backprojection. The order n of the Chebyshev polynomial is the counterpart of spatial frequency, and the factor $(n+1)$ is the counterpart of the factor $|\nu|$ that occurs in ordinary filtered backprojection.

Large object support To make further contact with filtered backprojection, consider what happens when R gets large. Intuitively, the inverse operator should become shift-invariant in this limit.

Recall the definition of the Chebyshev polynomial from (4.63):

$$U_n\left(\frac{p}{R}\right) \equiv \frac{\sin[(n+1)\cos^{-1}(p/R)]}{\sin[\cos^{-1}(p/R)]}\,. \tag{17.128}$$

If $R \gg p$, then

$$\cos^{-1}\left(\frac{p}{R}\right) \approx \frac{\pi}{2} - \frac{p}{R}\,. \tag{17.129}$$

Thus the denominator in (17.128) is approximately unity, and

$$U_n\left(\frac{p}{R}\right) \approx \sin\left[(n+1)\frac{\pi}{2}\right]\cos\left[(n+1)\frac{p}{R}\right] - \cos\left[(n+1)\frac{\pi}{2}\right]\sin\left[(n+1)\frac{p}{R}\right]\,. \tag{17.130}$$

With this approximation and a little trigonometry, the sum in (17.127) becomes

$$\sum_{n=0}^{\infty}(n+1)U_n\left(\frac{p'}{R}\right)U_n\left(\frac{p}{R}\right) = \sum_{n\ even}^{\infty}(n+1)\cos\left[(n+1)\frac{p}{R}\right]\cos\left[(n+1)\frac{p'}{R}\right]$$

$$+\sum_{n\ odd}^{\infty}(n+1)\sin\left[(n+1)\frac{p}{R}\right]\sin\left[(n+1)\frac{p'}{R}\right]\,. \tag{17.131}$$

If p and p' are both small compared to R, the sines and cosines vary slowly with n, and it is valid to replace the sums with integrals. If we define $2\pi\nu = (n+1)/R$ and note that n changes in steps of 2 in each sum, then $\Delta\nu = 1/(\pi R)$. With some more trig, the filter function is given in the limit by

$$h(p',p) = \lim_{R\to\infty} \frac{1}{2\pi^2 R^2} \sum_{n=0}^{\infty} (n+1)\, U_n\left(\frac{p'}{R}\right) U_n\left(\frac{p}{R}\right)$$

$$= \int_0^\infty \nu\, d\nu\, \cos[2\pi\nu(p'-p)] = \frac{1}{2}\int_{-\infty}^{\infty} |\nu|\, d\nu\, \exp[2\pi i\nu(p'-p)]\,. \tag{17.132}$$

As expected, the filter function is now a function of only $p' - p$. Moreover, we recognize the integral as the 1D Fourier transform of $|\nu|$, so it is precisely the filter function that always occurs in filtered backprojection. We have succeeded in reproducing the results from Sec. 4.4.3 by a very round-about route,[11] but along the way we also found an expression for the pseudoinverse of the 2D Radon transform with J equally spaced angles and finite object support.

17.2.3 Inversion of the 3D x-ray transform

The 3D x-ray transform is defined in (17.46), and an equivalent form is given in (17.47). As we discussed in Sec. 17.1.5, the main application of this transform is in cone-beam tomography, where the vector $\mathbf{r}$ in the transform is interpreted as the position of the vertex of the cone. In most data-collection geometries, this vertex moves along a path called the trajectory, and position along the trajectory can be specified by a scalar parameter τ. Thus the vertex position can be written as $\mathbf{r}_v(\tau)$, and the continuous data can be expressed by a slight modification of (17.46) and (17.47) as

$$g(\tau,\hat{\mathbf{s}}) = \int_0^\infty d\ell\; f[\mathbf{r}_v(\tau) - \hat{\mathbf{s}}\ell] = \int_S d^3r'\, f(\mathbf{r}') \frac{\delta\left[\hat{\mathbf{s}} - \frac{\mathbf{r}_v(\tau)-\mathbf{r}'}{|\mathbf{r}_v(\tau)-\mathbf{r}'|}\right]}{|\mathbf{r}_v(\tau) - \mathbf{r}'|^2}\,, \tag{17.133}$$

where S is the object support.

Our goal in this section is to discuss conditions under which the object $f(\mathbf{r}')$ can be recovered from the continuous, noise-free data $g(\tau,\hat{\mathbf{s}})$ and then to find explicit inversion formulas. Key to this effort will be the Hamaker formulas (17.48) and (17.51); these formulas establish a close relation between the 3D x-ray transform and the 3D Radon transform, and all known inversion formulas exploit this relation in some way.

Completeness conditions An excellent historical and mathematical account of data completeness in cone-beam tomography is presented by Defrise and Clack (1994). As these authors point out, the literature on this subject begins with the pioneering mathematical investigations of A. A. Kirillov (1961). If we translate his work into the language of cone-beam tomography, he concludes basically that any object can be recovered from cone-beam data if all planes in space pass through a trajectory point. This condition takes no account of finite object support and therefore leads to the impractical conclusion that the trajectory must be unbounded. B. D. Smith

[11] The sharp-eyed reader will note that there is an extra factor of $\frac{1}{2}$ in the final form of (17.132) that does not appear in Sec. 4.4.3. It arises since the integral over ϕ in (17.126) runs from 0 to 2π, while the corresponding integral in (4.155) runs from 0 to π.

(1985a) showed that any object of finite support could be reconstructed if all planes through the support contain at least one trajectory point.

Chen (1992) gave a completeness condition applicable when the object has finite support but we are interested in recovering the object only within some region of interest (ROI) contained in the support. This condition was improved by Tuy (1983) who showed that the object could be recovered within the ROI if all planes passing through the ROI contain at least one trajectory point. Another condition, given by Grangeat (1987, 1990), allows for the possibility that projections from some of the trajectory points are truncated by a finite detector size. We shall not make use of these alternative conditions here. Instead, we assume, following Smith, that all planes through the object contain at least one trajectory point.

General approach to inversion formulas Many papers on cone-beam reconstruction adhere to a four-step paradigm originated by Smith (1985b). The steps can be enumerated as:

1. Transform to the 3D Radon domain;

2. Reparameterize;

3. Filter;

4. Perform Radon backprojection.

Each of these steps can be formulated as a linear integral transform, so the overall sequence is another linear transform, mapping $g(\tau, \hat{\mathbf{s}})$ to a continuous object estimate $\hat{f}(\mathbf{r})$. For noise-free continuous data satisfying the completeness condition discussed above, we arrive at $\hat{f}(\mathbf{r}) = f(\mathbf{r})$. This inverse tomographic transform can then be used as a starting point to derive practical reconstruction algorithms to be applied to noisy, discrete data.

Step 1: Transformation to the Radon domain The first step is application of the Hamaker formula (17.48). Following the notation of Smith, we denote the result of this step as $G(\tau, \hat{\mathbf{n}})$ (where here the capital letter does *not* denote a Fourier transform) and write

$$G(\tau, \hat{\mathbf{n}}) = \int_{4\pi} d\Omega_{\hat{\mathbf{s}}}\; g(\tau, \hat{\mathbf{s}})\, h(\hat{\mathbf{s}} \cdot \hat{\mathbf{n}})\,. \tag{17.134}$$

As in Sec. 17.1.5 we require that $h(\cdot)$ be a function of one variable that is homogeneous of degree -2, *i.e.*, $h(\alpha x) = \frac{1}{\alpha^2} h(x)$. Candidate functions are the derivative of a delta function and the generalized function $-1/x^2$, which we know from (3.169) to be the 1D inverse Fourier transform of the ramp function $2\pi^2|\nu|$.

The usefulness of homogeneous functions of degree -2 becomes apparent when we substitute the final form of (17.133) into (17.134), obtaining

$$G(\tau, \hat{\mathbf{n}}) = \int_{4\pi} d\Omega_{\hat{\mathbf{s}}} \int_S d^3\mathbf{r}'\, f(\mathbf{r}')\, \frac{\delta\left[\hat{\mathbf{s}} - \frac{\mathbf{r}_v(\tau) - \mathbf{r}'}{|\mathbf{r}_v(\tau) - \mathbf{r}'|}\right]}{|\mathbf{r}_v(\tau) - \mathbf{r}'|^2}\, h(\hat{\mathbf{s}} \cdot \hat{\mathbf{n}})\,. \tag{17.135}$$

With a wave to Fubini, we interchange the order of integration and use the definition of the angular delta function, (2.155), to see that

$$G(\tau, \hat{\mathbf{n}}) = \int_S d^3\mathbf{r}'\, f(\mathbf{r}')\, \frac{h\left[\frac{\mathbf{r}_v(\tau)-\mathbf{r}'}{|\mathbf{r}_v(\tau)-\mathbf{r}'|}\cdot\hat{\mathbf{n}}\right]}{|\mathbf{r}_v(\tau)-\mathbf{r}'|^2}\,. \tag{17.136}$$

Now, because of the homogeneity, we can write

$$G(\tau, \hat{\mathbf{n}}) = \int_S d^3\mathbf{r}'\, f(\mathbf{r}')\, h\{[\mathbf{r}_v(\tau) - \mathbf{r}']\cdot\hat{\mathbf{n}}\}\,. \tag{17.137}$$

In particular, if $h(x)$ is the derivative of a delta function, then

$$G(\tau, \hat{\mathbf{n}}) = \left\{\frac{\partial}{\partial p}[\mathcal{R}_3\mathbf{f}](p, \hat{\mathbf{n}})\right\}_{p=\mathbf{r}_v(\tau)\cdot\hat{\mathbf{n}}}\,. \tag{17.138}$$

Thus we have related the cone-beam data to the derivative of the 3D Radon transform by using the homogeneity of the delta derivative to get rid of the $1/\mathbf{r}^2$ Jacobian factor inherent in cone-beam projections. We recall that the plane of integration for the Radon transform has normal $\hat{\mathbf{n}}$ and is a distance p from the origin; the vertex point must lie on this plane since $p = \mathbf{r}_v(\tau)\cdot\hat{\mathbf{n}}$.

We note also that (17.138) is essentially the dipole-sheet transform of the object (see Sec. 4.4.5), so recovery of the object from this stage amounts to implementing the inverse dipole-sheet transform. To do so, however, we must express $G(\tau, \hat{\mathbf{n}})$ in terms of the Radon parameter p rather than the trajectory parameter τ.

Step 2: Reparameterization If Smith's data-completeness condition is satisfied, it is always possible to find one or more values of τ such that $p = \mathbf{r}_v(\tau)\cdot\hat{\mathbf{n}}$ for any plane through the object defined by p and $\hat{\mathbf{n}}$. Many papers then simply define a new function $F(p, \hat{\mathbf{n}})$ (where the capital letter again does *not* denote Fourier transform) by writing

$$F(p, \hat{\mathbf{n}}) \equiv G(\tau, \hat{\mathbf{n}})\,, \tag{17.139}$$

with the side condition that τ signifies one of the solutions of $p = \mathbf{r}_v(\tau)\cdot\hat{\mathbf{n}}$. This step is often called *rebinning*, even though neither $F(p, \hat{\mathbf{n}})$ nor $G(\tau, \hat{\mathbf{n}})$ refers to binned data in any way.

Semantic issues aside, (17.139) presents both mathematical and practical difficulties. The practical problem is that several different values of τ correspond to the same p and $\hat{\mathbf{n}}$ if the plane intersects the trajectory in several places. In that case it is at best arbitrary to choose one such point, and when we consider noisy data there may be a noise advantage to averaging over the trajectory points. Mathematically, there are two difficulties. First, we would like to obtain an overall integral transformation that maps the data (a function of τ) to the final image, so we need an integral over τ. Second, it is not just the isolated point τ that is important with continuous data, but rather some infinitesimal neighborhood of that point; in other words, we must consider the Jacobian of the transformation from a function of τ to a function of p.

One way we might consider defining the required function of p is by

$$F(p, \hat{\mathbf{n}}) \stackrel{?}{=} \int d\tau\; G(\tau, \hat{\mathbf{n}})\, \delta[p - \mathbf{r}_v(\tau)\cdot\hat{\mathbf{n}}]\,. \tag{17.140}$$

By (2.33), the result of this integral is

$$F(p,\hat{\mathbf{n}}) = \sum_{k=1}^{K(p,\hat{\mathbf{n}})} \frac{G(\tau_k,\hat{\mathbf{n}})}{\left|\frac{\partial}{\partial\tau}\mathbf{r}_v(\tau)\cdot\hat{\mathbf{n}}\right|_{\tau=\tau_k}}, \tag{17.141}$$

where $\{\tau_k, k=1,...,K(p,\hat{\mathbf{n}})\}$ are the solutions to $p=\mathbf{r}_v(\tau)\cdot\hat{\mathbf{n}}$. This formula properly includes the Jacobians (the denominators in the summand), but it is awkward and unaesthetic.

A cleaner definition, and the one we shall adopt, is

$$F(p,\hat{\mathbf{n}}) \equiv \frac{1}{K(p,\hat{\mathbf{n}})}\int d\tau\; G(\tau,\hat{\mathbf{n}})\left|\frac{\partial}{\partial\tau}\mathbf{r}_v(\tau)\cdot\hat{\mathbf{n}}\right|\delta[p-\mathbf{r}_v(\tau)\cdot\hat{\mathbf{n}}]\,. \tag{17.142}$$

The integral now gives

$$F(p,\hat{\mathbf{n}}) = \frac{1}{K(p,\hat{\mathbf{n}})}\sum_{k=1}^{K(p,\hat{\mathbf{n}})} G(\tau_k,\hat{\mathbf{n}})\,. \tag{17.143}$$

Though (17.143) looks like (17.139) except for the averaging over equivalent trajectory points, we have the advantage from (17.142) that we now know how to express the reparameterization as an integral transform. The Jacobians have not been neglected but instead have been cancelled out by a judicious definition of the kernel of this transform.

Steps 3 and 4: Filtering and backprojection If we choose $h(x)$ as the derivative of a delta function, then (17.138) applies, and we see that

$$F(p,\hat{\mathbf{n}}) = \frac{\partial}{\partial p}[\mathcal{R}_3\mathbf{f}](p,\hat{\mathbf{n}})\,. \tag{17.144}$$

We know from (4.186) how to go from the derivative of the 3D Radon transform back to the original function; we need only differentiate again and backproject. 3D Radon backprojection can be conceptualized as smearing back over the plane of integration and integrating over all orientations of the plane. Since the plane is defined by $p=\mathbf{r}\cdot\hat{\mathbf{n}}$, backprojection is realized mathematically by substituting $\mathbf{r}\cdot\hat{\mathbf{n}}$ for p and then performing an angular integral. Thus we have [*cf.* (4.186)]

$$\hat{f}(\mathbf{r}) = -\frac{1}{4\pi^2}\int_{2\pi} d\Omega_{\hat{\mathbf{n}}}\left[\frac{\partial}{\partial p}F(p,\hat{\mathbf{n}})\right]_{p=\mathbf{r}\cdot\hat{\mathbf{n}}}. \tag{17.145}$$

If the data are given by (17.133), then $\hat{f}(\mathbf{r}) = f(\mathbf{r})$. This formula was first derived by Grangeat (1987).

Reconstruction operator Since we performed a sequence of linear transformations to get from the continuous data $g(\tau,\hat{\mathbf{s}})$ to the reconstruction $\hat{f}(\mathbf{r})$, there must be a single linear operator that performs the same task. In abstract form, there is an operator $\mathcal{O}$ such that $\hat{\mathbf{f}} = \mathcal{O}\mathbf{g}$. We shall now work out the kernel of that operator.

For generality, we let the filter used in Step 3 be denoted by $\tilde{h}(\cdot)$, which could be different from the filter $h(\cdot)$ used in Step 1. With this notation, (17.142) can be merged with the more general form of (17.145) to yield

$$\hat{f}(\mathbf{r}) = \int_{2\pi} d\Omega_{\hat{\mathbf{n}}} \int_{-\infty}^{\infty} dp'\, \tilde{h}(\mathbf{r}\cdot\hat{\mathbf{n}} - p') \frac{1}{K(p',\hat{\mathbf{n}})} \int d\tau\; G(\tau,\hat{\mathbf{n}}) \left| \frac{\partial}{\partial\tau}\mathbf{r}_v(\tau)\cdot\hat{\mathbf{n}} \right| \delta[p' - \mathbf{r}_v(\tau)\cdot\hat{\mathbf{n}}]$$

$$= \int_{2\pi} d\Omega_{\hat{\mathbf{n}}} \int d\tau\; \tilde{h}[\mathbf{r}\cdot\hat{\mathbf{n}} - \mathbf{r}_v(\tau)\cdot\hat{\mathbf{n}}] \frac{1}{K[\mathbf{r}_v(\tau)\cdot\hat{\mathbf{n}},\hat{\mathbf{n}}]} G(\tau,\hat{\mathbf{n}}) \left| \frac{\partial}{\partial\tau}\mathbf{r}_v(\tau)\cdot\hat{\mathbf{n}} \right| . \quad (17.146)$$

Inserting the definition of $G(\tau,\hat{\mathbf{n}})$ from (17.134) gives

$$\hat{f}(\mathbf{r}) = \int d\tau \int_{4\pi} d\Omega_{\hat{\mathbf{s}}} \int_{2\pi} d\Omega_{\hat{\mathbf{n}}}\; \tilde{h}[\mathbf{r}\cdot\hat{\mathbf{n}} - \mathbf{r}_v(\tau)\cdot\hat{\mathbf{n}}]\, h(\hat{\mathbf{s}}\cdot\hat{\mathbf{n}}) \frac{|\frac{\partial}{\partial\tau}\mathbf{r}_v(\tau)\cdot\hat{\mathbf{n}}|}{K[\mathbf{r}_v(\tau)\cdot\hat{\mathbf{n}},\hat{\mathbf{n}}]}\, g(\tau,\hat{\mathbf{s}}) . \quad (17.147)$$

Thus, if we write

$$\hat{f}(\mathbf{r}) = \int d\tau \int_{4\pi} d\Omega_{\hat{\mathbf{s}}}\; o(\mathbf{r};\tau,\hat{\mathbf{s}})\, g(\tau,\hat{\mathbf{s}}) , \quad (17.148)$$

the kernel must be given by

$$o(\mathbf{r};\tau,\hat{\mathbf{s}}) = \int_{2\pi} d\Omega_{\hat{\mathbf{n}}}\; \tilde{h}[\mathbf{r}\cdot\hat{\mathbf{n}} - \mathbf{r}_v(\tau)\cdot\hat{\mathbf{n}}]\, h(\hat{\mathbf{s}}\cdot\hat{\mathbf{n}}) \frac{|\frac{\partial}{\partial\tau}\mathbf{r}_v(\tau)\cdot\hat{\mathbf{n}}|}{K[\mathbf{r}_v(\tau)\cdot\hat{\mathbf{n}},\hat{\mathbf{n}}]} . \quad (17.149)$$

Choice of filter functions We still need to determine what filter functions can be used if we wish to have an exact inversion formula, where $\hat{f}(\mathbf{r}) = f(\mathbf{r})$. To this end, we look at the overall transformation from object to reconstruction. Writing this transformation as

$$\hat{f}(\mathbf{r}) = \int_{\infty} d^3\mathbf{r}'\, k(\mathbf{r},\mathbf{r}')\, f(\mathbf{r}') , \quad (17.150)$$

we see from (17.133) and (17.148) that

$$k(\mathbf{r},\mathbf{r}') = \int d\tau \int_{4\pi} d\Omega_{\hat{\mathbf{s}}}\; o(\mathbf{r};\tau,\hat{\mathbf{s}}) \frac{\delta\left[\hat{\mathbf{s}} - \frac{\mathbf{r}_v(\tau)-\mathbf{r}'}{|\mathbf{r}_v(\tau)-\mathbf{r}'|}\right]}{|\mathbf{r}_v(\tau)-\mathbf{r}'|^2} = \int d\tau\; \frac{o\left[\mathbf{r};\tau,\frac{\mathbf{r}_v(\tau)-\mathbf{r}'}{|\mathbf{r}_v(\tau)-\mathbf{r}'|}\right]}{|\mathbf{r}_v(\tau)-\mathbf{r}'|^2} . \quad (17.151)$$

Inserting (17.149) and using the homogeneity of $h(\cdot)$, we find

$$k(\mathbf{r},\mathbf{r}') = \int d\tau \int_{2\pi} d\Omega_{\hat{\mathbf{n}}}\; \tilde{h}[\mathbf{r}\cdot\hat{\mathbf{n}} - \mathbf{r}_v(\tau)\cdot\hat{\mathbf{n}}]\, h[\mathbf{r}_v(\tau)\cdot\hat{\mathbf{n}} - \mathbf{r}'\cdot\hat{\mathbf{n}}] \frac{|\frac{\partial}{\partial\tau}\mathbf{r}_v(\tau)\cdot\hat{\mathbf{n}}|}{K[\mathbf{r}_v(\tau)\cdot\hat{\mathbf{n}},\hat{\mathbf{n}}]} . \quad (17.152)$$

Now we make the change of variables $p = \mathbf{r}_v(\tau)\cdot\hat{\mathbf{n}}$. Recognizing the last factor in (17.152) as the Jacobian of this transformation, we see that

$$k(\mathbf{r},\mathbf{r}') = \int_{-\infty}^{\infty} dp \int_{2\pi} d\Omega_{\hat{\mathbf{n}}}\; \tilde{h}(\mathbf{r}\cdot\hat{\mathbf{n}} - p)\, h(p - \mathbf{r}'\cdot\hat{\mathbf{n}}) . \quad (17.153)$$

As an exercise, the reader can show that $k(\mathbf{r},\mathbf{r}') = \delta(\mathbf{r}-\mathbf{r}')$, yielding an exact inversion formula, if the Fourier transforms of the filter functions satisfy

$$H(\nu)\, \tilde{H}(\nu) = \nu^2 . \quad (17.154)$$

This condition is satisfied if $h(x) = \delta'(x)$ and $\tilde{h}(x) = -\delta'(x)/(4\pi^2)$, and it is this choice that gives (17.145). Another choice is ramp filters, $H(\nu) = \tilde{H}(\nu) = |\nu|$. Clack and Defrise (1994) have shown that the filters can also be certain linear combinations of derivatives and ramp filters. (Cross-terms in $\nu|\nu|$ cancel out by symmetry arguments.)

17.2.4 Inversion of attenuated transforms

We have encountered several tomographic transforms that include an exponential factor in the integrand. These *attenuated transforms* include the 2D attenuated Radon transform, the 2D exponential Radon transform and the 3D attenuated x-ray transform. (The 2D attenuated x-ray transform is the same as the 2D attenuated Radon transform, and the 3D attenuated and exponential Radon transforms are of little interest for reasons discussed in Sec. 17.1.6.)

As we saw in Sec. 17.1.6, the 2D exponential Radon transform, (17.58), applies when the attenuation coefficient is constant over a convex region. Inversion formulas for this case were found by Bellini *et al.* (1979) and Tretiak and Metz (1980). Over the next fifteen years, several other inversion formulas were found, and eventually Metz and Pan (1995) showed that all of them were special cases of a general formalism. (see also Pan and Metz, 1995).

Inversion formulas for the more general attenuated Radon transform, (17.54), were much more difficult to find, and many observers (including one author of this book) had thought they might not exist. Finally, however, an approach to a solution was suggested by Arbuzov (1998), and Novikov (2002a, 2002b) developed an explicit inversion formula. Implementations of this formula were presented by Kunyansky (2001) and Natterer (2001), and the latter author also presented a simpler derivation and an alternative inversion formula.

The key problem that remains unsolved at this writing is the attenuated x-ray transform. We know from Sec. 17.2.3 that inversion of the unattenuated x-ray transform is all about dealing with a weighting factor in the integrand, namely the factor $1/|\mathbf{r}_v(\tau) - \mathbf{r}'|^2$ in (17.133). Similarly, any attenuated transform brings in another weighting factor, the exponential attenuation factor. Real cone-beam data must necessarily involve both weighting — inverse-square and exponential attenuation — so it would be of some importance to devise continuous inversion formulas that include both.

Inversion of the 2D exponential Radon transform Because the inversion formulas for nonconstant attenuation are complex (in both senses of the word), we shall concentrate here on the 2D exponential Radon transform. For notational simplicity in what follows, we denote the operator for this transform as $\boldsymbol{\mathcal{R}}_\mu$ rather than $\boldsymbol{\mathcal{R}}_{2e,\mu}$ as used in (17.58); thus

$$g_\mu(p,\phi) = [\boldsymbol{\mathcal{R}}_\mu \mathbf{f}](p,\phi) = \int_\infty d^2r\, f(\mathbf{r})\, \delta(p - \mathbf{r}\cdot\hat{\mathbf{n}}) \exp(\mu \mathbf{r}\cdot\hat{\mathbf{n}}_\perp) \,. \tag{17.155}$$

It turns out to be useful to include $\boldsymbol{\mathcal{R}}_{-\mu}$ in our theory along with $\boldsymbol{\mathcal{R}}_\mu$. In particular, as we shall see below, the operator $\boldsymbol{\mathcal{R}}^\dagger_{-\mu}\boldsymbol{\mathcal{R}}_\mu$ turns out to be shift-invariant in spite of the fact that both the attenuating medium and the axis of rotation establish a preferred origin in object space. Moreover, the convolutional relation (4.194), which we derived in a signal-processing context in Sec. 4.4.6, generalizes to (Natterer, 1986)

$$\left(\boldsymbol{\mathcal{R}}^\dagger_{-\mu}\mathbf{h}\right) * \mathbf{f} = \boldsymbol{\mathcal{R}}^\dagger_{-\mu}[\mathbf{h} * (\boldsymbol{\mathcal{R}}_\mu \mathbf{f})] \,, \tag{17.156}$$

where $\mathbf{h}$ denotes an arbitrary function in data space, $h(p,\phi)$. Thus the asterisk on the left denotes 2D convolution but the one on the right denotes 1D convolution with respect to p for each ϕ. Derivation of (17.156) is left as an exercise.

Unfiltered backprojection Let us first examine the case $h(p,\phi) = \delta(p)$, so that the right-hand side of (17.156) becomes simply $\mathcal{R}^{\dagger}_{-\mu}\mathcal{R}_{\mu}\mathbf{f}$. Explicitly, from (17.155) and the definition of the adjoint [*cf.* (4.144)], we have

$$\left[\mathcal{R}^{\dagger}_{-\mu}\mathcal{R}_{\mu}\mathbf{f}\right](\mathbf{r}) = \int_{\infty} d^2r'\, f(\mathbf{r}')\int_0^{2\pi} d\phi\; \delta[(\mathbf{r}-\mathbf{r}')\cdot\hat{\mathbf{n}}]\exp[-\mu(\mathbf{r}-\mathbf{r}')\cdot\hat{\mathbf{n}}_{\perp}] = [\mathbf{f} * (\mathcal{R}^{\dagger}_{-\mu}\mathbf{h})](\mathbf{r})\,. \tag{17.157}$$

Since the integral over ϕ covers a full circle, we can without loss of generality choose $\mathbf{r}$ to lie along the x axis and write

$$\left[\mathcal{R}^{\dagger}_{-\mu}\mathbf{h}\right](\mathbf{r}) = \int_0^{2\pi} d\phi\; \delta(r\cos\phi)\exp(-\mu r\sin\phi)\,. \tag{17.158}$$

The integral can be performed with the help of (2.33), and we find

$$\left[\mathcal{R}^{\dagger}_{-\mu}\mathbf{h}\right](\mathbf{r}) = \frac{\cosh \mu r}{r}\,, \tag{17.159}$$

where $r = |\mathbf{r}|$. Note that this expression reduces to $1/r$ if $\mu \to 0$, and we know that $1/r$ is the PSF for unfiltered backprojection with the ordinary 2D Radon transform [see (4.167)].

We might think about performing a deconvolution in the 2D Fourier domain, but the exponential growth of $\cosh \mu r$ means that the Fourier transform of (17.159) cannot be defined in conventional terms, or even in terms of tempered distributions (see Sec. 3.3.4).

Exact filter function Another way to use (17.156) is to try to find the function $h(p,\phi)$ that yields a 2D delta function when backprojected with $\mathcal{R}^{\dagger}_{-\mu}$; that is,

$$\left[\mathcal{R}^{\dagger}_{-\mu}\mathbf{h}\right](\mathbf{r}) = \delta(\mathbf{r})\,. \tag{17.160}$$

With this function, (17.156) becomes

$$\mathbf{f} = \mathcal{R}^{\dagger}_{-\mu}[\mathbf{h} * (\mathcal{R}_{\mu}\mathbf{f})]\,. \tag{17.161}$$

If the filter function is independent of ϕ, it can be written as $h(p)$; the function $[\mathcal{R}^{\dagger}_{-\mu}\mathbf{h}](\mathbf{r})$ is rotationally symmetric in this case, and the left-hand side of (17.160) becomes [*cf.* (17.158)]

$$\left[\mathcal{R}^{\dagger}_{-\mu}\mathbf{h}\right](\mathbf{r}) = \int_0^{2\pi} d\phi\; h(r\cos\phi)\exp(-\mu r\sin\phi)\,. \tag{17.162}$$

Representing $h(p)$ in terms of its 1D Fourier transform $H(\nu)$, we obtain

$$\left[\mathcal{R}^{\dagger}_{-\mu}\mathbf{h}\right](\mathbf{r}) = \int_{-\infty}^{\infty} d\nu\; H(\nu)\int_0^{2\pi} d\phi\; \exp(-\mu r\sin\phi + 2\pi i r\cos\phi)\,. \tag{17.163}$$

A tabulated integral (Gradshteyn and Ryzhik, 1980, formula 3.937) yields

$$\left[\mathcal{R}^{\dagger}_{-\mu}\mathbf{h}\right](\mathbf{r}) = 2\pi\int_{-\infty}^{\infty} d\nu\; H(\nu)\,\mathrm{J}_0\left[2\pi r\sqrt{\nu^2 - (\mu/2\pi)^2}\right]\,. \tag{17.164}$$

This form immediately suggests the change of variables $\rho = \sqrt{\nu^2 - (\mu/2\pi)^2}$. If we take $H(\nu)$ to be zero for $|\nu| < \mu/2\pi$, then there is no worry about the argument of the Bessel function becoming imaginary. Since the integrand is even, we can write

$$\left[\mathcal{R}_{-\mu}^{\dagger}\mathbf{h}\right](\mathbf{r}) = 4\pi \int_0^{\infty} d\rho \; H\left[\sqrt{\rho^2 + (\mu/2\pi)^2}\right] \mathrm{J}_0(2\pi\rho r) \,. \tag{17.165}$$

We can now satisfy (17.160) almost by inspection. We know from (3.248) and (3.254) that

$$2\pi \int_0^{\infty} \rho\, d\rho \; \mathrm{J}_0(2\pi\rho r) = \delta(\mathbf{r})\,, \tag{17.166}$$

so (17.160) holds if we take

$$H(\nu) = \begin{cases} \frac{1}{2}|\nu| & \text{if} \quad |\nu| > \mu/2\pi \\ 0 & \text{if} \quad |\nu| \le \mu/2\pi \end{cases} \,. \tag{17.167}$$

This function, the familiar ramp filter with the center portion deleted, is plotted in Fig. 17.7. It was first derived by Tretiak and Metz (1980).

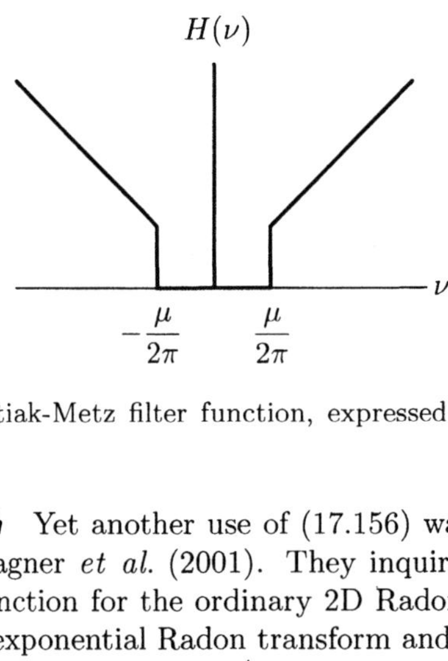

Fig. 17.7 The Tretiak-Metz filter function, expressed in the 1D frequency domain.

Neumann-series approach Yet another use of (17.156) was suggested, in a slightly different context, by Wagner *et al.* (2001). They inquired what would happen if one applied the filter function for the ordinary 2D Radon transform (*i.e.*, a ramp filter) to data from the exponential Radon transform and then followed it with the exponentially weighted backprojection $\mathcal{R}_{-\mu}^{\dagger}$. If we denote the ramp filter by $\mathbf{h}_0$ (the subscript indicating $\mu = 0$), then we can define

$$\hat{\mathbf{f}}_0 \equiv \mathcal{R}_{-\mu}^{\dagger}[\mathbf{h}_0 * \mathbf{g}_\mu]\,, \tag{17.168}$$

where $\mathbf{g}_\mu = \mathcal{R}_\mu \mathbf{f}$ is the (noise-free) exponential-Radon data.

Since (17.156) holds for any filter function, we see that

$$\hat{\mathbf{f}}_0 = \left(\mathcal{R}_{-\mu}^{\dagger}\mathbf{h}_0\right) * \mathbf{f} = \left[\left(\mathcal{R}_{-\mu}^{\dagger}\mathbf{h}_0\right) - \boldsymbol{\delta}\right] * \mathbf{f} + \mathbf{f} \equiv -\mathcal{K}\mathbf{f} + \mathbf{f}\,, \tag{17.169}$$

where $\boldsymbol{\delta}$ represents the 2D delta function $\delta(\mathbf{r})$ and $\mathcal{K}$ is the operator defined by the next-to-last term of (17.169).

Since (17.169) is in the form of (A.59), we can apply the Neumann series to write

$$\mathbf{f} = \hat{\mathbf{f}}_0 + \mathcal{K}\hat{\mathbf{f}}_0 + \mathcal{K}^2\hat{\mathbf{f}}_0 \ldots . \tag{17.170}$$

Thus, if the series converges, we can find the actual object $\mathbf{f}$ from $\hat{\mathbf{f}}_0$ in spite of having used the wrong filter function. Convergence is problematical since $\mathcal{R}_{-\mu}^{\dagger}$ is an unbounded operator if the object has infinite support, but Wagner *et al.* suggest that the problem can be avoided by assuming a sufficiently small support.

Fourier-series approach An elegant general approach to the exponential Radon transform was presented by Metz and Pan (1995). This approach begins by expanding $g_\mu(p,\phi)$ in an angular Fourier series as

$$g_\mu(p,\phi) = \sum_{k=-\infty}^{\infty} g_{\mu k}(p)\, e^{ik\phi} \tag{17.171}$$

and then further expands the coefficients in terms of their 1D Fourier transforms, so that

$$g_\mu(p,\phi) = \sum_{k=-\infty}^{\infty} e^{ik\phi} \int_{-\infty}^{\infty} d\nu\; G_{\mu k}(\nu)\, e^{2\pi i\nu p}\,. \tag{17.172}$$

Thus the projection data are represented in a hybrid manner by a Fourier transform on the p variable and a Fourier-series expansion in the angular variable ϕ. Explicitly,

$$G_{\mu k}(\nu) = \frac{1}{2\pi}\int_0^{2\pi} d\phi \int_{-\infty}^{\infty} dp\; g_\mu(p,\phi)\, e^{-ik\phi} e^{-2\pi i\nu p}\,. \tag{17.173}$$

We can also represent the Fourier *transform* of $f(\mathbf{r})$ in an angular Fourier *series* as

$$F(\boldsymbol{\rho}) = \sum_{k=-\infty}^{\infty} F_k(\rho)\exp(ik\theta_\rho)\,, \tag{17.174}$$

where the spatial frequency vector $\boldsymbol{\rho}$ has polar coordinates (ρ,θ_ρ).

Following Tretiak and Metz (1980) and Metz and Pan (1995), we now seek a relation between $G_{\mu k}(\nu)$ and $F_k(\rho)$. For consistent (noise-free) data, we can express $g_\mu(p,\phi)$ by (17.155) and represent $f(\mathbf{r})$ by its inverse Fourier transform, which is expressed by (17.174). Then (17.173) becomes

$$G_{\mu k}(\nu) = \frac{1}{2\pi}\int_0^{2\pi} d\phi \int_{-\infty}^{\infty} dp \int_{\infty} d^2r \int_{\infty} d^2\rho \sum_{k'=-\infty}^{\infty} F_{k'}(\rho)\exp(ik'\theta_\rho)\exp(2\pi i\boldsymbol{\rho}\cdot\mathbf{r})$$
$$\times\, \delta(p-\mathbf{r}\cdot\hat{\mathbf{n}})\exp(\mu\mathbf{r}\cdot\hat{\mathbf{n}}_\perp)\exp(-ik\phi)\exp(-2\pi i\nu p)\,. \tag{17.175}$$

The p integral can be performed by use of the delta function, and we recognize that $d^2r = r\,dr\,d\theta$ and $d^2\rho = \rho\,d\rho\,d\theta_\rho$, so we have

$$G_{\mu k}(\nu) = \frac{1}{2\pi}\int_0^{2\pi} d\phi \int_0^{\infty} r\,dr \int_0^{2\pi} d\theta \int_0^{\infty} \rho\,d\rho \int_0^{2\pi} d\theta_\rho \sum_{k'=-\infty}^{\infty} F_{k'}(\rho)\exp(ik'\theta_\rho)$$
$$\times\, \exp\left[2\pi i\rho r\cos(\theta-\theta_\rho)\right]\exp\left[\mu r\sin(\theta-\phi)\right]\exp(-ik\phi)\exp\left[-2\pi i\nu r\cos(\theta-\phi)\right]. \tag{17.176}$$

With the variable changes $\phi' = \theta-\phi$ and $\theta' = \theta-\theta_\rho$, the integral over θ_ρ yields $2\pi\,\delta_{kk'}$, so

$$G_{\mu k}(\nu) = \int_0^\infty r\,dr \int_0^\infty \rho\,d\rho\; F_k(\rho) \int_0^{2\pi} d\theta' \exp(2\pi i\rho r\cos\theta')\exp(-ik\theta')$$

$$\times \int_0^{2\pi} d\phi' \exp(\mu r\sin\phi')\exp(ik\phi')\exp(-2\pi i\nu r\cos\phi')\,. \tag{17.177}$$

Using two tabulated integrals (Gradshteyn and Ryzhik, 1980, formulas 3.715 and 3.937), we find

$$G_{\mu k}(\nu)=4\pi^2\left[\frac{\nu+\frac{\mu}{2\pi}}{\sqrt{\nu^2-(\frac{\mu}{2\pi})^2}}\right]^k \int_0^\infty r\,dr\int_0^\infty \rho\,d\rho\; F_k(\rho)\,\mathrm{J}_k(2\pi\rho r)\,\mathrm{J}_k\left[2\pi r\sqrt{\nu^2-\left(\frac{\mu}{2\pi}\right)^2}\right]. \tag{17.178}$$

Finally, we use the Fourier-Bessel theorem (Morse and Feshbach, 1953),

$$4\pi^2\int_0^\infty r\,dr\int_0^\infty \rho\,d\rho\; F_k(\rho)\,\mathrm{J}_k(2\pi\rho r)\,\mathrm{J}_k(2\pi\zeta r) = F_k(\zeta)\,, \tag{17.179}$$

to obtain (for $\nu \geq \frac{\mu}{2\pi}$)

$$G_{\mu k}(\nu) = \left[\frac{\frac{\mu}{2\pi}+\nu}{\sqrt{\nu^2-(\frac{\mu}{2\pi})^2}}\right]^k F_k\left[\sqrt{\nu^2-\left(\frac{\mu}{2\pi}\right)^2}\right]. \tag{17.180}$$

This is the desired relation between the mixed Fourier series-transform representation of the ERT data and the corresponding representation of the object. For $\mu = 0$, this relation is just a restatement of the central-slice theorem. An equivalent relation is

$$G_{\mu k}(-\nu) = \left[\frac{\frac{\mu}{2\pi}-\nu}{\sqrt{\nu^2-(\frac{\mu}{2\pi})^2}}\right]^k F_k\left[\sqrt{\nu^2-\left(\frac{\mu}{2\pi}\right)^2}\right]. \tag{17.181}$$

We can use either of these relations to find the object coefficients $F_k(\rho)$. From (17.180) and a change of variables, we find

$$F_k(\rho) = \left[\frac{\rho}{\frac{\mu}{2\pi}+\sqrt{\rho^2+(\frac{\mu}{2\pi})^2}}\right]^k G_{\mu k}\left[\sqrt{\rho^2+\left(\frac{\mu}{2\pi}\right)^2}\right], \tag{17.182}$$

and from (17.181) we find

$$F_k(\rho) = \left[\frac{\rho}{\frac{\mu}{2\pi}-\sqrt{\rho^2+(\frac{\mu}{2\pi})^2}}\right]^k G_{\mu k}\left[-\sqrt{\rho^2+\left(\frac{\mu}{2\pi}\right)^2}\right]. \tag{17.183}$$

With noise-free, continuous data, we can use either (17.182) or (17.183) to reconstruct the object exactly via (17.174) and an inverse Fourier transform.

Noisy, continuous data The derivation of (17.182) and (17.183) used the defining relation (17.155) for $g_\mu(p,\phi)$, and it was at that point that ideal, continuous, noise-free data were assumed. As we shall discuss in more detail in Secs. 17.2.5 and 17.2.6, real data are both noisy and discrete. It is of some theoretical interest, however, to imagine that the data are noisy but not discrete, in which case $g_\mu(p, \phi)$ is not given by (17.155) but instead by

$$g_\mu(p, \phi) = \int_\infty d^2r \; f(\mathbf{r})\, \delta(p - \mathbf{r} \cdot \hat{\mathbf{n}}) \exp\left(\mu \mathbf{r} \cdot \hat{\mathbf{n}}_\perp\right) + \Delta g(p, \phi)\,, \tag{17.184}$$

where $\Delta g(p, \phi)$ is some random process.

The immediate consequence of (17.184) is that $g_\mu(p, \phi)$ is not in consistency space (the range of $\mathcal{R}_\mu$) since there is no reason to believe that $\Delta g(p, \phi)$ is the projection of some object. Nevertheless, we can still apply all of the manipulations used above. That is, we can use the noisy, continuous data to compute the Fourier coefficients $G_{\mu k}(\nu)$ by means of (17.173), and we can then compute the right-hand sides of (17.182) and (17.183). What we cannot do, however, is claim that we have thereby found the actual object coefficient $F_k(\rho)$. Instead, we have found noisy estimates, denoted $\hat{F}_k^{(1)}(\rho)$ when (17.182) is used and $\hat{F}_k^{(2)}(\rho)$ when (17.183) is used. Because of the noise, these two estimates will not be identical.

Metz and Pan (1995) showed that different linear combinations of $\hat{F}_k^{(1)}(\rho)$ and $\hat{F}_k^{(2)}(\rho)$ reproduced different ERT inversion formulas previously found in the literature and also led to many new formulas. Clarkson (1999) showed how the different formulas correspond to different ways of projecting the inconsistent data onto consistency space.

17.2.5 Discretization of analytic reconstruction algorithms

So far in this chapter we have discussed the forward problem for SPECT in both CD and CC formulations, and we have discussed several analytic inversion formulas applicable to the CC case. We have not yet introduced any practical reconstruction algorithms that can be applied to discrete SPECT data, though general considerations on this issue were given in Chap. 15. Our goal in this section and the next is to show how the principles developed in Chap. 15 can be applied specifically to SPECT. The focus in this section is one-step linear algorithms obtained by discretizing analytic transforms. In Sec. 17.2.6 we turn to iterative algorithms, and in particular to formulating the requisite system matrix.

Discretization of analytic inverses In Sec. 15.2.4 we began with a CC operator $\mathcal{L}$ with a known left inverse $\mathcal{L}^{-1}$ and investigated how it could be applied to noisy, discrete data. The data model was assumed in (15.84) to be

$$\mathbf{g} = C \mathcal{D}_w \mathcal{L} \mathbf{f} + \mathbf{n}\,, \tag{17.185}$$

where C is a constant related to system sensitivity and exposure time, and $\mathcal{D}_w$ is a CD operator that acts on the continuous data and produces discrete measurements. The goal was to reconstruct the coefficients in some approximate object expansion; these coefficients (*e.g.*, pixel values) were assumed to be related to the object by (15.85):

$$\boldsymbol{\theta} = \mathcal{D}_\chi \mathbf{f} \qquad \text{or} \qquad \theta_n = \int_\infty d^2r \; \chi_n(\mathbf{r})\, f(\mathbf{r})\,. \tag{17.186}$$

For more discussion of this equation, see Sec. 7.1.3.

It was suggested in Sec. 15.2.4 that a reasonable (not necessarily optimal) way of estimating $\boldsymbol{\theta}$ is to define the matrix $\mathbf{O}$ by (15.88),

$$\mathbf{O} = \boldsymbol{\mathcal{D}}_{\chi} \boldsymbol{\mathcal{L}}^{-1} \boldsymbol{\mathcal{D}}_{w}^{\dagger}, \tag{17.187}$$

and use it to reconstruct the object coefficients as in (15.90):

$$\hat{\boldsymbol{\theta}} = \frac{1}{C} \mathbf{O} \mathbf{g}. \tag{17.188}$$

Discretization of the 2D inverse Radon transform To see how to apply these formulas to SPECT, consider first the situation where collimator blur and attenuation are neglected, so the irradiance on the detector plane is assumed to be described by the 2D Radon transform of the object, $\lambda(p, \phi)$. We further assume point sampling at uniform intervals in both p and ϕ, so we can index the discrete measurements with one index j for the ϕ variable and a second index k for the p variable. A convenient way to normalize the sampling operator $\boldsymbol{\mathcal{D}}_w$ is to take $w_{jk}(p, \phi) = [\Delta p \, \Delta \phi]^{\frac{1}{2}} \, \delta(p - p_k) \, \delta(\phi - \phi_j)$, where Δp and $\Delta \phi$ are the sampling intervals in p and ϕ, respectively. With these definitions, the mean value of the $(jk)^{th}$ measurement is given by

$$\overline{g}_{jk} = C' \int_{-\infty}^{\infty} dp \int_0^{\pi} d\phi \; \delta(p - p_k) \, \delta(\phi - \phi_j) \, \lambda(p, \phi) = C' \lambda(p_k, \phi_j) = C [\boldsymbol{\mathcal{D}}_w \boldsymbol{\mathcal{R}}_2 \mathbf{f}]_{jk} \,, \tag{17.189}$$

where C' is a constant related to the system and exposure time but independent of the definition of $\boldsymbol{\mathcal{D}}_w$, and $C \equiv C' [\Delta p \, \Delta \phi]^{-\frac{1}{2}}$.

With this data model and the inverse Radon transform as given in (4.161), the reconstruction matrix $\mathbf{O}$ has elements given by

$$\begin{aligned} O_{njk} &= \int_{\infty} d^2 r \; \chi_n(\mathbf{r}) \int_0^{\pi} d\phi \int_{-\infty}^{\infty} dp \; h(\mathbf{r} \cdot \hat{\mathbf{n}} - p) \, \delta(p - p_k) \, \delta(\phi - \phi_j) \\ &= \int_{\infty} d^2 r \; \chi_n(\mathbf{r}) \, h(\mathbf{r} \cdot \hat{\mathbf{n}}_j - p_k) \,, \end{aligned} \tag{17.190}$$

where $h(p)$ is the generalized function $-1/(2\pi^2 p^2)$, *i.e.*, the inverse Fourier transform of a ramp filter. Thus, with point sampling, we need only evaluate the shifted filter function at the sample points in data space, backproject, and then compute its scalar product in object space with (say) a pixel function.

By expressing $h(p)$ in terms of its Fourier transform, we can also write

$$\begin{aligned} O_{njk} &= \int_{\infty} d^2 r \; \chi_n(\mathbf{r}) \int_{-\infty}^{\infty} |\nu| \, d\nu \; \exp[2\pi i (\mathbf{r} \cdot \hat{\mathbf{n}}_j - p_k) \nu] \\ &= \int_{-\infty}^{\infty} |\nu| \, d\nu \; X_n^*(\hat{\mathbf{n}}_j \nu) \exp(-2\pi i \nu p_k) \,, \end{aligned} \tag{17.191}$$

where $X_n(\boldsymbol{\rho})$ is the 2D Fourier transform of $\chi_n(\mathbf{r})$. We can include apodization in this form by inserting into the integrand a factor $A(\nu)$ that goes to zero as ν increases [*cf.* (17.117)]. On the other hand, choice of $\chi_n(\mathbf{r})$ itself is a kind of apodization or regularization since $X_n(\hat{\mathbf{n}}_j \nu) \to 0$ as $\nu \to \infty$. The rolloff is more rapid if $\chi_n(\mathbf{r})$ is smoother or wider.

If we choose $\chi_n(\mathbf{r})$ to be the delta function $\delta(\mathbf{r}-\mathbf{r}_n)$, then $X_n(\hat{\mathbf{n}}_j\nu) = \exp(-2\pi i\mathbf{r}_n \cdot \hat{\mathbf{n}}_j\nu)$. In that case, it is imperative that we include an apodizing factor to control noise amplification, and the matrix elements are given by

$$O_{njk} = \int_{-\infty}^{\infty} |\nu|\, d\nu\, A(\nu) \exp(2\pi i\mathbf{r}_n \cdot \hat{\mathbf{n}}_j\nu) \exp(-2\pi i\nu p_k) = h_A(\mathbf{r}_n \cdot \hat{\mathbf{n}}_j - p_k)\,, \quad (17.192)$$

where $h_A(p) = \mathcal{F}_1^{-1}\{|\nu| A(\nu)\}$. Application of this matrix implements the commonly used filtered-backprojection algorithm.

Result of applying the discretized inverse If the data model (17.185) were accurate, $\hat{\boldsymbol{\theta}}$ would be expressed by [*cf.* (15.91)]

$$\hat{\boldsymbol{\theta}} = \mathcal{D}_\chi \mathcal{R}_2^{-1} \mathcal{D}_w^\dagger \mathcal{D}_w \mathcal{R}_2 \mathbf{f} + \frac{1}{C} \mathcal{D}_\chi \mathcal{R}_2^{-1} \mathcal{D}_w^\dagger \mathbf{n}\,. \quad (17.193)$$

The first term is the mean of $\boldsymbol{\theta}$ since the mean of $\mathbf{n}$ is, by definition, zero. To be explicit, we can write the mean of the nth component of $\boldsymbol{\theta}$ as

$$\bar{\hat{\theta}}_n = \int_\infty d^2r \int_\infty d^2r'\, \chi_n(\mathbf{r})\, p(\mathbf{r}, \mathbf{r}')\, f(\mathbf{r}')\,, \quad (17.194)$$

where $p(\mathbf{r}, \mathbf{r}')$ is the kernel of the CC operator $\mathcal{R}_2^{-1} \mathcal{D}_w^\dagger \mathcal{D}_w \mathcal{R}_2$, given by

$$p(\mathbf{r}, \mathbf{r}') = \Delta p\, \Delta\phi \sum_{j,k} h_A(\mathbf{r} \cdot \hat{\mathbf{n}}_j - p_k)\, \delta(p_k - \mathbf{r}' \cdot \hat{\mathbf{n}}_j)\,. \quad (17.195)$$

With the normalization we have adopted here, $p(\mathbf{r}, \mathbf{r}') \to \delta(\mathbf{r}-\mathbf{r}')$ in the limit that $A(\nu) \to 1$ and the number of samples in both p and ϕ goes to infinity so that the sums can be replaced by integrals. Thus, if our data model is correct and there is no noise, the discretized operator yields perfect reconstruction in that limit.

More realistically, however, we must recognize that the system operator $\mathcal{H}$ is not simply $C\mathcal{D}_w \mathcal{R}_2$; there are other effects such as collimator blur and attenuation, so we must write

$$\hat{\boldsymbol{\theta}} = \frac{1}{C} \mathcal{D}_\chi \mathcal{R}_2^{-1} \mathcal{D}_w^\dagger \mathcal{H} \mathbf{f} + \frac{1}{C} \mathcal{D}_\chi \mathcal{R}_2^{-1} \mathcal{D}_w^\dagger \mathbf{n}\,. \quad (17.196)$$

The general form (17.195) still applies, but now $p(\mathbf{r}, \mathbf{r}')$ is the kernel of the CC operator $\frac{1}{C} \mathcal{R}_2^{-1} \mathcal{D}_w^\dagger \mathcal{H}$. Explicitly,

$$p(\mathbf{r}, \mathbf{r}') = \frac{1}{C} \sum_{j,k} h_A(\mathbf{r} \cdot \hat{\mathbf{n}}_j - p_k)\, h_{jk}(\mathbf{r}')\,, \quad (17.197)$$

where $h_{jk}(\mathbf{r}')$ is what we usually call $h_m(\mathbf{r}')$, namely the kernel of the system operator $\mathcal{H}$. [The reader should not confuse $h_{jk}(\mathbf{r}')$ with the filter functions $h(p)$ or $h_A(p)$.]

Bias and estimability If $p(\mathbf{r}, \mathbf{r}')$ can be approximated by a delta function in (17.194), then the estimate of θ_n is unbiased. Unfortunately, this happy condition almost never occurs, for two reasons. First, there is no unbiased estimator for all true values of θ_n unless that parameter is estimable, as defined in Sec. 15.1.3, and we know from that section that it is estimable if and only if $\chi_n(\mathbf{r})$ can be written as a linear superposition of the sensitivity functions. For the data model of (17.185), that means that $\chi_n(\mathbf{r})$ is a superposition of line delta functions—not something one would ever be very interested in estimating. Second, even if θ_n is estimable, the specific estimate obtained by applying the discretized inverse Radon transform will still be biased in essentially every case since the discretized inverse Radon transform is not the pseudoinverse of the discretized Radon forward operator of (17.185), and that operator is not an accurate description of the real system in the first place. In short, filtered backprojection will inevitably lead to biased estimates of the parameters.

For any choice of $\chi_n(\mathbf{r})$ and any assumed object $f(\mathbf{r})$, one can numerically compute the bias in the estimate of θ_n, but the authors would discourage the readers from following that route, at least if $\chi_n(\mathbf{r})$ represents a pixel or voxel. As we have argued in Sec. 13.3.2, bias or mean-square error in pixel values is a virtually meaningless measure of image quality. Instead, the values computed by (17.188) should be regarded as merely a linear transformation of the raw data, and the transformed data will then be used by some observer to perform some practical task. We shall return in Sec. 17.3.4 to the question of how well the transformed vector $\hat{\boldsymbol{\theta}}$ captures the information content (in terms of task performance) of the data $\mathbf{g}$.

17.2.6 Matrices for iterative methods

The analytic inverses described above require discretization of a known inverse operator $\mathcal{L}^{-1}$; iterative methods, on the other hand, require discretization of an often poorly known forward operator $\mathcal{H}$. To discretize $\mathcal{L}^{-1}$ in (17.187), we needed two discretization operators, $\mathcal{D}_\chi$ and $\mathcal{D}_w$. The first served to map the continuous output of the inverse into the desired reconstruction functionals, while the second, in its adjoint form, served to convert the actual discrete data into a continuous form where we could apply $\mathcal{L}^{-1}$. For modeling a forward problem as a DD or matrix equation, the actual CD operator $\mathcal{H}$ already has the desired discrete output, so all we need is a discrete representation of the object. Methods for constructing such representations were introduced in Chap. 7 and used extensively in Chap. 15; we shall review this material briefly here and transcribe it into the notation of Sec. 17.1.1.

We represent the 3D object $f(\mathbf{r})$ approximately by

$$f_a(\mathbf{r}) = \sum_{\mathbf{n}=1}^{N} \theta_{\mathbf{n}} \phi_{\mathbf{n}}(\mathbf{r}), \tag{17.198}$$

where subscript a denotes approximate and subscript $\mathbf{n}$ is a multi-index, a 3D vector with integer components specifying location in a 3D array. The summation limits run from 1 to N on each component, so we are here considering an $N \times N \times N$ object representation, but the formalism is easily generalized to other arrays. It will be convenient to think of $\phi_{\mathbf{n}}(\mathbf{r})$ as a voxel function, which is uniform within a cube of side ϵ centered on point $\mathbf{r}_{\mathbf{n}}$, but of course the math is more general.

The approximate object function $f_a(\mathbf{r})$ can be regarded as a vector in an N^3-dimensional Hilbert space which we call representation space (see Sec. 7.1.2); when we take this view, (17.198) is written in operator form as

$$\mathbf{f}_a = \mathcal{D}_\phi^\dagger \boldsymbol{\theta}. \tag{17.199}$$

The system matrix $\mathbf{H}$ is defined by (7.307) or (15.13) as

$$\mathbf{H} \equiv \mathcal{H}\mathcal{D}_{\phi}^{\dagger}, \tag{17.200}$$

with elements given by (7.304) or (15.14). In the present notation, the elements are expressed as

$$H_{\mathbf{mn}} = \int_{\infty} d^3r \; h_{\mathbf{m}}(\mathbf{r})\, \phi_{\mathbf{n}}(\mathbf{r}), \tag{17.201}$$

where $\mathbf{m}$ is the 3D multi-index defined in Sec. 17.1.2; it specifies the 2D location on the detector face as well as the projection angle. This equation shows that a column of $\mathbf{H}$ is the image of $\phi_{\mathbf{n}}(\mathbf{r})$ for all projection directions. Note that the functions $\{\chi_{\mathbf{n}}(\mathbf{r})\}$ do not appear in $\mathbf{H}$ and that there is no need to introduce a data-space discretization operator $\mathcal{D}_w$ since the data are already discrete.

Matrices specific to SPECT The great advantage of iterative algorithms is that one can put all of one's effort into accurate modeling of the forward problem, with no worry about having to solve the corresponding inverse problem analytically. In the case of SPECT imaging with an Anger camera, accurate forward modeling requires accounting for attenuation and scatter of the radiation in the patient's body, radiometric factors (inverse-square and obliquity), depth-dependent blur by the collimator or pinhole, septal penetration, blurring associated with position estimation, and binning the position estimates into a discrete data vector $\mathbf{g}$. As we shall see below, these effects can be taken into account by theoretical analysis, direct measurement, Monte Carlo simulation or some combination of these methods.

All of these methods of determining the system matrix are facilitated by thinking of the matrix in terms of probabilities. We know that $H_{\mathbf{mn}}\theta_{\mathbf{n}}$ is the mean number of photons detected in bin $\mathbf{m}$ of the data array when the object strength in voxel $\mathbf{n}$ is $\theta_{\mathbf{n}}$. If we define $\theta_{\mathbf{n}}$ to be the mean number of photons emitted by voxel $\mathbf{n}$ during some exposure time, then $H_{\mathbf{mn}}$ is directly the probability that a photon emitted from voxel $\mathbf{n}$ is detected in bin $\mathbf{m}$.

This probabilistic viewpoint leads to several useful decompositions of the $\mathbf{H}$ matrix. For example, if a photon is recorded in bin $\mathbf{m}$, it may have arrived at the detector without any scattering, or it may have been scattered one or more times. Since these events are mutually exclusive, the probabilities add, and we can write

$$\mathbf{H} = \mathbf{H}^{(sc)} + \mathbf{H}^{(un)}, \tag{17.202}$$

where the superscript sc denotes *scattered* and un denotes *unscattered*. Both matrices are $M^2J \times N^3$ if we consider an $M \times M$ detector array stepped to J angles and an object decomposed via (17.198) into N^3 expansion functions such as voxels.

Analytic approaches to the $\mathbf{H}$ *matrix* The unscattered component of $\mathbf{H}$ was studied in detail in Sec. 16.2.2 for the case of a parallel-hole collimator, but in a CC formulation rather than the matrix formulation we seek here. The main result from that section was (16.82), which expressed the photon irradiance on the detector plane. We can adapt that expression to SPECT by adding an index j (specifying projection angle) to the photon irradiance $I_p(\mathbf{r})$ and to the collimator transmission $T(\mathbf{r}, \hat{\mathbf{s}})$. For simplicity, we drop the subscript p (which stood for *photon*), but, by some sort of perverted conservation law, we add a superscript as a reminder that we are dealing only with unscattered photons. Thus we rewrite (16.82) as

$$I_j^{(un)}(\mathbf{r}) = \frac{1}{4\pi}\int_{2\pi} d\Omega\, T_j(\mathbf{r},\hat{\mathbf{s}}) \int_0^\infty d\ell\, f(\mathfrak{r} - \hat{\mathbf{s}}\ell\,) \exp\left[-\int_0^\ell d\ell'\, \mu_{tot}(\mathfrak{r} - \hat{\mathbf{s}}\ell', \mathcal{E}_0)\right]. \tag{17.203}$$

The reader should recall our convention that the 2D vector $\mathbf{r}$ and the 3D vector $\mathfrak{r}$ refer to the same physical location.

The relation between the photon irradiance and the discrete detector output is discussed in detail in Sec. 16.2.3. For an Anger camera, the key result is (16.1047), which in the present notation becomes

$$\overline{g}_{\mathbf{m}}^{(un)} = \tau\eta(\mathcal{E}_0)\, P_{acc}(\mathcal{E}_0) \int_{\mathbf{m}} d^2\hat{r} \int_\infty d^2r\; \mathrm{pr}(\hat{\mathbf{r}}|\mathbf{r},\mathcal{E}_0)\, I_j^{(un)}(\mathbf{r}), \tag{17.204}$$

where the range of the $\hat{\mathbf{r}}$ integral is over the 2D extent of the detector pixel indexed by $\mathbf{m} = (m_x, m_y)$. Continuing our probabilistic interpretation, we note that $\int_{\mathbf{m}} d^2\hat{r}\; \mathrm{pr}(\hat{\mathbf{r}}|\mathbf{r},\mathcal{E}_0)$ is the probability that a photon of energy $\mathcal{E}_0$ striking the detector face at point $\mathbf{r}$ is estimated to fall in data bin $\mathbf{m}$, provided that it is absorbed [probability $\eta(\mathcal{E}_0)$] and that its estimated energy falls in the window [probability $P_{acc}(\mathcal{E}_0)$]. Combining (17.203) and (17.204) with (17.201) yields a complicated expression for the elements of $\mathbf{H}^{(un)}$:

$$H_{\mathbf{mn}}^{(un)} = \frac{\tau\,\eta(\mathcal{E}_0)\, P_{acc}(\mathcal{E}_0)}{4\pi} \int_{\mathbf{m}} d^2\hat{r} \int_\infty d^2r\; \mathrm{pr}(\hat{\mathbf{r}}|\mathbf{r},\mathcal{E}_0) \int_{2\pi} d\Omega\, T_j(\mathbf{r},\hat{\mathbf{s}})$$

$$\times \int_0^\infty d\ell\, \phi_{\mathbf{n}}(\mathfrak{r} - \hat{\mathbf{s}}\ell\,) \exp\left[-\int_0^\ell d\ell'\, \mu_{tot}(\mathfrak{r} - \hat{\mathbf{s}}\ell', \mathcal{E}_0)\right]. \tag{17.205}$$

We can put this expression into a more useful form by also adopting a discrete model for the detector. If we imagine that there is a pixel grid on the camera face, we can replace the integral over $\mathbf{r}$ with a discrete sum over points $\{\mathbf{r}_{\mathbf{m}'}\}$. For simplicity we assume that these points have the same spacing ϵ_d as in the final image array. Then we can define a block-diagonal $M^2J \times M^2J$ matrix $\mathbf{H}^{(det)}$ representing the detector; its elements are given by

$$H_{\mathfrak{m}\mathfrak{m}'}^{(det)} = \delta_{jj'}\, \eta(\mathcal{E}_0)\, P_{acc}(\mathcal{E}_0) \int_{\mathbf{m}} d^2\hat{r}\; \mathrm{pr}(\hat{\mathbf{r}}|\mathbf{r}_{\mathbf{m}'},\mathcal{E}_0), \tag{17.206}$$

where $\mathfrak{m} = (\mathbf{m}, j)$ and $\mathfrak{m}' = (\mathbf{m}', j')$. Thus $H_{\mathfrak{m}\mathfrak{m}'}^{(det)}$ is the probability that a photon striking the detector at point $\mathfrak{m}'$ is detected and assigned to bin $\mathfrak{m}$. This probability is assumed here to be independent of which projection is being measured — hence the block-diagonal form — but there could also be a dependence on j if different detectors were used for different projections.

Similarly, we can define an $M^2J \times N^3$ matrix $\mathbf{H}^{(un,prop)}$ representing propagation without scatter from the source location to the detector input; its elements are given by

$$H_{\mathfrak{m}'\mathbf{n}}^{(un,prop)} = \frac{\tau\,\epsilon_d^2}{4\pi}\int_{2\pi} d\Omega\, T_j(\mathbf{r}_{\mathbf{m}'},\hat{\mathbf{s}}) \int_0^\infty d\ell\, \phi_{\mathbf{n}}(\mathfrak{r}_{\mathbf{m}'} - \hat{\mathbf{s}}\ell\,) \exp\left[-\int_0^\ell d\ell'\, \mu_{tot}(\mathfrak{r}_{\mathbf{m}'} - \hat{\mathbf{s}}\ell', \mathcal{E}_0)\right]. \tag{17.207}$$

With these definitions, we see that

$$\mathbf{H}^{(un)} = \mathbf{H}^{(det)}\mathbf{H}^{(un,prop)} . \tag{17.208}$$

A further factorization is achieved if we assume that the attenuation coefficient $\mu_{tot}(\cdot)$ is slowly varying over a distance scale defined by the size of the voxel. If the voxel is small enough, then the function $\phi_{\mathbf{n}}(\mathbf{r}_{\mathbf{m}'} - \hat{\mathbf{s}}\ell)$ is zero unless $\mathbf{r}_{\mathbf{m}'} - \hat{\mathbf{s}}\ell \approx \mathbf{r}_{\mathbf{n}}$, so we can write

$$\exp\left[-\int_0^{\ell} d\ell' \, \mu_{tot}(\mathbf{r}_{\mathbf{m}'} - \hat{\mathbf{s}}\ell', \mathcal{E}_0)\right] \approx \exp\left[-\int_0^{|\mathbf{r}_{\mathbf{m}'}-\mathbf{r}_{\mathbf{n}}|} d\ell'' \, \mu_{tot}\left(\mathbf{r}_{\mathbf{n}} + \frac{\mathbf{r}_{\mathbf{m}'} - \mathbf{r}_{\mathbf{n}}}{|\mathbf{r}_{\mathbf{m}'} - \mathbf{r}_{\mathbf{n}}|}\ell'', \mathcal{E}_0\right)\right] , \tag{17.209}$$

where $\ell'' \equiv \ell - \ell'$, and we note that $\ell \approx |\mathbf{r}_{\mathbf{m}'} - \mathbf{r}_{\mathbf{n}}|$ within the current approximation. Since the right-hand side of (17.209) is independent of $\hat{\mathbf{s}}$ and ℓ, we can remove it from the integrals in (17.207) and write

$$H^{(un,prop)}_{\mathbf{m}'\mathbf{n}} = H^{(geom)}_{\mathbf{m}'\mathbf{n}} A_{\mathbf{m}'\mathbf{n}} , \tag{17.210}$$

where

$$H^{(geom)}_{\mathbf{m}'\mathbf{n}} = \frac{\tau \, \epsilon_d^2}{4\pi} \int_{2\pi} d\Omega \; T_j(\mathbf{r}_{\mathbf{m}'}, \hat{\mathbf{s}}) \int_0^{\infty} d\ell \; \phi_{\mathbf{n}}(\mathbf{r}_{\mathbf{m}'} - \hat{\mathbf{s}}\ell) \tag{17.211}$$

and

$$A_{\mathbf{m}'\mathbf{n}} = \exp\left[-\int_0^{|\mathbf{r}_{\mathbf{m}'}-\mathbf{r}_{\mathbf{n}}|} d\ell'' \, \mu_{tot}\left(\mathbf{r}_{\mathbf{n}} + \frac{\mathbf{r}_{\mathbf{m}'} - \mathbf{r}_{\mathbf{n}}}{|\mathbf{r}_{\mathbf{m}'} - \mathbf{r}_{\mathbf{n}}|}\ell'', \mathcal{E}_0\right)\right] . \tag{17.212}$$

The first factor in (17.210) represents geometric or straight-line propagation from the voxel location to the detector element without attenuation or scatter, and the second factor is the attenuation along this path in the small-voxel approximation. The element-by-element product in (17.210) is called a *Hadamard product* and denoted $\odot$ (see Sec. A.2.8), and we can write

$$\mathbf{H}^{(un,prop)} = \mathbf{H}^{(geom)} \odot \mathbf{A} , \qquad \mathbf{H}^{(un)} = \mathbf{H}^{(det)}\left[\mathbf{H}^{(geom)} \odot \mathbf{A}\right] . \tag{17.213}$$

We might also be able to assume that the attenuation factor is slowly varying over a scale defined by the spatial resolution of the detector or, in other words, that the attenuation of a photon from any point in the object to the point where the photon strikes the detector face is approximately the same as the attenuation to the point where the detector *thinks* the photon strikes it. If that is the case, we can replace $\mathbf{r}_{\mathbf{m}'}$ with $\mathbf{r}_{\mathbf{m}}$ in $A_{\mathbf{m}'\mathbf{n}}$, and we have

$$\mathbf{H}^{(un)} \approx \left[\mathbf{H}^{(det)}\mathbf{H}^{(geom)}\right] \odot \mathbf{A} . \tag{17.214}$$

The advantage of this form is that the product $\mathbf{H}^{(det)}\mathbf{H}^{(geom)}$ depends on the system, while $\mathbf{A}$ depends only on the particular patient being imaged. Thus, if we can characterize the system response in air, then a simple element-by-element correction yields a patient-specific $\mathbf{H}$ matrix, at least when scatter is neglected.

With scatter, the full $\mathbf{H}$ matrix is given by

$$\mathbf{H} = \mathbf{H}^{(sc)} + \left[\mathbf{H}^{(det)}\mathbf{H}^{(geom)}\right] \odot \mathbf{A} . \tag{17.215}$$

Now we have an additive patient-specific correction as well as a multiplicative one. The scatter matrix $\mathbf{H}^{(sc)}$ cannot, however, be decomposed as neatly as $\mathbf{H}^{(un)}$. The Hadamard factorization with an attenuation factor does not work since the scatter arriving at a point on the detector face may come from many different points in the object, each with its own attenuation factor. Moreover, the attenuation coefficient depends on photon energy $\mathcal{E}$, so different attenuation factors would have to be applied for different scattered photons to be rigorous.

To understand the roles of detector and propagation with scatter, we can define an energy-dependent scatter propagation matrix $\mathbf{H}^{(sc,prop)}(\mathcal{E})$ in such a way that $H_{\mathbf{m}'\mathbf{n}}^{(sc,prop)}(\mathcal{E})\Delta\mathcal{E}$ is the probability that a photon emitted from voxel $\mathbf{n}$ will undergo one or more scattering events and arrive at point $\mathbf{m}'$ on the discretized detector face with energy in the small interval $(\mathcal{E} - \frac{1}{2}\Delta\mathcal{E}, \mathcal{E} + \frac{1}{2}\Delta\mathcal{E})$. With this definition, the overall scatter matrix has elements given by

$$H_{\mathbf{mn}}^{(sc)} = \sum_{\mathbf{m}'} \int_0^{\mathcal{E}_0^-} d\mathcal{E}\ \eta(\mathcal{E})\, P_{acc}(\mathcal{E}) \int_{\mathbf{m}} d^2\hat{r}\ \mathrm{pr}(\hat{\mathbf{r}}|\mathbf{r}_{\mathbf{m}'}, \mathcal{E})\, H_{\mathbf{m}'\mathbf{n}}^{(sc,prop)}(\mathcal{E}) . \qquad (17.216)$$

Because of the integral over energy, factorization into purely detector-dependent and detector-independent factors is not possible. We may, however, be able to assume that $\mathrm{pr}(\hat{\mathbf{r}}|\mathbf{r}_{\mathbf{m}'}, \mathcal{E})$ is approximately independent of $\mathcal{E}$, at least for energies such that $P_{acc}(\mathcal{E}) \neq 0$. In that case, we can write

$$\mathbf{H}^{(sc)} = \mathbf{H}^{(est)}\, \mathbf{H}^{(sc,prop)} , \qquad (17.217)$$

where $\mathbf{H}^{(est)}$ is a matrix expressing the position-estimation process, with elements given by

$$H_{\mathbf{mm}'}^{(est)} = \delta_{jj'} \int_{\mathbf{m}} d^2\hat{r}\ \mathrm{pr}(\hat{\mathbf{r}}|\mathbf{r}_{\mathbf{m}'}) , \qquad (17.218)$$

and $\mathbf{H}^{(sc,prop)}$ (without the energy argument) is defined by

$$H_{\mathbf{m}'\mathbf{n}}^{(sc,prop)} = \int_0^{\mathcal{E}_0^-} d\mathcal{E}\ \eta(\mathcal{E}) P_{acc}(\mathcal{E}) H_{\mathbf{m}'\mathbf{n}}^{(sc,prop)}(\mathcal{E}) . \qquad (17.219)$$

Both of the factors in (17.217) involve characteristics of the detector, but the factorization is still useful since $\mathbf{H}^{(est)}$ expresses the spatial blur of the detector and $\mathbf{H}^{(sc,prop)}$ expresses its energy dependence. Moreover, $\mathbf{H}^{(est)}$ is independent of the patient being imaged, while $\mathbf{H}^{(sc,prop)}$ does depend on the patient.

Measurement of $\mathbf{H}$ As suggested in Sec. 7.4.1, it is possible to measure an $\mathbf{H}$ matrix by using a radioactive source with spatial distribution $\phi(\mathbf{r})$ and then systematically stepping it through the various positions $\{\mathbf{r}_{\mathbf{n}}\}$, creating the functions $\phi_{\mathbf{n}}(\mathbf{r}) = \phi(\mathbf{r} - \mathbf{r}_{\mathbf{n}})$. In principle, one could immerse this source in a medium with attenuating and scattering properties similar to those of the patient, thereby directly measuring $H_{\mathbf{mn}}$, but the practical difficulties in doing so hardly require enumeration.

A somewhat more practical method is to image the voxel source at various positions in air, which estimates $\mathbf{H}^{(det)}\mathbf{H}^{(geom)}$. The Hadamard product with a patient-specific attenuation matrix as in (17.214) then gives an estimate of $\mathbf{H}^{(un)}$ completely empirically.

This method is used routinely at the University of Arizona, but it produces startingly large matrices. For example, when the FASTSPECT system (Rowe *et al.*, 1993) is configured for imaging human brains, a volume of at least 3000 cm^3 must be mapped, requiring 375,000 values of $\mathbf{n}$ if 2 mm voxels are used. Since FASTSPECT uses 24 small, modular scintillation cameras, with the output of each discretized into a 64×64 array, each source location gives $24 \times 64 \times 64 \approx 100,000$ measurements, and the dimensions of $\mathbf{H}^{(un)}$ are thus around $100,000 \times 375,000$.

Sparseness Most iterative algorithms can be written to take advantage of sparse matrices. In SPECT, $\mathbf{H}^{(un)}$ is sparse since photons detected in bin $\mathbf{m}$ must have originated somewhere near the line of sight from that detector location back through the collimator or pinhole into the object space; most object voxels are far from this line of sight and cannot contribute to $g_{\mathbf{m}}$. Thus most of the elements of $\mathbf{H}^{(un)}$ are near zero and do not have to be stored. In FASTSPECT, for example, only about 2% of the elements are stored.

Scattered photons, on the other hand, can in principle make it from any object voxel to any data bin, so the sparseness is lost. Since iterative algorithms with sparse matrices execute faster than ones with filled matrices, there is a practical incentive to ignore scattering or to try to compensate for it somehow.

Approximate shift-invariance and interpolation of $\mathbf{H}$ In Sec. 16.2.2, we argued that the point response of a collimator was approximately shift-invariant for lateral translations of the point source [see (16.88)]. To be sure, this response depends on distance from the collimator face, but only by means of a slowly varying scale factor. Similarly, the point response of a pinhole aperture is also approximately shift-invariant laterally and slowly varying longitudinally. Moreover, the detector response is also approximately shift-invariant for most detectors. We can take advantage of these features to reduce the number of components of $\mathbf{H}^{(un)}$ that must be measured.

The trick is to decompose each column of $\mathbf{H}^{(un)}$ into its centroid on the detector and a spread about the centroid by writing

$$H_{\mathbf{mn}}^{(un)} = h_{j\mathbf{n}}[\mathbf{m} - \mathbf{m}_c(\mathbf{n}, j)] \, , \tag{17.220}$$

where the 2D index vector $\mathbf{m}_c(\mathbf{n}, j)$ denotes the centroid of the image of $\phi_{\mathbf{n}}(\mathbf{r})$ on the detector output for the jth projection angle. Because of the sparseness as discussed above, $h_{j\mathbf{n}}[\mathbf{m} - \mathbf{m}_c(\mathbf{n}, j)]$ is nonzero for only a small set of $\mathbf{m}$ near $\mathbf{m}_c(\mathbf{n}, j)$.

If $h_{j\mathbf{n}}(\mathbf{m})$ and $\mathbf{m}_c(\mathbf{n}, j)$ are slowly varying functions of source location $\mathbf{n}$, then we may not need to measure $\mathbf{H}_{\mathbf{mn}}^{(un)}$ for all $\mathbf{n}$; it might suffice to measure every other point or every third point in each of the three directions. The matrix elements for the unmeasured points can be recovered by interpolation of the centroids and averaging the blur functions $h_{j\mathbf{n}}(\mathbf{m})$ for the neighboring measured points. This interpolation can be done on the fly during iterative reconstruction, so it is not necessary to store the matrix elements for unmeasured points. If every other point is measured, there is a savings of a factor of 8 in measurement time and storage required, and if every third point is measured the savings is a factor of 27.

Monte-Carlo estimation of $\mathbf{H}$ The probabilistic interpretation of $\mathbf{H}$ suggests that it can be usefully estimated by Monte Carlo methods. As we saw in Sec. 10.4.5, the essence of a Monte Carlo calculation is simply to trace photons from a source to a detector in a computer. For Monte Carlo calculation of the $\mathbf{n}^{th}$ column of $\mathbf{H}$ with a voxel representation, the source in question is the $\mathbf{n}^{th}$ voxel, and tracing photons from this source is a computer simulation of the measurement process described above. One difference between the computer simulation and an actual measurement, however, is that we can choose to omit the effects of the detector in the former. That is, we can choose to estimate $\mathbf{H}^{(un,prop)}$ and $\mathbf{H}^{(sc,prop)}$ by Monte Carlo methods and then obtain the final $\mathbf{H}$ matrix by multiplying them by a sparse matrix characteristic of the detector.

The interpolation methods described above for measured $\mathbf{H}$ matrices apply also to Monte Carlo calculation, but now the savings are in computation time, not measurement time.

17.3 NOISE AND IMAGE QUALITY

In order to discuss image quality in SPECT, we need an understanding of the noise properties of the images, which means we must both characterize the noise in the raw data and analyze how that noise propagates through the data processing, including any preprocessing steps and the reconstruction algorithm itself. Basic tools for this analysis were developed in Chaps. 8, 11, 12 and 15, and here we shall apply them to SPECT.

In Sec. 17.3.1 we discuss the noise properties of the measured data and how it is modified before being used in image reconstruction. In Sec. 17.3.2, we discuss the effects of noise on reconstructed images, and in Sec. 17.3.3, we address the thorny topic of reconstruction artifacts. Finally, in Sec. 17.3.4 we apply what we have learned to objective assessment of image quality in SPECT.

17.3.1 Noise in the data

We know from the discussions in Chaps. 11 and 12 that the raw data in SPECT are strictly Poisson, in spite of such complicated phenomena as attenuation, scatter, depth-dependent blur and position estimation. That is, conditional on a particular object, the measurements $\{g_\mathrm{m}\}$ are independent of each other, and each is a Poisson random variable (see also Sec. 16.2.4).

The mean of g_m, however, is $[\mathcal{H}\mathbf{f}]_\mathrm{m}$, not $[\mathbf{H}\boldsymbol{\theta}]_\mathrm{m}$. As in (15.10), we can write

$$\mathbf{g} = \mathcal{H}\mathbf{f} + \mathbf{n}\,, \tag{17.221}$$

but in terms of the approximate representation (15.6) or (17.198), we can also write [*cf.* (15.11)]

$$\mathbf{g} = \mathcal{H}\mathbf{f}_a + \mathcal{H}\mathbf{f} - \mathcal{H}\mathbf{f}_a + \mathbf{n} \equiv \mathbf{H}\boldsymbol{\theta} + \boldsymbol{\epsilon}\,, \tag{17.222}$$

where the overall error $\boldsymbol{\epsilon}$ (modeling error plus noise) is given by

$$\boldsymbol{\epsilon} = \mathcal{H}\mathbf{f} - \mathcal{H}\mathbf{f}_a + \mathbf{n}\,. \tag{17.223}$$

There is no reason to believe that the modeling error is small compared to the Poisson noise. Indeed, the Poisson contribution to $\boldsymbol{\epsilon}$ can be made arbitrarily small by increasing the exposure time and hence the mean number of photons collected. (The modeling error can also be made arbitrarily small by increasing the number of expansion functions in (17.198), but this is seldom practical.)

Nevertheless, the covariance matrix of **g** is essentially unchanged by the modeling error, at least so long as object randomness is not considered. Without approximation, we have that

$$\left[\mathbf{K}_{\mathbf{g}|\mathbf{f}}\right]_{\mathbf{mm}'} = \left[\mathcal{H}\mathbf{f}\right]_{\mathbf{m}} \delta_{\mathbf{mm}'} , \tag{17.224}$$

where $\delta_{\mathbf{mm}'}$ is really the product of three Kronecker deltas, one for each of the two components of the detector index and one for the projection angles. Since $\mathbf{H}\boldsymbol{\theta}$ is usually a good approximation to $\mathcal{H}\mathbf{f}$, there is probably no great error in writing

$$\left[\mathbf{K}_{\mathbf{g}|\mathbf{f}}\right]_{\mathbf{mm}'} \approx \left[\mathbf{H}\boldsymbol{\theta}\right]_{\mathbf{m}} \delta_{\mathbf{mm}'} . \tag{17.225}$$

In particular, the modeling error does not introduce any correlations since $\mathcal{H}\mathbf{f}-\mathcal{H}\mathbf{f}_a$ is a nonrandom vector. We emphasize, however, that this is true only conditional on a particular object; when object randomness is considered, the covariance of the modeling error will not be diagonal.

Correction for detector imperfections Real detectors suffer from nonuniform response and possibly geometric distortion, and it is common practice to correct the raw projection data before applying a reconstruction algorithm. For example, nonuniformity in the flood response of the detector can be corrected by dividing by a stored flood image (see Sec. 7.2.1), and various interpolation algorithms can be used to correct distortion. These corrections make most reconstruction algorithms more accurate since the data after correction come closer to what was assumed in the derivation of the algorithm, but they also alter the noise properties of the data.

Simple flood division can have a surprisingly complicated effect on the covariance matrix of SPECT data. Since an experimental flood image itself suffers from Poisson noise, the corrected measurements are ratios of two Poisson random variables, which are not themselves a Poisson random variable. The covariance matrix for the corrected data in a single projection is still diagonal, but the diagonal elements are not equal to the mean. If, however, the same detector is used to record all of the projections by rotating it to different angles, then the fluctuations introduced by dividing by a common noisy flood are correlated from angle to angle, and the covariance matrix is no longer diagonal in the angular variables j and j'. Fortunately, this complication can be avoided by using a very large number of counts in the reference flood image.

Distortion in scintillation cameras of the Anger type arises from bias in the position estimation (see Sec. 12.3.6). With bias, detected events are not assigned to their correct locations on the camera face, even on average. If we apply a discretized analytic reconstruction to such distorted projection data, severe artifacts can result from the mismatch between the actual data characteristics and those assumed in derivation of the algorithm. Similarly, if we use an iterative algorithm with an analytic **H** matrix that takes no account of the distortion, then again artifacts ensue.

There are essentially three ways to avoid problems with camera distortion: (1) characterize it and incorporate it in the reconstruction algorithm; (2) avoid it in the first place by using unbiased position estimators; or (3) correct the data after measurement but before starting reconstruction. Options 1 and 2 require calibration steps that would be onerous in routine clinical applications. We can characterize the distortion by using a measured **H** matrix, but we saw in Sec. 17.2.6 that the resulting matrix is huge, and it can be used only in time-consuming iterative algorithms. Similarly, we know from Sec. 12.3.6 how to construct a maximum-likelihood position estimator, which is unbiased in the limit of a large number of optical photons per scintillation event, but it also requires an additional calibration step to determine the response functions of the individual photomultipliers.

Thus routine clinical practice currently uses some sort of distortion correction. The usual procedure is to record a high-count image of a radioactive grid and then interpolate the data so that the resulting image after interpolation is undistorted. When this same interpolation is applied to actual, noisy projection data, distortion is removed on average, but short-range correlations can be introduced. To the authors' knowledge, the effect of these correlations on the reconstructed images has not been studied.

Scatter correction Since real gamma-ray detectors do not have perfect energy discrimination, measured projection data will inevitably have contributions from both unscattered and scattered photons. Almost all reconstruction algorithms are derived on the assumption of no scatter, and severe quantitative errors can result when such algorithms are applied to real data. Much effort has been expended on ways of preprocessing the data to remove the effects of scatter on average, but much less attention has been paid to the effect of these corrections on statistical properties of the data.

One common way of removing scatter is to use two energy windows, one centered on the photopeak and one centered at a substantially lower energy, separated from the photopeak by more than the energy resolution of the detector. With isotopes that emit only a single gamma-ray energy, there is very little probability that an unscattered photon will contribute to the lower window, so the number of counts in that window can be used to form an estimate of the mean number of scattered photons contributing to a particular detector bin in a particular projection direction. This estimate can then be used along with some assumed scatter spectrum and knowledge of $P_{acc}(\mathcal{E})$ to compute an estimate of the scatter contribution in the photopeak window and subtract it off.

Since the number of counts in the lower window is a Poisson random variable, the corrected photopeak measurement is a weighted difference between Poisson random variables. This difference is not itself Poisson, but it is statistically independent from detector bin to detector bin and also from projection angle to projection angle. Thus the covariance matrix of the corrected data remains diagonal, but the diagonal elements are not equal to the mean.

Noise in the $\mathbf{H}$ *matrix* Up to here in this book, we have regarded the system matrix $\mathbf{H}$ as a deterministic quantity; we may not know it very well, but it isn't random. When we either measure it directly or compute it by Monte Carlo simulation, however, $\mathbf{H}$ becomes random in a sense: if we were to repeat the measurement or simulation, we would get a different matrix. On the other hand, if we use this random matrix in any reconstruction algorithm, it becomes a fixed characteristic of the algorithm. Errors of $\mathbf{H}$ arising from Poisson statistics in a measurement are, in principle, no different from modeling errors. They do not contribute to the covariance of reconstructed images if we think of that covariance in terms of repeated images, all reconstructed with the same $\mathbf{H}$, but they do, of course, affect objective measures of image quality.

17.3.2 Noise in reconstructed images

As we know from Secs. 15.2.6, 15.3.6 and 15.4.7, Poisson noise in the raw data can propagate in complicated ways through linear or nonlinear reconstruction algorithms. In this section we point out some of the complications that are specific to emission tomographic reconstruction.

Noise in one-step linear reconstructions The variance and covariance after application of a linear reconstruction operator $\mathbf{O}$ were derived in Sec. 15.2.6. Transcribing (15.120) and (15.121) into our multi-index notation and making the same approximation as in (17.225), we obtain

$$\left[\mathbf{K}_{\hat{\boldsymbol{\theta}}|\mathbf{f}}\right]_{\mathbf{nn}'} = \sum_{\mathbf{m}} [\mathcal{H}\mathbf{f}]_{\mathbf{m}} O_{\mathbf{nm}} O^*_{\mathbf{n}'\mathbf{m}} \approx \sum_{\mathbf{m}} [\mathbf{H}\boldsymbol{\theta}]_{\mathbf{m}} O_{\mathbf{nm}} O^*_{\mathbf{n}'\mathbf{m}} , \tag{17.226}$$

$$\mathrm{Var}\left\{\hat{\theta}_{\mathbf{n}}|\mathbf{f}\right\} = \sum_{\mathbf{m}} [\mathcal{H}\mathbf{f}]_{\mathbf{m}} |O_{\mathbf{nm}}|^2 \approx \sum_{\mathbf{m}} [\mathbf{H}\boldsymbol{\theta}]_{\mathbf{m}} |O_{\mathbf{nm}}|^2 . \tag{17.227}$$

The variance expressions can also be written as [*cf.* (15.122) and (15.123)]

$$\mathrm{Var}\left\{\hat{\theta}_{\mathbf{n}}|\mathbf{f}\right\} = \int_{\mathbf{S}_f} d^q r \, \aleph_{\mathbf{n}}(\mathbf{r})\, f(\mathbf{r}) \approx \sum_{\mathbf{n}'} \aleph_{\mathbf{nn}'} \boldsymbol{\theta}_{\mathbf{n}'} , \tag{17.228}$$

where $\aleph_{\mathbf{n}}(\mathbf{r})$ is the noise kernel, defined by

$$\aleph_{\mathbf{n}}(\mathbf{r}) \equiv \sum_{\mathbf{m}} |O_{\mathbf{nm}}|^2 h_{\mathbf{m}}(\mathbf{r}) , \tag{17.229}$$

and $\boldsymbol{\aleph}$ is a matrix approximation to the noise kernel, with elements given by

$$\aleph_{\mathbf{nn}'} \equiv \sum_{\mathbf{m}} |O_{\mathbf{nm}}|^2 H_{\mathbf{mn}'} . \tag{17.230}$$

The noise kernel thus makes it possible to see how noise at different locations in the reconstruction arises from different points in the object.

In evaluating these variance and covariance expressions, relatively crude approximations to $[\mathcal{H}\mathbf{f}]_{\mathbf{m}}$ or $[\mathbf{H}\boldsymbol{\theta}]_{\mathbf{m}}$ can often be used. Blurring of the data by the collimator or detector is relatively unimportant in (17.227), for example, since $|O_{\mathbf{nm}}|^2$ is nonnegative and hence itself a blurring operator. In addition, small or low-contrast features in the object can be ignored if they do not affect the mean data very much. For example, suppose the object is a cylindrical phantom of diameter D with uniform activity f_0 except for a small, hot sphere simulating a tumor. If the diameter of the tumor is D_t and its activity is f_t, then the tumor can be ignored for purposes of computing the noise (though not for purposes of computing the mean image) if $D_t f_t \ll D f_0$. In this case, most of the counts come from the large cylindrical region, so most of the noise comes from there as well.

Filtered backprojection To understand the implications of (17.226)–(17.230) for SPECT reconstructions, let us consider filtered backprojection with the sampled filter function of (17.192). That filter was derived for 2D reconstruction, but it is applicable to 3D rotating-camera SPECT with a parallel-hole collimator since the 3D volume can be reconstructed one 2D slice at a time (see Sec. 17.1.4).

The matrix elements of (17.192) are given by[12]

$$O_{\mathbf{n}jk} = \int_{-\infty}^{\infty} |\nu|\, d\nu\; A(\nu) \exp(2\pi i \mathbf{r_n} \cdot \hat{\mathbf{n}}_j \nu) \exp(-2\pi i \nu p_k) = h_A(\mathbf{r_n} \cdot \hat{\mathbf{n}}_j - p_k)\,, \quad (17.231)$$

where $h_A(p) = \mathcal{F}_1^{-1}\{|\nu|A(\nu)|\}$. This function (or actually, its negative) is plotted in Fig. 2.6b for one particular choice of $A(\nu)$, but other choices will give similar behavior: the filter function $h(p)$ will always have a central positive core and negative wings that fall off as $-1/p^2$.

With (17.229) and (17.231), the noise kernel for filtered backprojection is given by

$$\aleph_{\mathbf{n}}(\mathbf{r}) = \sum_j \sum_k h_A^2(\mathbf{r_n} \cdot \hat{\mathbf{n}}_j - p_k)\, h_{jk}(\mathbf{r})\,. \quad (17.232)$$

Similarly, the covariance matrix can be written as

$$\left[\mathbf{K}_{\hat{\boldsymbol{\theta}}|\mathbf{f}}\right]_{\mathbf{nn'}} = \sum_j \sum_k h_A(\mathbf{r_n} \cdot \hat{\mathbf{n}}_j - p_k)\, h_A(\mathbf{r_{n'}} \cdot \hat{\mathbf{n}}_j - p_k) [\boldsymbol{\mathcal{H}}\mathbf{f}]_{jk}\,. \quad (17.233)$$

As noted above, we do not need to be very precise in specifying the mean data $[\boldsymbol{\mathcal{H}}\mathbf{f}]_{jk}$ when computing the noise properties, so rough approximations to $h_{jk}(\mathbf{r})$ can be used. In particular, if we ignore attenuation and all blurring processes, we can write

$$h_{jk}(\mathbf{r}) \approx C\, \delta(p_k - \mathbf{r} \cdot \hat{\mathbf{n}}_j)\,, \quad (17.234)$$

where [*cf.* (16.98) and (16.104)]

$$C = \tau\, \eta(\mathcal{E}_0)\, P_{acc}(\mathcal{E}_0)\, \epsilon^2 \frac{\alpha_{pf} D_b^2}{16 L_b^2}\,, \quad (17.235)$$

with τ being the exposure time for a single projection and ϵ^2 the area of a bin in the camera output.

Continuous data Though the discrete expressions (17.232) and (17.233) should be used when accurate results are desired, considerable insight into the noise properties of filtered backprojection can be obtained by regarding the data as continuous and replacing the sums by integrals. First, we assume that very fine angular sampling is used so that the sum over j can be replaced by an integral over projection angle ϕ. With rotating-camera SPECT, this approximation can be realized in practice to an arbitrary accuracy since the camera can be stepped in fine angular increments with no difficulty. Second, we assume that the grid used to record the irradiance on the detector is fine compared to the large-scale structure of the mean data; as noted above, we do not need to worry much about small-scale structure in the mean data

[12] We are now using a mixed index notation, where the multi-index $\mathbf{n}$ specifies the voxel in the reconstruction and the scalar indices j and k specify rotation angle and 1D position on the detector, respectively. The reader should not confuse $\mathbf{n}$ with the unit vector $\hat{\mathbf{n}}_j$.

when calculating the noise. Taking the sampling intervals in p and ϕ as Δp and $\Delta\phi$, respectively, and assuming 360° rotation of the camera, we can rewrite (17.233) as

$$\left[\mathbf{K}_{\hat{\boldsymbol{\theta}}|\mathbf{f}}\right]_{\mathbf{nn}'} = \frac{1}{\Delta\phi\,\Delta p}\int_0^{2\pi} d\phi \int_{-\infty}^{\infty} dp\; h_A(\mathbf{r_n}\cdot\hat{\mathbf{n}}-p)\,h_A(\mathbf{r_{n'}}\cdot\hat{\mathbf{n}}-p)\,[\boldsymbol{\mathcal{H}}\mathbf{f}](p,\phi)\,. \quad (17.236)$$

With approximation (17.234), the covariance matrix becomes

$$\left[\mathbf{K}_{\hat{\boldsymbol{\theta}}|\mathbf{f}}\right]_{\mathbf{nn}'} = \frac{C}{\Delta\phi\,\Delta p}\int_\infty d^2r_o\; f(\mathbf{r}_o)\int_0^{2\pi} d\phi\; h_A[(\mathbf{r_n}-\mathbf{r}_o)\cdot\hat{\mathbf{n}}]\,h_A[(\mathbf{r_{n'}}-\mathbf{r}_o)\cdot\hat{\mathbf{n}}]\,. \quad (17.237)$$

The variance is obtained by setting $\mathbf{n} = \mathbf{n}'$, or

$$\mathrm{Var}\left\{\hat{\theta}_\mathbf{n}|\mathbf{f}\right\} = \frac{C}{\Delta\phi\,\Delta p}\int_\infty d^2r_o\; f(\mathbf{r}_o)\int_0^{2\pi} d\phi\; h_A^2[(\mathbf{r_n}-\mathbf{r}_o)\cdot\hat{\mathbf{n}}]\,, \quad (17.238)$$

so the noise kernel is given by

$$\aleph_\mathbf{n}(\mathbf{r}_o) = \frac{C}{\Delta\phi\,\Delta p}\int_0^{2\pi} d\phi\; h_A^2[(\mathbf{r_n}-\mathbf{r}_o)\cdot\hat{\mathbf{n}}]\,. \quad (17.239)$$

An immediate conclusion from (17.238) is that the mapping from the object to the variance distribution is shift-invariant under our assumptions of continuous data and simple Radon system kernel (17.234). It does not follow, however, that the noise is stationary; for a stationary random process, the variance must be independent of position, while here it is given as a convolution of the noise kernel with the object.

Another useful form for the noise kernel can be obtained by the change of variable $u = |\mathbf{r_n} - \mathbf{r}_o|\cos\phi'$, where ϕ' is the angle between $\mathbf{r_n} - \mathbf{r}_o$ and $\hat{\mathbf{n}}$. Then we have

$$\aleph_\mathbf{n}(\mathbf{r}_o) = \frac{2C}{\Delta\phi\,\Delta p}\int_{-|\mathbf{r_n}-\mathbf{r}_o|}^{|\mathbf{r_n}-\mathbf{r}_o|} du\;\frac{h_A^2(u)}{\sqrt{|\mathbf{r_n}-\mathbf{r}_o|^2-u^2}}\,. \quad (17.240)$$

To interpret this result, we recognize that $h_A^2(u)$ is nonnegative and falls off asymptotically as $1/u^4$, so it is highly concentrated near $u = 0$. The width of this function is of order $1/\nu_A$, where ν_A is the width of $A(\nu)$. When $|\mathbf{r_n} - \mathbf{r}_o| \gg 1/\nu_A$, we see that

$$\aleph_\mathbf{n}(\mathbf{r}_o) \approx \frac{2C}{\Delta\phi\,\Delta p}\frac{1}{|\mathbf{r_n}-\mathbf{r}_o|}\int_{-\infty}^{\infty} du\; h_A^2(u) = \frac{2C}{\Delta\phi\,\Delta p}\frac{1}{|\mathbf{r_n}-\mathbf{r}_o|}\int_{-\infty}^{\infty} d\nu\;\nu^2 A^2(\nu)\,, \quad (17.241)$$

where the last step follows from Parseval's theorem, (3.80). To be explicit, if we choose

$$A(\nu) = \exp\left(-\frac{\nu^2}{2\nu_0^2}\right), \quad (17.242)$$

then

$$\int_{-\infty}^{\infty} d\nu\;\nu^2 A^2(\nu) = \frac{\sqrt{\pi}}{2}\nu_0^3\,. \quad (17.243)$$

This result shows that the variance at a point in the reconstruction varies as the cube of the rolloff frequency of the filter function. Since the spatial resolution varies only linearly with ν_0, a two-fold increase in resolution is accompanied by an eight-fold increase in variance (or $2\sqrt{2}$-fold in standard deviation).

Disk object Additional features of the noise in filtered backprojection can be discerned by considering a uniform disk object. For simplicity, we continue to assume continuous data and the Radon kernel of (17.234).

For a disk object of radius R_d and activity f_0, (17.237) becomes

$$\left[\mathbf{K}_{\hat{\theta}|\mathbf{f}}\right]_{\mathbf{nn}'} = \frac{Cf_0}{\Delta\phi\,\Delta p}\int_0^{R_d} r_o\,dr_o \int_0^{2\pi} d\theta_o \int_0^{2\pi} d\phi\; h_A[(\mathbf{r_n}-\mathbf{r}_o)\cdot\hat{\mathbf{n}}]\,h_A[(\mathbf{r_{n'}}-\mathbf{r}_o)\cdot\hat{\mathbf{n}}]\,. \tag{17.244}$$

Representing the filter functions by their Fourier transforms and swapping integrals, we obtain

$$\left[\mathbf{K}_{\hat{\theta}|\mathbf{f}}\right]_{\mathbf{nn}'} = \frac{Cf_0}{\Delta\phi\,\Delta p}\int_{-\infty}^{\infty} d\nu\,|\nu|\;A(\nu)\int_{-\infty}^{\infty} d\nu'\,|\nu'|\;A(\nu')\int_0^{R_d} r_o\,dr_o\int_0^{2\pi} d\theta_o\int_0^{2\pi} d\phi$$

$$\times\,\exp\{2\pi i[(\mathbf{r_n}-\mathbf{r}_o)\cdot\hat{\mathbf{n}}\nu]\}\,\exp\{-2\pi i[(\mathbf{r_{n'}}-\mathbf{r}_o)\cdot\hat{\mathbf{n}}\nu']\}\,, \tag{17.245}$$

where the minus sign in the last exponential is allowed since $h_A(p)$ is real. The integrals over r_o and θ_o comprise the Fourier transform of the disk evaluated at $\hat{\mathbf{n}}(\nu-\nu')$, and the transform can be performed by (3.258). The integral over ϕ can be performed by (3.247), so we find

$$\left[\mathbf{K}_{\hat{\theta}|\mathbf{f}}\right]_{\mathbf{nn}'} = \frac{2\pi R_d C f_0}{\Delta\phi\,\Delta p}\int_{-\infty}^{\infty} d\nu\,|\nu|\;A(\nu)\int_{-\infty}^{\infty} d\nu'\,|\nu'|\;A(\nu')$$

$$\times\,\frac{\mathrm{J}_1[2\pi R_d(\nu-\nu')]}{\nu-\nu'}\,\mathrm{J}_0(2\pi|\mathbf{r_n}\nu-\mathbf{r_{n'}}\nu'|)\,. \tag{17.246}$$

As the disk radius goes to infinity, the besinc approaches a 1D delta function:

$$\lim_{R_d\to\infty}\frac{\mathrm{J}_1[2\pi R_d(\nu-\nu')]}{\nu-\nu'} = 2\,\delta(\nu-\nu')\,. \tag{17.247}$$

Thus

$$\left[\mathbf{K}_{\hat{\theta}|\mathbf{f}}\right]_{\mathbf{nn}'} \to \frac{4\pi R_d C f_0}{\Delta\phi\,\Delta p}\int_{-\infty}^{\infty} d\nu\,|\nu|^2\;A^2(\nu)\,\mathrm{J}_0(2\pi|\mathbf{r_n}-\mathbf{r_{n'}}|\nu)\,. \tag{17.248}$$

Thus the covariance *matrix* has the form of a continuous stationary autocovariance *function* sampled at $\mathbf{r_n}-\mathbf{r_{n'}}$. Since this function is rotationally symmetric, and since the 2D Fourier transform of a rotationally symmetric function is the Hankel transform [see (3.248)], we can also write

$$\left[\mathbf{K}_{\hat{\theta}|\mathbf{f}}\right]_{\mathbf{nn}'} \to \frac{2R_d C f_0}{\Delta\phi\,\Delta p}\mathcal{F}_2^{-1}\left\{|\nu|\,A^2(\nu)\right\}\big|_{r=|\mathbf{r_n}-\mathbf{r_{n'}}|}\,. \tag{17.249}$$

By the Wiener-Khinchin theorem (8.133), it follows that the noise power spectrum (before sampling on the reconstruction grid) is proportional to $|\nu|\,A^2(\nu)$. We must remember, however, that the noise is stationary in this sense only in the limit $R_d\to\infty$; for a finite disk the noise is not stationary and the noise power spectrum is not defined.

To see the lack of stationarity, we can rewrite (17.246) as

$$\left[\mathbf{K}_{\hat{\boldsymbol{\theta}}|\mathbf{f}}\right]_{\mathbf{nn}'} = \frac{2\pi R_d C f_0}{\Delta\phi\,\Delta p} \int_{-\infty}^{\infty} d\nu\,|\nu|\;A(\nu) \int_{-\infty}^{\infty} d\nu'\,|\nu'|\;A(\nu')$$

$$\times \frac{\mathrm{J}_1[2\pi R_d \Delta\nu]}{\Delta\nu}\,\mathrm{J}_0(2\pi|\overline{\mathbf{r}}\Delta\nu - \Delta\mathbf{r}\overline{\nu}|)\,, \tag{17.250}$$

where $\overline{\mathbf{r}} \equiv \frac{1}{2}(\mathbf{r_n} + \mathbf{r_{n'}})$, $\Delta\mathbf{r} \equiv \mathbf{r_n} - \mathbf{r_{n'}}$ and similarly for the frequency variables. Because of the presence of $\overline{\mathbf{r}}$, this covariance matrix is no longer obtained by sampling a stationary autocovariance function.

One qualitative observation from (17.250) is that the correlations tend to be directed radially away from the center of the disk since the second Bessel function has its maximum value when $\Delta\mathbf{r} = \overline{\mathbf{r}}\Delta\nu/\overline{\nu}$ or, in other words, when the vector $\Delta\mathbf{r}$ is parallel to $\overline{\mathbf{r}}$. This argument is confirmed by the reconstruction shown in Fig. 17.8 with its strong radial correlations.

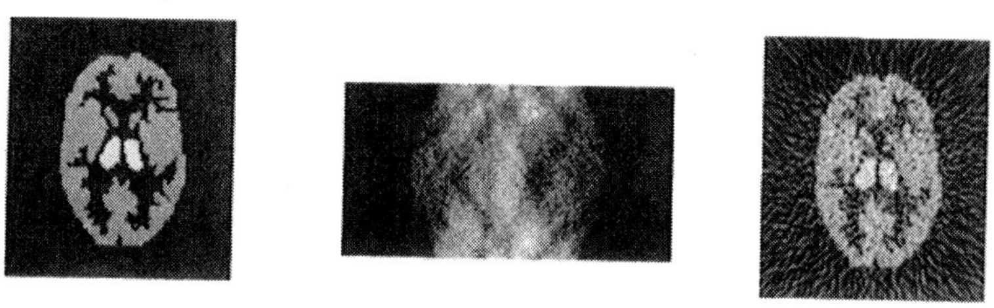

Object Sinogram Reconstruction

Fig. 17.8 Reconstruction of a brain phantom by filtered backprojection. (a) Simulated object. (b) Noisy sinogram. (c) Reconstruction. (Courtesy of Craig Abbey).

Another qualitative observation is that the variance is quite spread out, extending well beyond the edges of the object in Fig. 17.8. In essence, this comes about since the variance pattern is the object convolved with a noise kernel that varies approximately as $1/r_o$ as seen in (17.241). Moreover, because of this slow decay, the variance at the center of the disk is proportional to the radius of the disk, even when that radius is much greater than the spatial resolution of the system. In short, noise in filtered backprojection is very nonlocal.

The reader who wishes to study this problem further — and perhaps remove some of the approximations we made — is encouraged to use the stochastic Wigner distribution function. Numerical evaluation of the integrals will be required, but the results will explain the correlations seen in Fig. 17.8. Another very useful exercise is to apply the methods developed here to the case where attenuation in the object is modeled by the exponential Radon transform (see Sec. 17.1.6) and reconstruction is performed by the Tretiak-Metz algorithm.

Noise in nonlinear reconstructions As we saw in Chap. 15, there are two general approaches to nonlinear image reconstruction. One approach, called implicit estimation, is to specify a functional, often called an objective functional, that should be minimized by proper choice of the reconstruction vector $\hat{\boldsymbol{\theta}}$. The objective functional is frequently chosen so that the minimum is unique, and in that case it really doesn't matter, except in terms of computational effort, what specific algorithm is used to find the minimum. The other approach is to state a specific iterative algorithm, such as the MLEM (maximum-likelihood expectation maximization) algorithm of (15.305), and run it for some number of iterations. The algorithm need not be one for which convergence is guaranteed, and it need not be related to a specific objective functional.

Noise in implicit estimates was discussed in Sec. 15.3.6 under the assumption that the objective functional is differentiable (with respect to both the data and the reconstruction) at its minimum. This condition may not be satisfied if the functional includes a positivity constraint, but if it is, then the final reconstruction is approximately related to the data by a linear mapping as in (15.221). At this point the methods derived above for analyzing noise in one-step linear reconstruction can be applied, but the linear mapping depends on the object.

Noise in iterative algorithms can be analyzed by recursive methods described in Sec. 15.4.7. Again a differentiability condition is required, but now it is only the update rule that needs to be differentiable, not the functional at the final estimate. Many popular iterative algorithms, including the MLEM algorithm, satisfy this condition, and the recursive methods give highly accurate predictions of the mean and covariance as a function of iteration number (Wilson *et al.*, 1994; Wang and Gindi, 1997; Soares *et al.*, 1998).

Example: MLEM The MLEM algorithm, if ever run to completion, would maximize the likelihood $\mathrm{pr}(\mathbf{g}|\boldsymbol{\theta})$. As we discussed in Sec. 15.4.6, however, maximum likelihood is seldom a desirable end point in image reconstruction. Running the MLEM algorithm for a large number of iterations usually results in a virtually useless image, often one consisting of a few bright pixels like the night-sky reconstructions discussed in Sec. 15.3.5.

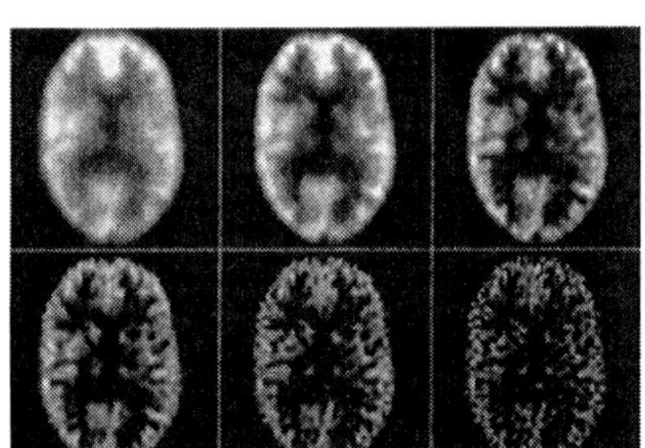

Fig. 17.9 Sequence of reconstructions of a brain phantom by the MLEM algorithm after 10, 20, 50, 100, 200, and 400 iterations (left to right, top to bottom). (Courtesy of D. W. Wilson.)

This behavior is seen in the sequence of reconstructions shown in Fig. 17.9. As the iteration proceeds, the image becomes sharper and noisier. When compared to the reconstruction obtained by filtered backprojection in Fig. 17.8, the MLEM images have a much more localized variance pattern, with no significant noise outside the boundaries of the disk. This point is reinforced by Fig. 17.10, where the mean image of a brain phantom is shown along with a spatial map of the variance. The variance pattern looks strikingly like the mean image, an effect that is predicted by the recursive approach to noise propagation discussed in Sec. 15.4.7. An intuitive way to understand the variance structure in MLEM, or any other reconstruction algorithm that enforces positivity, is to note that the constrained reconstruction cannot go negative. Thus when the mean is small, the variance must be small also.

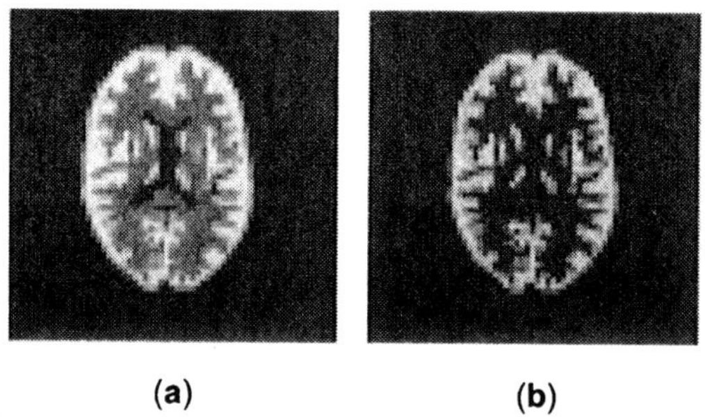

Fig. 17.10 Images illustrating the statistics of the MLEM algorithm. (a) Estimate of the mean image. (b) Estimate of the variance map. The object was a Hoffman brain phantom, and noise-free data were simulated for a SPECT system with 64 projection angles and 64 pixels per projection. Then 2,000 independent random data sets were generated by adding Poisson noise corresponding to an average of 100,000 counts. Reconstructions were performed for 100 iterations, and the 2,000 images were used to compute the sample mean and variance at each pixel, as shown. (Courtesy of D. W. Wilson.)

17.3.3 Artifacts

In addition to noise, there are also defects commonly referred to as *artifacts*[13] in reconstructed images. In general terms, artifacts are deviations between an object and its image, but we know from Sec. 7.1.4 that some deviations are inevitable; no digital image can ever exactly match a continuous object. Not all of these deviations, however, are called artifacts. We accept the fact that a digital image is blurred, perhaps noisy, and displayed on a discrete pixel grid, but if the image exhibits streaks or large areas of erroneous information, we refer to the defects as artifacts.

We can distinguish two general classes of artifact, depending on whether they are evident in the image of a point or require an object of large area. We shall call these two classes *point artifacts* and *areal artifacts*, respectively.

Qualitatively, a point artifact is any nonlocal or long-range structure in the overall system point response function (PRF), including the reconstruction algorithm. Thus blur and pixellization, being short range, are not artifacts, but streaks arising from inadequate angular sampling in tomography are. The scale that distinguishes local from nonlocal effects can be expressed in terms of the maximum spatial frequency passed by the system, including any reconstruction algorithm or post-processing filter used. If we denote this maximum frequency by ρ_{max}, then structures in the PRF extending beyond about $1/\rho_{max}$ will be termed nonlocal and hence artifactual.

[13]Etymologically, artifact (or artefact) derives from the Latin *arte factum*, something made with skill. However, it is *lack* of skill in designing or modeling the system that leads to artifacts. Even if we take the modern meaning of artifact as anything manmade, there is a logical problem since the reconstruction is made by a computer, not a human.

Areal artifacts can also be defined in terms of the system PRF, but they relate to spatial variations in the strength of the PRF. If we denote the PRF for the overall CD mapping from object function to discrete reconstructed image as $p_{\mathbf{n}}(\mathbf{r})$, then there are two distinct measures of the strength: the system sensitivity and the flood response. These terms are discussed in Secs. 7.2.1 and 7.3.1, but in brief system sensitivity is the sum of $p_{\mathbf{n}}(\mathbf{r})$ over $\mathbf{n}$ and flood uniformity is its integral over $\mathbf{r}$. Nonuniformities in either of these measures will be called areal artifacts.

Of course, nonconstant sensitivity or flood response can be problematical in direct imaging systems, but they can usually be corrected in those cases. For example, a measured flood response can be used to renormalize the image as in (7.109). With indirect imaging, on the other hand, we may not always know the cause of the areal artifact, or we may not have enough information to correct it. As an example, attenuation of the radiation in the patient's body in SPECT can lead to an areal artifact, and we cannot correct for it fully if we do not know the distribution of attenuation coefficient.

In reconstructed images, both kinds of artifact are joint characteristics of the imaging system and the reconstruction algorithm. They arise from null functions of the system and/or errors in system modeling, but their nature and their effect on objective measures of image quality depend on algorithmic issues such as nature and degree of regularization, number of iterations, and whether the algorithm enforces positivity.

Both kinds of artifact are deterministic rather than stochastic properties of an image. The term *noise artifacts* also appears in the literature, referring to long-range noise correlations, but we have already discussed that effect in Sec. 17.3.2.

Null functions To isolate the effect of null functions on artifacts, we ignore noise in the data, so that

$$\mathbf{g} = \mathcal{H}\mathbf{f}\,, \tag{17.251}$$

and we assume for now that we know $\mathcal{H}$ perfectly. If we use this $\mathcal{H}$ to compute a pseudoinverse estimate without adopting a discrete representation, say by the Backus-Gilbert method described in Sec. 15.2.2, then

$$\hat{\mathbf{f}} = \mathcal{H}^{+}\mathbf{g} = \mathcal{H}^{+}\mathcal{H}\mathbf{f} = \mathbf{f}_{meas} = \mathbf{f} - \mathbf{f}_{null}\,. \tag{17.252}$$

In many situations, including limited-angle tomography and aliasing, the term $\mathbf{f}_{null}$ is not spatially localized, so it produces an artifact. Note, however, that it is really the *lack* of null functions in the image that constitutes the artifact; the actual object contains null functions, and eliminating them from the image is what gives nonlocal structures.

Sampling and aliasing Sampling and aliasing were analyzed in Sec. 3.5, but from the viewpoint of bandlimited functions. The main result from that section is that a bandlimited function can be recovered exactly from its samples if the Nyquist condition is satisfied. The interpolating function turned out to be a sinc function of appropriate width. Given our definition of artifact, however, we should consider the point response function of the overall process (sampling and interpolation), and points are not bandlimited.

Suppose, for example, that we sample a 2D function $f(\mathbf{r})$ with a regular array of small detectors at locations $\mathbf{r} = \mathbf{m}\epsilon$, producing a data set $\{g_{\mathbf{m}}\}$. If $f(\mathbf{r}) = \delta(\mathbf{r} - \mathbf{r}')$, where $\mathbf{r}'$ is the point location, then at most one of the data values is nonzero. Calling that value $g_{\mathbf{m}'}$, we can write the result of sinc interpolation as [*cf.* (3.298)]

$$\hat{f}(\mathbf{r}) = \sum_{\mathbf{m}} g_{\mathbf{m}} \operatorname{sinc}\left(\frac{\mathbf{r} - \mathbf{m}\epsilon}{\epsilon}\right) = g_{\mathbf{m}'} \operatorname{sinc}\left(\frac{\mathbf{r} - \mathbf{m}'\epsilon}{\epsilon}\right). \tag{17.253}$$

The long tails of the sinc function are the artifacts. Similar long tails occur in tomographic reconstruction when the filter function has a sharp cutoff frequency.

Note that it was necessary to consider the interpolation or reconstruction step before the artifacts became evident. The CD PRF of the sampling step alone is compact since only a single detector element, $\mathbf{m} = \mathbf{m}'$, is activated by the point object. Other interpolation functions, such as triangle functions or Gaussians, would have produced more compact overall PRFs and weaker artifacts in the sense we have defined them.

Another way to look at aliasing is in terms of null functions. Any CD system has a null space, and any two objects that differ by a null function produce the same data and cannot be distinguished by the system, so we say they are aliased. Figure 3.8 gives an example of two cosine function that differ in frequency by the reciprocal of the sampling interval, so they agree exactly at the sample points. Neither of these functions is a null function of the sampling operator, but the difference between them is.

In tomography, aliasing occurs because there are a finite number of projections and a finite number of measurements per projection, but again two objects are aliased if they differ by a null function. This difference function is almost always spatially extended, so aliasing produces artifacts.

Aliasing and Fourier crosstalk A convenient way of quantifying the degree of aliasing is by the Fourier crosstalk matrix, introduced in Sec. 7.3.3 and applied to tomography in Sec. 17.1.3. Unlike other Fourier methods, the Fourier crosstalk matrix does not assume any form of shift-invariance, and it is motivated by CD system models rather than CC ones. The Fourier crosstalk matrix is an exact description of any CD system, fully equivalent to the integral operator $\mathcal{H}^\dagger\mathcal{H}$. Though it is an infinite matrix, its elements have simple physical interpretations. The diagonal elements tell us how well a particular Fourier-*series* coefficient is measured by the system. More importantly for the present discussion, however, the off-diagonal elements specify the degree of aliasing between two different Fourier components.

The idea of degree of aliasing is seldom discussed in treatments of direct imaging systems since it is not an issue for the idealized detector models usually used. If we consider an infinite detector array with uniformly spaced elements on a regular grid, then the null functions of the detector are easy to characterize. Two Fourier-transform kernels whose spatial frequencies differ by a multiple of the (vector) spatial frequency of the sampling grid produce the same data vectors (within a constant related to the size of the detector elements), so the null functions are differences between Fourier kernels. Loosely speaking, we say that the null functions are differences in plane waves, though of course we are discussing irradiance patterns and not waves at all.

In tomography, however, the null functions are not just differences between plane waves, and in fact they are usually quite difficult to determine. In the Fourier-series approach with the crosstalk matrix, we consider the action of the CD system on plane-wave components, but we use plane waves with finite support, and we do not pretend that these functions are either null functions or singular functions of the system. The degree of aliasing between two such components is measured by the cosine of the angle between the resulting vectors in the Hilbert data space [see (7.266)]. When these vectors are nearly parallel, then it is very difficult to distinguish the two Fourier-series components, and we say they are nearly aliased.

Examples of crosstalk matrices for cone-beam SPECT are given by Barrett and Gifford (1994). We shall have more to say about aliasing in SPECT and its relation to image quality in the next section.

Discretization errors Image reconstruction usually requires a discrete model of the object. As we know from Chap. 7, any such discretization must produce errors, but it not so obvious that these errors should be classified as artifacts. Discretization of an object function produces short-range errors and thus is not artifactual by itself, but when a discretized object is imaged (in the computer), then a kind of aliasing or moiré pattern can result, and that is artifactual.

This kind of aliasing is not described by the Fourier crosstalk matrix, which assumes that the object is a function of a continuous variable rather than a discrete array. Instead it is the result of applying a DD reconstruction operator to the data produced by a CD system.

For example, if the CD operator is the unregularized pseudoinverse of some matrix representation of the system operator, we know from Sec. 15.2.1 that the discrete reconstructed image in the absence of noise is given by [*cf.* (15.53)]

$$\hat{\boldsymbol{\theta}} = \mathbf{H}^{+}\mathbf{g} = \left[\boldsymbol{\mathcal{H}}\boldsymbol{\mathcal{D}}_{\phi}^{\dagger}\right]^{+}\boldsymbol{\mathcal{H}}\mathbf{f}\,. \tag{17.254}$$

By our definition, this image contains artifacts to the extent that the CD kernel of $\left[\boldsymbol{\mathcal{H}}\boldsymbol{\mathcal{D}}_{\phi}^{\dagger}\right]^{+}\boldsymbol{\mathcal{H}}$ has long tails or areal defects.

For comparison, we also know from Sec. 15.2.2 that it is possible to perform a reconstruction without adopting a discrete object representation or a system matrix. In that case, with no noise we get the reconstructed function,

$$\hat{\mathbf{f}} = \boldsymbol{\mathcal{H}}^{+}\boldsymbol{\mathcal{H}}\mathbf{f}. \tag{17.255}$$

We can sample this function for display purposes and smooth it in the process. If we use the same smoothing functions here as in the object discretization leading up to (17.254), the result is

$$\hat{\boldsymbol{\theta}} = \boldsymbol{\mathcal{D}}_{\phi}\boldsymbol{\mathcal{H}}^{+}\boldsymbol{\mathcal{H}}\mathbf{f}. \tag{17.256}$$

This time the reconstruction contains artifacts if the CC kernel of $\boldsymbol{\mathcal{D}}_{\phi}\boldsymbol{\mathcal{H}}^{+}\boldsymbol{\mathcal{H}}$ has long tails or areal defects. Since the operator $\boldsymbol{\mathcal{D}}_{\phi}$ is local and independent of location in the reconstructed image, it does not by itself introduce artifacts; any artifacts in (17.256) come from tails or areal defects of the CC operator $\boldsymbol{\mathcal{H}}^{+}\boldsymbol{\mathcal{H}}$. Recall from Sec. 15.2.1 that $\left[\boldsymbol{\mathcal{H}}\boldsymbol{\mathcal{D}}_{\phi}^{\dagger}\right]^{+}\boldsymbol{\mathcal{H}} \neq \boldsymbol{\mathcal{D}}_{\phi}\boldsymbol{\mathcal{H}}^{+}\boldsymbol{\mathcal{H}}$ except in special cases, so (17.254) can exhibit artifacts even if (17.256) does not. In physical terms, some of the artifacts from $\left[\boldsymbol{\mathcal{H}}\boldsymbol{\mathcal{D}}_{\phi}^{\dagger}\right]^{+}\boldsymbol{\mathcal{H}}$ are the result of a moiré effect between the discretization grid and the detector grid, but that effect is absent in $\boldsymbol{\mathcal{D}}_{\phi}\boldsymbol{\mathcal{H}}^{+}\boldsymbol{\mathcal{H}}$. On the other hand, this latter operator can also exhibit artifacts since $\boldsymbol{\mathcal{H}}$ is sampled in angle and in the detector plane.

Errors in system modeling Another broad class of artifacts arises from inaccurate modeling of the system. In general terms, we might collect data through an actual system described by a CD operator $\mathcal{H}$, but assume erroneously that the system was described by some other operator $\mathcal{H}_a$, where the subscript stands for *approximate* or *assumed*. If we form a matrix representation of the system, it will be given by $\mathbf{H} = \mathcal{H}_a \mathcal{D}_\phi^\dagger$ rather than $\mathcal{H}\mathcal{D}_\phi^\dagger$. Depending on the nature of the error in $\mathcal{H}$, a wide variety of artifacts can be produced.

Examples of physical system errors that might be ignored in the formation of the system matrix include alignment errors, such as tilt of the camera head or errors in locating the center of rotation, or detector distortion or nonuniformity. Any of these errors can produce long-range artifacts; for example, if the center of rotation is not specified correctly, it produces a characteristic streak artifact resembling a tuning fork around a point object.

Uncorrected detector nonuniformity can produce either streak or areal artifacts, depending on the nature of the nonuniformity. In general, if we define s_m as the sensitivity of the detector used in the mth measurement, we can write

$$g_m = s_m[\mathcal{H}\mathbf{f}]_m + n_m = \overline{s}[\mathcal{H}\mathbf{f}]_m + (s_m - \overline{s})[\mathcal{H}\mathbf{f}]_m + n_m \,, \tag{17.257}$$

where $\overline{s}$ is the average sensitivity of all detectors. This equation shows that the variations in sensitivity act, in a sense, as an object-dependent noise term. If the detector sensitivities are constant, this term is not stochastic, so this term does not enter into the data covariance matrix for a fixed object. It does, however, constitute a modeling error if we use an assumed $\mathbf{H}$ matrix that does not account for the detector sensitivity in the reconstruction.

To see how detector nonuniformity can lead to streak artifacts, suppose that a different array is used for each projection angle (as in some multi-camera systems developed at the University of Arizona) and that one of the arrays has a single bad element, say the one indexed by $\mathbf{m} = \mathbf{m}_0$. Then the image reconstructed by a matrix with elements $O_{\mathbf{nm}}$ might look relatively artifact-free except that it would contain the erroneous superimposed image $O_{\mathbf{nm}_0}$, which is just the DD PRF of the reconstruction operator. Since this PRF is highly nonlocal, the detector defect produces a streak artifact. If the same detector is used in all projections, as in any single-head rotating-camera system, then a bad element will show up in all projections and appear as a vertical line in the sinogram display of the data (see Sec. 4.4.1). Because of the sharp discontinuity in the data, a streak artifact will again result. On the other hand, slow variations in detector sensitivity will produce areal artifacts.

Motion artifacts Often an object moves or changes in some way during data acquisition, making it a spatio-temporal function, yet the reconstruction algorithm attempts to recover a purely spatial function. For example, during a SPECT scan, there is internal motion due to respiration, heart beats and peristalsis, and there may be an overall motion simply because the patient fidgets. Moreover, the radiotracer can redistribute during acquisition, but the reconstruction algorithm usually treats the object as static.

In some circumstances, temporal changes in the object do not cause artifacts. For example, if we collect all of the data with a static imaging system, so that each detector element is just a time integral of the radiation incident on it, then the data come from a time integral of the spatio-temporal object, and the algorithm seeks to recover this integrated object in some sense. In other words, there is motion blur, but no long-range structures that we would classify as artifacts.

If the detector system is also changing in time, however, complicated motion artifacts can arise. In rotating-camera SPECT, for example, a single camera is scanned to different projection angles. A fan-beam or cone-beam projection is acquired at each detector location, but if the object changes during the scan, then projections of *different* objects are acquired. Since the reconstruction algorithm is based on the assumption that the projections come from the same object, there are inconsistencies in the data that, in practice, lead to severe artifacts.

One might attempt to avoid this problem by doing a full spatio-temporal reconstruction, but the risk is that the motion artifacts would simply be replaced by sampling artifacts; if only M measurements can be made during the exposure time, then attempting to reconstruct the object at K time points means that only M/K of the measurements correspond to each temporal frame. The strategy may nevertheless succeed if there are strong temporal correlations.

Nonlinearities Many indirect imaging systems are really nonlinear but are modeled as linear to facilitate image reconstruction. In Sec. 16.1, for example, we discussed at length the nonlinear relationship between x-ray projection data and the object, specified as a distribution of the x-ray attenuation coefficient. In Sec. 16.1.7 we showed how the simple expedient of taking the logarithm of the data yields an approximate linearization, but we also discussed the residual errors. These errors can lead to areal artifacts in computed tomography if they are slowly varying, but also to streaks if a strongly absorbing point-like object is present.

In SPECT we do not have to deal with an inherently nonlinear relation between an object and its projections, but the detectors may exhibit nonlinearities at high count rates. As noted in Sec. 11.3.1, these nonlinearities result since the occurrence of one gamma-ray interaction can paralyze the detector and prevent the detection of another occurring soon after. This effect spoils the independence of the detected events and invalidates our Poisson data models, but even in the mean it constitutes a modeling error that can produce artifacts.

The general mathematical description for data produced by a nonlinear detector is

$$\mathbf{g} = \boldsymbol{\mathcal{N}}\{\boldsymbol{\mathcal{H}}\mathbf{f}\} = \boldsymbol{\mathcal{H}}\mathbf{f} + [\boldsymbol{\mathcal{N}} - \mathbf{I}]\{\boldsymbol{\mathcal{H}}\mathbf{f}\}\,, \tag{17.258}$$

where $\boldsymbol{\mathcal{N}}$ is a nonlinear operator. If $\boldsymbol{\mathcal{N}}$ is a monotonic point operator, as defined in Sec. 7.5.1, and if it is well characterized, then the effects can be corrected by applying $\boldsymbol{\mathcal{N}}^{-1}$ to the data before reconstruction. With Anger cameras, however, the nonlinearity may be neither local nor monotonic. A high count rate anywhere on the camera face can affect the entire detector, and the electronics might be paralyzable so that the observed count rate will first increase and then decrease with increasing photon flux. Such effects cannot be corrected by any data processing before reconstruction, so they are almost always ignored, with potential artifactual consequences.

Attenuation and scatter As we have discussed earlier in this chapter, attenuation and scattering of gamma rays in the patient's body are strong effects in SPECT, and they are a prime source of artifacts. Many methods for correcting the effects of attenuation and scatter have been described in the literature, but in every case they involve many assumptions and approximations. Nevertheless, they can be useful in reducing if not eliminating the artifacts. For a comprehensive review, see King *et al.* (2003).

Regularization and artifacts Regularization can cover up artifacts in some cases. If the null functions are predominantly high-frequency, then any regularization or post-reconstruction smoothing that eliminates high frequencies from the image also eliminates artifacts arising from null functions. Of course, the smoothing also blurs the image, but this blur can be local and hence not classified as an artifact. Similarly, artifacts related to the sidelobes of the sinc function with sharp-cutoff reconstruction filters are artifacts, but we know from Sec. 15.2.6 that they can be minimized or eliminated by apodization or regularization. Sometimes, however, the artifacts have significant low-frequency components, so regularization or smoothing does not help; an example would be the streak artifacts resulting from limited-angle tomography (Barrett *et al.*, 1991).

On the other hand, regularization can interact with other sources of artifacts and accentuate them. To see how, consider an iterative reconstruction with the least-squares data-agreement functional and Tikhonov regularization, as discussed in Sec. 15.3.3. If we neglect noise and discretization issues and presume perfect knowledge of $\mathcal{H}$, we know that the regularized least-squares solution is [*cf.* (15.156)]

$$\hat{\mathbf{f}}_\eta = \left[\mathcal{H}^\dagger\mathcal{H} + \eta\mathbf{I}\right]^{-1}\mathcal{H}^\dagger\mathbf{g} = \left[\mathcal{H}^\dagger\mathcal{H} + \eta\mathbf{I}\right]^{-1}\mathcal{H}^\dagger\mathcal{H}\mathbf{f}\,. \tag{17.259}$$

Now suppose that there are variations in detector sensitivity, so that

$$\mathcal{H} = \mathbf{s} \odot \mathcal{H}_0\,, \tag{17.260}$$

where $\mathbf{s}$ is a vector of sensitivity factors as in (17.257), $\odot$ denotes the Hadamard product, and $\mathcal{H}_0$ is an operator without sensitivity variations but otherwise identical to $\mathcal{H}$. With some abuse of notation, we can write

$$\hat{\mathbf{f}}_\eta = \left[\mathcal{H}_0^\dagger\mathbf{s}^2\mathcal{H}_0 + \eta\mathbf{I}\right]^{-1}\mathcal{H}_0^\dagger\mathbf{s}^2\mathcal{H}_0\mathbf{f}\,, \tag{17.261}$$

where $\mathcal{H}_0^\dagger\mathbf{s}^2\mathcal{H}_0\mathbf{f}$ is to be interpreted as $\mathcal{H}_0^\dagger[\mathbf{s} \odot \mathbf{s} \odot (\mathcal{H}_0\mathbf{f})]$.

Each sensitivity factor gets squared in forming $\mathcal{H}_0^\dagger\mathbf{s}^2\mathcal{H}_0$, but that would not have much effect on the image in the absence of regularization since

$$\lim_{\eta\to 0}\hat{\mathbf{f}}_\eta = \left[\mathcal{H}_0^\dagger\mathbf{s}^2\mathcal{H}_0\right]^{+}\mathcal{H}_0^\dagger\mathbf{s}^2\mathcal{H}_0\mathbf{f}\,. \tag{17.262}$$

In this limit, the pseudoinverse enforces strict agreement with the (noise-free) data, in the sense that $||\mathcal{H}_0(\hat{\mathbf{f}} - \mathbf{f})|| = 0$. Since the Hadamard product with a nonzero sensitivity factor introduces no new null functions, this implies $||\mathcal{H}(\hat{\mathbf{f}} - \mathbf{f})|| = 0$. Thus, in the limits of no noise and no regularization, the pseudoinverse reconstruction from $\mathcal{H}_0$ is artifact-free if the one from $\mathcal{H}$ is. In essence, the $\mathbf{s}^2$ factor in the forward operator is cancelled by the same factor in the pseudoinverse.

The situation is different, however, if the regularizing parameter does not approach zero. The effect of regularization is specifically to prevent complete agreement with the data, since agreement with noisy data produces noisy images. We are not considering noise here, but in the mean regularization prevents exact cancellation of the $\mathbf{s}^2$ factors and therefore makes the sensitivity variations visible in the reconstructed image, even if they are properly modeled.

One way to avoid this pitfall is not to regularize, instead using a very small value of η, but then to control noise by post-reconstruction smoothing of the image. Another possibility is to use a regularizer such as the gradient or Laplacian of $\mathbf{f}$ that is insensitive to the low spatial frequencies in the image. If the sensitivity is slowly varying, the effect cancels out even though high-frequency noise is well controlled.

Positivity and support constraints As we discussed in some detail in Chap. 15, prior information can often be put to very good use in indirect imaging. Though we may make assertions about smoothness and norms, what we really know about an object in most cases is just that it is nonnegative and has bounded support. In Sec. 15.4 we learned a number of methods for imposing positivity and support constraints.

As a general rule, constraints are useful to the degree that an unconstrained reconstruction would violate them. A well sampled tomographic system with a well characterized system operator and high-count data will produce a good image, with few negative values and little activity outside the true boundary of the object, even with algorithms that do not explicitly enforce positivity and support constraints. With poor sampling or an imperfectly characterized system operator, however, the images will contain artifacts. Streak artifacts often extend beyond the boundaries, and artifacts associated with null functions almost always contain negative values. In these cases, therefore, positivity and support constraints can be quite valuable.

We saw in Sec. 15.1.4 that positivity can place quantitative bounds on the magnitude of null functions, and that in turn places bounds on the level of artifacts in images obtained with algorithms that produce agreement with the data and also enforce positivity. Similar quantitative bounds can be ascribed to support constraints, but the details have not been worked out in the literature.

17.3.4 Image quality

In Sec. 16.2, we noted that a major limitation of planar nuclear medicine was the basic fact that it mapped a 3D object to a 2D image. For estimation tasks, attempts to quantify activity in a region of interest were thwarted by overlapping activity contributing to the projection of the ROI, and for detection tasks this same overlapping activity reduced the contrast and hence the lesion detectability. The clinical goals of ECT are therefore to obtain more accurate quantitation, improved contrast and hence improved lesion detectability.

The objective-assessment paradigm has been embraced enthusiastically by the SPECT community, and there is now a large literature on objective comparisons of different data-acquisition systems and algorithms. Ideal, Hotelling, channelized Hotelling and human observers have been used for detection and classification tasks, and variances, mean-square errors and Cramér-Rao bounds have been used for estimation tasks.

Our goal in this section is to give a broad overview of objective assessment of image quality as it has been applied to SPECT and to discuss some of the considerations in choosing tasks and observers.

SKE detectability in the raw data As in Sec. 16.2.5, we begin by considering the SKE/BKE detection task. This task can be performed by an observer given access to the raw, uncorrected data, to unreconstructed data after correction for distortion and/or scatter, or to reconstructed images. The observer can be the ideal observer, the Hotelling observer or even a human; for SKE/BKE tasks, humans can be trained to read sinogram displays of projection data.

The simplest of these combinations is the ideal observer acting on the raw, uncorrected data. The data are rigorously Poisson in that case, and the log-likelihood ratio was given in (13.131) or (16.117). To apply that formula to SPECT, we need only change the data index from $\mathbf{m}$ to $\mathfrak{m} = (\mathbf{m}, j)$, with j being the angular index as before; thus

$$\mathrm{SNR}_\lambda^2 = \frac{\left[\sum_{\mathfrak{m}}(\overline{g}_{2\mathfrak{m}} - \overline{g}_{1\mathfrak{m}})\ln(\overline{g}_{2\mathfrak{m}}/\overline{g}_{1\mathfrak{m}})\right]^2}{\frac{1}{2}\sum_{\mathfrak{m}}(\overline{g}_{2\mathfrak{m}} + \overline{g}_{1\mathfrak{m}})\ln^2(\overline{g}_{2\mathfrak{m}}/\overline{g}_{1\mathfrak{m}})} . \tag{17.263}$$

In SPECT the signal to be detected usually makes a small contribution to the mean data, in which case [*cf.* (13.135) or (16.119)]

$$\mathrm{SNR}_\lambda^2 \approx \sum_{\mathfrak{m}} \frac{s_{\mathfrak{m}}^2}{\overline{g}_{\mathfrak{m}}} . \tag{17.264}$$

At this point in Sec. 16.2.5, we considered a spatially compact signal and a slowly varying object, such that $\overline{g}_{\mathfrak{m}}$ was approximately the same for all pixels where the signal $s_{\mathfrak{m}}$ is nonzero. This approximation is less plausible in SPECT because of the angular variation; even if $\overline{g}_{\mathfrak{m}}$ is approximately the same for all detector elements in the vicinity of the signal in a single projection, it usually varies significantly from projection to projection.

One immediate conclusion from (17.263) or (17.264) is that it is desirable to have high uptake of the radiotracer in the lesion and low uptake (per unit volume) in the background; this is a consideration for the biochemist designing the tracers but not for the image scientist designing the system.

We can also use these formulas, as we did in Sec. 16.2.5 for the planar case, to study how the spatial resolution of the detector and collimator affect SNR^2. The conclusions will be the same: it is advantageous to improve the detector resolution, but a collimator with better resolution is detrimental to performance of this task since it comes at the expense of photon collection. In both planar nuclear medicine and SPECT, the optimum collimator for SKE/BKE lesion detection is no collimator at all. This conclusion should have no effect at all on the choice of collimators; it should affect our choice of task if the collimator is the element being evaluated.

SKE/BKE detectability in the reconstruction Though SKE/BKE tasks are not useful for collimator optimization, they can be used to study the effects of algorithmic parameters. In SPECT, the algorithm serves not only to provide a reconstructed image but also to control the noise (through choice of apodization or regularization) and to control artifacts by pre- or post-reconstruction data corrections and imposition of constraints. All of these measures influence SKE/BKE detectability by a human observer. They do not, however, influence the ideal observer, at least if we posit that the ideal observer is given access to the same input information that goes into the algorithm. To the extent that any of the data corrections or constraints are useful, the ideal observer — being ideal — will use them.

The Hotelling observer, on the other hand, does not have access to all possible information about the data; it knows only the mean vectors and the covariance matrices under the two hypotheses. We showed in Sec. 13.2.12 that the Hotelling observer was invariant to any linear, invertible transformation of the data, but it is not so obvious that Hotelling performance could not be improved by nonlinear algorithms.

MLEM algorithm and local NEQ To investigate the usefulness of the Hotelling observer in evaluating nonlinear algorithms, Donald Wilson (1994) used the MLEM algorithm on an SKE/BKE task and determined the Hotelling SNR as a function of the number of iterations. The MLEM algorithm was described in Sec. 15.4.6, and Wilson's methodology for computing the statistics of the images was described in Sec. 15.4.7.

The Hotelling SNR was computed by a discrete version of (13.266), which expresses the SNR for detection of a signal at location $\mathbf{r}_0$ in terms of the stochastic Wigner distribution function (WDF). For evaluating an iterative reconstruction algorithm, this formula can be written as

$$\left[\mathrm{SNR}_{Hot}^{(k)}(\mathbf{r}_0)\right]^2 = \int d^2\boldsymbol{\rho}\, \frac{\left|S^{(k)}(\boldsymbol{\rho})\right|^2}{W^{(k)}(\mathbf{r}_0, \boldsymbol{\rho})}, \tag{17.265}$$

where $S^{(k)}(\boldsymbol{\rho})$ is the Fourier transform of the mean signal in the reconstructed image at the kth iteration, and $W^{(k)}(\mathbf{r}_0, \boldsymbol{\rho})$ is the corresponding stochastic WDF. We use continuous notations for these quantities here, but in practice both $\mathbf{r}$ and $\boldsymbol{\rho}$ are measured on discrete grids and integrals are replaced by sums.

We know from Sec. 8.2.5 that the stochastic WDF can be interpreted as a local noise power spectrum (LNPS), describing the frequency content of the background texture in the vicinity of the signal location in the image. It would facilitate the analysis if we could also define a local modulation transfer function (LMTF), but it is not obvious how this can be done since the overall system — data acquisition plus reconstruction algorithm — is neither linear nor shift-invariant. Two assumptions are required. First, it is assumed, as above, that the signal in the raw data is weak compared to the background, so that the response to the signal can be represented as the linear term in a Taylor series. Then it is assumed that the signal is spatially compact and that the blurring by system and algorithm is slowly varying over the support of the signal.

With the weak-signal assumption, we can write the signal in the reconstruction at the kth iteration as

$$s^{(k)}(\mathbf{r}) \approx \int d^2r'\, p^{(k)}(\mathbf{r} - \mathbf{r}'; \mathbf{r}' |\, \mathbf{f}_b)\, f_s(\mathbf{r}'), \tag{17.266}$$

where $\mathbf{f}_s$ and $\mathbf{f}_b$ are the signal and background, respectively, in the object space, and $p^{(k)}(\mathbf{r} - \mathbf{r}'; \mathbf{r}' |\, \mathbf{f}_b)$ is the space-variant, object-dependent PSF for the overall system after k iterations. Analytic expressions for such a PSF can be devised (see Secs. 15.3.6 and 15.4.7), but it can also be determined experimentally in simulation studies by forming two noise-free data sets, one where the object is $\mathbf{f}_b$ and one where it is $\mathbf{f}_b$ plus a weak point source; after iterating for k steps, the difference between the two images is proportional to $p^{(k)}(\mathbf{r} - \mathbf{r}'; \mathbf{r}' |\, \mathbf{f}_b)$.

The second assumption takes advantage of the small spatial extent of the signal to write

$$s^{(k)}(\mathbf{r}) \approx \int d^2r'\, p^{(k)}(\mathbf{r}-\mathbf{r}';\mathbf{r}_0|\,\mathbf{f}_b)\, f_s(\mathbf{r}')\,, \tag{17.267}$$

where the signal part of the object function, $f_s(\mathbf{r}')$, is centered at $\mathbf{r}'=\mathbf{r}_0$. A Fourier transform yields

$$S^{(k)}(\boldsymbol{\rho}) = P^{(k)}(\boldsymbol{\rho}\,;\mathbf{r}_0|\,\mathbf{f}_b)\, F_s(\boldsymbol{\rho})\,, \tag{17.268}$$

where $P^{(k)}(\boldsymbol{\rho}\,;\mathbf{r}_0|\,\mathbf{f}_b)$ is a local transfer function, dependent on the known signal location and the known background object as well as the iteration number.

We can now rewrite (17.265) as

$$\left[\mathrm{SNR}^{(k)}_{Hot}(\mathbf{r}_0)\right]^2 = \left|P^{(k)}(0\,;\mathbf{r}_0|\,\mathbf{f}_b)\right|^2 \int d^2\rho\, \left|S^{(k)}(\boldsymbol{\rho})\right|^2 \mathrm{LNEQ}(\boldsymbol{\rho})\,, \tag{17.269}$$

where the *local noise-equivalent quanta* is defined by

$$\mathrm{LNEQ}(\boldsymbol{\rho}) \equiv \frac{[\mathrm{LMTF}(\boldsymbol{\rho})]^2}{\mathrm{LNPS}(\boldsymbol{\rho})} = \left|\frac{P^{(k)}(\boldsymbol{\rho};\mathbf{r}_0|\,\mathbf{f}_b)}{P^{(k)}(0;\mathbf{r}_0|\,\mathbf{f}_b)}\right|^2 \frac{1}{W^{(k)}(\mathbf{r}_0,\boldsymbol{\rho})}\,. \tag{17.270}$$

Thus we have manipulated the Hotelling SNR into a form that looks just like the one for linear, shift-invariant systems and stationary noise, yet we have not really made any of these assumptions. Because of the MLEM algorithm, the overall system is nonlinear, but it behaves linearly for weak signals; the nonlinear dependence on the background component remains. Similarly, the system is not shift-invariant and the noise is not stationary, but the signal is confined to a small spatial region and we neglect the variations of system response and noise properties over this region.

Wilson computed the LMTF, LNPS and LNEQ as a function of k for several objects and MLEM reconstructions. His results are shown in Fig. 17.11. As the iterations proceed, the algorithm recovers finer details and the LMTF gets broader. At the same time, however, the LNPS increases at all frequencies because of noise amplification; Eventually the LNPS approaches the characteristic ramp-like power spectrum that would be seen with filtered backprojection. If one looked only at LMTF and LNPS, it would not be obvious what effect the iterations were having on detectability, since both the signal and the noise are increasing at higher frequencies as k increases.

The LNEQ, however, tells an interesting story. Though both $[\mathrm{LMTF}]^2$ and LNEQ are increasing at all frequencies as k increases, their ratio is relatively unchanged, especially at higher spatial frequencies. Thus, for any signal that satisfies our assumptions, the Hotelling SNR is nearly invariant to iteration number. We could not reach this conclusion analytically since we know only that Hotelling performance is invariant to invertible, linear algorithms, and MLEM is nonlinear.

The conclusion to this point is that it doesn't matter very much how many iterations of MLEM are used, so long as the task is detection of a known, weak, compact signal in a known background and the observer is either ideal or Hotelling. To choose a stopping point for the MLEM algorithm, we must use a different task, a different observer, or both.

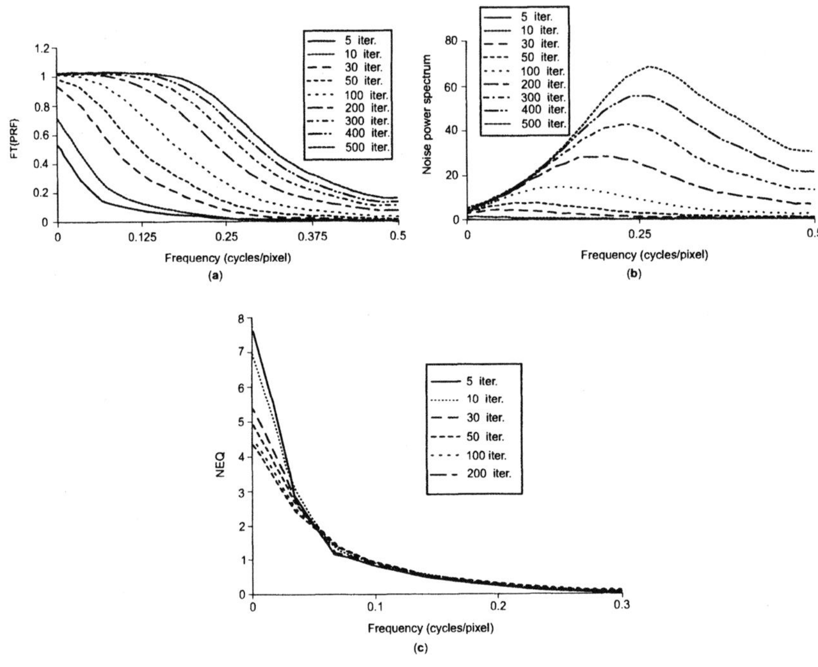

Fig. 17.11 Variation of local MTF, local noise power spectrum and local noise-equivalent quanta with iteration number in the MLEM algorithm. (Courtesy of D. W. Wilson.)

Effect of attenuation correction Another application of the Hotelling observer and SKE/BKE tasks was to the comparison of various linear reconstruction algorithms based on the exponential Radon transform (ERT). As we noted in Sec. 17.2.4, there are many different versions of the inverse ERT, all of which give the same mean image but which differ greatly in their noise characteristics.

The performance of the Hotelling observer on an SKE/BKE task was computed for several different inverse-ERT algorithms by Soares (1994, 1995). The task was detection of a small uniform signal disk superimposed on a larger background disk. The Hotelling SNR was computed on both the raw data and on the output of each of the algorithms. The key result was that all except one of the algorithms gave an SNR identical to that in the raw data. The exception was the Tretiak-Metz algorithm, described in Sec. 17.2.4, which showed much poorer performance. Since the Hotelling observer is invariant to linear, invertible algorithms, we must conclude that the Tretiak-Metz algorithm, though it is exact for continuous, noise-free data, is not invertible in the sense that discrete, noisy data could be recovered from the output of the algorithm. This conclusion is perhaps not surprising when one recalls that the filter function for this algorithm, shown in Fig. 17.7, sets a range of frequency components to zero.

This example illustrates one advantage of SKE/BKE tasks and the Hotelling observer: they can be used to identify algorithms that remove information relevant to a detection task from the data. However, the Hotelling observer provides no way of choosing among algorithms that do not remove information.

Random backgrounds We have seen that nonrandom, uniform backgrounds can give very misleading results in planar nuclear medicine and SPECT. We can introduce randomness in several ways, depending on whether the studies use physical phantoms, simulated images or real clinical images.

Physical phantoms are usually plastic containers filled with radioactive water and possibly various other plastic objects representing organs or lesions. The organs and lesions can be more or less radioactive than the water depending on what clinical application is under consideration. The background can be made random by using random levels of radioactivity in the water or organ phantoms or by moving the organs around in the water. Randomness on a smaller spatial scale can be incorporated by using many small vials with random activities or small nonradioactive objects immersed in the radioactive bath; see for example, Sain and Barrett (2003).

In simulation studies we have greater freedom in representing random backgrounds than is possible with physical phantoms. Just as in planar nuclear medicine (see Sec. 16.2.5), we can use the lumpy or clustered lumpy backgrounds described in Sec. 8.4.4, but these models have free parameters that must be chosen, and there can be concern that the conclusions of the image-quality studies might be affected by the choices. Kupinski *et al.* (2003) have outlined a method of estimating the parameters by using a relatively small number of actual images, provided they are taken through a well characterized imaging system.

Another possibility in simulation studies is to use high-resolution digitized anatomical images to define organ boundaries (see, *e.g.*, Gilland *et al.*, 1992) and then to randomly assign activity levels to each organ. To make the simulation more realistic, the statistics of these organ-to-organ activity variations can be determined from pharmacokinetic studies in animals or humans, and the internal variations within an organ can be simulated as above with a lumpy background.

Finally, for the ultimate in realism, clinical images can be used. For comparing reconstruction algorithms, the same raw projection data can be processed in several different ways, perhaps with or without attenuation correction or compensation for other system properties. If different collimators or other hardware variations are to be compared, the same patient must be imaged multiple times. Clinical images are out of the question for systematic optimization of the data-acquisition hardware, but they can be used to verify the results of a simulation-based optimization study.

Another practical issue with clinical images is obtaining an adequate number of cases with verified pathology. For this reason some investigators have used images of normal volunteers or verified normal patients and then added the lesions artificially. For example, Llacer (1993) used real PET projection data from normal subjects and added simulated projections of a lesion to generate data for abnormal cases. Chan *et al.* (1997) took this idea a step further by acquiring separate physical images of a small object representing a lesion and adding them to the normal patient data.

Random signals The signals in a SPECT detectability study can vary in location, size, shape and uptake of the tracer. In simulation studies, it is straightforward to generate any of these variations, but the main question is how to model the signal statistics.

Usually the main consideration in choosing the size and contrast of the lesion is statistical power rather than clinical realism. In comparing collimators, for example, it does little good to include large, high-contrast lesions that would be easily detected with any of the collimators under consideration, or small, low-contrast ones that are never visible. The lesion size and contrast should be chosen to challenge the systems and provide adequate statistical power in choosing one collimator over another. A good rule of thumb is that the lesion size and contrast should be chosen to yield an area under the ROC curve, for whatever observer is used, of about 0.85. Since this criterion cannot be met for all collimators simultaneously, it is advisable to use lesions of different sizes and contrasts, but there is little value in lumping them into a single binary decision task. Rather than asking whether a lesion of random size is present or absent, it is useful to conduct separate ROC studies with each lesion size. If it should turn out that collimator A is better for small lesions and collimator B is better for larger ones, that is valuable additional information that would be lost if different lesion sizes were randomly chosen in a single study. The only drawback is that more observer time is required for multiple ROC studies, but that is not much of an issue with model observers.

Similar considerations apply to signal location. In SPECT, detectability of a lesion is affected by its proximity to normal organ structures and by how deep it is within the patient's body, so it is useful to vary the signal location. It does not follow, however, that the signal location must be unknown to the observer. As discussed in Sec. 14.3.2, we can consider a sequence of SKE tasks and determine the variations of detectability with signal location. With numerical observers, we can even display these results as continuous maps of detectability as a function of position.

It is true, however, that signal size and location are random and unknown in clinical practice, and the practitioner cannot choose one collimator for superficial lesions and another for deep lesions; some kind of average figure of merit must be used. One approach that gives an overall figure of merit without sacrificing statistical power is to average the SKE detectability (SNR^2) over a clinically relevant range of lesion sizes and locations. An important question that has not been adequately studied is whether this average can give a different rank ordering of systems than would be obtained from a single ROC study with random locations.

Somewhat different considerations come into play when the reconstruction algorithm is being studied. For one thing, algorithmic variations tend to have less effect on detectability than do variations in the collimator, so there is less need to vary the signal contrast. The contrast can be chosen, say by means of a pilot study, and left at that value throughout a range of algorithmic variations.

Signal size is still important since almost any algorithm includes a regularization or smoothing parameter that varies the tradeoff between spatial resolution and noise, but again it is not necessary to vary the size randomly within a single ROC study. Should it turn out that one regularizing parameter is better for large lesions and another is better for small lesions, it would be possible in principle to allow the clinician to vary the regularization during the reading session.[14]

Randomness in signal location could be more important for algorithmic comparisons than for hardware optimization. In Chap. 14, we emphasized that the main function of an algorithm is to match the information in the raw data to the capabilities and limitations of a human observer. One limitation of humans is that

[14]Analysis of the socio-economic issues involving increased reading time vs. improved diagnosis is far too complicated for this book.

they are inefficient in searching an image for a lesion in a random, unknown location. Of course, ROC performance always degrades with location uncertainty, but the amount of degradation can depend on the reconstruction algorithm or display system. One algorithm could produce a lesion image of low contrast but also low noise, such that a human observer concentrating on a known location would have a good detectability. The same observer might have poor performance in a search task with that algorithm but could conceivably do better with a different algorithm that improved the contrast even at the expense of noise or resolution. The general question, which needs much further study, is whether the rank ordering of algorithms and displays is likely to be the same for SKE or SKEV (signal known exactly but variable) tasks as it is for search tasks.

Representative studies with human observers Psychophysical studies have become quite routine in nuclear medicine, and especially in SPECT. We cannot attempt a comprehensive review of this rapidly changing literature here, but we mention a few representative studies to indicate the general directions of the field. The studies listed under this heading involve just human observers without attempting to correlate the results to model observers; comparisons of humans and models as well as a few studies with model observers alone are discussed under the next heading.

Gilland *et al.* (1992) compared the filtered-backprojection (FBP) algorithm to MLEM using simulated SPECT images, and Llacer (1993) compared the same two algorithms for PET FDG (fluorinated deoxyglucose) imaging. In both cases MLEM was somewhat better, probably because it incorporates a positivity constraint and an accurate system model.

Gooley and Barrett (1992) compared a wide range of algorithms in a situation where even stronger prior information was available. They considered cardiac imaging with a blood-pool tracer that filled the left ventricle, and the task was to detect a wall-motion abnormality where a small region of the ventricle did not contract and hence showed up as a protrusion on an image obtained when the heart was maximally contracted (end systole). The strong prior information was that the activity within the ventricle was essentially uniform, so a slice through the object would show only two values, depending on whether the point was in the blood pool or not. Many different algorithms were evaluated; some of them assumed that the object was binary-valued and some did not. Algorithms that enforced this strong constraint led to much better human performance than those that did not, but there was little difference among algorithms that produced a continuous range of gray values. Amusingly, maximum entropy was minimum image quality for this task.

Numerous other human studies of cardiac imaging have been performed. LaCroix *et al.* (2000) evaluated iterative algorithms with attenuation correction for myocardial perfusion studies, where the tracer used went preferentially to perfused regions and hence the signal to be detected was a cold region called a perfusion defect. This study compared differences in defect detection between myocardial SPECT images reconstructed using conventional FBP without attenuation correction and those reconstructed using MLEM with nonuniform attenuation correction. It was found that MLEM with attenuation correction was superior, particularly for patients with large breasts or with a diaphragm raised to the level of the heart. If uncorrected, attenuation from these structures can produce cold regions similar to perfusion defects, hence false positives.

A similar study of the effects of attenuation correction in SPECT myocardial imaging was performed by Jang *et al.* (1998), and Sankaran *et al.* (2002) studied optimization of the regularization parameter and the detector-response compensation for the same application. Detector-response effects were studied with LROC analysis by Gifford *et al.* (2000b).

Representative studies with human and model observers There has been considerable research, summarized in Sec. 14.2.2, on validating anthropomorphic observers in general, but now there is also a substantial literature on this subject specific to nuclear medicine. When both human and model observers are used in the same paper and a high correlation between the results is found, it adds to our confidence in the predictive power of the models. The most investigated and most successful anthropomorphic models have been variants on the channelized Hotelling observer (CHO). Much of the rationale for the use of this observer in SPECT comes from work done at the University of Massachusetts, and we begin with a summary of that effort.

King *et al.* (1997) studied different data-acquisition and processing strategies for liver imaging. Both square, nonoverlapping (SQR) channels and difference-of-Gaussian (DOG) channels were used for the CHO. Since the goal was to see if the model observers predicted the human's rank ordering of the different acquisition strategies, the results were evaluated by Spearman's rank correlation coefficient. This coefficient turned out to be 0.94 for the SQR channels and 0.96 for the DOG channels if no scatter correction was used; poorer correlation was found with scatter correction. Using the same data, however, Gifford *et al.* (2000a) found much better correlation in the latter case by slightly modifying the parameters of the CHOs and by ensuring that the humans and CHOs operated on exactly the same images.

Gifford *et al.* (1999) compared the performance of human observers in an LROC (localization-ROC) experiment to the CHO on a simple ROC experiment. Of course the absolute performance of the humans, who had to contend with location uncertainty, was much worse than than that of the CHO on the SKE task, but surprisingly the rank ordering was excellent.

Other work by Gifford *et al.* (2002) studied the effects of subset size and number of iterations in block-iterative methods for tumor detection in thoracic SPECT. Again excellent agreement between human and CHO was found, with a Spearman rank correlation coefficient of 0.976.

Other groups have also successfully compared humans and CHOs. Wollenweber *et al.* (1999) studied defect detection in myocardial SPECT. Chen *et al.* (2002a) compared triple-head 360° vs. dual-head 180° acquisition, with and without attenuation correction. The CHO and human results both showed better detection performance in the 360° scan. The same group (Chen *et al.*, 2002b) also extended the CHO concept to 3D, giving the observers access to multiple tomographic slices in three different orientations. Both multi-slice and single-slice CHOs showed good correlation with humans.

A few papers have appeared (*e.g.*, Chan *et al.*, 1997, and Qi and Huesman, 2001) that use only model observers without checking the results against human studies. As the literature validating the models grows and a consensus on which model to use becomes firmly established, more such papers can be expected and the full value of model observers will be seen.

Computational issues with model observers The great advantage of model observers, of course, is that they permit a much wider exploration of system parameters than would be possible with psychophysical studies, but eventually computational limitations will become apparent. A notable step towards more efficient computation of CHO performance was taken by Bonetto *et al.* (2000). They used the approach of Fessler (1996) described in Sec. 15.3.6 to approximate the mean vector and covariance matrix of a MAP (maximum *a posteriori*) reconstruction. The results include the random effects of Poisson noise and background variations, but only to the extent that the prior used in the reconstruction algorithm matches the actual background randomness.

Artifacts and detectability In Sec. 17.3.3 we defined artifacts as nonlocal defects in the deterministic point response function of the overall imaging system, including the data acquisition and the reconstruction algorithm. Many artifacts can be characterized loosely as long tails on the PRF, but it is not so obvious why long tails are bad, in an objective sense. In fact, if we were considering an SKE/BKE detection task and an ideal observer, the long tails on the PRF would not degrade performance. The observer would know the PRF exactly and would be able to compute the resultant signal in the image in spite of the tails.

Thus, as with collimator optimization, we must consider more realistic tasks in order to see why the wide PRF is detrimental. Qualitatively, long tails are bad if the background is random since they couple random variations in the background to a distant signal location. A linear observer, whether Hotelling, CHO or human, will center a template on a known signal location but have a variance from background fluctuations that is much greater than it would be without the tails.

This effect can be seen neatly in the extensive work of Kenneth Hanson on tomography with sparse angular sampling (Hanson, 1988, 1989, 1990a, 1990b). He used typically just eight projections and an object consisting of a random collection of high-contrast discs. The signal to be detected was also a disc but with just one-tenth the amplitude of the background discs. With linear reconstructions algorithms, streak artifacts from the high-contrast discs were very evident and significantly reduced the detectability of the low-contrast disc. With nonlinear iterative algorithms such as ART (algebraic reconstruction technique), the artifacts were suppressed and the detectability improved. Hanson then went on to optimize the ART algorithm for task performance.

Much further work along these lines is needed to fully understand the role of artifacts in lesion detectability.

Estimation tasks In one sense, estimation tasks are the very essence of tomography. As we saw in Chap. 15, image reconstruction can be formulated in terms of estimation of the expansion coefficients in an approximate representation of the object. Usually a voxel representation is adopted, and the reconstruction is an attempt to estimate the integrals of the object over voxels, or voxel values for short.

It is common in the SPECT literature to try to use the accuracy of these voxel estimates as a measure of image quality. In this view, the goal of the imaging is not to perform a particular task but to recreate the object as accurately as possible. Fidelity rather than task performance is the measure of image quality. As we saw in Sec. 13.3.2, however, there is considerable ambiguity in defining image quality in terms of fidelity. We identified three different ways of defining the error between an object and an image and three different ways of averaging that error, so there are nine possible definitions of the mean-square error (MSE) between object and image.

Basically, this ambiguity arises because all real imaging systems have null functions. Infinitely many objects can give exactly the same mean data and hence the same values for any estimates derived from the data. If these objects have different values for their integrals over voxels, then there is no unambiguous way of defining the bias or MSE of the voxel estimates. As we phrased it in Sec. 15.1.3, voxel values are almost never estimable parameters.

From a task-based viewpoint, the inability to define the bias or MSE of voxel values is of no concern since the goal of imaging is never to determine even a single voxel value, much less a set of them. In clinical SPECT, we are often indeed interested in estimating quantitative parameters, but never just voxel values. Instead we might want to know the total uptake of a tracer in a region of interest in the brain or the change in volume of the left ventricle when the heart contracts. Unlike voxel values, these parameters are of direct medical interest, and mathematically they are much more likely to be estimable (see Sec. 15.1.3 for a definition). Thus estimation tasks, as we view them, are not about estimating a huge number of irrelevant and definitely unestimable parameters but rather a few highly relevant and approximately estimable ones.

Viewed this way, there is a large literature on estimation tasks in SPECT. The common term for an estimation task in that literature is *quantitation* or *quantification*.[15] The typical SPECT quantitation study uses a real or simulated phantom with a uniform background and immerses into it a spherical or cylindrical container with an activity higher than that of the background. Projection data are acquired and an image is formed by the authors' favorite reconstruction algorithm. Usually a fuzzy bump where the reconstructed gray values exceed those of the uniform background is seen in the image, and from this bump it is desired to estimate the activity in the container. A common approach is to define a region of interest the same size as the container and sum up the gray values in the region to get the activity estimate. The process can be repeated with many realizations of the raw data for the same object, and a bias, variance and mean-square error of this scalar parameter can be computed.

To put this discussion into the language of statistical estimation theory, the region-of-interest procedure yields a linear estimate with a predefined template. Unlike maximum-likelihood estimates it takes no account of the data statistics or the characteristics of the imaging system, and unlike Bayesian estimates it makes no assumptions about the prior distribution of parameter values. It is, nevertheless, a valid and common estimate, and its mean and variance can be assessed as any other estimate would be. If the container is large compared to the system resolution, the parameter is at least approximately estimable and a mean-square error can be defined and estimated.

[15]To *quantitate* means to measure or determine the quantity of something, especially with precision. To *quantify* can mean the same thing, but it often means assigning a quantity to something that has only quality. The reader can decide, on a scale of 1 to 10, which term is preferable in nuclear medicine.

Recommendations Rather than evaluating a system on how well an estimation task *is* performed by a suboptimal, *ad hoc* procedure, we can also evaluate it on how well the task *can be* performed by an optimal procedure. Our recommendation is to use the bias and variance of a maximum-likelihood estimator as the basic error characterizations for an estimation task, combining them into an MSE when a scalar figure of merit is desired. When there is difficulty in computing the variance of the ML estimate, Cramér-Rao bounds can be used.

These estimates and error measures can be computed either from the raw projection data or from the reconstructed image. The former is recommended if the objective is to evaluate the data-acquisition hardware in terms of an estimation task, while the latter is required if the algorithm is to be evaluated.

What limits task performance? In most discussions of estimator performance, the emphasis is on characteristics of the estimator. Maximum-likelihood estimation is preferred, for example, because it is known to be asymptotically unbiased, asymptotically efficient, and efficient in any case if an efficient estimator exists. These issues tell only a part of the story, however, when ML estimation is used in SPECT. The actual performance of the estimator is more likely to be limited by errors in modeling the imaging system than by intrinsic properties of the estimator.

If the input to the ML estimator is the raw projection data, then the ML estimator requires detailed knowledge of the image-forming process, including the attenuation and scatter in the patient's body and the space-variant blur of the collimator. It also requires knowledge of the data statistics, but in this case the data are pure Poisson. If, on the other hand, the input to the ML estimator is the reconstructed image rather than the raw data, then the algorithm should take into account attenuation, scatter and space-variant blur. In that case the deterministic description of the overall system might be assumed to be a simple blur function, but the noise on the input to the estimator is no longer Poisson (see Secs. 15.2.6, 15.3.6 and 15.4.7). In either case, errors in describing any aspect of the system will introduce a systematic error or bias that cannot be erased by any of the nice asymptotic properties of ML methods.

Practical examples Most of the practical work on evaluation and optimization of nuclear imaging systems for estimation tasks comes from the group at Harvard, which was previously mentioned in the context of planar imaging in Sec. 16.2.6. Some of their recent work deals with joint estimation-classification tasks, and some of it is ancillary work aimed at developing simulation tools, but we shall briefly survey it all here.

One particularly complicated clinical situation this group has studied uses tracers labelled with ^{67}Ga. This isotope has a complicated emission spectrum, with principal energies around 93, 185 and 300 keV, but also some higher-energy lines that cause severe difficulties with septal penetration. Moore and El Fakhri (2001) developed highly accurate and realistic Monte Carlo simulation methods in which the patient's anatomy is represented by segmented CT images of a commercial anthropomorphic torso phantom. With this tool, El Fakhri *et al.* (2002) studied the effect of the energy window settings on the performance of estimation and detection tasks. In other work, unpublished at this writing, these same authors have optimized the design of collimators for this application.

A related study (Kijewski *et al.*, 2001) of considerable practical importance applied similar ideas to analyzing the effects of the collimator and SPECT system geometry on simulated but realistic tasks related to diagnosis and management of Parkinson's disease. Cramér-Rao bounds on estimates of activity concentration and striatal volume were computed and related to the likelihood ratio for several binary classification tasks. The authors compared three commercial SPECT systems, two of relatively high spatial resolution and one of lower resolution. In all cases the higher-resolution systems performed better.

18

Coherent Imaging and Speckle

Coherent imaging modalities include holography, laser illuminators, radar and ultrasound. Images taken with these systems exhibit a phenomenon called *speckle*, which is not seen in our everyday experience with incoherent light. Whenever a diffuse object is illuminated with radiation that is spatially and temporally coherent, its image will show high-contrast, fine-scale structures not related in any obvious way to the characteristics of the object. This effect can be observed by shining a laser pointer on a piece of paper and inspecting the spot visually.[1] Even though the irradiance profile of the laser beam is smooth and featureless, the image on your retina will be highly noisy or "speckled."

Further experimentation with the laser pointer reveals other characteristics of speckle. It should be found that the size of the speckle blobs depends on the diameter of the pupil of the eye; if the experiment is performed in a darkened room where the pupil is dilated, the speckle blobs will be small, but if it is performed in bright sunlight or with a pinhole in front of the eye, the blobs will be larger. The contrast of the speckle patterns will, however, remain unchanged; unlike the kinds of noise considered in Chap. 12, speckle noise does not depend on the light level. If the laser beam is attenuated with a neutral-density filter or if a brighter laser is used, the contrast in the speckle pattern remains the same. Moreover, laser speckle is always in focus; the contrast does not depend on where you focus your eyes.

Our goals in this chapter are to understand these qualitative features of speckle noise and to give mathematical descriptions of them, and in the process to show how the mathematical methods of earlier chapters can be applied to coherent imaging. The treatment of diffraction theory and coherent imaging in Chap. 9 will be crucial, and some little-known aspects of random processes from Chap. 8 will find

[1]The reader is cautioned to observe only the reflected light and to obey Gaskill's rule of laser safety: "Never look down the laser beam with your remaining eye!"

application here. Of course, the principles and methodology of objective assessment of image quality, from Chaps. 13 and 14, will also be put to good use.

We begin in Sec. 18.1 by looking at the basic aspects of speckle with relatively little mathematics. We show why statistical models are applicable to speckle, and we state some well-known results for the univariate probability density function (PDF) and autocorrelation function of the irradiance in an observation plane with either imaging or nonimaging systems.

In Sec. 18.2 we analyze the statistical properties of speckle in detail. Our goal in this section is nothing less than a full infinite-dimensional description of the statistics of the irradiance in a simple nonimaging system. The main tool for this investigation will be the *characteristic functional*, introduced in Sec. 8.2.3.

In Sec. 18.3 we extend the analysis to imaging systems and relate it to the familiar continuous-to-discrete (CD) model for digital imaging systems that we have used often in this book. In this section our goal is characterize the multivariate (but finite-dimensional) statistics of the digital image as fully as possible.

In Sec. 18.4 we acknowledge that speckle is not the only source of randomness in coherent imaging systems. We consider also measurement noise and the stochastic nature of the objects being imaged, and we relate these effects to objective metrics of image quality.

In Secs. 18.2 – 18.4 we place considerable emphasis on object models that lead to Gaussian statistics on the fields in an image or observation plane, but in Sec. 18.5 we look at models that lead to decidedly non-Gaussian fields. In particular we consider objects that consist of randomly placed points, and we find a close connection with Poisson point processes as analyzed in Chap. 11.

In Sec. 18.6 we look specifically at 3D imaging systems such as medical ultrasound and radar where time of flight is used to encode the third dimension. The systems discussed there use amplitude-sensitive detectors to detect the scattered field, and the analysis is easier than for systems with irradiance sensitive detectors because the step of conversion of amplitude to irradiance is nonlinear. Both Gaussian and non-Gaussian speckle are treated.

18.1 BASIC CONCEPTS

Coherent light has a definite phase at each point on a surface, and in many situations that phase admits of simple mathematical descriptions. For example, a monochromatic plane wave has a phase that is constant on a plane, and a spherical wave has a phase that is constant on a sphere. If a coherent wave is reflected from a rough opaque surface or transmitted through a rough transparent surface such as a ground glass, it remains coherent but the phase is no longer simple, and in almost all situations it is unknown. The best we can say in general is that it varies rapidly and unpredictably from point to point. Speckle is the diffraction pattern from such an irregular, rapidly varying optical field. The fine structure results from interference of the light coming from different points on the rough object. The object roughness makes the phase complicated but does not make it vary randomly with time, so the light is still coherent and interference still occurs.

In this section we look qualitatively at some basic aspects of this phenomenon. In Sec. 18.1.1 we first discuss the role of statistics in describing what is essentially a deterministic problem. Then we review some basic results on the point statistics and correlation properties of speckle in a nonimaging context. In Sec. 18.1.2 we bring in a lens and a discrete detector array and begin to come to grips with just what it means to analyze speckle in an imaging system.

18.1.1 Elementary statistical considerations

If we knew the complex amplitude transmittance of a ground glass or the complex amplitude reflectance of a rough surface, we could compute the diffraction pattern by the methods developed in Chap. 9. Since we virtually never have that detailed information, we must resort to statistical methods. For example, we can consider a particular ground glass to be one realization of a spatial random process, and we can then investigate statistical properties of the diffraction pattern such as its mean, autocovariance function and various probability density functions. It is important to emphasize, however, that all statistical properties must be interpreted in terms of averages over ensembles of ground glasses, not as spatial or temporal averages. The statistical averages may turn out to be *functions* of space or time, but they are not *averages* over space or time. It can be quite misleading to invoke ergodicity in speckle problems.

Light emerging from rough surfaces One simplifying assumption that we can make regarding the statistical properties of the field emerging from a ground glass (or other rough object) is that the phase is completely random. If the thickness of a slab of glass changes by an amount Δh, then the phase of the light, relative to what it would be without the change in thickness, is $\Delta\phi = k(n-1)\,\Delta h$, where n is the refractive index of the glass and $k = 2\pi/\lambda$, with λ being the wavelength in free space. Thus if the height variations are random with a standard deviation of several wavelengths, the phases are approximately uniformly distributed over $(-\pi, \pi)$. Under this assumption a field of the form $A(\mathbf{r})\exp[i\phi(\mathbf{r})]$ is a zero-mean random process.

We can also say something about the autocorrelation function of this random process. Diffuse objects are rough on the scale of the wavelength of light — or else they wouldn't be diffuse. We know from Sec. 10.2.7 that the spatial autocorrelation function of an optical field is closely related to the angular distribution of the light. For example, a Lambertian surface has a radiance that is independent of angle and a radiant intensity proportional to the cosine of the angle of light propagation measured from the surface normal. We saw in (10.110) *et seq.* that this angular dependence is associated with a particular autocorrelation function, a sinc function with width approximately equal to the wavelength. We also mentioned — and left it to the reader to show — that any process that completely randomized the direction of light led to precisely this autocorrelation function and hence a Lambertian angular distribution.

Thus a natural statistical model of the light from a coherently illuminated diffuse object is that it is a 2D spatial random process with completely random phase and an autocorrelation function that is sharply peaked on the scale of a wavelength. This simple statement will take us quite far in the analysis of speckle.

Light on an observation plane Suppose the light from a coherently illuminated rough surface propagates through free space to an observation plane. We know from diffraction theory that all points on the surface contribute to the field at each point on the observation plane, and this observation allows us to draw further conclusions about the nature of speckle.

The elementary argument, to be made more precise in Sec. 18.2.4, is that many statistically independent points on the rough surface contribute to the field at each point on an observation plane. It then follows from the central-limit theorem that the complex field at each observation point is Gaussian. If the ground glass is rough enough that the phases of the light are uniformly distributed on $(-\pi, \pi)$, then the field at an observation point is a circular Gaussian, where the real and imaginary parts are i.i.d. Gaussians (see Sec. 8.3.6). In this case, we know that the the modulus of the complex field (square root of the irradiance) follows a Rayleigh law [see (C.140) and (8.231)], while the irradiance or squared modulus follows an exponential law [see (8.232)]. If the phase is not uniformly distributed on $(-\pi, \pi)$, there will be an undiffracted plane-wave component to the light, and the modulus will follow a Rician law [see (C.141)].

Speckle contrast When the irradiance obeys an exponential law, we can make a simple statement about the speckle contrast. We stated in Sec. C.5.3 (and the reader may verify) that the variance of an exponential law is equal to the square of the mean. Thus the standard deviation is equal to the mean, and the ratio of mean to standard deviation, often called a *signal-to-noise ratio*, is unity. This explains the observation, noted in the introduction of this chapter, that speckle contrast is independent of light level.

Blob size The probability laws discussed above give only the univariate statistics of the field or its modulus at a single point. A more complete statistical description requires accounting for the correlations in the field and irradiance. We shall now present a simple argument based on local spatial frequencies that will enable us to estimate the range of these correlations, roughly corresponding to the size of a speckle blob.

Consider a stationary ground glass in the plane $z = 0$ illuminated with a monochromatic plane wave. For now, no imaging system is used, and the diffraction pattern is observed in a parallel plane $z = z_0$. By Huygens' principle, each point on the ground glass emits a spherical wave, and since the light is coherent, each spherical wave interferes with every other spherical wave. The spherical wave coming from point $\mathbf{r}_1$ on the ground glass produces a wave at point $\mathbf{r}$ in the observation plane, a distance z from the ground glass, which is given in the Fresnel approximation by [*cf.* (9.94)]

$$u_z(\mathbf{r}) \propto \exp\left[i\pi \frac{|\mathbf{r} - \mathbf{r}_1|^2}{\lambda z} + i\phi(\mathbf{r}_1)\right], \tag{18.1}$$

where $\phi(\mathbf{r}_1)$ is the phase of the light at point $\mathbf{r}_1$. By (5.35), the 2D local spatial frequency of this wave at point $\mathbf{r}$ is given by

$$\boldsymbol{\rho}_1(\mathbf{r}) = \frac{1}{2\pi}\nabla\left[\frac{\pi}{\lambda z}|\mathbf{r} - \mathbf{r}_1|^2 + i\phi(\mathbf{r}_1)\right] = \frac{1}{\lambda z}(\mathbf{r} - \mathbf{r}_1), \tag{18.2}$$

where the subscript on $\boldsymbol{\rho}_1(\mathbf{r})$ indicates that it is associated with the wave emanating from $\mathbf{r}_1$. If we consider also a second point $\mathbf{r}_2$, then it generates a wave of local spatial frequency $\boldsymbol{\rho}_2(\mathbf{r})$, and the irradiance in the interference pattern between the two waves contains the difference frequencies[2] given by

$$\pm[\boldsymbol{\rho}_1(\mathbf{r}) - \boldsymbol{\rho}_2(\mathbf{r})] = \pm\frac{1}{\lambda z}(\mathbf{r}_2 - \mathbf{r}_1)\,. \tag{18.3}$$

Thus the largest spatial frequency in the interference pattern is determined by the largest separation of the two points on the ground glass, $|\mathbf{r}_2 - \mathbf{r}_1|_{max}$. As a rule of thumb, then, the correlation length or *blob size* of the speckle, denoted by ℓ_b, is the reciprocal of the maximum spatial frequency, or

$$\ell_b = \frac{\lambda z}{|\mathbf{r}_2 - \mathbf{r}_1|_{max}} = \frac{\lambda}{\Delta\theta}\,, \tag{18.4}$$

where $\Delta\theta$ is the angular subtense of the object from a point in the observation plane.

By considering the diffraction patterns of all pairs of points on the ground glass and averaging over the random phases, we can derive an expression for the autocorrelation function of the irradiance on the observation plane. This autocorrelation will be approximately zero when the distance between two observation points exceeds the blob size ℓ_b.

Multivariate statistics of the irradiance The simple arguments above yield the mean and autocorrelation function of the irradiance on the observation plane. If that irradiance were Gaussian, we would have a full statistical description since all properties of a Gaussian random process are determined by its mean and autocorrelation. Since an irradiance cannot be negative, however, it cannot be Gaussian, and we have just argued that its single-point PDF is exponential. To get a full description, therefore, we need all possible multi-point PDFs (see Sec. 8.2.2) or, equivalently, the characteristic functional of the process (see Sec. 8.2.3). Deriving this description is the major goal of Sec. 18.2.

18.1.2 Speckle in imaging

Consider a simple unit-magnification imaging system in which a lens of focal length f is placed a distance $2f$ from a coherently illuminated diffuse object and the image is observed a distance $2f$ behind the lens. To be definite, we can think of the object as a ground glass with a photographic transparency laid over it.

With the lens between the rough object and the observation plane, we can no longer say that all points on the object contribute to the field at each observation point; instead, only points within an area defined by the coherent point spread function (PSF) contribute. If the lens is diffraction-limited, then a large lens aperture means that a smaller area on the object contributes to each image point. If, on the other hand, the lens is aberrated, then a larger aperture may mean a *larger* PSF and *more* contributing area.

[2] One might wonder why there are no sum frequencies. The reason is that each wave has an unstated temporal factor with temporal frequency ν. The sum of the two spatial frequencies comes with a sum of the two temporal frequencies, and the difference of the spatial frequencies comes with the difference in the temporal frequencies. The factor $\exp[-2\pi i(\nu+\nu)t]$ averages to zero over any measurement time but $\exp[-2\pi i(\nu-\nu)t]$ does not.

If the contributing area contains many statistically independent regions on the ground glass, we can again invoke the central-limit theorem and claim that the field in the observation plane is Gaussian. If we also assume that the phase variations of the ground glass are approximately uniformly distributed on $(-\pi, \pi)$, the field is circular Gaussian.

The correlation length of this circular Gaussian is, however, no longer determined by the overall size of the ground glass; now the lens aperture controls the correlations. One way to see this point is to consider a very large ground glass with no photographic transparency. The field emerging from the ground glass is a circular Gaussian with a very short correlation length, and when it has propagated to the lens plane, it is still a circular Gaussian with a very short correlation length. According to (9.163), the lens multiplies the field by a quadratic phase, but adding a deterministic phase to one that is completely random leaves it completely random. Thus the field emerging from the lens is a circular Gaussian with a very short correlation length, but now it is nonzero only over the lens aperture. When it propagates to the image plane, it is yet another circular Gaussian, this time with a correlation length given approximately by (18.4), with $\Delta\theta$ now being the solid angle subtended by the lens aperture from a point in the image plane. Thus a smaller lens aperture means larger speckle blobs.

This argument depends critically on the field emerging from the ground glass having very short correlation length. At the opposite extreme, suppose the "ground" glass is smooth and slowly undulating. It may still have a total phase variation that is uniformly distributed over $(-\pi, \pi)$, but now its amplitude transmittance contains only relatively low spatial frequencies. We know from Sec. 9.2.1 that low spatial frequencies in the transmittance imply low deflection angles for the light. We can decompose the field into its plane-wave components, and most of those components will be within the pass-band of the lens if the lens aperture is large and the phase variations of the ground glass are slow. In that case, no speckle at all is observed; instead, the lens gives a faithful image of the phase object, reproducing its phase variations in the image plane. Speckle results when some of the light is deflected to larger angles and misses the lens.

The image detector Optical detectors respond to irradiance, so it is important to know the statistics of the irradiance. As we argued above, the irradiance at a point follows an exponential law if the field is circular Gaussian. This observation does not, however, take us very far towards the understanding of image statistics, for several reasons.

First, an optical detector integrates the irradiance over its area, and there is no guarantee that a single-point exponential law on the irradiance will lead to an exponential law on the detector output. To the contrary, if the detector area is large compared to the correlation length of the irradiance incident on the detector, we can again invoke the central-limit theorem and argue that the detector output is Gaussian, not exponential.

It is only when the detector is very small compared to this correlation length that an exponentially distributed output would be seen; observation of point statistics of the irradiance requires a point-like detector. Since, as noted above, the correlation length is controlled by the lens aperture, the point-detector approximation could be invalidated in any optical imaging system simply by stopping down the lens.

The second reason why the exponential law on the irradiance is inadequate for the analysis of imaging systems is that it is a univariate law, giving very little information about the multivariate statistics on the output of a detector array. It is tempting to assume that the outputs of different detector elements in an array are statistically independent. In that case the multivariate PDF would be a product of exponentials if each univariate PDF were exponential; this assumption is virtually never valid in practical systems. As we just saw, getting the exponential univariate PDF requires a detector small compared to the correlation length, but statistical independence requires that these point-like detectors be far apart compared to the correlation length. Practical detector arrays have contiguous detector elements with very little gap between them, so both conditions cannot be satisfied simultaneously.

Other noise sources The final reason for the inadequacy of the exponential law is that it describes only a single source of statistical variation. As we know from Chaps. 11 and 12, real optical detectors are also afflicted by Poisson noise from the discrete nature of photoelectric interactions and by Gaussian noise arising from electronic sources.

Moreover, the randomness of the objects being imaged must be considered in a full statistical description of the imaging process. In previous chapters we have often encountered doubly stochastic processes where the Poisson law was correct for a single object realization, but different objects gave different Poisson means. With coherent illumination and a photon-counting detector, we have a triply stochastic Poisson process. For a given object and speckle realization, the multivariate output of a discrete detector array is a Poisson random vector, but our statistical view of speckle considers an ensemble of ground glasses (or other fine-scale phase variations), and each of these realizations leads to a different mean vector for the Poisson distribution. If there is also an ensemble of objects (say photographic transparencies), each object produces its own PDF for the speckle distribution. The overall PDF is an average over Poissons given the irradiance distribution, an average over all irradiances given an object (but different ground glasses), and an average over objects. Moreover, if the output of the detector is processed by noisy electronics, both Gaussian and Poisson are present, and in that case we have a quadruply stochastic process.

Characteristic functions and functionals A full statistical description of a discrete image requires a huge multivariate PDF, or equivalently a huge multivariate characteristic function. In Chap. 8 we learned a lot about calculating characteristic functions for imaging systems, and we shall put that knowledge to good use in this chapter.

As we saw in Sec. 8.2.3, the continuous irradiance distribution incident on a detector array can be described by a characteristic *functional* (really an infinite-dimensional characteristic function), and we also saw that characteristic functionals can be propagated rather simply through linear imaging systems. The difficulty in the present context is that speckle is essentially a coherent phenomenon, so we must consider the imaging system as being linear in the field, but the detector responds to irradiance, so we need the characteristic functional not only for the field but also for its squared modulus. We shall see in Secs. 18.2 and 18.3 how to do this conversion.

Field-sensitive detection Analysis of speckle would be much easier if detectors responded to field rather than irradiance, and indeed some detectors do just that. For radio waves and microwaves we can build systems that respond directly to the field — otherwise your television set wouldn't work! As the temporal frequency of the wave increases, however, it becomes increasingly difficult to build electronics that will follow the rapid field variations, so field-sensitive detectors become less feasible, and they are essentially nonexistent in the optical frequency range.

An exception to this statement is *heterodyne* detectors. Since the 1920s all radio receivers have been built on the heterodyne principle in which the incoming modulated radio signal is mixed with a local oscillator signal in a nonlinear (ideally, square-law) detector. The nonlinearity produces sum and difference frequencies, and typically the difference-frequency signal is amplified further before final detection to extract the desired modulation signal.

Homodyne detection is the special case of heterodyne detection where the frequency of the local oscillator is the same as the center frequency of the incoming radio signal. If there were no modulation on that signal, the homodyne detector output would consist of two components, one at zero frequency and one at twice the original frequency. Both heterodyne and homodyne techniques can be applied to optical signals since the post-detection electronics needs only to be fast enough to respond to the difference signal, not the original optical signal.

The archetype optical homodyne system is holography. Consider an optical system with coherent illumination, and denote the field at the image plane by $u_{im}(\mathbf{r})$, and suppose we superimpose on it a local-oscillator wave (or *reference wave* in holographic jargon) $u_{ref}(\mathbf{r})$. If the two waves are completely coherent, then the total irradiance in the detector plane is given by

$$I(\mathbf{r}) = |u_{im}(\mathbf{r}) + u_{ref}(\mathbf{r})|^2 = |u_{im}(\mathbf{r})|^2 + |u_{ref}(\mathbf{r})|^2 + u_{im}^*(\mathbf{r})\, u_{ref}(\mathbf{r}) + u_{ref}^*(\mathbf{r})\, u_{im}(\mathbf{r})\,. \tag{18.5}$$

For homodyne detection, all four of these terms have the same temporal frequency, but they may have different spatial frequencies. For example, if the reference wave is a tilted plane wave with 2D spatial frequency $\boldsymbol{\rho}_{ref}$ in the detector plane, then the first two terms in (18.5) are independent of $\boldsymbol{\rho}_{ref}$, the third is centered at $\boldsymbol{\rho}_{ref}$ and the fourth is centered at $-\boldsymbol{\rho}_{ref}$. Because of this difference in the spatial-frequency spectra, the four terms can be separated in subsequent processing; holographic image reconstruction boils down to isolating the fourth term and extracting or recreating the field $u_{im}(\mathbf{r})$.

Heterodyne detection works essentially the same way, except now the four terms in (18.5) differ in their temporal frequencies. Post-detection bandpass filtering can isolate the fourth term, provided that the detector is fast enough to respond to the difference frequency.

With both homodyne or heterodyne detection, therefore, the combination of a square-law (irradiance-sensitive) detector and suitable post-detection processing results in a measured signal linear in the image field $u_{im}(\mathbf{r})$. Because the illumination is coherent in both cases, this field is speckled if it was the result of diffraction from a rough object. One useful approximation in analyzing the speckle statistics of homodyne and heterodyne systems, therefore, is to lump the square-law detector and post-detection processing into a single box in the system block diagram and assume that the output of the box is a linear functional of the input field, $u_{im}(\mathbf{r})$.

That this linearity is only an approximation can be seen by examining the contents of the box in more detail. If the box contains a discrete detector array, each element integrates the irradiance over its face, and the output of the array is a set of samples of the form $\int_m d^2r\ |u_{im}(\mathbf{r}) + u_{ref}(\mathbf{r})|^2$, where subscript m indicate integration over the face of the mth detector element. This integral is nonlinear in $u_{im}(\mathbf{r})$, and no amount of post-processing can fully undo the nonlinearity since the continuous field $u_{im}(\mathbf{r})$ cannot be reconstructed exactly from a finite set of samples. We might consider passing to the limit of an essentially infinite number of samples by using film as detector (which we almost always do in holography), but even then we have to assume that the film responds linearly to the irradiance.

Microwave and ultrasonic imaging The complications of heterodyne and homodyne detection are needed only because we cannot build detectors that are fast enough to respond to optical frequencies. This problem does not arise with radio waves and microwaves, where we can indeed build systems that are linear throughout. A microwave dish, for examples, integrates the field over its face and produces an output voltage that is a linear functional of the field. An array of microwave dishes is the analog of an optical detector array except that each element responds linearly to field rather than irradiance.

Most ultrasonic transducers also respond linearly to field, though in this case an acoustic or pressure field rather than an electromagnetic one. An ultrasonic transducer is made of a *piezoelectric* material in which a strain produces a voltage. To a good approximation, the total voltage from a transducer of finite area is proportional to the integral of the strain amplitude across that area. This strain distribution can be produced by an acoustic wave that has been reflected from a rough surface or a random volumetric distribution of scatterers, and in that case it suffers from speckle just as an optical wave does. An important difference, however, is that linear-systems theory can be used without approximation to analyze the speckle statistics in ultrasound. If the speckle at any stage can be assumed on the basis of central-limit arguments to be Gaussian, it remains Gaussian because any linear transformation of a Gaussian is a Gaussian (see Sec. 8.3.3).

Nevertheless, most papers on speckle in ultrasonic and microwave imaging do not use linear models, and most of them invoke Rayleigh or Rician rather than Gaussian statistics. The reason for this is that, even though the detectors respond linearly to the fields, the displayed image usually involves an envelope detection or demodulation step. To the extent that the detector array samples $u_{im}(\mathbf{r})$ finely, this post-detection processing is an attempt to compute $|u_{im}(\mathbf{r})|$ or $|u_{im}(\mathbf{r})|^2$. In the former case, Rayleigh statistics are relevant, and in the latter case, exponential statistics are applicable. In neither case, however, would the Rayleigh or exponential law provide the requisite information on the multivariate image statistics.

An additional complication in ultrasonic imaging, as usually practiced in medical sonography, is that pulsed signals are used to obtain spatial resolution along the direction perpendicular to the transducer face. In this case the system model must include both temporal and spatial aspects. Suitable models will be derived in Sec. 18.6.

18.2 SPECKLE IN A NONIMAGING SYSTEM

This section is devoted to the analysis of the simple system depicted in Fig. 18.1, where a ground glass is illuminated with a monochromatic plane wave and the diffraction pattern is observed on a plane parallel to the ground glass. Both the field and the irradiance in this diffraction pattern will be studied.

As noted in Sec. 18.1.1, we assume that the detailed properties of individual ground glasses are not known and not even of interest; our results are couched in terms of an ensemble of ground glasses. An individual ground glass is a sample function of a random process corresponding to this ensemble, and both the field and the irradiance in the observation plane are sample functions obtained by mathematical transformation of the ground-glass function.

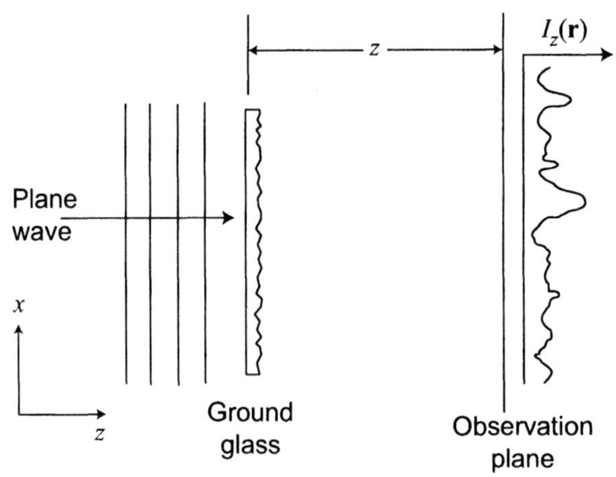

Fig. 18.1 Geometry for analyzing the speckle pattern in a nonimaging system.

Our objective is to learn as much as possible about the statistics of the field and irradiance random processes, while making as few assumptions as possible about the ground-glass random process. Since sample functions of a random process are vectors in an infinite-dimensional Hilbert space, we require an infinite-dimensional statistical description. We shall use the characteristic functional for this purpose; for important background information, see Secs. 8.2.3 and 8.3.5. Previous discussions of the statistical properties of speckle in finite-dimensional terms have been given by Dainty (1975, 1976, 1984), Goodman (1975, 1985) and Osche (2002). A previous treatment based on the infinite-dimensional characteristic functional is Zardecki and Delisle (1977).

18.2.1 Description of the ground glass

Descriptions of a ground glass (or other rough object) can be divided into three categories: physical, operational and mechanistic. A physical description focuses on how the ground glass is produced and what it looks like on a microscopic scale. An operational description asks only how the ground glass modifies a light field. A mechanistic description probes the interaction between the first two descriptions and asks how the physical characteristics relate to the operational ones. Any of these descriptions can be either deterministic or stochastic. A deterministic description gives a precise description of one particular piece of ground glass, and a stochastic description makes statements about ensemble averages.

We shall briefly survey all of these approaches but conclude at the end that we can be satisfied with a stochastic operational description, namely the random amplitude transmittance. The reader who agrees with this conclusion *a priori* can skip to Sec. 18.2.2.

What is a ground glass? Literally, a ground glass is a piece of glass that has been roughened by grinding it with an abrasive such as emery or corundum. Physically, its surface consists of irregular peaks and valleys. Operationally, it scatters light and appears diffuse to the eye. Much the same effect can be produced by etching the glass with hydrofluoric acid, so we may as well be discussing etched glass rather than ground glass. Operationally, we could also use a translucent material such as teflon or opal glass, which also scatter light and appear diffuse.

Surface-height profile If the light scattering takes place at a surface, as with a literally ground or etched glass, the physical quantity of interest is the profile of the rough surface. If the glass is in the x-y plane, its surface is described by $z = h(x, y)$, where $h(x, y)$ is usually referred to as the *surface-height profile* of the individual ground glass.

This profile can be measured by various devices called *profilometers*; for example, a fine stylus can be dragged over the surface and its deflection observed electrically or optically. Thus the surface-height profile $h(x, y)$, or $h(\mathbf{r})$ in our usual 2D vector notation, is an observable property of an individual ground glass. A common summary description of an individual profile is its standard deviation, often called the *RMS roughness*. Other measures obtainable by profilometry include the mean peak-to-valley distance and the histogram of observed heights.

If we consider an ensemble of ground glasses that have all been roughened in the same way, we can consider $h(\mathbf{r})$ to be a sample function of a random process. For a ground glass of finite size, this sample function is square-integrable and hence a vector in an $\mathbb{L}_2$ Hilbert space.

It is neither uncommon nor unreasonable to assume that the individual $h(\mathbf{r})$ is a sample function of a stationary random process, truncated by a rect function describing the overall support. We need not quibble over whether this statement should be interpreted as wide-sense or narrow-sense stationarity (see Sec. 8.2.4) since the same physical considerations justify both assumptions; we must assume only that the grinding process does the same things to the glass at all points. If one wanted to look in detail at this assumption, it would have to be admitted that the physical grinding process behaves differently at the edges of a specimen than at the center, so stationarity in either sense would be more applicable to etched glass than to ground glass, but this level of detail is far beyond our needs.

If we do make the assumption that an individual ground glass is a truncated sample from a stationary random process, the histogram of measured heights can be regarded as a histogram estimate of the true single-point PDF on the height, and the observed RMS roughness can be squared to get an estimate of the variance; both single-point PDF and variance of the random process are independent of position $\mathbf{r}$ under the assumption of stationarity. We can also use the profilometer data to get an estimate of the autocorrelation function of the process, and we might even take a discrete Fourier transform to estimate the power-spectral density. (See, however, the caveats in Sec. 8.2.7, and especially Fig. 8.1.)

Operational descriptions based on radiometry We have already met one operational description of a rough surface, the bidirectional transmittance distribution function

(BTDF) defined in Sec. 10.2.4. This description is appropriate to a ground glass since it is essentially a thin optical element with all of the scattering occurring at the surface. It is not a stochastic description since it involves only the average radiances of the incident and scattered light, and it is a very incomplete description since it says nothing about fluctuations around these averages.

Operational descriptions based on the field When the complex field is of interest, thin optical elements can be described deterministically by an *amplitude transmittance* $t(\mathbf{r})$, defined as in (9.153) by

$$u_+(\mathbf{r}) = u_-(\mathbf{r})\, t(\mathbf{r}) , \tag{18.6}$$

where $u_+(\mathbf{r})$ is the field just after the optical element, and $u_-(\mathbf{r})$ is the field incident on it. We can apply this idea to a ground glass simply by defining the transmittance $t_{GG}(\mathbf{r})$ as $u_+(\mathbf{r})/u_-(\mathbf{r})$. We don't really need to know how the ground glass does what it does, just how it modifies the field.

In this view, $t_{GG}(\mathbf{r})$ is one sample function drawn from the ensemble of ground glasses. The statistics of $t_{GG}(\mathbf{r})$ can be described just as with any other random process. Ideally we would like to know the characteristic functional, but as we shall see below we can quite often get by with much simpler statistical statements.

Mechanistic models A simple mechanistic model is the one used in Sec. 9.6.1 to describe the amplitude transmittance of a thin lens. As discussed in more detail in that section, a thin optical element modifies the phase of the light by an amount,

$$\phi(\mathbf{r}) = \frac{2\pi}{\lambda}(n_g - 1)\, h(\mathbf{r}) , \tag{18.7}$$

where λ is the free-space wavelength of the light and n_g is the refractive index of the glass. In this model, all of the light is presumed to pass through the element, and the amplitude transmittance is

$$t(\mathbf{r}) = \exp\left[i\phi(\mathbf{r})\right] . \tag{18.8}$$

Though a useful mental picture, (18.7) should not be taken seriously for rough surfaces. It ignores the fact that some light will be scattered or reflected in the backward direction, and the whole idea of phase modulation by optical elements stemmed from consideration of slabs of glass, not elements with structures comparable in size to a wavelength. Nevertheless, (18.7) does indicate that the phase variations result from surface-height variations, and it shows that very rough surfaces can produce very large phase changes.

Another mechanistic model, frequently used in SAR and reflection-mode ultrasound, assumes that the object consists of discrete point scatterers. Since each point is at a random location, it imparts a random phase to the wave. In its simplest form, this model takes the amplitude transmittance (or reflectance) to be a Poisson random process, but sometimes the scattering amplitude is also considered to be random. We shall have much more to say about point-scattering models in Sec. 18.5, but for now we shall think of the transmittance of the ground glass as a general random process, not necessarily a point process.

For further discussion of possible mechanistic models, see Zhao *et al.* (2001) and Osche (2002).

18.2.2 Some simplifying assumptions

We shall model the ground glass operationally as a pure phase object, with a random amplitude transmittance given by

$$t_{GG}(\mathbf{r}) = \exp[i\phi(\mathbf{r})]\, S(\mathbf{r})\,, \tag{18.9}$$

where $S(\mathbf{r})$ is a binary support function. The autocorrelation function of this complex random process is defined by

$$R_{GG}(\mathbf{r},\mathbf{r}') = \langle t_{GG}(\mathbf{r}) t^*_{GG}(\mathbf{r}')\rangle = S(\mathbf{r})\, S(\mathbf{r}')\, \langle \exp\{i[\phi(\mathbf{r}) - \phi(\mathbf{r}')]\}\rangle\,. \tag{18.10}$$

Two simplifying assumptions are commonly made with this model. One is that the ground glass is sufficiently rough that it completely randomizes the phases. In terms of the simple mechanistic model of (18.7), this amounts to assuming that $\sigma_h \gg \lambda$, where σ_h^2 is the variance of the surface-height profile. We can state this assumption in terms of the univariate or single-point PDF for the phase:

$$\mathrm{pr}[\phi(\mathbf{r})] = \frac{1}{2\pi}\,\mathrm{rect}\left[\frac{\phi(\mathbf{r})}{2\pi}\right] \qquad \text{(for all } \mathbf{r}). \tag{18.11}$$

It follows that

$$\langle t_{GG}(\mathbf{r})\rangle = 0\,, \tag{18.12}$$

where the average is over an ensemble of ground glasses. With this assumption, autocorrelation and autocovariance are the same thing since the mean is zero, and we denote the latter as $K_{GG}(\mathbf{r},\mathbf{r}')$.

The second useful assumption is that the autocovariance function is sharply peaked compared to other functions of interest, such as the point spread function of an imaging system. Thus, noting that $K_{GG}(\mathbf{r},\mathbf{r}) = 1$ within the support, we write

$$K_{GG}(\mathbf{r},\mathbf{r}') \approx S(\mathbf{r})\, \ell_c^2\, \delta(\mathbf{r}-\mathbf{r}')\,, \tag{18.13}$$

where ℓ_c is the *correlation length*,[3] defined by

$$\ell_c^2 \equiv \int_\infty d^2r'\, K_{GG}(\mathbf{r},\mathbf{r}')\,. \tag{18.14}$$

This integral is independent of $\mathbf{r}$ if ℓ_c is small compared to the support and $\mathbf{r}$ is not within ℓ_c of the border. In practice, we know from Sec. 10.2.7 that $\ell_c \approx \lambda$, where λ is the wavelength.

With this zero-mean complex Gaussian random process, it is tempting to assume further that circular Gaussian statistics apply (see Sec. 8.3.6), but that would be an oversimplification; for a pure phase function, $|t_{GG}(\mathbf{r})| = 1$, so Gaussian statistics are definitely not applicable. Nevertheless, we shall show via the central-limit theorem that the speckle field at an observation plane is very closely Gaussian.

[3]The reader should not confuse correlation length with coherence length. The latter term is often used to describe polychromatic radiation (see Sec. 9.7.4), but here we are considering monochromatic radiation where the coherence length is infinite.

Field emerging from the ground glass We assume for simplicity that the ground glass is illuminated with a unit-amplitude plane wave propagating in the z direction. The complex amplitude of this plane wave is $\exp(ikz)$, where $k = 2\pi/\lambda$. If we assume that the ground glass lies in the plane $z = 0$, then $u_-(\mathbf{r})$ in (18.6) is unity, and the field emerging from the ground glass is

$$u_0(\mathbf{r}) = t_{GG}(\mathbf{r}) \,. \tag{18.15}$$

Thus the field has exactly the same statistical properties as the ground glass itself with this illumination.

Propagation to the observation plane Our interest will be in the field in an observation plane parallel to the ground glass and a distance z away. We denote this field as $u_z(\mathbf{r})$ and the corresponding irradiance as $I_z(\mathbf{r}) \equiv |u_z(\mathbf{r})|^2$. We know from Sec. 9.4 how to compute $u_z(\mathbf{r})$ if we know $u_0(\mathbf{r})$. We can denote this relation abstractly as

$$\mathbf{u}_z = \boldsymbol{\mathcal{P}}_z \mathbf{u}_0 \,, \tag{18.16}$$

where $\boldsymbol{\mathcal{P}}_z$ is a propagation operator (not to be confused with a projection operator). Specifically, in the Fresnel approximation (see Sec. 9.4.6),

$$u_z(\mathbf{r}) = \frac{\exp(ikz)}{i\lambda z} \int_\infty d^2r_0 \; u_0(\mathbf{r}_0) \exp\left(i\pi \frac{|\mathbf{r} - \mathbf{r}_0|^2}{\lambda z} \right) . \tag{18.17}$$

If $t_{GG}(\mathbf{r})$ is a zero-mean random process, then so are $u_0(\mathbf{r})$ and $u_z(\mathbf{r})$. If, on the other hand, $\langle t_{GG}(\mathbf{r}) \rangle \neq 0$, then there is a nonrandom component to $u_z(\mathbf{r})$. If the support of the ground glass is large, this component is essentially an undiffracted plane wave. It is common terminology to say that the speckle is *fully developed* if $\langle u_0(\mathbf{r}) \rangle = 0$. If $\langle u_0(\mathbf{r}) \rangle \neq 0$ and there is an undiffracted plane-wave component, the speckle is said to be *partially developed.*

18.2.3 Propagation of characteristic functionals

In a statistical context, (18.16) and (18.17) apply to sample functions of random processes, and the theory of speckle boils down to analysis of the effects of various transformations on random processes. The transformation in (18.17) is a linear integral operator, but computation of the irradiance is a nonlinear point operation, and detection of the irradiance with a discrete detector array is a linear CD mapping. A useful tool for analyzing all of these transformations is the characteristic functional (see Secs. 8.2.3 and 8.3.5).

The characteristic functional for the transmittance of the ground glass is defined as

$$\Psi_{\mathbf{t}_{GG}}(\boldsymbol{\xi}) = \left\langle \exp\left[-i\pi \left(\boldsymbol{\xi}^\dagger \mathbf{t}_{GG} + \mathbf{t}_{GG}^\dagger \boldsymbol{\xi} \right) \right] \right\rangle , \tag{18.18}$$

where $\boldsymbol{\xi}$ is a vector in $\mathbb{L}_2(\mathbb{R}^2)$ corresponding to the complex-valued, square-integrable function $\xi(\mathbf{r})$, and $\mathbf{t}_{GG}$ corresponds to $t_{GG}(\mathbf{r})$. Similarly, the field emerging from the ground glass in the plane $z = 0$ has a characteristic functional given by

$$\Psi_{\mathbf{u}_0}(\boldsymbol{\xi}) = \left\langle \exp\left[-i\pi \left(\boldsymbol{\xi}^\dagger \mathbf{u}_0 + \mathbf{u}_0{}^\dagger \boldsymbol{\xi} \right) \right] \right\rangle . \tag{18.19}$$

The characteristic functional for the propagated field is given by

$$\Psi_{\mathbf{u}_z}(\boldsymbol{\xi}) = \left\langle \exp\left[-i\pi\left(\boldsymbol{\xi}^\dagger \mathbf{u}_z + \mathbf{u}_z{}^\dagger \boldsymbol{\xi}\right)\right]\right\rangle$$

$$= \left\langle \exp\left\{-i\pi\left[\boldsymbol{\xi}^\dagger(\boldsymbol{\mathcal{P}}_z\mathbf{u}_0) + (\boldsymbol{\mathcal{P}}_z\mathbf{u}_0)^\dagger\boldsymbol{\xi}\right]\right\}\right\rangle = \Psi_{\mathbf{u}_0}\left(\boldsymbol{\mathcal{P}}_z^\dagger\boldsymbol{\xi}\right). \tag{18.20}$$

In Sec. 18.2.4, we shall use the central-limit theorem to argue that $u_z(\mathbf{r})$ is a circular Gaussian random process under broad assumptions. When that conclusion holds, we can give an explicit form for $\Psi_{\mathbf{u}_z}(\boldsymbol{\xi})$; we know from (8.251) that it is given by

$$\Psi_{\mathbf{u}_z}(\boldsymbol{\xi}) = \exp\left(-\pi^2\boldsymbol{\xi}^\dagger\boldsymbol{\mathcal{K}}_{\mathbf{u}_z}\boldsymbol{\xi}\right), \tag{18.21}$$

where $\boldsymbol{\mathcal{K}}_{\mathbf{u}_z}$ is the autocovariance operator, an integral transform with the autocovariance function as its kernel. To be explicit, the quadratic form in (18.21) is a scalar given by

$$\boldsymbol{\xi}^\dagger\boldsymbol{\mathcal{K}}_{\mathbf{u}_z}\boldsymbol{\xi} = \int_\infty d^2r \int_\infty d^2r'\, \xi^*(\mathbf{r})\, K_{\mathbf{u}_z}(\mathbf{r},\mathbf{r}')\, \xi(\mathbf{r}'). \tag{18.22}$$

Since $\boldsymbol{\mathcal{K}}_{\mathbf{u}_z}$ is positive-definite, $\boldsymbol{\xi}^\dagger\boldsymbol{\mathcal{K}}_{\mathbf{u}_z}\boldsymbol{\xi}$ is real even though all factors in (18.22) may be complex.

We learned in Sec. 8.2.6 how to propagate autocovariance operators; with (8.147) and (18.13) we see that

$$\boldsymbol{\mathcal{K}}_{\mathbf{u}_z} = \boldsymbol{\mathcal{P}}_z\boldsymbol{\mathcal{K}}_{\mathbf{u}_0}\boldsymbol{\mathcal{P}}_z^\dagger \approx \ell_c^2\boldsymbol{\mathcal{P}}_z\boldsymbol{\mathcal{P}}_z^\dagger. \tag{18.23}$$

In the Fresnel approximation, the kernel of the operator $\boldsymbol{\mathcal{K}}_{\mathbf{u}_z}$ is[4]

$$K_{\mathbf{u}_z}(\mathbf{r},\mathbf{r}') = \frac{\ell_c^2}{\lambda^2 z^2}\int_A d^2r'' \exp\left(i\frac{\pi}{\lambda z}|\mathbf{r}-\mathbf{r}''|^2\right)\exp\left(-i\frac{\pi}{\lambda z}|\mathbf{r}'-\mathbf{r}''|^2\right)$$

$$= \frac{\ell_c^2}{\lambda^2 z^2}L^2\,\mathrm{sinc}\left(\frac{L}{\lambda z}|\mathbf{r}-\mathbf{r}'|\right), \tag{18.24}$$

where A denotes the area of the ground glass, of dimensions $L\times L$. The sinc function has a width of $\lambda z/L$, confirming the qualitative argument from Sec. 18.1.1 that the correlation length in the diffraction pattern is inversely proportional to the size of the ground glass.

18.2.4 Central-limit theorem

What follows is a fairly tedious derivation showing how the central-limit theorem applies to the present problem. The reader who is content with handwaving can skip this discussion and jump to Sec. 18.2.5.

The derivation here generally follows the development in Sec. 8.3.4, modified for random processes rather than random variables. A key difference, however, is

[4] A quadratic phase factor $\exp\left[i\frac{\pi}{\lambda z}(r^2 - r'^2)\right]$ has been dropped in (18.24) since it is near unity whenever the sinc function is appreciable.

that we do not add multiple random processes, but instead consider a single random process divided spatially into successively smaller component processes.

For generality we forego the zero-mean assumption and write

$$t_{GG}(\mathbf{r}) = \bar{t}_{GG} + \Delta t_{GG}(\mathbf{r}) \,, \tag{18.25}$$

and similarly for the fields. We assume that $\Delta t_{GG}(\mathbf{r})$ is rapidly varying so that the ground glass can be written as a sum of nonoverlapping regions with negligible statistical dependence. If the dimensions of the ground glass are $L \times L$ and those of the regions in question are $\epsilon \times \epsilon$, we write the random part of the field emerging from the ground glass as

$$\Delta u_0(\mathbf{r}) = \sum_{\mathbf{k}} \Delta u_0^{(\mathbf{k})}(\mathbf{r}) \,, \tag{18.26}$$

where $\mathbf{k}$ is a 2D multi-index of integer components. Explicitly,

$$\Delta u_0^{(\mathbf{k})}(\mathbf{r}) = \Delta u_0(\mathbf{r}) \operatorname{rect}\left(\frac{\mathbf{r} - \mathbf{k}\epsilon}{\epsilon}\right) \equiv w^{(\mathbf{k})}(\mathbf{r}) \, \Delta u_0(\mathbf{r}) \,, \tag{18.27}$$

where the components of $\mathbf{k}$ run from $-K$ to K, with $K = (L - \epsilon)/(2\epsilon)$ and K odd. Note that the sum of rect functions is unity within the $L \times L$ support of the ground glass. We assume that we can choose $\ell_c \ll \epsilon \ll L$, and assume further that ℓ_c is the characteristic length beyond which the random process is not just uncorrelated but statistically independent; thus (18.26) represents a decomposition of the field into statistically independent components.

Propagating to the plane z, we have

$$\Delta \mathbf{u}_z^{(\mathbf{k})} = \mathcal{P}_z \Delta \mathbf{u}_0^{(\mathbf{k})} \,. \tag{18.28}$$

The characteristic functional of $\Delta \mathbf{u}_z$ is

$$\begin{aligned} \Psi_{\Delta \mathbf{u}_z}(\boldsymbol{\xi}) &= \left\langle \exp\left[-i\pi \left(\boldsymbol{\xi}^\dagger \Delta \mathbf{u}_z + \Delta \mathbf{u}_z^\dagger \boldsymbol{\xi}\right)\right]\right\rangle \\ &= \left\langle \exp\left\{-i\pi \sum_{\mathbf{k}} \left[\boldsymbol{\xi}^\dagger \left(\mathcal{P}_z \Delta \mathbf{u}_0^{(\mathbf{k})}\right) + \left(\mathcal{P}_z \Delta \mathbf{u}_0^{(\mathbf{k})}\right)^\dagger \boldsymbol{\xi}\right]\right\}\right\rangle \\ &= \prod_{\mathbf{k}} \left\langle \exp\left\{-i\pi \left[\boldsymbol{\xi}^\dagger \left(\mathcal{P}_z \Delta \mathbf{u}_0^{(\mathbf{k})}\right) + \left(\mathcal{P}_z \Delta \mathbf{u}_0^{(\mathbf{k})}\right)^\dagger \boldsymbol{\xi}\right]\right\}\right\rangle = \prod_{\mathbf{k}} \Psi_{\Delta \mathbf{u}_0^{(\mathbf{k})}}\left(\mathcal{P}_z^\dagger \boldsymbol{\xi}\right) . \end{aligned} \tag{18.29}$$

If we were following the flow of Sec. 8.3.4 exactly, we would at this point assume that the the random processes $\Delta \mathbf{u}_0^{(\mathbf{k})}$ were independent and identically distributed, but in fact they are not identically distributed because of the spatial offsets. We can, however, expand $\Psi_{\Delta \mathbf{u}_0^{(\mathbf{k})}}(\boldsymbol{\xi})$ analogously to (8.205), yielding

$$\Psi_{\Delta \mathbf{u}_0^{(\mathbf{k})}}(\boldsymbol{\xi}) = 1 - \pi^2 \, \boldsymbol{\xi}^\dagger \left\langle \Delta \mathbf{u}_0^{(\mathbf{k})} \Delta \mathbf{u}_0^{(\mathbf{k})\dagger} \right\rangle \boldsymbol{\xi} + \ldots \,, \tag{18.30}$$

where we have made use of the facts that $\Delta \mathbf{u}_0^{(\mathbf{k})}$ has zero mean and $\boldsymbol{\xi}^\dagger \langle \Delta \mathbf{u}_0^{(\mathbf{k})} \Delta \mathbf{u}_0^{(\mathbf{k})\dagger} \rangle \boldsymbol{\xi}$ is real. We have also assumed that $\langle [\boldsymbol{\xi}^\dagger \mathcal{P}_z \Delta \mathbf{u}_0^{(\mathbf{k})}]^2 \rangle$ and its complex conjugate are negligible. We return to this assumption below; making it uncritically for now, we see that

$$\Psi_{\Delta \mathbf{u}_z}(\boldsymbol{\xi}) = \prod_{\mathbf{k}} \left[1 - \pi^2 \left(\mathcal{P}_z^\dagger \boldsymbol{\xi} \right)^\dagger \left\langle \Delta \mathbf{u}_0^{(\mathbf{k})} \Delta \mathbf{u}_0^{(\mathbf{k})\dagger} \right\rangle \mathcal{P}_z^\dagger \boldsymbol{\xi} + \ldots \right] . \tag{18.31}$$

At this point we invoke the fact that the field on the ground glass has a very short correlation length, much shorter than the region size ϵ and even shorter than a wavelength. The propagator $\mathcal{P}_z$, on the other hand, has a kernel that is slowly varying compared to a wavelength in the Fresnel approximation. Thus, for present purposes we can write

$$\left\langle \left[\Delta u_0^{(\mathbf{k})}(\mathbf{r}) \right] \left[\Delta u_0^{(\mathbf{k})}(\mathbf{r}') \right]^* \right\rangle \approx \ell_c^2 \, w^{(\mathbf{k})}(\mathbf{r}) \, \delta(\mathbf{r} - \mathbf{r}') \tag{18.32}$$

or, in operator form,

$$\left\langle \left[\Delta \mathbf{u}_0^{(\mathbf{k})} \right] \left[\Delta \mathbf{u}_0^{(\mathbf{k})} \right]^\dagger \right\rangle = \ell_c^2 \, \mathcal{W}^{(\mathbf{k})} , \tag{18.33}$$

with $\mathcal{W}^{(\mathbf{k})}$ being a multiplicative operator with kernel $w^{(\mathbf{k})}(\mathbf{r})\, \delta(\mathbf{r} - \mathbf{r}')$. Therefore,

$$\begin{aligned} \Psi_{\Delta \mathbf{u}_z}(\boldsymbol{\xi}) &= \prod_{\mathbf{k}} \left[1 - \pi^2 \ell_c^2 \left(\mathcal{P}_z^\dagger \boldsymbol{\xi} \right)^\dagger \mathcal{W}^{(\mathbf{k})} \left(\mathcal{P}_z^\dagger \, \boldsymbol{\xi} \right) + \ldots \right] \\ &= \prod_{\mathbf{k}} \left\{ 1 - \pi^2 \ell_c^2 \left(\boldsymbol{\xi}^\dagger \left[\mathcal{P}_z \mathcal{W}^{(\mathbf{k})} \mathcal{P}_z^\dagger \right] \boldsymbol{\xi} \right) + \ldots \right\} . \end{aligned} \tag{18.34}$$

Taking a logarithm yields

$$\ln\{\Psi_{\Delta \mathbf{u}_z}(\boldsymbol{\xi})\} = \sum_{\mathbf{k}} \ln \left\{ 1 - \pi^2 \ell_c^2 \left(\boldsymbol{\xi}^\dagger \left[\mathcal{P}_z \mathcal{W}^{(\mathbf{k})} \mathcal{P}_z^\dagger \right] \boldsymbol{\xi} \right) + \ldots \right\} . \tag{18.35}$$

The operator in square brackets here, $[\mathcal{P}_z \mathcal{W}^{(\mathbf{k})} \mathcal{P}_z^\dagger]$, has a kernel

$$\begin{aligned} \left[\mathcal{P}_z \mathcal{W}^{(\mathbf{k})} \mathcal{P}_z^\dagger \right] (\mathbf{r}, \mathbf{r}') &= \int_A d^2 r_1 \int_A d^2 r_2 \; p_z(\mathbf{r}, \mathbf{r}_1) \, w^{(\mathbf{k})}(\mathbf{r}_1) \, \delta(\mathbf{r}_1 - \mathbf{r}_2) \, p_z^*(\mathbf{r}_2, \mathbf{r}') \\ &= \int_A d^2 r_1 \; p_z(\mathbf{r}, \mathbf{r}_1) \, w^{(\mathbf{k})}(\mathbf{r}_1) \, p_z^*(\mathbf{r}_1, \mathbf{r}') . \end{aligned} \tag{18.36}$$

This integral varies as ϵ^2 and higher-order terms vary as higher powers. Thus, by the same arguments as in Sec. 8.3.4, we can drop the higher terms and expand the logarithm to obtain

$$\ln\{\Psi_{\Delta \mathbf{u}_z}(\boldsymbol{\xi})\} \sim - \sum_{\mathbf{k}} \pi^2 \ell_c^2 \boldsymbol{\xi}^\dagger \mathcal{P}_z \mathcal{W}^{(\mathbf{k})} \mathcal{P}_z^\dagger \boldsymbol{\xi} . \tag{18.37}$$

Finally, since $\sum_{\mathbf{k}} \mathcal{W}^{(\mathbf{k})}$ is just the unit operator, we get

$$\ln\{\Psi_{\Delta \mathbf{u}_z}(\boldsymbol{\xi})\} \sim -\pi^2 \ell_c^2 \boldsymbol{\xi}^\dagger \mathcal{P}_z \mathcal{P}_z^\dagger \boldsymbol{\xi} \tag{18.38}$$

or

$$\Psi_{\Delta \mathbf{u}_z}(\boldsymbol{\xi}) \sim \exp \left(-\pi^2 \ell_c^2 \boldsymbol{\xi}^\dagger \mathcal{P}_z \mathcal{P}_z^\dagger \boldsymbol{\xi} \right) . \tag{18.39}$$

The conclusion is that $\Delta u_z(\mathbf{r})$ is a circular Gaussian random process. If the mean transmittance of the ground glass is zero, then $u_z(\mathbf{r})$ is a also circular Gaussian. If the mean transmittance is not zero, we get

$$\Psi_{\mathbf{u}_z}(\boldsymbol{\xi}) \sim \exp \left[-i\pi \left(\boldsymbol{\xi}^\dagger \mathcal{P}_z \overline{\mathbf{u}}_0 + \overline{\mathbf{u}}_0^\dagger \mathcal{P}_z^\dagger \boldsymbol{\xi} \right) \right] \exp \left(-\pi^2 \ell_c^2 \boldsymbol{\xi}^\dagger \mathcal{P}_z \mathcal{P}_z^\dagger \boldsymbol{\xi} \right) . \tag{18.40}$$

A lapse of propriety Now let us return to (18.30), where we dropped $\langle[\boldsymbol{\xi}^\dagger \mathcal{P}_z \Delta \mathbf{u}_0^{(\mathbf{k})}]^2\rangle$ and its complex conjugate. This term can be written as

$$\left\langle \left[\boldsymbol{\xi}^\dagger \mathcal{P}_z \Delta \mathbf{u}_0^{(\mathbf{k})}\right]^2 \right\rangle = \boldsymbol{\xi}^\dagger \mathcal{P}_z \left\langle \left[\Delta \mathbf{u}_0^{(\mathbf{k})}\right] \left[\Delta \mathbf{u}_0^{(\mathbf{k})}\right]^t \right\rangle \mathcal{P}_z^t \boldsymbol{\xi}^{\dagger t} . \tag{18.41}$$

We could drop this term immediately if we could argue that $\Delta \mathbf{u}_0^{(\mathbf{k})}$ was a *proper* random process. Recall from Sec. 8.3.6 that a complex random vector or process $\mathbf{z}$ is proper if the mean $\langle \mathbf{z}\rangle$ and the *pseudocovariance* $\langle \mathbf{z}\mathbf{z}^t\rangle$ both vanish; in that case the second-order statistics are fully specified by the covariance defined as $\langle \mathbf{z}\mathbf{z}^\dagger\rangle$. A sufficient condition for $\mathbf{z}$ to be proper is that it be zero mean and that the real and imaginary parts of any component be *i.i.d.* Consideration of Fig. 18.2 shows, however, that $\Delta \mathbf{u}_0^{(\mathbf{k})}$ does not satisfy these conditions; its mean is zero by construction, but its real and imaginary parts are neither independent nor identically distributed. We therefore have to look further for the rationale for neglecting this term.

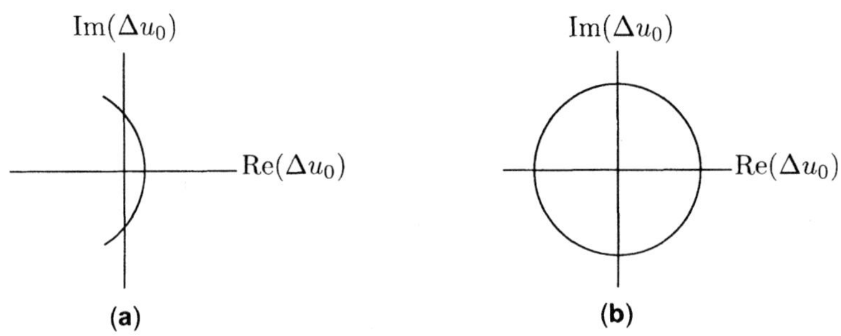

Fig. 18.2 (a) Diagram illustrating the statistical properties of the field in the plane $z = 0$ when it cannot be assumed that the phase of this field is uniformly distributed on $(0, 2\pi)$. It is assumed here that $\langle u_0(\mathbf{r})\rangle \neq 0$ and that $|u_0(\mathbf{r})| = 1$. Thus the allowed values of $\Delta u_0(\mathbf{r})$ lie on the arc shown, and by inspection the real and imaginary parts of $\Delta u_0(\mathbf{r})$ are neither independent nor identically distributed. (b) Similar diagram for the case where the phase is uniformly distributed. Now the mean is zero, $\Delta u_0(\mathbf{r}) = u_0(\mathbf{r})$, and the real and imaginary parts of $\Delta u_0(\mathbf{r})$ are i.i.d.

Since we have already assumed that the fields at two points on the ground glass are statistically independent unless the points are within ℓ_c, and that this distance is very small compared to the width of the kernel of $\mathcal{P}_z$, we can write the pseudocovariance function as [*cf.* (18.32)]

$$\left\langle \left[\Delta u_0^{(\mathbf{k})}(\mathbf{r})\right] \left[\Delta u_0^{(\mathbf{k})}(\mathbf{r}')\right] \right\rangle \approx a_c\, w^{(\mathbf{k})}(\mathbf{r})\, \delta(\mathbf{r} - \mathbf{r}') , \tag{18.42}$$

where $a_c \leq \ell_c^2$ because of partial phase cancellation. Writing out (18.41) in integral form and using (18.42) yields [*cf.* (18.36)]

$$\left\langle \left[\boldsymbol{\xi}^\dagger \mathcal{P}_z \Delta \mathbf{u}_0^{(\mathbf{k})}\right]^2 \right\rangle = a_c \int_\infty d^2r\; \xi^*(\mathbf{r}) \int_\infty d^2r'\, \xi^*(\mathbf{r}') \int_A d^2r_1\; p_z(\mathbf{r}, \mathbf{r}_1)\, w^{(\mathbf{k})}(\mathbf{r}_1)\, p_z(\mathbf{r}_1, \mathbf{r}') . \tag{18.43}$$

The important point to note here is the absence of a complex conjugate on either p_z factor, so there is no cancellation of quadratic phase factors. After performing steps analogous to (18.36) – (18.38) and inserting the Fresnel propagation kernel, we find that the $\mathbf{r}_1$ integral takes the form

$$\frac{1}{\lambda^2 z^2}\int_A d^2r_1 \exp\left[i\frac{\pi}{\lambda z}\left(|\mathbf{r}_1-\mathbf{r}|^2+|\mathbf{r}_1-\mathbf{r}'|^2\right)\right] . \tag{18.44}$$

Because of the oscillating factor of $\exp(2i\pi r_1^2/\lambda z)$, this integral will be very small (compared to A) if the overall area of the ground glass covers many Fresnel zones, or $L^2/\lambda z \gg 1$. This is the opposite of the Fraunhofer condition (see Sec. 9.5.2), so the pseudocovariance term has a small effect in the near field of the ground glass, in essence because the propagation introduces large phase variations.

In summary, we can overlook the impropriety if either the ground glass or the propagation fully randomizes the phase of the field.

18.2.5 Statistics of the irradiance

We now know how to write the characteristic functional for the field $u_z(\mathbf{r})$ in the observation plane. The next step is to study the statistics of the irradiance, $I_z(\mathbf{r}) \equiv |u_z(\mathbf{r})|^2$. Because of the nonlinear nature of the transformation from field to irradiance, there is no general rule for transforming the characteristic functionals, but we shall see that a useful expression is possible when the field is a circular Gaussian.

Our approach will be to consider first a finely sampled but discrete field pattern and compute its characteristic *function*, then pass to the limit of an infinitesimal sampling interval and get the characteristic *functional* for the irradiance. Point sampling causes no difficulty because the propagator $\mathcal{P}_z$ is a low-pass filter, blocking all 2D spatial frequencies of magnitude greater than the reciprocal of the wavelength (see Sec. 9.5.1).

We define an $N\times 1$ complex vector $\mathbf{u}_z$ with components $u_z(\mathbf{r}_n)$ and an $N\times 1$ real vector $\mathbf{I}_z$ with components given by $I_z(\mathbf{r}_n) = |u_z(\mathbf{r}_n)|^2$. The characteristic function (not functional, so we use ψ instead of Ψ) for $\mathbf{I}_z$ is defined by

$$\psi_{\mathbf{I}_z}(\boldsymbol{\zeta}) = \left\langle \exp(-2\pi i\boldsymbol{\zeta}^t\mathbf{I}_z)\right\rangle , \tag{18.45}$$

where $\boldsymbol{\zeta}$ is an $N\times 1$ real vector, and the average is over real and imaginary components of $\mathbf{u}_z$.

If the field in plane z is circular Gaussian, then

$$\begin{aligned}\psi_{\mathbf{I}_z}(\boldsymbol{\zeta}) &= \frac{1}{\pi^N\det(\mathbf{K}_{\mathbf{u}_z})}\int_\infty d^{2N}u_z \exp(-\mathbf{u}_z^\dagger\mathbf{K}_{\mathbf{u}_z}^{-1}\mathbf{u}_z)\exp(-2\pi i\boldsymbol{\zeta}^t\mathbf{I}_z) \\ &= \frac{1}{\pi^N\det(\mathbf{K}_{\mathbf{u}_z})}\int_\infty d^{2N}u_z \exp[-\mathbf{u}_z^\dagger(\mathbf{K}_{\mathbf{u}_z}^{-1}+2\pi i\mathbf{Z})\mathbf{u}_z] ,\end{aligned} \tag{18.46}$$

where $\mathbf{Z}$ is an $N\times N$ diagonal matrix with elements given by

$$Z_{mn} = \zeta_m\,\delta_{mn} . \tag{18.47}$$

Simultaneous diagonalization Key to evaluation of the integral in (18.46) is finding a representation where $\mathbf{K}_{\mathbf{u}_z}^{-1}$ and $\mathbf{Z}$ are simultaneously diagonal. We shall use the diagonalization procedure from Sec. 1.4.6; this procedure diagonalizes two Hermitian matrices $\mathbf{A}$ and $\mathbf{B}$, even if they do not commute, by three steps. The first step is to diagonalize $\mathbf{A}$ by a unitary transformation, then convert it to the unit matrix by a nonunitary *prewhitening* transform, finally diagonalize the transformed matrix $\mathbf{B}$ by another unitary transformation. The prewhitening step requires that $\mathbf{A}$ be nonsingular.

To apply this procedure to the problem at hand, we take $\mathbf{A} = \mathbf{K}_{\mathbf{u}_z}^{-1}$ and $\mathbf{B} = \mathbf{Z}$. Thus we first undiagonalize $\mathbf{Z}$, then rediagonalize it. We cannot, however, choose $\mathbf{A} = \mathbf{Z}$ since assuming that $\mathbf{Z}$ is nonsingular is the same as requiring all $\zeta_n \neq 0$; in fact, most of the interest in the characteristic function is at $\boldsymbol{\zeta} = \mathbf{0}$.

Following Sec. 1.4.6, we first do a Karhunen-Loève transformation (see also Sec. 8.2.7) on $\mathbf{K}_{\mathbf{u}_z}$:

$$\mathbf{K}_{\mathbf{u}_z}\boldsymbol{\Upsilon} = \boldsymbol{\Upsilon}\mathbf{M}\,, \qquad \boldsymbol{\Upsilon}^\dagger\mathbf{K}_{\mathbf{u}_z}\boldsymbol{\Upsilon} = \mathbf{M}\,, \qquad \boldsymbol{\Upsilon}^\dagger\mathbf{K}_{\mathbf{u}_z}^{-1}\boldsymbol{\Upsilon} = \mathbf{M}^{-1}\,, \tag{18.48}$$

where $\boldsymbol{\Upsilon}$ is a unitary matrix and $\mathbf{M}$ is a diagonal matrix with the eigenvalues of $\mathbf{K}_{\mathbf{u}_z}$ along the diagonal.

Next we prewhiten $\mathbf{K}_{\mathbf{u}_z}^{-1}$, obtaining

$$\mathbf{M}^{\frac{1}{2}}\boldsymbol{\Upsilon}^\dagger\mathbf{K}_{\mathbf{u}_z}^{-1}\boldsymbol{\Upsilon}\mathbf{M}^{\frac{1}{2}} = \mathbf{I}\,. \tag{18.49}$$

Finally, we rediagonalize $\mathbf{Z}$ to get

$$\mathbf{W}^\dagger\mathbf{Z}\mathbf{W} = \mathbf{D}\,, \tag{18.50}$$

where $\mathbf{D}$ is diagonal and

$$\mathbf{W} = \boldsymbol{\Upsilon}\mathbf{M}^{\frac{1}{2}}\boldsymbol{\Phi}\,. \tag{18.51}$$

Thus

$$\mathbf{W}^\dagger\mathbf{K}_{\mathbf{u}_z}^{-1}\mathbf{W} = \mathbf{I}\,, \qquad \mathbf{W}^\dagger\mathbf{Z}\mathbf{W} = \mathbf{D}\,. \tag{18.52}$$

Note that $\mathbf{W}^\dagger\mathbf{K}_{\mathbf{u}_z}^{-1}\mathbf{W} = \mathbf{I}$ does not imply that $\mathbf{W}^\dagger\mathbf{K}_{\mathbf{u}_z}\mathbf{W} = \mathbf{I}$ since $\mathbf{W}$ is not unitary.

Evaluation of the integral We can now perform the integral in (18.46) by introducing the change of variables:[5]

$$\mathbf{u}_z = \mathbf{W}\tilde{\mathbf{u}}_z\,. \tag{18.53}$$

The expression for the characteristic function now becomes[6]

$$\begin{aligned}\psi_{\mathbf{I}_z}(\boldsymbol{\zeta}) &= \frac{1}{\pi^N\det(\mathbf{K}_{\mathbf{u}_z})}\int_\infty d^{2N}u_z\,\exp[-\mathbf{u}_z^\dagger(\mathbf{K}_{\mathbf{u}_z}^{-1} + 2\pi i\mathbf{Z})\,\mathbf{u}_z] \\ &= \frac{|\det(\mathbf{W})|^2}{\pi^N\det(\mathbf{K}_{\mathbf{u}_z})}\int_\infty d^{2N}\tilde{u}_z\,\exp[-\tilde{\mathbf{u}}_z^\dagger(\mathbf{I} + 2\pi i\mathbf{D})\tilde{\mathbf{u}}_z]\,,\end{aligned} \tag{18.54}$$

where $\boldsymbol{\zeta}$ is now hidden within $\mathbf{D}$ (we'll retrieve it shortly), and $|\det(\mathbf{W})|^2$ is the Jacobian of the transformation. Since $\boldsymbol{\Phi}$ and $\boldsymbol{\Psi}$ are unitary, it follows from (A.78) and (A.83) that

$$|\det(\mathbf{W})|^2 = [\det(\mathbf{M}^{\frac{1}{2}})]^2 = \det(\mathbf{K})\,, \tag{18.55}$$

[5] Actually, this step is a bit tricky. With unitary transformations we would normally write $\tilde{\mathbf{u}} = \mathbf{W}^\dagger\mathbf{u}$, which is equivalent to (18.53) when $\mathbf{W}^\dagger = \mathbf{W}^{-1}$. To get the exponent into a form where our freshly diagonalized operators appear after a nonunitary transformation, however, we must define the transformation as in (18.53).

[6] The reader should not confuse the unit matrix $\mathbf{I}$ in this expression with the irradiance $I_z(\mathbf{r})$, which we write as $\mathbf{I}_z$ when we regard it as a vector in Hilbert space.

so the integral we need to evaluate is

$$\psi_{\mathbf{I}_z}(\boldsymbol{\zeta}) = \frac{1}{\pi^N} \int_\infty d^{2N}\tilde{u}_z \, \exp[-\tilde{\mathbf{u}}_z^\dagger (\mathbf{I} + 2\pi i \mathbf{D})\tilde{\mathbf{u}}_z] \,. \tag{18.56}$$

Since $\mathbf{D}$ is diagonal, this integral factors into a product of N 2D integrals of the form

$$\begin{aligned} I_n &= \int_{-\infty}^{\infty} dx_n \int_{-\infty}^{\infty} dy_n \, \exp[-a_n(x_n^2 + y_n^2)] \\ &= \int_{-\infty}^{\infty} dx_n \, \exp(-a_n \, x_n^2) \int_{-\infty}^{\infty} dy_n \, \exp(-a_n \, y_n^2) \,, \end{aligned} \tag{18.57}$$

where

$$a_n = 1 + 2\pi i D_{nn} \,. \tag{18.58}$$

Gaussians with complex coefficients have occurred several times previously in this book because of their role in Fresnel propagation problems (see Secs. 3.3.7, 4.3.2 and 9.4.6). By deforming the contour of integration as in (3.184), we find [*cf.* (3.185) with $\xi = 0$]

$$I_n = \frac{\pi}{a_n} \,, \tag{18.59}$$

and the original $2N$D integral in (18.46) is

$$I = \prod_{n=1}^{N} I_n = \prod_{n=1}^{N} \frac{\pi}{a_n} = \frac{\pi^N}{\det(\mathbf{I} + 2\pi i \mathbf{D})} \,. \tag{18.60}$$

Thus

$$\psi_{\mathbf{I}_z}(\boldsymbol{\zeta}) = \frac{1}{\det(\mathbf{I} + 2\pi i \mathbf{D})} = \frac{1}{\det(\mathbf{I} + 2\pi i \mathbf{W}^\dagger \mathbf{Z} \mathbf{W})} \,, \tag{18.61}$$

where, as a reminder,

$$\mathbf{W}^\dagger \mathbf{K}_{\mathbf{u}_z}^{-1} \mathbf{W} = \mathbf{I} \,, \qquad Z_{mn} = \zeta_n \, \delta_{mn} \,, \qquad [\mathbf{W}^\dagger \mathbf{Z} \mathbf{W}]_{mn} = D_n \, \delta_{mn} \,. \tag{18.62}$$

Equivalent forms By judiciously inserting a couple of unit matrices, written as $\mathbf{I} = \boldsymbol{\Psi}^\dagger \boldsymbol{\Psi}$, we get another expression for the characteristic function of the irradiance:

$$\psi_{\mathbf{I}_z}(\boldsymbol{\zeta}) = \left[\det\left(\mathbf{I} + 2\pi i \mathbf{U}^\dagger \mathbf{K}_{\mathbf{u}_z}^{\frac{1}{2}} \mathbf{Z} \mathbf{K}_{\mathbf{u}_z}^{\frac{1}{2}} \mathbf{U}\right)\right]^{-1} \,, \tag{18.63}$$

where $\mathbf{U} \equiv \boldsymbol{\Psi}\boldsymbol{\Phi}$ is the unitary operator that diagonalizes the Hermitian matrix $\mathbf{K}_{\mathbf{u}_z}^{\frac{1}{2}} \mathbf{Z} \mathbf{K}_{\mathbf{u}_z}^{\frac{1}{2}}$. Since determinants are unchanged by unitary transformations, (18.63) is equivalent to

$$\psi_{\mathbf{I}_z}(\boldsymbol{\zeta}) = \left[\det\left(\mathbf{I} + 2\pi i \mathbf{K}_{\mathbf{u}_z}^{\frac{1}{2}} \mathbf{Z} \mathbf{K}_{\mathbf{u}_z}^{\frac{1}{2}}\right)\right]^{-1} \,. \tag{18.64}$$

With some identities on determinants from App. A, we get

$$\det\left(\mathbf{I} + 2\pi i \mathbf{K}_{\mathbf{u}_z}^{\frac{1}{2}} \mathbf{Z} \mathbf{K}_{\mathbf{u}_z}^{\frac{1}{2}}\right) = \det\left[\mathbf{K}_{\mathbf{u}_z}^{\frac{1}{2}} \left(\mathbf{I} + 2\pi i \mathbf{K}_{\mathbf{u}_z}^{\frac{1}{2}} \mathbf{Z} \mathbf{K}_{\mathbf{u}_z}^{\frac{1}{2}}\right) \mathbf{K}_{\mathbf{u}_z}^{-\frac{1}{2}}\right] = \det(\mathbf{I} + 2\pi i \mathbf{K}_{\mathbf{u}_z} \mathbf{Z}) \,, \tag{18.65}$$

so

$$\psi_{\mathbf{I}_z}(\boldsymbol{\zeta}) = [\det(\mathbf{I} + 2\pi i \mathbf{K}_{\mathbf{u}_z} \mathbf{Z})]^{-1} \,. \tag{18.66}$$

This expression is consistent with the chi-squared characteristic function as given in (C.139). Specifically, if there are no correlations and $\mathbf{K}_{\mathbf{u}_z} = 2\sigma^2 \mathbf{I}$, then $\psi_{\mathbf{I}_z}(\boldsymbol{\zeta})$ is just a product of N factors like (C.139), each with 2 degrees of freedom.

Expansion in traces Determinants can be related to traces by (A.114), reproduced here for convenience:

$$-\ln[\det(\mathbf{I}-\mathbf{A})] = \operatorname{tr}\mathbf{A} + \tfrac{1}{2}\operatorname{tr}\mathbf{A}^2 + \tfrac{1}{3}\operatorname{tr}\mathbf{A}^3 + \ldots . \tag{18.67}$$

Using this relation with $\mathbf{A} = -2\pi i \mathbf{K}_{\mathbf{u}_z}\mathbf{Z}$, we get

$$\psi_{\mathbf{I}_z}(\boldsymbol{\zeta}) = \exp\left[-2\pi i \operatorname{tr}(\mathbf{K}_{\mathbf{u}_z}\mathbf{Z}) - 2\pi^2 \operatorname{tr}(\mathbf{K}_{\mathbf{u}_z}\mathbf{Z}\mathbf{K}_{\mathbf{u}_z}\mathbf{Z}) + \cdots\right] . \tag{18.68}$$

As noted in App. A, this expansion converges if all eigenvalues of $\mathbf{A}$ are < 1 in absolute value. Usually we want to use $\Psi_{\mathbf{I}_z}(\boldsymbol{\zeta})$ for calculating moments; convergence is guaranteed in that case since the moments all involve derivatives evaluated at the origin, and all eigenvalues of $\mathbf{A}$ approach zero as $\boldsymbol{\zeta} \to \mathbf{0}$.

Writing out the first term in the exponent of (18.68) in detail yields

$$\operatorname{tr}(\mathbf{KZ}) = \sum_{i,j} K_{ij} Z_{ji} = \sum_{i,j} K_{ij}\zeta_i\,\delta_{ji} = \sum_j K_{jj}\zeta_j . \tag{18.69}$$

Similarly, the second term is

$$\operatorname{tr}(\mathbf{KZKZ}) = \sum_{i,j,k\ell} K_{ij} Z_{jk} K_{k\ell} Z_{\ell i} = \sum_{j,k} \zeta_j K_{jk} K_{kj} \zeta_k = \sum_{j,k} \zeta_j |K_{jk}|^2 \zeta_k . \tag{18.70}$$

When these explicit expressions are used in (18.68), it is relatively straightforward to carry out the necessary derivatives and compute moments.

The continuous limit We can pass to the infinite-dimensional limit where vectors become functions by defining an integral operator $\boldsymbol{\mathcal{Z}}$ with kernel $\zeta(\mathbf{r})\,\delta(\mathbf{r}-\mathbf{r}')$.[7] Then the traces are interpreted as

$$\operatorname{tr}(\boldsymbol{\mathcal{K}}\boldsymbol{\mathcal{Z}}) = \int_\infty d^2r\ \zeta(\mathbf{r})\,K(\mathbf{r},\mathbf{r}) , \qquad \operatorname{tr}(\boldsymbol{\mathcal{K}}\boldsymbol{\mathcal{Z}}\boldsymbol{\mathcal{K}}\boldsymbol{\mathcal{Z}}) = \int_\infty d^2r \int_\infty d^2r'\,\zeta(\mathbf{r})\,|K(\mathbf{r},\mathbf{r}')|^2\,\zeta(\mathbf{r}') , \tag{18.71}$$

and similarly for higher-order terms.

At this stage we have an expression (albeit in terms of an infinite product of exponentials) for the characteristic functional of the irradiance for a nonimaging system. Specifically,

$$\Psi_{\mathbf{I}_z}(\boldsymbol{\zeta}) = \exp[-2\pi i \operatorname{tr}(\boldsymbol{\mathcal{K}}_{\mathbf{u}_z}\boldsymbol{\mathcal{Z}}) - 2\pi^2 \operatorname{tr}(\boldsymbol{\mathcal{K}}_{\mathbf{u}_z}\boldsymbol{\mathcal{Z}}\boldsymbol{\mathcal{K}}_{\mathbf{u}_z}\boldsymbol{\mathcal{Z}}) + \cdots] . \tag{18.72}$$

Equivalently, we can use the determinantal expression (18.66) and pass to the limit to obtain

$$\Psi_{\mathbf{I}_z}(\boldsymbol{\zeta}) = \frac{1}{\det(\boldsymbol{\mathcal{I}} + 2\pi i\,\boldsymbol{\mathcal{K}}_{\mathbf{u}_z}\boldsymbol{\mathcal{Z}})} , \tag{18.73}$$

provided we interpret the determinant of an integral operator as the infinite product of its eigenvalues. The factors in the product converge to unity as the eigenvalues of $\boldsymbol{\mathcal{K}}_{\mathbf{u}_z}$ approach zero, so the fact that the product is infinite causes no grief.

[7] This operator is the Hadamard product of the Hilbert-space vector $\boldsymbol{\zeta}$ and the unit operator $\boldsymbol{\mathcal{I}}$ (see Sec. A.2.8). We could have chosen to write it as $\boldsymbol{\mathcal{Z}} = \boldsymbol{\zeta} \odot \boldsymbol{\mathcal{I}}$, but expressing it in terms of its kernel may be clearer.

Single-point statistics We can check our results by computing the univariate statistics for the irradiance at a single spatial location. The univariate characteristic function is given by

$$\psi_{I_z(\mathbf{r}_0)}(\nu) = \langle \exp\left[-2\pi i\nu I_z(\mathbf{r}_0)\right]\rangle = \left\langle \exp\left[-2\pi i\nu \int_\infty d^2r \; I_z(\mathbf{r})\,\delta(\mathbf{r}-\mathbf{r}_0)\right]\right\rangle . \tag{18.74}$$

Integration against a delta function is permissible since the propagator is a low-pass filter and hence $I_z(\mathbf{r})$ is a function in a reproducing-kernel Hilbert space, specifically Paley-Wiener space (see Secs. 1.8 and 3.5). Thus we can say that $I_z(\mathbf{r}_0)$ is the scalar product[8] of $\delta(\mathbf{r}-\mathbf{r}_0)$ and the random process $I_z(\mathbf{r})$, and the univariate characteristic function is related to the characteristic functional of the process by

$$\psi_{I_z(\mathbf{r}_0)}(\nu) = \Psi_{\mathbf{I}_z}[\nu\,\delta(\mathbf{r}-\mathbf{r}_0)] . \tag{18.75}$$

Now $\zeta(\mathbf{r}) = \nu\,\delta(\mathbf{r}-\mathbf{r}_0)$ and ν is real, so

$$\begin{aligned} \mathrm{tr}(\boldsymbol{\mathcal{K}}_{\mathbf{u}_z}\boldsymbol{\mathcal{Z}}) &= \nu\, K_{\mathbf{u}_z}(\mathbf{r}_0,\mathbf{r}_0) , \\ \mathrm{tr}(\boldsymbol{\mathcal{K}}_{\mathbf{u}_z}\boldsymbol{\mathcal{Z}}\boldsymbol{\mathcal{K}}_{\mathbf{u}_z}\boldsymbol{\mathcal{Z}}) &= \nu^2|K_{\mathbf{u}_z}(\mathbf{r}_0,\mathbf{r}_0)|^2 = [\mathrm{tr}(\boldsymbol{\mathcal{K}}_{\mathbf{u}_z}\boldsymbol{\mathcal{Z}})]^2 , \text{ etc.} , \end{aligned} \tag{18.76}$$

and the desired univariate characteristic function is

$$\psi_{I_z(\mathbf{r}_0)}(\nu) = \exp[-2\pi i\nu\, K_{\mathbf{u}_z}(\mathbf{r}_0,\mathbf{r}_0) - 2\pi^2\nu^2[K_{\mathbf{u}_z}(\mathbf{r}_0,\mathbf{r}_0)]^2 + \cdots] . \tag{18.77}$$

The exponent is recognized as an expansion for the logarithm, $\ln(1+z) = z - \frac{1}{2}z^2 + \frac{1}{3}z^3 - \ldots$, valid for $|z| < 1$; hence

$$\psi_{I_z(\mathbf{r}_0)}(\nu) = \exp\{-\ln[1 + 2\pi i\nu\, K_{\mathbf{u}_z}(\mathbf{r}_0,\mathbf{r}_0)]\} , \tag{18.78}$$

so

$$\psi_{I_z(\mathbf{r}_0)}(\nu) = \frac{1}{1 + 2\pi i\nu\, K_{\mathbf{u}_z}(\mathbf{r}_0,\mathbf{r}_0)} , \tag{18.79}$$

which is the characteristic function for a chi-squared random variable with two degrees of freedom as given in (C.139), or equivalently for an exponential random variable as in (C.121).

Bivariate statistics If the irradiance at two points is of interest, the bivariate characteristic function is related to the characteristic functional for the random process by an extension of (18.75):

$$\psi_{I_z(\mathbf{r}_1),\,I_z(\mathbf{r}_2)}(\nu_1,\nu_2) = \Psi_{\mathbf{I}_z}[\nu_1\,\delta(\mathbf{r}-\mathbf{r}_1) + \nu_2\,\delta(\mathbf{r}-\mathbf{r}_2)] . \tag{18.80}$$

One way to proceed is to use (18.73) with $\boldsymbol{\mathcal{Z}}$ being the integral operator with kernel

$$Z(\mathbf{r},\mathbf{r}') = \zeta(\mathbf{r})\,\delta(\mathbf{r}-\mathbf{r}') = [\nu_1\,\delta(\mathbf{r}-\mathbf{r}_1) + \nu_2\,\delta(\mathbf{r}-\mathbf{r}_2)]\;\delta(\mathbf{r}-\mathbf{r}') . \tag{18.81}$$

To use (18.73), we first find the eigenvalues of $\boldsymbol{\mathcal{K}}_{\mathbf{u}_z}\boldsymbol{\mathcal{Z}}$; the eigenvalue equation is

$$\int_\infty d^2r''\,[\boldsymbol{\mathcal{K}}_{\mathbf{u}_z}\boldsymbol{\mathcal{Z}}](\mathbf{r},\mathbf{r}'')\,\phi(\mathbf{r}'') = \lambda\,\phi(\mathbf{r}) . \tag{18.82}$$

[8]The scalar product we intend is in $\mathbb{L}_2$, not the one for the reproducing-kernel Hilbert space. While the delta function is not in $\mathbb{L}_2$, we can realize the scalar product with a limiting argument.

Writing out the product $\mathcal{K}_{\mathbf{u}_z}\mathcal{Z}$ explicitly and using the kernel for $\mathcal{Z}$, we get

$$\int_\infty d^2r'' \int_\infty d^2r' \, K_{\mathbf{u}_z}(\mathbf{r},\mathbf{r}'') \, [\nu_1 \, \delta(\mathbf{r}''-\mathbf{r}_1) + \nu_2 \, \delta(\mathbf{r}''-\mathbf{r}_2)] \; \delta(\mathbf{r}''-\mathbf{r}') \, \phi(\mathbf{r}'')$$

$$= \nu_1 K_{\mathbf{u}_z}(\mathbf{r},\mathbf{r}_1) \, \phi(\mathbf{r}_1) + \nu_2 K_{\mathbf{u}_z}(\mathbf{r},\mathbf{r}_2) \, \phi(\mathbf{r}_2) = \lambda \, \phi(\mathbf{r}) \, . \tag{18.83}$$

It follows that $\phi(\mathbf{r})$ must have the form $\alpha K_{\mathbf{u}_z}(\mathbf{r},\mathbf{r}_1) + \beta K_{\mathbf{u}_z}(\mathbf{r},\mathbf{r}_2)$; there are only two linearly independent eigenfunctions and hence the rank of $\mathcal{K}_{\mathbf{u}_z}\mathcal{Z}$ is two in this case.

Since (18.83) must hold for all $\mathbf{r}$, including $\mathbf{r}_1$ and $\mathbf{r}_2$, we get two equations that can be written in matrix form as

$$\begin{bmatrix} \nu_1 K_{11} - \lambda & \nu_2 K_{12} \\ \nu_1 K_{21} & \nu_2 K_{22} - \lambda \end{bmatrix} \begin{bmatrix} \phi(\mathbf{r}_1) \\ \phi(\mathbf{r}_2) \end{bmatrix} = 0 \, , \tag{18.84}$$

where $K_{jk} \equiv K_{\mathbf{u}_z}(\mathbf{r}_j,\mathbf{r}_k)$. The two possible values for λ in this equation (call them λ_1 and λ_2) are found by setting the determinant of the 2×2 matrix to zero. Though the notation does not show it, these eigenvalues depend on ν_1 and ν_2.

We can use the eigenvalues of $\mathcal{K}_{\mathbf{u}_z}\mathcal{Z}$ to construct the eigenvalues of the operator in which we are really interested, $\mathcal{I} + 2\pi i\mathcal{K}_{\mathbf{u}_z}\mathcal{Z}$. Two of the eigenvalues of that operator are $1+2\pi i\lambda_1$ and $1+2\pi i\lambda_2$; there is also an infinite set of eigenvalues equal to unity, but they contribute nothing to the determinant in (18.73). After a bit of algebra we find

$$\psi_{I_z(\mathbf{r}_1),I_z(\mathbf{r}_2)}(\nu_1,\nu_2) = \frac{1}{(1+2\pi i\lambda_1)\,(1+2\pi i\lambda_2)}$$

$$= \frac{1}{[1+2\pi i\nu_1 \, K_{\mathbf{u}_z}(\mathbf{r}_1,\mathbf{r}_1)][1+2\pi i\nu_2 \, K_{\mathbf{u}_z}(\mathbf{r}_2,\mathbf{r}_2)] + 4\pi^2\nu_1\nu_2|K_{\mathbf{u}_z}(\mathbf{r}_1,\mathbf{r}_2)|^2} \, . \tag{18.85}$$

The corresponding bivariate PDF can be obtained by performing a 2D inverse Fourier transform; the result is given by Goodman (1975) in terms of I_0 Bessel functions. If $K_{\mathbf{u}_z}(\mathbf{r}_1,\mathbf{r}_2) = 0$, (18.85) reduces to the product of two chi-squared characteristic functions, each with two degrees of freedom, so the PDF reduces to the product of two exponentials.

18.3 SPECKLE IN AN IMAGING SYSTEM

Next we turn to speckle in optical imaging, adapting the analysis of Sec. 18.2 by adding a photographic transparency to serve as the object, a lens system to form the image, and a discrete detector array. The objective of the analysis is a statistical description, in the form of a characteristic function, for the detector output. In this section the only random effect we consider is speckle, but in Sec. 18.4 we shall extend the analysis further by bringing in measurement noise and object variation.

The system under consideration is described and analyzed deterministically in Sec. 18.3.1. The statistics are treated with the help of characteristic functionals in Sec. 18.3.2, and the effect of the image detector is discussed in Sec. 18.3.3.

18.3.1 The imaging system

The imaging system considered in this section is shown in Fig. 18.3. The object, consisting of a photographic transparency with amplitude transmittance t_{obj}, is placed directly over a ground glass, which is then illuminated with a normally incident plane wave as in Sec. 18.2. A thin lens is placed a distance $2f$ from the object, and a discrete detector array is placed in the focal plane, a distance $2f$. This system has a magnification of -1, but we shall avoid annoying minus signs by defining the axes in the image plane oppositely to those in the object plane; thus the x_0 direction in the object plane is pointing upward in Fig. 18.3, but the x direction in the image plane is pointing downward.

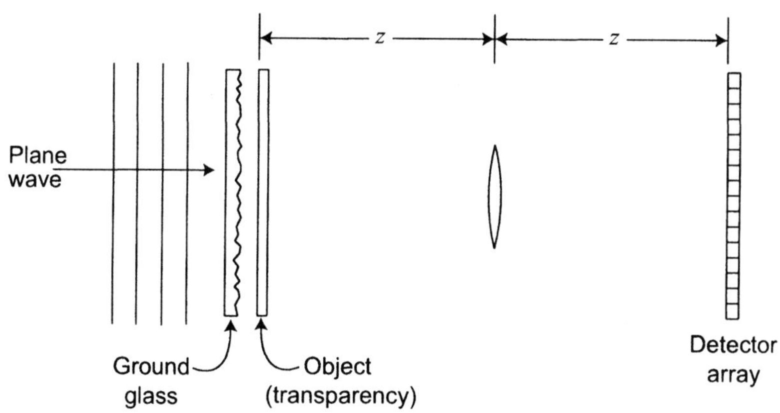

Fig. 18.3 Simple optical system with magnification of -1.

We can analyze this system in three stages. The first stage is the linear CC mapping from object field to image field, the second is a nonlinear point transformation from image-plane field to image-plane irradiance, and the third is a linear CD mapping from irradiance to the discrete detector outputs.

Coherent CC mapping The first stage is described by the integral transform

$$u_{im}(\mathbf{r}) = \int_\infty d^2r_0 \; p_{coh}(\mathbf{r} - \mathbf{r}_0; \mathbf{r}_0) \, u_0(\mathbf{r}_0) \, , \tag{18.86}$$

where $u_{im}(\mathbf{r})$ is the field in the image plane, $u_0(\mathbf{r}_0)$ is the field emerging from the object transparency in the plane $z = 0$, and $p_{coh}(\mathbf{r} - \mathbf{r}_0; \mathbf{r}_0)$ is the coherent point spread function (PSF). For the present problem, this PSF is given by (9.227); modified for our inverted coordinate system and $p = q = 2f$, that equation becomes

$$p_{coh}(\mathbf{r}; \mathbf{r}_0) = \frac{1}{(2\lambda f)^2} \, T_{pupil}\left(\frac{\mathbf{r}}{2\lambda f}; \mathbf{r}_0\right) \, , \tag{18.87}$$

where $T_{pupil}(\boldsymbol{\rho}; \mathbf{r}_0)$ is the Fourier transform with respect to the first variable of the 2D Fourier transform of the pupil function, $t_{pupil}(\mathbf{r}; \mathbf{r}_0)$. The second variable is needed here if we wish to include the effects of field-dependent aberrations; for more discussion of this point, see Secs. 9.6.3 and 9.7.1. If there are no field-dependent aberrations, both $T_{pupil}(\boldsymbol{\rho}; \mathbf{r}_0)$ and $p_{coh}(\mathbf{r}; \mathbf{r}_0)$ are independent of the second variable and (18.86) is a convolution.

We can write the mapping of (18.86) in abstract form by defining a coherent imaging operator $\boldsymbol{\mathcal{P}}_{coh}$, so that

$$\mathbf{u}_{im} = \boldsymbol{\mathcal{P}}_{coh} \mathbf{u}_0 \,. \tag{18.88}$$

Irradiance and CD mapping by the detector The second stage in the analysis is conversion of image-plane field to irradiance, stated simply as

$$I_{im}(\mathbf{r}) = |u_{im}(\mathbf{r})|^2 \,. \tag{18.89}$$

The third stage is the CD mapping by the detector array. For simplicity, we assume that the mean output of each detector element is proportional to the integral of the irradiance over its face, so

$$\overline{g}_m = \int_\infty d^2r \; w_m(\mathbf{r}) \, I_{im}(\mathbf{r}) \,, \tag{18.90}$$

where $w_m(\mathbf{r})$ is a constant responsivity factor times a rect function describing the spatial extent of the mth detector.

Object transparency The equations above relate the detector output to the field in the plane $z = 0$. To bring in the object transparency explicitly, we write

$$u_0(\mathbf{r}) = u_{GG}(\mathbf{r}) \, t_{obj}(\mathbf{r}) \,, \tag{18.91}$$

where $t_{obj}(\mathbf{r})$ is the complex amplitude transmission of the object, and $u_{GG}(\mathbf{r})$ is the field emerging from the ground glass and incident on the object; $u_{GG}(\mathbf{r})$ is identical to the amplitude transmission of the ground glass if the initial illumination is a unit-amplitude, normally incident plane wave.

We can put (18.91) into vector form by using the Hadamard notation introduced in Sec. A.2.8, with the convention that an element-by-element product of two vectors is denoted by simple juxtaposition. When the vectors in question are in an $\mathbb{L}_2$ Hilbert space, the juxtaposition of two vectors connotes the product of the corresponding functions, so (18.91) can be written as

$$\mathbf{u}_0 = \mathbf{u}_{GG} \, \mathbf{t}_{obj} \,. \tag{18.92}$$

The CC propagation rule, (18.88), now becomes

$$\mathbf{u}_{im} = \boldsymbol{\mathcal{P}}_{coh} \{ \mathbf{u}_{GG} \, \mathbf{t}_{obj} \} \,. \tag{18.93}$$

Where no confusion is likely to result, we may delete the braces and write $\mathbf{u}_{im} = \boldsymbol{\mathcal{P}}_{coh} \, \mathbf{u}_{GG} \, \mathbf{t}_{obj}$.

18.3.2 Propagation of characteristic functionals

The formulas in Sec. 18.3.1 describe the deterministic properties of the imaging system. To describe the statistical properties, we use characteristic functionals and basically retrace the steps of Sec. 18.3.1 with appropriate modifications.

Characteristic functional for the input field The characteristic functional for $\mathbf{u}_0$ can be written as

$$\Psi_{\mathbf{u}_0}(\boldsymbol{\xi}) = \left\langle \exp\left[-2\pi i \int_\infty d^2r\ \xi^*(\mathbf{r})\, u_{GG}(\mathbf{r})\, t_{obj}(\mathbf{r})\right]\right\rangle$$

$$= \left\langle \exp\left[-2\pi i \int_\infty d^2r\ [\xi(\mathbf{r})\, t^*_{obj}(\mathbf{r})]^*\ u_{GG}(\mathbf{r})\right]\right\rangle . \tag{18.94}$$

Thus

$$\Psi_{\mathbf{u}_0}(\boldsymbol{\xi}) = \Psi_{\mathbf{u}_{GG}}(\mathbf{t}^*_{obj}\boldsymbol{\xi}) . \tag{18.95}$$

When $\mathbf{t}_{obj}$ is moved to the other side of the scalar product in Hadamard notation, it acquires an asterisk rather than a dagger; regarded as an operator, the Hadamard product functions as scalar multiplication, and the adjoint of scalar multiplication by something is scalar multiplication by its complex conjugate.

Characteristic functional for the image field With (18.20) and (18.95), the characteristic functional for the field in the image plane becomes

$$\Psi_{\mathbf{u}_{im}}(\boldsymbol{\xi}) = \Psi_{\mathbf{u}_0}(\boldsymbol{\mathcal{P}}^\dagger_{coh}\boldsymbol{\xi}) = \Psi_{\mathbf{u}_{GG}}(\mathbf{t}^*_{obj}\boldsymbol{\mathcal{P}}^\dagger_{coh}\,\boldsymbol{\xi}) . \tag{18.96}$$

Thus if we know the characteristic functional for the ground glass (the only random element being considered here), we can transform it to the characteristic functional for the image field.

CLT revisited At this point in the derivation presented in Sec. 18.2, we invoked the central-limit theorem to argue that the field in the observation plane was circular Gaussian provided the ground glass fully randomized the input field. Key to the argument was the assumption that many independent regions on the ground glass contributed the field at any point in the observation plane.

That argument is less persuasive in an imaging context since the width of the coherent PSF defines the contributing region on the ground glass. For well corrected lenses of high numerical aperture, that width is of the order of a wavelength or two, and the correlation length is of order half a wavelength, so it is problematical to assume that the region defined by the PSF encompasses many independent regions. Moreover, even if the width of the PSF were large enough, we would still have to consider the effect of the object transparency. We cannot, for example, consider the object to be a small pinhole with width comparable to a wavelength.

In spite of these caveats, we shall assume here that the image-plane field is circular Gaussian, returning to the subject in Sec. 18.5 where we discuss object models that lead to non-Gaussian fields.

Field in the image plane With the assumption of a circular Gaussian form, the characteristic functional for the field in the image plane becomes

$$\Psi_{\mathbf{u}_{im}}(\boldsymbol{\xi}) = \exp(-\pi^2\boldsymbol{\xi}^\dagger\boldsymbol{\mathcal{K}}_{\mathbf{u}_{im}}\boldsymbol{\xi}) , \tag{18.97}$$

where

$$\boldsymbol{\mathcal{K}}_{\mathbf{u}_{im}} = \boldsymbol{\mathcal{P}}_{coh}\, \mathbf{t}_{obj}\, \langle \mathbf{u}_0\mathbf{u}_0^\dagger\rangle\, \mathbf{t}^*_{obj}\boldsymbol{\mathcal{P}}^\dagger_{coh} . \tag{18.98}$$

Thus, as with any circular Gaussian, all statistical properties of the image-plane field are determined solely by the autocorrelation function, which is also the autocovariance since the mean is zero.

If we assume that the correlation length of the field emerging from the ground glass is small compared to the width of the coherent PSF and that the object transparency is slowly varying, then we can represent the autocorrelation function of $u_0(\mathbf{r})$ by a delta function as in (18.13). In vector form, that means that $\langle \mathbf{u}_0 \mathbf{u}_0^\dagger \rangle = \ell_c^2 \, \mathbf{I}$, and we have

$$\mathcal{K}_{\mathbf{u}_{im}} = \ell_c^2 \, \mathcal{P}_{coh} \, \mathbf{t}_{obj} \, \mathbf{t}_{obj}^* \mathcal{P}_{coh}^\dagger = \ell_c^2 \, \mathcal{P}_{coh} \, |\mathbf{t}_{obj}|^2 \, \mathcal{P}_{coh}^\dagger \, , \tag{18.99}$$

where the last form uses a further embellishment on the Hadamard notation: $|\mathbf{t}_{obj}|^2$ is the $\mathbb{L}_2$ vector corresponding to the function $|t_{obj}(\mathbf{r})|^2$.

The autocorrelation function of the image-plane field is the kernel of the integral operator (18.99). Using (18.86), we see that

$$K_{\mathbf{u}_{im}}(\mathbf{r}, \mathbf{r}') = \ell_c^2 \int d^2 r_0 \; p_{coh}(\mathbf{r} - \mathbf{r}_0; \mathbf{r}_0) \, p_{coh}^*(\mathbf{r}' - \mathbf{r}_0; \mathbf{r}_0) \, |t_{obj}(\mathbf{r}_0)|^2 \, . \tag{18.100}$$

Irradiance in the image plane Though the mean field in the image plane is zero, the mean irradiance is not. In fact, the mean irradiance is obtained at once from (18.100) since

$$\langle I_{im}(\mathbf{r}) \rangle = \langle |u_{im}(\mathbf{r})|^2 \rangle = K_{\mathbf{u}_{im}}(\mathbf{r}, \mathbf{r}) = \ell_c^2 \int d^2 r_0 \; |p_{coh}(\mathbf{r} - \mathbf{r}_0; \mathbf{r}_0)|^2 \, |t_{obj}(\mathbf{r}_0)|^2 \, . \tag{18.101}$$

This expression agrees with (9.284), which gives the mean irradiance in an incoherent, quasimonochromatic imaging system. On average, a spatially incoherent source and a coherent source modulated with a fine ground glass give the same irradiance, though the meaning of mean is different. In incoherent imaging, the average is over an ensemble of source realizations, and in speckle problems it is over an ensemble of ground glasses. In the incoherent case, however, we can invoke ergodicity and replace the ensemble average with a time average, and the time average is what we observe with a temporally integrating detector (see Sec. 10.1.5). In the speckle problem, on the other hand, the irradiance is a time-independent speckle pattern and time averaging has no effect.

To study the statistics of the image-plane irradiance, we need its characteristic functional. Since we are assuming that the field is circular Gaussian, just as we did in Sec. 18.2.5, we can simply refer back to (18.73) and write

$$\Psi_{\mathbf{I}_{im}}(\boldsymbol{\zeta}) = \frac{1}{\det(\mathcal{I} + 2\pi i \, \mathcal{K}_{\mathbf{u}_{im}} \mathcal{Z})} \, , \tag{18.102}$$

where $\mathcal{Z}$ is the integral operator with kernel $\zeta(\mathbf{r}) \, \delta(\mathbf{r} - \mathbf{r}')$, and the determinant of an integral operator is interpreted as the product of its eigenvalues.

From this characteristic functional we can derive the univariate or single-point characteristic function and the corresponding univariate PDF just as we did in Sec. 18.2.5. From (18.79) we see that

$$\psi_{I_{im}(\mathbf{r})}(\nu) = \frac{1}{1 + 2\pi i \nu \, K_{\mathbf{u}_{im}}(\mathbf{r}, \mathbf{r})} \, , \tag{18.103}$$

from which it follows that $I_{im}(\mathbf{r})$ is a chi-squared random variable with two degrees of freedom, or equivalently an exponentially-distributed random variable (see

Secs. C.5.3 and C.5.5). The mean irradiance at point $\mathbf{r}$ is $K_{\mathbf{u}_{im}}(\mathbf{r},\mathbf{r})$, or simply $\langle|u_{im}(\mathbf{r})|^2\rangle$.

The bivariate characteristic function for the image-plane irradiance is given by an expression analogous to (18.85), with the substitution of (18.100) for the autocorrelation function.

The autocorrelation and autocovariance functions for the irradiance can be obtained several different ways. They can be computed from either the bivariate characteristic function (18.85) or the infinite-dimensional characteristic functional (18.102) by suitable differentiation. (In the latter case Fréchet derivatives, defined in Sec. 15.3.5, must be used, but they behave very much like ordinary derivatives.) A simpler route, however, is just to invoke the complex Gaussian moment theorem discussed in Sec. 8.3.6. By (8.250) we find

$$\begin{aligned}\langle I_{im}(\mathbf{r})\, I_{im}(\mathbf{r}')\rangle &= \\ \left\langle|u_{im}(\mathbf{r})|^2\,|u_{im}(\mathbf{r}')|^2\right\rangle &= K_{\mathbf{u}_{im}}(\mathbf{r},\mathbf{r})\,K_{\mathbf{u}_{im}}(\mathbf{r}',\mathbf{r}') + K_{\mathbf{u}_{im}}(\mathbf{r},\mathbf{r}')\,K_{\mathbf{u}_{im}}(\mathbf{r}',\mathbf{r}) \\ &= \langle I_{im}(\mathbf{r})\rangle\,\langle I_{im}(\mathbf{r}')\rangle + \left|K_{\mathbf{u}_{im}}(\mathbf{r},\mathbf{r}')\right|^2 . \qquad (18.104)\end{aligned}$$

Thus the autocovariance of the irradiance is just the squared modulus of the autocovariance of the field:

$$K_{\mathbf{I}_{im}}(\mathbf{r},\mathbf{r}') = \langle I_{im}(\mathbf{r})\, I_{im}(\mathbf{r}')\rangle - \langle I_{im}(\mathbf{r})\rangle\,\langle I_{im}(\mathbf{r}')\rangle = \left|K_{\mathbf{u}_{im}}(\mathbf{r},\mathbf{r}')\right|^2 . \qquad (18.105)$$

Stationarity, real and quasi To better understand the nature of the field covariance $K_{\mathbf{u}_{im}}(\mathbf{r},\mathbf{r}')$, we can write it out explicitly in terms of the pupil function of the system. As in Sec. 9.6.3, we write the pupil function as [*cf.* (9.181)]

$$t_{pupil}(\mathbf{r};\mathbf{r}_0) = \exp[ikW(\mathbf{r};\mathbf{r}_0)]\, t_{ap}(\mathbf{r}) , \qquad (18.106)$$

where $t_{ap}(\mathbf{r})$ is unity inside the clear aperture of the lens and zero outside, and the exponential factor accounts for the aberrations. With (18.87) and (18.100), we now have

$$\begin{aligned}K_{\mathbf{u}_{im}}(\mathbf{r},\mathbf{r}') &= \frac{\ell_c^2}{(2\lambda f)^4}\int_\infty d^2r_0\,|t_{obj}(\mathbf{r}_0)|^2\int_\infty d^2\tilde{r}\int_\infty d^2\tilde{r}'\, t_{ap}(\tilde{\mathbf{r}})\, t_{ap}(\tilde{\mathbf{r}}') \\ &\times \exp[ikW(\tilde{\mathbf{r}};\mathbf{r}_0)]\exp[-ikW(\tilde{\mathbf{r}}';\mathbf{r}_0)]\exp\left[-i\frac{2\pi}{2\lambda f}\tilde{\mathbf{r}}\cdot(\mathbf{r}-\mathbf{r}_0)\right]\exp\left[i\frac{2\pi}{2\lambda f}\tilde{\mathbf{r}}'\cdot(\mathbf{r}'-\mathbf{r}_0)\right] , \qquad (18.107)\end{aligned}$$

where all three integrals are formally over the 2D plane, but the integral over $\mathbf{r}_0$ is constrained to the object support by the factor $|t_{obj}(\mathbf{r}_0)|^2$ and the other two integrals are constrained to the clear aperture by the aperture transmittance factors. This is as far as we can go without further assumptions or approximations (we have already made the Fresnel approximation), but we shall consider several special cases.

First assume that the object transmittance $|t_{obj}(\mathbf{r}_0)|^2$ is a constant, which we set to unity for convenience, and assume further that the system, though aberrated, is shift invariant. This means that there can be field-independent aberrations such as spherical aberration and defocus but not field-dependent ones such as coma and astigmatism (see Sec. 9.6.4). Then $W(\tilde{\mathbf{r}};\mathbf{r}_0)$ is independent of $\mathbf{r}_0$, so we can write it as $W(\tilde{\mathbf{r}})$, and the $\mathbf{r}_0$ integral becomes

$$\int_\infty d^2r_0\,\exp\left[i\frac{2\pi}{2\lambda f}(\tilde{\mathbf{r}}-\tilde{\mathbf{r}}')\cdot\mathbf{r}_0\right] = (2\lambda f)^2\,\delta(\tilde{\mathbf{r}}-\tilde{\mathbf{r}}') . \qquad (18.108)$$

Now we can use this delta function to perform the integral over $\tilde{\mathbf{r}}'$ in (18.107), and we obtain

$$K_{\mathbf{u}_{im}}(\mathbf{r},\mathbf{r}') = \frac{\ell_c^2}{(2\lambda f)^2}\int_\infty d^2\tilde{r}\; t_{ap}(\tilde{\mathbf{r}})\exp\left[-i\frac{2\pi}{2\lambda f}\tilde{\mathbf{r}}\cdot(\mathbf{r}-\mathbf{r}')\right]. \tag{18.109}$$

Two key points are to be noted from this expression. First, the image-plane field is a stationary random process since its autocovariance function depends only on the difference $\mathbf{r}-\mathbf{r}'$. Second, the aberration factors have cancelled. Thus, by comparing (18.109) with (18.87), we see that the autocovariance function of the field in this case is the shift-invariant coherent PSF associated with an unaberrated lens of the same aperture as the actual aberrated one. To get these results, however, we had to assume that the object was a constant over an infinite field and that the optical system was shift-invariant over this field; neither of these assumptions is tenable.

To get a more realistic model, let us assume that the object is not constant but slowly varying, and that the aberrations are field dependent and also slowly varying. It is convenient to transform to sum and difference coordinates as in Sec. 8.2.4 or 13.2.13:

$$\bar{\mathbf{r}} = \tfrac{1}{2}(\mathbf{r}+\mathbf{r}'), \qquad \Delta\mathbf{r} = \mathbf{r}-\mathbf{r}', \tag{18.110}$$

so that

$$\mathbf{r} = \bar{\mathbf{r}} + \tfrac{1}{2}\Delta\mathbf{r}, \qquad \mathbf{r}' = \bar{\mathbf{r}} - \tfrac{1}{2}\Delta\mathbf{r}. \tag{18.111}$$

Note that the Jacobian of this transformation is unity. A similar transformation is used for $\tilde{\mathbf{r}}$ and $\tilde{\mathbf{r}}'$, and the autocovariance function of the field becomes

$$K_{\mathbf{u}_{im}}(\mathbf{r},\mathbf{r}') = K_{\mathbf{u}_{im}}(\bar{\mathbf{r}}+\tfrac{1}{2}\Delta\mathbf{r},\,\bar{\mathbf{r}}-\tfrac{1}{2}\Delta\mathbf{r}) \equiv \widetilde{K}_{\mathbf{u}_{im}}(\bar{\mathbf{r}},\Delta\mathbf{r}). \tag{18.112}$$

With these definitions, (18.107) takes the ungainly form,

$$\begin{aligned}\widetilde{K}_{\mathbf{u}_{im}}(\bar{\mathbf{r}},\Delta\mathbf{r}) &= \frac{\ell_c^2}{(2\lambda f)^4}\int_\infty d^2r_0\;|t_{obj}(\mathbf{r}_0)|^2\int_\infty d^2\bar{\tilde{r}}\int_\infty d^2\Delta\tilde{r}\\ &\times t_{ap}(\bar{\tilde{\mathbf{r}}}+\tfrac{1}{2}\Delta\tilde{\mathbf{r}})\,t_{ap}(\bar{\tilde{\mathbf{r}}}+\tfrac{1}{2}\Delta\tilde{\mathbf{r}})\exp[ikW(\bar{\tilde{\mathbf{r}}}+\tfrac{1}{2}\Delta\tilde{\mathbf{r}};\mathbf{r}_0)]\,\exp[-ikW(\bar{\tilde{\mathbf{r}}}-\tfrac{1}{2}\Delta\tilde{\mathbf{r}};\mathbf{r}_0)]\\ &\times\exp\left[-i\frac{2\pi}{2\lambda f}(\bar{\tilde{\mathbf{r}}}+\tfrac{1}{2}\Delta\tilde{\mathbf{r}})\cdot(\mathbf{r}-\mathbf{r}_0)\right]\cdot\exp\left[i\frac{2\pi}{2\lambda f}(\bar{\tilde{\mathbf{r}}}-\tfrac{1}{2}\Delta\tilde{\mathbf{r}})\cdot(\mathbf{r}'-\mathbf{r}_0)\right].\end{aligned} \tag{18.113}$$

As in Sec. 9.6.4 (and, indeed, in most of the optics literature), we neglect all aberrations beyond second order. That means that we can expand $W(\bar{\tilde{\mathbf{r}}}\pm\frac{1}{2}\Delta\tilde{\mathbf{r}};\mathbf{r}_0)$ in a Taylor series (with respect to its first argument) and truncate the series after the quadratic term. The terms constant and quadratic in $\Delta\tilde{\mathbf{r}}$ neatly cancel in (18.113), and the integral over that variable becomes

$$\begin{aligned}&\int_\infty d^2\Delta\tilde{r}\;t_{ap}(\bar{\tilde{\mathbf{r}}}+\tfrac{1}{2}\Delta\tilde{\mathbf{r}})\,t_{ap}(\bar{\tilde{\mathbf{r}}}-\tfrac{1}{2}\Delta\tilde{\mathbf{r}})\exp[ik\,\Delta\tilde{\mathbf{r}}\cdot\nabla W(\bar{\tilde{\mathbf{r}}};\mathbf{r}_0)]\exp\left[-i\frac{2\pi}{2\lambda f}\,\Delta\tilde{\mathbf{r}}\cdot(\bar{\mathbf{r}}-\mathbf{r}_0)\right]\\ &\approx t_{ap}(\bar{\tilde{\mathbf{r}}})\int_\infty d^2\Delta\tilde{r}\;\exp[ik\,\Delta\tilde{\mathbf{r}}\cdot\nabla W(\bar{\tilde{\mathbf{r}}};\mathbf{r}_0)]\exp\left[-i\frac{2\pi}{2\lambda f}\,\Delta\tilde{\mathbf{r}}\cdot(\bar{\mathbf{r}}-\mathbf{r}_0)\right]\\ &= (2\lambda f)^2\,t_{ap}(\bar{\tilde{\mathbf{r}}})\,\delta[\mathbf{r}_0-\bar{\mathbf{r}}+2f\,\nabla W(\bar{\tilde{\mathbf{r}}};\mathbf{r}_0)],\end{aligned} \tag{18.114}$$

where the approximation on the second line assumes that the aperture function has a large support and takes on only the values 0 and 1, while the last step involves some properties of delta functions. Recall from Sec. 9.6.6 that $2f\,\nabla W(\bar{\bar{\mathbf{r}}};\mathbf{r}_0)$ is the aberration-induced displacement of a ray originating from object point $\mathbf{r}_0$ and passing through point $\bar{\bar{\mathbf{r}}}$ in the pupil [*cf.* (9.210) and note that q in that equation is $2f$ in the present problem].

We can now use the delta function to perform the integral over $\mathbf{r}_0$. This delta function is not immediately in the form where we can invoke the sifting property of delta functions since its argument is a nonlinear function of $\mathbf{r}_0$ through the term $2f\,\nabla W(\bar{\bar{\mathbf{r}}};\mathbf{r}_0)$, but it is a reasonable approximation to replace $2f\,\nabla W(\bar{\bar{\mathbf{r}}};\mathbf{r}_0)$ with $2f\,\nabla W(\bar{\bar{\mathbf{r}}};\bar{\mathbf{r}})$ if the aberrations are slowly varying. (See Sec. 7.2.7 where we made a similar approximation in discussing shift-variant magnifiers.) If the object is also slowly varying over the scale of the aberration-induced displacement, we get, finally,

$$\begin{aligned}\widetilde{K}_{\mathbf{u}_{im}}(\bar{\mathbf{r}},\Delta\mathbf{r}) &\approx \frac{\ell_c^2}{(2\lambda f)^2}|t_{obj}(\bar{\mathbf{r}})|^2 \int_\infty d^2\bar{\bar{r}}\; t_{ap}(\bar{\bar{\mathbf{r}}}) \exp\left[-i\frac{2\pi}{2\lambda f}(\bar{\bar{\mathbf{r}}}\cdot\Delta\mathbf{r})\right] \\ &= \frac{\ell_c^2}{(2\lambda f)^2}|t_{obj}(\bar{\mathbf{r}})|^2\, T_{ap}\left(\frac{\Delta\mathbf{r}}{2\lambda f}\right) .\end{aligned} \tag{18.115}$$

Again the aberrations play no role; the correlation length under the present approximations is determined solely by the clear aperture of the lens.

In (18.115) we have cast the autocovariance function of the field into the quasistationary form originally encountered in Sec. 8.2.4 [*cf.* (8.119)]. In that section, however, we discussed only the second-order statistics of the random process. Here we can make a much stronger statement. Since all statistical properties of both the field and the irradiance are fully determined by the field autocovariance under the circular Gaussian assumption, (18.115) gives us an approximation through which we can compute the characteristic functionals of the field and irradiance and hence any desired higher-order moments of either.[9]

18.3.3 Effect of the detector

So far we have discussed the image-plane field and irradiance as continuous random processes. In practice the irradiance will be detected with one or more detectors of finite area. In this section we shall learn how to incorporate the detectors into our statistical descriptions of speckle under the assumption that the detection process itself is noise-free. In Sec. 18.4 we shall integrate the results of this section with what we know about measurement noise from previous chapters and get the complete statistics of the output of real, noisy, discrete detector arrays.

Single-element detectors We shall model a detector here as a device that integrates the irradiance over a finite area (see Sec. 10.1.5); real-world effects such as nonuniform response over this area and angular dependences are ignored. For now we consider a single detector element, but the analysis will be extended to arrays below.

[9]One might be tempted to take the approximation a step further and replace $T_{ap}(\Delta\mathbf{r}/2\lambda f)$ by a delta function, but this step is not warranted since (18.115) will be used in integrals where other factors might also be delta functions, as in (18.75). Moreover, we need the fact that the covariance operator is compact and hence has a denumerable set of eigenfunctions. We know from Chap. 1 that an integral operator is compact if its kernel satisfies the Hilbert-Schmidt condition, which $T_{ap}(\Delta\mathbf{r}/2\lambda f)$ does (over a finite support) but $\delta(\Delta\mathbf{r})$ does not.

As discussed qualitatively in Sec. 18.1.2, a key question is whether the detector area is smaller or larger than the speckle blob. If the detector is much smaller than the blob, integration over the detector area is basically point sampling, so the univariate PDF on the detector output (in the absence of measurement noise) will be essentially the single-point PDF of the irradiance field, which is an exponential under our circular-Gaussian assumptions. If the detector is much larger than the blob, on the other hand, we expect the PDF on the detector output to be Gaussian by dint of the central-limit theorem. We shall now learn how to make these qualitative observations quantitative.

Let g_{out} denote the output voltage from a single-element detector of area A_d. This voltage is related to the irradiance by

$$g_{out} = C \int_{A_d} d^2r \; I_{im}(\mathbf{r}) \,, \tag{18.116}$$

where C is the responsivity of the detector. It will prove useful to rewrite this equation as

$$g_{out} = \int_{\infty} d^2r \; w(\mathbf{r}) \, I_{im}(\mathbf{r}) = \mathbf{w}^\dagger \mathbf{I}_{im} \,, \tag{18.117}$$

where $w(\mathbf{r})$ takes on the value C within the area of the detector and zero otherwise, and $\mathbf{w}$ is the corresponding vector in Hilbert space.

The mapping in (18.117) is linear, and we know from Sec. 8.2.3 how to transform characteristic functionals through linear mappings. Specifically, from (8.95) and (18.117), we can write the characteristic *function* for g_{out} in terms of the characteristic *functional* for $\mathbf{I}_{im}$ as

$$\psi_{g_{out}}(\nu) = \left\langle \exp\left(-2\pi i \mathbf{w}^\dagger \mathbf{I}_{im} \, \nu\right)\right\rangle = \Psi_{\mathbf{I}_{im}}(\nu \mathbf{w}) \,. \tag{18.118}$$

To be explicit, we can use (18.102) to write

$$\psi_{g_{out}}(\nu) = \frac{1}{\det(\mathcal{I} + 2\pi i \nu \, \mathcal{K}_{\mathbf{u}_{im}} \mathcal{W})} \,, \tag{18.119}$$

where $\mathcal{W}$ is the integral operator with kernel $w(\mathbf{r})\,\delta(\mathbf{r}-\mathbf{r}')$. Alternatively, we can expand the determinant in traces as in (18.72) and write

$$\psi_{g_{out}}(\nu) = \exp[-2\pi i \nu \, \mathrm{tr}(\mathcal{K}_{\mathbf{u}_{im}} \mathcal{W}) - 2\pi^2 \nu^2 \, \mathrm{tr}(\mathcal{K}_{\mathbf{u}_{im}} \mathcal{W} \mathcal{K}_{\mathbf{u}_{im}} \mathcal{W}) + \cdots] \,. \tag{18.120}$$

This series will converge if ν is sufficiently small since the eigenvalues of $2\pi i \nu \, \mathcal{K}_{\mathbf{u}_{im}} \mathcal{W}$ are all proportional to ν.

Two limits To better understand these characteristic functions, we shall consider the limits of large and small detectors.

If the detector is very small compared to the correlation length of the speckle, we can approximate $w(\mathbf{r})$ by

$$w(\mathbf{r}) \approx C A_d \, \delta(\mathbf{r} - \mathbf{r}_d) \,, \tag{18.121}$$

where $\mathbf{r}_d$ is the detector location. Thus the kernel of $\mathcal{W}$ is $C A_d \, \delta(\mathbf{r} - \mathbf{r}_d)\, \delta(\mathbf{r} - \mathbf{r}')$, which can also be written as $C A_d \, \delta(\mathbf{r} - \mathbf{r}_d)\, \delta(\mathbf{r}' - \mathbf{r}_d)$. Since the kernel factors into a function of $\mathbf{r}$ times a function of $\mathbf{r}'$, it follows that the operator has rank one (see

Sec. 1.5.1), and since the constants are real, $\mathcal{W}$ is Hermitian. Moreover, a product of a rank-one operator with any other operator also has rank one, and rank is the number of nonzero eigenvalues for a Hermitian operator. Thus $\mathcal{K}_{\mathbf{u}_{im}}\mathcal{W}$ has only a single-nonzero eigenvalue. The reader can verify that this eigenvalue is $CA_d\,\overline{I}_{im}(\mathbf{r}_d)$ and that the corresponding eigenfunction is $\phi(\mathbf{r}) \propto K_{\mathbf{u}_{im}}(\mathbf{r},\mathbf{r}_d)$.

With this information, we can evaluate the determinant in (18.119). Recall that the determinant of an integral operator is defined as the product of its eigenvalues; the operator $\mathcal{I}+2\pi i\nu\,\mathcal{K}_{\mathbf{u}_{im}}\mathcal{W}$ has one eigenvalue equal to $1+2\pi i\,CA_d\,\nu\,\overline{I}_{im}(\mathbf{r}_d)$ and an infinite set of unity eigenvalues, so

$$\psi_{g_{out}}(\nu) = \frac{1}{1+2\pi i\,CA_d\,\nu\,\overline{I}_{im}(\mathbf{r}_d)}\,. \tag{18.122}$$

This is the characteristic function for an exponentially distributed scalar random variable of mean $CA_d\,\overline{I}_{im}(\mathbf{r}_d)$. It is no surprise that g_{out} is exponential since, with point sampling and no measurement noise, it merely reflects the single-point statistics of the irradiance.

Now consider the opposite limit of a large detector. This time the expansion in traces, (18.120), is more useful than the determinantal form. For simplicity we assume that the *mean* irradiance is approximately constant over the detector area (even though the detector may cover many speckle blobs). Then, with the help of (18.71) and (18.105),

$$\mathrm{tr}(\mathcal{K}_{\mathbf{u}_{im}}\mathcal{W}) = \int_\infty d^2r\; w(\mathbf{r})\,K_{\mathbf{u}_{im}}(\mathbf{r},\mathbf{r}) = CA_d\,K_{\mathbf{u}_{im}}(\mathbf{r}_d,\mathbf{r}_d) = CA_d\,\overline{I}_{im}(\mathbf{r}_d) = \overline{g}_{out}\,; \tag{18.122}$$

$$\begin{aligned}\mathrm{tr}(\mathcal{K}_{\mathbf{u}_{im}}\mathcal{W}\mathcal{K}_{\mathbf{u}_{im}}\mathcal{W}) &= \int_\infty d^2r \int_\infty d^2r'\; w(\mathbf{r})\,|K_{\mathbf{u}_{im}}(\mathbf{r},\mathbf{r}')|^2\,w(\mathbf{r}')\\ &= \int_\infty d^2r \int_\infty d^2r'\; w(\mathbf{r})\,K_{\mathbf{I}_{im}}(\mathbf{r},\mathbf{r}')\,w(\mathbf{r}')\,.\end{aligned} \tag{18.123}$$

To simplify this last form, let us assume that the irradiance is quasistationary over the detector area so that its autocovariance function can be factored as [*cf.* (18.112) and (18.115)][10]

$$K_{\mathbf{I}_{im}}(\mathbf{r},\mathbf{r}') = \widetilde{K}_{\mathbf{I}_{im}}(\overline{\mathbf{r}},\Delta\mathbf{r}) = \left[\overline{I}_{im}(\overline{\mathbf{r}})\right]^2 a(\Delta\mathbf{r})\,, \tag{18.124}$$

where $\overline{\mathbf{r}}$ and $\Delta\mathbf{r}$ are defined in (18.110), and the short-range factor is normalized so that $a(0) = 1$; we have also used the fact that the variance of the irradiance is the square of its mean for an exponential law. With this factorization, (18.123) becomes

$$\mathrm{tr}(\mathcal{K}_{\mathbf{u}_{im}}\mathcal{W}\mathcal{K}_{\mathbf{u}_{im}}\mathcal{W}) = \int_\infty d^2\overline{r} \int_\infty d^2\Delta r\; w(\overline{\mathbf{r}}+\tfrac{1}{2}\Delta\mathbf{r})\left[\overline{I}_{im}(\overline{\mathbf{r}})\right]^2 a(\Delta\mathbf{r})\,w(\overline{\mathbf{r}}-\tfrac{1}{2}\Delta\mathbf{r})\,. \tag{18.125}$$

[10]The reader should not confuse the two meanings of the overbar in this equation. When placed over a random variable as in $\overline{I}_{im}$, the overbar implies an ensemble average, but over a spatial variable as in $\overline{\mathbf{r}}$ it implies a spatial average position as in (18.110).

If the detector is large compared to the range of $a(\Delta\mathbf{r})$ (which is also the blob size), yet $\overline{I}_{im}(\overline{\mathbf{r}})$ is still approximately constant over the detector, we can write

$$\mathrm{tr}(\mathcal{K}_{\mathbf{u}_{im}}\mathcal{W}\mathcal{K}_{\mathbf{u}_{im}}\mathcal{W}) \approx C^2 A_d \left[\overline{I}_{im}(\mathbf{r}_d)\right]^2 \int_\infty d^2\Delta r \; a(\Delta\mathbf{r})\,. \tag{18.126}$$

The remaining integral defines the blob area A_b, and we can multiply and divide by the detector area to get

$$\mathrm{tr}(\mathcal{K}_{\mathbf{u}_{im}}\mathcal{W}\mathcal{K}_{\mathbf{u}_{im}}\mathcal{W}) \approx \frac{(CA_d)^2}{N_b}\left[\overline{I}_{im}(\mathbf{r}_d)\right]^2 = \frac{1}{N_b}(\overline{g}_{out})^2\,, \tag{18.127}$$

where N_b is the mean number of blobs in the detector area:

$$N_b \equiv \frac{A_d}{\int_\infty d^2\Delta r \; a(\Delta\mathbf{r})} = \frac{A_d}{A_b}\,. \tag{18.128}$$

Now we can insert (18.122) and (18.127) into (18.120) to obtain

$$\psi_{g_{out}}(\nu) \approx \exp\left[-2\pi i\,\overline{g}_{out}\,\nu - 2\pi^2\frac{(\overline{g}_{out})^2}{N_b}\,\nu^2 + \cdots\right]. \tag{18.129}$$

If the higher-order terms can be ignored, (18.129) becomes the characteristic function for a univariate normal random variable [*cf.* (C.116)]. This is the expected result since, roughly speaking, each blob represents a patch of the detector with irradiance independent of the irradiance in other patches. The detector sums the outputs from each patch, and hence the total output is normal by the central-limit theorem. It remains, however, to show that the higher-order terms are indeed negligible if the detector is large enough. In lieu of a formal proof, we present a heuristic argument.

The key is the second term in the exponent; since it has a negative, real coefficient, this term corresponds to a Gaussian factor in $\psi_{g_{out}}(\nu)$. As the detector gets larger, $N_b \propto A_d$ but $(\overline{g}_{out})^2 \propto A_d^2$, so the overall coefficient of ν^2 also grows linearly with A_d. Since this coefficient is in the exponent, the Gaussian factor rapidly gets smaller (for any given ν) as the detector gets larger. Conversely, a smaller ν suffices to make the Gaussian factor attain whatever level we deem negligible (say, 0.01). There must therefore be some A_d such that the Gaussian factor has died off to this level before the higher-order terms become appreciable.

The general case To determine the characteristic function and the PDF in the general case where the detector is neither very small nor very large, we must solve the eigenvalue problem for the operator $\mathcal{K}_{\mathbf{u}_{im}}\mathcal{W}$. If we denote the nth eigenvalue of this operator by $\lambda_{\mathcal{KW}}^{(n)}$, then (18.119) becomes

$$\psi_{g_{out}}(\nu) = \prod_{n=1}^{\infty}\left[1 + 2\pi i\nu\,\lambda_{\mathcal{KW}}^{(n)}\right]^{-1}. \tag{18.130}$$

If the eigenvalues are distinct, the corresponding univariate PDF is (Goodman, 1975):[11]

$$\mathrm{pr}(g_{out}) = \sum_{n=1}^{\infty} \frac{d_n}{\lambda_{\mathcal{KW}}^{(n)}} \exp\left(-\frac{g_{out}}{\lambda_{\mathcal{KW}}^{(n)}} \right), \qquad (g_{out} \geq 0), \tag{18.131}$$

where

$$d_n \doteq \prod_{\substack{m=1 \\ m \neq n}}^{\infty} \left(1 - \frac{\lambda_{\mathcal{KW}}^{(m)}}{\lambda_{\mathcal{KW}}^{(n)}} \right)^{-1}. \tag{18.132}$$

Thus the problem boils down to finding the eigenvalues, which usually must be done numerically.

One example where the eigenvalues can be found analytically uses the quasistationary covariance function of (18.115) and assumes that both the lens aperture and the detector response are described by 1D rect function (see Scribot, 1974 and Goodman, 1975). Thus the Fourier transform of the lens aperture is a sinc function, and the eigenvalue equation simplifies to a 1D problem of the form [*cf.* (4.64)],

$$const \cdot \int_{-\frac{1}{2}\epsilon}^{\frac{1}{2}\epsilon} dx' \,\mathrm{sinc}\left[B(x - x')\right] \phi_{\mathcal{KW}}^{(n)}(x') = \lambda_{\mathcal{KW}}^{(n)} \phi_{\mathcal{KW}}^{(n)}(x). \tag{18.133}$$

We know from Sec. 4.1.5 that the solutions to this equation are prolate spheroidal wavefunctions. The abrupt cutoff of the eigenvalue spectrum, as seen in Fig. 4.2, suggests other avenues for approximation, which the inquisitive reader may wish to follow.

SNR on the detector output If $N_b \gg 1$, then the characteristic function of the detector output is given by (18.129). Whether or not the higher-order terms are negligible, we can write that result as

$$\psi_{g_{out}}(\nu) = \exp\left[-2\pi i \, \overline{g}_{out} \, \nu - 2\pi^2 \sigma^2 \nu^2 + \cdots \right], \tag{18.134}$$

where σ^2 is the variance of the output signal, given by

$$\sigma^2 = \frac{(\overline{g}_{out})^2}{N_b}. \tag{18.135}$$

Thus the signal-to-noise ratio, defined here as the mean output divided by the standard deviation, is given by

$$\mathrm{SNR} = \sqrt{N_b}. \tag{18.136}$$

For a small detector, the SNR on the output is unity since the output statistics are just the point statistics of the irradiance, which is exponentially distributed and hence has a standard deviation equal to its mean. For a large detector, however, the SNR is improved since the speckle variations average out.

[11] The reader who wants to derive (18.131) should note that the eigenvalues are all nonnegative, so the poles are all in the upper half plane, and the contour can be closed on an infinite semicircle in the upper half plane since the exponential factor vanishes rapidly there.

Arrays It is relatively straightforward to extend the derivations above from a single detector to an array. The only difficulty is that we usually cannot assume that the elements in the array are either large or small compared to the blob size. As a numerical example, consider an array of 512 $\times$ 512 elements on a 1 cm^2 detector. The width of each element is thus about 2 μm. For an F/2 lens operating at a wavelength $\lambda = 0.5$ μm, the correlation length of the speckle (which is also the width of the diffraction-limited PSF), is about 1 μm, so the element size is comparable to the blob size.

Suppose that the $\mathbf{m}^{th}$ detector element in an array is located at position $\mathbf{r_m}$, where $\mathbf{m}$ is a 2D multi-index with integer components. Generalizing (18.117), the output of the detector element is given by

$$g_{\mathbf{m}} = \int_\infty d^2r\; w_{\mathbf{m}}(\mathbf{r})\, I_{im}(\mathbf{r})\,, \tag{18.137}$$

where

$$w_{\mathbf{m}}(\mathbf{r}) = C\,\mathrm{rect}\left(\frac{\mathbf{r}-\mathbf{r_m}}{\epsilon}\right). \tag{18.138}$$

Here C is the responsivity of the detector element and ϵ is its width. The set of all detector outputs is a vector $\mathbf{g}$, and (18.137) defines the discretization operator $\boldsymbol{\mathcal{D}}_w$ in:

$$\mathbf{g} = \boldsymbol{\mathcal{D}}_w\, \mathbf{I}_{im}\,. \tag{18.139}$$

Our objective is to understand the statistics of $\mathbf{g}$.

Multivariate characteristic function The generalizations of (18.118) and (18.119) are

$$\psi_{\mathbf{g}}(\boldsymbol{\xi}) = \left\langle \exp\left[-2\pi i\,(\boldsymbol{\mathcal{D}}_w \mathbf{I}_{im})^\dagger\, \boldsymbol{\xi}\right]\right\rangle = \Psi_{\mathbf{I}_{im}}(\boldsymbol{\mathcal{D}}_w^\dagger\, \boldsymbol{\xi}) = \frac{1}{\det(\boldsymbol{\mathcal{I}} + 2\pi i\, \boldsymbol{\mathcal{K}}_{\mathbf{u}_{im}} \boldsymbol{\mathcal{W}}_{\boldsymbol{\xi}})}\,, \tag{18.140}$$

where $\boldsymbol{\mathcal{W}}_{\boldsymbol{\xi}}$ is an integral operator with kernel

$$W_{\boldsymbol{\xi}}(\mathbf{r},\mathbf{r}') = [\boldsymbol{\mathcal{D}}_w^\dagger\, \boldsymbol{\xi}](\mathbf{r})\,\delta(\mathbf{r}-\mathbf{r}') = \sum_{\mathbf{m}} \xi_{\mathbf{m}} w_{\mathbf{m}}(\mathbf{r})\,\delta(\mathbf{r}-\mathbf{r}')\,. \tag{18.141}$$

The expansion in traces, (18.120), now becomes

$$\psi_{\mathbf{g}}(\boldsymbol{\xi}) = \exp[-2\pi i\,\mathrm{tr}(\boldsymbol{\mathcal{K}}_{\mathbf{u}_{im}} \boldsymbol{\mathcal{W}}_{\boldsymbol{\xi}}) - 2\pi^2\,\mathrm{tr}(\boldsymbol{\mathcal{K}}_{\mathbf{u}_{im}} \boldsymbol{\mathcal{W}}_{\boldsymbol{\xi}}\, \boldsymbol{\mathcal{K}}_{\mathbf{u}_{im}} \boldsymbol{\mathcal{W}}_{\boldsymbol{\xi}}) + \cdots]\,. \tag{18.142}$$

Convergence of this expansion requires that all eigenvalues of $2\pi i\, \boldsymbol{\mathcal{K}}_{\mathbf{u}_{im}} \boldsymbol{\mathcal{W}}_{\boldsymbol{\xi}}$ be < 1 in absolute value, which will occur if $|\boldsymbol{\xi}|$ is small enough.

The first two traces in the expansion are [*cf.* (18.71), (18.122) and (18.123)]

$$\mathrm{tr}(\boldsymbol{\mathcal{K}}_{\mathbf{u}_{im}} \boldsymbol{\mathcal{W}}_{\boldsymbol{\xi}}) = \sum_{\mathbf{m}} \xi_{\mathbf{m}} \int_\infty d^2r\; w_{\mathbf{m}}(\mathbf{r})\, K_{\mathbf{u}_{im}}(\mathbf{r},\mathbf{r})\,; \tag{18.143}$$

$$\mathrm{tr}(\boldsymbol{\mathcal{K}}_{\mathbf{u}_{im}} \boldsymbol{\mathcal{W}}_{\boldsymbol{\xi}}\, \boldsymbol{\mathcal{K}}_{\mathbf{u}_{im}} \boldsymbol{\mathcal{W}}_{\boldsymbol{\xi}}) = \sum_{\mathbf{m}} \xi_{\mathbf{m}} \sum_{\mathbf{m}'} \xi_{\mathbf{m}'} \int_\infty d^2r \int_\infty d^2r'\; w_{\mathbf{m}}(\mathbf{r})\, |K_{\mathbf{u}_{im}}(\mathbf{r},\mathbf{r}')|^2\, w_{\mathbf{m}'}(\mathbf{r}')\,. \tag{18.144}$$

Relation to the object Though (18.140) and (18.142) purport to specify statistical properties of an image, they do not explicitly show the effect of the object. To make that connection, we need to use the operator relation (18.99) or its kernel (18.100). Plugging (18.99) into (18.140) gives

$$\psi_{\mathbf{g}}(\boldsymbol{\xi}) = \frac{1}{\det(\mathcal{I} + 2\pi i\,\ell_c^2\,\mathcal{P}_{coh}\,|\mathbf{t}_{obj}|^2\,\mathcal{P}_{coh}^{\dagger}\,\mathcal{W}_{\boldsymbol{\xi}})}\,. \tag{18.145}$$

Note that only the squared modulus of the amplitude transmittance of the object appears. Recall that the object is a photographic transparency and that it is placed over a ground glass that completely randomizes the phase of the light; any additional phase modulation by the object transparency is irrelevant.

To acknowledge this point, and to put our results here into a notation compatible with previous chapters, we define

$$f(\mathbf{r}) \equiv |t_{obj}(\mathbf{r})|^2 \qquad \text{or} \qquad \mathbf{f} = |\mathbf{t}_{obj}|^2\,. \tag{18.146}$$

Thus the characteristic function is given by

$$\psi_{\mathbf{g}}(\boldsymbol{\xi}) = \frac{1}{\det\left(\mathcal{I} + 2\pi i\,\ell_c^2\,\mathcal{P}_{coh}\,\mathbf{f}\,\mathcal{P}_{coh}^{\dagger}\,\mathcal{W}_{\boldsymbol{\xi}}\right)}\,. \tag{18.147}$$

This characteristic function is conditional on a particular object, and we shall write it as $\psi_{\mathbf{g}|\mathbf{f}}(\boldsymbol{\xi})$ in Sec. 18.4.2 when we consider the object to be random.

Mean vector and covariance matrix The expansion in traces allows us just to read off the mean and covariance of $\mathbf{g}$. If we can write the characteristic function in the form

$$\psi_{\mathbf{g}}(\boldsymbol{\xi}) = \exp\left[-2\pi i \sum_{\mathbf{m}} b_{\mathbf{m}}\,\xi_{\mathbf{m}} - 2\pi^2 \sum_{\mathbf{m}}\sum_{\mathbf{m}'} \xi_{\mathbf{m}}\,A_{\mathbf{m}\mathbf{m}'}\xi_{\mathbf{m}'} + \cdots\right], \tag{18.148}$$

then we can show from (8.30) and (8.31) that $b_{\mathbf{m}}$ is the mean of $g_{\mathbf{m}}$ and $A_{\mathbf{m}\mathbf{m}'}$ is the $(\mathbf{m}, \mathbf{m}')$ element of the covariance matrix. Higher terms, which involve cubic and higher powers of $\xi_{\mathbf{m}}$, give no contribution to the mean or covariance after taking the requisite derivatives and setting all $\xi_{\mathbf{m}}$ to 0. Thus

$$\overline{g}_{\mathbf{m}} = \int_\infty d^2r\; w_{\mathbf{m}}(\mathbf{r})\,K_{\mathbf{u}_{im}}(\mathbf{r},\mathbf{r}) = \int_\infty d^2r\; w_{\mathbf{m}}(\mathbf{r})\overline{I}_{im}(\mathbf{r})\,; \tag{18.149}$$

$$\begin{aligned}[\mathbf{K}_{\mathbf{g}}]_{\mathbf{m}\mathbf{m}'} &= \int_\infty d^2r \int_\infty d^2r'\, w_{\mathbf{m}}(\mathbf{r})\,|K_{\mathbf{u}_{im}}(\mathbf{r},\mathbf{r}')|^2\, w_{\mathbf{m}'}(\mathbf{r}') \\ &= \int_\infty d^2r \int_\infty d^2r'\, w_{\mathbf{m}}(\mathbf{r})\,K_{\mathbf{I}_{im}}(\mathbf{r},\mathbf{r}')\, w_{\mathbf{m}'}(\mathbf{r}')\,. \end{aligned} \tag{18.150}$$

Neither of these results should be surprising; they are the usual expressions for propagating means and covariances through linear CD mappings.

Reduction to determinant of a matrix We began this section with the characteristic functional for the irradiance in the image plane and converted it to the characteristic function for the discrete data vector by using the rules for linear CD mappings, yet the result was still expressed as the determinant of an integral operator. We shall now show that it is possible to transform the characteristic function so that it involves the determinant of a matrix rather than an integral operator, at least for small detectors. The procedure is an extension of one we used in Sec. 18.2.5 to derive the bivariate point statistics of the irradiance.

A detector element is considered small for this purpose if its width ϵ is much less than the speckle blob size, which is approximately the width of the coherent PSF. In that case we can write [*cf.* (18.138)]

$$w_{\mathbf{m}}(\mathbf{r}) \approx C\epsilon^2\, \delta(\mathbf{r}-\mathbf{r}_{\mathbf{m}})\,, \tag{18.151}$$

and hence [*cf.* (18.138)]

$$W_{\boldsymbol{\xi}}(\mathbf{r},\mathbf{r}') \approx C\epsilon^2 \sum_{\mathbf{m}'} \xi_{\mathbf{m}'}\, \delta(\mathbf{r}-\mathbf{r}_{\mathbf{m}'})\, \delta(\mathbf{r}-\mathbf{r}')\,. \tag{18.152}$$

The key point is that the rank of $\mathcal{K}\mathcal{W}_{\boldsymbol{\xi}}$ (where we have deleted the subscript on $\mathcal{K}$ for notational simplicity) is less than or equal to the number of detector elements with this approximation; to see this, let $\phi^{(n)}_{\mathcal{KW}}(\mathbf{r})$ be an eigenfunction of $\mathcal{K}\mathcal{W}_{\boldsymbol{\xi}}$ with eigenvalue $\lambda^{(n)}_{\mathcal{KW}}$ and write

$$\left[\mathcal{K}\mathcal{W}_{\boldsymbol{\xi}}\phi^{(n)}_{\mathcal{KW}}\right](\mathbf{r}) \approx C\epsilon^2 \sum_{\mathbf{m}'} \xi_{\mathbf{m}'}\, K(\mathbf{r},\mathbf{r}_{\mathbf{m}'})\, \phi^{(n)}_{\mathcal{KW}}(\mathbf{r}_{\mathbf{m}'}) = \lambda^{(n)}_{\mathcal{KW}}\, \phi^{(n)}_{\mathcal{KW}}(\mathbf{r})\,. \tag{18.153}$$

Thus any eigenfunction of $\mathcal{K}\mathcal{W}_{\boldsymbol{\xi}}$ with nonzero eigenvalue is a linear combination of the functions $\{K(\mathbf{r},\mathbf{r}_{\mathbf{m}'})\}$. For an $M\times M$ detector array, there are M^2 functions in this set, and they are all linearly independent, so the rank of $\mathcal{K}\mathcal{W}_{\boldsymbol{\xi}}$ is $\leq M^2$, with equality if and only if all $\xi_{\mathbf{m}'} \neq 0$.

If we evaluate (18.153) at $\mathbf{r}=\mathbf{r}_{\mathbf{m}}$, we obtain

$$\left[C\epsilon^2\mathbf{K}\boldsymbol{\Xi} - \lambda^{(n)}_{\mathcal{KW}}\mathbf{I}\right]\boldsymbol{\Phi}^{(n)}_{\mathcal{KW}} = \mathbf{0}\,, \tag{18.154}$$

where $\boldsymbol{\Phi}^{(n)}_{\mathcal{KW}}$ is an $M^2\times 1$ vector with components $\{\phi^{(n)}_{\mathcal{KW}}(\mathbf{r}_{\mathbf{m}})\}$, and $\mathbf{K}$ and $\boldsymbol{\Xi}$ are $M^2\times M^2$ matrices defined by

$$[\mathbf{K}]_{\mathbf{mm}'} \equiv K(\mathbf{r}_{\mathbf{m}},\mathbf{r}_{\mathbf{m}'})\,, \qquad [\boldsymbol{\Xi}]_{\mathbf{mm}'} \equiv \xi_{\mathbf{m}}\, \delta_{\mathbf{mm}'}\,. \tag{18.155}$$

The eigenvalues are determined by solving

$$\det\left[C\epsilon^2\mathbf{K}\boldsymbol{\Xi} - \lambda^{(n)}_{\mathcal{KW}}\mathbf{I}\right] = 0\,. \tag{18.156}$$

Since the rank of the integral operator $\mathcal{K}\mathcal{W}_{\boldsymbol{\xi}}$ is $\leq M^2$, we can be assured that all nonzero eigenvalues will be found by solving this equation.

The original characteristic function from (18.140) is thus given by

$$\psi_{\mathbf{g}}(\boldsymbol{\xi}) = \frac{1}{\det(\mathcal{I}+2\pi i\,\mathcal{K}\mathcal{W}_{\boldsymbol{\xi}})} = \prod_{n=1}^{M^2} \frac{1}{1+2\pi i\, C\epsilon^2\lambda^{(n)}_{\mathcal{KW}}} = \frac{1}{\det(\mathbf{I}+2\pi i\, C\epsilon^2\mathbf{K}\boldsymbol{\Xi})}\,, \tag{18.157}$$

where $\mathbf{I}$ is an $M^2 \times M^2$ unit matrix.

The matrix-determinant form is particularly useful if we want to know the joint characteristic function for only a few detector outputs. For example, to get the bivariate statistics for any two $g_{\mathbf{m}}$, we need only two nonzero $\xi_{\mathbf{m}}$, so (18.156) requires evaluation of just a 2×2 determinant. We remind the reader, however, that the matrix form applies only with small detectors that provide essentially point sampling of the speckle irradiance.

18.4 NOISE AND IMAGE QUALITY

Though speckle is a form of noise, we still need to consider measurement noise (Gaussian or Poisson) as well as object randomness. In all, coherent imaging is triply (or even quadruply) stochastic: the image is a random vector conditional on the irradiance pattern, the irradiance pattern is a random process conditional on the object, and the object itself is another random process. If we consider both kinds of measurement noise — Gaussian electronic noise and Poisson photon-counting statistics — there are altogether four random processes in the problem.

Before trying to express these ideas mathematically, let us recall the meaning of randomness for each of these components. When we regard the object as random, we are envisioning drawing many objects from some ensemble and imaging each in turn. It may be difficult or even impossible to specify this ensemble, but in principle we are thinking in frequentist terms about sampling from an infinite set of possible objects. The objects discussed so far in this chapter have been photographic transparencies, and we imagine laying one after the other over an illuminated ground glass and imaging the emerging field. For each object, however, the irradiance pattern is also random because we can consider an ensemble of ground glasses; indeed, all statistical properties derived in Sec. 18.3 were for one object transparency and infinitely many ground glasses. Finally, if we use one object transparency and one ground glass but take repeated images, the data vector will be random because of photon statistics and electronic noise.

The goal of this section is to incorporate measurement noise and object randomness into our statistical analysis and then to use the results to discuss objective assessment of image quality in coherent imaging. We shall continue to consider a 2D imaging system where the object is a photographic transparency, but more general situations will be treated in Sec. 18.5.

18.4.1 Measurement noise

In Chap. 11 we presented a detailed account of Poisson random variables and processes, and in Chap. 12 we applied that knowledge to the analysis of photon noise in optical detectors. We also saw in Chap. 12, however, that numerous electronic noise processes could lead to Gaussian statistics instead of Poisson. We shall consider both of these kinds of noise in a detector array sensing a speckle distribution.

Photon noise As the starting point for discussing combined speckle and photon noise, we shall use (8.339), which expresses the transformation of the characteristic functional of a random process through a CD mapping, with the output of that mapping serving as the mean of a Poisson random vector. In the present discussion, the random process in question is the irradiance incident on the detector array, and the CD mapping is the discretization operator $\mathcal{D}_w$. Thus (8.339) becomes

$$\psi_{\mathbf{g}}(\boldsymbol{\xi}) = \Psi_{\mathbf{I}_{im}}[\mathcal{D}_w^\dagger \, \boldsymbol{\Gamma}(\boldsymbol{\xi})] \,, \tag{18.158}$$

where the operator $\boldsymbol{\Gamma}$ is defined by (8.338) as

$$[\boldsymbol{\Gamma}(\boldsymbol{\xi})]_{\mathbf{m}} = \frac{-1+\exp(-2\pi i\,\xi_{\mathbf{m}})}{-2\pi i} \,. \tag{18.159}$$

With this transformation, (18.140) becomes

$$\psi_{\mathbf{g}}(\boldsymbol{\xi}) = \frac{1}{\det\left(\mathcal{I} + 2\pi i\,\mathcal{K}_{\mathbf{u}_{im}} \mathcal{W}_{\boldsymbol{\Gamma}\boldsymbol{\xi}}\right)} = \frac{1}{\det\left(\mathcal{I} + 2\pi i\,\ell_c^2\,\mathcal{P}_{coh}\,\mathbf{f}\,\mathcal{P}_{coh}^\dagger\,\mathcal{W}_{\boldsymbol{\Gamma}\boldsymbol{\xi}}\right)} \,, \tag{18.160}$$

where $\mathcal{W}_{\boldsymbol{\Gamma}\boldsymbol{\xi}}$ is an integral operator with kernel [*cf.* (18.141)]

$$W_{\boldsymbol{\Gamma}\boldsymbol{\xi}}(\mathbf{r},\mathbf{r}') = \sum_{\mathbf{m}} \left[\frac{-1+\exp(-2\pi i\,\xi_{\mathbf{m}})}{-2\pi i}\right] w_{\mathbf{m}}(\mathbf{r})\,\delta(\mathbf{r}-\mathbf{r}') \,. \tag{18.161}$$

After a bit of manipulation, the kernel of the operator inside the determinant in (18.160) takes the form

$$\left[\mathcal{I} + 2\pi i\,\mathcal{K}_{\mathbf{u}_{im}} \mathcal{W}_{\boldsymbol{\Gamma}\boldsymbol{\xi}}\right](\mathbf{r},\mathbf{r}') = \delta(\mathbf{r}-\mathbf{r}') + K_{\mathbf{u}_{im}}(\mathbf{r},\mathbf{r}') \sum_{\mathbf{m}} [1-\exp(-2\pi i\,\xi_{\mathbf{m}})]\, w_{\mathbf{m}}(\mathbf{r}') \,. \tag{18.162}$$

The expansion in traces, (18.142), is still valid if it converges, but $\xi_{\mathbf{m}}$ must be replaced by $[\boldsymbol{\Gamma}(\boldsymbol{\xi})]_{\mathbf{m}}$ in (18.143) and (18.144); the result is

$$\begin{aligned} \psi_{\mathbf{g}}(\boldsymbol{\xi}) &= \exp\Big\{ -2\pi i \sum_{\mathbf{m}} [\boldsymbol{\Gamma}(\boldsymbol{\xi})]_{\mathbf{m}}\, \overline{g}_{\mathbf{m}} - 2\pi^2 \sum_{\mathbf{m},\mathbf{m}'} [\boldsymbol{\Gamma}(\boldsymbol{\xi})]_{\mathbf{m}} \left[\mathbf{K}_{\mathbf{g}}^{(nf)}\right]_{\mathbf{m}\mathbf{m}'} [\boldsymbol{\Gamma}(\boldsymbol{\xi})]_{\mathbf{m}'} + \cdots \Big\} \\ &= \exp\Big\{ \sum_{\mathbf{m}} [-1+\exp(-2\pi i\,\xi_{\mathbf{m}})]\, \overline{g}_{\mathbf{m}} \\ &\quad + \tfrac{1}{2} \sum_{\mathbf{m},\mathbf{m}'} [-1+\exp(-2\pi i\,\xi_{\mathbf{m}})] \left[\mathbf{K}_{\mathbf{g}}^{(nf)}\right]_{\mathbf{m}\mathbf{m}'} [-1+\exp(-2\pi i\,\xi_{\mathbf{m}'})] + \cdots \Big\} \,, \end{aligned} \tag{18.163}$$

where we have used (18.149) and (18.150). Recall, however, that (18.150) did not include the effects of Poisson noise, so the double integral in that equation is the covariance of $\mathbf{g}$ with speckle but without Poisson noise, hence the superscript (nf) for *noise-free*. With proper choice of the responsivity C, $g_{\mathbf{m}}$ is measured in units of detected events or counts, so $\overline{g}_{\mathbf{m}}$ can be interpreted as the mean number of counts in element $\mathbf{m}$ during the measurement time.

The multivariate characteristic function given in (18.160) or (18.163) contains all possible statistical information about the detector output when both speckle and photon noise are present. It applies to an arbitrary array of detector elements of arbitrary size. We shall now investigate several special cases and limits of this general result.

Mean vector and covariance matrix As we noted in Sec. 18.3.3, we can read off the mean and covariance of a random vector if we can express its characteristic function in the form (18.148). The characteristic function in (18.163) is not immediately in that form since the coefficients of successive terms are nonlinear functions of the frequency variables $\xi_{\mathbf{m}}$. We can, however, expand the exponentials to obtain

$$-1+\exp(-2\pi i\,\xi_{\mathbf{m}}) = -2\pi i\,\xi_{\mathbf{m}} - 2\pi^2\,\xi_{\mathbf{m}}^2 + \cdots . \tag{18.164}$$

Thus, through terms quadratic in the frequencies, we have

$$\psi_{\mathbf{g}}(\boldsymbol{\xi}) = \exp\left\{\sum_{\mathbf{m}}\left(-2\pi i\,\xi_{\mathbf{m}'} - 2\pi^2\,\xi_{\mathbf{m}}^2\right)\overline{g}_{\mathbf{m}} - 2\pi^2\sum_{\mathbf{m},\mathbf{m}'}\xi_{\mathbf{m}}\left[\mathbf{K}_{\mathbf{g}}^{(nf)}\right]_{\mathbf{mm}'}\xi_{\mathbf{m}'} + \cdots\right\}. \tag{18.165}$$

Comparison with (18.148) shows that $\overline{g}_{\mathbf{m}}$ is indeed the mean, and the covariance matrix is given by

$$\left[\mathbf{K}_{\mathbf{g}}\right]_{\mathbf{mm}'} = \overline{g}_{\mathbf{m}}\,\delta_{\mathbf{mm}'} + \left[\mathbf{K}_{\mathbf{g}}^{(nf)}\right]_{\mathbf{mm}'} . \tag{18.166}$$

This result is familiar from earlier chapters; it expresses the universal form for the covariance matrix of doubly stochastic Poisson noise, as first derived[12] in Sec. 11.2.2.

Large detector elements When we considered large detector elements with no Poisson noise in Sec. 18.3.3, we argued that the expansion in traces could be truncated after the second term in the exponent since that term was quadratic in the frequency variable and had a negative, real coefficient that increased with the detector area [see (18.129)]. As a result, the speckle statistics approached Gaussian. The coefficient of the first term in the exponent of the characteristic function was pure imaginary and gave the mean of the Gaussian.

With Poisson noise, all of this changes. It is no longer true that the coefficient of the first term is pure imaginary and that of the second term is pure real. In fact, both terms are complex and periodic in $\xi_{\mathbf{m}}$ with period 1, which tells us immediately that $g_{\mathbf{m}}$ can have only integer values. (We know from Sec. 3.3.8 that periodicity of a function restricts its Fourier transform to a discrete set of points.) Moreover, as noted above, we can no longer identify each term in the exponent with a specific power of $\boldsymbol{\xi}$; the $\boldsymbol{\Gamma}$ operator mixes the powers.

We can, however, terminate the series in the exponent of (18.163) with the single-sum term if the detector is large enough. The noise-free part of the covariance matrix approaches zero as N_b (the mean number of blobs across the detector) gets large [*cf.* (18.135)], so the double-sum term in (18.163) becomes negligible if N_b is large enough. In that case (18.163) becomes

$$\psi_{\mathbf{g}}(\boldsymbol{\xi}) = \exp\left\{\sum_{\mathbf{m}}[-1+\exp(-2\pi i\,\xi_{\mathbf{m}})]\,\overline{g}_{\mathbf{m}}\right\} = \prod_{\mathbf{m}}\exp\left\{[-1+\exp(-2\pi i\,\xi_{\mathbf{m}})]\,\overline{g}_{\mathbf{m}}\right\} . \tag{18.167}$$

This expression is recognized from (C.171) as the product of characteristic functions for Poisson random variables. Thus the speckle averages out and the detector

[12]In those previous results, we would have written $\overline{\overline{g}}_{\mathbf{m}}$ where we have $\overline{g}_{\mathbf{m}}$ here, but the first average, over the speckle irradiances, is already implicit in the definition of $\overline{g}_{\mathbf{m}}$ in (18.149). We shall reserve the double overbar for Sec. 18.4.2, where we consider the triply stochastic problem of combined speckle, Poisson noise and object randomness.

outputs revert to independent Poisson random variables as the detector gets large.

We encountered an analogous situation in Sec. 11.3.7 when we discussed photon-counting statistics with incoherent light [see especially the discussion below (11.128)]. In that case we also had two covariance terms, but one of them averaged out when we integrated over a long observation time, leaving us with just the Poisson component. The same thing happens with speckle except that the averaging is spatial. *Large detectors give Poisson statistics in spite of speckle.*

Gaussian limit of the Poisson If we insert (18.164) into (18.167), we get

$$\psi_{\mathbf{g}}(\boldsymbol{\xi}) = \prod_{\mathbf{m}} \exp\left[-2\pi i\, \xi_{\mathbf{m}}\, \overline{g}_{\mathbf{m}} - 2\pi^2\, \xi_{\mathbf{m}}^2\, \overline{g}_{\mathbf{m}} + \cdots\right] . \tag{18.168}$$

If $\overline{g}_{\mathbf{m}}$ is large enough, we can truncate this series after the first real term, the one quadratic in $\xi_{\mathbf{m}}$. Thus, if the detector elements are large and the mean counts per element are also large, the statistics approach independent Gaussian with the variance of each output equal to the mean.

Small detector elements, univariate statistics We can compute the univariate characteristic function for a single output $g_{\mathbf{m}}$ simply by setting $\xi_{\mathbf{m}'} = 0$ for all $\mathbf{m}' \neq \mathbf{m}$. Then (18.160) becomes

$$\psi_{g_{\mathbf{m}}}(\xi_{\mathbf{m}}) = \frac{1}{\det\left[\mathcal{I} + 2\pi i\, \mathcal{K}_{\mathbf{u}_{im}} \mathcal{W}_{\boldsymbol{\Gamma\xi}}^{\mathbf{m}}\right]} , \tag{18.169}$$

where $\mathcal{W}_{\boldsymbol{\Gamma\xi}}^{\mathbf{m}}$ is $\mathcal{W}_{\boldsymbol{\Gamma\xi}}$ with the summation sign in (18.161) omitted.

If the detector is much smaller than a blob and can be approximated with a delta function as in (18.121), the operator in the determinant has rank 1, and the determinant can be evaluated just as in Sec. 18.3.3. The univariate characteristic function becomes [*cf.* (18.122)]

$$\psi_{g_{\mathbf{m}}}(\xi_{\mathbf{m}}) = \frac{1}{1 + [1 - \exp(-2\pi i\, \xi_{\mathbf{m}})]\overline{g}_{\mathbf{m}}} = \frac{1}{1 + \overline{g}_{\mathbf{m}} - \overline{g}_{\mathbf{m}} \exp(-2\pi i\, \xi_{\mathbf{m}})} . \tag{18.170}$$

We can get the corresponding univariate PDF by taking an inverse Fourier transform. Temporarily dropping the subscript $\mathbf{m}$ for convenience, we must evaluate:

$$\mathrm{pr}(g) = \int_{-\infty}^{\infty} d\xi\, \frac{\exp(2\pi i g\, \xi)}{1 + \overline{g} - \overline{g} \exp(-2\pi i \xi)} . \tag{18.171}$$

Contour integration suggests itself. The exponential factor is analytic in the upper half plane and vanishes at infinity in that half plane (for $g > 0$). Thus we can close the contour with an arc of infinite radius crossing the positive imaginary axis. It remains to find the poles in the upper half plane and evaluate their residues.

The denominator vanishes when

$$\xi = n + \frac{i}{2\pi} \ln\left(\frac{1 + \overline{g}}{\overline{g}}\right) \equiv \xi_n , \qquad n \text{ integer} . \tag{18.172}$$

We can demonstrate that this singularity is a simple pole and find its residue by use of (B.47) and L'Hôpital's rule:

$$\lim_{\xi \to \xi_n} \frac{\xi - \xi_n}{1 + \overline{g} - \overline{g} \exp(-2\pi i \xi)} = \frac{1}{2\pi i \overline{g} \exp(-2\pi i \xi_n)} = \frac{1}{2\pi i (1 + \overline{g})} . \tag{18.173}$$

The residue of the integrand at the nth pole is thus

$$\text{res}_n = \frac{\exp(2\pi i \xi_n g)}{2\pi i(1+\overline{g})} . \tag{18.174}$$

There are an infinite number of poles within the contour; with (B.48) and a little algebra, the integral becomes

$$\oint_C d\xi \, \frac{\exp(2\pi i g\,\xi)}{1+\overline{g}-\overline{g}\exp(-2\pi i \xi)} = 2\pi i \sum_{n=-\infty}^{\infty} \text{res}_n = \sum_{n=-\infty}^{\infty} \exp(2\pi i n g) \frac{\overline{g}^{g}}{(1+\overline{g})^{1+g}} . \tag{18.175}$$

The sum will be recognized from (2.50) as a comb function, so

$$\text{pr}(g_{\mathbf{m}}) = \sum_{k=0}^{\infty} \frac{\overline{g}_{\mathbf{m}}^{k}}{(1+\overline{g}_{\mathbf{m}})^{1+k}} \, \delta(g_{\mathbf{m}} - k) \, , \tag{18.176}$$

where we have now reinstated the $\mathbf{m}$ subscript, and we have deleted all negative-k terms from the sum since $g_{\mathbf{m}}$ cannot be negative and hence $\delta(g_{\mathbf{m}} - k) = 0$ for $k < 0$. By (C.24), the coefficient of the delta function is the probability of occurrence of the discrete value $g_{\mathbf{m}} = k$, or

$$\Pr(g_{\mathbf{m}} = k) = \frac{\overline{g}_{\mathbf{m}}^{k}}{(1+\overline{g}_{\mathbf{m}})^{1+k}} . \tag{18.177}$$

This probability is the *Bose-Einstein law*, and we know from Sec. 11.1.4 that the Poisson transform of an exponential is a Bose-Einstein [see also (11.314)]. In the present problem, the irradiance at a single point is exponential, and the small detector samples the irradiance at a point and uses the result as the mean for a Poisson, with the resulting photon-counting distribution being, therefore, Bose-Einstein. It must be remembered, however, that this law applies only with small detectors and even then only to the univariate statistics.

Electronic noise If there is an array of noisy amplifiers or other noise sources, one for each detector element, then we can write

$$\mathbf{v} = \mathbf{g} + \mathbf{n} , \tag{18.178}$$

where $\mathbf{v}$ is the vector of amplifier output voltages and $\mathbf{n}$ is the corresponding vector of noise values. If $\mathbf{n}$ and $\mathbf{g}$ are statistically independent, we know from Sec. 8.5.3 that [*cf.* (8.335)]

$$\psi_{\mathbf{v}}(\boldsymbol{\xi}) = \psi_{\mathbf{n}}(\boldsymbol{\xi})\, \psi_{\mathbf{g}}(\boldsymbol{\xi}) . \tag{18.179}$$

If we neglect the Poisson noise and suppose that the electronic noise is the only noise process associated with the detection of a speckle pattern, then we have [*cf.* (18.140)]:

$$\psi_{\mathbf{v}}(\boldsymbol{\xi}) = \psi_{\mathbf{n}}(\boldsymbol{\xi})\, \Psi_{\mathbf{I}_{im}}(\boldsymbol{\mathcal{D}}_w^{\dagger}\, \boldsymbol{\xi}) . \tag{18.180}$$

To account for both electronic and Poisson noise, we write [*cf.* (18.158)]:

$$\psi_{\mathbf{v}}(\boldsymbol{\xi}) = \psi_{\mathbf{n}}(\boldsymbol{\xi})\, \Psi_{\mathbf{I}_{im}}[\boldsymbol{\mathcal{D}}_w^{\dagger}\, \boldsymbol{\Gamma}(\boldsymbol{\xi})] . \tag{18.181}$$

The usual model (see Chap. 12) is that the electronic noise $\mathbf{n}$ is zero-mean i.i.d. Gaussian, so

$$\psi_{\mathbf{n}}(\boldsymbol{\xi}) = \prod_{\mathbf{m}} \exp(-2\pi^2 \sigma_{el}^2 \xi_{\mathbf{m}}^2) \,. \tag{18.182}$$

If we use the expansion in traces to cast $\psi_{\mathbf{g}}(\boldsymbol{\xi})$ into the form (18.148), then

$$\psi_{\mathbf{v}}(\boldsymbol{\xi}) = \exp\left[-2\pi i \sum_{\mathbf{m}} \overline{g}_{\mathbf{m}}\, \xi_{\mathbf{m}} - 2\pi^2 \sum_{\mathbf{m}} \sum_{\mathbf{m}'} \xi_{\mathbf{m}} \,[\mathbf{K_v}]_{\mathbf{mm}'} \xi_{\mathbf{m}'} + \cdots \right], \tag{18.183}$$

where

$$\mathbf{K_v} = \mathbf{K_g} + \sigma_{el}^2 \mathbf{I} \,. \tag{18.184}$$

Thus, given the characteristic function in the standard form (18.148), we need only add a multiple of the unit matrix to the covariance in order to account for Gaussian electronic noise; there is no implication that the statistics of $\mathbf{g}$ itself are Gaussian.

18.4.2 Random objects

All of the statistical analysis to this point has been for a fixed object, but we know from Chaps. 13 and 14, that object variability is an important factor limiting task performance. In this section we shall see how to extend the analysis to include random objects.

Overall characteristic function Formally, the characteristic function of the data vector in the presence of speckle, Poisson noise and object variability is given by

$$\psi_{\mathbf{g}}(\boldsymbol{\xi}) = \left\langle \psi_{\mathbf{g}|\mathbf{f}}(\boldsymbol{\xi}) \right\rangle_{\mathbf{f}} = \left\langle \frac{1}{\det\left(\mathcal{I} + 2\pi i\, \ell_c^2\, \mathcal{P}_{coh}\, \mathbf{f} \mathcal{P}_{coh}^{\dagger}\, \mathcal{W}_{\Gamma\xi}\right)} \right\rangle_{\mathbf{f}} . \tag{18.185}$$

For purposes of computing moments, the expansion in traces (18.163) is useful:

$$\psi_{\mathbf{g}}(\boldsymbol{\xi}) = \left\langle \exp\left\{-2\pi i \sum_{\mathbf{m}} [\boldsymbol{\Gamma}(\boldsymbol{\xi})]_{\mathbf{m}}\, \overline{g}_{\mathbf{m}} - 2\pi^2 \sum_{\mathbf{m},\mathbf{m}'} [\boldsymbol{\Gamma}(\boldsymbol{\xi})]_{\mathbf{m}} \left[\mathbf{K}_{\mathbf{g}|\mathbf{f}}^{(nf)}\right]_{\mathbf{mm}'} [\boldsymbol{\Gamma}(\boldsymbol{\xi})]_{\mathbf{m}'} + \cdots \right\} \right\rangle_{\mathbf{f}}, \tag{18.186}$$

where the superscript (nf) has been added to indicate that this covariance matrix is computed without consideration of the Poisson noise, though of course it includes speckle; specifically, it is given by (18.150).

If we wanted also to incorporate electronic noise into the characteristic function, we would use (18.179) and (18.185) to write

$$\psi_{\mathbf{v}}(\boldsymbol{\xi}) = \psi_{\mathbf{n}}(\boldsymbol{\xi})\, \psi_{\mathbf{g}}(\boldsymbol{\xi}) = \psi_{\mathbf{n}}(\boldsymbol{\xi}) \left\langle \frac{1}{\det\left(\mathcal{I} + 2\pi i\, \ell_c^2\, \mathcal{P}_{coh}\, \mathbf{f} \mathcal{P}_{coh}^{\dagger}\, \mathcal{W}_{\Gamma\xi}\right)} \right\rangle_{\mathbf{f}} . \tag{18.187}$$

Since the electronic noise is assumed to be independent of the object, the factor $\psi_{\mathbf{n}}(\boldsymbol{\xi})$ is outside the average over $\mathbf{f}$. For simplicity, we shall omit any further discussion of electronic noise.

Mean and covariance In order to compute the overall mean vector and covariance matrix, we must carry out the expansion of the exponent in (18.186) through terms quadratic in the $\xi_{\mathbf{m}}$ and then compute first and second derivatives. Cubic and higher terms in the exponent will make no contribution to the mean and covariance.

The requisite expansion is [*cf.* (18.165)]

$$\psi_{\mathbf{g}}(\boldsymbol{\xi}) = \left\langle \exp\left\{ \sum_{\mathbf{m}} \left(-2\pi i\, \xi_{\mathbf{m}'} - 2\pi^2\, \xi_{\mathbf{m}}^2\right) \overline{g}_{\mathbf{m}} - 2\pi^2 \sum_{\mathbf{m},\mathbf{m}'} \xi_{\mathbf{m}} \left[\mathbf{K}_{\mathbf{g}}^{(nf)}\right]_{\mathbf{mm}'} \xi_{\mathbf{m}'} + \cdots \right\} \right\rangle_{\mathbf{f}}, \tag{18.188}$$

and the first two derivatives are

$$\frac{\partial}{\partial \xi_{\mathbf{m}}} \psi_{\mathbf{g}}(\boldsymbol{\xi}) = \left\langle \left[-2\pi i \overline{g}_{\mathbf{m}} - 4\pi^2 \xi_{\mathbf{m}} \overline{g}_{\mathbf{m}} - 4\pi^2 \sum_{\mathbf{m}'} \left[\mathbf{K}_{\mathbf{g}|\mathbf{f}}^{(nf)}\right]_{\mathbf{mm}'} \xi_{\mathbf{m}'} \right] \psi_{\mathbf{g}|\mathbf{f}}(\boldsymbol{\xi}) \right\rangle_{\mathbf{f}}; \tag{18.189}$$

$$\frac{\partial^2}{\partial \xi_{\mathbf{m}} \partial \xi_{\mathbf{m}'}} \psi_{\mathbf{g}}(\boldsymbol{\xi}) = -4\pi^2 \left\langle \left[\overline{g}_{\mathbf{m}}\, \delta_{\mathbf{mm}'} + \left[\mathbf{K}_{\mathbf{g}|\mathbf{f}}^{(nf)}\right]_{\mathbf{mm}'} + \overline{g}_{\mathbf{m}} \overline{g}_{\mathbf{m}'} \right] \psi_{\mathbf{g}|\mathbf{f}}(\boldsymbol{\xi}) \right\rangle_{\mathbf{f}}. \tag{18.190}$$

The overall mean is obtained by evaluating the first derivative at the origin:

$$\langle g_{\mathbf{m}} \rangle = -\frac{1}{2\pi i} \left[\frac{\partial}{\partial \xi_{\mathbf{m}}} \psi_{\mathbf{g}}(\boldsymbol{\xi}) \right]_{\mathbf{0}} = \langle \overline{g}_{\mathbf{m}} \rangle_{\mathbf{f}} \equiv \overline{\overline{g}}_{\mathbf{m}}. \tag{18.191}$$

The unsubscripted angle brackets on the left and the double overbar on the right imply an average over all random effects — speckle, Poisson noise and object variability — but the single overbar implies an average over only speckle and Poisson noise. Thus, as before, $\overline{g}_{\mathbf{m}}$ is the mean of $g_{\mathbf{m}}$ conditional on $\mathbf{f}$.

The second moment is

$$\langle g_{\mathbf{m}}\, g_{\mathbf{m}'} \rangle = \frac{1}{-4\pi^2} \left[\frac{\partial^2}{\partial \xi_{\mathbf{m}} \partial \xi_{\mathbf{m}'}} \psi_{\mathbf{g}}(\boldsymbol{\xi}) \right]_{\mathbf{0}} = \overline{\overline{g}}_{\mathbf{m}}\, \delta_{\mathbf{mm}'} + \left\langle \left[\mathbf{K}_{\mathbf{g}|\mathbf{f}}^{(nf)}\right]_{\mathbf{mm}'} \right\rangle_{\mathbf{f}} + \langle \overline{g}_{\mathbf{m}}\, \overline{g}_{\mathbf{m}'} \rangle_{\mathbf{f}}. \tag{18.192}$$

The overall covariance matrix is given by

$$[\mathbf{K}_{\mathbf{g}}]_{\mathbf{mm}'} \equiv \langle g_{\mathbf{m}}\, g_{\mathbf{m}'} \rangle - \overline{\overline{g}}_{\mathbf{m}}\, \overline{\overline{g}}_{\mathbf{m}'} = \overline{\overline{g}}_{\mathbf{m}}\, \delta_{\mathbf{mm}'} + \overline{K}_{\mathbf{mm}'}^{(nf)} + [\mathbf{K}_{\overline{\mathbf{g}}}]_{\mathbf{mm}'}, \tag{18.193}$$

where

$$\overline{K}_{\mathbf{mm}'}^{(nf)} \equiv \left\langle \left[\mathbf{K}_{\mathbf{g}|\mathbf{f}}^{(nf)}\right]_{\mathbf{mm}'} \right\rangle_{\mathbf{f}}, \tag{18.194}$$

and

$$[\mathbf{K}_{\overline{\mathbf{g}}}]_{\mathbf{mm}'} \equiv \langle \overline{g}_{\mathbf{m}}\, \overline{g}_{\mathbf{m}'} \rangle_{\mathbf{f}} - \overline{\overline{g}}_{\mathbf{m}}\, \overline{\overline{g}}_{\mathbf{m}'}. \tag{18.195}$$

Thus we now have three terms in the covariance matrix for this triply stochastic random process. The diagonal term $\overline{\overline{g}}_{\mathbf{m}}\, \delta_{\mathbf{mm}'}$ comes from the Poisson noise averaged over all objects and all speckle realizations. The "noise-free" term, (18.194), comes just from the speckle but is averaged over all objects.

The final term, (18.195), represents the variations of the object function as transferred through to the discrete data domain. It is identical to the covariance one would get for this same optical system operated with a long exposure time to average out the Poisson noise and a moving ground glass to average out the speckle. Equivalently, it is the covariance for the system used with an incoherent source and long exposure time, so that the only randomness left is the object variability.

18.4.3 Task performance

In this section we survey methods for computing observer performance on various tasks of interest. We rely heavily on Chap. 14, with the present goal being to see how the techniques described there can be applied specifically to speckle problems.

Discrimination between known objects The simplest of discrimination tasks is when the object must be either $\mathbf{f}_1$ or $\mathbf{f}_2$, both of which are nonrandom and known precisely to the observer. For example, $\mathbf{f}_1$ could refer to a known background and $\mathbf{f}_2$ to that same background plus a known signal, and in that case we refer to the task as SKE/BKE. The only noise source we consider here is speckle.

We have already examined one special case of this problem in Sec. 13.2.9. As an illustration of the ideal observer with non-Gaussian statistics, we considered a simplified (but not uncommon) model of speckle in which the measurements are statistically independent and exponentially distributed [see (13.136)]. We now know that this model is applicable only when the detector elements are small compared to the speckle blob size, yet separated by a distance large compared to the blob size. We found in that case that the log-likelihood ratio (13.137) yielded a linear discriminant, given in (13.217), though not the Hotelling discriminant, which for comparison was given in (13.218). For later reference, we rewrite (13.137) here in our multi-index notation as

$$\lambda(\mathbf{g}) = \sum_{\mathbf{m}} \left(\frac{1}{\overline{g}_{1\mathbf{m}}} - \frac{1}{\overline{g}_{2\mathbf{m}}} \right) g_{\mathbf{m}} \,. \tag{18.196}$$

We shall now revisit the SKE/BKE detection problem in speckle noise without these unrealistic assumptions. As in Sec. 13.2.12, we pose two questions: Is the log-likelihood ratio a linear discriminant? What is the optimal linear discriminant in the sense of maximizing the AUC? For background on these questions, see Sec. 13.2.12 and especially (13.219) – (13.226).

When is the log-likelihood ratio linear? The general answer to this question is that the log-likelihood ratio is a linear discriminant of the form [*cf.* (13.221)]

$$\lambda(\mathbf{g}) = \ln \left[\frac{\mathrm{pr}(\mathbf{g}|H_2)}{\mathrm{pr}(\mathbf{g}|H_1)} \right] = \mathbf{a}^t \mathbf{g} + c \tag{18.197}$$

if and only if the characteristic functions satisfy (13.226):

$$\psi_{\mathbf{g}|H_2}(\boldsymbol{\xi}) = const \cdot \psi_{\mathbf{g}|H_1}\left(\boldsymbol{\xi} + \frac{i}{2\pi}\mathbf{a} \right) . \tag{18.198}$$

If we can put the characteristic functions in this form, we can read off the linear discriminant $\mathbf{a}$ by inspection.

If the only noise present is speckle, then the characteristic function of the data under the jth hypothesis ($j = 1, 2$) is given from (18.147) as

$$\psi_{\mathbf{g}|H_j}(\boldsymbol{\xi}) = \frac{1}{\det(\mathcal{I} + 2\pi i\, \ell_c^2\, \mathcal{P}_{coh}\, \mathbf{f}_j\, \mathcal{P}_{coh}^\dagger\, \mathcal{W}_{\boldsymbol{\xi}})} \equiv \frac{1}{\det(\mathcal{I} + 2\pi i\, \mathcal{K}_j\, \mathcal{W}_{\boldsymbol{\xi}})} , \tag{18.199}$$

where $\mathcal{K}_j$ is the covariance operator for the image-plane field under hypothesis j.

The following manipulations will lead us in the direction of (18.198):

$$\psi_{\mathbf{g}|H_2}(\boldsymbol{\xi}) = \frac{1}{\det\left[\left(\mathcal{K}_2\mathcal{K}_1^{-1}\right)\left(\mathcal{K}_1\mathcal{K}_2^{-1} + 2\pi i\,\mathcal{K}_1\mathcal{W}_{\boldsymbol{\xi}}\right)\right]}$$

$$= \frac{1}{\det\left(\mathcal{K}_2\mathcal{K}_1^{-1}\right)\det\left(\mathcal{I} + 2\pi i\,\mathcal{K}_1\mathcal{W}_{\boldsymbol{\xi}} + \mathcal{K}_1\mathcal{K}_2^{-1} - \mathcal{I}\right)}$$

$$= const \cdot \frac{1}{\det\left\{\mathcal{I} + 2\pi i\,\mathcal{K}_1\left[\mathcal{W}_{\boldsymbol{\xi}} + \frac{i}{2\pi}\left(\mathcal{K}_1^{-1} - \mathcal{K}_2^{-1}\right)\right]\right\}}, \tag{18.200}$$

where, as a reminder [see (18.141)], $\mathcal{W}_{\boldsymbol{\xi}}$ has the kernel

$$[\mathcal{W}_{\boldsymbol{\xi}}](\mathbf{r}, \mathbf{r}') = \sum_{\mathbf{m}} \xi_{\mathbf{m}}\, w_{\mathbf{m}}(\mathbf{r})\,\delta(\mathbf{r} - \mathbf{r}'). \tag{18.201}$$

For comparison, the right-hand side of (18.198) can be written as

$$\psi_{\mathbf{g}|H_1}\left(\boldsymbol{\xi} + \frac{i}{2\pi}\mathbf{a}\right) = \frac{1}{\det\left\{\mathcal{I} + 2\pi i\,\mathcal{K}_1\left[\mathcal{W}_{\boldsymbol{\xi}} + \frac{i}{2\pi}\mathcal{W}_{\mathbf{a}}\right]\right\}}, \tag{18.202}$$

where

$$[\mathcal{W}_{\mathbf{a}}](\mathbf{r}, \mathbf{r}') = \sum_{\mathbf{m}} a_{\mathbf{m}}\, w_{\mathbf{m}}(\mathbf{r})\,\delta(\mathbf{r} - \mathbf{r}'). \tag{18.203}$$

Thus the log-likelihood ratio is a linear discriminant if

$$\det\left\{\mathcal{I} + 2\pi i\,\mathcal{K}_1\left[\mathcal{W}_{\boldsymbol{\xi}} + \frac{i}{2\pi}\mathcal{W}_{\mathbf{a}}\right]\right\} = \det\left\{\mathcal{I} + 2\pi i\,\mathcal{K}_1\left[\mathcal{W}_{\boldsymbol{\xi}} + \frac{i}{2\pi}\left(\mathcal{K}_1^{-1} - \mathcal{K}_2^{-1}\right)\right]\right\}. \tag{18.204}$$

If the detector elements are small compared to the speckle blob size, then the determinants of integral operators reduce to determinants of matrices as discussed in Sec. 18.3.3 [see (18.157)]. In that case the log-likelihood ratio is a linear discriminant if

$$\det\left\{\mathbf{I} + 2\pi i C\epsilon^2\mathbf{K}_1\left[\boldsymbol{\Xi} + \frac{i}{2\pi}\mathbf{A}\right]\right\} = \det\left\{\mathbf{I} + 2\pi i C\epsilon^2\mathbf{K}_1\left[\boldsymbol{\Xi} + \frac{i}{2\pi C\epsilon^2}\left(\mathbf{K}_1^{-1} - \mathbf{K}_2^{-1}\right)\right]\right\}, \tag{18.205}$$

where $\mathbf{K}_j$ and $\boldsymbol{\Xi}$ were defined in (18.155), and

$$[\mathbf{A}]_{\mathbf{mm}'} = a_{\mathbf{m}}\,\delta_{\mathbf{mm}'}. \tag{18.206}$$

Linear log-likelihoods: Example 1 As a first example of the use of this formalism, suppose that the detectors are small compared to a speckle blob, but sufficiently far apart that the measurements are uncorrelated. In that case we know already that the log-likelihood ratio is given by (18.196), but it is instructive to rederive that result from (18.205).

For uncorrelated measurements, we can write[13]

$$C\epsilon^2[\mathbf{K}_j]_{\mathbf{mm}'} = \overline{g}_{j\mathbf{m}}\,\delta_{\mathbf{mm}'}. \tag{18.207}$$

[13]The reader who was expecting a variance instead of a mean on the right-hand side of (18.207) should recall that $\mathbf{K}_j$ is the covariance of the field, not the covariance of the data. Thus $[\mathbf{K}_j]_{\mathbf{mm}} = \langle|u(\mathbf{r}_{\mathbf{m}})|^2\rangle_j$, and the factor of $C\epsilon^2$ accounts for integration across the detector face and conversion of the result to an output signal.

Thus (18.205) is satisfied if $\mathbf{A} = \left(\mathbf{K}_1^{-1} - \mathbf{K}_2^{-1}\right)/(C\epsilon^2)$, or

$$a_{\mathbf{m}} = \frac{1}{\overline{g}_{1\mathbf{m}}} - \frac{1}{\overline{g}_{2\mathbf{m}}}, \tag{18.208}$$

in agreement with (18.196).

Linear log-likelihoods: Example 2 Next we consider small detectors without assuming they are widely spaced compared to a blob size. In this case the measurements are correlated, and we cannot satisfy (18.205) by taking $\mathbf{A} \propto \mathbf{K}_1^{-1} - \mathbf{K}_2^{-1}$ since the former is diagonal by definition but the latter is not diagonal. We can, however, show that the determinants are equal even though the operators are not if we assume that the difference signal is slowly varying on the spatial scale of the correlations, so we can use the quasistationary approximation of (18.115).

To be specific, suppose the lens has a square aperture of side L_{ap}, so that (18.115) becomes

$$K_j(\mathbf{r}, \mathbf{r}') \approx \ell_c^2 f_j(\mathbf{r})\, \beta^2 \operatorname{sinc}[\beta(\mathbf{r} - \mathbf{r}')] , \tag{18.209}$$

where the sinc function with a vector argument is just a product of two 1D sinc functions, and

$$\beta \equiv \frac{L_{ap}}{2\lambda f} . \tag{18.210}$$

Note that the argument of $f_j(\cdot)$ is written as $\mathbf{r}$ in (18.209) and $\overline{\mathbf{r}}$ in (18.115); this difference is insignificant if $f_j(\mathbf{r})$ is slowly varying on the scale of the lens resolution, as it must be for (18.115) to be valid in the first place.

For detectors that are small compared to the blob size, the argument of the operator $\mathcal{K}_j \mathcal{W}_{\boldsymbol{\xi}}$ becomes

$$[\mathcal{K}_j \mathcal{W}_{\boldsymbol{\xi}}](\mathbf{r}, \mathbf{r}') \approx \ell_c^2 C\epsilon^2 \sum_{\mathbf{m}} \xi_{\mathbf{m}}\, f_j(\mathbf{r}_{\mathbf{m}})\, \beta^2 \operatorname{sinc}[\beta(\mathbf{r} - \mathbf{r}')]\, \delta(\mathbf{r}' - \mathbf{r}_{\mathbf{m}}) . \tag{18.211}$$

Applied to an arbitrary image-plane irradiance $I_{im}(\mathbf{r})$, this operator yields

$$[\mathcal{K}_j \mathcal{W}_{\boldsymbol{\xi}} \mathbf{I}_{im}](\mathbf{r}) = \ell_c^2 C\epsilon^2 \sum_{\mathbf{m}} \xi_{\mathbf{m}}\, f_j(\mathbf{r}_{\mathbf{m}})\, \beta^2 \operatorname{sinc}[\beta(\mathbf{r} - \mathbf{r}_{\mathbf{m}})]\; I_{im}(\mathbf{r}_{\mathbf{m}}) . \tag{18.212}$$

This expression is a linear combination of the sinc functions, so the range of $\mathcal{K}_j \mathcal{W}_{\boldsymbol{\xi}}$ is spanned by the set of functions $\{v_{\mathbf{m}}(\mathbf{r})\}$, where

$$v_{\mathbf{m}}(\mathbf{r}) \equiv \beta \operatorname{sinc}[\beta(\mathbf{r} - \mathbf{r}_{\mathbf{m}})] . \tag{18.213}$$

The orthonormality of this set,

$$(v_{\mathbf{m}}, v_{\mathbf{m}'}) \equiv \int_\infty d^2r\; v_{\mathbf{m}}(\mathbf{r})\, v_{\mathbf{m}'}(\mathbf{r}) = \delta_{\mathbf{m}\mathbf{m}'} , \tag{18.214}$$

can be verified by use of Parseval's theorem. (It helps to assume that $\beta\epsilon$ is an integer, though it suffices to say that $\beta\epsilon$ is large.)

Since $\operatorname{sinc}[\beta(\mathbf{r} - \mathbf{r}_{\mathbf{m}})]$ is reproduced by convolving it with $\beta^2 \operatorname{sinc}(\beta\mathbf{r})$, the range of $\mathcal{K}_j \mathcal{W}_{\boldsymbol{\xi}}$ is a finite-dimensional reproducing-kernel Hilbert space (see Sec. 1.8), which we shall denote by $\mathbb{K}$. The projector onto $\mathbb{K}$ is given by

$$\mathcal{P}_{\mathbb{K}} = \sum_{\mathbf{m}} \mathbf{v}_{\mathbf{m}} \mathbf{v}_{\mathbf{m}}^\dagger , \qquad [\mathcal{P}_{\mathbb{K}}](\mathbf{r}, \mathbf{r}') = \beta^2 \sum_{\mathbf{m}} \operatorname{sinc}[\beta(\mathbf{r} - \mathbf{r}_{\mathbf{m}})] \operatorname{sinc}[\beta(\mathbf{r}' - \mathbf{r}_{\mathbf{m}})] . \tag{18.215}$$

Thus (18.199) becomes

$$\psi_{\mathbf{g}|H_j}(\boldsymbol{\xi}) = \frac{1}{\det(\mathcal{I} + 2\pi i\, \mathcal{P}_{\mathbb{K}}\, \mathcal{K}_j\, \mathcal{W}_{\boldsymbol{\xi}})} = \frac{1}{\det(\mathcal{I} + 2\pi i\, \mathcal{P}_{\mathbb{K}}\, \mathcal{K}_j\, \mathcal{W}_{\boldsymbol{\xi}}\, \mathcal{P}_{\mathbb{K}})}, \tag{18.216}$$

where the last step follows from (A.85) and the fact that $\mathcal{P}_{\mathbb{K}}^2 = \mathcal{P}_{\mathbb{K}}$.

Retracing the steps leading up to (18.204), we see that the log-likelihood ratio is a linear discriminant in the present problem if

$$\mathcal{P}_{\mathbb{K}}\, \mathcal{K}_1 \mathcal{W}_{\mathbf{a}}\, \mathcal{P}_{\mathbb{K}} = \mathcal{P}_{\mathbb{K}}\, \mathcal{K}_1 \left[(\mathcal{P}_{\mathbb{K}}\, \mathcal{K}_1)^{-1} - (\mathcal{P}_{\mathbb{K}}\, \mathcal{K}_2)^{-1} \right] . \tag{18.217}$$

The kernel of the left-hand side of this equation is

$$[\mathcal{P}_{\mathbb{K}}\, \mathcal{K}_1 \mathcal{W}_{\mathbf{a}}\, \mathcal{P}_{\mathbb{K}}]\,(\mathbf{r}, \mathbf{r}') = \ell_c^2 C \epsilon^2 \beta^4 \sum_{\mathbf{m}} a_{\mathbf{m}} f_1(\mathbf{r_m})\, \mathrm{sinc}\,[\beta(\mathbf{r} - \mathbf{r_m})]\, \mathrm{sinc}\,[\beta(\mathbf{r}' - \mathbf{r_m})] \, . \tag{18.218}$$

To find the kernel of the right-hand side, note that

$$\begin{aligned}
&[\mathcal{P}_{\mathbb{K}}\, \mathcal{K}_j]\,(\mathbf{r}, \mathbf{r}') \\
&= \ell_c^2 C \epsilon^2 \beta^4 \sum_{\mathbf{m}} \mathrm{sinc}\,[\beta(\mathbf{r} - \mathbf{r_m})] \int d^2 r''\, \mathrm{sinc}\,[\beta(\mathbf{r}'' - \mathbf{r_m})]\, f_j(\mathbf{r}'')\, \mathrm{sinc}\,[\beta(\mathbf{r}'' - \mathbf{r}')] \\
&\approx \ell_c^2 C \epsilon^2 \beta^2 \sum_{\mathbf{m}} f_j(\mathbf{r_m})\, \mathrm{sinc}\,[\beta(\mathbf{r} - \mathbf{r_m})]\, \mathrm{sinc}\,[\beta(\mathbf{r}' - \mathbf{r_m})] \, .
\end{aligned} \tag{18.219}$$

It follows from the orthonormality of the basis functions that

$$\left[(\mathcal{P}_{\mathbb{K}}\, \mathcal{K}_j)^{-1} \right] (\mathbf{r}, \mathbf{r}') = \frac{\beta^2}{\ell_c^2 C \epsilon^2} \sum_{\mathbf{m}} \frac{1}{f_j(\mathbf{r_m})}\, \mathrm{sinc}\,[\beta(\mathbf{r} - \mathbf{r_m})]\, \mathrm{sinc}\,[\beta(\mathbf{r}' - \mathbf{r_m})] \, , \tag{18.220}$$

with the obvious assumption that $f_j(\mathbf{r_m}) \neq 0$.

With these kernels, (18.217) becomes

$$\sum_{\mathbf{m}} \left[\ell_c^2 C \epsilon^2 a_{\mathbf{m}} f_1(\mathbf{r_m}) - 1 + \frac{f_1(\mathbf{r_m})}{f_2(\mathbf{r_m})} \right] \mathrm{sinc}\,[\beta(\mathbf{r} - \mathbf{r_m})]\, \mathrm{sinc}\,[\beta(\mathbf{r}' - \mathbf{r_m})] = 0 \, . \tag{18.221}$$

Thus

$$a_{\mathbf{m}} = \frac{1}{\ell_c^2 C \epsilon^2} \left[\frac{1}{f_1(\mathbf{r_m})} - \frac{1}{f_2(\mathbf{r_m})} \right] , \tag{18.222}$$

which is equivalent to (18.208) since $\overline{g}_{j\mathbf{m}} = \ell_c^2 C \epsilon^2 f_j(\mathbf{r_m})$ for a slowly varying object.

Thus the log-likelihood ratio is obtained by a simple matched filter, but matched to the difference in the reciprocal objects rather than the difference object itself. If the object has low contrast, so that $f_2(\mathbf{r_m}) - f_1(\mathbf{r_m})$ is small and $f_1(\mathbf{r_m})$ and $f_1(\mathbf{r_m})$ are approximately constant over the signal region, then the log-likelihood is a simple non-prewhitening (NPW) matched filter. Perhaps surprisingly, no prewhitening is needed in spite of the correlations, basically because a sinc correlation corresponds to a flat power spectrum.

For examples of the use of NPW matched filters in SKE/BKE detection problems in speckle, see Smith *et al.* (1983) and Silverstein and O'Donnell (1988).

Gaussian approximations As we have just seen, it can be quite difficult to manipulate the expressions for speckle characteristic functions when the detectors are pointlike and there is no electronic or photon noise. Fortunately, the problems get easier, not harder, when some of these other effects are included, mainly because Gaussian statistics rapidly become applicable. We saw in Sec. 18.3.3 that the statistics of the detector output approach Gaussian when the detector area is much larger than the speckle blob size, essentially because the detector averages over many independent regions. In this same limit, as we saw in Sec. 18.4.1, the photon-counting statistics become Poisson, but Poissons are well approximated by Gaussians if the mean number of counts in an observation time is large. Finally, if we consider additive electronic noise as in Sec. 18.4.2, there is an additional Gaussian PDF to be convolved with the PDF of the detector output, making the overall PDF more nearly Gaussian.

Thus, the main use of the elegant characteristic functions presented here may be nothing more than providing expressions for the covariance matrix in many practical circumstances. Much further work is needed, however, to determine when more detailed statistical knowledge is useful in task performance.

Hotelling observer When the data statistics are Gaussian and the signal is weak enough that the covariance is the same under both hypotheses, we know that the ideal observer is the linear Hotelling observer, and the relevant figure of merit is the Hotelling trace or SNR. As we have seen in earlier chapters, however, we often use the Hotelling observer even when we cannot argue convincingly that the data are Gaussian. In the speckle context, that situation could arise with small detectors or short exposure times, and it almost certainly arises when object variability is considered.

If the BKE statistics are Gaussian but the objects demonstrate complicated, non-Gaussian statistics, then we must have recourse to sampling methods as discussed in Sec. 14.3.2. In applying these methods to speckle problems, the decomposition of the covariance matrix in (18.193) is useful since the Poisson term in that expression is diagonal and will usually be known analytically. Similarly, the second term in (18.193)—the so-called noise-free covariance—can often be analyzed by the methods of this chapter; though it is not diagonal, it will have only short-range correlations. That leaves the object-variability term to be estimated by sampling methods. There is nothing special about applying these methods to speckle problems, and they will not be discussed further here.

Channelized observer models As we know from Sec. 14.3.2, estimation of computation of the performance of the Hotelling observer (or optimal linear discriminant function) purely by sampling methods is difficult if the dimensionality of the data vector is large, as when the input to the discriminant is the output of a large detector array. Similarly, as we discussed in Sec. 14.3.3, computation of the performance of the ideal observer is difficult in high-dimensional problems since huge multivariate probability density functions are required. In both cases, dimensionality reduction or feature extraction is very helpful. Linear features derived from channels are particularly useful since the statistics of the channel outputs are readily related to the statistics of the data or even the underlying object.

In particular, (14.88) gives the transformation law for determining the characteristic function of a channel output vector when the characteristic functional for the objects is known and the system is linear. How this transformation rule is used in speckle problems depends on whether the detector integrates amplitude or irradiance. In the optical case, with irradiance-sensitive detectors, we can transform a characteristic function like (18.216) through the linear channels. With amplitude-sensitive detectors like the ones to be discussed in more detail in Sec. 18.6, we can go from the characteristic functional for the field distribution at the detector face all the way to the channel outputs.

In either case we can use the statistics of the channel outputs to compute the performance of channelized Hotelling or channelized ideal observers. Though this program has not yet been implemented, it appears to be an excellent way of taking advantage of the analytical results of this chapter.

Hybrid estimation-classification tasks As we discussed in Secs. 13.1.1 and 13.3.8, classification tasks are often performed with estimated object parameters as features. In many pattern-recognition problems, we are interested in small-scale features or texture of the object, but it is not obvious how coherent imagery can give us that kind of information. Indeed, as far as our analysis to this point goes, all rough objects are essentially equivalent. They all fully randomize the phase of the transmitted or reflected wave, and they all give rise to circular Gaussian statistics for the field.

Nevertheless, there is a huge literature on extracting texture information from coherent imagery, especially in the context of medical ultrasound or synthetic-aperture radar. The key to that endeavor is to recognize the limitations of the circular-Gaussian models. As we shall see in Sec. 18.5, many interesting new statistical distributions arise when we do not assume that the number of independent scattering elements within the area of the coherent PRF is large.

18.5 POINT-SCATTERING MODELS AND NON-GAUSSIAN SPECKLE

The development in this chapter so far has been based on the assumption that a ground glass completely randomizes the phase of a light wave and that the autocovariance function of the ground-glass transmittance is sharply peaked and can be approximated as a delta function. In other words, we have regarded the transmittance as a white-noise random process, and we haven't been very specific about the point statistics of that process. We were able to avoid being more specific about the ground glass since we invoked the central-limit theorem to argue that the field in an observation plane was always circular Gaussian. As we noted, however, that argument is suspect in an imaging system if the area of the PSF encompasses only a few correlation lengths of the ground glass.

In this section we shall adopt a different view of the object, assuming that it consists of a set of discrete point scatterers at random locations and possibly with random scattering amplitudes. A natural mathematical description of the object is thus a random point process, a topic we discussed at considerable length in Chap. 11. If we assume that the scattering points are statistically independent, we are led to describe the object as a Poisson point process, though necessarily a nonstationary one in interesting imaging situations. As in Chap. 11, however, there is also considerable interest in doubly stochastic point processes where the density of points is itself a random process.

Point-scattering models may be more applicable to reflection imaging than to transmission through a ground glass. Consider, for example, airborne imaging of a field of grass with a laser illuminator or microwave radar. Most of the radiation emitted from the aircraft will be forward scattered and eventually absorbed in the grass. Occasionally, however, a blade of grass will be oriented approximately normal to the incident direction and reflect the radiation back towards the imaging system in the aircraft. The object being imaged thus consists effectively of bright points or *glints*, and the amplitude and phase of each glint depend on the random position and orientation of the blade of grass.

Point-scattering models are also frequently used in medical ultrasound of soft organs. Red blood cells and liver lobules, for example, might be modeled as scattering points if they are small compared to the ultrasound wavelength. Phantoms consisting of small, dense points in gelatinous media can produce ultrasound speckle patterns that mimic well the speckle seen in actual tissue.

Our goal in this section is to recapitulate Sec. 18.3 but with point-scattering models in place of the delta-correlated ground glass assumed there. The result will be that the speckle field is not circular Gaussian, so we refer to the subject of this section as non-Gaussian speckle. Another common term is non-Rayleigh speckle, which refers to the fact that the square root of the irradiance does not follow a Rayleigh law, so the irradiance itself is not exponentially distributed.

Sec. 18.5.1 provides a general point-process description of the object field, states some statistical assumptions and derives the characteristic functional for the field. In Sec. 18.5.2 we propagate this object field to an image plane and determine its characteristic functional, but without the Gaussian approximation used in Sec. 18.3.2. Sec. 18.5.3 specializes the infinite-dimensional statistical description of the field in the previous section to univariate statistics of the field and irradiance at a single point.

18.5.1 Object fields and objects

All treatments of point-scattering models in the speckle literature express the object field as

$$u_{obj}(\mathbf{r}) = \sum_{n=1}^{N} a_n \exp(i\phi_n)\, \delta(\mathbf{r} - \mathbf{r}_n)\,, \tag{18.223}$$

but different physical situations require different mechanistic justifications of the model and different interpretation of the parameters. Key to all interpretations, however, is the realization that the delta function does not represent literally an object of infinitesimal width but merely something small compared to the resolution of the imaging system.

Consider, for example, a reflecting surface consisting of small flats or facets as shown in Fig. 18.4. For simplicity, we assume that all facets are parallel to the plane $z = 0$ and that the nth facet is a distance h_n from that reference plane. If this surface is illuminated by a unit-amplitude plane wave at normal incidence, then the wave propagates a distance $2h_n$ from $z = 0$ to the facet and back to $z = 0$. The phase shift relative to the phase of the incident wave at $z = 0$ is thus

$$\phi_n = \frac{2\pi}{\lambda} 2h_n\,, \tag{18.224}$$

where λ is the free-space wavelength of the radiation. Comparing this result to (18.7) we see that the phase shift in reflection is several times larger than in transmission for the same surface relief; in reflection the light travels twice as far, and in transmission the phase shift is reduced by a factor of $n_g - 1$ (with n_g being the refractive index of the glass, a number around 1.5) since phase is measured relative to what it would be in the absence of the ground glass.

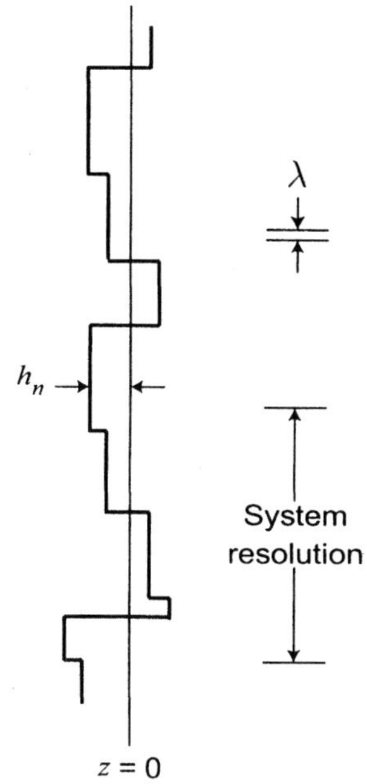

Fig. 18.4 Facet model for a reflecting surface.

If each facet is small compared to the resolution of the imaging system, then the field reflected from the facet can be treated as a spatial delta function. The amplitude a_n can then be interpreted as the amplitude reflectivity of the facet times its area. If the facets are at different angles, then a_n can also include the fraction of the light reflected towards the detector.

Another common situation where the point-scattering model of (18.223) is invoked is in scattering from aerosol particles in the atmosphere. In this case the phase shift arises, at least in part, from the random position of the aerosol particle in the illuminating field. We can write

$$u_{obj}(\mathbf{r}) = u_{inc}(\mathbf{r}) \sum_{n=1}^{N} a_n \, \delta(\mathbf{r} - \mathbf{r}_n) = \sum_{n=1}^{N} a_n \exp[i\Phi(\mathbf{r}_n)] \, \delta(\mathbf{r} - \mathbf{r}_n) \, , \qquad (18.225)$$

where we have assumed that the incident wave is described by the pure-phase function $u_{inc}(\mathbf{r}) = \exp[i\Phi(\mathbf{r})]$ and used a property of delta functions, (2.119). Thus the phase ϕ_n in this view is simply the phase of the illumination at the scattering point.

Defining the object Whatever the interpretation of the parameters in the object field, we must still specify just what we mean by "the object." One approach would be to identify $f(\mathbf{r})$ as the mean number of scattering points per unit area, which we denoted in Chap. 11 as $b(\mathbf{r})$. From (11.83), we know that $b(\mathbf{r})$ is the mean of the random sum of delta functions with unit weights:

$$b(\mathbf{r}) = \left\langle \sum_{n=1}^{N} \delta(\mathbf{r}-\mathbf{r}_n) \right\rangle . \tag{18.226}$$

In the present problem with complex weights, even if we define the object by $f(\mathbf{r}) = b(\mathbf{r})$, we cannot say that it is the mean of the object field; in fact, $\langle u_{obj}(\mathbf{r})\rangle = 0$ if the phases are uniformly distributed on $(-\pi, \pi)$.

An alternative is to associate the object with the squared modulus of the scattering amplitude. Then the object field would take the form,

$$u_{obj}(\mathbf{r}) = \sum_{n=1}^{N} \sqrt{f(\mathbf{r}_n)}\exp(i\phi_n)\,\delta(\mathbf{r}-\mathbf{r}_n) = \sqrt{f(\mathbf{r})}\sum_{n=1}^{N} \exp(i\phi_n)\,\delta(\mathbf{r}-\mathbf{r}_n) . \tag{18.227}$$

This definition is appealing since it expresses the object field as something associated with the object of interest times a random point process that we want to regard as noise; exactly this motivation led us in Sec. 18.3 to think of the object as a transparency placed over a ground glass. A disadvantage of (18.227), however, is that the object is independent of the density of scatterers, which might be of considerable interest.

We can combine the ideas behind (18.226) and (18.227) by defining the object as the mean density of scattering points times the average of the square of the scattering amplitude: $f(\mathbf{r}) \equiv b(\mathbf{r})\,\langle a_n^2\rangle$, where the expectation is defined with respect to a spatially varying PDF so it is a function of $\mathbf{r}$. As we shall see in Sec. 18.5.2, this quantity arises naturally when we study the statistics of the image field.

A more general way of modeling an object as a random point process is to give up on the scalar description and say that more than one characteristic of the object at each point is of interest, in which case the object is a vector-valued function (see Sec. 7.1.1). We could, for example, consider both the density of scattering points and the mean scattering amplitude as object characteristics of interest and define the components of an object vector by $f_1(\mathbf{r}) \equiv b(\mathbf{r})$ and $f_2(\mathbf{r}) \equiv \langle a_n\rangle$.

We shall not pursue the idea of a vector-valued object further, but instead focus on the scalar object field in what follows.

Statistical assumptions Since the object field is a complex random process, its characteristic functional is defined as in (18.19) by

$$\Psi_{\mathbf{u}_{obj}}(\boldsymbol{\xi}) \equiv \left\langle \exp\left[-i\pi\left(\boldsymbol{\xi}^\dagger \mathbf{u}_{obj} + \mathbf{u}_{obj}{}^\dagger \boldsymbol{\xi}\right)\right]\right\rangle , \tag{18.228}$$

where the average in its most general form is over $\{\mathbf{r}_n\}$, $\{a_n\}$, $\{\phi_n\}$, N itself, and the density $b(\mathbf{r})$. Once we have this characteristic functional, we can get the one for the image-plane field by use of (18.96), so our immediate task is to specify probability laws for the random variables and carry out the average in (18.228).

As usual in dealing with point processes, we assume that the points are statistically independent and identically distributed. If we assume further that the parameters of an individual point, $\mathbf{r}_n$, a_n and ϕ_n are independent of each other,[14]

[14]The assumption that ϕ_n is independent of $\mathbf{r}_n$ does not hold if the phase is determined simply by the position of the scattering point in the illumination field. In that case it might be valid to

the conditional density on these variables for a fixed N becomes

$$\mathrm{pr}\left[\{\mathbf{r}_n\}, \{a_n\}, \{\phi_n\}|N\right] = \prod_{n=1}^{N} \mathrm{pr}(\mathbf{r}_n)\, \mathrm{pr}(a_n)\, \mathrm{pr}(\phi_n)\,. \tag{18.229}$$

The joint density on all variables, including N, is thus

$$\mathrm{pr}\left[\{\mathbf{r}_n\}, \{a_n\}, \{\phi_n\}, N\right]$$

$$= \mathrm{pr}\left[\{\mathbf{r}_n\}, \{a_n\}, \{\phi_n\}|N\right] \mathrm{Pr}(N) = \mathrm{Pr}(N) \prod_{n=1}^{N} \mathrm{pr}(\mathbf{r}_n)\, \mathrm{pr}(a_n)\, \mathrm{pr}(\phi_n)\,. \tag{18.230}$$

To specify the problem fully, we need to give the component PDFs in this product. From (11.84), we know that $b(\mathbf{r})$, with proper normalization, is the PDF for the locations of the scattering points, so

$$\mathrm{pr}(\mathbf{r}_n) = \frac{b(\mathbf{r}_n)}{\int_{\mathbf{S}} d^2r'\, b(\mathbf{r}')} = \frac{1}{\overline{N}}\, b(\mathbf{r}_n)\,, \tag{18.231}$$

where $\mathbf{S}$ is the support of the object.

If we assume that the phases are fully randomized on $(-\pi, \pi)$, we can write

$$\mathrm{pr}(\phi_n) = \frac{1}{2\pi}\, \mathrm{rect}\left(\frac{\phi_n}{2\pi}\right)\,. \tag{18.232}$$

Various forms might be used for $\mathrm{pr}(a_n)$, but since a_n is the magnitude of a complex scattering amplitude, we must choose one that is defined for $0 \le a_n < \infty$; log-normal, gamma and K-Bessel densities are possibilities. If the object is a collection of identical particles with scattering cross section σ, we would take $\mathrm{pr}(a_n) = \delta(a_n - \sqrt{\sigma})$.

A natural choice for $\mathrm{Pr}(N)$ is a Poisson of mean $\overline{N}$, but for more generality we can treat $\overline{N}$ as random and write $\mathrm{Pr}(N)$ in the form of a Poisson transform as defined in (11.25):

$$\mathrm{Pr}(N) = \int_0^\infty d\overline{N}\, \mathrm{Pr}(N|\overline{N})\, \mathrm{pr}(\overline{N}) = \frac{1}{N!}\int_0^\infty d\overline{N}\, \exp(-\overline{N})\, \overline{N}^N\, \mathrm{pr}(\overline{N})\,. \tag{18.233}$$

Like $\mathrm{pr}(a_n)$, $\mathrm{pr}(\overline{N})$ must be defined on $(0, \infty)$. The Poisson is recovered by taking $\mathrm{pr}(\overline{N})$ as a delta function.

Characteristic functional for the object field Having assembled the needed probability laws, we turn next to computation of the characteristic functional for the object field. The development is similar to that in Sec. 11.3.10 but with the added complication of random amplitudes and phases.

In full generality, (18.228) can be written in the nested form,

$$\Psi_{\mathbf{u}_{obj}}(\boldsymbol{\xi}) \equiv \left\langle \left\langle \left\langle \exp\left[-i\pi\left(\boldsymbol{\xi}^\dagger \mathbf{u}_{obj} + \mathbf{u}_{obj}{}^\dagger \boldsymbol{\xi}\right)\right] \right\rangle_{\{a_n\},\{\phi_n\},\{\mathbf{r}_n\}|N} \right\rangle_{N|\overline{N}} \right\rangle_{\overline{N}}. \tag{18.234}$$

assume that the actual position of the point was $\mathbf{r}_n + \Delta\mathbf{r}_n$, where $\Delta\mathbf{r}_n$ was several wavelengths but nevertheless small compared to the system resolution. Then the independence of ϕ_n and $\mathbf{r}_n$ would be a reasonable model.

By use of the delta function in (18.223), we can rewrite the quantity being averaged as

$$\exp\left[-i\pi\left(\boldsymbol{\xi}^\dagger \mathbf{u}_{obj} + \mathbf{u}_{obj}{}^\dagger \boldsymbol{\xi}\right)\right] = \prod_{n=1}^{N} \exp\left\{-i\pi a_n \left[\exp(i\phi_n)\, \xi^*(\mathbf{r}_n) + \exp(-i\phi_n)\, \xi(\mathbf{r}_n)\right]\right\}$$

$$= \prod_{n=1}^{N} \exp\left\{-2\pi i a_n |\xi(\mathbf{r}_n)| \cos\left[\phi_n - \phi_\xi(\mathbf{r_n})\right]\right\} , \qquad (18.235)$$

where $\xi(\mathbf{r}_n) = |\xi(\mathbf{r}_n)| \exp[i\phi_\xi(\mathbf{r}_n)]$.

Since we are assuming that the point parameters are independent and identically distributed, the innermost expectation in (18.234) becomes

$$\left\langle \exp\left[-i\pi\left(\boldsymbol{\xi}^\dagger \mathbf{u}_{obj} + \mathbf{u}_{obj}{}^\dagger \boldsymbol{\xi}\right)\right] \right\rangle_{\{a_n\},\{\phi_n\},\{\mathbf{r}_n\}|N}$$

$$= \left[\left\langle \exp\left\{-2\pi i a_n |\xi(\mathbf{r}_n)| \cos\left[\phi_n - \phi_\xi(\mathbf{r_n})\right]\right\} \right\rangle_{a_n,\phi_n,\mathbf{r}_n}\right]^N . \qquad (18.236)$$

This step does not imply that the process is stationary in any sense, just that the individual scatterers are indistinguishable; in other words, $\mathrm{pr}(\mathbf{r}_n)$ is not necessarily constant over the support.

With (18.232) and a change of variables, the average over ϕ_n yields

$$\frac{1}{2\pi}\int_{-\pi}^{\pi} d\phi\, \exp(-2\pi i a_n |\xi(\mathbf{r}_n)| \cos\phi) = \mathrm{J}_0\left(2\pi a_n |\xi(\mathbf{r}_n)|\right) , \qquad (18.237)$$

where $\mathrm{J}_0(\,\cdot\,)$ is the zeroth-order Bessel function of the first kind. Note that the phase of $\xi(\mathbf{r}_n)$ has disappeared.

With (18.231) as the PDF on $\mathbf{r}_n$ and a generic PDF for a_n, we can now write

$$\left\langle \exp\left\{-2\pi i a_n |\xi(\mathbf{r}_n)| \cos\left[\phi_n - \phi_\xi(\mathbf{r_n})\right]\right\} \right\rangle_{a_n,\phi_n,\mathbf{r}_n}$$

$$= \frac{1}{\overline{\overline{N}}}\int_0^\infty da_n\, \mathrm{pr}(a_n) \int_{\mathbf{S}} d^2r_n\, b(\mathbf{r}_n)\, \mathrm{J}_0\left(2\pi a_n |\xi(\mathbf{r}_n)|\right) . \qquad (18.238)$$

Combining (18.234), (18.236) and (18.238), we obtain

$$\Psi_{\mathbf{u}_{obj}}(\boldsymbol{\xi}) = \left\langle \left\langle \left[\frac{1}{\overline{\overline{N}}}\int_0^\infty da_n\, \mathrm{pr}(a_n) \int_{\mathbf{S}} d^2r_n\, b(\mathbf{r}_n)\, \mathrm{J}_0\left(2\pi a_n |\xi(\mathbf{r}_n)|\right)\right]^N \right\rangle_{N|\overline{N}} \right\rangle_{\overline{N}} . \qquad (18.239)$$

The expectation over N for fixed $\overline{N}$ can be performed as in Sec. 11.3.10, with the result

$$\left\langle \left\{\frac{1}{\overline{\overline{N}}}\int_0^\infty da_n\, \mathrm{pr}(a_n) \int_{\mathbf{S}} d^2r_n\, b(\mathbf{r}_n)\, \mathrm{J}_0[2\pi a_n |\xi(\mathbf{r}_n)|]\right\}^N \right\rangle_{N|\overline{N}}$$

$$= \exp\left\{-\overline{N} + \int_0^\infty da_n\, \mathrm{pr}(a_n) \int_{\mathbf{S}} d^2r_n\, b(\mathbf{r}_n)\, \mathrm{J}_0[2\pi a_n |\xi(\mathbf{r}_n)|]\right\} . \qquad (18.240)$$

Since $\overline{N}$ is determined by the integral of the density of scatterers [see (18.231)], the remaining expectation is really over realizations of $b(\mathbf{r})$, so we can write

$$\Psi_{\mathbf{u}_{obj}}(\boldsymbol{\xi}) = \left\langle \exp\left\{ -\int_{\mathbf{S}} d^2r\, b(\mathbf{r}) + \int_0^\infty da_n\, \mathrm{pr}(a_n) \int_{\mathbf{S}} d^2r_n\, b(\mathbf{r}_n)\, \mathrm{J}_0[2\pi a_n |\xi(\mathbf{r}_n)|] \right\} \right\rangle_{\mathbf{b}} . \tag{18.241}$$

Sometimes we want to regard $b(\mathbf{r})$ as nonrandom; for example, if we define the object by $f(\mathbf{r}) = b(\mathbf{r})$ and consider just a single object, the average over $\mathbf{b}$ in (18.241) is not needed. In those cases, we shall use a conditional characteristic functional given by

$$\Psi_{\mathbf{u}_{obj}|\mathbf{b}}(\boldsymbol{\xi}) = \exp\left\{ -\int_{\mathbf{S}} d^2r\, b(\mathbf{r}) + \int_0^\infty da_n\, \mathrm{pr}(a_n) \int_{\mathbf{S}} d^2r_n\, b(\mathbf{r}_n)\, \mathrm{J}_0[2\pi a_n |\xi(\mathbf{r}_n)|] \right\}$$

$$= \exp\left\{ -\overline{N} + \int_{\mathbf{S}} d^2r_n\, b(\mathbf{r}_n) \langle \mathrm{J}_0[2\pi a_n |\xi(\mathbf{r}_n)|] \rangle_{a_n} \right\} . \tag{18.242}$$

This form should be compared to the characteristic functional for a Poisson point process with constant weights, given in (11.150) and rewritten here as

$$\Psi_{\mathbf{g}}(\mathbf{s}) = \exp\left\{ -\overline{N} + \int_{\mathbf{S}} d^2r_n\, b(\mathbf{r}_n) \exp[-2\pi i\, s(\mathbf{r}_n)] \right\} . \tag{18.243}$$

The key differences are the additional average over a_n in (18.242), the appearance there of the Bessel function rather than the complex exponential, and the use of a complex argument $\xi(\mathbf{r})$ in place of the real function $s(\mathbf{r})$ in (18.243). Curiously, (18.242) is pure real even though $\xi(\mathbf{r})$ is complex, and (18.243) is complex even though $s(\mathbf{r})$ is real.

18.5.2 Image fields

Since coherent imaging is a linear transformation of the object field to the image field, the characteristic functional for the latter can be found from a transformation rule like (18.96). For an arbitrary coherent imaging system with space-variant PSF $p_{coh}(\mathbf{r} - \mathbf{r}'; \mathbf{r}')$, we can write

$$u_{im}(\mathbf{r}) = \int_S d^2r'\, p_{coh}(\mathbf{r}-\mathbf{r}'; \mathbf{r}')\, u_{obj}(\mathbf{r}') = \sum_n a_n \exp(i\phi_n)\, p_{coh}(\mathbf{r}-\mathbf{r}_n; \mathbf{r}_n) . \tag{18.244}$$

In terms of the coherent imaging operator $\boldsymbol{\mathcal{P}}_{coh}$, we have

$$\mathbf{u}_{im} = \boldsymbol{\mathcal{P}}_{coh}\mathbf{u}_{obj} \qquad \text{or} \qquad u_{im}(\mathbf{r}) = [\boldsymbol{\mathcal{P}}_{coh}\mathbf{u}_{obj}](\mathbf{r}) . \tag{18.245}$$

With this operator and (18.96), we see that

$$\Psi_{\mathbf{u}_{im}}(\boldsymbol{\xi}) = \Psi_{\mathbf{u}_{obj}}(\boldsymbol{\mathcal{P}}^\dagger_{coh}\boldsymbol{\xi}) . \tag{18.246}$$

Explicitly, with (18.241),

$$\Psi_{\mathbf{u}_{im}}(\boldsymbol{\xi}) = \left\langle \exp\left\{ -\overline{N} + \int_0^\infty da_n\, \mathrm{pr}(a_n) \int_{\mathbf{S}} d^2r_n\, b(\mathbf{r}_n)\, \mathrm{J}_0\Big[2\pi a_n \big|[\boldsymbol{\mathcal{P}}^\dagger_{coh}\boldsymbol{\xi}](\mathbf{r}_n)\big|\Big] \right\} \right\rangle_{\mathbf{b}} . \tag{18.247}$$

This result is our most general statement of the statistics of the image-plane field. It is a modest extension of an expression obtained by Zardecki and Delisle (1977), who considered only free-space propagation and not an imaging system. In what follows we shall consider some additional approximations and special cases of this result.

The Gaussian limit In Sec. 18.2.4 we went through a lengthy derivation of the central-limit theorem for a coherently illuminated ground glass. The result was that the image-plane field approached a circular Gaussian random process if the ground glass had many independent regions within the area defined by the coherent PSF. A similar derivation for the present object model would show that the image field was circular Gaussian if the number of scattering points within the PSF area was large. Rather than presenting that derivation, however, we give here a heuristic argument which will suffice to determine the form of the circular Gaussian.

The starting point for the heuristic argument is that $\Psi_{\mathbf{u}_{im}}(\boldsymbol{\xi})$, like all characteristic functionals, must be unity when $\xi(\mathbf{r}) = 0$ for all $\mathbf{r}$. To see this point for (18.247), note that the Bessel function is identically one if $\xi(\mathbf{r}) = 0$ and hence $[\boldsymbol{\mathcal{P}}_{coh}\boldsymbol{\xi}](\mathbf{r}) = 0$. Then the integral over a_n yields unity and the one over $\mathbf{r}_n$ yields $\overline{N}$, neatly cancelling the other $\overline{N}$ in the exponent. If there are some regions for which $[\boldsymbol{\mathcal{P}}_{coh}\boldsymbol{\xi}](\mathbf{r}) \neq 0$, however, the cancellation is not exact, and in that case the exponential rapidly approaches zero as $\overline{N}$ gets large. Thus, as in our previous treatments of the central-limit theorem (see Secs. 8.3.4 and 18.2.4), the basic calculation is to expand the exponent in a power series and retain terms linear and quadratic in the frequency variable.

The requisite expansion for the present problem is

$$\mathrm{J}_0(z) = \sum_{k=0}^{\infty} \frac{\left(-\frac{1}{4}z^2\right)^k}{(k!)^2} = 1 - \tfrac{1}{4}z^2 + \cdots. \tag{18.248}$$

If we truncate this expansion, then the conditional characteristic functional of the image field for fixed $\mathbf{b}$, *i.e.*, (18.247) without the final average, becomes

$$\begin{aligned}\Psi_{\mathbf{u}_{im}|\mathbf{b}}(\boldsymbol{\xi}) &= \exp\left\{-\overline{N} + \int_0^{\infty} da_n \ \mathrm{pr}(a_n) \int_{\mathbf{S}} d^2 r_n \ b(\mathbf{r}_n)\left[1 - \pi^2 a_n^2 \left|[\boldsymbol{\mathcal{P}}_{coh}^{\dagger}\boldsymbol{\xi}](\mathbf{r}_n)\right|^2\right]\right\} \\ &= \exp\left\{-\pi^2 \langle a_n^2\rangle \int_{\mathbf{S}} d^2 r_n \ b(\mathbf{r}_n)\left|[\boldsymbol{\mathcal{P}}_{coh}^{\dagger}\boldsymbol{\xi}](\mathbf{r}_n)\right|^2\right\},\end{aligned} \tag{18.249}$$

where in the last step we have used the fact that $b(\mathbf{r}_n)$ is the mean number of points per unit area, so its integral over the whole field is just $\overline{N}$. The integral in the exponent can be written in detail as

$$\begin{aligned}&\int_{\mathbf{S}} d^2 r_n \ b(\mathbf{r}_n)\left|[\boldsymbol{\mathcal{P}}_{coh}^{\dagger}\boldsymbol{\xi}](\mathbf{r}_n)\right|^2 \\ &= \int_{\mathbf{S}} d^2 r \ \xi^*(\mathbf{r}) \int_{\mathbf{S}} d^2 r' \ \xi(\mathbf{r}') \int_{\mathbf{S}} d^2 r_n \ b(\mathbf{r}_n)\, p_{coh}(\mathbf{r}-\mathbf{r}_n;\mathbf{r}_n)\, p_{coh}^*(\mathbf{r}'-\mathbf{r}_n;\mathbf{r}_n) \,.\end{aligned} \tag{18.250}$$

Thus (18.249) becomes [*cf.* (18.97)]

$$\Psi_{\mathbf{u}_{im}|\mathbf{b}}(\boldsymbol{\xi}) = \exp\left(-\pi^2 \boldsymbol{\xi}^{\dagger} \boldsymbol{\mathcal{K}}_{\mathbf{u}_{im}} \boldsymbol{\xi}\right), \tag{18.251}$$

where $\mathcal{K}_{\mathbf{u}_{im}}$ is the autocovariance operator for the image field, with a kernel given by

$$K_{\mathbf{u}_{im}}(\mathbf{r},\mathbf{r}') = \langle a_n^2 \rangle \int_{\mathbf{S}} d^2 r_n \; b(\mathbf{r}_n)\, p_{coh}(\mathbf{r}-\mathbf{r}_n;\mathbf{r}_n)\, p_{coh}^*(\mathbf{r}'-\mathbf{r}_n;\mathbf{r}_n)\,. \qquad (18.252)$$

This result is identical with (18.100) if we replace $\ell_c^2\,|t_{obj}(\mathbf{r})|^2$ with $\langle a_n^2\rangle\, b(\mathbf{r})$. Thus a natural quantity to call an object is the mean density of scattering points times the average of the square of the scattering amplitude.

For a more formal discussion of the Gaussian limit, see Zardecki and Delisle (1977); they in turn cite Lord Rayleigh (1919).

Gaussian mixtures To get the full characteristic functional for the field in the limit of a large number of scatterers, we must still average (18.252) over realizations of the density $\mathbf{b}$. Since $\mathbf{u}_{im}$ has zero mean for each $\mathbf{b}$, we are averaging zero-mean Gaussian characteristic functionals with respect to variations in their covariance matrices. We encountered this situation in Sec. 8.4.3 where we argued that the output of high-pass or band-pass filters applied to many kinds of images could be viewed as mixtures of zero-mean Gaussians, and we showed that the resulting PDFs had long tails and cusps at the origin. A similar behavior can be expected for the field PDFs in speckle problems without the need for overt filtration since the speckle field naturally has zero mean.

In summary, if the density of scatters, $b(\mathbf{r})$, is nonrandom and large so that $\overline{N}_0 \to \infty$, then a circular Gaussian is expected for the field regardless of the statistics of a_n. If $b(\mathbf{r})$ and hence $\overline{N}_0$ are themselves random, however, then non-Gaussian fields can be expected. Moreover, as cautioned by Jakeman and Pusey (1976), large variations in a_n can lead to non-Gaussian behavior if $\overline{N}_0$ is large but not approaching infinity since then only a small fraction of the scatterers may contribute to the image field.

18.5.3 Univariate statistics of the image field and irradiance

If we evaluate the random process $u_{im}(\mathbf{r})$ at a particular point $\mathbf{r} = \mathbf{R}_0$, the resulting complex number is a random variable described by the univariate characteristic function $\psi_{u_{im}(\mathbf{R}_0)}(\nu)$ rather than the infinite-dimensional characteristic functional $\Psi_{\mathbf{u}_{im}}(\boldsymbol{\xi})$. We know from the discussion in Sec. 18.2.5, however, that the univariate characteristic function for $u_{im}(\mathbf{R}_0)$ is obtained from the characteristic functional by setting $\xi(\mathbf{r}) = \nu\,\delta(\mathbf{r}-\mathbf{R}_0)$ [*cf.* (18.75)]. Making this substitution in (18.247) and performing an integral with the delta function, we obtain

$$\psi_{u_{im}(\mathbf{R}_0)}(\nu)$$

$$= \left\langle \exp\left\{ -\overline{N} + \int_0^\infty da_n \; \mathrm{pr}(a_n) \int_{\mathbf{S}} d^2 r_n \; b(\mathbf{r}_n)\, \mathrm{J}_0[2\pi a_n |\nu\, p_{coh}(\mathbf{R}_0-\mathbf{r}_n;\mathbf{r}_n)|] \right\} \right\rangle_{\mathbf{b}} . \qquad (18.253)$$

We can get all moments of the field at point $\mathbf{R}_0$ from this equation, and we can also get all moments of the irradiance since $\langle [I_{im}(\mathbf{R}_0)]^n \rangle = \langle [u_{im}(\mathbf{R}_0) u_{im}^*(\mathbf{R}_0)]^n \rangle$. For ways of handling the complicated Fréchet derivatives involved in the moment calculations, see the appendix in Zardecki and Delisle (1977).

Micro-area approximation It is difficult to derive analytical results from the general expression (18.253), or even to gain qualitative insights, without further approximation. One common approach is the *micro-area approximation* suggested by Jakeman and Pusey in the 1970s. (For a review, see Jakeman and Tough, 1988.) Zardecki and Delisle (1977) go so far as to say that this model "appears to be the only feasible way of dealing with higher-order statistics."

The essence of the micro-area approximation is to ignore the detailed structure of the coherent PSF and replace $|p_{coh}(\mathbf{R}_0 - \mathbf{r}_n; \mathbf{r}_n)|$ in (18.253) by a uniform disc or cylinder function. The main statistical effect is then fluctuations of the number of point scatterers in the disc area.

To be more precise, we approximate the PSF as

$$|p_{coh}(\mathbf{R}_0 - \mathbf{r}_n; \mathbf{r}_n)| \approx |p_{coh}(\mathbf{0}; \mathbf{R}_0)| \operatorname{cyl}\left[\frac{|\mathbf{R}_0 - \mathbf{r}_n|}{D(\mathbf{R}_0)}\right], \tag{18.254}$$

where the cylinder function is defined in (3.257), and the diameter of the disc is determined such that

$$\int_{\mathbf{S}} d^2 r_n \; |p_{coh}(\mathbf{R}_0 - \mathbf{r}_n; \mathbf{r}_n)| \operatorname{cyl}\left[\frac{|\mathbf{R}_0 - \mathbf{r}_n|}{D(\mathbf{R}_0)}\right] = |p_{coh}(\mathbf{0}; \mathbf{R}_0)| \frac{\pi D^2(\mathbf{R}_0)}{4}. \tag{18.255}$$

Using this approximation and noting that $\mathrm{J}_0(0) = 1$, we can write the integral over $\mathbf{r}_n$ in (18.253) as

$$\int_{\mathbf{S}} d^2 r_n \; b(\mathbf{r}_n)\, \mathrm{J}_0[2\pi a_n |\nu\, p_{coh}(\mathbf{R}_0 - \mathbf{r}_n; \mathbf{r}_n)|]$$

$$\approx \int_{\mathbf{S}} d^2 r_n \; b(\mathbf{r}_n) + \{\mathrm{J}_0[2\pi a_n |\nu\, p_{coh}(\mathbf{0}; \mathbf{R}_0)|] - 1\} \int_{\mathbf{S}} d^2 r_n \; b(\mathbf{r}_n) \operatorname{cyl}\left[\frac{|\mathbf{R}_0 - \mathbf{r}_n|}{D(\mathbf{R}_0)}\right]$$

$$= \overline{N} + \overline{N_0}\{\mathrm{J}_0\,[2\pi a_n |\nu\, p_{coh}(\mathbf{0}; \mathbf{R}_0)|\,] - 1\}, \tag{18.256}$$

where

$$\overline{N} \equiv \int_{\mathbf{S}} d^2 r_n \; b(\mathbf{r}_n), \qquad \overline{N_0} \equiv \int_{\mathbf{S}} d^2 r_n \; b(\mathbf{r}_n) \operatorname{cyl}\left[\frac{|\mathbf{R}_0 - \mathbf{r}_n|}{D(\mathbf{R}_0)}\right]. \tag{18.257}$$

Thus $\overline{N}$ is the mean number of scatterers in the entire object support, and $\overline{N_0}$ is the mean number in the disc centered on point $\mathbf{R}_0$. Though the notation does not show it, $\overline{N_0}$ can be a function of $\mathbf{R}_0$ in general.

With (18.256), the univariate characteristic function of (18.253) becomes

$$\psi_{u_{im}(\mathbf{R}_0)}(\nu) = \left\langle \exp\left\{ -\overline{N_0} + \overline{N_0} \int_0^{\infty} da_n \; \mathrm{pr}(a_n)\, \mathrm{J}_0[2\pi a_n |\nu\, p_{coh}(\mathbf{0}; \mathbf{R}_0)|] \right\} \right\rangle_{\mathbf{b}}$$

$$= \left\langle \exp\left\{ -\overline{N_0} + \overline{N_0}\, \overline{\mathrm{J}_0[2\pi a_n |\nu\, p_{coh}(\mathbf{0}; \mathbf{R}_0)|]} \right\} \right\rangle_{\overline{N_0}}, \tag{18.258}$$

where the overbar on the Bessel function implies an average over the only remaining random variable in its argument, the scattering amplitude a_n. Note that the total number of scatterers, $\overline{N}$, no longer appears; it cannot since scatterers outside the disc defined by the PSF make no contribution to the field at the observation point $\mathbf{R}_0$. Moreover, the average over $\mathbf{b}$ is now equivalent to an average over $\overline{N_0}$, and the notation has been changed accordingly on the last line of (18.258).

The micro-area approximation can also be applied to multi-point statistics. To get the joint characteristic function for the field at K points, $\{\mathbf{r} = \mathbf{R}_k, k = 1, ..., K\}$, we set

$$\xi(\mathbf{r}) = \sum_{k=1}^{K} \nu_k \, \delta(\mathbf{r} - \mathbf{R}_k) \tag{18.259}$$

and then apply the micro-area approximation to each of the K disc areas. If the discs do not overlap, the scatterers in the individual micro-areas are statistically independent, and the joint characteristic function is a product of factors like (18.258). If they do overlap, we must take account of a sum within the argument of the Bessel function. For example, for $K = 2$ [*cf.* (18.253)],

$$\psi_{u_{im}(\mathbf{R}_1),u_{im}(\mathbf{R}_2)}(\nu_1, \nu_2)$$

$$= \left\langle \exp\left\{ -\overline{N} + \int_{\mathbf{S}} d^2 r_n \; b(\mathbf{r}_n) \, \mathrm{J}_0 \left[2\pi a_n \left| \sum_{k=1}^{2} \nu_k \, p_{coh}(\mathbf{0}; \mathbf{R}_k) \, \mathrm{cyl}\left(\frac{|\mathbf{R}_k - \mathbf{r}_n|}{D(\mathbf{R}_k)} \right) \right| \right] \right\} \right\rangle_{\mathbf{b}} . \tag{18.260}$$

There is no difficulty in principle in computing moments from this expression, but the required derivatives are tedious.

PDF of the irradiance The image-plane irradiance at a single point is related to the field at that point by $I_{im}(\mathbf{R}_0) = |u_{im}(\mathbf{R}_0)|^2$. The PDF for the irradiance can be found by the same procedure used in Sec. 8.3.6 to derive the exponential law for speckle irradiance, (8.232); we use (C.104) to transform the PDF on the real and imaginary parts of the field to the joint PDF on the irradiance and the phase angle of the field, then marginalize over the latter variable. Dropping the subscripts im and the spatial arguments temporarily for notational convenience, we write

$$\mathrm{pr}(I, \phi) = |J| \, \mathrm{pr}(u', u'') , \tag{18.261}$$

where prime and double-prime denote real and imaginary parts, respectively,

$$u' = \sqrt{I} \cos\phi , \qquad u'' = \sqrt{I} \sin\phi , \tag{18.262}$$

and J is a Jacobian given by

$$J = \det \begin{bmatrix} \frac{\partial u'}{\partial I} & \frac{\partial u''}{\partial I} \\ \frac{\partial u'}{\partial \phi} & \frac{\partial u''}{\partial \phi} \end{bmatrix} = \tfrac{1}{2} . \tag{18.263}$$

Thus

$$\mathrm{pr}(I, \phi) = \tfrac{1}{2} \, \mathrm{pr}(u', u'') \Big|_{u' = \sqrt{I}\cos\phi, u'' = \sqrt{I}\sin\phi} . \tag{18.264}$$

We can now express $\mathrm{pr}(u', u'')$ as the inverse Fourier transform of its characteristic function and take advantage of the fact that ϕ is uniformly distributed on $(0, 2\pi)$, so that

$$\mathrm{pr}(I, \phi) = \frac{1}{2\pi} \mathrm{pr}(I)$$

$$= \tfrac{1}{2} \int_{-\infty}^{\infty} d\nu' \int_{-\infty}^{\infty} d\nu'' \, \psi_u(\nu', \nu'') \exp\left[2\pi i \left(\nu' \sqrt{I} \cos\phi + \nu'' \sqrt{I} \sin\phi \right) \right] . \tag{18.265}$$

Since the characteristic function is rotationally symmetric in the complex plane, we can express the 2D inverse Fourier transform as a Hankel transform, yielding

$$\mathrm{pr}(I) = 2\pi^2 \int_0^\infty |\nu| d|\nu| \; \mathrm{J}_0\left(2\pi|\nu|\sqrt{I}\right) \psi_u(|\nu|) \,. \tag{18.266}$$

Using (18.258) and reinserting the subscripts and arguments, we obtain

$$\mathrm{pr}[I_{im}(\mathbf{R}_0)] = 2\pi^2 \int_0^\infty |\nu| d|\nu| \; \mathrm{J}_0\left[2\pi|\nu|\sqrt{I_{im}(\mathbf{R}_0)}\right]$$
$$\times \left\langle \exp\left\{-\overline{N}_0 + \overline{N}_0\, \overline{\mathrm{J}}_0[2\pi a_n|\nu\, p_{coh}(\mathbf{0};\mathbf{R}_0)|]\right\}\right\rangle_{\overline{N}_0} \,. \tag{18.267}$$

This form is valid in the micro-area approximation when the number of scatterers in the disc area centered at $\mathbf{R}_0$, denoted N_0, is a doubly stochastic Poisson random variable with random mean $\overline{N}_0$. An alternative and slightly more general form can be obtained by omitting the intermediate conditional average over N_0 for fixed $\overline{N}_0$; in that case we can write [*cf.* (18.240)]

$$\mathrm{pr}[I_{im}(\mathbf{R}_0)]$$
$$= 2\pi^2 \int_0^\infty |\nu| d|\nu| \; \mathrm{J}_0\left[2\pi|\nu|\sqrt{I_{im}(\mathbf{R}_0)}\right] \left\langle \left\{\overline{\mathrm{J}}_0\left[2\pi a_n|\nu\, p_{coh}(\mathbf{0};\mathbf{R}_0)|\right]\right\}^{N_0} \right\rangle_{N_0} \,. \tag{18.268}$$

This result (without the averages over a_n and N_0) was derived by Kluyver (1906) in response to a paper on random walks by Pearson (1905), and it was generalized to three dimensions by Rayleigh (1919). For a summary of these historical contributions, see Mardia (1972).

The Poisson case To evaluate (18.267) or (18.268) and get the PDF on the irradiance at a single point, we must specify PDFs for a_n and $\overline{N}_0$. The simplest case is when both $\overline{N}_0$ and a_n are nonrandom (and the latter has the same value for all n, so we can drop the subscript). With these assumptions, neither the average over $\overline{N}_0$ nor the one over a (signified by the overbar on $\overline{\mathrm{J}}_0$) is needed, and (18.268) becomes

$$\mathrm{pr}[I_{im}(\mathbf{R}_0)] = 2\pi^2 \sum_{N_0=0}^{\infty} \frac{\overline{N}_0^{N_0}}{N!}$$
$$\times \exp(-\overline{N}_0) \int_0^\infty |\nu| d|\nu| \; \mathrm{J}_0\left[2\pi|\nu|\sqrt{I_{im}(\mathbf{R}_0)}\right] \left\{\mathrm{J}_0[2\pi a|\nu\, p_{coh}(\mathbf{0};\mathbf{R}_0)|]\right\}^{N_0} \,. \tag{18.269}$$

To evaluate this integral, Lord Rayleigh (1919) suggested the approximation

$$[\mathrm{J}_0(x)]^{N_0} \approx \exp\left(-\tfrac{1}{4}N_0 x^2\right) \,, \tag{18.270}$$

which works fairly well even for N_0 as small as 3 or 4. The two sides of (18.270) agree for small x, and raising the Bessel function to a power causes it to fall off rapidly and suppresses the oscillatory sidelobes.

With Rayleigh's approximation, the integral in (18.269) is simply the Hankel transform of a Gaussian, and we find that

$$\mathrm{pr}[I_{im}(\mathbf{R}_0)] \approx \sum_{N_0=0}^{\infty} \frac{\overline{N}_0^{N_0}}{N_0!} \exp(-\overline{N}_0) \frac{1}{N_0\, a^2\, |p_{coh}(\mathbf{0};\mathbf{R}_0)|^2} \exp\left[-\frac{I_{im}(\mathbf{R}_0)}{N_0\, a^2\, |p_{coh}(\mathbf{0};\mathbf{R}_0)|^2}\right] . \tag{18.271}$$

We can go further if $\overline{N}_0$ is large. Then the Poisson probability is sharply peaked around $N_0 = \overline{N}_0$, and we obtain

$$\mathrm{pr}[I_{im}(\mathbf{R}_0)] \approx \frac{1}{\overline{I}_{im}(\mathbf{R}_0)} \exp\left[-\frac{I_{im}(\mathbf{R}_0)}{\overline{I}_{im}(\mathbf{R}_0)}\right], \tag{18.272}$$

where

$$\overline{I}_{im}(\mathbf{R}_0) = \overline{N}_0\, a^2\, |p_{coh}(\mathbf{0};\mathbf{R}_0)|^2 . \tag{18.273}$$

This exponential density should come as no surprise; when there are many scatterers with random phase, the real and imaginary parts of the field are i.i.d. normal by the central-limit theorem and hence the irradiance is exponentially distributed [see (8.232)].

Doubly stochastic Poissons and the K density More interesting results are obtained when $\overline{N}_0$ is random. Since $\overline{N}_0$ can, in principle, take any value between 0 and ∞, we need a PDF defined on the positive real line for $\mathrm{pr}(\overline{N}_0)$. Candidate densities include the exponential, Rayleigh and log-normal, but much attention in the speckle literature has centered on using a gamma distribution (see Sec. C.5.4). With its two free parameters α and β, the gamma law allows considerable flexibility in constructing approximate PDFs on $(0, \infty)$.

Jakeman and Pusey (1978) used the gamma law for $\mathrm{pr}(\overline{N}_0)$ in speckle statistics. Since the Poisson transform of a gamma is a negative binomial, their model is equivalent to saying that N_0 obeys a negative binomial. Specifically, if

$$\mathrm{pr}(\overline{N}_0) = \frac{\overline{N}_0^{\,\alpha-1} \exp\left(-\overline{N}_0/\beta\right)}{\beta^{\alpha}\,\Gamma(\alpha)}, \tag{18.274}$$

then (Saleh, 1978)[15]

$$\mathrm{Pr}(N_0) = \binom{N_0+\alpha-1}{N_0} \frac{\beta^{N_0}}{(1+\beta)^{N_0+\alpha}} . \tag{18.275}$$

The corresponding moment-generating function is (Abramowitz and Stegun, 1965, formula 26.1.23)

$$M_{N_0}(t) \equiv \langle \exp(N_0\, t)\rangle_{N_0} = \frac{1}{(1+\beta-\beta\, e^t)^{\alpha}} . \tag{18.276}$$

By differentiating the moment-generating function, we can show that the mean number of scatterers is $\alpha\beta$ and the variance is $\alpha\beta(1+\beta)$.

To apply the negative-binomial law to speckle, we use (18.268) in conjunction with the Rayleigh approximation (18.270). The relevant expectation is

$$\left\langle \{\mathrm{J}_0[2\pi a|\nu\, p_{coh}(\mathbf{0};\mathbf{R}_0)|]\}^{N_0} \right\rangle_{N_0} \approx \left\langle \exp\left[-\pi^2 N_0\, a^2 |p_{coh}(\mathbf{0};\mathbf{R}_0)|^2 |\nu|^2\right] \right\rangle_{N_0}$$

$$= \frac{1}{\{1+\beta-\beta\exp\left[-\pi^2 a^2 |p_{coh}(\mathbf{0};\mathbf{R}_0)|^2|\nu|^2\right]\}^{\alpha}} \approx \frac{1}{[1+\beta\,\pi^2 a^2 |p_{coh}(\mathbf{0};\mathbf{R}_0)|^2 |\nu|^2]^{\alpha}}, \tag{18.277}$$

[15]The negative binomial is often written as a binomial coefficient times $p^N(1-p)^{\alpha}$; to get the form in (18.274), we must identify p as $\beta/(1+\beta)$. If α is not an integer, the binomial coefficient must be expressed in terms of gamma functions.

where the penultimate step follows from (18.276). The last step is valid when $\alpha\beta \gg 1$, in which case only small values of $|\nu|^2$ are important; recall that $\alpha\beta$ is the mean of the original negative binomial distribution, so large $\alpha\beta$ is equivalent to large N_0, an assumption that was already used in the Rayleigh approximation.

To get an approximate expression for the univariate density on the irradiance, we substitute the last form of (18.277) into (18.268) and use a tabulated integral (Gradshteyn and Ryzhik, 1980, formula 6.565.4); the result is (Jakeman and Pusey, 1978; Jakeman, 1980)

$$\mathrm{pr}[I_{im}(\mathbf{R}_0)] = \frac{2}{\Gamma(\alpha)} \frac{1}{a^2 |p_{coh}(\mathbf{0};\mathbf{R}_0)|^2 \beta} \left[\frac{I_{im}(\mathbf{R}_0)}{a^2 |p_{coh}(\mathbf{0};\mathbf{R}_0)|^2 \beta} \right]^{\frac{\alpha-1}{2}} \mathrm{K}_{\alpha-1}\left[2\sqrt{\frac{I_{im}(\mathbf{R}_0)}{a^2 |p_{coh}(\mathbf{0};\mathbf{R}_0)|^2 \beta}} \right]. \tag{18.278}$$

This PDF, known as the *K distribution*, is defined in (C.142) and plotted in Fig. C.7. It reduces to the exponential distribution of (18.272) as $\alpha \to \infty$.

K-distributed scattering amplitudes As we have just shown, the K distribution arises in point-scattering speckle models when the scattering amplitudes are nonrandom but the number of scatterers in a micro-area obeys a negative-binomial distribution. In fact, it also arises when the number of scatterers is nonrandom but the amplitudes a_n themselves obey a K distribution.

Jakeman and Pusey (1976) considered the effect of scattering from the ocean surface the performance of microwave radars operating over the sea. They noted that the magnitude of the scattered signal is observed to follow Rayleigh statistics when a large area of the sea is illuminated, but with narrow beams and short pulses the micro-area to which the receiver is sensitive at any one time may have dimensions comparable to the longer wavelengths on the sea surface. The temporal signals from the clutter may then have a spiky "target-like" appearance, especially when the beam propagates nearly parallel to the ocean surface and the back-scatter is observed. Since these spikes or glints can give rise to false positives in a radar detection task, it is important to characterize the clutter statistics.

Jakeman and Pusey neglected the fluctuations in the number of scatterers in the micro-area and focused on fluctuations in the radar scattering cross section, which is included in the amplitude a_n. They provided arguments showing that a K distribution for a_n is both physically reasonable and analytically tractable. Specifically, they assumed that

$$\mathrm{pr}(a_n) = \frac{2b}{\Gamma(1+\nu)} \left(\frac{b a_n}{2} \right)^{\nu+1} \mathrm{K}_\nu(b a_n), \qquad \nu > -1, \tag{18.279}$$

where ν can be viewed as a skewness parameter; if $\nu \to -1$, this expression approaches a log-normal, and if $\nu \to \infty$ it approaches a Rayleigh density (Shankar, 1995).

With this density on the amplitudes and the assumption that N_0 is a nonrandom constant, Jakeman and Pusey (1976) find that (in our notation)

$$\mathrm{pr}[I_{im}(\mathbf{R}_0)] = \frac{b|p_{coh}(\mathbf{0};\mathbf{R}_0)|}{\sqrt{I_{im}(\mathbf{R}_0)}\,\Gamma(M)} \left[\frac{b}{2} \sqrt{\frac{I_{im}(\mathbf{R}_0)}{|p_{coh}(\mathbf{0};\mathbf{R}_0)|^2}} \right]^M \mathrm{K}_{M-1}\left[b\sqrt{\frac{I_{im}(\mathbf{R}_0)}{|p_{coh}(\mathbf{0};\mathbf{R}_0)|^2}} \right], \tag{18.280}$$

where

$$M = N_0(1+\nu)\,. \tag{18.281}$$

The remarkable similarity of (18.280) to (18.278) should be noted; when K distributions are observed experimentally, there may be no good way of deciding whether they arise from variations in N_0 or a_n. Nevertheless, the parameters of the distribution as estimated from empirical data might be useful features for pattern recognition (Joughin *et al.*, 1993).

The K distribution of scattering amplitudes has also been used to describe ultrasonic scattering from tissue (Shankar *et al.*, 1993; Narayanan *et al.*, 1994; Shankar, 1995; Molthen *et al.*, 1995; Molthen *et al.*, 1998).

Generalized K distributions Various generalizations of the K distribution have appeared in the literature. A common generalization is to consider weak scattering where the phases are not uniformly random and (18.232) does not hold. A useful model in that case is the *von Mises density*,

$$\mathrm{pr}(\phi_n) = \left[2\pi \mathrm{I}_0(\gamma)\right]^{-1} \exp[\gamma\cos(\phi_n)]\,\mathrm{rect}\left(\frac{\phi_n}{2\pi}\right), \tag{18.282}$$

where γ is a nonnegative constant and $\mathrm{I}_0(\,\cdot\,)$ is a modified Bessel function. This PDF is bell-shaped and peaked at $\phi_n = 0$; for large γ it is approximately Gaussian with a variance equal to $1/\gamma$, and it approaches $\delta(\phi_n)$ as $\gamma \to \infty$.

Using the von Mises PDF on the phases, taking the amplitudes a_n as nonrandom and assuming a negative-binomial distribution for N_0, Barakat (1986) derived a generalized PDF on the single-point irradiance and showed that it reduced to (18.278) as $\gamma \to 0$. He also showed that this PDF agreed well with experimental moments of atmospheric laser-scattering data obtained in central Florida.

Jakeman and Tough (1987) reinterpreted Barakat's model in terms of a random walk with a directional bias specified by the density of the phases. With the negative-binomial model for N_0, they showed that Barakat's generalized K density is obtained in the limit of large $\alpha\beta$, even without the specific von Mises density on the phases. They also extended Barakat's results to higher dimensions, along the way obtaining an n-dimensional generalization of the Rician density.

There has also been some attention to multivariate generalizations of the K distribution. A multivariate model accounting for polarization was developed by Yueh *et al.* (1989). Correlated K-distributed clutter was considered by Oliver (1985) and Marier (1995), and simulation methods were reported by Oliver and Tough (1986).

An interesting way of including spatial correlations in the K distribution was presented by Pentini *et al.* (1992). They considered a correlated circular Gaussian random process for the field, where, for example, the correlation function might be given by (18.100). They sampled the field on a grid of N points to get a correlated circular Gaussian random vector with PDF [*cf.* (8.245)]

$$\mathrm{pr}(\mathbf{u}|\tau) = \frac{1}{\pi^N \tau^N \det(\mathbf{K_u})} \exp\left(-\frac{1}{\tau}\mathbf{u}^\dagger \mathbf{K}_\mathbf{u}^{-1}\mathbf{u}\right), \tag{18.283}$$

where the nonnegative scalar τ can be interpreted as an overall power level or mean reflectivity of the object. If τ is a random variable independent of $\mathbf{u}$, then the overall density on $\mathbf{u}$ is a Gaussian mixture. Specifically, if τ follows a gamma law,

Pentini *et al.* show that $\mathrm{pr}(\mathbf{u})$ has the form

$$\mathrm{pr}(\mathbf{u}) = \int_0^\infty d\tau \; \mathrm{pr}(\mathbf{u}|\tau)\,\mathrm{pr}(\tau) \propto \left[\mathbf{u}^\dagger \mathbf{K}_{\mathbf{u}}^{-1}\mathbf{u}\right]^{\frac{\nu-N}{2}} \mathrm{K}_{\nu-N}\left(\mu\sqrt{\mathbf{u}^\dagger \mathbf{K}_{\mathbf{u}}^{-1}\mathbf{u}}\right), \qquad (18.284)$$

where ν and μ are constants related to the PDF on τ. Thus once again a K distribution arises, this time without recourse to a point-scattering model at all. The argument of the Bessel function is real (since $\mathbf{K}_{\mathbf{u}}^{-1}$ is Hermitian and positive-semidefinite), but the different components of $\mathbf{u}$ can have arbitrary correlations so long as they are consistent with the circular-Gaussian assumption.

Relation to Ricians Experimental histograms of $\sqrt{I_{im}(\mathbf{R}_0)}$ in coherent imaging are often well approximated by the Rician or Rice-Nakagami law defined in (C.141). Though originally derived by Rice for analysis of a known, nonrandom signal of amplitude A in Gaussian noise, this law is commonly used in speckle problems with A and σ^2 treated as adjustable parameters. This law would apply to speckle with interferometric or homodyne detection, but it does not seem to describe the non-interferometric systems we are analyzing in this chapter, so its mechanistic justification is not clear.

In one sense there is no need for a mechanistic justification of the Rician law; if it fits the data and if the parameters of the fit are useful features for pattern recognition, that is justification enough. Nevertheless, it is interesting to see how an approximate Rician might arise from different physical assumptions.

As we noted in Sec. 18.1.1, the Rician could occur in transmission imaging when the phase of the transmitted wave is not completely randomized, so that there is a nonrandom plane-wave component superimposed on the scattered field. Similarly, in reflection, something akin to a Rician could arise when the phases are not fully randomized, as in (18.282) with large γ. The generalized K distribution of Jakeman and Tough (1987) often approximates a Rician when expressed as a density on $\sqrt{I_{im}(\mathbf{R}_0)}$ rather than $I_{im}(\mathbf{R}_0)$.

The Rician density can also be obtained by assuming that there is a single strong point scatterer or specular reflector present, but there are two difficulties in this view. The first is the (often implicit) assumption that this strong scatterer, unlike the rest of the points in the object, is nonrandom. The second difficulty is that point scatterers outside the support of the coherent PSF of the imaging system make no contribution to the image-plane field at any given point; for the Rician density to be obtained exactly, therefore, one would have to assume that there is always exactly one point scatterer within the PSF no matter where one looks in the image.

Another model that leads to a Rician is an array of point scatterers with partial periodicity (Wagner *et al.*, 1986; Wagner *et al.*, 1987; Shankar, 1995). The wave diffracted from such an array consists of a coherent part, as if the array were perfectly periodic, and a diffuse part resulting from random displacement of the scatterers from the ideal lattice positions. The coherent part can serve as the nonrandom coherent signal envisioned in Rice's original work. Since the coherent component is the result of volume diffraction from a periodic structure, strong diffraction will occur only if the Bragg condition is satisfied (see Sec. 9.8.3). Thus the relative strength of the two components will depend not only on the degree of periodicity, but also on the directions of the incident and diffracted waves as well as the area (or volume) associated with the PSF. If one wished to test the validity of this model, the dependence of the empirical PDF could be studied as a function of these parameters.

Other densities for N_0 We have emphasized point-scattering models where N_0 follows a negative binomial law (or equivalently, $\overline{N}_0$ follows a gamma), but many other probability laws can be used for the number of scatterers in a micro-area. Delignon and Pieczynski (2002) summarize the possibilities and discuss ways of choosing among them in describing empirical data from synthetic-aperture radar.

Texture synthesis with speckle In Sec. 8.4 we discussed several parametric models for texture statistics. For example, the clustered lumpy background is a flexible model with a few free parameters that can be adjusted to mimic the textures seen in mammograms. With our current understanding of speckle statistics, it now appears feasible to use speckle fields as models of texture patterns that have nothing to do with coherent imaging. Suppose, for example, that we start with a nonstationary, doubly stochastic Poisson point process and image it — in a computer — through a coherent imaging system with a specified, possibly space-variant PSF. When we then form the irradiance — still in the computer — we will have a nonnegative random field that should be capable of mimicking many different natural textures. To make the synthesis realistic, we can adjust the parameters of the initial point process as well as those of the coherent PSF.

18.6 COHERENT RANGING

To this point we have considered speckle in 2D systems, but we have noted at several points that speckle is important in various forms of radar and ultrasound imaging. These methods have in common that they use amplitude-sensitive detectors, they are used to image 3D objects, and they use time of flight to encode the third dimension. In order to use time of flight, the radiation must be modulated in some way. The simplest option is just to pulse the radiation source, but much research has gone into chirp signals and other waveforms.

In spite of the modulation, radar and ultrasound systems are still coherent, so they exhibit speckle. In this section we shall analyze these systems both deterministically and stochastically and show how the speckle formalism developed earlier in the chapter can be extended to pulsed imaging.

We begin in Sec. 18.6.1 with a qualitative overview of the systems, treating radar and ultrasound in parallel to emphasize their similarities. The serious analytical effort begins in Sec. 18.6.2 by elucidating the linear operator that maps the 3D object to electrical signals from the amplitude-sensitive detectors. In Sec. 18.6.3 we develop a general characteristic functional to describe the statistics of speckle as manifest in the electrical signals. In Sec. 18.6.4 we consider issues in task performance and image quality.

18.6.1 System configurations

As shown in Fig. 18.5, the key elements in a radar or ultrasound system are an RF (radio-frequency) generator, a modulator, a transmitting element, a receiving element and a signal-processing system. Application of an electrical signal to the transmitting element produces a wave that propagates into space, reflects from some target and returns to the receiving element where it is converted back to an electrical signal. When the transmitting and receiving elements are identical as in Fig. 18.5 (or at least at essentially the same location), the system is said to be *monostatic*. When the transmitter and receiver are at different locations, the system is *bistatic*, and when multiple transmitters and/or receivers are used, the system is *multistatic*. Only monostatic systems will be discussed here.

The transmitted waves are highly directional or beam-like, so only targets within the beam receive radiation. If the targets are small, then they reradiate over a wide angular range, but only a small portion of the wave gets to the receiving element. For monostatic configurations, the receiver is sensitive to back-scattered radiation from the same beam-like region where the transmitted wave propagated.

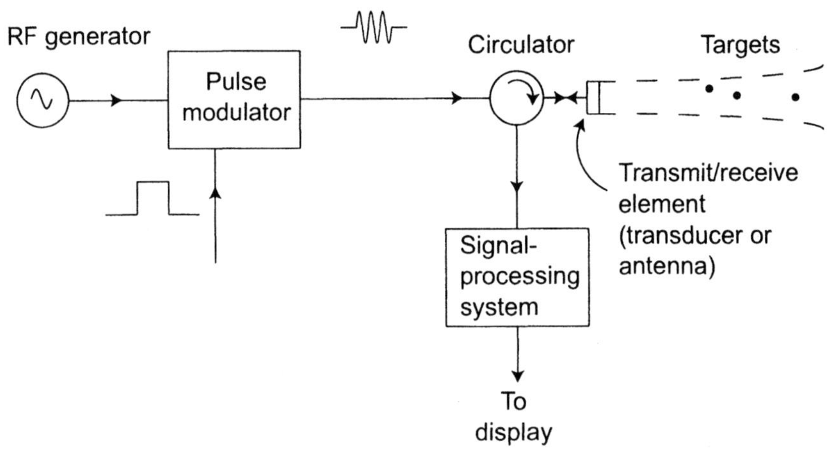

Fig. 18.5 A monostatic radar system.

Transducers We shall use the word *transducer* to describe the elements that convert the electrical signal to a propagating wave and back again. For radar the transducer is a microwave antenna, and for ultrasound it is a piezoelectric element.

The simplest (and most common) ultrasound transducer is a flat piezoelectric plate, usually in the form of a disc as shown in Fig. 18.6a. A high-frequency voltage is applied to electrodes on opposite sides of the disc, and the piezoelectric effect causes the material to vibrate at the same frequency. Maximum vibration amplitude is obtained when the thickness of the transducer is equal to half of the acoustic wavelength in the material.

If the vibrating transducer is placed in contact with a patient's body or other medium, it will induce a vibrational wave in the medium. For medical applications the vibrations are normal to the face of the transducer, so the induced wave is a compressional or density wave, similar to ordinary sound waves in air. Transverse or shear waves can be generated in some solid media, but they do not propagate in air, liquids or soft tissue. For medical use, the frequency of the wave is typically 5 – 10 MHz, corresponding to a wavelength in tissue of 0.15 – 0.3 mm.

Since all points on the transducer surface vibrate in phase, the generated wave has a planar wavefront over the surface of the disc. As it propagates, this plane wave diffracts, just as a monochromatic light wave would after passing through a disc aperture.

To avoid the diffraction spread and put maximum energy on a target at a known distance from the transducer, a curved ultrasound transducer can be used as shown in Fig. 18.6b. In this case the wave generated is a converging spherical wave, analogous to light after passing through a lens. In fact, actual acoustic lenses can be used with planar transducers to achieve the same result.

A microwave transducer is a reflecting dish illuminated with a feed horn as shown in Fig. 18.7. The dish is ideally parabolic but may be spherical in practice. If its radius of curvature is R, then it functions as a focusing element with focal length $f = R/2$. If the feed horn is one focal length from the dish, a plane electromagnetic wave is formed immediately after the reflection, and it then diffracts as it propagates just as the ultrasonic wave does. A converging spherical wave can be generated by moving the feed horn so that it is more than one focal length from the dish.

Radar systems operate over a wide variety of frequencies, from a few hundred MHz to tens of GHz. A common choice, referred to as *X band*, is 8 – 12 GHz. A 10 GHz radio wave in free space has a wavelength of 3 cm.

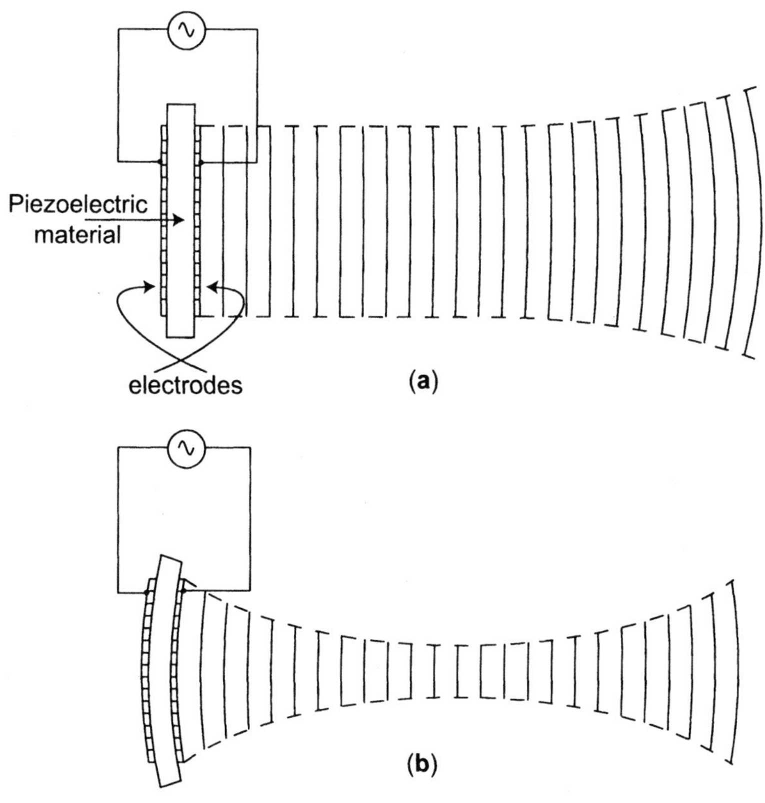

Fig. 18.6 Transducers for use in medical ultrasound. (a) Planar transducer produces a plane wave that eventually diffracts into a diverging spherical wave. (b) Curved transducer produces a converging spherical wave that comes to a focus at the center of curvature of the transducer.

Scanning methods If a pulsed RF signal is transmitted from a transducer in a given location, a wave packet propagates to a reflecting target, and part of the reflected wave returns to the transducer and is reconverted to an electrical signal. If the received electrical signal is displayed on an oscilloscope, an echo or *blip* is seen at a time corresponding to the round trip propagation time from transducer to target and back again. If there are many targets in the beam, there is an echo for each. The oscilloscope display is referred to as an *A-scan*, where A refers to the amplitude of the return signal. For our purposes, an A-scan can be regarded as a 1D image, with time delay corresponding to the range to the target.

To form a 2D image, the transducer can be tilted or rotated to produce beams in different directions. In radar it is common to rotate the antenna continuously, and in ultrasound the transducer is commonly swept over some angular arc, either mechanically or by hand. Conceptually it may be easier to think of the transducer as being pointed to a discrete set of positions, where a pulsed signal is transmitted and received at each position. The result is a 2D map of target locations in polar coordinates. Commonly, this information is displayed on a monitor with echo strength converted to brightness; this display is called a *B-scan* (B for brightness).

In a B-scan with different angular directions for the beam, information is obtained from the object along a fan of rays. In Chap. 17 we encountered fan-beam and cone-beam tomographic systems that also collect information along many rays passing through a single point. The key difference is that the tomographic systems collect *integrals* of some object property, while radar and ultrasound systems take advantage of time of flight to get information about where along the ray a target lies. Time of flight has been used in some PET systems to get similar information, and if the temporal resolution were adequate, PET images could be obtained without tomographic reconstruction. In practice, however, available gamma-ray detectors limit the time-of-flight resolution to about 10 cm, which is far from adequate for medical applications.

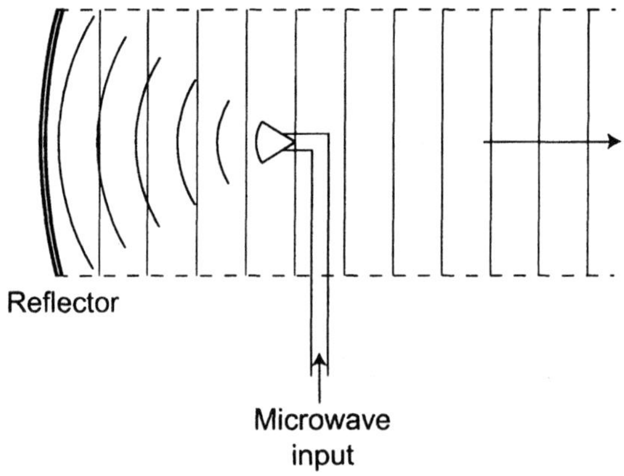

Fig. 18.7 Illustration of microwave transducer. The microwave input is located a distance $R/2$ from the reflector; the result is a plane wave.

Phased arrays Angular scanning of the beam is essential for obtaining a 2D image with a monostatic system, and focusing the beam can be useful to avoid diffraction spread. We know from Chap. 9 that an angular deviation corresponds to a 2D linear phase factor, $\exp(2\pi i \boldsymbol{\rho}_0 \cdot \mathbf{r})$ across a beam, and focusing corresponds to a quadratic phase factor, $\exp(-i\pi r^2/\lambda f)$. In essence, angular scanning and focusing are equivalent to imposing phase variations on the transducer.

The same phase variations can be imposed electronically rather than mechanically if the transducer is divided into many small elements that can be excited independently with different phases. The more elements used, the better the approximation to the desired continuous phase distribution but the more hardware needed to provide the requisite signals to each element. Ideally, the element spacing should be a half wavelength or less to give Nyquist sampling of any real (non-evanescent) plane-wave component of a general propagating beam (Macovski, 1979).

Phased arrays are commonly used in radar for rapid scanning, target tracking and beam forming. 1D phased arrays are also common in medical ultrasound. Rectangular elements are used, where one dimension, say L_x, of the rectangle is comparable to a wavelength but the other, L_y, is much larger than a wavelength. The result is a beam confined to be near the y-z plane but capable of being focused or scanned in that plane. Two-dimensional piezoelectric arrays with small square elements are also being developed; they permit arbitrary monostatic scanning and hence 3D imaging without mechanical motion.

Resolution As with any imaging system, spatial resolution is critical in pulsed ranging systems. We must distinguish lateral resolution (transverse to the beam) from longitudinal resolution (along the beam, in the range direction). Lateral resolution is computed just as in optics; a transducer of diameter D, for example, produces an angular divergence of λ/D in the Fraunhofer zone. Longitudinal resolution, on the other hand, is determined entirely by the modulation of the beam. For example, if we use a simple pulse of duration T, it produces a travelling wave packet of length vT, where v is the speed of propagation. The longitudinal resolution is thus about $vT/2$ (the factor of 2 arising since the delay is doubled for round-trip propagation).

To get good longitudinal resolution, therefore, we should use short pulses. The problem is that the pulse amplitude is usually constrained by engineering considerations such as electrical breakdown, so a shorter pulse has smaller energy, which makes it more difficult to detect weak echoes.

Pulse coding During World War II and since, many methods have been developed for getting good longitudinal resolution without sacrificing pulse energy and hence detectability (see Sec. 18.6.4). Klauder *et al.* (1960) classify the methods as chirp, bang and hiss. Bang refers to the use of simple pulse modulation as just discussed. Chirp refers to use of a temporal quadratic phase factor as the modulation. We know from Sec. 5.1.3 that a quadratic phase variation corresponds to a linear variation of local frequency with time, so the chirp is linear frequency modulation. Hiss refers to the use of noise as the modulation.

The principle of chirp radar is illustrated in Fig. 18.8. As shown there, the transmitted signal is a chirp where the lowest frequency is transmitted first. The received signal is sent through a dispersive delay line where the delay varies linearly with frequency and the lowest frequency is delayed the most. If the dispersion rate of the delay line is matched to the chirp of the signal, then all frequency components come out of the delay line at the same time, yielding a pulse of much higher amplitude and shorter duration. As we shall see in Sec. 18.6.2, the width of the compressed pulse is the reciprocal of the bandwidth of the chirp, while the total energy is determined by its width before compression. Thus chirps allow good detectability and also good longitudinal resolution.

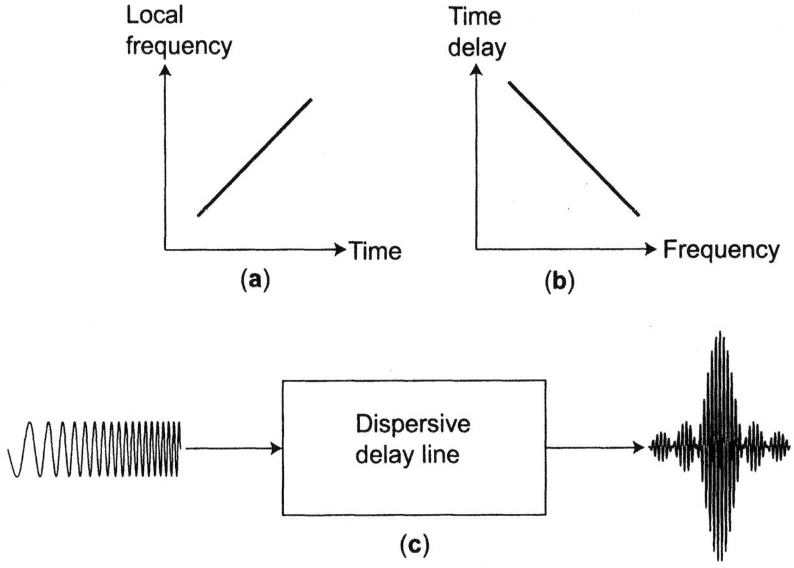

Fig. 18.8 Principle of chirp radar.

As we shall see in Sec. 18.6.2, the delay line is a matched filter for the chirp signal. That means that the delay line serves to cross-correlate the noisy received signal with a replica of the noise-free transmitted signal. If the target is a point, the noisy signal itself contains a replica of the transmitted signal, and the reason chirp radar works is that the correlation of a chirp with itself is sharply peaked, approaching a delta function as the width of the chirp increases.

Many other codes also have sharply peaked autocorrelations. The possibilities include nonlinear frequency modulation, frequency-jump chirp, step chirp and various phase-modulation patterns such as Barker codes. For details see Sullivan (2000).

Optical coherence tomography Although chirp signals are commonly used in modern radar, noise-modulated or hiss signals are not, having been superseded by sophisticated pulse codes. The principle of hiss radar is, however, the foundation for an optical ranging system called *optical coherence tomography* or *OCT*.

The basic system for OCT is shown in Fig. 18.9. The broadband optical source emits a wave that we can represent as $\tilde{u}(\mathbf{r}, t) \exp(-2\pi i \nu_0 t)$, where $\tilde{u}(\mathbf{r}, t)$ is a random process that can be regarded for present purposes as a noise modulation. The wave bounces off a target and returns to the system, where it is interfered with a delayed replica of itself. As we know from Sec. 9.7.4, interference occurs only when the path difference is less than the coherence length, given by $c/\Delta\nu$, where $\Delta\nu$ is the bandwidth of the source. Scanning in the range direction is accomplished by varying the delay, and fringes are observed only if the two path lengths match within the coherence length. Longitudinal or range resolution is thus $c/2\Delta\nu$ and can be improved, just as with a chirp, by increasing the bandwidth of the source.

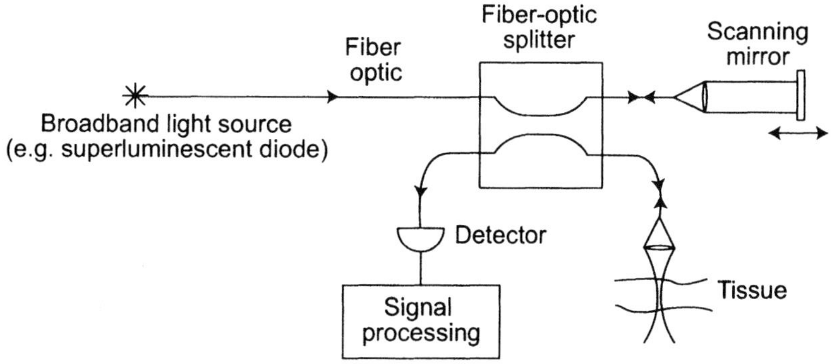

Fig. 18.9 Basic system for optical coherence tomography.

Doppler Often the targets in radar and medical ultrasound are in motion, and the reflected wave undergoes a Doppler shift. If the Doppler shift is measured, the system becomes a 4D imaging system, localizing the target in x, y, z and v_z, where x and y are transverse to the beam, z is along the beam and v_z is the component of $\mathbf{v}$ parallel to z. Resolution in x and y is determined by the transducer size and the wavelength; resolution in z is determined by the duration of the pulse after compression of other processing, and resolution in range is determined by our ability to measure the Doppler shift.

There is, however, a conflict between range resolution and Doppler resolution. If we choose a short pulse that contains only a few cycles of the center frequency, then we get good range resolution but poor Doppler resolution; since the pulse consists of a broad spectrum of frequencies, small frequency shift are hard to discern. Conversely, a long pulse gives good Doppler resolution. The use of a chirp does not help appreciably in this case, as we can see by looking again at Fig. 18.8a. A change in range z of a target causes a shift of the diagonal line in that figure along the horizontal axis, while a change in v_z causes a shift along the vertical axis. Except for what happens at the ends of the line, these two shifts are indiscernible, and range and Doppler shifts causes ambiguous variations in the received signal. The Woodward ambiguity function, introduced in Sec. 5.2.2, is a way of quantifying this ambiguity.

Synthetic-aperture radar Radar systems in aircraft or satellites can be used to get high-resolution images of the ground. The airborne antenna is generally small and a large area on the ground is illuminated, so these systems do not rely on λ/D to get the spatial resolution. Instead, many images are recorded as the aircraft flies along, and these images are coherent with respect to one another. They can, in effect, be made to interfere in the computer, and the result is that the resolution is determined by the distance the aircraft flies while a particular point on the ground is illuminated. Calling this distance L, we see that the angular resolution is λ/L rather than λ/D, so we have synthesized a much larger aperture, hence the name *synthetic-aperture radar* or *SAR*.[16]

[16]SAR is a second-order acronym; one of the letters stands for an acronym. Third-order acronyms are also possible; for example, ASF stands for Alaska SAR Facility.

The resolution properties of SAR images are quite counterintuitive. Consider a point target at range R from the aircraft. An antenna of diameter D illuminates a patch of length given by $R\lambda/D$, and this is also the length L of the synthesized aperture. After image reconstruction the linear resolution in the direction of the flight path is $R\lambda/L = (R\lambda)/(R\lambda/D)$ or simply D. Thus the resolution is the antenna size D, independent of both wavelength λ and range R. A smaller antenna is better since it illuminates a larger patch on the ground, and the size of the patch is the length of the synthesized aperture. Resolution in the range direction is determined by pulse width or pulse coding as with other forms of radar.

We shall not explicitly analyze SAR systems in what follows, but the formalism is general enough to include it. In particular, we shall consider multiple antenna configurations, which can include multiple positions along a flight path.

For excellent reviews of the mathematics of SAR, see Borden (1999) and Cheney (2001), and for a comprehensive treatment see Sullivan (2000).

18.6.2 Deterministic analysis

In this section we analyze a basic coherent ranging system deterministically; the corresponding stochastic analysis is presented in Sec. 18.6.3. Both analyses apply equally to microwave or ultrasound systems.

The system we shall analyze is monostatic, with the transducer lying in the plane $z = 0$ (or just behind it if the transducer is curved). The transducer is scanned either laterally by moving it in x and y, or angularly by tilting it physically or by controlling the linear phase factor in a phased array. In either case we use the discrete index j to denote the transducer configuration. For lateral scanning that index specifies the center coordinates of the transducer, (x_j, y_j), and for angular scanning it specifies the beam direction in polar coordinates, (θ_j, ϕ_j). For airborne SAR systems, j denotes the position of the antenna along the flight path, which can be taken as the x axis.

The transducer acts as a pulsed source, producing a beam-like wave packet in the space $z > 0$. The wave packet is reflected by inhomogeneities in the refractive index in the medium, and some of it returns to the transducer where it is reconverted to an electrical signal. For each j the transducer output is a function of time, so the collected data are denoted $g_j(t)$. We shall consider the refractive-index inhomogeneities to be functions of the continuous spatial variable $\mathbf{r} = (x, y, z)$ (with $z > 0$), so the system under consideration is a mixed CC-CD system in the terminology of Sec. 7.3.5. It is mixed in data space since $g_j(t)$ has both continuous and discrete indices, but it is continuous in object space since the object is a function of a continuous variable.

There are situations where the system is mixed in object space as well. If we consider polarized electromagnetic radiation, and the object has different scattering properties for the two polarizations, then the object is a 2D vector field and we need a discrete index on the object function. Similarly, in ultrasound the scattering is determined by both the object density and its compressibility, so again the object is a 2D vector field.

For simplicity, however, we shall forgo this generality and consider only scalar objects. Thus we are seeking an expression for a linear operator $\mathcal{L}$ that maps a function of (x, y, z) to a function of j and t. In order to make the analysis tractable, we assume weak scattering and apply the first Born approximation, as introduced in

Sec. 9.8.1. In that section, however, we assumed perfectly monochromatic radiation and solved the Helmholtz equation; here we must use the time-dependent wave equation in order to account for the pulse modulation.

The Born approximation in time-dependent problems We know from (9.23) that scalar waves in a homogeneous medium, where the speed of propagation is c_m, must satisfy

$$\left(\nabla^2 - \frac{1}{c_m^2}\frac{\partial^2}{\partial t^2}\right)u(\mathbf{r},t) = s(\mathbf{r},t)\,, \tag{18.285}$$

where $s(\mathbf{r},t)$ describes the radiation source. We shall be interested in solutions of this equation in situations where the speed of propagation varies with space and time. Perhaps surprisingly, the wave equation (18.285) still holds if we replace the constant c_m by the general function $c_m(\mathbf{r},t)$ so long as $u(\mathbf{r},t)$ varies much more rapidly with time than does $c_m(\mathbf{r},t)$; in other words, the Doppler shifts must be small compared to the mean frequency of the radiation. For simplicity, we assume in what follows that c_m is independent of time (thereby ruling out moving targets in radar).

To accommodate a speed of propagation that varies with $\mathbf{r}$, we define the refractive index as

$$n(\mathbf{r}) = \frac{c}{c_m(\mathbf{r})}\,, \tag{18.286}$$

where c is some reference speed, usually taken as the speed of light in vacuum for electromagnetic waves. For medical ultrasound, however, the reference speed is often taken as the speed of sound in water, about 1500 m/s; tissue is very similar to water acoustically, so the refractive index defined this way is near one.

Derivation of the Born approximation now parallels the treatment given in Sec. 9.8.1 for the monochromatic case. Equation (18.285) can be rewritten as

$$\left(\nabla^2 - \frac{1}{c^2}\frac{\partial^2}{\partial t^2}\right)u(\mathbf{r},t) = \frac{n^2(\mathbf{r})-1}{c^2}\frac{\partial^2}{\partial t^2}\,u(\mathbf{r},\mathbf{t}) + \mathbf{s}(\mathbf{r},\mathbf{t})\,. \tag{18.287}$$

We define $u_{inc}(\mathbf{r})$ as the field that would exist at point $(\mathbf{r},t)$ in the absence of the inhomogeneous index distribution. Specifically, as in (9.329), $u_{inc}(\mathbf{r},t)$ is the solution to

$$\left(\nabla^2 - \frac{1}{c^2}\frac{\partial^2}{\partial t^2}\right)u_{inc}(\mathbf{r},t) = s(\mathbf{r},t)\,. \tag{18.288}$$

We then write the total field as

$$u(\mathbf{r},t) = u_{inc}(\mathbf{r},t) + u_{sc}(\mathbf{r},t)\,, \tag{18.289}$$

where $u_{sc}(\mathbf{r},t) \equiv u(\mathbf{r},t) - u_{inc}(\mathbf{r},t)$ can be interpreted as the field scattered by the index inhomogeneities.

With these definitions, (18.287) in a source-free region ($z > 0$) becomes

$$\left(\nabla^2 - \frac{1}{c^2}\frac{\partial^2}{\partial t^2}\right)u_{sc}(\mathbf{r},t) = \frac{n^2(\mathbf{r})-1}{c^2}\frac{\partial^2}{\partial t^2}\,u(\mathbf{r},\mathbf{t})\,. \tag{18.290}$$

Notice that the unknown scattered field is hidden in $u(\mathbf{r},\mathbf{t})$ on the right.

The first Born approximation regards the scattered field as small compared to the incident field, so the wave equation (still in the source-free region) becomes

$$\left(\nabla^2 - \frac{1}{c^2}\frac{\partial^2}{\partial t^2}\right) u_{sc}(\mathbf{r},t) \approx \frac{n^2(\mathbf{r})-1}{c^2}\frac{\partial^2}{\partial t^2} u_{inc}(\mathbf{r},t) \approx k_0^2\left[1-n^2(\mathbf{r})\right] u_{inc}(\mathbf{r},\mathbf{t}), \tag{18.291}$$

where the last step follows if the incident field is narrowband with center frequency ν_0, and $k_0 \equiv 2\pi\nu_0/c$.

The equation for the scattered field, (18.291), is now identical to (18.288) except that the real source has been replaced by an effective source: the incident field modulated by the index inhomogeneities.

The object For purposes of deterministic analysis, we can identify "the object" as the scattering potential, defined as in (9.328) by

$$V(\mathbf{r}) \equiv k_0^2\left[1-n^2(\mathbf{r})\right]. \tag{18.292}$$

When we get to stochastic analysis in Sec. 18.6.3, we shall see that it is by no means so simple to say what the word "object" means.

The incident field Since multiple scattering is neglected in the first Born approximation, the incident field is found by solving the time-dependent wave equation, (18.288), with the transducer acting as a source but with no index inhomogeneities. All of the work in finding the solution was done in Sec. 9.3, and we can just appropriate the basic result here; from (9.66) we know that the incident field is given by

$$u_{inc}^{(j)}(\mathbf{r},t) = -\frac{1}{4\pi}\int_\infty d^3\mathbf{r}'\,\frac{1}{|\mathbf{r}-\mathbf{r}'|}s_j\left(\mathbf{r}',t-\frac{|\mathbf{r}-\mathbf{r}'|}{c}\right), \tag{18.293}$$

where $s_j(\mathbf{r},t)$ is the source function for the jth transducer configuration. In essence, the field at an observation point is the sum (integral) of the fields from all source points, delayed by the propagation time and reduced in amplitude by the $1/|\mathbf{r}-\mathbf{r}'|$ factor.

Spatially, the source function $s_j(\mathbf{r},t)$ contains a factor of $\delta(z)$ since the transducer lies in the plane $z=0$, so the integral in (18.293) is really over the x-y plane. Also, $s_j(\mathbf{r},t)$ vanishes when the point (x,y) lies outside the physical boundaries of the transducer in the jth configuration. The remaining function of (x,y) may include linear phase factors to describe tilt of the transducer or quadratic phase factors to describe focusing; either of these factors can be created mechanically, by tilting or curving the transducer, or electronically by programming the phase shifters in a phased array appropriately. For example, if

$$s_j(\mathbf{r}) \propto \delta(z)\exp\left[\frac{2\pi i}{\lambda_0}(\alpha_j x+\beta_j y) - \frac{\pi i}{\lambda_0 z_j}\left(x^2+y^2\right)\right], \tag{18.294}$$

it means that the beam is aimed in the direction specified by the direction cosines α_j and β_j (see Sec. 9.2.1) and focused in the plane $z=z_j$ (see Sec. 9.6.1).

Temporally, the source function consists of a carrier-frequency factor times a factor specifying the pulse modulation. If we assume that the carrier frequency and the pulse modulation are the same for every transducer configuration, we can write

$$s_j(\mathbf{r},t) = m(t)\exp(-2\pi i\nu_0 t)\, s_j(\mathbf{r}), \tag{18.295}$$

where $m(t)$ is the modulation function. Thus

$$u_{inc}^{(j)}(\mathbf{r},t) = -\frac{1}{4\pi}\exp(-2\pi i\nu_0 t)\int_\infty d^3\mathbf{r}'\,\frac{\exp(ik_0|\mathbf{r}-\mathbf{r}'|)}{|\mathbf{r}-\mathbf{r}'|}s_j(\mathbf{r}')\,m\left(t-\frac{|\mathbf{r}-\mathbf{r}'|}{c}\right), \tag{18.296}$$

where $k_0 \equiv 2\pi\nu_0/c$, and the real parts of all complex fields are to be taken. If $m(t)$ is a constant, (18.296) reduces to (9.67), which expresses the solution of the inhomogeneous Helmholtz equation with a prescribed source.

If $m(t)$ is slowly varying compared to the variation in propagation times from different points on the transducer to the scattering point $\mathbf{r}$, then we can approximate (18.296) by

$$u_{inc}^{(j)}(\mathbf{r},t) = -\frac{1}{4\pi}\exp(-2\pi i\nu_0 t)\,m\left(t-\frac{|\mathbf{r}-\mathbf{r}_{0j}|}{c}\right)\int_\infty d^3\mathbf{r}'\,\frac{\exp(ik_0|\mathbf{r}-\mathbf{r}'|)}{|\mathbf{r}-\mathbf{r}'|}\,s_j(\mathbf{r}'), \tag{18.297}$$

where $\mathbf{r}_{0j}$ is the center of the transducer in the jth configuration. In this form the result is identical to the Helmholtz expression, (9.67), except for an envelope function that moves along the beam at speed c. For later convenience we define

$$h_j(\mathbf{r}) \equiv -\frac{1}{4\pi}\int_\infty d^3\mathbf{r}'\,\frac{\exp(ik_0|\mathbf{r}-\mathbf{r}'|)}{|\mathbf{r}-\mathbf{r}'|}\,s_j(\mathbf{r}'). \tag{18.298}$$

Since the source is confined to the plane $z=0$, $h_j(\mathbf{r})$ is the monochromatic diffraction pattern of the transducer.

The scattered field The scattered field is also given by (18.293), but with the real source $s_j(\mathbf{r},t)$ replaced by the effective one:

$$\begin{aligned} u_{sc}^{(j)}(\mathbf{r}'',t) &= -\frac{1}{4\pi}\int_\infty d^3\mathbf{r}\,\frac{1}{|\mathbf{r}''-\mathbf{r}|}\,u_{inc}^{(j)}\left(\mathbf{r},t-\frac{|\mathbf{r}''-\mathbf{r}|}{c}\right)V(\mathbf{r}) \\ &= -\frac{1}{4\pi}\exp(-2\pi i\nu_0 t)\int_\infty d^3\mathbf{r}\,\frac{\exp(ik_0|\mathbf{r}''-\mathbf{r}|)}{|\mathbf{r}''-\mathbf{r}|}\,m\left(t-\frac{|\mathbf{r}-\mathbf{r}_{0j}|}{c}-\frac{|\mathbf{r}''-\mathbf{r}|}{c}\right)h_j(\mathbf{r})\,V(\mathbf{r}). \end{aligned} \tag{18.299}$$

Now the total delay of the envelope is the propagation time from the center of the transducer to the scattering point $\mathbf{r}$ to the observation point $\mathbf{r}''$.

The received signal When the scattered wave returns to the transducer, the reciprocity principle tells us that it undergoes the same weighting as on transmission; it is multiplied by a factor $s_j(\mathbf{r})$, and the resultant weighted amplitude is integrated across the transducer face (the sensitive area of the plane $z=0$). The output signal is thus

$$g_j(t) = C\int d^3\mathbf{r}''\,s_j(\mathbf{r}'')\,u_{sc}^{(j)}(\mathbf{r}'',t) \tag{18.300}$$

where C is a constant related to the responsivity of the transducer. With (18.299), we find

$$\begin{aligned} g_j(t) &= -\frac{C}{4\pi}\exp(-2\pi i\nu_0 t)\int_\infty d^3\mathbf{r}\,h_j(\mathbf{r})\,V(\mathbf{r}) \\ &\times\int d^3\mathbf{r}''\,s_j(\mathbf{r}'')\,\frac{\exp(ik_0|\mathbf{r}''-\mathbf{r}|)}{|\mathbf{r}''-\mathbf{r}|}\,m\left(t-\frac{|\mathbf{r}-\mathbf{r}_{0j}|}{c}-\frac{|\mathbf{r}''-\mathbf{r}|}{c}\right). \end{aligned} \tag{18.301}$$

Since the same transducer is used for receiving as for transmitting in a monostatic system, we can use the same approximation as in (18.297), approximating $|\mathbf{r}'' - \mathbf{r}|$ by $|\mathbf{r}_{0j} - \mathbf{r}|$ in the argument of $m(\,\cdot\,)$. We then obtain

$$g_j(t) = C \exp(-2\pi i \nu_0 t) \int_\infty d^3\mathbf{r}\, [h_j(\mathbf{r})]^2\, m\left(t - 2\frac{|\mathbf{r} - \mathbf{r}_{0j}|}{c}\right) V(\mathbf{r})\,. \tag{18.302}$$

We can write this result as

$$g_j(t) = \int d^3\mathbf{r}\; L_j(t, \mathbf{r})\, V(\mathbf{r}) \qquad \text{or} \qquad \mathbf{g} = \boldsymbol{\mathcal{L}}\mathbf{V}\,. \tag{18.303}$$

Thus the desired kernel for the linear operator $\boldsymbol{\mathcal{L}}$ is

$$L_j(t, \mathbf{r}) = C \exp(-2\pi i \nu_0 t)\, [h_j(\mathbf{r})]^2\, m\left(t - 2\frac{|\mathbf{r} - \mathbf{r}_{0j}|}{c}\right). \tag{18.304}$$

Note that the monochromatic diffraction pattern of the transducer, $h_j(\mathbf{r})$, appears as a square (not squared modulus) in the kernel. Any phase distortions in the transducer plane (deliberate or otherwise) get doubled because we go through the same transducer twice. Similarly, the time delay of the envelope is doubled since a round trip is required to get back to the transducer.

Electronic filtering There is almost always some electronic filtering applied to the transducer output. At the least, there is a bandpass filter to suppress noise outside the signal bandwidth, and if chirps or other coded pulses are used, the pulse compression or decoding is performed with an electrical filter. The filters are usually linear and shift invariant, so we can write the filter output as

$$\widetilde{g}_j(t) = \int_{-\infty}^{\infty} dt'\, g_j(t')\, q(t - t') \qquad \text{or} \qquad \widetilde{\mathbf{g}} = \boldsymbol{\mathcal{Q}}\mathbf{g}\,. \tag{18.305}$$

The overall mapping from $\mathbf{V}$ to the filter output is then given by

$$\widetilde{\mathbf{g}} = \boldsymbol{\mathcal{Q}}\boldsymbol{\mathcal{L}}\mathbf{V}\,, \tag{18.306}$$

where the kernel of $\boldsymbol{\mathcal{Q}}\boldsymbol{\mathcal{L}}$ is given by

$$[\boldsymbol{\mathcal{Q}}\boldsymbol{\mathcal{L}}]_j(t, \mathbf{r}) = \int_{-\infty}^{\infty} dt'\, L_j(t', \mathbf{r})\, q(t - t')\,. \tag{18.307}$$

With the help of (18.304), we obtain

$$[\boldsymbol{\mathcal{Q}}\boldsymbol{\mathcal{L}}]_j(t, \mathbf{r}) = C\, [h_j(\mathbf{r})]^2 \int_{-\infty}^{\infty} dt' \exp(-2\pi i \nu_0 t')\, m\,[t' - \tau_j(\mathbf{r})]\; q(t - t')\,, \tag{18.308}$$

where $\tau_j(\mathbf{r}) \equiv 2|\mathbf{r} - \mathbf{r}_{0j}|/c$.

Pulse compression To illustrate the behavior of the kernel of $\boldsymbol{\mathcal{Q}}\boldsymbol{\mathcal{L}}$, let us suppose that the modulation is a chirp, so that

$$m(t) = A \exp(-i\pi\beta t^2)\, \mathrm{rect}\left(\frac{t}{T}\right). \tag{18.309}$$

From (18.295), the source term in the wave equation is

$$s_j(\mathfrak{r},t) = A\exp(-2\pi i\nu_0 t - i\pi\beta t^2)\,\mathrm{rect}\left(\frac{t}{T}\right) s_j(\mathfrak{r})\,. \tag{18.310}$$

This signal extends over $-\frac{1}{2}T < t < \frac{1}{2}T$. The corresponding local frequency is obtained by differentiating the phase as in (5.27);[17] the result is

$$\nu(t) = -\frac{1}{2\pi}\frac{\partial}{\partial t}(-2\pi\nu_0 t - \pi\beta t^2) = \nu_0 + \beta t\,. \tag{18.311}$$

Thus the signal spectrum extends over $\nu_0 - \frac{1}{2}\beta T < \nu(t) < \nu_0 + \frac{1}{2}\beta T$. The bandwidth of the chirp is βT and the time-bandwidth product is βT^2.

Now suppose that the filter impulse response is the same as the transmitted signal except that β is changed to $-\beta$, which has the effect of converting an up-chirp (frequency increasing with time) to a down-chirp:

$$q(t) = \exp(-2\pi i\nu_0 t + i\pi\beta t^2)\,\mathrm{rect}\left(\frac{t}{T}\right) \equiv q_0(t)\exp(-2\pi i\nu_0 t)\,, \tag{18.312}$$

where the exponential factor tells us that an impulse applied to the filter will generate a bandpass signal centered on ν_0, and the factor $q_0(t)$ is the phase modulation of that signal. This impulse response can be realized by a dispersive delay line with a delay that varies linearly with frequency as in Fig. 18.8.

With (18.308) – (18.310) and some algebra, the kernel of $\mathcal{QL}$ becomes

$$[\mathcal{QL}]_j(t,\mathfrak{r}) = AC\,[h_j(\mathfrak{r})]^2 \exp\{-2\pi i\nu_0[t-\tau_j(\mathfrak{r})]\}\exp\{i\pi\beta[t-\tau_j(\mathfrak{r})]^2\}$$
$$\times \int_{-\infty}^{\infty} dt'' \exp\{-2\pi i\beta[t-\tau_j(\mathfrak{r})]t''\}\,\mathrm{rect}\left(\frac{t''}{T}\right)\mathrm{rect}\left[\frac{t-\tau_j(\mathfrak{r})-t''}{T}\right], \tag{18.313}$$

where $t'' = t' - \tau_j(\mathfrak{r})$. The important point is that the quadratic phase factors inside the integral have cancelled. The integral is now just the Fourier transform of the product of the two rects, which is another rect of width $T' = T - |t - \tau_j(\mathfrak{r})|$. The "frequency variable" in this Fourier transform is $\beta[t - \tau_j(\mathfrak{r})]$, and the transform is small if $\beta|t-\tau_j(\mathfrak{r})|T' \gg 1$. If the time-bandwidth product βT^2 is large compared to 1, we must have $|t - \tau_j(\mathfrak{r})| \ll T$, and hence

$$[\mathcal{QL}]_j(t,\mathfrak{r}) \approx AC\,[h_j(\mathfrak{r})]^2 \exp\{-2\pi i\nu_0[t-\tau_j(\mathfrak{r})]\}\,T\,\mathrm{sinc}\{\beta T[t-\tau_j(\mathfrak{r})]\}\,,$$
$$(\tau_j(\mathfrak{r}) \equiv 2|\mathfrak{r} - \mathfrak{r}_{0j}|/c)\,. \tag{18.314}$$

This kernel is just what we would have obtained by transmitting a pulse with $m(t) = AT\,\mathrm{sinc}(\beta T t)$ and doing no filtering. The actual transmitted pulse has been compressed from its initial width of T to a processed width of $1/(\beta T)$. The compression ratio CR is thus

$$CR \equiv \frac{\text{initial width}}{\text{width after processing}} = \frac{T}{(\beta T)^{-1}} = \beta T^2 = \text{time-bandwidth product}\,. \tag{18.315}$$

The factor of T in (18.314) will become important when we discuss noise. As the pulse is compressed, it is increased in amplitude, potentially aiding in the detection of signals buried in noise.

[17]There is a minus sign in (18.311) not seen in (5.27). The sign difference arises because we chose in Chap. 9 to use different sign conventions on spatial and temporal Fourier transforms in order to make the Fourier kernel $\exp[i(\mathbf{k}\cdot\mathfrak{r} - \omega t)]$ represent a plane wave travelling parallel to $\mathbf{k}$.

18.6.3 Statistical analysis

When we turn to the statistical aspects of coherent ranging systems, the key question concerns the nature of randomness and the role of probability. If we say, as we did in Sec. 18.6.2, that the object is the scattering potential **V**, and if we neglect measurement noise, then the conditional density $\text{pr}(\mathbf{g}|\mathbf{V})$ is merely $\delta(\mathbf{g} - \boldsymbol{\mathcal{L}}\mathbf{V})$ and any discussion of statistics is moot. On the other hand, we seldom know the details or fine structure of the scattering potential, and we might not be interested in them anyway. Statistical descriptions are a way of overcoming this lack of information, but it is incumbent on us to say what part of the scattering potential we think of as the object of interest and what part we regard as uninteresting and hence relegate to statistical summaries.

We avoided this issue in Sec. 18.2, where we considered free-space propagation of light from a ground glass, by saying that the ground glass was one realization of a random process. In that case all we were interested in was the statistical description of the diffracted light, and the implicit statistical ensemble was a set of ground glasses from which the one currently in the laser beam was drawn.

The problem became more acute in Sec. 18.3 when we added a lens to form an imaging system. We formulated the problem there by considering the object to be a photographic transparency $\mathbf{t}_{obj}$ placed over a coherently illuminated ground glass. We imagined a set of experiments where the ground glass was varied but the object transparency was not; thus $\text{pr}(\mathbf{g}|\mathbf{t}_{obj})$ was no longer a delta function in this interpretation, and we could regard $\mathbf{t}_{obj}$ or $|\mathbf{t}_{obj}|^2$ as the object.

Additional ways of defining an object were suggested in Sec. 18.5, where we considered distributions of point scatterers. One approach, evident in (18.226), was to regard the object $f(\mathbf{r})$ as the average density of point scatterers, where the average is over positions of the individual points but does not involve the phases. Another approach was to treat the object as an average energy reflectance, given by the density of scatterers times the average of the square of the scattering amplitude a_n.

These definitions can be extended to the 3D problems considered here, but they are not without their difficulties. When we want to perform a specific task, such as deciding whether a given speckle pattern was produced by an object from class 1 or class 2, it may be precisely the fine structure that is most important in the discrimination. In those cases we would not want to regard the fine structure as noise and lump it into $\text{pr}(\mathbf{g}|\mathbf{f})$, but instead would like to include it in $\text{pr}(\mathbf{f}|H_j)$, which is how we distinguish the classes in the first place. As we know from Chap. 13, the discrimination is ideally based on $\text{pr}(\mathbf{g}|H_j) = \int d\mathbf{f}\, \text{pr}(\mathbf{g}|\mathbf{f})\, \text{pr}(\mathbf{f}|H_j)$. Thus we would be better advised to use $\text{pr}(\mathbf{g}|\mathbf{f})$ to account for measurement noise and to reserve the speckle statistics for $\text{pr}(\mathbf{f}|H_j)$.

As a practical matter, however, this ideal-observer approach is beyond the current state of the art in object modeling. In practice, tissue characterization in ultrasound or identification of regions in SAR images of terrain is based on estimates of statistical parameters under assumptions of quasistationarity rather than a full consideration of $\text{pr}(\mathbf{f}|H_j)$. The estimates are treated as features, and the classifier is formulated in feature space rather than data space.

Dealing with reality As we formulated the coherent ranging problem in Sec. 18.6.2, the data consisted of a set of complex functions $\{g_j(t)\}$. Recall, however, that we are using the convention — virtually universal in wave propagation — that the real part of all complex quantities is understood. Thus the data are actually the set of real functions, $\{\text{Re}[g_j(t)]\}$. The functions that appear in the argument of the characteristic functional for the data are also real, and we denote them by $\{\xi_j(t)\}$, with the corresponding Hilbert-space vector denoted $\boldsymbol{\xi}$.

We define the characteristic functional for the real data as

$$\Psi_{\mathbf{g}}(\boldsymbol{\xi}) = \left\langle \exp\left\{-2\pi i \boldsymbol{\xi}^\dagger \,\text{Re}[\mathbf{g}]\right\}\right\rangle = \left\langle \exp\left\{-\pi i \boldsymbol{\xi}^\dagger [\mathbf{g} + \mathbf{g}^*]\right\}\right\rangle , \tag{18.316}$$

where, to be specific,

$$\boldsymbol{\xi}^\dagger \mathbf{g} \equiv \sum_{j=1}^{J} \int_{-\infty}^{\infty} dt \; \xi_j(t)\, g_j(t) , \qquad \boldsymbol{\xi}^\dagger \mathbf{g}^* \equiv \sum_{j=1}^{J} \int_{-\infty}^{\infty} dt \; \xi_j(t)\, g_j^*(t) , \tag{18.317}$$

with J being the total number of transducer configurations.

Since $\mathbf{g} = \boldsymbol{\mathcal{L}}\mathbf{V}$, we can also write

$$\boldsymbol{\xi}^\dagger \,\text{Re}[\mathbf{g}] = \tfrac{1}{2} \sum_{j=1}^{J} \int_{-\infty}^{\infty} dt \; \xi_j(t) \left\{ [\boldsymbol{\mathcal{L}}\mathbf{V}]_j(t) + [\boldsymbol{\mathcal{L}}\mathbf{V}]_j^*(t) \right\} . \tag{18.318}$$

The adjoint operator for this mixed CC-CD problem is defined by combining (1.43) and (1.45), yielding

$$[\boldsymbol{\mathcal{L}}^\dagger \boldsymbol{\xi}](\mathbf{r}) = \sum_{j=1}^{J} \int_{-\infty}^{\infty} dt \; \xi_j(t)\, L_j^*(t, \mathbf{r}) , \qquad [\boldsymbol{\mathcal{L}}^\dagger \boldsymbol{\xi}]^*(\mathbf{r}) = \sum_{j=1}^{J} \int_{-\infty}^{\infty} dt \; \xi_j(t)\, L_j(t, \mathbf{r}) . \tag{18.319}$$

Thus

$$\boldsymbol{\xi}^\dagger \,\text{Re}[\mathbf{g}] = \tfrac{1}{2} \int d^3 r \; \left\{ [\boldsymbol{\mathcal{L}}^\dagger \boldsymbol{\xi}]^*(\mathbf{r})\, V(\mathbf{r}) + V^*(\mathbf{r})\, [\boldsymbol{\mathcal{L}}^\dagger \boldsymbol{\xi}](\mathbf{r}) \right\} = \tfrac{1}{2} \left\{ [\boldsymbol{\mathcal{L}}^\dagger \boldsymbol{\xi}]^\dagger \mathbf{V} + \mathbf{V}^\dagger [\boldsymbol{\mathcal{L}}^\dagger \boldsymbol{\xi}] \right\} . \tag{18.320}$$

As usual, we define the characteristic functional of the complex scattering potential by

$$\Psi_{\mathbf{V}}(\boldsymbol{\zeta}) = \left\langle \exp\left[-i\pi \left(\boldsymbol{\zeta}^\dagger \mathbf{V} + \mathbf{V}^\dagger \boldsymbol{\zeta}\right)\right]\right\rangle , \tag{18.321}$$

where $\boldsymbol{\zeta}$ corresponds to the set of *complex* functions $\{\zeta_j(\mathbf{r})\}$, so that

$$\boldsymbol{\zeta}^\dagger \mathbf{V} \equiv \int d^3 r \; \boldsymbol{\zeta}^*(\mathbf{r})\, V(\mathbf{r}) . \tag{18.322}$$

But note from (18.320) that

$$\Psi_{\mathbf{g}}(\boldsymbol{\xi}) = \left\langle \exp\left\{-2\pi i \boldsymbol{\xi}^\dagger \,\text{Re}[\mathbf{g}]\right\}\right\rangle = \left\langle \exp\left\{-i\pi \left([\boldsymbol{\mathcal{L}}^\dagger \boldsymbol{\xi}]^\dagger \mathbf{V} + \mathbf{V}^\dagger \,[\boldsymbol{\mathcal{L}}^\dagger \boldsymbol{\xi}] \right)\right\}\right\rangle , \tag{18.323}$$

or, by comparison with (18.321),

$$\Psi_{\mathbf{g}}(\boldsymbol{\xi}) = \Psi_{\mathbf{V}}\left(\boldsymbol{\mathcal{L}}^\dagger \boldsymbol{\xi}\right) . \tag{18.324}$$

Thus the usual complex transformation rule works if we merely take $\boldsymbol{\xi}$ real, even though $\boldsymbol{\mathcal{L}}^\dagger \boldsymbol{\xi}$ is complex.

The Gaussian case For one time t and one detector configuration j, the data value is a weighted sum of contributions from object points in a *focal volume* defined by the kernel $L_j(t, \mathbf{r})$ defined in (18.304). As we discussed qualitatively in Sec. 18.6.1, the size of this volume is determined by both the duration of the modulation pulse $m(t)$ and by λ_0/D, where D is the diameter of the transducer and λ_0 is the free-space wavelength associated with the carrier frequency.

If the scattering potential in this focal volume can be divided into many statistically independent subvolumes, then it follows from the central-limit theorem that the received signal is Gaussian. Moreover, if we can argue that the phase of the reflected wave from each point in the volume is uniformly distributed over $(-\pi, \pi)$, then the statistics of the received signal are circular Gaussian. As we saw in (18.225), the phase of this wave arises at least in part from the random location of the scattering point in the illuminating beam. Thus a large focal volume leads us to the circular Gaussian model for two reasons: the volume contains more independent regions as it gets larger, and the phases vary more because the variations in round-trip propagation times to the transducer are greater.

If circular Gaussian statistics apply, all statistical properties of the received signal are determined by the autocovariance function of the scattering potential (which is also the autocorrelation function since circular Gaussians necessarily have zero mean). It follows from (8.147) that the autocovariance operator for $\mathbf{g}$ is related to that for $\mathbf{V}$ by

$$\boldsymbol{\mathcal{K}}_{\mathbf{g}} = \boldsymbol{\mathcal{L}}\boldsymbol{\mathcal{K}}_{\mathbf{V}}\boldsymbol{\mathcal{L}}^{\dagger}\,, \tag{18.325}$$

and the characteristic functional for the data becomes

$$\Psi_{\mathbf{g}}(\boldsymbol{\xi}) = \exp(-\pi^2 \boldsymbol{\xi}^{\dagger}\boldsymbol{\mathcal{L}}\boldsymbol{\mathcal{K}}_{\mathbf{V}}\boldsymbol{\mathcal{L}}^{\dagger}\boldsymbol{\xi})\,. \tag{18.326}$$

We can go a step further if the correlations of the scattering potential have very short range compared to the extent of the focal volume in all three dimensions. In that case we can approximate the autocovariance function (the kernel of $\boldsymbol{\mathcal{K}}_{\mathbf{V}}$) by

$$K_{\mathbf{V}}(\mathbf{r}, \mathbf{r}') = f(\mathbf{r})\,\delta(\mathbf{r} - \mathbf{r}')\,. \tag{18.327}$$

A physical interpretation of $f(\mathbf{r})$ will be given below.

With (18.327), (18.325) and (18.304), the autocovariance function for the data becomes

$$[\boldsymbol{\mathcal{K}}_{\mathbf{g}}]_{jj'}(t, t') = \int d^3r \int d^3r'\, L_j(t, \mathbf{r})\, f(\mathbf{r})\,\delta(\mathbf{r} - \mathbf{r}')\, L^*_{j'}(t', \mathbf{r}') = C^2 \exp[-2\pi i\nu_0(t - t')]$$

$$\times \int d^3r\; f(\mathbf{r})\,[h_j(\mathbf{r})]^2[h^*_{j'}(\mathbf{r})]^2\, m\!\left(t - 2\frac{|\mathbf{r} - \mathbf{r}_{0j}|}{c}\right) m^*\!\left(t' - 2\frac{|\mathbf{r} - \mathbf{r}_{0j'}|}{c}\right). \tag{18.328}$$

This expression shows that the data are correlated in time by an amount determined by the duration of the transmitted pulse; if $j = j'$ and $m(t)$ has duration T, then the integrand vanishes if $|t - t'| > T$. Signals from two different transducer configurations, $j \neq j'$, are correlated only if the two beams overlap.

Random point scatterers As in Sec. 18.5, it is often useful to think of the scattering in coherent ranging as coming from a collection of point objects at random locations. In this model the scattering potential is written

$$V(\mathbf{r}) = \sum_{n=1}^{N} a_n \exp(i\phi_n)\,\delta(\mathbf{r} - \mathbf{r}_n)\,. \tag{18.329}$$

A very similar expression, (18.223), appeared in Sec. 18.5 when we analyzed 2D coherent imaging, but the interpretation here is quite different; (18.223) could represent both the scatterers themselves and the scattered field since the illumination was just a plane wave. Here the illumination is a complicated spatio-temporal field.

Nevertheless, we can make good use of results from Sec. 18.5. If we assume that the phases ϕ_n are completely random and that the random positions $\mathfrak{r}_n$ are statistically independent but with a nonrandom density (per unit volume) $\mathit{b}(\mathfrak{r})$, then the derivation leading to (18.242) is still valid (for the scattering potential, not the field), and we see that

$$\Psi_{\mathbf{V}}(\boldsymbol{\zeta}) = \exp\left[-\overline{N} + \int_{\mathbf{S}} d^3\mathfrak{r}_n \; \mathit{b}(\mathfrak{r}_n) \left\langle \mathrm{J}_0[2\pi a_n |\zeta(\mathfrak{r}_n)|] \right\rangle_{a_n}\right] . \tag{18.330}$$

From this characteristic functional the reader can show [hint: use (18.248)] that the autocovariance function of $\mathbf{V}$ is given by

$$K_{\mathbf{V}}(\mathfrak{r}, \mathfrak{r}') = \mathit{b}(\mathfrak{r}) \left\langle a_n^2 \right\rangle \delta(\mathfrak{r} - \mathfrak{r}') . \tag{18.331}$$

Comparison with (18.327) then shows that

$$f(\mathfrak{r}) = \mathit{b}(\mathfrak{r}) \left\langle a_n^2 \right\rangle . \tag{18.332}$$

If we think of a_n as the square root of a backscattering cross section, then $f(\mathfrak{r})$ is the number of scatterers per unit volume times the average scattering cross section. Thus $f(\mathfrak{r})$ has dimensions of 1/length; since the delta function has dimension 1/length^3, $K_{\mathbf{V}}$ has dimensions 1/length^4, as it must since $\mathbf{V}$ has dimensions of 1/length^2 by (18.292).

We can also use (18.330) to analyze the statistics of the data. Application of (18.324) to (18.330) gives us immediately

$$\Psi_{\mathbf{g}}(\boldsymbol{\xi}) = \exp\left\{-\overline{N} + \int_{\mathbf{S}} d^3\mathfrak{r}_n \; \mathit{b}(\mathfrak{r}_n) \left\langle \mathrm{J}_0[2\pi a_n |[\boldsymbol{\mathcal{L}}^\dagger \boldsymbol{\xi}](\mathfrak{r}_n)|] \right\rangle_{a_n}\right\} . \tag{18.333}$$

When used with the definition of $\boldsymbol{\mathcal{L}}^\dagger$ from (18.319), this expression contains the complete statistical description of the data for the point-scattering model, provided the only noise present is speckle.

Electronic noise Radar and ultrasound signals are often quite weak, and post-detection electronic noise cannot be neglected. Poisson noise, on the other hand, does not arise since the energy of microwave photons and acoustic phonons is too low to liberate photoelectrons.

When electronic noise is included, the overall mapping from scattering potential to data is given by

$$\mathbf{g} = \boldsymbol{\mathcal{L}}\mathbf{V} + \mathbf{n} , \tag{18.334}$$

where $\mathbf{n} \Rightarrow n_j(t)$. The real part is still understood on $\boldsymbol{\mathcal{L}}\mathbf{V}$, but $\mathbf{n}$ is assumed real.

Since electronic noise is independent of signal level, $\mathbf{V}$ and $\mathbf{n}$ are statistically independent, and (18.324) becomes

$$\Psi_{\mathbf{g}}(\boldsymbol{\xi}) = \Psi_{\mathbf{n}}(\boldsymbol{\xi}) \, \Psi_{\mathbf{V}}(\boldsymbol{\mathcal{L}}^\dagger \boldsymbol{\xi}) . \tag{18.335}$$

As discussed in detail in Chap. 12, electronic noise is usually well modeled as a zero-mean Gaussian random process, and for simplicity we assume that its bandwidth is very large so we can take it as white noise with a constant power spectral density S_n. We assume also that the noise in different transducer configurations is uncorrelated, so the noise autocovariance function is

$$[\mathcal{K}_\mathbf{n}]_{jj'}(t,t') = S_n\, \delta_{jj'}\, \delta(t-t')\,, \qquad \mathcal{K}_\mathbf{n} = S_n \mathcal{I}, \tag{18.336}$$

where $\mathcal{I}$ is the unit operator in data space. With these assumptions, the noise characteristic functional is

$$\Psi_\mathbf{n}(\boldsymbol{\xi}) = \exp[-2\pi^2 S_n\, \boldsymbol{\xi}^\dagger \boldsymbol{\xi}] = \exp\left[-2\pi^2 S_n \sum_{j=1}^{J} \int_{-\infty}^{\infty} dt\ |\xi_j(t)|^2\right]. \tag{18.337}$$

By Parseval's theorem, an equivalent expression is

$$\Psi_\mathbf{n}(\boldsymbol{\xi}) = \exp\left[-2\pi^2 S_n \sum_{j=1}^{J} \int_{-\infty}^{\infty} d\nu\ |\Xi_j(\nu)|^2\right], \tag{18.338}$$

where $\Xi_j(\nu) \equiv \mathcal{F}_1\{\xi_j(t)\}$.

Filtering If we also include an electronic filtering step as in (18.305), the characteristic functional for the filter output is

$$\Psi_{\tilde{\mathbf{g}}}(\boldsymbol{\xi}) = \Psi_\mathbf{n}(\mathcal{Q}^\dagger \boldsymbol{\xi})\, \Psi_\mathbf{V}(\mathcal{L}^\dagger \mathcal{Q}^\dagger \boldsymbol{\xi})\,. \tag{18.339}$$

This characteristic functional is the complete statistical description of the noise in all of the coherent ranging systems we are considering in this section. It will form the basis for our discussion of task performance below.

For the case of white Gaussian noise, as in (18.337), the noise characteristic functional on the filter output becomes

$$\Psi_\mathbf{n}(\mathcal{Q}^\dagger \boldsymbol{\xi}) = \exp[-2\pi^2 S_n\, \boldsymbol{\xi}^\dagger \mathcal{Q}\mathcal{Q}^\dagger \boldsymbol{\xi}]\,. \tag{18.340}$$

In the Fourier domain [*cf.* (18.338)],

$$\Psi_\mathbf{n}(\mathcal{Q}^\dagger \boldsymbol{\xi}) = \exp\left[-2\pi^2 S_n \sum_{j=1}^{J} \int_{-\infty}^{\infty} d\nu\ |Q(\nu)\, \Xi_j(\nu)|^2\right], \tag{18.341}$$

where $Q(\nu)$ is the transfer function of the filter, $Q(\nu) \equiv \mathcal{F}_1\{q(t)\}$. The temporal autocovariance function on the filter output is thus [*cf.* (8.156)]

$$K_{\tilde{\mathbf{n}}}(t,t') = S_n[\mathcal{Q}\mathcal{Q}^\dagger](t,t') = S_n \int_{-\infty}^{\infty} d\nu\ |Q(\nu)|^2 \exp[-2\pi i\nu(t-t')]\,. \tag{18.342}$$

18.6.4 Task performance

The original task of radar is implicit in the word: radio detection and ranging. In early radar, spatial resolution was poor, and all targets could be considered to be points. The only thing that could be asked of the signal was whether a target was present and how far away it was. In the language of Sec. 13.3.9, this is a hybrid detection/estimation task.

As radar systems improved, it became feasible to distinguish one kind of target from another on the basis of the received signal; different aircraft, for example, had different *radar signatures*, so the task became classification instead of just detection. In addition, as Doppler capabilities were added (see Sec. 18.6.1), it became possible to estimate the speed of the target in the beam direction as well as the range, so the estimation part of the task became more complicated also.

Synthetic aperture radar and medical ultrasound add new possibilities in terms of task specification since they produce high-resolution images of objects regarded as functions of continuous variables, not just as point targets. From these images we can perform all of the same tasks that we would with any image: detection, classification, estimation of integrals of the object, mensuration, etc.

In this section we shall examine both detection and estimation tasks in coherent ranging systems. Both ideal and practical strategies for task performance will be discussed, and extensive use will be made of results from Chaps. 13 and 14.

Signal-detection tasks: What do we mean by "signal"? There are two distinct ways of defining the signal to be detected in 3D coherent ranging systems. One is simply to define the object in terms of the scattering potential $V(\mathbf{r})$ as we did in Sec. 18.6.2. We can then decompose the object into a background part $V_b(\mathbf{r})$ and a signal part $V_s(\mathbf{r})$ (see Sec. 18.5.3). The signal, the background or both can be regarded as random, depending on the detailed task specification. Even if we model $V_b(\mathbf{r})$ as a circular Gaussian, the data do not have zero mean under the signal-present hypothesis because $V_s(\mathbf{r})$ is not zero mean.

The other viewpoint is to think of the object as the density of scatterers, possibly weighted by the average of the square of the scattering amplitude as in (18.332). In this case a signal can be defined as a localized change in the density of scatterers, and the data do have zero mean if the phases are completely random.

SKE detection of a point target in free space We begin our discussion of signal detection in coherent ranging with the simplest of signals: a point object at a known location in free space. In terms of the scattering potential, the object is specified by

$$V_s(\mathbf{r}) = a_s\,\delta(\mathbf{r} - \mathbf{r}_0)\,. \tag{18.343}$$

Since we are considering free space, $V_b(\mathbf{r}) = 0$, and the mean data under the signal-present hypothesis are given by

$$\overline{g}_j(t) = [\boldsymbol{\mathcal{L}}\mathbf{V}_s]_j(t) = a_s L_j(t, \mathbf{r}_0) = C a_s \exp(-2\pi i \nu_0 t)\,[h_j(\mathbf{r}_0)]^2\, m\left(t - 2\frac{|\mathbf{r}_0 - \mathbf{r}_{0j}|}{c}\right), \tag{18.344}$$

where $L_j(t, \mathbf{r})$ is the kernel of $\boldsymbol{\mathcal{L}}$, and the last step has used (18.304).

Since the background is zero (hence nonrandom), the only noise is the Gaussian electronic noise. We know a lot about SKE detection in Gaussian noise from Chap. 13, and the only thing new here is that the data have both continuous (t) and discrete (j) indices. We expect the ideal observer to be a linear discriminant, but it is not obvious how to find the discriminant function in this mixed-data setting.

To find the ideal discriminant function, we use a result from Sec. 13.2.12. From (13.226) we know that the log-likelihood ratio $\lambda(\mathbf{g})$ is equivalent to a linear discriminant if and only if there exists a data-space vector $\mathbf{w}$ such that[18]

$$\Psi_{\mathbf{g}|H_2}(\boldsymbol{\xi}) \propto \Psi_{\mathbf{g}|H_1}\left(\boldsymbol{\xi} + \frac{i}{2\pi}\mathbf{w}\right). \tag{18.345}$$

Thus, in order for the log-likelihood ratio to be a linear discriminant, $\Psi_{\mathbf{g}|H_2}(\boldsymbol{\xi})$ must be proportional to $\Psi_{\mathbf{g}|H_1}(\boldsymbol{\xi})$ with each component shifted along the imaginary axis; the vector $\mathbf{w}$ is the ideal-observer discriminant function.

We apply this result here to the filtered data $\widetilde{\mathbf{g}}$ rather than the raw transducer output $\mathbf{g}$. The signal-present characteristic functional for a nonrandom $\mathbf{V}_s$ in free space is

$$\Psi_{\widetilde{\mathbf{g}}|H_2}(\boldsymbol{\xi}) = \left\langle \exp[-2\pi i(\mathcal{Q}\mathcal{L}\mathbf{V}_s + \mathcal{Q}\mathbf{n})^\dagger\boldsymbol{\xi}]\right\rangle_{\mathbf{n}} = \exp[-2\pi i(\mathcal{Q}\mathcal{L}\mathbf{V}_s)^\dagger\boldsymbol{\xi}]\Psi_{\mathbf{n}}(\mathcal{Q}^\dagger\boldsymbol{\xi}). \tag{18.346}$$

If we consider white Gaussian noise on the transducer output and use (18.340), we can write

$$\Psi_{\widetilde{\mathbf{g}}|H_2}(\boldsymbol{\xi}) = \exp[-2\pi i(\mathcal{Q}\mathcal{L}\mathbf{V}_s)^\dagger\boldsymbol{\xi}]\exp[-2\pi^2 S_n\boldsymbol{\xi}^\dagger\mathcal{Q}\mathcal{Q}^\dagger\boldsymbol{\xi}]. \tag{18.347}$$

The shifted version of the no-signal characteristic functional is

$$\Psi_{\widetilde{\mathbf{g}}|H_1}\left(\boldsymbol{\xi} + \frac{i}{2\pi}\mathbf{w}\right) = \exp\left[-2\pi^2 S_n\left(\boldsymbol{\xi} + \frac{i}{2\pi}\mathbf{w}\right)^\dagger \mathcal{Q}\mathcal{Q}^\dagger\left(\boldsymbol{\xi} + \frac{i}{2\pi}\mathbf{w}\right)\right]. \tag{18.348}$$

Recalling that $\widetilde{\mathbf{g}}$, $\boldsymbol{\xi}$ and $\mathbf{w}$ are all real since the actual data are real, and that $\mathcal{Q}\mathcal{L}\mathbf{V}_s$ is real by the convention that the real part is understood, we find that $\Psi_{\widetilde{\mathbf{g}}|H_2}(\boldsymbol{\xi}) \propto \Psi_{\widetilde{\mathbf{g}}|H_1}(\boldsymbol{\xi} + \frac{i}{2\pi}\mathbf{w})$ if

$$S_n\mathcal{Q}\mathcal{Q}^\dagger\mathbf{w} = \mathcal{Q}\mathcal{L}\mathbf{V}_s. \tag{18.349}$$

A formal solution to this equation is given by [*cf.* (1.194) and (1.199)]

$$\mathbf{w} = \frac{1}{S_n}[\mathcal{Q}\mathcal{Q}^\dagger]^+\mathcal{Q}\mathcal{L}\mathbf{V}_s = \frac{1}{S_n}\mathcal{Q}^{\dagger+}\mathcal{L}\mathbf{V}_s, \tag{18.350}$$

where the second step follows from (1.151). Other solutions exist, but they differ from this one by null functions of $\mathcal{Q}^\dagger$, which correspond to frequency components outside the filter passband and do not affect the detectability.

An explicit form for the discriminant function can be found by using the fact that $\mathcal{Q}$ and $\mathcal{Q}\mathcal{Q}^\dagger$ are 1D and shift-invariant. [Recall from Sec. 7.2.3 that the adjoint of convolution with $q(t)$ is correlation with $q^*(t)$.] With judicious insertion of 1D (temporal) Fourier operators, (18.350) becomes

$$\mathcal{F}_1\mathbf{w} = \frac{1}{S_n}\mathcal{F}_1\mathcal{Q}^{\dagger+}\mathcal{F}_1^{-1}\mathcal{F}_1\mathcal{L}\mathbf{V}_s. \tag{18.351}$$

[18]The derivation leading to (13.226) assumed a finite-dimensional data space, and $\Psi(\cdot)$ denoted a characteristic function rather than functional in that chapter, but the result is independent of the dimension of the data and therefore extends in the limit to continuous data and characteristic functionals.

If we write $q(t) = q_0(t)\exp(-2\pi i\nu_0 t)$ and $w_j(t) = w_{j0}(t)\exp(-2\pi i\nu_0 t)$, do the transforms and drop some irrelevant constants, (18.351) becomes

$$W_{j0}(\nu) \equiv \mathcal{F}_1\{w_{j0}(t)\} \propto [h_j(\mathbf{r}_0)]^2 \exp[-2\pi i\nu\tau_j(\mathbf{r}_0)]\,\frac{M(\nu)}{Q_0^*(\nu)}\,, \quad (\nu \text{ in filter passband})\,, \tag{18.352}$$

where capital letters denote Fourier transforms as usual, and $\tau_j(\mathbf{r}_0) \equiv 2|\mathbf{r}_0 - \mathbf{r}_{0j}|/c$. The 0 subscripts on W_{j0} and Q_0 indicate that the functions are translated back to zero frequency (*baseband* in radar parlance) before calculating the test statistic. In practice, this is done by homodyne detection in which a signal centered at frequency ν_0 is mixed with a local oscillator described in complex terms by $\exp(-2\pi i\nu_0 t)$ and only the low-frequency terms are retained.

The ideal-observer test statistic (an affine transformation of the log-likelihood ratio) is the scalar product in the mixed data space of $\mathbf{w}$ and $\widetilde{\mathbf{g}}$; by Parseval's theorem, it can be written as

$$t_{ideal}(\widetilde{\mathbf{g}}) = \mathbf{w}^\dagger \widetilde{\mathbf{g}} \propto \sum_{j=1}^{J} \int_{-\infty}^{\infty} d\nu\; W_{j0}^*(\nu)\, \widetilde{G}_{j0}(\nu)$$

$$\propto \sum_{j=1}^{J} [h_j^*(\mathbf{r}_0)]^2 \int_{-\nu_{max}}^{\nu_{max}} d\nu\; \frac{\widetilde{G}_{j0}(\nu)\, M^*(\nu)}{Q_0(\nu)} \exp[2\pi i\nu\tau_j(\mathbf{r}_0)]\,, \tag{18.353}$$

where the filter passband is $\nu_0 - \nu_{max} < \nu < \nu_0 + \nu_{max}$.

The second line in (18.353) tells us that the ideal observer will first undo the electrical filter by applying an inverse filter (thereby prewhitening the data temporally), then do a temporal matched filter on the signal with its original modulation. The phase factor in the integrand tells us that the observer is looking specifically for a signal delayed by the known amount $\tau_j(\mathbf{r}_0)$ in each of the received signals. The weighting with $[h_j^*(\mathbf{r}_0)]^2$ and the sum over j constitute a matched filter with respect to the discrete transducer configurations; no prewhitening is needed in this step since we assume [see (18.336)] that signals with different j are uncorrelated.

Point-target detectability and the radar equation Since the data are Gaussian and the discriminant function is linear, the ideal-observer test statistic obeys Gaussian statistics also. Thus the SNR of the test statistic as defined in (13.19) is identical to the detectability index d_A from (13.21), and either can be used to specify the performance. The reader can fill in the details, but it follows from (13.178) that SNR^2 for an ideal observer and a point target at $\mathbf{r} = \mathbf{r}_0$ is given by

$$\text{SNR}^2_{ideal}(\mathbf{r}_0) = \frac{1}{S_n}\mathbf{V}_s^\dagger \mathcal{L}^\dagger \mathcal{Q}^\dagger \mathcal{Q}^{\dagger +} \mathcal{L}\mathbf{V}_s = \frac{C^2|a_s|^2\mathcal{E}_t}{S_n}\sum_{j=1}^{J}|h_j(\mathbf{r}_0)|^4\,, \tag{18.354}$$

where $\mathcal{E}_t$ is the total energy in the transmitted signal, given by

$$\mathcal{E}_t \equiv \int_{-\nu_{max}}^{\nu_{max}} d\nu\; |M(\nu)|^2\,. \tag{18.355}$$

Each of the factors in (18.354) has a straightforward physical interpretation. We see that SNR^2 is inversely proportional to the power spectral density of the white noise going into the filter, but it is directly proportional to the signal strength as expressed by the cross section $|a_s|^2$. The transducer conversion factor C comes in quadratically since the noise is introduced after the transducer; thus an inefficient transducer reduces the signal but not the noise. The sum over j indicates that additional transducer configurations (or "looks" at the target) are beneficial to the degree that the target remains in the beam, as specified by the factor $|h_j(\mathbf{r})|^4$. The form of the transmitted signal comes into the SNR only as the total energy $\mathcal{E}_t$; the details of the modulation are irrelevant for detectability [*cf.* (13.120)]. In particular, note that the linear phase factor $\exp[2\pi i\nu\tau_j(\mathbf{r}_0)]$ seen in (18.353) has disappeared since the temporal matched filtering is a correlation, and the autocorrelation of $m(t-\tau)$ is the same as the autocorrelation of $m(t)$, namely a function peaked at $t=0$. (By contrast, the autoconvolution of $m(t-\tau)$ is peaked at $t=2\tau$.)

Since $h_j(\mathbf{r})$ is the amplitude in the diffraction pattern of the transducer, it falls off as $1/|\mathbf{r}_{0j}-\mathbf{r}_0|$ in the Fraunhofer region. Thus, for a monostatic system where all of the $\mathbf{r}_{0j}$ are the same,

$$\mathrm{SNR}^2_{ideal}(\mathbf{r}_0) \propto \frac{C^2|a_s|^2\mathcal{E}_t}{S_n}\,\frac{1}{|\mathbf{r}_{0j}-\mathbf{r}_0|^4}\,. \tag{18.356}$$

This expression is known as the *radar equation*. It shows that the point-target detectability falls off as the inverse fourth power of distance but that it can be increased by increasing either the energy of the transmitted signal or the transducer conversion efficiency. The latter option gets considerable attention in medical ultrasound.

SKE detection of extended targets in free space Much of what we learned above about ideal-observer detection for a point target in free space is readily extended to more complicated nonrandom targets, so long as we assume that the first Born approximation is valid. In particular, if there is no scattering background, so that $V_b(\mathbf{r})=0$ and the only noise is electronic, then the characteristic functionals of the filtered data under H_1 and H_2 are still given by (18.340) and (18.346), respectively. The ideal-observer test statistic is still linear, and the discriminant function is still given formally by (18.350).

To get a useful operational form for the test statistic, we use (18.303) – (18.305) to write [*cf.* (18.353)]

$$t_{ideal}(\mathbf{g}) \propto \sum_{j=1}^{J}\int d^3r\,[h_j^*(\mathbf{r})]^2\,V_s(\mathbf{r})\int_{-\nu_{max}}^{\nu_{max}} d\nu\,\frac{\widetilde{G}_{j0}(\nu)\,M^*(\nu)}{Q_0(\nu)}\exp[2\pi i\nu\tau_j(\mathbf{r})]\,. \tag{18.357}$$

As with point targets, the electrical filter is useless to the ideal observer, so the first step is to get rid of it by inverse-filtering (and hence temporally prewhitening) each received signal. Then the jth prewhitened signal is converted to a spatial function by setting $t=\tau_j(\mathbf{r})$, and this spatial signal is matched-filtered against the expected signal; no spatial prewhitening is required by dint of (18.336).

An expression for the detectability for an extended target will be given below when we allow both electronic noise and speckle.

SKE detection in speckle So far in this section, we have considered only targets in free space, but many problems in coherent ranging also involve a diffuse scattering background, so we have speckle as well as electronic noise. As we discussed in Sec. 18.6.3, the speckle may be well described by circular Gaussian statistics in some cases, but if the number of scatterers in a focal volume is small, non-Gaussian statistics may also arise.

In the Gaussian case, the data statistics are fully specified by the autocovariance operator, given by

$$\mathcal{K}_{\tilde{\mathbf{g}}} = \mathcal{Q}\mathcal{K}_{\mathbf{n}}\mathcal{Q}^{\dagger} + \mathcal{Q}\mathcal{L}\mathcal{K}_{\mathbf{V}}\mathcal{L}^{\dagger}\mathcal{Q}^{\dagger} . \tag{18.358}$$

An explicit form for the first term is given in (18.342), and one for the second term can be obtained from (18.328) by replacing $m(t)$ with $[q * m](t)$.

Since the noise is signal-independent and Gaussian, the ideal observer is a linear discriminant, and the discriminant function must satisfy [*cf.* (18.349)]

$$\left[S_n\mathcal{Q}\mathcal{Q}^{\dagger} + \mathcal{Q}\mathcal{L}\mathcal{K}_{\mathbf{V}}\mathcal{L}^{\dagger}\mathcal{Q}^{\dagger}\right]\mathbf{w} = \mathcal{Q}\mathcal{L}\mathbf{V}_s . \tag{18.359}$$

Without the speckle term in the covariance, we were able to get an explicit operational form for the discriminant, (18.352), by performing a temporal Fourier transform; because the electronic noise was stationary, Fourier transformation diagonalized the covariance operator. The speckle, on the other hand is not temporally stationary, so it is not obvious that a temporal Fourier transform will be useful in the present problem.

In similar situations in earlier chapters, we appealed to quasistationarity and assumed that the noise was locally stationary in the vicinity of the signal. For a signal centered at $\mathfrak{r} = \mathfrak{r}_0$, the essential approximation is to rewrite (18.327) as

$$K_{\mathbf{V}}(\mathfrak{r},\mathfrak{r}') = f_b(\mathfrak{r})\,\delta(\mathfrak{r}-\mathfrak{r}') \approx f_b(\mathfrak{r}_0)\,\delta(\mathfrak{r}-\mathfrak{r}') , \tag{18.360}$$

where $f_b(\mathfrak{r})$ is the background object, defined as in (18.332) by the mean density of scatterers times the mean scattering cross section. The assumption of spatial stationarity does not, however, immediately translate to temporal stationarity in the received signal; the covariance on $\mathbf{g}$ is given by [*cf.* (18.328)]

$$[\mathcal{K}_{\mathbf{g}}]_{jj'}(t,t')$$
$$\approx C^2 f_b(\mathfrak{r}_0)\exp[-2\pi i\nu_0(t-t')]\int d^3\mathfrak{r}\;[h_j(\mathfrak{r})]^2\,[h^*_{j'}(\mathfrak{r})]^2\,m[t-\tau_j(\mathfrak{r})]\,m^*[t'-\tau_{j'}(\mathfrak{r})] , \tag{18.361}$$

which is not just a function of $t - t'$. Some additional approximations are needed.

Since the transducers are assumed to lie in the plane $z = 0$, it is convenient to describe the response functions in terms of the 2D vector $\mathbf{r} = (x, y)$, so that $h_j(\mathfrak{r}) \equiv h_j(\mathbf{r}, z)$. In keeping with the quasistationary approximation (18.360), we can replace z with the signal coordinate z_0 in the response functions. The one place where we do not want to make this approximation, however, is in the argument of $m(\,\cdot\,)$; the modulation function determines the longitudinal resolution of the system, and we cannot assume that it is constant over the region of the signal. For propagation approximately parallel to the z axis, $\tau_j(\mathfrak{r}) \approx 2z/c$ (conveniently independent of j), so

$$[\mathcal{K}_{\mathbf{g}}]_{jj'}(t,t') \approx f_b(\mathfrak{r}_0)\,A_{jj'}\exp[-2\pi i\nu_0(t-t')]\int dz\; m\!\left(t-\frac{2z}{c}\right)m^*\!\left(t'-\frac{2z}{c}\right) , \tag{18.362}$$

where $A_{jj'}$ is a measure of the overlap of the beams from different transducer configurations:

$$A_{jj'} \equiv C^2 \int d^2r \, [h_j(\mathbf{r}, z_0)]^2 \left[h^*_{j'}(\mathbf{r}, z_0)\right]^2 . \qquad (18.363)$$

A change of variables $z' = z - ct/2$ yields

$$\int dz \, m\left(t - \frac{2z}{c}\right) m^*\left(t' - \frac{2z}{c}\right) = \int dz' \, m\left(-\frac{2z'}{c}\right) m^*\left(t' - t - \frac{2z'}{c}\right), \qquad (18.364)$$

showing that the covariance is now a function of only the time difference $t - t'$.

The remainder of the calculation proceeds by analogy with the case of SKE detection of a target in free space. By inserting Fourier operators appropriately, the reader can show that (18.359) is equivalent to

$$S_n|Q_0(\nu)|^2 \, W_{j0}(\nu) + \frac{c}{2} f_b(\mathbf{r}_0) \, |M(\nu)|^2 \, |Q_0(\nu)|^2 \sum_{j'=1}^{J} A_{jj'} W_{j'0}(\nu) = S_{j0}(\nu) , \qquad (18.365)$$

where $S_{j0}(\nu)$ is the signal transformed to the data domain by applying the operator $\mathcal{QL}$, Fourier-transforming, and shifting the frequency to baseband:

$$S_{j0}(\nu) \equiv [\mathcal{F}_1 \mathcal{QL} \mathbf{V}_s]_j \, (\nu - \nu_0) = CM(\nu) \, Q_0(\nu) \int d^3\mathbf{r} \, [h_j(\mathbf{r})]^2 \, V_s(\mathbf{r}) \exp[2\pi i \nu \tau_j(\mathbf{r})] . \qquad (18.366)$$

The solution of (18.365) can be written as

$$W_{j0}(\nu) = \frac{1}{S_n|Q_0(\nu)|^2} \sum_{j'=1}^{J} \left[\left(\mathbf{I} + \frac{c}{2S_n} f_b(\mathbf{r}_0) \, |M(\nu)|^2 \mathbf{A} \right)^{-1} \right]_{jj'} S_{j'0}(\nu) , \qquad (18.367)$$

where $\mathbf{A}$ is the $J \times J$ matrix with components $A_{jj'}$, and $\mathbf{I}$ is the $J \times J$ unit matrix.

Numerical evaluation of the inverse is not difficult since $\mathbf{A}$ is banded around the diagonal in practice; any reasonable scanning strategy will move the beam by an amount comparable to its width for each change in j since to move it in smaller increments would be to collect redundant data, while larger increments would risk missing the target altogether. Thus a small target will be seen by the system for only a few transducer configurations, and methods developed in Sec. 14.3.2 for dealing with nearly diagonal covariances, especially (14.39), are applicable here.

From (13.209), the ideal-observer detectability in this problem is

$$\mathrm{SNR}^2_{ideal} = \sum_{j=1}^{J} \int_{-\nu_{max}}^{\nu_{max}} d\nu \; S^*_{j0}(\nu) \, W_{j0}(\nu)$$

$$= \sum_{j=1}^{J} \sum_{j'=1}^{J} \int_{-\nu_{max}}^{\nu_{max}} d\nu \; \frac{1}{S_n|Q_0(\nu)|^2} \, S^*_{j0}(\nu) \left[\left(\mathbf{I} + \frac{c}{2S_n} f_b(\mathbf{r}_0) \, |M(\nu)|^2 \mathbf{A} \right)^{-1} \right]_{jj'} S_{j'0}(\nu) . \qquad (18.368)$$

Note that the characteristics of the electrical filter do not affect SNR^2_{ideal} (provided only that it doesn't cut off any signal frequencies) since $S^*_{j0}(\nu) S_{j'0}(\nu) \propto |Q_0(\nu)|^2$.

If $f_b(\mathbf{r}_0) \to 0$ (so there is no speckle) and we consider a point target, (18.368) reproduces (18.354). In the opposite case where $S_n \to 0$ and there is no electronic noise, we find

$$\mathrm{SNR}^2_{ideal} = \frac{2}{cf_b(\mathbf{r}_0)} \sum_{j=1}^{J} \sum_{j'=1}^{J} \int_{-\nu_{max}}^{\nu_{max}} d\nu \, \frac{1}{|M(\nu)|^2 \, |Q_0(\nu)|^2} \, S^*_{j0}(\nu) \left[\mathbf{A}^{-1}\right]_{jj'} S_{j'0}(\nu) \, . \tag{18.369}$$

The detectability in this limit is independent of both the electrical filter and the details of the modulation since $S^*_{j0}(\nu)S_{j'0}(\nu) \propto C^2|M(\nu)|^2 \, |Q_0(\nu)|^2$. It is even independent of the strength of the received wave since both the signal and the speckle scale the same way with the amplitude of the incident wave (contained in $|M(\nu)|^2$) and the transducer efficiency C^2 (contained in $\mathbf{A}$). The detectability does, however, depend on the details of the beam profile and the scanning strategy.

The SNR expressions in (18.368) and (18.369) simplify considerably if $\mathbf{A}$ is nearly diagonal, which can happen if the beam is translated by approximately its width on each step. When $\mathbf{A}$ is diagonal, (18.369) becomes[19]

$$\mathrm{SNR}^2_{ideal} = \frac{2}{cf_b(\mathbf{r}_0)} \sum_{j=1}^{J} \frac{\int_{-\nu_{max}}^{\nu_{max}} d\nu \, \left| \int d^3\mathbf{r} \, [h_j(\mathbf{r})]^2 \, V_s(\mathbf{r}) \exp[2\pi i\nu\tau_j(\mathbf{r})] \right|^2}{\int d^2r \, |h_j(\mathbf{r}, z_0)|^4} \, . \tag{18.370}$$

Since $\tau_j(\mathbf{r}) = 2z/c$ for propagation approximately parallel to the z axis, the numerator in (18.370) can be written

$$\int_{-\nu_{max}}^{\nu_{max}} d\nu \int d^2r \int dz \int d^2r' \int dz' \, [h^*_j(\mathbf{r}, z)]^2 \, V^*_s(\mathbf{r}, z)[h_j(\mathbf{r}', z')]^2 \, V_s(\mathbf{r}', z')$$
$$\times \exp\left[4\pi i\nu \frac{z'-z}{c}\right]$$
$$= 2\nu_{max} \int d^2r \int dz \int d^2r' \int dz' \, [h^*_j(\mathbf{r}, z)]^2 \, V^*_s(\mathbf{r}, z)[h_j(\mathbf{r}', z')]^2 \, V_s(\mathbf{r}', z')$$
$$\times \operatorname{sinc}\left[\frac{4}{c}\nu_{max}(z'-z)\right] . \tag{18.371}$$

We can go further if we assume that the signal to be detected is large and slowly varying in all three dimensions. First, if $V_s(\mathbf{r}, z)$ is slowly varying in the z direction compared to $\operatorname{sinc}\left[\frac{4}{c}\nu_{max}(z'-z)\right]$, we can write

$$2\nu_{max} \operatorname{sinc}\left[\frac{4}{c}\nu_{max}(z'-z)\right] \approx \frac{c}{2}\,\delta(z-z') \, , \tag{18.372}$$

and we can use this delta function to do the integral over z'. Similarly, if $V_s(\mathbf{r}, z)$ is slowly varying in x and y compared to the width of the beam, and the transducer is scanned laterally to successive positions $\mathbf{r}_j$, we can write

$$h^2_j(\mathbf{r}, z) = h^2_0(\mathbf{r} - \mathbf{r}_j, z) \approx \left[\int d^2r \, h^2_0(\mathbf{r}, z)\right] \delta(\mathbf{r} - \mathbf{r}_j) \, , \tag{18.373}$$

[19]Since the matrix $\mathbf{A}^{-1}$ arises from performing a prewhitening operation in j, neglecting the off-diagonal elements is justified even with beam overlap if we say we are using a non-prewhitening observer.

where $h_0(\mathbf{r}, z)$ is the response function for a transducer centered at $\mathbf{r}_j = \mathbf{0}$. Finally, we assume that the beam profile is nearly constant over the signal region, so $h_0(\mathbf{r}, z) \approx h_0(\mathbf{r}, z_0)$.

With all of these assumptions, the detectability takes the simple form,

$$\mathrm{SNR}^2_{ideal} \approx \frac{1}{f_b(\mathbf{r}_0)} \frac{\left|\int d^2r \; h_0^2(\mathbf{r}, z_0)\right|^2}{\int d^2r \; |h_0(\mathbf{r}, z_0)|^4} \sum_j \int dz \; |V_s(\mathbf{r}_j, z)|^2 . \tag{18.374}$$

We reiterate that this expression is for SKE detection of a broad, structureless signal described by the first Born approximation when there is no electronic noise and the background scattering that leads to speckle is spatially quasistationary. We also had to assume that the beam profiles do not change appreciably with z in the signal region, that the beams propagate approximately parallel to the z axis, and that the beam steps in x and y are approximately equal to the beam width so overlap can be neglected. If electronic noise is present or the assumptions about the beam profile are not valid, we have to use the more general expression given in (18.368), and probably evaluate it numerically.

SKE detection of increases in density of scatterers So far in this section we have considered the signal to be a known, nonrandom scattering potential $V_s(\mathbf{r})$. In many applications, however, we are interest in detecting an increased density of random scatterers. In medical ultrasound, for example, some tumors are evidenced by an increased backscattering or *echogenicity*, while cysts produce markedly weaker echo signals. Early work on detectability in this situation was carried out by Smith *et al.* (1983) and Wagner *et al.* (1983, 1986, 1987). More recent work on the subject was presented by Abbey *et al.* (2003) and Zemp *et al.* (2003).

A reasonable model for this problem is to treat the scattering potential as a zero-mean random process with different autocovariances under the two hypotheses (Abbey *et al.*, 2003). If we assume that the correlation length of the random process is small compared to the system resolution, then we can follow (18.327) and write

$$K_{\mathbf{V}1}(\mathbf{r}, \mathbf{r}') = f_b(\mathbf{r})\,\delta(\mathbf{r} - \mathbf{r}') , \qquad K_{\mathbf{V}2}(\mathbf{r}, \mathbf{r}') = [f_b(\mathbf{r}) + f_s(\mathbf{r})]\,\delta(\mathbf{r} - \mathbf{r}') , \tag{18.375}$$

where $K_{\mathbf{V}i}(\mathbf{r}, \mathbf{r}')$ is the autocovariance function under hypothesis H_i $(i = 1, 2)$, and $f_b(\mathbf{r})$ and $f_s(\mathbf{r})$ are defined similarly to (18.332) as the density of scatterers times the average cross section for background and signal, respectively. The corresponding autocovariance operators in the data domain are given by

$$\mathcal{K}_{\mathbf{g}\,i} = \mathcal{K}_{\mathbf{n}} + \mathcal{L}\mathcal{K}_{\mathbf{V}i}\mathcal{L}^\dagger . \tag{18.376}$$

If the mean number of scatterers in a resolution cell is large, then Gaussian statistics apply and the characteristic functionals for the data are given by (18.326). If the number of scatterers is small (but not too small), then the characteristic functional of (18.333) is applicable, but we shall address only the Gaussian case here.

In Sec. 13.2.10, we considered the problem of discrimination between signals in Gaussian noise with unequal covariance matrices. That analysis was for finite-dimensional random vectors, but the results extend readily to random processes; from (13.163), the log-likelihood ratio is the quadratic discriminant function,

$$\lambda(\mathbf{g}) = \tfrac{1}{2}\mathbf{g}^\dagger \left[\mathcal{K}_{\mathbf{g}1}^{-1} - \mathcal{K}_{\mathbf{g}2}^{-1}\right] \mathbf{g} . \tag{18.377}$$

The inverses exist because $\mathcal{K}_{\mathbf{n}}$ is a full-rank operator with our noise model, (18.336).

If we let $\Delta\mathcal{K}_{\mathbf{V}} \equiv \mathcal{K}_{\mathbf{V}2} - \mathcal{K}_{\mathbf{V}1}$ and make use of (18.376), we can write the operator in (18.377) as

$$\mathcal{K}_{\mathbf{g}1}^{-1} - \left(\mathcal{K}_{\mathbf{g}1} + \mathcal{L}\Delta\mathcal{K}_{\mathbf{V}}\mathcal{L}^{\dagger}\right)^{-1} = \mathcal{K}_{\mathbf{g}1}^{-1} - \left(\mathcal{I} + \mathcal{K}_{\mathbf{g}1}^{-1}\mathcal{L}\Delta\mathcal{K}_{\mathbf{V}}\mathcal{L}^{\dagger}\right)^{-1}\mathcal{K}_{\mathbf{g}1}^{-1}. \quad (18.378)$$

Following a suggestion by Abbey *et al.* (2003), we use the Neumann series (A.59) to obtain

$$\lambda(\mathbf{g}) = \tfrac{1}{2}\mathbf{g}^{\dagger}\mathcal{K}_{\mathbf{g}1}^{-1}\mathcal{L}\Delta\mathcal{K}_{\mathbf{V}}\mathcal{L}^{\dagger}\mathcal{K}_{\mathbf{g}1}^{-1}\mathbf{g} + \cdots. \quad (18.379)$$

The omitted terms are negligible if $\Delta\mathcal{K}_{\mathbf{V}}$ is small enough, so the truncated expansion should suffice if the signal contrast $f_s(\mathbf{r})/f_b(\mathbf{r})$ is very low. In the limit where the electronic noise goes to zero, we can also write

$$\lambda(\mathbf{g}) = \tfrac{1}{2}\mathbf{g}^{\dagger}\left[\mathcal{L}\mathcal{K}_{\mathbf{V}1}\mathcal{L}^{\dagger}\right]^{+}\mathcal{L}\Delta\mathcal{K}_{\mathbf{V}}\mathcal{L}^{\dagger}\left[\mathcal{L}\mathcal{K}_{\mathbf{V}1}\mathcal{L}^{\dagger}\right]^{+}\mathbf{g} + \cdots, \quad (18.380)$$

where $\left[\mathcal{L}\mathcal{K}_{\mathbf{V}1}\mathcal{L}^{\dagger}\right]^{+}$ is a Moore-Penrose pseudoinverse, defined according to (1.141) as[20]

$$\left[\mathcal{L}\mathcal{K}_{\mathbf{V}1}\mathcal{L}^{\dagger}\right]^{+} = \lim_{S_n \to 0}\left[S_n\mathcal{I} + \mathcal{L}\mathcal{K}_{\mathbf{V}1}\mathcal{L}^{\dagger}\right]^{-1}, \quad (18.381)$$

where S_n is the power spectral density of the white electronic noise (see (18.336)).

Various simplified forms of the log-likelihood ratio can be obtained by assuming quasistationarity ($f_b \approx$ constant) and assuming that the beam propagates approximately parallel to z with little change in lateral profile over the signal region. The calculations are similar to those presented above for detection of a nonrandom scattering potential and will not be detailed here.

Figures of merit for the ideal observer in this problem can be derived from likelihood-generating function $G(\beta)$ introduced in Sec. 13.2.7. We know from the discussion in that section that all properties of the likelihood ratio and its logarithm under both hypotheses are fully determined by $G(\beta)$. Moreover, the single number $G(0)$ is a very useful approximation to the SNR for the log-likelihood ratio (Clarkson and Barrett, 2000), and it can be used to set bounds on the ideal-observer AUC (Barrett *et al.*, 1998a, b; Shapiro, 1999; Clarkson, 2002).

One way of defining $G(0)$ is in terms of the Bhattacharyya distance; for a general M-dimensional random vector $\mathbf{g}$, we know from (13.97) that

$$G(0) = -4\log\left\{\int d^M g\,\left[\mathrm{pr}(\mathbf{g}|H_1)\,\mathrm{pr}(\mathbf{g}|H_2)\right]^{\frac{1}{2}}\right\}. \quad (18.382)$$

In the present problem, the data vector is an infinite-dimensional random process, but we can use (18.382) as if $\mathbf{g}$ were finite-dimensional and then let $M \to \infty$.

For a circular-Gaussian random vector, we know from (8.245) that

$$\mathrm{pr}(\mathbf{g}|H_i) = \frac{1}{\pi^N \det(\mathbf{K}_{\mathbf{g}i})}\exp\left(-\mathbf{g}^{\dagger}\mathbf{K}_{\mathbf{g}i}^{-1}\mathbf{g}\right). \quad (18.383)$$

[20]As written, (1.141) applies to an operator $\mathcal{H}$ that maps from one space to another. Here the operator $\mathcal{L}\mathcal{K}_{\mathbf{V}1}\mathcal{L}^{\dagger}$ is Hermitian and maps data space to itself, and (18.381) is an equivalent form in that case. The reader who doubts the validity of (18.381) is invited to check the Penrose equations.

Thus

$$G(0) = -4\log\left\{\frac{1}{\pi^N\sqrt{\det(\mathbf{K_{g1}})\det(\mathbf{K_{g2}})}}\int d^M g\, \exp\left[-\tfrac{1}{2}\mathbf{g}^\dagger\left(\mathbf{K}_{\mathbf{g}1}^{-1}+\mathbf{K}_{\mathbf{g}2}^{-1}\right)\mathbf{g}\right]\right\}$$

$$= 4\log\left\{\det\left(\mathbf{K}_{\mathbf{g}1}^{\frac{1}{2}}\right)\det\left(\mathbf{K}_{\mathbf{g}2}^{\frac{1}{2}}\right)\det\left(\tfrac{1}{2}\mathbf{K}_{\mathbf{g}1}^{-1}+\tfrac{1}{2}\mathbf{K}_{\mathbf{g}2}^{-1}\right)\right\}, \tag{18.384}$$

where in the second line we have used the normalization of $\text{pr}(\mathbf{g}|H_i)$ along with some properties of determinants from Sec. A.5; the square-root matrix is defined in Sec. A.8.3.

Now we define $\Delta\mathbf{K} \equiv \mathbf{K_{g2}} - \mathbf{K_{g1}}$ and use the following expansions:

$$\det\left[(\mathbf{K_{g1}}+\Delta\mathbf{K})^{\frac{1}{2}}\right] = \det\left[\mathbf{K}_{\mathbf{g}1}^{\frac{1}{2}}\left(\mathbf{I}+\tfrac{1}{2}\mathbf{K}_{\mathbf{g}1}^{-1}\Delta\mathbf{K}-\tfrac{1}{8}\mathbf{K}_{\mathbf{g}1}^{-1}\Delta\mathbf{K}\,\mathbf{K}_{\mathbf{g}1}^{-1}\Delta\mathbf{K}+\cdots\right)\right]; \tag{18.385}$$

$$\det\left[(\mathbf{K_{g1}}+\Delta\mathbf{K})^{-1}\right] = \det\left[\mathbf{K}_{\mathbf{g}1}^{-1}\left(\mathbf{I}-\mathbf{K}_{\mathbf{g}1}^{-1}\Delta\mathbf{K}+\mathbf{K}_{\mathbf{g}1}^{-1}\Delta\mathbf{K}\,\mathbf{K}_{\mathbf{g}1}^{-1}\Delta\mathbf{K}+\cdots\right)\right]. \tag{18.386}$$

To second order in $\Delta\mathbf{K}$, $G(0)$ is given by

$$G(0) \approx 4\log\left\{\det\left(\mathbf{I}+\tfrac{1}{8}\mathbf{K}_{\mathbf{g}1}^{-1}\Delta\mathbf{K}\,\mathbf{K}_{\mathbf{g}1}^{-1}\Delta\mathbf{K}\right)\right\} \approx \text{tr}\left[\mathbf{K}_{\mathbf{g}1}^{-1}\Delta\mathbf{K}\,\mathbf{K}_{\mathbf{g}1}^{-1}\Delta\mathbf{K}\right], \tag{18.387}$$

where the last step uses the series expansion (A.114) or (18.67), again terminated at second order in $\Delta\mathbf{K}$. The limit $M \to \infty$ is accomplished merely by replacing $\mathbf{K}$ with $\mathcal{K}$. The SNR on the log-likelihood ratio is then given by (13.96) as

$$\text{SNR}_\lambda^2 \approx 2G(0) \approx \text{tr}\left[\mathcal{K}_{\mathbf{g}1}^{-1}\Delta\mathcal{K}\,\mathcal{K}_{\mathbf{g}1}^{-1}\Delta\mathcal{K}\right]. \tag{18.388}$$

This expression is guaranteed to be real and nonnegative since it is equal to the sum of the squares of the eigenvalues of the Hermitian operator $\mathcal{K}_{\mathbf{g}1}^{-1}\Delta\mathcal{K}$.

Limits and special cases of this general result remain to be explored, and its relation to earlier work such as Wagner *et al.* (1983, 1986, 1987), Abbey *et al.* (2003) and Zemp *et al.* (2003) needs to be elucidated.

Known signals at unknown locations The SKE detection paradigm is unrealistic in radar and ultrasound because we never know in advance where the target is located. There are two ways we can consider incorporating this uncertainty into our analysis of detectability. The first is to consider a pure detection problem with random signal locations, and the second is to consider a joint detection-estimation problem. The difference is really in how we keep score. If the observer is given credit for a correct detection even it thinks the signal is at an incorrect location, then the problem is pure detection. In the joint detection-estimation problem, on the other hand, errors in position estimation are penalized. In the former case, figures of merit are related to the ROC curve, while in the latter case an LROC, FROC or AFROC curve can be used (see Sec. 14.2.3 and Figs. 14.4 and 14.5).

The likelihood ratio for an ideal observer attempting to detect a known signal at an unknown location in Gaussian noise was derived in Sec. 13.2.10. We saw in (13.161) that the optimal strategy in this case is to prewhiten the data, form the scalar product of the prewhitened data with the known signal, exponentiate the result and average over all possible signal locations; all information about signal location is lost in this averaging step. This strategy optimizes the area under the ROC curve with no penalty for incorrect signal location.

A common suboptimal strategy is the scanning matched filter or cross-correlator, where the noisy data are correlated with the known signal, essentially forming the scalar product for all possible shifts in a continuous fashion and omitting the exponentiation step. The final decision on signal-absent vs. signal-present is then made by choosing the shift for which the cross-correlator output is maximum and comparing the result to a threshold. For Gaussian noise, this strategy is equivalent to using the generalized likelihood ratio defined in (13.403). Various workers, including Nolte and Jaarsma (1967), Pelli (1985) and Wagner *et al.*, (1990b) have shown that this approach gives performance predictions very close to those of the optimal observer in the SNR ranges of experimental interest.

The joint detection-estimation problem was discussed in Sec. 13.3.9, where we formulated the problem as one of minimizing a cost function that involved both the detection and estimation aspects. With assumptions about the costs stated in that section, the optimal strategy is to first do the pure detection problem optimally by averaging the likelihood ratio over locations and then do MAP estimation if a signal-present decision is made.

In spite of the importance of joint detection-estimation problems in radar, ultrasound and many other areas, much further work is needed. In general terms, it is not yet clear how decisions about optimization of imaging systems might change depending on whether the task is formulated as pure detection or hybrid detection-estimation, and in the latter case how they might depend on the nature of the parameters to be estimated and their assumed probability densities. With particular reference to coherent imaging, it is not clear how the detectability is degraded by envelope detection and how it is affected by non-Gaussian statistics. It is also not yet known how to perform joint detection-estimation tasks optimally in non-Gaussian speckle.

19

Imaging in Fourier space

Imaging is fundamentally a way of obtaining information about the spatial distribution of some object characteristic, such as its transmittance, reflectance or radiant exitance. A direct-imaging system measures that distribution by sampling it over local, point-like spatial regions. Indirect imaging systems may start with nonlocal measurements but again the goal is to sample the spatial distribution in some sense.

In some indirect imaging systems, the initial measurements are essentially samples of the Fourier transform of the object distribution, rather than the desired distribution itself. Then the reconstruction step is some sort of inverse Fourier transform. Another way of thinking about such systems is that they are CD mappings in which the detector sensitivity functions approximate kernels of the Fourier transform; by contrast, direct imaging systems use sensitivity functions that approximate delta functions.

Fourier imaging systems are widespread. Magnetic resonance imaging (MRI) measures Fourier components of the induced magnetization in a patient's body, and some forms of synthetic-aperture radar sample the Fourier transform of the microwave reflectance of the terrain. Radioastronomy is basically the measurement of selected Fourier components of celestial sources of radio waves.

The whole field of Fourier optics (Goodman, 1968) is concerned with systems that use coherent light to study the Fourier transforms of object transmittances. In particular, certain kinds of holography yield measurements of the complex Fourier transforms of complex amplitude distributions.

Even tomographic systems such as SPECT (see Chap. 17) fit at least partially into the rubric of Fourier imaging; they measure line-integral projections of the object, but the central-slice theorem (Sec. 4.4.2) tells us that 1D Fourier transforms of the projections give us the 2D Fourier transform of the object.

In this chapter we shall illustrate the principles of Fourier imaging by considering two classes of measurement systems that directly acquire information about the object Fourier transform. The first class, called Fourier modulators and discussed in Sec. 19.1, consists of systems that use a sequence of masks that are related in a simple way to the Fourier kernel. The second class, covered in Sec. 19.2, consists of a variety of interferometers used mainly in astronomy. Our goal is to show how the tools developed in this book can be used to analyze and evaluate systems in both classes.

19.1 FOURIER MODULATORS

A Fourier modulator is a system that implements the Fourier integral literally. Computation of the Fourier transform of a function consists of two steps: modulating the function by multiplication with the Fourier kernel and integration of the result. Each step — modulation and integration — can be implemented in many different ways, depending on the physical nature of the object and the kind of radiation detector used. Here we consider mainly systems that measure the radiant exitance of some self-luminous object, though we shall also briefly discuss reflecting objects. Modulation will be by means of a series of masks placed directly over the object or over an intermediate image of the object.

The kinds of masks that can be used and the nature of the information they convey is surveyed in Sec. 19.1.1. Noise is included in the formalism in Sec. 19.1.2, and reconstruction of an object from a set of noisy samples of its Fourier transform is treated in Sec. 19.1.3. The key issue of image quality is discussed in Sec. 19.1.4.

19.1.1 Data acquisition

To pick a trivial starting point, the 2D Fourier transform is defined by

$$F(\boldsymbol{\rho}) = \int_\infty d^2r \; f(\mathbf{r}) \exp(-2\pi i \boldsymbol{\rho} \cdot \mathbf{r}) \, . \tag{19.1}$$

Suppose that we want to apply this definition to compute not the entire 2D distribution in the Fourier plane but rather its value at a single point, $\boldsymbol{\rho} = \boldsymbol{\rho}_0$. If we could synthesize a mask with the complex transmittance $\exp(-2\pi i \boldsymbol{\rho}_0 \cdot \mathbf{r})$, lay it over the object and integrate the resulting product, we would have the desired complex value, $F(\boldsymbol{\rho}_0)$.

There are several obvious problems with this approach. We may not know how to construct a complex mask, there may not be a good way of doing the complex integration, and even if we solve those problems, all that we have obtained is one point in the Fourier transform; to scan the entire Fourier plane we have to repeat the process with many different masks with different magnitudes and orientations of the spatial frequency vector $\boldsymbol{\rho}$. Undaunted, let us consider each of these problems in turn.

Complex masks Since $\exp(-2\pi i \boldsymbol{\rho}_0 \cdot \mathbf{r}) = \cos(2\pi \boldsymbol{\rho}_0 \cdot \mathbf{r}) - i \sin(2\pi \boldsymbol{\rho}_0 \cdot \mathbf{r})$, we can rewrite the definition of the Fourier transform as

$$F(\boldsymbol{\rho}_0) = \int_\infty d^2r \; f(\mathbf{r}) \cos(2\pi \boldsymbol{\rho}_0 \cdot \mathbf{r}) - i \int_\infty d^2r \; f(\mathbf{r}) \sin(2\pi \boldsymbol{\rho}_0 \cdot \mathbf{r}) \, . \tag{19.2}$$

Now we need two real masks instead of one complex one. Moreover, if $f(\mathbf{r})$ is real, we do not have to do a complex integration. Laying the cosine mask over the object and integrating the product gives the first integral in (19.2), and using the sine mask similarly gives the second integral; both integrals are real, and each yields one number. If we record the two real numbers separately in our lab notebook (or computer), we have the real and imaginary parts of $F(\boldsymbol{\rho}_0)$ for one particular $\boldsymbol{\rho}_0$.

Nonnegativity In many applications, $f(\mathbf{r})$ represents a radiant exitance, reflectance or other quantity that cannot go negative, and the relevant transmittance is a nonnegative real quantity between 0 and 1. We cannot synthesize a mask with transmittance $\cos(2\pi\boldsymbol{\rho}_0 \cdot \mathbf{r})$ but we can synthesize one with transmittance $\frac{1}{2}[1 \pm \cos(2\pi\boldsymbol{\rho}_0 \cdot \mathbf{r})]$. Thus it is advantageous to rewrite the definition of the Fourier transform once more as

$$F(\boldsymbol{\rho}_0) = \int_\infty d^2r \; f(\mathbf{r}) \, \tfrac{1}{2}[1 + \cos(2\pi\boldsymbol{\rho}_0 \cdot \mathbf{r})] - \int_\infty d^2r \; f(\mathbf{r}) \, \tfrac{1}{2}[1 - \cos(2\pi\boldsymbol{\rho}_0 \cdot \mathbf{r})]$$

$$- i \int_\infty d^2r \; f(\mathbf{r}) \, \tfrac{1}{2}[1 + \sin(2\pi\boldsymbol{\rho}_0 \cdot \mathbf{r})] + i \int_\infty d^2r \; f(\mathbf{r}) \, \tfrac{1}{2}[1 - \sin(2\pi\boldsymbol{\rho}_0 \cdot \mathbf{r})] \,. \qquad (19.3)$$

Now we have decomposed the complex integral into four real integrals, each of which has a nonnegative integrand. Each integral is obtained by laying the appropriate mask over the object and collecting all of the emerging light with a nonimaging detector. For each $\boldsymbol{\rho}_0$, we must measure four nonnegative real numbers in this way, but at least it is beginning to appear that there are no physical impediments to the measurement.

Reducing the number of masks The form of (19.3) is a bit misleading since it seems to imply that we need four masks for each spatial frequency. Suppose, however, that we have one mask with transmittance $\frac{1}{2}[1 + \cos(2\pi\boldsymbol{\rho}_0 \cdot \mathbf{r})]$. We can shift this mask to four positions, producing the other three required transmittances; for example, $\frac{1}{2}[1 + \sin(2\pi\boldsymbol{\rho}_0 \cdot \mathbf{r})]$ is obtained by shifting the original mask, $\frac{1}{2}[1 + \cos(2\pi\boldsymbol{\rho}_0 \cdot \mathbf{r})]$ a quarter period in the direction $\boldsymbol{\rho}_0$. Thus one physical mask serves for all four measurements required at one frequency $\boldsymbol{\rho}_0$.

Moreover, the same mask can be rotated to yield all of the spatial frequencies with the same magnitude as $\boldsymbol{\rho}_0$ but different orientations. Thus if we choose to scan the 2D Fourier plane in polar coordinates, one real, nonnegative mask will suffice for all of the complex measurements on a circle of radius ρ_0. Different masks will be needed for different radii in Fourier space.

Note also that it is not necessary to lay the mask directly on the object. It suffices to relay the object to an intermediate image plane and locate the mask there. If this relay system has variable magnification, one mask can serve for a range of scales of ρ.

Moiré masks One way to synthesize the needed masks is by the moiré effect.[1] Suppose we have two very fine masks, each of transmittance $\frac{1}{2}[1 + \cos(2\pi\boldsymbol{\rho}_c \cdot \mathbf{r})]$, where subscript c denotes *carrier frequency*. We shall assume that ρ_c is large compared to any spatial frequency present in the object. If we overlay the two masks, we get a low-frequency moiré pattern as shown in Fig. 19.1, and the two overlapping high-frequency masks serve as one of the low-frequency masks needed in the method

[1]The word moiré, referring to "watered silk," comes from the English *mohair*, which was apparently borrowed into French and then returned to English with the accent.

described above. The relative rotation angle between the two masks varies the magnitude of the spatial frequency of the pattern, and rotating both masks together varies its direction. Moreover, the phase of the moiré pattern can be shifted by shifting either mask laterally. The reader is invited to verify these contentions by making two copies of Fig. 19.1a on transparency paper and overlaying them, say on a light box or overhead projector.

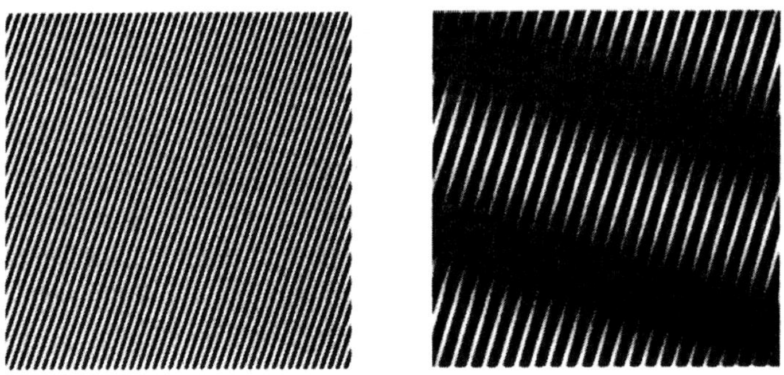

(a) (b)

Fig. 19.1 Illustration of the moiré pattern produced by overlaying two cosinusoidal patterns. (a) Single cosinusoidal pattern; (b) product of two identical patterns slightly rotated with respect to each other.

To explain the moiré effect mathematically, let us use a 2D rotation operator $\mathbf{R}(\theta)$ and write the transmittance of the overlaid masks as

$$T(\mathbf{r};\theta_1,\theta_2) = \tfrac{1}{2}\{1+\cos[2\pi\boldsymbol{\rho}_c\cdot\mathbf{R}(\theta_1)\mathbf{r}]\}\,\tfrac{1}{2}\{1+\cos[2\pi\boldsymbol{\rho}_c\cdot\mathbf{R}(\theta_2)\mathbf{r}]\}\,. \tag{19.4}$$

Since the rotation operator is unitary, $\mathbf{R}^\dagger(\theta) = \mathbf{R}^{-1}(\theta) = \mathbf{R}(-\theta)$, and we can use the definition of the adjoint to write

$$\begin{aligned} T(\mathbf{r};\theta_1,\theta_2) &= \tfrac{1}{2}\{1+\cos[2\pi\mathbf{R}(-\theta_1)\boldsymbol{\rho}_c\cdot\mathbf{r}]\}\,\tfrac{1}{2}\{1+\cos[2\pi\mathbf{R}(-\theta_2)\boldsymbol{\rho}_c\cdot\mathbf{r}] \\ &= \tfrac{1}{4}\{1+\cos[2\pi\boldsymbol{\rho}_{1,2}\cdot\mathbf{r}]\} + \text{high-frequency terms}\,, \end{aligned} \tag{19.5}$$

where

$$\boldsymbol{\rho}_{1,2} \equiv \mathbf{R}(-\theta_1)\boldsymbol{\rho}_c - \mathbf{R}(-\theta_2)\boldsymbol{\rho}_c\,. \tag{19.6}$$

As shown in Fig. 19.2, $\boldsymbol{\rho}_{1,2}$ is simply the vector difference of the two rotated spatial-frequency vectors.

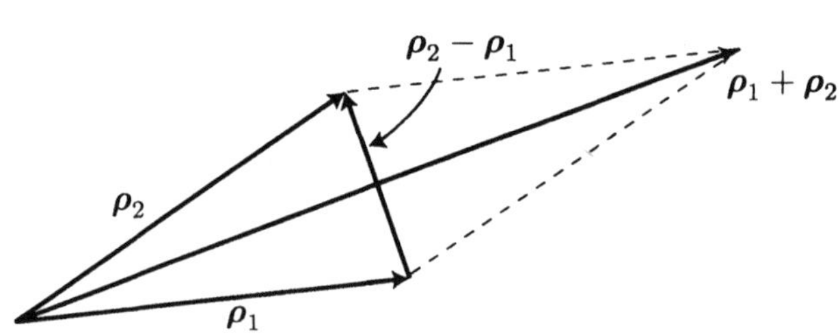

Fig. 19.2 Vector diagram illustrating the sum- and difference-frequency terms seen in the moiré pattern of Fig. 19.1b.

The high-frequency terms in (19.5) all have $\rho \geq \rho_c$ and can be neglected if the object Fourier transform is negligible at these frequencies. By the same argument, we could use square-wave patterns, and the harmonics of the carrier frequency can be neglected. Harmonics of the difference frequency, however, still remain.

The interesting difference-frequency term in (19.5) can be scanned over a portion of the Fourier plane by varying the two angles, θ_1 and θ_2. Specifically, if we allow θ_1 and θ_2 to cover $(0, 2\pi)$ but restrict $|\theta_1 - \theta_2| \leq \Delta\theta_{max}$, then we can reach any frequency that satisfies $\rho \leq 2\rho_c \sin \frac{1}{2}\Delta\theta_{max}$.

A possible moiré-Fourier imaging system takes the form shown in Fig. 19.3, where the moiré mask is placed in close proximity to the object, and the object is either self-luminous or a transparency placed over a diffuse light source. A single-element nonimaging detector collects some fraction of the emerging light. If we assume that the fraction of the light collected is the same for all points in the object plane, the mean detector output is given by

$$\overline{V}_{out} \propto \int_{\infty} d^2r \; f(\mathbf{r})\, T(\mathbf{r}; \theta_1, \theta_2)\,. \tag{19.7}$$

By shifting either mask laterally to get four terms as in (19.3) and by scanning the difference frequency by varying the two angles, we can sample the desired portion of the Fourier plane.

The configuration shown in Fig. 19.3 was first suggested by Mertz (1956). It can be used for any kind of radiation, and in fact it was used for gamma rays by Chou and Barrett (1978). In that case the masks were made of lead.

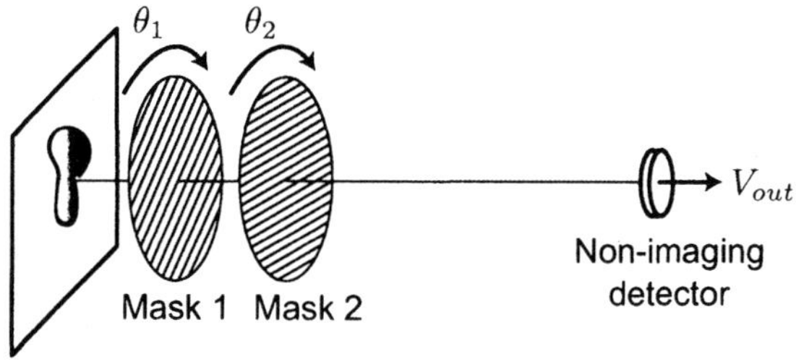

Fig. 19.3 Simple system for measuring Fourier components of a planar object. The object and the two masks are shown separated for clarity, but would actually be in close proximity to each other, with the nonimaging detector a larger distance away. Rotation of the two masks varies the spatial frequency of the moiré pattern and hence the Fourier component being measured.

Projecting fringe patterns If the object is reflecting rather than self-luminous or transmitting, we cannot place the mask in close proximity to it since it would then obscure the illumination. We could relay the object to an intermediate image plane and place the mask there, but we can also implement the mask by structuring the illumination itself.

Suppose $f(\mathbf{r})$ represents a reflectance, and let $I(\mathbf{r})$ be the illuminating irradiance. If we can arrange to make $I(\mathbf{r}) \propto \frac{1}{2}[1 + \cos(2\pi\boldsymbol{\rho}_0 \cdot \mathbf{r} + \phi)]$, then the total reflected light is given by one of the integrals in (19.3), with the phase ϕ specifying which one.

An easy way to get the desired irradiance profile is to interfere two plane waves. Suppose we produce two plane waves in a 3D space, with the complex amplitudes given by $u_1(\mathfrak{r}) = A\exp(i\mathbf{k}_1 \cdot \mathfrak{r})$ and $u_2(\mathfrak{r}) = A\exp(i\mathbf{k}_2 \cdot \mathfrak{r})$. If we let the two waves overlap on the plane $z = 0$, the resulting irradiance is[2]

$$\begin{aligned} I(\mathbf{r}) &= \left|u_1(\mathfrak{r}) + u_1(\mathfrak{r})\right|^2_{z=0} = 2|A|^2 \left\{1 + \cos[(\mathbf{k}_1 - \mathbf{k}_2)\cdot \mathfrak{r}]\right\}_{z=0} \\ &= 2|A|^2 \left[1 + \cos(2\pi \boldsymbol{\rho}_0 \cdot \mathbf{r})\right], \end{aligned} \tag{19.8}$$

where

$$\boldsymbol{\rho}_0 = \left(\frac{k_{1x} - k_{2x}}{2\pi}, \frac{k_{1y} - k_{2y}}{2\pi}\right). \tag{19.9}$$

Thus the 2D spatial frequency can be tuned by varying the **k** vectors of the two interfering plane waves.

Zone-plate moiré A Fresnel zone plate is a pattern of concentric annular zones, alternately clear and opaque, as illustrated in Fig. 19.4. Mathematically, a zone plate has a transmittance given by

$$T_{zp}(\mathbf{r}) = \tfrac{1}{2}\{1 + \mathrm{sgn}[\sin(\alpha r^2)]\}\, S(\mathbf{r}), \tag{19.10}$$

where sgn$[\cdot]$ is the signum function defined in Sec. 2.3.2, and $S(\mathbf{r})$ is a support function specifying the overall area of the zone plate. If $S(\mathbf{r})$ is a cylinder function [defined in (3.257)] centered on the origin, we refer to the zone plate as *on-axis*, but if it is a cylinder function displaced from the origin the zone plate is said to be *off-axis*. The transitions between clear and opaque zones occur at the zeros of the sine, or when $\alpha r_n^2 = n\pi$, $(n = 1, 2, 3, ...)$. Thus $r_1 = \sqrt{\pi/\alpha}$, and $r_n = r_1\sqrt{n}$. It follows that all zones of an on-axis zone plate have the same area, namely πr_1^2.

(a) (b)

Fig. 19.4 Fresnel zone plates. (a) On-axis; (b) off-axis.

[2]We employ the same convention for spatial vectors here as in earlier chapters. The gothic $\mathfrak{r}$ is a 3D vector and the Times Roman $\mathbf{r}$ is 2D, but they refer to the same physical point. Thus, if the point is on the plane $z = 0$, $\mathbf{r} = (x, y)$ and $\mathfrak{r} = (x, y, 0)$.

Overlaying two Fresnel zone plates yields straight-line moiré fringe patterns, as shown in Fig. 19.5. Different moiré spatial frequencies are generated by shifting one zone plate relative to the other, rather than by rotating the patterns as with the cosine masks discussed above. The reader may play with the zone-plate moiré effect by making two transparencies of Fig. 19.4.

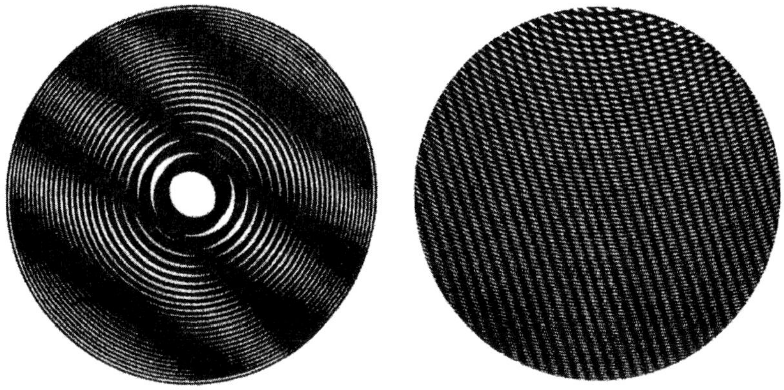

(a) (b)

Fig. 19.5 Moiré patterns produced by overlapping two identical zone plates with a slight relative shift. (a) On-axis zone plates; (b) off-axis zone plates.

The visual appearance of the moiré fringes in Fig. 19.5 is similar to that seen for cosine grating patterns in Fig. 19.1, but the math is quite different. Expanding $\mathrm{sgn}[\sin(\alpha r^2)]$ in a Fourier series in the variable r^2, we can write (Shulman, 1970; Barrett and Swindell, 1981, 1996)

$$T_{zp}(\mathbf{r}) = \left[\frac{1}{2} + \frac{1}{\pi i} \sum_{\substack{k=-\infty \\ (k \text{ odd})}}^{\infty} \frac{1}{k} \exp(-i\alpha k r^2) \right] S(\mathbf{r}) \,. \tag{19.11}$$

The net transmittance of two overlapping zone plates, one shifted by $\mathbf{r}_0$, is

$$T_{zp}(\mathbf{r})\, T_{zp}(\mathbf{r} - \mathbf{r}_0)$$

$$= \left[\frac{1}{2} + \frac{1}{\pi i} \sum_{\substack{k=-\infty \\ (k \text{ odd})}}^{\infty} \frac{1}{k} \exp(-i\alpha k r^2) \right] \left[\frac{1}{2} + \frac{1}{\pi i} \sum_{\substack{k'=-\infty \\ (k' \text{ odd})}}^{\infty} \frac{1}{k'} \exp(-i\alpha k' |\mathbf{r} - \mathbf{r}_0|^2) \right] S(\mathbf{r}, \mathbf{r}_0) \,, \tag{19.12}$$

where $S(\mathbf{r}, \mathbf{r}_0) \equiv S(\mathbf{r})\, S(\mathbf{r} - \mathbf{r}_0)$. The straight-line moiré fringes arise from terms with $k = -k'$. Focusing on $k = \pm 1$, we can write

$$T_{zp}(\mathbf{r})\, T_{zp}(\mathbf{r} - \mathbf{r}_0)$$

$$= \left\{ \tfrac{1}{4} + \tfrac{1}{\pi^2} \left[\exp(-i\alpha r^2) \exp(i\alpha |\mathbf{r} - \mathbf{r}_0|^2) + \exp(i\alpha r^2) \exp(-i\alpha |\mathbf{r} - \mathbf{r}_0|^2) \right] \right\} S(\mathbf{r}, \mathbf{r}_0) + \text{other terms} \,, \tag{19.13a}$$

which simplifies to

$$T_{zp}(\mathbf{r})\, T_{zp}(\mathbf{r} - \mathbf{r}_0) = \left[\tfrac{1}{4} + \tfrac{2}{\pi^2} \cos(2\alpha \mathbf{r} \cdot \mathbf{r}_0 - \alpha r_0^2) \right] S(\mathbf{r}, \mathbf{r}_0) + \text{ other terms} \,. \tag{19.13b}$$

Thus the fringes have a spatial frequency of $\alpha\mathbf{r}_0/\pi$ and a phase of αr_0^2. By varying the shift $\mathbf{r}_0$, we can generate a sequence of positive and negative sine and cosine moiré masks of varying frequency. The reader will appreciate this result more by actually observing the effect with transparencies made from Fig. 19.4.

Another way to understand (19.13) — and to make some statements about the neglected terms — is to use the concept of local spatial frequency, introduced in Sec. 5.1.3. We know from (5.35) that a pure phase function $f(\mathbf{r}) = \exp[i\Phi(\mathbf{r})]$ has a local frequency at point $\mathbf{r}$ given by $\boldsymbol{\rho}_{loc}(\mathbf{r}) = (2\pi)^{-1}\nabla\Phi(\mathbf{r})$. For Fresnel zone plates, each term in the expansion (19.11) is a pure phase function, and $\boldsymbol{\rho}_{loc}(\mathbf{r})$ for the kth term is given by $-(\alpha k/\pi)\mathbf{r}$ [*cf.* (5.37)]. When we overlap two zone plates as in (19.12), the moiré or difference-frequency term for $k = -k' = \pm 1$ is $\boldsymbol{\rho}_{loc}(\mathbf{r}) = \pm(\alpha/\pi)[\mathbf{r} - (\mathbf{r} - \mathbf{r}_0)] = \pm(\alpha/\pi)\mathbf{r}_0$. For $k = -k' \neq \pm 1$, we find $\boldsymbol{\rho}_{loc}(\mathbf{r}) = \pm(\alpha k/\pi)\mathbf{r}_0$, so these terms correspond to odd harmonics of the fundamental frequency $(\alpha/\pi)\mathbf{r}_0$; the moiré fringes are not pure cosines or sines for binary zone plates. Should we wish to eliminate the harmonic terms, we could use sinusoidal zone plates rather than binary ones. The reader is invited to retrace the analysis for this case.

Terms with $k \neq -k'$ are potentially more worrisome. Even with sinusoidal zone plates, we have the terms with $k = k' = \pm 1$, for which $\boldsymbol{\rho}_{loc}(\mathbf{r}) = \pm(\alpha/\pi)[\mathbf{r} + (\mathbf{r} - \mathbf{r}_0)] = \pm(\alpha/\pi)[2\mathbf{r} - \mathbf{r}_0]$. This sum-frequency term has a local frequency that depends linearly on $\mathbf{r}$, so it corresponds to a chirp in the moiré pattern [*cf.* (5.37)]. With this term present, we measure a linear combination of the Fourier and the Fresnel transform of the object (see Sec. 4.3.2).

To eliminate nettlesome sum-frequency terms, we can use off-axis zone plates. Suppose the support function $S(\mathbf{r})$ is given by $\text{cyl}[(\mathbf{r} - \mathbf{r}_c)/D]$, which describes an aperture of diameter D centered at $\mathbf{r} = \mathbf{r}_c$. For the kth term in the expansion, the local frequency at the center of the aperture is $(\alpha k/\pi)\mathbf{r}_c$. Thus, for $k = k' = \pm 1$ and for $r_0 \ll r_c$, the relevant local frequency is in the vicinity of $\pm(\alpha/\pi)2\mathbf{r}_c$; if this frequency is large compared to any spatial frequency in the object, the sum-frequency term is not important. This approximation is essentially the same as ignoring the carrier-frequency terms in (19.5), but with zone plates frequency is coupled to position by the local-frequency relation.

We shall return to the issue of neglected terms in Sec. 19.1.3 when we discuss reconstruction from Fourier samples.

Parallel data acquisition So far we have discussed systems that sample one Fourier component of the object at a time; getting another component requires replacing masks or performing some other mechanical motion. It would be desirable to have some way of measuring many different Fourier components simultaneously without mechanical motion.

With zone-plate moiré, the different masks are obtained by shifts of one zone plate with respect to the other, and the shifts can be accomplished by parallax rather than mechanically. Consider the system shown in Fig. 19.6. The object, assumed to be either self-luminous or an illuminated transparency, is in contact with the first zone plate. The second zone plate, half the scale of the first, is placed midway between the masked object and an observation plane. If we neglect diffraction,[3] light emerging from the first zone plate at point $\mathbf{r}$ and arriving at point

[3]This is the first mention of diffraction in this chapter; if the object and one or more masks are all in contact, diffraction is irrelevant.

$\mathbf{r}_d$ on the observation plane must have passed through the middle zone plate at the point $\mathbf{r}' = \frac{1}{2}(\mathbf{r}_d + \mathbf{r})$. (Note that $\mathbf{r}$, $\mathbf{r}'$ and $\mathbf{r}_d$ are 2D vectors in different planes.) If we denote the transmittance of the first zone plate by $T_{zp}(\mathbf{r})$, and that of the second (half-scale) zone plate by $T_{zp}(2\mathbf{r}')$, then the irradiance at point $\mathbf{r}_d$ is given by

$$I(\mathbf{r}_d) \propto \int_\infty d^2r\ f(\mathbf{r})\, T_{zp}(\mathbf{r})\, T_{zp}(\mathbf{r}_d + \mathbf{r})\,. \tag{19.14}$$

From this point on, we have exactly the same mathematics as in (19.12) and (19.13), but we see that the variable mechanical shift $\mathbf{r}_0$ has been replaced by the variable observation point $\mathbf{r}_d$. A detector array placed in the observation plane will sample many different shifts simultaneously. Since a shift $\mathbf{r}_d$ corresponds to a spatial frequency of $\alpha\mathbf{r}_d/\pi$ and a phase of αr_d^2 in the moiré pattern, a single readout of the detector array yields measurements of many different sine and cosine Fourier components.

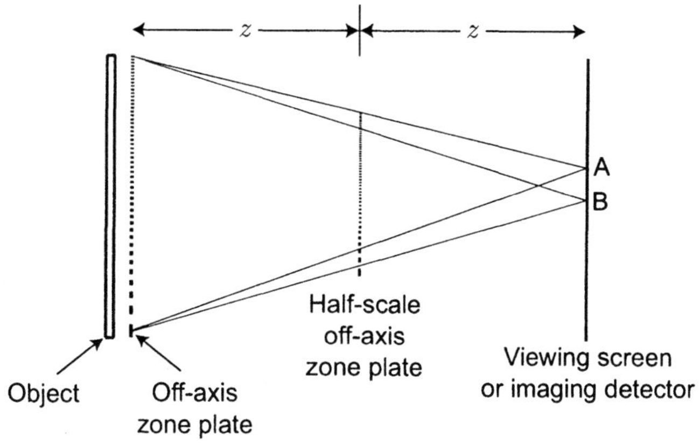

Fig. 19.6 Configuration for implementing the shift between two zone plates by parallax. Note the relative shift of the projections of the two zone plates from points A and B in the viewing plane.

Mertz Fresnel sandwich There is one minor problem with the configuration of Fig. 19.6. If we consider only the terms with $k = -k' = 1$, then

$$I(\mathbf{r}_d) \propto \int_\infty d^2r\ f(\mathbf{r}) \exp(i\alpha r^2) \exp(-i\alpha|\mathbf{r}_d + \mathbf{r}|^2) + \text{dull terms}$$

$$= \exp(-i\alpha r_d^2) \int_\infty d^2r\ f(\mathbf{r}) \exp(-2i\alpha\mathbf{r}_d \cdot \mathbf{r}) + \text{dull terms}\,. \tag{19.15}$$

This expression is the desired Fourier transform multiplied by a quadratic phase factor.

We could simply ignore the quadratic phase factor, since it is just a constant independent of $\mathbf{r}$ (hence outside the integral), but we can also cancel it out by adding one more zone plate as shown in Fig. 19.7. this third zone plate has the same scale as the first one, so it is described by the transmittance $T_{zp}(\mathbf{r}_d)$. The math gets very messy if we use the full expansion (19.11), but the essential point is that $T_{zp}(\mathbf{r}_d)$ contains a term in $\exp(+i\alpha r_d^2)$, which will serve to cancel the $\exp(-i\alpha r_d^2)$ in (19.15). Other terms are present as well, but we can argue that they are not so important if we use off-axis zone plates with sufficiently high center frequencies.

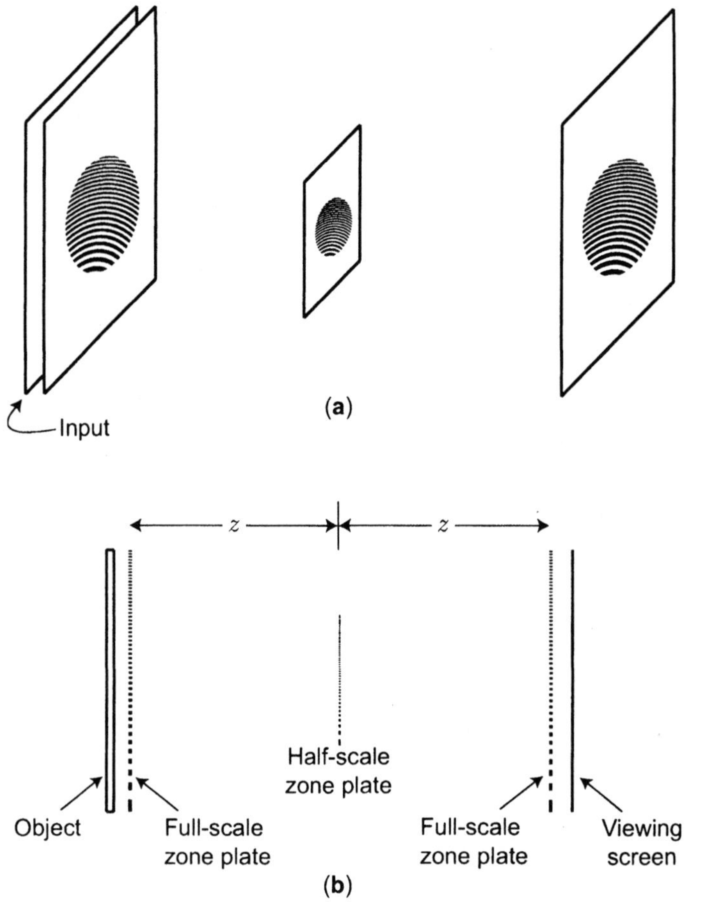

Fig. 19.7 Fresnel sandwich, obtained from the configuration of Fig. 19.6 by placing an additional zone plate in the output plane. (a) Layout. (b) schematic.

The three-zone-plate configuration was first presented in a delightful little book entitled *Transformations in Optics*, by Larry Mertz (1965). Mertz referred to Fig. 19.7 as a *Fresnel sandwich*, and he showed how it led to practical imaging applications as well as insightful analogies.

We have actually already met the Fresnel sandwich in earlier chapters. Recall from Sec. 4.3.3 that we can compute the Fourier transform of a function in a roundabout way if we multiply it by a chirp, convolve with the conjugate chirp, and multiply again by the original chirp. With judicious selection of terms, that is precisely what we are doing with the configuration of Fig. 19.7.

Another context in which we have met the same mathematics is in discussing coherent optical Fourier transformers in Sec. 9.7.2. We know from that section that the system shown in Fig. 19.8a performs a Fourier transformation on the electric field at its input. This system involves a section of free-space propagation, multiplication by the amplitude transmittance of a lens, and then another section of free-space propagation. In the Fresnel approximation, free-space propagation amounts to convolution with a chirp [see (9.94)], and transmission through a lens is equivalent to multiplication by a chirp [see (9.159)], so the system of Fig. 19.8a implements the convolve-multiply-convolve or CMC chirp-Fourier transform algorithm of Sec. 4.3.3. The Fresnel sandwich, on the other hand, implements the MCM (multiply-convolve-multiply) chirp-Fourier transform algorithm. A coherent optical system that also implements MCM is shown in Fig. 19.8b.

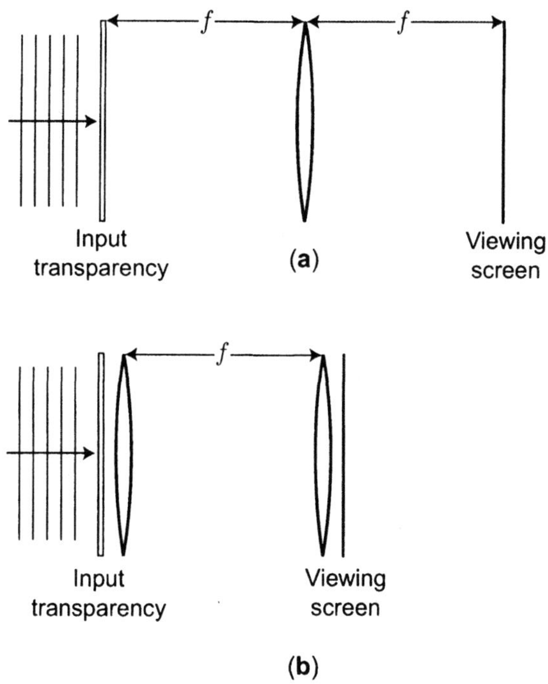

Fig. 19.8 Two coherent optical Fourier-transform systems. (a) System that implements the CMC chirp-Fourier transform algorithm; (b) system that implements the MCM algorithm, analogously to the Fresnel sandwich.

19.1.2 Noise

So far our analysis of Fourier imaging has neglected noise and treated the problem as CC. In reality we collect only a finite amount of noisy data, so a better description of a Fourier imaging system, as with any digital imaging system, is as a CD mapping. As detailed in Chap. 12, measurement noise can usually be modeled as either Gaussian or Poisson. In this section we shall discuss the effects of this noise on the discrete raw data and on estimates of the Fourier transform values derived from these data.

The data It will prove useful to denote each discrete measurement with two indices, one for the spatial frequency and one for the phase of the Fourier kernel. Thus we write

$$g_{mj} = C \int_{\infty} d^2r \; f(\mathbf{r})\, T_{mj}(\mathbf{r}) + n_{mj} \,, \tag{19.16}$$

where C is a radiometric constant and $T_{mj}(\mathbf{r})$ is the transmittance of one of the mask structures discussed in Sec. 19.1. For example, to describe the set of four integrals in (19.3), we take

$$T_{mj}(\mathbf{r}) = \tfrac{1}{2}[1 + \cos(2\pi \boldsymbol{\rho}_m \cdot \mathbf{r} + \phi_j)]\, S(\mathbf{r}) \,, \qquad \phi_j = (j-1)\tfrac{\pi}{2} \,, \qquad j = 0, 1, 2, 3 \,. \tag{19.17}$$

Thus (19.16) fits our usual CD mapping, $\mathbf{g} = \mathcal{H}\mathbf{f} + \mathbf{n}$, with the kernel given by

$$h_{mj}(\mathbf{r}) = \tfrac{1}{2} C[1 + \cos(2\pi \boldsymbol{\rho}_m \cdot \mathbf{r} + \phi_j)]\, S(\mathbf{r}) \,. \tag{19.18}$$

For both Gaussian and Poisson noise, the integral in (19.16) is the mean of g_{mj}, so $\langle n_{mj} \rangle = 0$ by definition.

Modified data: Points in the Fourier transform As a first step toward image reconstruction, we may choose to form an estimate of the complex Fourier transform value $F(\boldsymbol{\rho}_m)$. For the data described by (19.16)–(19.18), we can define these estimates in a natural way as

$$\hat{F}(\boldsymbol{\rho}_m) \equiv \frac{1}{C}\left[g_{m0} - g_{m2} - i g_{m1} + i g_{m3}\right] \,. \tag{19.19}$$

It can be verified that $\hat{F}(\boldsymbol{\rho}_m)$ is an unbiased estimate of $F(\boldsymbol{\rho}_m)$ and that it is the maximum-likelihood estimate in the case of i.i.d. Gaussian noise.

We can put (19.19) in matrix form by writing

$$\hat{F}(\boldsymbol{\rho}_m) = \sum_{m'j'} A_{mm'j'}\, g_{m'j'} \,. \tag{19.20}$$

By comparison with (19.19), we see that

$$A_{mm'j'} = \frac{1}{C}\, \delta_{mm'} [\delta_{j'0} - \delta_{j'2} - i\,\delta_{j'1} + i\,\delta_{j'3}] \,. \tag{19.21}$$

In operator form,

$$\hat{\mathbf{F}} = \mathbf{A}\mathbf{g} = \mathbf{A}\mathcal{H}\mathbf{f} + \mathbf{n} \,. \tag{19.22}$$

For some reconstruction algorithms, we may take $\hat{\mathbf{F}}$ as the data instead of $\mathbf{g}$.

Gaussian noise Consider the simple system of Fig. 19.3, and suppose the detector is afflicted with i.i.d. Gaussian noise such that each measurement has variance σ^2. Then the noise covariance matrix is given by

$$\mathbf{K_n} = \sigma^2 \mathbf{I} \,. \tag{19.23}$$

If we use four masks at each of M spatial frequencies, then $\mathbf{I}$ is the $4M \times 4M$ unit matrix.

If we consider only a single object, the noise covariance $\mathbf{K_n}$ is the same thing as the data covariance $\mathbf{K_g}$. We shall discuss the effect of object variability below, but for now consider only a single object.

For Gaussian noise and nonrandom objects, $\hat{F}(\boldsymbol{\rho}_m)$ is a linear combination of zero-mean Gaussian random variables, so it is itself a zero-mean Gaussian random variable. Since $\hat{F}(\boldsymbol{\rho}_m)$ is complex, however, we have to be careful in specifying its variance and covariance (see Sec. 8.3.6). The covariance matrix is defined by

$$\left[\mathbf{K}_{\hat{\mathbf{F}}}\right]_{mm'} \equiv \left\langle \Delta\hat{F}(\boldsymbol{\rho}_m)\, \Delta\hat{F}^*(\boldsymbol{\rho}_{m'}) \right\rangle , \tag{19.24}$$

where $\Delta\hat{F}(\boldsymbol{\rho}_m) \equiv \hat{F}(\boldsymbol{\rho}_m) - F(\boldsymbol{\rho}_m)$.

To evaluate this covariance matrix, note first that the real and imaginary parts of $\hat{F}(\boldsymbol{\rho}_m)$ are uncorrelated (and hence statistically independent since they are Gaussian). Moreover, $\hat{F}(\boldsymbol{\rho}_m)$ and $\hat{F}(\boldsymbol{\rho}_{m'})$ are uncorrelated (and independent) for $m \neq m'$. Thus

$$\left[\mathbf{K}_{\hat{\mathbf{F}}}\right]_{mm'} = \left\langle |\Delta\hat{F}(\boldsymbol{\rho}_m)|^2 \right\rangle \delta_{mm'} = \left\langle \left[\Delta\hat{F}_r(\boldsymbol{\rho}_m)\right]^2 + \left[\Delta\hat{F}_i(\boldsymbol{\rho}_m)\right]^2 \right\rangle \delta_{mm'} , \tag{19.25}$$

where subscripts r and i denote real and imaginary parts, respectively. Since the real and imaginary parts are i.i.d., we have, finally,

$$\left[\mathbf{K}_{\hat{\mathbf{F}}}\right]_{mm'} = \frac{4\sigma^2}{C^2}\,\delta_{mm'} , \tag{19.26}$$

or simply,

$$\mathbf{K}_{\hat{\mathbf{F}}} = \frac{4\sigma^2}{C^2}\mathbf{I} , \tag{19.27}$$

where now $\mathbf{I}$ is the $M \times M$ unit matrix.

Poisson noise If photon noise dominates and the objects are nonrandom, then each measurement is a Poisson random variable, and we can write the noise covariance matrix as [*cf.* (11.41)]

$$[\mathbf{K}_{\mathbf{n}}]_{mm'} = \overline{g}_m\,\delta_{mm'} . \tag{19.28}$$

The covariance on $\Delta\hat{F}(\boldsymbol{\rho}_m)$ in this case is found to be

$$\left[\mathbf{K}_{\hat{\mathbf{F}}}\right]_{mm'} = \frac{1}{C^2}\left[\overline{g}_{0m} + \overline{g}_{1m} + \overline{g}_{2m} + \overline{g}_{3m}\right]\delta_{mm'} . \tag{19.29}$$

Note that $\hat{F}$ is a linear combination of Poisson random variables but not itself Poisson.

Temporal noise issues Up to now we have regarded the measurement system as static; a mask is inserted, a measurement is taken, and then a new mask is inserted. It may, however, be more practical to use some continuous scanning mechanism. Then the required discrete measurements can be obtained by sampling the dynamic detector output at appropriate times.

When we consider such dynamic measurements, the noise must be treated as a temporal random process, and consideration must be given to the bandwidth of the electronics. Suppose, for simplicity, that the detector noise is a stationary, white, Gaussian random process, with power spectral density $S_n(\nu) = S_n = constant$. If we denote the temporal impulse response of the electronics by $p(t)$ and the corresponding transfer function by $P(\nu)$, then the power spectral density on the voltage at the output of the electronics is $S_V(\nu) = |P(\nu)|^2 S_n$ [see (8.156)]. Since the noise is stationary and the system is shift-invariant, all measurements have the same variance, namely [*cf.* (12.16)]

$$\sigma^2 = \int_{-\infty}^{\infty} d\nu \; S_V(\nu) = S_n \int_{-\infty}^{\infty} d\nu \; |P(\nu)|^2 = 2S_n |P(0)|^2 B \,, \tag{19.30}$$

where B is the *effective noise bandwidth* as defined in (12.18).

It is evident from (19.30) that we can reduce the noise variance by reducing the bandwidth, but doing so will also distort the desired signal[4] information. Suppose, for example, that we scan the interfering plane waves in such a way that the spatial frequency of the fringe pattern varies linearly with time,

$$\boldsymbol{\rho}(t) = \mathbf{a} + \mathbf{b}t \,. \tag{19.31}$$

If we assume for simplicity that the spatial phase of the fringe pattern is fixed at $\phi = \phi_j$, then the mean measurement $\overline{g}_{mj}$ is given by

$$\overline{g}_{mj} = C \int_{-\infty}^{\infty} dt \; p(t_m - t) \int_{\infty} d^2r \; \tfrac{1}{2}\{1 + \cos[2\pi(\mathbf{a} + \mathbf{b}t) \cdot \mathbf{r} + \phi_j]\} \, f(\mathbf{r}) \,. \tag{19.32}$$

The reader may show that

$$\overline{g}_{mj} = CP(0) \int_{\infty} d^2r \; \tfrac{1}{2} \left\{ 1 + \frac{|P(\mathbf{b} \cdot \mathbf{r})|}{P(0)} \cos[2\pi(\mathbf{a} + \mathbf{b}t_m) \cdot \mathbf{r} + \phi_j + \Psi_p(\mathbf{b} \cdot \mathbf{r})] \right\} f(\mathbf{r}) \,, \tag{19.33}$$

where $P(\nu) \equiv |P(\nu)| \exp[i\Psi_P(\nu)]$.

Two deleterious effects are seen in (19.33): the amplitude of the cosine modulation is reduced by the MTF of the electrical filter, and, perhaps more importantly, the phase of the modulation is shifted. Depending on the scan rate and the filter design, the phase shift could even convert the cosine to a sine, leading to an entirely erroneous estimate of the Fourier transform value. One might try to design the electrical filter so that $\Psi_P(\nu)$ is small at all relevant frequencies, but it cannot in principle be made zero; causal filters for which $p(t) = 0$ when $t < 0$ necessarily introduce phase shifts. For a simple RC filter, as discussed in Sec. 12.1.1, a $\pi/4$ phase shift occurs at the frequency for which $|P(\nu)|/P(0) = \frac{1}{2}$, and the limiting phase shift as $\nu \to \infty$ is $\pi/2$; more complicated filters can have larger phase shifts.

We are thus faced with a tradeoff: we want to choose the bandwidth of the filter small so as to reduce the noise, but not so small that it distorts the desired signal. Two general principles guide us in making this tradeoff. First, whatever distortion is incurred in the measurement should also be modeled in the reconstruction process; we should not blindly estimate Fourier transform values by (19.23) if (19.16) does not accurately describe the forward problem. Second, we should choose the bandwidth to optimize task performance. If the task depends on high spatial frequencies in the object, it may be better to use a larger bandwidth and accept more noise.

[4]We use the word "signal" here to mean "that which we are trying to measure," not "that which we are trying to discern in a detection task."

Object randomness So far we have discussed the contribution of measurement noise to the randomness in the data, but we know from previous chapters that object variability is also important in many cases. The two sources of randomness — measurement noise and object variability — make the data doubly stochastic, and there are two terms in the covariance matrix. The relevant theory is developed in Sec. 8.5.3, and we shall now apply it to the special case of imaging in Fourier space.

The general formulas we need are (8.347) and (8.348):

$$\mathbf{K_g} = \overline{\mathbf{K}}_\mathbf{n} + \mathbf{K}_{\overline{\mathbf{g}}}\,, \qquad \mathbf{K}_{\overline{\mathbf{g}}} = \mathcal{H}\mathcal{K}_\mathbf{f}\mathcal{H}^\dagger\,, \tag{19.34}$$

where $\overline{\mathbf{K}}_\mathbf{n}$ means the same thing as $\mathbf{K_n}$ for object-independent noise, but for Poisson noise it is given by (19.28) averaged over some ensemble of objects. The object randomness is described by $\mathbf{K}_{\overline{\mathbf{g}}}$, which is the autocovariance operator of the object, $\mathcal{K}_\mathbf{f}$, reflected into the data domain in the absence of measurement noise.

For the masks defined in (19.17) and real objects, we find

$$[\mathbf{K}_{\overline{\mathbf{g}}}]_{mj,m'j'} = \tfrac{1}{4}C^2 \int_\mathbf{S} d^2r \int_\mathbf{S} d^2r' \,\langle \Delta f(\mathbf{r})\, \Delta f(\mathbf{r}')\rangle \,[1+\cos(2\pi\boldsymbol{\rho}_m\cdot\mathbf{r}+\phi_j)]$$
$$\times\,[1+\cos(2\pi\boldsymbol{\rho}_{m'}\cdot\mathbf{r}'+\phi_{j'})]\,, \tag{19.35}$$

where $\Delta f(\mathbf{r}) \equiv f(\mathbf{r}) - \langle f(\mathbf{r})\rangle$. The integrals here are over the support of the masks, but since portions of the object outside of the masks never make any contribution to the data, we may as well think of $\mathbf{S}$ as the support of the objects. The expectation $\langle \Delta f(\mathbf{r})\, \Delta f(\mathbf{r}')\rangle$ is the autocovariance function of the object (*i.e.*, the kernel of $\mathcal{K}_\mathbf{f}$), which we denote $K_\mathbf{f}(\mathbf{r},\mathbf{r}')$.

With a little algebra, we find that the covariance matrix for the estimated Fourier samples is

$$\left[\mathbf{K}_{\hat{\mathbf{F}}}\right]_{mm'} = \frac{4\sigma^2}{C^2}\,\delta_{mm'} + \int_\mathbf{S} d^2r \int_\mathbf{S} d^2r'\, K_\mathbf{f}(\mathbf{r},\mathbf{r}')\exp[-2\pi i\,(\boldsymbol{\rho}_m\cdot\mathbf{r} - \boldsymbol{\rho}_{m'}\cdot\mathbf{r}')]\,. \tag{19.36}$$

Thus the object-variability part of the covariance matrix for the estimated Fourier samples is just a Fourier-transformed and sampled version of the autocovariance function of the object. One might be tempted to assume stationarity and simplify this expression further, but that impulse should be resisted in general. One situation in which at least quasistationarity can be justified is discussed in Sec. 19.1.4.

19.1.3 Reconstruction

Having collected a set of noisy measurements and characterized the noise, it remains to reconstruct an image of the object. We could simply refer back to Chap. 15 and say that all of the reconstruction algorithms introduced there are applicable, but this would ignore the fact that the data are related to the object by a well-understood transform, the Fourier transform. At the opposite extreme, we could reconstruct simply by applying an inverse Fourier transform (necessarily the inverse DFT since the data are discrete), but this would ignore many complications in the data-acquisition process.

Regular sampling in Fourier space We begin the discussion of image reconstruction by assuming that the sampled spatial frequencies (of a mask or projected fringe pattern) form a square grid of spacing $\Delta\rho$. In this case it is convenient to denote the spatial frequency as $\boldsymbol{\rho}_{\mathbf{k}} \equiv \mathbf{k}\Delta\rho$, where $\mathbf{k}$ is a 2D multi-index, $\mathbf{k} = (k_x, k_y)$, with integer components. Both components of $\mathbf{k}$ will be assumed to run from $-K$ to K, so the total number of sampled frequencies is $M = (2K+1)^2$.

We shall also assume that the sampled frequencies satisfy the Nyquist condition for sampling in frequency space, as discussed in Sec. 3.5.4. Specifically, if $f(\mathbf{r})$ vanishes outside of a square of side L, we require that $\Delta\rho \le 1/L$. Note that there is no requirement that $f(\mathbf{r})$ be bandlimited. We shall use the term *exact Nyquist sampling* to mean $\Delta\rho = 1/L$ and hence $\boldsymbol{\rho}_{\mathbf{k}} = \mathbf{k}/L$.

Backprojection In many inverse problems of the form $\mathbf{g} = \boldsymbol{\mathcal{H}}\mathbf{f} + \mathbf{n}$, it is useful to apply the adjoint or backprojection operator $\boldsymbol{\mathcal{H}}^\dagger$ to the data as a first step toward reconstruction. Consider data described by (19.16) – (19.18) with sampling on a regular grid. With the multi-index notation, backprojection is the superposition of mask functions weighted by the data values [*cf.* (1.45)]:

$$[\boldsymbol{\mathcal{H}}^\dagger \mathbf{g}](\mathbf{r}) = C \sum_{\mathbf{k}=-K}^{K} \sum_{j=0}^{3} g_{\mathbf{k}j} \, \tfrac{1}{2}[1 + \cos(2\pi\boldsymbol{\rho}_{\mathbf{k}} \cdot \mathbf{r} + \phi_j)] \,, \tag{19.37}$$

where the limits on the first sum imply that both components of $\mathbf{k}$ run over the specified range. Making use of (19.16) and (19.17), we find

$$[\boldsymbol{\mathcal{H}}^\dagger \mathbf{g}](\mathbf{r}) = C^2 \sum_{\mathbf{k}=-K}^{K} \sum_{j=0}^{3} \int_{\mathbf{S}} d^2r' \, f(\mathbf{r}') \, \tfrac{1}{2}[1+\cos(2\pi\boldsymbol{\rho}_{\mathbf{k}} \cdot \mathbf{r}' + \phi_j)] \, \tfrac{1}{2}[1+\cos(2\pi\boldsymbol{\rho}_{\mathbf{k}} \cdot \mathbf{r} + \phi_j)]$$

$$+ C \sum_{\mathbf{k}=-K}^{K} \sum_{j=0}^{3} n_{\mathbf{k}j} \, \tfrac{1}{2}[1 + \cos(2\pi\boldsymbol{\rho}_{\mathbf{k}} \cdot \mathbf{r} + \phi_j)] \,. \tag{19.38}$$

Interchanging the j sum and the integral and doing a bit of algebra yields[5]

$$\begin{aligned}
&[\boldsymbol{\mathcal{H}}^\dagger \mathbf{g}](\mathbf{r}) \\
&= C^2 \sum_{\mathbf{k}=-K}^{K} \left\{ \int_{\mathbf{S}} d^2r' \, f(\mathbf{r}') + \frac{1}{4} \int_{\mathbf{S}} d^2r' \, f(\mathbf{r}') \exp[2\pi i \boldsymbol{\rho}_{\mathbf{k}} \cdot (\mathbf{r} - \mathbf{r}')] \right. \\
&\qquad \left. + \frac{1}{4} \int_{\mathbf{S}} d^2r' \, f(\mathbf{r}') \exp[-2\pi i \boldsymbol{\rho}_{\mathbf{k}} \cdot (\mathbf{r} - \mathbf{r}')] \right\} \\
&\qquad + C \sum_{\mathbf{k}=-K}^{K} \sum_{j=0}^{3} n_{\mathbf{k}j} \, \tfrac{1}{2}[1 + \cos(2\pi\boldsymbol{\rho}_{\mathbf{k}} \cdot \mathbf{r} + \phi_j)] \,.
\end{aligned} \tag{19.39}$$

For exact Nyquist sampling, where $\boldsymbol{\rho}_{\mathbf{k}} = \mathbf{k}/L$, the integrals above are proportional to Fourier-series coefficients of the object, denoted $F_{\mathbf{k}}$ and defined by

$$F_{\mathbf{k}} \equiv \frac{1}{L^2} \int_{\mathbf{S}} d^2r' \, f(\mathbf{r}') \exp(-2\pi i \boldsymbol{\rho}_{\mathbf{k}} \cdot \mathbf{r}') = \frac{1}{L^2} F(\boldsymbol{\rho}_{\mathbf{k}}) \,, \tag{19.40}$$

[5] For those wishing to check the algebra, note that there are 4 terms in the sum over j and that $\sum_j \exp(\pm i\phi_j)$ and $\sum_j \exp(\pm 2i\phi_j)$ vanish.

where $F(\boldsymbol{\rho})$ is the ordinary 2D Fourier transform of the object (assumed to have support $\mathbf{S}$).

Making use of the symmetry $F_{\mathbf{k}} = F^*_{-\mathbf{k}}$, we find

$$[\mathcal{H}^\dagger \mathbf{g}](\mathbf{r}) = C^2L^2(2K+1)^2F_0 + \tfrac{1}{2}C^2L^2 \sum_{\mathbf{k}=-K}^{K} F_{\mathbf{k}} \exp(2\pi i \boldsymbol{\rho}_{\mathbf{k}} \cdot \mathbf{r})$$

$$+ C \sum_{\mathbf{k}=-K}^{K} \sum_{j=0}^{3} n_{\mathbf{k}j} \, \tfrac{1}{2}[1 + \cos(2\pi \boldsymbol{\rho}_{\mathbf{k}} \cdot \mathbf{r} + \phi_j)] \,. \tag{19.41}$$

The third term in this equation is the noise in the data transformed by $\mathcal{H}^\dagger$. Note that this transformation creates a continuous function of $\mathbf{r}$, so the transformed noise is a random process. The reader is invited to compute the autocovariance function of this random process.

The first term in (19.41) is not a function of $\mathbf{r}$, but it is object dependent since it is proportional to F_0, which is the average object value across the field of view. Moreover, the term can be very large because of the factor $(2K+1)^2$, which is the total number of sampled frequencies. The object is unknown, so we do not know the value of F_0, but we can estimate it from the data and subtract off the estimate; one way of doing so is presented below, but for now we shall just refer to the first term in (19.41), or any spatially constant term, as a DC term. Basically, the DC term in the backprojection comes from the DC term in the mask transmittances.

The second term in (19.41) is the desired reconstruction. We therefore define

$$\hat{f}(\mathbf{r}) \equiv \frac{1}{\frac{1}{2}C^2L^2}[\mathcal{H}^\dagger \mathbf{g}](\mathbf{r}) = \sum_{\mathbf{k}=-K}^{K} F_{\mathbf{k}} \exp(2\pi i \boldsymbol{\rho}_{\mathbf{k}} \cdot \mathbf{r}) + \text{DC term} + \text{noise} \,. \tag{19.42}$$

If the sum here ran over the infinite 2D grid of Fourier frequencies, it would equal the actual object $f(\mathbf{r})$. Instead it includes only the frequencies for which $-K \leq k_x$, $k_y \leq K$. Since the masks contain only these spatial frequencies, the functions $\{L^{-1} \exp(2\pi i \boldsymbol{\rho}_{\mathbf{k}} \cdot \mathbf{r})\}$ form a basis for the measurement space (no matter what set of frequencies is used). If $\boldsymbol{\rho}_{\mathbf{k}} = \mathbf{k}/L$ for some set of integer-valued multi-indices $\{\mathbf{k}\}$, this basis is orthonormal and the sum in (19.42) is exactly the measurement component of the object [*cf.* (1.165)]. Then we can write

$$\sum_{\mathbf{k}=-K}^{K} F_{\mathbf{k}} \exp(2\pi i \boldsymbol{\rho}_{\mathbf{k}} \cdot \mathbf{r}) = f_{meas}(\mathbf{r}) = [\mathcal{P}_{meas}\mathbf{f}](\mathbf{r}) \,, \tag{19.43}$$

where $\mathcal{P}_{meas}$ is the projector onto measurement space. Thus

$$\hat{f}(\mathbf{r}) = f_{meas}(\mathbf{r}) + \text{DC term} + \text{noise} \,. \tag{19.44}$$

The conclusion is that simple backprojection recovers the measurement component of the object except for noise and the DC term. This conclusion is, however, quite specific to the assumptions we made about the data. The masks had to be exactly described by (19.16) – (19.18), with pure cosinusoidal variations and no higher-order terms, and we had to assume exact Nyquist sampling on the same grid as used to define the Fourier series.

Backprojection from the Fourier domain Suppose we start not with the raw data $\mathbf{g}$ but rather with the estimated Fourier transform values $\hat{\mathbf{F}}$ as defined by (19.19) or (19.22). Backprojection in this case means applying the adjoint of the operator $\mathbf{A}\mathcal{H}$:

$$\hat{f}(\mathbf{r}) \equiv [\mathbf{A}\mathcal{H}]^\dagger \hat{\mathbf{F}} = [\mathbf{A}\mathcal{H}]^\dagger \mathbf{Ag} = \mathcal{H}^\dagger \mathbf{A}^\dagger \mathbf{Ag}. \tag{19.45}$$

As the reader may show,

$$\left[(\mathbf{A}\mathcal{H})^\dagger \mathbf{Ag}\right](\mathbf{r}) = \frac{1}{C}\sum_{\mathbf{k}} \exp(2\pi i \boldsymbol{\rho}_{\mathbf{k}} \cdot \mathbf{r})\,(g_{\mathbf{k}0} - g_{\mathbf{k}2} - i g_{\mathbf{k}1} + i g_{\mathbf{k}3})$$

$$= \sum_{\mathbf{k}} \exp(2\pi i \boldsymbol{\rho}_{\mathbf{k}} \cdot \mathbf{r})\,\hat{F}(\boldsymbol{\rho}_{\mathbf{k}}). \tag{19.46}$$

Thus backprojection from the Fourier estimates amounts to superposition of the complex conjugates of the Fourier kernels [see (1.45)] weighted by the estimates.

Since $\hat{F}(\boldsymbol{\rho}_{\mathbf{k}})$ is an unbiased estimator of $F(\boldsymbol{\rho}_{\mathbf{k}})$, we can write

$$\hat{f}(\mathbf{r}) = \sum_{\mathbf{k}} \exp(2\pi i \boldsymbol{\rho}_{\mathbf{k}} \cdot \mathbf{r})\,F(\boldsymbol{\rho}_{\mathbf{k}}) + \text{noise} = f_{meas}(\mathbf{r}) + \text{noise}. \tag{19.47}$$

There is no longer a DC term since we subtracted off the DC part of the mask transmittances in forming $\hat{F}(\boldsymbol{\rho}_{\mathbf{k}})$.

Irregular grids and pseudoinversion The sum in (19.42) or (19.47) can be identified with $f_{meas}(\mathbf{r})$ only if the Fourier basis functions are orthonormal, which is the case for sampling on a regular grid where $\boldsymbol{\rho}_{\mathbf{k}} = \mathbf{k}/L$.[6] For sampling on irregular grids, however, we do not recover the measurement component by simple backprojection. What is required is a pseudoinverse of $\mathcal{H}$ (or $\mathbf{A}\mathcal{H}$ when we start from the Fourier estimates), not just its adjoint; fortunately we know from Sec. 15.2.3 how to compute this pseudoinverse.

The basic trick used in Sec. 15.2.3 is broadly applicable. It amounts to using the identity (1.149) or (15.80):

$$\mathcal{H}^+ = \mathcal{H}^\dagger(\mathcal{H}\mathcal{H}^\dagger)^+. \tag{19.48}$$

In any CD problem, the operator $\mathcal{H}\mathcal{H}^\dagger$ is an $M \times M$ matrix (where M is the number of measurements). Its pseudoinverse can be found by standard methods if M is not too large, or the identity can be used as a starting point to develop iterative algorithms. Moreover, for Fourier samplers, $\mathcal{H}\mathcal{H}^\dagger$ is simply related by (15.79) to the Fourier transform of the support function.

Thus there is no great difficulty in dealing with irregular grids in Fourier samplers. We need to keep in mind, however, that simple backprojection does not recover the measurement component of the object in these cases.

[6]Strictly speaking, orthonormality also requires that the support function be square, but we can get around this problem with circular masks by considering a square that contains the circle. The circular support is then associated with the object rather than the basis function.

Discretization The reconstruction methods discussed so far in this section yield functions $\hat{f}(\mathbf{r})$, but of course any computer-implemented algorithm will produce only discrete vectors. We could simply sample the formulas above on a discrete grid to get $\hat{f}(\mathbf{r_n})$, but we can also adopt some approximate discrete representation for $f(\mathbf{r})$ at the outset and attempt to estimate the coefficients.

Following (7.24) or (15.6) we can write the approximate object representation as

$$f_a(\mathbf{r}) = \sum_{n=1}^{N} \theta_n \phi_n(\mathbf{r}) \,, \tag{19.49}$$

where $\{\phi_n(\mathbf{r})\}$ is any convenient set of expansion functions such as pixels. In operator notation, (19.49) becomes

$$\mathbf{f}_a = \mathcal{D}_\phi^\dagger \boldsymbol{\theta} \,, \tag{19.50}$$

where $\mathcal{D}_\phi$ is a CD discretization operator and $\mathcal{D}_\phi^\dagger$ is its adjoint (hence a DC operator). The data can be written as [*cf.* (15.10) – (15.14)]

$$\mathbf{g} = \mathcal{H}\mathbf{f} + \mathbf{n} = \mathbf{H}\boldsymbol{\theta} + \boldsymbol{\epsilon} \,, \tag{19.51}$$

where $\boldsymbol{\epsilon}$ includes both modeling error and noise, and the matrix $\mathbf{H}$ is given by

$$\mathbf{H} = \mathcal{H}\mathcal{D}_\phi^\dagger \,, \qquad H_{mn} = \int_{\mathbf{S}} d^q r \; h_m(\mathbf{r})\phi_n(\mathbf{r}) \,. \tag{19.52}$$

To apply this formalism to the present problem, we add a second data index j as in (19.16). If we take

$$\phi_n(\mathbf{r}) = \operatorname{rect}\left[\frac{\mathbf{r} - \mathbf{r}_n}{\epsilon}\right] \tag{19.53}$$

and use the system kernel given in (19.18), then we find

$$H_{mjn} = \frac{\epsilon^2 C}{2}[1 + \cos(2\pi\boldsymbol{\rho}_m \cdot \mathbf{r}_n + \phi_j)\operatorname{sinc}(\epsilon\boldsymbol{\rho}_m)] \,. \tag{19.54}$$

This matrix can now be used in any of the discrete reconstruction algorithms discussed in Chap. 15.

Use of the FFT Unlike many $\mathbf{H}$ matrices, the one given in (19.54) is not sparse; its elements are nonzero for almost any combination of m, j and n, so $4MN$ multiplies will be required to compute $\mathbf{H}\boldsymbol{\theta}$ at any step in an iterative reconstruction algorithm. If, however, we choose the frequencies $\{\boldsymbol{\rho}_m\}$ and the spatial points $\{\mathbf{r}_n\}$ to lie on regular grids of the same size, we can formulate the problem in terms of the DFT and take advantage of the FFT algorithm.

Reverting to multi-index notation, we let $\mathbf{r_n} = \mathbf{n}\epsilon$, where $\epsilon = L/N$ and $0 \le n_x$, $n_y \le N-1$, so an $N \times N$ grid is used for the object representation. To be able to use the same grid in the Fourier domain, we assume that measurements are taken for frequencies $\boldsymbol{\rho}_\mathbf{k} = \mathbf{k}\Delta\rho$, where $\Delta\rho = 1/L$ and $0 \le k_x, k_y \le N-1$; the use of only nonnegative integer indices causes no difficulty for real objects because of the symmetry relation (3.322).

With these grids, the matrix defined by (19.54) applied to an arbitrary $\boldsymbol{\theta}$ yields

$$[\mathbf{H}\boldsymbol{\theta}]_{\mathbf{k}j} = \frac{\epsilon^2 C}{2}\Theta_\mathbf{0} + \frac{\epsilon^2 C}{4}[\Theta_\mathbf{k}\exp(-i\phi_j) + \Theta_\mathbf{k}^*\exp(i\phi_j)] \,, \tag{19.55}$$

where $\boldsymbol{\Theta}$ is the 2D DFT of $\boldsymbol{\theta}$, defined by

$$\Theta_{\mathbf{k}} = \sum_{\mathbf{n}=0}^{N-1} \exp\left(-2\pi i \frac{\mathbf{k}\cdot\mathbf{n}}{N}\right) \theta_{\mathbf{n}} \,. \tag{19.56}$$

Thus the data vector associated with $\boldsymbol{\theta}$ is expressed in terms of DFTs and can be efficiently computed by means of the FFT algorithm. The reader may derive a similar expression for $\mathbf{H}^{\dagger}$ applied to an arbitrary $\mathbf{g}$. It is also of interest to consider what happens if we take an inverse DFT of the estimated Fourier components.

Iterative algorithms There are two problems with the reconstruction algorithms described so far: they are all based on the idealized system description of (19.16)-(19.18), and all of them can yield unphysical negative values. These difficulties can be circumvented by using an iterative algorithm with a positivity constraint, such as the EM algorithm described in Sec. 15.4.6.

Virtually any departure from the ideal system behavior can be incorporated into an iterative algorithm, provided only that the effect can be adequately modeled. For example, if the Fourier samples are obtained by one of the moiré methods described in Sec. 19.1, there is no need to neglect any of the higher-order terms; they can readily be incorporated into the $\mathbf{H}$ matrix, and the iterative algorithm will automatically compensate for them. Similarly, if there are phase shifts arising from the temporal response of the filter in a scanning system, they can be modeled as in (19.33) and again compensated by the algorithm.

The problem of negative values is inherent in the use of a finite set of spatial frequencies. If we represent a nonnegative function by a Fourier series and delete terms, the result will virtually always have negative values. Most iterative algorithms enforce a positivity constraint at each iteration and thereby correct this problem. With the EM algorithm, for example, the initial estimate is chosen to be nonnegative, and the multiplicative correction factor for the next estimate involves this estimate, the $\mathbf{H}$ matrix and the data [see (15.297)], all of which are nonnegative. Thus the second estimate is nonnegative, and by the same argument all subsequent estimates are nonnegative.

19.1.4 Image quality

The principles of task-based evaluation and optimization of imaging systems were enunciated in Chaps. 13 and 14. In this section we shall discuss the application of these principles to systems that measure Fourier components of an object, with particular attention to choice of the set of sampled frequencies $\{\boldsymbol{\rho}_n\}$.

SKE/BKE detection tasks We begin with SKE/BKE (signal known exactly, background known exactly) detection tasks. As we shall see, this task can lead to misleading conclusions about system optimization.

To be specific, consider the case of i.i.d. Gaussian noise and suppose that the data used for the detection task consist of Fourier components estimated according to (19.19). We know from (19.27) that the variance of each of these estimates is $4\sigma^2/C^2$, and it follows from (13.120) that the SNR for the ideal observer on an

SKE/BKE detection task is

$$\mathrm{SNR}_\lambda^2 = \frac{C^2}{4\sigma^2} \sum_{m=1}^{M} |S(\boldsymbol{\rho}_m)|^2 \,, \tag{19.57}$$

where $S(\boldsymbol{\rho})$ is the Fourier transform of the signal to be detected.

For most signals of interest, $|S(\boldsymbol{\rho})|$ is maximal at $\boldsymbol{\rho} = 0$, so the sum in (19.57) is maximized (for a fixed number of measurements) by choosing all $\boldsymbol{\rho}_m$ to be zero! In other words, we can do no better than measure the total integral of the object. Since the background is presumed to be known in all details, including its total integral, any increase in the integral of the object is an indication of the presence of a signal. In practice, however, we would never have this degree of prior information about the object, so we must choose a more realistic task for system optimization.

SKE/BKE discrimination tasks One way to avoid the incorrect conclusion that only the zero-frequency components are of interest is to consider discrimination between two different signals that have the same integral or DC value. This so-called *Rayleigh task* was discussed in Sec. 16.2.5. The ideal-observer SNR is still given by (13.120), but now the difference signal must be used in place of the signal in a detection task; (19.57) becomes

$$\mathrm{SNR}_\lambda^2 = \frac{C^2}{4\sigma^2} \sum_{m=1}^{M} |\Delta S(\boldsymbol{\rho}_m)|^2 \,. \tag{19.58}$$

There is some frequency $\boldsymbol{\rho}_0$ where $|\Delta S(\boldsymbol{\rho}_m)|$ is maximal, and (19.57) tells us that all samples should be taken at precisely that frequency. Again, this conclusion presumes an unrealistic amount of prior information and should not be taken seriously.

Random signals and backgrounds The best way to avoid the trap of assuming unrealistic prior information is to formulate the problem realistically in the first place, with unknown random signals and backgrounds. When we do so it becomes difficult to compute the ideal-observer SNR (see Sec. 14.3.3), and we must often revert to the Hotelling observer. Even that observer is complicated in the present problem, however, since it requires inversion of the covariance matrix given in (19.36). Methods for performing (or avoiding!) the inversion are discussed in Sec. 14.3.2, but an analytical solution would offer more insights.

We can make some headway analytically if we assume that we are interested in detecting a spatially localized signal with nonzero values only in the neighborhood of some point $\mathbf{r}_0$ and that the background statistics in this neighborhood are quasistationary (see Secs. 8.4.4 and 13.2.13). As in (13.259), the quasistationarity is best expressed in sum and difference coordinates; we define

$$K_{\mathbf{f}}(\mathbf{r}, \mathbf{r}') = K_{\mathbf{f}}(\overline{\mathbf{r}} + \tfrac{1}{2}\Delta\mathbf{r}, \overline{\mathbf{r}} - \tfrac{1}{2}\Delta\mathbf{r}) \equiv \widetilde{K}_{\mathbf{f}}(\overline{\mathbf{r}}, \Delta\mathbf{r}) \,, \tag{19.59}$$

where

$$\mathbf{r} \equiv \overline{\mathbf{r}} + \tfrac{1}{2}\Delta\mathbf{r} \,, \qquad \mathbf{r}' \equiv \overline{\mathbf{r}} - \tfrac{1}{2}\Delta\mathbf{r} \,. \tag{19.60}$$

For a signal localized near $\mathbf{r}_0$, only background variations in some neighborhood of $\overline{\mathbf{r}} = \mathbf{r}_0$ influence the detectability. If the background statistics in this neighborhood are quasistationary, it may be valid to approximate $\widetilde{K}_{\mathbf{f}}(\overline{\mathbf{r}}, \Delta\mathbf{r})$ by $\widetilde{K}_{\mathbf{f}}(\mathbf{r}_0, \Delta\mathbf{r})$. Moreover, it might be valid to assume that the background correlations are short range, so that $\widetilde{K}_{\mathbf{f}}(\mathbf{r}_0, \Delta\mathbf{r})$ drops to zero over a distance $|\Delta\mathbf{r}|$ small compared to the object support. With these assumptions, (19.36) becomes

$$\left[\mathbf{K}_{\hat{\mathbf{F}}}\right]_{mm'} = \frac{4\sigma^2}{C^2}\,\delta_{mm'}$$

$$+ \int_{\mathbf{S}} d^2\overline{r}\, \exp[-2\pi i\,(\boldsymbol{\rho}_m - \boldsymbol{\rho}_{m'}) \cdot \overline{\mathbf{r}}] \int_\infty d^2\Delta r\ \widetilde{K}_{\mathbf{f}}(\mathbf{r}_0, \Delta\mathbf{r}) \exp[-\pi i\,(\boldsymbol{\rho}_m + \boldsymbol{\rho}_{m'}) \cdot \Delta\mathbf{r}]\,. \tag{19.61}$$

In many cases of interest, the integral over $\overline{\mathbf{r}}$ is either approximately or exactly zero unless $m = m'$. For exact Nyquist sampling, that integral is exactly $L^2\delta_{mm'}$. For sampling on a sparser grid, the integral is small compared to L^2 if $|\boldsymbol{\rho}_m - \boldsymbol{\rho}_{m'}|L \gg 1$, where L is the width of the object support. Thus we can write

$$\left[\mathbf{K}_{\hat{\mathbf{F}}}\right]_{mm'} = \left[\frac{4\sigma^2}{C^2} + L^2 \int_\infty d^2\Delta r\ \widetilde{K}_{\mathbf{f}}(\mathbf{r}_0, \Delta\mathbf{r}) \exp\left(-2\pi i \boldsymbol{\rho}_m \cdot \Delta\mathbf{r}\right)\right] \delta_{mm'}\,. \tag{19.62}$$

The remaining integral is recognized as the stochastic Wigner distribution function [*cf.* (8.140)] for the zero-mean process $\Delta f(\mathbf{r})$, evaluated at $\mathbf{r} = \mathbf{r}_0$ and $\boldsymbol{\rho} = \boldsymbol{\rho}_m$. Denoting this Wigner distribution as $W_{\Delta f}(\mathbf{r}, \boldsymbol{\rho})$, we have

$$\left[\mathbf{K}_{\hat{\mathbf{F}}}^{-1}\right]_{mm'} = \left[\frac{4\sigma^2}{C^2} + L^2 W_{\Delta f}(\mathbf{r}_0, \boldsymbol{\rho}_m)\right]^{-1} \delta_{mm'}\,. \tag{19.63}$$

Thus the assumption of quasistationary noise leads to a diagonal (and therefore easily invertible) covariance matrix in this problem even though the system is far from shift-invariant.

With this inverse covariance, the Hotelling detectability satisfies

$$\mathrm{SNR}^2_{Hot}(\mathbf{r}_0) = \sum_{m=1}^{M} \frac{|S(\boldsymbol{\rho}_m)|^2}{\frac{4\sigma^2}{C^2} + L^2 W_{\Delta f}(\mathbf{r}_0, \boldsymbol{\rho}_m)}\,. \tag{19.64}$$

This expression reduces to (19.57) if there is no background variability. When there is background variability, it has a form similar to expressions derived in Chap. 13 [*e.g.*, (13.252) and (13.266)], but only within the validity of the quasistationary model.

One qualitative conclusion from (19.64) is that higher spatial frequencies are more useful with background variability than they are with a BKE task since the object Wigner distribution $W_{\Delta f}(\mathbf{r}_0, \boldsymbol{\rho})$ will tend to decrease with increasing frequency. The signal $|S(\boldsymbol{\rho})|^2$ also decreases with frequency in many problems, but if the spatial extent of the signal is smaller than the correlation length of the background, the ratio $|S(\boldsymbol{\rho})|^2/W_{\Delta f}(\mathbf{r}_0, \boldsymbol{\rho})$ increases with frequency, so higher frequencies provide more discrimination between signal and background. On the other hand, higher frequencies yield a smaller ratio of $|S(\boldsymbol{\rho})|^2$ to the frequency-independent measurement noise $4\sigma^2/C^2$. Numerical evaluation of (19.64) must be performed for quantitative comparisons of different sampling schemes, but no matrix inverses are needed because of the quasistationarity assumption.

The reader might be tempted to say, as we did in discussing (19.57) and (19.58), that there is some frequency $\boldsymbol{\rho}_0$ where the summand in (19.64) is maximal, so all samples should be taken at that one frequency. Recall, however, that we assumed in going from (19.61) to (19.62) that $\boldsymbol{\rho}_m \neq \boldsymbol{\rho}_{m'}$ if $m \neq m'$, and this assumption is certainly not true if all frequencies are the same.

Estimation tasks If an estimation task is to be performed, quantitative accuracy of the reconstruction is paramount. This accuracy depends on the choice of spatial frequencies, noise level and reconstruction algorithm, but in many cases the most important factor is the modeling of the forward problem. If, for example, we neglect higher-order terms in a moiré system, then we should assess their effect on estimates of the parameters of interest. If the noise level in the data is high, inaccurate modeling may make a small contribution to the mean-square error (MSE) in parameter estimates, but eventually, as exposure time is increased and the noise is reduced, modeling error will dominate.

We must remember, however, that MSE is applicable (at least in a non-Bayesian sense) only to estimable parameters (see Secs. 14.3.4 and 15.1.3). In the present problem, our basic measurements are unbiased estimates of the Fourier transform values $\{F(\boldsymbol{\rho}_n)\}$, so any linear combination of these values is estimable. If we wish to estimate the integral of the object over some region of interest (ROI), the template defining this region must be well approximated by a series of the form

$$t(\mathbf{r}) \approx \sum_n T_n \exp(2\pi i \boldsymbol{\rho}_n \cdot \mathbf{r}) \tag{19.65}$$

in order for the integral to be estimable. If this condition is not satisfied, different objects can give the same mean data but different true values for the ROI integral, and we would not know which object to select as defining the true value of the parameter. Some readers might opt for choosing a particular object, such as a uniform field, and defining an MSE with respect to this object, but in our view that would give a misleading picture of the estimation performance of the system. We recommend using estimable parameters for evaluating estimation performance whenever possible. For systems that measure Fourier samples, we should take enough samples for the approximation in (19.65) to be valid.

When we cannot satisfy approximation (19.65), the best we can do is use the ensemble mean-square error (EMSE) as defined in (13.288) in place of the MSE corresponding to a particular true value. To compute the EMSE exactly, we would need a PDF on the objects, but we can estimate it with a finite number of realistic simulated objects (see Sec. 14.3.4).

19.2 INTERFEROMETERS

Though often taught in elementary courses as distinct entities, interferometers and imaging systems are in fact virtually indistinguishable. All interferometers are imaging systems, and all imaging systems using coherent or partially coherent radiation involve interference.

Consider a common application of interferometry, testing the figure of a mirror during fabrication. The mirror under test is placed in one arm of an interferometer, and a reference surface such as a high-quality spherical mirror is placed in the other arm. The interferometer serves to overlap the two reflected waves, and a fringe pattern is formed on some detector. The key point for the present discussion, however, is that the reflected waves do not simply propagate to the detector; instead, they are imaged through some optical system (if only the human eye) so that the waves arriving at the detector are replicas of the waves emerging from the two mirrors. Then the fringe pattern is an image of the cosine of the phase difference between the waves.

In this example, the imaging system is direct, but there are also important indirect-imaging applications of interferometry. As a consequence of the van Cittert-Zernike theorem, data collected in interference experiments can be used to reconstruct an image of the illuminating source.

In this section we shall explore a variety of systems that can be viewed as either interferometers or imaging systems. In keeping with the spirit of the chapter, we use these systems to illustrate various mathematical points introduced earlier in the book.

We begin in Sec. 19.2.1 by discussing the classical double-slit experiment of Thomas Young as a way of introducing interferometers that use pairs of apertures. This section illustrates the concept of partial coherence and the van Cittert-Zernike theorem, and it shows how these principles can be adapted to imaging.

In Sec. 19.2.2 we give some now-familiar mathematical descriptions for noise in the raw data, and then we discuss the transition from noisy data to estimates of Fourier coefficients in interferometers.

Section 19.2.3 is a detailed treatment of a two-aperture interferometer of great historical and pedagogical significance, the Michelson stellar interferometer. Included is a discussion of how one goes from interferometric measurements to estimation of stellar diameters. Section 19.2.4 is a broad-brush look at some modern multiple-telescope systems that can operated as interferometers.

19.2.1 Young's double-slit experiment

Thomas Young (1773 – 1829) was an English physicist and physician who contributed an astonishing number of key concepts to modern science. He was the first to establish clearly the wave nature of light (thereby earning the animosity of English scientists devoted to Newton's corpuscular theory). He also showed that light consisted of transverse oscillations and laid the groundwork for the theory of polarization. He gave the first modern interpretation of energy, and he made key discoveries in elasticity and surface tension. In visual perception, he observed how the lens of the eye changes shape to focus at different distances, and he proposed that color vision involved only three color receptors. As if these physiological and physical contributions were not enough, he also translated the Rosetta stone and laid the foundation for modern Egyptology.

To an optics audience, Young's name is most familiar from the double-slit experiment illustrated in Fig. 19.9. There are several crucial deductions about the nature of light that can be made with this apparatus. First, when only one slit is open, a broad Fraunhofer diffraction pattern is seen, but when both slits are open simultaneously, a pattern of fine fringes appears. There are dark regions where less light is seen with both slits open. This observation demonstrates that light is a wave, capable of interference.

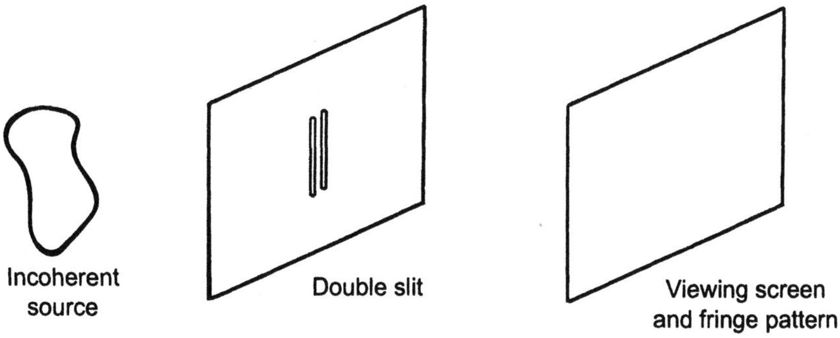

Fig. 19.9 Young's double-slit experiment.

Second, if we place polarizers over the slits, we find that the fringes disappear when the polarizations are orthogonal. This observation establishes the vector nature of the wave; the electric fields do not interfere if their dot product is zero.

Third, we can use the double-slit apparatus with modern position-sensitive, photon-counting detectors to better understand the quantum-mechanical nature of light and light-matter interactions. If we use a very weak light source, so that only one photon at a time is moving from the source to the detector plane, then each photon will impinge on the detector and be recorded at a single point. Nevertheless, after many photons have been recorded, the same fringe pattern is seen if both slits are open. If one slit is blocked, the fringes disappear, indicating that the picture of photons as localized particles is not adequate for describing this experiment.

This observation can be explained via quantum electrodynamics, where a probability amplitude is assigned to the path through each slit. The total probability for finding the photon at some point on the detector plane is the square of the sum of the probability amplitudes; since the amplitudes are complex, they can interfere, possibly making the total probability lower than it would be with only one slit open.

Alternatively, the observation can be explained by the semiclassical approach discussed in Sec. 10.1.4 where the field is treated classically but the atoms in the detector are treated quantum-mechanically. In this view, the interference is between the classical wave amplitudes emerging from the two slits; the irradiance on the detector plane, which is proportional to the square of the sum of the amplitudes, exhibits a classical fringe pattern. This irradiance then causes photoelectric interactions at random points on the detector surface. The observed pattern is a nonstationary Poisson random process where the probability density function on the positions of photoelectric interactions is proportional to the optical fluence (energy per unit area). Since fluence is irradiance times exposure time, it does not matter whether we use a weak light source (small irradiance) and long exposure time or a strong source and short time; we are not regarding the field as composed of photons, so there is no issue of how many photons are traversing the system simultaneously.

The experimental observation that is most important for the imaging applications of Young's double slit and other interferometers is that the visibility of the fringes is greatest for a quasimonochromatic point source, generally decreasing as the source size is increased. We know from Sec. 9.7 that the dependence of visibility on source distribution is a consequence of the spatial coherence of the field at the slit plane as described by the van Cittert-Zernike theorem. In fact, in that section we used a double-pinhole arrangement, Fig. 9.19, as a way of illustrating the concept of spatial coherence. The van Cittert-Zernike theorem was used as a description of the forward problem: given the source, compute the mutual coherence function and hence the fringe visibility. Here we shall discuss the inverse problem: given a set of fringe measurements taken with a double-pinhole or double-slit system, deduce properties of the source. As preliminaries, however, we shall discuss the fringes produced by point sources and extended sources, and we shall take another look at the meaning of the van Cittert-Zernike theorem and give a new and perhaps more intuitive derivation of it.

Monochromatic point source Consider first a monochromatic point source at 2D position $\mathbf{r}_s$ (where subscript s indicates *source*) in the plane $z = -z_0$. Two small pinholes are at 2D positions $\mathbf{r} = \mathbf{r}_1$ and $\mathbf{r} = \mathbf{r}_2$ in the plane $z = 0$, and the fringe pattern is observed in the plane $z = z_0$. The source emits a spherical wave that propagates to the pinhole plane, and each pinhole in turn emits a spherical wave that propagates to the observation plane. The interference of these two secondary spherical waves produces a straight-line fringe pattern.

We can analyze the fringe pattern by Fresnel diffraction theory as developed in Sec. 9.4.6. Within the Fresnel approximation, the field at the jth pinhole ($j = 1, 2$) is given by [*cf.* (9.94)]

$$u_0(\mathbf{r}_j) \propto \exp\left[i\frac{\pi}{\lambda z_0}\left|\mathbf{r}_j - \mathbf{r}_s\right|^2\right]. \tag{19.66}$$

If the pinholes are small compared to a wavelength, then a spherical wave from each propagates to the observation plane, and the field at point $\mathbf{r}$ is given by

$$u_{z_0}(\mathbf{r}) \propto \exp\left[i\frac{\pi}{\lambda z_0}\left(\left|\mathbf{r}_1 - \mathbf{r}_s\right|^2 + \left|\mathbf{r}_1 - \mathbf{r}\right|^2\right)\right] + \exp\left[i\frac{\pi}{\lambda z_0}\left(\left|\mathbf{r}_2 - \mathbf{r}_s\right|^2 + \left|\mathbf{r}_2 - \mathbf{r}\right|^2\right)\right]. \tag{19.67}$$

With a little algebra,[7] we find that the irradiance at point $\mathbf{r}$ in the observation plane is given by

$$I_{z_0}(\mathbf{r}) \propto \left|u_{z_0}(\mathbf{r})\right|^2 \propto 1 + \cos\left[\frac{2\pi}{\lambda z_0}(\mathbf{r}_2 - \mathbf{r}_1)\cdot(\mathbf{r}_s + \mathbf{r})\right]. \tag{19.68}$$

For fixed pinhole locations $\mathbf{r}_1$ and $\mathbf{r}_2$, this function describes the fringe pattern as a function of position $\mathbf{r}$ in the observation plane. Note that a maximum irradiance is seen at $\mathbf{r} = -\mathbf{r}_s$; the path length from the source through pinhole 1 to this point equals that through pinhole 2, so constructive interference occurs. We can say that the fringe pattern is centered at $\mathbf{r} = -\mathbf{r}_s$. Thus the lateral displacement or phase of the fringes is determined by the position $\mathbf{r}_s$ of the source point, and the fringe frequency $\boldsymbol{\rho}_f$ is determined by the pinhole spacing as

$$\boldsymbol{\rho}_f = \frac{1}{\lambda z_0}(\mathbf{r}_2 - \mathbf{r}_1). \tag{19.69}$$

The fringe visibility is 100% for a point source.

The relationship between (19.68) and (19.13) should not be overlooked. We know from (19.11) that a Fresnel zone plate is a sum of quadratic phase factors,

[7] We assume here that the origin of coordinates is halfway between the two pinholes, so that $r_1^2 = r_2^2$

just as spherical waves have complex amplitudes given by quadratic phase factors. When we take the product of two zone-plate transmittances as in (19.13), we get terms where the quadratic phases cancel, leaving phases that depend linearly on position, *i.e.*, straight-line moiré fringes. Similarly, when we interfere two spherical waves and compute the irradiance, we again get cancellation of quadratic phase terms so that straight-line interference fringes remain. The physics is very different in the two problems, but the math is quite similar.

Extended source: van Cittert-Zernike revisited Now consider a quasimonochromatic but spatially incoherent extended source. As discussed in Sec. 9.7.4, spatial incoherence means that the spatial autocorrelation function of the source is well approximated by a Dirac delta function, so there can be no interference between waves emanating from different source points. Instead, each point on the source produces its own fringe pattern, described by (19.68), and the total irradiance at point $\mathbf{r}$ is the sum (integral) of the irradiances from individual source points. If the radiant exitance of the source at point $\mathbf{r}_s$ is denoted $f(\mathbf{r}_s)$, we can write

$$I_{z_0}(\mathbf{r}) = C \int_\infty d^2r_s \; f(\mathbf{r}_s) \left\{1 + \cos[2\pi\boldsymbol{\rho}_f \cdot (\mathbf{r}_s + \mathbf{r})]\right\}, \tag{19.70}$$

where C is a constant containing geometric factors such as the distance z_0 and the pinhole diameters.

To put (19.70) into a more familiar form, we can define a dimensionless complex number γ by

$$\gamma \equiv \frac{\int_\infty d^2r_s \; f(\mathbf{r}_s) \exp[-2\pi i \boldsymbol{\rho}_f \cdot \mathbf{r}_s]}{\int_\infty d^2r_s \; f(\mathbf{r}_s)} = \frac{F(\boldsymbol{\rho}_f)}{F(0)}, \tag{19.71}$$

where $F(\boldsymbol{\rho})$ is the 2D Fourier transform of $f(\mathbf{r}_s)$, and, by the central-ordinate theorem, $F(0)$ is the integral of the object. With this definition, (19.70) becomes

$$I_{z_0}(\mathbf{r}) = CF(0)\left[1 + |\gamma| \cos\left(2\pi\boldsymbol{\rho}_f \cdot \mathbf{r} - \Phi_\gamma\right)\right], \tag{19.72}$$

where $\gamma \equiv |\gamma| \exp(i\Phi_\gamma)$.

As with a point source, the fringe frequency is determined solely by the pinhole positions, but now the fringe visibility $|\gamma|$ and phase Φ_γ are determined by the source distribution through (19.71). In fact, (19.71) is just a restatement of the van Cittert-Zernike theorem (9.317) when γ is recognized as the complex degree of coherence for points $\mathbf{r}_1$ and $\mathbf{r}_2$. By this theorem, the mutual coherence function is the Fourier transform of the source exitance, and by (9.259) the complex degree of coherence is a normalized version of the mutual coherence function.

Detector sensitivity functions To gain further insight into the meaning of the van Cittert-Zernike theorem, we can look at the sensitivity function associated with each detector element.

As with any linear CD system, we can write the mean measurement in the form

$$\overline{g}_{\mathbf{m}} = \int_\infty d^2r_s \; f(\mathbf{r}_s)\, h_{\mathbf{m}}(\mathbf{r}_s). \tag{19.73}$$

In the present problem, the detectors sample the continuous interference pattern, so we see from (19.68) and (19.69) that

$$h_{\mathbf{m}}(\mathbf{r}_s) \propto 1 + \cos[2\pi\boldsymbol{\rho}_f \cdot (\mathbf{r}_s + \mathbf{r}_{\mathbf{m}})]. \tag{19.74}$$

This form is familiar from Sec. 19.1.1 where we described several systems for measuring samples of the Fourier transform by modulating the object with a $(1 + \cos)$ pattern and integrating. In the double-pinhole setup, different detector elements correspond to cosine patterns all of the same frequency but with different phases. Thus the full array of measurements for one pinhole spacing can be used to estimate the complex Fourier transform for a single frequency.[8]

From pinholes to slits A practical difficulty with the system as described so far is that very small pinholes must be used in order for the waves emerging from them to be good approximations to spherical waves, as we have assumed. Small pinholes, however, collect very little light. One way to get more light is to follow the lead of Thomas Young and use slits rather than pinholes.

The analysis above is readily modified for slits. For a point source, straight-line fringes are still produced, and for a spatially incoherent source, the irradiances from different source points add linearly. Thus, for slits parallel to the y axis, (19.70) becomes

$$I_{z_0}(\mathbf{r}) = C' \int_\infty d^2 r_s \; f(\mathbf{r}_s) \left\{1 + \cos\left[2\pi \xi_f (x_s + x)\right]\right\}, \tag{19.75}$$

where C' is another radiometric constant (much larger than C), and the fringe frequency is given by [*cf.* (19.69)]

$$\xi_f = \frac{1}{\lambda z_0}(x_2 - x_1), \tag{19.76}$$

with x_1 and x_2 specifying the slit locations.

A consequence of (19.75) is that we get no information about $F(\xi, \eta)$ except for $\eta = 0$, *i.e.*, for $\boldsymbol{\rho}$ parallel to the x axis. Thus we need to rotate the object or the slits to obtain Fourier samples distributed over the 2D plane.

Another way to think about double-slit interferometry is in terms of the Radon transform and tomography (see Sec. 4.4). We know from the central-slice theorem, (4.150), that $F(\xi, 0)$ is the same as the 1D Fourier transform of the projection of the object along the y axis. If we sample ξ_f on a fine regular grid and extract the complex visibility for each sample, we can perform an inverse DFT to get an approximation to the projection $\lambda(p, \phi)$ for $\phi = 0$, and rotating the object or the slits will give other projections. Reconstruction can then proceed by any standard 2D tomographic algorithm such as filtered backprojection. Many algorithms, however, start by Fourier-transforming the projection data, and with double slits we directly collect data in Fourier space, so we may as well stay there.

19.2.2 Visibility estimation

As we have seen, a 2D fringe distribution is obtained for each pair of pinhole or slit locations, but it tells us relatively little about the object. The only useful information is contained in the visibility and phase of the fringes. If we can estimate γ from measured fringe data, we have an estimate of $F(\boldsymbol{\rho})$ for one particular $\boldsymbol{\rho}$,

[8] In fact, just four measurements would suffice if they were placed properly so that the phase $2\pi \boldsymbol{\rho}_f \cdot \mathbf{r}_m$ took on the values 0, $\pi/2$, π and $3\pi/2$, but it would be mechanically tricky to alter the detector locations for different pinhole spacings.

according to (19.71). To get additional Fourier components, we need additional pairs of pinhole locations.

Image reconstruction can thus be formulated as a two-step process. We can first estimate the Fourier components and then estimate the parameters in a discrete object representation. Alternatively, we can go directly from the raw data to a reconstructed image. In either case, it is necessary to characterize the noise in the data and how it affects the estimates of interest. In this section we look at visibility estimation in Gaussian and Poisson noise.

Visibility estimation — Gaussian data We assume a regular array of detectors in the observation plane. If each detector element is small compared to the period of the finest interference fringes used, the measurement by the detector at point $\mathbf{r_m}$ (where $\mathbf{m}$ is a 2D multi-index) is given by

$$g_{\mathbf{m}} = K\left[1 + |\gamma| \cos\left(2\pi \boldsymbol{\rho}_f \cdot \mathbf{r_m} - \Phi_\gamma\right)\right] + n_{\mathbf{m}} = \overline{g}_{\mathbf{m}} + n_{\mathbf{m}}\,, \tag{19.77}$$

where $n_{\mathbf{m}}$ is the noise in the measurement and K is a constant independent of $\mathbf{m}$ but proportional to the integral of the object. For simplicity we assume that the integral of the object can be estimated accurately from the fringe data or measured independently, so K will be regarded as known. Thus the quantities to be estimated are the two real parameters $|\gamma|$ and Φ_γ.

For i.i.d. normal noise, the log-likelihood for this problem is given by

$$\ln[\mathrm{pr}(\mathbf{g}\,|\,|\gamma|, \Phi_\gamma)] = const - \frac{1}{2\sigma^2}\sum_{\mathbf{m}}\{g_{\mathbf{m}} - K[1+|\gamma|\cos(2\pi\boldsymbol{\rho}_f \cdot \mathbf{r_m} - \Phi_\gamma)]\}^2\,. \tag{19.78}$$

It is convenient, however, to use γ and γ^* as the independent parameters, as discussed in Sec. A.9.4. Dropping the irrelevant constant, we can thus write the log-likelihood as

$$\ln\left[\mathrm{pr}(\mathbf{g}|\gamma,\gamma^*)\right]$$
$$= -\frac{1}{2\sigma^2}\sum_{\mathbf{m}}\left[g_{\mathbf{m}} - K - \tfrac{1}{2}K\gamma\exp(-2\pi i\boldsymbol{\rho}_f \cdot \mathbf{r_m}) - \tfrac{1}{2}K\gamma^*\exp(2\pi i\boldsymbol{\rho}_f \cdot \mathbf{r_m})\right]^2\,. \tag{19.79}$$

We need to maximize this expression with respect to γ and γ^*. Differentiating with respect to γ^* according to the differentiation rules of Sec. A.9.4, we see that

$$\frac{2\sigma^2}{K}\frac{\partial}{\partial\gamma^*}\ln\left[\mathrm{pr}(\mathbf{g}|\gamma)\right]$$
$$= \sum_{\mathbf{m}}\left[g_{\mathbf{m}} - K - \tfrac{1}{2}K\gamma\exp(-2\pi i\boldsymbol{\rho}_f \cdot \mathbf{r_m}) - \tfrac{1}{2}K\gamma^*\exp(2\pi i\boldsymbol{\rho}_f \cdot \mathbf{r_m})\right]\exp(2\pi i\boldsymbol{\rho}_f \cdot \mathbf{r_m})\,. \tag{19.80}$$

The derivative with respect to γ is just the complex conjugate of (19.80).

Equating (19.80) to zero gives

$$\sum_{\mathbf{m}}(g_{\mathbf{m}} - K)\exp(2\pi i\boldsymbol{\rho}_f \cdot \mathbf{r_m}) = \tfrac{1}{2}K\gamma\sum_{\mathbf{m}}1 + \tfrac{1}{2}K\gamma^*\sum_{\mathbf{m}}\exp(4\pi i\boldsymbol{\rho}_f \cdot \mathbf{r_m})\,. \tag{19.81}$$

We shall assume for simplicity that an integer number of fringes fit across the array, in which case the last sum is identically zero, Thus the maximum-likelihood estimate of the complex degree of coherence is given by (Walkup and Goodman, 1973)

$$\hat{\gamma} = \frac{2}{KM^2} \sum_{\mathbf{m}} (g_{\mathbf{m}} - K) \exp\left(2\pi i \boldsymbol{\rho}_f \cdot \mathbf{r}_{\mathbf{m}}\right) . \tag{19.82}$$

Computationally, $\hat{\gamma}$ is given by a discrete Fourier transform of the data, though only for a single frequency for each pair of pinhole locations. There is thus no advantage to the FFT algorithm, which is an efficient way of computing the DFT for a large set of frequencies.

Bias and variance Since M^2 is large, we can appeal to the asymptotic properties of ML estimators to argue that the estimate given by (19.82) is unbiased and efficient. It is instructive, however, to compute the bias and variance directly.

From (19.77) and (19.82), the mean of $\hat{\gamma}$ is

$$\langle \hat{\gamma} \rangle = \frac{2}{KM^2} \sum_{\mathbf{m}} (\overline{g}_{\mathbf{m}} - K) \exp(2\pi i \boldsymbol{\rho}_f \cdot \mathbf{r}_{\mathbf{m}})$$

$$= \frac{2}{KM^2} \sum_{\mathbf{m}} \left[\tfrac{K}{2}\gamma \exp(-2\pi i \boldsymbol{\rho}_f \cdot \mathbf{r}_{\mathbf{m}}) + \tfrac{K}{2}\gamma^* \exp(2\pi i \boldsymbol{\rho}_f \cdot \mathbf{r}_{\mathbf{m}})\right] \exp(2\pi i \boldsymbol{\rho}_f \cdot \mathbf{r}_{\mathbf{m}}) . \tag{19.83}$$

From the assumption made above that an integer number of fringes fit across the array, it follows that $\sum_{\mathbf{m}} \exp(4\pi i \boldsymbol{\rho}_f \cdot \mathbf{r}_{\mathbf{m}})$ is identically zero. Thus $\langle \hat{\gamma} \rangle = \gamma$, and the estimator is unbiased.

To study the variance, we write

$$\Delta\hat{\gamma} \equiv \hat{\gamma} - \gamma = \frac{2}{KM^2} \sum_{\mathbf{m}} \Delta g_{\mathbf{m}} \exp(2\pi i \boldsymbol{\rho}_f \cdot \mathbf{r}_{\mathbf{m}}) , \tag{19.84}$$

where $\Delta g_{\mathbf{m}} \equiv g_{\mathbf{m}} - \overline{g}_{\mathbf{m}}$. The zero-mean complex random variable $\Delta\hat{\gamma}$ has real and imaginary parts given, respectively, by

$$\Delta\hat{\gamma}_r = \frac{2}{KM^2} \sum_{\mathbf{m}} \Delta g_{\mathbf{m}} \cos(2\pi i \boldsymbol{\rho}_f \cdot \mathbf{r}_{\mathbf{m}}) ,$$

$$\Delta\hat{\gamma}_i = \frac{2}{KM^2} \sum_{\mathbf{m}} \Delta g_{\mathbf{m}} \sin(2\pi i \boldsymbol{\rho}_f \cdot \mathbf{r}_{\mathbf{m}}) . \tag{19.85}$$

Since the data covariance is given by

$$\langle \Delta g_{\mathbf{m}} \, \Delta g_{\mathbf{m}'} \rangle = \sigma^2 \, \delta_{\mathbf{m}\mathbf{m}'} , \tag{19.86}$$

we can show that

$$\text{Var}\left\{\Delta\hat{\gamma}_r\right\} = \text{Var}\left\{\Delta\hat{\gamma}_i\right\} = \frac{2\sigma^2}{K^2M^2} , \qquad \langle \Delta\hat{\gamma}_r \Delta\hat{\gamma}_i \rangle = 0 , \tag{19.87}$$

where we have used the fact that $\sum_{\mathbf{m}} \cos^2(2\pi i \boldsymbol{\rho}_f \cdot \mathbf{r}_{\mathbf{m}}) = \sum_{\mathbf{m}} \sin^2(2\pi i \boldsymbol{\rho}_f \cdot \mathbf{r}_{\mathbf{m}}) = \frac{1}{2}M^2$ but $\sum_{\mathbf{m}} \cos(2\pi i \boldsymbol{\rho}_f \cdot \mathbf{r}_{\mathbf{m}}) \sin(2\pi i \boldsymbol{\rho}_f \cdot \mathbf{r}_{\mathbf{m}}) = 0$ since an integer number of fringes fit across the array.

Thus $\Delta\hat{\gamma}_r$ and $\Delta\hat{\gamma}_i$ are uncorrelated and have the same variance. Moreover, each is a linear transformation of a Gaussian random variable, so they are i.i.d. Gaussian, and the complex quantity $\Delta\hat{\gamma}$ is circular Gaussian (see Sec. 8.3.6). The reader is invited to compute the Cramér-Rao bound for this problem and verify that $\hat{\gamma}$ is an efficient estimator.

Noise with photon-counting detectors So far in this section we have assumed that the interference pattern was detected by a 2D detector array in which the noise was i.i.d. normal. Now we consider photon-counting detectors.

As we showed classically in Sec. 11.3.7 and quantum-mechanically in Sec. 11.5.3, the photon statistics in the observation plane will virtually always be Poisson. All that is required is $\tau\,\Delta\nu \gg 1$, where $\Delta\nu$ is the spectral bandwidth of the source and τ is the observation time. We also know from Sec. 11.3 that all statistical properties of a Poisson random process are determined by the photon fluence $b(\mathbf{r})$. For narrowband radiation of mean frequency $\overline{\nu}$, photon fluence is fluence divided by the mean photon energy $h\overline{\nu}$, and fluence is irradiance times observation time, so

$$b(\mathbf{r}) = \tau \frac{I_{z_0}}{h\overline{\nu}} \,. \tag{19.88}$$

In spite of the term "photon fluence," this expression does not imply that the radiation actually consists of photons; it can be treated fully classically. Nevertheless, the pattern of photoelectric interactions in the observation plane will be a Poisson random process with fluence $\eta\, b(\mathbf{r})$, where η is the quantum efficiency. As discussed in Sec. 11.3.4, we can integrate this poisson random process over detector elements to get discrete Poisson data.

Though $\hat{\gamma}$ in (19.82) is the ML estimator only for i.i.d. Gaussian noise, it might be used with photon-counting detectors where the noise is Poisson. In that case we cannot be sure it is asymptotically efficient and unbiased. We need to examine the behavior of (19.82) for Poisson data when it is not ML, and we need to derive the ML estimator correctly under a Poisson model.

Using the Gaussian ML estimator with Poisson data Let us recompute the mean and variance of the Gaussian ML estimator but with Poisson data. Since the mean is the same for both Poisson and Gaussian data, it follows immediately from (19.83) that the estimator is unbiased no matter which data model is used. The definitions in (19.85) are still valid, and we just need to compute the variances and covariance of $\hat{\gamma}_r$ and $\hat{\gamma}_i$. In the Poisson case, the data covariance is given by

$$\langle \Delta g_{\mathbf{m}}\, \Delta g_{\mathbf{m}'} \rangle = \overline{g}_{\mathbf{m}}\, \delta_{\mathbf{m}\mathbf{m}'} \,, \tag{19.89}$$

and it follows that

$$\begin{aligned} &\mathrm{Var}\,\{\Delta\hat{\gamma}_r\} \\ &= \frac{4}{KM^4} \sum_{\mathbf{m}} \left[1 + \tfrac{1}{2}\gamma \exp(-2\pi i \boldsymbol{\rho}_f \cdot \mathbf{r}_{\mathbf{m}}) + \tfrac{1}{2}\gamma^* \exp(2\pi i \boldsymbol{\rho}_f \cdot \mathbf{r}_{\mathbf{m}})\right] \cos^2(2\pi \boldsymbol{\rho}_f \cdot \mathbf{r}_{\mathbf{m}}) \,. \end{aligned} \tag{19.90}$$

The spatial average across the array of the terms involving the complex exponentials is zero. The term varying just as $\cos^2(2\pi\boldsymbol{\rho}_f \cdot \mathbf{r}_{\mathbf{m}})$ sums to $\frac{1}{2}M^2$, as does the $\sin^2$ term in $\mathrm{Var}\,\{\Delta\hat{\gamma}_r\}$, and we are left with

$$\mathrm{Var}\,\{\Delta\hat{\gamma}_r\} = \mathrm{Var}\,\{\Delta\hat{\gamma}_i\} = \frac{2}{KM^2} \,, \qquad \langle \Delta\hat{\gamma}_r \Delta\hat{\gamma}_i \rangle = 0 \,. \tag{19.91}$$

This result differs from (19.87) in that the variance is proportional to $1/K$ rather than $1/K^2$. Since K is proportional to the source brightness and the exposure time, we see that the precision of the visibility estimates improves more rapidly with these parameters in the Gaussian case than in the Poisson case.

ML estimators for Poisson data The maximum-likelihood estimate of the complex visibility from Poisson data is given by

$$\hat{\gamma} = \underset{\gamma}{\text{argmax}} \; \{\ln[\text{pr}(\mathbf{g}|\gamma)]\} = \underset{\gamma}{\text{argmax}} \left\{ \sum_{\mathbf{m}} [g_{\mathbf{m}} \ln \overline{g}_{\mathbf{m}}(\gamma) - \overline{g}_{\mathbf{m}}(\gamma)] \right\}, \quad (19.92)$$

where

$$\overline{g}_{\mathbf{m}}(\gamma) = K \left[1 + |\gamma| \cos(2\pi \boldsymbol{\rho}_f \cdot \mathbf{r}_{\mathbf{m}} - \Phi_\gamma)\right] . \quad (19.93)$$

This equation must be solved iteratively, for example by an EM algorithm (see Sec. 15.4.6). Methods developed in Sec. 15.3.6 can be used to analyze the noise properties of the estimate.

From visibilities to image Having obtained visibility estimates, it remains to use them to reconstruct an image. Since many visibility measurements are needed, with many pairs of pinhole locations, we add an index to distinguish the different locations. For pinholes at $\mathbf{r}_1 = \mathbf{r}_{1n}$ and $\mathbf{r}_2 = \mathbf{r}_{2n}$, the fringe frequency is $\boldsymbol{\rho}_n = (\mathbf{r}_{2n} - \mathbf{r}_{1n})/(\lambda z_0)$, and the estimated object Fourier transform at this frequency is

$$\hat{F}(\boldsymbol{\rho}_n) = F(0)\, \hat{\gamma}(\boldsymbol{\rho}_n) . \quad (19.94)$$

Recall that we are assuming that $F(0)$ is measured independently, so estimation of γ directly yields an estimate of $F(\boldsymbol{\rho})$ at the fringe frequency. Note also that the index $\mathbf{m}$ has disappeared; even though we collect a 2D data set for each fringe frequency, we process it to extract a single complex number $\hat{F}(\boldsymbol{\rho}_n)$. The double-pinhole system is thus an alternative to the various systems treated in Sec. 19.1 for estimating values of the 2D Fourier transform of the object, and image reconstruction can be performed by any of the methods discussed in Sec. 19.1.3.

19.2.3 Michelson stellar interferometer

Albert A. Michelson (1842 – 1931) can rightfully be called the father of modern interferometry (though the French would nominate Fizeau, for reasons discussed in Sec. 19.2.4). A graduate of the U.S. Naval Academy, Michelson served as a science instructor there from 1875 to 1879, and during this time he began a series of measurements of the velocity of light. After a period of travel in Europe to study optics, he accepted a position in 1883 at the Case School of Applied Science in Cleveland. In collaboration with chemist Edward W. Morley, he constructed an exquisitely sensitive interferometer and used it in a famous experiment to measure the difference in the speed of light as the earth traversed the ether, the supposed medium for light propagation. In 1887 Michelson and Morley announced what has been called the most significant negative result in the history of science: they found no difference in the speed of light for propagation parallel or perpendicular to the earth's orbit. In modern language, the speed of light was independent of the reference frame. It was this critical observation that led Einstein to the special theory of relativity. In 1907 Michelson received the Nobel Prize in physics, the first awarded to an American.

Michelson's connection to imaging traces back to 1890 when he used an interferometer to study the moons of Jupiter (Michelson, 1890, 1891a), and soon thereafter he had given a clear statement of the role of fringe visibility (Michelson, 1891b). His biggest impact on astronomy, however, came in 1919 when he and Pease fitted a large telescope with an interferometer and used it to measure the diameter of the supergiant star Betelgeuse (Alpha Orionis) (Michelson and Pease, 1921).

The Michelson and Pease experiment is of interest in this chapter since it illustrates another technique for collecting data in Fourier space, and also because it is an example of an imaging system that relies on strong prior information, namely that stars are spherical. It is of enormous interest to the astronomy community because it allows high resolution without the expense of building a huge telescope.

Comprehensive reviews of interferometric imaging in astronomy are given by Tango and Twiss (1980), Roddier (1988), Shao and Colavita (1992), Lawson (2000), Quirrenbach (2001) and Saha (2002). A good popular discussion is given by Armstrong *et al.* (1995). Textbooks that provide useful background material include Hecht (1987), Lipson *et al.* (1995) and Born and Wolf (1999). A delightful collection of essays on many topics in optics, including the Michelson stellar interferometer, is Mansuripur (2002).

Acquisition geometry As shown schematically in Fig. 19.10a, Michelson and Pease placed two periscopes on a 6 m beam mounted in front of the 100 inch (2.54 m) Hooker telescope on Mt. Wilson. Each periscope consisted of two flat folding mirrors, and the outer mirror of each pair was movable. Two small apertures were used, each collecting light from one of the two periscopes. Then the primary mirror of the telescope was used to superimpose the star images from the two apertures, and fringes were observed visually. An equivalent optical system with a lens in place of the large spherical mirror is shown in Fig. 19.10b.

The setups shown in Fig. 19.10 can be regarded as modifications of Young's double-pinhole apparatus, with movable folding mirrors and their associated apertures in place of the movable pinholes. Unlike the double-pinhole experiment, however, the apertures are not assumed to be comparable in size to a wavelength. In the analysis of double pinholes, we assumed that the waves emerging from the pinholes were diverging spherical waves; in Fig. 19.10 the waves emerging from the apertures are close to plane waves, and the primary mirror or lens converts them to converging spherical waves which are imaged to Airy diffraction patterns in the focal plane. Thus, when both apertures are open, we see an Airy pattern modulated by fringes instead of a large-area fringe pattern as in the double-pinhole case [*cf.* (19.72)].

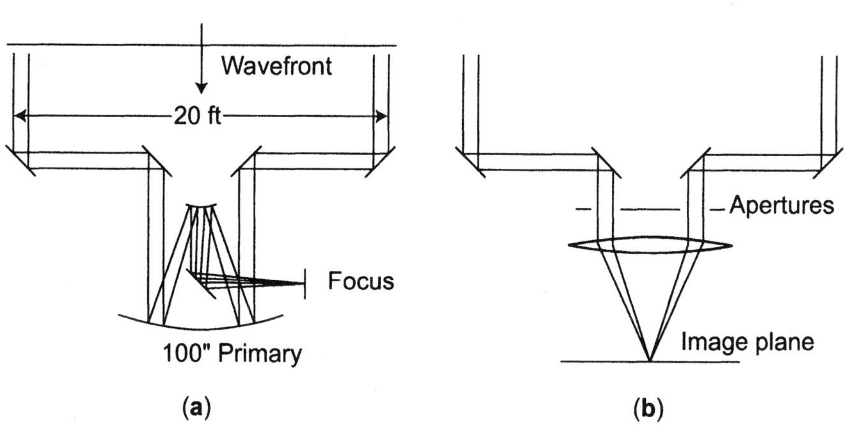

Fig. 19.10 The Michelson stellar interferometer. (a) Mirror system used by Michelson and Pease. (b) Equivalent lens system.

If the diameter of each aperture is D_{ap}, the angular size of the Airy pattern is approximately λ/D_{ap}; in practice this number is much larger than the angular diameter of the star. For example, a 20 cm aperture has an angular resolution of 2.5 μrad at a wavelength of 0.5 μm. By comparison, Betelgeuse has a diameter of 0.044 arc sec or 0.21 μrad.[9] One might think that Betelgeuse could be marginally resolved by using the full 100 inch aperture of the telescope, since its diffraction-limited resolution is approximately 0.2 μrad, but phase disturbances from the atmosphere limit the actual telescope resolution to about what one would get with a 20 cm aperture, around 2.5 μrad.

PSF analysis We can analyze the Michelson-Pease geometry just as we did with double pinholes, by first computing the irradiance image of a point source, then convolving it with an extended object distribution. We again assume that the object is spatially incoherent, so the image of an extended object is a linear superposition of the irradiance patterns from individual points. We shall treat the telescope as an ideal thin lens as in Fig. 19.10b, but essentially the same results apply to mirrors.

Because astronomical objects are essentially at an infinite distance from the imaging system, a monochromatic point source produces a plane wave described by $\exp(2\pi i k \hat{\boldsymbol{\alpha}}_0 \cdot \mathfrak{r})$, where $k = 2\pi/\lambda$, $\hat{\boldsymbol{\alpha}}_0 = (\alpha_{0x}, \alpha_{0y}, \alpha_{0z})$ is a 3D unit vector directed away from the source, and $\mathfrak{r} = (x, y, z)$ is a 3D position vector. In the plane $z = 0$, the wave can be written as $\exp(2\pi i \boldsymbol{\rho}_0 \cdot \mathbf{r})$, where $\mathbf{r} = (x, y)$ is a 2D position vector, and $\boldsymbol{\rho}_0 = (\alpha_{0x}/\lambda, \alpha_{0y}/\lambda)$ specifies the 2D spatial-frequency of the wave amplitude in that plane and hence the angular position of the source.

Let $\mathbf{r}_1$ and $\mathbf{r}_2$ be the 2D positions of the outer periscope mirrors in the plane $z = 0$, and let $\mathbf{r}_{a1}$ and $\mathbf{r}_{a2}$ be the 2D positions of the corresponding apertures (and hence also of the inner periscope mirrors). The vector distance $\mathbf{r}_1 - \mathbf{r}_2$ is called the *baseline*.

We choose the coordinate system so that the pupil lies in the plane $z = 0$, no matter how the telescope is aimed. Since the periscopes translate the fields laterally, the field in the pupil is given by

$$u^{(\delta)}_{pupil}(\mathbf{r}) \propto \exp[2\pi i \boldsymbol{\rho}_0 \cdot (\mathbf{r} + \mathbf{r}_1 - \mathbf{r}_{a1})] \operatorname{cyl}\left[\frac{\mathbf{r} - \mathbf{r}_{a1}}{D_{ap}}\right]$$

$$+ \exp[2\pi i \boldsymbol{\rho}_0 \cdot (\mathbf{r} + \mathbf{r}_2 - \mathbf{r}_{a2})] \operatorname{cyl}\left[\frac{\mathbf{r} - \mathbf{r}_{a2}}{D_{ap}}\right], \qquad (19.95)$$

where the cylinder function is defined by (3.257), and the superscript δ indicates that we are dealing with a point source.

In Sec. 9.6 we learned how to analyze the imaging properties of an ideal thin lens. The key was the Fresnel diffraction formula (9.98), which was used to express the field in the image plane as the Fourier transform of the field emerging from the lens times a quadratic phase factor. For an ideal lens in the Fresnel approximation, we saw in (9.159) that the lens itself introduces a quadratic phase factor that exactly cancels the quadratic phase factor in the Fresnel formula when the lens equation

[9] As a rule of thumb, 1 arc sec = 5 μrad; a more precise number is 4.848 μrad per arc sec.

(9.169) is satisfied. For the special case of a lens focused at infinity ($q = f$), the result is that [*cf.* (9.170) and (9.172)]

$$u_{im}(\mathbf{r}) \propto \exp\left(i\pi\frac{r^2}{\lambda f}\right)\left[\mathcal{F}_2\left\{u_{pupil}(\mathbf{r}')\right\}\right]_{\rho=\mathbf{r}/\lambda f}, \tag{19.96}$$

where $u_{pupil}(\mathbf{r}')$ is the field in the pupil (before it is modified by the amplitude transmittance of the lens). Thus the field in the image plane is essentially the Fourier transform of the field in the pupil.

Using (3.237), (3.238) and (3.259), we find

$$u_{im}^{(\delta)}(\mathbf{r}) \propto \exp\left(i\pi\frac{r^2}{\lambda f}\right)\frac{\pi D_{ap}^2}{4}\,\mathrm{besinc}\left(D_{ap}\left|\boldsymbol{\rho}_0 - \frac{\mathbf{r}}{\lambda f}\right|\right)$$
$$\times\left\{\exp\left[2\pi i\left(\boldsymbol{\rho}_0\cdot\mathbf{r}_1 - \frac{1}{\lambda f}\mathbf{r}\cdot\mathbf{r}_{a1}\right)\right] + \exp\left[2\pi i\left(\boldsymbol{\rho}_0\cdot\mathbf{r}_2 - \frac{1}{\lambda f}\mathbf{r}\cdot\mathbf{r}_{a2}\right)\right]\right\}. \tag{19.97}$$

The corresponding image-plane irradiance is

$$I_{im}^{(\delta)}(\mathbf{r}) = |u_{im}^{(\delta)}(\mathbf{r})|^2 \propto 2\left[\frac{\pi D_{ap}^2}{4}\,\mathrm{besinc}\left(D_{ap}\left|\boldsymbol{\rho}_0 - \frac{\mathbf{r}}{\lambda f}\right|\right)\right]^2$$
$$\times\left\{1 + \cos\left[2\pi\boldsymbol{\rho}_0\cdot(\mathbf{r}_1 - \mathbf{r}_2) - \frac{2\pi}{\lambda f}\mathbf{r}\cdot(\mathbf{r}_{a1} - \mathbf{r}_{a2})\right]\right\}. \tag{19.98}$$

The [besinc]^2 factor is the Airy pattern centered at $\mathbf{r} = \lambda f\boldsymbol{\rho}_0$, which is the geometric image of the point source. The $\{1 + \cos\}$ factor is the superimposed fringe pattern. The fringe frequency is determined by the aperture spacing $(\mathbf{r}_{a1} - \mathbf{r}_{a2})$; the baseline $(\mathbf{r}_1 - \mathbf{r}_2)$ affects the phase but not the frequency of the fringes. The fringes have 100% visibility for a point source.

Extended objects Above we considered a point source that radiates a plane wave with 2D spatial frequency $\boldsymbol{\rho}_0$ in the plane $z = 0$. An extended source produces a distribution of plane waves, hence a distribution of spatial frequencies. In keeping with common practice in astronomy, we shall use the direction cosines of the waves rather than spatial frequencies. Thus we specify the source by $f(\boldsymbol{\alpha}_s)$, where $\boldsymbol{\alpha}_s$ (without a hat since it is not a unit vector) is the 2D vector of direction cosines of the 3D wavevector, *i.e.*, $\boldsymbol{\alpha}_s = (\alpha_{sx}, \alpha_{sy})$; the third direction cosine is unneeded since $\alpha_{sx}^2 + \alpha_{sy}^2 + \alpha_{sz}^2 = 1$. Similarly, we use the 2D direction-cosine vector $\boldsymbol{\alpha} = (\alpha_x, \alpha_y)$ to specify position in the image plane. We shall assume that $f(\boldsymbol{\alpha}_s)$ is nonzero only for a very small range of α_x and α_y so that paraxial approximations apply. Thus we can relate angles to focal-plane coordinates by $\boldsymbol{\alpha} = \mathbf{r}/f$, and we can use infinite limits in integrals over α_x and α_y.

For an extended incoherent object, the image-plane irradiance (in angular units) is found by superimposing the contributions of each source point:

$$I_{im}(\boldsymbol{\alpha}) = C\int_\infty d^2\alpha_s\, f(\boldsymbol{\alpha}_s)\,\mathrm{besinc}^2\left(\frac{D_{ap}}{\lambda}|\boldsymbol{\alpha}_s - \boldsymbol{\alpha}|\right)$$
$$\times\left\{1 + \cos\left[\frac{2\pi}{\lambda}\boldsymbol{\alpha}_s\cdot(\mathbf{r}_1 - \mathbf{r}_2) - \frac{2\pi}{\lambda}\boldsymbol{\alpha}\cdot(\mathbf{r}_{a1} - \mathbf{r}_{a2})\right]\right\}, \tag{19.99}$$

where C is a radiometric constant that includes a factor of $(\pi D_{ap}^2/4)^2$.

Some limits of this result are interesting. First, if the periscopes are omitted so that $\mathbf{r}_1 = \mathbf{r}_{a1}$ and $\mathbf{r}_2 = \mathbf{r}_{a2}$, then the integral becomes a convolution; the image irradiance is the source distribution blurred by a PSF given by the besinc^2 modulated with a fringe pattern. This result is expected since the coherent PSF is just the Fourier transform of the pupil function, and the incoherent PSF is the squared-modulus of the coherent one (see Sec. 9.7.6).

Another limit that leads to a similar result, even with the periscopes, is when D_{ap} is small, as it was in the Michelson and Pease experiment. Small apertures reduce problems with atmospheric fluctuations and simplify the analysis of the output; only the latter advantage will be discussed here.

When the aperture is small, the width of the besinc function in (19.99) is large. If it is much larger than the angular diameter of the source and $f(\boldsymbol{\alpha}_s)$ is concentrated near $\boldsymbol{\alpha}_s = \boldsymbol{\alpha}_{s0}$, then we can approximate (19.99) as

$$I_{im}(\boldsymbol{\alpha}) = C \,\text{besinc}^2\left(\frac{D_{ap}}{\lambda}\left|\boldsymbol{\alpha}_{s0} - \boldsymbol{\alpha}\right|\right)$$

$$\times \int_\infty d^2\alpha_s\; f(\boldsymbol{\alpha}_s)\left\{1 + \cos\left[\frac{2\pi}{\lambda}\boldsymbol{\alpha}_s\cdot(\mathbf{r}_1 - \mathbf{r}_2) - \frac{2\pi}{\lambda}\boldsymbol{\alpha}\cdot(\mathbf{r}_{a1} - \mathbf{r}_{a2})\right]\right\}. \quad (19.100)$$

With this approximation, the same manipulations that led to (19.72) allow us to write

$$I_{im}(\boldsymbol{\alpha}) = CF(0)\,\text{besinc}^2\left(\frac{D_{ap}}{\lambda}\left|\boldsymbol{\alpha}_{s0} - \boldsymbol{\alpha}\right|\right)\left\{1 + |\gamma|\cos\left[\frac{2\pi}{\lambda}\boldsymbol{\alpha}\cdot(\mathbf{r}_{a1} - \mathbf{r}_{a2}) - \Phi_\gamma\right]\right\}, \quad (19.101)$$

where now

$$\gamma = \frac{F\left[(\mathbf{r}_1 - \mathbf{r}_2)/\lambda\right]}{F(0)}. \quad (19.102)$$

As expected, (19.101) shows that the image is an Airy pattern times a fringe pattern with a visibility determined by the object Fourier transform. A key point seen from (19.102) is that this Fourier transform is evaluated at a frequency determined by the baseline (spacing of the outer mirrors), even though this frequency would not be passed by the telescope itself. Subject only to practical constraints such as stability and phase shifts due to the atmosphere, arbitrarily high object frequencies can be measured.

Note that the argument of the numerator in (19.102) is dimensionless, as it must be since $F(\,\cdot\,)$ is the Fourier transform of a function of a direction-cosine vector, which is dimensionless. In general, we shall denote the frequency vector conjugate to $\boldsymbol{\alpha}$ as $\mathbf{u}$, with both vectors being dimensionless. Thus the numerator in (19.102) is $F(\mathbf{u})$ evaluated at $\mathbf{u} = (\mathbf{r}_1 - \mathbf{r}_2)/\lambda$. The reader should not confuse this frequency with the spatial frequency of the plane wave produced by a point on the source; the latter frequency is given by $\boldsymbol{\rho} = \boldsymbol{\alpha}/\lambda$. Thus, even though $\mathbf{u}$ and $\boldsymbol{\alpha}$ are both frequency variables in some sense, they are Fourier-conjugate variables in different domains. When we need to specify the components of $\mathbf{u}$, we shall follow common practice in astronomy and call them u and v; astronomers refer to the (angular) Fourier domain as the u-v plane.

Estimation of stellar diameter To a reasonable first approximation, stars are uniform discs. There may be some limb darkening since radiation travelling nearly tangent to the surface must traverse a longer path than radiation exiting nearly perpendicular to the surface, and there may be small surface features akin to solar flares, but if we ignore these details, the visibility function is given by

$$\gamma(\mathbf{u}) = \frac{F(\mathbf{u})}{F(0)} = \text{besinc}(uD_\alpha)\,, \tag{19.103}$$

where D_α is the angular diameter of the star and $u = |\mathbf{u}|$ is the magnitude of the angular frequency vector. This function has its first zero at $u = 1.22/D_\alpha$ and the second zero is at $2.23/D_\alpha$. It follows from (19.102) that the fringes vanish when

$$|\mathbf{r}_1 - \mathbf{r}_2| = 1.22\frac{\lambda}{D_\alpha} \qquad \text{or} \qquad |\mathbf{r}_1 - \mathbf{r}_2| = 2.23\frac{\lambda}{D_\alpha}\,. \tag{19.104}$$

For Betelgeuse at $\lambda = 0.5$ μm, the fringes vanish at mirror spacings of 2.9 and 5.3 m, so the 6 m beam used by Michelson and Pease was adequate to observe two nulls. They observed the fringes visually and adjusted the mirrors for zero visibility; the stellar diameter was then estimated simply from the corresponding mirror spacings.

A more modern approach would be to collect quantitative data at several mirror separations and use them to estimate the parameter D_α. A circuitous route would be to estimate the visibilities at each separation as discussed in Sec. 19.2.2 and then to fit a besinc to the result. Alternatively, the diameter could be estimated directly from the fringe data by ML methods. In either case, performance on this estimation task can be assessed by mean-square error (MSE) between the estimated and true diameters. Since there is only one parameter to be estimated, null functions are not an issue, and the objections to MSE raised in Sec. 14.1.2 do not apply.

Calculation of MSE is simplified if the noise level in the data is low. We know from Sec. 13.3.6 that ML estimators are asymptotically unbiased and efficient, which means that the MSE approaches the Cramér-Rao (CR) bound (Sec. 13.3.5) as the variance in the raw data (either fringe data or visibilities) approaches zero. Thus we can use the CR bound on the estimate of the diameter as a figure of merit for this problem.

To illustrate the procedure, suppose fringe data are recorded with a regularly spaced detector array for a single baseline. Suppose also that the star is exactly on the telescope axis so that $\boldsymbol{\alpha}_{s_0} = 0$, and that the aperture is small enough that (9.101) is applicable. The data, consisting of noisy samples of $I_{im}(\boldsymbol{\alpha})$, can then be written as[10]

$$g_m = K\,\text{besinc}^2\left(\frac{D_{ap}}{\lambda}|\boldsymbol{\alpha}_m|\right)\{1 + \text{besinc}(u_b D_\alpha)\cos(2\pi \mathbf{u}_f\cdot\boldsymbol{\alpha}_m)\} + n_m \equiv \overline{g}_m + n_m\,, \tag{19.105}$$

where K contains all of the radiometric factors [including aperture area and stellar brightness $F(0)$], $\mathbf{u}_f \equiv (\mathbf{r}_{a1} - \mathbf{r}_{a2})/\lambda$ is the fringe frequency in angular units, $\mathbf{u}_b \equiv (\mathbf{r}_1 - \mathbf{r}_2)/\lambda$ is the frequency associated with the baseline, and $u_b = |\mathbf{u}_b|$.

For i.i.d. Gaussian noise of variance σ^2, the CR bound is given by (13.372) as

$$\text{Var}(\hat{D}_\alpha) \geq \sigma^2 \left[\sum_m \left(\frac{d\overline{g}_m}{dD_\alpha}\right)^2\right]^{-1}\,. \tag{19.106}$$

[10]Note that the absolute-value signs on besinc$(u_b D_\alpha)$ have disappeared since the phase Φ_γ is either 0 or π, depending on whether besinc$(u_b D_\alpha)$ is positive or negative.

The parameter D_α appears only in the factor besinc$(u_b D_\alpha)$ in (19.105). The requisite derivative is given by Abramowitz and Stegun (1965, formula 9.1.30) as

$$\frac{d}{dD_\alpha}\,\text{besinc}(u_b D_\alpha) = -2\,\frac{\text{J}_2(\pi u_b D_\alpha)}{D_\alpha}\,, \tag{19.107}$$

where $\text{J}_2(\,\cdot\,)$ is a Bessel function. With this derivative, the final result for the CR bound is

$$\text{Var}(\hat{D}_\alpha) \geq \sigma^2 \left\{ \sum_m \left[2K\,\text{besinc}^2\left(D_{ap}\left|\boldsymbol{\alpha}_m\right|/\lambda\right) \frac{\text{J}_2(\pi u_b D_\alpha)}{D_\alpha} \cos\left(2\pi \mathbf{u}_f \cdot \boldsymbol{\alpha}_m\right) \right]^2 \right\}^{-1}. \tag{19.108}$$

One problem with this result is that a single baseline may not be sufficient to determine the stellar diameter unambiguously; if the visibility is near zero, several different D_α yield the same value for besinc$(u_b D_\alpha)$. If J different visibilities are measured for baselines $\mathbf{u}_b = \mathbf{u}_{bj}$, the Fisher informations for different baselines add, and (19.108) becomes

$$\text{Var}(\hat{D}_\alpha) \geq \sigma^2 \left\{ \sum_{j=1}^{J} \sum_m \left[2K\,\text{besinc}^2\left(\frac{D_{ap}\left|\boldsymbol{\alpha}_m\right|}{\lambda}\right) \frac{\text{J}_2(\pi u_{bj} D_\alpha)}{D_\alpha} \cos\left(2\pi \mathbf{u}_f \cdot \boldsymbol{\alpha}_m\right) \right]^2 \right\}^{-1}. \tag{19.109}$$

A CR bound such as (19.108) or (19.109) can be used to optimize the baselines and the aperture diameter for accuracy of the estimate of the stellar diameter. Note that a larger aperture gives more photons but fewer fringes since the width of the Airy pattern is less, so the optimum aperture diameter is not obvious without numerical evaluation. Note also that the accuracy depends on the actual stellar diameter.

If we knew an approximate value for the stellar diameter in advance and wanted to use only a single baseline, we could choose the optimal one by maximizing $\text{J}_2(\pi u_b D_\alpha)$ with respect to u_b. Since $\text{J}_2(x)$ is maximum at $x = 3.054$, the optimal baseline is $u_b = 0.97 D_\alpha^{-1}$; for comparison, zero visibility occurs when $u_b = 1.22 D_\alpha^{-1}$. More realistically, we might know only that we were looking for stellar diameters in some range, and we could average the right-hand side of (19.108) or (19.109) over that range to obtain a figure of merit for the system.

19.2.4 Interferometers with multiple telescopes

Building on the discussion of the Michelson stellar interferometer in the last section, we now consider other configurations that are used in modern astronomical interferometers. The systems treated here differ from the Michelson stellar interferometer in that two or more separate telescopes are used, as shown in Fig. 19.11 and 19.12, and in the use of optical delay lines. The delay lines may be afocal imaging systems, like the $4f$ system discussed in Sec. 9.7.2 but with movable mirrors to fold the path, or they may include fiber optics.

The delay lines are needed to compensate for the optical path difference (OPD) from a star to the two telescope apertures. As discussed in Sec. 9.7.4, interference cannot occur unless the total OPD is small compared to the coherence length of the source (or, equivalently, the difference in propagation times is small compared to the coherence time). Coherence time is the reciprocal of the spectral bandwidth of the source, so it can be increased by spectral filtering to reduce the bandwidth, but only at the expense of photons; a better solution is to equalize the paths with delay lines.

Delay lines are not necessary with a Michelson stellar interferometer. When the telescope is pointed exactly at the star of interest, the OPD to a point in the center of the image field is zero if the side mirrors are symmetrically located. As the telescope is scanned across the sky to look at different stars, the OPD remains zero since the side mirrors and apertures are attached to the primary mirror. In the configurations of Figs. 19.11 and 19.12, on the other hand, the OPD from a star to the telescope apertures varies as the telescopes are scanned, and the delay lines are needed to hold the total OPD constant.

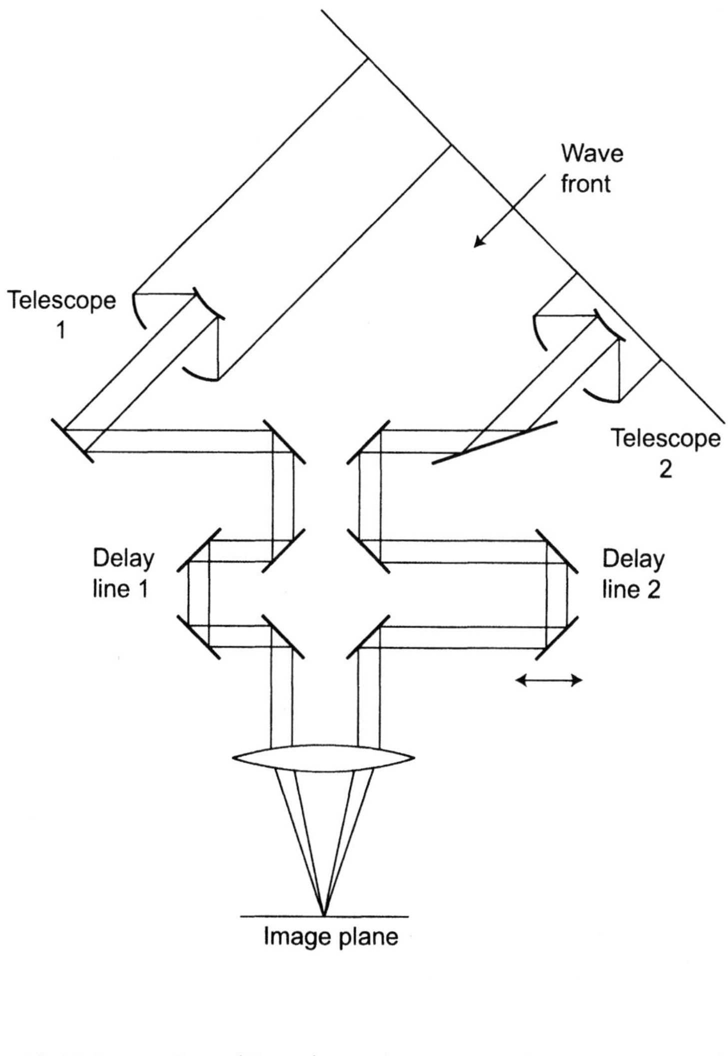

Fig. 19.11 Image-plane (Fizeau) interferometer with two telescopes (definitely not to scale). Details of the relay optics are omitted for clarity, but the image plane is common to the two telescopes.

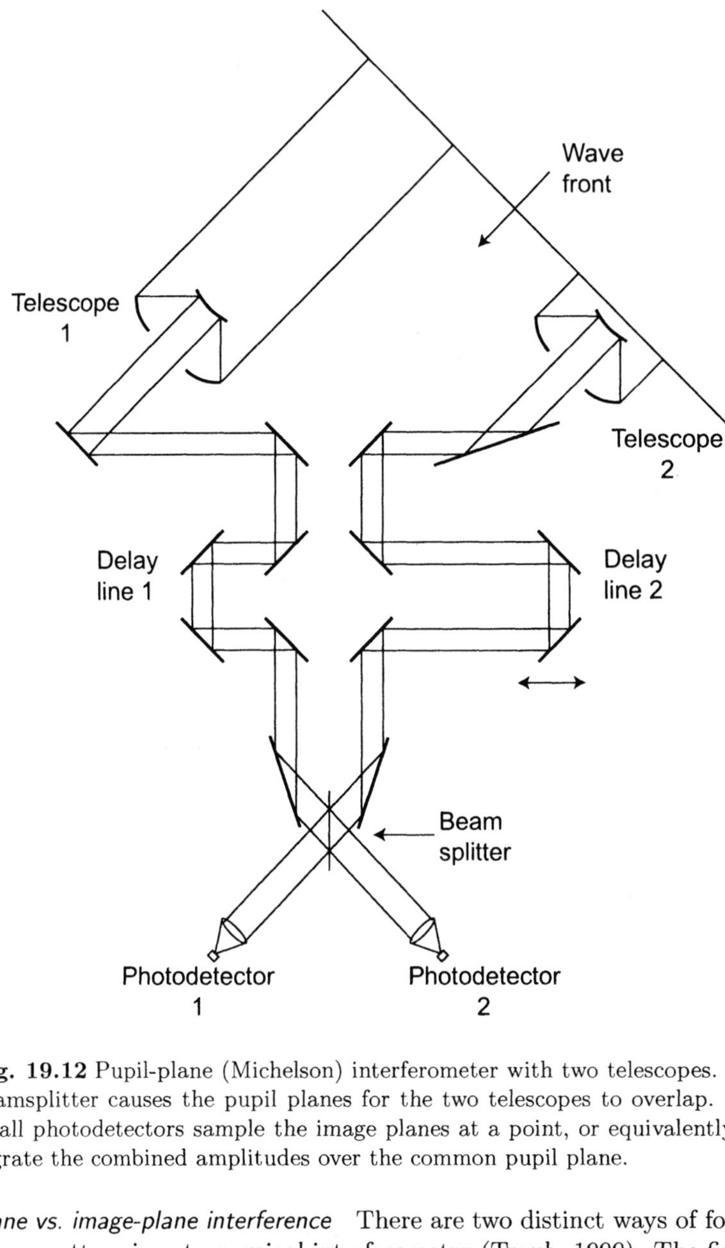

Fig. 19.12 Pupil-plane (Michelson) interferometer with two telescopes. The beamsplitter causes the pupil planes for the two telescopes to overlap. The small photodetectors sample the image planes at a point, or equivalently integrate the combined amplitudes over the common pupil plane.

Pupil-plane vs. image-plane interference There are two distinct ways of forming the interference pattern in astronomical interferometry (Traub, 1999). The first, shown in Fig. 19.11, is analogous to the Michelson stellar interferometer. The two beams are brought together in a common image plane, with a slight angle between them, and an image modulated by a fringe pattern is observed. This method is sometimes called *image-plane interferometry.*

The second method uses a beamsplitter to combine the beams with no angle between them; it can be regarded as the limit of the Michelson stellar interferometer as the distance between the apertures, $|\mathbf{r}_{a2} - \mathbf{r}_{a1}|$, goes to zero. Hence the spatial frequency of the fringe pattern goes to zero, but we can still observe fringes by varying the delay lines. Qualitatively, a point source produces identical Airy patterns from the two telescopes, and the phase difference between the two images oscillates with the time delay. In practice, the delay is continuously scanned, and the fringes are observed as temporal rather than spatial patterns. Since the relay optics and beamsplitter cause the images of the pupils of the two telescopes to overlap exactly, this method is called *pupil-plane interferometry*.

Fizeau and Michelson: History and semantics Armand-Hippolyte-Louis Fizeau (1819 – 1896) was a French physicist celebrated for his studies of the speed of light and for his early investigations into what we would now call partial coherence. By the 1860s, he was a member of the Académie des Sciences and a leading figure in French scientific circles.

In 1868 Fizeau suggested that a simple interferometer, constructed by placing two circular apertures over the pupil of a telescope, could be used to measure the diameters of astronomical bodies. A few years later Edouard Stéphan, director of the observatory at Marseilles, carried out the experiment suggested by Fizeau (Fizeau, 1873; Stéphan, 1874). Seventeen years elapsed before Michelson's first experiments in this field; in 1890 he measured the diameter of Jupiter, using essentially the same interferometer as suggested by Fizeau. An excellent short history of these developments is given by Lawson (2000). Lawson finds no evidence from the literature that Michelson was aware of the earlier French work until well after his experiments on Betelgeuse in 1920 – 1921, but he does speculate that Michelson might have had the opportunity to meet Fizeau during a trip to Paris in 1881.

The Fizeau interferometer is essentially the Michelson stellar interferometer without the periscope. Both instruments combine the beams in the image plane with a small angle between them, and both yield images modulated spatially with fringes. In current astronomical terminology, both are image-plane interferometers, and this method of beam combination is commonly referred to as the Fizeau mode or simply as a Fizeau interferometer.

On the other hand, the interferometer used by Michelson and Morley in their experiments on the speed of light combined the beams coaxially, with no angle between them, and allowed for scanning the mirrors temporally. Thus this instrument was a pupil-plane interferometer, and by extension all modern systems using coaxial beam combination and scanning mirrors are referred to as Michelson interferometers or are said to be operating in the Michelson mode. By this terminology, the Michelson stellar interferometer was a Fizeau interferometer, but many modern telescopes are Michelson interferometers.

Examples of Fizeau interferometers An example of a Fizeau interferometer is the Large Binocular Telescope (LBT), to be located on Mt. Graham in southeastern Arizona. Designed and constructed by a consortium of U.S., Italian and German institutions, the LBT is scheduled for completion in 2005. It consists of two 8.4 m, F/1.14 parabolic primaries on a single truss. It is equivalent in light-gathering power to a single 11.8 m telescope and in resolving power to a 22.8 m telescope. The F/15 secondary mirrors are adaptive, with individual feedback-controlled segments to compensate for atmospheric phase shifts. The system operates over a wavelength range of 0.4 to 400 μm in interferometric mode. One key advantage it has over other binocular telescopes is that it provides full coverage of the u-v plane.

An even more ambitious earth-based interferometer concept is called 20/20. Being designed by Roger Angel and collaborators at the University of Arizona, 20/20 will consist of two telescopes, each (incongruously) 21 m in diameter. The primary mirrors will consist of seven 8.4 m segments to be consistent with current mirror-fabrication facilities. The primaries will be F/0.7 and must therefore be markedly aspheric to get diffraction-limited performance. Atmospheric disturbances will be corrected to the diffraction limit of each telescope by 2.1 m adaptive secondary mirrors, also segmented to match the primaries. The two telescopes will move on a 100 m Dia. track in such a way that the baseline is always perpendicular to the line of sight to the source of interest, thus obviating the delay lines. Images will be combined in a Fizeau interferometer in order to get resolutions corresponding to baselines up to 100 m. In addition to its huge optical collection efficiency, 20/20 will have a field of view of about 30 arc sec, much larger than would be possible with conventional Michelson interferometers.

Both 20/20 and the LBT will be able to operate in a nulling mode where the light from a bright star will cancel out by destructive interference while light from a possible planet near the star will not be nulled. Since the beams must be combined coaxially to achieve the null, this mode (Bracewell and McPhie, 1979) is a variant on Michelson interferometry.

A spaceborne example of a Fizeau interferometer is the European Space Agency's DIVA satellite (Deutsches Interferometer für Vielkanalphotometrie und Astrometrie or, less nationalistically, Double Interferometer for Visual Astrometry). DIVA is essentially identical to the original Fizeau proposal as implemented by Stéphan. It uses a single telescope mirror with two 7.5 cm square masks separated (center to center) by 15 cm. It will be used to measure the positions, proper motions and parallaxes of all stars brighter than 15th magnitude.

Examples of Michelson interferometers An example of a Michelson interferometer is the NPOI (Navy Prototype Optical Interferometer) located on Anderson Mesa near Flagstaff, Arizona. Though NPOI consists of six telescopes, it can use pairwise Michelson recombination of the beams, and in this mode can be regarded as a set of two-aperture Michelson interferometers operating in parallel (Armstrong, *et al.*, 1998; Hummel, 2000). NPOI was the first long-baseline system to achieve interferometric beam combination in the visible region.

The VLT (Very Large Telescope) being constructed by the European Southern Observatory at Paranal, Chile, will consist of four 8.2 m telescopes, which can be operated separately or as an interferometer, plus three 1.8 m "outriggers" which will be used to get additional baselines in interferometry. When operated as an interferometer, this system is known as the VLTI (VLT Interferometer).

Many of key design decisions on the VLTI were made by constructing and testing a two-aperture system called VINCI (VLT INterferometer Commissioning Instrument). VINCI uses the Michelson mode with a fiberoptic beam combiner, an idea adapted from IOTA-FLUOR (Infrared-Optical Telescope Array, Fiberoptic Link Unit for Optical Recombination), a three-element Michelson interferometer on Mt. Hopkins in southern Arizona. In essence, a single-mode optical fiber is placed on the optical axis of each telescope, and a fiber coupler is used to combine the beams coherently.

There are many plans to put Michelson interferometers in space. A study team convened by the European Space Agency concluded that the Michelson concept was preferable to the Fizeau because of stability considerations, especially with very long baselines.

A space-based Michelson interferometer in the active design stage is called the Space Interferometry Mission (SIM). Scheduled for launch in 2009, SIM will consist of two 30 cm apertures on a fixed 10 m baseline. It requires 10 nm stabilization for astrometry and 1 nm stabilization for nulling, and it will provide 4 μas accuracy in position and parallax measurements of stars.

A follow-on mission would be called the Terrestrial Planet Finder (TPF). At this stage the acronym TPF denotes not a specific system but a set of concepts being explored for detecting earth-like planets around nearby stars. One possibility is a group of free-flying telescopes, each 3 – 4 m Dia., connected as an interferometer with baselines up to 1000 m.

Even more ambitious space interferometers known generically as Planet Imager and Life Finder are being actively discussed. Some of these concepts envision baselines as much as 1000 km. Though the challenge of providing nanometer stabilization over megameter distances is formidable to say the least, the resolution achievable by such systems would be spectacular. Daniel Goldin, former NASA director, has said that his dream is that his grandchildren will see images with Landsat resolution of a planet around another star. If the dream is ever to be achievable, it will require huge space-based interferometers.

EPILOGUE

The hardest part about writing this book was stopping. At the end of each chapter, we could see many other directions to explore, either segments of the recent literature that were not adequately represented or new problems for which we had assembled the tools but not yet begun to use them. Our inclination was to follow as many of these paths as possible, stretching our methodologies to their limits, but of course there would have been no end. We had to draw a line and send this tome to press. To give ourselves the courage to do so, we submit this modest epilogue, suggesting to the reader the paths we might have followed had we had several other lifetimes to do so, and we invite the reader to accompany us on the portions of the journey that we still hope to take.

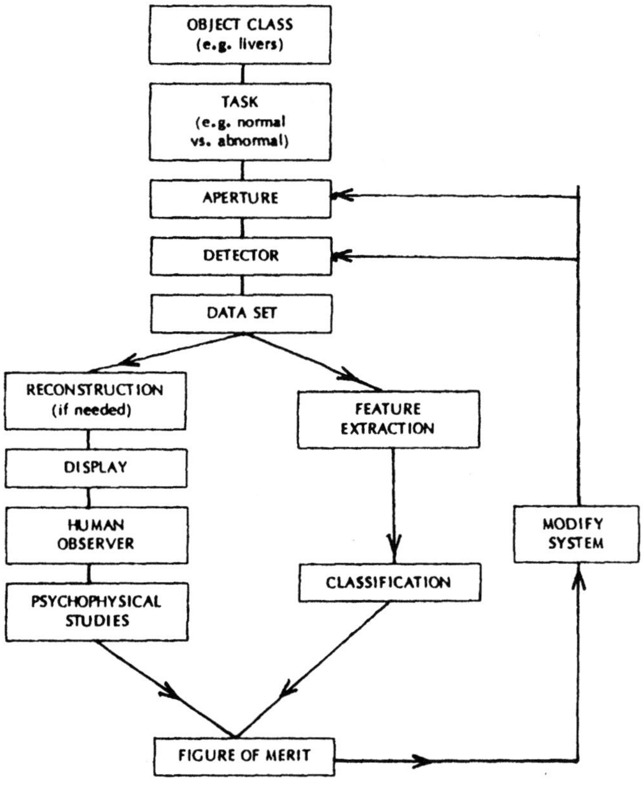

Fig. 1 Roadmap for the systematic optimization of gamma-ray imaging systems (from Myers *et al.*, 1986).

Systematic system optimization We begin the journey with a backward glance and a look at the roadmap to see where we are. In 1985, as the concept of task-based assessment of image quality was taking hold in the medical-imaging community, the diagram in Fig. 1 was presented at a conference at Georgetown University. This diagram embodies two basic principles that we have tried to stress in this book. First, it recognizes that meaningful measures of system quality must be based on the performance of specific observers on specific tasks of practical interest. Second, it envisions an imaging system as an integrated whole, with all components contributing to task performance and hence needing to be considered together.

A crucial feature of this diagram is the feedback path. After we have learned how to evaluate imaging systems, the next logical step is to incorporate the evaluation into a program of systematic, iterative optimization of task performance.

In the two decades since we first sketched this diagram, many researchers have contributed to the methodology implied in it. We now know much more about image formation, noise in imaging systems, image reconstruction algorithms and the properties of human and model observers.

The challenge for the future of image science is thus twofold: We must continue to refine our understanding of every block in diagrams like the above and we must think seriously about closing the loop and carrying out a full system optimization.

The search for excellence A basic difficulty in actually carrying out the plan suggested in the figure is that the specification of an imaging system is so complex. Consider the example of lens design, where the routine procedure is to assume some merit function and design the lens to optimize it. We would, of course, advocate that the merit function be chosen in relation to specific tasks, but no matter what merit function is used, the search for an optimum is difficult.

Even after the designer has selected the number and arrangement of the individual elements in the lens, there are still many parameters to be chosen, including interelement spacings, curvatures and indices of refraction. Iterative algorithms such as simulated annealing and genetic algorithms can be used to find a more-or-less global optimum in this parameter space, but there is no guarantee that some completely different arrangement of elements with a different set of free parameters might not perform better. Task-based assessment allows objective comparison of competing designs but does not obviate the role of human creativity in selecting the competitors in the first place.

If it doesn't fit, get a bigger hammer! Imaging systems acquire huge amounts of data and often require formidable data-processing resources even to produce one image. Since image quality is inherently a statistical concept, accurate evaluation of a single system requires a huge number of images, and systematic optimization requires, in effect, evaluation of a huge number of imaging systems.

Daunting though this prospect may seem, we can be encouraged by the rapid advances in computational capabilities. In its most expansive form, Moore's law says that every aspect of computer power (chip density, CPU speed, memory, data-transfer rates) will double every 18 months or so. For three decades shortsighted prognosticators have been predicting the end of this trend, but they have consistently been proven wrong. If the trend holds for another three decades, the gain in all facets of computer power will be about $2^{20} \simeq 10^6$. If we crudely benchmark a typical personal computer at the turn of the millennium (January 1, 2001, of course) as a single 1 GHZ processor with 100 MB of memory, we can envision the image scientist of 2031 having access to 1 PHz of processing speed and 100 TB of memory (T = tera = 10^{12}, P = peta = 10^{15}). These numbers might imply a huge number of processors and memory spread around the world, but they do not require an inordinate amount of technological optimism; the pessimistic view is that they will not be correct until 2041 or 2051.

What will image scientists of the not-so-distant future do with such computer power? Possibilities abound. Three-dimensional reconstructions on $1000 \times 1000 \times 1000$ grids will be routine, and rapid temporal sequences with hundreds or thousands of 3D frames will make high-resolution 4D imaging possible. Huge databases will be available for image archiving, and rapid access to the databases will facilitate automated image interpretation. Simulation tools will make the phrase "photographic realism" obsolete, and the authors of this book will quit chiding their colleagues for megalopinakophobia.

Most importantly, the computational arguments against meaningful system evaluation will disappear, and we can begin to close the loop and really optimize imaging systems. Along the way we have to increase our understanding of every component of the system, from radiation source to final observer, and we have to be creative in devising ways of improving them.

It's the data, stupid! A famous American politician told his campaign staff, "It's the economy, stupid!" thereby imploring them not to lose sight of the most significant factor in their endeavor. In imaging, the key factor is the data; robust data are as important to an image scientist as a robust economy is to a politician. In this analogy, image processing plays the role of spin — it can put a good face on a bad data set but cannot really overcome its limitations.

A data-driven approach to image science would obey three dicta: 1. Use all of the image data; 2. Get more image data; 3. Get more nonimaging data for use in conjunction with image data.

One might think that modern imaging does use all of the image data, but in fact there are information losses at several stages. As a simple example, a time exposure of a time-varying object discards potentially useful temporal data. A more complex example concerns position-sensitive photon-counting detectors where multiple sensor outputs are reduced to estimates of x and y coordinates for each photon. Information loss can occur in the estimation step and again when the estimates are binned into an image matrix. In fact, any binning or quantization of data is a potential source of information loss; moreover, we often use lossy algorithms for data compression. The challenge is to understand how serious these losses are, in terms of specific imaging tasks, and to devise ways of minimizing them.

Getting more imaging data may mean improving the spatial or temporal resolution of a system or it may mean acquiring information from different angular views, different wavelength bands or completely different imaging modalities. Often these advances come only after intensive research and considerable economic investment, so it is imperative to understand how they facilitate the intended application of the images.

Auxiliary nonimaging data are routinely used in conjunction with image data. A radiologist uses clinical indications and patient history along with radiographs in making a diagnosis. Analysts of landsat images use weather and climate data to aid in assessing the health of agricultural fields, and astronomers supplement their direct observations with data from celestial mechanics calculations and astrophysical simulations.

There are many other opportunities to integrate imaging and nonimaging data. The nonimaging data may concern the particular object being imaged or it may concern one or more classes of objects. In a medical context, for example, one might acquire further measurements on the particular patient being studied or one might compile a database of disease characteristics and use it to guide the acquisition and interpretation of medical images. The challenge is to learn what specific supplementary information is most useful to the goal of the imaging and to devise ways of acquiring it systematically and optimally.

How good does it have to be? It is not uncommon for the designer of an imaging system to ask a user what resolution is needed. The reply may well be that there is no need for resolution better than some value "because there is nothing to see" at a finer scale. Both the question and the answer are misguided.

The question is misguided since performance on any task will always improve with better data, either lower noise or finer resolution. The user of the imaging system or the marketing manager of the company that manufactures it might then say that the performance is "good enough" at some noise level and resolution, so there is no benefit in making it better. For example, a physician might be interested in detecting a tumor 1 cm in diameter and might feel that some particular system is adequate for this task. Pressed to say how he knows, the diligent physician might do an ROC study and get, say, 0.9 for the area under the curve. Is that good enough? Not for the patient whose tumor is missed! And even if economic cost-benefit arguments are adduced for not trying to improve the performance on this task, there will always be other tasks, such as detecting smaller lesions or distinguishing benign from malignant ones.

The issue is more subtle when we look at characteristics of individual components rather than the overall imaging systems. For example, if the point response function of a lens is much smaller than the size of a detector pixel, one might conclude that further improvements in the lens are not needed. Similarly, for decades it has been conventional wisdom in nuclear medicine that detector improvements are unnecessary since the limitation is the collimator.

In neither of these examples, however, does the conclusion hold up under close scrutiny. If a lens has better resolution than the detector with which it is used, the lens designer might choose to use a larger numerical aperture even though that measure increases the aberrations and degrades the lens contribution to the spatial resolution. A gain in light-collection efficiency could then be achieved with negligible loss in overall resolution, and new applications of the system in low light levels could become possible. Similarly, in the nuclear medicine example, improved detector resolution might indeed be useless with conventional parallel-hole collimators, but new approaches to image formation with multiple pinholes or coded apertures could take advantage of the improved detector capability.

It is not an overstatement to say that increased technological capability in any component of an imaging system will always lead to improved task performance when the design of the overall system is approached in an integrated and creative fashion.

Integrated computational imaging systems In 2001 the Optical Society of America conducted a topical meeting called ICIS — Integrated Computational Imaging Systems. The premise is that image processing and image acquisition are becoming increasingly indistinguishable. It is no longer the case that a hardware designer develops a camera and some kind of computer interface while a software specialist develops ways of processing and displaying the data. The data-acquisition system is almost always under computer control, and the control signals are frequently derived from the images themselves. Design of the hardware must take into account the needs and capabilities of the image analysis, and the nature of the processing can dictate what kind of data are acquired.

To a degree, this viewpoint is already used in the design of tomographic or other indirect imaging systems, where no image at all is obtained without processing, but there can still be a great symbiosis when image acquisition and processing are viewed as a whole and optimized in a coordinated way. As understanding of image quality advances, we can envision optimizing a system not for a class of objects but for a particular object; since we do not know what that object is before taking the image, we will have to modify and optimize the system during the image acquisition. The term *adaptive optics* can then take on a new meaning, adapting to the object being imaged and not just to corrupting influences such as the atmosphere.

To solve an inverse problem, concentrate on the forward problem Improved performance in an indirect imaging system can really come from only two sources: better data or better modeling of the data-acquisition system. The function of the reconstruction algorithm is only to narrow down the range of possible objects that are consistent with the data and the model of the system, and no algorithm can go beyond the limitations imposed by the data and the model.

To put this observation into practice, we need greatly improved models in many kinds of imaging. We need to move away from simplified system matrices with coarse voxels chosen purely for computational speed. In many applications we need accurate modeling of scattered and background radiation, and in imaging through turbid media we must move beyond the Born and Rytov approximations. We must avoid mathematical simplifications such as shift invariance and linearity when they are not justified. In optical systems, for example, we should account for shift-variant radiometric variations and off-axis aberrations as well as nonlinearities of the detector, and we should include the nonlinearities arising from partial coherence.

Achieving this level of accuracy in modeling will require careful system analysis and calibration and, of course, increased computer power, but it is the way to make progress in indirect imaging.

The last refuge Oscar Wilde said, "Consistency is the last refuge of the unimaginative." In inverse problems, however, consistency conditions are a largely untapped resource that can supplement incomplete or inaccurate data. As we saw in Chap. 15, consistency conditions can be derived by characterizing the range of an imaging operator, and in principle they can be used to reduce noise or correct for motion or other unknown characteristics of an imaging system.

Most known consistency conditions are based on continuous-to-continuous models of the imaging system, and most apply to forms of tomography where there are null functions of the adjoint operator. They thus apply to what one might call mathematical tomography rather than the real world of discrete, noisy data. A major theoretical challenge is to elucidate the relationship between mathematical tomography and real tomography and to understand what the continuous consistency conditions tell us about consistency of discrete data.

Tasks and observers A task-based assessment of image quality is useful in practice only if the task is meaningful. In Chaps. 16 – 19 we saw several examples where oversimplified tasks led to misleading conclusions about system design. Progress in objective assessment will require more complicated and more realistic tasks. A key question concerns extrapolation of conclusions from simple tasks to more complicated ones. For example, will a system optimized for detection of specified nonrandom signals also be optimal for tasks where the possible outcomes are not defined in advance?

As simulation approaches reality Image simulations are becoming ever more realistic, largely because of the entertainment industry. These words are being written shortly after the telecast of Super Bowl XXXVII, the American football championship. Graphics shown during that game cause one to wonder whether Superbowl LXVII will be played with real athletes or with simulations.

How can image scientists take advantage of techniques developed at Lucas Films and Dreamworks? Accurate depiction of the motion of animate objects will advance our understanding of dynamic imaging systems, and accurate modeling of variations in surface reflectance will aid in design and analysis of optical systems for viewing opaque objects.

Accurate simulation of fine-scale or textural variations requires a detailed understanding of the underlying object statistics and how they influence the image statistics. We need these statistical descriptions for several purposes, including evaluation of image quality, development of Bayesian reconstruction algorithms, image synthesis and pattern recognition.

New models of object textures must be devised, along with experimental methods for estimating the parameters of the models from real image data. Since objects are infinite-dimensional but only a finite number of image samples will be available, parsimonious low-dimensional descriptions must be sought, and ways of assessing whether they capture the essential features of the object variability must be developed.

Our perception of perception We also need to understand better the human perceptual process and to integrate it into the design and evaluation of imaging systems, including processing algorithms and display. The role of image reconstruction and enhancement is to match the raw image data to the human perceptual system, and effective design and meaningful assessment of these elements of the imaging chain require exploring the links between image science and cognitive science. Inclusion of frequency-selective channels and internal noise in our observer models is a first, halting, step to bring knowledge from visual perception into image science, but much more sophisticated models are possible. How are we to account for the strong nonlinearities of the visual system? What can we learn from the study of visual illusions that will affect the design of imaging systems?

Man vs. machine Computerized image analysis is becoming more powerful and commonplace. How do we optimize an image-acquisition system when the end user is a computer? Which tasks are best performed by the computer and which by humans? When humans perform well, can we decipher their strategy and build it into an algorithm? What is the tradeoff between computerized and computer-aided image analysis, and can we be quantitative about answering this question? For example, what is a meaningful performance measure for a computer algorithm that segments an image and presents it to humans in cartoon form? When does the cartoon enhance the ability of the human to grasp essential details (perform tasks) and when does it result in real information loss?

New technologies Although this book has dealt with the mathematics and physics of imaging, it has said very little about the technologies so essential to modern imaging. Some technological advances, such as optical and x-ray detector arrays, have been developed in direct response to the needs of imaging systems; others, such as optical data storage and computer displays, have been developed for more generic uses but have obvious applicability to imaging.

The greatest opportunity for creativity, however, comes from technologies developed in other fields that have no initially obvious connections to imaging. For example, lasers were an extension of masers, microwave amplifiers pursued initially for communications purposes. Superconducting magnets were developed long before their need in magnetic resonance imaging was apparent. And artificial radioisotopes were the result of wartime work on nuclear weapons, with no vision for their use in nuclear medicine. Yet in all of these cases the new technologies were quickly put to use in imaging systems.

Some emerging technologies with potential in imaging include femtosecond optical pulses, entangled photon states, diffractive and reflective x-ray optics, Josephson junctions and other superconducting devices, ultrasensitive seismometers and novel interferometers. Not to be overlooked are stunning advances in software in the areas of database management, artificial intelligence and data mining. A modern image scientist/technologist needs to be conversant with far more than the imaging literature and to remain alert to seemingly unrelated developments that can impact imaging.

New signals Science always searches for new ways of probing the universe, and increasingly the result of the probe is a multidimensional data set that can be manipulated and displayed as an image. One way to make progress in image science, or science in general, is to think up new things to map and new ways of probing them. Tables I and II in the Prologue should provide lots of hints in the search for new signals to image, but we can also look for new ways of applying ideas from image science to things we might not initially think of as images.

Sometimes ideas can come full circle, originating in another area, then being appropriated by image scientists, and finally being returned with embellishments to the original field. For example, in this book we have made considerable use of the Wigner distribution function (WDF) as a tool for signal and image analysis, although it originated in quantum mechanics. Now, however, there is considerable interest in imaging the quantum-mechanical Wigner function itself, and it turns out that tomographic reconstruction algorithms are the way to do it.

New dimensions We have long since passed the point where the word image implied a static, 2D construct. Modern imaging systems are almost always 3D ($x-y-z$ or $x-y-t$) or 4D ($x-y-z-t$), but there are many options for creative addition of yet other dimensions. The challenges lie in acquiring the data sets, doing high-dimensional image reconstruction and displaying the results.

To return to an example just given, the quantum-mechanical Wigner distribution function has been applied so far to 1D quantum states, such as the state of a single mode of the radiation field. If we want to measure the WDF of an

N-dimensional quantum state, it requires a $2N$-dimensional measurement/imaging system.

Similarly, if the quantity of interest is a statistical correlation function for a 2D field, it requires four variables for full specification, and correlations of 3D fields require six variables. There is no reason not to regard the correlation function, rather than the field itself, as the object to be imaged.

New pedagogies This book arose as an educational endeavor. It had its roots many years ago in a course on radiological imaging, and it eventually spawned several other courses on image science and noise. As the courses and the authors' educational and research interests evolved, it became apparent that the pedagogy of imaging was as challenging as the science and technology.

Students wanting to contribute at the cutting edge of image science must master a breadth of material comparable to — and perhaps even exceeding — the scope of mature disciplines such as chemistry and electrical engineering. They not only must study mathematics, statistics, physics and electronics, they must also understand how they interrelate and contribute to the whole gestalt of image science. They must be able to follow many diverse literatures and to pick and choose ideas and methodologies from them to solve their own problems in imaging. Like statisticians, they must appreciate the needs of their "clients," the end users of the images, but they must also be involved in optimizing the systems and developing the uses themselves.

Accommodating this breadth will require new departmental organizations, new interdisciplinary seminars and new professional societies. Academic departments and professional organizations devoted to optics and photonics will expand their base to include imaging, and umbrella interdisciplinary programs will arise to coordinate imaging activities across departments. Even social interactions among students and researchers perusing different aspects of the imaging elephant will help to broaden viewpoints and ease misconceptions.

We — the authors of this volume — have our own experiences in this educational process and our own thoughts about the critical educational needs in image science, and it is not a particularly difficult inverse problem to reconstruct those views from the contents of the book. We also realize, however, that optimization of the process is an ongoing challenge, where diverse views and imaginative approaches can be of immense value. We encourage a dialogue among educators and students in image science, and we hope that this book can stimulate that exchange in some small way. We look forward to observing the further development of image science as an academic discipline and making whatever contributions to it that we can.

APPENDIX A
Matrix Algebra

This appendix is a compendium of useful definitions and formulas related to matrices and vectors. The intent is to provide a ready reference rather than a tutorial. The formulas are presented here with little discussion and few derivations, though some of the topics are discussed more didactically in the main text of the book. Cross references to the main text are given where appropriate.

Good introductory treatments of matrices and discrete linear algebra are given by Eves (1966), Mirsky (1982), Pettofrezzo (1966), Strang (1980) and Usmani (1987). Excellent comprehensive texts are Golub and van Loan (1989) and Harville (1997); the latter is a particularly good match to the needs of this book. Lists of matrix identities are given in Siotani (1985), Pilz (1991) and Rade and Westergren (1990).

A.1 NOTATION AND TERMINOLOGY

An $M \times N$ matrix $\mathbf{A}$ is an ordered set of MN numbers arranged into an array with M rows and N columns. The number in the mth row and n^{th} column, denoted A_{mn}, is called the mn^{th} element of $\mathbf{A}$. The element can be either real or complex. The set of all $M \times N$ matrices with real elements is denoted $\mathbb{R}^{M \times N}$, while the set of all $M \times N$ matrices with complex elements is denoted $\mathbb{C}^{M \times N}$.

A *square* matrix is one with an equal number of rows and columns, $M = N$. An $N \times N$ square matrix is said to have *order* N or *degree* N. A *rectangular* matrix is one that has an unequal number of rows and columns ($M \neq N$). A *diagonal* matrix is a square matrix with $A_{mn} = 0$ if $m \neq n$. In other words, the nonzero elements are along the diagonal only. A square matrix $\mathbf{A}$ is said to be *upper triangular* if all of the elements below the diagonal are zero, *i.e.*, $A_{mn} = 0$ when $n < m$. Similarly, it is *lower triangular* if all elements above the diagonal are zero, *i.e.*, $A_{mn} = 0$ when $n > m$.

Two matrices are said to be equal if all elements of one are equal to the corresponding elements of the other. Thus

$$\mathbf{A} = \mathbf{B} \quad \text{if and only if} \quad A_{mn} = B_{mn} \quad \text{for all } m \text{ and } n\,. \tag{A.1}$$

If $\mathbf{A}$ is an $M \times N$ matrix and $\mathbf{A} = \mathbf{B}$, then $\mathbf{B}$ is also an $M \times N$ matrix.

An N-dimensional vector is an ordered set of N numbers. If these numbers are arranged in a column, the vector is referred to as a *column vector*, which can also be regarded as an $N \times 1$ matrix, with N rows and one column. In spite of this equivalence, vectors will be denoted with lower-case bold letters and matrices with upper-case ones. The word vector will be understood to mean column vector unless otherwise specified.

The *transpose* of an $M \times N$ matrix $\mathbf{A}$ is an $N \times M$ matrix denoted $\mathbf{A}^t$ obtained by interchanging rows and columns of $\mathbf{A}$. Thus

$$(\mathbf{A}^t)_{mn} = (\mathbf{A})_{nm} = A_{nm}\,. \tag{A.2}$$

The transpose of an $N \times 1$ column vector is a $1 \times N$ *row vector*, *i.e.*, a matrix with 1 row and N columns.

A *symmetric* matrix is a square matrix that is identical to its transpose, *i.e.*,

$$(\mathbf{A}^t)_{mn} = (\mathbf{A})_{mn} = A_{mn} = A_{nm} \quad \text{if } \mathbf{A} \text{ is symmetric}\,. \tag{A.3}$$

The *adjoint*[1] of an $M \times N$ matrix $\mathbf{A}$ is an $N \times M$ matrix denoted $\mathbf{A}^\dagger$, obtained by interchanging rows and columns of $\mathbf{A}$ and taking the complex conjugate of each element. Thus

$$(\mathbf{A}^\dagger)_{mn} = (\mathbf{A})^*_{nm} = A^*_{nm}\,. \tag{A.4}$$

Adjoint and transpose are synonymous if all elements of $\mathbf{A}$ are real. Note that $[\mathbf{A}^t]^t = \mathbf{A}$ and $[\mathbf{A}^\dagger]^\dagger = \mathbf{A}$.

A *Hermitian* matrix is a square matrix that is identical to its adjoint, *i.e.*,

$$(\mathbf{A}^\dagger)_{mn} = (\mathbf{A})_{mn} = A_{mn} = A^*_{nm} \quad \text{if } \mathbf{A} \text{ is Hermitian}\,. \tag{A.5}$$

A diagonal matrix is necessarily symmetric, and it is Hermitian if the diagonal elements are real. A real, Hermitian matrix is necessarily symmetric.

A square matrix $\mathbf{A}$ is said to be *skew-Hermitian* or *anti-Hermitian* if $\mathbf{A}^\dagger = -\mathbf{A}$. If $\mathbf{A}$ is skew-Hermitian, then $i\mathbf{A}$ is Hermitian.

A.2 BASIC ALGEBRAIC OPERATIONS

A.2.1 Addition and subtraction

Addition and subtraction of matrices are performed on an element-by-element basis. If we write

$$\mathbf{C} = \mathbf{A} \pm \mathbf{B}\,, \tag{A.6}$$

[1] Many older books use the term adjoint or *adjugate* to refer to a matrix of cofactors (to be defined in Sec. A.5.4). Our usage is common in the modern literature where a matrix is regarded as a linear operator.

it implies that

$$C_{mn} = A_{mn} \pm B_{mn} . \tag{A.7}$$

Subtraction is the inverse of addition, *i.e.*,

$$(\mathbf{A} + \mathbf{B}) - \mathbf{B} = \mathbf{A} . \tag{A.8}$$

Matrix addition is commutative and distributive:

$$\mathbf{A} + \mathbf{B} = \mathbf{B} + \mathbf{A} . \tag{A.9}$$

$$(\mathbf{A} + \mathbf{B}) + \mathbf{C} = \mathbf{A} + (\mathbf{B} + \mathbf{C}) . \tag{A.10}$$

The zero matrix, denoted $\mathbf{0}$, is a matrix of all zeros. It is the identity element for addition, so

$$\mathbf{A} + \mathbf{0} = \mathbf{A} \quad \text{and} \quad \mathbf{0} + \mathbf{A} = \mathbf{A} . \tag{A.11}$$

A.2.2 Scalar multiplication

A matrix $\mathbf{A}$ is said to be multiplied by a scalar α if every element of $\mathbf{A}$ is multiplied by α, *i.e.*,

$$[\alpha \mathbf{A}]_{mn} = \alpha A_{mn} , \tag{A.12}$$

where α and A_{mn} can both be complex. Scalar multiplication is commutative and associative:

$$\beta(\alpha \mathbf{A}) = \alpha(\beta \mathbf{A}) , \tag{A.13}$$

$$(\alpha + \beta)\mathbf{A} = \alpha \mathbf{A} + \beta \mathbf{A} , \tag{A.14}$$

where α and β are arbitrary real or complex numbers. Scalar multiplication is also associative with respect to addition:

$$\alpha(\mathbf{A} + \mathbf{B}) = \alpha \mathbf{A} + \alpha \mathbf{B} . \tag{A.15}$$

A.2.3 Matrix multiplication

In order to define a matrix-matrix product, which we write as

$$\mathbf{C} = \mathbf{A}\mathbf{B} , \tag{A.16}$$

the matrices must be *conformable.* That is, if $\mathbf{B}$ is a $P \times N$ matrix, then $\mathbf{A}$ must be an $M \times P$ matrix for some M. The product $\mathbf{C}$ is then an $M \times N$ matrix with elements given by

$$C_{mn} = \sum_{p=1}^{P} A_{mp} B_{pn} , \qquad m = 1, ..., M , \qquad n = 1, ..., N . \tag{A.17}$$

Matrix-vector multiplication is a special case of matrix-matrix multiplication. If $\mathbf{A}$ is an $M \times N$ matrix and $\mathbf{x}$ is an $N \times 1$ column vector, we can write

$$\mathbf{y} = \mathbf{A}\mathbf{x} , \tag{A.18}$$

where $\mathbf{y}$ is an $M \times 1$ column vector with elements given by

$$y_m = \sum_{n=1}^{N} A_{mn} x_n \,. \tag{A.19}$$

Matrix-matrix and matrix-vector multiplication are distributive and associative:

$$\mathbf{A}(\mathbf{B}+\mathbf{C}) = \mathbf{AB} + \mathbf{AC}\,; \tag{A.20}$$

$$\mathbf{A}(\mathbf{x}+\mathbf{y}) = \mathbf{Ax} + \mathbf{Ay}\,; \tag{A.21}$$

$$\mathbf{A}(\mathbf{BC}) = (\mathbf{AB})\mathbf{C}\,; \tag{A.22}$$

$$\mathbf{A}(\mathbf{Bx}) = (\mathbf{AB})\mathbf{x}\,. \tag{A.23}$$

On the other hand, matrix-matrix multiplication is not necessarily commutative. The statement $\mathbf{AB} = \mathbf{BA}$ makes no sense unless $\mathbf{A}$ and $\mathbf{B}$ are square matrices of the same size, and even then it is not true in general. If $\mathbf{A}$ and $\mathbf{B}$ are both $N \times N$ matrices and $\mathbf{AB} = \mathbf{BA}$, we say that $\mathbf{A}$ and $\mathbf{B}$ *commute*. Two diagonal $N \times N$ matrices necessarily commute.

The identity operator for matrix multiplication is the *unit matrix*, a square matrix with ones along the diagonal and zeros everywhere else. In an N-dimensional space, the unit matrix is denoted $\mathbf{I}_N$, though the subscript may be deleted if it is clear from context. The elements of $\mathbf{I}_N$ are given by

$$[\mathbf{I}_N]_{mn} = \delta_{mn}\,, \qquad m, n = 1, ..., N\,, \tag{A.24}$$

where δ_{mn} is the *Kronecker delta* symbol, which takes the value 1 if $m = n$ and 0 if $m \neq n$. The unit matrix $\mathbf{I}_N$ commutes with all $N \times N$ matrices.

A.2.4 Adjoints and transposes of products

From the definitions of matrix products and adjoints, it follows that

$$(c\mathbf{A})^\dagger = c^* \mathbf{A}^\dagger\,; \tag{A.25}$$

$$(\mathbf{AB})^\dagger = \mathbf{B}^\dagger \mathbf{A}^\dagger\,; \tag{A.26}$$

$$(\mathbf{A}_1\mathbf{A}_2...\mathbf{A}_k)^\dagger = \mathbf{A}_k^\dagger \mathbf{A}_{k-1}^\dagger ... \mathbf{A}_1^\dagger\,, \tag{A.27}$$

where c is any scalar and the matrices are assumed to be conformable.

Similar results apply to transposes:

$$(c\mathbf{A})^t = c\mathbf{A}^t\,; \tag{A.28}$$

$$(\mathbf{AB})^t = \mathbf{B}^t \mathbf{A}^t\,; \tag{A.29}$$

$$(\mathbf{A}_1\mathbf{A}_2...\mathbf{A}_k)^t = \mathbf{A}_k^t \mathbf{A}_{k-1}^t ... \mathbf{A}_1^t\,. \tag{A.30}$$

A.2.5 Inner product of two vectors

If $\mathbf{a}$ and $\mathbf{b}$ are vectors in an N-dimensional Euclidean space, their *inner product* or *scalar product* is denoted variously as $\mathbf{a} \cdot \mathbf{b}$, $(\mathbf{a}, \mathbf{b})$ or $\mathbf{a}^\dagger \mathbf{b}$. The latter notation is suggestive of matrix-matrix multiplication. If we regard $\mathbf{a}$ as an $N \times 1$ matrix with elements $A_{n1} = a_n$, then $\mathbf{a}^\dagger$ is its adjoint, a $1 \times N$ matrix or row vector with elements $A_{1n} = [A_{n1}]^* = a_n^*$. The product $\mathbf{a}^\dagger \mathbf{b}$ is then a 1×1 matrix or scalar, with the single element given by the usual rule for matrix-matrix multiplication,

$$\mathbf{a}^\dagger \mathbf{b} = \sum_{n=1}^{N} a_n^* b_n \, . \tag{A.31}$$

With this definition[2] of scalar product, the *norm* or length of a vector $\mathbf{a}$ is given by

$$||\mathbf{a}|| = \sqrt{\mathbf{a}^\dagger \mathbf{a}} = \sqrt{\sum_{n=1}^{N} |a_n|^2} \, . \tag{A.32}$$

If $\mathbf{b} = \mathbf{A}\mathbf{c}$ in (A.31), with $\mathbf{A}$ an $N \times M$ matrix and $\mathbf{c}$ an $M \times 1$ vector, we have

$$\mathbf{a}^\dagger \mathbf{A} \mathbf{c} = \sum_{n=1}^{N} a_n^* (\mathbf{A}\mathbf{c})_n = \sum_{n=1}^{N} \sum_{m=1}^{M} a_n^* A_{nm} c_m = \left[\mathbf{A}^\dagger \mathbf{a}\right]^\dagger \mathbf{c} \, . \tag{A.33}$$

For more discussion of inner products in a general Hilbert-space setting, see Chap. 1.

A.2.6 Outer product of two vectors

If $\mathbf{a}$ is an $N \times 1$ vector and $\mathbf{b}$ is $M \times 1$, their *outer product* or *tensor product* $\mathbf{b}\mathbf{a}^\dagger$ is an $M \times N$ matrix with components given by

$$\left[\mathbf{b}\mathbf{a}^\dagger\right]_{ij} = b_i a_j^* \, . \tag{A.34}$$

Similarly, $\mathbf{a}\mathbf{b}^\dagger$ is an $N \times M$ matrix with elements

$$\left[\mathbf{a}\mathbf{b}^\dagger\right]_{ij} = a_i b_j^* \, . \tag{A.35}$$

These two equations show that

$$\left[\mathbf{a}\mathbf{b}^\dagger\right]^\dagger = \mathbf{b}\mathbf{a}^\dagger \, , \tag{A.36}$$

in accord with (A.26).

Outer-product matrices can be used in matrix-matrix or matrix-vector products following the usual rules. For example, if $\mathbf{c}$ is an $M \times 1$ vector,

$$\mathbf{a}\mathbf{b}^\dagger \mathbf{c} = \mathbf{a}(\mathbf{b}^\dagger \mathbf{c}) = \mathbf{a} \sum_{m=1}^{M} b_m^* c_m \, , \tag{A.37}$$

[2]Equation (A.31) gives the Euclidean or $\mathbb{L}_2$ definition of scalar product; other possible definitions are discussed in Chap. 1, Sec. 1.1.4.

so $\mathbf{ab}^\dagger\mathbf{c}$ is an $N \times 1$ vector with elements

$$\left[\mathbf{ab}^\dagger\mathbf{c}\right]_n = a_n \sum_{m=1}^{M} b_m^* c_m \,. \tag{A.38}$$

Because of the associative laws, parentheses are superfluous in expressions like $\mathbf{ab}^\dagger\mathbf{c}$, which can be viewed as either the matrix $\mathbf{ab}^\dagger$ acting on the vector $\mathbf{c}$ or the scalar $\mathbf{b}^\dagger\mathbf{c}$ times the vector $\mathbf{a}$.

For more discussion of outer products, see Sec. 1.3.7 in Chap. 1.

A.2.7 Direct products

If $\mathbf{A}$ is an $M \times N$ matrix and $\mathbf{B}$ is a $P \times Q$ matrix, then the *direct product* or *Kronecker product* $\mathbf{A} \star \mathbf{B}$ is an $(MP) \times (NQ)$ matrix of the form

$$\mathbf{A} \star \mathbf{B} = \begin{bmatrix} a_{11}\mathbf{B} & a_{12}\mathbf{B} & \cdots & a_{1N}\mathbf{B} \\ a_{21}\mathbf{B} & a_{22}\mathbf{B} & \cdots & a_{2N}\mathbf{B} \\ \cdot & \cdot & \cdots & \cdot \\ \cdot & \cdot & \cdots & \cdot \\ a_{M1}\mathbf{B} & a_{M2}\mathbf{B} & \cdots & a_{MN}\mathbf{B} \end{bmatrix} . \tag{A.39}$$

The notation of (A.39) indicates that each element of the matrix $\mathbf{B}$ is multiplied by the scalar a_{11} and the resulting $P \times Q$ matrix is placed in the upper left position of $\mathbf{A} \star \mathbf{B}$, and similarly for the other elements indicated.

Elementary properties of the direct product are (Siotani *et al.*, 1985):

$$(c\mathbf{A}) \star \mathbf{B} = \mathbf{A} \star (c\mathbf{B}) = c(\mathbf{A} \star \mathbf{B}) \,; \tag{A.40}$$

$$(\mathbf{A} \star \mathbf{B}) \star \mathbf{C} = \mathbf{A} \star (\mathbf{B} \star \mathbf{C}) \,; \tag{A.41}$$

$$(\mathbf{A} \star \mathbf{B})^\dagger = \mathbf{A}^\dagger \star \mathbf{B}^\dagger \,; \tag{A.42}$$

$$(\mathbf{A} \star \mathbf{B})(\mathbf{C} \star \mathbf{D}) = (\mathbf{AC}) \star (\mathbf{BD}) \,; \tag{A.43}$$

$$(\mathbf{A} + \mathbf{B}) \star \mathbf{C} = (\mathbf{A} \star \mathbf{C}) + (\mathbf{B} \star \mathbf{C}) \,. \tag{A.44}$$

For reference, we list some relations between direct products and matrix inverses (see Sec. A.3), determinants (Sec. A.5) and traces (Sec. A.6):

$$(\mathbf{A} \star \mathbf{B})^{-1} = \mathbf{A}^{-1} \star \mathbf{B}^{-1} \qquad \text{for nonsingular } \mathbf{A} \text{ and } \mathbf{B}\,; \tag{A.45}$$

$$\det(\mathbf{A} \star \mathbf{B}) = [\det(\mathbf{A})]^P [\det(\mathbf{B})]^M \qquad \text{for } \mathbf{A}\colon M \times M \text{ and } \mathbf{B}\colon P \times P\,; \tag{A.46}$$

$$\mathrm{tr}\{\mathbf{A} \star \mathbf{B}\} = \mathrm{tr}\{\mathbf{A}\}\,\mathrm{tr}\{\mathbf{B}\} \,. \tag{A.47}$$

A.2.8 Hadamard products and other operations

If $\mathbf{A}$ and $\mathbf{B}$ are $M \times N$ matrices, their *Hadamard product* $\mathbf{A} \odot \mathbf{B}$ is another $M \times N$ matrix with elements given by

$$[\mathbf{A} \odot \mathbf{B}]_{mn} = A_{mn} B_{mn} \,. \tag{A.48}$$

The Hadamard product is thus an element-by-element product. The matrix of all ones, denoted $\mathbf{1}$, is the identity element for Hadamard multiplication. An exhaustive treatment of Hadamard products is given by Horn (1990).

We can extend the idea of Hadamard product to vectors just by regarding them as $N \times 1$ matrices. To be consistent with (A.48), we could denote the Hadamard product of two vectors as $\mathbf{a} \odot \mathbf{b}$, but it will often be convenient just to juxtapose the symbols and write $\mathbf{ab}$, defining this product by $[\mathbf{ab}]_n \equiv a_n b_n$. With matrices, the symbol $\odot$ serves to distinguish the Hadamard product from the ordinary matrix-matrix product, but with vectors the only products we have so far are the inner product $\mathbf{a}^\dagger \mathbf{b}$ and the outer product $\mathbf{ab}^\dagger$, so we are free to use $\mathbf{ab}$ for Hadamard product.

Many other operations can also be defined componentwise on vectors and matrices. For example, we can define an element-by-element ratio of two vectors by $[\mathbf{a}/\mathbf{b}]_n = a_n/b_n$, and a logarithm by $[\ln(\mathbf{a})]_n = \ln(a_n)$.

We can extend the concept further to functions, regarded as vectors in a Hilbert space (see Chap. 1). Thus, if the vector $\mathbf{a}$ denotes the function $a(x)$ and $\mathbf{b}$ denotes $b(x)$, then $\mathbf{ab}$ denotes $a(x)\,b(x)$ and $\ln(\mathbf{a})$ denotes $\ln[a(x)]$, so long as the resulting vectors are in the same Hilbert space as the original ones.

A.3 MATRIX INVERSION

A.3.1 Rank and invertibility

The *row rank* of a matrix is the number of linearly independent rows, while the *column rank* is the number of linearly independent columns. A basic theorem (Strang, 1980) states that the row rank and column rank are equal, so they are called simply the *rank*.

An $N \times N$ matrix $\mathbf{A}$ is called *invertible* or *nonsingular* if its rank $= N$. If that is the case, there exists a matrix $\mathbf{A}^{-1}$, called the *inverse* of $\mathbf{A}$, such that

$$\mathbf{A}^{-1}\mathbf{A} = \mathbf{A}\mathbf{A}^{-1} = \mathbf{I}_N \,. \tag{A.49}$$

By this definition, all rectangular matrices are singular. For rectangular matrices, no true inverse exists, but various *generalized inverses* or *pseudoinverses* can be defined. See Chap. 1 for a detailed discussion of this topic.

A simple example of a square but singular matrix is the outer product $\mathbf{ab}^\dagger$, where $\mathbf{a}$ and $\mathbf{b}$ are both $N \times 1$ vectors. This matrix has rank one since the ith row equals the jth row times the factor a_i^*/b_j^* [see (A.35)], so it is singular for $N > 1$.

See Sec. A.5.4 for a general expression for the inverse of a matrix in terms of determinants. Algorithms for computation of inverses can be found in Golub and van Loan (1989) or Press *et al.* (1992), but most readers will never need these details since standard packages such as Matlab and Mathematica can be used to find numerical or symbolic inverses.

A.3.2 General properties of matrix inverses

If we assume that all of the indicated inverses exist, the following general relations hold:

$$\left[\mathbf{A}^{-1}\right]^\dagger = \left[\mathbf{A}^\dagger\right]^{-1} ; \tag{A.50}$$

$$[c\mathbf{A}]^{-1} = \frac{1}{c}\mathbf{A}^{-1}, \qquad (c \text{ a scalar}); \tag{A.51}$$

$$[\mathbf{AB}]^{-1} = \mathbf{B}^{-1}\mathbf{A}^{-1}, \qquad (\mathbf{A} \text{ and } \mathbf{B} \text{ both } N \times N); \tag{A.52}$$

$$[\mathbf{A}^k]^{-1} = [\mathbf{A}^{-1}]^k, \qquad (k \text{ a positive integer}). \tag{A.53}$$

If a matrix $\mathbf{H}$ is Hermitian, its inverse $\mathbf{H}^{-1}$ is also Hermitian (if it exists).

A square matrix $\mathbf{U}$ is called *unitary* if its adjoint equals its inverse, *i.e.*, $\mathbf{U}^\dagger = \mathbf{U}^{-1}$. A unitary matrix with real elements satisfies $\mathbf{U}^\dagger = \mathbf{U}^t = \mathbf{U}^{-1}$ and is referred to as an *orthogonal* matrix. The columns of a unitary matrix form an orthonormal set, and the rows form another orthonormal set.

A.3.3 Inversion formulas for special cases

If $\mathbf{A}$ is a diagonal $N \times N$ matrix, its inverse is obtained by taking the reciprocal of each diagonal element, *i.e.*,

$$[\mathbf{A}^{-1}]_{mn} = \frac{1}{A_{nn}}\delta_{mn} \qquad \text{if } A_{mn} = A_{nn}\,\delta_{mn}. \tag{A.54}$$

Thus the unit matrix is its own inverse.

If $\mathbf{A}$ and $\mathbf{A} + \mathbf{pq}^\dagger$ are both nonsingular, with $\mathbf{p}$ and $\mathbf{q}$ being $N \times 1$ vectors and $\mathbf{A}$ being an $N \times N$ matrix, then

$$[\mathbf{A} + \mathbf{pq}^\dagger]^{-1} = \mathbf{A}^{-1} - \frac{\mathbf{A}^{-1}\mathbf{pq}^\dagger\mathbf{A}^{-1}}{1 + \mathbf{q}^\dagger\mathbf{A}^{-1}\mathbf{p}}. \tag{A.55}$$

A generalization of this result is the *matrix-inversion lemma*, also known as the *binomial inverse theorem* (Woodbury, 1950). Several forms of this useful theorem are (Tylavsky and Sohie, 1986):

$$[\mathbf{A} - \mathbf{UBV}]^{-1} = \mathbf{A}^{-1} + \mathbf{A}^{-1}\mathbf{U}\left[\mathbf{B}^{-1} - \mathbf{VA}^{-1}\mathbf{U}\right]^{-1}\mathbf{VA}^{-1} \tag{A.56a}$$

$$= \mathbf{A}^{-1} + \mathbf{A}^{-1}\mathbf{UB}\left[\mathbf{I} - \mathbf{U}^\dagger\mathbf{A}^{-1}\mathbf{UB}\right]^{-1}\mathbf{U}^\dagger\mathbf{A}^{-1} \qquad (\mathbf{U} = \mathbf{V}^\dagger) \tag{A.56b}$$

$$= \mathbf{A}^{-1} + \mathbf{A}^{-1}\mathbf{UB}\left[\mathbf{I} - \mathbf{VA}^{-1}\mathbf{UB}\right]^{-1}\mathbf{VA}^{-1} \tag{A.56c}$$

$$= \mathbf{A}^{-1} + \mathbf{A}^{-1}\mathbf{U}\left[\mathbf{I} - \mathbf{BVA}^{-1}\mathbf{U}\right]^{-1}\mathbf{BVA}^{-1} \tag{A.56d}$$

Form (A.56*a*) requires that $\mathbf{A}$ and $\mathbf{B}$ be nonsingular, while the other three forms require only that $\mathbf{A}$ be nonsingular. Of course, all forms require that the various matrices be conformable. An excellent discussion of these results is given by Tylavsky and Sohie (1986).

The inverse of partitioned matrices is also of interest. Given a nonsingular $N \times N$ matrix $\mathbf{A}$ and a nonsingular $M \times M$ matrix $\mathbf{B}$, we can form a partitioned $(M + N) \times (M + N)$ matrix with $N \times M$ and $M \times N$ submatrices of all zeros on the off-diagonals. The inverse of this partitioned matrix is given by

$$\begin{bmatrix} \mathbf{A} & \mathbf{0} \\ \mathbf{0} & \mathbf{B} \end{bmatrix}^{-1} = \begin{bmatrix} \mathbf{A}^{-1} & \mathbf{0} \\ \mathbf{0} & \mathbf{B}^{-1} \end{bmatrix}. \tag{A.57}$$

Finally, we consider the inverse of complex matrices. Let $\mathbf{C} = \mathbf{A} + i\mathbf{B}$, where $\mathbf{A}$ and $\mathbf{B}$ are real $N \times N$ matrices. Then $\mathbf{C}^{-1}$, if it exists, is given by (Rade and Westergren, 1990, p. 105)

$$[\mathbf{A} + i\mathbf{B}]^{-1} = \left[\mathbf{A} + \mathbf{B}\mathbf{A}^{-1}\mathbf{B}\right]^{-1} - i\mathbf{A}^{-1}\mathbf{B}\left[\mathbf{A} + \mathbf{B}\mathbf{A}^{-1}\mathbf{B}\right]^{-1}$$

$$= \mathbf{B}^{-1}\mathbf{A}[\mathbf{B} + \mathbf{A}\mathbf{B}^{-1}\mathbf{A}]^{-1} - i[\mathbf{B} + \mathbf{A}\mathbf{B}^{-1}\mathbf{A}]^{-1}, \tag{A.58}$$

where the first form requires the existence of $\mathbf{A}^{-1}$ and the second form requires $\mathbf{B}^{-1}$.

A.3.4 Neumann series

Consider a nonsingular $N \times N$ matrix of the form $\mathbf{I} - \mathbf{\Omega}$. Provided the series converges uniformly, we can express $[\mathbf{I} - \mathbf{\Omega}]^{-1}$ by a *Neumann series* (named for Carl Neumann, not John von),

$$[\mathbf{I} - \mathbf{\Omega}]^{-1} = \sum_{j=0}^{\infty} \mathbf{\Omega}^j . \tag{A.59}$$

This expansion is the operator generalization of the familiar expression for the sum of a geometric series:

$$\frac{1}{1-x} = \sum_{j=0}^{\infty} x^j , \tag{A.60}$$

where x is a scalar. Equation (A.60) is valid if the series converges uniformly, which requires that $|x| < 1$. The corresponding convergence requirement for (A.59) is that all eigenvalues of $\mathbf{\Omega}$ be less than unity in absolute value. (See Secs. 1.4.1 and 1.7.6 in Chap. 1.)

To prove that (A.59) is correct, we multiply the right-hand side by $[\mathbf{I} - \mathbf{\Omega}]$, obtaining

$$[\mathbf{I} - \mathbf{\Omega}] \sum_{j=0}^{\infty} \mathbf{\Omega}^j = \sum_{j=0}^{\infty} \mathbf{\Omega}^j - \sum_{j=0}^{\infty} \mathbf{\Omega}^{j+1}$$

$$= \mathbf{I} + \sum_{j=1}^{\infty} \mathbf{\Omega}^j - \sum_{j=0}^{\infty} \mathbf{\Omega}^{j+1} . \tag{A.61}$$

We now let $m = j - 1$ in the first sum, with the result

$$[\mathbf{I} - \mathbf{\Omega}] \sum_{j=0}^{\infty} \mathbf{\Omega}^j = \mathbf{I} + \sum_{m=0}^{\infty} \mathbf{\Omega}^{m+1} - \sum_{j=0}^{\infty} \mathbf{\Omega}^{j+1} = \mathbf{I} . \tag{A.62}$$

If we denote the series in (A.59) by $\mathbf{S}$, (A.62) shows that $[\mathbf{I} - \mathbf{\Omega}]\mathbf{S} = \mathbf{I}$. A similar calculation shows that $\mathbf{S}[\mathbf{I} - \mathbf{\Omega}] = \mathbf{I}$.

For more discussion of the Neumann series, see Chap. 1.

A.4 EIGENVECTORS AND EIGENVALUES

Section 1.4 in Chap. 1 is a detailed discussion of eigenvectors and eigenvalues, so only a brief survey is given here.

A.4.1 Basic concepts

The vector $\mathbf{u}$ is the *eigenvector* (German: characteristic vector) of a square $N \times N$ matrix $\mathbf{A}$ and the scalar λ is the corresponding *eigenvalue* if

$$\mathbf{A}\mathbf{u} = \lambda \mathbf{u}\,. \tag{A.63}$$

Equation (A.63) can also be written as $[\mathbf{A} - \lambda \mathbf{I}]\mathbf{u} = 0$. In this form it is recognized as a set of N homogeneous linear equations for N unknowns (the components of $\mathbf{u}$). As discussed in Sec. A.5.5, these equations have a nontrivial solution if and only if

$$\det(\mathbf{A} - \lambda \mathbf{I}) = 0\,, \tag{A.64}$$

where $\det(\cdot)$ denotes determinant (see Sec. A.5.1 for a definition).

When the determinant in (A.64) is expanded, a polynomial $P(\lambda)$ of degree N in the variable λ results, and (A.64) becomes[3]

$$P(\lambda) = \det[\mathbf{A} - \lambda \mathbf{I}] = (-1)^N \left[\lambda^N + a_1 \lambda^{N-1} + a_2 \lambda^{N-2} + ... + a_N\right]\,. \tag{A.65}$$

This equation is known as the *characteristic equation* for $\mathbf{A}$. The *Fundamental Theorem of Algebra* (Gellert *et al.*, 1977) assures us that the characteristic equation has N solutions (roots of the polynomial), though the roots are neither necessarily real nor necessarily distinct.

The characteristic equation can be solved (usually numerically) for each of these roots, which we shall denote as λ_n, $n = 1, ..., N$. Corresponding to each λ_n is an eigenvector $\mathbf{u}_n$.

A.4.2 Some general theorems

If a square matrix is diagonal ($A_{mn} = A_n\,\delta_{mn}$), its diagonal elements are immediately the eigenvalues, *i.e.*, $\lambda_n = A_{nn}$. The eigenvector $\mathbf{u}_n$ has a 1 in the nth position and 0 in all other positions, $[\mathbf{u}_n]_m = \delta_{nm}$. It is straightforward to show that (A.63) is satisfied with this λ_n and $\mathbf{u}_n$. Moreover, if the matrix is triangular, the diagonal elements are again the eigenvalues, though the eigenvectors are not so simple in this case.

Schur's lemma (Rade and Westergren, 1990) states that any square matrix can be reduced to upper triangular form by means of a unitary transformation. That is, it is possible to find a unitary matrix $\mathbf{U}$ such that

$$\mathbf{A}' = \mathbf{U}^\dagger \mathbf{A} \mathbf{U} \tag{A.66}$$

has only zero elements below the diagonal. Since the eigenvalues are unchanged by unitary transformation (see Chap. 1, Sec. 1.4.2), the diagonal elements of $\mathbf{A}'$ are also the eigenvalues of $\mathbf{A}$. Thus determination of the eigenvalues is equivalent to finding the unitary matrix $\mathbf{U}$ that reduces $\mathbf{A}$ to triangular form. Algorithms for this purpose are discussed in detail in Golub and van Loan (1989).

If a square matrix is approximately diagonal (*i.e.*, the off-diagonal elements are small), it is of interest to know how well the diagonal elements approximate the

[3]In (A.65), it can be shown (Eves, 1966, p. 200) that $a_N = \det(\mathbf{A})$ and $a_1 = \mathrm{tr}(\mathbf{A})$. Furthermore, all coefficients can be derived from traces of powers of $\mathbf{A}$ (Pettofrezzo, 1966, p. 84).

eigenvalues. *Gershgorin's theorem* (Golub and van Loan, 1989, p. 341) provides an answer to this question. Consider a disc in the complex z plane centered at $z = A_{nn}$. This disc consists of all points such that

$$|z - A_{nn}| \leq \sum_{m=1}^{N} |A_{nm}|(1 - \delta_{mn}), \tag{A.67}$$

where the factor $(1 - \delta_{mn})$ serves to omit the term $m = n$ from the sum. The disc in (A.67) is frequently called the nth *Gershgorin disc.* The theorem says that all eigenvalues of $\mathbf{A}$ must lie in the union of all N Gershgorin discs. If the matrix is diagonal, each disc has zero radius and the diagonal element is exactly the eigenvalue. Off-diagonal elements increase the radius of the discs and hence the uncertainty in the eigenvalues relative to the diagonal elements.

A square matrix is said to be *diagonally dominant* if

$$\sum_{m=1}^{N} |A_{nm}|(1 - \delta_{mn}) < A_{nn} \qquad \text{for all } n. \tag{A.68}$$

If this condition is satisfied, the matrix is invertible.

A.4.3 Hermitian matrices

Considerably stronger statements can be made about eigenvalues and eigenvectors if the matrix in question is Hermitian (or symmetric if all elements are real). A detailed discussion is given in Secs. 1.4.4 – 1.4.6 in Chap. 1, but the main results are listed here for reference.

(*i*) The eigenvalues of a Hermitian matrix are real;

(*ii*) Eigenvectors corresponding to different eigenvalues of a Hermitian matrix are orthogonal;

(*iii*) The N eigenvectors of an $N \times N$ Hermitian matrix are linearly independent and span the space N-dimensional Euclidean space $\mathbb{E}^N$ (see Sec. 1.1.2 in Chap. 1);

(*iv*) A Hermitian matrix can always be diagonalized by a unitary transformation;

(*v*) Two different Hermitian matrices $\mathbf{A}$ and $\mathbf{B}$ can be diagonalized by the *same* unitary transformation if and only if they commute, $\mathbf{AB} = \mathbf{BA}$.

(*vi*) For two different positive-definite Hermitian matrices $\mathbf{A}$ and $\mathbf{B}$ that do not necessarily commute, there exists a nonsingular matrix $\mathbf{W}$ such that $\mathbf{W}^\dagger\mathbf{AW} = \mathbf{I}$ and $\mathbf{W}^\dagger\mathbf{BW} = \mathbf{D}$, where $\mathbf{D}$ is diagonal. This matrix satisfies $\mathbf{BW} = \mathbf{AWD}$. (See Sec. A.8 for discussion of positive-definite matrices, and see Fukunaga (1990) or Sec. 1.4.6 for a discussion of the transformation.)

Note that point (*ii*) would allow two or more different eigenvectors not to be orthogonal if they share the same eigenvalue, though point (*iii*) ensures that they are linearly independent. If the vectors $\{\mathbf{e}_k, k = 1, ..., K\}$ are linearly independent, we can construct a set of mutually orthonormal vectors $\{\mathbf{u}_k, k = 1, ..., K\}$ by the process known as *Gram-Schmidt orthogonalization*. The algorithm is as follows:

$$(1)\ \mathbf{u}_1 = \frac{1}{||\mathbf{e}_1||}\mathbf{e}_1\ ;$$

$$(2)\ \mathbf{v}_2 = \mathbf{e}_2 - (\mathbf{e}_2, \mathbf{u}_1)\mathbf{u}_1\,, \qquad \mathbf{u}_2 = \frac{1}{||\mathbf{v}_2||}\mathbf{v}_2\ ;$$

$$\cdots\cdots$$

$$(K)\ \mathbf{v}_K = \mathbf{e}_K - (\mathbf{e}_K, \mathbf{u}_1)\mathbf{u}_1 - ... - (\mathbf{e}_K, \mathbf{u}_{K-1})\mathbf{u}_{K-1}\,, \quad \mathbf{u}_K = \frac{1}{||\mathbf{v}_K||}\mathbf{v}_K\ .$$

In step (2), the vector $\mathbf{e}_1$ is projected onto the normalized vector $\mathbf{u}_1$ and the projection is then subtracted from $\mathbf{e}_2$, guaranteeing that $\mathbf{v}_2$ is orthogonal to $\mathbf{u}_1$. After normalization, we have two orthonormal vectors $\mathbf{u}_1$ and $\mathbf{u}_2$. The process is then repeated, at each step constructing a new vector $\mathbf{v}_k$ orthogonal to all of the previous $\mathbf{u}_j$ for $j < k$, and then normalizing the result to get $\mathbf{u}_k$.

If the initial set $\{\mathbf{e}_k\}$ is a set of eigenvectors of the Hermitian matrix $\mathbf{H}$, and all of these eigenvectors have the same eigenvalue λ, then the new orthonormal vectors $\{\mathbf{u}_k\}$ are also eigenvectors and also all have the eigenvalue λ. Thus, if we assume that the Gram-Schmidt process has been applied if needed, we can be assured that the $\{\mathbf{u}_n, n = 1, ..., N\}$ form a complete orthonormal basis in $\mathbf{E}^N$. The orthonormality is expressed mathematically by

$$\mathbf{u}_m^\dagger \mathbf{u}_n = \delta_{nm}\,, \qquad n, m = 1, ..., N\,. \tag{A.69}$$

The completeness of the set $\{\mathbf{u}_n\}$ means that it can be used to represent the unit matrix in the form

$$\mathbf{I}_N = \sum_{n=1}^{N} \mathbf{u}_n \mathbf{u}_n^\dagger\,. \tag{A.70}$$

This relation is frequently referred to as the *closure* relation or the *resolution of the identity.*

The basis set consisting of the eigenvectors of a Hermitian matrix $\mathbf{H}$ can also be used to represent $\mathbf{H}$ itself as

$$\mathbf{H} = \sum_{n=1}^{N} \lambda_n \mathbf{u}_n \mathbf{u}_n^\dagger\,. \tag{A.71}$$

This representation, called the *spectral decomposition* of $\mathbf{H}$, is discussed in detail in Chap. 1 and exploited throughout the book.

A.5 DETERMINANTS

A.5.1 Definitions

If $\mathbf{A}$ is an $N \times N$ matrix with elements A_{mn}, then the *determinant* of $\mathbf{A}$, denoted $\det(\mathbf{A})$, is a scalar defined formally by

$$\det(\mathbf{A}) = \sum_{perm} (-1)^k A_{1s_1} A_{2s_2} ... A_{Ns_N}\,, \tag{A.72}$$

where the indices $s_1, s_2, ..., s_N$ are a permutation of $1, 2, ..., N$ and are hence distinct, and the sum runs over all possible permutations. Since there are $N!$ permutations, the sum has $N!$ terms in general (though some or all of them may be zero). The sign of each term is determined by the number k of inversions in the permutation. Cyclic permutations thus always occur with a plus sign.

The determinant can also be defined in terms of eigenvalues. If $\{\lambda_n, n = 1, ..., N\}$ are the eigenvalues of the $N \times N$ matrix $\mathbf{A}$, then

$$\det(\mathbf{A}) = \prod_{n=1}^{N} \lambda_n . \tag{A.73}$$

If $\mathbf{A}$ is singular, at least one of the eigenvalues is zero, and it follows from this equation that $\det(\mathbf{A}) = 0$.

A geometrical interpretation of determinant is that $|\det(\mathbf{A})|$ is the volume of the N-dimensional parallelopiped formed by the column (or row) vectors of $\mathbf{A}$. This result is familiar in ordinary 3D vector analysis. Three vectors $\mathbf{a}$, $\mathbf{b}$ and $\mathbf{c}$ define a parallelopiped with volume $|\mathbf{a} \cdot \mathbf{b} \times \mathbf{c}|$, where $\times$ denotes the usual vector cross product, and the triple scalar product $\mathbf{a} \cdot \mathbf{b} \times \mathbf{c}$ is computed as a determinant.

A.5.2 Special cases

If $\mathbf{A}$ is a 2×2 matrix, its determinant is given by

$$\det(\mathbf{A}) = A_{11}A_{22} - A_{12}A_{21} . \tag{A.74}$$

If $\mathbf{A}$ is a 3×3 matrix, its determinant is given by

$$\begin{aligned} \det(\mathbf{A}) = {} & A_{11}A_{22}A_{33} + A_{12}A_{23}A_{31} + A_{13}A_{21}A_{32} - A_{11}A_{23}A_{32} \\ & - A_{12}A_{21}A_{33} - A_{13}A_{22}A_{31} . \end{aligned} \tag{A.75}$$

If $\mathbf{A}$ is an $N \times N$ diagonal matrix,

$$\det(\mathbf{A}) = \Pi_{n=1}^{N} A_{nn} . \tag{A.76}$$

This equation is valid also if $\mathbf{A}$ is an upper or lower triangular matrix. An immediate consequence of (A.76) is that the determinant of the unit matrix is unity: $\det(\mathbf{I}) = 1$.

A.5.3 Properties of determinants

The determinant of the product of two $N \times N$ matrices is the product of their determinants:

$$\det(\mathbf{AB}) = \det(\mathbf{A}) \det(\mathbf{B}) . \tag{A.77}$$

Since the determinant of the unit matrix is unity, it follows that

$$\det(\mathbf{A}^{-1}) = \frac{1}{\det(\mathbf{A})} . \tag{A.78}$$

Directly from the definition, it follows that multiplying every element of an $N \times N$ matrix by a constant multiplies its determinant by the same constant to the Nth power:

$$\det(c\mathbf{A}) = c^N \det(\mathbf{A}) , \qquad (c \text{ a scalar}) . \tag{A.79}$$

Multiplying every element in a single row or column by c multiplies the determinant by c.

A determinant is unchanged if rows and columns are interchanged, so

$$\det(\mathbf{A}^t) = \det(\mathbf{A})\,. \tag{A.80}$$

The adjoint is formed by interchanging rows and columns and taking the complex conjugate of each element, so it follows that

$$\det(\mathbf{A}^\dagger) = [\det(\mathbf{A})]^*\,. \tag{A.81}$$

For a Hermitian matrix, $\mathbf{A}^\dagger = \mathbf{A}$ so $\det(\mathbf{A})$ is real. This result is in accord with (A.73) since, as proved in Chap. 1, the eigenvalues of a Hermitian matrix are real. If $\mathbf{U}$ is a unitary matrix, then $\mathbf{U}^{-1} = \mathbf{U}^\dagger$. By (A.78) and (A.81),

$$|\det(\mathbf{U})| = 1\,. \tag{A.82}$$

This result also follows from (A.73) since the eigenvalues of a unitary matrix are complex roots of unity.

An important consequence of (A.81) and (A.82) is that the determinant is unchanged by a unitary transformation:

$$\det(\mathbf{U}\mathbf{A}\mathbf{U}^\dagger) = \det(\mathbf{A})\,. \tag{A.83}$$

A unitary transformation is a special case of a similarity transformation. It is shown in Sec. 1.4.2 that similarity transformations do not alter eigenvalues. Therefore, with (A.73),

$$\det(\mathbf{S}\mathbf{A}\mathbf{S}^{-1}) = \det(\mathbf{A})\,, \tag{A.84}$$

where $\mathbf{S}$ is any nonsingular $N \times N$ matrix.

Additional symmetry properties of determinants are as follows:

(*i*) Interchange of two rows (columns) changes the sign of the determinant.

(*ii*) A determinant does not change if one row (column) is multiplied by a constant and added to another row (column).

(*iii*) A determinant equals zero if all elements of a row (column) are zero.

(*iv*) A determinant equals zero if one row (column) equals another row (column) times a constant.

It is often necessary to consider the determinant of the sum of two matrices. The following identity, discussed fully in Harville (1997), is quite useful:

$$\det(\mathbf{R} + \mathbf{STU}) = \det(\mathbf{R})\det(\mathbf{T})\det(\mathbf{T}^{-1} + \mathbf{U}\mathbf{R}^{-1}\mathbf{S})\,. \tag{A.85}$$

For many additional properties of determinants, see Muir (1960) and Andrews and Burge (1993), and for an early treatment of considerable historical interest, see Dodgson (1867). (Charles Lutwidge Dodgson, under the pseudonym Lewis Carroll, also wrote *Alice's Adventures in Wonderland.*)

A.5.4 Cofactors and inverses

If $\mathbf{A}$ is a nonsingular $N \times N$ matrix, the $(nm)^{th}$ element of its inverse is given by

$$\left[\mathbf{A}^{-1}\right]_{mn} = \frac{\mathrm{cof}_{nm}(\mathbf{A})}{\det(\mathbf{A})}, \tag{A.86}$$

where $\mathrm{cof}_{nm}(\mathbf{A})$ is the *cofactor*, defined as $(-1)^{n+m}$ times the determinant of the $(N-1) \times (N-1)$ matrix obtained by deleting the nth row and m^{th} column from $\mathbf{A}$. That determinant itself is called a *minor* of $\mathbf{A}$. It follows from (A.86) that the inverse does not exist if $\det(\mathbf{A}) = 0$.

The special case $N = 2$ yields

$$\begin{bmatrix} A_{11} & A_{12} \\ A_{21} & A_{22} \end{bmatrix}^{-1} = \frac{1}{A_{11}A_{22} - A_{12}A_{21}} \begin{bmatrix} A_{22} & -A_{12} \\ -A_{21} & A_{11} \end{bmatrix}. \tag{A.87}$$

From (A.74), the denominator is recognized as $\det(\mathbf{A})$.

Cofactors can also be used to evaluate determinants. The key result is

$$\det(\mathbf{A}) = \sum_{m=1}^{N} A_{nm}\, \mathrm{cof}_{nm}(\mathbf{A}), \tag{A.88}$$

where n denotes an arbitrary row. The utility of this result is that the determinant of an $N \times N$ matrix is expressed as a sum of determinants of $(N-1) \times (N-1)$ matrices. The process can be repeated $N-1$ times until only trivial determinants of 1×1 matrices need to be determined.

A.5.5 Cramer's rule

Consider a set of linear equations in the form

$$\mathbf{y} = \mathbf{A}\mathbf{x}, \tag{A.89}$$

where $\mathbf{A}$ is a nonsingular $N \times N$ matrix and $\mathbf{x}$ and $\mathbf{y}$ are $N \times 1$ column vectors. If $\mathbf{y}$ is known, the solution for $\mathbf{x}$ is given by $\mathbf{x} = \mathbf{A}^{-1}\mathbf{y}$, and the jth element of $\mathbf{x}$ can be found by *Cramer's rule*:

$$x_j = \frac{\det(\mathbf{A}_j)}{\det(\mathbf{A})}, \tag{A.90}$$

where $\mathbf{A}_j$ is the $N \times N$ matrix obtained by replacing the jth column of $\mathbf{A}$ with $\mathbf{y}$.

If all components of $\mathbf{y}$ are zero, the system of equations is said to be *homogeneous*. In that case $\det(\mathbf{A}_j)$ is identically zero (see Sec. A.5.3), so the only way the system $\mathbf{A}\mathbf{x} = \mathbf{0}$ can have a nontrivial solution for $\mathbf{x}$ is if $\det(\mathbf{A}) = 0$.

A.6 TRACES

A.6.1 Definitions

The *trace* or *spur* of a matrix is simply the sum of its diagonal elements:

$$\mathrm{tr}(\mathbf{A}) = \sum_{n=1}^{N} A_{nn}\,. \tag{A.91}$$

If $\mathbf{a}$ and $\mathbf{b}$ are both $N \times 1$ column vectors, the trace of their outer product is the same as the inner product:

$$\operatorname{tr}(\mathbf{a}\mathbf{b}^{\dagger}) = \mathbf{b}^{\dagger}\mathbf{a} = \sum_{n=1}^{N} b_n^* a_n \,. \tag{A.92}$$

A.6.2 Properties of traces

Some basic properties of the trace are as follows:

$$\operatorname{tr}(\mathbf{A}+\mathbf{B}) = \operatorname{tr}(\mathbf{A}) + \operatorname{tr}(\mathbf{B})\,; \tag{A.93}$$

$$\operatorname{tr}(\alpha\mathbf{A}) = \alpha \operatorname{tr}(\mathbf{A})\,, \tag{A.94}$$

where α is a scalar. If $\mathbf{A}$ is an $M \times N$ matrix and $\mathbf{B}$ is $N \times M$, then

$$\operatorname{tr}(\mathbf{AB}) = \operatorname{tr}(\mathbf{BA})\,. \tag{A.95}$$

The trace of a product of square matrices is invariant to cyclic permutation:

$$\operatorname{tr}(\mathbf{ABC}) = \operatorname{tr}(\mathbf{CAB}) = \operatorname{tr}(\mathbf{BCA})\,, \tag{A.96}$$

and similarly for products with any number of factors. It follows from (A.96) that the trace of a matrix is invariant to unitary transformations.

The trace of a square matrix is the sum of its eigenvalues (Pettofrezzo, 1966, p. 85):

$$\operatorname{tr}(\mathbf{A}) = \sum_{j=1}^{N} \lambda_j(\mathbf{A})\,, \tag{A.97}$$

where $\lambda_j(\mathbf{A})$ is the jth eigenvalue of $\mathbf{A}$. If the matrix can be diagonalized by a similarity transformation (as can all Hermitian matrices), (A.97) follows immediately from (A.96) since $\operatorname{tr}(\mathbf{S}^{-1}\mathbf{AS}) = \operatorname{tr}(\mathbf{A})$. The result in (A.97) is, however, more general; it holds for any square matrix.

Similarly, the trace of a square matrix $\mathbf{A}$ raised to the kth power is given by

$$\operatorname{tr}(\mathbf{A}^k) = \sum_{j=1}^{N} [\lambda_j(\mathbf{A})]^k \,, \qquad k \text{ a positive integer}\,. \tag{A.98}$$

A.7 FUNCTIONS OF MATRICES

Many different matrix functions can be defined as a power series, the general form of which is

$$f(\mathbf{A}) = \sum_{n=0}^{\infty} c_n \mathbf{A}^n \,. \tag{A.99}$$

The matrix $\mathbf{A}$ must be square, say $N \times N$, in order for its powers $\mathbf{A}^n$ to be defined, but it can be nonsymmetric and/or singular in what follows. If $\mathbf{A}$ has real elements and all coefficients c_n are real, $f(\mathbf{A})$ is a mapping from $\mathbb{R}^{N\times N} \to \mathbb{R}^{N\times N}$; if the elements can be complex, it is a mapping from $\mathbb{C}^{N\times N} \to \mathbb{C}^{N\times N}$.

One application of such a power-series function is the *Cayley-Hamilton theorem* (Eves, 1966), which states that every square matrix $\mathbf{A}$ satisfies its own characteristic equation. That is, $P(\mathbf{A}) = 0$, where $P(\mathbf{A})$ is the polynomial defined by (A.65) but with the scalar λ replaced by the matrix $\mathbf{A}$.

A.7.1 Matrix exponentials

The matrix exponential is defined by taking $c_n = 1/n!$ in (A.99), leading to

$$\exp(\mathbf{A}) = \sum_{n=0}^{\infty} \frac{1}{n!} \mathbf{A}^n . \tag{A.100}$$

If $\mathbf{A}$ and $\mathbf{B}$ commute, then

$$\exp(\mathbf{A}) \exp(\mathbf{B}) = \exp(\mathbf{B}) \exp(\mathbf{A}) = \exp(\mathbf{A} + \mathbf{B}) . \tag{A.101}$$

Somewhat more generally, if $\mathbf{A}$ and $\mathbf{B}$ both commute with their commutator, $\mathbf{AB} - \mathbf{BA}$, then

$$\exp\left[-\alpha(\mathbf{A} + \mathbf{B}) + \tfrac{1}{2}\alpha^2(\mathbf{AB} - \mathbf{BA})\right] = \exp(-\alpha\mathbf{A}) \exp(-\alpha\mathbf{B}) . \tag{A.102}$$

The matrix exponential $\exp(\alpha\mathbf{A})$ is never singular. In fact, its inverse is given by

$$[\exp(\alpha\mathbf{A})]^{-1} = \exp(-\alpha\mathbf{A}) . \tag{A.103}$$

All unitary matrices can be represented as

$$\mathbf{U} = \exp(i\mathbf{H}) , \tag{A.104}$$

where $\mathbf{H}$ is Hermitian. In this form, the adjoint of $\mathbf{U}$ is

$$\mathbf{U}^{\dagger} = \exp(-i\mathbf{H}) , \tag{A.105}$$

and the unitarity follows from (A.103).

A few additional properties of matrix exponentials are:

$$\det[\exp(\mathbf{A})] = \exp[\mathrm{tr}(\mathbf{A})] , \tag{A.106}$$

$$\mathbf{B}[\exp(\mathbf{A})]\mathbf{B}^{-1} = \exp(\mathbf{BAB}^{-1}) , \tag{A.107}$$

for any nonsingular matrix $\mathbf{B}$.

A.7.2 Trigonometric functions

Matrix sines and cosines are defined by

$$\sin(\mathbf{A}) = \mathbf{A} - \tfrac{1}{3!}\mathbf{A}^3 + \tfrac{1}{5!}\mathbf{A}^5 + \ldots ; \tag{A.108}$$

$$\cos(\mathbf{A}) = \mathbf{I} - \tfrac{1}{2!}\mathbf{A}^2 + \tfrac{1}{4!}\mathbf{A}^4 + \ldots . \tag{A.109}$$

It follows that

$$\exp(i\mathbf{A}) = \cos(\mathbf{A}) + i\sin(\mathbf{A}) . \tag{A.110}$$

If $\mathbf{A}$ and $\mathbf{B}$ commute, then

$$\cos(\mathbf{A} + \mathbf{B}) = \cos(\mathbf{A})\cos(\mathbf{B}) - \sin(\mathbf{A})\sin(\mathbf{B}) ; \tag{A.111}$$

$$\sin(\mathbf{A} + \mathbf{B}) = \sin(\mathbf{A})\cos(\mathbf{B}) + \cos(\mathbf{A})\sin(\mathbf{B}) . \tag{A.112}$$

A.7.3 Other functions

Several important results arise from a consideration of $\mathbf{I} - \mathbf{A}$, where $\mathbf{A}$ is an $N \times N$ matrix and $\mathbf{I}$ is the corresponding unit matrix. An extension of the Neumann series, discussed in Sec. A.3.4, shows that (Siotani *et al.*, 1985)

$$(\mathbf{I} - \mathbf{A})^{-m} = \sum_{n=0}^{\infty} \frac{\Gamma(m+n)}{\Gamma(m)n!} \mathbf{A}^n \qquad \text{for any real } m > 0\,, \tag{A.113}$$

if all eigenvalues of $\mathbf{A}$ are less than unity in absolute value.

The determinant of $\mathbf{I} - \mathbf{A}$ also has some interesting properties. For example, if all of the eigenvalues of $\mathbf{A}$ are less than unity in absolute value, then (Siotani *et al.*, 1985):

$$-\ln[\det(\mathbf{I} - \mathbf{A})] = \operatorname{tr}\mathbf{A} + \tfrac{1}{2}\operatorname{tr}\mathbf{A}^2 + \tfrac{1}{3}\operatorname{tr}\mathbf{A}^3 + \dots . \tag{A.114}$$

If $\mathbf{A}$ is a scalar (1×1 matrix), then determinant, trace and matrix are identical and (A.114) is the familiar expansion for the logarithm.

A.8 DEFINITE MATRICES AND QUADRATIC FORMS

A.8.1 Definitions

Given an $N \times N$ matrix $\mathbf{A}$ and an $N \times 1$ vector $\mathbf{x}$, we define a scalar *quadratic form* $Q_A(\mathbf{x})$ by

$$Q_A(\mathbf{x}) = \mathbf{x}^\dagger \mathbf{A} \mathbf{x}\,. \tag{A.115}$$

If the matrix $\mathbf{A}$ is Hermitian, $Q_A(\mathbf{x})$ is called a *Hermitian form.*

The matrix $\mathbf{A}$ is said to be *positive-definite* if its associated quadratic form $Q_A(\mathbf{x}) > 0$ for all $\mathbf{x}$ (except the trivial one with all elements zero). If $Q_A(\mathbf{x}) \geq 0$ for all $\mathbf{x}$, then $\mathbf{A}$ is said to be *positive-semidefinite* or *nonnegative-definite.* Similarly, $\mathbf{A}$ is *negative-definite* if $Q_A(\mathbf{x}) < 0$ for all nontrivial $\mathbf{x}$ and *negative-semidefinite* if $Q_A(\mathbf{x}) \leq 0$ for all $\mathbf{x}$. If none of these conditions prevail, then $\mathbf{A}$ is merely *indefinite.*

A.8.2 Conditions for definiteness

A necessary and sufficient condition for the $N \times N$ matrix $\mathbf{A}$ to be positive-definite is that all of the eigenvalues of $\mathbf{A}$ be greater than zero. By (A.73), this condition implies that $\det(\mathbf{A}) > 0$, but the converse does not necessarily hold; it is possible to have a positive determinant and an even number of negative eigenvalues. For example, a diagonal 3×3 matrix with $(1, -1, -1)$ along the diagonal has determinant $+1$ but is indefinite.

A stronger condition for definiteness is obtained by considering subdeterminants. A *minor* of order $N - k$ is obtained by deleting k rows and k columns ($k < N$) from the $N \times N$ matrix $\mathbf{A}$ and taking the determinant of the resulting $(N - k) \times (N - k)$ matrix. For example, we might delete the rows labelled by $n = n_1, n_2, ..., n_k$ and the columns labelled by $m = m_1, m_2, ..., m_k$. The minor is called a *principal minor* if the row and column labels are identical. A *leading principal minor* is one for which the row and column labels are successive integers from 1 to k. A necessary and sufficient condition for a Hermitian matrix to be positive-definite is that all of its leading principal minors be greater than zero (Eves, 1966). This condition rules out the 3×3 example given above; if we delete the first and second row and the first and second column, all that is left is the single element -1, and the corresponding principal minor is -1.

An important general theorem on positive-definite matrices states that:

> If $\mathbf{A}$ is a positive-definite matrix in $\mathbb{C}^{N\times N}$ and $\mathbf{B}$ is a matrix of rank K in $\mathbb{C}^{N\times K}$, then the $K \times K$ matrix $\mathbf{B}^\dagger\mathbf{AB}$ is also positive-definite.

This theorem is proved for real matrices on p. 140 of Golub and van Loan (1989), but their proof generalizes readily to the complex case.

An immediate consequence, which follows by setting $\mathbf{A} = \mathbf{I}$ in this theorem, is that any $K \times K$ matrix of the form $\mathbf{B}^\dagger\mathbf{B}$ is positive-definite if the $N \times K$ matrix $\mathbf{B}$ has rank K. Any matrix of the form $\mathbf{B}^\dagger\mathbf{B}$, without any restriction on $\mathbf{B}$, is at least positive-semidefinite. Moreover, it follows from (A.26) that $\mathbf{B}^\dagger\mathbf{B}$ is Hermitian. Thus, as soon as we can show that a matrix has the form $\mathbf{B}^\dagger\mathbf{B}$, we know that it is Hermitian and at least positive-semidefinite.

A stronger statement can be also made regarding Hermitian positive-definite matrices. Any such matrix $\mathbf{H}$ can be uniquely factored in the form

$$\mathbf{H} = \mathbf{G}\mathbf{G}^\dagger , \tag{A.116}$$

where $\mathbf{G}$ is square and lower triangular, with positive elements along the diagonal. This form, known as the *Cholesky factorization*, is at the heart of many numerical algorithms in linear algebra.

A.8.3 Square-root matrices

If the $N \times N$ Hermitian matrix $\mathbf{H}$ is positive-semidefinite, we can define its square root $\mathbf{H}^{\frac{1}{2}}$, an $N \times N$ Hermitian matrix that satisfies

$$\mathbf{H}^{\frac{1}{2}}\mathbf{H}^{\frac{1}{2}} = \mathbf{H}. \tag{A.117}$$

The easiest way to construct $\mathbf{H}^{\frac{1}{2}}$ is by means of the spectral decomposition (A.71). If we know the eigenvalues $\{\lambda_n\}$ and eigenvectors $\{\mathbf{u}_n\}$, then we can write the square-root matrix as

$$\mathbf{H}^{\frac{1}{2}} = \sum_{n=1}^{N} \sqrt{\lambda_n}\, \mathbf{u}_n \mathbf{u}_n^\dagger . \tag{A.118}$$

To show that (A.117) is satisfied by (A.118), we make use of the orthonormality condition (A.69).

If $\mathbf{H}$ is nonsingular, the inverse of the square-root matrix exists and is given by (Johnson and Wichern, 1988)

$$\mathbf{H}^{-\frac{1}{2}} = \sum_{n=1}^{N} \frac{1}{\sqrt{\lambda_n}} \mathbf{u}_n \mathbf{u}_n^\dagger . \tag{A.119}$$

A.9 DIFFERENTIATION FORMULAS

A.9.1 Derivatives and integrals of a matrix with respect to a real parameter

If $\mathbf{A}(t)$ is an $M \times N$ matrix with complex elements $A_{mn}(t)$ which depend on the real scalar parameter t, then we can define derivative and integral matrices by

$$\left[\frac{d}{dt}\mathbf{A}(t)\right]_{nm} = \frac{d}{dt}A_{nm}(t)\,; \tag{A.120}$$

$$\left[\int_a^b dt\ \mathbf{A}(t)\right]_{nm} = \int_a^b dt\ A_{nm}(t)\,. \tag{A.121}$$

The following manipulation rules for the derivative matrix can be verified:

$$\frac{d}{dt}[\mathbf{A}(t) + \mathbf{B}(t)] = \frac{d}{dt}\mathbf{A}(t) + \frac{d}{dt}\mathbf{B}(t)\,; \tag{A.122}$$

$$\frac{d}{dt}[\mathbf{A}(t)\,\mathbf{B}(t)] = \mathbf{A}(t)\frac{d}{dt}\mathbf{B}(t) + \left[\frac{d}{dt}\mathbf{A}(t)\right]\mathbf{B}(t)\,. \tag{A.123}$$

Note that the last term in (A.123) is not $\mathbf{B}d\mathbf{A}/dt$ unless the two matrices commute.

If $N = M$ and the inverse of $\mathbf{A}$ exists, it can also be differentiated (Rade and Westergren, 1990) as follows:

$$\frac{d}{dt}\mathbf{A}^{-1} = -\mathbf{A}^{-1}\frac{d\mathbf{A}}{dt}\mathbf{A}^{-1}\,. \tag{A.124}$$

A.9.2 Differentiation of a scalar-valued function with respect to a real vector

Let $\phi(\mathbf{x})$ be a complex differentiable scalar-valued function and $\mathbf{x}$ be a vector in $\mathbb{R}^N$. We define the *gradient* of $\phi(\mathbf{x})$, denoted $\partial\phi(\mathbf{x})/\partial\mathbf{x}$ or $\nabla\phi(\mathbf{x})$, as a vector in $\mathbb{R}^N$ such that its nth component is the partial derivative of $\phi(\mathbf{x})$ with respect to x_n, *i.e.*,

$$\left[\frac{\partial}{\partial\mathbf{x}}\phi(\mathbf{x})\right]_n = [\nabla\phi(\mathbf{x})]_n = \frac{\partial}{\partial x_n}\phi(\mathbf{x})\,. \tag{A.125}$$

From this definition it can be shown that

$$\frac{\partial}{\partial\mathbf{x}}\mathbf{a}^t\mathbf{x} = \mathbf{a}\,, \tag{A.126}$$

$$\frac{\partial}{\partial\mathbf{x}}\mathbf{x}^t\mathbf{A}\mathbf{x} = (\mathbf{A} + \mathbf{A}^t)\mathbf{x} = 2\mathbf{A}\mathbf{x} \qquad \text{if } \mathbf{A} \text{ is symmetric}\,, \tag{A.127}$$

where $\mathbf{a}$ is a constant $N \times 1$ vector, $\mathbf{A}$ is a constant $N \times N$ matrix, and both $\mathbf{a}$ and $\mathbf{A}$ can be complex.

We also define the derivative with respect to the transpose of a vector as the transpose of the derivative vector itself, so

$$\frac{\partial}{\partial\mathbf{x}^t}\phi(\mathbf{x}) = \left[\frac{\partial}{\partial\mathbf{x}}\phi(\mathbf{x})\right]^t\,. \tag{A.128}$$

If we regard $\partial\phi(\mathbf{x})/\partial\mathbf{x}$ as a column vector, then $\partial\phi(\mathbf{x})/\partial\mathbf{x}^t$ is a row vector.

The *Hessian matrix*, or matrix of second derivatives, is the $N \times N$ matrix with elements given by

$$\left[\frac{\partial^2}{\partial\mathbf{x}\partial\mathbf{x}^t}\phi(\mathbf{x})\right]_{mn} = \frac{\partial^2}{\partial x_m \partial x_n}\phi(\mathbf{x})\,. \tag{A.129}$$

This expression can be interpreted as the outer product (see Secs. A.2.6 and 1.3.7) of the column-vector operator $\partial/\partial\mathbf{x}$ with the row vector $\partial\phi(\mathbf{x})/\partial\mathbf{x}^t$. Some books use the notation ∇^2 for the Hessian, but that is too easily confused with the Laplacian; the Laplacian of a scalar function is a scalar, while the Hessian is a matrix. A better notation for the Hessian operator would be $\nabla\nabla^t$.

A.9.3 Differentiation with respect to real matrices

By analogy to (A.125), we define the derivative of the scalar function $\phi(\mathbf{A})$ with respect to the real $M \times N$ matrix $\mathbf{A}$ as the new $M \times N$ matrix $\partial\phi(\mathbf{A})/\partial\mathbf{A}$, with elements given by

$$\left[\frac{\partial\phi(\mathbf{A})}{\partial\mathbf{A}}\right]_{mn} = \frac{\partial\phi(\mathbf{A})}{\partial A_{mn}}\,. \tag{A.130}$$

An extensive list of identities for evaluating such matrix derivatives is given by Fukunaga (1990) and a very detailed treatment is given by Harville (1997). A few of the more interesting relations are listed here. All matrices are assumed to be real and conformable, but no other restrictions apply except those listed explicitly.

$$\frac{\partial}{\partial\mathbf{A}}\operatorname{tr}(\mathbf{BA}) = \mathbf{B}^t\,; \tag{A.131}$$

$$\frac{\partial}{\partial\mathbf{A}}\operatorname{tr}(\mathbf{UAVA}^t) = \mathbf{UAV} + \mathbf{U}^t\mathbf{AV}^t\,, \qquad (\mathbf{U} \text{ and } \mathbf{V} \text{ square})\,; \tag{A.132}$$

$$\frac{\partial}{\partial\mathbf{A}}\operatorname{tr}(\mathbf{A}^t\mathbf{BA}) = (\mathbf{B} + \mathbf{B}^t)\mathbf{A}\,, \qquad (\mathbf{B} \text{ square})\,; \tag{A.133}$$

$$\frac{\partial}{\partial A_{mn}}\det(\mathbf{A}) = \operatorname{cof}_{mn}(\mathbf{A})\,; \tag{A.134}$$

$$\frac{\partial\{\ln[\det(\mathbf{A})]\}}{\partial\mathbf{A}} = 2\mathbf{A}^{-1} - \operatorname{diag}(\mathbf{A}^{-1})\,, \qquad (\mathbf{A} \text{ symmetric, nonsingular})\,; \tag{A.135}$$

$$\frac{\partial[\det(\mathbf{A})]}{\partial\mathbf{A}} = \det(\mathbf{A})\left[2\mathbf{A}^{-1} - \operatorname{diag}(\mathbf{A}^{-1})\right]\,, \qquad (\mathbf{A} \text{ symmetric, nonsingular})\,, \tag{A.136}$$

where the notation $\operatorname{diag}(\mathbf{M})$ denotes the matrix $\mathbf{M}$ with its off-diagonal elements set to zero.

The following relations, taken from Fukunaga (1990), are useful in linear-discriminant analysis:

$$\frac{\partial}{\partial\mathbf{A}}\left\{\operatorname{tr}[(\mathbf{A}^t\mathbf{S}_2\mathbf{A})^{-1}(\mathbf{A}^t\mathbf{S}_1\mathbf{A})]\right\} = \frac{\partial}{\partial\mathbf{A}}\left\{\operatorname{tr}[(\mathbf{A}^t\mathbf{S}_1\mathbf{A})(\mathbf{A}^t\mathbf{S}_2\mathbf{A})^{-1}]\right\}$$

$$= -2\mathbf{S}_2\mathbf{A}(\mathbf{A}^t\mathbf{S}_2\mathbf{A})^{-1}(\mathbf{A}^t\mathbf{S}_1\mathbf{A})(\mathbf{A}^t\mathbf{S}_2\mathbf{A})^{-1} + 2\mathbf{S}_1\mathbf{A}(\mathbf{A}^t\mathbf{S}_2\mathbf{A})^{-1}\,. \tag{A.137}$$

A.9.4 Differentiation of a real function with respect to a complex scalar

In Secs. A.9.2 and A.9.3, the differentiation was performed with respect to vectors and matrices that were restricted to be real, but in optimization problems we often want to minimize a real, scalar-valued function of a complex $N \times 1$ vector $\mathbf{x}$. Minimization requires that the function be unchanged by an infinitesimal perturbation in either the real or imaginary part of any component of $\mathbf{x}$. Thus a total of $2N$ derivatives must vanish. In Sec. A.9.5 we shall develop techniques for dealing with such problems, but as a prelude in this section we take a look at one interpretation of differentiation with respect to a complex scalar x (soon to be one component of a complex vector).

We write the scalar x in terms of its real and imaginary parts, x' and x'' respectively, as

$$x = x' + ix'' . \tag{A.138}$$

The complex conjugate of x is given by

$$x^* = x' - ix'' . \tag{A.139}$$

The inverse relations are

$$x' = \tfrac{1}{2}(x + x^*) , \qquad x'' = \tfrac{1}{2i}(x - x^*) . \tag{A.140}$$

Next we define a complex derivative operator D by

$$D = \frac{\partial}{\partial x'} + i\frac{\partial}{\partial x''} . \tag{A.141}$$

It is important to realize that this operator takes individual partial derivatives with respect to x' and x'' and then combines them into a complex number. This is not the same as taking a total derivative with respect to the complex x, an operation discussed in detail in App. B. The operator D can be applied to real-valued or complex-valued functions of the complex x, while the usual complex derivative d/dz is defined only for complex-valued (in fact, analytic) functions.

The analogy of (A.141) to (A.125) should not be overlooked. If we think of x as a vector in the complex plane (see App. B) with components x' and x'' and corresponding unit vectors 1 and i, then D is a vector derivative like $\partial/\partial\mathbf{x}$ but with just two components.

From the definition of D, we have

$$D^* = \frac{\partial}{\partial x'} - i\frac{\partial}{\partial x''} ; \tag{A.142}$$

$$DD^* = \left(\frac{\partial}{\partial x'} + i\frac{\partial}{\partial x''}\right)\left(\frac{\partial}{\partial x'} - i\frac{\partial}{\partial x''}\right) = \frac{\partial^2}{\partial x'^2} + \frac{\partial^2}{\partial x''^2} . \tag{A.143}$$

Note that DD^* is the usual 2D Laplacian operator.

The following relations can be verified:

$$D[\phi(x) + \psi(x)] = D\phi(x) + D\psi(x) ; \tag{A.144}$$

$$D[\phi(x)\,\psi(x)] = \phi(x)\,D\phi(x) + \psi(x)\,D\phi(x) ; \tag{A.145}$$

$$D[\,|x|^n] = n|x|^{n-2}x . \tag{A.146}$$

These results are obvious extensions of familiar properties of real derivatives. On the other hand, if the operator D is applied to complex-valued functions, there are some surprises. For example,

$$Dx = \left(\frac{\partial}{\partial x'} + i\frac{\partial}{\partial x''}\right)(x' + ix'') = 0\,; \tag{A.147}$$

$$D^*x = \left(\frac{\partial}{\partial x'} - i\frac{\partial}{\partial x''}\right)(x' + ix'') = 2\,. \tag{A.148}$$

A symmetrized operator behaves more intuitively:

$$\tfrac{1}{2}(D + D^*)x = 1\,; \qquad \tfrac{1}{2}(D + D^*)x^n = nx^{n-1}\,. \tag{A.149}$$

To see the reasons for these peculiarities, we can regard (A.147) and (A.148) as specifying a coordinate transformation from (x', x'') to (x, x^*). The usual chain rule for differentiation then yields

$$\frac{\partial\phi(x',x'')}{\partial x} = \frac{\partial\phi(x',x'')}{\partial x'}\frac{\partial x'}{\partial x} + \frac{\partial\phi(x',x'')}{\partial x''}\frac{\partial x''}{\partial x} = \frac{1}{2}\frac{\partial\phi(x',x'')}{\partial x'} + \frac{1}{2i}\frac{\partial\phi(x',x'')}{\partial x''}\,. \tag{A.150}$$

Similarly,

$$\frac{\partial\phi(x',x'')}{\partial x}^* = \frac{\partial\phi(x',x'')}{\partial x'}\frac{\partial x'}{\partial x}^* + \frac{\partial\phi(x',x'')}{\partial x''}\frac{\partial x''}{\partial x}^* = \frac{1}{2}\frac{\partial\phi(x',x'')}{\partial x'} - \frac{1}{2i}\frac{\partial\phi(x',x'')}{\partial x''}\,. \tag{A.151}$$

If $\phi(x', x'')$ is real, $\partial\phi(x', x'')/\partial x^* = [\partial\phi(x', x'')/\partial x]^*$.

As an example, suppose

$$\phi(x', x'') = |x|^2 = xx^* = (x')^2 + (x'')^2\,. \tag{A.152}$$

Then, directly from (A.150), we have

$$\frac{\partial\phi(x',x'')}{\partial x} = \tfrac{1}{2}(2x') + \tfrac{1}{2i}(2x'') = x' - ix'' = x^*\,. \tag{A.153}$$

But note that this same result can be obtained more easily just by regarding x and x^* as independent variables, so that $\partial(xx^*)/\partial x = x^*$. By the same token,

$$\frac{\partial(xx^*)}{\partial x^*} = \tfrac{1}{2}(2x') - \tfrac{1}{2i}(2x'') = x' + ix'' = x\,. \tag{A.154}$$

In summary, we have the following interrelationships among various differential operators:

$$D = \left(\frac{\partial}{\partial x'} + i\frac{\partial}{\partial x''}\right) = 2\frac{\partial}{\partial x^*}\,; \tag{A.155}$$

$$D^* = \left(\frac{\partial}{\partial x'} - i\frac{\partial}{\partial x''}\right) = 2\frac{\partial}{\partial x}\,; \tag{A.156}$$

$$\tfrac{1}{2}(D + D^*) = \left(\frac{\partial}{\partial x^*} + \frac{\partial}{\partial x}\right) = \frac{\partial}{\partial x'}\,; \tag{A.157}$$

$$\tfrac{1}{2i}(D - D^*) = -i\left(\frac{\partial}{\partial x^*} - \frac{\partial}{\partial x}\right) = \frac{\partial}{\partial x''}\,. \tag{A.158}$$

The apparent paradox of (A.147) - (A.149) is now resolved. For example, (A.155) can be used to rewrite (A.147) as $2\partial x/\partial x^* = 0$, while (A.156) and (A.148) lead to $2\partial x/\partial x = 2$; both of these results are eminently reasonable if we regard x and x^* as independent variables.

A.9.5 Differentiation of a function with respect to a complex vector

We are now in a position to define derivatives with respect to a complex vector. Consider a scalar-valued function of N complex variables $x_1, ..., x_N$ or, equivalently, $2N$ real variables $x'_1, ..., x'_N$, $x''_1, ..., x''_N$. We can write this function as $\phi(\mathbf{x}', \mathbf{x}'')$, where $\mathbf{x}'$ is a vector of real parts and $\mathbf{x}''$ is a vector of imaginary parts, or simply as $\Phi(\mathbf{x})$, where $\mathbf{x} = \mathbf{x}' + i\mathbf{x}''$.

The vector derivatives with respect to real and imaginary parts are defined by

$$\left[\frac{\partial \Phi(\mathbf{x})}{\partial \mathbf{x}'}\right]_n = \frac{\partial \Phi(\mathbf{x})}{\partial x'_n}, \qquad \left[\frac{\partial \Phi(\mathbf{x})}{\partial \mathbf{x}''}\right]_n = \frac{\partial \Phi(\mathbf{x})}{\partial x''_n}. \tag{A.159}$$

Both $\partial\Phi(\mathbf{x})/\partial\mathbf{x}'$ and $\partial\Phi(\mathbf{x})/\partial\mathbf{x}''$ are $N \times 1$ column vectors.

Now we define the vector counterpart of D from (A.141) by

$$\nabla = \frac{\partial}{\partial \mathbf{x}'} + i\frac{\partial}{\partial \mathbf{x}''} \tag{A.160a}$$

or, equivalently,

$$[\nabla\Phi(\mathbf{x})]_n = \frac{\partial \Phi(\mathbf{x})}{\partial x'_n} + i\frac{\partial \Phi(\mathbf{x})}{\partial x''_n}. \tag{A.160b}$$

It is reasonable to call this operator ∇ since it is a natural generalization of the familiar real gradient operator. If we have a space with N real coordinates x_j, $j = 1, ..., N$, and associated orthonormal basis vectors $\mathbf{u}_j$, then

$$\nabla = \sum_{j=1}^{N} \mathbf{u}_j \frac{\partial}{\partial x_j}, \qquad (x_j, \mathbf{u}_j \text{ real}). \tag{A.161}$$

The operator defined by (A.160) has the same structure in $2N$ dimensions if we think of $\mathbf{u}_j$ and $i\mathbf{u}_j$ as comprising a set of $2N$ orthonormal unit vectors.

The vector counterpart of D^* is the adjoint of ∇, a row-vector operator defined such that

$$\nabla^\dagger = \frac{\partial}{\partial \mathbf{x}'^t} - i\frac{\partial}{\partial \mathbf{x}''^t}. \tag{A.162}$$

By extension of (A.155) and (A.156), we can also write

$$\nabla = 2\frac{\partial}{\partial \mathbf{x}^*} \quad \text{or} \quad [\nabla\Phi(\mathbf{x})]_n = 2\frac{\partial \Phi(\mathbf{x})}{\partial x_n^*}; \tag{A.163}$$

$$\nabla^\dagger = 2\frac{\partial}{\partial \mathbf{x}^t} \quad \text{or} \quad \left[\nabla^\dagger\Phi(\mathbf{x})\right]_n = 2\frac{\partial \Phi(\mathbf{x})}{\partial x_n}, \tag{A.164}$$

where $\partial/\partial x_n$ and $\partial/\partial x_n^*$ are to be interpreted according to (A.150) and (A.151).

The generalized Hessian is given by

$$\nabla\nabla^\dagger = \left(\frac{\partial}{\partial \mathbf{x}'} + i\frac{\partial}{\partial \mathbf{x}''}\right)\left(\frac{\partial}{\partial \mathbf{x}'^t} - i\frac{\partial}{\partial \mathbf{x}''^t}\right) \tag{A.165a}$$

or, equivalently,

$$\left[\nabla\nabla^\dagger\Phi(\mathbf{x})\right]_{nm} = 4\frac{\partial^2\Phi(\mathbf{x})}{\partial x_n^* \partial x_m}. \tag{A.165b}$$

Several useful results can be derived from these definitions. In what follows, $\mathbf{x}$ and $\mathbf{a}$ are $N \times 1$ column vectors and $\mathbf{A}$ is an $N \times N$ matrix; $\mathbf{A}$ and $\mathbf{a}$ may be complex but are independent of $\mathbf{x}$.

$$\nabla \mathbf{a}^\dagger \mathbf{x} = 0\,; \tag{A.166}$$

$$\nabla \mathbf{x}^\dagger \mathbf{a} = 2\mathbf{a}\,; \tag{A.167}$$

$$\nabla^\dagger \mathbf{x}^\dagger \mathbf{a} = 0\,; \tag{A.168}$$

$$\nabla^\dagger \mathbf{a}^\dagger \mathbf{x} = 2\mathbf{a}^\dagger\,; \tag{A.169}$$

$$\nabla \mathbf{x}^\dagger \mathbf{A}\mathbf{x} = 2\mathbf{A}\mathbf{x}\,; \tag{A.170}$$

$$\nabla^\dagger \mathbf{x}^\dagger \mathbf{A}\mathbf{x} = 2(\mathbf{x}^\dagger \mathbf{A})\,; \tag{A.171}$$

$$\nabla\nabla^\dagger \mathbf{x}^\dagger \mathbf{A}\mathbf{x} = 4\mathbf{A}\,. \tag{A.172}$$

A.10 TAYLOR EXPANSIONS

Many useful approximations are realized by expanding a scalar-valued function in a Taylor series and neglecting higher-order terms. The Taylor series for an analytic function of a single complex variable is discussed in App. B, Sec. B.3.4. Here we address the additional complications that arise when one expands a scalar-valued function of a real or complex vector. For background, we first review the Taylor expansion of a real function of a real scalar.

A.10.1 Real univariate Taylor series

If $\phi(x)$ is a real-valued function of a single real scalar and its first $K+1$ derivatives exist in an interval around $x = a$, then in this interval (Rade and Westergren, 1990)

$$\phi(x) = \sum_{k=0}^{K} \frac{\phi^{(k)}(a)}{k!}(x-a)^k + R_{K+1}\,, \tag{A.173}$$

where $\phi^{(k)}(x)$ is the kth derivative of $\phi(x)$, and R_{K+1} is a remainder term which satisfies

$$R_{K+1} = \left[\frac{\phi^{(K+1)}(b)}{(K+1)!}\right](x-a)^{K+1}\,, \tag{A.174}$$

for some b between a and x. If all derivatives exist in the interval, we can let $K \to \infty$, in which case $R_{K+1} \to 0$ and

$$\phi(x) = \sum_{k=0}^{\infty} \frac{\phi^{(k)}(a)}{k!}(x-a)^k\,. \tag{A.175}$$

This infinite series is known as the *Taylor series* for $\phi(x)$. The special case $a = 0$ yields the *Maclaurin series*.

Another useful form is obtained by a simple change of variables. Letting $a \to x'$ and $x - a \to a'$ (and then promptly dropping the primes), we have

$$\phi(x+a) = \sum_{k=0}^{\infty} \frac{\phi^{(k)}(x)}{k!} a^k\,, \tag{A.176}$$

provided, of course, that all of the derivatives exist. It is interesting to rewrite (A.176) in operator form as

$$\phi(x+a) = \exp\left[a\frac{d}{dx}\right]\phi(x)\,, \tag{A.177}$$

where the exponential differential operator is to be interpreted as

$$\exp\left[a\frac{d}{dx}\right] = \sum_{k=0}^{\infty}\frac{a^k}{k!}\frac{d^k}{dx^k}\,. \tag{A.178}$$

This operator is sometimes referred to as a *displacement operator* since it displaces the function $\phi(x)$ to $\phi(x+a)$.

A.10.2 Real multivariate Taylor series

One advantage of (A.178) is that it generalizes readily to the vector case. The multivariate Taylor expansion for a scalar-valued function of a real vector $\mathbf{x}$ is

$$\phi(\mathbf{x}+\mathbf{a}) = \exp\left(\mathbf{a}^t\frac{\partial}{\partial\mathbf{x}}\right)\phi(\mathbf{x}) = \phi(\mathbf{x}) + \left(\mathbf{a}^t\frac{\partial}{\partial\mathbf{x}}\right)\phi(\mathbf{x}) + \frac{1}{2}\left(\mathbf{a}^t\frac{\partial}{\partial\mathbf{x}}\right)\left(\mathbf{a}^t\frac{\partial}{\partial\mathbf{x}}\right)\phi(\mathbf{x}) + \ldots$$

$$= \phi(\mathbf{x}) + \mathbf{a}^t\frac{\partial\phi(\mathbf{x})}{\partial\mathbf{x}} + \tfrac{1}{2}\mathbf{a}^t\frac{\partial^2\phi(\mathbf{x})}{\partial\mathbf{x}\partial\mathbf{x}^t}\mathbf{a} + \ldots\,. \tag{A.179}$$

In component form, we have

$$\phi(\mathbf{x}+\mathbf{a}) = \phi(\mathbf{x}) + \sum_{i=1}^{N} a_i\frac{\partial}{\partial x_i}\phi(\mathbf{x}) + \frac{1}{2}\sum_{i=1}^{N}\sum_{j=1}^{N} a_i a_j\frac{\partial^2}{\partial x_i\partial x_j}\phi(\mathbf{x}) + \ldots\,. \tag{A.180}$$

For the real vectors being considered here, $\partial^2\phi(\mathbf{x})/\partial\mathbf{x}\partial\mathbf{x}^t$ is the Hessian matrix, and (A.179) shows that it is also the matrix of coefficients of second-order terms in the Taylor expansion.

In optimization problems, it is common to pick a direction specified by a unit vector $\hat{\mathbf{n}}$ and attempt to maximize or minimize a function $\phi(\mathbf{x})$ along this direction. Since the direction is fixed, the function can be represented by a *one-dimensional* Taylor series, even though it is a function of a vector (Gill *et al*, 1981, p. 53). The required expansion is a special case of (A.179):

$$\phi(\mathbf{x}+h\hat{\mathbf{n}}) = \exp\left(h\hat{\mathbf{n}}^t\frac{\partial}{\partial\mathbf{x}}\right)\phi(\mathbf{x})$$

$$= \phi(\mathbf{x}) + \hat{\mathbf{n}}^t\frac{\partial\phi(\mathbf{x})}{\partial\mathbf{x}}h + \tfrac{1}{2}\hat{\mathbf{n}}^t\frac{\partial^2\phi(\mathbf{x})}{\partial\mathbf{x}\partial\mathbf{x}^t}\hat{\mathbf{n}}h^2 + \ldots\,. \tag{A.181}$$

This result provides interpretations of the gradient and Hessian. For any direction $\hat{\mathbf{n}}$, $\hat{\mathbf{n}}^t[\partial\phi(\mathbf{x})/\partial\mathbf{x}]$ is the 1D derivative in that direction and $\hat{\mathbf{n}}^t[\partial^2\phi(\mathbf{x})/\partial\mathbf{x}\partial\mathbf{x}^t]\hat{\mathbf{n}}$ is the corresponding second derivative or curvature.

A.10.3 Complex univariate Taylor series

Consider a scalar-valued function of a complex scalar $x = x' + ix''$. One way to write a Taylor expansion for this function, denoted $\Phi(x)$ or $\phi(x', x'')$, is to consider it as a function of a real 2D vector with components x' and x''. Then (A.179) yields

$$\phi(x' + a', x'' + a'') = \exp\left(a'\frac{\partial}{\partial x'} + a''\frac{\partial}{\partial x''}\right)\phi(x', x'')$$

$$= \phi(x', x'') + \left(a'\frac{\partial}{\partial x'} + a''\frac{\partial}{\partial x''}\right)\phi(x', x'')$$

$$+ \frac{1}{2}\left(a'\frac{\partial}{\partial x'} + a''\frac{\partial}{\partial x''}\right)\left(a'\frac{\partial}{\partial x'} + a''\frac{\partial}{\partial x''}\right)\phi(x', x'') + \ldots . \tag{A.182}$$

This expansion works for any function of x' and x'' so long as all derivatives with respect to x' and x'' exist.

In terms of x and x^*, we can write this expansion as

$$\Phi(x + a) = \exp\left(a\frac{\partial}{\partial x} + a^*\frac{\partial}{\partial x^*}\right)\Phi(x) = \sum_{k=0}^{\infty}\frac{1}{k!}\left(a\frac{\partial}{\partial x} + a^*\frac{\partial}{\partial x^*}\right)^k\Phi(x), \tag{A.183}$$

which, by use of (A.155) and (A.156) can also be written

$$\Phi(x + a) = \exp\left[\tfrac{1}{2}(aD^* + a^*D)\right]\Phi(x). \tag{A.184}$$

A.10.4 Complex multivariate Taylor series

A scalar-valued function of an ND complex vector $\mathbf{x}$ can be regarded as a function $\phi(\mathbf{x}', \mathbf{x}'')$ of $2N$ variables, so (A.182) generalizes to

$$\phi(\mathbf{x}' + \mathbf{a}', \mathbf{x}'' + \mathbf{a}'') = \exp\left(\mathbf{a}'^t\frac{\partial}{\partial \mathbf{x}'} + \mathbf{a}''^t\frac{\partial}{\partial \mathbf{x}''}\right)\phi(\mathbf{x}', \mathbf{x}''). \tag{A.185}$$

Similarly, with the help of (A.160), the generalization of (A.184) is

$$\Phi(\mathbf{x} + \mathbf{a}) = \exp\left[\tfrac{1}{2}(\mathbf{a}\nabla^\dagger + \mathbf{a}^\dagger\nabla)\right]\Phi(\mathbf{x})$$

$$= \Phi(\mathbf{x}) + \left[\tfrac{1}{2}(\mathbf{a}\nabla^\dagger + \mathbf{a}^\dagger\nabla)\right]\Phi(\mathbf{x}) + \tfrac{1}{2}\left[\tfrac{1}{2}(\mathbf{a}\nabla^\dagger + \mathbf{a}^\dagger\nabla)\right]\left[\tfrac{1}{2}(\mathbf{a}\nabla^\dagger + \mathbf{a}^\dagger\nabla)\right]\Phi(\mathbf{x}) + \ldots . \tag{A.186}$$

A.11 MATRIX AND VECTOR INEQUALITIES

A.11.1 Classical inequalities

An important relation with many applications in this book is the Cauchy-Schwarz inequality (Johnson and Wichern, 1988, p. 63), which states that

$$|\mathbf{a}^\dagger\mathbf{b}|^2 \le (\mathbf{a}^\dagger\mathbf{a})(\mathbf{b}^\dagger\mathbf{b}). \tag{A.187}$$

The equality holds if and only if $\mathbf{b} = c\mathbf{a}$, where c is a scalar. One interpretation of this inequality can be seen by defining the angle between $\mathbf{a}$ and $\mathbf{b}$ by

$$\cos\theta_{ab} = \frac{\mathbf{a}^\dagger\mathbf{b}}{\sqrt{(\mathbf{a}^\dagger\mathbf{a})(\mathbf{b}^\dagger\mathbf{b})}}. \tag{A.188}$$

If $\mathbf{a}$ and $\mathbf{b}$ are real, the Cauchy-Schwarz inequality states that $\cos^2\theta_{ab} \le 1$, with equality holding if and only if $\mathbf{a}$ and $\mathbf{b}$ are parallel.

Additional useful inequalities arise when other vector norms are considered. The $\mathbb{L}_p$ norm is defined by

$$||\mathbf{x}||_p = \left[\sum_{n=1}^{N} |x_n|^p\right]^{1/p}. \tag{A.189}$$

In terms of this norm, *Minkowski's inequality* states that

$$||\mathbf{x}+\mathbf{y}||_p \le ||\mathbf{x}||_p + ||\mathbf{y}||_p, \tag{A.190}$$

and *Hölder's inequality* for Hadamard products (see Sec. A.2.8) states that

$$||\mathbf{x}\odot\mathbf{y}||_1 \le ||\mathbf{x}||_p||\mathbf{y}||_q, \qquad \frac{1}{p}+\frac{1}{q} = 1, \qquad p,q \ge 1. \tag{A.191}$$

A.11.2 Inequalities involving definite matrices

It is common in the literature to encounter statements such as $\mathbf{A} > \mathbf{B}$ or $\mathbf{A} \ge \mathbf{B}$, where $\mathbf{A}$ and $\mathbf{B}$ are two $N\times N$ matrices. These statements must be interpreted with care. Unlike the equality sign defined in (A.1), an inequality sign applied to matrices does *not* hold on an element-by-element basis. Rather, the statement $\mathbf{A} > \mathbf{B}$ means that for positive-definite matrices $\mathbf{A}$ and $\mathbf{B}$, $\mathbf{A}-\mathbf{B}$ is positive-definite, while $\mathbf{A} \ge \mathbf{B}$ means that $\mathbf{A}-\mathbf{B}$ is positive-semidefinite. The relation $\mathbf{A} > \mathbf{B}$ holds if and only if $\mathbf{x}^\dagger\mathbf{A}\mathbf{x} > \mathbf{x}^\dagger\mathbf{B}\mathbf{x}$ for all nonzero $\mathbf{x}$. Similar interpretations hold for $<$ and $\le$. This convention is called the *Loewner ordering* of the matrices $\mathbf{A}$ and $\mathbf{B}$.

With the Loewner convention, some very powerful inequalities can be adduced for definite matrices (Pilz, 1991). A few useful ones are given here.

If $\mathbf{A} > \mathbf{B}$, then

(*i*) $\mathbf{B}^{-1} > \mathbf{A}^{-1}$;
(*ii*) $\text{tr}(\mathbf{AC}) > \text{tr}(\mathbf{BC})$ for all positive-definite (and conformable) $\mathbf{C}$;
(*iii*) $\det(\mathbf{A}) > \det(\mathbf{B})$;
(*iv*) $\lambda_i(\mathbf{A}) > \lambda_i(\mathbf{B})$ for all i.

In these results, $>$ can be consistently replaced by $<$, $\le$ or $\ge$ and the results still hold. Inequality (*iv*) assumes that the eigenvalues are ordered so that $\lambda_1 \ge \lambda_2 \ge \lambda_3 \ldots$. Thus the entire eigenvalue spectrum of $\mathbf{A}$ must lie above that for $\mathbf{B}$ if $\mathbf{A} > \mathbf{B}$.

Another inequality involving a positive-definite matrix is the *extended Cauchy-Schwarz inequality*. It states that

$$|\mathbf{a}^\dagger\mathbf{b}|^2 \le (\mathbf{a}^\dagger\mathbf{K}\mathbf{a})(\mathbf{b}^\dagger\mathbf{K}^{-1}\mathbf{b}), \tag{A.192}$$

where $\mathbf{K}$ is positive-definite. Equality holds if and only if $\mathbf{b} = c\mathbf{K}\mathbf{a}$, where c is a scalar.

A.11.3 Matrix norms

The norm of an $N \times N$ matrix $\mathbf{A}$ can be defined as

$$||\mathbf{A}|| = \max_{\mathbf{x} \neq 0} \frac{||\mathbf{Ax}||}{||\mathbf{x}||}, \tag{A.193}$$

where $\mathbf{x}$ is an $N \times 1$ vector and $||\cdot||$ on the right denotes any desired vector norm. Thus it follows by definition that

$$||\mathbf{Ax}|| \leq ||\mathbf{A}||\,||\mathbf{x}||. \tag{A.194}$$

To be more precise about the operational form of the matrix norm, we must specify the associated vector norm to be used in (A.193). Three common choices for the latter are $\mathbb{L}_1$, $\mathbb{L}_\infty$ and $\mathbb{L}_2$, and the corresponding matrix norms are (Usmani, 1987)

$$||\mathbf{A}||_1 = \max_j \sum_i |A_{ij}|\ ; \tag{A.195}$$

$$||\mathbf{A}||_\infty = \max_i \sum_j |A_{ij}|\ ; \tag{A.196}$$

$$||\mathbf{A}||_2 = \sqrt{\text{maximum eigenvalue of } \mathbf{A}^\dagger\mathbf{A}}\ . \tag{A.197}$$

A common term for $||\mathbf{A}||_2$ is *spectral norm*. Each of these matrix norms leads to a useful inequality when substituted into (A.194).

A more familiar-looking matrix norm is the *usual norm*, defined via

$$||\mathbf{A}||^2 = \sum_{i,j} |A_{ij}|^2\,. \tag{A.198}$$

Associated with this norm is a scalar product of two matrices, defined as (Harville, 1997)

$$\mathbf{A}\cdot\mathbf{B} = \mathrm{tr}\{\mathbf{A}^\dagger\mathbf{B}\}\,. \tag{A.199}$$

In terms of this scalar product, the matrix version of the Cauchy-Schwarz inequality is (Harville, 1997)

$$|\mathbf{A}\cdot\mathbf{B}|^2 \leq ||\mathbf{A}||^2\,||\mathbf{B}||^2\,. \tag{A.200}$$

The equality holds if and only if $\mathbf{A} = \mathbf{0}$, $\mathbf{B} = \mathbf{0}$ or $\mathbf{A} = c\mathbf{B}$ for some scalar c.

APPENDIX B
Complex Variables

This appendix is intended as a brief survey of complex numbers, functions of a single complex variable and complex integration. Results needed in the main text are presented without proof and with relatively little discussion. For more details, the reader is referred to any of the many introductory texts on complex variables; particularly lucid ones are Carrier *et al.*, (1966), Churchill *et al.* (1974), Ahlfors (1979) and Wunsch (1994). More general texts with good sections on complex variables include Morse and Feshbach (1953), Arfken and Weber (1995) and Friedman (1991). Proofs of all statements made in this appendix can be found in these references.

B.1 COMPLEX ALGEBRA

B.1.1 What is a complex number?

A complex number is an ordered pair of real numbers obeying certain algebraic rules. If z_1 denotes the ordered pair (x_1, y_1) and z_2 denotes the pair (x_2, y_2), addition and multiplication are defined as follows:

$$\text{Addition:} \qquad z_3 = z_1 + z_2 \Rightarrow x_3 = x_1 + x_2\,, \qquad y_3 = y_1 + y_2 \tag{B.1}$$

$$\text{Multiplication:} \qquad z_3 = z_1 z_2 \Rightarrow x_3 = x_1 x_2 - y_1 y_2\,, \qquad y_3 = x_1 y_2 + x_2 y_1\,, \tag{B.2}$$

where $z_3 = (x_3, y_3)$. Subtraction and division are defined as the inverses of addition and multiplication, respectively. The usual algebraic laws (commutative, associative and distributive) are valid for complex numbers.

Equality of two complex numbers means equality for corresponding members of the pairs:

$$\text{Axiom of equality:} \qquad z_1 = z_2 \Rightarrow x_1 = x_2 \qquad \text{and} \qquad y_1 = y_2 \,. \tag{B.3}$$

In particular,

$$\text{Complex zero:} \qquad z_1 = 0 \Rightarrow x_1 = 0 \qquad \text{and} \qquad y_1 = 0 \,. \tag{B.4}$$

Thus 0 is understood to mean the ordered pair (0, 0).

B.1.2 Representations

A conventional representation of a complex number $z = (x, y)$ is:

$$z = x + iy \,, \tag{B.5}$$

where $i = \sqrt{-1}$. Since $\sqrt{-1}$ was once thought to be nonexistent,[1] y is referred to as the imaginary part of z, denoted $\text{Im}(z)$, while x is the real part or $\text{Re}(z)$.

The representation (B.5) is a compact way of summarizing the algebraic properties of complex numbers. It is easy to verify, for example, that the multiplication rule follows from (B.5) along with the usual rule for multiplying polynomials.

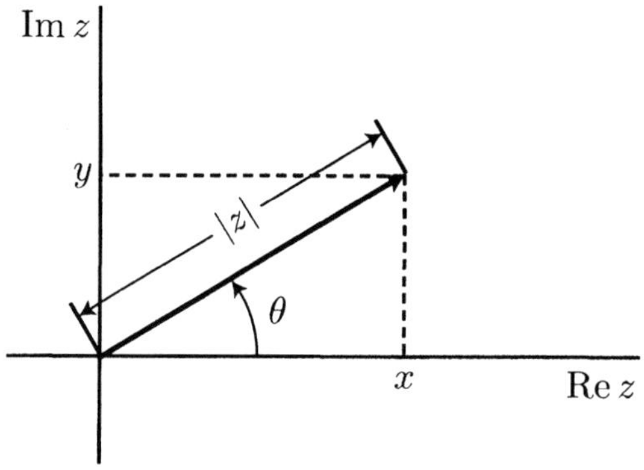

Fig. B.1 Representation of a complex number as a vector in a plane.

A graphical way of portraying a complex number is as a point in a plane (see Fig. B.1), with x and y as Cartesian coordinates (x, y). Equivalently, the complex number can be regarded as a 2D vector from the origin to the point (x, y). Though often attributed to Gauss, this representation apparently predated him (Boyer and Merzbach, 1989, p. 507).

The Argand or polar representation of z is

$$z = |z| \cos\theta + i|z| \sin\theta \,, \tag{B.6}$$

where $|z|$ and θ are known, respectively, as the *modulus* and *argument* of z. The notation $\theta = \arg(z)$ is common. It follows easily from (B.6) and Fig. B.1 that

$$|z|^2 = x^2 + y^2 \qquad \text{and} \qquad \arg(z) = \theta = \tan^{-1}\left(\frac{y}{x}\right) \,. \tag{B.7}$$

[1] Apparently the first to suggest the existence of an imaginary number was an obscure fifteenth-century French mathematician Nicolas Chuquet. For a fascinating history of the development of complex numbers, see Boyer and Merzbach (1989).

B.1.3 Complex exponentials and DeMoivre's theorem

The complex exponential $\exp(i\theta)$ can be defined in terms of the same Taylor series used for real variables. Recognizing that the Taylor series for $\exp(i\theta)$ equals the Taylor series for $\cos\theta$ plus i times the Taylor series for $\sin\theta$, we have

$$e^{i\theta} = \cos\theta + i\sin\theta\,. \tag{B.8}$$

With (B.8), representation (B.6) for a general complex number takes the more compact form,

$$z = |z|e^{i\theta}\,. \tag{B.9}$$

This form is especially handy for multiplication and division of complex numbers. If $z_1 = |z_1|\exp(i\theta_1)$ and $z_2 = |z_2|\exp(i\theta_2)$, then

$$z_1 z_2 = |z_1||z_2|\, e^{i(\theta_1+\theta_2)}\,, \qquad \frac{z_1}{z_2} = \frac{|z_1|}{|z_2|}\, e^{i(\theta_1-\theta_2)}\,. \tag{B.10}$$

From this result we can easily derive DeMoivre's theorem, which states that

$$z^n = [\,|z|(\cos\theta + i\sin\theta)]^n = \left[\,|z|\, e^{i\theta}\right]^n = |z|^n(\cos n\theta + i\sin n\theta)\,. \tag{B.11}$$

B.1.4 Complex conjugate

The complex conjugate of $z = x + iy$ is defined by

$$z^* = x - iy = |z|\, e^{-i\theta} = |z|(\cos\theta - i\sin\theta)\,. \tag{B.12}$$

The following properties follow from this definition:

$$(z^*)^* = z\,, \qquad zz^* = |z|^2\,, \qquad (z_1 \pm z_2)^* = z_1^* \pm z_2^*\,,$$
$$(z_1 z_2)^* = z_1^* z_2^*\,, \qquad \left(\frac{z_1}{z_2}\right)^* = \frac{z_1^*}{z_2^*}\,. \tag{B.13}$$

The real and imaginary parts of z are given in terms of z and z^* by

$$x = \tfrac{1}{2}(z + z^*)\,, \qquad y = \tfrac{1}{2i}(z - z^*)\,. \tag{B.14}$$

With (B.6), (B.8) and (B.14), we obtain the Euler formulas,

$$\cos\theta = \frac{e^{i\theta} + e^{-i\theta}}{2}\,, \qquad \sin\theta = \frac{e^{i\theta} - e^{-i\theta}}{2i}\,. \tag{B.15}$$

B.1.5 Roots of unity

An Nth root of unity is a number z such that $z^N = 1$, where N is an integer. If we restricted z to be real, the only solutions to this equation would be $z = 1$ for N odd or $z = \pm 1$ for N even. With complex numbers there are other possibilities. Consider the complex number W_N defined by

$$W_N = e^{-2\pi i/N}\,. \tag{B.16}$$

Raising W_N to the Nth power gives

$$(W_N)^N = (e^{-2\pi i/N})^N = e^{-2\pi i} = \cos 2\pi - i\sin 2\pi = 1\,. \tag{B.17}$$

Thus W_N is indeed an Nth root of unity, but it is not the only one. We can get other roots by raising W_N to other integer powers. To show that $[W_N]^k$ is an Nth root of unity for any integer k, note that

$$\left[(W_N)^k\right]^N = \left[(e^{-2\pi i/N})^k\right]^N = e^{-2\pi ik} = \cos 2\pi k - i\sin 2\pi k = 1\,. \tag{B.18}$$

The full set of Nth roots of unity is $\{\exp(-2\pi ik/N), k = 0, 1, ..., N-1\}$ since k and $k+N$ define the same complex number. There are always exactly N distinct N^{th} roots of unity.

B.2 FUNCTIONS OF A COMPLEX VARIABLE

A function $f(z)$ is a mapping from the complex number $z = x + iy$ to the complex number $f(z) = u + iv$, or equivalently from a point in the x-y plane to a point in the u-v plane. We thus write

$$f(z) = u(x,y) + iv(x,y)\,. \tag{B.19}$$

If a single point (x,y) maps to a single (u,v) or if two or more (x,y) points map to the same (u,v), we say the function is *single-valued.* If a single (x,y) maps to two or more (u,v), the function is *multiple-valued* or *multivalued.* Examples of single-valued functions include z^2, $1/z$ and $\cos z$, while examples of multivalued functions include $\sqrt{z}$, $\arg(z)$ and $\cos^{-1}(z)$.

B.2.1 Limits

The statement,

$$\lim_{z\to z_0} f(z) = w_0\,, \tag{B.20}$$

means that for every positive real number ϵ there exists a positive real number δ such that

$$|f(z) - w_0| < \epsilon \qquad \text{when} \qquad |z - z_0| < \delta \quad (z \neq z_0)\,. \tag{B.21}$$

In other words, $f(z)$ can be made arbitrarily close to w_0 by taking z sufficiently close to z_0, except possibly at the point z_0 itself.

A very important consequence of this definition is that when a limit of $f(z)$ exists at any point z_0, it must be unique. The value of the limit cannot depend on the direction of approach to z_0 in the complex plane. We can write

$$\lim_{z\to z_0} f(z) = \lim_{\substack{x\to x_0\\ y\to y_0}} u(x,y) + i \lim_{\substack{x\to x_0\\ y\to y_0}} v(x,y)\,, \tag{B.22}$$

and the limits on x and y can be taken in either order.

Consider two different functions $f(z)$ and $F(z)$ and suppose that both of them have limits as $z \to z_0$, *i.e.*,

$$\lim_{z\to z_0} f(z) = w_0\,, \qquad \lim_{z\to z_0} F(z) = W_0\,. \tag{B.23}$$

Then the following properties hold:

$$\lim_{z\to z_0} [f(z) + F(z)] = w_0 + W_0\,, \qquad \lim_{z\to z_0} [f(z)\,F(z)] = w_0 W_0\,,$$

$$\lim_{z\to z_0}\left[\frac{f(z)}{F(z)}\right]=\frac{w_0}{W_0} \qquad \text{if} \qquad W_0\neq 0\,. \tag{B.24}$$

If $P(z)$ is a polynomial of any finite order,

$$P(z)=a_0+a_1z+a_2z^2+\ldots+a_Nz^N\,, \tag{B.25}$$

then

$$\lim_{z\to z_0}P(z)=P(z_0)\,. \tag{B.26}$$

B.2.2 Continuity

A single-valued function $f(z)$ is said to be *continuous* at z_0 if and only if

$$f(z_0)\ \text{exists}\,, \qquad \lim_{z\to z_0}f(z)\ \text{exists}\,, \qquad \text{and} \qquad \lim_{z\to z_0}f(z)=f(z_0)\,. \tag{B.27}$$

These three conditions imply that $f(z)$ is defined in some neighborhood of z_0.

From the theorems above on limits, it can be shown that if two functions are continuous, their sum and product are continuous, and their quotient is continuous except for those values of z where the denominator vanishes. Every polynomial $P(z)$ is continuous at all z.

B.2.3 Derivatives

The derivative of a function of a complex variable is defined in the same manner as for a function of a real variable:

$$f'(z)=\lim_{\Delta z\to 0}\frac{f(z+\Delta z)-f(z)}{\Delta z}\,. \tag{B.28}$$

The function $f(z)$ is differentiable in any region where this limit exists. Since a limit must be unique if it exists, the derivative must be independent of the direction of Δz in the complex plane. An example of a function that is continuous but not differentiable (except at $z=0$) is $|z|^2$.

From the definition (B.28), the following rules for differentiation can be proved:

$$\frac{d}{dz}C=0\,, \qquad \frac{d}{dz}[Cf(z)]=C\left[\frac{df(z)}{dz}\right]\,, \qquad \frac{d}{dz}(z^n)=nz^{n-1}\,, \tag{B.29}$$

where C is a constant and n is a positive integer. If $f_1(z)$ and $f_2(z)$ are two functions whose derivatives exist, then

$$\frac{d}{dz}[f_1(z)+f_2(z)]=\frac{df_1(z)}{dz}+\frac{df_2(z)}{dz}\,,$$

$$\frac{d}{dz}[f_1(z)\,f_2(z)]=f_1(z)\frac{df_2(z)}{dz}+f_2(z)\frac{df_1(z)}{dz}\,,$$

$$\frac{d}{dz}\left[\frac{f_1(z)}{f_2(z)}\right]=\frac{f_2(z)\frac{df_1(z)}{dz}-f_1(z)\frac{df_2(z)}{dz}}{[f_2(z)]^2}\,, \qquad (f_2(z)\neq 0)\,. \tag{B.30}$$

B.2.4 Analytic functions

A function $f(z)$ is said to be *analytic* at z_0 if its derivative exists at every point in some neighborhood of z_0 (including the point z_0 itself). The terms *regular* and *holomorphic* are synonymous with analytic. If $f(z)$ is analytic at every point in the neighborhood of z_0 but not at z_0 itself, z_0 is called a *singular point* or *singularity* of $f(z)$. If $f(z)$ is analytic at some point z_0, then *all* derivatives of $f(z)$ can be shown to exist at that point.

A function $f(z)$ is analytic at ∞ if $f(1/z)$ is analytic at 0. A function that is analytic everywhere (except possibly at ∞) is called *entire*. For example, a polynomial is entire.

B.2.5 Cauchy-Riemann conditions

Analyticity of $f(z)$ imposes stringent requirements on its real and imaginary parts, as defined in (B.19). Let $u(x, y)$ and $v(x, y)$ and their partial derivatives with respect to x and y be single-valued and continuous in some neighborhood of a point $z_0 = x_0 + iy_0$. Then a necessary and sufficient condition that $f(z)$ be analytic at that point is that the *Cauchy-Riemann conditions* be satisfied:

$$\frac{\partial u(x,y)}{\partial x} = \frac{\partial v(x,y)}{\partial y}, \qquad \frac{\partial u(x,y)}{\partial y} = -\frac{\partial v(x,y)}{\partial x}. \tag{B.31}$$

As a corollary of the Cauchy-Riemann conditions, $u(x, y)$ and $v(x, y)$ are *harmonic functions*; *i.e.*, they satisfy the 2D Laplace equation:

$$\nabla^2 u(x,y) = \nabla^2 v(x,y) = 0. \tag{B.32}$$

Equation (B.32) shows that only certain functions can be used as the real or imaginary part of an analytic function, while (B.31) shows that the real and imaginary parts cannot be chosen independently.

As examples, z^n (n an integer) and $\cos z$ are entire (analytic for all z), but $|z|$ and z^* are not analytic anywhere.

B.2.6 Maxima, minima and zeros

For real functions of a real variable, maxima and minima are commonplace and easy to understand. The situation is very different with functions of complex variables. Since $f(z)$ is usually itself complex, it consists of two numbers $(u(z), v(z))$ which must be reduced to a single number if we are to even inquire about maxima and minima; the natural way to do this is to form the modulus $|f(z)|$. Motivated by our experience with real functions, we might then attempt to find the values of z for which $|f(z)|$ is a maximum or minimum. However, if $f(z)$ is an analytic function, an important theorem known as the *maximum-modulus theorem* states that no maximum can exist!

More precisely, if $f(z)$ is analytic within a circle of radius R centered on z_0, then

$$|f(z_0)| \leq M, \tag{B.33}$$

where M is the maximum value of $|f(z)|$ on that circle. Since $|f(z)|$ at any point z_0 is less than or equal to its value at some point on a circle surrounding z_0, the point z_0 cannot be a maximum of the modulus. If $f(z)$ is an entire function, we can extend the radius of the circle to infinity and conclude that $|f(z)|$ cannot have a maximum except at $z = \infty$.

As a corollary of the maximum-modulus theorem, if $f(z)$ is analytic and bounded [$|f(z)| \leq K$, K a constant] for all z, then $f(z)$ must be a constant. In other words, the only bounded, entire function is a constant. This result is known as *Liouville's theorem.*

An extension of the maximum-modulus theorem applies to all derivatives of an analytic function (which are themselves analytic). If $f(z)$ is analytic at z_0, then

$$|f^{(n)}(z_0)| \leq \frac{Mn!}{R^n}, \tag{B.34}$$

where M and R have the same meaning as above. Thus none of the derivatives of an analytic function can have maxima.

With real functions, the difference between maxima and minima is of little consequence; if $f(x)$ has a maximum at $x = x_0$, then $-f(x)$ has a minimum there. Complex functions are different. Though the modulus of an analytic function cannot have a maximum, it can indeed have a minimum. For example, if $f(z) = z^2$, $|f(z)|$ is minimum at $z = 0$. More generally, $|f(z)|$ takes on its minimum value of zero at all points where $f(z) = 0$.

However, the zeros of analytic functions also have some counterintuitive properties. They can occur only at isolated points and not along lines or areas of the complex plane. More precisely, unless a function is identically zero, about each point where the function is analytic there must be a neighborhood throughout which the function is nonzero, except possibly at the point itself. As a corollary, if $f(z)$ is analytic and zero over a finite line or area, it must be zero everywhere.

B.3 COMPLEX INTEGRATION

With real integrals, the value of a definite integral is fully specified by the function and the limits of integration. With complex integrals, however, the specific line or path of integration in the complex plane can also be important. When this path is a closed contour around some region R, we refer to the integral as a contour integral. Some of the most powerful results of complex analysis relate to contour integrals.

B.3.1 Definition of a line integral

By analogy to definite integrals of real functions, the integral of a complex function $f(z)$ along a line L from $z = \alpha$ to $z = \beta$ is defined by dividing the line into N small segments, adding up the contributions from each segment, and letting $N \to \infty$. If the jth segment extends from z_j to z_{j+1} and we define $\Delta_j z$ as $z_{j+1} - z_j$, then the integral is defined by

$$\int_L dz\; f(z) = \lim_{N\to\infty} \sum_{j=1}^{N} f(z_j')\, \Delta_j z\,, \tag{B.35}$$

where z_j' is any point on L between z_j and z_{j+1}. Equivalently, with (B.5) and (B.19), we can also write

$$\int_L dz\ f(z) = \int_L (dx + idy)\ [u(x,y) + iv(x,y)]\ , \tag{B.36}$$

where dx and dy are not independent but are constrained by the condition that (x, y) must lie on the line L.

B.3.2 Integrals of analytic functions

Though in general the line integral from $z = \alpha$ to $z = \beta$ depends on the specific line connecting the two points, an important exception occurs if $f(z)$ is analytic. If L and L' are two lines from α to β, R is the region between L and L' (including L and L' themselves), and $f(z)$ is analytic in R, then

$$\int_L dz\ f(z) = \int_{L'} dz\ f(z)\,. \tag{B.37}$$

This result gives us the freedom to deform the path of integration at will without changing the value of the integral, so long as the path remains entirely within the region of analyticity.

If the starting and ending points of a line integral are the same ($\alpha = \beta$), the path of integration is a closed loop referred to as a *contour*. By convention, contours are assumed to be traversed in a counter-clockwise direction unless otherwise stated.

A consequence of (B.37) known as the *Cauchy-Goursat theorem* says that the integral of an analytic function around a closed path is zero. That is, if $f(z)$ is analytic inside and on contour C, then

$$\oint_C dz\ f(z) = 0\,. \tag{B.38}$$

B.3.3 Integrals of z^n

Consider the integral of the function z^n around a closed contour C which encloses the origin. If $n \geq 0$, z^n is analytic and the integral is immediately zero by (B.38). If $n < 0$, on the other hand, z^n is singular at the origin and (B.38) is not applicable. We can, however, use (B.37) to deform the contour, allowing it to shrink to a circle of radius ϵ about the origin. On this new contour, $z = \epsilon\exp(i\theta)$ and $dz = izd\theta$, so we have

$$\oint_C dz\ z^n = i\epsilon^{n+1} \int_0^{2\pi} d\theta\ e^{i(n+1)\theta}\ . \tag{B.39}$$

The remaining integral is easily performed with the help of DeMoivre's theorem, (B.11), and we find

$$\oint_C dz\ z^n = \begin{cases} 2\pi i & \text{if } n = -1 \\ 0 & \text{otherwise} \end{cases}\ . \tag{B.40}$$

B.3.4 Cauchy integral formula

Suppose $f(z)$ is analytic inside and on contour C, and that the point z_0 is inside C. Then the celebrated *Cauchy integral formula* states that

$$\frac{1}{2\pi i}\oint_C dz\,\frac{f(z)}{z-z_0} = f(z_0)\,. \tag{B.41}$$

Without going through a formal proof, we can make this result plausible by using (B.37) to again justify shrinking the contour to a circle of radius ϵ around the point z_0. As $\epsilon \to 0$, the analytic function $f(z) \to f(z_0)$, a constant that can be taken out of the integral. Application of (B.40) then yields (B.41).

An extension of the Cauchy integral formula states that

$$\frac{n!}{2\pi i}\oint_C dz\,\frac{f(z)}{(z-z_0)^{n+1}} = f^{(n)}(z_0)\,, \tag{B.42}$$

where $f^{(n)}(z_0)$ denotes the nth derivative evaluated at z_0.

B.3.5 Series representations of complex functions

We have already seen that a polynomial of finite order is an analytic function. A *Taylor series* can be regarded as a polynomial of infinite order, and it too represents an analytic function. Specifically, if $f(z)$ is analytic at $z = z_0$, it can be expanded in a Taylor series of the form

$$f(z) = \sum_{n=0}^{\infty} a_n (z-z_0)^n\,, \tag{B.43}$$

where the coefficients are given by

$$a_n = \frac{1}{n!} f^{(n)}(z_0) = \frac{1}{2\pi i}\oint_C dz'\,\frac{f(z')}{(z'-z_0)^{n+1}}\,. \tag{B.44}$$

Any closed contour C can be used here, so long as $f(z)$ is analytic inside and on C. The Taylor series converges to $f(z)$ within a circle of radius R about z_0, where R is the distance from z_0 to the nearest singularity of $f(z)$. A *Maclaurin series* is the special case of a Taylor series when $z_0 = 0$.

If $f(z)$ is singular at z_0, it can still be represented as a power series in $z - z_0$, but negative powers are required. Suppose $f(z)$ is analytic in an annular region defined by two circles of radius R_1 and R_2 ($R_2 > R_1 > 0$) centered on z_0. Then a series representation of $f(z)$, called a *Laurent series*, is given by

$$f(z) = \sum_{n=-\infty}^{\infty} a_n (z-z_0)^n\,, \tag{B.45}$$

where the coefficients are

$$a_n = \frac{1}{2\pi i}\oint_C dz'\,\frac{f(z')}{(z'-z_0)^{n+1}}\,. \tag{B.46}$$

Here the contour C can be any closed contour in the annular region, and the Laurent series will converge to $f(z)$ in that region. If $f(z)$ has no other singularities besides the one at z_0 inside the inner circle, radius R_1 can approach zero, and the Laurent series will converge to $f(z)$ at all points inside the circle of radius R_2, save only z_0 itself.

B.3.6 Poles and residues

If the Laurent series for $f(z)$ around z_0 terminates at a maximum negative power of $-m$ (*i.e.*, $a_n = 0$ for $n < -m$) then the singularity at point z_0 is a *pole of order m*. A pole of order 1 is called a *simple pole*. If the series must be carried to $n = -\infty$, then z_0 is an *essential singularity*. If $a_n = 0$ for all $n < 0$, then the Laurent series reduces to the Taylor series and $f(z)$ is analytic at z_0. The coefficient a_{-1} is called the *residue* of $f(z)$ at $z = z_0$.

A function might have many poles, say at $z = z_j$, $j = 1, ..., J$. One conceptual way to calculate the residues is to expand $f(z)$ in J different Laurent series, one for each singularity, and read off the coefficient a_{-1} in each series. The problem with doing this in practice is that we end up with many coefficients that are of no interest. The following operational rule is a useful shortcut: To calculate the residue of $f(z)$ at a pole of order m at $z = z_j$, let $\phi(z) = (z - z_j)^m f(z)$. Then

$$\text{Residue of } f(z) \text{ at } z_j = \frac{\phi^{(m-1)}(z_j)}{(m-1)!} . \tag{B.47}$$

For a simple pole ($m = 1$), the residue at z_j is just the limit as $z \to z_j$ of $(z - z_j)f(z)$.

B.3.7 Residue theorem

Let C be a closed contour within and on which $f(z)$ is analytic except for a finite number of singular points at $z = z_j$, $j = 1, ..., J$. Then

$$\oint_C dz\ f(z) = 2\pi i \sum_j [\text{residue of } f(z) \text{ at } z = z_j] . \tag{B.48}$$

B.3.8 Use of residues to evaluate real integrals

One of the most important applications of complex integration is to perform *real* integrals. As an illustration, consider the integral

$$I = \int_{-\infty}^{\infty} dx\ \frac{\cos(\beta x)}{x^2 + \alpha^2} , \tag{B.49}$$

where α and β are real and positive; later α and β will be allowed to be negative. The integrand is real and the integral is along the real axis, but nevertheless contour integration is useful. To see this, note first that we can also write

$$I = \int_{-\infty}^{\infty} dx\ \frac{e^{i\beta x}}{x^2 + \alpha^2} , \tag{B.50}$$

since the integral of $\sin(\beta x)/(x^2 + \alpha^2)$ over the symmetric interval vanishes. We can now recognize (B.50) as the Fourier transform of $(x^2 + \alpha^2)^{-1}$ if we identify β as $-2\pi\xi$ (see Chap. 3).

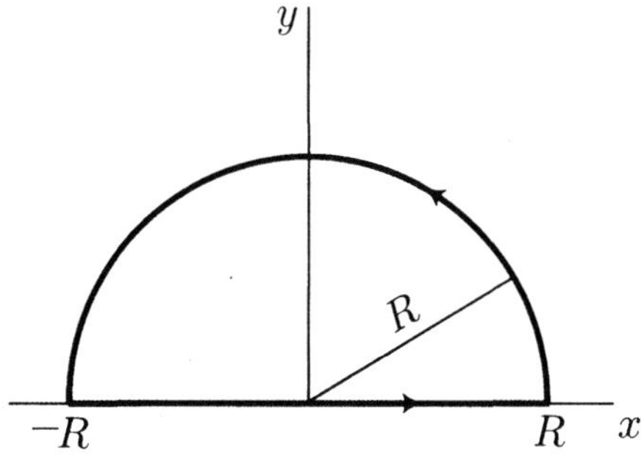

Fig. B.2 Contour used in evaluation of (B.50).

The next step is to convert the integral along the real axis into a contour integral in the complex plane. A suitable contour, shown in Fig. B.2, is one that runs from $-R$ to R along the real axis, then is closed by a counter-clockwise semicircle of radius R. Along the straight-line portion of C, $z = x$, and this portion of the contour integral approaches the desired integral I as $R \to \infty$. Along the semicircle, $z = R\exp(i\theta) = R\cos\theta + iR\sin\theta$, $dz = iR\exp(i\theta)d\theta$, and θ runs from 0 to π. We thus have

$$\oint_C dz\, \frac{e^{i\beta z}}{z^2+\alpha^2} = I + \lim_{R\to\infty} \int_0^{\pi} d\theta\, \frac{iR\, e^{i\theta}\, e^{i\beta R\cos\theta}\, e^{-\beta R\sin\theta}}{(R\cos\theta + iR\sin\theta)^2 + \alpha^2}\,. \tag{B.51}$$

Since both R and β are positive, the factor $\exp(-\beta R\sin\theta)$ vanishes exponentially as $R \to \infty$. Thus the limit in (B.51) is zero, and I is equal to the contour integral; all that remains is to evaluate that integral. To do so we write

$$I = \oint_C dz\, \frac{e^{i\beta z}}{z^2+\alpha^2} = \oint_C dz\, \frac{e^{i\beta z}}{(z+i\alpha)(z-i\alpha)}\,. \tag{B.52}$$

Since the exponential is an analytic function, the only singularities of the integrand are at the points $z = \pm i\alpha$, both of which are simple poles. The pole at $z = -i\alpha$ is outside the contour and does not affect the integral. By (B.47) the residue of the integrand at $z = i\alpha$ is $\exp[i\beta(i\alpha)]/2i\alpha$. Then (B.48) shows that

$$I = \frac{\pi}{\alpha}\, e^{-\beta\alpha}\,, \qquad \beta > 0\,, \qquad \alpha > 0\,. \tag{B.53}$$

If we wished to evaluate I for negative β or α, we could repeat the above calculation with a contour having a semicircle in the lower half of the complex plane. Alternatively, we can simply note from (B.49) that I is an even function of β and α. Either way, the general result for any real β and α is

$$I = \frac{\pi}{|\alpha|}\, e^{-|\beta\alpha|}\,, \qquad \beta\,, \qquad \alpha \text{ real}\,. \tag{B.54}$$

B.3.9 Cauchy principal value

In the example in Sec. B.3.8, we were interested in an integral along the real axis, and when we converted it to a contour integral, the poles were found to lie off the real axis. In other words, for real x the integrand remained finite. Sometimes, however, we are faced with an improper real integral with a singularity exactly on the path of integration. An example would be

$$I = \int_{-\infty}^{\infty} dx\, \frac{f(x)}{x}, \tag{B.55}$$

where $f(x)$ does not vanish at $x = 0$.

We would like to convert this integral to a contour integral of $f(z)/z$. If $f(z)$ vanishes sufficiently rapidly at ∞ in either the upper or lower half-plane, we can close the contour with an infinite semicircle as in Sec. B.3.8, but there is no general guidance on how to handle the singularity at $z = 0$. We might consider deforming the contour slightly to avoid the singularity, but it matters greatly whether we indent the contour above or below the pole (see Fig. B.3a and b). If the pole is outside the deformed contour, no matter how slightly, it does not contribute to the integral, but if it is inside, its full residue enters into (B.48).

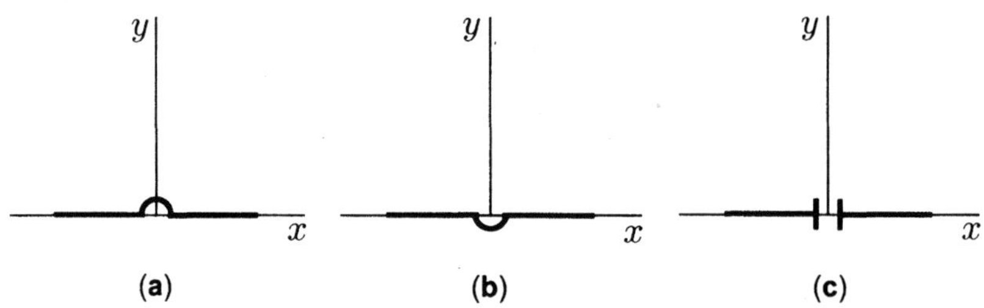

Fig. B.3 Possible contours that can be used for evaluation of (B.55). Only the portion of the contour on or near the real axis is shown; the contour will be closed by a semicircle in the upper or lower half-plane, depending on the behavior of $f(z)$. The contour in part c corresponds to the Cauchy Principal value. Choice among these contours must be made on physical rather than mathematical grounds.

Another option, often dictated by the physics or symmetry of a particular application, is to take the *Cauchy principal value* of the integral, defined for real integrals by

$$\mathcal{P}\int_{-\infty}^{\infty} dx\, \frac{f(x)}{x} = \lim_{\epsilon\to 0}\left\{\int_{-\infty}^{-\epsilon} dx\, \frac{f(x)}{x} + \int_{\epsilon}^{\infty} dx\, \frac{f(x)}{x}\right\}. \tag{B.56}$$

In terms of contour integrals, this definition is equivalent to using the interrupted contour of Fig. B.3c. Use of this contour is equivalent to counting *half* of the residue at any pole that lies exactly on the real axis, essentially averaging the results that would be obtained by the contours of Figs. B.3a and b.

To illustrate the use of the Cauchy principal value, let us compute the integral

$$I = \mathcal{P}\int_{-\infty}^{\infty} dx\, \frac{e^{i\beta x}}{x}, \tag{B.57}$$

which is the Fourier transform of the generalized function $\mathcal{P}\{1/x\}$ (see Sec. 3.3.7) if we let $\beta = -2\pi\xi$. As in Sec. B.3.8, we can close the contour with a semicircle in the upper half-plane if $\beta > 0$. Since the integrand vanishes exponentially along this semicircle, the contour integral equals the desired integral along the real axis. The only pole is at $z = 0$, exactly on the contour, and the residue of the integrand at that pole is 1. Since only half of that residue counts for a principal-value integral, we have

$$I = i\pi\,, \qquad \beta > 0\,. \tag{B.58}$$

The situation gets a bit more interesting if $\beta < 0$. In that case, we close the integral in the lower half-plane, but this means we are traversing the contour in a *clockwise* direction. Since all of our theorems are based on a counter-clockwise convention, we must reverse the direction, introducing a minus sign in the integral. Again the pole is on the contour and the residue is 1, so $I = -i\pi$ for $\beta < 0$. Combining these results and using the signum function defined in Sec. 2.3.2, we can write

$$I = i\pi\,\mathrm{sgn}(\beta)\,. \tag{B.59}$$

This same result could also have been obtained from (B.57) and (B.58) by noting that I is an odd function of β.

B.3.10 Summing a series

Just as complex integration can be used to perform real integrals, it can also be used to sum a real series. A very neat trick for this purpose is presented in Morse and Feshbach (1953), p. 413.

Suppose we are concerned with an infinite sum of the form

$$S = \sum_{n=-\infty}^{\infty} f(n)\,, \tag{B.60}$$

where we require $f(n)$ to be finite for all integers n and to vanish faster than $1/n$ as $n \to \infty$ in order for S to converge. We can convert $f(n)$ to a function of a complex variable simply by replacing n with z, and we assume that $|zf(z)|$ is bounded at infinity.

To evaluate S, we consider the contour integral

$$I = \pi \oint_C dz\; f(z) \cot \pi z\,. \tag{B.61}$$

The function $\pi \cot \pi z$ has simple poles of residue 1 at $z = 0, \pm 1, \pm 2, \ldots$, and it is bounded at infinity except along the real axis.

Now suppose we take the contour C as a full circle of radius R that does not intersect any poles of $\cot \pi z$. The length of this contour is $2\pi R$, so the integral must satisfy the inequality, $I \le 2\pi RM$, where M is the maximum value of the modulus of the integrand on the contour. However, since $|f(z)|$ vanishes faster than $1/R$ as $R \to \infty$ and the cotangent is bounded, the integral in (B.61) is, in fact, zero. Nevertheless, the residue theorem still holds, and we can also compute I from (B.48). As $R \to \infty$ all poles associated with either $f(z)$ or $\cot \pi z$ are enclosed in the contour and contribute to the integral. The poles of the cotangent are at $z = n$, while the poles of $f(z)$ are at $z = z_j$, $j = 1, \ldots, J$. By construction, the sum of the residues of the integrand at the poles of the cotangent is just S, so we have

$$I = 0 = 2\pi i \left[S + \sum_{j=1}^{J} (\text{residue of } \pi f(z) \cot(\pi z) \text{ at } z = z_j) \right], \tag{B.62}$$

where the sum runs over *poles of f(z) only*. Thus

$$S = \sum_{n=-\infty}^{\infty} f(n) = -\sum_{j=1}^{J} [\text{residue of } \pi f(z) \cot(\pi z) \text{ at } z = z_j] . \tag{B.63}$$

A similar formula for alternating series results from the use of the function $\pi \csc \pi z$, which has simple poles of residue $(-1)^n$ at $z = \pm n$. The counterpart of (B.63) is

$$\sum_{n=-\infty}^{\infty} (-1)^n f(n) = -\sum_{j=1}^{J} [\text{residue of } \pi f(z) \csc(\pi z) \text{ at } z = z_j] . \tag{B.64}$$

These two formulas are often very useful when $f(z)$ is a simple function with a few, easily calculated residues.

APPENDIX C
Probability

INTRODUCTION

A thorough grounding in probability theory is necessary in order to understand and model random variables and processes. Classification decisions and estimates based on random processes are themselves random; thus probability theory is required to describe the nature of such inferences. It is the purpose of this appendix to review the basic tools from probability theory used to describe one or two random variables; these tools are generalized to their vector forms in Chap. 8. In Sec. C.1 we start with a discussion of the calculus of probability, including the concepts of sample points and sample spaces, set theory, and the several basic approaches to probability. We then adopt an axiomatic approach to probability and consider the description of single and multiple random variables, and functions of such random variables (Secs. C.2, C.3, C.4). Some well-known probability laws for random variables are discussed in Secs. C.5 and C.6, and methods for generating random variables of a specified probability law are given in Sec. C.7.

The reader who is interested in reviewing these topics in greater depth is advised to consult one of the many general texts on these subjects, such as Johnson *et al.* (1992, 1994, 1995), Casella and Berger (1990), Davenport (1970) and Feller (1968). More sophisticated mathematical treatments are given by Shiryayev (1984) and Breiman (1992). A handy reference book is Evans *et al.* (1993).

C.1 Calculus of probability

Probability theory is a useful tool in the description of processes or experiments that have some uncertainty or randomness about them. In this appendix we will be concerned with descriptions of the ensemble statistics of the random variables in the experiment. Ensemble statistics, also referred to as population statistics, are hypothetical descriptions of the randomness of an experiment given perfect knowledge. This is to be contrasted with sample statistics, the description of data based on information derived from a finite number of random samples, which are not considered here.

In the next sections we provide a framework for describing experiments and their random outcomes using the language of set theory.

C.1.1 Outcomes, events, and spaces

We call a single run of an experiment a *trial.* At the conclusion of each trial, some experimental *outcome* or *sample* is recorded which has some randomness associated with it. Another trial would result in the recording of another realization of the experimental outcome. The randomness in the experiment means each sample may well take on a different, unpredictable value. For example, in a photon-counting experiment the outcome might be the detected number of counts. The outcomes of an experiment can be arbitrarily parsed into a chosen number of *events.* For example, the outcomes of a photon-counting experiment could be binned into three possible events: A_1: the measured number of counts n is less than N_0, A_2: n is equal to N_0, or A_3: n is greater than N_0. In this example, the events A_1, A_2, and A_3 are called *mutually exclusive* because the occurrence of one event on a trial precludes the others from occurring due to their disjoint nature. The set of all possible events for an experiment is termed the *event space* or *sample space*, denoted S. The event space we have set up for our photon-counting example is clearly arbitrary; we could just as well have designated each allowable value for n as an event. Each possible event associated with an experiment can be thought of as a sample point in the sample space. Every experiment has a *certain event*, though, which is the union of all events in the event space.

C.1.2 Concepts from set theory

Events and event spaces are central concepts in probability. We often find that we are interested not only in single events and event spaces for an experiment, but also in combinations of events and relationships between event spaces. It is natural to use the language of set theory for this purpose.

Let S denote the sample space of an experiment. Any possible outcome ζ of the experiment is said to be a member of the space S. This set membership is represented mathematically by $\zeta \in S$.

When an experiment has been performed and we say that some event A has occurred, this signifies that the outcome ζ of the experiment satisfies the conditions that specify event A. As in the photon-counting example of the previous section, some outcomes in S signify that event A occurred, while other outcomes signify that the event did not occur. Thus, any event can be regarded as a subset of the possible outcomes in the space S.

It is said that an event A is contained in another event B (or $A \subset B$) if every outcome that belongs to the subset defining the event A also belongs to the subset defining the event B. Equivalently, we say that A is a subset of B. For example, if A is the event that the number of photons detected in time t is less than N_1, and B is the event that the number of photons detected in the same interval is less than N_2, and $N_1 \leq N_2$, then A is contained in B.

If two events are defined such that A is contained in B and B is contained in A, then it follows that $A = B$. It is also easy to demonstrate that if A is contained in B and B is contained in the event C, then A is contained in C.

Some events are impossible. We cannot detect a negative number of photons, and physical constraints limit the maximum density of an image recorded on film. A subset that contains no outcomes is called the empty set and is denoted $\emptyset$.

Set theory operations In the sections below we present the basic relationships between various event combinations in the language of set theory. We shall build on these concepts in later sections of this appendix as we develop similar concepts regarding relationships between two or more random variables.

Union. The union of the events A and B is defined to be the set that contains all outcomes in either A or B or both; the union operation is a logical OR operation. The union operation is written $A \cup B$. Figure C.1 is a *Venn* diagram demonstrating the union operation. The following relationships involving the union of events are easily verified:

$$\text{Commutative law: } A \cup B = B \cup A. \tag{C.1a}$$

$$\text{Associative law: } A \cup (B \cup C) = (A \cup B) \cup C = A \cup B \cup C. \tag{C.1b}$$

$$\text{Union with itself: } A \cup A = A. \tag{C.1c}$$

$$\text{Union with empty set: } A \cup \emptyset = A. \tag{C.1d}$$

$$\text{Union with sample space: } A \cup S = S. \tag{C.1e}$$

Intersection. The intersection operation, denoted $A \cap B$, results in the set of outcomes common to both A and B. This operation is also depicted graphically in Fig. C.1. The intersection operation is the logical AND operation. The following relationships are easily shown to be true:

$$\text{Commutative law: } A \cap B = B \cap A. \tag{C.2a}$$

$$\text{Associative law: } (A \cap B) \cap C = A \cap (B \cap C) = A \cap B \cap C. \tag{C.2b}$$

$$\text{Distributive law: } A \cap (B \cup C) = (A \cap B) \cup (A \cap C). \tag{C.2c}$$

$$\text{Intersection with itself: } A \cap A = A. \tag{C.2d}$$

$$\text{Intersection with empty set: } A \cap \emptyset = \emptyset. \tag{C.2e}$$

$$\text{Intersection with sample space: } A \cap S = A. \tag{C.2f}$$

Now we can define mutual exclusivity in set theory terms: A and B are mutually exclusive or disjoint if $A \cap B = \emptyset$.

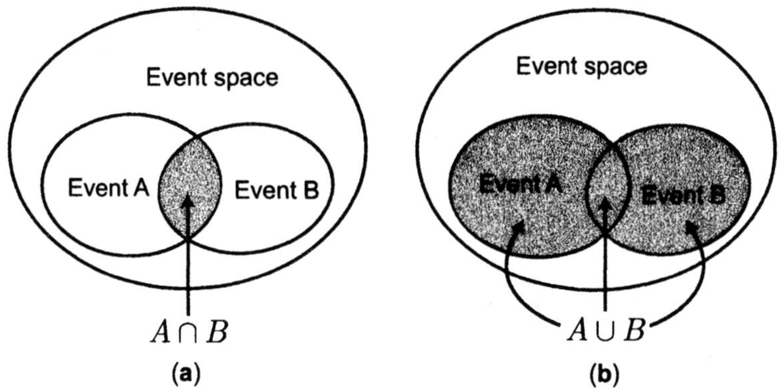

Fig. C.1 (a) A Venn diagram demonstrating the intersection operation as the shaded region where events A and B overlap. (b) A Venn diagram demonstrating the union operation; the shaded area is the union of event spaces A and B.

Complement. The complement of set A, denoted $\overline{A}$, is defined as the set consisting of all elements of S that are not in A. Then

$$\overline{\emptyset} = S\,; \tag{C.3a}$$

$$\overline{S} = \emptyset\,; \tag{C.3b}$$

$$A \cup \overline{A} = S\,; \tag{C.3c}$$

$$A \cap \overline{A} = \emptyset\,. \tag{C.3d}$$

De Morgan law. Using the Venn diagram of Fig. C.2, one can see that

$$\overline{(A \cup B)} = \overline{A} \cap \overline{B} \tag{C.4a}$$

and

$$\overline{(A \cap B)} = \overline{A} \cup \overline{B}\,. \tag{C.4b}$$

These relationships are useful because they tell us that in any set identity, when we replace unions with intersections, intersections with unions, and sets with their complements, the resulting identity also holds. We shall find use for the De Morgan law as we develop relationships between event probabilities in the sections below.

Partitions. A collection of events $A_1,\ A_2, ..., A_n$ is said to partition the sample space S if and only if

$$A_1 \cup A_2 \cup \cdots \cup A_n = S \tag{C.5a}$$

and

$$A_i \cap A_j = \emptyset \qquad \text{for} \qquad i \neq j\,. \tag{C.5b}$$

Fields We have defined events as collections of outcomes in an event space S. In the sections to follow we shall set the stage for assigning probabilities to events. We shall therefore restrict our attention to certain sets of events that are said to comprise a *field*. A field $\mathcal{F}$ is a set of events such that:

1. If $A_i \in \mathcal{F}$, then $\overline{A_i} \in \mathcal{F}$.

2. If $A_i \in \mathcal{F}$ and $A_j \in \mathcal{F}$, then $A_i \cup A_j \in \mathcal{F}$.

A *Borel field* $\mathcal{B}$ has the property that if the events $A_1, A_2, ..., A_k$ are contained in $\mathcal{B}$, then the events $\{A_1 \cap A_2 \cap \cdots \cap A_k\}$ and $\{A_1 \cup A_2 \cup \cdots \cup A_k\}$ are also contained in the field. One of the important properties of Borel fields is the assurance that whenever one considers set operations such as intersections and unions on the events, the resulting sets are also events for which probabilities can be assigned.

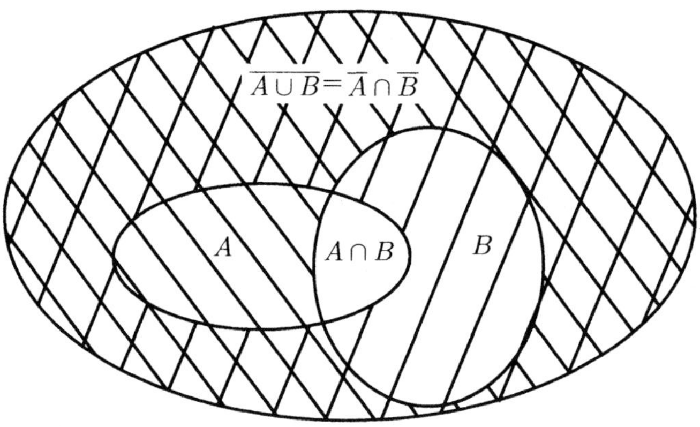

Fig. C.2 A Venn diagram illustrating the De Morgan law. The area with hatching in both directions is the complement of $A \cup B$. The region with hatching in either or both directions is the complement of $A \cap B$.

C.1.3 Definitions of probability

Each event A associated with an experiment has some probability of occurrence, denoted $\Pr(A)$. Much debate has centered on the issue of the assignment and interpretation of probability. As we describe in the Prologue, there are two major schools, known as the frequentists and Bayesians, with ardently defended positions regarding the assignment of probabilities. In the Prologue we discuss in some detail the controversy between these camps and our own philosophy toward the assignment of probabilities for problems in imaging. Here we briefly give several definitions of probability before presenting the axiomatic framework that will guide the rest of the appendix.

Relative frequency A useful definition of probability is based on the relative frequency of an experimental outcome, and a person who adopts this approach is referred to as a *frequentist*. In this school, probability is determined by repeatedly performing an experiment and counting the number of times the outcome actually occurs.

In frequentist parlance, if event A occurs $m(A)$ times in M trials, the relative frequency of the event is the number of occurrences of the event divided by the number of trials, $m(A)/M$. As the number of trials of an experiment approaches infinity, the probability of a particular event is the limit of the relative frequency of that event:

$$\Pr(A) = \lim_{M\to\infty} \frac{m(A)}{M}. \tag{C.6}$$

Those who propound this definition of probability emphasize its objectivity and the verifiability of the probability of some event A through experimental confirmation.

Ensemble definition In reality it is rarely practical to repeat an experiment a large number of times, and it is often impossible to duplicate the exact experimental conditions. An alternative conceptual approach is to postulate an infinite collection of identical experiments, called an *ensemble*, and to define probability as the frequency of occurrence of an event within the ensemble. Thus (C.6) still applies but with $m(A)$ interpreted as the number of members of the ensemble for which A occurs.

The ensemble definition of probability was first advanced by the physicist J. Willard Gibbs in the context of statistical mechanics. In that application, the ensemble is the set of possible states of a system of indistinguishable particles; an event is one possible set of state variables for the system. The famous Boltzmann-Gibbs distribution describes the equilibrium solution, that is, the distribution of energies for a system in thermal or mechanical equilibrium.

Classical definition The classical definition specifies a probability as the ratio of the number of favorable outcomes to the total number of possible outcomes for an experiment. For example, the roll of a six-sided die has six possible outcomes, three of which are favorable to an even event. The classicist would therefore conclude that the probability of an even event is $3/6 = 1/2$. Classical probability is distinct from the relative-frequency approach in that the favorable outcomes are counted in advance of the performing of an experiment. While classical probability is also considered to be objective, it is limited to situations where the experimental outcomes are equally likely, thereby rendering this definition useful only in situations where a symmetry is present between the various possible outcomes. In addition, the approach is not applicable when the number of outcomes is nondenumerable. Even when the number of outcomes is finite, one can imagine problems in specifying the possible outcomes and assuring their equal likelihood.

Subjective interpretation: Bayesian approach The Bayesian approach is founded on a subjective or personal interpretation of probability. In this approach probability is a measure of the weight of some belief. For example, a student believes he will get an A in a course based on some subjective assessment of his odds. This definition is used routinely by our judicial system when the jury is charged with determining whether or not a defendant is guilty beyond a reasonable doubt. Subjective probabilities can take on numerical values as well. A student can believe his probability of getting an A is 0.75, for example, based on information he has about the teacher and his own effort and ability. Note that while the frequentist interpretation of probability is in terms of real or hypothetical repeated trials of an experiment, a Bayesian's view departs from this picture radically. In the Bayesian world there is no consideration of a collection of experiments; probability is a statement of belief about a single event. This is an important distinction between these two schools of probabilistic thought.

One might think that the subjective nature of this definition, where probability is based largely on intuition, makes it difficult to codify into a rigorous mathematical framework. (In fact, many authors refer to this definition as the intuitive interpretation of probability.) However, authors such as Savage (1954), De Groot (1970), and de Finetti (1974, 1975) have developed axiomatic systems of probability where the probability functions are interpreted subjectively but are still required to satisfy axioms such as those found in the next section, or some that are very similar.

C.1.4 Axiomatic approach

Regardless of how one chooses to assign probabilities to events, it will be necessary for our purposes that algebraic manipulation of the event probabilities be possible. This requirement is satisfied by requiring the probabilities of events to satisfy a number of axioms. Probabilities that satisfy such a system of axioms are sometimes referred to as *mathematical probabilities.* Mathematical probabilities may stem from any of the definitions of probability given above.

The axiomatic approach to probability begins by defining a function $\Pr(\cdot)$ such that each event A is assigned a number $\Pr(A)$ called the probability of event A. This function must satisfy the following conditions:

I. $0 \leq \Pr(A) \leq 1$, or $\Pr(A)$ is nonnegative.

II. The probability of the certain event is 1, that is, $\Pr(S) = 1$.

III. If $A \cap B = \emptyset$, then $\Pr(A \cup B) = \Pr(A) + \Pr(B)$.

That is, the probability of occurrence of any of a set of mutually exclusive events is found by simply adding their individual probabilities.

The three properties given above are usually referred to as the Kolmogorov Axioms, after Andrei N. Kolmogorov (1903–1959), one of the fathers of modern probability theory. From the Kolmogorov axioms it is straightforward to derive other properties that event probabilities must obey, such as the following:

IV. $\Pr(\emptyset) = 0$.

V. $\Pr(\overline{A}) = 1 - \Pr(A)$.

VI. $\Pr(A \cup B) = \Pr(A) + \Pr(B) - \Pr(A \cap B)$.

These relationships can be useful when calculating the probabilities associated with possible outcomes of more complicated experiments. It is left to the reader as an exercise to show that the relative-frequency definition of probability (C.6) satisfies Kolmogorov's axioms.

C.1.5 Joint probability

We are often interested in the probability that two events A and B both occur. We refer to this event as the joint event $(A \cap B)$. The corresponding probability $\Pr(A \cap B)$, also denoted $\Pr(A, B)$, is the joint probability. The joint probability of two events A and B can be written as either $\Pr(A, B)$ or $\Pr(B, A)$ since the AND operation is commutative as in (C.2a), meaning event order is not important. Joint probabilities must satisfy the same basic axioms as the probabilities of the individual events, $\Pr(A)$ and $\Pr(B)$.

C.1.6 Conditional probability

Often we shall be interested in the following question: Given that event B has occurred, what is the probability that event A has occurred? In other words, we are interested in the probability of the event A *conditioned* on the occurrence of event B. We denote this probability $\Pr(A|B)$, where the vertical bar is read *given* or *conditioned on*. In the Venn diagram of Fig. C.1, we are interested in the probability that an outcome is a member of the set of outcomes defining event A, assuming that it is a member of the sample space defined by event B. When considering conditional probabilities, we are essentially redefining the sample space by restricting it to only those outcomes that are contained in the event hypothesized to have occurred, in this case event B.

The conditional probability $\Pr(A|B)$ is found by dividing the joint probability $\Pr(A, B)$ of events A and B by the probability that event B has occurred:

$$\Pr(A|B) = \frac{\Pr(A, B)}{\Pr(B)} . \tag{C.7}$$

If A and B are mutually exclusive, then

$$\Pr(A|B) = 0 . \tag{C.8}$$

On the other hand, if B implies A then

$$\Pr(A|B) = 1 . \tag{C.9}$$

Equation (C.7) can be generalized to more than two events as follows:

$$\Pr(A, B, C) = \Pr(A) \Pr(B|A) \Pr(C|A, B) \tag{C.10}$$

and

$$\Pr(A_1, A_2, ..., A_k) = \Pr(A_1) \Pr(A_2|A_1) \Pr(A_3|A_1, A_2) \cdots \Pr(A_k|A_1, A_2, ..., A_{k-1}) . \tag{C.11}$$

Independent events Suppose that events A and B are defined in a sample space S and both $\Pr(A)$ and $\Pr(B)$ are greater than zero. Events A and B are called *statistically independent* if $\Pr(A|B) = \Pr(A)$. It then follows from (C.7) that

$$\Pr(A, B) = \Pr(A) \Pr(B) \tag{C.12}$$

for statistically independent events.

Total probability If the set of events $\{B_i\}$ partitions the event space S [see (C.5)], and event A is also contained in S, then the total probability of event A can be found by summing the conditional probability of event A over all events $\{B_i\}$:

$$\Pr(A) = \sum_i \Pr(A|B_i)\Pr(B_i)\,. \tag{C.13}$$

C.1.7 Bayes' rule

Once the conditional probability $\Pr(A|B)$ has been defined by (C.7), it follows immediately that we can write the joint probability of events A and B as

$$\Pr(A, B) = \Pr(A|B)\Pr(B)\,. \tag{C.14}$$

We know from the properties of the intersection operation (C.2a) that $\Pr(A, B) = \Pr(B, A)$. Thus we can also express the joint probability in terms of the probability of event B conditioned on event A:

$$\Pr(A, B) = \Pr(B|A)\Pr(A)\,. \tag{C.15}$$

Equating (C.14) and (C.15) yields

$$\Pr(A|B)\Pr(B) = \Pr(B|A)\Pr(A)\,, \tag{C.16}$$

which leads to

$$\Pr(B|A) = \frac{\Pr(A|B)\Pr(B)}{\Pr(A)}\,. \tag{C.17}$$

More generally, if the set of events $\{B_i\}$ partition S then we can use the total probability rule of (C.13) to show that

$$\Pr(B_j|A) = \frac{\Pr(A|B_j)\Pr(B_j)}{\sum_i \Pr(A|B_i)\Pr(B_i)}\,. \tag{C.18}$$

Equation (C.18) is known formally as Bayes' rule, after Thomas Bayes (1704–1761), an English mathematician and philosopher. In Sec. C.4.3 below and throughout the text we shall say more about the applicability of Bayes' rule for drawing inferences about the occurrence of events using prior information and measured data.

C.2 SINGLE RANDOM VARIABLES

C.2.1 Definition of a random variable

In many scientific experiments the outcomes are numbers, such as the density of film at some spatial location or the number of photons that arrive at a detector within some time interval. The sample description space in the first example is the set of nonnegative real numbers, while in the second example it is confined to the set of nonnegative integers.

To use a numerical quantity to represent the outcome of an experiment, a mapping is needed from the original sample space S to the set of real numbers.[1] Suppose we have a sample space S with a probability function $\Pr(\cdot)$ defined on events in that sample space. A *random variable* x is a function that maps S onto the set of real numbers. That is, we observe $x = x'$ if and only if the outcome of the experiment is a $\zeta \in S$ such that $x(\zeta) = x'$. Then

$$\Pr(x = x') = \Pr\{\zeta \in S : x(\zeta) = x'\}. \tag{C.19}$$

Equation (C.19) is a formal definition of the probability function for the random variable x. In order for the probability function of (C.19) to satisfy the Kolmogorov Axioms, every event must be *measurable*. This requirement is met provided (Papoulis, 1991)

1. The set $\{x \leq x'\}$ is an event for any real number x';

2. The probabilities of the events $\{x = \infty\}$ and $\{x = -\infty\}$ equal 0.

A function that satisfies the above requirements is called measurable in the field $\mathcal{B}$ (see C.1.2).

As an example of a mapping to the set of real numbers, consider the experiment of tossing a coin. In this case there are two possible outcomes on each trial, either a head or a tail. A real-valued random variable can be associated with this experiment by assigning the numerical value 1 to the outcome if a head should result; if the outcome is a tail it is assigned the value 0. In this way the experimental outcomes are mapped to real values. Sometimes the outcome of an experiment is a numerical value that can be used as the random variable with no other mapping required. Such is the case, for example, if we are interested in the voltage across a resistor. The measured voltage V is the random variable of interest.

Notation Many books rigorously indicate random variables by using a special type font or by using capital letters (with the corresponding lower-case letter signifying a sample or a particular realization of the random variable). Though appealing in theory, this device runs into difficulty in practice.

One problem is that equations involving random variables are inherently ambiguous. If, for example, we write $y = ax + b$, where x and y have been identified as random variables and a and b are fixed parameters, the equation can mean that any particular realization of x, say $x = x_0$, is associated with a corresponding realization of y, namely $y = y_0 \equiv ax_0 + b$. Since this functional association holds for all realizations, it can as well be said to hold for the random variables themselves.

Another difficulty arises in imaging, as well as in many other physical applications, where it is not always clear whether a given quantity should be treated as a random variable. For example, an object to be imaged can be regarded either as fixed but unknown, as a spatial random process described by some known or unknown probability law, or as one sample function of such a random process. No matter which of these viewpoints we adopt, the equation describing the imaging

[1] We have deliberately restricted our discussion here to real random variables. Complex random variables are addressed in Chap. 8.

system has the same form (and the same ambiguity attaches to the image as to the object). If a particular operator equation relates an object to its image, it does so in the same way no matter whether we think of the object as nonrandom or as one sample function of a random process. And if we think of presenting an entire ensemble of random objects to the imaging system, the same operator can be used to compute the corresponding ensemble of images.[2]

Since it does not seem worthwhile to incorporate typographic distinctions that have no effect on the mathematical operations involved, we simply denote scalar random variables by italicized upper or lower case characters, just as for any other scalar mathematical variable. Vector random variables (discussed only briefly here but more thoroughly in Chap. 8) will be denoted by boldface lower-case letters, just as for any other vector. When it is necessary to distinguish a particular realization from the general random variable, the realization will be denoted either with a subscript or prime or by using a different case, but such situations will arise infrequently.

Discrete random variables In the coin-tossing example above, the random variable describing the outcome of the experiment took on isolated values, *i.e.*, either 0 or 1. Such a variable is called a *discrete random variable.* Formally, a random variable x is said to be discrete if x takes on only a finite number k of values $\{x_1, x_2, ..., x_k\}$ or, at most, a countably infinite set $\{x_i, i = 1, ..., \infty\}$ of such values. For a discrete random variable that can take on the values $\{x_i\}$, the probability of each outcome is denoted by $\Pr(x = x_i)$ or the shorthand $\Pr(x_i)$. It can be shown that

$$\sum_i \Pr(x = x_i) = 1\,, \tag{C.20}$$

where the sum is over the set of all possible values of i. This normalization is a consequence of the second Kolmogorov axiom.

Continuous random variables In some experiments the random variable of interest can take on a continuum of possible values, as in the example of the voltage across a resistor mentioned above. Variables of this type are known as *continuous random variables.*

A convenient way of describing the statistical properties of a continuous random variable is in terms of the *probability density function* (PDF), denoted $\mathrm{pr}(\cdot)$. This function is defined in terms of the probability that the random variable assumes a value within some specified small range of values. That is,

$$\mathrm{pr}(x_0) \equiv \lim_{\Delta x \to 0} \frac{1}{\Delta x} \Pr(x_0 - \tfrac{1}{2}\Delta x \le x \le x_0 + \tfrac{1}{2}\Delta x)\,. \tag{C.21}$$

In other words, for vanishingly small Δx, $\mathrm{pr}(x_0)\Delta x$ gives the probability that the random variable x takes on a value that is within $\pm\frac{1}{2}\Delta x$ of the particular value x_0.

The PDF can be used to calculate the probability that the random variable x falls in the finite interval (a, b) by dividing the interval into small, mutually exclusive elements of width Δx. Equation (C.21) then leads to

$$\Pr(a \le x \le b) = \int_a^b \mathrm{pr}(x')\, dx'\,. \tag{C.22}$$

[2] In this discussion we are not considering either measurement noise or randomness in $\boldsymbol{\mathcal{H}}$.

Since x is constrained to lie in the interval $(-\infty, \infty)$, it must be true that

$$\int_{-\infty}^{\infty} \mathrm{pr}(x')\, dx' = 1\,, \tag{C.23}$$

by Kolmogorov's second axiom.

C.2.2 Probability density functions

Although defined most naturally for continuous random variables, PDFs can be used also for discrete ones if we allow delta functions. Consider a random variable x that takes on the discrete set of values $\{x_i\}$. We can define a PDF for x by:

$$\mathrm{pr}(x) = \sum_i \mathrm{Pr}(x_i)\, \delta(x - x_i)\,, \tag{C.24}$$

where $\delta(x - x_i)$ is a Dirac delta function and, as before, $\mathrm{Pr}(x_i)$ is the *probability* (not density) that random variable x takes on the specific value x_i. We shall follow this typographic distinction consistently: lower-case $\mathrm{pr}(\cdot)$ will denote a PDF and capital $\mathrm{Pr}(\cdot)$ a probability.

To demonstrate the consistency of the definition in (C.24), we make use of some elementary properties of delta functions (see Chap. 2). Consider first a region $a < x \leq b$ that contains no allowed value of x. The probability that x falls in this region is given by the integral in (C.22). Since a delta function vanishes unless its argument is zero, which never happens over the postulated region, this integral is identically zero, as it must be since the region contains no allowed values.

Next consider a region of width 2ϵ around an allowed value, say x_j, and assume that ϵ is small enough that the region contains no other allowed values. From (C.22) and (C.24), the probability that x lies in this region is

$$\mathrm{Pr}(x_j - \epsilon < x < x_j + \epsilon) = \int_{x_j-\epsilon}^{x_j+\epsilon} dx' \sum_i \mathrm{Pr}(x_i)\, \delta(x' - x_i) = \mathrm{Pr}(x_j)\,, \tag{C.25}$$

where the last step used the sifting property of the delta function, (2.23). Thus (C.25) shows that the probability that x lies in the small region around x_j, as computed from the density (C.24), is the same as the probability that x is precisely equal to x_j.

Finally, consider a region that contains two allowed values, say x_j and x_k. The sifting property of delta functions now shows that the probability that x lies in this region is $\mathrm{Pr}(x_i) + \mathrm{Pr}(x_j)$, which, according to Kolmogorov Axiom III, is just the probability that x is either x_i or x_j.

Thus the delta-function density defined by (C.24) contains exactly the same information about the discrete random variable x as does the original probability law $\mathrm{Pr}(x_i)$.

C.2.3 Cumulative distribution function

Every measurable random variable x has an associated function called the cumulative distribution function of x, denoted $\mathrm{F}(x)$. The cumulative distribution function of the random variable x is defined for any real number c by

$$\mathrm{F}(c) = \mathrm{Pr}(x \leq c)\,. \tag{C.26}$$

That is, $\mathrm{F}(c)$ is the probability that the random variable x is less than or equal to c. That $\mathrm{F}(x)$ exists is a condition for a measurable random variable, according to the properties of such random variables given in Sec. C.2.1. The definition (C.26) holds true for all values of c, even though it encompasses values of c that may not be attainable by the random variable x. For example, the random variable may not be able to take on negative values. Then (C.26) would result in the finding that $\mathrm{F}(c) = 0$ for $c < 0$. Or, for a discrete random variable, the value of c may not be one of the values of x_i the variable can assume. Applying (C.26) in that instance would result in the sum of the probabilities $\Pr(x = x_i)$ for all $x_i < c$.

The cumulative distribution has the following properties:

$$\mathrm{F}(c_1) \leq \mathrm{F}(c_2) \qquad \text{if} \qquad c_1 < c_2\,, \tag{C.27a}$$

$$\lim_{c \to -\infty} \mathrm{F}(c) = 0\,, \tag{C.27b}$$

$$\lim_{c \to \infty} \mathrm{F}(c) = 1\,. \tag{C.27c}$$

Equation (C.27a) says that the distribution function is a nondecreasing function of c. This property holds since for every $c_1 < c_2$ the event $\{x \leq c_1\}$ is contained in the event $\{x \leq c_2\}$ and so the probability of the event $\{x \leq c_2\}$ must be at least as great. The proofs for (C.27b) and (C.27c) are based on the fact that x must be measurable, so it must take on some finite value $[\Pr(x = -\infty) = \Pr(x = \infty) = 0]$, and the probability of the certain event, $\Pr(x < \infty)$, is 1.

The cumulative distribution function is useful for describing the probability that a random variable lies within some interval. It can be shown that

$$\Pr(c_1 < x \leq c_2) = \mathrm{F}(c_2) - \mathrm{F}(c_1) \qquad \text{for all} \qquad c_1 < c_2\,. \tag{C.28}$$

We can distinguish among continuous and discrete random variables depending on the form of the cumulative distribution function, specifically, whether derivatives of $\mathrm{F}(x)$ exist for every value of x. A discrete random variable has a distribution function $\mathrm{F}(x)$ with a staircase form and discontinuities at the points x_i as shown in Fig. C.3. A continuous random variable has a continuous cumulative distribution function as shown in Fig. C.4.

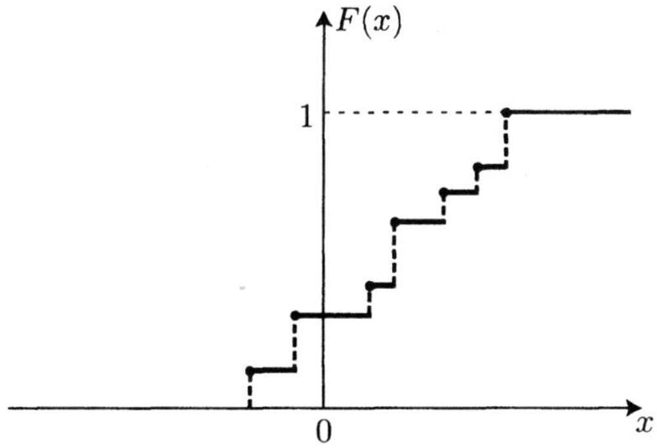

Fig. C.3 The cumulative distribution function of a discrete random variable.

For continuous random variables we can derive the probability that x takes on some particular value by allowing the interval in (C.28) to become arbitrarily small, leading to the definition of the PDF in terms of the derivative of the cumulative distribution function:

$$\mathrm{pr}(x) = \frac{d\,\mathrm{F}(x)}{dx} \, . \tag{C.29}$$

This, in turn, leads to the following result:

$$\mathrm{F}(x) = \int_{-\infty}^{x} \mathrm{pr}(x') \; dx' \, . \tag{C.30}$$

For discrete random variables we have an analogous expression for the cumulative distribution function:

$$\mathrm{F}(x) = \sum_{x_i \leq x} \mathrm{Pr}(x_i) \, . \tag{C.31}$$

Equations (C.29) and (C.30) also hold for the discrete case if we allow generalized functions. As we indicated below (C.28), the cumulative distribution function for a discrete random variable has a staircase form, which can be expressed as a sequence of step functions. If we make use of the fact that the derivative of a step function is a delta function (see Chap. 2), we see that taking the derivative of $\mathrm{F}(x)$ in (C.29) yields the delta functions of (C.24). Similarly, performing the integration operation of (C.30) on a series of delta functions gives back the staircase form for $\mathrm{F}(x)$.

From (C.29) - (C.31) we see that the cumulative distribution function contains the same information as the PDF. For continuous random variables, the PDF $\mathrm{pr}(x)$ can be thought of as a mass density along the real x axis. That is, $\mathrm{pr}(x)\,dx$ represents the mass in the interval $(x, x + \Delta x)$. In the discrete case, the impulses $\mathrm{Pr}(x_i)\,\delta(x - x_i)$ can be thought of as point masses of weight $\mathrm{Pr}(x_i)$ at locations x_i. We know from Kolmogorov's axioms that these masses are positive and that the sum of all mass along the real line is 1. The probability that the random variable x takes on a value in the interval (a, b) is the mass in that interval. The mass in the interval $(-\infty, x)$ is the distribution function $\mathrm{F}(x)$.

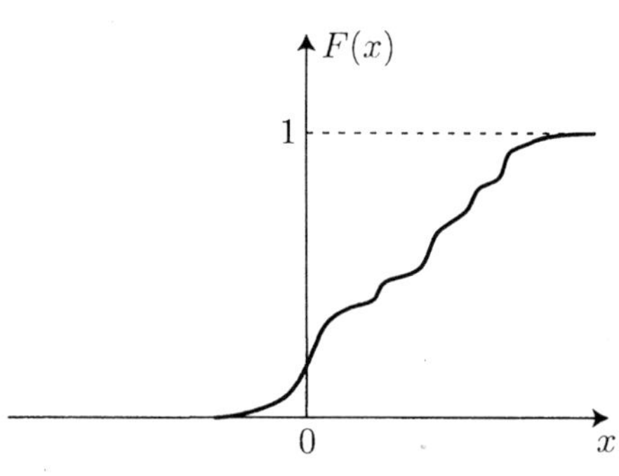

Fig. C.4 The cumulative distribution function of a continuous random variable.

C.2.4 Expectations

The PDF and cumulative distribution function are complete descriptions of the behavior of a random variable, but we are often interested in summary measures of this behavior. The most important single descriptor of a random variable is its *expected value*, which is also referred to as the *mean* of the random variable. The expected value of the random variable x will be written as $\mathrm{E}\{x\}$, $\langle x \rangle$ or $\overline{x}$; these notations will be used interchangeably.

A discrete random variable x has an expected value defined by

$$\mathrm{E}\{x\} = \sum_i x_i \, \mathrm{Pr}(x_i) \,, \tag{C.32}$$

where the sum is over all possible values of x_i. We see that the expected value is a weighted average of the possible values of x_i, where the weights are the probabilities that the random variable takes on each of those values. Note that, for a discrete random variable that takes on only integer values, its mean can take on any value in $(0, \infty)$.

A continuous random variable x has an expected value defined by

$$\mathrm{E}\{x\} = \int_{-\infty}^{\infty} x \, \mathrm{pr}(x) \, dx \,, \tag{C.33}$$

where $\mathrm{pr}(x)$ is the PDF for x. This definition applies to discrete random variables as well when (C.24) is used to define the PDF for the discrete random variable.

The *median* of a random variable, x_{med}, is defined as the value of x for which the cumulative distribution function $\mathrm{F}(x_{med}) = \mathrm{Pr}(x \leq x_{med})$ is equal to $1/2$. The value of the random variable is thus equally likely to be either greater than or less than the median.

A *mode* of a random variable, x_{mode}, is a value of x for which $\mathrm{pr}(x)$ takes on a local maximum. A *unimodal* PDF has a single maximum and thus a single mode. When $\mathrm{pr}(x)$ has an absolute maximum but also another local maximum at an equal or lower value, it is said to be *bimodal*. A probability density can have a locally flat region and therefore an infinite number of modes. Certain estimators, called maximum *a posteriori* estimators, make particular use of the mode of a PDF (see Chap. 13).

The mean of a random variable specifies the center of gravity of $\mathrm{pr}(x)$. Another important parameter is the *variance*, which is a measure of the spread of the random variable about its mean. The variance, σ^2, is defined by

$$\sigma^2 = \mathrm{Var}\{x\} = \mathrm{E}\{(x - \overline{x})^2\} = \int_{-\infty}^{\infty} (x - \overline{x})^2 \, \mathrm{pr}(x) \, dx \,. \tag{C.34}$$

Note that

$$\sigma^2 = \mathrm{E}\{x^2\} - \overline{x}^2 \,. \tag{C.35}$$

The positive square root σ is referred to as the *standard deviation*.

Chebyshev inequality Let x be an arbitrary random variable with mean $\overline{x}$ and finite variance σ^2. Then the Chebyshev inequality, given to us by the Russian mathematician P. L. Chebyshev (1821 – 1894), provides a bound on the probability that

x deviates from its mean by an amount $\delta > 0$. From the definition of variance we have

$$\sigma^2 \equiv \int_{-\infty}^{\infty} (x-\overline{x})^2 \operatorname{pr}(x)\, dx \geq \int_{|x-\overline{x}|\geq\delta} (x-\overline{x})^2 \operatorname{pr}(x)\, dx$$
$$\geq \delta^2 \int_{|x-\overline{x}|\geq\delta} \operatorname{pr}(x)\, dx = \delta^2 \Pr(\,|x-\overline{x}| \geq \delta)\,. \tag{C.36}$$

Thus

$$\Pr(|x-\overline{x}| \geq \delta) \leq \frac{\sigma^2}{\delta^2}\,. \tag{C.37}$$

Moments The behavior of a random variable can be further specified in terms of its *moments*, with the kth moment defined by

$$m_k = \mathrm{E}\left\{x^k\right\} = \int_{-\infty}^{\infty} x^k \operatorname{pr}(x)\, dx\,. \tag{C.38}$$

By the Schwarz inequality, $m_1^2 \leq m_2$ for any probability law.

Central moments The kth *central moment* μ_k of a random variable is defined by

$$\mu_k = \mathrm{E}\left\{(x-\overline{x})^k\right\} = \int_{-\infty}^{\infty} (x-\overline{x})^k \operatorname{pr}(x)\, dx\,. \tag{C.39}$$

Thus for any random variable $\mu_0 = 1$, $\mu_1 = 0$, and $\mu_2 = \sigma^2$.

The *skewness* of a random variable, given by

$$\text{Skewness} = \frac{\mu_3}{{\mu_2}^{3/2}}\,, \tag{C.40}$$

measures the lack of symmetry of the PDF. The *kurtosis* (Greek *kurtos*, convex, swelling) of a random variable, given by

$$\text{Kurtosis} = \frac{\mu_4}{\mu_2^2}\,, \tag{C.41}$$

measures the flatness or peakedness of the corresponding PDF. Skewness and kurtosis are sometimes used to describe the shapes of distributions for pattern recognition applications (Pratt, 1991).

Factorial moments Suppose n is a nonnegative, integer-valued random variable. The kth *factorial moment* of n is given by

$$c_k = \mathrm{E}\{n(n-1)\cdots(n-k+1)\} = \sum_{n=0}^{\infty} n(n-1)\cdots(n-k+1)\Pr(n) = \sum_{n=0}^{\infty} \frac{n!}{(n-k)!}\Pr(n)\,. \tag{C.42}$$

The first factorial moment c_1 is simply the mean m_1, and the second factorial moment is related to the mean and variance by

$$c_2 = \sigma^2 + m_1^2 - m_1\,. \tag{C.43}$$

One example of the applicability of the factorial moments in optics is in the description of the statistics of the number of photoelectrons emitted by a photosensitive surface upon which a light source is incident (Helstrom, 1995). Factorial moments come in when calculating the probability that more than a threshold number of counts are detected in a particular time interval.

Cumulants The *cumulants* (also called the *semi-invariants*) of a random variable are defined by

$$\exp\left[\sum_{k=1}^{\infty}\kappa_k\frac{t^k}{k!}\right] = \sum_{k=0}^{\infty} m_k\frac{t^k}{k!}\,, \tag{C.44}$$

where the m_k are the ordinary moments defined in (C.38). By expanding the left-hand side, and equating coefficients for equal powers of t, the reader can generate relationships between the cumulants and the ordinary moments. Of most interest, the first-order cumulant is the mean and the second-order cumulant is the variance of the random variable.

The cumulants are given their name because a random variable z formed by summation of a set of independent random variables $\{x_i\}$ will have cumulant κ_k equal to the sum of the cumulants of the underlying random variables for each value of k (Helstrom, 1995). The cumulants of the $\{x_i\}$ can thus be used to derive the PDF of z for cases where the PDFs of the underlying random variables are unknown. First the cumulants of the $\{x_i\}$ are determined (via a fitting algorithm); the sum of the cumulants gives the cumulants of z, which are then used to solve for the PDF of z.

C.3 FUNCTIONS OF A SINGLE RANDOM VARIABLE

We are often interested in a function of the outcome of an experiment, rather than the outcome itself. For example, we might want to know the square of the voltage across a resistor, where the voltage is itself a random variable. Probabilities can also be assigned to random variables derived from functions of random experimental outcomes.

C.3.1 Transformation of PDFs

Suppose x is a random variable with PDF $\mathrm{pr}(x)$. Any function $y = f(x)$ results in a new random variable whose probabilistic behavior depends on the form of $f(x)$ and the PDF of x. If the function $f(x)$ is monotonic, then it is one-to-one and onto, so that each x is mapped to only one y and each y can come from at most one x. Such functions are the easiest to work with, for they allow us to use a general formula for transforming the density function describing x to compute a density function for the random variable y. We must put one additional requirement on $f(x)$, and that is that $f^{-1}(y)$ must have a continuous derivative. The PDF for y is then determined by

$$\mathrm{pr}_y(y)\,dy = \mathrm{pr}_x(x)\,dx\,, \tag{C.45a}$$

or

$$\mathrm{pr}_y(y) = \mathrm{pr}_x[f^{-1}(y)]\left|\frac{d}{dy}f^{-1}(y)\right|, \tag{C.45b}$$

for all y in the range of $f(x)$. The absolute value appears in the second line because dx and dy in the first line are defined to be always positive. (We use subscripts on the density functions here to distinguish which function describes the behavior of x and which applies to y. Some books use such subscripts routinely. We shall use them only when confusion might otherwise arise as to which variable the density function describes.)

Equation (C.45) has applicability even if $f(x)$ is not monotonic, as long as $f(x)$ is monotonic over intervals for which (C.45) can be used. For example, suppose $y = x^2$ and we restrict y to be greater than 0. Then $f^{-1}(y) = \pm\sqrt{y}$, $|df^{-1}(y)/dy| = 1/(2\sqrt{y})$ and

$$\mathrm{pr}_y(y) = \frac{1}{2\sqrt{y}}\left[\mathrm{pr}_x(\sqrt{y}) + \mathrm{pr}_x(-\sqrt{y})\right] . \tag{C.46}$$

Thus the PDF of y is a sum representing the two monotonic intervals of $f(x)$. Many more examples on the use of (C.45) for transforming density functions can be found in Frieden (1991), Casella and Berger (1990) and Stark and Woods (1986).

C.3.2 Expected values

If y is a measurable function of the random variable x, the expected value of $y = f(x)$ can be computed from

$$\mathrm{E}\{y\} = \int_{-\infty}^{\infty} f(x)\,\mathrm{pr}(x)\,dx . \tag{C.47}$$

As a simple example, suppose $y = ax$. Then $\langle y\rangle = a\langle x\rangle$ and $\langle y^2\rangle = a^2\langle x^2\rangle$, so that the variance of y is related to the variance of x according to the relationship $\sigma_y^2 = a^2\sigma_x^2$. This is a specific consequence of the fact that the process of taking expectations is a linear operation. More generally, if $f_1(x)$ and $f_2(x)$ are two functions for which expectations exist, then

$$\mathrm{E}\{af_1(x) + bf_2(x) + c\} = a\mathrm{E}\{f_1(x)\} + b\mathrm{E}\{f_2(x)\} + c \tag{C.48}$$

for any constants a, b, and c.

Jensen's Inequality A function $f(x)$ of a nonrandom variable x is said to be convex in an interval I if, in that interval, $d^2f(x)/dx^2$ is greater than or equal to zero. Jensen's Inequality (Rade and Westergren, 1990) states that for a set of constants λ_i and a convex function $f(x)$,

$$f(x_1\lambda_1 + \cdots + x_n\lambda_n) \le \lambda_1 f(x_1) + \cdots + \lambda_n f(x_n) \tag{C.49}$$

for $\sum_{i=1}^{n}\lambda_i = 1$ and $\lambda_i \ge 0$.

Now suppose $f(x)$ is a convex function of a random variable x. Given the mean of the random variable, $\overline{x} = \mathrm{E}\{x\}$, Jensen's Inequality allows us to say the following about the mean of the function:

$$f(\mathrm{E}\{x\}) \le \mathrm{E}\{f(x)\} . \tag{C.50}$$

This expression can be verified by rewriting (C.47) using the elementary definition of integration and recognizing that the resulting factors, $\Delta x\,\mathrm{pr}(x_i)$, play the role of the λ_i in (C.49).

Taylor series A differentiable function of a random variable x can be expanded in a Taylor series about $\overline{x}$ to give

$$f(x) = \sum_k \frac{f^k(\overline{x})}{k!}(x - \overline{x})^k . \tag{C.51}$$

Taking the expectation of both sides and keeping only terms through second order, we find

$$\langle f(x)\rangle \approx f(\overline{x}) + \tfrac{1}{2}\sigma_x^2 f''(\overline{x})\,. \tag{C.52}$$

C.3.3 Generating functions

Characteristic function The characteristic function of the real-valued random variable x is defined by

$$\psi(\xi) \equiv \mathrm{E}\left\{e^{-2\pi i\xi x}\right\} = \int_{-\infty}^{\infty} dx\ \mathrm{pr}(x)\, e^{-2\pi i\xi x}\,. \tag{C.53}$$

It is evident from (C.53) that the characteristic function and the PDF form a Fourier transform pair. Thus the characteristic function uniquely determines the PDF. This relationship was called *Ein Schönes Theorem der Wahrscheinlichkeitsrechnung* by Gauss (1900). The PDF is written in terms of an inverse Fourier transform of $\psi(\xi)$ as

$$\mathrm{pr}(x) = \int_{-\infty}^{\infty} d\xi\ \psi(\xi)\, e^{2\pi i\xi x}\,. \tag{C.54}$$

Many authors choose to define the characteristic function as the expectation of $\exp(+2\pi i\xi t)$ or $\exp(+i\xi t)$. Since in imaging applications the random variable of interest is often the spatial position or time of an event, our definition (C.53) maintains the convention we use throughout this book: the forward Fourier transform maps the spatial or temporal domain to the spatial or temporal frequency domain.

The definition of $\psi(\xi)$ in terms of the expectation of $\exp(-2\pi i\xi x)$ leads naturally to the realization that moments of the random variable x can be derived through differentiation of $\psi(\xi)$:

$$\mathrm{E}\left\{x^k\right\} = (-2\pi i)^{-k}\, \frac{\partial^k}{\partial\xi^k}\psi(\xi)\bigg|_{\xi=0}\,. \tag{C.55}$$

This is the moment-generating property of the characteristic function. It is often much easier to perform the differentiation required in (C.55) than to do the integration of (C.38) in order to calculate the moments of x. As we shall see in Sec. C.4.5, the characteristic function can also be very useful in determining the PDF for sums of random variables.

Moment-generating function The *moment-generating function* $M(t)$ is related to the characteristic function by a simple change of variables. The moment-generating function of a random variable x is defined by

$$M(t) = \psi\left(-\frac{t}{2\pi i}\right) = \mathrm{E}\left\{e^{xt}\right\} = \int_{-\infty}^{\infty} dx\ \mathrm{pr}(x)\, e^{xt}\,. \tag{C.56}$$

Many authors define the moment-generating function and the characteristic function with the same sign in the exponential. The negative exponential in our definition of $\psi(\xi)$ results from having defined the characteristic function as the forward Fourier transform of the PDF (C.53).

As its name suggests, the moment-generating function can be used to compute moments:

$$\mathrm{E}\left\{x^k\right\} = \frac{\partial^k}{\partial t^k}M(t)\bigg|_{t=0}\,. \tag{C.57}$$

Thus the kth derivative of the moment-generating function, when evaluated at zero, allows the determination of the kth moment.

The moment-generating function can also be expressed in terms of a Taylor series in t about the origin:

$$M(t) = \sum_{k=0}^{\infty} \frac{1}{k!} \mathrm{E}\left\{x^k\right\} t^k . \tag{C.58}$$

The coefficients in the series can be seen to be proportional to the moments of the random variable x. This Taylor series representation is valid provided the series converges and all derivatives of $M(t)$ exist.

The primary use of the moment-generating function is to characterize PDFs. If the set of moments of a random variable x is unique (in that there is only one PDF with this set of moments) then the moment-generating function may be used to determine the probability density function for x (Helstrom, 1995). In general, however, two distribution functions with the same moments need not be the same. The reader interested in delving into this issue more deeply can find many related theorems and examples in Shiryayev (1984).

The moment-generating function is the two-sided Laplace transform (4.77) of the inverted PDF for x. That is,

$$M(t) = \int_{-\infty}^{\infty} dx \;\, \mathrm{pr}(-x)\, e^{-xt} = \boldsymbol{\mathcal{L}}\mathbf{a}_2 \left\{\mathrm{pr}(-x)\right\} . \tag{C.59}$$

Thus the inverse Laplace transform given in (4.78) can be used to derive the PDF on x given $M(t)$.

Cumulant-generating function The *cumulant-generating function* for a random variable x is related to the moment-generating function through a logarithm:[3]

$$S(t) = \ln M(-t) = \ln\langle e^{-xt}\rangle . \tag{C.60}$$

The cumulant-generating-function can be used to obtain the cumulants according to

$$\kappa_k = (-1)^k \left. \frac{\partial^k}{\partial t^k} S(t) \right|_{t=0} . \tag{C.61}$$

Probability-generating function We are often interested in nonnegative integer-valued random variables, for example when we are concerned with photon-counting statistics in detectors. Such variables are described by a set of probabilities $\mathrm{pr}(n)$ for each possible value of n. The characteristic function then takes on the form

$$\psi(\xi) = \langle e^{-2\pi i \xi n}\rangle = \sum_{n=0}^{\infty} \mathrm{Pr}(n)\, e^{-2\pi i \xi n} = \sum_{n=0}^{\infty} \mathrm{Pr}(n) \left(e^{-2\pi i \xi}\right)^n . \tag{C.62}$$

We see that the summation is in the form of a power series in $\exp(-2\pi i\xi)$. If we let the complex variable z replace $\exp(-2\pi i\xi)$ we can write

$$\langle z^n\rangle = \sum_{n=0}^{\infty} \mathrm{Pr}(n)\, z^n \equiv \Phi(z) . \tag{C.63}$$

[3]The minus sign in $\mathrm{pr}(\cdot)$ in (C.59) and $M(\cdot)$ in (C.60) arise from our definition of the moment-generating function, which differs from Helstrom in the sign of the exponent of (C.56).

This is the defining equation for $\Phi(z)$, the probability-generating function for the integer-valued random variable n. From (C.63) we see that $\Phi(z)$ resembles the z transform (4.88) of $\Pr(n)$, but with z^n in place of z^{-n}. Since (C.62) has the form of a complex Taylor series where the probabilities are the coefficients, comparison with (B.44) shows that

$$\Pr(n) = \frac{1}{n!}\frac{\partial^n}{\partial z^n}\Phi(z)\bigg|_{z=0} . \tag{C.64}$$

Hence the name probability-generating function.

Generation of factorial moments We again consider a nonnegative, integer-valued random variable n with a probability-generating function $\Phi(z)$. The factorial moments c_k [*cf.* (C.42)] can be obtained via differentiation of $\Phi(z)$:

$$c_k = \mathrm{E}\{n(n-1)(n-2)\cdots(n-k+1)\} = \sum_{n=0}^{\infty} n(n-1)(n-2)\cdots(n-k+1)\Pr(n)$$

$$= \frac{\partial^k}{\partial z^k}[\Phi(z)]\bigg|_{z=1} . \tag{C.65}$$

Thus $\Phi(z)$ can be used to generate factorial moments rather than probabilities by the simple expedient of evaluating the derivatives at 1 rather than 0. A function that generates factorial moments would logically be called the *factorial-moment-generating function*, but to avoid ugly hyphenation we shall refer to it simply as the FMGF.

C.3.4 Integrals of functions of random variables

The expected value of a random variable $y = f(x)$ is usually computable from a Riemann integral of the form (C.47). There is no difficulty in evaluating this integral when we can divide the range of integration into successively smaller intervals without concern for how the intervals are chosen. The rub comes when the summand is a function such that the value of the integral does in fact depend on how the subdivisions are chosen. For example, suppose $f(x)$ is a mapping of the form

$$f(x) = \begin{cases} 0 & \text{if } x \text{ is rational} \\ 1 & \text{if } x \text{ is irrational} \end{cases} . \tag{C.66}$$

We need to be able to integrate functions of this type in order to find the expectations of such events, but the process of integrating by cutting the x axis into smaller and smaller pieces and doing the sum will not give an answer independent of where the cuts fall. To remedy this difficulty we make use of the concepts of measure spaces and Lebesgue integration, as presented by Papoulis (1965).

The integral (C.47) can be rewritten, with the elementary definition of integration, as

$$\mathrm{E}\{f(x)\} = \int_{-\infty}^{\infty} dx\, f(x)\,\mathrm{pr}(x) = \lim_{\Delta x\to 0}\sum_{k=-\infty}^{\infty} \Delta x\, f(x_k)\,\mathrm{pr}(x_k)$$

$$= \lim_{\Delta x\to 0}\sum_{k=-\infty}^{\infty} f(x_k)\Pr(x_k - \tfrac{1}{2}\Delta x \le x \le x_k + \tfrac{1}{2}\Delta x) , \tag{C.67}$$

where $x_k = k\Delta x$, and we have used (C.21) in the last step. Equation (C.67) is a sum in terms of differential events $\{x_k - \frac{1}{2}\Delta x \le x \le x_k + \frac{1}{2}\Delta x\}$ that are mutually exclusive; their union is the certain event S. The expectation $\mathrm{E}\{f(x)\}$ is obtained by multiplying the probability of each differential event by the value of the function when that event is true, followed by summing over all such events. The resulting limit, obtained when $\Delta x \to 0$, is the Lebesgue integral in the space S:

$$\mathrm{E}\{f(x)\} = \int_S f(x)\, d\mathrm{F}\,, \tag{C.68}$$

where $\mathrm{F}(x)$ is the cumulative distribution function defined in Sec. C.2.3. Only events with nonzero measure in S (those that have finite probability) contribute to the expected value.

Lebesgue integration is one possible solution to the inadequacy of Riemann integration. Another approach is to use the theory of distributions, which is presented in Chap. 2. In that chapter we show that the derivative of a step function is a delta function. Then the definition of the PDF in terms of the derivative of the cumulative distribution function (C.29) holds even for discrete variables in a generalized sense.

C.3.5 Relationship between probability spaces and measure spaces

We have invoked terms from the theory of measure spaces as well as those from the realm of probability. Friedman (1991) presents the following table correlating terms used in the discussion of probability spaces to those of measure spaces:

Sample probability space	Normalized measure space
Elementary event	Element in space
Event	Measure space in A
Certain event	S
Impossible event	$\emptyset$
Probability of $A = \Pr(A)$	Measure of $A = \mu(A)$
Almost surely	Almost everywhere
Random variable x	Bounded measurable function x
Expected value of x	Lebesgue integral of x
Limit in probability	Limit in measure
Limit almost surely	Limit almost everywhere

C.4 TWO RANDOM VARIABLES

In this section we extend our discussion of random variables to include descriptions of the outcomes of two or more experiments, or two or more repetitions of the same experiment. Random variables considered two at a time can be considered to be two-dimensional random vectors, or bivariate random vectors. Chapter 8 extends the concepts of this section to higher dimensionality.

C.4.1 Joint probability

In Sec. C.1.5 we used the language of set theory to write the joint probability of two events. The concepts developed there require only minor notational changes for use in the description of random variables considered jointly. Suppose we are interested in an experiment involving two continuous random variables denoted x and y. Their joint PDF is written $\mathrm{pr}(x, y)$. This PDF is properly normalized if it satisfies

$$\int_{-\infty}^{\infty}\int_{-\infty}^{\infty} \mathrm{pr}(x', y')\, dx'dy' = 1\,. \tag{C.69}$$

The two-dimensional cumulative distribution function is written

$$\mathrm{F}(x, y) = \int_{-\infty}^{x}\int_{-\infty}^{y} \mathrm{pr}(x', y')\, dx'dy'\,. \tag{C.70}$$

The joint PDF is related to the two-dimensional cumulative distribution function by

$$\mathrm{pr}(x, y) = \frac{\partial^2}{\partial x \partial y}\mathrm{F}(x, y)\,, \tag{C.71}$$

the two-dimensional analogue of (C.29).

Similar expressions result for discrete random variables:

$$\mathrm{Pr}(x_i \text{ and } y_i) = \mathrm{Pr}(x_i, y_i) \tag{C.72}$$

is the joint probability that the discrete random variables x and y take on the values x_i and y_i. Proper normalization requires that

$$\sum_{x_i}\sum_{y_i} \mathrm{Pr}(x_i, y_i) = 1\,, \tag{C.73}$$

where the sums run over all possible values of x_i and y_i. The cumulative distribution function for a discrete two-dimensional probability is

$$\mathrm{F}(x, y) = \mathrm{Pr}(x_i \leq x, y_i \leq y) = \sum_{x_i \leq x}\sum_{y_i \leq y} \mathrm{Pr}(x_i, y_i)\,. \tag{C.74}$$

In the next sections, expressions will be given only for continuous random variables. As we have just seen, as well as earlier in Sec. C.2, expressions for discrete random variables can be obtained from their continuous counterparts by replacing the continuous integrals of (C.70) and (C.69) with sums over the allowed values of the discrete random variables.

C.4.2 Marginal and conditional probability

When an experiment involves two (or more) random variables, we often find ourselves interested in the statistics of one particular random variable, regardless of the behavior of the other. The *marginal probability* of random variable x is found by integrating the joint density of x and y over all possible values for y:

$$\mathrm{pr}(x) = \int_{-\infty}^{\infty} \mathrm{pr}(x, y)\, dy\,. \tag{C.75}$$

Marginal probabilities are used to compute probabilities or expectations of a single random variable exactly as discussed in Sec. C.2 and Sec. C.3.

Another important kind of probability is the *conditional probability*, as discussed earlier in Sec. C.1.6. The conditional probability density of x given y, written $\mathrm{pr}(x|y)$, is found by dividing the joint probability density $\mathrm{pr}(x,y)$ by the probability density of y:

$$\mathrm{pr}(x|y) = \frac{\mathrm{pr}(x,y)}{\mathrm{pr}(y)} . \tag{C.76}$$

The total probability of x is found by integrating the joint probability of x and y over all possible values of y:

$$\mathrm{pr}(x) = \int_{-\infty}^{\infty} \mathrm{pr}(x|y)\,\mathrm{pr}(y)\; dy . \tag{C.77}$$

Often we are interested in the PDF on a random variable x conditioned on a particular event A, written $\mathrm{pr}(x|A)$. More generally, if a set of events $\{A_i\}$ partitions the event space, then the total probability $\mathrm{pr}(x)$ is a weighted sum of the conditional probabilities:

$$\mathrm{pr}(x) = \sum_i \mathrm{pr}(x|A_i)\,\mathrm{Pr}(A_i) . \tag{C.78}$$

The resulting density function $\mathrm{pr}(x)$ is called a *mixture* PDF because it is a mixture or linear combination of probability densities.

Statistical independence Two random variables are said to be statistically independent if the value of one of them has no influence on the other. When two random variables are independent, it follows that their joint PDF takes the form

$$\mathrm{pr}(x,y) = \mathrm{pr}(x)\,\mathrm{pr}(y) . \tag{C.79}$$

It can be shown that the two-dimensional cumulative distribution function of two independent random variables also factors:

$$\mathrm{F}(x,y) = \mathrm{F}(x)\,\mathrm{F}(y) . \tag{C.80}$$

C.4.3 Bayes' rule

Bayes' rule was introduced in Sec. C.1.7 above. It can be obtained in terms of two continuous random variables by combining (C.76) and (C.77) to give

$$\mathrm{pr}(x|y) = \frac{\mathrm{pr}(y|x)\,\mathrm{pr}(x)}{\int_{-\infty}^{\infty} \mathrm{pr}(y|x)\,\mathrm{pr}(x)\; dx} . \tag{C.81}$$

C.4.4 Expectations, joint moments and covariance

The expected value of x conditioned on a particular value of y is obtained as follows:

$$\mathrm{E}\{x|y\} = \int_{-\infty}^{\infty} dx\; x\,\mathrm{pr}(x|y) . \tag{C.82}$$

The conditional expectation can be averaged over the variable on which it is conditioned to determine the expectation of the other random variable:

$$\mathrm{E}\{x\} = \int_{-\infty}^{\infty} dy \; \mathrm{E}\{x|y\} \, \mathrm{pr}(y) \, . \tag{C.83}$$

The variance of a single variable can be derived from conditional expectations as follows:

$$\begin{aligned}
\mathrm{Var}\{x\} &= \left\langle \left\langle (x - \langle\langle x\rangle_{x|y}\rangle_y)^2 \right\rangle_{x|y} \right\rangle_y \\
&= \left\langle \left\langle (x - \langle x\rangle_{x|y} + \langle x\rangle_{x|y} - \langle\langle x\rangle_{x|y}\rangle_y)^2 \right\rangle_{x|y} \right\rangle_y \\
&= \left\langle \left\langle (x - \langle x\rangle_{x|y})^2 \right\rangle_{x|y} \right\rangle_y + \left\langle \left\langle (\langle x\rangle_{x|y} - \langle\langle x\rangle_{x|y}\rangle_y)^2 \right\rangle_{x|y} \right\rangle_y \\
&= \mathrm{E}_y\{\mathrm{Var}\{x|y\}\} + \mathrm{Var}_y\{\mathrm{E}\{x|y\}\} \, ,
\end{aligned} \tag{C.84}$$

where subscripts have been added to the variance and expectation on the last line as a reminder that they apply to the y variable.

The *covariance* of two random variables is defined to be

$$\mathrm{Cov}\{x, y\} \equiv \mathrm{E}\{(x - \langle x\rangle)\,(y - \langle y\rangle)\} = \mathrm{E}\{xy\} - \langle x\rangle\langle y\rangle \, . \tag{C.85}$$

If x and y are independent, their covariance is 0.

The *correlation coefficient* ρ of two random variables is defined by

$$\rho = \frac{\mathrm{Cov}\{x, y\}}{\sqrt{\sigma_x^2 \sigma_y^2}} \, . \tag{C.86}$$

If x and y are independent, their correlation coefficient is 0 and x and y are said to be *uncorrelated.* However, the converse is not true; a correlation coefficient of 0 does not imply statistical independence.

Let x and y be independent random variables, and consider two functions $g(x)$ and $h(y)$, where $g(x)$ is a function only of x and $h(y)$ is a function only of y. Then

$$\mathrm{E}\{g(x)h(y)\} = \mathrm{E}\{g(x)\}\mathrm{E}\{h(y)\} \, . \tag{C.87}$$

This relationship can be verified using (C.79):

$$\begin{aligned}
\mathrm{E}\{g(x)h(y)\} &= \int_{-\infty}^{\infty}\int_{-\infty}^{\infty} \mathrm{pr}(x, y)\, g(x)\, h(y)\; dx\, dy \\
&= \int_{-\infty}^{\infty}\int_{-\infty}^{\infty} \mathrm{pr}(x)\, \mathrm{pr}(y)\, g(x)\, h(y)\; dx\, dy \\
&= \int_{-\infty}^{\infty} \mathrm{pr}(x)\, g(x)\; dx \int_{-\infty}^{\infty} \mathrm{pr}(y)\, h(y)\; dy = \mathrm{E}\{g(x)\}\mathrm{E}\{h(y)\} \, .
\end{aligned} \tag{C.88}$$

C.4.5 Functions of two random variables

In Sec. C.3.1 a method was presented for determining the PDF for a function $f(x)$ of the random variable x. We are often interested in functions of the form $z = f(x, y)$ and $h(u, v) = f(x, y)$. In the following sections we describe methods for determining the PDF of functions of two random variables.

One function of two random variables We start by considering a simple function z of two random variables x and y. The cumulative distribution of z, $\mathrm{F}_z(z')$, represents the probability that $f(x,y) = z$ is less than or equal to z'. There is some region Ω of the event space for which this condition is satisfied. That is, the event $\{z \leq z'\}$ is equal to the event $\{(x,y) \in \Omega\}$. Integration of the joint probability density function $\mathrm{pr}_{xy}(x,y)$ over the region Ω gives $\mathrm{F}_z(z)$:

$$F_z(z) = \iint_{(x,y)\in\Omega} \mathrm{pr}_{xy}(x,y)\, dx\, dy\,. \tag{C.89}$$

There will be instances where knowledge of the joint PDF $\mathrm{pr}_{xy}(x,y)$ and the form of $f(x,y)$ are enough to allow determination of $\mathrm{pr}_z(z)$. (In principle this is always true, although in practice cranking through the integral in (C.89) can be quite difficult.) Further on in this section we shall describe a second approach that provides an alternative solution strategy, which is through the use of characteristic functions. We shall first apply (C.89) to two special cases, where z is first the sum and then the product of two random variables. Our treatment follows that of Stark and Woods (1986).

Example 1: Functions of the form $z = x + y$ Suppose two independent random variables x and y are summed to form the random variable z. We next show that the PDF for z in this special case is given by the convolution of the PDFs for x and y.

The cumulative distribution function for z gives the probability that z is less than some value:

$$\begin{aligned}\mathrm{F}_z(z) &= \iint_{x+y\leq z} \mathrm{pr}_{xy}(x,y)\, dx\, dy = \int_{-\infty}^{\infty} dy \int_{-\infty}^{z-y} \mathrm{pr}_{xy}(x,y)\, dx \\ &= \int_{-\infty}^{\infty} [I_{xy}(z-y,y) - I_{xy}(-\infty,y)]\, dy\,,\end{aligned} \tag{C.90}$$

where $I_{xy}(x,y)$ is the indefinite integral

$$I_{xy}(x,y) = \int_{-\infty}^{x} \mathrm{pr}_{xy}(x',y)\, dx'\,. \tag{C.91}$$

Differentiation of $\mathrm{F}_z(z)$ yields the PDF of z:

$$\mathrm{pr}_z(z) = \frac{d\,\mathrm{F}_z(z)}{dz} = \int_{-\infty}^{\infty} \frac{d}{dz}[I_{xy}(z-y,y)]\, dy = \int_{-\infty}^{\infty} \mathrm{pr}_{xy}(z-y,y)\, dy\,. \tag{C.92}$$

This is a general result; we have not yet invoked the independence of x and y. The independence of x and y implies that the joint probability factors, giving

$$\mathrm{pr}_z(z) = \int_{-\infty}^{\infty} \mathrm{pr}_x(z-y)\,\mathrm{pr}_y(y)\, dy\,. \tag{C.93}$$

Integrals of this form, called convolutions, are discussed in more detail in Chap. 3. Equation (C.93) is expressed by saying that the PDF of a sum of independent random variables is the convolution of the density functions of the underlying random variables.

Whenever a random variable is formed from the sum of two or more random variables, the mean of the result is the sum of the means of the underlying random variables. That is, if $z = x+y$, then $\overline{z} = \overline{x}+\overline{y}$. Moreover, if the underlying variables are independent, the variances of the underlying variables add. Thus, for $z = x+y$ with x and y independent,

$$\sigma_z^2 = \sigma_x^2 + \sigma_y^2 \,. \tag{C.94}$$

Let us now extend the example to that of a weighted sum of random variables. Let $z = ax + by$, where the coefficients a and b are nonrandom and the x and y are random. The expected value of z is given by

$$\begin{aligned} \mathrm{E}\{z\} = \mathrm{E}\{ax+by\} &= \int_{-\infty}^{\infty} dx \int_{-\infty}^{\infty} dy\ (ax+by)\,\mathrm{pr}(x,y) \\ &= \int_{-\infty}^{\infty} dx \int_{-\infty}^{\infty} dy\ ax\,\mathrm{pr}(x,y) + \int_{-\infty}^{\infty} dx \int_{-\infty}^{\infty} dy\ by\,\mathrm{pr}(x,y)\,. \end{aligned} \tag{C.95}$$

Each integral in the last line of (C.95) results in a marginal probability by (C.75):

$$\mathrm{E}\{z\} = \int_{-\infty}^{\infty} dx\ ax\,\mathrm{pr}_x(x) + \int_{-\infty}^{\infty} dy\ by\,\mathrm{pr}_y(y) = a\overline{x} + b\overline{y}\,. \tag{C.96}$$

Thus for any random variable that is formed as a weighted sum of random variables, the expected value of the sum is given by the weighted sum of the means of the underlying random variables. While (C.94) requires that x and y be independent, (C.96) holds for any correlation relationship between x and y.

The process that allowed us to simplify (C.95) by integrating out all but one of the random variables from the joint probability is called *marginalization*. We will find this technique to be quite useful in the treatment of random processes in Chap. 8 and beyond.

Example 2: Functions of the form $z = xy$ To find the PDF for this case, we start as we did in the previous example by first determining the cumulative distribution function for z. Since $\mathrm{F}_z(z)$ represents the probability that z is less than some value, we need to determine the boundaries of the region Ω that defines the event $z = xy \le z'$. Region Ω is bounded by contours of constant z in $x-y$ space given by hyperbolae of the form $y = z/x$ (see Fig. C.5). The integral of (C.89) becomes

$$\mathrm{F}_z(z) = \int_0^{\infty} dy \int_{-\infty}^{z/y} \mathrm{pr}_{xy}(x,y)\ dx + \int_{-\infty}^{0} dy \int_{z/y}^{\infty} \mathrm{pr}_{xy}(x,y)\ dx \qquad \text{for} \qquad z \ge 0\,. \tag{C.97}$$

Let $I_{xy}(x,y)$ again be the indefinite integral defined in (C.91). Equation (C.97) can then be rewritten for $z \ge 0$ as

$$\mathrm{F}_z(z) = \int_0^{\infty} dy\ [I_{xy}(z/y,y) - I_{xy}(-\infty,y)] + \int_{-\infty}^{0} dy\ [I_{xy}(\infty,y) - I_{xy}(z/y,y)]\,. \tag{C.98}$$

Differentiation of $\mathrm{F}_z(z)$ yields the PDF for z:

$$\mathrm{pr}_z(z) = \frac{d\,\mathrm{F}_z(z)}{dz} = \int_{-\infty}^{\infty} \frac{1}{|y|}\,\mathrm{pr}_{xy}(z/y,y)\ dy, \qquad z \ge 0\,. \tag{C.99}$$

It can be shown that this expression is also valid for $z < 0$. If x and y are statistically independent, the joint probability in (C.99) factors and the density function for z becomes

$$\mathrm{pr}_z(z) = \int_{-\infty}^{\infty} \frac{1}{|y|}\,\mathrm{pr}_x(z/y)\,\mathrm{pr}_y(y)\,dy\,, \qquad z \geq 0\,. \tag{C.100}$$

This expression is similar to the convolution result of (C.93), except that the shift $z - y$ is replaced by the ratio z/y. Expressions of this sort are known as Mellin convolutions; Mellin transforms are treated in more detail in Sec. 4.2.2.

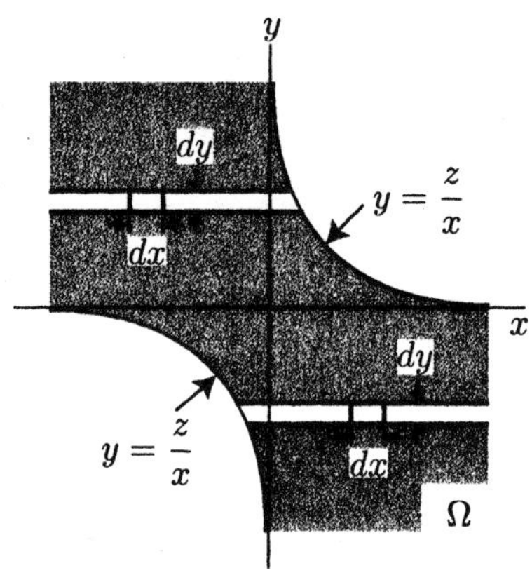

Fig. C.5 The boundaries of the region Ω that define the event $z = xy \leq z'$. Region Ω is bounded by contours of constant z in $x-y$ space given by hyperbolae of the form $y = z/x$.

Characteristic functions We return now to the subject of our first example. Consider again two independent random variables x and y with characteristic functions $\psi_x(\xi)$ and $\psi_y(\xi)$. The characteristic function of the random variable $z = x + y$ is given by

$$\psi_z(\xi) = \mathrm{E}\left\{e^{-2\pi i\xi z}\right\} = \mathrm{E}\left\{e^{-2\pi i\xi(x+y)}\right\} = \mathrm{E}\left\{e^{-2\pi i\xi x}\right\}\mathrm{E}\left\{e^{-2\pi i\xi y}\right\} = \psi_x(\xi)\psi_y(\xi)\,, \tag{C.101}$$

where we have made use of (C.87) to factor the expectation into functions of x and y. We found earlier that the density function for z was the convolution of the density functions for x and y. Equation (C.101) demonstrates that the characteristic function for z is the product of the characteristic functions for x and y. This relationship is as we should expect given that the characteristic function and the PDF form a Fourier transform pair (C.53). By virtue of the convolution theorem (3.132) we know that convolutions are transformed to simple products through Fourier transformation.

A similar derivation would show that if x and y are independent, with moment-generating functions $M_x(t)$ and $M_y(t)$, then the moment-generating function of the random variable $z = x + y$ is given by $M_z(t) = M_x(t)\,M_y(t)$.

Two functions of two random variables Suppose we have two continuous random variables, x and y, and two functions of those random variables $u = f_1(x, y)$ and $v = f_2(x, y)$. We are interested in the joint probability density $\mathrm{pr}_{uv}(u, v)$ of the new bivariate random vector (u, v). In this section we present a bivariate extension of the transformation of PDFs presented for functions of single random variables in Sec. C.3.1.

We assume that the functions $f_1(x,y)$ and $f_2(x,y)$ are differentiable and that they are also one-to-one and onto (see Sec. 1.3.4). Then the inverse mappings $x = h_1(u,v)$ and $y = h_2(u,v)$ are known to exist. The role played by the derivative in the univariate case [*cf.* (C.45)] is now played by the *Jacobian* of the transformation. The Jacobian, denoted J, is the determinant of the matrix of partial derivatives:

$$J = \det \begin{pmatrix} \frac{\partial x}{\partial u} & \frac{\partial x}{\partial v} \\ \frac{\partial y}{\partial u} & \frac{\partial y}{\partial v} \end{pmatrix} = \frac{\partial x}{\partial u}\frac{\partial y}{\partial v} - \frac{\partial x}{\partial v}\frac{\partial y}{\partial u}, \tag{C.102}$$

where

$$\frac{\partial x}{\partial u} = \frac{\partial h_1(u,v)}{\partial u} \tag{C.103a}$$

$$\frac{\partial x}{\partial v} = \frac{\partial h_1(u,v)}{\partial v} \tag{C.103b}$$

$$\frac{\partial y}{\partial u} = \frac{\partial h_2(u,v)}{\partial u} \tag{C.103c}$$

$$\frac{\partial y}{\partial v} = \frac{\partial h_2(u,v)}{\partial v}. \tag{C.103d}$$

Then, in direct analogy with (C.45), the joint PDF for (u,v) takes the form

$$\mathrm{pr}_{uv}(u,v) = \mathrm{pr}_{xy}[h_1(u,v), h_2(u,v)]\,|J|\,, \tag{C.104}$$

where J is assumed to be nonzero on the event space $\{x,y\}$.

C.5 CONTINUOUS PROBABILITY LAWS

C.5.1 Univariate normal

One of the most commonly encountered probability laws is the normal law, written in standard univariate form as

$$\mathrm{pr}(x) = \left(\frac{1}{2\pi\sigma^2}\right)^{\frac{1}{2}} \exp\left[-\frac{x^2}{2\sigma^2}\right], \quad -\infty < x < \infty. \tag{C.105}$$

This law is also commonly referred to as the Gaussian probability density, after the German mathematician and physicist Carl Friedrich Gauss (1777–1855). We shall use normal and Gaussian interchangeably. The normal law is used to model phenomena in almost all branches of science, including the distributions of height and weight in children, the IQ of a population, and the noise in some kinds of images. Its usefulness stems in part from the ease with which it can be manipulated, as well as its approximation to other distributions, as we shall see in later sections. Furthermore, under very weak assumptions it can be shown that a random variable formed by summing a set of other random variables is approximately normal (Sec. 8.3.4). Figure C.6 shows the familiar bell-shaped appearance of the normal distribution.

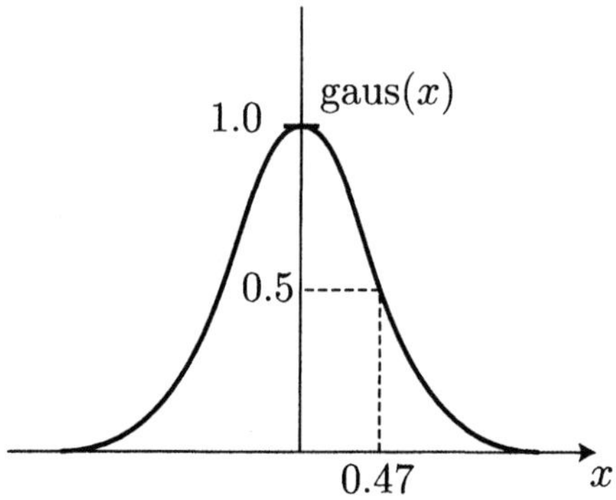

Fig. C.6 The standard normal distribution.

We shall first show that (C.105) is properly normalized. For normalization as a PDF we require

$$\left(\frac{1}{2\pi\sigma^2}\right)^{\frac{1}{2}} \int_{-\infty}^{\infty} \exp\left[-\frac{x^2}{2\sigma^2}\right] dx = 1 . \tag{C.106}$$

The integral shown above cannot be solved immediately since $\exp(-x^2)$ does not have an antiderivative. There is, however, a simple solution: the normalization integral can be manipulated into the form of a perfect differential by squaring and transforming to polar coordinates. The result is just the integral of an exponential, and we find immediately

$$\int_{-\infty}^{\infty} \exp\left[-\frac{x^2}{2\sigma^2}\right] dx = (2\pi\sigma^2)^{\frac{1}{2}} , \tag{C.107}$$

and so the PDF of (C.105) is indeed properly normalized.

Moments The mean of the PDF of (C.105) is zero because of the even symmetry and strong convergence of the integrand. The mean can be shifted to an arbitrary value $\overline{x}$ by writing

$$\mathrm{pr}(x) = \left(\frac{1}{2\pi\sigma^2}\right)^{\frac{1}{2}} \exp\left[-\frac{(x-\overline{x})^2}{2\sigma^2}\right] . \tag{C.108}$$

The symmetry now shifts to the point $\overline{x}$, making the first moment about $\overline{x}$ equal to zero. That $\overline{x}$ is indeed the first moment about the origin can be verified by letting $x' = x - \overline{x}$.

We calculate the second moment about the mean as

$$\langle (x-\overline{x})^2 \rangle = \left(\frac{1}{2\pi\sigma^2}\right)^{\frac{1}{2}} \int_{-\infty}^{\infty} dx\ (x-\overline{x})^2 \exp\left[-\frac{(x-\overline{x})^2}{2\sigma^2}\right] . \tag{C.109}$$

If we let $\beta = 1/2\sigma^2$ the definite integral can be identified as the derivative of the normalization integral; that is,

$$\int_{-\infty}^{\infty} dx\ (x-\overline{x})^2 \exp\left[-\frac{(x-\overline{x})^2}{2\sigma^2}\right] = -\frac{\partial}{\partial\beta} \int_{-\infty}^{\infty} dx\ \exp\left[-\beta(x-\overline{x})^2\right]$$

$$= -\frac{\partial}{\partial\beta}\left(\frac{\pi}{\beta}\right)^{\frac{1}{2}} = \tfrac{1}{2}\pi^{\frac{1}{2}}\beta^{-\frac{3}{2}} = \tfrac{1}{2}\pi^{\frac{1}{2}}\left(2\sigma^2\right)^{\frac{3}{2}} . \tag{C.110}$$

Multiplying by the normalization constant in front of the integral in (C.109), namely, $1/\sqrt{2\pi\sigma^2}$, we obtain

$$\left\langle (x-\overline{x})^2 \right\rangle = \sigma^2 . \tag{C.111}$$

We see therefore that the parameter σ^2 is the second moment of this density about the mean, *i.e.*, the variance. A random variable drawn from a normal distribution with parameters $\overline{x}$ and σ^2 is denoted $x \sim \mathcal{N}(\overline{x}, \sigma^2)$. A standard normal random variable has mean 0 and variance 1 and is denoted $x \sim \mathcal{N}(0,1)$.

By extending the previous steps we can show that all odd moments of this PDF about the mean are zero, and all even moments can be expressed in terms of σ^2. In particular, by applying $\partial/\partial\beta$ once more, we immediately find the fourth central moment:

$$\left\langle (x-\overline{x})^4 \right\rangle = 3\sigma^4 . \tag{C.112}$$

The kurtosis (C.41) of a Gaussian is thus 3. Distributions with kurtosis greater than 3 are more peaked than a Gaussian; they are referred to as *leptokurtic*. Kurtosis less than 3 yields a broader, flatter shape than that of a Gaussian; such distributions are referred to as *platykurtic*.

Cumulative distribution function Suppose we are given a normal random variable $x \sim \mathcal{N}(\overline{x}, \sigma^2)$ and we are interested in the probability that x is less than some value x'. The cumulative distribution contains this information:

$$\mathrm{F}(x') = \Pr(x < x') = \left(\frac{1}{2\pi\sigma^2}\right)^{\frac{1}{2}} \int_{-\infty}^{x'} \exp\left[-\frac{(x-\overline{x})^2}{2\sigma^2}\right] dx . \tag{C.113}$$

We convert this expression to one involving a standard normal by a transformation of variables. Let $y = (x-\overline{x})/\sigma$ and $y' = (x'-\overline{x})/\sigma$. Then

$$\mathrm{F}(x') = \frac{1}{\sqrt{2\pi}} \int_{-\infty}^{y'} \exp\left[-\tfrac{1}{2}y^2\right] dy$$

$$= \frac{1}{\sqrt{2\pi}} \int_{0}^{y'} \exp\left[-\tfrac{1}{2}y^2\right] dy - \frac{1}{\sqrt{2\pi}} \int_{0}^{-\infty} \exp\left[-\tfrac{1}{2}y^2\right] dy = \tfrac{1}{2}\,\mathrm{erf}\left(\frac{x'-\overline{x}}{\sigma\sqrt{2}}\right) + \frac{1}{2} , \tag{C.114}$$

where the function

$$\mathrm{erf}(a) \equiv \frac{2}{\sqrt{\pi}} \int_0^a \exp\left[-t^2\right] dt \tag{C.115}$$

is called the *error function*. Many probability and statistics books contain tables of error functions to facilitate the evaluation of interval probabilities of the form $\mathrm{pr}(a < x < b)$ for normal random variables. It should be noted that there is more than one definition in the literature for the error function.

Characteristic function We now derive the characteristic function of a normal random variable with mean equal to 0:

$$\psi(\xi) = \mathrm{E}\left\{e^{-2\pi i \xi x}\right\} = \frac{1}{\sqrt{2\pi\sigma^2}} \int_{-\infty}^{\infty} e^{-2\pi i \xi x} \exp\left[-\frac{x^2}{2\sigma^2}\right] dx$$

$$= \frac{1}{\sqrt{2\pi\sigma^2}} \boldsymbol{\mathcal{F}}\left\{\mathrm{gaus}\left(\frac{x}{\sqrt{2\pi\sigma^2}}\right)\right\} = \exp(-2\pi^2\xi^2\sigma^2) , \tag{C.116}$$

where we have made use of the Fourier relationships for Gaussians given in (3.177)–(3.180). Equation (C.116) can be generalized to a Gaussian random variable with mean $\overline{x}$ using the Fourier shift theorem (3.108).

C.5.2 Uniform distribution

A random variable is uniformly distributed if it has the following PDF:

$$\mathrm{pr}(x) = \begin{cases} 1/(b-a), & a < x \leq b \\ 0, & \text{elsewhere} \end{cases} . \tag{C.117}$$

The real parameters a and b define the limits of the possible outcome values of the random variable. This density function is also sometimes called the rectangular distribution. For shorthand, we shall denote a random variable that is drawn from a uniform distribution by $x \sim \mathcal{U}(a, b)$. By symmetry, the mean of the random variable is the midpoint: $\langle x \rangle = (b + a)/2$. It can also be shown that $\sigma^2 = (b - a)^2/12$. One optical application of this PDF is the description of a speckle source with a random phase uniformly distributed over 2π, so that $a = 0$, $b = 2\pi$, and $\mathrm{pr}(\theta) = 1/2\pi$.

C.5.3 Exponential distribution

The exponential probability law has many applications in imaging. It can be used to describe the random distribution of lifetimes of particles undergoing radioactive decay as well as the random intensity of a laser speckle pattern. A random variable x that has an exponential distribution has a PDF given by

$$\mathrm{pr}(x) = \begin{cases} \beta^{-1} e^{-x/\beta}, & x \geq 0 \\ 0, & x < 0 \end{cases} . \tag{C.118}$$

This density is properly normalized since

$$\int_0^\infty \frac{1}{\beta} e^{-x/\beta} \, dx = 1 . \tag{C.119}$$

It can be shown that the expected value of x is β and its variance is β^2.

The moment-generating function of an exponential random variable is

$$M(t) = \mathrm{E}\left\{e^{xt}\right\} = \frac{1}{\beta} \int_0^\infty e^{xt} e^{-x/\beta} \, dx = \frac{1}{1 - \beta t} . \tag{C.120}$$

A simple change of variables yields the characteristic function:

$$\psi(\xi) = \mathrm{E}\left\{e^{-2\pi i \xi x}\right\} = \frac{1}{1 + 2\pi i \xi \beta} . \tag{C.121}$$

C.5.4 Gamma and beta distributions

Gamma distribution The general form of the gamma probability density is written

$$\mathrm{pr}(x) = \frac{x^{\alpha-1} e^{-x/\beta}}{\beta^\alpha \Gamma(\alpha)} , \quad 0 \leq x < \infty , \tag{C.122}$$

where $\Gamma(\alpha)$ is the usual gamma function; for integer α, $\Gamma(\alpha) = (\alpha - 1)!$, but more generally $\Gamma(\alpha)$ is defined by

$$\Gamma(\alpha) = \int_0^\infty dx\, x^{\alpha-1} e^{-x}\,, \tag{C.123}$$

where $\alpha > 0$. A random variable that follows this law is said to be drawn from the $\Gamma(\alpha, \beta)$ law. The parameter α primarily influences the shape of the density function, while β primarily affects the scale of the PDF. It can be shown that the gamma density function is properly normalized since

$$\int_0^\infty x^{\alpha-1} e^{-x/\beta}\, dx = \beta^\alpha \Gamma(\alpha)\,. \tag{C.124}$$

The mean of a gamma distribution is given by

$$\mathrm{E}\{x\} = \frac{1}{\beta^\alpha \Gamma(\alpha)} \int_0^\infty x\, x^{\alpha-1} e^{-x/\beta}\, dx = \frac{1}{\beta^\alpha \Gamma(\alpha)} \beta^{\alpha+1} \Gamma(\alpha + 1) = \alpha\beta\,, \tag{C.125}$$

where we have made use of the normalization of the gamma distribution [*cf.* (C.123)] and definition (C.122). A similar derivation of $\mathrm{E}\{x^2\}$ would involve a $\Gamma(\alpha + 2, \beta)$ distribution, with the final result that the variance of a gamma-distributed random variable is $\alpha\beta^2$.

The moment-generating function of a gamma-distributed random variable is given by

$$M(t) = \frac{1}{\beta^\alpha \Gamma(\alpha)} \int_0^\infty e^{xt} x^{\alpha-1} e^{-x/\beta}\, dx = \frac{1}{\beta^\alpha \Gamma(\alpha)} \Gamma(\alpha) \left[\frac{\beta}{1 - \beta t}\right]^\alpha = \left[\frac{1}{1 - \beta t}\right]^\alpha, \tag{C.126}$$

where we again have made use of (C.124). The characteristic function is thus given by

$$\psi(\xi) = \mathrm{E}\left\{e^{-2\pi i \xi x}\right\} = \left[\frac{1}{1 + 2\pi i \xi \beta}\right]^\alpha. \tag{C.127}$$

It is left as an exercise for the reader to show that the gamma probability law describes a random variable formed by summing independent random variables that are each exponentially distributed. (*Hint*: Show that the product of N exponential moment-generating functions, each with exponential parameter β, gives the moment-generating function of a gamma distribution with $\alpha = N$.)

Beta distribution The beta function $B(\alpha, \beta)$, given by

$$B(\alpha, \beta) = \int_0^1 x^{\alpha-1} (1 - x)^{\beta-1}\, dx\,, \tag{C.128}$$

can be used to form a family of distributions known as the beta family:

$$\mathrm{pr}(x) = \frac{1}{B(\alpha, \beta)} x^{\alpha-1} (1 - x)^{\beta-1}\,, \qquad 0 < x < 1\,, \qquad \alpha > 0\,, \qquad \beta > 0\,. \tag{C.129}$$

From the definition of $B(\alpha, \beta)$ it is simple to show that the beta density function is properly normalized. The beta probability law is useful for describing continuous random variables restricted to lie between 0 and 1. For this reason it is often used to describe proportions. It reduces to the uniform law $\mathcal{U}(0,1)$ when $\alpha = \beta = 1$.

The beta function is related to the gamma function in the following way:

$$B(\alpha, \beta) = \frac{\Gamma(\alpha)\Gamma(\beta)}{\Gamma(\alpha+\beta)} . \tag{C.130}$$

Thus it is often possible to make use of the properties of gamma functions when manipulating beta-distributed random variables.

The form of the beta probability law makes calculation of its moments particularly easy:

$$\mathrm{E}\{x^n\} = \frac{1}{B(\alpha,\beta)} \int_0^1 x^n x^{\alpha-1}(1-x)^{\beta-1}\, dx = \frac{1}{B(\alpha,\beta)} B(\alpha+n, \beta) \tag{C.131}$$

by the definition of the beta function (C.128).

In particular,

$$\langle x \rangle = \frac{\alpha}{\alpha+\beta} \quad \text{and} \quad \mathrm{Var}\,\{x\} = \frac{\alpha\beta}{(\alpha+\beta)^2(\alpha+\beta+1)} . \tag{C.132}$$

C.5.5 Chi-squared random variables

Suppose we are given a standard normal random variable x. A new variable y, defined by $y = x^2$, is described by the chi-squared probability law. The form of this PDF for y can be determined using the law of transformation of PDFs of random variables (C.45):

$$\mathrm{pr}_y(y) = \mathrm{pr}_x(y) \left| \frac{dx}{dy} \right| = \frac{1}{\sqrt{2\pi y}}\, e^{-\frac{1}{2}y} . \tag{C.133}$$

More generally, if a random variable z is given by a sum of independent random variables,

$$z = \sum_{i=1}^{n} x_i^2 , \tag{C.134}$$

where each of the $x_i \sim \mathcal{N}(0,1)$, then

$$\mathrm{pr}(z) = \frac{z^{(n-2)/2}\, e^{-z/2}}{2^{n/2}\,\Gamma\left(\frac{n}{2}\right)} , \tag{C.135}$$

where we have made use of the fact that $\Gamma(\frac{1}{2}) = \sqrt{\pi}$. This probability density is referred to as *chi-squared*, or χ^2, with n *degrees of freedom*. We shall denote a random variable that is drawn from a chi-squared distribution of this form by $z \sim \chi^2_n$. It can be seen that the χ^2 distribution is a special case of the gamma distribution, with $\alpha = n/2$ and $\beta = 2$. The mean and variance of the chi-squared distribution can be determined from the results for the gamma distribution:

$$\mathrm{E}\{z\} = n \quad \text{and} \quad \mathrm{Var}(z) = 2n . \tag{C.136}$$

Note that in the special case where $n = 2$, or $z = x^2 + y^2$, (C.135) reduces to the exponential law.

To derive the characteristic function, consider again the simple case where $y = x^2$. The characteristic function is the Fourier transform of the PDF for y [*cf.* (C.133)]:

$$\psi_y(\xi) = \frac{1}{\sqrt{2\pi}} \int_0^\infty \exp(-2\pi i \xi y) \exp\left(-\frac{y}{2}\right) y^{-1/2} \, dy \,. \tag{C.137}$$

With the change of variables $x = \sqrt{y}$, this becomes

$$\psi_y(\xi) = \frac{1}{\sqrt{2\pi}} \int_0^\infty \exp(-2\pi i \xi x^2) \exp\left(\frac{-x^2}{2}\right) 2 \, dx$$

$$= \sqrt{\frac{2}{\pi}} \int_0^\infty \exp[-\tfrac{1}{2}x^2(1 + 4\pi i \xi)] \, dx = (1 + 4\pi i \xi)^{-1/2} \,. \tag{C.138}$$

The characteristic function of z, where z is again the sum of the squares of n mutually independent standard-normal random variables, is a generalization of (C.138):

$$\psi_z(\xi) = (1 + 4\pi i \xi)^{-n/2} \,. \tag{C.139}$$

The chi-squared law is used in describing the statistics of the irradiance for coherent imaging systems.

C.5.6 Rayleigh density function

The *generalized Rayleigh* PDF describes the behavior of a random variable $z = \sqrt{\sum_{i=1}^n x_i^2}$, where the $\{x_i\}$ are i.i.d. drawn from a $\mathcal{N}(0, \sigma^2)$ PDF. This density function describes the distribution of Euclidian distances from the origin to the point in an n-dimensional space defined by the $\{x_i\}$.

The *Rayleigh density function* refers to the special 2D case, where $z = \sqrt{x^2 + y^2}$, and x and y are independent, zero-mean Gaussian random variables. The form of the PDF is given by

$$\mathrm{pr}(z) = \frac{z}{\sigma^2} \exp\left(-\frac{z^2}{2\sigma^2}\right), \qquad z \geq 0 \,, \tag{C.140}$$

where σ^2 is the variance of both x and y.

The PDF given in (C.140) describes the magnitude of a complex random variable z with real part x and imaginary part y. The Rayleigh PDF is encountered in phase-insensitive detection systems, where x and y are the real and imaginary parts of a random field. Lord Rayleigh (1880) originally derived this PDF while considering the random-walk problem as it relates to scattered fields in acoustics. Johnson *et al.* (1994) give many important developments and applications of the Rayleigh distribution. One use for the Rayleigh law is for describing the statistics of the modulus of the complex field on an observation plane illuminated by light from a coherently illuminated rough surface.

C.5.7 Rician density function

In this case the random variable is formed by $z = \sqrt{x^2 + y^2}$, only now $x \sim \mathcal{N}(A, \sigma^2)$ while $y \sim \mathcal{N}(0, \sigma^2)$. The resulting probability density for z can be shown to be

$$\mathrm{pr}(z) = \frac{z}{\sigma^2} \exp\left[-\frac{1}{2}\frac{A^2 + z^2}{\sigma^2}\right] I_0\left(\frac{zA}{\sigma^2}\right), \qquad z \geq 0 \,, \tag{C.141}$$

where $I_0(\cdot)$ is the zero-order modified Bessel function of the first kind. The PDF of (C.141) is called the *Rician* density function or sometimes the Rice-Nakagami density function (Rice, 1944, 1945). Rice derived this PDF in the analysis of a sinusoidal signal current of known amplitude in the presence of a noise current. Irradiance, or the squared modulus of a field, follows an exponential law [see (8.232)]. An imaging application of the Rician law arises in the characterization of the statistics of the modulus of the complex field for coherent systems without complete phase randomization.

C.5.8 K distribution

The K distribution is a generalization of the Rayleigh and Rician laws. Whereas the Rayleigh and Rician distributions result from the summation of Gaussian random variables, the K distribution is the solution when the underlying random variables follow a K distribution. The K distribution is written

$$\mathrm{pr}(z) = \frac{2b}{\Gamma(N)}\left[\frac{bz}{2}\right]^N \mathrm{K}_{N-1}(bz)\,, \qquad z \geq 0\,, \tag{C.142}$$

where $b \geq 0$, $N > 0$, and $\mathrm{K}_{N-1}(\cdot)$ is a modified Bessel function of the second kind with order $N-1$. A family of K distributions is shown in Fig. C.7.

In a speckle problem, N is the number of reflectors and b is a scale parameter. As $N \to \infty$ the K distribution approaches a Rayleigh distribution.

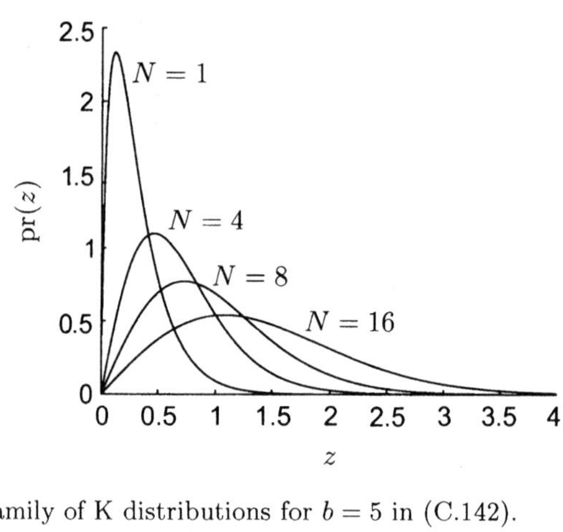

Fig. C.7 A family of K distributions for $b = 5$ in (C.142).

C.5.9 Log-normal probability law

If the logarithm of a random variable obeys a univariate normal law, the variable itself is said to obey a log-normal law. Consider a random variable z defined by

$$z = Ae^x\,, \tag{C.143}$$

or

$$x = \ln(z/A)\,. \tag{C.144}$$

The variable z is described by a log-normal probability law if x is normal. Note that a log-normal random variable is not the log of a normal one, but rather a random variable whose log is normal. In fact, this distribution might better be called the *antilognormal* distribution since it describes the behavior of an exponential, that is, antilogarithmic function of a random variable.

To find the density for z, we assume $x \sim \mathcal{N}(\overline{x}, \sigma^2)$ and use the transformation rule (C.45). The result is

$$\mathrm{pr}(z) = \frac{1}{\sqrt{2\pi}\,\sigma z} \exp\left[-\frac{(\ln z - \ln A - \overline{x})^2}{2\sigma^2}\right], \qquad (z > 0). \tag{C.145}$$

This density has three free parameters, A, $\overline{x}$ and σ^2. In terms of these parameters, the mean and variance of z are given by

$$\overline{z} = A \exp\left[\overline{x} + \tfrac{1}{2}\sigma^2\right], \qquad \mathrm{Var}\{z\} = A^2 \exp(2\overline{x}) \left[e^{2\sigma^2} - e^{\sigma^2}\right]. \tag{C.146}$$

The log-normal probability law often comes into play when a random variable z is the product of independent random variables:

$$z = \prod_{i=1}^{N} x_i\,. \tag{C.147}$$

Thus

$$\ln z = \sum_{i=1}^{N} \ln x_i\,. \tag{C.148}$$

Since $\ln x_i$ and $\ln x_j$ are statistically independent if x_i and x_j are, the right-hand side of (C.148) is the sum of N independent random variables. By the central-limit theorem (discussed in Chap. 8), the sum is approximately normal, so z is approximately a log-normal random variable.

The log-normal law can arise in the description of the pixel PDFs of reconstructions from tomographic data.

C.5.10 Distributions with infinite moments

There are probability laws that are important in optics that are not so easily manipulated to determine underlying moments and moment-generating functions. The next subsections describe three such interesting distributions.

$1/x$ distribution The $1/x$ probability law has the form

$$\mathrm{pr}(x) = \frac{K}{x}, \qquad 0 < a \le x \le b\,. \tag{C.149}$$

The normalization constant K can easily be determined:

$$K^{-1} = \int_a^b \frac{1}{x}\, dx = \ln(b) - \ln(a)\,. \tag{C.150}$$

The mean and variance of x are given by

$$\mathrm{E}\{x\} = \int_a^b K\, dx = K(b - a) \tag{C.151}$$

and

$$\mathrm{Var}\{x\} = \mathrm{E}\left\{x^2\right\} - [\mathrm{E}\{x\}]^2 = \frac{K}{2}\left(b^2 - a^2\right) - K^2\left(b - a\right)^2 . \tag{C.152}$$

Note that the variance $\to \infty$ if $a \to 0$ or $b \to \infty$.

The $1/x$ distribution with $a = 0$ and $b = \infty$ is sometimes used in Bayesian inference problems when it is desirable to have a probability law that is independent of scale or units. Jeffreys (1961) advocated this distribution as a noninformative prior. The fact that the PDF cannot be normalized for $a = 0$ turns out not to be important, and $1/x$ is referred to as an *improper prior.*

Cauchy probability law A *Cauchy* random variable is described by the following PDF:

$$\mathrm{pr}(x) = \frac{a}{\pi}\frac{1}{a^2 + x^2}, \qquad -\infty < x < \infty . \tag{C.153}$$

This probability law is named for Auguste Louis Cauchy (1789 – 1857), a French mathematician whose writings contributed greatly to many fields, including astronomy, optics, hydrodynamics, and function theory.

It is straightforward to verify that the Cauchy law is properly normalized. However, consider a Cauchy random variable with parameter $a = 1$. The evaluation of the expected value of this random variable leads to the integral

$$\mathrm{E}\{x\} = \int_{-\infty}^{\infty} x \frac{1}{\pi}\frac{1}{1+x^2}\, dx . \tag{C.154}$$

This integral does not strictly exist since it is not absolutely convergent, but it is reasonable to interpret it as a Cauchy principal value (see Sec. B.3.9). With this interpretation,

$$\mathrm{E}\{x\} = \mathcal{P}\int_{-\infty}^{\infty} x \frac{1}{\pi}\frac{1}{1+x^2}\, dx = \lim_{c\to\infty}\left[\int_{-c}^{c} x \frac{1}{\pi}\frac{1}{1+x^2}\, dx\right] = 0 . \tag{C.155}$$

Thus, with the principal-value interpretation, the mean is zero as expected from the symmetry of $\mathrm{pr}(x)$.

The variance of a Cauchy random variable does not exist because

$$\mathrm{E}\left\{x^2\right\} = \int_{-\infty}^{\infty} x^2 \frac{1}{\pi}\frac{1}{1+x^2}\, dx \tag{C.156}$$

fails to converge in any sense.

The characteristic function of a Cauchy random variable is given by

$$\psi(\xi) = \exp(-2\pi a|\xi|) . \tag{C.157}$$

The Cauchy law is useful for describing the distribution of light incident along a line below a point source. It can also be shown that the ratio of two independent random variables both drawn from standard normal distributions is a Cauchy random variable.

Lévy distributions Paul Lévy (1886 – 1971), a French mathematician, considered a generalized form of the central-limit theorem when he investigated the conditions under which a random variable $y = \sum_i^N x_i$, where all the x_i have distributions from the same family, has a distribution from the same family. Distributions for which this occurs are called *Lévy*, *Lévy-stable*, or simply *stable* distributions.

The PDF of a Lévy distribution has no closed form in general. The moment-generating function of a Lévy distribution is given by (Feller, 1971)

$$M(\xi) = \exp(-b|\xi|^q)\,, \tag{C.158}$$

where q is the Lévy index $(0 < q \leq 2)$ and b is a width parameter. When $q = 2$, the Lévy PDF is a Gaussian, and when $q = 1$ it is a Cauchy distribution. For $q < 2$ the tails of the density decrease sufficiently slowly that all moments beyond the first are divergent. Nevertheless, the width of the distribution can be described, and the parameter b performs this function. It follows from (C.101) that the sum of two independent Lévy random variables of index q and width b is another Lévy random variable of index q and width $2b$.

The name of the Italian sociologist Vilfredo Pareto is often connected to the Lévy distribution, hence Pareto-Lévy or Paretian distributions, owing to Pareto's investigation of income distributions using stable distributions. Mandelbrot (1960, 1963) made extensive use of stable distributions for analyzing financial time-series data. Lévy distributions have also been used to describe animal foraging paths and weather and earthquake patterns. In imaging, Lévy distributions are used to describe the outputs of bandpass or high-pass filters (see Sec. 8.4). Numerical routines are available for the simulation of Lévy distributions (Mantegna, 1994).

C.6 DISCRETE PROBABILITY LAWS

C.6.1 Bernoulli trials and binomial statistics

Bernoulli random variables In addition to the coin-tossing example given in Sec. C.2.1, there are many other examples of experiments where the outcome can be classified as either a success or failure and mapped to a random variable x accordingly. A success (*e.g.*, a head in the coin-tossing experiment) accords the value 1 to x, while a failure (a tail in the coin-tossing example) results in x being assigned the value 0. A random variable that is mapped to the set $\{0, 1\}$ according to the random outcome of an experiment is said to be a *Bernoulli* random variable. The probability law for a Bernoulli random variable is given by

$$\Pr(x = 0) = 1 - p$$

$$\Pr(x = 1) = p\,, \tag{C.159}$$

where p, the probability of a success, must be between 0 and 1. The mean and variance of a Bernoulli random variable are given by

$$\langle x \rangle = p \qquad \text{and} \qquad \sigma^2 = p(1-p)\,. \tag{C.160}$$

Binomial law Suppose we perform a sequence of experiments, with the outcome of each individual experiment scored as either success or failure. The probability of a success on each trial is p $(0 \leq p \leq 1)$ and the probability of a failure on each trial is $q = 1 - p$. That is, the outcome of each trial is a Bernoulli random variable. The probability of n successes in N trials is then described by the *binomial* law:

$$\Pr(n) = \binom{N}{n} p^n q^{N-n}, \tag{C.161}$$

where the binomial coefficient,

$$\binom{N}{n} \equiv \frac{N!}{n!(N-n)!}, \tag{C.162}$$

is often expressed verbally as N choose n since it gives the number of ways n objects can be chosen from a set of N indistinguishable objects. We shall denote a random variable that is drawn from a binomial probability distribution by $n \sim \mathcal{B}(p, N)$.

The mean and variance of a binomial random variable are given by

$$\langle n \rangle = Np \qquad \text{and} \qquad \sigma^2 = Np(1-p). \tag{C.163}$$

The binomial law describes selection processes in imaging. For example, the probability that a photon will be absorbed (or not) when it strikes a detector can be modeled by a binomial law.

Multinomial law The binomial law governs the probability of outcomes for a two-alternative Bernoulli trial experiment. When there are M distinct possible outcomes on each trial (for example, the tossing of a six-sided die gives $M = 6$), the statistics of the experiment are given by a multinomial law. Let N be the number of trials in the experiment, and let p_i be the probability of outcome i for each trial, where i runs from 1 to M. The number of times outcome i occurs is denoted n_i. The joint probability of the number of times each possible outcome occurs in N trials is given by:

$$\Pr(n_1, n_2, ..., n_M) = \frac{N!}{n_1! n_2! \cdots n_M!} p_1^{n_1} p_2^{n_2} \cdots p_M^{n_M} = N! \prod_{i=1}^{M} \frac{p_i^{n_i}}{n_i!}, \tag{C.164}$$

where $N = n_1 + n_2 + \cdots + n_M$ and $\sum_i p_i = 1$.

C.6.2 Poisson distribution

As we shall see in considerable detail in Chaps. 11 and 12, the Poisson probability law plays a fundamental role in problems where discrete events are counted. In optics and imaging, the events will often be photoelectric absorption processes. The randomness in the number of such processes, loosely referred to as *photon noise* or *shot noise*, is often the dominant noise in practical imaging systems.

Here we present a compendium of some of the important mathematical properties of Poisson random variables. A more complete discussion of these properties and the physical context in which they arise is given in Chap. 11.

Probability law, mean and variance A discrete random variable N with a sample space given by the natural numbers (zero and the positive integers) is said to be Poisson-distributed or to obey the Poisson probability law if

$$\Pr(N = n) = \frac{e^{-\lambda} \lambda^n}{n!}, \qquad n = 0, 1, 2, \ldots. \tag{C.165}$$

Note that this probability law is specified entirely by the single number λ, often called *the parameter* of the Poisson distribution.

As elsewhere in this appendix, we shall use the shorthand $\Pr(n)$ for $\Pr(N = n)$ where no confusion can arise. When it is necessary to indicate the parameter explicitly, we shall write the probability as $\Pr(n|\lambda)$, a notation that will prove advantageous in many applications where λ can be a random variable.

The parameter λ is also the mean value of N. To show this, we write

$$\langle N\rangle = \sum_{n=0}^{\infty} n\Pr(n) = \exp(-\lambda)\sum_{n=1}^{\infty} n\frac{\lambda^n}{n!}\,. \tag{C.166}$$

The change of limits is valid since 0! is defined to be one and hence the $n = 0$ term in the sum is zero. Letting $m = n-1$ and recognizing that $n/n! = 1/(n-1)! = 1/m!$, we find

$$\langle N\rangle = \lambda\exp(-\lambda)\sum_{m=0}^{\infty}\frac{\lambda^m}{m!} = \lambda\,. \tag{C.167}$$

Because of this result, we shall often use the symbol $\overline{N}$ instead of λ for the parameter of the Poisson distribution.

Higher moments must necessarily be expressible in terms of $\overline{N}$ (or λ) since that is the only parameter in $\Pr(n)$. By a derivation similar to the one just given for $\langle N\rangle$, we can show that the second moment is given by

$$\langle N^2\rangle = \overline{N}^2 + \overline{N}\,. \tag{C.168}$$

This result implies that the variance is equal to the mean, *i.e.*,

$$\mathrm{Var}\{N\} = \left\langle\left[N-\overline{N}\right]^2\right\rangle = \overline{N}\,. \tag{C.169}$$

The equality of the mean and the variance is a hallmark of the Poisson distribution. Only pure numbers (dimensionless quantities) can have this property since otherwise the dimensions of the variance would be the square of those of the mean. The Poisson law applies only to integer-valued random variables, and these integers have no dimensions attached to them.

Characteristic function and moment-generating function The moment-generating function for the Poisson distribution is given by

$$M(t) = \langle\exp(tN)\rangle = \exp\left(-\overline{N}\right)\sum_{N=0}^{\infty}\frac{\left(\overline{N}e^t\right)^N}{N!} = \exp\left[\overline{N}(e^t-1)\right]\,. \tag{C.170}$$

By a simple change of variables, the characteristic function is given by

$$\psi(\xi) = \langle\exp(-2\pi i\xi N)\rangle = \exp\left[\overline{N}\left(e^{-2\pi i\xi}-1\right)\right]\,. \tag{C.171}$$

From either of these functions, any desired moment can be computed. The first and second moments have already been given, and the third and fourth are:

$$\langle N^3\rangle = \overline{N}^3 + 3\overline{N}^2 + \overline{N}\,; \tag{C.172}$$

$$\langle N^4\rangle = \overline{N}^4 + 6\overline{N}^3 + 7\overline{N}^2 + \overline{N}\,. \tag{C.173}$$

This sequence can be continued by use of the recursion relation (Metz, 1969):

$$\langle N^k \rangle = \overline{N} \sum_{n=0}^{k-1} \binom{k-1}{n} \langle N^n \rangle . \tag{C.174}$$

A few of the central moments are:

$$\langle (N - \overline{N})^3 \rangle = \overline{N} ; \tag{C.175}$$

$$\langle (N - \overline{N})^4 \rangle = 3\overline{N}^2 + \overline{N} ; \tag{C.176}$$

$$\langle (N - \overline{N})^5 \rangle = 10\overline{N}^2 + \overline{N} ; \tag{C.177}$$

$$\langle (N - \overline{N})^6 \rangle = 15\overline{N}^3 + 5\overline{N}^2 + \overline{N} . \tag{C.178}$$

The factorial moments of a Poisson are particularly simple. The kth factorial moment is given by (Kotz *et al.*, 1986)

$$c_k = \left\langle N^{(k)} \right\rangle = \langle N(N-1) \cdots (N-k+1) \rangle = \overline{N}^k . \tag{C.179}$$

Recursive calculation of Poisson probabilities Because of the factorial, it can be difficult to compute $\Pr(n)$ directly for large n. The following recursive relation provides a solution to this problem:

$$\ln[\Pr(n+1)] = \ln[\Pr(n)] + \ln \overline{N} - \ln(n+1) . \tag{C.180}$$

Knowing $\ln[\Pr(n)]$, one can thus find $\ln[\Pr(n+1)]$ by adding a numerically well-behaved correction term $\ln \overline{N} - \ln(n+1)$. An equivalent relation is

$$(n+1) \Pr(n+1) = \overline{N} \Pr(n) . \tag{C.181}$$

Linear combinations of Poisson random variables The following theorems are given in Haight (1967) and Kotz *et al.* (1986): If N_1 and N_2 are independent Poisson random variables with means $\overline{N}_1$ and $\overline{N}_2$, respectively, then $N_1 + N_2$ is a Poisson random variable with mean $\overline{N}_1 + \overline{N}_2$. Note, however, that $N_1 - N_2$ is not a Poisson random variable since its variance is not equal to its mean; the variance of $N_1 - N_2$ is $\overline{N}_1 + \overline{N}_2$ if N_1 and N_2 are independent and Poisson, but the mean is $\overline{N}_1 - \overline{N}_2$. A linear combination $\alpha N_1 + \beta N_2$ of two independent Poisson random variables is a Poisson random variable if and only if $\alpha = \beta = 1$.

Conversely, if N_1 and N_2 are independent random variables and $N_1 + N_2$ is a Poisson random variable, then N_1 and N_2 must each be Poisson-distributed.

Approximations to the Poisson law A Poisson random variable takes on only integer values, but it is nevertheless sometimes convenient to approximate $\Pr(n)$ by a continuous function of n. Several such approximations are given in Haight (1967) and Abramowitz and Stegun (1965), but the most useful approach is to approximate a Poisson by a Gaussian. To obtain this approximation, we need only replace x with n, $\overline{x}$ with $\overline{N}$, and σ_x^2 with $\overline{N}$ in the general univariate normal density (C.105). The result is

$$\Pr(n) \simeq \left(2\pi\overline{N}\right)^{-1/2} \exp\left[-\left(n - \overline{N}\right)^2 / 2\overline{N}\right] . \tag{C.182}$$

This expression is an excellent approximation if $\overline{N}$ is greater than about 10, but is often usable even for $\overline{N}$ as small as 3.

C.6.3 Bose-Einstein probability law

The *Bose-Einstein* probability law is important in optics applications involving photocount statistics for thermal radiation. In Chap. 11 the Bose-Einstein law is shown to be the appropriate description for the distribution of photocounts resulting from the process of detecting an incident intensity that obeys an exponential probability law. The Bose-Einstein distribution is written

$$\Pr(n) = \frac{\overline{N}^n}{\left(1+\overline{N}\right)^{n+1}}, \qquad n = 0, 1, 2, \ldots, \tag{C.183}$$

where $\overline{N}$ is the mean of the Bose-Einstein random variable. To verify the normalization of this expression we note that

$$\sum_{n=0}^{\infty} \Pr(n) = \sum_{n=0}^{\infty} \frac{1}{1+\overline{N}} \left[\frac{\overline{N}}{1+\overline{N}}\right]^n . \tag{C.184}$$

The well-known relationship for a geometric series,

$$\sum_{k=0}^{\infty} a x^k = \frac{a}{1-x}, \tag{C.185}$$

can then be used to show that the sum of (C.184) is equal to 1. From this simple derivation it should be no surprise that the Bose-Einstein law is often referred to as the *geometric* distribution.

The Bose-Einstein distribution can be rewritten as

$$\begin{aligned} \Pr(n) &= \exp\left(n \ln \overline{N}\right) \exp\left[-(n+1)\ln\left(1+\overline{N}\right)\right] \\ &= \exp\left\{n\left[\ln \overline{N} - \ln\left(1+\overline{N}\right)\right] - \ln\left(1+\overline{N}\right)\right\} \\ &= A e^{-Cn}, \end{aligned} \tag{C.186}$$

where $A = 1/(1+\overline{N})$ and $C = \ln(1+\overline{N}) - \ln \overline{N}$. We see from (C.186) that the probability law for n is an exponential that has been discretely sampled because of the discrete nature of n.

Determination of the mean and variance of the distribution involve similar manipulations of geometric series. The variance of the Bose-Einstein law can be shown to be written in terms of the mean as:

$$\text{Var}\{N\} = \overline{N}\left(1+\overline{N}\right). \tag{C.187}$$

The characteristic function for the Bose-Einstein distribution is given by (Saleh, 1978)

$$\psi(\xi) = \left(1+\overline{N} - \overline{N}e^{-2\pi i \xi}\right)^{-1}. \tag{C.188}$$

The factorial moments of this distribution are given by (Goodman, 1985)

$$c_k = \text{E}\{n(n-1)\cdots(n-k+1)\} = k!\,\overline{N}^k. \tag{C.189}$$

C.6.4 Negative binomial law

A random variable that follows the negative binomial law is described by

$$\Pr(n) = \binom{n+N-1}{N-1} p^n (1-p)^N \qquad 0 \leq n < \infty\,, \tag{C.190}$$

where $0 \leq p \leq 1$.

The mean and variance are given by

$$\langle n \rangle = \frac{Np}{1-p} \qquad \text{and} \qquad \sigma^2 = \frac{Np}{(1-p)^2}\,. \tag{C.191}$$

The negative binomial law describes the number of failures n that occur before reaching a fixed number of successes N in a set of Bernoulli trials, where there are two possible outcomes, the outcomes are independent across trials, and the probability of failure p on each trial is the same. The negative binomial also can arise in doubly stochastic Poisson processes, where there is a Poisson random variable whose underlying parameter is itself drawn from a Gamma distribution.

C.7 Sampling methods

Above we have introduced a variety of analytical expressions for probabilities and probability density functions. Often we need to draw samples from one of these distributions on a computer, but most digital random-number generators provide only random variables uniformly distributed from 0 to 1. Fortunately there are methods by which such random-number generators can be used to generate random variables with arbitrary densities; we shall describe two of them here.

C.7.1 Rejection method

Suppose first that we wish to generate a sample from a PDF $\mathrm{pr}_u(u)$, where u is a continuous random variable defined on the interval (0, 1); an example would be the beta density defined in (C.129). In the rejection method, we generate a pair of random numbers, x and y, both uniformly distributed on (0, 1), where x is a candidate for the desired random number u. We accept the candidate and assign $u = x$ if $\mathrm{pr}_u(x) < y$. We can think of x and y as defining a point in an x-y plane; we accept only points that lie below the curve given by $y = \mathrm{pr}_u(x)$. Points corresponding to low PDF have a low probability of being accepted.

To see that the accepted variables follow the desired PDF, we note that the probability of an accepted variable x falling in a vanishingly small interval $(u - \frac{1}{2}\Delta u, u + \frac{1}{2}\Delta u)$ is the probability that a variable in this range is proposed in the first place times the probability that it is accepted. Since the proposals have PDF $\mathrm{pr}_x(x) = 1$, the probability of a proposal in $(u - \frac{1}{2}\Delta u, u + \frac{1}{2}\Delta u)$ is simply Δu, and the probability of acceptance is $\mathrm{pr}_u(u)$. Thus the probability of an accepted x being in $(u - \frac{1}{2}\Delta u, u + \frac{1}{2}\Delta u)$ is $\mathrm{pr}_u(u)\Delta u$, so it follows from (C.21) that the accepted variables are distributed according to $\mathrm{pr}_u(u)$.

The rejection method can be extended to any continuous random variable u defined on a finite interval (a, b) by drawing x from the uniform distribution on

$(0, 1)$, defining a new random number $x' = a + (b - 1)x$, and then proceeding as above with x' in place of x.

If the range of u is infinite, the rejection method can be applied only approximately. If $\text{pr}_u(u)$ falls off rapidly as $u \to \pm\infty$, it may be acceptable to truncate the PDF to a finite range and apply the rejection method. Of course, values outside this range will never be generated, but that may be acceptable in some cases. As the range gets wider, the approximation gets better but more of the proposals are rejected.

C.7.2 Cumulative-distribution method

This method requires knowledge of the cumulative distribution function (CDF) as defined by (C.26) or (C.30). We denote the CDF for a continuous random variable u as $F_u(\cdot)$, and we observe that this function has a unique inverse $F_u^{-1}(\cdot)$ since the CDF is monotonically increasing. We can draw a random variable t from a uniform density on $(0, 1)$ and define a new random variable x by

$$x = F_u^{-1}(t)\,. \tag{C.192}$$

The PDF on x is given from (C.45a) as

$$\text{pr}_x(x) = \text{pr}_t(t)\frac{dt}{dx}\,. \tag{C.193}$$

(No absolute-value signs are needed since $dt/dx \geq 0$.) Since $\text{pr}_t(t) = 1$, we see that

$$\text{pr}_x(x) = \frac{d}{dx}F_u(x) = \frac{d}{dx}\int_{-\infty}^{x} du\ \text{pr}_u(u) = \text{pr}_u(x)\,. \tag{C.194}$$

Thus the variable x generated by this procedure indeed has the desired PDF.

Bibliography

Aarsvold, J. N. (1993), Multiple-pinhole coded-aperture tomography: A model and analysis, Ph.D. Dissertation, University of Arizona, Tucson, AZ.

Abbey, C. K. and Barrett, H. H. (1995), Linear iterative reconstruction algorithms: Study of observer performance, in *Proc. 14th Intl. Conf. on Info. Proc. in Med. Imag.* (Bizais, Y., Barillot, C. and DiPaola, R., Eds.), Kluwer Academic, Dordrecht, pp. 65–76.

Abbey, C. K. and Denny, J. L. (1996), The Barankin bound: Instability in certain estimation problems, *Proc. SPIE* **2708**, 53–60.

Abbey, C. K., Clarkson, E., Barrett, H. H., Müller, S. P. and Rybicky, F. J. (1998), A method for approximating the density of maximum likelihood and maximum *a posteriori* estimates under a Gaussian noise model, *Med. Image Analysis*, **2**:4, 1–9.

Abbey, C. K., Eckstein, M. P. and Bochud, F. O. (1999), Estimation of human-observer templates in two-alternative forced-choice experiments, *Proc. SPIE* **3663**, 284–295.

Abbey, C. K. and Bochud, F. O. (2000), Modeling visual detection tasks in correlated image noise with linear model observers, in *Handbook of Medical Imaging*, Vol. 1: *Physics and Psychophysics* (Beutel, J., Kundel, H. and Van Metter, R., Eds.), SPIE Press, Bellingham, WA, pp. 629–654.

Abbey, C. K. and Barrett, H. H. (2001), Human and model-observer performance in ramp-spectrum noise: Effects of regularization and object variability, *J. Opt. Soc. Am. A* **18**, 473–488.

Abbey, C. K. and Eckstein, M. P. (2001), Maximum-likelihood and maximum-a posteriori estimates of human-observer templates, *Proc. SPIE* **4324**, 114–122.

Abbey, C. K., Zemp, R. I. and Insana, M. F. (2003), Pre-envelope deconvolution for increased lesion detection efficiency in ultrasonic imaging, *Proc. SPIE* **5034**, 280–288.

Abramowitz, M. and Stegun, I. A. (1965), *Handbook of Mathematical Functions*, Dover, New York.

Ahlfors, L. V. (1979), *Complex Analysis*, McGraw-Hill, New York.

Ahumada, A. J., Jr. and Watson, A. B. (1985), Equivalent-noise model for contrast detection and discrimination, *J. Opt. Soc. Am.* **2**, 1133–1139.

Ahumada, A. J., Jr. and Null, C. H. (1993), Image quality: A multidimensional problem, in *Digital Images and Human Vision* (Watson, A. B., Ed.), MIT Press, Cambridge, MA, pp. 141–148.

Alazraki, N. P. and Mishkin, F. S., Eds. (1988), *Fundamentals of Nuclear Medicine*, 2nd edition, Society of Nuclear Medicine, New York.

Albert, A. (1972), *Regression and the Moore-Penrose Pseudoinverse*, Academic Press, New York.

Alieva, T., Lopez, V., Lopez, G. A. and Almeida, L. B. (1994), The fractional Fourier transform in optical propagation problems, *J. Mod. Opt.* **41**, 1037–1044.

Alpert, B. (1993), A class of bases in $\mathbb{L}^2$ for the sparse representation of integral operators, *SIAM J. Math. Anal.* **24**, 246–262.

American College of Radiology (ACR) (1998), Breast Imaging Reporting and Data System (BI-RADS), Third Edition, Reston, Va.

Anand, V. B. (1993), *Computer Graphics and Geometric Modeling for Engineers*, Wiley, New York.

Anderson, T. M., Jr., Mintzer, R. A., Hoffer, P. B., Lusted, L. B., Smith, V. C. and Pokorny, J. (1973), Nuclear image transmission by picturephone: Evaluation by ROC curve method, *Invest. Radiol.* **8**, 244–250.

Anderson, T. W. (1971), *An Introduction to Multivariate Statistical Analysis* John Wiley & Sons, New York.

Andrews, H. C. and Hunt, B. R. (1977), *Digital Image Restoration*, Prentice-Hall, Englewood Cliffs, NJ.

Andrews, G. E. and Burge, W. H. (1993), Determinant identities, *Pacific J. Math.* **158**, 1–14.

Anger, H. O. (1958), Scintillation camera, *Rev. Sci. Instrum.* **29**, 27–33.

Anger, H. O. (1964), Scintillation camera with multichannel collimators, *J. Nucl. Med.* **5**, 515–531.

Arbuzov, E. V., Bukhgeim, A. L. and Kazantsev, S. G. (1998), Two-dimensional tomography problems and the theory of A-analytic functions, *Siberian Adv. Math.* **8**, 1–20.

Arfken, G. B. and Weber, H. J. (1995), *Mathematical Methods for Physicists*, 4th ed., Academic Press, San Diego, CA.

Armstrong, J. T., Hutter, D. J., Johnston, K. J. and Mozurkewich, D. (May 1995), Stellar optical interferometry in the 1990s, *Phys. Today* **48**:5, 42–49.

Armstrong, J. T., Mozurkewich, D., Rickard, L. J., Hutter, D. J., Benson, J. A., Bowers, P. F., Elias II, N. M., Hummel, C. A., Johnston, K. J., Buscher, D. F., Clark III, J. H., Ha, L., Ling, L.-C., White, N. M. and Simon, R. S. (1998), The Navy prototype optical interferometer, *Astrophys. J.* **496**, 550–571.

Armstrong, M. A. (1988), *Groups and Symmetry*, Springer-Verlag, New York.

Arridge, S. R., Schweiger, M., Hiraoka, M. and Delpy, D. T. (1993), A finite element approach for modelling photon transport in tissue, *Med. Phys.* **20**:2, 299–309.

Ashcroft, N. W. and Mermin, N. D. (1976), *Solid State Physics*, W. B. Saunders, Philadelphia, PA.

Backus, G. and Gilbert, F. (1967), Numerical applications of a formalism for geophysical inverse problems, *Geophys. J. Roy. Astr. Soc.* **13**, 247–276.

Backus, G. and Gilbert, F. (1968), The resolving power of gross earth data, *Geophys. J. Roy. Astr. Soc.* **16**, 169–205.

Bak, P., Tang, C. and Wiesenfeld, K. (1988), Self-organized criticality, *Phys. Rev. A* **38**, 364–374.

Baltes, H. P., Geist, J. and Walther, A. (1978), Radiometry and coherence, in *Inverse Source Problems in Optics* (Baltes, H. P., Ed.), Springer-Verlag, Berlin, Chap. 5.

Bamber, D. (1975), The area above the ordinal dominance graph and the area below the receiver operating graph, *J. Math. Psych.* **12**, 387–415.

Banvard, R. A., Ed. (1996), The Visible Human Project Conference Proceedings, October 7-8, 1996, National Institutes of Health, William H. Natcher Conference Center, Bethesda, Maryland, USA.

Banvard, R. A., Pinciroli, F. and Cerveri, P., Eds. (1998), The Second Visible Human Project Conference Proceedings, October 1-2, 1998, National Institutes of Health, William H. Natcher Conference Center, Bethesda, Maryland, USA.

Banvard, R. A., Ed. (2000), The Third Visible Human Project Conference Proceedings, October 5-6, 2000, National Institutes of Health, William H. Natcher Conference Center, Bethesda, Maryland, USA.

Barakat, R. (1976), Sums of independent lognormally distributed random variables. *J. Opt. Soc. Am.* **66**, 211–216.

Barakat, R. (1986), Weak-scatter generalization of the K-density function with application to laser scattering in atmospheric turbulence, *J. Opt. Soc. Am. A* **3**:4, 401–409.

Barankin, E. W. (1949), Locally best unbiased estimates, *Ann. Math. Stat.* **20**, 477–501.

Barlow, H. B. (1956), Retinal noise and absolute threeshold, *J. Opt. Soc. Am.* **45**, 634–639. Reprinted in Cohn, T. E., Ed. (1993), Collected Works in Optics, Vol. 3: *Visual Detection*, Optical Society of America, Washington, DC, pp. 47–52.

Barlow, H. B. (1978), The efficiency of detecting changes in density in random dot patterns, *Vision Res.* **18**, 637–650.

Barlow, H. B. and Reeves, B. C. (1979), The versatility and absolute efficiency of detecting mirror symmetry in random dot displays, *Vision Res.* **19**, 783–793.

Barlow, H. B. (1989), Unsupervised learning, *Neural Comput.* **1**, 295–311.

Barrett, H. H. and Swindell, W. (1981), *Radiological Imaging: Theory of Image Formation, Detection, and Processing*, Vols. I and II, Academic Press, New York.

Barrett, H. H. (1982), Dipole-sheet transform, *J. Opt. Soc. Am.* **72**, 468–475.

Barrett, H. H. (1984a), The Radon transform and its applications, in *Progress in Optics*, Vol. 21 (Wolf, E., Ed.), Elsevier Science, Amsterdam.

Barrett, H. H. (1984b), Three-dimensional image reconstruction from planar projections, with application to optical data processing, in *Transformations in Optical Signal Processing* (Rhodes, W. T., Fienup, J. R. and Saleh, B. E. A., Eds.), SPIE, Bellingham, WA.

Barrett, H. H., Rolland, J. P., Wagner, R. F. and Myers, K. J. (1989), Detection and discrimination of known signals in inhomogeneous, random backgrounds, *Proc. SPIE* **1090**, 176–182.

Barrett, H. H. (1990), Objective assessment of image quality: Effect of quantum noise and object variability, *J. Opt. Soc. Am. A* **12**, 834–852.

Barrett, H. H., Aarsvold, J. N. and Roney, T. J. (1991), Null functions and eigenfunctions: tools for the analysis of imaging systems, in *Information Processing in Medical Imaging* (Ortendahl, D. A. and Llacer, J., Eds.), Wiley-Liss, New York, pp. 211–226.

Barrett, H. H., Gooley, T. A., Girodias, K. A., Rolland, J. P., White T. A. and Yao, J. (1992), Linear discriminants and image quality, *Image Vision Comp.* **10**, 451–460.

Barrett, H. H., Yao, J., Rolland, J. P. and Myers, K. J. (1993), Model observers for assessment of image quality, *Proc. Natl. Acad. Sci. USA* **90**, 9758–9765.

Barrett, H. H., Wilson, D. W. and Tsui, B. M. W. (1994), Noise properties of the EM algorithm: I. Theory, *Phys. Med. Biol.* **39**, 833–846.

Barrett, H. H. and Gifford, H. C. (1994), Cone-beam tomography with discrete data sets, *Phys. Med. Biol.* **39**, 451–476.

Barrett, H. H., Denny, J. L., Wagner, R. F. and Myers, K. J. (1995), Objective assessment of image quality. II. Fisher information, Fourier crosstalk, and figures of merit for task performance, *J. Opt. Soc. Am. A* **12**, 834–852.

Barrett, H. H., Denny, J. L., Gifford, H. C., Abbey, C. K., Wagner, R. F. and Myers, K. J. (1996), Generalized NEQ: Fourier analysis where you would least expect to find it, *Proc. SPIE* **2708**, 41–52.

Barrett, H. H. and Swindell, W. (1996), *Radiological Imaging: Theory of Image Formation, Detection and Processing* (paperback edition, print on demand), Academic Press, New York.

Barrett, H. H. and Abbey, C. K. (1997), Bayesian detection of random signals on random backgrounds, in *Information Processing in Medical Imaging* (Duncan, J. and Gindi, G., Eds.), Springer Lecture Notes in Computer Science, **1234**, Springer-Verlag, New York, pp. 155–166.

Barrett, H. H., Wagner, R. F. and Myers, K. J. (1997a), Correlated point processes in radiological imaging, *Proc. SPIE* **3032**, 110–124.

Barrett, H. H., White, T. and Parra, L. C. (1997b), List-mode likelihood, *J. Opt. Soc. Am. A* **14**, 2914–2923.

Barrett, H. H., Abbey, C. K. and Clarkson, E. (1998a), Some unlikely properties of the likelihood ratio and its logarithm, *Proc. SPIE* **3340**, 65–77.

Barrett, H. H., Abbey, C. K. and Clarkson, E. (1998b), Objective assessment of image quality. III. ROC metrics, ideal observers, and likelihood-generating functions, *J. Opt. Soc. Am. A* **15**, 1520–1535.

Barrett, H. H., Abbey, C. K., Gallas, B. and Eckstein, M. (1998c), Stabilized estimates of Hotelling-observer detection performance in patient-structured noise, *Proc. SPIE* **3340**, 27–43.

Barrett, H. H., Gallas, B., Clarkson, E. and Clough, A. (1998d), Scattered radiation in nuclear medicine: A case study in the Boltzmann transport equation, in *Computational Radiology and Imaging: Therapy and Diagnosis* (Borgers, C. and Natterer, F., Eds.), Springer-Verlag, New York.

Barrett, H. H., Myers, K. J., Gallas, B., Clarkson, E. and Zhang, H. (2001), Megalopinakophobia: Its symptoms and cures, *Proc. SPIE* **4320**, 299–307.

Barten, P. G. J. (1987), The SQRI method: a new method for the evaluation of visible resolution on a display, *Proc. Soc. Info. Display* **28**, 253–262.

Barten, P. G. J. (1990), Subjective image quality of high-definition television pictures, *Proc. Soc. Info. Display* **31**, 239–243.

Barten, P. G. J. (1992), Physical model for the contrast sensitivity of the human eye, *Proc. SPIE* **1666**, 57–72.

Barten, P. G. J. (1993), Spatio-temporal model for the contrast sensitivity of the human eye and its temporal aspects, *Proc. SPIE* **1913**, 2–14.

Bastiaans, M. J. (1978), The Wigner distribution function applied to optical signals and systems, *Opt. Commun.* **25**, 26–30.

Bastiaans, M. J. (1979a), Transport equation for the Wigner distribution function, *Opt. Acta* **26**, 1265–1272.

Bastiaans, M. J. (1979b), Transport equation for the Wigner distribution function in an inhomogeneous and dispersive medium, *Opt. Acta* **26**, 1333–1344.

Bastiaans, M. (1981), Signal description by means of a local frequency spectrum, *Proc. SPIE* **373**, 49–62.

Bastiaans, M. (1994), Gabor's signal expansion and the Zak transform, *Appl. Opt.* **33**, 5241–5255.

Bates, R. H. T. and McDonnell, M. J. (1986), *Image Restoration and Reconstruction*, No. 16 in Oxford Engineering Science Series, Oxford University Press, New York.

Bayes, T. (1764), An essay towards solving a problem in the doctrine of chances, *Philos. Trans. Roy. Soc.* **53**, 370–418. Reprinted in *Biometrika* **45**, 296–315, 1958.

Beam, C., Layde, P. M. and Sullivan, D. C. (1996), Variability in the interpretation of screening mammograms by US radiologists, *Arch. Intern. Med.* **156**, 209–213.

Beck, R. N. (1964a), A theory of radioisotope scanning systems, in *Medical Radioisotope Scanning*, Vol. 1, IAEA, Vienna, pp. 35–56.

Beck, R. N. (1964b), A theory of radioisotope scanning systems, in *Medical Radioisotope Scanning*, Vol. 1, IAEA, Vienna, pp. 211–231.

Beck, R. N. (1968a), The scanning system as a whole: General considerations, in *Fundamental Problems in Scanning* (Gottschalk, A. and Beck, R. N., Eds.), Thomas, Springfield, IL, Chap. 3.

Beck, R. N. (1968b), Collimation of gamma rays, in *Fundamental Problems in Scanning* (Gottschalk, A. and Beck, R. N., Eds.), Thomas, Springfield, IL, Chap. 6.

Begg, C. B. and Greenes, R. A. (1983), Assessment of diagnostic tests when disease verification is subject to selection bias, *Biometrics* **39**, 207–215.

Begg, C. B. (1987), Biases in the assessment of diagnostic tests, *Stat. Med.* **6** 411–423.

Begg, C. B. and McNeil, B. J. (1988), Assessment of radiologic tests: Control of bias and other design considerations, *Radiology* **167**, 565–569.

Begg, C. B. and Metz, C. E. (1990), Consensus diagnoses and "Gold Standards," *Med. Decis. Making* **10**, 29–30.

Beiden, S. V., Wagner, R. F. and Campbell, G. (2000a), Components-of-variance models and multiple-bootstrap experiments: A alternative method for random-effects, receiver operating characteristic analysis, *Acad. Radiol.* **7**, 341–349.

Beiden, S. V., Campbell, G., Meier, K. L. and Wagner, R. F. (2000b), On the problem of ROC analysis without truth: The EM algorithm and the information matrix, *Proc. SPIE* **3981**, 126–134.

Beiden, S. V., Wagner, R. F., Campbell, G., Metz, C. E. and Jiang, Y. (2001a), Components-of-variance models for random-effects ROC analysis: The case of unequal variance structures across modalities, *Acad. Radiol.* **8**, 605–615.

Beiden, S. V., Wagner, R. F., Campbell, G. and Chan, H.-P. (2001b), Analysis of uncertainties in estimates of components of variance in multivariate ROC analysis, *Acad. Radiol.* **8**, 616–622.

Bell, A. J. and Sejnowski, T. J. (1997), The 'independent components' of natural scenes are edge filters, *Vision Res.* **37**:23, 3327–3338.

Bell, E. T. (1937), *Men of Mathematics*, Simon & Schuster, New York.

Bellini, S., Piacentini, M., Cafforio, C. and Rocca, F. (1979), Compensation of tissue absorption in emission tomography, *IEEE Trans. Acoust. Speech Signal Process.* **27**, 213–2188.

Bellman, R. (1995), *Introduction to Matrix Analysis*, Society for Industrial and Applied Mathematics, Philadelphia, PA.

Ben-Israel, A. and Greville, T. N. E. (1974), *Generalized Inverses: Theory and Application*, Wiley, New York.

Berbaum, K. S., Franken, E. A., Dorfman, D. D. and Caldwell, R. T. (2000), Proper ROC analysis and joint ROC analysis of the satisfaction of search effect in chest radiology, *Acad. Radiol.* **7**, 945–958.

Bernamont, J. (1937), Fluctuations in the resistance of thin films, *Proc. Phys. Soc.* **49**, 138–139.

Bertero, M., DeMol, C. and Pike, E. R. (1985), Linear inverse problems with discrete data: I. General formulation and singular system analysis, *Inverse Probl.* **1**, 301–330.

Bertero, M. (1989), Linear inverse and ill-posed problems, in *Advances in Electronics and Electron Physics*, Vol. 75 (Hawkes, P. W., Ed.), Academic Press, New York.

Bertero, M. and Boccacci, P. (1998), *Introduction to Inverse Problems in Imaging*, Institute of Physics, Bristol.

Bertero, M., DeMol, C. and Pike, E. R. (1988), Linear inverse problems with discrete data: II. Stability and regularization, *Inverse Probl.* **4**, 573–594.

Besag, J. (1973), Spatial interaction and the statistical analysis of lattice systems (with discussion), *J. Roy. Stat. Soc. B* **36**, 192–236.

Besag, J., Green, P., Higdon, D. and Mengersen, K. (1995), Bayesian computation and stochastic systems, *Stat. Sci.* **10**, 3–66.

Bhattacharyya, A. (1943), On a measure of divergence between two statistical populations defined by their probability distributions, *Bull. Calcutta Math. Soc.* **35**, 99–110.

Bhattacharyya, A. (1946), On some analogues of the amount of information and their use in statistical estimation, *Sankhia* **8**, 1–14.

Bhattacharyya, A. (1947), On some analogues of the amount of information and their use in statistical estimation, *Sankhia* **8**, 201–218.

Bhattacharyya, A. (1948), On some analogues of the amount of information and their use in statistical estimation, *Sankhia* **8**, 315–328.

Bialynicki-Birula, I. (1998), Exponential localization of photons, *Phys. Rev. Lett.* **80**, 5247–5250.

Big Bird (1969), The letter E, Sesame Street, Corporation for Public Broadcasting, New York.

Bjorken, J. D. and Drell, S. D. (1964), *Relativistic Quantum Mechanics,* McGraw-Hill, New York.

Blackwell, H. R. (1946), Contrast thresholds of the human eye, *J. Opt. Soc. Am.* **36**, 624–643.

Blackwell, K. T. (1998), The effect of white and filtered noise on contrast detection thresholds, *Vision Res.* **38**, 267–280.

Blake, R., Cool, S. J. and Crawford, M. L. J. (1974), Visual resolution in the cat, *Vision Res.* **14**, 1211–1217.

Blakemore, C. and Campbell, F. W. (1969), On the existence of neurons in the human visual system selectively sensitive to orientation and size of retinal images, *J. Physiol.* **203**, 237–260.

Blume, H. and Hemminger, B. M. (1997), Image presentation in digital radiology: Perspectives on the emerging DICOM display function standard and its application, *Radiographics* **17**:3, 769–777.

Bochud, F. O., Verdun, F. R., Hessler, C. and Valley, J. F. (1995), Detectability on radiological images: The effect of the anatomical noise, *Proc. SPIE* **2436**, 156–164.

Bochud, F. O., Abbey, C. K. and Eckstein, M. P. (1999a), Statistical texture synthesis of mammographic images with clustered lumpy backgrounds, *Opt. Express* **4**, 193–199.

Bochud, F. O., Abbey, C. K. and Eckstein, M. P. (1999b), Further investigation of the effect of phase spectrum on visual detection in structured backgrounds, *Proc. SPIE* **3663**, 273–281.

Boff, K. R., Kaufman, L. and Thomas, J. P. (1986), *Handbook of Perception and Human Performance,* Wiley, New York.

Bonetto, P., Qi, J. and Leahy, R. M. (2000), Covariance approximation for fast and accurate computation of channelized Hotelling observer statistics, *IEEE Trans. Nucl. Sci.* **47**, 1567–1572.

Bookstein, F. L. (1991), *Morphometric Tools for Landmark Data: Geometry and Biology*, Cambridge University Press, Cambridge.

Borden, B. (1999), *Radar Imaging of Airborne Targets: A Primer for Applied Mathematicians and Scientists*, Institute of Physics, Bristol and Philadelphia.

Born, M. and Wolf, E. (1999), *Principles of Optics*, 7th (expanded) ed., Cambridge University Press, Cambridge.

Boullion, T. L. and Odell, P. (1971), *Generalized Inverse Matrices*, Wiley, New York.

Box, E. P. and Tiao, G. C. (1992), *Bayesian Inference in Statistical Analysis*, reprint edition, Wiley-Interscience, New York.

Boyer, C. B. and Merzbach, U. C. (1989), *A History of Mathematics*, Wiley, New York.

Bracewell, R. (1965), *The Fourier Transform and Its Applications*, McGraw-Hill, New York.

Bracewell, R. N. and McPhie R. H. (1979), Searching for non-solar planets, *Icarus* **38**:1, 136–147.

Bregman, L. M. (1965), Finding the common point of convex sets by the method of successive projections, *Dokl. Akad. Nauk* **162**, 487–490.

Breiman, L. (1992), *Probability*, Society for Industrial and Applied Mathematics, Philadelphia, PA.

Brigham, E. O. (1974), *The Fast Fourier Transform*, Prentice-Hall, Englewood Cliffs, NJ.

Brillouin, L. (1956), *Science and Information Theory*, Academic Press, New York.

Brown, D. G., Insana, M. F. and Tapiovaara, M. (1995), Detection performance of the ideal decision function and its Maclaurin expansion: Signal position unknown, *J. Acoust. Soc. Am.* **97**, 379–398.

Bube, R. H. (1960), *Photoconductivity of solids*, Wiley, New York.

Bunch, P. C., Hamilton, J. F., Sanderson, G. K. and Simmons, A. H. (1978), A free response approach to the measurement and characterization of radiographic-observer performance, *J. Appl. Photogr. Eng.* **4**, 166–171.

Bunch, P. C., Hamilton, J. F., Sanderson, G. K. and Simmons, A. H. (1997), A free response approach to the measurement and characterization of radiographic observer performance, *Proc. SPIE* **127**, 124–135.

Buonocore, M. H., Brody, W. R. and Macovski, A. (1981), A natural pixel decomposition for two-dimensional image reconstruction, *IEEE Trans. Bio-Med. Eng.* **28**, 69–78.

Burgess, A. (1992), *A Mouthful of Air*, William Morrow, New York.

Burgess, A. E., Humphrey, K. and Wagner, R. F. (1979), Detection of bars and discs in quantum noise, *Proc. SPIE* **173**, 34–40.

Burgess, A. E., Wagner, R. F., Jennings, R. J. and Barlow, H. B. (1981), Efficiency of human visual discrimination, *Science* **214**, 93–94.

Burgess, A. E. and Ghandeharian, H. (1984a), Visual signal detection. I. Ability to use phase information, *J. Opt. Soc. Am. A* **1**, 900–905.

Burgess, A. E. and Ghandeharian, H. (1984b), Visual signal detection II. Signal-location identification, *J. Opt. Soc. Am. A* **1**, 906–910.

Burgess, A. E. (1985a), Visual signal detection. III. On Bayesian use of prior knowledge and cross correlation, *J. Opt. Soc. Am. A* **2**, 1498–1507.

Burgess, A. E. (1985b), Statistical efficiency of perceptual decisions, *Proc. SPIE* **454**, 18–26.

Burgess, A. E. and Colborne, B. (1988), Visual signal detection IV: Observer inconsistency, *J. Opt. Soc. Am. A* **5**, 617–627.

Burgess, A. E. (1994), Statistically defined backgrounds: Performance of a modified non-prewhitening observer model, *J. Opt. Soc. Am. A* **11**, 1237–1242.

Burgess, A. E. (1995), Comparison of receiver operating characteristic and forced choice observer performance measurement methods, *Med. Phys.* **22**, 643–655.

Burgess, A. E., Li, X. and Abbey, C. K. (1997), Visual signal detectability with two noise components: Anomalous masking effects, *J. Opt. Soc. Am. A* **14**, 2420–2442.

Burgess, A. E. (1999), Visual signal detection with two-component noise: Low-pass spectrum effects, *J. Opt. Soc. Am. A* **16**, 694–704.

Burgess, A. E. (2001), Evaluation of detection model performance in power-law noise, *Proc. SPIE* **4324**, 123–132.

Burgess, R. E. (1959), Homophase and heterophase fluctuations in semiconducting crystals, *Discuss. Faraday Soc.* **21**:1, 51–158.

Burnashev, M. V. (1998), On one useful inequality in the testing of hypotheses, *IEEE Trans. Info. Th.* **44**, 1668–1670.

Butzer, P. L. (1983), A survey of the Whittaker-Shannon sampling theorem and some of its extensions, *J. Math. Res. Exposition* **3**, 185–212.

Buvat, I., Laffont, S., Le Cloirec, J., Bourguet, P. and Di Paola, R. (2001), Importance of the choice of the collimator for the detection of small lesions in scintimammography: A phantom study, *Phys. Med. Biol.* **46**, 1343–1356.

Byrne, C. L. (1993), Iterative image reconstruction algorithms based on cross-entropy minimization, *IEEE Trans. Image Process.* **2**, 96–103.

Byrne, C. L. (1995), *Erratum and addendum* to Iterative image reconstruction algorithms based on cross entropy minimization, *IEEE Trans. Image Process.* **4**, 226–227.

Caelli, T. and Moraglia, G. (1986), On the detection of signals embedded in natural scenes, *Percept. Psychophys.* **39**, 87–95.

Callen, H. B. and Welton, T. A. (1951), Irreversibility and generalized noise, *Phys. Rev.* **83**, 34–40.

Caloyannides, M. A. (1974), Microcycle spectral estimates of 1/f noise in semiconductors, *J. Appl. Phys.* **45**, 307–316.

Campbell, F. W. and Kulikowski, J. J. (1966), Orientational selectivity of the human visual system, *J. Physiol.* **187**, 437–445.

Campbell, F. W. and Robson, J. G. (1968), Application of Fourier analysis to the visibility of gratings, *J. Physiol.* **197**, 551–566.

Campbell, F. W., Nachmias, J. and Jukes, J. (1970), Spatial frequency discrimination in human vision, *J. Opt. Soc. Am.* **60**, 555–559.

Campbell, F. W., Johnstone, J. R. and Ross, J. (1981), An explanation for the visibility of low frequency gratings, *Vision Res.* **21**, 723–730.

Campbell, G., Douglas, M. A., Bailey, J. J. (1988). Nonparametric comparison of two tests of cardiac function on the same patient population using the entire ROC curve, *Proc. Of Computers in Cardiology*, Computer Society of the IEEE, 267-270.

Campbell, G. A. and Foster, R. M. (1948), *Fourier Integrals for Practical Applications*, Van Nostrand, Princeton, NJ.

Campbell, S. L. and Meyer, C. D. (1979), *Generalized Inverses of Linear Transformations*, Pitman, London.

Capp, M. P. (1981), Radiological imaging 2000 AD, *Radiology* **138**, 541–550.

Caradus, S. R. (1978), *Generalized Inverses and Operator Theory*, Queen's University, Kingston, Ontario.

Cargill, E. B. (1989), A mathematical liver model and its application to system optimization and texture analysis, Ph.D. Dissertation, University of Arizona, Tucson, AZ.

Carlson, C. R. and Cohen, R. (1980), A simple psychophysical model for predicting the visibility of displayed information, *Proc. Soc. Info. Display* **21**, 229–245.

Carrier, G. F., Krook, M. and Pearson, C. E. (1966), *Functions of a Complex Variable: Theory and Technique*, McGraw-Hill, New York.

Carslaw, H. S. (1930), *Introduction to the Theory of Fourier's Series and Integrals*, 3rd rev. ed., Dover, New York.

Casasent, D. and Psaltis, D. (1976), Scale-invariant optical correlation using Mellin transforms, *Opt. Commun.* **17**, 59–63.

Casella, G. and Berger, R. L. (1990), *Statistical Inference*, Duxbury Press, Belmont, CA.

Casella, G. and George, E. (1992), Explaining the Gibbs Sampler, *Am. Stat.* **46**, 167–174.

Cavanaugh, P. (1978), Size and position invariance in the visual system, *Perception* **7**, 167–177.

Censor, Y. (1981), Row-action methods for huge and sparse systems and their applications, *SIAM Rev.* **23**, 444–466.

Censor, Y. and Zenios, S. A. (1997), Parallel Optimization: Theory, Algorithms and Applications, Oxford University Press, New York.

Cercignani, C. (1998), *Ludwig Boltzmann: The Man Who Trusted Atoms*, Oxford University Press, New York.

Chakraborty, D. P. (1989), Maximum likelihood analysis of free-response receiver operating characteristic (FROC) data, *Med. Phys.* **16**, 561–568.

Chakraborty, D. P. and Winter, L. H. L. (1990), Free-response methodology: Alternate analysis and a new observer-performance experiment, *Radiology* **174**, 873–881.

Chakraborty, D. P. and Eckert, M. P. (1995), Quantitative versus subjective evaluation of mammography accreditation phantom images, *Med. Phys.* **22**, 133–143.

Chakraborty, D. P. (2000), The FROC, AFROC and DROC variants of the ROC analysis, in *Handbook of Medical Imaging*, Vol. 1: *Physics and Psychophysics* (Beutel, J., Kundel, H. and Van Metter, R., Eds.), SPIE Press, Bellingham, WA, pp. 771–796.

Chakraborty, D. P., Liu, X., O'Shea, M. and Toto, L. C. (2000), A quantitative method for visual phantom image quality evaluation, *Proc. SPIE* **3981**, 24–33.

Champeney, D. C. (1987), *A Handbook of Fourier Theorems*, Cambridge University Press, Cambridge.

Chan, M., Leahy, R., Mumcuoglu, E. and Cherry, S. (1997), Comparing lesion detection performances of PET image reconstruction algorithms: A case study, *IEEE Trans. Nucl. Sci.* **44**:4, 1558–1563.

Chandresekhar, S. (1960), *Radiative Transfer*, Dover, New York.

Chen, J. (1992), A theoretical framework of regional cone-beam tomography, *IEEE Trans. Med. Imaging* **11**, 342–350.

Chen, M., Peter, J., Jaszczak, R. J., Gilland, D. R., Bowsher, J. E., Tournai, M. P. and Metzler, S. D. (2002a), Observer studies of cardiac lesion detectability with triple-head 360° versus dual-head 180° SPECT acquisition using simulated projection data, *IEEE Trans. Nucl. Sci.* **49**, 655–660.

Chen, M., Bowsher, J. E., Baydush, A. H., Gilland, K. L., DeLong, D. M. and Jaszczak, R. J. (2002b), Using the Hotelling observer on multi-slice and multi-view simulated SPECT myocardial perfusion images, *IEEE Trans. Nucl. Sci.* **49**, 661–667.

Chib, S. and Greenberg, E. (1995), Understanding the Metropolis-Hastings algorithm, *Am. Stat.* **49**, 327–335.

Cheney, M. (2001), A mathematical tutorial on synthetic aperture radar, *SIAM Rev.* **43**, 301–312.

Chou, C. and Barrett, H. H. (1978), Gamma-ray imaging in Fourier space, *Opt. Lett.* **3**:5, 187–189.

Chubb, C. and Landy, M. S. (1991), Orthogonal distribution analysis: A new approach to the study of texture perception, in *Computational Models of Visual Processing* (Landy, M. and Movshon, J. A., Eds.), MIT Press, Cambridge, MA.

Chui, C. K. (1992), *An Introduction to Wavelets*, Academic Press, New York.

Churchill, R. V., Brown, J. W. and Verkey, R. F. (1974), *Complex Variables and Applications*, 3rd ed., McGraw-Hill, New York.

Claasen, T. A. C. M. and Mecklenbräuker, W. F. G. (1980), The Wigner distribution function — a tool for time-frequency signal analysis, *Philips J. Res.* **35**, 217–250.

Clack, R. and Defrise, M. (1994), Cone beam reconstruction by the use of Radon transform intermediate functions, *J. Opt. Soc. Am. A* **11**, 580-585.

Clarkson, E. and Barrett, H. H. (1997), A bound on null functions for digital imaging systems with positivity constraints, *Opt. Lett.* **22**, 814–815.

Clarkson, E. and Barrett, H. H. (1998a), Bounds on null functions of linear digital imaging systems, *J. Opt. Soc. Am. A* **15**, 1355–1360.

Clarkson, E. and Barrett, H. H. (1998b), Symmetry properties of an imaging system and consistency conditions in image space, *Phys. Med. Biol.* **43**, 1039–1048.

Clarkson, E. (1999), Projections onto the range of the exponential Radon transform and reconstruction algorithms, *Inverse Probl.* **15**, 563–571.

Clarkson, E. and Barrett, H. H. (2000), Approximations to ideal-observer performance on signal-detection tasks, *Appl. Opt.* **39**, 1783–1793.

Clarkson, E. and Barrett, H. H. (2001), High-pass filters give histograms with positive kurtosis, *Opt. Lett.* **26**, 1253–1255.

Clarkson, E. (2002), Bounds on the area under the receiver operating characteristic curve for the ideal observer, *J. Opt. Soc. Am. A* **19**, 1963–1968.

Clarkson, E. and Barrett, H. H. (2002), Statistical decision theory and tumor detection, in *Image Processing Techniques for Tumor Detection* (Strickland, R., Ed.), Dekker, New York, Chap. 4.

Clarkson, E., Kupinski, M. A. and Barrett, H. H. (2002), Transformation of characteristic functionals through imaging systems, *Opt. Express* **10**:13, 536–539.

Clarkson, E., Kupinski, M. A. and Hoppin, J. A. (2003), Assessing the accuracy of estimates of the likelihood ratio, *Proc. SPIE* **5034**, 135–143.

Clinthorne, N. H., Rogers, W. L., Shao, L., and Koral, K. F. (1987), A hybrid maximum likelihood position computer for scintillation cameras, *IEEE Trans. Nucl. Sci.* **34**, 97-101.

Clough, A. and Barrett, H. H. (1983), Attenuated Radon and Abel transforms, *J. Opt. Soc. Am.* **73**, 1590–1595.

Cohen, A. and d'Ales, J. P. (1995), Nonlinear approximation of stochastic processes, in *Wavelets and Statistics* (Antoniadis, A. and Oppenheim, G., Eds.), Springer-Verlag, New York.

Cohen, A. and d'Ales, J. P. (1997), Nonlinear approximation of random functions, *SIAM J. Appl. Math.* **57**, 518–540.

Cohen, G., Wagner, L. K., Amtey, S. R. and DiBianca, F. A. (1981), Contrast-detail-dose and dose efficiency analysis of a scanning digital and a screen-film-grid radiographic system, *Med. Phys.* **8**, 358–367.

Cohen, L. (1995), *Time-frequency Analysis*, Prentice-Hall, Englewood Cliffs, NJ.

Cohen-Tannoudji, C., Diu, B. and Laloë, F. (1977), *Quantum Mechanics*, Vol. I, Wiley, New York.

Cohen-Tannoudji, C., Dupont-Roc, J. and Grynberg, G. (1989), *Photons and Atoms: Introduction to Quantum Electrodynamics*, Wiley-Interscience, New York.

Cohn, T. E., Ed. (1993), Collected Works in Optics, Vol. 3: *Visual Detection*, Optical Society of America, Washington, DC.

Collins, S. A. (1970), Lens-system diffraction integral written in terms of matrix optics, *J. Opt. Soc. Am.* **60**:9, 1168–1179.

Combettes, P. L. (1993), The foundations of set theoretic estimation, *Proc. IEEE* **81**, 182–208.

Comon, P. (1994), Independent component analysis, A new concept? *Signal Process.* Special issue on Higher-Order Statistics **36**:3, 287–314.

Cook, L. T., Insana, M. F., McFadden, M. A., Hall, T. J. and Cox, G. G. (1995), Contrast-detail analysis of image degradation due to lossy compression, *Med. Phys.* **22**, 715–721.

Cooley, J. W. and Tukey, J. W. (1965), An algorithm for the machine calculation of complex Fourier series, *Math. Comput.* **19**:90, 297–301.

Cornfield, I. P., Fomin, S. V. and Sinai, Y. G. (1982), *Ergodic Theory*, Springer-Verlag, Berlin.

Cornsweet, T. N. (1970), *Visual Perception*, Academic Press, New York.

Courant, R. and Hilbert, D. (1989), *Methods of Mathematical Physics*, Vol. I, Classics ed., Wiley, New York.

Cox, D. R. and Hinkley, D. V. (1994), *Theoretical Statistics*, Chapman & Hall, London.

Cramér, H. (1946), *Mathematical Methods of Statistics*, Princeton University Press, Princeton, NJ.

Cross, G. and Jain, A. (1983), Markov random field texture models, *IEEE Trans. PAMI* **5**:1, 25–39.

Csiszár, I. (1991), Why least-squares and maximum entropy? An axiomatic approach to inference for linear inverse problems, *Ann. Stat.* **19**:4, 2032–2066.

Cunningham, D. R., Laramore, R. D. and Barrett, E. (1976), Detection in Image Dependent Noise, *IEEE Trans. Inform. Theory* **IT-22**, 603–610.

Dainty, J. C. and Shaw, R. (1974), *Image Science: Principles, Analysis, and Evaluation of Photographic-Type Imaging Processes*, Academic Press, London.

Dainty, J. C., Ed. (1975), *Laser Speckle and Related Phenomena*, Springer-Verlag, Berlin.

Dainty, J. C. (1976), The statistics of speckle patterns, *Progress in Optics* **14**, 3–46.

Dainty, J. C. (1984), *Laser Speckle and Related Phenomena* (2nd ed.), Springer-Verlag, Berlin.

Daly, S. (1993), The visible differences predictor: An algorithm for the assessment of image fidelity, in *Digital Images and Human Vision* (Watson, A. B., Ed.), MIT Press, Cambridge, MA, pp. 179–206.

Danielson, G. C. and Lanczos, C. (1942), Some improvements in practical Fourier analysis and their application to x-ray scattering from liquids, *J. Franklin I.* **233**, 365–380, 435–452.

Daubechies, I. (1988), Orthonormal bases of compactly supported wavelets, *Commun. Pure Appl. Math.* **41**, 909–996.

Daubechies, I. (1992), *Ten Lectures on Wavelets*, Society for Industrial and Applied Mathematics, Philadelphia, PA.

Daugman, J. D. (1985), Uncertainty relation for resolution in space, spatial frequency, and orientation optimized by two-dimensional visual cortical filters, *J. Opt. Soc. Am. A* **2**, 1160–1169.

Daugman, J. D. (1988), Complete discrete 2D Gabor transform by neural networks for image analysis and compression, *IEEE Trans. Acoust. Speech Signal Processing* **36**, 1169–1179.

Davenport, W. B., Jr. and Root, W. L. (1958), *An Introduction to the Theory of Random Signals and Noise*, McGraw-Hill, New York. Reprinted by IEEE Press, New York, 1987.

Davenport, W. B., Jr. (1970), *Probability and Random Processes: An Introduction for Applied Scientists and Engineers*, McGraw-Hill, New York.

Davis, P. J. (1979), *Circulant Matrices*, Wiley, New York.

Davison, M. E. and Grunbaum, F. A. (1981), Tomographic reconstructions with arbitrary directions, *Commun. Pure Appl. Math.* **34**, 77–120

Deans, S. R. (1983), *The Radon Transform and Some of Its Applications*, Wiley-Interscience, New York.

DeBelder, M., Bollen, R. and Duville, R. (1971), A new approach to the evaluation of radiographic systems, *J. Photogr. Sci.* **19**, 126–131.

de Boor, C. (1978), *A Practical Guide to Splines*, Springer-Verlag, New York.

DeBroglie, L. (1924), Recherches sur la théorie des Quanta, Ph.D. Dissertation, University of Paris. English translation can be found in Phase waves of Louis deBroglie, *Am. J. Phys.* **40**:9, 1315–1320, September 1972.

de Bruijn, N. G. (1973), A theory of generalized functions, with applications to Wigner distribution and Weyl correspondence, *Nieuw Archief voor Wiskunde* **21**:3, 205–280.

de Finetti, B. (1974, 1975), *Theory of Probability*, Vols. 1 and 2, Wiley, Chichester.

Defrise, M. and Clack, R. (1994), A cone-beam reconstruction algorithm using shift-variant filtering and cone-beam backprojection, *IEEE Trans. Med. Imag. 13*, 186–195.

De Groot, M. H. (1970), *Optimal Statistical Decisions*, McGraw-Hill, New York.

de Hevesy, G. (1962), *Adventures in Radioisotope Research*, Pergamon Press, New York.

Delignon, Y. and Pieczynski, W. (2002), Modeling non-Rayleigh speckle distribution in SAR images, *IEEE Trans. on Geoscience and Remote Sensing* **40**:6, 1430–1435.

DeLong, E. R., DeLong, D. M. and Clarke-Pearson, D. L. (1988), Comparing the areas of two or more correlated receiver operating characteristic curves: A non-parametric approach, *Biometrics* **44**, 837–845.

Dempster, A. P., Laird, N. M. and Rubin, D. B. (1977), Maximum likelihood from incomplete data via the EM algorithm, *J. Roy. Stat. Soc. Ser. B* **39**, 1–38.

DePalma, J. J. and Lowry, E. M. (1962), Sine-wave response of the visual system, *J. Opt. Soc. Am.* **228**, 328–335.

Dereniak, E. L. and Crowe, D. G. (1984), *Optical Radiation Detectors*, Wiley, New York.

Descartes, (1637), *La Dioptrique.*

DeValois, R. L., Albrecht, D. G. and Thorell, L. G. (1982), Spatial frequency selectivity of cells in the macaque visual cortex, *Vision Res.* **22**, 545–559.

Diaconis, P. and Freedman, D. (1981), On the statistics of vision: The Julesz conjecture, *J. Math. Psychol.* **24**, 112–118.

Dodgson, C. L. (1867), *An Elementary Treatise on Determinants, with Their Application to Simultaneous Linear Equations and Algebraical Geometry*, Macmillan, London.

Doob, J. L. (1953), *Stochastic Processes*, Wiley, New York.

Dorf, R. C. (2000), The Electrical Engineering Handbook, 2nd ed., CRC Press, Boca Raton, FL.

Dorfman, D. D. and Alf, E., Jr. (1968), Maximum likelihood estimation of parameters of signal detection theory — a direct solution, *Psychometrika* **33**, 117–124.

Dorfman, D. D. and Alf, E. (1969), Maximum likelihood estimation of parameters of signal detection theory and determination of confidence intervals — rating method data, *J. Math. Psych.* **6**, 487–496.

Dorfman, D. D., Berbaum, K. S. and Metz, C. E. (1992), ROC rating analysis: Generalization to the population of readers and cases with the jackknife method, *Invest. Radiol.* **27**, 723–731.

Dorfman, D. D. and Metz, C. E. (1995), Multi-reader multi-case ROC analysis: Comments on Begg's commentary, *Acad. Radiol.* **2**:Suppl. 1, S76.

Dorfman, D. D., Berbaum, K. S., Metz, C. E., Lenth, R. V., Hanley, J. A. and Dagga, H. A. (1996), Proper ROC analysis: The bigamma model, *Acad. Radiol.* **4**, 138–149.

Dorfman, D. D., Berbaum, K. S., Lenth, R. V., Chen, Y.-F. and Donaghy, B. A. (1998), Monte Carlo validation of a multireader method for receiver operating characteristic discrete rating data: Factorial experimental design, *Acad. Radiol.* **5**, 591–602.

Dorfman, D. D., Berbaum, K. S. and Brandser, E. A. (2000a), A contaminated binormal model for ROC data. Part I. Some interesting examples of binormal degeneracy, *Acad. Radiol.* **7**, 420–426.

Dorfman, D. D. and Berbaum, K. S. (2000b), A contaminated binormal model for ROC data. Part II. A formal model, *Acad. Radiol.* **7**, 427–437.

Dorfman, D. D. and Berbaum, K. S. (2000c), A contaminated binormal model for ROC data. Part III. Initial evaluation with detection ROC data, *Acad. Radiol.* **7**, 438–447.

Dorsch, R. G. (1995), Fractional Fourier transforms of variable order based on a modular lens system, *Appl. Opt.* **34**, 6016–6020.

D'Orsi, C. J. and Swets, J. A. (1995), Variability in the interpretation of mammograms (letter), *N. Engl. J. Med.* **332**, 1172.

Dryden, I. L. and Mardia, K. V. (1998), *Statistical Shape Analysis*, Wiley, New York.

Dubin, D. A. and Hennings, M. A. (1990), *Quantum Mechanics, Algebras and Distributions*, Longman Scientific and Technical Publications, Harlow, United Kingdom.

Duda, R. O., Hart, P. E. and Stork, D. G. (2001), *Pattern Classification*, Wiley, New York.

Duderstat, J. J. and Martin, W. R (1979), *Transport Theory*, Wiley-Interscience, New York.

Duflo, M. and Moore, C. C. (1976), On the regular representation of a nonunimodular locally compact group, *J. Funct. Anal.* **21**, 209–243.

Durnin, J. (1987), Exact solutions for nondiffracting beams. I. The scalar theory, *J. Opt. Soc. Am. A* **4**:4, 651–654.

Duta, N., Sonka, M. and Jain, A. K. (1999), Learning shape models from examples using automatic shape clustering and procrustes analysis, in *Information Processing in Medical Imaging: Proceedings of the Sixteenth Conference* (Kuba, A., Samal, M. and Todd-Pokvopek, A., Eds.), Springer-Verlag, New York, pp. 370–375.

Easton, R. L., Jr. and Barrett, H. H. (1987), Tomographic transformations in optical signal processing, in *Optical Signal Processing* (Horner, J., Ed.), Academic Press, San Diego, pp. 335–386.

Eckstein, M. P. and Whiting, J. S. (1995), Lesion detection in structured noise, *Acad. Radiol.* **2**, 249–253.

Eckstein, M. P. and Whiting, J. S. (1996), Visual signal detection in structured backgrounds. I. Effect of number of possible spatial locations and signal contrast, *J. Opt. Soc. Am. A* **13**, 1777–1787.

Eckstein, M. P., Ahumada, A. J. and Watson, A. B. (1997), Visual signal detection in structured backgrounds. II. Effect of contrast gain control, background variations and white noise, *J. Opt. Soc. Am. A* **14**, 2406–2419.

Eckstein, M. P., Abbey, C. K., Bochud, F. O., Bartroff, J. L. and Whiting, J. S. (1999), The effect of image compression in model and human performance, *Proc. SPIE* **3663**, 243–252.

Eckstein, M. P., Abbey, C. K. and Bartroff, M. P. (2000a), Model observer optimization of JPEG image compression, *Proc. SPIE* **3981**, 106–115.

Eckstein, M. P., Abbey, C. K. and Bochud, F. O. (2000b), A practical guide to model observers for visual detection in synthetic and natural noisy images, in *Handbook of Medical Imaging*, Vol. 1: *Physics and Psychophysics* (Beutel, J., Kundel, H. and Van Metter, R., Eds.), SPIE Press, Bellingham, WA, pp. 593–628.

Eckstein, M. P. and Abbey, C. K. (2001), Model observers for signal known statistically tasks, *Proc. SPIE* **4324**, 91–102.

Eckstein, M. P., Pham, B. and Abbey, C. K. (2002), The effect of image compression for model and human observers in signal known statistically tasks, *Proc. SPIE* **4320**, 13–24

Edwards, A. W. F. (1974), The history of likelihood, *Int. Stat. Rev.* **42**, 4–15.

Edwards, D. C., Kupinski, M. A., Nishikawa, R. M. and Metz, C. E. (2000), Estimation of linear observer templates in the presence of multi-peaked Gaussian noise through 2AFC experiments, *Proc. SPIE* **3981**, 86–96.

Egan, J. P., Greenberg, G. Z. and Schulman, A. I. (1961), Operating characteristics, signal detection, and the method of free response, *J. Acoust. Soc. Am.* **33**, 993–1007.

El Fakhri, G., Buvat, I., Benali, H., Todd-Pokropek, A. and Di Paola R. (2000), Relative impact of scatter, collimator response, attenuation, and finite spatial resolution corrections in cardiac SPECT, *J. Nucl. Med.* **41**, 1400–1408.

El Fakhri, G., Moore, S. C. and Kijewski, M. F. (2002), Optimization of Ga-67 imaging for detection and estimation tasks: Dependence of imaging performance on spectral acquisition parameters, *Med. Phys.* **29**, 1859–1866.

Elmore, J. G., Wells, C. K., Lee, C. H., Howard, D. H. and Feinstein, A. R. (1994), Variability in radiologists' interpretations of mammograms, *N. Engl. J. Med.* **331**, 1493–1499.

Engl, H. W., Hanke, M. and Neubauer, A. (1996), *Regularization of Inverse Problems*, Kluwer Academic, Dordrecht, Netherlands.

Erdélyi, A. (1954), *Tables of Integral Transforms*, Vol. 1, McGraw-Hill, New York.

Eskin, J. D. (1997), Semiconductor gamma-ray imaging detectors for nuclear medicine, Ph.D. Dissertation, University of Arizona, Tucson, AZ.

Eskin, J. D., Barrett, H. H. and Barber, H. B. (1999), Signals induced in semiconductor gamma-ray imaging detectors, *J. Appl. Phys.* **85**, 647–659.

Evans, M., Hastings, N. and Peacock, B. (1993), *Statistical Distributions*, 2nd ed., Wiley, New York.

Eves, H. (1966), *Elementary Matrix Theory*, Dover, New York.

Fano, U. (1947), Ionization yield of radiations. II. The fluctuations of the number of ions, *Phys. Rev.* **72**, 26–29.

Fante, R. L. (1981), Relationship between radiative-transport theory and Maxwell's equations in dielectric media, *J. Opt. Soc. Am.* **71**, 460–468.

Farrell, J., Trontelj, H., Rosenberg, C. and Wiseman, J. (1991), Perceptual metrics for monochrome image compression, *Soc. Info. Display Digest* **22**, 631–634.

Fässler, A. and Stiefel, E. (1992), *Group Theoretical Methods and Their Applications*, English translation by Wong, B. D., Birkhäuser, Boston.

Feller, W. (1968), *An Introduction to Probability Theory and Its Applications*, Vol. 1, 3rd ed., Wiley, New York.

Feller, W. (1971), *An Introduction to Probability Theory and Its Applications*, Vol. 2, 2nd ed., Wiley, New York.

Fellgett, P. B. and Linfoot, E. H. (1955), On the assessment of optical images, *Philos. Trans. Roy. Soc.* **A247**, 369–407.

Ferwerda, H. A. (1999), The radiative transfer equation for scattering media with a spatially varying refractive index, *J. Opt. A. Pure Appl. Opt.* **1**, L1–L2.

Fessler, J. A. (1994), Penalized weighted least-squares image reconstruction for positron emission tomography, *IEEE Trans. Med. Imaging* **13**:2, 290–300.

Fessler, J. A. (1996), Mean and variance of implicitly defined biased estimators (such as penalized maximum likelihood): Application to tomography, *IEEE Trans. Image Process.* **5**, 493–506.

Fessler, J. A., Clinthorne, N. H. and Rogers, W. L. (1993), On complete-data spaces for PET reconstruction algorithms, *IEEE Trans. Nucl. Sci.* **40**, 1055–1061.

Fessler, J. A. and Hero, A. O. (1994), Space-alternating generalized EM algorithm, *IEEE Trans. Signal Process.* **42**, 2664–2677.

Fessler, J. A. and Rogers, W. L. (1996), Spatial resolution properties of penalized-likelihood image reconstruction methods: Space-invariant tomographs, *IEEE Trans. Image Process.* **5**:9, 1346–1358.

Fick, A. (1855), Über Diffusion, *Poggendorf 's Annalen der Physik und Chemie* **94**, 59–86.

Fiddy, M. A. (1992), Linearized and approximate methods for inversion of scattered field data, in *Inverse Problems in Scattering and Imaging* (Pike, E. R. and Bertero, M., Eds.), Adam Hilger, Bristol, Chap. 3.

Field, D. J. (1987), Relations between the statistics of natural images and the response properties of cortical cells, *J. Opt. Soc. Am. A* **4**:12, 2379–2394.

Field, D. J. (1994), What is the goal of sensory coding?, *Neural Comput.* **6**, 559–601.

Fiete, R. D., Barrett, H. H., Smith, W. E. and Myers, K. J. (1987), The Hotelling Trace Criterion and its correlation with human observer performance, *J. Opt. Soc. Am. A* **4**, 945–953.

Fiorentini, A., Jeanne, M. and di Francia, G. T. (1955), Measurement of differential threshold in the presence of spatial illumination gradient, *Atti. Fond. Georgio Ronchi* **12**, 371–379.

Fisher, R. A. (1922), On the mathematical foundations of theoretical statistics, *Philos. Trans. Roy. Soc. Lond.* **222**, 309–368. Reprinted as #10 in Fisher, R. A. (1950), *Contributions to Mathematical Statistics*, Wiley, New York.

Fisher, R. A. (1925), Theory of statistical estimation, *Proc. Cambridge Philos. Soc.* **22**, 700–725.

Fisher, R. A. (1934), Two new properties of mathematical likelihood, *Proc. Roy. Soc. Lond.* **144**, 285–307.

Fisher, R. A. (1935), The logic of inductive inference, *J. Roy. Stat. Soc.* **98**, 39–54.

Fisher, R. A. (1936), The use of multiple measurements in taxonomic problems, *Ann. of Eugenics* **7**, 179–188. Reprinted in Fisher, R. A. (1950), *Contributions to Mathematical Statistics*, Wiley, New York.

Flammer, C. (1957), *Spheroidal Wave Functions*, Stanford University Press, Palo Alto, CA.

Flandrin, P. and Martin, W. (1997), The Wigner-Ville spectrum of nonstationary random signals, in *The Wigner Distribution — Theory and Applications in Signal Processing* (Mecklenbräuker, W. and Hlawatsch, F., Eds.), Elsevier, Amsterdam, The Netherlands.

Foley, J. D., van Dam, A., Feiner, S. K. and Hughes, J. F. (1990), *Computer Graphics: Principles and Practice*, Addison-Wesley, Reading, MA.

Fox, A. G. and Li, T. (1961), Resonant modes in a maser interferometer, *Bell Syst. Tech. J.* **46**, 453–488.

Friberg, A. T. (1991), Propagation of a generalized radiance in paraxial optical systems, *J. Opt. Soc. Am. A* **30**:18, 2443–2446.

Frieden, B. R. (1983), *Probability, Statistical Optics, and Data Testing: A Problem Solving Approach*, Springer-Verlag, New York.

Frieden, B. R. (1991), *Probability, Statistical Optics, and Data Testing: A Problem Solving Approach*, 2nd ed., Springer-Verlag, New York.

Friedman, B. (1991), *Lectures on Applications-Oriented Mathematics*, Wiley-Interscience, New York.

Fryback, D. G. and Thornbury, J. R. (1991), The efficacy of diagnostic imaging, *Med. Decis. Making* **11**, 88–94.

Fukunaga, K. (1990), *Statistical Pattern Recognition*, 2nd ed., Academic Press, San Diego.

Furenlid, L. R., Clarkson, E., Marks, D. G. and Barrett, H. H. (2000), Spatial pileup considerations for pixellated gamma-ray detectors, *IEEE Trans. Nucl. Sci.* **47**, 1399–1402.

Gallas, B. D. (2001), Signal detection in lumpy backgrounds, Ph.D. Dissertation, University of Arizona, Tucson, AZ.

Gallas, B. D. (2003), Variance of the channelized Hotelling observer from a finite number of trainers and testers, *Proc. SPIE* **5034**, 100–111.

Gallas, B. D. and Barrett, H. H. (2003), Validating the use of channels to estimate the ideal linear observer, *J. Opt. Soc. Am. A***20**, 1725-1739.

Garland, L. H. (1949), On the scientific evaluation of diagnostic procedures, *Radiology* **52**, 309–327.

Garra, B. S., Insana, M. F., Shawker, T. H., Wagner, R. F., Bradford, M. and Russel, M. A. (1989), Quantitative ultrasonic detection and classification of liver disease: Comparison with human observer performance, *Invest. Radiol.* **24**, 196–203.

Gaskill, J. D. (1978), *Linear Systems, Fourier Transforms, and Optics*, Wiley, New York.

Gauss, C. F. (1823), *Theory of the Combination of Observations Least Subject to Errors*, English translation by Stewart, G. W., SIAM, Philadelphia, PA, 1995.

Gauss, C. F. (1900), Schönes Theorem der Wahrscheinlichkeitsrechnung, *Werke* **8**, 88. Koniglichen Gesellenschaft der Wissenschaften zu Göttingen, Teubner, Leipzig.

Geisler, W. S. and Davila, K. D. (1985), Ideal discriminators in spatical vision: Two-point stimuli, *J. Opt. Soc. Am. A* **2**, 1483–1497.

Gel'fand, I. M. (1961), *Lectures on Linear Algebra* (translated by Shenitzer, A.), Dover, New York.

Gellert, W., Küstner, H., Hellwich, M. and Kästner, H. (1977), *The VNR Concise Encyclopedia of Mathematics*, Van Nostrand Reinhold, New York.

Geman, S. and Geman, D. (1984), Stochastic relaxation, Gibbs distributions, and the Bayesian restoration of images, *IEEE Trans. Pattern Anal. Mach. Intell.* **6**, 721–741.

Geman, S. and McClure, D. (1985), Bayesian image analysis: An application to single photon emission tomography, in *Proceedings of the Statistical Computing Section*, pp. 12–18, Washington, DC.

Gerchberg, R. W. (1974), Super-resolution through error energy reduction, *Opt. Acta* **12**:9, 709–720.

Gersho, A. and Gray, R. M. (1992), *Vector Quantization and Signal Compression*, Kluwer Academic, Boston.

Geyer, C. J. and Thompson, E. A. (1992), Constrained Monte Carlo maximum likelihood for dependent data, *J. Roy. Stat. Soc. Ser. B* **54**, 657–699.

Gifford, H. (1997), Theory and application of Fourier crosstalk: An evaluator for digital-system design, Ph.D. Dissertation, University of Arizona, Tucson, AZ.

Gifford, H. C., Wells, R. G. and King, M. A. (1999), A comparison of human observer LROC and numerical observer ROC for tumor detection in SPECT images, *IEEE Trans. Nucl. Sci.* **46**, 1032–1037.

Gifford, H. C., King, M. A., de Vries, D. J. and Soares, E. J. (2000a), Channelized Hotelling and human observer correlation for lesion detection in hepatic SPECT imaging, *J. Nucl. Med.* **41**, 514–521.

Gifford, H. C., King, M. A., Wells, R. G., Hawkins, W. G., Narayanan, M. V. and Pretorius, P. H. (2000b), LROC analysis of detector-response compensation in SPECT, *IEEE Trans. Med. Imaging* **19**, 463–473.

Gifford, H. C. and King, M. A. (2001), Case-sampling in LROC: A Monte Carlo analysis, *Proc. SPIE* **4324**, 143–150.

Gifford, H. C., King, M. A., Narayanan, M. V., Pretorius, P. H., Smyczynski, M. S. and Wells, R. G. (2002), Effect of block-iterative acceleration on GA-67 tumor detection in thoracic SPECT, *IEEE Trans. Nucl. Sci.* **49**, 50–552.

Giger, M. L. and Doi, K. (1987), Effect of pixel size on detectability of low-contrast signals in digital radiography, *J. Opt. Soc. Am. A* **4**, 966–975.

Gilks, W. R., Richardson, S. and Spiegelhalter, D. J. (1996), *Markov Chain Monte Carlo in Practice*, Chapman & Hall, London.

Gill, P. E., Murray, W. and Wright, M. H. (1981), *Practical Optimization*, Academic Press, London.

Gilland, D. R., Tsui, B. M. W., Metz, C. E., Jaszczak, R. J. and Perry, J. R. (1992), An evaluation of maximum likelihood-expectation maximization reconstruction for SPECT by ROC analysis, *J. Nucl. Med.* **33**, 451–457.

Ginsburg, A. P., Evans, D. W., Sekuler, R. and Harp, S. A. (1982), Contrast sensitivity predicts pilots' performance in aircraft simulators, *Am. J. Optomet. Physiol. Opt.* **59**, 105–109.

Ginsburg, A. P. and Evans, D. W. (1984), Rapid measurement of contrast sensitivity using new contrast sensitivity vision test chart system: Initial population data, in *Proceedings of the Human Factors Society*, 28th Annual Meeting (Alluisi, M. J., De Groot, S. and Alluisi, E. A., Eds.), pp. 123–127.

Girod, B. (1993), What's wrong with mean-squared error? in *Digital Images and Human Vision* (Watson, A. B., Ed.), MIT Press, Cambridge, MA, pp. 207–220.

Glasko, V. B. (1988), *Inverse Problems of Mathematical Physics*, American Institute of Physics, New York.

Glassner, A. S. (1995), *Principles of Digital Image Synthesis*, Vol. 1 and 2, Morgan Kaufmann, San Francisco.

Glauber, R. J. (1963), Coherent and incoherent states of the radiation field, *Phys. Rev.* **131**, 2766–2788.

Glauber, R. J. (1965), in *Quantum Optics and Electronics* (Les Houches Summer School of Theoretical Physics, University of Grenoble), (DeWitt, C., Blandin, A. and Cohen-Tannoudji, C., Eds.), Gordon and Breach, New York.

Gmitro, A. F., Gindi, G. R., Barrett, H. H. and Easton, R. L. (1983), Two-dimensional image processing by one-dimensional filtering of projection data, *Proc. SPIE* **388**, 132–139.

Gohberg, I. and Goldberg, S. (1981), *Basic Operator Theory*, Birkhäuser, Boston.

Gold, M. R., Siegel, J. E., Russell, L. B. and Weinstein, M. C. (1996), *Cost-effectiveness in Health and Medicine,* Oxford University Press, New York.

Golub, G. H. and Van Loan, C. F. (1989), *Matrix Computations*, 2nd ed., Johns Hopkins University Press, Baltimore.

Gonzales, R. C. and Wintz, P. (1977), *Digital Image Processing*, Addison-Wesley, Reading, MA.

Good, W. F., Sumkin, J. H., Dash, N., Johns, C. M., Zuley, M. L., Rockette, H. E. and Gur, D. (1999), Observer sensitivity to small differences: A multipoint rank-order experiment, *Am. J. Radiol.* **173**:2, 275–278.

Goodenough, D. J. (1975), Objective measures related to ROC curves, *Proc. SPIE* **47**, 134–141.

Goodman, J. W. (1968), *Introduction to Fourier Optics*, McGraw-Hill, New York.

Goodman, J. W. (1975), Statistical Properties of Laser Speckle Patterns, in *Laser Speckle and Related Phenomena*, Springer-Verlag, Heidelberg, Chap. 2, pp. 9–75.

Goodman, J. W. (1985), *Statistical Optics*, Wiley, New York.

Goodman, N. R. (1963), Statistical analysis based on a certain complex Gaussian distribution, *Ann. Math. Stat.* **34**, 152–176.

Gooley, T. A. and Barrett, H. H. (1992), Evaluation of the statistical methods of image reconstruction through ROC analysis, *IEEE Trans. Med. Imaging*, **11**:2, 276–283.

Gordon, R., Bender, R. and Herman, G. (1970), Algebraic reconstruction technique (ART) for three-dimensional electron microscopy and X-ray photography, *J. Theor. Biol.* **29**, 471–481.

Gori, F. (1994), Why is the Fresnel transform so little known?, in *Current Trends in Optics* (Dainty, C., Ed.), Academic Press, London, Chap. 10.

Gorman, J. D. and Hero, A. O. (1990), Lower bounds for parametric estimation with constraints, *IEEE Trans. Inform. Theory* **36**, 1285–1301.

Gouy, G. L. (1886), Sur le mouvement lumineux, *J. Phys. Paris* **5**, 354–362.

Gradshteyn, I. S. and Ryzhik, I. M. (1980), *Table of Integrals, Series and Products*, 4th ed., Academic Press, Orlando, FL.

Graham, N. and Nachmias, J. (1971), Detection of grating patterns containing two spatial frequencies: A comparison of single-channel and multiple-channels models, *Vision Res.* **11**, 251–259.

Grangeat, P. (1987), Analyse d'un Système d'Imagerie 3D par Reconstruction à partir de Radiographies X en Géométrie Conique, Ph.D. Dissertation, École Nationale Supérieure des Télécommunications, Paris.

Grangeat, P. (1990), Mathematical framework of cone beam 3d reconstruction via the first derivative of the radon transform, in *Mathematical Methods in Tomography* (Herman, G. T., Louis, A. K. and Natterer, F., Eds), Springer-Verlag, New York.

Gray, R. M. and Macovski, A. (1976), Maximum a posteriori estimation of position in scintillation cameras, *IEEE Trans. Nucl. Sci.* **NS-23**:1, 849–852.

Green, D. M. and Swets, J. A. (1966), *Signal Detection Theory and Psychophysics*, Wiley, New York. Reprinted, with corrections, 1974, Krieger, Huntington, NY.

Grey, D. R. and Morgan, B. J. T. (1972), Some aspects of ROC curve-fitting: Normal and logistic models, *J. Math. Psych.* **9**, 128–139.

Groetsch, C. W. (1977), *Generalized Inverses of Linear Operators: Representations and Approximations*, Marcel Dekker, New York.

Guignard, P. A. (1982), A comparative method based on ROC analysis for the quantitation of observer performance in scintigraphy, *Phys. Med. Biol.* **27**, 1163–1176.

Gull, S. F. and Skilling J. (1984), Maximum entropy method in image processing, *Proc. IEEE* **131-F**, 649–659.

Gunter, D. L. (1996), Collimator Characteristics and Design, in Henken, R. E. (1996) *Nuclear Medicine*, Mosby Year Book, St. Louis, Chap. 8.

Gur, D., Rubin, D. A. and Kart, B. H. (1997), Forced choice and ordinal discrete rating assessment of image quality: A comparison, *J. Dig. Imaging* **10**, 103–107.

Haight, F. A. (1967), *Handbook of the Poisson Distribution*, Wiley, New York.

Hakimmashhadi, H. (1988), Discrete Fourier transform and FFT, in *Signal Processing Handbook* (Chen, C. H., Ed.), Marcel Dekker, New York, Chap. 3.

Hald, A. (1998), *A History of Mathematical Statistics from 1750 to 1930*, Wiley, New York.

Halperin, B. (1962), The product of projection operators, *Acta Sci. Math.* **23**, 96–99.

Halpern, E. J., Albert, M., Krieger, A. M., Metz, C. E. and Maidement, A. D. (1996), Comparison of receiver operating characteristic curves on the basis of optimal operating points, *Acad. Radiol.* **3**, 245–253.

Halter, M. (1976), On the spatial breadth of spatial frequency channels in human visual detection, Ph.D. Dissertation, University of California, Berkeley, CA.

Hamaker, C., Smith, K. T., Solmon, D. C. and Wagner, S. L. (1980), The divergent beam x-ray transform, *Rocky Mountain J. Math.* **10**, 253–283.

Hamaker, C. and Solmon, D. C. (1978), The angles between the null spaces of X-rays, *J. Math. Anal. Appl.* **62**, 1–23.

Hamermesh, M. (1989), *Group Theory and Its Application to Physical Problems*, Dover, New York.

Hanley, J. A. (1988), The robustness of the "binormal" assumptions used in fitting ROC curves, *Med. Decis. Making* **8**, 197–203.

Hanley, J. A. (1989), Receiver operating characteristic (ROC) methodology: The state of the art, *Crit. Revs. Diag. Imaging* **29**, 307–335.

Hanley J. A. and McNeil, B. J. (1982), The meaning and use of the area under a receiver operating characteristic (ROC) curve, *Radiology* **143**, 29–36.

Hanley, J. A. and McNeil, B. J. (1983), A method of comparing the areas under ROC curves derived from the same cases, *Radiology* **148**, 839–843.

Hanson, K. M. (1977), Detectability in the presence of computed tomographic reconstruction noise, *Proc. SPIE* **127**, 304–312.

Hanson, K. M. and Wecksung, G. W. (1985), Local basis-function approach to computed tomography, *Appl. Opt.* **24**, 4028–4039.

Hanson, K. M. (1988), POPART — Performance optimized algebraic reconstruction technique, *Proc. SPIE* **1001**, 318–325.
Hanson, K. M. (1989), Optimization for object localization of the constrained algebraic reconstruction technique, *Proc. SPIE* **1090**, 146–153.
Hanson, K. M. (1990a), Optimization of the constrained algebraic reconstruction technique for a variety of visual tasks, in *Information Processing in Medical Imaging: Proceedings of the Eleventh Conference* (Ortendahl, D. A. and Llacer, J., Eds.), Wiley-Liss, New York, pp. 45–57.
Hanson, K. M. (1990b), Method of evaluating image recovery algorithms based on task performance, *J. Opt. Soc. Am. A* **7**, 1294–1304.
Harris, J. L. (1964), Resolving power and decision theory, *J. Opt. Soc. Am.* **54**, 606–611.
Hartline, H. K. (1934), Intensity and duration in the excitation of single photoreceptor units, *J. Cellular Compar. Physiol.* **5**, 229–247.
Hartline, H. K. (1940), The receptive fields of optic nerve fibers, *Am. J. Physiol.* **130**, 690–699.
Hartline, H. K. (1949), Inhibition of activity of visual receptors by illuminating nearby retinal areas in the limulus eye, *Federat. Proc.* **8**, 69.
Hartline, H. K. and Ratliff, F. (1957), Inhibitory interactions of receptor units in the eye of Limulus, *J. Gen. Physiol.* **40**, 351–376.
Harville, D. A. (1997), Matrix Algebra From a Statistician's Perspective, Springer-Verlag, New York.
Hastings, W. K. (1970), Monte Carlo sampling methods using Markov chains and their applications, *Biometrika* **57**, 97–109.
Hayes, J. (1992), Fast Fourier transforms and their applications, in *Applied Optics and Optical Engineering* (Shannon, R. R. and Wyant, J. C., Eds.), Academic Press, Orlando, FL, Chap. 2.
Haynor, D. (1997), Bayes and medical imaging: it's time to make priors a priority, *Proc. SPIE* **3034**, 2–9.
Hebert, T. and Leahy, R. (1989), A generalized EM algorithm for 3-d Bayesian reconstruction from Poisson data using Gibbs priors, *IEEE Trans. Med. Imaging* **8**, 194–202.
Hecht, E. (1987), *Optics*, 2nd ed., Addison Wesley, Reading, MA.
Hecht, S., Shlaer, S. and Pirenne, M. H. (1942), Energy, quanta and vision, *J. Gen. Physiol.* **25**, 819–840. Reprinted in Cohn, T. E., Ed. (1993), Collected Works in Optics, Vol. 3: *Visual Detection*, Optical Society of America, Washington, DC, pp. 6–27.
Heeger, D. J. and Bergen, J. R. (1995), Pyramid-based texture analysis/synthesis, *Computer Graphics Proc.* **29**, 229–238.
Heideman, M. T., Johnson, D. H. and Burrus, C. S. (1985), Gauss and the history of the FFT, *Arch. Hist. Exact Sci.* **34**, 265–277. Also in *IEEE Acoust. Speech Signal Process. Mag.* **1**, 14–21.
Heine, J. J., Deans, S. R. and Clarke, L. P. (1999), Multiresolution probability analysis of random fields, *J. Opt. Soc. Am. A* **16**, 6–16.
Helgason, S. (1980), The Radon transform, *Progress in Mathematics* **5**, Birkhäuser, Boston, MA. 2nd ed., Springer Verlag (August 1999).
Helstrom, C. W. (1964), The detection and resolution of optical signals, *IEEE Trans. Info. Theory* **52**, 275–287.

Helstrom, C. W. (1967), Image restoration by the method of least squares, *J. Opt. Soc. Am.* **57**, 297–303.

Helstrom, C. W. (1995), *Elements of Signal Detection and Estimation*, Prentice-Hall, Englewood Cliffs, NJ.

Henkelman, R. M., Kay, I. and Bronskill, M. J. (1990), Receiver operator characteristic (ROC) analysis without truth, *Med. Decis. Making* **10**, 24–29.

Herman, G. T. and Chan, M. (1995), Bayesian image reconstruction with image-modeling priors, in *Proc. 1995 IEEE Workshop on Nonlinear Signal and Image Processing*, pp. 94–97.

Hero, A. O. and Fessler, J. A. (1994), A recursive algorithm for computing Cramer-Rao-type bounds on estimator covariance, *IEEE Trans. Inform. Theory* **40**:4, 1205–1210.

Hero, A. O., Fessler, J. A. and Usman, M. (1996), Exploring estimator bias-variance tradeoffs using the uniform CR bound, *IEEE Trans. Signal Process.* **44**, 2026–2041.

Herzberger, M. (1980), *Modern Geometrical Optics*, Robert E. Krieger, Huntington, NY. Reprint of 1958 edition, Interscience Publishers.

Hestenes, M. R. (1975), Pseudoinverses and conjugate gradients, *CACM* **18**:1, 40–43.

Hestenes, M. R. and Stiefel, E. (1952), Methods of conjugate gradients for solving linear systems, *J. Res. Nat. Bur. Stand.* **49**, 409–436.

Hoeschen, C., Fessel, A., Buhr, E. and Döhring, W. (2000), Determination of the x-ray intensity pattern in mammography with very high spatial resolution, *Proc. SPIE* **3977**, 220–230.

Hoffman, H. M., Irwin, A., Prayaga, R. and Murray, M. (1996), Virtual anatomy from the visible man: Creating tools for medical education, in *First Visible Human Conference Proceedings* (Banvard, R., Ed.), National Library of Medicine, Bethesda MD.

Hoffman, K. P. and Stone, J. (1971), Conduction velocity of afferents to cat visual cortex: A correlation with cortical receptive field properties, *Brain Res.* **32**, 460–466.

Hooge, F. N. (1969), 1/f noise is no surface effect, *Phys. Lett.* **29A**:3, 139–140.

Hoppin, J. W., Kupinski, M. A., Kastis, G. A., Clarkson, E. and Barrett, H. H. (2002), Objective Comparison of Quantitative Imaging Modalities Without the Use of a Gold Standard, *IEEE Trans. Med. Imaging* **21**:5, 441–449.

Horn, R. A. (1990), The Hadamard product, *Proc. Symp. Appl. Math* **40**, 87–169.

Hotelling, H. (1931), The generalization of Student's ratio, *Ann. Math. Stat.* **2**, 360–378.

Hubel, D. H. and Wiesel, T. N. (1962), Receptive fields, binocular interaction and functional architecture in the cat's visual cortex, *J. Physiol.* **160**, 106–154.

Hubel, D. H. and Wiesel, T. N. (1968), Receptive fields and functional architecture of monkey striate cortex, *J. Physiol.* **195**, 215-243.

Hubel, D. H. and Wiesel, T. N. (1974), Uniformity of monkey striate cortex: A parallel relationship between field size, scatter, and magnification factor, *J. Comp. Neural.* **158**, 295–306.

Huck, F. O., Fales, C. L. and Rahman, Z. (1997), *Visual Communication Theory: An Information Theory Approach*, Kluwer Academic, Boston.

Hudson, H. M. and Larkin, R. S. (1994), Accelerated image reconstrucion using ordered subsets of projection data, *IEEE Trans. Med. Imaging* **13**, 601–609.

Hultgren, B. O. (1990), Subjective quality factor revisited, *Proc. SPIE* **1249**, pp. 12–23.

Hummel, C. A. (2000), The practice of interferometry with NPOI, *Proc. SPIE* **4006**, 584.

Image Resolution Assessment and Reporting Standards (IRARS) Committee (1996), *Civil NIIRS Reference Guide.*

Image Resolution Assessment and Reporting Standards (IRARS) Committee (1996), *Civil NIIRS Reference Guide, Appendix III, History of NIIRS.*

International Commission on Radiation Units and Measurements (ICRU) (1996), *Medical Imaging — The Assessment of Image Quality*, Report 54, ICRU, Bethesda, MD.

Ishida, M., Doi, K., Loo, L.-N., Metz, C. E. and Lehr, J. L. (1984), Digital image processing: Effect on detectability of simulated low-contrast radiographic patterns, *Radiology* **150**, 569–575.

Ishimaru, A. (1978), *Wave Propagation and Scattering in Random Media*, Vol. 1, Academic Press, New York.

Jackson, J. D. (1998), *Classical Electrodynamics*, 3rd ed., Wiley, New York.

Jakeman, E. (1980), On the statistics of K-distributed noise, *J. Phys. A: Math. Gen.* **13**, 31–48.

Jakeman, E. and Pusey, P. N. (1976), A Model for Non-Rayleigh Sea Echo, *IEEE Trans. Antennas Propagation* **AP-24**:6, 806–814.

Jakeman, E. and Pusey, P. N. (1978), Significance of K distribution in scattering experiments, *Phys. Rev. Lett.* **40**, 546–550.

Jakeman, E. and Tough, R. J. A. (1987), Generalized K distributions: a statistical model for weak scattering, *J. Opt. Soc. Am. A* **4**, 1764–1772.

Jakeman, E. and Tough, R. J. A. (1988), Non-Gaussian Models for the Statistics of Scattered Waves, *Advances in Physics* **37**:5, 471–529.

James, G. and Liebeck, M. (1993), *Representations and Characters of Groups*, Cambridge University Press, Cambridge.

Jang, S., Jaszczak, R. J., Tsui, B. M. W., Metz, C. E., Gilland, D. R., Turkington, T. G. and Coleman, R. E. (1998), ROC evaluation of the lesion detectability with and without non-uniform attenuation compensation in SPECT myocardial perfusion imaging: A phantom study, *IEEE Trans. Nucl. Sci.* **45**, 2080–2088.

Jeffreys, H. (1939), *Theory of Probability*, Oxford University Press, London.

Jeffreys, H. (1961), *Theory of Probability*, 3rd ed., Oxford University Press, London.

Jeffreys, H. (1973), *Scientific Inference*, 3rd ed., Cambridge University Press, New York.

Jerri, A. J. (1977), The Shannon sampling theorem — Its various extensions and applications: A tutorial review, *Proc. IEEE* **65**, 1565–1596.

Jerri, A. J. (1986), *The Sampling Expansion — A Detailed Bibliography*, Monograph, Clarkson University, Potsdam, NY.

Jerri, A. J. (1992), *Integral and Discrete Transforms with Applications and Error Analysis*, Marcel Dekker, New York.

Jiang, Y., Metz, C. E. and Nishikawa, R. M. (1996), A receiver operating characteristic partial area index for highly sensitive diagnostic tests, *Radiology* **201**, 745–750.

Johns, H. E. and Cunningham, J. R. (1983), *The Physics of Radiology*, 4th ed., Charles C. Thomas, Springfield, IL.

Johnson, G. E. (1994), Constructions of particular random processes, *Proc. IEEE* **82**, 270–285.

Johnson, J. B. (1925), The Schottky effect in low frequency circuits, *Phys. Rev.* **26**, 71.

Johnson, J. B. (1928), Thermal agitation of electricity in semiconductors, *Phys. Rev.* **32**, 97.

Johnson, N. L. and Kotz, S. (1969), *Discrete Distributions*, Houghton-Mifflin, Boston.

Johnson, N. L. and Kotz, S. (1970), *Continuous Univariate Distributions*, Vol. 2, Houghton-Mifflin, Boston.

Johnson, N. L., Kotz, S. and Kemp, A. W. (1992), *Univariate Discrete Distributions,* 2nd ed., Wiley, New York.

Johnson, N. L., Kotz, S. and Balakrishnan, N. (1994), *Continuous Univariate Distributions*, Vol. 1, 2nd ed., Wiley, New York.

Johnson, N. L., Kotz, S. and Balakrishnan, N. (1995), *Continuous Univariate Distributions*, Vol. 2, 2nd ed., Wiley, New York.

Johnson, R. A. and Wichern, D. W. (1988), *Applied Multivariate Statisical Analysis*, 2nd ed., Prentice-Hall, Englewood Cliffs, NJ.

Joughin, I. R., Percival, D. B. and Winebrenner, D. P. (1993), Maximum Likelihood Estimation of K Distribution Parameters for SAR Data, *IEEE Trans. on Geoscience and Remote Sensing* **31**:5, 989–999.

Judy, P. F., Swensson, R. G. and Szulc, M. (1981), Lesion detection and signal-to-noise ratio in CT, *Med. Phys.* **8**, 13–23.

Judy, P. F. and Swensson, R. G. (1985), Detectability of lesions of various sizes on CT images, *Proc. SPIE* **535**, 38–42.

Judy, P. F., Kijewski, M. F. and Swensson, R. G. (1997), Observer detection performance loss: Target size uncertainty, *Proc. SPIE* **3036**, 39–47.

Julesz, B. (1962), Visual pattern discrimination, *IRE Trans. Inform. Theory* **8**, 84–92.

Julesz, B. (1981), Textons, the elements of texture perception, and their interactions, *Nature* **290**, 91–97.

Jung, R. and Kornhuber, H., Eds. (1961), *Neurophysiologie und Psychophysik des visuellen systems*, Springer-Verlag, Heidelberg.

Kaas, J. H. (1978), The organization of the visual cortex, in *Sensory Systems of Primates* (Noback, C. R., Ed.), Plenum, New York, pp. 151–179.

Kaczmarz, S. (1937), Angenäherte Auflosung von Systemen linearer Gleichungen, *Bull. Int. Acad. Pol. Sci. Lett.* **35**, 355–357.

Kaiser, G. (1994), *A Friendly Guide to Wavelets*, Birkhäuser, Boston.

Kaluza, R. (1996), *The Life of Stefan Banach. Through a Reporter's Eyes*, Birkhäuser, Cambridge, MA.

Kanatani, K. (1990), *Group-Theoretical Methods in Image Understanding*, Springer-Verlag, Berlin.

Kanwal, R. P. (1983), *Generalized Functions*, Academic Press, New York.

Karush, W. (1939), *Minima of Functions of Several Variables with Inequalities as Side Constraints*, Master's Thesis, University of Chicago, Chicago, IL.

Kasdin, N. J. (1995), Discrete simulation of colored noise and stochastic processes and $1/f^{\alpha}$ power law noise generation, *Proc. IEEE* **83**, 802–827.

Kay, J. (1994), Statistical models for PET and SPECT data, *Stat. Methods Med. Res.* **3**, 5–21.

Kay, S. M. (1998), *Fundamentals of Statistical Signal Processing*, Vol. II: *Detection Theory*, Prentice-Hall, Upper Saddle River, NJ.

Kelly, D. H. (1977), Visual contrast sensitivity, *Opt. Acta* **24**, 107–129.

Kendall, D. G., Barden, D., Carne, T. K. and Le, H. (1999), *Shape and Shape Theory*, Wiley, Chichester.

Kendall, M. G. and Stuart, A. (1979), *The Advanced Theory of Statistics*, Vol. 2, 4th ed., Griffin, London.

Kepler, J. (1604), Ad Vitellionem Paralipomena, quibus Astronomiae Pars Optica traditur...de modo visionis, humorum oculi usu, contra Opticos & Anatomicos, C. Marnius & Heirs of J. Aubrius: Frankfurt.

Keshner, M. S. (1982), 1/f noise, *Proc. IEEE* **70**:3, 212–218.

Khinchin, A. I. (1949), *Statistical Mechanics*, Dover, New York.

Kijewski, M. F., Swensson, R. G. and Judy, P. F. (1989), Analysis of rating data from multiple-alternative tasks, *J. Math. Psych.* **33**, 428–451.

Kijewski, M. F., Müller, S. P. and Moore, S. C. (1992), The Barankin bound: A model of detection with location uncertainty, *Proc. SPIE* **1768**, 153–160.

Kijewski, M. F., Moore, S. C., Jadvar, H., Zimmerman, R. E. and Müller, S. P. (2001), Effects of SPECT collimation and system geometry on classification tasks related to Parkinson's disease, *IEEE Trans. Nucl. Sci.* **48**, 734–738.

King, J. L., Britton, C. A., Gur, D., Rockette, H. E. and Davis, P. L. (1993), On the validity of the continuous and discrete confidence rating scales in receiver operating characteristic studies, *Invest. Radiol.* **28**, 962–963.

King, M. A., DeVries, D. J. and Soares, E. J. (1997), Comparison of the channelized Hotelling and human observers for lesion detection in hepatic SPECT imaging, *Proc. SPIE* **3036**, 14–20.

King, M. A., Glick, S. J., Pretorius, P. H., Wells, R. G., Gifford, H. C., Narayanan, M. V. and Farncombe, T. (2003), Attenuation, scatter, and spatial resolution compensation in SPECT, in *Emission Tomography: The Fundamentals of PET and SPECT* (Wernick, M. N. and Aarsvold, J. N., Eds.), Academic Press, to appear.

Kingston, R. H. (1995), *Optical Sources, Detectors, and Systems*, Academic Press, San Diego.

Kirillov, A. A. (1961), On a problem of I. M. Gel'fand, *Sov. Math. Dokl.* **2**, 268–269 (English translation).

Kirkpatrick, S., Gelatt, C. D., Jr. and Vecchi, M. P. (1983), Optimization by simulated annealing, *Science* **220**, 671–680.

Kirkpatrick, S., Gelatt, C. D., Jr. and Vecchi, M. P. (1984), Optimization by simulated annealing: Quantitative study, *J. Stat. Phys.* **34**, 975–986.

Kirsch, A. (1996), *An Introduction to the Mathematical Theory of Inverse Problems*, Springer-Verlag, New York.

Klauder, J. R., Price, A. C., Darlington, S. and Albersheim, W. J. (1960), The theory and design of chirp radar, *Bell Syst. Tech. J.* **39**, 745–808.

Klauder, J. R. (1985), A coherent-state primer, in *Coherent States — Applications in Physics and Mathematical Physics* (Klauder, J. R. and Skagerstam, B., Eds.), World Scientific Publishing, Singapore.

Kluyver, J. C. (1905/1906), A Local Probability Problem, Proceedings Section of Science, *K. Akad. Van Wet. te Amsterdam* **8**, 341–350.

Knoll, G. F. (1999), *Radiation Detection and Measurement*, 3rd ed., Wiley, New York.

Kogan, S. (1996), *Electronic Noise and Fluctuations in Solids*, Cambridge University Press, Cambridge.

Kogelnik, H. and Li, T. (1966), Laser beams and resonators, *Proc. IEEE* **54**, 1312–1329.

Kolmogorov, A. N. (1950), *Foundations of the Theory of Probability*, Chelsea, New York.

Körner, T. W. (1988), *Fourier Analysis*, Cambridge University Press, Cambridge.

Kotz, S., Johnson, N. L. and Read, C. B., Eds. (1982), *Encyclopedia of Statistical Sciences*, Vols. 1 and 2, Wiley, New York.

Kotz, S., Johnson, N. L. and Read, C. B., Eds. (1986), *Encyclopedia of Statistical Sciences*, Vol. 7, Wiley, New York.

Kotz, S., Johnson, N. L. and Read, C. B., Eds. (1988), *Encyclopedia of Statistical Sciences*, Vol. 8, Wiley, New York.

Krane, K. (1983), *Modern Physics*, Wiley, Toronto.

Kraniauskas, P. (1994), A plain man's guide to the FFT, *IEEE Signal Proc. Mag.* **11**:2, 24–35.

Krantz, D. H. (1969), Threshold theories of signal detection, *Psychol. Rev.* **76**, 308–324. Reprinted in Cohn, T. E., Ed. (1993), Collected Works in Optics, Vol. 3: *Visual Detection*, Optical Society of America, Washington, DC, pp. 265–281.

Kreysig, E. (1978), *Introductory Functional Analysis with Applications*, Wiley, New York.

Kronauer, R. E. and Zeevi, Y. Y. (1985), Reorganization and diversification of signals in vision, *IEEE Trans. Syst. Man Cyb.* **15**, 91–101.

Krupinski, E. A., Johnson, J. P., Roehrig, H., Lubin, J. and Engstrom, M. (2003), Human vision model to predict observer performance: Detection of microcalcifications as a function of monitor phosphor, *Proc. SPIE* **5034**, 20–24.

Kuffler, S.W. (1953), Discharge patterns and functional organization of mammalian retina, *J. Neurophysiol.* **16**, 37–68.

Kuhn, H. W. and Tucker, A. W. (1951), Nonlinear programming, in *Proceedings of the Second Berkeley Symposium on Mathematical Statistics and Probability* (Neyman, J., Ed.), University of California Press, Berkeley, CA, pp. 481–492.

Kundel, H. L. and Polansky, M. (1997), Mixture distribution and receiver operating characteristic analysis of bedside chest imaging with screen-film and computed radiography, *Acad. Radiol.* **4**, 1–7.

Kundel, H. L. and Polansky, M. (1998), Comparing observer performance with mixture distribution analysis when there is no external gold standard, *Proc. SPIE* **3340**, 78–84.

Kundel, H. L., Polansky, M. and Phelan, M. (2001), Evaluating imaging systems in the absence of truth: A comparison of ROC and mixture distribution analysis in computer aided diagnosis in mammography, *Proc. SPIE* **4324**, 153–158.

Kunyansky, L. A. (2001), A new SPECT reconstruction algorithm based on the Novikov explicit inversion formula, *Inverse Probl.* **17**, 293–306.

Kupinski, M. A., Hoppin, J. W., Clarkson, E. and Barrett, H. H. (2002), Estimation in medical imaging without a gold standard, *Acad. Radiol.* **9**, 290–297.

Kupinski, M. A., Clarkson, E., Hoppin, J., Chen, L. and Barrett, H. H. (2003a), Experimental determination of object statistics from noisy images, *J. Opt. Soc. Am. A* **20**, 421–429.

Kupinski, M. A., Hoppin, J., Clarkson, E. and Barrett, H. H. (2003b), Ideal-observer computation using Markov-chain Monte Carlo, *J. Opt. Soc. Am. A* **20**, 430–438.

Kupinski, M. A., Clarkson, E., Gross, K. and Hoppin, J. W. (2003c), Optimizing imaging hardware for estimation tasks, *Proc. SPIE* **5034**, 309–313.

LaCroix, K. J., Tsui, B. M. W., Frey, E. C. and Jaszczak, R. J. (2000), Receiver operating characteristic evaluation of iterative reconstruction with attenuation correction in Tc-99m-sestamibi myocardial SPECT images, *J. Nucl. Med.* **41**, 502–513.

Lam, E. Y. and Goodman, J. W. (2000), A mathematical analysis of the DCT coefficient distributions for images, *IEEE Trans. Image Process.* **9**:10, 1661–1666.

Landesman, B. T. and Barrett, H. H. (1988), Gaussian amplitude functions that are exact solutions to the scalar Helmholtz equation, *J. Opt. Soc. Am. A* **5**, 1610–1619.

Landweber, L. (1951), An iteration formula for Fredholm integral equations of the first kind, *Am. J. Math.* **73**, 615–625.

Lang, S. (1993), *Real and Functional Analysis*, 3rd ed., Springer-Verlag, New York.

Lange, K. and Carson, R. (1984), EM reconstruction algorithms for emission and transmission tomography, *J. Comput. Assist. Tomogr.* **8**, 306–316.

Lathi, B. P. (1992), *Linear Systems and Signals*, Berkeley-Cambridge Press, Carmichael, CA.

Lawson, C. L. and Hanson, R. J. (1995), *Solving Least Squares Problems*, Society for Industrial and Applied Mathematics, Philadelphia, PA.

Lawson, P. R., Ed. (2000), Notes on the history of stellar interferometry, in *Principles of Long Baseline Stellar Interferometry*, Course notes from the 1999 Michelson Summer School, JPL Publication 00-009 07/00.

Lawson, W. (1971), Electro-optical system evaluation, in *Photoelectronic Imaging Devices* (Biberman, L. M. and Nudelman, S., Eds.), Plenum Press, New York.

Leahy, R., Mosher, J., Spencer, M., Huang, M., Lewine, J. (1998), A study of dipole localization accuracy for MEG and EEG using a human skull phantom, *Electroencephalography and Clinical Neurophysiology* **107**, 159-173.

Legge, G. E. and Foley, J. M. (1980), Contrast masking in human vision, *J. Opt. Soc. Am.* **70**, 1458–1471.

Lehmann, E. L. (1991), *Theory of Point Estimation,* Wadsworth & Brooks, Pacific Grove, CA.

Levi, A. and Stark, H. (1987), Restoration from phase and magnitude by generalized projections, in *Image Recovery: Theory and Application* (Stark, H., Ed.), Academic Press, San Diego.

Lewis, E. E. and Miller, W. F., Jr. (1984), *Computational Methods of Neutron Transport*, Wiley-Interscience, New York.

Lewitt, R. M. (1990), Multidimensional digital image representations using generalized Kaiser-Bessel window functions, *J. Opt. Soc. Am. A* **7**, 1834–1846.

Lewitt, R. M. (1992), Alternatives to voxels for image representation in iterative reconstruction algorithms, *Phys. Med. Biol.* **37**, 705–716.

Lighthill, M. J. (1958), *An Introduction to Fourier Analysis and Generalized Functions*, Cambridge University Press, Cambridge.

Linfoot, E. H. (1955), Information theory and optical images, *J. Opt. Soc. Am.* **45**, 808–819.

Lipson, S. G., Lipson, H. and Tannhauser, D. S. (1995), *Optical Physics*, Cambridge University Press, Cambridge.

Lipster, R. S. and Shiryayev, A. N. (1977), *Statistics of Random Processes I: General Theory* (translated by Aries, A. B.), Springer-Verlag, New York.

Ljungberg, M., Strand, S. E. and King, M. A. (1998), *Monte-Carlo Calculations in Nuclear Medicine*, IOP Publishing, Bristol and Philadelphia.

Llacer, J. and Veklerov, E. (1989), Feasible images and practical stopping rule for iterative algorithms in emission tomography, *IEEE Trans. Med. Imaging* **8**, 186–193.

Llacer, J. (1993), Results of a clinical receiver operating characteristic study combining filtered backprojection and maximum likelihood estimator images in FDG-PET studies, *J. Nucl. Med.* **34**, 1198–1203.

Lloyd, C. and Beaton, R. (1990), Design of a spatio-chromatic human vision model for evaluating full-color display systems, *Proc. SPIE* **1249**, 23–37.

Loève, M. (1963), *Probability Theory*, 3rd ed., D. Van Nostrand Company, New York.

Logan, B. F. and Shepp, L. A. (1975), Optimal reconstruction of a function from its projections, *Duke Math. J.* **42**:4, 645–659.

Lohmann, A. W. (1993), Image rotation, Wigner rotation, and the fractional Fourier transform, *J. Opt. Soc. Am. A* **10**, 2181–2186.

Lohmann, A. W. and Mendlovic, D. (1994), Image formation of a self Fourier object, *Appl. Opt.* **33**, 153–157.

Lombard, F. J. and Martin, F. (1961), Statistics of electron multiplication, *Rev. Sci. Instrum.* **31**, 200–201.

Lomont, J. S. (1959), *Applications of Finite Groups*, Academic Press, New York.

Longhurst, R. S. (1973), *Geometrical and Physical Optics*, 3rd ed., Longman Group, London.

Loo, L.-N., Doi, K. and Metz, C. E. (1984), A comparison of physical image quality indices and observer performance in the radiographic detection of nylon beads, *Phys. Med. Biol.* **29**, 837–856.

Loo, L.-N., Doi, K. and Metz, C. E. (1985), Investigation of basic imaging properties in digital radiography. 4. Effect of unsharp masking on the detectability of simple patterns, *Med. Phys.* **12**, 209–214.

Lorch, E. R. (1962), *Spectral Theory*, Oxford University Press, New York.

Loudon, R. (1973), *The Quantum Theory of Light*, Clarendon Press, Oxford.

Lubin, J. (1993), The use of psychophysical data and models in the analysis of display system performance, in *Digital Images and Human Vision* (Watson, A. B., Ed.), MIT Press, Cambridge, MA, pp. 163–178.

Lubin, J. (1995), A visual discrimination model for imaging system design and evaluation, in *Visual Models for Target Detection and Recognition* (Peli, E., Ed.), World Scientific, Singapore, pp. 245–283.

Lucy, L. B. (1974), An iterative technique for the rectification of observed distributions, *Astron. J.* **79**, 745–754.

Ludwig, D. (1966), The Radon transform on Euclidean space, *Commun. Pure Appl. Math.* **19**, 49–81.

Lukyanchikova, N. B. (1996), *Noise Research in Semiconductor Physics* (Jones, B. K., Ed.), Gordon and Breach Science, Amsterdam.

Lusted, L. B. (1968), *Introduction to Medical Decision Making,* Thomas, Springfield, IL.

Lusted, L. B. (1971), Signal detectability and medical decision-making, *Science* **171**, 1217–1219.

Ma, G. and Hall, W. J. (1993), Confidence bands for receiver operating characteristic curves, *Med. Decis. Making* **13**, 191–197.

Maffei, L. and Fiorentini, A. (1973), The visual cortex as a spatial frequency analyser, *Vision Res.* **13**, 1255–1267.

Magnus, W. and Oberhettinger, F. (1949), *Formulas and Theorems of the Special Functions of Mathematical Physics*, Chelsea Publishing, New York.

Mahajan, V. N. (1981), Zernike annular polynomials for imaging systems with annular pupils, *J. Opt. Soc. Am.* **71**, 75–84.

Mallat, S. (1989), A theory for multiresolution signal decomposition: The wavelet representation, *IEEE PAMI* **11**, 674–693.

Mallat, S. (1999), *A Wavelet Tour of Signal Processing*, 2nd ed., Academic Press, New York.

Maloney, R. K., Mitchison, G. J. and Barlow, H. B. (1987), Limit to the detection of Glass patterns in the presence of noise, *J. Opt. Soc. Am. A* **4**, 2336–2341.

Mammone, R. (1987), Image restoration using linear programming, in *Image Recovery: Theory and Applications*, (Stark, H., Ed.), Academic Press, Orlando, Chap. 4.

Mandel, L. (1959), Image fluctuations in cascade intensifiers, *Brit. J. Appl. Phys.* **10**, 233–234.

Mandel, L. and Wolf, E. (1976), Spectral coherence and the concept of cross-spectral purity, *J. Opt. Soc. Am.* **66**, 529–535.

Mandel, L. and Wolf, E. (1995), *Optical Coherence and Quantum Optics,* Cambridge University Press, Cambridge.

Mandelbrot, B. B. (1960), The Pareto-Lévy law and the distribution of income, *Int. Econ. Rev.* **1**, 79–106.

Mandelbrot, B. B. (1963), Variation of certain speculative prices, *J. Business* **36**, 394–419.

Mandelbrot, B. (1999), *Multifractals and 1/f Noise: Wild Self-Affinity in Physics*, Springer-Verlag, New York.

Mangasaria, O. L. (1994), *Nonlinear Programming*, Society for Industrial and Applied Mathematics, Philadelphia, PA.

Mansuripur, M. (2002), *Classical Optics and its Applications*, Cambridge University Press, Cambridge.

Mantegna, R. N. (1994), Fast, accurate algorithm for the numerical simulation of Levy stable stochastic processes, *Phys. Rev. E* **49**, 4677–4683.

Marcelja, S. (1980), Mathematical description of the response of simple cortical cells, *J. Opt. Soc. Am.* **70**, 1297–1300.

Marchand, E. W. and Wolf, E. (1974a), Radiometry with sources of any state of coherence, *J. Opt. Soc. Am.* **64**, 1219–1226.

Marchand, E. W. and Wolf, E. (1974b), Walther's definitions of generalized radiance, *J. Opt. Soc. Am.* **64**, 1273–1274.

Mardia, K. V. (1972), *Statistics of Directional Data*, Academic Press, London.

Mardia, K. V., Kent, J. T. and Bibby, J. M. (1979), *Multivariate Analysis*, Academic Press, New York.

Margenau, H. and Murphy, G. M. (1956), *The Mathematics of Chemistry and Physics*, D. Van Nostrand, New York.

Marier, L. J. (1995), Correlated K-distributed clutter generation for radar detection and track, *IEEE Trans. Aerospace and Electronic Systems* **31**, 568–580.

Marks, D. G., Barber, H. B., Barrett, H. H., Tueller, J. and Woolfenden, J. M. (1999), Improving performance of a CdZnTe imaging array by mapping the detector with gamma rays, *Nucl. Instrum. Methods Phys. Res. A* **428**, 102–112.

Marks, D. (2000), Estimation methods for semiconductor gamma-ray detectors, Ph.D. Dissertation, University of Arizona, Tucson, AZ.

Marks, R. J. (1991), *Introduction to Shannon Sampling and Interpolation Theory*, Springer-Verlag, New York.

Massof, R. W. and Emmel, T. C. (1987), Criterion-free parameter-free distribution-independent index of diagnostic test performance, *Appl. Opt.* **26**, 1395–1408.

Matteuci, M. (1970), *History of the Motor Car*, Crown, New York.

Maurer, S. B. (July, 1995), SIAM news.

Mayer, M. J. and Tyler, C. W. (1986), Invariance of the slope of the psychometric function with spatial summation, *J. Opt. Soc. Am. A* **3**, 1166–1172.

McAulay, R. J. and Hofstetter, E. M. (1971), Barankin bounds on parameter estimation, *IEEE Trans. Inform. Theory* **17**, 669–676.

McClish, D. K. (1989), Analyzing a portion of the ROC curve, *Med. Decis. Making* **9**, 190–195.

McClish, D. K. (1990), Determining a range of false-positive rates for which ROC curves differ, *Med. Decis. Making* **10**, 283–287.

McColl, J. H., Holmes, A. P. and Ford, I. (1994), Statistical methods in neuroimaging with particular application to emission tomography, *Stat. Methods Med. Res.* **3**, 63–86.

McLachlan, G. J. and Krishnan T. (1997), *The EM Algorithm and Extensions*, Wiley, New York.

McNeil, B. J. and Adelstein, S. J. (1976), Determining the value of diagnostic and screening tests, *J. Nucl. Med.* **17**, 439–448.

McNeil, B. J. and Hanley, J. A. (1984), Statistical approaches to the analysis of receiver operating characteristic (ROC) curves, *Med. Decis. Making* **4**, 137–150.

Mecklenbräuker, W. and Hlawatsch, F., Eds. (1997), *The Wigner Distribution — Theory and Applications in Signal Processing*, Elsevier, Amsterdam, The Netherlands.

Melsa, J. L. and Cohn, D. L. (1978), *Decision and Estimation Theory*, McGraw-Hill, New York.

Mendlovic, D., Ozaktas, H. M. and A. W. Lohmann (1994) Self Fourier functions and fractional Fourier Transforms, *Opt. Commun.* **105**, 36–38.

Mertz, L. (1956), Optical Fourier synthesizer, *J. Opt. Soc. Am.* **46**, 548 – 551.

Mertz, L. (1965), *Transformations in Optics*, Wiley, New York.

Messiah, A. (1961), *Quantum Mechanics*, Vol. I, North-Holland Publishing, Amsterdam.

Messiah, A. (1962), *Quantum Mechanics*, Vol. II, North-Holland Publishing, Amsterdam.

Metropolis, N., Rosenbluth, A., Rosenbluth, M., Teller, A. and Teller, E. (1953), Equations of state calculations by fast computing machines, *J. Chem. Phys.* **21**, 1087–1091.

Metz, C. E. (1969), A mathematical investigation of radioisotope scan image processing, Ph.D. Dissertation, University of Pennsylvania, Philadelphia, PA.

Metz, C. E., Goodenough, D. J. and Rossmann, K. (1973), Evaluation of receiver operating characteristic curve data in terms of information theory, with applications in radiography, *Radiology* **109**, 297–303.

Metz, C. E. and Pizer, S. M. (1971), Nonstationary and nonlinear scintigram processing, Paper presented at the 2nd International Conference on Data Handling and Image Processing in Scintigraphy, Hanover, West Germany.

Metz, C. E., Starr, S. J. and Lusted, L. B. (1976), Quantitative evaluation of visual detection performance in medicine: ROC-analysis and determination of diagnostic benefit, in *Medical Images: Formation, Perception and Measurement* (Hay, G. A., Ed.), Wiley, New York, pp. 220–241.

Metz, C. E. (1978), Basic principles of ROC analysis, *Semin. Nucl. Med.* **8**, 283–298.

Metz, C. E. and Kronman, H. B. (1980), Statistical significance tests for binormal ROC curves, *J. Math. Psych.* **22**, 218–243.

Metz, C. E., Wang, P.-L. and Kronman, H. B. (1984), A new approach for testing the significance of differences between ROC curves measured from correlated data, in *Information Processing in Medical Imaging* (Deconinck, F., Ed.), Martinus Nijhoff, The Hague, pp. 432–445.

Metz, C. E. (1986a), Statistical analysis of ROC data in evaluating diagnostic performance, In *Multiple Regression Analysis: Applications in the Health Sciences* (Herbert, D. and Myers, R., Eds.), American Institute of Physics, New York, pp. 365–384.

Metz, C. E. (1986b), ROC methodology in radiologic imaging, *Invest. Radiol.* **21**, 720–733.

Metz, C. E. (1989), Some practical issues of experimental design and data analysis in radiological ROC studies, *Invest. Radiol.* **24**, 234–245.

Metz, C. E and Shen, J.-H. (1992), Gains in accuracy from replicated readings of diagnostic images: Prediction and assessment in terms of ROC analysis, *Med. Decis. Making* **12**, 60–75.

Metz, C. E. (1993), Quantification of failure to demonstrate statistical significance: The usefulness of confidence intervals, *Invest. Radiol.* **28**, 59–63.

Metz, C. E. and Pan, X. (1995), A unified analysis of exact methods of inverting the 2D exponential Radon transform, with implications for noise control in SPECT, *IEEE Trans. Med. Imaging* **14**, 643–658.

Metz, C. E., Herman, B. A. and Roe, C. A. (1998a), Statistical comparison of two ROC curve estimates obtained from partially-paired datasets, *Med. Decis. Making* **18**, 110–121.

Metz, C. E., Herman, B. A. and Shen, J-H. (1998b), Maximum-likelihood estimation of ROC curves from continuously-distributed data, *Stat. Med.* **17**, 1033–1053.

Metz, C. E. (1999), Fundamental ROC analysis, in *Handbook of Medical Imaging*, Vol. 1: *Progress in Medical Physics and Psychophysics* (Beutel, J., Kundel, H. L. and Van Metter, R. L., Eds.), SPIE Press, Bellingham, WA.

Metz, C. E. and Pan, X. (1999), "Proper" binormal ROC curves: Theory and maximum-likelihood estimation, *J. Math. Psych.* **43**, 1–33.

Metz, C. E. (2000), Fundamental ROC analysis, in *Handbook of Medical Imaging*, Vol. 1: *Physics and Psychophysics* (Beutel, J., Kundel, H. and Van Metter, R., Eds.), SPIE Press, Bellingham, WA, pp. 751–769.

Meyer, Y. (1993), *Wavelets: Algorithms and Applications*, Society for Industrial and Applied Mathematics, Philadelphia, PA.

Meystre, P. and Sargent, M. (1990), *Elements of Quantum Optics*, Springer-Verlag, Berlin.

Michelson, A. A. (1890), On the application of interference methods to astronomical measurements, Plates I and II, *Philos. Mag.* **30**, 1–20.

Michelson, A. A. (1891a), Measurement of Jupiter's satellites by interference, *Nature* **45**, 160–161.

Michelson, A. A. (1891b), Visibility of interference fringes in the focus of a telescope, *Philos. Mag.* **31**, 256–259.

Michelson, A. A. and Pease, F. G. (1921), Measurement of the diameter of Alpha Orionis with the interferometer, *Astrophys. J.* **53**, 249–259.

Middleton, D. (1960), *An Introduction to Statistical Communication Theory*, McGraw-Hill, New York. Reprinted as IEEE Press Classic Reissue, 1996.

Middleton, D. (1996), *An Introduction to Statistical Communication Theory*, IEEE, New York.

Miller, M. I., Snyder, D. L. and Miller, T. R. (1985), Maximum-likelihood reconstruction for single-photon emission computed-tomography, *IEEE Trans. Nucl. Sci.* **NS32**, 769–778.

Milster, T. D., Selberg, L. A., Barrett, H. H., Easton, R. L., Rossi, G. R., Arendt, J. and Simpson, R. G. (1984), A modular scintillation camera for use in nuclear medicine, *IEEE Trans. Nucl. Sci.* **31**, 578–580.

Milster, T. D., Selberg, L. A., Barrett, H. H., Landesman, A. L. and Seacatt, R. H., III (1985), Digital position estimation for the modular scintillation camera, *IEEE Trans. Nucl. Sci.* **32**, 748–752.

Milster, T. D., Aarsvold, J. N., Barrett, H. H., Landesman, A. L., Mar, L. S., Patton, D. D., Roney, T. J., Rowe, R. K. and Seacat, R. H., III (1990), A full-field modular gamma camera, *J. Nucl. Med.* **31**:5, 632–639.

Mirsky, L. (1982), *An Introduction to Linear Algebra*, Dover, New York.

Molthen, R. C., Shankar, P. M. and Reid, J. M. (1995), Tissue characterization in ultrasonic B scans using non-Rayleigh statistics, *Ultrasound in Medicine and Biology* **21**, 161–170.

Molthen, R. C., Shankar P. M., Reid J. M., Forsberg F., Halpern, E. J., Piccoli, C.. W. and Goldberg, B. B. (1998), Comparisons of the Rayleigh and K-distribution models using in vivo breast and liver tissue, *Ultrasound in Medicine and Biology* **24**, 93–100.

Montroll, E. W. and Shlesinger, M. F. (1982), On 1/f noise and other distributions with long tails, *Proc. Natl. Acad. Sci. USA* **79**, 3380–3383.

Moore, E. H. (1920), On the reciprocal of the general algebraic matrix (Abstract), *Bull. Am. Math. Soc.* **26**, 394–395.

Moore, S. C. and El Fakhri, G. (2001), Realistic Monte Carlo simulation of Ga-67 SPECT imaging, *IEEE Trans. Nucl. Sci.* **48**, 720–724.

Morozov, V. A. (1993), *Regularization Methods for Ill-posed Problems*, CRC Press, Boca Raton, FL.

Morse, P. M. and Feshbach, H. (1953), *Methods of Theoretical Physics*, McGraw-Hill, New York.

Mostafavi, H. and Sakrison, D. (1976), Structure and properties of a single channel in the human visual system, *Vision Res.* **16**, 957–968.

Movshon, J. A., Thompson, I. D. and Tolhurst, D. J. (1978), Spatial summation in the respective fields of simple cells in the cat's striate cortex, *J. Physiol.* **283**, 53–77.

Muir, T. A (1960), *Treatise on the Theory of Determinants*, Dover, New York.

Müller, S. P., Polak, J. F., Kijewski, M. F. and Holman, B. L. (1986), Collimator selection for SPECT brain imaging: The advantage of high resolution, *J. Nucl. Med.* **27**, 1729–1738.

Müller, S. P., Kijewski, M. F., Moore, S. C. and Holman, B. L. (1990) Maximum-likelihood estimation — a model for optimal quantitation in nuclear medicine, *J. Nucl. Med.* **31**, 1693–1701.

Müller, S. P., Kijewski, M. F. and Moore, S. C., (1995), The efficiency of ML estimates at low SNR can be inferred from the probability distribution of the estimates, *J. Nucl. Med.* **36**, 119.

Myers, K. J., Barrett, H. H., Borgstron, M. C., Patton, D. D. and Seeley, G. W. (1985), Effect of noise correlation on detectability of disk signals in medical imaging, *J. Opt. Soc. Am. A* **2**, 1752–1759.

Myers, K. J., Barrett, H. H., Borgstrom, M. C., Cargill, E. B., Clough, A. V., Fiete, R. D., Milster, T. D., Patton, D. D., Paxman, R. G., Seeley, G. W., Smith, W. E. and Stempski, M. O. (1986), A systematic approach to the design of diagnostic systems for nuclear medicine, in *Information Processing in Medical Imaging: Proceedings of the Ninth Conference* (Bacharach, S. L., Ed.), Martinus Nijhoff, Dordrecht, The Netherlands, pp. 431–444.

Myers, K. J. and Barrett, H. H. (1987), Addition of a channel mechanism to the ideal-observer model, *J. Opt. Soc. Am. A* **4**, 2447–2457.

Myers, K. J., Rolland, J. P., Barrett, H. H. and Wagner, R. F. (1990), Aperture optimization for emission imaging: Effect of a spatially varying background, *J. Opt. Soc. Am. A* **7**, 1279–1293.

Myers, K. J., Wagner, R. F. and Hanson, K. M. (1993), Rayleigh task performance in tomographic reconstructions: Comparison of human and machine performance, *Proc. SPIE* **1898**, 628–637.

Nachmias, J. (1981), On the psychometric function for contrast detection, *Vision Res.* **21**, 215–223.

Nachmias, J. and Steinman, R. M. (1963), Study of absolute visual detection by the rating-scale method, *J. Opt. Soc. Am.* **53**, 1206–1213. Reprinted in Cohn, T. E., Ed. (1993), Collected Works in Optics, Vol. 3: *Visual Detection*, Optical Society of America, Washington, DC, pp. 59–66.

Naimark, M. A. and Stern A. I. (1982), *Theory of Group Representations*, Springer-Verlag, New York.

Narayanan, M. V., Shankar, P. M. and Reid, J. M. (1994), Non-Rayleigh statistics of ultrasound backscattered signals, *IEEE Trans. Ultrasonics Ferroelectrics Frequency Control* **41**, 845–851.

Nashed, M. Z. (1976), *Generalized Inverses and Applications*, Academic Press, New York.

National Electrical Manufacturers Association (NEMA) (2001), *Digital Imaging and Communications in Medicine (DICOM) Part 14: Grayscale Standard Display Function*, NEMA, Rosslyn, VA.

Natterer, F. (1986), *The Mathematics of Computerized Tomography*, Wiley, New York. Reprinted in SIAM series, Classics in Applied Mathematics, Society of Industrial and Applied Mathematics, Philadelphia, PA, 2001.

Natterer F. (2001), Inversion of the attenuated Radon transform, *Inverse Probl.* **17**, 113–119.

Natterer, F. and Wübbeling, F. (2001), *Mathematical Methods in Image Reconstruction*, Society of Industrial and Applied Mathematics, Philadelphia, PA.

Nazarathy, M., Hardy, A. and Shamir, J. (1986), Misaligned first-order optics: Canonical operator theory, *J. Opt. Soc. Am. A* **3**, 1360–1369.

Nazarathy, M. and Shamir, J. (1982a), First order optics — A canonical representation — lossless systems, *J. Opt. Soc. Am.* **72**, 356–364.

Nazarathy, M. and Shamir, J. (1982b), First-order optics — operator representation for systems with loss or gain, *J. Opt. Soc. Am.* **72**, 1398–1408.

Neelamkavil, F. (1994), *Computer Simulation and Modelling*, Wiley, New York.

Neeser, F. D. and Massey, J. L. (1993), Proper complex random processes with application to information theory, *IEEE Trans. Inform. Theory* **39**, 1293–1302.

Nicodemus, F. (1963a), Radiance, *Am. J. Phys.* **31**, 368–377.

Nicodemus, F. (1963b), *Geometrical Considerations and Nomenclature for Reflectance*, Monograph No. 160, National Bureau of Standards, Washington, DC.

Nicodemus, F. (1973), Normalization in radiometry, *Appl. Opt.* **12**, 2960–2973.

Nieto-Vesperinas, M. (1991), *Scattering and Diffraction in Physical Optics*, Wiley-Interscience, New York.

Nolte, L. W. and Jaarsma, D. (1967), More on the detection of one of M orthogonal signals, *J. Acoust. Soc. Am.* **41**, 497–505.

North, D. O. (1943), Analysis of factors which determine signal-noise discrimination in pulsed carrier systems, *RCA Tech. Rept.* **PTR-6C**, RCA. Reprinted in *Proc. IEEE* **51**, 1016–1027, 1963

Novikov, R. G. (2002a), An inversion formula for the attenuated x-ray transformation, *Ark. Mat.* **40**, 145-167.

Novikov, R. G. (2002b), On the range characterization for the two-dimensional attenuated x-ray transformation, *Inverse Probl.* **18**, 677–700

Nyquist, H. (1928a), Certain topics in telegraph transmission theory, *AIEE Trans.* **47**, 617–644.

Nyquist, H. (1928b), Thermal agitation of electric charge in conductors, *Phys. Rev.* **32**, 110–113.

Oberhettinger, F. (1972), *Tables of Bessel Transforms*, Springer-Verlag, Berlin.

Oberhettinger, F. (1973), *Tables of Laplace Transforms*, Springer-Verlag, Berlin.

Oberhettinger, F. (1974), *Tables of Mellin Transforms*, Springer-Verlag, Berlin.

Obuchowski, N. A. (1995), Multireader, multimodality receiver operating characteristic curve studies: Hypothesis testing and sample size estimation using an analysis of variance approach with dependent observations, *Acad. Radiol.* **2**:Suppl. 1, S22.

Obuchowski, N. A. (1998), Sample size calculations in studies of test accuracy, *Stat. Methods Med. Res.* **7**, 371–392.

Ogawa, H. (1988), An operator pseudo-inversion lemma, *SIAM J. Appl. Math.* **48**, 1527–1531.

Ogawa, H. and Oja, E. (1986), Projection filter, Wiener filter and Karhunen-Loève subspaces in digital image restoration, *J. Math. Anal. Appl.* **114**, 37–51.

Ohara, K., Chan, H.-P., Doi, K., Giger, M. L. and Fujita, H. (1986), Investigation of basic imaging properties in digital radiography. 8. Detection of simulated low-contrast objects in digital subtraction angiographic images, *Med. Phys.* **13**, 304–311.

Oliver, C. J. (1985), Correlated K-distributed clutter models, *Opt. Acta* **32**, 1515–1547.

Oliver, C. J. and Tough, R. J. (1986), On the simulation of correlated K-distributed random clutter, *Opt. Acta* **33**, 223–250.

Olshausen, B. A. and Field, D. J. (1996), Emergence of simple-cell receptive field properties by learning a sparse code for natural images, *Nature* **381**, 607–609.

Oppenheim, A. V. and Lim, J. S. (1981), The importance of phase in signals, *Proc. IEEE* **69**, 529–541.

Oppenheim, A. V. and Schafer, R. W. (1975), *Digital Signal Processing*, Prentice-Hall, Englewood Cliffs, NJ.

Oppenheim, A. V. and Schafer, R. W. (1989), *Discrete-Time Signal Processing*, Prentice-Hall, Englewood Cliffs, NJ.

Oppenheim, A. V. and Willsky, A. S. (1983), *Signals and Systems*, Prentice-Hall, Englewood Cliffs, NJ.

O'Rourke, J. (1998), *Computational Geometry in C*, 2nd ed., Cambridge University Press, Cambridge.

Osche, G. R. (2002), *Optical Detection Theory for Laser Applications*, Wiley, Hoboken, NJ.

O'Sullivan, J. A., Blahut, R. E. and Snyder, D. L. (1998), Information-theoretic image formation, *IEEE Trans. Inform. Theory* **44**, 2094–2123.

Owsley, C., Sekuler, R. and Siemsen, D. (1983), Contrast sensitivity throughout adulthood, *Vision Res.* **23**, 689–699.

Ozaktas, H. M., Barshan, B., Mendlovic, D. and Onural, L. (1994), Convolution, filtering and multiplexing in fractional Fourier domain and their relation to chirp and wavelet transforms, *J. Opt. Soc. Am. A* **11**, 547–559.

Ozaktas, H. M. and Mendlovic, D. (1994), The fractional Fourier transform as a tool for analyzing beam propagation and spherical mirror resonators, *Opt. Lett.* **19**, 1678–1680.

Paley, R. E. A. C. and Wiener, N. (1934), *Fourier Transforms in the Complex Domain*, Vol. 19, American Mathematical Society Colloquium Publications, New York.

Palmer, J. M. (1994), The measurement of transmission, absorption, emission and reflection, in *Handbook of Optics*, Vol. 2 (Bass, M., Ed.), McGraw-Hill, New York, Chap. 25.

Pan, X. and Metz, C. (1995), Analysis of noise properties of a class of exact method of inverting the 2-D exponential Radon transform, *IEEE Trans. Med. Imaging* **14**, 659–668.

Pan, X. and Metz, C. E. (1997), The "proper" binormal model: Parametric ROC curve estimation with degenerate data, *Acad. Radiol.* **4**, 380–389.

Papoulis, A. (1962), *The Fourier Integral and Its Applications*, McGraw-Hill, New York.

Papoulis, A. (1965), *Probability, Random Variables, and Stochastic Processes*, McGraw-Hill, New York.

Papoulis, A. (1968), *Systems and Transforms with Applications in Optics*, McGraw-Hill, New York.

Papoulis, A. (1975), A new algorithm in spectral analysis and band-limited extrapolation, *IEEE Trans. Circuits Syst.* **22**, 735–742.

Papoulis, A. (1991), *Probability, Random Variables, and Stochastic Processes*, 3rd ed., McGraw-Hill, New York.

Papoulis, A. (1994), Pulse compression, fiber communications and diffraction: A unified approach, *J. Opt. Soc. Am. A* **11**, 3–13.

Park, S., Kupinski, M. A., Clarkson, E. and Barrett, H. H. (2003), Ideal-observer performance under signal and background uncertainty, in *Information Processing in Medical Imaging: Proceedings of the Eighteenth Conference* (Taylor, C. and Noble, A., Eds.), in press.

Parra, L. and Barrett, H. H. (1998), List-mode likelihood: EM algorithm and image quality estimation demonstrated on 2-D PET, *IEEE Trans. Med. Imaging* **17**:2, 228–235.

Patton, D. D. (1978), Introduction to clinical decision making, *Semin. Nucl. Med.* **8**, 273–282.

Patton, D. D. (2000), The father of nuclear medicine: Establishing paternity, *J. Nucl. Med.* **41**, 26N-30N.

Patton, D. D. and Woolfenden, J. M. (1989), A utility-based model for comparing the cost-effectiveness of diagnostic studies, *Invest. Radiol.* **24**, 263–271.

Pearson, K. (1894), Contributions to the mathematical theory of evolution, *Philos. Trans. Roy. Soc. Ser. A.* **185**, 71–110.

Pearson, K. (1905), The Problem of the Random Walk, *Nature* **LXXII**, 294–342.

Pelli, D. (1981), Effects of visual noise, Ph.D. Dissertation, Cambridge University, Cambridge, England.

Pelli, D. (1985), Uncertainty explains many aspects of visual contrast detection and discrimination, *J. Opt. Soc. Am. A* **2**, 1508–1530.

Penrose, R. (1955), A generalized inverse for matrices, *Proc. Camb. Philol. Soc.* **51**, 406–413.

Pentini, R. A., Farina, A. and Zirilli, F. (1992), Radar detection of targets located in a coherent K distributed clutter background, *IEEE Proc.* **139**, 239–245.

Percival, D. B. and Walden, A. T. (1993), *Spectral Analysis for Physical Applications*, University Press, Cambridge, UK.

Pettofrezzo, A. J. (1966), *Matrices and Transformations*, Dover, New York.

Peterson, W. W., Birdsall, T. G. and Fox, W. C. (1954), The theory of signal detectability, *Trans. IRE* **PGIT 4**, 171–212.

Phillips, G. C. and Wilson, H. R. (1984), Orientation bandwidths of spatial mechanisms measured by masking, *J. Opt. Soc. Am. A* **1**:2, 226–232.

Pierre, D. A. (1986), *Optimization Theory with Applications*, Paperback edition, Dover, New York.

Pike, E. R. and Bertero, M., Eds. (1992), *Inverse Problems in Scattering and Imaging*, Adam Hilger, Bristol, United Kingdom.

Pilz, J. (1991), *Bayesian Estimation and Experimental Design in Linear Regression Models*, Wiley, Chichester, England.

Pineda, A. R., Barrett, H. H. and Arridge, S. R. (2000), Spatially varying detectability for optical tomography, *Proc. SPIE* **3977**, 77–83.

Pineda, A. R. and Barrett, H. H. (2001), What does DQE say about lesion detectability in digital radiography?, *Proc. SPIE* **4320**, 561–569.

Pineda, A. R. and Barrett, H. H. (2004a), Figures of merit for digital radiography. I. Flat background and deterministic blurring, *Med. Phys.* **31**, 348-358.

Pineda, A. R. and Barrett, H. H. (2004b), Figures of merit for digital radiography. II. Finite number of secondaries, structured and random backgrounds, *Med. Phys.* **31**, 359-367.

Pirenne, M. H. (1943), Binocular and uniocular threshold of vision, *Nature* **152**, 698–699. Reprinted in Cohn, T. E., Ed. (1993), Collected Works in Optics, Vol. 3: *Visual Detection*, Optical Society of America, Washington, DC, pp. 383–384.

Pizer, S. (1981), Intensity mappings to linearize display devices, *Computer Graphics and Image Processing* **17**:3, 262–268.

Polansky, M. (2000), Agreement and accuracy mixture distribution analysis, in *Handbook of Medical Imaging*, Vol. 1: *Physics and Psychophysics* (Beutel, J., Kundel, H. and Van Metter, R., Eds.), SPIE Press, Bellingham, WA, 797–835.

Politte, D. G. and Snyder, D. L. (1988), The use of constraints to eliminate artifacts in maximum-likelihood image estimation for emission tomography, *IEEE Trans. Nucl. Med.* **35**:1, 608–610.

Pollehn, H. and Roehrig, H. (1970), Effect of noise on the modulation transfer function of the visual channel, *J. Opt. Soc. Am.* **60**, 842–848.

Possolo, A., Ed. (1991), *Spatial Statistics and Imaging*, Institute of Mathematical Statistics, Hayward, CA.

Poularikas, A. D., Ed. (1996), *The Transforms and Applications Handbook*, CRC Press, Boca Raton, FL.

Pratt, W. K. (1991), *Digital Image Processing*, 2nd ed., Wiley, New York.

Preparata, F. and Shamos, M. (1985), *Computational Geometry: An Introduction*, Springer-Verlag, New York.

Press, S. J. (1989), *Bayesian Statistics: Principles, Models and Applications*, Wiley, New York.

Press, W. H., Teukolsky, S. A., Vetterling, W. T. and Flannery, B. P. (1992), *Numerical Recipes: The Art of Scientific Computing*, Cambridge University Press, Cambridge.

Pretorius, P. H., King, M. A., Tsui, B. M. W., LaCroix, K. J. and Xia, W. (1999), A mathematical model of motion of the heart for use in generating source and attenuation maps for simulating emission imaging, *Med. Phys.* **26**, 2323–2332.

Pretorius, P. H., Xia, W., King, M. A., Tsui, B. M. W., Pan, T. S. and Villegas, B. J. (1997), Evaluation of right and left ventricular volume and ejection fraction using a mathematical cardiac torso phantom for gated pool SPECT, *J. Nucl. Med.* **38**:10, 1528–1534.

Qi, J. and Huesman, R. H. (2001), Theoretical study of lesion detectability of MAP reconstruction using computer observers, *IEEE Trans. Med. Imaging* **20**, 815–822.

Quenouille, M. H. (1956), Notes on bias in estimation, *Biometrika* **43**, 353–360.

Quirrenbach, A. (2001), Optical interferometry, *Ann. Rev. Astron. Astrophys.* **39**, 353–401.

Rabbani, M. and Jones, P. W. (1991), *Digital Image Compression Techniques*, Tutorial Texts in Optical Engineering, Vol. TT7, SPIE Optical Engineering Press, Bellingham, WA.

Rabbani, M., Shaw, R. and van Metter, R. (1987), Detective quantum efficiency of imaging systems with amplifying and scattering mechanisms, *J. Opt. Soc. Am. A* **4**, 895–901.

Rade, L. and Westergren, B. (1990), *Beta Mathematics Handbook*, 2nd ed., CRC Press, Boca Raton, FL.

Rade, L. and Westergren, B. (1992), *Beta Mathematics Handbook*, 3rd ed., CRC Press, Sweden.

Radon, J. (1917), Ber. Verh. Sächs. Akad. Wiss. Leipzig, *Math. Phys. K1.* **69**, 262–277. Translated in Deans (1983).

Ramirez, R. W. (1985), *The FFT Fundamentals and Concepts*, Prentice-Hall, Englewood Cliffs, NJ.

Ramo, S. (1939), Current induced by electron motion, *Proc. IRE* **27**, 584–585.

Ransohoff, D. F. and Feinstein, A. R. (1978), Problems of spectrum and bias in evaluating the efficacy of diagnostic tests, *New Engl. J. Med.* **299**, 926–930.

Rao, C. R. (1945), Information and accuracy attainable in the estimation of statistical parameters, *Bull. Calcutta Math. Soc.* **37**, 81–91.

Ratliff, F. (1965), *Mach Bands: Quantitative Studies on Neural Networks in the Retina*, Hodden Day, San Francisco.

Rayleigh, J. W. S. (1880), On the resultant of a large number of vibrations of the same pitch and of arbitrary phase, *Philos. Mag., 5th Ser.* **10**, 73–78.

Rayleigh, J. W. S. (1903), On the spectrum of an irregular disturbance, *Philos. Mag.* **5**, 238–243.

Rayleigh, J. W. S. (1919), On the Problem of Random Vibrations, and of Random Flights in One, Two, or Three Dimensions, *The London, Edinburgh, and Dublin Philosophical Magazine and Journal of Science* **37**:220, 321–347.

Reed, C. (1996), *Hilbert*, Springer-Verlag, New York.

Reed, I. S. (1962), On the moment theorem for complex Gaussian processes, *IRE Trans. Inform. Theory* **IT-8**, 194–195.

Regan, D. (2000), *Human Perception of Objects: Early Visual Processing of Spatial Form Defined by Luminance, Color, Texture, Motion, and Binocular Disparity*, Sinauer Associates, Sunderland, MA.

Reif, F. (1965), *Fundamentals of Statistical and Thermal Physics*, McGraw-Hill, New York.

Reimer, L. (1985), *Scanning Electron Microscopy: Physics of Image Formation and Microanalysis*, Springer-Verlag, Berlin.

Revesz, G., Kundel, H. L. and Graber, M. A. (1974), The influence of structured noise on the detection of radiologic abnormalities, *Invest. Radiol.* **9**, 479–486.

Revesz, G., Kundel, H. L. and Bonitatibus, M. (1983), The effect of verification on the assessment of imaging techniques, *Invest. Radiol.* **18**, 194–198.

Rice, S. O. (1944), Mathematical analysis of random noise, *Bell Syst. Tech. J.* **23**, 282–332.

Rice, S. O. (1945), Mathematical analysis of random noise, *Bell Syst. Tech. J.* **24**, 46–156.

Richards, J. I. and Youn, H. K. (1990), *Theory of Distributions: A Nontechnical Introduction*, Cambridge University Press, Cambridge.

Richardson, W. H. (1972), Bayesian-based iterative method of image restoration, *J. Opt. Soc. Am.* **62**, 55–59.

Riskin, H. (1984), *The Fokker-Planck Equation: Methods of Solution and Applications*, Springer-Verlag, Berlin.

Robert, C. P. and Casella, G. (1999), *Monte Carlo Statistical Methods*, Springer-Verlag, New York.

Robson, J. G. (1966), Spatial and temporal contrast sensitivity functions of the visual system, *J. Opt. Soc. Am.* A**56**, 1141–1142.

Rockette, H. E., Obuchowski, N., Metz, C. E. and Gur, D. (1990), Statistical issues in ROC curve analysis, *Proc. SPIE* **1234**, 111–119.

Rockette, H. E., Gur, D. and Metz, C. E. (1992), The use of continuous and discrete confidence judgments in receiver operating characteristic studies of diagnostic imaging techniques, *Invest. Radiol.* **27**, 169–172.

Rockette, H. E., King, J. L., Medina, J. L., Eisen, H. B., Brown, M. L. and Gur, D. (1995), Imaging systems evaluation: Effect of subtle cases on the design and analysis of receiver operating characteristic studies, *Am. J. Radiol.* **165**, 679–683.

Rockette, H. E., Johns, C. M., Weissman, J. L., Holbert, J. M., Sumkin, J. H., King, J. L. and Gur, D. (1997), Relationship of subjective ratings of image quality and observer performance, *Proc. SPIE* **3036**, 152–159.

Rockette H. E., King, J. L., Thaete, F. L., Fuhrman, C. R., Slifko, R. M. and Gur, D. (1998), Selection of subtle cases for observer-performance studies: The importance of knowing the true diagnosis, *Acad. Radiol.* **5**, 86–92.

Rockette, H. E., Li, W., Brown, M. L., Britton, C. A., Towers, J. T. and Gur, D. (2001), Statistical test to assess rank-order imaging studies, *Acad. Radiol.* **24**, 24–30.

Roddier, F. (1988), Interferometric imaging in optical astronomy, *Phys. Rep.* **170**:2, 97–166.

Roe, C. A. and Metz, C. E. (1997a), The Dorfman-Berbaum-Metz method for statistical analysis of multi-reader, multi-modality ROC data: Validation by computer simulation, *Acad. Radiol.* **4**, 298–303.

Roe, C. A. and Metz, C. E. (1997b), Variance-component modeling in the analysis of receiver operating characteristic index estimates, *Acad. Radiol.* **4**, 587–600.

Rogala, E. W. (1999), Task-based assessment of a proposed phase-shifting interferometer/ellipsometer, Ph.D. Dissertation, University of Arizona, Tucson, AZ.

Rogala, E. W. and Barrett, H. H. (1997), Phase-shifting interferometry and maximum-likelihood estimation theory, *Appl. Opt.* **36**, 8871–8876.

Rogala, E. W. and Barrett, H. H. (1998a), Phase-shifting interferometry and maximum-likelihood estimation theory II: A generalized solution, *Appl. Opt.* **37**, 7253–7258.

Rogala, E. W. and Barrett, H. H. (1998b), A phase-shifting interferometer/ellipsometer capable of measuring the complex index of refraction and the surface height profile of a test surface, *J. Opt. Soc. Am. A* **15**, 538–548.

Rogala, E. W. and Barrett, H. H. (1998c), Assessing and optimizing the performance of a phase-shifting interferometer capable of measuring the complex index and profile of a test surface, *J. Opt. Soc. Am. A* **15**, 1670–1685.

Rogers, A. E. E., Hinteregger, H. F., Whitney, A. R., Counselman, C. C., Shapiro, I. I., Wittels, J. J., Klemperer, W. K., Warnock, W. W., Clark, T. A., Hutton, L. K., Marandino, G. E., Ronnang, B. O., Rydbeck, O. E. H., Niell, A. E. (1974), The structure of radio sources 3C 273B and 3C 84 deduced from the "closure" phases and visibility amplitudes observed with three-element interferometers, *Astrophys. J.* **193**, 294.

Rogers, D. F. (1998), *Procedural Elements for Computer Graphics*, 2nd ed., WCB/McGraw-Hill, New York.

Rogers, D. F. and Adams, J. A. (1990), *Mathematical Elements for Computer Graphics*, 2nd ed., WCB/McGraw-Hill, New York.

Rogers, W. L., Han, K. S., Jones, L. W. and Beierwaltes, W. H. (1972), Application of a Fresnel zone plate to gamma-ray imaging, *J. Nucl. Med.* **13**, 612.

Rogers, W. L., Koral K. F., Mayans, R., Leonard, P. F., Thrall, J. H., Brady, T. J. and Keyes, J. W. (1980), Coded-aperture imaging of the heart, *J. Nucl. Med.* **21**, 371–378.

Roggeman, M. C. and Welsh, B. (1996), *Imaging Through Turbulence*, CRC Press, Boca Raton, FL.

Rolland, J. P., Barrett, H. H. and Seeley, G. W. (1989), Quantitative study of deconvolution and display mappings for long-tailed point-spread functions, *Proc. SPIE* **1092**, 17–21.

Rolland, J. P. and Barrett, H. H. (1992), Effect of random background inhomogeneity on observer detection performance, *J. Opt. Soc. Am. A* **9**:5, 649–658.

Rolland, J. P. and Strickland, R. (1997), An approach to the synthesis of biological tissue, *Opt. Express* **1**:13, 414–423.

Rolland, J. P., Goon, A. and Yu, L. (1998), Synthesis of textured complex backgrounds, *Opt. Eng.* **37**:7, 2055–2063.

Rolland, J. P. (2000), Synthesizing anatomical images for image understanding, in *Handbook of Medical Imaging*, Vol. II: *Progress in Medical Physics and Psychophysics* (Beutel, J., Van Metter, R. and Kundel, H. L., Eds.), SPIE Press, Bellingham, WA, Chap. 13.

Rose, A. (1948), The sensitivity performance of the human eye on an absolute scale, *J. Opt. Soc. Am.* **38**, 196–208.

Ross, S. M. (1993), *Introduction to Probability Models*, Academic Press, New York.

Rowe, R. K., Aarsvold, J. N., Barrett, H. H., Chen, J. C., Klein, W. P., Moore, B. A., Pang, I. W., Patton, D. D. and White, T. A. (1993), A stationary hemispherical SPECT imager for three-dimensional brain imaging, *J. Nucl. Med.* **34**, 474–480.

Ruck, D. W., Rogers, S. K., Kabrisky, M., Oxley, M. E. and Suter, B. W. (1990), The multilayer perceptron as an approximation to a Bayes optimal discriminant function, *IEEE Trans. Neural Networks* **1**, 296–298.

Russell, L. B., Gold, M. R., Siegel, J. E., Daniels, N. and Weinstein, M. C. (1996), The role of cost-effectiveness analysis in health and medicine, *JAMA* **276**, 1172–1177.

Sabatier, P. C., Ed. (1987), *Inverse Problems: An Interdisciplinary Study*, Suppl. 19 to *Advances in Electronics and Electron Physics*, Academic Press, London.

Sabatier, P. C., Ed. (1991), *Inverse Methods in Action*, Springer-Verlag, Berlin.

Sachs, M., Nachmias, J. and Robson, J. (1971), Spatial-frequency channels in human vision, *J. Opt. Soc. Am.* **61**, 1176–1186.

Saha, S. K. (2002), Modern optical astronomy: Technology and impact of interferometry, *Rev. Mod. Phys.* **74**, 550–600

Sain, J. D. and Barrett, H. H. (2003), Performance evaluation of a modular gamma camera using detectability index, *J. Nucl. Med.* **44**:1, 58–66.

Saito, H., Tanaka, K., Fukuda, Y. and Oyamada, H. (1988), Analysis of discontinuity in visual contours in area 19 of the cat, *J. Neurosci.* **8**, 1131–1143.

Saleh, B. E. A., (1978a), *Photoelectron Statistics*, Springer-Verlag, Berlin.

Saleh, B. E. A. (1978b), Bilinear processing of 1-D signals by use of linear 2-D coherent optical processors (TE), *Appl. Opt.* **17**:21, 3408–3411.

Saleh, B. E. A. and Freeman, M. O. (1987), Optical transformations, in *Optical Signal Processing* (Horner, J., Ed.), Adademic Press, San Diego, CA.

Saleh, B. E. A. and Teich, M. C. (1982), Multiplied Poisson noise in pulse, particle and photon detection, *Proc. IEEE* **70**, 229–245.

Saleh, B. E. A. and Teich, M. C. (1991), *Fundamentals of Photonics*, Wiley, New York.

Sandberg, I. W. (1963), On the properties of systems that distort signals, *Bell Syst. Tech. J.* **42**, 2033–2047.

Sankaran, S., Frey, E. C., Gilland, K. L. and Tsui, B. M. W. (2002), Optimum compensation method and filter cutoff frequency in myocardial SPECT: A human observer study, *J. Nucl. Med.* **43**, 432–438.

Sarfatt, J. (1963), *Nuovo Cimento* **27**, 1119.

Sargent, M., Scully, M. O. and Lamb, W. E. (1974), *Laser Physics*, Addison-Wesley, Reading, MA.

Savage, L. J. (1954), *The Foundations of Statistics*, Wiley, New York.

Sayood, K. (2000), *Introduction to Data Compression*, 2nd ed., Morgan-Kaufman, San Francisco.

Scales, L. E. (1985), *Introduction to Non-Linear Optimization*, Springer, New York.

Scharf, L. L. (1991), *Statistical Signal Processing: Detection, Estimation, and Time-Series Analysis*, Addison-Wesley, Reading, MA.

Schensted, I. (1976), *A Course on the Application of Group Theory to Quantum Mechanics*, NEO Press, Peaks Island, ME.

Schetzen, M. (1980), *The Volterra and Wiener Theories of Nonlinear Systems*, Wiley, New York.

Schottky, W. (1926), Small-shot effect and flicker effect, *Phys. Rev.* **28**, 74–103.

Schroeder, M. R. (1991), *Fractals, Chaos, Power Laws: Minutes from an Infinite Paradise*, W. H. Freeman, New York.

Schuster, A. (1894), On interference phenomena, *Philos. Mag.* **37**, 509–545.

Schuster, A. (1904), *The Theory of Optics*, London.

Schuster, A. (1906), The periodogram and its optical analogy, *Proc. Roy. Soc.* **77**, 136–140.

Schwartz, L. (1950), *Théorie des Distributions*, Hermann Cie., Paris.

Scribot, A. A. (1974), First-order probability density functions of speckle measured with a finite aperture, *Opt. Commun.* **11**, 238–241.

Scully, M. O. and Lamb, W. E. (1969), Quantum theory of an optical maser. III. Theory of photoelectron counting statistics, *Phys. Rev.* **179**:2, 368–374.

Seger, O. (1993), *Model Building and Restoration with Applications in Confocal Microscopy*, Department of Electrical Engineering, Linköping University, Linköping, Sweden.

Seltzer, S. E., Swensson, R. G., Judy, P. F. and Nawfel, R. D. (1988), Size discrimination in computed tomographic images: Effects of feature contrast and display window, *Invest. Radiol.* **23**, 455–462.

Sezan, M. I. (1992), An overview of convex projections theory and its applications to image recovery problems, *Ultramicroscopy* **40**, 55–67.

Shankar, P. M., Reid, J. M., Ortega, H., Piccoli, C. W., Goldberg, B. B. (1993), Use of non-Rayleigh statistics for the identification of tumors in ultrasonic B scans of the breast, *IEEE Trans. Med. Imaging* **12**, 687–692.

Shankar, P. M. (1995), A model for ultrasonic scattering from tissues based on the K distribution, *Phys. Med. Biol.* **40**, 1633–1649.

Shannon, C. E. (1948), A mathematical theory of communication, *Bell Syst. Tech. J.* **27**, 379–423.

Shao, M. and Colavita, M. M. (1992), Long-baseline optical and infrared stellar interferometry, *Ann. Rev. Astron. Astrophys.* **30**, 457–498.

Shapiro, J. H. (1999), Bounds on the area under the ROC curve, *J. Opt. Soc. Am. A* **16**, 53–57.

Shapley, R. and Enroth-Cugell, C. (1985), Visual adaptation and retinal gain control, *Progr. Retinal Res.* **3**, 263–346.

Shaw, R. (1963), The equivalent quantum efficiency of the photographic process, *J. Photogr. Sci.* **11**, 199–204.

Shaw, R. (1978), Evaluating the efficiency of imaging processes, *Rep. Prog. Phys.* **41**, 1103–1155.

Shepard, R. N., Romney, A. K. and Nerlove, S. B. (1972), *Multidimensional Scaling*, Vol. I, *Theory*, Seminar Press, New York.

Shepp, L. A. and Logan, B. F. (1974), The Fourier reconstruction of a head section, *IEEE Trans. Nucl. Sci.* **21**, 21–43.

Shepp, L. A. and Vardi, Y. (1982), Maximum likelihood reconstruction for emission tomography, *IEEE Trans. Med. Imaging* **1**, 113–122.

Shiryayev, A. N. (1984), *Probability* translated by Boas, R. P., Springer-Verlag, New York.

Shockley, W. (1938), Currents to conductors induced by a moving point charge, *J. Appl. Phys.* **9**, 635–636.

Shockley, W. and Pierce, J. R. (1938), A theory of noise for electron multipliers, *Proc. IRE* **26**, 321–332.

Shore, J. E. and Johnson, R. W., (1990), Axiomatic derivation of the principle of maximum entropy and the principle of minimum cross-entropy, *IEEE Trans. Inform. Theory* **26**, 26–37.

Shulman, A. R. (1970), *Optical Data Processing*, Wiley, New York.

Sibul, L. H. (1995), Application of reproducing and invariance properties of wavelet and Fourier-Wigner transforms, *Proc. SPIE* **2569**, 418–428.

Siegman, A. E. (1986), *Lasers*, University Science Books, Mill Valley, CA.

Sillion, F. X. and Puech, C. (1994), *Radiosity and Global Illumination*, Morgan Kaufmann, San Francisco.

Silverman, B. W. (1986), *Density Estimation for Statistics and Data Analysis*, Chapman & Hall, London.

Silverstein, S. D. and O'Donnell, M. (1988), Theory of frequency and temporal compounding in coherent imaging: Speckle suppression and image resolution, *J. Opt. Soc. Am. A* **5**, 104–113.

Simoncelli, E. and Adelson, E. (1989), Nonseparable QMF pyramids, *Proc. SPIE* **1199**, 1242–1246.

Simpson, A. J. and Fitter, M. J. (1973), What is the best index of detectability? *Psych. Bull.* **80**, 481–488.

Siotani, M., Hayakawa, T. and Fujikoshi, Y. (1985), *Modern Multivariate Statistical Analysis: A Graduate Course and Text*, American Sciences Press, Columbus, OH.

Skilling, J. (1988), The axioms of maximum entropy, in *Maximum-Entropy and Bayesian Methods in Science and Engineering*, Vol. 1 (Erickson, G. J. and Smith, C. R., Eds.), Kluwer, Dordrecht.

Slepian, D. and Pollak, H. O. (1961), Prolate spheroidal wave functions, Fourier analysis and uncertainty – I, *Bell Syst. Tech. J.* **40**, 43–64.

Slepian, D. (1976), On bandwidth, *Proc. IEEE* **64**, 292–300.

Small, C. G. (1996), *The Statistical Theory of Shape*, Springer, New York.

Smith, B. D. (1985a), Image reconstruction from cone-beam projections: Necessary and sufficient conditions and reconstruction methods, *IEEE Trans. Med. Imaging* **MI-4**, 14–28.

Smith, B. D. (1985b), Derivation of the extended fan-beam formula, *IEEE Trans. Med. Imaging* **MI-4**, 177–184.

Smith, L. (1984), *Linear Algebra*, 2nd ed., Springer-Verlag, New York.

Smith, L. P. (1953), *Mathematical Methods for Scientists and Engineers*, Dover, New York.

Smith, S. W., Wagner, R. F., Sandrik, J. M. and Lopez, H. (1983), Low contrast detectability and contrast/detail analysis in medical ultrasound, *IEEE Trans. Sonics Ultrasonics* **30**, 164–173.

Smith, W. E., Barrett, H. H. and Paxman, R. G. (1983), Reconstruction of objects from coded images by simulated annealing, *Opt. Lett.* **8**:4, 199–201.

Smith, W. E. and Barrett, H. H. (1986), Hotelling trace criterion as a figure of merit for the optimization of imaging systems, *J. Opt. Soc. Am. A* **3**, 717–723.

Snyder, D. L. and Miller, M. I. (1991), *Random Point Processes in Time and Space*, 2nd ed., Springer-Verlag, New York.

Soares, E. J. (1994), Attenuation, noise and image quality in single-photon emission computed tomography, Ph.D. Dissertation, University of Arizona, Tucson, AZ.

Soares, E. J., Barrett, H. H. and Abbey, C. K. (1995), Noise characterization and objective image-quality assessment of SPECT imaging, in *Information Processing in Medical Imaging* (Bizais, Y., Barillot, C. and Di Paola, R., Eds.), Kluwer Academic, Boston, pp. 353–354.

Soares, E. J., Byrne, C. L. and Glick, S. J. (1998), Noise characterization of block-iterative reconstruction algorithms: I. Theory, *IEEE Trans. Med. Imaging* **19**:4, 261–270.

Sokolnikoff, I. S. and Redheffer, R. M. (1958), *Mathematics of Physics and Modern Engineering*, McGraw-Hill, New York.

Spiro, I., Jones, R. C. and Wark, D. Q. (1965), Atmospheric transmission concepts, symbols, units and nomenclature, *Infrared Phys.* **5**, 11–36.

Stakgold, I. (1967), *Boundary Value Problems of Mathematical Physics*, Vol. I, Macmillan, New York.

Stakgold, I. (1979), *Green's Functions and Boundary Value Problems*, Wiley, New York.

Stark, H., Ed. (1987), *Image Recovery: Theory and Application*, Academic Press, San Diego.

Stark, H. and Sezan, M. I. (1994), Image processing using projection methods, in *Real-Time Optical Information Processing* (Javidi, B. and Horner, J. L., Eds.), Academic Press, San Diego, pp. 185–232.

Stark, H. and Woods, J. W. (1986), *Probability, Random Processes, and Estimation Theory for Engineers*, Prentice-Hall, Englewood Cliffs, NJ.

Starr, S. J., Metz, C. E., Lusted, L. B. and Goodenough, D. J. (1975), Visual detection and localization of radiographic images, *Radiology* **116**, 533–538.

Stavroudis, O. N. (1972), *The Optics of Rays, Wavefronts and Caustics*, Academic Press, New York.

Stayman, J. W. and Fessler, J. A. (2000), Regularization for uniform spatial resolution properties in penalized-likelihood image reconstruction, *IEEE Trans. Med. Imaging* **19**:6, 601–615.

Stéphan, E. (1874), Sur l'extreme petitesse du diamtre apparent des etoiles fixes, *Comptes Rendus de l'Acaemie des Sciences* **78**, 1008–1012.

Stoica, P. and Marzetta, T. L. (2001), Parameter estimation problems with singular information matrices, *IEEE Trans. Signal Proc.* **49**, 87–89.

Stoisiek, M. and Wolf, D. (1976), Recent investigations on the stationarity of $1/f$ noise, *J. Appl. Phys.* **47**:1, 362.

Strang, G. (1980), *Linear Algebra and Its Applications*, 2nd ed., Academic Press, Orlando, FL.

Strang, G. and Fix, G. J. (1973), *An Analysis of the Finite Element Method*, Prentice-Hall, Englewood Cliffs, NJ.

Strichartz, R. (1994), *A Guide to Distribution Theory and Fourier Transforms*, CRC Press, Boca Raton, FL.

Stromeyer, C. F., III and Julesz, B. (1972), Spatial-frequency masking in vision: Critical bands and spread of masking, *J. Opt. Soc. Am.* **62**, 1221–1232.

Stromeyer, C. F., III and Klein, S. (1975), Evidence against narrow-band spatial frquency channels in human vision: The detectability of frequency modulated gratings, *Vision Res.* **15**, 899–910.

Sullivan, R. J. (2000), *Microwave Radar: Imaging and Advanced Concepts*, Artech House, Norwood, MA.

Surdin, M. (1939), Fluctuations de courant thermionique at le "flicker effect," *J. Phys. Radium* **10**, 188.

Swank, R. K. (1973), Absorption and noise in x-ray phosphors, *J. Appl. Phys.* **44**, 4199–4203.

Swensson, R. M. and Green, D. M. (1977), On the relations between random walk models for two-choice response times, *J. Math. Psychol.* **15**, 282–291.

Swensson, R. G. and Judy, P. F. (1981), Detection of noisy visual targets: Models for the effects of spatial uncertainty and signal-to-noise ratio, *Percept. Psychophys.* **29**, 521–534.

Swensson, R. G. (1996), Unified measurement of observer performance in detecting and localizing target objects on images, *Med. Phys.* **23**, 1709–1725.

Swensson, R. G., King, J. L., Good, W. F. and Gur, D. (2000), Using a constrained formulation based on probability summation to fit receiver operating characteristic (ROC) curves, *Proc. SPIE* **3981**, 145–153.

Swets, J. A., Tanner, W. P., Jr. and Birdsall, T. G. (1961), Decision processes in perception, *Psych. Rev.* **68**, 301–340.

Swets, J. A., Ed. (1964), *Signal Detection and Recognition by Human Observers*, Wiley, New York. Reprinted by Peninsula Publishing, Los Altos, CA, 1988.

Swets, J. A. (1979), ROC analysis applied to the evaluation of medical imaging techniques, *Invest. Radiol.* **4**, 109–120.

Swets, J. A., Pickett, R. M., Whitehead, S. F., Getty, D. J., Schnur, J. A., Swets, J. B. and Freeman, B. A. (1979), Assessment of diagnostic technologies, *Science* **105**, 753–759.

Swets, J. A. and Pickett, R. M. (1982), *Evaluation of Diagnostic Systems: Methods from Signal Detection Theory*, Academic Press, New York.

Swets, J. A. (1986), Form of empirical ROCs in discrimination and diagnostic tasks: Implications for theory and measurement of performance, *Psychol. Bull.* **99**, 181–198.

Swets, J. A. (1988), Measuring the accuracy of diagnostic systems, *Science* **240**, 1285–1293.

Szilard, L. (1929), Über die Entropieverminderung in einem thermodynamischen System bei Eingriffen intelligenter Wesen, *Z. Physik* **53**, 840–856.

Tan, P. and Drossos, C. (1975), Invariance properties of maximum likelihood estimators, *Math. Mag.* **Jan.-Feb.**, 37–41.

Tandon, J. L. and Bilger, H. R. (1976), 1/f noise as a non-stationary process: Experimental evidences and some analytical conditions, *J. Appl. Phys* **47**, 1697–1701.

Tanner, W. P., Jr. and Swets, J. A. (1954), A decision-making theory of visual detection, *Psychol. Rev.* **61**, 401–409. Reprinted in Cohn, T. E., Ed. (1993), Collected Works in Optics, Vol. 3: *Visual Detection*, Optical Society of America, Washington, DC, pp. 34–42.

Tapiovaara, M. (1990), Ideal observer and absolute efficiency of detecting mirror symmetry in random images, *J. Opt. Soc. Am. A* **7**, 2245–2253.

Tatarskii, V. I. (1983), The Wigner representation of quantum mechanics, *Sov. Phys. Uspekhi* **26**:4, 311–327.

Teich, M. C. and Saleh, B. E. A. (1988), Photon bunching and antibunching, in *Progress in Optics*, Vol. 26 (Wolf, E., Ed.), Elsevier Science, Amsterdam.

Teich, M. C. and Saleh, B. E. A. (1990), Squeezed and antibunched light, *Phys. Today* **43**, 26–34.

Ter Haar, D. (1955), Foundations of statistical mechanics, *Rev. Mod. Phys.* **27**, 289–338.

Thomas, J. P. (1985), Effect of static-noise and grating masks on detection and identification of grating targets, *J. Opt. Soc. Am. A* **2**, 1586–1592.

Thompson, M. L. and Zucchini, W. (1989), On the statistical analysis of ROC curves, *Stat. Med.* **8**, 1277–1290.

Thomson, G. P. (1965), Electronic waves, in *Nobel Lectures: Physics 1921–1941*, Elsevier, Amsterdam, pp. 386–394; 397–403 .

Tikhonov, A. N. (1943), On stability of inverse problems, *Dokl. Akad. Nauk USSR* **39**:5, 195–198.

Tikhonov, A. N. and Arsenin, V. Ya. (1979), *Methods for Solving Ill-posed Problems*, Nauka, MoscoW.

Tikochinsky, Y., Tishby, N. Z. and Levine, R. D. (1984), Consistent inference of probabilities for reproducible experiments, *Phys. Rev. Lett.* **52**, 1357–1369.

Tinkham, M. (1964), *Group Theory and Quantum Mechanics*, McGraw-Hill, New York.

Titchmarsh, E. C. (1948), *Introduction to the Theory of Fourier Integrals*, 2nd ed., Oxford University Press, London.

Toledano, A. and Gatsonis, C. A. (1995), Regression analysis of correlated receiver operating characteristic data, *Acad. Radiol.* **2**:Suppl. 1, S30–S36.

Toledano, A. Y. and Gatsonis, C. (1996), Ordinal regression methodology for ROC curves derived from correlated data, *Stat. Med.* **15**, 1807–1826.

Toledano, A. Y. and Gatsonis, C. (1999), GEEs for ordinal categorical data: Arbitrary patterns of missing responses and missingness in a key covariate, *Biometrics* **22**, 488–496.

Tolhurst, D. J. (1973), Separate channels for the analysis of the shape and movement of a moving visual stimulus, *J. Phys.* **231**, 385–402.

Tolhurst, D. J. (1975), Sustained and transient channels in human vision, *Vision Res.* **15**, 1151–1155.

Tolhurst, D. J., Movshon, J. A. and Dean, A. F. (1983), The statistical reliability of signals in single neurons in cat and monkey visual cortex, *Vision Res.* **23**, 775–785.

Torgerson, W. S. (1958), *Theory and Methods of Scaling*, Wiley, New York.

Tosteson, A. and Begg, C. (1988), A general regression methodology for ROC curve estimation, *Med. Decis. Making* **8**, 204–215.

Towers, J. D., Holbert, J. M., Britton, C. A., Costello, P., Sciullli, R. and Gur, D. (2000), Multipoint rank-order study methodology: Observer issues, *Invest. Radiol.* **35**:2, 125–130.

Traub, W. A. (1999), Beam combination and fringe measurement, in *Principles of Long Baseline Stellar Interferometry*, Course notes from the 1999 Michelson Summer School (Lawson, P. R., Ed.), JPL Publication 00-009 07/00, Ch. 3.

Tretiak, O. and Metz, C. (1980), The exponential Radon transform, *SIAM J. Appl. Math.* **39**, 341–354.

Tsui, B. M. W., Metz, C. E., Atkins, F. B., Starr, S. J. and Beck, R. N. (1978), A comparison of optimum spatial resolution in nuclear imaging based on statistical theory and on observer performance, *Phys. Med. Biol.* **23**:4, 654–676.

Tsui, B. M. W., Metz, C. E. and Beck, R. N. (1983), Optimum detector spatial resolution for discriminating between tumour uptake distributions in scintigraphy, *Phys. Med. Biol.* **28**:7, 775–788.

Tsui, B. M. W., Terry, J. A. and Gullberg, G. T. (1993), Evaluation of cardiac cone-beam SPECT using observer performance experiments and ROC analysis, *Invest. Radiol.* **28**:12, 1101–1112.

Tsui, B. M. W., Gunter, D. L., Beck, R. N. and Patton, J. A. (1996), Physics of Collimator Design, in Sandler, M. P., Coleman, R. E., Wackers, F. J. Th., Patton, J. A., Gottschalk, A. and Hoffer, P. B., Eds. *Diagnostic Nuclear Medicine* (3rd ed.), Vol. 1, Williams & Wilkins, Baltimore, MD, pp. 67–80.

Turner, D. A. (1978), An intuitive approach to receiver operating characteristic curve analysis, *J. Nucl. Med.* **19**, 213–220.

Tuy, H. K. (1983), An inversion formula for cone-beam reconstruction, *SIAM J. Appl. Math.* **43**, 546–552.

Tylavsky, D. J. and Sohie, G. R. L. (1986), Generalization of the matrix inversion lemma, *Proc. IEEE* **74**:7, 1050–1052.

Usmani, R. A. (1987), *Applied Linear Algebra*, Marcel Dekker, New York.

van der Ziel, A. (1950), On the noise spectra of semi-conductor noise and of flicker effect, *Physica* **16**:4, 359–375.

van der Ziel, A. (1988), Unified presentation of 1/f noise in electron devices: Fundamental 1/f noise sources, *Proc. IEEE* **76**:3, 233–258.

Van de Walle, R., Barrett, H. H., Myers, K. J., Altbach, M. I., Desplanques, B., Gmitro, A. F., Cornelis, J. and Lemahieu, I. (2000), Reconstruction of MR images from data acquired on a general non-regular grid by pseudoinverse calculation, *IEEE Trans. Med. Imaging* **19**:12, 1160–1167.

van Laarhoven, P. J. M. and Aarts, E. H. L. (1987), *Simulated Annealing: Theory and Applications*, D. Reidel Publishing, Dordrecht, Holland.

van Meter, D. and Middleton, D. S. O. (1954), Modern statistical approaches to reception in communication theory, *IRE Trans.* **PGIT-4**, 119–145.

Van Nes, F. L. and Bouman, M. A. (1967), Spatial modulation transfer in the human eye, *J. Opt. Soc. Am.* **57**, 401–406.

Van Trees, H. L. (1968), *Detection, Estimation, and Modulation Theory, Part I*, Wiley, New York.

Van Vleck, J. H. and Middleton, D. (1946), A theoretical comparison of the visual, aural, and meter reception of pulsed signals in the presence of noise, *J. Appl. Phys.* **17**, 940–971.

Ville, J. (1948), Théorie et applications de la notion de signal analytique, *Cables Transm.* **2A1**, 61–74.

von Neumann, J. (1950), *Functional Operators*, Vol. II: *The Geometry of Orthogonal Spaces*, Annals of Mathematics Studies 22 (Griffiths, P. A., Mather, J. N. and Stein, E. M., Eds.), Princeton University Press, Princeton, NJ.

Voss, R. F. and Clarke, J. (1976), Flicker (1/f) noise: Equilibrium temperature and resistance fluctuations, *Phys. Rev.* **B 13**, 556–573.

Wagner, J.-M., Noo, F. and Clackdoyle, R. (2001), 3D Image reconstruction from exponential X-ray projections using Neumann series, in *Conference Record of the 2001 IEEE International Conference on Acoustics, Speech, and Signal Processing* (IEEE Signal Processing Society, Ed.), pp. 2017–2020.

Wagner, R. F. (1978), Decision theory and the signal-to-noise ratio of Otto Schade, *Photogr. Sci. Eng.* **22**, 41–46.

Wagner, R. F., Brown, D. G. and Pastel, M. S. (1979), Application of information theory to the assessment of computed tomography, *Med. Phys.* **6**, 83–94.

Wagner, R. F., Brown, D. B. and Metz, C. E. (1981), On the multiplex advantage of coded source/aperture photon imaging, *Proc. SPIE* **314**, 72–76.

Wagner R. F., Smith S. W., Sandrik J. M. and Lopez, H. (1983), Statistics of speckle in ultrasound B scans, *IEEE Trans. Sonics Ultrasonics* **30**, 156–163.

Wagner, R. F. and Brown, D. G. (1985), Unified SNR analysis of medical imaging systems, *Phys. Med. Biol.* **30**:6, 489–518.

Wagner, R. F., Insana, M. F. and Brown, D. G. (1986), Unified approach to the detection and classification of speckle texture in diagnostic ultrasound, *Opt. Eng.* **25**, 738–742.

Wagner, R. F., Insana, M. F. and Brown, D. G. (1987), Statistical properties of radio frequency and envelope-detected signals with applications to medical ultrasound, *J. Opt. Soc. Am.* **4**, 910–922.

Wagner, R. F., Myers, K. J., Brown, D. G., Tapiovaara, M. J. and Burgess, A. E. (1989), Higher-order tasks: Human vs. machine performance, *Proc. SPIE* **1090**, 183–194.

Wagner, R. F., Insana, M. F., Brown, D. G., Garra, B. S. and Jennings, R. J. (1990a), Texture discrimination: Radiologist, machine, and man, in *Vision: Coding and Efficiency* (Blakemore, C., Ed.), Cambridge University Press, Cambridge, pp. 310–318.

Wagner, R. F., Myers, K. J., Burgess, A. E., Brown, D. G. and Tapiovaara, M. J. (1990b), Maximum a posteriori detection: Figures of merit for detection under uncertainty, *Proc. SPIE* **1232**, 195–204.

Wagner, R. F., Myers, K. J. and Hanson, K. M. (1992), Task performance on constrained reconstructions: Human observer performance compared with sub-optimal Bayesian performance, *Proc. SPIE* **1652**, 352–362.

Wagner, R. F., Chan, H.-P., Sahiner, B., Petrick, N. and Mossoba, J. T. (1997) Finite-sample effects and resampling plans: Applications to linear classifiers and computer-aided diagnosis, *Proc. SPIE* **3034**, 467–477.

Wagner, R. F., Beiden, S. V. and Metz, C. E. (2001), Continuous vs. categorical data for ROC analysis: Some quantitative considerations, *Acad. Radiol.* **8**, 328–334.

Wahba, G. (1990), *Spline Models for Observational Data*, Society for Industrial and Applied Mathematics, Philadelphia, PA.

Walker, J. S. (1991), *Fast Fourier Transforms*, CRC Press, Boca Raton, FL.

Walkup, J. F. and Goodman, J. W. (1973), Limitations of fringe-parameter estimation at low light levels, *J. Opt. Soc. Am.* **63**, 470–478.

Walter, G. G. (1994), *Wavelets and Other Orthogonal Systems with Applications*, CRC Press, Boca Raton, FL.

Walther, A. (1968), Radiometry and coherence, *J. Opt. Soc. Am.* **58**, 1256–1259.

Walther, A. (1973), Radiometry and coherence, *J. Opt. Soc. Am.* **63**, 1622–1623.

Walther, A. (1978), Propagation of the generalized radiance through lenses, *J. Opt. Soc. Am.* **68**, 1606–1610.

Wang, W. and Gindi, G. (1997), Noise analysis of MAP-EM algorithms for emission tomography, *Phys. Med. Biol.* **42**, 2215–2232.

Watson, A. B. and Robson, J. G. (1981), Discrimination at threshold: Labelled detectors in human vision, *Vision Res.* **21**, 1115–1122.

Watson, A. B. (1983), Detection and recognition of simple spatial forms, in *Physical and Biological Processing of Images* (Braddick, O. J. and Sleigh, A. C., Eds.), Springer-Verlag, New York, pp. 100–114. Watson, A. B. and Pelli, D. G. (1983), QUEST: A Bayesian adaptive psychometric method, *Perception and Psychophysics* **33**(2), 113-120.

Watson, A. B. (1987), The cortex transform: Rapid computation of simulated neural images, *Compt. Vision Graphics Image Proc.* **39**, 311–327.

Watson, A. B. and Ahumada, A. J. (1989), A hexagonal orthogonal-oriented pyramid as a model of image representation in the visual cortex, *IEEE Trans. Bio-Med. Eng.* **36**, 97–106.

Watson, A. B. and Fitzhugh, A. (1990), The method of constant stimuli is inefficient, *Perception and Psychophysics* **47**(1), 87-91.

Weinberg, A. M. and Wigner, E. P. (1958), *The Physical Theory of Neutron Chain Reactors*, University of Chicago Press, Chicago, IL.

Weinert, H. L., Ed. (1983), *Reproducing Kernel Hilbert Spaces: Application in Statistical Signal Processing*, Hutchinson Ross, Stroudsburg, PA.

Weinstein, M. C. and Fineberg, H. V. (1980), *Clinical Decision Analysis*, W. B. Saunders, Philadelphia, PA.

Wentzell, A. D. (1981), *A Course in the Theory of Stochastic Processes* (translated by Chomet, S.), McGraw-Hill, New York.

West, B. J. and Shlesinger, M. F. (1989), On the ubiquity of 1/f noise, *Int. J. Mod. Phys. B* **3**:6, 795–819.

West, B. J. and Shlesinger, M. F. (1990), The noise in natural phenomena, *Am. Sci.* **78**, 40–45.

Weyl, H. (1952), *Symmetry*, Princeton University Press, Princeton, NJ.

Whalen, A. D. (1971), *Detection of Signals in Noise*, Academic Press, San Diego, CA.

Whitrow, G. J. (1965), entry on Joseph Fourier, in *Encyclopedia Brittanica* (Benton, W., Ed.), Encyclopedia Brittanica, Chicago, IL.

Wieand, S., Gail, M. H., James, B. R. and James, K. L. (1989), A family of nonparametric statistics for comparing diagnostic markers with paired or unpaired data, *Biometrika* **76**, 585–592.

Wiener, N. (1930), Generalized harmonic analysis, *Acta Math.* **55**, 117–258.

Wiener, N. (1942), *Extrapolation, Interpolation and Smoothing of Stationary Time Series*, MIT Press, Cambridge, MA.

Wigner, E. P. (1932), On the quantum correction for thermo-dynamic equilibrium, *Phys. Rev.* **40**, 749-759.

Wilson, D. W. (1994), Noise and resolution properties of FB and ML-EM reconstructed SPECT images, Ph.D. Dissertation, University of North Carolina, Chapel Hill, NC.

Wilson, H. and Bergen, J. (1979), A four mechanism model for threshold spatial vision, *Vision Res.* **19**, 19–32.

Wilson, H. R. and Gelb, D. J. (1984), Modified line-element theory for spatial-frequency and width discrimination, *J. Opt. Soc. Am. A* **1**, 124–131.

Wilson, D. W., Tsui, B. M. W. and Barrett, H. H. (1994), Noise properties of the EM algorithm: II Monte Carlo simulations, *Phys. Med. Biol.* **39**, 847–872.

Wolf, E. (1976), New theory of radiative energy transfer in free electromagnetic fields, *Phys. Rev. D* **13**, 869–886.

Wolf, E. (1978), Coherence and radiometry, *J. Opt. Soc. Am.* **68**, 6–17.

Wolf, M. (1980), Signal-to-noise ratio and the detection of detail in non-white noise, *Proc. SPIE* **24**, 99–103.

Wollenweber, S. D., Tsui, B. M. W., Lalush, D. S., Frey, E. C., Lacroix, K. J. and Gullberg, G. T. (1999), Comparison of Hotelling observer models and human observers in defect detection from myocardial SPECT imaging, *IEEE Trans. Nucl. Sci.* **46**, 2098–2103.

Woodbury, M. A. (1950), *Inverting Modified Matrices*, Memo. Rep. 42, Statistical Research Group, Princeton, NJ.

Wooding, R. A. (1956), The multivariate distribution of complex normal variables, *Biometrika* **43**, 212–215.

Wünsch, A. D. (1994), *Complex Variables with Applications*, 2nd ed., Addison-Wesley, Reading, MA.

Yao, J. and Barrett, H. H. (1992), Predicting human performance by a channelized Hotelling observer model, *Proc. SPIE* **1768**, 161–168.

Yaroslavsky, L. P. (1985), *Digital Picture Processing: An Introduction*, Springer-Verlag, Berlin.

Youla, D. C. (1978), Generalized image restoration by the method of alternating orthogonal projections, *IEEE Trans. Circuits Syst.* **CAS-25**, 694–702.

Youla, D. C. and Webb, H. (1982), Image restoration by the method of convex projections: Part 1 – Theory, *IEEE Trans. Med. Imaging* **1**, 81–94.

Young, T. Y. and Fu, K. S. (1986), *Handbook of Pattern Recognition and Image Processing*, Academic Press, Orlando, FL.

Yueh, S. H., Kong, J. A., Jao, K. A., Shin, R. T. and Novak, L. M. (1989), K-distribution and polarimetricterrain radar clutter, *J. Electromagnetic Waves Appls.* **3**:8, 747–768.

Zadeh, L. A. and Ragazzini, J. R. (1952), Optimal filters for the detection of signals in noise, *Proc. IRE* **40**, 1223–1231.

Zardecki, A. and Delisle, C. (1977), Higher order statistics of light scattered by a random phase screen, *Opt. Acta* **24**, 241–259.

Zayed, A. I. (1993), *Advances in Shannon's Sampling Theory*, CRC Press, Boca Raton, FL.

Zemanian, A. H. (1965), *Distribution Theory and Transform Analysis*, Dover, New York.

Zemanian, A. H. (1987), *Generalized Integral Transforms*, Dover, New York.

Zemp, R. I., Abbey, C. K. and Insana, M. F. (2003), Generalized NEQ for assessment of ultrasound image quality, *Proc. SPIE* **5030**, 391–402 .

Zernike, F. (1934), Diffraction theory of the knife-edge test and its improved form, the phase contrast method, *Mon. Not. R. Astron. Soc.* **94**, 377–384.

Zetzsche, C. and Hauske, G. (1989), Multiple channel model for the prediction of subjective image quality, in *Human Vision, Visual Processing, and Digital Display* (Rogowitz, B. E., Ed.), *Proc. SPIE* **1077**, 209–216.

Zhang, H. (2001), Signal detection in medical imaging, Ph.D. Dissertation, University of Arizona, Tucson, AZ.

Zhang, H., Clarkson, E. and Barrett, H. H. (2001a), Feature-extraction method based on the ideal observer, *Proc. SPIE* **4322**, 440–447.

Zhang, H., Clarkson, E. and Barrett, H. H. (2001b), Nonlinear discriminant analysis, *Proc. SPIE* **4322**, 448–455.

Zhao, Y.-P., Wang, G.-C. and Lu, T.-M. (2001), *Characterization of Amorphous and Crystalline Rough Surfaces — Principles and Applications, Experimental Methods in the Physical Sciences*, Vol. 37, Academic Press, New York.

Zhou, X. H. and Gatsonis, C. A. (1996), A simple method for comparing correlated ROC curves using incomplete data, *Stat. Med.* **15**, 1687–1693.

Zhu, S. C., Wu, Y. and Mumford D. (1998), Filters, random fields and maximum entropy (FRAME), *Int. J. Comput. Vision* **27**, 1–20.

Zubairy, M. and Wolf, E. (1977), Exact equations for radiative transfer for energy and momentum in free electromagnetic fields, *Opt. Commun.* **20**, 321–324.

Zubal, I. G., Harrell, C. R., Smith, E. O., Rattner, Z., Gindi, G. and Hoffer, P. B. (1994), Computerized 3-dimensional segmented human anatomy, *Med. Phys.* **21**, 299–302.

Zweig, H. J. (1965), Detective quantum efficiency of photodetectors with some amplifying mechanism, *J. Opt. Soc. Am.* **52**, 525–528.

Zweig, M. H. and Campbell, G. (1993), Receiver-operating characteristic (ROC) plots: A fundamental evaluation tool in clinical medicine, *Clin. Chem.* **39**, 561–577. Erratum published in *Clin. Chem.* **39**, 1589 (1993).

Index